P9-CJE-699

3/18/87 B.S.

WITHDRAWN
JUN 0 5 2024
DAVID O. McKAY LIBRARY
BYU-IDAHO

Building Design and Construction Handbook

Other McGraw-Hill Handbooks of Interest

Baumeister • Marks' Standard Handbook for Mechanical Engineers
Brady and Clauser • Materials Handbook
Brater • Handbook of Hydraulics
Callender • Time-Saver Standards for Architectural Design Data
Conover • Grounds Maintenance Handbook
Considine • Energy Technology Handbook
Crocker and King • Piping Handbook
Croft, Carr, and Watt • American Electricians' Handbook
Davis and Sorensen • Handbook of Applied Hydraulics
Emerick • Handbook of Mechanical Specifications for Buildings and Plants
Emerick • Heating Handbook
Emerick • Troubleshooters' Handbook for Mechanical Systems
Fink and Beaty • Standard Handbook for Electrical Engineers
Foster • Handbook of Municipal Administration and Engineering
Gieck • Engineering Formulas
Harris • Handbook of Noise Control
Harris and Crede • Shock and Vibration Handbook
Havers and Stubbs • Handbook of Heavy Construction
Heyel • The Foreman's Handbook
Hicks • Standard Handbook of Engineering Calculations
Higgins and Morrow • Maintenance Engineering Handbook
King and Brater • Handbook of Hydraulics
La Londe and Janes • Concrete Engineering Handbook
Leonards • Foundation Engineering
Lund • Industrial Pollution Control Handbook
Manas • National Plumbing Code Handbook
Mantell • Engineering Materials Handbook
Merritt • Standard Handbook for Civil Engineers
Merritt • Structural Steel Designers' Handbook
Myers • Handbook of Ocean and Underwater Engineering
O'Brien • Contractor's Management Handbook
Peckner and Bernstein • Handbook of Stainless Steels
Perry • Engineering Manual
Rossnagel • Handbook of Rigging
Smeaton • Switchgear and Control Handbook
Stanair • Plant Engineering Handbook
Streeter • Handbook of Fluid Dynamics
Timber Engineering Co. • Timber Design and Construction Handbook
Tuma • Engineering Mathematics Handbook
Tuma • Handbook of Physical Calculations
Tuma • Technology Mathematics Handbook
Urquhart • Civil Engineering Handbook
Waddell • Concrete Construction Handbook
Woods • Highway Engineering Handbook

Building Design and Construction Handbook

Fourth Edition

Frederick S. Merritt

Consulting Engineer
West Palm Beach, Fla.

McGraw-Hill Book Company

New York   St. Louis   San Francisco   Auckland
Bogotá   Hamburg   Johannesburg   London   Madrid
Mexico   Montreal   New Delhi   Panama   Paris
São Paulo   Singapore   Sydney   Tokyo   Toronto

Library of Congress Cataloging in Publication Data
Main entry under title:
Building design and construction handbook.

   Rev. ed. of Building construction handbook. 3rd ed. 1975.
   Includes bibliographies and index.
   1. Building—Handbooks, manuals, etc.
I. Merritt, Frederick S.
TH151.B825   1982   690   81-15602
ISBN 0-07-041521-8   AACR2

Copyright © 1982, 1975, 1965, 1958 by McGraw-Hill, Inc. All rights reserved. Printed in the United States of America. Except as permitted under the United States Copyright Act of 1976, no part of this publication may be reproduced or distributed in any form or by any means, or stored in a data base or retrieval system, without the prior written permission of the publisher.

    4567890   KPKP   89876

ISBN 0-07-041521-8

The editors for this book were Harold B. Crawford and Geraldine Fahey, the designer was Mark E. Safran, and the production supervisor was Sally Fliess. It was set in Century Schoolbook by University Graphics, Inc.

Printed and bound by The Kingsport Press.

Contents

Contributors xxiii
Preface xxv

1. Building Systems
by Frederick S. Merritt 1-1

 1-1 Principles of Architecture 1-1
 1-2 Systems 1-2
 1-3 Traditional Design Procedures 1-4
 1-4 Traditional Construction Procedures 1-5
 1-5 Role of the Client in Design and Construction 1-7
 1-6 Building Costs 1-8
 1-7 Major Building Systems 1-9
 1-8 Value Engineering 1-20
 1-9 Systems Design 1-26
 1-10 Building Codes 1-33
 1-11 Zoning Codes 1-35
 1-12 Other Regulations 1-36
 1-13 Systems Design by Team 1-37
 1-14 Application of Systems Design 1-37

2. Business, Art, and Profession of Architecture
by William E. Diamond II 2-1

 2-1 Business of the Architect 2-1
 2-2 Art of the Architect 2-3
 2-3 Profession of the Architect 2-4

Basic and Traditional Services 2-6

 2-4 Selection of Architect 2-6
 2-5 Building Program 2-7
 2-6 Design Stage 2-10
 2-7 Contract Documents 2-12
 2-8 Architect's Role during Construction 2-14
 2-9 Liability 2-16
 2-10 Architect and Consultants 2-17
 2-11 Professional Engineers 2-17
 2-12 Design by Team 2-17

Additional Services 2-18

 2-13 Nontraditional Services 2-18

Major Conditions Affecting Architectural Practice 2-19

 2-14 Economic Factors 2-19
 2-15 Changing Construction Market 2-19
 2-16 Systems 2-19
 2-17 Multiple-Contract Construction 2-21
 2-18 Accelerated Design and Construction 2-22
 2-19 Professional Construction Managers 2-23

Construction without Professional Services 2-24

 2-20 Package Builders 2-24
 2-21 Home Builders 2-24
 2-22 Design-and-Build Companies 2-24
 2-23 Turnkey Developers 2-25

3. Protection of Structures and Occupants against Hazards
by Frederick S. Merritt 3-1

- 3-1 Design Life of Buildings 3-2
- 3-2 Safety Factors 3-2

Wind Protection 3-2

- 3-3 Wind Characteristics 3-3
- 3-4 Wind Pressures and Suctions 3-3
- 3-5 Failure Modes 3-8
- 3-6 Limitation of Drift 3-9

Protection against Earthquakes 3-13

- 3-7 Earthquake Characteristics 3-13
- 3-8 Seismic Scales 3-14
- 3-9 Effects of Ground Conditions 3-14
- 3-10 Seismic Loads 3-15
- 3-11 Aseismic Design 3-16

Protection against Water 3-16

- 3-12 Floodproofing 3-16
- 3-13 Waterproofing 3-17

Protection against Fire 3-29

- 3-14 Fire-Protection Standards 3-29
- 3-15 Fire-Protection Concepts 3-31
- 3-16 Fire Loads and Resistance Ratings 3-31
- 3-17 Fire Barriers 3-32
- 3-18 Fire-Resistance Classification of Buildings 3-34
- 3-19 Extinguishment of Fires 3-35
- 3-20 Fire Detection 3-39
- 3-21 Smoke and Heat Venting 3-41
- 3-22 Emergency Egress 3-43
- 3-23 Building Operation in Emergencies 3-46
- 3-24 Safety during Construction 3-47

Lightning Protection 3-48

- 3-25 Characteristics of Lightning 3-48
- 3-26 Methods for Protecting against Lightning 3-49

Protection against Intruders 3-50

- 3-27 Security Measures 3-50

4. Building Materials
by Albert G. H. Dietz 4-1

Cementitious Materials 4-1

- 4-1 Portland Cements 4-2
- 4-2 Air-Entraining Portland Cements 4-3
- 4-3 Aluminous Cements 4-4
- 4-4 Natural Cements 4-4
- 4-5 Hydraulic Limes 4-4
- 4-6 Quicklimes 4-5
- 4-7 Mason's Hydrated Lime 4-5
- 4-8 Finishing Hydrated Limes 4-6
- 4-9 Low-Temperature Gypsum Derivatives 4-6
- 4-10 Oxychloride Cements 4-6
- 4-11 Masonry Cements 4-7

Aggregates 4-7

- 4-12 Coarse Aggregates 4-7
- 4-13 Fine Aggregates 4-8
- 4-14 Heavyweight Aggregates 4-8
- 4-15 Lightweight Aggregates 4-8

Mortars and Concretes 4-9

- 4-16 Mortars 4-9
- 4-17 Normal Concrete 4-10
- 4-18 Air-Entraining Concretes 4-13
- 4-19 Lightweight Concretes 4-13
- 4-20 Heavyweight Concretes 4-14
- 4-21 Concrete Masonry Units 4-14

Burned-Clay Units 4-16

- 4-22 Brick 4-16
- 4-23 Structural Clay Tile 4-17
- 4-24 Ceramic Tiles 4-17
- 4-25 Architectural Terra Cotta 4-17

Building Stones 4-18

- 4-26 Strength of Stone 4-19
- 4-27 Specific Gravity, Porosity, and Absorption of Stone 4-19
- 4-28 Permeability of Stone 4-19
- 4-29 Thermal Expansion of Stone 4-19
- 4-30 Freezing and Thawing of Stone 4-19

Gypsum Products 4-21

- 4-31 Gypsumboard 4-21
- 4-32 Gypsum Lath 4-22
- 4-33 Gypsum Sheathing Board 4-22

4-34 Gypsum Partition Tile or Block 4-22
4-35 Gypsum Plank 4-22

Glass and Glass Block 4-22

4-36 Glass for General Use 4-22
4-37 Glass Block 4-23

Asbestos-Cement Products 4-23

4-38 Properties of Asbestos-Cement 4-23

Wood 4-25

4-39 Mechanical Properties of Wood 4-25
4-40 Hygroscopic Properties of Wood 4-29
4-41 Commercial Grades of Wood 4-29
4-42 Destroyers and Preservatives 4-31
4-43 Glues and Adhesives 4-33
4-44 Plywood and Other Fabricated Wood Boards 4-34
4-45 Wood Bibliography 4-35

Ferrous Metals 4-35

4-46 Steels 4-36
4-47 Irons 4-36
4-48 Standard Steels 4-36
4-49 Heat Treatment and Hardening of Steels 4-37
4-50 Steel Alloys 4-40
4-51 Stainless Steels 4-41
4-52 Welding Ferrous Metals 4-45
4-53 Corrosion of Iron and Steel 4-46
4-54 Ferrous Metals Bibliography 4-47

Aluminum and Aluminum-Based Alloys 4-47

4-55 Aluminum-Alloy Designations 4-48
4-56 Clad Aluminum 4-48
4-57 Corrosion Resistance and Corrosion Protection of Aluminum 4-48
4-58 Finishes for Aluminum 4-51
4-59 Welding and Brazing of Aluminum 4-51
4-60 Aluminum Bibliography 4-52

Copper and Copper-Based Alloys 4-52

4-61 Copper 4-52
4-62 Plain Brass 4-56
4-63 Leaded Brass 4-56
4-64 Tin Brass 4-57
4-65 Nickel Silvers 4-57
4-66 Cupronickel 4-57
4-67 Silicon Bronze 4-57
4-68 Aluminum Bronze 4-58
4-69 Tin Bronze 4-58
4-70 Copper Bibliography 4-58

Lead and Lead-Based Alloys 4-58

4-71 Applications of Lead 4-58
4-72 Lead Bibliography 4-59

Nickel and Nickel-Based Alloys 4-59

4-73 Properties of Nickel and Its Alloys 4-59
4-74 Nickel Bibliography 4-60

Plastics 4-60

4-75 General Properties of Plastics 4-60
4-76 Fillers and Plasticizers 4-61
4-77 Molding and Fabricating Methods for Plastics 4-61
4-78 Thermosetting Plastics 4-62
4-79 Thermoplastic Resins 4-64
4-80 Cellulose Derivatives 4-66

Combinations of Plastics and Other Materials 4-67

4-81 High-Pressure Laminates 4-67
4-82 Reinforced Plastics 4-67
4-83 Polymeric Mortar, Concrete, and Grout 4-68
4-84 Plastics Bibliography 4-70

Porcelain-Enameled Products 4-70

4-85 Porcelain Enamel on Metal 4-70
4-86 Porcelain Bibliography 4-71

Rubber 4-71

4-87 Synthetic Rubbers 4-71
4-88 Properties of Rubber 4-72
4-89 Laminated Rubber 4-73
4-90 Rubber Bibliography 4-74

Asphalt and Bituminous Products 4-74

4-91 Dampproofing and Waterproofing Asphalts 4-74
4-92 Roofing Asphalts and Pitches 4-74
4-93 Roofing Felts 4-74
4-94 Roll Roofing 4-75
4-95 Asphalt Shingles 4-75
4-96 Asphalt Mastics and Grouts 4-75
4-97 Bituminous Pavements 4-75
4-98 Asphalt Bibliography 4-75

Joint Seals 4-75

4-99 Calking Compounds 4-76
4-100 Sealants 4-76
4-101 Gaskets 4-77
4-102 Joint Seals Bibliography 4-77

Coatings 4-77

4-103 Vehicles or Binders 4-78
4-104 Pigments for Paints 4-79
4-105 Resins for Paints 4-79
4-106 Coatings Bibliography 4-80

5. Structural Theory
by Frederick S. Merritt 5-1

Stress and Strain 5-2

5-1 Types of Loads 5-2
5-2 Design Loads 5-3
5-3 Static Equilibrium 5-11
5-4 Unit Stress and Strain 5-12
5-5 Hooke's Law 5-13
5-6 Constant Unit Stress 5-13
5-7 Poisson's Ratio 5-15
5-8 Thermal Stresses 5-15
5-9 Strain Energy 5-15

Stresses at a Point 5-17

5-10 Stress Notation 5-17
5-11 Stress Components 5-17
5-12 Two-Dimensional Stress 5-18
5-13 Principal Stresses 5-18
5-14 Maximum Shearing Stress 5-19
5-15 Mohr's Circle 5-19

Straight Beams 5-20

5-16 Types of Beams 5-20
5-17 Reactions 5-21
5-18 Internal Forces 5-21
5-19 Shear Diagrams 5-22
5-20 Bending-Moment Diagrams 5-23
5-21 Moments in Uniformly Loaded Beams 5-23
5-22 Shear-Moment Relationship 5-25
5-23 Moving Loads and Influence Lines 5-25
5-24 Maximum Bending Moment 5-26
5-25 Bending Stresses in a Beam 5-27
5-26 Moment of Inertia 5-27
5-27 Section Modulus 5-29
5-28 Shearing Stresses in a Beam 5-29
5-29 Combined Shear and Bending Stress 5-30
5-30 Stresses in the Plastic Region 5-30
5-31 Beam Deflections 5-30
5-32 Combined Axial and Bending Loads 5-33
5-33 Eccentric Loading 5-33
5-34 Unsymmetrical Bending 5-37
5-35 Beams with Unsymmetrical Sections 5-38

Curved Beams 5-39

5-36 Stresses in Curved Beams 5-39
5-37 Curved Beams with Rectangular Sections 5-40
5-38 Curved Beams with Circular Sections 5-40
5-39 Curved I or T Sections 5-40
5-40 Axial and Bending Loads on Curved Beams 5-41
5-41 Slope and Deflection of Curved Beams 5-42

Graphic-Statics Fundamentals 5-42

5-42 Parallelogram of Forces 5-42
5-43 Resolution of Forces 5-43
5-44 Force Polygons 5-43
5-45 Equilibrium Polygons 5-43
5-46 Beam and Truss Reactions by Graphics 5-44
5-47 Shear and Moment Diagrams by Graphics 5-45

Roof Trusses 5-46

5-48 Method of Sections 5-47
5-49 Method of Joints 5-48
5-50 Bow's Notation 5-49
5-51 Graphical Analysis of Trusses 5-49

General Tools for Structural Analysis 5-50

5-52 Virtual Work 5-50
5-53 Castigliano's Theorems 5-52
5-54 Method of Least Work 5-52
5-55 Dummy Unit-Load Method 5-53
5-56 Truss Deflections 5-54
5-57 Reciprocal Theorem 5-56

Continuous Beams and Frames 5-57

5-58 General Method of Analysis for Continuous Members 5-57
5-59 Sign Convention 5-59
5-60 Fixed-End Moments 5-59
5-61 Fixed-End Stiffness 5-61
5-62 Fixed-End Carry-Over Factor 5-62
5-63 Moment Distribution by Converging Approximations 5-62
5-64 Continuous Frames 5-64
5-65 Moment-Influence Factors 5-66
5-66 Deflection of Supports 5-66
5-67 Procedure for Sidesway 5-67
5-68 Single-Cycle Moment Distribution 5-70
5-69 Method for Checking Moment Distribution 5-74
5-70 Plane Grid Frameworks 5-76

Buckling of Columns 5-82

5-71 Stable Equilibrium 5-83
5-72 Unstable Equilibrium 5-83
5-73 Effect of End Conditions 5-84
5-74 Inelastic Buckling 5-84
5-75 Column Curves 5-84
5-76 Local Buckling 5-85
5-77 Behavior of Actual Columns 5-85

Torsion 5-86

5-78 Shear Center 5-86
5-79 Stresses due to Torsion 5-86

Structural Dynamics 5-88

5-80 Properties of Materials under Dynamic Loading 5-88
5-81 Natural Period of Vibration 5-89
5-82 Impact and Sudden Loads 5-95
5-83 Dynamic Analysis of Simple Structures 5-98
5-84 Resonance and Damping 5-104
5-85 Approximate Design for Dynamic Loading 5-106
5-86 Approximate Earthquake Analysis 5-108
5-87 Reduction of Vibration 5-114

Stresses in Arches 5-115

5-88 Three-Hinged Arches 5-116
5-89 Two-Hinged Arches 5-116
5-90 Fixed Arches 5-118
5-91 Stresses in Arch Ribs 5-120

Membrane and Thin-Shell Structures 5-121

5-92 Thin-Shell Analysis 5-121
5-93 Membrane Theory for Thin Shells 5-121
5-94 Membrane Forces in Shells of General Shape 5-123
5-95 Membrane Forces in a Hyperbolic Paraboloid Shell 5-124
5-96 Elliptical-Paraboloid Membrane Forces 5-125
5-97 Membrane Forces in Shells of Revolution 5-125
5-98 Bending Theory for Thin Shells 5-128
5-99 Stresses in Thin Shells 5-129

Air-Stabilized Structures 5-130

5-100 Pneumatic Construction 5-130
5-101 Membrane Stresses 5-132

Folded Plates 5-132

5-102 Folded-Plate Theory 5-133
5-103 Yitzhaki Method for Folded Plates 5-133

Cable-Supported Structures 5-140

5-104 Simple Cables 5-141
5-105 Cable Systems 5-146

x   Contents

Rapid Approximate Analysis of
Multistory Frames   5-153

5-106   Portal Method   5-154
5-107   Cantilever Method   5-155

Ultimate Strength of Ductile
Flexible Members   5-155

5-108   Theory of Plastic Behavior   5-156
5-109   Upper and Lower Bounds for Ultimate Loads   5-157

Finite-Element Methods   5-160

5-110   Force-Displacement Relations   5-161
5-111   Matrix Force (Flexibility) Method   5-162
5-112   Matrix Displacement (Stiffness) Method   5-164
5-113   Element Flexibility and Stiffness Matrices   5-166

6. Soil Mechanics and Foundations
by Charles P. Gupton   6-1

6-1   Physical Properties of Soils   6-1
6-2   States of Matter Affecting Soil Behavior   6-2
6-3   Soil Moisture   6-3
6-4   Resistance of Soils to Pressure   6-3
6-5   Soil Properties Important in Engineering   6-4

Identification, Sampling, and
Testing of Soils   6-5

6-6   Soil Identification   6-5
6-7   Subsurface Exploration   6-10
6-8   Laboratory Tests   6-15

Soil Improvement   6-15

6-9   Man-Made Fills   6-16
6-10   Densification   6-18
6-11   Surcharges   6-19
6-12   Grouting   6-21
6-13   Drainage   6-21

Foundation Design   6-21

6-14   Steps in Foundation Design   6-21
6-15   Bearing and Load Capacity of Foundations   6-22

6-16   Settlement of Foundations   6-25
6-17   Spread Footings   6-30
6-18   Combined Foundations and Mats   6-36
6-19   Pile Footings   6-37
6-20   Load Capacity of Piles   6-42
6-21   Installation of Piles   6-45
6-22   Open Shafts for Foundations   6-46
6-23   Slurry Trench Walls   6-47
6-24   Drilled Shafts   6-47

Earth Pressure on Walls   6-49

6-25   Active Earth Pressure   6-49
6-26   Passive Earth Pressure   6-51
6-27   At-Rest Pressure   6-52
6-28   Pressures on Underground Structures   6-52

Sheeting and Bracing of
Excavations   6-52

6-29   Sheeting Design Pressures   6-52
6-30   Bracing of Sheetpiles   6-53
6-31   Anchorage of Sheeting   6-54
6-32   Sheeting Removal   6-55

Causes of Foundation Failures   6-56

6-33   Failures due to Heaving   6-56
6-34   Settlement Failures   6-56
6-35   Bibliography   6-57

7. Wood Construction
by Maurice J. Rhude   7-1

7-1   Basic Characteristics and How to Use Them   7-2
7-2   Sectional Properties of Lumber and Timber   7-3
7-3   Basic and Allowable Stresses for Timber   7-3
7-4   Structural Grading of Wood   7-9
7-5   Moisture Content of Wood   7-14
7-6   Checking in Timbers   7-15
7-7   Allowable Unit Stresses and Modifications for Structural Lumber   7-18
7-8   Lateral Support of Wood Framing   7-22
7-9   Combined Stresses in Timber   7-24

7-10	Effect of Shrinkage or Swelling on Slope of Curved Members 7-25	8-13	Open-Web Joist Floors 8-18
		8-14	Cellular-Steel Floors 8-19
		8-15	Composite Concrete-Steel Beams 8-20
7-11	Fabrication of Structural Timber 7-25		
		8-16	Effect of Intermediate Beams on Costs 8-20
7-12	Fabrication of Glued-Laminated Timbers 7-26		
		8-17	Other Floor Systems 8-20
7-13	Preservative Treatments for Wood 7-28	8-18	Roof Systems 8-21

7-14 Resistance of Wood to Chemical Attack 7-30
7-15 Designing for Fire Safety 7-30
7-16 Mechanical Fastenings 7-32
7-17 Glued Fastenings 7-46
7-18 Wood Columns 7-48
7-19 Design of Timber Joists 7-50
7-20 Design of Timber Beams 7-51
7-21 Deflection and Camber of Timber Beams 7-56
7-22 Minimum Roof Slopes 7-56
7-23 Design of Wood Trusses 7-57
7-24 Design of Timber Arches 7-59
7-25 Wood Framing for Small Houses 7-61
7-26 Wood Foundations 7-64
7-27 Timber Decking 7-66
7-28 Plywood 7-68
7-29 Composite Construction 7-82
7-30 Pole Construction 7-83
7-31 Timber Erection 7-84

8. Structural-Steel Construction
by Henry J. Stetina 8-1

8-1 Structural-Steel Shapes 8-1
8-2 Tolerances for Structural Shapes 8-2
8-3 Cambered Beams 8-2
8-4 In-Service Steel 8-3
8-5 Structural-Steel Specifications 8-3

Steel-Framing Systems 8-5

8-6 Wall-Bearing Framing 8-5
8-7 Skeleton Framing 8-7
8-8 Long-Span Steel Framing 8-8
8-9 Steel and Concrete Framing 8-15

Floor and Roof Systems 8-16

8-10 Factors Affecting Floor Design 8-16
8-11 Concrete-Arch Floors 8-17
8-12 Concrete-Pan System 8-17

Allowable Design Stresses 8-21

8-19 Allowable Tension in Steel 8-21
8-20 Allowable Shear in Steel 8-22
8-21 Allowable Compression in Steel 8-23
8-22 Effective Column Length 8-24
8-23 Web Crippling 8-26
8-24 Pipe and Tubular Columns 8-28
8-25 Bending 8-28
8-26 Bearing 8-33
8-27 Combined Axial Compression and Bending 8-34
8-28 Combined Axial Tension and Bending 8-35
8-29 Design of Beam Sections for Torsion 8-35
8-30 Wind and Seismic Stresses 8-36
8-31 Cyclic Loaded Members 8-36
8-32 Stresses in Welds 8-37
8-33 Stresses for Rivets and Bolts 8-39
8-34 Combined Shear and Tension on Rivets and Bolts 8-40
8-35 Allowable Stresses for Composite Design 8-41
8-36 Criteria for Plate Girders 8-45
8-37 Design of Trusses 8-49

Dimensional and Deflection Limitations 8-50

8-38 Slenderness Ratios 8-50
8-39 Width-Thickness Ratios 8-50
8-40 Limits on Deflections 8-50

Plastic Design of Steel 8-54

8-41 Plastic-Design Criteria 8-54
8-42 Comparison of Elastic and Plastic Design Methods 8-57

Bracing 8-58

- 8-43 Column Bracing 8-58
- 8-44 Beam Bracing 8-59
- 8-45 Capacity of Bracing 8-61
- 8-46 Lateral Forces on Building Frames 8-61
- 8-47 Bracing Tall Buildings 8-62
- 8-48 Shear Walls 8-64
- 8-49 Bracing Industrial-Type Buildings 8-65
- 8-50 Bracing Crane Structures 8-67
- 8-51 Design of X Bracing 8-67
- 8-52 Bracing Rigid Frames 8-67

Fasteners 8-67

- 8-53 Hot-Driven Rivets 8-68
- 8-54 Unfinished Bolts 8-68
- 8-55 High-Strength Bolts 8-69
- 8-56 Other Bolt-Type Fasteners 8-71
- 8-57 Bolt Holes 8-71
- 8-58 Welding 8-72
- 8-59 Combinations of Fasteners 8-74
- 8-60 Fastener Symbols 8-75
- 8-61 Erection Clearances for Fasteners 8-76

Connections 8-77

- 8-62 Stresses in Bearing-Type Connections 8-77
- 8-63 Stresses in Friction-Type Connections 8-81
- 8-64 Stresses in Welded Connections 8-82
- 8-65 Types of Beam Connections 8-84
- 8-66 Bolted Framed Connections 8-84
- 8-67 Bolted Seated Connections 8-85
- 8-68 Welded Framed Connections 8-86
- 8-69 Welded-Seat Connections 8-87
- 8-70 End-Plate Connections 8-87
- 8-71 Special Connections 8-89
- 8-72 Simple, Rigid and Semirigid Connections 8-90
- 8-73 Column Splices 8-93
- 8-74 Beam Splices 8-96

Steel Erection 8-96

- 8-75 Erection Equipment 8-97
- 8-76 Clearance for Erecting Beams 8-97
- 8-77 Erection Sequence 8-99
- 8-78 Field-Welding Procedures 8-99
- 8-79 Field Tolerances 8-100
- 8-80 Adjusting Lintels 8-102

Painting 8-103

- 8-81 Corrosion of Steel 8-103
- 8-82 Painting Steel Structures 8-104
- 8-83 Paint Systems 8-105
- 8-84 Field Painting Steel 8-105
- 8-85 Steel in Contact with Concrete 8-106

Fire Protection for Structural Steel 8-107

- 8-86 Need for Fire Protection 8-107
- 8-87 Effect of Heat on Steel 8-107
- 8-88 Fire Protection of Exterior Members 8-108
- 8-89 Materials for Improving Fire Resistance 8-108
- 8-90 Pierced Ceilings and Floors 8-111
- 8-91 Fire-Resistance Ratings 8-112
- 8-92 Bibliography 8-113

9. Cold-Formed Steel Construction
by Don S. Wolford 9-1

Open-Web Steel Joists 9-1

- 9-1 Joist Design and Fabrication 9-2
- 9-2 Design of Open-Web Floor Joists 9-4
- 9-3 Extended Ends and Ceiling Extensions 9-5
- 9-4 Bridging and Anchors for Open-Web Joists 9-6
- 9-5 Formwork for Joist Construction 9-6
- 9-6 Slab Reinforcement in Steel-Joist Construction 9-6
- 9-7 Open-Web Joists in Roof Construction 9-6
- 9-8 Fire Resistance of Open-Web Joist Construction 9-7

Cold-Formed Shapes 9-7

9-9 Material for Cold-Formed Shapes 9-8
9-10 Utilization of the Cold Work of Forming 9-12
9-11 Types of Cold-Formed Sections 9-13
9-12 Design Principles for Cold-Formed Sections 9-14
9-13 Structural Behavior of Flat Compression Elements 9-16
9-14 Unit Stresses for Cold-Formed Shapes 9-20
9-15 Design of Unstiffened Elements Subject to Compression 9-21
9-16 Stiffened Elements Subject to Compression 9-23
9-17 Light-Gage Members with Unusually Wide Flanges on Short Spans 9-27
9-18 Maximum Flat-Width Ratios for Light-Gage Elements 9-27
9-19 Laterally Unsupported Light-Gage Beams 9-28
9-20 Web Stresses in Cold-Formed Shapes 9-29
9-21 Cold-Formed Steel Columns 9-31
9-22 Combined Axial and Bending Stresses 9-38
9-23 Cylindrical Tubular Members 9-38
9-24 Corrugated Sheets 9-39
9-25 Stainless Steel Structural Design 9-41
9-26 Electric-Arc Welding of Cold-Formed Steel 9-42
9-27 Electric Resistance Welding of Cold-Formed Steel 9-45
9-28 Maximum Spacing of Welds 9-47
9-29 Bolting Cold-Formed Members 9-49
9-30 Riveting of Cold-Formed Shapes 9-51
9-31 Self-Tapping Screws for Joining Light-Gage Components 9-52
9-32 Special Fasteners for Light-Gage Steel 9-53

Steel Roof Deck 9-53

9-33 Types of Steel Roof Deck 9-54
9-34 Materials for Steel Roof Deck 9-54
9-35 Load-Carrying Capacity of Steel Roof Deck 9-58
9-36 Details and Accessories for Steel Roof Deck 9-58
9-37 Insulation of Steel Roof Deck 9-59
9-38 Fire Resistance of Steel Roof Deck 9-59
9-39 Other Forms of Steel Roof Decking 9-59

Cellular Steel Floors 9-61

9-40 Advantages of Cellular Decking 9-61
9-41 Types of Cellular Floor Units 9-62
9-42 Materials for Cellular Decking 9-63
9-43 Structural Design of Cellular Floors 9-64
9-44 Details and Accessories for Cellular Steel Floors 9-64
9-45 Fire Resistance of Cellular Floor Construction 9-65

Other Forms of Lightweight Steel Construction 9-66

9-46 Building Systems 9-66

10. Concrete Construction *by Paul F. Rice and Edward S. Hoffman* 10-1

Concrete and Its Ingredients 10-1

10-1 Cements 10-1
10-2 Aggregates 10-3
10-3 Proportioning Concrete Mixes 10-3
10-4 Yield Calculations 10-4
10-5 Properties and Tests of Fresh (Plastic) Concrete 10-5
10-6 Properties and Tests of Hardened Concrete 10-6
10-7 Measuring and Mixing Concrete Ingredients 10-9
10-8 Admixtures 10-10

Quality Control 10-12

10-9 Mix Design 10-12
10-10 Check Tests of Materials 10-14
10-11 At the Mixing Plant—Yield Adjustments 10-14

xiv   Contents

10-12   At the Placing Point—Slump Adjustments   10-15
10-13   Strength Tests   10-15
10-14   Test Evaluation   10-17

Formwork   10-18

10-15   Responsibility for Formwork   10-18
10-16   Materials and Accessories for Forms   10-19
10-17   Pressure of Fresh Concrete on Vertical Forms   10-19
10-18   Design Loads for Horizontal Forms   10-19
10-19   Lateral Bracing for Shoring   10-20
10-20   Form Removal and Reshoring   10-21
10-21   Special Forms   10-21
10-22   Inspection of Formwork   10-21

Reinforcement   10-22

10-23   Reinforcing Bars   10-22
10-24   Welded-Wire Fabric   10-24
10-25   Prestressing Steel   10-25
10-26   Fabrication and Placing of Rebars   10-25
10-27   Bar Supports   10-27
10-28   Inspection of Reinforcement   10-27

Concrete Placement   10-28

10-29   Methods of Placing   10-28
10-30   Excess Water   10-29
10-31   Consolidation   10-29
10-32   Concreting Vertical Elements   10-30
10-33   Concreting Horizontal Elements   10-30
10-34   Bonding to Hardened Concrete   10-31
10-35   Heavy-Duty Floor Finishes   10-31
10-36   Concreting in Cold Weather   10-32
10-37   Concreting in Hot Weather   10-32
10-38   Curing Concrete   10-32
10-39   Joints in Concrete   10-33
10-40   Inspection of Concrete Placement   10-35

Analysis of Concrete Structures   10-35

10-41   Analyses of One-Way Floor and Roof Systems   10-35
10-42   Two-Way Slab Frames   10-37
10-43   Special Analyses   10-37

Structural Design of Flexural Members   10-38

10-44   Strength Design with Load Factors   10-38
10-45   Allowable-Stress Design at Service Loads (Alternate Design Method)   10-39
10-46   Strength Design for Flexure   10-41
10-47   Shear in Flexural Members   10-45
10-48   Torsion in Concrete Members   10-46
10-49   Development, Anchorage, and Splices of Reinforcement   10-48
10-50   Crack Control   10-55
10-51   Deflection of Concrete Beams and Slabs   10-57

One-Way Reinforced-Concrete Slabs   10-60

10-52   Analysis and Design of One-Way Slabs   10-60
10-53   Embedded Pipes in One-Way Slabs   10-62

One-Way Concrete-Joist Construction   10-62

10-54   Standard Sizes of Joists   10-63
10-55   Design of Joist Construction   10-64
10-56   Reinforcement of Joists for Flexure   10-65
10-57   Shear in Joists   10-66

Two-Way Construction   10-67

10-58   Analysis and Design of Flat Plates   10-67
10-59   Stiffness in Two-Way Construction   10-70
10-60   Bar Lengths and Details for Flat Plates   10-71
10-61   Flat Slabs   10-72

10-62 Two-Way Slabs on Beams 10-75
10-63 Estimating Guide for Two-Way Construction 10-75

Beams 10-76

10-64 Definitions of Flexural Members 10-76
10-65 Flexural Reinforcement 10-76
10-66 Reinforcement for Shear and Flexure 10-79
10-67 Reinforcement for Torsion and Shear 10-80
10-68 Crack Control in Beams 10-82

Walls 10-82

10-69 Bearings Walls 10-82
10-70 Nonbearing Walls 10-84
10-71 Cantilever Retaining Walls 10-85
10-72 Counterfort Retaining Walls 10-86
10-73 Retaining Walls Supported on Four Sides 10-87

Foundations 10-88

10-74 Types of Foundations 10-88
10-75 General Design Principles for Footings 10-89
10-76 Spread Footings for Walls 10-91
10-77 Spread Footings for Individual Columns 10-92
10-78 Combined Spread Footings 10-93
10-79 Strap Footings 10-95
10-80 Mat Foundations 10-96
10-81 Pile Foundations 10-97
10-82 Drilled-Pier Foundations 10-97

Columns 10-99

10-83 Basic Assumptions for Strength Design of Columns 10-99
10-84 Design Requirements for Columns 10-102
10-85 Design for Minimum Bending Moment 10-103
10-86 Column Ties and Tie Patterns 10-104
10-87 Biaxial Bending of Columns 10-106
10-88 Slenderness Effects on Concrete Columns 10-107
10-89 Economy in Column Design 10-108

Special Construction 10-109

10-90 Deep Beams 10-109
10-91 Shear Walls 10-110
10-92 Reinforced Concrete Arches 10-112
10-93 Reinforced Concrete Thin Shells 10-113
10-94 Concrete Folded Plates 10-115
10-95 Slabs on Grade 10-115
10-96 Seismic-Resistant Concrete Construction 10-116
10-97 Composite Flexural Members 10-117

Precast Concrete Members 10-118

10-98 Design Methods for Precast Members 10-118
10-99 Reinforcement Cover in Precast Members 10-118
10-100 Tolerances for Precast Construction 10-119
10-101 Accelerated Curing 10-119
10-102 Precast Floor and Roof Systems 10-119
10-103 Precast Ribbed Slabs, Folded Plates, and Shells 10-119
10-104 Wall Panels 10-120
10-105 Lift-Slabs 10-121

Prestressed Concrete Construction 10-122

10-106 Basic Principles of Prestressed Concrete 10-122
10-107 Losses in Prestress 10-123
10-108 Allowable Stresses at Service Loads 10-124
10-109 Design Procedure for Prestressed-Concrete Beams 10-125
10-110 Flexural-Strength Design of Prestressed Concrete 10-126
10-111 Shear-Strength Design of Prestressed Concrete 10-127
10-112 Bond, Development, and Grouting of Tendons 10-129
10-113 Application and Measurement of Prestress 10-130
10-114 Concrete Cover in Prestressed Members 10-130

11. Engineered Brick Construction
by Alan H. Yorkdale 11-1

11-1 Design Standards 11-1
11-2 Definitions of Design Terms 11-2
11-3 Materials for Engineered Brick Construction 11-3
11-4 Determination of Masonry Compression Strength 11-3
11-5 Allowable Stresses for Brick Construction 11-4
11-6 Allowable Loads for Walls and Columns 11-6
11-7 Shear-Wall Action 11-7
11-8 Diaphragms 11-9
11-9 Torque Distribution to Shear Walls 11-11
11-10 Example of Torque Distribution to Shear Walls 11-11
11-11 Relative Deflections of Shear Walls 11-12
11-12 Diaphragm-Deflection Limitations 11-13
11-13 Shear-Wall Rigidity 11-14
11-14 Effects of Shear-Wall Arrangements 11-16
11-15 Coupled Shear Walls 11-17
11-16 Eccentrically Loaded Shear Walls 11-18
11-17 Allowable Stresses for Shear Walls 11-18
11-18 Bearing Walls 11-19
11-19 Floor-Wall Connections 11-19
11-20 Example—End Shear Wall, Concrete Joist Floors 11-20
11-21 Example—End Shear Wall, Concrete Plank Floors 11-26
11-22 Example—Interior Shear Wall, Concrete Joist Floors 11-27
11-23 Example—Interior Shear Wall, Concrete Plank Floors 11-32
11-24 Bibliography 11-33

12. Walls, Partitions, and Interior Finishes
by Frederick S. Merritt 12-1

Masonry Walls 12-1

12-1 Masonry Definitions 12-2
12-2 Quality of Materials for Masonry 12-4
12-3 Mortar 12-5
12-4 Allowable Stresses in Masonry 12-6
12-5 Construction of Masonry 12-6
12-6 Bond between Wythes in Masonry Walls 12-9
12-7 Grouted Masonry 12-11
12-8 Flashing in Masonry Walls 12-12
12-9 Repair of Leaky Joints 12-13
12-10 Lateral Support for Masonry Walls 12-14
12-11 Masonry-Thickness Requirements 12-15
12-12 Corbeling 12-18
12-13 Provisions for Dimensional Changes after Construction 12-18
12-14 Glass Block 12-20
12-15 Openings, Chases, and Recesses in Masonry Walls 12-21
12-16 Support Conditions for Walls 12-22
12-17 Chimneys and Fireplaces 12-22

Stud Walls 12-23

12-18 Stud-Wall Construction 12-23
12-19 Sheathing 12-24

Curtain Walls 12-25

12-20 Functional Requirements for Curtain Walls 12-25
12-21 Wood Facades 12-26
12-22 Wall Shingles 12-26
12-23 Stucco 12-26
12-24 Metal, Plastic, and Asbestos-Cement Siding 12-27
12-25 Precast-Concrete or Metal and Glass Facings 12-28
12-26 Sandwich Panels 12-28

Partitions 12-30

12-27 Types of Partitions 12-30
12-28 Structural Requirements of Partitions 12-31

Ceramic-Tile Construction 12-31

12-29 Types of Ceramic Tile 12-32
12-30 Tile Installation Methods 12-32

Panel Finishes 12-36

12-31 Plywood Finishes 12-36
12-32 Asbestos-Cement Boards 12-36
12-33 Fiber and Pulp Boards 12-36

Plaster and Gypsumboard 12-36

12-34	Plaster and Gypsumboard Construction Terms	12-37
12-35	Plaster Finishes	12-43
12-36	Gypsum Bases for Plaster	12-45
12-37	Metal Lath	12-46
12-38	Masonry Bases for Plaster	12-48
12-39	Plaster Base Coats	12-49
12-40	Finish Plaster Coats	12-50
12-41	Gypsumboard Finishes	12-51
12-42	General Application Procedures for Gypsumboard	12-53
12-43	Single-Ply Gypsumboard Construction	12-55
12-44	Multi-Ply Gypsumboard Construction	12-57
12-45	Finishing Procedures for Gypsumboard	12-59
12-46	Isolation and Control Joints	12-60

Floor Coverings 12-63

12-47	Asphalt Tiles	12-63
12-48	Cork Tiles	12-63
12-49	Vinyl Flooring	12-64
12-50	Vinyl-Asbestos Tiles	12-64
12-51	Backed Vinyl	12-64
12-52	Rubber	12-65
12-53	Linoleum	12-65
12-54	Installation of Thin Coverings	12-65
12-55	Carpets	12-66
12-56	Terrazzo	12-68
12-57	Concrete Floors	12-69
12-58	Wood Floors	12-69
12-59	Industrial Floors	12-69
12-60	Conductive Flooring	12-70
12-61	Specifications and Standards for Flooring	12-71

13. Windows
by Philip M. Grennan 13-1

13-1	How Much Window Area?	13-1
13-2	Window Materials	13-2
13-3	Wood Windows	13-2
13-4	Definitions for Wood Windows	13-2
13-5	Steel Windows	13-5
13-6	Aluminum Windows	13-5
13-7	Stainless-Steel and Bronze Windows	13-6
13-8	Window Sizes	13-6
13-9	Window Types	13-6
13-10	Windows in Wall-Panel Construction	13-14
13-11	Mechanical Operators for Windows	13-14
13-12	Weather Stripping and Storm Sash	13-15
13-13	Window Glass	13-15
13-14	Glazing Compounds	13-19
13-15	Factory-Sealed Double Glazing	13-20
13-16	Gaskets for Glazing	13-21

14. Doors *by Frederick S. Merritt* 14-1

Ordinary Doors 14-1

14-1	Exterior Doors	14-2
14-2	Traffic Flow and Safety	14-3
14-3	Fire and Smokestop Doors	14-4
14-4	Door Materials	14-6
14-5	Structural Requirements for Openings and Doors	14-7
14-6	Swinging Doors	14-8
14-7	Ordinary Horizontally Sliding Doors	14-10
14-8	Ordinary Vertically Sliding Doors	14-10
14-9	Revolving Doors	14-11

Special-Purpose Doors 14-12

14-10	Large Horizontally Sliding Doors	14-13
14-11	Large Vertically Sliding Doors	14-14
14-12	Large Swinging Doors	14-14
14-13	Horizontally Hinged Doors	14-14
14-14	Radiation-Shielding Doors	14-15

15. Roof Coverings
by J. C. Gudas 15-1

15-1	Roof Slopes	15-1
15-2	Bituminous and Plastic Roofing Materials	15-2
15-3	Types of Roof Coverings	15-3
15-4	Selection of Roof Coverings	15-4

Single-Unit Coverings 15-5

15-5 Built-Up Roofs 15-5
15-6 Dependence of Built-Up Roofing on Roof Decks 15-6
15-7 Types of Roof Decks 15-7
15-8 Vapor Barriers for Built-Up Roofing 15-8
15-9 Insulation with Built-Up Roofing 15-9
15-10 Built-Up Roofing Failures 15-10
15-11 Temporary Roof Coverings 15-10
15-12 Good-Practice Recommendations for Built-Up Roofing 15-11
15-13 Inverted Built-Up Roofing System 15-12
15-14 Foamed-in-Place Polyurethane Roof System 15-13
15-15 Metal Roof Coverings 15-13
15-16 Cold-Process Roof-Covering Systems 15-15
15-17 Elastomeric Roof Coverings 15-16
15-18 Single-Ply Roofing Systems 15-17
15-19 Corrugated Asbestos-Cement Sheets 15-19
15-20 Rigid Vinyl Roofing 15-20
15-21 Venting Lightweight-Concrete Roofs 15-20
15-22 Roof Moisture-Detection Surveys 15-21
15-23 Air- and Tension-Supported Roofs 15-21

Multiunit Roof Coverings 15-22

15-24 Asphalt Shingles 15-22
15-25 Wood Shingles 15-24
15-26 Asbestos-Cement Shingles 15-25
15-27 Roofing Tiles 15-26
15-28 Roofing Slate 15-27
15-29 Roof Finishing Details 15-27

16. Builders' Hardware
by Richard A. Hudnut 16-1

16-1 Finishing and Rough Hardware 16-1
16-2 When to Select Hardware 16-2
16-3 Effects of Codes and Regulations on Hardware 16-2
16-4 Function of Hardware Consultants 16-2
16-5 Factors Affecting Selection of Hardware 16-3
16-6 Template and Nontemplate Hardware 16-3

Finishing Hardware 16-3

16-7 Applicable Standards for Finishing Hardware 16-3
16-8 Hinges and Butts 16-4
16-9 How to Select Hinges 16-6
16-10 Door-Closing Devices 16-8
16-11 Locks and Latches 16-10
16-12 Hand of Doors 16-13
16-13 Casement Sash Operators 16-14
16-14 Sliding Hardware 16-14
16-15 Folding-Door Hardware 16-14
16-16 Locking Bolts (Doors and Windows) 16-14

Rough Hardware 16-14

16-17 Window Hardware 16-14
16-18 Inserts, Anchors, and Hangers 16-15
16-19 Nails 16-16
16-20 Screws 16-18
16-21 Welded Studs 16-18
16-22 Powder-Driven Studs 16-20
16-23 Miscellaneous Anchoring Devices 16-21
16-24 Bolts 16-21

17. Acoustics *by Lyle F. Yerges* 17-1

17-1 Sound Production and Transmission 17-1
17-2 Nomenclature for Analysis of Sound 17-2
17-3 Elastic Media 17-3
17-4 Hearing 17-4
17-5 Force and Energy Concepts 17-6
17-6 Units and Dimensions for Sound Measurements 17-6
17-7 Sound Measurements and Measurement Scales 17-7
17-8 Sound and Vibration Control 17-8
17-9 Airborne Sound Transmission 17-8
17-10 Bypassing of Sound Barriers 17-11

Plaster and Gypsumboard 12-36

12-34 Plaster and Gypsumboard Construction Terms 12-37
12-35 Plaster Finishes 12-43
12-36 Gypsum Bases for Plaster 12-45
12-37 Metal Lath 12-46
12-38 Masonry Bases for Plaster 12-48
12-39 Plaster Base Coats 12-49
12-40 Finish Plaster Coats 12-50
12-41 Gypsumboard Finishes 12-51
12-42 General Application Procedures for Gypsumboard 12-53
12-43 Single-Ply Gypsumboard Construction 12-55
12-44 Multi-Ply Gypsumboard Construction 12-57
12-45 Finishing Procedures for Gypsumboard 12-59
12-46 Isolation and Control Joints 12-60

Floor Coverings 12-63

12-47 Asphalt Tiles 12-63
12-48 Cork Tiles 12-63
12-49 Vinyl Flooring 12-64
12-50 Vinyl-Asbestos Tiles 12-64
12-51 Backed Vinyl 12-64
12-52 Rubber 12-65
12-53 Linoleum 12-65
12-54 Installation of Thin Coverings 12-65
12-55 Carpets 12-66
12-56 Terrazzo 12-68
12-57 Concrete Floors 12-69
12-58 Wood Floors 12-69
12-59 Industrial Floors 12-69
12-60 Conductive Flooring 12-70
12-61 Specifications and Standards for Flooring 12-71

13. Windows
by Philip M. Grennan 13-1

13-1 How Much Window Area? 13-1
13-2 Window Materials 13-2
13-3 Wood Windows 13-2
13-4 Definitions for Wood Windows 13-2
13-5 Steel Windows 13-5
13-6 Aluminum Windows 13-5
13-7 Stainless-Steel and Bronze Windows 13-6
13-8 Window Sizes 13-6
13-9 Window Types 13-6
13-10 Windows in Wall-Panel Construction 13-14
13-11 Mechanical Operators for Windows 13-14
13-12 Weather Stripping and Storm Sash 13-15
13-13 Window Glass 13-15
13-14 Glazing Compounds 13-19
13-15 Factory-Sealed Double Glazing 13-20
13-16 Gaskets for Glazing 13-21

14. Doors *by Frederick S. Merritt* 14-1

Ordinary Doors 14-1

14-1 Exterior Doors 14-2
14-2 Traffic Flow and Safety 14-3
14-3 Fire and Smokestop Doors 14-4
14-4 Door Materials 14-6
14-5 Structural Requirements for Openings and Doors 14-7
14-6 Swinging Doors 14-8
14-7 Ordinary Horizontally Sliding Doors 14-10
14-8 Ordinary Vertically Sliding Doors 14-10
14-9 Revolving Doors 14-11

Special-Purpose Doors 14-12

14-10 Large Horizontally Sliding Doors 14-13
14-11 Large Vertically Sliding Doors 14-14
14-12 Large Swinging Doors 14-14
14-13 Horizontally Hinged Doors 14-14
14-14 Radiation-Shielding Doors 14-15

15. Roof Coverings
by J. C. Gudas 15-1

15-1 Roof Slopes 15-1
15-2 Bituminous and Plastic Roofing Materials 15-2
15-3 Types of Roof Coverings 15-3
15-4 Selection of Roof Coverings 15-4

Single-Unit Coverings 15-5

15-5 Built-Up Roofs 15-5
15-6 Dependence of Built-Up Roofing on Roof Decks 15-6
15-7 Types of Roof Decks 15-7
15-8 Vapor Barriers for Built-Up Roofing 15-8
15-9 Insulation with Built-Up Roofing 15-9
15-10 Built-Up Roofing Failures 15-10
15-11 Temporary Roof Coverings 15-10
15-12 Good-Practice Recommendations for Built-Up Roofing 15-11
15-13 Inverted Built-Up Roofing System 15-12
15-14 Foamed-in-Place Polyurethane Roof System 15-13
15-15 Metal Roof Coverings 15-13
15-16 Cold-Process Roof-Covering Systems 15-15
15-17 Elastomeric Roof Coverings 15-16
15-18 Single-Ply Roofing Systems 15-17
15-19 Corrugated Asbestos-Cement Sheets 15-19
15-20 Rigid Vinyl Roofing 15-20
15-21 Venting Lightweight-Concrete Roofs 15-20
15-22 Roof Moisture-Detection Surveys 15-21
15-23 Air- and Tension-Supported Roofs 15-21

Multiunit Roof Coverings 15-22

15-24 Asphalt Shingles 15-22
15-25 Wood Shingles 15-24
15-26 Asbestos-Cement Shingles 15-25
15-27 Roofing Tiles 15-26
15-28 Roofing Slate 15-27
15-29 Roof Finishing Details 15-27

16. Builders' Hardware by Richard A. Hudnut 16-1

16-1 Finishing and Rough Hardware 16-1
16-2 When to Select Hardware 16-2
16-3 Effects of Codes and Regulations on Hardware 16-2
16-4 Function of Hardware Consultants 16-2
16-5 Factors Affecting Selection of Hardware 16-3
16-6 Template and Nontemplate Hardware 16-3

Finishing Hardware 16-3

16-7 Applicable Standards for Finishing Hardware 16-3
16-8 Hinges and Butts 16-4
16-9 How to Select Hinges 16-6
16-10 Door-Closing Devices 16-8
16-11 Locks and Latches 16-10
16-12 Hand of Doors 16-13
16-13 Casement Sash Operators 16-14
16-14 Sliding Hardware 16-14
16-15 Folding-Door Hardware 16-14
16-16 Locking Bolts (Doors and Windows) 16-14

Rough Hardware 16-14

16-17 Window Hardware 16-14
16-18 Inserts, Anchors, and Hangers 16-15
16-19 Nails 16-16
16-20 Screws 16-18
16-21 Welded Studs 16-18
16-22 Powder-Driven Studs 16-20
16-23 Miscellaneous Anchoring Devices 16-21
16-24 Bolts 16-21

17. Acoustics by Lyle F. Yerges 17-1

17-1 Sound Production and Transmission 17-1
17-2 Nomenclature for Analysis of Sound 17-2
17-3 Elastic Media 17-3
17-4 Hearing 17-4
17-5 Force and Energy Concepts 17-6
17-6 Units and Dimensions for Sound Measurements 17-6
17-7 Sound Measurements and Measurement Scales 17-7
17-8 Sound and Vibration Control 17-8
17-9 Airborne Sound Transmission 17-8
17-10 Bypassing of Sound Barriers 17-11

17-11	Structure-Borne Sound Transmission 17-12		18-23	Air-Distribution Temperature for Cooling 18-72
17-12	Damping of Vibrations 17-12		18-24	Condensers 18-72
17-13	Sound Absorption 17-13		18-25	Compressor-Motor Units 18-74
17-14	Control of Reflection and Reverberation 17-14		18-26	Cooling Equipment—Central Plant Packaged Units 18-75
17-15	Installation of Absorbents 17-17		18-27	Zoning 18-76
17-16	Performance Data 17-18		18-28	Packaged Air-Conditioning Units 18-76
17-17	Acoustical Criteria 17-20		18-29	Absorption Units for Cooling 18-78
17-18	Helpful Hints for Noise Control 17-22		18-30	Ducts for Air Conditioning 18-78
17-19	Bibliography 17-24		18-31	Built-Up Air-Conditioning Units 18-80

18. Heating, Ventilation, and Air Conditioning
by Frank C. Yanocha 18-1

18-1	Definitions of Terms for Heating, Ventilation and Air Conditioning (HVAC) 18-2
18-2	Heat and Humidity 18-7
18-3	Heat Flow and Human Comfort 18-15
18-4	Thermal Insulation 18-22
18-5	Prevention of Damage from Condensation 18-27
18-6	Ventilation 18-30
18-7	Heat Losses 18-37
18-8	Heat Gains 18-40

Methods of Heating Buildings 18-43

18-9	General Procedure for Sizing a Heating Plant 18-43
18-10	Heating-Load-Calculation Example 18-45
18-11	Warm-Air Heating 18-46
18-12	Hot-Water Heating Systems 18-51
18-13	Steam-Heating Systems 18-54
18-14	Unit Heaters 18-57
18-15	Radiant Heating 18-58
18-16	Snow Melting 18-61
18-17	Radiators and Convectors 18-63
18-18	Heat Pumps 18-63
18-19	Solar Heating 18-64
18-20	Integrated Systems 18-66

Methods of Cooling and Air Conditioning 18-67

18-21	Sizing an Air-Conditioning Plant 18-67
18-22	Refrigeration Cycles 18-70

18-32	All-Air Systems 18-81
18-33	Air-Water Systems 18-85
18-34	Control Systems for Air Conditioning 18-86
18-35	Heating and Air Conditioning 18-88
18-36	Industrial Air Conditioning 18-90
18-37	Chemical Cooling 18-90
18-38	Year-Round Air Conditioning 18-91

19. Plumbing—Water-Supply, Sprinkler, and Wastewater Systems *by Tyler G. Hicks* 19-1

19-1	Health Requirements for Plumbing 19-1

Water Supply 19-2

19-2	Water Quality 19-2
19-3	Water Treatment 19-4
19-4	Water Quantity and Pressures 19-5
19-5	Water Distribution in Buildings 19-6
19-6	Plumbing Fixtures 19-10
19-7	Water Demand and Fixture Units 19-11
19-8	Water-Pipe Sizing 19-15
19-9	Domestic-Water Heaters 19-20

Wastewater Piping 19-21

19-10	Wastewater Disposal 19-22
19-11	Sewers 19-23
19-12	Wastewater System Elements 19-25

19-13	Waste-Pipe Materials 19-27		21-3	Color Rendering of Light 21-8
19-14	Layout of Waste Piping 19-28		21-4	Quantity of Light 21-9
19-15	Interceptors 19-28		21-5	Lighting Methods 21-10
19-16	Piping for Industrial Wastes 19-29		21-6	Daylight 21-12
19-17	Rainwater Drainage 19-29		21-7	Characteristics of Lamps 21-13
19-18	Waste-Pipe Sizing 19-32		21-8	Characteristics of Lighting Fixtures 21-20
19-19	Venting 19-32		21-9	Systems Design of Lighting 21-23
19-20	Plumbing-System Inspection and Tests 19-35		21-10	Bibliography 21-25

Gas Piping 19-36

19-21	Gas Supply 19-36
19-22	Gas-Pipe Sizes 19-36
19-23	Estimating Gas Consumption 19-36
19-24	Gas-Pipe Materials 19-37
19-25	Testing Gas Piping 19-38

Sprinkler Systems 19-38

19-26	Sprinkler System Standards and Approvals 19-38
19-27	Types of Sprinkler Systems 19-39
19-28	Sprinkler Piping 19-45
19-29	Water Supplies for Sprinklers 19-47
19-30	Standpipes 19-48
19-31	Central-Station Supervisory Systems 19-49

20. Electric Power
by Charles J. Wurmfeld 20-1

20-1	Direct-Current Systems 20-2
20-2	Alternating-Current Systems 20-4
20-3	Electrical Loads 20-10
20-4	Electrical Conductors and Raceways 20-10
20-5	Power System Apparatus 20-13
20-6	Communication Systems 20-21
20-7	Electrical Distribution in Buildings 20-21
20-8	Circuit and Conductor Calculations 20-27

21. Lighting *by Jonathan S. Fink* 21-1

21-1	Light and Sight 21-2
21-2	Quality of Light 21-5

22. Vertical Circulation
by Frederick S. Merritt 22-1

22-1	Ramps 22-2
22-2	Stairs 22-4
22-3	Emergency Egress Stairways 22-7
22-4	Wood Stairs 22-8
22-5	Steel Stairs 22-8
22-6	Concrete Stairs 22-9
22-7	Escalators 22-9
22-8	Elevator Installations 22-13
22-9	Definitions of Elevator Terms 22-15
22-10	Elevator Hoistways 22-18
22-11	Elevator Cars 22-20
22-12	Electric Elevators 22-23
22-13	Hydraulic Elevators 22-28
22-14	Planning for Passenger Elevators 22-29
22-15	Hospital Elevators 22-34
22-16	Planning for Freight Elevators 22-34
22-17	Dumbwaiters 22-37
22-18	Vertical Conveyors and Pneumatic Tubes 22-37
22-19	Mail Chutes 22-38

23. Surveying for Buildings
by Reginald S. Brackett 23-1

23-1	Units of Measurement 23-1
23-2	Survey of General Topography 23-2
23-3	Survey of Nearby Buildings 23-6
23-4	Establishment of Elevations before Construction 23-7
23-5	General Layout for Construction 23-7
23-6	Surveying Instruments 23-11

24. Estimating Building Construction Costs
by E. D. Lowell 24-1

- 24-1 Estimating before, during, and after Design 24-1
- 24-2 Types of Estimates 24-3
- 24-3 Preliminary Steps in Estimating 24-5
- 24-4 Quantity Takeoffs 24-6
- 24-5 Unit Prices 24-8
- 24-6 Allowances 24-10
- 24-7 Estimate Details 24-11
- 24-8 Evaluation of General Cost Items 24-11
- 24-9 Computer Estimating 24-14
- 24-10 Pipe Estimating by Computer 24-14
- 24-11 Cost Estimate and Construction Schedule 24-15
- 24-12 Presentation of the Estimate 24-17
- 24-13 Reference Material 24-20
- 24-14 Cost Indices 24-20

25. Construction Management
by Robert F. Borg 25-1

- 25-1 Types of Construction Companies 25-1
- 25-2 Construction Company Organization 25-3
- 25-3 Contractor's Business Consultants 25-4
- 25-4 Sources of Business 25-5
- 25-5 What Constitutes the Contract Documents? 25-6
- 25-6 Major Concerns with Building Codes 25-7
- 25-7 Estimating, Bidding, and Costs 25-9
- 25-8 Types of Bids and Contracts 25-10
- 25-9 Professional Construction Managers 25-11
- 25-10 Contract Administration 25-12
- 25-11 Subcontracts 25-13
- 25-12 Purchase Orders 25-17
- 25-13 Insurance 25-21
- 25-14 Construction Contract Bonds 25-30
- 25-15 Scheduling and Expediting 25-32
- 25-16 Fast Tracking 25-36
- 25-17 Project Management 25-37
- 25-18 Field Supervision 25-40
- 25-19 Who Pays for the Unexpected? 25-40
- 25-20 Changes, Extras, and Claims 25-43
- 25-21 Arbitration and Mediation 25-45
- 25-22 Trade Payment Breakdowns and Payments 25-47
- 25-23 Cost Records 25-50
- 25-24 Accounting Methods 25-51
- 25-25 Safety 25-54
- 25-26 Community Relations 25-55
- 25-27 Relations with Public Agencies in Executing Construction Operations 25-56
- 25-28 Labor Relations 25-57
- 25-29 Social and Environmental Concerns in Construction 25-58
- 25-30 Systems Building 25-59
- 25-31 Basics of Successful Management 25-62

26. Specifications
by Joseph F. Ebenhoeh, Jr. 26-1

- 26-1 Role of Specifications in Construction 26-1
- 26-2 Types of Specifications 26-2
- 26-3 Components of Specifications 26-4
- 26-4 Bidding Requirements 26-4
- 26-5 General Conditions and Special Conditions 26-6
- 26-6 Contract Forms 26-7
- 26-7 Technical Specifications 26-7
- 26-8 Automated Specifications 26-9
- 26-9 Covers and Binding for Specifications 26-11
- 26-10 Construction Contracts 26-11
- 26-11 Revision of Contract Documents 26-13
- 26-12 Aids for Specification Writers 26-13
- 26-13 Trade Associations and Societies 26-14

Appendix Factors for Conversion to the Metric System (SI) of Units
by Frederick S. Merritt A-1

Index follows Appendix

Contributors

ROBERT F. BORG President, Kreisler Borg Florman Construction Company, Scarsdale, N.Y. *(Construction Management)*

REGINALD S. BRACKETT, Bay Shore, N.Y. *(Surveying for Buildings)*

WILLIAM E. DIAMOND II, AIA, The Woodlands, Tex. *(Business, Art, and Profession of Architecture)*

ALBERT G. H. DIETZ Professor, Building Engineering, School of Architecture and Planning, Massachusetts Institute of Technology, Cambridge, Mass. *(Building Materials)*

JOSEPH F. EBENHOEH, JR. Chief, Architectural Specifications Division, Albert Kahn Associates, Inc., Detroit, Mich. *(Specifications)*

JONATHAN S. FINK Electrical Engineer, Sanders & Thomas, Inc., Pottstown, Pa. *(Lighting)*

PHILLIP M. GRENNAN Consulting Engineer, Rockville Centre, N.Y. *(Windows)*

J. C. GUDAS Editor-Publisher, American Roofer and Building Improvement Contractor, Downers Grove, Ill. *(Roof Coverings)*

CHARLES P. GUPTON Partner, Dames & Moore, Boca Raton, Fla. *(Soil Mechanics and Foundations)*

TYLER G. HICKS International Engineering Associates, New York, N.Y. *(Plumbing—Water-Supply, Sprinkler, and Wastewater Systems)*

EDWARD S. HOFFMAN Vice President, Klein and Hoffman, Inc., Chicago, Ill. *(Concrete Construction)*

RICHARD A. HUDNUT Product Standards Coordinator, Builders Hardware Manufacturers Association, New York, N.Y. *(Builders' Hardware)*

E. D. LOWELL formerly Chief Estimator, Kaiser Engineers, Oakland, Calif. *(Estimating Building Construction Costs)*

FREDERICK S. MERRITT Consulting Engineer, West Palm Beach, Fla. *(Building Systems. Protection of Structures and Occupants Against Hazards. Structural Theory. Walls, Partitions, and Interior Finishes. Doors. Vertical Circulation.)*

MAURICE J. RHUDE President, Sentinel Structures, Inc., Peshtigo, Wisc. *(Wood Construction)*

PAUL F. RICE Vice President, Engineering, Concrete Reinforcing Steel Institute, Chicago, Ill. *(Concrete Construction)*

HENRY J. STETINA Consulting Engineer, Jenkintown, Pa. *(Structural-Steel Construction)*

DON S. WOLFORD Consulting Engineer, Middletown, Ohio *(Cold-Formed Steel Construction)*

CHARLES J. WURMFELD Consulting Engineer, Wurmfeld Associates, P.C., New York, N.Y. *(Electric Power)*

FRANK C. YANOCHA Chief Mechanical Engineer, HVAC, Sanders & Thomas, Inc., Pottstown, Pa. *(Heating, Ventilation, and Air Conditioning)*

LYLE F. YERGES Consulting Engineer, Downers Grove, Ill. *(Acoustics)*

ALAN H. YORKDALE Vice President, Engineering and Research, Brick Institute of America, McLean, Va. *(Engineered Brick Construction)*

Preface

For this fourth edition, the title of the *Building Construction Handbook* has been expanded to *Building Design and Construction Handbook* to indicate the broader scope of the book. The new edition may well be considered a new book. It contains several new sections, much additional subject matter in sections that have been retained from earlier editions, and substantial revisions of those sections to treat new developments.

Many of the revisions were necessitated by changes in the specifications for design and construction with commonly used construction materials. Other changes and additions of new information were made necessary by growing public concern with energy conservation, environmental impact of new construction, and rising building construction and operating costs. To alleviate these concerns, new concepts for design and construction are required, and introduction of such concepts was one of the objectives of this edition of the handbook.

In preparing this edition, the objectives of previous editions, which gained ready acceptance for them, were retained—basically to provide in a single volume a compendium of the best of current building design and construction practices, information that would be of the greatest usefulness to those who have to make decisions affecting selection of building materials and construction methods. Emphasis is placed on fundamental principles and practical applications of them, with special attention to simplified procedures. Frequent reference is made to other sources where additional authoritative, detailed information may be obtained.

Some of the new sections in the fourth edition were added to present new developments in design and construction. Other new sections resulted from contributions of new authors, who brought a fresh viewpoint, new ideas, and additional information to the treatment of their subjects. In addition, the other authors took advantage of the opportunity created by publication of a new edition to update and improve almost all parts of the book and to add material on other new developments. Although obsolete and less-important information in the earlier editions have been deleted to make space for the additions, there has been a substantial increase in the number of pages in the book.

The opening section of the fourth edition, Section 1, Building Systems, is new. Serving as an introduction to the other sections of the book, it briefly describes building components, design procedures and the construction process, subjects that are treated in detail in the following sections. But even more importantly, the first section presents a new concept of building design and construction—systems design—to help designers and builders meet the challenge of producing better buildings with lower life-cycle costs.

Section 3, Protection of Structures and Occupants against Hazards, also is new. It describes the effects on buildings of wind, earthquakes, fire, lightning, water, and flooding and indicates how buildings can be constructed to withstand them. This section also presents the principles of and methods for protecting against intruders and for resisting lateral forces that act on buildings—a subject that was not treated in one place in previous editions of the book. Nevertheless, as before, structural analysis of lateral-force-resisting framing is covered in another section (Section 5), and Sections 7 through 11 describe how structural theory may be applied to design for various structural materials.

Another new section is Section 21, Lighting. It represents a considerable expansion, by a new author, of a very brief treatment of this subject in the section Electric Power and Lighting in previous editions. Devotion of a separate section to lighting presents an opportunity not only to provide much additional information on artificial illumination but also to add an informative treatment of daylighting, which had not been covered in previous editions of the book.

Section 6, Soil Mechanics and Foundations; Section 9, Cold-Formed Steel Construction; and Section 18, Heating, Ventilation, and Air Conditioning have been rewritten by new authors to cover recent developments in those fields. Also, Sections 7, 8, and 10 have been revised to reflect the changes that have occurred in design specifications for wood, steel, and concrete and in construction practices in recent years. In addition, systems for enclosing buildings are treated in this edition in contiguous sections (12 through 15), and Section 12 integrates the treatment of walls, partitions, and interior finishes. Furthermore, Section 19, Water-Supply, Plumbing, Sprinkler, and Wastewater Systems, and Section 20, Electric Power, have been expanded to give fuller treatment of basic principles and functioning of system components.

To avoid cumbersome language, terms such as "workmanship" and the pronoun "he" are used to refer to both men and women.

In preparing this reference work, the contributors drew heavily on numerous sources of information. Many of these are credited or given as references throughout the book, but space limitations preclude mentioning them all. The editor and the contributors wish to acknowledge their indebtedness to these sources and to express their gratitude.

The editor is especially grateful to the contributors, not only because he appreciates the great value of their contributions but also because he is keenly aware of their considerable sacrifices in taking time to prepare their sections.

We all hope that you will find this edition even more useful than the previous ones and that application of the information it contains will stimulate greater production of cost-effective, energy-efficient and environmentally sound buildings.

Frederick S. Merritt

Building Design and Construction Handbook

Section 1

Building Systems

Frederick S. Merritt
Consulting Engineer
West Palm Beach, Fla.

Sociological changes, new developments in industry and commerce, inflationary economies of nations, and advances in building technology place an ever-increasing burden on building designers and constructors. They need more and more knowledge and skill to cope with the demands placed on them.

The public continually demands more complex buildings than in the past. They must serve more purposes, last longer, and require less maintenance and repair. As in the past, they must look pretty. Yet, both building construction and operating costs must be kept within acceptable limits or new construction will cease.

To meet this challenge successfully, continual improvements in building design and construction must be made. Building designers and constructors should be alert to these advances and learn how to apply them skillfully.

One advance of note is the adaptation to building design of operations research, or systems design, developed around the middle of the twentieth century and originally applied with noteworthy results to design of machines and electronic equipment. In the past, design of a new building was mainly an imitation of the design of an existing building. Innovations were often developed fortuituously and by intuition and were rare occurrences. In contrast, systems design encourages innovation. It is a precise procedure that guides creativity toward the best decisions. As a result, it can play a significant role in meeting the challenges posed by increasing building complexity and costs. The basic principles of systems design are presented in this section.

1-1. Principles of Architecture.

A building is an assemblage that is firmly attached to the ground and that provides total or nearly total shelter for machines, processing equipment, performance of human activities, storage of human possessions, or any combination of these.

Building design is the process of providing all information necessary for construction of a building that will meet its owner's requirements and also satisfy public health, welfare, and safety requirements. **Architecture** is the art and science of building design. **Building construction** is the process of assembling materials to form a building.

Building design may be legally executed only by persons deemed competent to do so by the state in which the building is to be constructed. Competency is determined on the basis of education, experience, and ability to pass a written test of design skills.

Architects are persons legally permitted to practice architecture. **Engineers** are experts in specific scientific disciplines and are legally permitted to design parts of buildings; in some cases, complete buildings. In some states, persons licensed as **building designers** are permitted to design certain types of buildings.

Building construction is generally performed by laborers and craftsmen engaged for the purpose by an individual or organization, called a **contractor.** The contractor signs an agreement, or contract, with the building owner under which the contractor agrees to construct a specific building on a specified site and the owner agrees to pay for the materials and services provided. Often, building construction is restricted only to contractors granted a license for the purpose by the community in which the building is to be constructed.

In design of a building, architects should be guided by the following principles:

1. The building should be constructed to serve purposes specified by the client.

2. The design should be constructable by known techniques and with available labor and equipment, within an acceptable time.

3. The building should be capable of withstanding the elements and normal usage for a reasonably long period of time.

4. Both inside and outside, the building should be visually pleasing.

5. No part of the building should pose a hazard to the safety or health of its occupants under normal usage, and the building should provide for safe evacuation or refuge in emergencies.

6. The building should provide the degree of shelter from the elements and of control of the interior environment—air, temperature, humidity, light, and acoustics—specified by the client and not less than the minimums required for safety and health of the occupants.

7. The building should be constructed to minimize adverse environmental impact on its neighbors.

8. Operation of the building should consume a minimum of energy while permitting the structure to serve its purposes.

9. The sum of costs of construction, operation, maintenance, repair, and anticipated future alterations should be kept within the limit specified by the client.

To provide all the information necessary for the construction of a building, the ultimate objective of design is production of **drawings,** or **plans,** showing what is to be constructed, **specifications** stating what materials and equipment are to be incorporated in the building, and a **construction contract** between the client and a contractor. Designers also should observe construction of the building while it is in process. This should be done not only to assist the client in ensuring that the building is being constructed in accordance with plans and specifications but also to obtain information that will be useful in design of future buildings.

See also Sec. 2.

1-2. Systems.

Systems design comprises a rational, orderly series of steps that leads to the best decision for a given set of conditions. The procedure requires:

Analysis of a building as a system

Synthesis, or selection of components, to form a system that meets specific objectives while subject to constraints, or variables controllable by designers

Appraisal of system performance, including comparisons with alternative systems

Feedback to analysis and synthesis of information obtained in system evaluation, to improve the design

The prime advantage of the procedure is that, through comparisons of alternatives and data feedback to the design process, systems design converges on an optimum, or best, system for the given conditions. Another advantage is that the procedure enables designers to clarify the requirements for the building being designed. Still another advantage is that the procedure provides a common basis of understanding and promotes cooperation between the specialists in various aspects of building design.

For a building to be treated as a system, as required in systems design, it is necessary to know what a system is and what its basic characteristics are.

A system is an assemblage formed to satisfy specific objectives and subject to constraints and restrictions and consisting of two or more components that are interrelated and compatible, each component being essential to the required performance of the system.

Because the components are required to be interrelated, operation, or even the mere existence, of one component affects in some way the performance of other components. Also, the required performance of the system as a whole, as well as the constraints on the system, imposes restrictions on each component.

A building meets the preceding requirements. By definition, it is an assemblage (Art. 1-1). It is constructed to serve specific purposes. It is subject to constraints while doing so, inasmuch as designers can control properties of the system by selection of components (Art. 1-9). Building components, such as walls, floors, roofs, windows, and doors, are interrelated and compatible with each other. The existence of any of these components affects to some extent the performance of the others. And the required performance of the building as a whole imposes restrictions on the components. Consequently, a building has the basic characteristics of a system, and systems-design procedures should be applicable to it.

Systems Analysis. A group of components of a system may also be a system. Such a group is called a **subsystem.** It, too, may be designed as a system, but its goal must be to assist the system of which it is a component to meet its objectives. Similarly, a group of components of a subsystem may also be a system. That group is called a **subsubsystem.**

For brevity, the major subsystems of a building are referred to as systems in this book.

In a complex system, such as a building, subsystems and other components may be combined in a variety of ways to form different systems. For the purposes of building design, the major systems are usually defined in accordance with the construction trades that will assemble them, for example, structural framing, plumbing, electrical systems, and heating, ventilation, and air conditioning.

In systems analysis, a system is resolved into its basic components. Subsystems are determined. Then, the system is investigated to determine the nature, interaction, and performance of the system as a whole. The investigation should answer such questions as:

What does each component (or subsystem) do?
What does the component do it to?
How does the component serve its function?
What else does the component do?

Why does the component do the things it does?
What must the component really do?
Can it be eliminated because it is not essential or because another component can assume its tasks?

See also Art. 1-8.

1-3. Traditional Design Procedures.

Systems design of buildings requires a different approach to design and construction than that used in traditional design (Art. 1-9). Because traditional design and construction procedures are still widely used, however, it is desirable to incorporate as much of those procedures in systems design as is feasible without destroying its effectiveness. This will make the transition from traditional design to systems design easier. Also, those trained in systems design of buildings will then be capable of practicing in traditional ways, if necessary.

There are several variations of traditional design and construction. These are described throughout this book. For the purpose of illustrating how they may be modified for systems design, however, one widely used variation, which will be called basic traditional design and construction, is described in the following and in Art. 1-4.

In the basic traditional design procedure, design usually starts when a client recognizes the need for and economic feasibility of a building and engages an architect, a professional with a broad background in building design. The architect, in turn, engages consulting engineers and other consultants.

For most buildings, structural, mechanical, and electrical consulting engineers are required. A structural engineer is a specialist trained in the application of scientific principles to design of load-bearing walls, floors, roofs, foundations, and skeleton framing needed for the support of buildings and building components. A mechanical engineer is a specialist trained in the application of scientific principles to design of plumbing, elevators, escalators, horizontal walkways, dumbwaiters, conveyors, installed machinery, and heating, ventilation, and air conditioning. An electrical engineer is a specialist trained in the application of scientific principles to design of electric circuits, electric controls and safety devices, electric motors and generators, electric lighting, and other electric equipment.

For buildings on a large site, the architect may engage a landscape architect as a consultant. For a concert hall, an acoustics consultant may be engaged; for a hospital, a hospital specialist; for a school, a school specialist.

The architect does the overall planning of the building and incorporates the output of the consultants into the contract documents. The architect determines what internal and external spaces the client needs, the sizes of these spaces, their relative locations, and their interconnections. The results of this planning are shown in floor plans, which also diagram the internal flow, or circulation, of people and supplies. Major responsibilities of the architect are enhancement of the appearance inside and outside of the building and keeping adverse environmental impact of the structure to a minimum. The exterior of the building is shown in drawings, called elevations. The location and orientation of the building is shown in a site plan. The architect also prepares the specifications for the building. These describe in detail the materials and equipment to be installed in the structure. In addition, the architect, usually with the aid of an attorney engaged by the client, prepares the construction contract.

The basic traditional design procedure is executed in several stages. In the first stage, the architect develops a **program,** or list of the client's requirements. In the second stage, or **conceptual phase,** the architect translates requirements into spaces, relates the spaces and makes sketches, called schematics, to illustrate the concepts. When sufficient information is obtained on the size and general construction of the building, a rough estimate is made of construction cost. If this cost does not exceed the cost budgeted by the client for construction, the next stage, **design development,** proceeds. In this stage, the architect and consultants work out more details and show the results in preliminary construction drawings and outline specifications. A preliminary cost estimate utilizing the greater amount of information on the building now available is then prepared. If this cost does not exceed the client's budget, the final stage, the **contract documents phase,** starts. It culminates in production of working, or construction, drawings and specifications, which are incorporated in the contract between the client and a builder and therefore become legal documents. Before the documents are completed, however, a final cost estimate is prepared. If the cost exceeds the client's budget, the design is revised to achieve the necessary cost reduction.

After the estimated cost is brought within the budget and the client has approved the contract documents, the architect helps the owner in obtaining bids from contractors or in negotiating a construction price with a qualified contractor. For private work, construction not performed for a governmental agency, the owner generally awards the construction contract to a contractor, called a **general contractor.** Assigned the responsibility for construction of the building, this contractor may perform some, all, or none of the work. Usually, much of the work is let out to specialists, called **subcontractors.** For public work, there may be a legal requirement that bids be taken and the contract awarded to the lowest responsible bidder. Sometimes also, separate contracts have to be awarded for the major specialties, such as mechanical and electrical trades, and to a general contractor, who is assigned responsibility for coordinating the work of the trades and performance of the work. (See also Art. 1.4.)

Building design should provide for both normal and emergency conditions. The latter includes fire, explosion, power cutoffs, hurricanes, and earthquakes. The design should include facilities for handicapped persons.

See also Sec. 2.

1-4. Traditional Construction Procedures.

As mentioned in Art. 1-3, construction is performed by contractors. While they would like to satisfy the owner and the building designers, contractors have the main objective of making a profit. Hence, their initial task is to prepare a bid price based on an accurate estimate of construction costs. This requires development of a concept for performance of the work and a construction time schedule. After a contract has been awarded, contractors must furnish and pay for all materials, equipment, power, labor, and supervision required for construction. The owner compensates the contractors for construction costs and services.

A **general contractor** assumes overall responsibility for construction of a building. The contractor engages **subcontractors** who take responsibility for the work of the various trades required for construction. For example, a plumbing contractor installs the plumbing, an electrical contractor installs the electrical system, a steel erector erects structural steel, and an elevator contractor installs elevators. Their contracts are with the general contractor, and they are paid by the general contractor.

Sometimes, in addition to a general contractor, the owner contracts separately with specialty contractors, such as electrical and mechanical contractors, who perform a substantial amount of the work required for a building. Such contractors are called **prime contractors.** Their work is scheduled and coordinated by the general contractor, but they are paid directly by the owner.

Sometimes also, the owner awards a contract to an organization for both the design and construction of a building. Such organizations are called **turnkey contractors.** One variation of this type of contract is employed by developers of groups of one-family homes or low-rise apartment buildings. The **homebuilder** designs and constructs the dwellings, but the design is substantially completed before owners purchase the homes.

Administration of the construction procedure often is difficult. Consequently, some owners seek assistance from an expert, called a **professional construction manager,** with extensive construction experience, who receives a fee for his services. The construction manager negotiates with general contractors and helps select one to construct the building. The manager usually also supervises selection of subcontractors. During construction, he helps control costs, expedite equipment and material deliveries, and keep the work on schedule (see Art. 25-9). In some cases, instead, the owner may prefer to engage a **construction program manager,** to assist in administering both design and construction.

Construction contractors employ labor that may or may not be unionized. Unionized craftsmen are members of unions that are organized by construction trades, such as carpenter, plumber, and electrician unions. Union members will perform only the work assigned to their trade. On the job, groups of workers are supervised by **foremen,** all of whom report to a **superintendent.**

During construction, all work should be inspected. For the purpose, the owner, often through the architect and consultants, engages inspectors. The field inspectors may be placed under the control of an owner's representative, who may be titled *clerk of the works, architect's superintendent, engineer's superintendent,* or *resident engineer.* The inspectors have the responsibility of ensuring that construction meets the requirements of the contract documents and is performed under safe conditions. Such inspections may be made at frequent intervals.

In addition, inspections also are made by representatives of one or more governmental agencies. They have the responsibility of ensuring that construction meets legal requirements and have little or no concern with detailed conformance with the contract documents. Such legal inspections are made periodically or at the end of certain stages of construction. One agency that will make frequent inspections is the local or state building department, whichever has jurisdiction. The purpose of these inspections is to ensure conformance with the local or state building code.

Following is a description of the basic traditional construction procedure for a multistory building:

After the award of a construction contract to a general contractor, the owner may ask the contractor to start work before signing of the contract by giving the contractor a *letter of intent* or after signing of the contract by issuing a *written notice to proceed.* The contractor then obtains construction permits, as required, from governmental agencies, such as the local building, water, sewer, and highway departments.

The general contractor plans and schedules construction operations in detail and mobilizes equipment and personnel for the project. Subcontractors are notified of the contract award and issued letters of intent or awarded subcontracts, then are given, at appropriate times, notice to proceed.

Before construction starts, the general contractor orders a survey to be made of adjacent structures and terrain, both for the record and to become knowledgeable of local conditions. A survey is then made to lay out construction.

Field offices for the contractor are erected on or near the site. If desirable for safety reasons to protect passersby, the contractor erects a fence around the site and an overhead protective cover, called a bridge. Structures required to be removed from the site are demolished and the debris is carted away.

Next, the site is prepared to receive the building. This work may involve grading the top surface to bring it to the proper elevations, excavating to required depths for basement and foundations, and shifting of utility piping. For deep excavations, earth sides are braced and the bottom is drained.

Major construction starts with the placement of foundations, on which the building rests. This is followed by the erection of load-bearing walls and structural framing. Depending on the height of the building, ladders, stairs, or elevators may be installed to enable construction personnel to travel from floor to floor and eventually to the roof. Also, hoists may be installed to lift materials to upper levels. If needed, temporary flooring may be placed for use of personnel.

As the building rises, pipes, ducts, and electric conduit and wiring are installed. Then, permanent floors, exterior walls, and windows are constructed. At the appropriate time, permanent elevators are installed. If required, fireproofing is placed for steel framing. Next, fixed partitions are built and the roof and its covering, or roofing, are put in place.

Finishing operations follow. These include installation of the following: ceilings; tile; wallboard; wall paneling; plumbing fixtures; heating furnaces; air-conditioning equipment; heating and cooling devices for rooms; escalators; floor coverings; window glass; movable partitions; doors; finishing hardware; electrical equipment and apparatus, including lighting fixtures, switches, outlets, transformers, and controls; and other items called for in the drawings and specifications. Field offices, fences, bridges, and other temporary construction must be removed from the site. Utilities, such as gas, electricity, and water, are hooked up to the building. The site is landscaped and paved. Finally, the building interior is painted and cleaned.

The owner's representatives then give the building a final inspection. If they find that the structure conforms with the contract documents, the owner accepts the project and gives the general contractor final payment on issuance by the building department of a certificate of occupancy, which indicates that the completed building meets building-code requirements.

See also Secs. 23 to 26.

1-5. Role of the Client in Design and Construction.

Article 1-4 points out that administration of building construction is difficult, as a result of which some clients, or owners, engage a construction manager or construction program manager to act as the owner's authorizing agent and project overseer. The reasons for the complexity of construction administration can be seen from an examination of the owner's role before and during construction.

After the owner recognizes the need for a new building, the owner establishes project goals and determines the economic feasibility of the project. If it appears to be feasible, the owner develops a building program (list of requirements), budget, and time schedule for construction. Next, preliminary arrangements are made to finance construction. Then, the owner selects a construction program manager or an architect for design of the building. Later, a construction manager may be chosen, if desired.

The architect may seek from the owner approval of the various consultants that will be needed for design. If a site for the building has not been attained at this stage, the architect can assist in site selection. When a suitable site has been found, the owner purchases it and arranges for surveys and subsurface explorations to provide information for locating the building, access, foundation design and construction, and landscaping. It is advisable at this stage for the owner to start developing harmonious relations with the community in which the building will be erected.

During design, the owner assists with critical design decisions; approves schematic drawings, rough cost estimates, preliminary drawings, outline specifications, preliminary cost estimate, contract documents, and final cost estimate; pays designers' fees in installments as design progresses; and obtains a construction loan. Then, the owner awards the general contract for construction and orders construction to start. Also, the owner takes out liability, property, and other desirable insurance.

At the start of construction, the owner arranges for construction permits. As construction proceeds, the owner's representatives inspect the work to ensure compliance with the contract documents. Also, the owner pays contractors in accordance with the terms of the contract. Finally, the owner approves and accepts the completed project.

One variation of the preceding procedure is useful when time available for construction is short. It is called **phased, or fast-track, construction.** In this variation, the owner engages a construction manager and a general contractor before design has been completed, to get an early start on construction. Work then proceeds on some parts of the building while other parts are still being designed. For example, excavation and foundation construction are carried out while design of the structural framing is being finished. The structural framing is erected, while heating, ventilation, and air-conditioning, electrical, plumbing, wall, and finishing details are being developed. For tall buildings, the lower portion can be constructed, while the upper part is still being designed. For large, low-rise buildings, one section can be built, while another is under design.

1-6. Building Costs.

Construction cost of a building usually is a dominant design concern. One reason is that if construction cost exceeds the owner's budget, the owner may cancel the project. Another reason is that costs, such as property taxes and insurance, that occur after completion of the building often are proportional to the initial cost. Hence, owners usually try to keep that cost low. Designing a building to minimize construction cost, however, may not be in the owner's best interests. There are many other costs that the owner incurs during the anticipated life of the building that should be taken into account.

Before construction of a building starts, the owner generally has to make a sizable investment in the project. The major portion of this expenditure usually goes for purchase of the site and building design. Remaining preconstruction costs include those for feasibility studies, site selection and evaluation, surveys, and program definition.

The major portion of the construction cost is the sum of the payments to the general contractor and prime contractors. Remaining construction costs usually consist of interest on the construction loan, permit fees, and costs of materials, equipment, and labor not covered by the construction contracts.

The **initial cost** to the owner is the sum of preconstruction, construction, and occupancy costs. The latter covers costs of moving possessions into the building and start-up of utility services, such as water, gas, electricity, and telephone.

After the building is occupied, the owner incurs costs for operation and maintenance of the buildings. Such costs are a consequence of decisions made during building design.

Often, postconstruction costs are permitted to be high so that initial costs can be kept low. For example, operating the building may be expensive because the design makes artificial lighting necessary when daylight could have been made available or because extra heating and air conditioning are necessary because of inadequate insulation of walls and roof. As another example, maintenance may be expensive because of the difficulty of changing electric lamps or because cleaning the building is time-consuming and laborious. In addition, frequent repairs may be needed because of poor choice of materials during design. Hence, operation and maintenance costs over a specific period of time, say 10 or 20 years, should be taken into account in optimizing the design of a building.

Life-cycle cost is the sum of initial, operating, and maintenance costs. Generally, it is life-cycle cost that should be minimized in building design rather than construction cost. This would enable the owner to receive the greatest return on the investment in the building.

Nevertheless, a client usually establishes a construction budget independent of life-cycle cost. This often is necessary because the client does not have adequate capital for an optimum building and places too low a limit on construction cost. The client hopes to have sufficient capital later to pay for the higher operating and maintenance costs or for replacement of undesirable building materials and installed equipment. Sometimes, the client establishes a low construction budget because the client's goal is a quick profit on early sale of the building, in which case the client has little or no concern with future high operating and maintenance costs for the building. For these reasons, construction cost frequently is a dominant concern in design.

See also Sec. 24.

1-7. Major Building Systems.

The simplest building system consists of only two components. One component is a floor, a flat, horizontal surface on which human activities can take place. The other component is an enclosure that extends over the floor and generally also around it to provide shelter from the weather for human activities.

The ground may serve as the floor in primitive buildings. In better buildings, however, the floor may be a structural deck laid on the ground or supported above ground on structural members, such as the joist and walls in Fig. 1-1. Use of a deck and structural members adds at least two different types of components, or two subsystems, to the simplest building system. Also, often, the enclosure over the floor requires supports, such as the rafter and walls in Fig. 1-1, and the walls, in turn, are seated on foundations in the ground. Additionally, **footings** are required at the base of the foundations to spread the load over a large area of the ground, to prevent the building from sinking (Fig. 1-2a). Consequently, even slight improvements in a primitive building introduce numerous additional components, or subsystems, into a building.

More advanced buildings consist of numerous subsystems, which are referred to as systems in this book when they are major components. Major subsystems generally include structural framing and foundations, enclosure systems, plumbing, lighting,

Fig. 1-1 Vertical section through one-story building with basement shows locations of some major components. *(Reprinted with permission from F. S. Merritt, "Building Engineering and Systems Design," Van Nostrand Reinhold, New York.)*

acoustics, safety systems, vertical-circulation elements, electric power and signal systems, and heating, ventilation, and air conditioning (HVAC).

Structural System. The portion of a building that extends above the ground level outside it is called the **superstructure.** The portion below the outside ground level is called the **substructure.** The parts of the substructure that distribute building loads to the ground are known as **foundations.**

Foundations may take the form of walls. When the ground under the building is excavated for a cellar, or basement, the foundation walls have the additional task of retaining the earth along the outside of the building (Fig. 1-1). The superstructure in such cases is erected atop the foundation walls.

The footing under a wall (Fig. 1-2a) is called a **continuous spread footing.** A slender structural member, such as a column (Fig. 1-2b), usually is seated on an **individual spread footing.** When the soil is so weak, however, that the spread footings for columns become very large, it often is economical to combine the footings into a single footing under the whole building. Such a footing is called a **raft,** or **mat, footing** or a **floating foundation.** For very weak soils, it generally is necessary to support the foundations on **piles** (Fig. 1-2c). These are slender structural members that are hammered or otherwise driven through the weak soil, often until the tips seat on rock or a strong layer of soil.

Fig. 1-2 Commonly used foundations: (a) foundation wall on continuous footing; (b) individual spread footing for a column; (c) pile footing for a column.

The foundation system must be designed to transmit the loads from the superstructure structural system directly to the ground in such a manner that settlement of the completed building as the soil deflects will be within acceptable limits. The superstructure structural system, in turn, should be designed to transmit its loads to the foundation system in the manner anticipated in the design of the foundations. (See also Sec. 6.)

In most buildings, the superstructure structural system consists of floor and roof decks, horizontal members that support them, and vertical members that support the other components.

The horizontal members are generally known as **beams,** but they also are called by different names in specific applications. For example:

Joists are closely spaced to carry light loads.
Stringers support stairs.
Headers support structural members around openings in floors, roofs, and walls.
Purlins are placed horizontally to carry level roof decks.
Rafters are placed on an incline to carry sloping roof decks.
Girts are light horizontal members that span between columns to support walls.
Lintels are light horizontal beams that support walls at floor levels in multistory buildings or that carry the part of walls above openings for doors and windows.
Girders may be heavily loaded beams or horizontal members that support other beams (Fig. 1-3).
Spandrels carry exterior walls and support edges of floors and roofs in multistory buildings.
Trusses serve the same purposes as girders but consist of slender horizontal, vertical and inclined components with large open spaces between them. The spaces are triangular in shape. Light beams similarly formed are called **open-web joists** (Fig. 1-6d).

Fig. 1-3 Structural-steel skeleton framing for a multistory building. *(Courtesy of the American Institute of Steel Construction.)*

Floor and roof decks or the beams that support them are usually seated on load-bearing walls or carried by columns, which carry the load downward. (The horizontal members also may be suspended on hangers, which transmit the load to other horizontal members at a higher level.) The system comprising decks, beams, and bearing walls is known as load-bearing construction (Fig. 1-1). The system composed of decks, beams, and columns is known as skeleton framing (Fig. 1-3).

Both types of systems must be designed to transmit to the foundations vertical (gravity) loads, vertical components of inclined loads, horizontal (lateral) loads, and horizontal components of inclined loads. Vertical walls and columns have the appropriate alignment for carrying vertical loads downward. But acting alone, these structural members are inadequate for resisting lateral forces.

One way to provide lateral stability is to incorporate in the system diagonal members, called **bracing** (Fig. 1-3). Bracing, columns and beams then work together to carry the lateral loads downward. Another way is to rigidly connect beams to columns to prevent a change in the angle between the beams and columns, thus making them work together as a **rigid frame** to resist lateral movement. Still another way is to provide long walls, known as **shear walls,** in two perpendicular directions. Lateral forces on the building can be resolved into forces in each of these directions. The walls then act like vertical beams **(cantilevers)** in transmitting the forces to the foundations. (See also Arts. 3-5 and 3-6.)

Because of the importance of the structural system, the structural members should be protected against damage, especially from fire. For fire protection, bracing may be encased in fire-resistant floors, roofs, or walls. Similarly, columns may be encased in walls, and beams may be encased in floors. Or a fire-resistant material, such as concrete, mineral fiber, or plaster, may be used to box in the structural members (Fig. 1-6c).

See also Secs. 7 to 10.

Systems for Enclosing Buildings. Buildings are enclosed for privacy, to exclude wind, rain, and snow from the interior, and to control interior temperature and humidity. A single-enclosure type of system is one that extends continuously from the ground to enclose the floor. Simple examples are cone-like tepees and dome igloos. A multiple-enclosure type of system consists of a horizontal or inclined top covering, called a **roof** (Fig. 1-1), and vertical or inclined side enclosures called **walls**.

Roofs may have any of a wide variety of shapes. A specific shape may be selected because of appearance, need for attic space under the roof, requirements for height between roof and floor below, desire for minimum enclosed volume, structural economy,

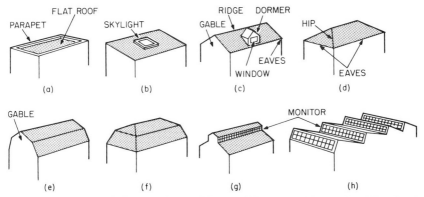

Fig. 1-4 Roofs composed of plane surfaces: (a) flat roof, (b) shed roof; (c) pitched roof; (d) hipped roof; (e) gambrel roof; (f) mansard roof; (g) monitored roof; (h) sawtooth roof. *(Reprinted with permission from F. S. Merritt, "Building Engineering and Systems Design," Van Nostrand Reinhold, New York.)*

or requirements for drainage of rainwater and shedding of snow. While roofs are sometimes given curved surfaces, more often roofs are composed of one or more plane surfaces. Some commonly used types are shown in Fig. 1-4.

The flat roof shown in Fig. 1-4a is nearly horizontal but has a slight pitch for drainage purposes. A more sloped roof is called a shed roof (Fig. 1-4b). A pitched roof (Fig. 1-4c) is formed by a combination of two inclined planes. Four inclined planes may be combined to form either a hipped roof (Fig. 1-4d) or a gambrel roof (Fig. 1-4e). A mansard roof (Fig. 1-4f) is similar to a hipped roof but, composed of additional planes, encloses a larger volume underneath. Any of the preceding roofs may have glazed openings, called **skylights** (Fig. 1-4b), for daylighting the building interior. The roofs shown in Fig. 1-4c to f are often used to enclose attic space. Windows may be set in **dormers** that project from a sloped roof (Fig. 1-4c).

Monitored roofs are sometimes used for daylighting and ventilating the interior. A **monitor** is a row of windows installed vertically, or nearly so, above a roof (Fig. 1-4g). Figure 1-4h illustrates a variation of a monitored roof that is called a sawtooth roof.

The basic element in a roof is a thin, waterproof covering, called **roofing** (Sec. 15).

Because it is thin, it is usually supported on **sheathing,** a thin layer, or **roof deck,** a thick layer, which in turn, is carried on structural members, such as beams or trusses. The roof or space below should contain thermal insulation (Fig. 1-6c and d).

Exterior walls enclose a building below the roof. The basic element in the walls is a strong, durable, water-resistant facing. For added strength or lateral stability, this facing may be supplemented on the inner side by a backing or sheathing (Fig. 1-5a). For esthetic purposes, an interior facing usually is placed on the inner side of the backing. A layer of insulation should be incorporated in walls to resist passage of heat.

Generally, walls may be built of unit masonry, panels, framing, or a combination of these materials.

Unit masonry consists of small units, such as clay brick, concrete block, glass block or clay tile, held together by a cement such as mortar. Figure 1-5a shows a wall built of concrete blocks.

Panel walls consist of units much larger than unit masonry. Made of metal, concrete, glass, plastics, or preassembled bricks, a panel may extend from foundation to roof in single-story buildings, or from floor to floor or from window header in one story to window sill of floor above in multistory buildings. Large panels may incorporate one or more windows. Figure 1-5c shows a concrete panel with a window.

Framed walls consist of slender, vertical, closely spaced structural members, tied together with horizontal members at top and bottom, and interior and exterior facings.

Fig. 1-5 Types of exterior-wall construction: (a) concrete-block wall; (b) wood-framed wall; (c) precast-concrete curtain wall.

1-14 Building Systems

Thermal insulation may be placed between the components. Figure 1-5b shows a wood-framed exterior wall.

Combination walls are constructed of several different materials. Metal, brick, concrete, or clay tile may be used as the exterior facing because of strength, durability, and water and fire resistance. These materials, however, are relatively expensive. Consequently, the exterior facing is made thin and backed up with a less expensive material. For example, brick may be used as an exterior facing with wood framing or concrete block as the backup.

Exterior walls may be classified as curtain walls or bearing walls. **Curtain walls**

(a)

(b)

serve primarily as an enclosure. Supported by the structural system, such walls need be strong enough to carry only their own weight and wind pressure on the exterior face. **Bearing walls,** in contrast, serve not only as an enclosure but also to transmit to the foundations loads from other building components, such as beams, floors, roofs, and other walls (Fig. 1-5a and b). (See also Secs. 11 and 12.)

Openings are provided in exterior walls for a variety of purposes, but mainly for windows and doors. Where openings occur, structural support must be provided over them to carry the weight of the wall above and any other loads on that portion of the wall. Usually, a beam called a lintel is placed over openings in masonry walls (Fig. 1-5a) and a beam called a top header is set over openings in wood-framed walls.

A **window** usually consists of transparent glass or plastics (**glazing**) held in place by light framing, called **sash.** The window is fitted into a frame secured to the walls (Fig. 1-5a). For sliding windows, the frame carries guides in which the sash slides. For swinging windows, stops against which the window closes are built into the frame.

Fig. 1-6 Examples of floor-ceiling and roof-ceiling systems. (a) Concrete structural slab carries hollow, steel deck, concrete fill, and flexible-tile flooring. (b) Acoustical-tile ceiling incorporating a lighting fixture with provisions for air distribution is suspended below a floor. (c) Insulated roof and steel beams are sprayed with mineral fiber for fire protection. (d) Insulated roof and open-web joists are protected by a fire-rated suspended ceiling.

Hardware is provided to enable the window to function as required. For movable windows, the hardware includes grips for moving them, locks, hinges for swinging windows, and sash balances and pulleys for vertically sliding windows.

The main purposes of windows are to illuminate the building interior with daylight, to ventilate the interior, and to give occupants a view of the outside. For retail stores, windows may have the major purpose of giving passersby a view of items displayed inside. (See also Sec. 13.)

Doors are installed in exterior walls to give access to or from the interior or to prevent such access. For similar reasons, doors are also provided in interior walls and partitions. Thus, a door may be part of a system for enclosing a building or a component of a system for enclosing interior spaces.

Systems for Enclosing Interior Spaces. The interior of a building usually is compartmented into spaces or rooms by horizontal dividers (floor-ceiling or roof ceiling systems) and vertical dividers (interior walls and partitions). (The term partitions is generally applied to non-load-bearing walls.)

Floor-Ceiling Systems. The basic element of a floor is a load-carrying deck. For protection against wear, esthetic reasons, foot comfort, or noise control, a **floor covering** often is placed over the deck, which then may be referred to as a **subfloor.** Figure 1-6a shows a concrete subfloor with a flexible-tile floor covering. A hollow-cold-formed steel deck is incorporated in the subfloor to house electric wiring.

In some cases, a subfloor may be strong and stiff enough to span, unaided, long distances between supports provided for it. In other cases, the subfloor is closely supported on beams. The subfloor in Fig. 1-6a, for example, is shown constructed integrally with concrete beams, which carry the loads from the subfloor to bearing walls or columns.

The underside of a floor or roof and of beams supporting it, including decorative treatment when applied to that side, is called a **ceiling.** Often, however, a separate ceiling is suspended below a floor or roof for esthetic or other reasons. Figure 1-6b shows such a ceiling. It is formed with acoustical panels and incorporates a lighting fixture and air-conditioning inlets and outlets.

Metal and wood subfloors and beams require fire protection. Figure 1-6c shows a roof and its steel beams protected on the underside by a sprayed-on mineral fiber. Figure 1-6d shows a roof and open-web steel joists protected on the underside by a continuous, suspended, fire-resistant ceiling. As an alternative to encasement in or shielding by a fire-resistant material, wood may be made fire-resistant by treatment with a fire-retardent chemical.

Fire Ratings. Tests have been made, usually in conformance with "Standard Methods of Tests of Building Construction and Materials," ASTM E119, developed by the American Society for Testing and Materials, to determine the length of time specific assemblies of materials can withstand a standard fire, specified in E119. On the basis of test results, each construction is assigned a fire rating, which gives the time in hours that the assembly can withstand the fire. Fire ratings for various types of construction may be obtained from local, state, or model building codes or the "Fire Resistance Design Manual," published by the Gypsum Association, 1603 Orrington Ave., Evanston, Ill. 60201.

Interior Walls and Partitions. Interior space dividers do not have to withstand such severe conditions as do exterior walls. For instance, they are not exposed to rain, snow, and solar radiation. Bearing walls, however, must be strong enough to transmit to supports below them the loads to which they are subjected. Usually, such interior walls extend vertically from the roof to the foundations of a building and carry floors and roof. The basic element of a bearing wall may be a solid core, as shown in Fig. 1-7d, or closely spaced vertical framing **(studs),** as shown in Fig. 1-7b.

Non-load-bearing partitions do not support floors or roof. Hence, partitions may be

Fig. 1-7 Types of partitions: (*a*) non-load-bearing; (*b*) gypsumboard on metal studs; (*c*) gypsumboard face panels laminated to a gypsum core panel; (*d*) concrete bearing wall, floors, and beams. *(Reprinted with permission from F. S. Merritt, "Building Engineering and Systems Design," Van Nostrand Reinhold, New York.)*

made of such thin materials as sheet metal (Fig. 1-7*a*), brittle materials as glass (Fig. 1-7*a*), or weak materials as gypsum (Fig. 1-7*c*). Light framing may be used to hold these materials in place. Because they are non-load-bearing, partitions may be built and installed to be easily shifted or to be foldable, like a horizontally sliding door. (See also Sec. 12.)

Wall Finishes. Walls are usually given a facing that meets specific architectural requirements for the spaces enclosed. Such requirements include durability under indoor conditions, ease of maintenance, attractive appearance, fire resistance, water resistance, and acoustic properties appropriate to the occupancy of the space enclosed. The finish may be the treated surface of the exposed wall material, such as the smooth, painted face of a sheet-metal panel, or a separate material, such as plaster, gypsumboard, plywood, or wallpaper. (See also Sec. 12.)

Doors. Openings are provided in interior walls and partitions to permit passage of people and equipment from one space to another. Doors are installed in the openings to control passage and also to provide privacy.

Usually, a door frame is set around the perimeter of the opening to hold the door in place (Fig. 1-8). Depending on the purpose of the door, size, and other factors, the door may be hinged to the frame at top, bottom, or either side. Or the door may be constructed to slide vertically or horizontally or to rotate about a vertical axis in the center of the opening (revolving door). (See also Sec. 14.)

Hardware is provided to enable the door to function as required. For example, hinges are provided for swinging doors, and guides are installed for sliding doors. Locks or latches are placed in or on doors to prevent them from being opened. Knobs or pulls are attached to doors for hand control.

Builders' Hardware. This is a general term applied to fastenings and devices, such as nails, screws, locks, hinges, and pulleys. These items generally are classified as either finishing hardware or rough hardware (Sec. 16).

Plumbing. The major systems for conveyance of liquids and gases in pipes within a building are

Fig. 1-8 Example of door and frame.

classified as plumbing. Plumbing pipes usually are connected to others that extend outside the building to a supply source, such as a public water main or utility gas main, or to a disposal means, such as a sewer. For health, safety, and other reasons, pipes of different types of plumbing systems must not be interconnected, and care must be taken to prevent flow from one system to another.

The major purposes of plumbing are: (1) to convey water and heating gas, if desired, from sources outside a building to points inside where the fluid or gas is needed, and (2) to collect waste water and storm water in the building, on the roof, or elsewhere on the site and convey the liquid to sewers outside the building.

For these purposes, plumbing requires fixtures for collecting discharged water and wastes; pipes for supply and disposal; valves for controlling flow; drains, and other accessories. For more details, see Sec. 19.

Heating, Ventilation, and Cooling (HVAC). Part of the environmental control systems within buildings, along with lighting and sound control, HVAC is often necessary for the health and comfort of building occupants. Sometimes, however, HVAC may be needed for manufacturing processes, product storage, or operation of equipment, such as computers. HVAC usually is used to control temperature, humidity, air movement, and air quality in the interior of buildings.

Ventilation is required to supply clean air for breathing, to furnish air for operation of combustion equipment, and to remove contaminated air. Ventilation, however, also can be used for temperature control by bringing outside air into a building when there is a desirable temperature difference between that air and the interior air.

The simplest way to ventilate is to open windows. When this is not practicable, mechanical ventilation is necessary. This method employs fans to draw outside air into the building and distribute the air, often through ducts, to interior spaces. The method, however, can usually be used only in mild weather. To maintain comfort conditions in the interior, the fresh air may have to be heated in cold weather and cooled in hot weather.

Heating and cooling of a building interior may be accomplished in any of a multitude of ways. Various methods are described in Sec. 18.

Lighting. For health, safety, and comfort of occupants, a building interior should be provided with an adequate quantity of light, good quality of illumination, and proper color of light. The required illumination may be supplied by natural or artificial means.

Daylight is the source of natural illumination. It enters a building through fenestration, such as windows in the exterior walls or monitors or skylights on the roof.

Artificial illumination can be accomplished in primitive fashion by burning of candles or of oil or gas in lamps. Generally, however, artificial illumination is more conveniently obtained through consumption of electrical energy in incandescent, fluorescent, electroluminescent, or other electric lamps. In such cases, the light source is housed in a **luminaire,** or **lighting fixture.** More details are given in Sec. 21.

Acoustics. The science of sound, its production, transmission, and effects are applied in building design for sound and vibration control.

A major objective of acoustics is provision of an environment that enhances communication in the building interior, whether the sound is created by speech or music. This is accomplished by installation of enclosures with appropriate acoustic properties around sound sources and receivers. Another important objective is reduction or elimination of noise—unwanted sound—from building interiors. This may be accomplished by elimination of the noise at the source, by installation of sound barriers, or by placing sound-absorbing materials on the surfaces of enclosures.

Still another objective is reduction or elimination of vibrations that can annoy occupants, produce noise by rattling loose objects, or crack or break parts or contents of a building. The most effective means of preventing undesirable vibrations is correction of the source. Otherwise, the source should be isolated from the building structure and potential transmission paths should be interrupted with carefully designed discontinuities.

For more details, see Sec. 17.

Electric Power and Signal Systems. Electric power is generally bought from a nearby utility and often supplemented for emergency purposes by power from batteries or a generating plant on the site. Purchased power is brought from the power lines connected to the generating source to an entrance control point and a meter in the building. From there, conductors distribute the electricity throughout the building to outlets where the power can be tapped for lighting, heating, and operating electric devices.

Two interrelated types of electrical systems are usually provided within a building. One type is used for communications, including telephone, television, background music, paging, signal and alarm systems. The second type serves the other electrical needs of the building and its occupants.

In addition to conductors and outlets, an electrical system also incorporates devices and apparatus for controlling electric voltage and current. Because electricity can be hazardous, the system must be designed and installed to prevent injury to occupants and damage to building components.

For more details, see Sec. 20.

Vertical-Circulation Elements. In multistory buildings, provision must be made for movement of people, supplies, and equipment between the various levels. This may be accomplished with ramps, stairs, escalators, elevators, dumbwaiters, vertical conveyors, pneumatic tubes, mail chutes, or belt conveyors. Some of the mechanical equipment, however, may not be used for conveyance of people.

A **ramp**, or sloping floor, is often used for movement of people and vehicles in such buildings as stadiums and garages. In most buildings, however, stairs are installed because they can be placed on a steeper slope and therefore occupy less space than ramps. Nevertheless, access to at least one entrance to a building should be provided by a ramp for use by handicapped persons.

A **stairway** consists of a series of steps and landings. Each **step** consists of a horizontal platform, or **tread,** and a vertical separation or enclosure, called a **riser** (Fig. 1-9a). Railings are placed along the sides of the stairway and floor openings for safety

Fig. 1-9 Vertical-circulation elements: (a) stairs; (b) electric traction elevator; (c) hydraulic elevator.

reasons. Also, structural members may be provided to support the stairs and the floor edges. Often, in addition, the stairway must be enclosed for fire protection.

Escalators, or moving stairs, are installed in such buildings as department stores and transportation terminals, or in the lower stories of office buildings and hotels, where there is heavy pedestrian traffic between floors. Such powered stairs consist basically of a conveyor belt with steps attached; an electric motor for moving the belt, and steps, controls, and structural supports.

Elevators are installed to provide speedier vertical transportation, especially in tall buildings. Transportation is provided in an enclosed car that moves along guides, usually within a fire-resistant vertical shaft but sometimes unenclosed along the exterior of a building. The shaft, or the exterior wall, has openings, protected by doors, at each floor to provide access to the elevator car. The car may be suspended on and moved by cables (Fig. 1-9b) or set atop a piston moved by hydraulic pressure (Fig. 1-9c).

More information on vertical-circulation elements is given in Sec. 22.

(F. S. Merritt, "Building Engineering and Systems Design," Van Nostrand Reinhold, New York.)

1-8. Value Engineering.

As indicated in Art. 1-3, the client in the initial design phase develops a program, or list of requirements. The goal of the designers is to select a system that meets these requirements. Before the designers do this, however, it is advisable for them to question whether the requirements represent the client's actual needs. Can the criteria and standards affecting the design be made less stringent? After the program has been revised to answer these questions, the designers select a system. Next, it is advisable for the designers to question whether the system provides the best value at the lowest cost. Value engineering is a useful procedure for answering this question and selecting a better alternative if the answer indicates this is desirable.

Value engineering is the application of the scientific method to the study of values of systems. The major objective of value engineering in building design and construction is reduction of initial and life-cycle costs (Art. 1-6). Thus, value engineering has one of the objectives of systems design, in which the overall goal is production of an optimum building, and should be incorporated in the systems-design procedure.

The **scientific method,** which is incorporated in the definitions of value engineering and systems design, consists of the following steps:

1. Collection of data and observations of natural phenomena
2. Formulation of an hypothesis capable of predicting future observations
3. Testing of the hypothesis to verify the accuracy of its predictions and abandonment or improvement of the hypothesis if it is inaccurate

Those who conduct or administer value studies are often called **value engineers** or **value analysts.** They generally are organized into an interdisciplinary team for value studies for a specific project. Sometimes, however, an individual, such as an experienced contractor, performs value engineering services for the client for a fee or a percentage of savings achieved by the services.

Value Analysis. Value is a measure of benefits anticipated from a system or from the contribution of a component to system performance. This measure must be capable of serving as a guide in a choice between alternatives in evaluations of system performance. Because generally in comparisons of systems only relative values need be considered, value takes into account both advantages and disadvantages, the former being considered positive and the latter negative. It is therefore possible in comparisons of systems that the value of a component of a system may be negative and subtracts from the overall performance of the system.

System evaluations would be relatively easy if a monetary value could always be placed on performance. Then, benefits and costs could be compared directly. Value, however, often must be based on a subjective decision of the client. For example, how much extra is an owner willing to pay for beauty, prestige, or better community relations? Will the owner accept gloom, glare, draftiness, or noise for a savings in cost? Consequently, other values than monetary must be considered in value analysis. Such considerations require determination of the relative importance of the client's requirements and weighting of values accordingly.

Value analysis is the part of the value-engineering procedure devoted to investigation of the relation between costs and values of components and systems and alternatives to these. The objective is to provide a rational guide for selection of the lowest-cost system that meets the client's actual needs.

Measurement Scales. For the purposes of value analysis, it is essential that characteristics of a component or system on which a value is to be placed be distinguishable. An analyst should be able to assign different numbers, not necessarily monetary, to values that are different. These numbers may be ordinates of any one of the following four measurement scales: ratio, interval, ordinal, nominal.

Ratio Scale. This scale has the property that, if any characteristic of a system is assigned a value number k, any characteristic that is n times as large must be assigned a value number nk. Absence of the characteristic is assigned the value zero. This type of scale is commonly used in engineering, especially in cost comparisons. For example, if a value of $10,000 is assigned to system A and of $5000 to system B, then A is said to cost twice as much as B.

Interval Scale. This scale has the property that equal intervals between assigned values represent equal differences in the characteristic being measured. The scale zero is assigned arbitrarily. The Celsius scale of temperature measurement is a good example of an interval scale. Zero is arbitrarily established as the temperature at which water freezes; the zero value does not indicate absence of heat. The boiling point of water is arbitrarily assigned the value of 100. The scale between 0 and 100 is then divided into 100 equal intervals called degrees (°C). Despite the arbitrariness of the selection of the zero point, the scale is useful in heat measurement. For example, changing the temperature of an object from 40°C to 60°C, an increase of 20°C, requires twice as much heat as changing the temperature from 45°C to 55°C, an increase of 10°C.

Ordinal Scale. This scale has the property that the magnitude of a value number assigned to a characteristic indicates whether a system has more, or less, of the characteristic than another system has or is the same with respect to that characteristic. For example, in a comparison of the privacy afforded by different types of partitions, each may be assigned a number that ranks it in accordance with the degree of privacy that it provides. Partitions with better privacy are given larger numbers. Ordinal scales are commonly used when values must be based on subjective judgments of nonquantifiable differences between systems.

Nominal Scale. This scale has the property that the value numbers assigned to a characteristic of systems being compared merely indicate whether the systems differ in this characteristic. But no value can be assigned to the difference. This type of scale is often used to indicate the presence or absence of a characteristic or component. For example, the absence of a means of access to equipment for maintenance may be represented by zero or a blank space, whereas the presence of such access may be denoted by 1 or X.

Weighting. In practice, construction cost usually is only one factor, perhaps the only one with a monetary value, of several factors that must be evaluated in a comparison of systems. In some cases, some of the other characteristics of the system may be more important to the owner than cost. Under such circumstances, the comparison may be made by use of an ordinal scale for ranking each characteristic and then weighting the rankings in accordance with the importance of the characteristic to the owner.

As an example of the use of this procedure, calculations for comparison of two partitions are shown in Table 1-1. Alternative 1 is an all-metal partition and alternative 2 is made of glass and metal.

In Table 1-1, characteristics of concern in the comparison are listed in the first column. The numbers in the second column indicate the relative importance of each char-

TABLE 1-1 Comparison of Alternative Partitions*

		Alternatives			
		1 All metal		2 Glass and metal	
Characteristics	Relative importance	Relative value	Weighted value	Relative value	Weighted value
Construction cost	8	10	80	8	64
Appearance	9	7	63	9	81
Sound transmission	5	5	25	4	20
Privacy	3	10	30	2	6
Visibility	10	0	0	8	80
Movability	2	8	16	8	16
Power outlets	4	0	0	0	0
Durability	10	9	90	9	90
Low maintenance	8	7	56	5	40
Total weighted values			360		397
Cost			$12,000		$15,000
Ratio of values to cost			0.0300		0.0265

*Reprinted with permission from F. S. Merritt, "Building Engineering and Systems Design," Van Nostrand Reinhold Company, New York.

acteristic to the owner: 1 denotes lowest priority and 10 highest priority. These are the weights. In addition, each of the partitions is ranked on an ordinal scale, with 10 as the highest value, in accordance with the degree to which it possesses each characteristic. These rankings are listed as relative values in Table 1-1. For construction cost, for instance, the metal partition is assigned a relative value of 10 and the glass-metal partition a value of 8, because the metal partition costs a little less than the other one. In contrast, the glass-metal partition is given a relative value of 8 for visibility, because the upper portion is transparent, whereas the metal partition has a value of zero, because it is opaque.

To complete the comparison, the weight of each characteristic is multiplied by the relative value of the characteristic for each partition and entered in Table 1-1 as a weighted value. For construction cost, for example, the weighted values are $8 \times 10 = 80$ for the metal partition and $8 \times 8 = 64$ for the glass-metal partition. The weighted values for each partition are then added, yielding 360 for alternative 1 and 397 for alternative 2. While this indicates that the glass-metal partition is better, it may not be the best for the money. To determine whether it is, the weighted value for each partition is divided by its cost, yielding 0.0300 for the metal partition and 0.0265 for the other. Thus, the metal partition appears to offer more value for the money and would be recommended.

Economic Comparisons. In a choice between alternative systems, only the difference between system values are significant and need to be compared.

Suppose, for example, the economic effect of adding 1 in of thermal insulation to a building is to be investigated. In a comparison, it is not necessary to compute the total cost of the building with and without the insulation. Generally, the value analyst need only subtract the added cost of 1 in of insulation from the decrease in HVAC cost to obtain the net saving or cost increase resulting from addition of insulation. A net saving would encourage addition of insulation. Thus, a decision can be reached without the complex computation of total building cost.

In evaluating systems, value engineers must take into account not only initial and life-cycle costs but also the return the client wishes to make on the investment in the building. Generally, a client would like not only to maximize profit, the difference between revenue from use of the building and total costs, but also ensure that the rate of return, the ratio of profit to investment, is larger than all of the following:

Rate of return expected from the type of business
Interest rate for borrowed money
Rate for government bonds or notes
Rate for highly rated corporate bonds

The client is concerned with interest rates because all costs represent money that must be borrowed or that could otherwise be invested at a current interest rate. The client also has to be concerned with time, measured from the date at which an investment is made, because interest cost increases with time. Therefore, in economic comparisons of systems, interest rates and time must be taken into account. (Effects of monetary inflation can be taken into account in much the same way as interest.)

An economic comparison usually requires evaluation of initial capital investments, salvage values after several years, annual disbursements and annual revenues. Because each element in such a comparison may have associated with it an expected useful life different from that of the other elements, the different types of costs and revenues must be made commensurable by reduction to a common basis. This is commonly done by either:

1. Converting all costs and revenues to equivalent uniform annual costs and income
2. Converting all costs and revenues to present worth of all costs and revenues at time zero

Present worth is the money that, invested at time zero, would yield at later times required costs and revenues at a specified interest rate. In economic comparisons, the conversions should be based on a rate of return on investment that is attractive to the client. It should not be less than the interest rate the client would have to pay if the amount of the investment had to be borrowed. For this reason, the desired rate of return is called interest rate in conversions. Calculations also should be based on actual or reasonable estimates of time periods. Salvage values, for instance, should be taken as the expected return on sale or trade-in of an item after a specific number of years that it has been in service. Interest may be considered compounded annually.

Future Value. Based on the preceding assumptions, a sum invested at time zero increases in time to

$$S = P(1 + i)^n \qquad (1\text{-}1)$$

where S = future amount of money, equivalent to P, at the end of n periods of time with interest i
i = interest rate

n = number of interest periods, years
P = sum of money invested at time zero = present worth of S

Present Worth. Solution of Eq. (1-1) for P yields the present worth of a sum of money S at a future date:

$$P = S(1 + i)^{-n} \tag{1-2}$$

The present worth of payments R made annually for n years is

$$P = R\frac{1 - (1 + i)^{-n}}{i} \tag{1-3}$$

The present worth of the payments R continued indefinitely can be obtained from Eq. (1-3) by making n infinitely large:

$$P = \frac{R}{i} \tag{1-4}$$

Capital Recovery. A capital investment P at time zero can be recovered in n years by making annual payments of

$$R = P\frac{i}{1 - (1 + i)^{-n}} = P\left[\frac{i}{(1 + i)^n - 1} + i\right] \tag{1-5}$$

When an item has salvage value V after n years, capital recovery R can be computed from Eq. (1-5) by subtraction of the present worth of the salvage value from the capital investment P.

$$R = [P - V(1 + i)^{-n}]\left[\frac{i}{(1 + i)^n - 1} + i\right] \tag{1-6}$$

Example. To illustrate the use of these formulas, an economic comparison is made in the following for two air-conditioning units being considered for an office building. Costs are estimated as follows:

	Unit 1	Unit 2
Initial cost	$300,000	$500,000
Life, years	10	20
Salvage value	$50,000	$100,000
Annual costs	$30,000	$20,000

Cost of operation, maintenance, repairs, property taxes and insurance are included in the annual costs. The present-worth method is used for the comparison, with interest rate $i = 8\%$.

Conversion of all costs and revenues to present worth must be based on a common service life, although the two units have different service lives, 10 and 20 years, respectively. For the purpose of the conversion, it may be assumed that replacement assets will repeat the investment and annual costs predicted for the initial asset. (Future values, however, should be corrected for monetary inflation.) In some cases, it is convenient to select for the common service life the least common multiple of the lives of the units being compared. In other cases, it may be more convenient to assume that the investment and annual costs continue indefinitely. The present worth of such annual costs is called **capitalized cost.**

For this example, a common service life of 20 years, the least common multiple of 10 and 20, is selected. Hence, it is assumed that unit 1 will be replaced at the end of the tenth period at a cost of $300,000 less the salvage value. Similarly, the replacement unit will be assumed to have the same salvage value after 20 years.

TABLE 1-2 Example Comparison of Two Air-Conditioning Units

	Unit 1	Unit 2
Initial investment	$300,000	$500,000
Present worth of replacement cost in 10 years $P - V$ at 8% interest [Eq. (1-2)]	115,800	115,800
Present worth of annual costs for 20 years at 8% interest [Eq. (1-3)]	294,540	196,360
Present worth of all costs	710,340	696.360
Revenue:		
Present worth of salvage value after 20 years at 8% interest [Eq. (1-2)]	10,730	21,450
Net cost:		
Present worth of net cost in 20 years at 8% interest	$699,610	$674,910

The calculations in Table 1-2 indicate that the present worth of the net cost of unit 2 is less than that for unit 1. If cost were the sole consideration, purchase of unit 2 would be recommended.

Value Analysis Procedure. In building design, value analysis generally starts with a building system or subsystem proposed by the architect and consultants. The client or the client's representative appoints an interdisciplinary team to study the system or subsystem and either recommend its use or propose a more economical alternative. The team coordinator sets goals and priorities for the study and may appoint task groups to study parts of the building in accordance with the priorities. The value analysts should follow a systematic, scientific procedure for accomplishing all the necessary tasks that comprise a value analysis. The procedure should provide an expedient format for recording the study as it progresses, assure that consideration has been given to all information, some of which may have been overlooked in development of the proposed system, and logically resolve the analysis into components that can be planned, scheduled, budgeted, and appraised.

The greatest cost reduction can be achieved by analysis of every component of a building. This, however, is not practical, because of the short time usually available for the study and because the cost of the study increases with time. Hence, it is advisable that the study concentrate on those building systems (or subsystems) whose cost is a relatively large percentage of the total building (or system) cost, because those components have possibilities for substantial cost reduction.

During the initial phase of value analysis, the analysts should obtain a complete understanding of the building and its major systems by rigorously reviewing the program, proposed design and all other pertinent information. They should also define the functions, or purposes, of each building component to be studied and estimate the cost of accomplishing the functions. Thus, the analysts should perform a systems analysis, as indicated in Art. 1-2, answer the questions listed in Art. 1-2 for the items to be studied, and estimate the initial and life-cycle costs of the items.

In the second phase of value analysis, the analysts should question the cost-effectiveness of each component to be studied. Also, by use of imagination and creative techniques, they should generate several alternative means for accomplishing the required functions of the component. Then, in addition to answers to the questions in Art. 1-2, the analysts should obtain answers to the following questions:

Do the original design and each alternative meet performance requirements?
What does each cost installed and over the life cycle?

Will it be available when needed? Will skilled labor be available?
Can any components be eliminated?
What other components will be affected by adoption of an alternative? What will the resulting changes in the other components cost? Will there be a net saving in cost?

In investigating the possibility of elimination of a component, the analysts also should see if any part of it can be eliminated, if two parts or more can be combined into one, and if the number of different sizes and types of an element can be reduced. If costs might be increased by use of a nonstandard or unavailable item, the analysts should consider substitution of a more appropriate alternative. In addition, consideration should be given to simplification of construction or installation of components and to ease of maintenance and repair.

In the following phase of value analysis, the analysts should critically evaluate the original design and alternatives. The ultimate goal should be recommendation of the original design or an alternative, whichever offers the greatest value and cost-savings potential. The analysts also should submit estimated costs for the original design and the alternative.

In the final phase, the analysts should prepare and submit to the client or the client's representative who appointed them a written report on the study and resulting recommendations. Also, they should submit a workbook containing detailed backup information.

Value engineering should start during the conceptual phase of design. Then, it has the greatest impact on cost control and no cost is involved in making design changes. During later design phases, design changes involve some cost, especially when substitution of major subsystems is involved, but the cost is nowhere near as great as when changes are made during construction. Such changes should be avoided if possible. Value engineering, however, should be applied to the project specifications and construction contract. Correction of unnecessary and overconservative specifications and contract provisions offers considerable potential for cost reduction.

(E. D. Heller, "Value Management: Value Engineering and Cost Reduction," Addison-Wesley, Reading, Mass.; L. D. Miles, "Techniques of Value Analysis and Engineering," McGraw-Hill Book Co., New York; A. Mudge, "Value Engineering," McGraw-Hill Book Company, New York; M. C. Macedo, P. V. Dobrow and J. J. O'Rourke, "Value Management for Construction," John Wiley & Sons, Inc., New York.)

1-9. Systems Design.

The basic traditional design procedure (Art. 1-3), which has been widely used for many years, and commonly used variations of it have resulted in many excellent buildings. It needs improvement, however, because clients cannot be certain that its use gives the best value for the money or that the required performance could not have been attained at lower cost. The uncertainty arises because historically:

    1. Actual construction costs often exceed low bids or negotiated prices, because of design changes during construction; unanticipated delays during construction, which increase costs; and unforeseen conditions, such as unexpectedly poor subsurface conditions that make excavation and foundation construction more expensive.

    2. Construction, operation, or maintenance costs are higher than estimated, because of design mistakes or omissions.

    3. Separation of design and construction into different specialties leads to underestimated or overestimated construction costs and antagonistic relations between designers and builders.

4. Construction costs are kept within the client's budget at the expense of later higher operating, maintenance, and repair costs.

5. Coordination of the output of architects and consultants is not sufficiently close for production of an optimum building for the client's actual needs.

One objective of systems design is to correct these defects. This can be done while retaining the desirable features of traditional procedures, such as development of building design in stages, with progressively more accurate cost estimates and frequent client review. Systems design therefore should at least do the following:

1. Question the cost effectiveness of proposed building components and stimulate generation of lower-cost alternatives that achieve the required performance. This can be done by incorporating value engineering in systems design.

2. More closely coordinate the work of various design specialists and engage building construction and operation experts to assist in design.

3. Take into account both initial and life-cycle costs.

4. Employ techniques that will reduce the number of design mistakes and omissions that are not discovered until after construction starts.

Systems Design Procedure. Article 1-2 defines systems and explains that systems design comprises a rational, orderly series of steps that leads to the best decision for a given set of conditions. Article 1-2 also lists the basic components of the procedure as analysis, synthesis, appraisal, and feedback. Following is a more formal definition:

Systems design is the application of the scientific method to selection and assembly of components or subsystems to form the optimum system to attain specified goals and objectives while subject to given constraints and restrictions.

The scientific method is defined in Art. 1-8. Goals, objectives, and constraints are discussed later.

Systems design of buildings, in addition to correcting defects in traditional design, must provide answers to the following questions:

1. What does the client actually want the building to accomplish (goals, objectives, and associated criteria)?

2. What conditions exist, or will exist after construction, that are beyond the designers' control?

3. What requirements for the building or conditions affecting system performance does design control (constraints and associated standards)?

4. What performance requirements and time and cost criteria can the client and designers use to appraise system performance?

Collection of information necessary for design of the building starts at the inception of design and may continue through the contract documents phase. Data collection is an essential part of systems design but because it is continuous throughout design it is not listed as one of the basic steps.

For illustrative purposes, the systems design procedure is shown resolved into nine basic steps in Fig. 1-10. Because value analysis is applied in step 5, steps 4 through 8 covering synthesis, analysis, and appraisal may be repeated several times. Each iteration should bring the design closer to the optimum.

In preparation for step 1, the designers should secure a building program and information on existing conditions that will affect building design. In step 1, the designers use the available information to define goals to be met by the system.

Goals. These state what the building is to accomplish, how it will affect the environment and other systems, and how other systems and the environment will affect the

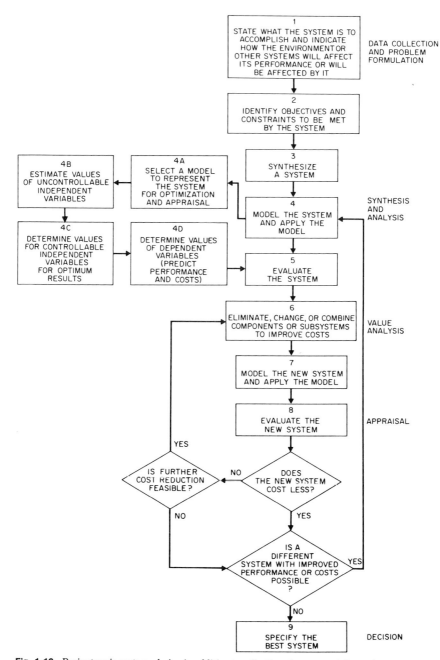

Fig. 1-10 Basic steps in systems design in addition to collection of necessary information.

building. Goals should be generalized but brief statements, encompassing all the design objectives. They should be sufficiently specific, however, to guide generation of initial and alternative designs and control selection of the best alternative.

A simple example of a goal is: Design a branch post-office building with 100 employees to be constructed on a site owned by the client. The building should harmonize with neighboring structures. Design must be completed within 90 days and construction within 1 year. Construction cost is not to exceed $500,000.

When systems design is applied to a subsystem, goals serve the same purpose as for a system. They indicate the required function of the subsystem and how it affects and is affected by other systems.

Objectives. With the goals known, the designers can advance to step 2 and define the system objectives. These are similar to goals but supply in detail the requirements that the system must satisfy to attain the goals.

In listing objectives, the designers may start with broad generalizations that they later develop at more detailed levels to guide design of the system. Some objectives, such as minimization of initial costs, life-cycle costs and construction time, should be listed. Other objectives that apply to the design of almost every building, such as the health, safety, and welfare objectives of the building, zoning, and Occupational Safety and Health Administration regulations, are too numerous to list and may be adopted by reference. Objectives should be sufficiently specific to guide the planning of building interior spaces and selection of specific characteristics for the building and its components: appearance, strength, durability, stiffness, operational efficiency, maintenance, and fire resistance. Also, objectives should specify the degree of control needed for operation of systems provided to meet the other objectives.

At least one criterion must be associated with each objective. The criterion is a range of values within which the performance of the system must lie for the objective to be met. The criterion should be capable of serving as a guide in evaluations of alternative systems. For example, for fire resistance of a wall, the criterion might be 2-hr fire rating.

In addition to establishing criteria, the designers should weight the objectives in accordance with the relative importance of the objectives to the client (Art. 1-8). These weights should also serve as guides in comparisons of alternatives.

System Constraints. In step 2 of systems design, the designers should also define constraints on the system. Constraints are restrictions on the values of design variables that represent properties of the system and are controllable by the designers. Designers are seldom completely free to choose any values desired for controllable variables because of various restrictions, which may be legal ones such as building or zoning code requirements, or may be economic, physical, chemical, temporal, psychological, sociological, or esthetic requirements. Such restrictions may fix the values of the controllable variables or establish a range in which they must lie.

At least one standard must be associated with each constraint. A standard is a value or range of values governing a property of the system. The standard specifying a fixed value may be a minimum or maximum value.

For example, a designer may be seeking to determine the thickness of a load-bearing brick wall. The local building code may state that such a wall may not be less than 8 in thick. This requirement is a minimum standard. The designer may then select a wall thickness of 8 in or more. The requirements of other systems, however, may indicate that the wall thickness may not exceed 16 in. This is a maximum standard. Furthermore, bricks may be available only in nominal widths of 4 in. Hence, the constraints limit the values of the controllable variable, in this case wall thickness, to 8, 12, or 16 in.

Synthesis. In step 3, the designers must conceive at least one system that satisfies the objectives and constraints. For this, they rely on their past experience, knowledge, imagination, and creative skills and on advice from consultants, including value engineers, construction experts, and experienced operators of the type of facilities to be designed.

In addition, the designers should select systems that are cost-effective and can be erected speedily. To save design time in selection of a system, the designers should investigate alternative systems in a logical sequence for potential for achieving optimum results. The following is a possible sequence:

1. Selection of an available industrialized building, a system that is preassembled in a factory. Such a system is likely to be low cost, because of the use of mass-production techniques and factory wages, which usually are lower than those for field personnel. Also, the quality of materials and construction may be better than for custom-built structures, because of assembly under controlled conditions and close supervision.

2. Design of an industrialized building (if the client needs several of the same type of structure).

3. Assembling a building with prefabricated components or systems. This type of construction is similar to that used for industrialized buildings except that the components preassembled are much smaller parts of the building system.

4. Specification of as many prefabricated and standard components as feasible. Standard components are off-the-shelf items, readily available from building supply companies.

5. Repetition of the same component as many times as possible. This may permit mass production of some nonstandard components. Also, repetition may speed construction, because field personnel will work faster as they become familiar with the components.

6. Design of components for erection so that building trades will be employed on the site continuously. Work that compels one trade to wait for completion of work by another trade delays construction and is costly.

Models. In step 4, the designers should represent the system by a model that will enable them to analyze the system and evaluate its performance. The model should be simple, consistent with the role for which it is selected, for practical reasons. The cost of formulating and using the model should be negligible compared with the cost of assembling and testing the actual system.

For every input to a system, there must be a known, corresponding input to the model such that the responses (output) of the model to that input are determinable and correspond to the response of the system to its input. The correlation may be approximate but nevertheless close enough to serve the purposes for which the model is to be used. For example, for cost estimates during the conceptual phase of design, use may be made of a cost model that yields only reasonable guesses of construction costs. The cost model used in the contract documents phase, however, should be accurate.

Models may be classified as iconic, symbolic or analog. The **iconic type** may be the actual system or a part of it or merely bear a physical resemblance to the actual system. This type is often used for physical tests of performance, such as load or wind-tunnel tests or adjustments of controls. **Symbolic models** represent by symbols the input and output of a system and are usually amenable to mathematical analysis of a system. They enable relationships to be generally, yet compactly, expressed, are less costly to develop and use than other types of models, and are easy to manipulate. **Analog models** are real systems but with physical properties different from those of the actual system. Examples include dial watches for measuring time, thermometers for measuring heat changes, slide rules for multiplying numbers, flow of electric current for measuring heat flow through a metal plate, and soap membranes for measuring torsion in an elastic shaft.

Variables representing input and properties of a system may be considered independent variables. These are of two types:

1. Variables that the designers can control or constraints: $x_1, x_2, x_3, \ldots$
2. Variables that are uncontrollable: $y_1, y_2, y_3, \ldots$

Variables representing system output or performance may be considered dependent variables: $z_1, z_2, z_3, \ldots$

The dependent variables are functions of the independent variables. These functions also contain parameters, which can be adjusted in value to calibrate the model to the behavior of the actual system.

Step 4 of systems design then may be resolved into four steps, as indicated in Fig. 1-10:

1. Select and calibrate a model to represent the system for optimization and appraisal.
2. Estimate values for the uncontrollable, independent variables.
3. Determine values for the controllable variables.
4. Determine the output or performance of the system from the relationship of dependent and independent variables by use of the model.

Cost Models. As an example of the use of models in systems design, consider the following cost models:

$$C = Ap \tag{1-7}$$

where C = construction cost of building
A = floor area, ft^2, in the building
p = unit construction cost, dollars per square foot

This is a symbolic model applicable only in the early stages of design when systems and subsystems are specified only in general form. Both A and p are estimated, usually on the basis of past experience with similar types of buildings.

$$C = \Sigma\, A_i p_i \tag{1-8}$$

where A_i = convenient unit of measurement for the ith system
p_i = cost per unit for the ith system

This symbolic cost model is suitable for estimating building construction cost in preliminary design stages after types of major systems have been selected. Equation (1-8) gives the cost as the sum of the cost of the major systems, to which should be added the estimated costs of other systems and contractor's overhead and profit. A_i may be taken as floor or wall area, square feet, pounds of steel, cubic yards of concrete, or any other applicable parameter for which the unit cost may be reasonably accurately estimated.

$$C = \Sigma\, A_j p_j \tag{1-9}$$

where A_j = convenient unit of measurement for the jth subsystem
p_j = cost per unit for the jth subsystem

This symbolic model may be used in the design development phase and later after components of the major systems have been selected and greater accuracy of the cost estimate is feasible. Equation (1-9) gives the construction cost as the sum of the costs of all the subsystems, to which should be added contractor's overhead and profit.

For more information on cost estimating, see Sec. 24.

Optimization. The objective of systems design is to select the single best system for a given set of conditions, a process known as optimization. When more than one property of the system is to be optimized or when there is a single characteristic to be optimized but it is nonquantifiable, an optimum solution may or may not exist. If it does exist, it may have to be found by trial and error with a model or by methods such as those described in Art. 1-8.

When one characteristic, such as construction cost, of a system is to be optimized, the criterion may be expressed as

$$\text{Optimize } z_r = f_r(x_1, x_2, x_3, \ldots y_1, y_2, y_3, \ldots) \tag{1-10}$$

where z_r = dependent variable to be maximized or minimized
x = controllable variable, identified by a subscript
y = uncontrollable variable, identified by a subscript
f_r = objective function

Generally, however, there are restrictions on values of the independent variables. These restrictions may be expressed as

$$\begin{aligned} f_1(x_1, x_2, x_3, \ldots y_1, y_2, y_3, \ldots) &\geq 0 \\ f_2(x_1, x_2, x_3, \ldots y_1, y_2, y_3, \ldots) &\geq 0 \\ &\cdots\cdots\cdots\cdots\cdots\cdots\cdots\cdots\cdots \\ f_n(x_1, x_2, x_3, \ldots y_1, y_2, y_3, \ldots) &\geq 0 \end{aligned} \tag{1-11}$$

Simultaneous solution of Eqs. (1-10) and (1-11) yields the optimum values of the variables. The solution may be obtained by use of such techniques as calculus, linear programming, or dynamic programming depending on the nature of the variables and the characteristics of the equations.

Direct application of Eqs. (1-10) and (1-11) to a whole building, its systems and its larger subsystems usually is impractical, because of the large number of variables and the complexity of their relationships. Hence optimization generally has to be attained in a different way, generally by such methods as suboptimization or simulation.

Systems with large numbers of variables may sometimes be optimized by a process called **simulation,** which involves trial and error with the actual system or a model. In simulation, the properties of the system or model are adjusted with a specific input or range of inputs to the system, and outputs or performance are measured until an optimum result is obtained. When the variables are quantifiable and models are used, the solution usually can be expedited by use of high-speed electronic computers. The actual system may be used when it is available and accessible and changes in it will have little or no effect on construction costs. For example, after installation of air ducts, an air-conditioning system may be operated for a variety of conditions to determine the optimum damper position for control of airflow for each condition.

Suboptimization is a trial-and-error process in which designers try to optimize a system by first optimizing its subsystems. It is suitable when components influence each other in series. For example, consider a structural system consisting only of roof, columns, and footings. The roof has a known load (input), exclusive of its own weight. Design of the roof affects the columns and footings, because its output equals the load on the columns. Design of the columns loads only the footings. Design of the footings, however, has no effect on any of the other structural components. Therefore, the structural components are in series and they may be designed by suboptimization to obtain the minimum construction cost or least weight of the system.

Suboptimization of the system may be achieved by first optimizing the footings; for example, designing the lowest-cost footings. Next, the design of both the columns and the footings should be optimized. (Optimization of the columns alone will not yield an optimum structural system, because of the effect of the column weight on the footings.) Finally, roof, columns, and footings together should be optimized. (Optimization of the roof alone will not yield an optimum structural system, because of the effect of its weight on columns and footings. A low-cost roof may be very heavy, requiring costly columns and footings, whereas the cost of a lightweight roof may be so high as to offset any savings from less-expensive columns and footings. An alternative roof may provide optimum results.)

Appraisal. In step 5 of systems design, the designers should evaluate the results obtained in step 4, modeling the system and applying the model. The designers should verify that construction and life-cycle costs will be acceptable to the client and that the proposed system satisfies all objectives and constraints.

During the preceding steps, value analysis may have been applied to parts of the building. In step 6, however, value analysis should be applied to the whole building system. This process may result in changes only to parts of the system, producing a new system, or several alternatives to the original design may be proposed. In steps 7 and 8, therefore, the new systems, or at least those with good prospects, should be modeled and evaluated. During and after this process, completely different alternatives may be conceived. As a result, steps 4 through 8 should be repeated for the new concepts. Finally, in step 9, the best of the systems studied should be selected.

(R. J. Aguilar, "Systems Analysis and Design in Engineering, Architecture, Construction and Planning," Prentice-Hall, Inc., Englewood Cliffs, N.J.; R. L. Ackoff and M. W. Saseini, "Fundamentals of Operations Research," John Wiley & Sons, Inc., New York; K. I. Majid, "Optimum Design of Structures," Halsted Press/Wiley, New York; E. J. McCormick, "Human Factors in Engineering," McGraw-Hill Book Company, New York; F. S. Merritt, "Building Engineering and Systems Design," Van Nostrand Reinhold, New York; R. DeNeufville and J. H. Stafford, "Systems Analysis for Engineers and Managers," McGraw-Hill Book Company, New York; L. Spunt, "Optimum Structural Design," Prentice-Hall, Englewood Cliffs, N.J.)

1-10. Building Codes.

Many of the restrictions encountered in building design are imposed by legal regulations. While all must be met, those in building codes are the most significant because they affect almost every part of a building.

Building codes are established under the police powers of a state to protect the health, welfare, and safety of communities. A code is administered by a building official of the municipality or state that adopts it by legislation. Development of a local code may be guided by a national model code, such as those promulgated by the International Conference of Building Officials, Inc., Building Officials and Code Administrators International, Inc., Southern Building Code Congress International, Inc., and National Conference of States on Building Codes and Standards.

In general, building-code requirements are the minimum needed for public protection. Design of a building must satisfy these requirements. Often, however, architects and engineers must design more conservatively, to meet the client's needs, produce a more efficient building system, or take into account conditions not covered fully by code provisions.

Construction drawings for a building should be submitted to the building-code administrator before construction starts. If the building will meet code requirements, the administrator issues a building permit, on receipt of which the contractor may commence building. During construction, the administrator sends inspectors periodically to inspect the work. If they discover a violation, they may issue an order to remove it or they may halt construction, depending on the seriousness of the violation. On completion of construction, if the work conforms to code requirements, the administrator issues to the owner a certificate of occupancy.

Forms of Codes. Codes often are classified as specifications type or performance type. A specification-type code names specific materials for specific uses and specifies minimum or maximum dimensions, for example, "a brick wall may not be less than 6 in thick." A performance-type code, in contrast, specifies required performance of a construction but leaves materials, methods, and dimensions for the designers to choose. Performance-type codes are generally preferred, because they give designers greater design freedom in meeting clients' needs, while satisfying the intent of the code. Most codes, however, are neither strictly specifications nor performance type but rather a mixture of the two. The reason for this is that insufficient information is currently available for preparation of an entire enforceable performance code.

The organization of building codes varies with locality. Generally, however, they consist of two parts, one dealing with administration and enforcement and the other specifying requirements for design and construction in detail.

Part 1 usually covers licenses, permits, fees, certificates of occupancy, safety, projections beyond street lines, alterations, maintenance, applications, approval of drawings, stop-work orders, and posting of buildings to indicate permissible live loads and occupant loads.

Part 2 gives requirements for structural components, lighting, HVAC, plumbing, gas piping and fixtures, elevators and escalators, electrical distribution, stairs, corridors, walls, doors, and windows. This part also defines and sets limits on occupancy and construction-type classifications. In addition, the second part contains provisions for safety of public and property during construction operations and for fire protection and means of egress after the building is occupied.

Many of the preceding requirements are *adopted by reference* in the code from nationally recognized standards or codes of practice. These may be promulgated by agencies of the Federal government or by such organizations as the American National Standards Institute, American Society for Testing and Materials, American Institute of Steel Construction, American Concrete Institute, and American Institute of Timber Construction.

Code Classifications of Buildings. Building codes usually classify a building in accordance with the fire zone in which it is located, the type of occupancy, and the type of construction, which is an indication of the fire protection offered.

The **fire zone** in which a building is located may be determined from the community's fire-district zoning map. The building code specifies the types of construction and occupancy groups permitted or prohibited in each fire zone.

The **occupancy group** to which a building official assigns a building depends on the use to which the building is put. Typical classifications include one- and two-story dwellings; apartment buildings, hotels, dormitories; industrial buildings with noncombustible, combustible, or hazardous contents; schools; hospitals and nursing homes; and places of assembly, such as theaters, concert halls, auditoriums, and stadiums.

Type of construction of a building is determined, in general, by the fire ratings assigned to its components. A code usually establishes two major categories: combustible and noncombustible construction. The combustible type may be subdivided in accordance with the fire protection afforded major structural components and the rate at which they will burn; for example, heavy timber construction is considered slow-burning. The noncombustible type may be subdivided in accordance with the fire-resistive characteristics of components.

Building codes may limit the floor area that may be enclosed at any level within walls of specified fire resistance. The limitations depend on occupancy group and type of construction. The purpose is to delay or prevent spread of fire over large portions of the building. For the same reason, building codes also may restrict building height and number of stories. In addition, to permit rapid and orderly egress in emergencies, such as fire, codes limit the occupant load, or number of persons allowed in a building or room. In accordance with permitted occupant loads, codes indicate the number of exits of adequate capacity and fire protection that must be provided.

1-11. Zoning Codes.

Like building codes, zoning codes are established under the police powers of the state, to protect the health, welfare, and safety of the public. Zoning, however, primarily regulates land use by controlling types of occupancy of buildings, building height, and density and activity of population in specific parts of a jurisdiction.

Zoning codes are usually developed by a planning commission and administered by the commission or a building department. Land-use controls adopted by the local planning commission for current application are indicated on a zoning map. It divides the jurisdiction into districts, shows the type of occupancy, such as commercial, industrial, or residential, permitted in each district and notes limitations on building height and bulk and on population density in each district.

The planning commission usually also prepares a master plan as a guide to the growth of the jurisdiction. A future land-use plan is an important part of the master plan. The commission's objective is to steer changes in the zoning map in the direction of the future land-use plan. The commission, however, is not required to adhere rigidly to the plans for the future. As conditions warrant, the commission may grant variances from any of the regulations.

In addition, the planning commission may establish land subdivision regulations, to control development of large parcels of land. While the local zoning map specifies minimum lot area for a building and minimum frontage a lot may have along a street, subdivision regulations, in contrast, specify the level of improvements to be installed in new land-development projects. These regulations contain criteria for location, grade, width, and type of pavement of streets, length of blocks, open spaces to be provided, and right of way for utilities.

A jurisdiction may also be divided into fire zones in accordance with population density and probable degree of danger from fire. The fire-zone map indicates the limitations on types of construction that the zoning map would otherwise permit.

In the vicinity of airports, zoning may be applied to maintain obstruction-free approach zones for aircraft and to provide noise-attenuating distances around the airports. Airport zoning limits building heights in accordance with distance from the airport.

Control of Building Height. Zoning places limitations on building dimensions to limit population density and to protect the rights of occupants of existing buildings to light, air, and esthetic surroundings. Various zoning ordinances achieve these objectives in a variety of ways, including establishment of a specific maximum height or number of stories, limitation of height in accordance with street width, setting miminums for distances of buildings from lot lines, or relating total floor area in a building to the lot area or to the area of the lot occupied by a building. Applications of some of these limitations are illustrated in Fig. 1-11.

Figure 1-11a shows a case where zoning prohibits buildings from exceeding 12 stories or 150 ft in height. Figure 1-11b illustrates a case where zoning relates building height to street width. In this case, for the specific street width, zoning permits a building to be erected along the lot boundary to a height of six stories or 85 ft. Greater heights are permitted, however, so long as the building does not penetrate *sky-exposure planes*. For the case shown in Fig. 1-11b, these planes start at the lot line at the 85-ft height and incline inward at a slope of 3:1. Some zoning codes will permit the upper part of the building to penetrate the planes if the floor area of the tower at any level does not exceed 40% of the lot area and the ratio of floor area to lot area (**floor-area ratio**) of the whole building does not exceed 15. To maximize the floor area in the building and maintain verticality of exterior walls, designers usually set back the upper parts of a building in a series of steps (Fig. 1-11b).

1-36 Building Systems

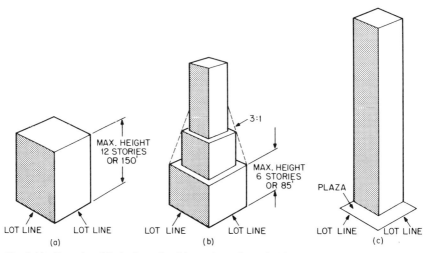

Fig. 1-11 Examples of limitations placed by zoning codes on building height: (a) height limitations for buildings constructed along lot boundaries; (b) setbacks required by a 3:1 sky exposure plane; (c) height of a sheer tower occupying only part of a lot is limited by the total floor area permitted. *(Reprinted with permission from F. S. Merritt, "Building Engineering and Systems Design," Van Nostrand Reinhold, New York.)*

Some zoning ordinances, however, permit an alternative that many designers prefer. If the building is set back from the lot lines at the base to provide a street-level plaza, which is a convenience to the public and reduces building bulk, zoning permits the building to be erected as a sheer tower (Fig. 1-11c). The code may set a maximum floor-area ratio of 15 or 18, depending on whether the floor area at any level of the tower does not exceed 50 or 40%, respectively, of the lot area.

1-12. Other Regulations.

In addition to building and zoning codes, building design and construction must comply with many other regulations. These include those of the local or state health, labor, and fire departments; local utility companies; and local departments of highways, streets, sewers, and water. These agencies may require that drawings for the building be submitted for review and that a permit be granted before construction starts.

Also, building construction and conditions in buildings after completion must comply with regulations of the U.S. Occupational Safety and Health Administration (OSHA) based on the Occupational Safety and Health Act originally passed by Congress in 1970. There is, however, no provision in this law for reviewing building plans before construction starts. OSHA usually inspects buildings only after an accident occurs or a complaint has been received. Therefore, building owners, designers and contractors should be familiar with OSHA requirements and enforce compliance with them.

Other government agencies also issue regulations affecting buildings. For example, materials used in military construction must conform with Federal specifications. For residential construction requiring government-insured mortgages, design must comply with standards of the Federal Housing Administration (FHA "Minimum Property Standards") or the Veterans Administration.

1-13. Systems Design by Team.

For efficient and successful execution of systems design of buildings, a design organization superior to that used for traditional design (Art. 1-3) is highly desirable. For systems design, the various specialists required should form a building team, to contribute their skills in concert.

One reason why the specialists should work closely together is that in systems design account must be taken of the effects of each component on the performance of the building and of the interaction of building components. Another reason is that for cost effectiveness, unnecessary components should be eliminated and, where possible, two or more components should be combined. When the components are the responsibility of different specialists, these tasks can be accomplished with facility only when the specialists are in direct and immediate communication.

In addition to the design consultants required for traditional design, the building team should be staffed with value engineers, cost estimators, construction experts, and building operators and users experienced in operation of the type of building to be constructed. Because of the diversity of skills present on such a team, it is highly probable that all ramifications of a decision will be considered and chances for mistakes and omissions will be reduced.

(W.W. Caudill, "Architecture by Team," and F. S. Merritt, "Building Engineering and Systems Design," Van Nostrand Reinhold, New York.)

1-14. Application of Systems Design.

Systems design may be used profitably in all phases of building design. Systems design, however, is most advantageous in the early design stages. One system may be substituted for another, and components may be eliminated or combined in those stages with little or no cost.

Systems design should be preferably applied in the contract documents stage only to the details being worked out then. Major changes are likely to be costly. Value analysis, though, should be applied to the specifications and construction contract, because such studies may achieve significant cost savings.

Systems design should be applied in the construction stage only when design is required because of changes necessary in plans and specifications at that time. Time available at that stage, however, may not be sufficient for thorough studies. Nevertheless, value analysis should be applied to the extent feasible.

(F. S. Merritt, "Building Engineering and Systems Design," Van Nostrand Reinhold, New York.)

Section 2

Business, Art, and Profession of Architecture

William E. Diamond II, AIA

The practice of architecture is a business, an art, and a profession. An architect, however, is primarily a professional. He is a qualified expert, licensed to practice. His functions are discharged primarily for the welfare of other than his own interests. His profession, when it reaches its highest potential, is not only artful in a figurative sense, but is a literally and historically accepted art form. Functioning within an economic frame of reference, he is inevitably required to exercise good business sense both in the delivery of his service and in his own behalf.

2-1. Business of the Architect.

An architect is a businessman as well as a professional. His business, large or small, has many of the requirements of any business. To perform, it must be profitable.

When a business happens to be the profession of architecture, the profit requirement takes on special significance. Architects must obtain and manage the resources required to perform services well. Without good business sense and proper business practices, an architect may find himself restricted in his ability to provide the best services to his clients. An architect who settles for too little fee, who does not keep his office costs in order, or who does not manage his own processes well is constrained in his ability to be a good professional. It is a wise client who understands and incorporates similar concerns in selection of an architect and fee negotiation.

Architects have a responsiblity to employees to provide the most favorable working conditions and compensation the marketplace will permit. If they fail to obtain or manage fees properly, they will not be able to discharge that responsibility.

Obtaining Commissions and Making Agreements. Architects obtain commissions through reputation, personal contacts, and respect generated through their prac-

NOTE: Minor revisions were made to this section for the 4th edition by Frederick S. Merritt, Consulting Engineer.

tices and community and personal activities. The agreements reached in securing work are always centered around the architectural services to be provided and the fees required to cover those services.

The major concern of an architect at the start of a project is to arrive at an accurate assessment of the scope of the work to be performed. The nature of the project, the degree of professional involvement, and the skills required must be considered in arriving at an equitable fee arrangement. Types of fees that may be used are:

Percentage of construction cost of project.
Fee plus reimbursement of expenses.
Multiple of direct personnel expense.
Stipulated or lump sum.
Salary.

For a typical project requiring standard services, the **percentage-of-construction-cost fee** is a safe standard. Years of experience with the relationship between the scope of architectural services required for various sizes of standard construction contracts provide a basis for such rule-of-thumb fee agreements.

For projects, where atypical work or services are required, other arrangements are more suitable. For example, for projects where the scope of work is indefinite, a **cost-plus fee** is often best. It permits work to proceed on an as-authorized basis, without undue gambling for either party to the contract. Under such an agreement, the architect is reimbursed for his costs, and also receives an agreed-on fee for each unit of effort he expends on the project. Special studies, consultations, investigations, and unusual design efforts are often performed under such an arrangement.

For atypical projects where the scope can be clearly defined, a **lump-sum fee** is often appropriate. In all cases, however, if the architect is a good businessman, he will know his own costs and will be able to project accurately the scope of work required to accomplish fixed tasks. He should take care, for the protection of his client, his staff, and his own interests, that he does not settle for fees that do not cover his own costs adequately. Otherwise, the client's interests will suffer, and his own financial stability will be undermined.

Fee agreements should be accompanied by a well-defined understanding in the form of a written agreement between architect and owner.

Architect As a Manager. The architect chosen for a project manages several aspects of the project simultaneously. He manages his own internal resources, he manages the processes that deliver service to the owner, and through that service he manages the programs of owner needs through the facility process to the creation of a built environment. The requirement that the architect be a capable businessman is, therefore, far-reaching. Furthermore, changes in the construction industry place increasing demands on this aspect of architecture.

The need for good business sense, and a thorough knowledge of the architect's costs, is reinforced by the need to manage his fees throughout the project. Allocation, commitment, and monitoring of the expenditure of resources are of critical importance to the success or failure of every project. Only when these are properly done can the correct services, the right advice, the appropriate design, and the best documents be delivered to clients.

Once the resources required to deliver services are insured, the architect must provide management skills to see that these services are timely, well coordinated, accurate, and well related to the client's needs. This is especially important on large projects, in large offices, or when dealing with the architect's employees and consultants. The best talent must be secured, appropriately organized, directed, and coordinated to insure that the project receives well-integrated and well-directed professional service.

Finally, the object of the entire endeavor is to produce the facility the owner needs, within his budget and on schedule. The architect's resources and the services provided are in many ways only a tool in managing the construction process for the benefit of the owner. The architect's management of materials and technology and his relationships with the owner and contractors will account in good measure for the success of the project.

Running an Architectural Practice. As a businessman, an architect is faced with acquiring a staff, advancing outstanding staff members, and removing unacceptable workers. He must keep records of his business expenses, file tax returns, provide employee benefits, distribute and account for profits, and keep accurate cost records for project planning and to satisfy government requirements. He must meet legal requirements for practice as an individual or as a corporation. In many of these areas, he is assisted by experts. It is impossible for an architect to practice effectively or successfully without a thorough understanding and complete concern for the business of architecture.

2-2. Art of the Architect.

It is the art of the architect that is the most visible aspect of his practice; yet, it is the most difficult to discuss objectively. Historically, architecture has been regarded as one of the fine arts, and has been defined as the art of building well. No course in art history would be complete without a thorough discussion of the edifices of the past. But not all architects are artists, and few buildings can be termed architecture in the strict sense of the word.

While some architects are regarded as artistic, the general public is not likely to apply the term *artist* to the architect who lives down the street. This is in part due to a narrow attitude about art and a broad view of architects and their services.

In the Renaissance period of history, many great artists were equally skilled in various arts. Painting, sculpture, architecture, physics and anatomy were such arts. Leonardo da Vinci is a well-known example of this generality. Since da Vinci's time, knowledge has broadened and specialization has taken over. The arts have been separated from the sciences: one considered objective, the other subjective.

Art and the artist have become more specialized as well. The fine artist is not part of the daily experience of the general public. His work is the subject of special shows and exhibits. Few homes or buildings are graced with original works of a true artist. Buildings on the other hand have become more technical, more scientific in their design and construction.

This discussion is not intended to imply that art has diminished in the concern of architects. They are no less concerned today about the esthetic, subjective aspects of the environment they create than before. But their concerns have become broader. They no longer practice only for a wealthy few. They no longer design only churches, palaces, and monuments. They are no longer restricted by a very limited technology or relatively unrestricted by budget. Finally, the construction process is no longer solely the work of highly skilled independent artisans and craftsmen.

With all these changes, the art of the architect has also changed. Architects no longer are able to lavish human labor on endless surface areas. Social and economic changes have dictated that no one man's productive life may be totally consumed by the carving or setting of tile for a few square yards of wall. Buildings may not be constructed from basic materials worked with great skill. Instead, mass-produced materials and equipment are fitted together artfully to form the built environment. Facilities are constructed for all elements in the society, with an ever-widening and changing technology, and within stringent budgetary requirements.

Architecture is simpler, cleaner, yet more diverse, and in most ways just as artful as in the past. This does not mean that every building can be called architecture in the strictest sense, any more than in the past. Too many builders are building too much. It does mean, however, that a broader cross section of the built environment receives the services of architects, and thus some esthetic concern. Often, the results are less than perfect to the purist; yet it cannot be said that the art is gone from architecture. To the contrary, the best work of the best architects is technically superb and still a work of art in the traditional sense.

So architecture is still an art; an art, however, inseparable from technology. The divisions of art and science have not been and cannot be imposed on architecture.

2-3. Profession of the Architect.

The term "professional" is broadly applied in regular usage. It is often used to imply a high degree of skill and capability in a given field of endeavor. It also may indicate that the person to whom the term is applied discharges the functions fairly and with great concern for the interests of the people and communities with whom he or she deals. Finally, professional in its narrowest usage differentiates a group of vocational pursuits that are other than business or labor.

Architects are professionals in the latter sense, through choice, education, experience, and license. When they discharge responsibilities well, they lay claim to the term in the fullest sense of the word. Their responsibilities to be professional are not only traditional, but also a legal requirement.

Licensed Architects. The most basic professional credential of an architect is his license in each state in which he practices. Licensing is intended to insure that adequate skills and knowledge are applied to protect the health, safety, and welfare of those affected by the practice of the profession, and to make certain that anyone who represents himself as an architect may be relied on to observe basic requirements for design and construction of buildings. Licensing insures that an owner seeking services, that the public occupying a building, and that architects in practice are all fairly protected from the unqualified, the incompetent, and the unethical.

Obtaining a license to practice places special obligations on architects, to the public, to clients, and to the architectural profession. Consequently, candidates, to receive their license, must meet educational and experience requirements. They must be recommended by other architects under whose supervision they have worked. They must pass a series of examinations that test their abilities in functional design and basic engineering, their knowledge of construction and professional practice, and their understanding of the requirements and conduct of the profession.

A typical architect receives 5 to 6 years of college-level education, and obtains either a professional degree at the Bachelor's level or a Master's degree. This is typically combined with 3 to 4 years of experience before he or she is eligible to take license examinations. After this period, it is usual for the architect to serve an additional period of internship in a firm other than his or her own before offering services directly to the public.

Professional Ethics. The American Institute of Architects has formulated the following basic principles for guidance of architects:

Advice and counsel constitute the service of the profession. Given in verbal, written, or graphic form, they are normally rendered in order that buildings with their equipment and the areas about them, in addition to being well suited to their purposes, well planned for health, safety, and efficient operation and economical maintenance, and soundly constructed of materials and by methods most appropriate and economical for their particular uses, shall have a beauty and distinction that lift them above the commonplace. It is the purpose of the profession of architecture to render such services from the beginning to the completion of a project.

The fulfillment of that purpose is advanced every time the architect renders the highest quality of service he is capable of giving. Particularly, his drawings, specifications, and other documents should be complete, definite, and clear concerning his intentions, the scope of the contractor's work, the material and methods of construction to be employed, and the conditions under which the construction is to be completed and the work paid for. The relation of the architect to his client depends on good faith. The architect should explain the exact nature and extent of his services and the conditional character of estimates made before final drawings and specifications are complete.

The contractor depends on the architect to guard his interests as well as those of the client. The architect should condemn workmanship and materials that are not in conformity with the contract documents, but it is also his duty to give every reasonable aid toward a complete understanding of those documents so that mistakes may be avoided. He should not call on a contractor to compensate for oversights and errors in the contract documents without additional consideration. An exchange of information between architects and those who supply and handle building materials is commended and encouraged.

The architect, in his investments and in his business relations outside his profession, must be free from financial or personal interests that tend to weaken or discredit his standing as an unprejudiced and honest adviser, free to act in his client's best interests. The use of free engineering services offered by manufacturers, jobbers of building materials, appliances, and equipment, or contractors may be accompanied by an obligation that can become detrimental to the best interest of the owner.

The architect may offer his services to anyone on the generally accepted basis of commission, salary, or fee, as architect, consultant, adviser, or assistant, provided he rigidly maintains his professional integrity, disinterestedness, and freedom to act.

He should refrain from associating himself with, or allowing the use of his name by, any enterprise of questionable character.

Architects should work together through their professional organizations to promote the welfare of the physical environment, and should share in the interchange of technical information and experience.

The architect should seek opportunities to be of service in civic affairs, and, to the best of his ability, advance the safety, health, and well-being of the community in which he resides by promoting appreciation of good design, the value of good construction, and the proper placement of facilities and the adequate development of the areas about them.

The architect should take action to advance the interests of his employees, providing suitable working conditions for them, requiring them to render competent and efficient services, and paying them adequate and just compensation therefor. The architect should encourage and sponsor those who are entering the profession, assisting them to a full understanding of the functions, duties, and responsibilities of the architectural profession.

Every architect should contribute toward justice, courtesy, and sincerity in his profession. In the conduct of his practice, he must maintain a wholly professional attitude toward those he serves, toward those who assist him in his practice, toward his fellow architects, and toward the members of other professions. His daily performance should command respect to the extent that the profession will benefit from the example he sets both to other professionals and to the public in general.

Standards of Behavior. (Established by the American Institute of Architects and mandatory for its membership, AIA Document No. 330.) An architect is remunerated for his services solely by his professional commission, salary, or fee, and is debarred from any other source of compensation in connection with the works and duties entrusted to him.

An architect shall not falsely or maliciously injure the professional reputation, prospects, or business of a fellow architect.

An architect who has been engaged or retained as professional adviser in a competition, cannot, if the competition is abandoned, be employed as architect for this project.

An architect shall avoid exaggerated, misleading, or paid publicity. He shall not take part, nor give assistance, in obtaining advertisements or other support toward meeting the expense of any publication illustrating his works, nor shall he permit others to solicit such advertising or other support in his name.

An architect shall represent truthfully and clearly to his prospective client or employer his qualifications and capabilities to perform services.

An architect shall above all serve and promote the public interest in the effort to improve human environment, and he shall act in a manner to bring honor and dignity to the profession of architecture. He shall conform to the registration law governing the practice of architecture in any jurisdiction in which he practices.

BASIC AND TRADITIONAL SERVICES

A typical project begins at the moment a need for facilities is identified by a potential owner. This is true whether the facility will be a modest home or a massive multimillion-dollar complex.

Among his first acts, the owner selects an architect. Establishment of a basic program or statement of requirements, along with budgets and time schedules, follow close behind. With these in hand, the intense work of the architect begins. He designs a building, estimates its cost, and reviews his work with the owner. On approval of the design, the architect incorporates his design into contract documents to be bid on by contractors. He presides over bidding, assists in selection of a bidder, and aids in establishing contracts between owner and contractor. Finally, the architect oversees construction, interprets contract documents when necessary, and assists the contractor in his dealings with the owner. At the conclusion of the work, the architect tries to ensure that all contracts have been completed, and in accordance with plans and specifications, and that all contractual obligations have been satisfied by both parties to the contract.

2-4. Selection of Architect.

Proper selection of an architect for a project is critical to the project, and should not be mishandled or performed on the wrong basis. The services of an architect are expertise, advice, and counsel; therefore, the owner's selection process should be directed at determining the following factors:

Professionalism. The ability of the architect to perform his duties fairly and in a professional manner is of major importance. His reputation and actions during the selection process are particularly important indicators. Those who simply appear concerned about obtaining the commission without arriving at a full understanding of the owner's needs and demonstrating the ability to perform the necessary services should be avoided. The standing in the community and reputation among previous clients of the architect to be selected are important.

Overall Abilities. While licensing laws establish reliable minimum standards of competence, ability varies. Professional ability is the only competition for work that is to the benefit of the owner. In this area, previous work is the guide (unless a design competition is held).

Background and Experience. Of importance in selecting the appropriate professional is any peculiarity of the commission. Some architects may have special talents or greater experience in certain areas. In some cases, knowledge or understanding of local or special conditions may be critical to the project.

One of the most common mistakes made by owners is solicitation of proposed fees as a basis of selection. Most reputable architects, believing the procedure to be unprofessional and counter to the best interest of the client, will not compete on the basis of fees. Because the direct costs of most architects are the same, differences in proposed fees reflect varying degrees or qualities of service. Since the value of a finished project is determined more by the services the architect provides than by any other single factor, it is critical to the owner that the degree or quality of service be based on a clear and detailed understanding of the commission. A wise owner selects an architect on the basis of his abilities. Negotiations for fees should be undertaken after the architect has been selected and after detailed discussions with the architect. In consideration of total project value, this is the only sensible procedure and may be relied on to produce the most efficient services.

In the final analysis, the value of architectural services to an owner depends on the effectiveness of the architect. If the right services are obtained, the benefit can be expected to be maximized. These services are the direct result of the owner's selection processes.

2-5. Building Program.

The program, representing the results of early basic, critical decisions, is a statement of requirements for the facility to be built and is associated with a budget. The architect for the project often is called on to assist the owner in defining his needs in physical terms. (This assistance is not normally included in typical fee curves used as a guide.) Functions are translated into length, width, and height, and become rooms. Related functions become areas of buildings. These are not designed solutions, but requirements for design to satisfy.

For complex buildings, experts or consultants in facility planning may be engaged. Hospitals, schools, and industrial installations are only a few of such complex cases. For these, the requirements of a complex set of physical and functional needs must be reconciled into clear requirements for a building design.

In many cases, a basic view of the kind of construction is developed during programming, simple or complex, unusual or typical. Areas of special concern are often identified for early work. For example, unusual environmental requirements may dictate an intense effort in the area of mechanical equipment.

During this early stage, the architect is called on to determine which requirements beyond the direct needs of the client must be satisfied. Codes and zoning requirements are the most common such issues (see Arts. 1-10 and 1-11). The architect must early identify restrictions and requirements that may affect the design and construction of the building.

Occupational Safety and Health Laws. Under federal and state laws, minimum standards of health and safety have been established for practically all occupations. The requirements are imposed mainly on employers and, in that they affect physical conditions, these regulations are an important consideration in the design of buildings where workers are employed. Such standards include lighting requirements, toilet-fixture counts, ventilation, and noise levels, to mention only a few.

Zoning ordinances and building codes are principally enforced by two methods, granting of building permits and inspection of construction work in progress, leading to issuance of certificates of occupancy when completed buildings satisfy all regulations. These allow examination of building plans and work under way by local officials charged with administration of these regulations. Conformance is assured through review, discussion, agreement, inspection, and permit granting. With federal regulations, however, the procedure is different. While advice is available, official review of plans does not occur. Enforcement is often undertaken only after worker complaint. An inspection is then made. If the complaint is upheld, the employer may be liable to a fine without sufficient time to correct the deficiency. While inspectors may be reasonable, the procedure places the employer, who is often the building owner, and the architect in a difficult position. Because official review and consent are not possible in assessing the specific application of the various regulations, and because of the legal liability inherent in a violation of federal occupational and safety laws, the architect and the owner must exercise great care to comply with those regulations.

Environmental Impact. Projects funded in whole or part by the Federal government or some states are subject to regulations that require careful consideration of environmental consequences of the projects. For maintenance of good public relations, such consideration is also advisable for privately financed projects. The reason for this is that it was public concern over environmental damage caused by past construction that led to passage of the National Environment Policy Act of 1969 (NEPA).

This Act requires all agencies of the Federal government to include, in every recommendation or report on proposals for legislation and other major Federal actions significantly affecting the quality of human environment, a detailed statement by the responsible official on:

1. The environmental impact of the proposed action.
2. Any adverse environmental effects that cannot be avoided should the proposals be implemented.
3. Alternatives to the proposed action.
4. The relationship between local short-term uses of the environment and the maintenance and enhancement of long-term productivity.
5. Any irreversible and irretrievable commitments of resources that would be involved in the proposed action should it be implemented.

The NEPA requirement to file an environmental impact statement (EIS) must be fulfilled by the Federal agency responsible for the project rather than the recipient of the Federal funds. Nevertheless, the project's architects or engineers may be required to supply the necessary data for support of the EIS.

Courts have decided that the NEPA applies to any Federal project or any project involving Federal funds. The funding need not be direct. In one case, for example, a court ruled that before guaranteeing repayment of a loan for construction of an apartment building, the Department of Housing and Urban Development should have submitted an EIS, because the structure will change the character of the neighborhood. It will concentrate population in the area, may increase vehicular traffic and will result in a loss of view from certain neighboring property.

The NEPA also states that in order to carry out the policy set forth in the act, it is the continuing responsiblility of the Federal government to use all practicable means, consistent with other essential considerations of national policy, to improve and coordinate federal plans, functions, programs and resources to achieve the following:

1. Fulfill the responsibilities of each generation as trustee of the environment for succeeding generations.
2. Assure for all Americans safe, healthful, productive and esthetically and culturally pleasing surroundings.
3. Attain the widest range of beneficial uses of the environment without degradation, risk to health or safety or other undesirable and unintended consequences.
4. Preserve important historic, cultural and natural aspects of our national heritage, and maintain, wherever possible, an environment that supports diversity and variety of individual choice.
5. Achieve a balance between population and resource use that will permit high standards of living and a wide sharing of life's amenities.
6. Enhance the quality of renewable resources and approach the maximum attainable recycling of depletable resources.

Consequently, an assessment of a project's environmental impact should provide a description in detail of and data on the following:

1. Design of the project.
2. Its environmental setting.
3. Probable environmental impact of the project on:
   a. Significant historic, scientific, or archaeological places or structures.
   b. Ecological, recreational, or scenic resources.
   c. Vegetation, wildlife, land features, or bodies of water.
   d. Transportation and congestion.
   e. Quality of air and water, esthetics of the vicinity, and noise levels.
   f. Development of potential hazards, such as fire, floods, erosion, or earthquakes.
   g. Community welfare and growth.
   h. Displacement of residences, businesses, or social or recreational facilities.
4. Measures for minimizing adverse effects of the project.
5. Alternatives to the project, including comparisons of project costs and benefits with those of the alternatives and including costs and benefits of not constructing the project.
6. Resources that will be irretrievably lost, especially fuels and energy.

Preparation of such an EIS requires that an interdisciplinary approach be adopted, that due consideration be given to unquantifiable values, that all reasonable alternatives to the project be analyzed, that public notice of the study be promulgated, and that the public be apprised of comments made by Federal agencies on the EIS.

While building designers should recognize the importance of the client's needs and desires, a primary concern of design should be the health, welfare and safety of building users, whether they be occupants, visitors, or the client's employees. The good of the community should be another major concern, not only for moral reasons but also because ignoring it may have adverse economic and political effects on the client. Consequently, assessment of environmental impact is important for all projects, and measures should be taken to eliminate or at least minimize adverse effects.

Energy Conservation. In response to the national need for energy conservation and in recognition of the high consumption of energy in buildings, the Department of Energy gave a grant to the American Society of Heating, Refrigerating, and Air-Conditioning Engineers (ASHRAE) for development of a national energy conservation standard for new buildings. The resulting standard, ASHRAE 90-75, promulgated in 1975, established thermal design requirements for exterior walls and roofs. It was incorporated in many building codes.

Seeking greater energy-use reduction, Congress passed the Energy Conservation Standards for New Buildings Act of 1976. The Act mandated the development of energy performance standards for new buildings (BEPS). Accordingly, the Department of Energy began development of such standards, for adoption by Federal government agencies and state and local building codes. BEPS consists of three fundamental elements:

1. **Energy budget levels** for different classifications of buildings in different climates, expressed as an annual rate of energy consumption (British thermal units per square foot per year).
2. A method for applying these energy budget levels to a specific building design to obtain a specific annual rate of energy consumption, or **design energy budget.** for the proposed building.
3. A method for calculating the estimated annual rate of energy consumption, or **design energy consumption,** of the proposed building.

The design energy consumption may not exceed the design energy budget of a new building.

Even without these regulations, energy conservation for buildings makes good sense, for a reduction in energy use also reduces building operating costs. It is worthwhile, therefore, to spend more on a building initially to save energy over its service life—at least up to the point where the amortized annual value of the increased investment equals the annual savings in energy costs.

As a consequence, life-cycle cost, which is the sum of initial, operating, and maintenance costs, should be given preference over initial cost in establishment of a cost budget for a proposed building.

(F. S. Merritt, "Building Engineering and Systems Design," Van Nostrand Reinhold Company, New York.)

2-6. Design Stage.

Once the basic requirements of a project have been established, including the owner's requirements, budget, and schedules, as well as zoning, building codes, and other environmental patterns, design of a building can begin. The central objective of the design process is to arrive at the proposed structure that will most appropriately meet the owner's needs for physical space. The requirements may be unusual in some respects, or relatively straightforward. In all cases, schedules and costs are major considerations.

The design process is a complex interaction of the requirements of the owner; the experience, knowledge, and ability of the architect; and the site and surrounding environment of the proposed structure. The process is one of making tentative decisions in response to parts of the problem, and testing for validity in comparison with solutions to other parts and other requirements (Table 2-1). The final solution must be a well-integrated whole, and not just an assembly of individual answers to various portions of the problem.

Owner review of the proposed solution from the earliest decisions to a complete schematic design is essential. The proposals of the architect must be examined to determine that the owner's needs have been properly understood and integrated, and that budgets and other requirements are being maintained.

Costs. The initial requirements of the owner must include some budget figure. The key word is **budget**. A budget is a value judgment or willingness to pay. An owner places a value on his new facility that he is willing to pay; this is his budget figure.

The architect prepares cost estimates. His early estimates are based on a rough approximation of the cost of a facility not yet designed but which can approximate the owner's requirements. The key word is *estimate*.

TABLE 2-1 Summary of Design Phase Activities

Conference with client to outline project, its general purposes, general plan, feasibility, location, general type of construction and equipment; to discuss the probable time required to build, and to estimate approximate costs
Development of the background of the project as a basis and format for contract documents
Visits to the site and siting studies
Diagrammatic and blocked-out studies of elements of the project (schematics)
Examination of laws, ordinances, codes, standards, rules, and regulations of governmental authorities
On approval of schematics by the client, preparation of preliminary drawings—plans, sections, elevations (design development)
Preparation of outline specifications, describing construction, materials, equipment, and basic mechanical and electrical systems
Preparation of preliminary estimates of the costs of the work
Detailing, preparation of contract documents, and final cost estimate

The comparison with budget is important. When problems regarding cost occur, it is generally the result of a situation in which the owner finds his willingness to pay, or budget, considerably lower than the architect's estimate of probable cost for meeting project requirements. On the assumption that the owner's requirements are clearly understood by the architect, the owner may face a critical set of decisions:

Budget the same funds for reduced or modified requirements?
Increase the budget?
Proceed with the objective that detail design efficiency will make the comparison more favorable?
Cancel the project?

Some combination of the first three actions is the usual response. It is important, however, for both the owner and the architect to understand clearly that the setting of requirements and establishing of budgets is the ultimate responsibility of the owner. Designing the facility and accurate cost estimating are the responsibility of the architect.

Most disputes arising between an architect and owner usually flow out of a failure to respect these responsibilities by one or both parties. The architect must design the building to meet owner requirements and not to satisfy his own desires. He has a responsibility to keep the owner informed as to the probable cost implications of each change in requirements, and at each stage in the design process. The owner, on the other hand, should understand that his requirements must be clearly communicated and that changes in the project will affect costs.

The architect, though responsible for making accurate estimates, cannot guarantee his estimates. While he can anticipate, he cannot control the cost of materials, equipment, labor, or other conditions in the construction market that affect costs. He can be expected to be responsible for designing efficiently and within the budget. He can also be expected to bring considerable design and management leverage to bear on costs.

Schematic Designs. The schematic design stage should be completed with site plans and floor plans worked out, basic building massing determined, facades laid out in a general way, and basic materials and the physical systems (structural, mechanical, electrical) selected. A formal submission to the owner is normally made at completion of the schematic phase, complete with a review of estimated construction costs and items that will be incorporated in the contract documents and that are peculiar to the owner or the project.

Design Development. On approval by the owner of the schematic design, the project moves into the design development phase. The emphasis moves from concern for overall relationships, functions, and massing to more technical concerns for the constructability of the building. While technology is never beyond the consideration of the architect, it is during this stage that he concentrates on the impact of technology on the overall solution. The architect's esthetic concerns are not reduced during this period; rather, they are shifted from function, mass, and space to surface and detail.

It is possible, during the design development stage, that changes will occur in the basic design. For example, the design development of technology, materials, or building systems may lead to a problem or to an opportunity previously unforeseen in earlier steps. This is, however, an exception rather than a general rule. The main purpose of design development is to bring the facility into clearer focus and to a higher level of resolution.

The basic structural, mechanical, and electrical systems will be generally engineered and laid out, typical exterior walls will be designed, typical construction details worked out, and the junctions of typical materials considered. While the facility will not yet be completely detailed, much of the uncertainty will be resolved.

Outline Specifications. Drawings describing the facility in some detail are prepared, and outline specifications established. These specifications are not written for contract purposes, but to record for review the basic decisions that will be incorporated in the contract specifications. The outline specifications tell the owner what materials and special conditions of construction and special requirements of the contract are proposed for the work.

At this point, the architect prepares another formal cost estimate. He is in a position to refine his former estimate to a great degree, and can expect that it will have a much higher degree of accuracy than that for schematics. (See also Art. 2-7.)

2-7. Contract Documents.

With a fully approved design (Art. 2-6) and the authorization of the owner, the architect next prepares contract documents, including working drawings. The main elements of contract documents are:

Drawings
Specifications
General conditions
Supplementary conditions
Owner-contractor agreement

(See also Art. 25-5 and Sec. 26.)

While drawings are prepared to permit a contractor to construct the building, the phase during which these drawings are made has a greater significance. The architect, at this phase, is preparing documents as a basis for a legal contract between the owner and one or more contractors. These documents center around the drawings, and the task of preparing the drawings consumes most of the architect's resources for the phase.

But there are other critical parts of the total effort. Detailed specifications will also be prepared. In addition, the basic legal conditions (general and supplementary) will be determined. Aside from defining the parties to the contract, and their relationships, these documents describe such concerns as the rights, duties, and responsibilities of all parties during the construction period, insurance requirements, methods of payments, procedures for settling disputes, and ways of dissolving the contract. Additionally, requirements peculiar to the owner or the facility are spelled out. Provisions for such items as site utilities, temporary structures, and temporary heat for winter construction may be detailed in various locations in the documents.

Simultaneously with creation of the legal documents, fine detailing of the building proceeds and the last level of design decision making occurs. Every condition of material, surface, joint, and construction must be considered and defined so as to be clear to the contractor. The drawings and specifications must be complete in defining the building.

The contract documents are the basis for bidding, construction contract, and construction. The architect's responsibilities to the owner for clarity and accuracy are thus extended to cover the contractor, who will work from the documents. These documents must be clear, correct in every detail, and without conflicting information or requirements.

Contractors who bid, contract, and work from the architect's documents rely on his abilities. Any mistakes the architect may make can prove costly to a contractor as well as to the owner.

Final Cost Estimate. On completion of the contract documents, the architect prepares his final and most accurate estimate of construction cost. It is intended to confirm previous estimates, incorporate more detailed knowledge of the building, reflect owner-requested changes, and assess the construction market at the time of bidding. (See also Sec. 24.)

Bidding and Contract Award. With authorization of the owner, the project is put on the construction market. In some cases, a single contractor may be asked to negotiate a contract directly. This is particularly appropriate for small work or where unique abilities or knowledge peculiar to a specific contractor are important.

For private work, a selected list of prequalified bidders may be invited to bid competitively for the work. The list is usually prepared with contractor reputation, financial stability, and construction capabilities in mind.

For public work, bidding is open to all licensed contractors. In this case, bids are requested by advertisement, and plans and specifications are made generally available to all bidders.

Under all conditions, bidding must be handled fairly and equitably. All contractors must be treated even-handedly, and given the same information at about the same time. Bids must be received simultaneously, and must be in precisely the same form for fair comparison. The low bidder may not always be the successful bidder, but the criteria of selection must be uniformly and fairly applied without partiality. The preceding activities are summarized in Table 2-2.

TABLE 2-2 Summary of Contract Documents, Bid, and Award

(Prepared as the basis for taking bids and establishing a contract between owner and contractor)
Working drawings:
   Architectural drawings
   Structural drawings
   Mechanical drawings
   Electrical drawings
   Site development and other special engineering drawings
Specifications:
   Type and quality of materials
   Desired finishes
   Manner of construction
General conditions:
   Responsibilities of parties to the contract
   Insurance requirements, bonds, methods of payment, and other conditions supplementary to drawings and specifications
Approvals:
   Assistance of the architect in obtaining approval of governmental agencies, when required
Bidding:
   Architect furnishes stipulated number of copies of plans and specifications to the client (and to contractors for bidding, if directed to do so)
   Architect may issue "addenda" to change, modify, or clarify the contract documents to be bid
   Architect may provide assistance in preparation and letting of contracts for construction
   Prequalification of bidders and advertisement of bids
   Proposal form for bid submission, to insure fair competition
   For public works, "notice to contractors," usually advertised in an official publication, containing:
      Brief description of building
      Where proposals will be received
      Date and time for receipt of proposals
      Where contract documents may be examined and obtained
      Deposit required for securing copies of drawings and specifications
   "Instructions to bidders" with requirements regarding submission of bids
   For all public and some private work, a certified check or bidder's bond is required, to insure that the successful bidder will sign a contract for the bid price
   When the architect presides over the bid opening, he must insure simultaneous receipt of bids and proper tabulation for impartial comparison
Award of contract:
   On review of all bids, the architect is generally responsible for the recommendation of award and preparation of the contract between client and successful bidder

2-8. Architect's Role during Construction.

Construction of a building results from fulfillment of a contract (or contracts) between an owner and one or more contractors. The architect's responsibility is to ensure fair performance of this contract by both parties. His prime functions in this regard are interpretation of drawings and specifications and establishment and enforcement of standards of acceptability for the work.

Administration. Traditional duties of the architect during construction include administration of the contract as the agent of the owner. The architect seeks to ensure that the contractor fulfills the terms of the contract documents. The architect, however, also has a serious responsibility to the contractor, which requires interpretation of contract documents fairly and without prejudice to the contractor and assisting the contractor in obtaining payment.

The architect does not traditionally become involved in the conduct of the work or determine the methods of construction. He does, however, have a general responsibility to see that the construction site is maintained in a safe condition and that all terms of the contract are observed. Among his diverse concerns, he has a responsibility to the owner to see that the contractor discharges his financial obligations to subcontractors and suppliers during execution of the contract, so that no financial irregularities complicate the project.

At the outset of construction, the architect has many duties. Among them, he advises the owner of his responsibilities in regard to property insurance, to protect the property of the contractor (including the work) from fire or "acts of God." The architect's dealings with the contractor usually center around approval of subcontractor lists, contractor's schedule of values, and proposed progress schedule to determine the value of in-place work as a basis of payment.

Construction Inspection. Once work is under way, the architect inspects construction, as necessary, to check progress of work, assure compliance with contract requirements, and guide the contractor in interpretation of the contract documents for full performance of his contract.

The architect's supervision should cover:

Checking of bench marks for building location.
Examination of soil for footings after excavation. (Tests may be necessary.)
Inspection of connections with public utilities.
Determination of need for additional subsoil drainage.
Verification of accuracy of placing base plates for steel columns.
Checking concrete forms for alignment and bracing.
Inspection of concrete reinforcing and electrical conduits, and of location of sleeves.
Check on quality of concrete.
Determination of adequacy of weather protection.
Check of wiring and piping for conformance with contract documents.
Check of partition layout and plastering grounds.
Check of finish materials for proper installation and compliance with contract documents.
Similar checks on other items required by plans and specifications.

Shop-Drawing Checking. Throughout the construction period, there are important specific activities the architect may perform that keep the project running smoothly. To assure that the intent of design is carried out, and to assist the contractor in interpreting the documents, the architect checks shop drawings submitted by contractors. His checking must be performed with care and dispatch, to prevent upset of construction schedules. Where the contractor is unclear either in his shop drawings or in any other area, the architect may prepare full-size detail drawings or other supplementary drawings necessary to assure compliance with contract requirements.

Payments to Contractors. From a financial point of view also, the architect has serious responsibilities. He issues certificates to actuate owner payment to the contractor as prescribed in the agreement (usually on a monthly basis). In ascertaining the amount to be paid under any certificate, the architect should assure that:

Value of work done has been accurately ascertained.

Amount of the retained percentage has been deducted (usually 10%).

Total amount of change orders approved by the owner to date, and the correct total of previous payments are shown.

No reasons exist for withholding payment; that is, there is no defective work, no evidence indicating the probable filing of claims, no failure of the contractor to make payments to subcontractors or suppliers, no reasonable doubt that the contract can be completed for the balance unpaid, no notice of damage to another contractor.

On determination that the work has been completed, the architect issues a "certificate of payment" to the contractor, sending him the contractor's and owner's copies. The contractor sends the owner's copy to the owner for payment. If required, the contractor should certify that bills have been paid and that his payment has been received.

Valuation of work forming the basis of a certificate of payment should be regarded as one of the architect's more serious duties. Failure to correctly value the work will do some injury to the owner or the contractor. In past court cases, where a loss has fallen on an owner or contractor because of an inaccurate valuation, the architect has been held liable for the full amount of the loss.

Change Orders. During the contract period, reason may arise to make changes in the work. These are accomplished by issuing documents to the contractor describing the desired changes. Such changes may include owner deletion or addition of equipment, adjustments for unforeseen site conditions, or errors or omissions in the documents.

Once the proposed change has been reviewed by the contractor, he makes a proposal for the work. The architect checks the extra cost or credit, and issues bulletins and change orders, as necessary, for any modification of contract.

Issuance of orders for changes in the contract, however, must be authorized in writing by the owner. (See also Arts. 25-25 and 25-26.)

Project Completion Activities. As the project nears completion, the architect's attention is turned to wrapping up work and agreements in an orderly and well-coordinated manner. This is accomplished through final acceptance of the work and issuance of the final certificates. The normal procedure is as follows:

1. The contractor sends the architect a statement claiming substantial completion.

2. If the architect concurs, he sends a statement of substantial completion to the owner with copies to the contractor.

3. A semifinal inspection is made by the architect in the presence of the owner's project representative.

4. The contractor submits to the architect for approval all required guarantees, certificates of inspection, and bonds.

5. Lists of required corrections are made, one for each trade, with provision for indicating satisfactory correction by checking or punching.

6. Final inspection is made by the architect and the owner.

7. If everything is acceptable, a certificate of completion is sent to the contractor with copies to the owner.

8. All required operating instructions are submitted to the architect for approval.

9. Complete keying schedule with master keys, submaster keys, room keys, and any special keying is delivered to the architect for approval.

10. The contractor submits an affidavit that all bills have been paid, with any exceptions noted and covered by releases of liens.

11. The architect sends owner's and contractor's copies of final certificate of payment to the contractor.

12. The owner makes final payment and notifies the architect.

13. The architect sends the owner permit to occupy, with copies to the contractor.

Before issuing a final or semifinal certificate of payment, the architect should be sure that:

Degree of completion of work called for by the contract has been reached.

Any period of time required to elapse between the issuance of the semifinal and final certificate for payment has elapsed.

Owner's interests have been protected in respect to the filing of liens, usually by a release of liens and affidavit.

Accounts between owner and contractor have been adjusted, including original contract sum, additions, and deductions as included on change orders, cash allowances, deductions for uncorrected work, and deduction for liquidated damages.

Any written guarantees, such as those for roofing, have been deposited with the architect or owner, as well as certificate of occupancy and all certificates of inspection, such as for boilers or electrical work, or by public authorities.

2-9. Liability.

In general, an architect, in discharging basic and traditional services for a project, is liable for losses that would not have occurred had he used *reasonable care and skill*. Legally, this means the degree of care and skill currently furnished by members of the profession generally, rather than by its most outstanding members.

This, however, does not mean that an architect can expect to get by with just average performance. His reputation rides on his performance, and reputation is the most important means of obtaining commissions. As a professional, every architect must act in the best interests of his client. It is not an accident that only through discharging this responsibility an architect ensures his own success.

The construction process entails commitment of sizable resources by owners and contractors. The architect is instrumental in the proper commitment of these funds. His main function is to protect the owner. The architect does not, however, guarantee the work of the contractor.

In general, the contract drawings serve as a record of the instructions given by the architect. If losses result from the contractor's failure to follow these instructions, the liability is clearly the contractor's rather than the architect's. If, on the other hand, losses are due to faulty design, as indicated by the contract documents, the liability is clearly the architect's.

In the case of foundations, it is generally impracticable after completion of a building either to determine whether the contract documents have been followed or to check the soundness of the design. It is difficult to uncover the work, and because of lack of data regarding the character of the soil, checking the design would have little meaning. For these reasons, settlement of foundations is usually taken as prima facie evidence of neglect or lack of skill on the part of the architect. The architect, therefore, is generally held to be liable for damage resulting from a foundation failure.

Another definite liability of the architect is for any losses that result from errors in his certificates for payment to the contractor.

Any architect may delegate authority to his employees. But by doing so, he does not escape responsibility for their acts. As a professional and a businessman, he must be expected to choose competent employees and associates and to supervise their work closely.

2-10. Architect and Consultants.

In the performance of any services, an architect may retain consultants to supplement his own skills, to deliver a more complete and well-coordinated service to his clients. Such consultants include structural, mechanical, and electrical engineers and specialists in specific building types.

In all cases, the architect's legal responsibilities to the owner remain firm. He is fully responsible for the services he delivers. The consultants, in turn, are responsible to the architect. Following this principle, the architect is responsible to his clients for an error in structural engineering made by his structural consultant. The architect may claim damages in a separate action if he is sued, but the responsibility to his client is his alone. Consequently, it is wise for the architect and his clients to evaluate the expertise the architect may have in supervising others before retaining consultants in other areas of responsibility.

2-11. Professional Engineers.

Integral wtih the design of any facility is a series of systems concerning which the architect has basic knowledge and some design skill. These systems are the structural, mechanical, electrical, and site-paving and drainage systems. On small, simple projects, the architect may choose to design the entire project himself. On larger projects, however, he usually relies on the design capabilities of specialists in these engineering areas.

In some architect offices, these skills are retained in much the same fashion that an owner retains an architect, through a professional services contract. In other, generally larger offices, the staff may include professionals in one or more of these disciplines.

By whatever means they are employed for a project, these specialists play an indispensable role in design and construction. They are expected to perform in just as professional a fashion as the architects with whom they collaborate.

The major distinctions between architects and engineers run along generalist and specialist lines. The generalists are ultimately responsible for the work. It is for this reason that an architect is generally employed by an owner. On some special projects, however, where one of the engineering specialties becomes the predominant feature (dams, power plants, sewage treatment, certain research or industrial installations), an owner may select an engineering professional to assume responsibility for design and construction. On other projects, it is the unique and imaginative contribution of the engineer that may make the most significant total impact. The overall strength of a dynamic, exposed structure, the sophistication of complex lighting systems, or the quiet efficiency of a well-designed mechanical system may prove to be the major source of owner pride in a facility. In any circumstance, the responsibilities of the professional engineer for competence and contribution are just as great and important to the project as those of the architect.

2-12. Design by Team.

For many reasons, building design and construction tend to become increasingly complex and sophisticated, and sizes of typical projects to become larger. As a result, design and construction processes tend to become too difficult to be controlled by a single individual and require instead the efforts of a closely integrated team of diverse professionals, each of whom is preeminent in an area of expertise and identifies with the total project.

Design by team places extraordinary demands on the team members. These members may be architects, engineers, construction specialists, financial experts, planners, statisticians, building-type consultants, owner representatives, or other experts. Three rules govern their performance:

1. The field of expertise of each team member must be distinct and clear. The member must be sufficiently demanding and expert in his field to command respect of his teammates, and he must be given complete and final professional authority in his field.

2. Team members must be concerned more about the total project than about their own field of expertise. They must be willing to contribute to all areas of the work. They must establish criteria for their own fields based on total-project requirements rather than just on the standards of their disciplines. They must be sensitive to the fields of the other professionals.

3. Team members must be able to identify conditions where project demands and standards of their disciplines coincide (an opportunity), and to reconcile the conflicts (a process of design optimization).

Team members generally find that they must tackle some design problems individually and others in concert with one or more teammates. The most difficult problems are often those that must be solved by a group. If team effort is successful and creative in developing solutions, a higher-quality project than the contribution of a single designer is produced. But if the team members develop solutions by compromise, the project is "designed by committee," with poor results likely.

Team design, therefore, has the potential of rising above the work of a single designer. It also has the potential to go the other way. The outcome depends on the quality of the professionals employed and the manner in which they are organized and operate. Leadership, in particular, can play an important role in developing the potential of the team.

See also Art. 1-13.

ADDITIONAL SERVICES

Beyond traditional services, architects often are called on to extend themselves into areas where, because of basic knowledge, skill, and experience, they may provide unique professional services to clients. Such services may be delivered as part of basic services, or on a completely separate basis.

2-13. Nontraditional Services.

Most architects may provide certain of these services. Site selection, program definition, and various kinds of feasibility studies are good examples. In other instances, particular experience, a specialized practice, or a unique association with other professionals may prepare a practitioner to provide services that could not normally be obtained from an architect. Hospital consultation (with a specialist in health facilities) and computer scheduling and cost control (from a computer-oriented architectural firm) are examples in this area.

The architect should be a generalist, problem solver, businessman, and manager. It is these characteristics, in combination with in-depth experience in some specific areas, that often lead to selection of an architect for performance of hybrid tasks where intra- or interdisciplinary tradeoffs are required to determine broad issues, make complex decisions, or produce overall plans.

Architects have been called on to plan hospitals, manage large construction projects, inventory national resources, select sites, and plan universities. It is important to understand, however, that the training and licensing of an architect do not generally qualify all architects for such work. It is instead the unique combination of generalist background with specialized experience that produces such a consultant.

In whatever area an architect undertakes to provide service, whether through his own skills or through the skills of subcontractors, his responsibilities to perform competently as a professional and in a legal sense are undiminished.

MAJOR CONDITIONS AFFECTING ARCHITECTURAL PRACTICE

Changes set in motion with the industrial revolution continue to bring major shifts in construction practices. With these changes, the architect and his services also change.

2-14. Economic Factors.

A building owner views his facilities in economic terms first and foremost. As real estate, facilities have investment value that must be related to cost, return, and other economic issues. The trend toward more owner sophistication in economic decision making regarding facilities will continue. This means that the clients architects serve will become increasingly cost conscious and will expect these concerns to be dealt with more expertly. For the architect, this is a challenge, both as a professional and a businessman, to develop greater skill in cost control through design and involvement with construction.

Costs of material and labor in the construction industry rise continuously. This means that new facilities are usually more expensive to construct than in the past, often not only from an absolute point of view but relative to everything else. For a large project of several years' duration, costs may rise substantially while construction proceeds. A contractor bidding a lump sum for a multimillion-dollar project with the expectation of providing materials and labor several years ahead and at costs that cannot be predicted must cover such risks. He cannot bid "tight." He must arrange to receive funds from the contract to cover such risks. The contract sum always includes these costs.

2-15. Changing Construction Market.

As costs rise, skills employed in the construction industry have tended to decline. Skilled craftsmen capable of artistic interpretation have become scarce and, in effect, have been replaced by laborers who perform rote construction chores.

At the same time, the trend has been for large general contractors to become removed from direct involvement in current technology or in actual performance of construction work and to become more involved as managers of the process. Those involved in technology and its changes are suppliers, manufacturers, and subcontractors, who may have no direct relationship to the owner and the architect from a contractual point of view.

These changes in construction in a time of increasing economic pressure on owners, especially owners of large projects, place greater stress on architects to manage the costs and technology associated with construction but reduce leverage via traditional processes.

2-16. Systems.

Development of systems methods based on computer technology has impacted the design and construction of facilities. Construction scheduling, cost-control techniques, accounting, information-retrieval systems, graphics, program modeling, and simulation—all computer-assisted—are often used in the facility production process. These techniques and the capabilities they make possible have an enormous impact on facility planning, design, and construction.

Building Systems. Architects currently seldom detail stone and wood trim for buildings. Instead, they select products that incorporate the details most suitable for their buildings. Manufacturers or producers design such items along with their products. Windows, doors, wall surfaces, toilet enclosures, plumbing accessories, masonry units are usually designed not by an architect for a project but by a manufacturer. Mass production, high labor costs, and development of manufacturing processes and materials heretofore not available have contributed to these changes in architectural practice.

Similarly, engineers can select more comprehensive products. Air-conditioning systems of increasingly larger sizes are becoming available with a greater variety of equipment and controls. Larger structural systems are available on a preengineered basis. Integrated ceiling systems, including lighting, air conditioning, ceiling surface, and ceiling structure, with the ability to adapt to wall systems, are produced by several manufacturers on a total-concept basis.

Also, contractors and suppliers in mechanical, electrical, and structural areas are becoming more specialized.

All these factors have contributed to development of a phenomenon known as *building systems*. A building is usually thought of as an assembly of systems, such as structural; exterior skin; heating, ventilating, and air conditioning; plumbing; power, signal, and lighting; walls; and ceilings. In many instances, these physical systems conform to the area of one, or of a group of, contractors, building tradesmen, or suppliers who function in a discrete set of construction activities.

Building systems have always existed within facilities. For instance, structural systems and wall systems have occurred in all buildings. But these distinctions have become less academic and more real. The continuing increase in complexity of buildings; in specialization of design roles; and in specialization and sophistication of manufacturers, suppliers, and contractors has made the boundaries between the various systems within buildings more clearly defined. In most cases, the subcontracting of work to a general contractor and various specialist contractors has recognized and followed these lines. Consequently, many building systems can be obtained as a package, or each one can be viewed as a clearly separate portion of the total building. (See also Sec. 1.)

Systems Building. The existence of well-defined and generally accepted building systems has led to another level in the systems process. Contracts are sometimes awarded by an owner to a contractor for one or more of the building systems. This level of recognition of the existence of discrete parts of the total facility is indicative of a phenomenon called *systems building*.

Systems building, while relying on the definitions of building systems, is not "hardware" oriented. It is instead a management process that manipulates the design and construction of the facility through the design and installation of the various building systems.

It is possible, for example, to buy a structural system of known capacity and dimensions, a preengineered air-conditioning system, a completely designed exterior-wall system, and a ceiling and interior-partitioning system, and to combine them uniquely with other parts of the building designed in a more traditional way. By this method, it is possible to place on a unique foundation, as a total facility, a building that has not been designed or built before.

Systems building is applicable to a number of circumstances. Commercial, industrial, and governmental owners often have a continuing building program that requires repetitive erection of a single-facility prototype in various locations. Such owners are not interested in a unique design and construction effort for substantially the same facility each time a new site is acquired. The process consumes needless time. Furthermore, economic efficiency in design and construction cannot be achieved with such a traditional one-at-a-time process. Under the general heading of systems building, a broad range of options are open to such owners (Table 2-3).

Design of a unique building or building part is not always necessary. Preengineered or premanufactured buildings or building parts are available. Schools, for example, often

TABLE 2-3 System Buildings Options for Facility Acquisition

Specific design for an individual building (traditional approach)
Specific systems designed for a group building program
Application or assembly of predesigned systems, or preconstructed or commercially available systems, buildings, or components

acquire additional classroom space by the purchase and integration of such structures or components into a larger complex of complete buildings.

Facilities may also be acquired via specific systems designed for a group building program with direct participation from systems suppliers. (Such a program would call for erection of a large number of basically similar buildings over a period of time.)

An extreme example of systems building might have facilities acquired through the lease and assembly of preconstructed buildings and building systems components. This approach could meet temporary and urgent needs. Furthermore, in a second phase, leased structures could be replaced in phased erection of either uniquely designed or system-designed permanent structures. The options and combinations associated with systems building are endless.

Thus, systems building generates choices other than one-at-a-time design for one-at-a-time, on-site construction. In the examples cited, a fairly broad definition of systems building and building systems has been utilized. The most significant efforts in this regard have been in more comprehensive product design, systems design for special building programs, and, in the factory, assembly of building components and totally integrated building parts. Housing, schools, and hotel design and construction have been in the forefront of systems building. (See also Art. 25-30.)

(F. S. Merritt, "Building Engineering and Systems Design," Van Nostrand Reinhold Company, New York; "Metal Building Systems," Metal Building Dealers Assn., 1406 Third National Building, Dayton, Ohio 45402.)

2-17. Multiple-Contract Construction.

Under traditional procedures, a general contract is awarded to a single contractor for the total project. He, in turn, assumes complete responsibility for the total work and is obligated to deliver the completed project for the contract price. To accomplish this, he establishes subcontracts between himself and other contractors and suppliers who provide most of the actual facility.

Under multiple-contract construction, however, prime contracts are established directly between an owner and the suppliers and subcontractors who might normally deal with a general contractor.

Prime contracts directly between the owner and the contractor or supplier provide greater flexibility in design of a facility. Design of some parts of a building can be combined with purchase and installation of predesigned and premanufactured items, and with products that are especially designed for a specific application by the manufacturer or supplier. Thus, the owner and architect can obtain the most competitive and appropriate preengineered systems, or develop and acquire unique products through direct contact with suppliers. Traditionally designed and constructed portions of the building can also be contracted for separately in one or more packages.

The interface between multiple contracts must be carefully defined and managed. One method of achieving this goal is assignment of the contracts to a general contractor once the unique design or most appropriate system has been obtained. Other methods are discussed in Art. 2-19.

The award of multiple contracts for a single facility may occur simultaneously. Reasons for simultaneous multiple contracts include advantages of dealing directly with the system supplier and financial leverage. Award of several contracts of moderate size instead of one large one has a number of beneficial effects:

Bidding is more diverse. A large project may be beyond the financial capability of many otherwise qualified contractors. With the total project reduced to several relatively small contracts, more contractors can bid competitively. Where the contracts are written to follow trade or system specialties, bids can be taken on each of the parts of the facility, and total project cost assembled from a combination of low bids. A general contract for the same project, in contrast, would provide a single low bid that would not necessarily be composed of the lowest cumulative total of all subcontract prices.

With budgets, design, estimates, contract documents, and bids established for several parts of the project, instead of one large lump sum, it is easier to make design tradeoffs throughout the entire process to control overall costs. If the structural contract is over the budget, it is feasible to reduce the cost of carpeting and drapes to compensate and thus maintain the overall project budget. This can also be accomplished under a single release contract. But the complete contract structure imposed on an entire project, and more detailed bidding information, make tight cost control a direct and strong result of multiple contracts.

Multiple contracts are not always awarded simultaneously in the same manner as if the total project were covered by a single contract. When some staggering of contracts occurs, the object is usually to reduce the total time required to design and construct the project. (See also Arts. 2-18 and 25-8.)

2-18. Accelerated Design and Construction.

The culmination of most of the procedures discussed in this section, accelerated design and construction rely on the traditional services of architects and require both systems concepts and multiple contracts. The traditional process of design and construction and the roles and responsibilities of the various parties need not be profoundly changed.

In the traditional process, the entire facility moves stage by stage through the entire process (programming, design, design development, contract documents, bid and award of contracts, construction, and acceptance of completed project).

With any form of accelerated design and delivery, these phases remain substantially the same. But the various building systems or subsystems move through them at different times and result in multiple contracts released at different times.

For any project, basic building siting is determined early in the design process. Therefore, at an early date in design, a contract can be let for rough site work.

Similarly, basic structural decisions can be made before all the details of the building are established, and permit early award of foundation and structural contracts. Under such circumstances, construction can be initiated very early in the design process, instead of at the end of a lengthy design and contract preparation period. Months and even years can be taken out of the traditional project schedule. When the systems building techniques discussed in Art. 2-16 are employed, purchase of preengineered, commercially available building systems can be integrated into the accelerated design and construction process, reducing time even more drastically.

The major requirements for a project in which design and construction occur simultaneously are threefold:

1. Accurate cost estimating to maintain project budgets.
2. Full understanding of the construction process so that design decisions and contract documents for each building system or subsystem are completed in a fashion that meets the requirements of an ongoing construction process.
3. Tight management of the construction process with feedback into the design process to maintain clear definition of the required contract packages. Overall project cost control and overall project construction responsibilities, including management of the interface of independent prime contracts, must also be undertaken by a construction manager.

Often, the major purpose of accelerated design and construction is to reduce the effect of rapidly increasing construction costs during the extended project design and construction period. For large projects extending several years, contractors and subcontractors quote costs for materials and labor that will be installed several years later. In most cases, the costs associated with such work are uncertain. Bidding for such work, especially when it is of large magnitude, must be very conservative. Accelerated design and construction, however, bring all the financial benefits of shortened project duration and reduce the effect of cost escalation. Also, bidding can be closer to the actual costs, thus reducing bidding risk to the contractor. The combination of early bidding, shortened contract duration, reduced escalation, smaller packages, and greater number of bidders can produce large savings in overall construction costs.

A major objection to accelerated design and construction is that project construction is initiated prior to obtaining bids for the total project and prior to complete assurance that the total project budget can be maintained. In this regard, the reliability of estimating becomes even more critical. It is the experience of most owners and architects involved with multiple contracts, however, that such contracts, bid one at a time, can be readily compared with a total budget breakdown and thus provide safeguards against budget overruns. The ability to design, bid, and negotiate each contract as a separate entity provides optimum cost control. (See also Art. 25-16.)

For accelerated design and construction programs to work effectively, a new role is required. This is the role of the professional construction manager (Art. 2-19).

2-19. Professional Construction Managers.

While a construction manager is necessary on any multiple-contract project, he is most fully utilized in accelerated design and construction. He is a professional who does not contract for performing the actual work. Rather, he provides services during the design and construction process. He must provide the direction and strategy necessary for composing contract packages for award. He must schedule the construction processes and communicate the requirements that continuing construction places on the design process. He must manage the processes of construction in the field. He must provide the leadership required to solve field problems and coordinate the work of the several contractors.

Since all contractors contract directly with the owner, the construction manager has no direct financial interest in the work. He acts only as the agent of the owner in the same professional capacity as the architect.

The construction manager may be a qualified architect, engineer, or contractor. It is not safe to assume, however, that all architects, engineers, or contractors possess the thorough knowledge of design and construction and the management skills necessary to discharge this unique role successfully. No matter who plays the role, however, a close team relationship between design and construction management leaders is critical. The process is continuous and interacting. Its success depends on the correct flow of information, decisions, and documents.

Use of construction management is not restricted to accelerated design and construction programs. On any multiple-contract project, for example, the owner may elect to employ a construction manager or management agency to maintain coordination between the work of the several contractors and to assume responsibility for certain project-wide construction tasks such as site maintenance. Regardless of the type of project or method of awarding contracts, the construction manager must be expert in construction and in management techniques. He must be retained on a professional basis with his fees earned as a result of his service to the owner and not a result of construction performed. (See also Arts. 25-1 and 25-9.)

CONSTRUCTION WITHOUT PROFESSIONAL SERVICES

There are numerous occasions on which an owner does not choose to employ an architect as his agent. Such instances include the small-builder house and employment of a design-build firm for more substantial facilities. When the owner employs a professional as his agent, he secures an expert in design and construction committed to acquiring the most suitable facility for the owner's needs and budget. When an owner acquires facilities without use of such services, he deals directly with those whose business it is to provide a built facility for profit. With the exception of some residential construction, an architect is involved in the process. His client, however, is the package builder and not the owner.

2-20. Package Builders.

There are several kinds of package builders. Their main attraction is that they will, for a fixed and often guaranteed cost, deliver to the owner a designed and constructed facility in a single package. Often, such a builder will acquire property, arrange financing, and include the entire project under one contract. Such services vary in degree of control and participation that the owner is afforded. In some cases, the owner may be offered one or more stock prototypes. In other cases, the design of the facility may be specific for the building to be erected. (See also Arts. 2-21 to 2-23, 25-1, and 25-8.)

2-21. Home Builders.

Residential builders typically offer several models with which their forces are familiar. The potential owner is offered a few choices at reasonable prices. He must shop for the model that best suits his budget and his needs.

There are local, capable, and sensitive builders who take pride in providing the best quality at a price the market will bear. Such builders may or may not employ an architect. In either case, they are to be sought out and valued by the owner seeking a modest builder home.

2-22. Design-and-Build Companies.

Designer-builders offer to design and construct a facility for a fixed lump-sum price, as a total contract. They will bid competitively to provide this service or will provide free design services prior to owner commitment to the project and as a basis for negotiation.

Their design work is not primarily aimed at cost-performance tradeoffs, but at reduced cost for acceptable quality.

The design-build approach to facilities is best employed when an owner requires relatively straightforward facilities and does not require the ability to participate in detailed decision making regarding the various building systems and materials. This does not mean that he has no control over these items. On the contrary, he is often permitted a wide range of selection. But the range of choices is affected by the fixed-cost restraints imposed by the designer-builder and accepted by the owner. When the facilities required are within the range of relatively standard industry-wide prototypes, this restriction may have little significance.

A common misconception regarding design-build is that poor-quality work inevitably results. While there is a general benefit to the builder for reductions in material and labor costs, the more reputable designer-builder may be relied on to deliver a building well within industry standards.

Facilities where higher quality systems, more sensitive design needs, or atypical technical requirements occur deserve the services of an independent design professional. The designer-builder or package builders, however, may offer a desirable service for more straightforward projects. (See also Arts. 25-1 and 25-8.)

2-23. Turnkey Developers.

The term *turnkey* is often used interchangeably with design-build (Art. 2-22). The major distinction in meaning is that turnkey implies more complete independence on the part of the turnkey builder.

In turnkey projects, the major input of the owner is a document outlining his requirements. Bids are usually taken wherein the turnkey developer submits design and cost proposals to meet these requirements. Once a developer is selected, the owner sells his property to the developer or authorizes its purchase from a third party under option. From this time on, the owner has little or no participation in the project. Instead, he guarantees that he will buy the project back from the developer, as a package, at completion.

As with design-build companies, the professional designer is on the developer team and is not an agent of the owner. Once overall requirements are established, detail decisions are made by the turnkey developer. The owner deals only with the final product. This means he has less control than with design-build or other similar procedures.

Section 3

Protection of Structures and Occupants against Hazards

Frederick S. Merritt
Consulting Engineer,
West Palm Beach, Fla.

Studies of building failures provide information that building designers should use to prevent similar catastrophes. Many of the lessons learned from failures have led to establishment of safety rules in building codes. These rules, however, generally are minimum requirements and apply to ordinary structures. Building designers, therefore, should use judgment in applying code requirements and should adopt more stringent design criteria where conditions dictate.

Such conditions are especially likely to exist for buildings in extreme climates or in areas exposed to natural hazards, such as high winds, earthquakes, floods, landslides, and lightning. Stricter criteria should also be used for buildings that are tall and narrow, are low but very large, have irregular or unusual shapes, house hazardous materials, or are of novel construction. Furthermore, building codes may not contain provisions for some hazards against which building designers nevertheless should provide protection. Examples of such hazards are vandalism, trespass, and burglary. In addition, designers should anticipate conditions that may exist in buildings in emergencies and provide refuge for occupants or safe evacuation routes.

Building designers also should use judgment in determining the degree of protection to be provided against specific hazards. Costs of protection should be commensurate with probable losses from an incident. In many cases, for example, it is uneconomical to construct a building that will be immune to extreme earthquakes, high winds of tornadoes, arson, bombs, burst dams, or professional burglars. Full protection, however, should always be provided against hazards with a high probability of occurrence accompanied by personal injuries or high property losses. Such hazards include hurricanes and gales, fire, and vandals.

3-2 Protection of Structures and Occupants against Hazards

3-1. Design Life of Buildings.

For natural phenomena, design criteria may be based on the probability of occurrence of extreme conditions, as determined from statistical studies of weather in specific localities. These probabilities are often expressed as mean recurrence intervals.

A **mean recurrence interval** of an extreme condition is the average time, in years, between occurrences of a condition equal to or worse than the specified extreme condition. For example, the mean recurrence interval of a wind of 60 mi/hr or more may be recorded for Los Angeles as 50 years. Thus, after a building has been erected in Los Angeles, chances are that in the next 50 years it will be subjected only once to a wind of 60 mi/hr or more. Consequently, if the building was assumed to have a 50-year life, designers might logically design it basically for a 60-mi/hr wind, with a safety factor included in the design to protect against low-probability faster winds. Mean recurrence intervals are the basis for minimum design loads for high winds, snowfall, and earthquake in "Building Code Requirements for Minimum Design Loads in Buildings and Other Structures," ANSI A58.1, American National Standards Institute, 1430 Broadway, New York, N.Y. 10018.

3-2. Safety Factors.

Design of buildings for both normal and emergency conditions should always incorporate a safety factor against failure. The magnitude of the safety factor should be selected in accordance with the importance of a building, the extent of personal injury or property loss that may result if a failure occurs, and the degree of uncertainty as to the magnitude or nature of loads and the properties and behavior of building components.

As usually incorporated in building codes, a safety factor for quantifiable system variables is a number greater than unity. The factor may be applied in either of two ways.

One way is to relate the maximum permissible load, or demand, on a system under service conditions to design capacity. This system property is calculated by dividing by the safety factor the ultimate capacity, or capacity at failure, for sustaining that type of load. For example, suppose a structural member assigned a safety factor of 2 can carry 1000 lb before failure occurs. The design capacity then is $1000/2 = 500$ lb.

The second way in which codes apply safety factors is to relate the ultimate capacity of a system to a design load. This load is calculated by multiplying the maximum load under service conditions by a safety factor, often referred to as a **load factor**. For example, suppose a structural member assigned a load factor of 2 is required to carry a service load of 500 lb. Then, the member should be designed to have a capacity for sustaining a design load of $500 \times 2 = 1000$ lb, without failing.

While both methods achieve the objective of providing reserve capacity against unexpected conditions, use of load factors offers the advantage of greater flexibility in design of a system for a combination of different loadings, because a different load factor can be assigned to each type of loading in accordance with probability of occurrence and effects of other uncertainties.

Safety factors for various building systems are discussed in following sections of the book. This section presents general design principles for protection of buildings and occupants against high winds, earthquakes, water, fire, lightning, and intruders.

WIND PROTECTION

For practical design, wind and earthquake may be treated as horizontal, or lateral, loads. Although wind and seismic loads may have vertical components, these generally are small and readily resisted by columns and bearing walls.

The variation with height of the magnitude of a wind load for a multistory building differs from that of a seismic load. Nevertheless, provisions for resisting either type of load are similar.

In areas where the probability of either a strong earthquake or a high wind is small, it is nevertheless advisable to provide in buildings considerable resistance to both types of load. In many cases, such resistance can be incorporated with little or no increase in costs over designs that ignore either high wind or seismic resistance.

3-3. Wind Characteristics.

Because wind loads are considered horizontal forces, wind pressure, for design purposes, should be assumed to be applied to the gross area of the vertical projection of that portion of the building above the average level of the adjoining ground. Although the loads are assumed to be horizontal, they may nevertheless apply either inward pressures or suctions to inclined and horizontal surfaces. In any case, wind loads should be considered to act normal to the exposed building surfaces. Furthermore, wind should be considered to be likely to come from any direction, unless it is known for a specific locality that extreme winds may come only from one direction. As a consequence of this assumption, each wall of a rectangular building should be considered in design to be subject to the maximum wind load.

Winds generally strike a building in gusts. Consequently, the building is subjected to dynamic loading. Nevertheless, except for unusually tall or narrow buildings, it is common practice to treat wind as a static loading, even though wind pressures are not constant.

Estimation of design wind pressures is complicated by several factors. One factor is the effect of natural and man-made obstructions along the ground. Another factor is the variation of wind velocity with height above ground. Still another factor complicating wind-pressure calculation is the effect of building or building-component shape or geometry (relationship of height or width to length) on pressures. For important buildings, it is advisable to base design wind pressures on the results of wind tunnel tests of a model of a building, neighboring buildings, and nearby terrain.

3-4. Wind Pressures and Suctions.

Pressures are considered positive when they tend to push a building component toward the building interior. They are treated as negative for suctions or uplifts, which tend to pull components outward.

Figure 3-1a illustrates wind flow over the sloping roof of a low building. For roofs with inclines up to about 30°, the wind may create an uplift over the entire roof (Fig. 3-1b). Also, as shown in Fig. 3-1b and c, the pressure on the external face of the windward wall is positive and on the leeward wall, negative (suction). If there are openings in the walls, the wind will impose internal pressures on the walls, floors, and roof. The net pressure on any building component, therefore, is the vector sum of the pressures acting on opposite faces of the component.

Because of the wind characteristics described in Art. 3-3 and the dependence of wind pressures on building geometry, considerable uncertainty exists as to the magnitude, direction, and duration of the maximum wind loads that may be imposed on any portion of a specific building. Consequently, numerous assumptions, based to some extent on statistical evidence, generally are made to determine design wind loads for buildings. Minimum requirements for wind loads are presented in local and model building codes and in "Building Code Requirements for Minimum Design Loads in Buildings and Other Structures," ANSI A58.1, American National Standards Institute.

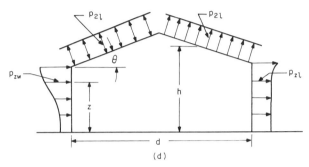

Fig. 3-1 Effects of wind on a low building with pitched roof. (*a*) Airflow at the building. (*b*) Wind applies inward pressure against the windward wall, suction on the leeward wall, and uplift over all of a roof with slight slopes. (*c*) With a steep roof, inward pressure acts on the windward side of the roof and uplift only on the leeward side. (*d*) Pressure distribution along walls and roof assumed for design of wind bracing of building.

Codes usually permit design wind loads to be determined either by mathematical calculations in accordance with an analytical procedure specified in the code or by wind-tunnel tests. Such tests are advisable for structures with unusual shapes, unusual response to lateral loading, or location where channeling effects or buffeting in the wake of upwind obstructions are likely to occur. Tests also are desirable where wind records are not available or when more accurate information is needed. ANSI A58.1 requires that the following conditions be met in execution of wind-tunnel tests:

1. The tests should be conducted on a model scaled with respect to dimensions, mass distribution, stiffness, and damping of the proposed structure.

2. Air motion should be modeled to account for variation of wind speed with elevation.

3. The geometric scale of the model should not be greater than 3 times that of the longitudinal component of turbulence.

4. Instruments used should have response characteristics consistent with the required accuracy of measurements to be recorded.

5. Account should be taken of the dependence of forces and pressures on the Reynolds number of the air motion.

In the analytical methods specified by building codes, maximum wind speeds observed in a region are converted to velocity pressures. These are then multiplied by various factors, to take into account building, site, and wind characteristics, to obtain design static wind loads. Bear in mind, however, that, in general, code requirements are applicable to pressures considerably smaller than those created by tornadoes, which may have wind speeds up to 600 mi/hr.

Wind Pressures and Suctions   3-5

▨ REGION WITH MINIMUM BASIC WIND SPEED OF 70 MPH
■ ROUGH TERRAIN OR OCEAN PROMONTORIES WITH HIGH WINDS

Fig. 3-2 Contours indicate variation of basic wind speed, mi/hr, in 48 contiguous states of the United States and in Alaska. Basic wind speed may be taken as 80 mi/hr for Hawaii and 95 mi/hr for Puerto Rico. *(Based on data in ANSI A58.1, American National Standards Institute.)*

Velocity pressures due to wind to be used in building design vary with type of terrain, distance above ground level, importance of building, likelihood of hurricanes, and basic wind speed recorded near the building site. The wind pressures are assumed to act horizontally on the building area projected on a vertical plane normal to the wind direction.

The basic wind speed used in design is the fastest-mile wind speed recorded at a height of 10 m (32.8 ft) above open, level terrain with a 50-year mean recurrence interval.

Unusual wind conditions often occur over rough terrain and around ocean promontories. Basic wind speeds applicable to such regions should be selected with the aid of meteorologists and the application of extreme-value statistical analysis to anemometer readings taken at or near the site of the proposed building. Generally, however, basic wind velocities, mi/hr, may be assumed for the United States as indicated in Fig. 3-2.

For design purposes, wind pressures should be determined in accordance with the degree to which terrain surrounding the building to be designed exposes it to the wind. ANSI A58.1 establishes four classes of exposure:

Exposure A applies to centers of large cities, where for at least one-half mile upwind from the building the majority of structures are over 70 ft high and lower buildings extend at least one more mile upwind.

Exposure B applies to wooded or suburban terrain or to urban areas with closely spaced buildings mostly less than 70 ft high, where such conditions prevail upwind for a distance from the building of at least 1500 ft or 10 times the building height.

Exposure C exists for flat, open country or exposed terrain with obstructions less than 30 ft high.

Exposure D applies to flat, coastal areas directly exposed to wind blowing over a large expanse of water with a shoreline at a distance from the building of not more than 1500 ft or 10 times the building height.

For design purposes also, the following formulas may be used to determine, for heights z (in feet) greater than 15 ft above ground, a pressure coefficient K for converting wind speeds to pressures.

For Exposure A, for heights up to 1500 ft above ground level,

$$K = 0.000517 \left(\frac{z}{32.8} \right)^{2/3} \qquad (3\text{-}1)$$

For z less than 15 ft, $K = 0.00031$.

For Exposure B, for heights up to 1200 ft above ground level,

$$K = 0.00133 \left(\frac{z}{32.8} \right)^{4/9} \qquad (3\text{-}2)$$

For z less than 15 ft, $K = 0.00095$.

For Exposure C, for heights up to 900 ft above ground level,

$$K = 0.00256 \left(\frac{z}{32.8} \right)^{2/7} \qquad (3\text{-}3)$$

For z less than 15 ft, $K = 0.0020$.

For Exposure D, for heights up to 700 ft above ground level,

$$K = 0.00357 \left(\frac{z}{32.8} \right)^{1/5} \qquad (3\text{-}4)$$

For z less than 15 ft, $K = 0.0031$.

For ordinary buildings not subject to hurricanes, the velocity pressure q_z, psf, at height z may be calculated from

$$q_z = KV^2 \qquad (3\text{-}5)$$

where V = basic wind speed, mi/hr, but not less than 70 mi/hr.

For important buildings, such as hospitals and communication buildings, for buildings sensitive to wind, such as slender skyscrapers, and for buildings presenting a high degree of hazard to life and property, such as auditoriums, q_z computed from Eq. (3-5) should be increased 7%.

To allow for hurricanes, q_z should be increased 5% for ordinary buildings and by 11% for important, wind-sensitive or high-risk buildings along coastlines. These increases may be assumed to reduce uniformly with distance from the shore to zero for ordinary buildings and 5% for the more important or sensitive buildings at points 100 mi inland.

Wind pressures on low buildings are different at a specific elevation from those on tall buildings. Hence, building codes may give different formulas for pressures for the two types of construction. In any case, however, design wind pressure should be a minimum of 10 psf.

Multistory Buildings. For design of the main wind-force resisting system of ordinary, rectangular, multistory buildings, the design pressure at any height z, ft, above ground may be computed for windward walls from

$$p_{zw} = G_o C_{pw} q_z \qquad (3\text{-}6)$$

where p_{zw} = design wind pressure, psf, on windward wall
G_o = gust response factor
C_{pw} = external pressure coefficient for windward wall
q_z = velocity pressure computed from Eq. (3-5) and modified for hurricanes and building importance, risks, and wind sensitivity

Wind Pressures and Suctions

For roofs and leeward walls, the design pressure at elevation z is

$$p_{zl} = G_o C_p q_h \qquad (3\text{-}7)$$

where p_{zl} = design pressure, psf, on roof or leeward wall
C_p = external pressure coefficient for roof or leeward wall
q_h = velocity pressure at mean roof height h (see Fig. 3-1d)

In these equations, the gust response factor may be taken approximately as

$$G_o = 0.65 + \frac{8.58D}{(h/30)^n} \geq 1 \qquad (3\text{-}8)$$

where D = 0.16 for Exposure A, 0.10 for Exposure B, 0.07 for Exposure C and 0.05 for Exposure D
n = ⅓ for Exposure A, ⅖ for Exposure B, ½ for Exposure C and 0.1 for Exposure D
h = mean roof height, ft

Based on the results of wind-tunnel tests, ANSI A58.1 recommends that the external pressure coefficient for windward walls C_{pw} be taken equal to 0.8. For leeward walls, subjected to suction, C_p depends on the ratio of the depth d to width b of the building and may be assumed as follows:

$$d/b = \quad 1 \text{ or less} \quad 2 \quad 4 \text{ or more}$$
$$C_p = \quad -0.5 \quad -0.3 \quad -0.2$$

The negative sign indicates suction. Table 3-1 lists values of C_p for pressures on roofs.

TABLE 3-1 External Pressure Coefficients C_p for Roofs*

Flat roofs						−0.7
Wind parallel to ridge of sloping roof						
h/b or $h/d \leq 2.5$						−0.7
h/b or $h/d > 2.5$						−0.8
Wind perpendicular to ridge of sloping roof, at angle θ with horizontal						
Leeward side						−0.7
Windward side						

	Slope of roof θ, deg					
h/d	10	20	30	40	50	60 or more
0.3 or less	0.2	0.2	0.3	0.4	0.5	
0.5	−0.9	−0.75	−0.2	0.3	0.5	0.01θ
1.0	−0.9	−0.75	−0.2	0.3	0.5	
1.5 or more	−0.9	−0.9	−0.9	0.35	0.21	

*h = height of building, ft; d = depth, ft, of building in direction of wind; b = width, ft, of building transverse to wind.
Based on data in ANSI A58.1-1981.

Flexible Buildings. These are structures with a fundamental natural frequency less than 1 Hz or with a ratio of height to least horizontal dimension (measured at midheight for buildings with tapers or setbacks) exceeding 5. For such buildings, the main wind-force resisting system should be designed for a pressure on windward walls at any height z, ft, above ground computed from

$$p_{zw} = G_f C_{pw} q_z \qquad (3\text{-}9)$$

where G_f = gust response factor determined by analysis of the system taking into account its dynamic properties. For leeward walls of flexible buidlings,

$$p_{zl} = G_f C_p q_h \tag{3-10}$$

(In ANSI A58.1-1981, G_f is represented by $\overline{G}_3$. Requiring a knowledge of the fundamental frequency, structural damping characteristics, and type of exposure of the building, the formula for G_f is complicated, but computations may be simplified somewhat by use of tables and charts in the standard.)

One-Story Buildings. For design of the main wind-force resisting system of rectangular, one-story buildings, the design pressure at any height z, ft, above ground may be computed for windward walls from

$$p_{zw} = (G_o C_p + C_{p1}) q_z \tag{3-11}$$

where C_{p1} = 0.75 if the percentage of openings in one wall exceeds that of other walls by 10% or more
= 0.25 for all other cases

For roofs and leeward walls, the design pressure at elevation z is

$$p_{zl} = G_o C_p q_h - C_{p2} q_z \tag{3-12}$$

where C_{p2} = +0.75 or −0.25 if the percentage of openings in one wall exceeds that of other walls by 10% or more
= ±0.25 for all other cases

(Positive signs indicate pressures acting toward a wall; negative signs indicate pressures acting away from the wall.)

Design Guides. For a rough estimate of basic design wind pressures, use Table 5-1c, Art. 5-2. For more accurate design formulas and procedures, consult ANSI A58.1. The standard also presents formulas and describes methods for computing wind pressures on building components, arched roofs, roofs over open buildings or frameworks, chimneys, tanks, trussed towers, tower guys, overhangs, and balconies.

3-5. Failure Modes.

Consideration of the ways in which winds may damage or destroy buildings suggests provisions that should be made to prevent failures. Past experience with building damage by winds indicates buildings are likely to fail by overturning; sliding; separation of components; excessive sway, or drift; or structural collapse.

Subjected to lateral forces W, a building may act as a rigid body and overturn. It would tend to rotate about the edge of its base on the leeward side (Fig. 3-3a). Overturning is resisted by the weight of the building M with a lever arm e measured from the axis of rotation. Building codes usually require that

$$Me \geq 1.5 Wh \tag{3-13}$$

where Wh is the overturning moment.

Resistance to overturning may be increased by securely anchoring buildings to foundations. When this is done, the weight of earth atop the footings and pressing against foundation walls may be included with the weight of the building.

In addition to the danger of overturning, there is the risk of a building being pushed laterally by high winds. Sliding is resisted by friction at the base of the footings and earth pressure against foundation walls (Fig. 3-3b). (Consideration should be given to the possibility that soil that is highly resistant to building movement when dry may become weak when wet.) Another danger is that a building may be pushed by wind off the foundations (Fig. 3-3c). Consequently, to prevent this, a building should be firmly anchored to its foundations.

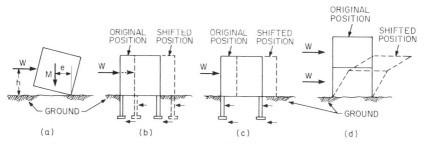

Fig. 3-3 Some ways in which wind may destroy a building: (*a*) overturning; (*b*) sliding through the ground; (*c*) sliding off the foundations; (*d*) excessive sidesway, or drift.

Buildings also may be damaged by separation of other components from each other. Therefore, it is essential that all connections between structural members and between other components and their supports be capable of resisting design wind loads. The possibility of separation of components by uplift or suction should not be overlooked. Such pressures can slide a roof laterally or lift it from its supports, tear roof coverings, rip off architectural projections, and suck out windows.

Another hazard is drift (sway) or collapse without overturning or sliding. Excessive drift when the wind rocks a building can cause occupant discomfort, induce failure of structural components by fatigue, or lead to complete collapse of the structure. The main resistance to drift usually is provided by structural components, such as beams, columns, and walls that are also assigned the task of supporting gravity loads. Some means must be provided to transmit wind or seismic loads from these members to the foundations and thence to the ground. Otherwise, the building may topple like a house of cards (Fig. 3-3*d*).

3-6. Limitation of Drift.

There are no generally accepted criteria for maximum permissible lateral deflections of buildings. Some building codes limit drift of any story of a building to a maximum of 0.25% of the story height for wind and 0.50% of the story height for earthquake loads. Buildings less than 14 stories high and with height-width ratios of 2.5 or less are exempt from these limitations. The severer limitation of drift caused by wind loads is applied principally because they are likely to occur more frequently than earthquakes and will produce motions that will last much longer.

Three basic methods are commonly used, separately or in combination with each other, to prevent collapse of buildings under lateral loads, limit drift and transmit the loads to the foundations. These methods are illustrated in Fig. 3-4. One method is to incorporate shear walls in a building. A shear wall is a vertical cantilever with high resistance to horizontal loads parallel to its length (Fig. 3-4*a*). A pair of perpendicular walls can resist wind from any direction, because any wind load can be resolved into components in the planes of the walls (Fig. 3-4*b*).

A second method of providing resistance to lateral loads is to incorporate diagonal structural members to carry lateral forces to the ground (Fig. 3-4*c*). (The diagonals in Fig. 3-4*c* are called X bracing. Other types of bracing are illustrated in Fig. 3-7.) Under lateral loads, the braced bays of a building act like cantilever vertical trusses. The arrows in Fig. 3-4*c* show the paths taken by wind forces from points of application to the ground. Note that the lateral loads impose downward axial forces on the leeward columns, causing compression, and uplift on the windward columns, causing tension.

A third method of providing resistance to lateral loads is to integrate the beams, or girders, and columns into rigid frames (Fig. 3-4*d*). In a rigid frame, connections between

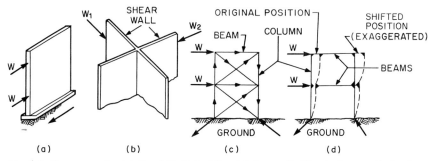

Fig. 3-4 Some ways of restricting drift of a building: (a) shear wall; (b) pair of perpendicular shear walls; (c) diagonal bracing; (d) rigid frames.

horizontal and vertical components prevent any change of angle between the members under loads. (Drift can occur only if beams and columns bend.) Such joints are often referred to as rigid, moment, or wind connections. They prevent the frame from collapsing in the manner shown in Fig. 3-3d until the loads are so large that the strength of the members and connections is exhausted. Note that in a rigid frame, leeward columns are subjected to bending and axial compression and windward columns are subjected to bending and axial tension.

In addition to using one or more of the preceding methods, designers can reduce drift by proper shaping of buildings, arrangements of structural components, and selection of members with adequate dimensions and geometry to withstand changes in dimensions. Shape is important because low, squat buildings have less sideway than tall, narrow buildings, and buildings with circular or square floor plans have less sideway than narrow, rectangular buildings with the same floor area per story.

Low Buildings. Figure 3-5a illustrates the application of diagonal bracing to a low, industrial-type building. Bracing in the plane of the roof acts with the rafters, ridge beam, and an edge roof beam as an inclined truss, which resists wind pressures on the roof. Each truss transmits the wind load to the ends of the building. Diagonals in the end walls transmit the load to the foundations. Wind pressure on the end walls is resisted by diagonal bracing in the end panels of the longitudinal walls. Wind pressure on the longitudinal walls, like wind on the roof, is transmitted to the end walls.

Fig. 3-5 Bracing of low buildings: (a) diagonal bracing in roofs and walls; (b) isolated pairs of shear walls in a T pattern; (c) service-core enclosure used as shear walls; (d) shear walls at ends of building and rigid frames in the perpendicular direction.

For large buildings, rigid frames are both structurally efficient and economic.

Alternatively, for multistory buildings, shear walls may be used. Figure 3-5b shows shear walls arranged in the shape of a T in plan, to resist wind from any direction. Figure 3-5c illustrates the use of walls enclosing stairwells and elevator shafts as shear walls. In apartment buildings, closet enclosures also can serve as shear walls if designed for the purpose. Figure 3-5d shows shear walls placed at the ends of a building to resist wind on its longitudinal walls. Wind on the shear walls, in turn, is resisted by girders and columns in the longitudinal direction acting as rigid frames. (See also Arts. 11-7 to 11-19.)

Tall Buildings. For low buildings, structural members sized for gravity loads may require little or no enlargement to also carry stresses due to lateral loads. For tall buildings, however, structural members often have to be larger than sizes necessary only for gravity loads. With increase in height, structural material requirements increase rapidly. Therefore, for tall buildings, designers should select wind-bracing systems with high structural efficiency to keep material requirements to a minimum.

While shear walls, diagonal bracing, and rigid frames can be used even for very tall buildings, simple framing arrangements, such as planar systems, are not so efficient in high structures as more sophisticated framing. For example, shear walls or rigid frames in planes parallel to the lateral forces (Fig. 3-6a) may sway considerably at the top if the building is tall (more than 30 stories) and slender. Resistance to drift may be improved, however, if the shear walls are arranged in the form of a tube within the building (Fig. 3-6b). (The space within the tube can be utilized for stairs, elevators, and other services. This space is often referred to as the **service core**.) The cantilevered tube is much more efficient in resisting lateral forces than a series of planar, parallel shear walls containing the same amount of material. Similarly, rigid frames and diagonal bracing may be arranged in the form of an internal tube to improve resistance to lateral forces.

The larger the size of the cantilevered tube for a given height, the greater will be its resistance to drift. For maximum efficiency of a simple tube, it can be arranged to enclose the entire building (Fig. 3-6c). For the purpose, bracing or a rigid frame may be erected behind or in the exterior wall, or the exterior wall itself may be designed to act as a perforated tube. Floors act as horizontal diaphragms to brace the tube and distribute the lateral forces to it.

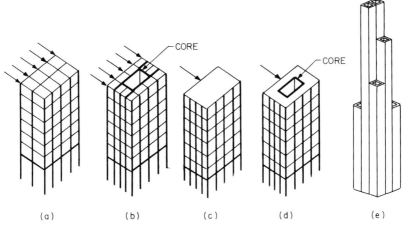

Fig. 3-6 Bracing of tall buildings: (a) diagonal bracing, rigid frames or shear walls placed in planes (bents) parallel to the lateral forces; (b) interior tube enclosing service core; (c) perforated tube enclosing the building; (d) tube within a tube; (e) bundled tubes.

For very tall buildings, when greater strength and drift resistance are needed than can be provided by a simple tube, the tube around the exterior may be augmented by an internal tube (Fig. 3-6d), or by other arrangements of interior bracing, such as shear walls attached and perpendicular to the exterior tube. As an alternative, a very tall building may be composed of several interconnected small tubes, which act together in resisting lateral forces (Fig. 3-6e). Known as bundled tubes, this type of framing offers greater flexibility in floor-area reduction at various levels than a tube-within-tube type, because the tubes in a bundle can differ in height.

Diagonal bracing is more efficient in resisting drift than the other methods, because the structural members carry the lateral loads to the foundations as axial forces, as shown in Fig. 3-4c, rather than as a combination of bending, shear, and axial forces. Generally, the bracing is arranged to form trusses composed of triangular configurations, because of the stability of such arrangements. The joints between members comprising a triangle cannot move relative to each other unless the length of the members changes. Figure 3-7a illustrates the use of X bracing in the center bay of a multistory building to form a vertical cantilever truss to resist lateral forces.

Other forms of bracing, however, may be used as an alternative to reduce material requirements. Figure 3-7b shows how a single diagonal can be used in the center bay to form a vertical truss. In large bays, however, the length of the diagonal may become too long for structural efficiency. Hence, two or more diagonals may be inserted in the bay instead, as shown in Fig. 3-7c to e. The type of bracing in Fig. 3-7c is known as K bracing; that in Fig. 3-7d, as V bracing; and that in Fig. 3-7e, as inverted V bracing. The V type, however, has the disadvantage of restricting deflection of the beams to which the diagonals are attached and thus compelling the diagonals to carry gravity loads applied to the beams.

The bracing shown in Fig. 3-7a to e has the disadvantage of obstructing the bay and interfering with placement of walls, doors, passageways, and, for bracing along the building exterior, placement of windows. Accordingly, the inverted V type often is converted to knee bracing, short diagonals placed near beam-to-column joints. When knee bracing also is architecturally objectionable because of interference with room arrangements, an alternative form of wind bracing, such as rigid frames or shear walls, has to be adopted.

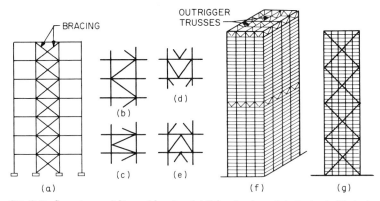

Fig. 3-7 Some types of diagonal bracing: (a) X bracing in an interior bent; (b) single diagonal; (c) K bracing; (d) V bracing; (e) inverted V bracing; (f) horizontal trusses at the roof and intermediate levels to restrict drift; (g) X bracing on the exterior of a building.

Trusses also can be placed horizontally to stiffen buildings for less drift. For example, Fig. 3-7f shows a building with wind bracing provided basically by an internal vertical cantilever tube. A set of horizontal trusses at the roof and a similar set at an intermediate level tie the tube to the exterior columns. The trusses reduce the drift at the top of the building by utilizing bending resistance of the columns. A belt of horizontal trusses around the building exterior at the roof and the intermediate level also helps resist drift of the building by utilizing bending resistance of the exterior columns.

When not considered architecturally objectionable, diagonal bracing may be placed on the building exterior to form a braced tube. The bracing may also serve as columns to transmit floor and roof loads to the ground. Figure 3-7g shows how multistory X bracing has been used to create a braced tube for a skyscraper.

See also Art. 3-11 and Secs. 7 through 11.

(Council on Tall Buildings and Urban Habitat, "Planning and Design of Tall Buildings," Vols. SC, SB, and CB, American Society of Civil Engineers, New York; F. S. Merritt, "Structural Steel Designers' Handbook," McGraw-Hill Book Company, New York; E. Simiu and R. H. Scanlon, "Wind Effects on Structures," John Wiley & Sons, Inc., New York.)

PROTECTION AGAINST EARTHQUAKES

Buildings should be designed to withstand minor earthquakes without damage, because they may occur almost everywhere. For major earthquakes, it may not be economical to prevent all damage but collapse should be precluded.

Because an earthquake and a high wind are not likely to occur simultaneously, building codes usually do not require that buildings be designed for a combination of large seismic and wind loads. Thus, designers may assume that the full strength of wind bracing is also available to resist drift caused by earthquakes.

The methods of protecting against high winds described in Art. 3-6 may also be used for protecting against earthquakes. Shaking of buildings produced by temblors, however, is likely to be much severer than that caused by winds. Consequently, additional precautions must be taken to protect against earthquakes. Because such protective measures will also be useful in resisting unexpectedly high winds, such as those from tornadoes, however, it is advisable to apply aseismic design principles to all buildings.

These principles require that collapse be avoided, oscillations of buildings damped, and damage to both structural and nonstructural components minimized. Nonstructural components are especially liable to damage from large drift. For example, walls are likely to be stiffer than structural framing and therefore subject to greater seismic forces. The walls, as a result, may crack or collapse. Also, they may interfere with planned actions of structural components and cause additional damage. Consequently, aseismic design of buildings should make allowance for large drift, for example, by providing gaps between adjoining buildings and between adjoining building components not required to be rigidly connected together and by permitting sliding of such components. Thus, partitions and windows should be free to move in their frames so that no damage will occur when an earthquake wracks the frames.

3-7. Earthquake Characteristics.

Earthquakes are produced by sudden release of tremendous amounts of energy within the earth by a sudden movement at a point called the **hypocenter.** (The point on the surface of the earth directly above the hypocenter is called the **epicenter.**) The resulting

3-14 Protection of Structures and Occupants against Hazards

shock sends out longitudinal and transverse vibrations in all directions, both through the earth's crust and along the surface, and at different velocities. Consequently, the shock waves arrive at distant points at different times.

As a result, the first sign of the advent of an earthquake at a distant point is likely to be faint surface vibration of short duration as the first longitudinal waves arrive at the point. Then, severe shocks of longer duration occur there, as other waves arrive.

Movement at any point of the earth's surface during a temblor may be recorded with seismographs and plotted as seismograms, which show the variation with time of displacements. Seismograms of past earthquakes indicate that seismic wave forms are very complex.

Ground accelerations are also very important, because they are related to the inertial forces that act on building components during an earthquake. Accelerations are recorded in accelerograms, which are a plot of the variation with time of components of the ground accelerations. Newton's law relates acceleration to force:

$$F = Ma = \frac{W}{g} a \tag{3-14}$$

where F = force, lb
M = mass accelerated
a = acceleration of the mass, ft/s^2
W = weight of building component accelerated, lb
g = acceleration due to gravity = 32.2 ft/s^2

3-8. Seismic Scales.

For study of the behavior of buildings in past earthquakes and application of the information collected to contemporary aseismic design, it is useful to have some quantitative means for comparing earthquake severity. Two scales, the Modified Mercalli and the Richter, are commonly used in the United States.

The Modified Mercalli scale compares earthquake intensity by assigning values to human perceptions of the severity of oscillations and extent of damage to buildings. The scale has 12 divisions. The severer the reported oscillations and damage, the higher is the number assigned to the earthquake intensity (Table 3-2).

The Richter scale assigns numbers M to earthquake intensity in accordance with the amount of energy released, as measured by the maximum amplitude of ground motion:

$$M = \log A - 1.73 \log \frac{100}{D} \tag{3-15}$$

where M = earthquake magnitude 100 km from epicenter
A = maximum amplitude of ground motion, micrometers
D = distance, km, from epicenter to point where A is measured

The larger the ground displacement at a given location, the higher the value of the number assigned on the Richter scale. A Richter magnitude of 8 corresponds approximately to a Modified Mercalli intensity of XI, and for smaller intensities, Richter scale digits are about one unit less than corresponding Mercalli Roman numerals.

3-9. Effects of Ground Conditions.

The amplitude of ground motion at a specific location during an earthquake depends not only on distance from the epicenter but also on the types of soil at the location. (Some soils suffer a loss of strength in an earthquake and allow large, uneven foundation set-

TABLE 3-2 Modified Mercalli Intensity Scale (Abridged)

Intensity	Definition
I	Detected only by sensitive instruments.
II	Felt by a few persons at rest, especially on upper floors. Delicate suspended objects may swing.
III	Felt noticeably indoors; not always recognized as an earthquake. Standing automobiles rock slightly. Vibration similar to that caused by a passing truck.
IV	Felt indoors by many, outdoors by few; at night some awaken. Windows, dishes, doors rattle. Standing automobiles rock noticeably.
V	Felt by nearly everyone. Some breakage of plaster, windows, and dishes. Tall objects disturbed.
VI	Felt by all; many frightened and run outdoors. Falling plaster and damaged chimneys.
VII	Everyone runs outdoors. Damage of buildings negligible to slight, depending on quality of construction. Noticeable to drivers of automobiles.
VIII	Damage slight to considerable in substantial buildings, great in poorly constructed structures. Walls thrown out of frames; walls, chimneys, monuments fall; sand and mud ejected.
IX	Considerable damage to well-designed structures; structures shifted off foundations; buildings thrown out of plumb; underground pipes damaged. Ground cracked conspicuously.
X	Many masonry and frame structures destroyed; rails bent; water splashed over banks; landslides; ground cracked.
XI	Bridges destroyed; rails bent greatly; most masonry structures destroyed; underground service pipes out of commission; landslides; broad fissures in ground.
XII	Total damage. Waves seen in surface level; lines of sight and level distorted; objects thrown into air.

tlements, which cause severe property damage.) Ground motion usually is much larger in alluvial soils (sands or clays deposited by flowing water) than in rocky areas or diluvial soils (material deposited by glaciers). Reclaimed land or earth fills generally undergo even greater displacements than alluvial soils. Consequently, in selection of sites for buildings in zones where severe earthquakes are highly probable during the life of the buildings, preference should be given to sites with hard ground or rock to considerable depth, with sand and gravel as a less desirable alternative and clay as a poor choice.

3-10. Seismic Loads.

During an earthquake, the ground may move horizontally in any direction and up and down, shifting the building foundations correspondingly. Inertial forces, or seismic loads, on the building resist the displacements. Major damage usually is caused by the horizontal components of these loads, inasmuch as vertical structural members generally have adequate strength to resist the vertical components. Hence, as for wind loads, buildings should be designed to resist the maximum probable horizontal component applied in any direction.

Seismic forces vary rapidly with time. Therefore, they impose a dynamic loading on buildings. Calculation of the building response to such loading is complex (Art. 5-85) and is usually carried out only for important buildings that are very tall and slender. For other types of buildings, building codes generally permit use of an alternative static loading for which structural analysis is much simpler (Art. 5-86).

3-16   Protection of Structures and Occupants against Hazards

3-11. Aseismic Design.

The basic methods for providing wind resistance—shear walls, diagonal bracing, and rigid frames (Art. 3-6)—are also suitable for resisting seismic loads. Ductile rigid frames, however, are preferred because of large energy-absorbing capacity. Building codes encourage their use by permitting them to be designed for smaller seismic loads than those required for shear walls and diagonal bracing. (Ductility is a property that enables a structural member to undergo considerable deformation without failing. The more a member deforms, the more energy it can absorb and therefore the greater is the resistance it can offer to dynamic loads.)

Designers usually utilize floors and roofs, acting as horizontal diaphragms, to transmit lateral forces to the resisting structural members. Horizontal bracing, however, may be used instead. Where openings occur in floors and roofs, for example, for floors and elevators, structural framing should be provided around the opening to bypass the lateral forces.

As for wind loads, the weight of the building and of earth adjoining foundations are the only forces available to prevent the horizontal loads from overturning the building. [See Eq. (3-13) in Art. 3-5.] Also, as for wind loads, the roof should be firmly anchored to the superstructure framing, which, in turn, should be securely attached to the foundations. Furthermore, individual footings, especially pile and caisson footings, should be tied to each other to prevent relative movement.

Building codes often limit the drift per story under the equivalent static seismic loading mentioned previously to a maximum of 0.5% of the story height. Connections and intersections of curtain walls and partitions with each other or with the structural framing should allow for a relative movement of at least twice the calculated drift in each story. Such allowances for displacement may be larger than those normally required for dimensional changes caused by temperature variations.

(ICBO "Uniform Building Code," International Conference of Building Officials, Inc., 5360 South Workman Mill Road, Whittier, Cal. 90601; F. S. Merritt, "Structural Steel Designers' Handbook," McGraw-Hill Book Company, New York; N. M. Newmark and E. Rosenblueth, "Fundamentals of Earthquake Engineering," Prentice-Hall, Englewood Cliffs, N.J.)

PROTECTION AGAINST WATER

Whether thrust against and into a building by a flood, driven into the interior by a heavy rain, leaking from plumbing, or seeping through the exterior enclosure, water can cause costly damage to a building. Consequently, designers should protect buildings and their contents against water damage.

Protective measures may be divided into two classes: floodproofing and waterproofing. Floodproofing provides protection against flowing surface water, commonly caused by a river overflowing its banks. Waterproofing provides protection against penetration through the exterior enclosure of buildings of groundwater, rainwater, and melting snow.

3-12. Floodproofing.

A flood occurs when a river rises above an elevation, called flood stage, and is not prevented by enclosures from causing damage beyond its banks. Buildings constructed in a flood plain, an area that can be inundated by a flood, should be protected against a flood with a mean recurrence interval of 100 years. Maps showing flood-hazard areas in the United States can be obtained from the Federal Insurance Administrator, Department of Housing and Urban Development, who administers the National Flood Insurance Program. Minimum criteria for floodproofing are given in National Flood Insurance Rules and Regulations (Federal Register, Vol. 41, No. 207, Oct. 26, 1976).

Major objectives of floodproofing are to protect fully buildings and contents from damage from a 100-year flood, reduce losses from more devastating floods, and lower flood insurance premiums. Floodproofing, however, would be unnecessary if buildings were not constructed in flood plains. Building in a flood plain should be avoided unless the risk to life is acceptable and construction there can be economically and socially justified.

Some sites in flood plains possess some ground high enough to avoid flood damage. If such sites must be used, buildings should be clustered on the high areas. Where such areas are not available, it may be feasible to build up an earth fill, with embankments protected against erosion by water, to raise structures above flood levels. Preferably, such structures should not have basements, because they would require costly protection against water pressure.

An alternative to elevating a building on fill is raising it on stilts (columns in an unenclosed space). In that case, utilities and other services should be protected against damage from flood flows. The space at ground level between the stilts may be used for parking automobiles, if the risk of water damage to them is acceptable or if they will be removed before flood waters reach the site.

Buildings that cannot be elevated above flood stage should be furnished with an impervious exterior. Windows should be above flood stage, and doors should seal tightly against their frames. Doors and other openings may also be protected with a flood shield, such as a wall. Openings in the wall for access to the building may be protected with a movable flood shield, which for normal conditions can be stored out of sight and then positioned in the wall opening when a flood is imminent.

To prevent water damage to essential services for buildings in flood plains, important mechanical and electrical equipment should be located above flood level. Also, auxiliary electric generators to provide some emergency power are desirable. In addition, pumps should be installed to eject water that leaks into the building. Furthermore, unless a building is to be evacuated in case of flood, an emergency water supply should be stored in a tank above flood level, and sewerage should be provided with cutoff valves to prevent backflow.

3-13. Waterproofing.

In addition to protecting buildings against floods, designers also should adopt measures that prevent groundwater, rainwater, snow, or melted snow from penetrating into the interior through the exterior enclosure. Water may leak through cracks, expansion joints or other openings in walls and roofs, or through cracks around windows and doors. Also, water may seep through solid but porous exterior materials, such as masonry. Leakage generally may be prevented by use of weatherstripping around windows and doors, impervious waterstops in joints, or calking of cracks and other openings. Methods of preventing seepage, however, depend on the types of materials used in the exterior enclosure.

Definitions of Terms Related to Water Resistance

Permeability. Quality or state of permitting passage of water and water vapor into, through, and from pores and interstices, without causing rupture or displacement.

---

*Excerpted with minor revisions from Sec. 12 of the third edition of this handbook, authored by Cyrus C. Fishburn, formerly with the Division of Building Technology, National Bureau of Standards.

3-18 Protection of Structures and Occupants against Hazards

Terms used in this section to describe the permeability of materials, coatings, structural elements, and structures follow in decreasing order of permeability:

Pervious or Leaky. Cracks, crevices, leaks, or holes larger than capillary pores, which permit a flow or leakage of water, are present. The material may or may not contain capillary pores.
Water resistant. Capillary pores exist that permit passage of water and water vapor, but there are few or no openings larger than capillaries that permit leakage of significant amounts of water.
Water repellent. Not "wetted" by water; hence, not capable of transmitting water by capillary forces alone. However, the material may allow transmission of water under pressure and may be permeable to water vapor.
Waterproof. No openings are present that permit leakage or passage of water and water vapor; the material is impervious to water and water vapor, whether under pressure or not.

These terms also describe the permeability of a surface coating or a treatment against water penetration, and they refer to the permeability of materials, structural members, and structures whether or not they have been coated or treated.

Permeability of Concrete and Masonry. Concrete contains many interconnected voids and openings of various sizes and shapes, most of which are of capillary dimensions. If the larger voids and openings are few in number and not directly connected with each other, there will be little or no water penetration by leakage and the concrete may be said to be water-resistant.

Concrete in contact with water not under pressure ordinarily will absorb it. The water is drawn into the concrete by the surface tension of the liquid in the wetted capillaries.

Water-resistant concrete for buildings should be a properly cured, dense, rich concrete containing durable, well-graded aggregate. The water content of the concrete should be as low as is compatible with workability and ease of placing and handling. Resistance of concrete to penetration of water may be improved, however, by incorporation of a water-repellent admixture in the mix during manufacture. (See also Sec. 10.)

Water-repellent concrete is permeable to water vapor. If a vapor-pressure gradient is present, moisture may penetrate from the exposed face to an inner face. The concrete is not made waterproof (in the full meaning of the term) by the use of an integral water repellent. Note also that water repellents may not make concrete impermeable to penetration of water under pressure. They may, however, reduce absorption of water by the concrete.

Most masonry units also will absorb water. Some are highly pervious under pressure. The mortar commonly used in masonry will absorb water too but usually contains few openings permitting leakage.

Masonry walls may leak at the joints between the mortar and the units, however. Except in single-leaf walls of highly pervious units, leakage at the joints results from failure to fill them with mortar and poor bond between the masonry unit and mortar. As with concrete, rate of capillary penetration through masonry walls is small compared with the possible rate of leakage.

Capillary penetration of moisture through above-grade walls that resist leakage of wind-driven rain is usually of minor importance. Such penetration of moisture into well-ventilated subgrade structures may also be of minor importance if the moisture is readily evaporated. However, long-continued capillary penetration into some deep, confined subgrade interiors frequently results in an increase in relative humidity, a decrease in evaporation rate, and objectionable dampness.

Roof Drainage. Many roof failures have been caused by excessive water accumulation. In most cases, the overload that caused failure was not anticipated in design of those roofs, because the designers expected rainwater to run off the roof. But because of inadequate drainage, the water ponded instead.

On flat roofs, ponding of rainwater causes structural members to deflect. The resulting bowing of the roof surface permits more rainwater to accumulate, and the additional weight of this water causes additional bowing and collection of even more water. This process can lead to roof collapse. Similar conditions also can occur in the valleys of sloping roofs.

To avoid water accumulation, roofs should be sloped toward drains and pipes that have adequate capacity to conduct water away from the roofs, in accordance with local plumbing codes. Minimum roof slope for drainage should be at least ¼ in/ft, but larger slopes are advisable.

The primary drainage system should be supplemented by a secondary drainage system at a higher level to prevent ponding on the roof above that level. The overflow drains should be at least as large as the primary drains and should be connected to drain pipes independent of the primary system. The roof and its structural members should be capable of sustaining the weight of all rainwater that could accumulate on the roof if part or all of the primary drainage system should become blocked.

Drainage for Subgrade Structures. Subgrade structures located above groundwater level in drained soil may be in contact with water and wet soil for periods of indefinite duration after long-continued rains and spring thaws. Drainage of surface and subsurface water, however, may greatly reduce the time during which the walls and floor of a structure are subjected to water, may prevent leakage through openings resulting from poor workmanship and reduce the capillary penetration of water into the structure. If subsurface water cannot be removed by drainage, the structure must be made waterproof or highly water-resistant.

Surface water may be diverted by grading the ground surface away from the walls and by carrying the runoff from roofs away from the building. The slope of the ground surface should be at least ¼ in/ft for a minimum distance of 10 ft from the walls. Runoff from high ground adjacent to the structure should also be diverted.

Proper subsurface drainage of ground water away from basement walls and floors requires a drain of adequate size, sloped continuously, and, where necessary, carried around corners of the building without breaking continuity. The drain should lead to a storm sewer or to a lower elevation that will not be flooded and permit water to back up in the drain.

Drain tile should have a minimum diameter of 6 in and should be laid in gravel or other kind of porous bed at least 6 in below the basement floor. The open joints between the tile should be covered with a wire screen or building paper to prevent clogging of the drain with fine material. Gravel should be laid above the tile, filling the excavation to an elevation well

Fig. 3-8 Drainage at the bottom of a foundation wall.

above the top of the footing. Where considerable water may be expected in heavy soil, the gravel fill should be carried up nearly to the ground surface and should extend from the wall a distance of at least 12 in (Fig. 3-8).

Concrete Floors on Ground. These should preferably not be constructed in low-lying areas that are wet from ground water or periodically flooded with surface water. The ground should slope away from the floor. The level of the finished floor should be at least 6 in above grade. Further protection against ground moisture and possible flooding

3-20 Protection of Structures and Occupants against Hazards

of the slab from heavy surface runoffs may be obtained with subsurface drains located at the elevation of the wall footings.

All organic material and topsoil of poor bearing value should be removed in preparation of the subgrade, which should have a uniform bearing value to prevent unequal settlement of the floor slab. Backfill should be tamped and compacted in layers not exceeding 6 in in depth.

Where the subgrade is well-drained, as where subsurface drains are used or are unnecessary, floor slabs of residences should be insulated either by placing a granular fill over the subgrade or by use of a lightweight-aggregate concrete slab covered with a wearing surface of gravel or stone concrete. The granular fill, if used, should have a minimum thickness of 5 in and may consist of coarse slag, gravel, or crushed stone, preferably of 1-in minimum size. A layer of 3-, 4-, or 6-in-thick hollow masonry building units is preferred to gravel fill for insulation and provides a smooth, level bearing surface.

Moisture from the ground may be absorbed by the floor slab. Floor coverings, such as oil-base paints, linoleum, and asphalt tile, acting as a vapor barrier over the slab, may be damaged as a result. If such floor coverings are used and where a complete barrier against the rise of moisture from the ground is desired, a two-ply bituminous membrane waterproofing should be placed beneath the slab and over the insulating concrete or granular fill (Fig. 3-9). The top of the lightweight-aggregate concrete, if used, should be troweled or brushed to a smooth level surface for the membrane. The top of the granular fill should be covered with a grout coating, similarly finished. (The grout coat, ½ to 1 in thick, may consist of a 1:3 or a 1:4 mix by volume of portland cement and sand. Some ⅜- or ½-in maximum-sized coarse aggregate may be added to the grout if desired.) After the top surface of the insulating concrete or grout coating has hardened and dried, it should be mopped with hot asphalt or coal-tar pitch and covered before cooling with a lapped layer of 15-lb bituminous saturated felt. The first ply of felt then should be mopped with hot bitumen and a second ply of felt laid and mopped on its top surface. Care should be exercised not to puncture the membrane, which should preferably be covered with a coating of mortar, immediately after its completion. If properly laid and protected from damage, the membrane may be considered to be a waterproof barrier.

Where there is no possible danger of water reaching the underside of the floor, a single layer of 55-lb smooth-surface asphalt roll roofing or a similar product of 55-lb minimum weight may be substituted for the membrane. Joints between the sheets should be lapped and sealed with bituminous mastic. Great care should be taken to prevent puncturing of the roofing layer during concreting operations. When so installed, asphalt roll roofing

Fig. 3-9 Insulated concrete slab on ground with membrane dampproofing.

provides a low-cost and adequate barrier against the movement of excessive amounts of moisture by capillarity and in the form of vapor.

If a lightweight-aggregate insulating concrete slab is used, it is placed directly on the subgrade. The slab should have a minimum thickness of 4 in and a minimum compressive strength of 1500 psi. The coarse aggregate should not exceed 1 in in size. If the insulating slab is to be covered with a membrane, the slab should be permitted to dry after curing and before the membrane is placed. The wearing surface applied over the membrane should be at least 2½ in and preferably 3 in thick of 3000 psi concrete and should be reinforced with welded wire fabric weighing at least 40 lb per square. A 6 × 6 mesh of No. 6 wire is suitable. If the concrete wearing slab is to be covered with an oil-base paint or a floor covering that is a vapor barrier, the slab should be in a thoroughly dry condition before it is covered.

Basement Floors. Where a basement is to be used in drained soils as living quarters or for the storage of things that may be damaged by moisture, the floor should be insulated and should preferably contain the membrane waterproofing previously described. In general, the design and construction of such basement floors are similar to those of floors on ground.

If passage of moisture from the ground into the basement is unimportant or can be satisfactorily controlled by air conditioning or ventilation, the waterproof membrane need not be used. The concrete slab should have a minimum thickness of 4 in and need not be reinforced, but should be laid on a granular fill or other insulation placed on a carefully prepared subgrade. The concrete in the slab should have a minimum compressive strength of 2000 psi and may contain an integral water repellent.

A basement floor below the water table will be subjected to hydrostatic upward pressures. The floor should be made heavy enough to counteract the uplift.

A bituminous-filled joint between the basement walls and a floor over drained soil will prevent leakage into the basement of any water that may occasionally accumulate under the slab. Space for the joint may be provided by use of beveled siding strips, which are removed after the concrete has hardened. After the slab is properly cured, it and the wall surface should be in as dry a condition as is practicable before the joint is filled to ensure a good bond of the filler and to reduce the effects of slab shrinkage on the permeability of the joint. Hot asphalt or coal-tar pitch may be placed in the joint after its sides have been primed with a suitable bitumen.

Monolithic Concrete Basement Walls. These should have a minimum thickness of 6 in. Where insulation is desirable, as where the basement is used for living quarters, lightweight aggregate, such as those prepared by calcining or sintering blast-furnace slag, clay, or shale that meet the requirements of American Society for Testing and Materials Standard C330, may be used in the concrete. The concrete should have a minimum compressive strength of 2000 psi.

For the forms in which concrete for basement walls is cast, form ties of an internal-disconnecting type are preferable to twisted-wire ties. Entrance holes for the form ties should be sealed with mortar after the forms are removed. If twisted-wire ties are used, they should be cut a minimum distance of 1½ in inside the face of the wall and the holes filled with mortar.

The resistance of the wall to capillary penetration of water in temporary contact with the wall face may be increased by the use of a water-repellent admixture. The water repellent may also be used in the concrete at and just above grade to reduce the capillary rise of moisture from the ground into the superstructure walls.

Where it is desirable to make the wall resistant to passage of water vapor from the outside and to increase its resistance to capillary penetration of water, the exterior wall face may be treated with an impervious coating. The continuity and the resultant effectiveness in resisting moisture penetration of such a coating is dependent on the smoothness and regularity of the concrete surface and on the skill and technique used in applying the coating to the dry concrete surface. Some bituminous coatings that may be used are listed below in increasing order of their resistance to moisture penetration:

Spray- or brush-applied asphalt emulsions.
Spray- or brush-applied bituminous cutbacks.
Trowel coatings of bitumen with organic solvent, applied cold.
Hot-applied asphalt or coal-tar pitch, preceded by application of a suitable primer.

Cementitious brush-applied paints and grouts and trowel coatings of a mortar increase moisture resistance of monolithic concrete, especially if such coatings contain a water repellent. However, in properly drained soil, such coatings may not be justified unless needed to prevent leakage of water through openings in the concrete resulting from segregation of the aggregate and bad workmanship in casting the walls. The trowel coatings may also be used to level irregular wall surfaces in preparation for the application of a bituminous coating.

Basement Walls of Masonry Units. Water-resistant basement walls of masonry units should be carefully constructed of durable materials to prevent leakage and damage due to frost and other weathering exposure. Frost action is most severe at the grade line and may result in structural damage and leakage of water. Where wetting followed by sudden severe freezing may occur, the masonry units should meet the requirements of the following specifications:

Building brick (solid masonry units made from clay or shale), American Society for Testing and Materials Standard C62, Grade SW.
Facing brick (solid masonry units made from clay or shale), ASTM Standard C216, Grade SW.
Structural clay load-bearing wall tile, ASTM Standard C34, Grade LBX.
Hollow load-bearing concrete masonry units, ASTM Standard C90, Grade N.

For such exposure conditions, the mortar should be a Type S mortar (Table 12-1) having a minimum compressive strength of 1800 psi when tested in accordance with the requirements of ASTM Standard C270. For milder freezing exposures and where the walls may be subjected to some lateral pressure from the earth, the mortar should have a minimum compressive strength of 1000 psi.

Leakage through an expansion joint in a concrete or masonry foundation wall may be prevented by insertion of a waterstop in the joint. Waterstops should be of the bellows type, made of 16-oz copper sheet, which should extend a minimum distance of 6 in on either side of the joint. The sheet should be embedded between wythes of masonry units or faced with a 2-in thick cover of mortar reinforced with welded-wire fabric. The outside face of the expansion joint should be filled flush with the wall face with bituminous plastic cement conforming with the requirements of Federal Specification SS-C-153.

Rise of moisture, by capillarity, from the ground into the superstructure walls may be greatly retarded by use of an integral water-repellent admixture in the mortar. The water-repellent mortar may be used in several courses of masonry located at and just above grade.

The use of shotcrete or trowel-applied mortar coatings, ¾ in or more in thickness, to the outside faces of both monolithic concrete and unit-masonry walls greatly increases their resistance to penetration of moisture. Such coatings cover and seal construction joints and other vulnerable joints in the walls against leakage. When applied in a thickness of 2 in or more, they may be reinforced with welded-wire fabric to reduce the incidence of large shrinkage cracks in the coating. However, the cementitious coatings do not protect the walls against leakage if the walls, and subsequently the coatings, are badly cracked as a result of unequal foundation settlement, excessive drying shrinkage, and thermal changes.

Two trowel coats of a mortar containing 1 part portland cement to 3 parts sand by volume should be applied to the outside faces of basement walls built of hollow masonry units. One trowel coat may suffice on the outside of all-brick and of brick-faced walls.

The wall surface and the top of the wall footing should be cleansed of dirt and soil, and the masonry should be thoroughly wetted with water. While still damp, the surface should be covered with a thin scrubbed-on coating of portland cement tempered to the consistency of thick cream. Before this prepared surface has dried, a ⅜-in-thick trowel-applied coating or mortar should be placed on the wall and over the top of the footing; a fillet of mortar may be placed at the juncture of the wall and footing.

Where a second coat of mortar is to be applied, as on hollow masonry units, the first coat should be scratched to provide a rough bonding surface. The second coat should be applied at least 1 day after the first, and the coatings should be cured and kept damp by wetting for at least 3 days. A water-repellent admixture in the mortar used for the second or finish coat will reduce the rate of capillary penetration of water through the walls. If a bituminous coating is not to be used, the mortar coating should be kept damp until the backfill is placed.

Thin, impervious coatings may be applied to the plaster coating if resistance to penetration of water vapor is desired. The plaster coating should be dry and clean before the bituminous coating is applied over the surfaces of the wall and the top of the footing. Unless backed with a tier of masonry or a self-supporting layer of concrete, the bituminous coatings may fail, in time, to bond properly to the inner face of the masonry. They should not be used on the inner faces of the walls unless properly supported in position.

Impervious Membranes. These are waterproof barriers providing protection against penetration of water under hydrostatic pressure and water vapor. To resist hydrostatic pressure, the membrane should be made continuous in the walls and floor of a basement. It also should be protected from damage during building operations and should be laid by experienced workers under competent supervision. It usually consists of three or more alternate layers of hot, mopped-on asphalt or coal-tar pitch and plies of bituminous saturated felt or woven cotton fabric. The number of moppings exceeds the number of plies by one.

An alternative is a $\frac{1}{16}$- to $\frac{1}{8}$-in-thick layer of butyl rubber or rubberized asphalt and plastic film, secured with a compatible adhesive. In installation, manufacturers' recommendations should be carefully followed.

Bituminous saturated cotton fabric is stronger and is more extensible than bituminous saturated felt but is more expensive and more difficult to lay. At least one or two of the plies in a membrane should be of saturated cotton fabric to provide strength, ductility, and extensibility to the membrane. Where vibration, temperature changes, and other conditions conducive to displacement and volume changes in the basement are to be expected, the relative number of fabric plies may be increased.

The minimum weight of bituminous saturated felt used in a membrane should be 13 lb per 100 ft^2. The minimum weight of bituminous saturated woven cotton fabric should be 10 oz/yd^2.

Although a membrane is held rigidly in place, it is advisable to apply a suitable primer over the surfaces receiving the membrane and to aid in the application of the first mopped-on coat of hot asphalt or coal-tar pitch.

Materials used in the membrane should meet the requirements of the following current American Society for Testing and Materials standards:

Creosote primer for coal-tar pitch—D43.
Primer for asphalt—D41.
Coal-tar pitch—D450.
Asphalt—D449.
Woven cotton fabric, bituminous saturated—D173.

Coal-tar saturated felt—D227.
Asphalt saturated felt—D226.

The number of plies of saturated felt or fabric should be increased with increase in the hydrostatic head to which the membrane is to be subjected. Five plies is the maximum commonly used in building construction, but 10 or more plies have been recommended for pressure heads of 35 ft or greater. The thickness of the membrane crossing the wall footings at the base of the wall should be no greater than necessary, to keep very small the possible settlement of the wall due to plastic flow in the membrane materials.

The amount of primer to be used may be about 1 gal per 100 ft^2. The amount of bitumen per mopping should be at least 4½ gal per 100 ft^2. The thickness of the first and last moppings is usually slightly greater than the thickness of the moppings between the plies.

The surfaces to which the membrane is to be applied should be smooth, dry, and at a temperature above freezing. Air temperature should be not less than 50°F. The temperature of coal-tar pitch should not exceed 300°F and asphalt, 350°F.

If the concrete and masonry surfaces are not sufficiently dry, they will not readily absorb the priming coat, and the first mopping of bitumen will be accompanied by bubbling and escape of steam. Should this occur, application of the membrane should be stopped and the bitumen already applied to damp surfaces should be removed.

The membrane should be built up ply by ply, the strips of fabric or felt being laid immediately after each bed has been hot-mopped. One of several methods of laying the plies is the solid or continuous method, in which a portion of each strip of fabric or felt is in contact with the supporting base of the membrane. The lap of succeeding plies or strips over each other is dependent on the width of the roll and the number of plies. In any membrane there should be some lap of the top or final ply over the first, initial ply. The American Railway Engineering Association requires a minimum distance of 2 in for this lap (Fig. 3-10). End laps should be staggered at least 24 in, and the laps between succeeding rolls should be at least 12 in.

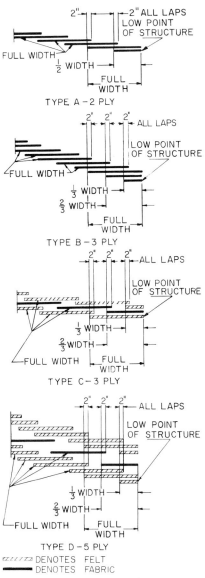

Fig. 3-10 Lapping of membranes recommended by the American Railway Engineering Association.

For floors, the membrane should be placed over a concrete base or subfloor whose top surface is troweled smooth and which is level with the tops of the wall footings. The membrane should be started at the outside face of one wall and extend over the wall footing, which may be keyed. It should cover the floor and tops of other footings to the outside faces of the other walls, forming a continuous horizontal waterproof barrier. The plies should project from the edges of the floor membrane and lap into the wall membrane.

The loose ends of felt and fabric must be protected; one method is to fasten them to a temporary vertical wood form about 2 ft high, placed just outside the wall face. Immediately after the floor membrane has been laid, its surface should be protected and covered with a layer of portland-cement concrete, at least 2 in thick.

For walls, the installed membrane should be protected against damage and held in position by protection board or a facing of brick, tile, or concrete block. A brick facing should have a minimum thickness of 2½ in. Facings of asphalt plank, asphalt block, or mortar require considerable support from the membrane itself and give protection against abrasion of the membrane from lateral forces only. Protection against downward forces such as may be produced by settlement of the backfill is given only by the self-supporting masonry walls.

The kind of protective facing may have some bearing on the method of constructing the membrane. The membrane may be applied to the exterior face of the wall after its construction, or it may be applied to the back of the protective facing before the main wall is built. The first of these methods is known as the outside application; the second is known as the inside application (Fig. 3-11).

Fig. 3-11 Two methods of applying waterproofing membranes to walls: (*a*) outside application (membrane applied to the wall, then protected by the backing); (*b*) inside application (membrane applied to the backing, after which the wall is built).

For the inside application, a protective facing of considerable stiffness against lateral forces must be built, especially if the wall and its membrane are to be used as a form for the casting of a main wall of monolithic concrete. The inner face of the protecting wall must be smooth or else leveled with mortar to provide a suitable base for the membrane. The completed membrane should be covered with a ⅜-in-thick layer of mortar to protect it from damage during construction of the main wall.

Application of wall membranes should be started at the bottom of one end of the wall and the strips of fabric or felt laid vertically. Preparation of the surfaces and laying of the membrane proceed much as they do with floor membranes. The surfaces to which the membrane is attached must be dry and smooth, which may require that the faces of masonry walls be leveled with a thin coat of grout or mortar. The plies of the wall membrane should be lapped into those of the floor membrane.

If the outside method of application is used and the membrane is faced with masonry, the narrow space between the units and the membrane should be filled with mortar as the units are laid. The membrane may be terminated at the grade line by a return into the superstructure wall facing.

Waterstops in joints in walls and floors containing a bituminous membrane should be the metal-bellows type. The membrane should be placed on the exposed face of the joint and it may project into the joint, following the general outline of the bellows.

The protective facing for the membrane should be broken at the expansion joint and the space between the membrane and the line of the facing filled with a bituminous plastic cement.

Details at pipe sleeves running through the membrane must be carefully prepared. The membrane should be reinforced with additional plies and may be calked at the sleeve. Steam and hot-water lines should be insulated to prevent damage to the membrane.

Above-Grade Water-Resistant Walls. The rate of moisture penetration through capillaries in above-grade walls is low and usually of minor importance. However, such walls should not permit leakage of wind-driven rain through openings larger than those of capillary dimension.

Walls of cellular concrete, made of aggregates containing no fines, are highly permeable and require surface coatings of grout or mortar to prevent leakage of wind-driven rain.

Precast-concrete or metal panels are usually made of dense, highly water-resistant materials. However, walls made of these panels are vulnerable to leakage at the joints. In such construction, edges of the panels may be recessed and the interior of vertical joints filled with grout or other sealant after the panels are aligned.

Calking compound is commonly used as a facing for the joints. Experience has shown that calking compounds often weather badly; their use as a joint facing creates a maintenance problem and does not prevent leakage of wind-driven rain after a few years' exposure.

The amount of movement to be expected in the vertical joints between panels is a function of the panel dimensions and the seasonal fluctuation in temperature and, for concrete, the moisture content of the concrete. For panel construction, it may be more feasible to use an interlocking water-resistant joint. For concrete, the joint may be faced on the weather side with mortar and backed with either a compressible premolded strip or calking.

Brick walls 4 in or more in thickness can be made highly water-resistant. The measures that need to be taken to ensure there will be no leakage of wind-driven rain through brick facings are not extensive and do not require the use of materials other than those commonly used in masonry walls. The main factors that need to be controlled are the rate of suction of the brick at the time of laying and filling of all joints with mortar (Art. 12-5).

In general, the greater the number of brick leaves, or wythes, in a wall, the more water-resistant the wall.

Walls of hollow masonry units are usually highly permeable, and brick-faced walls backed with hollow masonry units are greatly dependent upon the water resistance of the brick facing to prevent leakage of wind-driven rain. For exterior concrete masonry walls without facings of brick, protection against leakage may be obtained by facing the walls with a cementitious coating of paint, stucco, or shotcrete.

For walls of rough-textured units, a portland cement–sand grout provides a highly water-resistant coating. The cement may be either white or gray.

Factory-made portland-cement paints containing a minimum of 65%, and preferably 80%, portland cement may also be used as a base coat on concrete masonry. Application of the paint should conform with the requirements of American Concrete Institute Standard 616. The paints, stuccos, and shotcrete should be applied to dampened surfaces. Shotcrete should conform with the requirements of ACI Standard 506.

Pneumatically applied coatings thinner than those discussed above and which contain durable, low-absorptive aggregates are also highly water-resistant.

Cavity walls, particularly brickfaced cavity walls, may be made highly resistant to leakage through the wall facing. However, as usually constructed, facings are highly permeable, and the leakage is trapped in the cavity and diverted to the outside of the wall through conveniently located weep holes. This requires that the inner tier of the cavity be protected against the leakage by adequate flashings, and weep holes should be placed at the bottom of the cavities and over all wall openings. The weep holes may be formed by the use of sash-cord head joints or ⅜-in-diameter rubber tubing, withdrawn after the wall is completed.

Flashings should preferably be hot-rolled copper sheet of 10-oz minimum weight. They should be lapped at the ends and sealed either by solder or with bituminous plastic cement. Mortar should not be permitted to drop into the flashings and prevent the weep holes from functioning.

Prevention of Cracking. Shrinkage of concrete masonry because of drying and a drop in temperature may result in cracking of a wall and its cementitious facing. Such cracks readily permit leakage of wind-driven rain. The chief factor reducing incidence of shrinkage cracking is the use of dry block. When laid in the wall, the block should have a low moisture content, preferably one that is in equilibrium with the driest condition to which the wall will be exposed.

The block should also have a low potential shrinkage. See moisture-content requirements in ASTM C90 and method of test for drying shrinkage of concrete block in ASTM C426.

Formation of large shrinkage cracks may be controlled by use of steel reinforcement in the horizontal joints of the masonry and above and below wall openings. Where there may be a considerable seasonal fluctuation in temperature and moisture content of the wall, high-tensile-strength, deformed-wire joint reinforcement should be placed in at least 50% of all bed joints in the wall.

Use of control joints faced with calking compound has also been recommended to control shrinkage cracking; however, this practice is marked by frequent failures to keep the joints sealed against leakage of rain. Steel joint reinforcement strengthens a concrete masonry wall, whereas control joints weaken it, and the calking in the joints requires considerable maintenance.

Water-Resistant Surface Treatments for Above-Grade Walls. Experience has shown that leakage of wind-driven rain through masonry walls, particularly those of brick, ordinarily cannot be stopped by use of an inexpensive surface treatment or coating that will not alter the appearance of the wall. Such protective devices either have a low service life or fail to stop all leakage.

Both organic and cementitious pigmented coating materials, properly applied as a continuous coating over the exposed face of the wall, do stop leakage. Many of the organic pigmented coatings are vapor barriers and are therefore unsuitable for use on the outside, "cold" face of most buildings. If vapor barriers are used on the cold face of the wall, it is advisable to use a better vapor barrier on the warm face to reduce condensation in the wall and behind the exterior coating.

Coatings for masonry may be divided into four groups, as follows: (1) colorless coating materials; (2) cementitious coatings; (3) pigmented organic coatings; and (4) bituminous coatings.

Colorless Coating Materials. The colorless "waterproofings" are often claimed to stop leakage of wind-driven rain through permeable masonry walls. Solutions of oils, paraffin wax, sodium silicate, chlorinated rubber, silicone resins, and salts of fatty acids have been applied to highly permeable test walls and have been tested at the National Bureau of Standards under exposure conditions simulating a wind-driven rain. Most of these solutions contained not more than 10% of solid matter. These treatments reduced the rate of leakage but did not stop all leakage through the walls. The test data show that colorless coating materials applied to permeable walls of brick or concrete masonry may not provide adequate protection against leakage of wind-driven rain.

Solutions containing oils and waxes tended to seal the pores exposed in the faces of the mortar joints and masonry units, thereby acting more or less as vapor barriers, but did not seal the larger openings, particularly those in the joints.

Silicone water-repellent solutions greatly reduced leakage through the walls as long as the treated wall faces remained water-repellent. After an exposure period of 2 or 3 hr, the rate of leakage gradually increased as the water repellency of the wall face diminished.

Coatings of the water-repellent, breather type, such as silicone and "soap" solutions, may be of value in reducing absorption of moisture into the wall surface. They may be of special benefit in reducing the soiling and disfiguration of stucco facings and light-colored masonry surfaces. They may be applied to precast-concrete panels to reduce volume changes that may otherwise result from changes in moisture content of the concretes. However, it should be noted that a water-repellent treatment applied to the surface may cause water, trapped in the masonry, to evaporate beneath the surface instead of at the surface. If the masonry is not water-resistant and contains a considerable amount of soluble salts, as evidenced by efflorescence, application of a water repellent may cause salts to be deposited beneath the surface, thereby causing spalling of the masonry. The water repellents therefore should be applied only to walls having water-resistant joints. Furthermore, application of a colorless material makes the treated face of the masonry water-repellent and may prevent the proper bonding of a cementitious coating that could otherwise be used to stop leakage.

Cementitious Coatings. Coatings of portland-cement paints, grouts, and stuccos and of pneumatically applied mortars are highly water-resistant. They are preferred above all other types of surface coatings for use as water-resistant base coatings on above-grade concrete masonry. They may also be applied to the exposed faces of brick masonry walls that have not been built to be water-resistant.

The cementitious coatings absorb moisture and are of the breather type, permitting passage of water vapor. Addition of water repellents to these coatings does not greatly affect their water resistance but does reduce the soiling of the surface from the absorption of dirt-laden water. If more than one coating is applied, as in a two-coat paint or stucco facing job, the repellent is preferably added only to the finish coat, thus avoiding the difficulty of bonding a cementitious coating to a water-repellent surface.

The technique used in applying the cementitious coatings is highly important. The backing should be thoroughly dampened. Paints and grouts should be scrubbed into place with stiff fiber brushes and the coatings should be properly cured by wetting. Properly applied, the grouts are highly durable; some grout coatings applied to concrete masonry test walls were found to be as water-resistant after 10 years out-of-doors exposure as when first applied to the walls.

Pigmented Organic Coatings. These include textured coatings, mastic coatings, conventional paints, and aqueous dispersions. The thick-textured and mastic coatings are usually spray-applied but may be applied by trowel. Conventional paints and aqueous dispersions are usually applied by brush or spray. Most of these coatings are vapor barriers but some textured coatings, conventional paints, and aqueous dispersions are breathers. Except for the aqueous dispersions, all the coatings are recommended for use with a primer.

Applied as a continuous coating, without pinholes, the pigmented organic coatings are highly water-resistant. They are most effective when applied over a smooth backing. When they are applied with paintbrush or spray by conventional methods to rough-textured walls, it is difficult to level the surface and to obtain a continuous water-resistant coating free from holes. A scrubbed-on cementitious grout used as a base coat on such walls will prevent leakage through the masonry without the use of a pigmented organic coating.

The pigmented organic coatings are highly decorative but may not be so water-resistant, economical, or durable as the cementitious coatings.

Bituminous Coatings. Bituminous cutbacks, emulsions, and plastic cements are usually vapor barriers and are sometimes applied as "dampproofers" on the inside faces of masonry walls. Plaster is often applied directly over these coatings, the bond of the plaster to the masonry being only of a mechanical nature. Tests show that bituminous coatings applied to the inside faces of highly permeable masonry walls, not plastered, will readily blister and permit leakage of water through the coating. It is advisable not to depend on such coatings to prevent the leakage of wind-driven rain unless they are incorporated in the masonry or held in place with a rigid self-sustaining backing.

Even though the walls are resistant to wind-driven rain, but are treated on their inner faces with a bituminous coating, water may be condensed on the warm side of the coating and damage to the plaster may result, whether the walls are furred or not. However, the bituminous coating may be of benefit as a vapor barrier in furred walls, if no condensation occurs on the warm side.

See also Secs. 10 and 12.

("Manual for Railway Engineering," Chap. 29, "Waterproofing," American Railway Engineering Association, 59 E. Van Buren St., Chicago, Ill. 60605; "Manual of Concrete Practice," Vol. 3, "Admixtures for Concrete," "A Guide to the Use of Waterproofing, Dampproofing, Protective, and Decorative Barrier Systems for Concrete," and "Guide to Joint Sealants for Concrete Structures," American Concrete Institute, P.O. Box 19150, Redford Station, Detroit, Mich. 48219.)

PROTECTION AGAINST FIRE

There are two distinct aspects of fire protection: life safety and property protection. Although providing for one aspect generally results in some protection for the other, the two goals are not mutually inclusive. A program that provides for prompt notification and evacuation of occupants meets the objectives for life safety, but provides no protection for property. Conversely, it is possible that adequate property protection might not be sufficient for protection of life.

Absolute safety from fire is not attainable. It is not possible to eliminate all combustible materials or all potential ignition sources. Thus, in most cases, an adequate fire protection plan must assume that unwanted fires will occur despite the best efforts to prevent them. Means must be provided to minimize the losses caused by the fires that do occur.

The first obligation of designers is to meet legal requirements while providing the facilities required by the client. In particular, the requirements of the applicable building code must be met. The building code will contain fire safety requirements, or it will specify some recognized standard by reference. Many owners will also require that their own insurance carrier be consulted—to obtain the most favorable insurance rate, if for no other reason.

3-14. Fire-Protection Standards.

The standards most widely adopted are those published by the National Fire Protection Association (NFPA), 470 Atlantic Ave., Boston, Mass. 02210. The NFPA "National Fire Codes" comprise 16 volumes containing more than 200 standards, updated annually.

3-30 Protection of Structures and Occupants against Hazards

(These are also available separately.) The standards are supplemented by the NFPA "Fire Protection Handbook," which contains comprehensive and detailed discussion of fire problems and much valuable statistical and engineering data.

Underwriters Laboratories, Inc. (UL), 207 East Ohio St., Chicago, Ill. 60611, publishes testing laboratory approvals of devices and systems in its "Fire Protection Equipment List," updated annually and by bimonthly supplements. The publication outlines the tests that devices and systems must pass to be listed. The UL "Building Materials List" describes and lists building materials, ceiling-floor assemblies, wall and partition assemblies, beam and column protection, interior finish materials, and other pertinent data. UL also publishes lists of "Accident Equipment," "Electrical Equipment," "Electrical Construction Materials," "Hazardous Location Equipment," "Gas and Oil Equipment," and others.

Separate standards for application to properties insured by the Factory Mutual System are published by the Factory Mutual Engineering Corporation (FM), Norwood, Mass. 02062. The "Handbook of Industrial Loss Prevention," prepared by the FM staff and published by McGraw-Hill Book Company, New York, treats many industrial fire problems in detail. FM also publishes a list of devices and systems it has tested and approved.

The General Services Administration, acting for the Federal government, has developed many requirements that must be considered, if applicable. Also, the Federal government encourages cities to adopt some uniform code.

The Federal Occupational Safety and Health Act of 1970 (OSHA) sets standards for protecting the health and safety of nearly all employees. It is not necessary that a business be engaged in interstate commerce for the law to apply. OSHA defines **employer** as "a person engaged in a business affecting commerce who has employees, but does not include the United States or any State or political subdivision of a State."

An employer is required to "furnish to each of his employees employment and a place of employment which are free from recognized hazards that are causing or are likely to cause death or serious physical harm to his employees." He is also required to "comply with occupational safety and health standards promulgated under the Act." Regulation and enforcement are the responsibility of the Department of Labor.

There are also other OSHA requirements, such as record keeping. Every employer must permit Labor Department investigators to enter his establishment "without delay and at reasonable times."

Part 1910 of the Act contains all the standards that have been adopted. Subpart E, "Means of Egress," adopts NFPA 101, "Life Safety Code" as the standard. Subpart L, "Fire Protection," adopts the appropriate NFPA standard, where any is given.

Building codes consist of a set of rules aimed at providing reasonable safety to the community, to occupants of buildings, and to the buildings themselves. The codes may adopt the standards mentioned previously and other standards concerned with fire protection by reference or adapt them to the specific requirements of the community. In the absence of a municipal or state building code, designers may apply the provisions of the Uniform Building Code, promulgated by the International Conference of Building Officials, or other national model code.

Many states have codes for safety to life in commercial and industrial buildings, administered by the Department of Labor, the State Fire Marshal's Office, the State Education Department, or the Health Department. Some of these requirements are drastic and must always be considered.

Obtaining optimum protection for life and property can require consultation with the owner's insurance carrier, municipal officials, and the fire department. If the situation is complicated enough, it can require consultation with a specialist in all phases of fire protection and prevention. In theory, municipal building codes are designed for life safety and for protection of the public, whereas insurance-oriented codes (except for NFPA 101, "Life Safety Code") are designed to minimize property fire loss. Since about 70% of any building code is concerned with fire protection, there are many circumstances that can best be resolved by a fire protection consultant.

3-15. Fire-Protection Concepts.

Although fires in buildings can be avoided, they nevertheless occur. Some of the reasons for this are human error, arson, faulty electrical equipment, poor maintenance of heating equipment, and natural causes, such as lightning. Consequently, buildings should be designed to minimize the probability of a fire and to protect life and limit property damage if a fire should occur. The minimum steps that should be taken for the purpose are as follows:

1. Limit potential fire loads, with respect to both combustibility and ability to generate smoke and toxic gases.

2. Provide means for prompt detection of fires, with warnings to occupants who may be affected and notification of the presence of fire to fire fighters.

3. Communication of instructions to occupants as to procedures to adopt for safety, such as to staying in place, proceeding to a designated refuge area, or evacuating the building.

4. Provide means for early extinguishment of any fire that may occur, primarily by automatic sprinklers but also by trained fire fighters.

5. Make available also for fire fighting an adequate water supply, appropriate chemicals, adequate-size piping, conveniently located valves on the piping, hoses, pumps, and other equipment necessary.

6. Prevent spread of fire from building to building, either through adequate separation or by enclosure of the building with incombustible materials.

7. Partition the interior of the building with fire barriers, or divisions, to confine a fire to a limited space.

8. Enclose with protective materials structural components that may be damaged by fire (fireproofing).

9. Provide refuge areas for occupants and safe evacuation routes to outdoors.

10. Provide means for removal of heat and smoke from the building as rapidly as possible without exposing occupants to these hazards, with the air-conditioning system, if one is present, assisting the removal by venting the building and by pressurizing smokeproof towers, elevator shafts, and other exits.

11. For large buildings, install standby equipment for operation in emergencies of electrical systems and elevators.

These steps are discussed in the following articles.

3-16. Fire Loads and Resistance Ratings.

The nature and potential magnitude of fire in a building are directly related to the amount and physical arrangement of combustibles present, as contents of the building or as materials used in its construction. Because of this, all codes classify buildings by **occupancy** and **construction**, because these features are related to the amount of combustibles.

The total amount of combustibles is called the **fire load** of the building. Fire load is expressed in pounds per square foot (psf) of floor area, with an assumed calorific value of 7000 to 8000 Btu/lb. (This Btu content applies to organic materials similar to wood and paper. Where other materials are present in large proportion, the weights must be

3-32 Protection of Structures and Occupants against Hazards

adjusted accordingly. For example, for petroleum products, fats, waxes, alcohol, and similar materials, the weights are taken at twice their actual weights, because of the Btu content.)

National Bureau of Standards burnout tests presented in NBS Report BMS92 indicate a relation between fire load and fire severity as shown in Table 3-3.

TABLE 3-3 Relation between Weight of Combustibles and Fire Severity*

Average Weight of Combustibles, Psf	Equivalent Fire Severity, Hr
5	½
7½	¾
10	1
15	1½
20	2
30	3
40	4½
50	6
60	7½

* Based on National Bureau of Standards Report BMS92, "Classifications of Building Constructions," U.S. Government Printing Office, Washington, D.C. 20402.

The temperatures used in standard fire tests of building components are indicated by the internationally recognized time-temperature curve shown in Fig. 3-12. Fire resistance of construction materials, determined by standard fire tests, is expressed in hours. The Underwriters Laboratories "Building Materials List" tabulates fire ratings for materials and assemblies it has tested.

Every building code specifies required fire-resistance ratings for structural members, exterior walls, fire divisions, fire separations, ceiling-floor assemblies, and any other constructions for which a fire rating is necessary. (Fire protection for structural steel is discussed in Arts. 8-86 to 8-91. Design for fire resistance of open-web steel joist construction is treated in Art. 9-8, of steel roof deck in Art. 9-38, and of cellular floor construction in Art. 9-45. Design for fire safety with wood construction is covered in Art. 7-15.)

Building codes also specify the ratings required for interior finish of walls, ceilings, and floors. These are classified as to flame spread, fuel contributed, and smoke developed, determined in standard tests performed according to ASTM E84 or ASTM E119. Ratings of specific products are given in the UL "Building Materials List."

3-17. Fire Barriers.

A major consideration in building design is safety of the community. Hence, buildings should be designed to control fires so that they will not spread from building to building.

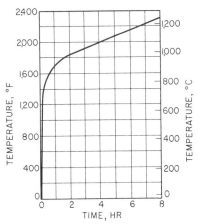

Fig. 3-12 Time-temperature curve for standard fire test.

One way that building codes try to achieve this objective is to establish fire zones or fire limits that restrict types of construction or occupancy that can be used. Additional zoning regulations establish minimum distances between buildings. Another way to achieve the objective is to specify the types of construction that can be used for enclosing the exterior of buildings. The distance between adjoining buildings, fire rating, and stability when exposed to fire of exterior walls, windows, and doors, and percent of window area are some of the factors taken into account in building codes for determination of the construction classification of a building.

To prevent spread of fire from roof to roof, building codes also often require that exterior walls extend as a parapet at least 3 ft above the roof level. Parapets also are useful in shielding fire fighters who may be hosing a fire from roofs of buildings adjoining the one on fire. In addition, buildings should be topped with roof coverings that are fire-resistant.

Fire Divisions. To prevent spread of fire vertically in building interiors, building codes generally require that floor-ceiling and roof-ceiling assemblies be fire-resistant. The fire rating of such assemblies is one of the factors considered in determination of the construction classification of a building. Also, openings in floors and roofs should be fire-protected, although building codes do not usually require this for one-story or two-story dwellings. For the purpose, an opening, such as that for a stairway, may be protected with a fire-resistant enclosure and fire doors. In particular, stairways and escalator and elevator shafts should be enclosed, not only to prevent spread of fire but also to provide a protected means of egress from the building for occupants and of approach to the fire source by fire fighters.

To prevent spread of fire horizontally in building interiors, it is desirable to partition interiors with fire divisions. A **fire division** is any construction with the fire-resistance rating and structural stability under fire conditions required for the type of occupancy and construction of the building to bar the spread of fire between adjoining buildings or between parts of the same building on opposite sides of the division. A fire division may be an exterior wall, fire window, fire door, fire wall, ceiling, or firestop.

A fire wall should be built of incombustible material, have a fire rating of at least 4 hr, and extend continuously from foundations to roof. Also, the wall should have sufficient structural stability in a fire to allow collapse of construction on either side without the wall collapsing. Building codes restrict the size of openings that may be provided in a fire wall and require the openings to be fire-protected (Art. 14-3).

To prevent spread of fire through hollow spaces, such spaces should be firestopped. A **firestop** is a solid or compact, tight closure set in a hollow, concealed space in a building to retard spread of flames, smoke, or hot gases. All partitions and walls should be firestopped at every floor level, at the top-story ceiling level, and at the level of support for roofs. Also, very large unoccupied attics should be subdivided by firestops into areas of 3000 ft^2 or less. Similarly, any large concealed space between a ceiling and floor or roof should be subdivided. For the purpose, firestops extending the full depth of the space should be placed along the line of supports of structural members and elsewhere, if necessary, to enclose areas not exceeding 1000 ft^2 when situated between a floor and ceiling or 3000 ft^2 when located between a ceiling and roof.

Openings between floors for pipes, ducts, wiring, and other services should be sealed with the equal of positive firestops. Partitions between each floor and a suspended ceiling above are not generally required to be extended to the slab above unless this is necessary for required compartmentation. But smoke stops should be provided at reasonable intervals to prevent passage of smoke to noninvolved areas.

Height and Area Restrictions Limitations on heights and floor areas included between fire walls in any story of a building are given in every building code and are directly related to occupancy and construction. From the standpoint of fire protection, these provisions are chiefly concerned with safety to life, and endeavor to ensure this

3-34 Protection of Structures and Occupants against Hazards

through requirements determining minimum number of exits, proper location of exits, and maximum travel distance (hence escape time) necessary to reach a place of refuge. The limitations are also aimed at limiting the size of fires.

Unlimited height and area are permitted for the most highly fire-resistant type of construction. Permissible heights and areas are decreased with decrease in fire resistance of construction. Area permitted between fire walls in any story reduces to 6000 ft^2 for a one-story, wood-frame building.

Installation of automatic sprinklers increases permissible heights and areas in all classes, except those allowed unlimited heights and areas.

Permissible unlimited heights and areas in fire-resistive buildings considered generally satisfactory in the past may actually not be safe. A series of fires involving loss of life and considerable property damage opened the fire safety of such construction to question. As a result, some cities have made more stringent the building-code regulations applicable to high-rise buildings.

Many building codes prohibit floor areas of unlimited size unless the building is sprinklered. Without automatic sprinklers, floor areas must be subdivided into fire-wall-protected areas of from 7500 to 15,000 ft^2 and the enclosing fire walls must have 1- or 2-hr fire ratings, depending on occupancy and construction.

3-18. Fire-Resistance Classification of Buildings.

Although building codes classify buildings by occupancy and construction, there is no universal standard for number of classes of either occupancy or construction. The National Building Code (NBC), sponsored by the National Conference of States on Building Codes and Standards, lists nine occupancy classes and seven construction classes. The Building Officials and Code Administrators International, Inc. (BOCA) code lists 17 occupancy classes and 10 construction classes. The New York City code lists 10

TABLE 3-4 Approximate Fire Loads for Various Occupancies*

Occupancy Class	Typical Average Fire Load Including Floors and Trim, Psf
Assembly	10.0
Business	12.6
Educational	7.6
High hazard	†
Industrial	25.0
Institutional	5.7
Mercantile	15–20
Residential	8.8
Storage	30.0

* From National Bureau of Standards Report BMS92, "Classifications of Building Constructions," U.S. Government Printing Office, Washington, D.C. 20402.

† Special provisions are made for this class, and hazards are treated for the specific conditions encountered, which might not necessarily be in proportion to the actual fire load.

occupancy classes with a total of 19 subclasses, and 2 construction classes with a total of 10 subclasses. The nine occupancy classes of the National Building Code are tabulated with approximate fire loads in Table 3-4. This table should be used only as a guide. For a specific project refer to the applicable local code. Codes do not relate life-safety hazards to the actual fire load, but deal with them through requirements for exit arrangements, interior finishes, and ventilation.

The seven construction classes of the National Building Code are:

1. Fire-resistive construction—Type A.
2. Fire-resistive construction—Type B.
3. Protected noncombustible construction.
3. Unprotected noncombustible construction.
5. Heavy-timber construction.
6. Ordinary construction.
7. Wood-frame construction.

The required fire resistance varies from 4 hr for exterior bearing walls and interior columns in the Fire Resistive—Type A class to 1 hr for walls and none for columns in the wood-frame construction class.

Class of construction affects fire-protection-system design through requirements that structural members as well as contents of buildings be protected.

3-19. Extinguishment of Fires.

Design of all buildings should include provisions for prompt extinguishment of fires. Apparatus installed for the purpose should take into account the nature and amount of combustible and smoke-producing materials that may be involved in a fire. Such apparatus may range from small, hand-held extinguishers that discharge a stream of water to retard small fires until fire fighters arrive to hoses attached to a large, pressurized water supply and automatic fire sprinklers. Also desirable are fire and smoke detectors and a protective signaling system that sounds an alarm to alert building occupants and calls fire fighters.

Classes of Fires. For convenience in defining effectiveness of extinguishing media, Underwriters Laboratories, Inc., has developed a classification that separates combustible materials into four types:

1. *Class A fires* involve ordinary combustibles and are readily extinguishable by water or cooling, or by coating with a suitable chemical powder.
2. *Class B fires* involve flammable liquids where smothering is effective and where a cooling agent must be applied with care.
3. *Class C fires* are those in live electrical equipment where the extinguishing agent must be nonconductive. Since a continuing electrical malfunction will keep the fire source active, circuit protection must operate, after which an electrically conductive agent can be used with safety.
4. *Class D fires* involve metals that burn, such as magnesium, sodium, and powdered aluminum. Special powders are necessary for such fires, as well as special training for operators. These fires should never be attacked by untrained personnel.

Automatic Sprinklers. The most widely used apparatus for fire protection in buildings is the automatic sprinkler system. In one or more forms, automatic sprinklers are effective protection in all nine of the National Building Code occupancy classes (Art. 3-18). Special treatment and use of additional extinguishing agents, though, may be required in many high-hazard, industrial, and storage occupancies.

Sprinkler systems are suitable for extinguishing all Class A fires and, in many cases, also Class B and C fires. For Class B fires, a sealed (fusible) head system may be used if the flammable liquid is in containers or is not present in large quantity. Sprinklers have a good record for extinguishing fires in garages, for example. An oil-spill fire can be extinguished or contained when the water is applied in the form of spray, as from a sprinkler head. When an oil spill or process-pipe rupture can release flammable liquid under pressure, an open-head (deluge) system may be required to apply a large volume of water quickly and to keep surrounding equipment cool.

For Class C fires, water can be applied to live electrical equipment if it is done in the form of a nonconducting foglike spray. This is usually the most economical way to protect outdoor oil-filled transformers and oil circuit breakers.

Fire protection should be based on complete coverage of the building by the sprinkler system. Partial coverage is rarely advisable, because extinguishing capacity is based on detecting and extinguishing fires in their incipiency, and the system must be available at all times in all places. Systems are not designed to cope with fires that have gained headway after starting in unsprinklered areas.

Basically, a sprinkler system consists of a network of piping installed at the ceiling or roof and supplied with water from a suitable source. On the piping at systematic intervals are placed heat-sensitive heads, which discharge water when a predetermined temperature is reached at any head. A gate valve is installed in the main supply, and drains are provided. An alarm can be connected to the system so that local and remote signals can be given when the water flows.

Standpipes. Hoses supplied with water from standpipes are the usual means of manual application of water to interior building fires. Standpipes are usually designed for this use by the fire department, but they can be used by building fire fighters also.

Standpipes are necessary in buildings higher than those that ground-based fire department equipment can handle effectively. The National Building Code requires standpipes in buildings higher than 50 ft. The New York City building-code requirement starts at six stories or 75 ft, except where there are more than 10,000 ft^2 per floor, in which case they are required for two or more stories.

See also Sec. 19.

Chemical Extinguishment. Fires involving some materials may not be readily extinguished with water alone. When such materials may be present in a building, provision should be made for application of appropriate chemicals.

Foamed chemicals, mostly masses of air- or gas-filled bubbles, formed by chemical or mechanical means, may be used to control fires in flammable liquids. Foam is most useful in controlling fires in flammable liquids with low flash points and low specific gravity, such as gasoline. The mass of bubbles forms a cohesive blanket that extinguishes fire by excluding air and cooling the surface.

Foam clings to horizontal surfaces and can also be used on vertical surfaces of process vessels to insulate and cool. It is useful on fuel-spill fires, to extinguish and confine the vapors.

For fire involving water-soluble liquids, such as alcohol, a special foam concentrate must be used. Foam is not suitable for use on fires involving compressed gases, such as propane, nor is it practical on live electrical equipment. Because of the water content, foam cannot be used on fires involving burning metals, such as sodium, which reacts with water. It is not effective on oxygen-containing materials.

Three distinct types of foam are suitable for fire control: chemical foam, air foam (mechanical foam), and high-expansion foam.

Chemical foam was the first foam developed for fire fighting. It is formed by the reaction of water with two chemical powders, usually sodium bicarbonate and aluminum sulfate. The reaction forms carbon dioxide, which is the content of the bubbles. This foam is the most viscous and tenacious of the foams. It forms a relatively tough blanket, resistant to mechanical or heat disruption. The volume of expansion may be as much as 10 times that of the water used in the solution.

Chemical foam is sensitive to the temperature at which it is formed, and the chemicals tend to deteriorate during long storage periods. It is not capable of being transported through long pipe lines. For these reasons, it is not used as much as other foams. National Fire Protection Association standard NEPA 11 covers chemical foam.

Air foam (mechanical foam) is made by mechanical mixing of water and a protein-based chemical concentrate. There are several methods of combining the components, but essentially the foam concentrate is induced into a flowing stream of water through a metering orifice and a suitable device, such as a venturi. The volume of foam generated is from 16 to 33 times the volume of water used. Several kinds of mixing apparatus are available, choice depending on volume required, availability of water, type of hazard, and characteristics of the protected area or equipment.

Air foam can be conducted through pipes and discharged through a fixed chamber mounted in a bulk fuel storage tank, or it can be conducted through hoses and discharged manually through special nozzles. This foam can also be distributed through a sprinkler system of special design to cover small equipment, such as process vessels, or in multi-system applications, over an entire airplane hangar. The standard for use and installation of air foam is NFPA 11, and for foam-water sprinkler systems, NFPA 16.

High-expansion foam was developed for use in coal mines, where its extremely high expansion rate allowed it to be generated quickly in sufficient volume to fill mine galleries and reach inaccessible fires. This foam can be generated in volumes of from 100 to 1000 times the volume of water used, with the latter expansion in most general use. The foam is formed by passage of air through a screen constantly wetted by a solution of chemical concentrate, usually with a detergent base. The foam can be conducted to a fire area by ducts, either fixed or portable, and can be applied manually by small portable generators. Standard for equipment and use of high-expansion foam is NFPA 11A.

High-expansion foam is useful for extinguishing fires by totally flooding indoor confined spaces, as well as for local application to specific areas. It extinguishes by displacing air from the fire and by the heat-absorbing effect of converting the foam water content into steam. The foam forms an insulating barrier for exposed equipment or building components.

High-expansion foam is more fragile than chemical or air foam. Also, it is not generally reliable when used outdoors where it is subject to wind currents. High-expansion foam is not toxic, but it has the effect of disorienting people who may be trapped in it.

Carbon dioxide is useful as an extinguishing agent, particularly on surface fires, such as those involving flammable liquids in confined spaces. It is nonconductive and is effective on live electrical equipment. Because carbon dioxide requires no clean-up, it is desirable on equipment such as gasoline or diesel engines. The gas can be used on Class A fires. But when a fire is deep-seated, an extended discharge period is required to avoid rekindling.

Carbon dioxide provides its own pressure for discharge and distribution and is non-reactive with most common industrial materials. Because its density is 1½ times that of air, carbon dioxide tends to drop and to build up from the base of a fire. Extinguishment of a fire is effected by reduction of the oxygen concentration surrounding a fire.

Carbon dioxide may be applied to concentrated areas or machines by hand-held equipment, either carried or wheeled. Or the gas may be used to flood totally a room containing a hazard. The minimum concentrations for total flooding for fires involving some commercial liquids are listed in "Standard on Carbon-Dioxide Extinguishing Systems," NFPA 12.

Carbon dioxide is not effective on fires involving burning metals, such as magnesium, nor is it effective on oxygen-containing materials, such as nitrocellulose. Hazard to personnel is involved to the extent that a concentration of 9% will cause suffocation in a few minutes, and concentrations of 20% can be fatal. When used in areas where personnel are present, a time delay before discharge is necessary to permit evacuation.

For use in total flooding systems, carbon dioxide is available in either high-pressure or low-pressure equipment. Generally, it is more economical to use low-pressure equipment for large volumes, although there is no division point applicable in all cases.

Halon 1301 is one of a series of halogenated hydrocarbons, bromotrifluoromethane ($CBrF_3$), used with varying degrees of effectiveness as fire-extinguishing agents. Its use, in general, is the same as that of carbon dioxide. It also needs no clean-up. Its action is quicker than carbon dioxide, however, because Halon 1301 does not have to displace the air from around the fire.

At room temperatures, Halon 1301 is a colorless, odorless gas with a density 5 times that of air. The gas is stored in pressure vessels, similar to those for carbon dioxide. These vessels can be manifolded for large systems. Because the vapor pressure varies widely with temperature of Halon 1301, the containers are superpressurized with dry nitrogen to 600 psig or 360 psig.

Action of the gas as an extinguisher is considered to be that of a chemical inhibitor, interfering with the chain reaction necessary to maintain combustion. The action is almost instantaneous. Required concentrations are substantially less than those for carbon dioxide.

Because of the low concentration provided, and the density of Halon 1301, the discharge system must provide a thorough mixing of the agent with the air surrounding the fire. Flame extinguishment is very rapid. This characteristic is aided by quick detection of fires, indicating the need for an automatically actuated system.

If rapid egress is possible, there should be no prohibition on use of Halon 1301 in normally occupied areas in concentrations less than 7%. It is considered by Underwriters Laboratories to be about half as toxic as carbon tetrachloride, which was used for many years as a fire-extinguishing agent. The decomposition products of Halon 1301, however, are irritating, and prolonged exposure of the gas to temperatures above 900°F will cause the formation of toxic fluorides and bromides.

Dry chemical extinguishing agents were used originally to extinguish Class B fires and were described as a dry chemical extinguishing agent. It consisted of a sodium bicarbonate base with additives to prevent caking and to improve fluid flow characteristics. Later, multipurpose dry chemicals effective on Class A, B, and C fires were developed. These chemicals are distinctly different from the dry powder extinguishing agents used on combustible metals described below.

Dry chemicals are effective on surface fires, especially on flammable liquids. When used on Class A fires, they do not penetrate into the burning material. So when a fire involves porous or loosely packed material, water is used as a backup. The major effect of dry chemicals is due almost entirely to ability to break the chain reaction of combustion. A minor effect of smothering is obtained on Class A fires.

Fires that are likely to rekindle are not effectively controlled by dry chemicals. When these chemicals are applied to machinery or equipment at high temperatures, caking can cause some difficulty in cleaning up after the fire.

Dry chemicals can be discharged in local applications by hand-held extinguishers, wheeled portable equipment, or nozzles on hose lines. These chemicals can also be used for extinguishing fires by total flooding, when they are distributed through a piped system with special discharge nozzles. The expellant gas is usually dry nitrogen.

Dry powder extinguishing agents are powders effective in putting out combustible-metal fires. There is no universal extinguisher that can be used on all fires involving combustible metals. Such fires should never be fought by untrained personnel.

There are several proprietary agents effective on several metals, but none should be used without proper attention to the manufacturer's instructions and the specific metal involved. For requirements affecting handling and processing of combustible metals, reference should be made to National Fire Protection Association standards NFPA 48 and 652 for magnesium, NFPA 481 for titanium, NFPA 482M for zirconium, and NFPA 65 and 651 for aluminum.

3-20. Fire Detection.

Every fire-extinguishing activity must start with detection. To assist in this, many types of automatic detectors are available, with a wide range of sensitivity. Also, a variety of operations can be performed by the detection system. It can initiate an alarm, local or remote, visual or audible; notify a central station; actuate an extinguishing system; start or stop fans or processes, or perform any other operation capable of automatic control.

There are five general types of detectors, each employing a different physical means of operation. The types are designated *fixed-temperature, rate-of-rise, photoelectric, combustion-products,* and *ultraviolet* or *infrared* detectors.

A wide variety of detectors has been tested and reported on by Underwriters Laboratories, Inc. See Art. 3-14.

Fixed-Temperature Detectors. In its approval of any detection device, UL specifies the maximum distance between detectors to be used for area coverage. This spacing should not be used without competent judgment. In arriving at the permitted spacing for any device, UL judges the response time in comparison with that of automatic sprinkler heads spaced at 10-ft intervals. Thus, if a device is more sensitive than a sprinkler head, the permitted spacing is increased until the response times are nearly equal. If greater sensitivity is desired, the spacing must be reduced.

With fixed-temperature devices, there is a thermal lag between the time the ambient temperature reaches rated temperature and the device itself reaches that temperature. For thermostats having a rating of 135°F, the ambient temperature can reach 206°F.

Disk thermostats are the cheapest and most widely used detectors. The most common type employs the principle of unequal thermal expansion in a bimetallic assembly to operate a snap-action disk at a preset temperature, to close electrical contacts. These thermostats are compact. The disk, ½ in in diameter, is mounted on a plastic base 1¾ in in diameter. The thermostats are self-resetting, the contacts being disconnected when normal temperature is restored.

Thermostatic cable consists of two sheathed wires separated by a heat-sensitive coating which melts at high temperature, allowing the wires to contact each other. The assembly is covered by a protective sheath. When any section has functioned, it must be replaced.

Continuous detector tubing is a more versatile assembly. This detector consists of a small-diameter Inconel tube, of almost any length, containing a central wire, separated from the tube by a thermistor element. At elevated temperatures, the resistance of the thermistor drops to a point where a current passes between the wire and the tube. The current can be monitored, and in this way temperature changes over a wide range, up to 1000°F, can be detected. The detector can be assembled to locate temperature changes of different magnitudes over the same length of detector. It is self-restoring when normal temperature is restored. This detector is useful for industrial applications, as well as for fire detection.

Fusible links are the same devices used in sprinkler heads and are made to operate in the same temperature range. Melting or breaking at a specific temperature, they are used to restrain operation of a fire door, electrical switch, or similar mechanical function, such as operation of dampers. Their sensitivity is substantially reduced when installed at a distance below a ceiling or other heat-collecting obstruction.

Rate-of-Rise Detectors. Detectors and detector systems are said to operate on the rate-of-rise principle when they function on a rapid increase in temperature, whether the initial temperature is high or low. The devices are designed to operate when temperature rises at a specified number of degrees, usually 10 or 15°F, per minute. They are not affected by normal temperature increases and are not subject to thermal lag, as are fixed-temperature devices.

A rate-of-rise system widely used to operate deluge and preaction sprinkler systems employs a hollow copper bulb connected to a suitably encased diaphragm by small-diameter copper tubing. Increased ambient temperature raises the temperature and hence the pressure of air in the bulb. The increased pressure is transmitted to the diaphragm through the tubing. The expanded diaphragm starts a lever train that performs any function required. A calibrated vent, however, permits slowly developed air pressure to escape. A limited number of bulbs can be placed on a tubing circuit, but any number of circuits can be used to operate a single system.

A related system uses copper tubing as the detector, with a specified minimum length exposed to a given protected area. Pressure from heated air is transmitted to one or two suitably encased diaphragms, which are connected to electrical contacts.

A self-contained pneumatic detector employs an air chamber about the size and shape of half a tennis ball. The ball is mounted on a plastic base containing a flexible diaphragm connected to electrical contacts. Pressure from heated air in the chamber bulges the diaphragm directly, and closes or opens electrical contacts. A fixed-temperature element is also incorporated in the device.

Rate-of-rise operation also is provided by a device consisting of a sealed, stainless-steel cylinder ⅝ in in diameter, 3 in long, enclosing two struts in compression, with electrical contacts. Rapid increase in temperature causes the cylinder to expand faster than the struts, relieving the compression and causing the contacts to close. Fixed-temperature operation is obtained at a predetermined maximum temperature, when all components have expanded to a point where the contacts close.

Similar action is obtained with a different device used to operate a pilot air or water line controlling a deluge system. An aluminum tube 1½ in in diameter, 19 in long, encloses an air or water valve. Rapid rise in temperature reaches the outer case first, expanding it. The expanding case pulls a rod, opening the valve. Fixed-temperature operation is obtained by installing a standard sprinkler head in an opening provided in the base.

A rate-of-rise detector employing thermocouples is used by one manufacturer. A parabolic mirror directs radiant or convected heat to the thermopile for maximum effect.

Photoelectric Detectors. These indicate a fire condition by detecting the smoke. Sensitivity can be adjusted to operate when obscuration is as low as 0.4% per ft. In these devices, a light source is directed so that it does not impinge on a photoelectric cell. When sufficient smoke particles are concentrated in the chamber, their reflected light reaches the cell, changing its resistance and initiating a signal.

These detectors are particularly useful when a potential fire is likely to generate a substantial amount of smoke before appreciable heat and flame erupt. A fixed-temperature, snap-action disk is usually included in the assembly.

Combustion-Products Detectors. Two physically different means, designated ionization type and resistance-bridge type, are used to operate combustion-products detectors.

The **ionization type,** most generally used, employs ionization of gases by alpha particles emitted by a small quantity of radium or americum. The detector contains two ionization chambers, one sealed and the other open to the atmosphere, in electrical balance with a cold-cathode tube or transistorized amplifier. When sufficient combustion products enter the open chamber, the electrical balance is upset, and the resulting current operates a relay.

The **resistance-bridge type** of detector operates when combustion products change the impedance of an electric bridge grid circuit deposited on a glass plate.

Combustion-products detectors are designed for extreme early warning, and are most useful when it is desirable to have warning of impending combustion when combustion products are still invisible. These devices are sensitive in some degree to air currents, temperature, and humidity, and should not be used without consultation with competent designers.

Flame Detectors. These discriminate between visible light and the light produced by combustion reactions. Ultraviolet detectors are responsive to flame having wavelengths up to 2850 Å. The effective distance between flame and detectors is about 10 ft for a 5-in-diam pan of gasoline, but a 12-in-square pan fire can be detected at 30 ft.

Infrared detectors are also designed to detect flame. These are not designated by range of wavelength because of the many similar sources at and above the infrared range. To identify the radiation as a fire, infrared detectors usually employ the characteristic flame flicker, and have a built-in time delay to eliminate accidental similar phenomena.

3-21. Smoke and Heat Venting.

In extinguishment of any building fire, the heat-absorption capacity of water is the principal medium of reducing the heat release from the fire. When, however, a fire is well-developed, the smoke and heat must be released from confinement to make the fire approachable for final manual action. If smoke and heat venting is not provided in the building design, holes must be opened in the roof or building sides by the fire department. In many cases, it has been impossible to do this, with total property losses resulting.

Large-area, one-story buildings can be provided with venting by use of monitors, or a distribution of smaller vents. Multistory buildings present many problems, particularly since life safety is the principal consideration in these buildings.

Ventilation facilities should be provided in addition to the protection afforded by automatic sprinklers and hose stations.

Large One-Story Buildings. For manufacturing purposes, low buildings are frequently required to be many hundreds of feet in each horizontal dimension. Lack of automatic sprinklers in such buildings has proven to be disastrous where adequate smoke and heat venting has not been provided. Owners generally will not permit fire division walls, because they interfere with movement and processing of materials. With the whole content of a building subject to the same fire, fire protection and venting are essential to prevent large losses in windowless buildings, underground structures, and buildings housing hazardous operations.

There is no accepted formula for determining the exact requirements for smoke and heat venting. Establishment of guide lines is the nearest approach that has been made to venting design, and these must be adapted to the case at hand. Consideration must be given to quantity, shape, size, and combustibility of contents.

Venting Ratios. It is generally accepted that the ratio of effective vent opening to floor area should be:

Low-heat-release contents	1:150
Moderate-heat-release contents	1:100
High-heat-release contents	1:30–1:50

Venting can be accomplished by use of monitors, continuous vents, unit-type vents, or sawtooth skylights. In moderate-sized buildings exterior-wall windows may be used if they are near the eaves.

Monitors must be provided with operable panels or other effective means of providing openings at the required time.

Continuous gravity vents are continuous narrow slots provided with a weather hood above. Movable shutters can be provided and should be equipped to open automatically in a fire condition.

Vent Spacing. Unit-type vents are readily adapted to flat roofs, and can be installed in any required number, size, and spacing. They are made in sizes from 4×4 ft to 10×10 ft, with a variety of frame types and means of automatic opening. In arriving at the number and size of vents, preference should be given to a large number of small vents, rather than a few large vents. Because it is desirable to have a vent as near as possible to any location where a fire can start, a limit should be placed on the distance between units.

The generally accepted maximum distance between vents is:

Low-heat-release contents	150 ft
Moderate-heat-release contents	120 ft
High-heat-release contents	75–100 ft

Releasing Methods. Roof vents should be automatically operated by means that do not require electric power. They also should be capable of being manually operated. Roof vents approved by Underwriters Laboratories, Inc. are available from a number of manufacturers.

Refer to National Fire Protection Association standard NFPA 204 in designing vents for large, one-story buildings. Tests conducted prior to publication of NFPA 231C indicated that a sprinkler system designed for adequate density of water application will eliminate the need for roof vents, but the designers would be well advised to consider the probable speed of fire and smoke development in making a final decision. NFPA 231C covers the rack storage of materials as high as 20 ft.

High-Rise Buildings. Building codes vary in their definition of high-rise buildings, but the intent is to define buildings in which fires cannot be fought successfully by ground-based equipment and personnel. Thus, ordinarily, high-rise means buildings 100 ft or more high. In design for smoke and heat venting, however, any multistory building presents the same problems.

Because smoke inhalation has been the cause of nearly all fatalities in high-rise buildings, some building codes require that a smoke venting system be installed and made to function independently of the air-conditioning system. Also, smoke detectors must be provided to actuate exhaust fans and at the same time warn the fire department and the building's control center. The control center must have two-way voice communication, selectively, with all floors and be capable of issuing instructions for occupant movement to a place of safety.

Because the top story is the only one that can be vented through the roof, all other stories must have the smoke conducted through upper stories to discharge safely above the roof. A separate smoke shaft extending through all upper stories will provide this means. It should be provided with an exhaust fan and should be connected to return-air ducts with suitable damper control of smoke movement, so that smoke from any story can be directed into the shaft. The fan and dampers should be actuated by smoke detectors installed in suitable locations at each inlet to return-air ducts. Operation of smoke detectors also should start the smoke-vent-shaft fan and stop supply-air flow. Central-station supervision (Art. 3-23) should be provided for monitoring smoke-detector operation. Manual override controls should be installed in a location accessible under all conditions.

Windows with fixed sash should be provided with means for emergency opening by the fire department.

Pressurizing stair towers to prevent the entrance of smoke is highly desirable but difficult to accomplish. Most standpipe connections are usually located in stair towers, and it is necessary to open the door to the fire floor to advance the hose stream toward the fire. A more desirable arrangement would be to locate the riser in the stair tower, if required by code, and place the hose valve adjacent to the door to the tower. Some codes permit this, and it is adaptable to existing buildings.

3-22. Emergency Egress.

In addition to providing means for early detection of fire, preventing its spread, and extinguishing it speedily, building designers should also provide the appropriate number, sizes, and arrangements of exits to permit quick evacuation of occupants if fire or other conditions dangerous to life occur. Buildings should be designed to preclude development of panic in emergencies, especially in confined areas where large numbers of persons may assemble. Hence, the arrangement of exit facilities should permit occupants to move freely toward exits that they can see clearly and that can be reached by safe, unobstructed, uncongested paths. Redundancy is highly desirable; there should be more than one path to safety, so that loss of a single path will not prevent escape of occupants from a danger area. The paths should be accessible to and usable by handicapped persons, including those in wheelchairs, if they may be occupants.

Building codes generally contain requirements for safe, emergency egress from buildings. Such requirements also are concisely presented in the "Life Safety Code" of the National Fire Protection Association.

Egress Components. Many building codes define an exit as a safe means of egress from the interior of a building to an open exterior space beyond the reach of a building fire or give an equivalent definition. Other codes consider an exterior door or a stairway leading to access to such a door to be an exit. To prevent misunderstandings, the "Life Safety Code" defines a means of egress composed of three parts.

Accordingly, a **means of egress** is a continuous, unobstructed path for evacuees from any point in a building to a public way. Its three parts are:

Exit access—that portion that leads to an entrance to an exit.

Exit—the portion that is separated from all other building spaces by construction or equipment required to provide a protected path to the exit discharge.

Exit discharge—the portion that connects the termination of an exit to a public way.

Means of egress may be provided by exterior and interior doors and enclosed horizontal and vertical passageways, including stairs and escalators. (Elevators and exterior fire escapes are not generally recognized as reliable means of egress in a fire.) Exit access includes the space from which evacuation starts and passageways and doors that must be traversed to reach an exit.

Types of Exits. Building codes generally recognize the following as acceptable exits when they meet the codes' safety requirements:

Corridors—enclosed horizontal or slightly inclined public passageways, which lead from interior spaces toward an exit discharge. Minimum floor-to-ceiling height permitted is generally 7 ft 6 in, but 7 ft may be used for short stretches. Minimum width depends on type of occupancy (Table 3-6). Codes may require subdivision of corridors into lengths not exceeding 300 ft for educational buildings and 150 ft for institutional buildings. Subdivision should be accomplished with noncombustible partitions incorporating smokestop doors. In addition, codes may require the corridor enclosures to have a fire rating of 1 or 2 hr.

Exit passageways—horizontal extensions of vertical passageways. Minimum floor-to-ceiling height is the same as for corridors. Width should be at least that of the vertical passageways. Codes may require passageway enclosures to have a 2-hr fire rating. A street-floor lobby may serve as an exit passageway if it is sufficiently wide to accommodate the probable number of evacuees from all contributing spaces at the lobby level.

Exit doors—doors providing access to streets or to stairs or exit passageways. Those at stairs or passageways should have a fire rating of at least ¾ hr (Art. 14-3).

Horizontal exit—passageway to a refuge area. The exit may be a fire door through a wall with a 2-hr fire rating, a balcony providing a path around a fire barrier, or a bridge or tunnel between two buildings. Doors in fire barriers with 3- or 4-hr fire ratings should have a 1½-hr rated door on each face of the fire division. Walls permitted to have a lower fire rating may incorporate a single door with a rating of at least 1½ hr. Balconies, bridges, and tunnels should be at least as wide as the doors providing access to them, and enclosures or sides of these passageways should have a fire rating of 2 hr or more. Exterior-wall openings below or within 30 ft of an open bridge or balcony should have at least ¾-hr fire protection.

Interior stairs—stairs that are inside a building and that serve as an exit. Except in one-story or two-story low-hazard buildings, such stairs should be built of noncombustible materials. Stairway enclosures generally should have a 2-hr fire rating. Building codes, however, may exempt low dwellings from this requirement.

Exterior stairs—stairs that are open to the outdoors and that serve as an exit to ground level. Height of such stairs is often limited to 75 ft or six stories. The stairs should be protected by a fire-resistant roof and should be built of noncombustible materials. Wall openings within 10 ft of the stairs should have ¾-hr fire protection.

Smokeproof tower—a continuous fire-resistant enclosure protecting a stairway from fire or smoke in a building. At every floor, a passageway should be provided by vestibules or balconies directly open to the outdoors and at least 40 in wide. Tower enclosures should have a 2-hr fire rating. Access to the vestibules or balconies and entrances to the tower should be provided by doorways at least 40 in wide, protected by self-closing fire doors.

Escalators—moving stairs. Building codes may permit their use as exits if they meet the safety requirements of interior stairs and if they move in the direction of exit travel or stop gradually when an automatic fire-detection system signals a fire.

Moving walks—horizontal or inclined conveyor belts for passengers. Building codes may permit their use as exits if they meet the safety requirements for exit passageways and if they move in the direction of exit travel or stop gradually when an automatic fire-detection system signals a fire.

Refuge Areas. A refuge area is a space protected against fire and smoke. When located within a building, the refuge should be at about the same level as the areas served and separated from them by construction with at least a 2-hr fire rating. Access to the refuge areas should be protected by fire doors with a fire rating of 1½ hr or more.

A refuge area should be large enough to shelter comfortably its own occupants plus those from other spaces served. The minimum floor area required may be calculated by allowing 3 ft^2 of unobstructed space for each ambulatory person and 30 ft^2 per person for hospital or nursing-home patients. Each refuge area should be provided with at least one horizontal or vertical exit, such as a stairway, and in locations more than 11 stories above grade, with at least one elevator.

Location of Exits. Building codes usually require a building to have at least two means of egress from every floor. Exits should be remote from each other, to reduce the chance that both will be blocked in an emergency.

All exit access facilities and exits should be located so as to be clearly visible to building occupants or signs should be installed to indicate the direction of travel to the exits.

Signs marking the locations of exits should be illuminated with at least 5 ft-c of light. Floors of means of egress should be illuminated with at least 1 ft-c of artificial light whenever the building is occupied.

If an open floor area does not have direct access to an exit, a protected, continuous passageway should be provided directly to an exit. The passageway should be kept open at all times. Occupants using the passageway should not have to pass any high-hazard areas not fully shielded.

To ensure that occupants will have sufficient escape time in emergencies, building codes limit the travel distance from the most remote point in any room or space to a door that opens to an outdoor space, stairway, or exit passageway. The maximum travel dis-

TABLE 3-5 Typical Occupant Load Requirements for Buildings

Occupancy	Net floor area per occupant, ft^2
Auditoriums	7
Billiard rooms	50
Bowling alleys	50
Classrooms	20
Dance floors	7
Dining spaces (nonresidential)	12
Exhibition spaces	10
Garages and open parking structures	250
Gymnasiums	15
Habitable rooms	200
Industrial shops	200
In schools	50
Institutional sleeping rooms	120
Kindergartens	35
Kitchens (nonresidential)	200
Laboratories	50
Preparation rooms	100
Libraries	25
Locker rooms	12
Offices	100
Passenger terminals or platforms	1.5C*
Sales areas (retail)	
First floor or basement	30
Other floors	60
Seating areas (audience) in places of assembly	
Fixed seats	D†
Movable seats	10
Skating rinks	15
Stages	S‡
Storage rooms	300

*C = capacity of all passenger vehicles that can be unloaded simultaneously.
†D = number of seats or occupants for which space is to be used.
‡S = 75 persons per unit of width of exit openings serving a stage directly, or one person per 15 ft^2 of performing area plus one person per 50 ft^2 of remaining area plus number of seats that may be placed for an audience on stage.

tance permitted depends on the type of occupancy and whether the space is sprinklered. For example, for corridors not protected by sprinklers, maximum permitted length may range from 100 ft for storage and institutional buildings to 150 ft for residential, mercantile, and industrial occupancies. With sprinkler protection, permitted length may range from 150 ft for high-hazard and storage buildings to 300 ft for commercial buildings, with 200 ft usually permitted for other types of occupancies.

Building codes also may prohibit or limit the lengths of passageways or courts that lead to a dead end. For example, a corridor that does not terminate at an exit is prohibited in high-hazard buildings. For assembly, educational, and institutional buildings, the maximum corridor length to a dead end may not exceed 30 ft, whereas the maximum such length is 40 ft for residential buildings and 50 ft for all other occupancies, except high-hazard.

Required Exit Capacity. Minimum width of a passageway for normal use is 36 in. This is large enough to accommodate one-way travel for persons on crutches or in wheelchairs. For two-way travel, a 60-in width is necessary. (A corridor, however, need not be 60 in wide for its full length, if 60 × 60-in passing spaces, alcoves, or corridor intersections are provided at short intervals.) Building codes, however, may require greater widths to permit rapid passage of the anticipated number of evacuees in emergencies. This number depends on a factor called the occupant load.

Occupant load of a building space is the maximum number of persons that may be in the space at any time. Building codes may specify the minimum permitted capacity of exits in terms of occupant load, given as net floor area, square feet, per person, for various types of occupancy (Table 3-5). The number of occupants permitted in a space served by the exits then can be calculated by dividing the floor area, square feet, by the specified occupant load.

The occupant load of any space should include the occupant load of other spaces if the occupants have to pass through that space to reach an exit.

With the occupant load known, the required width for an exit or an exit door can be determined by dividing the occupant load on the exit by the capacity of the exit.

Capacities of exits and access facilities generally are measured in units of width of 22 in, and the number of persons per unit of width is determined by the type of occupancy. Thus, the number of units of exit width for a doorway is found by dividing by 22 the clear width of the doorway when the door is in the open position. (Projections of stops and hinge stiles may be disregarded.) Fractions of a unit of width less than 12 in should not be credited to door capacity. If, however, 12 in or more is added to a multiple of 22 in, one-half unit of width can be credited. Building codes indicate the capacities in persons per unit of width that may be assumed for various means of egress. Recommendations of the "Life Safety Code" of the National Fire Protection Association, 470 Atlantic Ave., Boston, Mass. 02210, are summarized in Table 3-6.

TABLE 3-6 Capacities, Persons per Unit of Width, for Means of Egress

Level egress components, including doors	100
Stairway	60
Ramps 44 in or more wide, slope not more than 10%	100
Narrower or steeper ramps	
Up	60
Down	100

3-23. Building Operation in Emergencies.

For buildings that will be occupied by large numbers of persons, provision should be made for continuation of services essential to safe, rapid evacuation of occupants in event of fire or other emergencies and for assisting safe movement of fire fighters, medical personnel, or other aides.

Standby electric power, for example, should be available in all buildings to replace the basic power source if it should fail. The standby system should be equipped with a generator that will start automatically when normal power is cut off. The emergency power supply should be capable of operating all emergency electric equipment at full power within 1 min of failure of normal service. Such equipment includes lights for exits, elevators for fire fighters' use, escalators and moving walks designated as exits, exhaust fans and pressurizing blowers, communication systems, fire detectors, and controls needed for fire fighting and life safety during evacuation of occupants.

In high-rise buildings, at least one elevator should be available for control by fire fighters and to give them access to any floor from the street-floor lobby. Also, elevator controls should be designed to preclude elevators from stopping automatically at floors affected by fire.

Supervision of emergency operations can be efficiently provided by personnel at a control center placed in a protected area. This center may include an electronic computer, supplemented by personnel performing scheduled maintenance, and should be capable of continuously monitoring alarms, gate valves on automatic fire sprinklers, temperatures, air and water pressures, and perform other pertinent functions. Also, the center should be capable in emergencies of holding two-way conversations with occupants and notifying police and fire departments of the nature of the emergencies. In addition, provision should be made for the control center to dispatch investigators to sources of potential trouble or send maintenance personnel to make emergency repairs when necessary. Standards for such installations are NFPA 72A, "Local Protective Signaling Systems," NFPA 72B, "Auxiliary Protective Signaling Systems," NFPA 72C, "Remote Station Protective Signaling Systems," and NFPA 72D, "Proprietary Protective Signaling Systems."

For economical building operation, the emergency control center may be made part of a control center used for normal building operation and maintenance. Thus, the control center may normally control HVAC to conserve energy, turn lights on and off, and schedule building maintenance and repair. When an emergency occurs, emergency control should be activated in accordance with prepared plans for handling each type of emergency.

The control center need not be located within the building to be supervised nor operated by in-house personnel. Instead, an external central station may provide the necessary supervision. Such services are available in most cities and are arranged by contract, usually with an installation charge and an annual maintenance charge. Requirements for such systems are in National Fire Protection Association standard NFPA 71.

3-24. Safety during Construction.

Most building codes provide specific measures that must be taken for fire protection during construction of buildings. But when they do not, fundamental fire-safety precautions must be taken. Even those structures that will, when completed, be noncombustible contain quantities of forming and packing materials that present a serious fire hazard.

Multistory buildings should be provided with access stairways and, if applicable, an elevator for fire department use. Stairs and elevator should follow as closely as possible the upward progress of the structure and be available within one floor of actual building height. In buildings requiring standpipes, the risers should be placed in service as soon as possible, and as close to the construction floor as practicable. Where there is danger of freezing, the water supply can consist of a Siamese connection for fire department use.

In large-area buildings, required fire walls should be constructed as soon as possible. Competent watchman service also should be provided.

3-48 Protection of Structures and Occupants against Hazards

The greatest source of fires during construction is portable heaters. Only the safest kind should be used, and these safeguarded in every practical way. Fuel supplies should be isolated and kept to a minimum.

Welding operations also are a source of fires. They should be regulated in accordance with building-code requirements.

Control of tobacco smoking is impossible during building construction, so control of combustible materials is necessary. Good housekeeping should be provided, and all combustible materials not necessary for the work should be removed as soon as possible.

Construction offices and shanties should be equipped with adequate portable extinguishers. So should each floor in a multistory building.

LIGHTNING PROTECTION

Lightning, a high-voltage, high-current electrical discharge between clouds and the ground, may strike and destroy life and property anywhere thunderstorms have occurred in the past. Buildings and their occupants, however, can be protected against this hazard by installation of a special electrical system. Because an incomplete or poor installation can cause worse damage or injuries than no protection at all, a lightning-protection system should be designed and installed by experts.

As an addition to other electrical systems required for a building, a lightning-protection system increases the construction cost of a building. A building owner therefore has to decide whether potential losses justify the added expenditure. In doing so, the owner should take into account the importance of the building, danger to occupants, value and nature of building contents, type of construction, proximity of other structures or trees, type of terrain, height of building, number of days per year during which thunderstorms may occur, costs of disruption of business or other activities and the effects of loss of essential services, such as electrical and communication systems. (Buildings housing flammable or explosive materials generally should have lightning protection.) Also, the owner should compare the cost of insurance to cover losses with the cost of the protection system.

3-25. Characteristics of Lightning.

Lightning strikes are associated with thunderstorms. In such storms, the base of the clouds generally develops a negative electrical charge, which induces a positive charge in the earth directly below. As the clouds move, the positive charges, being attracted by the negative charges, follow along the surface of the earth and climb up buildings, antennas, trees, power transmission towers, and other conducting or semiconducting objects along the path. The potential between clouds and earth may build up to 10^6 to 10^9 V. When the voltage becomes great enough to overcome the electrical resistance of the air between the clouds and the ground or an object on it, current flows in the form of a lightning flash. Thus, the probability of a building being struck by lightning depends not only on the frequency of occurrence of thunderstorms but also on building height relative to nearby objects and the intensity of cloud charges.

Destruction at the earth's surface may result not only at points hit by lightning directly but also by electrostatic induction at points several feet away. Also, lightning striking a tall object may flash to a nearby object that offers a suitable path to the ground.

Lightning often shatters nonconductors or sets them on fire if they are combustible. Conductors struck may melt. Living things may be burned or electrocuted. Also, lightning may induce overvoltages in electrical power lines, sending electrical charges along the lines in both directions from the stricken point to ground. Direct-stroke overvoltages may range up to several million volts and several hundred thousand amperes. Induced strokes, which occur more frequently, may be on the order of several hundred thousand volts with currents up to 2000 A. Such overvoltages may not only damage electric equipment connected to the power lines but also buildings served by them. Consequently, lightning protection is necessary for outdoor conductors as well as for buildings.

3-26. Methods for Protecting against Lightning.

Objectives of lightning protection are life safety, prevention of property damage, and maintenance of essential services, such as electrical and communication systems. Lightning protection usually requires installation of electrical conductors that extend from points above the roof of a building to the ground, for the purpose of conducting to the ground lightning that would otherwise strike the building. Such an installation, however, possesses the potential hazard that, if not done properly, lightning may flash from the lightning conductors to other building components. Hence, the system must ensure that the lightning discharge is diverted away from the building and its contents. Lightning protection systems should conform to the standards of the American National Standards Institute, National Fire Protection Association (NFPA 78, "Lightning Protection Code") and Underwriters Laboratories (UL 96A, "Master Labeled Lightning-Protection Systems").

The key element in diverting lightning away from a building is an air terminal or lightning rod, a conductor that projects into the air at least 12 in above the roof. Air terminals should be spaced at intervals not exceeding 25 ft. Alternatively, a continuous wire conductor or a grid of such conductors may be placed along the highest points of a roof. If the tallest object on a roof is a metal mast, it can act as an air terminal. A metal roof also can serve as an air terminal, but only if all joints are made electrically continuous by soldering, welding, or interlocking. Arranged to provide a cone of protection over the entire building, all the air terminals should be connected by conductors to each other and, by the same or other conductors, to the ground along at least two separated paths.

For roof and down conductors, copper, copper-clad steel, galvanized steel or a metal alloy that is as resistant to corrosion as copper may be used. (A solid copper conductor should be at least ¼ in in diameter.) Direct connections between dissimilar metals should be avoided to prevent corrosion. Metal objects and non-current-carrying components of electrical systems should be kept at least 6 ft away from the lightning conductors or should be bonded to the nearest lightning conductor. Sharp bends in the conductors are not desirable. If a 90° bend must be used, the conductor should be firmly anchored, because the high current in a lightning stroke will tend to straighten the bend. If the conductor has a U bend, the high current may induce an electric arc to leap across the loop while also exerting forces to straighten out the bend.

In steel-frame buildings, the steel frame can be used as a down conductor. In such cases, the top of the frame should be electrically connected to air terminals and the base should be electrically connected to grounding electrodes. Similarly, the reinforcing steel of a reinforced concrete building can be used as down conductors if the reinforcing steel is bonded together from foundations to roof.

The final and equally important elements of a lightning-protection system are grounding electrodes and the earth itself. The type and dimensions of the grounds, or grounding electrodes, depends on the electrical resistance, or resistivity, of the earth, which can be measured by technicians equipped with suitable instruments. The objective of the grounding installation, which should be electrically bonded to the down conductors, should be an earth-system resistance of 10 Ω or less. Underground water pipes can serve as grounds if they are available. If not, long metal rods can be driven into the ground to serve as electrodes. Where earth resistivity is poor, an extensive system of buried wires may be required.

(J. L. Marshall, "Lightning Protection," John Wiley & Sons, Inc., New York.)

PROTECTION AGAINST INTRUDERS

Prevention of illegal entry into buildings by professional criminals determined to break in is not practical. Hence, the prime objective of security measures is to make illegal entry difficult. If this is done, it will take an intruder longer to gain entry or will compel the intruder to make noise, thus increasing the chances of detection and apprehension. Other objectives of security measures are detection of break-in attempts and intruders, alarming intruders so that they leave the premises before they cause a loss or injury, and alerting building occupants and the police of the break-in attempt. Also, an objective is to safeguard valuable assets by placing them in a guarded, locked, secure enclosure with access limited only to approved personnel.

Some communities have established ordinances setting minimum requirements for security and incorporated them in the building code. (Communities that have done this include Los Angeles, Oakland, and Concord in California; Indianapolis, Ind.; Trenton, N.J.; Arlington Heights Village, Ill.; Arlington County, Va.; and Prince George's County, Md.) Provisions of these codes cover security measures for doors and windows and associated hardware, accessible transoms, roof openings, safes, lighting of parking lots, and intrusion-detection devices. For buildings requiring unusual security measures, owners and designers should obtain the advice of a security expert.

3-27. Security Measures.

Basic security for a building is provided by commonly used walls and roofs with openings protected by doors with key-operated locks or windows with latches. The degree of protection required for a building and its occupants beyond basic security and privacy needs depends on the costs of insurance and security measures relative to potential losses from burglary and vandalism.

For a small building not housing small items of great value (these can be placed in a safety deposit box in a bank), devices for detecting break-in attempts are generally the most practical means for augmenting basic security. Bells, buzzers, or sirens should be installed to sound an alarm and notify the police when an intruder tries to enter the locked building or a security area.

For a large building or a building requiring tight security, defense should be provided in depth. Depending on the value of assets to be protected, protection should start at the boundary of the property, with fences, gates, controlled access, guard patrols, exterior illumination, alarms, or remote surveillance by closed-circuit television. This defense should be backed up by similar measures at the perimeter of the building and by security locks and latches on doors and windows. Openings other than doorways or windows should be barred or made too small for human entry and screened. Within the building, valuables should be housed in locked rooms or a thick, steel safe, with controlled access to those areas.

For most types of occupancy, control at the entrance often may be provided by a receptionist who records names of visitors and persons visited, notifies the latter and can advise the police of disturbances. When necessary, the receptionist can be augmented by a guard at the control point or in a security center and, in very large or high-rise buildings, by a roving guard available for emergencies. If a large security force is needed, facilities should be provided in the building for an office for the security administrator and staff, photographic identification, and squad room and lockers—all in or adjoining a security center.

The security center may be equipped with or connected to electronic devices that do the following:

1. Detect a break-in attempt and sound an alarm.
2. Identify the point of intrusion.
3. Turn on lights.
4. Display the intruder on closed-circuit television and record observations on videotape.
5. Notify the police.
6. Limit entry to specific spaces only to approved personnel and only at permitted times.
7. Change locks automatically.

In addition, the center may be provided with emergency reporting systems, security guard tour reporting systems, fire detection and protection systems, including supervision of automatic fire sprinklers, HVAC controls, and supervision of other life safety measures.

(P. S. Hopf, "Handbook of Building Security Planning and Design," McGraw-Hill Book Company, New York.)

Section 4

Building Materials

Albert G. H. Dietz
Massachusetts Institute of Technology

This section describes the basic materials used in building construction and describes their common applications. For discussion purposes, materials used in similar applications are grouped and discussed in sequence, for example, masonry materials, wood, metals, plastics, etc.

Masonry materials, in the widest sense of the term, include all the inorganic, nonmetallic materials of construction. These encompass a wide range of materials from insulating blocks of lightweight concrete to slate tile roofs. Only the most important of these will be presented in this section, and attention will be focused as far as possible on those characteristics of the materials that affect their utilization, strength, durability, volume change, etc.

CEMENTITIOUS MATERIALS

Cementitious materials include the many and varied products that may be mixed with water or another liquid to form a paste that may or may not have aggregate added to it. The paste (mortar or concrete if aggregate is used) is temporarily plastic and may be molded or deformed. But later, it hardens or sets to a rigid mass.

There are many varieties of cements and numerous ways of classification. One of the simplest classifications is by the chemical constituent that is responsible for the setting or hardening of the cement. On this basis, the silicate and aluminate cements, wherein the setting agents are calcium silicates and aluminates, constitute the most important group of modern cements. Included in this group are the portland, aluminous, and natural cements.

Limes, wherein the hardening is due to the conversion of hydroxides to carbonates, were formerly widely used as the sole cementitious material, but their slow setting and hardening are not too compatible with modern requirements. Hence, their principal

NOTE: Sections covering masonry materials were written by James A. Murray (deceased), former Associate Professor of Materials, Massachusetts Institute of Technology, for the first edition and updated by Professor Dietz for subsequent editions.

function today is to plasticize the otherwise harsh cements and add resilience to mortars and stuccoes. Use of limes is beneficial in that their slow setting promotes healing, the recementing of hairline cracks.

Another class of cements is composed of calcined gypsum and its related products. The gypsum cements are widely used in interior plaster and for fabrication of boards and blocks; but the solubility of gypsum prevents its use in construction exposed to any but extremely dry climates.

Oxychloride cements constitute a class of specialty cements of unusual properties. Their cost prohibits their general use in competition with the cheaper cements; but for special uses, such as the production of sparkproof floors, they cannot be equaled.

Masonry cements or mortar cements are widely used because of their convenience. While they are, in general, mixtures of one or more of the above-mentioned cements with some admixes, they deserve special consideration because of their economies.

4-1. Portland Cements.

Portland cements are the most common of the modern hydraulic cements. They are made by blending a carefully proportioned mixture of calcareous and argillaceous materials. The mixture is burned in a rotary kiln at a temperature of about 2700°F to form hard nodulized pellets, called clinker. The clinker is ground with a retarder (usually rock gypsum) to a fine powder, which constitutes the portland cement.

Since cements are rarely used in construction without addition of aggregates to form mortars or concretes, properties of the end products are of much greater importance to the engineer than the properties of the cement from which they are made. Properties of mortars and concretes are determined to a large extent by the amount and grading of aggregate, amount of mixing water, and use of admixtures.

One property of mortars and concretes—durability—is greatly affected by the chemical composition of the cement used. It is therefore pertinent to consider the composition of several types of cement and the effect of composition on the use of the cement.

The composition of portland cement is now usually expressed in terms of the potential phase compounds present, the tricalcium silicate (C_3S), dicalcium silicate (C_2S), tricalcium aluminate (C_3A), and tetracalcium aluminum ferrite (C_4AF). Portland cements are made in eight types, distinction between the types being based upon both physical and chemical requirements. Some requirements of the American Society for Testing and Materials Specification for Portland Cement (C150-78) are given in Table 4-1. Other requirements of the specification, such as autoclave expansion, time of set, and magnesia content, which are constant for all types of cement, are omitted from the table.

Type I, general-purpose cement, is the one commonly used for many structural purposes. Chemical requirements for this type of cement are limited to magnesia and sulfur trioxide contents and loss on ignition, since the cement is adequately defined by its physical characteristics.

Type II is a modified cement for use in general concrete where a moderate exposure to sulfate attack may be anticipated or where a moderate heat of hydration is required. These characteristics are attained by placing limitations on the C_3S and C_3A content of the cement. Type II cement gains strength a little more slowly than Type I but ultimately will achieve equal strength. It is generally available in most sections of the country and is preferred by some engineers over Type I for general construction.

Type III cement attains high early strength. In 3 days, strength of concrete made with it is practically equal to that made with Type I or Type II cement at 28 days. This high early strength is attained by finer grinding (although no minimum is placed upon the fineness by specification) and by increasing the C_3S and C_3A content of the cement. Type III cement, however, has high heat evolution and therefore should not be used in

TABLE 4-1 Chemical and Physical Requirements for Portland Cement

Type: Name:	I and IA General- purpose	II and IIA Modi- fied	III and IIIA High early	IV Low- heat	V Sulfate- resisting
C_3S, max. %				35	
C_2S, min. %				40	
C_3A, max. %		8	15	7	5
Fineness, specific surface, m^2/kg, avg. min, by turbidimeter	160	160		160	160
Compressive strength, psi, mortar cubes of 1 part cement and 2.75 parts graded standard sand after:					
1 day					
Standard			1800		
Air-entraining			1450		
3 days					
Standard	1800	1500	3500		1200
Air-entraining	1450	1200	2800		
7 days					
Standard	2800	2500		1000	2200
Air-entraining	2250	2000			
28 days					
Standard				2500	3000

large masses. It also has poor sulfate resistance. Type III cement is not always available out of building materials dealers' stocks but may be obtained by them from the cement manufacturer on short notice.

Type IV is a low-heat cement, which has been developed for mass concrete construction. Normal Type I cement, if used in large masses that cannot lose heat by radiation, will liberate enough heat during the hydration of the cement to raise the temperature of the concrete as much as 50 or 60°F. This results in a relatively large increase in dimensions while the concrete is still soft and plastic. Later, as the concrete cools after hardening, shrinkage causes cracks to develop, weakening the concrete and affording points of attack for aggressive solutions. The potential-phase compounds which make the largest contribution to the heat of hydration are C_3S and C_3A; so the amounts of these that are permitted to be present are limited. Since these compounds also produce the early strength of cement, the limitation results in a cement that gains strength relatively slowly. This is of little importance, however, in the mass concrete for which this type of cement is designed.

Type V is a special portland cement intended for use when high sulfate resistance is required. Its resistance to sulfate attack is attained through the limitation on the C_3A content. It is particularly suitable for structures subject to attack by liquors containing sulfates, such as sea water and some other natural waters.

Both Type IV and Type V cements are specialty cements; they are not normally available from dealer's stock but are usually obtainable for use on a large job if arrangements are made with the cement manufacturer in advance.

4-2. Air-Entraining Portland Cements.

For use in the manufacture of air-entraining concrete, agents may be added to the cement by the manufacturer, thereby producing "air-entraining portland cements." These cements are available as Types IA, IIA, and IIIA.

Most cements will exceed the requirements of ASTM specification C150 as shown in Table 4-1 by a comfortable margin.

4-3. Aluminous Cements.

Aluminous cements are prepared by fusing a mixture of aluminous and calcareous materials (usually bauxite and limestone) and grinding the resultant product to a fine powder. These cements are characterized by their rapid-hardening properties and the high strength developed at early ages. Table 4-2 (adapted from F. M. Lea "Chemistry of Cement and Concrete," St. Martin's Press, Inc., New York) shows the relative strengths of 4-in cubes of 1:2:4 concrete made with normal portland, high-early-strength portland, and aluminous cements.

TABLE 4-2 Relative Strengths of Concretes Made from Portland and Aluminous Cements

Days	Compressive strength, psi		
	Normal portland	High-early portland	Aluminous
1	460	790	5,710
3	1,640	2,260	7,330
7	2,680	3,300	7,670
28	4,150	4,920	8,520
56	4,570	5,410	8,950

Since a large amount of heat is liberated with rapidity by aluminous cement during hydration, care must be taken not to use the cement in places where this heat cannot be dissipated. It is usually not desirable to place aluminous-cement concretes in lifts of over 12 in, otherwise the temperature rise may cause serious weakening of the concrete.

Aluminous cements are much more resistant to the action of sulfate waters than are portland cements. They also appear to be much more resistant to attack by water containing aggressive carbon dioxide or weak mineral acids than the silicate cements. Their principal use is in concretes where advantage may be taken of their very high early strength or of their sulfate resistance, and where the extra cost of the cement is not an important factor.

4-4. Natural Cements.

Natural cements are formed by calcining a naturally occurring mixture of calcareous and argillaceous substances at a temperature below that at which sintering takes place. The American Society for Testing and Materials Specification for Natural Cement (C10) requires that the temperature shall be no higher than is necessary to drive off the carbonic acid gas. Since natural cements are derived from naturally occurring materials and no particular effort is made to adjust the composition, both the composition and properties vary rather widely. Some natural cements may be almost the equivalent of portland cement in properties; others are much weaker. Natural cements are principally used now in masonry mortars and as an admixture in portland-cement concretes.

4-5. Hydraulic Limes.

Hydraulic lime is made by calcining a limestone containing silica and alumina to a temperature short of incipient fusion so as to form sufficient free lime (CaO) to permit hydration and at the same time leaving unhydrated sufficient calcium silicates to give the dry powder its hydraulic properties (see American Society for Testing and Materials Specification C141).

Because of the low silicate and high lime contents, hydraulic limes are relatively weak. They find their principal use in masonry mortars.

4-6. Quicklimes.

When limestone is heated to a temperature in excess of 1700°F, the carbon dioxide content is driven off and the remaining solid product is quicklime. It consists essentially of calcium and magnesium oxides plus impurities such as silica, iron, and aluminum oxides. The impurities are usually limited to less than 5%. If they exceed 10%, the product may be a hydraulic lime.

Two classes of quicklime are recognized, high-calcium and dolomitic. A high-calcium quicklime usually is considered as one containing less than 5% magnesium oxide. A dolomitic quicklime usually contains from 35 to 40% magnesium oxide. A few quicklimes are found that contain from 5 to 35% magnesium oxide and are called magnesian limes.

The outstanding characteristic of quicklime is its ability to slake with water. When quicklime is mixed with from 2 to 3 times its weight of water, a chemical reaction takes place: the calcium oxide combines with water to form calcium hydroxide, and sufficient heat is evolved to bring the entire mass to a boil. The resulting product is a suspension of finely divided calcium hydroxide (and magnesium hydroxide or oxide if dolomitic lime is used) in water. On cooling, the semifluid mass stiffens to a putty of such consistency that it may be shoveled or carried in a hod. This slaked quicklime putty, when cooled and preferably screened, is the material used in construction. Quicklime should always be thoroughly slaked.

The yield of putty will vary, depending upon the type of quicklime, its degree of burning, slaking conditions, etc., and will usually be from 70 to 100 ft^3 of putty per ton of quicklime. The principal use of the putty is in masonry mortars, where it is particularly valuable because of the high degree of plasticity or workability it imparts to the mortar. It is used at times as an admixture in concrete for the purpose of improving workability. It also is used in some localities as finish-coat plaster where full advantage may be taken of its high plasticity.

4-7. Mason's Hydrated Lime.

Hydrated limes are prepared from quicklimes by addition of a limited amount of water. After hydration ceases to evolve heat, the resulting product is a fine, dry powder. It is then classified by air-classification methods to remove undesirable oversize particles and packaged in 50-lb sacks. It is always a factory-made product, whereas quicklime putty is almost always a job-slaked product.

Mason's hydrated limes are those hydrates suitable for use in mortars, base-coat plasters, and concrete. They necessarily follow the composition of the quicklime. High-calcium hydrates are composed primarily of calcium hydroxide. Normal dolomitic hydrates are composed of calcium hydroxide plus magnesium oxide.

Plasticity of mortars made from normal mason's hydrated limes (Type N) is fair. It is better than that attained with most cements but not nearly so high as that of mortars made with an equivalent amount of slaked quicklime putty.

The normal process of hydration of a dolomitic quicklime at atmospheric pressure results in the hydration of the calcium fraction only, leaving the magnesium oxide portion substantially unchanged chemically. When dolomitic quicklime is hydrated under pressure, the magnesium oxide is converted to magnesium hydroxide. This results in the so-called "special" hydrates (Type S), which not only have their magnesia contents substantially completely hydrated but also have a high degree of plasticity immediately on wetting with water. Mortars made from Type S hydrates are more workable than those made from Type N hydrates. In fact, Type S hydrates are nearly as workable as those made from slaked quicklime putties. The user of this type of hydrate may therefore have the convenience of a bagged product and a high degree of workability without having the trouble and possible hazard of slaking quicklime.

4-8. Finishing Hydrated Limes.

Finishing hydrated limes are particularly suitable for use in the finishing coat of plaster. They are characterized by a high degree of whiteness and of plasticity. Practically all finishing hydrated limes are produced in the Toledo district of Ohio from dolomitic limestone. The normal hydrate is composed of calcium hydroxide and magnesium oxide. When first wetted, it is no more plastic than Type N mason's hydrates. It differs from the latter, however, in that, on soaking overnight, the finishing hydrated lime develops a very high degree of plasticity, whereas the mason's hydrate shows relatively little improvement in plasticity on soaking.

Finishing hydrated lime is also made by the pressure method, which forms a Type S hydrate. This material, similar in characteristics (except color) to the Type S mason's hydrate, is also used for finish-coat plaster but does not have quite as satisfactory characteristics for plastering as does Type N finishing hydrate.

4-9. Low-Temperature Gypsum Derivatives.

When gypsum rock ($CaSO_4 \cdot 2H_2O$) is heated to a relatively low temperature, about 130°C, three-fourths of the water of crystallization is driven off. The resulting product is known by various names such as hemihydrate, calcined gypsum, and first-settle stucco. Its common name, however, is **plaster of paris**. It is a fine powder, usually white. When mixed with water, it sets rapidly and attains its strength on drying. While it will set under water, it does not gain strength and ultimately, on continued water exposure, will disintegrate.

Plaster of paris, with set retarded or unretarded, is used as a molding plaster for preparing ornamental plaster objects or as gaging plaster, which is used with finishing hydrated lime to form the smooth white-coat finish on plaster walls. The unretarded plaster of paris is used by manufacturers to make gypsum block, tile, and gypsumboard (wallboard, lath, backerboard, coreboard, etc.).

When plaster of paris is retarded and mixed with fiber such as sisal, it is marketed under the name of hardwall plaster or cement plaster. (The latter name is misleading, since it does not contain any portland cement.) Hardwall plaster, mixed with water and with from 2 to 3 parts of sand by weight, is widely used for base-coat plastering. In some cases wood fiber is used in place of sand, making a "wood-fibered" plaster.

Special effects are obtained by combining hardwall plaster with the correct type of aggregate. With perlite or vermiculite aggregate, a lightweight plaster is obtained.

Gypsum plasters, in general, have a strong set, gain their full strength when dry, do not have abnormal volume changes, and have excellent fire-resistance characteristics. They are not well adapted, however, for use under continued damp conditions or intermittent wet conditions.

4-10. Oxychloride Cements.

Lightly calcined magnesium oxide mixed with a solution of magnesium chloride forms a cement known as magnesium oxychloride cement, or **Sorel cement**. It is particularly useful in making flooring compositions in which it is mixed with colored aggregates. Floors made of oxychloride cement are sparkproof and are more resilient than floors of concrete.

Oxychloride cement has very strong bonding power and may be used with greater quantities of aggregate than are possible with portland cement because of its higher bonding power. It also bonds well with wood and is used in making partition block or tile with wood shavings or sawdust as aggregate. It is moderately resistant to water but should not be used under continually wet conditions.

4-11. Masonry Cements.

Masonry cements, or—as they are sometimes called—mortar cements, are intended to be mixed with sand and used for setting brick, tile, stone, etc. They may be any one of the hydraulic cements already discussed or mixtures of them in any proportion.

Many commercial masonry cements are mixtures of portland cement and pulverized limestone, often containing as much as 50 or 60% limestone. They are sold in bags containing from 70 to 80 lb, each bag nominally containing a cubic foot. Price per bag is commonly less than that of portland cement, but because of the use of the lighter bag, cost per ton is higher than that of portland cement.

Since there is no specification as to chemical content and physical requirements of masonry cement, specifications are quite liberal; some manufacturers vary the composition widely, depending upon competition, weather conditions, or availability of materials. Resulting mortars may vary widely in properties.

AGGREGATES

Aggregate may be defined as an inert material which may be bound together by a matrix into a conglomerated mass.

As such, the term aggregate includes the normal sands, gravels, and crushed stone used for making mortar, plaster, and concrete. It also includes the lightweight materials, such as slag, vermiculite, and pumice, used to prepare lightweight plasters and concretes. It does not include finely divided or pulverized stone or sand sometimes used with cement that is more properly termed an admixture or filler.

As in the case of cements, properties of aggregates are of less importance than properties of mortars or concretes made from aggregates. The most important property of an aggregate is its soundness, or its dimensional stability. Concrete may be unsound if made with a sound aggregate if not properly proportioned, handled, and placed; but no amount of precaution will enable sound concrete to be made with unsound aggregate.

Cleanliness, strength, and gradation are important in any aggregate. For most purposes, an aggregate should be free of clay, silt, organic matter, and salts. Such cleanliness is usually readily obtainable if facilities are available for washing the aggregate; but bank sands and gravels may contain clay and loam unless special precautions are taken. Strength is important but is infrequently measured, reliance being placed to a large extent upon the performance of existing structures as being evidence of adequate strength. Grading, while important, is under the control of the producer, who can usually adjust his operation to meet the grading requirements of any normal specification.

4-12. Coarse Aggregates.

Coarse aggregates comprise that portion of the aggregate predominantly retained on a No. 4 sieve (4.75 mm). Normal coarse aggregate is composed of gravel or crushed stone and for structural concrete is usually required to be fairly uniformly graded from a maximum size of about 2 in down to $\frac{3}{16}$ in. It should, of course, be clean and strong as well as sound.

ASTM Standard C125 defines gravel as (1) coarse aggregate resulting from natural disintegration and abrasion of rock or processing of weakly bound conglomerate, or (2) that portion of an aggregate retained on a No. 4 sieve. The first definition applies to an entire aggregate, in a natural condition or after processing; the second to a portion of an aggregate.

Crushed stone is "the product resulting from the artificial crushing of rocks, boulders or large cobblestones, substantially all faces of which have resulted from the crushing operation." (See ASTM Standard C125.) As a result of its processing, it is angular in contour. It should be roughly cubical in shape for use in concrete but frequently contains flat or needle-shaped particles that may be a source of weakness in the concrete. Crushed stone is usually of a single rock species and is not so heterogeneous as gravel. It may contain unsound impurities of the same characteristics as found in gravel.

Blast-furnace slag is used as a coarse aggregate in some localities where it is economically available. It is lighter in weight than gravel and crushed stone but normally is not considered a lightweight aggregate.

4-13. Fine Aggregates.

Fine aggregate is that portion of aggregate finer than a No. 4 sieve. It is usually graded fairly uniformly from No. 4 to No. 100 in size. Unless otherwise mentioned, fine aggregate is usually considered to be sand, the product of the natural disintegration and abrasion of rock or processing of completely friable sandstone. The fine product obtained from crushing of stone is known as "stone sand," while that from crushing of slag is known as "slag sand." Sand for use in concrete must contain negligible amounts of clay, coal, and lignite, material finer than a No. 220 sieve, shale, alkali, mica, coated grains, soft and flaky particles, and organic impurities.

4-14. Heavyweight Aggregates.

These include magnetite, with a specific gravity δ of 4.3; barite, $\delta = 4.2$; limonite, $\delta = 3.8$; ferrophosphorus, $\delta = 6.3$; and steel shot or punchings, $\delta = 7.6$. Such heavyweight aggregates may be used instead of gravel or crushed stone to produce a dense concrete; for example, for shielding of nuclear reactors.

4-15. Lightweight Aggregates.

Lightweight aggregates include perlite, exfoliated vermiculite, pumice, scoria, tuff, blast furnace slag, fly ash, diatomite, and expanded clay, shale, and slate. When mixed with cement, the resulting concrete also is light in weight and has low thermal conductivity. Strength of the concrete will be almost inversely proportional to the weight of the aggregate, although comparisons are not always easily made because of the difficulty of making concrete mixes that have this type of aggregate as the sole variable.

ASTM Specifications C330, C331, and C332 require that lightweight aggregates for structural and insulating concrete and concrete masonry units be composed predominantly of cellular and granular inorganic material. For structural concrete and masonry, dry loose weight should not exceed 70 lb/ft^3 for fine aggregates, 55 lb/ft^3 for coarse aggregates, and 65 lb/ft^3 for combined fine and coarse aggregates. C332 divides aggregates for insulating concretes into two groups. Group I contains aggregates, such as perlite and vermiculite, prepared by expanding products. Dry loose weight of the perlite should be between 7.5 and 12 lb/ft^3, and of the vermiculite, between 5.5 and 10 lb/ft^3. These aggregates produce concretes weighing from 15 to 50 lb/ft^3, the maximum thermal conductivity of which may be 1.50. Group II contains aggregates, such as pumice, scoria, and tuff, prepared by processing natural materials, and aggregates prepared by expanding, calcining, or sintering products. For Group II, dry loose weight should not exceed 70 lb/ft^3 for fine aggregates, 55 lb/ft^3 for coarse aggregates, and 65 lb/ft^3 for combined fine and coarse aggregates. The aggregates generally produce concretes weighing from 45 to 90 lb/ft^3, the maximum thermal conductivity of which may be 3.00.

MORTARS AND CONCRETES

4-16. Mortars.

Mortars are composed of a cementitious material, fine aggregate, sand, and water. They are used for bedding unit masonry, for plasters and stuccoes, and with the addition of coarse aggregate, for concrete. In this article, consideration will be given primarily to those mortars used for unit masonry and plasters.

Properties of mortars vary greatly, being dependent on the properties of the cementitious material used, ratio of cementitious material to sand, characteristics and grading of the sand, and ratio of water to solids.

Packaging and Proportioning. Mortars are usually proportioned by volume. A common specification is that not more than 3 ft^3 of sand shall be used with 1 ft^3 of cementitious material. Difficulty is sometimes encountered, however, in determining just how much material constitutes a cubic foot: a bag of cement (94 lb) by agreement is called a cubic foot in proportioning mortars or concretes, but an actual cubic foot of lime putty may be used in proportioning mortars. Since hydrated limes are sold in 50-lb bags, each of which makes somewhat more than a cubic foot of putty, weights of 40, 42, and 45 lb of hydrated lime have been used as a cubic foot in laboratory studies; but on the job, a bag is frequently used as a cubic foot. Masonry cements are sold in bags containing 70 to 80 lb, and again a bag is considered a cubic foot.

Workability is an important property of mortars, particularly of those used in conjunction with unit masonry of high absorption. Workability is controlled by the character of the cement and amount of sand. For example, a mortar made from 3 parts sand and 1 part slaked lime putty will be more workable than one made from 2 parts sand and 1 part portland cement. But the 3:1 mortar has lower strength. By proper selection or mixing of cementitious materials, a satisfactory compromise may usually be obtained, producing a mortar of adequate strength and workability.

Water retention—the ratio of flow after one-minute standard suction to the flow before suction—is used as an index of the workability of mortars. A high value of water retention is considered desirable for most purposes. There is, however, a wide variation in water retention of mortars made with varying proportions of cement and lime and with varying limes. ASTM C270 requires mortar mixed to an initial flow of 100 to 115, as determined by the method of ASTM C109, to have a flow after suction of at least 75%.

Strength of mortar is frequently used as a specification requirement, even though it has little relation to the strength of masonry. (See, for example, ASTM Specifications C270 and C476.) The strength of mortar is affected primarily by the amount of cement in the matrix. Other factors of importance are the ratio of sand to cementing material, curing conditions, and age when tested.

Volume change of mortars constitutes another important property. Normal volume change (as distinguished from unsoundness) may be considered as the shrinkage during early hardening, shrinkage on drying, expansion on wetting, and changes due to temperature.

After drying, mortars expand again when wetted. Alternate wetting and drying produces alternate expansion and contraction, which apparently continues indefinitely with portland-cement mortars.

Coefficient of Expansion. The coefficients of thermal expansion of several mortars, reported in "Volume Changes in Brick Masonry Materials," *Journal of Research of the National Bureau of Standards,* Vol. 6, p. 1003, ranged from 0.38×10^{-5} to 0.60×10^{-5} for masonry-cement mortars; from 0.41×10^{-5} to 0.53×10^{-5} for lime mortars, and from 0.42×10^{-5} to 0.61×10^{-5} for cement mortars. Composition of the cementitious material apparently has little effect upon the coefficient of thermal expansion of a mortar.

High-Bond Mortars. When polymeric materials, such as polyvinylidene chloride, are added to mortar, greatly increased bonding, compressive, and shear strength result. To obtain this high strength, the other materials, including sand, water, Type I or III portland cement, and a workability additive, such as pulverized ground limestone or marble dust, must be of quality equal to that of the ingredients of standard mortar. The high strength of the mortar makes it possible to apply appreciable bending and tensile stresses to masonry, resulting in thinner walls and making possible the prelaying of single-wythe panels that can be hoisted into place.

4-17. Normal Concrete.

Concrete is a mixture of portland cement, fine aggregate, coarse aggregate, and water, which is temporarily fluid and capable of being poured, cast, or molded, but which later sets and hardens to form a solid mass. Normal concrete is here considered to be concrete made with sand as fine aggregate and with conventional aggregates, such as gravel or crushed stone, and to which no air-entrainment admixture has been added. It is to be distinguished from air-entrained concrete, which has some air deliberately added, from lightweight concretes made with special lightweight aggregates or with a large air content, from heavyweight concretes made with special heavyweight aggregates, from fiber-reinforced concretes incorporating short steel or glass fibers, and from polymer concretes made or impregnated with polymers.

The ultimate consumer of concrete is interested in three major properties of concrete—adequate strength, adequate durability, and minimum cost. In the attainment of these three objectives, the concrete designer may wish to vary the proportions of ingredients widely. The principal variations that may be imposed are:

1. The water–cement ratio
2. The cement–aggregate ratio
3. The size of coarse aggregate
4. The ratio of fine aggregate to coarse aggregate
5. The type of cement

With varying ingredients, concretes may be made with strengths as low as 1500 psi or more than 8000 psi at 28 days. Durability may vary from almost complete durability under the most severe exposure to none at all unless the concrete is protected from weather. Material costs may vary from very expensive to very cheap.

Workability. This property of fluid concrete is of great importance. It involves the concepts of fluidity, cohesiveness, adhesiveness, plasticity, and probably others. A definition sometimes used is "ability to flow without segregation." Workability is enhanced by high cement factors and high sand-to-coarse-aggregate ratios; but these factors also involve high cost and may involve undesirable, high-volume change. The concrete designer must strike a balance between all factors involved and must frequently allow less workability than he wants in order to obtain concrete that is satisfactory from some other aspect.

Density of fluid concrete made with conventional aggregates will approximate 150 lb/ft³. This value may be used in calculating strength of formwork and density of set concrete.

Setting and hardening of concrete are caused by a chemical reaction between the constituents of the cement and the water. This reaction is accompanied by the evolution of considerable heat. Normally, in structural concrete, this heat is removed by conduction, radiation, or convection about as rapidly as it is generated; but in large masses, the heat cannot be dissipated rapidly and may be sufficient to raise the temperature of the mass by 50 to 60°F. In such cases, Type IV cement may be used to keep the temperature rise as low as possible and thereby prevent delayed cracking.

Strength of concrete is conventionally measured in the United States by testing concrete cylinders 6 in in diameter and 12 in high in compression.

The prime factor affecting strength is the water-cement (W/C) ratio. This is commonly expressed as a ratio by weight but may be expressed as gallons of water per sack of cement. Table 4-3 shows the relation between the W/C ratio and 28-day compressive strength of normal concrete.

The strength is affected by the age of the concrete when tested, as shown in Table 4-4.

Strength of concrete is affected by the cement content, as shown in Table 4-5.

Strength of concrete is diminished by curing at low temperatures, as is shown in Table 4-6.

TABLE 4-3 Relation between W/C* Ratio and Compressive Strength

Net W/C by Weight	28-day Strength, Psi
0.40	5,500
0.42	5,200
0.44	5,000
0.46	4,800
0.49	4,400
0.53	4,000
0.58	3,600
0.62	3,300
0.67	2,900
0.71	2,600
0.75	2,400

* Water-cement.

TABLE 4-4 Relation between Age and Compressive Strength, Psi

Age	Type I cement	Type II cement	Type III cement
3 days	2,100	2,100	3,400
7 days	3,200	3,000	4,400
28 days	4,500	4,500	5,500
90 days	5,000	5,400	6,100
1 year	5,600	5,800	6,700

Other factors have a bearing on the strength of a particular concrete but may usually be compensated for in mix design. Strength of a concrete made with one aggregate may be somewhat less than that made with a different aggregate, but a slight decrease in water-cement ratio of the former concrete may make the two identical. Similarly, variations in strength due to grading of aggregate and to the type or brand of cement may be compensated in the mix design.

Volume Change. This property is of importance equal to that of strength. Volume

TABLE 4-5 Relation between Cement Content and Strength of Concrete

Cement Content, Lb per Cu Yd	Avg 28-day Strength, Psi
350	2,000
400	2,700
450	3,300
500	3,800
550	4,300
600	4,800
650	5,200

TABLE 4-6 Effect of Curing Temperature on Relative Strength of Concrete

Curing Temp, °F	Relative 28-day Strength 70°F = 100
40	78
55	98
70	100
85	101
100	103
115	105

changes that occur while concrete is in a plastic state are normally not of great importance in mass concrete, but in reinforced concrete they may be. The result of settlement, sedimentation, or bleeding may be to cause the concrete to settle away from the underside of the reinforcement, thereby impairing bond of the concrete to the reinforcing steel. This may be avoided by using a workable nonbleeding concrete, obtainable by using less water, finer cement, pozzolanic admixtures, or air entrainment.

Volume changes that occur after hardening are apt to result in cracking, which weakens the concrete and makes it susceptible to early disintegration by attack by frost or by aggressive solutions.

Concrete is subject to expansion and contraction with wetting and drying. The drying shrinkage of concrete is slightly more than ½ in per 100 ft. This, in magnitude, is equivalent to a thermal change of about 83°F.

Coefficient of Expansion. The thermal coefficient varies with the cement content, water content, type of aggregate, and other factors, but averages 0.0000055 (in/in)/°F.

Modulus of Elasticity. The stress-strain curve for concrete is curved over its length. The portion from zero load to about 40% of the ultimate load, however, is sufficiently straight that a secant modulus of elasticity may be determined for this portion. The modulus of elasticity, calculated in this manner, is about 1000 times the compressive strength in pounds per square inch.

Creep. When concrete is subjected to a load, it initially undergoes an elastic deformation, the magnitude of which is proportional to the load and the elasticity of the concrete. If load application continues, however, deformation increases gradually with time. This deformation under load is termed creep or plastic flow. The creep continues for an indefinite time, measurements having been made for periods in excess of 10 years. It proceeds at a continuously diminishing rate and approaches some limiting value. The amount of creep at this limiting value appears to be from 1 to 3 times the amount of the elastic deformation. Usually, more than 50% of the ultimate creep takes place within the first 3 months after loading. Upon unloading, an immediate elastic recovery takes place, followed by a plastic recovery of lesser amount than the creep on first loading.

(ACI 211.1, "Recommended Practice for Selecting Proportions for Normal and Heavyweight Concrete," American Concrete Institute.)

4-18. Air-Entraining Concretes.

All normal concrete contains some air, usually between 0.5 and 1.0%. By the addition of a small amount of an air-entraining agent, the amount of air contained in the concrete may be materially increased to 10% or more (by volume) if desired. This extra air is present in the form of small, discrete air bubbles, which have a marked effect upon the workability, durability, and strength of the concrete. Resistance to repeated cycles of freezing and thawing, in particular, is much greater than that of normal concrete.

Since addition of air decreases strength of concrete, it is necessary to compensate for this decrease by decreasing the water–cement ratio or decreasing unit water content, or both. Even with compensating adjustments of this nature, it is usually impossible to maintain the design strength of the concrete if more than 6% air is present. Consequently, air contents are usually specified to be about 4%, with 6% air as a maximum.

Air-entrained concrete has a lower unit weight than normal concrete. A concrete that would weigh 150 lb/ft^3 without air will weigh only 144 with 4% entrained air.

Addition of air to concrete makes a great improvement in workability. If air is added to a normal concrete of 2-in slump without compensating adjustments, an air-entrained concrete with a slump of 6 to 8 in will result. By removal of some of the water, thereby increasing the water-cement ratio, the slump may be brought back to 2 in, but the resulting air-entrained concrete is still much more workable and more cohesive and has less bleeding and segregation than the normal concrete of equivalent slump. This improvement in workability persists, even if some of the sand is withdrawn, as is common practice.

The other properties of air-entrained concrete, such as rate of development of strength, volume change, and elasticity, appear to be identical with those of normal concrete, if allowance is made for the weakening effect of the air and the additional workability attained through its incorporation.

4-19. Lightweight Concretes.

Appreciable savings in weight and accompanying savings in steel costs and other costs are often obtainable through the use of lightweight concretes. Often, these are concretes wherein the normal heavy aggregates have been replaced with lightweight aggregates (see Art. 4-15).

Lighter concretes, in general, have higher absorptions, lower strengths, lower elastic moduli, lower thermal conductivity, and higher shrinkage than heavier concretes. The designer of lightweight concrete must base his design on specific tests of the particular aggregate under consideration rather than upon average data obtainable from tables and charts.

Other types of lightweight concretes are produced by various methods.

Expanded Concretes. One of these methods is to mix a concrete or mortar with a treated aluminum powder. The reaction of the aluminum with calcium hydroxide liberated by the cement evolves hydrogen, which causes the plastic concrete to expand. The expansion may be as much as 100% of the unexpanded volume. The cement sets at about the time the expansion is completed. Hence the set concrete remains in the expanded form. Concretes weighing from 75 to 100 lb/ft^3 are produced in this manner. The upper surface of the concrete may be irregular if there has been irregularity in distribution of the aluminum powder.

Foamed Concrete. Lightweight concrete can be made by mixing cement and fine filler, such as pulverized stone, silica flour, or fly ash, with a stabilized foam. By proper proportioning of the volumes of foam and solids, concretes weighing as little as 10 lb/ft^3 have been produced. These materials have very low strength but are well adapted for production of insulated and fire-resistant roof slabs and other members where strength is of little importance.

Nailable Concrete. Concrete into which nails can be driven and which will hold the nails firmly is produced by mixing equal volumes of cement, sand, and pine sawdust with enough water to give a slump of 1 to 2 in. This concrete, after moist curing for 2 days and drying for 1 day, has excellent holding power for nails.

(ACI 211.2, "Recommended Practice for Selecting Proportions for Structural Lightweight Concrete," American Concrete Institute.)

4-20. Heavyweight Concretes.

Concretes made with heavyweight aggregates are used for shielding and structural purposes in construction of nuclear reactors and other structures exposed to high-intensity radiation (see Art. 4-14). Heavyweight aggregates are used for making shielding concretes because absorption of such radiation is proportional to density, and, consequently, these aggregates have greater capacity for absorption than those ordinarily used for normal concrete. With such aggregates, concrete weighing up to about 385 lb/ft^3 can be produced.

Concrete made with limonite or magnetite can develop densities of 210 to 224 lb/ft^3 and compressive strengths of 3200 to 5700 psi. With barite, concrete may weigh 230 lb/ft^3 and have a strength of 6000 psi. With steel punchings and sheared bars as coarse aggregate and steel shot as fine aggregate, densities of 250 to 288 lb/ft^3 and strengths of about 5600 psi can be attained. Generally, grading of aggregates and mix proportions are similar to those used for normal concrete.

Mixing and placing operations, however, are more difficult than for normal concrete, because of segregation. To avoid this, heavyweight concretes may be grouted into place. These concretes usually do not have good resistance to weathering or abrasion.

(ACI 211.1, "Recommended Practice for Selecting Proportions for Normal and Heavyweight Concrete," American Concrete Institute.)

4-21. Concrete Masonry Units.

A wide variety of manufactured products are produced from concrete and used in building construction. These include such items as concrete brick, concrete block or tile, concrete floor and roof slabs, precast wall panels, precast beams, and cast stone. These items are made both from normal dense concrete mixes and from mixes with lightweight aggregates. Concrete blocks are made with holes through them to reduce their weight and to enable masons to grip them.

Nominal size (actual dimensions plus width of mortar joint) of hollow concrete block usually is 8 × 8 × 16 in. Solid blocks often are available with nominal size of 4 × 8 × 16 in or 4 × 2½ × 8 in.

Properties of the units vary tremendously—from strong, dense load-bearing units used under exposed conditions to light, relatively weak, insulating units used for roof and fire-resistant construction.

Many types of concrete units have not been covered by adequate standard specifications. For these units, reliance must be placed upon the manufacturer's specifications. Requirements for strength and absorption of concrete brick and block established by the American Society for Testing and Materials for Type I (moisture-controlled) and Type II (non-moisture-controlled) units are summarized in Table 4-7.

Manufactured concrete units have the advantage (or sometimes disadvantage) that curing is under the control of the manufacturer. Many methods of curing are used, from simply stacking the units in a more or less exposed location to curing under high-pressure

TABLE 4-7 Summary of ASTM Specification Requirements for Concrete Masonry Units

	Compressive strength, min, psi		Moisture content for Type I units, max, % of total absorption (average of 5 units)			Moisture absorption, max, lb/ft³ (average of 5 units)		
			Avg annual relative humidity, %			Oven-dry weight of concrete, lb/ft³		
	Avg of 5 units	Individual min	Over 75	75 to 50	Under 50	125 or more	105 to 125	Under 105
Concrete building brick, ASTM C55:								
N-I, N-II (high strength severe exposures)	3500	3000				10	13	15
S-I, S-II (general use, moderate exposures)	2500	2000				13	15	18
Linear shrinkage, %:								
0.03 or less			45	40	35			
0.03 to 0.45			40	35	30			
Over 0.045			35	30	25			
Solid, load-bearing units, ASTM C145:								
N-I, N-II (unprotected exterior walls below grade or above grade exposed to frost)	1800	1500				13	15	18
S-I, S-II (protected exterior walls below grade or above grade exposed to frost)	1200	1000						20*
Linear shrinkage, %: (Same as for brick)								
Hollow, load-bearing units, ASTM C90:								
N-I, N-II (general use)	1000	800				13	15	18
S-I, S-II (above grade, weather protected)	700	600						20*
Linear shrinkage, %: (Same as for brick)								
Hollow, non-load-bearing units, ASTM C129	600	500						
Linear shrinkage, %: (Same as for brick)								

*For units weighing less than 85 lb/ft³.

steam. The latter method appears to have considerable merit in reducing ultimate shrinkage of the block. Shrinkage may be as small as ¼ to ⅜ in per 100 ft for concrete units cured with high-pressure steam ("Shrinkage and Curing in Masonry Units," *Journal of the American Concrete Institute,* Vol. 23, p. 393). These values are about one-half as great as those obtained with normal atmospheric curing. Tests for moisture movement in blocks cured with high-pressure and high-temperature steam indicated expansions of from ¼ to ½ in per 100 ft after saturation of previously dried specimens.

BURNED-CLAY UNITS

Use of burned-clay structural units dates from prehistoric times. Hence durability of well-burned units has been adequately established through centuries of exposure in all types of climate.

Modern burned-clay units are made in a wide variety of sizes, shapes, colors, and textures to suit the requirements of modern architecture. They include such widely diverse units as common and face brick; hollow clay tile in numerous shapes, sizes, and designs for special purposes; ceramic tile for decorative and sanitary finishes, and architectural terra cotta for ornamentation.

Properties of burned-clay units vary with the type of clay or shale used as raw material, method of fabrication of the units, and temperature of burning. As a consequence, some units, such as salmon brick, are underburned, highly porous, and of poor strength. But others are almost glass hard, have been pressed and burned to almost eliminate porosity, and are very strong. Between these extremes lie most of the units used for construction.

4-22. Brick.

Brick have been made in a wide range of sizes and shapes, from the old Greek brick, which was practically a 23-in cube of 12,650 in^3 volume, to the small Belgian brick, approximately 1¾ × 3⅝ × 4½ in with a total volume of only 27 in.^{3} The present common nominal sizes in the United States are 4 or 6 in thick by 2⅔ or 4 in high by 8 or 12 in long. Actual dimensions are smaller, usually by the amount of the width of the mortar joint. Current specification requirements for strength and absorption of building brick are given in Table 4-8 (see ASTM Specifications C62 and C216). Strength and absorption of brick from different producers vary widely.

TABLE 4-8 Physical Requirements for Clay or Shale Solid Brick

Grade	Compressive strength, flat, min, psi		Water absorption, 5-hr boil, max—%		Saturation* coefficient, max—%	
	Avg of 5	Individual	Avg of 5	Individual	Avg of 5	Individual
SW—Severe weathering	3,000	2,500	17.0	20.0	0.78	0.80
MW—Moderate weathering	2,500	2,200	22.0	25.0	0.88	0.90
NW—No exposure	1,500	1,250	No limit	No limit	No limit	No limit

* Ratio of 24-hr cold absorption to 5-hr boil absorption.

Thermal expansion of brick reported in "Thermal Expansion of Clay Building Brick," *Journal of Research of the National Bureau of Standards,* Vol. 27, p. 197, ranged from 0.0000017 per °F for a sample of fire-clay brick to 0.0000069 per °F for a surface-clay brick. The average coefficient for 80 specimens of surface-clay brick was 0.0000033; that for 41 specimens of shale brick was 0.0000034. A variation of from 0.0000036 to 0.0000085 was reported in "Volume Changes in Brick Masonry Materials," *Journal of Research of the National Bureau of Standards,* Vol. 6, p. 1003. Wetting tests of brick indicated expansions varying from 0.0005 to 0.025%.

The thermal conductivity of dry brick as measured by several investigators ranges from 1.29 to 3.79 Btu/(hr) (ft^2) (°F) (in). The values are increased by wetting.

4-23. Structural Clay Tile.

Structural clay tiles are hollow burned-clay masonry units with parallel cells. Such units have a multitude of uses: as a facing tile for interior and exterior unplastered walls, partitions, or columns; as load-bearing tile in masonry constructions designed to carry superimposed loads; as partition tile for interior partitions carrying no superimposed load; as fireproofing tile for protection of structural members against fire; as furring tile for lining the inside of exterior walls; as floor tile in floor and roof construction; and as header tiles, which are designed to provide recesses for header units in brick or stone-faced walls. Units are available with the following ranges in nominal dimensions: 8 to 16 in in length, 4 in for facing tile to 12 in for load-bearing tile in height, and 2 in for facing tile to 12 in for load-bearing tile in thickness.

Two general types of tile are available—side-construction tile, designed to receive its principal stress at right angles to the axis of the cells, and end-construction tile designed to receive its principal stress parallel to the axis of the cells.

Tiles are also available in a number of surface finishes, such as opaque glazed tile, clear ceramic-glazed tile, nonlustrous glazed tile, and scored, combed, or roughened finishes designed to receive mortar, plaster, or stucco.

Requirements of the appropriate ASTM specifications for absorption and strength of several types of tile are given in Table 4-9 (see ASTM Specifications C34, C56, C57, C212, and C126 for details pertaining to size, color, texture, defects, etc.). Strength and absorption of tile made from similar clays but from different sources and manufacturers vary widely.

The modulus of elasticity of tile may range from 1,620,000 to 6,059,000 psi ("Strength, Absorption and Freezing Resistance of Hollow Building Tile," *Journal of the American Ceramic Society,* Vol. 7, No. 3, p. 189).

4-24. Ceramic Tiles.

Ceramic tile is a burned-clay product used primarily for decorative and sanitary effects. It is composed of a clay body on which is superimposed a decorative glaze.

The tiles are usually flat but vary in size from about ½ in square to more than 6 in. Their shape is also widely variable—squares, rectangles, and hexagons are the predominating forms, to which must be added coved moldings and other decorative forms. These tiles are not dependent on the color of the clay for their final color, since they are usually glazed. Hence, they are available in a complete color gradation from pure whites through pastels of varying hue to deep solid colors and jet blacks.

Properties of the base vary somewhat. In particular, absorption ranges from almost zero to about 15%. The glaze is required to be impervious to liquids and should not stain, crack, or craze.

Ceramic tiles are applied on a solid backing by means of a mortar or adhesive. They are usually applied with the thinnest possible mortar joint; consequently accuracy of dimensions is of greatest importance. Since color, size, and shape of tile are important, selection of tile should be based on the current literature of the manufacturer.

4-25. Architectural Terra Cotta.

The term "terra cotta" has been applied for centuries to decorative molded-clay objects whose properties are similar to brick. The molded shapes are fired in a manner similar to brick.

Terra cotta is frequently glazed to produce a desired color or finish. This introduces the problem of cracking or crazing of the glaze, particularly over large areas.

Structural properties of terra cotta are similar to those of brick.

TABLE 4-9 Physical Requirement Specification for Structural Clay Tile

	Absorption, % (1 hr boiling)		Compressive strength, psi (based on gross area)			
			End-construction tile		Side-construction tile	
Type and grade	Avg of 5 tests	Individual max	Min avg of 5 tests	Individual min	Min avg of 5 tests	Individual min
Load-bearing (ASTM C34):						
LBX	16	19	1400	1000	700	500
LB	25	28	1000	700	700	500
Non-load-bearing (ASTM C56):						
NB		28				
Floor tile (ASTM C57):						
FT 1		25	3200	2250	1600	1100
FT 2		25	2000	1400	1200	850
Facing tile (ASTM C212):						
FTX	9 (max)	11				
FTS	16 (max)	19				
Standard			1400	1000	700	500
Special duty			2500	2000	1200	1000
Glazed units (ASTM C126)			3000	2500	2000	1500

LBX. Tile suitable for general use in masonry construction and adapted for use in masonry exposed to weathering. They may also be considered suitable for direct application of stucco.

LB. Tile suitable for general use in masonry where not exposed to frost action, or in exposed masonry where protected with a facing of 3 in or more of stone, brick, terra cotta, or other masonry.

NB. Non-load-bearing tile made from surface clay, shale, or fired clay.

FT 1 and FT 2. Tile suitable for use in flat or segmental panels or in combination tile and concrete ribbed-slab construction.

FTX. Smooth-face tile suitable for general use in exposed exterior and interior masonry walls and partitions, and adapted for use where tiles low in absorption, easily cleaned, and resistant to staining are required and where a high degree of mechanical perfection, narrow color range, and minimum variation in face dimensions are required.

FTS. Smooth or rough-texture face tile suitable for general use in exposed exterior and interior masonry walls and partitions and adapted for use where tile of moderate absorption, moderate variation in face dimensions, and medium color range may be used, and where minor defects in surface finish, including small handling chips, are not objectionable.

Standard. Tile suitable for general use in exterior or interior masonry walls and partitions.

Special duty. Tile suitable for general use in exterior or interior masonry walls and partitions and designed to have superior resistance to impact and moisture transmission, and to support greater lateral and compressive loads than standard tile construction.

Glazed units. Ceramic-glazed structural clay tile with a glossy or satin-mat finish of either an opaque or clear glaze, produced by the application of a coating prior to firing and subsequently made vitreous by firing.

BUILDING STONES

Principal building stones generally used in the United States are limestones, marbles, granites, and sandstones. Other stones such as serpentine and quartzite are used locally but to a much lesser extent. Stone, in general, makes an excellent building material, if properly selected on the basis of experience; but the cost may be relatively high.

Properties of stone depend on what nature has provided. Therefore, the designer does not have the choice of properties and color available in some of the manufactured building units. The most the stone producer can do for purchasers is to avoid quarrying certain stone beds that have been proved by experience to have poor strength or poor durability.

4-26. Strength of Stone.

Data on the strength of building stones are presented in Table 4-10, summarized from *U.S. National Bureau of Standards Technical Papers,* No. 123, B. S. Vol. 12; No. 305, Vol. 20, p. 191; No. 349, Vol. 21, p. 497; *Journal of Research of the National Bureau of Standards,* Vol. 11, p. 635; Vol. 25, p. 161). The data in Table 4-10 pertain to dried specimens. Strength of saturated specimens may be either greater or less than that of completely dry specimens. In a comparison of 90 specimens of marble tested in compression, both dry and saturated, it was found that the strength of 69 saturated marbles was less than that of the same marbles when dry by amounts ranging from 0.3 to 46.0% and averaging 9.7%. On the other hand, 21 marbles showed an increase in strength on soaking; the increase ranged from 0.04 to 35.4% and averaged 7.6%.

The modulus of rupture of dry slate is given in Table 4-10 as ranging from 6000 to 15,000 psi. Similar slates, tested wet, gave moduli ranging from 4700 to 12,300 psi. The ratio of wet modulus to dry modulus varied from 0.42 to 1.12 and averaged 0.73.

4-27. Specific Gravity, Porosity, and Absorption of Stone.

Data on the true specific gravity, bulk specific gravity, unit weights, porosity, and absorption of various stones have been obtained by Kessler and his associates. Some of these data are given in Table 4-11.

4-28. Permeablilty of Stone.

Permeability of stone varies with type of stone, thickness, and driving pressure that forces water through the stone. Table 4-12 presents data for the more common stones at three different pressures, as reported in "Permeability of Stone," *U.S. National Bureau of Standards Technical Papers,* No. 305, Vol. 20, p. 191. The units of measurement of permeability are cubic inches of water that will flow through a square foot of a specimen ½ in thick in 1 hr.

4-29. Thermal Expansion of Stone.

Data on thermal expansion of building stones as given in Table 4-13 show that limestones have a wide range of expansion as compared with granites and slates.

Marble loses strength after repeated heating and cooling. A marble that had an original strength of 9174 psi had a strength after 50 heatings to 150°C of 8998 psi—a loss of 1.9%. After 100 heatings to 150°C, the strength was only 8507 psi, or a loss of 7.3%. The latter loss in strength was identical with that obtained on freezing and thawing the same marble for 30 cycles. Also, marble retains a permanent expansion after repeated heating ("Thermal Expansion of Rock at High Temperatures," *Transactions of the Royal Society of Canada,* Vol. 4, sec. 3, p. 19). For specimens 20 cm in length, total expansions of over 2 mm and permanent expansions of over 1 mm were measured.

4-30. Freezing and Thawing of Stone.

In freezing and thawing tests of 89 different marbles ("Physical and Chemical Tests of Commercial Marbles of U.S.," *U.S. National Bureau of Standards Technical Papers,* No.

TABLE 4-10 Strength Characteristics of Commercial Building Stones

Stone	Compressive strength, psi, range	Modulus of rupture, psi, range	Shear strength, psi, range	Tensile strength, psi, range	Elastic modulus, psi, range	Toughness Range	Toughness Avg	Wear resistance Range	Wear resistance Avg
Granite	7,700–60,000	1,430–5,190	2,000–4,800	600–1,000	5,700,000–8,200,000	8–27	13	43.9–87.9	60.8
Marble	8,000–50,000	600–4,900	1,300–6,500	150–2,300	7,200,000–14,500,000	2–23	6	6.7–41.7	18.9
Limestone	2,600–28,000	500–2,000	800–4,580	280–890	1,500,000–12,400,000	5–20	7	1.3–24.1	8.4
Sandstone	5,000–20,000	700–2,300	300–3,000	280–500	1,900,000–7,700,000	2–35	10	1.6–29.0	13.3
Quartzite	16,000–45,000					5–30	15		
Serpentine	11,000–28,000	1,300–11,000		800–1,600	4,800,000–9,600,000			13.3–111.4	46.9
Basalt	28,000–67,000					5–40	20		
Diorite	16,000–35,000					6–38	23		
Syenite	14,000–28,000								
Slate		6,000–15,000	2,000–3,600	3,000–4,300	9,800,000–18,000,000	10–56		5.6–11.7	7.7
Diabase						6–50	19		
Building limestone						3–8	4.4		

TABLE 4-11 Specific Gravity and Porosity of Commercial Building Stones

Stone	Specific gravity		Unit weight, lb per cu ft	Porosity, %	Absorption, %	
	True	Bulk			By weight	By volume
Granite.....	2.599–3.080	2.60–3.04	157–187	0.4–3.8	0.02–0.58	0.4–1.8
Marble.....	2.718–2.879	2.64–2.86	165–179	0.4–2.1	0.01–0.45	0.04–1.2
Limestone...	2.700–2.860	1.87–2.69	117–175	1.1–31.0	....	6–15
Slate	2.771–2.898	2.74–2.89	168–180	0.1–1.7	0.00–1.63	0.3–2.0
Basalt......	....	2.9–3.2				
Soapstone...	....	2.8–3.0				
Gneiss	....	2.7–3.0				
Serpentine...	....	2.5–2.8	158–183			
Sandstone...	....	2.2–2.7	119–168	1.9–27.3	....	6–18
Quartzite....	....	....	165–170	1.5–2.9		

TABLE 4-12 Permeability of Commercial Building Stones

[in^3/(ft^2)(hr) for 1/2-in thickness]

Pressure, psi	1.2	50	100
Granite	0.06–0.08	0.11	0.28
Slate	0.006–0.008	0.08–0.11	0.11
Marble	0.06–0.35	1.3–16.8	0.9–28.0
Limestone	0.36–2.24	4.2–44.8	9.0–109
Sandstone	4.2–174.0	51.2	221

TABLE 4-13 Coefficient of Thermal Expansion of Commercial Building Stones

Stone	Range of Coefficient
Limestone	$(4.2–22) \times 10^{-6}$
Marble	$(3.6–16) \times 10^{-6}$
Sandstone	$(5.0–12) \times 10^{-6}$
Slate	$(9.4–12) \times 10^{-6}$
Granite..........	$(6.3–9) \times 10^{-6}$

123, Vol. 12), after 30 cycles, 66 marbles showed loss of strength ranging from 1.2 to 62.1% and averaging 12.3% loss. The other 23 marbles showed increases in strength ranging from 0.5 to 43.9% and averaging 11.2% increase.

Weight change was also determined in this investigation to afford another index of durability. Of 86 possible comparisons after 30 cycles of freezing and thawing, 16 showed no change in weight, 64 showed decreases in weight ranging from 0.01 to 0.28% and averaging 0.04% loss, while 6 showed increases in weight ranging from 0.01 to 0.08% and averaging 0.04%.

GYPSUM PRODUCTS

4-31. Gypsumboard.

This product consists of a core of set gypsum surfaced with specially manufactured paper firmly bonded to the core. It is designed to be used without addition of plaster for walls, ceilings, or partitions and provides a surface suitable to receive either paint or paper (see also Sec. 12). Gypsumboard is extensively used in "dry-wall" construction, where plaster is eliminated. It is also available with one surface covered with aluminum or other heat-reflecting type of foil, or with imitation wood-grain or other patterns on the exposed surface so that no additional decoration is required.

4-32. Gypsum Lath.

Gypsum lath is similar to gypsumboard in that it consists of a core of set gypsum surfaced with paper. The paper for gypsumboard, however, is produced so that it is ready to receive paint or paper, while that for gypsum lath is specially designed or treated so that plaster will bond tightly to the paper. In addition, some lath provides perforations or other mechanical key devices to assist in holding the plaster firmly on the lath. It is also available with reflective foil backing (see also Sec. 12).

Veneer plasters, special proprietary compositions for thin plaster surfaces, are best applied over veneer plaster base, similar to gypsum lath, but produced to accommodate the veneer plaster compositions. Both gypsum lath and veneer base are made as regular, X-rated (fire-retardant), and insulating (foil-backed) types.

4-33. Gypsum Sheathing Board.

Gypsum sheathing boards are similar in construction to gypsumboard (Art. 4-31), except that they are provided with a water-repellent paper surface. They are commonly made in only ½-in thickness, and with a nominal width of 24 or 48 in. They are made with either square edges or with V tongue-and-groove edges. Sheathing boards also are available with a water-repellent core.

4-34. Gypsum Partition Tile or Block.

Gypsum tiles or blocks are used for non-load-bearing partition walls and for protection of columns, elevator shafts, etc., against fire. They have been essentially replaced by dry-wall systems.

4-35. Gypsum Plank.

A precast gypsum product used particularly for roof construction is composed of a core of gypsum cast in the form of a plank, with wire-fabric reinforcement and usually with tongue-and-groove metal edges and ends. The planks are available in two thicknesses—a 2-in plank, which is 15 in wide and 10 ft long, and a 3-in plank which is 12 in wide and 30 in long. (See ASTM C377.)

GLASS AND GLASS BLOCK

4-36. Glass for General Use.

Glass is so widely used for decorative and utilitarian purposes in modern construction that it would require an encyclopedia to list all the varieties available. Clear glass for windows and doors is made in varying thicknesses or strengths, also in double layers to obtain additional thermal insulation. Safety glass, laminated from sheets of glass and plastic, or made with embedded wire reinforcement, is available for locations where breakage might be hazardous. For ornamental work, glass is available in a wide range of textures, colors, finishes, and shapes. See also Art. 13-13.

For many of its uses, glass presents no problem to the materials engineer. It is used in relatively small sheets or sections, which are frequently free-floating in their supports; therefore, expansion is not a problem. Most kinds of glass are highly resistant to weathering so that durability of the glass itself is not a problem, although the weathertightness of the frame may be a serious problem. Much glass is used in interiors for esthetic reasons where it is not subject to the stresses imposed upon many of the other masonry materials. An exception to this general condition is found in the case of glass block.

4-37. Glass Block.

Glass blocks are made by first pressing or shaping half blocks to the desired form, then fusing the half blocks to form a complete block. A block is usually 3⅞ in thick and 5¾, 7¾, or 11¾ in square. The center of the block is hollow and is under a partial vacuum, which adds to the insulating properties of the block. Corner and radial blocks are also available to produce desired architectural effects.

Glass block is commonly laid up in a cement or a cement-lime mortar. Since there is no absorption by the block to facilitate bond of mortar, various devices are employed to obtain a mechanical bond. One such device is to coat the sides of the block with a plastic and embed therein particles of sand. The difficulty in obtaining permanent and complete bond sometimes leads to the opening up of mortar joints; a wall of glass block, exposed to the weather, may leak badly in a rainstorm unless unusual precautions have been taken during the setting of the block to obtain full and complete bond.

Glass blocks have a coefficient of thermal expansion that is from 1½ to 2 times that of other masonry. For this reason, large areas of block may expand against solid masonry and develop sufficient stress so that the block will crack. Manufacturers usually recommend an expansion joint every 10 ft or so, to prevent building up of pressure sufficient to crack the block. With adequate protection against expansion and with good workmanship, or with walls built in protected locations, glass-block walls are ornamental, sanitary, excellent light transmitters, and have rather low thermal conductivity.

ASBESTOS-CEMENT PRODUCTS

Asbestos-cement products are formed from a mixture of portland cement and asbestos fiber that has been wetted and then pressed into a board or sheet form. Organic fiber is added in some cases to promote resiliency and ease of machining. Curing agents, water-repellent admixtures, pigments, mineral granules, and mineral fillers may or may not be added, depending upon the particular end use of the product and its desired appearance.

4-38. Properties of Asbestos-Cement.

Since these materials are used in sheet form and frequently under exposed weather conditions, the properties of most importance are flexural strength, deflection under load while wet, and absorption. The requirements of ASTM Specifications C220, C221, C222, and C223 for asbestos-cement products are given in Table 4-14.

Type U, or general-utility flat sheets, are suitable for exterior and interior use, having sufficient strength for general utility and construction purposes and suitable where high density, glossy finish, and lowest absorption are not essential. Sheets of this type are easy to cut and work, are bendable, and need not be drilled for nailing when thickness is ¼ in or less (ASTM Specification C220).

Type F, or flexible, high-strength flat sheets, are suitable for exterior and interior use where a combination of strength and density, flexibility, smooth surface, and low absorp-

TABLE 4-14 Specification Requirements for Asbestos-Cement Products

	Thickness, in	Size of test specimen, in	Length of test span, in	Average breaking load, lb	Average deflection, in at maximum load		Max. water absorption, % by weight
					Parallel with length	Parallel with width	
Flat sheets							
Type F	1/8	6 × 12	10	20	0.45	0.35	25
	3/16	6 × 12	10	50	0.30	0.20	25
	1/4	6 × 12	10	90	0.20	0.15	25
	3/8	6 × 12	10	190	0.12	0.10	25
	1/2	6 × 12	10	360	0.08	0.06	25
	5/8	6 × 12	10	560	0.05	0.03	25
Type U	3/16	6 × 12	10	35	0.20	0.15	30
	1/4	6 × 12	10	65	0.15	0.10	30
	3/8	6 × 12	10	145	0.10	0.08	30
	1/2	6 × 12	10	260	0.07	0.05	30
	5/8	6 × 12	10	400	0.04	0.02	30
Corrugated sheets							
Type A		12.6 × 36	30	125*			
Type B		12.6 × 36	30	315*			
Type C		12.6 × 36	30	675*			
Type D		12.6 × 36	30	1050*			
Type E		12.6 × 36	30	1740*			
Roofing shingles and siding							
3/16 in	0.15 min						
Strong direction		6 × 12	10	27 min†	0.15 min		
Weak direction		6 × 12	10	22 min†	0.15 min		
1/4 in	0.255 min						
Strong direction		6 × 12	10	60 min‡	0.10 min		
Weak direction		6 × 12	10	50 min‡	0.10 min		
3/8 in	0.345						
Strong direction		6 × 12	10	140 min§	0.065 min		
Weak direction		6 × 12	10	115 min§	0.065 min		

*Bending moment, ft-lb/ft, for centrally applied load.
†Subtract 2 lb for siding.
‡Subtract 5 lb for siding.
§Subtract 10 lb for siding.

tion are desired. Sheets of this type are readily cuttable and workable and need not be drilled for fastening when thickness is ¼ in or less (ASTM C220).

Corrugated sheets are designed to provide weather-exposed surfaces for roofs, walls, and other elements of buildings and other structures, and for decorative and other purposes (ASTM C221). Corrugated sheets are available as Types A and B, generally used for roofs and walls, and Types C, D, and E, generally used for bulkheading applications.

Roofing shingles are designed to provide weather-exposed surfaces on roofs of buildings (ASTM C222). They are available in a number of shapes and patterns.

Siding shingles and clapboards are designed to provide weather-exposed sidewall surfaces of buildings (ASTM C223).

WOOD

Wood is a building material made from trees. It is a natural polymer composed of cells in the shape of long, thin tubes with tapered ends. The cell wall consists of cellulose crystals, which are bonded together by a complex amorphous lignin composed of carbohydrates. Most of the cells in a tree trunk are oriented vertically. Consequently, properties of wood in the direction of cell axes, usually referred to as longitudinal, or parallel to grain, differ from those in the other (radial or circumferential) directions, or across the grain.

4-39. Mechanical Properties of Wood.

Because of its structure, wood has different strength properties parallel and perpendicular to the grain. Tensile, bending, and compressive strengths are greatest parallel to the grain and least across the grain, whereas shear strength is least parallel to the grain and greatest across the grain. Except in plywood, the shearing strength of wood is usually governed by the parallel-to-grain direction.

The compressive strength of wood at an angle other than parallel or perpendicular to the grain is given by the following formula:

$$C_\theta = \frac{C_1 C_2}{C_1 \sin^2 \theta + C_2 \cos^2 \theta} \qquad (4\text{-}1)$$

in which C_θ is the strength at the desired angle θ with the grain, C_1 is the compressive strength parallel to grain, and C_2 is the compressive strength perpendicular to the grain.

Increasing moisture content reduces all strength properties except impact bending, in which green wood is stronger than dry wood. The differences are brought out in Table 4-15. In practice, no differentiation is made between the strength of green and dry wood in engineering timbers, because of seasoning defects that may occur in timbers as they dry and because large timbers normally are put into service without having been dried. This is not true of laminated timber, in which dry wood must be employed to obtain good glued joints. For laminated timber, higher stresses can be employed than for ordinary lumber. In general, compression and bending parallel to the grain are affected most severely by moisture, whereas modulus of elasticity, shear, and tensile strength are affected less. In practice, tensile strength parallel to the grain is taken equal to the bending strength of wood.

In Table 4-15 are summarized also the principal mechanical properties of the most important American commercial species.

Values given in the table are average ultimate strengths. To obtain working stresses from these, the following must be considered: (1) Individual pieces may vary 25% above and below the average. (2) Values given are for standard tests that are completed in a few minutes. Over a period of years, however, wood may fail under a continuous load about $\tfrac{9}{16}$ that sustained in a standard test. (3) The modulus of rupture of a standard 2-in-deep flexural-test specimen is greater than that of a deep beam. In deriving working stresses, therefore, variability, probable duration of load, and size are considered, and reduction factors are applied to the average ultimate strengths to provide basic stresses, or working stresses, for blemishless lumber. These stresses are still further reduced to account for such blemishes as knots, wane, slope of grain, shakes, and checks, to provide working stresses for classes of commercial engineering timbers. (See Sec. 7 for engineering design in timber.)

**TABLE 4-15 Strength of Some Commercially Important Woods Grown in the United States*
(Results of Tests on Small, Clear Specimens†)**

Commercial name of species	Specific gravity	Modulus of Rupture, ksi	Modulus of Elasticity, ksi	Compression parallel to grain, maximum crushing strength, psi	Compression perpendicular to grain, fiber stress at proportional limit, psi	Shear parallel to grain, maximum shearing strength, psi	Side hardness, load perpendicular to grain, lb
Ash, white	0.55	9.6	1440	3,990	670	1380	960
	0.60	15.4	1740	7,410	1160	1950	1320
Beech, American	0.56	8.6	1380	3,550	540	1290	850
	0.64	14.9	1720	7,300	1010	2010	1300
Birch, yellow	0.55	8.3	1500	3,380	430	1110	780
	0.62	16.6	2010	8,170	970	1880	1260
Cedar, western red	0.31	5.2	940	2,770	240	770	260
	0.32	7.5	1110	4,560	460	990	350
Chestnut, American	0.40	5.6	930	2,470	310	800	420
	0.43	8.6	1230	5,230	620	1080	540
Cypress, bald	0.42	6.6	1180	3,580	400	810	390
	0.46	10.6	1440	6,360	730	1000	510
Douglas fir, coast	0.45	7.7	1560	3,780	380	900	500
	0.48	12.4	1950	7,240	800	1130	710
Douglas fir, interior, west	0.46	7.7	1510	3,870	420	940	510
	0.50	12.6	1820	7,440	760	1290	660
Elm, American	0.46	7.2	1110	2,910	360	1000	620
	0.50	11.8	1340	5,520	690	1510	830
Hemlock, eastern	0.38	6.4	1070	3,080	360	850	400
	0.40	8.9	1200	5,410	650	1060	500
Hemlock, western	0.42	6.6	1310	3,360	280	860	410
	0.45	11.3	1640	7,110	550	1250	540

Species							
Hickory, pecan	0.60	9.8	1370	3,990	780	1480	1310
	0.66	13.7	1730	7,850	1720	2080	1820
Locust, black	0.66	13.8	1850	6,800	1160	1760	1570
	0.69	19.4	2050	10,180	1830	2480	1700
Larch, western	0.48	4.9	960	3,760	400	870	510
	0.52	13.1	1870	7,640	930	1360	830
Maple, sugar	0.56	9.4	1550	4,020	640	1460	970
	0.63	15.8	1830	7,830	1470	2330	1450
Oak, northern red	0.56	8.3	1350	3,440	610	1210	1000
	0.63	14.3	1820	6,760	1010	1780	1290
Oak, white	0.60	8.3	1250	3,560	670	1250	1060
	0.68	15.2	1780	7,440	1070	2000	1360
Pine, shortleaf	0.47	7.4	1390	3,530	350	910	440
	0.51	13.1	1750	7,270	820	1390	690
Pine, longleaf	0.54	8.5	1590	4,320	480	1040	590
	0.59	14.5	1980	8,470	960	1510	870
Pine, sugar	0.34	4.9	1030	2,460	210	720	270
	0.36	8.2	1190	4,460	500	1130	380
Pine, western white	0.35	4.7	1190	2,430	190	680	260
	0.38	9.7	1460	5,040	470	1040	420
Yellow poplar	0.38	6.0	1220	2,660	270	790	440
	0.42	10.1	1580	5,540	500	1190	540
Redwood, old growth	0.38	7.5	1180	4,200	420	800	410
	0.40	10.0	1340	6,150	700	940	480
Spruce, white	0.37	5.6	1070	2,570	240	690	320
	0.40	9.8	1340	5,470	460	1080	480
Tupelo, black	0.46	7.0	1030	3,040	480	1100	640
	0.50	9.6	1200	5,520	930	1340	810

*From U.S. Forest Products Laboratory.
†Values in first line are for green material. Values in second line are adjusted to 12% moisture content.

TABLE 4-16 Shrinkage of Commercially Important Woods Grown in the United States

Species	Shrinkage (% of dimension when green) when oven-dried from green to 0% moisture (test values)		
	Radial	Tangential	Volumetric
Ash, commercial white	4.9	7.8	13.3
Basswood	6.6	9.3	15.8
Beech	5.5	11.9	17.2
Birch, yellow	7.3	9.5	16.8
Cedar:			
Eastern red	3.1	4.7	7.8
Northern white	2.2	4.9	7.2
Port Orford	4.6	6.9	10.1
Western red	2.4	5.0	6.8
Cherry, black	3.7	7.1	11.5
Chestnut	3.4	6.7	11.6
Douglas fir:			
Coast region	4.8	7.6	12.4
Interior, north	3.8	6.9	10.7
Interior, west	4.8	7.5	11.8
Elm:			
American	4.2	7.2	14.6
Rock	4.8	8.1	14.9
Hemlock:			
Eastern	3.0	6.8	9.7
Western	4.2	7.8	12.4
Hickory:			
Pecan	4.9	8.9	13.6
Shag bark	7.0	10.5	16.7
Larch, western	4.5	9.1	14.0
Maple:			
Red	4.0	8.2	12.6
Sugar	4.8	9.9	14.7
Oak:			
Red, northern	4.0	8.6	13.7
White	5.6	10.5	16.3
Pine:			
Loblolly	4.8	7.4	12.3
Longleaf	5.1	7.5	12.2
Eastern white	2.1	6.1	8.2
Ponderosa	3.9	6.2	9.7
Shortleaf	4.6	7.7	12.3
Sugar	2.9	5.6	7.9
Western white	4.1	7.4	11.8
Poplar, yellow	4.6	8.2	12.7
Redwood, old growth	2.6	4.4	6.8
Spruce:			
Black	4.1	6.8	11.3
Engelmann	3.8	7.1	11.0
Tupelo:			
Black	5.1	8.7	14.4
Water	4.2	7.6	12.5
Walnut, black	5.5	7.8	14.4

*From U.S. Forest Products Laboratory.

4-40. Hygroscopic Properties of Wood.

Because of its nature, wood tends to absorb moisture from the air when the relative humidity is high, and to lose it when the relative humidity is low. Moisture imbibed into the cell walls causes the wood to shrink and swell as the moisture content changes with the relative humidity of the surrounding air. The maximum amount of imbibed moisture the cell walls can hold is known as the fiber-saturation point, and for most species is in the vicinity of 25 to 30% of the oven-dry weight of the wood. Free water held in the cell cavities above the fiber-saturation point has no effect upon shrinkage or other properties of the wood. Changes in moisture content below the fiber-saturation point cause negligible shrinkage or swelling along the grain, and such shrinkage and swelling are normally ignored; but across the grain, considerable shrinkage and swelling occur in both the radial and tangential direction. Tangential shrinkage (as in flat-cut material) is normally approximately 50% greater than radial shrinkage (as in edge-grain material).

Table 4-16 gives average-shrinkage values in the radial and tangential directions for a number of commercial species as they dry from the fiber-saturation point (or green condition) to oven-dry. For intermediate changes in moisture content, shrinkage is proportional; e.g., if the fiber-saturation point is assumed to be 28% (average), a change in moisture content from 21 to 14% would cause shrinkage one-quarter as great as the shrinkage from 28% to zero.

Table 4-17 gives measured average moisture contents of wood in buildings in various parts of the United States.

TABLE 4-17 Measured Moisture-Content Values for Various Wood Items*

Use of lumber	Moisture content (% of weight of oven-dry wood) for					
	Dry Southwestern states		Damp Southern coastal states		Remainder of the United States	
	Avg	Individual pieces	Avg	Individual pieces	Avg	Individual pieces
Interior finish woodwork and softwood flooring.............................	6	4–9	11	8–13	8	5–10
Hardwood flooring.....................	6	5–8	10	9–12	7	6–9
Siding, exterior trim, sheathing, and framing.............................	9	7–12	12	9–14	12	9–14

* From U.S. Forest Products Laboratory.

4-41. Commercial Grades of Wood.

Lumber is graded by the various associations of lumber manufacturers having jurisdiction over various species. Two principal sets of grading rules are employed: (1) for softwoods, and (2) for hardwoods.

Softwoods. Softwood lumber is classified as dry, moisture content 19% or less; and green, moisture content above 19%.

According to the American Softwood Lumber Standard, softwoods are classified according to use as:

Yard lumber: lumber of grades, sizes, and patterns generally intended for ordinary construction and general building purposes.

Structural lumber: lumber 2 in or more in nominal thickness and width for use where working stresses are required.

Factory and shop lumber: lumber produced or selected primarily for manufacturing purposes.

Softwoods are classified according to extent of manufacture as:

Rough lumber: lumber that has not been dressed (surfaced) but has been sawed, edged, and trimmed.

Dressed (surfaced) lumber: lumber that has been dressed by a planing machine (for the purpose of attaining smoothness of surface and uniformity of size) on one side (S1S), two sides (S2S), one edge (S1E), two edges (S2E), or a combination of sides and edges (S1S1E, S1S2, S2S1E, S4S).

Worked lumber: lumber that, in addition to being dressed, has been matched, shiplapped or patterned:

Matched lumber: lumber that has been worked with a tongue on one edge of each piece and a groove on the opposite edge.

Shiplapped lumber: lumber that has been worked or rabbeted on both edges, to permit formation of a close-lapped joint.

Patterned lumber: lumber that is shaped to a pattern or to a molded form.

Softwoods are also classified according to nominal size:

Boards: lumber less than 2 in in nominal thickness and 2 in or more in nominal width. Boards less than 6 in in nominal width may be classified as strips.

Dimension: lumber from 2 in to, but not including, 5 in in nominal thickness, and 2 in or more in nominal width. Dimension may be classified as framing, joists, planks, rafters, studs, small timbers, etc.

Timbers: lumber 5 in or more nominally in least dimension. Timber may be classified as beams, stringers, posts, caps, sills, girders, purlins, etc.

Actual sizes of lumber are less than the nominal sizes, because of shrinkage and dressing. In general, dimensions of dry boards, dimension lumber, and timber less than 2 in wide or thick are $\frac{1}{4}$ in less than nominal; from 2 to 7 in wide or thick, $\frac{1}{2}$ in less, and above 6 in wide or thick, $\frac{3}{4}$ in less. Green lumber less than 2 in wide or thick is $\frac{1}{32}$ in more than dry; from 2 to 4 in wide or thick, $\frac{1}{16}$ in more, 5 and 6 in wide or thick, $\frac{1}{8}$ in more, and 8 in or above in width and thickness, $\frac{1}{4}$ in more than dry lumber. (See Table 7-2.) There are exceptions, however.

Yard lumber is classified on the basis of quality as:

Appearance: lumber of good appearance and finishing qualities, often called *select*.
    Suitable for natural finishes.
        Practically clear.
        Generally clear and of high quality.
    Suitable for paint finishes.
        Adapted to high-quality paint finishes.
        Intermediate between high-finishing grades and common grades, and partaking somewhat of the nature of both.

Common: lumber suitable for general construction and utility purposes, often given various commercial designations.

    For standard construction use.

        Suitable for better-type construction purposes.

        Well adapted for good standard construction.

        Designed for low-cost temporary construction.

    For less exacting purposes.

        Low quality, but usable.

Structural lumber is assigned modulus of elasticity values and working stresses in bending, compression parallel to grain, compression perpendicular to grain, and horizontal shear in accordance with ASTM procedures. These values take into account such factors as sizes and locations of knots, slope of grain, wane, and shakes or checks, as well as such other pertinent features as rate of growth and proportions of summerwood.

Factory and shop lumber is graded with reference to its use for doors and sash, or on the basis of characteristics affecting its use for general cut-up purposes, or on the basis of size of cutting. The grade of factory and shop lumber is determined by the percentage of the area of each board or plank available in cuttings of specified or of given minimum sizes and qualities. The grade of factory and shop lumber is determined from the poor face, although the quality of both sides of each cutting must be considered.

Hardwoods. Because of the great diversity of applications for hardwood both in and outside the construction industry, hardwood grading rules are based upon the proportion of a given piece that can be cut into smaller pieces of material clear on one or both sides and not less than a specified size. Grade classifications are therefore based on the amount of clear usable lumber in a piece.

Special grading rules of interest in the construction industry cover hardwood interior trim and moldings, in which one face must be practically free of imperfections and in which Grade A may further limit the amount of sapwood as well as stain. Hardwood dimension rules, in addition, cover clears, which must be clear both faces; clear one face; paint quality, which can be covered with paint; core, which must be sound on both faces and suitable for cores of glued-up panels; and sound, which is a general-utility grade.

Hardwood flooring is graded under two separate sets of rules: (1) for maple, birch, and beech; and (2) for red and white oak and pecan. In both sets of rules, color and quality classifications range from top-quality to the lower utility grades. Oak may be further subclassified as quarter-sawed and plain-sawed. In all grades, top-quality material must be uniform in color, whereas other grades place no limitation on color.

Shingles are graded under special rules, usually into three classes: Numbers 1, 2, and 3. Number 1 must be all edge grain and strictly clear, containing no sapwood. Numbers 2 and 3 must be clear to a distance far enough away from the butt to be well covered by the next course of shingles.

4-42. Destroyers and Preservatives.

The principal destroyers of wood are decay, caused by fungus, and attack by a number of animal organisms of which termites, carpenter ants, grubs of a wide variety of beetles, teredo, and limnoria are the principal offenders. In addition, fire annually causes widespread destruction of wood structures.

Decay will not occur if wood is kept well ventilated and air-dry or, conversely, if it is kept continuously submerged so that air is excluded.

Most termites in the United States are subterranean and require contact with the soil. The drywood and dampwood termites found along the southern fringes of the country and along the west coast, however, do not require direct soil contact and are more difficult to control.

Teredo, limnoria, and other water-borne wood destroyers are found only in salt or brackish waters.

Various wood species vary in natural durability and resistance to decay and insect attack. The sapwood of all species is relatively vulnerable; only the heartwood can be considered to be resistant. Table 4-18 lists the common species in accordance with heartwood resistance. Such a list is only approximate, and individual pieces deviate considerably.

Preservatives employed to combat the various destructive agencies may be subdivided into oily, water-soluble salts, and solvent-soluble organic materials. The principal

TABLE 4-18 Resistance to Decay of Heartwood of Domestic Woods

Resistant or very resistant	Moderately resistant	Slightly or nonresistant
Baldcypress (old growth)*	Baldcypress (young growth)*	Alder
Catalpa	Douglas fir	Ashes
Cedars	Honeylocust	Aspens
Cherry, black	Larch, western	Basswood
Chestnut	Oak, swamp chestnut	Beech
Cypress, Arizona	Pine, eastern white*	Birches
Junipers	Southern pine:	Buckeye
Locust, black†	Longleaf*	Butternut
Mesquite	Slash*	Cottonwood
Mulberry, red†	Tamarack	Elms
Oak:		Hackberry
Bur		Hemlocks
Chestnut		Hickories
Gambel		Magnolia
Oregon white		Maples
Post		Oak (red and black species)
White		Pines (other than longleaf, slash, and eastern white)
Osage orange†		Poplars
Redwood		Spruces
Sassafras		Sweetgum
Walnut, black		True firs (western and eastern)
Yew, Pacific†		Willows
		Yellow poplar

*The southern and eastern pines and baldcypress are now largely second growth with a large proportion of sapwood. Consequently, substantial quantities of heartwood lumber of these species are not available.
†These woods have exceptionally high decay resistance.
From U.S. Forest Products Laboratory.

oily preservatives are coal-tar creosote and creosote mixed with petroleum. The most commonly employed water-soluble salts are acid copper chromate, chromated copper arsenate and arsenite, copper naphthenate, chromated zinc chloride, and other materials that are often sold under various proprietary names. The principal solvent-soluble organic materials are chlorinated phenols, such as pentachlorphenol.

Preservatives may be applied in a variety of ways, including brushing and dipping, but for maximum treatment, pressure is required to provide deep side-grain penetration. Butts of poles and other parts are sometimes placed in a hot boiling creosote or salt solution, and after the water in the wood has been converted to steam, they are quickly transferred to a cold vat of the same preservative. As the steam condenses, it produces a partial vacuum, which draws the preservative fairly deeply into the surface.

Pressure treatments may be classified as full-cell and empty-cell. In the full-cell treatment, a partial vacuum is first drawn in the pressure-treating tank to withdraw most of the air in the cells of the wood. The preservative is then let in without breaking the vacuum, after which pressure is applied to the hot solution. After treatment is completed, the individual cells are presumably filled with preservative. In the empty-cell method, no initial vacuum is drawn, but the preservative is pumped in under pressure against the back pressure of the compressed air in the wood. When the pressure is released, the air in the wood expands and forces out excess preservative, leaving only a coating of preservative on the cell walls.

Retentions of preservative depend on the application. For teredo-infestation, full-cell creosote treatment to refusal may be specified, ranging from 16 to 20 lb per cubic foot of wood. For ordinary decay conditions and resistance to termites and other destroyers of a similar nature, the empty-cell method may be employed with retentions in the vicinity of 6 to 8 lb of creosote per cubic foot of wood. Salt retentions generally range in the vicinity of 1½ to 3 lb of dry salt retained per cubic foot of wood.

Solvent-soluble organic materials, such as pentachlorphenol, are commonly employed for the treatment of sash and door parts to impart greater resistance to decay. This is commonly done by simply dipping the parts in the solution and then allowing them to dry. As the organic solvent evaporates, it leaves the water-insoluble preservative behind in the wood.

These organic materials are also employed for general preservative treatment, including fence posts and structural lumber. The water-soluble salts and solvent-soluble organic materials leave the wood clean and paintable. Creosote in general cannot be painted over, although partial success can be achieved with top-quality aluminum-flake pigment paints.

Treatment against fire consists generally of applying salts containing ammonium and phosphates, of which monoammonium phosphate and diammonium phosphate are widely employed. At retentions of 3 to 5 lb of dry salt per cubic foot, the wood does not support its own combustion, and the afterglow when fire is removed is short. A variety of surface treatments is also available, most of which depend on the formation of a blanket of inert-gas bubbles over the surface of the wood in the presence of flame or other sources of heat. The blanket of bubbles insulates the wood beneath and retards combustion.

4-43. Glues and Adhesives.

A variety of adhesives is now available for use with wood, depending on the final application. The older adhesives include animal glue, casein glue, and a variety of vegetable glues, of which soybean is today the most important. Animal glues provide strong, tough, easily made joints, which, however, are not moisture-resistant. Casein mixed with cold water, when properly formulated, provides highly moisture-resistant glue joints, although they cannot be called waterproof. The vegetable glues have good dry strength but are not moisture-resistant.

The principal high-strength glues today are synthetic resins, of which phenol formaldehyde, urea formaldehyde, resorcinol formaldehyde, and melamine formaldehyde are the most important. Phenol, resorcinol, and melamine provide glue joints that are completely waterproof and will not separate when properly made even on boiling. Urea formaldehyde provides a glue joint of high moisture resistance, although not quite so good as the other three. Phenol and melamine require application of heat, as well as pressure, to cure the adhesive. Urea and resorcinol, however, can be formulated to be mixed with water at ordinary temperatures and hardened without application of heat above room temperature. Waterproof plywood is commonly made in hot-plate presses with phenolic or melamine adhesive. Resorcinol is employed where heat cannot be applied, as in a variety of assembly operations and the manufacture of laminated parts like ships' keels, which must have the maximum in waterproof qualities. Epoxide resins provide strong joints.

An emulsion of polyvinyl acetate serves as a general-purpose adhesive, for general assembly operations where maximum strength and heat or moisture resistance are not required. This emulsion is merely applied to the surfaces to be bonded, after which they are pressed together and the adhesive is allowed to harden.

4-44. Plywood and Other Fabricated Wood Boards.

As ordinarily made, plywood consists of thin sheets, or veneers, of wood glued together. The grain is oriented at right angles in adjacent plies. To obtain plywood with balance—that is, which will not warp, shrink, or twist unduly—the plies must be carefully selected and arranged to be mirror images of each other with respect to the central plane. The outside plies or faces are parallel to each other and are of species that have the same shrinkage characteristics. The same holds true of the cross bands. As a consequence, plywood has an odd number of plies, the minimum being three.

Principal advantages of plywood over lumber are its more nearly equal strength properties in length and width, greater resistance to checking, greatly reduced shrinkage and swelling, and resistance to splitting.

The approach to equalization of strength of plywood in the various directions is obtained at the expense of strength in the parallel-to-grain direction; i.e., plywood is not so strong in the direction parallel to its face plies as lumber is parallel to the grain. But plywood is considerably stronger in the direction perpendicular to its face plies than wood is perpendicular to the grain. Furthermore, the shearing strength of plywood in a plane perpendicular to the plane of the plywood is very much greater than that of ordinary wood parallel to the grain. In a direction parallel to the plane of the plywood, however, the shearing strength of plywood is less than that of ordinary wood parallel to the grain, because in this direction rolling shear occurs in the plywood; i.e., the fibers in one ply tend to roll rather than to slide.

Depending on whether plywood is to be used for general utility or for decorative purposes, the veneers employed may be cut by peeling from the log, by slicing, or today very rarely, by sawing. Sawing and slicing give the greatest freedom and versatility in the selection of grain. Peeling provides the greatest volume and the most rapid production, because logs are merely rotated against a flat knife and the veneer is peeled off in a long continuous sheet.

Plywood is classified as interior or exterior, depending on the type of adhesive employed. Interior-grade plywood must have a reasonable degree of moisture resistance but is not considered to be waterproof. Exterior-grade plywood must be completely waterproof and capable of withstanding immersion in water or prolonged exposure to outdoor conditions.

In addition to these classifications, plywood is further subclassified in a variety of ways depending on the quality of the surface ply. Top-quality is clear on one or both faces, except for occasional patches. Lower qualities permit sound defects, such as knots and similar blemishes, which do not detract from the general utility of the plywood but detract from its finished appearance.

Particle Board. Wood chips, sawdust, and flakes are pressed with a binder (urea-formaldehyde or phenol-formaldehyde) to form boards (sheathing, underlayment, core-stock), having uniform strength and low shrinkage in the plane of the board.

Hardboard. Wood chips (exploded by high-pressure steam into wood fibers) and lignin are pressed to form boards of various densities. Additives may add weather resistance and other properties.

4-45. Wood Bibliography.

Forest Products Laboratory, Forest Service, U.S. Department of Agriculture: "Wood Handbook," Government Printing Office, Washington, D.C.

National Hardwood Lumber Association, Chicago, Ill.: "Rules for the Measurement and Inspection of Hardwood Lumber, Cypress, Veneer, and Thin Lumber."

National Forest Products Association, Washington, D.C.: "National Design Specifications for Stress-grade Lumber and Its Fastenings."

U.S. Department of Commerce, National Bureau of Standards, Washington, D.C.: American Softwood Lumber Standard, Voluntary Practice Standard PS20; Douglas Fir Plywood, Commercial Standard CS 45; Hardwood Plywood, Commercial Standard CS 35.

Western Wood Products Association, Portland, Ore.: "Western Woods Use Book."

FERROUS METALS

The iron-carbon equilibrium diagram in Fig. 4-1 shows that, under equilibrium conditions (slow cooling) if not more than 2.0% carbon is present, a solid solution of carbon in gamma iron exists at elevated temperatures. This is called austenite. If the carbon content is less than 0.8%, cooling below the A_3 temperature line causes transformation

Fig. 4-1 Iron-carbon diagram.

of some of the austenite to ferrite, which is substantially pure alpha iron (containing less than 0.01% carbon in solution). Still further cooling to below the A_1 line causes the remaining austenite to transform to pearlite—the eutectoid mixture of fine plates, or lamellas, of ferrite and cementite (iron carbide) whose iridescent appearance under the microscope gives it its name.

If the carbon content is 0.8%, no transformation on cooling the austenite occurs until the A_1 temperature is reached. At that point, all the austenite transforms to pearlite, with its typical "thumbprint" microstructure.

At carbon contents between 0.80 and 2.0%, cooling below the A_{cm} temperature line causes iron carbide, or cementite, to form in the temperature range between A_{cm} and $A_{1,3}$. Below $A_{1,3}$, the remaining austenite transforms to pearlite.

4-46. Steels.

The plain iron-carbon metals with less than 0.8% carbon content consist of ferrite and pearlite and provide the low-carbon (0.06 to 0.30%), medium-carbon (0.30 to 0.50%), and high-carbon (0.50 to 0.80%) steels called hypoeutectoid steels. The higher-carbon or hypereutectoid tool steels contain 0.8 to 2.0% carbon and consist of pearlite and cementite. The eutectoid steels occurring in the vicinity of 0.8% carbon are essentially all pearlite.

4-47. Irons.

Metals containing substantially no carbon (several hundredths of 1%) are called irons, of which wrought iron, electrolytic iron, and "ingot" iron are examples.

Wrought iron, whether made by the traditional puddling method or by mixing very low carbon iron and slag, contains a substantial amount of slag. Because it contains very little carbon, it is soft, ductile, and tough and, like low-carbon ferrous metals generally, is relatively resistant to corrosion. It is easily worked. When broken, it shows a fibrous fracture because of the slag inclusions. "Ingot" iron is a very low carbon iron containing no slag, which is also soft, ductile, and tough.

Above 2.0% carbon content is the region of the cast irons. Above the $A_{1,3}$ temperature, austenite, the eutectic ledeburite and cementite occur; below the $A_{1,3}$ temperature, the austenite transforms to pearlite, and a similar transformation of the ledeburite occurs.

When the silicon content is kept low, and the metal is cooled rapidly, *white cast iron* results. It is hard and brittle because of the high cementite content. White cast iron as such has little use; but when it is reheated and held a long time in the vicinity of the transformation temperature, then cooled slowly, the cementite decomposes to ferrite and nodular or temper carbon. The result is black-heart *malleable iron.* If the carbon is removed during malleabilization, white-heart malleable iron results.

If the silicon content is raised, and the metal is cooled relatively slowly, *gray cast iron* results. It contains cementite, pearlite, ferrite, and some free carbon, which gives it its gray color. Gray iron is considerably softer and tougher than white cast iron and is generally used for castings of all kinds. Often, it is alloyed with elements like nickel, chromium, copper, and molybdenum.

At 5.0% carbon, the end product is hard, brittle iron carbide or cementite.

4-48. Standard Steels.

The American Iron and Steel Institute and the Society of Automotive Engineers have designated standard compositions for various steels including plain carbon steels and

TABLE 4-19 Standard Steels for Hot-Rolled Bars

(Basic Open-Hearth and Acid Bessemer Carbon Steels)

SAE and AISI No.	Chemical composition limits, %			
	Carbon	Manganese	Max phosphorus	Max sulfur
1008	0.10 max	0.30/0.50	0.040	0.050
1010	0.08/0.13	0.30/0.60	0.040	0.050
1015	0.13/0.18	0.30/0.60	0.040	0.050
1020	0.18/0.23	0.30/0.60	0.040	0.050
1025	0.22/0.28	0.30/0.60	0.040	0.050
1030	0.28/0.34	0.60/0.90	0.040	0.050
1040	0.37/0.44	0.60/0.90	0.040	0.050
1050	0.48/0.55	0.60/0.90	0.040	0.050
1070	0.65/0.75	0.60/0.90	0.040	0.050
1084	0.80/0.93	0.60/0.90	0.040	0.050
1095	0.90/1.03	0.30/0.50	0.040	0.050

alloy steels. AISI and SAE numbers and compositions for several representative hot-rolled carbon-steel bars are given in Table 4-19.

Most of the steel used for construction is low- to medium-carbon, relatively mild, tough, and strong, fairly easy to work by cutting, punching, riveting, and welding. Table 4-20 summarizes the most important carbon steels, cast irons, wrought iron, and low-alloy steels used in construction as specified by the American Society for Testing and Materials.

Prestressed concrete imposes special requirements for reinforcing steel. It must be of high strength with a high yield point and minimum creep in the working range. Tables 4-20 and 4-21 give ASTM specification requirements for wires and strands.

4-49. Heat Treatment and Hardening of Steels.

Heat-treated and hardened steels are sometimes required in building operations. The most familiar heat treatment is annealing, a reheating operation in which the metal is usually heated to the austenitic range and cooled slowly to obtain the softest, most ductile state. Cold working is often preceded by annealing. Annealing may be only partial, just sufficient to relieve internal stresses that might cause deformation or cracking, but not enough to reduce markedly the increased strength and yield point brought about by the cold working, for example.

Heat treatments are best exemplified by isothermal transformation curves, also called S curves and TTT (transformation-temperature-time) curves. Figure 4-2 is typical. It shows that above A_{e3} austenite is stable, between A_{e1} and A_{e3} some austenite transforms to ferrite, but below A_{e1} various transformations occur at various rates depending on temperature. A specimen cooled to above the "knee" at 900°F shows proeutectoid transformation to ferrite (between the first two curves) followed by transformation to pearlite (ferrite plus carbide) in the second zone, until transformation is complete. At the knee, transformation is extremely rapid, and the proeutectoid formation of ferrite disappears. A specimen cooled rapidly to below the knee and held at a lower temperature transforms completely to ferrite plus carbide in a strong, hard form called bainite. If cooled quickly to below the M_s temperature and held, the transformation product is the very hard, strong, but relatively brittle form known as martensite.

In plain carbon steels, the knee of the curve is extremely critical because of the short

TABLE 4-20 ASTM Requirements for Structural, Reinforcing, and Fastening Steels

	ASTM specification	Tensile strength, min, ksi	Yield point, min, ksi	Elongation in 8 in, min, %	Elongation in 2 in, min, %	Bend test, ratio of bend diameter, in, to specimen thickness, in*				
						0–¾	¾–1	1–1½	1½–2	Over 2
Structural steel	A36	58–80	36	20	23	½	1	1½	2½	3
High-strength, low-alloy, structural Mn-V steel	A441	60–70	40–50	18	21	1	1½	2	2½	3
High-strength, low-alloy, structural steel	A242	63–70	42–50	18	21	1	1½	2	2½	3
High-strength, low-alloy columbium-vanadium steels	A572	60–80	42–65	20–15	24–17	Depends on grade*				
High-strength, low-alloy structural steel	A588	63–70	42–50	18	21	1	1½	2	2½	3
High-yield-strength, quenched and tempered alloy steel	A514	110–130	90–100		17–18	2	2	3	4	4
Structural steel	A529	60–85	42	19		1				
High-strength steel bolts	A325	105	81		14					
High-strength, alloy steel bolts	A490	150–170	115–130		14					
Carbon-steel sheet and strip	A570	49–65	30–50	19–10	25–11	Depends on grade*				

Material	ASTM designation	Tensile strength (ksi)	Yield strength (ksi)	Elongation (%)	180° bend test; ratio of pin diameter to specimen diameter
Bolts and nuts, machine	A307	60–100		18	
Sheetpiling	A328	70	38.5	17	
Cast steel, 65-35, annealed	A27	70	38.5	17	2
Reinforcing steel for concrete:					
Billet-steel bars	A615				
Grade 40		70	40	7–11	Under No. 6: 4; Nos. 6, 7, 8, 9, 10, 11: 5
Grade 60		90	60	7–9	Under No. 6: 4; No. 6: 5; Nos. 7, 8: 6; Nos. 9, 10, 11: 8
Rail-steel bars	A616				
Grade 50		80	50	5–6	Under No. 8: 6; Nos. 9, 10, 11: 8†
Grade 60		90	60	4.5–6	Under No. 8: 6; Nos. 9, 10, 11: 8†
Cold-drawn wire	A82	70–80	56–70		W7 or under: 1 over W7: 2
Seven-wire prestressing strand	A416				
Grade 250		250	‡	3.5 in 24-in gage length	
Grade 270		270	‡	3.5 in 24-in gage length	

*Optional. See ASTM A6, "General Requirements for Rolled Steel Plates, Shapes, Sheet Piling, and Bars for Structural Use."

†90° bend for No. 11 bars.

‡Yield strength should be at least 85% of the minimum breaking strength.

TABLE 4-21 ASTM Requirements for Prestressing Wire

Material	ASTM specification No.	Tensile strength, ksi	Minimum yield point, ksi
Stress-relieved strand for prestressed concrete:	A416		
Grade 250		250	85% of breaking
Grade 270		270	strength, 1% extension
Stress-relieved wire for prestressed concrete:	A421		
BA		235–240	199.75–204, 1% extension
WA		235–250	199.75–212.5, 1% extension

transformation time. It may be impossible to cool the centers of heavy parts fast enough to get below the knee, and hardenability is therefore limited. The addition of alloying elements almost always shifts the knee to the right, that is, makes transformation more sluggish and therefore increases the hardenability of the metal. Many alloys, such as several stainless steels, remain austenitic at ordinary temperatures; others can easily be made martensitic throughout, but still others merely become ferritic and cannot be hardened.

Often, a hard surface is required on a soft, tough core. Two principal casehardening methods are employed. For **case carburizing,** the low- to medium-carbon steel is packed in carbonaceous materials and heated to the austenite range. Carbon diffuses into the surface, providing a hard, high-carbon case when the part is cooled. For **nitriding,** the part is exposed to ammonia gas or a cyanide at moderately elevated temperatures. Extremely hard nitrides are formed in the case and provide a hard surface. Cast iron is frequently cast against metal inserts in the molds. The cast metal is chilled by the inserts and forms white cast iron, whereas the rest of the more slowly cooled metal becomes gray cast iron.

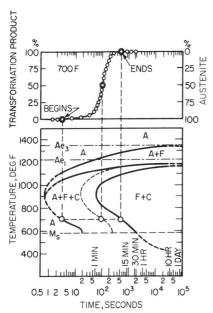

Fig. 4-2 Typical isothermal transformation curve at a single temperature (top) and complete isothermal transformation diagram (bottom), showing effects of heat treatment. *("Metals Handbook," American Society for Metals.)*

4-50. Steel Alloys.

Plain carbon steels can be given a great range of properties by heat treatment and by working; but addition of alloying elements greatly extends those properties or makes the heat-treating operations easier and simpler. For example, combined high tensile strength and toughness, corrosion resistance, high-speed cutting, and many other specialized purposes require alloy steels. However, the most important effect of alloying is the influence on hardenability. The most important alloying elements from the standpoint of building, and their principal effects, are summarized below:

Aluminum restricts grain growth during heat treatment and promotes surface hardening by nitriding.

Chromium is a hardener, promotes corrosion resistance (see Art. 4-51), and promotes wear resistance.

Copper promotes resistance to atmospheric corrosion and is sometimes combined with molybdenum for this purpose in low-carbon steels and irons. It strengthens steel and increases the yield point without unduly changing elongation or reduction of area.

Manganese in low concentrations promotes hardenability and nondeforming, nonshrinking characteristics for tool steels. In high concentrations, the steel is austenitic under ordinary conditions, is extremely tough, and work-hardens readily. It is therefore used for teeth of power-shovel dippers, railroad frogs, rock crushers, and similar applications.

Molybdenum is usually associated with other elements, especially chromium and nickel. It increases corrosion resistance, raises tensile strength and elastic limit without reducing ductility, promotes casehardening and improves impact resistance.

Nickel boosts tensile strength and yield point without reducing ductility; increases low-temperature toughness, whereas ordinary carbon steels become brittle; promotes casehardening; and in high concentrations improves corrosion resistance under severe conditions. It is often used with chromium (see Art.4-51). **Invar** contains 36% nickel.

Silicon strengthens low-alloy steels; improves oxidation resistance; with low carbon yields transformer steel, because of low hysteresis loss and high permeability; in high concentrations provides hard, brittle castings, resistant to corrosive chemicals, useful in plumbing lines for chemical laboratories.

Sulfur promotes free machining, especially in mild steels.

Titanium prevents intergranular corrosion of stainless steels by preventing grain-boundary depletion of chromium during such operations as welding and heat treatment.

Tungsten, vanadium, and **cobalt** are all used in high-speed tool steels, because they promote hardness and abrasion resistance. Tungsten and cobalt also increase high-temperature hardness.

The principal effects of alloying elements are summarized in Table 4-22.

4-51. Stainless Steels.

Stainless steels of primary interest in building are the wrought stainless steels of the austenitic type. The austenitic stainless steels contain both chromium and nickel. Total content of alloy metals is not less than 23%, with chromium not less than 16% and nickel not less than 7%. Table 4-23 summarizes the composition and principal properties of these metals.

Austenitic stainless steels are tough, strong, and shock-resistant, but work-harden readily; so some difficulty on this score may be experienced with cold working and machining. These steels can be welded readily but may have to be stabilized (e.g., AISI Types 321 and 347) against carbide precipitation and intergranular corrosion due to welding unless special precautions are taken. These steels have the best high-temperature strength and resistance to scaling of all the stainless steels.

Types 303 and 304 are the familiar 18-8 stainless steels widely used for building applications. These and Types 302 and 316 are the most commonly employed stainless steels. Where maximum resistance to corrosion is required, such as resistance to pitting by seawater and chemicals, the molybdenum-containing Types 316 and 317 are best.

TABLE 4-22 Effects of Alloying Elements in Steel

Element	Solid solubility		Influence on ferrite	Influence on austenite (hardenability)	Influence exerted through carbide		Principal functions
	In gamma iron	In alpha iron			Carbide-forming tendency	Action during tempering	
Aluminum (Al)	1.1% (increased by C)	36%	Hardens considerably by solid solution	Increases hardenability mildly, if dissolved in austenite	Negative (graphitizes)	....	1. Deoxides efficiently 2. Restricts grain growth (by forming dispersed oxides or nitrides) 3. Alloying element in nitriding steel
Chromium (Cr)	12.8% (20% with 0.5% C)	Unlimited	Hardens slightly; increases corrosion resistance	Increases hardenability moderately	Greater than Mn; less than W	Mildly resists softening	1. Increases resistance to corrosion and oxidation 2. Increases hardenability 3. Adds some strength at high temperatures 4. Resists abrasion and wear (with high carbon)
Cobalt (Co)	Unlimited	75%	Hardens considerably by solid solution	Decreases hardenability as dissolved	Similar to Fe	Sustains hardness by solid solution	1. Contributes to red hardness by hardening ferrite
Manganese (Mn)	Unlimited	3%	Hardens markedly; reduces plasticity somewhat	Increases hardenability moderately	Greater than Fe; less than Cr	Very little, in usual percentages	1. Counteracts brittleness from the sulfur 2. Increases hardenability inexpensively
Molybdenum (Mo)	3% ± (8% with 0.3% C)	37.5% (less with lowered temp)	Provides age-hardening system in high Mo-Fe alloys	Increases hardenability strongly (Mo > Cr)	Strong; greater than Cr	Opposes softening by secondary hardening	1. Raises grain-coarsening temperature of austenite 2. Deepens hardening 3. Counteracts tendency toward temper brittleness 4. Raises hot and creep strength, red hardness 5. Enhances corrosion resistance in stainless steel 6. Forms abrasion-resisting particles
Nickel (Ni)	Unlimited	10% (irrespective of carbon content)	Strengthens and toughens by solid solution	Increases hardenability mildly, but tends to retain austenite with higher carbon	Negative (graphitizes)	Very little in small percentages	1. Strengthens unquenched or annealed steels 2. Toughens pearlitic-ferritic steels (especially at low temperature) 3. Renders high-chromium iron alloys austenitic

TABLE 4-22 Effects of Alloying Elements in Steel (Continued)

Element	Solid solubility		Influence on ferrite	Influence on austenite (hardenability)	Influence exerted through carbide		Principal functions
	In gamma iron	In alpha iron			Carbide-forming tendency	Action during tempering	
Phosphorus, (P)	0.5%	2.8% (irrespective of carbon content)	Hardens strongly by solid solution	Increases hardenability	Nil	....	1. Strengthens low-carbon steel 2. Increases resistance to corrosion 3. Improves machinability in free-cutting steels
Silicon (Si)	2% ± (9% with 0.35% C)	18.5% (not much changed by carbon)	Hardens with loss in plasticity (Mn < Si < P)	Increases hardenability moderately	Negative (graphitizes)	Sustains hardness by solid solution	1. Used as general-purpose deoxidizer 2. Alloying element for electrical and magnetic sheet 3. Improves oxidation resistance 4. Increases hardenability of steel carrying nongraphitizing elements 5. Strengthens low-alloy steels

Titanium (Ti)	0.75% (1% ± with 0.20% C)	6% ± (less with lowered temp)	Provides age-hardening system in high Ti-Fe alloys	Probably increases hardenability very strongly as dissolved. The carbide effects reduce hardenability	Greatest known (2% Ti renders 0.50% carbon steel unhardenable)	Persistent carbides probably unaffected. Some secondary hardening	1. Fixes carbon in inert particles
							  a. Reduces martensitic hardness and hardenability in medium-chromium steels
							  b. Prevents formation of austenite in high-chromium steels
							  c. Prevents localized depletion of chromium in stainless steel during long heating

Tungsten (W)	6% (11% with 0.25% C)	33% (less with lowered temp)	Provides age-hardening system in high W-Fe alloys	Increases hardenability strongly in small amounts	Strong	Opposes softening by secondary hardening	1. Forms hard, abrasion-resistant particles in tool steels 2. Promotes hardness and strength at elevated temperature
Vanadium (V)	1% (4% with 0.20% C)	Unlimited	Hardens moderately by solid solution	Increases hardenability very strongly, as dissolved	Very strong (V < Ti or Cb)	Max for secondary hardening	1. Elevates coarsening temperature of austenite (promotes fine grain) 2. Increases hardenability (when dissolved) 3. Resists tempering and causes marked secondary hardening

* "Metals Handbook," American Society for Metals.

TABLE 4-23 Stainless Steels—ASTM A276-78

ASTM No.	Carbon, max %	Chromium, %	Nickel, %	Other elements, max %	Tensile strength, ksi A†	Tensile strength, ksi B‡	Yield point, ksi A†	Yield point, ksi B‡	Elongation, % in 2 in. A†	Elongation, % in 2 in. B‡	Reduction area, % A†	Drawing or stamping	Remarks* General properties	Welding properties
302	0.15	17–19	8–10	Mn 2, Si 1	75	95–125	30	45–100	40	12–28	50	Very good	A readily fabricated stainless steel for decorative or corrosion-resistant use	Very good; tough welds
302B	0.15	17–19	8–10	Si 2–3, Mn 2	75		30		40		50		Silicon added to increase resistance to scaling at high temperatures	Fairly good. May crack; not recommended
304	0.08	18–20	8–10.5	Mn 2, Si 1	75	95–125	30	45–100	40	12–28	50		A low-carbon 18-8 steel, weldable with less danger of intercrystalline corrosion	
308	0.08	19–21	10–12	Mn 2, Si 1	75		30		40		50		For use when corrosion resistance greater than that of 18-8 is needed	
309	0.20	22–24	12–15	Mn 2, Si 1	75		30		40		50	Good	For use at elevated temperature, combining high scaling resistance and good strength	Good
309S	0.08	22–24	12–15	Mn 2, Si 1	75		30						Low carbon permits welded fabrication with a minimum of carbide precipitation	
310	0.25	24–26	19–22	Mn 2, Si 1	75		30		40		50	Good	Similar to 25-12 stainless, with higher nickel content for greater stability at welding temperatures	Good
316	0.08	16–18	10–14	Mo 2–3, Mn 2, Si 1	75	95–125	30	45–100	40	12–28	50	Good	Superior resistance to chemical corrosion	Very good
321	0.08	17–19	9–12	Ti 5 × C min, Mn 2, Si 1	75		30		40		50	Good	An 18-8 type, stabilized against intercrystalline corrosion at elevated temperatures	Very good
347	0.08	17–19	9–13	Mn 2, Si 1, Cb-Ta 10 × C, min	75		30		40		50	Good	A stabilized 18-8 steel for service at elevated temperatures	Very good

*Metals Handbook, American Society for Metals. †A, annealed. ‡B, cold-finished.

For resistance to ordinary atmospheric corrosion, some of the martensitic and ferritic stainless steels, containing 15 to 20% chromium and no nickel, are employed. The martensitic steels, in general, range from about 12 to 18% chromium and from 0.08 to 1.10% carbon. Their response to heat treatment is similar to that of the plain carbon steels. When chromium content ranges from 15 to 30% and carbon content is below 0.35%, the steels are ferritic and nonhardenable. The high-chromium steels are resistant to oxidizing corrosion and are useful in chemical plants.

4-52. Welding Ferrous Metals.

General welding characteristics of the various types of ferrous metals are as follows:

Wrought iron is ideally forged but may be welded by other methods if the base metal is thoroughly fused. Slag melts first and may confuse unwary operators.

Low-carbon iron and steels (0.30%C or less) are readily welded and require no preheating or subsequent annealing unless residual stresses are to be removed.

Medium-carbon steels (0.30 to 0.50%C) can be welded by the various fusion processes. In some cases, especially in steel with more than 0.40% carbon, preheating and subsequent heat treatment may be necessary.

High-carbon steels (0.50 to 0.90%C) are more difficult to weld and, especially in arc welding, may have to be preheated to at least 500°F and subsequently heated between 1200 and 1450°F. For gas welding, a carburizing flame is often used. Care must be taken not to destroy the heat treatment to which high-carbon steels may have been subjected.

Tool steels (0.80 to 1.50%C) are difficult to weld. Preheating, postannealing, heat treatment, special welding rods, and great care are necessary for successful welding.

Welding of structural steels is governed by the American Welding Society "Structural Welding Code," AWS D1.1, the American Institute of Steel Construction Specification for the Design, Fabrication and Erection of Structural Steel for Buildings, or a local building code. AWS D1.1 specifies tests to be used in qualifying welders and types of welds. The AISC Specification and many building codes require, in general, that only qualified welds be used and that they be made only by qualified welders.

Structural steels to be welded should be of a weldable grade, such as A36, A441, A572, A588, or A514. These steels may be welded by shielded metal arc, submerged arc, gas metal arc, flux-cored arc, electroslag, electrogas, or stud-welding processes.

Shielded-metal-arc welding fuses parts to be joined by the heat of an electric arc struck between a coated metal electrode and the material being joined, or base metal. The electrode supplies filler material for making the weld, gas for shielding the molten metal from the air, and flux for refining this metal.

Submerged-arc welding fuses the parts to be joined by the heat of an electric arc struck between a bare metal electrode and base metal. The weld is shielded from the air by flux. The electrode or a supplementary welding rod supplies metal filler for making the weld.

Gas-metal-arc welding produces fusion by the heat of an electric arc struck between a filler-metal electrode and base metal, while the molten metal is shielded by a gas or mixture of gas and flux. For structural steels, the gas may be argon, argon with oxygen, or carbon dioxide.

Electroslag welding uses a molten slag to melt filler metal and surfaces of the base metal and thus make a weld. The slag, electrically conductive, is maintained molten by its resistance to an electric current that flows between an electrode and the base metal. The process is suitable only for welding in the vertical position. Moving, water-cooled shoes are used to contain and shape the weld surface. The slag shields the molten metal.

Electrogas welding is similar to the electroslag process. The electrogas process, however, maintains an electric arc continuously, uses an inert gas for shielding, and the electrode provides flux.

Stud welding is used to fuse metal studs or similar parts to other steel parts by the heat of an electric arc. A welding gun is usually used to establish and control the arc, and to apply pressure to the parts to be joined. At the end to be welded, the stud is equipped with a ceramic ferrule, which contains flux and which also partly shields the weld when molten.

Preheating before welding reduces the risk of brittle failure. Initially, its main effect is to lower the temperature gradient between the weld and adjoining base metal. This makes cracking during cooling less likely and gives entrapped hydrogen, a possible source of embrittlement, a chance to escape. A later effect of preheating is improved ductility and notch toughness of base and weld metals and lower transition temperature of weld. When, however, welding processes that deposit weld metal low in hydrogen are used and suitable moisture control is maintained, the need for preheat can be eliminated. Such processes include use of low-hydrogen electrodes and inert-arc and submerged-arc welding.

Weldability of structural steels is influenced by their chemical content. Carbon, manganese, silicon, nickel, chromium, and copper, for example, tend to have an adverse effect, whereas molybdenum and vanadium may be beneficial. To relate the influence of chemical content on structural steel properties to weldability, the use of a carbon equivalent has been proposed. One formula suggested is

$$C_{eq} = C + \frac{Mn}{4} + \frac{Si}{4} \tag{4-2}$$

where C = carbon content, %
Mn = manganese content, %
Si = silicon content, %

Another proposed formula includes more elements:

$$C_{eq} = C + \frac{Mn}{6} + \frac{Ni}{20} + \frac{Cr}{10} - \frac{Mo}{50} - \frac{V}{10} + \frac{Cu}{40} \tag{4-3}$$

where Ni = nickel content, %
Cr = chromium content, %
Mo = molybdenum content, %
V = vanadium content, %
Cu = copper content, %

Carbon equivalent appears to be related to the maximum rate at which a weld and adjacent base metal may be cooled after welding without underbead cracking occurring. The higher the carbon equivalent, the lower will be the allowable cooling rate. Also, the higher the carbon equivalent, the more important use of low-hydrogen electrodes and preheating becomes. (O. W. Blodgett, "Design of Welded Structures," The James F. Lincoln Arc Welding Foundation, Cleveland, Ohio.)

4-53. Corrosion of Iron and Steel.

Corrosion of ferrous metals is caused by the tendency of iron (anode) to go into solution in water as ferrous hydroxide and displace hydrogen, which in turn combines with dissolved oxygen to form more water. At the same time, the dissolved ferrous hydroxide is converted by more oxygen to the insoluble ferric hydroxide, thereby allowing more iron to go into solution. Corrosion, therefore, requires liquid water (as in damp air) and oxygen (which is normally present dissolved in the water).

Protection against corrosion takes a variety of forms:

Deaeration. If oxygen is removed from water, corrosion stops. In hot-water heating systems, therefore, no fresh water should be added. Boiler feedwater is sometimes deaerated to retard corrosion.

Coatings

1. **Paints.** Most paints are based on oxidizing oil and a variety of pigments, of which red lead, oxides of iron, lead sulfate, zinc sulfate, graphite, aluminum, and various hydrocarbons are a few. No one paint is best for all applications. Other paints are coatings of asphalt and tar.

2. **Metallic.** Zinc applied by hot dipping (**galvanizing**) or hot powder (**sherardizing**), hot tin dip, hot aluminum dip, and electrolytic plates of tin, copper, nickel, chromium, cadmium, and zinc. A mixture of lead and tin is called **terneplate.** Zinc is anodic to iron and protects, even after the coating is broken, by sacrificial protection. Tin and copper are cathodic and protect as long as the coating is unbroken but may hasten corrosion by pitting and other localized action once the coating is pierced.

3. **Chemical.** Insoluble phosphates, such as iron or zinc phosphate, are formed on the surface of the metal by treatment with phosphate solutions. These have some protective action and also form good bases for paints. Black oxide coatings are formed by treating the surface with various strong salt solutions. This coating is good for indoor use but has limited life outdoors. It provides a good base for rust-inhibiting oils.

Cathodic Protection. As corrosion proceeds, electric currents are produced as the metal at the anode goes into solution. If a sufficient countercurrent is produced, the metal at the anode will not dissolve. This is accomplished in various ways, such as connecting the iron to a more active metal like magnesium (rods suspended in domestic water heaters) or connecting the part to be protected to buried scrap iron and providing an external current source such as a battery or rectified current from a power line (protection of buried pipe lines).

4-54. Ferrous Metals Bibliography.

American Iron and Steel Institute, 1000 16th St., N.W., Washington, D.C. 20036:
"Carbon Steels, Chemical Composition Limits," "Constructional Alloys, Chemical Composition Limits"; "Steel Products Manuals."
American Society for Testing and Materials, Philadelphia, Pa.: "Standards."
American Society for Metals, Cleveland, Ohio: "Metals Handbook."

ALUMINUM AND ALUMINUM-BASED ALLOYS

Pure aluminum and aluminum alloys are used in buildings in various forms. High-purity aluminum (at least 99% pure) is soft and ductile but weak. It has excellent corrosion resistance and is used in buildings for such applications as bright foil for heat insulation, roofing, flashing, gutters and downspouts, exterior and interior architectural trim, and as pigment in aluminum-based paints. Its high heat conductivity recommends it for cooking utensils. The electrical conductivity of the electrical grade is 61% of that of pure copper on an equal-volume basis and 201% on an equal-weight basis.

Aluminum alloys are generally harder and stronger than the pure metal. Furthermore, pure aluminum is difficult to cast satisfactorily, whereas many of the alloys are readily cast.

4-55. Aluminum-Alloy Designations.

The alloys may be classified: (1) as cast and wrought, and (2) as heat-treatable and non-heat-treatable. Alloys are heat treatable if the dissolved constituents are less soluble in the solid state at ordinary temperatures than at elevated temperatures, thereby making age-hardening possible. When heat treated to obtain complete solution, the product may be unstable and tend to age spontaneously. It may also be treated to produce stable tempers of varying degree. Cold working or strain hardening is also possible, and combinations of tempering and strain hardening can also be obtained.

Because of these various possible combinations, a system of letter and number designations has been worked out by the producers of aluminum and aluminum alloys to indicate the compositions and the tempers of the various metals. Wrought alloys are designated by a four-digit index system. 1xxx is for 99.00% aluminum minimum. The last two digits indicate the minimum aluminum percentage. The second digit represents impurity limits. (EC is a special designation for electrical conductors.) 2xxx to 8xxx represent alloy groups in which the first number indicates the principal alloying constituent, and the last two digits are identifying numbers in the group. The second digit indicates modification of the basic alloy. The alloy groups are

Copper	2xxx
Manganese	3xxx
Silicon	4xxx
Magnesium	5xxx
Magnesium and silicon	6xxx
Zinc	7xxx
Other elements	8xxx
Unused series	9xxx

For cast alloys, a similar designation system is used. The first two digits identify the alloy or its purity. The last digit, preceded by a decimal point, indicates the form of the material; for example, casting or ingot. Casting alloys may be sand or permanent-mold alloys.

Among the wrought alloys, the letters F, O, H, W, and T indicate various basic temper designations. These letters in turn may be followed by numerals to indicate various degrees of treatment. Temper designations are summarized in Table 4-24.

The structural alloys generally employed in building fall in the 2xxx, 5xxx, and 6xxx categories. Architectural alloys often used include 3xxx, 5xxx, and 6xxx groups.

4-56. Clad Aluminum.

Pure aluminum is generally more corrosion resistant than its alloys. Furthermore, its various forms—pure and alloy—have different solution potentials; that is, they are anodic or cathodic to each other, depending on their relative solution potentials. A number of alloys are therefore made with centers or "cores" of aluminum alloy, overlaid with layers of metal, either pure aluminum or alloys, which are anodic to the core. If galvanic corrosion conditions are encountered, the cladding metal protects the core sacrificially.

4-57. Corrosion Resistance and Corrosion Protection of Aluminum.

Although aluminum ranks high in the electromotive series of the metals, it is highly corrosion resistant because of the tough, transparent, tenacious film of aluminum oxide that

TABLE 4-24 Basic Temper Designations for Wrought Aluminum Alloys*

-F **As fabricated.** This designation applies to the products of shaping processes in which no special control over thermal conditions or strain-hardening is employed. For wrought products, there are no mechanical property limits.

-O **Annealed.** This designation applies to wrought products annealed to obtain the lowest-strength temper, and to cast products annealed to improve ductility and dimensional stability.

-H **Strain hardened (wrought products only).** This designation applies to products that have their strength increased by strain hardening, with or without supplementary thermal treatments to produce some reduction in strengths. The H is always followed by two or more digits.

-W **Solution heat treated.** An unstable temper applicable only to alloys that spontaneously age at room temperature after solution heat treatment. This designation is specific only when the period of natural aging is indicated: for example, W 1/2 hr.

-T **Thermally treated to produce stable tempers other than F, O, or H.** This designation applies to products that are thermally treated, with or without supplementary strain hardening, to produce stable tempers. The T is always followed by one or more digits:

Basic Tempers. The first digit following the H or T indicates the specific combination of basic operations, as follows:

-H1 **Strain hardened only.** This designation applies to products that are strain hardened to obtain the desired strength without supplementary thermal treatment. The number following this designation indicates the degree of strain hardening.

-H2 **Strain hardened and partially annealed.** This designation applies to products that are strain hardened more than the desired final amount and then reduced in strength to the desired level by partial annealing. For alloys that age-soften at room temperature, the H2 tempers have the same minimum ultimate tensile strength as the corresponding H3 tempers. For other alloys, the H2 tempers have the same minimum ultimate tensile strength as the corresponding H1 tempers and slightly higher elongation. The number following this designation indicates the degree of strain hardening remaining after the product has been partially annealed.

-H3 **Strain hardened and stabilized.** This designation applies to products that are strain hardened and whose mechanical properties are stabilized by a low-temperature thermal treatment that causes slightly lowered tensile strength and improved ductility. H3 is applicable only to those alloys that, unless stabilized, gradually age-soften at room temperature. The number following this designation indicates the degree of strain hardening before the stabilization treatment.

Degree of strain hardening. The digit after H1, H2, and H3 indicates degree of strain hardening. Zero represents annealing, while 8 indicates tempering to produce an ultimate tensile strength equivalent to that achieved by a cold reduction (temperature during reduction not to exceed 120°F) of about 75% following a full anneal. Tempers producing ultimate tensile strengths at equal intervals between that of the 0 temper and that of the 8 temper are designated by 1 to 7. Numeral 9 designates tempers that produce a minimum ultimate tensile strength exceeding that of the 8 temper by 2.0 ksi or more.

-T1 **Cooled from an elevated-temperature shaping process and naturally aged to a substantially stable condition.** Applies to products not cold worked after cooling from an elevated-temperature shaping process, or in which the effects of cold work in flattening or straightening the material may not be recognized in mechanical-property limits.

-T2 **Cooled from an elevated-temperature shaping process, cold worked, and naturally aged to a substantially stable condition.** Applies to products cold worked to improve strength after cooling from an elevated-temperature shaping process, or in which the effects of cold work in flattening or straightening the material is recognized in mechanical-property limits.

-T3 **Solution heat treated, cold worked, and naturally aged to a substantially stable condition.** Applies to products that are cold worked after solution heat treatment, or for which the effects of cold work in flattening or straightening the material are recognized in mechanical-property limits.

TABLE 4-24 Basic Temper Designations for Wrought Aluminum Alloys* (*Continued*)

-T4 **Solution heat treated and naturally aged to a substantially stable condition.** Applies to products not cold worked after solution heat treatment, or in which the effects of cold work in flattening or straightening the material are recognized in mechanical-property limits.

-T5 **Cooled from an elevated-temperature shaping process and then artificially aged.** Applied to products that are not cold worked after cooling from an elevated-temperature shaping process, or in which the effects of cold work in flattening or straightening the material may not be recognized in mechanical-property limits.

-T6 **Solution heat treated and then artificially aged.** This designation applies to products that are not cold worked after solution heat treatment, or for which the effects of cold work in flattening or straightening the material may not be recognized in mechanical-property limits.

-T7 **Solution heat treated and then stabilized.** Applies to products that are stabilized after solution heat treatment to carry them beyond the point of maximum strength to provide control of some special characteristics.

-T8 **Solution heat treated, cold worked, and then artificially aged.** This designation applies to products that are cold worked to improve strength, or for which the effects of cold work in flattening or straightening the material are recognized in mechanical-property limits.

-T9 **Solution heat treated, artificially aged, and then cold worked.** This designation applies to products that are cold worked to improve strength.

-T10 **Cooled from an elevated-temperature shaping process, cold worked, and then artificially aged.** This designation applies to products that are artificially aged after cooling from an elevated-temperature shaping process, such as casting or extrusion, and then cold worked to further improve strength.

Treatment Variations. Additional digits, larger than zero, may be added to designations T1 through T10 to indicate a variation in treatment that significantly alters the characteristics of the product.

*Aluminum Association.

rapidly forms on any exposed surface. It is this corrosion resistance that recommends aluminum for building applications. For most exposures, including industrial and seacoast atmospheres, the alloys normally recommended are adequate, particularly if used in usual thicknesses and if mild pitting is not objectionable.

Pure aluminum is the most corrosion resistant of all and is used alone or as cladding on strong-alloy cores where maximum resistance is wanted. Of the alloys, those containing magnesium, manganese, chromium, or magnesium and silicon in the form of $MgSi_2$ are highly resistant to corrosion. The alloys containing substantial proportions of copper are more susceptible to corrosion, depending markedly on the heat treatment.

Certain precautions should be taken in building. Aluminum is subject to attack by alkalies, and it should therefore be protected from contact with wet concrete, mortar, and plaster. Clear methacrylate lacquers or strippable plastic coatings are recommended for interiors, and methacrylate lacquer for exterior protection during construction. Strong alkaline and acid cleaners should be avoided and muriatic acid should not be used on masonry surfaces adjacent to aluminum. If aluminum must be contiguous to concrete and mortar outdoors, or where it will be wet, it should be insulated from direct contact by asphalts, bitumens, felts, or other means. As is true of other metals, atmospheric-deposited dirt must be removed to maintain good appearance.

Electrolytic action between aluminum and less active metals should be avoided, because the aluminum then becomes anodic. If aluminum must be in touch with other metals, the faying surfaces should be insulated by painting with asphaltic or similar paints, or by gasketing. Steel rivets and bolts, for example, should be insulated. Drainage from copper-alloy surfaces onto aluminum must be avoided. Frequently, steel surfaces can be galvanized or cadmium-coated where contact is expected with aluminum. The zinc or cadmium coating is anodic to the aluminum and helps to protect it.

4-58. Finishes for Aluminum.

Almost all finishes used on aluminum may be divided into three major categories in the system recommended by The Aluminum Association: mechanical finishes, chemical finishes, and coatings. The last may be subdivided into anodic coatings, resinous and other organic coatings, vitreous coatings, electroplated and other metallic coatings, and laminated coatings.

In The Aluminum Association system, mechanical and chemical finishes are designated by M and C, respectively, and each of the five classes of coating is also designated by a letter. The various finishes in each category are designated by two-digit numbers after a letter. The principal finishes are summarized in Table 4-25.

TABLE 4-25 Finishes for Aluminum and Aluminum Alloys

Type of finish	Designation*
Mechanical finishes:	
As fabricated	M1Y
Buffed	M2Y
Directional textured	M3Y
Nondirectional textured	M4Y
Chemical finishes:	
Nonetched cleaned	C1Y
Etched	C2Y
Brightened	C3Y
Chemical conversion coatings	C4Y
Coatings:	
Anodic	
General	A1Y
Protective and decorative (less than 0.4 mil thick)	A2Y
Architectural Class II (0.4–0.7 mil thick)	A3Y
Architectural Class I (0.7 mil or more thick)	A4Y
Resinous and other organic coatings	R1Y
Vitreous coatings	V1Y
Electroplated and other metallic coatings	E1Y
Laminated coatings	L1Y

*Y represents digits (0, 1, 2, ... 0) or X (to be specified) that describe the surface, such as specular, satin, matte, degreased, clear anodizing or type of coating.

4-59. Welding and Brazing of Aluminum.

Weldability and brazing properties of aluminum alloys depend heavily on their composition and heat treatment. Most of the wrought alloys can be brazed and welded, but sometimes only by special processes. The strength of some alloys depends on heat treatment after welding. Alloys heat treated and artificially aged are susceptible to loss of strength at the weld, because weld metal is essentially cast. For this reason, high-strength structural alloys are commonly fabricated by riveting or bolting, rather than by welding.

Brazing is done by furnace, torch, or dip methods. Successful brazing is done with special fluxes.

4-60. Aluminum Bibliography.

The Aluminum Association, 818 Connecticut Ave., N.W., Washington, D.C. 20006: "Aluminum Standards and Data," and "Designation Systems for Aluminum Finishes."

Van Lancker, M.: "Metallurgy of Aluminum Alloys," John Wiley & Sons, Inc., New York.

COPPER AND COPPER-BASED ALLOYS

Copper and its alloys are widely used in the building industry for a large variety of purposes, particularly applications requiring corrosion resistance, high electrical conductivity, strength, ductility, impact resistance, fatigue resistance, or other special characteristics possessed by copper or its alloys. Some of the special characteristics of importance to building are ability to be formed into complex shapes, appearance, and high thermal conductivity, although many of the alloys have low thermal conductivity and low electrical conductivity as compared with the pure metal.

In the following paragraphs, brief comments will be made regarding the properties, typical compositions, and uses of the principal copper and copper-based alloys of interest in building. The principal properties are summarized in Table 4-26.

4-61. Copper.

The excellent corrosion resistance of copper makes it suitable for such applications as roofing, flashing, cornices, gutters, downspouts, leaders, fly screens, and similar applications. For roofing and flashing, soft-annealed copper is employed, because it is ductile and can easily be bent into various shapes. For gutters, leaders, downspouts, and similar applications, cold-rolled hard copper is employed, because its greater hardness and stiffness permit it to stand without large numbers of intermediate supports.

Copper and copper-based alloys, particularly the brasses, are employed for water pipe in buildings, because of their corrosion resistance. Electrolytic tough-pitch copper is usually employed for electrical conductors, but for maximum electrical conductivity and weldability, oxygen-free high-conductivity copper is used.

When arsenic is added to copper, it appears to form a tenacious adherent film, which is particularly resistant to pitting corrosion. Phosphorus is a powerful deoxidizer and is particularly useful for copper to be used for refrigerator tubing and other applications where flaring, flanging, and spinning are required. Arsenic and phosphorus both reduce the electrical conductivity of the copper.

For flashing, copper is frequently coated with lead to avoid the green patina formed on copper that is sometimes objectionable when it is washed down over adjacent surfaces, such as ornamental stone. The patina is formed particularly in industrial atmospheres. In rural atmospheres, where industrial gases are absent, the copper normally turns to a deep brown color.

Principal types of copper and typical uses are (see also Table 4-26):

Electrolytic tough pitch (99.90% copper) is used for electrical conductors—bus bars, commutators, etc.; building products—roofing, gutters, etc.; process equipment—kettles, vats, distillery equipment; forgings. General properties are high electrical conductivity, high thermal conductivity, and excellent working ability.

Deoxidized (99.90% copper and 0.025% phosphorus) is used, in tube form, for water and refrigeration service, oil burners, etc.; in sheet and plate form, for welded construction. General properties include higher forming and bending qualities than electrolytic copper. They are preferred for coppersmithing and welding (because of resistance to embrittlement at high temperatures).

TABLE 4-26 Copper and Copper Alloys*

Alloy name	Electrolytic tough pitch		Deoxidized		Commercial bronze, 90%		Red brass, 85%		Cartridge brass, 70%		Muntz metal		High-leaded brass		Forging brass		Architectural bronze	
Working properties:																		
1. Cold-working	1. Excellent		1. Excellent		1. Excellent		1. Excellent		1. Excellent		1. Fair		1. Poor		1. Fair		1. Very poor	
2. Hot-working	2. Excellent		2. Excellent		2. Excellent		2. Good		2. Good		2. Excellent		2. Poor		2. Excellent		2. Excellent	
3. Machining	3. Fair		3. Poor		3. Poor		3. Poor		3. Fair		3. Good		3. Excellent		3. Good		3. Good	
4. Welding†	4. Deoxidized copper preferred		4. ‡		4. ‡		4. ‡		4. ‡		4. ‡		4. Not recommended		4. Not recommended		4. Not recommended	
5. Soldering	5. Excellent		5. Excellent		5. Excellent		5. Excellent		5. Excellent		5. Excellent		5. Excellent		5. Good		5. Excellent	
6. Polishing	6. Excellent		6. Excellent		6. Excellent		6. Excellent		6. Excellent		6. Excellent		6. Excellent		6. Excellent		6. Excellent	
Young's modulus of elasticity, psi (000,000 omitted)	17		17		17		17		16		15		15		15		14	
Avg coefficient of thermal expansion, per °F (68 to 570°F)	0.0000098		0.0000098		0.0000102		0.0000104		0.0000111		0.0000116		0.0000113		0.0000115		0.0000116	
Electrical conductivity, % IACS at 68°F	101		85 annealed		44		37		28		28		26		27		28	
Thermal conductivity, Btu/(ft²)(ft)(hr)(/°F), at 68°F	226		196		109		92		70		71		67		69		71	
	Hard	Soft	Hard	Soft	Hard	Soft	Hard	Soft	Hard	Soft	Hard	Soft	Hard	Soft	Hard	Soft¶	Hard	Soft¶
Tensile strength, 1000 psi:																		
Sheet	50	32	50	32	61	37	70	40	76	47	80	54	74	49		52		60
Rod	48	32	55	32		38	70	40	70	48		45		52				
Tube									78	47								
Elongation, % in 2 in:																		
Sheet	6	45	6	45	5	45	5	47	8	62	10	45	7	52		45		30
Rod	16	55	8	45		50	8	55	30	65								
Tube									8	65								
Yield strength, 1000 psi: 0.5% extension under load:																		
Sheet	47	8	45	10	54	10	57	12	63	15	60	21	60	17		20		20
Rod	45	8	50	10		12	58	12	52	16								
Tube									64	15								
Rockwell hardness:																		
Sheet	50B	40F	50B	40F	70B	53F	77B	59F	82B	64F	85B	80F	80B	68F		78F		65B
Rod	47B	40F	60B	40F		57F	77B	60F	80B	65F								
Tube									82B	64F								

4-53

TABLE 4-26 Copper and Copper Alloys* (Continued)

Alloy name	Admiralty	Manganese bronze (A)	Nickel silver, 18% (A), deep drawing	Cupronickel 10%	Cupronickel, 30%	Silicon bronze (high-silicon bronze, A)	Aluminum silicon bronze	Phosphor bronze, 5% (A)	Aluminum bronze, 5%
Working properties:									
1. Cold-working	1. Good	1. Poor	1. Excellent	1. Excellent	1. Excellent	1. Good	1. Poor	1. Excellent	1. Good
2. Hot-working	2. Poor	2. Excellent	2. Fair	2. Fair	2. Fair	2. Excellent	2. Excellent	2. Poor	2. Good
3. Machining	3. Fair	3. Good	3. Fair	3. Fair	3. Fair	3. Fair	3. Excellent	3. Fair	3. Fair
4. Welding†	4. ‡	4. ‡	4. ‡	4. ‡	4. ‡	4. ‡	4. ‡	4. ‡	4. ‡
5. Soldering	5. Good	5. Excellent	5. Excellent	5. Excellent	5. Excellent	5. Excellent	5. Fair	5. Excellent	5. Good
6. Polishing	6. Good	6. Excellent	6. Excellent	6. Excellent	6. Good	6. Good	6. Excellent	6. Excellent	6. Good
Young's modulus of elasticity, psi (000,000 omitted)	16	15	18	18	22	17	14	16	17
Avg. coefficient of thermal expansion, per °F (68 to 570°F)	0.0000112	0.0000118	0.0000090	0.0000095	0.0000090	0.0000100	0.0000100	0.0000099	0.0000099
Electrical conductivity, % IACS at 68°F	25	24	6	9	4.6	7	7	15	18
Thermal conductivity, Btu/(ft²)(ft)(hr)(°F), at 68°F	64	61	19	22	17	21	26	40	48

	Hard	Soft	Hard	Soft	Hard	Soft	Hard	Soft	Hard	Soft	Hard§	Soft	Hard§	Soft	Hard	Soft	Hard	Soft	Soft
Tensile strength, 1000 psi:																			
Sheet	100			55	85	58	65		40	82	54	94	60		80	100	47		
Rod		53	80					60	44	80	60					75			50
Tube																			
Elongation, % in 2 in:																			
Sheet	3		19	33	3	40		30	4	35	8		60		20	4	64		
Rod		65														25			60
Tube							10	42	5	45									
Yield strength, 1000 psi:																			
0.5% extension under load:																			
Sheet	80		56		74	25		15	76	10	58	25		43	80	19			
Rod		22		22			57	16	75	25						65			19
Tube																			
Rockwell hardness:																			
Sheet	95B		90B	65B	87B	85F	70	20	85B	40B	93B	82F		80B	95B		73F		
Rod		75F							65F	80F					80B				65F
Tube								72B											

*Revere Copper and Brass, Inc.
†By fusion methods; conductivity of the coppers makes resistance welding (spot and seam) impractical. Coppers, however, may be resistance-brazed by a patented method.
Basis of rating: sheet, hard—0.040 in stock, previously rolled 4 B & S numbers hard (37.5% area reduction).
Basis of rating: rod, hard—1 in and under diameter, previously drawn through 25 to 35% area reduction.
‡For welding—gas-shielded-arc processes preferred.
¶Shapes.
§Hard, for rod, bolt temper, extra hard.

4-62. Plain Brass.

A considerable range of brasses is obtainable for a large variety of end uses. The high ductility and malleability of the copper-zinc alloys, or brasses, make them suitable for operations like deep drawing, bending, and swaging. They have a wide range of colors. They are generally less expensive than the high-copper alloys.

Grain size of the metal has a marked effect upon its mechanical properties. For deep drawing and other heavy working operations, a large grain size is required, but for highly finished polished surfaces, the grain size must be small.

Like copper, brass is hardened by cold working. Hardnesses are sometimes expressed as quarter hard, half hard, hard, extra hard, spring, and extra spring, corresponding to reductions in cross section during cold working ranging from approximately 11 to 69%. Hardness is strongly influenced by alloy composition, original grain size, and form (strip, rod, tube, wire).

Brass compositions range from higher copper content to zinc contents as high as 40% or more. Brasses with less than 36% zinc are plain alpha solid solutions; but Muntz metal, with 40% zinc, contains both alpha and beta phases.

The principal plain brasses of interest in building, and their properties are (see also Table 4-26):

Commercial bronze, 90% (90.0% copper, 10.0% zinc). Typical uses are forgings, screws, weatherstripping, and stamped hardware. General properties include excellent cold working and high ductility.

Red brass, 85% (85.0% copper, 15.0% zinc). Typcial uses are dials, hardware, etched parts, automobile radiators, and tube and pipe for plumbing. General properties are higher strength and ductility than copper, and excellent corrosion resistance.

Cartridge brass, 70% (70.0% copper, 30.0% zinc). Typical uses are deep drawing, stamping, spinning, etching, rolling—for practically all fabricating processes—cartridge cases, pins, rivets, eyelets, heating units, lamp bodies and reflectors, electrical sockets, drawn shapes, etc. General properties are best combination of ductility and strength of any brass, and excellent cold-working properties.

Muntz metal (60.0% copper, 40.0% zinc). Typical uses are sheet form, perforated metal, architectural work, condenser tubes, valve stems, and brazing rods. General properties are high strength combined with low ductility.

4-63. Leaded Brass.

Lead is added to brass to improve its machinability, particularly in such applications as automatic screw machines where a freely chipping metal is required. Leaded brasses cannot easily be cold-worked by such operations as flaring, upsetting, or cold heading. Several leaded brasses of importance in the building field are the following (see also Table 4-26):

High-leaded brass (64.0% copper, 34.0% zinc, 2.0% lead). Typical uses are engraving plates, machined parts, instruments (professional and scientific), nameplates, keys, lock parts, and tumblers. General properties are free machining and good blanking.

Forging brass (60.0% copper, 38.0% zinc, 2.0% lead). Typical uses are hot forging, hardware, and plumbing goods. General properties are extreme plasticity when hot and a combination of good corrosion resistance with excellent mechanical properties.

Architectural bronze (56.5% copper, 41.25% zinc, 2.25% lead). Typical uses are handrails, decorative moldings, grilles, revolving door parts, miscellaneous architectural trim, industrial extruded shapes (hinges, lock bodies, automotive parts). General properties are excellent forging and free-machining properties.

4-64. Tin Brass.

Tin is added to a variety of basic brasses to obtain hardness, strength, and other properties that would otherwise not be available. Two important alloys are (see also Table 4-26):

Admiralty (71.0% copper, 28.0% zinc, 1.0% tin, 0.05% arsenic). Typical uses are condenser and heat-exchanger plates and tubes, steam-power-plant equipment, chemical and process equipment, and marine uses. General properties are excellent corrosion resistance, combined with strength and ductility.

Manganese bronze (58.5% copper, 39.0% zinc, 1.4% iron, 1.0% tin, 0.1% manganese). Typical uses are forgings, condenser plates, valve stems, and coal screens. General properties are high strength combined with excellent wear resistance.

4-65. Nickel Silvers.

These are alloys of copper, nickel, and zinc. Depending on the composition, they range in color from a definite to slight pink cast through yellow, green, whitish green, whitish blue, to blue. A wide range of nickel silvers is made, of which only one typical composition will be described (see also Table 4-26). Those that fall in the combined alpha-beta phase of metals are readily hot-worked and therefore are fabricated without difficulty into such intricate shapes as plumbing fixtures, stair rails, architectural shapes, and escalator parts. Lead may be added to improve machining.

Nickel silver, 18% (A) (65.0% copper, 17.0% zinc, 18.0% nickel). Typical uses are hardware, architectural panels, lighting, electrical and plumbing fixtures. General properties are high resistance to corrosion and tarnish, malleable, and ductile. Color: silver-blue-white.

4-66. Cupronickel.

Copper and nickel are alloyed in a variety of compositions of which the high-copper alloys are called the cupronickels. Typical commercial types of cupronickel contain 10% or 30% nickel (Table 4-26):

Cupronickel, 10% (88.5% copper, 10% nickel, 1.5% iron). Recommended for applications requiring corrosion resistance, especially to salt water, as in tubing for condensers, heat exchangers, and formed sheets.

Cupronickel, 30% (70.0% copper, 30.0% nickel). Typical uses are condenser tubes and plates, tanks, vats, vessels, process equipment, automotive parts, meters, refrigerator pump valves. General properties are high strength and ductility and resistance to corrosion and erosion. Color: white-silver.

4-67. Silicon Bronze.

These are high-copper alloys containing percentages of silicon ranging from about 1% to slightly more than 3%. In addition, they generally contain one or more of the four elements, tin, manganese, zinc, and iron. A typical one is high-silicon bronze, type A (see also Table 4-26):

High-silicon bronze, A (96.0% copper, 3.0% silicon, 1.0% managanese). Typical uses are tanks—pressure vessels, vats; weatherstrips, forgings. General properties are corrosion resistance of copper and mechanical properties of mild steel.

4-68. Aluminum Bronze.

Like aluminum, these bronzes form an aluminum oxide skin on the surface, which materially improves resistance to corrosion, particularly under acid conditions. Since the color of the 5% aluminum bronze is similar to that of 18-carat gold, it is used for costume jewelry and other decorative purposes. Aluminum-silicon bronzes (Table 4-26) are used in applications requiring high tensile properties in combination with good corrosion resistance in such parts as valves, stems, air pumps, condenser bolts, and similar applications. Their wear-resisting properties are good; consequently, they are used in slide liners and bushings.

4-69. Tin Bronze.

Originally and historically, the bronzes were all alloys of copper and tin. Today, the term bronze is generally applied to engineering metals having high mechanical properties and the term brass to other metals. The commercial wrought bronzes do not usually contain more than 10% tin because the metal becomes extremely hard and brittle. When phosphorus is added as a deoxidizer, to obtain sound dense castings, the alloys are known as phosphor bronzes. The two most commonly used tin bronzes contain 5 or 8% tin. Both have excellent cold-working properties (Table 4-26).

4-70. Copper Bibliography.

American Society for Testing and Materials, Philadelphia Pa.: "Standards."
"Alloy Data," Copper Development Assn., 405 Lexington Ave, New York, N.Y. 10017

LEAD AND LEAD-BASED ALLOYS

Lead is used primarily for its corrosion resistance. Lead roofs 2000 years old are still intact. Exposure tests indicate corrosion penetrations ranging from less than 0.0001 in to less than 0.0003 in in 10 years in atmospheres ranging from mild rural to severe industrial and seacoast locations. Sheet lead is therefore used for roofing, flashing, spandrels, gutters, and downspouts.

4-71. Applications of Lead.

Because the green patina found on copper may wash away sufficiently to stain the surrounding structure, lead-coated copper is frequently employed. ASTM B101-78 covers two classes, as listed in Table 4-27.

Lead pipe for the transport of drinking water should be used with care. Distilled and very soft waters slowly dissolve lead and may cause cumulative lead poisoning. Hard

TABLE 4-27 Classes of Lead-Coated Copper

Class	Weight of lead coating, lb per 100 ft^2, total, coated both sides	
	Min	Max
Class A standard (general utility)	12	15
Class B heavy	20	30

Properties of Nickel and Its Alloys   4-59

waters apparently deposit a protective coating on the wall of the pipe and little or no lead is subsequently dissolved in the water.

Principal alloying elements used with building leads are antimony (for hardness and strength) and tin. But copper, arsenic, bismuth, nickel, zinc, silver, iron, and manganese are also added in varying proportions.

Soft solders consist of varying percentages of lead and tin. For greater hardness, antimony is added, and for higher-temperature solders, silver is added in small amounts. ASTM Standard B32 specifies properties of soft solders.

Low-melting alloys and many bearing metals are alloys of lead, bismuth, tin, cadmium, and other metals including silver, zinc, indium, and antimony. The fusible links used in sprinkler heads and fire-door closures, made of such alloys, have a low melting point, usually lower than the boiling point of water. Yield (softening) temperatures range from 73 to 160°F and melting points from about 80 to 480°F, depending on the composition.

4-72. Lead Bibliography.

American Society for Metals, Cleveland, Ohio: "Metals Handbook."
American Society for Testing and Materials, Philadelphia, Pa.: "Standards."

NICKEL AND NICKEL-BASED ALLOYS

Nickel is used mostly as an alloying element with other metals, but it finds use in its own right, largely as electroplate or as cladding metal. Among the principal high-nickel alloys are Monel and Inconel. The nominal compositions of these metals are given in Table 4-28.

4-73. Properties of Nickel and Its Alloys.

Nickel is resistant to alkaline corrosion under nonoxidizing conditions but is corroded by oxidizing acids and oxidizing salts. It is resistant to fatty acids, other mildly acid

TABLE 4-28 Composition of Nickel Alloys

Content	Nickel alloy, low-carbon N02201	Nickel alloy N02200	Monel N04400	Inconel N06600	70–30 cupro-nickel C71500	90–10 cupro-nickel C70600
	ASTM B160	ASTM B160	ASTM B127	ASTM B168	ASTM B171	ASTM B171
Carbon	0.02	0.15	0.3	0.15 max		
Manganese	0.35	0.35	2.00 max	1.0 max	1.0 max	1.0 max
Sulfur	0.01	0.01	0.024 max	0.015 max		
Silicon	0.35	0.35	0.5	0.5 max		
Chromium				14–17		
Nickel	99 min	99 min	63–70	72 min	29–33	9–11
Copper	0.25	0.25	Remainder	0.5 max	65 min	86.5 min
Iron	0.40 max	0.40 max	2.5 max	6–10	0.40–1.0	1.0–1.8
Lead					0.05 max	0.05 max
Zinc					1.0	1.0

conditions, such as food processing and beverages, and resists oxidation at temperatures as high as 1600°F.

Monel is widely used in kitchen equipment. It is better than nickel in reducing conditions like warm unaerated acids, and better than copper under oxidizing conditions, such as aerated acids, alkalies, and salt solutions. It is widely used for handling chlorides of many kinds.

Inconel is almost completely resistant to corrosion by food products, pharmaceuticals, biologicals, and dilute organic acids. It is superior to nickel and Monel in resisting oxidizing acid salts like chromates and nitrates but is not resistant to ferric, cupric, or mercuric chlorides. It resists scaling and oxidation in air and furnace atmospheres at temperatures up to 2000°F.

4-74. Nickel Bibliography.

International Nickel Co., New York: "Nickel and Nickel Alloys."

Hoerson, Albert, Jr.: "Nonferrous-clad Plate Steels," Chap. 13 in A. G. H. Dietz, "Composite Engineering Laminates," M.I.T. Press, Cambridge, Mass.

PLASTICS

The synonymous terms plastics and synthetic resins denote synthetic organic high polymers, all of which are plastic at some stage in their manufacture. Plastics fall into two large categories—thermoplastic and thermosetting materials.

4-75. General Properties of Plastics.

Thermoplastics may be softened by heating and hardened by cooling any number of times. Thermosetting materials are either originally soft or liquid, or they soften once upon heating; but upon further heating, they harden permanently. Some thermosetting materials harden by an interlinking mechanism in which water or other by-product is given off, by a process called condensation; but others, like the unsaturated polyesters, harden by a direct interlinking of the basic molecules without release of a by-product.

Most plastics are modified with plasticizers, fillers, or other ingredients. Consequently, each base material forms the nucleus for a large number of products having a wide variety of properties. This section can only indicate generally the range of properties to be expected.

Because plastics are quite different in their composition and structure from other materials, such as metals, their behavior under stress and under other conditions is likely to be different from other materials. Just as steel and lead are markedly different and are used for different applications, so the various plastics materials—some hard and brittle, others soft and extensible—must be designed on different bases and used in different ways. Some plastics show no yield point, because they fail before a yield point can be reached. Others have a moderately high elastic range, followed by a highly plastic range. Still others are highly extensible and are employed at stresses far beyond the yield point.

More than many other materials, plastics are sensitive to temperature and to the rate and time of application of load. How these parameters influence the properties is indicated in a general way in Fig. 4-3, which shows that for many plastics an increase in temperature, increase in plasticizer content, and decrease in rate of load application mean an increase in strain to fracture, accompanied by a decrease in maximum stress. This viscoelastic behavior, combining elastic and viscous or plastic reaction to stress, is unlike the behavior of materials which are traditionally considered to behave only elastically.

4-76. Fillers and Plasticizers.

Fillers are commonly added, particularly to the thermosetting plastics, to alter their basic characteristics. For example, wood flour converts a hard, brittle resin, difficult to handle, into a cheaper, more easily molded material for general purposes. Asbestos fibers provide better heat resistance; mica gives better electrical properties; and a variety of fibrous materials, such as chopped fibers, chopped fabric, and chopped tire cords, increase the strength and impact properties.

Plasticizers are added to many thermoplastics, primarily to transform hard and rigid materials into a variety of forms having varying degrees of softness, flexibility, and strength. In addition, dyes or pigments, stabilizers, and other products may be added.

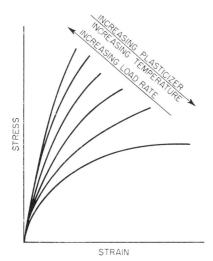

Fig. 4-3 Stress-strain diagram shows the influence of temperature, plasticizer, and rate of loading on behavior of plastics.

4-77. Molding and Fabricating Methods for Plastics.

Both thermosetting and thermoplastic molding materials are formed into final shape by a variety of molding and fabricating methods:

Thermosetting materials are commonly formed by placing molding powder or molded preform in heated dies and compressing under heat and pressure into the final infusible shape. Or they are formed by forcing heat-softened material into a heated die for final forming into the hard infusible shape.

Thermoplastics are commonly formed by injection molding; that is, by forcing soft, hot plastic into a cold die, where it hardens by cooling. Continuous profiles of thermoplastic materials are made by extrusion. Thermoplastic sheets, especially transparent acrylics, are frequently formed into final shape by heating and then blowing to final form under compressed air or by drawing a partial vacuum against the softened sheet.

Foamed plastics are employed for thermal insulation in refrigerators, buildings, and many other applications. In buildings, plastics are either prefoamed into slabs, blocks, or other appropriate shapes, or they are foamed in place.

Prefoamed materials, such as polystyrene, are made by adding a blowing agent and extruding the mixture under pressure and at elevated temperatures. As the material emerges from the extruder, it expands into a large "log" that can be cut into desired shapes. The cells are "closed"; that is, they are not interconnecting and are quite impermeable.

Foamed-in-place plastics are made with pellets or liquids. The pellets, made, for example, of polystyrene, are poured into the space to be occupied, such as a mold, and heated, whereupon they expand and occupy the space. The resulting mass may be permeable between pellets. Liquid-based foams, exemplified by polyurethane, are made by mixing liquid ingredients and immediately casting the mixture into the space to be occupied. A quick reaction results in a foam that rises and hardens by a thermosetting reaction. When blown with fluorocarbon gases, such foams have exceptionally low thermal conductivities.

4-62 Building Materials

All the plastics can be machined, if proper allowance is made for the properties of the materials.

Plastics are often combined with sheet or mat stocks, such as paper, cotton muslin, glass fabric, glass filament mats, nylon fabric, and other fabrics, to provide laminated materials in which the properties of the combined plastic and sheet stock are quite different from the properties of either constituent by itself. Two principal varieties of laminates are commonly made: (1) High-pressure laminates employing condensation-type thermosetting materials, which are formed at elevated temperatures and pressures. (2) Reinforced plastics employing unsaturated polyesters and epoxies, from which no by-products are given off, and consequently, either low pressures or none at all may be required to form combinations of these materials with a variety of reinforcing agents, like glass fabric or mat.

In the following articles, the principal varieties of plastics are briefly described and their principal fields of application indicated. In Table 4-29, ranges of values to be expected of specimens made and tested under specific conditions, usually those of the American Society for Testing and Materials, are given. The method of fabrication and specific application of a particular plastic may severely modify these properties.

4-78. Thermosetting Plastics.

General properties of thermosetting plastics are described in Art. 4-75. Following are properties of several thermosetting plastics used in buildings:

Phenol Formaldehyde. These materials provide the greatest variety of thermosetting molded plastic articles. They are used for chemical, decorative, electrical, mechanical, and thermal applications of all kinds. Hard and rigid, they change slightly, if at all, on aging indoors but, on outdoor exposure, lose their bright surface gloss. However, the outdoor-exposure characteristics of the more durable formulations are otherwise generally good. Phenol formaldehydes have good electrical properties, do not burn readily, and do not support combustion. They are strong, light in weight, and generally pleasant to the eye and touch, although light colors by and large are not obtainable because of the fairly dark-brown basic color of the resin. They have low water absorption and good resistance to attack by most commonly found chemicals.

Epoxy and Polyester Casting Resins. These are used for a large variety of purposes. For example, electronic parts with delicate components are sometimes cast completely in these materials to give them complete and continuous support, and resistance to thermal and mechanical shock. Some varieties must be cured at elevated temperatures; others can be formulated to be cured at room temperatures. One of the outstanding attributes of the epoxies is their excellent adhesion to a variety of materials, including such metals as copper, brass, steel, and aluminum.

Polyester Molding Materials. When compounded with fibers, particularly glass fibers, or with various mineral fillers, including clay, the polyesters can be formulated into putties or premixes that are easily compression- or transfer-molded into parts having high impact resistance.

Melamine Formaldehyde. These materials are unaffected by common organic solvents, greases, and oils, as well as most weak acids and alkalies. Their water absorption is low. They are insensitive to heat and are highly flame-resistant, depending on the filler. Electrical properties are particularly good, especially resistance to arcing. Unfilled materials are highly translucent and have unlimited color possibilities. Principal fillers are alpha cellulose for general-purpose compounding; minerals to improve electrical properties, particularly at elevated temperatures; chopped fabric to afford high shock resistance and flexural strength; and cellulose, mainly for electrical purposes.

TABLE 4-29 Selected Properties of Plastics*

Property	ASTM test method	Acrylonitrile butadiene styrene	Acrylic PMMA	Cellulosics	Epoxies	Fluoroplastics	Melamine formaldehyde	Nylon polyamide	Phenol formaldehyde
Tensile strength, psi	D638	2,500–11,000	7,000–11,000	2,000–9,000	2,500–20,000	2,000–12,000	5,000–13,000	10,000–28,000	3,500–18,000
Elongation, %	D638	3–100	2–10	5–800	0.5–70	8–500	0.6–1.0	2–300	0.2–4.0
Compressive strength, psi	D695	4,500–14,000	12,000–18,000	2,100–8,000	1,000–40,000	1,700–10,000	20,000–45,000	12,000–29,400	12,000–34,000
Flexural strength, psi	D790	4,000–15,500	13,000–19,000	2,000–16,000	1,000–30,000	2,000–11,000	9,000–23,000	5,000–41,000	5,500–60,000
Tensile elastic modulus, ksi	D638	130–740	380–450	50–250	1–3,000	50–1,200	1,100–2,400	100–1,450	800–3,300
Compressive elastic modulus, ksi	D695	130–450	370–460			60–120		180–370	
Flexural elastic modulus, ksi	D790	130–710	420–460	90–300	500–4,500	80–950	1,100	140–1,300	800–3,300
Izod impact, ft·lb/in, notch	D256A	2.5–12	0.3–0.5	0.4–8.5	0.15–10.0	2.5–No break	0.2–18	0.6–No break	0.2–18.0
Hardness, Rockwell	D785	R75–115 M85	M85–105	R10–125	M55–112	D50–80	M115–125	R81–120 M78–101	M95–120
Thermal expansion, 10^{-6} in/(in)(°C)	D696	21–130	50–90	80–200	5.5–100	10–100	15–45	15–100	8–75
Thermal conductivity, 10^{-4} cal·cm/(s)(cm²)(°C)	C177	4.5–8.0	4.0–6.0	3.8–8.0	4.0–35.0	4.7–10.0	6.5–11.5	5.1–9.6	3.5–14
24-hr water absorption, 1/8-in thickness, %	D570	0.20–0.60	<0.01–0.13	0.8–6.5				0.2–2.0	0.03–1.2

Property	Polycarbonate	Polyesters, thermosetting	Polyethylene	Poly-propylene	Polystyrene and copolymers	Silicones	Urethanes	Urea-formaldehyde	Vinyl and copolymers
Tensile strength, psi	7,000–19,000	500–50,000	600–9,000	4,000–15,000	2,800–18,000	350–6,500	175–11,000	5,500–13,000	1,500–12,000
Elongation, %	3–110	<1–310	1.5–900	2–600	1–100	100–700	3–1,100	<1	2–450
Compressive strength, psi	11,000–18,000	13,000–50,000	2,700–7,000	3,500–9,800	4,000–19,000	Flexible–20,000	20,000	25,000–45,000	900–22,000
Flexural strength, psi	13,500–23,000	6,000–80,000	2,000–11,000	5,000–22,000	4,200–20,000	Flexible–14,000	700–19,000	10,000–18,000	4,200–17,000
Tensile elastic modulus, ksi	345–1,250	500–4,500	14–500	100–1,500	220–1,200		10–350	1,000–1,500	50–600
Compressive elastic modulus, ksi	350–1,300		8–150				10–100		
Flexural elastic modulus	300–1,100	1,000–3,000	8–350	130–1,000	240–1,150	Flexible–2,500	10–610	1,300–1,600	380–750
Izod impact, ft·lb/in, notch	1.2–No break	0.2–30.0	0.5–No break	0.4–20	0.25–3.4	0.25–No break	0.4–flexible	0.25–0.40	0.3–5.6
Hardness, Rockwell	R117, M70–92	50–80 (Barcol)	R50–75	R50–111 M50–85	M10–90	Soft–M90	30–35 (Barcol)	M110–120	R117–122 M50–85
Thermal Expansion, 10^{-6} in/(in)(°C)	22–68	15–100	48–220	27–100	36–101	20–800	100–200	22–36	64–190
Thermal conductivity, 10^{-4} cal·cm/(s)(cm²)(°C)	4.6–9.0	15–25	8–12	2.8–8.8	2.0–6.6	3.5–7.5	1.7–7.4	7–10	3–3.7
24-hr water absorption, 1/8-in thickness, %	0.14	0.01–2.5	<0.01–0.13	<0.01–0.06					0.01–3.0

*Modern Plastics Encyclopedia.

4-64   Building Materials

Polyurethane. This plastic is used in several ways in building. As thermal insulation, it is used in the form of foam, either prefoamed or foamed in place. The latter is particularly useful in irregular spaces. When blown with fluorocarbons, the foam has an exceptionally low K-factor and is, therefore, widely used in thin-walled refrigerators. Other uses include field-applied or baked-on clear or colored coatings and finishes for floors, walls, furniture, and casework generally. The rubbery form is employed for sprayed or troweled-on roofing, and for gaskets and calking compounds.

Urea Formaldehyde. Like the melamines, these offer unlimited translucent to opaque color possibilities, light-fastness, good mechanical and electrical properties, and resistance to organic solvents as well as mild acids and alkalies. Although there is no swelling or change in appearance, the water absorption of urea formaldehyde is relatively high, and it is therefore not recommended for applications involving long exposure to water. Occasional exposure to water is without deleterious effect. Strength properties are good, although special shock-resistant grades are not made.

Silicones. Unlike other plastics, silicones are based on silicon rather than carbon. As a consequence, their inertness and durability under a wide variety of conditions are outstanding. As compared with the phenolics, their mechanical properties are poor, and consequently glass fibers are added. Molding is more difficult than with other thermosetting materials. Unlike most other resins, they may be used in continuous operations at 400°F; they have very low water absorption; their dielectric properties are excellent over an extremely wide variety of chemical attack; and under outdoor conditions their durability is particularly outstanding. In liquid solutions, silicones are used to impart moisture resistance to masonry walls and to fabrics. They also form the basis for a variety of paints and other coatings capable of maintaining flexibility and inertness to attack at high temperaures in the presence of ultraviolet sunlight and ozone. Silicone rubbers maintain their flexibility at much lower temperatures than other rubbers.

4-79. Thermoplastic Resins.

Materials under this heading in general can be softened by heating and hardened by cooling.

Acrylics. In the form of large transparent sheets, these are used in aircraft enclosures and building construction. Although not so hard as glass, they have perfect clarity and transparency. Among the most resistant of the transparent plastics to sunlight and outdoor weathering, they possess an optimum combination of flexibility and sufficient rigidity with resistance to shattering. A wide variety of transparent, translucent, and opaque colors can be produced. The sheets are readily formed to complex shapes. They are used for such applications as transparent windows, outdoor and indoor signs, parts of lighting equipment, decorative and functional automotive parts, reflectors, household-appliance parts, and similar applications. They can be used as large sheets, molded from molding powders, or cast from the liquid monomer.

Acrylonitrile-Butadiene-Styrene (ABS). This three-way copolymer provides a family of tough, hard, chemically resistant resins with many grades and varieties, depending on variations in constituents. The greatest use is for pipes and fittings, especially drain-waste-vent (DWV). Other uses include buried sewer and water lines, mine pipe, well casings, conduit, and appliance housings.

Polyethylene. In its unmodified form, this is a flexible, waxy, translucent plastic. It maintains flexibility at very low temperatures, in contrast with many other thermoplastic materials.

Polyethylene may be provided as low-density, or standard, or as high-density, or linear material. High-density polyethylene has greater strength and stiffness, withstands somewhat higher temperatures, and has a more sharply defined softening temperature range. The heat-distortion point of the low-density polyethylenes is low; these plastics are not recommended for uses above 150°F. However, high-density materials have higher heat-distortion temperatures. Some may be heated to temperatures above 212°F. But the heat-distortion point may rise well above 250°F for plastics irradiated with high-energy beams. Unlike most plastics, this material is partly crystalline. It is highly inert to solvents and corrosive chemicals of all kinds at ordinary temperatures. Usually low moisture permeability and absorption are combined with excellent electrical properties. Its density is lower than that of any other commercially available nonporous plastic. When compounded with black pigment, its weathering properties are good. It is widely used as a primary insulating material on wire and cable and has been used as a replacement for the lead jacket in communication cables and other cables. It is widely used also as thin flexible film for packaging, particularly of food, and as corrosionproof lining for tanks and other chemical equipment.

Polypropylene. This polyolefin is similar in many ways to its counterpart, polyethylene, but is generally harder, stronger, and more temperature-resistant. It finds a great many uses, among them piping and complete water cisterns for water closets in plumbing systems.

Polycarbonate. Excellent transparency, high impact resistance and good resistance to weathering combine to recommend this plastic for safety glazing and for general illumination and shatter-resistant fixtures. It is available in large, clear, tinted, and opaque sheets that can be formed into shells, domes, globes, and other forms. It can be processed by standard molding methods.

Polytetrafluoroethylene. When fluorine replaces hydrogen, the resulting plastic is a highly crystalline linear-type polymer, unique among organic compounds in its chemical inertness and resistance to change at high and low temperatures. Its electrical properties are excellent. Its outstanding property is extreme resistance to attack by corrosive agents and solvents of all kinds. At temperatures well above 500°F, it can be held for long periods with practically no change in properties, except loss in tensile strength. Service temperatures are generally maintained below 480°F. It is not embrittled at low temperatures, and its films remain flexible at temperatures below -100°F. Polytetrafluoroethylene is used in buildings in which resistance to extreme conditions or low friction is desired. In steam lines, for example, supporting pads of this plastic permit the lines to slide easily over the pads. The temperatures involved have little or no effect. Other low-friction applications include, for example, seats for trusses. Mechanical properties are only moderately high, and reinforcement may be necessary to prevent creep and squeeze-out under heavy loads. These fluorocarbons are difficult to wet; consequently, they are often used as parting agents, or where sticky materials must be handled.

Polyvinylfluoride. This has much of the superior inertness to chemical and weathering attack typical of the fluorocarbons. Among other uses, it is used as thin-film overlays for building boards to be exposed outdoors.

Polyvinyl Formal and Polyvinyl Butyral. Polyvinyl formal resins are principally used as a base for tough, water-resistant insulating enamel for electric wire. Polyvinyl butyral is the tough interlayer in safety glass. In its cross-linked and plasticized form, polyvinyl butyral is extensively used in coating fabrics for raincoats, upholstery, and other heavy-duty moisture-resistant applications.

Vinyl Chloride Polymers and Copolymers. These materials vary from hard and rigid to highly flexible. Polyvinyl chloride is naturally hard and rigid but can be plasticized to any required degree of flexibility as in raincoats and shower curtains. Copolymers, including vinyl chloride plus vinyl acetate, are naturally flexible without plasticizers. Nonrigid vinyl plastics are widely used as insulation and jacketing for electric wire and cable because of their electrical properties and their resistance to oil and water. Thin films are used for rainwear and similar applications, whereas heavy-gage films and sheets are widely used for upholstery. Vinyl chlorides are used for floor coverings in the form of tile and sheet because of their abrasion resistance and relatively low water absorption. The rigid materials are used for tubing, pipe, and many other applications where their resistance to corrosion and action of many chemicals, especially acids and alkalies, recommends them. They are attacked by a variety of organic solvents, however. Like all thermoplastics, they soften at elevated temperatures.

Vinylidene Chloride. This material is highly resistant to most inorganic chemicals and to organic solvents generally. It is impervious to water on prolonged immersion, and its films are highly resistant to moisture-vapor transmission. It can be sterilized, if not under load, in boiling water. Mechanical properties are good. It is not recommended for uses involving high-speed impact, shock resistance, or flexibility at subfreezing temperatures.

Polystyrene. Polystyrene formulations constitute a large and important segment of the entire field of thermoplastic materials. Numerous modified polystyrenes provide a relatively wide range of properties. Polystyrene is one of the lightest of the presently available commercial plastics. It is relatively inexpensive, easily molded, has good dimensional stability, and good stability at low temperatures; it is brilliantly clear when transparent and has an infinite range of colors. Water absorption is negligible even after long immersion. Electrical characteristics are excellent. It is resistant to most corrosive chemicals, such as acids, and to a variety of organic solvents, although it is attacked by others. Polystyrenes as a class are considerably more brittle and less extensible than many other thermoplastic materials, but these properties are markedly improved in copolymers. Under some conditions, they have a tendency to develop fine cracks, known as craze marks, on exposure, particularly outdoors. This is true of many other thermoplastics, especially when highly stressed.

Nylon. Molded nylon is used in increasing quantities for impact and high resistance to abrasion. It is employed in small gears, cams, and other machine parts, because even when unlubricated they are highly resistant to wear. Its chemical resistance, except to phenols and mineral acids, is excellent. Extruded nylon is coated onto electric wire, cable, and rope for abrasion resistance. Applications like hammerheads indicate its impact resistance.

4-80. Cellulose Derivatives.

Cellulose is a naturally occurring high polymer found in all woody plant tissue and in such materials as cotton. It can be modified by chemical processes into a variety of thermoplastic materials, which, in turn, may be still further modified with plasticizers, fillers, and other additives to provide a wide variety of properties.

Cellulose Acetate. This is the basis of safety film developed to overcome the highly flammable nature of cellulose nitrate, the first plastic to be commercially available. Starting as film, sheet, or molding powder, cellulose acetate is made into a variety of items such as transparent packages and a large variety of general-purpose items. Depending on the plasticizer content, it may be hard and rigid or soft and flexible. Moisture absorption of this and all other cellulosics is relatively high, and they are therefore not recommended for long-continued outdoor exposure. But cellulose acetate film, reinforced with metal mesh, is widely used for temporary enclosures of buildings during construction.

Cellulose Acetate Butyrate. The butyrate copolymer is inherently softer and more flexible than cellulose acetate and consequently requires less plasticizer to achieve a given degree of softness and flexibility. It is made in the form of clear transparent sheet and film, or in the form of molding powders, which can be molded by standard injection-molding procedures into a wide variety of applications. Like the other cellulosics, this material is inherently tough and has good impact resistance. It has infinite colorability, like the other cellulosics. Cellulose acetate butyrate tubing is used for such applications as irrigation and gas lines.

Cellulose Nitrate. One of the toughest of the plastics, cellulose nitrate is widely used for tool handles and similar applications requiring high impact strength. The high flammability requires great caution, particularly in the form of film. Most commercial photographic film is cellulose nitrate as opposed to safety film. Cellulose nitrate is the basis of most of the widely used commercial lacquers for furniture and similar items.

COMBINATIONS OF PLASTICS AND OTHER MATERIALS

Plastics often are used as part of a composite construction with other materials. The composites may be in the form of laminates, matrix systems, sandwich structures, or combinations of these.

4-81. High-Pressure Laminates.

Laminated thermosetting products consist of fibrous sheet materials combined with a thermosetting resin, usually phenol formaldehyde or melamine formaldehyde. The commonly used sheet materials are paper, cotton fabric, asbestos paper or fabric, nylon fabric, and glass fabric. The usual form is flat sheet, but a variety of rolled tubes and rods is made.

Decorative Laminates. These high-pressure laminates consist of a base of phenolic resin-impregnated kraft paper over which a decorative overlay, such as printed paper, is applied. Over all this is laid a thin sheet of melamine resin. When the entire assemblage is pressed in a hot-plate press at elevated temperatures and pressures, the various layers are fused together and the melamine provides a completely transparent finish, resistant to alcohol, water, and common solvents. This material is widely used for tabletops, counter fronts, wainscots, and similar building applications. It is customarily bonded to a core of plywood to develop the necessary thickness and strength. In this case, a backup sheet consisting of phenolic resin and paper alone, without the decorative surface, is employed to provide balance to the entire sandwich.

4-82. Reinforced Plastics.

These are commonly made with phenolic, polyester, and epoxide resins combined with various types of reinforcing agents, of which glass fibers in the form of mats or fabrics are the most common. Because little or no pressure is required to form large complex parts, rather simple molds can be employed for the manufacture of such things as boat hulls and similar large parts. In buildings, reinforced plastics have been rather widely used in the form of corrugated sheet for skylights and side lighting of buildings, and as molded shells, concrete forms, sandwiches, and similar applications.

These materials may be formulated to cure at ordinary temperatures, or they may require moderate temperatures to cure the resins. Customarily, parts are made by laying up successive layers of the glass fabric or the glass mat and applying the liquid resin to them. The entire combination is allowed to harden at ordinary temperatures, or it is placed in a heated chamber for final hardening. It may be placed inside a rubber bag and a vacuum drawn to apply moderate pressure, or it may be placed between a pair of matching molds and cured under moderate pressure in the molds.

The high impact resistance of these materials combined with good strength properties and good durability recommends them for building applications. When the quantity of reinforcing agent is kept relatively low, a high degree of translucence may be achieved, although it is less than that of the acrylics and the other transparent thermoplastic materials.

Table 4-30 presents some engineering properties of several types of reinforced plastics.

Fabrics for Air-Supported Roofs. Principal requirements for fabrics and coatings for air-supported structures are high strip tensile strength in both warp and fill directions, high tear resistance, good coating adhesion, maximum weathering resistance, maximum joint strength, good flexing resistance, and good flame resistance. Translucency may or may not be important, depending on the application. The most commonly used fabrics are nylon, polyester, and glass. Neoprene and Hypalon have commonly been employed for military and other applications where opacity is desired. For translucent application, vinyl chloride and fluorocarbon polymers are more common. Careful analysis of loads and stresses, especially dynamic wind loads, and means of joining sections and attaching to anchorage is required.

4-83. Polymeric Mortar, Concrete, and Grout.

Properties of portland-cement products may be improved by addition of monomeric or polymeric materials.

Mortars. Styrene-butadiene or polyvinylidene chloride added to masonry mortar increases all strength properties, but most markedly tensile strength (about 100%), flexural strength (about 100%) and shear strength (about 400 to 500%). Compressive strength increases less (about 25 to 75%), and the elastic modulus may increase up to several hundred percent, or it may be reduced. Because of the increase in mortar strength, brick walls laid up in a single wythe may be used instead of double wythe, and brick panels can be prelaid and hoisted into place. Some corrosiveness and darkening of the mortar with age, however, have been reported.

Concrete. Polymers may be added to concrete by impregnating standard concrete, by adding monomers or emulsified polymers to fresh concrete during mixing, or by replacing the portland cement with a polymer. For impregnation, such monomers as methyl methacrylate, styrene, acrylonitrile and tert-butyl styrene are impregnated into dried and evacuated concrete and polymerized in place by heat, chemical action, or high-energy radiation. Concrete road surfaces may be dried with high-temperature heat in an enclosure and soaked with a monomer that is subsequently cured by heating. Compressive strength of concrete has been increased by impregnation 200 to 400%, elastic modulus 200 to 300%, with the stress-strain curve linear almost to failure, creep greatly reduced, and resistance to sulfate attack and freezing and thawing greatly increased. Thermal coefficient of expansion may be increased about 2.5%.

For mixing with fresh concrete, monomers or polymer latexes of methyl methacrylate, polyester-styrene, epoxy-styrene, furans, styrene-butadiene, and vinylidene chloride have been employed. Results have ranged from disappointing to doubling of compressive and tensile strengths. The major advantage is that standard mixing and placing can be employed, instead of the post-impregnation procedures needed for impregnated concrete. Polymer-modified concrete has been used for restoring and resurfacing deteriorated road surfaces.

In **polymer concrete,** such polymers as methyl methacrylate and unsaturated polyester replace the usual portland cement. Resulting strength properties are similar to those of impregnated concrete. Unlike standard concrete, polymer concrete can be cured in hours or minutes, depending on temperature, but commonly overnight at room temperature, and is ready for use immediately. A common application is in wall panels with 1-in-thick facings of glass-fiber-reinforced polymer concrete surrounding a 1-in-thick core of plastic foam. Polymer concrete is used for rapid repair of holes in concrete roads.

TABLE 4-30 General Properties of Some Reinforced Plastics*

Property	Polyester		Epoxy			Phenolic		
	Glass fabric	Glass mat	Nonwoven continuous glass filament	Nonwoven noncontinuous glass filament	Glass fabric	Glass fabric	Cellulose paper	Cotton fabric
Specific gravity	1.5–2.1	1.4–2.1	1.75–2.00	1.50–1.85	1.70–2.00	1.55–1.95	1.28–1.40	1.30–1.40
Tensile strength, psi	18,000–65,000	10,000–34,000	11,000–21,500	18,000–35,000	35,000–85,000	20,000–50,000	8,000–20,000	7,000–16,000
Tensile elastic modulus, ksi	1,000–3,400	1,000–2,200	1,500–2,500	1,500–2,500	2,000–4,000	2,000–3,700	800–2,000	500–1,500
Compressive strength, psi	20,000–60,000	20,000–45,000	9,000–12,000	12,000–25,000	35,000–80,000	34,000–75,000	20,000–40,000	30,000–44,000
Compressive elastic modulus, ksi	3,000–4,000	1,500–2,500	1,200–2,500	1,200–2,500	2,000–4,000	2,000–3,800	800	600
Flexural strength, psi	12,500–90,000	15,000–40,000	12,000–21,000	20,000–25,000	40,000–105,000	16,000–80,000	10,500–30,000	14,000–30,000
Flexural elastic modulus, ksi	600–3,500	1,000–2,500	1,500–2,500	1,500–2,500	2,000–4,500	2,000–4,000	800–1,500	800–1,000
Shear strength, psi	12,000–23,000	10,000–20,000	3,000–6,000		14,000–25,000	17,200–24,000	6,000–14,000	7,000–18,000
Izod impact strength, ft·lb/in, notch	19–35	4.5–28.0			5.5–25.0	4.0–18.0	0.3–1.10	0.8–3.00
Hardness, Rockwell	M100–110	M70–110	M100–108		M105–120	M100–110	M70–120	M90–115
24-hr water absorption, 1/8-in thickness, %	0.15–2.50	0.1–1.3	0.04–0.30		0.04–0.30	0.12–2.70	0.2–6.0	0.7–4.4
Thermal conductivity, 10^{-4} cal·cm/(sec)(cm²)(°C)		1.7–8.7			7.1–7.4	1.4–1.8	1.8–7.0	2.1–7.0
Thermal expansion, 10^{-6} in/(in)(°C)		10–42	9–25		10–15	15–25	14–34	18–32

*Modern Plastics Encyclopedia.

Slab Toppings. At least partly because of excellent adhesion, epoxies are formulated with sand and other fillers to provide surfacing materials for concrete. Unlike standard concrete topping, epoxy-based surfacing materials can be thin. They are especially useful for smoothing uneven, irregular surfaces. The epoxy cures quickly, allowing use of the surface in a short time.

Grout. Cracked concrete can be repaired with an epoxy grout. The grout is forced into cracks under pressure for deep penetration. Because of its good bonding strength, the epoxy grout can largely restore strength, while, at the same time, sealing the crack against penetration by liquids.

4-84. Plastics Bibliography.

American Concrete Institute, SP-58, "Polymers in Concrete-International Symposium."
American Society of Civil Engineers, "Structural Plastics Design Manual."
American Society for Testing and Materials, "Standards."
"Modern Plastics Encyclopedia," Plastics Catalog Corp., New York.
Dietz, A. G. H.: "Plastics for Architects and Engineers," M.I.T. Press, Cambridge, Mass.
Harper, C., "Handbook of Plastics and Elastomers," McGraw-Hill Book Company, New York.
Skeist, I.: "Plastics in Building," Van Nostrand Reinhold Company, New York.

PORCELAIN-ENAMELED PRODUCTS

Porcelain-enameled metal is used for indoor and outdoor applications because of its hardness, durability, washability, and color possibilities. For building purposes, porcelain enamel is applied to sheet metal and cast iron, the former for a variety of purposes including trim, plumbing, and kitchen fixtures, and the latter almost entirely for plumbing fixtures. Most sheet metal used for porcelain enameling is steel—low in carbon, manganese, and other elements. Aluminum is also used for vitreous enamel.

4-85. Porcelain Enamel on Metal.

Low-temperature softening glasses must be employed, especially with sheet metal, to avoid the warping and distortion that would occur at high temperatures. To obtain lower softening temperatures than would be attainable with high-silica glasses, boron is commonly added. Fluorine may replace some of the oxygen, and lead may also be added to produce easy-flowing brilliant enamels; but lead presents an occupational health hazard.

Composition of the enamel is carefully controlled to provide a coefficient of thermal expansion as near that of the base metal as possible. If the coefficient of the enamel is greater than that of the metal, cracking and crazing are likely to occur, but if the coefficient of the enamel is slightly less, it is lightly compressed upon cooling, a desirable condition because glass is strong in compression.

To obtain good adhesion between enamel and metal, one of the so-called transition elements used in glass formulation must be employed. Cobalt is favored. Apparently, the transition elements promote growth of iron crystals from base metal into the enamel, encourage formation of an adherent oxide coating on the iron, which fuses to the enamel, or develop polar chemical bonds between metal and glass.

Usually, white or colored opaque enamels are desired. Opacity is promoted by mixing in, but not dissolving, finely divided materials possessing refractive indexes widely different from the glass. Tin oxide, formerly widely used, has been largely displaced by less expensive and more effective titanium and zirconium compounds. Clay adds to opacity. Various oxides are included to impart color.

Most enameling consists of a ground coat and one or two cover coats fired on at slightly lower temperatures; but one-coat enameling of somewhat inferior quality can be accomplished by first treating the iron surface with soluble nickel salts.

The usual high-soda glasses used to obtain low-temperature softening enamels are not highly acid-resistant and therefore stain readily and deeply when iron-containing water drips on them. Enamels highly resistant to severe staining conditions must be considerably harder; i.e., have higher softening temperatures and therefore require special techniques to avoid warping and distorting of the metal base.

Interiors of refrigerators are often made of procelain-enameled steel sheets for resistance to staining by spilled foods, whereas the exteriors are commonly baked-on synthetic-resin finishes.

4-86. Porcelain Bibliography.

Norton, F. H.: "Elements of Ceramics," Addison-Wesley Publishing Company, Cambridge, Mass., 1952.

Kingery, W. D., H. K. Bowen, and D. R. Uhlmann: "Introduction to Ceramics," John Wiley & Sons, Inc., New York.

RUBBER

Rubber for construction purposes is both natural and synthetic. Natural rubber, often called crude rubber in its unvulcanized form, is composed of large complex molecules of isoprene.

4-87. Synthetic Rubbers.

The principal synthetic rubbers are the following:

GR-S is the one most nearly like crude rubber and is the product of styrene and butadiene copolymerization. It is the most widely used of the synthetic rubbers. It is not oil-resistant but is widely used for tires and similar applications.

Nitril is a copolymer of acrylonitrile and butadiene. Its excellent resistance to oils and solvents makes it useful for fuel and solvent hoses, hydraulic-equipment parts, and similar applications.

Butyl is made by the copolymerization of isobutylene with a small proportion of isoprene or butadiene. It has the lowest gas permeability of all the rubbers and consequently is widely used for making inner tubes for tires and other applications in which gases must be held with a minimum of diffusion. It is used for gaskets in buildings.

Neoprene is made by the polymerization of chloroprene. It has very good mechanical properties and is particularly resistant to sunlight, heat, aging, and oil; it is therefore used for making machine belts, gaskets, oil hose, insulation on wire cable, and other applications to be used for outdoor exposure, such as roofing, and gaskets for building and glazing.

Sulfide rubbers—the polysulfides of high molecular weight—have rubbery properties, and articles made from them, such as hose and tank linings and glazing compounds, exhibit good resistance to solvents, oils, ozone, low temperature, and outdoor exposure.

Silicone rubber, which also is discussed in Art. 4-78, when made in rubbery consistency forms a material exhibiting exceptional inertness and temperature resistance. It is therefore used in making gaskets, electrical insulation, and similar products that maintain their properties at both high and low temperatures.

Additional elastomers, which are sometimes grouped with the rubbers, include polyethylene, cyclized rubber, plasticized polyvinyl chloride, and polybutene.

4-88. Properties of Rubber.

In the raw state, rubbers are generally quite plastic, especially when warm, have relatively low strength, are attacked by various solvents, and often can be dissolved to form cements. These characteristics are necessary for processing, assembling, and forming but are not consistent with the strength, heat stability, and elasticity required in a finished product. The latter desirable properties are obtained by vulcanization, a chemical change involving the interlinking of the rubber molecules and requiring incorporation of sulfur, zinc oxide, organic accelerators, and other ingredients in the raw rubber prior to heat treatment.

Other compounding ingredients do not enter into the chemical reaction but add considerably to the mechanical properties of finished rubber. This is particularly true of carbon black. Antioxidants retard the rate of deterioration of rubber on exposure to light and oxygen. The great variety of materials entering into various rubber compounds therefore provides a wide range of properties. In addition, many rubber products are laminated structures of rubber compounds combined with materials like fabric and metals.

The two principal categories of rubber are hard and soft. Hard rubber has a high degree of hardness and rigidity produced by vulcanization with high proportions of sulfur (ranging as high as 30 to 50% of the rubber). In soft rubber, however, sulfur may range as low as 1 to 5% and usually not more than 10%.

Principal properties of rubber compounds of interest in building are the following:

Age Resistance. Rubber normally oxidizes slowly on exposure to air at ordinary temperatures. Oxidation is accelerated by heat and ozone. Hard rubber ages less rapidly than soft, and most of the synthetics age at somewhat the same or greatly retarded rates compared with natural rubber.

Compressibility. With the exception of sponge rubber, completely confined rubber, as in a tight container, is virtually impossible to compress.

Elongation. Soft vulcanized compounds may stretch as much as 1000% of the original length, whereas the elongation of hard rubber ranges usually between 1 and 50%. Even after these great elongations, rubber will return practically to its original length.

Energy Absorption. The energy-storing ability of rubber is about 150 times that of spring-tempered steel. The resilient energy-storing capacity of rubber is 14,600 ft-lb/lb as compared with 95.3 for spring steel.

Expansion and Contraction. The volumetric changes caused by temperature are generally greater for rubber than for metals. Soft vulcanized-rubber compounds, for example, have thermal coefficients of expansion ranging from 0.00011 to 0.00005, whereas hard vulcanized-rubber compounds range from 0.00004 to 0.000015, as compared with coefficients of aluminum of 0.000012, glass of 0.000005, and steel of 0.000007.

Friction. The coefficient of friction between soft rubber and dry steel may exceed unity. But when the surfaces are wet, the friction coefficient drops radically and may become as low as 0.02 in water-lubricated rubber bearings.

Heat Resistance. Vulcanized-natural-rubber compounds are normally limited in use to temperatures of 150 to 200°F, whereas some of the synthetic-rubber compounds like the silicones may be used at temperatures as high as 450°F.

Light Resistance. The natural, GR-S, and nitryl rubber compounds under tension are likely to crack when exposed to sunlight. Other rubbers are highly resistant to similar attack. Sunlight discolors hard rubber somewhat and reduces its surface electrical resistivity.

Gas Permeability. The polysulfide and butyl compounds have very low rates of permeation by air, hydrogen, helium, and carbon dioxide, in contrast with the natural and GR-S compounds, whose gas permeability is relatively high. A butyl inner tube loses air only about 10% as rapidly as a natural rubber tube.

Strength. The tensile strength based on original cross section ranges from 300 to 4500 psi for soft-rubber stocks, and from about 1000 to 10,000 psi for hard rubber. Under compression, soft rubber merely distorts, whereas true hard rubber can be subjected to 10,000 to 15,000 psi before distorting markedly.

4-89. Laminated Rubber.

Rubber is often combined with various textiles, fabrics, filaments, and metal wire to obtain strength, stability, abrasion resistance, and flexibility. Among the laminated materials are the following:

V Belts. These consist of a combination of fabric and rubber, frequently combined with reinforcing grommets of cotton, rayon, steel, or other high-strength material extending around the central portion.

Flat Rubber Belting. This laminate is a combination of several plies of cotton fabric or cord, all bonded together by a soft-rubber compound.

Conveyor Belts. These, in effect, are moving highways used for transporting such material as crushed rock, dirt, sand, gravel, slag, and similar materials. When the belt operates at a steep angle, it is equipped with buckets or similar devices and becomes an elevator belt. A typical conveyor belt consists of cotton duct plies alternated with thin rubber plies; the assembly is wrapped in a rubber cover, and all elements are united into a single structure by vulcanization. A conveyor belt to withstand extreme conditions is made with some textile or metal cords instead of the woven fabric. Some conveyor belts are especially arranged to assume a trough form and made to stretch less than similar all-fabric belts.

Rubber-Lined Pipes, Tanks, and Similar Equipment. The lining materials include all the natural and synthetic rubbers in various degrees of hardness, depending on the application. Frequently, latex rubber is deposited directly from the latex solution onto the metal surface to be covered. The deposited layer is subsequently vulcanized. Rubber linings can be bonded to ordinary steel, stainless steel, brass, aluminum, concrete, and wood. Adhesion to aluminum is inferior to adhesion to steel. Covering for brass must be compounded according to the composition of the metal.

Rubber Hose. Nearly all rubber hose is laminated and composed of layers of rubber combined with reinforcing materials like cotton duck, textile cords, and metal wire. The modern laminated rubber hose can be made with a large variety of structures. Typical hose consists of an inner rubber lining, a number of intermediate layers consisting of braided cord or cotton duck impregnated with rubber, and outside that, several more layers of fabric, spirally wound cord, spirally wound metal, or in some cases, spirally wound flat steel ribbon. Outside of all this is another layer of rubber to provide resistance to abrasion. Hose for transporting oil, water, wet concrete under pressure, and for dredging purposes is made of heavy-duty laminated rubber.

Vibration Insulators. These usually consist of a layer of soft rubber bonded between two layers of metal. Another type of insulator consists of a rubber tube or cylinder vulcanized to two concentric metal tubes, the rubber being deflected in shear. A variant of this consists of a cylinder of soft rubber vulcanized to a tubular or solid steel core and a steel outer shell, the entire combination being placed in torsion to act as a spring. Heavy-duty mounts of this type are employed on trucks, buses, and other applications calling for rugged construction.

4.90. Rubber Bibliography.

American Society for Testing and Materials, Philadelphia, Pa.: "Standards."
Gobel, E. F.: "Rubber Springs Design," John Wiley & Sons, Inc., New York.

ASPHALT AND BITUMINOUS PRODUCTS

Asphalt, because of its water-resistant qualities and good durability, is used for many building applications to exclude water, to provide a cushion against vibration and expansion, and in similar applications.

4-91. Dampproofing and Waterproofing Asphalts.

For dampproofing (mopped-on coating only) and waterproofing (built-up coating of one or more plies), three types of asphalt are recognized (ASTM Specification D449):

Type A, an easily flowing, soft, adhesive, "self-healing" material for use underground or wherever similar moderate temperatures are found.

Type B, a less susceptible asphalt for use aboveground where temperatures do not exceed 125°F.

Type C, for use aboveground where exposed on vertical surfaces to direct sunlight or other areas where temperatures exceed 125°F.

Softening ranges are, respectively, 115 to 145°F, 145 to 170°F, and 180 to 200°F.

4-92. Roofing Asphalts and Pitches.

For constructing built-up roofing, four grades of asphalt are recognized (ASTM D312): Type 1, for inclines up to ½ in/ft; Type II, for inclines up to 1½ in/ft; Type III, for inclines up to 3 in/ft; and Type IV, suited for inclines up to 6 in/ft, generally in areas with relatively high year-round temperatures. Types I through IV may be either smooth or surfaced with slag or gravel. Softening ranges are 135 to 150°F, 158 to 176°F, 180 to 200°F, and 210 to 225°F, respectively.

Coal-tar pitches for roofing, dampproofing, and waterproofing are of three types (ASTM D450): Type 1, for built-up roofing systems; Type 2, for dampproofing and membrane waterproofing systems: Type 3, for built-up roofing, but containing less volatiles than Type 1. Softening ranges are 126 to 140°F, 106 to 126°F, and 133 to 147°F, respectively.

4-93. Roofing Felts.

For built-up waterproofing and roofing, three types of membranes are employed: felt (ASTM D226, D227), asbestos felt (ASTM D250), and cotton fabrics (ASTM D173). Felts are felted sheets of vegetable or animal fibers or mixtures of both, saturated with asphalt or with coal tar conforming to ASTM D312 and D450.

Standard asphalt felts weigh 15, 20, or 30 lb per square (100 ft^2), and standard coal-tar felts weigh 13 lb per square.

Asbestos felts, weighing 15 or 30 lb per square, are asphalt-saturated felts containing at least 85% asbestos fiber.

Cotton fabrics are open-weave materials weighing at least 3½ oz/yd^2 before saturation, with thread counts of 24 to 32 per inch. The saturants are either asphalts or coal tars. The saturated fabric must weigh at least 10 oz/yd^2.

4-94. Roll Roofing.

Asphalt roll roofing, shingles, and siding consist basically of roofing felt, first uniformly impregnated with hot asphaltic saturant and then coated on each side with at least one layer of a hot asphaltic coating and compounded with a water-insoluble mineral filler. The bottom or reverse side, in each instance, is covered with some suitable material, like powdered mica, to prevent sticking in the package or roll.

Granule-surfaced roll roofing (ASTM D249) is covered uniformly on the weather side with crushed mineral granules, such as slate. Minimum weight of the finished roofing must be 81 to 83 lb per square (100 ft^2), and the granular coating must weigh at least 18.5 lb per square.

Roll roofing (ASTM 224), surfaced with powdered talc or mica, is made in two grades, 39.8 and 54.6 lb per square, of which at least 18 lb must be the surfacing material.

4-95. Asphalt Shingles.

There are two standard types: Type I, uniform thickness; Type II, thick butt (ASTM D225). Average weights must be 95 lb per square (100 ft^2). For Type I, the weather-side coating must weigh 23.0 lb per square; for Type II, 30.0 lb per square. The material in these shingles is similar to granule-surfaced roll roofing.

4-96. Asphalt Mastics and Grouts.

Asphalt mastics used for waterproofing floors and similar structures, but not intended for pavement, consist of mixtures of asphalt cement, mineral filler, and mineral aggregate, which can be heated at about 400°F to a sufficiently soft condition to be poured and troweled into place. The raw ingredients may be mixed on the job or may be premixed, formed into cakes, and merely heated on the job (ASTM D491).

Bituminous grouts are suitable for waterproofing above or below ground level as protective coatings. They also can be used for membrane waterproofing or for bedding and filling the joints of brickwork. Either asphaltic or coal-tar pitch materials of dampproofing and waterproofing grade are used, together with mineral aggregates as coarse as sand.

4-97. Bituminous Pavements.

Asphalts for pavement (ASTM D946) contain petroleum asphalt cement, derived by the distillation of asphaltic petroleum. Various grades are designated as 40-50, 60-70, 85-100, 120-150, and 200-300, depending upon the depth of penetration of a standard needle in a standard test (ASTM D5).

Emulsions range from low to high viscosity and quick- to slow-setting (ASTM D977).

4-98. Asphalt Bibliography.

American Society for Testing and Materials, Philadelphia, Pa.: "Standards."

JOINT SEALS

Calking compounds, sealants, and gaskets are employed to seal the points of contact between similar and dissimilar building materials that cannot otherwise be made completely tight. Such points include glazing, the joints between windows and walls, the many joints occurring in the increasing use of panelized construction, the copings of parapets, and similar spots.

The requirements of a good joint seal are: (1) good adhesion to or tight contact with the surrounding materials, (2) good cohesive strength, (3) elasticity to allow for compression and extension as surrounding materials retract or approach each other because of changes in moisture content or temperature, (4) good durability or the ability to maintain their properties over a long period of time without marked deterioration, and (5) no staining of surrounding materials such as stone.

4-99. Calking Compounds.

These traditional sealers are used mostly with traditional materials such as masonry, with relatively small windows, and at other points where motion of building components is relatively small. They are typically composed of bodied linseed or soy oil, or both, combined with asbestos fiber, calcium carbonate (ground marble or limestone), tinting pigments, a gelling agent, drier, and mineral spirits (thinners).

Two types are commonly employed, gun grade and knife grade. Gun grades are viscous semiliquids suitable for application by hand or air-operated calking guns. Knife grades are stiffer and are applied by knife, spatula, or mason's pointing tools.

Because calking compounds are based on drying oils that eventually harden in contact with the air, the best joints are generally thick and deep, with a relatively small portion exposed to the air. The exposed surface is expected to form a tough protective skin for the soft mass underneath, which in turn provides the cohesiveness, adhesiveness and elasticity required. Thin shallow beads cannot be expected to have the durability of thick joints with small exposed surface areas.

4-100. Sealants.

For joints and other points where large movements of building components are expected, rubbery, elastomeric materials may be used as sealants. Whereas traditional calking compounds should not be used where movements of more than 5% or at most 10% are expected, larger movements, typically 10 to 25%, can be accommodated by the rubbery sealants.

Some rubbery sealants consist of two components, mixed just before application. Polymerization occurs, leading to conversion of the viscous material to a rubbery consistency. The working time or pot life before this occurs varies, depending upon formulation and temperature, from a fraction of an hour to several hours or a day. Other formulations are single-component and require no mixing. They harden upon exposure to moisture in the air.

Various curing agents, accelerators, plasticizers, fillers, thickeners, and other agents may be added, depending on the basic material and the end-use requirements.

Among the polymeric materials employed are:

Acrylics: solvent-release type, water-release type, latex
Butyls: skinning and nonskinning
Polysulfide: two-part and one-part
Silicone: one-part
Polyurethane: two-part and one-part
Chlorosulfonated polyethylene: one-part
Polyurethane-polyepoxide: two-part

Characteristics of the preceding formulations vary. Hence, the proper choice of materials depends upon the application. A sealant with the appropriate hardness, extensibility, useful temperature ranges, expected life, dirt pickup, staining, colorability, rate of cure to tack-free condition, toxicity, resistance to ultraviolet light, and other attributes should be chosen for the specific end use.

In many joints, such as those between building panels, it is necessary to provide backup; that is, a foundation against which the compound can be applied. This serves to limit the thickness of the joint, to provide the proper ratio of thickness to width, and to force the compound into intimate contact with the substrate, thereby promoting adhesion. For the purpose, any of various compressible materials, such as polyethylene or polyurethane rope, or oakum, may be employed.

To promote adhesion to the substrate, various primers may be needed. (To prevent adhesion of the compound to parts of the substrate where adhesion is not wanted, any of various liquid and tape bond-breakers may be employed.) Generally, good adhesion requires dry, clean surfaces free of grease and other deleterious materials.

4-101. Gaskets.

Joint seals described in Arts. 4-99 and 4-100 are formed in place; that is, soft masses are put into the joints and conform to their geometry. A gasket, on the other hand, is preformed and placed into a joint whose geometry must conform with the gasket in such a way as to seal the joint by compression of the gasket. Gaskets, however, are cured under shop-controlled conditions, whereas sealants cure under variable and not always favorable field conditions.

Rubbery materials most commonly employed for gaskets are cellular or noncellular (dense) neoprene, EPDM (ethylene-propylene polymers and terpolymers) and polyvinylchloride polymers.

Gaskets are generally compression types or lock-strip *(zipper)* types. The former are forced into the joint and remain tight by being kept under compression. With lock-strip gaskets, a groove in the gasket permits a lip to be opened and admit glass or other panel, after which a strip is forced into the groove, tightening the gasket in place. If the strip is separable from the gasket, its composition is often harder than the gasket itself.

See also Art. 13-19.

For setting large sheets of glass and similar units, setting or supporting spacer blocks of rubber are often combined with gaskets of materials such as vulcanized synthetic rubber and are finally sealed with the elastomeric rubber-based sealants or glazing compounds.

4-102. Joint Seals Bibliography.

"Building Seals and Sealants," STP 606, American Society for Testing and Materials, Philadelphia, Pa.

Cook, J. P.: "Construction Sealants and Adhesives," John Wiley & Sons, Inc., New York.

Damusis, A.: "Sealants," Van Nostrand Reinhold Company, New York.

COATINGS

Protective and decorative coatings generally employed in building are the following:

Oil Paint. Drying-oil vehicles or binders plus opaque and extender pigments.
Water Paint. Pigments plus vehicles based on water, casein, protein, oil emulsions, and rubber or resin latexes, separately or in combination.
Calcimine. Water and glue, with or without casein, plus powdered calcium carbonate and any desired colored pigments.

Varnish. Transparent combination of drying oil and natural or synthetic resins.
Enamel. Varnish vehicle plus pigments.
Lacquer. Synthetic-resin film former, usually nitrocellulose, plus plasticizers, volatile solvents, and other resins.
Shellac. Exudations of the lac insect, dissolved in alcohol.
Japan. Solutions of metallic salts in drying oils, or varnishes containing asphalt and opaque pigments.
Aluminum Paint. Fine metallic aluminum flakes suspended in drying oil plus resin, or in nitrocellulose.

4-103. Vehicles or Binders.

Following are descriptions of the most commonly used vehicles and binders for paint:

Natural drying oils. Drying oils harden by absorbing oxygen. The most important natural oils are linseed from flax seed (for many years the standard paint vehicle), tung oil (faster drying, good compatibility with varnish), oiticica oil (similar to tung), safflower (best nonyellowing oil), soybean (flexible films), dehydrated castor (good adhesion, fast drying), and fish oil (considered inferior but cheap).

Alkyds. These, the most widely used paint vehicles, are synthetic resins that are modified with various vegetable oils to produce clear resins that are harder than natural oils. Properties of the film depend on relative proportions of oil and resin. The film is both air drying and heat hardening.

Latexes. Latex paints are based on emulsions of various polymers including acrylics, polyvinyl acetate, styrene-butadiene, polyvinyl chloride, and rubber. They are easy to apply, dry quickly, have no solvent odor, and application tools are easily cleaned with soap and water. The films adhere well to various surfaces, have good color retention, and have varying degrees of flexibility.

Epoxy and epoxy-polyester. Catalyzed two-part, all-epoxy coatings are formed by addition of a catalyst to the liquid epoxy just before application (pot life a few minutes to a day). Films are as hard as many baked-on coatings and are resistant to solvents and traffic. Oil-modified epoxy esters, in contrast, harden on oxidation without a catalyst. They are less hard and chemically resistant than catalyzed epoxies, but dry fast and are easily applied. Epoxy-polyesters mixed just before use, produce smooth finishes suitable for many interior surfaces and are chemically resistant.

Polyurethanes. These produce especially abrasion-resistant, fast-hardening coatings. Two-component formulations, of variable pot life, are mixed just before use. One-component formulations cure by evaporation and reaction with moisture in air (30 to 90% relative humidity). Oils and alkyds may be added.

Vinyl solutions. Solutions of polyvinyl chloride and vinyl esters dry rapidly and are built up by successive, sprayed thin coatings. They characteristically have low gloss, high flexibility, and inertness to water but are sensitive to some solvents. Adhesion may be a problem. Weather resistance is excellent.

Dryers. These are catalysts that hasten the hardening of drying oils. Most dryers are salts of heavy metals, especially cobalt, manganese, and lead, to which salts of zinc and calcium may be added. Iron salts, usable only in dark coatings, accelerate hardening at high temperatures. Dryers are normally added to paints to hasten hardening, but they must not be used too liberally or they cause rapid deterioration of the oil by overoxidation.

Thinners. These are volatile constituents added to coatings to promote their spreading qualities by reducing viscosity. They should not react with the other constituents and should evaporate completely. Commonly used thinners are turpentine and mineral spirits, i.e., derivatives of petroleum and coal tar.

4-104. Pigments for Paints.

Pigments may be classified as white and colored, or as opaque and extender pigments. The hiding power of pigments depends on the difference in index of refraction of the pigment and the surrounding medium—usually the vehicle of a protective coating. In opaque pigments, these indexes are markedly different from those of the vehicles (oil or other); in extender pigments, they are nearly the same. The comparative hiding efficiencies of various pigments must be evaluated on the basis of hiding power per pound and cost per pound.

Principal white pigments, in descending order of relative hiding power per pound, are approximately as follows: rutile titanium dioxide, anatase titanium dioxide, zinc sulfide, titanium-calcium, titanium-barium, zinc sulfide–barium, titanated lithopone, lithopone, antimony oxide, zinc oxide.

Zinc oxide is widely used by itself or in combination with other pigments. Its color is unaffected by many industrial and chemical atmospheres. It imparts gloss and reduces chalking but tends to crack and alligator instead.

Zinc sulfide is a highly opaque pigment widely used in combination with other pigments.

Titanium dioxide and extended titanium pigments have high opacity and generally excellent properties. Various forms of the pigments have different properties. For example, anatase titanium dioxide promotes chalking, whereas rutile inhibits it.

Colored pigments for building use are largely inorganic materials, especially for outdoor use, where the brilliant but fugitive organic pigments soon fade. The principal inorganic colored pigments are:

Metallic. Aluminum flake or ground particle, copper bronze, gold leaf, zinc dust.

Black. Carbon black, lampblack, graphite, blue basic lead sulfate (gray), vegetable black, and animal blacks.

Earth colors. Yellow ocher, raw and burnt umber, raw and burnt sienna; reds and maroons.

Blue. Ultramarine, iron ferrocyanide (Prussian, Chinese, Milori).

Brown. Mixed ferrous and ferric oxide.

Green. Chromium oxide, hydrated chromium oxide, chrome greens.

Orange. Molybdated chrome orange.

Red. Iron oxide, cadmium red, vermilion.

Yellow. Zinc chromate, cadmium yellows, hydrated iron oxide.

Extender pigments are added to extend the opaque pigments, increase durability, provide better spreading characteristics, and reduce cost. The principal extender pigments are silica, china clay, talc, mica, barium sulfate, calcium sulfate, calcium carbonate, and such materials as magnesium oxide, magnesium carbonate, barium carbonate, and others used for specific purposes.

4-105. Resins for Paints.

Natural and synthetic resins are used in a large variety of air-drying and baked finishes. The natural resins include both fossil resins, which are harder and usually superior in quality, and recent resins tapped from a variety of resin-exuding trees. The most important fossil resins are amber (semiprecious jewelry), Kauri, Congo, Boea Manila, and Pontianak. Recent resins include Damar, East India, Batu, Manila, and rosin. Shellac, the product of the lac insect, may be considered to be in this class of resins.

The synthetic resins, in addition to the ones discussed in Art. 4-103, are increasing in importance, especially for applications requiring maximum durability. Among them are phenol formaldehyde, melamine formaldehyde, urea formaldehyde, silicones, fluorocarbons and cellulose acetate-butyrate.

Phenolics in varnishes are used for outdoor and other severe applications on wood and metals. They are especially durable when baked.

Melamine and urea find their way into a large variety of industrial finishes, such as automobile and refrigerator finishes.

Silicones are used when higher temperatures are encountered than can be borne by the other finishes.

Fluorocarbons are costly but provide high-performance coatings, industrial siding, and curtain walls with excellent gloss retention, stain resistance and weather resistance.

Cellulose acetate-butyrate provides shop-applied, high-gloss finishes.

4-106. Coatings Bibliography.

Banov, A., "Paints and Coatings Handbook," Structures Publishing Company, Farmington, Mich.

Burns, R. M., and Bradley, W., "Protective Coatings for Metals," Van Nostrand Reinhold Company, New York.

Martens, C. R., "The Technology of Paints, Varnishes and Lacquers," Van Nostrand Reinhold Company, New York.

Section 5

Structural Theory

Frederick S. Merritt
Consulting Engineer
West Palm Beach, Fla.

Structural theory establishes symbolic (mathematical) models for describing the behavior of structures. These models, when verified by laboratory and field tests and observations of structures under service conditions, form the basis for analysis and design methods for structures.

Structural design is the application of structural theory to ensure that buildings and other structures are built to support all loads and resist all constraining forces that may be reasonably expected to be imposed on them during their expected service life, without hazard to occupants or users and preferably without dangerous deformations, excessive sidesway (drift), or annoying vibrations (see Sec. 3). In addition, good design requires that this objective be achieved economically.

Provision should be made in application of structural theory to design for abnormal as well as normal service conditions. Abnormal conditions may arise as a result of accidents, fire, explosions, tornadoes, severer-than-anticipated earthquakes, floods, and inadvertent or even deliberate overloading of building components. Under such conditions, parts of a building may be damaged. The structural system, however, should be so designed that the damage will be limited in extent and undamaged portions of the building will remain stable. For the purpose, structural elements should be proportioned and arranged to form a stable system under normal service conditions; in addition, the system should have sufficient continuity and ductility, or energy-absorption capacity, so that if any small portion of it should sustain damage, other parts will transfer loads (at least until repairs can be made) to remaining structural components capable of transmitting the loads to the ground.

(D. M. Schultz, F. F. P. Burnett and M. Fintel, "A Design Approach to General Structural Integrity," in "Design and Construction of Large-Panel Concrete Structures," U.S. Department of Housing and Urban Development, 1977; E. V. Leyendecker and B. R. Ellingwood, "Design Methods for Reducing the Risk of Progressive Collapse in Buildings," NBS Buildings Science Series 98, National Bureau of Standards, 1977.)

STRESS AND STRAIN

Loads are the external forces acting on a structure. Stresses are the internal forces that resist them. Depending on the manner in which the loads are applied, they tend to deform the structure and its components—tensile forces tend to stretch, compressive forces to squeeze together, torsional forces to twist, and shearing forces to slide parts of the structure past each other.

5-1 Types of Loads.

External loads on a structure may be classified in several different ways. In one classification, they may be considered as static or dynamic.

Static loads are forces that are applied slowly and then remain nearly constant. One example is the weight, or dead load, of a floor or roof system.

Dynamic loads vary with time. They include repeated and impact loads.

Repeated loads are forces that are applied a number of times, causing a variation in the magnitude, and sometimes also in the sense, of the internal forces. A good example is a bridge crane.

Impact loads are forces that require a structure or its components to absorb energy in a short interval of time. An example is the dropping of a heavy weight on a floor slab, or the shock wave from an explosion striking the walls and roof of a building.

External forces may also be classified as distributed and concentrated.

Uniformly distributed loads are forces that are, or for practical purposes may be considered, constant over a surface area of the supporting member. Dead weight of a rolled-steel I beam is a good example.

Concentrated loads are forces that have such a small contact area as to be negligible compared with the entire surface area of the supporting member. A beam supported on a girder, for example, may be considered, for all practical purposes, a concentrated load on the girder.

Another common classification for external forces labels them axial, eccentric, and torsional.

An axial load is a force whose resultant passes through the centroid of a section under consideration and is perpendicular to the plane of the section.

An eccentric load is a force perpendicular to the plane of the section under consideration but not passing through the centroid of the section, thus bending the supporting member (see Arts. 5-33, 5-35, and 5-79).

Torsional loads are forces that are offset from the shear center of the section under consideration and are inclined to or in the plane of the section, thus twisting the supporting member (see Art. 5-79).

Also, building codes classify loads in accordance with the nature of the source. For example:

Dead loads include materials, equipment, constructions, or other elements of weight supported in, on, or by a building, including its own weight, that are intended to remain permanently in place.

Live loads include all occupants, materials, equipment, constructions, or other elements of weight supported in, on, or by a building and that will or are likely to be moved or relocated during the expected life of the building.

Impact loads are a fraction of the live loads used to account for additional stresses and deflections resulting from movement of the live loads.

Wind loads are maximum forces that may be applied to a building by wind in a mean recurrence interval, or a set of forces that will produce equivalent stresses.

Snow loads are maximum forces that may be applied by snow accumulation in a mean recurrence interval.

Seismic loads are forces that produce maximum stresses or deformations in a building during an earthquake.

5-2. Design Loads.

In designing structural members, designers should use whichever is larger of the following:

1. Loadings specified in the local or state building code.
2. Probable maximum loads, based not only on current site conditions and original usage of proposed building spaces but also on possible future events. Loads that are of uncertain magnitude and that may be treated as statistical variables should be selected in accordance with a specific probability that the chosen magnitudes will not be exceeded during the life of the building or in accordance with the corresponding mean recurrence interval. The mean recurrence interval generally used for ordinary permanent buildings is 50 years. The interval, however, may be set at 25 years for structures with no occupants or offering negligible risk to life, or at 100 years for permanent buildings with a high degree of sensitivity to the loads and an unusually high degree of hazard to life and property in case of failure.

In the absence of a local or state building code, designers can be guided by loads specified in a national model building code or by the following data:

Loads applied to structural members may consist of the following, alone or in combination: dead, live, impact, earth pressure, hydrostatic pressure, snow, ice, rain, wind, or earthquake loads; constraining forces, such as those resulting from restriction of thermal, shrinkage, or moisture-change movements; or forces caused by displacements or deformations of members, such as those caused by creep, plastic flow, differential settlement, or sidesway (drift).

Dead Loads. Actual weights of materials and installed equipment should be used. See Tables 5-1d and 5-2.

Live Loads. These may be concentrated or distributed loads and should be considered placed on the building to produce maximum effects on the structural member being designed. Minimum live loads to be used in building design are listed in Table 5-1. These include an allowance for impact, except as noted in the footnote of Table 5-1b.

Partitions generally are considered to be live loads, because they may be installed at any time, almost anywhere, to subdivide interior spaces, or may be shifted from original places to other places in the future. Consequently, unless a floor is designed for a large live load, for example, 80 lb/ft^2, the weight of partitions should be added to other live loads, whether or not partitions are shown on the working drawings for building construction.

Because of the low probability that a large floor area contributing load to a specific structural member will be completely loaded with maximum design live loads, building codes generally permit these loads to be reduced for certain types of occupancy. Usually, however, codes do not permit any reduction for places of public assembly, dwellings, garages for trucks and buses, or one-way slabs. For areas with a minimum required live load exceeding 100 lb/ft^2 and for passenger-car garages, live loads on columns supporting

TABLE 5-1   Minimum Design Live and Wind Loads

a. Uniformly distributed live loads, lb/ft², impact included[a]

Occupancy or use	Load	Occupancy or use	Load
Assembly spaces:		Marquees	75
Auditoriums[b] with fixed seats	60	Morgue	125
Auditoriums[b] with movable seats	100	Office buildings:	
Ballrooms and dance halls	100	Corridors above first floor	80
Bowling alleys, poolrooms,		Files	125
similar recreational areas	75	Offices	50
Conference and card rooms	50	Penal institutions:	
Dining rooms, restaurants	100	Cell blocks	40
Drill rooms	150	Corridors	100
Grandstand and reviewing-stand		Residential:	
seating areas	100	Dormitories:	
Gymnasiums	100	Nonpartitioned	60
Lobbies, first-floor	100	Partitioned	40
Roof gardens, terraces	100	Dwellings, multifamily:	
Skating rinks	100	Apartments	40
Bakeries	150	Corridors	80
Balconies (exterior)	100	Hotels:	
Up to 100 ft² on one- and		Guest rooms, private corridors	40
two-family houses	60	Public corridors	80
Bowling alleys, alleys only	40	Housing, one- and two-family:	
Broadcasting studios	100	First floor	40
Catwalks	30	Storage attics	80
Corridors:		Uninhabitable attics	20
Areas of public assembly, first-		Upper floors, habitable attics	30
floor lobbies	100	Schools:	
Other floors same as occupancy		Classrooms	40
served, except as indicated		Corridors	80
elsewhere in this table		Shops with light equipment	60
Fire escapes:		Stairs and exitways	100
Multifamily housing	40	Handrails, vertical and horizontal	
Others	100	thrust, lb/lin ft	50
Garages:		Storage warehouse:	
Passenger cars	50	Heavy	250
Trucks and buses		Light	125
Hospitals:		Stores:	
Operating rooms, laboratories,		Retail:	
service areas	60	Basement and first floor	100
Patients' rooms, wards,		Upper floors	75
personnel areas	40	Wholesale	100
Kitchens other than domestic	150	Telephone equipment rooms	80
Laboratories, scientific	100	Theaters:	
Libraries:		Aisles, corridors, lobbies	100
Corridors above first floor	80	Dressing rooms	40
Reading rooms	60	Projection rooms	100
Stack rooms, books and shelving		Stage floors	150
at 65 lb/ft³, but at least	150	Toilet areas	40
Manufacturing and repair areas:			
Heavy	250		
Light	125		

[a]See Eqs. (5-1) and (5-2).
[b]Including churches, schools, theaters, courthouses, and lecture halls.
[c]Use American Association of State Highway and Transportation Officials highway lane loadings.

TABLE 5-1 Minimum Design Live and Wind Loads *(Continued)*

b. Concentrated live loads[d]

Location	Load, lb
Elevator machine room grating (on 4-in² area)	300
Finish, light floor-plate construction (on 1-in² area)	200
Garages:	
Passenger cars:	
Manual parking (on 20-in² area)	2,000
Mechanical parking (no slab), per wheel	1,500
Trucks, buses (on 20-in² area), per wheel	16,000
Office floors (on area 2.5 ft square)	2,000
Roof-truss panel point over garage, manufacturing, or storage floors	2,000
Scuttles, skylight ribs, and accessible ceilings (on area 2.5 ft square)	200
Sidewalks (on area 2.5 ft square)	8,000
Stair treads (on 4-in² area at center of tread)	300

[d]Use instead of uniformly distributed live load, except for roof trusses, if concentrated loads produce greater stresses or deflections. Add impact factor for machinery and moving loads: 100% for elevators, 20% for light machines, 50% for reciprocating machines, 33% for floor or balcony hangers. For craneways, add a vertical force equal to 25% of maximum wheel load; a lateral force equal to 10% of the weight of trolley and lifted load, at the top of each rail; and a longitudinal force equal to 10% of maximum wheel loads, acting at top of rail.

c. Wind pressures for design of framing, vertical walls, and windows of ordinary rectangular buildings, lb/ft²[e]

Height zone, ft above curb	Exposures								
	A^f	B^g	C^h	A^f	B^g	C^h	A^f	B^g	C^h
	Coastal areas, N.W. and S.E. United States[i]			Northern and central United States[j]			Other parts of United States[k]		
0–50	20	40	65	15	25	40	15	20	35
51–100	30	50	75	20	35	50	15	25	40
101–300	40	65	85	30	45	60	20	35	45
301–600	65	85	105	40	55	70	35	45	55
Over 600	85	100	120	60	70	80	45	55	65

[e]For winds with 50-year recurrence interval. For computation of more exact wind pressures, see Art. 3-4 and ANSI Standard A58.1.
[f]Centers of large cities and very rough, hilly terrain.
[g]Suburban areas, towns, city outskirts, wooded areas, and rolling terrain.
[h]Flat, open country; open, flat coastal belts, and grassland.
[i]110-mi/hr basic wind speed.
[j]90-mi/hr basic wind speed.
[k]80-mi/hr basic wind speed.

d. Minimum design loads for materials

Material	Load, lb/ft³	Material	Load, lb/ft³
Aluminum, cast	165	Gravel, dry	104
Bituminous products:		Gypsum, loose	70
Asphalt	81	Ice	57.2
Petroleum, gasoline	42	Iron, cast	450
Pitch	69	Lead	710
Tar	75	Lime, hydrated, loose	32

TABLE 5-1 Minimum Design Live and Wind Loads *(Continued)*

d. Minimum design loads for materials *(continued)*

Material	Load, lb/ft³	Material	Load, lb/ft³
Brass, cast	534	Lime, hydrated, compacted	45
Bronze, 8 to 14% tin	509	Magnesium alloys	112
Cement, portland, loose	90	Mortar, hardened:	
Cement, portland, set	183	Cement	130
Cinders, dry, in bulk	45	Lime	110
Coal, anthracite, piled	52	Riprap (not submerged):	
Coal, bituminous or lignite, piled	47	Limestone	83
Coal, peat, dry, piled	23	Sandstone	90
Charcoal	12	Sand, clean and dry	90
Copper	556	Sand, river, dry	106
Cork, compressed	14.4	Silver	656
Earth (not submerged):		Steel	490
Clay, dry	63	Stone, quarried, piled:	
Clay, damp	110	Basalt, granite, gneiss	96
Clay and gravel, dry	100	Limestone, marble, quartz	95
Silt, moist, loose	78	Sandstone	82
Silt, moist, packed	96	Shale, slate	92
Sand and gravel, dry, loose	100	Tin, cast	459
Sand and gravel, dry, packed	100	Water, fresh	62.4
Sand and gravel, wet	120	Water, sea	64
Gold, solid	1205	Zinc	450

more than one floor may be decreased 20%. Except for the preceding cases, a reduced live load L, lb/ft², may be computed from

$$L = \left(0.25 + \frac{15}{\sqrt{A_I}}\right) L_o \quad (5\text{-}1)$$

where L_o = unreduced live load, lb/ft² (see Table 5-1*a*)

A_I = influence area, or floor area over which the influence surface for structural effects is significantly different from zero

= area of four surrounding bays for an interior column, plus similar areas from supported floors above, if any

= area of two adjoining bays for an interior girder or for an edge column, plus similar areas from supported floors above, if any

= area of one bay for an edge girder or for a corner column, plus similar areas from supported floors above, if any

The reduced live load L, however, should not be less than $0.5L_o$ for members supporting one floor or $0.4L_o$ for members supporting two or more floors.

Roofs used for promenades should be designed for a minimum live load of 60 lb/ft², and those used for gardens or assembly, for 100 psf. Ordinary roofs should be designed for a minimum live load L, lb/ft², computed from

$$L = 20R_1R_2 \geq 12 \quad (5\text{-}2)$$

where $R_1 = 1.2 - 0.001A_t$ but not less than 0.6 nor more than 1.0

A_t = tributary area, ft², for structural member being designed

$R_2 = 1.2 - 0.05r$ but not less than 0.6 nor more than 1.0

r = rise of roof in 12 in for a pitched roof or 32 times the ratio of rise to span for an arch or dome

This minimum live load need not be combined with snow load for design of a roof but should be designed for the larger of the two.

Subgrade Pressures. Walls below grade should be designed for lateral soil pressures and the hydrostatic pressure of subgrade water, plus the load from surcharges at ground level. Design pressures should take into account the reduced weight of soil because of buoyancy when water is present. In design of floors at or below grade, uplift due to hydrostatic pressures on the underside should be considered.

Wind Loads. Characteristics of wind loads and minimum design wind loads are discussed in Arts. 3-3 and 3-4. For a rough estimate, the data in Table 5-1c may be used as a guide.

Snow, Ice, and Rain Loads. These, in effect, are nonuniformly distributed, vertical, live loads that are imposed by nature and hence are generally uncertain in magnitude and duration. They may occur alone or in combination. Design snow loads preferably should be determined for the site of the proposed building with the advice of meteorologists and application of extreme-value statistical analysis to rain and snow records for the locality.

Rain loads depend on drainage and may become large enough to cause roof failure when drainage is blocked (see Art. 3-13).

Ice loads are created when snow melts, then freezes, or when rain follows a snow storm and freezes. These loads should be considered in determining the design snow load. Snow loads may consist of pure snow or a mixture of snow, ice, and water.

Design snow loads on roofs may be assumed to be proportional to the maximum ground snow load p_g, lb/ft², measured in the vicinity of the building, with a 50-year mean recurrence interval. Determination of the constant of proportionality should take into account:

1. Appropriate mean recurrence interval.
2. Roof exposure. Wind may blow snow off the roof or onto the roof from nearby higher roofs or create nonuniform distribution of snow.
3. Roof thermal conditions. Heat escaping through the roof melts the snow. If the water can drain off, the snow load decreases. Also, for sloped roofs, if they are warm, there is a tendency for snow to slide off. Insulated roofs, however, restrict heat loss from the interior and therefore are subjected to larger snow loads.
4. Type of occupancy and uses of building. More conservative loading should be used for public-assembly buildings, because of the risk of great loss of life and injury to occupants if overloads should cause the roof to collapse.
5. Roof slope. The steeper a roof, the greater is the likelihood of good drainage and that snow will slide off.

In addition, roof design should take into account not only the design snow load uniformly distributed over the whole roof area but also possible unbalanced loading, because of snow being blown off part of the roof, and possible snow drifts piling up over a portion of the roof.

For **flat roofs,** in the absence of building-code requirements, the basic snow load when the ground snow load p_g is 20 lb/ft² or less may be taken as

$$p_{\min} = p_g \qquad (5\text{-}3a)$$

When p_g is between 20 and 25 lb/ft², the minimum allowable design load is $p_{min} = 20$ lb/ft², and when p_g exceeds 25 lb/ft², the basic snow load may be taken as

$$p_f = 0.8 p_g \qquad (5\text{-}3b)$$

TABLE 5-2 Minimum Design Dead Loads

	lb/ft²
Walls	
Clay brick	
High-absorption, per 4-in wythe	34
Medium absorption, per 4-in wythe	39
Low-absorption, per 4-in wythe	46
Sand-lime brick, per 4-in wythe	38
Concrete brick	
4-in, with heavy aggregate	46
4-in, with light aggregate	33
Concrete block, hollow	
8-in, with heavy aggregate	55
8-in, with light aggregate	35
12-in, with heavy aggregate	85
12-in, with light aggregate	55
Clay tile, loadbearing	
4-in	24
8-in	42
12-in	58
Clay tile, nonloadbearing	
2-in	11
4-in	18
8-in	34
Furring tile	
1 1/2-in	8
2-in	10
Glass block, 4-in	18
Gypsum block, hollow	
2-in	9.5
4-in	12.5
6-in	18.5

	lb/ft²
Floor Finishes	
Asphalt block, 2-in	24
Cement, 1-in	12
Ceramic or quarry tile, 1-in	12
Hardwood flooring, 7/8-in	4
Plywood subflooring, 1/2-in	1.5
Resilient flooring, such as asphalt tile and linoleum	2
Slate, 1-in	15
Softwood subflooring, per in of thickness	3
Terrazzo, 1-in	12
Wood block, 3-in	4

	lb/ft²	
Wood joists, double wood floor, joist size	12-in spacing	16-in spacing
2 × 6	6	5
2 × 8	6	6
2 × 10	7	6
2 × 12	8	7
3 × 6	7	6
3 × 8	8	7
3 × 10	9	8
3 × 12	11	9
3 × 14	12	10

	lb/ft²
Concrete Slabs	
Stone aggregate, reinforced, per in of thickness	12.5
Cinder, reinforced, per in of thickness	9.25
Lightweight aggregate, reinforced, per in of thickness	9

Masonry	lb/ft³
Cast-stone masonry	144
Concrete, stone aggregate, reinforced	150
Ashlar:	
Granite	165
Limestone, crystalline	165
Limestone, oölitic	135
Marble	173
Sandstone	144

Roof and Wall Coverings	lb/ft²
Asbestos-cement corrugated or shingles	4
Asphalt shingles	2
Composition:	
3-ply ready roofing	1
4-ply felt and gravel	5.5
5-ply felt and gravel	6
Copper or tin	1
Corrugated iron	2
Sheathing (gypsum), 1/2-in	2
Sheathing (wood), per in thickness	3
Slate, 3/16-in	7
Wood shingles	3

Waterproofing	lb/ft²
Five-ply membrane	5

Ceilings	lb/ft²
Plaster (on tile or concrete)	5
Suspended metal lath and gypsum plaster	10
Suspended metal lath and cement plaster	15
Suspended steel channel supports	2
Gypsumboard per 1/4-in thickness	1.1

Floor Fill	lb/ft²
Cinders, no cement, per in of thickness	5
Cinders, with cement, per in of thickness	9
Sand, per in of thickness	8

Partitions	lb/ft²
Plaster on masonry	
Gypsum, with sand, per in of thickness	8.5
Gypsum, with lightweight aggregate, per in	4
Cement, with sand, per in of thickness	10
Cement, with lightweight aggregate, per in	5
Plaster, 2-in solid	20
Metal studs	
Plastered two sides	18
Gypsumboard each side	6
Wood studs, 2 × 4-in	
Unplastered	3
Plastered one side	11
Plastered two sides	19
Gypsumboard each side	7

Glass	lb/ft²
Single-strength	1.2
Double-strength	1.6
Plate, 1/8-in	1.6

Insulation	lb/ft²
Cork, per in of thickness	1.0
Foamed glass, per in of thickness	0.8
Glass-fiber bats, per in of thickness	0.06
Polystyrene, per in of thickness	0.2
Urethane	0.17
Vermiculite, loose fill, per in of thickness	0.05

where p_f = design snow load, lb/ft², for a flat roof that may have unheated space underneath and that may be located where the wind cannot be relied on to blow snow off, because of nearby higher structures or trees

p_g = ground snow load, lb/ft²

For roofs sheltered from the wind, increase p_f computed from Eq. (5-3b) by 20%, and for windy sites, reduce p_f 10%. For a poorly insulated roof with heated space underneath, decrease p_f by 30%.

Increase p_f 10% for large office buildings and public-assembly buildings, such as auditoriums, schools, factories. Increase p_f 20% for essential buildings, such as hospitals, communication buildings, police and fire stations, power plants, and for structures housing expensive objects or equipment. Decrease p_f 20% for structures with low human occupancy, such as farm buildings.

The **ground snow load** p_g should be determined from an analysis of snow depths recorded at or near the site of the proposed building. For a rough estimate in the absence of building-code requirements, p_g may be taken as follows for the United States, except for mountainous regions:

0–5 lb/ft²—southern states from about latitude N32° southward

10–15 lb/ft²—Pacific coast between latitudes N32° and N40° and other states between latitudes N32° and N37°

20–30 lb/ft²—Pacific coast from latitude N40° northward and other states between latitudes N37° and N40°

40–50 lb/ft²—north Atlantic and central states between latitudes N40° and N43°

60–80 lb/ft²—northern New England between latitudes N43° and N45° and central states from N43° northward

80–120 lb/ft²—Maine above latitude N45°

For **sloping roofs,** the snow load depends on whether the roof will be warm or cold. In either case, the load may be assumed to be zero for roofs making an angle θ of 70° or more with the horizontal. Also, for any slope, the load need not be taken greater than p_f given by Eq. (5-3b). For slopes θ, deg, between 0° and 70°, the snow load, lb/ft², acting vertically on the projection of the roof on a horizontal plane, may be computed for warm roofs from

$$p_s = \left(\frac{70 - \theta}{40}\right) p_f \leq p_f \qquad (5\text{-}4a)$$

and for cold roofs from

$$p_s = \left(\frac{70 - \theta}{25}\right) p_f \leq p_f \qquad (5\text{-}4b)$$

Hip and gable roofs should be designed for the condition of the whole roof loaded with p_s, and also with the windward side unloaded and the leeward side carrying $1.5p_s$.

For **curved roofs,** the snow load on the portion that is steeper than 70° may be taken as zero. For the less-steep portion, the load p_s may be computed as for a sloped roof, with θ taken as the angle with the horizontal of a line from the crown to points on the roof where the slope starts to exceed 70°. Curved roofs should be designed with the whole area fully loaded with p_s. They also should be designed for the case of snow only on the leeward side, with the load varying uniformly from $0.5p_s$ at the crown to $2p_s$ at points where the roof slope starts to exceed 30° and then decreasing to zero at points where the slope starts to exceed 70°.

Multiple folded-plate, sawtooth, and barrel-vault roofs similarly should be designed for unbalanced loads increasing from $0.5p_s$ at ridges to $3p_s$ in valleys.

Snow drifts may accumulate adjacent to higher roofs that are less than 20 ft horizontally away. The reason for this is that wind may blow snow from the higher roof onto the lower roof. Drifts also may accumulate at projections above roofs, such as at parapets, solar collectors, and penthouse walls. Drift loads accordingly should be taken into account when:

1. The ground snow load p_g exceeds 10 lb/ft².
2. A higher roof exists (or may be built in the future) within 20 ft of the building, if the height differential, ft, exceeds $1.2p_f/\gamma$, where p_f is computed from Eq. (5-3b) and γ is the snow density, lb/ft³.
3. A projection extends a distance, ft, exceeding $1.2p_f/\gamma$ above the roof and is more than 15 ft long.

In computation of drift loads, the snow density γ, lb/ft³, may be taken as follows:

$$p_g = 11\text{--}30 \quad 31\text{--}60 \quad 60 \text{ or more}$$
$$\gamma = 15 \quad 20 \quad 25$$

The drift may be assumed to be a triangular prism with maximum height, located adjacent to a higher roof or along a projection, taken as $h_d = 2p_g/\gamma$, modified by factors for risk and exposure, described for flat roofs. Width of the prism should be at least 10 ft and may be taken as $3h_d$ for projections up to 50 ft long and as $4h_d$ for projections more than 50 ft long. Accordingly, the load varies uniformly with distance from a projection, from $h_d\gamma$ at the projection to zero. For drifts due to snow load from a higher roof at a horizontal distance S, ft, away horizontally ($S \leq 20$ ft), the maximum drift intensity may be taken as $h_d\gamma(20 - S)/20$.

Rain-Snow Load Combination. In roof design, account should be taken of the combination of the design snow load with a temporary water load from an intense rainstorm, including the effects of roof deflection on ponding. The added water load depends on the drainage characteristics of the roof, which, in turn, depend on the roof slope. For a flat roof, the rain surcharge may be taken as 8 lb/ft² for slopes less ¼ in/ft and as 5 lb/ft² for steeper slopes, except where the minimum allowable design snow load p_{min} exceeds p_f computed from Eq. (5-3b). In such cases, these water surcharges may be reduced by $p_{min} - p_f$.

For methods of computing more accurate snow loads, see the latest edition of "Building Code Requirements for Minimum Design Loads in Buildings," ANSI A58.1, American National Standards Institute.

(W. Tobiasson and R. Redfield, "Snow Loads for the United States," Part II, and S. C. Colbeck, "Snow Loads Resulting from Rain on Snow," U.S. Army Cold Regions Research and Engineering Laboratory, Hanover, N.H.)

Seismic Loads. These are the result of horizontal and vertical movements imposed on a building by earth vibrations during an earthquake. Changing accelerations of the building mass during the temblor create changing inertial forces, which are assumed in building design to act as seismic loads at the various floor and roof levels in proportion to the portion of the building mass at those levels. Because analysis of building response to such dynamic loading generally is very complex, building codes permit, for design of ordinary buildings, substitution of equivalent static loading for the dynamic loading (see Art. 5-86).

5-3. Static Equilibrium.

If a structure and its components are so supported that, after a very small deformation occurs, no further motion is possible, they are said to be in equilibrium. Under such circumstances, internal forces, or stresses, exactly counteract the loads.

Several useful conclusions may be drawn from the state of static equilibrium: Since there is no translatory motion, the sum of the external forces must be zero; and since there is no rotation, the sum of the moments of the external forces about any point must be zero.

For the same reason, if we consider any portion of the structure and the loads on it, the sum of the external and internal forces on the boundaries of that section must be zero. Also, the sum of the moments of these forces must be zero.

In Fig. 5-1, for example, the sum of the forces R_L and R_R needed to support the roof truss is equal to the 20-kip load on the truss (1 kip = 1 kilopound = 1000 lb = 0.5 ton).

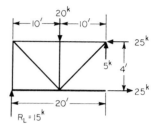

Fig. 5-1 Truss in equilibrium under load. Upward-acting forces equal those acting downward.

Fig. 5-2 Portion of a truss is held in equilibrium by stresses in its components.

Also, the sum of the moments of the external forces is zero about any point; about the right end, for instance, it is $40 \times 15 - 30 \times 20 = 600 - 600$.

In Fig. 5-2 is shown the portion of the truss to the left of section AA. The internal forces at the cut members balance the external load and hold this piece of the truss in equilibrium.

Generally, it is convenient to decompose the forces acting on a structure into components parallel to a set of perpendicular axes that will simplify computations. For example, for forces in a single plane—a condition commonly encountered in building design—the most useful technique is to resolve all forces into horizontal and vertical components. Then, for a structure in equilibrium, if H represents the horizontal components, V the vertical components, and M the moments of the components about any point in the plane,

$$\Sigma H = 0 \quad \Sigma V = 0 \quad \text{and} \quad \Sigma M = 0 \tag{5-5}$$

These three equations may be used to evaluate three unknowns in any nonconcurrent coplanar force system, such as the roof truss in Figs. 5-1 and 5-2. They may determine the magnitude of three forces for which the direction and point of application already are known, or the magnitude, direction, and point of application of a single force.

Suppose, for the truss in Fig. 5-1, the reactions at the supports are to be computed. Taking moments about the right end and equating to zero yields $40\,R_L - 30 \times 20 = 0$, from which left reaction $R_L = 600/40 = 15$ kips. Equating the sum of the vertical forces to zero gives $20 - 15 - R_R = 0$, from which the right reaction $R_R = 5$ kips.

5.4. Unit Stress and Strain.

For convenience in evaluating the load-carrying ability of a member of a structure, it is customary to express its strength in terms of the maximum unit stress, or internal force per unit of area, that it can sustain without suffering structural damage. Then, application of a safety factor to the maximum unit stress determines a unit stress that should not be exceeded when design loads are applied to the member. That unit stress is known as the allowable stress or working stress.

To ascertain whether a structural member has adequate load-carrying capacity, the designer generally has to compute the maximum unit stress produced by design loads in the member for each type of internal force—tensile, compressive, or shearing—and compare it with the corresponding allowable unit stress.

When the loading is such that the unit stress is constant over a section under consideration, the stress may be obtained by dividing the force by the area of the section. But in general, the unit stress varies from point to point. In that case, the unit stress at any point in the section is the limiting value of the ratio of the internal force on any small area to that area, as the area is taken smaller and smaller.

Sometimes in the design of a structure, unit stress may not be the prime consideration. The designer may be more interested in limiting the deformation or strain.

Deformation in any direction is the total change in the dimension of a member in that direction.

Unit strain in any direction is the deformation per unit of length in that direction.

When the loading is such that the unit strain is constant over a portion of a member, it may be obtained by dividing the deformation by the original length of that portion. In general, however, the unit strain varies from point to point in a member. Like a varying unit stress, it represents the limiting value of a ratio.

5-5. Hooke's Law.

For many materials, unit strain is proportional to unit stress, until a certain stress, the **proportional limit,** is exceeded. Known as Hooke's law, this relationship may be written as

$$f = E\epsilon \quad \text{or} \quad \epsilon = \frac{f}{E} \tag{5-6}$$

where f = unit stress
ϵ = unit strain
E = modulus of elasticity

Hence, when the unit stress and modulus of elasticity of a material are known, the unit strain can be computed. Conversely, when the unit strain has been found, the unit stress can be calculated.

When a member is loaded and the unit stress does not exceed the proportional limit, the member will return to its original dimensions when the load is removed. The **elastic limit** is the largest unit stress that can be developed without a permanent deformation remaining after removal of the load.

Some materials possess one or two **yield points.** These are unit stresses in the region of which there appears to be an increase in strain with no increase or a small decrease in strain. Thus, the materials exhibit plastic deformation. For materials that do not have a well-defined yield point, the offset yield strength is used as a measure of the beginning of plastic deformation.

The **offset yield strength,** or **proof stress** as it is sometimes referred to, is defined as the unit stress corresponding to a permanent deformation, usually 0.01% (0.0001 in/in) or 0.20% (0.002 in/in).

5-6. Constant Unit Stress.

The simplest cases of stress and strain are those in which the unit stress and strain are constant. Stresses due to an axial tension or compression load or a centrally applied shearing force are examples; also an evenly applied bearing load. These loading conditions are illustrated in Figs. 5-3 to 5-6.

For the axial tension and compression loadings, we take a section normal to the centroidal axis (and to the applied forces). For the shearing load, the section is taken along a plane of sliding. And for the bearing load, it is chosen through the plane of contact between the two members.

Fig. 5-3 Tension member. **Fig. 5-4** Compression member. **Fig. 5-5** Bracket in shear. **Fig. 5-6** Bearing load.

Since for these loading conditions, the unit stress is constant across the section, the equation of equilibrium may be written

$$P = Af \tag{5-7}$$

where P = load
f = a tensile, compressive, shearing, or bearing unit stress
A = cross-sectional area for tensile or compressive forces, or area on which sliding may occur for shearing forces, or contact area for bearing loads

For torsional stresses, see Art. 5-79.

The unit strain for the axial tensile and compressive loads is given by the equation

$$\epsilon = \frac{e}{L} \tag{5-8a}$$

where ϵ = unit strain
e = total lengthening or shortening of the member
L = original length of the member

Applying Hooke's law and Eq. (5-8a) to Eq. (5-7) yields a convenient formula for the deformation:

$$e = \frac{PL}{AE} \tag{5-8b}$$

where P = load on the member
A = its cross-sectional area
E = modulus of elasticity of the material

[Since long compression members tend to buckle, Eqs. (5-7) and (5-8) are applicable only to short members.]

While tension and compression strains represent a simple stretching or shortening of a member, shearing strain represents a distortion due to a small rotation. The load on the small rectangular portion of the member in Fig. 5-5 tends to distort it into a parallelogram. The unit shearing strain is the change in the right angle, measured in radians.

Modulus of rigidity, or shearing modulus of elasticity, is defined by

$$G = \frac{v}{\gamma} \tag{5-9a}$$

where G = modulus of rigidity
v = unit shearing stress
γ = unit shearing strain

It is related to the modulus of elasticity in tension and compression E by the equation

$$G = \frac{E}{2(1 + \mu)} \tag{5-9b}$$

where μ is a constant known as Poisson's ratio.

5-7. Poisson's Ratio.

Within the elastic limit, when a material is subjected to axial loads, it deforms not only longitudinally but also laterally. Under tension, the cross section of a member decreases, and under compression, it increases. The ratio of the unit lateral strain to the unit longitudinal strain is called Poisson's ratio.

For many materials, this ratio can be taken equal to 0.25. For structural steel, it is usually assumed to be 0.3.

Assume, for example, that a steel hanger with an area of 2 in^2 carries a 40-kip (40,000-lb) load. The unit stress is 40,000/2, or 20,000 psi. The unit tensile strain, taking the modulus of elasticity of the steel as 30,000,000 psi, is 20,000/30,000,000, or 0.00067 in/in. With Poisson's ratio as 0.3, the unit lateral strain is -0.3×0.00067, or a shortening of 0.00020 in/in.

5-8. Thermal Stresses.

When the temperature of a body changes, its dimensions also change. Forces are required to prevent such dimensional changes, and stresses are set up in the body by these forces.

If α is the coefficient of expansion of the material and T the change in temperature, the unit strain in a bar restrained by external forces from expanding or contracting is

$$\epsilon = \alpha T \tag{5-10a}$$

According to Hooke's law, the stress f in the bar is

$$f = E\alpha T \tag{5-10b}$$

where E = modulus of elasticity.

5-9. Strain Energy.

When a bar is stressed, energy is stored in it. If a bar supporting a load P undergoes a deformation e the energy stored in it is

$$U = \tfrac{1}{2}Pe \tag{5-11a}$$

This equation assumes the load was applied gradually and the bar is not stressed beyond the proportional limit. It represents the area under the load-deformation curve up to the load P. Applying Eqs. (5-6) and (5-7) to Eq. (5-11a) gives another useful equation for energy:

$$U = \frac{f^2}{2E} AL \tag{5-11b}$$

where f = unit stress
E = modulus of elasticity of the material
A = cross-sectional area
L = length of the bar

Since AL is the volume of the bar, the term $f^2/2E$ indicates the energy stored per unit of volume. It represents the area under the stress-strain curve up to the stress f. Its value when the bar is stressed to the proportional limit is called the **modulus of resilience.** This modulus is a measure of the capacity of the material to absorb energy without danger of being permanently deformed and is of importance in designing members to resist energy loads.

Equation (5-11a) is a general equation that holds true when the principle of superposition applies (the total deformation produced by a system of forces is equal to the sum of the elongations produced by each force). In the general sense, P in Eq. (5-11a) represents any group of statically interdependent forces that can be completely defined by one symbol, and e is the corresponding deformation.

The strain-energy equation can be written as a function of either the load or the deformation.

For axial tension or compression:

$$U = \frac{P^2 L}{2AE} \qquad U = \frac{AEe^2}{2L} \qquad (5\text{-}11c)$$

where P = axial load
e = total elongation or shortening
L = length of the member
A = cross-sectional area
E = modulus of elasticity

For pure shear:

$$U = \frac{V^2 L}{2AG} \qquad U = \frac{AGe^2}{2L} \qquad (5\text{-}11d)$$

where V = shearing load
e = shearing deformation
L = length over which deformation takes place
A = shearing area
G = shearing modulus

For torsion:

$$U = \frac{T^2 L}{2JG} \qquad U = \frac{JG\phi^2}{2L} \qquad (5\text{-}11e)$$

where T = torque
ϕ = angle of twist
L = length of shaft
J = polar moment of inertia of the cross section
G = shearing modulus

For pure bending (constant moment):

$$U = \frac{M^2 L}{2EI} \qquad U = \frac{EI\theta^2}{2L} \qquad (5\text{-}11f)$$

where M = bending moment
θ = angle of rotation of one end of the beam with respect to the other
L = length of beam
I = moment of inertia of the cross section
E = modulus of elasticity

For beams carrying transverse loads, the strain energy is the sum of the energy for bending and that for shear.

See also Art. 5-55.

STRESSES AT A POINT

Tensile and compressive stresses are sometimes referred to also as **normal stresses,** because they act normal to the cross section. Under this concept, tensile stresses are considered as positive normal stresses and compressive stresses as negative.

5-10. Stress Notation.

Suppose a member of a structure is acted upon by forces in all directions. For convenience, let us establish a reference set of perpendicular coordinate x, y, and z axes. Now let us take at some point in the member a small cube with sides parallel to the coordinate axes. The notations commonly used for the components of stress acting on the sides of this element and the directions assumed as positive are shown in Fig. 5-7.

For example, for the sides of the element perpendicular to the z axis, the normal component of stress is denoted by f_z. The shearing stress v is resolved into two components and requires two subscript letters for a complete description. The first letter indicates the direction of the normal to the plane under consideration. The second letter indicates the direction of the component of the stress. Considering the sides perpendicular to the z axis, the shear component in the x direction is labeled v_{zx} and that in the y direction v_{zy}.

5-11. Stress Components.

If, for the small cube in Fig. 5-7, moments of the forces acting on it are taken about the x axis, considering the cube's dimensions as dx, dy, and dz, the equation of equilibrium requires that

$$v_{zy}\, dx\, dy\, dz = v_{yz}\, dx\, dy\, dz$$

(Forces are taken equal to the product of the area of the face and the stress at the center.) Two similar equations can be written for moments taken about the y axis and the z axis. These equations show that

$$v_{xy} = v_{yx} \qquad v_{zx} = v_{xz} \qquad \text{and} \qquad v_{zy} = v_{yz} \qquad (5\text{-}12)$$

In words, the components of shearing stress on two perpendicular faces and acting normal to the intersection of the faces are equal.

Consequently, to describe the stresses acting on the coordinate planes through a point, only six quantities need be known. These stress components are f_x, f_y, f_z, $v_{xy} = v_{yx}$, $v_{yz} = v_{zy}$, and $v_{zx} = v_{xz}$.

If the cube in Fig. 5-7 is acted on only by normal stresses f_x, f_y, and f_z, from Hooke's law and the application of Poisson's ratio, the unit strains in the x, y, and z directions, in accordance with Arts. 5-5 and 5-6, are, respectively,

$$\begin{aligned}\epsilon_x &= \frac{1}{E}[f_x - \mu(f_y + f_z)] \\ \epsilon_y &= \frac{1}{E}[f_y - \mu(f_x + f_z)] \\ \epsilon_z &= \frac{1}{E}[f_z - \mu(f_x + f_y)]\end{aligned} \qquad (5\text{-}13)$$

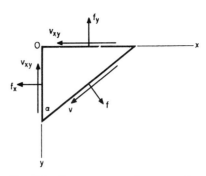

Fig. 5-7 Stresses in a rectangular coordinate system.

Fig. 5-8 Stresses at a point on a plane inclined to the axes.

where μ = Poisson's ratio. If only shearing stresses act on the cube in Fig. 5-7, the distortion of the angle between edges parallel to any two coordinate axes depends only on shearing-stress components parallel to those axes. Thus, the unit shearing strains are (see Art. 5-6)

$$\gamma_{xy} = \frac{1}{G} v_{xy} \qquad \gamma_{yz} = \frac{1}{G} v_{yz} \quad \text{and} \quad \gamma_{zx} = \frac{1}{G} v_{zx} \qquad (5\text{-}14)$$

5-12. Two-Dimensional Stress.

When the six components of stress necessary to describe the stresses at a point are known (Art. 5-11), the stress on any inclined plane through the same point can be determined. For the case of two-dimensional stress, only three stress components need be known.

Assume, for example, that at a point O in a stressed plate, the components f_x, f_y, and v_{xy} are known (Fig. 5-8). To find the stresses for any plane through the z axis, take a plane parallel to it close to O. This plane and the coordinate planes form a triangular prism. Then, if α is the angle the normal to the plane makes with the x axis, the normal and shearing stresses on the inclined plane, obtained by application of the equations of equilibrium, are

$$f = f_x \cos^2 \alpha + f_y \sin^2 \alpha + 2v_{xy} \sin \alpha \cos \alpha \qquad (5\text{-}15)$$
$$v = v_{xy}(\cos^2 \alpha - \sin^2 \alpha) + (f_y - f_x) \sin \alpha \cos \alpha \qquad (5\text{-}16)$$

5-13. Principal Stresses.

A plane through a point on which stresses act may be assigned a direction for which the normal stress is a maximum or a minimum. There are two such positions, perpendicular to each other. And on those planes, there are no shearing stresses.

The directions in which the normal stresses become maximum or minimum are called principal directions and the corresponding normal stresses principal stresses.

To find the principal directions, set the value of v given by Eq. (5-16) equal to zero. The resulting equation is

$$\tan 2\alpha = \frac{2v_{xy}}{f_x - f_y} \qquad (5\text{-}17)$$

If the x and y axes are taken in the principal directions, v_{xy} is zero. Consequently, Eqs. (5-15) and (5-16) may be simplified to

$$f = f_x \cos^2 \alpha + f_y \sin^2 \alpha \qquad (5\text{-}18)$$
$$v = \tfrac{1}{2} \sin 2\alpha (f_y - f_x) \qquad (5\text{-}19)$$

where f and v are, respectively, the normal and shearing stress on a plane at an angle α with the principal planes and f_x and f_y are the principal stresses.

Pure Shear. If on any two perpendicular planes only shearing stresses act, the state of stress at the point is called pure shear or simple shear. Under such conditions, the principal directions bisect the angles between the planes on which these shearing stresses occur. The principal stresses are equal in magnitude to the unit shearing stresses.

5-14. Maximum Shearing Stress.

The maximum unit shearing stress occurs on each of two planes that bisect the angles between the planes on which the principal stresses act. The maximum shear is equal to one-half the algebraic difference of the principal stresses:

$$\max v = \frac{f_1 - f_2}{2} \qquad (5\text{-}20)$$

where f_1 is the maximum principal stress and f_2 the minimum.

5-15. Mohr's Circle.

The relationship between stresses at a point may be represented conveniently on Mohr's circle (Fig. 5-9). In this diagram, normal stress f and shear stress v are taken as coordinates. Then, for each plane through the point, there will correspond a point on the circle, whose coordinates are the values of f and v for the plane.

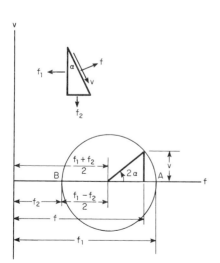

Fig. 5-9 Mohr's circle for stresses at a point—constructed from known principal stresses.

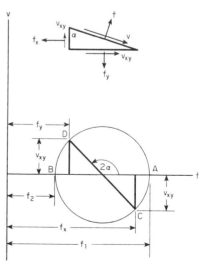

Fig. 5-10 Stress circle constructed from two known normal positive stresses, f_x and f_y, and a shear v_{xy}.

To construct the circle given the principal stresses, mark off the principal stresses f_1 and f_2 on the f axis (points A and B in Fig. 5-9). Tensile stresses are measured to the right of the v axis and compressive stresses to the left. Construct a circle with its center on the f axis and passing through the two points representing the principal stresses. This is the Mohr's circle for the given stresses at the point under consideration.

Suppose now, we wish to find the stresses on a plane at an angle α to the plane of f_1. If a radius is drawn making an angle 2α with the f axis, the coordinates of its intersection with the circle represent the normal and shearing stresses acting on the plane.

Mohr's circle can also be plotted when the principal stresses are not known but the stresses f_x, f_y, and v_{xy}, on any two perpendicular planes, are. The procedure is to plot the two points representing these known stresses with respect to the f and v axes (points C and D in Fig. 5-10). The line joining these points is a diameter of Mohr's circle. Constructing the circle on this diameter, we find the principal stresses at the intersection with the f axis (points A and B in Fig. 5-10).

For more details on the relationship of stresses and strains at a point, see Timoshenko and Goodier, "Theory of Elasticity," McGraw-Hill Book Company, New York.

STRAIGHT BEAMS

Beams are the horizontal members used to support vertically applied loads across an opening. In a more general sense, they are structural members that external loads tend to bend, or curve.

5-16. Types of Beams.

There are many ways in which beams may be supported. Some of the more common methods are shown in Figs. 5-11 to 5-16.

The beam in Fig. 5-11 is called a simply supported, or **simple beam**. It has supports near its ends, which restrain it only against vertical movement. The ends of the beam are free to rotate. When the loads have a horizontal component, or when change in length of the beam due to temperature may be important, the supports may also have to prevent horizontal motion. In that case, horizontal restraint at one support is generally sufficient.

The distance between the supports is called the **span**. The load carried by each support is called a **reaction**.

The beam in Fig. 5-12 is a **cantilever**. It has only one support, which restrains it

Fig. 5-11 Simple beam.

Fig. 5-12 Cantilever beam.

Fig. 5-13 Beam with one end fixed.

Fig. 5-14 Fixed-end beam.

Fig. 5-15 Beam with overhangs.

Fig. 5-16 Continuous beam.

from rotating or moving horizontally or vertically at that end. Such a support is called a **fixed end.**

If a simple support is placed under the free end of the cantilever, the beam in Fig. 5-13 results. It has one end fixed, one end simply supported.

The beam in Fig. 5-14 has both ends fixed. No rotation or vertical movement can occur at either end. In actual practice, a fully fixed end can seldom be obtained. Some rotation of the beam ends generally is permitted. Most support conditions are intermediate between those for a simple beam and those for a fixed-end beam.

In Fig. 5-15 is shown a beam that overhangs both its simple supports. The overhangs have a free end, like a cantilever, but the supports permit rotation.

When a beam extends over several supports, it is called a **continuous beam** (Fig. 5-16).

Reactions for the beams in Figs. 5-11, 5-12, and 5-15 may be found from the equations of equilibrium. They are classified as **statically determinate beams** for that reason.

The equations of equilibrium, however, are not sufficient to determine the reactions of the beams in Figs. 5-13, 5-14, and 5-16. For those beams, there are more unknowns than equations. Additional equations must be obtained on the basis of deformations permitted; on the knowledge, for example, that a fixed end permits no rotation. Such beams are classified as **statically indeterminate.** Methods for finding the stresses in that type of beam are given in Arts. 5-55, 5-57 to 5-69, and 5-110 to 5-113.

5-17. Reactions.

As an example of the application of the equations of equilibrium (Art. 5-3) to the determination of the reactions of a statically determinate beam, we shall compute the reactions of the 60-ft-long beam with overhangs in Fig. 5-17. This beam carries a uniform load of 200 lb/lin ft over its entire length and several concentrated loads. The supports are 36 ft apart.

Fig. 5-17 Beam with overhangs loaded both uniformly and with concentrated loads.

To find reaction R_1, we take moments about R_2 and equate the sum of the moments to zero (clockwise rotation is considered positive, counterclockwise, negative):

$$-2000 \times 48 + 36R_1 - 4000 \times 30 - 6000 \times 18 + 3000 \times 12 - 200 \times 60 \times 18 = 0$$
$$R_1 = 14{,}000 \text{ lb}$$

In this calculation, the moment of the uniform load was found by taking the moment of its resultant, which acts at the center of the beam.

To find R_2, we can either take moments about R_1 or use the equation $\Sigma V = 0$. It is generally preferable to apply the moment equation and use the other equation as a check.

$$3000 \times 48 - 36R_2 + 6000 \times 18 + 4000 \times 6 - 2000 \times 12 + 200 \times 60 \times 18 = 0$$
$$R_2 = 13{,}000 \text{ lb}$$

As a check, we note that the sum of the reactions must equal the total applied load:

$$14{,}000 + 13{,}000 = 2000 + 4000 + 6000 + 3000 + 12{,}000$$
$$27{,}000 = 27{,}000$$

5-18. Internal Forces.

Since a beam is in equilibrium under the forces applied to it, it is evident that at every section internal forces are acting to prevent motion. For example, suppose we cut the

5-22 Structural Theory

Fig. 5-18 Portions of a beam are held in equilibrium by internal stresses.

beam in Fig. 5-17 vertically just to the right of its center. If we total the external forces, including the reaction, to the left of this cut (see Fig. 5-18a), we find there is an unbalanced downward load of 4000 lb. Evidently, at the cut section, an upward-acting internal force of 4000 lb must be present to maintain equilibrium. Again, if we take moments of the external forces about the section, we find an unbalanced moment of 54,000 ft-lb. So there must be an internal moment of 54,000 ft-lb acting to maintain equilibrium.

This internal, or resisting, moment is produced by a couple consisting of a force C acting on the top part of the beam and an equal but opposite force T acting on the bottom part (Fig. 5-18b). The top force is the resultant of compressive stresses acting over the upper portion of the beam, and the bottom force is the resultant of tensile stresses acting over the bottom part. The surface at which the stresses change from compression to tension—where the stress is zero—is called the **neutral surface**.

5-19. Shear Diagrams.

The unbalanced external vertical force at a section is called the shear. It is equal to the algebraic sum of the forces that lie on either side of the section. Upward-acting forces on the left of the section are considered positive, downward forces negative; signs are reversed for forces on the right.

A diagram in which the shear at every point along the length of a beam is plotted as an ordinate is called a shear diagram. The shear diagram for the beam in Fig. 5-17 is shown in Fig. 5-19b.

Fig. 5-19 Shear diagram for the beam with loads shown in Fig. 5-17.

The diagram was plotted starting from the left end. The 2000-lb load was plotted downward to a convenient scale. Then, the shear at the next concentrated load—the left support—was determined. This equals $-2000 - 200 \times 12$, or -4400 lb. In passing from just to the left of the support to a point just to the right, however, the shear changes by the magnitude of the reaction. Hence, on the right-hand side of the left support the shear is $-4400 + 14,000$, or 9600 lb. At the next concentrated load, the shear is $9600 - 200 \times 6$, or 8400 lb. In passing the 4000-lb load, however, the shear changes to $8400 - 4000$, or 4400 lb. Proceeding in this manner to the right end of the beam, we terminate with a shear of 3000 lb, equal to the load on the free end there.

It should be noted that the shear diagram for a uniform load is a straight line sloping downward to the right (see Fig. 5-21). Therefore, the shear diagram was completed by connecting the plotted points with straight lines.

Shear diagrams for commonly encountered loading conditions are given in Figs. 5-30 to 5-41.

5-20. Bending-Moment Diagrams.

The unbalanced moment of the external forces about a vertical section through a beam is called the bending moment. It is equal to the algebraic sum of the moments about the section of the external forces that lie on one side of the section. Clockwise moments are considered positive, counterclockwise moments negative, when the forces considered lie on the left of the section. Thus, when the bending moment is positive, the bottom of the beam is in tension.

A diagram in which the bending moment at every point along the length of a beam is plotted as an ordinate is called a bending-moment diagram.

Figure 5-20c is the bending-moment diagram for the beam loaded with concentrated loads only in Fig. 5-20a. The bending moment at the supports for this simply supported beam obviously is zero. Between the supports and the first load, the bending moment is proportional to the distance from the support, since it is equal to the reaction times the distance from the support. Hence the bending-moment diagram for this portion of the beam is a sloping straight line.

The bending moment under the 6000-lb load in Fig. 5-20a considering only the force to the left is 7000×10, or 70,000 ft-lb. The bending-moment diagram, then, between the left support and the first concentrated load is a straight line rising from zero at the left end of the beam to 70,000, plotted to a convenient scale, under the 6000-lb load.

The bending moment under the 9000-lb load, considering the forces on the left of it, is $7000 \times 20 - 6000 \times 10$, or 80,000 ft-lb. (It could have been more easily obtained by considering only the force on the right, reversing the sign convention: $8000 \times 10 = 80,000$ ft-lb.) Since there are no loads between the two concentrated loads, the bending-moment diagram between the two sections is a straight line.

If the bending moment and shear are known at any section of a beam, the bending moment at any other section may be computed, providing there are no unknown forces between the two sections. The rule is:

The bending moment at any section of a beam is equal to the bending moment at any section to the left, plus the shear at that section times the distance between sections, minus the moments of intervening loads. If the section with known moment and shear is on the right, the sign convention must be reversed.

For example, the bending moment under the 9000-lb load in Fig. 5-20a could also have been obtained from the moment under the 6000-lb load and the shear to the right of the 6000-lb load given in the shear diagram (Fig. 5-20b). Thus, $80,000 = 70,000 + 1000 \times 10$. If there had been any other loads between the two concentrated loads, the moment of these loads about the section under the 9000-lb load would have been subtracted.

Bending-moment diagrams for commonly encountered loading conditions are given in Figs. 5-30 to 5-41. These may be combined to obtain bending moments for other loads.

5-21. Moments in Uniformly Loaded Beams.

When a beam carries a uniform load, the bending-moment diagram does not consist of straight lines. Consider, for example, the beam in Fig. 5-21a, which carries a uniform

Fig. 5-20 Shear and moment diagrams for a beam with concentrated loads.

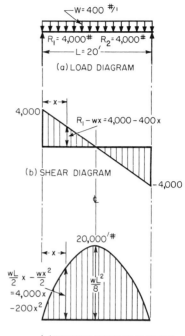

Fig. 5-21 Shear and moment diagrams for a uniformly loaded beam.

load over its entire length. As shown in Fig. 5-21c, the bending-moment diagram for this beam is a parabola.

The reactions at both ends of a simply supported uniformly loaded beam are both equal to $wL/2 = W/2$, where w is the uniform load in pounds per linear foot, $W = wL$ is the total load on the beam, and L is the span.

The shear at any distance x from the left support is $R_1 - wx = wL/2 - wx$ (see Fig. 5-21b). Equating this expression to zero, we find that there is no shear at the center of the beam.

The bending moment at any distance x from the left support is

$$M = R_1 x - wx\left(\frac{x}{2}\right) = \frac{wLx}{2} - \frac{wx^2}{2} = \frac{w}{2}x(L-x) \qquad (5\text{-}21)$$

Hence:

The bending moment at any section of a simply supported, uniformly loaded beam is equal to one-half the product of the load per linear foot and the distances to the section from both supports.

The maximum value of the bending moment occurs at the center of the beam. It is equal to $wL^2/8 = WL/8$.

5-22. Shear-Moment Relationship.

The slope of the bending-moment curve for any point on a beam is equal to the shear at that point; i.e.,

$$V = \frac{dM}{dx} \tag{5-22a}$$

Since maximum bending moment occurs when the slope changes sign, or passes through zero, maximum moment (positive or negative) occurs at the point of zero shear.

After integration, Eq. (5-22a) may also be written

$$M_1 - M_2 = \int_{x_2}^{x_1} V\, dx \tag{5-22b}$$

This equation indicates that the change in bending moment between any two sections of a beam is equal to the area of the shear diagram between ordinates at the two sections.

5-23. Moving Loads and Influence Lines.

One of the most helpful devices for solving problems involving variable or moving loads is an influence line. Whereas shear and moment diagrams evaluate the effect of loads at all sections of a structure, an influence line indicates the effect at a given section of a unit load placed at any point on the structure.

For example, to plot the influence line for bending moment at some point A on a beam, a unit load is applied at some point B. The bending moment at A due to the unit load at B is plotted as an ordinate to a convenient scale at B. The same procedure is followed at every point along the beam and a curve is drawn through the points thus obtained.

Actually, the unit load need not be placed at every point. The equation of the influence line can be determined by placing the load at an arbitrary point and computing the bending moment in general terms. (See also Art. 5-57.)

Suppose we wish to draw the influence line for reaction at A for a simple beam AB (Fig. 5-22a). We place a unit load at an arbitrary distance of xL from B. The reaction at A due to this load is $1 \cdot xL/L = x$. Then, $R_A = x$ is the equation of the influence line. It represents a straight line sloping upward from zero at B to unity at A (Fig. 5-22a). In other words, as the unit load moves across the beam, the reaction at A increases from zero to unity in proportion to the distance of the load from B.

Figure 5-22b shows the influence line for bending moment at the center of a beam. It resembles in appearance the bending-moment diagram for a load at the center of the beam, but its significance is entirely different. Each ordinate gives the moment at midspan for a load at the corresponding location. It indicates that, if a unit load is placed at a distance xL from one end, it produces a bending moment of $½\, xL$ at the center of the span.

Figure 5-22c shows the influence line for shear at the quarter point of a beam. When the load is to the right of the quarter point, the shear is positive and equal to the left reaction. When the load is to the left, the shear is negative and equal to the right reaction.

The diagram indicates that, to produce maximum shear at the quarter point, loads should be placed only to the right of the quarter point, with the largest load at the quarter point, if possible. For a uniform load, maximum shear results when the load extends from the right end of the beam to the quarter point.

Suppose, for example, that the beam is a crane girder with a span of 60 ft. The wheel loads are 20 and 10 kips, respectively, and are spaced 5 ft apart. For maximum shear at

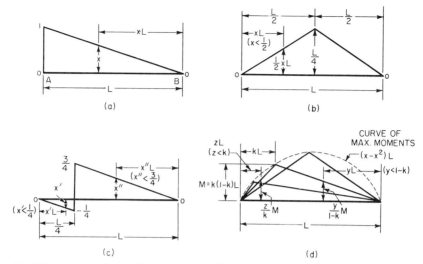

Fig. 5-22 Influence lines for (a) reaction at A; (b) midspan bending moment; (c) quarter-point shear; and (d) bending moments at several points in a beam.

the quarter point, the wheels should be placed with the 20-kip wheel at that point and the 10-kip wheel to the right of it. The corresponding ordinates of the influence line (Fig. 5-22c) are ¾ and ⁴⁰⁄₄₅ × ¾. Hence, the maximum shear is 20 × ¾ + 10 × ⁴⁰⁄₄₅ × ¾ = 21.7 kips.

Figure 5-22d shows influence lines for bending moment at several points on a beam. It is noteworthy that the apexes of the diagrams fall on a parabola, as shown by the dashed line. This indicates that the maximum moment produced at any given section by a single concentrated load moving across a beam occurs when the load is at that section. The magnitude of the maximum moment increases when the section is moved toward midspan, in accordance with the equation shown in Fig. 5-22d for the parabola.

5-24. Maximum Bending Moment.

When there is more than one load on the span, the influence line is useful in developing a criterion for determining the position of the loads for which the bending moment is a maximum at a given section.

Maximum bending moment will occur at a section C of a simple beam as loads move across it when one of the loads is at C. The proper load to place at C is the one for which the expression $W_a/a - W_b/b$ (Fig. 5-23) changes sign as that load passes from one side of C to the other.

When several loads move across a simple beam, the maximum bending moment produced in the beam may be near but not necessarily at midspan. To find the maximum moment, first determine the position of the loads for maximum moment at midspan. Then shift the loads until the load P_2 that was at the center of the beam is as far from midspan as the resultant of all the loads on the span is on the other side of midspan (Fig. 5-24). Maximum moment will occur under P_2.

When other loads move on or off the span during the shift of P_2 away from midspan, it may be necessary to investigate the moment under one of the other loads when it and the resultant are equidistant from midspan.

Moment of Inertia   5-27

Fig. 5-23 Moving loads on a simple beam AB placed for maximum bending moment at C.

Fig. 5-24 Moving loads placed to impose the largest maximum bending moment in a simple beam.

5-25. Bending Stresses in a Beam.

To derive the commonly used flexure formula for computing the bending stresses in a beam, we have to make the following assumptions.

   1. The unit stress in any plane parallel to the neutral surface of a beam is proportional to the unit strain of the plane.
   2. The modulus of elasticity in tension is the same as that in compression.
   3. The total and unit axial strain in any plane parallel to the neutral surface are both proportional to the distance of that plane from the neutral surface. (Cross sections that are plane before bending remain plane after bending. This requires that all planes have the same length before bending; thus, that the beam be straight.)
   4. The loads act in a plane containing the centroidal axis of the beam and are perpendicular to that axis. Furthermore, the neutral surface is perpendicular to the plane of the loads. Thus, the plane of the loads must contain an axis of symmetry of each cross section of the beam. (The flexure formula does not apply to a beam loaded unsymmetrically. See Arts. 5-34 and 5-35.)
   5. The beam is proportioned to preclude prior failure or serious deformation by torsion, local buckling, shear, or any cause other than bending.

Equating the bending moment to the resisting moment due to the internal stresses at any section of a beam yields

$$M = \frac{fI}{c} \qquad (5\text{-}23)$$

Fig. 5-25 Unit stresses on a beam section caused by bending of the beam.

M is the bending moment at the section, f is the normal unit stress in a plane at a distance c from the neutral axis (Fig. 5-25), and I is the moment of inertia of the cross section with respect to the neutral axis. If f is given in pounds per square inch (psi), I in in^4, and c in inches, then M will be in inch-pounds. Generally, c is taken as the distance to the outermost fiber. See also Arts. 5-26 and 5-27.

5-26. Moment of Inertia.

The neutral axis in a symmetrical beam is coincidental with the centroidal axis; i.e., at any section the neutral axis is so located that

$$\int y \, dA = 0 \qquad (5\text{-}24)$$

5-28   Structural Theory

where dA is a differential area parallel to the axis (Fig. 5-25), y is its distance from the axis, and the summation is taken over the entire cross section.

Moment of inertia with respect to the neutral axis is given by

$$I = \int y^2 \, dA \tag{5-25}$$

Values of I for several common types of cross section are given in Fig. 5-26. Values for structural-steel sections are presented in "Steel Construction Manual," American Insti-

Fig. 5-26 Geometric properties of a section.

tute of Steel Construction, Chicago, Ill. When the moments of inertia of other types of sections are needed, they can be computed directly by application of Eq. (5-25) or by breaking the section up into components for which the moment of inertia is known.

If I is the moment of inertia about the neutral axis, A the cross-sectional area, and d the distance between that axis and a parallel axis in the plane of the cross section, then the moment of inertia about the parallel axis is

$$I' = I + Ad^2 \qquad (5\text{-}26)$$

With this equation, the known moment of inertia of a component of a section about the neutral axis of the component can be transferred to the neutral axis of the complete section. Then, summing up the transferred moments of inertia for all the components yields the moment of inertia of the complete section.

When the moments of inertia of an area with respect to any two rectangular axes are known, the moment of inertia with respect to any other axis passing through the point of intersection of the two axes may be obtained through the use of Mohr's circle, as for stresses (Fig. 5-10). In this analog, I_x corresponds with f_x, I_y with f_y, and the **product of inertia** I_{xy} with v_{xy} (Art. 5-15).

$$I_{xy} = \int xy \, dA \qquad (5\text{-}27)$$

The two perpendicular axes through a point about which the moments of inertia are a maximum and a minimum are called the principal axes. The products of inertia are zero for the principal axes.

5-27. Section Modulus.

The ratio $S = I/c$ in Eq. (5-23) is called the section modulus. I is the moment of inertia of the cross section about the neutral axis and c the distance from the neutral axis to the outermost fiber. Values of S for common types of sections are given in Fig. 5-26.

5-28. Shearing Stresses in a Beam.

Fig. 5-27 Unit shearing stresses on a beam section.

The vertical shear at any section of a beam is resisted by nonuniformly distributed, vertical unit stresses (Fig. 5-27). At every point in the section, there is also a horizontal unit stress, which is equal in magnitude to the vertical unit shearing stress there [see Eq. (5-12)].

At any distance y' from the neutral axis, both the horizontal and vertical shearing unit stresses are equal to

$$v = \frac{V}{It} A' \overline{y} \qquad (5\text{-}28)$$

where V = vertical shear at the cross section
t = thickness of beam at distance y' from neutral axis
I = moment of inertia about neutral axis
A' = area between the outermost fiber and the fiber for which the shearing stress is being computed
$\overline{y}$ = distance of center of gravity of this area from the neutral axis (Fig. 5-27)

For a rectangular beam with width b and depth d, the maximum shearing stress occurs at mid-depth. Its magnitude is

5-30 Structural Theory

$$v = \frac{12V}{bd^3b}\frac{bd^2}{8} = \frac{3}{2}\frac{V}{bd}$$

That is, the maximum shear stress is 50% greater than the average shear stress on the section. Similarly, for a circular beam, the maximum is one-third greater than the average. For an I beam, however, the maximum shearing stress in the web is not appreciably greater than the average for the web section alone, if it is assumed that the flanges take no shear.

5-29. Combined Shear and Bending Stress.

For deep beams on short spans and beams made of low-tensile-strength materials, it is sometimes necessary to determine the maximum stress f' on an inclined plane caused by a combination of shear and bending stress—v and f, respectively. This stress f', which may be either tension or compression, is greater than the normal stress. Its value may be obtained by application of Mohr's circle (Art. 5-15), as indicated in Fig. 5-10, but with $f_y = 0$, and is

$$f' = \frac{f}{2} + \sqrt{v^2 + \left(\frac{f}{2}\right)^2} \qquad (5\text{-}29)$$

5-30. Stresses in the Plastic Range.

When the bending stresses exceed the proportional limit and stress no longer is proportional to strain, the distribution of bending stresses over the cross section ceases to be linear. The outer fibers deform with little change in stress, while the fibers not stressed beyond the proportional limit continue to take more stress as the load on the beam increases. (See also Art. 5-108.)

Modulus of rupture is defined as the stress computed from the flexure formula [Eq. (5-23)] for the maximum bending moment a beam sustains at failure. It is not the actual maximum unit stress in the beam but is sometimes used to compare the strength of beams.

5-31. Beam Deflections.

When a beam is loaded, it deflects. The new position of its longitudinal centroidal axis is called the **elastic curve.**

At any point of the elastic curve, the radius of curvature is given by

$$R = \frac{EI}{M} \qquad (5\text{-}30)$$

where M = bending moment at the point
E = modulus of elasticity
I = moment of inertia of the cross section about the neutral axis

Since the slope dy/dx of the curve is small, its square may be neglected, so that, for all practical purposes, $1/R$ may be taken equal to d^2y/dx^2, where y is the deflection of a point on the curve at a distance x from the origin of coordinates. Hence, Eq. (5-30) may be rewritten

$$M = EI\frac{d^2y}{dx^2} \qquad (5\text{-}31)$$

To obtain the slope and deflection of a beam, this equation may be integrated, with M expressed as a function of x. Constants introduced during the integration must be evaluated in terms of known points and slopes of the elastic curve.

Equation (5-31), in turn, may be rewritten after one integration as

$$\theta_B - \theta_A = \int_A^B \frac{M}{EI} dx \tag{5-32}$$

in which θ_A and θ_B are the slopes of the elastic curve at any two points A and B. If the slope is zero at one of the points, the integral in Eq. (5-32) gives the slope of the elastic curve at the other. It should be noted that the integral represents the area of the bending-moment diagram between A and B with each ordinate divided by EI.

The **tangential deviation** t of a point on the elastic curve is the distance of this point, measured in a direction perpendicular to the original position of the beam, from a tangent drawn at some other point on the elastic curve.

$$t_B - t_A = \int_A^B \frac{Mx}{EI} dx \tag{5-33}$$

Equation (5-33) indicates that the tangential deviation of any point with respect to a second point on the elastic curve equals the moment about the first point of the M/EI diagram between the two points. The **moment-area method** for determining the deflection of beams is a technique in which Eqs. (5-32) and (5-33) are utilized.

Suppose, for example, the deflection at midspan is to be computed for a beam of uniform cross section with a concentrated load at the center (Fig. 5-28).

Since the deflection at midspan for this loading is the maximum for the span, the slope of the elastic curve at the center of the beam is zero; i.e., the tangent is parallel to the undeflected position of the beam. Hence, the deviation of either support from the midspan tangent is equal to the deflection at the center of the beam. Then, by the moment-area theorem [Eq. (5-33)], the deflection y_c is given by the moment about either support of the area of the M/EI diagram included between an ordinate at the center of the beam and that support.

$$y_c = \frac{1}{2} \frac{PL}{4EI} \frac{L}{2} \frac{2}{3} \frac{L}{2} = \frac{PL^3}{48EI}$$

Suppose, now, the deflection y at any point D at a distance xL from the left support (Fig. 5-28) is to be determined. Referring to the sketch, we note that the distance DE from the undeflected position of D to the tangent to the elastic curve at support A is given by

$$y + t_{AD} = xt_{AB}$$

Fig. 5-28 Elastic curve for a simple beam.

where t_{AD} is the tangential deviation of D from the tangent at A and t_{AB} is the tangential deviation of B from that tangent. This equation, which is perfectly general for the deflection of any point of a simple beam, no matter how loaded, may be rewritten to give the deflection directly:

$$y = xt_{AB} - t_{AD} \tag{5-34}$$

But t_{AB} is the moment of the area of the M/EI diagram for the whole beam about support B. And t_{AD} is the moment about D of the area of the M/EI diagram included between ordinates at A and D. Hence

$$y = x\frac{1}{2}\frac{PL}{4EI}\frac{L}{2}\left(\frac{2}{3}+\frac{1}{3}\right)L - \frac{1}{2}\frac{PLx}{2EI}xL\frac{xL}{3} = \frac{PL^3}{48EI}x(3-4x^2)$$

It is also noteworthy that, since the tangential deviations are very small distances, the slope of the elastic curve at A is given by

$$\theta_A = \frac{t_{AB}}{L} \tag{5-35}$$

This holds, in general, for all simple beams regardless of the type of loading.

The procedure followed in applying Eq. (5-34) to the deflection of the loaded beam in Fig. 5-28 is equivalent to finding the bending moment at D with the M/EI diagram serving as the load diagram. The technique of applying the M/EI diagram as a load and determining the deflection as a bending moment is known as the **conjugate-beam method**.

The conjugate beam must have the same length as the given beam; it must be in equilibrium with the M/EI load and the reactions produced by the load; and the bending moment at any section must be equal to the deflection of the given beam at the corresponding section. The last requirement is equivalent to requiring that the shear at any section of the conjugate beam with the M/EI load be equal to the slope of the elastic curve at the corresponding section of the given beam. Figure 5-29 shows the conjugates for various types of beams.

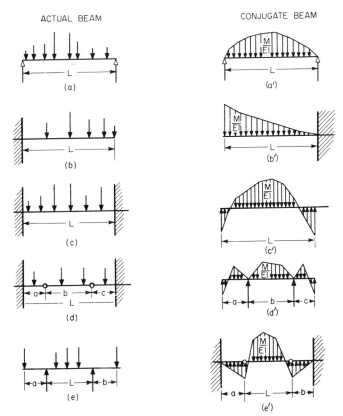

Fig. 5-29 Conjugate beams.

Deflections for several types of loading on simple beams are given in Figs. 5-30 to 5-35 and for overhanging beams and cantilevers in Figs. 5-36 to 5-41.

When a beam carries a number of loads of different types, the most convenient method of computing its deflection generally is to find the deflections separately for the uniform and concentrated loads and add them up.

For several concentrated loads, the easiest solution is to apply the reciprocal theorem (Art. 5-57). According to this theorem, if a concentrated load is applied to a beam at a point A, the deflection it produces at point B is equal to the deflection at A for the same load applied at $B(d_{AB} = d_{BA})$.

Suppose, for example, the midspan deflection is to be computed. Then, assume each load in turn applied at the center of the beam and compute the deflection at the point where it originally was applied from the equation of the elastic curve given in Fig. 5-33. The sum of these deflections is the total midspan deflection.

Another method for computing deflections of beams is presented in Art. 5-55. This method may also be applied to determining the deflection of a beam due to shear.

5-32. Combined Axial and Bending Loads.

For stiff beams, subjected to both transverse and axial loading, the stresses are given by the principle of superposition if the deflection due to bending may be neglected without serious error. That is, the total stress is given with sufficient accuracy at any section by the sum of the axial stress and the bending stresses. The maximum stress equals

$$f = \frac{P}{A} + \frac{Mc}{I} \tag{5-36}$$

where P = axial load
A = cross-sectional area
M = maximum bending moment
c = distance from neutral axis to outermost surface at the section where maximum moment occurs
I = moment of inertia of cross section about neutral axis at that section

When the deflection due to bending is large and the axial load produces bending stresses that cannot be neglected, the maximum stress is given by

$$f = \frac{P}{A} + (M + Pd)\frac{c}{I} \tag{5-37}$$

where d is the deflection of the beam. For axial compression, the moment Pd should be given the same sign as M, and for tension, the opposite sign, but the minimum value of $M + Pd$ is zero. The deflection d for axial compression and bending can be obtained by applying Eq. (5-31). (S. Timoshenko and J. M. Gere, "Theory of Elastic Stability," McGraw-Hill Book Company, New York; Friedrich Bleich, "Buckling Strength of Metal Structures," McGraw-Hill Book Company, New York.) However, it may be closely approximated by

$$d = \frac{d_o}{1 - (P/P_c)} \tag{5-38}$$

where d_o = deflection for the transverse loading alone
P_c = the critical buckling load $\pi^2 EI/L^2$ (see Art. 5-72)

5-33. Eccentric Loading.

An eccentric longitudinal load in the plane of symmetry produces a bending moment Pe where e is the distance of the load from the centroidal axis. The total unit stress is the sum of the stress due to this moment and the stress due to P applied as an axial load:

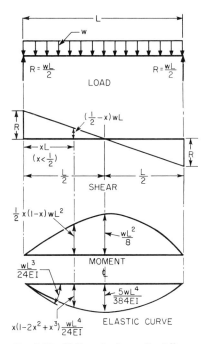

Fig. 5-30 Uniform load over the full span of a simple beam.

Fig. 5-31 Uniform load over part of a simple beam.

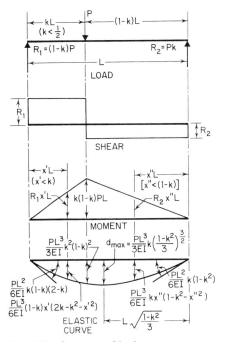

Fig. 5-32 Concentrated load at any point of a simple beam.

Fig. 5-33 Concentrated load at midspan of a simple beam.

Fig. 5-34 Two equal concentrated loads on a simple beam.

Fig. 5-35 Several equal concentrated loads on a simple beam.

Fig. 5-36 Concentrated load at the end of a beam overhang.

Fig. 5-37 Concentrated load at the end of a cantilever.

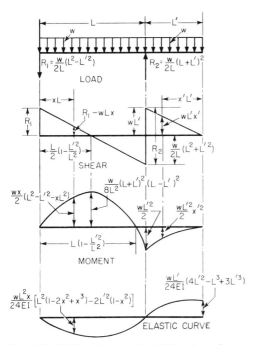

Fig. 5-38 Uniform load over the full length of a beam with overhang.

Fig. 5-39 Uniform load over the full length of a cantilever.

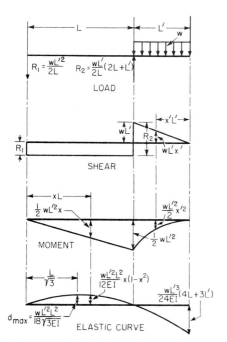

Fig. 5-40 Uniform load on a beam overhang.

Fig. 5-41 Triangular loading on a cantilever.

$$f = \frac{P}{A} \pm \frac{Pec}{I} = \frac{P}{A}\left(1 \pm \frac{ec}{r^2}\right) \tag{5-39}$$

where A = cross-sectional area
c = distance from neutral axis to outermost fiber
I = moment of inertia of cross section about neutral axis
r = **radius of gyration,** which is equal to $\sqrt{I/A}$

Figure 5-26 gives values of the radius of gyration for some commonly used cross sections.

If there is to be no tension on the cross section, e should not exceed r^2/c. For a rectangular section with width b and depth d, the eccentricity, therefore, should be less than $b/6$ and $d/6$; i.e., the load should not be applied outside the middle third. For a circular cross section with diameter D the eccentricity should not exceed $D/8$.

When the eccentric longitudinal load produces a deflection too large to be neglected in computing the bending stress, account must be taken of the additional bending moment Pd, where d is the deflection. This deflection may be computed by employing Eq. (5-31) or closely approximated by

$$d = \frac{4eP/P_c}{\pi(1 - P/P_c)} \tag{5-40}$$

P_c is the critical buckling load, $\pi^2 EI/L^2$ (see Art. 5-72).

If the load P does not lie in a plane containing an axis of symmetry, it produces bending about the two principal axes through the centroid of the cross section. The stresses are given by

$$f = \frac{P}{A} \pm \frac{Pe_x c_x}{I_y} \pm \frac{Pe_y c_y}{I_x} \tag{5-41}$$

where A = cross-sectional area
e_x = eccentricity with respect to principal axis YY
e_y = eccentricity with respect to principal axis XX
c_x = distance from YY to outermost fiber
c_y = distance from XX to outermost fiber
I_x = moment of inertia about XX
I_y = moment of inertia about YY

The principal axes are the two perpendicular axes through the centroid for which the moments of inertia are a maximum or a minimum and for which the products of inertia are zero.

5-34. Unsymmetrical Bending.

Bending caused by loads that do not lie in a plane containing a principal axis of each cross section of a beam is called unsymmetrical bending. If the bending axis of the beam lies in the plane of the loads, to preclude torsion (see Art. 5-25), and if the loads are perpendicular to the bending axis, to preclude axial components, the stress at any point in a cross section is given by

$$f = \frac{M_x y}{I_x} \pm \frac{M_y x}{I_y} \tag{5-42}$$

where M_x = bending moment about principal axis XX
M_y = bending moment about principal axis YY
x = distance from point for which stress is to be computed to YY axis
y = distance from point to XX axis
I_x = moment of inertia of the cross section about XX
I_y = moment of inertia about YY

5-38 Structural Theory

If the plane of the loads makes an angle θ with a principal plane, the neutral surface will form an angle α with the other principal plane such that

$$\tan \alpha = \frac{I_x}{I_y} \tan \theta \tag{5-43}$$

5-35. Beams with Unsymmetrical Sections.

In the derivation of the flexure formula $f = Mc/I$ (Art. 5-25), the assumption is made that the beam bends, without twisting, in the plane of the loads and that the neutral surface is perpendicular to the plane of the loads. These assumptions are correct for beams with cross sections symmetrical about two axes when the plane of the loads contains one of these axes. They are not necessarily true for beams that are not doubly symmetrical. The reason is that in beams that are doubly symmetrical the bending axis coincides with the centroidal axis, whereas in unsymmetrical sections the two axes may be separate. In the latter case, if the plane of the loads contains the centroidal axis but not the bending axis, the beam will be subjected to both bending and torsion.

The **bending axis** may be defined as the longitudinal line in a beam through which transverse loads must pass to preclude the beam's twisting as it bends. The point in each section through which the bending axis passes is called the **shear center,** or center of twist. The shear center is also the center of rotation of the section in pure torsion (Art. 5-78). Its location depends on the dimensions of the section.

If a beam has an axis of symmetry, the shear center lies on it. In doubly symmetrical beams, the shear center lies at the intersection of the two axes of symmetry and hence coincides with the centroid.

For any section composed of two narrow rectangles, such as a T beam or an angle, the shear center may be taken as the intersection of the longitudinal center lines of the rectangles.

For a channel section with one axis of symmetry, the shear center is outside the section at a distance from the centroid equal to $e(1 + h^2A/4I)$, where e is the distance from the centroid to the center of the web, h is the depth of the channel, A the cross-sectional area, and I the moment of inertia about the axis of symmetry. (The web lies between the shear center and the centroid.)

Fig. 5-42 Bending stresses in a curved beam.

Locations of shear centers for several other sections are given in Friedrich Bleich, "Buckling Strength of Metal Structures," Chap. III, McGraw-Hill Book Company, New York.

CURVED BEAMS

Structural members, such as arches, crane hooks, chain links, and frames of some machines, that have considerable initial curvature in the plane of loading are called curved beams. The flexure formula of Art. 5-25, $f = Mc/I$, cannot be applied to them with any reasonable degree of accuracy unless the depth of the beam is small compared with the radius of curvature.

Unlike the condition in straight beams, unit strains in curved beams are not proportional to the distance from the neutral surface, and the centroidal axis does not coincide with the neutral axis. Hence the stress distribution on a section is not linear but more like the distribution shown in Fig. 5-42c.

5-36. Stresses in Curved Beams.

Just as for straight beams, the assumption that plane sections before bending remain plane after bending generally holds for curved beams. So the total strains are proportional to the distance from the neutral axis. But since the fibers are initially of unequal length, the unit strains are a more complex function of this distance. In Fig. 5-42a, for example, the bending couples have rotated section AB of the curved beam into section $A'B'$ through an angle $\Delta d\theta$. If ϵ_o is the unit strain at the centroidal axis and ω is the angular unit strain $\Delta d\theta/d\theta$, then the unit strain at a distance y from the centroidal axis (measured positive in the direction of the center of curvature) is

$$\epsilon = \frac{DD'}{DD_o} = \frac{\epsilon_o R\, d\theta - y\Delta\, d\theta}{(R - y)\, d\theta} = \epsilon_o - (\omega - \epsilon_o)\frac{y}{R - y} \tag{5-44}$$

where R = radius of curvature of centroidal axis.

Equation (5-44) can be expressed in terms of the bending moment if we take advantage of the fact that the sum of the tensile and compressive forces on the section must be zero and the moment of these forces must be equal to the bending moment M. These two equations yield

$$\epsilon_o = \frac{M}{ARE} \quad \text{and} \quad \omega = \frac{M}{ARE}\left(1 + \frac{AR^2}{I'}\right) \tag{5-45}$$

where A is the cross-sectional area, E the modulus of elasticity, and

$$I' = \int \frac{y^2\, dA}{1 - y/R} = \int y^2\left(1 + \frac{y}{R} + \frac{y^2}{R^2} + \cdots\right) dA \tag{5-46}$$

It should be noted that I' is very nearly equal to the moment of inertia I about the centroidal axis when the depth of the section is small compared with R, so that the maximum ratio of y to R is small compared with unity. M is positive when it decreases the radius of curvature.

Since the stress $f = E\epsilon$, we obtain the stresses in the curved beam from Eq. (5-44) by multiplying it by E and substituting ϵ_o and ω from Eq. (5-45):

$$f = \frac{M}{A\,R} - \frac{My}{I'}\frac{1}{1 - y/R} \tag{5-47}$$

The distance y_o of the neutral axis from the centroidal axis (Fig. 5-42) may be obtained from Eq. (5-47) by setting $f = 0$:

$$y_o = \frac{I'R}{I' + AR^2} \tag{5-48}$$

Since y_o is positive, the neutral axis shifts toward the center of curvature.

5-37. Curved Beams with Rectangular Sections.

Taking the first three terms of the series in Eq. (5-46) and integrating, we obtain for a rectangular cross section

$$I' = I\left(1 + \frac{3}{5}\frac{c^2}{R^2}\right) \tag{5-49}$$

where I is the moment of inertia about the centroidal axis and c is one-half the depth of the beam. The stresses in the outer fibers are given by

$$\begin{aligned} f_{max} &= -\frac{Mc}{I}\left(\frac{I}{I'}\frac{R}{R-c} - \frac{c}{3R}\right) \\ f_{min} &= +\frac{Mc}{I}\left(\frac{I}{I'}\frac{R}{R+c} + \frac{c}{3R}\right) \end{aligned} \tag{5-50}$$

It is noteworthy that the factor by which Mc/I is multiplied in Eq. (5-50) is a function of c/R and independent of the other dimensions of the beam.

5-38. Curved Beams with Circular Sections.

Integrating the series in Eq. (5-46) for a circle and taking the first three terms of the result, we obtain

$$I' = I\left(1 + \frac{1}{2}\frac{c^2}{R^2}\right) \tag{5-51}$$

where I = moment of inertia about centroidal axis
c = one-half depth of beam
The stresses in the outer fibers are given by

$$\begin{aligned} f_{max} &= -\frac{Mc}{I}\left(\frac{I}{I'}\frac{R}{R-c} - \frac{c}{4R}\right) \\ f_{min} &= +\frac{Mc}{I}\left(\frac{I}{I'}\frac{R}{R+c} + \frac{c}{4R}\right) \end{aligned} \tag{5-52}$$

For this section, too, the factor by which Mc/I is multiplied is a function of c/R and independent of the other dimensions of the beam.

5-39. Curved I or T Beams.

If Eq. (5-47) is applied to I or T beams or tubular members, it may indicate circumferential flange stresses that are much lower than will actually occur. The error is due to the fact that the outer edges of the flanges deflect radially. The effect is equivalent to having only part of the flanges active in resisting bending stresses. Also, accompanying the flange deflections, there are transverse bending stresses in the flanges. At the junction with the web, these reach a maximum, which may be greater than the maximum circumferential stress. Furthermore, there are radial stresses (normal stresses acting in the direction of the radius of curvature) in the web that also may have maximum values greater than the maximum circumferential stress.

A good approximation to the stresses in I or T beams is as follows: for circumferential stresses, Eq. (5-47) may be used with a modified cross section, which is obtained by using a reduced flange width. The reduction is calculated from $b' = \alpha b$, where b is the length of the portion of the flange projecting on either side from the web, b' is the corrected length, and α is a correction factor determined from equations developed by H. Bleich. α is a function of b^2/rt, where t is the flange thickness and r the radius of the center of the flange:

$b^2/rt =$	0.5	0.7	1.0	1.5	2	3	4	5
$\alpha =$	0.9	0.6	0.7	0.6	0.5	0.4	0.37	0.33

When the parameter b^2/rt is greater than 1.0, the maximum transverse bending stress is approximately equal to 1.7 times the stress obtained at the center of the flange from Eq. (5-47) applied to the modified section. When the parameter equals 0.7, that stress should be multiplied by 1.5, and when it equals 0.4, the factor is 1.0. In Eq. (5-47), I' for I beams may be taken for this calculation approximately equal to

$$I' = I\left(1 + \frac{c^2}{R^2}\right) \tag{5-53}$$

where I = moment of inertia of modified section about its centroidal axis
R = radius of curvature of centroidal axis
c = distance from centroidal axis to center of the more sharply curved flange

Because of the high stress factor, it is advisable to stiffen or brace curved I-beam flanges.

The maximum radial stress will occur at the junction of web and flange of I beams. If the moment is negative, that is, if the loads tend to flatten out the beam, the radial stress is tensile, and there is a tendency for the more sharply curved flange to pull away from the web. An approximate value of this maximum stress is

$$f_r = -\frac{A_f}{A}\frac{M}{t_w c_g r'} \tag{5-54}$$

where f_r = radial stress at junction of flange and web of a symmetrical I beam
A_f = area of one flange
A = total cross-sectional area
M = bending moment
t_w = thickness of web
c_g = distance from centroidal axis to center of flange
r' = radius of curvature of inner face of more sharply curved flange

(A. P. Boresi, O. Sidebottom, F. B. Seely, and J. O. Smith, "Advanced Mechanics of Materials," John Wiley & Sons, Inc., New York.)

5-40. Axial and Bending Loads on Curved Beams.

If a curved beam carries an axial load P as well as bending loads, the maximum unit stress is

$$f = \frac{P}{A} \pm \frac{Mc}{I}K \tag{5-55}$$

where K is a correction factor for the curvature [see Eqs. (5-50) and (5-52)]. The sign of M is taken positive in this equation when it increases the curvature, and P is positive when it is a tensile force, negative when compressive.

5-41. Slope and Deflection of Curved Beams.

If we consider two sections of a curved beam separated by a differential distance ds (Fig. 5-42), the change in angle $\Delta\,d\theta$ between the sections caused by a bending moment M and an axial load P may be obtained from Eq. (5-45), noting that $d\theta = ds/R$.

$$\Delta\,d\theta = \frac{M\,ds}{EI'}\left(1 + \frac{I'}{A\,R^2}\right) + \frac{P\,ds}{A\,RE} \tag{5-56}$$

where E is the modulus of elasticity, A the cross-sectional area, R the radius of curvature of the centroidal axis, and I' is defined by Eq. (5-45) [see also Eqs. (5-49), (5-51), and (5-53)].

If P is a tensile force, the length of the centroidal axis increases by

$$\Delta\,ds = \frac{P\,ds}{A\,E} + \frac{M\,ds}{A\,RE} \tag{5-57}$$

The effect of curvature on shearing deformations for most practical applications is negligible.

For shallow sections (depth of section less than about one-tenth the span), the effect of axial forces on deformations may be neglected. Also, unless the radius of curvature is very small compared with the depth, the effect of curvature may be ignored. Hence, for most practical applications, Eq. (5-56) may be used in the simplified form:

$$\Delta\,d\theta = \frac{M\,ds}{EI} \tag{5-58}$$

For deeper beams, the action of axial forces, as well as bending moments, should be taken into account; but unless the curvature is sharp, its effect on deformations may be neglected. So only Eq. (5-58) and the first term in Eq. (5-57) need be used. (S. Timoshenko and D. H. Young, "Theory of Structures," McGraw-Hill Book Company, New York.) See also Arts. 5-88 to 5-91.

GRAPHIC-STATICS FUNDAMENTALS

A force may be represented by a straight line of fixed length. The length of line to a given scale represents the magnitude of the force. The position of the line parallels the line of action of the force. And an arrowhead on the line indicates the direction in which the force acts.

Forces are concurrent when their lines of action meet. If they lie in the same plane, they are coplanar.

5-42. Parallelogram of Forces.

The resultant of several forces is a single force that would produce the same effect on a rigid body. The resultant of two concurrent forces is determined by the parallelogram law:

If a parallelogram is constructed with two forces as sides, the diagonal represents the resultant of the forces (Fig. 5-43a).

The **resultant** is said to be equal to the sum of the forces, sum here meaning, of course, addition by the parallelogram law. Subtraction is carried out in the same manner as addition, but the direction of the force to be subtracted is reversed.

If the direction of the resultant is reversed, it becomes the **equilibrant,** a single force that will hold the two given forces in equilibrium.

Fig. 5-43 Addition of forces by (a) parallelogram law; (b) triangle construction; (c) polygon construction.

5-43. Resolution of Forces.

To resolve a force into two components, a parallelogram is drawn with the force as a diagonal. The sides of the parallelogram represent the components. The procedure is: (1) Draw the given force. (2) From both ends of the force draw lines parallel to the directions in which the components act. (3) Draw the components along the parallels through the origin of the given force to the intersections with the parallels through the other end. Thus, in Fig. 5-43a, P_1 and P_2 are the components in directions OA and OB of the force represented by OC.

5-44. Force Polygons.

Examination of Fig. 5-43a indicates that a step can be saved in adding the two forces. The same resultant could be obtained by drawing only the upper half of the parallelogram. Hence, to add two forces, draw the first force; then draw the second force beginning at the end of the first one. The resultant is the force drawn from the origin of the first force to the end of the second force, as shown in Fig. 5-43b. Again, the equilibrant is the resultant with direction reversed.

From this diagram, an important conclusion can be drawn: **If three forces meeting at a point are in equilibrium, they will form a closed force triangle.**

The conclusions reached for addition of two forces can be generalized for several concurrent forces: To add several forces, $P_1, P_2, P_3, \ldots, P_n$, draw P_2 from the end of P_1, P_3 from the end of P_2, etc. The force required to close the force polygon is the resultant (Fig. 5-43c).

If a group of concurrent forces are in equilibrium, they will form a closed force polygon.

5-45. Equilibrium Polygons.

When the forces are coplanar but not concurrent, the force polygon will yield the magnitude and direction of the resultant but not its point of application. To complete the solution, the easiest method generally is to employ an auxiliary force polygon, called an equilibrium, or funicular (string), polygon. Sides of this polygon represent the lines of action of certain components of the given forces; more specifically, they take the configuration of a weightless string holding the forces in equilibrium.

5-44   Structural Theory

In Fig. 5-44a are shown three forces P_1, P_2, and P_3. The magnitude and direction of their resultant R are determined from the force polygon in Fig. 5-44b. To construct the equilibrium polygon for determining the position of the resultant, select a point O in or near the force polygon. From O, which is called a **pole**, draw rays OA, OB, OC, and OD to the extremities of the forces (Fig. 5-44b). Note that consecutive pairs of rays form closed triangles with the given forces, and therefore, they are components of the forces in the directions of the rays. Furthermore, each of the rays OB and OC, drawn to a vertex of the polygon formed by two given forces, represents, in turn, two forces with the same line of action but acting in opposite directions to help hold the given forces in equilibrium. Ray OA is a component not only of P_1 but also of R; similarly, OD is a component of both P_3 and R, and with OA holds R in equilibrium.

Fig. 5-44 Force and equilibrium polygons for a system of forces.

If in Fig. 5-44a, where the given forces are shown in their relative positions, lines are drawn from each force in the direction of the rays forming their components, they will form a funicular polygon. For example, starting with a convenient point on P_1 (Fig. 5-44a), draw a line parallel to OB until it intersects P_2. This represents the line of action of both force BO, which helps hold P_1 in equilibrium, and OB, which helps hold P_2 in equilibrium. Through the intersection with P_2, draw a line parallel to OC until it intersects P_3. Through this last intersection, draw a line parallel to DO, one component of R, and through the starting point on P_1, draw a line parallel to OA, the second component of R, completing the funicular polygon. These two lines intersect on the line of action of R.

5-46. Beam and Truss Reactions by Graphics.

The equilibrium polygon can also be used to find the reactions of simple beams and trusses supporting vertical loads. First, a force polygon is constructed to obtain the magnitude and direction of the resultant of the loads, which is equal in magnitude but opposite in direction to the sum of the reactions. Second, rays are drawn to the vertexes of the polygon from a conveniently located pole. These rays are used to construct all but one side of an equilibrium polygon. The closing side is the common line of action of two equal but opposite forces that act with a pair of rays already drawn to hold the reactions in equilibrium. Therefore, draw a line through the pole parallel to the closing side. The intersection with the resultant separates it into two forces, which are equal to the reactions sought.

For example, suppose the reactions were to be obtained graphically for the beam (or truss) in Fig. 5-45a. As a first step, the force polygon $ABCD$ is constructed, a pole selected, and rays drawn to the extremities of the forces (Fig. 5-45b). Since the loads are parallel, the force polygon is a straight line. The sum of the reactions, in this case, is equal to the length of the line AD.

Next, the equilibrium polygon is constructed. Starting with a convenient point on the line of action of R_1, draw a line oa parallel to ray OA in Fig. 5-45b and locate its intersection with the line of action of P_1 (Fig. 5-45d). Through this intersection, draw ob, a line parallel to OB, to the intersection with P_2, then a line oc parallel to OC, and finally a line od parallel to OD, which terminates on the line of action of R_2. Draw the closing line oe of the polygon between this last intersection and the starting point of the polygon. The last step is to draw through the pole (Fig. 5-45b) OE, a line parallel to oe, the closing

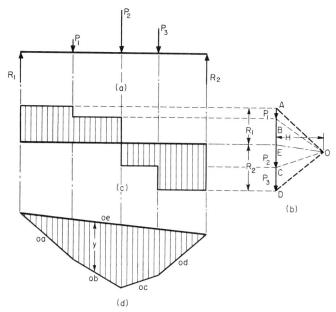

Fig. 5-45 Shear and moment diagrams obtained by graphic statics for (a) loaded beam; (b) force polygon; (c) shear diagram; (d) equilibrium polygon and bending-moment diagram.

line of the equilibrium polygon, cutting the force-polygon resultant at E. Then, $DE = R_2$, and $EA = R_1$. (See Art. 5-17 for an analytical method of computing reactions.)

5-47. Shear and Moment Diagrams by Graphics.

The shear at any section of a beam is equal to the algebraic sum of the loads and reactions on the left of the section, upward-acting forces being considered positive, downward forces negative. If the forces are arranged in the proper order, the shear diagram may be obtained directly from the force polygon after the reactions have been determined.

For example, the shear diagram for the beam in Fig. 5-45a can be easily obtained from the force polygon $ABCDE$ in Fig. 5-45b. The zero axis is a line parallel to the beam through E. As indicated in Fig. 5-45c, the ordinates of the shear diagram are laid off, starting with R_1 along the line of action of the left reaction, by drawing lines parallel to the zero axis through the extremities of the forces in the force polygon. (See Art. 5-19 for an analytical method of determining shears.)

The moment of a force about a point can be obtained from the equilibrium and force polygons. In the equilibrium diagram, draw a line parallel to the force through the given point. Measure the intercept of this line between the two adjacent funicular-polygon sides (extended if necessary) that originate at the given force. The moment is the product of this intercept by the distance of the force-polygon pole from the force. The intercept should be measured to the same linear scale as the beam and load positions, and the pole distance to the same scale as the forces in the force polygon.

As a consequence of this relationship between the sides of the funicular polygon, each ordinate (parallel to the forces) multiplied by the pole distance is equal to the bending moment at the corresponding section of the beam or truss. In Fig. 5-45d, for example, the equilibrium polygon, to scale, is the bending-moment diagram for the beam in Fig. 5-45a. At any section, the moment equals the ordinate y multiplied by the pole distance H. (See Art. 5-20 for an analytical method.)

ROOF TRUSSES

A truss is a coplanar system of structural members joined together at their ends to form a stable framework. Neglecting small changes in the lengths of the members due to loads, the relative positions of the joints cannot change.

Three bars pinned together to form a triangle represent the simplest type of truss. Some of the more common types of roof trusses are shown in Fig. 5-46.

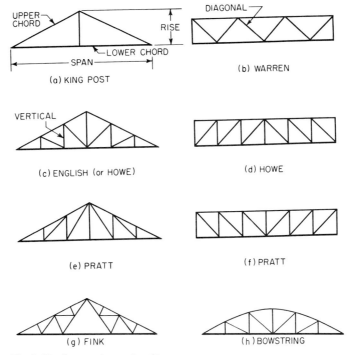

Fig. 5-46 Common types of roof trusses.

The top members are called the upper chord; the bottom members, the lower chord; and the verticals and diagonals, web members.

The purpose of roof trusses is to act like big beams, to support the roof covering over long spans. They not only have to carry their own weight and the weight of the roofing and roof beams, or purlins, but cranes, wind loads, snow loads, suspended ceilings, and equipment, and a live load to take care of construction, maintenance, and repair loading. These loads are applied at the intersection of the members, or panel points, so that the members will be subjected principally to axial stresses.

5-48. Method of Sections.

A convenient method of determining the stresses in truss members is to isolate a portion of the truss by a section so chosen as to cut only as many members with unknown stresses as can be evaluated by the laws of equilibrium applied to that portion of the truss. The stresses in the cut members are treated as external forces. Compressive forces act toward the panel point and tensile forces away from the joint.

Suppose, for example, we wish to find the stress in chord AB of the truss in Fig. 5-47a. We can take a vertical section XX close to panel point A. This cuts not only AB but AD and ED as well. The external 10-kip (10,000-lb) loading and 25-kip reaction at the left are held in equilibrium by the compressive force C in AB, tensile force T in ED, and tensile force S in AD (Fig. 5-47b). The simplest way to find C is to take moments about D, the point of intersection of S and T, eliminating these unknowns from the calculation.

$$-9C + 36 \times 25 - 24 \times 10 - 12 \times 10 = 0$$

from which C is found to be 60 kips.

Similarly, to find the stress in ED, the simplest way is to take moments about A, the point of intersection of S and C:

$$-9T + 24 \times 25 - 12 \times 10 = 0$$

from which T is found to be 53.3 kips.

On the other hand, the stress in AD can be easily determined by two methods. One takes advantage of the fact that AB and ED are horizontal members, requiring AD to carry the full vertical shear at section XX. Hence we know that the vertical component V of $S = 25 - 10 - 10 = 5$ kips. Multiplying V by sec θ (Fig. 5-47b), which is equal to the ratio of the length of AD to the rise of the truss ($^{15}\!/_9$), S is found to be 8.3 kips. The

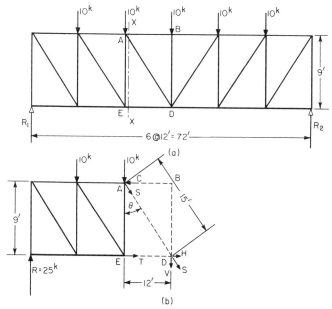

Fig. 5-47 Truss stresses determined by method of sections.

5-48 Structural Theory

second method—presented because it is useful when the chords are not horizontal—is to resolve S into horizontal and vertical components at D and take moments about E. Since both T and the horizontal component of S pass through E, they do not appear in the computations, and C already has been computed. Equating the sum of the moments to zero gives $V = 5$, as before.

5-49. Method of Joints.

Another useful method for determining the stresses in truss members is to select sections that isolate the joints one at a time and then apply the laws of equilibrium to each. Considering the stresses in the cut members as external forces, the sum of the horizontal components of the forces acting at a joint must be zero, and so must be the sum of the vertical components. Since the lines of action of all the forces are known, we can therefore compute two unknown magnitudes at each joint by this method. The procedure is to start at a joint that has only two unknowns (generally at the support) and then, as stresses in members are determined, analyze successive joints.

Let us, for illustration, apply the method to joint 1 of the truss in Fig. 5-48a. Equating the sum of the vertical components to zero, we find that the vertical component of the top chord must be equal and opposite to the reaction, 12 kips (12,000 lb). The stress in the top chord at this joint, then, must be a compression equal to $12 \times {}^{30}/_{18} = 20$ kips.

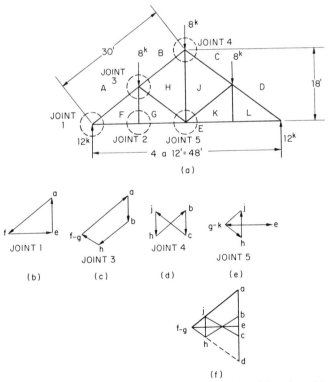

Fig. 5-48 Method of joints applied to a roof truss (a) to (e), and graphical solution with a Maxwell diagram (f).

From the fact that the sum of the horizontal components must be zero, we find that the stress in the bottom chord at the joint must be equal and opposite to the horizontal component of the top chord. Hence the stress in the bottom chord must be a tension equal to $20 \times \frac{24}{30} = 16$ kips.

Moving to joint 2, we note that, with no vertical loads at the joint, the stress in the vertical is zero. Also, the stress is the same in both bottom chord members at the joint, since the sum of the horizontal components must be zero.

Joint 3 now contains only two unknown stresses. Denoting the truss members and the loads by the letters placed on opposite sides of them, as indicated in Fig. 5-48a, the unknown stresses are S_{BH} and S_{HG}. The laws of equilibrium enable us to write the following two equations, one for the vertical components and the second for the horizontal components:

$$\Sigma V = 0.6 S_{FA} - 8 - 0.6 S_{BH} + 0.6 S_{HG} = 0$$
$$\Sigma H = 0.8 S_{FA} - 0.8 S_{BH} - 0.8 S_{HG} = 0$$

Both unknown stresses are assumed to be compressive; i.e., acting toward the joint. The stress in the vertical does not appear in these equations, because it was already determined to be zero. The stress in FA, S_{FA}, was found from analysis of joint 1 to be 20 kips. Simultaneous solution of the two equations yields $S_{HG} = 6.7$ kips and $S_{BH} = 13.3$ kips. (If these stresses had come out with a negative sign, it would have indicated that the original assumption of their directions was incorrect; they would, in that case, be tensile forces instead of compressive forces.)

5-50. Bow's Notation.

The method of designating the loads and stresses used in the previous paragraph is the basis of Bow's notation. Capital letters are placed in the spaces between truss members and between forces. Each member and load is then designated by the letters on opposite sides of it. For example, in Fig. 5-48a, the upper chord members are AF, BH, CJ, and DL. The loads are AB, BC, and CD, and the reactions are EA and DE. Stresses in the members generally are designated by the same letters but in lower case.

5-51. Graphical Analysis of Trusses.

The method of joints may also be used with graphical techniques. Figure 5-48b, c, d, and e shows how the polygon of force may be applied at each joint to determine the two unknown stresses. Since each joint is in equilibrium, the loads and the stresses in each member must form a closed polygon. The known forces are constructed first. Then, from the origin and the end point, respectively, a line is drawn parallel to the line of action of each of the two unknown stresses. Their intersection determines the magnitude of the unknown stresses.

This solution, as well as the preceding ones, presumes that the reactions are known. They may be computed analytically or graphically, with load and funicular polygons, as explained in Arts. 5-17 and 5-46.

Examination of Fig. 5-48b, c, d, and e indicates that each stress occurs in two force polygons. Hence the graphical solution can be shortened by combining the polygons. The combination of the various polygons for all the joints into one stress diagram is known as a **Maxwell diagram**.

The procedure consists of first constructing the force polygon for the loads and reactions and then applying the method of joints, employing each stress as it is found for finding the stresses in the next joint. To make it easy to determine whether the stresses are compression or tension, loads and reactions should be plotted in the force polygon in the order in which they are passed in going clockwise around the truss. Similarly, in drawing the force polygon for each joint, the forces should be plotted in a clockwise direction around the joint. If these rules are followed, the order of the space letters indicates the direction of the forces—compressive stresses act toward the joint, tensile stresses away from the joint. Going around joint 1 clockwise, for example, we find the reaction EA as an upward-acting vertical force ea (e to a) in Fig. 5-48b; the top-chord stress af acting toward the joint (a to f), and the bottom-chord stress fe acting away from the joint (f to e). Hence af is compressive, fe tensile.

To construct a Maxwell diagram for the truss in Fig. 5-48a, we lay off the loads and reactions in clockwise order (Fig. 5-48f). The force polygon $abcde$ is a straight line because all the forces are vertical. To solve joint 1, a line is drawn through point a in Fig. 5-48f parallel to AF and a second line through point e parallel to FE. Their intersection is point f. Going around this triangle in the same order as the loads and members are encountered in traveling clockwise around joint 1, we find ea, the reaction, as an upward force; going from a to f, we move toward the joint, indicating that af is a compressive stress; and going from f to e, we move away from the joint, indicating that fe is a tensile stress.

At joint 2, the stress in the vertical is zero. Hence point g coincides with point f.

To solve joint 3, we start with the known stress af and proceed clockwise around the joint. To complete the force polygon, we draw a line through point b parallel to BH and a line through g parallel to GH. Their intersection locates point h. Similarly, the force polygon for joint 4 is completed by drawing a line through c parallel to CJ and a line through h parallel to JH.

Wind loads on a truss with a sloping top chord are assumed to act normal to the roof. In that case, the load polygon will be an inclined line or a true polygon. The reactions are computed generally on the assumption either that both are parallel to the resultant of the wind loads or that one end of the truss is free to move horizontally and therefore will not resist the horizontal components of the loads. The stress diagram is plotted in the same manner as for vertical loads after the reactions are found.

Some trusses are complex and require special methods of analysis. For methods of solving these, see Norris et al., "Elementary Structural Analysis," McGraw-Hill Book Company, New York.

GENERAL TOOLS FOR STRUCTURAL ANALYSIS

For some types of structures, the equilibrium equations are not sufficient to determine the reactions or the internal stresses. These structures are called statically **indeterminate.**

For the analysis of such structures, additional equations must be written on the basis of a knowledge of the elastic deformations. Hence methods of analysis that enable deformations to be evaluated in terms of unknown forces or stresses are important for the solution of problems involving statically indeterminate structures. Some of these methods, like the method of virtual work, are also useful in solving complicated problems involving statically determinate systems.

5-52. Virtual Work.

A virtual displacement is an imaginary small displacement of a particle consistent with the constraints upon it. Thus, at one support of a simply supported beam, the virtual displacement could be an infinitesimal rotation $d\theta$ of that end but not a vertical movement. However, if the support is replaced by a force, then a vertical virtual displacement may be applied to the beam at that end.

Virtual work is the product of the distance a particle moves during a virtual displacement by the component in the direction of the displacement of a force acting on the particle. If the displacement and the force are in opposite directions, the virtual work is negative. When the displacement is normal to the force, no work is done.

Suppose a rigid body is acted upon by a system of forces with a resultant R. Given a virtual displacement ds at an angle α with R, the body will have virtual work done on it equal to $R \cos \alpha \, ds$. (No work is done by internal forces. They act in pairs of equal magnitude but opposite direction, and the virtual work done by one force of a pair is equal but opposite in sign to the work done by the other force.) If the body is in equilibrium under the action of the forces, then $R = 0$ and the virtual work also is zero.

Thus, the principle of virtual work may be stated: **If a rigid body in equilibrium is given a virtual displacement, the sum of the virtual work of the forces acting on it must be zero.**

As an example of how the principle may be used to find a reaction of a statically determinate beam, consider the simple beam in Fig. 5-49a, for which the reaction R is to be determined. First, replace the support by an unknown force R. Next, move that end of the beam upward a small amount dy as in Fig. 5-49b. The displacement under the load P will be $x \, dy/L$, upward. Then, by the principle of virtual work, $R \, dy - Px \, dy/L = 0$, from which $R = Px/L$.

Fig. 5-49 Principle of virtual work applied to determination of a simple-beam reaction (a) and (b) and to the reaction of a beam with a suspended span (c) and (d).

The principle may also be used to find the reaction R of the more complex beam in Fig. 5-49c. The first step again is to replace the support by an unknown force R. Next, apply a virtual downward displacement dy at hinge A (Fig. 5-49d). The displacement under the load P will be $x \, dy/c$, and at the reaction R will be $a \, dy/(a + b)$. According to the principle of virtual work, $-Ra \, dy/(a + b) + Px \, dy/c = 0$, from which $R = Px(a + b)/ac$. In this type of problem, the method has the advantage that only one reaction need be considered at a time and internal forces are not involved.

When an elastic body is deformed, the virtual work done by the internal forces is equal to the corresponding increment of the strain energy dU, in accordance with the principle of virtual work.

Assume a constrained elastic body acted upon by forces $P_1, P_2, \ldots$, for which the corresponding deformations are $e_1, e_2 \ldots$. Then, $\Sigma P_n \, de_n = dU$. The increment of the strain energy due to the increments of the deformations is given by

$$dU = \frac{\partial U}{\partial e_1} de_1 + \frac{\partial U}{\partial e_2} de_2 + \cdots$$

In solving a specific problem, a virtual displacement that is most convenient in simplifying the solution should be chosen. Suppose, for example, a virtual displacement is selected that affects only the deformation e_n corresponding to the load P_n, other deformations being unchanged. Then, the principle of virtual work requires that

$$P_n \, de_n = \frac{\partial U}{\partial e_n} de_n$$

This is equivalent to

$$\frac{\partial U}{\partial e_n} = P_n \qquad (5\text{-}59)$$

which states that the partial derivative of the strain energy with respect to any specific deformation gives the corresponding force.

Suppose, for example, the stress in the vertical bar in Fig. 5-50 is to be determined. All bars are made of the same material and have the same cross section. If the vertical bar stretches an amount e under the load P, the inclined bars will each stretch an amount $e \cos \alpha$. The strain energy in the system is [from Eq. (5-11c)]

$$U = \frac{AE}{2L}(e^2 + 2e^2 \cos^3 \alpha)$$

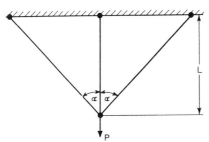

Fig. 5-50 Indeterminate truss.

and the partial derivative of this with respect to e must be equal to P; that is

$$P = \frac{AE}{2L}(2e + 4e \cos^3 \alpha)$$

$$= \frac{AEe}{L}(1 + 2 \cos^3 \alpha)$$

Noting that the force in the vertical bar equals AEe/L, we find from the above equation that the required stress equals $P/(1 + 2 \cos^3 \alpha)$.

5-53. Castigliano's Theorems.

It can also be shown that, if the strain energy is expressed as a function of statically independent forces, the partial derivative of the strain energy with respect to one of the forces gives the deformation corresponding to that force. (See Timoshenko and Young, "Theory of Structures," McGraw-Hill Book Company, New York.)

$$\frac{\partial U}{\partial P_n} = e_n \qquad (5\text{-}60)$$

This is known as Castigliano's first theorem. (His second theorem is the principle of least work.)

5-54. Method of Least Work.

If displacement of a structure is prevented, as at a support, the partial derivative of the strain energy with respect to that supporting force must be zero, according to Castigliano's first theorem. This establishes his second theorem:

The strain energy in a statically indeterminate structure is the minimum consistent with equilibrium.

As an example of the use of the method of least work, we shall solve again for the stress in the vertical bar in Fig. 5-50. Calling this stress X, we note that the stress in each of the inclined bars must be $(P - X)/2 \cos \alpha$. With the aid of Eq. (5-11c), we can express the strain energy in the system in terms of X as

$$U = \frac{X^2 L}{2AE} + \frac{(P - X)^2 L}{4AE \cos^3 \alpha}$$

Hence, the internal work in the system will be a minimum when

$$\frac{\partial U}{\partial X} = \frac{XL}{AE} - \frac{(P - X)L}{2AE \cos^3 \alpha} = 0$$

Solving for X gives the stress in the vertical bar as $P/(1 + 2 \cos^3 \alpha)$, as before (Art. 5-52).

5-55. Dummy Unit-Load Method.

In Art. 5-9, the strain energy for pure bending was given as $U = M^2 L/2EI$ in Eq. (5-11f). To find the strain energy due to bending stress in a beam, we can apply this equation to a differential length dx of the beam and integrate over the entire span. Thus,

$$U = \int_0^L \frac{M^2\, dx}{2EI} \tag{5-61}$$

If M represents the bending moment due to a generalized force P, the partial derivative of the strain energy with respect to P is the deformation d corresponding to P. Differentiating Eq. (5-61) under the integral sign gives

$$d = \int_0^L \frac{M}{EI} \frac{\partial M}{\partial P}\, dx \tag{5-62}$$

The partial derivative in this equation is the rate of change of bending moment with the load P. It is equal to the bending moment m produced by a unit generalized load applied at the point where the deformation is to be measured and in the direction of the deformation. Hence, Eq. (5-62) can also be written

$$d = \int_0^L \frac{Mm}{EI}\, dx \tag{5-63}$$

To find the vertical deflection of a beam, we apply a dummy unit vertical load at the point where the deflection is to be measured and substitute the bending moments due to this load and the actual loading in Eq. (5-63). Similarly, to compute a rotation, we apply a dummy unit moment.

As a simple example, let us apply the dummy unit-load method to the determination of the deflection at the center of a simply supported, uniformly loaded beam of constant moment of inertia (Fig. 5-51a). As indicated in Fig. 5-51b, the bending moment at a distance x from one end is $(wL/2)x - (w/2)x^2$. If we apply a dummy unit load vertically at the center of the beam (Fig. 5-51c), where the vertical deflection is to be determined, the moment at x is $x/2$, as indicated in Fig. 5-51d. Substituting in Eq. (5-63) gives

$$d = 2 \int_0^{L/2} \left(\frac{wL}{2} x - \frac{w}{2} x^2 \right) \frac{x}{2} \frac{dx}{EI} = \frac{5wL^4}{384 EI}$$

Fig. 5-51 Dummy unit-load method applied to a uniformly loaded beam (*a*) to find midspan deflection; (*b*) moment diagram for the uniform load; (*c*) unit load at midspan; (*d*) moment diagram for the unit load.

Fig. 5-52 End rotation due to end moment: (*a*) by dummy unit-load method; (*b*) moment diagram for end moment; (*c*) unit moment applied at beam end; (*d*) moment diagram for unit moment.

As another example, let us apply the method to finding the end rotation at one end of a simply supported, prismatic beam produced by a moment applied at the other end. In other words, the problem is to find the end rotation at B, θ_B, in Fig. 5-52a, due to M_A. As indicated in Fig. 5-52b, the bending moment at a distance x from B caused by M_A is $M_A x/L$. If we applied a dummy unit moment at B (Fig. 5-52c), it would produce a moment at x of $(L - x)/L$ (Fig. 5-52d). Substituting in Eq. (5-63) gives

$$\theta_B = \int_0^L M_A \frac{x}{L} \frac{L-x}{L} \frac{dx}{EI} = \frac{M_A L}{6EI}$$

To determine the deflection of a beam caused by shear, Castigliano's theorems can be applied to the strain energy in shear

$$V = \int\int \frac{v^2}{2G} dA\, dx$$

where v = shearing unit stress
G = modulus of rigidity
A = cross-sectional area

5-56. Truss Deflections.

The dummy unit-load method may also be adapted for the determination of the deformation of trusses. As indicated by Eq. (5-11c), the strain energy in a truss is given by

$$U = \sum \frac{S^2 L}{2AE} \qquad (5\text{-}64)$$

which represents the sum of the strain energy for all the members of the truss. S is the stress in each member caused by the loads. Applying Castigliano's first theorem and differentiating inside the summation sign yield the deformation:

$$d = \sum \frac{SL}{AE} \frac{\partial S}{\partial P} \qquad (5\text{-}65)$$

The partial derivative in this equation is the rate of change of axial stress with the load P. It is equal to the axial stress u in each bar of the truss produced by a unit load applied at the point where the deformation is to be measured and in the direction of the deformation. Consequently, Eq. (5-65) can also be written

$$d = \sum \frac{SuL}{AE} \qquad (5\text{-}66)$$

To find the deflection of a truss, apply a dummy unit vertical load at the panel point where the deflection is to be measured and substitute in Eq. (5-66) the stresses in each member of the truss due to this load and the actual loading. Similarly, to find the rotation of any joint, apply a dummy unit moment at the joint, compute the stresses in each member of the truss, and substitute in Eq. (5-66). When it is necessary to determine the relative movement of two panel points, apply dummy unit loads in opposite directions at those points.

It is worth noting that members that are not stressed by the actual loads or the dummy loads do not enter into the calculation of a deformation.

As an example of the application of Eq. (5-66), let us compute the deflection of the truss in Fig. 5-53. The stresses due to the 20-kip load at each panel point are shown in Fig. 5-53a, and the ratio of length of members in inches to their cross-sectional area in

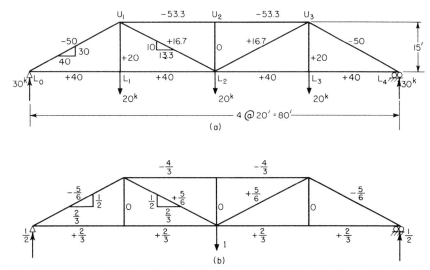

Fig. 5-53 Dummy unit-load method applied to a loaded truss: (a) to find midspan deflection; (b) unit load applied at midspan.

square inches is given in Table 5-3. We apply a dummy unit vertical load at L_2, where the deflection is required. Stresses u due to this load are shown in Fig. 5-53b and Table 5-3.

The computations for the deflection are given in Table 5-3. Members not stressed by the 20-kip loads or the dummy unit load are not included. Taking advantage of the symmetry of the truss, we tabulate the values for only half the truss and double the sum.

TABLE 5-3 Deflection of a Truss

Member	L/A	S	u	SuL/A
L_0L_2	160	+40	$+\tfrac{2}{3}$	4,267
L_0U_1	75	−50	$-\tfrac{5}{6}$	3,125
U_1U_2	60	−53.3	$-\tfrac{4}{3}$	4,267
U_1L_2	150	+16.7	$+\tfrac{5}{6}$	2,083
				13,742

$$d = \frac{SuL}{AE} = \frac{2 \times 13{,}742{,}000}{30{,}000{,}000} = 0.916 \text{ in.}$$

Also, to reduce the amount of calculation, we do not include the modulus of elasticity E, which is equal to 30,000,000, until the very last step, since it is the same for all members.

5-57. Reciprocal Theorem.

Consider a structure loaded by a group of independent forces A, and suppose that a second group of forces B are added. The work done by the forces A acting over the displacements due to B will be W_{AB}.

Now, suppose the forces B had been on the structure first, and then load A had been applied. The work done by the forces B acting over the displacements due to A will be W_{BA}.

The reciprocal theorem states that $W_{AB} = W_{BA}$.

Some very useful conclusions can be drawn from this equation. For example, there is the reciprocal deflection relationship: **The deflection at a point A due to a load at B is equal to the deflection at B due to the same load applied at A. Also, the rotation at A due to a load (or moment) at B is equal to the rotation at B due to the same load (or moment) applied at A.**

Another consequence is that deflection curves may also be influence lines to some scale for reactions, shears, moments, or deflections **(Muller-Breslau principle).** (Influence lines are defined in Art. 5-23.) For example, suppose the influence line for a reaction is to be found; that is, we wish to plot the reaction R as a unit load moves over the structure, which may be statically indeterminate. For the loading condition A, we analyze the structure with a unit load on it at a distance x from some reference point. For loading condition B, we apply a dummy unit vertical load upward at the place where the reaction is to be determined, deflecting the structure off the support. At a distance x from the reference point, the displacement is d_{xR} and over the support the displacement is d_{RR}. Hence $W_{AB} = -1(d_{xR}) + Rd_{RR}$. On the other hand, W_{BA} is zero, since loading condition A provides no displacement for the dummy unit load at the support in condition B. Consequently, from the reciprocal theorem,

$$R = \frac{d_{xR}}{d_{RR}}$$

Since d_{RR} is a constant, R is proportional to d_{xR}. Hence the influence line for a reaction can be obtained from the deflection curve resulting from a displacement of the support

(Fig. 5-54). The magnitude of the reaction is obtained by dividing each ordinate of the deflection curve by the displacement of the support.

Similarly, the influence line for shear can be obtained from the deflection curve produced by cutting the structure and shifting the cut ends vertically at the point for which the influence line is desired (Fig. 5-55).

Fig. 5-54 Reaction-influence line for a continuous beam.

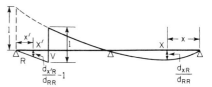

Fig. 5-55 Shear-influence line for a continuous beam.

Fig. 5-56 Moment-influence line for a continuous beam.

Fig. 5-57 Deflection-influence line for a continuous beam.

The influence line for bending moment can be obtained from the deflection curve produced by cutting the structure and rotating the cut ends at the point for which the influence line is desired (Fig. 5-56).

And finally, it may be noted that the deflection curve for a load of unity at some point of a structure is also the influence line for deflection at that point (Fig. 5-57).

CONTINUOUS BEAMS AND FRAMES

Fixed-end beams, continuous beams, continuous trusses, and rigid frames are statically indeterminate. The equations of equilibrium are not sufficient for the determination of all the unknown forces and moments. Additional equations based on a knowledge of the deformation of the member are required.

Hence, while the bending moments in a simply supported beam are determined only by the loads and the span, bending moments in a statically indeterminate member are also a function of the geometry, cross-sectional dimensions, and modulus of elasticity.

5-58. General Method of Analysis for Continuous Members.

Continuous beams and frames consist of members that can be treated as simple beams, the ends of which are prevented by moments from rotating freely. Member LR in the continuous beam $ALRBC$ in Fig. 5-58a, for example, can be isolated, as shown in Fig. 5-58b, and the elastic restraints at the ends replaced by couples M_L and M_R. In this way, LR is converted into a simply supported beam acted upon by transverse loads and end moments.

Fig. 5-58 Any span of a continuous beam (a) can be treated as a simple beam, as shown in (b) and (c). In (c), the moment diagram is resolved into basic components.

The bending-moment diagram for LR is shown at the left in Fig. 5-58c. Treating LR as a simple beam, we can break this diagram down into three simple components, as shown at the right of the equals sign in Fig. 5-58c. Thus, the bending moment at any section equals the simple-beam moment due to the transverse loads, plus the simple-beam moment due to the end moment at L, plus the simple-beam moment due to the end moment at R.

Obviously, once M_L and M_R have been determined, the shears may be computed by taking moments about any section. Similarly, if the reactions or shear are known, the bending moments can be calculated.

A general method for determining the elastic forces and moments exerted by redundant supports and members is as follows: Remove as many redundant supports or members as necessary to make the structure statically determinate. Compute for the actual loads the deflections or rotations of the statically determinate structure in the direction of the forces and couples exerted by the removed supports and members. Then, in terms of these forces and couples, compute the corresponding deflections or rotations the forces and couples produce in the statically determinate structure (see Arts. 5-55 and 5-31). Finally, for each redundant support or member write equations that give the known rotations or deflections of the original structure in terms of the deformations of the statically determinate structure.

For example, one method of finding the reactions of the continuous beam $ALRBC$ in Fig. 5-58a is to remove supports L, R, and B temporarily. The beam is now simply supported between A and C, and the reactions and moments can be computed from the laws of equilibrium. Beam AC deflects at points L, R, and B, whereas we know that continuous beam $ALRBC$ is prevented from deflecting at these points by the supports there. This information enables us to write three equations in terms of the three unknown reactions that were eliminated to make the beam statically determinate.

So we first compute the deflections d_1, d_2, d_3 of simple beam AC at L, R, and B due to the loads. Then, in terms of the unknown reactions R_1 at L, R_2 at R, R_3 at B, we compute the deflection at L when AC is loaded with only the three reactions. This deflection is the sum $-y_{11}R_1 - y_{12}R_2 - y_{13}R_3$, where y_{ij} represents the deflection at support i due to a unit load at support j. The first equation is obtained by noting that the sum of this deflection and the deflection at L due to the loads must be zero. Similarly, equations can be written making the total deflection at R and B, successively, equal to zero. In this way, three equations with three unknowns are obtained, which can be solved simultaneously to yield the required reactions.

$$d_1 = y_{11}R_1 + y_{12}R_2 + y_{13}R_3$$
$$d_2 = y_{21}R_1 + y_{22}R_2 + y_{23}R_3 \qquad (5\text{-}67)$$
$$d_3 = y_{31}R_1 + y_{32}R_2 + y_{33}R_3$$

See also Arts. 5-110 to 5-113.

5-59. Sign Convention.

For moment distribution, the following sign convention is most convenient: A moment acting at an end of a member or at a joint is positive if it tends to rotate the joint clockwise, negative if it tends to rotate the joint counterclockwise. Hence, in Fig. 5-58, M_R is positive and M_L is negative.

Similarly, the angular rotation at the end of a member is positive if in a clockwise direction, negative if counterclockwise. Thus, a positive end moment produces a positive end rotation in a simple beam.

For ease in visualizing the shape of the elastic curve under the action of loads and end moments, bending-moment diagrams should be plotted on the tension side of each member. Hence, if an end moment is represented by a curved arrow, the arrow will point in the direction in which the moment is to be plotted.

5-60. Fixed-End Moments.

A beam so restrained at its ends that no rotation is produced there by the loads is called a fixed-end beam, and the end moments are called fixed-end moments. Fixed-end moments may be expressed as the product of a coefficient and WL, where W is the total load on the span L. The coefficient is independent of the properties of other members of the structure. Thus, any member can be isolated from the rest of the structure and its fixed-end moments computed.

Fixed-end moments may be determined conveniently by the moment-area method or the conjugate-beam method (Art. 5-31).

Fixed-end moments for several common types of loading on beams of constant moment of inertia (prismatic beams) are given in Figs. 5-59 to 5-62. Also, the curves in

Fig. 5-59 Moments for concentrated load at any point of a fixed-end beam.

Fig. 5-60 Moments for a uniform load on a fixed-end beam.

5-60   Structural Theory

Fig. 5-61 Moments for two equal concentrated loads on a fixed-end beam.

Fig. 5-62 Moments for several equal concentrated loads on a fixed-end beam.

Fig. 5-64 enable fixed-end moments to be computed easily for any type of loading on a prismatic beam. Before the curves can be entered, however, certain characteristics of the loading must be calculated. These include $\bar{x}L$, the location of the center of gravity of the loading with respect to one of the loads; $G^2 = \Sigma b_n^2 P_n/W$, where $b_n L$ is the distance from each load P_n to the center of gravity of the loading (taken positive to the right); and $S^3 = \Sigma b_n^3 P_n/W$. (See Case 9, Fig. 5-63.) These values are given in Fig. 5-63 for some common types of loading.

The curves in Fig. 5-64 are entered with the location a of the center of gravity with respect to the left end of the span. At the intersection with the proper G curve, proceed

Fig. 5-63 Characteristics of loadings.

horizontally to the left to the intersection with the proper S line, then vertically to the horizontal scale indicating the coefficient m by which to multiply WL to obtain the fixed-end moment. The curves solve the equations:

$$m_L = \frac{M_L^F}{WL} = G^2[1 - 3(1 - a)] + a(1 - a)^2 + S^3 \qquad (5\text{-}68a)$$

$$m_R = \frac{M_r^F}{WL} = G^2(1 - 3a) + a^2(1 - a) - S^3 \qquad (5\text{-}68b)$$

where M_L^F is the fixed-end moment at the left support and M_R^F at the right support.

As an example of the use of the curves, find the fixed-end moments in a prismatic beam of 20-ft span carrying a triangular loading of 100 kips, similar to the loading shown

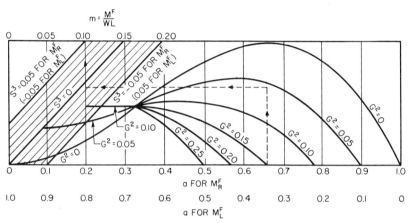

Fig. 5-64 Chart for fixed-end moments due to any type of loading.

in Case 4, Fig. 5-63, distributed over the entire span, with the maximum intensity at the right support.

Case 4 gives the characteristics of the loading: $y = 1$; the center of gravity is $0.33L$ from the right support, so $a = 0.667$; $G^2 = \frac{1}{18} = 0.056$; and $S^3 = -\frac{1}{135} = -0.007$. To find M_R^F, we enter Fig. 5-64 with $a = 0.67$ on the upper scale at the bottom of the diagram, and proceed vertically to the estimated location of the intersection of the coordinate with the $G^2 = 0.06$ curve. Then, we move horizontally to the intersection with the line for $S^3 = -0.007$, as indicated by the dash line in Fig. 5-64. Referring to the scale at the top of the diagram, we find the coefficient m_R to be 0.10. Similarly, with $a = 0.67$ on the lowest scale, we find the coefficient m_L to be 0.07. Hence, the fixed-end moment at the right support is $0.10 \times 100 \times 20 = 200$ ft-kips, and at the left support $-0.07 \times 100 \times 20 = -140$ ft-kips.

5-61. Fixed-End Stiffness.

To correct a fixed-end moment to obtain the end moment for the actual conditions of end restraint in a continuous structure, the end of the member must be permitted to rotate. The amount it will rotate depends on its stiffness, or resistance to rotation.

The fixed-end stiffness of a beam is defined as the moment required to produce a rotation of unity at the end where it is applied, while the other end is fixed against rotation. It is represented by K_R^F in Fig. 5-65.

For prismatic beams, the fixed-end stiffnesses for both ends are equal to $4EI/L$, where E is the modulus of elasticity, I the moment of inertia of the cross section about the centroidal axis, and L the span (generally taken center to center of supports). When deformations are not required to be calculated, only the relative value of K^F for each member need be known; hence, only the ratio of I to L has to be computed. (For prismatic beams with a hinge at one end, the actual stiffness is $3EI/L$, or three-fourths the fixed-end stiffness.)

(a) ELASTIC CURVE

(b) MOMENT DIAGRAM

Fig. 5-65 Fixed-end stiffness.

For beams of variable moment of inertia, the fixed-end stiffness may be calculated by methods presented later in this section or obtained from prepared tables. See Art. 5-62.

5-62. Fixed-End Carry-Over Factor.

When a moment is applied at one end of a beam, a resisting moment is induced at the far end if that end is restrained against rotation (Fig. 5-65). The ratio of the resisting moment at a fixed end to the applied moment is called the fixed-end carry-over factor C^F.

For prismatic beams, the fixed-end carry-over factor toward either end is 0.5. It should be noted that the applied moment and the resisting moment have the same sign (Fig. 5-65); i.e., if the applied moment acts in a clockwise direction, the carry-over moment also acts clockwise.

For beams of variable moment of inertia, the fixed-end carry-over factor may be calculated by methods presented later in this section or obtained from tables such as those in the "Handbook of Frame Constants," EB034D, Portland Cement Association, Skokie, Ill. 60077, and J. M. Gere, "Moment Distribution Factors for Beams of Tapered I-section," American Institute of Steel Construction, Chicago, Ill.

5-63. Moment Distribution by Converging Approximations.

The frame in Fig. 5-66 consists of four prismatic members rigidly connected together at O and fixed at ends A, B, C, and D. If an external moment U is applied at O, the sum of the end moments in each member at O must be equal to U. Furthermore, all members must rotate at O through the same angle θ, since they are assumed to be rigidly connected there. Hence, by the definition of fixed-end stiffness, the proportion of U induced in the end of each member at O is equal to the ratio of the stiffness of that member to the sum of the stiffnesses of all the members at the joint.

Suppose a moment of 100 ft-kips is applied at O, as indicated in Fig. 5-66b. The relative stiffness (or I/L) is assumed as shown in the circle on each member. The distribution factors for the moment at O are computed from the stiffnesses and shown in the boxes. For example, the distribution factor for OA equals its stiffness divided by the sum of the stiffnesses of all the members at the joint: $3/(3 + 2 + 4 + 1) = 0.3$. Hence, the moment induced in OA at O is $0.3 \times 100 = 30$ ft-kips. Similarly, OB gets 10 ft-kips, OC 40 ft-kips, and OD 20 ft-kips.

Because the far ends of these members are fixed, one-half of these moments are carried over to them. Thus $M_{AO} = 0.5 \times 30 = 15$; $M_{BO} = 0.5 \times 10 = 5$; $M_{CO} = 0.5 \times 40 = 20$; and $M_{DO} = 0.5 \times 20 = 10$.

Most structures consist of frames similar to the one in Fig. 5-66, or even simpler, joined together. Though the ends of the members are not fixed, the technique employed for the frame in Fig. 5-66b can be applied to find end moments in such continuous structures.

Before the general method is presented, one short cut is worth noting. Advantage can be taken when a member has a hinged end to reduce the work of distributing moments. This is done by using the true stiffness of a member instead of the fixed-end stiffness. (For a prismatic beam with one end hinged, the stiffness is three-fourths the fixed-end stiffness; for a beam with variable I, it is equal to the fixed-end stiffness times $1 - C_L^F C_R^F$.) Naturally, the carry-over factor toward the hinge is zero.

(a) ELASTIC CURVE FOR UNBALANCED MOMENT AT JOINT O

When a joint is neither fixed nor pinned but is restrained by elastic members connected there, moments can be distributed by a series of converging approximations. All joints are locked against rotation. As a result, the loads will create fixed-end moments at the ends of every loaded member. At each joint, a moment equal to the algebraic sum of the fixed-end moments there is required to hold it fixed. Then, one joint is unlocked at a time by applying a moment equal but opposite in sign to the moment that was needed to prevent rotation. The unlocking moment must be distributed to the members at the joint in proportion to their fixed-end stiffnesses and the distributed moments carried over to the far ends.

(b) STIFFNESSES AND DISTRIBUTION FACTORS FOR A FRAME

Fig. 5-66 Bending moments in a simple frame.

After all joints have been released at least once, it generally will be necessary to repeat the process—sometimes several times— before the corrections to the fixed-end moments become negligible. To reduce the number of cycles, the unlocking of joints should start with those having the greatest unbalanced moments.

Suppose the end moments are to be found for the continuous beam $ABCD$ in Fig. 5-67. The I/L values for all spans are equal; therefore, the relative fixed-end stiffness for all members is unity. However, since A is a hinged end, the computation can be shortened by using the actual relative stiffness, which is ¾. Relative stiffnesses for all members are shown in the circle on each member. The distribution factors are shown in boxes at each joint.

The computation starts with determination of fixed-end moments for each member (Art. 5-60). These are assumed to have been found and are given on the first line in Fig. 5-67. The greatest unbalanced moment is found from inspection to be at hinged end A; so this joint is unlocked first. Since there are no other members at the joint, the full unlocking moment of $+400$ is distributed to AB at A and one-half of this is carried over to B. The unbalance at B now is $+400 - 480$ plus the carry-over of $+200$ from A, or a

5-64 Structural Theory

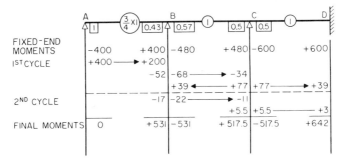

Fig. 5-67 Moment distribution by converging approximations.

total of +120. Hence, a moment of −120 must be applied and distributed to the members at B by multiplying by the distribution factors in the corresponding boxes.

The net moment at B could be found now by adding the entries for each member at the joint. However, it generally is more convenient to delay the summation until the last cycle of distribution has been completed.

The moment distributed to BA need not be carried over to A, because the carry-over factor toward the hinged end is zero. However, half the moment distributed to BC is carried over to C.

Similarly, joint C is unlocked and half the distributed moments carried over to B and D, respectively. Joint D should not be unlocked, since it actually is a fixed end. Thus, the first cycle of moment distribution has been completed.

The second cycle is carried out in the same manner. Joint B is released, and the distributed moment in BC is carried over to C. Finally, C is unlocked, to complete the cycle. Adding the entries for the end of each member yields the final moments.

5-64. Continuous Frames.

In design of continuous frames, one objective is to find the maximum end moments and interior moments produced by the worst combination of loading. For maximum moment at the end of a beam, live load should be placed on that beam and on the beam adjoining the end for which the moment is to be computed. Spans adjoining these two should be assumed to be carrying only dead load.

For maximum midspan moments, the beam under consideration should be fully loaded, but adjoining spans should be assumed to be carrying only dead load.

The work involved in distributing moments due to dead and live loads in continuous frames in buildings can be greatly simplified by isolating each floor. The tops of the upper columns and the bottoms of the lower columns can be assumed fixed. Furthermore, the computations can be condensed considerably by following the procedure recommended in "Continuity in Concrete Building Frames," EB033D, Portland Cement Association, Skokie, Ill. 60077, and indicated in Fig. 5-68.

Figure 5-68 presents the complete calculation for maximum end and midspan moments in four floor beams AB, BC, CD, and DE. Building columns are assumed to be fixed at the story above and below. None of the beam or column sections is known to begin with; so as a start, all members will be assumed to have a fixed-end stiffness of unity, as indicated on the first line of the calculation.

On the second line, the distribution factors for each end of the beams are shown, calculated from the stiffnesses (Arts. 5-61 and 5-63). Column stiffnesses are not shown, because column moments will not be computed until moment distribution to the beams

	A		B			C			D		E
1. RELATIVE STIFFNESS		$K^F=1$			$K^F=1$			$K^F=1$			$K^F=1$	
2. DISTRIBUTION FACTOR	1/3		1/4	1/4		1/4	1/4		1/4	1/4		1/3
3. F.E.M. DEAD LOAD	—		+91	−37		+37	−70		+70	−59		—
4. F.E.M. TOTAL LOAD	−172	+99	+172	−78	+73	+78	−147	+85	+147	−126	+63	+126
5. CARRY-OVER	−17 ←	+11	+29	−1 →	+1	−2	−11 →	+7	+14	−21 ←	+13	+7
6. ADDITION	−189	+18	+201	−79	−1	+76	−158	+9	+161	−147	+5	+133
7. DISTRIBUTION	+63		−30	−30		+21	+21		−4	−4		−44
8. MAX. MOMENTS	−126	+128	+171	−109	+73	+97	−137	+101	+157	−151	+81	+89

Fig. 5-68 Bending moments in a continuous frame.

has been completed. Then the sum of the column moments at each joint may be easily computed, since they are the moments needed to make the sum of the end moments at the joint equal to zero. The sum of the column moments at each joint can then be distributed to each column there in proportion to its stiffness. In this example, each column will get one-half the sum of the column moments.

Fixed-end moments at each beam end for dead load are shown on the third line, just above the heavy line, and fixed-end moments for live plus dead load on the fourth line. Corresponding midspan moments for the fixed-end condition also are shown on the fourth line and, like the end moments, will be corrected to yield actual midspan moments.

For maximum end moment at A, beam AB must be fully loaded, but BC should carry dead load only. Holding A fixed, we first unlock joint B, which has a total-load fixed-end moment of $+172$ in BA and a dead-load fixed-end moment of -37 in BC. The releasing moment required, therefore, is $-(172 - 37)$, or -135. When B is released, a moment of $-135 \times \frac{1}{4}$ is distributed to BA. One-half of this is carried over to A, or $-135 \times \frac{1}{4} \times \frac{1}{2} = -17$. This value is entered as the carry-over at A on the fifth line in Fig. 5-68. Joint B is then relocked.

At A, for which we are computing the maximum moment, we have a total-load fixed-end moment of -172 and a carry-over of -17, making the total -189, shown on the sixth line. To release A, a moment of $+189$ must be applied to the joint. Of this, $189 \times \frac{1}{3}$, or 63, is distributed to AB, as indicated on the seventh line of the calculation. Finally, the maximum moment at A is found by adding lines 6 and 7: $-189 + 63 = -126$.

For maximum moment at B, both AB and BC must be fully loaded, but CD should carry only dead load. We begin the determination of the moment at B by first releasing joints A and C, for which the corresponding carry-over moments at BA and BC are $+29$ and $-(+78 - 70) \times \frac{1}{4} \times \frac{1}{2} = -1$, shown on the fifth line in Fig. 5-68. These bring the total fixed-end moments in BA and BC to $+201$ and -79, respectively. The releasing moment required is $-(201 - 79) = -122$. Multiplying this by the distribution factors for BA and BC when joint B is released, we find the distributed moments, -30, entered on line 7. The maximum end moments finally are obtained by adding lines 6 and 7: $+171$ at BA and -109 at BC. Maximum moments at C, D, and E are computed and entered in Fig. 5-68 in a similar manner. This procedure is equivalent to two cycles of moment distribution.

The computation of maximum midspan moments in Fig. 5-68 is based on the assumption that in each beam the midspan moment is the sum of the simple-beam midspan moment and one-half the algebraic difference of the final end moments (the span carries full load but adjacent spans only dead load). Instead of starting with the simple-beam moment, however, we begin with the midspan moment for the fixed-end condition and apply two corrections. In each span, these corrections are equal to the carry-over moments entered on line 5 for the two ends of the beam multiplied by a factor.

For beams with variable moment of inertia, the factor is $\pm\frac{1}{2}[(1/C^F) + D - 1]$ where C^F is the fixed-end carry-over factor toward the end for which the correction factor is being computed and D is the distribution factor for that end. The plus sign is used for correcting the carry-over at the right end of a beam, and the minus sign for the carry-over at the left end. For prismatic beams, the correction factor becomes $\pm\frac{1}{2}(1 + D)$.

For example, to find the corrections to the midspan moment in AB, we first multiply the carry-over at A on line 5, -17, by $-\frac{1}{2}(1 + \frac{1}{3})$. The correction, $+11$, is also entered on the fifth line. Then, we multiply the carry-over at B, $+29$, by $+\frac{1}{2}(1 + \frac{1}{4})$ and enter the correction, $+18$, on line 6. The final midspan moment is the sum of lines 4, 5, and 6: $+99 + 11 + 18 = +128$. Other midspan moments in Fig. 5-68 are obtained in a similar manner.

5-65. Moment-Influence Factors.

In certain types of framing, particularly those in which different types of loading conditions must be investigated, it may be convenient to find maximum end moments from a table of moment-influence factors. This table is made up by listing for the end of each member in the structure the moment induced in that end when a moment (for convenience, $+1000$) is applied to every joint successively. Once this table has been prepared, no additional moment distribution is necessary for computing the end moments due to any loading condition.

For a specific loading pattern, the moment at any beam end M_{AB} may be obtained from the moment-influence table by multiplying the entries under AB for the various joints by the actual unbalanced moments at those joints divided by 1000, and summing (see also Art. 5-67 and Table 5-4).

5-66. Deflection of Supports.

For some framing, it is convenient to know the effect of a deflection of a support normal to the original position of a continuous beam. But the moment-distribution method is based on the assumption that such movement of a support does not occur. However, the method can be modified to evaluate the end moments resulting from a support movement.

The procedure is to distribute moments, as usual assuming no deflection at the supports. This implies that additional external forces are exerted at the supports to prevent deflection. These forces can be computed. Then, equal and opposite forces are applied to the structure to produce the final configuration, and the moments that they produce are distributed as usual. These moments added to those obtained with undeflected supports yield the final moments.

To apply this procedure, it is first necessary to know the fixed-end moments for a beam with supports at different levels. In Fig. 5-69a, the right end of a beam with span L is at a height d above the left end. To find the fixed-end moments, we first consider the left end hinged, as in Fig. 5-69b. Noting that a line connecting the two supports makes an angle approximately equal to d/L (its tangent) with the original position of the beam, we apply a moment at the hinged end to produce an end rotation there equal to d/L. By the definition of stiffness, this moment equals $K_L^F d/L$. The carry-over to the right end is C_R^F times this.

By the law of reciprocal deflections (Art. 5-57), the fixed-end moment at the right end of a beam caused by a rotation of the other end is equal to the fixed-end moment at the left end of the beam caused by the same rotation at the right end. Therefore, the carry-

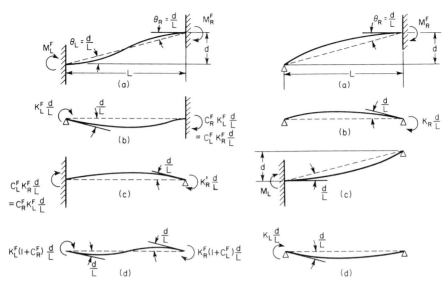

Fig. 5-69 Bending moments due to displacement of a support of a fixed-end beam.

Fig. 5-70 Moments due to displacement of a support of a beam with one end fixed, one end hinged.

over moment for the right end in Fig. 5-69b is also equal to $C_L^F K_R^F/L$ (see Fig. 5-69c). By adding the end moments for the loading conditions in Fig. 5-69b and c, we obtain the end moments in Fig. 5-69d, which is equivalent to the deflected beam in Fig. 5-69a:

$$M_L^F = K_L^F (1 + C_R^F) \frac{d}{L} \qquad (5\text{-}69)$$

$$M_R^F = K_R^F (1 + C_L^F) \frac{d}{L} \qquad (5\text{-}70)$$

In a similar manner, the fixed-end moment can be found for a beam with one end hinged and the supports at different levels (Fig. 5-70):

$$M^F = K \frac{d}{L} \qquad (5\text{-}71)$$

where K is the actual stiffness for the end of the beam that is fixed. For prismatic beams, this value is three-fourths that for fixed-end stiffness; for beams of variable moments of inertia, it is equal to the fixed-end stiffness times $(1 - C_L^F C_R^F)$.

5-67. Procedure for Sidesway.

Computation of moments due to sidesway, or drift, in rigid frames is conveniently executed by the following method:

1. Apply forces to the structure to prevent sidesway while the fixed-end moments due to loads are distributed.
2. Compute the moments due to these forces.
3. Combine the moments obtained in Steps 1 and 2 to eliminate the effect of the forces that prevented sidesway.

Suppose the rigid frame in Fig. 5-71 is subjected to a 2000-lb horizontal load acting to the right at the level of beam BC. The first step is to compute the moment-influence factors (Table 5-4) by applying moments of $+1000$ at joints B and C, assuming sidesway prevented. Since there are no intermediate loads on the beams and columns, the only fixed-end moments that need be considered are those in the columns resulting from lateral deflection of the frame caused by the horizontal load.

This deflection, however, is not known initially. So we assume an arbitrary deflection, which produces a fixed-end moment of $-1000M$ at the top of column CD. M is an unknown constant to be determined from the fact that the sum of the shears in the deflected columns must be equal to the 2000-lb load. The same deflection also produces a moment of $-1000M$ at the bottom of CD [see Eqs. (5-69) and (5-70)].

From the geometry of the structure, we furthermore note that the deflection of B relative to A is equal to the deflection of C relative to D. Then, according to Eqs. (5-69) and (5-70), the fixed-end moments in the columns are proportional to the stiffnesses of the columns and hence are equal in AB to $-1000M \times \% = -3000M$. The column fixed-end moments are entered in the first line of Table 5-5, which is called a moment-collection table.

Fig. 5-71 Rigid frame.

In the deflected position of the frame, joints B and C are unlocked. First, we apply a releasing moment of $+3000M$ at B and distribute it by multiplying by 3 the entries in the column marked "$+1000$ at B" in Table 5-4. Similarly, a releasing moment of $+1000M$ is applied at C and distributed with the aid of Table 5-4. The distributed moments are entered in the second and third lines of Table 5-5. The final moments are the sum of the fixed-end moments and the distributed moments and are given in the fifth line.

Isolating each column and taking moments about one end, we find that the overturning moment due to the shear is equal to the sum of the end moments. We have one such equation for each column. Adding these equations, noting that the sum of the shears equals 2000 lb, we obtain

$$-M(2052 + 1104 + 789 + 895) = -2000 \times 20$$

from which we find $M = 8.26$. This value is substituted in the sidesway totals in Table 5-5 to yield the end moments for the 2000-lb horizontal load.

TABLE 5-4 Moment-Influence Factors for Fig. 5-71

Member	$+1{,}000$ at B	$+1{,}000$ at C
AB	351	−105
BA	702	−210
BC	298	210
CB	70	579
CD	−70	421
DC	−35	210

Suppose now a vertical load of 4000 lb is applied to BC of the rigid frame in Fig. 5-71, 5 ft from B. Tables 5-4 and 5-5 can again be used to determine the end moments with a minimum of labor:

The fixed-end moment at B, with sideway prevented, is $-12,800$, and at C $+3200$. With the joints locked, the frame is permitted to move laterally an arbitrary amount, so that in addition to the fixed-end moments due to the 4000-lb load, column fixed-end moments of $-3000M$ at B and $-1000\,M$ at C are induced. Table 5-5 already indicates

TABLE 5-5 Moment-Collection Table for Fig. 5-71

Remarks	AB +	AB −	BA +	BA −	BC +	BC −	CB +	CB −	CD +	CD −	DC +	DC −
Sideway, FEM		3,000M		3,000M						1,000M		1,000M
B moments	1,053M		2,106M		894M		210M			210M		105M
C moments		105M		210M	210M		579M		421M		210M	
Partial sum	1,053M	3,105M	2,106M	3,210M	1,104M		789M		421M	1,210M	210M	1,105M
Totals		2,052M		1,104M	1,104M		789M			789M		895M
For 2,000-lb load		17,000		9,100	9,100		6,500			6,500		7,400
4,000-lb load, FEM						12,800	3,200					
B moments	4,490		8,980		3,820		897			897		448
C moments	336		672			672		1,853	1,347		672	
Partial sum	4,826		9,652		3,820	13,472	4,097	1,853		2,244		1,120
No-sideway sum	4,826		9,652			9,652	2,244			2,244		1,120
Sideway M		4,710		2,540	2,540		1,810			1,810		2,060
Totals	120		7,110		7,110		4,050			4,050		3,180

the effect of relieving these column moments by unlocking joints B and C. We now have to superimpose the effect of releasing joints B and C to relieve the fixed-end moments for the vertical load. This we can do with the aid of Table 5-4. The distribution is shown in the lower part of Table 5-5. The sums of the fixed-end moments and distributed moments for the 4000-lb load are shown on the line "No-sideway sum."

The unknown M can be evaluated from the fact that the sum of the horizontal forces acting on the columns must be zero. This is equivalent to requiring that the sum of the column end moments equals zero:

$$-M(2052 + 1104 + 789 + 895) + 4826 + 9652 - 2244 - 1120 = 0$$

from which $M = 2.30$. This value is substituted in the sideway total in Table 5-5 to yield the sideway moments for the 4000-lb load. The addition of these moments to the totals for no sideway yields the final moments.

This procedure enables one-story bents with straight beams to be analyzed with the necessity of solving only one equation with one unknown regardless of the number of bays. If the frame is several stories high, the procedure can be applied to each story. Since an arbitrary horizontal deflection is introduced at each floor or roof level, there are as many unknowns and equations as there are stories.

The procedure is more difficult to apply to bents with curved or polygonal members between the columns. The effect of the change in the horizontal projection of the curved or polygonal portion of the bent must be included in the calculations. In many cases, it may be easier to analyze the bent as a curved beam (arch).

(A. Kleinlogel, "Rigid Frame Formulas," Frederick Ungar Publishing Co., New York.)

5-68. Single-Cycle Moment Distribution.

In the method of moment distribution by converging approximations, all joints but the one being unlocked are considered fixed. In distributing moments, the stiffness and carry-over factors used are based on this assumption. However, if actual stiffnesses and carry-over factors are employed, moments can be distributed throughout continuous frames in a single cycle.

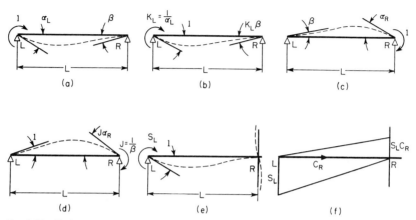

Fig. 5-72 End rotations of a simple beam (a) to (d) and of a beam that is part of a continuous frame (e) and (f).

Formulas for actual stiffnesses and carry-over factors can be written in several simple forms. The equations given in the following text were chosen to permit the use of existing tables for beams of variable moment of inertia that are based on fixed-end stiffnesses and fixed-end carry-over factors, such as the "Handbook of Frame Constants," EB034D, Portland Cement Association, Skokie, Ill.

Considerable simplification of the formulas results if they are based on the simple-beam stiffness of members of continuous frames. This value can always be obtained from tables of fixed-end properties by multiplying the fixed-end stiffness given in the tables by $(1 - C_L^F C_R^F)$, in which C_L^F is the fixed-end carry-over factor to the left and C_R^F is the fixed-end carry-over factor to the right.

To derive the basic constants needed, we apply a unit moment to one end of a member, considering it simply supported (Fig. 5-72a). The end rotation at the support where the moment is applied is α, and at the far end, the rotation is β. By the dummy-load method (Art. 5-55), if x is measured from the β end,

$$\alpha = \frac{1}{L^2} \int_0^L \frac{x^2}{EI_x} dx \qquad (5\text{-}72)$$

$$\beta = \frac{1}{L^2} \int_0^L \frac{x(L-x)}{EI_x} dx \qquad (5\text{-}73)$$

in which I_x = moment of inertia at a section a distance of x from the β end
E = modulus of elasticity

Simple-beam stiffness K of the member is the moment required to produce a rotation of unity at the end where it is applied (Fig. 5-72b). Hence, at each end of a member, $K = 1/\alpha$.

For prismatic beams, K has the same value for both ends and is equal to $3EI/L$. For haunched beams, K for each end can be obtained from tables for fixed-end stiffnesses, as mentioned previously, or by numerical integration of Eq. (5-72).

While the value of α, and consequently of K, is different at opposite ends of an unsymmetrical beam, the value of β is the same for both ends, in accordance with the law of reciprocal deflections (compare Fig. 5-72a and c). This is also evident from Eq. (5-73), where $L - x$ can be substituted for x without changing the value of the integral.

Now, if we apply a moment J at one end of a simple beam to produce a rotation of unity at the other end (Fig. 5-72d), this moment will be equal to $1/\beta$ and will have the same value regardless of the end at which it is applied. As can be shown, K/J is equal to the **fixed-end carry-over factor.**

J is equal to $6EI/L$ for prismatic beams. For haunched beams, it can be computed by numerical integration of Eq. (5-73).

Actual stiffness S of the end of an unloaded span is the moment producing a rotation of unity at the end where it is applied when the other end of the beam is restrained against rotation by other members of the structure (Fig. 5-72e).

The bending-moment diagram for a moment S_L applied at the left end of a member of a continuous frame is shown in Fig. 5-72f. As indicated, the moment carried over to the far end is $S_L C_R$, where C_R is the carry-over factor to the right. At L, the rotation produced by S_L alone is S_L/K_L, and by $S_L C_R$ alone is $-S_L C_R/J$. The sum of these angles must equal unity by definition of stiffness:

$$\frac{S_L}{K_L} - \frac{S_L C_R}{J} = 1$$

Solving for S_L and noting that $K_L/J = C_L^F$, the fixed-end carry-over factor to the left, we find the formula for the stiffness of the left end of a member:

$$S_L = \frac{K_L}{1 - C_L^F C_R} \qquad (5\text{-}74)$$

For the right end:

$$S_R = \frac{K_R}{1 - C_R^F C_L} \qquad (5\text{-}75)$$

For prismatic beams, the stiffness formulas reduce to

$$S_L = \frac{K}{1 - C_R/2} \qquad S_R = \frac{K}{1 - C_L/2} \qquad (5\text{-}76)$$

where $K = 3EI/L$.

When the far end of a prismatic beam is fully fixed against rotation, the carry-over factor equals ½. Hence, the fixed-end stiffness equals $4K/3$. This indicates that the effect of partial restraint on prismatic beams is to vary the stiffness between K for no restraint and $1.33K$ for full restraint. Because of this small variation, an estimate of the actual stiffness of a beam may be sufficiently accurate in many cases.

Restraint R at the end of an unloaded beam in a continuous frame is the moment applied at that end to produce a unit rotation in all the members of the joint. Since the sum of the moments at the joint must be zero, R must be equal to the sum of the stiffnesses of the adjacent ends of the members connected to the given beam at that joint.

Furthermore, the moment induced in any of these other members bears the same ratio to the applied moment as the stiffness of the member does to the restraint. Consequently, *end moments are distributed at a joint in proportion to the stiffnesses of the members.*

Actual carry-over factors can be computed by modifying the fixed-end carry-over factors. In Fig. 5-72e and f, by definition of restraint, rotation at joint R is $-S_L C_R/R_R$, which must be equal to the rotation of the beam at R caused by the moments at L and R. The rotation due to $S_L C_R$ alone is equal to $S_L C_R/K_R$, and the rotation due to S_L alone is $-S_L/J$. Hence $-S_L C_R/R_R = S_L C_R/K_R - S_L/J$. Solving for C_R and noting that $K_R/J = C_R^F$, the fixed-end carry-over factor to the right, we find the actual carry-over factor to the right:

$$C_R = \frac{C_R^F}{1 + K_R/R_R} \quad (5\text{-}77)$$

Similarly, the actual carry-over factor to the left is

$$C_L = \frac{C_L^F}{1 + K_L/R_L} \quad (5\text{-}78)$$

Approximate Carry-Over Factors. In analyzing a continuous beam, we generally know the carry-over factors toward the ends of the first and last spans. Starting with these values, we can calculate the rest of the carry-over factors and the stiffnesses of the members. However, in many frames, there are no end conditions known in advance. To analyze these structures, we must assume several carry-over factors.

This will not complicate the analysis, because in many cases it will be found unnecessary to correct the values of C based on assumed carry-over factors of preceding spans. The reason is that C is not very sensitive to the restraint at far ends of adjacent members. When carry-over factors are estimated, the greatest accuracy will be attained if the choice of assumed values is restricted to members subject to the greatest restraint.

A very good approximation to the carry-over factor for prismatic beams may be obtained from the following formula, which is based on the assumption that far ends of adjacent members are subject to equal restraint:

$$C = \frac{\Sigma K - K}{2(\Sigma K - \delta K)} \quad (5\text{-}79)$$

where ΣK = sum of K values of all members at joint toward which carry-over factor is acting

K = simple-beam stiffness of member for which carry-over factor is being computed

δ = a factor that varies from zero for no restraint to ¼ for full restraint at far ends of connecting members

Since δ varies within such narrow limits, it affects C very little.

Estimation Example. To illustrate the estimation and calculation of carry-over factors, the carry-over factors and stiffness in the clockwise direction will be computed for the frame in Fig. 5-73a. Relative I/L, or K values, are given in the circles.

A start will be made by estimating C_{AB}. Taking $\delta = ⅛$, we apply Eq. (5-79) at B with $K = 3$ and $\Sigma K = 5$ and find $C_{AB} = 0.216$, as shown in Fig. 5-73a. We can now compute the stiffness S_{AB} employing Eq. (5-76): $S_{AB} = 3/(1 - 0.108) = 3.37$. Noting that $R_{AD} = S_{AB}$, we can use the exact formula [Eq. (5-78)] to obtain the carry-over factor for DA: $C_{DA} = 0.5/(1 + 6/3.37) = 0.180$. Continuing around the frame in this manner, we return to C_{AB} and recalculate it by Eq. (5-78), obtaining 0.221. This differs only slightly from the estimated value. The change in C_{DA} caused by the new value of C_{AB} is negligible.

If a bending moment of 1000 ft-lb were introduced at A in AB, it would induce a moment of 1000 C_{AB} = 221 ft-lb at B; 221 × 0.321 = 71 ft-lb at C; 71 × 0.344 = 24 ft-lb at D; 24 × 0.180 = 4 ft-lb at A, etc.

Beam Analysis. To show how moments in a continuous beam would be computed, the end moments will be determined for the beam in Fig. 5-73b, which is identical with

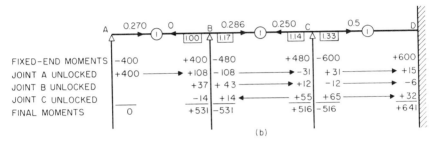

Fig. 5-73 Moments in (a) a quadrangular frame and (b) a continuous beam.

the one in Fig. 5-67 for which moments were obtained by converging approximations. Relative I/L, or K values, are shown in the circles on each span. Since A is a hinged end, $C_{BA} = 0$. $S_{BA} = 1/(1 - 0) = 1$. Since there is only one member joined to BC at B, $R_{BC} = S_{BA}$, and $C_{CB} = 0.5/(1 + 1/1) = 0.250$. With this value, we compute $S_{CB} = 1.14$. To obtain the carry-over factors for the opposite direction, we start with $C_{CD} = 0.5$, since we know D is a fixed end. This enables us to compute $S_{CD} = 1.33$ and the remainder of the beam constants.

The fixed-end moments are given on the first line of calculations in Fig. 5-73b. We start the distribution by unlocking A by applying a releasing moment of $+400$. Since A is unrestrained, the full 400 is given to A and $400 \times 0.270 = 108$ is carried over to B. If several members had been connected to AB at B, this moment, with sign changed, would be distributed to them in proportion to their stiffnesses. But since only one member is connected at B, the moment at BC is -108. Next, $-108 \times 0.286 = -31$ is carried over to C. Finally, a moment of $+15$ is carried to D.

Then, joint B is unlocked. The unbalanced moment of $400 - 480 = -80$ is counteracted with a moment of $+80$, which is distributed to BA and BC in proportion to their stiffnesses (shown in the boxes at the joint). BC, for example gets

$$80 \times \frac{1.17}{1.17 + 1.00} = 43$$

The carry-over to C is $43 \times 0.286 = 12$, and to D, -6.

Similarly, the unbalanced moment at C is counteracted and distributed to CB and CD in proportion to the stiffnesses shown in the boxes at C, then carried over to B and D. The final moments are the sum of the fixed-end and distributed moments.

Special Cases. Advantage can be taken of certain properties of loads and structures to save work in distribution by using carry-over factors as the ratio of end moments in loaded members. For example, suppose it is obvious, from symmetry of loading and structure, that there will be no end rotation at an interior support. The part of the structure on one side of this support can be isolated and the moments distributed only in this part, with the carry-over factor toward the support taken as C^F.

Again, suppose it is evident that the final end moments at opposite ends of a span must be equal in magnitude and sign. Isolate the structure on each side of this beam and distribute moments only in each part, with the carry-over factor for this span taken as 1.

5-69. Method for Checking Moment Distribution.

End moments computed for a continuous structure must be in accordance with both the laws of equilibrium and the requirements of continuity. At each joint, therefore, the sum of the moments must be equal to zero (or to an external moment applied there), and the end of every member connected there must rotate through the same angle. It is a simple matter to determine whether the sum of the moments is zero, but further calculation is needed to prove that the moments yield the same rotation for the end of each member at a joint. The following method not only will indicate that the requirements of continuity are satisfied but also will tend to correct automatically any mistakes that may have been made in computing the end moments.

Consider a joint O made up of several members OA, OB, OC, etc. The members are assumed to be loaded and the calculation of end moments to have started with fixed-end moments. For any one of the members, say OA, the end rotation at O for the fixed-end condition is

$$0 = \frac{M^F_{OA}}{K_{OA}} - \frac{M^F_{AO}}{J_{OA}} - \phi_{OA} \qquad (5\text{-}80)$$

where M^F_{OA} = fixed-end moment at O
M^F_{AO} = fixed-end moment at A
K_{OA} = simple-beam stiffness at O
J_{OA} = moment required at A to produce a unit rotation at O when the span is considered simply supported
ϕ_{OA} = simple-beam end rotation at O caused by loads

For the final end moments, the rotation at O is

$$\theta = \frac{M_{OA}}{K_{OA}} - \frac{M_{OA}}{J_{OA}} - \phi_{OA} \qquad (5\text{-}81)$$

Subtracting Eq. (5-80) from Eq. (5-81) and multiplying by K_{OA} yields

$$K_{OA}\theta = M_{OA} - M^F_{OA} - C^F_{AO} M'_{OA} \qquad (5\text{-}82)$$

in which the fixed-end carry-over factor toward O, C^F_{AO}, has been substituted for K_{OA}/J_{OA}, and M'_{OA} for $M_{AO} - M^F_{AO}$. An expression analogous to Eq. (5-82) can be written for each of the other members at O. Adding these to Eq. (5-82), we obtain

$$\theta \Sigma K_O = \Sigma M_O - \Sigma M^F_O - \Sigma C^F_O M'_O \qquad (5\text{-}83)$$

Equating the value of θ obtained from Eqs. (5-82) and (5-83) and solving for M_{OA}, we determine the equations that will check the joint for continuity:

$$M_{OA} = M^F_{OA} + C^F_{AO} M'_{OA} - m_{OA} \qquad (5\text{-}84)$$

$$m_{OA} = \frac{K_{OA}}{\Sigma K_O} (-\Sigma M_O + \Sigma M^F_O + \Sigma C^F_O M'_O) \qquad (5\text{-}85)$$

Method for Checking Moment Distribution 5-75

Similar equations can be written for the other members at O by substituting the proper letter for A in the subscripts.

If the calculations based on these equations are carried out in table form, the equations prove to be surprisingly simple (see Tables 5-6 and 5-7).

For prismatic beams, the terms $C^F M'$ become ½ the change in the fixed-end moment at the far end of each member at a joint.

Examples. Suppose we want to check the beam in Fig. 5-73b. Each joint and the ends of the members connected are listed in Table 5-6, and a column is provided for the summation of the various terms for each joint. K values are given on line 1, the end moments to be checked on line 2, and the fixed-end moments on line 3. On line 4 is entered one-half the difference obtained when the fixed-end moment is subtracted from the final moment at the far end of each member. $-m$ is placed on line 5 and the corrected end moment on line 6. The $-m$ values are obtained from the summation columns by adding line 2 to the negative of the sum of lines 3 and 4 and distributing the result to the members of the joint in proportion to the K values. The corrected moment M is the sum of lines 3, 4, and 5.

TABLE 5-6 Moment-Distribution Check

(No deflection at supports)

Line No.	Term	Joint A		Joint B			Joint C			Joint D	
		AB	Sum	BA	BC	Sum	CB	CD	Sum	DC	Sum
1	K	1	1	1	1	2	1	1	2	1	∞
2	M	0	0	+600	−600	0	+450	−450	0	+700	
3	M^F	−400	−400	+400	−480	−80	+480	−600	−120	+600	
				First Cycle							
4	½M'	+100	+100	+200	−15	+185	−60	+50	−10	+75	
5	$-m$	+300	+300	−53	−52	−105	+65	+65	+130	0	
6	M	0	0	+547	−547	0	+485	−485	0	+675	
				Second Cycle							
7	½M'	.....	.....	+200	+2	+202	−33	+37	+4	+57	
8	$-m$	.....	.....	−61	−61	−122	+58	+58	+116	0	
9	M	.....	.....	+539	−539	0	+505	−505	0	+657	

Assume that a mistake was made in computing the end moments for Fig. 5-73b, giving the results shown on line 2 of Table 5-6 for the fixed-end moments on line 3. The correct moments can be obtained as follows:

At joint B, the sum of the incorrect moments is zero, as shown in the summation column on line 2. The sum of the fixed-end moments at B is -80, as indicated on line 3. For BA, ½ M' is obtained from lines 2 and 3 of the column for AB: ½ × (0 + 400) = +200, which is entered on line 4. The line 4 entry for BC is obtained from ½ × (450 − 480) = −15. The sum of the line 4 values at B is therefore 200 − 15 = +185. Entered in the summation column, this is then added to the summation value on line 3, the sign is changed and the number on line 2 (in this case zero) added to the sum, giving −105, which is noted in the summation column on line 5. The values for BA and BC on line 5 are obtained by multiplying −105 by the ratio of the K value of each member to the sum of the K values at the joint; that is, for BA, $-m = -105 \times \frac{1}{2} = -53$. The corrected moment for BA, the sum of lines 3, 4, and 5, is $+400 + 200 - 53 = +547$. The other corrected moments are found in the same way and are shown on line 6.

The new end moments differ considerably from the moments on line 2, indicating that those moments were incorrect. A comparison with Fig. 5-73b shows that the new moments are nearer to the correct answer than those on line 2. Even closer results can be obtained by repeating the calculations using the moments on line 6 as demonstrated in the second cycle in Table 5-6.

Equations (5-80) to (5-85) can be generalized to include the effect of the movement d of a support in a direction normal to the initial position of a span of length L:

$$K_{OA}\theta = M_{OA} - M^F_{OA} - C^F_{AO} M'_{OA} + K_{OA}\frac{d}{L} \tag{5-86}$$

$$M_{OA} = M^F_{OA} + C^F_{AO} M'_{OA} - K_{OA}\frac{d}{L} - m_{OA} \tag{5-87a}$$

$$m_{OA} = \frac{K_{OA}}{\Sigma K_O}\left(-\Sigma M_O + \Sigma M^F_O + \Sigma C^F_O M'_O - \Sigma K_O \frac{d}{L}\right) \tag{5-87b}$$

For each span with a support movement, the term Kd/L can be obtained from the fixed-end moment due to this deflection alone; for OA, for example, by multiplying moment M^F_{OA} by $(1 - C^F_{AO}C^F_{OA})/(1 + C^F_{OA})$. For prismatic beams, this factor reduces to $\frac{1}{2}$. Equation (5-86) is called a **slope-deflection equation**.

TABLE 5-7 Moment-Distribution Check
(Frame with sidesway)

Line No.	Term	Joint A		Joint B			Joint C			Joint D	
		AB	Sum	BA	BC	Sum	CB	CD	Sum	DC	Sum
1	K	6	∞	6	3	9	3	2	5	2	∞
2	M	+120	...	+7,110	-7,110	0	+4,050	-4,050	0	-3,180	
3	M^F	0	...	0	-12,800	-12,800	+3,200	0	+3,200	0	
4	$-Kd/L$	-3,450	...	-3,450	0	-3,450	0	-1,150	-1,150	-1,150	
5	$+\frac{1}{2}M'$	+3,555	...	+60	+425	+485	+2,845	-1,590	+1,255	-2,025	
6	$-m$	0	...	+10,510	+5,255	+15,765	-1,983	-1,322	-3,305	0	
7	M	+105	...	+7,120	-7,120	0	+4,062	-4,062	0	-3,175	

In Table 5-7, the solution for the frame in Fig. 5-71 is checked for the condition in which a 4,000-lb vertical load was placed 5 ft from B on span BC. The computations are similar to those in Table 5-6, except that the terms $-Kd/L$ are included for the columns to account for the sidesway. These values are obtained from the sidesway fixed-end moments in Table 5-5. For BA, for example, $Kd/L = \frac{1}{2} \times 3000M$, with $M = 2.30$, as found in the solution. The check indicates that the original solution was sufficiently accurate for a slide-rule computation. If line 7 had contained a different set of moments, the shears would have had to be investigated again. A second cycle could be carried out by distributing the unbalance to the columns to obtain new Kd/L values.

5-70. Plane Grid Frameworks.[*]

Grid framing comprises two or more sets of parallel girders or trusses interconnected at their intersections, or nodes. In plane grid framing, the girders lie in a single plane, usu-

---

[*]Reprinted with permission from F. S. Merritt, "Structural Steel Designers' Handbook," McGraw-Hill Book Company, New York.

ally horizontal. Grid framing, however, may also be composed of space trusses. In that case, they usually are referred to as double-layer grid framing. Methods of analyzing single- and double-layer framing generally are similar. This article therefore will illustrate the technique with the simpler plane framing and with girders, instead of plane trusses. Loading will be taken as vertical. Girders will be assumed continuous at all nodes, except terminals.

Girders may be arranged in numerous ways for plane grid framing. Figure 5-74 shows some ways of placing two sets of girders. The grid in Fig. 5-74a consists of orthogonal

Fig. 5-74 Orthogonal grids. (a) Girders on short spacing. (b) Girders with wide spacing and beams between. (c) Girders placed diagonally.

sets laid perpendicular to boundary girders. Columns may be placed at the corners, along the boundaries, or at interior nodes. In the following analysis, for illustrative purposes, columns will be assumed only at the corners, and interior girders will be assumed simply supported on the boundary girders. With wider spacing of interior girders, the arrangement shown in Fig. 5-74b may be preferable. With beams in alternate bays spanning in perpendicular directions, loads are uniformly distributed to the girders. Alternatively, the interior girders may be set parallel to the main diagonals, as indicated in Fig. 5-74c. The method of analysis for this case is much the same as for girders perpendicular to boundary members. The structure, however, need not be rectangular or square, nor need the interior members be limited to two sets of girders.

Many methods have been used successfully to obtain exact or nearly exact solutions for grid framing, which may be highly indeterminate. These include consistent deflections, finite differences, moment distribution or slope deflection, flat plate analogy, and model analysis. This article will be limited to illustrating the use of the method of consistent deflections.

In this method, each set of girders is temporarily separated from the other sets. Unknown loads satisfying equilibrium conditions then are applied to each set. Equations are obtained by expressing node deflections in terms of the loads and equating the deflection at each node of one set to the deflection of the same node in another set. Simultaneous solution of the equations yields the unknown loads on each set. With these now known, bending moments, shears, and deflections of all the girders can be computed by conventional methods.

For a simply supported grid, the unknowns generally can be selected and the equations formulated so that there are one unknown and one equation for each interior node. The number of equations required, however, can be drastically reduced if the framing is made symmetrical about perpendicular axes and the loading is symmetrical or antisymmetrical. For symmetrical grids subjected to unsymmetrical loading, the amount of work involved in analysis often can be decreased by resolving loads into symmetrical and antisymmetrical components. Figure 5-75 shows how this can be done for a single load unsymmetrically located on a grid. The analysis requires the solution of four sets of simultaneous equations, and addition of the results, but there are fewer equations in each set than for unsymmetrical loading. The number of unknowns may be further decreased when the proportion of a load at a node to be assigned to a girder at that node can be determined by inspection or simple computation. For example, for a square orthogonal grid, each girder at the central node carries half the load there when the grid loading is symmetrical or antisymmetrical.

For analysis of simply supported grid girders, influence coefficients for deflection at any point induced by a unit load are useful. They may be computed from the following formulas.

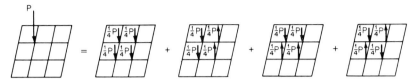

Fig. 5-75 Resolution of a load into symmetrical and antisymmetrical components.

The deflection at a distance xL from one support of a girder produced by a concentrated load P at a distance kL from that support (Fig. 5-76) is given by

$$\delta = \frac{PL^3}{6EI} x(1-k)(2k - k^2 - x^2) \qquad 0 \leq x \leq k \tag{5-88}$$

$$\delta = \frac{PL^3}{6EI} k(1-x)(2x - x^2 - k^2) \qquad k \leq x \leq 1 \tag{5-89}$$

where L = span of simply supported girder
E = modulus of elasticity of the girder material
I = moment of inertia of girder cross section

For deflections, the elastic curve is also the influence curve, when $P = 1$. Hence, the influence coefficient for any point of the girder may be written

$$\delta' = \frac{L^3}{EI} [x,k] \tag{5-90}$$

where

$$[x,k] = \begin{cases} \dfrac{x}{6}(1-k)(2k - k^2 - x^2) & 0 \leq x \leq k \\ \dfrac{k}{6}(1-x)(2x - x^2 - k^2) & k \leq x \leq 1 \end{cases} \tag{5-91}$$

The deflection at a distance xL from one support of the girder produced by concentrated loads P at distances kL and $(1-k)L$ from that support (Fig. 5-77) is given by

$$\delta = \frac{PL^3}{EI} (x,k) \tag{5-92}$$

where

$$(x,k) = \begin{cases} \dfrac{x}{6}(3k - 3k^2 - x^2) & 0 \leq x \leq k \\ \dfrac{k}{6}(3x - 3x^2 - k^2) & k \leq x \leq \dfrac{1}{2} \end{cases} \tag{5-93}$$

The deflection at a distance xL from one support of the girder produced by a downward concentrated load P at distance kL from the support and an upward concentrated load P at a distance $(1-k)L$ from the support (antisymmetric loading, Fig. 5-78) is given by

$$\delta = \frac{PL^3}{EI} \{x,k\} \tag{5-94}$$

where

$$\{x,k\} = \begin{cases} \dfrac{x}{6}(1-2k)(k-k^2-x^2) & 0 \le x \le k \\ \dfrac{k}{6}(1-2x)(x-x^2-k^2) & k \le x \le \dfrac{1}{2} \end{cases} \quad (5\text{-}95)$$

For convenience in analysis, the loading carried by the grid framing is converted into concentrated loads at the nodes. Suppose, for example, that a grid consists of two sets of parallel girders as in Fig. 5-74, and the load at interior node r is P_r. Then, it is convenient to assume that one girder at the node is subjected to an unknown force X_r there, and the other girder therefore carries a force $P_r - X_r$ at the node. With one set of girders detached from the other set, the deflections produced by these forces can be determined with the aid of Eqs. (5-88) to (5-95).

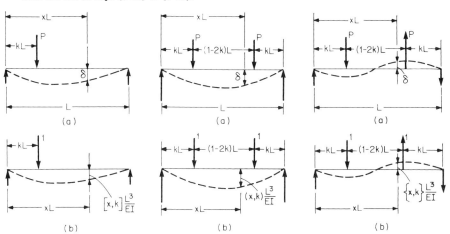

Fig. 5-76 Single load on a beam. (a) Deflection curve. (b) Curve of influence coefficients for deflection at distance xL from left support.

Fig. 5-77 Two equal downward loads symmetrically placed on a beam: (a) deflection curve; (b) curve of influence coefficients.

Fig. 5-78 Equal upward and downward loads symmetrically placed on a beam: (a) deflection curve; (b) curve of influence coefficients.

Square Grid. A simple example will be used to illustrate the application of the method of consistent deflections. Assume an orthogonal grid within a square boundary (Fig. 5-79a). There are $n = 4$ equal spaces of width h between girders. Columns are located at the corners A, B, C, and D. All girders have a span $nh = 4h$ and are simply supported at their terminals, though continuous at interior nodes. To simplify the example, all girders are assumed to have equal and constant moment of inertia I. Interior nodes carry a concentrated load P. Exterior nodes, except corners, are subjected to a load $P/2$.

Because of symmetry, only five different nodes need be considered. These are numbered from 1 to 5 in Fig. 5-79a, for identification. By inspection, loads P at nodes 1 and 3 can be distributed equally to the girders spanning in the x and y directions. Thus, when the two sets of parallel girders are considered separated, girder 4-4 in the x direction carries a load of $P/2$ at midspan (Fig. 5-79b). Similarly, girder 5-5 in the y direction carries loads of $P/2$ at the quarter points (Fig. 5-79c).

Let X_2 be the load acting on girder 4-4 (x direction) at node 2 (Fig. 5-79b). Then, $P - X_2$ acts on girder 5-5 (y direction) at midspan (Fig. 5-79c). The reaction R of girders 4-4 and 5-5 are loads on the boundary girders (Fig. 5-79d).

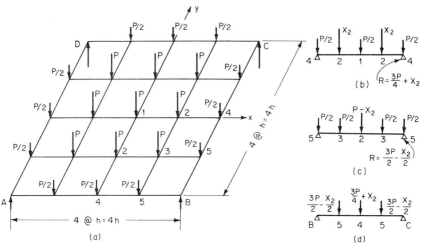

Fig. 5-79 Orthogonal grid: (a) loads distributed to joints; (b) loads on midspan girder; (c) loads on quarter-point girder; (d) loads on boundary girder.

Because of symmetry, X_2 is the only unknown in this example. Only one equation is needed to determine it.

To obtain this equation, equate the vertical displacement of girder 4-4 (x direction) at node 2 to the vertical displacement of girder 5-5 (y direction) at node 2. The displacement of girder 4-4 equals its deflection plus the deflection of node 4 on BC. Similarly, the displacement of girder 5-5 equals its deflection plus the deflection of node 5 on AB or its equivalent BC.

When use is made of Eqs. (5-90) and (5-92), the deflection of girder 4-4 (x direction) at node 2 equals

$$\delta_2 = \frac{n^3 h^3}{EI}\left\{\left[\frac{1}{4},\frac{1}{2}\right]\frac{P}{2} + \left(\frac{1}{4},\frac{1}{4}\right)X_2\right\} + \delta_4 \qquad (5\text{-}96a)$$

where δ_4 is the deflection of BC at node 4. By Eq. (5-91), $[\frac{1}{4},\frac{1}{2}] = (\frac{1}{48})(\frac{11}{16})$. By Eq. (5-93), $(\frac{1}{4},\frac{1}{4}) = \frac{1}{48}$. Hence,

$$\delta_2 = \frac{n^3 h^3}{48 EI}\left(\frac{11}{32}P + X_2\right) + \delta_4 \qquad (5\text{-}96b)$$

For the loading shown in Fig. 5-79d,

$$\delta_4 = \frac{n^3 h^3}{EI}\left\{\left[\frac{1}{2},\frac{1}{2}\right]\left(\frac{3P}{4} + X_2\right) + \left(\frac{1}{2},\frac{1}{4}\right)\left(\frac{3P}{2} - \frac{X_2}{2}\right)\right\} \qquad (5\text{-}97a)$$

By Eq. (5-91), $[\frac{1}{2},\frac{1}{2}] = \frac{1}{48}$. By Eq. (5-93), $(\frac{1}{2},\frac{1}{4}) = (\frac{1}{48})(\frac{11}{8})$. Hence, Eq. (5-97a) becomes

$$\delta_4 = \frac{n^3 h^3}{48 EI}\left(\frac{45}{16}P + \frac{5}{16}X_2\right) \qquad (5\text{-}97b)$$

Similarly, the deflection of girder 5-5 (y direction) at node 2 equals

$$\delta_2 = \frac{n^3 h^3}{EI}\left\{\left[\frac{1}{2},\frac{1}{2}\right](P - X_2) + \left(\frac{1}{2},\frac{1}{4}\right)\frac{P}{2}\right\} + \delta_5$$

$$= \frac{n^3 h^3}{48 EI}\left(\frac{27}{16}P - X_2\right) + \delta_5 \qquad (5\text{-}98)$$

For the loading shown in Fig. 5-79d,

$$\delta_5 = \frac{n^3h^3}{EI}\left\{\left[\frac{1}{4},\frac{1}{2}\right]\left(\frac{3P}{4}+X_2\right)+\left(\frac{1}{4},\frac{1}{4}\right)\left(\frac{3P}{2}-\frac{X_2}{2}\right)\right\}$$
$$= \frac{n^3h^3}{48EI}\left(\frac{129}{64}P+\frac{3}{16}X_2\right) \quad (5\text{-}99)$$

The needed equation for determining X_2 is obtained by equating the right-hand side of Eqs. (5-96b) and (5-98) and substituting δ_4 and δ_5 given by Eqs. (5-97b) and (5-99). The result, after division of both sides of the equation by $n^3h^3/48EI$, is

$$\frac{11}{32}P + X_2 + \frac{45}{16}P + \frac{5}{16}X_2 = \frac{27}{16}P - X_2 + \frac{129}{64}P + \frac{3}{16}X_2 \quad (5\text{-}100)$$

Solution of the equation yields

$$X_2 = \frac{35P}{136} = 0.257P \quad \text{and} \quad P - X_2 = \frac{101P}{136} = 0.743P$$

With these forces known, the bending moments, shears, and deflections of the girders can be computed by conventional methods.

Rectangular Grid. To examine a more general case of symmetrical framing, consider the orthogonal grid with rectangular boundaries in Fig. 5-80a. In the x direction, there are n spaces of width h. In the y direction, there are m spaces of width k. Only members symmetrically placed in the grid are the same size. Interior nodes carry a concentrated load P. Exterior nodes, except corners, carry $P/2$. Columns are located at the corners. For identification, nodes are numbered in one quadrant. Since the loading, as well as the framing, is symmetrical, corresponding nodes in the other quadrants may be given corresponding numbers.

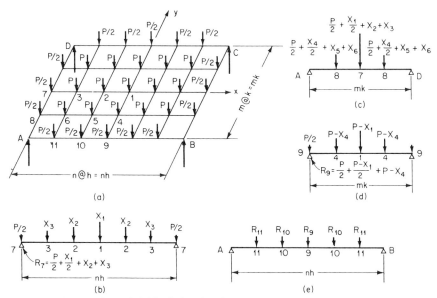

Fig. 5-80 Rectangular grid: (a) loads distributed to joints; (b) loads on longer midspan girder; (c) loads on shorter boundary girder; (d) loads on shorter midspan girder; (e) loads on longer boundary girder.

5-82   Structural Theory

At any interior node r, let X_r be the load carried by the girder spanning in the x direction. Then, $P - X_r$ is the load at that node applied to the girder spanning in the y direction. For this example, therefore, there are six unknowns X_r, because r ranges from 1 to 6. Six equations are needed for determination of X_r. They may be obtained by the method of consistent deflections. At each interior node, the vertical displacement of the x-direction girder is equated to the vertical displacement of the y-direction girder, as in the case of the square grid.

To indicate the procedure for obtaining these equations, the equation for node 1 in Fig. 5-80a will be developed. When use is made of Eqs. (5-90) and (5-92), the deflection of girder 7-7 at node 1 (Fig. 5-80b) equals

$$\delta_1 = \frac{n^3 h^3}{EI_7} \left\{ \left[\frac{1}{2}, \frac{1}{2} \right] X_1 + \left(\frac{1}{2}, \frac{1}{3} \right) X_2 + \left(\frac{1}{2}, \frac{1}{6} \right) X_3 \right\} + \delta_7 \quad (5\text{-}101)$$

where I_7 = moment of inertia of girder 7-7
δ_7 = deflection of girder AD at node 7

Girder AD carries the reactions of the interior girders spanning in the x direction (Fig. 5-80c).

$$\delta_7 = \frac{m^3 k^3}{EI_{AD}} \left\{ \left[\frac{1}{2}, \frac{1}{2} \right] \left(\frac{P}{2} + \frac{X_1}{2} + X_2 + X_3 \right) \right.$$
$$\left. + \left(\frac{1}{2}, \frac{1}{4} \right) \left(\frac{P}{2} + \frac{X_4}{2} + X_5 + X_6 \right) \right\} \quad (5\text{-}102)$$

where I_{AD} = moment of inertia of girder AD. Similarly, the deflection of girder 9-9 at node 1 (Fig. 5-80d) equals

$$\delta_1 = \frac{m^3 k^3}{EI_9} \left\{ \left[\frac{1}{2}, \frac{1}{2} \right] (P - X_1) + \left(\frac{1}{2}, \frac{1}{4} \right) (P - X_4) \right\} + \delta_9 \quad (5\text{-}103)$$

where I_9 = moment of inertia of girder 9-9
δ_9 = deflection of girder AB at node 9

Girder AB carries the reactions of the interior girders spanning in the y direction (Fig. 5-80c).

$$\delta_9 = \frac{n^3 h^3}{EI_{AB}} \left\{ \left[\frac{1}{2}, \frac{1}{2} \right] \left(\frac{P}{2} + \frac{P - X_1}{2} + P - X_4 \right) \right.$$
$$+ \left(\frac{1}{2}, \frac{1}{3} \right) \left(\frac{P}{2} + \frac{P - X_2}{2} + P - X_5 \right)$$
$$\left. + \left(\frac{1}{2}, \frac{1}{6} \right) \left(\frac{P}{2} + \frac{P - X_3}{2} + P - X_6 \right) \right\} \quad (5\text{-}104)$$

where I_{AB} = moment of inertia of girder AB. The equation for vertical displacement at node 1 is obtained by equating the right-hand side of Eqs. (5-101) and (5-103) and substituting δ_7 and δ_9 given by Eqs. (5-102) and (5-104).

After similar equations have been developed for the other five interior nodes, the six equations are solved simultaneously for the unknown forces X_r. When these have been determined, moments, shears, and deflections for the girders can be computed by conventional methods.

BUCKLING OF COLUMNS

Columns are compression members whose cross-sectional dimensions are relatively small compared with their length in the direction of the compressive force. Failure of such members occurs because of instability when a certain axial load (called the critical or **Euler load**) is equaled or exceeded. The member may bend, or buckle, suddenly and collapse.

Hence the strength P of a column is not determined by the unit stress in Eq. (5-7) ($P = Af$) but by the maximum load it can carry without becoming unstable. The condition of instability is characterized by disproportionately large increases in lateral deformation with slight increase in axial load. Instability may occur in slender columns before the unit stress reaches the elastic limit.

5-71. Stable Equilibrium.

Consider, for example, an axially loaded column with ends unrestrained against rotation, shown in Fig. 5-81. If the member is initially perfectly straight, it will remain straight as long as the load P is less than the critical load P_c. If a small transverse force is applied, the column will deflect, but it will return to the straight position when this force is removed. Thus, when P is less than P_c, internal and external forces are in stable equilibrium.

5-72. Unstable Equilibrium.

If $P = P_c$ and a small transverse force is applied, the column again will deflect, but this time, when the force is removed, the column will remain in the bent position (dashed line in Fig. 5-81). The equation of this elastic curve can be obtained from Eq. (5-31):

$$EI\frac{d^2y}{dx^2} = -P_c y \qquad (5\text{-}105)$$

in which E = modulus of elasticity
I = least moment of inertia
y = deflection of the bent member from the straight position at a distance x from one end

Fig. 5-81 Buckling of a long column.

This assumes, of course, that the stresses are within the elastic limit. Solution of Eq. (5-105) gives the smallest value of the Euler load as

$$P_c = \frac{\pi^2 EI}{L^2} \qquad (5\text{-}106)$$

Equation (5-106) indicates that there is a definite finite magnitude of an axial load that will hold a column in equilibrium in the bent position when the stresses are below the elastic limit. Repeated application and removal of small transverse forces or small increases in axial load above this critical load will cause the member to fail by buckling. Internal and external forces are in a state of unstable equilibrium.

It is noteworthy that the Euler load, which determines the load-carrying capacity of a column, depends on the stiffness of the member, as expressed by the modulus of elasticity, rather than on the strength of the material of which it is made.

By dividing both sides of Eq. (5-106) by the cross-sectional area A and substituting r^2 for I/A (r is the radius of gyration of the section), we can write the solution of Eq. (5-105) in terms of the average unit stress on the cross section:

$$\frac{P_c}{A} = \frac{\pi^2 E}{(L/r)^2} \qquad (5\text{-}107)$$

This holds only for the elastic range of buckling; i.e., for values of the slenderness ratio L/r above a certain limiting value that depends on the properties of the material. For inelastic buckling, see Art. 5-74.

5-73. Effect of End Conditions.

Equation (5-107) was derived on the assumption that the ends of the column are free to rotate. It can be generalized, however, to take into account the effect of end conditions:

$$\frac{P_c}{A} = \frac{\pi^2 E}{(kL/r)^2} \qquad (5\text{-}108)$$

where k is the factor that depends on the end conditions. For a pin-ended column, $k = 1$; for a column with both ends fixed, $k = \frac{1}{2}$; for a column with one end fixed and one end pinned, k is about 0.7; and for a column with one end fixed and one end free from all restraint, $k = 2$.

5-74. Inelastic Buckling.

Equations (5-106) and (5-108), having been derived from Eq. (5-105), the differential equation for the elastic curve, are based on the assumption that the critical average stress is below the elastic limit when the state of unstable equilibrium is reached. In members with slenderness ratio L/r below a certain limiting value, however, the elastic limit is exceeded before the column buckles. As the axial load approaches the critical load, the modulus of elasticity varies with the stress. Hence Eqs. (5-106) and (5-108), based on the assumption that E is a constant, do not hold for these short columns.

After extensive testing and analysis, prevalent engineering opinion favors the Engesser equation for metals in the inelastic range:

$$\frac{P_t}{A} = \frac{\pi^2 E_t}{(kL/r)^2} \qquad (5\text{-}109)$$

This differs from Eqs. (5-106) to (5-108) only in that the tangent modulus E_t (the actual slope of the stress-strain curve for the stress P_t/A) replaces the modulus of elasticity E in the elastic range. P_t is the smallest axial load for which two equilibrium positions are possible, the straight position and a deflected position.

5-75. Column Curves.

Curves obtained by plotting the critical stress for various values of the slenderness ratio are called column curves. For axially loaded, initially straight columns, the column curve consists of two parts: (1) the Euler critical values, and (2) the Engesser, or tangent-modulus critical values.

The latter are greatly affected by the shape of the stress-strain curve for the material of which the column is made, as shown in Fig. 5-82. The stress-strain curve for a material, such as an aluminum alloy or high-strength steel, which does not have a sharply defined yield point, is shown in Fig. 5-82a. The corresponding column curve is drawn in Fig. 5-82b. In contrast, Fig. 5-82c presents the stress-strain curve for structural steel, with a sharply defined point, and Fig. 5-82d the related column curve. This curve becomes horizontal as the critical stress approaches the yield strength of the material and the tangent modulus becomes zero, whereas the column curve in Fig. 5-82b continues to rise with decreasing values of the slenderness ratio.

Fig. 5-82 Column curves: (a) stress-strain curve for a material that does not have a sharply defined yield point; (b) column curve for this material; (c) stress-strain curve for a material with a sharply defined yield point; (d) column curve for that material.

Examination of Fig. 5-82b also indicates that slender columns, which fall in the elastic range, where the column curve has a large slope, are very sensitive to variations in the factor k, which represents the effect of end conditions. On the other hand, in the inelastic range, where the column curve is relatively flat, the critical stress is relatively insensitive to changes in k. Hence the effect of end conditions on the stability of a column is of much greater significance for long columns than for short columns.

5-76. Local Buckling.

A column may not only fail by buckling of the member as a whole but as an alternative, by buckling of one of its components. Hence, when members like I beams, channels, and angles are used as columns or when sections are built up of plates, the possibility of the critical load on a component (leg, half flange, web, lattice bar) being less than the critical load on the column as a whole should be investigated.

Similarly, the possibility of buckling of the compression flange or the web of a beam should be looked into.

Local buckling, however, does not always result in a reduction in the load-carrying capacity of a column. Sometimes, it results in a redistribution of the stresses enabling the member to carry additional load.

5-77. Behavior of Actual Columns.

For many reasons, columns in structures behave differently from the ideal column assumed in deriving Eqs. (5-106) and (5-109). A major consideration is the effect of accidental imperfections, such as nonhomogeneity of materials, initial crookedness, and unintentional eccentricities of the axial load, since neither field nor shopwork can be perfect. These and the effects of residual stresses usually are taken into account by a proper choice of safety factor.

There are other significant conditions, however, that must be considered in any design rule: continuity in frame structures and eccentricity of the axial load. Continuity affects column action in two ways. The restraint at column ends determines the value of k, and bending moments are transmitted to the column by adjoining structural members.

Because of the deviation of the behavior of actual columns from the ideal, columns generally are designed by empirical formulas. Separate equations usually are given for short columns, intermediate columns, and long columns. For specific materials—steel, concrete, timber—these formulas are given in Secs. 7 to 10.

For more details on column action, see F. Bleich, "Buckling Strength of Metal Structures," McGraw-Hill Book Company, New York, 1952; and S. Timoshenko and J. M. Gere, "Theory of Elastic Stability," McGraw-Hill Book Company, New York, 1961.

TORSION

Forces that cause a member to twist about a longitudinal axis are called torsional loads. Simple torsion is produced only by a couple, or moment, in a plane perpendicular to the axis.

If a couple lies in a nonperpendicular plane, it can be resolved into a torsional moment, in a plane perpendicular to the axis, and bending moments, in planes through the axis.

5-78. Shear Center.

The point in each section of a member normal to the axis about which the section twists is called the shear center. (The location of the shear center in commonly used shapes is given in Art. 5-35.) If the loads on a beam, for example, do not pass through the shear center, they cause the beam to twist.

5-79. Stresses Due to Torsion.

Simple torsion is resisted by internal shearing stresses. These can be resolved into radial and tangential shearing stresses, which being normal to each other also are equal (see Art. 5-11). Furthermore, on planes that bisect the angles between the planes on which the shearing stresses act, there also occur compressive and tensile stresses. The magnitude of these normal stresses is equal to that of the shear. Therefore, when torsional loading is combined with other types of loading, the maximum stresses occur on inclined planes and can be computed by the methods explained in Arts. 5-12 and 5-15.

If a circular shaft (hollow or solid) is twisted, a section that is plane before twisting remains plane after twisting. Within the proportional limit, the shearing stress at any point in a transverse section varies with the distance from the center of the section. The maximum shear occurs at the circumference and is given by

$$v = \frac{Tr}{J} \qquad (5\text{-}110)$$

where T = torsional moment
R = radius of section
J = polar moment of inertia

Polar moment of inertia is defined by

$$J = \int \rho^2 \, dA \qquad (5\text{-}111)$$

where ρ = radius from shear center to any point in section
dA = differential area at the point

In general, J equals the sum of the moments of inertia about any two perpendicular axes through the shear center. For a solid circular section, $J = \pi r^4/2$.

Within the proportional limit, the angular twist for a circular bar is given by

$$\theta = \frac{TL}{GJ} \tag{5-112}$$

where L = length of bar
G = shearing modulus of elasticity (see Art. 5-6)

If a shaft is not circular, a plane transverse section before twisting does not remain plane after twisting. The resulting warping increases the shearing stresses in some parts of the section and decreases them in others, compared with the shearing stresses that would occur if the section remained plane. Consequently, shearing stresses in a noncircular section are not proportional to distances from the shear center. In elliptical and rectangular sections, for example, maximum shear occurs on the circumference at a point nearest the shear center.

For a solid rectangular section, this maximum may be expressed in the following form:

$$v = \frac{T}{kb^2 d} \tag{5-113}$$

where b = short side of the rectangle
d = long side
k = a constant depending on the ratio of these sides:

$d/b =$	1.0	1.5	2.0	2.5	3	4	5	10	∞
$k =$	0.208	0.231	0.246	0.258	0.267	0.282	0.291	0.312	0.333

(S. Timoshenko and J. N. Goodier, "Theory of Elasticity," McGraw-Hill Book Company, New York.)

If a thin-shell hollow tube is twisted, the shearing force per unit of length on a cross section (shear flow) is given approximately by

$$H = \frac{T}{2A} \tag{5-114}$$

where T = torsional moment
A = area enclosed by mean perimeter of wall of tube

And the unit shearing stress is given approximately by

$$v = \frac{H}{t} = \frac{T}{2At} \tag{5-115}$$

where t = thickness of tube. For a rectangular tube with sides of unequal thickness, the total shear flow can be computed from Eq. (5-114), and the shearing stress along each side from Eq. (5-115).

For a narrow rectangular section, the maximum shear is very nearly equal to

$$v = \frac{T}{\frac{1}{3} b^2 d} \tag{5-116}$$

where b = short dimension of the rectangle
d = long side

This formula can also be used to find the maximum shearing due to torsion in members, such as I beams and channels, made up of thin rectangular components. Let $J =$

⅓Σb^3d, where b is the thickness of each rectangular component and d the corresponding length. Then, the maximum shear is given approximately by

$$v = \frac{Tb}{J} \tag{5-117}$$

In this equation, b is the thickness of the web or the flange. The maximum shearing stress is at the center of one of the long sides of the rectangular part that has the greatest thickness.

(A. P. Boresi, O. Sidebottom, F. B. Seely and J. O. Smith, "Advanced Mechanics of Materials," 3rd Ed., John Wiley & Sons, Inc., New York.)

STRUCTURAL DYNAMICS

Article 5-1 notes that loads can be classified as static or dynamic and that the distinguishing characteristic is the rate of application of load. If a load is applied slowly, it may be considered static. Since dynamic loads may produce stresses and deformations considerably larger than those caused by static loads of the same magnitude, it is important to know reasonably accurately what is meant by slowly.

A useful definition can be given in terms of the natural period of vibration of the structure or member to which the load is applied. If the time in which a load rises from zero to its maximum value is more than double the natural period, the load may be treated as static. Loads applied more rapidly may be dynamic. Structural analysis and design for such loads are considerably different from and more complex than those for static loads.

In general, exact dynamic analysis is possible only for relatively simple structures, and only when both the variation of load and resistance with time are a convenient mathematical function. Therefore, in practice, adoption of approximate methods that permit rapid analysis and design is advisable. And usually, because of uncertainties in loads and structural resistance, computations need not be carried out with more than a few significant figures, to be consistent with known conditions.

5-80. Properties of Materials under Dynamic Loading.

In general, mechanical properties of structural materials improve with increasing rate of load application. For low-carbon steel, for example, yield strength, ultimate strength, and ductility rise with increasing rate of strain. Modulus of elasticity in the elastic range, however, is unchanged. For concrete, the dynamic ultimate strength in compression may be much greater than the static strength.

Since the improvement depends on the material and the rate of strain, values to use in dynamic analysis and design should be determined by tests approximating the loading conditions anticipated.

Under many repetitions of loading, though, a member or connection between members may fail because of "fatigue" at a stress smaller than the yield point of the material. In general, there is little apparent deformation at the start of a fatigue failure. A crack forms at a point of high stress concentration. As the stress is repeated, the crack slowly spreads, until the member ruptures without measurable yielding. Though the material may be ductile, the fracture looks brittle.

Some materials (generally those with a well-defined yield point) have what is known as an **endurance limit.** This is the maximum unit stress that can be repeated, through a definite range, an indefinite number of times without causing structural damage. Generally, when no range is specified, the endurance limit is intended for a cycle in which the stress is varied between tension and compression stresses of equal value. For a different range, if f is the endurance limit, f_y the yield point, and r the ratio of the minimum stress to the maximum, then the relationship between endurance stresses is given approximately by

$$f_{max} = \frac{2f}{(1 - r) + (f/f_y)(1 + r)} \tag{5-118}$$

A range of stress may be resolved into two components—a steady, or mean, stress and an alternating stress. The endurance limit sometimes is defined as the maximum value of the alternating stress that can be superimposed on the steady stress an indefinitely large number of times without causing fracture. If f is the endurance limit for completely reversed stresses, s the steady unit stress, and f_u the ultimate tensile stress, then the alternating stress may be obtained from a relationship of the type

$$f_a = f\left(1 - \frac{s^n}{f_u}\right) \tag{5-119}$$

where n lies between 1 and 2, depending on the material.

Design of members to resist repeated loading cannot be executed with the certainty with which members can be designed to resist static loading. Stress concentrations may be present for a wide variety of reasons, and it is not practicable to calculate their intensities. But sometimes it is possible to improve the fatigue strength of a material or to reduce the magnitude of a stress concentration below the minimum value that will cause fatigue failure.

In general, avoid design details that cause severe stress concentrations or poor stress distribution. Provide gradual changes in section. Eliminate sharp corners and notches. Do not use details that create high localized constraint. Locate unavoidable stress raisers at points where fatigue conditions are the least severe. Place connections at points where stress is low and fatigue conditions are not severe. Provide structures with multiple load paths or redundant members, so that a fatigue crack in any one of the several primary members is not likely to cause collapse of the entire structure.

Fatigue strength of a material may be improved by cold-working the material in the region of stress concentration, by thermal processes, or by prestressing it in such a way as to introduce favorable internal stresses. Where fatigue stresses are unusually severe, special materials may have to be selected with high energy absorption and notch toughness.

(J. H. Faupel, "Engineering Design," John Wiley & Sons, Inc., New York; C. H. Norris et al., "Structural Design for Dynamic Loads," McGraw-Hill Book Company, New York; W. H. Munse, "Fatigue of Welded Steel Structures," Welding Research Council, 345 East 47th Street, New York, N.Y. 10017.)

5-81. Natural Period of Vibration.

A preliminary step in dynamic analysis and design is determination of this period. It can be computed in many ways, including by application of the laws of conservation of energy and momentum or Newton's second law of motion, $F = M(dv/dt)$, where F is force, M mass, v velocity, and t time. But in general, an exact solution is possible only for simple structures. Therefore, it is general practice to seek an approximate—but not necessarily inexact—solution by analyzing an idealized representation of the actual member or structure. Setting up this model and interpreting the solution require judgment of a high order.

Natural period of vibration is the time required for a structure to go through one cycle of free vibration, that is, vibration after the disturbance causing the motion has ceased.

5-90   Structural Theory

To compute the natural period, the actual structure may be conveniently represented by a system of masses and massless springs, with additional resistances provided to account for energy losses due to friction, hysteresis, and other forms of damping. In simple cases, the masses may be set equal to the actual masses; otherwise, equivalent masses may have to be computed (Art. 5-85). The spring constants are the ratios of forces to deflections.

For example, a single mass on a spring (Fig. 5-83b) may represent a simply supported beam with mass that may be considered negligible compared with the load W at midspan

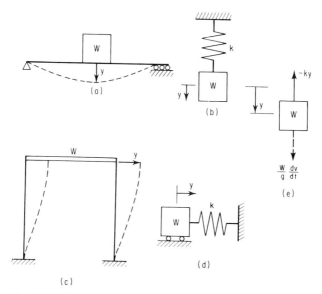

Fig. 5-83 Mass on a weightless spring (b) or (d) may represent the motion of (a) a beam or (c) a rigid frame in free vibration.

(Fig. 5-83a). The spring constant k should be set equal to the load that produces a unit deflection at midspan; thus, $k = 48EI/L^3$, where E is the modulus of elasticity, psi; I the moment of inertia, in^4; and L the span, in, of the beam. The idealized mass equals W/g, where g is the acceleration due to gravity, 386 in/s^2.

Also, a single mass on a spring (Fig. 5-83d) may represent the rigid frame in Fig. 5-83c. In that case, $k = 2 \times 12EI/h^3$, where I is the moment of inertia, in^4, of each column and h the column height, in. The idealized mass equals the sum of the masses on the girder and the girder mass. (Weight of columns and walls is assumed negligible.)

The spring and mass in Fig. 5-83b and d form a one-degree system. The **degree of a system** is determined by the least number of coordinates needed to define the positions of its components. In Fig. 5-83, only the coordinate y is needed to locate the mass and determine the state of the spring. In a two-degree system, such as one comprising two masses connected to each other and to the ground by springs and capable of movement in only one direction, two coordinates are required to locate the masses.

If the mass with weight W, lb, in Fig. 5-83 is isolated, as shown in Fig. 5-83e it will be in dynamic equilibrium under the action of the spring force $-ky$ and the inertia force $(d^2y/dt^2)(W/g)$. Hence, the equation of motion is

$$\frac{W}{g}\frac{d^2y}{dt^2} + ky = 0 \tag{5-120}$$

where y = displacement of mass, in, measured from rest position. Equation (5-120) may be written in the more convenient form

$$\frac{d^2y}{dt^2} + \frac{kg}{W}y = \frac{d^2y}{dt^2} + \omega^2 y = 0 \tag{5-121}$$

The solution is

$$y = A \sin \omega t + B \cos \omega t \tag{5-122}$$

where A and B are constants to be determined from initial conditions of the system, and

$$\omega = \sqrt{\frac{kg}{W}} \tag{5-123}$$

is the natural circular frequency, rad/s.

The motion defined by Eq. (5-122) is harmonic. Its natural period, s, is

$$T = \frac{2\pi}{\omega} = 2\pi \sqrt{\frac{W}{gk}} \tag{5-124}$$

Its natural frequency, Hz, is

$$f = \frac{1}{T} = \frac{1}{2\pi} \sqrt{\frac{kg}{W}} \tag{5-125}$$

If, at time $t = 0$, the mass has an initial displacement y_0 and velocity v_0, substitution in Eq. (5-122) yields $A = v_0/\omega$ and $B = y_0$. Hence, at any time t, the mass is completely located by

$$y = \frac{v_0}{\omega} \sin \omega t + y_0 \cos \omega t \tag{5-126}$$

The stress in the spring can be computed from the displacement y.

Vibrations of Lumped Masses. In multiple-degree systems, an independent differential equation of motion can be written for each degree of freedom. Thus, in an N-degree system with N masses, weighing $W_1, W_2, \ldots, W_N$ lb, and N^2 springs with constants k_{rj} ($r = 1, 2, \ldots, N; j = 1, 2, \ldots, N$), there are N equations of the form

$$\frac{W_r}{g}\frac{d^2y_r}{dt^2} + \sum_{j=1}^{N} k_{rj}y_j = 0 \qquad r = 1, 2, \ldots, N \tag{5-127}$$

Simultaneous solution of these equations reveals that the motion of each mass can be resolved into N harmonic components. They are called the fundamental, second, third, etc., harmonics. Each set of harmonics for all the masses is called a **normal mode** of vibration.

There are as many normal modes in a system as degrees of freedom. Under certain circumstances, the system could vibrate freely in any one of these modes. During any such vibration, the ratio of displacement of any two of the masses remains constant. Hence, the solutions of Eqs. (5-127) take the form

$$y_r = \sum_{n=1}^{N} a_{rn} \sin \omega_n(t + \tau_n) \tag{5-128}$$

where a_{rn} and τ_n are constants to be determined from the initial conditions of the system and ω_n is the natural circular frequency for each normal mode.

To determine ω_n, set $y_1 = A_1 \sin \omega t$; $y_2 = A_2 \sin \omega t \ldots$ Then, substitute these and their second derivatives in Eqs. (5-127). After dividing each equation by $\sin \omega t$, the following N equations result:

$$\left(k_{11} - \frac{W_1}{g}\omega^2\right) A_1 + k_{12}A_2 + \ldots + k_{1N}A_N = 0$$

$$k_{21}A_1 + \left(k_{22} - \frac{W_2}{g}\omega^2\right) A_2 + \ldots + k_{2N}A_N = 0 \qquad (5\text{-}129)$$

$$\cdots\cdots\cdots\cdots\cdots\cdots\cdots\cdots\cdots\cdots\cdots\cdots\cdots\cdots\cdots\cdots$$

$$k_{N1}A_1 + k_{N2}A_2 + \ldots + \left(k_{NN} - \frac{W_N}{g}\omega^2\right) A_N = 0$$

If there are to be nontrivial solutions for the amplitudes $A_1, A_2, \ldots, A_N$, the determinant of their coefficients must be zero. Thus,

$$\begin{vmatrix} k_{11} - \dfrac{W_1}{g}\omega^2 & k_{12} & \ldots & k_{1N} \\ k_{21} & k_{22} - \dfrac{W_2}{g}\omega^2 & \ldots & k_{2N} \\ \cdots & \cdots & \cdots & \cdots \\ k_{N1} & k_{N2} & \ldots & k_{NN} - \dfrac{W_N}{g}\omega^2 \end{vmatrix} = 0 \qquad (5\text{-}130)$$

Solution of this equation for ω yields one real root for each normal mode. And the natural period for each normal mode can be obtained from Eq. (5-124).

If ω for a normal mode now is substituted in Eqs. (5-129), the amplitudes $A_1, A_2, \ldots, A_N$ for that mode can be computed in terms of an arbitrary value, usually unity, assigned to one of them. The resulting set of modal amplitudes defines the **characteristic shape** for that mode.

The normal modes are mutually orthogonal; that is,

$$\sum_{r=1}^{N} W_r A_{rn} A_{rm} = 0 \qquad (5\text{-}131)$$

where W_r is the rth mass out of a total of N, A represents the characteristic amplitude of a normal mode, and n and m identify any two normal modes. Also, for a total of S springs

$$\sum_{s=1}^{S} k_s y_{sn} y_{sm} = 0 \qquad (5\text{-}132)$$

where k_s is the constant for the sth spring and y represents the spring distortion.

When there are many degrees of freedom, this procedure for analyzing free vibration becomes very lengthy. In such cases, it may be preferable to solve Eqs. (5-129) by numerical, trial-and-error procedures, such as the Stodola-Vianello method. In that method, the solution converges first on the highest or lowest mode. Then, the other modes are determined by the same procedure after elimination of one of the equations by use of Eq. (5-131). The procedure requires assumption of a characteristic shape, a set of amplitudes A_{r1}. These are substituted in one of Eqs. (5-129) to obtain a first approximation of ω^2. With this value and with $A_{N1} = 1$, the remaining $N - 1$ equations are solved to obtain a new set of A_{r1}. Then, the procedure is repeated until assumed and final characteristic amplitudes agree.

Because even this procedure is very lengthy for many degrees of freedom, the Rayleigh approximate method may be used to compute the fundamental mode. The frequency obtained by this method, however, may be a little on the high side.

The Rayleigh method also starts with an assumed set of characteristic amplitudes A_{r1} and depends for its success on the small error in natural frequency produced by a relatively larger error in the shape assumption. Next, relative inertia forces acting at each mass are computed: $F_r = W_r A_{r1}/A_{N1}$, where A_{N1} is the assumed displacement at one of the masses. These forces are applied to the system as a static load and displacements B_{r1} due to them calculated. Then, the natural frequency can be obtained from

$$\omega^2 = \frac{g \sum_{r=1}^{N} F_r B_{r1}}{\sum_{r=1}^{N} W_r B_{r1}^2} \tag{5-133}$$

where g is the acceleration due to gravity, 386 in/s². For greater accuracy, the computation can be repeated with B_{r1} as the assumed characteristic amplitudes.

When the Rayleigh method is applied to beams, the characteristic shape assumed initially may be chosen conveniently as the deflection curve for static loading.

The Rayleigh method may be extended to determination of higher modes by the Schmidt orthogonalization procedure, which adjusts assumed deflection curves to satisfy Eq. (5-131). The procedure is to assume a shape, remove components associated with lower modes, then use the Rayleigh method for the residual deflection curve. The computation will converge on the next higher mode. The method is shorter than the Stodola-Vianello procedure when only a few modes are needed.

For example, suppose the characteristic amplitudes A_{r1} for the fundamental mode have been obtained and the natural frequency for the second mode is to be computed. Assume a value for the relative deflection of the rth mass A_{r2}. Then, the shape with the fundamental mode removed will be defined by the displacements

$$a_{r2} = A_{r2} - c_1 A_{r1} \tag{5-134}$$

where c_1 is the participation factor for the first mode.

$$c_1 = \frac{\sum_{r=1}^{N} W_r A_{r2} A_{r1}}{\sum_{r=1}^{N} W_r A_{r1}^2} \tag{5-135}$$

Substitute a_{r2} for B_{r1} in Eq. (5-133) to find the second-mode frequency and, from deflections produced by $F_r = W_r a_{r2}$, an improved shape. (For more rapid convergence, A_{r2} should be selected to make c_1 small.) The procedure should be repeated, starting with the new shape.

For the third mode, assume deflections A_{r3} and remove the first two modes:

$$a_{r3} = A_{r3} - c_1 A_{r1} - c_2 A_{r2} \tag{5-136}$$

The participation factors are determined from

$$c_1 = \frac{\sum_{r=1}^{N} W_r A_{r3} A_{r1}}{\sum_{r=1}^{N} W_r A_{r1}^2} \qquad c_2 = \frac{\sum_{r=1}^{N} W_r A_{r3} A_{r2}}{\sum_{r=1}^{N} W_r A_{r2}^2} \tag{5-137}$$

Use a_{r3} to find an improved shape and the third-mode frequency.

Vibrations of Distributed Masses. For some structures with mass distributed throughout, it sometimes is easier to solve the dynamic equations based on distributed mass than the equations based on equivalent lumped masses. A distributed mass has an infinite number of degrees of freedom and normal modes. Every particle in it can be considered a lumped mass on springs connected to other particles. Usually, however, only the fundamental mode is significant, though sometimes the second and third modes must be taken into account.

For example, suppose a beam weighs w lb/lin ft and has a modulus of elasticity E, psi, and moment of inertia I, in^4. Let y be the deflection at a distance x from one end. Then, the equation of motion is

$$EI \frac{\partial^4 y}{\partial x^4} + \frac{w}{g} \frac{\partial^2 y}{\partial t^2} = 0 \tag{5-138}$$

(This equation ignores the effects of shear and rotational inertia.) The deflection y_n for each mode, to satisfy the equation, must be the product of a harmonic function of time $f_n(t)$ and of the characteristic shape $Y_n(x)$, a function of x with undetermined amplitude. The solution is

$$f_n(t) = c_1 \sin \omega_n t + c_2 \cos \omega_n t \tag{5-139}$$

where ω_n is the natural circular frequency and n indicates the mode, and

$$Y_n(x) = A_n \sin \beta_n x + B_n \cos \beta_n x + C_n \sinh \beta_n x + D_n \cosh \beta_n x \tag{5-140}$$

where

$$\beta_n = \sqrt[4]{\frac{w \omega_n^2}{EIg}} \tag{5-141}$$

For a simple beam, the boundary (support) conditions for all values of time t are $y = 0$ and bending moment $M = EI\, \partial^2 y/\partial x^2 = 0$. Hence, at $x = 0$ and $x = L$, the span length, $Y_n(x) = 0$ and $d^2 Y_n/dx^2 = 0$. These conditions require that

$$B_n = C_n = D_n = 0 \qquad \beta_n = \frac{n\pi}{L}$$

to satisfy Eq. (5-140). Hence, according to Eq. (5-141), the natural circular frequency for a simply supported beam is

$$\omega_n = \frac{n^2 \pi^2}{L^2} \sqrt{\frac{EIg}{w}} \tag{5-142}$$

The characteristic shape is defined by

$$Y_n(x) = \sin \frac{n\pi x}{L} \tag{5-143}$$

The constants c_1 and c_2 in Eq. (5-139) are determined by the initial conditions of the disturbance. Thus, the total deflection, by superposition of modes, is

$$y = \sum_{n=1}^{\infty} A_n(t) \sin \frac{n\pi x}{L} \tag{5-144}$$

where $A_n(t)$ is determined by the load (see Art. 5-83).

Equations (5-139) to (5-141) apply to spans with any type of end restraints. Figure 5-84 shows the characteristic shape and gives constants for determination of natural circular frequency ω and natural period T for the first four modes of cantilever, simply supported, fixed-end, and fixed-hinged beams. To obtain ω, select the appropriate constant from Fig. 5-84 and multiply it by $\sqrt{EI/wL^4}$. To get T, divide the appropriate constant by $\sqrt{EI/wL^4}$.

To determine the characteristic shapes and natural periods for beams with variable cross section and mass, use the Rayleigh method. Convert the beam into a lumped-mass

Fig. 5-84 Coefficients for computing natural circular frequencies and natural periods of vibrations of prismatic beams.

system by dividing the span into elements and assuming the mass of each element to be concentrated at its center. Also, compute all quantities, such as deflection and bending moment, at the center of each element. Start with an assumed characteristic shape and apply Eq. (5-140).

Methods are available for dynamic analysis of continuous beams. (R. Clough and J. Penzien, "Dynamics of Structures," McGraw-Hill Book Company, New York; D. G. Fertis and E. C. Zobel, "Transverse Vibration Theory," The Ronald Press Company, New York.) But even for beams with constant cross section, these procedures are very lengthy. Generally, approximate solutions are preferable.

(J. M. Biggs, "Introduction to Structural Dynamics," McGraw-Hill Book Company, New York; N. M. Newmark and E. Rosenblueth, "Fundamentals of Earthquake Engineering," Prentice-Hall, Englewood Cliffs, N.J.)

5-82. Impact and Sudden Loads.

Under impact, there is an abrupt exchange or absorption of energy and drastic change in velocity. Stresses caused in the colliding members may be several times larger than stresses produced by the same weights applied statically.

An approximation of impact stresses in the elastic range can be made by neglecting the inertia of the body struck and the effect of wave propagation and assuming that the kinetic energy is converted completely into strain energy in that body. Consider a prismatic bar subjected to an axial impact load in tension. The energy absorbed per unit of volume when the bar is stressed to the proportional limit is called the **modulus of resilience**. It is given by $f_y^2/2E$, where f_y is the yield stress and E the modulus of elasticity, both in psi.

Below the proportional limit, the unit stress, psi, due to an axial load U, in-lb, is

$$f = \sqrt{\frac{2UE}{AL}} \qquad (5\text{-}145)$$

where A is the cross-sectional area, in^2, and L the length of bar, in. This equation indicates that, for a given unit stress, energy absorption of a member may be improved by increasing its length or area. Sharp changes in cross section should be avoided, however, because of associated high stress concentrations. Also, uneven distribution of stress in a member due to changes in section should be avoided. For example, if part of a member is given twice the diameter of another part, the stress under axial load in the larger portion is one-fourth that in the smaller. Since the energy absorbed is proportional to the square of the stress, the energy taken per unit of volume by the larger portion is therefore only one-sixteenth that absorbed by the smaller. So despite the increase in volume caused by doubling of the diameter, the larger portion absorbs much less energy than the smaller. Thus, energy absorption would be larger with a uniform stress distribution throughout the length of the member.

Impact on Short Members. If a static axial load W would produce a tensile stress f' in the bar and an elongation e', in, then the axial stress produced in a short member when W falls a distance h, in, is

$$f = f' + f' \sqrt{1 + \frac{2h}{e'}} \qquad (5\text{-}146)$$

if f is within the proportional limit. The elongation due to this impact load is

$$e = e' + e' \sqrt{1 + \frac{2h}{e'}} \qquad (5\text{-}147)$$

These equations indicate that the stress and deformation due to an energy load may be considerably larger than those produced by the same weight applied gradually.

The same equations hold for a beam with constant cross section struck by a weight at midspan, except that f and f' represent stresses at midspan and e and e', midspan deflections.

According to Eqs. (5-146) and (5-147), a sudden load ($h = 0$) causes twice the stress and twice the deflection as the same load applied gradually.

Impact on Long Members. For very long members, the effect of wave propagation should be taken into account. Impact is not transmitted instantly to all parts of the struck body. At first, remote parts remain undisturbed, while particles struck accelerate rapidly to the velocity of the colliding body. The deformations produced move through the struck body in the form of elastic waves. The waves travel with a constant velocity, ft/s,

$$c = 68.1 \sqrt{\frac{E}{\rho}} \qquad (5\text{-}148)$$

where E = modulus of elasticity, psi
ρ = density of the struck body, lb/ft^3

If an impact imparts a velocity v, ft/s, to the particles at one end of a prismatic bar, the stress, psi, at that end is

$$f = E\frac{v}{c} = 0.0147v \sqrt{E\rho} = 0.000216\rho cv \qquad (5\text{-}149)$$

if f is in the elastic range. In a compression wave, the velocity of the particles is in the direction of the wave. In a tension wave, the velocity of the particles is in the direction opposite the wave.

In the plastic range, Eqs. (5-148) and (5-149) hold, but with E as the tangent modulus of elasticity. Hence, c is not a constant and the shape of the stress wave changes as it moves. The elastic portion of the stress wave moves faster than the wave in the plastic range. Where they overlap, the stress and irrecoverable strain are constant.

(The impact theory is based on an assumption difficult to realize in practice—that contact takes place simultaneously over the entire end of the bar.)

At the free end of a bar, a compressive stress wave is reflected as an equal tension wave, and a tension wave as an equal compression wave. The velocity of the particles there equals $2v$.

At a fixed end of a bar, a stress wave is reflected unchanged. The velocity of the particles there is zero, but the stress is doubled, because of the superposition of the two equal stresses on reflection.

For a bar with a fixed end struck at the other end by a moving mass weighing W_m lb, the initial compressive stress, psi, is

$$f_o = 0.0147 v_o \sqrt{E\rho} \tag{5-150}$$

where v_o is the initial velocity of the particles, ft/s, at the impacted end of the bar and E and ρ, the modulus of elasticity, psi, and density, lb/ft³, of the bar. As the velocity of W_m decreases, so does the pressure on the bar. Hence, decreasing compressive stresses follow the wave front. At any time $t < 2L/c$, where L is the length of the bar, in, the stress at the struck end is

$$f = f_o e^{-2\alpha t/\tau} \tag{5-151}$$

where $e = 2.71828$, α is the ratio of W_b, the weight of the bar, to W_m, and $\tau = 2L/c$.

When $t = \tau$, the wave front with stress f_o arrives back at the struck end, assumed still to be in contact with the mass. Since the velocity of the mass cannot change suddenly, the wave will be reflected as from a fixed end. During the second interval, $\tau < t < 2\tau$, the compressive stress is the sum of two waves moving away from the struck end and one moving toward this end.

Maximum stress from impact occurs at the fixed end. For α greater than 0.2, this stress is

$$f = 2f_o(1 + e^{-2\alpha}) \tag{5-152}$$

For smaller values of α, it is given approximately by

$$f = f_o \left(1 + \sqrt{\frac{1}{\alpha}} \right) \tag{5-153}$$

Duration of impact, time it takes for the impact stress at the struck end to drop to zero, is approximately

$$T = \frac{\pi L}{c \sqrt{\alpha}} \tag{5-154}$$

for small values of α.

When W_m is the weight of a falling body, velocity at impact is $\sqrt{2gh}$, when it falls a distance h, in. Substitution in Eq. (5-150) yields $f_o = \sqrt{2EhW_b/AL}$, since $W_b = \rho AL$ is the weight of the bar. Putting $W_b = \alpha W_m$; $W_m/A = f'$, the stress produced by W_m when applied gradually, and $E = f'L/e'$, where e' is the elongation for the static load, gives $f_o = f' \sqrt{2h\alpha/e'}$. Then, for values of α smaller than 0.2, the maximum stress, from Eq. (5-153), is

$$f = f' \left(\sqrt{\frac{2h\alpha}{e'}} + \sqrt{\frac{2h}{e'}} \right) \tag{5-155}$$

For larger values of α, the stress wave due to gravity acting on W_m during impact should be added to Eq. (5-152). Thus, for α larger than 0.2,

$$f = 2f'(1 - e^{-2\alpha}) + 2f'\sqrt{\frac{2h\alpha}{e'}}(1 + e^{-2\alpha}) \tag{5-156}$$

Equations (5-155) and (5-156) correspond to Eq. (5-146), which was developed without wave effects being taken into account. For a sudden load, $h = 0$, Eq. (5-156) gives for the maximum stress $2f'(1 - e^{-2\alpha})$, not quite double the static stress, the result indicated by Eq. (5-146). (See also Art. 5-83.)

(S. Timoshenko and J. N. Goodier, "Theory of Elasticity," McGraw-Hill Book Company, New York; S. Timoshenko and D. H. Young, "Engineering Mechanics," McGraw-Hill Book Company, New York; J. H. Faupel, "Engineering Design," John Wiley & Sons, Inc., New York.)

5-83. Dynamic Analysis of Simple Structures.

Articles 5-80 to 5-82 present a theoretic basis for analysis of structures under dynamic loads. As noted in Art. 5-81, an approximate solution based on an idealized representation of an actual member or structure is advisable for dynamic analysis and design. Generally, the actual structure may be conveniently represented by a system of masses and massless springs, with additional resistances to account for damping. In simple cases, the masses may be set equal to the actual masses; otherwise, equivalent masses may be substituted for the actual masses (Art. 5-85). The spring constants are the ratios of forces to deflections (see Art. 5-81).

Usually, for structural purposes, the data sought are the maximum stresses in the springs and their maximum displacements and the time of occurrence of the maximums. This time is generally computed in terms of the natural period of vibration of the member or structure, or in terms of the duration of the load. Maximum displacement may be calculated in terms of the deflection that would result if the load were applied gradually.

Fig. 5-85 One-degree system acted on by a force varying with time.

The term D by which the static deflection e', spring forces, and stresses are multiplied to obtain the dynamic effects is called the **dynamic load factor.** Thus, the dynamic displacement is

$$y = De' \tag{5-157}$$

And the maximum displacement y_m is determined by the maximum dynamic load factor D_m, which occurs at time t_m.

One-Degree Systems. Consider the one-degree-of-freedom system in Fig. 5-85a. It may represent a weightless beam with a mass weighing W lb applied at midspan and subjected to a varying force $F_o f(t)$, or a rigid frame with a mass weighing W lb at girder level and subjected to this force. The force is represented by an arbitrarily chosen constant force F_o times $f(t)$, a function of time.

If the system is not damped, the equation of motion in the elastic range is

$$\frac{W}{g}\frac{d^2y}{dt^2} + ky = F_o f(t) \tag{5-158}$$

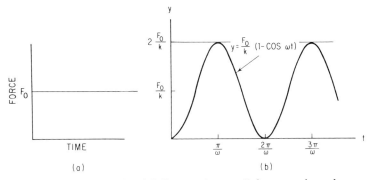

Fig. 5-86 Harmonic motion. (a) Constant force applied to an undamped one-degree system, such as the one in Fig. 5-85a. (b) Displacements vary with time like a cosine curve.

where k is the spring constant and g the acceleration due to gravity, 386 in/s². The solution consists of two parts. The first, called the complementary solution, is obtained by setting $f(t) = 0$. This solution is given by Eq. (5-122). To it must be added the second part, the particular solution, which satisfies Eq. (5-158).

The general solution of Eq. (5-158), arrived at by treating an element of the force-time curve (Fig. 5-85b) as an impulse, is

$$y = y_o \cos \omega t + \frac{v_o}{\omega} \sin \omega t + e'\omega \int_0^t f(\tau) \sin \omega(t - \tau) \, d\tau \qquad (5\text{-}159)$$

where y = displacement of mass from equilibrium position, in
y_o = initial displacement of mass ($t = 0$), in
$\omega = \sqrt{kg/W}$ = natural circular frequency of free vibration
k = spring constant = force producing unit deflection, lb/in
v_o = initial velocity of mass, in/s
$e' = F_o/k$ = displacement under static load, in

A closed solution is possible if the integral can be evaluated.

Assume, for example, the mass is subjected to a suddenly applied force F_o that remains constant (Fig. 5-86a). If y_o and v_o are initially zero, the displacement y of the mass at any time t can be obtained from the integral in Eq. (5-159) by setting $f(\tau) = 1$:

$$y = e'\omega \int_0^t \sin \omega(t - \tau) \, d\tau = e'(1 - \cos \omega t) \qquad (5\text{-}160)$$

This equation indicates that the dynamic load factor $D = 1 - \cos \omega t$. It has a maximum value $D_m = 2$ when $t = \pi/\omega$. Figure 5-86b shows the variation of displacement with time. For values of D_m for some other types of forces, see Table 5-8.

Multidegree Systems. A multidegree lumped-mass system may be analyzed by the modal method after the natural frequencies of the normal modes have been determined (Art. 5-81). This method is restricted to linearly elastic systems in which the forces applied to the masses have the same variation with time. For other cases, numerical analysis must be used.

In the modal method, each normal mode is treated as an independent one-degree system. For each degree of the system, there is one normal mode. A natural frequency and a characteristic shape are associated with each mode. In each mode, the ratio of the displacements of any two masses is constant with time. These ratios define the characteristic shape. The modal equation of motion for each mode is

$$\frac{d^2 A_n}{dt^2} + \omega_n^2 A_n = \frac{gf(t) \sum_{r=1}^{j} F_r \phi_{rn}}{\sum_{r=1}^{j} W_r \phi_{rn}^2} \qquad (5\text{-}161)$$

where A_n = displacement in the nth mode of an arbitararily selected mass
ω_n = natural frequency of the nth mode
$F_r f(t)$ = varying force applied to the rth mass
W_r = weight of the rth mass
j = number of masses in the system
ϕ_{rn} = ratio of the displacement in the nth mode of the rth mass to A_n
g = acceleration due to gravity

We define the modal static deflection as

$$A_n' = \frac{g \sum_{r=1}^{j} F_r \phi_{rn}}{\omega_n^2 \sum_{r=1}^{j} W_r \phi_{rn}^2} \qquad (5\text{-}162)$$

Then, the response for each mode is given by

$$A_n = D_n A_n' \qquad (5\text{-}163)$$

where D_n = dynamic load factor.

Since D_n depends only on ω_n and the variation of force with time $f(t)$, solutions for D_n obtained for one-degree systems also apply to multidegree systems. Thus, the dynamic load factor for each mode may be obtained from Table 5-8. The total deflection at any point is the sum of the displacements for each mode, $\Sigma A_n \phi_{rn}$, at that point.

Beams. The response of beams to dynamic forces can be determined in a similar way. The modal static deflection is defined by

$$A_n' = \frac{\int_0^L p(x) \phi_n(x) \, dx}{\omega_n^2 \dfrac{w}{g} \int_0^L \phi_n^2(x) \, dx} \qquad (5\text{-}164)$$

where $p(x)$ = load distribution on the span [$p(x)f(t)$ is the varying force]
$\phi_n(x)$ = characteristic shape of the nth mode (see Art. 5-81)
L = span length
w = uniformly distributed weight on the span

The response of the beam then is given by Eq. (5-163), and the dynamic deflection is the sum of the modal components, $\Sigma A_n \phi_n(x)$.

Nonlinear Responses. When the structure does not react linearly to loads, the equations of motion can be solved by numerical analysis if resistance is a unique function of displacement. Sometimes, the behavior of the structure can be represented by an idealized resistance-displacement diagram that makes possible a solution in closed form. Figure 5-87a shows such a diagram.

Elastic-Plastic Responses. Resistance is assumed linear ($R = ky$) in Fig. 5-87a until a maximum R_m is reached. After that, R remains equal to R_m for increases in y substantially larger than the displacement y_e at the elastic limit. Thus, some portions of the structure deform into the plastic range. Figure 5-87a, therefore, may be used for

TABLE 5-8 Maximum Response and Occurrence Time for Short-Time Loads*

Rectangular Pulses (Fig. 5-88a)

t_d/T	Elastic response		$y_m/y_e = 1.5$		$y_m/y_e = 2.0$		$y_m/y_e = 5.0$		$y_m/y_e = 10.0$	
	D_m	t_m/T	R_m/F_o	t_m/T	R_m/F_o	t_m/T	R_m/F_o	t_m/T	R_m/F_o	t_m/T
0.1	0.6	0.3	0.5	0.3	0.4	0.4	0.2	0.5	0.2	0.7
0.2	1.2	0.4	0.8	0.4	0.6	0.4	0.4	0.5	0.3	0.8
0.3	1.6	0.4	1.2	0.4	1.0	0.5	0.5	0.6	0.4	0.8
0.4	1.9	0.5	1.4	0.5	1.1	0.5	0.7	0.7	0.5	0.9
0.5	2.0	0.5	1.5	0.6	1.2	0.6	0.8	0.8	0.6	0.9
1.0	2.0	0.5	1.5	0.6	1.3	0.8	1.1	1.1	0.9	1.3
10.0	2.0	0.5	1.5	0.6	1.3	0.8	1.1	1.7	1.1	3.0

Triangular Loads (Fig. 5-88b)

t_d/T	Elastic response		$y_m/y_e = 1.5$		$y_m/y_e = 2.0$		$y_m/y_e = 5.0$		$y_m/y_e = 10.0$	
	D_m	t_m/T	R_m/F_o	t_m/T	R_m/F_o	t_m/T	R_m/F_o	t_m/T	R_m/F_o	t_m/T
0.1	0.3	0.3	0.2	0.3	0.2	0.4	0.1	0.5		
0.2	0.6	0.3	0.4	0.4	0.3	0.4	0.2	0.6	0.1	0.8
0.3	0.8	0.4	0.6	0.4	0.5	0.5	0.3	0.7	0.2	0.9
0.4	1.0	0.4	0.7	0.4	0.6	0.5	0.4	0.7	0.3	0.9
0.5	1.2	0.4	0.9	0.4	0.7	0.6	0.4	0.7	0.3	0.9
1.0	1.5	0.5	1.1	0.5	1.0	0.6	0.7	0.8	0.5	1.0
5.0	2.0	0.5	1.4	0.6	1.3	0.6	1.0	1.4	0.9	2.0
10.0	2.0	0.5	1.4	0.7	1.3	0.7	1.1	1.5	1.0	2.5

Constant Force with Short Rise Times (Fig. 5-88c)

t_r/T	Elastic response		$y_m/y_e = 1.5$		$y_m/y_e = 2.0$		$y_m/y_e = 5.0$		$y_m/y_e = 10.0$	
	D_m	t_m/T	R_m/F_o	t_m/T	R_m/F_o	t_m/T	R_m/F_o	t_m/T	R_m/F_o	t_m/T
0.1	2.0	0.5	1.5	0.8	1.3	1.0	1.1	1.7	1.1	3.5
0.5	1.7	0.8	1.3	1.0	1.2	1.1	1.1	2.6	1.0	6.0
1.0	1.0	1.0	1.0	1.0	1.0	1.0	1.0	1.0	1.0	1.0
1.5	1.2	1.4	1.0	2.2	1.0	3.0	1.0	3.0		
2.0	1.0	2.0	1.0	2.0	1.0	2.0	1.0	2.0		

Triangular Pulses (Fig. 5-88d)

t_d/T	Elastic response		$y_m/y_e = 1.5$		$y_m/y_e = 2.0$		$y_m/y_e = 5.0$		$y_m/y_e = 10.0$	
	D_m	t_m/T	R_m/F_o	t_m/T	R_m/F_o	t_m/F	R_m/F_o	t_m/F	R_m/F_o	t_m/F
0.1	0.3	0.3	0.2	0.3	0.2	0.4				
0.2	0.6	0.4	0.5	0.4	0.4	0.4	0.2	0.6	0.2	0.6
0.3	1.0	0.4	0.7	0.4	0.5	0.4	0.3	0.6	0.2	0.9
0.4	1.2	0.5	0.8	0.5	0.7	0.5	0.4	0.7	0.3	0.9
0.5	1.3	0.5	0.9	0.6	0.8	0.6	0.5	0.7	0.3	1.0
1.0	1.6	0.7	1.1	0.7	1.0	0.8	0.7	1.0	0.5	1.3
5.0	1.1	2.7	1.0	3.0	0.9	3.1	0.8	3.5	0.8	3.9
10.0	1.0	5.0	1.0	5.5	0.9	6.0	0.9	6.3	0.8	7.0

* Source: U.S. Army Corps of Engineers, "Design of Structures to Resist the Effects of Atomic Weapons," Manual EM 1110-345-415.

Fig. 5-87 Response in the plastic range of a one-degree system with resistance characteristics indicated in (a) to a constant force (b) is shown in (c).

ductile structures only rarely subjected to severe dynamic loads. When this diagram can be used for designing such structures, more economical designs can be produced than for structures limited to the elastic range, because of the high energy-absorption capacity of structures in the plastic range.

For a one-degree system, Eq. (5-158) can be used as the equation of motion for the initial sloping part of the diagram (elastic range). For the second stage, $y_e < y < y_m$, where y_m is the maximum displacement, the equation is

$$\frac{W}{g}\frac{d^2y}{dt^2} + R_m = F_o f(t) \qquad (5\text{-}165)$$

For the unloading stage, $y < y_m$, the equation is

$$\frac{W}{g}\frac{d^2y}{dt^2} + R_m - k(y_m - y) = F_o f(t) \qquad (5\text{-}166)$$

Suppose, for example, the one-degree undamped system in Fig. 5-85a behaves in accordance with the bilinear resistance function of Fig. 5-87a and is subjected to a suddenly applied constant load (Fig. 5-87b). With zero initial displacement and velocity, the response in the first stage ($y < y_e$), according to Eq. (5-160), is

$$y = e'(1 - \cos \omega t_1) \qquad (5\text{-}167)$$

$$\frac{dy}{dt} = e'\omega \sin \omega t_1 \qquad (5\text{-}168)$$

Equation (5-160) also indicates that y_e will be reached at a time t_e such that $\cos \omega t_e = 1 - y_e/e'$.

For convenience, let $t_2 = t - t_e$ be the time in the second stage; thus, $t_2 = 0$ at the start of that stage. Since the condition of the system at that time is the same as at the end of the first stage, the initial displacement is y_e and the initial velocity $e'\omega \sin \omega t_e$.

The equation of motion of the second stage is

$$\frac{W}{g}\frac{d^2y}{dt^2} + R_m = F_o \qquad (5\text{-}169)$$

The solution, taking into account initial conditions after integrating, for $y_e < y < y_m$ is

$$y = \frac{g}{2W}(F_o - R_m)t_2^2 + e'\omega t_2 \sin \omega t_e + y_e \qquad (5\text{-}170)$$

Maximum displacement occurs at the time

$$t_m = \frac{W\omega e'}{g(R_m - F_o)} \sin \omega t_e \qquad (5\text{-}171)$$

and can be obtained by substituting t_m in Eq. (5-170).

The third stage, unloading after y_m has been reached, can be determined from Eq. (5-166) and conditions at the end of the second stage. The response, however, is more easily found by noting that the third stage consists of an elastic, harmonic residual vibration. In this stage the amplitude of vibration is $(R_m - F_o)/k$, since this is the distance between the neutral position and maximum displacement, and in the neutral position the spring force equals F_o. Hence, the response can be obtained directly from Eq. (5-160) by substituting $y_m - (R_m - F_o)/k$ for e', because the neutral position, located at $y = y_m - (R_m - F_o)/k$, occurs when $\omega t_3 = \pi/2$, where $t_3 = t - t_e - t_m$. The solution is

$$y = y_m - \frac{R_m - F_o}{k} + \frac{R_m - F_o}{k} \cos \omega t_3 \qquad (5\text{-}172)$$

Response in the three stages is shown in Fig. 5-87c. In that diagram, however, to represent a typical case, the coordinates have been made nondimensional by expressing y in terms of y_e and the time in terms of T, the natural period of vibration.

Figure 5-88 shows some idealized force-time functions. When actual conditions approximate these, maximum displacement and time of its occurrence may be obtained from Table 5-8, which was calculated on the assumption of a bilinear resistance-displacement relationship (Fig. 5-87a). For each type of load, as load duration or time of rise varies, Table 5-8 lists the dynamic load factors in the elastic range and the ratio R_m/F_o

= y_e/e' in the plastic range. Values in the plastic range are given for several values of y_m/y_e.

(J. M. Biggs, "Introduction to Structural Dynamics," and R. Clough and J. Penzien, "Dynamics of Structures," McGraw-Hill Book Company, New York; D. G. Fertis and E. C. Zobel, "Transverse Vibration Theory," The Ronald Press Company, New York; N. M. Newmark and E. Rosenblueth, "Fundamentals of Earthquake Engineering," Prentice-Hall, Englewood Cliffs, N.J.)

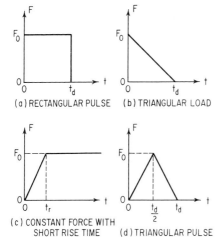

Fig. 5-88 Some idealized force-time relationships.

5-84. Resonance and Damping.

Damping in structures, resulting from friction and other causes, resists motion imposed by dynamic loads. Generally, the effect is to decrease the amplitude and lengthen the period of vibrations. If damping is large enough, vibration may be eliminated.

When maximum stress and displacement are the prime concern, damping may not be of great significance for short-time loads. These maximums usually occur under such loads at the first peak of response, and damping, unless unusually large, has little effect in a short period of time. But under conditions close to resonance, damping has considerable effect.

Resonance is the condition of a vibrating system under a varying load such that the amplitude of successive vibrations increases. Unless limited by damping or changes in the condition of the system, amplitudes may become very large.

Two forms of damping generally are assumed in structural analysis, viscous or constant (Coulomb). For viscous damping, the damping force is taken proportional to the velocity but opposite in direction. For Coulomb damping, the damping force is assumed constant and opposed in direction to the velocity.

Viscous Damping. For a one-degree system (Arts. 5-81 to 5-83), the equation of motion for a mass weighing W lb and subjected to a force F varying with time but opposed by viscous damping is

$$\frac{W}{g}\frac{d^2y}{dt^2} + ky = F - c\frac{dy}{dt} \qquad (5\text{-}173)$$

where y = displacement of the mass from equilibrium position, in
k = spring constant, lb/in
t = time, s
c = coefficient of viscous damping
g = acceleration due to gravity = 386 in/s²

Let us set $\beta = cg/2W$ and consider those cases in which $\beta < \omega$, the natural circular frequency [Eq. (5-123)], to eliminate unusually high damping (overdamping). Then, for initial displacement y_o and velocity v_o, the solution of Eq. (5-173) with $F = 0$ is

$$y = e^{-\beta t}\left(\frac{v_o + \beta y_o}{\omega_d}\sin \omega_d t + y_o \cos \omega_d t\right) \qquad (5\text{-}174)$$

where $\omega_d = \sqrt{\omega^2 - \beta^2}$ and $e = 2.71828$. Equation (5-174) represents a decaying harmonic motion with β controlling the rate of decay and ω_d the natural frequency of the damped system.

When $\beta = \omega$

$$y = e^{-\omega t}[v_o t + (1 + \omega t)y_o] \qquad (5\text{-}175)$$

which indicates that the motion is not vibratory. Damping producing this condition is called critical, and, from the definition of β, the critical coefficient is

$$c_d = \frac{2W\beta}{g} = \frac{2W\omega}{g} = 2\sqrt{\frac{kW}{g}} \qquad (5\text{-}176)$$

Damping sometimes is expressed as a percent of critical (β as a percent of ω).

For small amounts of viscous damping, the damped natural frequency is approximately equal to the undamped natural frequency minus $\frac{1}{2}\beta^2/\omega$. For example, for 10% critical damping ($\beta = 0.1\omega$), $\omega_d = \omega[1 - \frac{1}{2}(0.1)^2] = 0.995\omega$. Hence, the decrease in natural frequency due to damping generally can be ignored.

Damping sometimes is measured by **logarithmic decrement,** the logarithm of the ratio of two consecutive peak amplitudes during free vibration.

$$\text{Logarithmic decrement} = \frac{2\pi\beta}{\omega} \qquad (5\text{-}177)$$

For example, for 10% critical damping, the logarithmic decrement equals 0.2π. Hence, the ratio of a peak to the following peak amplitude is $e^{0.2\pi} = 1.87$.

The complete solution of Eq. (5-173) with initial displacement y_o and velocity v_o is

$$y = e^{-\beta t}\left(\frac{v_o + \beta y_o}{\omega_d}\sin \omega_d t + y_o \cos \omega_d t\right)$$
$$+ e'\frac{\omega^2}{\omega_d}\int_0^t f(\tau)e^{-\beta(t-\tau)}\sin \omega_d(t - \tau)\, d\tau \qquad (5\text{-}178)$$

where e' is the deflection that the applied force would produce under static loading. Equation (5-178) is identical to Eq. (5-159) when $\beta = 0$.

Unbalanced rotating parts of machines produce pulsating forces that may be represented by functions of the form $F_o \sin \alpha t$. If such a force is applied to an undamped one-degree system, Eq. (5-159) indicates that if the system starts at rest the response will be

$$y = \frac{F_o g}{W}\left(\frac{1/\omega^2}{1 - \alpha^2/\omega^2}\right)\left(\sin \alpha t - \frac{\alpha}{\omega}\sin \omega t\right) \qquad (5\text{-}179)$$

And since the static deflection would be $F_o/k = F_o g/W\omega^2$, the dynamic load factor is

$$D = \frac{1}{1 - \alpha^2/\omega^2}\left(\sin \alpha t - \frac{\alpha}{\omega}\sin \omega t\right) \qquad (5\text{-}180)$$

If α is small relative to ω, maximum D is nearly unity; thus, the system is practically statically loaded. If α is very large compared with ω, D is very small; thus, the mass cannot follow the rapid fluctuations in load and remains practically stationary. Therefore, when α differs appreciably from ω, the effects of unbalanced rotating parts are not too serious. But if $\alpha = \omega$, resonance occurs; D increases with time. Hence, to prevent structural damage, measures must be taken to correct the unbalanced parts to change α, or to change the natural frequency of the vibrating mass, or damping must be provided.

The response as given by Eq. (5-179) consists of two parts, the free vibration and the forced part. When damping is present, the free vibration is of the form of Eq. (5-174) and is rapidly damped out. Hence, the free part is called the **transient response,** and the forced part, the **steady-state response**. The maximum value of the dynamic load factor for the steady-state response D_m is called the **dynamic magnification factor**. It is given by

$$D_m = \frac{1}{\sqrt{(1 - \alpha^2/\omega^2)^2 + (2\beta\alpha/\omega^2)^2}} \qquad (5\text{-}181)$$

With damping, then, the peak values of D_m occur when $\alpha = \omega\sqrt{1 - \beta^2/\omega^2}$ and are approximately equal to $\omega/2\beta$. For example, for 10% critical damping,

$$D_m = \frac{\omega}{0.2\omega} = 5$$

So even small amounts of damping significantly limit the response at resonance.

Coulomb Damping. For a one-degree system with Coulomb damping, the equation of motion for free vibration is

$$\frac{W}{g}\frac{d^2y}{dt^2} + ky = \pm F_f \qquad (5\text{-}182)$$

where F_f is the constant friction force and the positive sign applies when the velocity is negative. If initial displacement is y_o and initial velocity is zero, the response in the first half cycle, with negative velocity, is

$$y = \left(y_o - \frac{F_f}{k}\right)\cos \omega t + \frac{F_f}{k} \qquad (5\text{-}183)$$

equivalent to a system with a suddenly applied constant force. For the second half cycle, with positive velocity, the response is

$$y = \left(-y_o + 3\frac{F_f}{k}\right)\cos \omega \left(t - \frac{\pi}{\omega}\right) - \frac{F_f}{k} \qquad (5\text{-}184)$$

If the solution is continued with the sign of F_f changing in each half cycle, the results will indicate that the amplitude of positive peaks is given by $y_o - 4nF_f/k$, where n is the number of complete cycles, and the response will be completely damped out when $t = ky_oT/4F_f$, where T is the natural period of vibration, or $2\pi/\omega$.

Analysis of the steady-state response with Coulomb damping is complicated by the possibility of frequent cessation of motion.

(S. Timoshenko, D. H. Young, and W. Weaver, "Vibration Problems in Engineering," 4th ed., John Wiley & Sons, Inc., New York; D. D. Barkan, "Dynamics of Bases and Foundations," McGraw-Hill Book Company; W. C. Hurty and M. F. Rubinstein, "Dynamics of Structures," Prentice-Hall, Englewood Cliffs, N.J.)

5-85. Approximate Design for Dynamic Loading.

Complex analysis and design methods seldom are justified for structures subject to dynamic loading because of lack of sufficient information on loading, damping, resistance to deformation, and other factors. In general, it is advisable to represent the actual structure and loading by idealized systems that permit a solution in closed form (see Arts. 5-80 to 5-84).

Whenever possible, represent the actual structure by a one-degree system consisting of an equivalent mass with massless spring. For structures with distributed mass, simplify the analysis in the elastic range by computing the response only for one or a few of the normal modes. In the plastic range, treat each stage—elastic, elastic-plastic, and plastic—as completely independent; for example, a fixed-end beam may be treated, when in the elastic-plastic stage, as a simply supported beam.

Choose the parameters of the equivalent system to make the deflection at a critical point, such as the location of the concentrated mass, the same as it would be in the actual structure. Stresses in the actual structure should be computed from the deflections in the equivalent system.

Compute an assumed shape factor ϕ for the system from the shape taken by the actual structure under static application of the loads. For example, for a simple beam in the elastic range with concentrated load at midspan, ϕ may be chosen, for $x < L/2$, as $(Cx/L^3)(3L^2 - 4x^2)$, the shape under static loading, and C may be set equal to 1 to make ϕ equal to 1 when $x = L/2$. For plastic conditions (hinge at midspan), ϕ may be taken as Cx/L, and C set equal to 2, to make $\phi = 1$ when $x = L/2$.

For a structure with concentrated forces, let W_r be the weight of the rth mass, ϕ_r the value of ϕ for a specific mode at the location of that mass, and F_r the dynamic force acting on W_r. Then, the equivalent weight of the idealized system is

$$W_e = \sum_{r=1}^{j} W_r \phi_r^2 \tag{5-185}$$

where j is the number of masses. The equivalent force is

$$F_e = \sum_{r=1}^{j} F_r \phi_r \tag{5-186}$$

For a structure with continuous mass, the equivalent weight is

$$W_e = \int w \phi^2 \, dx \tag{5-187}$$

where w is the weight in lb/lin ft. The equivalent force is

$$F_e = \int q \phi \, dx \tag{5-188}$$

for a distributed load q, lb/lin ft.

The resistance of a member or structure is the internal force tending to restore it to its unloaded static position. For most structures, a bilinear resistance function, with slope k up to the elastic limit and zero slope in the plastic range (Fig. 5-87a), may be assumed. For a given distribution of dynamic load, maximum resistance of the idealized system may be taken as the total load with that distribution that the structure can support statically. Similarly, stiffness is numerically equal to the total load with the given distribution that would cause a unit deflection at the point where the deflections in the actual structure and idealized system are equal. Hence, the equivalent resistance and stiffness are in the same ratio to the actual as the equivalent forces to the actual forces.

Let k be the actual spring constant, g acceleration due to gravity, 386 in/s^2, and

$$W' = \frac{W_e}{F_e} \Sigma F \tag{5-189}$$

where ΣF represents the actual total load. Then, the equation of motion of an equivalent one-degree system is

$$\frac{d^2 y}{dt^2} + \omega^2 y = g \frac{\Sigma F}{W'} \tag{5-190}$$

and the natural circular frequency is

$$\omega = \sqrt{\frac{kg}{W'}} \tag{5-191}$$

The natural period of vibration equals $2\pi/\omega$. Equations (5-190) and (5-191) have the same form as Eqs. (5-121), (5-123), and (5-158). Consequently, the response can be computed as indicated in Arts. 5-81 to 5-83.

Whenever possible, select a load-time function for ΣF to permit use of a known solution, such as those in Table 5-8.

For preliminary design of a one-degree system loaded into the plastic range by a suddenly applied force that remains substantially constant up to the time of maximum response, the following approximation may be used for that response:

$$y_m = \frac{y_e}{2(1 - F_o/R_m)} \qquad (5\text{-}192)$$

where y_e is the displacement at the elastic limit, F_o the average value of the force, and R_m the maximum resistance of the system. This equation indicates that for purely elastic response, R_m must be twice F_o; whereas, if y_m is permitted to be large, R_m may be made nearly equal to F_o, with greater economy of material.

For preliminary design of a one-degree system subjected to a sudden load with duration t_d less than 20% of the natural period of the system, the following approximation can be used for the maximum response:

$$y_m = \frac{1}{2} y_e \left[\left(\frac{F_o}{R_m} \omega t_d \right)^2 + 1 \right] \qquad (5\text{-}193)$$

where F_o is the maximum value of the load and ω the natural frequency. This equation also indicates that the larger y_m is permitted to be, the smaller R_m need be.

For a beam, the spring force of the equivalent system is not the actual force, or reaction, at the supports. The real reactions should be determined from the dynamic equilibrium of the complete beam. This calculation should include the inertia force, with distribution identical with the assumed deflected shape of the beam. For example, for a simply supported beam with uniform load, the dynamic reaction in the elastic range is $0.39R + 0.11F$, where R is the resistance, which varies with time, and $F = qL$ is the load. For a concentrated load F at midspan, the dynamic reaction is $0.78R - 0.28F$. And for concentrated loads $F/2$ at each third point, it is $0.62R - 0.12F$. (Note that the sum of the coefficients equals 0.50, since the dynamic-reaction equations must hold for static loading, when $R = F$.) These expressions also can be used for fixed-end beams without significant error. If high accuracy is not required, they also can be used for the plastic range.

5-86. Approximate Earthquake Analysis.

Structures often are designed to resist the dynamic forces of earthquakes by use of equivalent static loads. Many building codes specify such a method. (See also Arts. 3-7 to 3-11.)

For example, in the Uniform Building Code (1979), computation of static, horizontal forces for aseismic design requires that the base shear, or total horizontal force developed at grade, kips, be determined first from

$$V = ZIKCSW \qquad (5\text{-}194)$$

Z provides for variation in design forces with changes in the probability of seismic intensity with zones. $Z = \frac{1}{8}$ for zone 0, $Z = \frac{3}{16}$ for zone 1, $Z = \frac{3}{8}$ for zone 2, $Z = \frac{3}{4}$ for zone 3, and $Z = 1$ for zone 4 (Fig. 5-89).

W, in general, is the total dead load, kips. For storage and warehouse occupancies, however, W should be taken as the dead load plus 25% of the live load. There are other cases, however, where designers would find it prudent and realistic to include a portion of the live load in W.

I is a factor that depends on the type of building occupancy. For buildings in which 300 or more persons may assemble in one room, $I = 1.25$. For essential buildings, such

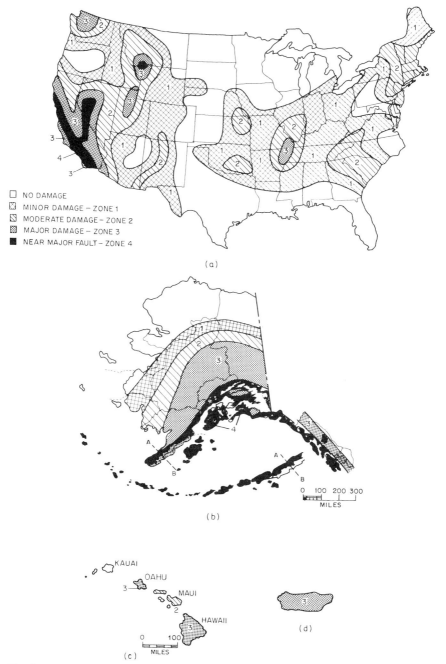

Fig. 5-89 Zones of probable seismic intensity in (*a*) mainland United States; (*b*) Alaska; (*c*) Hawaii; (*d*) Puerto Rico.

as hospitals, fire and police stations, and emergency centers, $I = 1.5$. For other occupancies, I may be taken as unity.

C is called the seismic coefficient.

$$C = \frac{1}{15\sqrt{T}} \quad (5\text{-}195)$$

T is the fundamental elastic period of vibration of the structure, s. For a building as a whole, C need not exceed 0.12.

T may be computed from

$$T = \frac{0.05 h_n}{\sqrt{D}} \quad (5\text{-}196)$$

where h_n = height, ft, of uppermost level above base
D = plan dimension, ft, in direction parallel to applied forces

More appropriate values of T may be used if they can be substantiated by technical data. For example, T may be determined from Eq. (5-133). If the lateral-force-resisting system is a moment-resisting space frame designed to withstand all the lateral forces, and is not enclosed or adjoined by more rigid elements, T may be taken as $0.1N$, where N = number of stories above grade.

Space frames, as defined in the Uniform Building Code, are three-dimensional structural systems composed of interconnected members, other than bearing walls, and laterally supported to function as a complete, self-contained unit, with or without the aid of horizontal diaphragms or floor bracing systems. Such vertical, load-carrying frames may be considered moment-resisting if the members and joints are capable of resisting design lateral forces by bending moments.

S is a coefficient equal to or greater than unity that accounts for forces resulting from soil-structure interaction. Some building codes may permit S to be chosen as follows:

S = 1 for rock or similar material characterized by a shear wave velocity exceeding 2500 ft/s or for stiff soil, such as stable deposits of sand, gravel, or stiff clay, overlying rock at a depth of less than 200 ft

  = 1.2 for stable deposits of sand, gravel, or stiff clay overlying rock at a depth greater than 200 ft

  = 1.5 for sands or weak to medium-stiff clays 30 ft or more deep (the clay strata may incorporate layers of sand or gravel)

For sites where soil characteristics are not known in detail or do not fit any of the preceding descriptions, S may be taken as 1.5.

In any case, the product CS need not be taken more than 0.14 in Eq. (5-194).

The value of S depends on the characteristic site period T_s, which may be determined from geotechnical tests in accordance with the engineering standard of the International Conference of Building Officials, Inc., "Determination of the Characteristic Site Period, T_s." This period should be taken as the value within the range of probable site periods that is nearest to the fundamental elastic period of vibration T of the building, but not less than 0.3 s.

For $T/T_s \leq 1$,

$$S = 1 + \frac{T}{T_s} - 0.5\left(\frac{T}{T_s}\right)^2 \geq 1 \quad (5\text{-}197)$$

For $T/T_s > 1$,

$$S = 1.2 + 0.6\frac{T}{T_s} - 0.3\left(\frac{T}{T_s}\right)^2 \geq 1 \quad (5\text{-}198)$$

For $T_s \geq 2.5$, S should be computed with $T_s = 2.5$.

K takes into account the potential for inelastic energy absorption in moment-resisting frames. It also recognizes the redundancy of framing, or second line of defense, present in most complete frames, whether or not designed to resist lateral loads. Buildings that do not possess at least a complete vertical, load-carrying space frame are penalized by assignment of a high K. Table 5-9 lists suggested values for K.

TABLE 5-9 K for Aseismic Design of Buildings and Other Structures*

Framing system	K
Building framing systems other than those listed below	1.00
Building with a box system	1.33
Buildings with a ductile, moment-resisting space frame and shear walls or braced frames designed so that:	
Frames and shear walls resist the total lateral force in proportion to their rigidities, taking into account interaction of shear walls and frames	
Shear walls or braced frames acting independently of the moment-resisting space frame resist the total required lateral force	
The ductile, moment-resisting space frame can resist at least 25% of the required lateral force	0.80
Buildings with a ductile, moment-resisting space frame with capacity to resist the total required lateral force	0.67
Elevated tanks plus full contents, on four or more cross-braced legs and not supported by a building†	2.50
Structures other than buildings and those listed in Table 5-10	2.00
Buildings with vertical resisting elements of unreinforced masonry‡	4.00

*Where prescribed wind loads produce higher stresses, those loads should be used instead of the seismic loads.

†The minimum value of KC in Eq. (5-194) is 0.12, and the maximum value need not exceed 0.25. Framing should be designed for an accidental horizontal torsion due to the lateral shear acting with an eccentricity of 5% of the maximum dimension in plan at the level of the shear. For tanks on the ground or connected to or part of a building, see Table 5-10.

‡Such construction should not be used in zones 3 and 4.

Box systems, for which $K = 1.33$ or 4.00, employ shear walls or braced frames subjected primarily to axial stresses to resist the required lateral forces. Such systems are considered to lack a complete vertical, load-carrying space frame.

Note, however, that a building with shear walls or braced frames may be assigned $K = 0.80$ if it also has a ductile moment-resisting space frame capable of resisting at least 25% of the lateral force. In that case, the shear walls, acting independently of the space frame, must be able to withstand the total required lateral force.

A **ductile, moment-resisting space frame** of structural steel is a space frame capable of developing plastic hinges at connections of beams and girders to columns. In seismic zones 3 and 4, each such connection should be capable of developing the full plastic capacity of the beam or girder at the connection, unless analysis or tests indicate that a smaller capacity would be adequate. To ensure adequate ductility where beam flange area is reduced, for example, by bolt holes, plastic hinges should not be permitted to occur at such locations unless the ratio of ultimate strength to yield strength of the steel exceeds 1.5. In addition, special precautions must be taken to prevent buckling; for example, width-thickness and depth-thickness ratios must satisfy plastic-design criteria. And the effective lengths used in determining the slenderness ratios of columns should ignore any assistance provided by shear walls and braced frames in resisting lateral forces. Furthermore, the capacity of butt welds in tension should be verified by nondestructive testing.

For ductile, moment-resisting space frames of reinforced concrete, the Uniform Building Code places stringent requirements on member sizes, amount and location of reinforcement, concrete strength, anchorage of reinforcement, and beam-column joints.

A building whose sole resistance to lateral forces is provided by a ductile, moment-resisting space frame may be assigned $K = 0.67$. Before assigning a K value of 1.00 or less, however, designers should ascertain that action or failure of more-rigid elements surrounding or adjoining the frame will not impair its ability to carry vertical or lateral loads.

In seismic zones 2, 3, and 4, structural members of frames that are not part of the lateral-force-resisting system should have the capacity to carry vertical loads and bending moment due to $3/K$ times the distortions resulting from the forces calculated from Eq. (5-194). Similarly, for buildings in seismic zones 3 and 4 and those in zone 2 with I greater than unity, all members of braced frames should be designed for 1.25 times the forces induced by the loads calculated from Eq. (5-194).

Buildings often may qualify for more than one K value, for example, when they have different framing systems in perpendicular directions. Note, however, that if $K = 1.33$ is required in one direction, it also will be required in the other.

For elevated tanks, independent of buildings, the relatively high value $K = 3$ is assigned because of their relatively poor performance in earthquakes. This high value is especially desirable for water tanks because of their importance to communities in case of fire after an earthquake.

Structures other than buildings are assigned $K = 2$. This relatively high value reflects a general lack of redundancy, so that little yielding can be tolerated without failure. This value also assumes absence of nonstructural elements that can contribute materially to damping.

Parts of Structures. Building components and their anchorages to the main structural system should be designed for seismic forces given by

$$F_p = ZIC_p W_p \qquad (5\text{-}199)$$

(This equation has no bearing on determination of lateral forces to be applied to the building as a whole.)

Values of the seismic coefficient C_p are given in Table 5-10. W_p, in general, is the weight of the component. I is the occupancy factor used for the building. For anchorage of equipment required for life-safety systems, however, I should be taken as 1.5.

The Uniform Building Code also requires that floor and roof diaphragms and horizontal bracing systems for seismic loads be designed to resist horizontal forces computed from

$$F_{px} = w_{px} \sum_{i=x}^{n} F_i \bigg/ \sum_{i=x}^{n} w_i \leq 0.30 ZI w_{px} \quad \text{but not less than } 0.14 ZI w_{px} \qquad (5\text{-}200)$$

where F_{px} = seismic force on the diaphragm or horizontal bracing system at floor or roof level x
F_i = lateral force applied to level i, as computed from Eq. (5-202)
n = number of levels in structure
w_i = portion of dead load (plus 25% of floor live load in storage and warehouse occupancies) assigned to level i
w_{px} = weight of diaphragm or bracing at level x, plus live load as for w_i, and weight of tributary elements
I = occupancy importance factor, as used in Eq. (5-194)

If lateral forces are to be transferred from vertical structural components, such as columns and walls, above a diaphragm or bracing system to lower vertical structural members resisting lateral forces, the transferred forces should be added to those deter-

TABLE 5-10 C_p for Aseismic Design of Components of Buildings and Other Structures

Components	Direction of force	C_p
Exterior bearing and nonbearing walls, interior bearing walls and partitions, interior nonbearing walls and partitions over 10 ft high, masonry or concrete fences over 6 ft high*	Normal to flat surface	0.30
Cantilever parapet and other cantilever walls, except retaining walls; chimneys and stacks	Normal to flat surface	0.80
Exterior and interior ornamentations and appendages	Any direction	0.80
When connected to or part of a building: penthouses; anchorage and supports for chimneys, stacks and tanks; storage racks more than 8 ft high; equipment and machinery	Any direction	0.30
Framing for suspended ceilings in seismic zones 2, 3 and 4†	Any direction	0.30
Connections for exterior panels or other elements attached to or enclosing the exterior‡	Any direction	1.33
Connections for prefabricated structural elements other than walls, with force applied at center of gravity of assemblage	Any horizontal direction	0.30

*Deflection of walls or partitions under a force of 5 psf applied perpendicular to them shall not exceed 1/240 the span for walls with brittle finishes or 1/120 the span for walls with flexible finishes.

Precast, nonbearing, nonshear wall panels or other elements attached to or enclosing the exterior shall accommodate movements of the structure resulting from lateral forces or temperature changes. Connections and panel joints shall allow for a relative movement between stories of at least 1/2-in or twice the story drift under wind or $3/K$ times the calculated elastic story displacement under seismic loads, whichever is greater.

†Use ceiling weight of at least 4 lb/ft², including lighting fixtures and other equipment supported by ceiling.

‡Connections should have sufficient ductility and rotation capacity to preclude fracture of concrete or brittle failures at or near welds. Movement in the plane of a panel for story drift may be provided by sliding connections with slotted or oversize holes, or by bending of steel. Use $I = 1$ in Eq. (5-199) for an entire connector assembly.

mined from Eq. (5-200). Such transfers may be necessitated by offsets in placement of vertical members in successive stories or by changes in the stiffness of those members.

Distribution of Seismic Loads. Seismic forces are assumed to act at each floor level on vertical planar frames, or bents, or on shear walls extending in the direction of the loads. The seismic loads at each level should be distributed over the floor or roof area in accordance with the distribution of mass on that level.

Total horizontal shear at any level should be distributed to the bents or shear walls of the lateral-force distribution system in proportion to their rigidities. The distribution should, however, take into account the rigidities of horizontal bracing and diaphragms (floors and roofs). In lightly loaded structures, for example, diaphragms may be sufficiently flexible to permit independent action of the lateral-force-resisting bents. A strong temblor could cause severe distress in frames and diaphragms if relative rigidities were not properly evaluated.

The total horizontal shear, kips, on a structure is V, given by Eq. (5-194). The Uniform Building Code recommends that the portion of V to be assigned to the top of a multistory structure be computed from

$$F_t = 0.07TV \qquad (5\text{-}201)$$

where T = fundamental elastic period of vibration of the structure in the direction of the lateral force, s. F_t need not be more than $0.25V$ and may be taken as zero when $T \leq 0.7$ sec. Equation (5-201) recognizes the influence of higher modes of vibration as well as deviations from straight-line deflection patterns, particularly in tall buildings with relatively small dimensions in plan.

The code also recommends that the seismic force F_x, to be assigned to any level at a height h_x, ft, above the ground be calculated from

$$F_x = (V - F_t) \frac{w_x h_x}{\sum_{i=1}^{n} w_i h_i} \qquad (5\text{-}202)$$

where w_x = portion of W located at or assigned to level x
h_x = height, ft, of level x above ground level
w_i = portion of W located at or assigned to level i
h_i = height, ft, of level i above ground level
n = number of levels in structure

Seismic force distribution for buildings or structural frames with irregular shapes should be determined by dynamic analysis.

Overturning. The equivalent static lateral forces applied to a building at various levels induce overturning moments. At any level, the overturning moment equals the sum of the products of each force and its height above that level. The overturning moments acting on the base of the structure and in each story are resisted by axial forces in vertical elements and footings.

At any level, the increment in the design overturning moment should be distributed to the resisting elements in the same proportion as the distribution of shears to those elements. Where a vertical resisting element is discontinued, the overturning moment at that level should be carried down as loads to the foundation.

(J. M. Biggs, "Introduction to Structural Dynamics," and R. Clough and J. Penzien, "Dynamics of Structures," McGraw-Hill Book Company, New York; E. Rosenblueth, "Design of Earthquake-Resistant Structures," Halsted/Wiley, Somerset, N.J.; N. M. Newmark and E. Rosenblueth, "Fundamentals of Earthquake Engineering," Prentice-Hall, Englewood Cliffs, N.J.; S. Okamoto, "Introduction to Earthquake Engineering," John Wiley & Sons, Inc., New York.)

5-87. Reduction of Vibrations.

In general, vibrations in a building are objectionable from the viewpoint of physical comfort, noise, protection of delicate apparatus, and danger of fatigue failures (Figs. 5-90 and 5-91). The simplest ways to eliminate them are to isolate vibrating machinery, install supports that transmit a minimum of energy, install shock absorbers, or insert mufflers and screens to absorb noise (see Sec. 17).

Sometimes, machines with rotating parts can be made to vibrate less by balancing the rotating parts to reduce the exciting force or the speed can be adjusted to prevent resonance. Also, the masses of the supports can be adjusted to change the natural frequency so that resonance is avoided.

The surfaces of a foundation subjected to vibration should be shaped to preclude the possibility of the vibrations being reflected from them back into the interior in a direction in which a stress buildup can occur. Such a foundation should be made of a homogeneous material, since objectionable stresses may occur in the region of an inhomogeneity. The foundation should be massive—weight comparable with that of the machine it supports—and should be isolated by an air gap or insulating material from the rest of the building.

("Vibration in Buildings," Building Research Station Digest No. 78, Her Majesty's Stationery Office, London, June, 1955; D. D. Barkan, "Dynamics of Bases and Foundations," McGraw-Hill Book Company, New York.)

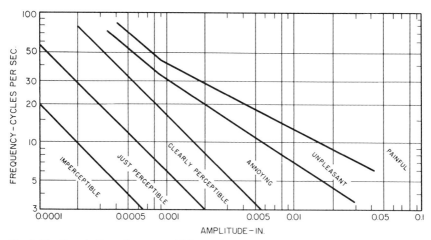

Fig. 5-90 Chart indicates human sensitivity to vibrations. *(British Research Station Digest No. 78.)*

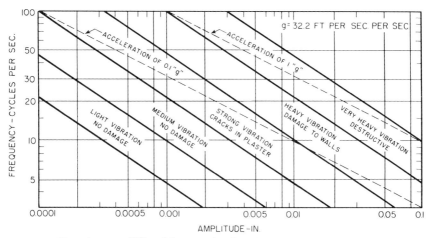

Fig. 5-91 Chart shows possibility of damage to buildings from vibration. *(British Research Station Digest No. 78.)*

STRESSES IN ARCHES

An arch is a curved beam, the radius of curvature of which is very large relative to the depth of the section. It differs from a straight beam in that: (1) loads induce both bending and direct compressive stresses in an arch; (2) arch reactions have horizontal components even though loads are all vertical; and (3) deflections have horizontal as well as vertical components (see also Arts. 5-36 to 5-41).

The necessity of resisting the horizontal components of the reactions is an important consideration in arch design. Sometimes these forces are taken by tie rods between the supports, sometimes by heavy abutments or buttresses.

Arches may be built with fixed ends, as can straight beams, or with hinges at the supports. They may also be built with a hinge at the crown.

5-88. Three-Hinged Arches.

An arch with a hinge at the crown as well as at both supports (Fig. 5-92) is statically determinate. There are four unknowns—two horizontal and two vertical components of the reactions—but four equations based on the laws of equilibrium are available: (1) The sum of the horizontal forces must be zero. (2) The sum of the moments about the left support must be zero. (3) The sum of the moments about the right support must be zero. (4) The bending moment at the crown hinge must be zero (not to be confused with the sum of the moments about the crown, which also must be equal to zero but which would not lead to an independent equation for the solution of the reactions).

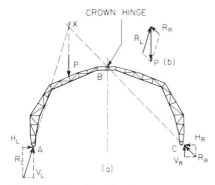

Fig. 5-92 Three-hinged arch.

Stresses and reactions in three-hinged arches can be determined graphically by utilizing the principles presented in Arts. 5-43 to 5-47 and 5-51 and taking advantage of the fact that the bending moment at the crown hinge is zero. For example, in Fig. 5-92a, a concentrated load P is applied to segment AB of the arch. Then, since the bending moment at B must be zero, the line of action of the reaction at C must pass through the crown hinge. It intersects the line of action of P at X. The line of action of the reaction at A must also pass through X. Since P is equal to the sum of the reactions, and since the directions of the reactions have thus been determined, the magnitude of the reactions can be measured from a parallelogram of forces (Fig. 5-92b). When the reactions have been found, the stresses can be computed from the laws of statics (see Art. 5-91) or, in the case of a trussed arch, determined graphically.

5-89. Two-Hinged Arches.

When an arch has hinges at the supports only (Fig. 5-93), it is statically indeterminate, and some knowledge of its deformations is required to determine the reactions. One procedure is to assume that one of the supports is on rollers. This makes the arch statically determinate. The reactions and the horizontal movement of the support are computed for this condition (Fig. 5-93b). Then, the magnitude of the horizontal force required to return the movable support to its original position is calculated (Fig. 5-93c). The reactions for the two-hinged arch are finally found by superimposing the first set of reactions on the second (Fig. 5-93d).

For example, if δx is the horizontal movement of the support due to the loads, and if $\delta x'$ is the horizontal movement of the support due to a unit horizontal force applied to the support, then

$$\delta x + H \delta x' = 0 \tag{5-203}$$

$$H = -\frac{\delta x}{\delta x'} \tag{5-204}$$

where H is the unknown horizontal reaction. (When a tie rod is used to take the thrust, the right-hand side of the first equation is not zero, but the elongation of the rod,

HL/AE.) The dummy unit-load method [Eq. (5-63)] can be used to compute δx and $\delta x'$:

$$\delta x = \int_A^B \frac{My}{EI} ds - \int_A^B \frac{N\,dx}{AE} \tag{5-205}$$

where M = moment at any section resulting from loads
N = normal thrust on cross section
A = cross-sectional area of arch
y = ordinate of section measured from A as origin, when B is on rollers
I = moment of inertia of section
E = modulus of elasticity
ds = differential length along axis of arch
dx = differential length along horizontal

$$\delta x' = -\int_A^B \frac{y^2}{EI} ds - \int_A^B \frac{\cos \alpha \, dx}{AE} \tag{5-206}$$

where α = the angle the tangent to axis at the section makes with horizontal. Unless the thrust is very large and would be responsible for large strains in the direction of the arch axis, the second term on the right-hand side of Eq. (5-205) can usually be ignored.

In most cases, integration is impracticable. The integrals generally must be evaluated by approximate methods. The arch axis is divided into a convenient number of sections and the functions under the integral sign evaluated for each section. The sum is approximately equal to the integral. Thus, for the usual two-hinged arch,

$$H = \frac{\sum_A^B (My\,\Delta s/EI)}{\sum_A^B (y^2\,\Delta s/EI) + \sum_A^B (\cos \alpha\,\Delta x/AE)} \tag{5-207}$$

Fig. 5-93 Two-hinged arch.

5-90. Fixed Arches.

An arch is considered fixed when rotation is prevented at the supports (Fig. 5-94). Such an arch is statically indeterminate; there are six reaction components and only three equations are available from conditions of equilibrium. Three more equations must be obtained from a knowledge of the deformations.

One way to determine the reactions is to consider the arch cut at the crown, forming two cantilevers. First, the horizontal and vertical deflections and rotation produced at the end of each half arch by the loads are computed. Next, the deflection components and rotation at those ends are determined for unit vertical force, unit horizontal force, and unit moment applied separately at the ends. These deformations, multiplied, respectively, by V, the unknown shear; H, the unknown horizontal thrust at the crown; and M, the unknown moment there, yield the deformation caused by the unknown forces at the crown. Adding these deformations algebraically to the corresponding one produced by the loads gives the net movement of the free end of each half arch. Since these ends must deflect and rotate the same amount to maintain continuity, three equations can be obtained in this manner for the determination of V, H, and M. The various deflections can be computed by the dummy unit-load method [Eq. (5-63)], as demonstrated in Art. 5-89 for two-hinged arches.

The solution of the equations, however, can be simplified considerably if the center of coordinates is shifted to the elastic center of the arch and the coordinate axes are properly oriented. If the unknown forces and moments V, H, and M are determined at the elastic center, each equation will contain only one unknown. When the unknowns at the elastic center are determined, the shears, thrusts, and moments at any point on the arch can be found by application of the laws of equilibrium. (See Art. 5-91.)

Determination of the location of the elastic center of an arch is equivalent to finding

Fig. 5-94 Fixed-end arch.

the center of gravity of an area. Instead of an increment of area dA, however, an increment of length ds multiplied by a width $1/EI$ must be used. (E is the modulus of elasticity, I the moment of inertia.) Since, in general, integrals are difficult or impracticable to evaluate, the arch axis usually is divided into a convenient number of elements of length Δs and numerical integration is used, as described in Art. 5-89. Then, if the origin of coordinates is temporarily chosen at A, the left support of the arch in Fig. 5-94b, and if x' is the horizontal distance to a point on the arch and y' the vertical distance, the coordinates of the elastic center are

$$X = \frac{\sum_{A}^{B}(x'\,\Delta s/EI)}{\sum_{A}^{B}(\Delta s/EI)} \qquad Y = \frac{\sum_{A}^{B}(y'\,\Delta s/EI)}{\sum_{A}^{B}(\Delta s/EI)} \tag{5-208}$$

If the arch is symmetrical about the crown, the elastic center lies on a normal to the tangent at the crown. In that case, there is a savings in calculations by taking the origin of the temporary coordinate system at the crown and measuring coordinates parallel to the tangent and the normal. To determine Y, the distance of the elastic center from the crown, Eq. (5-208) can be used with the summations limited to the half arch between crown and either support. For a symmetrical arch also, the final coordinate system should be chosen parallel to the tangent and normal to the crown.

After the elastic center has been located, the origin of a new coordinate system should be taken at the center. For convenience, the new coordinates x_1, y_1 may be taken parallel to those in the temporary system. Then, for an unsymmetrical arch, the final coordinate axes should be chosen so that the x axis makes an angle α, measured clockwise, with the x_1 axis such that

$$\tan 2\alpha = \frac{2\sum_{A}^{B}(x_1 y_1\,\Delta s/EI)}{\sum_{A}^{B}(x_1^2\,\Delta s/EI) - \sum_{A}^{B}(y_1^2\,\Delta s/EI)} \tag{5-209}$$

The unknown forces H and V at the elastic center should be taken parallel, respectively, to the final x and y axes.

For a coordinate system with origin at the elastic center and axes oriented to satisfy Eq. (5-209),

$$H = \frac{\sum_{A}^{B}(M'y\,\Delta s/EI)}{\sum_{A}^{B}(y^2\,\Delta s/EI)}$$

$$V = \frac{\sum_{A}^{B}(M'x\,\Delta s/EI)}{\sum_{A}^{B}(x^2\,\Delta s/EI)} \tag{5-210}$$

$$M = \frac{\sum_{A}^{B}(M'\,\Delta s/EI)}{\sum_{A}^{B}(\Delta s/EI)}$$

where M' is the average bending moment on each element due to loads. To account for the effect of an increase in temperature t, add $EctL$ to the numerator of H, where c is the coefficient of expansion and L the distance between abutments.

(S. Timoshenko and D. H. Young, "Theory of Structures," McGraw-Hill Book Company, New York; S. F. Borg and J. J. Gennaro, "Modern Structural Analysis," Van Nostrand Reinhold Company, Inc., New York.)

5-91. Stresses in Arch Ribs.

When the reactions have been found for an arch (Arts. 5-88 to 5-90), the principal forces acting on any cross section can be found by applying the equations of equilibrium. For example, consider the portion of a fixed arch in Fig. 5-95, where the forces at the elastic center H_e, V_e, and M_e are known and the forces acting at point X are to be found. The load P, H_e, and V_e may be resolved into components parallel to the axial thrust N and the shear S at X, as indicated in Fig. 5-95. Then, by equating the sum of the forces in each direction to zero, we get

$$N = V_e \sin \theta_x + H_e \cos \theta_x + P \sin (\theta_x - \theta)$$
$$S = V_e \cos \theta_x - H_e \sin \theta_x + P \cos (\theta_x - \theta)$$
(5-211)

And by taking moments about X and equating to zero, we obtain

$$M = V_e x + H_e y - M_e + Pa \cos \theta + Pb \sin \theta \qquad (5\text{-}212)$$

The shearing unit stress on the arch cross section at X can be determined from S with the aid of Eq. (5-28). The normal unit stresses can be calculated from N and M with the aid of Eq. (5-36).

In designing an arch, it may be necessary to compute certain secondary stresses, in addition to those caused by live, dead, wind, and snow loads. Among the secondary stresses to be considered are those due to temperature changes, rib shortening due to thrust or shrinkage, deformation of tie rods, and unequal settlement of footings. The procedure is the same as for loads on the arch, with the deformations producing the secondary stresses substituted for or treated the same as the deformations due to loads.

Fig. 5-95 Portion of an arch, including the elastic center.

MEMBRANE AND THIN-SHELL STRUCTURES

A structural membrane or shell is a curved surface structure. Usually, it is capable of transmitting loads in more than two directions to supports. It is highly efficient structurally when it is so shaped, proportioned, and supported that it transmits the loads without bending or twisting.

A membrane or a shell is defined by its middle surface, halfway between its extrados, or outer surface, and intrados, or inner surface. Thus, depending on the geometry of the middle surface, it might be a type of dome, barrel arch, cone, or hyperbolic paraboloid. Its thickness is the distance, normal to the middle surface, between extrados and intrados.

5-92. Thin-Shell Analysis.

A thin shell is a shell with a thickness relatively small compared to its other dimensions. But it should not be so thin that deformations would be large compared with the thickness.

The shell should also satisfy the following conditions: Shearing stresses normal to the middle surface are negligible. Points on a normal to the middle surface before it is deformed lie on a straight line after deformation. And this line is normal to the deformed middle surface.

Calculation of the stresses in a thin shell generally is carried out in two major steps, both usually involving the solution of differential equations. In the first, bending and torsion are neglected (membrane theory, Art. 5-93). In the second step, corrections are made to the previous solution by superimposing the bending and shear stresses that are necessary to satisfy boundary conditions (bending theory, Art. 5-98).

Ribbed Shells. For long-span construction, thin shells often are stiffened at intervals by ribs. Usually, the construction is such that the shells transmit some of the load imposed on them to the ribs, which then perform structurally as more than just stiffeners. Stress and strain distributions in shells and ribs consequently are complicated by the interaction between shells and ribs. The shells restrain the ribs, and the ribs restrain the shells. Hence, ribbed shells usually are analyzed by approximate methods based on reasonable assumptions.

For example, for a cylindrical shell with circumferential ribs, the ribs act like arches. For an approximate analysis, the ribbed shell therefore may be assumed to be composed of a set of arched ribs with the thin shell between the ribs acting in the circumferential direction as flanges of the arches. In the longitudinal direction, it may be assumed that the shell transfers load to the ribs in flexure. Designers may adjust the results of a computation based on such assumptions to correct for a variety of conditions, such as the effects of free edges of the shell, long distances between ribs, relative flexibility of ribs and shell, and characteristics of the structural materials.

5-93. Membrane Theory for Thin Shells.

Thin shells usually are designed so that normal shears, bending moments and torsion are very small, except in relatively small portions. In the membrane theory, these stresses are ignored.

Despite the neglected stresses, the remaining stresses are in equilibrium, except possibly at boundaries, supports, and discontinuities. At any interior point, the number of equilibrium conditions equals the number of unknowns. Thus, in the membrane theory, a thin shell is statically determinate.

The membrane theory does not hold for concentrated loads normal to the middle surface, except possibly at a peak or valley. The theory does not apply where boundary conditions are incompatible with equilibrium. And it is inexact where there is geometric incompatibility at the boundaries. The last is a common condition, but the error is very small if the shell is not very flat. Usually, disturbances of membrane equilibrium due to incompatibility with deformations at boundaries, supports, or discontinuities are appreciable only in a narrow region about each source of disturbance. Much larger disturbances result from incompatibility with equilibrium conditions.

To secure the high structural efficiency of a thin shell, select a shape, proportions, and supports for the specific design conditions that come as close as possible to satisfying the membrane theory. Keep the thickness constant; if it must change, use a gradual taper. Avoid concentrated and abruptly changing loads. Change curvature gradually. Keep discontinuities to a minimum. Provide reactions that are tangent to the middle surface. At boundaries, ensure, to the extent possible, compatibility of shell deformations with deformations of adjoining members, or at least keep restraints to a minimum. Make certain that reactions along boundaries are equal in magnitude and direction to the shell forces there.

Means usually adopted to satisfy these requirements at boundaries and supports are illustrated in Fig. 5-96. In Fig. 5-96a, the slope of the support and provision for movement normal to the middle surface ensure a reaction tangent to the middle surface. In Fig. 5-96b, a stiff rib, or ring girder, resists unbalanced shears and transmits normal forces to columns below. The enlarged view of the ring girder in Fig. 5-96c shows gradual thickening of the shell to reduce the abruptness of the change in section. The stiffening ring at the lantern in Fig. 5-96d, extending around the opening at the crown, projects above the middle surface, for compatibility of strains, and connects through a transition curve with the shell; often, the rim need merely be thickened when the edge is upturned, and the ring can be omitted. In Fig. 5-96e, the boundary of the shell is a stiffened edge. In Fig. 5-96f, a scalloped shell provides gradual tapering for transmitting the loads to

Fig. 5-96 Special provisions made at supports and boundaries of thin shells to meet requirements of the membrane theory include: (a) a device to ensure a reaction tangent to the middle surface; (b) stiffened edges, such as the ring girder at the base of a dome; (c) gradually increased shell thickness at a stiffening member; (d) a transition curve at changes in section; (e) a stiffening edge obtained by thickening the shell; (f) scalloped edges; (g) a flared support.

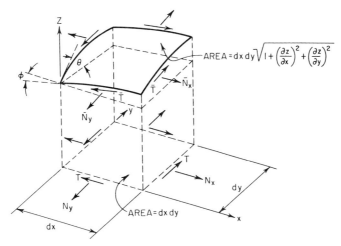

Fig. 5-97 Shears and normal forces acting on an element of the middle surface of a thin shell in the membrane theory.

the supports, at the same time providing access to the shell enclosure. And in Fig. 5-96g, a column is flared widely at the top to support a thin shell at an interior point.

Even when the conditions for geometric compatibility are not satisfactory, the membrane theory is a useful approximation. Furthermore, it yields a particular solution to the differential equations of the bending theory.

5-94. Membrane Forces in Shells of General Shape.

By applying the equilibrium conditions to a small element of the middle surface, one obtains the membrane forces. The sides of the element are acted upon by normal forces and shears, as indicated in Fig. 5-97. For simplification, the equilibrium equations are usually expressed in terms of the projection of these forces on the xy plane. Thus:

$$\overline{N}_x = N_x \frac{\cos \phi}{\cos \theta} \qquad \overline{N}_y = N_y \frac{\cos \theta}{\cos \phi} \qquad \overline{T} = T \tag{5-213}$$

The forces act per unit of length of the edges of the element and over the full thickness of the shell, and hence are called unit forces. The subscripts indicate the coordinate axis they parallel.

Assume that the loading on the element per unit area in the xy plane is given by its components X, Y, and Z. Then, for equilibrium in the direction of the three coordinate axes:

$$\frac{\partial N_x}{\partial x} + \frac{\partial T}{\partial y} + X = 0 \qquad \frac{\partial N_y}{\partial y} + \frac{\partial T}{\partial x} + Y = 0$$

$$\frac{\partial}{\partial x}\left(N_x \frac{\partial z}{\partial x} + T \frac{\partial z}{\partial y}\right) + \frac{\partial}{\partial y}\left(N_y \frac{\partial z}{\partial y} + T \frac{\partial z}{\partial x}\right) + Z = 0 \tag{5-214}$$

where $\tan \phi = \partial z/\partial x$ and $\tan \theta = \partial z/\partial y$. These equations can be reduced to a single equation by introduction of a stress function $F(x,y)$ such that

$$N_x = \frac{\partial^2 F}{\partial y^2} - \int X\, dx \qquad N_y = \frac{\partial^2 F}{\partial x^2} - \int Y\, dy \qquad T = -\frac{\partial^2 F}{\partial x\, \partial y} \tag{5-215}$$

where the lower limits of the integrals are constants and the upper limits are x and y, respectively.

The equilibrium equation for the general case of a thin shell then is

$$\frac{\partial^2 F}{\partial x^2}\frac{\partial^2 z}{\partial y^2} - 2\frac{\partial^2 F}{\partial x\,\partial y}\frac{\partial^2 z}{\partial x\,\partial y} + \frac{\partial^2 F}{\partial y^2}\frac{\partial^2 z}{\partial x^2} = P \quad (5\text{-}216)$$

$$P = -Z + X\frac{\partial z}{\partial x} + Y\frac{\partial z}{\partial y} + \frac{\partial^2 z}{\partial x^2}\int X\,dx + \frac{\partial^2 z}{\partial y^2}\int Y\,dy$$

Hence, the membrane forces are determined when this equation is solved for F. (S. Timoshenko and S. Woinowsky-Krieger, "Theory of Plates and Shells," McGraw-Hill Book Company, New York.)

5-95. Membrane Forces in a Hyperbolic-Paraboloid Shell.

Equation (5-216) can be used to determine the membrane forces in a thin shell with a hyperbolic-paraboloid middle surface (Fig. 5-98). Assume, for example, that the equation of the middle surface is

$$z = cxy \quad (5\text{-}217)$$

where c is a constant, and the loading on the shell (Fig. 5-99) is vertical, say, a snow load, $-Z_s$. Then, $P = -Z = Z_s$. (Note that when x is a constant or y is a constant, the intercept with the surface is a straight line, though the surface is curved.)

Differentiation of the surface equation gives

$$\frac{\partial^2 z}{\partial x^2} = \frac{\partial^2 z}{\partial y^2} = 0 \qquad \frac{\partial^2 z}{\partial x\,\partial y} = c$$

Substitution of these derivatives in Eqs. (5-216) and (5-215) yields

$$-2\frac{\partial^2 F}{\partial x\,\partial y}c = 2Tc = -Z = Z_s$$

from which results

$$T = \frac{Z_s}{2c} \quad (5\text{-}218)$$

Furthermore, N_x can be any function of y, and N_y any function of x.

Suppose the shell to be stiffened along its edges by a diaphragm with negligible resis-

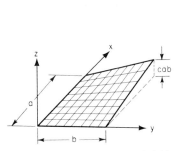

Fig. 5-98 Hyperbolic-paraboloid thin shell.

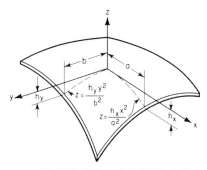

Fig. 5-99 Elliptical-paraboloid thin shell.

tance normal to its plane. Then, $N_x = \overline{N}_x = 0$ and $N_y = \overline{N}_y = 0$, since the edges are free of normal forces.

Assume now the load to be the weight of the shell, a constant $-Z_w$ per unit area of the surface. The area of the horizontal projection of an element of the surface equals $(1/(1 + c^2x^2 + c^2y^2)^{1/2}$ times the area of the element. Then, the horizontal projections of the unit forces are

$$T = \frac{Z_w}{2c}(1 + c^2x^2 + c^2y^2)^{1/2}$$

$$N_x = -\frac{Z_w y}{2} \log_e [cx + (1 + c^2x^2 + c^2y^2)^{1/2}] + f_1(y) \quad (5\text{-}219)$$

$$N_y = -\frac{Z_w x}{2} \log_e [cy + (1 + c^2x^2 + c^2y^2)^{1/2}] + f_2(x)$$

where $f_1(y)$ is an arbitrary function of y and $f_2(x)$ an arbitrary function of x; logarithms are taken to the base e. The actual unit forces $\overline{N}_x$ and $\overline{N}_y$ can be calculated by applying Eq. (5-213) to Eq. (5-219), noting that $\tan \phi = -cy$ and $\tan \theta = -cx$. (S. Timoshenko and S. Woinowsky-Krieger, "Theory of Plates and Shells," 2d ed., McGraw-Hill Book Company, New York.)

5-96. Elliptical-Paraboloid Membrane Forces.

Equations (5-216) and (5-215) also can be used to obtain membrane forces in an elliptical paraboloid (parabolas in vertical planes, ellipses in horizontal)

$$z = \frac{h_x x^2}{a^2} + \frac{h_y y^2}{b^2} \quad (5\text{-}220)$$

rectangular in plan (Fig. 5-99). The solution usually is given as a Fourier series, summed for $n = 1, 3, 5, \ldots, \infty$.

Suppose the shell to be subjected to a uniform load p over its horizontal projection. Assume that forces normal to the boundaries vanish. Then, the unit forces, acting per unit of length over the thickness of the shell, are

$$\overline{N}_x = -\frac{pb^2}{2h_x}\left(\frac{a^4 + 4h_x^2 x^2}{b^4 + 4h_y^2 y^2}\right)^{1/2}\left[1 + \frac{4}{\pi}\sum_n (-1)^{(n+1)/2} \frac{\cosh(n\pi x/c)}{n \cosh(n\pi a/c)} \cos \frac{n\pi y}{2b}\right]$$

$$\overline{N}_y = \frac{2pa^4}{\pi h_x b^2}\left(\frac{b^4 + 4h_y^2 y^2}{a^4 + 4h_x^2 x^2}\right)^{1/2}\sum_n (-1)^{(n+1)/2} \frac{\cosh(n\pi x/c)}{n \cosh(n\pi a/c)} \cos \frac{n\pi y}{2b} \quad (5\text{-}221)$$

$$T = \frac{2pa^2}{\pi h_x}\sum_n (-1)^{(n+1)/2} \frac{\sinh(n\pi x/c)}{n \cosh(n\pi a/c)} \sin \frac{n\pi y}{2b}$$

where $c = 2a(h_y/h_x)^{1/2}$.

The series for T does not converge at the corners. Actually, bending moments and torsion will exist in those regions, counteracting the huge membrane shears.

5-97. Membrane Forces in Shells of Revolution.

In a shell of revolution, the middle surface results from the rotation of a plane curve about an axis, called the "shell axis." A plane through the axis cuts the surface in a meridian. A plane normal to the axis cuts the surface in a circle, called a parallel (Fig. 5-100a).

The membrane forces in a shell of revolution usually are taken parallel to the x, y, z, axes of a cartesian coordinate system at each point of the middle surface. The x axis is

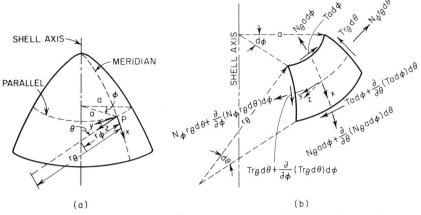

Fig. 5-100 Shell of revolution: (*a*) coordinate system used in analysis; (*b*) unit forces acting on an element.

tangent to the meridian, the *y* axis is tangent to the parallel, and the *z* axis lies in the direction of the normal to the surface (Fig. 5-100*b*).

A given point *P* on the surface is determined by the angle θ between the shell axis and the normal through *P*, and by the angle ϕ between the radius through *P* of the parallel on which it lies and a fixed, reference direction. Let r_θ be the radius of curvature of the meridian, and let r_ϕ, the length of the shell normal between *P* and the shell axis, be the radius of curvature of the normal section at *P*. If *a* is the radius of the parallel through *P*, then $a = r_\phi \sin \theta$.

Consider the element of the middle surface of the shell in Fig. 5-100*b*. Its edges are subjected to unit normal and shear forces, acting per unit of length of edge over the thickness of the shell. Assume that the loading on the element per unit of area is given by its *X*, *Y*, *Z* components. Then, the equations of equilibrium are

$$\frac{\partial}{\partial \theta}(N_\theta r_\phi \sin \theta) + \frac{\partial T}{\partial \phi} r_\theta - N_\phi r_\theta \cos \theta + X r_\theta r_\phi \sin \theta = 0$$

$$\frac{\partial N_\phi}{\partial \phi} r_\theta + \frac{\partial}{\partial \theta}(T r_\phi \sin \theta) + T r_\theta \cos \theta + Y r_\theta r_\phi \sin \theta = 0 \quad (5\text{-}222)$$

$$N_\theta r_\phi + N_\phi r_\theta + Z r_\theta r_\phi = 0$$

Symmetrical Loading. When the loads also are symmetrical about the shell axis, Eqs. (5-222) take a simpler form. If *a* is the radius of a parallel and *R* the resultant of the total vertical load on the shell above that parallel, then the first equation can be written in the alternative form:

$$N_\theta = -\frac{R}{2\pi a} \sin \theta = -\frac{R}{2\pi r_\phi} \sin^2 \theta \quad (5\text{-}223)$$

Substitution of this in the third equation yields

$$N_\phi = \frac{R}{2\pi r_\theta} \sin^2 \theta - Z r_\phi \quad (5\text{-}224)$$

Because of rotational symmetry

$$T = 0 \quad (5\text{-}225)$$

Spherical Domes. For a spherical shell, $r_\theta = r_\phi = r$. If the load p is uniform over the horizontal projection of the shell, then Eqs. (5-223) to (225) indicate that the unit shear $T = 0$; with $R = \pi a^2 p$, the unit meridional thrust $N_\theta = -pr/2$; and the unit hoop force $N_\phi = -(pr/2)\cos 2\theta$. Thus, there is a constant meridional compression throughout the shell. The hoop forces are compressive in the upper half of the shell and tensile in the lower half, vanishing at $\theta = 45°$.

Again for a spherical dome, if the load w is uniform over the area of the shell, then Eqs. (5-223) to (225) give $T = 0$; with $R = 2\pi r^2(1 - \cos\theta)w$, the unit meridional thrust $N_\theta = -wr/(1 + \cos\theta)$; and unit hoop force $N_\phi = wr\{[1/(1 + \cos\theta)] - \cos\theta\}$. In this case, the compression along the meridian increases with θ. The hoop forces are compressive in the upper part of the shell, become zero at $\theta = 51°\,51'$ and turn to tension for larger values of θ. Usually, at the lower dome boundary, these tensile hoop forces are resisted by a ring girder, and since shell and girder under the membrane theory will have different strains, bending stresses will be imposed on the shell.

When there is an opening around the crown of the dome, as in Fig. 5-96b, the upper edge may be thickened or reinforced with a ring girder to resist the hoop forces. If $2\theta_0$ is the angle of the opening and P the vertical load per unit of length of the compression ring, then for a uniform load w over the area of the shell

$$N_\theta = -wr\frac{\cos\theta_0 - \cos\theta}{\sin^2\theta} - P\frac{\sin\theta_0}{\sin^2\theta}$$

$$N_\phi = wr\left(\frac{\cos\theta_0 - \cos\theta}{\sin^2\theta} - \cos\theta\right) + P\frac{\sin\theta_0}{\sin^2\theta} \quad (5\text{-}226)$$

Conical Shells. For a conical shell, parallel circles are located by the coordinate s, measured from the vertex along a generator, instead of the angle θ, and the unit meridional force is denoted by N_s (Fig. 5-101). The equilibrium equations are

$$\frac{\partial}{\partial s}(N_s s) + \frac{\partial T}{\partial \phi}\frac{1}{\cos\theta} - N_\phi + Xs = 0$$

$$\frac{\partial N_\phi}{\partial \phi}\frac{1}{\cos\theta} + \frac{\partial}{\partial s}(Ts) + T + Ys = 0 \quad (5\text{-}227)$$

$$N_\phi + Zs\cot\theta = 0$$

For a vertical load p uniform over the horizontal projection of a conical shell, Eqs. (5-227) give

$$T = 0 \quad N_s = -\frac{ps}{2}\cot\theta \quad \text{and} \quad N_\phi = -ps\frac{\cos^3\theta}{\sin\theta} \quad (5\text{-}228)$$

Suppose that the cone has an opening at the top at $s = s_0$, and the shell is subjected to a vertical uniform load w over its area. Then,

$$T = 0 \quad N_s = -\frac{s^2 - s_0^2}{2s}\frac{w}{\sin\theta} \quad \text{and} \quad N_\phi = -ws\frac{\cos^2\theta}{\sin\theta} \quad (5\text{-}229)$$

For a horizontal load q uniform over the vertical projection of the cone with lantern,

$$T = -q\frac{s^3 - s_0^3}{3s^2}\sin\phi$$

$$N_s = -\frac{qs}{2}\left[\cos\theta - \frac{1}{3\cos\theta} - \frac{s_0^2}{s^2}\left(\cos\theta - \frac{1}{\cos\theta}\right) - \frac{s_0^3}{s^3}\frac{2}{3\cos\theta}\right]\cos\phi \quad (5\text{-}230)$$

$$N_\phi = -qs\cos\theta\cos\phi$$

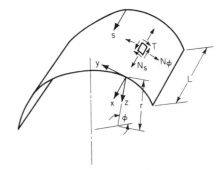

Fig. 5-101 Conical thin shell.　　**Fig. 5-102** Cylindrical thin shell.

Barrel Arches. For a cylindrical shell with horizontal axis, a point on the surface is located by its distance s along a generator from one end and by the angle ϕ which the radius of curvature at the point makes with the horizontal (Fig. 5-102). The equilibrium equations are

$$\frac{\partial N_s}{\partial s}r + \frac{\partial T}{\partial \phi} + Xr = 0 \qquad \frac{\partial N_\phi}{\partial \phi} + \frac{\partial T}{\partial s}r + Yr = 0 \qquad N_\phi + Zr = 0 \qquad (5\text{-}231)$$

The third equation indicates that $N_\phi = 0$ when $Z = 0$. Hence, for uniform vertical loads over the horizontal projection of the shell area, vertical reactions vanish along the longitudinal edges of a cylindrical shell if the tangent to the directrix at the edges is vertical. Shells with a semicircular, semielliptical, or cycloidal directrix have this property.

Equations (5-231) can be used to demonstrate an unusual property of parabolic vaults. The equation of the parabola can be rewritten from the common form $u = v^2/2r_0$, where r_0 is the radius of curvature at the vertex ($\phi = \pi/2$), into the more convenient form $r = r_0/\sin^3 \phi$. Now, for a vertical load p uniform over the horizontal projection, $X = 0$, $Y = -p \sin \phi \cos \phi$, and $Z = p \sin^2 \phi$. Then, solution of Eqs. (5-231) yields $T = 0$ and $N_s = 0$, so that loads pass to the supports only in the ϕ direction. $N_\phi = -pr_0/\sin \phi$. Thus, under this type of a loading, a parabolic vault behaves like a parabolic arch. The same results are obtained with a catenary-shaped vault under vertical loading uniform over the shell area.

For a semicircular barrel arch, r is a constant. If the vertical load w is uniform over the shell area, $X = 0$, $Y = w \cos \phi$, and $Z = w \sin \phi$. Then, Eqs. (5-231) indicate that $T = -w(L - 2s) \cos \phi$, $N_s = -w(s/r)(L - s) \sin \phi$, and $N_\phi = -wr \sin \phi$. No supports are needed along the longitudinal edges, since N_ϕ is zero there, but provision must be made to resist the shear. Such provision, usually in the form of an edge member, generally creates bending stresses due to incompatibility of deformations at the shell boundary. The shears transmit the loads to the edges $s = 0$ and $s = L$, where supports must be provided.

Other Shapes. For membrane stresses in a variety of shells, see W. Flügge, "Stresses in Shells," Springer-Verlag, New York; D. P. Billington, "Thin Shell Concrete Structures," McGraw-Hill Book Company, New York; and S. Timoshenko and S. Woinowski-Krieger, "Theory of Plates and Shells" McGraw-Hill Book Company, New York.

5-98. Bending Theory for Thin Shells.

When equilibrium conditions are not satisfied or incompatible deformations exist at boundaries, bending and torsion stresses arise in the shell. Sometimes, the design of the shell and its supports can be modified to reduce or eliminate these stresses (Art. 5-92).

When the design cannot eliminate them, provision must be made for the shell to resist them.

But even for the simplest types of shells and loading, the stresses are difficult to compute. In bending theory, a thin shell is statically indeterminate; deformation conditions must supplement equilibrium conditions in setting up differential equations for determining the unknown forces and moments. Solution of the resulting equations may be tedious and time-consuming, if indeed solution is possible.

In practice, therefore, shell design relies heavily on the designer's experience and judgment. The designer should consider the type of shell, material of which it is made, and support and boundary conditions, and then decide whether to apply a bending theory in full, use an approximate bending theory, or make a rough estimate of the effects of bending and torsion. (Note that where the effects of a disturbance are large, these change the normal forces and shears computed by the membrane theory.) For concrete domes, for example, the usual procedure is to use as a support a deep, thick girder or a heavily reinforced or prestressed tension ring, and the shell is gradually thickened in the vicinity of this support (Fig. 5-96c).

Circular barrel arches, with ratio of radius to distance between supporting arch ribs less than 0.25, may be designed as beams with curved cross section. Secondary stresses, however, must be taken into account. These include stresses due to volume change of rib and shell, rib shortening, unequal settlement of footings, and temperature differentials between surfaces.

Bending theory for cylinders and domes is given in W. Flügge, "Stresses in Shells," Springer-Verlag, New York; D. P. Billington, "Thin Shell Concrete Structures," McGraw-Hill Book Company, New York; S. Timoshenko and S. Woinowsky-Krieger, "Theory of Plates and Shells," McGraw-Hill Book Company, New York; "Design of Cylindrical Concrete Shell Roofs," Manual of Practice No. 31, American Society of Civil Engineers.

5-99. Stresses in Thin Shells.

The results of the membrane and bending theories are expressed in terms of unit forces and unit moments, acting per unit of length over the thickness of the shell. To compute the unit stresses from these forces and moments, usual practice is to assume normal forces and shears to be uniformly distributed over the shell thickness and bending stresses to be linearly distributed.

Then, normal stresses can be computed from equations of the form

$$f_x = \frac{N_x}{t} + \frac{M_x}{t^3/12} z \qquad (5\text{-}232)$$

where z = distance from middle surface
t = shell thickness
M_z = unit bending moment about axis parallel to direction of unit normal force N_x

Similarly, shearing stresses produced by central shears and twisting moments may be calculated from equations of the form

$$v_{xy} = \frac{T}{t} \pm \frac{D}{t^3/12} z \qquad (5\text{-}233)$$

where D = twisting moment. Normal shearing stresses may be computed on the assumption of a parabolic stress distribution over the shell thickness:

$$v_{xz} = \frac{V}{t^3/6} \left(\frac{t^2}{4} - z^2 \right) \qquad (5\text{-}234)$$

where V = unit shear force normal to middle surface

5-130   Structural Theory

For axes rotated with respect to those used in the thin-shell analysis, use Eqs. (5-15) and (5-16) to transform stresses or unit forces and moments from the given to the new axes.

AIR-STABILIZED STRUCTURES

A true membrane is able to withstand tension but is completely unable to resist bending. Although it is highly efficient structurally, like a shell, a membrane must be much thinner than a shell and therefore can be made of a very lightweight material, such as fabric, with considerable reduction in dead load compared with other types of construction. Such a thin material, however, would buckle if subjected to compression. Consequently, a true membrane, when loaded, deflects and assumes a shape that enables it to develop tensile stresses that resist the loads.

Membranes may be used for the roof of a building or as a complete exterior enclosure. One way to utilize a membrane for these purposes is to hang it with initial tension between appropriate supports. For example, a tent may be formed by supporting fabric atop one or more tall posts and anchoring the outer edges of the stretched fabric to the ground. As another example, a dish-shaped roof may be constructed by stretching a membrane and anchoring it to the inner surface of a ring girder. In both examples, loads induce only tensile stresses in the membrane. The stresses may be computed from the laws of equilibrium, because a membrane is statically determinate.

Another way to utilize a membrane as an enclosure or roof is to pretension the membrane to enable it to carry compressive loads. For the purpose, forces may be applied, and retained as long as needed, around the edges or over the surface of the membrane to induce tensile stresses that are larger than the largest compressive stresses that loads will impose. As a result, compression due to loads will only reduce the prestress, and the membrane will always be subjected only to tensile stresses.

5-100. Pneumatic Construction.

A common method of pretensioning a membrane enclosure is to pressurize the interior with air. Sufficient pressure is applied to counteract dead loads, so that the membrane actually floats in space. Slight additional pressurization is also used to offset wind and other anticipated loads. Made of lightweight materials, a membrane thus can span large distances economically. This type of construction, however, has the disadvantage that energy is continuously required for operation of air compressors to maintain interior air at a higher pressure than that outdoors.

Pressure differentials used in practice are not large. They often range between 0.02 and 0.04 psi (3 and 5 psf). Air must be continually supplied, because of leakage. While there may be some leakage of air through the membrane, more important sources of air loss are the entrances and exits to the structure. Air locks and revolving doors, however, can reduce these losses.

An air-stabilized enclosure, in effect, is a membrane bag held in place by small pressure differentials applied by environmental energy. Such a structure is analogous to a soap film. The shape of a bubble is determined by surface-tension forces. The membrane is stressed equally in all directions at every point. Consequently, the film forms shapes with minimum surface area, frequently spherical. Because of the stress distribution, any shape that can be obtained with soap films is feasible for an air-stabilized enclosure. Figure 5-103c shows a configuration formed by a conglomeration of bubbles as an illustration of a shape that can be adopted for an air-stabilized structure.

In practice, shapes of air-stabilized structures often resemble those used for thin-shell enclosures. For example, spherical domes (Fig. 5-103a) are frequently constructed with a membrane. Also, membranes are sometimes shaped as semicircular cylinders with quarter-sphere ends (Fig. 5-103b).

Air-stabilized enclosures may be classified as air-inflated, air-supported, or hybrid structures, depending on the type of support.

Fig. 5-103 Some shapes for air-supported structures. *(Reprinted with permission from F. S. Merritt, "Building Engineering and Systems Design," Van Nostrand Reinhold Company, New York.)*

Air-inflated enclosures are completely supported by pressurized air entrapped within membranes. There are two main types, inflated-rib structures and inflated dual-wall structures.

In inflated-rib construction, the membrane enclosure is supported by a framework of air-pressurized tubes, which serve much like arch ribs in thin-shell construction (Art. 5-92). The principle of their action is demonstrated by a water hose. A flexible hose, when empty, collapses under its own weight on short spans or under loads normal to its length; but it stiffens when filled with water. The water pressure tensions the hose walls and enables them to withstand compressive stresses.

In inflated dual-walled construction, pressurized air is trapped between two concentric membranes (Fig. 5-104). The shape of the inner membrane is maintained by suspending it from the outer one. Because of the large volume of air compressed between the membranes, this type of construction can span longer distances than can inflated-rib structures.

Because of the variation of air pressure with changes in temperature, provision must be made for adjustment of the pressure of the compressed air in air-inflated structures. Air must be vented to relieve excessive pressures, to prevent overtensioning of the membranes. Also, air must be added to compensate for pressure drops, to prevent collapse.

Air-supported enclosures consist of a single membrane supported by the difference between internal air pressure and external atmospheric pressure (Fig. 5-105). The pressure differential deflects the membrane outward, inducing tensile stresses in it, thus enabling it to withstand compressive forces. To resist the uplift, the construction must be securely anchored to the ground. Also, the membrane must be completely sealed around its perimeter to prevent air leakage.

Hybrid structures consist of one of the preceding types of pneumatic construction augmented by light metal framing, such as cables. The framing may be merely a safety

Fig. 5-104 Inflated dual-wall structure.

Fig. 5-105 Air-supported structure.

measure to support the membrane if pressure should be lost or a means of shaping the membrane when it is stretched. Under normal conditions, air pressure against the membrane reduces the load on the framing from heavy wind and snow loads.

5-101. Membrane Stresses.

Air-supported structures are generally spherical or cylindrical because of the supporting uniform pressure.

When a spherical membrane with radius R, in, is subjected to a uniform radial internal pressure, p, psi, the internal unit tensile force, lb/in, in any direction, is given by

$$T = \frac{pR}{2} \tag{5-235}$$

In a cylindrical membrane, the internal unit tensile force, lb/in, in the circumferential direction is given by

$$T = pR \tag{5-236}$$

where R = radius, in, of the cylinder. The longitudinal membrane stress depends on the conditions at the cylinder ends. For example, with immovable end enclosures, the longitudinal stress would be small. If, however, the end enclosure is flexible, a tension about half that given by Eq. (5-236) would be imposed on the membrane in the longitudinal direction.

Unit stress in the membrane can be computed by dividing the unit force by the thickness, in, of the membrane.

(R. N. Dent, "Principles of Pneumatic Architecture," John Wiley & Sons, Inc., New York.)

FOLDED PLATES

A folded-plate structure consists of a series of thin planar elements, or flat plates, connected to one another along their edges. Usually used on long spans, especially for roofs, folded plates derive their economy from the girder action of the plates and the mutual support they give one another.

Longitudinally, the plates may be continuous over their supports. Transversely, there may be several plates in each bay (Fig. 5-106). At the edges, or folds, they may be capable of transmitting both moment and shear or only shear.

Fig. 5-106 Folded-plate structure.

5-102. Folded-Plate Theory.

A folded-plate structure has a two-way action in transmitting loads to its supports. Transversely, the elements act as slabs spanning between plates on either side. The plates then act as girders in carrying the load from the slabs longitudinally to supports, which must be capable of resisting both horizontal and vertical forces.

If the plates are hinged along their edges, the design of the structure is relatively simple. Some simplification also is possible if the plates, though having integral edges, are steeply sloped or if the span is sufficiently long with respect to other dimensions that beam theory applies. But there are no criteria for determining when such simplification is possible with acceptable accuracy. In general, a reasonably accurate analysis of folded-plate stresses is advisable.

Several good methods are available (D. Yitzhaki, "The Design of Prismatic and Cylindrical Shell Roofs," North Holland Publishing Company, Amsterdam; "Phase I Report on Folded-plate Construction," Proceedings Paper 3741, *Journal of the Structural Division, American Society of Civil Engineers,* December, 1963; and A. L. Parme and J. A. Sbarounis, "Direct Solution of Folded Plate Concrete Roofs," EB021D, Portland Cement Association, Skokie, Ill.). They all take into account the effects of plate deflections on the slabs and usually make the following assumptions:

The material is elastic, isotropic, and homogeneous. The longitudinal distribution of all loads on all plates is the same. The plates carry loads transversely only by bending normal to their planes and longitudinally only by bending within their planes. Longitudinal stresses vary linearly over the depth of each plate. Supporting members, such as diaphragms, frames, and beams, are infinitely stiff in their own planes and completely flexible normal to their own planes. Plates have no torsional stiffness normal to their own planes. Displacements due to forces other than bending moments are negligible.

Regardless of the method selected, the computations are rather involved; so it is wise to carry out the work by computer or, when done manually, in a well-organized table. The Yitzhaki method (Art. 5-103) offers some advantages over others in that the calculations can be tabulated, it is relatively simple, it requires the solution of no more simultaneous equations than one for each edge for simply supported plates, it is flexible, and it can easily be generalized to cover a variety of conditions.

5-103. Yitzhaki Method for Folded Plates.

Based on the assumptions and general procedure given in Art. 5-102, the Yitzhaki method deals with the slab and plate systems that comprise a folded-plate structure in two ways. In the first, a unit width of slab is considered continuous over supports immovable in the direction of the load (Fig. 5-107b). The strip usually is taken where the longitudinal plate stresses are a maximum. Secondly, the slab reactions are taken as loads on the plates, which now are assumed to be hinged along the edges (Fig. 5-107c). Thus, the slab reactions cause angle changes in the plates at each fold. Continuity is restored by applying to the plates an unknown moment at each edge. The moments can be determined from the fact that at each edge the sum of the angle changes due to the loads and to the unknown moments must equal zero.

The angle changes due to the unknown moments have two components. One is the angle change at each slab end, now hinged to an adjoining slab, in the transverse strip of unit width. The second is the angle change due to deflection of the plates. The method assumes that the angle change at each fold varies in the same way longitudinally as the angle changes along the other folds.

The method will be demonstrated for the symmetrical folded-plate structure in Fig. 5-107a. To simplify the explanation, the tabulated calculations will be broken up into

5-134 Structural Theory

Fig. 5-107 Folded plate is analyzed by first considering a transverse strip (a) as a continuous slab on supports that do not settle (b). Then, the slabs are assumed hinged (c), and acted upon by the reactions computed in the first step and by unknown moments to correct for this assumption. In the longitudinal direction, the plates act as deep girders (e) with shears along the edges, positive directions shown (f). Slab reactions are resolved into plate forces, parallel to the planes of the plates (d).

several steps, though, in practice, the table is suitable for continuous calculation from start to finish. Dimensions and geometrical data are given in Table 5-11.

In the table, h is the depth in inches, t the thickness in inches, a the horizontal projection of the depth in inches, and A the cross-sectional area of a plate in square inches. ϕ is the angle each plate makes with the horizontal, in degrees. W is the vertical load in pounds per linear foot of plate and thus also the total load in pounds on a 12-in-wide slab.

TABLE 5-11 Geometric Data for Symmetric Folded Plate

Plate	h, in.	t, in.	a, in.	ϕ, deg	$\cos \phi$	$\tan \phi$	A, sq in.	W, lb per ft
1 and 6	48	7	0	90	0	.....	326	350
2 and 5	108	3	93.5	30	0.866	0.577	324	337.5
3 and 4	108	3	106.4	10	0.985	0.1763	324	337.5

Step 1. Compute the loads on a 12-in-wide transverse strip at midspan. For the edge beam, $W = 350$ lb/lin ft. For a load of 37.5 lb/ft^2 on the plates, each slab in the strip carries 337.5 lb.

Step 2. Consider the strip as a continuous slab supported at the folds (Fig. 5-107b), and compute the bending moments by moment distribution. To take advantage of the symmetrically loaded, symmetrical structure, edge 3 is taken as fixed, and end moments are distributed only in the left half of the strip. Also, since edge 1 is free to rotate, the relative stiffness of plate 2 is taken as three-fourths its actual relative stiffness to eliminate the need for distributing moments to edge 1. The computations are shown in Table 5-12.

TABLE 5-12 Slab on Rigid Supports

Property	Plate 1		Plate 2		Plate 3	
	Edge 0	Edge 1	Edge 1	Edge 2	Edge 2	Edge 3
Stiffness...............	...	...	...	$\tfrac{3}{4} \times 1$	1	Fixed
Ratio..................	...	...	...	0.428	0.572	0
M^F, in.-lb............	0	0	0	3,945	$-2{,}996$	2,996
				-406	-543	-272
M, in.-lb..............	0	0	0	3,539	$-3{,}539$	2,724

Step 3. From the end moments M found in Step 2, compute slab reactions and plate loads. Reactions (positive upward) at the nth edge are

$$R_n = V_n + V_{n+1} + \frac{M_{n-1} + M_n}{a_n} - \frac{M_n + M_{n+1}}{a_{n+1}} \tag{5-237}$$

where V_n, V_{n+1} = shears at both sides of edge n
M_n = moment at edge n, in-lb
M_{n-1} = moment at edge $(n-1)$, in-lb
M_{n+1} = moment at edge $(n+1)$, in-lb

The calculations are shown in Table 5-13.

TABLE 5-13 Slab Reactions and Plate Loads

Property	Plate 1		Plate 2		Plate 3	
	Edge 0	Edge 1	Edge 1	Edge 2	Edge 2	Edge 3
Shears..................	0	350	168.8	168.8	168.8	337.5
$\pm \Sigma M/a_n$...........	0	0	-37.8	37.8	7.7	-15.4
Reaction...............	0	481		383.1		322.1
k......................	...	...			0.401	0.3526
R/k...................	...	...			957	913
From R_{n-1}............		0		0	-972	
From R_n...............		481		1,105	927	
Plate load.............		481		1,105	-45	
M, ft-lb................		254,000		584,000	$-23{,}800$	

5-136 Structural Theory

For the vertical beam, the shear is zero at edge 0, and $W = 350$ at edge 1. For the plates, the shear is $W/2 = 168.8$ at edges 1 and 2, and $W = 337.5$ at edge 3 to include the load from plate 4. These values are shown on the first line of the table. The entries on the second line are obtained by dividing the sum of the end moments in each plate by a, the horizontal projection of the plate. Entries on the right for each edge are written with sign changed. The reactions are obtained by adding the values for each edge on the first and second lines of Table 5-13.

Let $k = \tan \phi_n - \tan \phi_{n+1}$, where ϕ is positive as shown in Fig. 107a. Then, the load (positive downward) on the nth plate is

$$P_n = \frac{R_n}{k_n \cos \phi_n} - \frac{R_{n-1}}{k_{n-1} \cos \phi_n} \tag{5-238}$$

(Figure 5-107d shows the resolution of forces at edge n; $n - 1$ is similar.) Equation (5-238) does not apply for the case of a vertical reaction on a vertical plate, for R/k is the horizontal component of the reaction.

In Table 5-13, the two components of Eq. (5-238) are shown on separate lines after the values of R/k, the second component being entered first, with sign changed. The components are summed to obtain the plate loads in pounds per linear foot.

Step 4. Calculate the midspan (maximum) bending moment in each plate. In this example, each plate is a simple beam and $M = PL^2/8$, ft-lb, where L is the span in feet. The moments for each plate are entered in Table 5-13 after the plate loads. These moments are assumed to vary parabolically along the edges (Fig. 5-107e).

Step 5. Determine the free-edge longitudinal stresses at midspan. In each plate, these can be computed from

$$f_{n-1} = \frac{72M}{Ah} \qquad f_n = -\frac{72M}{Ah} \tag{5-239}$$

where f is the stress in psi, M the moment in ft-lb from Step 4, and tension is taken as positive, compression as negative.

Step 6. Apply a shear to adjoining edges to equalize the stresses there. Compute the adjusted stresses by converging approximations, similar to moment distribution. To do this, distribute the unbalanced stress at each edge in proportion to the reciprocals of the areas of the plates, and use a carry-over factor of $-\frac{1}{2}$ to distribute the stress to a far edge. Edge 0, being a free edge, requires no distribution of the stress there. Edge 3, because of symmetry, may be treated the same, and distribution need be carried out only in the left half of the structure. The calculations are shown in Table 5-14.

TABLE 5-14 Plate Stresses for Slab Loads

Property	Plate 1		Plate 2		Plate 3	
	Edge 0	Edge 1	Edge 1	Edge 2	Edge 2	Edge 3
$1/A$ ratio	.....	0.491	0.509	0.5	0.5	
Free-edge f	1,133	−1,133	1,200	−1,200	−49	49
Distribution	−573	1,145	−1,188	594	−279	139
			−139	278		
	34	−68	71	−36	−18	9
			−9	18		
	2	−5	4	−2	−1	
				1		
Adjusted f	596	−61	−61	−347	−347	197

Step 7. Compute the midspan edge deflections. In general, the vertical component δ, in inches, can be computed from

$$\frac{E}{L^2}\delta_n = \frac{15}{k_n}\left(\frac{f_{n-1} - f_n}{a_n} - \frac{f_n - f_{n+1}}{a_{n+1}}\right) \quad (5\text{-}240)$$

where E = modulus of elasticity, psi
$k = \tan \phi_n - \tan \phi_{n+1}$, as in Step 3
The factor E/L^2 is retained for convenience; it is eliminated by dividing the simultaneous angle equations by it. For a vertical plate, the vertical deflection is given by

$$\frac{E}{L^2}\delta_n = \frac{15(f_{n-1} - f_n)}{h_n} \quad (5\text{-}241)$$

The calculations are shown in Table 5-15.

TABLE 5-15 Edge Rotations for Plate Loads

Property	Plate 1		Plate 2		Plate 3	
	Edge 0	Edge 1	Edge 1	Edge 2	Edge 2	Edge 3
$\pm \Delta f/(a \text{ or } h)$	.....	13.70	3.06	5.11	−5.11	−5.11
$E\delta_n/L^2$	.....	205.5		305.6		−435
$\mp E\Delta\delta/L^2 a$	.....	......	......	1.07	6.95	−13.90
$E\theta_P/L^2$	.....	......	......	8.02		−13.90

On the first line of the table are entered the results of dividing the difference of the stresses at the edges of each plate by the depth for the vertical edge beam and by the horizontal projection of the depth for the other plates. The results for the top edge of a plate are entered with sign changed. The second line, giving $E\delta_n/L^2$, is obtained from the first line by multiplying by 15 for the edge beam or by adding the entries for each plate, and then multiplying by $15/k_n$.

Step 8. Compute the midspan angle change θ_P at each edge. This can be determined from

$$\frac{E}{L^2}\theta_P = -\frac{\delta_{n-1} - \delta_n}{a_n} + \frac{\delta_n - \delta_{n+1}}{a_{n+1}} \quad (5\text{-}242)$$

by continuing the calculations of Step 7 in Table 5-15. The third line of the table is obtained by subtracting the successive entries on the second line and then dividing by the horizontal projection of the plate. The left-hand entry at each edge is written with sign changed. The fourth line, giving $E\theta_P/L^2$, is the sum of the third-line entries for each edge.

Step 9. To correct the edge rotations with a symmetrical loading, apply an unknown moment of $+1,000m_n \sin(\pi x/L)$, in-lb (positive when clockwise) to plate n at edge n and $-1000m_n \sin(\pi x/L)$ to its counterpart, plate n' at edge n'. Also, apply $-1000m_n \sin(\pi x/L)$ to plate $(n + 1)$ at edge n and $+1000m_n \sin(\pi x/L)$ to its counterpart; x is the distance along an edge from the end of a plate. (The sine function is assumed to make the loading vary longitudinally in approximately the same manner as the deflections.) At midspan, the absolute value of these moments is $1000m_n$.

The 12-in-wide transverse strip at midspan, hinged at the supports, will then be subjected at the supports to moments of $1000m_n$. Compute the rotations thus caused in the slabs from

$$\frac{E}{L^2}\theta''_{n-1} = \frac{166.7 h_n m_n}{L^2 t_n^3}$$

$$\frac{E}{L^2}\theta''_n = \frac{333.3 m_n}{L^2}\left(\frac{h_n}{t_n^3} + \frac{h_{n+1}}{t_{n+1}^3}\right) \quad (5\text{-}243)$$

$$\frac{E}{L^2}\theta''_{n+1} = \frac{166.7 h_{n+1} m_n}{L^2 t_{n+1}^3}$$

The results of this calculation are shown in Table 5-16.

TABLE 5-16 Edge Rotations, Reactions, and Plate Loads for Unknown Moments

	Plate 1		Plate 2		Plate 3	
	Edge 0	Edge 1	Edge 1	Edge 2	Edge 2	Edge 3
$E\theta''/L^2$ for m_2......	...	...	...	0.631m_2		0.316m_2
$E\theta''/L^2$ for m_3......	...	...	...	0.158m_3		0.631m_3
R for m_2............	...	−10.70m_2		20.08m_2		−18.76m_2
R for m_3............	...	...		−9.38m_3		18.76m_3
k..................	...	...		0.401		0.3526
R/k for m_2.........	...	...		50.3m_2		−53.2m_2
R/k for m_3.........	...	...		−23.4m_3		53.2m_3
P for m_2............	−10.70m_2		58.1m_2		−105.2m_2	
P for m_3............	0		−27.1m_3		77.8m_3	
M for m_2...........	−4,580m_2		24,840m_2		−45,100m_2	
M for m_3...........	0		−11,620m_3		33,300m_3	

Because edge 1, in effect, is hinged in the actual structure, the correction moments need be applied only at edges 2 and 3. And the angle changes in the slabs due to these moments need be computed only at those edges. Results are given in terms of m_2, the moment at edge 2, and m_3, the moment at edge 3.

Step 10. Compute the slab reactions and plate loads due to the unknown moments. The reactions are

$$R_{n-1} = -\frac{1,000 m_n}{a_n} \quad R_n = 1,000 m_n\left(\frac{1}{a_n} + \frac{1}{a_{n+1}}\right) \quad R_{n+1} = -\frac{1,000 m_n}{a_{n+1}} \quad (5\text{-}244)$$

The plate loads are

$$P_n = \frac{1}{\cos \phi_n}\left(\frac{R_n}{k_n} - \frac{R_{n-1}}{k_{n-1}}\right) \quad (5\text{-}245)$$

The calculations are shown in Table 5-16.

Step 11. Assume that the loading on each plate is $P_n \sin(\pi x/L)$ (Fig. 5-107e), and calculate the midspan (maximum) bending moment. For a simple beam,

$$M = PL^2/\pi^2$$

The moments also are shown in Table 5-16 in terms of m_2 and m_3.

Step 12. Using Eq. (5-239), compute the free-edge longitudinal stresses at midspan. Then, as in Step 6, apply a shear at each edge to equalize the stresses. Determine the adjusted stresses by converging approximations. The results of this calculation are shown in Table 5-17.

TABLE 5-17 Plate Stresses for Unknown Moments

Stress	Plate 1		Plate 2		Plate 3	
	Edge 0	Edge 1	Edge 1	Edge 2	Edge 2	Edge 3
Free-edge f, for m_2	$-20.5m_2$	$20.5m_2$	$51.1m_2$	$-51.1m_2$	$-92.6m_2$	$92.6m_2$
Adjusted f, for m_2	$-31.2m_2$	$41.9m_2$	$41.9m_2$	$-66.3m_2$	$-66.3m_2$	$79.5m_2$
Free-edge f, for m_3			$-23.8m_3$	$23.8m_3$	$68.5m_3$	$-68.5m_3$
Adjusted f, for m_3	$9.2m_3$	$-18.4m_3$	$-18.4m_3$	$41.4m_3$	$41.4m_3$	$-54.9m_3$

Step 13. Compute the vertical component of the edge deflections at midspan from

$$\frac{E}{L^2}\delta_n = \frac{144}{\pi^2 k_n}\left(\frac{f_{n-1} - f_n}{a_n} - \frac{f_n - f_{n+1}}{a_{n+1}}\right) \tag{5-246}$$

or for a vertical plate from

$$\frac{E}{L^2}\delta_n = \frac{144(f_{n-1} - f_n)}{\pi^2 h_n} \tag{5-247}$$

The results of this calculation are shown in Table 5-18. The procedure is the same as Step 7.

Step 14. Using Eq. (5-242), determine the midspan angle change θ' at each edge. The computations are given in Table 5-18 and follow the procedure of Step 8.

TABLE 5-18 Edge Rotations for Unknown Moments

Rotation	Plate 1		Plate 2		Plate 3	
	Edge 0	Edge 1	Edge 1	Edge 2	Edge 2	Edge 3
$\pm\Delta f/(a$ or $h)$ for m_2		$-1.522m_2$	$1.155m_2$	$1.368m_2$	$-1.368m_2$	$-1.368m_2$
$E\delta/L^2$, for m_2		$-22.2m_2$		$92.0m_2$		$-113.2m_2$
$\mp E\Delta\delta/L^2 a$, for m_2				$1.222m_2$	$1.927m_2$	$-3.854m_2$
$E\theta'/L^2$, for m_2				$3.149m_2$		$-3.854m_2$
$\pm\Delta f/(a$ or $h)$ for m_3		$0.574m_3$	$-0.640m_3$	$-0.904m_3$	$0.904m_3$	$0.904m_3$
$E\delta/L^2$, for m_3		$8.4m_3$		$-56.2m_3$		$74.8m_3$
$\mp E\Delta\delta/L^2 a$, for m_3				$-0.691m_3$	$-1.230m_3$	$2.460m_3$
$E\theta'/L^2$, for m_3				$-1.921m_3$		$2.460m_3$

Step 15. At each edge, set up an equation by putting the sum of the angle changes equal to zero. Thus, after division through by E/L^2: $\theta_P + \theta'' + \Sigma\theta' = 0$. Solve these simultaneous equations for the unknown moments.

The sum of the angles for edge 2 is

$$8.02 + 0.631m_2 + 0.158m_3 + 3.149m_2 - 1.921m_3 = 0$$

The sum of the angles for edge 3 is

$$-13.90 + 0.316m_2 + 0.631m_3 - 3.854m_2 + 2.460m_3 = 0$$

The solution is $m_2 = -0.040$ and $m_3 = 4.45$.

Step 16. Determine the actual reactions, loads, stresses, and deflections by substituting for m_2 and m_3 the values just found. For example, the final longitudinal stress at edge 0 is 596 − 31.2(−0.040) + 9.2(4.45) = 638 psi. Similarly, the final longitudinal stress at edge 1 is −145 psi; at edge 2, −160 psi, and at edge 3, −50 psi.

Step 17. Compute the shear stresses. The shear stress at edge n in pounds (Fig. 5-107f) is

$$T_n = T_{n-1} - \frac{f_{n-1} + f_n}{2} A_n \qquad (5\text{-}248)$$

In the example, $T_0 = 0$, so the shears at the edges can be obtained successively, since the stresses f are known. The calculations are shown in Table 5-19.

TABLE 5-19 Shearing Stresses

Stress	Plate 1		Plate 2		Plate 3	
	Edge 0	Edge 1	Edge 1	Edge 2	Edge 2	Edge 3
$-(f_{n-1}+f_n)A_n/2$	...	−83,500		49,500		34,000
Midspan edge shear, lb	0	−83,500		−34,000		0
Edge shear, psi, $x = 0$	0	−61.2	−143	−58	−58	0
$0.75PL/A$	69.9		148.3		46.0	
0.25 edge shear	0	−15.3	−35.8	−14.5	−14.5	0
Middepth shear, psi	54.6		98.0		31.5	

For a uniformly loaded folded plate, the shear stress S, psi, at any point on an edge n is approximately

$$S = \frac{2T_{\max}}{3Lt}\left(\frac{1}{2} - \frac{x}{L}\right) \qquad (5\text{-}249)$$

with a maximum at plate ends of

$$S_{\max} = \frac{T_{\max}}{3Lt} \qquad (5\text{-}250)$$

The edge shears computed with Eq. (5-250) are given in Table 5-19.

The shear stress, psi, at middepth (not always a maximum) is

$$v_n = \left(\frac{3P_n L}{2A_n} + \frac{S_{n-1} + S_n}{2}\right)\left(\frac{1}{2} - \frac{x}{L}\right) \qquad (5\text{-}251)$$

and has its largest value at $x = 0$:

$$v_{\max} = \frac{0.75 P_n L}{A_n} + \frac{S_{n-1} + S_n}{4} \qquad (5\text{-}252)$$

The middepth shears also are given in Table 5-19.

For more details, see D. Yitzhaki and Max Reiss, "Analysis of Folded Plates, Proceedings Paper 3303, *Journal of the Structural Division,* American Society of Civil Engineers, October, 1962.

CABLE-SUPPORTED STRUCTURES*

A cable is a linear structural member, like a bar of a truss. The dimensions of a cable relative to its length, however, are so small that it cannot withstand bending or compres-

---

*Reprinted with permission from F. S. Merritt, "Structural Steel Designers' Handbook," McGraw-Hill Book Company, New York.

sion. Consequently, under loads at an angle to its longitudinal axis, a cable sags and assumes a shape that enables it to develop tensile stresses that resist the loads.

Structural efficiency results from two cable characteristics: (1) uniformity of tensile stresses over the cable cross section, and (2) usually, small variation of tension along the longitudinal axis. Hence, it is economical to use materials with very high tensile strength for cables.

Cables sometimes are used in building construction as an alternative to such tension members as hangers, ties, or tension chords of trusses. For example, cables are used in

Fig. 5-108 Two types of cable-stayed girder construction for roofs.

a form of long-span cantilever-truss construction in which a horizontal roof girder is supported at one end by a column and near the other end by a cable that extends diagonally upward to the top of a vertical mast above the column support (cable-stayed-girder construction, Fig. 5-108). Cable stress can be computed for this case from the laws of equilibrium.

Cables also may be used in building construction instead of girders, trusses, or membranes to support roofs. For the purpose, cables may be arranged in numerous ways. It is consequently impractical to treat in detail in this book any but the simplest types of such applications of cables. Instead, general procedures for analyzing cable-supported structures are presented in the following.

5-104. Simple Cables.

An ideal cable has no resistance to bending. Thus, in analysis of a cable in equilibrium, not only is the sum of the moments about any point equal to zero but so is the bending moment at any point. Consequently, the equilibrium shape of the cable corresponds to the funicular, or bending-moment, diagram for the loading (Fig. 5-109a). As a result, the tensile force at any point of the cable is tangent there to the cable curve.

The point of maximum sag of a cable coincides with the point of zero shear. (Sag in this case should be measured parallel to the direction of the shear forces.)

Fig. 5-109 Simple cable: (a) cable with a uniformly distributed load; (b) cable with supports at different levels.

Stresses in a cable are a function of the deformed shape. Equations needed for analysis, therefore, usually are nonlinear. Also, in general, stresses and deformations cannot be obtained accurately by superimposition of loads. A common procedure in analysis is to obtain a solution in steps by using linear equations to approximate the nonlinear ones and by starting with the initial geometry to obtain better estimates of the final geometry.

For convenience in analysis, the cable tension, directed along the cable curve, usually is resolved into two components. Often, it is advantageous to resolve the tension T into a horizontal component H and a vertical component V (Fig. 5-109b). Under vertical loading then, the horizontal component is constant along the cable. Maximum tension occurs at the support. V is zero at the point of maximum sag.

For a general, distributed vertical load q, the cable must satisfy the second-order linear differential equation

$$Hy'' = q \qquad (5\text{-}253)$$

where y = rise of cable at distance x from low point (Fig. 5-109b)
$y'' = d^2y/dx^2$

Catenary. Weight of a cable of constant cross section represents a vertical loading that is uniformly distributed along the length of cable. Under such a loading, a cable takes the shape of a catenary.

Take the origin of coordinates at the low point C and measure distance s along the cable from C (Fig. 5-109b). If q_o is the load per unit length of cable, Eq. (5-253) becomes

$$Hy'' = \frac{q_o \, ds}{dx} = q_o \sqrt{1 + y'^2} \qquad (5\text{-}254)$$

where $y' = dy/dx$. Solving for y' gives the slope at any point of the cable

$$y' = \frac{\sinh q_o x}{H} = \frac{q_o x}{H} + \frac{1}{3!}\left(\frac{q_o x}{H}\right)^3 + \cdots \qquad (5\text{-}255)$$

A second integration then yields the equation for the cable shape, which is called a catenary.

$$y = \frac{H}{q_o}\left(\cosh \frac{q_o x}{H} - 1\right) = \frac{q_o}{H}\frac{x^2}{2!} + \left(\frac{q_o}{H}\right)^3 \frac{x^4}{4!} + \cdots \qquad (5\text{-}256)$$

If only the first term of the series expansion is used, the cable equation represents a parabola. Because the parabolic equation usually is easier to handle, a catenary often is approximated by a parabola.

For a catenary, length of arc measured from the low point is

$$s = \frac{H}{q_o}\sinh \frac{q_o x}{H} = x + \frac{1}{3!}\left(\frac{q_o}{H}\right)^2 x^3 + \cdots \qquad (5\text{-}257)$$

Tension at any point is

$$T = \sqrt{H^2 + q_o^2 s^2} = H + q_o y \qquad (5\text{-}258)$$

The distance from the low point C to the left support L is

$$a = \frac{H}{q_o}\cosh^{-1}\left(\frac{q_o}{H} f_L + 1\right) \qquad (5\text{-}259)$$

where f_L = vertical distance from C to L. The distance from C to the right support R is

$$b = \frac{H}{q_o}\cosh^{-1}\left(\frac{q_o}{H} f_R + 1\right) \qquad (5\text{-}260)$$

where f_R = vertical distance from C to R.

Given the sags of a catenary f_L and f_R under a distributed vertical load q_o, the horizontal component of cable tension H may be computed from

$$\frac{q_o l}{H} = \cosh^{-1}\left(\frac{q_o f_L}{H} + 1\right) + \cosh^{-1}\left(\frac{q_o f_R}{H} + 1\right) \qquad (5\text{-}261)$$

where l = span, or horizontal distance between supports L and $R = a + b$. This equation usually is solved by trial. A first estimate of H for substitution in the right-hand side of the equation may be obtained by approximating the catenary by a parabola. Vertical components of the reactions at the supports can be computed from

$$R_L = H \sinh \frac{q_o a}{H} \qquad R_R = H \sinh \frac{q_o b}{H} \qquad (5\text{-}262)$$

Parabola. Uniform vertical live loads and uniform vertical dead loads other than cable weight generally may be treated as distributed uniformly over the horizontal projection of the cable. Under such loadings, a cable takes the shape of a parabola.

Take the origin of coordinates at the low point C (Fig. 5-109b). If w_o is the load per foot horizontally, Eq. (5-253) becomes

$$Hy'' = w_o \qquad (5\text{-}263)$$

Integration gives the slope at any point of the cable

$$y' = \frac{w_o x}{H} \qquad (5\text{-}264)$$

A second integration then yields the parabolic equation for the cable shape

$$y = \frac{w_o x^2}{2H} \qquad (5\text{-}265)$$

The distance from the low point C to the left support L is

$$a = \frac{l}{2} - \frac{Hh}{w_o l} \qquad (5\text{-}266)$$

where l = span, or horizontal distance between supports L and $R = a + b$
h = vertical distance between supports
The distance from the low point C to the right support R is

$$b = \frac{l}{2} + \frac{Hh}{w_o l} \qquad (5\text{-}267)$$

When supports are not at the same level, the horizontal component of cable tension H may be computed from

$$H = \frac{w_o l^2}{h^2}\left(f_R - \frac{h}{2} \pm \sqrt{f_L f_R}\right) = \frac{w_o l^2}{8f} \qquad (5\text{-}268)$$

where f_L = vertical distance from C to L
f_R = vertical distance from C to R
f = sag of cable measured vertically from chord LR midway between supports (at $x = Hh/w_o l$)
As indicated in Fig. 5-109b,

$$f = f_L + \frac{h}{2} - y_M \qquad (5\text{-}269)$$

where $y_M = Hh^2/2w_o l^2$. The minus sign should be used in Eq. (5-268) when low point C is between supports. If the vertex of the parabola is not between L and R, the plus sign should be used.

The vertical components of the reactions at the supports can be computed from

$$V_L = w_o a = \frac{w_o l}{2} - \frac{Hh}{l} \qquad V_R = w_o b = \frac{w_o l}{2} + \frac{Hh}{l} \qquad (5\text{-}270)$$

Tension at any point is

$$T = \sqrt{H^2 + w_o^2 x^2} \qquad (5\text{-}271)$$

Length of parabolic arc RC is

$$L_{RC} = \frac{b}{2} \sqrt{1 + \left(\frac{w_o b}{H}\right)^2} + \frac{H}{2w_o} \sinh \frac{w_o b}{H} = b + \frac{1}{6}\left(\frac{w_o}{H}\right)^2 b^3 + \cdots \qquad (5\text{-}272)$$

Length of parabolic arc LC is

$$L_{LC} = \frac{a}{2} \sqrt{1 + \left(\frac{w_o a}{H}\right)^2} + \frac{H}{2w_o} \sinh \frac{w_o a}{H} = a + \frac{1}{6}\left(\frac{w_o}{H}\right)^2 a^3 + \cdots \qquad (5\text{-}273)$$

When supports are at the same level, $f_L = f_R = f$, $h = 0$, and $a = b = l/2$. The horizontal component of cable tension H may be computed from

$$H = \frac{w_o l^2}{8f} \qquad (5\text{-}274)$$

The vertical components of the reactions at the supports are

$$V_L = V_R = \frac{w_o l}{2} \qquad (5\text{-}275)$$

Maximum tension occurs at the supports and equals

$$T_L = T_R = \frac{w_o l}{2} \sqrt{1 + \frac{l^2}{16f^2}} \qquad (5\text{-}276)$$

Length of cable between supports is

$$\begin{aligned} L &= \frac{1}{2} \sqrt{1 + \left(\frac{w_o l}{2H}\right)^2} + \frac{H}{w_o} \sinh \frac{w_o l}{2H} \\ &= l\left(1 + \frac{8}{3}\frac{f^2}{l^2} - \frac{32}{5}\frac{f^4}{l^4} + \frac{256}{7}\frac{f^6}{l^6} + \cdots \right) \end{aligned} \qquad (5\text{-}277)$$

If additional uniformly distributed load is applied to a parabolic cable, the change in sag is approximately

$$\Delta f = \frac{15}{16}\frac{l}{f}\frac{\Delta L}{5 - 24f^2/l^2} \qquad (5\text{-}278)$$

For a rise in temperature t, the change in sag is about

$$\Delta f = \frac{15}{16}\frac{l^2 ct}{f(5 - 24f^2/l^2)}\left(1 + \frac{8}{3}\frac{f^2}{l^2}\right) \qquad (5\text{-}279)$$

where c = coefficient of thermal expansion.

Elastic elongation of a parabolic cable is approximately

$$\Delta L = \frac{Hl}{AE}\left(1 + \frac{16}{3}\frac{f^2}{l^2}\right) \qquad (5\text{-}280)$$

where A = cross-sectional area of cable
 E = modulus of elasticity of cable steel
 H = horizontal component of tension in cable

If the corresponding change in sag is small, so that the effect on H is negligible, this change may be computed from

$$\Delta f = \frac{15}{16} \frac{Hl^2}{AEf} \frac{1 + 16f^2/3l^2}{5 - 24f^2/l^2} \tag{5-281}$$

For the general case of vertical dead load on a cable, the initial shape of the cable is given by

$$y_D = \frac{M_D}{H_D} \tag{5-282}$$

where M_D = dead-load bending moment that would be produced by the load in a simple beam
H_D = horizontal component of tension due to dead load

For the general case of vertical live load on the cable, the final shape of the cable is given by

$$y_D + \delta = \frac{M_D + M_L}{H_D + H_L} \tag{5-283}$$

where δ = vertical deflection of cable due to live load
M_L = live-load bending moment that would be produced by the live load in a simple beam
H_L = increment in horizontal component of tension due to live load

Subtraction of Eq. (5-282) from Eq. (5-283) yields

$$\delta = \frac{M_L - H_L y_D}{H_D + H_L} \tag{5-284}$$

If the cable is assumed to take a parabolic shape, a close approximation to H_L may be obtained from

$$\frac{H_L}{AE} K = \frac{w_D}{H_D} \int_0^l \delta \, dx - \frac{1}{2} \int_0^l \delta'' \delta \, dx \tag{5-285}$$

$$K = l\left[\frac{1}{4}\left(\frac{5}{2} + \frac{16f^2}{l^2}\right)\sqrt{1 + \frac{16f^2}{l^2}} + \frac{3l}{32f}\log_e\left(\frac{4f}{l} + \sqrt{1 + \frac{16f^2}{l^2}}\right)\right] \tag{5-286}$$

where $\delta'' = d^2\delta/dx^2$.

If elastic elongation and δ'' can be ignored, Eq. (5-285) simplifies to

$$H_L = \frac{\int_0^l M_L \, dx}{\int_0^l y_D \, dx} = \frac{3}{2fl} \int_0^l M_L \, dx \tag{5-287}$$

Thus, for a load uniformly distributed horizontally w_L,

$$\int_0^l M_L \, dx = \frac{w_L l^3}{12} \tag{5-288}$$

and the increase in the horizontal component of tension due to live load is

$$H_L = \frac{3}{2fl} \frac{w_L l^3}{12} = \frac{w_L l^2}{8f} = \frac{w_L l^2}{8} \frac{8H_D}{w_D l^2} = \frac{w_L}{w_D} H_D \tag{5-289}$$

When a more accurate solution is desired, the value of H_L obtained from Eq. (5-289) can be used for an initial trial in solving Eqs. (5-284) and (5-285).

(S. P. Timoshenko and D. H. Young, "Theory of Structures," and F. S. Merritt,

"Structural Steel Designers' Handbook," McGraw-Hill Book Company, New York; W. T. O'Brien and A. J. Francis, "Cable Movements under Two-dimensional Loads," *Journal of the Structural Division, ASCE,* Vol. 90, No. ST3, *Proceedings Paper* 3929, June 1964, pp. 89–123; W. T. O'Brien, "General Solution of Suspended Cable Problems," *Journal of the Structural Division, ASCE,* Vol. 93, No. ST1, *Proceedings Paper* 5085, February, 1967, pp. 1–26; W. T. O'Brien, "Behavior of Loaded Cable Systems," *Journal of the Structural Division, ASCE,* Vol. 94, No. ST10, *Proceedings Paper* 6162, October 1968, pp. 2281–2302; G. R. Buchanan, "Two-dimensional Cable Analysis," *Journal of the Structural Division, ASCE,* Vol. 96, No. ST7, *Proceedings Paper* 7436, July 1970, pp. 1581–1587.)

5-105. Cable Systems.

Analysis of simple cables is described in Art. 5-104. Cables, however, may be assembled into many types of systems. One important reason for such systems is that roofs to be supported are two- or three-dimensional. Consequently, three-dimensional cable arrangements often are advantageous. Another important reason is that cable systems can be designed to offer much higher resistance to vibrations than simple cables do.

Like simple cables, cable systems behave nonlinearly. Thus, accurate analysis is difficult, tedious, and time-consuming. As a result, many designers use approximate methods that appear to have successfully withstood the test of time. Because of the numerous types of systems and the complexity of analysis, only general procedures will be outlined in this article.

Cable systems may be stiffened or unstiffened. Stiffened systems, usually used for suspension bridges, are rarely used in buildings. This article will deal only with unstiffened systems, that is, systems where loads are carried to supports only by cables.

Often, unstiffened systems may be classified as a network or as a cable truss, or double-layered plane system.

Networks consist of two or three sets of parallel cables intersecting at an angle (Fig. 5-110). The cables are fastened together at their intersections.

Cable trusses consist of pairs of cables, generally in a vertical plane. One cable of each pair is concave downward, the other concave upward (Fig. 5-111).

Cable Trusses. Both cables of a cable truss are initially tensioned, or prestressed, to a predetermined shape, usually parabolic. The prestress is made large enough that any compression that may be induced in a cable by loads only reduces the tension in the

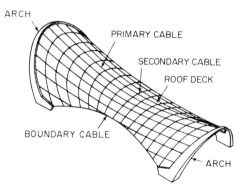

Fig. 5-110 Cable network.

cable; thus, compressive stresses cannot occur. The relative vertical position of the cables is maintained by verticals, or spreaders, or by diagonals. Diagonals in the truss plane do not appear to increase significantly the stiffness of a cable truss.

Figure 5-111 shows four different arrangements of the cables, with spreaders, in a cable truss. The intersecting types (Fig. 5-111*b* and *c*) usually are stiffer than the others, for given size cables and given sag and rise.

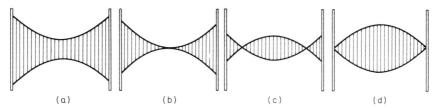

Fig. 5-111 Planar cable systems: (*a*) completely separated cables; (*b*) cables intersecting at midspan; (*c*) crossing cables; (*d*) cables meeting at supports.

For supporting roofs, cable trusses often are placed radially at regular intervals (Fig. 5-112). Around the perimeter of the roof, the horizontal component of the tension usually is resisted by a circular or elliptical compression ring. To avoid a joint with a jumble of cables at the center, the cables usually are also connected to a tension ring circumscribing the center.

Properly prestressed, such double-layer cable systems offer high resistance to vibrations. Wind or other dynamic forces difficult or impossible to anticipate may cause resonance to occur in a single cable, unless damping is provided. The probability of resonance occurring may be reduced by increasing the dead load on a single cable. But this is not economical, because the size of cable and supports usually must be increased as well. Besides, the tactic may not succeed, because future loads may be outside the design range. Damping, however, may be achieved economically with interconnected cables under different tensions, for example, with cable trusses or networks.

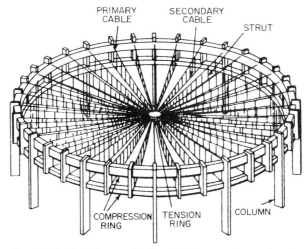

Fig. 5-112 Cable trusses placed radially to support a round roof.

The cable that is concave downward (Fig. 5-111) usually is considered the load-carrying cable. If the prestress in that cable exceeds that in the other cable, the natural frequencies of vibration of both cables will always differ for any value of live load. To avoid resonance, the difference between the frequencies of the cables should increase with increase in load. Thus, the two cables will tend to assume different shapes under specific dynamic loads. As a consequence, the resulting flow of energy from one cable to the other will dampen the vibrations of both cables.

Natural frequency, cycles per second, of each cable may be estimated from

$$\omega_n = \frac{n\pi}{l}\sqrt{\frac{Tg}{w}} \tag{5-290}$$

where n = integer, 1 for the fundamental mode of vibration, 2 for the second mode, . . .
l = span of cable, ft
w = load on cable, kips per ft
g = acceleration due to gravity = 32.2 ft/s²
T = cable tension, kips

The spreaders of a cable truss impose the condition that under a given load the change in sag of the cables must be equal. But the changes in tension of the two cables may not be equal. If the ratio of sag to span f/l is small (less than about 0.1), Eq. (5-281) indicates that, for a parabolic cable, the change in tension is given approximately by

$$\Delta H = \frac{16}{3}\frac{AEf}{l^2}\Delta f \tag{5-291}$$

where Δf = change in sag
A = cross-sectional area of cable
E = modulus of elasticity of cable steel

Double cables interconnected with struts may be analyzed as discrete or continuous systems. For a discrete system, the spreaders are treated as individual members. For a continuous system, the spreaders are replaced by a continuous diaphragm that ensures that the changes in sag and rise of cables remain equal under changes in load. Similarly, for analysis of a cable network, the cables, when treated as a continuous system, may be replaced by a continuous membrane.

Double Cable with Equivalent Diaphragm. To illustrate the procedure for a cable truss treated as a continuous system, the type shown in Fig. 5-111d will be analyzed. The bottom cable will be assumed to be the load-carrying cable. Both cables are prestressed and are assumed to be parabolic. The horizontal component H_{iu} of the initial tension in the upper cable is given. The resulting rise is f_u, and the weight of cables and spreaders is taken as w_c. Span is l.

The horizontal component of the prestress in the bottom cable H_{ib} can be determined by equating the bending moment in the system at midspan to zero.

$$H_{ib} = \frac{f_u}{f_b}H_{iu} + \frac{w_c l^2}{8f_b} = \frac{(w_c + w_i)l^2}{8f_b} \tag{5-292}$$

where f_b = sag of lower cable
w_i = uniformly distributed load exerted by diaphragm on each cable when cables are parabolic

Setting the bending moment at the high point of the upper cable equal to zero yields

$$w_i = \frac{8H_{iu}f_u}{l^2} \tag{5-293}$$

Thus, the lower cable carries a uniform downward load $w_c + w_i$, while the upper cable is subjected to a distributed upward force w_i.

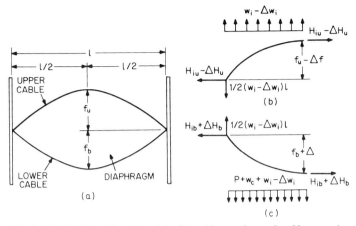

Fig. 5-113 Double-cable system: (*a*) cables with spreaders replaced by an equivalent diaphragm; (*b*) forces acting on the top cable; (*c*) forces acting on the bottom cable.

Suppose a load p uniformly distributed horizontally is now applied to the system (Fig. 5-113*a*). This load may be dead load or dead load plus live load. It will decrease the tension in the upper cable by ΔH_u and the rise by Δf (Fig. 5-113*b*). Correspondingly, the tension in the lower cable will increase by ΔH_b and the sag by Δf (Fig. 5-113*c*). The force exerted by the diaphragm on each cable will decrease by Δw_i.

The changes in tension may be computed from Eq. (5-291). Also, application of this equation to the bending-moment equations for the midpoints of each cable and simultaneous solution of the resulting pair of equations yields the changes in sag and diaphragm force. The change in sag may be estimated from

$$\Delta f = \frac{1}{H_{iu} + H_{ib} + (A_u f_u^2 + A_b f_b^2)16E/3l^2} \frac{pl^2}{8} \tag{5-294}$$

where A_u = cross-sectional area of upper cable
A_b = cross-sectional area of lower cable

The decrease in uniformly distributed diaphragm force is given approximately by

$$\Delta w_i = \frac{(H_{iu} + 16A_u E f_u^2/3l^2)p}{H_{iu} + H_{ib} + (A_u f_u^2 + A_b f_b^2)16E/3l^2} \tag{5-295}$$

And the change in load on the lower cable is nearly

$$p - \Delta w_i = \frac{(H_{ib} + 16A_b E f_b^2/3l^2)p}{H_{iu} + H_{ib} + (A_u f_u^2 + A_b f_b^2)16E/3l^2} \tag{5-296}$$

In Eqs. (5-294) to (5-295), the initial tensions H_{iu} and H_{ib} generally are relatively small compared with the other terms and can be neglected. If then $f_u = f_b$, as is often the case, Eq. (5-295) simplifies to

$$\Delta w_i = \frac{A_u}{A_u + A_b} p \tag{5-297}$$

and Eq. (5-296) becomes

$$p - \Delta w_i = \frac{A_b}{A_u + A_b} p \tag{5-298}$$

The horizontal component of tension in the upper cable for load p may be computed from

$$H_u = H_{iu} - \Delta H_u = \frac{w_i - \Delta w_i}{w_i} H_{iu} \tag{5-299}$$

The maximum vertical component of tension in the upper cable is

$$V_u = \frac{(w_i - \Delta w_i)l}{2} \tag{5-300}$$

The horizontal component of tension in the lower cable may be computed from

$$H_b = H_{ib} + \Delta H_b = \frac{w_c + w_i + p - \Delta w_i}{w_c + w_i} H_{ib} \tag{5-301}$$

The maximum vertical component of tension in the lower cable is

$$V_b = \frac{(w_c + w_i + p - \Delta w_i)l}{2} \tag{5-302}$$

In general, in analysis of cable systems, terms of second-order magnitude may be neglected. But changes in geometry should not be ignored.

Double Cable with Discrete Spreaders. Treatment of a cable truss as a discrete system may be much the same as that for a cable network considered a discrete system. For loads applied to the cables between joints, or nodes, the cable segments between nodes are assumed parabolic. The equations given in Art. 5-104 may be used to determine the forces in the segments and the forces applied at the nodes. Equilibrium equations then can be written for the forces at each joint.

These equations, however, generally are not sufficient for determination of the forces acting in the cable system. These forces also depend on the deformed shape of the network. They may be determined from equations for each joint that take into account both equilibrium and displacement conditions.

For a cable truss (Fig. 5-114a), prestressed initially into parabolic shapes, the forces in the cables and spreaders can be found from equilibrium conditions, as indicated in Fig. 5-114b to e. With the horizontal component of the initial tension in the upper cable H_{iu} given, the prestress in the segment to the right of the high point of that cable (joint 1, Fig. 5-114b) is $T_{iu1} = H_{iu}/\cos \alpha_{Ru1}$. The vertical component of this tension equals $W_{i1} - W_{cu1}$, where W_{i1} is the force exerted by the spreader and W_{cu1} is the load on joint 1 due to the weight of the upper cable. [If the cable is symmetrical about the high point, the vertical component of tension in the cable segment is $(W_{i1} - W_{cu1})/2$.] The direction cosine of the cable segment cos α_{Ru1} is determined by the geometry of the upper cable after it is prestressed. Hence, W_{i1} can be readily computed when H_{iu} is known.

With W_{i1} determined, the initial tension in the lower cable at its low point (joint 3, Fig. 5-114d) can be found from equilibrium requirements in similar fashion and by setting the bending moment at the low point equal to zero. Similarly, the cable and spreader forces at adjoining joints (joints 2 and 4, Fig. 5-114c and e) can be determined.

Suppose now vertical loads are applied to the system. They can be resolved into concentrated vertical loads acting at the nodes, such as the load P at a typical joint O_b of the bottom cable, shown in Fig. 5-115b. The equations of Art. 5-104 can be used for the purpose. The loads will cause vertical displacements δ of all the joints. The spreaders, however, ensure that the vertical displacement of each upper-cable node equals that of the lower-cable node below. A displacement equation can be formulated for each joint of the system. This equation can be obtained by treating a cable truss as a special case of a cable network.

Cable Systems   5-151

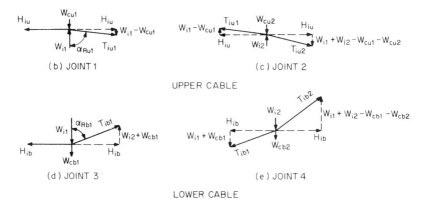

Fig. 5-114 Double-cable system: (a) cables with discrete spreaders; (b) to (e) forces acting at joints 1 to 4.

Cable Networks. A cable network (Fig. 5-110), as explained earlier, consists of interconnected cables. Let joint 0 in Fig. 5-115a represent a typical joint in a cable network and 1, 2, 3, ... adjoining joints. Cable segments 01, 02, 03, ... intersect at 0. 0 is selected as the origin of a three-dimensional coordinate system.

In general, a typical cable segment $0r$ will have direction cosines $\cos \alpha_{rx}$ with respect

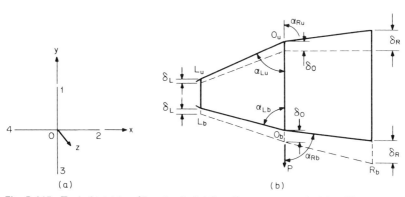

Fig. 5-115 Typical joint in cable network. (a) Coordinate system at the joint. (b) Displacement of the cables by a load at a joint in the network.

to the x axis, $\cos \alpha_{ry}$ with respect to the y axis, and $\cos \alpha_{rz}$ with respect to the z axis. A load P at 0 can be resolved into components P_x parallel to the x axis, P_y parallel to the y axis, and P_z parallel to the z axis. Similarly, the displacement of any joint r can be resolved into components δ_{rx}, δ_{ry}, and δ_{rz}. For convenience, let

$$\Delta_x = \delta_{rx} - \delta_{0x} \qquad \Delta_y = \delta_{ry} - \delta_{0y} \qquad \Delta_z = \delta_{rz} - \delta_{0z} \qquad (5\text{-}303)$$

For a cable-network joint, in general, then, where n cable segments interconnect, three equations can be established:

$$\sum_{r=1}^{n} \left[\frac{EA_r}{l_r} \cos \alpha_{rz} (\Delta_x \cos \alpha_{rx} + \Delta_y \cos \alpha_{ry} + \Delta_z \cos \alpha_{rz}) \right] + P_z = 0 \qquad (5\text{-}304a)$$

$$\sum_{r=1}^{n} \left[\frac{EA_r}{l_r} \cos \alpha_{ry} (\Delta_x \cos \alpha_{rx} + \Delta_y \cos \alpha_{ry} + \Delta_z \cos \alpha_{rz}) \right] + P_y = 0 \qquad (5\text{-}304b)$$

$$\sum_{r=1}^{n} \left[\frac{EA_r}{l_r} \cos \alpha_{rx} (\Delta_x \cos \alpha_{rx} + \Delta_y \cos \alpha_{ry} + \Delta_z \cos \alpha_{rz}) \right] + P_x = 0 \qquad (5\text{-}304c)$$

where E = modulus of elasticity of cable steel
A_r = cross-sectional area of cable segment $0r$
l_r = length of chord from 0 to r

These equations are based on the assumption that the ratio of rise or sag to span of the cables is small, and that, for any cable segment, initial tension T_i can be considered negligible compared with EA.

For a cable truss, $n = 2$ for a typical joint. If only vertical loading is applied, only Eq. (5-304a) is needed. At typical joints O_u of the upper cable and O_b of the bottom cable (Fig. 5-115b), the vertical displacement is denoted by δ_o. The displacements of the joints L_u and L_b on the left of O_u and O_b are indicated by δ_L. Those of the joints R_u and R_b on the right of O_u and O_b are represented by δ_R. Then, for joint O_u, Eq. (5-304a) becomes

$$\frac{EA_{Lu}}{l_{Lu}} \cos^2 \alpha_{Lu} (\delta_L - \delta_O) + \frac{EA_{Ru}}{l_{Ru}} \cos^2 \alpha_{Ru} (\delta_R - \delta_O) = W_i - \Delta W_i - W_{cu} \qquad (5\text{-}305)$$

where W_i = force exerted by spreader at O_u and O_b before application of P
ΔW_i = change in spreader force due to P
W_{cu} = load at O_u from weight of upper cable
A_{Lu} = cross-sectional area of upper-cable segment on the left of O_u
l_{Lu} = length of chord from O_u to L_u
A_{Ru} = cross-sectional area of upper-cable segment on the right of O_u
l_{Ru} = length of chord from O_u to R_u

For joint O_b, Eq. (5-304a) becomes, on replacement of subscript u by b,

$$\frac{EA_{Lb}}{l_{Lb}} \cos^2 \alpha_{Lb} (\delta_L - \delta_O) + \frac{EA_{Rb}}{l_{Rb}} \cos^2 \alpha_{Rb} (\delta_r - \delta_O) = -P - W_i + \Delta W_i - W_{cb} \qquad (5\text{-}306)$$

where W_{cb} = load at O_b due to weight of lower cable and spreader.

Thus, for a cable truss with m joints in each cable, there are m unknown vertical displacements δ and m unknown changes in spreader force ΔW_i. Equations (5-305) and (5-306), applied to upper and lower nodes, respectively, provide $2m$ equations. Simultaneous solution of these equations yields the displacements and forces needed to complete the analysis.

The direction cosines in Eqs. (5-304) to (5-306), however, should be those for the displaced cable segments. If the direction cosines of the original geometry of a cable network are used in these equations, the computed deflections will be larger than the true deflections, because cables become stiffer as sag increases. The computed displacements, however, may be used to obtain revised direction cosines. The equations may then be solved again to yield corrected displacements. The process can be repeated as many times as necessary for convergence.

For cable networks in general, convergence can often be speeded by computing the direction cosines for the third cycle of solution with node displacements that are obtained by averaging the displacements at each node computed in the first two cycles.

(F. S. Merritt, "Structural Steel Designers' Handbook," McGraw-Hill Book Company, New York; H. Mollman, "Analysis of Plane Prestressed Cable Structures," *Journal of the Structural Division, ASCE,* Vol. 96, No. ST10, *Proceedings Paper* 7598, October 1970, pp. 2059–2082; D. P. Greenberg, "Inelastic Analysis of Suspension Roof Structures," *Journal of the Structural Division, ASCE,* Vol. 96, No. ST5, *Proceedings Paper* 7284, May 1970, pp. 905–930; H. Tottenham and P. G. Williams, "Cable Net: Continuous System Analysis," *Journal of the Engineering Mechanics Division, ASCE,* Vol. 96, No. EM3, *Proceedings Paper* 7347, June 1970, pp. 277–293; A. Siev, "A General Analysis of Prestressed Nets," *Publications, International Association for Bridge and Structural Engineering,* Vol. 23, pp. 283–292, Zurich, Switzerland, 1963; A. Siev, "Stress Analysis of Prestressed Suspended Roofs," *Journal of the Structural Division, ASCE,* Vol. 90, No. ST4, *Proceedings Paper* 4008, August 1964, pp. 103–121; C. H. Thornton and C. Birnstiel, "Three-dimensional Suspension Structures," *Journal of the Structural Division, ASCE,* Vol. 93, No. ST2, *Proceedings Paper* 5196, April 1967, pp. 247–270.)

RAPID APPROXIMATE ANALYSIS OF MULTISTORY FRAMES

Buildings must be designed to resist horizontal forces as well as vertical loads. In tall buildings, the lateral forces must be given particular attention, because if they are not properly provided for, they can cause collapse of the structure (Art. 3-5).

Horizontal forces are generally taken care of with X bracing, shear walls, or wind connections. **X bracing,** usually adopted for low industrial-type buildings, transmits the forces to the ground by truss action. **Shear walls** are vertical slabs capable of transmitting the forces to the ground without shearing or buckling; they generally are so deep in the direction of the lateral loads that bending stresses are easily handled. **Wind connections** are a means of establishing continuity between girders and columns. The structure is restrained against lateral deformation by the resistance to rotation of the members at the connections. (See also Art. 3-6.)

The continuous rigid frames thus formed can be analyzed by the methods of Arts. 5-63 to 5-68, 5-80 to 5-85, or 5-110 to 5-113. Sometimes, however, approximate methods are used.

It is noteworthy that for most buildings even the "exact" methods are not exact. In the first place, the forces acting are not static loads, but generally dynamic; they are uncertain in intensity, direction, and duration. Earthquake forces, usually assumed as a percentage of the weight of the building above each level, act at the base of the structure, not at each floor level as is assumed in design, and accelerations at each level vary nearly linearly with distance above the base. Also, at the beginning of a design, the sizes of members are not known, so the exact resistance to lateral deformation cannot be calculated. Furthermore, floors, walls, and partitions help resist the lateral forces in a very uncertain way. Hence, the approximate methods described in the following articles often are useful in analyzing tall buildings subjected to lateral loads.

5-106. Portal Method.

Since an exact analysis is impossible, most designers prefer a wind-analysis method based on reasonable assumptions and requiring a minimum of calculations. One such method is the so-called "portal method."

It is based on the assumptions that points of inflection (zero bending moment) occur at the midpoints of all members and that exterior columns take half as much shear as do interior columns. These assumptions enable all moments and shears throughout the building frame to be computed by the laws of equilibrium.

Consider, for example, the roof level (Fig. 5-116a) of a tall building. A wind load of 600 lb is assumed to act along the top line of girders. To apply the portal method, we cut the building along a section through the inflection points of the top-story columns, which are assumed to be at the column midpoints, 6 ft down from the top of the building. We need now consider only the portion of the structure above this section.

Since the exterior columns take only half as much shear as do the interior columns, they each receive 100 lb, and the two interior columns, 200 lb. The moments at the tops of the columns equal these shears times the distance to the inflection point. The wall end of the end girder carries a moment equal to the moment in the column. (At the floor level below, as indicated in Fig. 5-116b, that end of the end girder carries a moment equal to the sum of the column moments.) Since the inflection point is at the midpoint of the girder, the moment at the inner end of the girder must be the same as at the outer end. The moment in the adjoining girder can be found by subtracting this moment from the column moment, because the sum of the moments at the joint must be zero. (At the floor level below, as shown in Fig. 5-116b, the moment in the interior girder is found by subtracting the moment in the exterior girder from the sum of the column moments.)

Fig. 5-116 Wind stresses in a tall building.

Girder shears then can be computed by dividing girder moments by the half span. When these shears have been found, column loads can be easily computed from the fact that the sum of the vertical loads must be zero, by taking a section around each joint through column and girder inflection points. As a check, it should be noted that the column loads produce a moment that must be equal to the moments of the wind loads above the section for which the column loads were computed. For the roof level (Fig. 5-116a), for example, $-50 \times 24 + 100 \times 48 = 600 \times 6$.

5-107. Cantilever Method.

Another wind-analysis procedure that is sometimes employed is the cantilever method. Basic assumptions here are that inflection points are at the midpoints of all members and that direct stresses in the columns vary as the distances of the columns from the center of gravity of the bent. The assumptions are sufficient to enable shears and moments in the frame to be determined from the laws of equilibrium.

For multistory buildings with height-to-width ratio of 4 or more, the Spurr modification is recommended ("Welded Tier Buildings," U.S. Steel Corp.). In this method, the moments of inertia of the girders at each level are made proportional to the girder shears.

The results obtained from the cantilever method generally will be different from those obtained by the portal method. In general, neither solution is correct, but the answers provide a reasonable estimate of the resistance to be provided against lateral deformation. (See also *Transactions of the ASCE*, Vol. 105, pp. 1713–1739, 1940.)

ULTIMATE STRENGTH OF DUCTILE FLEXURAL MEMBERS

When an elastic material, such as structural steel, is loaded in tension with a gradually increasing load, stresses are proportional to strains up to the proportional limit (near the yield point). If the material, like steel, also is ductile, then it continues to carry load beyond the yield point, though strains increase rapidly with little increase in load (Fig. 5-117a).

Fig. 5-117 Stress-strain relationship for a ductile material generally is similar to the curve shown in (a). To simplify plastic analysis, the portion of (a) enclosed by the dash lines is approximated by the curve in (b), which extends to the range where strain hardening begins.

Similarly, a beam made of an elastic material continues to carry more load after the stresses in the outer surfaces reach the yield point. However, the stresses will no longer vary with distance from the neutral axis, so the flexure formula [Eq. (5-23)] no longer holds. However, if simplifying assumptions are made, approximating the stress-strain relationship beyond the elastic limit, the load-carrying capacity of the beam can be computed with satisfactory accuracy.

5-108. Theory of Plastic Behavior.

For a ductile material, the idealized stress-strain relationship in Fig. 5-117b may be assumed. Stress is proportional to strain until the yield-point stress f_y is reached, after which strain increases at a constant stress.

For a beam of this material, the following assumptions will also be made: plane sections remain plane, strains thus being proportional to distance from the neutral axis; properties of the material in tension are the same as those in compression; its fibers behave the same in flexure as in tension; and deformations remain small.

Strain distribution across the cross section of a rectangular beam, based on these assumptions, is shown in Fig. 5-118a. At the yield point, the unit strain is ϵ_y and the curvature ϕ_y, as indicated in (1). In (2), the strain has increased several times, but the section still remains plane. Finally, at failure, (3), the strains are very large and nearly constant across upper and lower halves of the section.

Corresponding stress distributions are shown in Fig. 5-118b. At the yield point, (1), stresses vary linearly and the maximum is f_y. With increase in load, more and more fibers reach the yield point, and the stress distribution becomes nearly constant, as indicated in (2). Finally, at failure, (3), the stresses are constant across the top and bottom parts of the section and equal to the yield-point stress.

(a) STRAIN DISTRIBUTION

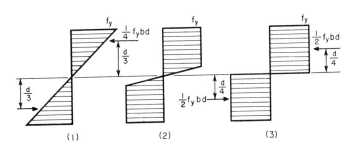

(b) STRESS DISTRIBUTION

Fig. 5-118 Strain distribution is shown in (a) and stress distribution in (b) for a cross section of a beam as it is loaded beyond the yield point, for the idealized stress-strain relationship in Fig. 5-117b: stage (1) shows the condition at the yield point of the outer fibers; (2) after yielding starts; (3) at ultimate load.

The resisting moment at failure for a rectangular beam can be computed from the stress diagram for stage 3. If b is the width of the member and d its depth, then the ultimate moment for a rectangular beam is

$$M_P = \frac{bd^2}{4} f_y \tag{5-307}$$

Since the resisting moment at stage 1 is $M_y = f_y bd^2/6$, the beam carries 50% more moment before failure than when the yield-point stress is first reached in the outer fibers.

A circular section has an M_P/M_y ratio of about 1.7, while a diamond section has a ratio of 2. The average wide-flange rolled-steel beam has a ratio of about 1.14.

Plastic Hinges. The relationship between moment and curvature in a beam can be assumed to be similar to the stress-strain relationship in Fig. 5-117b. Curvature ϕ varies linearly with moment until $M_y = M_P$ is reached, after which ϕ increases indefinitely at constant moment. That is, a plastic hinge forms.

Moment Redistribution. This ability of a ductile beam to form plastic hinges enables a fixed-end or continuous beam to carry more load after M_P occurs at a section, because a redistribution of moments takes place. Consider, for example, a uniformly loaded, fixed-end, prismatic beam. In the elastic range, the end moments are $M_L = M_R = WL/12$, while the midspan moment M_C is $WL/24$. The load when the yield point is reached in the outer fibers at the beam ends is $W_y = 12M_y/L$. Under this load the moment capacity of the ends of the beam is nearly exhausted; plastic hinges form there when the moment equals M_P. As load is increased, the ends then rotate under constant moment and the beam deflects like a simply supported beam. The moment at midspan increases until the moment capacity at that section is exhausted and a plastic hinge forms. The load causing that condition is the ultimate load W_u since, with three hinges in the span, a link mechanism is formed and the member continues to deform at constant load. At the time the third hinge is formed, the moments at ends and center are all equal to M_P. Therefore, for equilibrium, $2M_P = W_u L/8$, from which $W_u = 16M_P/L$. Since for the idealized moment-curvature relationship, M_P was assumed equal to M_y, the carrying capacity due to redistribution of moments is 33% greater than W_y.

5-109. Upper and Lower Bounds for Ultimate Loads.

Methods for computing the ultimate strength of continuous beams and frames may be based on two theorems that fix upper and lower limits for load-carrying capacity.

Upper-Bound Theorem. A load computed on the basis of an assumed link mechanism will always be greater than or at best equal to the ultimate load.

Lower-Bound Theorem. The load corresponding to an equilibrium condition with arbitrarily assumed values for the redundants is smaller than or at best equal to the ultimate loading—provided that everywhere moments do not exceed M_P.

Equilibrium Method. This method, based on the lower-bound theorem, usually is easier for simple cases. The steps involved are:

1. Select redundants that if removed would leave the structure determinate.
2. Draw the moment diagram for the determinate structure.
3. Sketch the moment diagram for an arbitrary value of each redundant.
4. Combine the moment diagrams, forming enough peaks so that the structure will act as a link mechanism if plastic hinges are formed at those points.
5. Compute the value of the redundants from the equations of equilibrium, assuming that at the peaks $M = M_P$.
6. Check to see that there are sufficient plastic hinges to form a mechanism and that M is everywhere less than or equal to M_P.

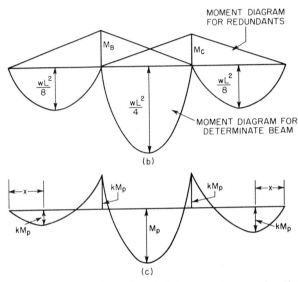

Fig. 5-119 Continuous beam shown in (a) carries twice as much uniform load in the center span as in the side span. In (b) are shown the moment diagrams for this loading condition with redundants removed and for the redundants. The two moment diagrams are combined in (c), producing peaks at which plastic hinges are assumed to form.

Consider, for example, the continuous beam $ABCD$ in Fig. 5-119a with three equal spans, uniformly loaded, the center span carrying double the load of the end span. Assume that the plastic moment for the end spans is k times that for the center span ($k < 1$). For what value of k will the ultimate strength be the same for all spans?

Figure 5-119b shows the moment diagram for the beam made determinate by ignoring the moments at B and C, and the moment diagram for end moments M_B and M_C applied to the determinate beam. Figure 5-119c shows the combined moment diagram. If plastic hinges form at all the peaks, a link mechanism will exist.

Since at B and C the joints can develop only the strength of the weakest beam, a plastic hinge will form when $M_B = M_C = kM_p$. A plastic hinge also will form at the center of span BC when the midspan moment is M_P. For equilibrium to be maintained, this will occur when $M_P = wL^2/4 - \tfrac{1}{2}M_B - \tfrac{1}{2}M_C = wL^2/4 - kM_P$, from which

$$M_P = \frac{wL^2}{4(1+k)} \tag{5-308}$$

Maximum moment occurs in spans AB and CD where $x = L/2 - M/wL$—or if $M = kM_P$, where $x = L/2 - kM_P/wL$. A plastic hinge will form at this point when the moment equals kM_P. For equilibrium, therefore,

$$kM_P = \frac{w}{2}x(L-x) - \frac{x}{L}kM_P = \frac{w}{2}\left(\frac{L}{2} - \frac{kM_P}{wL}\right)\left(\frac{L}{2} + \frac{kM_P}{wL}\right) - \left(\frac{1}{2} - \frac{kM_P}{wL^2}\right)kM_P$$

This leads to the quadratic equation:

$$\frac{k^2 M_P^2}{wL^2} - 3kM_P + \frac{wL^2}{4} = 0 \tag{5-309}$$

Substitution of M_P from Eq. (5-308) in Eq. (5-309) yields

$$7k^2 + 4k = 4 \quad \text{or} \quad k(k + \tfrac{4}{7}) = \tfrac{4}{7}$$

from which $k = 0.523$. And the ultimate load is

$$wL = \frac{4M_P(1+k)}{L} = 6.1\frac{M_P}{L} \tag{5-310}$$

Mechanism Method. This method, based on the upper-bound theorem, requires the following steps:

1. Determine the location of possible hinges (points of maximum moment).
2. Pick combinations of hinges to form possible link mechanisms. Certain types are classed as elementary, such as the mechanism formed when hinges are created at the ends and center of a fixed-end or continuous beam, or when hinges form at top and bottom of a column subjected to lateral load, or when hinges are formed at a joint in the members framing into it, permitting the joint to rotate freely. If a rigid frame has n possible plastic hinges and x redundant forces or moments, then there are $n - x$ independent equilibrium equations and $n - x$ elementary mechanisms. Both elementary mechanisms and possible combinations of them must be investigated.
3. Apply a virtual displacement to each possible mechanism in turn and compute the internal and external work (Art. 5-52).
4. From the equality of internal and external work, compute the critical load for each mechanism. The mechanism with the lowest critical load is the most probable, and its load is the ultimate, or limit load.
5. Make an equilibrium check to ascertain that moments everywhere are less than or equal to M_P.

As an example of an application of the mechanism method, let us find the ultimate load for the rigid frame of constant section throughout in Fig. 5-120a. Assume that the vertical load at midspan is equal to 1.5 times the lateral load.

Maximum moments can occur at five points—$A, B, C, D,$ and E—and plastic hinges may form there. Since this structure has three redundants, the number of elementary mechanisms is $5 - 3 = 2$. These are shown in Fig. 5-120b and c. A combination of these mechanisms is shown in Fig. 5-120d.

Let us first investigate the beam alone, with plastic hinges at $B, E,$ and C. As indicated in Fig. 5-120e, apply a virtual rotation θ to BE at B. The center deflection, then, is $\theta L/2$. Since C is at a distance of $L/2$ from E, the rotation at C must be θ and at E, 2θ. The external work, therefore, is $1.5P$ times the midspan deflection, $\theta L/2$, which equals $\tfrac{3}{4}\theta PL$. The internal work is the sum of the work at each hinge, or $M_P\theta + 2M_P\theta + M_P\theta = 4M_P\theta$. Equating internal and external work: $\tfrac{3}{4}\theta PL = 4M_P\theta$, from which $P = 5.3M_P/L$.

Next, let us investigate the frame mechanism, with plastic hinges at $A, B, C,$ and D. As shown in Fig. 5-120f, apply a virtual rotation θ to AB. The point of application of the lateral load P will then move a distance $\theta L/2$, and the external work will be $\theta PL/2$. The

Fig. 5-120 Ultimate-load possibilities for a rigid frame of constant cross section and with fixed-end bases.

internal work, being the sum of the work at the hinges, will be $4M_P\theta$. Equating internal and external work, $\theta PL/2 = 4M_P\theta$, from which $P = 8M_P/L$.

Finally, let us compute the critical load for the combination beam-frame mechanism in Fig. 5-120g. Again, apply a virtual rotation θ to AB, moving B horizontally and E vertically a distance of $\theta L/2$. The external work, therefore, equals

$$\frac{\theta PL}{2} + \frac{3}{4}\theta PL = \frac{5\theta PL}{4}$$

The internal work is the sum of the work at hinges $A, E, C,$ and D and is equal to $M_P\theta + 2M_P\theta + M_P\theta = 6M_P\theta$. Equating internal and external work, $5\theta PL/4 = 6M_P\theta$, from which $P = 4.8M_P/L$.

The combination mechanism has the lowest critical load. Now an equilibrium check must be made to ensure that the moments everywhere in the frame are less than M_P. If they are, then the combination mechanism is the correct solution.

Consider first the length EC of the beam as a free body. Taking moments about E, noting that the moments at E and C equal M_P, the shear at C is computed to be $4M_P/L$. Therefore, the shear at B is $1.5 \times 4.8M_P/L - 4M_P/L = 3.2M_P/L$. By taking moments about B, with BE considered as a free body, the moment at B is found to be $0.6M_P$. Similar treatment of the columns as free bodies indicates that nowhere is the moment greater than M_P. Therefore, the combination mechanism is the correct solution, and the ultimate load for the frame is $4.8M_P/L$ laterally and $7.2M_P/L$ vertically at midspan.

(R. O. Disque, "Applied Plastic Design in Steel," Van Nostrand Reinhold Company, New York; "Plastic Design in Steel—A Guide and Commentary, M & R No. 41, American Society of Civil Engineers, New York; B. G. Neal, "Plastic Methods of Structural Analysis," John Wiley & Sons, Inc., New York.)

FINITE-ELEMENT METHODS

From the basic principles given in preceding articles, systematic procedures have been developed for determining the behavior of a structure from a knowledge of the behavior under load of its components. In these methods, called finite-element methods, a structural system is considered an assembly of a finite number of finite-size components, or elements. These are assumed to be connected to each other only at discrete points, called nodes. From the characteristics of the elements, such as their stiffness or flexibility, the characteristics of the whole system can be derived. With these known, internal stresses and strains throughout can be computed.

Choice of elements to be used depends on the type of structure. For example, for a truss with joints considered hinged, a natural choice of element would be a bar, subjected only to axial forces. For a rigid frame, the elements might be beams subjected to bending and axial forces, or to bending, axial forces, and torsion. For a thin plate or shell, elements might be triangles or rectangles, connected at vertices. For three-dimensional structures, elements might be beams, bars, tetrahedrons, cubes, or rings.

For many structures, because of the number of finite elements and nodes, analysis by a finite-element method requires mathematical treatment of large amounts of data and solution of numerous simultaneous equations. For this purpose, the use of high-speed computers is advisable.

The mathematics of such analyses is usually simpler and more compact when the data are handled in matrix form. Matrix notation is especially convenient in indicating the solution of simultaneous linear equations. For example, suppose a set of equations is represented in matrix notation by $\mathbf{AX} = \mathbf{B}$. Multiplication of both sides of the equation by inverse $\mathbf{A}^{-1}$ yields $\mathbf{A}^{-1}\mathbf{AX} = \mathbf{A}^{-1}\mathbf{B}$. Since $\mathbf{A}^{-1}\mathbf{A} = \mathbf{I}$, the identity matrix, and $\mathbf{IX} = \mathbf{X}$, the solution of the equations is given by $\mathbf{X} = \mathbf{A}^{-1}\mathbf{B}$. (F. S. Merritt, "Modern Mathematical Methods for Engineers," McGraw-Hill Book Company, New York.)

The methods used for analyzing structures generally may be classified as force (flexibility) or displacement (stiffness) methods.

In analysis of statically indeterminate structures by force methods, forces are chosen as redundants, or unknowns. The choice is made in such a way that equilibrium is satisfied. These forces are then determined from the solution of equations that ensure compatibility of all displacements of elements at each node. After the redundants have been computed, stresses and strains throughout the structure can be found from equilibrium equations and stress-strain relations.

In displacement methods, displacements are chosen as unknowns. The choice is made in such a way that geometric compatibility is satisfied. These displacements are then determined from the solution of equations that ensure that forces acting at each node are in equilibrium. After the unknowns have been computed, stresses and strains throughout the structure can be found from equilibrium equations and stress-strain relations.

In choosing a method, the following should be kept in mind: In force methods, the number of unknowns equals the degree of indeterminancy. In displacement methods, the number of unknowns equals the degrees of freedom of displacement at nodes. The fewer the unknowns, the fewer the calculations required.

5-110. Force-Displacement Relations.

Assume a right-handed cartesian coordinate system, with axes x, y, z. Assume also at each node of a structure to be analyzed a system of base unit vectors, $\mathbf{e}_1$ in the direction of the x axis, $\mathbf{e}_2$ in the direction of the y axis, and $\mathbf{e}_3$ in the direction of the z axis. Forces and moments acting at a node are resolved into components in the directions of the base vectors. Then, the forces and moments at the node may be represented by the vector $P_i\mathbf{e}_i$, where P_i is the magnitude of the force or moment acting in the direction of $\mathbf{e}_i$. This vector, in turn, may be conveniently represented by a column matrix $\mathbf{P}$. Similarly, the displacements—translations and rotation—of the node may be represented by the vector

$\Delta_i \mathbf{e}_i$, where Δ_i is the magnitude of the displacement acting in the direction of $\mathbf{e}_i$. This vector, in turn, may be represented by a column matrix Δ.

For compactness, and because, in structural analysis, similar operations are performed on all nodal forces, all the loads, including moments, acting on all the nodes may be combined into a single column matrix $\mathbf{P}$. Similarly, all the nodal displacements may be represented by a single column matrix Δ.

If the independent loads and displacements are listed in the appropriate order in each matrix, a matrix equation relating them can be written:

$$\Delta = \mathbf{FP} \tag{5-311}$$

For an elastic structure, $\mathbf{F}$ is the flexibility matrix. It is square (has the same number of rows as columns) and symmetric ($\mathbf{F} = \mathbf{F}^T$). A typical element F_{ij} of $\mathbf{F}$ gives the deflection of a node in the direction of displacement Δ_i when a unit force acts at the same or another node in the direction of force P_j. The jth column of $\mathbf{F}$, therefore, contains all the nodal displacements when one force P_j is set equal to unity and all other independent forces are zero.

Multiplication of both sides of Eq. (5-311) by $\mathbf{F}^{-1}$ yields $\mathbf{F}^{-1}\Delta = \mathbf{F}^{-1}\mathbf{FP} = \mathbf{IP} = \mathbf{P}$, or

$$\mathbf{P} = \mathbf{K}\Delta \tag{5-312}$$

where $\mathbf{K}$ = inverse of flexibility matrix = $\mathbf{F}^{-1}$.

For an elastic structure, $\mathbf{K}$ is the stiffness matrix. It too is square and symmetric. A typical element K_{ij} of $\mathbf{K}$ gives the force at a node, in the direction of load P_i, that results when the same or another node is given a unit displacement in the direction of displacement Δ_j. The jth column of $\mathbf{K}$, therefore, contains the nodal forces that produce a unit displacement of the node at which Δ_j occurs and in the direction of Δ_j, but no other nodal displacements throughout the structure.

For some structures, the flexibility or stiffness matrix can be constructed from the definition. In other cases, the matrix may have to be synthesized from the flexibility or stiffness matrices of finite elements comprising the structures. The procedures are explained in the following articles.

5-111. Matrix Force (Flexibility) Method.

In this article, the structure is assumed to consist of n finite elements connected only at the nodes. The flexibility matrices of the elements are assumed known. Right-handed cartesian coordinate axes, x, y, z, are chosen for the structure, with base unit vectors $\mathbf{e}_i$ parallel to these axes at each mode, as in Art. 5-110. The flexibility matrix for the whole structure can be constructed from the element flexibility matrices, as will be demonstrated.

For the ith element, nodal forces and displacements are related by

$$\delta_i = \mathbf{f}_i \mathbf{S}_i \qquad i = 1, 2, \ldots n \tag{5-313}$$

where δ_i = matrix of displacements of the nodes of the ith element in directions of base vectors

$\mathbf{f}_i$ = flexibility matrix of the ith element (Art. 5-113)

$\mathbf{S}_i$ = matrix of forces, moments, torques acting at nodes of the ith element in directions of base vectors

For compactness, this relationship between nodal displacements and forces for each element can be combined into a single matrix equation applicable to all the elements:

$$\delta = \mathbf{fS} \tag{5-314}$$

where δ = matrix of all nodal displacements for all elements
S = matrix of all forces at the nodes of all elements

$$\mathbf{f} = \begin{bmatrix} \mathbf{f}_1 & 0 & \cdots & 0 \\ 0 & \mathbf{f}_2 & \cdots & 0 \\ \cdots & \cdots & \cdots & \cdots \\ 0 & 0 & \cdots & \mathbf{f}_n \end{bmatrix} \tag{5-315}$$

Element forces $\mathbf{S}$ can be determined from the loads $\mathbf{P}$ at the nodes to meet equilibrium requirements at the nodes:

$$\mathbf{S} = \mathbf{a}_0 \mathbf{P} \tag{5-316}$$

where $\mathbf{a}_0$ is a matrix of influence coefficients. The jth column of $\mathbf{a}_0$ contains the element forces when one load $P_j = 1$, and all other loads are set equal to zero.

From Eqs. (5-314) and (5-316),

$$\delta = \mathbf{f} \mathbf{a}_0 \mathbf{P} \tag{5-317}$$

From energy relations, it can be shown that

$$\Delta = \mathbf{a}_0^T \delta \tag{5-318}$$

where Δ = matrix of displacements at the nodes of the structure
$\mathbf{a}_0^T$ = transpose of $\mathbf{a}_0$ = matrix $\mathbf{a}_0$ with rows and columns interchanged
Finally, from Eqs. (5-317) and (5-318) comes the relationship that permits computation of the flexibility matrix of the whole structure from the flexibility matrices of its elements:

$$\Delta = \mathbf{a}_0^T \mathbf{f} \mathbf{a}_0 \mathbf{P} \tag{5-319}$$

Since also, by Eq. (5-311), $\Delta = \mathbf{FP}$,

$$\mathbf{F} = \mathbf{a}_0^T \mathbf{f} \mathbf{a}_0 \tag{5-320}$$

Equations (5-311) and (5-313) to (5-320) can be used to compute all forces and displacements in a statically determinate structure.

If the structure is statically indeterminate, however, $\mathbf{a}_0$ cannot be developed from the equilibrium equations alone. Forces and displacements, though, can be determined from the following equations:

First, the structure is reduced, made statically determinate by replacing some constraints with redundant forces, represented by a matrix $\mathbf{X}$. All element forces $\mathbf{S}$ are related to loads $\mathbf{P}$ at the nodes of the reduced structure by

$$\mathbf{S} = \mathbf{a}_0 \mathbf{P} + \mathbf{a}_1 \mathbf{X} \tag{5-321}$$

where $\mathbf{a}_0$ = matrix of influence coefficients for loads acting on reduced structure
$\mathbf{a}_1$ = matrix of influence coefficients for redundants acting on reduced structure
The ith column of $\mathbf{a}_1$ contains the element forces resulting from a unit load applied in the direction of $\mathbf{X}_i$ at the node where $\mathbf{X}_i$ acts. Substitution of Eq. (5-321) in Eq. (5-314) yields

$$\delta = \mathbf{f}\mathbf{S} = \mathbf{f}(\mathbf{a}_0 \mathbf{P} + \mathbf{a}_1 \mathbf{X}) \tag{5-322}$$

Generally, the displacements associated with redundants are zero. Hence, by Eq. (5-318),

$$\mathbf{a}_1^T \delta = \mathbf{0} \tag{5-323}$$

Solving for the redundants from Eqs. (5-322) and (5-323) yields

$$\mathbf{X} = -\mathbf{D}^{-1} \mathbf{D}_0 \mathbf{P} \tag{5-324}$$

where $\mathbf{D}_0 = \mathbf{a}_1^T \mathbf{f} \mathbf{a}_0$
$\mathbf{D}^{-1}$ = inverse of matrix $\mathbf{D}$
$\mathbf{D} = \mathbf{a}_1^T \mathbf{f} \mathbf{a}_1$

Substitution of the known values of the redundants in Eq. (5-321) permits computation of the element forces from

$$\mathbf{S} = (\mathbf{a}_0 - \mathbf{a}_1 \mathbf{D}^{-1} \mathbf{D}_0)\mathbf{P} \tag{5-325}$$

Methods of developing flexibility matrices of finite elements are given in Art. 5-113.

Loads between Nodes. When loads act along a beam, they should be replaced by equivalent loads at the nodes—simple-beam reactions and fixed-end moments, both with signs reversed. The equivalent forces should be used in constructing $\mathbf{a}_0$ for Eqs. (5-316) and (5-321) for determining redundants, but the fixed-end moments should not be included in $\mathbf{a}_0$ in Eq. (5-325) for computing element forces. When the element forces are obtained from Eq. (5-330) or (5-340), the final element forces then are determined by adding the fixed-end moments and simple-beam reactions to the solution of Eq. (5-330) or (5-340).

(H. I. Laursen, "Structural Analysis," and O. C. Zienkiewicz, "The Finite Element Method," McGraw-Hill Book Company, New York; C. S. Desai, "Elementary Finite Element Method," and R. H. Gallagher, "Finite Element Analysis," Prentice-Hall, Englewood Cliffs, N.J.)

5-112. Matrix Displacement (Stiffness) Method.

In this article, the structure is assumed to consist of n finite elements connected only at the nodes. The stiffness matrices of the elements are assumed known. Right-handed cartesian coordinate axes x, y, z, are chosen for the structure with base unit vectors $\mathbf{e}_i$ parallel to these axes at each node, as in Art. 5-110. Forces and displacements are resolved into components along $\mathbf{e}_i$. The stiffness matrix for the whole structure can be constructed from the element stiffness matrices, as will be demonstrated.

For the ith element, nodal forces and displacements are related by

$$\mathbf{S}_i = \mathbf{k}_i \delta_i \qquad i = 1, 2, \ldots n \tag{5-326}$$

where $\mathbf{S}_i$ = matrix of forces, including moments and torques, acting at the nodes of the ith element
$\mathbf{k}_i$ = stiffness matrix of the ith element (Art. 5-113)
δ_i = matrix of displacements of the nodes of the ith element

For compactness, this relationship between nodal displacements and forces for each element can be combined into a single matrix equation applicable to all the elements:

$$\mathbf{S} = \mathbf{k}\delta \tag{5-327}$$

where $\mathbf{S}$ = matrix of all forces acting at the nodes of all elements
δ = matrix of all nodal displacements for all elements

$$\mathbf{k} = \begin{bmatrix} \mathbf{k}_1 & 0 & \ldots & 0 \\ 0 & \mathbf{k}_2 & \ldots & 0 \\ \cdot\cdot & \cdot\cdot & \ldots & \cdot\cdot \\ 0 & 0 & \ldots & \mathbf{k}_n \end{bmatrix} \tag{5-328}$$

Element nodal displacements δ can be determined from the displacements Δ of the nodes of the structure to ensure geometric compatibility:

$$\delta = \mathbf{b}_0 \Delta \tag{5-329}$$

where $\mathbf{b}_0$ is a matrix of influence coefficients. The jth column of $\mathbf{b}_0$ contains the element nodal displacements when the node where Δ_j occurs is given a unit displacement in the direction of Δ_j, and no other nodes are displaced.

From Eqs. (5-327) and (5-329)

$$\mathbf{S} = \mathbf{k}\mathbf{b}_0\Delta \qquad (5\text{-}330)$$

From energy relationships, it can be shown that

$$\mathbf{P} = \mathbf{b}_0^T\mathbf{S} \qquad (5\text{-}331)$$

where $\mathbf{P}$ = matrix of all the loads acting at the nodes of the structure
$\mathbf{b}_0^T$ = transpose of $\mathbf{b}_0$ = matrix $\mathbf{b}_0$ with rows and columns interchanged

Finally, from Eqs. (5-330) and (5-331) comes the relationship that permits computation of the stiffness matrix of the whole structure from the stiffness matrices of its elements:

$$\mathbf{P} = \mathbf{b}_0^T\mathbf{k}\mathbf{b}_0\Delta \qquad (5\text{-}332)$$

Since also, by Eq. (5-312), $\mathbf{P} = \mathbf{K}\Delta$,

$$\mathbf{K} = \mathbf{b}_0^T\mathbf{k}\mathbf{b}_0 \qquad (5\text{-}333)$$

With the stiffness matrix known, the nodal displacements can be computed from the nodal loads. Multiplication of both sides of Eq. (5-312) by $\mathbf{K}^{-1}$ gives

$$\Delta = \mathbf{K}^{-1}\mathbf{P} \qquad (5\text{-}334)$$

This permits member forces to be computed from Eq. (5-330).

For structures with a large number of nodal displacements, inversion of the stiffness matrix $\mathbf{K}$ directly may be impracticable. The size of $\mathbf{K}$, however, often can be reduced by taking advantage initially of conditions at nodes where loads do not act in the directions of permissible displacements.

Let $\mathbf{X}$ be the matrix of the unknown displacements at those nodes. Then, as in Eq. (5-329),

$$\delta = \mathbf{b}_0\Delta + \mathbf{b}_1\mathbf{X} \qquad (5\text{-}335)$$

where $\mathbf{b}_0$ = matrix of influence coefficients for displacements of nodes where loads act in directions of displacements
$\mathbf{b}_1$ = matrix of influence coefficients for displacements of nodes where loads do not act in directions of displacements

The jth column of $\mathbf{b}_1$ contains the element nodal displacements when the node where X_j occurs is given a unit displacement in the direction of X_j, and no other nodes are displaced. Substitution of Eq. (5-335) in Eq. (5-327) yields

$$\mathbf{S} = \mathbf{k}(\mathbf{b}_0\Delta + \mathbf{b}_1\mathbf{X}) \qquad (5\text{-}336)$$

Since $P_j = 0$ for those displacements not associated with loads, Eq. (5-331) indicates that

$$\mathbf{b}_1^T\mathbf{S} = 0 \qquad (5\text{-}337)$$

Solving for the unknown displacements from Eq. (5-336) and (5-337) gives

$$\mathbf{X} = -\mathbf{B}^{-1}\mathbf{B}_0\Delta \qquad (5\text{-}338)$$

where $\mathbf{B}_0 = \mathbf{b}_1^T\mathbf{k}\mathbf{b}_0$
$\mathbf{B}^{-1}$ = inverse of matrix $\mathbf{B}$
$\mathbf{B} = \mathbf{b}_1^T\mathbf{k}\mathbf{b}_1$

Substitution for **X** in Eq. (5-335) gives the element nodal displacements:

$$\delta = (\mathbf{b}_0 - \mathbf{b}_1\mathbf{B}^{-1}\mathbf{B}_0)\Delta = \mathbf{b}_2\Delta \tag{5-339}$$

where $\mathbf{b}_2 = \mathbf{b}_0 - \mathbf{b}_1\mathbf{B}^{-1}\mathbf{B}_0$. From Eqs. (5-336), (5-338) and (5-339), the element member forces are given by

$$\mathbf{S} = \mathbf{k}\mathbf{b}_2\Delta \tag{5-340}$$

And from Eq. (5-331),

$$\mathbf{P} = \mathbf{b}_0^T\mathbf{k}\mathbf{b}_2\Delta \tag{5-341}$$

Since also, by Eq. (5-312), $\mathbf{P} = \mathbf{K}\Delta$, the stiffness matrix for the whole structure is given by

$$\mathbf{K} = \mathbf{b}_0^T\mathbf{k}\mathbf{b}_2 \tag{5-342}$$

This matrix may be considerably smaller than that given by Eq. (5-333). As before, nodal displacements can be computed from Eq. (5-334) after **K** has been inverted.

Methods of developing stiffness matrices of finite elements are described in Art. 5-113.

(See bibliography at the end of Art. 5-111.)

5-113. Element Flexibility and Stiffness Matrices.

As indicated in Arts. 5-111 and 5-112, the relationship between independent forces and displacements at nodes of finite elements comprising a structure is determined by flexibility matrices $\mathbf{f}_i$ [Eq. (5-313)] or stiffness matrices $\mathbf{k}_i$ [Eq. (5-326)] of the elements. In some cases, the components of these matrices can be developed from the defining equations.

The jth column of a flexibility matrix of a finite element, for which we will use the symbol **f** without subscript in this article, contains all the nodal displacements of the element when one force S_j is set equal to unity and all other independent forces are set equal to zero.

The jth column of a stiffness matrix of a finite element, for which we will use the symbol **k** without subscript in this article, consists of the forces acting at the nodes of the element to produce a unit displacement of the node at which δ_j occurs and in the direction of δ_j but no other nodal displacements of the element.

Bars with Axial Stress Only. As an example of the use of the definitions of flexibility and stiffness, consider the simple case of an elastic bar under tension applied by axial forces P_i and P_j at nodes i and j, respectively (Fig. 5-121). The bar might be the finite element of a truss, such as a diagonal or a hanger. Connections to other members are made at nodes i and j, which can transmit only forces in the directions i to j or j to i.

Fig. 5-121 Elastic bar in tension.

For equilibrium, $P_i = P_j = P$. Displacement of node j relative to node i is e. From Eq. (5-5), $e = PL/AE$, where L is the initial length of the bar, A the bar cross-sectional area, and E the modulus of elasticity. Setting $P = 1$ yields the flexibility of the bar,

$$f = \frac{L}{AE} \tag{5-343}$$

Setting $e = 1$ gives the stiffness of the bar,

$$k = \frac{AE}{L} \tag{5-344}$$

Beams with Bending Only. As another example of the use of the definition to determine element flexibility and stiffness matrices, consider the simple case of an elastic beam in bending applied by moments M_i and M_j at nodes i and j, respectively (Fig. 5-122a). The beam might be a finite element of a rigid frame. Connections to other members are made at nodes i and j, which can transmit moments and forces normal to the beam.

Nodal displacements of the element can be sufficiently described by rotations θ_i and θ_j relative to the straight line between nodes i and j. For equilibrium, forces $V_j = -V_i$

Fig. 5-122 Beam subjected to end moments and shears.

normal to the beam are required at nodes j and i, respectively, and $V_j = (M_i + M_j)/L$, where L is the span of the beam. Thus, M_i and M_j are the only independent forces acting. Hence, Eqs. (5-313) and (5-326) can be written for this element as

$$\theta = \begin{bmatrix} \theta_i \\ \theta_j \end{bmatrix} = \mathbf{f} \begin{bmatrix} M_i \\ M_j \end{bmatrix} = \mathbf{fM} \tag{5-345}$$

$$\mathbf{M} = \begin{bmatrix} M_i \\ M_j \end{bmatrix} = \mathbf{k} \begin{bmatrix} \theta_i \\ \theta_j \end{bmatrix} = \mathbf{k}\theta \tag{5-346}$$

The flexibility matrix $\mathbf{f}$ then will be a 2×2 matrix. The first column can be obtained by setting $M_i = 1$ and $M_j = 0$ (Fig. 5-122b). The resulting angular rotations are given by Eqs. (5-72) and (5-73). For a beam with constant moment of inertia I and modulus of elasticity E, the rotations are $\alpha = L/3EI$ and $\beta = -L/6EI$. Similarly, the second column can be developed by setting $M_i = 0$ and $M_j = 1$.

The flexibility matrix for a beam in bending then is

$$\mathbf{f} = \begin{bmatrix} \dfrac{L}{3EI} & -\dfrac{L}{6EI} \\ -\dfrac{L}{6EI} & \dfrac{L}{3EI} \end{bmatrix} = \dfrac{L}{6EI} \begin{bmatrix} 2 & -1 \\ -1 & 2 \end{bmatrix} \tag{5-347}$$

The stiffness matrix, obtained in a similar manner or by inversion of $\mathbf{f}$, is

$$\mathbf{k} = \begin{bmatrix} \dfrac{4EI}{L} & \dfrac{2EI}{L} \\ \dfrac{2EI}{L} & \dfrac{4EI}{L} \end{bmatrix} = \dfrac{2EI}{L} \begin{bmatrix} 2 & 1 \\ 1 & 2 \end{bmatrix} \tag{5-348}$$

Beams Subjected to Bending and Axial Forces. For a beam subjected to nodal moments M_i and M_j and axial forces P, flexibility and stiffness are represented by 3×3 matrices. The load-displacement relations for a beam of span L, constant moment of inertia I, modulus of elasticity E, and cross-sectional area A are given by

$$\begin{bmatrix} \theta_i \\ \theta_j \\ e \end{bmatrix} = \mathbf{f} \begin{bmatrix} M_i \\ M_j \\ P \end{bmatrix} \qquad \begin{bmatrix} M_i \\ M_j \\ P \end{bmatrix} = \mathbf{k} \begin{bmatrix} \theta_i \\ \theta_j \\ e \end{bmatrix} \tag{5-349}$$

In this case, the flexibility matrix is

$$\mathbf{f} = \frac{L}{6EI} \begin{bmatrix} 2 & -1 & 0 \\ -1 & 2 & 0 \\ 0 & 0 & \eta \end{bmatrix} \quad (5\text{-}350)$$

where $\eta = 6I/A$, and the stiffness matrix is

$$\mathbf{k} = \frac{EI}{L} \begin{bmatrix} 4 & 2 & 0 \\ 2 & 4 & 0 \\ 0 & 0 & \psi \end{bmatrix} \quad (5\text{-}351)$$

where $\psi = A/I$.

General Methods for Computing Flexibility. Each component f_{ij} of a flexibility matrix equals a displacement at a node caused by a unit force at the same or another node. When the matrix for a finite element cannot readily be constructed from this definition, other means must be used.

For some elements, the components of the flexibility matrix may be obtained by integrating over the whole volume of the element

$$f_{ij} = \int_V \sigma^T \epsilon \, dV \quad (5\text{-}352)$$

where σ^T = transpose of the matrix of the stress produced in the element by a unit force at the node where displacement f_{ij} occurs and in the direction of that displacement

ϵ = matrix of the unit strains produced in the element by a unit force at the same or another node

dV = differential volume

For use in the integral, note that for a three-dimensional, homogeneous, elastic material with modulus of elasticity E, shearing modulus G, and Poisson's ratio μ, stresses σ and unit strains ϵ in the direction of x, y, z Cartesian coordinate axes are related by

$$\epsilon = \phi \sigma \quad (5\text{-}353)$$

$$\epsilon^T = [\epsilon_x \quad \epsilon_y \quad \epsilon_z \quad \gamma_{xy} \quad \gamma_{xz} \quad \gamma_{yz}] \quad (5\text{-}354a)$$

$$\sigma^T = [\sigma_x \quad \sigma_y \quad \sigma_z \quad \tau_{xy} \quad \tau_{xz} \quad \tau_{yz}] \quad (5\text{-}354b)$$

$$\phi = \frac{1}{E} \begin{bmatrix} 1 & -\mu & -\mu & 0 & 0 & 0 \\ -\mu & 1 & -\mu & 0 & 0 & 0 \\ -\mu & -\mu & 1 & 0 & 0 & 0 \\ 0 & 0 & 0 & \dfrac{E}{G} & 0 & 0 \\ 0 & 0 & 0 & 0 & \dfrac{E}{G} & 0 \\ 0 & 0 & 0 & 0 & 0 & \dfrac{E}{G} \end{bmatrix} \quad (5\text{-}354c)$$

See Arts. 5-10 to 5-12.

For some elements, the components of the flexibility matrix may be obtained by differentiating twice the strain energy U in the element due to forces at the nodes:

$$f_{ij} = \frac{\partial^2 U}{\partial S_i \partial S_j} \quad (5\text{-}355)$$

where S_i = force acting at the node where displacement f_{ij} occurs in the direction of that displacement

S_j = force acting at the same or another node

Flexibility of Thin Triangular Element. For an elastic, triangular element with thickness t very small compared with other dimensions, the terms in Eq. (5-353) become

$$\epsilon^T = [\epsilon_x \quad \epsilon_y \quad \gamma_{xy}] \qquad (5\text{-}356a)$$

$$\sigma^T = [\sigma_x \quad \sigma_y \quad \tau_{xy}] \qquad (5\text{-}356b)$$

$$\phi = \frac{1}{E}\begin{bmatrix} 1 & -\mu & 0 \\ -\mu & 1 & 0 \\ 0 & 0 & \dfrac{E}{G} \end{bmatrix} \qquad (5\text{-}356c)$$

See Arts. 5-10 to 5-12.

Assume that forces S_i act only at the vertices (nodes) of the triangle and in the directions of the x and y axes. Of the six possible forces that can act on the element, only the three forces S_1, S_2, S_3, shown in Fig. 5-123, are independent. The other three can be computed from these from equilibrium conditions. The independent forces can be formed into a column matrix **S** for loads on the element. This matrix is related to the nodal displacements δ of the element by Eq. (5-314), $\delta = \mathbf{f}\mathbf{S}$, where **f** is the flexibility matrix of the element. In this case, δ also is a column vector with three components. Thus, **f** is a 3×3 matrix.

A key assumption will now be made: *Stresses in the element are constant.*

This places constraints on the displacements and forces of the structure composed of such elements. As a result, equilibrium or geometric compatibility may be violated, and the idealized structure may

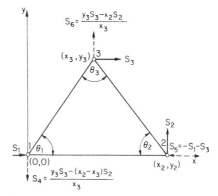

Fig. 5-123 Forces acting at nodes of a thin triangular element.

then be stiffer or more flexible than the actual structure. The approximation may be improved by such refinements as selecting an element with more nodes (triangle with nodes at vertices and midpoints of sides) or a smaller element. (The former choice usually requires fewer calculations for about the same degree of accuracy.)

For constant σ_x, σ_y, and τ_{xy}, the stresses and nodal forces are related in terms of the nodal coordinates by

$$\sigma = \alpha\mathbf{S} \qquad (5\text{-}357)$$

$$\alpha = \frac{2}{t}\begin{bmatrix} -\dfrac{1}{y_3} & 0 & \dfrac{x_3 - x_2}{x_2 y_3} \\ 0 & -\dfrac{1}{x_3} & \dfrac{y_3}{x_2 x_3} \\ 0 & 0 & \dfrac{1}{x_2} \end{bmatrix} \qquad (5\text{-}358)$$

From energy relations, it can be shown that for integration over the volume of the element

$$\int_V \epsilon^T \sigma \, dV = \delta^T \mathbf{S} \qquad (5\text{-}359)$$

Substituting Eq. (5-357) in Eq. (5-359) and transposing matrices result in the relation between strains and nodal displacements:

5-170 Structural Theory

$$\delta = \int_V \alpha^T \epsilon \, dV \qquad (5\text{-}360)$$

By Eqs. (5-353) and (5-357), $\epsilon = \phi\sigma = \phi\alpha\mathbf{S}$. Hence, Eq. (5-360) can be written

$$\delta = \left[\int_V \alpha^T \phi \alpha \, dV\right] \mathbf{S} \qquad (5\text{-}361)$$

But also, $\delta = \mathbf{fS}$. Hence, the flexibility matrix is given by

$$\mathbf{f} = \int_V \alpha^T \phi \alpha \, dV = \int_V \alpha^T \phi \alpha t \, dA = \frac{x_2 y_3}{2} t \, \alpha^T \phi \alpha \qquad (5\text{-}362)$$

Substitution for α from Eq. (5-358) and for ϕ from Eq. (5-356c) gives, with $E/G = 2(1 + \mu)$,

$$\mathbf{f} = \frac{2}{Et} \begin{bmatrix} \dfrac{x_2}{y_3} & -\dfrac{\mu x_2}{x_3} & \dfrac{x_2 - x_3}{y_3} + \dfrac{\mu y_3}{x_3} \\ -\dfrac{\mu x_2}{x_3} & \dfrac{x_2 y_3}{x_3^2} & \dfrac{\mu(x_3 - x_2)}{x_3} - \dfrac{y_3^2}{x_3^2} \\ \dfrac{x_2 - x_3}{y_3} + \dfrac{\mu y_3}{x_3} & \dfrac{\mu(x_3 - x_2)}{x_3} - \dfrac{y_3^2}{x_3^2} & \dfrac{(x_3 - x_2)^2}{x_2 y_3} + \dfrac{2 y_3 (\mu x_2 + x_3)}{x_2 x_3} + \dfrac{y_3^3}{x_2 x_3^2} \end{bmatrix}$$

$$(5\text{-}363)$$

Flexibility Matrix for Rotated Axes. If the orientation of the coordinate axes for Eq. (5-363) differs from that of the axes chosen for the whole structure, the flexibility matrix and nodal forces and displacements must be transformed accordingly. Assume that the triangular element is rotated counterclockwise through an angle θ relative to the coordinate axes of the structure. Then, new nodal forces $\mathbf{S}$ in the directions of those axes and the original nodal forces

$$\mathbf{S}_o = [S_1 \quad S_4 \quad S_2 \quad S_5 \quad S_3 \quad S_6]^T$$

are related by

$$\mathbf{S}_o = \mathbf{RS} \qquad (5\text{-}364)$$

where $\mathbf{R}$ is a diagonal matrix with elements

$$\mathbf{R}_{ij} = \begin{bmatrix} \cos\theta & \sin\theta \\ -\sin\theta & \cos\theta \end{bmatrix} \text{ for } i = j \qquad \mathbf{R}_{ij} = \mathbf{0} \text{ for } i \neq j \qquad (5\text{-}365)$$

From Eq. (5-364), the relation between original nodal forces and new nodal forces can be expressed as

$$[S_1 \quad S_2 \quad S_3]_o^T = \mathbf{r}[S_1 \quad S_2 \quad S_3]^T \qquad (5\text{-}366)$$

The transformed flexibility matrix $\mathbf{f}$ then is given in terms of the flexibility matrix $\mathbf{f}_o$ of Eq. (5-363) by

$$\mathbf{f} = \mathbf{r}^T \mathbf{f}_o \mathbf{r} \qquad (5\text{-}367)$$

General Methods for Computing Stiffness. Each component k_{ij} of a stiffness matrix equals a force at a node when a unit displacement occurs only at the same or another node. When the matrix for a finite element cannot readily be developed from this definition, other means must be used.

For some elements, the components of the stiffness matrix can be obtained by integrating over the whole volume of the element

$$k_{ij} = \int_V \epsilon^T \sigma \, dV \qquad (5\text{-}368)$$

Element Flexibilty and Stiffness Matrices 5-171

where ϵ^T = transpose of the matrix of unit strains produced in the element when the node at which force k_{ij} acts is given a unit displacement in the direction of the force while no other displacements occur

σ = matrix of the stresses in the element when a unit displacement occurs at the same or another node while no other displacements occur

dV = differential volume

See Arts. 5-10 to 5-12.

For some elements, the components of the stiffness matrix can be obtained by differentiating the strain energy U in the element because of nodal displacements:

$$k_{ij} = \frac{\partial^2 U}{\partial \delta_i \partial \delta_j} \qquad (5\text{-}369)$$

where δ_i = displacement of the node where k_{ij} acts, in the direction of k_{ij}
δ_i = displacement of the same or another node

Stiffness of Thin Triangular Element. For an elastic triangular element with modulus of elasticity E, shearing modulus G, Poisson's ratio μ, and thickness t very small relative to other dimensions, stresses σ are related to strains ϵ by

$$\sigma = \kappa \epsilon \qquad (5\text{-}370)$$

$$\sigma^T = [\sigma_x \; \sigma_y \; \tau_{xy}] \qquad (5\text{-}371a)$$

$$\epsilon^T = [\epsilon_x \; \epsilon_y \; \gamma_{xy}] \qquad (5\text{-}371b)$$

$$\kappa = \frac{E}{1-\mu^2} \begin{bmatrix} 1 & \mu & 0 \\ \mu & 1 & 0 \\ 0 & 0 & \dfrac{1-\mu}{2} \end{bmatrix} \qquad (5\text{-}371c)$$

See Arts. 5-10 to 5-12.

Assume that forces S_i act only at the vertices (nodes) of the triangle and in the directions of the x, y axes. A displacement δ_i in the direction of each axis can occur at each node, as shown in Fig. 5-124. Hence, six nodal displacements can be chosen.

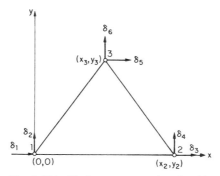

Fig. 5-124 Displacements at nodes of a thin triangular element.

They can be formed into a displacement matrix δ. This matrix is related to the matrix **S** of nodal forces by Eq. (5-326), $\mathbf{S} = \mathbf{k}\delta$, where **k** is the stiffness matrix of the element.

A key assumption will now be made: *Unit strains in the element are constant.* (This places constraints on the displacements and forces, as discussed in connection with the development of the flexibility matrix of a triangular element.) Because

5-172   Structural Theory

$$\epsilon^T = \begin{bmatrix} \dfrac{\partial \delta_x}{\partial x} & \dfrac{\partial \delta_y}{\partial y} & \dfrac{\partial \delta_x}{\partial y} + \dfrac{\partial \delta_y}{\partial x} \end{bmatrix} \quad (5\text{-}372)$$

where ϵ^T = transpose of the matrix of unit strains in the element
δ_x = component of displacement within the element in the direction of x axis
δ_y = component of displacement within the element in the direction of y axis
this assumption is equivalent to requiring that displacements vary linearly within the element. In matrix form, this relation can be written:

$$\begin{bmatrix} \delta_x \\ \delta_y \end{bmatrix} = \mathbf{Xc} = \begin{bmatrix} 1 & x & y & 0 & 0 & 0 \\ 0 & 0 & 0 & 1 & x & y \end{bmatrix} [c_1 c_2 c_3 c_4 c_5 c_6]^T \quad (5\text{-}373)$$

where c_i = constant. The strains then, by Eq. (5-372), are given by

$$\epsilon = \mathbf{Cc} = \begin{bmatrix} 0 & 1 & 0 & 0 & 0 & 0 \\ 0 & 0 & 0 & 0 & 0 & 1 \\ 0 & 0 & 1 & 0 & 1 & 0 \end{bmatrix} [c_1 c_2 c_3 c_4 c_5 c_6]^T \quad (5\text{-}374)$$

Nodal displacements can be obtained from Eq. (5-373) on substitution of the nodal coordinates (Fig. 5-124):

$$\delta = \begin{bmatrix} 1 & 0 & 0 & 0 & 0 & 0 \\ 0 & 0 & 0 & 1 & 0 & 0 \\ 1 & x_2 & 0 & 0 & 0 & 0 \\ 0 & 0 & 0 & 1 & x_2 & 0 \\ 1 & x_3 & y_3 & 0 & 0 & 0 \\ 0 & 0 & 0 & 1 & x_3 & y_3 \end{bmatrix} \begin{bmatrix} c_1 \\ c_2 \\ c_3 \\ c_4 \\ c_5 \\ c_6 \end{bmatrix} \quad (5\text{-}375)$$

Solving for the constants yields

$$\mathbf{c} = \frac{1}{x_2 y_3} \begin{bmatrix} x_2 x_3 & 0 & 0 & 0 & 0 & 0 \\ -y_3 & 0 & 0 & y_3 & 0 & 0 \\ x_3 - x_2 & 0 & -x_3 & 0 & x_2 & 0 \\ 0 & x_2 y_3 & 0 & 0 & 0 & 0 \\ 0 & -y_3 & 0 & y_3 & 0 & 0 \\ 0 & x_3 - x_2 & 0 & -x_3 & 0 & x_2 \end{bmatrix} \begin{bmatrix} \delta_1 \\ \delta_2 \\ \delta_3 \\ \delta_4 \\ \delta_5 \\ \delta_6 \end{bmatrix} \quad (5\text{-}376)$$

Substitution of Eq. (5-376) in Eq. (5-374) develops the relation between unit strains and nodal displacements:

$$\epsilon = \beta\delta \quad (5\text{-}377)$$

$$\beta = \frac{1}{x_2 y_3} \begin{bmatrix} -y_3 & 0 & 0 & y_3 & 0 & 0 \\ 0 & x_3 - x_2 & 0 & -x_3 & 0 & x_2 \\ x_3 - x_2 & -y_3 & -x_3 & y_3 & x_2 & 0 \end{bmatrix} \quad (5\text{-}378)$$

From energy relations, it can be shown that for integration over the whole volume of the element

$$\int_V \sigma^T \epsilon \, dV = \mathbf{S}^T \delta \quad (5\text{-}379)$$

Substituting Eq. (5-377) in (5-379) and transposing matrices results in the relation between stresses and nodal forces:

$$\mathbf{S} = \int_V \beta^T \sigma \, dV \quad (5\text{-}380)$$

By Eqs. (5-370) and (5-377), $\sigma = \kappa\epsilon = \kappa\beta\delta$. Hence, Eq. (5-380) can be written as

$$\mathbf{S} = \left[\int_V \beta^T \kappa \beta \, dV \right] \delta \quad (5\text{-}381)$$

But also, $\mathbf{S} = \mathbf{k}\delta$. Consequently, according to Eq. (5-381), the stiffness matrix for the triangular element is given by

$$\mathbf{k} = \int_V \boldsymbol{\beta}^T \boldsymbol{\kappa} \boldsymbol{\beta} \, dV = \int_V \boldsymbol{\beta}^T \boldsymbol{\kappa} \boldsymbol{\beta} t \, dA = \frac{x_2 y_3}{2} t \boldsymbol{\beta}^T \boldsymbol{\kappa} \boldsymbol{\beta} \tag{5-382}$$

Substitution for β from Eq. (5-378) and for κ from Eq. (5-371c) gives

$$\mathbf{k} = \frac{Et}{2(1-\mu^2)\,x_2 y_3} \begin{bmatrix} y_3^2 + x_{23}^2\lambda & x_{23}y_3(\mu+\lambda) & -y_3^2 + x_3 x_{23}\lambda \\ x_{23}y_3(\mu+\lambda) & x_{23}^2 + y_3^2\lambda & -y_3(x_{23}\mu - x_3\lambda) \\ -y_3^2 + x_3 x_{23}\lambda & -y_3(x_{23}\mu - x_3\lambda) & y_3^2 + x_3^2\lambda \\ y_3(x_3\mu - x_{23}\lambda) & x_3 x_{23} - y_3^2\lambda & -x_3 y_3(\mu+\lambda) \\ -x_2 x_{23}\lambda & -x_2 y_3\lambda & -x_2 x_3\lambda \\ -x_2 y_3\mu & -x_2 x_{23} & x_2 y_3\mu \end{bmatrix}$$

$$\begin{matrix} y_3(x_3\mu - x_{23}\lambda) & -x_2 x_{23}\lambda & -x_2 y_3\mu \\ x_3 x_{23} - y_3^2\lambda & -x_2 y_3\lambda & -x_2 x_{23} \\ -x_3 y_3(\mu+\lambda) & -x_2 x_3\lambda & x_2 y_3\mu \\ x_3^2 + y_3^2\lambda & x_2 y_3\lambda & -x_2 x_3 \\ x_2 y_3\lambda & x_2^2\lambda & 0 \\ -x_2 x_3 & 0 & x_2^2 \end{matrix} \tag{5-383}$$

where $\lambda = (1-\mu)/2$
$x_{23} = x_2 - x_3$

Stiffness Matrix for Rotated Axes. If the orientation of the coordinate axes for Eq. (5-383) differs from that of the axes chosen for the whole structure, the stiffness matrix and nodal displacements must be transformed accordingly. Assume that the triangular element is rotated counterclockwise through an angle θ relative to the coordinate axes of the structure. Then, new nodal displacements δ in the direction of those axes are related to the original nodal displacements δ_o used in determining Eq. (5-383) by

$$\delta_o = \mathbf{R}\delta \tag{5-384}$$

where $\mathbf{R}$ is the diagonal matrix with elements given by Eq. (5-365). The transformed stiffness matrix $\mathbf{k}$ then is given in terms of the stiffness matrix $\mathbf{k}_o$ of Eq. (5-383) by

$$\mathbf{k} = \mathbf{R}^T \mathbf{k}_o \mathbf{R} \tag{5-385}$$

(See bibliography at the end of Art. 5-111.)

Section 6

Soil Mechanics and Foundations

Charles P. Gupton
Partner, Dames & Moore

Foundations constitute the building subsystems that are used to transmit building loads directly to the earth below. Foundations must be built to distribute the loads over contiguous soils so that they and the underlying materials will have sufficient strength and rigidity to carry the loads without excessive deformations. Because of the interactions of soils and foundations, the characteristics of the soils underlying a building strongly influence selection of types and sizes of foundations. The foundations, in turn, may significantly affect design of the superstructure, the project construction time, and, consequently, building construction costs. For production of a safe, cost-effective building, therefore, a knowledge of soil mechanics and foundation design and construction is essential.

The study of soils, their properties and their behavior for engineering purposes is the province of soil mechanics. This section deals with the applications of soil mechanics to design and construction of foundations for buildings. Structural design of concrete foundations is treated in Arts. 10-74 to 10-82.

6-1. Physical Properties of Soils.

The geologist defines a soil as an altered rock. The engineer defines a soil as the material that supports or loads a structure at its base.

Natural soil materials may be divided into four classifications: sands and gravels, silts, clays, and organics. Sands and gravels are granular, nonplastic materials. Clays, composed of much smaller particles, exhibit plasticity and are cohesive. Silts are intermediate in grain size, typically behave as granular materials but may be slightly plastic. Organic materials are principally vegetable refuse.

The origin and mode of deposition of soils sheds much light on their nature and variability in the field. There are two origins of soils: residual and sedimentary. Residual soils form in situ by the chemical weathering of rocks and, having never been physically disturbed, retain the minor geologic features of the parent material. (Changes from soil to rock in the field are usually gradational.) Sedimentary soils are those transported and deposited by the action of rivers, seas, glaciers, and wind. The mode of deposition often controls grain size, its variation, and the stratigraphy and uniformity of the soil units.

The engineer needs to know, for complete identification of a soil: the size of grains, gradation, particle shapes, grain orientation, chemical makeup, and colloidal and dust fractions. Even then, the physical properties can be made to vary over a large range with slight chemical additives or electrochemical controls.

Where surface properties are important, grain shape becomes at least equal in influence to size gradation. Normally, an important characteristic is the relative positioning of the grains within the soil body, since this controls the resistance to internal displacement and is at least a qualitative measure of shear and compression strengths.

There have been many attempts to codify all soils into classes of similar and recognizable properties. As more information is collected on soil properties, the systems of classification become more elaborate and complicated. One difficulty is the attempt to use similar classifications for different uses; for instance, a system applicable to highway design has less value when the problem is basically that of building foundations.

6-2. States of Matter Affecting Soil Behavior.

Soils can exhibit solid, viscous, plastic, or liquid action; if the true state can be predicted, the structural design of foundations can be prepared accordingly.

By contrast, solids are materials having constant density, elasticity, and internal resistance, little affected by normal temperature changes, moisture variations, and vibration below seismic values. Deformation by shearing takes place along two sets of parallel planes, with the angle between the sets constant for any material and independent of the nature or intensity of the external forces inducing shearing strain. The harder and more brittle the material, the more this angle differs from a right angle.

If tension is taken as negative compression, Hartmann's law states that the acute angle formed by the shear planes is bisected by the axis of maximum compression, and the obtuse angle by the axis of minimum compression (usually tension). Shearing planes do not originate simultaneously and are not uniformly distributed.

The angle of shear is related to the ultimate compression and tension stresses by Mohr's formula:

$$\cos \theta = \frac{f_1 - f_2}{f_1 + f_2} \tag{6-1}$$

i.e., the cosine of the angle of shear equals the ratio of the difference of the normal stresses to the sum. When lateral pressures are applied externally simultaneously with loads, the angle of shear increases; i.e., the material becomes less brittle.

These basic properties of solids can be used in designs affecting soils only if the soils remain solid. When changes in conditions modify soil structures so that they are not solid, these properties are nonexistent, and a new set of rules governs the actions. Most soils will act like solids, but only up to a certain loading. This limit loading depends on many external factors, such as flow of moisture, temperature, vibration, age, and sometimes the rate of loading.

No marked subdivision exists between liquid, plastic, and viscous action. These three states have the common property that volume changes are difficult to accomplish, but shape changes occur continuously. They differ in the amount of force necessary to start motion. There is a minimum value necessary for the plastic or viscous states, while a negligible amount will start motion in the liquid state. Upon cessation of the force, a plastic material will stop motion, but liquid and viscous materials will continue indefinitely, or until counteracting forces come into play. Usually, division between solid and plastic states is determined by the percentage of moisture in the soil. This percentage, however, is not a constant but decreases with increase in pressure on the material. Furthermore, the entire relationship between moisture percentage and change of state can be deranged by adding chemicals to the water.

Liquids retained in a vessel are almost incompressible. In water-bearing soils, therefore, if movement or loss of water can be prevented, volume change and settlement will be avoided. Water loss can be prevented by both physical and chemical changes in the nature of the water.

6-3. Soil Moisture.

Water is usually present in a soil as a film on its particles or as a liquid between particles, or pore water. If the water content of soil is chiefly film or adsorbed moisture, the mass will not act as a liquid. All solids tend to adsorb or condense upon their surfaces any liquids (and gases) with which they come into contact.

The kind of ion, or metal element, in the chemical makeup of a solid has a great influence on how much water can be adsorbed. Ion-exchange procedures for soil stabilization and percolation control are therefore an important part of soil mechanics.

Water films are tougher than pore water. (Terzaghi, in 1920, stated that water films less than two-millionths of an inch thick are semisolid in action; they will not boil or freeze at normal temperatures.) Consequently, saturated soil freezes much more easily than damp soil, and ice crystals will grow by sucking free moisture out of the pores. A sudden thaw will then release large concentrations of water, often with drastic results. When liquids evaporate, they first change to films, requiring a large thermal increment for the change from film to vapor. Thus, the effect of temperature on soil action may be explained by the reduction of film thickness with rise of temperature.

The presence of moisture in a soil is essential for controlled compaction. Compaction of soils artificially or by natural causes is best carried out under fairly definite moisture contents, since redistribution of soil grains to a closely packed volume cannot be accomplished without sufficient moisture to coat each particle. The film acts as a lubricant to permit easier relative movements and by its capillary tension holds the grains in position. Obviously, finer grains require more water for best stabilization than coarser materials.

6-4. Resistance of Soils to Pressure.

Before 1640, Galileo distinguished between solids, semifluids, and fluids. He stated that semifluids, unlike fluids, maintain their condition when heaped up, and if a hollow or cavity is made, agitation causes the filling up of the hollow, whereas in solids, the hollow does not fill up. This is an early description of the property known as the natural slope of granular materials—an easily observed condition for clean, dry sands, but a variable slope for soils containing clays with varying moisture percentages. The angle of natural repose should not be confused with the angle of internal friction, although many writers have followed Woltmann, who, in translating Coulomb's papers, made that mistake.

The fundamental laws of friction were applied to soils by Coulomb. He recognized that the resistance along a surface of failure within a soil is a function of both the load per unit area and the surface of contact.

Resistance of soils to deformation is controlled mainly by their shear resistance. Shear resistance in soils is derived from the sum of two components, friction and cohesion. Frictional resistance develops from the irregularities of grain-to-grain contacts and is proportional to the normal force between the particles. Cohesion, the soil's maximum tensile strength, is a result of the attractive forces operating between grains in close contact and is independent of the normal pressure. Such true cohesion is relatively rare. Some frictional behavior is normal for most soils.

6-5. Soil Properties Important in Engineering.

The normally important soil properties are:

Density. The amount of solid material in a unit volume is called the dry density. For granular soils and fibrous organics, the dry density is the most important factor controlling their engineering properties. One such property is the state, or degree, of compaction which usually is expressed in terms of relative density, or ratio (as percentage) of the difference between dry density of the natural soil and its minimum dry density to the difference between its maximum and minimum dry densities. During construction of engineering fills, however, the degree of compaction is usually specified and measured as the ratio of the actual in-place dry density to the maximum dry density obtained in a laboratory moisture-density relationship test (ASTM D1557 or D698).

Internal Friction. Coulomb pure friction corresponds to simple shear in the theory of elasticity. Internal friction is usually expressed geometrically as the angle of internal friction ϕ, where $\tan \phi = f$, the coefficient of friction. The frictional component of shear T_{max} in a soil mass thus is equal to $N \tan \phi$, where N is the normal force acting on the mass.

Values of ϕ range from about 28° for loose sands and nonplastic silts to about 48° for dense sands and gravels. The value increases with increasing density, angularity, and gradation; decreases if mica is present; is relatively unaffected by rate of loading and grain size; and may be either increased or decreased under repeated or cyclic loading. Many designers use T_{max} as the total shear resistance—an assumption also made in most earth-pressure formulas.

Cohesion. This is the maximum tensile strength of the soil. It results from a complicated interaction of many factors, such as colloidal adhesion of the grain covering, capillary tension of the moisture films, electrostatic attraction of charged surfaces, drainage conditions, and stress history. True cohesion exists only in clays with edge-to-face contact between particles. Apparent cohesion can occur in nonplastic fine-grained soils under partial saturation conditions. The value of cohesion to be used in design depends directly on the drainage conditions under the imposed loading as well as on the method of testing for its determination and must be evaluated carefully.

(G. A. Leonards, "Foundation Engineering," McGraw-Hill Book Company, New York; T. W. Lambe and R. J. Whitman, "Soil Mechanics," John Wiley & Sons, Inc., New York.)

Compressibility. This property defines the stress-strain characteristics of the soil. Application of added stresses to a soil mass results in volume changes and associated displacements. Such displacements at the foundation level lead to foundation settlements. Limitation of settlements to allowable values often controls foundation design, particularly for granular soils.

For granular soils, compressibility is expressed in terms of Young's modulus E, taken usually as the secant modulus of the stress-strain curve from a standard triaxial test. The modulus decreases with increasing axial stress and increases with increasing confining pressure and with repeated loading.

Compressibility of saturated clays is expressed as the compression index C_c, together with a determination of the maximum past pressure to which they have been subjected. Both values are determined from standard laboratory one-dimensional consolidation tests (ASTM D2435). C_c is the change in void ratio per logarithmic cycle of stress and is a function of past stress history. For practical applications, its value in the stress range of interest must be known.

Permeability. This is the ability of the soil mass to conduct fluid flow under a unit of hydraulic gradient. In foundation design, only permeability under saturated conditions usually is needed. Permeabilities of most soil formations are highly variable and strongly sensitive to relatively small variations in the soil mass.

Controlled by the size and continuity of the pore space and consequently by the grain size, permeability typically is an anisotropic property, usually being greater in the horizontal direction than in the vertical. Permeability values for the various soil classifications vary by a factor of more than 10 million, as determined directly by laboratory or field permeability tests or indirectly by consolidation tests and grain-size analyses. Field pumping tests with observation wells provide the most reliable values.

Other Properties. There are other minor soil properties that are sometimes of importance. The vegetable-free refuse content of a soil, for example, may affect the fixity of any properties induced by treatment. For instance, soils very heavy in rotted vegetation, containing tannic acids, are not suitable for cement stabilization. As another example, soils with large limestone-dust content can be weakened by water flow through the mass or disintegrated by percolating sewage and other waste liquors.

IDENTIFICATION, SAMPLING, AND TESTING OF SOILS

To permit the application of previous experience to new soil conditions, a standard system of soil identification is necessary. For the purpose, classification of a soil usually is based on physical properties determined in accordance with standardized testing procedures. Testing for soil properties or reactions of soils to applied loads involves both laboratory and field procedures.

6-6. Soil Identification.

Field investigations to identify soils can be made by surface surveying, aerial surveying, or geophysical or subsurface exploratory analysis. Complete knowledge of the geological structure of an area permits definite identification from a surface reconnaissance. Tied together with a mineralogical classification of the surface layers, the reconnaissance can at least recognize structures of some soils. The surveys, however, cannot alone determine soil behavior unless identical conditions have previously been encountered.

Geological and agronomic soil maps and detailed reports useful for this purpose are available for most populated areas in the United States. They are issued by the U.S. Department of Agriculture, U.S. Geological Survey, and corresponding state offices. Old surveys are of great value in locating original shore lines and stream courses, as well as existence of surface-grade changes.

A complete site inspection is a necessary adjunct to the data collected from maps and surveys and will often clarify the question of uniformity. Furthermore, inspection of neighboring structures will point up some possible difficulties.

Aerial surveying has been developed to the extent where very rapid determination of soil types can be made at low cost over very large areas. Stereoptic photographic data, correlated with standard charts, identify soil types from color, texture, drainage characteristics, and ground cover.

Soil Classification. The most widely accepted soil classification system is the Unified Soil Classification (Table 6-1). It sets definite criteria for naming soils and lists them in fixed divisions based on grain size and laboratory tests of physical characteristics.

TABLE 6-1 Unified Soil Classification Including Identification and Description[a]

Major division	Group symbol	Typical name	Field identification procedures[b]	Laboratory classification criteria[c]
A. Coarse-grained soils (more than half of material larger than No. 200 sieve)[d]				
1. Gravels (more than half of coarse fraction larger than No. 4 sieve)[e]				
Clean gravels (little or no fines)	GW	Well-graded gravels, gravel-sand mixtures, little or no fines	Wide range in grain sizes and substantial amounts of all intermediate particle sizes	$D_{60}/D_{10} > 4$ $1 < D_{30}^2/D_{10}D_{60} < 3$ D_{10}, D_{30}, D_{60} = sizes corresponding to 10, 30, and 60% on grain-size curve
	GP	Poorly graded gravels or gravel-sand mixtures, little or no fines	Predominantly one size, or a range of sizes with some intermediate sizes missing	Not meeting all gradation requirements for GW
Gravels with fines (appreciable amount of fines)	GM	Silty gravels, gravel-sand-silt mixtures	Nonplastic fines or fines with low plasticity (see ML soils)	Atterberg limits below A line or PI < 4
	GC	Clayey gravels, gravel-sand-clay mixtures	Plastic fines (see CL soils)	Atterberg limits above A line with PI > 7	Soils above A line with $4 < PI < 7$ are borderline cases, require use of dual symbols
2. Sands (more than half of coarse fraction smaller than No. 4 sieve)[e]				
Clean sands (little or no fines)	SW	Well-graded sands, gravelly sands, little or no fines	Wide range in grain sizes and substantial amounts of all intermediate particle sizes	$D_{60}/D_{10} > 6$ $1 < D_{30}^2/D_{10}D_{60} < 3$
	SP	Poorly graded sands or gravelly sands, little or no fines	Predominantly one size, or a range of sizes with some intermediate sizes missing	Not meeting all gradation requirements for SW
Sands with fines (appreciable amount of fines)	SM	Silty sands, sand-silt mixtures	Nonplastic fines or fines with low plasticity (see ML soils)	Atterberg limits above A line or PI < 4
	SC	Clayey sands, sand-clay mixtures	Plastic fines (see CL soils)	Atterberg limits above A line with PI > 7	Soils with Atterberg limits above A line while $4 < PI < 7$ are borderline cases, require use of dual symbols

Information required for describing coarse-grained soils:

For undisturbed soils, add information on stratification, degree of compactness, cementation, moisture conditions, and drainage characteristics. Give typical name; indicate approximate percentage of sand and gravel; maximum size, angularity, surface condition, and hardness of the coarse grains; local or geological name and other pertinent descriptive information; and symbol in parentheses. Example: *Silty sand*, gravelly; about 20% hard, angular gravel particles, ½ in. maximum size; rounded and subangular sand grains, coarse to fine; about 15% nonplastic fines with low dry strength; well compacted and moist in place; alluvial sand; (SM).

Major division	Group symbol	Typical names	Identification procedure[f]			Laboratory classification criteria[e]
			Dry strength (crushing characteristics)	Dilatancy (reaction to shaking)	Toughness (consistency near PL)	
B. Fine-grained soils (more than half of material smaller than No. 200 sieve)[d]						
Silts and clays with liquid limit less than 50	ML	Inorganic silts and very fine sands, rock flour, silty or clayey fine sands, or clayey silts with slight plasticity	None to slight	Quick to slow	None	Plasticity chart for laboratory classification of fine-grained soils compares them at equal liquid limit. Toughness and dry strength increase with increasing plasticity index (PI)
	CL	Inorganic clays of low to medium plasticity, gravelly clays, sandy clays, silty clays, lean clays	Medium to high	None to very slow	Medium	
	OL	Organic silts and organic silty clays of low plasticity	Slight to medium	Slow	Slight	
Silts and clays with liquid limit more than 50	MH	Inorganic silts, micaceous or diatomaceous fine sandy or silty soils, elastic silts	Slight to medium	Slow to none	Slight to medium	
	CH	Inorganic clays of high plasticity, fat clays	None to very high	None	High	
	OH	Organic clays of medium to high plasticity	Medium to high	None to very slow	Slight to medium	
C. Highly organic soils						
	Pt	Peat and other highly organic soils	Readily identified by color, odor, spongy feel, and often by fibrous texture			

TABLE 6-1 Unified Soil Classification Including Identification and Description *(Continued)*

Field identification procedures for fine-grained soils or fractions*g*:

Dilatancy (reaction to shaking)	Dry strength (crushing characteristics)	Toughness (consistency near PL)
After removing particles larger than No. 40 sieve, prepare a pat of moist soil with a volume of about ½ cu in. Add enough water if necessary to make the soil soft but not sticky. Place the pat in the open palm of one hand and shake horizontally, striking vigorously against the other hand several times. A positive reaction consists of the appearance of water on the surface of the pat, which changes to a livery consistency and becomes glossy. When the sample is squeezed between the fingers, the water and gloss disappear from the surface, the pat stiffens, and finally it cracks or crumbles. The rapidity of appearance of water during shaking and of its disappearance during squeezing assist in identifying the character of the fines in a soil. Very fine clean sands give the quickest and most distinct reaction, whereas a plastic clay has no reaction. Inorganic silts, such as a typical rock flour, show a moderately quick reaction.	After removing particles larger than No. 40 sieve, mold a pat of soil to the consistency of putty, adding water if necessary. Allow the pat to dry completely by oven, sun, or air drying, then test its strength by breaking and crumbling between the fingers. This strength is a measure of character and quantity of the colloidal fraction contained in the soil. The dry strength increases with increasing plasticity. High dry strength is characteristic of clays of the CH group. A typical inorganic silt possesses only very slight dry strength. Silty fine sands and silts have about the same slight dry strength but can be distinguished by the feel when powdering the dried specimen. Fine sand feels gritty, whereas a typical silt has the smooth feel of flour.	After particles larger than the No. 40 sieve are removed, a specimen of soil about ½ cu in. in size is molded to the consistency of putty. If it is too dry, water must be added. If it is too sticky, the specimen should be spread out in a thin layer and allowed to lose some moisture by evaporation. Then, the specimen is rolled out by hand on a smooth surface or between the palms into a thread about $\frac{1}{8}$ in. in diameter. The thread is then folded and rerolled repeatedly. During this manipulation, the moisture content is gradually reduced and the specimen stiffens, finally loses its plasticity, and crumbles when the plastic limit (PL) is reached. After the thread crumbles, the pieces should be lumped together and a slight kneading action continued until the lump crumbles. The tougher the thread near the PL and the stiffer the lump when it finally crumbles, the more potent is the colloidal clay fraction in the soil. Weakness of the thread at the PL and quick loss of coherence of the lump below the PL indicate either organic clay of low plasticity or materials such as kaolin-type clays and organic clays that occur below the *A* line. Highly organic clays have a very weak and spongy feel at PL.

Information required for describing fine-grained soils:

For undisturbed soils, add information on structure, stratification, consistency in undisturbed and remolded states, moisture, and drainage conditions. Give typical name; indicate degree and character of plasticity; amount and maximum size of coarse grains; color in wet condition; odor, if any; local or geological name and other pertinent descriptive information; and symbol in parentheses. Example: *Clayey silt*, brown; slightly plastic; small percentage of fine sand; numerous vertical root holes; firm and dry in place; loess; (ML).

^{a} Adapted from recommendations of Corps of Engineers and U.S. Bureau of Reclamation. All sieve sizes United States standard.
^{b} Excluding particles larger than 3 in. and basing fractions on estimated weights.
^{c} Use grain-size curve in identifying the fractions as given under field identification.
For coarse-grained soils, determine percentage of gravel and sand from grain-size curve. Depending on percentage of fines (fractions smaller than No. 200 sieve), coarse-grained soils are classified as follows:
   Less than 5% fines . . . . GW, GP, SW, SP
   More than 12% fines . . . . GM, GC, SM, SC
   5% to 12% fines. . . . . Borderline cases requiring use of dual symbols
Soils possessing characteristics of two groups are designated by combinations of group symbols; for example, GW-GC indicates a well-graded, gravel-sand mixture with clay binder.
^{d} The No. 200 sieve size is about the smallest particle visible to the naked eye.
^{e} For visual classification, the ¼-in. size may be used as equivalent to the No. 4 sieve size.
^{f} Applicable to fractions smaller than No. 40 sieve.
^{g} These procedures are to be performed on the minus 40-sieve-size particles (about 1/64 in.). Simply remove by hand the coarse particles that interfere with the tests.
For field classification purposes, screening is not intended.

6-9

6-7. Subsurface Exploration.

This is the field phase of soils analysis and substructure design, and its importance cannot be overemphasized. Inadequate, inaccurate or misleading information at this stage is the most common cause of uneconomical foundation and earthworks designs, and of their failures. The key word is exploration. The purpose of this work is to determine—by exploratory techniques—the nature of the site subsurface conditions and their impact on design. The work must therefore be planned and executed so as to reveal the nature of the soils and not simply conducted in a perfunctory manner. The type and extent of exploration techniques, in-situ testing, and sampling methods thus should be chosen in accordance with the unknowns associated with the site, geologic hazards that may be reasonably expected, the loading intensity imposed by the structure, and the amount of settlement of the building that can be tolerated.

The intensity and methodology of the exploration effort will vary widely. (No standard can be applied, nor should one be.) In heavily developed areas where the subsurface conditions and engineering characteristics are well known through prior work for other structures, an appropriate investigation might well consist of a limited, confirming-type program, even though the structure under design may be a major facility. In contrast, a lightly loaded structure in a remote area with poor subsurface conditions may require an extensive investigation. Some building codes specify exploration programs, *but these are minimum criteria*. Blind adherence to their requirements does not constitute prudent engineering practice. The best guideline is: the foundation engineer should be reasonably certain that no *major unknowns* exist at the site and should know the controlling characteristics of the subsurface materials.

In accordance with this general philosophy, it is often advisable to conduct subsurface investigations in two or more phases, where successive phases provide increasing detail. Start with a few borings, or other widely spaced explorations. From these, ascertain the stratigraphy and general soil properties. Then, plan the second phase to fill in blind spots, to confirm uniformity or predictability, or to delineate and define anomalies. A third-level detailed effort is usually required only to further define anomalies or to conduct special testing imposed by the requirements of the particular facility under design.

Remote-Sensing Techniques. These methods provide indirect evidence of the subsurface materials. They do not indicate directly the engineering properties but can give depths to strata contacts in many instances and may permit qualitative evaluations of the materials.

Aerial surveys are appropriate where large areas are to be explored. Analyses of conventional aerial stereoscopic photographs; thermal and false-color, infrared imagery; multispectral satellite imagery; or side-looking aerial radar can disclose the surface topography and drainage, linear features that reflect geologic structure, type of surface soil and often the type of underlying rock. These techniques are particularly useful in locating filled-in sinkholes in karst regions, which are characterized by closely spaced sinks.

Geophysical exploration provides information quickly and is economical for supplementing data from borings. The geophysical techniques include seismic reflection and refraction, uphole and crosshole seismic tests, electrical resistivity, microgravimetric measurements, and acoustical subbottom profiling.

Seismic exploration of subsoils is a carry-over from practices standardized in oil-field surveys to determine discontinuities in soil structure. The principles involved are the well-known characteristics of sound-wave transmission, reflection, and refraction in passing through materials of different densities. The method charts arrival time at various points of sound waves induced by explosion or hammer blows.

A similar technique uses the variation in electric conductivity of various soil densities and at discontinuities in layer contacts. The microgravimetric method employs very small variations in the gravity field to locate subsurface voids or changes in rock type. Subbottom profiling is used offshore to provide a continuous record, from seismic reflection, of the subbottom stratigraphic contacts.

In many cases, geophysical methods can give a good picture of certain soil properties, depths of layers, and depths to rock. They cannot be expected to give more than an average or statistical picture of conditions or to give definite identification of the kinds of soils. Information from these methods, however, is an excellent guide for programming a complete exploratory investigation and may be invaluable in providing coverage between widely spaced boreholes.

In-Situ Exploration and Testing Techniques. A wide variety of techniques are available for in-situ investigations of subsurface conditions. Several of these methods have been standardized, and many are utilized worldwide.

Test Pits and Trenches. These provide direct visual observation and hand sampling of soils but are limited to practical depths of 10 ft or less. Trenches are advantageous for locating surface contacts between steeply dipping strata. Test pits are an economical and rapid means of obtaining subsurface information but should not be used alone unless the materials lying below the pit are known or are of no concern. The soils in pits and trenches may be sampled by manually pushing thin-walled tubes into the bottom or sides.

Probes. These are drilled or driven soundings, without samples, used to locate top of rock or to obtain cuttings for soil identification. Driven probes give good indication of soil properties. Rock probes can also indicate relative rock soundness by measuring the rate of drill penetration under constant down pressure.

Borings. The traditional method of exploring the subsurface is to drill and examine holes and the material removed from them. Many foundation construction difficulties, however, often of a very expensive nature, result from either unreliable methods of sampling borings or from too much reliance on extrapolation of the results. Consequently, expert engineering judgment should be employed in planning and executing soil explorations with borings.

A boring can be expected to depict truly only the conditions at the boring location. The information obtained there may or may not be indicative of the conditions between borings. Hence, choosing the number and location of borings calls for good judgment.

Many municipal and state building codes specify a minimum number of borings for a specific size of site. The information obtained from these borings, however, may not be adequate for foundation design and construction. Often, after the code-required borings have been completed, the boring rig is removed from the site; but during excavation or foundation construction, discontinuities of layers are discovered. At that stage, the cost of additional borings is high. But the borings should be made. Without exception, their cost is a small part of either the wastage in overdesign to offset lack of sufficient information or the cost of corrective revisions that will be necessary during foundation construction.

The usual plan for making supplementary borings is to spot an additional boring midway between a pair of borings that shows inconsistent conditions. Then, if the new boring is not a reasonable average of the two, another boring may be placed between the pair with the greatest variation. More borings should be added, as indicated by the results of the preceding explorations. In this way, the location of changes in condition can be closely fixed and overconservative estimates eliminated.

Borings are advanced with continuous flight augers for rotating drill bits. A circulating fluid is used during boring to remove the cuttings. To avoid contamination of soil samples taken from the borings, it is usually advantageous to use the rotary-drilling method with bentonitic drilling mud as the circulating fluid. Regardless of the method used, great care should be taken that the fluid level inside the borehole never falls below the water table, even during the interval that the drilling tools are being removed from the hole.

During the drilling and sampling of borings, a log should be kept by an engineer, geologist, or trained technician of the hole depth where samples are taken or tests made, depth to strata changes and to the water table, and the results of any in-situ tests, such as the standard penetration test (ASTM D1586). Additionally, the logs should include a complete description of all materials encountered, as judged from the cuttings and samples. Where the borehole encounters rock, the log should note the core recovery, in length and in percentage of the distance drilled into the rock; any weathered seams or solution features; and any joints and their inclination. For many rock formations, determination and notation on the log of the rock quality designation (RQD) is a good indication of rock soundness. A typical boring log for soil is shown in Fig. 6-1.

Borings provide a means for both sampling and testing the subsurface soils. The most common borehole test is the Standard Penetration Test or SPT (ASTM D1586). In this test, a split-barrel sampler, or spoon, with 2-in outside diameter is driven into the borehole bottom by a 140-lb hammer falling 30 in. The number of blows required to drive the spoon 12 in, the N value in blows per foot, is recorded. N is an indication of the density of granular soils and may reflect the consistency of nonsensitive clays. The SPT provides a moderately disturbed soil sample, but it is suitable for observation, classification, and index testing. The blow count, however, is subject to many influences during drilling and sampling and in the materials penetrated. Therefore, N should be carefully evaluated. (See V. F. B. de Mello, "The Standard Penetration Test," Proceedings of the Fourth Panamerican Conference on Soil Mechanics and Foundation Engineering, Vol. 1, 1971, San Juan, Puerto Rico, and J. Schmerttmann, "Measurement of In-situ Shear Strength," Conference on In-situ Measurement of Soil Properties, Vol. II, 1975, American Society of Civil Engineers.)

Other borehole tests include the vane shear, for measuring the in-situ shear strength of clays and organic soils, and the pressure meter, in which an expanding vertical membrane inside a borehole measures the stiffness and strength of the surrounding soil.

Soil and rock sampling devices should be selected to maximize recovery and limit disturbances of the sample to the degree needed. Sound, unweathered rock should be sampled with core barrels fitted with diamond bits. Several sizes are available, from AX bits to 4.5-in-O.D. bits. The NX bit is most often used, but the larger sizes normally give better recovery in weathered, closely jointed, or porous rock. Recovery can also usually be improved by use of multiple-tube core barrels. A typical rock boring log is shown in Fig. 6-2.

The SPT barrel is a relatively good sampler for most soils, although the sample is moderately disturbed. In firm-to-stiff clays, relatively undisturbed samples can be obtained with hydraulically pushed, thin-wall Shelby tubes. In very soft clays and loose sands, a piston sampler gives better results. Dense soils and severely weathered rocks may be sampled with Dennison or Pitcher samplers.

Soundings. A sounding with a cone penetrometer is an effective supplementary technique for use in conjunction with samples from borings. The hydraulically or mechanically advanced Dutch cone utilizes a conical point to measure penetration resistance and a follower sleeve to measure frictional resistance. These two values are good indicators of strength and stiffness and relatively reliable indicators of soil type.

Cone penetrometers may also be advanced dynamically by driving with a hammer, as in the SPT. Obtained in this manner, the test results are analogous to SPT blow counts and can be used for the same purposes when appropriate correction factors are applied.

Fig. 6-1 Typical soil boring log.

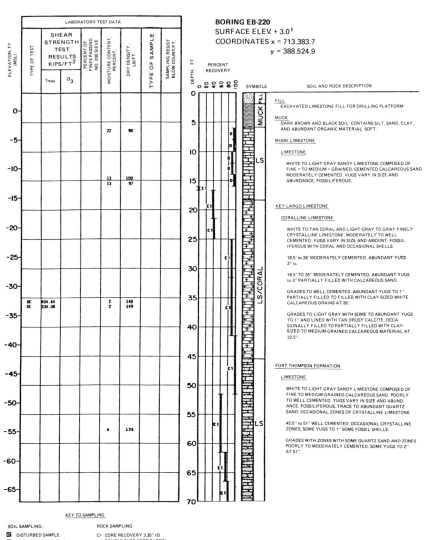

Fig. 6-2 Typical rock boring log.

6-8. Laboratory Tests.

Basically laboratory tests give more precise values of engineering properties than do interpretations of simple field tests—provided the samples are truly representative of the subsurface conditions. Laboratory testing of soils has developed into a complicated maze of interrelated tests, with many forms of apparatus. A summary of these tests is published periodically by the American Society for Testing and Materials (see STP 479, "Special Procedures for Testing Soil and Rock for Engineering Purposes"). Many of the test techniques are applicable only to certain restricted soil groups and give misleading data if otherwise used.

If settlement-sensitive structures, major earthworks, dams, or steep slopes are to be founded on soft or uncertain soils, laboratory tests should be performed on representative samples. Laboratory tests, however, are expensive and time-consuming. Therefore, except on large projects, usually few such tests are made. Consequently, the samples for testing should be carefully selected.

Index tests, such as Atterberg limits, density, and grain-size distribution, are used to classify and characterize soils, indicate general engineering characteristics of soils, determine suitability of soils as fill material, and estimate susceptibility of soils to ground-improvement techniques. It is common to make at least several of these tests on each stratum of interest.

Compressibility tests, such as consolidation and triaxial compression tests, provide values for estimating settlements under load. Consolidation tests are performed on plastic soils and triaxial compression tests, on granular materials. Consolidation tests also give time-rate of settlements, change in permeability with consolidation and the maximum past pressure to which the soil has been subjected. Triaxial compression tests are also used to evaluate the stiffness of sands, which increases with increasing confining pressure, and the strength gain of clays under consolidation.

Laboratory strength tests measure soil strength for determination of bearing capacity, resistance to lateral earth pressure, and slope stability. Strengths of sands are measured by triaxial and direct shear tests. Laboratory vane-shear, unconfined-compression and triaxial-compression tests are applicable to cohesive soils.

In strength testing, it is necessary that drainage conditions during the tests simulate as closely as possible those that will prevail in the soil stratum at the time, and for the structure, under consideration. All the tests *except triaxial* are essentially undrained and approximate rapid field loading, such as that for most building foundations and excavations. Triaxial tests permit testing under undrained, consolidated undrained, and consolidated drained conditions. Tests under drained conditions would be used for such applications as checking stability under an embankment after excess pore pressures have dissipated.

If loose, saturated sands may be subjected to earthquake loads, the strength of the sands under cyclic loading and their liquefaction potential may be determined in the laboratory by cyclic triaxial tests.

SOIL IMPROVEMENT

Soil as an engineering material differs from stone, timber, and other natural products in that soil can be altered to conform to desired characteristics. Soil improvement is an age-old practice for building on marginal ground and is often used in modern geotechnical-engineering practice. Improvement may be achieved by in-situ methods or by construction of man-made fills. In either case, the objectives are increased bearing capacity or decreased settlements, or both. Many techniques, including densification, surcharging, grading, and fill construction, have been developed and become widely accepted. They are largely responsible for the increased and economical usage of marginal sites.

6.9 Man-Made Fills.

The term *fill* is used in this section to mean those earth materials that are placed primarily to level or to raise the ground surface, in contrast with retaining structures, such as earth dams and dikes. Most of the general principles presented in the following, however, apply to both types of earthwork.

Nearly all building sites require some man-made fills, even if for nothing more than the subbase for floor slabs or pavements. Yet, many undesirable conditions, including improper compaction, volume changes, and unanticipated settlements due to weights of fills, often are associated with fills. To prevent such conditions from occurring, fills should be viewed as structural elements of a project and designed accordingly, not haphazardly. Materials and their gradation, placement, degree of compaction, and occasionally their thickness should be selected so as to properly support the loads anticipated.

There are two basic types of fills: those placed in dry by conventional earthmoving equipment and techniques, and those placed wet by hydraulic dredges. The latter type is used mainly for filling behind bulkheads or for large-scale areal fills.

A wide range of materials and grain sizes are suitable for fills for most purposes, but all organic matter and refuse should be prohibited. Economics usually requires that the source of fill materials be as close to the site as possible, yet may eliminate some materials. For example, economics may not allow the drying of saturated fine-grained soils. For most fills, maximum particle size in the 18 in below foundations, slabs, or the surface should be limited to 3 in.

The most common test for determining the gross suitability of a soil as fill and for providing a standard for compaction is the moisture-density relationship test (ASTM D698 and D1557), often called the Proctor test. Figure 6-3 shows typical Proctor test results for sands, silts, and clays, as well as the 100% saturation or zero-air-voids line. Several such laboratory tests should be performed on the borrow material, and the standard moisture-density relationship should be established. The peak of the curve indicates the maximum density obtainable in the laboratory by the test method, as well as the optimum moisture content.

The two ASTM tests represent different levels of compactive effort. In the field, however, much higher compactive efforts may be employed than those used in the laboratory. Thus, a different moisture-density relationship may exist on the site. The Proctor test results should, therefore, not be regarded as an inherent property of the material. Nevertheless, the curves indicate the sensitivity to moisture content and the degree of field control that may be required to obtain the specified density.

Fill Compaction. The required degree of compaction for a fill is expressed normally as a minimum percentage of the maximum dry density obtained in the laboratory test, to be accomplished within a stipulated moisture range. Typically, densities representing 90 to 100% of the maximum density, at moisture contents within 2 to 4% of the optimum moisture content, are specified. (ASTM D1557 is used as the standard when high bearing capacity and low compressibility are needed; ASTM D698 should be used where the requirements are less, such as beneath parking lots.) Minimum densities of 90 to 95% of the maximum density are appropriate for most fills; 100% compaction is often required beneath roadways, footings, or other highly loaded areas.

Note that field densities can be greater than 100% of the maximum laboratory value. Furthermore, with greater compactive effort, such densities can be achieved at moisture contents off the laboratory curve. Care should be taken, however, that fine-grained

Fig. 6-3 Typical moisture-density relationships for soils as obtained from laboratory tests made in accordance with ASTM D1557-64T. Method C of this standard was used to determine the curve for sand and gravel; method A was used for all other curves.

materials not be overcompacted on the dry side of optimum, because they may swell and soften significantly on saturation.

Conventional fills are placed in lifts of 8 to 24 in. Each lift should be compacted before placement of the next lift. Actual compaction achieved may be determined by field density tests performed on each lift. For the purpose, wet density and moisture content should be measured and the dry density computed. Field densities may be obtained by the sand-cone (ASTM D1556) or balloon volume-meter (ASTM D2167) method, from an undisturbed sample or with the nuclear moisture-density meter. For most applications, one field density test for each 4000 to 10,000 ft^2 of lift surface is adequate.

On large projects where heavy compaction equipment is used, lift thicknesses of 18 to 24 in are acceptable. For most projects, however, lift thicknesses should be restricted to 8 to 12 in.

Hydraulically placed fills composed of dredged soils usually need not be compacted as they are being placed; many never are compacted. Segregation of the silt and clay fractions of the dredged material is common and generally not harmful; but accumulation of these fines in pockets adjacent to bulkheads or beneath structures should be avoided. Proper use of internal dikes, weirs, and decanting techniques can prevent such accumulations.

Bear in mind that fills are very heavy dead loads and may severely stress even deeply buried strata. A 1-ft depth of compacted fill is equal in load to about 1.5 stories of an ordinary office building. Consequently, the following undesirable conditions may occur: If a building straddles a cut-fill boundary line, there is a potential for sharply delineated differential settlement. Deep hydraulic fills may cause surface subsidence of several feet. Pile-supported structures with slabs on grade over deep fills may, as the fill settles, be damaged because of slab, utility, and entranceway movement relative to the structure. Tieback rods that are used for bulkheads and pass through a hydraulic fill may be loaded by subsequent fill settlement. To avoid this, the rods should be sleeved through the fill.

6-10. Densification.

Any of several different techniques, usually involving some form of vibration, may be used for densification, the in-situ compaction of primarily granular soils to increase their density. Applicability of these methods is sensitive to grain size, as shown in Fig. 6-4. Consequently, grain-size distribution must be carefully considered in selection of the densification method.

Clean sands can be readily densified to depths of about 6 ft by rolling the surface with a heavy, vibratory, steel-drum roller. The vibration frequency is to some extent adjustable but is generally most effective in the range of 25 to 30 Hz. (Little densification will be achieved below 6 ft, and the top foot may be actually loosened.)

Vibroflotation and Terra-Probe increase densities of sands by the multiple insertions of vibrating probes. The cylindrical voids created are filled with imported sand. Insertions are generally made in clusters, with typical spacing of about 4.5 ft, where building columns will be erected. Relative densities of 85% or higher are obtainable throughout the depth of insertion, which may exceed 40 ft. These techniques however, will not work if fines content exceeds about 15% or if organic matter is present in colloidal form in amounts greater than about 5% by weight.

Compaction piles are an alternative used to densify sands and permit the subsequent use of shallow foundations. Such piles may be of any material but commonly are wood or a sand replacement (sand pile) in which a shell is driven and the resultant hole filled with sand. The volume displaced by the piles and the vibration accompanying pile driving lead to densification of the surrounding soils. Typically, the foundation structural element does not bear directly on the compaction pile, but rather on the densified mass. Compaction piles would normally be used under the same structural requirements and subsurface conditions as Vibroflotation or the Terra-Probe technique.

Another technique for large-scale areal densification is dynamic compaction, as developed by Techniques Louis Menard, involving dropping of heavy weights from great heights on the ground surface. Weights may range from 10 to 40 tons. Dropping heights may be as large as 100 ft, and impact spacings up to 60 ft on center may be used. Multiple drops are made at each location, and several such passes across the area are required.

Fig. 6-4 Applicability of soil-improvement techniques in accordance with grain size of soils.

This technique is capable of densifying the widest range of grain sizes and materials, as shown in Fig. 6-4.

With 15-ton weights falling 80 ft, loose sands can be compacted to dense or very dense states to depths of 25 to 35 ft. With 40-ton weights, the effective depths can be extended to 40 to 50 ft, with further increase in density. With sands and nonplastic silts, enforced ground settlements on the order of 2 ft can be achieved.

The technique accomplishes densification as if the site had been subjected to numerous miniearthquakes, with the compaction resulting from partial liquefaction (where the soil is saturated) and the passage of the wave train. Elevated pore pressures are produced in saturated masses, and successive passes of pounding must await the dissipation of these pressures if the following pass is to be effective in producing further compaction. Figure 6-5 presents the results of a large-scale dynamic compaction effort in Bangladesh, shown as a plot of mean Standard Penetration Test values, before and after dynamic compaction, as a function of depth.

6-11. Surcharges.

Sometimes, suitable foundation materials are underlain by soft, compressible clays that would permit unacceptable settlements. In such cases, the site often can be made usable

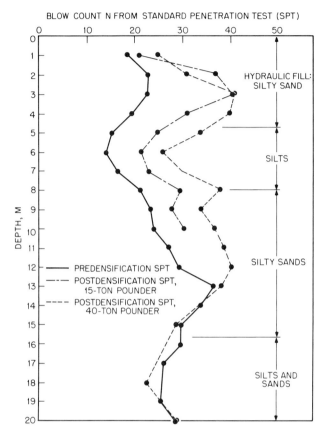

Fig. 6-5 Results of Standard Penetration Test (SPT) measurements of densification achieved with dynamic compaction.

by surcharging the surface. The intent of surcharging is to preload the ground and preconsolidate the clays, thus canceling the settlements that would otherwise have occurred under the structures. A concurrent objective may be to increase the strength of the clay.

In practice, the process is straightforward if the clay is overlain by sufficiently strong soils that bearing capacity is not a concern. The area to be improved is loaded with loose, dumped earth until the weight is equivalent to the load that will later be imposed by the permanent structure. (If highly plastic clays or thick layers with little internal drainage are present, sand drains may have to be inserted to achieve consolidation within a reasonable time.) Settlements of the original ground surface and of the clay layer should be closely monitored during and after placement of the surcharge. The fill should remain until settlement has about stopped. Then the surcharge may be removed and the structures erected. If surcharging has been done properly, the structures should experience no further settlement due to primary consolidation. Potential settlements due to secondary compression, however, should be evaluated, particularly if the soft soils are highly organic.

The technique is effective and most beneficial for large areas. The need for low-cost temporary fill and the long times occasionally required for settlement are the principal constraints.

6-12. Grouting.

The injection of cementitious materials or chemical agents into soil and rock masses improves them by either filling voids or causing compaction of the masses. Such grouting often is used to fill openings in rock masses, typically carbonate rocks, to restrict seepage or prevent structural collapse. Special materials and methods have been developed for specific purposes. For example, injection of quick-setting gels is effective in stabilizing loose sands, and high-pressure cement grouting can produce localized compaction of loose sands.

6-13. Drainage.

Permanent lowering of a near-surface water table in fine sands and silts often improves the near-surface soils markedly, particularly for such applications as roads, parking areas, and low-rise residential construction. Drainage is effective because the strength of soils decreases, in general, with an increase in amount and pressure of pore water.

In granular soils, the water table can be lowered with pumps and vertical wells, and the removed water conducted to off-site disposal basins or sewers. For silts, which are difficult to drain by such means, electroosmosis, or electrical drainage, may be used. This method employs the principle that water flows to the cathode when a direct current passes through saturated soils. The water may be pumped out at the cathode.

Measures should be taken to maintain the lowered water table. Surface and subsurface flows should be intercepted and conducted away from the site. For the purpose, the ground surface should be graded to control the flow of surface runoff and intercepting drains should be installed in the ground. Such drains, laid approximately along contours, are especially useful in stabilizing earth slopes.

FOUNDATION DESIGN

The selection of foundation design criteria—type of foundation and its depths and allowable load or bearing pressure—is often an iterative process. To provide proper support, all foundations must satisfy two requirements simultaneously: adequate bearing capacity and structurally tolerable settlements. Although related, the two requirements usually cannot be automatically met simultaneously. A foundation that fails in bearing also settles excessively, but one with normally adequate bearing capacity may also settle excessively. The two factors, bearing or load capacity and settlement, then must be examined and the foundation design based on the limiting condition.

6-14. Steps in Foundation Design.

In practice, the general procedure for foundation design is:

1. Determine the ultimate bearing capacity of the one or more types of foundation that are feasible, given the subsurface conditions and structural requirements.

2. Reduce the ultimate bearing capacities computed by a factor of safety of 2 to 3. Use the higher factor of safety where there is less certainty about the subsurface conditions.

3. Compute the settlements that may occur for a foundation with the reduced allowable bearing capacity and the given structural loadings.

4. If settlements are structurally acceptable, compute relative costs of the satisfactory foundation types, on a comparative basis such as cost per ton of column load or cost per square foot of building area. These costs should include those for all structural elements of the foundation system, such as caps for piles and any ground improvement works required, and any unusual costs such as dewatering. Consider the time required for construction. All other things being equal, choose the least-cost system.

5. If settlements are unacceptable for all types considered, then examine other alternatives, such as ground improvement, building relocation, reduced bearing pressures or loads, different bearing depths, and revision of superstructure. Repeat steps 3 and 4 until a safe design offering the best economy is selected.

6-15. Bearing or Load Capacity of Foundations.

Bearing or load capacity is a characteristic of a foundation-soil system, not an intrinsic quality of the soil only. Soils of different type differ in bearing capacity, but also for a specific soil, bearing capacity may vary with the type, shape, size, and depth of the foundation element applying pressure.

There are two basic types of foundations, shallow and deep. Also, there are several varieties of each type.

Shallow foundations include spread, continuous and strapped footings, mats, and compensated rafts. Deep foundations include drilled caissons, or piers, and many varieties of driven and cast-in-place concrete piles.

Note that it is not sufficient to base foundation design on bearing capacity of soil and foundation alone. The design should also limit settlement (see Art. 6-16).

Shallow Foundations. The ultimate bearing capacity q_{ult}, psf, of a soil under an infinitely long strip foundation is given by

$$q_{ult} = cN_c + qN_q + \frac{\gamma B}{2} N_\gamma \qquad (6\text{-}2)$$

where c = cohesion of the soil (see Art. 6-5)
q = overburden pressure, psf, at level of foundation base
γ = unit weight of soil, lb/ft^3
B = foundation width, ft
N_c, N_q, N_γ = bearing-capacity factors

Variation of the bearing-capacity factors with ϕ, angle of internal friction of the soil, is shown in Fig. 6-6.

The basic equation presumes an infinitely long strip foundation of width B and should be corrected for other shapes. Correction factors by which the bearing-capacity factors should be multiplied are given in Table 6-2, in which L = footing length.

The derivation of Eq. (6-2) presumes the soils to be homogeneous throughout the

TABLE 6-2 Shape Corrections for Bearing-Capacity Factors of Shallow Foundations[*]

	Correction factor		
Shape of foundation	N_c	N_q	N_γ
Strip	1.00	1.00	1.00
Rectangle	$1 + (B/L)(N_q/N_c)$	$1 + (B/L)\tan\phi$	$1 - 0.4(B/L)$
Circle and square	$1 + (N_q/N_c)$	$1 + \tan\phi$	0.60

[*]After E. E. De Beer, as modified by A. S. Vesić, "Bearing Capacity of Shallow Foundations," in "Foundation Engineering Handbook," H. F. Winterkorn and H. Fang, editors, Van Nostrand Reinhold Company, New York.

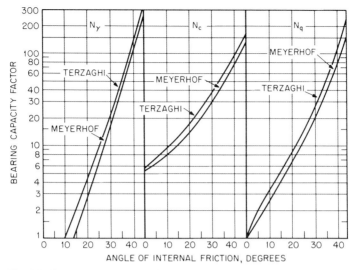

Fig. 6-6 Bearing-capacity factors determined by Terzaghi and Meyerhof for use in Eq. (6-2).

stressed zone, which is seldom the case. Consequently, adjustments may be required for departures from homogeneity. For stratified clay soils, for example, N_c may be adjusted as indicated in Fig. 6-7. In sands, if there is a moderate variation in strength, it is safe to use Eq. (6-2), but with bearing-capacity factors representing a weighted average strength.

For strongly varied soil profiles or interlayered sands and clays, the bearing capacity of each layer should be determined. This should be done by assuming the foundation bears on each layer successively but at the contact pressure for the depth below the bottom of the foundation of the top of the layer. Examples of pressure distributions below the foundation that may be assumed are shown in Figs. 6-8 and 6-9.

Deep Foundations. The load capacity of deep-footing foundations can be evaluated analytically in basically the same manner as described for shallow foundations. In the case of driven piles, there are additional methods for determining the load capacity that are based on pile resistance to driving.

Piles. A relatively large number of static equations have been proposed for computing the capacity of piles. Terzaghi and Peck proposed the formula

$$P_u = Q_{ip} + 2\pi r f_s L_f \qquad (6\text{-}3)$$

where P_u = ultimate load on pile, kips
Q_{ip} = point bearing capacity, kips, of pile, computed with the same equations as for shallow foundations, except that deep-foundation bearing-capacity factors are used (see J. E. Bowles, "Foundation Analysis and Design," McGraw-Hill Book Company, New York)
r = radius, ft, of pile
L_f = embedded length, ft, of the pile
f_s = friction, psf, along the pile (ranges from 0.5 to 2 ksf for cohesionless soils and may be taken as the cohesion for clays)

Fig. 6-7 Bearing-capacity factor N_c for stratified clay plots as a function of the ratio of the undrained shear strength c_2 of the lower layer to that of the upper layer c_1, and of the ratio of the depth D of the top of the lower layer below the foundation to the width B of the foundation.

(See K. Terzaghi and R. B. Peck, "Soil Mechanics in Engineering Practice," John Wiley & Sons, Inc., New York. A more rigorous formula, preferred for tapered piles, is given in R. J. Nordlund, "Bearing Capacity of Piles in Cohesionless Soils," Journal of the Soil Mechanics and Foundations Division, May 1963, American Society of Civil Engineers.)

The load capacity of driven piles is often estimated from their resistance to driving. The most commonly used driving formula is the *Modified Engineering News Formula*

$$P = \frac{2.5E}{S + 0.1} \frac{W_r + n^2 W_p}{W_r + W_p} \qquad (6\text{-}4)$$

where P = pile design load, kips
  E = manufacturer's pilehammer rated energy, ft-kips
  S = average set or tip penetration, inches per blow for the last five blows for gravity hammers and the last 20 blows for other types of hammers
  W_r = weight of ram, kips
  W_p = weight of pile, kips
  n = coefficient of restitution of cap blocks or cushions or butt of pile

While generally satisfactory in most applications, the formula gives results such that the safety factor may vary widely. Pile-driving formulas should never be used blindly, without references to pile-driving experience in the area of the site or to load tests.

Piles generally are driven in groups, rather than singly. Consequently, the group efficiency should be checked against the performance of a single pile. For piles supported by friction with the soil (**friction piles**), the perimeter friction of the group normally would be less than the sum of the friction forces on the individual piles. (In sands, densification due to driving a pile cluster increases the supporting friction forces.) For piles supported at the tip (**bearing piles**), the group efficiency becomes even more important with respect to the capacity of pile foundations to resist uplift loads.

Drilled Caissons. The bearing capacity of drilled caissons or piers is computed as if they were deep footings, except that account may be taken of the friction between shaft and soil. End-bearing should be computed as for footings, but with deep-foundation bearing-capacity factors (see J. E. Bowles, "Foundation Analysis and Design," McGraw-Hill Book Company, New York.) Adhesion or friction developed along the caisson shaft adds to the total bearing capacity and may be significant where the construction technique promotes good contact of the caisson concrete with relatively undisturbed soil or rock around the periphery of the shaft. In clays, the adhesion of soil to the shaft may approach 75% of the soil cohesion. (A value of $0.45c$, with a limit of 2000 psf, is recommended by A. W. Skempton in "Cast-in-situ Bored Piles in London Clay," *Geotechnique*, Vol. 9, No. 4.) In sands, the friction should be assumed to carry the weight of the shaft concrete only. In sound rock, the adhesion will essentially equal the concrete bond, whereas somewhat higher values may be realized if the rock is porous and is intruded by the concrete. In relatively weak porous limestones in south Florida, load tests demonstrated shaft design adhesion values of 8 to 10 tons/ft^2.

6-16. Settlement of Foundations.

All foundations settle. All of the materials used or soils encountered in ordinary construction are, to some degree, compressible under the range of stresses induced by most structures. The amount of settlement that can be accommodated safely by a structure is a function of the uniformity of the movements, the settlement rate, and the time of occurrence of settlement relative to the construction sequence, as well as the sensitivity of the structure as governed by column spacing, type of construction, and spacing of expansion joints.

In general, nuisance- or damage-causing settlements are those that create differential movements between building components and that occur after construction has been completed. Everything else being equal, differential movements become increasingly difficult to tolerate as column spacing decreases.

Cast-in-place concrete structures are typically far more sensitive to such movements than are steel buildings. For most steel structures, a deflected beam slope of 1:300 is about the steepest acceptable, whereas, for rigid concrete structures the deflected beam slope should not exceed about 1:800 to 1:1000.

Estimation of foundation settlements requires a knowledge of the stress distribution within the strata being analyzed and their compressibilities.

For most applications, stresses may be computed by the pressure-bulb concept with the methods of either Boussinesq or Westergaard. For thick deposits, use the Boussinesq distribution shown in Fig. 6-8; for thinly stratified soils, use use the Westergaard approach shown in Fig. 6-9. These charts indicate the stresses q beneath a single foundation unit that applies a pressure at its base of q_o.

Most facilities, however, involve not only multiple foundation units of different sizes, but also floor slabs, perhaps fills, and other elements that contribute to the induced stresses. The stresses used for settlement calculation should include the overlapping and contributory stresses that may arise from these multiple loads.

Cohesive Soils. Foundation settlement due to clays arises from the phenomenon called consolidation. Evaluation of the settlement of clays requires knowledge of the compressibility as well as the prior stress history of the deposit. Analysis of these parameters can be complex. (See, for example, T. W. Lambe and R. J. Whitman, "Soil Mechanics," John Wiley & Sons, Inc., New York.)

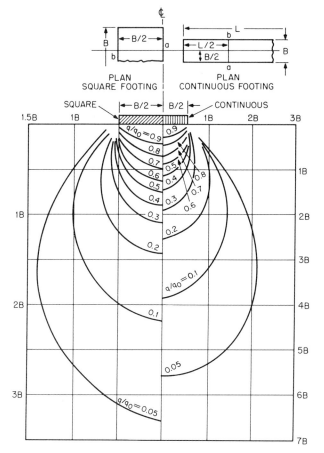

Fig. 6-8 Boussinesq stress distribution under a square footing with side B and under a continuous footing with width B.

Settlement due to clay may be estimated from

$$S_o = \frac{C_c}{1 + e_o} H \log \frac{P_o + \Delta P}{P_o} \tag{6-5}$$

where S_o = total settlement
C_c = compression index, taken as the slope of the virgin compression curve
e_o = initial void ratio, or ratio of void volume to volume of solids in a soil mass
H = layer thickness, or one-half the layer thickness if it is drained top and bottom
P_o = preconsolidation pressure
ΔP = induced additional pressure

Values of C_c and P_o are obtained from laboratory consolidation, or odometer, test results plotted for void ratio vs. log P or for unit strain vs. log P, in which case the slope of the virgin curve C_r is called the compression ratio

$$C_r = \frac{C_c}{1 + e_o} \tag{6-6}$$

The portion of the curve obtained with continually increasing P is called the **virgin compression curve**. A different curve is obtained for unloading and for reloading.

Several techniques have been proposed for determination of the maximum past pressure P_o. The most commonly used are those developed by Casagrande and Burmeister, as shown in Fig. 6-10. With the Casagrande construction, to determine P_o, draw a tangent at the point of sharpest curvature (minimum radius of curvature) of the consolidation curve, a horizontal line at the point of tangency, and a line bisecting the angle between the two. Extend the steeply sloped, nearly straight portion of the virgin curve backward, and its intersection with the bisecting line is taken as the maximum past pressure.

The Burmeister technique requires a small unload-reload cycle just beyond the point at which the consolidation curve begins to bend downward sharply. A characteristic triangle is formed between the rebound and reload curves, as shown in Fig. 6-10. Shift the rebound curve upward, parallel to itself, to a point where a geometrically similar triangle of the same vertical height is found. The pressure at that point represents the maximum past pressure. For other techniques, see G. A. Leonards, "Foundation Engineering," McGraw-Hill Book Company, New York, and H. F. Winterkorn and H. Fang, "Foundation Engineering Handbook," Van Nostrand Reinhold Company, New York.

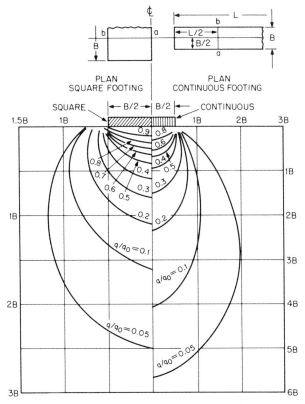

Fig. 6-9 Westergaard stress distribution under a square footing with side B and under a continuous footing with width B.

Fig. 6-10 Determination of maximum past pressure in clay from the curve relating consolidation pressure and void ratio: (a) method devised by Casagrande; (b) Burmeister's method.

The rate of settlement is usually an important consideration where clays are involved. The time, months (or years), to end of primary consolidation is given by

$$t = \frac{H^2}{C_v} \qquad (6\text{-}7)$$

where C_v = coefficient of consolidation, ft² per month (or year)
 H = length, ft, of longest drainage path to an outlet, for example, the ground surface or a drained pervious stratum

C_v may be determined from the time-compression curve of a single load increment during the laboratory test. Two graphical methods are commonly used, the square-root-of-time and the log-time fitting methods. (See preceding references.)

The coefficient of consolidation as determined by laboratory techniques is very sensitive to sample disturbance, testing techniques, and a number of other factors. Hence, there may be substantial error in the estimate of the time required to complete the primary consolidation of clays.

Some very plastic clays and most organic soils are subject to further compression beyond the end of primary consolidation. This creep-like characteristic is termed secondary compression and may represent a significant portion of the total settlement. For evaluation techniques and typical values, see T. W. Lambe and R. J. Whitman, "Soil Mechanics," John Wiley & Sons, Inc., New York.

Granular Soils. The bearing pressure of foundations on granular soils is usually governed by allowable settlements, rather than by bearing capacity. Settlements of foundations on these soils are estimated by empirical techniques utilizing field test data and by methods based on elastic theory.

One method is based on the standard penetration test. In this method, the allowable soil pressure q_{all}, ksf, for settlement of 1 in is computed from

$$q_{\text{all}} = 0.72(N - 3)\left(\frac{B+1}{2B}\right)^2 K_d \qquad (6\text{-}8a)$$

where N = representative standard penetration resistance, blows per foot, at depth B below foundation = $N'[8/(p + 2)]$ when corrected for overburden pressure
 N' = actual blows per foot
 p = effective overburden pressure, ksf
 B = footing width, ft
 K_d = depth factor = $1 + 0.2D/B \geq 1.2$
 D = depth of foundation base, ft

Allowable bearing capacity may also be estimated from the cone penetration test as

$$q_{\text{all}} = q_c/15 \qquad B \leq 4 \text{ ft} \qquad (6\text{-}8b)$$
$$q_{\text{all}} = q_c/25 \qquad B > 4 \text{ ft} \qquad (6\text{-}8c)$$

where q_c = cone penetration resistance, ksf.

Immediate Settlement of Compressible Layers. Immediate settlement, ft, of a foundation may be computed, on the basis of the theory of elasticity, from

$$S = \mu_0 \mu_1 q B \left(\frac{1 - \mu^2}{E_s}\right) \qquad (6\text{-}9)$$

where μ_0 and μ_1 = depth and shape factors obtainable from Fig. 6-11
 μ = Poisson's ratio for the soil
 q = pressure at base of foundation, ksf
 E_s = elastic modulus, ksf

S should be computed for each compressible soil layer under the foundation, and the settlements should be added to yield the total immediate settlement. For soils subject to

Fig. 6-11 Shape and depth factors for use in Eq. (6-9). *(After N. Janbu.)*

consolidation, total settlement equals the sum of immediate settlement and consolidation settlement.

While E_s may be estimated from the undrained shear strength S_u, determined from a triaxial compression test, it may range from $250S_u$ to $500S_u$. Because of this wide variation, especially in view of the likelihood of sample disturbance, it is advisable to determine E_s from field tests, such as the Standard Penetration Test, Cone Penetration Test (CPT), or plate loading tests (see Art. 6-17). For example, E_s, ksf, may be estimated from SPT data by the following, for sand

$$E_s = 10(N + 15) \tag{6-10}$$

and for clayey sands,

$$E_s = 6(N + 5) \tag{6-11}$$

where N = resistance, blows per foot. From CPT data, E_s may be approximated by $3q_c$ for sand and by $2q_c$ for clay, where q_c is the cone penetration resistance, ksf.

6-17. Spread Footings.

The most economical foundation for a load is one that transmits the load directly to the soil. In consequence, a spread footing is the most frequently used type of foundation for an isolated column.

Allowable Bearing Pressures. Approximate allowable soil bearing pressures, without tests, for various soils and rocks are given in Table 6-3 for normal conditions. These basic bearing pressures may be increased when the base of the footing is embedded beyond normal depth. Rock values may be increased by 10% for each foot of embedment beyond 4 ft in fully confined conditions, but the values may not exceed twice the basic values.

In any case, bearing pressures should be limited to values such that the proposed construction will be safe against failure of the soil under 100% overload.

Plate Loading Tests. Use of load tests for determining bearing values is as old as the art of foundation design. The techniques described centuries ago have only been changed by the use of equipment to prepare the test area and to handle the static loads.

The standard method for determining the bearing capacity of soil for spread footings is given in ASTM D1194. In this test, loads are applied to the soil through a steel plate by jacks. The jacks may react against a loaded truck, the weight of iron castings set on a frame, or a beam held in place by screw anchors embedded in the ground. The plate is loaded incrementally, and settlements are recorded for each increment of load. From these data, the load-settlement curve is plotted. Loads should be applied until a total settlement of 1 in occurs, the capacity of the loading apparatus is reached or a bearing failure occurs.

In this test, practical limits on capability for applying load require small plates, usually much smaller than actual footings. As a consequence, the test results cannot be directly translated to footing design. A major reason for this is that bearing capacity in sands and settlements of both sands and clays are related to footing size [see Eq. (6-2)]. Therefore, to scale the results of plate loading tests to footings, tests should be performed on plates of two different sizes, the smallest being not less than 1 ft square. The results may be used to verify bearing capacity estimated by other means or to obtain the coefficients of Eq. (6-2) that are unknown.

In clays, the bearing capacity q is independent of footing size. Thus, q_f for a footing equals q_p determined by the plate loading test.

TABLE 6-3 Allowable Bearing Pressures

Soil material	tons/ft^2	Notes
Unweathered sound rock	60	No adverse seam structure
Medium rock	40	
Intermediate rock	20	
Weathered, seamy, or porous rock	2 to 8	
Hardpan	12	Well cemented
Hardpan	8	Poorly cemented
Gravel soils	10	Compact, well graded
Gravel soils	8	Compact with more than 10% gravel
Gravel soils	6	Loose, poorly graded
Gravel soils	4	Loose, mostly sand
Sand soils	3 to 6	Dense
Fine sand	2 to 4	Dense
Clay soils	5	Hard
Clay soils	2	Medium stiff
Silt soils	3	Dense
Silt soils	1½	Medium dense
Compacted fills		Compacted to 90 to 95% of maximum density (ASTM D1557)
Fills and soft soils	2 to 4	By field or laboratory test only

In sand, the bearing capacity, according to Eq. (6-1a), is proportional to footing width. Thus,

$$q_f = q_p \frac{B_f}{B_p} \qquad (6\text{-}12)$$

where B_f = footing width, ft
B_p = plate width, ft

The settlement S_f of footings may be estimated from the settlement S_p obtained in a plate loading test from the following. In clays,

$$S_f = S_p \frac{B_f}{B_p} \qquad (6\text{-}13a)$$

and in sands,

$$S_f = S_p \left(\frac{2B_f}{B_f + B_p} \right)^2 \qquad (6\text{-}13b)$$

Bear in mind that results from tests with small plates are highly influenced by local irregularities beneath the plates that may not similarly influence footings. Furthermore, and more importantly, plate loading tests are of short duration, are applicable only to immediate settlements, and do not normally reflect consolidation of underlying clay deposits that may be stressed by the footings. The results of the tests, therefore, may be misleading.

Because of their inherent limitations, plate loading tests are seldom used alone to establish design criteria. Instead, soil-strength parameters and compressibility values are either determined or interpreted from field and laboratory tests; then, bearing capacities and settlements are computed and the limiting bearing capacity is chosen. (See Arts. 6-15 and 6-16.)

Cause of Settlement of Footings. Since there are on record many cases of foundation failures but very few footing failures, it must be concluded that the ground on which the footings rest is the usual cause of the trouble.

The displacement of the load carried can be one or several of the following movements:

1. Compression of the footing material elastically.
2. Deflection of the footing unit elastically.
3. Compression of the soil directly below the footing elastically.
4. Vertical yield of the soil directly below the footing because of horizontal transfer of stress, with consequent horizontal strain—elastic and plastic—possibly not equally in all directions.
5. Vertical yield of the soil because of change in character of the soil, such as change of water content under pressure, change in volume at contact planes of different materials, change in chemical constituents of soil or rock.

It should be kept in mind that items 3, 4, and 5 may be caused not only by the direct loading but also by loadings on adjacent footings. Also, the loading at the bottom of a footing is not uniform over its entire area.

The action of loads on soil may be conceived as a series of interrelated phases, chiefly:

1. The load compresses the soil directly below it.
2. The compressed soil bulges laterally.
3. The bulge compresses the side soil areas.
4. The side soil areas resist and distort, sometimes showing definite surface bulges.

Amounts of Settlement. From data that have been accumulated, it can be concluded that:

1. In soft rock, hardpan, and dense glacial deposits, settlements of from ¼ to ⅜ in must be expected and will occur almost immediately under loads up to 10 tons/ft².
2. In dry sands, settlements of ½ to ¾ in must be expected under loads from 3 to 4 tons/ft² and will occur almost immediately.
3. In wet sands, settlement of ¼ to ½ in will occur immediately under loads from 2 to 3 tons/ft² and will increase to ¾ in in a period of months, if the sand is drained.
4. In stiff clays mixed with gravel or sand, settlements under loads of 3 to 4 tons/ft² will be ½ in immediately and may increase to about 2 in in about 2 years.
5. In soft clays, settlements under even small loads may be substantial and will continue with time. Settlements of 6 to 9 in are not unusual.

These settlements are for small footings; large footings will settle more. The settlement of large areas in granular soils will not be uniform, chiefly because the base pressures are not uniform. Furthermore, edge shear on a footing provides some resistance to settlement. The settlement of large areas in plastic soils will not be uniform, because the base pressures are not uniform and the effect of plastic flow is much less in the interior regions.

Footing Loadings. The load to be used for footing design must be at least the dead load, including the weight of all fixed partitions, furniture, finishes, etc. However, it is doubtful whether the live load has any effect on settlements, except in such structures as grandstands, where the live-load increment may be over 150 psf for a few minutes sporadically.

Where building codes are enforced, the loads used in foundation design usually are fixed by a code. Two requirements are common:

1. Footings should be designed to sustain full dead load and a reduced summation of live loads on all floors, the percentage of reduction increasing with the number of stories, but seldom exceeding 50%. Pressures under the footings are limited by listed maximum unit pressures for various classes of soils.
2. Footings should be designed as in 1, but the areas of the footings should be proportioned for equal soil pressure under dead load only.

In method 2, unit base pressures under dead plus live load are greater for interior footings than for wall footings. Since wall footings are usually rectangular—and if at street lines, on soil less affected by adjacent loads—the opposite condition would be more conductive to equal settlements. In method 1, since the computed live loads seldom, if ever, exist, the actual base pressures on the interior footings are less than those on the exterior footings.

Effect of Pressure Distribution on Footing Design. The rigid solution for the stresses in a footing for a pressure distribution on the base that is a function of the settlement is many times indeterminate. As compared with a uniform distribution, the effect of the variable distribution on the computed bending moments in a single loaded footing (whether square, round, or strip shape) may be large, but the effects on computed shear, as well as unit shear and bond stresses in the concrete, may be disregarded from the points of both safety and economy. Where bending moment controls the design, as well as in combination footings of either the mat, cantilever, or combined types, the departure from uniform distribution may affect both safety and economy.

Why, then, have there not been many structural failures of footings constructed from designs based on the assumption of uniform distribution? The usual explanations point to the surplus strength in the materials (factors of safety) as well as the nonexisting live loads included in the design. However, since unequal settlements probably occur, even though the total base pressures may be less than computed, another explanation is necessary. It is found in the theory of "limit design": in a closed system of forces and restraints, local failure cannot occur, because the continuity of the mass causes a transfer of load to the spots which can best take the load.

Footing Design. Despite the uncertainties of pressure distribution under footings, an approximate solution for some simple footings may be developed, the method being necessary only where large bending moments control the design.

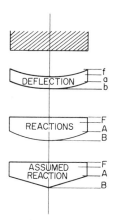

Fig. 6-12 Soil reactions on a strip footing.

The simplest case is a strip footing carrying a uniform continuous line load. The total load per foot w on the footing is equal to the summation of unit pressure below the footing. The shape of the loading curve is assumed similar to the deflection curve of the footing. The maximum unit pressure is $(F + A + B)$; the minimum is $(F + A)$, where F is the unit weight of the footing, A the minimum pressure, exclusive of footing weight, and B the maximum change in pressure from a uniform distribution due to defection of the footing (Fig. 6-12).

First, compute the maximum deflection of the footing as a cantilever, assuming the load w uniformly distributed: $b = wL^4/8EI$, where L is the cantilever length, E the modulus of elasticity, and I the moment of inertia based on an assumed section, footing weight being disregarded temporarily. Next, assume that the ratio of this deflection to the maximum settlement is the same as the ratio of the pressure B to maximum soil pressure. Draw the resulting base-pressure curve, taking into account the fact that the area between the pressure curve and the zero axis must equal the total load. Assume that the deflection curve is a straight line to simplify the computation (assumed reaction diagram, Fig. 6-12). Then, repeat the process using the new pressure curve in place of the assumed uniform distribution. Generally, the first approximation is close enough for economy.

Taking a specific example to illustrate the method, we shall compute the pressures for a concrete strip footing 10 ft wide and 20 in deep, which carries a wall load of 60,000 lb/lin ft. The soil is such that the average pressure, 6000 psf, will cause a settlement of ¼ in.

Under uniform distribution, w is 6000 psf; E for concrete will be taken as 2,000,000 psi; I for a 1-ft length of footing is $12 \times 20^3/12$, or 8000 in^4.

$$b = \frac{wL^4}{8EI} = \frac{6000 \times 5^4 \times 12^3}{8 \times 2,000,000 \times 8000} = 0.05 \text{ in}$$

By proportion,

$$B = \frac{0.05}{0.25} \times 6000 = 1200 \text{ psf}$$

Since the area of the pressure curve must equal 60,000 lb, i.e.,

$$\frac{L}{2}(2A + B) = P$$

$$A = \frac{P}{L} - \frac{B}{2} = 6000 - \frac{1200}{2} = 5400 \text{ psf}$$

Footing weight of 250 psf must be added to these pressures. Base pressures then vary from 5650 psf at the edge to 6850 psf at the center.

For use in the next step, note that deflection of a cantilever under triangular load, with zero at the free end, is $WL^3/15EI$ where W is the total load. Assume the pressures just computed as the load. The new approximate deflection then is $12^3[5400 \times 5^4/8EI + 1200 \times 5^4/(15 \times 2EI)]$, or practically 0.05 in, differing less than 5% from the first approximation. Moments and shears can be computed from the pressures already obtained, with sufficient accuracy.

For a square footing with a point load, similar assumptions can be made. In addition, assume that the deflection of a corner is $wL^4/8EI$ and that of the middle of a side, two-thirds as much, so that the deflected footing is dish-shaped.

For a footing 10×10 ft by 30 in thick carrying a load of 600,000 lb on a soil that settles ¼ in under the average load, 6000 psf, the deflections under a uniform 6000 psf load are computed to be 0.050 and 0.033 in at the corners and middle of the sides, respectively. The corresponding vertical ordinates of the dish-shaped loading curve are $0.05 \times 6,000/0.25 = 1200$ and 800 lb (departure from uniform distribution). Consider half a quarter section, carrying an eighth of the load, or 75,000 lb on an area of 12.5 ft² and assume straight-line pressure variation from center to corners and to center of edges. The "loss" in bearing value, represented by a pyramidal volume with 5-ft altitude, is $(800 + 1200)/2 \times ⅔ \times 5 = 8300$ lb. Unit pressure at the center of footing, therefore, is $(75,000 + 8300)/12.5 = 6700$ psf. At the middle of the edge, the pressure is $6700 - 800 = 5900$ psf. And at the corner, the pressure is $6700 - 1200 = 5500$ psf. To these units are to be added the weight of the footing, or 375 psf. Moments can then be computed from this distribution, and will show some saving over the assumption of uniform distribution.

Helpful Hints for Footing Design. Several practical considerations may be used as a guide in design and construction:

1. The more rigid the footing, the more uniform the pressure on the base.
2. Nonuniform distribution may be disregarded in footings resting on soils having a bearing value over 3 tons/ft².
3. On similar soils, with the same loads per unit area, square footings settle more than rectangular footings.
4. On the same soils, for the same shaped footings under the same average load intensity, the larger the footing, the greater the settlement.
5. Footings on the exterior of a building, not adjacent to other loads, will settle less than those affected by adjacent loads.
6. Footings will probably settle again when adjacent soil areas are loaded.
7. Plate loading tests are reliable only when site conditions are duplicated exactly.
8. Flexible footings may be designed for smaller bending moments than are given by the assumption of uniform distribution.
9. Footing failures are seldom structural failures.
10. Uniform settlements of buildings can be obtained by taking into consideration: actual loadings to be expected; greater soil bearing value of exterior locations; provision for slip joints for settlement due to adjacent future loadings; effect of difference in levels of footings in close proximity; changes in portions of the soil area due to water, weathering, and vibration; and differences in bearing value of differently shaped footings.
11. Maximum soil pressure under centers of footings is no criterion of expected settlements, because of the distribution of load due to rigidity.
12. Bearing pressures of soil and expected settlements deduced from load tests in the field or laboratory tests on undisturbed samples are of little value if the conditions in the construction work are not considered.
13. Exposure of soils during excavation may modify the soil structure and change all the characteristics.
14. All structures will settle, and provision must be made for the expected settlement by keeping them free of adjacent structures that have moved into a stable position.

15. Lateral resistance in soils must not be disturbed or reduced below that necessary to balance the lateral stresses resulting from the added loadings.

6-18. Combined Foundations and Mats.

In many cases, it is not possible or economical to provide a footing centrally located under each column and wall load. There are two corrective procedures normally used for such cases: combination of loads for one footing or imposition of balancing loads to neutralize the eccentric relative positions of load and footing. Two or more loads may be supported on a single footing, and in some locations an entire building may be constructed on a mat, or raft. Balancing of loads can be accomplished by means of structural beams acting as levers (often called "pan handles" or "pump handles") between column loads, or by imposition of counterbalancing moment resistance in the form of rigid frames.

Entire building areas may be covered by a single mat, where soils have extremely low bearing values at desired settlements and where the weight of the structure must sometimes be partly or wholly balanced by removal of soil weight to avoid soil failure.

Design of Combined Footings. For isolated footings the form of soil-resistance distribution may not be a determining factor in the economy of structural design, since the bending moment of the pressures about the centrally located column usually does not control the design of either the concrete or the reinforcing. The opposite condition holds, however, when the superimposed loads are not single and centrally located. And since very little is known concerning the true distribution of soil resistance at the contact face between soil and concrete, the usual design procedure is to assume uniform distribution, with sufficient soil resistance to counterbalance eccentricity of loading. However, there are available numerical computer solutions that can model the variable loading and deflection.

Once a design of this type is prepared, analysis of the deflections resulting from a uniform distribution can be used to revise the shape of the soil-reaction curve. The resulting moments will usually be found less than when computed for the uniform assumption.

For this computation, the soil-reaction modulus, or relationship between gross deformation of the soil and bearing value, must be known. This can be taken from the results of laboratory or field tests, but the effect of edge shears should be considered.

When more than one load is supported on a footing, the center of gravity of the loads should coincide with the geometric center of the footing. In addition, axes of symmetry through the column loadings should coincide with axes dividing the footing into equal areas. Since column loads are seldom of constant value, and since the variation in uncertain loadings (live load, impact, wind, etc.) is different even for adjacent columns, these requirements can only be approximately met.

When the soil is highly sensitive to load changes, combined footings should be avoided, if possible, since irreversible tipping may cause structural damage. A simple example is when you stand on a plank on a soft soil and shift your weight from one leg to another.

For combined footings, a study must be made of different possible load combinations, and the shape of the footing should be adjusted so that there will be pressure over the entire base area at all times. Any curling of the footing, especially in water-bearing soils, will permit softening of part of the base area. Unequal settlements must result.

Mat Footings. The application of the **Moseley principle** of least resistance—if a system of reactions statically stable can be shown to exist to counteract all external forces, then the structure will be safe if designed for such reactions—permits some economical modifications of the assumption of uniform base pressures. This is especially true where the footings are extended to provide support for interior walls in addition to adjacent columns. One assumption is to allot to an area symmetrically located about each column the load of that column, the area being determined by the full soil bearing value. In that way the soil resistances are in the form of rectangular loads with blank spaces between, and the computed bending moments are considerably reduced. For large mats, the same assumption will result in strips of loading under the lines joining the columns. Some mats are constructed on such strips, with thinner slabs closing the "no-load" gaps.

One danger in the construction of mats, or rafts, often overlooked, should be noted. Since they are usually built on soft, saturated soils, where moisture removal causes shrinkage and settlement, the absorption of moisture by the concrete mat and dissipation by evaporation within cellar spaces must be prevented. For all qualities of concrete, such moisture travel can be appreciable, since the temperature gradient will always be upward. A waterproof layer preventing moisture loss is a wise precaution.

Strapped Footings. Used to support two columns, strapped footings consist of a pad under each column with a strap between them. Where soil resistance is assumed only under the footings the strap connections (levers), often built on disturbed soil, or even a flexible form liner, are more economical than combined footings. The usual assumptions are that base pressure is uniform under each pad, or bearing area, and that a rigid beam connects the two column loads, so that the center of gravity of the bearing areas coincides with the center of gravity of the column loads.

6-19. Pile Foundations.

A pile is defined as a structural unit introduced into the ground to transmit loads to lower soil strata or to alter the physical properties of the ground. It is of such shape, size, and length that the supporting material immediately underlying the base of the unit cannot be manually inspected. This definition does not limit the type of materials making up the pile, the manner of insertion into the ground, or the loads to be transferred to lower strata. However, where manual inspection of the bottom is possible, as in shallow pits or sufficiently wide, deep caissons, these types of construction are not classified as piles.

Pile Caps. Usually, structural members, such as columns and bearing walls, have to be supported on a group of closely spaced piles. In such cases, the load is transmitted from the structural members to the piles through a thick concrete footing, or cap, in which the pile tops are embedded.

Pile caps are designed as structural members with no allowance for any load support on the soil directly below.

Load Transfer from Piles. Piles transfer load to lower strata by frictional resistance along the surface in contact with the soil and by direct bearing at or near the bottom of the pile. The portion of the load carried by friction and by end bearing is known between limits.

The maximum frictional resistance between soil and pile surface is a definite limit to the amount of load that can be transferred by friction. The confined character of the soil considerably increases its bearing value and a compressed volume of soil at the pile tip acts as an integral part of the base, of unknown total area.

Soil Limitations on Piles. Structural piles can be relied upon within the limitations normally set on the materials of which they are made. Maximum pile loading is more dependent on the nature of the soils into which the pile is driven than on the materials comprising the pile.

Structures on piles subject to lateral movements should be braced by inclined, or batter, piles or designed to take the load in bending in the piles.

Piles installed in soils subject to future subsidence and consolidation must be investigated for possible future loads when the consolidation occurs and the weight of these upper layers hangs on the piles. (Several suggestions offered to avoid this negative skin friction have been tried but found wanting.) Bear in mind that the maximum negative skin friction is equal to the surface frictional resistance and consider this load as part of the total design load.

Pile Embedment in Soil. Piles require a definite grip in resistant soil. It is a wise precaution to insist on at least 10 ft of embedment in natural soil of sufficient stiffness to warrant a presumptive bearing value of at least 2 tons/ft^2. The embedment should be below grade or such elevation as will not be affected by future excavations.

Pile Bracing. Each pile should be braced against lateral displacement by rigid connection to at least two other piles in radial directions not less than 60° apart. Where fewer than three piles are in a group, or where extra lateral stability is made necessary by the nature of loads or softness of soil layers, concrete ties or braces should be provided between caps. Such bracing should be considered as struts, and all reinforcing bars should be anchored in the caps to develop full tension value. A continuous reinforced-concrete mat 6 in or more in thickness, supported by and anchored to the pile caps, or in which piles are embedded at least 3 in, may be used as the bracing tie, if the slab does not depend on the soil for the direct support of its own weight and loads.

Pile Spacing. Inasmuch as the purpose of piles is to transfer loads into soil layers, pile spacing should be a function of the load to be carried and the capacity of the soil to take the load.

Piles bearing on rock or penetrating into rock should have a minimum spacing, center to center, of twice the average diameter, or 1.75 times the diagonal of the pile, but not less than 24 in. This spacing will not overload the rock, even if the bearing piles are fully loaded and no load relief occurs from skin friction.

All other piles, transferring a major part of the load to the soil by skin friction, should have a minimum spacing center to center of twice the average diameter, or 1.75 times the diagonal of the pile, but not less than 30 in. For a pile group consisting of more than four piles, such minimum spacing should be increased by 10% for each interior pile, to a maximum of 40%.

Where lot-line conditions, space restrictions, or obstructions prevent proper spacing of piles, the carrying capacity should be reduced for those piles that are too close to each other or less than half the required spacing from a lot line. The percentage reduction of load-carrying capacity of each pile should be taken as one-half of the percentage reduction in required spacing. If, for example, a pile is 24 in from a lot line instead of 30 in, the percentage reduction of spacing is 20%; so pile capacity is reduced 20%/2, or 10%. Such reduction in values is sometimes economical where closer spacing permits the placing of a pile group below the columns and eliminates eccentricity straps or combined footings.

Minimum Sizes of Piles. Usual restrictions on pile dimensions are as follows: The pile point should be at least 6 in in diameter. Piles of uniform section should have a minimum diameter of 8 in or, if not circular, a minimum thickness of 7½ in. For tapered piles, the butt at cutoff should be at least 8 in in diameter.

Actually, if piles can satisfy stress restrictions as short columns and withstand stresses during driving, there is no justification for dimensional limitations. Generally, piles of minimum size are suitable only for end-bearing piles, because they have too little surface to develop sufficient skin friction.

Timber Piles. Continuously kept wet, timber piles are quite permanent. But a narrow band of alternately wet and dry condition, sometimes caused by seasonal water-level fluctuations, will soon be apparent in pile failures.

Where timber piles may be partly above water level, the pile should be pressure-treated to a final retention of not less than 12 lb of creosote per cubic foot of wood. The tops of all creosoted piles as cut off should be below ground level and the cutoff section should be treated with three coats of hot creosote oil. Creosoted piles above ground level require some replenishment of preservative with age, as well as protection against fire damage.

Almost every kind of timber can be used for a pile. Some species take the shock of driving better than others, and the species have different static compression values. Prime-grade sticks cannot be expected to be sold for piles; yet it is reasonable to insist on timber cut above the ground swell and free from decay, from unsound or grouped knots, from wind shakes, and short reversed bends. The maximum diameter of any sound knot should be one-third the diameter of the pile section where the knot occurs, but not more than 4 in in the lower half of the pile length or more than 5 in elsewhere. All knots should be trimmed flush with the body of the pile, and ends should be squared with the axis.

Piles should have reasonably uniform taper throughout the length and should be so straight that a line joining the centers of point and butt should not depart from the body of the pile. All dimensions should be measured inside the bark, which may be left on the pile if untreated but must be removed before treatment with any preservative. The diameter at any section is the average of the maximum and minimum dimensions; piles not exactly circular in section should not be rejected.

No timber pile should have a point less than 6 in in diameter, except for temporary use. Where the point is reinforced with a steel shoe, the minimum dimension is at the upper end of such shoe. No untreated timber piles should be used unless the cutoff or top level of the pile is below permanent water-table level. This level must not be assumed higher than the invert level of any sewer, drain, or subsurface structure, existing or planned, in the immediate vicinity, or higher than the water level at the site resulting from the lowest drawdown of wells or sumps.

Of the various timber species, cedar, western hemlock, Norway pine, spruce, or similar kinds are restricted to an average unit compression at any section of 600 psi; cypress, Douglas fir, hickory, oak, southern pine, or similar kinds are allowed a value of 800 psi. A maximum load of 20 tons is commonly allowed on timber piles with 6-in points, and 25 tons where the point is 8 in or more, subject, of course, to compliance with indicated or tested driving resistance and maximum stresses.

Timber piles are seldom used as end bearing piles, but there is no real objection to such use, if the piles are properly installed.

Steel H Piles. Where heavy resistances exist in layers that overlie the bearing depths, steel H piles will often take the punishment of the heavy driving much better than any other type. After insertion in the ground, the soil gripped between the web and inside faces of the flanges becomes an integral part of the pile, so that frictional resistance is measured along the surface of the enclosing rectangle and not along the metal surface of the section.

To prevent local crippling during driving, the section should have flange projections not exceeding 14 times the minimum thickness of metal in either web or flange and with total flange width not less than 85% of the depth of the section. All metal thicknesses should be at least 0.40 in.

Other structural-steel sections, or combinations of sections, having flange widths and depths of at least 10 in and metal thickness of at least ½ in may also be used—usually in combination with steel sheeting. Most rolling mills issue special sections for use as piles.

The load-carrying value of a steel pile is usually restricted to a maximum unit stress of 12,600 psi at any cross section but not more than 35% of the yield stress.

Concrete Piles. Precast piles are most suitable for large projects. The necessity for accurate length determination and the added cost of cutoffs, plus the high cost of equipment to make and handle the piles, tend to make precast piles more expensive than the various cast-in-place types.

Precast piles must be reinforced for local stresses during lifting and driving and must be cured to full strength before use. Prestressing is economical for precast piles, since the cost of ordinary reinforcing is at least half the total cost of a precast pile before driving.

Cast-in-place concrete piles come in many variations, in which concrete fills premolded cavities in the soil. The cavity for a pile can be formed by removal of the soil or by forcing the volume displaced into the adjacent soil, as is done by timber or precast-concrete piles. The soil cavity can be made by auger or casing carried to a desired depth and filled with concrete. Volume displacement is accomplished by driving a thin, metal, closed-end shell stiffened with a retractable mandrel or a heavier cylindrical metal shell, which is retracted as the concrete filling is installed.

Retracted shells are usually heavy steel pipe, sometimes with reinforced cutting edges and with driving rings welded at the top to provide a grip for retraction.

In some types, the concrete is compressed by a vibrating or drop hammer and tends to squeeze out laterally below the casing to form annular expansions. These greatly increase the bearing value of the pile.

One advantage of cast-in-place pipe or thin-shell piles is the inspector's ability to examine the cavity before placement of the concrete, to check depth of penetration and variation from plumb or desired batter position.

Usually, the empirical resistance taken as a measure of pile value is indicated by the downward movement during casing driving, before any concrete is placed.

For all cast-in-place piles, care must be exercised to clean the cavity of all foreign matter and to fill the entire volume with concrete. If water enters the shell, tremie tubes may be necessary for casting the concrete.

Structural concrete piles are designed as short concrete columns. A maximum unit stress of 25% of the 28-day cylinder strength in compression is permitted for working-stress design or appropriate load factors, an assumed 5% eccentricity of loading, and $\phi = 0.70$ ($\phi = 0.75$ where a permanent steel casing of ¼-in or greater wall thickness is used) may be applied for ultimate-strength design. Reinforcing steel is evaluated as for reinforced concrete columns. Metal shells in which concrete is cast and which acts with the concrete as part of a pile may, with a ⅙-in allowance for loss in thickness by corrosion, be stressed to 35% of the yield stress.

Concrete Displacement Piles. The Franki extrusion pile requires a steel shell to be set into a pit. The bottom is filled with gravel or dry concrete to a depth of two or three pile diameters, and the fill then is pushed into the soil by dropping a heavy ram inside the shell. The concrete packs and grips the shell sufficiently to pull it along through almost any type of soil. When the shell reaches the desired depth, rapid impact with the ram breaks up the plug and extrudes it into the soil in the shape of an inverted mushroom (Fig. 6-13). Concrete is added before the plug material entirely clears the shell. As the additional concrete is extruded, the shell is withdrawn.

The result is a very rough-surfaced concrete cylinder with a number of annular fins projecting into the soil. At the bottom is a mushroom-shaped expansion. Under load tests, these piles have carried much more load than is indicated by the area of the mushroom multiplied by the normally presumptive bearing capacity of the soil. These piles are suitable for granular soils and develop load capacities of 100 to 150 tons.

Drilled Injection Piles. These are small-diameter concrete shafts. This type of pile is constructed by first drilling into the soil with an auger, then following with simultaneous extraction of the auger and filling of the cavity with a cement grout that is injected through the hollow shaft of the auger. Piles of this type have been used successfully as friction piles. The method also has been used for for building continuous bulkhead cutoffs, similar to tight sheeting.

Reused Piles. Reuse of piles, whether where structures have been demolished or for new structures, should be restricted to places where complete data are available for

Fig. 6-13 Steps in producing a concrete displacement pile. *(Reprinted with permission from F. S. Merritt, "Standard Handbook for Civil Engineers," McGraw-Hill Book Company, New York.)*

length and driving conditions of each pile. Where a group of piles is to be reused for new loadings, their supporting capacity should be restricted to about 75% of rated capacity, unless they are first load-tested. Additional piles for the same structure should be of similar type and about the same length as those being reused.

Pile Splices. Piles are often made up of special sections of structural elements. Adjoining sections, if desired, may be composed of different materials. The load capacity of such composite piles is usually set equal to that of the weakest section. Splices should be designed to hold the sections together during driving as well as to resist the later effects from lateral loads or the driving of adjacent piles.

Timber-pile sections are usually spliced with sleeves—short, tight-fitting steel pipes into which the ends of the piles are driven. Sometimes drift pins or through bolts are used to hold the assembly together. Steel rolled-section or pipe piles are often spliced with a full-strength weld. A precast-concrete pile may be extended with prefabricated steel splices or by exposing and lapping or welding the longitudinal reinforcing steel and that of the extension section, then concreting the extension.

6-20. Load Capacity of Piles.

All types of piles must be able to safely support the maximum combination of the following loads:

1. All dead loads and the weight of the pile cap and superimposed loads.
2. All live loads, reduced as permitted by building code because of floor area or number of stories in the building.
3. Lateral force and moment reactions, including the effect of eccentricity, if any, between the column load and the center of gravity of the pile group.
4. That amount of the vertical, lateral, and moment reactions resulting from wind and seismic loads in excess of one-third of the respective reactions totaled from 1, 2, and 3.

The maximum load permitted on a vertical pile is the allowable pile load, as determined by formula, test, or computation, applied concentrically in the direction of its axis. (See Art. 6-15.)

Lateral-load capacity, deflection, and pile bending stresses should be determined by rational analysis. (See H. F. Winterkorn and H. Fang, "Foundation Engineering Handbook," Van Nostrand Reinhold Company, New York.) Lateral loads that cause allowable deflections are usually on the order of 3 to 5 tons per pile. Very stiff soils and stiff piles, however, permit substantially greater loads. Where piles are subjected to both axial and lateral loads, the compressive and bending stresses are additive and should be checked to see that the allowable pile stress is not exceeded.

If batter piles are used to take lateral forces, the resultant loads along the axis of the pile should not exceed the allowable pile value. Also, allowable unit stresses in any pile section must not be exceeded.

The total pile load is considered as existing in all end-bearing piles without relief. For a friction pile, the entire load is assumed to exist at the section located at two-thirds the embedded length of the pile, measured from the top, unless the rate of load transfer is estimated or determined otherwise.

Pile Load Tests. The load capacity of piles is often determined directly by full-scale load tests. One reason for this is that even where capacity has been estimated by static or dynamic techniques, most building codes require verification by load tests.

When the allowable pile load is to be determined by load tests, subsurface investigations must first prove that the conditions in the test area are fairly uniform. The test piles should be driven close to points where borings were made, to establish a correlation with borehole observations. A complete pile-driving log record should be kept. Driving should be carried to resistances indicated by dynamic formulas for the desired load capacity or to tip elevations indicated by static analyses.

There are four basic types of pile load tests. The test most often used, and required by most building codes, is the "Standard Test." In this method, the pile is loaded in eight increments, usually to twice the design load. Each increment is held until the settlement rate is 0.01 in/hr or for 2 hr, whichever is less. The final load is held for 24 to 48 hr.

The quick maintained-load method uses a constant time interval per load increment and thus minimizes the time-dependent phenomena, which may produce misleading results in the Standard Test. In the quick test, load increments are held for a constant interval of 5 to 15 min and about 25 load increments are applied. A better definition of the load-deflection curve results. One quick test method is given is ASTM D1143.

Another quick method is the constant-rate-of-penetration test, in which the pile is forced to settle at a constant rate. Tests indicate that this test may provide superior definition of the load-deflection curve.

Occasionally, test piles are subjected to cyclic loading by introduction of unloading-reloading cycles. Tests indicate, however, that this method sheds little light on tip movement (supposedly the principal reason for its use), and, in general, is inferior to the quick tests.

The interpretation of pile load tests to determine failure and design loads is a matter deserving close attention. Building codes often stipulate the criteria by which the failure load should be determined but usually also provide alternative methods or permit some flexibility. Some codes, for example, set the maximum allowable load as one-half that load causing a net settlement of 0.005 in/ton, a gross settlement of 1 in, or a disproportionate increase in settlement.

Failure-load determination techniques that incorporate time-dependent phenomena or that may be altered by the simple changing of scales on a graphic plot may yield results that are misleading. The preferred methods for determining failure load employ graphical techniques. For the purpose, in one method, the failure load is defined as the load corresponding to the butt movement that exceeds the elastic deformation by 0.15 in plus a factor equal to the pile diameter, in, divided by 120. This technique is intended for application to results from quick tests.

In another method, the failure load is obtained from the load-settlement curve. In this case, the failure load is defined as the load corresponding to the intersection of the elastic line of the pile and a line with a slope of 0.05 in/ton drawn tangent to the load settlement curve.

Tip movement may be assumed to occur under the load corresponding to the point where the load-settlement curve becomes tangent to the elastic line of the pile.

In the subsequent driving of piles in the area in which piles were tested, all piles of the same type, driven with the same equipment and showing the same or greater resistance to penetration of the soil as was shown by the piles tested should be allowed the average capacity obtained from the load tests. Any piles that stop at higher levels and are one-third or more shorter than the average length of the test piles should be reevaluated.

Bearing Piles. Piles driven to end bearing in rock or in extremely tough and dense layers directly overlying bedrock are called bearing piles. They may be of any type of material but are usually steel or precast concrete, to withstand the heavy driving. (If timber piles are used, extreme care should be exercised that the piles are not overdriven.) Steel piles may be either rolled H sections or concrete-filled pipes.

The load capacities of bearing piles may be estimated from driving formulas or static equations, or determined by test loads. If a pile is driven to essential refusal, the allowable load is governed by the structural capacity of the pile. Typical load capacities of rolled section, heavy pipes, and precast concrete range from 100 to 200 tons. (See also Art. 6-15.)

Special pile forms may be used to obtain a very high allowable load. One frequently used is known by its proprietary name of **drilled-in caisson.** It is formed with a heavy steel pipe with a reinforced cutting edge, which is driven into bedrock and then cleaned out. A socket is then drilled or chopped into the rock at the bottom of the pipe. Next, the pipe is filled with concrete. For higher load capacity, the pile may be reinforced as a concrete column or with a structural-steel shape that extends into the socket (Fig. 6-14).

The load-carrying capacity is the structural capacity of the pile as a column, but may not exceed the load transfer from the socket to the rock. This load is the sum of the allowable bearing value on the bottom of the socket plus the bond of the concrete to the rock along the surface area of contact in the depth of the socket. Not economical for

Fig. 6-14 Typical vertical section through a drilled-in caisson for (*a*) concrete superstructure and (*b*) steel superstructure. (*c*) Typical horizontal section. *(Reprinted by permission from F. S. Merritt, "Standard Handbook for Civil Engineers," McGraw-Hill Book Company, New York.)*

small loads, such piles have been used to sustain loads of 3000 tons, requiring reinforcing with the heaviest rolled-steel column sections with plates welded to them.

The high cost of the steel pipe shell warrants retraction during concreting. Often, when the shell is to be retracted, a thin shell liner is inserted to act as a form for the concrete.

Friction Piles. Piles not driven to a bearing on hard ground or rock are known as friction piles and are assumed to have little or no end bearing. When the piles stop in soils underlain by compressible layers, a careful analysis of the imposed loading on such layers is necessary, to determine what settlements are to be expected from consolidation. Usually, piles evaluated by impact formula only are restricted to a 30-ton value, with further restriction to 25 tons or less for wood piles.

Steel tubes 18 in in diameter have been used as friction piles for 100 and 200 tons, but in some cases, it was evident that the soils could not take such load concentration in groups of four to six piles and that resistances dissipated with time, even when the piles were not loaded.

Bearing value of friction piles is affected by vibrations transferred to the piles. Tests under 500 to 850 vibrations per minute showed additional settlements in the range of 0.2 to 0.5 in in silt and sand.

Pile Deterioration. Any analysis of load transfer and determination of pile load-carrying value is based on the premise that the pile structure will not change. That danger comes from many directions; the pile may rust away or rot, and exterior conditions, such as soil acidity, fluctuation in water levels, and loss of lateral support in the soil encasing the piles must be carefully considered and precautionary measures taken.

The rusting of steel surfaces in the ground (the average corrosion loss is 0.003 in/year) cannot harm the load value of a pile, unless the mill scale becomes loose and surface friction is seriously affected. The life of steel H and sheet piling is far beyond the normal life of a building.

6.21. Installation of Piles.

Equipment and methods for installing piles should be such as to produce satisfactory piles in proper position and alignment. This involves the proper choice of equipment for the size and weight of pile, expected load value, and type of soil to be penetrated.

Hammers. Piles usually are driven into place with hammers powered by gravity (very slow), compressed air, steam, or diesel fuel. Generally, heavier piles require heavier hammers. Use of wave equation analyses facilitates proper matching of hammer to pile. Piles also may be driven with vibratory hammers.

Experience with wood piles indicates that a lighter hammer than required by some building codes should be used to reduce breakage of piles, certainly in areas where soil layers change suddenly in resistance to penetration. Although the usual argument that a big spike should not be driven with a tack hammer is true, so is the opposite statement that a long slender finishing nail should not be driven with a sledge.

A pile-driving rig may comprise a rig specially built for pile driving or a specially equipped crane. A specialized rig may consist of a framework and platform for supporting the equipment needed in pile driving, which may include a hammer, boilers, winches, and engine. If a crane is used, it should be equipped with a boom for handling pile leads and the hammer, and with a drum or winch head with a line to the head of the boom for positioning the piles. For underwater driving, an enclosed steam hammer may be used, with a steam-exhaust hose extending above the water surface. Water may be kept out of the casing by maintenance of air pressure at about 0.5 psi in the lower part of the hammer housing. As an alternative, the hammer may be kept above water by insertion of a follower, or pile extension, at the top of the pile.

Jetting Piles into Place. Other methods of pile insertion include the water-jet loosening of the soil. This allows the pile to drop by gravity or under the weight of a hammer resting on the butt. Jetting, however, is not necessary or advisable in soils other than fairly coarse, dense sands. Even there, it must be remembered that jetting is excavation of soil; it has been so decided by the New York State Supreme Court, and all legal restrictions concerning excavations along a property line are available protections for the adjacent owner.

Jetting should be stopped at least 3 ft above the final pile-tip elevation, and the piles should then be driven at least 3 ft to the required resistance. The condition of adjacent completed piles should be checked, and if loss of resistance is found, all piles should be redriven.

Jacking. Piles also may be installed with jacks reacting against static loads (in the form of an existing structural mass or a temporary setup of weights). This method is usually limited to addition of piles during underpinning operations or within existing buildings with inadequate headroom for a pile-driving rig.

Tolerances for Piles. In design of a cap, each pile is considered a point load concentrated at the center of the cutoff cross section. However, only in ideal soil conditions and with perfect equipment will the piles all be inserted vertically and exactly in the planned position so that this assumption is true. Some practical rules of acceptance of piles not perfectly positioned must control every pile job:

A tolerance of 3 in in any direction from the designated location should be permitted without reduction in load capacity—provided that no pile in the group will then carry more than 110% of its normal permitted value, even though the average loadings of the piles are much less. The distribution of column loads into the individual piles should be determined either analytically or graphically from a plotted graph of the surveyed locations of the cutoff centers.

Where any pile is overloaded, either additional balancing piles should be added to the group, or structural levers should be added for load redistribution or connection with adjacent groups.

Computation of actual pile loads can be simplified by plotting the pile and column locations to scale, then locating the center of gravity of the piles graphically and using the line between this point and the center of the column loads as an axis of symmetry. Multiplying pile values by the offsets of all piles from the normal axis drawn through the center of gravity of the piles gives a total moment equal to the eccentricity moment. Assume a linear variation for reduction or increase of pile value with distance from the axis. If the maximum pile value is just beyond the 110% limit, a second approximation should be tried, relocating the center of gravity of the piles on a weighted basis, using the values from the first solution. This will slightly reduce the eccentricity and may be sufficient to permit approval of the pile group.

Where piles are installed out of plumb, a tolerance of 2% of the pile length, measured by plumb bob inside a casing pile or by extending the exposed section of solid piles, is permissible. A greater variation requires special provision to balance the resultant horizontal and vertical forces, using ties to adjacent column caps where necessary. For hollow-pipe piles, which are driven out of plumb or where obstructions have caused deviations from the intended line, the exact shape of the pile can be determined by the use of a magnetic shape-measuring instrument, or an electronic plumb bob. After the deflection curve of a pile has been plotted, the reduction in allowable load, if any, can be determined.

6-22. Open Shafts for Foundations.

Before the invention of heavy construction equipment, the only way that a foundation could be carried to considerable depths through soft soils was by slow, manual excavation within a protective construction. To a large degree, this development was a carry-over from experience gathered in well-sinking operations. (A parallel development in more recent times is the drilled-caisson procedure modeled on oil-well-exploration methods. See Art. 6-24.) The sheeted and braced deep-water well logically became a construction shaft for a deep footing. The ancient masonry ring enclosure in a well, which was successively underpinned as the excavation progressed, as developed in many parts of North Africa, the Babylonian area of influence, and in India, is still copied, with only slight modifications, in very deep shafts.

Shafts may be constructed to great depths, even through saturated soft soil, by the following methods:

1. Excavation, followed by sheeting the desired outline with braced horizontal or vertical members of wood, steel, or precast concrete.

2. Excavation, following the insertion of vertical sheeting, introducing bracing at the desired depths, when reached. This type of construction usually is called a **cofferdam.**

3. Method 2 sometimes is modified by setting up a completely braced shell, or **caisson,** which sinks from its weight (aided, if necessary, by temporary surcharge loadings) as the excavation undercuts the shell.

4. Pneumatic caissons are used for excavations carried to depths at which it would be impossible with ordinary caissons to control the flow of water or soil through the bottom. They are capped to provide a top seal and work is done under artificially applied air pressure, sufficient to counterbalance hydrostatic pressure. Entrance to the working chamber is through double-gated locks.

5. Resistance to sinking of a caisson limits the depth to about 100 ft, so the **nested caisson** was introduced. Lower sections of this type of caisson fit within the completed ring and can be jacked down against the resistance of the upper section. With this procedure, a three-section caisson has reached 400-ft depths.

6. Similar depths can be reached by freezing the ground around the volume to be excavated.

6-23. Slurry Trench Walls.

Use of a bentonite-slurry-filled trench dug to or into rock, and later displacement of the slurry by tremie concrete, is often an economical method for foundation-wall construction (Fig. 6-15). Walls, usually a minimum of 24 in thick, can be constructed by this method in any soil condition. Reinforcing cages may be inserted into the slurry before concreting. Good bond between the steel and the concrete has resulted.

Reuse of the slurry, a specially homogenized machine mix, necessitates dividing the length of the wall into sections. On some projects, leakage problems have caused trouble. In the slurry-wall constructions for the Budapest subway, the entire inside face was covered with ⅜-in steel plates welded to a continuous sheet to prevent leakage. On other jobs, grouting of the joints was found necessary.

Application of the slurry method to enclosure of large, deep cellars, such as those of the World Trade Center in New York, has shown that the method can compete economically with normal cofferdam construction. For such large areas, interior bracing frames are too expensive. Anchors drilled into the exterior rock are usually provided to replace the temporary bracing.

6-24. Drilled Shafts.

As a development of power drilling for water and oil wells, there is a group of drilled-shaft procedures for obtaining economical exposures of suitable bearing layers. Development of this type of foundation has progressed rapidly since about 1960 and now finds wide acceptance.

Several types of bits and drilling rigs are available. The drilling rigs are heavy equipment utilizing square Kelly bars, often telescoping, and capable of high torques and considerable downpressure. The most commonly used bits are the flight auger, bucket auger, and underreaming bit. The flight and bucket augers advance the shaft, remove loosened material to the surface, and produce a cylindrical shaft of constant diameter. The underream bit is used to enlarge the base by undercutting the shaft and forming what is known as the *bell*. The flight augers, equipped with specially hardened teeth, are capable of cutting many of the sedimentary rocks as well as weathered igneous and metamorphic rocks.

Shaft diameters range from 2 to about 10 ft, with bell diameters ranging up to about 15 ft. The heavy equipment now available permits shaft construction to depths approaching 150 ft.

Drilled caissons are very efficient foundations where high-strength bearing materials can be reached. Satisfactory performance, however, requires not only an adequate bearing material but also that the hole be cleaned before concreting and that there be no mud inclusions or voids in the concrete. Because drilling in unstable ground can encounter severe difficulties, design of a caisson foundation should always be preceded by a detailed evaluation of constructability.

Fig. 6-15 Slurry-trench method for constructing a continuous concrete wall: (a) excavating one section; (b) concreting one section while another is excavated. (Reprinted by permission from F. S. Merritt, "Standard Handbook for Civil Engineers," McGraw-Hill Book Company, New York.)

EARTH PRESSURE ON WALLS

In foundation construction, three types of earth pressure are commonly encountered:

Active earth pressure. This exerts horizontal and vertical load components against any structure that impedes the tendency of the earth to fall, slide, or creep into its natural state of equilibrium but yields slightly under the pressure.
Passive earth pressure. This pressure is mobilized when a structure tends to compress the earth.
At-rest pressure. This is a horizontal pressure exerted by soil against a nonyielding vertical surface.

Each of these pressures is dependent on many of the physical properties of the soil, as well as the relative rigidity of the soil and structure. The most important properties of the soil appear to be density, angle of internal friction in sands, and cohesion and overconsolidation ratio in clays. Some typical values for various soil types are given in Table 6-4.

TABLE 6-4 Typical Soil Densities and Internal Resistance

Soil type	Average weight, w, lb per cu ft	Coefficient of internal resistance, $\tan \phi$
Soft flowing mud.........	105–120	0.18
Wet fine sand............	110–120	0.27–0.58
Dry sand................	90–110	0.47–0.70
Gravel..................	120–135	0.58–0.84
Compact loam...........	90–110	0.58–1.00
Loose loam..............	75–90	0.27–0.58
Clay....................	95–120	0.18–1.00
Cinders.................	35–45	0.47–1.00
Coke....................	40–50	0.58–1.00
Anthracite coal..........	52	0.58
Bituminous coal.........	50	0.70
Ashes...................	40	0.84
Wheat..................	50	0.53

6.25. Active Earth Pressure.

The horizontal component of active earth pressure for any material acting against an ordinary wall is closely given by the general wedge theory for the case of a vertical wall and fill with a substantially horizontal top surface. This wall is assumed to be backfilled by usual construction methods and capable of a small rotational movement (of the magnitude of 0.001 its height), to mobilize the internal friction of the backfill. This rotation is equivalent to an outward movement of ¼ in at the top of a 20-ft wall under earth pressure.

The horizontal component of lateral pressure, P_a, psf, at depth h is given for vertical walls and horizontal fills by

$$P_a = whK_a - 2c\sqrt{K_a} \qquad (6\text{-}14)$$

where w = average unit weight, lb/ft³, of the soil
h = depth, ft
c = cohesion of the soil, psf
$K_a = \tan^2(45 - \phi/2)$
ϕ = angle of internal friction, deg

Table 6-5 gives values of K_a for various values of ϕ.

TABLE 6-5 Active and Passive Pressure Coefficients

Values of $K_a = \tan^2\left(45° - \dfrac{\phi}{2}\right) = \dfrac{1 - \sin\phi}{1 + \sin\phi}$

and of $K_p = \tan^2\left(45° + \dfrac{\phi}{2}\right) = \dfrac{1 + \sin\phi}{1 - \sin\phi}$

ϕ	$\tan\phi$	K_a	K_p
0	0	1.00	1.00
10	0.176	0.70	1.42
20	0.364	0.49	2.04
25	0.466	0.41	2.47
30	0.577	0.33	3.00
35	0.700	0.27	3.69
40	0.839	0.22	4.40
45	1.000	0.17	5.83
50	1.192	0.13	7.55
60	1.732	0.07	13.90
70	2.748	0.03	32.40
80	5.671	0.01	132.20
90	∞	0	∞

The general wedge-theory formulas (taking into account friction along the surface of a wall) may also be used for evaluation of the horizontal component for all other conditions of wall batters and sloping fills. However, comparison with experimental results indicates that the results are somewhat too small for negative surcharges and somewhat too large for positive surcharges, but the differences are no greater than 10%. It is quite accurate enough, taking into account the uncertainties of conditions as to actual slope, to use a table of ratios, referred to as the simplified case of vertical wall and horizontal fill (see Table 6-6).

The vertical component of lateral pressure, in all cases, is such that the resultant pressure forms an angle with the normal to the back of the wall equal to the angle of wall friction. However, under no condition can this angle exceed the angle of internal friction of the fill.

The pressure of soils that, because of lack of drainage and because of their nature, may become fluid at any time—whether such fluid material is widespread or only a narrow layer against the wall—is the same as hydrostatic pressure of a liquid having the same density as that soil.

The pressure of submerged soils is given by Eq. (6-14) with the weight of material reduced by buoyancy (for the solid fraction of the soil only) and the coefficient of internal friction evaluated for the submerged condition; in addition, full hydrostatic pressure of water must be included. For granular materials, submergence affects the coefficients of internal and wall friction very little.

TABLE 6-6 Lateral Pressure Ratios for General Conditions

For Horizontal Fill

Wall Slope*	%
+1:6	107
+1:12	103
Vertical	100
−1:12	95
−1:6	90

For Vertical Walls

$\phi =$	40°	30°	20°
Fill Slope†		%	
+10°	123	110	109
0	100	100	100
−10°	74	83	88

NOTE: Percentages apply to values given by Eq. (6-14) for the case of vertical wall and horizontal fill and for a specific value of ϕ. In computing the above values, the angle of friction between the wall and fill is assumed equal to ϕ, but not greater than 30°.

* + represents a slope away from the fill; − represents a slope toward the fill.

† + represents a slope uphill away from the wall; − represents a downhill slope.

The pressure of fills during saturation and prior to complete submergence and the pressure during drainage periods are affected by the rapidity of water movement. Drainage produces a slight temporary decrease in pressure from normal. Submergence produces an expansion of the fill with consequent increase in pressure. Such variations will not occur if adequate provision is made for drainage; if such provision is not made, the soil may become submerged and the pressure should be computed as for a submerged soil.

Surface loading increases the lateral pressure on the wall. For a uniform surcharge, the lateral pressure on the wall is increased by the surcharge pressure multiplied by the appropriate earth-pressure coefficient.

The pressure in pits and bins is not given by the wedge theory unless a correction for sidewall friction is made, in which case actual field observations of pressures are closely checked.

6-26. Passive Earth Pressure.

The horizontal component P_p, psf, of passive pressure at depth h may be computed from

$$P = whK_p + 2c \sqrt{K_p} \tag{6-15}$$

where w = average unit weight, lb/ft³, of the soil
h = depth, ft
$K_p = \tan^2(45 + \phi/2)$
c = cohesion of the soil, psf
ϕ = angle of internal friction, deg

6-27. At-Rest Pressure.

Earth pressures of soil at rest develop on vertical walls that are nonyielding. Such pressures represent an intermediate case between the limiting cases of active and passive pressures. Pressures at rest are calculated as for active or passive conditions from Eq. (6-14) or (6-15), except that the coefficient K_o is employed.

$$K_o = 1 - \sin \phi \tag{6-16}$$

for cohesionless soils and

$$K_o = 0.95 - \sin \phi \tag{6-17}$$

for normally consolidated clays.

For clays, K_o will increase with overconsolidation ratio (OCR), the ratio of preconsolidation pressure to overburden pressure, and may approach a value of 3 for very high values of OCR.

6-28. Pressures on Underground Structures.

The vertical load reaction due to soil above a buried structure has been determined chiefly on culverts, although there are considerable data available from tunnel-construction experience. The relative rigidity or flexibility of surrounding earth and the intruding culvert, tunnel, or foundation has considerable influence on whether the load to be carried is greater than, equal to, or less than the weight of the soil directly above. (A summary of general relationships is given in several reports of the Iowa State College Engineering Experiment Station, Ames, Iowa, where a succession of fine work has been done on this subject from 1910 to the present.) Rigidity in the backfill has the same effect.

SHEETING AND BRACING OF EXCAVATIONS

Sheeting and bracing should be carefully designed to hold excavation banks in safe position during construction. Nevertheless, trouble due to bank movement or collapse develops more often than necessary, because of inadequate design. Legal responsibility for preventing damage to adjacent property, public streets, or buried utility services during and after foundation construction is usually placed on the builder. Because sheeting and bracing are temporary, however, the tendency of builders is to minimize materials and labor and "take a chance." The results usually are expensive reconstruction, loss of valuable time, and lengthy litigation. Some cities require filing of a carefully prepared design for temporary sheeting.

6-29. Sheeting Design Pressures.

Sheeting without special anchorage or buttresses can be used in several forms or combinations of shapes. Any formula used for stresses in sheeting should be based on assumptions of the shape that the sheeting takes under load.

Measurement of strut loads and sheeting pressures, both in the field and in the laboratory, led to the pressure diagrams in Figure 6-16. As can be seen, these diagrams result in pressures greater than the Coulomb pressure [Eq. (6-14)] on the upper part of the wall, because of the nature of the sheeting deflection during construction.

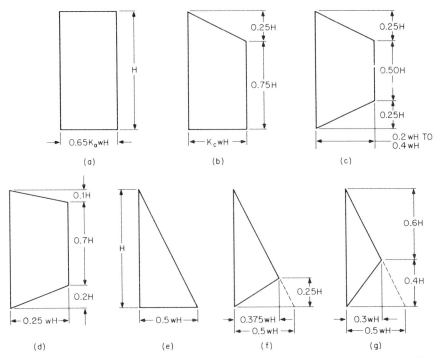

Fig. 6-16 Diagrams representing lateral earth pressure against braced sheeting. As recommended by R. B. Peck and K. Terzaghi (a) for sand, with $K_a = (1 - \sin \phi)/(1 + \sin \phi)$; (b) for soft to medium clay, with $K_c = 1 - 4mc$, where c is the cohesion of the clay and $m = 1$ for most clays but can be as low as 0.4 for unstable clays; and (c) for stiff, fissured clay. As recommended by G. Tschebotarioff for (d) sand, (e) soft clays, (f) medium clay, and (g) stiff clay.

6-30. Bracing of Sheetpiles.

Sheetpiling can be provided with additional supports by either bracing installed inside the cut or by anchoring the top to buried *dead men* located outside (Fig. 6-17). Where space is available, the latter procedure is more economical. A dead man may be a battered pile, or a buried mass of concrete, or a short, continuous sheetpile wall to which are attached adjustable lengths of cables or rods. (See also Art. 6-31.)

The reactions to be provided at each level of support may be determined from the pressure diagrams in Fig. 6-16. The actual distribution of pressure on the sheeting depends on the flexibility of the sheeting, which, in turn, is affected by the relative stiffness of its supports. A liberal allowance for support at each level is therefore necessary, and the total of all the resistances should be considerably above the total pressure.

Normally, water pressure need not be considered, because the excavation must be kept dry, and any hydrostatic pressures on the back of the sheeting can be easily relieved. If water can accumulate in back of the sheeting, however, such pressure must be considered. Bear in mind also that saturation will reduce considerably the soil-anchorage reaction at the bottom of the sheeting and increase the upper reactions. Saturation of the bottom will also require deeper penetration of the sheeting, to avoid blowouts or boils

Fig. 6-17 Types of braced sheeting: (a) horizontal wood lagging between soldier piles, braced with wales and rakers; (b) continuous sheetpile wall braced with wales and rakers; (c) sheetpiles tied back to a concrete dead man; (d) soldier piles tied back with tensioned steel bars or wire strands; (e) parallel sheetpile walls with cross-lot bracing; (f) sheetpile cofferdam with cross-lot bracing in perpendicular directions. *(Reprinted with permission from F. S. Merritt, "Building Engineering and Systems Design," Van Nostrand Reinhold Company, New York.)*

along the sheeting, either of which will prove disastrous. Drainage is a cheap investment for safety.

Bracing against previously completed footings (Fig. 6-17a) is permissible only if the footings are properly braced against undisturbed soil or have enough lateral strength. In evaluation of lateral strength, the frictional resistance on the base of footings (disregarding any cohesive forces) may be added to the passive resistance of the soil against vertical faces.

6-31. Anchorage of Sheeting.

Where exterior buried anchorages are used for sheeting, the pullout resistance of an anchor varies with the type of soil and size of the anchor and its location relative to the bracing. For long, continuous dead men, the allowable anchor pull p, kips, may be computed from

$$P = (P_p - P_a)\frac{L}{\text{SF}} \quad (6\text{-}18)$$

where P_p = passive earth force, kips/lin ft, on anchor
P_a = active earth force, kips/lin ft, on anchor
L = length, ft, of anchor
SF = factor of safety (minimum of 2 is recommended)

For short dead men, W. C. Teng has proposed the following. For granular soils,

$$A = L(P_p - P_a) + \tfrac{1}{3} K_o w (K_p + K_a) d_2^3 \tan \phi \qquad (6\text{-}19)$$

where d_2 = depth to bottom of dead man, ft
A = ultimate anchor capacity, kips
K_a, K_p = pressure coefficients given in Table 6-5
K_o = pressure coefficient given by Eq. (6-16)

For cohesive soils,

$$A = L\left[(P_p - P_a) + q_u d_2\right] \qquad (6\text{-}20)$$

where q_u = unconfined compressive strength, psf.

(W. C. Teng, "Foundation Design," Prentice-Hall, Englewood Cliffs, N.J.)

Where possible, anchors should be located so that the failure wedge of the anchor does not intercept the failure wedge of the bulkhead. If the anchor lies entirely within the bulkhead failure zone, it will be ineffective; if the two failure zones do not intercept, the anchor will not load the wall and will be completely effective. For the intermediate case, the allowable anchor pull should be reduced to

$$N_p = [(P_p - P_a) - (P'_p - P'_a)] \frac{L}{\text{SF}} \qquad (6\text{-}21)$$

where P'_p, P'_a = the forces acting upon the vertical surface extending from the point of intersection of the two failure zones, kips/lin ft.

Tendon Anchors. With the techniques developed for posttensioned concrete, sheeting may be anchored by means of tendons inserted in diagonally drilled holes outside the sheeted excavation, tensioned against the wales to desired reactions, and grouted (Fig. 6-17d). This type of anchorage may be used instead of interior bracing frames. The grouted end of the tendon must be located beyond any possible slip plane in the soil body. Usually, the anchorage is in rock; some successful operations, however, have anchored the tendons or high-strength rods in soil. Tendons are allowed the same stress values as in posttensioned concrete. They should be pretested to 10 to 15% above the desired reaction before the holding anchorage is locked on the wales. Special sloping reaction blocks are needed on the wales, and are usually made of double beams, or else, the wales are set on a slope to provide a right-angle bearing surface for the tendon blocks.

Legal clearance from adjacent property owners and public lands is necessary to permit the subsurface encroachment by such anchorage.

Where excavations are necessarily kept open for long periods, say more than a year, some provision may be necessary to avoid possible corrosion loss of the metal in the tendons. The usual wet environment, often with stray electric currents, may cause some serious loss in original strength of the anchorages. Fully grouting the drilled holes in rock or the cased holes in earth after the tendons have been tensioned can reduce the hazard of corrosion.

6-32. Sheeting Removal.

Sheeting should not be removed until all backfill is properly consolidated. Some additional fill and compaction are necessary as sheeting is removed to avoid internal slip and possible damage to adjacent pavements.

CAUSES OF FOUNDATION FAILURES

Structures that show unwanted vertical or lateral movements as a result of foundation settlements, heaves, or displacements are examples where the foundation design is a failure. Rarely does a structural failure occur in the footing; the usual trouble is in the soil and a consequence of the assumption that movements either will not occur or will be uniform.

6-33. Failures due to Heaving.

That many structures are affected by the heave of fine-grained soils is now common knowledge. Certain bentonitic clay soils undergo large volume changes with seasonal variation in moisture content—especially when lightly loaded. The phenomenon is further complicated by the shading effect, both under and toward the northerly side of the buildings, causing an inequality in the drying out of the soil.

Tidal intrusion of water along shore lines has caused upward movements of light frame structures.

Evaporation of moisture taken out by trees adjacent to foundations has caused sufficient volume change to tip and crack walls.

6-34. Settlement Failures.

Compact, hard, dry clays do not remain in that condition if water is present. Consolidation of soil layers from drainage induced by sewer and other subsurface trenches, as well as the normal slower consolidation from normal pressure of overburden, can cause tipping of foundations. Even pile foundations have been known to settle and move laterally where unconsolidated layers—usually not uniform in thickness—start to shrink after the trigger action of pile driving or change in loading conditions.

A very common failure is the dragging down of an existing wall by new construction built in intimate contact; an open gap or sliding plane must be provided between any previously settled-in-place structure and new work. A similar trouble is the sympathetic settlement of existing structures on a stable foundation underlain by soft strata, which are compressed and bent by new loading, even if the two structures do not touch.

Settlements can be easily observed, and the trouble can often be recognized from the shape of the masonry cracks. Every design should be examined to see where and how foundation movements can occur, and design modifications should be made to eliminate the effects on other structures and to equalize movements, in amount and direction, in the new structure.

For example, a large structure designed by the Los Angeles County Flood Control District was to be founded partly on natural shale and partly on a blended compacted fill. The formula for the blending of fills was determined by mixing decomposed granite and broken shale to obtain the same shear strength as the natural shale. The computed expected settlement was 0.014 ft. Under 90% of the design load, the completed structure showed an actual settlement of 0.012 ft. Procedures of this type are available and should be used to avoid failures of the foundation design.

Settlements are possible for pile foundations or even footings on rock, and any assumption made that no settlement will occur can only lead to trouble. Pile foundations will settle as much as the supporting soils under the distributed loads.

Certain shale and siltstone will disintegrate under continuous loading if they become saturated. Unless bottoms of footings are sealed, percolations of ground or bleed water may soften such rocks significantly.

Settlements produced by dissolving of limestone by percolating ground water can be especially damaging. Such movements are usually catastrophic and must be avoided. Before a structure is built over limestone, there should be adequate indications that such activity is highly unlikely during the life of the facility.

Backfill for basement walls with soils that shrink on dehydration—often caused by suction of the soil water into the concrete—show voids in the fill immediately adjacent to the wall. Such conditions have been known to be the cause of complete collapse of a wall immediately after a heavy rainfall, when enough water entered the voids to provide full hydrostatic pressure, for which the wall was not designed.

Localized footing failures within buildings are often caused by the pipe trenches dug after the footings are completed, without regard to relative depths of excavations. Similar troubles often occur when new trunk sewers, or even transit tunnels, are built in public streets without proper underpinning of all footings above the surface of influence from the new excavation. Such construction will also affect the normal groundwater table, and together with pumping for industrial and air-conditioning requirements, has in the past lowered the water level sufficiently to cause rotting of untreated wood piles. This difficulty has been so prevalent that many cities now restrict the use of untreated piles for any area where groundwater level may not remain stable. Proximity of natural water areas is no proof of such stability; the groundwater level has been shown to be 35 ft below sea level at only short distances from bulkhead lines where industrial pumping was carried to excess.

A serious foundation failure of a heavy factory bearing on compact fine sand was caused by the sinking of some wells within the building and pumping process water from the subsoil. The underlying soil layers were reduced in thickness, and a dish-shaped floor resulted.

Local overload on ground-floor slabs, transmitted to some footings, explains the odd-shaped roofs on many single-story warehouses.

All these troubles can be controlled, but only by recognition in design that foundations will settle if soil support is altered and that every new foundation will settle as load is applied.

6-35. Bibliography.

O. Andersland and D. Anderson, "Geotechnical Engineering for Cold Regions," McGraw-Hill Book Company, New York.

J. E. Bowles, "Foundation Analysis and Design," McGraw-Hill Book Company, New York.

G. A. Fletcher and V. A. Smoots, "Construction Guide for Soils and Foundations," John Wiley & Sons, Inc., New York.

J. K. Mitchell, "Fundamentals of Soil Behavior," John Wiley & Sons, Inc., New York.

R. B. Peck, W. E. Hanson and T. H. Thornburn, "Foundation Engineering," John Wiley & Sons, Inc., New York.

W. H. Perloff and W. Baron, "Soil Mechanics: Principles and Applications," John Wiley & Sons, Inc., New York.

F. E. Richart, Jr.; R. Hall, Jr., and R. D. Woods, "Vibrations of Soils and Foundations," Prentice-Hall, Englewood Cliffs, N.J.

K. Terzaghi and R. B. Peck, "Soil Mechanics in Engineering Practice," John Wiley & Sons, Inc., New York.

G. Tschebotarioff, "Foundations, Retaining and Earth Structures," McGraw-Hill Book Company, New York.

J. J. Tuma and M. Abdel-Hady, "Engineering Soil Mechanics," Prentice-Hall, Englewood Cliffs, N.J.

H. F. Winterkorn and H. Fang, "Foundation Engineering Handbook," Van Nostrand Reinhold Company, New York.

Section 7

Wood Construction

Maurice J. Rhude
President, Sentinel Structures, Inc.
Peshtigo, Wis.

Wood is the greatest renewable resource in the building materials field. More trees can always be grown to provide the wood needed for construction. Trees often are planted with a purpose, stands are improved, and the trees are then harvested with a particular end product in mind.

As a consequence of its origin, wood as a building material has inherent characteristics with which users should be familiar. For example, although cut simultaneously from trees growing side by side in a forest, two boards of the same species and size most likely do not have the same strength. The task of describing this nonhomogeneous material, with its variable biological nature, is not easy. But it can be described accurately, and much better than was possible in the past, for research has provided much useful information on wood properties and behavior in structures.

Research has shown, for example, that a compression grade cannot be used, without modification, for the tension side of a deep bending member. Also, a bending grade cannot be used, unless modified, for the tension side of a deep bending member, or for a tension member. Experience indicates that typical growth characteristics are more detrimental to tensile strength than to compressive strength. Furthermore, research has made possible better estimates of engineering qualities of wood. No longer is it necessary to use only visual inspection, keyed to averages, for estimating the engineering qualities of a piece of wood.

Practice, not engineering design, has been the criterion in the past in home building. As a result, the strength of wood often was not fully utilized. With a better understanding of wood now possible, the availability of sound structural design criteria, and development of economical manufacturing processes, greater and more efficient use is being made of wood in home building.

Improvements in adhesives also have contributed to the betterment of wood construction. In particular, the laminating process, employing adhesives to build up thin boards into deep timbers, improves on nature. Not only are stronger structural members thus made available, but also higher grades of lumber can be placed in regions of greatest stress, and lower grades in regions of lower stress, for overall economy. Despite variations in strength of wood, lumber can be transformed into glued-laminated timbers of predictable strength and with very little variability in strength.

7-1. Basic Characteristics and How to Use Them.

Wood differs in several significant ways from other building materials. Its cellular structure is responsible, to a considerable degree, for this. Because of this structure, structural properties depend on orientation. While most structural materials are essentially isotropic, with nearly equal properties in all directions, wood has three principal grain directions—longitudinal, radial, and tangential. (Loading in the longitudinal direction is referred to as parallel to the grain, whereas transverse loading is considered across the grain.) Parallel to the grain, wood possesses high strength and stiffness. Across the grain, strength is much lower. (In tension, wood stressed parallel to the grain is 25 to 40 times stronger than when stressed across the grain. In compression, wood loaded parallel to the grain is 6 to 10 times stronger than when loaded perpendicular to the grain.) Furthermore, a wood member has three moduli of elasticity, with a ratio of largest to smallest as large as 150:1.

Wood undergoes dimensional changes from causes different from those for dimensional changes in most other structural materials. For instance, thermal expansion of wood is so small as to be unimportant in ordinary usage. Significant dimensional changes, however, occur because of gain or loss in moisture. Swelling and shrinkage from this cause vary in the three grain directions; size changes about 6 to 16% tangentially, 3 to 7% radially, but only 0.1 to 0.3% longitudinally.

Wood offers numerous advantages nevertheless in construction applications—beauty, versatility, durability, workability, low cost per pound, high strength-to-weight ratio, good electrical insulation, low thermal conductance, and excellent strength at low temperatures. It is resistant to many chemicals that are highly corrosive to other materials. It has high shock-absorption capacity. It can withstand large overloads of short time duration. It has good wearing qualities, particularly on its end grain. It can be bent easily to sharp curvature. A wide range of finishes can be applied for decorative or protective purposes. Wood can be used in both wet and dry applications. Preservative treatments are available for use when necessary, as are fire retardants. Also, there is a choice of a wide range of species with a wide range of properties.

In addition, a wide variety of wood framing systems is available. The intended use of a structure, geographical location, configuration required, cost, and many other factors determine the framing system to be used for a particular project.

Design Recommendations. The following recommendations aim at achieving economical designs with wood framing:

Use standard sizes and grades of lumber. Consider using standardized structural components, whether lumber, stock glued beams, or complex framing designed for structural adequacy, efficiency, and economy.

Use standard details wherever possible. Avoid specially designed and manufactured connecting hardware.

Use as simple and as few joints as possible. Place splices, when required, in areas of lowest stress. Do not locate splices where bending moments are large, thus avoiding design, erection, and fabrication difficulties.

Avoid unnecessary variations in cross section of members along their length.

Use identical member designs repeatedly throughout a structure, whenever practicable. Keep the number of different arrangements to a minimum.

Consider using roof profiles that favorably influence the type and amount of load on the structure.

Specify allowable design stresses rather than the lumber grade or combination of grades to be used.

Select an adhesive suitable for the service conditions, but do not overspecify. For example, waterproof resin adhesives need not be used where less expensive water-resistant adhesives will do the job.

Use lumber treated with preservatives where service conditions dictate. Such treatment need not be used where decay hazards do not exist. Fire-retardant treatments may be used to meet a specific flame-spread rating for interior finish, but are not necessary for large-cross-sectional members that are widely spaced and already a low fire risk.

Instead of long, simple spans, consider using continuous or suspended spans or simple spans with overhangs.

Select an appearance grade best suited to the project. Do not specify premium appearance grade for all members if it is not required.

Table 7-1 may be used as a guide to economical span ranges for roof and floor framing in buildings.

7-2. Sectional Properties of Lumber and Timber.

Details regarding dressed sizes of various species of wood are given in the grading rules of agencies that formulate and maintain such rules. Dressed sizes in Table 7-2 are from the American Softwood Lumber Standard, Voluntary Product Standard PS20-70. These sizes are generally available, but it is good practice to consult suppliers before specifying sizes not commonly used to find out what sizes are on hand or can be readily secured.

Table 7-3 lists properties of sections of solid-sawn lumber and timber.

Weight and Specific Gravity. Specific gravity is a reliable indicator of fiber content. Also, specific gravity and the strength and stiffness of solid wood or laminated products are interrelated. See Table 7-4 for weights and specific gravities of several commercial lumber species.

Glued-Laminated Timber. Standard finished sizes of structural glued-laminated timber should be used to the extent that conditions permit. These standard finished sizes are based on lumber sizes given in Voluntary Product Standard PS20-70. Other finished sizes may be used to meet the size requirements of a design, or to meet other special requirements.

Nominal 2-in-thick lumber, surfaced to 1½ in before gluing, is used to laminate straight members and curved members having radii of curvature within the bending-radius limitations for the species. (Formerly, a net thickness of 1⅝ in was common for nominal 2-in lumber and may be used, dependent on its availability.) Nominal 1-in-thick lumber, surfaced to ¾ in before gluing, may be used for laminating curved members when the bending radius is too short to permit use of nominal 2-in-thick laminations, if the bending-radius limitations for the species are observed. Other lamination thicknesses may be used to meet special curving requirements.

Table 7-5 lists properties of sections of glued-laminated timber.

7-3. Basic and Allowable Stresses for Timber.

Testing of a species to determine average strength properties should be carried out from either of two viewpoints:

1. Tests should be made on specimens of large size containing defects. Practically all structural uses involve members of this character.

2. Tests should be made on small, clear specimens to provide fundamental data. Factors to account for the influence of various characteristics may be applied to establish the strength of structural members.

TABLE 7-1 Economical Span Range for Framing Members

Framing member	Economical span range, ft	Usual spacing, ft
Roof beams (generally used where a flat or low-pitched roof is desired):		
Simple span:		
Constant depth		
Solid-sawn	0-40	4-20
Glued-laminated	20-100	8-24
Tapered	25-100	8-24
Double tapered (pitched beams)	25-100	8-24
Curved beams	25-100	8-24
Simple beam with overhangs (usually more economical than simple span when span is over 40 ft):		
Solid-sawn	24	4-20
Glued-laminated	10-90	8-24
Continuous span:		
Solid-sawn	10-50	4-20
Glued-laminated	10-50	8-24
Arches (three-hinged for relatively high-rise applications and two-hinged for relatively low-rise applications):		
Three-hinged:		
Gothic	40-90	8-24
Tudor	30-120	8-24
A-frame	20-160	8-24
Three-centered	40-250	8-24
Parabolic	40-250	8-24
Radial	40-250	8-24
Two-hinged:		
Radial	50-200	8-24
Parabolic	50-200	8-24
Trusses (provide openings for passage of wires, piping, etc.):		
Flat or parallel chord	50-150	12-20
Triangular or pitched	50-90	12-20
Bowstring	50-200	14-24
Tied arches (where no ceiling is desired and where a long clear span is desired with low rise):		
Tied segment	50-100	8-20
Buttressed segment	50-200	14-24
Domes	50-350	8-24
Simple-span floor beams:		
Solid sawn	6-20	4-12
Glued laminated	6-40	4-16
Continuous floor beams	25-40	4-16
Roof sheathing and decking:		
1-in. sheathing	1-4	
2-in. sheathing	6-10	
3-in. roof deck	8-15	
4-in. roof deck	12-20	
Plywood sheathing	1-4	
Sheathing on roof joists	1.33-2	
Plank floor decking (floor and ceiling in one):		
Edge to edge	4-16	
Wide face to wide face	4-16	

TABLE 7-2 Nominal and Minimum Dressed Sizes of Boards, Dimension, and Timbers

Item	Thickness, in.			Face width, in.		
	Nominal	Minimum dressed		Nominal	Minimum dressed	
		Dry*	Green†		Dry*	Green†
Boards	1	3/4	25/32	2	1 1/2	1 9/16
	1 1/4	1	1 1/32	3	2 1/2	2 9/16
	1 1/2	1 1/4	1 9/32	4	3 1/2	3 9/16
				5	4 1/2	4 5/8
				6	5 1/2	5 5/8
				7	6 1/2	6 5/8
				8	7 1/4	7 1/2
				9	8 1/4	8 1/2
				10	9 1/4	9 1/2
				11	10 1/4	10 1/2
				12	11 1/4	11 1/2
				14	13 1/4	13 1/2
				16	15 1/4	15 1/2
Dimension	2	1 1/2	1 9/16	2	1 1/2	1 9/16
	2 1/2	2	2 1/16	3	2 1/2	2 9/16
	3	2 1/2	2 9/16	4	3 1/2	3 9/16
	3 1/2	3	3 1/16	5	4 1/2	4 5/8
				6	5 1/2	5 5/8
				8	7 1/4	7 1/2
				10	9 1/4	9 1/2
				12	11 1/4	11 1/2
				14	13 1/4	13 1/2
				16	15 1/4	15 1/2
	4	3 1/2	3 9/16	2	1 1/2	1 9/16
	4 1/2	4	4 1/16	3	2 1/2	2 9/16
				4	3 1/2	3 9/16
				5	4 1/2	4 5/8
				6	5 1/2	5 5/8
				8	7 1/4	7 1/2
				10	9 1/4	9 1/2
				12	11 1/4	11 1/2
				14	....	13 1/2
				16	....	15 1/2
Timbers	5 and thicker	...	1/2 in. less	5 and wider	....	1/2 in. less

* Dry lumber is defined as lumber seasoned to a moisture content of 19% or less.

† Green lumber is defined as lumber having a moisture content in excess of 19%.

TABLE 7-3 Properties of Sections of Solid-Sawn Wood

Nominal size, in.	Standard dressed size, in. (S4S)	Area of section sq in.	Moment of inertia in.^{4}	Section modulus, in.^{3}	Weight, lb per lin ft, of piece when weight of wood, lb per cu ft, equals					
					25	30	35	40	45	50
1 × 3	¾ × 2½	1.875	0.977	0.781	0.326	0.391	0.456	0.521	0.586	0.651
1 × 4	¾ × 3½	2.625	2.680	1.531	0.456	0.547	0.638	0.729	0.820	0.911
1 × 6	¾ × 5½	4.125	10.398	3.781	0.716	0.859	1.003	1.146	1.289	1.432
1 × 8	¾ × 7¼	5.438	23.817	6.570	0.944	1.133	1.322	1.510	1.699	1.888
1 × 10	¾ × 9¼	6.938	49.466	10.695	1.204	1.445	1.686	1.927	2.168	2.409
1 × 12	¾ × 11¼	8.438	88.989	15.820	1.465	1.758	2.051	2.344	2.637	2.930
2 × 3*	1½ × 2½	3.750	1.953	1.563	0.651	0.781	0.911	1.042	1.172	1.302
2 × 4	1½ × 3½	5.250	5.359	3.063	0.911	1.094	1.276	1.458	1.641	1.823
2 × 6	1½ × 5½	8.250	20.797	7.563	1.432	1.719	2.005	2.292	2.578	2.865
2 × 8	1½ × 7¼	10.875	47.635	13.141	1.888	2.266	2.643	3.021	3.398	3.776
2 × 10	1½ × 9¼	13.875	98.932	21.391	2.409	2.891	3.372	3.854	4.336	4.818
2 × 12	1½ × 11¼	16.875	177.979	31.641	2.930	3.516	4.102	4.688	5.273	5.859
2 × 14	1½ × 13¼	19.875	290.775	43.891	3.451	4.141	4.831	5.521	6.211	6.901
3 × 1	2½ × ¾	1.875	0.088	0.234	0.326	0.391	0.456	0.521	0.586	0.651
3 × 2	2½ × 1½	3.750	0.703	0.938	0.651	0.781	0.911	1.042	1.172	1.302
3 × 4	2½ × 3½	8.750	8.932	5.104	1.519	1.823	2.127	2.431	2.734	3.038
3 × 6	2½ × 5½	13.750	34.661	12.604	2.387	2.865	3.342	3.819	4.297	4.774
3 × 8	2½ × 7¼	18.125	79.391	21.901	3.147	3.776	4.405	5.035	5.664	6.293
3 × 10	2½ × 9¼	23.125	164.886	35.651	4.015	4.818	5.621	6.424	7.227	8.030
3 × 12	2½ × 11¼	28.125	296.631	52.734	4.883	5.859	6.836	7.813	8.789	9.766
3 × 14	2½ × 13¼	33.125	484.625	73.151	5.751	6.901	8.051	9.201	10.352	11.502
3 × 16	2½ × 15¼	38.125	738.870	96.901	6.619	7.943	9.266	10.590	11.914	13.238
4 × 1	3½ × ¾	2.625	0.123	0.328	0.456	0.547	0.638	0.729	0.820	0.911
4 × 2	3½ × 1½	5.250	0.984	1.313	0.911	1.094	1.276	1.458	1.641	1.823
4 × 3	3½ × 2½	8.750	4.557	3.646	1.519	1.823	2.127	2.431	2.734	3.038
4 × 4	3½ × 3½	12.250	12.505	7.146	2.127	2.552	2.977	3.403	3.828	4.253
4 × 6	3½ × 5½	19.250	48.526	17.646	3.342	4.010	4.679	5.347	6.016	6.684
4 × 8	3½ × 7¼	25.375	111.148	30.661	4.405	5.286	6.168	7.049	7.930	8.811
4 × 10	3½ × 9¼	32.375	230.840	49.911	5.621	6.745	7.869	8.993	10.117	11.241
4 × 12	3½ × 11¼	39.375	415.283	73.828	6.836	8.203	9.570	10.938	12.305	13.672
4 × 14	3½ × 13¼	47.250	717.609	106.313	8.203	9.844	11.484	13.125	14.766	16.406
4 × 16	3½ × 15¼	54.250	1,086.130	140.146	9.418	11.302	13.186	15.069	16.953	18.837
6 × 1	5½ × ¾	4.125	0.193	0.516	0.716	0.859	1.003	1.146	1.289	1.432
6 × 2	5½ × 1½	8.250	1.547	2.063	1.432	1.719	2.005	2.292	2.578	2.865
6 × 3	5½ × 2½	13.750	7.161	5.729	2.387	2.865	3.342	3.819	4.297	4.774
6 × 4	5½ × 3½	19.250	19.651	11.229	3.342	4.010	4.679	5.347	6.016	6.684
6 × 6	5½ × 5½	30.250	76.255	27.729	5.252	6.302	7.352	8.403	9.453	10.503
6 × 8	5½ × 7½	41.250	193.359	51.563	7.161	8.594	10.026	11.458	12.891	14.323
6 × 10	5½ × 9½	52.250	392.963	82.729	9.071	10.885	12.700	14.514	16.328	18.142
6 × 12	5½ × 11½	63.250	697.068	121.229	10.981	13.177	15.373	17.569	19.766	21.962
6 × 14	5½ × 13½	74.250	1,127.672	167.063	12.891	15.469	18.047	20.625	23.203	25.781
6 × 16	5½ × 15½	85.250	1,706.776	220.229	14.800	17.760	20.720	23.681	26.641	29.601
6 × 18	5½ × 17½	96.250	2,456.380	280.729	16.710	20.052	23.394	26.736	30.078	33.420
6 × 20	5½ × 19½	107.250	3,398.484	348.563	18.620	22.344	26.068	29.792	33.516	37.240
6 × 22	5½ × 21½	118.250	4,555.086	423.729	20.530	24.635	28.741	32.847	36.953	41.059
6 × 24	5½ × 23½	129.250	5,948.191	506.229	22.439	26.927	31.415	35.903	40.391	44.878
8 × 1	7¼ × ¾	5.438	0.255	0.680	0.944	1.133	1.322	1.510	1.699	1.888
8 × 2	7¼ × 1½	10.875	2.039	2.719	1.888	2.266	2.643	3.021	3.398	3.776
8 × 3	7¼ × 2½	18.125	9.440	7.552	3.147	3.776	4.405	5.035	5.664	6.293
8 × 4	7¼ × 3½	25.375	25.904	14.802	4.405	5.286	6.168	7.049	7.930	8.811
8 × 6	7½ × 5½	41.250	103.984	37.813	7.161	8.594	10.026	11.458	12.891	14.323
8 × 8	7½ × 7½	56.250	263.672	70.313	9.766	11.719	13.672	15.625	17.578	19.531
8 × 10	7½ × 9½	71.250	535.859	112.813	12.370	14.844	17.318	19.792	22.266	24.740
8 × 12	7½ × 11½	86.250	950.547	165.313	14.974	17.969	20.964	23.958	26.953	29.948
8 × 14	7½ × 13½	101.250	1,537.734	227.813	17.578	21.094	24.609	28.125	31.641	35.156
8 × 16	7½ × 15½	116.250	2,327.422	300.313	20.182	24.219	28.255	32.292	36.328	40.365
8 × 18	7½ × 17½	131.250	3,349.609	382.813	22.786	27.344	31.901	36.458	41.016	45.573
8 × 20	7½ × 19½	146.250	4,634.297	475.313	25.391	30.469	35.547	40.625	45.703	50.781
8 × 22	7½ × 21½	161.250	6,211.484	577.813	27.995	33.594	39.193	44.792	50.391	55.990
8 × 24	7½ × 23½	176.250	8,111.172	690.313	30.599	36.719	42.839	48.958	55.078	61.198
10 × 1	9¼ × ¾	6.938	0.325	0.867	1.204	1.445	1.686	1.927	2.168	2.409
10 × 2	9¼ × 1½	13.875	2.602	3.469	2.409	2.891	3.372	3.854	4.336	4.818
10 × 3	9¼ × 2½	23.125	12.044	9.635	4.015	4.818	5.621	6.424	7.227	8.030
10 × 4	9¼ × 3½	32.375	33.049	18.885	5.621	6.745	7.869	8.993	10.117	11.241
10 × 6	9½ × 5½	52.250	131.714	47.896	9.071	10.885	12.700	14.514	16.328	18.142
10 × 8	9½ × 7½	71.250	333.984	89.063	12.370	14.844	17.318	19.792	22.266	24.740
10 × 10	9½ × 9½	90.250	678.755	142.896	15.668	18.802	21.936	25.069	28.203	31.337
10 × 12	9½ × 11½	109.250	1,204.026	209.396	18.967	22.760	26.554	30.347	34.141	37.934
10 × 14	9½ × 13½	128.250	1,947.797	288.563	22.266	26.719	31.172	35.625	40.078	44.531
10 × 16	9½ × 15½	147.250	2,948.068	380.396	25.564	30.677	35.790	40.903	46.016	51.128
10 × 18	9½ × 17½	166.250	4,242.836	484.896	28.863	34.635	40.408	46.181	51.953	57.726
10 × 20	9½ × 19½	185.250	5,870.109	602.063	32.161	38.594	45.026	51.458	57.891	64.323
10 × 22	9½ × 21½	204.250	7,867.479	731.896	35.460	42.552	49.644	56.736	63.828	70.920
10 × 24	9½ × 23½	223.250	10,274.148	874.396	38.759	46.510	54.262	62.014	69.766	77.517
12 × 1	11¼ × ¾	8.438	0.396	1.055	1.465	1.758	2.051	2.344	2.637	2.930
12 × 2	11¼ × 1½	16.875	3.164	4.219	2.930	3.516	4.102	4.688	5.273	5.859
12 × 3	11¼ × 2½	28.125	14.648	11.719	4.883	5.859	6.836	7.813	8.789	9.766
12 × 4	11¼ × 3½	39.375	40.195	22.969	6.836	8.203	9.570	10.938	12.305	13.672
12 × 6	11½ × 5½	63.250	159.443	57.979	10.981	13.177	15.373	17.569	19.766	21.962
12 × 8	11½ × 7½	86.250	404.297	107.813	14.974	17.969	20.964	23.958	26.953	29.948
12 × 10	11½ × 9½	109.250	821.651	172.979	18.967	22.760	26.554	30.347	34.141	37.934
12 × 12	11½ × 11½	132.250	1,457.505	253.479	22.960	27.552	32.144	36.736	41.328	45.920
12 × 14	11½ × 13½	155.250	2,357.859	349.313	26.953	32.344	37.734	43.125	48.516	53.906

Nominal size, in.	Standard dressed size, in. (S4S)	Area of section, sq in.	Moment of inertia in.[4]	Section modulus, in.[3]	Weight, lb per lin ft, of piece when weight of wood, lb per cu ft, equals					
					25	30	35	40	45	50
12 × 16	11½ × 15½	178.250	5,568.713	460.479	30.946	37.135	43.325	49.514	55.703	61.892
12 × 18	11½ × 17½	201.250	5,136.066	586.979	34.939	41.927	48.915	55.903	62.891	69.878
12 × 20	11½ × 19½	224.250	7,105.922	728.813	38.932	46.719	54.505	62.292	70.078	77.865
12 × 22	11½ × 21½	247.250	9,524.273	885.979	42.925	51.510	60.095	.68.681	77.266	85.851
12 × 24	11½ × 23½	270.250	12,437.129	1,058.479	46.918	56.302	65.686	75.069	84.453	93.837
14 × 2	13¼ × 1½	19.875	3.727	4.969	3.451	4.141	4.831	5.521	6.211	6.901
14 × 3	13¼ × 2½	33.125	17.253	13.802	5.751	6.901	8.051	9.201	10.352	11.502
14 × 4	13¼ × 3½	46.375	47.34	27.052	8.047	9.657	11.266	12.877	14.485	16.094
14 × 6	13½ × 5½	74.250	187.172	68.063	12.891	15.469	18.047	20.625	23.203	25.781
14 × 8	13½ × 7½	101.250	474.609	126.563	17.578	21.094	24.609	28.125	31.641	35.156
14 × 10	13½ × 9½	128.250	964.547	203.063	22.266	26.719	31.172	35.625	40.078	44.531
14 × 12	13½ × 11½	155.250	1,710.984	297.563	26.953	32.344	37.734	43.125	48.516	53.906
14 × 14	13½ × 13½	209.250	4,189.359	540.563	36.328	43.594	50.859	58.125	65.391	72.656
14 × 18	13½ × 17½	236.250	6,029.297	689.063	41.016	49.219	57.422	65.625	73.828	82.031
14 × 20	13½ × 19½	263.250	8,341.734	855.563	45.703	54.844	63.984	73.125	82.266	91.406
14 × 22	13½ × 21½	290.250	11,180.672	1,040.063	50.391	60.469	70.547	80.625	90.703	100.781
14 × 24	13½ × 23½	317.250	14,600.109	1,242.563	55.078	66.094	77.109	88.125	99.141	110.156
16 × 2	15¼ × 1½	38.125	19.857	15.885	6.619	7.944	9.267	10.592	11.915	13.240
16 × 4	15¼ × 3½	53.375	54.487	31.135	9.267	11.121	12.975	14.828	16.682	18.536
16 × 6	15¼ × 5½	85.250	214.901	78.146	14.800	17.760	20.720	23.681	26.641	29.601
16 × 8	15¼ × 7½	116.250	544.922	145.313	20.182	24.219	28.255	32.292	36.328	40.365
16 × 10	15¼ × 9½	147.250	1,107.443	233.146	25.564	30.677	35.790	40.903	46.016	51.128
16 × 12	15¼ × 11½	178.250	1,964.463	341.646	30.946	37.135	43.325	49.514	55.703	61.892
16 × 14	15¼ × 13½	209.250	3,177.984	470.813	36.328	43.594	50.859	58.125	65.391	72.656
16 × 16	15¼ × 15½	240.250	4,810.004	620.646	41.710	50.052	58.394	66.736	75.078	83.420
16 × 18	15¼ × 17½	271.250	6,922.523	791.146	47.092	56.510	65.929	75.347	84.766	94.184
16 × 20	15¼ × 19½	302.250	9,577.547	982.313	52.474	62.969	73.464	83.958	94.453	104.948
16 × 22	15¼ × 21½	333.250	12,837.066	1,194.146	57.856	69.427	80.998	92.569	104.141	115.712
16 × 24	15¼ × 23½	364.250	16,763.086	1,426.646	63.238	75.885	88.533	101.181	113.828	126.476
18 × 6	17½ × 5½	96.250	242.630	88.229	16.710	20.052	23.394	26.736	30.078	33.420
18 × 8	17½ × 7½	131.250	615.234	164.063	22.786	27.344	31.901	36.458	41.016	45.573
18 × 10	17½ × 9½	166.250	1,250.338	263.229	28.863	34.635	40.408	46.181	51.953	57.726
18 × 12	17½ × 11½	201.250	2,217.943	385.729	34.939	41.927	48.915	55.903	62.891	69.878
18 × 14	17½ × 13½	236.250	3,588.047	531.563	41.016	49.219	57.422	65.625	73.828	82.031
18 × 16	17½ × 15½	271.250	5,430.648	700.729	47.092	56.510	65.929	75.347	84.766	94.184
18 × 18	17½ × 17½	306.250	7,815.754	893.229	53.168	63.802	74.436	85.069	95.703	106.337
18 × 20	17½ × 19½	341.250	10,813.359	1,109.063	59.245	71.094	82.943	94.792	106.641	118.490
18 × 22	17½ × 21½	376.250	14,493.461	1,348.229	65.321	78.385	91.450	104.514	117.578	130.642
18 × 24	17½ × 23½	411.250	18,926.066	1,610.729	71.398	85.677	99.957	114.236	128.516	142.795
20 × 6	19½ × 5½	107.250	270.359	98.313	18.620	22.344	26.068	29.792	33.516	37.240
20 × 8	19½ × 7½	146.250	685.547	182.813	25.391	30.469	35.547	40.625	45.703	50.781
20 × 10	19½ × 9½	185.250	1,393.234	293.313	32.161	38.594	45.026	51.458	57.891	64.323
20 × 12	19½ × 11½	224.250	2,471.422	429.813	38.932	46.719	54.505	62.292	70.078	77.865
20 × 14	19½ × 13½	263.250	3,998.109	592.313	45.703	54.844	63.984	73.125	82.266	91.406
20 × 16	19½ × 15½	302.250	6,051.297	780.813	52.474	62.969	73.464	83.958	94.453	104.948
20 × 18	19½ × 17½	341.250	8,708.984	995.313	59.245	71.094	82.943	94.792	106.641	118.490
20 × 20	19½ × 19½	380.250	12,049.172	1,235.813	66.016	79.219	92.422	105.625	118.829	132.031
20 × 22	19½ × 21½	419.250	16,149.859	1,502.313	72.786	87.344	101.901	116.458	131.016	145.573
20 × 24	19½ × 23½	458.250	21,089.047	1,794.813	79.557	95.469	111.380	127.292	243.203	159.115
22 × 6	21½ × 5½	118.250	298.088	108.396	20.530	24.635	28.741	32.847	36.953	41.059
22 × 8	21½ × 7½	161.250	755.859	201.563	27.995	33.594	39.193	44.792	50.391	55.990
22 × 10	21½ × 9½	204.250	1,536.130	323.396	35.460	42.552	49.644	56.736	63.828	70.920
22 × 12	21½ × 11½	247.250	2,724.901	473.896	42.925	51.510	60.095	68.681	77.266	85.851
22 × 14	21½ × 13½	290.250	4,408.172	653.063	50.391	60.469	70.547	80.625	90.703	100.781
22 × 16	21½ × 15½	333.250	6,671.941	860.896	57.856	69.427	80.998	92.569	104.141	115.712
22 × 18	21½ × 17½	376.250	9,602.211	1,097.396	65.321	78.385	91.450	104.514	117.578	130.642
22 × 20	21½ × 19½	419.250	13,284.984	1,362.563	72.786	87.344	101.901	116.458	131.016	145.573
22 × 22	21½ × 21½	462.250	17,806.254	1,656.396	80.252	96.302	112.352	128.403	144.453	160.503
22 × 24	21½ × 23½	505.250	23,252.023	1,978.896	87.717	105.260	122.804	140.347	157.891	175.434
24 × 6	23½ × 5½	129.250	325.818	118.479	22.439	26.927	31.415	35.903	40.391	44.878
24 × 8	23½ × 7½	176.250	826.172	220.313	30.599	36.719	42.839	48.958	55.078	61.198
24 × 10	23½ × 9½	223.250	1,679.026	353.479	38.759	46.510	54.262	62.014	69.766	77.517
24 × 12	23½ × 11½	270.250	2,978.380	517.979	46.918	56.302	65.686	75.069	84.453	93.837
24 × 14	23½ × 13½	317.250	4,818.234	713.813	55.078	66.094	77.109	88.125	99.141	110.156
24 × 16	23½ × 15½	364.250	7,292.586	940.979	63.238	75.885	88.533	101.181	113.828	126.476
24 × 18	23½ × 17½	411.250	10,495.441	1,199.479	71.398	85.677	99.957	114.236	128.516	142.795
24 × 20	23½ × 19½	458.250	14,520.797	1,489.313	79.557	95.469	111.380	127.292	143.203	159.115
24 × 22	23½ × 21½	505.250	19,462.648	1,810.479	87.717	105.260	122.804	140.347	157.891	175.434
24 × 24	23½ × 23½	552.250	25,415.004	2,162.979	95.877	115.052	134.227	153.403	172.578	191.753

* For lumber surfaced 1⅝ in. thick, instead of 1½ in., the area, moment of inertia, and section modulus may be increased 8.33%.

TABLE 7-4 Weights and Specific Gravities of Commercial Lumber Species

Species	Specific gravity based on oven-dry weight and volume at 12% moisture content	Weight, lb per cu ft		Adjusting factor for each 1% change in moisture content	Moisture content when green (avg), %	Specific gravity based on oven-dry weight and volume when green	Weight when green, lb per cu ft
		At 12% moisture content	At 20% moisture content				
Softwoods:							
Cedar:							
Alaska	0.44	31.1	32.4	0.170	38	0.42	35.5
Incense	0.37	25.0	26.4	0.183	108	0.35	42.5
Port Orford	0.42	29.6	31.0	0.175	43	0.40	35.0
Western red	0.33	23.0	24.1	0.137	37	0.31	26.4
Cypress, southern	0.46	32.1	33.4	0.167	91	0.42	45.3
Douglas fir:							
Coast region	0.48	33.8	35.2	0.170	38	0.45	38.2
Inland region	0.44	31.4	32.5	0.137	48	0.41	36.3
Rocky Mountain	0.43	30.0	31.4	0.179	38	0.40	34.6
Fir, white	0.37	26.3	27.3	0.129	115	0.35	39.6
Hemlock:							
Eastern	0.40	28.6	29.8	0.150	111	0.38	43.4
Western	0.42	29.2	30.2	0.129	74	0.38	37.2
Larch, western	0.55	38.9	40.2	0.170	58	0.51	46.7
Pine:							
Eastern white	0.35	24.9	26.2	0.167	73	0.34	35.1
Lodgepole	0.41	28.8	29.9	0.142	65	0.38	36.3
Norway	0.44	31.0	32.1	0.142	92	0.41	42.3
Ponderosa	0.40	28.1	29.4	0.162	91	0.38	40.9
Southern shortleaf	0.51	35.2	36.5	0.154	81	0.46	45.9
Southern longleaf	0.58	41.1	42.5	0.179	63	0.54	50.2
Sugar	0.36	25.5	26.8	0.162	137	0.35	45.8
Western white	0.38	27.6	28.6	0.129	54	9.36	33.0
Redwood	0.40	28.1	29.5	0.175	112	0.38	45.6
Spruce:							
Engelmann	0.34	23.7	24.7	0.129	80	0.32	32.5
Sitka	0.40	27.7	28.8	0.145	42	0.37	32.0
White	0.40	29.1	29.9	0.104	50	0.37	33.0
Hardwoods:							
Ash, white	0.60	42.2	43.6	0.175	42	0.55	47.4
Beech, American	0.64	43.8	45.1	0.162	54	0.56	50.6
Birch:							
Sweet	0.65	46.7	48.1	0.175	53	0.60	53.8
Yellow	0.62	43.0	44.1	0.142	67	0.55	50.8
Elm, rock	0.63	43.6	45.2	0.208	48	0.57	50.9
Gum	0.52	36.0	37.1	0.133	115	0.46	49.7
Hickory:							
Pecan	0.66	45.9	47.6	0.212	63	0.60	56.7
Shagbark	0.72	50.8	51.8	0.129	60	0.64	57.0
Maple, sugar	0.63	44.0	45.3	0.154	58	0.56	51.1
Oak:							
Red	0.63	43.2	44.7	0.187	80	0.56	56.0
White	0.68	46.3	47.6	0.167	68	0.60	55.6
Poplar, yellow	0.42	29.8	31.0	0.150	83	0.40	40.5

Tests made in accordance with the first viewpoint have the disadvantage that the results apply only to the particular combination of characteristics existing in the test specimens. To determine the strength corresponding to other combinations requires additional tests; thus, an endless testing program is necessary. The second viewpoint permits establishment of fundamental strength properties for each species, and application of general rules to cover the specific conditions involved in a particular case.

It is this second viewpoint that has been generally accepted. When a species has been adequately investigated under this concept, there should be no need for further tests on that species unless new conditions arise. ASTM Standard D143, "Standard Methods of Testing Small, Clear Specimens of Timber," gives the procedure for determination of fundamental data on wood species.

Basic stresses are essentially unit stresses applicable to clear and straight-grained defect-free material. These stresses, derived from the results of tests on small, clear specimens of green wood, include an adjustment for variability of material, length of loading period, and factor of safety. They are considerably less than the average of the species. They require only an adjustment for grade to become allowable unit stresses.

Allowable unit stresses are computed for a particular grade by reducing the basic stress according to the limitations on defects for that grade. The basic stress is multiplied by a strength ratio to obtain an allowable stress. This strength ratio represents that proportion of the strength of a defect-free piece that remains after taking into account the effect of strength-reducing features.

The principal factors entering into the establishment of allowable unit stress for each species include inherent strength of wood, reduction in strength due to natural growth characteristics permitted in the grade, effect of long-time loading, variability of individual species, possibility of some slight overloading, characteristics of the species, size of member and related influence of seasoning, and factor of safety. The effect of these factors is a strength value for practical-use conditions lower than the average taken from tests on small, clear specimens.

Basic stresses for laminated timbers under wet service conditions are the same as for solid timbers; i.e., these stresses are based on the strength of wood in the green condition. When moisture content in a member will be low throughout its service, a second set of higher basic stresses, based on the higher strength of dry material, may be used. Technical Bulletin 479, U.S. Department of Agriculture, "Strength and Related Properties of Woods Grown in the United States," presents test results on small, clear, and straight-grained wood species in the green state and in the 12%-moisture-content, air-dry condition. Technical Bulletin 1069, U.S. Department of Agriculture, "Fabrication and Design of Glued-laminated Structural Members," gives the basic stresses for clear, solid-sawn members and glued-laminated timbers in the wet condition, and basic stresses for clear, glued-laminated timbers in the dry condition.

Allowable unit stresses for commercial species of lumber, widely accepted for wood construction in economical and efficient designs, are given in "National Design Specification for Wood Construction," National Forest Products Association. Allowable unit stresses for glued-laminated timber are given in the American Institute of Timber Construction "Timber Construction Manual," John Wiley & Sons, Inc., and in the AITC "Standard Specifications for Structural Glued-laminated Timber," AITC 117.

See also Arts. 7-5 to 7-7 and 7-9.

7-4. Structural Grading of Wood.

Strength properties of wood are intimately related to moisture content and specific gravity. Therefore, data on strength properties, unaccompanied by corresponding data on these physical properties, would be of little value.

TABLE 7-5 Properties of Sections of Glued-Laminated Timber

2¾-in. Width

No. of laminations 1½-in.	No. of laminations ¾-in.	d	C_F	A	S	I	Vol.
2	4	3.00	1.00	6.8	3.4	5.1	0.05
	5	3.75	1.00	8.4	5.3	9.9	0.06
3	6	4.50	1.00	10.1	7.6	17.1	0.07
	7	5.25	1.00	11.8	10.3	27.1	0.08
4	8	6.00	1.00	13.5	13.5	40.5	0.09
	9	6.75	1.00	15.2	17.1	57.7	0.11
5	10	7.50	1.00	16.9	21.1	79.1	0.12
	11	8.25	1.01	18.6	25.5	105.3	0.13
6	12	9.00	1.00	20.2	30.4	136.7	0.14
	13	9.75	1.00	21.9	35.6	173.8	0.15
7	14	10.50	1.00	23.6	41.3	217.0	0.16
	15	11.25	1.00	25.3	47.5	267.0	0.18
8	16	12.00	1.00	27.0	54.0	324.0	0.19
	17	12.75	0.99	28.7	61.0	388.6	0.20
9	18	13.50	0.99	30.4	68.3	461.3	0.21
	19	14.25	0.98	32.1	76.1	542.6	0.22
10	20	15.00	0.98	33.8	84.4	632.8	0.23

3⅛-in. Width

No. of laminations 1½-in.	No. of laminations ¾-in.	d	C_F	A	S	I	Vol.
2	4	3.00	1.00	9.4	4.7	7.0	0.06
	5	3.75	1.00	11.7	7.3	13.7	0.08
3	6	4.50	1.00	14.1	10.5	23.7	0.10
	7	5.25	1.00	16.4	14.4	37.7	0.11
4	8	6.00	1.00	18.8	18.8	56.3	0.13
	9	6.75	1.00	21.1	23.7	80.1	0.15
5	10	7.50	1.00	23.4	29.3	109.9	0.16
	11	8.25	1.01	25.8	35.4	146.2	0.18
6	12	9.00	1.00	28.1	42.2	189.8	0.20
	13	9.75	1.00	30.5	49.5	241.4	0.21
7	14	10.50	1.00	32.8	57.4	301.5	0.23
	15	11.25	1.00	35.2	65.9	370.8	0.24
8	16	12.00	1.00	37.5	75.0	450.0	0.26
	17	12.75	0.99	39.8	84.7	539.8	0.28
9	18	13.50	0.99	42.2	94.9	640.7	0.29
	19	14.25	0.98	44.5	105.8	753.6	0.31
10	20	15.00	0.98	46.9	117.2	878.9	0.33
	21	15.75	0.97	49.2	129.2	1,017.4	0.34
11	22	16.50	0.97	51.6	141.8	1,169.8	0.36
	23	17.25	0.96	53.9	155.0	1,336.7	0.37
12	24	18.00	0.96	56.3	168.8	1,518.8	0.39
	25	18.75	0.95	58.6	183.1	1,716.6	0.41
13	26	19.50	0.95	60.9	198.0	1,931.0	0.42
	27	20.25	0.94	63.3	213.6	2,162.4	0.44

6¾-in. Width

No. of laminations 1½-in.	No. of laminations ¾-in.	d	C_F	A	S	I	Vol.
4	8	6.00	1.00	40.5	40.5	121.5	0.28
	9	6.75	1.00	45.6	51.3	173.0	0.32
5	10	7.50	1.00	50.6	63.3	237.5	0.35
	11	8.25	1.00	55.7	76.6	315.9	0.39
6	12	9.00	1.00	60.8	91.1	410.1	0.42
	13	9.75	1.00	65.8	106.9	521.4	0.46
7	14	10.50	1.00	70.9	124.0	651.2	0.49
	15	11.25	1.00	75.9	142.4	800.9	0.53
8	16	12.00	1.00	81.0	162.0	972.0	0.56
	17	12.75	0.99	86.1	182.9	1,165.9	0.60
9	18	13.50	0.99	91.1	205.0	1,384.0	0.63
	19	14.25	0.98	96.2	228.4	1,627.7	0.67
10	20	15.00	0.98	101.3	253.1	1,898.4	0.70
	21	15.75	0.97	106.3	279.1	2,197.7	0.74
11	22	16.50	0.97	111.4	306.3	2,526.8	0.77
	23	17.25	0.96	116.4	334.8	2,887.3	0.81
12	24	18.00	0.96	121.5	364.5	3,280.5	0.84
	25	18.75	0.95	126.6	395.5	3,707.9	0.88
13	26	19.50	0.95	131.6	427.8	4,170.9	0.91
	27	20.25	0.94	136.7	461.3	4,670.9	0.95
14	28	21.00	0.94	141.8	496.1	5,209.3	0.98
	29	21.75	0.94	146.8	532.2	5,787.6	1.02
15	30	22.50	0.93	151.9	569.5	6,407.2	1.05
	31	23.25	0.93	156.9	608.1	7,069.5	1.09
16	32	24.00	0.93	162.0	648.0	7,776.0	1.12
	33	24.75	0.92	167.1	689.1	8,528.0	1.16
17	34	25.50	0.92	172.1	731.5	9,327.0	1.20
	35	26.25	0.92	177.2	775.2	10,174.4	1.23
18	36	27.00	0.91	182.3	820.1	11,071.7	1.27
	37	27.75	0.91	187.3	866.3	12,020.2	1.30
19	38	28.50	0.91	192.4	913.8	13,021.4	1.34
	39	29.25	0.91	197.4	962.5	14,076.7	1.37
20	40	30.00	0.90	202.5	1,012.5	15,187.5	1.41
	41	30.75	0.90	207.6	1,063.8	16,355.3	1.44
21	42	31.50	0.90	212.6	1,116.3	17,581.4	1.48
	43	32.25	0.90	217.7	1,170.1	18,867.4	1.51
22	44	33.00	0.89	222.8	1,225.1	20,214.6	1.55
	45	33.75	0.89	227.8	1,281.4	21,624.4	1.58
23	46	34.50	0.89	232.9	1,339.0	23,098.3	1.62
	47	35.25	0.89	237.9	1,397.9	24,637.7	1.65
24	48	36.00	0.88	243.0	1,458.0	26,244.0	1.69
	49	36.75	0.88	248.1	1,519.4	27,918.7	1.72
25	50	37.50	0.88	253.1	1,582.0	29,663.1	1.76
	51	38.25	0.88	258.2	1,645.9	31,478.7	1.79

8¾-in. Width

No. of laminations 1½-in.	No. of laminations ¾-in.	d	C_F	A	S	I	Vol.
6	12	9.00	1.00	78.8	118.1	531.6	0.55
	13	9.75	1.00	85.3	138.6	675.8	0.59
7	14	10.50	1.00	91.9	160.8	844.1	0.64
	15	11.25	1.00	98.4	184.6	1,038.2	0.68
8	16	12.00	1.00	105.0	210.0	1,260.0	0.73
	17	12.75	0.99	111.6	237.1	1,511.3	0.77
9	18	13.50	0.99	118.1	265.8	1,794.0	0.82
	19	14.25	0.98	124.7	296.1	2,109.9	0.87
10	20	15.00	0.98	131.3	328.1	2,460.9	0.91
	21	15.75	0.97	137.8	361.8	2,848.8	0.96
11	22	16.50	0.97	144.4	397.0	3,275.5	1.00
	23	17.25	0.96	150.9	433.9	3,742.8	1.05
12	24	18.00	0.96	157.5	472.5	4,252.5	1.09
	25	18.75	0.95	164.1	512.7	4,806.5	1.14
13	26	19.50	0.95	170.6	554.5	5,406.7	1.18
	27	20.25	0.94	177.2	598.0	6,054.8	1.23
14	28	21.00	0.94	183.8	643.1	6,752.8	1.28
	29	21.75	0.94	190.3	689.9	7,502.5	1.32
15	30	22.50	0.93	196.9	738.3	8,305.7	1.37
	31	23.25	0.93	203.4	788.3	9,164.2	1.41
16	32	24.00	0.93	210.0	840.0	10,080.0	1.46
	33	24.75	0.92	216.6	893.3	11,054.8	1.50
17	34	25.50	0.92	223.1	948.3	12,090.6	1.55
	35	26.25	0.92	229.7	1,004.9	13,189.1	1.59
18	36	27.00	0.91	236.3	1,063.1	14,352.2	1.64
	37	27.75	0.91	242.8	1,123.0	15,581.7	1.69
19	38	28.50	0.91	249.4	1,184.5	16,879.6	1.73
	39	29.25	0.91	255.9	1,247.7	18,247.5	1.78
20	40	30.00	0.90	262.5	1,312.5	19,687.5	1.82
	41	30.75	0.90	269.1	1,378.9	21,201.3	1.87
21	42	31.50	0.90	275.6	1,447.0	22,790.7	1.91
	43	32.25	0.90	282.2	1,516.8	24,457.7	1.96
22	44	33.00	0.89	288.8	1,588.1	26,204.1	2.00
	45	33.75	0.89	295.3	1,661.1	28,031.6	2.05
23	46	34.50	0.89	301.9	1,735.8	29,942.2	2.10
	47	35.25	0.89	308.4	1,812.1	31,937.7	2.14
24	48	36.00	0.88	315.0	1,890.0	34,020.0	2.19
	49	36.75	0.88	321.6	1,969.6	36,190.9	2.23
25	50	37.50	0.88	328.1	2,050.8	38,452.2	2.28
	51	38.25	0.88	334.7	2,133.6	40,805.7	2.32
26	52	39.00	0.88	341.3	2,218.1	43,253.4	2.37
	53	39.75	0.88	347.8	2,304.3	45,797.5	2.42
27	54	40.50	0.87	354.4	2,392.0	48,438.6	2.46
	55	41.25	0.87	360.9	2,481.5	51,179.8	2.51

7-10

5 1/8-in. Width

14	28	21.00	0.94	65.6	229.7	2,411.7	0.46		26	52	39.00	0.88	263.3	1,711.6	33,366.9	1.83		28	56	42.00	0.87	367.5	2,572.5	54,022.5	2.60
	29	21.75	0.94	68.0	246.4	2,679.5	0.47			53	39.75	0.88	268.3	1,777.6	35,329.2	1.86			57	42.75	0.87	374.1	2,665.2	56,968.6	2.55
15	30	22.50	0.93	70.3	263.7	2,966.3	0.49		27	54	40.50	0.87	273.4	1,845.3	37,367.0	1.90		29	58	43.50	0.87	380.6	2,759.5	60,019.8	2.64
	31	23.25	0.93	72.7	281.5	3,272.9	0.50			55	41.25	0.87	278.4	1,914.3	39,481.6	1.93			59	44.25	0.87	387.2	2,855.5	63,178.1	2.69
16	32	24.00	0.93	75.0	300.0	3,600.0	0.52		28	56	42.00	0.87	283.5	1,984.5	41,674.5	1.97		30	60	45.00	0.86	393.8	2,953.1	66,445.3	2.73
										57	42.75	0.87	288.6	2,056.0	43,947.2	2.00			61	45.75	0.86	400.3	3,052.4	69,823.3	2.78
3	6	4.50	1.00	23.1	17.3	38.9	0.16		29	58	43.50	0.87	293.6	2,128.8	46,301.0	2.04		31	62	46.50	0.86	406.9	3,153.3	73,313.8	2.83
	7	5.25	1.00	26.9	23.5	61.8	0.19			59	44.25	0.87	298.7	2,202.8	48,737.4	2.07			63	47.25	0.86	413.4	3,255.8	76,918.8	2.87
4	8	6.00	1.00	30.8	30.8	92.3	0.21		30	60	45.00	0.86	303.8	2,278.1	51,257.8	2.11		32	64	48.00	0.86	420.0	3,360.0	80,640.0	2.92
	9	6.75	1.00	34.6	38.9	131.3	0.24			61	45.75	0.86	308.8	2,354.7	53,863.7	2.14			65	48.75	0.86	426.6	3,465.8	84,479.4	2.96
5	10	7.50	1.00	38.4	48.0	180.2	0.27		31	62	46.50	0.86	313.9	2,432.5	56,556.4	2.18		33	66	49.50	0.85	433.1	3,573.3	88,438.7	3.01
	11	8.25	1.00	42.3	58.1	239.6	0.29			63	47.25	0.86	318.9	2,511.6	59,337.3	2.21			67	50.25	0.85	439.7	3,682.4	92,519.9	3.05
6	12	9.00	1.00	46.1	69.2	311.3	0.32		32	64	48.00	0.86	324.0	2,592.0	62,208.0	2.25		34	68	51.00	0.85	446.3	3,793.1	96,724.7	3.10
	13	9.75	1.00	50.0	81.2	395.8	0.35												69	51.75	0.85	452.8	3,905.5	101,055.0	3.14
7	14	10.50	1.00	53.8	94.2	494.4	0.37											35	70	52.50	0.85	459.4	4,019.5	105,512.7	3.19
	15	11.25	1.00	57.7	108.1	608.1	0.40												71	53.25	0.85	465.9	4,135.2	110,099.6	3.24
8	16	12.00	1.00	61.5	123.0	738.0	0.43											36	72	54.00	0.85	472.5	4,252.5	114,817.5	3.28
	17	12.75	0.99	65.3	138.9	885.2	0.45												73	54.75	0.85	479.1	4,371.4	119,668.3	3.33
9	18	13.50	0.99	69.2	155.7	1,050.8	0.48											37	74	55.50	0.84	485.6	4,492.0	124,653.9	3.37
	19	14.25	0.99	73.0	173.4	1,235.8	0.51												75	56.25	0.84	492.2	4,614.3	129,776.0	3.42
10	20	15.00	0.98	76.9	192.2	1,441.4	0.53											38	76	57.00	0.84	498.8	4,738.1	135,036.6	3.46
	21	15.75	0.98	80.7	211.9	1,668.6	0.56												77	57.75	0.84	505.3	4,863.6	140,437.4	3.51
11	22	16.50	0.97	84.6	232.5	1,918.5	0.59											39	78	58.50	0.84	511.9	4,990.8	145,980.4	3.55
	23	17.25	0.97	88.4	254.2	2,192.2	0.61												79	59.25	0.84	518.4	5,119.6	151,667.3	3.60
12	24	18.00	0.96	92.3	276.8	2,490.8	0.64											40	80	60.00	0.84	525.0	5,250.0	157,500.0	3.65
	25	18.75	0.96	96.1	300.3	2,815.2	0.67												81	60.75	0.84	531.6	5,382.1	163,480.4	3.69
13	26	19.50	0.95	99.9	324.8	3,166.8	0.69											41	82	61.50	0.83	538.1	5,515.8	169,610.3	3.74
	27	20.25	0.95	103.8	350.3	3,546.4	0.72												83	62.25	0.83	544.7	5,651.1	175,891.5	3.78
14	28	21.00	0.94	107.6	376.7	3,955.2	0.75											42	84	63.00	0.83	551.3	5,788.1	182,326.0	3.83
	29	21.75	0.94	111.5	404.1	4,394.3	0.77																		
15	30	22.50	0.93	115.3	432.4	4,864.7	0.80																		
	31	23.25	0.93	119.2	461.7	5,367.6	0.83																		
16	32	24.00	0.93	123.0	492.0	5,904.0	0.85																		
	33	24.75	0.92	126.8	523.2	6,475.0	0.88																		
17	34	25.50	0.92	130.7	555.4	7,081.6	0.91																		
	35	26.25	0.92	134.5	588.6	7,725.0	0.93																		
18	36	27.00	0.91	138.4	622.7	8,406.3	0.96																		
	37	27.75	0.91	142.2	657.8	9,126.4	0.99																		
19	38	28.50	0.91	146.1	693.8	9,886.6	1.01																		
	39	29.25	0.91	149.9	730.8	10,687.8	1.04																		
20	40	30.00	0.90	153.8	768.8	11,531.5	1.07																		
	41	30.75	0.90	157.6	807.7	12,417.9	1.09																		
21	42	31.50	0.90	161.4	847.5	13,348.9	1.12																		
	43	32.25	0.90	165.3	888.4	14,325.2	1.15																		
22	44	33.00	0.89	169.1	930.2	15,348.1	1.17																		
	45	33.75	0.89	173.0	972.9	16,418.5	1.20																		
23	46	34.50	0.89	176.8	1,016.7	17,537.6	1.23																		
	47	35.25	0.89	180.7	1,061.4	18,706.4	1.25																		
24	48	36.00	0.88	184.5	1,107.0	19,926.0	1.28																		

7-11

TABLE 7-5 Properties of Sections of Glued-Laminated Timber* (Continued)

No. of laminations 1½-in.	No. of laminations ¾-in.	d	C_F	A	S	I	Vol.	No. of laminations 1½-in.	No. of laminations ¾-in.	d	C_F	A	S	I	Vol.	No. of laminations 1½-in.	No. of laminations ¾-in.	d	C_F	A	S	I	Vol.
				10¾-in. Width								12¼-in. Width								14¼-in. Width			
7	14	10.50	1.00	112.9	197.5	1,037.0	0.78	8	16	12.00	1.00	147.0	294.0	1,764.0	1.02	9	18	13.50	0.99	192.4	432.8	2,921.7	1.34
	15	11.25	1.00	120.9	226.8	1,275.5	0.84		17	12.75	0.99	156.2	331.9	2,115.8	1.08		19	14.25	0.98	203.1	482.3	3,436.2	1.41
8	16	12.00	1.00	129.0	258.0	1,548.0	0.90	9	18	13.50	0.99	165.4	372.1	2,511.6	1.15	10	20	15.00	0.98	213.8	534.4	4,007.8	1.48
	17	12.75	0.99	137.1	291.3	1,856.8	0.95		19	14.25	0.98	174.6	414.6	2,953.8	1.21		21	15.75	0.97	224.4	589.1	4,639.5	1.56
9	18	13.50	0.99	145.1	326.5	2,204.1	1.01	10	20	15.00	0.98	183.8	459.4	3,445.3	1.28	11	22	16.50	0.97	235.1	646.6	5,334.4	1.63
	19	14.25	0.98	153.2	363.8	2,592.2	1.06		21	15.75	0.97	192.9	506.4	3,988.6	1.34		23	17.25	0.96	245.8	706.7	6,095.5	1.71
10	20	15.00	0.98	161.3	403.1	3,023.4	1.12	11	22	16.50	0.97	202.1	555.8	4,585.7	1.40	12	24	18.00	0.96	256.5	769.5	6,925.5	1.78
	21	15.75	0.97	169.3	444.4	3,500.0	1.18		23	17.25	0.96	211.3	607.5	5,239.7	1.47		25	18.75	0.95	267.2	835.0	7,827.8	1.86
11	22	16.50	0.97	177.4	487.8	4,024.2	1.23	12	24	18.00	0.96	220.5	661.5	5,953.5	1.53	13	26	19.50	0.95	277.9	903.1	8,805.2	1.93
	23	17.25	0.96	185.4	533.1	4,598.3	1.29		25	18.75	0.95	229.7	717.7	6,728.9	1.60		27	20.25	0.94	288.6	973.9	9,860.7	2.00
12	24	18.00	0.96	193.5	580.5	5,224.5	1.34	13	26	19.50	0.95	238.9	776.3	7,569.4	1.66	14	28	21.00	0.94	299.3	1,047.4	10,997.4	2.08
	25	18.75	0.95	201.6	629.9	5,905.2	1.40		27	20.25	0.94	248.1	837.2	8,476.5	1.72		29	21.75	0.94	309.9	1,123.5	12,218.3	2.15
13	26	19.50	0.95	209.6	681.3	6,642.5	1.46	14	28	21.00	0.94	257.2	900.4	9,453.9	1.79	15	30	22.50	0.93	320.6	1,202.3	13,526.4	2.23
	27	20.25	0.94	217.7	734.7	7,438.8	1.51		29	21.75	0.94	266.4	965.8	10,503.1	1.85		31	23.25	0.93	331.3	1,283.8	14,924.6	2.30
14	28	21.00	0.94	225.8	790.1	8,296.3	1.57	15	30	22.50	0.93	275.6	1,033.6	11,627.9	1.91	16	32	24.00	0.93	342.0	1,368.0	16,416.0	2.38
	29	21.75	0.94	233.8	847.6	9,217.3	1.62		31	23.25	0.93	284.8	1,103.6	12,829.5	1.97		33	24.75	0.92	352.7	1,454.8	18,003.6	2.45
15	30	22.50	0.93	241.9	907.0	10,204.1	1.68	16	32	24.00	0.93	294.0	1,176.0	14,112.0	2.04	17	34	25.50	0.92	363.4	1,544.3	19,690.4	2.52
	31	23.25	0.93	249.9	968.3	11,258.6	1.74		33	24.75	0.92	303.2	1,250.6	15,476.8	2.10		35	26.25	0.92	374.1	1,636.5	21,479.4	2.60
16	32	24.00	0.93	258.0	1,032.0	12,384.0	1.79	17	34	25.50	0.92	312.4	1,327.6	16,926.8	2.17	18	36	27.00	0.91	384.8	1,731.4	23,373.6	2.67
	33	24.75	0.92	266.1	1,097.5	13,581.7	1.85		35	26.25	0.92	321.6	1,406.8	18,464.1	2.23		37	27.75	0.91	395.4	1,828.9	25,376.0	2.75
17	34	25.50	0.92	274.1	1,165.0	14,854.1	1.90	18	36	27.00	0.91	330.8	1,488.4	20,093.1	2.30	19	38	28.50	0.91	406.1	1,929.1	27,489.6	2.82
	35	26.25	0.92	282.2	1,234.6	16,203.7	1.96		37	27.75	0.91	339.9	1,572.2	21,813.7	2.36		39	29.25	0.91	416.8	2,032.0	29,717.4	2.89
18	36	27.00	0.91	290.3	1,306.1	17,632.7	2.02	19	38	28.50	0.91	349.1	1,658.3	23,631.4	2.42	20	40	30.00	0.90	427.5	2,137.5	32,062.5	2.97
	37	27.75	0.91	298.3	1,379.7	19,143.3	2.07		39	29.25	0.91	358.3	1,746.7	25,545.7	2.49		41	30.75	0.90	438.2	2,245.7	34,527.8	3.04
19	38	28.50	0.91	306.4	1,455.3	20,737.8	2.13	20	40	30.00	0.90	367.5	1,837.5	27,562.5	2.55	21	42	31.50	0.90	448.9	2,356.6	37,116.4	3.12
	39	29.25	0.91	314.4	1,532.9	22,418.4	2.18		41	30.75	0.90	376.7	1,930.5	29,680.8	2.62		43	32.25	0.90	459.6	2,470.1	39,831.1	3.19
20	40	30.00	0.90	322.5	1,612.5	24,187.3	2.24	21	42	31.50	0.90	385.9	2,025.8	31,907.0	2.68	22	44	33.00	0.89	470.3	2,586.4	42,675.2	3.27
	41	30.75	0.90	330.6	1,694.1	26,047.3	2.30		43	32.25	0.90	395.1	2,123.4	34,239.7	2.74		45	33.75	0.89	480.9	2,705.3	45,651.5	3.34
21	42	31.50	0.90	338.6	1,777.8	28,000.1	2.35	22	44	33.00	0.89	404.2	2,223.4	36,685.7	2.81	23	46	34.50	0.89	491.6	2,826.8	48,763.1	3.41
	43	32.25	0.90	346.7	1,863.4	30,047.5	2.41		45	33.75	0.89	413.4	2,325.6	39,244.4	2.87		47	35.25	0.89	502.3	2,951.1	52,012.9	3.49
22	44	33.00	0.89	354.8	1,951.1	32,193.8	2.46	23	46	34.50	0.89	422.6	2,430.1	41,919.1	2.93	24	48	36.00	0.88	513.0	3,078.0	55,404.0	3.56
	45	33.75	0.89	362.8	2,040.8	34,438.8	2.52		47	35.25	0.89	431.8	2,536.9	44,712.7	3.00		49	36.75	0.88	523.7	3,207.6	58,939.4	3.64
23	46	34.50	0.89	370.9	2,132.5	36,786.2	2.58	24	48	36.00	0.88	441.0	2,646.0	47,628.0	3.06	25	50	37.50	0.88	534.4	3,339.8	62,622.1	3.71
	47	35.25	0.89	378.9	2,226.3	39,237.8	2.63		49	36.75	0.88	450.2	2,757.4	50,667.0	3.13		51	38.25	0.88	545.1	3,474.8	66,455.0	3.79
24	48	36.00	0.88	387.0	2,322.0	41,796.0	2.69	25	50	37.50	0.88	459.4	2,871.1	53,833.0	3.19	26	52	39.00	0.88	555.8	3,612.4	70,441.8	3.86
	49	36.75	0.88	395.1	2,419.8	44,463.1	2.74		51	38.25	0.88	468.6	2,987.1	57,127.8	3.25		53	39.75	0.87	566.4	3,752.6	74,583.9	3.93
25	50	37.50	0.88	403.1	2,519.5	47,241.2	2.80	26	52	39.00	0.88	477.8	3,105.4	60,554.8	3.32	27	54	40.50	0.87	577.1	3,895.5	78,885.3	4.01
	51	38.25	0.88	411.2	2,621.3	50,132.8	2.86		53	39.75	0.87	486.9	3,226.0	64,115.8	3.38		55	41.25	0.87	587.8	4,041.2	83,350.0	4.08
26	52	39.00	0.88	419.3	2,725.1	53,139.9	2.91	27	54	40.50	0.87	496.1	3,348.8	67,814.1	3.45	28	56	42.00	0.87	598.5	4,189.5	87,979.5	4.16
	53	39.75	0.87	427.3	2,830.9	56,265.0	2.97		55	41.25	0.87	505.3	3,474.0	71,651.5	3.51		57	42.75	0.87	609.2	4,340.5	92,777.4	4.23
27	54	40.50	0.87	435.4	2,938.8	59,510.3	3.02	28	56	42.00	0.87	514.5	3,601.5	75,631.6	3.57	29	58	43.50	0.87	619.9	4,494.1	97,746.5	4.30
	55	41.25	0.87	443.5	3,048.6	62,878.1	3.08		57	42.75	0.87	523.7	3,731.3	79,755.5	3.64		59	44.25	0.86	630.6	4,650.4	102,890.1	4.38
28	56	42.00	0.87	451.5	3,160.5	66,370.5	3.14	29	58	43.50	0.87	532.9	3,863.9	84,027.7	3.70	30	60	45.00	0.86	641.3	4,809.4	108,211.0	4.45
	57	42.75	0.87	459.6	3,274.4	69,989.9	3.19		59	44.25	0.86	542.1	3,999.7	88,449.1	3.76		61	45.75	0.86	651.9	4,971.0	113,712.2	4.53
29	58	43.50	0.87	467.6	3,390.3	73,738.6	3.25	30	60	45.00	0.86	551.2	4,134.4	93,023.4	3.83	31	62	46.50	0.86	662.6	5,135.3	119,396.7	4.60
	59	44.25	0.86	475.7	3,508.2	77,618.8	3.30		61	45.75	0.86	560.4	4,273.3	97,752.2	3.89		63	47.25	0.86	673.3	5,302.3	125,267.7	4.68
30	60	45.00	0.86	483.8	3,628.1	81,632.8	3.36	31	62	46.50	0.86	569.6	4,414.6	102,639.3	3.96	32	64	48.00	0.86	684.0	5,472.0	131,328.0	4.75
	61	45.75	0.86	491.8	3,750.1	85,782.9	3.41		63	47.25	0.86	578.8	4,558.1	107,685.9	4.02		65	48.75	0.85	694.7	5,644.3	137,580.7	4.82
31	62	46.50	0.86	499.9	3,874.0	90,071.2	3.47	32	64	48.00	0.86	588.0	4,704.0	112,896.0	4.08	33	66	49.50	0.85	705.4	5,819.3	144,028.8	4.90
	63	47.25	0.86	507.9	4,000.0	94,500.2	3.53		65	48.75	0.85	597.2	4,852.1	118,270.7	4.15		67	50.25	0.85	716.1	5,997.0	150,675.2	4.97
32	64	48.00	0.86	516.0	4,128.0	99,072.0	3.58	33	66	49.50	0.85	606.4	5,002.6	123,814.2	4.21	34	68	51.00	0.85	726.8	6,177.4	157,523.1	5.05
	65	48.75	0.85	524.1	4,258.0	103,789.0	3.64		67	50.25	0.85	615.6	5,155.3	129,527.4	4.27		69	51.75	0.85	737.4	6,360.4	164,575.3	5.12

No.	d	C_F	A	S	I	Vol.	No.	d	C_F	A	S	I	Vol.	No.	d	C_F	A	S	I	Vol.
33	49.50	0.85	532.1	4,390.0	108,653.3	3.70		51.00	0.85	624.8	5,310.8	135,414.6	4.34		52.50	0.85	748.1	6,546.1	171,835.3	5.20
	50.25	0.85	540.2	4,524.1	113,667.3	3.75		51.75	0.85	633.9	5,467.7	141,476.6	4.40		53.25	0.85	758.8	6,734.5	179,305.0	5.27
34	51.00	0.85	548.3	4,660.2	118,833.2	3.81		52.50	0.85	643.1	5,627.3	147,717.8	4.47		54.00	0.85	769.5	6,925.5	186,998.5	5.34
	51.75	0.85	556.3	4,798.2	124,153.3	3.86		53.25	0.85	652.3	5,789.3	154,138.9	4.53		54.75	0.84	780.2	7,119.2	194,888.4	5.42
35	52.50	0.85	564.4	4,938.3	129,629.9	3.92		54.00	0.85	661.5	5,953.5	160,744.5	4.59		55.50	0.84	790.9	7,315.6	203,007.7	5.49
	53.25	0.85	572.4	5,080.4	135,265.2	3.98		54.75	0.85	670.7	6,120.0	167,535.1	4.66		56.25	0.84	801.6	7,514.6	211,349.5	5.57
36	54.00	0.85	580.5	5,224.5	141,061.6	4.03		55.50	0.84	679.9	6,288.8	174,515.4	4.72		57.00	0.84	812.3	7,716.4	219,916.7	5.64
	54.75	0.85	588.6	5,370.6	147,021.1	4.09		56.25	0.84	689.1	6,459.9	181,685.8	4.79		57.75	0.84	822.9	7,920.8	228,712.4	5.71
37	55.50	0.85	596.6	5,518.8	153,146.2	4.14		57.00	0.84	698.2	6,633.4	189,051.2	4.85		58.50	0.84	833.6	8,127.8	237,739.5	5.79
	56.25	0.85	604.7	5,668.9	159,439.1	4.20		57.75	0.84	707.4	6,809.1	196,611.7	4.91		59.25	0.84	844.3	8,337.6	247,001.0	5.86
38	57.00	0.85	612.8	5,821.1	165,902.1	4.26		58.50	0.84	716.6	6,987.1	204,372.5	4.98		60.00	0.84	855.0	8,550.0	256,500.0	5.94
	57.75	0.84	620.8	5,975.3	172,537.4	4.31		59.25	0.84	725.8	7,167.4	212,333.5	5.04		60.75	0.84	865.7	8,765.1	266,239.5	6.01
39	58.50	0.85	628.9	6,131.5	179,347.3	4.37		60.00	0.84	735.0	7,350.0	220,500.0	5.10		61.50	0.84	876.4	8,982.8	276,222.5	6.09
	59.25	0.84	636.9	6,289.8	186,334.1	4.42		60.75	0.84	744.2	7,534.9	228,871.8	5.17		62.25	0.84	887.1	9,203.3	286,451.9	6.16
40	60.00	0.84	645.0	6,450.0	193,500.0	4.48		61.50	0.84	753.4	7,722.1	237,454.4	5.23		63.00	0.84	897.8	9,426.9	296,950.8	6.23
	60.75	0.84	653.1	6,612.3	200,847.4	4.54		62.25	0.84	762.6	7,911.6	246,247.3	5.30		63.75	0.83	908.4	9,652.1	307,662.3	6.31
41	61.50	0.84	661.1	6,776.5	208,378.4	4.59		63.00	0.84	771.8	8,103.4	255,256.3	5.36		64.50	0.83	919.1	9,880.6	318,649.2	6.38
	62.25	0.84	669.2	6,942.8	216,095.3	4.65		63.75	0.83	781.0	8,297.4	264,480.7	5.42		65.25	0.83	929.8	10,111.7	329,894.6	6.46
42	63.00	0.84	677.3	7,111.1	224,000.5	4.70		64.50	0.83	790.1	8,493.8	273,926.5	5.49		66.00	0.83	940.5	10,345.5	341,401.5	6.53
	63.75	0.84	685.3	7,281.4	232,096.1	4.76		65.25	0.83	799.3	8,692.5	283,592.7	5.55		66.75	0.83	951.2	10,582.0	353,173.0	6.61
43	64.50	0.83	693.4	7,453.8	240,384.5	4.87		66.00	0.83	808.5	8,893.5	293,485.5	5.61		67.50	0.83	961.9	10,821.1	365,137.0	6.68
	65.25	0.83	701.4	7,628.5	248,867.9	4.87		66.75	0.83	817.7	9,096.7	303,603.8	5.68		68.25	0.83	972.6	11,062.9	377,521.5	6.75
44	66.00	0.83	709.5	7,804.5	257,548.5	4.93		67.50	0.83	826.9	9,302.3	313,954.1	5.74		69.00	0.83	983.3	11,307.4	390,104.5	6.83
	66.75	0.83	717.6	7,982.9	266,428.8	4.98		68.25	0.83	836.1	9,510.2	324,534.9	5.81		69.75	0.83	993.9	11,554.5	402,964.0	6.90
45	67.50	0.83	725.6	8,163.3	275,510.8	5.04		69.00	0.83	845.2	9,720.4	335,352.9	5.87		70.50	0.82	1,004.6	11,804.3	416,103.1	6.98
	68.25	0.83	733.7	8,345.7	284,796.9	5.10		69.75	0.82	854.4	9,932.8	346,406.5	5.93		71.25	0.82	1,015.3	12,056.8	429,524.8	7.05
46	69.00	0.83	741.8	8,530.1	294,289.3	5.15		70.50	0.82	863.6	10,147.6	357,702.7	6.00		72.00	0.82	1,026.0	12,312.0	443,232.0	7.12
	69.75	0.82	749.8	8,716.6	303,990.5	5.21		71.25	0.82	872.8	10,364.6	369,239.4	6.06		72.75	0.82	1,036.7	12,569.8	457,227.8	7.20
47	70.50	0.82	757.9	8,905.3	313,902.4	5.26		72.00	0.82	882.0	10,584.0	381,024.0	6.12		73.50	0.82	1,047.4	12,830.3	471,515.1	7.27
	71.25	0.82	765.9	9,095.5	324,027.5	5.32		72.75	0.82	891.2	10,805.6	393,054.2	6.19		74.25	0.82	1,058.1	13,093.5	486,097.1	7.35
48	72.00	0.82	774.0	9,288.0	334,368.0	5.38		73.50	0.82	900.4	11,029.4	405,337.6	6.25		75.00	0.82	1,068.8	13,359.4	500,976.6	7.42
	72.75	0.82	782.1	9,482.5	344,926.3	5.43		74.25	0.82	909.6	11,255.8	417,871.6	6.32		75.75	0.81	1,079.4	13,627.9	516,156.7	7.50
49	73.50	0.82	790.1	9,679.0	355,704.8	5.49		75.00	0.81	918.8	11,484.4	430,664.1	6.38		76.50	0.81	1,090.1	13,899.1	531,640.5	7.57
	74.25	0.82	798.2	9,877.6	366,704.8	5.54		75.75	0.81	927.9	11,715.2	443,712.2	6.44		77.25	0.81	1,100.8	14,173.0	547,430.7	7.64
50	75.00	0.82	806.3	10,078.1	377,929.7	5.60		76.50	0.81	937.1	11,948.3	457,024.7	6.51		78.00	0.81	1,111.5	14,449.5	563,330.6	7.72
								77.25	0.81	946.3	12,183.7	470,596.7	6.57		78.75	0.81	1,122.2	14,728.7	579,943.1	7.79
								78.00	0.81	955.5	12,421.5	484,438.5	6.64		79.50	0.81	1,132.9	15,010.6	596,671.2	7.87
								78.75	0.81	964.7	12,661.5	498,545.9	6.70		80.25	0.81	1,143.6	15,295.1	613,718.0	7.94
								79.50	0.81	973.9	12,903.8	512,927.8	6.76		81.00	0.81	1,154.3	15,582.4	631,086.2	8.02
								80.25	0.81	983.1	13,148.4	527,580.3	6.83		81.75	0.80	1,164.9	15,872.3	648,779.2	8.09
								81.00	0.81	992.2	13,395.4	542,512.7	6.89		82.50	0.80	1,175.6	16,164.8	666,799.8	8.16
								81.75	0.81	1,001.4	13,644.5	557,720.6	6.95		83.25	0.80	1,186.3	16,460.1	685,151.2	8.24
								82.50	0.81	1,010.6	13,896.1	573,213.9	7.02		84.00	0.80	1,197.0	16,758.0	703,836.1	8.31
								83.25	0.81	1,019.8	14,149.9	588,987.6	7.08		84.75	0.80	1,207.7	17,058.6	722,857.7	8.39
								84.00	0.81	1,029.0	14,406.0	605,052.0	7.15		85.50	0.80	1,218.4	17,361.8	742,218.8	8.46
															86.25	0.80	1,229.1	17,667.8	761,922.8	8.54
															87.00	0.80	1,239.8	17,976.4	781,972.3	8.61
															87.75	0.80	1,250.4	18,287.6	802,370.6	8.68
															88.50	0.80	1,261.1	18,601.6	823,120.6	8.76
															89.25	0.80	1,271.8	18,918.2	844,225.2	8.83
															90.00	0.80	1,282.5	19,237.5	865,687.6	8.91
															90.75	0.80	1,293.2	19,559.5	887,510.6	8.98
															91.50	0.80	1,303.9	19,884.1	909,697.3	9.05
															92.25	0.80	1,314.6	20,211.4	932,250.8	9.13
															93.00	0.80	1,325.3	20,541.4	955,174.0	9.20
															93.75	0.80	1,335.9	20,874.0	978,470.0	9.28
															94.50	0.80	1,346.6	21,209.3	1,002,141.6	9.35
															95.25	0.79	1,357.3	21,547.3	1,026,192.0	9.43
															96.00	0.79	1,368.0	21,888.0	1,050,624.2	9.50

*d = section depth, in.; C_F = size factor; A = cross-sectional area, sq in.; S = section modulus, in.^{3}; I = moment of inertia, in.^{4}; Vol. = volume, cu ft per lin ft.

7-14 Wood Construction

The strength of wood is actually affected by many other factors, such as rate of loading, duration of load, temperature, direction of grain, and position of growth rings. Strength is also influenced by such inherent growth characteristics as knots, cross grain, shakes, and checks.

Analysis and integration of available data have yielded a comprehensive set of simple principles for grading structural timber (Tentative Methods for Establishing Structural Grades of Lumber, ASTM D245).

The same characteristics, such as knots and cross grain, that reduce the strength of solid timber also affect the strength of laminated members. However, there are additional factors, peculiar to laminated wood, that must be considered: Effect on strength of bending members is less from knots located at the neutral plane of the beam, a region of low stress. Strength of a bending member with low-grade laminations can be improved by substituting a few high-grade laminations at the top and bottom of the member. Dispersement of knots in laminated members has a beneficial effect on strength. With sufficient knowledge of the occurrence of knots within a grade, mathematical estimates of this effect may be established for members containing various numbers of laminations.

Allowable design stresses taking these factors into account are higher than for solid timbers of comparable grade. But cross-grain limitations must be more restrictive than for solid timbers, to justify these higher allowable stresses.

7-5. Moisture Content of Wood.

Wood is unlike most other structural materials in regard to the causes of its dimensional changes. These are primarily from gain or loss of moisture, not change in temperature. It is for this reason that expansion joints are seldom required for wood structures to permit movement with temperature changes. It partly accounts for the fact that wood structures can withstand extreme temperatures without collapse.

A newly felled tree is green (contains moisture). In removing the greater part of this water, seasoning first allows free water to leave the cavities in the wood. A point is reached where these cavities contain only air, and the cell walls still are full of moisture. The moisture content at which this occurs, the fiber-saturation point, varies from 25 to 30% of the weight of the oven-dry wood.

During removal of the free water, the wood remains constant in size and in most properties (weight decreases). Once the fiber-saturation point has been passed, shrinkage of the wood begins as the cell walls lose water. Shrinkage continues nearly linearly down to zero moisture content (Table 7-6). (There are, however, complicating factors, such as the effects of timber size and relative rates of moisture movement in three directions, longitudinal, radial, and tangential to the growth rings.) Eventually, the wood assumes a condition of equilibrium, with the final moisture content dependent on the relative humidity and temperature of the ambient air. Wood swells when it absorbs moisture, up to the fiber-saturation point. The relationship of wood moisture content, temperature, and relative humidity can actually define an environment (Fig. 7-1).

This explanation has been simplified. Outdoors, rain, frost, wind, and sun can act directly on the wood. Within buildings, poor environmental conditions may be created for wood by localized heating, cooling, or ventilation. The conditions of service must be sufficiently well known to be specifiable. Then, the proper design stress can be assigned to wood and the most suitable adhesive selected.

Dry Condition of Use. Design stresses—Allowable unit stresses for dry condition of use are applicable for normal loading when the wood moisture content in service is less than 16%, as in most covered structures.

Adhesive selection—Dry-use adhesives are those that perform satisfactorily when the moisture content of wood does not exceed 16% for repeated or prolonged periods of service, and are to be used only when these conditions exist.

Wet Condition of Use. Design stresses—Allowable unit stresses for wet condition of use are applicable for normal loading when the moisture content in service is 16% or more. This may occur in members not covered or in covered locations of high relative humidity.

Adhesive selection—Wet-use adhesives will perform satisfactorily for all conditions, including exposure to weather, marine use, and where pressure treatments are used, whether before or after gluing. Such adhesives are required when the moisture content exceeds 16% for repeated or prolonged periods of service.

7-6. Checking in Timbers.

Separation of grain, or checking, is the result of rapid lowering of surface moisture content combined with a difference in moisture content between inner and outer portions of the piece. As wood loses moisture to the surrounding atmosphere, the outer cells of the member lose at a more rapid rate than the inner cells. As the outer cells try to shrink, they are restrained by the inner portion of the member. The more rapid the drying, the greater will be the differential in shrinkage between outer and inner fibers, and the greater the shrinkage stresses. Splits may develop.

Checks affect the horizontal shear strength of timber. A large reduction factor is applied to test values in establishing allowable unit stresses, in recognition of stress concentrations at the ends of checks. Allowable unit stresses for horizontal shear are adjusted for the amount of checking permissible in the various stress grades at the time of the grading. Since strength properties of wood increase with dryness, checks may enlarge with increasing dryness after shipment, without appreciably reducing shear strength.

Cross-grain checks and splits that tend to run out the side of a piece, or excessive checks and splits that tend to enter connection areas, may be serious and may require servicing. Provisions for controlling the effects of checking in connection areas may be incorporated in design details.

To avoid excessive splitting between rows of bolts caused by shrinkage during seasoning of solid-sawn timbers, the rows should not be spaced more than 5 in apart, or a saw kerf, terminating in a bored hole, should be provided between the lines of bolts. Whenever possible, maximum end distances for connections should be specified to minimize the effect of checks running into the joint area. Some designers require stitch bolts in members, with multiple connections loaded at an angle to the grain. Stitch bolts, kept tight, will reinforce pieces where checking is excessive.

One of the principal advantages of glued-laminated timber construction is relative freedom from checking. Seasoning checks may, however, occur in laminated members for the same reasons that they exist in solid-sawn members. When laminated members are glued within the range of moisture contents set in AITC 103, "Standard for Structural Glued-laminated Timber," they will approximate the moisture content in normal-use conditions, thereby minimizing checking. Moisture content of the lumber at the time of gluing is thus of great importance to the control of checking in service. However, rapid changes in moisture content of large wood sections after gluing will result in shrinkage or swelling of the wood, and during shrinking, checking may develop in both glued joints and wood.

Differentials in shrinkage rates of individual laminations tend to concentrate shrinkage stresses at or near the glue line. For this reason, when checking occurs, it is usually at or near glue lines. The presence of wood-fiber separation indicates adequate glue bonds, and not delamination.

In general, checks have very little effect on the strength of glued-laminated members. Laminations in such members are thin enough to season readily in kiln drying without developing checks. Since checks lie in a radial plane, and the majority of laminations are

TABLE 7-6 Shrinkage Values of Wood Based on Dimensions When Green

Species	Dried to 20% MC*			Dried to 6% MC†			Dried to 0% MC		
	Radial, %	Tangential, %	Volumetric, %	Radial, %	Tangential, %	Volumetric, %	Radial, %	Tangential, %	Volumetric, %
Softwoods:‡									
Cedar:									
Alaska	0.9	2.0	3.1	2.2	4.8	7.4	2.8	6.0	9.2
Incense	1.1	1.7	2.5	2.6	4.2	6.1	3.3	5.2	7.6
Port Orford	1.5	2.3	3.4	3.7	5.5	8.1	4.6	6.9	10.1
Western red	0.8	1.7	2.3	1.9	4.0	5.4	2.4	5.0	6.8
Cypress, southern	1.3	2.1	3.5	3.0	5.0	8.4	3.8	6.2	10.5
Douglas fir:									
Coast region	1.7	2.6	3.9	4.0	6.2	9.4	5.0	7.8	11.8
Inland region	1.4	2.5	3.6	3.3	6.1	8.7	4.1	7.6	10.9
Rocky Mountain	1.2	2.1	3.5	2.9	5.0	8.5	3.6	6.2	10.6
Fir, white	1.1	2.4	3.3	2.6	5.7	7.8	3.2	7.1	9.8
Hemlock:									
Eastern	1.0	2.3	3.2	2.4	5.4	7.8	3.0	6.8	9.7
Western	1.4	2.6	4.0	3.4	6.3	9.5	4.3	7.9	11.9
Larch, western	1.4	2.7	4.4	3.4	6.5	10.6	4.2	8.1	13.2
Pine:									
Eastern white	0.8	2.0	2.7	1.8	4.8	6.6	2.3	6.0	8.2
Lodgepole	1.5	2.2	3.8	3.6	5.4	9.2	4.5	6.7	11.5
Norway	1.5	2.4	3.8	3.7	5.8	9.2	4.6	7.2	11.5
Ponderosa	1.3	2.1	3.2	3.1	5.0	7.7	3.9	6.3	9.6
Southern (avg.)	1.6	2.6	4.1	4.0	6.1	9.8	5.0	7.6	12.2
Sugar	1.0	1.9	2.6	2.3	4.5	6.3	2.9	5.6	7.9
Western white	1.4	2.5	3.9	3.3	5.9	9.4	4.1	7.4	11.8
Redwood (old growth)	0.9	1.5	2.3	2.1	3.5	5.4	2.6	4.4	6.8
Spruce:									
Engelmann	1.1	2.2	3.5	2.7	5.3	8.3	3.4	6.6	10.4
Sitka	1.4	2.5	3.8	3.4	6.0	9.2	4.3	7.5	11.5
Hardwoods:‡									
Ash, white	1.6	2.6	4.5	3.8	6.2	10.7	4.8	7.8	13.4
Beech, American	1.7	3.7	5.4	4.1	8.8	13.0	5.1	11.0	16.3
Birch:									
Sweet	2.2	2.8	5.2	5.2	6.8	12.5	6.5	8.5	15.6
Yellow	2.4	3.1	5.6	5.8	7.4	13.4	7.2	9.2	16.7
Elm, rock	1.6	2.7	4.7	3.8	6.5	11.3	4.8	8.1	14.1
Gum, red	1.7	3.3	5.0	4.2	7.9	12.0	5.2	9.9	15.0
Hickory:									
Pecan§	1.6	3.0	4.5	3.9	7.1	10.9	4.9	8.9	13.6
True	2.5	3.8	6.0	6.0	9.0	14.3	7.5	11.3	17.9
Maple, hard	1.6	3.2	5.0	3.9	7.6	11.9	4.9	9.5	14.9
Oak:									
Red	1.3	2.7	4.5	3.2	6.6	10.8	4.0	8.2	13.5
White	1.8	3.0	5.3	4.2	7.2	12.6	5.3	9.0	15.8
Poplar, yellow	1.3	2.4	4.1	3.2	5.7	9.8	4.0	7.1	12.3

* MC = moisture content, as a percent of weight of oven-dry wood. These shrinkage values have been taken as one-third the shrinkage to the oven-dry condition as given in the last three columns of this table.

† These shrinkage values have been taken as four-fifths of the shrinkage to the oven-dry condition as given in the last three columns of this table.

‡ The total longitudinal shrinkage of normal species from fiber saturation to oven-dry condition is minor. It usually ranges from 0.17 to 0.3% of the green dimension.

§ Average of butternut hickory, nutmeg hickory, water hickory, and pecan.

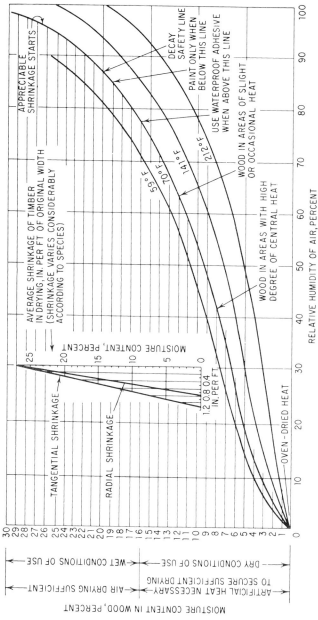

Fig. 7-1 Approximate relationship of wood equilibrium moisture content and temperature and relative humidity is shown by the curves. The triangular diagram indicates the effect of wood moisture content on shrinkage.

7-17

essentially flat grain, checks are so positioned in horizontally laminated members that they will not materially affect shear strength. When members are designed with laminations vertical (with wide face parallel to the direction of load application), and when checks may affect the shear strength, the effect of checks may be evaluated in the same manner as for checks in solid-sawn members.

Seasoning checks in bending members affect only the horizontal shear strength. They are usually not of structural importance unless the checks are significant in depth and occur in the midheight of the member near the support, and then only if shear governs the design of the members. The reduction in shear strength is nearly directly proportional to the ratio of depth of check to width of beam. Checks in columns are not of structural importance unless the check develops into a split, thereby increasing the slenderness ratio of the columns.

Minor checking may be disregarded, since there is ample factor of safety in allowable unit stresses. The final decision as to whether shrinkage checks are detrimental to the strength requirements of any particular design or structural member should be made by a competent engineer experienced in timber construction.

7-7. Allowable Unit Stresses and Modifications for Structural Lumber.

Articles 7-3 to 7-6 discuss conditions for modifications of basic stresses to determine allowable unit stresses for wood. A wide range of allowable unit stresses for many species is given in the "National Design Specification for Wood Construction" (NDS). They apply to solid-sawn lumber under normal duration of load and under continuously dry service conditions, such as those in most covered structures. For continuously wet-use service conditions, apply the appropriate factor from Table 7-7. NDS may be obtained from the National Forest Products Association, 1619 Massachusetts Ave., N.W., Washington, D.C. 20036.

The modifications for duration of load, temperature, treatment, and size are the same as for glued-laminated timber.

Stresses for Structural Glued-Laminated Timber. The allowable unit stresses given in the AITC "Timber Construction Manual" and "Standard Specification for Structural Glued-laminated Timber," AITC 117, are for normal conditions of loading and the assumptions of uniform loading, horizontal laminating, and a 12-in-deep member with a span-to-depth ratio of 21:1. Tables are given for dry-use and for wet-use conditions. AITC specifications may be obtained from the American Institute of Timber Construction, 333 W. Hampden Ave., Englewood, Colo. 80110.

Allowable unit stresses for dry-use conditions are applicable when the moisture content in service is less than 16%, as in most covered structures. Allowable unit stresses for wet-use conditions are applicable when the moisture content in service is 16% or more,

TABLE 7-7 Allowable-Stress Modification Factors for Moisture in Solid-Sawn Lumber

Service condition—moisture content exceeds 19%	Allowable unit stress					
	Extreme fiber in bending F_b	Tension parallel to grain F_t	Compression parallel to grain F_c	Compression perpendicular to grain $F_{c\perp}$	Horizontal shear F_v	Modulus of elasticity E
Lumber 2 to 4 in thick	0.86	0.84	0.70	0.67	0.97	0.97
Lumber 5 in or more thick	1.00	1.00	0.91	0.67	1.00	1.00

as may occur in exterior or submerged construction and in some structures housing wet processes or otherwise having constantly high relative humidities.

Allowable stresses for vertically laminated members made of combination grades of lumber are the weighted average of the lumber grades.

Requirements for limiting cross grain, type, and location of end joints, and certain manufacturing and other requirements, must be met for these allowable-unit-stress combinations to apply.

Species other than those referenced in these specifications may be used if allowable unit stresses are established for them in accordance with the provisions of U.S. Product Standard PS 56-73.

For laminated bending members 16¼ in or more in depth, the outermost tension-side laminations representing 5% of the total depth of the member should have additional grade restrictions, as described in AITC 117, "Standard Specification for Structural Glued-laminated Timber of Douglas Fir, Southern Pine, Western Larch, and California Redwood."

Use of hardwoods in glued-laminated timbers is covered by AITC 119, "Standard Specification for Hardwood Glued-laminated Timber."

Lumber that is E-rated, as well as visually graded and then positioned in laminated timber, is covered by AITC 120, "Standard Specification for Structural Glued-laminated Timber Using E-rated and Visually Graded Lumber of Douglas Fir, Southern Pine, Hemlock, and Lodgepole Pine." The lumber is sorted into E grades by measuring the modulus of elasticity of each piece and positioning the lumber in the member according to stiffness, but still grading visually so as to meet the visual grade requirements.

Modifications for Duration of Load. Wood can absorb overloads of considerable magnitude for short periods; thus, allowable unit stresses are adjusted accordingly. The elastic limit and ultimate strength are higher under short-time loading. Wood members under continuous loading for years will fail at loads one-half to three-fourths as great as would be required to produce failure in a static-bending test when the maximum load is reached in a few minutes.

Normal load duration contemplates fully stressing a member to the allowable unit stress by the application of the full design load for a duration of about 10 years (either continuously or cumulatively).

When a member is fully stressed by maximum design loads for long-term loading conditions (greater than 10 years, either continuously or cumulatively), the allowable unit stresses are reduced to 90% of the tabulated values.

When duration of full design load (either continuously or cumulatively) does not exceed 10 years, tabulated allowable unit stresses should be increased as follows: 15% for 2-month duration, as for snow; 25% for 7-day duration; 33⅓% for wind or earthquake; and 100% for impact. These increases are not cumulative. The allowable unit stress for normal loading may be used without regard to impact, if the stress induced by impact does not exceed the allowable unit stress for normal loading. These adjustments do not apply to modulus of elasticity, except when it is used to determine allowable unit loads for columns.

Modifications for Temperature. Tests show that wood increases in strength as temperature is lowered below normal. Tests conducted at about —300°F indicate that the important strength properties of dry wood in bending and compression, including stiffness and shock resistance, are much higher at extremely low temperatures.

Some reduction of the allowable unit stresses may be necessary for members subjected to elevated temperatures for repeated or prolonged periods. This adjustment is especially desirable where high temperature is associated with high moisture content.

Temperature effect on strength is immediate. Its magnitude depends on the moisture content of the wood and, when temperature is raised, on the duration of exposure.

Between 0 and 70°F, the static strength of dry wood (12% moisture content) roughly increases from its strength at 70°F about ⅓ to ½% for each 1°F decrease in temperature. Between 70 and 150°F, the strength decreases at about the same rate for each 1°F increase in temperature. The change is greater for higher wood moisture content.

After exposure to temperatures not much above normal for a short time under ordinary atmospheric conditions, the wood, when temperature is reduced to normal, may recover essentially all its original strength. Experiments indicate that air-dry wood can probably be exposed to temperatures up to nearly 150°F for a year or more without a significant permanent loss in most strength properties. But its strength while at such temperatures will be temporarily lower than at normal temperature.

When wood is exposed to temperatures of 150°F or more for extended periods of time, it will be permanently weakened. The nonrecoverable strength loss depends on a number of factors, including moisture content and temperature of the wood, heating medium, and time of exposure. To some extent, the loss depends on the species and size of the piece.

Glued-laminated members are normally cured at temperatures of less than 150°F. Therefore, no reduction in allowable unit stresses because of temperature effect is necessary for curing.

Adhesives used under standard specifications for structural glued-laminated members, for example, casein, resorcinol-resin, phenol-resin, and melamine-resin adhesives, are not affected substantially by temperatures up to those that char wood. Use of adhesives that deteriorate at low temperatures is not permitted by standard specifications for structural glued-laminated timber. Low temperatures appear to have no significant effect on the strength of glued joints.

Modifications for Pressure-Applied Treatments. The allowable stresses given for wood also apply to wood treated with a preservative when this treatment is in accordance with American Wood Preservers Association (AWPA) standard specifications. These limit pressure and temperature. Investigations have indicated that, in general, any weakening of timber as a result of preservative treatment is caused almost entirely by subjecting the wood to temperatures and pressures above the AWPA limits.

Highly acidic salts, such as zinc chloride, tend to hydrolyze wood if they are present in appreciable concentrations. Fortunately, the concentrations used in wood preservative treatments are sufficiently small that strength properties other than impact resistance are not greatly affected under normal service conditions. A significant loss in impact strength may occur, however, if higher concentrations are used.

None of the other common salt preservatives is likely to form solutions as highly acidic as those of zinc chloride. So, in most cases, their effect on the strength of the wood can be disregarded.

In wood treated with highly acidic salts, such as zinc chloride, moisture is the controlling factor in corrosion of fastenings. Therefore, wood treated with highly acidic salts is not recommended for use under highly humid conditions.

The effects on strength of all treatments, preservative and fire-retardant, should be investigated, to assure that adjustments in allowable unit stresses are made when required.

("Manual of Recommended Practice," American Wood Preservers Association.)

Modifications for Size Factor. When the depth of a rectangular beam exceeds 12 in, the tabulated unit stress in bending F_b should be reduced by multiplication by a size factor C_F.

$$C_F = \left(\frac{12}{d}\right)^{1/9} \tag{7-1}$$

where d = depth of member, in. Table 7-5 lists values of C_F for various cross-sectional sizes.

TABLE 7-8 Change in Size Factor C_F

Span-depth ratio	Change, %	Loading condition for simply supported beams	Change, %
7	+6.2	Single concentrated load	+7.8
14	+2.3	Uniform load	0
21	0	Third-point load	−3.2
28	−1.6		
35	−2.8		

Equation (7-1) applies to bending members satisfying the following basic assumptions: simply supported beam, uniformly distributed load, and span-depth ratio $L/d = 21$. C_F may thus be applied with reasonable accuracy to beams usually used in buildings. Where greater accuracy is required for other loading conditions or span-depth ratios, the percentage changes given in Table 7-8 may be applied directly to the size factor given by Eq. (7-1). Straight-line interpolation may be used for L/d ratios other than those listed in the table.

Modifications for Radial Tension or Compression. The radial stress induced by a bending moment in a member of constant cross section may be computed from

$$f_r = \frac{3M}{2Rbd} \qquad (7\text{-}2)$$

where M = bending moment, in-lb
R = radius of curvature at centerline of member, in
b = width of cross section, in
d = depth of cross section, in

Equation (7-2) can also be used to estimate the stresses in a member with varying cross section. Information on more exact procedures for calculating radial stresses in curved members with varying cross section can be obtained from the American Institute of Timber Construction.

When M is in the direction tending to decrease curvature (increase the radius), tensile stresses occur across the grain. For this condition, the allowable tensile stress across the grain is limited to one-third the allowable unit stress in horizontal shear for southern pine for all load conditions, and for Douglas fir and larch for wind or earthquake loadings. The limit is 15 psi for Douglas fir and larch for other types of loading. These values are subject to modification for duration of load. If these values are exceeded, mechanical reinforcement sufficient to resist all radial tensile stresses is required.

When M is in the direction tending to increase curvature (decrease the radius), the stress is compressive across the grain. For this condition, the allowable stress is limited to that for compression perpendicular to grain for all species covered by AITC 117.

Modifications for Curvature Factor. For the curved portion of members, the allowable unit stress in bending should be modified by multiplication by the following curvature factor:

$$C_c = 1 - 2000 \left(\frac{t}{R}\right)^2 \qquad (7\text{-}3)$$

where t = thickness of lamination, in
R = radius of curvature of lamination, in

t/R should not exceed $\frac{1}{100}$ for hardwoods and southern pine, or $\frac{1}{125}$ for softwoods other than southern pine. The curvature factor should not be applied to stress in the straight portion of an assembly, regardless of curvature elsewhere.

The recommended minimum radii of curvature for curved, structural glued-laminated members are 9 ft 4 in for ¾-in laminations, and 27 ft 6 in for 1½-in laminations. Other radii of curvature may be used with these thicknesses, and other radius-thickness combinations may be used.

Certain species can be bent to sharper radii, but the designer should determine the availability of such sharply curved members before specifying them.

Modifications for Lateral Stability. The tabulated allowable bending unit stresses are applicable to members that are adequately braced. When deep and slender members not adequately braced are used, allowable bending unit stresses must be reduced. For the purpose, a slenderness factor should be applied, as indicated in the AITC "Timber Construction Manual." (See also Art. 7-8.)

The reduction in bending stresses determined by applying the slenderness factor should not be combined with a reduction in stress because of the application of the size factor. In no case may the allowable bending unit stress used in design exceed the stress determined by applying a size factor.

7-8. Lateral Support of Wood Framing.

To prevent beams and compression members from buckling, they may have to be braced laterally. Need for such bracing, and required spacing, depend on the unsupported length and cross-sectional dimensions of members.

When buckling occurs, a member deflects in the direction of its least dimension b, unless prevented by bracing. (In a beam, b usually is taken as the width.) But if bracing precludes buckling in that direction, deflection can occur in the direction of the perpendicular dimension d. Thus, it is logical that unsupported length L, b, and d play important roles in rules for lateral support, or in formulas for reducing allowable stresses for buckling.

For glued-laminated beams, design for lateral stability is based on a function of Ld/b^2. For solid-sawn beams of rectangular cross section, maximum depth-width ratios should satisfy the approximate rules, based on nominal dimensions, summarized in Table 7-9. When the beams are adequately braced laterally, the depth of the member below the brace may be taken as the width. For glued-laminated arches, maximum depth-width ratios should satisfy the approximate rules, based on actual dimensions, in Table 7-10.

Where joists frame into arches or the compression chords of trusses and provide adequate lateral bracing, the depth, rather than the width, of arch or truss chord may be taken as b. The joists should be erected with upper edges at least ½ in above the supporting member, but low enough to provide adequate lateral support. The depth of arch or compression chord also may be taken as b, where joists or planks are placed on top of the arches or chords and securely fastened to them and blocking is firmly attached between the joists.

For glued-laminated beams, no lateral support is required when the depth does not exceed the width. In that case also, the allowable bending stress does not have to be adjusted for lateral instability. Similarly, if continuous support prevents lateral movement of the compression flange, lateral buckling cannot occur and the allowable stress need not be reduced.

When the depth of a glued-laminated beam exceeds the width, bracing must be provided at supports. This bracing must be so placed as to prevent rotation of the beam in a plane perpendicular to its longitudinal axis. And unless the compression flange is braced at sufficiently close intervals between the supports, the allowable stress should be adjusted for lateral buckling. All other modifications of the stresses, except for size factor, should be made.

TABLE 7-9 Approximate Lateral-Support Rules for Sawn Beams*

Depth-width ratio (nominal dimensions)	Rule
2 or less	No lateral support required
3 to 4	Hold ends in position with full-depth blocking, bridging, hangers, or other structural members
5	Hold ends in position and compression edge in line, e.g., with direct connection of sheathing, decking, or joists
6	Hold ends in position and compression edge in line, as for 5 to 1, and provide adequate bridging or blocking at intervals not exceeding 8 ft
7	Hold ends in position and both edges firmly in line

If a beam is subject to both flexure and compression parallel to grain, the ratio may be as much as 5:1 if one edge is held firmly in line, e.g., by rafters (or roof joists) and diagonal sheathing. If the dead load is sufficient to induce tension on the underside of the rafters, the ratio for the beam may be 6:1.

*From "National Design Specification for Wood Construction," National Forest Products Association.

When the buckling factor

$$C_s = \sqrt{\frac{L_e d}{b^2}} \tag{7-4}$$

where L_e is effective length, in, and b and d also are in inches, does not exceed 10, the allowable bending stress does not have to be adjusted for lateral buckling. Such beams are classified as short beams.

For intermediate beams, for which C_s exceeds 10 but is less than

$$C_k = \sqrt{\frac{3E}{5F_b}} \tag{7-5}$$

the allowable bending stress should be determined from

$$F'_b = F_b \left[1 - \frac{1}{3}\left(\frac{C_s}{C_k}\right)^4 \right] \tag{7-6}$$

where E = modulus of elasticity, psi

F_b = allowable bending stress, psi, adjusted, except for size factor

For long beams, for which C_s exceeds C_k but is less than 50, the allowable bending stress should be determined from

$$F'_b = \frac{0.40E}{(C_s)^2} \tag{7-7}$$

In no case should C_s exceed 50.

The effective length L_e for Eq. (7-4) is given in terms of unsupported length of beam in Table 7-11. Unsupported length is the distance between supports or the length of a

TABLE 7-10 Approximate Lateral-Support Rules for Glued-Laminated Arches

Depth-width ratio (Actual Dimensions)	Rule
5 or less	No lateral support required
6	Brace one edge at frequent intervals

TABLE 7-11 Ratio of Effective Length to Unsupported Length of Glued-Laminated Beams

Simple beam, load concentrated at center	1.61
Simple beam, uniformly distributed load	1.92
Simple beam, equal end moments	1.84
Cantilever beam, load concentrated at unsupported end	1.69
Cantilever beam, uniformly distributed load	1.06
Simple or cantilever beam, any load (conservative value)	1.92

cantilever when the beam is laterally braced at the supports to prevent rotation and adequate bracing is not installed elsewhere in the span. When both rotational and lateral displacement are also prevented at intermediate points, the unsupported length may be taken as the distance between points of lateral support. If the compression edge is supported throughout the length of the beam and adequate bracing is installed at the supports, the unsupported length is zero.

Acceptable methods of providing adequate bracing at supports include anchoring the bottom of a beam to a pilaster and the top of the beam to a parapet; for a wall-bearing roof beam, fastening the roof diaphragm to the supporting wall or installing a girt between beams at the top of the wall; for beams on wood columns, providing rod bracing.

For continuous lateral support of a compression flange, composite action is essential between deck elements, so that sheathing or deck acts as a diaphragm. One example is a plywood deck with edge nailing. With plank decking, nails attaching the plank to the beams must form couples, to resist rotation. In addition, the planks must be nailed to each other, for diaphragm action. Adequate lateral support is not provided when only one nail is used per plank and no nails between planks.

(American Institute of Timber Construction, "Timber Construction Manual," John Wiley & Sons, Inc., New York; "National Design Specification," National Forest Products Association; "Western Woods Use Book," Western Wood Products Association, 1500 Yeon Building, Portland Ore. 97204.)

7-9. Combined Stresses in Timber.

Allowable unit stresses given in the "National Design Specification for Wood Construction" apply directly to bending, horizontal shear, tension parallel or perpendicular to grain, and compression parallel or perpendicular to grain. For combined axial tension or compression in short columns ($L/d \leq 11$) P/A and bending stress M/S, stresses are limited by

$$\frac{P/A}{F} + \frac{M/S}{F_b} \leq 1 \tag{7-8a}$$

where F = allowable tension stress parallel to grain or allowable compression stress parallel to grain adjusted for unsupported length, psi
F_b = allowable bending stress, psi

For combined axial compression and bending in members with $L/d > 11$, JP/A should be subtracted from F_b in Eq. (7-8a), where

$$J = \frac{L/d - 11}{K - 11} \leq 1 \tag{7-8b}$$

and K is a factor that depends on column cross-sectional properties [see Eqs. (7-12) and (7-15)].

When a bending stress f_x, shear stress f_{xy}, and compression or tension stress f_y perpendicular to the grain exist simultaneously, the stresses should satisfy the Norris interaction formula:

$$\frac{f_x^2}{F_x^2} + \frac{f_y^2}{F_y^2} + \frac{f_{xy}^2}{F_{xy}^2} \leq 1 \tag{7-9}$$

where F_x = allowable bending stress F_b modified for duration of loading but not for size factor
F_y = allowable stress in compression or tension perpendicular to grain $F_{c\perp}$ or $F_{b\perp}$ modified for duration of loading
F_{xy} = allowable horizontal shear stress F_v modified for duration of loading

The usual design formulas do not apply to sharply curved glued-laminated beams. Nor do they hold where laminations of more than one species, each with markedly different modulus of elasticity, comprise the same member.

7-10. Effect of Shrinkage or Swelling on Shape of Curved Members.

Wood shrinks or swells across the grain, but has practically no dimensional change along the grain. Radial swelling causes a decrease in the angle between the ends of a curved member; radial shrinkage causes an increase in this angle.

Such effects may be of great importance in three-hinged arches that become horizontal, or nearly so, at the crest of a roof. Shrinkage, increasing the relative end rotations, may cause a depression at the crest and create drainage problems. For such arches, therefore, consideration must be given to moisture content of the member at time of fabrication and in service, and to the change in end angles that results from change in moisture content and shrinkage across the grain.

7-11. Fabrication of Structural Timber.

Fabrication consists of boring, cutting, sawing, trimming, dapping, routing, planing, and otherwise shaping, framing, and finishing wood units, sawn or laminated, including plywood, to fit them for particular places in a final structure. Whether fabrication is performed in shop or field, the product must exhibit a high quality of workmanship.

Jigs, patterns, templates, stops, or other suitable means should be used for all complicated and multiple assemblies to ensure accuracy, uniformity, and control of all dimensions. All tolerances in cutting, drilling, and framing must comply with good practice and applicable specifications and controls. At the time of fabrication, tolerances must not exceed those listed below, unless they are not critical and not required for proper performance. Specific jobs, however, may require closer tolerances.

Location of Fastenings. Spacing and location of all fastenings within a joint should be in accordance with the shop drawings and specifications, with a maximum permissible tolerance of $\pm \frac{1}{16}$ in. The fabrication of members assembled at any joint should be such that the fastenings are properly fitted.

Bolt-Hole Sizes. Bolt holes in all fabricated structural timber, when loaded as a structural joint, should be $\frac{1}{16}$ in larger in diameter than bolt diameter for $\frac{1}{2}$-in and larger-diameter bolts, and $\frac{1}{32}$ in larger for smaller-diameter bolts. Larger clearances may be required for other bolts, such as anchor bolts and tension rods.

Holes and Grooves. Holes for stress-carrying bolts, connector grooves, and connector daps must be smooth and true within $\frac{1}{16}$ in per 12 in of depth. The width of a split-ring connector groove should be within $+0.02$ in of and not less than the thickness of the corresponding cross section of the ring. The shape of ring grooves must conform generally to the cross-sectional shape of the ring. Departure from these requirements may be allowed when supported by test data. Drills and other cutting tools should be set to conform to the size, shape, and depth of holes, grooves, daps, etc., specified in the "National Design Specification for Wood Construction," National Forest Products Association.

Lengths. Members should be cut within $\pm \frac{1}{16}$ in of the indicated dimension when they are up to 20 ft long, and $\pm \frac{1}{16}$ in per 20 ft of specified length when they are over 20 ft long. Where length dimensions are not specified or critical, these tolerances may be waived.

End Cuts. Unless otherwise specified, all trimmed square ends should be square within $\frac{1}{16}$ in per foot of depth and width. Square or sloped ends to be loaded in compression should be cut to provide contact over substantially the complete surface.

7-12. Fabrication of Glued-Laminated Timbers.

Structural glued-laminated timber is made by bonding layers of lumber together with adhesive so that the grain direction of all laminations is essentially parallel. Narrow boards may be edge-glued, and short boards may be end-glued. The resultant wide and long laminations are then face-glued into large, shop-grown timbers.

Recommended practice calls for lumber of nominal 1- and 2-in thicknesses for laminating. The lumber is dressed to ¾- and 1½-in thicknesses, respectively, just before gluing. The thinner laminations are generally used in curved members.

Constant-depth members normally are a multiple of the thickness of the lamination stock used. Variable-depth members, because of tapering or special assembly techniques, may not be exact multiples of these lamination thicknesses.

Nominal width of stock, in	3	4	6	8	10	12	14	16
Standard member finished width, in	2 1/4	3 1/8	5 1/8	6 3/4	8 3/4	10 3/4	12 1/4	14 1/4

Standard widths are most economical, since they represent the maximum width of board normally obtained from the stock used in laminating.

When members wider than the stock available are required, laminations may consist of two boards side by side. These edge joints must be staggered, vertically in horizontally laminated beams (load acting normal to wide faces of laminations), and horizontally in vertically laminated beams (load acting normal to the edge of laminations). In horizontally laminated beams, edge joints need not be edge-glued. Edge gluing is required in vertically laminated beams.

Edge and face gluings are the simplest to make, end gluings the most difficult. Ends are also the most difficult surfaces to machine into scarfs or finger joints.

A plane sloping scarf (Fig. 7-2), in which the tapered surfaces of laminations are glued together, can develop 85 to 90% of the strength of an unscarfed, clear, straight-grained control specimen. Finger joints (Fig. 7-3) are less wasteful of lumber. Quality can be adequately controlled in machine cutting and in high-frequency gluing. A combination of thin tip, flat slope on the side of the individual fingers, and a narrow pitch is desired.

The length of fingers should be kept short for savings of lumber, but long for maximum strength.

The usefulness of structural glued-laminated timbers is determined by the lumber used and glue joint produced. Certain com-

Fig. 7-2 Plane sloping scarf.

Fig. 7-3 Finger joints: (a) fingers formed by cuts perpendicular to the wide face of the board; (b) fingers formed by cuts perpendicular to the edges.

binations of adhesive, treatment, and wood species do not produce the same quality of glue bond as other combinations, although the same gluing procedures are used. Thus, a combination must be supported by adequate experience with a laminator's gluing procedure.

The only adhesives currently recommended for wet-use and preservative-treated lumber, whether gluing is done before or after treatment, are resorcinol and phenol-resorcinol resins. The prime adhesive for dry-use structural laminating is casein. Melamine and melamine-urea blends are used in smaller amounts for high-frequency curing of end gluings.

Glued joints are cured with heat by several methods. R.F. (high-frequency) curing of glue lines is used for end joints and for limited-size members where there are repetitive gluings of the same cross section. Low-voltage resistance heating, where current is passed through a strip of metal to raise the temperature of a glue line, is used for attaching thin facing pieces. The metal may be left in the glue line as an integral part of the completed member. Printed electric circuits, in conjunction with adhesive films, and adhesive films impregnated on paper or on each side of a metal conductor placed in the glue line are other alternatives.

Preheating the wood to ensure reactivity of the applied adhesive has limited application in structural laminating. The method requires adhesive application as a wet or dry film simultaneously to all laminations, and then rapid handling of multiple laminations.

Curing the adhesive at room temperature has many advantages. Since wood is an excellent insulator, a long time is required for elevated ambient temperature to reach inner glue lines of a large assembly. With room-temperature curing, equipment needed to heat the glue line is not required, and the possibility of injury to the wood from high temperature is avoided.

7-13. Preservative Treatments for Wood.

Wood-destroying fungi must have air, suitable moisture, and favorable temperatures to develop and grow in wood. Submerge wood permanently and totally in water, or keep the moisture content below 18 to 20%, or hold temperatures below 40°F or above 110°F, and wood remains permanently sound. If wood moisture content is kept below the fiber-saturation point (25 to 30%) when the wood is untreated, decay is greatly retarded. Below 18 to 20% moisture content, decay is completely inhibited.

If wood cannot be kept dry, a wood preservative, properly applied, must be used.

Wood members are permanent without treatment if located in enclosed buildings where good roof coverage, proper roof maintenance, good joint details, adequate flashing, good ventilation, and a well-drained site assure moisture content of the wood continuously below 20%. Also, in arid or semiarid regions where climatic conditions are such that the equilibrium moisture content seldom exceeds 20%, and then only for short periods, wood members are permanent without treatment.

Where wood is in contact with the ground or with water, where there is air and the wood may be alternately wet and dry, a preservative treatment, applied by a pressure process, is necessary to obtain an adequate service life. In enclosed buildings where moisture given off by wet-process operations maintains equilibrium moisture contents in the wood above 20%, wood structural members must be preservatively treated. So must wood exposed outdoors without protective roof covering and where the wood moisture content can go above 18 to 20% for repeated or prolonged periods.

Where wood structural members are subject to condensation by being in contact with masonry, preservative treatment is necessary.

To obtain preservatively treated glued-laminated timber, lumber may be treated before gluing, and the members then glued to the desired size and shape. Or the already-glued and machined members may be subjected to certain treatments. When laminated members do not lend themselves to treatment because of size and shape, gluing of treated laminations is the only method of producing adequately treated members.

There are problems in gluing some treated woods. Certain combinations of adhesive, treatment, and wood species are compatible; other combinations are not. All adhesives of the same type do not produce bonds of equal quality for a particular wood species and preservative. The bonding of treated wood is dependent on the concentration of preservative on the surface at the time of gluing, and on the chemical effects of the preservative on the adhesive. In general, longer curing times or higher curing temperatures are needed for treated than for untreated wood to obtain comparable adhesive bonds. Assembly times for the two types of wood are different as well.

Each type of preservative and method of treatment has certain advantages. The preservative to be used depends on the service expected of the member for the specific conditions of exposure. The minimum retentions shown in Table 7-12 may be increased where severe climatic or exposure conditions are involved.

Creosote and creosote solutions have low volatility. They are practically insoluble in water, and thus are most suitable for severe exposure, contact with ground or water, and where painting is not a requirement or a creosote odor is not objectionable.

Oil-borne chemicals are organic compounds dissolved in a suitable petroleum carrier oil, and are suitable for outdoor exposure or where leaching may be a factor, or where painting is not required. Depending on the type of oil used, they may result in relatively

TABLE 7-12 Recommended Minimum Retentions of Preservatives, lb/ft³*

Preservatives	Sawn and laminated timbers		Laminations		Sawn and laminated timbers		Laminations	
	Western woods†	Southern pine	Western woods†	Southern pine	Western woods†	Southern pine	Western woods†	Southern pine
	Ground contact				Above ground			
Creosote or creosote solutions:								
Creosote	12	12	12	12	6	6	6	6
Creosote—coal-tar solution	12	12	NR‡	12	6	6	NR‡	6
Creosote-petroleum solution	12	NR‡	12	NR‡	6	NR‡	6	NR‡
Oil-borne chemicals:								
Pentachlorophenol (5% in specified petroleum oil)	0.6	0.6	0.6	0.6	0.3§	0.3	0.3	0.3
Penta (water-repellent, moderate decay hazard)	NR‡	NR‡	NR‡	NR‡	0.2	0.2	0.25	0.25
Water-borne inorganic salts:								
Chromated zinc arsenate (Boliden salt)	NR‡	NR‡	NR‡	NR‡	0.50	0.50	0.45	0.45
Acid copper chromate (Celcure)	NR‡	NR‡	0.50	0.50	0.25	0.25	0.25	0.25
Ammoniacal copper arsenite (Chemonite)	0.40	0.40	0.40	0.40	0.25	0.25	0.25	0.25
Chromated zinc chloride	NR‡	NR‡	NR‡	NR‡	0.50	0.50	0.50	0.50
Copperized chromated zinc arsenate	NR‡	NR‡	NR‡	NR‡	0.45	0.45	0.45	0.45
Chromated copper arsenate (Greensalt, Erdalith)	0.60	0.60	0.60	0.60	0.25	0.25	0.25	0.25
Fluor chrome arsenate phenol (Tanalith, Wolman salt, Osmossar, Osmosalt)	NR‡	NR‡	NR‡	NR‡	0.25	0.25	0.25	0.25

*See latest edition of AITC 109 Treating Standard for Structural Timber Framing, American Institute of Timber Construction or AWPA Standards C2 and C28, American Wood Preservers Association.
†Douglas fir, western hemlock, western larch.
‡NR—not recommended
§0.5 lb for timber less than 5 in thick.

7-29

clean surfaces. While there is a slight odor from such treatment, it is usually not objectionable.

Water-borne inorganic salts are dissolved in water or aqua ammonia, which evaporates after treatment and leaves the chemicals in the wood. The strength of solutions varies to provide net retention of dry salt required. These salts are suitable where clean and odorless surfaces are required. The surfaces are paintable after proper seasoning.

When treating before gluing is required, water-borne salts, or oil-borne chemicals in mineral spirits, or AWPA P9 volatile solvent are recommended. When treatment before gluing is not required or desired, creosote, creosote solutions, or oil-borne chemicals are recommended.

7-14. Resistance of Wood to Chemical Attack.

Wood is superior to many building materials in resistance to mild acids, particularly at ordinary temperatures. It has excellent resistance to most organic acids, notably acetic. However, wood is seldom used in contact with solutions that are more than weakly alkaline. Oxidizing chemicals and solutions of iron salts, in combination with damp conditions, should be avoided.

Wood is composed of roughly 50 to 70% cellulose, 25 to 30% lignin, and 5% extractives with less than 2% protein. Acids such as acetic, formic, lactic, and boric do not ionize sufficiently at room temperature to attack cellulose, and thus do not harm wood.

When the pH of aqueous solutions of weak acids is 2 or more, the rate of hydrolysis of cellulose is small and dependent on the temperature. A rough approximation of this temperature effect is that, for every 20°F increase, the rate of hydrolysis doubles. Acids with pH values above 2, or bases with pH below 10, have little weakening effect on wood at room temperature, if the duration of exposure is moderate.

7-15. Designing for Fire Safety.

Maximum protection of the occupants of a building and of the property itself can be achieved by taking advantage of the fire-endurance properties of wood in large cross sections, and by close attention to details that make a building fire-safe. Building materials alone, building features alone, or detection and fire-extinguishing equipment alone cannot provide maximum safety from fire in buildings. A proper combination of these three will provide the necessary degree of protection for the occupants and the property.

The following should be investigated:

Degree of protection needed, as dictated by the occupancy or operations taking place in the building.

Number, size, type (such as direct to the outside), accessibility of exitways (particularly stairways), and their distance from each other.

Installation of automatic alarm and sprinkler systems.

Separation of areas in which hazardous processes or operations take place, such as boiler rooms and workshops.

Enclosure of stairwells and use of self-closing fire doors.

Fire stopping and elimination, or proper protection, of concealed spaces.

Interior finishes to assure surfaces that will not spread flame at hazardous rates.

Roof venting equipment or provision of draft curtains where walls might interfere with production operations.

When exposed to fire, wood forms a self-insulating surface layer of char, which provides its own fire protection. Even though the surface chars, the undamaged wood beneath retains its strength and will support loads in accordance with the capacity of the uncharred section. Heavy-timber members have often retained their structural integrity through long periods of fire exposure and remained serviceable after refinishing of the charred surfaces. This fire endurance and excellent performance of heavy timber are attributable to the size of the wood members, and to the slow rate at which the charring penetrates.

The structural framing of a building, which is the criterion for classifying a building as combustible or noncombustible, has little to do with the hazard from fire to the building occupants. Most fires start in the building contents and create conditions that render the inside of the structure uninhabitable long before the structural framing becomes involved in the fire. Thus, whether the building is of combustible or noncombustible classification has little bearing on the potential hazard to the occupants. However, once the fire starts in the contents, the material of which the building is constructed can be of significant help in facilitating evacuation, fire fighting, and property protection.

The most important protection factors for occupants, firemen fighting the fire, and the property, as well as adjacent exposed property, are prompt detection of the fire, immediate alarm, and rapid extinguishment of the fire. Firemen do not fear fire fighting in buildings of heavy-timber construction as in buildings of many other types of construction. They need not fear sudden collapse without warning; they usually have adequate time, because of the slow-burning characteristics of the timber, to ventilate the building and fight the fire from within the building or on top.

With size of member of particular importance to fire endurance of wood members, building codes specify minimum dimensions for structural members, and classify buildings with wood framing as heavy-timber construction, ordinary construction, or wood-frame construction.

Heavy-timber construction is the type in which fire resistance is attained by placing limitations on the minimum size, thickness, or composition of all load-carrying wood members; by avoidance of concealed spaces under floors and roofs; by use of approved fastenings, construction details, and adhesives; and by providing the required degree of fire resistance in exterior and interior walls. (See AITC 108, Heavy Timber Construction, American Institute of Timber Construction.)

Ordinary construction has exterior masonry walls and wood-framing members of sizes smaller than heavy-timber sizes.

Wood-frame construction has wood-framed walls and structural framing of sizes smaller than heavy-timber sizes.

Depending on the occupancy of a building or hazard of operations within it, a building of frame or ordinary construction may have its members covered with fire-resistive coverings. The interior finish on exposed surfaces of rooms, corridors, and stairways is important from the standpoint of its tendency to ignite, flame, and spread fire from one location to another. The fact that wood is combustible does not mean that it will spread flame at a hazardous rate. Most codes exclude the exposed wood surfaces of heavy-timber structural members from flame-spread requirements, because such wood is difficult to ignite and, even with an external source of heat, such as burning contents, is resistant to spread of flame.

Fire-retardant chemicals may be impregnated in wood with recommended retentions to lower the rate of surface flame spread and make the wood self-extinguishing if the external source of heat is removed. After proper surface preparation, the surface is paintable. Such treatments are accepted under several specifications, including Federal and military. They are recommended only for interior or dry-use service conditions or locations protected against leaching. These treatments are sometimes used to meet a specific flame-spread rating for interior finish, or as an alternate to noncombustible secondary members and decking meeting the requirements of Underwriters Laboratories, Inc., NM 501 or NM 502, nonmetallic roof-deck assemblies in otherwise heavy-timber construction.

7-16. Mechanical Fastenings.

Nails, spikes, screws, lags, bolts, and timber connectors, such as shear plates and split rings, are used for connections in wood construction. Joint-design data have been established by experience and tests, because determination of stress distribution in wood and metal fasteners is complicated.

Allowable loads or stresses and methods of design for bolts, connectors, and other fasteners used in one-piece sawn members also are applicable to laminated members.

Problems can arise, however, if a deep-arch base section is bolted to the shoe attached to the foundation by widely separated bolts. A decrease in wood moisture content and shrinkage will set up considerable tensile stress perpendicular to the grain, and splitting may occur. If the moisture content at erection is the same as that to be reached in service, or if the bolt holes in the shoe are slotted to permit bolt movement, the tendency to split will be reduced.

Fasteners subject to corrosion or chemical attack should be protected by painting, galvanizing, or plating. In highly corrosive atmospheres, such as in chemical plants, metal fasteners and connections should be galvanized or made of stainless steel. Consideration may be given to covering connections with hot tar or pitch. In such extreme conditions, lumber should be at or below equilibrium moisture content at fabrication to reduce shrinkage. Such shrinkage could open avenues of attack for the corrosive atmosphere.

Iron salts are frequently very acidic and show hydrolytic action on wood in the presence of free water. This accounts for softening and discoloration of wood observed around corroded nails. This action is especially pronounced in acidic woods, such as oak, and in woods containing considerable tannin and related compounds, such as redwood. It can be eliminated, however, by using zinc-coated, aluminum, or copper nails.

Nails and Spikes. Hardened deformed-shank nails and spikes are made of high-carbon-steel wire and are headed, pointed, annularly or helically threaded, and heat-treated and tempered, to provide greater strength than common wire nails and spikes. But the same loads as given for common wire nails and spikes or the corresponding

TABLE 7-13 Nail and Spike Dimensions

Nails			Spikes		
Penny-weight	Length, in.	Wire diam, in.	Penny-weight	Length, in.	Wire diam, in.
6d	2	0.113	10d	3	0.192
8d	2½	0.131	12d	3¼	0.192
10d	3	0.148	16d	3½	0.207
12d	3¼	0.148	20d	4	0.225
16d	3½	0.162	30d	4½	0.244
20d	4	0.192	40d	5	0.263
30d	4½	0.207	50d	5½	0.283
40d	5	0.225	60d	6	0.283
50d	5½	0.244	5/16	7	0.312
60d	6	0.263	3/8	8½	0.375

lengths are used with a few exceptions. The common nail and spike dimensions are shown in Table 7-13.

Nails should not be driven closer together than half their length, unless driven in prebored holes. Nor should nails be closer to an edge than one-quarter their length. When one structural member is joined to another, penetration of nails into the second or farther timber should be at least half the length of the nails. Holes for nails, when necessary to prevent splitting, should be bored with a diameter less than that of the nail. If this is done, the same allowable load as for the same-size fastener with a bored hole applies in both withdrawal and lateral resistance.

Nails or spikes should not be loaded in withdrawal from the end grain of wood. Also, nails inserted parallel to the grain should not be used to resist tensile stresses parallel to the grain.

Safe lateral loads and safe resistance to withdrawal of common wire nails are given in Table 7-14, and of spikes in Table 7-15. The tables provide a safety factor of about 6 against failure. Joint slippage would be objectionable long before ultimate load is reached.

For lateral loads, if a nail or spike penetrates the piece receiving the point for a distance less than 11 times the nail or spike diameter, the allowable load is determined by straight-line interpolation between full load at 11 diameters, as given in the tables, and zero load for zero penetration. But the minimum penetration in the piece receiving the point must be at least 3⅜ diameters. Allowable lateral loads for nails or spikes driven into end grain are two-thirds the tabulated values. If a nail or spike is driven into unseasoned wood that will remain wet or will be loaded before seasoning, the allowable load is 75% of the tabulated values, except that for hardened, deformed-shank nails the full load may be used. Where properly designed metal side plates are used, the tabulated allowable loads may be increased 25%.

TABLE 7-14 Strength of Wire Nails

Size of nail, pennyweight	6d	8d	10d	12d	16d	20d	30d	40d	50d	60d
Length of nail, in	2	2 1/2	3	3 1/4	3 1/2	4	4 1/2	5	5 1/2	6
Safe lateral strength, lb (inserted perpendicular to the grain of the wood, penetrating 11 diameters), Douglas fir or southern pine	63	78	94	94	108	139	155	176	199	223
Safe resistance to withdrawal, lb/lin in of penetration into the member receiving the point (inserted perpendicular to the grain of the wood):										
Douglas fir (dense)	29	34	38	38	42	49	53	58	63	67
Southern pine (dense)	35	41	46	46	50	59	64	70	76	81

TABLE 7-15 Strength of Spikes

Size of spike, pennyweight	10d	12d	16d	20d	30d	40d	50d	60d	5/16 in.	3/8 in.
Length of spike, in	3	3 1/2	3 1/2	4	4 1/2	5	5 1/2	6	7	8 1/2
Safe lateral strength, lb (inserted perpendicular to the grain of the wood, penetrating 11 diameters), Douglas fir and southern pine	139	139	155	176	199	223	248	248	288	37 9
Safe resistance to withdrawal, lb/lin in of penetration into the member receiving the point (inserted perpendicular to the grain of the wood):										
Douglas fir (dense)	49	49	53	58	63	67	73	73	80	96
Southern pine (dense)	59	59	64	70	76	81	88	88	97	116

Wood Screws. The common types of wood screws have flat, oval, or round heads. The flat-head screw is commonly used if a flush surface is desired. Oval and round-headed screws are used for appearance or when countersinking is objectionable.

Wood screws should not be loaded in withdrawal from end grain. They should be inserted perpendicular to the grain by turning into predrilled holes, and should not be started or driven with a hammer. Spacings, end distances, and side distances must be such as to prevent splitting.

Table 7-16 gives the allowable loads for lateral resistance at any angle of load to grain, when the wood screw is inserted perpendicular to the grain (into the side grain of the main member) and a wood side piece is used. Embedment must be seven times the shank diameter into the member holding the point. For less penetration, reduce loads in proportion; penetration, however, must not be less than 4 times the shank diameter.

Table 7-16 also gives allowable withdrawal loads, lb/in, of penetration of the threaded portion of a screw into the member holding the point when the screw is inserted perpendicular to the grain of the wood.

When metal side plates rather than wood side pieces are used, the allowable lateral load, at any angle of load to grain, may be increased 25%. Allowable lateral loads when loads act perpendicular to the grain and the screw is inserted into end grain are two-thirds of those shown.

For Douglas fir and southern pine, the lead hole for a screw loaded in withdrawal should have a diameter of about 70% of the root diameter of the screw. For lateral resistance, the part of the hole receiving the shank should be about seven-eighths the diameter of the shank, and that for the threaded portion should be about seven-eighths the diameter of the screw at the root of the thread.

The loads given in Table 7-16 are for wood screws in seasoned lumber, for joints used indoors or in a location always dry. When joints are exposed to the weather, use 75%, and where joints are always wet, 67% of the tabulated loads. For lumber pressure-impregnated with fire-retardant chemicals, use 75% of the tabulated loads.

Lag Screws or Lag Bolts. Lag screws are commonly used because of their convenience, particularly where it would be difficult to fasten a bolt or where a nut on the surface would be objectionable. They range, usually, from about 0.2 to 1.0 in in diameter, and from 1 to 16 in in length. The threaded portion ranges from ¾ in for 1- and 1¼-in-long lag screws to half the length for all lengths greater than 10 in.

Lag screws, like wood screws, require prebored holes of the proper size. The lead hole for the shank should have the same diameter as the shank. The lead-hole diameter for the threaded portion varies with the density of the wood species. For Douglas fir and southern pine, the hole for the threaded portion should be 60 to 75% of the shank diameter. The smaller percentage applies to lag screws of smaller diameters. Lead holes slightly larger than those recommended for maximum efficiency should be used with lag screws of excessive length.

In determining withdrawal resistance, the allowable tensile strength of the lag screw at the net or root section should not be exceeded. Penetration of the threaded portion to a distance of about 7 times the shank diameter in the denser species and 10 to 12 times the shank diameter in the less dense species will develop approximately the ultimate tensile strength of a lag screw.

The resistance of a lag screw to withdrawal from end grain is about three-quarters that from side grain.

Table 7-17 gives the allowable normal-duration lateral and withdrawal loads for Douglas fir and southern pine. When lag screws are used with metal plates, the allowable lateral loads parallel to the grain are 25% higher than with wood side plates. No increase is allowed for load applied perpendicular to the grain.

Lag screws should preferably not be driven into end grain, because splitting may develop under lateral load. If lag screws are so used, however, the allowable loads should be taken as two-thirds the lateral resistance of lag screws in side grain with loads acting perpendicular to the grain.

Spacings, edge and end distances, and net section for lag-screw joints should be the same as those for joints with bolts of a diameter equal to the shank diameter of the lag screw.

For more than one lag screw the total allowable load equals the sum of the loads permitted for each lag screw, provided that spacings, end distances, and edge distances are sufficient to develop the full strength of each lag screw.

The allowable loads in Table 7-17 are for lag screws in lumber seasoned to a moisture content about equal to that which it will eventually have in service. For lumber installed unseasoned and seasoned in place, the full allowable lag-screw loads may be used for a joint having a single lag screw and loaded parallel or perpendicular to grain; or a single row of lag screws loaded parallel to grain; or multiple rows of lag screws loaded parallel to grain with separate splice plates for each row. For other types of lag-screw joints, the allowable loads are 40% of the tabulated loads. For lumber partly seasoned when fabricated, proportionate intermediate loads between 100 and 40% may be used.

TABLE 7-16 Strength of Wood Screws

Gage of screw	6	7	8	9	10	12	14	16	18	20	24
Diam of screw, in	0.138	0.151	0.164	0.177	0.190	0.216	0.242	0.268	0.294	0.320	0.372
Allowable lateral load, lb (normal duration), Douglas fir or southern pine	75	90	106	124	143	185	232	284	342	406	548
Allowable withdrawal load, lb/in of penetration (normal duration):											
Douglas fir	102	112	121	131	140	160	179	199	217	236	275
Southern pine	119	130	141	152	163	186	208	230	253	275	320

TABLE 7-17 Strength of Lag Screws

Diam of lag screw, in	1/4		5/16		3/8		7/16		1/2		9/16		5/8		3/4		7/8		1		1 1/8		1 1/4	
Allowable withdrawal load, lb/in, of penetration of the threaded part into the member holding the point (normal duration):																								
Douglas fir	232		274		314		353		390		426		461		528		593		656		716		775	
Southern pine	260		307		352		385		437		477		516		592		664		734		802		868	

Allowable lateral load, lb per lag screw, in single shear with 1/2-in metal side plates (Douglas fir or southern pine) for normal duration

Length of lag screw in main members, in	1/4 Par-allel to grain	1/4 Per-pen-dic-ular to grain	5/16 Par-allel to grain	5/16 Per-pen-dic-ular to grain	3/8 Par-allel to grain	3/8 Per-pen-dic-ular to grain	7/16 Par-allel to grain	7/16 Per-pen-dic-ular to grain	1/2 Par-allel to grain	1/2 Per-pen-dic-ular to grain	9/16 Par-allel to grain	9/16 Per-pen-dic-ular to grain	5/8 Par-allel to grain	5/8 Per-pen-dic-ular to grain	3/4 Par-allel to grain	3/4 Per-pen-dic-ular to grain	7/8 Par-allel to grain	7/8 Per-pen-dic-ular to grain	1 Par-allel to grain	1 Per-pen-dic-ular to grain	1 1/8 Par-allel to grain	1 1/8 Per-pen-dic-ular to grain	1 1/4 Par-allel to grain	1 1/4 Per-pen-dic-ular to grain
3	210	160																						
4	235	185	265	180	320	195																		
5			355	240	480	290	370	210	415	215														
6			375	255	535	325	575	320	625	325	455	225	490	235										
7			400	270	545	330	710	405	850	440	680	340	1005	355	525									
8					555	340	735	415	945	490	930	460	1250	480	650									
9							750	425	970	505	1095	540	1460	600	810									
10							760	430	985	510	1210	600	1480	700	935	1190								
11									990	515	1240	615	1500	720	950	1480	1035	2490						
12											1250	620	1510	725	965	1840	1200	2880	1485	3710				
13													1540	740	970	2130	1230	2960	1550	3880	1960	4900		
14																2160	1240	2990	1560	3900	1970	4920		
15																2190	1250	3000	1570	3930	1980	4950	6060	2420
16																2220	1260	3030	1580	3960	1990	4980	6110	2450
																			1580	3960	2000	5000	6150	2460

For lumber pressure-impregnated with fire-retardant chemicals, allowable loads for lag screws are the same as those for unseasoned lumber. Where joints are to be exposed to weather, use 75%, and where joints are always wet, 67% of the tabulated allowable loads.

Bolts. Standard machine bolts with square heads and nuts are used extensively in wood construction. Spiral-shaped dowels are also used at times to hold two pieces of wood together; they are used to resist checking and splitting in railroad ties and other solid-sawn timbers.

Holes for bolts should always be prebored and should have a diameter that permits the bolt to be driven easily ($\frac{1}{32}$ to $\frac{1}{16}$ in larger than the bolt). Careful centering of holes in main members and splice plates is necessary.

Center-to-center distance along the grain between bolts acting parallel to the grain should be at least 4 times the bolt diameter. For a tension joint, distance from end of wood to center of nearest bolt should be at least 7 times the bolt diameter for softwoods, and 5 times for hardwoods. For compression joints, the end distance should be at least 4 times the bolt diameter for both softwoods and hardwoods. If closer spacings are used, the loads allowed should be reduced proportionately.

Also for bolts acting parallel to the grain, the distance from the center of a bolt to the edge of the wood should be at least $1\frac{1}{2}$ times the bolt diameter. Usually, however, the edge distance is set at half the distance between bolt rows. In any event, the area at the critical section through the joint must be sufficient to keep unit stresses in the wood within the allowable.

The critical section is that section at right angles to the direction of the load that gives maximum stress in the member over the net area remaining after bolt holes at the section are deducted. For parallel-to-grain loads, the net area at a critical section should be at least 100% for hardwoods, and 80% for softwoods, of the total area in bearing under all the bolts in the joint.

For parallel- or perpendicular-to-grain loads, spacing between rows paralleling a member should not exceed 5 in unless separate splice plates are used for each row.

For bolts bearing perpendicular to the grain, center-to-center spacing across the grain should be at least 4 times the bolt diameter if wood side plates are used. But if the design load is less than the bolt-bearing capacity of the plates, the spacing may be reduced proportionately. For metal side plates, the spacing only need be sufficient to permit tightening of the nuts. Distance from end of wood to center of bolt should be at least four diameters. So should the distance between center of bolt and edge toward which the load is acting. The edge distance at the opposite edge is relatively unimportant.

A load applied to only one end of a bolt perpendicular to its axis may be taken as half the symmetrical two-end load.

Allowable bolt loads parallel and perpendicular to the grain are the same for dense Douglas fir and southern pine and are given in Table 7-18.

These basic loads are for permanent loads. For short-duration loads, basic loads may be modified as noted in Art. 7-7.

Groups of Bolts. When bolts are properly spaced and aligned, the allowable load on a group of bolts may be taken as the sum of the individual load capacities.

Timber Connectors. These are metal devices used with bolts for producing joints with fewer bolts without reduction in strength. Several types of connectors are available. Usually, they are either steel rings that are placed in grooves in adjoining members to prevent relative movement or metal plates embedded in the faces of adjoining timbers. The bolts are used with these connectors to prevent the timbers from separating. The load is transmitted across the joint through the connectors.

Split rings (Fig. 7-4a) are the most efficient device for joining wood to wood. They are placed in circular grooves cut by a hand tool in the contact surfaces. About half the depth

TABLE 7-18 Allowable Load, lb, on One Bolt Loaded in Double Shear and Bearing on Dense Douglas Fir or Southern Pine*

Diameter of bolt, in D	Length of bolt l, in, in main member											
	1 1/2		2		2 1/2		3		3 1/2		4	
	Parallel to grain P	Perpendicular to grain Q	Parallel to grain P	Perpendicular to grain Q	Parallel to grain P	Perpendicular to grain Q	Parallel to grain P	Perpendicular to grain Q	Parallel to grain P	Perpendicular to grain Q	Parallel to grain P	Perpendicular to grain Q
1/2	1100	500	1370	670	1480	840	1490	1010	1490	1140	1490	1180
5/8	1380	570	1820	760	2140	950	2290	1140	2320	1330	2330	1510
3/4	1660	630	2210	840	2710	1060	3080	1270	3280	1480	3340	1690
7/8	1940	700	2580	930	3210	1160	3770	1390	4190	1630	4450	1860
1	2220	760	2960	1010	3680	1270	4390	1520	5000	1770	5470	2030
1 1/4												
1 1/2												

Diameter of bolt, in D	Length of bolt l, in, in main member											
	4 1/2		5 1/2		7 1/2		9 1/2		11 1/2		13 1/2	
	Parallel to grain P	Perpendicular to grain Q	Parallel to grain P	Perpendicular to grain Q	Parallel to grain P	Perpendicular to grain Q	Parallel to grain P	Perpendicular to grain Q	Parallel to grain P	Perpendicular to grain Q	Parallel to grain P	Perpendicular to grain Q
1/2	2330	1640	2330	1650	2330	1480						
5/8	3350	1900	3350	2200	3350	2130						
3/4	4530	2090	4570	2550	4560	2840	3350	1920				
7/8	5770	2280	5930	2790	5950	3550	4570	2660	4560	1980		
1	7980	2670	8940	3260	9310	4450	5950	3460	5950	3240	5960	2410
1 1/4							9300	5210	9300	5110	9300	4860
1 1/2							13410	6480	13410	7200	13400	7070

*For three-member joints with side members at least half as thick as the main member. When steel plates are used for side members, increase tabulated values for parallel-to-grain loading 25%, but do not increase values for perpendicular-to-grain loading. Length of bolt l is measured in main member. For a two-member joint, use the smaller of the following: one-half the tabulated value for the thicker member or one-half the tabulated value for a piece twice the thickness of the thinner member. For a joint with four or more members of equal thickness, use for each shear plane one-half the tabulated value for the thickness of the corresponding member. If the members are not equally thick, resolve the joint into contiguous three-member joints. Determine the design load for each of these from the table and assign one-half to each shear plane. When a shear plane is assigned two different values, use the smaller.

Fig. 7-4 Timber connectors: (*a*) split ring; (*b*) wood members connected with split ring and bolt; (*c*) shear plate; (*d*) steel plate connected to wood member with shear plate and bolt; (*e*) wood members connected with a pair of shear plates and bolt. *(Reprinted with permission from F. S. Merritt, "Building Engineering and Systems Design," Van Nostrand Reinhold Company, New York.)*

of each ring is in each of the two members in contact (Fig. 7-4*b*). A bolt hole is drilled through the center of the core encircled by the groove.

A split ring has a tongue and groove split to permit simultaneous bearing of the inner surface of the ring against the core, and the outer surface of the ring against the outer wall of the groove. The ring is beveled for ease of assembly. Rings are manufactured in 2½- and 4-in diameter.

Shear plates are intended for wood-to-steel connections (Fig. 7-4*c* and *d*). But when used in pairs, they may be used for wood-to-wood connections (Fig. 7-4*e*), replacing split rings. Set with one plate in each member at the contact surface, they enable the members to slide easily into position during fabrication of the joint, thus reducing the labor needed to make the connection. Shear plates are placed in precut daps and are completely embedded in the timber, flush with the surface. As with split rings, the role of the bolt through each plate is to prevent the components of the joint from separating; loads are transmitted across the joint through the plates. They come in 2⅝- and 4-in diameter.

Shear plates are useful in demountable structures. They may be installed in the members immediately after fabrication and held in position by nails.

Toothed rings and spike grids sometimes are used for special applications. But split rings and shear plates are the prime connectors for timber construction.

TABLE 7-19 Allowable Loads for One Split Ring and Bolt in Single Shear

(Normal loading)

Split-ring diam, in.	Bolt diam, in.	No. of faces of piece with connectors on same bolt	Thickness (net) of lumber, in.	Loaded parallel to grain		Loaded perpendicular to grain		
				Min edge distance, in.	Allowable load per connector and bolt, lb	Edge distance, in.		Allowable load per connector and bolt, lb
						Unloaded edge, min	Loaded edge	
2½	½	1	1 min	1¾	2,270	1¾	1¾ min	1,350
							2¾ or more	1,620
			1½ or more	1¾	2,730	1¾	1¾ min	1,620
							2¾ or more	1,940
		2	1½ min	1¾	2,100	1¾	1¾ min	1,250
							2¾ or more	1,500
			2 or more	1¾	2,730	1¾	1¾ min	1,620
							2¾ or more	1,940
4	¾	1	1 min	2¾	3,510	2¾	2¾ min	2,030
							3¾ or more	2,440
			1½ or more	2¾	5,160	2¾	2¾ min	2,990
							3¾ or more	3,590
		2	1½ min	2¾	3,520	2¾	2¾ min	2,040
							3¾ or more	2,450
			2	2¾	4,250	2¾	2¾ min	2,470
							3¾ or more	2,960
			2½	2¾	5,000	2¾	2¾ min	2,900
							3¾ or more	3,480
			3 or more	2¾	5,260	2¾	2¾ min	3,050
							3¾ or more	3,660

Table 7-19 gives safe working loads for split rings and bolts. Table 7-20 gives safe working loads for shear plates and bolts.

In design for wind forces acting alone or with dead and live loads, the safe working loads for connectors may be increased by one-third. But the number and size of connectors should not be less than those required for dead and live loads alone.

Impact may be disregarded up to the following percentages of the static effect of the live load producing the impact:

Connector	Impact allowance, %
Split ring, any size, bearing in any direction	100
Shear plate, any size, bearing parallel to grain	66⅔
Shear plate, any size, bearing perpendicular to grain	100

One-half of any impact load that remains after disregarding the percentages indicated should be included with the dead and live loads in designing the joint.

The above procedures for increasing the allowable loads on connectors for suddenly applied and short-duration loads do not reduce the actual factor of safety of the joint. They are recommended because of the favorable behavior of wood under such forces. Different values are allowed for different types and sizes of connectors and directions of bearing because of variations in the extent to which distortion of the metal, as well as the strength of the wood, affects the ultimate strength of the joint.

TABLE 7-20 Allowable Loads for One Shear Plate and Bolt in Single Shear

(Normal Load Duration and Wood Side Plates*)

Shear-plate diam, in	Bolt diam, in	No. of faces of piece with connectors on same bolt	Thickness (net) of lumber, in	Loaded parallel to grain		Loaded perpendicular to grain		
				Min edge distance, in	Allowable load per connector and bolt, lb	Edge distance, in		Allowable load per connector and bolt, lb
						Unloaded edge, min	Loaded edge	
2 5/8	3/4	1	1 1/2 min	1 3/4	2670	1 3/4	1 3/4 min	1550
							2 3/4 or more	1860
		2	1 1/2 min	1 3/4	2080	1 3/4	1 3/4 min	1210
							2 3/4 or more	1450
			2	1 3/4	2730	1 3/4	1 3/4 min	1590
							2 3/4 or more	1910
			2 1/2 or more	1 3/4	2860	1 3/4	1 3/4 min	1660
							2 3/4 or more	1990
4	3/4 or 7/8	1	1 1/2 min	2 3/4	3750	2 3/4	2 3/4 min	2180
							3 3/4 or more	2620
			1 3/4 or more	2 3/4	4360	2 3/4	2 3/4 min	2530
							3 3/4 or more	3040
		2	1 3/4 min	2 3/4	2910	2 3/4	2 3/4 min	1680
							3 3/4 or more	2020
			2	2 3/4	3240	2 3/4	2 3/4 min	1880
							3 3/4 or more	2260
			2 1/2	2 3/4	3690	2 3/4	2 3/4 min	2140
							3 3/4 or more	2550
			3	2 3/4	4140	2 3/4	2 3/4 min	2400
							3 3/4 or more	2880
			3 1/2 or more	2 3/4	4320	2 3/4	2 3/4 min	2500
							3 3/4 or more	3000

*Tabulated loads also apply to metal side plates, except that for 4-in shear plates the parallel-to-grain (not perpendicular) loads for wood side plates shall be increased 11%; but loads shall not exceed: for all loadings, except wind, 2900 lb for 2 5/8-in shear plates; 4970 and 6720 lb for 4-in shear plates with 3/4- and 7/8-in bolts, respectively; or for wind loading, 3870, 6630, and 9020 lb, respectively. If bolt threads are in bearing on the shear plate, reduce the preceding values by one-ninth.

Metal side plates, when used, shall be designed in accordance with accepted metal practices. For A36 steel, the following unit stresses, psi, are suggested for all loadings except wind: net section in tension, 22,000; shear, 14,500; bearing, 33,000. For wind, these values may be increased one-third. If bolt threads are in bearing, reduce the preceding shear and bearing values by one-ninth.

The tabulated loads apply to seasoned timbers used where they will remain dry. If the timbers will be more or less continuously damp or wet in use, two-thirds of the tabulated values should be used.

Safe working loads for split rings and shear plates for angles between 0° (parallel to grain) and 90° (perpendicular to grain) may be obtained from the Scholten nomograph (Fig. 7-5). This determines the bearing strength of wood at various angles to the grain. The chart is a graphical solution of the Hankinson formula:

$$N = \frac{PQ}{P \sin^2 \theta + Q \cos^2 \theta} \tag{7-10}$$

where N, P, and Q are, respectively, the allowable load, lb, or stress, psi, at inclination θ with the direction of grain, parallel to grain, and perpendicular to grain.

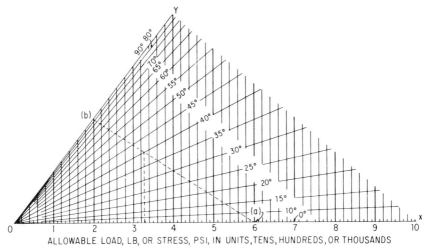

Fig. 7-5 Scholten nomograph for determining allowable bolt load on or bearing stress in wood at various angles to the grain.

For example, given $P = 6000$ lb and $Q = 2000$ lb, find the allowable load at an angle of 40° with the grain. Connect 6 on the 0° line (point a) with the intersection of a vertical line through 2 and the 90° line (point b). $N = 3280$ lb is found directly below the intersections of line ab with the 40° line.

Safe working loads are based on the assumption that the wood at the joint is clear and relatively free from checks, shakes, and splits. If knots are present in the longitudinal projection of the net section within a distance from the critical section of half the diameter of the connector, the area of the knot should be subtracted from the area of the critical section. It is assumed that slope of the grain at the joint does not exceed 1 in 10.

The stress, whether tension of compression, in the net area, the area remaining at the critical section after subtracting the projected area of the connectors and the bolt from the full cross-sectional area of the member, should not exceed the safe stress of clear wood in compression parallel to the grain.

Tables 7-19 and 7-20 list the least thickness of member that should be used with the various sizes of connectors. The loads listed for the greatest thickness of member with each type and size of connector unit are the maximums to be used for all thicker material. Loads for members with thicknesses between those listed may be obtained by interpolation.

Fig. 7-6 Measurement of spacing, edge distance, and end distance for connectors.

TABLE 7-21 Connector Spacings and End Distances

(With Corresponding Percentages of Tabulated Loads)

Split ring diam, in	Shear plate diam, in	Spacing parallel to grain		Spacing perpendicular to grain		End distance		
		Spacing, in	Percent of tabulated load	Minimum, in	Percent of tabulated load	In tension members, in	In compression members, in	Percent of tabulated load*
				Parallel to grain loading				
2 1/2	2 5/8	6 3/4	100	3 1/2 min.	100	5 1/2	4	100
2 1/2	2 5/8	3 1/2 min	50	3 1/2 min.	100	2 3/4 min	2 1/2 min	62.5
4	4	9	100	5 min	100	7	5 1/2	100
4	4	5 min	50	5 min	100	3 1/2 min	3 1/4 min	62.5
				Perpendicular to grain loading				
2 1/2	2 5/8	3 1/2 min		4 1/4	100	5 1/2	5 1/2	100
2 1/2	2 5/8	3 1/2 min		3 1/2 min	50	2 3/4 min	2 3/4 min	62.5
4	4	5 min		6	100	7	7	100
4	4	5 min		5 min	50	3 1/2 min	3 1/2 min	62.5

*No reduction in end distance is permitted for compression members loaded parallel to grain.
From NFPA "National Design Specification for Wood Construction."

TABLE 7-22 Modification Factors K for Joints with Connectors, Bolts, and Laterally Loaded Lag Screws and Wood Side Plates

A_1/A_2*	A_1, in²*†	\multicolumn{11}{c}{Number of fasteners in a row}										
		2	3	4	5	6	7	8	9	10	11	12
0.5‡	<12	1.00	0.92	0.84	0.76	0.68	0.61	0.55	0.49	0.43	0.38	0.34
	12–19	1.00	0.95	0.88	0.82	0.75	0.68	0.62	0.57	0.52	0.48	0.43
	>19–28	1.00	0.97	0.93	0.88	0.82	0.77	0.71	0.67	0.63	0.59	0.55
	>28–40	1.00	0.98	0.96	0.92	0.87	0.83	0.79	0.75	0.71	0.69	0.66
	>40–64	1.00	1.00	0.97	0.94	0.90	0.86	0.83	0.79	0.76	0.74	0.72
	>64	1.00	1.00	0.98	0.95	0.91	0.88	0.85	0.82	0.80	0.78	0.76
1.0‡	<12	1.00	0.97	0.92	0.85	0.78	0.71	0.65	0.59	0.54	0.49	0.44
	12–19	1.00	0.98	0.94	0.89	0.84	0.78	0.72	0.66	0.61	0.56	0.51
	>19–28	1.00	1.00	0.97	0.93	0.89	0.85	0.80	0.76	0.72	0.68	0.64
	>28–40	1.00	1.00	0.99	0.96	0.92	0.89	0.86	0.83	0.80	0.78	0.75
	>40–64	1.00	1.00	1.00	0.97	0.94	0.91	0.88	0.85	0.84	0.82	0.80
	>64	1.00	1.00	1.00	0.99	0.96	0.93	0.91	0.88	0.87	0.86	0.85

*A_1 = cross-sectional area of main members before boring or grooving.
 A_2 = sum of the cross-sectional areas of side members before boring or grooving.
†When A_1/A_2 exceeds 1.0, use A_2 instead of A_1.
‡When A_1/A_2 exceeds 1.0, use A_2/A_1.
For A_1/A_2 between 0 and 1.0, interpolate or extrapolate from the tabulated values.
From NFPA "National Design Specification for Wood Construction."

Rows of Fasteners. For connectors in a multiple joint, there are several rules: Spacings, edge distances, and end distances should be measured as indicated in Fig. 7-6. The edge distances given in Tables 7-19 and 7-20 are the minimum permitted. End distances and appropriate percentages of tabulated allowable loads are listed in Table 7-21. For end distances intermediate between the minimum and that required for full load, the allowable load may be determined by interpolation. Table 7-21 also lists spacings and appropriate percentages of tabulated design loads. For spacings intermediate between the minimum and that required for full load, the allowable load may be determined by interpolation.

The total allowable load for a joint with two or more connectors may be taken as the sum of the allowable loads for the individual connectors, but for any row of fasteners of the same size and type, the load should not exceed P_r computed from

$$P_r = P_s K \qquad (7\text{-}11)$$

where P_s = sum of the allowable loads, lb, for individual fasteners in the row
 K = modification factor, as obtained from Table 7-22 or 7-23

Placement of connectors in joints with members at right angles to each other is subject to the limitations of either member. Since rules for alignment, spacing, and edge and end distance of connectors for all conceivable directions of applied load would be complicated, designers must rely on a sense of proportion and adequacy in applying the above rules to conditions of loading outside the specific limitations mentioned.

Anchor Bolts. To attach columns or arch bases to concrete foundations, anchor bolts are embedded in the concrete, with sufficient projection to permit placement of angles or shoes bolted to the wood. Sometimes, instead of anchor bolts, steel straps are embedded in the concrete with a portion projecting above for bolt attachment to the wood members.

Washers. Bolt heads and nuts bearing on wood require metal washers to protect the wood and to distribute the pressure across the surface of the wood. Washers may be

TABLE 7-23 Modification Factors *K* for Joints with Connectors, Bolts, and Laterally Loaded Lag Screws and Metal Side Plates

A_1/A_2*	A_1, in²*	\multicolumn{11}{c}{Number of fasteners in a row}
		2	3	4	5	6	7	8	9	10	11	12
2–12	25–39	1.00	0.94	0.87	0.80	0.73	0.67	0.61	0.56	0.51	0.46	0.42
	40–64	1.00	0.96	0.92	0.87	0.81	0.75	0.70	0.66	0.62	0.58	0.55
	65–119	1.00	0.98	0.95	0.91	0.87	0.82	0.78	0.75	0.72	0.69	0.66
	120–199	1.00	0.99	0.97	0.95	0.92	0.89	0.86	0.84	0.81	0.79	0.78
12–18	40–64	1.00	0.98	0.94	0.90	0.85	0.80	0.75	0.70	0.67	0.62	0.58
	65–119	1.00	0.99	0.96	0.93	0.90	0.86	0.82	0.79	0.75	0.72	0.69
	120–199	1.00	1.00	0.98	0.96	0.94	0.92	0.89	0.96	0.83	0.80	0.78
	200	1.00	1.00	1.00	0.98	0.97	0.95	0.93	0.91	0.90	0.88	0.87
18–24	40–64	1.00	1.00	0.96	0.93	0.89	0.84	0.79	0.74	0.69	0.64	0.59
	65–119	1.00	1.00	0.97	0.94	0.92	0.89	0.86	0.83	0.80	0.76	0.73
	120–199	1.00	1.00	0.99	0.98	0.96	0.94	0.92	0.90	0.88	0.86	0.85
	200	1.00	1.00	1.00	1.00	0.98	0.96	0.95	0.93	0.92	0.92	0.91
24–30	40–64	1.00	0.98	0.94	0.90	0.85	0.80	0.74	0.69	0.65	0.61	0.58
	65–119	1.00	0.99	0.97	0.93	0.90	0.86	0.82	0.79	0.76	0.73	0.71
	120–199	1.00	1.00	0.98	0.96	0.94	0.92	0.89	0.87	0.85	0.83	0.81
	200	1.00	1.00	0.99	0.98	0.97	0.95	0.93	0.92	0.90	0.89	0.89
30–35	40–64	1.00	0.96	0.92	0.86	0.80	0.74	0.68	0.64	0.60	0.57	0.55
	65–119	1.00	0.98	0.95	0.90	0.86	0.81	0.76	0.72	0.68	0.65	0.62
	120–199	1.00	0.99	0.97	0.95	0.92	0.88	0.85	0.82	0.80	0.78	0.77
	200	1.00	1.00	0.98	0.97	0.95	0.93	0.90	0.89	0.87	0.86	0.85
35–42	40–64	1.00	0.95	0.89	0.82	0.75	0.69	0.63	0.58	0.53	0.49	0.46
	65–119	1.00	0.97	0.93	0.88	0.82	0.77	0.71	0.67	0.63	0.59	0.56
	120–199	1.00	0.98	0.96	0.93	0.89	0.85	0.81	0.78	0.76	0.73	0.71
	200	1.00	0.99	0.98	0.96	0.93	0.90	0.87	0.84	0.82	0.80	0.78

*A_1 = cross-sectional area of main member before boring or grooving.
A_2 = sum of cross-sectional areas of metal side plates before drilling.
From NFPA "National Design Specification for Wood Construction."

cast, malleable, cut, round-plate, or square-plate. When subjected to salt air or salt water, they should be galvanized or given some type of effective coating. Ordinarily, washers are dipped in red lead and oil before installation.

Setscrews should never be used against a wood surface. It may be possible, with the aid of proper washers, to spread the load of the setscrew over sufficient surface area of the wood that the compression strength perpendicular to grain is not exceeded.

Table 7-24 gives washer sizes capable of developing the capacity of A307 bolts.

Tie Rods. To resist the horizontal thrust of arches not buttressed, tie rods are required. The tie rods may be installed at ceiling height or below the floor. Table 7-25 gives the allowable loads on bars used as tie rods.

Hangers. Standard and special hangers are used extensively in timber construction. Stock hangers are available from a number of manufacturers, but most hangers are of special design. Where appearance is of prime importance, concealed hangers are frequently selected.

(National Design Specification for Wood Construction, National Forest Products Association, 1619 Massachusetts Ave., N.W., Washington, D.C. 20026; "Design Manual for TECO Timber Connector Construction," Timber Engineering Co., 1619 Massachu-

TABLE 7-24 Allowable Loads for Bolts and Washers

Rod or bolt		Plate washers†		Cut washers			
Diam, in.	Tensile capacity,* lb	Side of square, in.	Thickness, in.	Outside diam, in.	Hole diam, in.	Thickness, in.	Max load,‡ lb
3/8	1,550	1 3/4	3/16	1	7/16	5/64	290
7/16	2,100	2 1/8	1/4	1 1/4	1/2	5/64	460
1/2	2,750	2 3/8	1/4	1 3/8	9/16	7/64	550
5/8	4,300	3	5/16	1 3/4	11/16	9/64	910
3/4	6,190	3 3/4	3/8	2	13/16	5/32	1,100
7/8	8,420	4 3/8	1/2	2 1/4	13/16	11/64	1,100
1	11,000	5	1/2	2 1/2	1 1/16	11/64	1,800
1 1/8	13,920	5 5/8	5/8	2 3/4	1 1/4	11/64	2,100
1 1/4	17,180	6 3/8	3/4	3	1 3/8	11/64	2,500
1 3/8	20,800	7	3/4	3 1/4	1 1/2	3/16	2,900
1 1/2	24,700	7 3/4	7/8	3 1/2	1 5/8	3/16	3,400
1 5/8	29,000	8 3/8	7/8	3 3/4	1 3/4	3/16	3,900
1 3/4	33,700	9	1	4	1 7/8	3/16	4,400
1 7/8	38,700	9 3/4	1	4 1/4	2	3/16	5,000
2	44,000	10 1/4	1 1/8	4 1/2	2 1/8	3/16	5,500
2 1/8	49,700	11	1 1/8				
2 1/4	55,700	11 3/4	1 1/4	4 3/4	2 3/8	7/32	6,000
2 3/8	62,000	12 1/2	1 1/4				
2 1/2	68,700	13	1 3/8	5	2 5/8	15/64	6,500
2 5/8	75,800	13 3/4	1 1/2				
2 3/4	83,200	14 1/2	1 1/2	5 1/4	2 7/8	1/4	6,800
2 7/8	90,900	15 1/4	1 5/8				
3	99,000	15 3/4	1 5/8	5 1/2	3 1/8	9/32	7,200

* Based on ASTM A36 steel and A307 bolts.
† Size required to develop capacity of rod or bolt in accordance with note ‡.
‡ Based on allowable strength of wood in compression perpendicular to grain of 450 psi.

setts Ave., N.W., Washington, D.C. 20036; "Western Woods Use Book," Western Wood Products Association, 1500 Yeon Building, Portland, Ore. 97204.)

7-17. Glued Fastenings.

Glued joints are generally made between two pieces of wood where the grain directions are parallel (as between the laminations of a beam or arch). Or such joints may be between solid-sawn or laminated timber and plywood, where the face grain of the plywood is either parallel or at right angles to the grain direction of the timber.

It is only in special cases that lumber may be glued with the grain direction of adjacent pieces at an angle. When the angle is large, dimensional changes from changes in wood moisture content set up large stresses in the glued joint. Consequently, the strength of the joint may be considerably reduced over a period of time. Exact data are not available, however, on the magnitude of this expected strength reduction.

In joints connected with plywood gusset plates, this shrinkage differential is minimized, because plywood swells and shrinks much less than does solid wood.

Glued joints can be made between end-grain surfaces. But they are seldom strong enough to meet the requirements of even ordinary service. Seldom is it possible to

TABLE 7-25 Allowable Tension on Round and Square Upset Bars

UNC and 4UN Class-2A Thread

Diam d or side s, in.	Round bars			Square bars		
	Capacity,* lb	Upset		Capacity,* lb	Upset	
		Diam D, in.	Length L, in.		Diam D, in.	Length L, in.
$3/4$	9,700	1	4	12,400	$1\,1/8$	4
$7/8$	13,200	$1\,1/8$	4	16,900	$1\,1/4$	4
1	17,300	$1\,3/8$	4	22,000	$1\,1/2$	4
$1\,1/8$	21,800	$1\,1/2$	4	27,800	$1\,3/4$	4
$1\,1/4$	27,000	$1\,3/4$	4	34,400	2	$4\,1/2$
$1\,3/8$	32,700	$1\,3/4$	4	41,600	2	$4\,1/2$
$1\,1/2$	38,900	2	$4\,1/2$	49,500	$2\,1/4$	5
$1\,5/8$	45,600	$2\,1/4$	5	58,100	$2\,1/2$	$5\,1/2$
$1\,3/4$	53,000	$2\,1/4$	5	67,400	$2\,1/2$	$5\,1/2$
$1\,7/8$	60,800	$2\,1/8$	$5\,1/2$	77,300	$2\,3/4$	$5\,1/2$
2	69,100	$2\,1/2$	$5\,1/2$	88,000	$2\,3/4$	$5\,1/2$
$2\,1/8$	78,000	$2\,3/4$	$5\,1/2$	99,300	3	6
$2\,1/4$	87,500	$2\,3/4$	$5\,1/2$	111,400	$3\,1/4$	$6\,1/2$
$2\,3/8$	97,500	3	6	124,100	$3\,1/4$	$6\,1/2$
$2\,1/2$	108,000	$3\,1/4$	$6\,1/2$	137,500	$3\,1/2$	7
$2\,5/8$	119,100	$3\,1/4$	$6\,1/2$	151,600	$3\,3/4$	7

* Based on ASTM A36 steel.

develop more than 25% of the tensile strength of the wood in such butt joints. It is for this reason that plane sloping scarfs of relatively flat slope (Fig. 7-2), or finger joints with thin tips and flat slope on the sides of the individual fingers (Fig. 7-3), are used to develop a high proportion of the strength of the wood.

Joints of end grain to side grain are also difficult to glue properly. When subjected to severe stresses as a result of unequal dimensional changes in the members due to changes in moisture content, joints suffer from severely reduced strength.

For the preceding reasons, joints between end-grain surfaces and between end-grain and side-grain surfaces should not be used if the joints are expected to carry load.

For joints made with wood of different species, the allowable shear stress for parallel-grain bonding is equal to the allowable shear stress parallel to the grain for the weaker species in the joint. This assumes uniform stress distribution in the joint. When grain direction is not parallel, the allowable shear stress on the glued area between the two pieces may be estimated from the Scholten nomograph (Fig. 7-5).

Adhesives used for fabricating structural glued-laminated timbers also are satisfactory for other glued joints. In selecting an adhesive, consideration should be given to initial and service-condition wood moisture content.

[AITC 103, Standard for Structural Glued-laminated Timber, American Institute of Timber Construction, 833 W. Hampden Ave., Englewood, Colo. 80110; Federal Specification MMM-A-125, Adhesive, Casein-type, Water- and Mold-resistant, General Services Administration, Washington, D.C. 20405; Military Specification MIL-A-397B, Adhesive, Room-temperature and Intermediate-temperature Setting Resin (Phenol, Resorcinol, and Melamine Base), and Military Specification MIL A-5534A, Adhesive, High-temperature Setting Resin (Phenol, Melamine, and Resorcinol Base), U.S. Naval Supply Depot, Philadelphia, Pa. 19120.]

7-18. Wood Columns.

Wood compression members may be a solid piece of timber (Fig. 7-7a) or laminated (Fig. 7-7b) or built up of spaced members (Fig. 7-7c). The latter comprise two or more wood compression members with parallel longitudinal axes. The members are separated at ends and midpoints by blocking, and joined to the end blocking with connectors with adequate shear resistance.

Short Columns. Columns with a ratio of unsupported length L, in, to least dimension d, in, less than 11 fail by crushing. The allowable concentric load for such members equals the cross-sectional area, in^2, times F_c, the allowable compression stress, psi, parallel to grain for the species, adjusted for service conditions and duration of load.

Long and Intermediate Columns. When the slenderness ratio L/d exceeds 11, wood columns generally fail by buckling. In that case, the allowable stress is determined from formulas that yield values less than F_c. The computed allowable stress must be adjusted for duration of load.

For square or rectangular, solid-sawn or glued-laminated columns with L/d larger than 11 but smaller than K computed from

$$K = 0.671 \sqrt{\frac{E}{F_c}} \tag{7-12}$$

Fig. 7-7 Behavior of wood columns depends on length-thickness or length-depth ratios: (a) solid timber column; (b) glued-laminated column; (c) spaced column.

the allowable stress F_c' should not exceed

$$F_c' = \left[1 - \frac{1}{3}\left(\frac{L/d}{K}\right)^4\right] F_c \leq F_c \qquad (7\text{-}13)$$

where F_c = allowable compression stress parallel to grain
E = modulus of elasticity of the wood, adjusted for duration of load

For long rectangular, solid-sawn or glued-laminated columns ($L/d \geq K$), the allowable unit stress may not exceed F_c or

$$F_c' = \frac{0.30E}{(L/d)^2} \qquad (7\text{-}14)$$

In no case may L/d exceed 50. The formula was derived for pin-end conditions, but may also be used for square-cut ends.

For round columns, the allowable stress may not exceed that for a square column of the same cross-sectional area.

For tapered columns, d may be taken as the sum of the least dimension plus one-third the difference between this and the maximum thickness parallel to this dimension.

For spaced columns, L/d' should not exceed 80, L/d'' should not exceed 50, and L'/d' should not exceed 40 (see Fig. 7-7c). Allowable stresses depend on the fixity condition, a or b, as defined in Fig. 7-7c. The factor K for spaced columns is defined by

$$K = 0.671 \sqrt{\frac{C_x E}{F_c}} \qquad (7\text{-}15)$$

where C_x = 2.5 for fixity condition a
= 3.0 for fixity condition b

For spaced columns with components having L'/d' larger than 11 but smaller than K, the allowable stress should not exceed F_c' computed from Eq. (7-13) with K determined from Eq. (7-15).

For long components of spaced columns ($L'/d' \geq K$), the allowable stress should not exceed

$$F_c' = \frac{0.30 C_x E}{(L'/d')^2} \qquad (7\text{-}16)$$

Each member of a spaced column should be designed separately on the basis of its L'/d'. The allowable load on each equals its cross-sectional area, in², times its allowable stress, psi, adjusted for duration of load. The sum of the allowable loads on the individual members should equal or exceed the total load on the column.

When a single spacer block is placed in the middle tenth of a spaced column, connectors are not required. They should be used for multiple spaced blocks. The distance between two adjacent blocks may not exceed half the distance between centers of connectors in the end blocks in opposite ends of the column.

For all types of columns, the distance between adequate bracing, including beams and struts, should be used as L in determining L/d. The largest L/d for the column or any component, whether it be computed for a major or minor axis, should be used in calculating the allowable unit stress.

Bending plus Compression. For combined axial and bending stress, Eq. (7-8a) or (7-8b) governs. Bending caused by transverse loads, wind loads, or eccentric loads, or any combination of these, should be taken into account.

The critical section of columns supporting trusses frequently exists at the connection of knee brace to column. Where no knee brace is used, or the column supports a beam,

Fig. 7-8 Typical anchorages of wood column to base: (*a*) column anchored to concrete base with U strap; (*b*) anchorage with steel angles; (*c*) anchorage with a welded box shoe.

the critical section for moment usually occurs at the bottom of truss or beam. Then a rigid connection must be provided to resist moment, or adequate diagonal bracing must be provided to carry wind loads into a support.

Figure 7-8 shows typical column base anchorages, and Fig. 7-9 typical beam-to-column connections (AITC 104, Typical Construction Details, American Institute of Timber Construction).

(American Institute of Timber Construction, "Timber Construction Manual," John Wiley & Sons, Inc., New York; "Western Woods Use Book," Western Wood Products Association, 1500 Yeon Building, Portland, Ore. 97204; "Wood Structural Design Data," National Forest Products Association, 1619 Massachusetts Ave, Washington, D.C. 20036.)

7-19. Design of Timber Joists.

Joists are relatively narrow beams, usually spaced 12 to 24 in c to c. They generally are topped with sheathing and braced with diaphragms or cross bridging at intervals up to 10 ft. For joist spacings of 16 to 24 in c to c, 1-in sheathing usually is required. For spacings over 24 in, 2 in or more of wood decking is necessary.

Standard beam formulas for bending, shear, and deflection may be used to determine joist sizes. Connections shown in Figs. 7-9 and 7-10 may be used for joists.

See also Art. 7-20

Fig. 7-9 Typical wood beam and girder connections to columns: (a) girder to steel column; (b) girder to wood column; (c) beam to pipe column; (d) beam to wood column, with steel strap welded to steel side plates; (e) beam to wood column, with a T plate; (f) beam-column connection with spiral dowel and shear plates.

7-20. Design of Timber Beams.

Standard beam formulas for bending, shear, and deflection may be used to determine beam sizes. Ordinarily, deflection governs design; but for short, heavily loaded beams, shear is likely to control.

Fig. 7-11 shows the types of beams commonly produced in timber. Straight and single- and double-tapered straight beams can be furnished solid-sawn or glued-laminated. The curved members can be furnished only glued-laminated. Beam names describe the top and bottom surfaces of the beam. The first part describes the top surface, the word following the hyphen the bottom. Sawn surfaces on the tension side of a beam should be avoided.

Table 7-26 gives the load-carrying capacity for various cross-sectional sizes of glued-laminated, simply supported beams.

Example. With the following data, design a straight, glued-laminated beam, simply supported and uniformly loaded: span, 28 ft; spacing, 9 ft c to c; live load, 30 lb/ft^2; dead load, 5 lb/ft^2 for deck and 7.5 lb/ft^2 for roofing. Allowable bending stress of combination grade is 2400 psi, with modulus of elasticity $E = 1,800,000$ psi. Deflection limitation is $L/180$, where L is the span, ft. Assume the beam is laterally supported by the deck throughout its length.

With a 15% increase for short-duration loading, the allowable bending stress F_b becomes 2760 psi, and the allowable horizontal shear F_v, 230 psi.

Fig. 7-10 Beam connections: (*a*), (*b*) wood beam anchored on wall with steel angles; (*c*) with welded assembly; (*d*) beam anchored directly with bolt; (*e*) beam supported on girder with bent-strap hanger; (*f*) similar support for purlins; (*g*) saddle connects beam to girder (suitable for one-sided connection); (*h*), (*i*) connections with concealed hangers; (*j*), (*k*) connections with steel angles.

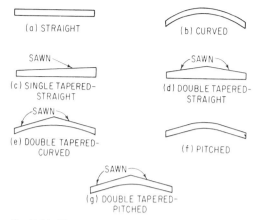

Fig. 7-11 Types of glued-laminated beams.

Assume the beam will weigh 22.5 lb/lin ft, averaging 2.5 lb/ft². Then, the total uniform load comes to 45 psf. So the beam carries $w = 45 \times 9 = 405$ lb/lin ft.

The end shear $V = wL/2$ and the maximum shearing stress $= 3V/2 = 3wL/4$. Hence, the required area, in², for horizontal shear is

$$A = \frac{3wL}{4F_v} = \frac{wL}{306.7} = \frac{405 \times 28}{306.7} = 37.0 \text{ in}^2$$

The required section modulus, in³, is

$$S = \frac{1.5wL^2}{F_b} = \frac{1.5 \times 405 \times 28^2}{2760} = 172.6 \text{ in}^3$$

If $D = 180$, the reciprocal of the deflection limitation, then the deflection equals $5 \times 1728wL^4/384EI \leq 12L/D$, where I is the moment of inertia of the beam cross section, in⁴. Hence, to control deflection, the moment of inertia must be at least

$$I = \frac{1.875DwL^3}{E} = \frac{1.875 \times 180 \times 405 \times 28^3}{1,800,000} = 1688 \text{ in}^4$$

Assume that the beam will be fabricated with 1½-in laminations. From Table 7-5, the most economical section satisfying all three criteria is 5¼ × 16½, with $A = 84.6$, $S = 232.5$, and $I = 1918.5$. But it has a size factor of 0.97 %. So the allowable bending stress must be reduced to $2760 \times 0.97 = 2677$ psi. And the required section modulus must be increased accordingly to $172.6/0.97 = 178$. The selected section still is adequate.

Suspended-Span Construction. Cantilever systems may be comprised of any of the various types and combinations of beam illustrated in Fig. 7-12. Cantilever systems permit longer spans or larger loads for a given size member than do simple-span systems, if member size is not controlled by compression perpendicular to grain at the supports or by horizontal shear. Substantial design economies can be effected by decreasing the depths of the members in the suspended portions of a cantilever system.

For economy, the negative bending moment at the supports of a cantilevered beam should be equal in magnitude to the positive moment.

Consideration must be given to deflection and camber in cantilevered multiple spans. When possible, roofs should be sloped the equivalent of ¼ in per foot of horizontal distance between the level of drains and the high point of the roof to eliminate water pockets, or provisions should be made to ensure that accumulation of water does not produce

TABLE 7-26 Load-Carrying Capacity of Simple-Span Laminated Beams*

Span, ft	Spacing, ft	Roof beam total load-carrying capacity						Floor beams total load
		30 psf	35 psf	40 psf	45 psf	50 psf	55 psf	50 psf
8	4	3⅛ × 4½	3⅛ × 4½	3⅛ × 6	3⅛ × 6	3⅛ × 6	3⅛ × 6	3⅛ × 6
	6	3⅛ × 4½	3⅛ × 4½	3⅛ × 6	3⅛ × 6	3⅛ × 6	3⅛ × 6	3⅛ × 6
	8	3⅛ × 4½	3⅛ × 4½	3⅛ × 6	3⅛ × 6	3⅛ × 6	3⅛ × 6	3⅛ × 7½
10	4	3⅛ × 4½	3⅛ × 4½	3⅛ × 6	3⅛ × 6	3⅛ × 6	3⅛ × 6	3⅛ × 7½
	6	3⅛ × 4½	3⅛ × 6	3⅛ × 6	3⅛ × 6	3⅛ × 6	3⅛ × 7½	3⅛ × 7½
	8	3⅛ × 6	3⅛ × 6	3⅛ × 7½	3⅛ × 7½	3⅛ × 7½	3⅛ × 7½	3⅛ × 9
	10	3⅛ × 6	3⅛ × 7½	3⅛ × 7½	3⅛ × 7½	3⅛ × 7½	3⅛ × 7½	3⅛ × 9
12	6	3⅛ × 6	3⅛ × 6	3⅛ × 7½	3⅛ × 7½	3⅛ × 7½	3⅛ × 7½	3⅛ × 9
	8	3⅛ × 6	3⅛ × 7½	3⅛ × 9	3⅛ × 9	3⅛ × 9	3⅛ × 9	3⅛ × 10½
	10	3⅛ × 7½	3⅛ × 7½	3⅛ × 9	3⅛ × 9	3⅛ × 9	3⅛ × 10½	3⅛ × 10½
	12	3⅛ × 7½	3⅛ × 9	3⅛ × 9	3⅛ × 9	3⅛ × 10½	3⅛ × 10½	3⅛ × 12
14	8	3⅛ × 7½	3½ × 9	3⅛ × 9	3⅛ × 9	3⅛ × 10½	3⅛ × 10½	3⅛ × 12
	10	3⅛ × 9	3⅛ × 9	3⅛ × 10½	3⅛ × 10½	3⅛ × 10½	3⅛ × 12	3⅛ × 12
	12	3⅛ × 9	3⅛ × 10½	3⅛ × 10½	3⅛ × 10½	3⅛ × 12	3⅛ × 12	3⅛ × 13½
	14	3⅛ × 10½	3⅛ × 10½	3⅛ × 12	3⅛ × 12	3⅛ × 12	3⅛ × 13½	3⅛ × 13½
16	8	3⅛ × 9	3⅛ × 9	3⅛ × 10½	3⅛ × 10½	3⅛ × 12	3⅛ × 12	3⅛ × 13½
	12	3⅛ × 10½	3⅛ × 12	3⅛ × 12	3⅛ × 12	3⅛ × 13½	3⅛ × 13½	3⅛ × 15
	14	3⅛ × 12	3⅛ × 12	3⅛ × 13½	3⅛ × 13½	3⅛ × 15	3⅛ × 15	3⅛ × 15
	16	3⅛ × 12	3⅛ × 13½	3⅛ × 13½	3⅛ × 15	3⅛ × 15	3⅛ × 16½	3⅛ × 15
18	8	3⅛ × 9	3⅛ × 10½	3⅛ × 12	3⅛ × 12	3⅛ × 12	3⅛ × 13½	3⅛ × 15
	12	3⅛ × 12	3⅛ × 12	3⅛ × 13½	3⅛ × 13½	3⅛ × 15	3⅛ × 16½	3⅛ × 16½
	16	3⅛ × 13½	3⅛ × 15	3⅛ × 15	3⅛ × 16½	5⅛ × 13½	5⅛ × 13½	5⅛ × 15
	18	3⅛ × 15	3⅛ × 15	3⅛ × 16½	3⅛ × 18	5⅛ × 15	5⅛ × 15	5⅛ × 15
20	8	3⅛ × 12	3⅛ × 12	3⅛ × 13½	3⅛ × 13½	3⅛ × 13½	3⅛ × 15	3⅛ × 16½
	12	3⅛ × 13½	3⅛ × 13½	3⅛ × 15	3⅛ × 16½	3⅛ × 16½	5⅛ × 13½	5⅛ × 15
	16	3⅛ × 15	3⅛ × 16½	3⅛ × 18	3⅛ × 18	5⅛ × 15	5⅛ × 16½	5⅛ × 18
	18	3⅛ × 16½	3⅛ × 16½	3⅛ × 18	5⅛ × 15	5⅛ × 16½	5⅛ × 16½	5⅛ × 18
22	8	3⅛ × 13½	3⅛ × 13½	3⅛ × 13½	3⅛ × 15	3⅛ × 15	3⅛ × 16½	5⅛ × 15
	12	3⅛ × 15	3⅛ × 15	3⅛ × 16½	3⅛ × 18	3⅛ × 18	5⅛ × 15	5⅛ × 16½
	16	3⅛ × 16½	3⅛ × 18	5⅛ × 15	5⅛ × 16½	5⅛ × 16½	5⅛ × 18	5⅛ × 19½
	18	3⅛ × 18	5⅛ × 15	5⅛ × 16½	5⅛ × 16½	5⅛ × 18	5⅛ × 18	5⅛ × 19½
24	8	3⅛ × 13½	3⅛ × 15	3⅛ × 15	3⅛ × 16½	3⅛ × 16½	3⅛ × 18	5⅛ × 16½
	12	3⅛ × 16½	3⅛ × 16½	3⅛ × 18	3⅛ × 18	5⅛ × 16½	5⅛ × 16½	5⅛ × 18
	16	3⅛ × 18	5⅛ × 16½	5⅛ × 16½	5⅛ × 18	5⅛ × 18	5⅛ × 19½	5⅛ × 21
	18	5⅛ × 15	5⅛ × 16½	5⅛ × 18	5⅛ × 18	5⅛ × 19½	5⅛ × 21	5⅛ × 21
26	8	3⅛ × 15	3⅛ × 16½	3⅛ × 16½	3⅛ × 16½	3⅛ × 18	5⅛ × 16½	5⅛ × 18
	12	3⅛ × 18	3⅛ × 18	5⅛ × 16½	5⅛ × 16½	5⅛ × 18	5⅛ × 18	5⅛ × 19½
	16	5⅛ × 16½	5⅛ × 16½	5⅛ × 18	5⅛ × 18	5⅛ × 19½	5⅛ × 21	5⅛ × 22½
	18	5⅛ × 16½	5⅛ × 18	5⅛ × 18	5⅛ × 19½	5⅛ × 21	5⅛ × 21	5⅛ × 22½
28	8	3⅛ × 16½	3⅛ × 16½	3⅛ × 16½	3⅛ × 18	5⅛ × 16½	5⅛ × 16½	5⅛ × 19½
	12	3⅛ × 18	5⅛ × 16½	5⅛ × 18	5⅛ × 18	5⅛ × 18	5⅛ × 19½	5⅛ × 21
	16	5⅛ × 18	5⅛ × 18	5⅛ × 19½	5⅛ × 19½	5⅛ × 21	5⅛ × 22½	5⅛ × 24
	18	5⅛ × 18	5⅛ × 18	5⅛ × 19½	5⅛ × 21	5⅛ × 22½	5⅛ × 24	5⅛ × 24
30	8	3⅛ × 18	3⅛ × 18	5⅛ × 16½	5⅛ × 16½	5⅛ × 18	5⅛ × 18	5⅛ × 21
	12	5⅛ × 16½	5⅛ × 18	5⅛ × 18	5⅛ × 19½	5⅛ × 19½	5⅛ × 21	5⅛ × 22½
	16	5⅛ × 18	5⅛ × 19½	5⅛ × 19½	5⅛ × 21	5⅛ × 22½	5⅛ × 24	5⅛ × 25½
	18	5⅛ × 19½	5⅛ × 21	5⅛ × 21	5⅛ × 22½	5⅛ × 24	5⅛ × 25½	5⅛ × 27
32	8	3⅛ × 18	5⅛ × 16½	5⅛ × 18	5⅛ × 18	5⅛ × 18	5⅛ × 19½	5⅛ × 21
	12	5⅛ × 18	5⅛ × 19½	5⅛ × 19½	5⅛ × 21	5⅛ × 21	5⅛ × 22½	5⅛ × 24
	16	5⅛ × 19½	5⅛ × 21	5⅛ × 22½	5⅛ × 22½	5⅛ × 24	5⅛ × 25½	5⅛ × 27
	18	5⅛ × 21	5⅛ × 21	5⅛ × 22½	5⅛ × 24	5⅛ × 25½	5⅛ × 27	5⅛ × 28½
34	8	5⅛ × 16½	5⅛ × 18	5⅛ × 18	5⅛ × 19½	5⅛ × 19½	5⅛ × 21	5⅛ × 22½
	12	5⅛ × 19½	5⅛ × 19½	5⅛ × 21	5⅛ × 21	5⅛ × 22½	5⅛ × 24	5⅛ × 25½
	16	5⅛ × 21	5⅛ × 22½	5⅛ × 22½	5⅛ × 24	5⅛ × 25½	5⅛ × 27	5⅛ × 28½
	18	5⅛ × 22½	5⅛ × 22½	5⅛ × 24	5⅛ × 25½	5⅛ × 27	5⅛ × 28½	5⅛ × 28½
36	12	5⅛ × 19½	5⅛ × 21	5⅛ × 22½	5⅛ × 22½	5⅛ × 24	5⅛ × 25½	6¾ × 25½
	16	5⅛ × 22½	5⅛ × 24	5⅛ × 24	5⅛ × 25½	5⅛ × 27	6¾ × 27	6¾ × 27
	18	5⅛ × 22½	5⅛ × 24	5⅛ × 25½	5⅛ × 28½	5⅛ × 30	6¾ × 27	6¾ × 28½
	20	5⅛ × 24	5⅛ × 25½	5⅛ × 27	5⅛ × 30	6¾ × 27	6¾ × 28½	6¾ × 30
38	12	5⅛ × 21	5⅛ × 22½	5⅛ × 24	5⅛ × 24	5⅛ × 25½	5⅛ × 27	6¾ × 27
	16	5⅛ × 24	5⅛ × 24	5⅛ × 25½	5⅛ × 25½	5⅛ × 28½	5⅛ × 30	6¾ × 28½
	18	5⅛ × 24	5⅛ × 25½	5⅛ × 27	5⅛ × 30	6¾ × 27	6¾ × 28½	6¾ × 30
	20	5⅛ × 25½	5⅛ × 27	5⅛ × 28½	6¾ × 27	6¾ × 28½	6¾ × 30	6¾ × 31½

Span, ft	Spacing, ft	Roof beams total load-carrying capacity						Floor beams total load
		30 psf	35 psf	40 psf	45 psf	50 psf	55 psf	50 psf
40	12	5⅛ × 22½	5⅛ × 24	5⅛ × 24	5⅛ × 25½	5⅛ × 27	6¾ × 25½	6¾ × 28½
	16	5⅛ × 24	5⅛ × 25½	5⅛ × 27	5⅛ × 28½	6¾ × 27	6¾ × 28½	6¾ × 31½
	18	5⅛ × 25½	5⅛ × 27	5⅛ × 28½	6¾ × 27	6¾ × 28½	6¾ × 30	6¾ × 31½
	20	5⅛ × 27	5⅛ × 28½	6¾ × 27	6¾ × 28½	6¾ × 30	6¾ × 31½	6¾ × 33
42	12	5⅛ × 24	5⅛ × 24	5⅛ × 25½	5⅛ × 27	6¾ × 25½	6¾ × 25½	6¾ × 30
	16	5⅛ × 25½	5⅛ × 27	5⅛ × 28½	5⅛ × 30	6¾ × 28½	6¾ × 28½	6¾ × 33
	18	5⅛ × 27	5⅛ × 28½	5⅛ × 30	6¾ × 28½	6¾ × 30	6¾ × 31½	6¾ × 33
	20	5⅛ × 28½	5⅛ × 30	6¾ × 28½	6¾ × 30	6¾ × 31½	6¾ × 33	6¾ × 34½
44	12	5⅛ × 24	5⅛ × 25½	5⅛ × 27	5⅛ × 27	6¾ × 25½	6¾ × 27	6¾ × 31½
	16	5⅛ × 27	5⅛ × 28½	5⅛ × 30	6¾ × 28½	6¾ × 30	6⅜ × 31½	6¾ × 33
	18	5⅛ × 28½	5⅛ × 30	6¾ × 28½	6¾ × 30	6¾ × 31½	6¾ × 33	6¾ × 34½
	20	5⅛ × 30	6¾ × 27	6¾ × 30	6¾ × 30	6¾ × 33	6¾ × 34½	6¾ × 36
46	12	5⅛ × 25½	5⅛ × 27	5⅛ × 28½	6¾ × 25½	6¾ × 27	6¾ × 28½	6¾ × 31½
	16	5⅛ × 28½	5⅛ × 30	5⅛ × 28½	6¾ × 30	6¾ × 31½	6¾ × 33	6¾ × 36
	18	5⅛ × 28½	6¾ × 28½	6¾ × 30	6¾ × 31½	6¾ × 33	6¾ × 34½	6¾ × 36
	20	5⅛ × 30	6¾ × 28½	6¾ × 31½	6¾ × 33	6¾ × 34½	6¾ × 36	8¾ × 34½
48	12	5⅛ × 27	5⅛ × 28½	5⅛ × 30	5⅛ × 30	6¾ × 28½	6¾ × 30	6¾ × 33
	16	5⅛ × 30	6¾ × 28½	6¾ × 30	6¾ × 30	6¾ × 31½	6¾ × 34½	6¾ × 37½
	18	5⅛ × 30	6¾ × 30	6¾ × 30	6¾ × 33	6¾ × 34½	6¾ × 36	8¾ × 34½
	20	6¾ × 28½	6¾ × 30	6¾ × 31½	6¾ × 34½	6¾ × 36	6¾ × 37½	8¾ × 36
50	12	5⅛ × 28½	5⅛ × 28½	6¾ × 30	6¾ × 28½	6¾ × 30	6¾ × 31½	8¾ × 34½
	16	5⅛ × 30	6¾ × 30	6¾ × 30	6¾ × 31½	6¾ × 33	6¾ × 36	8¾ × 34½
	18	6¾ × 28½	6¾ × 30	6¾ × 31½	6¾ × 34½	6¾ × 34½	8¾ × 33	8¾ × 36
	20	6¾ × 30	6¾ × 31½	6¾ × 33	6¾ × 36	6¾ × 37½	8¾ × 34½	8¾ × 37½
52	12	5⅛ × 28½	5⅛ × 30	6¾ × 28½	6¾ × 30	6¾ × 31½	6¾ × 31½	6¾ × 36
	16	6¾ × 28½	6¾ × 30	6¾ × 31½	6¾ × 33	6¾ × 34½	6¾ × 37½	8¾ × 36
	18	6¾ × 30	6¾ × 31½	6¾ × 33	6¾ × 34½	6¾ × 37½	6¾ × 39	8¾ × 37½
	20	6¾ × 31½	6¾ × 33	6¾ × 34½	6¾ × 37½	6¾ × 39	8¾ × 36	8¾ × 39
54	12	5⅛ × 30	6¾ × 28½	6¾ × 30	6¾ × 31½	6¾ × 33	6¾ × 33	8¾ × 37½
	16	6¾ × 30	6¾ × 31½	6¾ × 33	6¾ × 34½	6¾ × 36	6¾ × 37½	8¾ × 37½
	18	6¾ × 31½	6¾ × 33	6¾ × 34½	6¾ × 36	6¾ × 39	8¾ × 36	8¾ × 39
	20	6¾ × 33	6¾ × 34½	6¾ × 36	6¾ × 39	8¾ × 36	8¾ × 37½	8¾ × 40½
56	12	6¾ × 28½	6¾ × 30	6¾ × 31½	6¾ × 33	6¾ × 33	6¾ × 34½	8¾ × 36
	16	6¾ × 31½	6¾ × 33	6¾ × 34½	6¾ × 36	6¾ × 37½	6¾ × 34½	8¾ × 39
	18	6¾ × 33	6¾ × 34½	6¾ × 36	6¾ × 37½	6¾ × 34½	6¾ × 37½	8¾ × 40½
	20	6¾ × 33	6¾ × 36	6¾ × 37½	8¾ × 34½	6¾ × 37½	8¾ × 39	8¾ × 42
58	12	6¾ × 30	6¾ × 31½	6¾ × 31½	6¾ × 33	6¾ × 34½	6¾ × 36	8¾ × 37½
	16	6¾ × 31½	6¾ × 34½	6¾ × 36	6¾ × 37½	8¾ × 39	8¾ × 36	8¾ × 40½
	18	6¾ × 33	6¾ × 34½	6¾ × 37½	6¾ × 39	8¾ × 36	8¾ × 39	8¾ × 42
	20	6¾ × 34½	6¾ × 36	6¾ × 39	8¾ × 36	8¾ × 39	8¾ × 40½	8¾ × 43½
60	12	6¾ × 30	6¾ × 31½	6¾ × 33	6¾ × 34½	6¾ × 36	6¾ × 37½	8¾ × 39
	16	6¾ × 33	6¾ × 33	6¾ × 36	6¾ × 36	6¾ × 36	6¾ × 37½	8¾ × 42
	18	6¾ × 34½	6¾ × 36	6¾ × 39	6¾ × 36	8¾ × 37½	8¾ × 39	8¾ × 43½
	20	6¾ × 36	6¾ × 37½	8¾ × 36	8¾ × 37½	8¾ × 40½	8¾ × 42	8¾ × 45

* This table applies to straight, simply supported, laminated timber beams. Other beam support systems may be employed to meet varying design conditions.
1. Roofs should have a minimum slope of ¼ in. per ft to eliminate water ponding.
2. Beam weight must be subtracted from total load carrying capacity. Floor beams are designed for uniform loads of 40-psf live load and 10-psf dead load.
3. Allowable stresses:
Bending stress, F_b = 2,400 psi (reduced by size factor).
Shear stress F_v = 165 psi.
Modulus of elasticity, E = 1,800,000 psi.
For roof beams, F_b and F_v were increased 15% for short duration of loading.
4. Deflection limits:
Roof beams—1/180 span for total load.
Floor beams—1/360 span for 40-psf live load only.
For preliminary design purposes only.
For more complete design information, see the AITC "Timber Construction Manual."

Fig. 7-12 Cantilevered-beam systems. A is a single cantilever. B is a suspended beam. C has a double cantilever, and D is a beam with one end suspended.

greater deflection and live loads than anticipated. Unbalanced loading conditions should be investigated for maximum bending moment, deflection, and stability.

(American Institute of Timber Construction, "Timber Construction Manual," John Wiley & Sons, Inc., New York; "Western Woods Use Book," Western Wood Products Association, 1500 Yeon Building, Portland, Ore. 97204; "Wood Structural Design Data," National Forest Products Association, 1619 Massachusetts Ave., N.W., Washington, D.C. 20036.)

7-21. Deflection and Camber of Timber Beams.

The design of many structural systems, particularly those with long span, is governed by deflection. Strength calculations based on allowable stresses alone may result in excessive deflection. Limitations on deflection increase member stiffness.

Table 7-27 gives recommended deflection limits, as a fraction of the beam span, for timber beams. The limitation applies to live load or total load, whichever governs.

Glued-laminated beams are cambered by fabricating them with a curvature opposite in direction to that corresponding to deflections under load. Camber does not, however, increase stiffness. Table 7-28 lists recommended minimum cambers for glued-laminated timber beams.

7-22. Minimum Roof Slopes.

Flat roofs have collapsed during rainstorms even though they were adequately designed on the basis of allowable stresses and definite deflection limitations. The failures were

TABLE 7-27   Recommended Roof-Beam Deflection Limitations, in*
(In Terms of Span, l, in)

Use classification	Live load only	Dead load plus live load
Roof beams:		
Industrial	$l/180$	$l/120$
Commercial and institutional:		
Without plaster ceiling............	$l/240$	$l/180$
With plaster ceiling..............	$l/360$	$l/240$
Floor beams:		
Ordinary usage†	$l/360$	$l/240$
Highway bridge stringers..............	$l/200$ to $l/300$	
Railway bridge stringers	$l/300$ to $l/400$	

* Camber and Deflection, AITC 102, Appendix B, American Institute of Timber Construction.

† Ordinary usage classification is intended for construction in which walking comfort, minimized plaster cracking, and elimination of objectionable springiness are of prime importance. For special uses, such as beams supporting vibrating machinery or carrying moving loads, more severe limitations may be required.

TABLE 7-28   Recommended Minimum Camber for Glued-Laminated Timber Beams*

Roof beams†	1½ times dead-load deflection
Floor beams‡	1½ times dead-load deflection
Bridge beams: §	
Long span	2 times dead-load deflection
Short span	2 times dead load plus ½ applied-load deflection

* Camber and Deflection, AITC 102, Appendix B, American Institute of Timber Construction.

† The minimum camber of 1½ times dead-load deflection will produce a nearly level member under dead load alone after plastic deformation has occurred. Additional camber is usually provided to improve appearance or provide necessary roof drainage.

‡ The minimum camber of 1½ times dead-load deflection will produce a nearly level member under dead load alone after plastic deformation has occurred. On long spans, a level ceiling may not be desirable because of the optical illusion that the ceiling sags. For warehouse or similar floors where live load may remain for long periods, additional camber should be provided to give a level floor under the permanently applied load.

§ Bridge members are normally cambered for dead load only on multiple spans to obtain acceptable riding qualities.

---

caused by ponding of water as increasing deflections permitted more and more water to collect.

Roof beams should have a continuous upward slope equivalent to ¼ in/ft between a drain and the high point of a roof, in addition to minimum recommended camber (Table 7-28), to avoid ponding. When flat roofs have insufficient slope for drainage (less than ¼ in/ft), the stiffness of supporting members should be such that a 5-lb/ft² load will cause no more than ½-in deflection.

Because of ponding, snow loads or water trapped by gravel stops, parapet walls, or ice dams magnify stresses and deflections from existing roof loads by

$$C_p = \frac{1}{1 - W'L^3/\pi^4 EI} \tag{7-17}$$

where C_p = factor for multiplying stresses and deflections under existing loads to determine stresses and deflections under existing loads plus ponding

W' = weight of 1 in of water on roof area supported by beam, lb

L = span of beam, in

E = modulus of elasticity of beam material, psi

I = moment of inertia of beam, in⁴

(Kuenzi and Bohannan, "Increases in Deflection and Stresses Caused by Ponding of Water on Roofs," Forest Products Laboratory, Madison, Wis.)

7-23. Design of Wood Trusses.

Type of truss and arrangement of members may be chosen to suit the shape of structure, the loads, and stresses involved. The types most commonly built are bowstring, flat or parallel chord, pitched, triangular or A type, camelback, and scissors (Fig. 7-13). For most construction other than houses, trusses usually are spaced 12 to 20 ft apart. For houses, very light trusses generally are erected 16 to 24 in c to c.

Joints are critical in the design of a truss. Use of a specific truss type is often governed by joint considerations.

Chords and webs may be single-leaf (or monochord), double-leaf, or multiple-leaf members. Monochord trusses and trusses with double-leaf chords and single-leaf web system are common. Web members may be attached to the sides of the chords, or web members may be in the same plane as the chords and attached with straps or gussets.

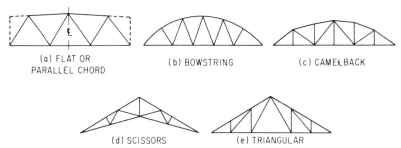

Fig. 7-13 Types of wood trusses.

Individual truss members may be solid-sawn, glued-laminated, or mechanically laminated. Glued-laminated chords and solid-sawn web members are usually used. Steel rods or other steel shapes may be used as members of timber trusses if they meet design and service requirements.

The bowstring truss is by far the most popular. Spans of 100 to 200 ft are common, with single or two-piece top and bottom chords of glued-laminated timber, webs of solid-sawn timber, and metal heel plates, chord plates, and web-to-chord connections. This system is lightweight for the loads that it can carry. It can be shop or field assembled. Attention to the top chord, bottom chord, and heel connections is of prime importance, since they are the major stress-carrying components. Since the top chord is nearly the shape of an ideal arch, stresses in chords are almost uniform throughout a bowstring truss; web stresses are low under uniformly distributed loads.

Parallel-chord trusses, with slightly sloping top chords and level bottom chords, are used less often, because chord stresses are not uniform along their length and web stresses are high. Hence, different cross sections are required for successive chords, and web members and web-to-chord connections are heavy. Eccentric joints and tension stresses across the grain should be avoided in truss construction, whenever possible, but particularly in parallel-chord trusses.

Triangular trusses and the more ornamental camelback and scissors trusses are used for shorter spans. They usually have solid-sawn members for both chords and webs where degree of seasoning of timbers, hardware, and connections are of considerable importance.

Truss Joints. For joints, split-ring and shear-plate connectors are generally most economical. Sometimes, when small trusses are field-fabricated, only bolted joints are used. However, grooving tools for connectors can also be used effectively in the field.

Framing between Trusses. Longitudinal sway bracing perpendicular to the plane of the truss is usually provided by solid-sawn X bracing. Lateral wind bracing may be provided by end walls or intermediate walls, or both. The roof system and horizontal bracing should be capable of transferring the wind load to the walls. Knee braces between trusses and columns are often used to provide resistance to lateral loads.

Horizontal framing between trusses consists of struts between trusses at bottom-chord level and diagonal tie rods, often of steel with turnbuckles for adjustment.

Bowstring-Truss Framing. Table 7-29 gives typical bowstring-truss dimensions based on uniform loading conditions on the top chord, as normally imposed by roof joists.

For ordinary roof loads and spacing of 16 to 24 ft, vertical X bracing is required every 30 to 40 ft of chord length. This bracing is placed in alternate bays. Horizontal T-strut bracing should be placed from lower chord to lower chord in the same line as the vertical X bracing for the complete length of the building. If a ceiling is framed into the lower chords, these struts may be omitted.

TABLE 7-29 Bowstring-Truss Dimensions

Span range, ft	No. of panels	Avg truss height*	Avg arc length†
53–57	6	8 ft 7 in.	60 ft 8 in.
58–62	...	9 ft 3 in.	65 ft 11 in.
63–67	...	9 ft 11 in.	71 ft 5 in.
68–72	8	10 ft 7 in.	76 ft 5 in.
73–77	...	11 ft 3 in.	81 ft 8 in.
78–82	...	11 ft 11 in.	86 ft 10 in.
83–87	...	12 ft 7 in.	92 ft 1 in.
88–92	...	13 ft 3 in.	97 ft 4 in.
93–97	...	13 ft 11½ in.	102 ft 7 in.
98–102	...	14 ft 7½ in.	107 ft 10 in.
103–107	10	15 ft 5 in.	113 ft 2 in.
108–112	...	16 ft 1 in.	118 ft 5 in.
113–117	...	16 ft 9 in.	123 ft 8 in.
118–122	...	17 ft 5 in.	128 ft 11 in.
123–127	...	18 ft 1 in.	134 ft 2 in.
128–132	12	18 ft 11 in.	139 ft 6 in.
133–137	...	19 ft 7 in.	144 ft 9 in.

* The vertical distance from top of truss heel bearing plate to top of chord at midspan.
† Measured along the top of the top chord, wood to wood. Dimensions shown are for the longest of the span range. For smaller spans, lengths are proportionately shorter.

Joists, spaced 12 to 24 in c to c, usually rest on the top chords of the trusses and are secured there by toenailing. They may also be placed on ledgers attached to the sides of the upper chords or set in metal hangers, thus lowering the roof line.

Purlins, large-cross-sectional joists spaced 4 to 8 ft c to c, are often set on top of the top chords, where they are butted end to end and secured to the chords with clip angles. Purlins may also be set between the top chords on metal purlin hangers. Roof sheathing, 1-in. D&M (dressed and matched) or ⅜- to ½-in thick plywood, is laid directly on joists. For purlin construction, 2-in D&M roof sheathing is normally used.

("Design Manual for Teco Timber Connector Construction," Timber Engineering Co., 1619 Massachusetts Ave., N.W., Washington, D.C. 20036; AITC 102, Appendix A, "Trusses and Bracing," American Institute of Timber Construction, 333 W. Hampden Ave., Englewood, Colo., 80110.)

7-24. Design of Timber Arches.

Arches may be two-hinged, with hinges at each base, or three-hinged, with a hinge at the crown. Figure 7-14 presents typical forms of arches.

Tudor arches are gabled rigid frames with curved haunches. Columns and pitched roof beam on each side of the crown usually are one piece of glued-laminated timber. This type of arch is frequently used in church construction with a high rise.

Preliminary sizes of key sections of a Tudor arch, for use in a detailed arch analysis, may be determined for uniform vertical loads as follows (for steep roofs, the effect of wind loads must be investigated): Assume an allowable bending stress F_b = 2200 psi, after adjustment for short-duration loading, curvature, and size factor; horizontal shear

Fig. 7-14 Types of wood arches.

stress $F_v = 230$ psi; live plus dead load of 45 lb/ft²; span $L = 60$ ft; rise from base to crown $T = 25$ ft; arch spacing = 16 ft; and haunch radius = 7 ft for ¾-in laminations (Fig. 7-15a).

The total uniform vertical load $w = 45 \times 16 = 720$ lb/lin ft. Compute the horizontal thrust:

$$H = \frac{wL^2}{8T} = \frac{720 \times 60^2}{8 \times 25} = 12{,}960 \text{ lb}$$

For this shearing force, determine the base depth d for an assumed width b of 6¾ in.

$$d = \frac{3H}{2F_v b} = \frac{3 \times 12{,}960}{2 \times 230 \times 6.75} = 12.6$$

Use $d = 12\frac{1}{2}$ in. Next compute the bending moment M at the critical section, which is a distance D from the base hinge. To find D, subtract the dimension given in Table 7-30 for a 3:12 pitch from the wall height: $17.50 - 5.47 = 12.03$ ft.

$$M = 12HD = 12 \times 12{,}960 \times 12.03 = 1{,}871{,}000 \text{ in-lb}$$

The section modulus required at the critical section then is

$$S = \frac{M}{F_b} = \frac{1{,}871{,}000}{2200} = 850 \text{ in}^3$$

From Table 7-5, for ¾-in laminations, select the most economical section with this section modulus: 6¾ × 27¾, with $S = 866$. Make the upper tangent the same size as the lower.

TABLE 7-30 Distance of Critical Section below Working Point
(*WP* in Fig. 7-15)

Pitch	1:12	2:12	3:12	4:12	5:12	6:12	7:12	8:12	9:12
Distance, ft	6.43	5.94	5.47	5.05	4.67	4.32	4.03	3.75	3.5

Finally, determine the depth d at the crown. This depth should at least equal the width, and preferably, it should be one-third larger. Hence,

$$d = 1.33 \times 6.75 = 9 \text{ in}$$

This may be increased for architectural reasons or to equal depth of purlins. Figure 7-15b summarizes the preliminary dimensions of the arch.

Segmented arches are fabricated with overlapping lumber segments, nailed or glued-laminated. They generally are three-hinged, and they may be tied or buttressed. If an arch is tied, the tie rods, which resist the horizontal thrust, may be above the ceiling or below grade, and simple connections may be used where the arch is supported on masonry walls, concrete piers, or columns (Fig. 7-16).

Segmented arches are economical because of the ease of fabricating them and simplicity of field erection. Field splice joints are minimized; generally there is only one simple connection, at the crown (Fig. 7-17c). Except for extremely long spans, they are shipped in only two pieces. Erected, they need not be concealed by false ceilings, as may be necessary with trusses. And the cross section is large enough for segmented arches to be classified as heavy-timber construction.

Figure 7-18 shows typical moment connections for arches.

7-25. Wood Framing for Small Houses.

Fig. 7-15 Tudor arch.

Though skeleton framing may be used for one- and two-family dwellings, such structures up to three stories high generally are built with load-bearing walls. When wood framing is used, the walls are conventionally built with slender studs spaced 16 in c to c. Similarly, joists and rafters, which are supported on the walls and partitions, are usually also spaced 16 in c to c. Facings, such as sheathing and wallboard, and decking, floor underlayment, and roof sheathing are generally available in appropriate sizes for attachment to studs, joists, and rafters with that spacing. Sometimes, however, builders find use of 24-in spacing with thicker attached materials more economical.

Wood studs are usually set in walls and partitions with wide faces perpendicular to the face of the wall or partition. The studs are nailed at the bottom to bear on a horizontal plank, called the bottom or sole plate, and at the top to a pair of horizontal planks, called the top plate. These plates often are the same size as the studs. Joists or rafters may be supported on the top plate or on a header, called a ribband, supported in a cutout in the studs (Fig. 7-19).

Studs may be braced against racking by diagonals or horizontal blocking and facing materials, such as plywood or gypsum sheathing.

Fig. 7-16 Bases for segmented wood arches: (a) and (b) tie rod anchored to arch shoe; (c) hinge anchorage for large arch; (d) welded arch shoe.

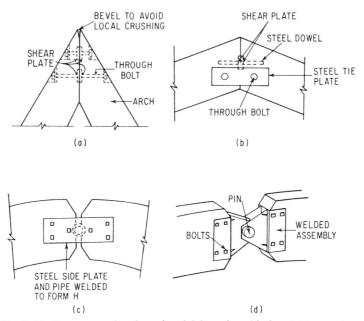

Fig. 7-17 Crown connections for arches: (a) for arches with slope 4:12 or greater, connection consists of pairs of back-to-back shear plates with through bolts or threaded rods counterbored into the arch; (b) for arches with flatter slopes, shear plates centered on a dowel may be used in conjunction with tie plates and through bolts; (c) and (d) hinge at crown.

With 2 × 4 or 3 × 4-in studs, 16 in c to c, walls may be as tall as 14 ft; with 2 × 6-in studs, up to 20 ft high. Angles at corners where stud walls or partitions meet should be framed solid so that no lath may extend from one room to another.

Three types of wood bearing-wall construction generally are used in the United States: braced frame or eastern, balloon frame, and platform frame or western.

Braced-Timber Framing. This type originated in New England when the settlers were required to cut framing members from logs. It is the oldest type but it is also the strongest and most rigid.

In modern adaptations, sometimes called the **combination frame,** a braced-timber exterior frame consists of a wood sill resting on the foundation wall; comparatively heavy posts at all corners and some intermediate points, seated on the sill; horizontal members, called girts, at second-floor level—"dropped girts" supporting ends of joists and "raised girts" paralleling the joists; a plate across the top of the frame, supporting roof rafters; and studs, 16 in c to c between posts, extending from sill to girts and from girts to plate. Diagonal bracing is set between posts and sill, girts and plate. Posts are tenoned into sills, and girts are connected to posts with mortise-and-tenon joints.

Rough boards, ⅞ in thick, generally are used to sheathe braced frames. Set horizontally, each board should be attached with at least two nails to each stud (three nails if boards are more than 6 in wide).

Balloon Framing. This type (Fig. 7-19a), built almost entirely of 2-in lumber, is distinguished by two-story-high studs, to the faces of which joists at all levels are nailed. Ends of the upper-story joists rest on a false girt, or "ribband." It differs from the braced frame also because of the omission of diagonal brac-

Fig. 7-18 Moment connections for an arch: (a) and (b) connections with top and bottom steel plates; (c) connections with side plates.

ing; instead the frame is stiffened by diagonal boards or sheathing. Fire stops—horizontal 2 × 4s—are placed between the studs at about midheight and at floor level in each story.

Diagonal boards should be nailed to each stud with two or four nails. (Three nails are no better than two under racking action, because the center nail acts as a pivot and adds nothing to the strength.) Plywood or gypsum sheathing, applied in large sheets, also imparts considerable rigidity to the frame. It should be nailed with 8d common nails, 6 in c to c, for ⅝-in material and with 6d nails, 2½ in c to c, for ¼-in material.

Platform Framing. This type (Fig. 7-19b and c), also known as western framing, is characterized by one-story-high walls supported by the floors at each level. At the bot-

7-64 Wood Construction

Fig. 7-19 Framing for small houses: (a) balloon; (b) and (c) platform or western; (d) detail of balloon framing at second floor.

tom, joists rest on the sill and frame into headers of the same depth, which extend around the periphery of the building. Generally, before the superstructure is erected, the first-floor subflooring is laid on the joists to serve as a working platform. Then, a sole plate of the same size as the studs is set on top of the subfloor to form a base for the verticals. The studs extend to a plate at the next floor level. (Some builders assemble the wall framing in a horizontal position, then tilt it into place.) Joists are seated on the plate and frame into a header. After that, subflooring is placed over the joists, extending to the edge of the frame, another sole is laid, and the cycle is repeated. For this type of framing also, plywood or gypsum sheathing, well nailed, is highly desirable.

7-26. Wood Foundations.

Plywood and lumber walls are an alternative to concrete for foundation walls for one-story and multistory houses and other light-frame buildings. Main components of a wood foundation wall are plywood, ⅝ in or more thick, and wood studs, spaced 12 in or more on centers, both pressure-treated with preservative (Fig. 7-20). The advantages of a wood foundation over concrete are faster construction, because there is no delay due to the wait for concrete or unit masonry to cure, and easier interior finishing, because wood foundations provide nailable studs for the usual finishes.

Plywood should be an exterior type or an interior type that has been bonded with

Fig. 7-20 Wood foundation.

exterior glue, both types manufactured to meet the requirements of U.S. Product Standard PS 1 for Construction and Industrial Plywood or one of the American Plywood Association's performance-engineered proprietary specifications. Lumber should be grade marked by an approved inspection agency, should be capable of accepting pressure preservative treatment, and should be of a species for which allowable unit stresses are given in the NFPA "National Design Specification for Wood Construction."

Treatment for the plywood and lumber involves impregnating into the wood under heat and pressure ammoniacal copper arsenate or chromated copper arsenate (A, B, or C). Salt retention should be at least 0.60 lb/ft^3 of wood, 50% more than building codes usually require for ground-contact applications. After impregnation, plywood should be dried to a moisture content of 18% or less, and lumber, to 19% or less. Portions of the wall more than 8 in above the ground, however, need not be pressure treated. If any of the materials have to be cut after treatment, the cut edges, unless they will be 8 in or more above grade, should be field treated with the preservative used in the original treatment but with a minimum concentration of 3% in solution. To minimize end cutting and field treatment of footing plates, these members may be extended past the corners of the foundation.

Only corrosion-resistant fasteners should be used in wood foundations. Plywood may be attached to the studs with Type 304 or 316 stainless-steel nails, hot-dipped or hot-tumbled galvanized nails, or silicon-bronze or copper fasteners. Lumber-to-lumber fasteners, whether above or below grade, should be hot-dipped or hot-tumbled galvanized nails, except that stainless-steel fasteners should be used for knee-wall assemblies.

To keep the interior of the foundation walls dry, it is advisable to enclose the exterior with a waterproofing membrane, such as polyethylene film, and to ensure good drainage. For the latter purpose, the ground surface should be sloped away from the building at ½ in/ft for a distance of at least 6 ft and gravel should be placed against the walls and under the cellar floor.

("The All-Weather Wood Foundation System: Design, Fabrication and Installation Manual," National Forest Products Association, 1619 Massachusetts Ave., Washington, D.C. 20036; "All-Weather Wood Foundation System," American Plywood Association, P.O. Box 11700, Tacoma, Wash. 98411; "Under-Floor Plenum Manual," National Association of Home Builders, 15th and M St., Washington, D.C. 20005.)

7-27. Timber Decking.

Wood decking used for floor and roof construction may consist of solid-sawn planks with nominal thickness of 2, 3, or 4 in. Or it may be panelized or laminated. Panelized decking is made up of splined panels, usually about 2 ft wide.

Mechanically laminated decks are erected by setting square-edge dimension lumber on edge and fastening wide face to wide face. If side nails are used, they should be long enough to penetrate 2½ lamination thicknesses for load transfer. Where supports are 4 ft c to c or less, side nails should be spaced not more than 30 in c to c. They should be staggered one-third of the spacing in adjacent laminations. When supports are more than 4 ft c to c, nails should be spaced about 18 in c to c alternately near top and bottom edges. These nails also should be staggered one-third of the spacing in adjacent laminations. Two side nails should be used at each end of butt-joined pieces. In addition, laminations should be toenailed to supports with nails 4 in (20d) or longer. When supports are 4 ft c to c or less, alternate laminations should be nailed to alternate supports; for longer spans, alternate laminations at least should be nailed to every support.

For glued-laminated decking, two or more pieces of lumber are laminated into a single decking member, usually with 2- to 4-in nominal thickness.

Solid-sawn decking usually is fabricated with edges tongued and grooved, shiplap, or groove cut for splines, for transfer of vertical load between pieces. The decking may be end-matched, square end, or end-grooved for splines. As indicated in Fig. 7-21, the decking may be arranged in various patterns over supports.

In Type 1, the pieces are simply supported. Type 2 has a controlled random layup. Type 3 contains intermixed cantilevers. Type 4 consists of a combination of simple-span and two-span continuous pieces. Type 5 is two-span continuous.

In Types 1, 4, and 5, end joints bear on supports. For this reason, these types are recommended for thin decking, such as 2-in.

Type 3, with intermixed cantilevers, and Type 2, with controlled random layup, are used for deck continuous over three or more spans. These types permit some of the end joints to be located between supports. Hence, provision must be made for stress transfer at those joints. Tongue-and-groove edges, wood splines on each edge of the course, horizontal spikes between courses, and end matching or metal end splines may be used to transfer shear and bending stresses.

In Type 2, the distance between end joints in adjacent courses should be at least 2 ft for 2-in deck and 4 ft for 3- and 4-in deck. Joints approximately lined up (within 6 in of being in line) should be separated by at least two courses. All pieces should rest on at least one support. And not more than one end joint should fall between supports in each course.

In Type 3, every third course is simple span. Pieces in other courses cantilever over

Timber Decking   7-67

Fig. 7-21 Typical arrangement patterns for heavy-timber decking.

supports and end joints fall at alternate quarter or third points of the spans. Each piece rests on at least one support.

To restrain laterally supporting members of 2-in deck in Types 2 and 3, the pieces in the first and second courses and in every seventh course should bear on at least two supports. End joints in the first course should not occur on the same supports as end joints in the second course unless some construction, such as plywood overlayment, provides continuity. Nail end distance should be sufficient to develop the lateral nail strength required.

Heavy-timber decking is laid with wide faces bearing on the supports. Each piece must be nailed to each support. Each end at a support should be nailed to it. For 2-in decking a 3½-in (16d) toe and face nail should be used in each 6-in-wide piece at supports, and three nails for wider pieces. Tongue-and-groove decking generally is also toenailed through the tongue. For 3-in decking, each piece should be toenailed with one 4-in (20d) spike and face-nailed with one 5-in (40d) spike at each support. For 4-in decking, each piece should be toenailed at each support with one 5-in (40d) nail and face-nailed there with one 6-in (60d) spike.

Courses of 3- and 4-in double tongue-and-groove decking should be spiked to each other with 8½-in spikes not more than 30 in apart. One spike should not be more than 10 in from each end of each piece. The spikes should be driven through predrilled holes. Two-inch decking is not fastened together horizontally with spikes.

7-68 Wood Construction

Deck design usually is governed by maximum permissible deflection in end spans. But each design should be checked for bending stress. Deflection, in, may be calculated for a 12-in-wide section from

$$\delta = \frac{wL^4}{K_1 EI} \tag{7-18}$$

where w = uniform load, lb/ft^2
L = span, ft
E = modulus of elasticity, psi
I = moment of inertia, in^4
K_1 = constant, obtained from Table 7-31

Bending stress, psi, for a 12-in-wide section may be computed from

$$f_b = \frac{wL^2}{K_2 S} \tag{7-19}$$

where S = section modulus, in^3. Use the full cross section of the deck for Types 1, 4, and 5; use two-thirds for Types 2 and 3

K_2 = constant, obtained from Table 7-31

(AITC 112, Standard for Heavy Timber Roof Decking, and AITC 118, Standard for 2-in. Nominal Thickness Lumber Roof Decking for Structural Applications, American Institute of Timber Construction, 333 W. Hampden Ave., Englewood, Colo., 80110; AITC "Timber Construction Manual," John Wiley & Sons, Inc., New York.)

TABLE 7-31 Constants for Computing Deck Deflections and Bending Stresses

Deck type	K_1	K_2
1	0.0445	0.667
2	0.0579	0.555
3	0.0608	0.555
4	0.0631	0.667
5	0.1071	0.667

7-28. Plywood.

Plywood is made of softwood layers glued together; the grain of each layer is perpendicular to the grain of adjoining layers. Softwood plywood is manufactured in accordance with U.S. Product Standard PS1-74 for Construction and Industrial Plywood. Plywood for exterior use should be made with an adhesive that is fully waterproof and suitable for wet service conditions and severe exposure.

PS1-74 classifies the softwoods according to stiffness:

Group 1. Douglas fir from Washington, Oregon, California, Idaho, Montana, Wyoming, British Columbia, and Alberta; western larch; southern pine (loblolly, longleaf, shortleaf, slash); yellow birch; American beech; tan oak.

Group 2. Port Orford cedar; Douglas fir from Nevada, Utah, Colorado, Arizona, and New Mexico; fir (California red, grand, noble, Pacific silver, white); western hemlock; red and white lauan; western white pine, red pine; black maple; yellow poplar; red and Sitka spruce.

Group 3. Red alder; Alaska cedar; jack, lodgepole, spruce and ponderosa pine; paper birch; subalpine fir; eastern hemlock; bigleaf maple; redwood; black, Engelmann and white spruce.

Group 4. Incense and western red cedar, sugar and eastern white pine, eastern and black (western poplar) cottonwood, cativo, paper birch, and bigtooth and quaking aspen.

Group 5. Balsam fir, basswood, and balsam poplar.

Construction plywood is graded under the standard in accordance with two basic systems. One system covers engineered grades, the other appearance grades.

Engineered grades consist largely of unsanded sheathing panels designated C-D Interior or C-C Exterior. The latter is bonded with exterior glue. Either of these grades may be classified as Structural I or Structural II, both of which are made with exterior glue and are subject to other requirements, such as limitations on knot size and repairs of defects. Structural I is made only of Group 1 species and is stiffer than other grades. Structural II may be made of Group 1, 2, or 3, or any combination of these species. Structural I and II are suitable for such applications as box beams, gusset plates, stressed-skin panels, and folded-plate roofs.

Appearance grades, except for Plyform, are designated by panel thickness, veneer classification of face and back veneers, and species group of the veneers. For Plyform, class designates a mix of species.

Table 7-32 describes the various grades of plywood and indicates how they are generally used. The table also indicates the stress-grade level to be used for determination of stresses from Table 7-35.

The standard classifies plywood made for use as concrete forms in two grades. Plyform (B-B) Class I is limited to Group 1 species on face and back, with limitations on inner plies. Plyform (B-B) Class II permits Group 1, 2, or 3 for face and back, with limitations on inner plies. High-density overlay should be specified for both classes when highly smooth, grain-free concrete surfaces and maximum reuses are required. The bending strength of Plyform Class I is greater than that of Class II. Grades other than Plyform, however, may be used for forms.

Manufacturers stamp plywood to indicate group, grade, mill number, product standard, thickness, exposure durability classification, and Plyform class. They also include a span rating. This consists of a pair of numbers separated by a slant bar, on unsanded construction grades. The number on the left indicates the maximum recommended spacing, in, of supports when the panel is used for roof decking. The number on the right gives the maximum recommended spacing, in, of supports when the panel is used for subflooring. If a zero is stamped on the right, the panel should not be used for subflooring. Table 7-34 relates species group to plywood thickness and span rating.

Appearance grades of plywood also are identified by a stamp. The group number indicates the species group of the veneer used on face and back. Letter symbols describe these veneer grades used in plywood:

N—Special-order natural-finish veneer. Selected all heartwood or all sapwood. Free of open defects. Allows some repairs.

A—Smooth and paintable. Neatly made repairs permissible. Also used for natural finish in less-demanding applications.

B—Solid-surface veneer. Circular repair plugs and tight knots permitted. Can be painted.

C—Minimum veneer permitted in exterior-type plywood. Knotholes up to 1 in in diameter. (Occasional knotholes ½ in larger permitted if total width of all knots and knotholes within a specified section does not exceed certain limits.) Limited splits and tight knots up to 1½ in are permitted.

C plugged—Improved C veneer with splits limited to ⅛ in in width and knotholes and borer holes limited to ¼ × ½ in.

D—Used only in interior-type plywood for inner plies and backs. Permits limited splits and knots and knotholes up to 2½ in (½ in larger under specified conditions).

TABLE 7-32 Applications of Plywood Grades and Guide to Use of Tables 7-33 and 7-35

Plywood grade	Description and use	Veneer grade Face	Veneer grade Back	Veneer grade Inner	Common thicknesses, in	Grade stress level (Table 7-35)	Species group	Section property table
(a) Interior-type plywood								
C-D INT-APA	Unsanded sheathing grade for wall, roof, subflooring, and industrial applications such as pallets and for engineering design, with proper stresses. Also available with intermediate and exterior glue.* For permanent exposure to weather or moisture only, exterior-type plywood is suitable.	C	D	D	5/16, 3/8, 1/2, 5/8, 3/4	S-3*	See Table 7-34	Table 7-33a (unsanded)
Structural I C-D INT-APA or Structural II C-D INT-APA†	Plywood grades to use where strength properties are of maximum importance, such as plywood-lumber components. Made with exterior glue only. Structural I is made from all Group 1 woods. Structural II allows Group 3 woods.	C	D	D	5/16, 3/8, 1/2, 5/8, 3/4	S-2	For Structural I, use Group 1. For Structural II, use Group 3.	Table 7-33b (unsanded)
Underlayment INT-APA	For underlayment or combination subfloor-underlayment under resilient floor coverings. Available with exterior glue. Touch-sanded. Available with tongue and groove.	C plugged	D	C & D	1/2, 19/32, 5/8 23/32, 3/4	S-3*	As Specified	Table 7-33a (touch-sanded)
C-D Plugged INT-APA	For built-ins, wall and ceiling tile backing, *not* for underlayment. Available with exterior glue. Touch-sanded.	C plugged	D	D	1/2, 19/32, 5/8, 23/32, 3/4	S-3*	As Specified	Table 7-33a (touch-sanded)
Structural I or II† Underlayment or C-D plugged	For higher strength requirements for underlayment or built-ins. Structural I constructed from all Group 1 woods. Made with exterior glue only.	C plugged	D	C & D	1/2, 19/32, 5/8, 23/32, 3/4	S-2	For Structural I, use Group 1. For Structural II, use Group 3.	Table 7-33b (touch-sanded)
2.4.1 INT-APA	Combination subfloor-underlayment. Quality floor base. Available with exterior glue, most often touch-sanded. Available with tongue and groove.	C plugged	D	C & D	1 1/8	S-3*	Group 1	Table 7-33a
Appearance grades	Generally applied where a high quality surface is required. Includes N-N, N-A, N-B, N-D, A-A, A-B, A-D, B-B, and B-D INT-APA Grades.	B or better	D or better	D	1/4, 3/8, 1/2, 5/8, 3/4	S-3*	As Specified	Table 7-33a (sanded)
(b) Exterior-type plywood								
C-C EXT-APA	Unsanded sheathing grade with waterproof glue bond for wall, roof, subfloor, and industrial applications such as pallet bins.	C	C	C	5/16, 3/8, 1/2, 5/8, 3/4	S-1	See Table 7-34	Table 7-33a (unsanded)

7-70

Grade designation	Description	Face	Back	Inner plies	Thickness (in)	Surface & stress	Species group	Table
Structural I C-C EXT-APA or Structural II C-C EXT-APA†	"Structural" is a modifier for this unsanded sheathing grade. For engineering applications in construction and industry where full exterior-type panels are required. Structural I is made from Group 1 woods only.	C	C	C	5/16, 3/8, 1/2, 5/8, 3/4	S-1	For Structural I, use Group 1. For Structural II, use Group 3.	Table 7-33b (unsanded)
Underlayment EXT-APA and C-C plugged EXT-APA	Underlayment for combination subfloor underlayment or two-layer floor under resilient floor coverings where severe moisture conditions may exist. Also for controlled atmosphere rooms and many industrial applications. Touch-sanded. Available with tongue and groove.	C plugged	C	C	1/2, 19/32, 23/32, 5/8, 3/4	S-2	As Specified	Table 7-33a (touch-sanded)
Structural I or II† underlayment EXT-APA or C-C plugged EXT-APA	For higher strength underlayment where severe moisture conditions may exist. All Group 1 construction in Structural I. Structural II allows Group 3 woods.	C plugged	C	C	1/2, 19/32, 23/32, 5/8, 3/4	S-2	For Structural I, use Group 1. For Structural II, use Group 3	Table 7-33b (touch-sanded)
B-B Plyform Class I or II†	Concrete-form grade with high reuse factor. Sanded both sides, mill-oiled unless otherwise specified. Available in HDO. For refined design information on this special-use panel see APA publication "Plywood for Concrete Forming" (form V345). Design using values from this specification will result in a conservative design.	B	B	C	5/8, 3/4	S-2	For Class I, use Group 1. For Class II, use Group 3.	Table 7-33a (sanded)
Marine EXT-APA	Superior exterior-type plywood made only with Douglas Fir or Western Larch. Special solid-core construction. Available with MDO or HDO face. Ideal for boat hull construction.	A or B	B	B	1/4, 3/8, 1/2, 5/8, 3/4	A face & back use S-1. B face or back use S-2	Group 1	Table 7-33b (sanded)
Appearance Grades	Generally applied where a high quality surface is required. Includes AA, A-B, A-C, B-B, B-C, HDO and MDO EXT-APA. Appearance grades may be modified to Structural I. For such designation use Group 1 stresses and Table 7-34b (sanded) section properties.	B or better	C or better	C	1/4, 3/8, 1/2, 5/8, 3/4	A or C face and back use S-1. B face or back use S-2	As Specified	Table 7-33a (sanded)

*When exterior glue is specified, i.e., "interior with exterior glue," stress level 2 (S-2) should be used.
†Check local suppliers for availability of Structural II and Plyform Class II grades.
From "Plywood Design Specification," American Plywood Association.

7-71

TABLE 7-33 Effective Section Properties for Plywood[*]

(a) Face plies of species group different from that of inner plies (includes all product standard grades except those noted in part b)

Nominal thickness, in	Approximate weight, lb/ft²	Effective thickness for shear, in	Stress applied parallel to face grain					Stress applied perpendicular to face grain			
			Area A, in²/ft	Moment of inertia I, in⁴/ft	Eff. section modulus KS, in³/ft	Rolling shear constant Ib/Q, in²/ft	Area A, in²/ft	Moment of inertia I, in⁴/ft	Eff. section modulus KS, in³/ft	Rolling shear constant Ib/Q, in²/ft
Unsanded panels										
5/16-U	1.0	0.283	1.914	0.025	0.124	2.568	0.660	0.001	0.023	—
3/8-U	1.1	0.293	1.866	0.041	0.162	3.108	0.799	0.002	0.033	—
1/2-U	1.5	0.316	2.500	0.086	0.247	4.189	1.076	0.005	0.057	2.585
5/8-U	1.8	0.336	2.951	0.154	0.379	5.270	1.354	0.011	0.095	3.252
3/4-U	2.2	0.467	3.403	0.243	0.501	6.823	1.632	0.036	0.232	3.717
7/8-U	2.6	0.757	4.109	0.344	0.681	7.174	2.925	0.162	0.542	5.097
1-U	3.0	0.859	3.916	0.493	0.859	9.244	3.611	0.210	0.660	6.997
1 1/8-U	3.3	0.877	4.621	0.676	1.047	10.008	3.464	0.307	0.821	8.483
Sanded panels										
1/4-S	0.8	0.304	1.680	0.013	0.092	2.175	0.681	0.001	0.020	—
3/8-S	1.1	0.313	1.680	0.038	0.176	3.389	1.181	0.004	0.056	—
1/2-S	1.5	0.450	1.947	0.077	0.266	4.834	1.281	0.018	0.150	3.099
5/8-S	1.8	0.472	2.280	0.129	0.356	6.293	1.627	0.045	0.234	3.922
3/4-S	2.2	0.589	2.884	0.197	0.452	7.881	2.104	0.093	0.387	4.842
7/8-S	2.6	0.608	2.942	0.278	0.547	8.225	3.199	0.157	0.542	5.698
1-S	3.0	0.846	3.776	0.423	0.730	8.882	3.537	0.253	0.744	7.644
1 1/8-S	3.3	0.865	3.854	0.548	0.840	9.883	3.673	0.360	0.918	9.032
Touch-sanded panels										
1/2-T	1.5	0.346	2.698	0.083	0.271	4.252	1.159	0.006	0.061	2.746
19/32-T	1.7	0.491	2.618	0.123	0.337	5.403	1.610	0.019	0.150	3.220
5/8-T	1.8	0.497	2.728	0.141	0.364	5.719	1.715	0.023	0.170	3.419
23/32-T	2.1	0.503	3.181	0.196	0.447	6.600	2.014	0.035	0.226	3.659
3/4-T	2.2	0.509	3.297	0.220	0.477	6.917	2.125	0.041	0.251	3.847
(2·4·1)1 1/8-T	3.3	0.855	4.592	0.653	0.995	9.933	4.120	0.283	0.763	7.452

(b) All plies from same species group (includes Structural I and Marine)

Unsanded panels								
5/16-U	1.0	0.356	2.375	0.025	0.144	2.567	1.188	0.002	0.029	—
3/8-U	1.1	0.371	2.226	0.041	0.195	3.107	1.438	0.003	0.043	—
1/2-U	1.5	0.403	2.906	0.091	0.318	4.188	1.938	0.007	0.077	2.574
5/8-U	1.8	0.434	3.464	0.155	0.433	5.268	2.438	0.015	0.122	3.238
3/4-U	2.2	0.606	3.672	0.247	0.573	6.817	2.938	0.059	0.334	3.697
7/8-U	2.6	0.776	4.388	0.346	0.690	6.948	3.510	0.192	0.584	5.086
1-U	3.0	1.088	5.200	0.529	0.922	8.512	6.500	0.366	0.970	6.986
1 1/8-U	3.3	1.119	6.654	0.751	1.164	9.061	5.542	0.503	1.131	8.675
Sanded panels								
1/4-S	0.8	0.342	1.680	0.013	0.092	2.172	1.226	0.001	0.027	—
3/8-S	1.1	0.373	1.680	0.038	0.177	3.382	2.126	0.007	0.078	—
1/2-S	1.5	0.545	1.947	0.078	0.271	4.816	2.305	0.030	0.217	3.076
5/8-S	1.8	0.576	2.280	0.131	0.361	6.261	2.929	0.077	0.343	3.887
3/4-S	2.2	0.748	3.848	0.202	0.464	7.926	3.787	0.162	0.570	4.812
7/8-S	2.6	0.778	3.952	0.288	0.569	7.539	5.759	0.275	0.798	5.671
1-S	3.0	1.091	5.215	0.479	0.827	7.978	6.367	0.445	1.098	7.639
1 1/8-S	3.3	1.121	5.593	0.623	0.955	8.840	6.611	0.634	1.356	9.031
Touch-sanded panels								
1/2-T	1.5	0.403	2.698	0.084	0.282	4.246	2.086	0.008	0.082	2.720
19/32-T	1.7	0.567	3.127	0.124	0.349	5.390	2.899	0.030	0.212	3.183
5/8-T	1.8	0.575	3.267	0.144	0.378	5.704	3.086	0.037	0.242	3.383
23/32-T	2.1	0.598	3.337	0.201	0.469	6.582	3.625	0.057	0.322	3.596
3/4-T	2.2	0.606	3.435	0.226	0.503	6.900	3.825	0.067	0.359	3.786

*From "Plywood Design Specification," American Plywood Association.

TABLE 7-34 Species Group as Determined by Span Rating for Plywood

Thickness, in	Span Rating†						
	12/0	16/0	20/0	24/0	32/16	42/20	48/24
5/16	4	3	1				
3/8		4	3	1			
1/2				4	1		
5/8					3	1	
3/4						3	1
7/8						4	3

*Use allowable stress for species group tabulated for plywood thickness and span rating.

†30/12, 5/8-in, and 36/16, 3/4-in panels sometimes are available. Check local supplier.

From "Plywood Design Specification," American Plywood Association.

Table 7-33 lists effective section properties per foot of width of plywood panels, for use with the combinations of grades and species given in Table 7-32. The section properties should be used with the allowable stresses tabulated in Table 7-35 for the species group of the face plies of the panels. With the section properties given in Table 7-33, plywood panels, despite their anisotropic properties, may be designed as if they are homogeneous, orthotropic plates. The section properties depend mainly on the characteristics of the face plies, including the direction of grain relative to direction of stress, and are independent of the number of plies in the panels. (For a standard 4 × 8-ft panel, the face-grain direction is parallel to the 8-ft edge.) Different section properties are given in the table for unsanded panels, such as sheathing grades C-C and C-D, than for sanded panels, such as A and B face plies, or touch-sanded panels, such as underlayment, overlaid, or plugged grades.

In general, the panel thickness to use in stress calculations should be the nominal thicknesses tabulated in Table 7-33. For computation of shear through the thickness of the panel, however, the effective thickness for shear listed in the third column of Table 7-33 should be used. This is a smaller value, to compensate for the lesser effectiveness of inner plies in mixed-species panels. The additional shear resistance afforded by the glue in the panels, however, is also taken into account.

The moments of inertia I in Table 7-33 should be used only for stiffness calculations with the moduli of elasticity E tabulated in Table 7-35, but the I values should not be used for computation of bending stresses. Instead, for flexural stresses, the effective section moduli KS given in Table 7-33 should be used. The reason for this is that in calculation of these moduli the outermost tension ply has been ignored. Stressed perpendicular to the face grain, this ply adds little strength to the panel. (The KS values listed apply only to panels used with faces perpendicular to the loads.)

Table 7-33 also gives the rolling-shear constant, Ib/Q, in²/ft. Multiplication of this constant by the allowable rolling shear stress F_s, given in Table 7-35, yields the shear in the plane of the plies, or the rolling shear, V, lb/ft.

The allowable stresses in Table 7-35 apply only for normal loading conditions and should be modified for other conditions, as is done for lumber (Art. 7-7), except for panel size. The allowable bending stresses F_b, as given in Table 7-35, apply only for sections 24 in or more wide. For plywood strips more than 12 in but less than 24 in wide, F_b values from the table should be reduced 25%, and for strips 12 in or less wide, the reduction should be 50%.

TABLE 7-35 Allowable Unit Stresses for Plywood, psi*

Type of stress	Species group of face ply	Grade stress level†				
		S-1		S-2		S-3
		Wet‡	Dry§	Wet‡	Dry§	Dry only§
Extreme fiber stress in bending, F_b. Tension in plane of plies, F_t. Face grain parallel or perpendicular to span. (At 45° to face grain use 1/6 F_t)	1	1,430	2,000	1,190	1,650	1,650
	2, 3	980	1,400	820	1,200	1,200
	4	940	1,330	780	1,110	1,110
Compression in plane of plies, F_c parallel or perpendicular to face grain. (At 45° to face grain use 1/3 F_c)	1	970	1,640	900	1,540	1,540
	2	730	1,200	680	1,100	1,100
	3	610	1,060	580	990	990
	4	610	1,000	580	950	950
Shear in plane perpendicular to plies, F_v parallel or perpendicular to face grain. (At 45° to face grain use 2 F_v)	1	205	250	205	250	210
	2, 3	160	185	160	185	160
	4	145	175	145	175	155
Shear, rolling, in the plane of plies, F_s parallel or perpendicular to face grain. (At 45° to face grain use 1 1/3 F_s)	Marine and Structural I	63	75	63	75	
	Structural II and 2·4·1	49	56	49	56	55
	All other	44	53	44	53	48
Modulus of rigidity or shear modulus, G, in plane of plies	1	70,000	90,000	70,000	90,000	82,000
	2	60,000	75,000	60,000	75,000	68,000
	3	50,000	60,000	50,000	60,000	55,000
	4	45,000	50,000	45,000	50,000	45,000
Bearing (on face). Perpendicular to plane of plies, $F_{c\perp}$	1	210	340	210	340	340
	2, 3	135	210	135	210	210
	4	105	160	105	160	160
Modulus of elasticity E in bending in plane of plies. Face grain parallel or perpendicular to span	1	1,500,000	1,800,000	1,500,000	1,800,000	1,800,000
	2	1,300,000	1,500,000	1,300,000	1,500,000	1,500,000
	3	1,100,000	1,200,000	1,100,000	1,200,000	1,200,000
	4	900,000	1,000,000	900,000	1,000,000	1,000,000

*Conforming to U.S. Product Standard PS 1-74 for Construction and Industrial Plywood. Normal load basis.
†See Table 7-32 for applicable grades of plywood. To qualify for stress level S-1, gluelines must be exterior and only veneer grades N, A, and C are allowed in either face or back. For stress level S-2, gluelines must be exterior and veneer grade B, C-Plugged, and D are allowed on the face or back. Stress level S-3 includes all panels with interior or intermediate glue lines.
‡Moisture content in service 16% or more. Use exterior-type plywood for M.C. > 18%.
§Moisture content in service less than 16%.
From "Plywood Design Specification," American Plywood Association.

Example of Use of Tables 7-32 to 7-35. Determine, for stress applied parallel to face grain, the section properties, allowable bending stress F_b, rolling-shear constant, and modulus of elasticity E for a sheathing grade, 32/16 C-D INT-APA panel.

Table 7-34 indicates that panels with a 32/16 span rating are available with thicknesses of ½ and ⅝ in. If the ½-in thickness is chosen, Table 7-34 requires that species group 1 be used. From Table 7-34, for stress applied parallel to grain, $I = 0.086$ in^4/ft, $KS = 0.247$ in^3/ft, and the rolling-shear constant is 4.189 in^2/ft. Table 7-32 indicates that, for the C-D INT-APA panel, stress level 3 (S-3) should be used. Finally, from Table 7-35, for S-3 and species group 1, $F_b = 1650$ psi, $F_s = 48$ psi, and $E = 1,800,000$ psi.

TABLE 7-36 Plywood Roof Sheathing and Subflooring Recommendations[a]

a. Roof sheathing laid with face grain perpendicular to joists[b]

Panel span rating	Plywood thickness, in	Maximum span, in[c]	Unsupported edge— max, length, in[d]	Allowable roof loads,[e] lb/ft² Spacing of supports, in, c to c									
				12	16	20	24	30	32	36	42	48	60	72
12/0	5/16	12	12	100 (130)										
16/0	5/16, 3/8	16	16	130 (170)	55 (75)									
20/0	5/16, 3/8	20	20		85 (110)	45 (55)								
24/0	3/8, 1/2	24	24		150 (160)	75 (100)	45 (60)							
30/12	5/8	30	26			145 (165)	85 (110)	40 (55)						
32/16	1/2, 5/8	32	28				90 (105)	45 (60)	40 (50)					
36/16	3/4	36	30				125 (145)	65 (85)	55 (70)	35 (50)				

7-76

Identification Index	Thickness (in)	Max Span (in)	Max Span (in)						
42/20	5/8, 3/4, 7/8	42	32	80 (105)	65 (90)	45 (60)	35 (40)			25 (30)
48/24	3/4, 7/8	48	36		105 (115)	75 (90)	55 (55)	40 (40)	45 (45)	30 (30)
2-4-1	1 1/8	72	48			160 (160)	95 (95)	70 (70)	45 (45)	25 (35)
1 1/8 in G 1 and 2	1 1/8	72	48			145 (145)	85 (85)	65 (65)	40 (40)	30 (30)
1 1/4 in G 3 and 4	1 1/4	72	48			160 (165)	95 (95)	75 (75)	45 (45)	25 (35)

[a] American Plywood Association.
[b] Applies to Standard, Structural I and II, and C-C grades only. Plywood continuous over two or more spans.
For applications where the roofing is to be guaranteed by a performance bond, recommendations may differ somewhat from these values. Contact American Plywood Association for bonded roof recommendations.
Use 2-in (6d) common, smooth, ring-shank, or spiral-thread nails for 1/2-in thick or less, and 2 1/2-in (8d) common, smooth, ring-shank, or spiral thread for plywood 1 in thick or less (if ring-shank or spiral-thread nails have the same diameter as common). Use 2 1/2-in (8d) ring-shank or spiral-thread or 3-in (10d) common, smooth-shank nails for 2-4-1, 1 1/8-, and 1 1/4-in panels. Space nails 6 in at panel edges and 12 in at intermediate supports, except that where spans are 48 in or more, nails shall be 6 in c to c at all supports.
[c] The spans shall not be exceeded for any load conditions.
[d] Provide adequate blocking, tongue-and-grooved edges, or other suitable edge support such as PlyClips when spans exceed indicated value. Use two PlyClips for 48-in or greater spans, and one for lesser spans.
[e] Uniform load deflection limitation: $\frac{1}{180}$ the span under live load plus dead load, $\frac{1}{240}$ under live load only. Allowable live load shown without parentheses and allowable total load shown within parentheses.
Allowable loads were established by laboratory test and calculations on the assumption of evenly distributed loads. Figures shown are not applicable for concentrated loads.

TABLE 7-36 Plywood Roof Sheathing and Subflooring Recommendations (*Continued*)

b. Allowable shears in plywood roof diaphragms for wind or seismic loads, lb/ft[f] (plywood and framing assumed already designed for perpendicular loads)

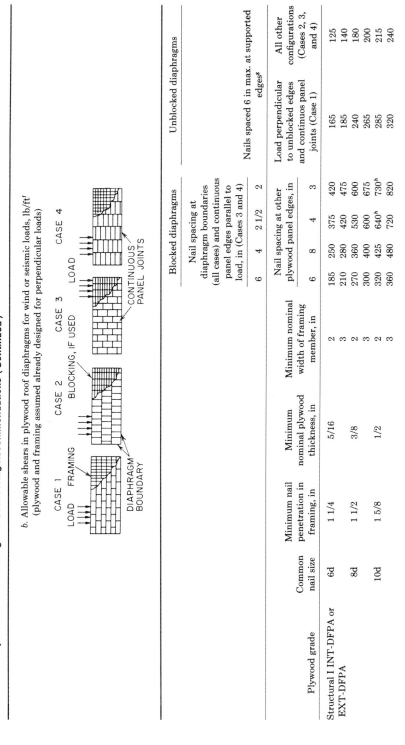

Plywood grade	Common nail size	Minimum nail penetration in framing, in	Minimum nominal plywood thickness, in	Minimum nominal width of framing member, in	Blocked diaphragms								Unblocked diaphragms	
					Nail spacing at diaphragm boundaries (all cases) and continuous panel edges parallel to load, in (Cases 3 and 4)				Nail spacing at other plywood panel edges, in				Nails spaced 6 in max. at supported edges[g]	
					6	4	2 1/2	2	6	8	4	3	Load perpendicular to unblocked edges and continuos panel joints (Case 1)	All other configurations (Cases 2, 3, and 4)
Structural I INT-DFPA or EXT-DFPA	6d	1 1/4	5/16	2	185	250	375	420	185	250	375	420	165	125
				3	210	280	420	475	210	280	420	475	185	140
	8d	1 1/2	3/8	2	270	360	530	600	270	360	530	600	240	180
				3	300	400	600	675	300	400	600	675	265	200
	10d	1 5/8	1/2	2	320	425	640[h]	730[h]	320	425	640[h]	730[h]	285	215
				3	360	480	720	820	360	480	720	820	320	240

7-78

C-C EXT-DFPA	6d	5/16	2	170	225	335	380	110
Structural II INT-DFPA,			3	190	250	380	430	125
Standard C-D INT-		3/8	2	185	250	375	420	125
DFPA, sheathing and	8d	3/8	3	210	280	420	475	140
other DFPA grades			2	240	320	480	545	160
except Species Group 5		1/2	3	270	360	540	610	180
		1/2	2	270	360	530	600	180
			3	300	400	600	675	200
	10d	1/2	2	290	385	575^h	655^h	190
			3	325	430	650	735	215
		5/8	2	320	425	640^h	730^h	215
			3	360	480	720	820	240

^{f}Design for diaphragm stresses depends on direction of continuous panel joints with reference to load, not on direction of long dimensions of plywood sheet. Continuous framing may be in either direction.

^{g}Space nails 12 in on center along intermediate framing members.

^{h}Reduce tabulated allowable shears 10% when boundary members provide less than 3-in nominal nailing surface.

c. Plywood subflooring and subfloor underlayment with face grain perpendicular to joists^i

Panel span rating	Plywood thickness, in	Maximum span,^{j} in	Nail size and type	Nail spacing, in	
				Panel edges	Intermediate
30/12	5/8	12^k	2 1/2-in (8d) common	6	10
32/16	1/2, 5/8	16^l	2 1/2-in (8d) common^m	6	10
36/16	3/4	16^l	2 1/2-in (8d) common	6	10
42/20	5/8, 3/4, 7/8	20^l	2 1/2-in (8d) common	6	10
48/24	3/4, 7/8	24	2 1/2-in (8d) common	6	10
2-4-1					
1 1/2-in Groups 1 and 2	1 1/8	48	3-in (10d) common	6	6
1 1/4-in Groups 3 and 4	1 1/4	48	3-in (10d) common	6	6

^{i}These values apply for Structural I and II, Standard sheathing, and C-C Exterior grades only, for application of ^{25}/$_{32}$-in wood-strip flooring or separate underlayment layer. Plywood continuous over two or more spans.

^{j}Edges shall be tongued and grooved, or supported with blocking, unless underlayment is installed, or finish floor is ^{25}/$_{32}$-in wood strip. Spans limited to values shown because of possible effect of concentrated loads. At indicated maximum spans, panels will support uniform loads of at least 65 lb/ft^2. For spans of 24 in or less, panels will support loads of at least 100 lb/ft^2.

^{k}May be 16 in if ^{25}/$_{32}$-in wood-strip flooring is installed at right angles to joists.

^{l}May be 24 in if ^{25}/$_{32}$-in wood-strip flooring is installed at right angles to joists.

^{m}2-in (6d) common nail permitted if plywood is ½ in thick.

TABLE 7-37 Plywood Wall Sheathing and Siding Recommendations[a]

a. Plywood wall sheathing[b]

Panel span rating	Panel thickness, in	Maximum stud spacing, in		Nail size[c]	Nail spacing, in	
		Exterior covering nailed to:			Panel edges (when over framing)	Intermediate (each stud)
		Stud	Sheathing			
12/0, 16/0, 20/0	3/16	16	16[d]	2-in (6d)	6	12
16/0, 20/0, 24/0	3/8	24	24[d]	2-in (6d)	6	12
24/0, 32/16	1/2	24	24	2-in (6d)	6	12

[a]American Plywood Association.
[b]The plywood may be installed horizontally or vertically. When plywood sheathing is used, building paper and diagonal wall bracing can be omitted. Plywood should be continuous over two or more spans.
[c]Common, annular, spiral-threaded, or T nails of the same diameter as common nails [0.113 in for 2-in-long (6d) nails] may be used. Staples also are permitted, but at reduced spacing.
[d]When siding such as shingles are nailed only to the sheathing, apply the plywood with the face grain perpendicular to the supports.

b. Plywood siding

Application	Plywood thickness, in	Max spacing of supports, in c to c	Nail length and type[e]	Nail spacing, in	
				Panel edges	Intermediate
Panel siding[f]	3/8[g]	16	2-in (6d) casing or siding	6	12
	1/2	24	2-in (6d) casing or siding	6	12
	5/8 or thicker	24	2 1/2-in (8d) casing or siding	6	12
Lap siding	3/8[h]	16	2-in (6d) casing or siding	One nail per stud along bottom edge	4 at vertical joint; 8 at studs (siding wider than 12 in)
	1/2	20	2 1/2-in (8d) casing or siding		
	5/8	24	2 1/2-in (8d) casing or siding		
Bevel siding	9/16 min butt	16	2-in (6d) casing or siding	Same as lap siding	
Soffits or ceilings	3/8	24	2-in (6d) casing or siding	6 (or one nail at each support)	12
	5/8	48			12

[e]Use galvanized, aluminum, or other noncorrosive nails. Nails may be color-coated.
[f]Battens, if used, can be applied with 2 1/2-in-long (8d) noncorrosive casing nails spaced 12 in c to c (staggered).
[g]When separate sheathing is applied, 3/8-in panel siding may be used over supports 24 in on centers, 1/4-in over supports 16 in c to c.
[h]When separate sheathing is applied, 5/16-in overlaid plywood lap siding may be used over supports 16 in c to c.

c. Allowable shears in plywood walls, lb/lin in[i]

Plywood species	Min nominal plywood thickness, in	Plywood sheathing direct to framing				Plywood sheathing over 1/2-in gypsum sheathing				Plywood siding direct to framing			Plywood siding on 1/2-in gypsum sheathing	
		Nail spacing, in, at plywood panel edges[j]				Nail spacing, in, at plywood panel edges[k]				Nail spacing, in, at plywood panel edges[l]			Nail spacing, in, at plywood panel edges[n]	
		6	4	2 1/2	2	6	4	2 1/2	2	6	4		6	4
Structural I INT-DFPA, or EXT-DFPA	5/16	200	300	450	510	200	300	450	510	160	240		160	240
	3/8	280	430	640	730	280	430	640[q]	730[q]	200[m]	300[o]		200[o]	300[o]
	1/2, 5/8	340	510	770[q]	870[q]					200	300		200	300
Structural II INT-DFPA, C-C EXT-DFPA,	5/16	180	270	400	450	180	270	400	450	140	210		140	210
	3/8	260	380	570	640	260	380	570[q]	640[q]	160[m]	240[o]		160[o]	240[o]
Standard C-D INT-DFPA, DFPA panel siding[p]	1/2, 5/8	310	460	690[q]	770[q]					160	240		160	240

[i]All panel edges should be backed with framing, which should be 2-in nominal or wider. The plywood may be installed horizontally or vertically. Space nails 12 in c to c along intermediate framing members. Drive nails at least 1 1/2 in into framing.
[j]Smooth, bright, common, or galvanized box nails—use 2-in (6d) for 5/16-in plywood, 2 1/22-in (8d) for 3/8-in, and 3-in (10d) for 1/2-in.
[k]Smooth, bright, common, or galvanized box nails—use 2 1/2-in (8d) for 5/16-in plywood and 3-in (10d) for 3/8-in.
[l]Galvanized casing nails—use 2-in (6d) or 2 1/2-in[m] (8d) for 3/8-in plywood and 2 1/2-in (8d) for 1/2- and 5/8-in. For siding attached with galvanized box nails, use shears for sheathing with the same size nail. For 3/8- and 1/2-in plywood, nails at panel edges should penetrate the full plywood thickness.
[m]Shear value for 2 1/2-in (8d) nails.
[n]Galvanized casing nails—use 2 1/2-in (8d) or 3-in (10d) for 3/8-in plywood and 3-in (10d) for 1/2- and 5/8-in. For siding attached with galvanized box nails, use shears for sheathing with the same size nail. For 3/8- and 1/2-in plywood, nails at panel edges should penetrate the full plywood thickness.
[o]Shear value for 3-in (10d) nails.
[p]Shears in this group may be increased 20% when using plywood one or more sizes thicker than minimum.
[q]Reduce 10% when boundary members provide less than 3-in nominal nailing surface.

Allowable Loads. The allowable uniform load w_b, psf, in bending for a plywood panel loaded perpendicular to its face may be computed from

$$w_b = \frac{K_b F_b KS}{L^2} \tag{7-20}$$

where L = span, ft, center to center of supports
K_b = 8 for single-span or two-span (equal-span) panels
= 10 for three-span panels with equal spans

The allowable uniform load w_s, psf, for rolling shear in a panel subjected to bending may be computed from

$$w_s = \frac{K_s F_s (\text{lb per } Q)}{L - b} \tag{7-21}$$

where b = width of support, ft
K_s = 2 for single-span panels
= 1.6 for two-span panels with equal spans
= 1.67 for three-span panels with equal spans

Table 7-36 lists allowable loads and nail requirements for plywood roof sheathing. It also gives maximum support spacing and nail requirements for plywood subflooring. Table 7-37 presents similar data for plywood wall sheathing.

Deflections. Bending deflections δ_b, in, including shear deflections, may be computed from

$$\delta_b = \frac{w(L + k)^4}{EI} K_d \tag{7-22}$$

where k = 0.25 in for 2-in (nominal width) framing
= 0.625 in for 4-in (nominal width) framing
w = uniform load on panel, psf
K_d = 1.875 for single-span panels
= 0.778 for two-span panels with equal spans $L + k$
= 0.991 for three-span panels with equal spans $L + k$

For computation of shear deflections separately, as would be required for short spans (span-thickness ratios of about 20 or less), increase E obtained from Table 7-35 by 10% and use for all span conditions:

$$\delta_s = \frac{wt^2(L - b)^2}{EI} C \tag{7-23}$$

where t = nominal panel thickness, in
C = 1.134 for panels placed with face grain perpendicular to supports
= 0.567 for panels placed with face grain parallel to supports

Table 7-38 gives the allowable uniform load to ensure that the deflection of a plywood floor will not exceed $l/180$, where l is the joist spacing, in.

("The Plywood How to Book," "Plywood Construction Guide for Residential Building," "Plywood Construction Systems for Commercial and Industrial Buildings," "Plywood Concrete Forms," "Plywood Design Specification," "Plywood in Apartments," "Plywood Truss Designs," "Plywood Components," American Plywood Association, 1119 A St., Tacoma, Wash. 98401.)

7-29. Composite Construction.

The combination of concrete and steel is a prime example of composite construction. Steel serves a function, resisting tension, not handled as well by concrete. Wood and concrete do not complement each other in the same manner, since the strength proper-

TABLE 7-38 Safe Uniform Floor Load for $l/180$ Deflection*

Joist spacing l, in.	Plywood thickness, in.	Safe load, psf
12	3/8	326
	1/2	435
	5/8	543
	3/4	653
16	3/8	193
	1/2	327
	5/8	408
	3/4	490
24	3/8	57
	1/2	111
	5/8	217
	3/4	311

* Exterior-type Douglas fir, C-C grade, applied with the face grain across supports.

ties of these materials do not differ so greatly. Nevertheless, some use has been made of concrete and wood in bridge decks. There are some materials, however, which, when combined with wood, yield improved constructions.

For example, an open-web joist has been fabricated with wood top and bottom chords and metal web members. Also, as an example, glued-laminated wood beams have been prestressed with high-strength steel strands. These were placed longitudinally in the tension zone of the beams and tensioned to apply a bending moment opposite to those that will be imposed in service. Results indicate that such beams are, on the average, about one-third stronger than nonprestressed beams, and variation in strength properties is about half (B. Bohannan, "Prestressed Laminated Wood Beams," U.S. Forest Service Research Paper, January, 1964, U.S. Forest Products Laboratory, Madison, Wis.). In another experiment, steel plates under tension were glued to the outer, tensile laminations of wood beams. Such beams were, on the average, 25% stiffer and 75% stronger then nonprestressed beams, and variability in strength properties decreased 70% (J. Peterson, "Wood Beams Prestressed with Bonded Tension Elements," *Journal of the Structural Division,* No. ST1, February, 1965, American Society of Civil Engineers).

7-30. Pole Construction.

Wood poles are used for various types of construction, including flagpoles, utility poles, and framing for buildings. These employ preservatively treated round poles set into the ground as columns. The ground furnishes vertical and horizontal support and prevents rotation at the base (Table 7-39).

Isolated poles, such as flagpoles or signs, may be designed with lateral bearing values equal to twice these tabulated values.

In buildings, a bracing system can be provided at the top of the poles to reduce bending moments at the base and to distribute loads. Design of buildings supported by poles without bracing requires good knowledge of soil conditions, to eliminate excessive deflection or sidesway.

Bearing values under the base of poles should be checked. For backfilling the holes, well-tamped native soil, sand, or gravel may be satisfactory. But concrete or soil cement

Wood Construction

TABLE 7-39   Allowable Lateral Passive Soil Pressure

Class of material	Allowable stress per foot of depth below natural grade, psf	Max allowable stress, psf
Good: Compact, well-graded sand and gravel, hard clay, well-graded fine and coarse sand (all drained so water will not stand)	400	8,000
Average: Compact fine sand, medium clay, compact sandy loam, loose sand and gravel (all drained so water will not stand)	200	2,500
Poor: Soft clay, clay loam, poorly compacted sand, clays containing large amounts of silt (water stands during wet season)	100	1,500

is more effective. They can reduce the required depth of embedment and improve bearing capacity by increasing the skin-friction area of the pole. Skin friction is effective in reducing uplift due to wind.

To increase bearing capacity under the base end of poles for buildings, concrete footings often are used. They should be designed to withstand the punching shear of the poles and bending moments. Thickness of concrete footings should be at least 12 in. Consideration should be given to use of concrete footings even in firm soils, such as hard dry clay, coarse firm sand, or gravel.

Calculation of required depth of embedment in soil of poles subject to lateral loads generally is impractical without many simplifying assumptions. While an approximate analysis can be made, the depth of embedment should be checked by tests, or at least against experience in the same type of soil.

Table 7-40 gives allowable stresses for treated poles, Table 7-41 lists standard dimensions for Douglas fir and southern pine poles, and Table 7-42 gives safe concentric column loads.

7-31. Timber Erection.

Erection of timber framing requires experienced crews and adequate lifting equipment to protect life and property and to assure that the framing is properly assembled and not damaged during handling.

Each shipment of timber should be checked for tally and evidence of damage. Before erection starts, plan dimensions should be verified in the field. The accuracy and adequacy of abutments, foundations, piers, and anchor bolts should be determined. And the erector must see that all supports and anchors are complete, accessible, and free from obstructions.

Jobsite Storage.   If wood members must be stored at the site, they should be placed where they do not create a hazard to other trades or to the members themselves. All framing, and especially glued-laminated members, stored at the site should be set above the ground on appropriate blocking. The members should be separated with strips, so that air may circulate around all sides of each member. The top and all sides of each storage pile should be covered with a moisture-resistant covering that provides protection from the elements, dirt, and jobsite debris. (Do not use clear polyethylene films,

TABLE 7-40 Stress Values for Treated Wood Poles*

Species	Modulus of rupture, extreme fiber in bending, psi†	Extreme fiber in bending F_b, psi‡	Modulus of elasticity E, psi‡	Compression parallel to grain F_c, psi‡
Cedar, Northern white	4000	1540	600,000	740
Cedar, Western red	6000	1850	900,000	1030
Douglas fir	8000	2700	1,500,000	1360
Hemlock, Western	7400	2380	1,300,000	1250
Larch, Western	8400	2940	1,500,000	1500
Pine, jack	6600	2100	1,100,000	1100
Pine, lodgepole	6600	1820	1,100,000	980
Pine, ponderosa	6000	1710	1,000,000	920
Pine, red or Norway	6600	2100	1,300,000	1020
Pine, Southern	8000	2740	1,500,000	1360

*Air-dried prior to treatment.
†Based on *American National Standard Specifications and Dimensions for Wood Poles*, ANSI 05.1-1972.
‡Based on ASTM D 2899-70T, "Tentative Method for Establishing Design Stresses for Round Timber Piles."

TABLE 7-41 Dimensions of Douglas-Fir and Southern-Pine Poles*

Class	1	2	3	4	5	6	7	9	10
Min circumference at top, in.	27	25	23	21	19	17	15	15	12

Length of pole, ft	Ground-line distance from butt,† ft	Min circumference at 6 ft from butt, in.							
20	4	31.0	29.0	27.0	25.0	23.0	21.0	19.5	17.5	14.0
25	5	33.5	31.5	29.5	27.5	25.5	23.0	21.5	19.5	15.0
30	5½	36.5	34.0	32.0	29.5	27.5	25.0	23.5	20.5	
35	6	39.0	36.5	34.0	31.5	29.0	27.0	25.0		
40	6	41.0	38.5	36.0	33.5	31.0	28.5	26.5		
45	6½	43.0	40.5	37.5	35.0	32.5	30.0	28.0		
50	7	45.0	42.0	39.0	36.5	34.0	31.5	29.0		
55	7½	46.5	43.5	40.5	38.0	35.0	32.5			
60	8	48.0	45.0	42.0	39.0	36.0	33.5			
65	8½	49.5	46.5	43.5	40.5	37.5				
70	9	51.0	48.0	45.0	41.5	38.5				
75	9½	52.5	49.0	46.0	43.0					
80	10	54.0	50.5	47.0	44.0					
85	10½	55.0	51.5	48.0						
90	11	56.0	53.0	49.0						
95	11	57.0	54.0	50.0						
100	11	58.5	55.0	51.0						
105	12	59.5	56.0	52.0						
110	12	60.5	57.0	53.0						
115	12	61.5	58.0							
120	12	62.5	59.0							
125	12	63.5	59.5							

* See the latest edition of Standard Specifications and Dimensions for Wood Poles, ANSI 05.1.
† The figures in this column are intended for use only when a definition of ground line is necessary to apply requirements relating to scars, straightness, etc.

TABLE 7-42 Safe Concentric Column Loads on Douglas-Fir and Southern-Pine Poles, lb*

Top diam, in	8	7	6	5
Pole class	2	3	5	6
Unsupported pole length, ft (above ground line):				
0	68,500	52,500	38,500	26,000
10	68,500	51,000	28,500	14,000
12	61,500	36,500	20,500	10,000
14	46,000	27,500	15,500	8,000
16	36,000	22,000	12,500	6,500
18	29,500	17,500	10,000	5,000
20	24,500	15,000	8,500	4,500
25	16,500	10,000	6,000	
30	12,500	7,500		
35	10,000	6,000		

*See the latest edition of Standard Specifications and Dimensions for Wood Poles, ANSI 05.1.

since wood members may be bleached by sunlight.) Individual wrappings should be slit or punctured on the lower side to permit drainage of water that accumulates inside the wrapping.

Glued-laminated members of Premium and Architectural Appearance (and Industrial Appearance in some cases) are usually shipped with a protective wrapping of water-resistant paper. While this paper does not provide complete freedom from contact with water, experience has shown that protective wrapping is necessary to ensure proper appearance after erection. Though used specifically for protection in transit, the paper should remain intact until the roof covering is in place. It may be necessary, however, to remove the paper from isolated areas to make connections from one member to another. If temporarily removed, the paper should be replaced and should remain in position until all the wrapping is removed.

At the site, to prevent surface marring and damage to wood members, the following precautions should be taken:

Lift members or roll them on dollies or rollers out of railroad cars. Unload trucks by hand or crane. Do not dump, drag, or drop members.

During unloading with lifting equipment, use fabric or plastic belts, or other slings that will not mar the wood. If chains or cables are used, provide protective blocking or padding.

Guard against soiling, dirt, footprints, abrasions, or injury to shaped edges or sharp corners.

Equipment. Adequate equipment of proper load-handling capacity, with control for movement and placing of members, should be used for all operations. It should be of such nature as to ensure safe and expedient placement of the material. Cranes and other mechanical devices must have sufficient controls that beams, columns, arches, or other elements can be eased into position with precision. Slings, ropes, cables, or other securing devices must not damage the materials being placed.

The erector should determine the weights and balance points of the framing members before lifting begins, so that proper equipment and lifting methods may be employed. When long-span timber trusses are raised from a flat to a vertical position preparatory to lifting, stresses entirely different from design stresses may be introduced. The magnitude and distribution of these stresses depend on such factors as weight, dimensions, and type of truss. A competent rigger will consider these factors in determining how much suspension and stiffening, if any, is required and where it should be located.

Accessibility. Adequate space should be available at the site for temporary storage of materials from time of delivery to the site to time of erection. Material-handling equipment should have an unobstructed path from jobsite storage to point of erection. Whether erection must proceed from inside the building area or can be done from outside will determine the location of the area required for operation of the equipment. Other trades should leave the erection area clear until all members are in place and are either properly braced by temporary bracing or are permanently braced in the building system.

Assembly and Subassembly. Whether done in a shop or on the ground or in the air in the field, assembly and subassembly are dependent on the structural system and the various connections involved.

Care should be taken with match-marking on custom materials. Assembly must be in accordance with the approved shop drawings. Any additional drilling or dapping, as well as the installation of all field connections, must be done in a workmanlike manner.

Trusses are usually shipped partly or completely disassembled. They are assembled on the ground at the site before erection. Arches, which are generally shipped in half sections, may be assembled on the ground or connections may be made after the half arches are in position. When trusses and arches are assembled on the ground at the site, assembly should be on level blocking to permit connections to be properly fitted and securely tightened without damage. End compression joints should be brought into full bearing and compression plates installed where intended.

Before erection, the assembly should be checked for prescribed overall dimensions, prescribed camber, and accuracy of anchorage connections. Erection should be planned and executed in such a way that the close fit and neat appearance of joints and the structure as a whole will not be impaired.

Field Welding. Where field welding is required, the work should be done by a qualified welder in accordance with job plans and specifications, approved shop drawings, and specifications of the American Institute of Steel Construction and the American Welding Society.

Cutting and Fitting. All connections should fit snugly in accordance with job plans and specifications and approved shop drawings. All cutting, framing, and boring should be done in accordance with good shop practices. Any field cutting, dapping, or drilling should be done in a workmanlike manner, with due consideration given to final use and appearance.

Bracing. Structural elements should be placed to provide restraint or support, or both, to ensure that the complete assembly will form a stable structure. This bracing may extend longitudinally and transversely. It may comprise sway, cross, vertical, diagonal, and like members that resist wind, earthquake, erection, acceleration, braking, and other forces. And it may consist of knee braces, cables, rods, struts, ties, shores, diaphragms, rigid frames, and other similar components in combinations.

Bracing may be temporary or permanent. Permanent bracing, required as an integral part of the completed structure, is shown on the architectural or engineering plans and usually is also referred to in the job specifications. Temporary construction bracing is required to stabilize or hold in place permanent structural elements during erection until other permanent members that will serve the purpose are fastened in place. This bracing is the responsibility of the erector and is normally furnished and erected by him. It should be attached so that children and other casual visitors cannot remove it or prevent it from serving as intended. Protective corners and other protective devices should be installed to prevent members from being damaged by the bracing.

In timber-truss construction, temporary bracing can be used to plumb trusses during erection and hold them in place until they receive the rafters and roof sheathing. The major portion of temporary bracing for trusses is left in place, because it is designed to brace the completed structure against lateral forces.

Failures during erection occur occasionally and regardless of construction material used. The blame can usually be placed on insufficient or improperly located temporary erection guys or braces, overloading with construction materials, or an externally applied force sufficient to render temporary erection bracing ineffective.

Structural members of wood must be stiff, as well as strong. They must also be properly guyed or laterally braced, both during erection and permanently in the completed structure. Large rectangular cross sections of glued-laminated timber have relatively high lateral strength and resistance to torsional stresses during erection. However, the erector must never assume that a wood arch, beam, or column cannot buckle during handling or erection.

Specifications often require that:

1. Temporary bracing shall be provided to hold members in position until the structure is complete.

2. Temporary bracing shall be provided to maintain alignment and prevent displacement of all structural members until completion of all walls and decks.

3. The erector should provide adequate temporary bracing and take care not to overload any part of the structure during erection.

While the magnitude of the restraining force that should be provided by a cable guy or brace cannot be precisely determined, general experience indicates that a brace is adequate if it supplies a restraining force equal to 2% of the applied load on a column or of the force in the compression flange of a beam. It does not take much force to hold a member in line; but once it gets out of alignment, the force then necessary to hold it is substantial.

Section 8

Structural-Steel Construction

Henry J. Stetina
Consulting Engineer, Jenkintown, Pa.,
Formerly Senior Regional Engineer,
American Institute of Steel Construction

Structural-steel construction embraces only the range of hot-rolled steel sections (or shapes) and plates, of thicknesses of ⅛ in or more, together with such fittings as rivets, bolts, bracing rods, and turnbuckles.

The shop operation of cutting steel plates and shapes to final size, punching, drilling, and assembling the components into finished members ready for shipment is called fabricating. Most companies comprising the structural-steel fabricating industry not only take contracts to furnish the fabricating services but also install or erect the steel—a branch of the industry called *erecting*.

The skilled laborers employed in the field to actually place and connect the steel in position are called *ironworkers*.

To promote uniformity in bidding practices, the American Institute of Steel Construction has adopted a more specific, although lengthy, definition for structural steel in the AISC "Code of Standard Practice." It is well for the owner and his engineer to understand fully just what the fabricator will supply when the latter submits a price for furnishing "structural steel." When it is desired that bids should include any other material, such as ornamental iron, open-web joists, and miscellaneous metalwork, a statement to that effect should appear in the bidding invitation.

8-1. Structural-Steel Shapes.

Steel mills have a standard classification for the many products they make, one of which is *structural shapes (heavy)*. By definition this classification takes in all shapes having at least one cross-sectional dimension of 3 in or more. Shapes of lesser size are classified as *structural shapes (light)* or, more specifically, bars.

Shapes are identified by their cross-sectional characteristics—angles, channels, beams, columns, tees, pipe, tubing, and piles. For convenience, structural shapes are simply identified by letter symbols according to the following schedule.

8-2 Structural-Steel Construction

Section	Symbol
Wide-flange shapes	W
Standard I shapes	S
Bearing-pile shapes	HP
Similar shapes that cannot be grouped in W, S, or HP	M
Structural tees cut from W, S, or M shapes	WT, ST, MT
American standard channels	C
All other channel shapes	MC
Angles	L

The industry recommended standard (adopted 1970) for indicating a specific size of beam or column-type shape on designs, purchase orders, shop drawings, etc., specifies listing of symbol, depth, and weight, in that order. For example, W14 × 30 identifies a wide-flange shape with nominal depth of 14 in and weight of 30 lb/lin ft. The ×, read as "by," is merely a separation.

Similarly, plates are billed in the order of symbol (PL), followed by thickness and width, thus: PL¾ × 18.

Each shape has its particular functional use, but the workhorse of building construction is the wide-flange W section. For all practical purposes, W shapes have parallel flange surfaces. The profile of a W shape of a given nominal depth and weight available from different producers is essentially the same, except for the size of fillets between web and flanges.

8-2. Tolerances for Structural Shapes.

Mills are granted a tolerance because of variations peculiar to working of hot steel and wear of equipment. Limitations for such variations are established by Specification A6 of the American Society for Testing and Materials.

Wide-flange beams or columns, for example, may vary in depth by as much as ½ in, i.e., ¼ in over and under the nominal depth. The designer should always keep this in mind. Fillers, shims, and extra weld metal installed during erection may not be desirable, but often they are the only practical solution to dimensional variations from nominal.

Cocked flanges on column members are particularly troublesome to the erector, for it is not until the steel is erected in the field that the full extent of mill variations becomes evident. This is particularly true for a long series of spans or bays, where the accumulating effect of dimensional variation of many columns may require major adjustment. Fortunately, the average variation usually is negligible and nominal erection clearance allowed for by the fabricator will suffice.

Mill tolerances also apply to beams ordered from the mills cut to length. Where close tolerance may be desired, as sometimes required for welded connections, it may be necessary to order the beams long and then finish the ends in the fabricating shop to precise dimensions.

8-3. Cambered Beams.

Frequently, designers want long-span beams slightly arched (cambered) to offset deflection under load and to prevent a flat or saggy appearance. Such beams may be procured from the mills, the required camber being applied to cold steel. The AISC "Manual of Steel Construction" gives the maximum cambers that mills can obtain and their prediction of the minimum cambers likely to remain permanent. Smaller cambers than these minimums may be specified, but their permanency cannot be guaranteed. It should be observed that nearly all beams will have some camber as permitted by the tolerance for straightness, and advantage may be taken of such camber in shop fabrication.

A method of cambering, not dependent on mill facilities, is to employ heat. In welded construction, it is commonplace to flame-straighten members that have become distorted. By the same procedure, it is possible to distort or camber a beam to desired dimensions.

8-4. In-Service Steel.

Over the years, there have been numerous changes in the size of beam and column sections. The reasons are many: efficiency, popularity, stocking, standardization, simplification, and conservation. Occasionally, an engineer is called on to investigate an existing building, possibly because of increased floor loadings or because an additional story or two is to be added. Whatever the reason, he may be faced with the task of identifying sections that are no longer rolled and for which physical properties are not readily obtainable. To aid the designer in this problem, the American Institute of Steel Construction published in 1953 a complete compilation of steel and wrought-iron beams and columns that were rolled in this country during the period 1873–1952. Knowing the year of construction, consequently the approximate date of mill rolling, will enable the engineer to establish grade of the steel material and thus determine the yield and working stresses. (AISC, 400 N. Michigan Ave., Chicago, Ill. 60611.)

8-5. Structural-Steel Specifications.

Uniform quality is an outstanding characteristic of structural steel. This high degree of uniformity, regardless of the source of the steel, is achieved through universal acceptance of American Society for Testing and Materials specifications. There is no significant metallurgical or physical difference between the products rolled by United States structural mills.

In contrast to the single all-purpose structural steel (former ASTM A7) in general use during the first half of the twentieth century, many structural steels are now recommended by the American Institute of Steel Construction for building construction. While several are restricted to particular products, such as pipe, tubing, and strip steel for open-web steel joists, most of the specifications pertain to material for the plates and shapes generally used for principal members in steel structures. The most important characteristics of these steels are given in Table 8-1.

Weldability. All steels listed in Table 8-1 are of good welding quality.

Steel Strengths. In general, all steels of greater strength than A36 are more costly to buy from mills, but this does not mean they are too expensive to use. Their advantage of greater strength and possibly simpler fabrication often results in the most economical construction. Nevertheless, the desire for one grade of structural steel that is widely available and generally economical has made A36 the most commonly used steel.

Table 8-1 lists several steels with more than one level of tensile strength and yield stress, the levels being dependent on thickness of material. The listed thicknesses are precise for plates and nearly correct for shapes. To obtain the precise values for shapes, refer to the AISC "Steel Construction Manual" or to mill catalogs.

Weathering Steels. Strictly from a consideration of cost efficiency, corrosion-resistant A242 and A588 steels are not economical for ordinary usage. Their use in buildings, however, may be justified where steel will be exposed and perhaps left unpainted. On exposure to ordinary atmospheric conditions, these steels, called weathering steels, take on a thin rust film that inhibits further corrosion. Thus, they have been used without paint for building frameworks exposed to the elements.

TABLE 8-1 Characteristics of Structural Steels

ASTM specification	Thickness, in.	Minimum tensile strength, ksi	Minimum yield stress,* ksi	Relative corrosion resistance†
Carbon Steels				
A36	To 8 in. incl.	58–80¶	36	1‡
A529	To ½ in. incl.	60–85¶	42	1
High-strength, Low-alloy Steels				
A441	To ¾ incl.	70	50	2
	Over ¾ to 1½	67	46	2
	Over 1½ to 4 incl.	63	42	2
	Over 4 to 8 incl.	60	40	2
A572	Gr 42: to 4 incl.	60	42	1
	Gr 45: to 1½ incl.	60	45	1
	Gr 50: to 1½ incl.	65	50	1
	Gr 55: to 1½ incl.	70	55	1
	Gr 60: to 1 incl.	75	60	1
	Gr 65: to ½ incl.	80	65	1
A242	To ¾ incl.	70	50	4–8
	Over ¾ to 1½	67	46	4–8
	Over 1½ to 4 incl.	63	42	4–8
A588	To 4 incl.	70	50	4
	Over 4 to 5	67	46	4
	Over 5 to 8 incl.	63	42	4
Heat-treated Low-alloy Steels				
A514	To ¾ incl.	115–135	100	1–4
	Over ¾ to 2½	115–135	100	1–4
	Over 2½ to 4 incl.	105–135	90	1–4

* Yield stress or yield strength, whichever shows in the stress-strain curve.
† Relative to carbon steels low in copper.
‡ A36 steel with 0.20% copper has a relative corrosion resistance of 2.
¶ Minimum tensile strength may not exceed the higher value.

Steel Identification. Because of the many steel types and strengths in use, ASTM specifications require each piece of hot-rolled steel to be properly identified with vital information, including heat numbers of the ingots. The American Institute of Steel Construction "Specification for the Design, Fabrication and Erection of Structural Steel for Buildings" requires fabricators to be prepared to demonstrate, by written procedure and by actual practice, the visible identification of all main stress-carrying elements at least through shop assembly. Steel identification includes ASTM designation, heat numbers (if required), and mill test reports when specially ordered.

Structural steels usually are referred to by their corresponding ASTM designation. For example, A36 steel is steel produced to meet requirements of ASTM A36.

AISC Specification. The AISC "Specification for the Design, Fabrication and Erection of Structural Steel for Buildings," first promulgated in 1923, is widely used throughout the United States. The AISC revises the specification periodically to keep abreast of professional development research findings, and availability of new materials.

AISC also issues the following supplementary specifications: "Specification for Architecturally Exposed Structural Steel," and in conjunction with the Steel Joist Institute, "Standard Specifications for Open Web Steel Joists, H-Series" and "Standard Specifications for Longspan Steel Joists, LH-Series, and Deep Longspan Steel Joists, DLH-Series" (see Arts. 9-1 to 9-8).

STEEL-FRAMING SYSTEMS

Steel construction may be classified into three broad categories: wall bearing, skeleton framing, and long-span construction. Depending on the functional requirements of the building itself and the materials used in constructing its roof, floors, and walls, one or more of these methods of framing may be employed in a single structure.

8-6. Wall-Bearing Framing.

Probably the oldest and commonest type of framing, wall-bearing framing (not to be confused with bearing-wall construction), occurs whenever a wall of a building, interior or exterior, is used to support ends of main structural elements carrying roof or floor loads. The walls must be strong enough to carry the reaction from the supported members and thick enough to ensure stability against any horizontal forces that may be imposed. Such construction often is limited to relatively low structures, because load-bearing walls become massive in tall structures. Nevertheless, a wall-bearing system may be advantageous for tall buildings when designed with reinforcing steel (Sec. 11).

Fig. 8-1 Lintels supporting masonry.

8-6 Structural-Steel Construction

A common application of wall-bearing construction may be found in many one-family homes. A steel beam, usually 8 or 10 in deep, is used to carry the interior walls and floor loads across the basement with no intermediate supports, the ends of the beam being supported on the foundation walls. The relatively shallow beam depth affords maximum headroom for the span. In some cases, the spans may be so large that an intermediate support becomes necessary to minimize deflection—usually a steel pipe column serves this purpose.

Another example of wall-bearing framing is the member used to support masonry over windows, doors, and other openings in a wall. Such members, called *lintels*, may be a steel angle section (commonly used for brick walls in residences) or, on longer spans and for heavier walls, a fabricated assembly. A variety of frequently used types is shown in Fig. 8-1. In types (*b*), (*c*), and (*e*), a continuous plate is used to close the bottom, or soffit, of the lintel, and to join the load-carrying beams and channels into a single shipping unit. The gap between the toes of the channel flanges in type (*d*) may be covered by a door frame or window trim, to be installed later. Pipe and bolt separators are used to hold the two channels together to form a single member for handling. (See also Art. 8-80.)

Bearing Plates. Because of low allowable pressures on masonry, bearing plates (sometimes called masonry plates) are usually required under the ends of all beams that rest on masonry walls, as illustrated in Fig. 8-2. Even when the pressure on the wall under a member is such that an area no greater than the contact portion of the member itself is required, wall plates are sometimes prescribed, if the member is of such weight that it must be set by the steel erector. The plates, shipped loose and in advance of steel erection, are then set by the mason to provide a satisfactory seat at the proper elevation.

Anchors. The beams are usually anchored to the masonry. Of the two common methods shown in the American Institute of Steel Construction "Manual of Steel Construction," **government anchors** as illustrated in Fig. 8-2 have been preferred.

Nonresidential Uses. Another common application for the wall-bearing system is in one-story commercial and light industrial-type construction. The masonry side walls support the roof system, which may be rolled beams, open-web joists, or light trusses. Clear spans of moderate size are usually economical, but for longer spans (probably over 40 ft), wall thickness and size of buttresses (pilasters) must be built to certain specified minimum proportions commensurate with the span—a requirement of building codes to assure stability. Therefore, the economical aspect should be carefully investigated. It may cost less to introduce steel columns and keep wall size to the minimum permissible. On the other hand, it may be feasible to reduce the span by introducing intermediate columns and still retain the wall-bearing system for the outer end reactions.

Planning for Erection. One disadvantage of wall-bearing construction needs emphasizing: Before steel can be set by the ironworkers, the masonry must be built up to the proper elevation to receive it. When these elevations vary, as is the case at the end of a pitched or arched roof, then it may be necessary to proceed in alternate stages, progress of erection being interrupted by the work that must be performed by the masons, and vice

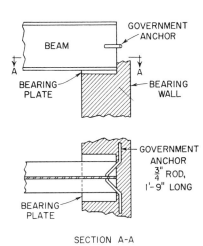

Fig. 8-2 Wall-bearing beam.

versa. The necessary timing to avoid delays is seldom obtained. A few columns or an additional rigid frame at the end of a building may cost less than calling trades to a job on some intermittent and expensive arrangement. Remember, too, that labor-union regulations may prevent the trades from handling any material other than that belonging to their own craft. An economical rule may well be: Lay out the work so that the erector and his ironworkers can place and connect all the steelwork in one continuous operation.

(Bibliography, Art. 8-92.)

8-7. Skeleton Framing.

In skeleton framing all the gravity loadings of the structure, including the walls, are supported by the steel framework. Such walls are termed **nonbearing** or **curtain walls**. This system made the skyscraper possible. Steel, being so much stronger than all forms of masonry, is capable of sustaining far greater load in a given space, thus obstructing less of the floor area in performing its function.

With columns properly spaced to provide support for the beams spanning between them, there is no limit to the floor and roof area that can be constructed with this type of framing, merely by duplicating the details for a single bay. Erected tier upon tier, this type of framing can be built to any desired height. Fabricators refer to this type of construction as "beam and column"; a typical arrangement is illustrated in Fig. 8-3.

The spandrel beams, marked B1 in Fig. 8-3, are located in or under the wall so as to reduce eccentricity caused by wall loads. Figure 8-4 shows two methods for connecting to the spandrel beam the shelf angle that supports the outer course of masonry over window openings 6 ft or more in width. In order that the masonry contractor may proceed expeditiously with his work, these shelf angles must be in alignment with the face of the building and at the proper elevation to match a masonry joint. The connection of the angles to the spandrel beams is made by bolting; shims are provided to make the adjustments for line and elevation (Art. 8-80).

Figure 8-4a illustrates a typical connection arrangement when the outstanding leg of the shelf angle is about 3 in or less below the bottom flange of the spandrel beam; Fig. 8-4b illustrates the corresponding arrangement when the outstanding leg of the shelf angle is more than about 3 in below the bottom flange of the spandrel beam. In the cases represented by Fig. 8-4b, the shelf angles are usually shipped attached to the spandrel beam; if the distance from the bottom flange to the horizontal leg of the shelf angle is greater than 10 in, a hanger may be required.

In some cases, as over door openings, the accurate adjustment features provided by Fig. 8-4a and b may not be needed. It may then be more economical to simplify the

Fig. 8-3 Typical steel beam-and-column framing, shown in plan.

8-8 Structural-Steel Construction

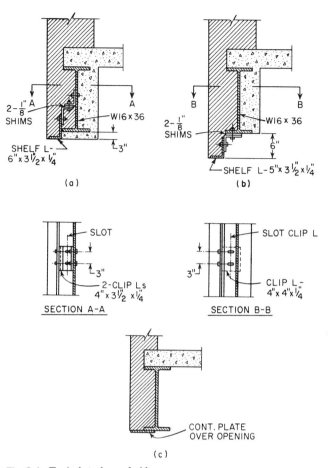

Fig. 8-4 Typical steel spandrel beams.

detail, as shown in Fig. 8-4c. The elevation and alignment will then conform to the permissible tolerances associated with the steel framework.

8-8. Long-Span Steel Framing.

Large industrial buildings, auditoriums, gymnasiums, theaters, hangars, and exposition buildings require much greater clear distance between supports than can be supplied by beam and column framing. When the clear distance is greater than can be spanned with rolled beams, several alternatives are available. These may be classified as *girders, simple trusses, arches, rigid frames, cantilever-suspension spans,* and various types of space frames, such as *folded plates, curvilinear grids, thin-shell domes, two-way trusses, and cable networks.*

Girders are the usual choice where depths are limited, as over large unobstructed areas in the lower floors of tall buildings, where column loads from floors above must be carried across the clear area. Sometimes, when greater strength is required than is available in rolled beams, cover plates are added to the flanges (Fig. 8-5a) to provide the additional strength.

When depths exceed the limit for rolled beams, i.e., for spans exceeding about 67 ft (based on the assumption of a depth-span ratio of 1:22 with 36 in deep Ws), the girder must be built up from plates and shapes. Welded girders usually are preferred to the old-type conventional riveted girders (Fig. 8-5b), composed of web plate, angles, and cover plates. Where this type may still be specified, all components are fastened with high-strength bolts rather than with rivets.

Welded girders generally are composed of three plates (Fig. 8-5c). This type offers the most opportunity for simple fabrication, efficient use of material, and least weight. Top and bottom flange plates may be of different size (Fig. 8-5d), an arrangement advantageous to composite construction, which integrates a concrete floor slab with the girder flange, to function together.

Heavy girders may use cover-plated tee sections (Fig. 8-5e). Where lateral loads are a factor, as in the case of girders supporting cranes, a channel may be fastened to the top flange (Fig. 8-5f). In exceptionally heavy construction, it is not unusual to use a pair of girders diaphragmed together to share the load (Fig. 8-5g).

Fig. 8-5 Typical built-up girders.

The availability of high-strength, weldable steels resulted in development of **hybrid girders.** For example, a high-strength steel, say A572 Grade 50, whose yield stress is 50 ksi, may be used in a girder for the most highly stressed flanges, and the lower-priced A36 steel, whose yield stress is 36 ksi, may be used for lightly stressed flanges and web plate and detail material. The AISC Specification requires that the top and bottom flanges at any cross section have the same cross-sectional area, and that the steel in these flanges be of the same grade. The allowable bending stress may be slightly less than for conventional homogeneous girders of the high-strength steel, to compensate for possible overstress in the web at the junction with the flanges. Hybrid girders are efficient and economical for heavy loading and long spans and, consequently, are frequently employed in bridgework.

Trusses. When depth limits permit, a more economical way of spanning long distances is with trusses, for both floor and roof construction. Because of their greater depth, trusses usually provide a greater stiffness against deflection when compared pound for pound with the corresponding rolled beam or plate girder than otherwise would be required. Six general types of trusses frequently used in building frames are shown in Fig. 8-6, together with modifications that can be made to suit particular conditions.

Trusses in Fig. 8-6a to d and k may be used as the principal supporting members in floor and roof framing; types e to j serve a similar function in the framing of symmetrical roofs having a pronounced pitch. As shown, types a to d have a top chord that is not quite parallel to the bottom chord. Such an arrangement is used to provide for drainage of flat roofs. Most of the connections of the roof beams **(purlins),** which these trusses support, can be identical, which would not be the case if the top chord were dead level and the elevation of the purlins varied. When used in floors, truss types a to d have parallel chords.

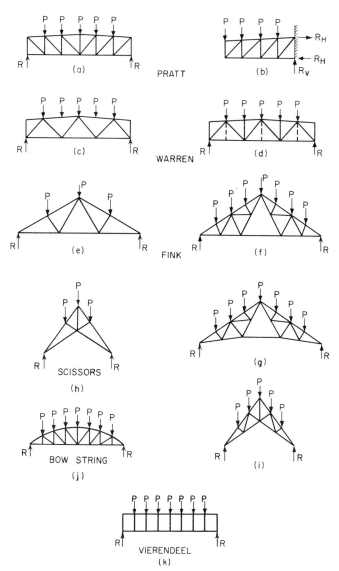

Fig. 8-6 Types of steel trusses.

Properly proportioned, bow-string trusses (Fig. 8-6j) have the unique characteristic that the stress in their web members is relatively small. The top chord, which usually is formed in the arc of a circle, is stressed in compression, and the bottom chord in tension. In spite of the relatively expensive operation of forming the top chord, this type of truss has proved very popular in roof framing on spans of moderate lengths up to about 100 ft.

The Vierendeel truss (Fig. 8-6k) generally is shop welded to develop full rigidity of connections between the vertical and chords. It is useful where absence of diagonals is desirable to permit passage between the verticals.

Trusses also may be used for long spans, as three-dimensional trusses (space frames) or as grids. In two-way grids, one set of parallel lines of trusses is intersected at 90° by another set of trusses so that the verticals are common to both sets. Because of the rigid connections at the intersections, loads are distributed nearly equally to all trusses. Reduced truss depth and weight savings are among the apparent advantages of such grids.

Long-span joists are light trusses closely spaced to support floors and flat roofs. They conform to standard specifications (Art. 8-5) and to standard loading tables. Both Pratt and Warren types are used, the shape of chords and webs varying with the fabricator. Yet, all joists with the same designation have the same guaranteed load-supporting capacity. The standard loading tables list allowable loads for joists up to 72 in deep and with clear span up to 144 ft. The joists may have parallel or sloping chords or other configuration.

Truss Applications. Cross sections through a number of buildings having roof trusses of the general type just discussed are shown diagrammatically in Fig. 8-7. Cross section (*a*) might be that of a storage building or a light industrial building. A Fink truss has been used to provide a substantial roof slope. Roofs of this type are often designed

Fig. 8-7 Some examples of structures with trussed roofs.

8-12 Structural-Steel Construction

to carry little loading, if any, except that produced by wind and snow, since the contents of the building are supported on the ground floor. For light construction, the roof and exterior wall covering may consist of thin, cold-formed metal or asbestos-cement sheets, lapped at their seams so as to shed rain. Lighting and ventilation, in addition to that provided by windows in the vertical side walls, frequently are furnished by means of sash installed in the vertical side of a continuous monitor, framing for which is indicated by the dotted lines in the sketch.

Cross section (*b*) shows a scissors truss supporting the high roof over the nave of a church. This type of truss is used only when the roof pitch is steep, as in ecclesiastical architecture.

A modified Warren truss, shown in cross section (*c*), might be one of the main supporting roof members over an auditorium, gymnasium, theater, or other assembly-type

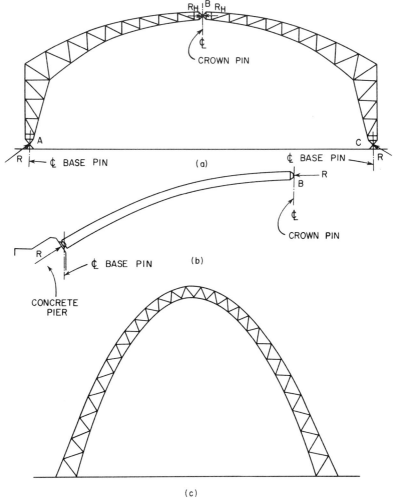

Fig. 8-8 Steel arches: (*a*) and (*b*) three-hinged; (*c*) fixed.

building where large, unobstructed floor space is required. Similar trusses, including modified Pratt, are used in the roofs of large garages, terminal buildings, and airplane hangars, for spans ranging from about 80 up to 500 ft.

The Pratt truss (Fig. 8-7d) is frequently used in industrial buildings, while (e) depicts a type of framing often used where overhead traveling cranes handle heavy loads from one point on the ground to another.

Arches. When very large clear spans are needed, the bent framing required to support walls and roof may take the form of solid or open-web arches, of the kind shown in Fig. 8-8. A notable feature of bents (a) and (b) is the heavy steel pins at points A, B, and C, connecting the two halves of the arch together at the crown and supporting them at the foundation. These pins are designed to carry all the reaction from each half arch, and to function in shear and bearing much as a single rivet is assumed to perform when loaded in double shear.

Use of hinge pins offers two advantages in long-span frames of the type shown in Fig. 8-8. In the first place, they simplify design calculations. Second, they simplify erection. All the careful fitting can be done and strong connections required to develop the needed strength at the ends of the arch can be made in the shop, instead of high above ground in the field. When these heavy members have been raised in the field about in their final position, the upper end of each arch is adjusted, upward or downward, by means of jacks near the free end of the arch. When the holes in the pin plates line up exactly, the crown pin is slipped in place and secured against falling out by the attachment of keeper plates. The arch is then ready to carry its loading. Bents of the type shown in Fig. 8-8a and b are referred to as **three-hinged arches.**

When ground conditions are favorable and foundations are properly designed, and if the loads to be carried are relatively light, as, for example, for a large gymnasium, a **hingeless arch** similar to the one shown diagrammatically in Fig. 8-8c may offer advantage in overall economy.

In many cases, the arches shown in Fig. 6-8a and b are designed without the pins at B *(two-hinged arch)*. Then, the section at B must be capable of carrying the moment and shear present. Therefore, the section at B may be heavier than for the three-hinged arch, and erection will be more exacting for correct closure.

Rigid Frames. These are another type of long-span bent. In design, the stiffness afforded by beam-to-column connections is carefully evaluated and counted on in the design to relieve some of the bending moment that otherwise would be assumed as occurring with maximum intensity at midspan. Typical examples of rigid frame bents are shown in Fig. 8-9. When completely assembled in place in the field, the frames are fully continuous throughout their entire length and height. A distinguishing characteristic of rigid frames is the absence of pins or hinges at the crown, or midspan.

In principle, single-span rigid-frame bents are either two-hinged or hingeless arches—the latter, when the column bases are fully restrained by large rigid foundations, to which they are attached by a connection capable of transmitting moment as well as shear. Since such foundations may not be economical, if at all possible, because of soil conditions, the usual practice is to consider the bents hinged at each reaction. However, this does not imply the necessity of expensive pin details; in most cases, sufficient rotation of the column base can be obtained with the ordinary flat-ended base detail and a single line of anchor bolts placed perpendicular to the span on the column center line. Many designers prefer to obtain a hinge effect by concentrating the column load on a narrow bar, as shown in Fig. 8-9c; this refinement is worthwhile in larger spans.

Regardless of how the frame is hinged, there is a problem in resisting the horizontal shear that the rigid frame imparts to the foundation. For small spans and light thrusts, it may be feasible to depend on the foundation to resist lateral displacement. However, more positive performance and also reduction in costs are usually obtained by connecting

Fig. 8-9 Steel rigid frames: (*a*) single bent; (*b*) continuous frame with underfloor tie; (*c*) connection of tie to a column; (*d*) two-hinged.

opposite columns of a frame with tie rods, as illustrated in Fig. 8-9*b*, thus eliminating these horizontal forces from the foundation.

For ties on small spans, it may be possible to utilize the reinforcing bars in the floor slab or floor beams, by simply connecting them to the column bases. On larger spans, it is advisable to use tie rods and turnbuckles, the latter affording the opportunity to prestress the ties and thus compensate for elastic elongation of the rods when stressed. Prestressing the rod during erection to 50% of its value has been recommended for some major installations; but, of course, the foundations should be checked for resisting some portion of the thrust.

Single-story, welded rigid frames often are chosen where exposed steelwork is desired for such structures as churches, gymnasiums, auditoriums, bowling alleys, and shopping centers, because of attractive appearance and economy. Columns may be tapered, girders may vary in depth linearly or parabolically, haunches (knees) may be curved, field joints may be made inconspicuous, and stiffeners may simply be plates.

Field Splices. One problem associated with long-span construction is that of locating field splices compatible with the maximum sizes of members that can be shipped and erected. Field splices in frames are generally located at or near the point of counterflexure, thus reducing the splicing material to a minimum. In general, the maximum height for shipping by truck is 8 ft, by rail 10 ft. Greater overall depths are possible, but these should always be checked with the carrier; they vary with clearances under bridges and through tunnels.

Individual shipping pieces must be stiff enough to be handled without buckling or other injury, light enough to be lifted by the raising equipment, and capable of erection without interference from other parts of the framework. This suggests a study of the entire frame to ensure orderly erection, and to make provisions for temporary bracing of the members, to prevent jackknifing, and for temporary guying of the frame, to obtain proper alignment.

Hung-Span Beams. In some large one-story buildings, an arrangement of cantilever-suspension (hung) spans (Fig. 8-10) has proved economical and highly efficient. This layout was made so as to obtain equal maximum moments, both negative and positive, for the condition of uniform load on all spans. A minimum of three spans is required; that is, a combination of two end spans (A) and one interior span (B). The connection

Fig. 8-10 Hung-span steel construction.

at the end of the cantilever (point D) must be designed as a shear connection only. If the connection is capable of transmitting moment as well as shear, it will change the design to one of continuity and the dimensions in Fig. 8-10 will not apply. This scheme of cantilever and suspended spans is not necessarily limited to one-story buildings.

As a rule, interior columns are separate elements in each story. Therefore, horizontal forces on the building must be taken solely by the exterior columns (Bibliography, Art. 8-92).

8-9. Steel and Concrete Framing.

In another type of framing system, a partial use of structural steel has an important role, namely, *combination framing* of reinforced concrete and structural steel.

Composite or combination construction actually occurs whenever concrete is made to assist steel framing in carrying loads. The term composite, however, often is used for the specific cases in which concrete slabs act together with flexural members (Art. 8-15).

Reinforced-concrete columns of conventional materials when employed in tall buildings and for large spans become excessively large. One method of avoiding this objectionable condition is to use high-strength concrete and high-strength reinforcing bars. Another is to use a structural-steel column core. In principle, the column load is carried by both the steel column and the concrete that surrounds the steel shape; building codes usually contain an appropriate formula for this condition.

A number of systems employ a combination of concrete and steel in various ways. One method features steel columns supporting a concrete floor system by means of a steel shearhead connected to the columns at each floor level. The shallow grillage is embedded in the floor slab, thus obtaining a smooth ceiling without drops or capitals.

Another combination system is the **lift-slab** method. In this system, the floor slabs are cast one on top of another at ground level. Jacks, placed on the permanent steel columns, raise the slabs, one by one, to their final elevation, where they are made secure to the columns. When fireproofing is required, the columns may be boxed in with any one of many noncombustible materials available for that purpose. The merit of this system is the elimination of formwork and shoring that are essential in conventional reinforced-concrete construction.

For high-rise buildings, structural-steel framing often is used around a central, load-bearing, concrete core, which contains elevators, stairways, and services. The thick walls of the core, whose tubular configuration may be circular, square, or rectangular, are designed to resist all the wind forces as well as gravity loads. Sometimes, the surrounding steel framing is cantilevered from the core, or the perimeter members are hung from trusses or girders atop the core and possibly also, in very tall buildings, at midheight of the core. (Bibliography, Art. 8-92.)

FLOOR AND ROOF SYSTEMS

In most types of buildings, floor and roof systems are so intimately related to the structural frame that the two must be considered together in the development of a steel design. Both are important, but floors usually predominate in tier-type structures.

8-10. Factors Affecting Floor Design.

Selection of a suitable and economical floor system for a steel-frame building involves many considerations: load-carrying capacity, durability, fire resistance, dead weight, overall depth, facility for installing power, light, and telephones, facility for installing air conditioning, sound transmission, appearance, maintenance, and construction time.

Building codes specify minimum design live loads for buildings. In the absence of a code regulation, one may use "Minimum Design Loads in Buildings and Other Structures," American Standard A58.1, American National Standards Institute. See also Art. 5-2. Floors should be designed to support the actual loading or these minimum loads, whichever is larger. Most floors can be designed to carry any given load. However, in some instances, a building code may place a maximum load limit on particular floor systems without regard to calculated capacity.

Resistance to lateral forces should not be disregarded, especially in areas of seismic disturbances or for perimeter windbents. In designs for such conditions floors may be employed as horizontal diaphragms to distribute lateral forces to walls or framing designed to transmit them to the ground.

Durability becomes a major consideration when a floor is subject to loads other than static or moderately kinetic types of forces. For example, a light joist system may be just the floor for an apartment or an office building but may be questionable for a manufacturing establishment where a floor must resist heavy impact and severe vibrations. Shallow floor systems deflect more than deep floors; the system selected should not permit excessive or objectionable deflections.

Fire resistance and fire rating are very important factors, because building codes, in the interest of public safety, specify the degree of resistance that must be provided. Many floor systems are rated by the codes or by fire underwriters for purposes of satisfying code requirements or basing insurance rates (see Arts. 8-86 to 8-91).

The dead weight of the floor system, including the framing, is an important factor affecting economy of construction. For one thing, substantial saving in the weight and cost of a steel frame may result with lightweight floor systems. In addition, low dead weight may also reduce foundation costs.

Sometimes, the depth of a floor system is important. For example, the height of a building may be limited for a particular type of fire-resistant construction or by zoning laws. The thickness of the floor may well be the determining factor limiting the number of stories that can be built. Also, the economy of a deep floor is partly offset by the increase in height of walls, columns, pipes, etc.

Another important consideration, particularly for office buildings and similar-type occupancies, is the need for furnishing an economical and flexible electrical wiring system. With the accent on movable partitions and ever-changing office arrangements, the readiness and ease with which telephones, desk lights, and electric-powered business machines can be relocated are of major importance. Therefore, the floor system that by its makeup provides large void spaces or cells for concealing wiring possesses a distinct advantage over competitive types of solid construction. Likewise, the problem of recessed lighting in the ceiling may disclose an advantage in one system over another. Furthermore, air conditioning and ventilation are a must in new office buildings; a competitive necessity in older structures, if at all possible. So location of ducts and method of attachment warrant study of several floor systems for optimum results.

Sound transmission and acoustical treatments are other factors that need to be evaluated. A wealth of data is available in reports published by the National Bureau of Standards. In general, floor systems of sandwich type with air spaces between layers will afford better resistance to sound transmission than solid systems, which do not interrupt sound waves. Although the ideal soundproof floor is impractical, costwise, there are several reasonably satisfactory systems. Much depends on type of occupancy, floor coverings, and ceiling finish—acoustical plaster or tile.

Appearance and maintenance are also weighed by the designer and the owner. A smooth, neat ceiling is usually a prerequisite for residential occupancy, a less expensive finish may be deemed satisfactory for an institutional building.

Speed of construction is essential. Contractors are alert to the systems that enable the following-up trades to work immediately behind the erector and with unimpeded efficiency.

In general, the following constructions are commonly used in conjunction with steel framing; concrete arch, concrete joists (removable pan), steel joists, cellular steel and composite concrete-steel beams.

8-11. Concrete-Arch Floors.

The type of floor often used in office and industrial buildings consists of a reinforced-concrete slab, about 4 in thick, supported on steel beams spaced about 8 ft apart. These beams are designated secondary or intermediate, whereas the beams or girders framing into columns and supporting secondary beams are the primary members (Fig. 8-3). It is called an "arch" floor because it evolved from earlier construction with thick brick and tile floors that were truly flat-top arches; some of the early reinforced-concrete floors also were cast on arch-shaped forms.

The arch system is so much heavier than the contemporary lightweight systems that it has lost most of its earlier appeal for light-occupancy construction. It ranks high, however, where heavy loads, durability, and rigidity are key factors. It is without competition whenever building codes restrict the use of other types. The system works well with the steel frame because all formwork can be supported directly, or by wire suspension, on the floor beams, thus obviating vertical shoring.

8-12. Concrete-Pan System.

Concrete floors cast on removable metal forms or pans, which form the joists, are frequently used with steel girders in certain areas. Since the joists span the distance between columns, intermediate steel beams are not needed (Fig. 8-11). This floor generally weighs less than the arch system but still considerably more than the lightest types.

There are a number of variations of the concrete-joist system, such as the "grid" or "waffle" system, where the floor is cast on small, square, removable pans, or domes, so that the finished product becomes a two-way joist system. Other systems employ permanent filler blocks—usually a lightweight tile. Some of these variations fall in the heaviest floor classification; also the majority require substantial forms and shoring.

TYPICAL PANEL

FIRE-RESISTIVE CEILING

ENLARGED SECTION A

Fig. 8-11 Concrete joist floor.

8-13. Open-Web Joist Floors.

The lightest floor system in common use is the open-web joist construction shown in Fig. 8-12. It is popular for all types of light occupancies, principally because of initial low cost.

Many types of open-web joists are on the market. Some employ bars in their makeup, while others are entirely of rolled shapes; they all conform to standards and good-practice specifications promulgated by the Steel Joist Institute and the American Institute of Steel Construction (see Art. 8-5). All joists conform to the standard loading tables and carry the same size designation so that the designer need only indicate on his drawings the standard marking without reference to manufacturer, just as he would for a steel beam or column section.

Satisfactory joist construction is assured by adhering to SJI and AISC recommendations. Joists generally are spaced 2 ft c to c; they should be adequately braced (with bridging) during construction to prevent rotation or buckling; and to avoid "springy" floors, they should be carefully selected to provide sufficient depth.

One of the main appeals of this system is the elimination of falsework. Joists are easily handled, erected, and connected to supporting beams—usually by tack welding. Temporary coverage and working platforms are quickly placed. The open space between joists, and through the webs, may be utilized for ducts, cables, light fixtures, and piping. A thin floor slab is cast on steel lath, corrugated-steel sheets, or wire-reinforced paper lath laid on top of the joists. A plaster ceiling may be suspended or attached directly to the bottom flange of the joist.

Lightweight beams, or so-called "junior" beams, are also used in the same manner as

Fig. 8-12 Open-web steel joist.

the open-web joists, and with the same advantages and economy, except that the solid webs do not allow so much freedom in installation of utilities. Beams may be spaced according to their safe load capacity; 3- and 4-ft spacings are common. As a type, therefore, the lightweight-steel-beam floor is intermediate between concrete arches and open-web joists.

8-14. Cellular-Steel Floors.

Light-gage steel decking is frequently used in modern office buildings. One type is illustrated in Fig. 8-13. Other manufacturers make similar cellular metal decks, the primary difference being in the shape of the cells. Often, decking with half cells is used. These are open ended on the bottom, but flat sheet close those cells, usually 5 to 6 ft apart, that incorporate services. Sometimes, cells are enlarged laterally to transmit air for air conditioning.

Two outstanding advantages held for cellular floors are rapidity of erection and the ease with which present and future connections can be made to telephone, light, and power wiring, each cell serving as a conduit. Each deck unit becomes a working platform immediately on erection, thus enabling the several finishing trades to follow right behind the steel erector.

Although the cost of the steel deck system may be larger than that of other floor systems, the cost differential can be narrowed to competitive position when equal consideration for electrical facility is imposed on the other systems; e.g., the addition of 4 in of concrete fill to cover embedded electrical conduit on top of a concrete flat-slab floor.

In earlier floors of this type, the steel decking was assumed to be structurally independent. In that case, the concrete fill served only to provide fire resistance and a level floor. Most modern deckings are bonded or locked to the concrete, so that the two materials act as a unit. Therefore, the metal can be made thinner or the spans made longer. Usually, only top-quality stone concrete (ASTM C33 aggregates) is used, although lightweight concrete made with ASTM C330 aggregates is an acceptable alternative.

Usage of cellular deck in composite construction is facilitated by economical attachment of shear connectors to both the decking and underlying beams. For example, when welded studs are used, a welding gun automatically fastens the studs through two layers of hot-dipped galvanized decking to the unpainted top flanges of the steel beams. This construction is similar to composite concrete-steel beams (Art. 8-15).

The total floor weight of cellular steel construction is low, comparable to open-web steel joists. Weight savings of about 50% are obtained in comparison with concrete slabs; 30% savings in overall weight of the building. However, costwise, a big factor in a high-

Fig. 8-13 Cellular-steel floor.

labor-rate area is the elimination of costly formwork needed for concrete slabs, since the decking serves as the form.

Fire resistance for any required rating is contributed by the fill on top of the cells and by the ceiling below (Fig. 8-13). Generally, removable panels for which no fire rating is claimed are preferred for suspended ceilings, and fireproofing materials are applied directly to the under side of the metal deck and all exposed surfaces of steel floor beams, a technique often called spray-on fireproofing (Art. 8-89).

8-15. Composite Concrete-Steel Beams.

In composite construction, the structural concrete slab is made to assist the steel beams in supporting loads. Hence, the concrete must be bonded to the steel to ensure shear transfer. When the steel beams are completely encased in the concrete, the natural bond is considered capable of resisting horizontal shear. But that bond generally is disregarded when only the top flange is in contact with the concrete, and the slab acts much like a cover plate on the beam. (See also Art. 8-35.) Instead, shear connectors are used to resist the horizontal shear. Commonly used connectors are welded studs, hooked or headed, and short lengths of channels.

Usually, composite construction is most efficient for heavy loading, long spans, large beam spacing, and restricted depths. Because the concrete serves much like a cover plate, lighter steel beams may be used for given loads, and deflections are smaller than for noncomposite construction.

8-16. Effect of Intermediate Beams on Costs.

The joist systems, either steel or concrete, require no intermediate support, since they are obtainable in lengths to meet the normal bay dimensions in tier building construction. On the other hand, the concrete arch and cellular-steel floors are usually designed with one or two intermediate beams within the panel. The elimination of secondary beams does not necessarily mean overall economy just because the structural-steel contract is less. These beams are simple to fabricate and erect and allow much duplication. An analysis of contract prices shows that the cost per ton of secondary beams will average 20% under the cost per ton for the whole steel structure; or viewed another way, the omission of secondary beams increases the price per ton on the balance of the steelwork by $3\frac{1}{2}$% on the average. This fact should be taken into account when making a cost analysis of several systems.

8-17. Other Floor Systems.

Aside from the basic floor systems of Arts. 8-11 to 8-15, there are numerous adaptations and proprietary systems, such as these:

Cofar (corrugated form and reinforcing), where concrete reinforcement is shop-welded to corrugated sheets, which serves as the form for the concrete slab. Clear spans up to 14 ft are possible.

Smooth ceilings, where the floor slab is supported with the aid of short cantilever steel beams, or grids, rigidly connected to the columns and embedded in the slab—a system that eliminates beams between columns.

Dox, a system of precast-concrete blocks tied together in the shop with steel rods to form beams or slabs.

Flexicore, a prefabricated, light, concrete floor plank, achieving weight saving through the use of circular, longitudinal, hollow spaces.

Battledeck or steel-plate floor, the concrete slab of the arch system being replaced with a steel plate. It requires close spacing of secondary beams.

In addition, there are available a number of precast floor planks (or tiles) of concrete or gypsum, some featuring lightweight aggregates.

8-18. Roof Systems.

These are similar in many respects to the floor types discussed in Arts. 8-11 to 8-17; in fact, for flat-top tier buildings, the roof may be just another floor. However, when roof loads are smaller than floor loads, as is usually the case, it may be economical to lighten the roof construction. For example, steel joists may be spaced farther apart. Where roof decking is used, the spacing of the joists is determined by the load-carrying ability of the applied decking and of the joists.

Most of the considerations discussed previously for floors also are applicable to roof systems. In addition, however, due thought should be given to weather resistance, heat conductance and insulation, moisture absorption and vapor barriers, and especially to maintenance.

Many roof systems are distinctive as compared with the floor types; for example, the corrugated sheet-metal roofing commonly employed on many types of industrial or mill buildings. The sheets rest on small beams, channels, or joists, called *purlins,* which in turn are supported by trusses. Similar members on the sidewalls are called *girts.*

ALLOWABLE DESIGN STRESSES

Structural-steel members are designed for any one, or any combination, of five possible stress conditions: bending, shearing, web crippling, axial tension, and axial compression. There are other conditions that must be investigated under special conditions: local buckling, excessive deflection, and torsion.

When elastic theory is used, design is based on allowable unit stresses, usually those given in local building codes. Since several grades of steel are available, it is expedient to relate all allowable unit stresses to the minimum yield stress specified for each grade.

When plastic theory is used, design is based on the ultimate strength of members. A safety factor, comparable to that established for elastic design, is applied to the design load to determine the ultimate-load capacity required of a member.

Formulas and allowable stresses in the following articles follow recommendations of the American Institute of Steel Construction "Specification for the Design, Fabrication and Erection of Structural Steel for Buildings," 1978. An accompanying commentary provides background and clarifies the intent of the specification. Both are incorporated in the AISC "Manual of Steel Construction."

8-19. Allowable Tension in Steel.

Unit tensile stress F_t on the gross area should not exceed $0.60F_y$, where F_y is the minimum yield stress of the steel member (see Table 8-2). Nor should F_t exceed $0.50F_u$, where F_u is the minimum tensile strength of the steel member, when the allowable stress is applied to the net area of a member connected with fasteners requiring holes. However, if the fastener is a large pin, as used to connect eyebars, pin plates, etc., F_t is limited to $0.45F_y$ on the net area. Therefore, for the popular A36 steel, the allowable tension stresses for gross and net areas are 22.0 and 29.0 ksi, respectively, and in the case of pin plates, 16.2 ksi.

TABLE 8-2 Allowable Tension on Gross Area

F_y, ksi	F_t, ksi	F_y, ksi	F_t, ksi
36.0	22.0	60.0	36.0
42.0	25.2	65.0	39.0
45.0	27.0	90.0	54.0
50.0	30.0	100.0	60.0
55.0	33.0		

In determination of the net area of a tension member, the width of a hole for a rivet or bolt should be taken $\frac{1}{16}$ in larger than the nominal dimension of the hole normal to the direction of applied stress. Although for a tension member without holes, the gross section should be taken normal to the direction of applied stress, for a member with holes, the net section should be chosen as the one with the smallest area that passes through any chain of holes across the width of the member. Thus, the net section may pass through a chain of holes lying in a plane normal to the direction of applied stress or through holes along a diagonal or zigzag line.

Net section for a member with a chain of holes extending along a diagonal or zigzag line is the product of the net width and thickness. To determine net width, deduct from the gross width the sum of the diameters of all the holes in the chain, then add, for each gage space in the chain, the quantity

$$\frac{s^2}{4g}$$

where s = longitudinal spacing (**pitch,** in) of any two consecutive holes
g = transverse spacing (**gage,** in) of the same two holes

The critical net section of the member is obtained from that chain with the least net width.

When a member axially stressed in tension is subjected to nonuniform transfer of load because of connections through bolts or rivets to only some of the elements of the cross section, as in the case of a W, M, or S shape connected solely by rivets or bolts through the flanges, the net area should be reduced as follows: 10% if the flange width is at least two-thirds the beam depth and at least three fasteners lie along the line of stress; 10% also for structural tees cut from such shapes; 15% for any of the preceding shapes that do not meet those criteria and for other shapes that have at least three fasteners in line of stress; and 25% for all members with only two fasteners in the line of stress.

8-20. Allowable Shear in Steel.

Unit stress in shear on the gross sections should not be greater than $F_v = 0.40F_y$ where F_y is the minimum yield point of the steel (Table 8-3).

TABLE 8-3 Allowable Shear on Gross Section

F_y, ksi	F_v, ksi	F_y, ksi	F_v, ksi
36.0	14.5	60.0	24.0
42.0	17.0	65.0	26.0
45.0	18.0	90.0	36.0
50.0	20.0	100.0	40.0
55.0	22.0		

Unit shear on a beam web may be computed by dividing the total shear at a section by the web area, taken as the product of web thickness and overall beam depth. Except when the web has cutouts or holes, beams never fail because of high shear. Failure will be a form of buckling because of the complex pattern of stresses.

Beams with web-angle end connections and coped top flange should be checked for shear on the critical plane through the holes in the web. In this case, the allowable unit shear stress should be taken as $F_v = 0.30F_u$, where F_u is the minimum tensile strength of the steel.

A special case occurs when a web lies in a plane common to intersecting members; for example, the knee of a rigid frame. Then, shear stresses generally are high. Such webs, in elastic design, should be reinforced when the web thickness is less than $32M/A_{bc}F_y$, where M is the algebraic sum of clockwise and counterclockwise moments (in ft-kips) applied on opposite sides of the connection boundary, and A_{bc} is the planar area of the connection web, in^2 (approximately the product of the depth of the member introducing the moment and the depth of the intersecting member). In plastic design, this thickness is determined from $23M_p/A_{bc}F_y$, where M_p is the plastic moment, or M times a load factor of 1.70. In this case, the total web shear produced by the factored loading should not exceed the web area (depth times thickness) capacity in shear. Otherwise, the web must be reinforced with diagonal stiffeners or a doubler plate.

For deep girder webs, allowable shear is reduced. The reduction depends on the ratio of clear web depth between flanges to web thickness and an aspect ratio of stiffener spacing to web depth. In practice this reduction does not apply when the ratio of web depth to thickness is less than $380/\sqrt{F_y}$. Table 10 in Appendix A of the 1978 AISC "Specification for the Design, Fabrication and Erection of Structural Steel for Buildings" gives the allowable shear stresses in plate girder webs.

8-21. Allowable Compression in Steel.

The allowable compressive stress on the gross section of axially loaded members is given by formulas determined by the effective slenderness ratios Kl/r of the members, where Kl is effective length (Art. 8-22) and r is the least radius of gyration. A critical value, designated C_c, occurs at the slenderness ratio corresponding to the maximum stress for

TABLE 8-4 Slenderness Ratio at Maximum Stress for Elastic Buckling Failure

F_y, ksi	C_c	F_y, ksi	C_c
36.0	126.1	60.0	97.7
42.0	116.7	65.0	93.8
45.0	112.8	90.0	79.8
50.0	107.0	100.0	75.7
55.0	102.0		

elastic buckling failure (Table 8-4). This is illustrated in Fig. 8-14. An important fact to note: when Kl/r exceeds $C_c = 126.1$, the allowable compressive stress is the same for A36 and all higher-strength steels.

$$C_c = \sqrt{2\pi^2 E/F_y} \qquad (8\text{-}1)$$

where E = the modulus of elasticity of the steel = 29,000 ksi
F_y = the specified minimum yield stress, ksi

When Kl/r for any unbraced segment is less than C_c, the allowable compressive stress, ksi, is

$$F_a = \frac{[1 - (Kl/r)^2/2C_c^2]F_y}{\text{FS}} \tag{8-1a}$$

where FS is the safety factor, which varies from 1.67 when $Kl/r = 0$ to 1.92 when $Kl/r = C_c$.

$$\text{FS} = \frac{5}{3} + \frac{3Kl/r}{8C_c} - \frac{(Kl/r)^3}{8C_c^3} \tag{8-2}$$

When Kl/r is greater than C_c:

$$F_a = \frac{12\pi^2 E}{23(Kl/r)^2} = \frac{149{,}000}{(Kl/r)^2} \tag{8-1b}$$

This is the Euler column formula for elastic buckling with a constant safety factor of 1.92 applied.

Increased stresses are permitted for bracing and secondary members with l/r greater than 120. (K is taken as unity.) For such members, the allowable compressive stress is

$$F_{as} = \frac{F_a}{1.6 - l/200r} \tag{8-3}$$

where F_a is given by Eq. (8-1a) or (8-1b). The higher stress is justified by the relative unimportance of these members and the greater restraint likely at their end connections. The full unbraced length should always be used for l.

Tables giving allowable stresses for the entire range of Kl/r appear in the AISC "Manual of Steel Construction." Approximate values may be obtained from Fig. 8-14.

8-22. Effective Column Length.

The proper application of the column formulas, Eqs. (8-1a) and (8-1b) in Art. 8-21, is dependent on judicious selection of K. This term is defined as the ratio of effective column length to actual unbraced length.

For a pin-ended column with translation of the ends prevented, $K = 1$. But in general, K may be greater or less than unity. For example, consider the columns in the frame in Fig 8-15. They are dependent entirely on their own stiffness for stability against sidesway. If enough axial load is applied to them, their effective length will exceed their actual length. But if the frame were braced to prevent sidesway, the effective length would be less than the actual length because of the resistance to end rotation provided by the girder.

Theoretical values of K for six idealized conditions in which joint rotation and translation are either fully realized or nonexistent are given in Fig. 8-16. Also noted are values recommended by the Column Research Council for use in design when these conditions are approximated. Since joint fixity is seldom fully achieved, slightly higher design values than theoretical are given for fixed-end columns.

Specifications do not provide criteria for sidesway resistance under vertical loading, because it is impossible to evaluate accurately the contribution to stiffness of the various components of a building. Instead, specifications cite the general conditions that have proven to be adequate.

Constructions that inhibit sidesway in building frames include substantial masonry walls, interior shear walls; braced towers and shafts; floors and roofs providing dia-

Fig. 8-14 Allowable stresses for axial compression.

phragm action; that is, stiff enough to brace the columns to shear walls or bracing systems; frames designed primarily to resist large side loadings or to limit horizontal deflection; and diagonal X bracing in the planes of the frames. Compression members in trusses are considered to be restrained against translation at connections. Generally, for all these constructions, K may be taken as unity, but a value less than one is permitted if proven by analysis.

When resistance to sidesway depends solely on the stiffness of the frames; for example, in tier buildings with light curtain walls or with wide column spacing, and with no diagonal bracing systems or shear walls, the designer may use any of several proposed rational methods for determining K. A quick estimate, however, can be made by using the alignment chart in the AISC "Manual." The effective length Kl of compression members, in such cases, should not be less than the actual unbraced length.

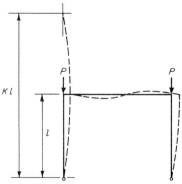

Fig. 8-15 Configurations of members of rigid frame caused by sidesway.

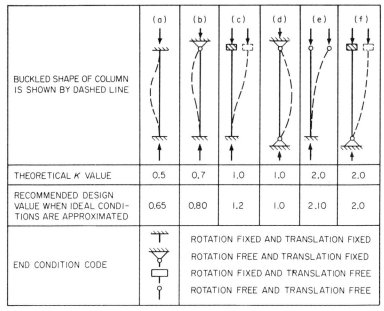

Fig. 8-16 Values of K for idealized conditions.

Fig. 8-17 Web crippling in a simple beam. Critical web section is assumed at fillet.

8-23. Web Crippling.

Webs of rolled beams and welded plate girders should be so proportioned that the compressive stress, ksi, at the web toe of the fillets does not exceed

$$F_a = 0.75 F_y$$

where F_y = specified minimum yield stress, ksi.

Web failure probably would be in the form of buckling caused by concentrated loading, either at an interior load or at the supports. The capacity of the web to transmit the forces safely should be checked.

Load Distribution. Loads are resisted not only by the part of the web directly under them but also by the parts immediately adjacent. A 45° distribution usually is assumed, as indicated in Fig. 8-17 for two common conditions. The distance k is determined by the point where the fillet of the flange joins the web; it is tabulated in the beam tables of the AISC "Manual of Steel Construction." F_a is applicable to the horizontal web strip of length $b + k$ at the end support or $b + 2k$ under an interior load. Bearing stiffeners are required when F_a is exceeded.

Bearing atop Webs. The sum of the compression stresses resulting from loads bearing directly on or through a flange on the compression edge of a plate-girder web should not exceed the following:

When the flange is restrained against rotation, the allowable compressive stress, ksi, is

$$F_a = \left[5.5 + \frac{4}{(a/h)^2} \right] \frac{10{,}000}{(h/t)^2} \tag{8-4}$$

Fig. 8-18 Web crippling in column at joint with welded beam.

When the flange is not restrained against rotation,

$$F_a = \left[2 + \frac{4}{(a/h)^2}\right] \frac{10{,}000}{(h/t)^2} \tag{8-5}$$

where a = clear distance between transverse stiffeners, in
h = clear distance between flanges, in
t = web thickness, in

The load may be considered distributed over a web length equal to the panel length (distance between vertical stiffeners) or girder depth, whichever is less.

Web Stiffeners on Columns. The web of a column may also be subject to crippling by the thrust from the compression flange of a rigidly connected beam, as shown at point a in Fig. 8-18. Likewise, to ensure full development of the beam plastic moment, the column flange opposite the tensile thrust at point b may require stiffening.

Web stiffeners having a combined cross-sectional area A_{st}, in^2, are required on the column whenever A_{st} computed from Eq. (8-6) is positive.

$$A_{st} = \frac{P - F_{yc}t(t_b + 5k)}{F_{ys}} \tag{8-6}$$

where t = thickness of column web, in
t_b = thickness, in, of beam flange delivering concentrated load
F_{yc} = column steel yield stress, ksi
F_{ys} = stiffener steel yield stress, ksi
P = computed force delivered by beam flange or connection plate multiplied by ⅝ when force is a result of dead and live loads, or by ⅔ when it is a result of wind or earthquake forces, kips
k = distance from face of column to edge of fillet on rolled sections (use equivalent for welded sections)

Regardless of the preceding requirement, a single or double stiffener is needed opposite the compression force delivered to the column at point a when

$$d_c > \frac{4100\, t^3 \sqrt{F_{yc}}}{P} \tag{8-7}$$

where d_c = clear distance, in, between column flanges (clear of fillets). Also, a pair of stiffeners is needed opposite the tension force at point b when

$$t_f < 0.4 \sqrt{\frac{P}{F_{yc}}} \tag{8-8}$$

where t_f = thickness of column flange, in.

The thickness of a stiffener should not be less than one-half the thickness of the beam flange or plate that delivers force P to the column. Stiffener width should not be less than one-third of the flange or plate width.

8-24. Pipe and Tubular Columns.

Pipe columns are particularly useful in low-height buildings where light loads prevail, and where attractive architectural effects are derived from exposed steel. Theoretically, the circular section is ideal because the slenderness ratio, l/r, is the same in every direction; however, the use of pipe columns for multistory buildings is limited by the cost of making beam connections.

Pipe meeting the requirements of American Society for Testing and Materials Specification A53, Types E and S, Grade B, is comparable to A36 steel, with F_y = 36 ksi. It comes in three weight classifications: standard, extra strong, and double extra strong, and in diameters ranging up to 26 in.

Several mills produce square and rectangular tubing in sizes ranging from 3 × 2 and 2 × 2 to 12 × 8 and 10 × 10 in, with wall thickness up to ⅝ in. These flat-sided shapes afford easier connections than pipes, not only for connecting beams but also for such items as window and door frames.

The main strength properties of several grades of steel used for pipe and tubular sections are summarized in Table 8-5.

TABLE 8-5 Characteristics of Pipe and Tubular Steels

ASTM spec.	Grade	Product	Min tensile strength, ksi	Min yield stress, ksi
A53	B	Pipe	60.0	35.0*
A500	A	Round	45.0	33.0
	A	Shaped	45.0	39.0
	B	Round	58.0	42.0
	B	Shaped	58.0	46.0
A501	...	All tubing	58.0	36.0
A618	I	All tubing	70.0	50.0
	II	All tubing	70.0	50.0
	III	All tubing	65.0	50.0

* Use 36.0 for purpose of design.

The concentric load capacity of pipes and tubes is determined from the column formulas of Art. 8-21. It is also given in safe-load tables in the AISC "Manual of Steel Construction." The formulas and tables are based on the assumption that the pipes and tubes are not filled with concrete.

8-25. Bending.

Allowable bending stresses for structural-steel members depend on their cross-sectional geometry. Beams classified as **compact** are allowed a bending stress, ksi, $F_b = 0.66 F_y$ for the extreme surfaces in both tension and compression, where F_y is the specified yield stress, ksi. Such members have an axis of symmetry in the plane of loading, their compression flange is adequately braced to prevent lateral displacement, and they develop their full plastic moment (section modulus times yield stress) before buckling.

Compactness Requirements. To qualify as compact, members must meet the following conditions:

1. The flanges must be continuously connected to the web or webs.
2. The width-thickness ratio of unstiffened projecting elements of the compression flange must not exceed $65.0/\sqrt{F_y}$. For computation of this ratio, width b equals one-half the full flange width of I-shaped sections, or the distance from the free edge to the first row of fasteners (or welds) for projecting plates, or the full width of legs of angles, flanges of zees or channels, or tee stems. See also Art. 8-41.
3. The depth-thickness ratio d/t_w of webs must not exceed $640(1 - 3.74 f_a/F_y)/\sqrt{F_y}$ when f_a, the computed axial stress, is equal to or less than $0.16 F_y$, or $257/\sqrt{F_y}$ when $f_a > 0.16 F_y$.
4. The width-thickness ratio of stiffened compression flange plates in box sections and that part of the cover plates for beams and built-up members that is included between longitudinal lines of rivets, bolts, or welds must not exceed $190/\sqrt{F_y}$.
5. For the compression flange of members not box shaped to be considered supported, the unbraced length between lateral supports should not exceed $76.0 b_f/\sqrt{F_y}$ or $20,000 \, A_f/F_y d$, where b_f is the flange width, A_f the flange area, and d the web depth.
6. The unbraced length for rectangular box-shaped members with depth not more than 6 times the width and with flange thickness not more than 2 times the web thickness must not exceed $(1950 + 1200 \, M_1/M_2) b/F_y$. The unbraced length in such cases, however, need not be less than $1200 b/F_y$. M_1 is the smaller and M_2 the larger of bending moments at points of lateral support.
7. The diameter-thickness ratio of hollow circular steel sections must not exceed $3300/F_y$.

Stresses for Compact Beams. Most sections used in building framing, including practically all rolled W shapes of A36 steel and most of those with $F_y = 50$ ksi, comply with these requirements for compactness, as illustrated in Fig. 8-19. Such sections, therefore, are designed with $F_b = 0.66 F_y$. Excluded from qualifying are hybrid girders, tapered girders, and sections made from A514 steel.

Braced sections that meet the requirements for compactness, and are continuous over their supports or rigidly framed to columns, are also permitted a redistribution of the design moments. Negative gravity-load moments at supports may be reduced 10%. But then, the maximum positive moment must be increased by 10% of the average negative moments. This moment redistribution does not apply to cantilevers, hybrid girders, or members made of A514 steel.

Stresses for Noncompact Beams. Many other beam-type members, including nearly compact sections that do not meet all seven requirements, are accorded allowable bending stresses, some higher and some considerably lower than $0.66 f_y$, depending on such conditions as shape factor, direction of loading, inherent resistance to torsion or buckling, and external lateral support. The common conditions and applicable allowable bending stresses are summarized in Fig. 8-20. In the formulas,

l = distance, in, between cross sections braced against twist or lateral displacement of the compression flange
r_t = radius of gyration, in, of a section comprising the compression flange plus one-third of the compression web area, taken about an axis in the plane of the web
A_f = area of the compression flange, in^2

Fig. 8-19 Requirements for laterally supported compact sections.

42.0	45.0	50.0	55.0	60.0	65.0
10.0	9.7	9.2	8.8	8.4	8.1
$98.8 - 8.8 f_a$	$95.4 - 7.9 f_a$	$90.5 - 6.8 f_a$	$86.3 - 5.9 f_a$	$82.6 - 5.2 f_a$	$79.4 - 4.6 f_a$
39.7	38.3	36.3	34.7	33.2	31.9
29.3	28.3	26.9	25.6	24.5	23.6
$11.7 b_f$	$11.3 b_f$	$10.7 b_f$	$10.2 b_f$	$9.8 b_f$	$9.4 b_f$
$476 \frac{A_f}{d}$	$444 \frac{A_f}{d}$	$400 \frac{A_f}{d}$	$364 \frac{A_f}{d}$	$333 \frac{A_f}{d}$	$308 \frac{A_f}{d}$
—	—	—	—	—	
$28.6 b$	$26.7 b$	$24.0 b$	$21.8 b$	$20.0 b$	$18.5 b$

SECTION	REQUIREMENTS	F_b
LOAD, I-shape, SYMMETRICAL ABOUT BOTH AXES (b_f, t_f)	MEETS COMPACTNESS 1 & 2*	$0.75\,F_y$ FOR BOTH TENSION AND COMPRESSION
	MEETS COMPACTNESS 1*† $$\frac{65.0}{\sqrt{F_y}} < \frac{b_f}{2t_f} < \frac{95.0}{\sqrt{F_y}}$$	$F_y\left[1.075 - 0.005\left(\dfrac{b_f}{2t_f}\right)\sqrt{F_y}\right]$
T-shape, SYM. ABOUT (b, t_f)	MEETS COMPACTNESS 1,3,4,5*† $$\frac{65.0}{\sqrt{F_y}} < \frac{b}{t_f} < \frac{95.0}{\sqrt{F_y}}$$	$F_y\left[0.79 - 0.002\left(\dfrac{b}{t_f}\right)\sqrt{F_y}\right]$
LOAD on SOLID SQUARES AND ROUNDS, SOLID FLAT ELEMENTS BENT ABOUT WEAKER AXIS	SOLID SQUARES AND ROUNDS SOLID FLAT ELEMENTS BENT ABOUT WEAKER AXIS	$0.75\,F_y$ FOR BOTH TENSION AND COMPRESSION
UNBRACED LENGTH l, box section (b, t_f), LATERAL SUPPORT	MEETS COMPACTNESS 1,3,6 $$\frac{b}{t_f} \leq \frac{238}{\sqrt{F_y}} \text{ OR } \leq \frac{317}{\sqrt{F_y}} \text{ IF}$$ PERFORATED WITH ACCESS HOLES MIN. $l = \dfrac{1{,}200\,b}{F_y}$	$0.60\,F_y$ FOR BOTH TENSION AND COMPRESSION
TENSION FLANGE FLEXURAL SECTIONS NOT COVERED ABOVE		$0.60\,F_y$
UNBRACED LENGTH l, LOAD, LATERAL SUPPORT, FOR CHANNELS USE ONLY ⟶	$\dfrac{b}{t_f} \leq \dfrac{95.0}{\sqrt{F_y}}$ WHEN $\dfrac{319}{\sqrt{F_y}} \leq \dfrac{l}{r_t} \leq \dfrac{714}{\sqrt{F_y}}$ WHEN $\dfrac{l}{r_t} \geq \dfrac{714}{\sqrt{F_y}}$ OR FOR SOLID RECTANGULAR FLANGE WITH AREA NOT LESS THAN TENSION FLANGE †	FOR COMPRESSION, USE LARGER OF FOLLOWING: $F_y\left[\dfrac{2}{3} - \dfrac{F_y(l/r_t)^2}{1{,}530{,}000}\right] \leq 0.60\,F_y$ $\dfrac{170{,}000}{(l/r_t)^2} \leq 0.60\,F_y$ $\dfrac{12{,}000\,A_f}{l\,d} \leq 0.60\,F_y$
COMPRESSION FLANGE FOR FLEXURAL SECTIONS NOT COVERED ABOVE	$\dfrac{b}{t_f} \leq \dfrac{95.0}{\sqrt{F_y}}$ $l \leq \dfrac{152\,b}{\sqrt{F_y}}$ FOR MAJOR-AXIS BENDING	$0.60\,F_y$

* EXCEPT A514 STEEL † EXCEPT HYBRID GIRDERS

NOTE: FOR I-SHAPED SECTIONS $b_f = 2b$

Fig. 8-20 Allowable bending stresses for sections not qualifying as compact.

TABLE 8-6 Allowable Bending Stresses, ksi

F_y	$0.60F_y$	$0.66F_y$	$0.75F_y$
36.0	22.0	24.0	27.0
42.0	25.2	27.7	31.5
45.0	27.0	29.7	33.8
50.0	30.0	33.0	37.5
55.0	33.0	36.3	41.3
60.0	36.0	39.6	45.0
65.0	39.0	42.9	48.8
90.0	54.0		
100.0	60.0		

Lateral Support. In computation of allowable bending stresses in compression for beams with distance between lateral supports exceeding requirements, a range sometimes called laterally unsupported, the AISC formulas contain a moment factor C_b in recognition of the beneficial effect of internal moments, both in magnitude and direction, at the points of support. For the purpose of this summary, however, the moment factor has been taken as unity and the formulas simplified in Fig. 8-20. The formulas are exact for the case in which the bending moment at any point within an unbraced length is larger than that at both ends of this length. They are conservative for all other cases. Where more refined values are desired, reference should be made to the AISC Specification.

Limits on Width-Thickness Ratios. For flexural members in which the width-thickness ratios of compression elements exceed the limits given in Fig. 8-20 and which are usually lightly stressed, appropriate allowable bending stresses are suggested in "Slender Compression Elements," Appendix C, AISC "Specification for Design, Fabrication and Erection of Structural Steel for Buildings."

For additional discussion of lateral support, see Art. 8-44; also, additional information on width-thickness ratios of compression elements is given in Fig. 8-27.

The allowable bending stresses F_b, ksi, for values often used for various grades of steel are listed in Table 8-6.

8-26. Bearing.

For bearing on contact surfaces, such as milled, sawed, or accurately finished joints, bearing of pins in finished holes, and ends of fitted bearing stiffeners, the allowable stress is

$$F_p = 0.90F_y \qquad (8\text{-}9)$$

where F_y is the specified minimum yield stress of the steel. When the parts in contact have different yield stresses, use the smaller F_y (Table 8-7).

TABLE 8-7 Allowable Bearing on Milled Surfaces, ksi

F_y	F_p	F_y	F_p
36.0	32.4	60.0	54.0
42.0	37.8	65.0	58.5
45.0	40.5	90.0	81.0
50.0	45.0	100.0	90.0
55.0	49.5		

TABLE 8-8 Allowable Bearing on Expansion Rollers and Rockers, kips/lin in

F_y	F_p	F_y	F_p
36.0	0.76d	60.0	1.55d
42.0	0.96d	65.0	1.72d
45.0	1.06d	90.0	2.54d
50.0	1.22d	100.0	2.87d
55.0	1.39d		

TABLE 8-9 Allowable Bearing on Masonry, ksi

On sandstone and limestone	0.40
On brick in cement mortar	0.25
On the full area of concrete	$0.35f'_c$
On less than full concrete area	$0.35f'_c \sqrt{A_2/A_1} \leq 0.7f'_c$

where f'_c = specified compressive strength, ksi, of the concrete
A_1 = bearing area
A_2 = concrete area

The allowable bearing stress on expansion rollers and rockers, in-lb/lin in, is

$$F_p = \frac{F_y - 13}{20} 0.66d \tag{8-10}$$

where d is the diameter of roller or rocker, in (Table 8-8).

Allowable bearing stresses on masonry usually can be obtained from a local or state building code, whichever governs. In the absence of such regulations, however, the values in Table 8-9 may be used.

8-27. Combined Axial Compression and Bending.

When a member is subjected to bending, the neutral axis is displaced from its original position a maximum distance Δ. If it also carries an axial load P, secondary bending moments, equal to the product of the load and the displacements of the axis, result. This often is referred as the P-Δ effect.

When the computed axial stress, f_a, is less than 15% of F_a, the stress that would be permitted if axial force alone were present, the influence of the secondary moment is so small that it may be neglected and a straight-line interaction formula may be used. Thus, when $f_a/F_a \leq 0.15$:

$$\frac{f_a}{F_a} + \frac{f_{bx}}{F_{bx}} + \frac{f_{by}}{F_{by}} \leq 1.0 \tag{8-11}$$

where subscripts x and y indicate, respectively, the major and minor axes of bending (if bending is about only one axis, then the term for the other axis is omitted), and

f_b = computed compressive bending stress, ksi, at point under consideration
F_b = compressive bending stress, ksi, that is allowed if bending alone existed

When $f_a/F_a > 0.15$, the effect of the secondary bending moment should be taken into account and the member proportioned to satisfy Eqs. (8-12) and 8-13):

$$\frac{f_a}{F_a} + \frac{C_{mx}f_{bx}}{[1 - f_a/F'_{ex}]F_{bx}} + \frac{C_{my}f_{by}}{[1 - f_a/F'_{ey}]F_{by}} \leq 1.0 \tag{8-12}$$

$$\frac{f_a}{0.60F_y} + \frac{f_{bx}}{F_{bx}} + \frac{f_{by}}{F_{by}} \leq 1.0 \tag{8-13}$$

where, as before subscripts x and y indicate axes of bending

$$F'_e = \frac{12\pi^2 E}{23(Kl_b/r_b)^2}$$

E = modulus of elasticity, 29,000 ksi
l_b = actual unbraced length, in, in the plane of bending
r_b = corresponding radius of gyration, in
K = effective-length factor in the plane of bending
C_m = reduction factor whose value is determined from the following conditions:

1. For compression members in frames subject to joint translation (sidesway), C_m = 0.85.

2. For restrained compression members in frames braced against joint translation and not subject to transverse loading between their supports in the plane of bending, C_m = $0.6 - 0.4 M_1/M_2$, but not less than 0.4, where M_1/M_2 is the ratio of the smaller to larger moments at the ends of that portion of the member unbraced in the plane of bending under consideration. M_1/M_2 is positive when the member is bent in reverse curvature, and negative when it is bent in single curvature.

3. For compression members in frames braced against joint translation in the plane of loading and subjected to transverse loading between their supports, the value of C_m may be determined by rational analysis. Instead, however, C_m may be taken as 0.85 for members whose ends are restrained, and 1.0 for ends unrestrained.

In wind and seismic design F'_e may be increased one-third, conforming with the increases covered in Art. 8-30.

Additional information, including illustrations of the foregoing three conditions for determining the value of C_m, is given in the "Commentary on the AISC Specification for the Design, Fabrication and Erection of Structural Steel for Buildings."

8-28. Combined Axial Tension and Bending.

Members subject to both axial tension and bending stresses should be proportioned to satisfy Eq. (8-11), with f_b and F_b, respectively, as the computed and allowable bending tensile stress. But the compressive bending stresses must not exceed the values given in Art. 8-25.

8-29. Design of Beam Sections for Torsion.

This is a special type of load application, since in normal practice eccentric loads on beams are counterbalanced to the point where slight eccentricities may be neglected. For example, spandrel beams supporting a heavy masonry wall may not be concentric with the load, thus inducing torsional stresses, but these will largely be canceled out by the equally eccentric loads of the floor, partitions, attached beams, and similar restraints. For this reason, one seldom finds any ill effects from torsional stresses.

It is during the construction phase that torsion may be in evidence, usually the result of faulty construction procedure. In Fig. 8-21 are illustrated some of the bad practices that have caused trouble in the field: when forms for concrete slabs are hung on one edge of a beam (usually the light secondary beam) the weight of the wet concrete may be sufficient to twist the beam. Figure 8-21a shows the correct method, which reduces torsional effect to the minimum. Likewise for spandrels, the floor ties, if any, forms, or the

8-36   Structural-Steel Construction

Fig. 8-21 Steel beams subjected to torsion—good and bad practices.

slab itself should be placed prior to the construction of the eccentric wall (Fig. 8-21b). Connectors for heavy roofing sheets when located on one side of the purlin may distort the section; the condition should be corrected by staggering, as indicated in Fig. 8-21c.

Equations for computing torsion stresses are given in Arts. 5-78 and 5-79. Also, see Bibliography, Art. 8-92.

8-30. Wind and Seismic Stresses.

For wind or earthquake forces, acting alone or in combination with the design dead and live loads, allowable stresses may be increased one-third. But the resultant section should not be less than that required for dead and live loads alone without this one-third increase in stress. The increase is allowed because wind and seismic forces are of short duration.

8-31. Cyclic Loaded Members.

Relatively few structural members in a building are ever subjected to large, repeated variations of stress or stress reversals (tension to compression, and vice versa) that could cause fatigue damage to the steel. Members need not be investigated for this possibility unless the number of stress cycles exceeds 20,000, which is nearly equivalent to two applications every day for 25 years.

Wind and seismic stresses of any appreciable magnitude occur too infrequently to cause fatigue damage. On the other hand, members such as cranes, crane girders, and supports for heavy machinery (e.g., presses and forges) are subjected to frequent stress variations. Allowable stresses given in the previous articles for static loadings have to be reduced to offset the lower resistance of steel to fatigue conditions. The subject is treated in considerable detail, both for members and their connections, including suggested maximum stress ranges, in "Fatigue," Appendix B, AISC Specification for the Design, Fabrication and Erection of Structural Steel for Buildings. An interesting point to note is that all steels (except for one condition favorable to A514 steel) are treated alike; i.e., fatigue damage is independent of steel strength. (See also, F. S. Merritt, "Structural Steel Designers' Handbook," McGraw-Hill Book Company, New York.)

8-32. Stresses for Welds.

Allowable stresses for welds joining structural steel parts depend on the type of welded joint, strength of electrode, and strength of base material. Commonly used welded joints are fillet and groove. The latter is classified in accordance with penetration, as either complete or partial. Welded joints are illustrated in Fig. 8-39, p. 8-73.

When proper electrodes are used for the grade of base steel, the allowable stresses for welded groove joints summarized in Table 8-10 may be used.

TABLE 8-10 Allowable Stresses for Groove Welds

Type of groove weld	Kind of stress	Direction of stress	Allowable stress
Complete penetration	Tension or compression	Parallel to axis of weld	Same as base metal
	Tension or compression	Normal to effective throat	Same as base metal
	Shear	In plane of effective throat	30% of F_t of weld metal, but shear stress on base metal may not exceed $0.40F_y$*
Partial penetration	Tension or compression	Parallel to axis of weld†	Same as base metal
	Tension	Normal to effective throat	30% of F_t of weld metal, but tensile stress on base metal may not exceed $0.60F_y$*
	Compression	Normal to effective throat	Same as base metal
	Shear	Parallel to axis of weld	30% of F_t of weld metal, but shear stress on base metal may not exceed $0.40F_y$*

*F_t = minimum tensile strength, ksi, of weld metal.
  F_y = specified minimum yield stress, ksi, of base metal.
†When fillet welds and partial-penetration groove welds are used to connect elements, such as girder flanges to webs, the tensile or compressive stresses in these elements, parallel to the weld axes, may be disregarded.

TABLE 8-11 Allowable Shear Stresses for Fillet Welds

Shear F_v	Electrode	Matching base steel
18.0 ksi	AWS A5.1, E60XX electrodes AWS A5.17, F6X-EXXX flux-electrode combination AWS A5.20, E60T-X electrodes	A500 Grade A
21.0 ksi	AWS A5.1 or A5.5, E70XX electrodes AWS A5.17, F7X-EXXX flux-electrode combination AWS A5.18, E70S-X or E70U-1 electrodes AWS A5.20, E70T-X electrodes	A36, A53 Grade B A242 A441, A500 Grade B A501, A529, A572 Grades 42 to 60, A588
24.0 ksi	AWS A5.5, E80XX electrodes Grade 80 submerged arc, gas metal-arc or flux-cored arc weld metal	A572 Grade 65
27.0 ksi	AWS A5.5, E90XX electrodes Grade 90 submerged arc, gas metal-arc or flux-cored arc weld metal	A514 over 2½ in. thick
30.0 ksi	AWS A5.5, E100XX electrodes Grade 100 submerged arc, gas metal-arc or flux-cored arc weld metal	A514 over 2½ in. thick
33.0 ksi	AWS A5.5, E110XX electrodes Grade 110 submerged arc, gas metal-arc or flux-cored arc weld metal	A514 2½ in. and less in thickness

The most significant characteristic of fillet-welded joints (illustrated in Fig. 8-39) is that all forces, regardless of direction, are resolved as shear on the effective throat. When fillet welds are used to join elements such as girder flanges to a web, they are designed for horizontal shear without regard to the tensile or compressive stresses in the elements. Allowable shear stresses for fillet welds based upon the strengths of electrodes and base steels are given in Table 8-11.

Allowable shear stress on the effective area of groove or fillet welds is limited to 0.3 the nominal strength of weld metal, ksi, except that the stress on base metal cannot exceed $0.40F_y$, where F_y is the yield strength, ksi, of the base metal. With the exception of allowable tensile stress for a complete-penetration groove weld, it is permissible to use weld metal whose strength is less than indicated as "matching" in Table 8-11. In general, weld metal one strength stronger than "matching," as given in Table 8-11, is permitted for all welds.

Specifications for all welding electrodes, promulgated by the American Welding Society (AWS), are identified as A5.1, A5.5, A5.17, etc., depending on the welding process (Art. 8-58). Electrodes for manual arc welding, often called stick electrodes, are designated by the letter E followed by four or five digits. The first two or three digits designate the strength level; thus, E70XX means electrodes having a minimum tensile strength of 70.0 ksi. Allowable shear stress on the deposited weld metal is taken as 0.30 times the electrode strength classification; thus, 0.30 times 70 for an E70 results in an allowable stress of 21.0 ksi. The remaining digits provide information on the intended usage, such as the particular welding positions and types of electrode coating.

(AISC "Specification for the Design, Fabrication and Erection of Structural Steel for Buildings," "Structural Welding Code," AWS D1.1, American Welding Society, 2501 N.W. 7th St., Miami, Fla. 33125.)

8.33. Stresses for Rivets and Bolts.

Allowable tension and shear stresses on rivets, bolts, and threaded parts (kips per square inch of area of rivets before driving or unthreaded body area of bolts and threaded parts, except as noted) are given in Table 8-12.

The allowable bearing stress on the projected area of bolts in bearing-type connections (see Art. 8-62) and on rivets is given by $F_p = 1.5F_u$, where F_u is the specified minimum tensile strength of the connected material (not the fastener) (Table 8-13).

Since connections do not fail in bearing, use of bearing stresses serves only as an index of the efficiency of net sections. The same index is valid for joints assembled with rivets or bolts, regardless of fastener shear strength or the presence or absence of threads in the bearing area. But bearing stress is not restricted in friction-type connections assembled with A325 and A490 bolts. Similarly, no distinction is made between bearing stresses for single and double, or enclosed, bearing.

TABLE 8-12 Allowable Tension and Shear Stresses, ksi, on Rivets, Bolts, and Threaded Parts

			Shear F_v		
			Friction-type connections		
Fastener	Tension $F_t{}^a$	Bearing type connections	Standard-size holes[b]	Over-size and short-slotted holes[b]	Long-slotted holes[b]
A502, Grade 1, rivets	23.0	17.5			
A502, Grade 2, rivets	29.0	22.0			
Threaded parts with threads not excluded from shear planes	$0.33F_u{}^c$	$0.17F_u{}^c$			
Theaded parts with threads excluded from shear planes	$0.33F_u{}^c$	$0.22F_u{}^c$			
A307 bolts	20.0	10.0[d]			
A325 bolts					
Type F[e]	44.0		17.5[f]	15.0[f]	12.5[f]
Type N[e]	44.0	21.0			
Type X[e]	44.0	30.0			
A490 bolts					
Type F[e]	54.0		22.0[f]	19.0[f]	16.0[f]
Type N[e]	54.0	28.0			
Type X[e]	54.0	40.0			

[a]Static loading only. For fatigue conditions, see the AISC "Specification for the Design, Fabrication and Erection of Structural Steel for Buildings."
[b]For definitions of hole sizes, see Art. 8-57.
[c]F_u = minimum tensile strength, ksi, of fastener.
[d]Threads permitted in shear planes.
[e]F = friction-type connection.
N = connection where threads are excluded from shear planes.
X = connection where threads are excluded from shear planes.
[f]The allowable shear stress is based on the assumption that contact surfaces are free of paint, loose mill scale, grease, etc. Higher stresses may be used, depending on surface preparation and finishes. See Appendix E of the 1978 AISC Specifications mentioned above.

TABLE 8-13 Allowable Bearing, ksi, on Projected Area of Bolts and Rivets

F_u	F_p	F_u	F_p
45.0	67.5	67.0	101.0
55.0	82.5	70.0	105.0
58.0*	87.0	75.0	113.0
60.0	90.0	80.0	120.0
63.0	94.5	100.0	150.0
65.0	97.5	110.0	165.0

*For A36 Steel

8-34. Combined Shear and Tension on Rivets and Bolts.

Tests indicate that allowable stress for concurrent shear and tension on fasteners can be obtained from an interaction formula representing an ellipse (Fig. 8-22). But for simplication, use of three straight lines is recommended. This permits substantial concurrent stresses without a reduction. For example, as shown in Fig. 8-22 for A325 bolts in a bearing-type connection with threads excluded from shear planes, if a bolt is fully stressed in tension at the allowable stress of 44 ksi, it can also be stressed in shear up to 7.9 ksi. If it is stressed in shear at the allowable stress of 30 ksi, it can also be stressed in tension up to 13 ksi. Between these limiting conditions, the combined stresses are limited to those determined by the diagonal straight line. The ellipse, straight lines, and limits are different for different types of fasteners.

For combined shear and tension, with shear stress f_v produced by the same force as that producing the tension, but not exceeding the values in Table 8-12, the tension stress f_t in fasteners in bearing-type joints should not exceed the values indicated in Table 8-14.

For friction-type joints subjected to combined shear and tension, the allowable shear stress given in Table 8-12 should be reduced. This should be done by multiplying that stress by $1 - f_t A_b/T_b$, where f_t is the average tensile stress, ksi, due to the direct load; A_b is the nominal bolt area, in^2, and T_b is the specified pretension load, kips, of the bolt. (For wind and seismic loads, the reduced allowable shear should be increased by one-third.) When the tensile stress is caused by a bending moment generated in the plane of the web of a beam, as in the case of typical framed-beam connections (Art. 8-66), the shear component is offset by increased compression below the beam axis. Thus, actually, no shear exists on the fasteners in tension.

Fig. 8-22 Allowable combined stress in A325 bolts in bearing-type connections, with threads excluded from shear plane.

TABLE 8-14 Allowable Tension F_t, ksi, for Fasteners Subjected to Combined Shear and Tension in Bearing-Type Connections*

Fastener	Threads not excluded from shear planes	Threads excluded from shear planes
A502, Grade 1, rivets	$30† - 1.3f_v \leq 23†$	
A502, Grade 2, rivets	$38† - 1.3f_v \leq 29†$	
A307 bolts	$26† - 1.8f_v \leq 20†$	
Threaded parts	$0.43F_u† - 1.8f_v \leq 0.33F_u†$	$0.43F_u† - 1.4f_v \leq 0.33F_u†$
A325 bolts	$55† - 1.8f_v \leq 44†$	$55† - 1.4f_v \leq 44†$
A490 bolts	$68† - 1.8f_v \leq 54†$	$68† - 1.4f_v \leq 54†$

*f_v = shear stress, ksi, produced by forces applied to the connected parts.
F_u = minimum tensile strength, ksi, of fasteners.
†Should be increased by one-third when the forces include wind or seismic loads.

8-35. Allowable Stresses for Composite Design.

In composite construction, steel beams are so interconnected with the concrete slab they support that beam and slab act together to resist compression. In one type (Fig. 8-23c), the concrete completely encases the steel beam; in another type (Fig. 8-23a and b), the slab rests on the top flange of the beam, and shear is transferred mechanically between the steel and concrete.

Encased Beams. Two design methods are approved for the encased type. In one method, stresses are computed on the assumption that the steel beam alone supports all the dead load applied prior to concrete hardening (unless the beam is temporarily

Fig. 8-23 Steel-concrete composite beam construction: (a) and (b) with welded-stud shear connectors; (c) with encasement in concrete.

shored), and the composite beam supports the remaining dead and live loads. Then, for positive bending moments, the total stress, ksi, on the steel-beam bottom flange is

$$f_b = \frac{M_D}{S} + \frac{M_L}{S_t} \leq 0.66 F_y \qquad (8\text{-}14)$$

where F_y = specified yield stress of the steel, ksi
M_D = dead-load bending moment in-kips
M_L = live-load bending moment, in-kips
S = section modulus of steel beam, in^3
S_t = section modulus of transformed section, in^3. To obtain the transformed equivalent steel area, divide the effective concrete area by the modular ratio n (modulus of elasticity of steel divided by modulus of elasticity of concrete). In computation of effective concrete area, use effective width of concrete slab (Fig. 8-23a and b)

The stress $0.66F_y$ is allowed because the steel beam is restrained against lateral buckling.

The second method stems from a "shortcut" provision contained in many building codes. This provision simply permits higher bending stresses in beams encased in concrete. For example,

$$f_b = \frac{M_D + M_L}{S} \leq 0.76 F_y \qquad (8\text{-}15)$$

Of course, this higher stress would not be realized, because of composite action.

Beams with Shear Connectors. For composite construction where shear connectors transfer shear between slab and beam, the design is based on behavior at ultimate load. It assumes that all loads are resisted by the composite section, even if shores are not used during construction to support the steel beam until the concrete gains strength. For this case, the computed stress in the bottom flange for positive bending moment is

$$f_b = \frac{M_D + M_L}{S_t} \leq 0.66 F_y \qquad (8\text{-}16)$$

where S_t = section modulus, in^3, of transformed section of composite beam. Figure 8-23a and b give the width of slab assumed effective in composite action. To prevent overstressing the bottom flange of the steel beam when temporary shoring is omitted, a limitation is placed on the value of S_t used in computation of f_b with Eq. (8-16):

$$S_t \leq \left(1.35 + 0.35 \frac{M_L}{M_D}\right) S_s \qquad (8\text{-}17)$$

where M_D = moment, in-kips, due to loads applied prior to concrete hardening (75% cured)
M_L = moment, in-kips, due to remaining dead and live loads
S_s = section modulus, in^3, of steel beam alone relative to bottom flange

Shear on Connectors. Shear connectors usually are studs or channels. The total horizontal shear to be taken by the connectors between the point of maximum positive moment and each end of a simple beam, or the point of counterflexure in a continuous beam, is the smaller of the values obtained from Eqs. (8-18) and (8-19).

$$V_h = \frac{0.85 f'_c A_c}{2} \qquad (8\text{-}18)$$

$$V_h = \frac{A_s F_y}{2} \qquad (8\text{-}19a)$$

where f'_c = specified strength of concrete, ksi
A_c = actual area of effective concrete flange, as indicated in Fig. 8-23a and b, in^2
A_s = area of steel beam, in^2

In continuous composite beams, where shear connectors are installed in negative-moment regions, the longitudinal reinforcing steel in the concrete slab may be considered to act compositely with the steel beam in those regions. In such cases, the total horizontal shear to be resisted by the shear connectors between an interior support and each adjacent inflection point is

$$V_h = \frac{S_{sr} F_{yr}}{2} \qquad (8\text{-}19b)$$

where A_{sr} = total area, in², of longitudinal reinforcing steel within the effective width of the concrete slab at the interior support
F_{yr} = specified yield stress of the reinforcing steel, ksi

These formulas represent the horizontal shear at ultimate load divided by 2 to approximate conditions at working load.

Number of Connectors. The minimum number of connectors N_1, spaced uniformly between the point of maximum moment and adjacent points of zero moment, is V_h/q, where q is the allowable shear load on a single connector, as given in Table 8-15. Values in this table, however, are applicable only to concrete made with aggregates conforming to ASTM C33. For concrete made with rotary-kiln-produced aggregates conforming to ASTM C330 and with concrete weight of 90 pcf or more, the allowable shear load for one connector is obtained by multiplying the values in Table 8-15 by the factors in Table 8-16.

If a concentrated load occurs between the points of maximum and zero moments, the minimum number of connectors required between the concentrated load and the point of zero moment is given by

$$N_2 = \frac{V_h}{q} \left(\frac{S_t M_c/M - S_s}{S_t - S_s} \right) \qquad (8\text{-}20)$$

where M = maximum moment, in-kips
M_c = moment, in-kips, at concentrated load $< M$
S_s = section modulus, in³, of steel beam relative to bottom flange
S_t = section modulus, in³, of transformed section of composite beam relative to bottom flange but not to exceed S_t computed from Eq. (8-17)

TABLE 8-15 Allowable Horizontal-Shear Loads, *q*, for Connectors, kips

(Applicable Only to Concrete Made with ASTM C33 Aggregates)

Connector	$f'_c = 3.0$	$f'_c = 3.5$	$f'_c \geq 4.0$
1/2-in dia. × 2-in hooked or headed stud*	5.1	5.5	5.9
5/8-in dia. × 2 1/2-in hooked or headed stud*	8.0	8.6	9.2
3/4-in dia. × 3-in hooked or headed stud*	11.5	12.5	13.3
7/8-in dia. × 3 1/2-in hooked or headed stud*	15.6	16.8	18.0
3-in channel, 4.1 lb	4.3w†	4.7w†	5.0w†
4-in channel, 5.4 lb	4.6w†	5.0w†	5.3w†
5-in channel, 6.7 lb	4.9w†	5.5w†	5.6w†

*Length given is minimum.
†w = length of channel, in.

TABLE 8-16 Shear-Load Factors for Connectors in Lightweight Concrete

Air dry weight, pcf, of concrete	90	95	100	105	110	115	120
Factors for $f'_c \leq 4.0$ ksi	0.73	0.76	0.78	0.81	0.83	0.86	0.88
Factors for $f'_c \geq 5.0$ ksi	0.82	0.85	0.87	0.91	0.93	0.96	0.99

Fig. 8-24 Steel-concrete composite beam construction with formed steel decking: (a) ribs parallel to beam; (b) ribs transverse to beam [refer to (a) for applicable requirements].

The allowable shear loads for connectors incorporate a safety factor of about 2.5 applied to ultimate load for the commonly used concrete strengths. Not to be confused with shear values for fasteners given in Art. 8-33, the allowable shear loads for connectors are applicable only with Eqs. (8-18) and (8-19).

Connector Details. Shear connectors, except those installed in ribs of formed steel decks, should have a minimum lateral concrete cover of 1 in. The diameter of a stud connector, unless located directly over the beam web, is limited to 2.5 times the thickness of the beam flange to which it is welded. Minimum center-to-center stud spacing is 6 diameters along the longitudinal axis, 4 diameters transversely. Studs may be spaced uniformly, rather than in proportion to horizontal shear, inasmuch as tests show a redistribution of shear under high loads similar to the stress redistribution in large riveted and bolted joints. Maximum spacing is 8 times the slab thickness.

Slab Compressive Stresses. These seldom are critical, but they should be investigated, especially when the slab lies on only one side of the steel beam (Fig. 8-23b). Usually, the effective width of the slab is determined by the criterion: width may be assumed up to 8 times the thickness. Thickening the slab, particularly when there is a cover plate on the beam tension flange, generally is economical.

Formed Steel Decking. Concrete slabs are frequently cast on permanent steel decking whose configuration may be described as ribbed, corrugated, cellular, or blended cellular (see Sec. 9). Two distinct composite-design constructions are inherent: ribs parallel or ribs perpendicular to the supporting beams or girders. (See Fig. 8-24.) The design method described for composite concrete-slab and steel-beam construction is also applicable for systems utilizing formed steel decking, except for some modification recommended by the AISC "Specification for the Design, Fabrication and Erection of Structural Steel for Buildings" and as illustrated in Fig. 8-24.

The allowable horizontal shear loads given in Tables 8-15 and 8-16 may have to be adjusted for use with formed steel decking. For decking with ribs parallel to supports (Fig. 8-24a), the allowable loads should be reduced when w/h is less than 1.5 by multiplying the tabulated values by

$$0.6 \left(\frac{w}{h}\right)\left(\frac{H}{h} - 1\right) \leq 1 \quad (8\text{-}21a)$$

where w = average width of concrete rib, in
h = nominal rib height, in
H = length of stud after welding, in, but not more than $(h + 3)$ for computations

For decking with ribs perpendicular to supports, the reduction factor is:

$$\left(\frac{0.85}{\sqrt{N}}\right)\left(\frac{w}{h}\right)\left(\frac{H}{h} - 1\right) \leq 1 \quad (8\text{-}21b)$$

where N = number of studs on a beam and in one rib, but three studs are the maximum that may be considered effective.

8-36. Criteria for Plate Girders.

A plate girder is an I shape whose main components are plates or plates and angles. Basically, it consists of a web plate attached along each edge normal to and along the central axis of a flange (Fig. 8-25). Usually, the web is welded to the flanges with continuous fillet welds. The connection, however, may also be made with rivets or bolts through angles.

Fig. 8-25 Plate girders: (a) welded; (b) bolted.

8-46 Structural-Steel Construction

TABLE 8-17 Limiting Depth-Thickness Ratios for Plate-girder Webs

F_y, ksi	h/t Eq. (8-22a)	h/t Eq. (8-22b)	F_y, ksi	h/t Eq. (8-22a)	h/t Eq. (8-22b)
36.0	322	333	60.0	207	258
42.0	282	309	65.0	192	248
45.0	266	298	90.0	143	211
50.0	243	283	100.0	130	200
55.0	223	270			

Allowable stresses for tension, shear, compression, bending, and bearing are the same for plate girders as those assigned in Arts. 8-19, 8-20, 8-23, 8-25 and 8-26. But reductions in allowable stress are required under some conditions, and there are limitations on the proportions of girder components.

Web Depth-Thickness Limits. The ratio of the clear distance h between flanges, in, to web thickness t, in, is limited by

$$\frac{h}{t} \leq \frac{14{,}000}{\sqrt{F_y(F_y + 16.5)}} \qquad (8\text{-}22a)$$

where F_y is the specified yield stress of the compression flange steel, ksi (Table 8-17). When, however, transverse stiffeners are provided at spacings not exceeding 1.5 times the girder depth, the limit on h/t is increased to

$$\frac{h}{t} \leq \frac{2{,}000}{\sqrt{F_y}} \qquad (8\text{-}22b)$$

General Design Method. Plate girders may be proportioned to resist bending on the assumption that the moment of inertia of the gross cross section is effective. No deductions need be made for fastener holes, unless the holes reduce the gross area of either flange by more than 15%. When they do, the excess should be deducted.

Hybrid girders, which have higher-strength steel in the flanges than in the web, may also be proportioned by the moment of inertia of the gross section when they are not subjected to an axial force greater than 15% of the product of yield stress of the flange steel and the area of the gross section. At any given section, the flanges must have the same cross-sectional area and be made of the same grade of steel.

The allowable compressive bending stress F_b must be reduced from that given in Art. 8-25 where h/t exceeds $760/\sqrt{F_b}$. For greater values of this ratio, the allowable compressive bending stress, except for hybrid girders, becomes

$$F'_b \leq F_b \left[1 - 0.0005 \frac{A_w}{A_f} \left(\frac{h}{t} - \frac{760}{\sqrt{F_b}} \right) \right] \qquad (8\text{-}23)$$

where A_w = the web area, in²

A_f = the compression flange area, in²

For hybrid girders, not only is the allowable compressive bending stress limited to that given by Eq. (8-23), but also the maximum stress in either flange may not exceed

$$F'_b \leq F_b \left[\frac{12 + (A_w/A_f)(3\alpha - \alpha^3)}{12 + 2(A_w/A_f)} \right] \qquad (8\text{-}24)$$

where a = ratio of web yield stress to flange yield stress.

Flange Limitations. The projecting elements of the compression flange must comply with the limitations for b/t given in Art. 8-39. The area of cover plates, where used, should not exceed 0.70 times the total flange area. Partial-length cover plates (Fig. 8-25b) should extend beyond the theoretical cutoff point a sufficient distance to develop their share of bending stresses at the cutoff point. Preferably for welded-plate girders, the flange should consist of a series of plates, which may differ in thickness and width, joined end to end with complete-penetration groove welds (Fig. 8-25a).

Bearing Stiffeners. These are required on girder webs at unframed ends. They may also be needed at concentrated loads, including supports. Set in pairs, bearing stiffeners may be angles or plates placed on opposite sides of the web, usually normal to the bending axis. Angles are attached with one leg against the web. Plates are welded perpendicular to the web. The stiffeners should have close bearing against the flanges through which they receive their loads, and should extend nearly to the edges of the flanges.

These stiffeners are designed as columns, with allowable stresses as given in Art. 8-21. The column section is assumed to consist of a pair of stiffeners and a strip of girder web with width 25 times web thickness for interior stiffeners and 12 times web thickness at ends. In computing the effective slenderness ratio Kl/r, use an effective length Kl of at least 0.75 the length of the stiffeners.

Intermediate Stiffeners. With properly spaced transverse stiffeners strong enough to act as compression members, a plate-girder web can carry loads far in excess of its buckling load. The girder acts, in effect, like a Pratt truss, with the stiffeners as struts and the web forming fields of diagonal tension. The following formulas for stiffeners are based on this behavior. Like bearing stiffeners, intermediate stiffeners are placed to project normal to the web and the bending axis, but they may consist of a single angle or plate. They may be stopped short of the tension flange a distance up to 4 times the web thickness. If the compression flange is a rectangular plate, single stiffeners must be attached to it to prevent the plate from twisting. When lateral bracing is attached to stiffeners, they must be connected to the compression flange to transmit at least 1% of the total flange stress, except when the flange consists only of angles.

The total shear force, kips, divided by the web area, in², for any panel between stiffeners should not exceed the allowable shear F_v given by Eqs. (8-25a) and (8-25b).

Except for hybrid girders, when C_v is less than unity:

$$F_v = \frac{F_y}{2.89}\left[C_v + \frac{1 - C_v}{1.15\sqrt{1 + (a/h)^2}}\right] \leq 0.4 F_y \qquad (8\text{-}25a)$$

For hybrid girders or when C_v is more than unity or when intermediate stiffeners are omitted:

$$F_v = \frac{F_y C_v}{2.89} \leq 0.4 F_y \qquad (8\text{-}25b)$$

where a = clear distance between transverse stiffeners, in
h = clear distance between flanges within an unstiffened segment, in

$$C_v = \frac{45{,}000 k}{F_y (h/t)^2} \text{ when } C_v \text{ is less than 0.8}$$

$$= \frac{190}{h/t}\sqrt{\frac{k}{F_y}} \text{ when } C_v \text{ is more than 0.8}$$

t = web thickness, in
$k = 5.34 + 4/(a/h)^2$ when $a/h > 1$
$ = 4 + 5.34/(a/h)^2$ when $a/h < 1$

Stiffeners for an end panel or for any panel containing large holes and for adjacent panels should be so spaced that the largest average web shear f_v in the panel does not exceed the allowable shear given by Eq. (8-25b).

Intermediate stiffeners are not required when h/t is less than 260 and f_v is less than the allowable stress given by Eq. (8-25b). When these criteria are not satisfied, stiffeners

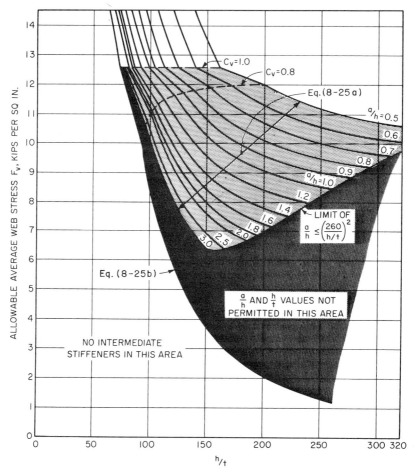

Fig. 8-26 Chart shows relationship between allowable web shears in plate girders, with $F_y = 36$, and web thickness, distance between flanges and stiffener spacing.

should be spaced so that the applicable allowable shear, Eq. (8-25a) or (8-25b), is not exceeded, and in addition, so that a/h is not more than $[260/(h/t)]^2$ or 3.

Solution of the preceding formulas for stiffener spacing requires assumptions of dimensions and trials. The calculations can be facilitated by using tables in the AISC "Manual of Steel Construction." Also, Fig. (8-26) permits rapid selection of the most efficient stiffener arrangement, for webs of A36 steel. Similar charts can be drawn for other steels.

If the tension field concept is to apply to plate girder design, care is necessary to ensure that the intermediate stiffeners function as struts. When these stiffeners are spaced to satisfy Eq. (8-25a), their gross area, in² (total area if in pairs) should be at least

$$A_{st} = \frac{1 - C_v}{2} \left[\frac{a}{h} - \frac{(a/h)^2}{\sqrt{1 + (a/h)^2}} \right] Y Dht \qquad (8\text{-}26)$$

where Y = ratio of yield stress of web steel to yield stress of stiffener steel
D = 1.0 for stiffeners in pairs
  = 1.8 for single-angle stiffeners
  = 2.4 for single-plate stiffeners

When the greatest shear stress f_v in a panel is less than F_v determined from Eq. (8-25a), the gross area of the stiffeners may be reduced in the ratio f_v/F_v.

The moment of inertia of a stiffener or pair of stiffeners, about the web axis, should be at least $(h/50)^4$. The connection of these stiffeners to the web should be capable of developing shear, in kips per lineal inch of single stiffener or pair, of at least

$$f_{vs} = h \sqrt{\left(\frac{F_y}{340}\right)^3} \qquad (8\text{-}27)$$

where F_v is the yield stress of the web steel (Table 8-18). This shear also may be reduced in the ratio f_v/F_v, as above.

TABLE 8-18 Required Shear Capacity of Intermediate-Stiffener Connections to Girder Web

F_y, ksi	f_{vs}, kips per lin in.	F_y, ksi	f_{vs}, kips per lin in.
36.0	0.034h	60.0	0.074h
42.0	0.043h	65.0	0.084h
45.0	0.048h	90.0	0.136h
50.0	0.056h	100.0	0.160h
55.0	0.065h		

Combined Stresses in Web. A check should be made for combined shear and bending in the web where the tensile bending stress is approximately equal to the maximum permissible. When f_v, the shear force at the section divided by the web area, is greater than that permitted by Eq. (8-25b), the tensile bending stress in the web should be limited to no more than $0.6F_y$ or $F_y(0.825 - 0.375f_v/F_v)$, where F_v is the allowable web shear given by Eq. (8-25a) or (8-25b). For girders with A514 steel flanges and webs, when the flange bending stress is more than 75% of the allowable, the allowable shear stress in the web should not exceed that given by Eq. (8-25b).

Also, the compressive stresses in the web should be checked (see Art. 8-23).

8-37. Design of Trusses.

A truss is a framework arranged so as to form a series of rigid triangles, the bending moments being translated into axial stresses. The assumptions that these stresses are applied along the center of gravity of each member and that these stress lines meet at a point are not always precisely true. Steel angles, for example, in bolted construction are located so that gage lines, not gravity axes, are on the stress lines. Naturally, for wide angles with two gage lines, the line closest to the gravity axis is centered on the stress line to minimize eccentricity. Secondary stresses caused by such eccentricity should be avoided or minimized.

As a general rule, members framing into trusses are located at **panel points**—the intersections of three or more members. When loads are introduced between panel points, the truss members subjected to them must be designed for combined axial stress and bending.

8-50   Structural-Steel Construction

The design methods for trusses are the same as the basic methods of designing for tension, compression, and combined stresses (Arts. 8-19, 8-21, 8-27, and 8-28) plus connections at the panel points.

DIMENSIONAL AND DEFLECTION LIMITATIONS

Allowable stresses are based on certain minimum sizes of structural members and their elements that make possible full development of strength before premature buckling occurs. The higher the allowable stresses the more stringent must be the dimensional restrictions to preclude buckling or excessive deflections.

8-38. Slenderness Ratios.

Column slenderness ratios and effective lengths are defined and discussed in Arts. 8-21 and 8-22. The American Institute of Steel Construction "Specification for the Design, Fabrication and Erection of Structural Steel for Buildings" limits the effective slenderness ratio Kl/r to 200 for columns, struts, and truss members, where K is the ratio of effective length to actual unbraced length l, and r is the least radius of gyration.

A practical rule also establishes limiting slenderness ratios l/r for tension members:

For main members	240
For bracing and secondary members	300

But this does not apply to rods or other tension members that are drawn up tight (prestressed) during erection. The purpose of the rule is to avoid objectionable slapping or vibration in long, slender members.

8-39. Width-Thickness Ratios.

The AISC specifies several restricting ratios for compression members. One set applies to projecting elements subjected to axial compression or compression due to bending. Another set applies to compression elements supported along two edges.

Figure 8-27 lists maximum width-thickness ratios, b/t, for commonly used elements and grades of steel. Tests show that when b/t of elements normal to the direction of compressive stress does not exceed these limits, the member may be stressed close to the yield stress without failure by local buckling. Because the allowable stress increases with F_y, the specified yield stress of the steel, width-thickness ratios are less for higher-strength steels.

These b/t ratios should not be confused with the width-thickness ratios described in Art. 8-25. There, more restrictive conditions are set in defining compact sections qualified for higher allowable stresses.

Limits on width-thickness ratios for plastic designs are given in Art. 8-41.

8-40. Limits on Deflections.

The AISC "Specification for the Design, Fabrication and Erection of Structural Steel for Buildings" restricts the maximum live-load deflection of beams and girders supporting plastered ceilings to $\frac{1}{360}$ of the span. This rule is intended to prevent cracking of monolithic plaster; therefore, it is not necessarily applicable to members with unfinished ceilings or ceilings of other materials.

All other deflection criteria are left to the designer; it is impracticable to cover with a simple specification all the possible variations in loads, occupancies, and tolerable amounts of movement.

Minimum Depth-Span Ratios. As a guide, Table 8-19 lists suggested minimum depth-span ratios for various loading conditions and yield strengths of steel up to $F_y = 50.0$ ksi. These may be useful for estimating or making an initial design selection. Since maximum deflection is a straight-line function of maximum bending stress f_b and therefore is nearly proportional to F_y, a beam of steel with $F_y = 100.0$ ksi would have to be twice the depth of a beam of steel with $F_y = 50.0$ ksi when each is stressed to allowable values and has the same maximum deflection.

Vibration of large floor areas that are usually free of physical dampeners such as partitions may occur in buildings such as shopping centers and department stores, where pedestrian traffic is heavy. The minimum depth-span ratios suggested for "heavy pedestrian traffic" are intended to provide an acceptable solution.

One rule of thumb, often used to determine beam depth quickly, is that the depth, in, equals one-half the span length, ft. This, however, is accurate only for the case in which $F_b = 24.0$ ksi, which, in general, holds for A36 steel.

Ponding. Beams for flat roofs may require a special investigation to assure stability against water accumulation, commonly called ponding, unless there is adequate provision for drainage during heavy rainfall. The AISC "Specification for the Design, Fabrication and Erection of Structural Steel for Buildings" gives these criteria for stable roofs:

$$C_p + 0.9C_s \leq 0.25 \tag{8-28}$$

$$I_d \geq \frac{25S^4}{10^6}$$

where $C_p = 32L_sL_p^4/10^7\,I_p$
$C_s = 32SL_s^4/10^7\,I_s$
L_p = column spacing in direction of girder, ft (length of primary members)
L_s = column spacing perpendicular to direction of girder, ft (length of secondary member)
S = spacing of secondary members, ft
I_p = moment of inertia for primary members, in^4
I_s = moment of inertia for secondary member, in^4. Where a steel deck is supported on primary members, it is considered the secondary member. Use $0.85I_s$ for joists and trusses
I_d = moment of inertia of a steel deck supported on secondary members, in^4/ft

Uniform-Load Deflections. For the common case of a uniformly loaded simple beam loaded to the maximum allowable bending stress, the deflection in inches may be computed from

$$\delta = \frac{5}{24}\frac{F_b l}{Ed/l} \tag{8-29}$$

where F_b = the allowable bending stress, ksi
l = the span, in
E = 29,000 ksi
d/l = the depth-span ratio

Camber. Trusses of 80-ft or greater span should be cambered to offset dead-load deflections. Crane girders 75 ft or more in span should be cambered for deflection under dead load plus one-half live load.

$F_y =$	36.0	42.0	45.0	50.0	55.0	60.0	65.0	90.0	100.0
$\frac{76}{\sqrt{F_y}}$	12.7	11.7	11.3	10.7	10.2	9.8	9.4	8.0	7.6
$\frac{95}{\sqrt{F_y}}$	15.8	14.7	14.2	13.4	12.8	12.3	11.8	10.0	9.5

MAXIMUM $\frac{b}{t}$

SINGLE ANGLE

DOUBLE ANGLE

GIRDER WEB / STIFFENER

DOUBLE ANGLE

WELDED GIRDERS, BEAM AND TEE FLANGES

COVER PLATES

STEM OF TEES	$\dfrac{127}{\sqrt{F_y}}$	21.2	19.6	18.9	18.0	17.1	16.4	15.8	13.4	12.7
COLUMN WEBS AND DIAPHRAGMS	$\dfrac{253}{\sqrt{F_y}}$	42.2	39.0	37.7	35.8	34.1	32.7	31.4	26.7	25.3
COVER PLATES SOLID	$\dfrac{238}{\sqrt{F_y}}$	39.7	36.7	35.5	33.7	32.1	30.7	29.5	25.1	23.8
PERFORATED	$\dfrac{317}{\sqrt{F_y}}$	52.8	48.9	47.3	44.8	42.7	40.9	39.3	33.4	31.7

Fig. 8-27 Maximum width-thickness ratios for compression elements.

TABLE 8-19 Suggested Minimum Depth-Span Ratios for Beams

Specific beam condition	F_y, ksi				Deflection at	
	36.0	42.0	45.0	50.0	$0.60F_y$	$0.66F_y$
Heavy shock or vibration	$\dfrac{1}{18}$	$\dfrac{1}{15.5}$	$\dfrac{1}{14.5}$	$\dfrac{1}{13}$	$\dfrac{l}{357}$	$\dfrac{l}{324}$
Heavy pedestrian traffic	$\dfrac{1}{20}$	$\dfrac{1}{17}$	$\dfrac{1}{16}$	$\dfrac{1}{14.5}$	$\dfrac{l}{320}$	$\dfrac{l}{291}$
Normal loading	$\dfrac{1}{22}$	$\dfrac{1}{19}$	$\dfrac{1}{18}$	$\dfrac{1}{16}$	$\dfrac{l}{290}$	$\dfrac{l}{264}$
Beams for flat roofs*	$\dfrac{1}{25}$	$\dfrac{1}{21.5}$	$\dfrac{1}{20}$	$\dfrac{1}{18}$	$\dfrac{l}{258}$	$\dfrac{l}{232}$
Roof purlins, except for flat roofs*	$\dfrac{1}{28}$	$\dfrac{1}{24}$	$\dfrac{1}{22}$	$\dfrac{1}{20}$	$\dfrac{l}{232}$	$\dfrac{l}{210}$

* Investigate for stability against ponding. l = span of beam.

PLASTIC DESIGN OF STEEL

8-41. Plastic-Design Criteria.

The AISC "Specification for the Design, Fabrication and Erection of Structural Steel for Buildings" permits plastic design for simple or continuous beams and braced or unbraced planar rigid frames. The design method is applicable to all structural steels listed in Table 8-1 for which F_y does not exceed 65.0 ksi.

Plastic design is not recommended for parts of structures subjected to fatigue-producing loads, such as continuous crane-runway girders, although rigid frame supports for them may be plastic designed.

Load Factors. A distinguishing characteristic of plastic design is the absence of controlling allowable stresses, which are basic to elastic design. Instead, conventional safety factors are maintained through use of load factors by which the design loads are multiplied to determine the required capacity at ultimate load. (See example, Art. 8-42.)

The AISC Specification recommends two load factors:

1.70 for gravity loads (dead plus live)
1.30 for gravity loads plus wind or seismic forces

Braced Frames. A moderate-height multistory framing system is considered braced when securely attached shear walls (interior and perimeter), floor slabs, and roof decks function together with the steel framework (Art. 8-48).

For taller buildings, the steel frame by itself must be braced to maintain stability under the effects of both gravity and horizontal loads. This is achieved either by a system of diagonal bracing or by a moment-resisting frame in which the beam-to-column connections are rigid (Art. 8-47). The members (columns, beams, diagonals) comprising the vertical bracing system can be analyzed as a simply connected vertical-cantilever truss for frame buckling and lateral stability. But in that case, the axial force in a member caused by the factored loading may not exceed $0.85 P_y$, where P_y is the axial-load capacity at yield, the yield stress times the area of the member. Girders and beams in the bracing system must also satisfy Eq. (8-30) for beam-columns, with P_{cr} taken as the maximum axial strength of the beam based on actual l/r between braced points in the plane of bending.

TABLE 8-20 Maximum Width-Thickness Ratios for Plastic Design

	Yield stress F_y						
	36.0	42.0	45.0	50.0	55.0	60.0	65.0
b/t_f* for flanges of W beams and similar built-up single-web sections	8.5	8.0	7.4	7.0	6.6	6.3	6.0
b/t* for flange plates in box sections	31.7	29.3	28.3	26.9	25.6	24.5	23.6
d/t_w* for webs of members subjected to bending:							
When $P/P_y \geq 0.27$†	42.8	39.6	38.3	36.3	34.7	33.2	31.9
When $P/P_y = 0$‡	68.5	63.5	61.3	58.2	55.5	53.2	51.1

* b = projection, in., of compression flange of W and similar sections or width, in. (distance between longitudinal lines of fasteners or welds) of compression flange of box sections
t_f = thickness, in., of that flange in W and similar beams
t = thickness, in., of that flange in box sections
d = depth, in., of web
t_w = thickness, in., of web
† Depth-thickness ratios based on $257/\sqrt{F_y}$. P = applied axial load, kips, and P_y = yield stress, ksi, times cross-sectional area of member.
‡ $d/t_w = (1 - 1.4P/P_y)412/\sqrt{F_y}$ when $0 \leq P/P_y \leq 0.27$; use straight-line interpolation between values in table for $P/P_y = 0$ and $P/P_y \geq 0.27$.

Unbraced Frames. Unbraced multistory frames are permitted when the effect of frame instability and column axial deformation is considered in the analysis. Such frames must be stable for both of the aforementioned factored loadings. In this case, the axial column force may not exceed $0.75P_y$.

Limits on Width-Thickness Ratios. In plastic design, the limits on width-thickness ratios for compression elements are more restrictive than corresponding ratios for elastic design (Art. 8-39). The latter ratios ensure safety against local buckling of compression elements up to the yield stress, whereas the ratios for plastic design are set to ensure adequate hinge rotation at ultimate loads and throughout the plastic range. The maximum width-thickness ratios permitted for compression elements are listed in Table 8-20.

Columns. The maximum strength of an axially loaded column, kips, is given by

$$P_{cr} = 1.7 \, AF_a \qquad (8\text{-}30)$$

where A = gross area of column, in²
F_a = allowable stress, ksi, given by Eq. (8-1a or b)
The factored axial load, kips, may not exceed P_{cr}.

For combined axial force P and moment M, a column must satisfy the following interaction formulas:

$$\frac{P}{P_{cr}} + \frac{C_m M}{M_m(1 - P/P_e)} \leq 1.0 \qquad (8\text{-}31a)$$

$$\frac{P}{P_y} + \frac{M}{1.18 M_p} \leq 1.0 \qquad M \leq M_p \qquad (8\text{-}31b)$$

where M_p = plastic moment, in kips = $F_y Z$
F_y = steel yield stress, ksi
Z = plastic section modulus, in³, of column
P = applied axial load, kips
P_e = $(^{23}/_{12})AF'_e$

F_e' = stress defined for Eq. (8-12)
M = maximum applied moment, in kips, produced by applied loads
C_m = coefficient defined for Eq. (8-12)
M_m = critical moment in absence of axial load, in-kips
   = M_p for column braced in weak direction
   = $M_p[1.07 - (l/r_y)\sqrt{F_y}/3160] \leq M_p$ for column not braced in weak direction. Additional information is given in the "Commentary on the AISC Specification"
l/r_y = slenderness ratio about weak axis

See also following discussion of lateral bracing.

Shear. Webs of beams, girders, columns and large connections (e.g., rigid-frame knees) should be proportioned for shear so that

$$V_u \leq 0.55 F_y t d \qquad (8\text{-}32)$$

where V_u = shear, kips, for the factored loads
d = depth of member, in
t = thickness of web, in

Should the computed shear exceed V_u, the web may be reinforced with stiffeners or doubler plates.

Web Stiffeners. These are required at points of load application where a plastic hinge would form. Conditions governing web thickness and the need for stiffeners for members subjected to concentrated loads delivered by flanges of a member framing into it are summarized in Art. 8-23. The rules are applicable to both elastic and plastic design methods.

Beams. The plastic bending capacity of a beam is given by

$$M_p = Z F_y \qquad (8\text{-}33)$$

where M_p = moment, in-kips, at which a plastic hinge would form
F_y = specified minimum yield stress, ksi
Z = plastic section modulus, in^3, about beam axis under consideration. The AISC "Manual of Steel Construction" lists Z for both the x and y axes of all rolled-beam sections

Lateral Bracing. An important consideration in plastic design of structural members is the adequacy of lateral bracing. To ensure behavior as predicted by plastic theory, lateral bracing must be provided to resist lateral and torsional displacements at plastic-hinge locations. Also, the laterally unsupported distance l_{cr} between each hinge and an adjacent braced point must satisfy the criteria:

$$\frac{l_{cr}}{r_y} = \frac{1{,}375}{F_y} + 25 \qquad 1.0 > \frac{M}{M_p} > -0.5 \qquad (8\text{-}34a)$$

$$\frac{l_{cr}}{r_y} = \frac{1{,}375}{F_y} \qquad -0.5 \geq \frac{M}{M_p} > -1.0 \qquad (8\text{-}34b)$$

where r_y = radius of gyration, in, of member about its weak axis
M = lesser of the moments at ends of unbraced segment, in-kips
M_p = moment, in-kips, at which a hinge would form at end of member
M/M_p = end-moment ratio, positive when segment is bent in reverse curvature and negative when bent in single curvature

Members built into a masonry wall, with web perpendicular to the wall, are considered laterally supported.

Fasteners and Welds. Capacities of high-strength bolts, in both bearing and friction-type connections, A307 bolts, rivets, and welds for resisting forces computed for factored loading may be taken equal to 1.70 times the capacities determined with allowable unit stresses used for elastic design.

In general, other aspects of plastic design follow the criteria recommended for elastic design. Additional information is given in the AISC Specification and Commentary. (See also Bibliography, Art. 8-92.)

8-42. Comparison of Elastic and Plastic Design Methods.

To illustrate the fundamental difference between elastic design, based on allowable stresses, and plastic design, based on load factors, the following simple example is presented in which a laterally supported beam with both ends fixed is designed by both methods.

Problem: Given total bending moment $WL/8 = 234$ ft-kips, where W is the total uniform gravity loading, kips, and L is the span, ft. Assume both ends of the beam are fixed. Select a compact beam section of A36 steel.

Solution by Elastic Theory: The maximum positive moment is

$$\frac{WL}{24} = \frac{1}{3} \times \frac{WL}{8} = \frac{234}{3} = 78 \text{ ft-kips}$$

The maximum negative moment is $WL/12 = \frac{2}{3} \times WL/8 = 156$ ft-kips (Fig. 8-28).

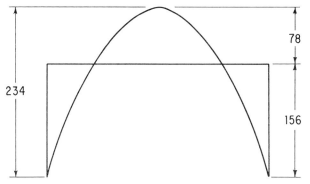

Fig. 8-28 Bending-moment diagram for a fixed-end beam with $WL/8 = 234$.

As noted in Art. 8-25, a limited redistribution of moments is permitted by the AISC specification. The negative end moments may be reduced 10% if the positive moment is increased by 10% of the average end moments. Thus, the adjusted end negative moments are $156 - 0.1 \times 156 = 140.4$, and the adjusted positive moment is $78 + 15.6 = 93.6$, because both end moments are equal. The negative moment, 140.4 ft-kips, governs. So with the allowable bending stress, $F_b = 24$ kips/in^2, the required elastic section modulus is

$$S_x = 140.4 \times 12/24 = 70.2$$

Because a W16 × 40 has $S_x = 64.6$, whereas a W16 × 45 has $S_x = 72.5$, use the W16 × 45.

Solution by Plastic Theory. Assume plastic hinges to form at both ends and at the midpoint of the beam. The plastic moment M_p is the same at all those points and thus must be equal to half the total, or simple-beam, bending moment. With a load factor of

1.70, $M_p = 1.70 \times \frac{1}{2} \times WL/8 = 1.70 \times 117$. Then, the required plastic section modulus is

$$Z_x = \frac{M_p}{F_y} = 1.70 \times 117 \times \frac{12}{36} = 66.3$$

Because a W16 × 36 has $Z_x = 63.9$ and a W16 × 40 has $Z_x = 72.7$, and other sections with a plastic modulus between those values are heavier, use the W16 × 40. The weight saving compared with elastic design is 11%.

BRACING

Bracing as it applies to steel structures includes secondary members incorporated into the system of main members to serve these principal functions:

1. Slender compression members, such as columns, beams, and truss elements are braced, or laterally supported, so as to restrain the tendency to buckle in a direction normal to the stress path. The rigidity, or resistance to buckling, of an individual member is determined from its length and certain physical properties of its cross section. Economy and size usually determine whether bracing is to be employed.

2. Since most structures are assemblies of vertical and horizontal members forming rectangular (or square) panels, they possess little inherent rigidity. Consequently, additional rigidity must be supplied by a secondary system of members or by rigid or semirigid joints between members. This is particularly necessary when the framework is subject to lateral loads, such as wind, earthquakes, and moving loads. Exempt from this second functional need for bracing are trusses, which are basically an arrangement of triangles possessing in their planes an inherent ideal rigidity both individually and collectively.

3. There frequently is a need for bracing to resist erection loads and to align or prevent overturning, in a direction normal to their planes, of trusses, bents, or frames during erection. Such bracing may be temporary; however, usually bracing needed for erection is also useful in supplying rigidity to the structure and therefore is permanently incorporated into the building. For example, braces that tie together adjoining trusses and prevent their overturning during erection are useful to prevent sway—even though the swaying forces may not be calculable.

8-43. Column Bracing.

Interior columns of a multistory building are seldom braced between floor connections. Bracing of any kind generally interferes with occupancy requirements and architectural considerations. Since the slenderness ratio l/r in the weak direction usually controls column size, greatest economy is achieved by using only wide-flange column sections or similar built-up sections.

It is frequently possible to reduce the size of wall columns by introducing knee braces or struts in the plane of the wall, or by taking advantage of deep spandrels or girts that may be otherwise required. Thus the slenderness ratio for the weak and strong axis can be brought into approximate balance. The saving in column weight may not always be justified; one must take into account the weight of additional bracing and cost of extra details.

Column bracing is prevalent in industrial buildings because greater vertical clearances necessitate longer columns. Tall slender columns may be braced about both axes to obtain an efficient design.

Undoubtedly, heavy masonry walls afford substantial lateral support to steel columns embedded wholly or partly in the wall. The general practice, however, is to disregard this assistance.

An important factor in determining column bracing is the allowable stress for the column section (Art. 8-21). Column formulas for obtaining this stress are based on the ratio of two variables, effective length Kl and the physical property called radius of gyration r. For bracing for plastic design, see Art. 8-41.

The question of when to brace (to reduce the unsupported length and thus slenderness ratio) is largely a matter of economics and architectural arrangements; thus no general answer can be given.

Composite Columns. Concrete encasement (when used) is usually credited with partial lateral support. This is reflected in an increase in allowable load-carrying capacity. For example, the American Concrete Institute Building Code (ACI 318) permits the strength of concrete columns with a structural-steel core (composite compression members) to be computed for the same limiting conditions applicable to ordinary reinforced-concrete members. Thus, for factored loads, the capacity of the column is determined by the yield stress F_y of the structural steel, the compressive strength of the concrete $(0.85f'_c)$, the yield stress of longitudinal reinforcing steel, critical buckling load, and slenderness ratio. But the minimum thickness of the concrete encasement around the core should be at least 1½ in or $b\sqrt{F_y/3E_s}$ for each face of width b, or $h\sqrt{F_y/8E_s}$ for circular sections with diameter h. The concrete should have a strength f'_c of at least 2.5 ksi. Design yield stress for the structural steel core may not exceed 50 ksi. The concrete encasement must be spiral bound or laterally tied. In addition, longitudinal reinforcing bars with a total cross-sectional area between 1 and 8% of the net concrete section must be placed within the ties or spiral. (These bars may be taken into account in computing the area and moment of inertia of the core for spiral-bound columns, but not the moment of inertia for tied columns.) Then, the radius of gyration, in, of the composite member may be taken as

$$r = \sqrt{\frac{0.2E_c I_g + E_s I_t}{0.2E_c A_g + E_s A_t}} \qquad (8\text{-}35)$$

where E_c = modulus of elasticity of concrete, ksi
E_s = modulus of elasticity of structural steel core, ksi
I_g = moment of inertia, in⁴, of gross concrete section about centroidal axis, neglecting core and reinforcement
I_t = moment of inertia, in⁴, of structural steel about centroidal axis of member cross section
A_g = gross area of composite column, in²
A_t = area of structural steel, in²

In computation of the buckling load $P_c = \pi^2 EI/(Kl)^2$, where K is the effective length factor and l the unbraced length, in., EI of the composite section may be taken as

$$EI = \frac{0.2E_c I_g + E_s I_t}{1 + \beta_d} \qquad (8\text{-}36)$$

where β_d = ratio of maximum design dead-load moment to maximum design total-load moment, always positive.

8-44. Beam Bracing.

Economy in size of member dictates whether laterally unsupported beams should have additional lateral support between end supports. Lateral support at intermediate points should be considered whenever the allowable stress obtained from the reduction formulas for large l/r_t given in Fig. 8-20 falls below some margin, say 25%, of the stress allowed for the fully braced condition. There are cases, however, where stresses as low as 4.0 ksi have been justified, because intermediate lateral support was impractical.

The question often arises: When is a steel beam laterally supported? There is no fixed rule in specifications (nor any intended in this discussion) because the answer requires application of sound judgment based on experience. Tests and studies that have been made indicate that it takes rather small forces to balance the lateral thrusts of initial buckling.

In Fig. 8-29 are illustrated some of the common situations encountered in present-day practice. In general, **positive lateral support is provided by:**

(*a*) and (*b*)   All types of cast-in-place concrete slabs (questionable for vibrating loads and loads hung on bottom flange).

(*c*)   Metal and steel plate decks, welded connections.

(*d*)   Wood decks nailed securely to nailers bolted to beam.

(*e*) and (*f*)   Beam flange tied or braced to strut system, either as shown in (*e*) or by means of cantilever tees, as shown in (*f*); however, struts should be adequate to resist rotation.

Fig. 8-29 Lateral support for beams.

(g) Purlins used as struts, with tees acting as cantilevers (common in rigid frames and arches). If plate stiffeners are used, purlins should be connected to them with high-strength bolts to ensure rigidity.

(h) Open-web joists tack-welded (or the equivalent) to the beams, but the joists themselves must be braced together (bridging), and the flooring so engaged with the flanges that the joists, in turn, are adequately supported laterally.

(i) Purlins connected close to the compression flange.

(k) Tees (part of cast-in-place gypsum construction) welded to the beams.

Doubtful lateral support is provided by:

(j) Purlins seated on beam webs, where the seats are distant from the critical flange

(l) Precast slabs not adequately fastened to the compression flange.

The reduction formulas for large l/r_t given in Fig. 8-20 do not apply to steel beams fully encased in concrete, even though no other lateral support is provided.

Introducing a secondary member to cut down the unsupported length does not necessarily result in adequate lateral support. The assumed thrust, or rather the resistive capacity of the member, must be traced through the system to ascertain its effectiveness. For example the system in Fig. 8-30a may be free to deflect laterally as shown. This can be prevented by a rigid floor system that acts as a diaphragm, or in the absence of a floor, it may be necessary to X-brace the system as shown in Fig. 8-30b.

8-45. Capacity of Bracing.

For an ideally straight, exactly concentrically loaded beam or column, only a small force may be needed to be supplied by an intermediate brace intended to reduce the unbraced length of a column or the unsupported length of the compression flange of a beam. But there is no generally accepted method of calculating that force.

The only function of a brace is to provide a node point in the buckled configuration. Since no force is considered present at the node, rigidity is the only requirement for the brace. But actual members do contain nonuniform residual stresses and slight initial crookedness and may be slightly misaligned, and these eccentricities create a thrust that must be resisted by the brace.

Fig. 8-30 Lateral bracing systems (a) without and (b) with X bracing.

A rule used by some designers that has proved satisfactory is to design the brace for 2% of the axial load of columns, or 2% of the total compressive stress in beam flanges. Studies and experimental evidence indicate that this rule is conservative.

8-46. Lateral Forces on Building Frames.

Design of bracing to resist forces induced by wind, seismic disturbances, and moving loads, such as those caused by cranes, is not unlike, in principle, design of members that support vertical dead and live loads. These lateral forces are readily calculable. They are collected at points of application and then distributed through the structural system and delivered to the ground. Wind loads, for example, are collected at each floor level and distributed to the columns that are selected to participate in the system. Such loads are cumulative; that is, columns resisting wind shears must support at any floor level all the wind loads on the floors above the one in consideration.

8-47. Bracing Tall Buildings.

If the steel frame of the multistory building in Fig. 8-31a is subjected to lateral wind load, it will distort as shown in Fig. 8-31b, if the connections of columns and beams are of the standard type, for which rigidity (resistance to rotation) is nil. One can visualize this readily by assuming each joint is connected with a single pin. Naturally, the simplest method to prevent this distortion is to insert diagonal members—triangles being inherently rigid, even if all the members forming the triangles are pin-connected.

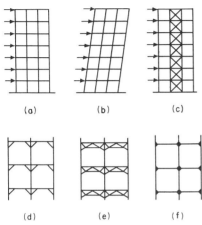

Fig. 8-31 Wind bracing for multistory buildings.

Braced Bents. Bracing of the type in Fig. 8-31c, called X bracing, is both efficient and economical. Unfortunately, X bracing is usually impracticable because of interference with doors, windows, and clearance between floor and ceiling. Architects of modern office buildings require clear floor areas; this offers flexibility of space use, with movable partitions. But about the only place for X bracing in this type of building is in the elevator shaft, fire tower, or wherever a windowless wall is required. As a result, additional bracing must be supplied by other methods. On the other hand, X bracing is used extensively for bracing industrial buildings of the shed or mill types.

Moment-Resisting Frames. Designers, however, have a choice of several alternative methods. Knee braces, shown in Fig. 8-31d, or portal frames, shown in Fig. 8-31e, may be used in outer walls, where they are likely to interfere only with windows. For buildings with window walls, the bracing often used is the bracket type (Fig. 8-31f). It simply develops the end connection for the calculated wind moment. Connections vary in type, depending on size of members, magnitude of wind moment, and compactness needed to comply with floor-to-ceiling clearances.

Figure 8-32 illustrates a number of bracket-type wind-braced connections. The minimum type, represented in Fig. 8-32a, consists of angles top and bottom; they are ample for moderate-height buildings. Usually the outstanding leg (against the column) is of a size that permits only one gage line; a second line of fasteners would not be effective because of the eccentricity. When greater moment resistance is needed, the type shown in Fig. 8-32b should be considered. This is the type that has become rather conventional in field-bolted construction. Figure 8-32c illustrates the maximum size with beam stubs having flange widths that permit additional gage lines, as shown. It is thus possible on larger wide-flange columns to obtain 16 fasteners in the stub-to-column connection.

The resisting moment of a given connection varies with the distance between centroids of the top and bottom connection piece. To increase this distance, thus increasing the moment, an auxiliary beam may be introduced as shown in Fig. 8-32d, if it does not create an interference.

All the foregoing types may be of welded construction, rather than bolted. In fact, it

Fig. 8-32 Typical wind connections.

is not unusual to find mixtures of both because of the fabricator's decision to shop-bolt and field-weld, or vice versa. Welding, however, has much to offer in simplifying details and saving weight, as illustrated in Fig. 8-32e, f, and g. The last represents the ultimate efficiency with respect to weight saving, and furthermore, it eliminates interfering details.

Deep wing brackets (Fig. 8-32h and i) are sometimes used for wall beams and spandrels designed to take wind stresses. Such deep brackets are, of course, acceptable for interior beam bracing whenever the brackets do not interfere with required clearances.

Not all beams need be wind-braced in tall buildings. Usually the wind is concentrated on certain column lines, called **bents,** and the forces are carried through the bents to the ground. For example, in a wing of a building, it is possible to concentrate the wind load on the outermost bent. To do so may require a stiff floor or diaphragmlike system capable of distributing the wind loads laterally. One-half these loads may be transmitted to the outer bent, and one-half to the main building to which the wing connects.

Braced bents are invariably necessary across the narrow dimension of a building. The question arises as to the amount of bracing required in the long dimension, since wind of equal unit intensity is assumed to act on all exposed faces of structures. In buildings of square or near square proportions, it is likely that braced bents will be provided in both directions. In buildings having a relatively long dimension, as compared with width, the need for bracing diminishes. In fact, in many instances, wind loads are distributed over so many columns that the inherent rigidity of the whole system is sufficient to preclude the necessity of additional bracing.

8-64   Structural-Steel Construction

Why all the emphasis on beam-to-column connections and not on column-to-column joints, since both joints are subjected to the same wind forces? Columns are compression members and as such transmit their loads, from section above to section below, by direct bearing between finished ends. It is not likely, in the average building, for the tensile stresses induced by wind loads ever to exceed the compressive pressure due to dead loads. Consequently, there is no theoretical need for bracing a column joint. Actually, however, column joints are connected together with nominal splice plates for practical considerations—to tie the columns during erection and to obtain vertical alignment.

This does not mean that one may always ignore the adequacy of column splices. In lightly loaded structures, or in exceptionally tall but narrow buildings, it is possible for the horizontal wind forces to cause a net uplift in the windward column because of the overturning action. The commonly used column splices should then be checked for their capacity to resist the maximum net tensile stresses caused in the column flanges. This computation and possible heaving up of the splice material may not be thought of as bracing; yet, in principle, the column joint is being "wind-braced" in a manner similar to the wind-braced floor-beam connections.

Tubular Framing. Designs of some very tall buildings employ a structural concept that departs from the foregoing conventional method of wind-bracing steel frames. In these buildings, the perimeter columns and spandrel beams form a theoretical, vertical-cantilever hollow tube that resists the entire lateral load. The floors, through diaphragm action, transfer the lateral forces to the exterior columns. These members are closely spaced and either X-braced or rigidly connected to the spandrel beams to approximate an idealized solid-surface tube. Interior columns merely support their share of the gravity loads. The elimination of floorbeam-to-column moment connections and standardization of floorbeam sizes, freed from accumulative lateral forces, which are characteristic of conventional design, offer substantial savings.

8-48. Shear Walls.

It is well known that masonry walls enveloping the steel frame, interior masonry walls, and perhaps some of the stiffer partitions can resist a substantial amount of lateral load. Rigid floor systems participate by distributing the shears induced at each floor level to the columns and walls. Yet, it is common design practice to carry wind loads on the steel frame, little or no credit being given to the substantial resistance rendered by the floors and walls. In the past, some engineers deviated from this conservatism by assigning a portion of the wind loads to the floors and walls; but even then, the steel frame carried the major share. The present trend in multistory building design, particularly for offices, is to walls of glass, thin metallic curtain walls, lightweight floors, and removable partitions. This construction imposes on the steel frame almost complete responsibility for transmittal of wind loads to the ground. Consequently, the importance of windbracing tall steel structures should not diminish.

In fact, in tall, slender, partitioned buildings, such as hotels and apartments, the problem of preventing or reducing to tolerable limits the cracking of rigid-type partitions has been related to the wracking action of the frame caused by excessive deflection. One solution that may be used for exceptionally slender frames (those most likely to deflect excessively) is to supplement the normal bracing of the steel frame with shear walls. Acting as vertical cantilevers in resisting lateral forces, these walls, often constructed of reinforced concrete, may be arranged much like structural shapes, such as plates, channels, Ts, Is, or Hs. (See also Art. 3-6 and Sec. 11.) Walls needed for fire towers, elevator shafts, divisional walls, etc., may be extended and reinforced to serve as shear walls, and may relieve the steel frame of cumbersome bracing or uneconomical proportions.

Fig. 8-33 Relative stiffness of bents depends on restraints on columns.

Fig. 8-34 Bracing in factory transmits wind loads to the ground.

8-49. Bracing Industrial-Type Buildings.

Bracing of low industrial buildings for horizontal forces presents fewer difficulties than bracing of multistory buildings, because the designer is virtually free to select the most efficient bracing without regard to architectural considerations or interferences. For this reason, conventional X bracing is widely used—but not exclusively. Knee braces, struts, and sway frames are used where needed.

Wind forces acting upon the frame shown in Fig. 8-33a, with hinged joints at the top and bottom of supporting columns, would cause collapse as indicated in Fig. 8-33b. In practice, the joints would not be hinged. However, a minimum-type connection at the truss connection and a conventional column base with anchor bolts located on the axis transverse to the frame would approximate this theoretical consideration of hinged joints. Therefore, the structure requires bracing capable of preventing collapse and undue deflection.

In the usual case, the connection between truss and column will be stiffened by means of knee braces (Fig. 8-33c). The rigidity so obtained may be supplemented by providing partial rigidity at the column base by simply locating the anchor bolts in the plane of the bent.

In buildings containing overhead cranes, the knee brace may interfere with crane operation. Then, the interference may be eliminated by fully anchoring the column base so that the column may function as a vertical cantilever (Fig. 8-33d).

The method often used for very heavy industrial buildings is to obtain substantial rigidity at both ends of the column so that the behavior under lateral load will resemble the condition illustrated in Fig. 8-33e. In both (d) and (e), the footings must be designed for such moments.

A common assumption in wind distribution for the type of light mill building shown in Fig. 8-34 is that the windward columns take a large share of the load acting on the side of the building and deliver the load directly to the ground. The remaining wind load

8-66   Structural-Steel Construction

on the side is delivered by the same columns to the roof systems, where the load joins with the wind forces imposed directly on the roof surface. Then, by means of diagonal X bracing, working in conjunction with the struts and top chords of the trusses, the load is carried to the eave struts, thence to the gables and, through diagonal bracing, to the foundations.

Because wind may blow in all directions, the building also must be braced for the wind load on the gables. This bracing becomes less important as the building increases in length and conceivably could be omitted in exceptionally long structures. The stress path is not unlike that assumed for the transverse wind forces. The load generated on the ends is picked up by the roof system and side framing, delivered to the eaves, and then transmitted by the diagonals in the end sidewall bays to the foundation.

No distribution rule for bracing is intended in this discussion; bracing can be designed many different ways. Whereas the foregoing method would be sufficient for a small building, a more elaborate treatment is suggested for larger structures.

(a) VIEW SHOWING BRACING IN PLANE OF ROOF AND IN WALLS

(b) BRACING IN PLANE OF BOTTOM CHORD OF TRUSSES

(c) TYPICAL SWAY FRAME

Fig. 8-35 Braced bays in one-story building transmit wind loads to the ground.

Braced bays, or towers, are usually favored in the better-braced structures, such as in Fig. 8-35. There, a pair of transverse bents are connected together with X bracing in the plane of the columns, plane of truss bottom chords, plane of truss top chords, and by means of struts and sway frames. It is assumed that each such tower can carry the wind load from adjacent bents, the number depending on assumed rigidities, size, span, and also on sound judgment. Usually every third or fourth bent should become a braced bay. Participation of bents adjoining the braced bay can be assured by insertion of bracing designated "intermediate" in Fig. 8-35b. This bracing is of greater importance when knee braces between trusses and columns cannot be used. When maximum lateral stiffness of intermediate bents is desired, it can be obtained by extending the X bracing across the span; this is shown with broken lines in Fig. 8-35b.

Buildings with flat or low-pitched roofs, shown in Fig. 8-7d and e, require little bracing because the trusses are framed into the columns. These columns are designed for the heavy moments induced by wind pressure against the building side. The bracing that would be provided, at most, would consist of X bracing in the plane of the bottom chords for purpose of alignment during erection and a line or two of sway frames for longitudinal rigidity. Alignment bracing is left in the structure since it affords a secondary system for distributing wind loads.

8-50. Bracing Crane Structures.

All buildings that house overhead cranes should be braced for the thrusts induced by sideway and longitudinal motions of the cranes. Bracing used for wind or erection may, of course, be assumed to sustain the lateral crane loadings. These forces are usually concentrated on one bent. Therefore, normal good practice dictates that adjoining bents share in the distribution. Most effective is a system of X bracing located in the plane of the bottom chords of the roof trusses.

In addition, the bottom chords should be investigated for possible compression, although the chords normally are tension members. A heavily loaded crane is apt to draw the columns together, conceivably exerting a greater compression stress than the actual tension stress obtainable under condition of zero snow load. This may indicate the need for intermediate bracing of the bottom chord.

8-51. Design of X Bracing.

X bracing may consist of rods (probably the least expensive) or stiffer shapes, such as angles. Rods are unsuitable for compression, whereas an angle may be able to take some compression and thereby aid another diagonal member in tension. Generally, the rod or the angle is designed to take all the stress in tension.

One objection to rods is that they may rattle, particularly in buildings subjected to vibrations. Angle bracing is often larger than need be for the calculated stress. The American Institute of Steel Construction, to assure satisfactory sevice under normal operation, requires that tension members (except rods) have a minimum slenderness ratio l/r of 300.

8-52. Bracing Rigid Frames.

Rigid frames of the type shown in Fig. 8-9a have enjoyed popular usage for gymnasiums, auditoriums, mess halls, and with increasing frequency, industrial buildings. The stiff knees at the junction of the column with the rafter imparts excellent transverse rigidity. Each bent is capable of delivering its share of wind load directly to the footings. Nevertheless, some bracing is advisable, particularly for resisting wind loads against the end of the building. Most designers emphasize the importance of an adequate eave strut; it usually is arranged so as to brace the inside flange (compression) of the frame knee, the connection being located at the midpoint of the transition between column and rafter segments of the frame. Intermediate X bracing in the plane of the rafters usually is omitted.

FASTENERS

There are three basic types of fasteners: rivets, bolts, and welds. Rivets, formerly widely used for main connections, are now seldom used in new construction.

Many variables affect the selection of a fastener. Included among these are economy of fabrication and of erection, availability of equipment, inspection problems, labor supply, and such design considerations as fatigue, size and type of connections, continuity of framing, reuse, and maintenance.

Most structural shops are sufficiently versatile to furnish all types of fasteners, but some are limited. For example, a shop arranged for all-welded construction may be short on punching and reaming capacity to furnish bolted construction.

8-68   Structural-Steel Construction

It is not uncommon for steel framing to be connected with such combinations as shop and field welds, or shop welds and field bolts. Even for field fasteners, there may be a combination of welds for the main beam-to-column connections and bolts for the small fill-in beams. The variables affecting the decisions to use these fasteners and combinations of them are too numerous and often too controversial to permit establishment of fixed rules.

8-53. Hot-Driven Rivets.

Use of field riveting declined decades ago when trained riveters became scarce. Shop riveting, however, continued until the increasing volume of welded construction raised the cost of infrequent riveting to the point where it usually became more economical to substitute high-strength bolts, although the purchase price of bolts is higher. As a consequence, few structural shops now have the necessary equipment for riveted construction.

Available rivets in sizes up to 1½-in diam conform to ASTM Specification A502, Grades 1 or 2. Grade 1 carbon-steel rivets are intended for use with the lower-strength carbon steels, while Grade 2 manganese-carbon rivets are more efficient for use with the higher-strength carbon and low-alloy, high-strength structural steels. Holes for rivets should not be more than $\frac{1}{16}$ in larger than the nominal rivet diameter.

Allowable stresses for rivets are given in Table 8-12.

8-54. Unfinished Bolts.

Known in construction circles by several names—ordinary, common, machine, or rough—unfinished bolts are characterized chiefly by the rough appearance of the shank. They are covered by ASTM Specification A307. They fit into holes $\frac{1}{16}$ in larger in diameter than the nominal bolt diameter. Symbols for bolts are shown in Fig. 8-36.

Fig. 8-36 Symbols for shop and field bolts.

Allowable stresses are given in Tables 8-12 and 8-14. The low allowable stress in shear takes into account the possibility of threads in shear planes. Thus, washers used to extend the bolt body are unnecessary.

One advantage of unfinished bolts is the ease of making a connection; only a wrench is required. On large jobs, however, erectors find they can tighten bolts more economically with a pneumatic-powered impact wrench. Power tightening probably yields greater uniformity of tension in the bolts and makes for a better-balanced connection.

While some old building codes restrict unfinished bolts to minor applications, such as small, secondary (or intermediate) beams in floor panels and in certain parts of one-story, shed-type buildings, the AISC, with a basis of many years of experience, permits A307 bolts for main connections on structures of substantial size. For example, these bolts may be used for beam and girder connections to columns in buildings up to 125 ft in height.

There is an economic relation between the strength of a fastener and that of the base material. So while A307 may be economical for connecting steel with a 36-ksi yield point, this type of bolt may not be economical with 50-ksi yield-point steel. The number of fasteners to develop the latter becomes excessive and perhaps impractical due to size of detail material.

A307 bolts should always be considered for use, even in an otherwise all-welded building, for minimum-type connections, such as for purlins, girts, struts, etc.

See also Art. 8-57.

Locking Devices for Bolts. Unfinished bolts (ASTM A307) and interference-body-type bolts (Art. 8-56) usually come with American Standard threads and nuts. Properly tightened, connections with these bolts give satisfactory service under static loads. But when the connections are subjected to vibration or heavy dynamic loads, a locking device is desirable to prevent the nut from loosening.

Locking devices may be classified according to the method employed: special threads, special nuts, special washers, and what may be described as field methods. Instead of conventional threads, bolts may be supplied with a patented self-locking thread called Dardelet. Sometimes, locking features are built into the nuts. Patented devices, the Automatic-Nut, Union-Nut, and Pal-Nut, are among the common ones. Washers may be split rings or specially toothed. Field methods generally used include checking, or distorting, the threads by jamming them with a chisel or locking by tack welding the nuts.

8-55. High-Strength Bolts.

Development of high-strength bolts is vested in the Research Council on Riveted and Bolted Structural Joints of the Engineering Foundation. Its "Specification for Structural Steel Joints Using A325 or A490 Bolts" was adopted by the American Institute for Steel Construction. Bolts conforming to ASTM A449 are acceptable, but their usage is restricted to bearing-type connections (Fig. 8-37) requiring bolt diameters greater than 1½ in. Furthermore, when they are required to be tightened to more than 50% of their

Fig. 8-37 Two main types of connections with high-strength bolts. Although, in general, no paint is permitted on faying surfaces in friction-type connections, the following exceptions are allowed: scored galvanized coatings, inorganic zinc-rich paint, metallized zinc or aluminum coatings.

8-70 Structural-Steel Construction

specified minimum tensile strength, hardened steel washers should be installed under the heads.

When high-strength bolts are used in a connection, they are highly tensioned by tightening of the nuts and thus tightly clamp together the parts of the connection.

For convenient computation of load capacity, the clamping force and resulting friction are resolved as shear. Bearing between bolt body and connected material is not a factor until loads become large enough to cause slippage between the parts of the connection. The bolts are assumed to functon in shear following joint slippage into full bearing.

The clamping and bearing actions lead to the dual concept: friction-type connections and bearing-type connections. For the latter, the allowable shear depends on the cross-sectional bolt area at the shear plane. Hence, two shear values are assigned, one for the full body area and one for the reduced area at the threads (Fig. 8-37). Allowable stresses are given in Tables 8-12 to 8-14.

TABLE 8-21 Washer Requirements for High-Strength Bolts

Method of tensioning	A325 bolts	A490 bolts	
		Base material $F_y < 40.0$	Base material $F_y > 40.0$
Calibrated wrench	One washer under turned element	Two washers	One washer under turned element
Turn-of-the-nut	None	Two washers	One washer under turned element
Both methods, slotted and over-sized holes	Two washers	Two washers	Two washers

Washers. Washer requirements, depending on method of installation and diameter of holes, are summarized in Table 8-21.

Identification. There is no difference in appearance of high-strength bolts intended for either friction or bearing-type connections. To aid installers and inspectors in identifying the several available grades of steel, bolts and nuts are manufactured with permanent markings (Fig. 8-38). Symbols for bolts for use on drawings are shown in Fig. 8-36.

Bolt Tightening. Specifications require all high-strength bolts to be tightened to 70% of their specified minimum tensile strength, which is nearly equal to the proof load (specified lower bound to the proportional limit) for A325 bolts, and within 10% of the proof load for A490 bolts. Tightening above these minimum tensile values does not damage the bolts, but it is prudent to avoid excessive uncontrolled tightening. The required minimum tension, kips, for bolt sizes is given in Table 8-22. There are three methods for tightening bolts to assure the prescribed tensioning:

Fig. 8-38 Identification markings on heads and nuts of high-strength bolts.

Turn-of-Nut. By means of a manual or powered wrench, the head or nut is turned from an initial snug-tight position. The amount of rotation, varying from one-third to a full turn, depends on the ratio of bolt length (underside of head to end of point) to bolt diameter and on the disposition (normal or sloped not more than 1:20 with respect to the bolt axis) of the outer surfaces of bolted parts. Required rotations are tabulated in the "Specification for Structural Steel Joints Using A325 or A490 Bolts."

Calibrated Wrench. By means of a powered wrench with automatic cutoff and calibration on the job. Control and test are accomplished with a hydraulic device equipped with a gage that registers the tensile stress developed.

Direct Tension Indicator. Special indicators are permitted on satisfactory demonstration of performance. One example is a hardened steel washer with protrusions on one face. The flattening that occurs on bolt tightening is measured and correlated with the induced tension.

See also Art. 8-57.

TABLE 8-22 Minimum Tightening Tension, kips, for High-Strength Bolts

Dia, in.	A325	A490
$5/8$	19	24
$3/4$	28	35
$7/8$	39	49
1	51	64
$1\,1/8$	56	80
$1\,1/4$	71	102
$1\,3/8$	85	121
$1\,1/2$	103	148

8-56. Other Bolt-Type Fasteners.

Interference body or bearing-type bolts are characterized by a ribbed or interrupted-ribbed shank and a button-shaped head; otherwise, including strength, they are similar to the regular A325 high-strength bolts. The extreme diameter of the shank is slightly larger than the diameter of the bolt hole. Consequently, the tips of the ribs or knurlings will groove the side of the hole, assuring a tight fit. One useful application has been in high television towers, where minimum-slippage joints are desired with no more installation effort than manual tightening with a spud wrench. Nuts may be secured with lock washers, self-locking nuts, or Dardelet self-locking threads. The main disadvantage of interference body bolts is the need for accurate matching of truly concentric holes in the members being joined; reaming sometimes is necessary.

Huckbolts are grooved (not threaded) and have an extension on the end of the shank. When the bolt is in the hole, a hydraulic machine, similar to a bolting or riveting gun, engages the extension. The machine pulls on the bolt to develop a high clamping force, then swages a collar into the grooved shank and snaps off the extension, all in one quick operation.

8-57. Bolt Holes.

For bearing-type connections, the diameter of a standard hole may not exceed the nominal bolt diameter by more than $1/16$ in. For friction-type connections, the diameter of a hole may be larger than the nominal bolt diameter by $3/16$ in for bolts up to $7/8$ in in diameter, by ¼ in for 1-in bolts, and by ⁵⁄₁₆ in for larger bolts. (Such holes are called oversized.) Hardened washers should be placed over oversized holes in an outer ply.

Slotted holes are usually provided for fasteners in connections to permit expansion or adjustment. Holes that depart only slightly from round are referred to as short-slotted holes. Elongated holes are called long-slotted holes. For either type of hole, the slots may be formed without regard to direction of loading in friction-type connections; but in bearing-type connections, slot length should be placed normal to the direction of loading. Also, for either type of hole, the width should not exceed the bolt diameter by more than ¹⁄₁₆ in.

For short-slotted holes, the length should not exceed the nominal bolt diameter by more than ¼ in for bolts up to ⅞ in in diameter, by more than ⁵⁄₁₆ in for 1-in bolts, and by more than ⅜ in for larger bolts. Washers (hardened when used with high-strength bolts) should be placed over short-slotted holes in an outer ply.

Long-slotted holes may have a length up to 2½ times the bolt diameter. They may be used in only one of the connected parts at an individual faying surface. When the slot is in an outer ply, plate washers or a continuous bar with standard holes should be installed to cover the entire slot. Washers or bars used with high-strength bolts should be at least ⁵⁄₁₆ in thick.

8-58. Welding.

The method of joining steel parts by fusion at high temperatures is widely used in steel construction, especially in fabricating shops, where conditions favorable to quality production are easily controlled. Field welding, however, depends on availability of skilled welders and reliability of inspection services, and therefore field bolting often is used instead.

An important advantage of welding is economy of material. Welding reduces the amount of detail connection material, especially in continuous and rigid-frame designs, where welded connections are relatively simple. In addition to economy, welding is quiet and therefore highly desirable for use in congested areas.

Any of several welding processes may be used: manual shielded metal-arc, submerged-arc, flux-cored arc, gas-metal arc, electrogas, and electroslag. They are not all interchangeable, however; each has its advantageous applications.

Many building codes accept the recommendations of the American Welding Society ("Structural Welding Code," AWS D1.1). The American Institute of Steel Construction "Specification for the Design, Fabrication and Erection of Structural Steel for Buildings" embodies many of this code's salient requirements. Symbols for welds are shown in Fig. 8-40.

Weld Types. Practically all welds used for connecting structural steel are of either of two types: fillet or groove.

Figure 8-39*a* and *b* illustrates a typical fillet weld. As stated in Art. 8-32, all stresses on fillet welds are resolved as shear on the effective throat. The normal throat dimension, as indicated in Fig. 8-39*a* and *b,* is the effective throat for all welding processes, except the submerged-arc method. The deep penetration characteristic of the latter process is recognized by increasing the effective throat dimension, as shown in Fig. 8-39*c.* Allowable stresses for fillet welds are given in Table 8-11.

Groove welds (Fig. 8-39*d, e,* and *f*) are classified in accordance with depth of solid weld metal as either complete or partial penetration. Most grooves, such as those in Fig. 8-39*d* and *e,* are made complete-penetration welds by the workmanship requirements: use backup strips (Figs. 8-55 and 8-56) or remove slag inclusions and imperfections (step called back-gouging) on the unshielded side of the root weld. The partial-penetration groove weld shown in Fig. 8-39*f* is typical of the type of weld used for box-type members

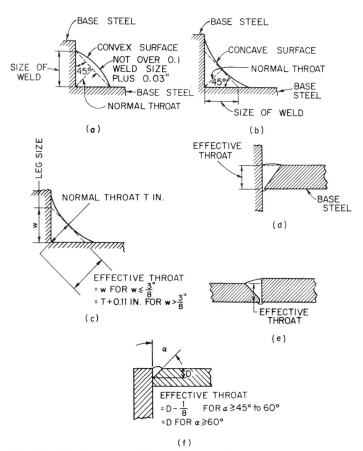

Fig. 8-39 Effective throats of fillet and groove welds.

and column splice joints. Effective throat is dependent upon the welding process, welding position, and the chamfer angle α. The indicated effective throat (Fig. 8-39f) is proper for the shielded-metal-arc processes and for all welding positions. Allowable stresses for groove welds are given in Table 8-10.

Base-Metal Temperatures. An important requirement in the production of quality welds is the temperature of base metal. Minimum preheat and interpass temperature as specified by the AWS and AISC standards must be obtained within 3 in of the welded joint before welding starts and then maintained until completion. Table 8-23 gives the temperature requirements based on thickness (thickest part of joint) and welding process for several structural steels. When base metal temperature is below 32°F, it must be preheated to at least 70°F and maintained at that temperature during welding. No welding is permitted when ambient temperature is below 0°F. Additional information, including temperature requirements for other structural steels, is given in AWS D1.1 and the AISC Specification.

Another quality-oriented requirement applicable to fillet welds is minimum leg size, depending on thickness of steel (Table 8-24). The thicker part connected governs, except that the weld size need not exceed the thickness of the thinner part. This rule is intended

TABLE 8-23 Minimum Preheat and Interpass Temperatures for Base Metal to Be Welded

Steel*	Shielded-metal-arc welding with other than low-hydrogen electrodes		Shielded-metal-arc welding with low-hydrogen electrodes, gas-metal-arc, and flux-cored arc welding	
	Thickness, in	Temp, °F	Thickness, in	Temp, °F
A36	To 3/4 in incl.	32	To 3/4 in incl.	32
	Over 3/4 to 1 1/2	150	Over 3/4 to 1 1/2	50
	Over 1 1/2 to 2 1/2	225	Over 1 1/2 to 2 1/2	150
	Over 2 1/2	300	Over 2 1/2	225
A242	Not permitted		To 3/4 in incl.	32
A441			Over 3/4 to 1 1/2	70
A588			Over 1 1/2 to 2 1/2	150
A572 to $F_y = 50$			Over 2 1/2	225

*For temperatures for other steels, see AWS D1.1, "Structural Welding Code," American Welding Society.

TABLE 8-24 Minimum Sizes* of Fillet and Partial-Penetration Welds

Base-metal Thickness, In.	Weld Size, In.
To 1/4 incl.	1/8
Over 1/4 to 1/2	3/16
Over 1/2 to 3/4	1/4
Over 3/4 to 1 1/2	5/16
Over 1 1/2 to 2 1/4	3/8
Over 2 1/4 to 6	1/2
Over 6	5/8

* Leg dimension for fillet welds; minimum effective throat for partial-penetration groove welds.

to minimize the effects of restraint resulting from rapid cooling due to disproportionate mass relationships.

8-59. Combinations of Fasteners.

The American Institute of Steel Construction "Specification for the Design, Fabrication and Erection of Structural Steel for Buildings" distinguishes between existing and new framing in setting conditions for connections of fasteners:

In new work, A307 bolts or high-strength bolts in bearing-type connections should not be considered as sharing the stress with welds. If welds are used, they should be designed to carry the entire stress in the connection. But when one leg of a connection angle is connected with one type of fastener and the other leg with a different type, this rule does not apply. The stress is transferred into the angle by one method, then taken out by another. Such a connection is commonly used, since one fastening method may be selected for shop work and a different method for field connection. But high-strength bolts may share stress with welds on the same leg in friction-type joints (Art. 8-55) if the bolts are tightened before the welds are made.

For connections in existing frames, existing rivets and high-strength bolts may be used for carrying stresses from existing dead loads, and welds may be provided for addi-

8-60. Fastener Symbols.

Fasteners are indicated on design, shop, and field erection drawings by notes and symbols. A simple note may suffice for bolts; for example: "⅞-in A325 bolts, except as noted." Welds require more explicit information, since their location is not so obvious as that of holes for bolts.

Symbols are standard throughout the industry. Figure 8-36 shows the symbols for bolts, Fig. 8-40 the symbols for welds. The welding symbols (Fig. 8-40a) together with the information key (Fig. 8-40b) are from the American Welding Society "Symbols for Welding and Nondestructive Testing, AWS A2.4.

Fig. 8-40 Welding symbols.

8-61. Erection Clearances for Fasteners.

All types of fasteners require clearance for proper installation in both shop and field. Shop connections seldom are a problem, since each member is easily manipulated and wholly under the control of the shop and drafting-room personnel. Field connections, however, require careful planning, because connections can be made only after all members are erected and aligned in final position. It becomes the responsibility of the shop draftsmen to arrange connection details in such manner that adequate clearance is provided in the field.

Clearances are necessary for two reasons: to permit entry, as in the case of bolts entering into holes, and to permit tightening of bolts with pneumatic bolting machines, and movement of "stick" electrodes in welding. Approximate clearances for installing high-strength bolts with powered impact wrenches are given in Fig. 8-41.

Somewhat less standardized, but equally important, are the clearances for welded joints. Ample room should be provided for the operator to observe the weld as it is deposited and to manipulate the welding electrode. In many respects, more latitude is afforded to the welder since he can shorten the electrode (12 to 18 in initially) and even prebend it to accommodate a tight situation.

In general, the preferred position of the electrode for fillet welding in the horizontal position is in a plane forming an angle of 30° with the vertical side of the weld. Frequently, this angle must be varied to clear a projecting part, as illustrated in Fig. 8-42a and b. Satisfactory results can be obtained when the weld root is visible to the operator and when the clear distance to a projection that may interfere with the electrode is not less than one-half the projection height ($y/2$ in Fig. 8-42b).

Still another example of welding clearance is illustrated in Fig. 8-42c. There the objective is to weld an end-connection angle to a beam web, which is in a horizontal

Fig. 8-41 Approximate clearances for installing high-strength bolts with powered impact wrenches.

Fig. 8-42 Welding clearances required for electrodes: (*a*) and (*b*) near projecting flanges; (*c*) at end-connection angles.

plane, as it would lie on shop skids. The 20° angle in the plan view is the minimum for satisfactory welding with a straight electrode. Assuming a ½-in setback for the end of the beam and a ⅜-in outside diameter of the electrode, the clearance should not be less than 1¼ in for an angle whose width *W* is 3 in, or less than 1⅝ in where *W* equals 4 in.

Although both of the foregoing examples deal with shop welding, the beams being laid in positions most favorable to the work, the limiting positions of the electrode are equally applicable to comparable situations that may arise with field connections.

CONNECTIONS

Design of connections and splices is a specialized art of the structural-steel industry known as *detailing*. Usually, designers simply indicate the type and size of fastener and type of connection; for example, "⅞-in A325 bolts in bearing-type joints, connections framed." For beams, the designer may furnish the reactions; if not, then the detailer would determine the reactions from the uniform-load capacity (tabulated in the AISC "Manual of Steel Construction"), giving due consideration to the effect of large concentrated loads near the connection.

For wind-resisting connections in multistory buildings, the wind moments are usually given in the design drawings and perhaps a sketch of a typical connection showing type of wind bracing desired.

Complete development of each connection becomes the responsibility of the fabricator's engineering organization.

8-62. Stresses in Bearing-Type Connections.

When some slip, although very small, may occur between connected parts, the fasteners are assumed to function in shear. The presence of paint on contact surfaces is therefore

Fig. 8-43 Bolted connection in shear and bearing: (*a*) with bolt in single shear; (*b*) with bolt in double shear (two shearing planes).

of no consequence. Fasteners may be rivets, A307 bolts or high-strength bolts, or any other similar fastener not dependent on development of friction on the contact surfaces.

Single shear occurs when opposing forces act on a fastener as shown in Fig. 8-43*a*, the plates tending to slide on their contact surfaces. The body of the fastener resists this tendency; a state of shear then exists over the cross-sectional area of the fastener.

Double shear takes place whenever three or more plates act on a fastener as illustrated in Fig. 8-43*b*. There are two or more parallel shearing surfaces (one on each side of the middle plate in Fig. 8-43*b*). Accordingly, the shear strength of the fastener is measured by its ability to resist two or more single shears.

Bearing on Base Metal. This is a factor to consider; but, as stated in Art. 8-33, calculation of bearing stresses in most joints is useful only as an index of efficiency of the net section of tension members.

Edge Distances. The AISC "Specification for the Design, Fabrication and Erection of Structural Steel for Buildings" recommends minimum edge distances, center of hole to edge of connected part, as given in Table 8-25. In addition, the edge distance, in, when in the direction of force should not be less than $2P/F_u t$, where P is the force, kips, transmitted by one fastener to the part for which the edge distance is applicable; F_u is the specified minimum tensile strength of the part (not the fastener), ksi, and t is the thickness of the part, in.

A special rule applies to beams with framed connections that are usually designed for the shear due to beam reactions. The edge distance for the beam web, with standard-size

TABLE 8-25. Minimum Edge Distance for Punched, Reamed, or Drilled Holes, In

Fastener diameter, in.	At sheared edges	At rolled edges of plates, shapes or bars or gas-cut edges†
1/2	7/8	3/4
5/8	1 1/8	7/8
3/4	1 1/4	1
7/8	1 1/2 *	1 1/8
1	1 3/4 *	1 1/4
1 1/8	2	1 1/2
1 1/4	2 1/4	1 5/8
Over 1 1/4	1 3/4 × diameter	1 1/4 × diameter

* These may be 1 1/4 in. at the ends of beam connection angles.

† All edge distances in this column may be reduced 1/8 in. when the hole is at a point where stress does not exceed 25% of the maximum allowed stress in the element.

holes, should be not less than $2P_R/F_u t$, where P_R is the beam reaction per bolt, kips. This rule, however, need not be applied when the bearing stress transmitted by the fastener does not exceed $0.90F_u$.

The maximum distance from the center of a fastener to the nearest edge of parts in contact should not exceed 6 in or 12 times the part thickness.

Minimum Spacing. The AISC specification also requires that the minimum distance between centers of rivet and bolt holes not be less than 2⅔ times the rivet or bolt diameter. But at least three diameters is desirable. Additionally, the hole spacing, in, when along the line of force, should not be less than $2P/F_u t - d/2$, where P, F_u, and t are as previously defined for edge distance and d = nominal diameter of fastener, in. Since this rule is for standard-size holes, appropriate adjustments should be made for oversized and slotted holes. In no case should the clear distance between holes be less than the fastener diameter.

Eccentric Loading. Stress distribution is not always as simple as for the joint in Fig. 8-43a where the fastener is directly in the line of stress. Sometimes, the load is applied eccentrically, as shown in Fig. 8-44. For such connections, tests show that use of actual eccentricity to compute the maximum force on the extreme fastener is unduly

Fig. 8-44 Eccentrically loaded fastener groups: (a) with bolts in shear; (b) with bolts in tension and shear.

conservative because of plastic behavior and clamping force generated by the fastener. Hence, it is permissible to reduce the actual eccentricity to a more realistic "effective" eccentricity.

For fasteners equally spaced on a single gage line, the effective eccentricity in inches is given by

$$l_{\text{eff}} = l - \frac{1 + 2n}{4} \qquad (8\text{-}37)$$

where l = the actual eccentricity
n = the number of fasteners

For the bracket in Fig. 8-44b, the reduction applied to l_1 is $(1 + 2 \times 6)/4 = 3.25$ in.

For fasteners on two or more gage lines

$$l_{\text{eff}} = l - \frac{1 + n}{2} \qquad (8\text{-}38)$$

where n is the number of fasteners per gage line. For the bracket in Fig. 8-44a, the reduction is $(1 + 4)/2 = 2.5$ in.

In Fig. 8-44a, the load P can be resolved into an axial force and a moment: Assume two equal and opposite forces acting through the center of gravity of the fasteners, both forces being equal to and parallel to P. Then, for equal distribution on the fasteners, the shear on each fastener caused by the force acting in the direction of P is $f_v = P/n$, where n is the number of fasteners.

The other force forms a couple with P. The shear stress f_e due to the couple is proportional to the distance from the center of gravity and acts perpendicular to the line from the fastener to the center. In determining f_e, it is convenient to first express it in terms of x, the force due to the moment Pl_{eff} on an imaginary fastener at unit distance from the center. For a fastener at a distance a from the center, $f_e = ax$, and the resisting moment is $f_e a = a^2 x$. The sum of the moments equals Pl_{eff}. This equation enables x to be evaluated and hence, the various values of f_e. The resultant R of f_e and f_v can then be found; a graphical solution usually is sufficiently accurate. The stress so obtained must not exceed the allowable value of the fastener in shear (Arts. 8-33 and 8-34).

For example, in Fig. 8-44a, $f_v = P/8$. The sum of the moments is

$$4a_1^2 x + 4a_2^2 x = Pl_{\text{eff}}$$

$$x = \frac{Pl_{\text{eff}}}{4a_1^2 + 4a_2^2}$$

Then, $f_e = a_2 x$ for the most distant fastener, and R can be found graphically as indicated in Fig. 8-44a.

Tension and Shear. For fastener group B in Fig. 8-44b, use actual eccentricity l_2 since these fasteners are subjected to combined tension and shear. Here too, the load P can be resolved into an axial shear force through the fasteners and a couple. Then, the stress on each fastener caused by the axial shear is P/n, where n is the number of fasteners. The tensile forces on the fasteners vary with distance from the center of rotation of the fastener group.

A simple method, erring on the safe side, for computing the resistance moment of group B fasteners assumes that the center of rotation coincides with the neutral axis of the group. It also assumes that the total bearing pressure below the neutral axis equals the sum of the tensile forces on the fasteners above the axis. Then, with these assumptions, the tensile force on the fastener farthest from the neutral axis is

$$f_t = \frac{d_{\max} Pl_2}{\Sigma A d^2} \qquad (8\text{-}39)$$

where d = distance of each fastener from the neutral axis
d_{max} = distance from neutral axis of farthest fastener
A = nominal area of each fastener

The maximum resultant stresses f_t and $f_v = P/n$ are then plotted as an ellipse and R is determined graphically. The allowable stress is given as the tensile stress F_t as a function of the computed shear stress f_v. (In Table 8-14, allowable stresses are given for the ellipse approximated by three straight lines.)

Note that the tensile stress of the applied load is not additive to the internal tension (pretension) generated in the fastener on installation. On the other hand, the AISC Specification does require the addition to the applied load of tensile stresses resulting from prying action, depending on the relative stiffness of fasteners and connection material. Prying force Q (Fig. 8-45b) may vary from negligible to a substantial part of the total tension in the fastener. A method for computing this force is given in the AISC "Manual of Steel Construction."

The old method for checking the bending strength of connection material ignored the effect of prying action. It simply assumed bending moment equal to P/n times e (Fig. 8-45). This procedure may be used for noncritical applications.

Fig. 8-45 Fasteners in tension. Prying action on the connection causes a moment $M = Pe/n$ on either side, where P = applied load, e its eccentricity, as shown above, and n the number of fasteners resisting the moment.

8-63. Stresses in Friction-Type Connections.

Design of this type of connection assumes that the fastener, under high initial tensioning, develops frictional resistance between the connected parts, preventing slippage despite external load. Properly installed rivets and A307 bolts provide some friction, but since it is not dependable it is ignored. High-strength steel bolts tightened nearly to their yield strengths, however, develop substantial, reliable friction. No slippage will occur at design loads if the contact surfaces are clean and free of paint or have only scored galvanized coatings, inorganic zinc-rich paint, or metallized zinc or aluminum coatings.

The AISC "Specification for the Design, Fabrication and Erection of Structural Steel for Buildings" lists allowable "shear" stresses for high-strength bolts in friction-type connections. Though there actually is no shear on the bolt shank, the shear concept is convenient for measuring bolt capacity.

Since most joints in building construction can tolerate tiny slippage, bearing-type joints, which are allowed much higher stresses for the same high-strength bolts when the threads are not in shear planes, may, for reasons of economy, lessen the use of friction-type joints.

The capacity of a friction-type connection does not depend on the bearing of the bolts against the sides of their holes. Hence, general specification requirements for protection against high bearing stresses or bending in the bolts may be ignored.

If the fasteners B in Fig. 8-44b are in a friction-type connection, the bolts above the neutral axis will lose part of their clamping force; but this is offset by a compressive force below the neutral axis. Consequently, there is no overall loss in frictional resistance to slippage.

When it is apparent that there may be a loss of friction (which occurs in some type of brackets and hangers subject to tension and shear) and slip under load cannot be tolerated, the working value in shear should be reduced in proportion to the ratio of residual tension to initial tension.

Friction-type connections subjected to eccentric loading, such as that illustrated in Fig. 8-44, are analyzed in the same manner as bearing-type connections (Art. 8-62).

8-64. Stresses in Welded Connections.

Welds are of two general types, fillet (Fig. 8-46a) and groove (Fig. 8-46b), with allowable stresses dependent on grade of weld and base steels. Since all forces on a fillet weld are resisted as shear on the effective throat (Art. 8-58), the strength of connections resisting direct tension, compression and shear are easily computed on the basis that a kip of fillet shear resists a kip of the applied forces. Many connections, some of which are shown in Fig. 8-47, are not that simple because of eccentricity of applied force with respect to the fillets. In designing such joints it is customary to take into account the actual eccentricity.

The underlying design principles for eccentric welded connections are similar to those for eccentric riveted or bolted connections (Art. 8-62). Consider the welded bracket in Fig. 8-48. The first step is to compute the center of gravity of the weld group. Then, the load P can be resolved into an equal and parallel load through the center of gravity and a couple. The load through the center of gravity is resisted by a uniform shear on the welds; for example, if the welds are all the same size, shear per linear inch is $f_v = P/n$

(a) FILLET WELD (b) GROOVE WELDS

Fig. 8-46 Two main types of weld.

Fig. 8-47 Typical eccentric welded connections.

where n is the total linear inches of weld. The moment Pl of the couple is resisted by the moment of the weld group. The maximum stress, which occurs on the weld element farthest from the center of gravity, may be expressed as $f_e = Pl/S$, where S is the polar section modulus of the weld group.

To find S, first compute the moments of inertia I_X of the welds about the XX axis and I_Y about the perpendicular YY axis. (If the welds are all the same size, their lengths, rather than their relative shear capacities, can be conveniently used in all moment calculations.) The polar moment of inertia $J = I_X + I_Y$, and the polar section modulus $S = J/a$, where a is the distance from the center of gravity to the farthest weld element. The resultant R of f_v and f_e, which acts normal to the line from the center of gravity to the weld element for which the stress is being determined, should not exceed the capacity of the weld element (Art. 8-32).

Fig. 8-48 Stresses on welds caused by eccentricity.

8-65. Types of Beam Connections.

In general, all beam connections are classified as either *framed* or *seated*. In the framed type, the beam is connected to the supporting member with fittings (short angles are common) attached to the beam web. With seated connections, the ends of the beam rest on a ledge or seat, in much the same manner as if the beam rested on a wall.

8-66. Bolted Framed Connections.

When a beam is connected to a support, a column or a girder, with web connection angles, the joint is termed "framed." Each connection should be designed for the end reaction of the beam, and type, size and strength of the fasteners, and bearing strength of base materials should be taken into account. To speed design, the AISC "Manual of Steel Construction" lists a complete range of suitable connections with capacities depending on these variables. Typical connections for beam or channels ranging in depth from 3 to 30 in are shown in Fig. 8-49.

To provide sufficient stability and stiffness, the length of connection angles should be at least half the clear depth of beam web.

For economy, select the minimum connection adequate for the load. For example, assume an 18-in beam is to be connected. The AISC Manual lists three- and four-row connections in addition to the five-row type shown in Fig. 8-49. Total shear capacity

REGULAR   HEAVY

Fig. 8-49 Typical bolted framed connections.

ranges from a low of 26.5 kips for ¾-in-diam A307 bolts in a three-row regular connection to a high of 263.0 kips for 1-in-diam A325 bolts in a five-row heavy connection, bearing type. This wide choice does not mean that all types of fasteners should be used on a project, but simply that the tabulated data cover many possibilities, enabling an economical selection. Naturally, one type of fastener should be used throughout, if practical; but shop and field fasteners may be different.

Bearing stresses on beam webs should be checked against allowable stresses (Art. 8-33), except for friction-type connections, in which bearing is not a factor. Sometimes, the shear capacity of the field fasteners in bearing-type connections may be limited by bearing on thin webs, particularly where beams frame into opposite sides of a web. This could occur where beams frame into column or girder webs.

One side of a framed connection usually is shop connected, the other side field connected. The capacity of the connection is the smaller of the capacities of the shop or field group of fasteners.

In the absence of specific instructions in the bidding information, the fabricator should select the most economical connection. Deeper and stiffer connections, if desired by the designer, should be clearly specified.

Fig. 8-50 Typical bolted seated connections: (a) stiffened seat, (b) unstiffened seat.

8-67. Bolted Seated Connections.

Sizes, capacities, and other data for seated connections for beams, shown in Fig. 8-50, are tabulated in the AISC "Manual of Steel Construction." Two types are available, stiffened seats (Fig. 8-50a) and unstiffened seats (Fig. 8-50b).

Unstiffened Seats. Capacity is limited by the bending strength of the outstanding, horizontal leg of the seat angle. A 4-in leg 1 in thick generally is the practical limit. An angle of A36 steel with these dimensions has a top capacity of 60.5 kips for beams of A36 steel, and 78.4 kips when $F_y = 50$ ksi for the beam steel. Therefore, for larger end reactions, stiffened seats are recommended.

The actual capacity of an unstiffened connection will be the lesser of the bending strength of the seat angle, the shear resistance of the fasteners in the vertical leg, or the bearing strength of the beam web. (See also Art. 8-23 for web crippling stresses.) Data in the AISC Manual make unnecessary the tedious computations of balancing the seat-angle bending strength and beam-web bearing.

The nominal setback from the support of the beam to be seated is ½ in. But tables for seated connections assume ¾ in to allow for mill underrun of beam length.

Stiffened Seats. These may be obtained with either one or two stiffener angles, depending on the load to be supported. As a rule, stiffeners with outstanding legs having a width less than 5 in are not connected together; in fact, they may be separated, to line up the angle gage line (recommended centerline of fasteners) with that of the column.

The capacity of a stiffened seat is the lesser of the bearing strength of the fitted angle stiffeners or the shear resistance of the fasteners in the vertical legs. Crippling strength of the beam web usually is not the deciding factor, because of ample seat area. When legs larger than 5 in wide are required, eccentricity should be considered, in accordance with the technique given in Art. 8-62. The center of the beam reaction may be taken at the midpoint of the outstanding leg.

Advantages. Seated connections offer some significant advantages. For economical fabrication, the beams merely are punched and are free from shop-fastened details. They pass from the punching machine to the paint shed, after which they are ready for delivery. In erection, the seat provides an immediate support for the beam while the erector aligns the connecting hole. The top angle is used to prevent accidental rotation of the beam. For framing into column webs, seated connections allow more erection clearance

8-86   Structural-Steel Construction

for entering the trough formed by column flanges than do framed connections. A framed beam usually is detailed to within $\frac{1}{16}$ in of the column web. This provides about $\frac{1}{8}$ in total clearance, whereas a seated beam is cut about $\frac{1}{2}$ in short of the column web, yielding a total clearance of about 1 in. Then, too, each seated connection is wholly independent, whereas for framed beams on opposite sides of a web, there is the problem of aligning the holes common to each connection.

Frequently, the angles for framed connections are shop attached to columns. Sometimes, one angle may be shipped loose to permit erection. This detail, however, cannot be used for connecting to column webs, because the column flanges may obstruct entering or tightening of bolts. In this case, a seated connection has a distinct advantage.

8-68. Welded Framed Connections.

The AISC "Manual of Steel Construction" tabulates sizes and capacities of angle connections for beams for three conditions: all welded, both legs (Fig. 8-51); web leg shop welded, outstanding leg for hole-type fastener; and web leg for hole-type fastener installed in shop, outstanding leg field welded. Tables are based on E70 electrodes. Thus, the connections made with A36 steel are suitable for beams of both carbon and high-strength structural steels.

Eccentricity of load with respect to the weld patterns causes stresses in the welds that must be considered in addition to direct shear. Assumed forces, eccentricities, and induced stresses are shown in Fig. 8-51b. Stresses are computed as in the example in Art.

Fig. 8-51 Welded framed connection on beam web: (a) weld locations along connection angles; (b) forces on welds.

8-64, based on vector analysis that characterizes elastic design. The capacity of welds A or B that is smaller will govern design.

If ultimate strength (plastic design) of such connections is considered, many of the tabulated "elastic" capacities are more conservative than necessary. Although AISC deemed it prudent to retain the "elastic" values for the weld patterns, recognition was given to research results on plastic behavior by reducing the minimum beam-web thickness required when welds A are on opposite sides of the web. As a result, welded framed connections are now applicable to a larger range of rolled beams than strict elastic design would permit.

Shear stresses in the supporting web for welds B should also be investigated, particularly when beams frame on opposite side of the web.

8-69. Welded-Seat Connections.

Also tabulated in the AISC "Manual of Steel Construction," welded-seat connections (Fig. 8-52) are the welded counterparts of bolted-seat connections for beams (Art. 8-67). As for welded framed connections (Art. 8-68), the load capacities for seats, taking into account the eccentricity of loading on welds, are computed by "elastic" vector analysis. Assumptions and the stresses involved are shown in Fig. 8-52c.

An unstiffened seat angle of A36 steel has a maximum capacity of 60.5 kips for supporting beams of A36 steel, and 78.4 kips for steel with F_y = 50 ksi (Fig. 8-52a). For heavier loads, a stiffened seat (Fig. 8-52b) should be used.

Stiffened seats may be a beam stub, a tee section, or two plates welded together to form a tee. Thickness of the stiffener (vertical element) depends on the strengths of beam and seat materials. For a seat of A36 steel, stiffener thickness should be at least that of the supported beam web when the web is A36 steel, and 1.4 times thicker for web steel with F_y = 50 ksi.

When stiffened seats are on line on opposite sides of a supporting web of A36 steel, the weld size made with E70 electrodes should not exceed one-half the web thickness, and for web steel with F_y = 50 ksi, two-thirds the web thickness.

Although top or side lug angles will hold the beam in place in erection, it often is advisable to use temporary erection bolts to attach the bottom beam flange to the seat. Usually, such bolts may remain after the beam flange is welded to the seat.

8-70. End-Plate Connections.

The art of welding makes feasible connections that were not possible with older-type fasteners, e.g., end-plate connections (Fig. 8-53).

Of the several variations, only the flexible type (Fig. 8-53a) has been "standardized" with tabulated data in the AISC "Manual of Steel Construction." Flexibility is assured by making the end plate ¼ in thick wherever possible (never more than ⅜ in). Such connections in tests exhibit rotations similar to those for framed connections.

The weld connecting the end plate to the beam web is designed for shear. There is no eccentricity. Weld size and capacity are limited by the shear strength of the beam web adjoining the weld. Effective length of weld is reduced by twice the weld size to allow for possible deficiencies at the ends.

As can be observed, this type of connection requires accurate cutting of the beam to length. Also the end plates must be squarely positioned so as to compensate for mill and shop tolerances.

The end plate connection is easily adapted for resisting beam moments (Fig. 8-53b, c, and d). One deterrent, however, to its use for tall buildings where column flanges are massive and end plates thick is that the rigidity of the parts may prevent drawing the

Fig. 8-52 Welded-seat connections: (a) unstiffened seat; (b) stiffened seat; (c) stresses in welds.

Fig. 8-53 End-plate connections.

Special Connections   8-89

Fig. 8-54 Examples of special connections.

surfaces into tight contact. Consequently, it may not be easy to make such connections accommodate normal mill and shop tolerances.

8-71. Special Connections.

In some structural frameworks, there may be connections in which a standard type (Arts. 8-66 to 8-70) cannot be used. Beam centers may be offset from column centers, or intersection angles may differ from 90°, for example.

For some skewed connections the departure from the perpendicular may be taken care of by slightly bending the framing angles. When the practical limit for bent angles is exceeded, bent plates may be used (Fig. 8-54a).

Special one-sided angle connections, as shown in Fig. 8-54b, are generally acceptable for light beams. When such connections are used, the eccentricity of the fastener group in the outstanding leg should be taken into account. Length l may be reduced to the effective eccentricity (Art. 8-62).

Spandrel and similar beams lined up with a column flange may be conveniently connected to it with a plate (Fig. 8-54c and d). The fasteners joining the plate to the beam web should be capable of resisting the moment for the full lever arm l for the connection in Fig. 8-54c. For beams on both sides of the column with equal reactions, the moments balance out. But the case of live load on one beam only must be considered. And bear in mind the necessity of supporting the beam reaction as near as possible to the column center to relieve the column of bending stresses.

When spandrels and girts are offset from the column, a Z-type connection (Fig. 8-54e) may be used. The eccentricity for beam-web fasteners should be taken as l_1, for column-flange fasteners as l_2, and for fasteners joining the two connection angles as l_3 when l_3 exceeds 2½ in; smaller values of l_3 may be considered negligible.

8-72. Simple, Rigid, and Semirigid Connections.

Moment connections are capable of transferring the forces in beam flanges to the column. This moment transfer, when specified, must be provided for in addition to and usually independent of the shear connection needed to support the beam reaction. Framed, seated, and end-plate connections (Arts. 8-66 to 8-70) are examples of shear connections. Those shown in Fig. 8-32 are moment connections. In Fig. 8-32a to g, flange stresses are developed independently of the shear connections, whereas in h and i, the forces are combined and the entire connection resolved as a unit.

Moment connections may be classified according to their design function: those resisting moment due to lateral forces on the structure, and those needed to develop continuity, with or without resistance to lateral forces.

The connections generally are designed for the computed bending moment, which often is less than the beam's capacity to resist moment. A maximum connection is obtained, however, when the beam flange is developed for its maximum allowable stress.

The ability of a connection to resist moment depends on the elastic behavior of the parts. For example, the light lug angle shown connected to the top flange of the beam in Fig. 8-55b is not designed for moment and accordingly affords negligible resistance to rotation. In contrast, full rigidity is expected of the direct welded flange-to-column connection in Fig. 8-56a. The degree of fixity, therefore, is an important factor in design of moment connections.

Fixity of End Connections. Specifications recognize three types of end connections: simple, rigid, and semirigid. The type designated *simple* (unrestrained) is intended to support beams and girders for shear only and leave the ends free to rotate under load. The type designated *rigid* (known also as rigid-frame, continuous, restrained frame) aims at not only carrying the shear but also providing sufficient rigidity to hold virtually unchanged the original angles between members connected. *Semirigid,* as the name implies, assumes that the connections of beams and girders possess a dependable and known moment capacity intermediate in degree between the simple and rigid types. Figure 8-57 illustrates these three types together with the uniform-load moments obtained with each type.

Although no definite relative rigidities have been established, it is generally conceded that the simple or flexible type could vary from zero to 15% (some researchers recommend 20%) end restraint and that the rigid type could vary from 90 to 100%. The semirigid types lie between 15 and 90%, the precise value assumed in the design being largely dependent on experimental analysis. These percentages of rigidity represent the ratio of the moment developed by the connection, with no column rotation, to the moment developed by a fully rigid connection under the same conditions, multiplied by 100.

Framed and seated connections offer little or no restraint. In addition, several other

Fig. 8-55 Methods of constructing flexible welded connections.

arrangements come within the scope of simple-type connections, although they appear to offer greater resistance to end rotations. For example, in Fig. 8-55a, a top plate may be used instead of an angle for lateral support, the plate being so designed that plastic deformation may occur in the narrow unwelded portion. Naturally, the plate offers greater resistance to beam rotation than a light angle, but it can provide sufficient flexibility that the connection can be classified as a simple type. Plate and welds at both ends are proportioned for about 25% of the beam moment capacity. The plate is shaped so that the metal across the least width is at yield stress when the stresses in the wide portion, in the butt welds, and in the fillet welds are at allowable working values. The unwelded length is then made from 20 to 50% greater than the least width to assure ductile yielding. This detail can also be developed as an effective moment-type connection.

Another flexible type is the direct web connection in Fig. 8-55b. Figured for shear loads only, the welds are located on the lower part of the web, where the rotational effect of the beam under load is the least, a likely condition when the beam rests on erection seats and the axis of rotation centers about the seat rather than about the neutral axis.

Tests indicate that considerable flexibility also can be obtained with a properly proportioned welded top-plate detail as shown in Fig. 8-55c without narrowing it as in Fig. 8-55a. This detail is usually confined to wind-braced simple-beam designs. The top plate is designed for the wind moment on the joint, at the increased stresses permitted for wind loads.

The problem of superimposing wind bracing on what is otherwise a clear-cut simple beam with flexible connections is a complex one. Some compromise is usually effected

8-92 Structural-Steel Construction

between theory and actual design practice. Two alternates usually are permitted by building codes:

1. Connections designed to resist assumed wind moments should be adequate to resist the moments induced by the gravity loading and the wind loading, at specified increased unit stresses.

2. Connections designed to resist assumed wind moments should be so designed that larger moments, induced by gravity loading under the actual condition of restraint, will be relieved by deformation of the connection material.

Obviously, these options envisage some nonelastic, but self-limiting, deformation of the structural-steel parts. Innumerable wind-braced buildings of riveted, bolted, or welded construction have been designed on this assumption of plastic behavior and have proved satisfactory in service.

Fully rigid, bolted beam end connections are not often used because of the awkward, bulky details, which, if not interfering with architectural clearances, are often so costly to design and fabricate as to negate the economy gained by using smaller beam sections. In appearance, they resemble the type shown in Fig. 8-32 for wind bracing; they are developed for the full moment-resisting capacity of the beam.

Much easier to accomplish and more efficient are welded rigid connections (Fig. 8-56). They may be connected simply by butt-welding the beam flanges to the columns—the "direct" connection shown in Fig. 8-56a and b. Others may prefer the "indirect" method, with top plates, because this detail permits ordinary mill tolerance for beam length. Welding of plates to stiffen the column flanges, when necessary, is also relatively simple (Art. 8-23).

Fig. 8-56 Methods of constructing rigid welded connections.

Fig. 8-57 Effect of rigidity of end connections on end moments.

In lieu of the erection seat angle in Fig. 8-56b, a patented, forged hook-and-eye device, known as Saxe erection units, may be used. The eye, or seat, is shop welded to the column, and the hook, or clip, is shop welded to the under side of the beam bottom flange. For deep beams, a similar unit may be located on the top flange to prevent accidental turning over of the beams. Saxe units are capable of supporting normal erection loads and dead weight of members; but their contribution to the strength of the connection is ignored in computing resistance to shear.

A comparison of fixities intermediate between full rigidity and zero restraint in Fig. 8-57 reveals an optimum condition attainable with 75% rigidity; end and center-span moments are equal, each being $WL/16$, or one-half the simple-beam moment. The saving in weight of beam is quite apparent.

Perhaps the deterrent to a broader usage of semirigid connections has been the proviso contained in specifications: "permitted only upon evidence that the connections to be used are capable of resisting definite moments without overstress of the fasteners." As a safeguard, the proportioning of the beam joined by such connections is predicated upon no greater degree of end restraint than the minimum known to be effected by the connection. Suggested practice, based on research with welded connections, is to design the end connections for 75% rigidity but to provide a beam sized for the moment that would result from 50% restraint; i.e., $WL/12$. ("Report of Tests of Welded Top Plate and Seat Building Connections," *The Welding Journal,* Research Supplement 146S–165S, 1944.) The type of welded connection in Fig. 8-55c, when designed for the intended rigidity, is generally acceptable.

End-plate connections (Fig. 8-53) are another means of achieving negligible, partial, and full restraint.

8-73. Column Splices.

Column-to-column connections are usually determined by the change in section. In general, a change is made at every second floor level, where a shop or field splice is located. From an erection viewpoint, as well as for fabrication and shipment, splices at every

Fig. 8-58 Bolted column splices with friction-type connections.

third floor may be more economical because of the reduced number of pieces to handle. Naturally, this advantage is partly offset by extra weight of column material, because the column size is determined by loads on the lowest story of each tier, there being an excess of section for the story or two above.

Splices are located just above floor-beam connections, usually about 2 to 3 ft above the floor. Because column stresses are transferred from column to column by bearing, the splice plates are of nominal size, commensurate with the need for safe erection and bending moments the joint may be subjected to during erection. From the viewpoint of moment resistance, a conventional column splice develops perhaps 20% of the moment capacity of the column.

Although column splices are not standardized, as are beam connections, they are generally uniform throughout the industry. (Splices shown in the Appendix, "Structural Steel Detailing," American Institute of Steel Construction, may be used by the fabricator or erector unless the designer indicates his preference.) Figure 8-58 illustrates the common types of column splices made with high-strength bolts. In Fig. 8-58a and b, the upper column bears directly on the lower column; filler plates are supplied in (b) when the differences in depth of the two columns are greater than can be absorbed by erection clearance.

As a rule, some erection clearance should be provided. When columns of the same nominal depth are spliced, it is customary to supply a ⅛-in fill under each splice plate on the lower column or, as an alternate, to leave the bolt holes open on the top gage line below the finished joint until the upper shaft is erected. The latter procedure permits the erector to spring the plates apart to facilitate entry of the upper column.

When the upper column is of such dimension that its finished end does not wholly bear on the lower column, one of two methods must be followed: In Fig. 8-58c, stresses in a portion of the upper column not bearing on the lower column are transferred by

Column Splices 8-95

Fig. 8-59 Welded column splices.

means of flange plates that are finished to bear on the lower column. These bearing plates must be attached with sufficient single-shear bolts to develop the load transmitted through bearing on the finished surface.

When the difference in column size is pronounced, the practice is to use a horizontal bearing plate as shown in Fig. 8-58d. These plates, known as **butt plates,** may be attached to either shaft with tack welds or clip angles. Usually it is attached to the upper shaft, because a plate on the lower shaft may interfere with erection of the beams that frame into the column web.

Somewhat similar are welded column splices. In Fig. 8-59a, a common case, holes for erection purposes are generally supplied in the splice plates and column flanges as shown. Some fabricators, however, prefer to avoid drilling and punching of thick pieces, and use instead clip angles welded on the inside flanges of the columns, one pair at diagonally opposite corners, or some similar arrangement. Figure 8-59b and c corresponds to the bolted splices in Fig. 8-58c and d. The shop and field welds for the welded butt plate in Fig. 8-59c may be reversed, to provide erection clearance for beams seated just below the splice. The erection clip angles would then be shop welded to the underside of the butt plate, and the field holes would pierce the column web.

The butt-weld splice in Fig. 8-59d is the most efficient from the standpoint of material saving. The depth of the bevel as given in the illustration is for the usual column splice, in which moment is unimportant. However, should the joint be subjected to considerable moment, the bevel may be deepened; but a ⅛-in minimum shoulder should remain for the purpose of landing and plumbing the column. For full moment capacity, a complete-penetration welded joint would be required.

8-74. Beam Splices.

These are required in rigid frames, suspended-span construction, and continuous beams. Such splices are usually located at points of counterflexure or at points where moments are relatively small. Therefore, splices are of moderate size. Flanges and web may be spliced with plates or butt welded.

Fig. 8-60 Welded beam splices.

For one reason or another it is sometimes expedient to make a long beam from two short lengths. A welded joint usually is selected, because the beams can be joined together without splice plates and without loss of section because of bolt holes. Also, from the viewpoint of appearance, the welded joint is hardly discernible.

Usually, the joint must be 100% efficient, to develop the full section. Figure 8-60 illustrates such a detail. The back side of the initial weld is gouged or chipped out; access holes in the beam webs facilitate proper edge preparation and depositing of the weld metal in the flange area in line with the web. Such holes are usually left open, because plugs would add undesirable residual stresses to the joint.

STEEL ERECTION

A clear understanding of what the fabricator furnishes or does not furnish to the erector, particularly on fabrication contracts that may call for delivery only, is all-important—and most fabricated steel is purchased on delivery basis only.

Purchasing structural steel has been greatly simplified by the industry's "Code of Standard Practice for Buildings and Bridges"—an American Institute of Steel Construction publication. A provision in the construction contract making the code part of the contract is often used, since it establishes a commonly accepted and well-defined line of demarcation between what is, and what is not, to be furnished under the contract. Lacking such a provision, the contract, to avoid later misunderstandings, must enumerate in considerable detail what is expected of both parties to the contract.

Under the code—and unless otherwise specifically called for in the contract documents—such items as steel sash, corrugated-iron roofing or siding, and open-web steel joists, and similar items, even if made of steel and shown on the contract design drawings, are not included in the category "structural steel." Also, such items as door frames are excluded, even when made of structural shapes, if they are not fastened to the structure in such way as to comply with "constituting part of the steel framing." On the other hand, loose lintels shown on design plans or in separate schedules are included.

According to the code, a fabricator furnishes with "structural steel," to be erected by someone else, the field bolts required for fastening the steel. The fabricator, however, does not furnish the following items unless specified in the invitation to bid: shims, fitting-up bolts, drift pins, temporary cables, welding electrodes, or thin leveling plates for column bases.

The code also defines the erection practices. For example, the erector does not paint field boltheads and nuts, field welds, or touch up abrasions in the shop coat, or perform any other field painting unless required in specifications accompanying the invitation to bid.

8-75. Erection Equipment.

If there is a universal piece of erection equipment, it is the crane. Mounted on wheels or tractor treads, it is extremely mobile, both on the job and in moving from job to job. Practically all buildings are erected with this efficient raising device. The exception, of course, is the skyscraper whose height exceeds the reach of the crane. Operating on ground level, cranes have been used to erect buildings of about 20 stories, the maximum height being dependent on the length of the boom and width of building.

Cranes may also be mounted on locomotive cars. Their use in building construction is limited to heavy plant construction serviced by railroads. By laying the rails early in the construction schedule, the erector can bring in the heavy steelwork on railway cars and use a heavy-duty locomotive crane for erection.

The guy derrick is a widely used raising device for erection of tall buildings. Its principal asset is the ease by which it may be "jumped" from tier to tier as erection proceeds upward. The boom and mast reverse position; each in turn serves to lift up the other. It requires about 2 hr to make a two-story jump.

Stiff-leg derricks and gin poles are two other rigs sometimes used, usually in the role of auxiliaries to cranes or guy derricks. Gin poles are the most elementary—simply a guyed boom. The base must be secure because of the danger of kicking out. The device is useful for the raising of incidental materials, for dismantling and lowering of larger rigs, and for erection of steel on light construction where the services of a crane are unwarranted.

Stiff-leg derricks are most efficient where they may be set up to remain for long periods of time. They have been used to erect multistory buildings but are not in popular favor because of the long time required to jump from tier to tier. Among the principal uses for stiff legs are (1) unloading steel from railroad cars for transfer to trucks, (2) storage and sorting, and (3) when placed on a flat roof, raising steel to roof level, where it may be sorted and placed within reach of a guy derrick.

Less time for "jumping" the raising equipment is needed for cranes mounted on steel box-type towers, about three stories high, that are seated on interior elevator wells or similar shafts for erecting steel. These tower cranes are simply jacked upward hydraulically or raised by cables, with the previously erected steelwork serving as supports. In another method, a stiff-leg derrick is mounted on a trussed platform, spanning two or more columns, and so powered that it can creep up the erected exterior columns. In addition to the advantage of faster jumps, these methods permit steel erection to proceed as soon as the higher working level is reached.

8-76. Clearance for Erecting Beams.

Clearances to permit tightening bolts and welding are discussed in Art. 8-61. In addition, designers also must provide sufficient field clearance for all members so as to permit erection without interference with members previously erected. The shop draftsman should always arrange the details so that the members can be swung into their final position without shifting the members to which they connect from their final positions. The following examples illustrate the conditions most frequently encountered in building work:

In framed beam connections (Fig. 8-61) the slightly shorter distance out-to-out of

Fig. 8-61 Erection clearance for beams.

Fig. 8-62 Alternate method for providing erection clearance.

Fig. 8-63 Clearance for beam seated on column-web connections.

connection angles ($B - \frac{1}{8}$ in), as compared with the face-to-face distance between supporting members, is usually sufficient to allow forcing the beam into position. Occasionally, however, because the beam is relatively short, or because heavy connection angles with wide outstanding legs are required, the diagonal distance A may exceed the clearance distance B. If so, the connection for one end must be shipped bolted to the framed beam to permit its removal during erection.

An alternate solution is to permanently fasten one connection angle of each pair to the web of the supporting beam, temporarily bolting the other angle to the same web for shipment, as shown in Fig. 8-62. The beam should be investigated for the clearance in swinging past permanently bolted connection angles. Attention must also be paid to possible interference of stiffeners in swinging the beam into place when the supporting member is a plate girder.

Another example is that of a beam seated on column-web connections (Fig. 8-63). The first step is to remove the top angles and shims temporarily. Then, while hanging from the derrick sling, the beam is tilted until its ends clear the edges of the column flanges, after which it is rotated back into a horizontal position and landed on the seats. The greatest diagonal length G of the beam should be about $\frac{1}{8}$ in less than the face-to-face distance F between column webs. It must also be such as to clear any obstruction above; e.g., G must be equal to or less than C, or the obstructing detail must be shipped bolted for temporary removal. To allow for possible overrun, the ordered length L of the beam should be less than the detailing length E by at least the amount of the permitted cutting tolerance.

Frequently, the obstruction above the beam connection may be the details of a column splice. As stated in Art. 8-73, it may be necessary to attach the splice material on the lower end of the upper shaft, if erection of the beam precedes erection of the column in the tier above.

8-77. Erection Sequence.

The order in which steel is to be fabricated and delivered to the site should be planned in advance so as not to conflict with the erector's methods or his construction schedule. For example, if steel is to be erected with derricks, the approximate locations at which the derricks will be placed will determine the shipping installments, or sections, into which the frame as a whole must be segregated for orderly shipment. When installments are delivered to the site at predetermined locations, proper planning will eliminate unnecessary rehandling. Information should be conveyed to the drafting room so that the shipping installments can be indicated on the erection plans and installments identified on the shipping lists.

In erection of multistory buildings with guy derricks, the practice is to hoist and place all columns first, spandrel beams and wall bracing next, and interior beams with filler beams last. More specifically, erection commences with bays most distant from the derrick and progresses toward the derrick, until it is closed in. Then, the derrick is jumped to the top and the process is repeated for the next tier. Usually, the top of the tier is planked over to obtain a working platform for the erectors and also to afford protection for the trades working below. However, before the derrick is jumped, the corner panels are plumbed; similarly when panels are erected across the building, cables are stretched to plumb the structure.

There is an established sequence for completing the connections. The raising gang connects members together with temporary fitting-up bolts. The number of bolts is kept to a minimum, just enough to draw the joint up tight and take care of the stresses caused by dead weight, wind, and erection forces. Permanent connections are made as soon as alignment is within tolerance limits. Usually, permanent bolting or welding follows on the heels of the raising gang. Sometimes, the latter moves faster than the gang making the permanent connections, in which case it may be prudent to skip every other floor, thus obtaining permanent connections as close as possible to the derrick—a matter of safe practice.

Some erectors prefer to use permanent high-strength (A325 and A490) bolts for temporary fitting up. Because bolts used for fit-up are not tightened to specified minimum tension, they may be left in place and later tightened as required for permanent installation.

8-78. Field-Welding Procedures.

The main function of a welding sequence is to control distortion due primarily to the effects of welding heat. In general, a large input of heat in a short time tends to produce the greatest distortion. Therefore, it is always advisable, for large joints, to weld in stages, with sufficient time between each stage to assure complete dispersal of heat, except for heat needed to satisfy interpass-temperature requirements (Art. 8-58). Equally important, and perhaps more efficient from the erector's viewpoint, are those methods that balance the heat input in such a manner that the distortional effects tend to cancel out.

Welding on one flange of a column tends to leave the column curled toward the welded side cooling, because of shrinkage stresses. A better practice for beams connecting to both sides of a column is to weld the opposite connections simultaneously. Thus the shrinkage of each flange is kept in balance and the column remains plumb.

If simultaneous welding is not feasible, then the procedure is to weld in stages. About 60% of the required weld might be applied on the first beam, then the joint on the opposite flange might be completely welded, and finally, welding on the first beam would be completed. Procedures such as this will go far to reduce distortion.

Experience has shown that it is good practice to commence welding at or near the center of a building and work outward. Columns should be checked frequently for vertical alignment, because shrinkage in the welds tends to shorten the distance between columns. Even though the dimensional change at each joint may be very small, it can accumulate to an objectionable amount in a long row of columns. One way to reduce the distortion is to allow for shrinkage at each joint, say, $\frac{1}{16}$ in for a 20-ft bay, by tilting or spreading the columns. Thus, a spread of $\frac{1}{8}$ in for the two ends of a beam with flanges butt welded to the columns may be built in at the fabricating shop; for example, by increasing the spacing of erection-bolt holes in the beam bottom flange. Control in the field, however, is maintained by guy wires until all joints are welded.

Shortening of bays can become acute in a column row in which beams connect to column flanges, because the shrinkage shortening could possibly combine with the mill underrun in column depths. Occasionally, in addition to spreading the columns, it may be necessary to correct the condition by adding filler plates or building out with weld metal.

Some designers of large welded structures prefer to detail the welding sequence for each joint. For example, on one project, the procedure for the joint shown in Fig. 8-64 called for four distinct operations, or stages: first, the top 6 in of the shear weld on the vertical connection was made; second, the weld on the top flange; third, the bottom-flange weld; and fourth, the remaining weld of the vertical connection. The metal was allowed to return to normal temperature before starting each stage. One advantage of this procedure is the prestressing benefits obtained in the connecting welds. Tensile stresses are developed in the bottom-flange weld on cooling; compressive stresses of equal magnitude consequently are produced in the top flange. Since these stresses are opposite to those caused by floor loads, welding stresses are useful in supporting the floor loads. Although this by-product assistance may be worthwhile, there are no accepted methods for resolving the alleged benefits into design economy.

Fig. 8-64 Indication of sequence in welding connections.

Multistory structures erected with a guy derrick supported on the steelwork as it rises will be subjected by erection loads to stresses and strains. The resulting deformations should be considered in formulating a field-welding sequence.

8-79. Field Tolerances.

Dimensional variations in the field often are a consequence of permissible variations in rolling of steel and in shop fabrication. Limits for mill variations are prescribed in the American Society for Testing and Materials Standard Specification A6, "General Requirements for Delivery of Rolled Steel Plates, Shapes, Sheet Piling, and Bars for

Structural Use." For example, wide-flange beams are considered straight, vertically or laterally, if they are within ⅛ in for each 10 ft of length. Similarly, columns are straight if the deviation is within ⅛ in per 10 ft, with a maximum deviation of ⅜ in.

It is standard practice to compensate in shop details for certain mill variations. The adjustments are made in the field, usually with clearances and shims.

Shop-fabrication tolerance for straightness of columns and other compression members often is expressed as a ratio, 1:1000, between points of lateral support. (This should

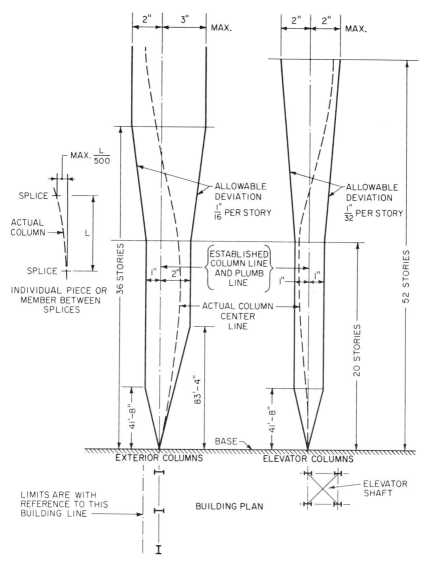

Fig. 8-65 Permissible deviations from plumb for columns. Limits shown are based on the assumption that the center of the column base coincides with the established column line.

8-102   Structural-Steel Construction

be recognized as approximately the equivalent of ⅛ in per 10 ft, and since such members rarely exceed 30 ft in length, between lateral supports, the ⅜-in maximum deviation prevails.) Length of fabricated beams have a tolerance of 1/16 in up to 30 ft and ⅛ in over 30 ft. Length of columns finished to bear on their ends have a tolerance of 1/32 in.

Erected beams are considered level and aligned if the deviation does not exceed 1:500. Similarly, columns are plumb and aligned if the deviation of individual pieces, between splices in the usual multistory building, does not exceed 1:500. The total or accumulative displacement for multistory columns cannot exceed the limits prescribed in the American Institute of Steel Construction "Code of Standard Practice." For convenience, these are indicated in Fig. 8-65. Control is placed only on the exterior columns and those in the elevator shaft.

Field measurements to determine whether columns are plumb should always be made at night or on cloudy days, never in sunshine. Solar heat induces differential thermal stresses, which cause the structure to curl away from the sun by an amount that renders plumbing measurements useless.

If beam flanges are to be field welded (Fig. 8-56a) and the shear connection is a high-strength-bolted, friction-type joint, the holes should be made oversize or horizontal slotted (Art. 8-55), thus providing some built-in adjustment to accommodate mill and shop tolerances for beams and columns.

Similarly, for beams with framed connections (Fig. 8-49 and 8-51) that will be field bolted to columns, allowance should be made in the details for finger-type shims, to be used where needed for column alignment.

Because of several variables, bearing of column joints is seldom in perfect contact across the entire cross-sectional area. The AISC recommends acceptance if gaps between the bearing surfaces do not exceed 1/16 in. Should a gap exceed 1/16 in and an engineering investigation shows need for more contact area, the gap may be filled with mild steel shims.

Tolerance for placing of machinery directly on top of several beams is another problem occasionally encountered in the field. The elevation of beam flanges will vary because of permissible variations for mill rolling, fabrication, and erection. This should be anticipated and adequate shims provided for field adjustments.

8-80. Adjusting Lintels.

Lintels supported on the steel frame (sometimes called shelf angles) may be permanently fastened in the shop to the supporting spandrel beam, or they may be attached so as to allow adjustment in the field (see Fig. 8-4, Art. 8-7). In the former case, the final position is solely dependent on the alignment obtained for the spandrel itself, whereas for the latter, lintels may be adjusted to line and grade independently of the spandrel. Field adjustment is the general rule for all multistory structures. Horizontal alignment is obtained by using slotted holes in the connection clip angles; vertical elevation (grade) is obtained with shims.

When walls are of masonry construction, a reasonable amount of variation in the position of lintels may be absorbed without much effort by masons. So the erector can adjust the lintels immediately following the permanent fastening of the spandrels to the columns. This procedure is ideal for the steel erector, because it allows him to complete his contract without costly delays and without interference with other trades. Subsequent minor variations in the position of the lintels, because of deflection or torsional rotation of the spandrel when subjected to dead weight of the floor slab, are usually absorbed without necessitating further lintel adjustment.

However, with lightweight curtain walls, the position of the lintels is important, because large paneled areas afford less latitude for variation. As a rule, the steel erector is unable to adjust the lintels to the desired accuracy at the time the main framework is erected. If he has contracted to do the adjusting, he must wait until the construction engineer establishes the correct lines and grades. In the usual case, floor slabs are concreted immediately after the steelwork is inspected and accepted. The floor grades then determined become the base to which the lintels can be adjusted. At about the same time, the wall contractor has his scaffolds in place, and by keeping pace with wall construction, the steel erector, working from the wall scaffolds, adjusts the lintels.

In some cases, the plans call for concrete encasement of the spandrel beams, in which case concreting is accomplished with the floor slab. Naturally, the construction engineer should see that the adjustment features provided for the lintels are not frozen in the concrete. One suggestion is to box around the details, thus avoiding chopping out concrete. In some cases, it may be possible to avoid the condition entirely by locating the connection below the concrete encasement, where the adjustment is always accessible.

The whole operation of lintel adjustment is one of coordination between the several trades; that this be carried out in an orderly fashion is the duty of the construction engineer. Furthermore, the desired procedure should be carefully spelled out in the job specifications so that erection costs can be estimated fairly.

Particularly irksome to the construction engineer is the lintel located some distance below the spandrel and supported on flexible, light steel hangers. This detail can be troublesome because it has no capacity to resist torsion. Avoid by developing the lintel and spandrel to act together as a single member.

PAINTING

Protection of steel surfaces has been, since the day steel was first used, a vexing problem for the engineers, paint manufacturers, and maintenance personnel. Over the years, there have been many developments, the result of numerous studies and research activities. However, it was not until 1950 that a concerted attempt was made to correlate all available data. Then, the Steel Structures Painting Council was organized, and an effort was made to determine and outline the best methods developed up to the present time, issue specifications covering practical and economical methods of surface preparation and painting steel structures, and engage in further research aimed at reducing or preventing steel corrosion.

Results are published in the "Steel Structures Painting Manual." This work is in two volumes—Vol. I, "Good Painting Practice," and Vol. II, "Systems and Specifications" (Steel Structures Painting Council, 4400 Fifth Avenue, Pittsburgh Pa. 15213). Each of the Council's paint systems covers the method of cleaning surfaces, types of paint to be used, number of coats to be applied, and techniques to be used in their applications. Each surface treatment and paint system is identified by uniform nomenclature, e.g., Paint System Specification SSPC-PS7.00-64T, which happens to be the identity of the minimum-type protection as furnished for most buildings.

8-81. Corrosion of Steel.

Ordinarily, steel corrodes in the presence of both oxygen and water, but corrosion rarely takes place in the absence of either. For instance, steel does not corrode in dry air, and corrosion is negligible when the relative humidity is below 70%, the critical humidity at normal temperature. Likewise, steel does not corrode in water that has been effectively deaerated. Therefore, the corrosion of structural steel is not a serious problem, except where water and oxygen are in abundance and where these primary prerequisites are supplemented with corrosive chemicals such as soluble salts, acids, cleaning compounds, and welding fluxes.

8-104　Structural-Steel Construction

In ideal dry atmospheres, a thin transparent film of iron oxide forms. This layer of ferric oxide is actually beneficial, since it protects the steel from further oxidation.

When exposed to water and oxygen in generous amounts, steel corrodes at an average rate of roughly 5 mils loss of surface metal per year. If the surface is comparatively dry, the rate drops to about ½ mil per year after the first year, the usual case in typical industrial atmospheres. Excessively high corrosion rates occur only in the presence of electrolytes or corrosive chemicals; usually, this condition is found in localized areas of a building.

Mill scale, the thick layer of iron oxides that forms on steel during the rolling operations, is beneficial as a protective coating, provided it is intact and firmly adheres to the steel. In the mild environments generally encountered in most buildings, mill scale that adheres tightly after weathering and handling offers no difficulty. In buildings exposed to high humidity and corrosive gases, broken mill scale may be detrimental to both the steel and the paint. Through electrochemical action, corrosion sets in along the edges of the cracks in the mill scale and in time loosens the scale, carrying away the paint.

Galvanic corrosion takes place when dissimilar metals are connected together. Noble metals such as copper and nickel should not be connected to structural steel with steel fasteners, since the galvanic action destroys the fasteners. On the other hand, these metals may be used for the fasteners, because the galvanic action is distributed over a large area and consequently little or no harm is done. When dissimilar metals are to be in contact, the contacting surfaces should be insulated; paint is usually satisfactory.

8-82. Painting Steel Structures.

Evidence obtained from dismantled old buildings and from frames exposed during renovation indicates that corrosion does not occur where steel surfaces are protected from the atmosphere. Where severe rusting was found and attributed to leakage of water, presence or absence of shop paint had no significant influence. Consequently, the American Institute of Steel Construction "Specification for the Design, Fabrication and Erection of Structural Steel for Buildings" exempts from one-coat shop paint, at one time mandatory, all steel framing that is concealed by interior finishing materials—ceilings, fireproofing partitions, walls, and floors.

Structures may be grouped as follows: (1) those that need no paint, shop or field; (2) those in which interior steelwork will be exposed, probably field painted; (3) those fully exposed to the elements. Thus, shop paint is required only as a primer coat before a required coat of field paint.

Group (1) could include such structures as apartment buildings, hotels, dormitories, office buildings, stores, and schools, where the steelwork is enclosed by other materials. The practice of omitting the shop and field paint for these structures, however, may not be widely accepted because of tradition and the slowness of building-code modernization. Furthermore, despite the economic benefit of paint omission, clean, brightly painted steel during construction has some publicity value.

In group (2) are warehouses, industrial plants, parking decks, supermarkets, one-story schools, inside swimming pools, rinks, and arenas, all structures shielded from the elements but with steel exposed in the interior. Field paint may be required for corrosion protection or appearance or both. The severity of the corrosion environment depends on type of occupancy, exposure, and climatic conditions. The paint system should be carefully selected for optimum effectiveness.

In group (3) are those structures exposed at all times to the weather: crane runways, fire escapes, towers, exposed exterior columns, etc. When made of carbon steel, the members will be painted after erection and therefore should be primed with shop paint. The paint system selected should be the most durable one for the atmospheric conditions at the site. For corrosion-resistant steels, such as those meeting ASTM Specification A242 and A588, field painting may be unnecessary. On exposure, these steels acquire a relatively hard coat of oxide, which shields the surface from progressive rusting. The color, russet brown, has architectural appeal.

8-83. Paint Systems.

The Steel Structures Painting Council has correlated surface preparations and primer, intermediate, and finish coats of paints into systems, each designed for a common service condition ("Steel Structures Painting Manual"). In addition, the Council publishes specifications for each system and individual specifications for surface preparations and paints. Methods for surface cleaning include solvent, hand-tool, power-tool, pickling, flame, and several blast techniques.

Surface preparation is directly related to the type of paints. In general, a slow-drying paint containing oil and rust-inhibitive pigments and one possessing good wetting ability may be applied on steel nominally cleaned. On the other hand, a fast-drying paint with poor wetting characteristics requires exceptionally good surface cleaning, usually entailing complete removal of mill scale. Therefore, in specifying a particular paint, the engineer should include the type of surface preparation, to prevent an improper surface condition from reducing the effectiveness of an expensive paint.

Paint selection and surface preparation are a matter of economics. For example, while blast-cleaned surfaces are conceded to be the best paint foundation for lasting results, the high cost is not always justified. Nevertheless, the Council specifies a minimum surface preparation by a blast cleaning process for such paints as alkyd, phenolic, vinyl, coal tar, epoxy, and zinc-rich.

As an aid for defining and evaluating the various surface preparations, taking into account the initial condition of the surface, an international visual standard is available and may be used. A booklet of realistic color photographs for this purpose can be obtained from the Council or American Society for Testing and Materials. The applicable standard and acceptance criteria are given in "Quality Criteria and Inspection Standards," published by the American Institute of Steel Construction.

The Council stresses the relationship between the prime coat (shop paint) and the finish coats. A primer that is proper for a particular type of field paint could be an unsatisfactory base for another type of field paint. Since there are numerous paint formulations, refer to Council publications when faced with a painting condition more demanding than ordinary.

In the absence of specific contract requirements for painting, the practice described in the American Institute of Steel Construction "Specification for the Design, Fabrication and Erection of Structural Steel for Buildings" may be followed. This method may be considered "nominal." The steel is brushed, by hand or power, to remove loose mill scale, loose rust, weld slag, flux deposit, dirt, and foreign matter. Oil and grease spots are solvent cleaned. The shop coat is a commercial-quality paint applied by brushing, dipping, roller coating, flow coating, or spraying to a 2-mil thickness. It affords only short-time protection. Therefore, finished steel that may be in ground storage for long periods or otherwise exposed to excessively corrosive conditions may exhibit some paint failure by the time it is erected, a condition beyond the control of the fabricator. Where such conditions can be anticipated, as for example, an overseas shipment, the engineer should select the most effective paint system.

8-84. Field-Painting Steel.

There is some question as to justification for protecting steelwork embedded in masonry or in contact with exterior masonry walls built according to good workmanship standards but not impervious to moisture. For example, in many instances, the masonry backing for a 4-in brick wall is omitted to make way for column flanges. Very definitely, a 4-in wall will not prevent penetration of water. In many cases, also, though a gap is provided between a wall and steelwork, mortar drippings fall into the space and form bridges over which water may pass, to attack the steel. The net effect is premature failure of both wall and steel. Walls have been shattered—sheared through the brick—by the powerful expansion of rust formations. The preventatives are: (1) coating the steel with suitable paint and (2) good wall construction (see Sec. 12).

A typical building code reads: "Special precautions shall be taken to protect the outer surfaces of steel columns located in exterior walls against corrosion, by painting such surfaces with waterproof paints, by the use of mastic, or by other methods or waterproofing approved by the building inspector."

In most structures an asphalt-type paint is used for column-flange protection. The proviso is sometimes extended to include lintels and spandrels, since the danger of corrosion is similar, depending on the closeness and contact with the wall. However, with the latter members, it is often judicious to supplement the paint with flashing, either metallic or fabric. A typical illustration, taken from an actual apartment-building design, is shown in Fig. 8-66.

In general, building codes differ on field paint; either paint is stipulated or the code is silent. From a practical viewpoint, the question of field painting cannot be properly resolved with a single broad rule. For an enclosed building in which the structural members are enveloped, for example, a field coat is sheer wastage, except for exterior steel members in contact with walls. On the other hand, exposed steel subject to high-humidity atmospheres and to exceptionally corrosive gases and contaminants may need two or three field coats.

Fig. 8-66 Flashing at spandrel and lintels.

Manufacturing buildings should always be closely scrutinized, bearing in mind that original conditions are not always permanent. As manufacturing processes change, so do the corrosive environments stimulated by new methods. It is well to prepare for the most adverse eventuality.

Special attention should be given to steel surfaces that become inaccessible, e.g., tops of purlins in contact with roof surfaces. A three-coat job of particularly suitable paint may pay off in the long run, even though it delays placement of the roof covering.

8-85. Steel in Contact with Concrete.

According to the "Steel Structures Painting Manual," Vol. I, "Good Painting Practice" (Steel Structures Painting Council, Pittsburgh, Pa.):

1. Steel that is embedded in concrete for reinforcing should not be painted. Design considerations require strong bond between the reinforcing and the concrete so that the stress is distributed; painting of such steel does not supply sufficient bond. If the concrete is properly made and of sufficient thickness over the metal, the steel will not corrode.

2. Steel that is encased with exposed lightweight concrete that is porous should be painted with at least one coat of good-quality rust-inhibitive primer. When conditions are severe, or humidity is high, two or more coats of paint should be applied, since the concrete may accelerate corrosion.

3. When steel is enclosed in concrete of high density or low porosity, and when the concrete is at least 2 in to 3 in thick, painting is not necessary, since the concrete will protect the steel.

4. Steel in partial contact with concrete is generally not painted. This creates an undesirable condition, for water may seep into the crack between the steel and the concrete, causing corrosion. A sufficient volume of rust may be built up, spalling the concrete. The only remedy is to chip or leave a groove in the concrete at the edge next to the steel and seal the crack with an alkali-resistant calking compound (such as bituminous cement).

5. Steel should not be encased in concrete that contains cinders, since the acidic condition will cause corrosion of the steel.

FIRE PROTECTION FOR STRUCTURAL STEEL

Structural steel is a noncombustible material. It is therefore satisfactory for use without protective coverage in many types of buildings where noncombustibility is sufficient, from the viewpoint of either building ordinances or owner's preference. When structural steel is used in this fashion, it is described as "exposed" or "unprotected." Naturally, unprotected steel may be selected wherever building codes permit combustible construction; the property of noncombustibility makes it more attractive than competitive materials not possessing this property.

Exposed or unprotected structural steel is commonly used for industrial-type buildings, hangars, auditoriums, stadiums, warehouses, parking garages, billboards, towers, and low stores, schools, and hospitals. In most cases, these structures contain little combustible material. In others, where the contents are highly combustible, sprinkler or deluge systems may be incorporated to protect the steelwork.

8-86. Need for Fire Protection of Steel.

Steel building frames and floor systems must be covered with fire-resistant materials in certain buildings to reduce the chance of fire damage. These structures may be tall buildings, such as offices, apartments, and hotels, or low-height buildings, such as warehouses, where there is a large amount of combustible content. The buildings may be located in congested areas, where the spread of fire is a strong possibility. So for public safety, as well as to prevent property loss, building codes regulate the amount of fire resistance that must be provided.

The following are some of the factors that enter into the determination of minimum fire resistance for a specific structure: height, floor area, type of occupancy (a measure of combustible contents), fire-fighting apparatus, sprinkler systems, and location in a community (fire zone), which is a measure of hazard to adjoining properties.

8-87. Effect of Heat on Steel.

A moderate rise in temperature of structural steel, say up to 500°F, is beneficial in that the strength is about 10% greater than the normal value. Above 500°F, strength falls off,

8-108   Structural-Steel Construction

until at 700°F it is approximately equal to the normal temperature strength. At a temperature of 1000°F, the compressive strength of steel is about the same as the maximum allowable working stress in columns.

Unprotected steel members have a rating of about 15 min, based on fire tests of columns with cross-sectional areas of about 10 in^2. Heavier columns, possessing greater mass for dissipation of heat, afford greater resistance—20 min perhaps. Columns with reentrant space between flanges filled with concrete, but otherwise exposed, have likewise been tested; where the total area of the solid cross section approximates 36 in^2, the resistance is 30 min, and where the area is 64 in^2, the resistance is 1 hr.

The average coefficient of expansion for structural steel between the temperatures of 100 and 1200°F is given by the formula

$$C = 0.0000061 + 0.0000000019t \tag{8-40}$$

in which C = coefficient of expansion per °F
t = temperature, °F

Below 100°F, the average coefficient of expansion is taken as 0.0000065.

The modulus of elasticity of structural steel, about 29,000 ksi at room temperature, decreases linearly to 25,000 ksi at 900°F. Then, it drops at an increasing rate at higher temperatures.

8-88. Fire Protection of Exterior Members.

Steel members, such as spandrel beams and columns, on the exterior of a building may sometimes be left exposed or may be protected in an economical manner from fire damage, whereas interior steel members of the same building may be required to be protected with more expensive insulating materials, as discussed in Art. 8-89. Standard fire tests for determining fire-endurance ratings of exterior steel members are not available. But from many tests, data have been obtained that provide a basis for analytical, thermodynamic methods for fire-safe design. (See for example, "Fire-Safe Structural Steel—A Design Guide," American Iron and Steel Institute, 1000 16th St., N.W., Washington, D.C. 20036.)

The tests indicate that an exterior steel spandrel beam with its interior side protected by fire-resistant construction need only have its flanges fire protected. This may be simply done by application of fireproofing, such as sprayed-on mineral fibers, to the upper surface of the top flange and the under surface of the bottom flange. In addition, incombustible flame-impingement shields should enclose the flanges to deflect flames that may be emitted through windows. The shields, for example, may be made of ¼-in-thick weathering steel. This construction prevents the temperature of the spandrel beam from reaching a critical level.

Exposed-steel columns on the outside of a building may be made fire safe by placement at adequate distances from the windows. Such columns may also be located closer to the building when placed on the side of windows at such distances that the steel is protected by the building walls against flame impingement. Thermodynamic analysis can indicate whether or not the chosen locations are fire safe.

8-89. Materials for Improving Fire Resistance.

Structural steel may be protected with any of many materials—brick, stone, concrete, gypsumboard, gypsum block, sprayed-on mineral fibers, and various fire-resistant plasters.

Concrete insulation serves well for column protection, in that it gives additional stability to the steel section. Also, it is useful where abrasion resistance is needed. Concrete,

Materials for Improving Fire Resistance   8-109

Fig. 8-67 Fire protection of steel columns by encasement with (a) concrete, (b) plaster on gypsum lath, (c) plaster on metal lath, (d) furring and gypsumboard, (e) gypsumboard without furring, and (f) gypsum block and plaster.

however, is not an efficient insulating medium compared with fire-resistant plasters. Normally, it is placed completely around the columns, beams, or girders, with all reentrant spaces filled solid (Fig. 8-67a). Although this procedure contributes to the stability of columns and effects composite action in beams and slabs, it has the disadvantage of imposing great weight on the steel frame and foundations. For instance, full protection of a W12 column with stone concrete weighs about 355 psf, whereas plaster protection weighs about 40 psf, and lightweight concretes made with such aggregates as perlite, vermiculite, expanded shale, expanded slag, pumice, pumicite and sintered flyash weigh less than 100 psf.

Considerable progress has been made in the use of lightweight plasters with aggregates possessing good insulating properties. Two aggregates used extensively are perlite and vermiculite; they replace sand in the sanded-gypsum plaster mix. A 1-in thickness weighs about 4 psf, whereas the same thickness of sanded-gypsum plaster weighs about 10 psf.

Typical details of lightweight plaster protection for columns are shown in Fig. 8-67b and c. Generally, vermiculite and perlite plaster thicknesses of 1 to 1¾ in afford protection of 3 and 4 hr, depending on construction details. Good alternatives include gypsum board (Fig. 8-67d and e) or gypsum block (Fig. 8-67f).

For buildings where rough usage is expected, a hard, dense insulating material such as concrete, brick, or tile would be the logical selection for fire protection.

For many buildings, finished ceilings are mandatory. It is therefore logical to employ the ceiling for protecting roof and floor framing. All types of gypsum plasters are used extensively for this dual purpose. Figure 8-68 illustrates typical installations. For 2-hr floors, ordinary sand-gypsum plaster ¾ in thick is sufficient. Three- and four-hour floors

Fig. 8-68 Fire protection for floor framing with incombustible floor construction: (a) section showing suspended plaster ceiling; (b) attached plaster ceiling; (c) furred plaster ceiling; (d) suspended ceiling with lay-in, fire-rated acoustic panels; (e) detail of panel support in (d); (f) detail showing fire protection around recessed lighting; (g) detail showing fire protection around air-conditioning duct and grille.

may be obtained with perlite gypsum and vermiculite gypsum in the thickness range of ¾ to 1 in.

Instead of plastered ceilings, use may be made of fire-rated dry ceilings, acoustic tiles, or drop (lay-in) panels (Fig. 8-68d and e).

Another alternative is to spray the structural steel mechanically (where it is not protected with concrete) with plasters of gypsum, perlite, or vermiculite, proprietary cementitious mixtures, or mineral fibers not deemed a health hazard during spraying (Fig. 8-69). In such cases, the fire-resistance rating of the structural system is independent of the ceiling. Therefore, the ceiling need not be of fire-rated construction. Drop panels, if used, need not be secured to their suspended supports.

Fig. 8-69 Typical fire protection with sprayed material.

Still another sprayed-on material is the intumescent fire-retardant coating, essentially a paint. Tested in conformance with ASTM Specification E119, a $\frac{3}{16}$-in-thick coat applied to a steel column has been rated 1 hr, ½-in-thick coating 2 hr. As applied, the coating has a hard, durable finish, but at high temperatures, it puffs to many times its original thickness, thus forming an effective insulating blanket. Thus, it serves the dual need for excellent appearance and fire protection.

Aside from dual functioning of ceiling materials, the partitions, walls, etc., being of incombustible material, also protect the structural steel, often with no additional assistance. Fireproofing costs, therefore, may be made a relatively minor expense in the overall costs of a building through dual use of materials.

8-90. Pierced Ceilings and Floors.

Some buildings require recessed light fixtures and air-conditioning ducts, thus interrupting the continuity of fire-resistive ceilings. A rule that evolved from early standard fire tests permitted 100 in^2 of openings for noncombustible pipes, ducts, and electrical fixtures in each 100 ft^2 of ceiling area.

It has since been demonstrated, with over 100 fire tests that included electrical fixtures and ducts, that the fire-resistance integrity of ceilings is not impaired when, in general:

Recessed light fixtures, 2 by 4 ft, set in protective boxes, occupy no more than 25% of the gross ceiling area.

Air-duct openings, 30 in maximum in any direction, are spaced so as not to occupy more than 576 in^2 of each 100 ft^2 of gross ceiling area. They must be protected with fusible-link dampers against spread of smoke and heat.

These conclusions are not always applicable. Reports of fire tests of specific floor systems should be consulted.

A serious infringement of the fire rating of a floor system could occur when pipes, conduit, or other items pierce the floor slab, a practice called "poke-through." Failure to calk the openings with insulating material results in a lowering of fire ratings from hours to a few minutes.

8-91. Fire-Resistance Ratings.

Most standard fire tests on structural-steel members and assemblies have been conducted at one of two places—the National Bureau of Standards, Washington, D.C., or the Underwriters' Laboratories, Northbrook, Ill. Fire-testing laboratories also are available at Ohio State University, Columbus, Ohio, and the University of California, Berkeley, Calif. Laboratory test reports form the basis for establishing ratings. Summaries of these tests, together with tabulation of recognized ratings, are published by a number of organizations listed below. The trade associations, for the most part, limit their ratings to those constructions employing the material they represent.

The American Insurance Association (formerly The National Board of Fire Underwriters), 85 John St., New York, N.Y. 10038.

The National Bureau of Standards, Washington, D.C. 20234.

Gypsum Association, 1603 Orrington Ave., Evanston, Ill. 60201.

Metal Lath/Steel Framing Association, 221 North LaSalle St., Chicago, Ill. 60601

Perlite Institute, 45 West 45th St., New York, N.Y. 10036.

American Iron and Steel Institute, 1000 16th St., N.W., Washington, D.C. 20036.

American Institute of Steel Construction, 400 N. Michigan Ave., Chicago, Ill. 60611.

8-92. Bibliography.

Title	Author	Publisher
Designing Fire Protection for Steel Columns		AISI*
Designing Fire Protection for Steel Trusses		AISI*
Fire-Safe Structural Steel		AISI*
Iron and Steel Beams, 1873–1952		AISC†
Guide to Shop Painting of Structural Steel		AISC†
Manual of Steel Construction		AISC†
Structural Steel Detailing		AISC†
Fundamentals of Welding		AWS‡
Structural Welding Code, D1.1		AWS‡
Structural Steel Design	J. Bowles	McGraw-Hill
Structural Steel Design Data Manual	J. Bowles	McGraw-Hill
Analysis and Design of Steel Buildings	S. W. Crawley and R. M. Dillon	Wiley
Concepts in Building Firesafety	M. D. Egan	Wiley
Structural Engineering Handbook	E. Gaylord and C. Gaylord	McGraw-Hill
Guide to Stability Design Criteria for Metal Structures	B. G. Johnston	Wiley
Steel Design for Structural Engineers	B. O. Kuzmanovich and N. Nicholas	Prentice-Hall
Structural Steel Designers' Handbook	F. S. Merritt	McGraw-Hill
Standard Details for Fire-Resistive Building Construction	L. Przetak	McGraw-Hill
Structural Design Guide to AISC Specification for Buildings	P. F. Rice and E. S. Hoffman	Van Nostrand Reinhold

*American Iron and Steel Institute, 1000 16th St., N.W., Washington, D.C. 20036.
†American Institute of Steel Construction, 400 N. Michigan Ave., Chicago, Ill. 60611.
‡American Welding Society, 2501 N.W. 7th St., Miami, Fla. 33125.

Section 9

Cold-Formed Steel Construction

Don S. Wolford
Consulting Engineer
Middletown, Ohio

The term cold-formed steel construction, as used in this section, refers to structural components that are made of steel but are produced from other material than hot-rolled structural steel shapes and plates. This section deals with fabricated components made from basic forms, such as bars, sheet and strip, and include:

Open-web steel joists, which are usually fabricated from relatively small bars, bar-size angles, and shapes formed from flat-rolled material.

Sheet- and strip-steel structural members produced either by roll-forming or by bending- or press-brake operations.

Wall panels and floor and roof deck made of steel sheets shaped to provide both surface coverage and load-carrying capacity.

Alone or in combination with structural steel, cold-formed components are used in light-occupancy structures and in custom-built and factory-produced buildings for all kinds of occupancies, from small and sometimes temporary structures to massive high-rise buildings.

This section describes a few forms that have become well established, and the general principles underlying their structural design and use.

OPEN-WEB STEEL JOISTS

Under "Standard Specifications for Open-web Steel Joists, H-Series," adopted jointly by the Steel Joist Institute (SJI) and the American Institute of Steel Construction (AISC), open-web steel joists are relatively small, parallel-chord steel trusses. They are suitable for direct support of floors and roof decks in buildings, when designed in accordance with those specifications and the standard load tables that accompany them.

9-2 Cold-Formed Steel Construction

Fig. 9-1 Some examples of open-web steel-joist floor construction.

Longspan steel joists and joist girders, also covered by specifications adopted by SJI and AISC, are essentially structural steel trusses designed for somewhat heavier duty and substantially longer spans than regular open-web steel joists. They are not included in the following discussion.

As usually employed in floor construction, open-web steel joists are covered by a slab of concrete, 2 to 2½ in thick, cast on permanent forms (centering). A wall-bearing, open-web steel-joist floor assembly, with ceiling, is illustrated in Fig. 9-1b. Use of open-web steel joists in roofs is discussed in Art. 9-7.

The joists usually are supported either on structural-steel framing (Fig. 9-1a) or masonry bearing walls. When used with structural steel, the joists are preferably welded to the supporting framework, but may be bolted or clipped. For wall-bearing applications, wall anchors are usually specified, as shown in Fig. 9-1b.

In addition to light weight, one of the advantages of open-web steel joist construction is that the opening between chords provides space for electrical work, ducts, and piping.

9-1. Joist Design and Fabrication.

Standardization of open-web steel joists under SJI-AISC specifications mentioned previously consists essentially of definition of product, specification of design basis, and specifications covering bridging and certain other details. Exact forms of the members and web systems, and methods of manufacture, are determined by individual producers.

A number of proprietary designs have been developed. A few types are shown in Fig. 9-2. For the regular underslung end construction, the standard end depth is 2½ in, as shown. Square-end joists, which do not have underslung ends, have been supplied for special purposes.

SJI publishes load tables for standardized joists in depths of 8 to 30 in in 2-in increments. There are several different weights for each joist depth, except for the 8-in, which has only one standard weight.

H-Series Joists. The H series is designed at a basic working stress of 30 ksi, based on a minimum specified yield strength of chords of 50 ksi. This strength is obtained by using either high-strength, low-alloy steel, or cold-formed sections of carbon steel with yield strength increased by the forming operation. Webs may be made of A36 steel and designed for an allowable tensile stress of 22 ksi.

Specifications. Detailed rules for the structural design of open-web steel joists are given in SJI-AISC "Standard Specifications for Open-web Steel Joists, H-Series," Steel Joist Institute, 1205 48th Ave. North, Myrtle Beach, SC 29577. (See also Art. 9-2.)

Composite Construction. Although standard open-web steel joists are designed as simply supported, self-contained structural members, other joists have been designed with the top chord intended to act in a composite manner with the concrete slab that covers it. Composite action may be obtained by shaping the chord to provide shear anchorage within the slab, or by other means. Detailed information may be obtained from manufacturers' catalogs.

Fabrication. Open-web steel joists are different in one important respect from the fabricated structural-steel framing members commonly used in building constructions: the joists are usually manufactured by production-line methods with special equipment

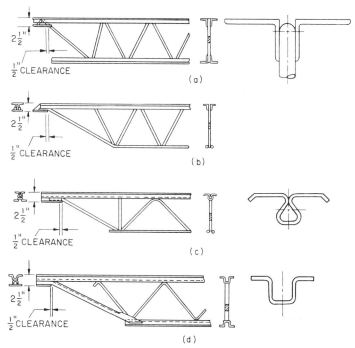

Fig. 9-2 Some examples of open-web steel joists.

designed for quantity production. Component parts are generally joined together by either resistance or electric-arc welding.

Shop Paint. Open-web steel joists are given a prime coat of paint prior to shipment. The SJI-AISC specifications call for shop paint that conforms to or meets the minimum performance requirements of Steel Structures Painting Council Specification 15-68T for Type I (red oxide) or Type II (asphalt), or of Federal Specifications T-TP-636 (red oxide).

9-2. Design of Open-Web Joist Floors.

Open-web joists are designed primarily to be used under uniformly distributed loading and at substantially uniform spacing. They can safely carry concentrated loads, however, if proper consideration is given to the effect of such loads. For example, good practice requires that heavy concentrated loads be applied at joist panel points. (The weight of an ordinary partition running crosswise of the joists usually is considered distributed by the floor slab to the extent that it will not cause appreciable local bending in the top chords of the joists.) But the joists must be selected to resist the bending moments, shears, and end reactions due to such loads.

Joists at Openings. Open-web joists are not designed to be used as individual framing members, although special joists have been so used under special conditions.

Fig. 9-3 Header construction with open-web joists at floor openings.

Relatively small openings between joists may usually be framed with angle, channel, or Z-shaped headers supported on adjacent joists (Fig. 9-3). Larger openings should be framed in structural steel.

Headers should preferably be located so that they are supported at panel points of the trimmer joists. In any case in which that is not practicable and in which the header reaction exceeds 0.4 kip, consideration must be given to the bending stresses induced in the top chords of the trimmer joists by the concentrated loads from the header, or struts may be inserted in the joists at the points at which the header is supported ("Recommended Code of Standard Practice for Open-web and Longspan Steel Joists," Steel Joist Institute, 1703 Parham Road, Richmond, Va. 23229).

Joists at Heavy Loads. Open-web steel joists can be doubled and even tripled where necessary to support heavy loads.

Design Methods. The method of selecting joist sizes for any floor depends on whether the effect of any cross partitions or other concentrated loads must be considered. Under uniform loading only, joist sizes and spacings are most conveniently selected from a table of safe loads. The Steel Joist Institute standard load tables provide that equal concentrated top-chord loads, as in bulb-tee roof construction, may be considered uniformly distributed if the spacing of the concentrations along the joists does not exceed

Fig. 9-4 Framing plan for part of a floor of open-web steel-joist construction.

33 in. Where other concentrated loads or nonuniform loads exist, calculate bending moments, end reactions, and shears, and select joists accordingly. Any case in which there may be local bending of either chord should be investigated by appropriate methods.

The deflection of open-web steel joists can be computed in the same way as for other types of beams. In computing moment of inertia of the cross section, however, the chords only should be considered, and the web system should be neglected. The resulting moment of inertia should be reduced 15% to account for that part of the deflection caused by web-member strain.

Effective moments of inertia of standard SJI-AISC joists are published by the Steel Joist Institute (Art. 9-1) or are available from manufacturers.

Depth-Span Limitations. The SJI-AISC specifications provide that the clear span of a joist shall not exceed 24 times its depth, except that for floors the clear span of an H-series joist shall not exceed 20 times its depth.

Figure 9-4 shows a portion of a framing plan for a typical open-web joist floor.

9-3. Extended Ends and Ceiling Extensions.

Plastered ceilings attached directly to regular open-web joists are usually supported at the underslung ends by means of *ceiling extensions.* These may consist of an extension of the lower chord of the joist past the end bottom-chord panel point (Figs. 9-1 and 9-5a), or they may consist of special attachments, depending on the length of the extension and the detailed design of the joist. For details, in a particular case, refer to the manufacturers' catalogs.

Fig. 9-5 Open-web steel joist with (a) ceiling extension, (b) extended end. *(Steel Joist Institute.)*

Open-web joists are frequently provided with extended ends overhanging the support (Fig. 9-5b). Depending on the length of the extension and the detailed design of the joist, the extended end may consist of a simple extension of the top chord of the joist, separate fabricated attachments, or a combination of both. The Steel Joist Institute publishes load tables for extended ends based on attachments that are separately fabricated from angle or channel sections. Since details may vary among different manufacturers, the individual manufacturer's practice should be followed in a particular case.

9-4. Bridging and Anchors for Open-Web Joists.

When properly made and installed, open-web steel joists are capable of withstanding all usual service requirements. The joists are, however, very flexible in the lateral direction. They are not intended to be used without proper lateral support. Consequently, bridging (bracing) should be installed between joists as soon as possible after joists have been placed and before application of construction loads. The principal function of bridging is to provide lateral support during the construction period.

Bridging details vary with different manufacturers. The most commonly used type is composed of rods or angles fastened to the top and bottom chords of the joists in a vertical plane.

SJI-AISC specifications for open-web joists contain provisions for the number of rows of bridging for each chord size.

Care must be exercised even after bridging has been installed, to avoid undue concentrations of construction loads.

Connections. It is important that masonry anchors be used at walls on wall-bearing joists. Where the joists rest on steel beams, the joists should be welded, bolted, or clipped to the beams. The standard specifications of the Steel Joist Institute should be followed in all such details.

9-5. Formwork for Joist Construction.

Formwork (centering) for concrete floors and roofs with open-web joists usually consists of a decking of corrugated steel sheets. Especially produced for this purpose and marketed under various tradenames, the sheets are usually placed on top of the joists and form a permanent part of the construction. Expanded metal rib lath or paper-backed welded-wire fabric has also been used.

Corrugated sheets can be fastened with self-tapping screws or welded to the joists. SJI-AISC specifications for open-web joists require that each attachment to the top chords of the joists be capable of resisting a lateral force of at least 0.3 kip. Spacing of attachments should not exceed 36 in along the top chord.

9-6. Slab Reinforcement in Steel Joist Construction.

When the usual cast-in-place floor slab is used, it is customary to install reinforcing bars in two perpendicular directions or to use welded-wire fabric. No other reinforcement is usually considered necessary.

9-7. Open-Web Joists in Roof Construction.

Open-web joists frequently are used as purlins to support roofs at spacings determined by the load-carrying capacity of the decking. The joists should be properly braced. For steep roofs, sag rods should be provided for lateral support, as for structural-steel purlins.

Open-web steel joists have also been used as rafters (inclined beams) to support sloping roofs.

When open-web joists are used for supporting flat roofs, consideration should be given to the problem of ponding, or increase in load due to the accumulation of rainwater in concavities caused by deflection of the roof structure. SJI-AISC specifications for open-web steel joists require that, unless a roof surface is provided with sufficient slope toward points of free drainage or adequate individual drains to prevent accumulation of rainwater, the roof system should be investigated to ensure stability under ponding conditions.

A roof slope of ¼ in/ft is usually considered sufficient for normal conditions of free drainage. For a roof wholly or partly surrounded by parapets or other walls, and provided with individual drains, it is important that the drains be properly designed and maintained to prevent build-up of live load beyond safe limits. Such build-up, including excessive snow accumulation, can prove more damaging than the ponding that may be caused by deflection of the roof system.

The stability investigation may be made in the same way, and follow the same criteria, as for structural steel (Art. 8-40). Where open-web steel joists rest on relatively rigid supports, such as masonry bearing walls, an approximate solution for the required moment of inertia of the joists, derived from Eq. (8-28) with C_s considered negligible for bearing walls, is given by

$$I_{\min} = 0.0000128 SL^4 \qquad (9\text{-}1)$$

where $I_{\min}$ = minimum allowable moment of inertia of each joist, in^4
S = spacing of joists, ft
L = span of joists, ft

According to AISC, the ponding deflection contributed by a metal deck is usually such a small part of the total ponding deflection of a roof panel that it is sufficient to limit the moment of inertia of the deck, per foot of width normal to its span, to 0.000025 times the fourth power of its span length. ("Commentary on AISC Specification," 1978.) In any close or doubtful case, the stability against ponding of the entire roof structure should be investigated, as for structural steel.

For a detailed treatment of steel joist roofs under ponding loads, refer to "Structural Design of Steel Joist Roofs to Resist Ponding Loads," *Steel Joist Institute Technical Digest*, No. 3, May 1971.

9-8. Fire Resistance of Open-Web Steel Joist Construction.

Any required degree of fire resistance can be obtained for steel joist construction with proper protection. Numerous fire-resistance tests have been made of floor and roof assemblies that include steel joists. Complete lists of such tests are published by the American Insurance Association, Factory Mutual System, and the Underwriters Laboratories, Inc. (See also "Fire Resistance Classification of Building Constructions," National Bureau of Standards, Report BMS 92.)

Table 9-1 lists a few typical ratings for steel joist construction. Some additional ratings for open-web steel joists supporting steel roof deck are listed in Table 9-15.

COLD-FORMED SHAPES

Cold-formed shapes usually imply relatively small, thin sections made by bending sheet or strip steel in roll-forming machines, press brakes, or bending brakes. Because of the relative ease and simplicity of the bending operation and the comparatively low cost of forming rolls and dies, the cold-forming process lends itself well to the manufacture of special shapes for special purposes and makes it possible to use thin material shaped for maximum stiffness.

The use of cold-formed shapes for ornamental and other non-load-carrying purposes

TABLE 9-1 Fire Ratings for Some Floor-Ceiling Assemblies with Open-Web Steel Joists

1-hr fire resistance
  Concrete slab: 2 in or more thick, reinforced
  Ceiling: 1/2-in-thick gypsumboard*
  Steel joists: No. 3† chord or heavier
1 1/2-hr fire resistance
  Concrete slab: 2 in or more thick, reinforced
  Ceiling: 5/8-in-thick gypsumboard*
  Steel joists: No. 4† chord or heavier
2-hr fire resistance
  Concrete slab: 2 1/2 in or more thick, reinforced
  Ceiling: 1/2-in-thick gypsumboard*
  Steel joists: No. 3† chord or heavier
3-hr fire resistance
  Concrete slab: 2 1/2 in or more thick, reinforced
  Ceiling: 5/8-in-thick gypsumboard*
  Steel joists: No. 3† chord or heavier
4-hr fire resistance
  Concrete slab: 2 1/2 in or more thick, reinforced
  Steel joists: No. 5† chord or heavier
  Furring channels: 3/4-in cold-rolled.
  Ceiling: 3/4-in vermiculite plaster aggregate in fibered gypsum scratch and brown coats or unfibered brown coat, with metal lath.

*Application of up to 1/8 in of plaster to gypsumboard is permissible.
†Chord size is given by the last digit of the joist designation.

is commonplace. Door and window frames, metal-partition work, non-load-bearing studs, facing, and all kinds of ornamental sheet-metal work employ such shapes. The following deals with cold-formed shapes used for structural purposes in the framing of buildings.

There is no standard series of cold-formed structural sections, such as those for hot-rolled shapes, although groups of such sections have been designed ("Cold-formed Steel Design Manual," American Iron and Steel Institute). For the most part, however, cold-formed structural shapes are designed to serve a particular purpose. The general approach of the designer is therefore similar to that involved in the design of built-up structural sections, namely, to fit a particular application.

As a general rule, cold-formed shapes will cost more per pound than hot-rolled sections. They will accordingly be found to be economical under the following circumstances:

1. Where their use permits a substantial reduction in weight relative to comparable hot-rolled sections. This will occur usually where relatively light loads are to be supported over short spans, or where stiffness rather than strength is the controlling factor in the design.

2. In a special case in which a suitable combination of standard rolled shapes would be heavy and uneconomical.

3. Where the quantities required are too small to justify the investment in equipment necessary to produce a suitable hot-rolled section.

4. In dual-purpose panel work, where both strength and coverage are desired.

9-9. Material for Cold-Formed Shapes.

Cold-formed shapes are usually made from hot-rolled sheet or strip steel. Cold-rolled material, that is, steel that has been cold-reduced to the desired thickness, is used for the thinner gages or where, for any reason, the surface finish, mechanical properties, or

TABLE 9-2 Classification by Size of Flat-Rolled Carbon Steel

a. Hot-rolled

Width, in.	Thicknesses, in.			
	0.2300 and thicker	0.2299–0.2031	0.2030–0.1800	0.1799–0.0449
To 3½ incl.	Bar	Bar	Strip	Strip[a]
Over 3½ to 6 incl.	Bar	Bar	Strip	Strip[b]
Over 6 to 8 incl	Bar	Strip	Strip	Strip
Over 8 to 12 incl	Plate[c]	Strip	Strip	Strip
Over 12 to 48 incl	Plate[d]	Sheet	Sheet	Sheet
Over 48	Plate[d]	Plate[d]	Plate[d]	Sheet

b. Cold-rolled

Width, in.	Thicknesses, in.		
	0.2500 and thicker	0.2499–0.0142	0.0141 and thinner
To 12, incl	Bar	Strip[e,f]	Strip[e]
Over 12 to 23¹⁵⁄₁₆, incl . .	Sheet[g]	Sheet[g]	Strip[h]
Over 23¹⁵⁄₁₆	Sheet	Sheet	Black plate[i]

[a] 0.0255-in. minimum thickness.
[b] 0.0344-in. minimum thickness.
[c] Strip, up to and including 0.5000-in. thickness, when ordered in coils.
[d] Sheet, up to and including 0.5000-in. thickness, when ordered in coils.
[e] Except that when the width is greater than the thickness, with a maximum width of ½ in. and a cross-sectional area not exceeding 0.05 sq in., and the material has rolled or prepared edges, it is classified as flat wire.
[f] Sheet, when slit from wider coils and supplied with cut edge (only), in thicknesses 0.0142 to 0.0821 and widths 2 to 12 in., inclusive, and carbon content 0.25% maximum by ladle analysis.
[g] May be classified as strip when a special edge, a special finish, or single-strand rolling is specified or required.
[h] Also classified as black plate[i], depending on detailed specifications for edge, finish, analysis, and other features.
[i] Black plate is a cold-rolled, uncoated tin-mill product that is supplied in relatively thin gages.

closer tolerances that result from cold-reducing are desired. Manufacture of cold-formed shapes from plates for use in building construction is possible but done infrequently.

Plate, Sheet, or Strip? The commercial distinction between steel plates, sheet, and strip is principally a matter of thickness and width of material, although in some sizes it depends on whether the material is furnished in flat form or in coils, whether it is carbon or alloy steel, and, particularly for cold-rolled material, on surface finish, type of edge, temper or heat treatment, chemical composition, and method of production. Although the manufacturers' classification of flat-rolled steel products by size is subject to change from time to time, and may vary slightly among different producers, that given in Table 9-2 for carbon steel is representative.

Carbon steel is generally used, although high-strength, low-alloy steel may be used where strength or corrosion resistance justify it. Stainless steel may be used for exposed work.

Mechanical Properties. Material to be used for structural purposes generally conforms to one of the standard specifications of the American Society for Testing and Materials (ASTM). Table 9-3 lists the ASTM specifications for structural-quality carbon

TABLE 9-3 Principal Mechanical Properties of Structural Quality Sheet, Strip, and Plate Steel

ASTM designation	Material	Grade	Minimum yield point, ksi	Minimum tensile strength, ksi		Minimum elongation, % in 2 in	Bend test, 180°, ratio of inside diameter to thickness
				Hot rolled	Cold rolled		
A446	Galvanized sheet steel, zinc-coated by the hot-dip process, structural quality	A	33		45	20	1 1/2
		B	37		52	18	2
		C	40		55	16	2 1/2
		D	50		65	12	†
		E	80		82		†
		F	50		70	12	1
A570	Hot-rolled sheet and strip, carbon steel	30	30	49		*	1 1/2
		36	36	53		*	2
		40	40	55		*	2 1/2
		45	45	60		*	3
		50	50	65		*	
A606	Hot-rolled and cold-rolled sheet and strip, high-strength, low-alloy steel	Cut lengths	50	70		22	1
		Coils	45	65		22	1
		Annealed or normalized	45	65		22	1
		Cold rolled	45		65	22	1
						HR§ CR§	
A607	Hot-rolled and cold-rolled, high-strength, low-alloy columbium or vanadium steels, sheet and strip, cut lengths or coils	45	45	60	60	25 22	1
		50	50	65	65	22 20	1
		55	55	70	70	20 18	1 1/2
		60	60	75	75	18 16	2
		65	65	80	80	16 15	2 1/2
		70	70	85	85	14 14	3
A611	Cold-rolled sheet, structural carbon-steel sheet, cut lengths or coils	A	25		42	26	0
		B	30		45	24	1
		C	33		48	22	1 1/2

Spec	Description	Grade	Yield point min, ksi	Tensile strength, ksi		Elong in 8 in, %	Elong in 2 in, %
A36	Structural steel (plates only)		40	58–80		20	2
A242	High-strength, low-alloy structural steel (plates 3/4 in and under)	D	80	70	52	23	†
		E	36		82	†	‡
A441	High-strength, low-alloy structural manganese vanadium steel (plates 3/4 in and under)		50	70		†	‡
A529	Structural steel with 42 ksi minimum yield point (1/2 in maximum thickness) (plates only)		42	60–85		†	‡
A572	High-strength, low-alloy columbium-vanadium steels of structural quality (plates only)	42	42	60		24	‡
		50	50	65		21	‡
		60	60	75		18	‡
		65	65	80		17	‡
A588	High-strength, low-alloy structural steel with 50 ksi minimum yield point to 4 in thick (plates only)	A	50	70		21	‡
		B	50	70		21	‡
		C	50	70		21	‡
		D	50	70		21	‡
		E	50	70		21	‡
		F	50	70		21	‡
		G	50	70		21	‡
		H	50	70		21	‡
		J	50	70		21	‡

*Varies, see specification.
†Not specified or required.
‡S14 bend test.
§HR = hot rolled; CR = cold rolled.

and low-alloy sheet and strip, and the principal mechanical properties of the material described by them.

Stainless Steel Applications. Stainless-steel cold-formed shapes, although not ordinarily used in floor and roof framing, are widely used in exposed components, such as stairs; railings and balustrades; doors and windows; mullions; fascias, curtain walls and panel work; and other applications in which a maximum degree of corrosion resistance, retention of appearance and luster, and compatibility with other materials are primary considerations. Stainless-steel sheet and strip are available in several types and grades, with different strength levels and different degrees of formability, and in a wide range of finishes.

Information useful in design of stainless steel cold-formed members can be obtained from the "Stainless-Steel Cold-Formed Structural Design Manual," American Iron and Steel Institute, 1000 16th St., N.W., Washington, D.C. 20036. The manual is applicable to material covered by ASTM A666, "Austenitic Stainless Steel, Sheet, Strip, Plate and Flat Bars for Structural Applications," which contains requirements for 201, 202, 301, 302, 304 and 316 types of stainless steels. Further information on these steels as well as on stainless steels covered by ASTM A167, A176 and A412 may be obtained from AISI also and from the International Nickel Company, Inc. See also Art. 9-25.

Coatings. Material for cold-formed shapes may be either *black* (uncoated) or galvanized. Because of its higher cost, galvanized material is used only where exposure conditions warrant paying for the increased protection afforded against corrosion.

Low-carbon sheets coated with vitreous enamel are frequently used for facing purposes, but not as a rule to perform a load-carrying function in buildings.

Selection of Grade. The choice of a grade of material, within a given class or specification, usually depends on the severity of the forming operation required to make the desired shape, strength desired (see also Art. 9-10), weldability requirements, and the economics of the situation. Grade C of ASTM A611, with a specified minimum yield point of 33 ksi, has long been popular for structural use. Some manufacturers, however, use higher-strength grades to good advantage.

Gage Numbers. Thickness of cold-formed shapes is sometimes expressed as the manufacturers' standard gage number of the material from which the shapes are formed. *Use of decimal parts of an inch, instead of gage numbers, however, is preferable.* The relationships between gage number, weight, and thickness for uncoated and galvanized steel sheets are given in Table 9-4 for even gages.

9-10. Utilization of the Cold Work of Forming.

When strength alone, particularly yield strength, is an all-important consideration in selecting a material or grade for cold-formed shapes, it is sometimes possible to take advantage of the strength increase that results from cold-working the material during the forming operation and thus use a lower-strength, more workable, and possibly more economical grade than would otherwise be required. This increase in strength is ordinarily most noticeable in relatively stocky, compact sections produced in the thicker gages. Cold-formed chord sections for open-web steel joists (Figs. 9-2c and d) are good examples. Overall average yield strengths of more than 150% of the minimum specified yield strength of the plain material are obtained in such sections.

The strengthening effect of the forming operation varies across the section, but is most pronounced at the bends and corners of a cold-formed section. Accordingly, the overall strength of shapes in which bends and corners constitute a relatively large percentage of the whole section is increased more than that of shapes having a relatively large proportion of wide, flat elements that are not heavily worked in forming. For the

TABLE 9-4 Gages, Weights, and Thicknesses of Sheets

Steel manufacturer's standard gage No.	Weight, psf	Equivalent sheet thickness, in*	Galvanized sheet gage No.	Weight, psf	Thickness equivalent,† in
4	9.3750	0.2242			
6	8.1250	0.1943			
8	6.8750	0.1644	8	7.03125	0.1681
10	5.6250	0.1345	10	5.78125	0.1382
12	4.3750	0.1046	12	4.53125	0.1084
14	3.1250	0.0747	14	3.28125	0.0785
16	2.5000	0.0598	16	2.65625	0.0635
18	2.0000	0.0478	18	2.15625	0.0516
20	1.5000	0.0359	20	1.65625	0.0396
22	1.2500	0.0299	22	1.40625	0.0336
24	1.0000	0.0239	24	1.15625	0.0276
26	0.7500	0.0179	26	0.90625	0.0217
28	0.6250	0.0149	28	0.78125	0.0187
30	0.5000	0.0120	30	0.65625	0.0157
32	0.40625	0.0097	32	0.56250	0.0134
34	0.34375	0.0082			
36	0.28125	0.0067			
38	0.25000	0.0060			

*Thickness equivalents of steel are based on 0.023912 in/(lb-ft^2) (reciprocal of 41.820 psf per inch of thickness, although the density of steel is ordinarily taken as 489.6 lb/ft^3, 0.2833 lb/in^3, or 40.80 psf per inch of thickness). The density is adjusted because sheet weights are calculated for specified widths and lengths of sheets, with all shearing tolerances on the over side, and also because sheets are somewhat thicker at the center than at the edges. The adjustment yields a close approximation of the relationship between weight and thickness. ("Steel Products Manual, Carbon Steel Sheets," American Iron and Steel Institute.)

†Total thickness, in, including zinc coating. To obtain base metal thickness, deduct 0.0015 in per ounce coating class, or refer to ASTM A446.

latter type of shapes, the strength of the plain, unformed sheet or strip may be a controlling factor in the selection of a grade of material.

Full-section tests constitute a relatively simple, straightforward way of determining as-formed strength. They are particularly applicable to sections that do not contain any elements that may be subject to local buckling; that is, sections for which the form factor Q (defined in Art. 9-21) is unity. However, each case has to be considered individually in determining the extent to which cold-forming will produce an increase in utilizable strength. For further information, refer to the American Iron and Steel Institute "Specification for the Design of Cold-Formed Steel Structural Members" and its "Commentary."

9-11. Types of Cold-Formed Sections.

Many cold-formed shapes used for structural purposes are similar in their general configuration to hot-rolled structural sections. Channels, angles, and zees can be roll-formed in a single operation from one piece of material. I sections are usually made by welding two channels back to back or by welding two angles to a channel. All sections of this kind may be made with either plain flanges as in Fig. 9-6a to d, j, and m, or with flanges stiffened by lips at outer edges, as in Fig. 9-6e to h, k, and n.

In addition to these sections, which follow somewhat conventional lines and have their counterparts in hot-rolled structural sections, the flexibility of the forming process makes it relatively easy to obtain inverted U, or hat-shaped, sections and open box sections (Fig. 9-6o to q). These sections are very stiff in a lateral direction and can be used without lateral support where other more conventional types of sections would fail because of lateral instability.

Other special shapes are illustrated in Fig. 9-7. Some of these are nonstructural in nature; others are used for special structural purposes.

Figure 9-8 shows a few cold-formed stainless steel sections.

An important characteristic of cold-formed shapes is that thickness of section is substantially uniform. (A slight reduction in thickness may occur at bends but may be ignored for computing weights and section properties.) This means that, for a specified gage, the amount of flange material in a section, such as a channel, is almost entirely a function of the width of the section—except where additional flange has been obtained by doubling the material back on itself. Another distinguishing feature of cold-formed sections is that the corners are rounded on both the inside and outside of the bend, since the shapes are formed by bending flat material.

Sharp corners, such as can be obtained with hot-rolled structural channels, angles, and zees, cannot be obtained in cold-formed shapes, except with very soft material and in a coining or upsetting operation, as distinguished from a simple bending operation. This is not customary in the manufacture of structural cold-formed sections; and in proportioning such sections, the inside radius of bends should never be less, and should preferably be 33 to 100% greater, than specified for the relatively narrow ASTM bend-test specimens.

Deck and panel sections, such as are used for floors, roofs, and walls, are as a rule considerably wider, relative to their depth, than are the structural framing members shown in Figs. 9-6 to 9-8. Such shapes are illustrated and discussed in Arts. 9-33 to 9-43.

9-12. Design Principles for Cold-Formed Sections.

The structural behavior of cold-formed shapes follows the same laws of structural mechanics as does that of conventional structural-steel shapes and plates. Thus, design procedures commonly used in the selection of hot-rolled shapes are generally applicable to cold-formed sections. Although only a portion of a section, in some instances, may be considered structurally effective, computation of the structural properties of the effective portion follows conventional procedure.

Fig. 9-6 Typical cold-formed steel structural sections.

The uniform thickness of most cold-formed sections, and the fact that the widths of the various elements comprising such a section are usually large relative to the thickness, make it possible to consider, in computing structural properties (moment of inertia, section modulus, etc.), that such properties vary directly as the first power of the thickness. So in most cases, section properties can be approximated by first assuming that the sec-

Fig. 9-7 Miscellaneous cold-formed shapes. *(Bethlehem Steel Corp.)*

tion is made up of a series of line elements, omitting the thickness dimension, and then multiplying the result by the thickness to obtain the final value. With this method, the final multiplier is always the first power of the thickness, and first-power quantities such as radius of gyration and those locating the centroid of the section do not involve the thickness dimension. The assumption that the area, moment of inertia, and section modulus vary directly as the first power of the thickness is particularly useful in determining the required thickness of a section after the widths of the various elements that comprise it have been fixed. Although the method is sufficiently accurate for most practical purposes, it is well, particularly where the section is fairly thick relative to the widths of its elements, to check the final result through an exact method of computation.

Fig. 9-8 Cold-formed stainless steel sections. *(The International Nickel Co., Inc.)*

Properties of thin elements are given in Table 9-5.

One of the distinguishing characteristics of light-gage cold-formed sections is that they are usually composed of elements that are relatively wide and thin, and as a result, attention must be given to certain modes of structural behavior customarily neglected in dealing with heavier sections, such as hot-rolled structural shapes.

When wide thin elements are in axial compression—as in the case of a beam flange or a part of a column—they tend to buckle elastically at stresses below the yield point. This local elastic buckling is not to be confused with the general buckling that occurs in the failure of a long column or of a laterally unsupported beam. Rather local buckling represents failure of a single element of a section, and conceivably may be relatively unrelated to buckling of the entire member. In addition, there are other factors—such as shear lag, which gives rise to nonuniform stress distribution; torsional instability, which may be considerably more pronounced in thin sections than in thicker ones, requiring more attention to bracing; and other related structural phenomena customarily ignored in conventional structural work that sometimes must be considered with thin material. Means of taking care of these factors in ordinary structural design are described in the American Iron and Steel Institute (AISI) "Specification for the Design of Cold-Formed Steel Structural Members."

9-13. Structural Behavior of Flat Compression Elements.

It is convenient, in discussing buckling of plates or flat compression elements in beams, columns, or other structural members, to use **flat-width ratio** w/t, the ratio of the width

TABLE 9-5 Properties of Area and Line Elements

Area

$A = bt$
$I_x = \dfrac{b^3 t}{12}$
$I_y = \dfrac{bt^3}{12}$

Rectangle

Line

$A_l = b$
$I_{lx} = \dfrac{b^3}{12}$
$I_{ly} = 0$

$I_1 = \dfrac{bt}{12}(b^2 \sin^2 \theta + t^2 \cos^2 \theta)$
$I_2 = \dfrac{bt}{12}(b^2 \sin^2 \phi + t^2 \cos^2 \phi)$

Inclined Rectangle

$I_1 = \dfrac{Ln^2}{12}$
$I_2 = \dfrac{Lm^2}{12}$

$\beta = 2\alpha$

$A = \beta t \left(\dfrac{r_1 + r_2}{2}\right)$
$\bar{y} = \dfrac{2}{3A}(r_1^3 - r_2^3)\sin\dfrac{\beta}{2}$
$I_x = \tfrac{1}{8}(r_1^4 - r_2^4)(\beta + \sin\beta) - A\bar{y}^2$
$I_y = \tfrac{1}{8}(r_1^4 - r_2^4)(\beta - \sin\beta)$

$\beta = 2\alpha$
$L = R\beta$
$\bar{y} = 2R\,\dfrac{\sin\dfrac{\beta}{2}}{\beta}$
$I_x = R^3\left(\dfrac{\beta + \sin\beta}{2} - \dfrac{4\sin^2\dfrac{\beta}{2}}{\beta}\right)$
$I_y = \tfrac{1}{2}R^3(\beta - \sin\beta)$

Circular Arc

$A = \dfrac{\pi}{4}t(r_1 + r_2)$
$\quad = 0.7854t(r_1 + r_2)$
$\bar{y} = \dfrac{4}{3\pi}\dfrac{(r_1^3 - r_2^3)}{t(r_1 + r_2)}$
$\quad = 0.424\dfrac{(r_1^3 - r_2^3)}{t(r_1 + r_2)}$
$I_x = \dfrac{\pi}{16}(r_1^4 - r_2^4) - A\bar{y}^2$
$\quad = 0.1964(r_1^4 - r_2^4) - A\bar{y}^2$

$L = \dfrac{\pi R}{2}$
$\bar{y} = \dfrac{2}{\pi}R = 0.637R$
$I_x = 0.1488R^3$

For small radii:

r_2	A	$\bar{y}$	I_x	
$2t$	$3.927t^2$	$1.613t$	$2,549t^4$	
$1.5t$	$3.142t^2$	$1.300t$	$1.369t^4$	90° Circular Corner
t	$2.356t^2$	$0.990t$	$0.635t^4$	
$0.75t$	$1.963t^2$	$0.838t$	$0.400t^4$	
$0.5t$	$1.571t^2$	$0.690t$	$0.235t^4$	

w of a single flat element, in, exclusive of edge fillets, to the thickness t of the element, in (Fig. 9-9).

For structural design computations, flat compression elements of cold-formed structural members can be divided into two kinds—stiffened elements and unstiffened elements. **Stiffened compression elements** are flat compression elements; i.e., plane compression flanges of flexural members and plane webs and flanges of compression members, of which *both* edges parallel to the direction of stress are stiffened by a web, flange, stiffening lip, or the like (AISI "Specification for the Design of Cold-formed Steel Structural Members"). A flat element stiffened at only *one* edge parallel to the direction of stress is called an **unstiffened element**. If the sections shown in Fig. 9-9 are used as compression members, the webs are considered stiffened compression elements. The lips that stiffen the outer edges of the flanges are, in turn, unstiffened elements. Any section composed of a number of plane portions, or elements, can be broken down into a combination of stiffened and unstiffened elements.

Fig. 9-9 Compression elements.

The following requirements apply only to carbon and low-alloy steels.

Stiffener Requirements. According to the AISI Specification, in order that a compression element may qualify as a stiffened compression element, its edge stiffeners should comply with the following:

$$I_{min} = 1.83t^4 \sqrt{\left(\frac{w}{t}\right)^2 - \frac{4000}{F_y}} \quad \text{but not less than } 9.2t^4 \qquad (9\text{-}2)$$

where w/t = flat-width ratio of stiffened element
I_{min} = minimum allowable moment of inertia of stiffener (of any shape) about its own centroidal axis parallel to stiffened element, in^4
F_y = specified minimum yield point of the material, ksi

Where the stiffener consists of a simple lip bent at right angles to the stiffened element, the required overall depth d, in, of such a lip may be determined with satisfactory accuracy from the following formula:

$$d = 2.8t \sqrt[6]{\left(\frac{w}{t}\right)^2 - \frac{4000}{F_y}} \qquad (9\text{-}3)$$

but not less than $4.8t$. A simple lip should not be used as an edge stiffener for any element having a flat-width ratio greater than 60.

The values of I_{min} and d from Eqs. (9-2) and (9-3) are plotted in Fig. 9-10 for $F_y = 33$ and $F_y = 50$ ksi. The influence of F_y is negligible for w/t ratios above about 25 or 30. For practical reasons, in the thinner gages, the depth d of a simple lip will frequently be specified larger than the required minimum.

Multiple-Stiffened Elements. The discussion of stiffened compression elements in this section deals primarily with simple elements, such as those shown in Figs. 9-6 to 9-9. Multiple-stiffened elements, i.e., elements that have longitudinal stiffeners placed between webs or between a web and an edge, are sometimes encountered, particularly in deck and panel work. For methods of computing the effectiveness of such elements and for requirements for stiffeners for such elements, see the AISI "Specification for the Design of Cold-formed Steel Structural Members."

Buckling of Unstiffened Elements. Whether local buckling of a flat compression element must be considered depends on its flat-width ratio. For unstiffened elements, local buckling does not have to be considered unless the flat-width ratio of the elements exceeds $63.3/\sqrt{F_y}$. (See also Arts. 9-15 and 9-18.)

Buckling of Stiffened Elements. For stiffened elements, which behave in buckling in a substantially different manner from unstiffened elements, the flat-width ratio

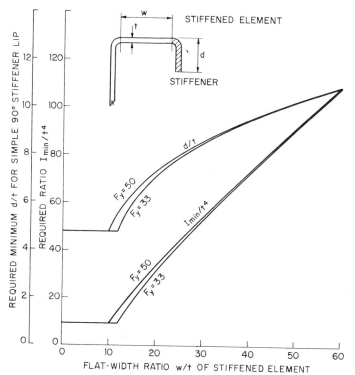

Fig. 9-10 Curves for determining minimum dimensions of edge stiffeners for stiffened compression elements.

beyond which allowance must be made for local buckling varies with the actual unit stress in the element. The effect of local buckling may be ignored for stiffened elements with flat-width ratios equal to or less than $(w/t)_{lim}$ in Eqs. (9-4) and (9-5). (These equations do not apply to the sides of square and rectangular structural tubing, for which a more liberal treatment is permissible. See the AISI Specification.)

For safe-load determination, i.e., in computing effective area and section modulus:

$$\left(\frac{w}{t}\right)_{lim} = \frac{171}{\sqrt{f}} \qquad (9\text{-}4)$$

where f = computed unit stress, ksi, in the element based upon effective width.

Equation (9-4) is based on a safety factor of about 1.67 against yield stress at the outer fiber of the section. For any other safety factor m, multiply the right-handed side of Eq. (9-4) by $\sqrt{1.67/m}$.

For deflection determinations, i.e., in computing moment of inertia to be used in deflection calculations or other calculations involving stiffness:

$$\left(\frac{w}{t}\right)_{\lim} = \frac{221}{\sqrt{f}} \tag{9-5}$$

See also Arts. 9-16 and 9-18.

Beam and Column Design. As long as the proportions of cold-formed sections are such that there are no unstiffened compression elements having net flat-width ratios greater than $63.3/\sqrt{F_y}$, and there are no stiffened compression elements having flat-width ratios greater than the limits established by Eq. (9-4), they can be treated, in their structural use as beams and columns, in the same general way as ordinary structural shapes, and without any modification to take into account local buckling. The maximum widths of flat compression elements that may be so used may be determined from the charts of Figs. 9-11 and 9-14.

When the above limits are exceeded, the treatment in Arts. 9-15 and 9-16 is recommended. (AISI, "Cold-formed Steel Design Manual.")

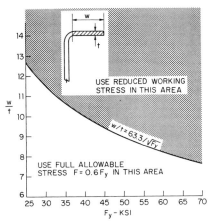

Fig. 9-11 Curve for determining maximum net width of unstiffened compression elements without stress reduction.

Curves and Corners. The theoretical buckling strength of a curved plate carrying axial compression transverse to the direction of curvature is many times that of a flat plate, even if the curvature is very slight. Hence, ordinary bends and corners in cold-formed shapes may be relied upon to carry any stress up to the full yield strength of the material. The possibility of local buckling in transversely curved plates in compression need not be of concern unless the curve is very shallow relative to the width and thickness of the element. Such cases are rarely encountered in load-carrying elements in buildings. For doubtful cases that have not been proven by experience or tests, refer to aircraft practice or to appropriate tests dealing with elastic stability. The discussion of compression elements in this section deals only with flat elements.

9-14. Unit Stresses for Cold-Formed Shapes.

The basic working stress F, ksi, in tension and bending for cold-formed sections made of sheet and strip carbon or low-alloy steels is usually specified as $F = 0.60F_y$, where F_y is the specified minimum yield point of the material, ksi. This corresponds to a safety factor of 1.67, applied to the yield point. As discussed in Art. 9-15, this basic stress must be reduced in dealing with wide unstiffened elements in compression to provide against local buckling.

An increase of 33⅓% in all allowable stresses is customary for combined wind, earthquake, and gravity loads, or in members subject only to the effect of wind and earthquake forces. (D. S. Ellifritt, "The Mysterious One-third Stress Increase," *Engineering Journal*, fourth quarter, 1977, American Institute of Steel Construction.)

9-15. Design of Unstiffened Elements Subject to Compression.

Unstiffened elements are defined in Art. 9-13. The following requirements apply only to carbon and low-alloy steels.

Compute section properties of the full section of compression members in the conventional manner, and use the basic working stress if the flat-width ratio w/t of any unstiffened element does not exceed $63.3/\sqrt{F_y}$, where F_y is the yield point, ksi, of the steel. (See Fig. 9-11.)

If w/t is greater than $63.3/\sqrt{F_y}$, the AISI "Specification for the Design of Cold-formed Steel Structural Members" prescribes that the allowable compression stress be reduced to a value F_c, ksi, given by Eqs. (9-6) to (9-8).

For the usual case in which F_y is 33 ksi or more, Eqs. (9-6) and (9-7) are applicable when the flat-width ratio w/t is less than 25.

For $63.3/\sqrt{F_y} < w/t \leq 144/\sqrt{F_y}$,

$$F_c = F_y \left[0.767 - 0.00264 \frac{w}{t} \sqrt{F_y} \right] \quad (9\text{-}6)$$

For $144/\sqrt{F_y} < w/t \leq 25$,

$$F_c = \frac{8000}{(w/t)^2} \quad (9\text{-}7)$$

In the relatively infrequent case in which F_y is less than 33 ksi, use straight-line interpolation between $F_c = F$ (basic working stress) at $w/t = 63.3\sqrt{F_y}$, and $F_c = 12.8$ ksi at $w/t = 25$.

For values of w/t from 25 to 60, inclusive, for all grades of steel,

$$F_c = 19.8 - 0.28 \frac{w}{t} \quad (9\text{-}8)$$

except that for angle struts (Eq. (9-7) applies).

These equations are plotted in Fig. 9-12 for yield points of 33 and 50 ksi. Note that Eq. (9-6) plots as a pair of straight lines drawn between $F_c = F$ when $w/t = 63.3/\sqrt{F_y}$,

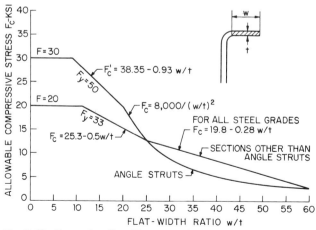

Fig. 9-12 Curves for allowable compressive stress in unstiffened elements for $F_y = 33$ ksi and $F_y = 50$ ksi.

TABLE 9-6 Allowable Compression Stresses in Unstiffened Elements

1	2	3	4	5	6
	For $w/t \leq 63.3/\sqrt{F_y}$		For $w/t \leq 144/\sqrt{F_y}$ or 25, whichever is smaller		
F_y, ksi	$63.3/\sqrt{F_y}$	$F_c = 0.60F_y$*	$144/\sqrt{F_y} \leq 25$	F_c†	For $w/t > 144/\sqrt{F_y}$ or 25
25	12.7	15	25	12.8	If w/t is greater than $144/\sqrt{F_y}$
30	11.5	18	25	12.8	(column 4) but less than 25,
33	11.0	20	25	12.8	$F_c = 8000/(w/t)^2$
36‡	10.6	22	24	13.9	
37	10.4	22	23.7	14.2	
40	10.0	24	22.8	15.4	
42	9.8	25	22.2	16.2	
45	9.4	27	21.5	17.3	For all flat-width ratios greater
50	9.0	30	20.4	19.2	than 25, use Eqs. (9-7) and
55	8.5	33	19.4	21.8	(9-8), as applicable, to
60	8.2	36	18.6	23.1	obtain F_c
65	7.8	39	17.9	25.0	
70	7.6	42	17.2	27.0	
75	7.3	45	16.6	29.0	
80§	7.1	48	16.1	30.9	

*Rounded to nearest whole ksi.
†$F_c = 8000/(w/t)^2$ for $w/t = 144/\sqrt{F_y}$ or 25. For w/t values between those in columns 2 and 4, F_c may be obtained by straight-line interpolation between columns 3 and 5.
‡Material of this yield point is not listed in the ASTM sheet and strip specifications summarized in Table 9-3. It is included here because it is a popular standard grade of structural-steel plates, shapes, and bars as covered by ASTM A36.
§Sheet steel with yield strength as high as 80 ksi may be used, with allowable working unit stress of 0.6 F_y, but deflection and stiffness may be limitations in designs to which such steel is applied.

and $F_c = 8000/(w/t)^2$ when $w/t = 144/\sqrt{F_y}$. Note also that, when $w/t \geq 25$, F_c is the same curve for all grades of steel.

These limiting values and the corresponding allowable compression stresses are given in Table 9-6. Straight-line interpolation may be used between them.

Selection of Flat-Width Ratios. For maximum economy of material in the use of unstiffened elements in compression, w/t should be held to a maximum of $63.3/\sqrt{F_y}$, so that the full basic working unit stress may be used. With Eqs. (9-6) to (9-8), the maximum capacity of an element (unit stress times cross-sectional area) occurs at about $w/t = 35$ for F_y below about 40 ksi, and at $w/t = 144/\sqrt{F_y}$ for higher-strength grades. The reason for this is that as w/t increases beyond these values the buckling stress decreases faster than the area increases.

Example. As an example of the application of the equations, consider the double-channel section shown in Fig. 9-13, which is to be used as a simply supported beam about axis xx with the top flange in compression. The two legs of that flange are unstiffened compression elements. The section has a section modulus of about $27t$, where t is the thickness of the material, in. It is to be made of ASTM A611, Grade C, uncoated steel, $F_y = 33$ ksi.

If the thickness is 0.120 in, then the flat-width ratio of the unstiffened top flange is $10.4 < (63.3/\sqrt{33} = 11.0)$. Thus, the section can be used at the basic working

Fig. 9-13 Double-channel beam.

stress of 20 ksi under gravity loads. Its section modulus is 27 × 0.120 = 3.2 in³, and its resisting moment is 3.2 × 20/12 = 5.3 ft-kips.

If the thickness is 0.060 in, then the flat-width ratio of the unstiffened compression flange becomes 20.8 < (144/$\sqrt{33}$ = 25). According to Eq. (9-6) or by interpolation in Table 9-6, the allowable stress must be reduced to 14.9 ksi. The section modulus is 27 × 0.060 = 1.6 in³, and the resisting moment of the section is 1.6 × 14.9/12 = 2.0 ft-kips.

If the thickness is further reduced to 0.03 in, then the flat-width ratio of the unstiffened compression elements becomes 41.7, and the allowable stress about 8 ksi. The section modulus is reduced to 0.81 in³, and the resisting moment is only 0.54 ft-kips.

The above example illustrates the desirability of holding w/t for unstiffened compression elements to a maximum value of $63.3/\sqrt{F_y}$, $w/t = 11$ in this case, unless larger values are required for reasons other than load-carrying capacity alone.

9-16. Stiffened Elements Subject to Compression.

As discussed in Art. 9-13, for carbon and low-alloy steels, local buckling of stiffened elements in compression cannot be ignored when the flat-width ratio w/t exceeds $171/\sqrt{f}$ for safe-load determination (Fig. 9-14), or $221/\sqrt{f}$ for deflection calculations. In such cases, compute section properties based on an *effective width* b, in, of each stiffened compression element (Fig. 9-15). Determine b by means of Eqs. (9-9) and (9-10).

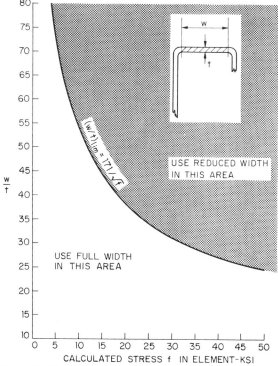

Fig. 9-14 Curve for determining maximum net width for full effectiveness of stiffened compression elements for load-carrying calculations.

Fig. 9-15 Effective width of stiffened compression elements.

For safe-load determinations, i.e., in computing effective area and section modulus

$$\frac{b}{t} = \frac{253}{\sqrt{f}}\left[1 - \frac{55.3}{(w/t)\sqrt{f}}\right] \tag{9-9}$$

where f = unit stress, ksi, in the element computed on basis of reduced section
  w = width of the element, in
  t = thickness, in

For deflection determinations, i.e., in computing moment of inertia to be used in deflection calculations or in other calculations involving stiffness,

$$\frac{b}{t} = \frac{326}{\sqrt{f}}\left[1 - \frac{71.3}{(w/t)\sqrt{f}}\right] \tag{9-10}$$

with f, w, and t the same as for Eq. (9-9). [Equations (9-9) and (9-10) do not apply to the sides of square or rectangular structural tubing, for which a more liberal treatment is permissible. See the AISI "Specification for the Design of Cold-formed Steel Structural Members."]

Stiffened Elements Subject to Compression 9-25

As shown in Fig. 9-15, that portion of the element not considered to be effective is assumed to be located symmetrically about the centerline of the element.

The curves of Figs. 9-16 and 9-17 were plotted from Eqs. (9-9) and (9-10), and may be used to determine the ratio b/t for different values of w/t and unit stress.

The effective width b in Eqs. (9-9) and (9-10) is dependent on the unit stress f, and since reduced-section properties are a function of effective width, the reverse is also true. In the general case of flexural members, therefore, determination of effective width by these formulas requires successive approximations: First, a unit stress is assumed for the compression flange. With that stress, b is computed. Next, the section modulus and unit stress in the compression flange are calculated for the reduced section. Then the process is repeated if the effective width so computed should be significantly different from that for the stress first assumed.

However, this somewhat cumbersome procedure can be avoided and the correct value of b/t obtained directly from the formulas when f is known or is held to a specified maximum allowable value (such as 20 ksi for flexural members of Grade C uncoated steel), and the neutral axis of the section is closer to the tension flange than the compression flange so that the stress in the compression flange always controls. The latter condition holds for symmetrical channels, zees, and I sections used as flexural members about their major axes, such as (e), (f), (k), and (n) in Fig. 9-6. For hat and box sections, of the proportions shown in (o) and (p) in Fig. 9-6, or unsymmetrical channels, zees, and I sections, the error introduced by basing the effective width of the compression flange on an f value equal to the basic working stress F will generally be negligible, even though the neutral axis is above the geometric centerline. For wide, inverted, pan-shaped sec-

Fig. 9-16 Curves for determining effective width of stiffened compression elements for safe-load calculations (not applicable to sides of square and rectangular tubing).

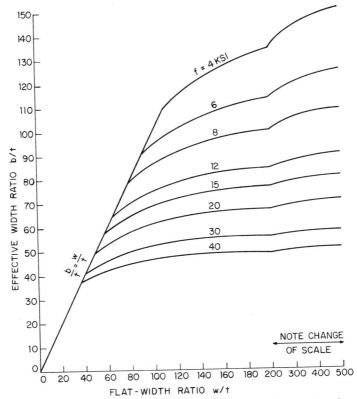

Fig. 9-17 Curves for determining effective width of stiffened compression elements for deflection calculations (not applicable to sides of square and rectangular tubing).

tions, such as deck and panel sections, a somewhat more accurate determination will frequently prove desirable; but extreme accuracy in determining the effective width of the compression flange is not essential, since relatively large variations in effective width will have relatively little effect on section modulus and moment of inertia. Even for sections of this kind, a good first approximation can be made by basing b/t on the basic working stress F, and two approximations will almost always suffice.

In computing moment of inertia for deflection or stiffness calculations, properties of the full unreduced section can be used without significant error when w/t of the compression elements exceeds $(w/t)_{\text{lim}}$ but does not exceed about 60 or 70.

Example. As an example of effective-width determination, load-carrying capacity is to be computed for the hat section in Fig. 9-18a, which is to be made of ASTM A611, Grade C, uncoated steel. This shape is to be used as a simply supported beam with the top flange in compression, at a basic working stress of 20 ksi.

The top flange, 3 in wide, is a stiffened compression element. If the thickness is ⅛ in, the flat-width ratio is 24. For $f = 20$ ksi, $(w/t)_{\text{lim}}$, according to Eq. (9-4), is 38.2 > 24. Therefore, properties of the section may be computed in the usual manner, with the entire section considered structurally effective.

If, however, the thickness is ¹⁄₁₆ in, the flat-width ratio becomes 48, and Eq. (9-9) applies. For this value of w/t and $f = 20$ ksi, Eq. (9-9) gives $b/t = 42$. Thus, only 87½%

of the top flange is considered effective. The neutral axis of the section will lie below the horizontal centerline; so the compression-flange stress will control. Since that stress is limited to 20 ksi, the effective width may be accurately determined from Eq. (9-9) for 20 ksi.

For the section in Fig. 9-18b, in which the horizontal centroidal axis will be much nearer the compression than the tension flange, the stress in the tension flange controls.

Fig. 9-18 Hat sections.

Determination of the unit stress in the compression flange and the effective width of that flange requires a trial-and-error procedure.

9-17. Light-Gage Members with Unusually Wide Flanges on Short Spans.

The phenomenon known as *shear lag*, resulting in nonuniform stress distribution in the flanges of a flexural member, requires recognition only in comparatively extreme cases of concentrated loads on spans that are short relative to the width of the member. The effects of shear lag can be taken care of by an effective-width procedure applicable to both tension and compression flanges. The AISI "Specification for the Design of Cold-formed Steel Structural Members" provides methods for handling this situation. This specification should be consulted in any case of concentrated loading in which the span of a beam is less than 30 times the half width of flange for I-beam and similar sections or 30 times half the distance between webs for box or U-type sections.

9-18. Maximum Flat-Width Ratios for Light-Gage Elements.

When the flat-width ratio exceeds about 30 for an unstiffened element and 250 for a stiffened element of carbon or low-alloy steel, noticeable buckling of the element is likely to develop at relatively low stresses. Present-day practice is to permit buckles to develop in the sheet and to take advantage of what is known as the post-buckling strength of the section. The effective-width formulas [Eqs. (9-9) and (9-10)] are based on this practice. However, to avoid excessive deformations, overall flat-width ratios, disregarding intermediate stiffeners and based on the actual thickness of the element, should not exceed the following:

Stiffened compression element having one longitudinal edge connected to a web or flange, the other to a simple right-angle lip	60
Stiffened compression element having both edges stiffened by other stiffening means than a simple right-angle lip	90
Stiffened compression element with both longitudinal edges connected to a web or flange element such as in a hat, U, or box type of section	500
Unstiffened compression elements	60

In special kinds of panel work, where the flat compression element may be stiffened by insulation board or other similar material fastened to it with an adhesive, elastic buckling does not occur with the same freedom as without such stiffening. The extent to which the stiffening effect of collateral materials may be utilized has to be determined by tests in each case. It will further depend on the extent to which the adhesive can be relied on to furnish continuous permanent bond between the two materials, and whether or not such bond is likely to be affected by fire or other external actions. In general, it is unwise to rely on collateral materials to prevent elastic buckling in the compression element.

9-19. Laterally Unsupported Light-Gage Beams.

In the relatively infrequent cases in which light-gage cold-formed sections used as beams are not laterally supported at frequent intervals, the unit stress must be reduced to avoid failure from lateral instability. The amount of reduction depends on the shape and proportions of the section and the spacing of lateral supports. For details, see the AISI "Specification for the Design of Cold-formed Steel Structural Members."

Because of the torsional flexibility of light-gage channel and zee sections, their use as beams without lateral support is not recommended. In the usual case in which one flange of such a section is connected to a deck or sheathing material, bracing of the other flange may or may not be needed to prevent twisting of the member, depending on: (1) whether or not the deck or sheathing material and its connection to the beam will effectively restrain the connected beam flange from lateral deflection and twist, (2) the dimensions of the member and the span, and (3) whether the unbraced flange is in compression or tension. In any doubtful case, a load test should be made to determine whether additional bracing is necessary. When bracing against twist is required, means of proportioning such bracing can be found in the AISI Specification. Bracing details will depend on the particular form of the sections involved. Bridging, such as is used with open-web steel-joist construction (Art. 9-4), can be effectively used.

In any case in which laterally unsupported beams must be used, or where lateral buckling of a flexural member is likely to be a problem, consideration should be given to use of relatively bulky sections having two webs, such as hat or box sections, as illustrated in Fig. 9-6o and p. Sections for which the moment of inertia about the vertical y axis is equal to or greater than that about the horizontal x axis do not fail because of lateral instability; but as the ratio I_y/I_x decreases, the section becomes more susceptible to failure by lateral buckling. Thus, double-web sections almost always are considerably more stable laterally than single-web sections of normal proportions. Hat sections (Fig. 9-6o) are especially suitable where lateral stiffness is necessary, and when the flat-width ratios of the individual elements are low enough to be fully effective structurally (Art. 9-13), design of the section is simple and straightforward. Sections of the kind shown in Fig. 9-7b having a top width of 2 in, depth of 2½ in, and bottom flanges ⅝ in wide have been successfully used under floor loads without any lateral support on spans up to 7 ft, or 42 times the top width.

Even with hat sections, extreme proportions should be avoided. With very high narrow sections made of thin material, lateral instability may result from lack of stiffness in the webs, although the flat-width ratio of the compression flange may be low. It is desirable also to keep the flat-width ratio of the bottom flanges relatively low to avoid the necessity of using a low allowable stress (Art. 9-15) when the section is to be used as a continuous beam and the bottom flanges are in compression over the interior supports.

In pan construction, consisting of inverted hat sections or U sections, in which the narrow flanges are in compression, the lateral stability of the compression flanges depends on the rigidity of the webs and the bottom, or tension, flange. An approximate

TABLE 9-7 Allowable Shear Stresses in Webs, ksi, and Maximum Depth-Thickness Ratios*

1	2	3	4	5	6
	For $F_v \leq [0.4F_y =$ $(2/3)F]$		For F_v when $h/t \leq (547/\sqrt{F_y} = 424/\sqrt{F})$		
F_y	$0.4F_y$	Max. h/t	F_v	$h/t = 547/\sqrt{F_y}$	
25	10	76.0	6.9	109.4	For h/t less than the values in
30	12	69.4	8.3	99.9	column 3, use F_v from
33	13.2	66.2	9.2	95.2	column 2
37	14.8	62.5	10.3	89.9	
40	16	60.1	11.1	86.5	For h/t between the values
42	16.8	58.6	11.7	84.4	in columns 3 and 5, $F_v =$
45	18	56.7	12.5	81.5	$152\sqrt{F_y}/(h/t) = 118\sqrt{F}/(h/t)$
50	20	53.7	13.9	77.4	
55	22	51.2	15.3	73.8	For h/t greater than the values in
60	24	49.1	16.7	70.6	columns 5, $F_v = 83{,}200/(h/t)^2$
65	26	47.1	18.1	67.9	
70	28	45.4	19.4	65.4	
75	30	43.8	20.8	63.2	
80	32	42.5	22.2	61.2	

*F_v = allowable shear stress, ksi
F_y = minimum specified yield point, ksi
F = basic working stress in tension and bending, ksi
h = clear distance between flanges, in
t = thickness of web, in. Where the web consists of two sheets, as in the case of two channels fastened back to back to form an I section, each sheet should be considered a separate web carrying its share of the shear
The values of F_v in this table are based upon the basic working stress values F_c from Table 9-6.

method of handling this case analytically is given in the AISI "Cold-formed Steel Design Manual," or tests may be made.

9-20. Web Stresses in Cold-Formed Shapes.

Allowable shear stresses on the gross section of webs of cold-formed carbon or low-alloy steel shapes may be assigned the same values, relative to yield point, as for structural steel. Since web stiffeners are not ordinarily used in such shapes, except possibly at reactions and points of concentrated loading, allowable shear stresses may be expressed in somewhat simpler terms than for structural-steel girders, where web stiffeners are customary. Shear stresses allowable under the AISI "Specification for the Design of Cold-formed Steel Structural Members" are summarized in Table 9-7. For any case in which intermediate stiffeners are used, as in plate girders, web design may be handled in the same way as for structural steel.

The use of unstiffened webs in which the depth-thickness ratio h/t exceeds 150 is not recommended unless stiffeners are provided at reactions and under concentrated loads. In such cases, h/t should not exceed 200.

Combined Shear and Bending. The buckling effect of combined shear and bending stresses in beam webs will rarely be critical. However, it may be when high h/t values and high-strength steel are being used, with high working stresses, and then only where high shear and high bending stresses occur at the same section. An interaction formula for investigating such a case can be found in the AISI Specification.

TABLE 9-8 Maximum Reaction or Concentrated Load on Single Unreinforced Webs of Cold-Formed Steel Sections*

[ASTM A611, Grade C, Uncoated Steel (F_y = 33 ksi), Corner Radii = Thickness of Material]

Gage and thickness	H, in	Max. end reaction or concentrated load on end of cantilever P_{max} (1), kips				Max. interior reaction or concentrated load P_{max} (2), kips			
		Length of bearing, B, in				Length of bearing, B, in			
		1	2	4	8	1	2	4	8
No. 18, 0.0478 in	1	0.40				0.78			
	2	0.38	0.53			0.74	0.83		
	4	0.34	0.45	0.67		0.68	0.75	0.90	
	6	0.29	0.36	0.50	0.77	0.61	0.67	0.78	1.00
No. 16, 0.0598 in	1	0.57				1.18			
	2	0.55	0.75			1.15	1.26		
	4	0.51	0.68	0.99		1.08	1.18	1.37	
	6	0.47	0.59	0.83		1.00	1.08	1.25	
	8	0.42	0.50	0.66	0.95	0.92	0.99	1.12	1.38
No. 14, 0.0747 in	2	0.82	1.08			1.79	1.92		
	4	0.77	1.00	1.43		1.69	1.83	2.09	
	6	0.73	0.91	1.28		1.60	1.72	1.96	
	8	0.68	0.82	1.10	1.64	1.51	1.61	1.81	2.20
	10	0.64	0.73	0.93	1.29	1.42	1.50	1.67	1.98
No. 12, 0.1046 in	4	1.42	1.78	2.47		3.34	3.55	3.95	
	6	1.38	1.69	2.31		3.22	3.41	3.79	
	8	1.33	1.60	2.19	3.21	3.10	3.27	3.61	4.29
	10	1.28	1.51	1.96	2.85	2.98	3.13	3.43	4.04
	12	1.24	1.42	1.78	2.46	2.85	2.99	3.26	3.80
No. 10, 0.1345 in	4	2.25	2.73	3.62		5.54	5.82	6.40	
	6	2.20	2.64	3.52		5.39	5.65	6.16	
	8	2.16	2.55	3.34	4.81	5.24	5.48	5.96	6.85
	10	2.11	2.46	3.16	4.46	5.08	5.31	5.75	6.64
	12	2.06	2.37	2.98	4.11	4.93	5.14	5.54	6.36

*Values in this table were computed from the following equations:

TABLE 9-8 Maximum Reaction or Concentrated Load on Single Unreinforced Webs of Cold-Formed Steel Sections (*Continued*)

$$P_{max}(1) = 0.1t^2 \left(980 + 42\frac{B}{t} - 0.22\frac{B}{t}\frac{h}{t} - 0.11\frac{h}{t} \right) \quad (9\text{-}11)$$

$$P_{max}(2) = 0.1t^2 \left(3050 + 23\frac{B}{t} - 0.09\frac{B}{t}\frac{h}{t} - 5\frac{h}{t} \right) \quad (9\text{-}12)$$

where t = web thickness
  h = clear distance between flanges
  B = length of bearing
  H = overall depth

In solution of these equations, B should not be assigned any values greater than h.

Equation (9-12) and corresponding values in the table for interior reactions and concentrated loads apply only where the distance x exceeds $1.5h$. Otherwise, Eq. (9-11) and corresponding values for end reactions and concentrated loads on ends of cantilevers govern.

Web Crippling. Empirical formulas for design against web crippling also can be found in this specification. Those formulas are summarized in Tables 9-8 and 9-9 for ASTM A611 Grade C uncoated steel and for sections where inside-corner radii are equal to the thickness of the material. Care must be taken not to extend the formulas beyond the tabular ranges of B/t. Table 9-8 applies to all cases of unreinforced webs, including hat sections as well as channels and zees. Table 9-9 applies only where the web is backed up by an opposing web or portion of a web, such as in an I section made by welding two angles to a channel.

For other grades of steel or other corner radii adjust the tabular values as indicated by Eqs. (9-15a) and (9-15b):

Multiply the values for $P_{max}(1)$ in Table 9-8 by the quantity

$$k \left(1.15 - 0.15\frac{r}{t} \right)(1.33 - 0.33k) \quad (9\text{-}15a)$$

and the values for $P_{max}(2)$ in Table 9-8 by the quantity

$$k \left(1.06 - 0.06\frac{r}{t} \right)(1.22 - 0.22k) \quad (9\text{-}15b)$$

Multiply the values in Table 9-9 by the ratio of 20 to the basic working stress, ksi, or by $33/F_y$, where F_y is the yield point of the steel, ksi.

In Eqs. (9-15a) and (9-15b), k = specified minimum yield point, ksi, divided by 33; t = web thickness, in; r = inside corner radius, in, if not greater than $4t$. For larger corner radii the web crippling strength should be determined by test.

9-21. Cold-Formed Steel Columns.

When cold-formed carbon or low-alloy steel sections are used as compression members, conventional procedures may be followed without modification, except when a section falls into either of the following categories:

Class I. Sections containing elements that exceed the limits $w/t = 63.3/\sqrt{F_y}$ for unstiffened elements, and $w/t = 171/\sqrt{F} = 221/\sqrt{F_y}$ for stiffened elements, where F = basic working stress, ksi, and F_y = yield point, ksi, of the steel. When either of these limits is exceeded, requiring that provision be made against failure by local buckling, the reduced strength of the section can be taken into account by introducing a form factor or buckling factor into the column design formula. This is done merely by using QF_y instead of F_y in the basic formula, where the factor Q is determined as outlined later.

TABLE 9-9 Maximum Reactions and Concentrated Loads Bearing on Restrained Webs of Cold-Formed Steel*

[ASTM A611, Grade C, Uncoated Steel (F_y = 33), Corner Radii = Thickness of Material]

	End bearing, P_{max} (3)				Interior bearing, P_{max} (4)			
	Length of bearing, B, in				Length of bearing, B, in			
Gage and thickness	1	2	4	8	1	2	4	8
No. 18, 0.0478 in	0.53	0.61	0.72		1.01	1.22	1.51	
No. 16, 0.0598 in	0.80	0.91	1.08		1.50	1.79	2.20	
No. 14, 0.0747 in	1.20	1.37	1.59		2.22	2.63	3.21	
No. 12, 0.1046 in	2.24	2.51	2.88	3.40	4.06	4.73	5.69	7.04
No. 10, 0.1345 in	3.60	3.98	4.51	5.28	6.38	7.38	8.77	10.74

*Values in this table were computed from the following equations:

$$P_{max}(3) = 20t^2 \left(7.4 + 0.93 \sqrt{\frac{B}{t}} \right) \tag{9-13}$$

$$P_{max}(4) = 20t^2 \left(11.1 + 2.41 \sqrt{\frac{B}{t}} \right) \tag{9-14}$$

where t = web thickness (each web sheet)
h = clear distance between flanges
Values are in kips for one web thickness t. Multiply by 2 for double-thickness webs.
Values of P_{max} (4) apply only where the distance x is greater than $1.5h$. Otherwise, use P_{max} (3).
The effective length of bearing B for substitution in the above formulas should not be taken as greater than h, and tabular values of P for values of B greater than h should not be used.

Class II. Open sections whose shear center and centroid do not coincide, and which are not braced against twisting. Under these conditions, because of low torsional resistance, failure of a thin-walled section may occur by torsional-flexural buckling instead of by simple column action. Hence, the torsional-flexural strength should be investigated as outlined later. This precaution applies to such sections as channels, hat sections, angles, tees, unsymmetrical I sections, and any other singly symmetrical or nonsymmetrical sections, if twisting can occur. It is not necessary for doubly symmetrical shapes, such as symmetrical I sections; closed sections, such as box or tubular sections; point-symmetric shapes such as equal-flange zee sections; or for any section which is held against twisting throughout its length, as by the attachment of wall sheathing.

Bear in mind, however, that members braced against twisting in the finished structure may not be so braced during construction.

The safe-load formulas recommended by the AISI "Specification for the Design of Cold-formed Steel Structural Members" for concentrically loaded columns of carbon or low-alloy steel are as follows:

A. Sections that do not fall into either Class I or II, i.e., sections not subject to either local buckling or twisting, and are at least 0.09 in thick:

Same as for structural steel (see Arts. 8-21 and 8-22)

B. All sections that are less than 0.09 in thick and all sections, regardless of thickness, that fall into Class I but do not fall into Class II, i.e., sections not subject to twisting but containing elements that may be subject to local buckling ($Q < 1.0$):

For $Kl/r < (\sqrt{2\pi^2 E/QF_y} = 763/\sqrt{QF_y})$,

$$\frac{P}{A} = 0.522 Q F_y - \left(\frac{QF_y}{1{,}494}\right)^2 \left(\frac{Kl}{r}\right)^2 \tag{9-16}$$

where P = total allowable axial load, kips
A = full, unreduced cross-sectional area of member, in^2
F_y = specified minimum yield point, ksi
Kl = effective unsupported length of member, in. (For determination of the effective length factor K, follow structural-steel practice or refer to the AISI "Specification for the Design of Cold-formed Steel Structural Members" and its "Commentary.")
r = radius of gyration of full, unreduced cross section, in
Q = a factor determined as described later

For $Kl/r \geq 763/\sqrt{QF_y}$,

$$\frac{P}{A} = \frac{12\pi^2 E}{23(Kl/r)^2} = \frac{151{,}900}{(Kl/r)^2} \tag{9-17}$$

as for structural steel columns.

Equation (9-16) represents a family of curves beginning at $P/A = 0.522QF_y$ for $Kl/r = 0$ and tangent to the Euler curve of Eq. (9-17) at $Kl/r = 763/\sqrt{QF_y}$, $P/A = 0.261QF_y$ (Fig. 9-19). An infinite number of curves may be drawn, each for a different value of QF_y. Equation (9-16) reduces to Eq. (9-18) for ASTM A611, Grade C, uncoated steel ($F_y = 33$ ksi), and to Eq. (9-19) for a high-strength low-alloy steel for which $F_y = 50$ ksi.

For $F_y = 33$ and $Kl/r < 133/\sqrt{Q}$,

$$\frac{P}{A} = 17.2 Q - 0.000488 Q^2 \left(\frac{Kl}{r}\right)^2 \tag{9-18}$$

For $F_y = 50$ ksi and $Kl/r < 108/\sqrt{Q}$,

$$\frac{P}{A} = 26.1 Q - 0.00112 Q^2 \left(\frac{Kl}{r}\right)^2 \tag{9-19}$$

For Kl/r greater than the limiting values for Eqs. (9-18) and (9-19), for all grades of steel, and all values of Q, use Eq. (9-17).

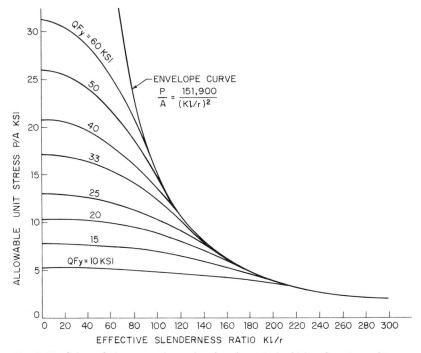

Fig. 9-19 Column design curves for sections less than 0.09 in thick and sections subject to local buckling ($Q < 1$) but not twisting (group B sections).

C. Sections that fall into Class II:
1. For singly symmetrical shapes (shapes symmetrical about only one axis) that do not also fall into Class I; that is, shapes for which $Q = 1.0$, the AISI Specification requires that P/A be the smaller of the following:
   a. P/A determined as for sections in group A or B, as applicable.
   b. A function of the torsional-flexural buckling stress σ_{TFO}:
   If $\sigma_{TFO} > 0.5F_y$,

$$\frac{P}{A} = 0.522F_y - \frac{F_y^2}{7.67\sigma_{TFO}} \qquad (9\text{-}20)$$

If $\sigma_{TFO} \leq 0.5F_y$,

$$\frac{P}{A} = 0.522\sigma_{TFO} \qquad (9\text{-}21)$$

Determination of σ_{TFO} is a complex procedure that involves the location of the shear center, the polar radius of gyration about the shear center, and the torsion and warping constants of the section. For thin-walled shapes of uniform thickness, these quantities are not mathematically difficult to determine, but the process is likely to prove tedious and laborious unless computerized. But the mere fact that a section or component falls in Class II does not necessarily mean that the torsional-flexural mode will control, i.e., that Eqs. (9-20) and (9-21) will yield a lower allowable stress than that given by the regular column formula. For a few common shapes of normal proportions, a preliminary appraisal of whether the torsional-flexural buckling stress may control the design can be made by means of the following approximate but conservative criteria (adapted from "Torsional-Flexural Buckling, Elastic and Inelastic, of Cold-Formed Thin-Walled Columns," by Alexander Chajes, Pen J. Fang, and George Winter, *Cornell Engineering Research Bulletin* 66-1, Cornell University, August, 1966, and the AISI "Cold-Formed Steel Design Manual"):

For plain equal-leg angles (Fig. 9-20a), torsional-flexural buckling need not be investigated if

$$\frac{t}{a^2} Kl \geq 1.1 \tag{9-22}$$

where t = thickness of material, in
a = leg width, in, as shown in Fig. 9-20a
Kl = effective length, in, between twist-restraining attachments (for open, singly symmetrical or nonsymmetrical components of built-up sections, the effective length between battens, diaphragms, or other bracing)

For equal-leg lipped angles for which $c/a \leq 0.4$ (Fig. 9-20b), torsional-flexural buckling need not be investigated if

$$\frac{t}{a^2} Kl \geq 1.1 + \frac{3c}{a} \tag{9-23}$$

For symmetrical channels and hat sections, see Figs. 9-21 to 9-23.

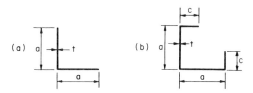

Fig. 9-20 Equal-legged angles.

Torsional-flexural buckling should be investigated if $(t/a^2)Kl$ is less than the values given by Eqs. (9-22) and (9-23), or falls below and to the right of the curves in Figs. 9-21 to 9-23; if, for lipped angles (Fig. 9-20b), c/a is greater than 0.4; or if, for hat sections (Fig. 9-23), c/a is less than 0.2 or greater than 0.8. More precise methods of determining the failure mode, and methods of calculating the torsional-flexural buckling stress, are available in the AISI Specification and Design Manual.

2. For singly symmetrical shapes that also fall into Class I, i.e., sections for which Q is less than 1.0, either replace F_y by QF_y in Eqs. (9-20) and (9-21), or use tests to determine the critical load. General rules relating to such tests are given in the AISI Specification.
3. For nonsymmetric shapes, i.e., shapes that are not symmetric about any point or axis, the AISI Specification recommends either analysis or tests. A general method of analysis is given in the AISI Design Manual.

Note that torsional-flexural buckling must be considered for sections in group C only if a member or component is not restrained against twisting over any appreciable proportion of its length. The amount of restraint required to prevent twisting is small, perhaps of the same order of magnitude as that required to brace a column or to keep a wall stud from buckling in the plane of the wall. In any unproven or doubtful case, however, tests are advisable.

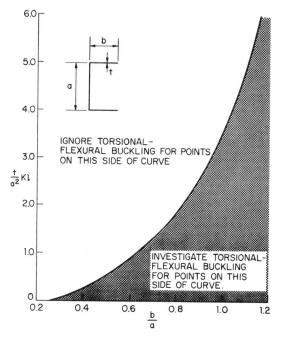

Fig. 9-21 Failure mode for plain channels under concentric compressive load.

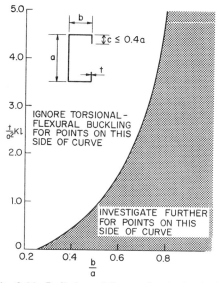

Fig. 9-22 Preliminary failure-mode criterion for lipped channels under concentric compressive load for $c/a \leq 0.4$.

For bracing and secondary members with l/r greater than 120, the AISI Specification recommends that the allowable P/A value determined as for sections in group A, B, or C be increased to

$$\left(\frac{P}{A}\right)_s = \frac{P/A}{1.3 - (l/r)/400} \quad (9\text{-}24)$$

where l is the actual unbraced length of the member. (The effective-length factor K is taken as unity for secondary members.) Kl/r should not exceed 200, except that during construction a value of 300 may be temporarily allowed.

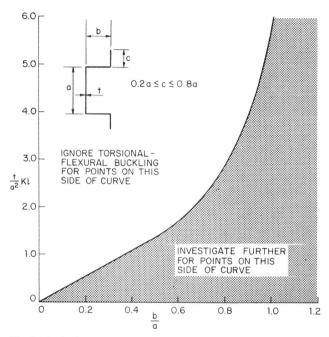

Fig. 9-23 Preliminary failure-mode criterion for hat sections under concentric compressive load for values of c/a between 0.2 and 0.8.

Determination of Q. The form factor, or buckling factor, Q can never be greater than 1.0. For sections that do not contain any elements that exceed the lower w/t limits set for Class I, $Q = 1.0$ and disappears from the equation. For sections that do contain such elements, Q is determined as follows:

a. For members composed entirely of stiffened elements, Q is the ratio between the effective design area, as determined from the effective design widths of such elements, and the full or gross area of the cross section. The effective design area used in determining Q is to be based on the basic design stress F allowed in tension and bending.

b. For members composed entirely of unstiffened elements, Q is the ratio between the allowable compression stress for the weakest element of the cross section (the element having the largest flat-width ratio) and the basic design stress.

c. For members composed of both stiffened and unstiffened elements, the factor Q is to be the product of a stress factor Q_s computed as outlined in b above and an area factor

Q_a computed as outlined in a above. However, the stress on which Q_a is to be based shall be that value of the unit stress f_c used in computing Q_s; and the effective area to be used in computing Q_a shall include the full area of all unstiffened elements.

9-22. Combined Axial and Bending Stresses.

For eccentrically loaded members of carbon or low-alloy steel in groups A and B of Art. 9-21, i.e., members not subject to torsional-flexural buckling, the procedure for dealing with combined axial and flexural loads is substantially the same as for structural steel, and the same types of interaction equations may be used. The AISI "Specification for the Design of Cold-formed Steel Structural Members" prescribes the following criteria for compressive axial loading:

$$\frac{f_a}{F_{a1}} + \frac{C_{mx}f_{bx}}{\left(1 - \frac{f_a}{F'_{ex}}\right)F_{bx}} + \frac{C_{my}f_{by}}{\left(1 - \frac{f_a}{F'_{ey}}\right)F_{by}} \leq 1.0 \tag{9-25}$$

$$\frac{f_a}{F_{ao}} + \frac{f_{bx}}{F_{b1x}} + \frac{f_{by}}{F_{b1y}} \leq 1.0 \tag{9-26}$$

When $f_a/F_{a1} \leq 0.15$, Eq. (9-27) may be used instead of Eqs. (9-25) and (9-26).

$$\frac{f_a}{F_{a1}} + \frac{f_{bx}}{F_{bx}} + \frac{f_{by}}{F_{by}} \leq 1.0 \tag{9-27}$$

where F_{a1} = allowable axial stress P/A under concentric loading only
F_{ao} = allowable axial stress under concentric loading when effective column length $Kl = 0$
F_{b1x}, F_{b1y} = allowable bending stress in compression, about axis x and y, respectively when the possibility of lateral buckling is excluded. For sections containing unstiffened compression elements, F_{b1x} and F_{b1y} are limited to the allowable stresses in such elements as recommended in Art. 9-15
f_a = computed stress under concentric loading only
f_{bx}, f_{by} = computed bending stresses about axis x and y, respectively
C_m = reduction factor defined in Art. 8-27

Note that for members with unstiffened compression elements the bending stresses f_{bx} and f_{by} are to be computed on the basis of the effective width of such elements, as discussed in Art. 9-16.

For handling combined axial and bending loads on eccentrically loaded members in group C (Class II of Art. 9-21, i.e., members that may be subject to failure by torsional-flexural buckling), the procedure is more complex. The AISI "Cold-Formed Steel Design Manual" should be consulted in such cases.

9-23. Cylindrical Tubular Members.

Closed structural members are effective in resisting torques, and round tubes are the most efficient in this respect. When carbon or low-alloy steel tubes are used in compression or bending, local buckling will not occur at stresses below the yield point if

$$D/t \leq 3300/F_y \tag{9-28}$$

where D = mean diameter of tube, in
t = base metal thickness, in
F_y = yield point, ksi

Compression unit stress should not exceed the basic design stress F. Between $D/t = 3300/F_y$ and $D/t = 13,000/F_y$ the following design stress should be used:

$$F_r = 662t/D + 0.399F_y \qquad (9\text{-}29)$$

The allowable stress P/A for compression members, however, must not exceed values obtained from Eq. (9-16) or (9-17) with Q taken as unity.

The preceding recommendations for round tubes presume no utilization of strengthening through cold work of forming. However, welded steel tubing is ordinarily used with considerable amounts of unrelieved cold work, which tends to strengthen the steel. For more information on this subject, see D. S. Wolford and M. J. Rebholz, "Beam and Column Tests of Welded Steel Tubing with Design Recommendations," Bulletin No. 233, October 1958, American Society for Testing and Materials. The general subject of cylinders and tubes is broadly reviewed by D. R. Sherman in "Tentative Criteria for Structural Application of Steel Tubing and Pipe," Publication SP604-876-7.5M-MP, August 1976, American Iron and Steel Institute.

9-24. Corrugated Sheets.

These are the oldest type of cold-formed steel structural members. They have been used since 1784, when Henry Cort introduced sheet rolling in England.

The commonest form of corrugated sheet, the arc-and-tangent type, has the basic cross section shown in Fig. 9-24a. Its section properties are readily calculated with factors taken from Fig. 9-24b to f and substituted in the following formulas.

The area, in², of the corrugated sheet may be determined from

$$A = \lambda bt \qquad (9\text{-}30)$$

where b = width of sheet, in
t = sheet thickness, in
$\lambda = \dfrac{(2/K + \alpha)\sin\alpha + (1 - 2\alpha/K)\cos\alpha - 1}{1 - \cos\alpha}$ (See Fig. 9-24d)
K = pitch-depth ratio of a corrugation = p/d
p = pitch, in, of corrugation
d = depth, in, of corrugation
α = tangent angle, radians, or angle of web with respect to the neutral axis of the sheet cross section

The moment of inertia, in⁴, of the corrugated sheet may be obtained from

$$I = C_5 bt^3 + C_6 bd^2 t \qquad (9\text{-}31)$$

where $C_5 = \dfrac{q(6\alpha + \sin 2\alpha - 8\sin\alpha) + 4\sin\alpha + K\cos\alpha}{12K}$ (See Fig. 9-24b)

$C_6 = \dfrac{1}{K}\left[q^3\left(6\alpha + \sin 2\alpha - 8\sin\alpha - \dfrac{4}{3}\tan^3\alpha \sin^2\alpha \right) \right.$

$+ q^2(4\sin\alpha + K\tan^3\alpha \sin\alpha - 4\alpha)$

$\left. + q\left(\alpha - \dfrac{1}{4}K^2\tan^3\alpha\right) + \dfrac{K^3\tan^2\alpha}{48\cos\alpha}\right]$ (See Fig. 9-24c)

$q = \dfrac{r}{d} = \dfrac{K\tan\alpha - 2}{4(\sec\alpha - 1)}$ (See Fig. 9-24e)

The section modulus of the corrugated sheet may be computed from

$$S = \dfrac{2I}{d + t} \qquad (9\text{-}32)$$

9-40 Cold-Formed Steel Construction

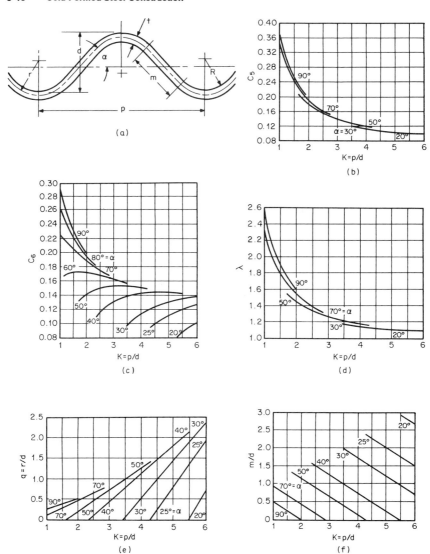

Fig. 9-24 Factors for determining section properties of the arc-and-tangent type of corrugated steel sheet shown in (a).

The radius of gyration, in, is given by

$$\rho = \sqrt{\frac{I}{A}} \qquad (9\text{-}33a)$$

and the tangent length-depth ratio is

$$\frac{m}{d} = \frac{\sin \alpha}{1 - \cos \alpha} - \frac{K}{2} \qquad (9\text{-}33b)$$

(See Fig. 9-24f.)

Example. Consider a corrugated sheet with a 6-in pitch, 2-in depth, inside radius R of 1⅛ in, and thickness t of 0.135 in. The mean radius r is then $1.125 + 0.135/2 = 1.192$ in; $q = r/d = 1.192/2 = 0.596$ in, and $K = p/d = \frac{6}{2} = 3$. From Fig. 9-24$e$, angle α is found to be nearly 45°. For $p/d = 3$ and $\alpha = 45°$, Fig. 9-24b, c, d, and f indicate that $C_5 = 0.14$, $C_6 = 0.145$, $\lambda = 1.24$, and $m/d = 0.93$. Section properties per inch of corrugated width are then computed as follows:

From Eq. (9-30):

$$A = 1.24 \times 1 \times 0.135 = 0.167 \text{ in}^2$$

From Eq. (9-31):

$$I = 0.14 \times 1(0.135)^3 + 0.145 \times 1(2)^2 0.135 = 0.0786 \text{ in}^4$$

From Eq. (9-32):

$$S = \frac{2 \times 0.0786}{2 + 0.135} = 0.0736 \text{ in}^3$$

From Eq. (9-33a):

$$\rho = \sqrt{\frac{0.0786}{0.167}} = 0.686 \text{ in}$$

and from Eq. (9-33b):

$$m = 0.93 \times 2 = 1.86 \text{ in}$$

I, S, and A for corrugated sheets with width b are obtained by multiplying the per-inch values by b.

Unit Stresses. The allowable unit bending stress F_r, ksi, at extreme fibers of corrugated sections of carbon or low-alloy steel may be taken as $0.6F_y$, if r/t does not exceed $1,650/F_y$. For $1,650/F_y \leq r/t < 6,500/F_y$,

$$F_r = 331t/r + 0.399F_y \qquad (9\text{-}34)$$

where F_y = specified minimum yield point of the steel, ksi.

Section properties of corrugated sheets with cross sections composed of flat elements may be computed with the linear method given in Art. 9-12, by combining properties of the various elements as given in Table 9-5. (See also "Sectional Properties of Corrugated Sheets Determined by Formula," *Civil Engineering*, February 1954.)

9-25. Stainless Steel Structural Design.

Cold-formed stainless steel structural members require different design approaches from those presented in Arts. 9-9 through 9-22 for cold-formed structural members of carbon and low-alloy steels, unless the stainless steel is a ferritic type, largely alloyed with chromium, and exhibits a sharp-yielding stress-strain curve.

Because of excellent corrosion resistance, the stainless steels of most interest for structural members, especially in buildings, are the austenitic types, largely alloyed with both chromium and nickel. These alloys have stress-strain curves that do not show sharp yield points but instead are rounded, and the proportional limits are quite low compared with the yield strengths. The "AISI Specification for the Design of Cold-Formed Stainless Steel Structural Members" presents design treatments paralleling those of Arts. 9-9 through 9-22 for carbon and low-alloy steels. (For detailed information on austenitic grades of stainless steels, see ASTM A666, "Austenitic Stainless Steel, Sheet, Strip, Plate and Flat Bar for Structural Applications." See also W. Yu, "Cold-Formed Steel Structures," McGraw-Hill Book Company, New York.)

9-26. Electric-Arc Welding of Cold-Formed Steel.

Cold-formed components of flat-rolled steel are readily joined by both fusion- and resistance-welding methods. In the field, fusion welding is particularly useful for fastening decks, walls, and roofs of cold-formed panels to structural framing. Initial design should anticipate joining by welding rather than merely substitute for other fasteners like rivets, bolts, or screws. This dictum calls for arrangement of parts that ensures accessibility for welding and good fit-up, with the objectives of concealing any unsightly welds that may be made and avoiding notches that could cause fatigue cracks and brittle fracture.

Steel surfaces for welding should be bare and free of oil, grease, paint, scale, rust, and dirt. Paint should be applied only after the welding operation. Metallic-coated steels, such as those galvanized (zinc-coated), aluminized, or terne-coated (lead-tin alloy), may be welded using techniques developed for coated steels. ("Welding Handbook," American Welding Society, 2501 N.W. 7th St., Miami, Fla. 33125 and O. W. Blodgett, "Design of Weldments," James F. Lincoln Arc Welding Foundation, Cleveland, Ohio 44117.)

Arc welding may be done in the shop or the field. Factors favoring arc welding are portability and versatility of equipment as well as freedom in joint design. Only one side of a joint need be accessible, and overlap of parts is not required if joint fit-up is good. As with structural steel, the welding procedure to use depends on material composition, work conditions, and service requirements.

When thin material is welded, care should be taken to minimize distortion and to avoid burning through the material because of the use of too much current. Distortion can be reduced by taking the following precautions:

Avoid overwelding. Match weld sizes to service requirements. Do not use electrodes that are too large relative to the thickness of material being welded. Design joints to minimize shrinking, warping, and twisting. Use jigs and fixtures to hold weldments of thin materials, to control distortion. Also, use tooling designed as heat sinks to provide restraining forces. Try to counteract the magnitudes and directions of predicted distortions by preangling of the weldment parts. For butt welds and other edge-to-edge welds in thin material, properly align the parts being joined and provide firm backup.

Types of arc-welding processes for light-gage steel are as follows:

Shielded-metal-arc welding (manual stick electrode) is the most commonly used process, because of its versatility, but skilled operators are required. The welds can be made in any position. Vertical and overhead welding, however, should be avoided when possible.

Gas-metal-arc welding requires use of special equipment to feed a continuous spool of bar or flux-cored wire into the arc. A shielding gas, such as argon or carbon dioxide, is used to protect the arc zone from oxidation and nitrogen pickup. The process is relatively fast and close control can be maintained over the deposit. The process is not applicable to materials less than $\frac{1}{32}$ in thick but is used extensively for thicker steels.

Gas-tungsten-arc welding operates by maintaining an arc between a nonconsumable tungsten electrode and the work. Close weld control is possible. The process is not used for high-production fabrication, except in specialized applications, because of the relatively high cost.

Arc-Spot Welding. One form of arc-spot welding, an adaptation of gas-metal-arc welding, requires a special welding torch and timer. The welding torch is positioned on a lapped joint to burn a hole through the top sheet of the joint. Also, the torch melts some filler wire to fill the hole, thereby fusing the two parts together. Only one-side access to the joint is necessary. This process is useful when field welding must be done by inexperienced welders.

Another form of arc-spot welding is an application of gas-tungsten-arc welding. The heat of the arc melts a spot completely through one sheet of a lapped joint and partly through the second sheet. When the arc is cut off, fusion results without addition of filler metal.

The common types of arc-welded joint configurations are as follows:

Arc-spot welds (puddle welds). These result when coalescence proceeds from the surface of one sheet into one or more other sheets of a lapped joint without formation of a hole in any sheet (Fig. 9-25a).

Fig. 9-25 Common types of arc-welded joints: (a) arc-spot (round puddle) weld; (b) arc-seam (oblong puddle) weld; (c) flare-bevel weld; (d) flare-vee weld; (e) square groove weld; (f) fillet weld.

Arc-seam welds. These are basically the same as arc-spot welds but are made linearly without slots in the sheets (Fig. 9-25b). Both types of welds are useful in attaching cold-formed steel decks and panels to supporting framing.

Flare-bevel and flare-vee welds. These are made on the outsides of curved edges of bends in cold-formed members (Fig. 9-25c and d).

Square-groove welds. These are rarely used for thin steel (Fig. 9-25e).

Fillet welds. These are made along the edges of sheets in lapped joints and may be longitudinal or transverse to the direction of loading (Fig. 9-25f).

Load Capacities of Welds. As in conventional structural welding, it is general practice to require that the deposited filler metal have a tensile strength at least equal to that of the members being joined. For members of unequal-strength steels, the weld material should be at least as strong as the weaker material.

Failures in welded sheet-steel connections are generally quite complicated. They often occur in a combination of basic modes accompanied by large out-of-plane inelastic deformation.

Fillet-welded joints usually fail by tearing of a sheet next to the weld. Failure is the

9-44 Cold-Formed Steel Construction

Fig. 9-26 Typical failure modes of welded joints: (*a*) sheet tear at transverse fillet weld; (*b*) sheet tear at longitudinal fillet weld; (*c*) transverse sheet tear at flare-bevel weld; (*d*) longitudinal sheet tear at flare-bevel weld; (*e*) shearing of round puddle weld; (*f*) sheet tear at round puddle weld; (*g*) sheet tear at and buckling of round puddle weld.

result of applied tension in transverse welds (Fig. 9-26*a*) or of applied shear in longitudinal welds (Fig. 9-26*b*).

Cold-formed channels welded with flare-bevel welds tend to tear a sheet along the weld (Fig. 9-26*c* and *d*).

Arc-spot welds (round puddle welds) fail by weld shear (Fig. 9-26*e*), transverse or longitudinal sheet tearing (Fig. 9-26*f*), tearing and buckling (Fig. 9-26*g*), or a combination of these modes.

Design of arc-welded joints of sheet metal is treated in the "Specification for Welding Sheet Steel in Structures," AWS D1.3, American Welding Society. Maximum allowable load capacities of arc-welded joints of sheet metal, including cold-formed members 0.180 in or less in thickness, are as follows:

Groove Welds in Butt Joints. Properly matched groove welds (not shown in Fig. 9-26) can be expected to develop the full strength of the base metal. The allowable unit stress for matching electrode and base-metal combinations for a groove weld in a butt joint, welded from one or both sides, equals that of the lower-strength base metal in the connection, if an effective throat equal to the thickness of the material is consistently obtained.

Arc-Spot (Puddle) Welds. The allowable load capacity P, kips, is as follows:
For $d/t < 240/\sqrt{F_y}$,

$$P = 2.2 t d_a F_t \qquad (9\text{-}35)$$

where F_y = specified minimum yield point of sheet steel, ksi
d = diameter, in, of outer surface of arc-spot weld or width of seam weld
t = thickness, in, of sheet steel, exclusive of coatings, for a single sheet or combined thicknesses of sheet steel for a double sheet
d_a = average diameter, in, of arc-spot weld or average width of seam weld
    = $d - t$ for a single sheet
    = $d - 2t$ for a double sheet
F_t = permissible unit tensile stress, ksi, in sheet steel = $0.40 F_{ult}$
F_{ult} = specified minimum ultimate tensile strength, ksi, of the sheet steel

For $d/t \geq 240/\sqrt{F_y}$,

$$P = 1.4 t d_a F_t \qquad (9\text{-}36)$$

In no case, however, should the allowable load exceed

$$P = \frac{\pi}{4} d_e^2 F_w \qquad (9\text{-}37)$$

where d_e = effective diameter, in, of arc-spot weld or effective width of arc-seam weld at fused surface = $0.7d - 1.5t$
F_w = permissible unit shear stress, ksi, in weld metal = $0.30 F_{xx}$
F_{xx} = strength level, ksi, of the weld metal designated by the AWS electrode classification

For the preceding equations to be applicable, weld washers made of sheet steel at least 0.060 in thick should be used to contain arc-spot welds of sheet steel thinner than 0.028 in.

Arc-Seam Welds. The allowable load capacity P, kips, is as follows:
For $d/t < 240/\sqrt{F_y}$,

$$P = (L + 3.2d_a)tF_s \qquad (9\text{-}38)$$

where L = length, in, of seam weld, exclusive of circular ends
F_s = permissible unit shear stress, ksi, in sheet metal = $0.30 F_{ult}$
For $d/t \geq 240/\sqrt{F_y}$,

$$P = (L + 2d_a)tF_s) \qquad (9\text{-}39)$$

In no case, however, should the allowable load exceed

$$P = \left(\frac{\pi}{4} d_e^2 + Ld_e/2\right) F_w \qquad (9\text{-}40)$$

For allowable-capacity calculation purposes, L should not be taken larger than $3d_a$.

Fillet Welds Transverse to Direction of Loading. The allowable load capacity, kips, is

$$P = LtF_t \qquad (9\text{-}41)$$

Fillet Welds Longitudinal to Direction of Loading. The allowable load capacity, kips, is a follows: For $L \geq 2.5$ in,

$$P = LtF_s \qquad (9\text{-}42a)$$

For $L < 2.5$ in,

$$P = 2t^{1.2}L^{1/3}F_s \qquad (9\text{-}42b)$$

Flare-Bevel Groove Welds Loaded Transverse to Weld Axis. The allowable load capacity, kips, is

$$P = 0.80tLF_t \qquad (9\text{-}43)$$

Flare-Bevel Groove Welds Loaded Longitudinal to Weld Axis. The allowable load capacity, kips, is as follows: For single shear,

$$P = tLF_s \qquad (9\text{-}44a)$$

For double shear,

$$P = 2tLF_s \qquad (9\text{-}44b)$$

9.27. Electric Resistance Welding of Cold-Formed Steel.

Resistance welding is defined as a group of welding processes wherein coalescence is produced by the heat obtained from resistance of the work to flow of electric current in a circuit of which the work is a part, and by the application of pressure ("Welding Handbook," American Welding Society). Because of the size of the equipment required, resistance welding is essentially a shop process. Speed and low cost per weld are factors favoring its selection.

Almost all resistance welding processes require a lap-type joint. The amount of contacting overlap varies from ⅜ to 1 in, depending on sheet thickness. Access to both sides of the joint is normally required. Adequate clearance must be provided for electrodes and welder arms.

Spot welding is the most common resistance welding process. The work is held between electrodes through which an electric current passes. A weld is formed at the interface between the pieces being joined, and it consists of a cast-steel nugget. The nugget has a diameter about equal to that of the electrode face and should penetrate 60 to 80% of each sheet thickness.

For structural design purposes, spot welding can be treated in the same way as rivets, except that no reduction in net section resulting from holes need be made. Table 9-10

TABLE 9-10 Resistance Welding Data for Uncoated Low-Carbon Steel

Thickness t of thinnest outside piece, in	Min OD of electrode D, in	Min contacting overlap, in	Min weld spacing c to c, in	Approx diam of fused zone, in	Min shear capacity per weld,* kips	Diam of projection D, in
			Plain spot welding			
0.021	3/8	7/16	3/8	0.13	0.32	
0.031	3/8	7/16	1/2	0.16	0.57	
0.040	1/2	1/2	3/4	0.19	0.92	
0.050	1/2	9/16	7/8	0.22	1.35	
0.062	1/2	5/8	1	0.25	1.85	
0.078	5/8	11/16	1 1/4	0.29	2.70	
0.094	5/8	3/4	1 1/2	0.31	3.45	
0.109	5/8	13/16	1 5/8	0.32	4.15	
0.125	7/8	7/8	1 3/4	0.33	5.00	
			Pulsation welding			
1/8	1	7/8		3/8	5	
3/16	1 1/4	1 1/8		9/16	10	
1/4	9/16	1 3/8		3/4	15	
			Projection welding			
0.125		11/16	9/16	0.338	4.8	0.281
0140		3/4	5/8	7/16	6.0	0.312
0.156		13/16	11/16	1/2	7.5	0.343
0.171		7/8	3/4	9/16	8.5	0.375
0.187		15/16	13/16	9/16	10.0	0.406
0.203		1	7/8	5/8	12	0.437
0.250		1 1/4	1 1/16	11/16	15	0.531

*Allowable shear strengths may be taken as one-fourth the tabulated capacities.

gives the essential information for design purposes for uncoated material based on "Recommended Practices for Resistance Welding," American Welding Society, 1966. Maximum allowable loads per weld for design purposes are based on shear strengths of welds observed in tests after application of a safety factor of 2.5 to the lower bounds of the data. Note that the thickest material listed in Table 9-10 for plain spot welding is ⅛ in. Thicker material can be resistance welded by projection or by pulsation methods if high-capacity spot welders for thicker material than ⅛ in are not available.

Projection welding is a form of spot welding in which the effects of current and pressure are intensified by concentrating them on small areas of projections embossed in the sheet to be welded. Thus, satisfactory resistance welds can be made on material thicker than ⅛ in with spot welders that are ordinarily limited to thinner stocks.

Pulsation welding, better called **multiple-impulse welding**, makes spot welds with more than one impulse of current, a maneuver that makes some spot welders useful for materials thicker than ⅛ in. The choice between projection welding and impulse welding depends on the work being produced, volume of output, and the equipment available.

Spot welding of higher-strength steels than those covered by the data in Table 9-10 may require special welding conditions if the higher shear strengths of the high-strength steels are to be developed.

All steels used for spot welding should be free of scale. Therefore, either hot-rolled and pickled steels or cold-rolled steels are usually specified.

Steels containing more than 0.15% carbon are not as readily spot welded as lower-carbon steels, unless special techniques are used to ensure ductile spot welds. Moreover, high-carbon steels, such as ASTM A446, Grade D, which may have carbon content as high as 0.40% (by heat analysis), are not recommended for resistance welding. Designers should resort to other means of joining such steels.

In the detailing of spot-welded joints, it is necessary to maintain sufficient overlaps to ensure consistent weld strengths and minimum distortion at joints. The minimum weld spacings specified in Table 9-10 should be observed, or else shunting of current to previously made nearby welds may reduce current in welds being made and result in poor welds. Joint design should provide sufficient clearance between the electrodes and the work to prevent short-circuiting of the current needed to make satisfactory spot welds. For information on spot welding of coated steels, see "Recommended Practices for Resistance Welding of Coated Low-Carbon Steel," American Welding Society.

9.28. Maximum Spacing of Welds.

In addition to observing requirements for minimum spacing of welds it is also necessary to observe certain precautions with respect to the maximum spacing that may be permitted. These are as follows for carbon and low-alloy steels:

For Shear Transfer. The spacing should not exceed that required to transmit the shear between the connected parts. This refers, of course, to the primary function of the welds to act in the same way as rivets in structural connections.

For Column Action. The AISI "Specification for the Design of Cold-formed Steel Structural Members" recommends that, to avoid failure of a compression cover plate or sheet by column action (cylindrical buckling) between welds, weld spacing in line of stress, in, should not exceed $200t/\sqrt{f}$, where t is the thickness, in, of the cover plate, and f is the design stress, ksi, in the cover plate.

To Prevent Local Plate Buckling. The AISI Specification prescribes that the spacing, in line of stress, of welds connecting a compression cover plate or sheet to a stiffener or other element should not exceed three times the flat width of the narrowest unstiffened compression element in that portion of the cover plate that is tributary to the connections. But the spacing need not be less than the following, unless closer spacing is required to transmit shear or prevent cylindrical buckling:

Value of F_c permitted in unstiffened element	Minimum required spacing of connections, in
$0.54F_y$ or less	$228t/\sqrt{F_y}$
Greater than $0.54F_y$	$190t/\sqrt{F_y}$

where F_y = specified minimum yield point of the steel, ksi
 t = thickness, in, of cover plate or sheet
 F_c = allowable compressive stress in the unstiffened element, ksi
This reduces to the following, for F_y = 33 ksi and F_y = 50 ksi:

	Minimum required spacing, in, unless closer spacing is required to transmit shear or prevent cylindrical buckling	
Allowable F_c, nearest ksi	$F_y = 33$	$F_y = 50$
18 or less	$40t$	$32t$
Greater than 18, not more than 20	$33t$	$32t$
Greater than 20, not more than 27		$32t$
Greater than 27		$27t$

For intermittent fillet welds parallel to the direction of stress, the spacing should be taken as the clear distance, in, between welds plus ½ in. In all other cases, the spacing should be taken as the center-to-center distance between connections.

None of the above requirements regarding weld spacing applies to cover sheets that act only as sheathing material and are not considered load-carrying elements.

Between Two Channels Forming an I. The AISI Specification also contains recommendations for the maximum longitudinal spacing of welds connecting two channels to form an I section (Figs. 9-6m and 9-6n). For compression members, it is recommended that spacing, in, not exceed

$$s_{max} = \frac{Lr_{cy}}{2r_y} \quad (9\text{-}45)$$

where r_{cy} is the radius of gyration, in, of one channel about its centroidal axis parallel to web, and L/r_y is the slenderness ratio of the built-up member about the axis that controls its load-carrying capacity. This axis may be the minor axis for free-standing columns, or the major axis where that axis controls, as in the case of braced wall studs.

For flexural members, the AISI provisions are somewhat more complicated, and involve the span of the beam, the tensile strength of each connection, the distance between rows of connections, the intensity of the load on the beam, and the location of the shear center of each channel. In most cases, any reasonably close spacing of the welds will be satisfactory, but it is always advisable to check the spacing against the AISI criteria.

TABLE 9-11 Allowable Shear Stresses for Various Bolt Steels in Bearing-Type Connections

ASTM designation	Type of steel	Shear on gross cross section of bolts, ksi	
		Threads not excluded	Threads excluded
A307	Low-carbon	10.0	10.0
A325	High strength	15.0	22.0
A354	Quenched and tempered alloy	18.0	27.0
A449	Quenched and tempered	18.0	27.0
A490	Quenched and tempered alloy	22.5	32.0

9-29. Bolting Cold-Formed Steel Members.

Bolting is often convenient in cold-formed steel construction. Bolts, nuts, and washers should generally conform to the requirements of the ASTM designations of Table 9-11, which lists allowable shear values for bearing-type connections.

The clear distance between bolts set in rows parallel to the direction of load and also the distance from any bolt center to the end of a sheet or other boundary toward which the joint load is directed should conform to the following limitations:

When $F_u/F_y \geq 1.15$,

$$e_{min} = \frac{P}{0.5F_u t} \quad (9\text{-}46)$$

When $F_u/F_y < 1.15$,

$$e_{min} = \frac{P}{0.45F_u t} \quad (9\text{-}47)$$

where e_{min} = minimum clear distance, in, between bolts in line of stress or from any bolt center to a sheet boundary toward which joint load is directed
P = force transmitted by a bolt, kips
t = thickness, in, of thinnest connected part
F_u = specified minimum ultimate tensile strength of steel, ksi
F_y = specified minimum tensile yield point of steel, ksi

The minimum distance between centers of bolt holes should provide sufficient clearance for bolt heads, nuts, washers, and the wrench. This distance, however, should not be less than 3 times the nominal bolt diameter. Moreover, the distance from the center of any bolt to the end or other boundary of the connecting members should be at least 1.5 times the bolt diameter.

The tensile stress on the net section of a bolted connection F_t, ksi, should not exceed $0.60F_y$ nor the following stresses when $F_u \geq 1.35F_y$: When $t \geq \frac{3}{16}$ in,

$$F_t = 0.50F_u \quad (9\text{-}48)$$

When $t < \frac{3}{16}$ in for double-shear connections, with washers under both the bolt head and the nut,

$$F_t = 0.50F_u(1 - 0.9r + 3rd/s) \leq 0.50F_u \quad (9\text{-}49)$$

When $t < \frac{3}{16}$ in for single-shear connections, with washers under both the bolt head and the nut,

$$F_t = 0.45F_u(1 - 0.9r + 3rd/s) \leq 0.45F_u \quad (9\text{-}50)$$

where r = force transmitted by the bolt or bolts at the section under consideration divided by the tensile force in the member at that section (if r is less than 0.2, it may be taken equal to zero)

d = bolt diameter, in

s = spacing, in, of bolts measured perpendicular to line of stress (for a single bolt, s = width of sheet)

When $F_u < 1.35F_y$, substitute $F_u/1.35$ for F_u in Eqs. (9-48) to (9-50). The bearing stress on the area dt should not exceed the values given in Table 9-12.

TABLE 9-12 Allowable Bearing Stresses for Bolted Connections with Washers under Both Bolt Head and Nut

Thickness of connected part, in	Type of joint	F_u/F_y for connected part	Allowable bearing stress F_p, ksi
3/16 or more	All types	No limit	$1.50F_u$
Less than 3/16 but not less than 0.024	Inside sheet of double-shear connection	1.15 or more Less than 1.15	$1.50F_u$ $1.35F_u$
	Single shear and outside sheets of double-shear connection	No limit	$1.35F_u$

The American Iron and Steel Institute "Specifications for the Design of Cold-formed Steel Structural Members," 1980, is the source of the preceding provisions. It also presents allowable tension and bearing stresses for bolted joints without washers.

Example. Consider the upper sheet in each of the two-bolt arrangements shown in Fig. 9-27. In both cases, thickness of sheet t is less than 3/16 in, and F_u is more than $1.35F_y$. Bolts are ⅝ in in diameter. In Case 1, the bolts are arranged in a single transverse row, and the force transmitted by that row is equal to the total load T. The force in the sheet at the bolts is also considered to be T. Therefore, $r = T/T = 1$. With $s = 2$ in and bolt diameter $d = ⅝$ in, $d/s = 0.31$. The allowable net-section stress is, therefore, by Eq. (9-50),

$$F_t = 0.45F_u(1.0 - 0.9 \times 1 + 3 \times 1 \times 0.31) = 0.45F_u \times 1.03$$

Since the quantity in the parentheses is greater than 1 and $F_u > 1.35F_y$, the allowable net-section stress is $0.45F_u$, or 100% of the basic allowable stress F.

In Case 2, the two bolts are arranged in a single line in the direction of the applied load. The force being transmitted by the bolt at each section (1-1 and 2-2) is $T/2$. For the upper sheet, the force at section 1-1 is the total force T. At section 1-1, therefore, $r = (T/2)/T = ½$. In this case, the spacing s is the total sheet width, 4 in. Hence, $d/s = (⅝)/4 = 0.16$. The allowable net-section stress in the upper sheet at section 1-1 is, therefore, by Eq. (9-50),

$$F_t = 0.45F_u(1.0 - 0.9 \times ½ + 3 \times ½ \times 0.16) = 0.45F_u \times 0.79 = 0.36\,F_u$$

Since F_t is less than $0.45F_u$, the allowable net-section stress is 79% of the basic allowable stress F. At section 2-2, for the upper sheet, $r = 1$ because both the force transmitted by the bolt and the force in the sheet at that section equal $T/2$. The allowable stress at that section given by Eq. (9-50) is $0.45F_u \times 0.58 = 0.26F_u$, but this low stress has no significance, because the force in the upper sheet is only 50% of the total force T. For the lower sheet in Case 2, the allowable stresses at sections 1-1 and 2-2 are those computed for the upper sheet at sections 2-2 and 1-1, respectively.

Fig. 9-27 Bolted connections with two bolts.

Despite the stress reduction required in Case 2, the total force T that can be transferred through this joint on the basis of net-section tensile stress is about the same for both cases. This will not necessarily be true for other sizes and joint arrangements.

Let us assume now that $\frac{3}{16}$-in-thick steel of ASTM A611, Grade C, is used as the joined materials of Cases 1 and 2. For Grade C, $F_y = 33$ ksi, $F_u = 48$ ksi, and $F_u > 1.35 F_y$. The allowable tension force for Case 1 is

$$P = 0.45 \times 48[(4 - 2 \times \tfrac{5}{8}) \times \tfrac{3}{16}] = 11.14 \text{ kips}$$

The allowable tension force for Case 2 is

$$P = 0.36 \times 48[(4 - \tfrac{5}{8}) \times \tfrac{3}{16}] = 10.94 \text{ kips}$$

The minimum clear distance for Case 1 is

$$e_{\min} = 11.14/(0.5 \times 48 \times \tfrac{3}{16}) = 2.48 \text{ in}$$

The minimum clear distance for Case 2 is

$$e_{\min} = 10.94/(0.5 \times 48 \times \tfrac{3}{16}) = 2.43 \text{ in}$$

The shear stress on the $\tfrac{5}{8}$-in-diam bolts in Case 1 would be

$$f_v = \frac{11.14}{(\tfrac{5}{8})^2 \times 0.7854 \times 2} = 18.16 \text{ ksi}$$

Table 9-12 gives an allowable shear stress of 22.0 ksi for A325 bolts, with threads excluded, which is more than needed for the load for Case 1.

The shear stress on the $\tfrac{5}{8}$-in-diam bolts in Case 2 would be

$$f_v = \frac{10.94}{(\tfrac{5}{8})^2 \times 0.7854 \times 2} = 17.83 \text{ ksi.}$$

Table 9-12 gives an allowable shear stress of 22.0 ksi for A325 bolts, threads excluded, which is satisfactory for the load for Case 2.

For further information on bolted joints, see "Specification for Structural Joints Using ASTM A325 or A490 Bolts" as approved by the Research Council on Riveted and Bolted Structural Joints.

9-30. Riveting of Cold-Formed Shapes.

Although widely practiced in aircraft work and traditional applications such as drainage products, riveting seldom is used for load-carrying cold-formed members in buildings.

9-52 Cold-Formed Steel Construction

When riveting is used for such purposes, ordinary structural practice may be followed as a guide in strength calculations, but, for maximum rivet spacing, rules for maximum allowable spacing of spot welds (Art. 9-28) should be observed. The provisions of Art. 9-29 for net-section stresses should also be observed, unless the results of carefully made tests or experience with a particular application dictate otherwise.

Care must be taken in making use of the results of tests of joints in thin material, because of the variations in strength that can result from variations in the texture of the surfaces being joined, hole clearance, and tightness after driving.

9-31. Self-Tapping Screws for Joining Light-Gage Components.

Self-tapping screws that are hardened so that their threads form or cut mating threads in one or both of the parts being connected are frequently used for making field joints in light-gage steel construction. Such screws provide a rapid and efficient means of making light-duty connections. The screws are especially useful for such purposes as fastening sheet-metal siding and roofing to structural steel; making attachments at joints, side laps, and closures in siding, roofing, and decking; fastening collateral materials to steel framing; and fastening steel studs to sill plates or channel tracks. The screws also may be used for fastening bridging to steel joists and studs, fastening corrugated decking to steel joists, and similar connections to secondary members.

A few types of self-tapping screws are illustrated in Fig. 9-28. Other types are available. There are many different head styles—slotted, recessed, hexagonal, flat, round, etc. Some types, called *Sems,* are supplied with preassembled washers under the heads. Other types are supplied with neoprene washers for making watertight joints in roofing.

All the types of screws shown in Fig. 9-28 require prepunched or predrilled holes. Self-drilling screws, which have a twist drill point that drills a hole of the proper size

	SHEET METAL			STRUCTURAL STEEL	CASTINGS		FORGINGS
	STAINLESS STEEL	STEEL, BRASS, ALUMINUM, ETC.			NON-FERROUS: ALUMINUM, MAGNESIUM, BRONZE, ETC.	FERROUS: GRAY IRON, STEEL, MALLEABLE IRON, ETC.	STEEL, BRASS, BRONZE, ETC.
	0.015" TO 0.050" THICK	0.015" TO 0.050" THICK	0.050" TO 0.200" THICK	0.200" TO 0.500" THICK			
THREAD-FORMING SCREWS TYPE AB	▨	▨	▨	▨			
TYPE B	▨	▨	▨	▨	▨		
THREAD-CUTTING SCREWS TYPE F	▨	▨	▨	▨	▨	▨	▨
DRIVE SCREWS TYPE U			▨	▨	▨	▨	▨

Fig. 9-28 Self-tapping screws. *(Extracted with permission from "P-K Selector Guide," USM Corporation, Parker-Kalon Division.)* For other types, head styles, detailed dimensions, and information on self-drilling screws, refer to manufacturers' literature.

just ahead of threading, are especially suited for field work, because they eliminate a separate punching or drilling operation. Another type of self-drilling screw, capable of being used in relatively thin gages of material in situations where the parts being joined can be firmly clamped together, has a very sharp point that pierces the material until the threads engage.

Torsional-strength requirements for self-tapping screws have been standardized under American National Standards Institute B18.6.4, "Slotted and Recessed Head Tapping Screws and Metallic Drive Screws." Safe loads in shear and tension on such screws can vary considerably, depending on type of screw and head, tightening torque, and details of the assembly. They may be used for structural load-carrying purposes on the basis of tests of the type of assembly involved, manufacturers' recommendations, or experience with a particular application.

Essential body dimensions for some types of self-tapping screws are given in Table 9-13. Complete details of these and other types, and recommended hole sizes, may be found in ANSI B18.6.4, and in manufacturers' publications.

TABLE 9-13 Average Diameters of Self-Tapping Screws, in*

Number or size, in	Types AB and B (Fig. 9.28)		Type F† (Fig. 9-28)	Type U (Fig. 9-28)
	Outside	Root	Outside	Outside
No. 4	0.112	0.084	0.110	0.114
No. 6	0.137	0.102	0.136	0.138
No. 8	0.164	0.119	0.161	0.165
No. 10	0.186	0.138	0.187	0.180
No. 12	0.212	0.161	0.213	0.209
No. 14‡ or 1/4	0.243	0.189	0.247	0.239‡
5/16	0.312	0.240	0.309	0.312
3/8§	0.376	0.304	0.371	0.375

*Averages of standard maximum and minimum dimensions adopted under ANSI B18.6.4-1966.

† Type F has threads of machine-screw type approximating the Unified Thread Form (ANSI B1.1-1960). The figures shown are averages of those for two different thread pitches for each size of screw.

‡ Size No. 14 for Type U.

§ Does not apply to Type AB.

9-32. Special Fasteners for Light-Gage Steel.

Special fasteners such as tubular rivets, blind rivets (capable of being driven from one side only), special bolts used for "blind insertion," special studs, lock nuts, and the like, and even metal stitching, which is an outgrowth of the common office stapling device, are all used for special applications. When such a fastener is required, refer to manufacturers' catalogs, and base any structural strength that may be attributed to it on the results of carefully made tests or the manufacturers' recommendations.

STEEL ROOF DECK

The term steel roof deck usually refers to relatively long, ribbed steel sheets with overlapping or interlocking edges designed to serve primarily for the support of roof loads. A steel roof deck assembly, with insulation and roofing, is illustrated in Fig. 9-29.

Fig. 9-29 Roof-deck assembly.

9-33. Types of Steel Roof Deck.

Most steel decking intended primarily to be used as roof deck has relatively narrow longitudinal ribs from 1½ to 2 in deep and spaced about 6 in c to c, as illustrated in Fig. 9-30. Some manufacturers produce shallower sections, with ribs from ⅞ to 1 in deep and spaced about 4 in c to c. Others produce deeper sections, with 8-in rib spacing.

In addition to ribbed sections of the general type illustrated in Fig. 9-30, some manufacturers produce long-span roof-deck sections, from 3 to 7½ in deep, some types of which are shown in Fig. 9-31. The cellular floor decks described in Arts. 9-40 to 9-42 may also be used in roofs.

The flat top portions of roof-deck sections may be either plain, as illustrated, or may have one or more longitudinal flutes, as shown for the long-span section in Fig. 9-31a; or they may have other patterns formed in them. Details of roof-deck sections vary with different manufacturers.

Fig. 9-30 Ribbed roof-deck section. Other widths and profiles also are available.

Roof-deck sections are also available with perforations in the ribs or top sheets to form acoustical ceilings. When the ribs are perforated (Fig. 9-32), they are placed in the normal position, with the flat portion up, and the rib space is filled with sound-absorbing material. If the top sheet is perforated, the deck is placed with ribs upstanding, and the space between them is filled with acoustical material. Closed cellular sections with perforated bottom sheets may also be used for acoustical ceilings in roof construction. (See also Art. 9-42 and Fig. 9-37.)

9-34. Materials for Steel Roof Deck.

The usual material for steel roof deck is structural-quality sheet steel, either black or galvanized, conforming to Grade C of ASTM A611 for uncoated material, or to Grade A of A446 for galvanized material. Some manufacturers, however, produce decking from higher-strength grades also.

Common thicknesses for sections of the general type shown in Fig. 9-30 are given in Table 9-14. Some manufacturers, however, offer these sections in heavier material. The very shallow, short-span sections with ribs spaced 4 in c to c are usually even thinner. The deep, long-span sections of Fig. 9-31 are furnished in thicknesses up to about 0.1 in.

Steel deck made from uncoated material is given a prime coat of paint or other protective material before shipment. Galvanized material may or may not be painted.

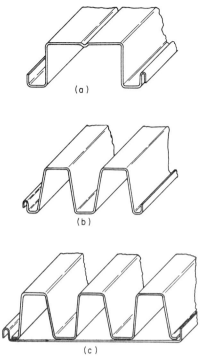

Fig. 9-31 Long-span roof-deck sections. See Figs. 9-35 and 9-37 for other sections used for long-span roof decks.

Fig. 9-32 Roof-deck sections with perforated ribs for acoustical control. *(H. H. Robertson Co.)* Cellular units with perforated bottom sheets (Fig. 9-37) also are used as roof deck with acoustic ceilings.

TABLE 9-14 Allowable Total (Dead plus Live) Uniform Loads, psf, on Steel Roof Deck*

Deck type	Span condition	Design thickness, in	Span—c to c joists or purlins, ft-in									
			4-0	4-6	5-0	5-6	6-0	6-6	7-0	7-6	8-0	8-6	9-0
NR 22	SIMPLE	0.0295	74	58	47								
NR 20		0.0358	90	72	58	48	40						
NR 18		0.0474	121	95	77	64	54	46					
NR 22	2-SPAN	0.0295	80	64	51	42							
NR 20		0.0358	96	76	62	51	43						
NR 18		0.0474	124	98	79	66	55	47	40				
NR 22	3 OR MORE	0.0295	100	79	64	53	45						
NR 20		0.0358	120	95	77	64	53	46					
NR 18		0.0474	155	122	99	82	69	59	51	44			
			4-0	4-6	5-0	5-6	6-0	6-6	7-0	7-6	8-0	8-6	9-0
IR 22	SIMPLE	0.0295	86	68	55	45							
IR 20		0.0358	106	83	68	56	47	40					
IR 18		0.0474	141	112	90	75	63	54	46	40			
IR 22	2-SPAN	0.0295	93	74	60	49	41						
IR 20		0.0358	112	88	72	59	50	42					
IR 18		0.0474	145	114	93	76	64	55	47	41			
IR 22	3 OR MORE	0.0295	116	92	74	62	52	44					
IR 20		0.0358	140	110	89	74	62	53	46	40			
IR 18		0.0474	181	143	116	96	80	68	59	51	45	40	

NARROW RIB DECK TYPE NR
RIBS APPROX. 6" C-C
MAX 1"
MIN 3/8"
1 1/2" MIN

INTERMEDIATE RIB DECK TYPE IR
RIBS APPROX. 6" C-C
MAX 1 3/4"
MIN 1"
1 1/2" MIN

WIDE RIB DECK
TYPE WR
RIBS APPROX.
6" C-C
MAX 2 1/2"
MIN 1 3/4"
1 1/2" MIN

			5-0	5-6	6-0	6-6	7-0	7-6	8-0	8-6	9-0	9-6	10-0
SIMPLE	WR 22	0.0295	89	70	56	46							
	WR 20	0.0358	112	87	69	56	47	40					
	WR 18	0.0474	154	118	94	76	63	53	45				
2-SPAN	WR 22	0.0295	98	81	68	58	50	43		43			
	WR 20	0.0358	125	103	87	74	64	56	49	57	51	45	40
	WR 18	0.0474	165	136	115	98	84	73	64				
3 OR MORE	WR 22	0.0295	122	101	85	70	58	49	42	45			
	WR 20	0.0358	156	129	108	87	72	60	52	59	51	45	40
	WR 18	0.0474	207	171	143	120	98	81	69				

*Load tables were calculated with sectional properties for minimum thicknesses of 0.028, 0.034, and 0.045 in, corresponding respectively to design thicknesses of 0.0295, 0.0358, and 0.0474 in, exclusive of coating on base metal.
Loads shown in tables are uniformly distributed total (dead plus live) loads, psf. Loads in shaded areas are governed by live-load deflection not in excess of 1/240 × span. The dead load included is 10 psf. All other loads are governed by the allowable flexural stress limit of 20 ksi for a 33-ksi minimum yield point.
Rib-width limitations shown are taken at the theoretical intersection points of flange.
Span length assumes c-to-c spacing of supports. Tabulated loads shall not be increased by assuming clear-span dimensions.
Bending moment formulas used for flexural stress limitation are: for simply supported and two-span decking, $M = wl^2/8$; for decking with three continuous spans or more, $M = wl^2/10$.
Deflection formulas for deflection limitation are: For simply supported decking, $\Delta = 5wl^4/384EI$; for two- and three-span decking, $\Delta = 3wl^4/384EI$.
Normal installations covered by these tables do not require midspan fasteners for spans of 5 ft or less.
From "Design Manual for Floor Decks and Roof Decks," Steel Deck Institute, 1978.

Weight of the roof-deck sections illustrated in Fig. 9-30 varies, depending on the detailed design and dimensions of the section. With 0.0358-in-thick material, sections 1½ in deep will weigh between 2 and 3 psf. For structural design calculations, 3.0 psf is suggested for 0.0474-in-thick, 2.5 psf for 0.0358-in-thick, and 2.0 psf for 0.0295-in-thick deck. The weight of the deeper types varies from less than 2 psf to more than 11 psf, depending on design and thickness.

9-35. Load-Carrying Capacity of Steel Roof Deck.

Standard load tables published by the Steel Deck Institute (SDI) apply only to decking of the general configuration shown in Fig. 9-30, with 6-in rib spacing, 1½-in minimum depth of ribs, three different rib widths, and three different thicknesses of material. (See Table 9-14.) Allowable uniformly distributed loads vary from 40 psf to well over 200 psf, depending on width of rib, span condition (simple or continuous), length of span c to c of supports, and whether allowable stress or deflection controls. Live-load deflection is limited to $\frac{1}{240}$ times the span. Safe-load and deflection calculations are based on the American Iron and Steel Institute "Specification for the Design of Cold-formed Steel Structural Members," following the procedures outlined in Arts. 9-12 to 9-32.

The SDI specifications require that steel deck units be anchored to the supporting framework to resist the following gross uplift forces, from which the dead load of the roof-deck construction may be deducted:

45 psf for eave overhangs
30 psf for all other roof areas

The SDI specifications further provide: "Suspended ceilings, light fixtures, ducts or other utilities shall not be supported by the steel deck."

Safe loads and useful spans for sections deeper than those covered by the SDI load table are substantially greater than those shown in Table 9-14, and for the deepest long-span sections, tabulated spans are as long as 35 ft for roof loads. For detailed information regarding such sections, refer to manufacturers' publications.

In addition to their normal function as roof panels to resist gravity loading, steel roof-deck assemblies can be used as shear diaphragms under lateral loads, such as wind and seismic forces. When steel roof deck is used for this purpose, special attention must be given to attachments between panels and attachments of panels to building frames. Detailed information is available from manufacturers. (See also A. H. Nilson, "Shear Diaphragms of Light-Gage Steel," *Journal of the Structural Division,* Vol. 86, No. ST11, November 1960, American Society of Civil Engineers.)

9-36. Details and Accessories for Steel Roof Deck.

As illustrated in Fig. 9-30, all roof-deck sections have lapping or interlocking edges. Most sections are designed so that the ends lap (Fig. 9-33a). Special filler or cover pieces are sometimes provided instead of telescoping end laps.

Special auxiliary shapes to fit various roofing details are furnished by most manufacturers of steel roof deck. These usually include side closure plates, ridge and valley plates, and cant strips. Although details and dimensions will vary with different manufacturers, the general shapes of these accessories are as shown in Fig. 9-33.

Steel roof decks are usually fastened to structural-steel supports by electric arc welding or by self-tapping screws. The Steel Deck Institute recommends that adjacent sections spanning more than 5 ft be fastened together at midspan. For further information see the "Design Manual for Floor Decks and Roof Decks," Steel Deck Institute, P. O.

Fig. 9-33 Roof-deck details.

Box 3812, St. Louis, Mo. 63122. In a particular case, the details to be used depend on the circumstances surrounding the job and on the recommendations of the manufacturer.

9-37. Insulation of Steel Roof Deck.

Although insulation is not ordinarily supplied by steel roof deck manufacturers, it is more or less standard practice to install insulation between the roof deck and the roof covering, as illustrated in Fig. 9-29. The Steel Deck Institute "Design Manual for Floor Decks and Roof Decks" contains the following recommendations with respect to insulation:

> All steel roof decks shall be covered with a material of sufficient insulating value as to prevent condensation under normal occupancy conditions. Insulation shall be adequately attached to the steel deck by adhesives or mechanical fasteners. Insulation materials shall be protected from the elements at all times during storage and installation.

9-38. Fire Resistance of Steel Roof Deck.

The Steel Deck Institute "Design Manual for Floor Decks and Roof Decks," 1978, contains the fire-resistance ratings given in Table 9-15. Additional fire ratings are available.

9-39. Other Forms of Steel Roof Decking.

Forms of steel roof decking other than those described in Arts. 9-33 and 9-34 are also available. Corrugated sheets are an example. In addition to the traditional sinusoidal

9-60 Cold-Formed Steel Construction

cross section, corrugated sheets are available in a variety of other cross-sectional configurations, sizes, and load-carrying capacities. Some of these have special venting slots, for use with gypsum roofs. Others have protrusions or indentations formed in them, or cross wires welded to them, to provide anchorage in concrete fill and function in a composite manner with the concrete. Some are made from full-hard material such as Grade E of ASTM A611 and A446, with yield points of 80 ksi and higher.

TABLE 9-15 Fire-Resistance Ratings, Steel Roof Deck Construction

Hours	Roof construction	Insulation	Underside protection	Authority
2	Min 1 1/2-in-deep steel deck on steel joists, 6 ft c to c max	Min 1-in-thick, UL-listed mineral-fiber board	7/8-in-thick, lightweight-aggregate, gypsum plaster on metal lath	UL Design* P404
	Min 1 1/2-in-deep steel deck on steel joists, 5 ft 6 in c to c max	Min 3/4-in-thick, UL-listed glass-fiber insulation board	1-in-thick vermiculite, gypsum plaster on metal-lath suspended ceiling	UL Design P409
1 1/2	Min 1 1/2-in-deep steel deck on steel beams	1-in-thick, UL-listed mineral-fiber insulation board applied in two layers of ½-in board, with joints staggered	Min 1 3/8 to 1 1/2-in-thick, direct-applied, sprayed vermiculite plaster, UL listed	UL Design P703
1	Min 1 1/2-in-deep steel deck on steel beams	1-in-thick, UL-listed mineral-fiber insulation board	Min 1 3/8 to 1 1/2-in-thick, direct-applied, sprayed vermiculite plaster, UL listed	UL Design P701
1	Min 7/8-in deep steel deck on steel joists, 4 ft c to c max	1- to 3-in-thick, UL-listed mineral-fiber insulation board	5/8-in-thick, ceramic, acoustical, lay-in boards in ceiling grid	UL Designs P211 and P210
1	Min 1 3/8-in-deep steel deck on 10-in steel joists, 5 ft 8 in c to c max	1-in-thick, listed mineral-fiber insulation board	5/8-in-thick, listed, acoustical, lay-in boards in listed ceiling grid	FM Design† FC37-1 hr

*Fire Resistance Index, Underwriters' Laboratories, Inc., January, 1972.
†Factory Mutual.

Although ribbed roof-deck sections are usually placed with the flat portion up, as shown in Figs. 9-29 and 9-30, they may also be used with the ribs upstanding without the acoustical treatment mentioned in Art. 9-33. Roofs of this general kind have found numerous applications, especially for light industrial, agricultural, and relatively small storage structures. Sections for this purpose may be conventional, or they may have very wide, flat portions between upstanding ribs.

Manufacturers' literature contains a wealth of detailed information regarding all types of roof-deck sections and their applications.

CELLULAR STEEL FLOORS

Cellular steel floors are often used in many kinds of structures, including massive highrise buildings for institutional, business, and mercantile occupancies.

9-40. Advantages of Cellular Decking.

One of the principal advantages of cellular construction (Fig. 9-34) is that, in addition to acting as load-carrying components, the decking provides spaces for power and telephone

Fig. 9-34 Cellular steel floor assemblies. (*a*) With cells used for wiring and distribution of conditioned air. *(H. H. Robertson Co.)* (*b*) With shear connectors for composite action. *(Inland-Ryerson Construction Products Co.)*

wiring. With proper outlet fittings, changes in electrical services can be made quickly and easily. Cells of the proper size can also function as air ducts. In addition, use of cellular steel construction can result in a substantial saving in weight, particularly important in high-rise buildings.

From the standpoint of the contractor, cellular decking units are advantageous, because they can be quickly installed. Following structural-steel erection very closely, they can provide an immediate working platform for all trades.

Cellular steel floors are not to be considered, necessarily, light-load construction. They are capable of carrying the heaviest loads likely to be encountered in any commercial building.

9-41. Types of Cellular Floor Units.

Many different designs of cellular steel panels are available. Outline cross sections of some popular types are shown in Fig. 9-35. Cellular long-span roof deck sections of the type illustrated in Fig. 9-31c are also suitable for use in floor construction, although they have been generally superseded by those shown in Fig. 9-35.

Fig. 9-35 Cellular steel floor sections. These are shown with and without indentations or embossments to indicate a few basic outlines. For other profiles, widths, and depths, see manufacturers' publications.

Cellular steel units may be designed to function as self-contained structural components, carrying the entire floor load with the concrete acting only as fill, or they may be designed to function as a composite member with the concrete slab. Cellular units designed for composite action have indentations or protrusions (embossments) formed in them, as may be seen in Fig. 9-34b, to provide mechanical anchorage with the concrete. Most of the units used in new construction are of the composite type.

Where the utmost in flexibility of electrical services is not essential, cellular sections are frequently alternated with open sections that do not have the bottom sheet (Fig. 9-34b), forming what is known as a *blend* system. Figure 9-36 shows a cross section of a portion of a blend system that combines closed cells for electrical raceways, closed cells for air ducts, and open sections.

Fig. 9-36 Portion of blend system combining electrical cells, air cells, and open units, shown through one pattern of indentations and embossments providing for composite action with the concrete slab. Also shown are welded studs, which are placed at beams supporting the decking, for composite action beween beams and slab. *(H. H. Robertson Co.)* See also Fig. 9-34.

The bottom sheets of closed cells can be perforated and acoustic pads inserted to provide acoustic ceilings. Some cells are available with specially designed pads to provide acoustic control and at the same time permit air to escape through the perforations, so that the cell acts as an air-diffusion chamber. The ceiling cavities in the deeper open sections (without the bottom sheet) may also be used to house fixtures for recessed lighting (Fig. 9-37).

9-42. Materials for Cellular Decking.

Cellular steel floor and roof panels usually are made of structural-quality sheet steel conforming to the requirements for Grade C of ASTM A611 (uncoated) or Grade A of A446 (galvanized), both with a minimum specified yield point of 33 ksi. Thicknesses range from 0.0299 to 0.1046 in. Combinations of thicknesses may be used, for example, 0.0478-in-thick top and 0.0359-in-thick bottom sheet.

The thickness of concrete over the tops of cellular units varies from about 2½ in up, depending on job requirements and on whether the concrete functions simply as fill material or as a structural component, composite with the steel units. For composite

Fig. 9-37 Cellular steel floor panel combining an acoustical ceiling, air-diffusion chamber, and recessed lighting.

action, concrete having a compressive strength of 3000 psi is commonly used, although concrete of higher strength is sometimes specified. It may be either stone concrete weighing about 145 lb/ft^3, or lightweight concrete, weighing about 110 lb/ft^3. (Concrete containing chloride salts should not be used with galvanized units.) Shrinkage reinforcement in the concrete usually consists of welded wire fabric. Wire gage depends on slab thickness.

9-43. Structural Design of Cellular Floors.

Strength and deflection calculations of cellular floor units are usually based upon the AISI "Specification for the Design of Cold-formed Steel Structural Members," using the procedures discussed in Arts. 9-12 through 9-28. Structural design for composite action follows conventional practice. It may be based on either shored or unshored construction.

Usual spans for the popular 1½- and 3-in-deep sections of the kind shown in Fig. 9-35a and b range from about 7 to 16 ft, depending on depth and thickness of section, thickness of concrete, and whether the design is noncomposite, unshored composite, or shored composite. The deeper sections, Fig. 9-35d, for example, can span up to 32 ft under floor loadings when used in shored composite construction.

As for other forms of construction, deflection limitations will sometimes control the design. These limitations depend to some extent on local building-code requirements. The traditional figure of $\frac{1}{360}$ times the span is a common limit for live-load deflection. Regardless of deflection considerations, it is generally considered good practice to limit the span of plain (noncomposite) floor units to 25 times the total depth of the floor (the distance from the bottom of the floor units to the top of the structural concrete slab), and to limit the span of composite cellular floor construction to 32 times that depth.

In addition to normal floor-load service, cellular steel floors can also function as shear diaphragms under lateral loadings, such as wind or seismic forces.

Steel floor units usually are attached to supporting framework by electric-arc welding. Where the units are continuous over one or more supports, attachment to the intermediate supports is usually specified. Intermittent positive attachment along overlapping edges is achieved by welding. Some units are made with continuously interlocking joints along the edges.

Welded-stud shear connectors can be installed in the troughs of cellular units, at steel beams, to provide for composite steel-concrete design of the beams, and incidentally to fasten the deck to the beams. Some cellular units are available with wide troughs designed to accommodate pairs of studs lengthwise of the beams (Fig. 9-34b). Units are also available with markedly trapezoidal troughs for additional concrete when required for shear transfer (Fig. 9-36).

Each manufacturer publishes specifications, load tables, and detailed recommendations for the use of the company's products. Manufacturers' recommendations should always be followed in designing any cellular-floor installation.

9-44. Details and Accessories for Cellular Steel Floors.

Structural details of cellular units differ considerably among manufacturers. These details include the exact configurations and dimensions of the sections, the design of the edge laps or interlocks, and the pattern of the indentations or embossments that are formed in some units for composite action. Structural accessories required for all designs include closure pieces, installed at the ends of cell runs; flashing or cover strips at columns, beams and girders parallel to the cells, and elsewhere where the cellular construction is interrupted; and sealing material at the joints of abutting cells.

In addition to these structural accessories, other components include trench headers

(Fig. 9-34a) and crossover cells for wiring; header ducts and inlet fittings for air cells (usually a part of the air-circulation system); shelf tees for supporting the translucent cover sheets of open units housing recessed lighting fixtures; hanger tabs for suspended ceilings; and other items required to produce a complete floor-ceiling system. Nonstructural accessories are not necessarily furnished by the manufacturer of the floor units. Some of them may be included in the electrical and air-conditioning contracts.

9-45. Fire Resistance of Cellular Floor Construction.

Any required degree of fire protection for cellular steel floor and roof assemblies can be obtained with concrete topping and either rated membrane ceilings or direct application compounds (spray-on fireproofing). Under certain provisos with respect to the kind and thickness of concrete topping, cellular floors without any underside protection have been awarded fire-resistance ratings up to and including 3 hr (Table 9-16). A 2-hr rating requires less thickness of concrete than a 3-hr, and is sufficient to satisfy the fire-resistance requirements prescribed by practically all building codes for floor-ceiling assemblies for any occupancy in which cellular floors are likely to be used. Underside protection, however, is usually required for units underneath trench headers (Fig. 9-34a) and for the supporting steelwork. Under some conditions, it may prove economical to use a

TABLE 9-16 Fire-Resistance Ratings for Cellular Steel Floors without Underside Protection but with Restrained Assemblies

Hours	Floor units of galvanized steel	Concrete cover over top of units*	UL design No.†
3	1 1/2- to 3-in-deep cellular or open	4 3/16 in (lightweight)	D902 (225-3 hr)
2	1 1/2- to 3-in-deep cellular or open	3 1/4 in (lightweight)	D840 (267-2 hr)
	1 1/2- to 3-in-deep cellular or open	4 1/2 in (stone)	D902 (300-2 hr)
	4 1/2- to 7 1/2-in-deep cellular	3 in or 4 in	D903 (304-2 hr)
	4 1/2- to 7 1/2-in-deep cellular, with perforated bottom sheets	3 1/4 in or 4 in	D903 (304-2 hr)
	4 1/2- to 7 1/2-in deep open	3 1/2 in or 4 1/2 in	D903 (304-2 hr)
1 1/2	1 1/2- to 3-in-deep cellular or open	4 in (stone)	D902 (32-1 1/2 hr)
	4 1/2- to 7 1/2-in-deep cellular or open	3 in or 3 1/2 in	D903 (33-1 1/2 hr)
	4 1/2- to 7 1/2-in-deep cellular, solid bottom sheets	3 1/4 in (stone)	D903 (33-1 1/2 hr)
1	1 1/2- to 3-in-deep cellular or open	3 1/2 in (stone)	D902 (59-1 hr)
	4 1/2- to 7 1/2-in-deep cellular, solid or perforated bottom sheets	2 1/2 in (lightweight)	D903 (60-1 hr)
	4 1/2- to 7 1/2-in-deep open	2 3/4 in (lightweight) or 3 in (stone)	D903 (60-1 hr)

*Where two or three different types of units (solid bottom, perforated, or open) in a blend system require different thicknesses of concrete for the desired fire rating, the greatest concrete thickness required for any type of unit in the combination should be used throughout. Unless otherwise indicated, concrete may be lightweight or stone.

†Fire Resistance Index, Underwriters' Laboratories, Inc., Jan., 1979, Pre-1972 numbers are in parentheses.

minimum of concrete cover and protect the entire underside with a spray-on compound. This depends on the economics of each job.

The listing in Table 9-16 is not intended to be all-inclusive. For other ratings, for ratings with underside protection, and for other information, see the Underwriters Laboratories Fire Resistance Directory. Additional details of fire tests and fire resistance ratings for cellular and noncellular steel floor-ceiling assemblies are available from the American Insurance Assocation, Factory Mutual Research Corp., and other fire-testing agencies, as well as from the manufacturers of steel floor and roof decking.

OTHER FORMS OF LIGHTWEIGHT STEEL CONSTRUCTION

9-46. Building Systems.

Open-web sections of various kinds, unitary cold-formed framing members, and panels may be combined in a variety of ways to form complete building systems. A number of such systems are available, most of them including sheet-steel curtain-wall construction. (See Sec. 12.) They may or may not include structural-steel components. The systems are suitable for a variety of occupancies—industrial, commercial, educational, agricultural, and residential.

Modular construction, in which more or less standardized modules are put together to form buildings of various sizes, uses many different types of steel components.

Most of these building systems are of a proprietary nature. Refer to manufacturers' literature for detailed information in this rapidly changing field.

(Metal Building Dealers Association and Metal Building Manufacturers Association, "Metal Building Systems," MBMA, 1230 Keith Building, Cleveland, Ohio 44115.)

Section 10

Concrete Construction

Paul F. Rice
Vice President, Engineering
Concrete Reinforcing Steel Institute, Chicago, Ill.

Edward S. Hoffman
Vice President
Klein and Hoffman, Inc., Chicago, Ill.

Economical, durable construction with concrete requires a thorough knowledge of its properties and behavior in service, of approved design procedures, and of recommended field practices. Not only is such knowledge necessary to avoid disappointing results, especially when concrete is manufactured and formed on the building site, but also to obtain maximum benefits from its unique properties.

To provide the needed information, several organizations promulgate standards, specifications, recommended practices, and reports. Where these organizations are referred to in this section, they will be designated by their initials: PCA for Portland Cement Association, ACI for American Concrete Institute, CRSI for Concrete Reinforcing Steel Institute, and ASTM for American Society for Testing and Materials. In particular, for brevity, "Building Code Requirements for Reinforced Concrete," ACI 318-77, will be referred to either as ACI 318-77 or the ACI Building Code.

This code contains the following basic definitions:

Concrete is a mixture of portland cement, fine aggregate, coarse aggregate and water.

Admixture is a material other than portland cement, aggregate, or water that is added to concrete to modify its properties.

In this section, unless indicated otherwise, these definitions apply to the terms *concrete* and *admixture*.

CONCRETE AND ITS INGREDIENTS

10-1. Cements.

Articles 4-1 and 4-2 describe the various types of portland cement and give the chemical and physical requirements for them: Portland cements (ASTM C150) or air-entraining portland cements (ASTM C175) are available in types I to V and IA to IIIA, respectively, for use under different service conditions. Portland-type cements also include portland blast-furnace slag cement (ASTM C205) and portland-pozzolan cement (ASTM C340), which may be employed in concrete under the ACI Building Code, and may therefore be included under its definition of concrete. Proprietary "shrinkage-compensating" portland-type cements are also available for general use in concrete.

Although all the preceding cements can be employed for concrete, they are *not* interchangeable. (See Table 4-1.) Note that both tensile and compressive strengths vary considerably, at early ages in particular, even for the five types of basic portland cement. Consequently, although contract specifications for concrete strength are usually based on a standard 28-day age for the concrete, the proportions of ingredients required differ for each type. For the usual building project, where the load-strength relationship is likely to be critical at a point in strength gain equivalent to 7-day standard curing (Fig. 10-1), substitution of a different type (sometimes brand) of cement without reproportioning the mix may be dangerous.

The accepted specifications (ASTM) for cements do not regulate cement temperature nor color. Nevertheless, in hot-weather concreting, the temperature of the fresh concrete and therefore of its constituents must be controlled. Cement temperatures above 170°F are not recommended ("Recommended Practice for Hot Weather Concreting," ACI 305).

For exposed architectural concrete, not intended to be painted, control of color is desirable. For uniform color, the water-cement ratio and cement content must be kept constant, because they have significant effects on concrete color. Bear in mind that because of variations in the proportions of natural materials used, cements from different sources differ markedly in color. A change in brand of cement therefore can cause a change in color. Color differences also provide a convenient check for substitution of types (or brands) of cement different from those used in trial batches made to establish proportions to be employed for a building.

Fig. 10-1 Typical strength-gain rate with standard curing of non-air-entrained concrete with water-cement ratio of 0.50 and Chicago-area aggregates. *(Courtesy Materials Service Corp., Chicago, Ill.)*

10-2. Aggregates.

Articles 4-12 to 4-15 describe the characteristics of aggregates that are generally acceptable for concrete. Only material conforming to specifications for normal-weight aggregate (ASTM C330) or lightweight aggregate for structural concrete (ASTM C330) is accepted under the ACI Building Code without special tests. When an aggregate for which no experience record is available is considered for use, the modulus of elasticity and shrinkage as well as the compressive strength should be determined from trial batches of concrete made with the aggregate. In some localities, aggregates acceptable under C33 or C330 may impart abnormally low ratios of modulus of elasticity to strength (E_c/f'_c) or high shrinkage to concrete. Such aggregates should not be used.

10-3. Proportioning Concrete Mixes.

Principles for proportioning concrete to achieve a prescribed compressive strength after a given age under standard curing are simple.

1. The strength of a hardened concrete mix depends on the water-cement ratio. The water and cement form a paste. If the paste is made with more water, it becomes weaker (Fig. 10-2).

2. The ideal minimum amount of cement paste is that which will coat all aggregate particles and fill all voids.

3. For practical purposes, fresh concrete must possess workability sufficient for the placement conditions. For a given strength and with given materials, the cost of the mix increases as the workability increases. Additional workability is provided by more fine aggregate and more water, but more cement must also be added to keep the same water-cement ratio.

Because of the variations in material constituents, temperature, and workability required at job sites, theoretical approaches for determining ideal mix proportions usually do not give satisfactory results on the job. Most concrete therefore is proportioned empirically, in accordance with results from trial batches made with the materials to be used on the job. Small adjustments in the initial basic mix may be made as a job progresses; the frequency of such adjustments usually depends on the degree of quality control.

When new materials or exceptional quality control will be employed, the trial-batch method is the most reliable and efficient procedure for establishing proportions.

In determination of a concrete mix, a series of trial batches (or past field experience) is used to establish a curve relating the water-cement ratio to the strength and ingredient proportions of concrete, including admixtures if specified, for the range of desired strengths and workability (slump). Each point on the curve should represent the average of test results on at least three specimens, and the curve should be determined by at least three points. Depending on anticipated quality control, a demonstrated or expected coefficient of variation or standard deviation is assumed for determination of minimum average strength of test specimens (Art. 10-9). Mix proportions are selected from the curve to produce this average strength.

For any large project, significant savings can be made through use of quality control to reduce the overdesign otherwise required by a building code (law). When the owner's specifications include a minimum cement content, however, much of the economic incentive for the use of quality control is lost. See Fig. 10-3 for typical water-cement ratios.

Note that separate procedures are required for selecting proportions when lightweight aggregates are used, because their water-absorption properties differ from those of normal-weight aggregates.

Fig. 10-2 Variation of 28-day compressive strength of normal-weight concrete with water-cement ratio. Solid lines for air-entrained and non-air-entrained concretes indicate average results of tests by Materials Service Corp., Chicago, Ill. Dashed lines indicate relationship given in the ACI Building Code (ACI 318-77) for maximum permissible water-cement ratio and specified 28-day strengths.

Fig. 10-3 Variation of 28-day compressive strength of concrete with type of aggregate and water-cement ratio, except that strengths in excess of 7000 psi were determined at 56 days. All mixes contained a water-reducing agent and 100 lb/yd^3 of fly ash, and were non-air-entrained. In calculation of the water-cement ratio, two-thirds of the weight of the fly ash was added to the cement content.

("Building Code Requirements for Reinforced Concrete," ACI 318-77; "Recommended Practice for Selecting Proportions for Normal and Heavyweight Concrete," ACI 211.1; "Recommended Practice for Evaluation of Strength Test Results of Concrete," ACI 214; "Recommended Practice for Selecting Proportions for Structural Lightweight Concrete," ACI 211.2, American Concrete Institute.)

10-4. Yield Calculation.

Questions often arise between concrete suppliers and buyers regarding "yield," or volume of concrete supplied. A major reason for this is that often the actual yield may be less than the yield calculated from the volumes of ingredients. For example, if the mix temperature varies, less air may be entrained; or if the sand becomes drier and no corrections in batch weights are made, the yield will be under that calculated.

If the specific gravity (sp. gr.) and absorption (abs.) of the aggregates have been determined in advance, accurate yield calculations can be performed as often as necessary to adjust the yield for control of the concrete.

Example.

Yield of Non-Air-Entrained Concrete. The following material properties were recorded for materials used in trial batches: fine aggregate (sand) sp. gr. = 2.65, abs. = 1%; coarse aggregate (gravel) sp. gr. = 2.70, abs. = 0.5%; and cement, sp. gr. = 3.15 (typical). These properties are not expected to change significantly as long as the aggregates used are from the same source. The basic mix proportions for 1 yd^3 of concrete, selected from the trial batches are:

Cement: 564 lb (6 bags)
Surface-dry sand: 1170 lb
Surface-dry gravel: 2000 lb
Free water: 300 lb/yd^3 (36 gal/yd^3)

Check the yield.

$$\text{Cement volume} = \frac{564}{3.15 \times 62.4} = 2.87 \text{ ft}^3$$

$$\text{Water volume} = 300/62.4 = 4.81 \text{ ft}^3$$

$$\text{Sand volume} = \frac{1170}{2.65 \times 62.4} = 7.08 \text{ ft}^3$$

$$\text{Gravel volume} = \frac{2000}{2.70 \times 62.4} = \underline{11.87 \text{ ft}^3}$$

$$\text{Total volume of solid constituents} = 26.63 \text{ ft}^3$$

Volume of entrapped air = 27 − 26.63 = 0.37 ft^3 (1.4%)
Total weight, lb/yd^3 = 564 + 300 + 1170 + 2000 = 4034
Total weight, lb/ft^3 = 4034/27 = 149.4
Weight of standard 6 × 12 in cylinder (0.1963 ft^3) = 29.3 lb

These results indicate that some rapid field checks should be made. Total weight, lb, divided by the total volume, yd^3, reported on the trip tickets for truck mixers should be about 4000 on this job, unless a different slump was ordered and the proportions adjusted accordingly. If the specified slump for the basic mix was to be reduced, weight, lb/yd^3, should be increased, because less water and cement would be used and the cement paste (water plus cement) weighs 864/7.68 = 113 lb/ft^3 < 149.4 lb/ft^3. If the same batch weights are used for all deliveries, and the slump varies erratically, the yield also will vary. For the same batch weights, a lower slump is associated with underyield, a higher slump with overyield. With a higher slump, overyield batches are likely to be understrength, because some of the aggregate has been replaced by water.

The basic mix proportions in terms of weights may be based on surface-dry aggregates or on oven-dry aggregates. The surface-dry proportions are somewhat more convenient, since absorption then need not be considered in calculation of free water. Damp sand and gravel carry about 5 and 1% free water, respectively. The total weight of this free water should be deducted from the basic mix weight of water (300 lb/yd^3 in the example) to obtain the weight of water to be added to the cement and aggregates. The weight of water in the damp aggregates also should be added to the weights of the sand and gravel to obtain actual batch weights, as reported on truck-mixer delivery tickets.

10.5. Properties and Tests of Fresh (Plastic) Concrete.

About 2½ gal of water can be chemically combined with each 94-lb sack of cement for full hydration and maximum strength. Water in excess of this amount will be required, however, to provide necessary workability.

Although concrete technologists define and measure workability and consistency separately and in various ways, the practical user specifies only one—slump (technically a measure of consistency). The practical user regards workability requirements simply as provision of sufficient water to permit concrete to be placed and consolidated without honeycomb or excessive water rise; to make concrete "pumpable" if it is to be placed by pumps; and for slabs, to provide a surface that can be finished properly. These workability requirements vary with the job and the placing, vibration, and finishing equipment used.

Slump is tested in the field very quickly. An open-ended, 12-in-high, truncated metal cone is filled in three equal-volume increments and each increment is consolidated separately, all according to a strict standard procedure (ASTM C143, "Slump of Portland Cement Concrete"). Slump is the sag of the concrete, in, after the cone is removed. The slump should be measured to the nearest ¼ in, which is about the limit of accuracy reproducible by expert inspectors.

Unless the test is performed exactly in accordance with the standard procedure, the results are not comparable and therefore are useless.

The slump test is invalidated if: the operator fails to anchor the cone down by standing on the base wings; the test is performed on a wobbly base, such as forms carrying traffic or a piece of metal on loose pebbles; the cone is not filled by inserting material in small amounts all around the perimeter, or filled and tamped in three equal increments; the top two layers are tamped deeper than their depth plus about 1 in; the top is pressed down to level it; the sample has been transported and permitted to segregate without remixing; unspecified operations, such as tapping the cone, occur; the cone is not lifted up smoothly in one movement; the cone tips over because of filling from one side or pulling the cone to one side; or if the measurement of slump is not made to the center vertical axis of the cone.

Various penetration tests are quicker and more suitable for untrained personnel than the standard slump test. In each case, the penetration of an object into a flat surface of fresh concrete is measured and related to slump. These tests include use of the patented "Kelley ball" (ASTM C360, "Ball Penetration in Fresh Portland Cement Concrete") and a simple, standard tamping rod with a bullet nose marked with equivalent inches of slump.

Air Content. A field test frequently required measures the air entrapped and entrained in fresh concrete. Various devices (air meters) that are available give quick, convenient results. In the basic methods, the volume of a sample is measured, then the air content is removed or reduced under pressure, and finally the remaining volume is measured. The difference between initial and final volume is the air content. (See ASTM C138, C173, and C231.)

Cement Content. Tests on fresh concrete sometimes are employed to determine the amount of cement present in a batch. Although performed more easily than tests on hardened concrete, tests on fresh concrete nevertheless are too difficult for routine use and usually require mobile laboratory equipment.

For yield tests on fresh concrete, see Arts. 10-10 to 10-12.

10.6. Properties and Tests of Hardened Concrete.

The principal properties of concrete with which designers are concerned and symbols commonly used for some of these properties are:

f'_c = specified compressive strength, psi, determined in accordance with ASTM C39 from standard 6×12 in cylinders under standard laboratory curing. Unless otherwise specified, f'_c is based on tests on cylinders 28 days old

E_c = modulus of elasticity, psi, determined in accordance with ASTM C469; usually assumed as $E_c = \omega^{1.5}(33)\sqrt{f'_c}$, or for normal-weight concrete (about 145 lb/ft³), $E_c = 57,000\sqrt{f'_c}$

ω = weight, lb/ft³, determined in accordance with ASTM C138 or C567

f_t = direct tensile strength, psi

f_{ct} = average splitting tensile strength, psi, of lightweight-aggregate concretes determined by the split cylinder test (ASTM C496)

f_r = modulus of rupture, psi, the tensile strength at the extreme fiber in bending (commonly used for pavement design) determined in accordance with ASTM C78

Other properties, frequently important for particular conditions are: durability to resist freezing and thawing when wet and with deicers, color, surface hardness, impact hardness, abrasion resistance, shrinkage, behavior at high temperatures (about 500°F), insulation value at ordinary ambient temperatures, insulation at the high temperatures of a standard fire test, fatigue resistance, and for arctic construction, behavior at cold temperatures (-60 to -75°F). For most of the research on these properties, specially devised tests were employed, usually to duplicate or simulate the conditions of service anticipated. (See "Index to Proceedings of the American Concrete Institute.")

In addition to the formal testing procedures specified by ASTM and the special procedures described in the research references, some practical auxiliary tests, precautions in evaluating tests, and observations that may aid the user in practical applications follow.

Compressive Strength, f'_c. The standard test (ASTM C39) is used to establish the quality of concrete, as delivered, for conformance to specifications. Tests of companion field-cured cylinders measure the effectiveness of the curing (Art. 10-13).

Core tests (ASTM C42) of the hardened concrete in place, if they give strengths higher than the specified f'_c or an agreed-on percentage of f'_c (often 85%), can be used for acceptance of material, placing, consolidation, and curing. If the cores taken for these tests show unsatisfactory strength but companion cores given accelerated additional curing show strengths above the specified f'_c, these tests establish acceptance of the material, placing, and consolidation, and indicate the remedy, more curing, for the low in-place strengths.

For high-strength concretes, say above 5000 psi, care should be taken that the capping material is also high strength. Better still, the ends of the cylinders should be ground to plane.

Indirect testing for compressive strength includes surface-hardness tests (impact hammer). Properly calibrated, these tests can be employed to evaluate field curing. (See also Art. 10-13.)

Modulus of Elasticity E_c. This property is used in all design, but it is seldom determined by test, and almost never as a regular routine test. For important projects, it is best to secure this information at least once, during the tests on the trial batches at the various curing ages. An accurate value will be useful in prescribing camber or avoiding unusual deflections. An exact value of E_c is invaluable for long-span, thin-shell construction, where deflections can be large and must be predicted accurately for proper construction and timing removal of forms. Figure 10-4 is a plot of $E_c = \omega^{1.5}(33)\sqrt{f'_c}$ the formula given in the ACI Building Code.

Tensile Strength. The standard splitting test is a measure of almost pure uniform tension f_{ct}. The beam test (Fig. 10-5a) measures bending tension f_r on extreme surfaces (Fig. 10-5b), calculated for an assumed perfectly elastic, triangular stress distribution.

The split-cylinder test (Fig. 10-5c) is used for structural design. It is not sensitive to minor flaws or the surface condition of the specimen. The most important application of the splitting test is in establishment of design values for reinforcing-steel development length, shear in concrete, and deflection of structural lightweight-aggregate concretes.

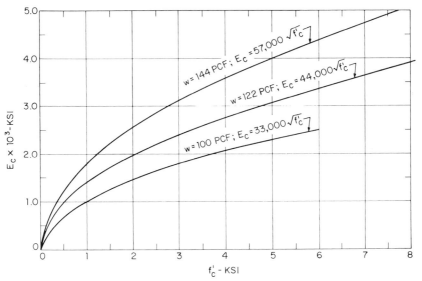

Fig. 10-4 Relationship between modulus of elasticity in compression and 28-day compressive strength of concrete. (ACI 318-77.)

Fig. 10-5 Test methods for tensile strength of concrete: (a) beam test determines modulus of rupture f_r; (b) stress distribution assumed for calculation of f_r; (c) split-cylinder test measures internal tension f_{ct}; (d) stress distribution assumed for f_{ct}.

The values f_{ct} (Fig. 10-5d) and f_r bear some relationship to each other, but are not interchangeable. The beam test is very sensitive, especially to flaws on the surface of maximum tension and to the effect of drying-shrinkage differentials, even between the first and last of a group of specimens tested on the same day. The value f_r is widely used in pavement design, where all testing is performed in the same laboratory and results are then comparable.

Special Properties. Frequently, concrete may be employed for some special purpose for which special properties are more important than those commonly considered. Sometimes, it may be of great importance to enhance one of the ordinary properties. These special applications often become apparent as new developments using new materials or as improvements using the basic materials. The partial list of special properties is constantly expanding—abrasion and impact resistance (heavy-duty floor surfacings), heat resistance (chimney stacks and jet engine dynamometer cells), light weight (concrete canoes), super-high-compressive strength, over 10 ksi (high-rise columns), waterproof concrete, resistance to chemical attack (bridge decks, chemical industry floors, etc.), increased tensile strength (highway resurfacing, precast products, etc.), no-shrink or expansive concrete (grouting under base plates), etc. Some of these special properties are achieved with admixtures (see Art. 10-8). Some utilize special cements (high-alumina cement for heat resistance or shrinkage-compensating cement for no-shrink concrete). Some utilize special aggregates (lightweight aggregate, steel fiber, asbestos fiber, glass fiber, and special heavy aggregate). (See "State-of-the-Art Report on Fiber Reinforced Concrete," ACI 544.1 R-73.) Some special properties—increased compressive and tensile strength, waterproofing, and improved chemical resistance are achieved with polymers, either as admixtures or surface treatment of hardened concrete. (See "Polymers in Concrete," ACI 548 R-77.)

10-7. Measuring and Mixing Concrete Ingredients.

Methods of measuring the quantities and mixing the ingredients for concrete, and the equipment available, vary greatly. For very small projects where mixing is performed on the site, the materials are usually batched by volume. Under these conditions, accurate proportioning is very difficult. To achieve a reasonable minimum quality of concrete, it is usually less expensive to prescribe an excess of cement than to employ quality control. The same conditions make use of air-entraining cement preferable to separate admixtures. This practical approach is preferable also for very small projects to be supplied with ready-mixed concrete. Economy with excess cement will be achieved whenever volume is so small that the cost of an additional sack of cement per cubic yard is less than the cost of a single compression test.

For engineered construction, some measure of quality control is always employed. In general, all measurements of materials including the cement and water should be by weight. The ACI Building Code provides a sliding scale of overdesign for concrete mixes that is inversely proportional to the degree of quality control provided. In the sense used here, such overdesign is the difference between the specified f'_c and the actual average strength as measured by tests.

Mixing and delivery of structural concrete may be performed by a wide variety of equipment and procedures:

Site mixed, for delivery by chute, pump, truck, conveyor, or rail dump cars. (Mixing procedure for normal-aggregate concretes and lightweight-aggregate concretes to be pumped are usually different, because the greater absorption of some lightweight aggregates must be satisfied before pumping.)

Central-plant mixed, for delivery in either open dump trucks or mixer trucks.

Central-plant batching (weighing and measuring), for mixing and delivery by truck ("dry-batched" ready mix).

Complete portable mixing plants are available and are commonly used for large building or paving projects distant from established sources of supply.

Generally, drum mixers are used. For special purposes, various other types of mixers are required. These special types include countercurrent mixers, in which the blades revolve opposite to the turning of the drum, usually about a vertical axis, for mixing very dry, harsh, nonplastic mixes. Such mixes are required for concrete masonry or heavy-duty floor toppings. Dry-batch mixers are used for dry shotcrete (sprayed concrete), where water and the dry-mixed cement and aggregate are blended between the nozzle of the gun and impact at the point of placing.

("Recommended Practice for Measuring, Mixing, Transporting and Placing Concrete," ACI 304.)

10-8. Admixtures.

These are materials other than portland cement, aggregate, or water that are added to concrete to modify its properties.

Air Entrainment. Air-entraining admixtures (ASTM C260) may be interground as additives with the cement at the mill or added separately at the concrete mixing plant, or both. Where quality control is provided, it is preferable to add such admixtures at the concrete plant so that the resulting air content can be controlled for changes in temperature, sand, or job requirements.

Use of entrained air is recommended for all concrete exposed to weathering or deterioration from aggressive chemicals. The ACI Building Code requires air entrainment for all concrete subject to freezing temperatures while wet. Detailed recommendations for air content are available in "Recommended Practice for Selecting Proportions for Normal and Heavyweight Concrete," ACI 211.1, and "Recommended Practice for Selecting Proportions for Structural Lightweight Concrete," ACI 211.2.

One common misconception relative to air entrainment is the fear that it has a deleterious effect on concrete strength. Air entrainment, however, improves workability. This will usually permit some reduction in water content. For lean, low-strength mixes, the improved workability permits a relatively large reduction in water content, sand content, and water-cement ratio, which tends to increase concrete strength. The resulting strength gain offsets the strength-reducing effect of the air itself, and a net increase in concrete strength is achieved. For rich, high-strength mixes, the relative reduction in the water-cement ratio is lower and a small net decrease in strength results, about on the same order of the air content (4 to 7%). The improved durability and reduction of segregation in handling, because of the entrained air, usually make air entrainment desirable, however, in all concrete except extremely high-strength mixtures, such as for lower-story interior columns or heavy-duty interior floor toppings for industrial wear.

Accelerators. Calcium chloride for accelerating the rate of strength gain in concrete (ASTM D98) is perhaps the oldest application of admixtures. Old specifications for winter concreting or masonry work commonly required use of a maximum of 1 to 3% $CaCl_2$ by weight of cement for all concrete. Proprietary admixtures now available may include accelerators, but not necessarily $CaCl_2$. The usual objective for use of an accelerator is to reduce curing time by developing 28-day strengths in 7 days approximately (ASTM C494).

In spite of users' familiarity with $CaCl_2$, a number of misconceptions about its effect persist. It has been sold (sometimes under proprietary names) as an accelerator, a cement replacement, an "antifreeze," a "waterproofer," and a "hardener." It is simply an accelerator; any improvement in other respects is pure serendipity. Recent discoveries, however, indicate possible corrosion damage from indiscriminate use of chloride-containing material in concrete exposed to stray currents, containing dissimilar metals, containing prestressing steel subject to stress corrosion, or exposed to severe wet freezing or salt water. For further information, see "Guide for Use of Admixtures in Concrete," ACI 212.2R.

Retarders. Unless proper precautions are taken, hot-weather concreting may cause "flash set," plastic shrinkage, "cold joints," or strength loss. Admixtures that provide controlled delay in the set of a concrete mix without reducing the rate of strength gain during subsequent curing offer an inexpensive solution to many hot-weather concreting problems. These (proprietary) admixtures are usually combined with water-reducing admixtures that more than offset the loss in curing time due to delayed set (ASTM C494). See "Hot Weather Concreting," ACI 305R, for further details on retarders, methods of cooling concrete materials, and limiting temperatures for hot-weather concreting.

Waterproofing. A number of substances, such as stearates and oils, have been employed as masonry-mortar and concrete admixtures for "waterproofing." Indiscriminate use of such materials in concrete without extremely good quality control usually results in disappointment. The various water-repellent admixtures are intended to prevent capillarity, but most severe leakage in concrete occurs at honeycombs, cold joints, cracks, and other noncapillary defects. Concrete containing water-repellent admixtures also requires extremely careful continuous curing, since it will be difficult to rewet after initial drying.

Waterproof concrete can be achieved by use of high-strength concrete with a low-water-cement ratio, air entrained to reduce segregation, designed to minimize crack width, and with good quality control and inspection during the mixing, placing, and curing operations. Surface coatings can be used to improve resistance to water penetration of vertical or horizontal surfaces. For detailed information on surface treatments, see "Guide to Durable Concrete," ACI 201.2R.

Cement Replacement. The term "cement replacement" is frequently misused in reference to chemical admixtures intended as accelerators or water reducers. Strictly, a cement replacement is a finely ground material, usually weakly cementitious (pozzolanic), which combines into a cementlike paste replacing some of the cement paste to fill voids between the aggregates. The most common applications of these admixtures are for low-heat, low-strength mass concrete or for concrete masonry. In the former, they fill voids and reduce the heat of hydration; in the latter, they fill voids and help to develop the proper consistency to be self-standing as the machine head is lifted in the forming process. Materials commonly used are fly ash (Fig. 10-3), hydraulic lime, natural cement and pozzolans. Fly ash and other pozzolans used as admixtures should conform to ASTM C618, "Specifications for Fly Ash and Raw or Calcined Natural Pozzolans for Use in Portland Cement Concrete."

Special-Purpose Admixtures. The list of materials employed from earliest times as admixtures for various purposes includes almost everything from human blood to synthetic coloring agents.

Admixtures for coloring concrete are available in all colors. The oldest and cheapest is perhaps carbon black.

Admixtures causing expansion for use in sealing cracks or under machine bases, etc., include powdered aluminum and finely ground iron.

Special admixtures are available for use where the natural aggregate is alkali reactive, to neutralize this reaction.

Proprietary admixtures are available that increase the tensile strength or bond strength of concrete. They are useful for making repairs to concrete surfaces.

For special problems requiring concrete with unusual properties, detailed recommendations of "Guide for Use of Admixtures in Concrete," ACI 212.2R, and references it contains, may be helpful.

For all these special purposes, a thorough investigation of admixtures proposed is recommended. Tests should be made on samples containing various proportions for colored concrete. Strength and durability tests should be made on concrete to be exposed to sunlight, freezing, salt, or any other job condition expected, and special tests should be made for any special properties required, as a minimum precaution.

QUALITY CONTROL

10-9. Mix Design.

Concrete mixes are designed with the aid of information obtained from trial batches or field experience with the materials to be used. In either case, the proportions of ingredients must be selected to produce, for at least three test specimens, an average strength f_{cr} greater than the specified strength f'_c.

The required excess, $f_{cr} - f'_c$, depends on the standard deviation σ expected. Strength data for determining the standard deviation can be considered suitable if they represent either a group of at least 30 consecutive tests representing materials and conditions of control similar to those expected or the statistical average for two groups totaling 30 or more tests. The tests used to establish standard deviation should represent concrete produced to meet a specified strength within 1000 psi of that specified for the work proposed.

$$\sigma = \sqrt{\frac{(x_1 - \overline{x})^2 + (x_2 - \overline{x})^2 + (x_3 - \overline{x})^2 + \cdots + (x_n - \overline{x})^2}{n}} \qquad (10\text{-}1)$$

where $x_1, x_2, \ldots, x_n$ = strength, psi, obtained in test of first, second, $\ldots$, nth sample, respectively
n = number of tests
$\overline{x}$ = average strength, psi of n cylinders

Coefficient of variation is the standard deviation expressed as a percentage of the average strength. ("Recommended Practice for Evaluation of Strength Test Results of Concrete," ACI 214.)

The strength used as a basis for selecting proportions of a mix should exceed the required f'_c by at least the amount indicated in Table 10-1.

The values for f_{cr} in Table 10-1 are the larger of the values calculated from Eqs. (10-2) and (10-3).

$$f_{cr} = f'_c + 1.343\sigma \qquad (10\text{-}2)$$
$$f_{cr} = f'_c + 2.326\sigma - 500 \qquad (10\text{-}3)$$

For an established supplier of concrete, it is very important to be able to document the value of σ. This value is based on a statistical analysis in which Eq. (10-1) is applied to at least 30 consecutive tests. These tests must represent similar materials and conditions of control not stricter than those to be applied to the proposed project. The lower the value of σ obtained from the tests, the closer the average strength is permitted to be to the specified strength. A supplier is thus furnished an economic incentive, lower cement content, to develop a record of good control (low σ). A supplier who does maintain such a record can, in addition, avoid the expense of trial batches.

Where no such production record exists, for example, when new sources of cement or aggregate are supplied to an established plant, to a new facility, such as a portable plant on the site, or for the first attempt at a specified strength f'_c more than 1000 psi above previous specified strengths, trial batches must be used as a basis for selecting initial proportions.

TABLE 10-1 Recommended Average Strengths of Test Cylinders for Selecting Proportions for Concrete Mixes

Range of standard deviation σ, psi	Average strength f_{cr}, psi
Under 300	$f'_c + 400$
300–400	$f'_c + 550$
400–500	$f'_c + 700$
500–600	$f'_c + 900$
Over 600	$f'_c + 1,200$

If the specified $f'_c \leq 5000$ psi for non-air-entrained concrete or $f'_c \leq 4500$ psi for air-entrained normal-weight concrete, trial batches need not be employed. Table 10-2 lists maximum water-cement ratios that may be used to select *initial* proportions for normal-weight concrete. These ratios are expected to provide $f_{cr} - f'_c \geq 1200$ psi and are economical only for small projects where the cost of trial batches is not justified.

Note that the term "initial proportions" includes cement. The initial proportions can be used during progress of a project only as long as the strength-test results justify them. The process of quality control of concrete for a project requires maintenance of a running average of strength-test results and changes in the proportions whenever the actual degree of control (standard deviation σ) varies from that assumed for the initial proportioning. Equations (10-2) and (10-3) are applied for this analysis. With contract specifications based on the 1977 ACI Building Code, no minimum cement content is required; so good control during a long-time project is rewarded by permission to use a lower cement content than would be permitted with inferior control.

Regardless of the method (field experience, trial batches, or maximum water-cement ratio) used, the basic initial proportions must be based on mixes with both air content and slump at the maximum permitted by the specifications.

Other ACI Building Code requirements for mix design are:

TABLE 10-2 Maximum Water-Cement Ratios for Specified Concrete Strengths*

Specified compressive strength f'_c, psi†	Non-air-entrained concrete		Air-entrained concrete	
	Absolute ratio by weight	U.S. gal per 94-lb bag of cement	Absolute ratio by weight	U.S. gal per 94-lb bag of cement
2500	0.67	7.6	0.54	6.1
3000	0.58	6.6	0.46	5.2
3500	0.51	5.8	0.40	4.5
4000	0.44	5.0	0.35	4.0
4500	0.38	4.3	‡	‡
5000	‡	‡	‡	‡

*As specified by ACI 318-77, Table 4.5.
†28-day strengths for cements meeting strength limits of ASTM C150 Types I, IA, II, or IIA, and 7-day strengths for Types III or IIIA. With most materials, the water-cement ratios shown will provide average strengths greater than indicated.
‡For strengths above 4500 psi (non-air-entrained concrete) and 4000 psi (air-entrained concrete), proportions should be established by field experience or laboratory trial batches.

TABLE 10-3 Required Air Entrainment in Concrete Exposed to Freezing and Thawing while Wet*

Nominal maximum size of coarse aggregate, in	Total air content, vol %
3/8	6–10
1/2	5–9
3/4	4–8
1	3.5–6.5
1 1/2	3–6
2	2.5–5.5
3	1.5–4.5

*From ACI 318-77, Table 4.6.1.

1. Concrete exposed to freezing and thawing while wet should have air entrained within the limits in Table 10-3, and the water-cement ratio by weight should not exceed 0.53. If lightweight aggregate is used, f'_c should be at least 3000 psi.

2. For watertight, normal-weight concrete, maximum water-cement ratios by weight are 0.50 for fresh water and 0.45 for seawater. With lightweight aggregate, minimum $f'_c = 3750$ psi for concrete exposed to fresh water and $f'_c = 4000$ psi for seawater.

Although the Code does not distinguish between a "concrete production facility" with in-house control and an independent concrete laboratory control service, the distinction is important. Very large suppliers have in-house professional quality control. Most smaller suppliers do not. Where the records of one of the latter might indicate a large standard deviation, but an independent quality-control service is utilized, the standard deviation used to select $f_{cr} - f'_c$ should be based on the proven record of the control agency. Ideally, the overdesign should be based, in these cases, on the record of the control agency operating in the concrete plant used.

10-10. Check Tests of Materials.

Without follow-up field control, all the statistical theory involved in mix proportioning becomes an academic exercise.

The complete description of initial proportions should include: cement analysis and source; specific gravity, absorption, proportions of each standard sieve size; fineness modulus; and organic tests for fine and coarse aggregates used, as well as their weights and maximum nominal sizes.

If the source of any aggregate is changed, new trial batches should be made. A cement analysis should be obtained for each new shipment of cement.

The aggregate gradings and organic content should be checked at least daily, or for each 150 yd³. The moisture content (or slump) should be checked continuously for all aggregates, and suitable adjustments should be made in batch weights. When the limits of ASTM C33 or C330 for grading or organic content are exceeded, proper materials should be secured and new mix proportions developed, or until these measures can be effected, concrete production may continue on an emergency basis but with a penalty of additional cement.

10-11. At the Mixing Plant—Yield Adjustments.

Well-equipped concrete producers have continuous measuring devices to record changes in moisture carried in the aggregates or changes in total free water in the contents of the mixer. The same measurements, however, may be easily made manually by quality-control personnel.

To illustrate: for the example in Art. 10-4, the surface-dry basic mix is cement, 564 lb; water, 300 lb; sand, 1170 lb; and gravel, 2000 lb. Absorption is 1% for the sand and 0.5% for the gravel. If the sand carries 5.5% and the gravel 1.0% total water by weight, the added free water becomes:

Sand: 1170 (0.055 − 0.01) = 53 lb
Gravel: 2000 (0.010 − 0.005) = 10 lb

Batch weights adjusted for yield become:

Cement: 564 lb
Water: 300 − 53 − 10 = 237 lb
Sand: 1170 + 53 = 1223 lb
Gravel: 2000 + 10 = 2010 lb

Note that the corrective adjustment includes adding to aggregate weights as well as deducting water weight. Otherwise, the yield will be low, and slump (slightly) increased. The yield would be low by about

$$\frac{53 + 10}{2.65 \times 62.4} = 0.381 \text{ ft}^3/\text{yd}^3 = 1.4\%$$

10-12. At the Placing Point—Slump Adjustments.

With good quality control, no water is permitted on the mixing truck. If the slump is too low (or too high) on arrival at the site, additional cement must be added. If the slump is too low (the usual complaint), additional water and cement in the prescribed water-cement ratio can also be added. After such additions, the contents must be thoroughly mixed, 2 to 3 min at high speed. Because placing-point adjustments are inconvenient and costly, telephone or radio communication with the supply plant is desirable so that most such adjustments may be made conveniently at the plant.

Commonly, a lesser degree of control is accepted in which the truck carries water, the driver is on the honor system not to add water without written authorization from a responsible agent at the site, and the authorization as well as the amounts added are recorded on the record (trip ticket) of batch weights.

Note: If site adjustments are made, test samples for strength-test specimens should be taken only after all site adjustments. For concrete in critical areas, such as lower-floor columns in high-rise buildings, strictest quality control is recommended.

10-13. Strength Tests.

Generally, concrete quality is measured by the specified compressive strength f'_c of 6 × 12 in cylinders after 28 days of laboratory curing.

Conventional Tests. The tests made for this purpose but performed after various periods of field curing are typically specified for tests of curing adequacy. For lightweight-aggregate concretes only, the same type of laboratory-cured test specimen is tested for tensile splitting strength f_{ct} to establish design values for deflection, development of reinforcing steel, and shear. Applicable ASTM specifications for these tests:

C21, "Making and Curing Concrete Compressive and Flexural Strength Test Specimens in the Field."

C39, "Test for Compressive Strength of Molded Concrete Cylinders."

C496, "Test for Splitting Tensile Strength of Molded Concrete Cylinders."

The specifications for standard methods and procedures of testing give general directions within which the field procedures can be adjusted to job conditions. One difficulty arises when the specimens are made in the field from samples taken at the job site. During the first 48 hr after molding, the specimens are very sensitive to damage and variations from standard laboratory curing conditions, which can significantly reduce the strength-test results. Yet, job conditions may preclude sampling, molding, and field storage on the same spot.

If the fresh-concrete sample must be transported more than about 100 ft to the point of molding cylinders, some segregation occurs. Consequently, the concrete sample should be remixed to restore its original condition. After the molds for test cylinders have been filled, if the specimens are moved, high-slump specimens segregate in the molds; low-slump specimens in the usual paper or plastic mold are often squeezed out of shape or separated into starting cracks. Such accidental damage varies with slump, temperature, time of set and molding, and degree of carelessness.

If the specimen cylinders are left on the job site, they must be protected against drying and accidental impact from construction traffic. If a workman stumbles over a specimen less than 3 days old, it should be inspected for damage. The best practice is to provide a small, insulated, dampproofed, locked box on the site in which specimens can be cast, covered, and provided with 60 to 80°F temperature and 100% humidity for 24 to 72 hr. Then, they can be transported and subjected to standard laboratory curing conditions at the testing laboratory. When transported, the cylinders should be packed and handled like fresh eggs, since loose rattling will have about an equivalent effect in starting incipient cracks.

Similarly, conditions for field-cured cylinders must be created as nearly like those of the concrete in place as possible. Also, absolute protection against impact or other damage must be provided. Because most concrete in place will be in much larger elements than a test cylinder, most of the in-place concrete will benefit more from retained heat of hydration (Fig. 10-6). This effect decreases rapidly, because the rate of heat development is greatest initially. To ensure similar curing conditions, field-cured test cylinders should be stored for the first 24 hr in the field curing box with the companion cylinders for laboratory curing. After this initial curing, the field-cured cylinders should be stored near the concrete they represent and cured under the same conditions.

Exceptions to this initial curing practice arise when the elements cast are of dimensions comparable to those of the cylinders, or the elements cast are not protected from drying or low temperatures, including freezing, or test cylinders are cured *inside* the elements they represent (patented system).

These simple, seemingly overmeticulous precautions will eliminate most of the unnecessary, expensive, job-delaying controversies over low tests. Both contractor and owner are justifiably annoyed when costly later tests on hardened concrete, after an even more costly job delay, indicate that the original fresh-concrete test specimens were defective and not the building concrete.

Special Tests. Many other strength tests or tests for special qualities are occasionally employed for special purposes. Those most often encountered in concrete building construction are strength tests on drilled cores and sawed beams (ASTM C42); impact tests (ASTM C805), e.g., Schmidt hammer; pullout tests (ASTM C900); penetration tests (ASTM C803); determination of modulus of elasticity during the standard compression test; and deflection measurements on a finished building element under load (Chapter 20, ACI 318-77). (See also "Commentary on ACI 318-77" and the "Manual of Concrete Inspection," (ACI SP-2.)

Newer methods for evaluating in-situ strength of concrete include the following: Methods, such as the one in which test cylinders are field-cured inside the in-situ concrete, measure compressive strength directly, refined even to measuring it in a desired

Fig. 10-6 Effect of curing temperature on strength-gain rate of concrete, with 28-day strength as basis.

direction. Others actually measure other properties, such as penetration, impact, or pullout, which are indirect measures of compressive strength, but may be employed because the property they measure is itself important. For example, in cantilevered form construction where forms for each new lift are bolted into the previous lift, pullout results may be more meaningful than standard compression tests. (See "Testing Hardened Concrete," ACI Monograph No. 9, 1976.) Most of the in-situ tests may also be classified as accelerated tests, although not all accelerated tests are performed in situ.

Because construction time is continually becoming a more important factor in overall construction economy, the standard 28-day strength becomes less significant. For example, the final strength at completion of a high-rise project requiring high-strength concrete in lower-story columns is often specified at 90 days. At the other extreme, a floor system may be loaded by the forms and concrete for the floor above in as little as two days. These conditions demand accelerated testing. (See "Use of Accelerated Strength Testing," ACI 214.1R-81.)

10-14. Test Evaluation.

On small jobs, the results of tests on concrete after the conventional 28 days of curing may be valuable only as a record. In these cases, the evaluation is limited to three options: (1) accept results, (2) remove and replace faulty concrete, or (3) conduct further tests to confirm option (1) or (2) or for limited acceptance at a lower-quality rating. The same comment can be applied to a specific element of a large project. If the element supports 28 days' additional construction above, the consequences of these decisions are expensive.

Samples sufficient for at least five strength tests of each class of concrete should be taken at least once each day, or once for each 150 yd³ of concrete or each 5000 ft² of surface area placed. Each strength test should be the average for two cylinders from the same sample. The strength level of the concrete can be considered satisfactory if the averages of all sets of three consecutive strength-test results equal or exceed the specified strength f'_c and no individual strength-test result falls below f'_c by more than 500 psi.

If individual tests of laboratory-cured specimens produce strengths more than 500 psi below f'_c, steps should be taken to assure that the load-carrying capacity of the structure is not jeopardized. Three cores should be taken for each case of a cylinder test more than 500 psi below f'_c. If the concrete in the structure will be dry under service conditions, the cores should be air-dried (temperature 60 to 80°F, relative humidity less than 60%) for 7 days before the tests and should be tested dry. If the concrete in the structure will be more than superficially wet under service conditions, the cores should be immersed in water for at least 48 hr and tested wet.

Regardless of the age on which specified design strength f'_c is based, large projects of long duration offer the opportunity for adjustment of mix proportions during the project. If a running average of test results and deviations from the average is maintained, then, with good control, the standard deviation achieved may be reduced significantly below the usually conservative, initially assumed standard deviation. In that case, a saving in cement may be realized from an adjustment corresponding to the improved standard deviation. If control is poor, the owner must be protected by an increase in cement. Specifications that rule out either adjustment are likely to result in less attention to quality control.

FORMWORK

For a recommended overall basis for specifications and procedures, see "Recommended Practice for Concrete Formwork," ACI 347. For materials, details, etc., for builders, see "Formwork for Concrete," ACI SP-4.

10-15. Responsibility for Formwork.

The exact legal determination of responsibilities for formwork failures among owner, architect, engineer, general contractor, subcontractors, or suppliers can be determined only by a court decision based on the complete contractual arrangements undertaken for a specific project.

Generally accepted practice makes the following rough division of responsibilities:

Safety. The general contractor is responsible for the design, construction, and safety of formwork. Subcontractors or material suppliers may subsequently be held responsible to the general contractor. The term "safety" here includes prevention of any type of formwork failure. The damage caused by a failure always includes the expense of the form itself, and may also include personal injury or damage to the completed portions of a structure. Safety also includes protection of all personnel on the site from personal injury during construction. Only the supervisor of the work can control the workmanship in assembly and the rate of casting on which form safety ultimately depends.

Structural Adequacy of the Finished Concrete. The structural engineer is responsible for the design of the concrete structure. The reason for specification requirements that the architect or engineer approve the order and time of form removal, shoring, and reshoring is to ensure proper structural behavior during such removal and to prevent overloading of recently constructed concrete below or damage to the concrete from which forms are removed prematurely. The architect or engineer should require approval for locations of construction joints not shown on plans to ensure proper transfer of shear and other forces through these joints. Specifications should also require that debris be cleaned from form material and the bottom of vertical element forms, and that form-release agents used be compatible with appearance requirements and future finishes to be applied. None of these considerations, however, involves the safety of the formwork per se.

10-16. Materials and Accessories for Forms.

When a particular design or desired finish imposes special requirements, and only then, the engineers' specifications should incorporate these requirements and preferably require sample panels for approval of finish and texture. Under competitive bidding, best bids are secured when the bidders are free to use ingenuity and their available materials ("Formwork for Concrete," ACI SP-4).

10-17. Pressure of Fresh Concrete on Vertical Forms.

This may be estimated from

$$p = 150 + 9000 \frac{R}{T} \tag{10-4}$$

where p = lateral pressure, psf
R = rate of filling, ft/hr
T = temperature of concrete, °F

See Fig. 10-7a.

For columns, the maximum pressure p_{max} is 3000 psf or $150h$, whichever is less, where h = height, ft, of fresh concrete above the point of pressure. For walls where R does not exceed 7 ft/hr, p_{max} = 2000 psf or $150h$, whichever is less.

For walls with rate of placement $R > 7$,

$$p = 150 + \frac{43,400}{T} + 2800 \frac{R}{T} \tag{10-5}$$

where p_{max} = 2000 psf or $150h$, whichever is less. See Fig. 10-7b.

The calculated form pressures should be increased if concrete weight exceeds 150 psf, cements are used that are slower setting than standard portland cement, slump is more than 4 in (not recommended), retarders are used to slow set, or the concrete is revibrated full depth, or forms are externally vibrated. Under these conditions, a safe design assumes that the concrete is a fluid with weight w and $p_{max} = wh$ for the full height of placement.

10-18. Design Loads for Horizontal Forms.

Best practice is to consider all known vertical loads, including the formwork itself, plus concrete, and to add an allowance for live load. This allowance, including workmen, runways, and equipment, should be at least 50 psf. When concrete will be distributed from overhead by a bucket or by powered buggies, an additional allowance of at least 25 psf for impact load should be added. Note that the weight of a loaded power buggy dropping off a runway, or an entire bucket full of concrete dropped at one spot, is not considered and might exceed designs based on 50- or 75-psf live load. Formwork should be designed alternatively, with continuity, to accept such spot overloads and distribute them to various unloaded areas, or with independently braced units to restrict a spot overload to a spot failure. The first alternative is preferable.

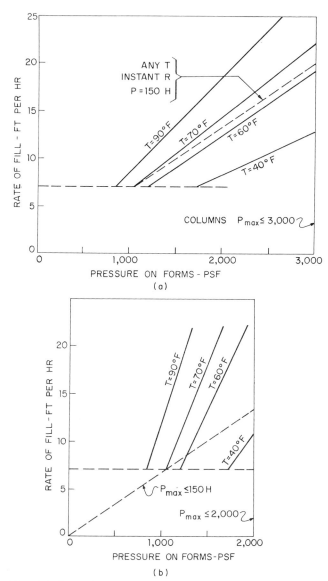

Fig. 10-7 Internal pressures exerted by concrete on formwork: (a) column forms; (b) wall forms.

10-19. Lateral Bracing for Shoring.

Most failures of large formwork are "progressive," vertically through several floors, or horizontally, as each successive line of shoring collapses like a house of cards. To eliminate all possibility of a large costly failure, the overall formwork shoring system should be reviewed before construction to avoid the usual "house-of-cards" design for vertical loads only. Although it is not always possible to foresee exact sources or amounts of lateral forces, shoring for a floor system should be braced to resist at least 100 lb/lin ft acting horizontally upon any of the edges, or a total lateral force on any edge equal to 2% of the total dead loads on the floor, whichever is larger.

Wall forms should be braced to resist local building-code wind pressures, plus at least 100 lb/lin ft at the top in either direction. The recommendation applies to basement wall forms even though wind may be less, because of the high risk of personal injury in the usual restricted areas for form watchers and other workmen.

10-20. Form Removal and Reshoring.

Much friction between contractors' and owners' representatives is created because of misunderstanding of the requirements for form removal and reshoring. The contractor is concerned with a fast turnover of form reuse for economy (with safety), whereas the owner wants quality, continued curing for maximum in-place strength, and an adequate strength and modulus of elasticity to minimize initial deflection and cracking. Both want a satisfactory surface.

Satisfactory solutions for all concerned consist of the use of high-early-strength concrete or accelerated curing, or substitution of a means of curing protection other than formwork. The use of field-cured cylinders (Arts. 10-6 and 10-13) in conjunction with appropriate nondestructive in-place strength tests (Art. 10-13) enables owner and contractor representatives to measure the rate of curing to determine the earliest time for safe form removal.

Reshoring or ingenious formwork design that keeps shores separate from surface forms, such as "flying forms" that are attached to the concrete columns (Strickland Shoring System, Jacksonville, Fla.), permits early stripping without premature stress on the concrete. Properly performed, reshoring is ideal from the contractors' viewpoint. But the design of reshores several stories in depth becomes very complex. The loads delivered to supporting floors are very difficult to predict and often require a higher order of structural analysis than that of the original design of the finished structure. To evaluate these loads, knowledge is required of the modulus of elasticity E_c of each floor (different), properties of the shores (complicated in some systems by splices), and the initial stress in the shores, which is dependent on how hard the wedges are driven or the number of turns of screw jacks, etc. ("Formwork for Concrete," ACI SP-4). When stay-in-place shores are used, reshoring is simpler (because variations in initial stress, which depend on workmanship, are eliminated), and a vertically progressive failure can be averted.

One indirect measure is to read deflections of successive floors at each stage. With accurate measurements of E_c, load per floor can then be estimated by structural theory. A more direct measure (seldom used) is strain measurement on the shores, usable with metal shores only. On large important projects, where formwork cost and cost of failure justify such expense, both types of measurement can be employed.

10-21. Special Forms.

Special formwork may be required for uncommon structures, such as folded plates, shells, arches, and posttensioned-in-place designs, or for special methods of construction, such as slip forming with the form rising on the finished concrete or with the finished concrete descending as excavation progresses, permanent forms of any type, preplaced-grouted-aggregate concreting, underwater concreting, and combinations of precast and cast-in-place concreting.

10-22. Inspection of Formwork.

Inspection of formwork for a building is a service usually performed by the architect, engineer, or both, for the owner and, occasionally, directly by employees of the owner.

10-22 Concrete Construction

Formwork should be inspected before the reinforcing steel is in place to ensure that the dimensions and location of the concrete conform to design drawings (Art. 10-15). This inspection would, however, be negligent if deficiencies in the areas of contractor responsibility were not noted also.

(See "Recommended Practice for Concrete Formwork," ACI 347, and "Formwork for Concrete," ACI SP-4, for construction check lists, and "Manual of Concrete Inspection," ACI SP-2.)

REINFORCEMENT

10-23. Reinforcing Bars.

The term *deformed steel bars for concrete reinforcement* is commonly shortened to *rebars*. The short form will be used in this section.

Standard United States Grade 60 rebars are available in 11 standard sizes, designated on design drawings and specifications by a size number (Table 10-4).

Standard billet-steel rebars are produced in two strength grades: 40 and 60. Axle-steel rebars come in Grades 40 and 60, and rail-steel rebars in Grades 50 and 60. The grade number indicates yield strength, ksi.

The standard grade for building construction is Grade 60 ("Manual of Standard Practice," Concrete Reinforcing Steel Institute, 933 N. Plum Grove Road, Shaumberg, Ill. 60195). It is also the standard grade on which the 1977 ACI Building Code requirements are based. Most of the United States rebar production is billet steel Grade 60. ASTM specifications for Grade 40 rebars provide for only four commonly available bar sizes, Nos. 3 through 6.

Bending recommendations for Grade 60 rebars conforming to both ACI Building Code and industry practice are given in Table 10-5.

Standard rebars are not produced to any weldability requirements. (ASTM A706 is a specification for special low-alloy steel rebars suitable for welding, generally without preheat, and for seismic-resistant "ductile frames." Such rebars are never carried in stock and have seldom been used. Specification of A706 should be considered only for very large projects where its special properties would justify its premium price.) The user intending any welding should follow procedures recommended by the American Welding Society (AWS D1.4), which are required by the ACI Building Code.

The standard specifications for rebars in the United States are ASTM A615 (billet

TABLE 10-4 Standard United States Rebar Sizes*

Bar size no.	Weight, lb per ft	Nominal diameter, in.	Cross-sectional area, sq in.	Nominal perimeter, in.
3	0.376	0.375	0.11	1.178
4	0.668	0.500	0.20	1.571
5	1.043	0.625	0.31	1.963
6	1.502	0.750	0.44	2.356
7	2.044	0.875	0.60	2.749
8	2.670	1.000	0.79	3.142
9	3.400	1.128	1.00	3.544
10	4.303	1.270	1.27	3.990
11	5.313	1.410	1.56	4.430
14	7.650	1.693	2.25	5.320
18	13.600	2.257	4.00	7.090

*Courtesy Concrete Reinforcing Steel Institute.

TABLE 10-5 Standard Hooks*

	Recommended end hooks—all grades, in or ft-in			
Bar size no.	180° hooks			90° hooks
	D†	A or G	J	A or G
3	2 1/4	5	3	6
4	3	6	4	8
5	3 3/4	7	5	10
6	4 1/2	8	6	1-0
7	5 1/4	10	7	1-2
8	6	11	8	1-4
9	9 1/2	1-3	11 3/4	1-7
10	10 3/4	1-5	1-1 1/4	1-10
11	12	1-7	1-2 3/4	2-0
14	18 1/4	2-3	1-9 3/4	2-7
18	24	3-0	2-4 1/2	3-5

†D = finished bend diameter, in.

Stirrups (ties similar)

Stirrup and tie hook dimensions, in— Grades 40-50-60 ksi

Bar size no.	D, in	90° hook	135° hook	
		Hook A or G	Hook A or G	H, approx.
3	1 1/2	4	4	2 1/2
4	2	4 1/2	4 1/2	3
5	2 1/2	6	5 1/2	3 3/4
6	4 1/2	1-0	7 3/4	4 1/2
7	5 1/4	1-2	9	5 1/4
8	6	1-4	10 1/4	6

135° Seismic stirrup/tie hook dimensions, (ties similar) in—Grades 40-50-60 ksi

Bar size no.	D, in	135° hook	
		Hook A or G	H, approx.
3	1 1/2	5	3 1/2
4	2	6 1/2	4 1/2
5	2 1/2	8	5 1/2
6	4 1/2	10 3/4	6 1/2
7	5 1/4	1-0 1/2	7 3/4
8	6	1-2 1/4	9

*Notes:
1. All specific sizes recommended by CRSI in this table meet minimum requirements of ACI 318-77.
2. 180° hook J dimension (sizes 10, 11, 14, and 18) and A or G dimension (Nos. 14 and 18) have been revised to reflect recent test research using ASTM/ACI bend-test criteria as a minimum.
3. Tables for stirrup and tie hook dimensions have been expanded to include sizes 6, 7, and 8 to reflect current design practices.

Courtesy of the Concrete Reinforcing Steel Institute.

steel), A616 (rail steel), and A617 (axle steel). Use the latest issue of these specifications, because properties involving yield strength and bendability are upgraded periodically.

10-24. Welded-Wire Fabric (WWF).

Welded-wire fabric is an orthogonal grid made with two kinds of cold-drawn wire: smooth (or plain) or deformed. The wires can be spaced in each direction of the grid as desired, but for buildings, usually at 12 in maximum. Sizes of wires available in each type, with standard and former designations, are shown in Table 10-6.

Welded-wire fabric usually is designated WWF on drawings. Sizes of WWF are designated by spacing followed by wire sizes; for example, WWF 6 × 12, W11/W7, which indicates smooth wires, size W11, spaced at 6 in, and size W7, spaced at 12 in. WWF 6 × 12, D-11/D-7 indicates deformed wires of the same nominal size and spacing.

All WWF can be designed for Grade 60 material. Some of the smaller sizes are

TABLE 10-6 Standard Wire Sizes for Reinforcement

Deformed wire (A496) size	Plain wire (A82) size	Nominal dia, in.	Nominal area, sq in.	AS & W Steel-wire gages			
				Dia, in.	AS & W gage	Dia, in.	Area, sq in.
D-31	W31	0.628	0.310	$1/2$	...	0.5000	0.1964
D-30	W30	0.618	0.300	...	7/0	0.4900	0.1886
D-29	...	0.608	0.290	$15/32$	...	0.4688	0.1726
D-28	W28	0.597	0.280	...	6/0	0.4615	0.1673
D-27	...	0.586	0.270	$7/16$	...	0.4375	0.1503
D-26	W26	0.575	0.260	...	5/0	0.4305	0.1456
D-25	...	0.564	0.250	$13/32$	...	0.4062	0.1296
D-24	W24	0.553	0.240	...	4/0	0.3938	0.1218
D-23	...	0.541	0.230	$3/8$	...	0.3750	0.1104
D-22	W22	0.529	0.220	...	3/0	0.3625	0.1032
D-21	...	0.517	0.210	$11/32$	...	0.3438	0.09281
D-20	W20	0.504	0.200	...	2/0	0.3310	0.08605
D-19	...	0.491	0.190	$5/16$	...	0.3125	0.07670
D-18	W18	0.478	0.180	...	0	0.3065	0.07378
D-17	...	0.465	0.170	...	1	0.2830	0.06290
D-16	W16	0.451	0.160	$9/32$	...	0.2812	0.06213
D-15	...	0.437	0.150	...	2	0.2625	0.05412
D-14	W14	0.422	0.140	$1/4$	...	0.2500	0.04909
D-13	...	0.406	0.130	...	3	0.2437	0.04664
D-12	W12	0.390	0.120	...	4	0.2253	0.03987
D-11	...	0.374	0.110	$7/32$	...	0.2188	0.03758
D-10	W10	0.356	0.100	...	5	0.2070	0.03365
D-9	...	0.338	0.090	...	6	0.1920	0.02895
D-8	W8	0.319	0.080	$3/16$	...	0.1875	0.02761
D-7	W7	0.298	0.070	...	7	0.1770	0.02461
D-6	W6	0.276	0.060	...	8	0.1620	0.02061
...	W5.5	0.264	0.055	$5/32$	...	0.1562	0.01918
D-5	W5	0.252	0.050	...	9	0.1483	0.01727
...	W4.5	0.240	0.045	...	10	0.1350	0.01431
D-4	W4	0.225	0.040	$1/8$	...	0.1250	0.01227
...	W3.5	0.211	0.035	...	11	0.1205	0.01140
D-3	W3	0.195	0.030	...	12	0.1055	0.008715
...	W2.5	0.178	0.025	$3/32$	...	0.0938	0.006903
D-2	W2	0.159	0.020	...	13	0.0915	0.006576
...	W1.5	0.138	0.015	...	14	0.0800	0.005027
D-1	W1	0.113	0.010	...	15	0.0720	0.004072
...	W0.5	0.080	0.005	...	16	0.0625	0.003068

Fabrication and Placing of Rebars   10-25

required to develop higher yield strengths measured at 0.2% offset, but limitations in the ACI Building Code on strain restrict general use to yield strength of 60 ksi.

The standard specifications for wire in the United States are:

ASTM A82, Plain Wire
ASTM A496, Deformed Wire
ASTM A185, Plain Wire, WWF
ASTM A497, Deformed Wire, WWF

10-25. Prestressing Steel.

Cold-drawn high-strength wires, singly or stranded, with ultimate tensile strengths up to 270 ksi, and high-strength, alloy-steel bars, with ultimate tensile strengths up to 160 ksi, are used in prestressing. The applicable specifications for wire and strands are:

ASTM A416, Uncoated Seven-Wire Stress-Relieved Strand.
ASTM A421, Uncoated Stress-Relieved Wire.

The ACI Building Code also permits use of other types of strand or wire, including galvanized wire or strand, that meet the minimum requirement of ASTM A416 and A421.

Single wires or strands are used for plant-made pre-tensioned, prestressed members. Posttensioned prestressing may be performed with the member in place, on a site fabricating area, or in a plant. In posttensioned applications, single wires are usually grouped into parallel-wire tendons. Posttensioned tendons may also consist of solid bars or strands.

High-strength, alloy-steel bars for posttensioning must be proof-stressed during manufacture to 85% of the minimum guaranteed tensile strength, and subsequently given stress-relieving heat treatment. Final physical properties tested on full sections must conform to the following minimum requirements:

Yield strength (0.2% offset): 0.85 f_{pu} (specified ultimate).
Elongation in 20 diameters at rupture: 4%.
Reduction of area at rupture: 20%.

10-26. Fabrication and Placing of Rebars.

Fabrication of rebars consists of cutting to length and required bending. The preparation of shop drawings and (field) placing drawings is termed *detailing.* Ordinarily, the rebar supplier details, fabricates, and delivers to the site, as required. In the far-western states, the rebar supplier also ordinarily places the bars. In the New York City area, fabrication is performed on the site by the same (union) workmen who place the reinforcement. (See "Details and Detailing of Concrete Reinforcement," ACI 315.)

Fabrication Tolerances. The generally accepted industry tolerances on fabrication are the same regardless of varying contractual arrangements. (See Figs. 10-8, 10-9, and 10-10.) ("Manual of Standard Practice," CRSI.)

Erection. For construction on small sites, such as high-rise buildings in metropolitan areas, delivery of materials is a major problem. Reinforcement required for each area to be concreted at one time is usually delivered separately. Usually, the only available space for storage of this steel is the formwork in place. Under such conditions, unloading time becomes important.

The bars for each detail length, bar size, or mark number are wired into *bundles* for delivery. A *lift* may consist of one or more bundles grouped together for loading or unloading. The maximum weight of a single lift for unloading is set by the job crane capacity. The maximum weight of a *shop lift* for loading is usually far larger, and so shop

Fig. 10-8 Tolerances in fabrication of bar sizes Nos. 3 to 11: (a) cutting tolerances; (b) tolerances for standard hooks.

Fig. 10-9 Special fabrication tolerances: (a) end deviation at a cut; (b) maximum gap at erected end-bearing splices.

Fig. 10-10 Shipping limitations: (a) height limit; (b) length and height limits.

lifts may consist of several separately bundled *field lifts*. If the field equipment limits a field lift to less than 3000 lb, the usual minimum, special arrangements for building are required. Where site storage is provided, the most economical unloading without an immediately available crane is by dumping or rolling bundles off the side. Unloading arrangements should be agreed on in advance, so that loading can be carried out in the proper order and bars bundled appropriately. ("Placing Reinforcing Bars," CRSI.)

Placement Tolerances. The ACI Building Code prescribes rebar placement tolerances applicable simultaneously to effective depth d and to concrete cover in all flexural members, walls, and columns as follows:

Where d is 8 in or less, $\pm \frac{3}{8}$ in; more than 8 in, $\pm \frac{1}{2}$ in. These tolerances may not reduce cover more than one-third of that specified. The tolerance on longitudinal position of bends or ends of bars is ± 2 in. At discontinuous ends, the specified cover cannot be reduced more than $\frac{1}{2}$ in.

Bundling. Rebars may be placed in concrete members singly or in bundles (up to four No. 11 or smaller bars per bundle). This practice reduces rebar congestion or the need for several layers of single, parallel bars in girders. For columns, it eliminates many interior ties and permits use of No. 11 or smaller bars where small quantities of No. 14 or 18 are not readily available.

Only straight bars should be bundled ordinarily. Exceptions are bars with end hooks, usually at staggered locations, so that the bars are not bent as a bundle ("Placing Reinforcing Bars," CRSI).

A bundle is assembled by wiring the separate bars tightly in contact. If they are preassembled, placement in forms of long bundles requires a crane. Because cutoffs or splices of bars within a bundle must be staggered, it will often be necessary to form the bundle in place.

Bending and Welding Limitations. The ACI Code contains the following restrictions:

All bars must be bent without heating, except as permitted by the engineer.

Bars partly embedded in concrete may not be bent without permission of the engineer.

No welding of crossing bars (tack welding) is permitted without the approval of the engineer.

For unusual bends, heating may be permitted because bars bend more easily when heated. They are also less likely to crack if bent at less than the minimum radii permitted. If not embedded in thin sections of concrete, heating the bars to temperatures between 1200 and 1400°F facilitates bending, usually without damage to the bars or splitting of the concrete. If partly embedded bars are to be bent, heating controlled within these limits, plus the provision of a round fulcrum for the bend to avoid a sharp kink in the bar, are essential.

Tack welding creates a *metallurgical notch effect,* seriously weakening the bars. If different size bars are tacked together, the notch effect is aggravated in the larger bar. Tack welding therefore should never be permitted at a point where bars are to be fully stressed, and never for the assembly of ties or spirals to column verticals or stirrups to main beam bars.

When large, preassembled reinforcement units are desired, the engineer can plan the tack welding necessary as a supplement to wire ties at points of low stress or to added bars not required in the design.

10-27. Bar Supports.

Bar supports are commercially available in four general types of material: wire, precast concrete, molded plastic, and asbestos concrete. Complete specifications for the various types of wire and precast-concrete bar supports, as well as recommended maximum spacings and details for use, are given in the CRSI "Manual of Standard Practice," and "Wire Bar Supports," Product Standard, U.S. Department of Commerce.

Wire bar supports are generally available in the United States in four classes of rust resistance: A, plain; B, with plastic-covered feet; C, stainless or with stainless-steel wire feet; and D, special stainless; listed in the order of increasing cost. Precast-block bar supports are commercially available in the far-western states in three types: plain block, block with embedded tie wire for bottom-bar support, and block with a hole for the leg of a vertical bar for top- and bottom-bar support.

Various proprietary molded plastic bar supports are available and have been used successfully in the United States. There is, however, no generally established set of standard types or specifications for the plastic material. British standards warn the user that the coefficient of temperature expansion for the molded plastic materials used can differ from that of concrete by as much as 10 to 1. Investigation of this property is advisable before use of plastic supports in concrete that will be exposed to high variations in temperature.

10-28. Inspection of Reinforcement.

This involves approval of rebar material for conformance to the physical properties required, such as ASTM specifications for the strength grade specified; approval of the bar details and shop and placing drawings; approval of fabrication to meet the approved details within the prescribed tolerances; and approval of rebar placing.

Approvals of rebar material may be made on the basis of mill tests performed by the manufacturer for each heat from which the bars used originated. If samples are to be taken for independent strength tests, measurements of deformations, bending tests, and minimum weight, the routine samples may be best secured at the mill or the fabrication shop before fabrication. Occasionally, samples for check tests are taken in the field; but in this case, provision should be made for extra lengths of bars to be shipped and for schedules for the completion of such tests before the material is required for placing. Sampling at the point of fabrication, before fabrication, is recommended.

Inspection of fabrication and placement is usually most conveniently performed in the field, where gross errors would require correction in any event.

Under the ACI Building Code, the bars should be free of oil, paint, form coatings, and mud when placed. Rust or mill scale sufficiently loose to damage the bond is normally dislodged in handling.

If heavily rusted bars (which may result from improper storage for a long time exposed to rusting conditions) are discovered at the time of placing, a quick field test of suitability requires only scales, a wire brush, and calipers. In this test, a measured length of the bar is wire-brushed manually and weighed. If less than 94% of the nominal weight remains, or if the height of the deformations is deficient, the rust is deemed excessive. In either case, the material may then be rejected or penalized as structurally inadequate. Where space permits placing additional bars to make up the structural deficiency (in bond value or weight), as in walls and slabs, this solution is preferred, because construction delay then is avoided. Where job specifications impose requirements on rust more severe than the structural requirements of the ACI Code, for example, for decorative surfaces exposed to weather, the inspection should employ the special criteria required.

CONCRETE PLACEMENT

The principles governing proper placement of concrete are:

Segregation must be avoided during all operations between the mixer and the point of placement, including final consolidation and finishing.

The concrete must be thoroughly consolidated, worked solidly around all embedded items, and should fill all angles and corners of the forms.

Where fresh concrete is placed against or on hardened concrete, a good bond must be developed.

Unconfined concrete must not be placed under water.

The temperature of fresh concrete must be controlled from the time of mixing through final placement, and protected after placement.

("Mixing and Placing," ACI 218-77 and "Commentary"; "Specifications for Structural Concrete for Buildings," ACI 301; "Recommended Practice for Concrete Floor and Slab Construction," ACI 302.)

10-29. Methods of Placing.

Concrete may be conveyed from a mixer to point of placement by any of a variety of methods and equipment, if properly transported to avoid segregation. Selection of the most appropriate technique for economy depends on job conditions, especially job size, equipment, and the contractor's experience. In building construction, concrete usually is placed with hand- or power-operated buggies; drop-bottom buckets with a crane; inclined chutes; flexible and rigid pipe by pumping; *shotcrete,* in which either dry materials and water are sprayed separately or mixed concrete is shot against the forms; and for underwater placing, tremie chutes (closed flexible tubes). For mass-concrete construction, side-dump cars on narrow-gage track or belt conveyors may be used. For pavement, concrete may be placed by bucket from the swinging boom of a paving mixer, directly by dump truck or mixer truck, or indirectly by trucks into a spreader.

A special method of placing concrete suitable for a number of unusual conditions consists of grout-filling preplaced coarse aggregate. This method is particularly useful for underwater concreting, because grout, introduced into the aggregate through a vertical pipe gradually lifted, displaces the water, which is lighter than the grout. Because of bearing contact of the aggregate, less than usual overall shrinkage is also achieved.

10-30. Excess Water.

Even within the specified limits on slump and water-cement ratio, excess water must be avoided. In this context, excess water is present for the conditions of placing if evidence of water rise (vertical segregation) or water flow (horizontal segregation) occurs. Excess water also tends to aggravate surface defects by increased leakage through form openings. The result may be honeycomb, sandstreaks, variations in color, or soft spots at the surface.

In vertical formwork, water rise causes weak planes between each layer deposited. In addition to the deleterious structural effect, such planes, when hardened, contain voids through which water may pass.

In horizontal elements, such as floor slabs, excess water rises and forms a weak laitance layer at the top. This layer suffers from low strength, low abrasion resistance, high shrinkage, and generally poor quality.

10-31. Consolidation.

The purpose of consolidation is to eliminate voids of entrapped air and to ensure intimate complete contact of the concrete with the surfaces of the forms and the reinforcement. Intense vibration, however, may also reduce the volume of desirable entrained air; but this reduction can be compensated by adjustment of the mix proportions.

Powered internal vibrators are usually used to achieve consolidation. For thin slabs, however, high-quality, low-slump concrete can be effectively consolidated, without excess water, by mechanical surface vibrators. For precast elements in rigid, watertight forms, external vibration (of the form itself) is highly effective. External vibration is also effective with in-place forms, but should not be used unless the formwork is specially designed for the temporary increase in internal pressures to full fluid head plus the impact of the vibrator ("Formwork for Concrete," ACI SP-4).

Except in certain paving operations, vibration of the reinforcement should be avoided. Although it is effective, the necessary control to prevent overvibration is difficult. Also, when concrete is placed in several lifts or layers, vibration of vertical rebars passing into partly set concrete below may be harmful. Note, however, that revibration of concrete before the final set, under controlled conditions, can improve concrete strength markedly and reduce surface voids (bugholes). This technique is too difficult to control for general use on field-cast vertical elements, but it is very effective in finishing slabs with powered vibrating equipment.

Manual spading is most efficient for removal of entrapped air at form surfaces. This method is particularly effective where smooth impermeable form material is used and the surface is inward sloping.

On the usual building project, different conditions of placement are usually encountered that make it desirable to provide for various combinations of the techniques described. One precaution generally applicable is that the vibrators not be used to move the concrete laterally.

("Recommended Practice for Consolidation of Concrete," ACI 309.)

10-32. Concreting Vertical Elements.

The interior of columns is usually congested; it contains a large volume of reinforcing steel compared with the volume of concrete, and has a large height compared with its cross-sectional dimensions. Therefore, though columns should be continuously cast, the concrete should be placed in 2- to 4-ft-deep increments and consolidated with internal vibrators. These should be lifted after each increment has been vibrated. If delay occurs in concrete supply before a column has been completed, every effort should be made to avoid a cold joint. When the remainder of the column is cast, the first increment should be small, and should be vibrated to penetrate the previous portion slightly.

In all columns and reinforced narrow walls, concrete placing should begin with 2 to 4 in of grout. Otherwise, loose stone will collect at the bottom and form honeycomb. This grout should be proportioned for about the same slump as the concrete or slightly more, but at the same or lower water-cement ratio. (Some engineers prefer to start vertical placement with a mix having the same proportions of water, cement, and fine aggregate, but with one-half the quantity of coarse aggregate, as in the design mix, and to place a starting layer 6 to 12 in deep.)

When concrete is placed for walls, the only practicable means to avoid segregation is to place no more than a 24-in layer in one pass. Each layer should be vibrated separately and kept nearly level.

For walls deeper than 4 ft, concrete should be placed through vertical, flexible trunks or chutes located about 8 ft apart. The trunks may be flexible or rigid, and come in sections so that they can be lifted as the level of concrete in place rises. The concrete should not fall free, from the end of the trunk, more than 4 ft or segregation will occur, with the coarse aggregate ricocheting off the forms to lodge on one side. Successive layers after the initial layer should be penetrated by internal vibrators for a depth of about 4 to 6 in to ensure complete integration at the surface of each layer. Deeper penetration can be beneficial (revibration), but control under variable job conditions is too uncertain for recommendation of this practice for general use.

The results of poor placement in walls are frequently observed: sloping layer lines; honeycombs, leaking, if water is present; and, if cores are taken at successive heights, up to a 50% reduction in strength from bottom to top. Some precautions necessary to avoid these ill effects are:

Place concrete in level layers through closely spaced trunks or chutes.
Do not place concrete full depth at each placing point.
Do not move concrete laterally with vibrators.
For deep, long walls, reduce the slump for upper layers 2 to 3 in below the slump for the starting layer.
On any delay between placing of layers, vibrate the concrete thoroughly at the interface.

If concreting must be suspended between planned horizontal construction joints, level off the layer cast, remove any laitance and excess water, and form a straight, level construction joint, if possible, with a small cleat attached to the form on the exposed face (see also Art. 10-39).

10-33. Concreting Horizontal Elements.

Concrete placement in horizontal elements follows the same general principles outlined in Art. 10-32. Where the surface will be covered and protected against abrasion and weather, few special precautions are needed.

For concrete slabs, careless placing methods result in horizontal segregation, with desired properties in the wrong location, the top consisting of excess water and fines with low abrasion and weather resistance, and high shrinkage. For a good surface in a one-course slab, low-slump concrete and a minimum of vibration and finishing are desirable. Immediate screeding with a power-vibrated screed is helpful in distributing low-slump, high-quality concrete. No further finishing should be undertaken until free water, if any, disappears. A powered, rotary tamping float can smooth very-low-slump concrete at this stage. Final troweling should be delayed, if necessary, until the surface can support the weight of the finisher.

When concrete is placed for deep beams that are monolithic with a slab, the beam should be filled first. Then, a short delay for settlement should ensue before slab concrete is cast. Vibration through the top slab should penetrate the beam concrete sufficiently to ensure thorough combination.

When a slab is cast, successive batches of concrete should be placed on the edge of previous batches, to maintain progressive filling without segregation. For slabs with sloping surfaces, concrete placing should usually begin at the lower edge.

For thin shells in steeply sloping areas, placing should proceed downslope. Slump should be adjusted and finishing coordinated to prevent restraint by horizontal bars from causing plastic cracking in the fresh concrete.

10-34. Bonding to Hardened Concrete.

The surface of hardened concrete should be rough and clean where it is to be bonded with fresh concrete.

Vertical surfaces of planed joints may be prepared easily by wire brushing them, before complete curing, to expose the coarse aggregate. (The timing can be extended, if desired, by using a surface retarder on the bulkhead form.) For surfaces fully cured without earlier preparation, sandblasting, bush hammering, or acid washes (thoroughly rinsed off) are effective means of preparation for bonding new concrete. (See also Art. 10-33.)

Horizontal surfaces of previously cast concrete, for example, of walls, are similarly prepared. Care should be taken to remove all laitance and to expose sound concrete and coarse aggregate. (See also Art. 10-32. For two-course floors, see Art. 10-35.)

10-35. Heavy-Duty Floor Finishes.

Floor surfaces highly resistant to abrasion and impact are required for many industrial and commercial uses. Such surfaces are usually built as two-course construction, with a base or structural slab topped by a wearing surface. The two courses may be cast integrally or with the heavy-duty surface applied as a separate topping.

In the first process, which is less costly, ordinary structural concrete is placed and screeded to nearly the full depth of the floor. The wearing surface concrete, made with special abrasion-resistant aggregate, emery, iron filings, etc., then is mixed, spread to the desired depth, and troweled before final set of the concrete below.

The second method requires surface preparation of the base slab, by stiff brooming before final set to roughen the surface and thorough washing before the separate heavy-duty topping is cast. For the second method, the topping is a very dry (zero-slump) concrete, made with ⅜-in maximum-size special aggregate. This topping should be designed for a minimum strength, $f'_c = 6000$ psi. It must be tamped into place with powered tampers or rotary floats. (*Note:* If test cylinders are to be made from this topping, standard methods of consolidation will not produce a proper test; tamping similar in effect to that applied to the floor itself is necessary.) One precaution vital to the separate topping method is that the temperatures of topping and base slab must be kept compatible.

("Recommended Practice for Concrete Floor and Slab Construction," ACI 302.)

10-36. Concreting in Cold Weather.

Frozen materials should never be used. Concrete should not be cast on a frozen subgrade, and ice must be removed from forms before concreting. Concrete allowed to freeze wet, before or during early curing, may be seriously damaged. Furthermore, temperatures should be kept above 40°F for any appreciable curing (strength gain).

Concrete suppliers are equipped to heat materials and to deliver concrete at controlled temperatures in cold weather. These services should be utilized.

In very cold weather, for thin sections used in building, the freshly cast concrete must be enclosed and provided with temporary heat. For more massive sections or in moderately cold weather, it is usually less expensive to provide insulated forms or insulated coverings to retain the initial heat and subsequent heat of hydration generated in the concrete during initial curing.

The curing time required depends on the temperature maintained and whether regular or high-early-strength concrete is employed. High-early-strength concrete may be achieved with accelerating admixtures (Art. 10-8) or with high-early-strength cement (Types III or IIIA) or by a lower water-cement ratio, to produce the required 28-day strength in about 7 days.

An important precaution in employing heated enclosures is to supply heat without drying the concrete or releasing carbon dioxide fumes. Exposure of fresh concrete to drying or fumes results in chalky surfaces. Another precaution is to avoid rapid temperature changes of the concrete surfaces when heating is discontinued. The heat supply should be reduced gradually, and the enclosure left in place to permit cooling to ambient temperatures gradually, usually over a period of at least 24 hr.

("Cold-Weather Concreting," ACI 306R.)

10-37. Concreting in Hot Weather.

Mixing and placing concrete at a high temperature may cause *flash set* in the mixer, during placing, or before finishing can be completed. Also, loss of strength can result from casting hot concrete.

In practice, most concrete is cast at about 70 ± 20°F. Research on the effects of casting temperature shows highest strengths for concrete cast at 40°F and significant but practically unimportant increasing loss of strength from 40°F to 90°F. For higher temperatures, the loss of strength becomes important. So does increased shrinkage. The increased shrinkage is attributable not only to the high temperature, but also to the increased water content required for a desired slump as temperature increases. See Fig. 10-6.

For ordinary building applications, concrete suppliers control temperatures of concrete by cooling the aggregates and, when necessary, by supplying part of the mixing water as crushed ice. In very hot weather, these precautions plus sectional casting, to permit escape of the heat of hydration, may be required for massive foundation mats. Retarding admixtures are also used with good effect to reduce slump loss during placing and finishing.

("Hot-Weather Concreting," ACI 305R.)

10-38. Curing Concrete.

Curing of concrete consists of the processes, natural and artificially created, that affect the extent and rate of hydration of the cement.

Many concrete structures are cured without artificial protection of any kind. They are allowed to harden while exposed to sun, wind, and rain. This type of curing is unreliable, because water may evaporate from the surface.

Various means are employed to cure concrete by controlling its moisture content or its temperature. In practice, curing consists of conserving the moisture within newly placed concrete by furnishing additional moisture to replenish water lost by evaporation. Usually, little attention is paid to temperature, except in winter curing and steam curing.

More effective curing is beneficial in that it makes the concrete more watertight and increases the strength.

Methods for curing may be classified as:

1. Those that supply water throughout the early hydration process and tend to maintain a uniform temperature. These methods include ponding, sprinkling, and application of wet burlap or cotton mats, wet earth, sawdust, hay, or straw.

2. Those designed to prevent loss of water but having little influence on maintaining a uniform temperature. These methods include waterproof paper and impermeable membranes. The latter is usually a clear or bituminous compound sprayed on the concrete to fill the pores and thus prevent evaporation. A fugitive dye in the colorless compound aids the spraying and inspection.

A white pigment that gives infrared reflectance can be used in a curing compound to keep concrete surfaces cooler when exposed to the sun.

The criterion for judging the adequacy of field curing provided in the ACI Building Code is that the field-cured test cylinders produce 85% of the strengths developed by companion laboratory-cured cylinders at the age for which strength is specified.

10-39. Joints in Concrete.

Several types of joints may occur or be formed in concrete structures:

Construction joints are formed when fresh concrete is placed against hardened concrete.

Expansion joints are provided in long components to relieve compressive stresses that would otherwise result from a temperature rise.

Contraction joints (control joints) are provided to permit concrete to contract during a drop in temperature and to permit drying shrinkage without resulting uncontrolled random cracking.

Contraction joints should be located at places where concrete is likely to crack because of temperature changes or shrinkage. The joints should be inserted where there are thickness changes and offsets. Ordinarily, joints should be spaced 30 ft c to c or less in exposed structures, such as retaining walls.

To avoid unsightly cracks due to shrinkage, a dummy-type contraction joint is frequently used (Fig. 10-11). When contraction takes place, a crack occurs at this deliber-

Fig. 10-11 Control joints for restraining shrinkage and temperature cracks: (*a*) slab on grade; (*b*) wall.

ately made plane of weakness. In this way, the crack is made to occur in a straight line easily sealed.

Control joints may also consist of a 2- or 3-ft gap left in a long wall or slab, with the reinforcement from both ends lapped in the gap. Several weeks after the wall or slab has been concreted, the gap is filled with concrete. By that time, most of the shrinkage has taken place. A checkerboard pattern of slab construction can be used for the same reason.

In expansion joints, a filler is usually provided to separate the two parts of the structure. This filler should be a compressible substance, such as corkboard or premolded mastic. The filler should have properties such that it will not be squeezed out of the joint, will not slump when heated by the sun, and will not stain the surface of the concrete.

To be waterproof, a joint must be sealed. For this purpose, copper flashing may be used. It is usually embedded in the concrete on both sides of the joint, and folded into the joint so that the joint may open without rupturing the metal. The flashing must be strong enough to hold its position when the concrete is cast.

Proprietary flexible water stops and polysulphide calking compounds may also be used as sealers.

Open expansion joints are sometimes employed for interior locations where the opening is not objectionable. When exposed to water from above, as in parking decks, open joints may be provided with a gutter below to drain away water.

The engineer should show all necessary vertical and horizontal joints on design drawings. All pertinent details affecting reinforcement, water stops, and sealers should also be shown.

Construction joints should be designed and located if possible at sections of minimum shear. These sections will usually be at the center of beams and slabs, where the bending moment is highest. They should be located where it is most convenient to stop work. The construction joint is often keyed for shearing strength.

If it is not possible to concrete an entire floor in one operation, vertical joints preferably should be located in the center of a span. Horizontal joints are usually provided between columns and floor; columns are concreted first, then the entire floor system.

Various types of construction joints are shown in Fig. 10-12. The numbers on each section refer to the sequence of placing concrete.

If the joint is horizontal, as in Fig. 10-12a, water may be trapped in the key of the joint. If the joint is vertical, the key is easily formed by nailing a wood strip to the inside of the forms. A raised key, as in Fig. 10-12b, makes formwork difficult for horizontal joints.

In the horizontal joint in Fig. 10-12c, the key is made by setting precast-concrete blocks into the concrete at intermittent intervals. The key in Fig. 10-12d is good if the shear acts in the directions shown.

The V-shaped key in Fig. 10-12e can be made manually in the wet concrete for horizontal joints.

The key is eliminated in Fig. 10-12f, reliance being placed on friction on the roughened surface. This method may be used if the shears are small, or if there are large compressive forces or sufficient reinforcement across the joint.

See also Arts. 10-32 to 10-34.

Fig. 10-12 Types of construction joints. Circled numbers indicate order of casting.

10-40. Inspection of Concrete Placement.

Concrete should be inspected for the owner before, during, and after casting. Before concrete is placed, the formwork must be free of ice and debris and properly coated with bond-breaker oil. The rebars must be in place, properly supported to bear any traffic they will receive during concrete placing. Conduit, inserts, and other items to be embedded must be in position, fixed against displacement. Construction personnel should be available, usually carpenters, bar setters, and other trades, if piping or electrical conduit is to be embedded, to act as form watchers and to reset any rebars, conduit, or piping displaced.

As concrete is cast, the slump of the concrete must be observed and regulated within prescribed limits, or the specified strengths based on the expected slump may be reduced. If the inspector of placing is also responsible for sampling and making cylinders, he should test slump, entrained air, temperatures, and unit weights during concreting. He should control any field adjustment of slump and added water and cement. The inspector should also ascertain that handling, placing, and finishing procedures that have been agreed on in advance are properly followed, to avoid segregated concrete. In addition, he should ensure that any emergency construction joints made necessary by stoppage of concrete supply, rain, or other delays are properly located and made in accordance with procedures specified or approved by the engineer.

Inspection is complete only when concrete is cast, finished, protected for curing, and attains full strength.

(ACI "Manual of Concrete Inspection.")

ANALYSIS OF CONCRETE STRUCTURES

Under the ACI Building Code, concrete structures generally may be analyzed by elastic theory. When specific limiting conditions are met, certain approximate methods are permitted. For some cases, the Code recommends an empirical method.

10-41. Analyses of One-Way Floor and Roof Systems.

The ACI Building Code permits an approximate analysis for continuous systems in ordinary building if:

Components are not prestressed.
In successive spans, the ratio of the larger span to the smaller does not exceed 1.20.
The spans carry only uniform loads.
The ratio of live load to dead load does not exceed 3.

This analysis determines the maximum moments and shears at faces of supports, and the midspan moments representing envelope values for the respective loading combinations. In this method, moments are computed from

$$M = CwL_n^2 \qquad (10\text{-}6)$$

where C = coefficient, given in Fig. 10-13
w = uniform load
L_n = clear span for positive moment or shear and the average of adjacent clear spans for negative moment

For an elastic ("exact") analysis, the spans L of members that are not built integrally with their supports should be taken as the clear span plus the depth of slab or beam but need not exceed the distance between centers of supports. For spans of continuous

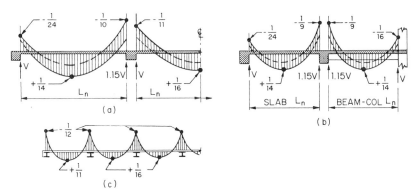

Fig. 10-13 Coefficients C for calculation of bending moments from $M = CwL_n^2$ in approximate analysis of beams and one-way slabs with uniform load w. For shears, $V = 0.5wL_n$. (a) More than two spans. (b) Two-span slab or beam. (c) Slabs—all spans, $L_n \leq 10$ ft.

frames, spans should be taken as the distance between centers of supports. For solid or ribbed slabs with clear spans not exceeding 10 ft, if built integrally with their supports, spans may be taken as the clear distance between supports.

If an elastic analysis is performed for continuous flexural members for each loading combination expected, calculated moments may be redistributed if the ratio ρ of tension-reinforcement area to effective concrete area or ratio of $\rho - \rho'$, where ρ' is the compression-reinforcement ratio, to the balanced-reinforcement ratio ρ_b, lie within the limits given in the ACI Code. When ρ or $\rho - \rho'$ is not more than $0.5\rho_b$, percentage

$$\gamma = 20\left(1 - \frac{\rho - \rho'}{\rho_b}\right) \tag{10-7}$$

Positive moments should be changed accordingly (increased by this percentage when negative moments are decreased) to reflect the moment redistribution property of under-reinforced concrete in flexure.

For example, suppose a 20-ft interior span of a continuous slab with equal spans is made of concrete with a strength f'_c of 4 ksi and reinforced with bars having a yield strength f_y of 60 ksi. Dead load and live load are both 0.100 ksf. The design moments are determined as follows:

Maximum negative moments occur at the supports when this span and adjacent spans carry both dead and live loads. Call this loading Case 1. For Case 1 then, maximum negative moment equals

$$M = -(0.100 + 0.100)(20)^2/12 = -6.67 \text{ ft-kips/ft}$$

The corresponding positive moment at midspan is 3.33 ft-kips/ft.

Maximum positive moment in the interior span occurs when it carries full load but adjacent spans support only dead loads. Call this loading Case 2. For Case 2, then, the negative moment is $-(6.67 - 3.33/2) = -5.00$ ft-kips/ft, and the maximum positive moment is 5.00 ft-kips/ft. Figure 10-14a shows the maximum moments.

For the concrete and reinforcement properties given, the balanced-reinforcement ratio is, by Eq. (10-26),

$$\rho_b = \frac{0.85\beta_1 f'_c}{f_y} \frac{87}{87 + f_y} = 0.0285$$

Assume now that reinforcement ratios for the top steel and bottom steel are 0.00267 and

Fig. 10-14 Bending moments in an interior 20-ft span of a continuous one-way slab: (a) moments for Case 1 (this and adjacent spans fully loaded) and Case 2 (this span fully loaded but adjacent spans with only dead load); (b) Case 1 moments after redistribution.

0.002, respectively. If alternate bottom bars extend into the supports, $\rho' = 0.001$. Substitution in Eq. (10-7) gives for the redistribution percentage

$$\gamma = 20\left(1 - \frac{0.00267 - 0.001}{0.0285}\right) = 18.8\%$$

The negative moment (Case 1) therefore can be decreased to $M = -6.67(1 - 0.188) = -5.41$ ft-kips/ft. The corresponding positive moment at midspan is $10 - 5.41 = 4.59$ kips/ft (Fig. 10-14b).

For Case 2 loading, if the negative moment is increased 18.8%, it becomes $-5.94 > 5.41$ ft-kips/ft. Therefore, the slab should be designed for the moments shown in Fig. 10-14b.

10-42. Two-Way Slab Frames.

The ACI Building Code prescribes an *equivalent column* concept for use in analysis of two-way slab systems. This concept permits a three-dimensional (space-frame) analysis in which the "equivalent column" combines the flexibility (reciprocal of stiffness) of the real column and the torsional flexibility of the slabs or beams attached to the column at right angles to the direction of the bending moment under consideration. The method, applicable for all ratios of successive spans and of dead to live load, is an elastic ("exact") analysis called the "equivalent frame method."

An approximate procedure, the "direct design method," is also permitted (within limits of load and span). This method constitutes the direct solution of a one-cycle moment distribution. (See also Art. 10-58.)

(P. F. Rice and E. S. Hoffman, "Structural Design Guide to the ACI Building Code," Van Nostrand Reinhold Company, New York.)

10-43. Special Analyses.

Space limitations preclude more than a brief listing of some of the special analyses required for various special types of reinforced concrete construction and selected basic references for detailed information. Further references to applicable research are available in each of the basic references.

Seismic-loading-resistant ductile frames: ACI 318-77; ACI Detailing Manual, 1980.

High-rise construction, frames, shear walls, frames plus shear walls, and tube concept: "Planning and Design of Tall Buildings," Vols. SC, CL, and CB, American Society of Civil Engineers.

Nuclear containment structures: ASME-ACI Code for Nuclear Containment Structures, ACI 359; also, ACI 349 and 349R.

Sanitary engineering structures: "Concrete Sanitary Engineering Structures," ACI 350R.

Bridges: "Analysis and Design of Reinforced Concrete Bridge Structures," ACI 343R.

It should be noted that the ACI Building Code specifically provides for the acceptance of analyses by computer or model testing to supplement the manual calculations when required by building officials.

STRUCTURAL DESIGN OF FLEXURAL MEMBERS

10-44. Strength Design with Load Factors.

Safe, economical strength design of reinforced concrete structures requires that their ultimate-load-carrying capacity be predictable or known. The safe, or service-load-carrying capacity can then be determined by dividing the ultimate-load-carrying capacity by a factor of safety.

The ACI Building Code provides for strength design of concrete members by use of **design loads** (actual and specified loads multiplied by load factors). Design axial forces, shears, and moments in members are determined as if the structure were elastic. Strength-design theory is then used to design critical sections for these axial forces, shears, and moments.

Strength design of reinforced concrete flexural members (Art. 10-64) may be based on the following assumptions and applicable conditions of equilibrium and compatibility of strains:

1. Strain in the reinforcing steel and the concrete is directly proportional to the distance from the neutral axis (Fig. 10-15).

2. The maximum usable strain at the extreme concrete compression fiber equals 0.003 in/in.

3. When the strain, in/in, in reinforcing steel is less than f_y/E_s, where f_y = yield strength of the steel and E_s = its modulus of elasticity, the steel stress, psi, equals 29,000,000 times the steel strain. After the steel yield strength has been reached, the stress remains constant at f_y, though the strain increases.

4. Except for prestressed concrete (Art. 10-106) or plain concrete, the tensile strength of the concrete is negligible in flexure.

5. The shape of the concrete compressive stress distribution may be assumed to be a rectangle, trapezoid, parabola, or any other shape in substantial agreement with comprehensive strength tests.

6. For a rectangular stress block, the compressive stress in the concrete should be

Fig. 10-15 Stress and strain in a rectangular reinforced-concrete beam, reinforced for tension only, at ultimate load: (a) section; (b) strain; (c) stress.

taken as $0.85f'_c$. This stress may be assumed constant from the surface of maximum compressive strain to a depth of $a = \beta_1 c$, where c is the distance to the neutral axis (Fig. 10-15). For $f'_c = 4000$ psi, $\beta_1 = 0.85$. For greater concrete strengths, β_1 should be reduced 0.05 for each 1000 psi in excess of 4000, but β_1 should not be taken less than 0.65.

(See also Art. 10-83 for columns.)

The ACI Code requires that the strength of a member based on strength design theory include a **capacity reduction factor** ϕ to provide for small adverse variations in materials, workmanship, and dimensions individually within acceptable tolerances. The degree of ductility, importance of the member, and the accuracy with which the member's strength can be predicted were considered in assigning values to ϕ:

ϕ should be taken as 0.90 for flexure and axial tension; 0.85 for shear and torsion; 0.70 for bearing on concrete; 0.65 for bending in plain concrete; and for axial compression combined with bending, 0.75 for members with spiral reinforcement, and 0.70 for other members.

For combinations of loads, a structure and its members should have the following strength U:

Dead load D and live load L, plus their internal moments and forces:

$$U = 1.4D + 1.7L \qquad (10\text{-}8)$$

Wind load W:

$$U = 0.75(1.4D + 1.7L + 1.7W) \qquad (10\text{-}9)$$

When D and L reduce the effects of W:

$$U = 0.9D + 1.3W \qquad (10\text{-}10)$$

Earthquake forces E:

$$U = 0.75(1.4D + 1.7L + 1.87E) \qquad (10\text{-}11)$$

When D and L reduce the effects of E:

$$U = 0.9D + 1.43E \qquad (10\text{-}12)$$

Lateral earth pressure H:

$$U = 1.4D + 1.7L + 1.7H \qquad (10\text{-}13)$$

When D and L reduce the effects of H:

$$U = 0.9D + 1.7H \qquad (10\text{-}14)$$

Lateral pressure F from liquids:

$$U = 1.4D + 1.7L + 1.4F \qquad (10\text{-}15)$$

Impact effects, if any, should be included with the live load L.

Where the effects of differential settlement, creep, shrinkage, or temperature change can be significant, they should be included with the dead load D, and the strength should not be less than $1.4D$ or

$$U = 0.75(1.4D + 1.7L) \qquad (10\text{-}16)$$

10-45. Allowable-Stress Design at Service Loads (Alternate Design Method).

Nonprestressed, reinforced-concrete flexural members (Art. 10-64) may be designed for flexure by the alternate design method of ACI 318-77 (working-stress design). In this method, members are designed to carry service loads (load factors and ϕ are taken as unity) under the straight-line (elastic) theory of stress and strain. (Because of creep in the concrete, only stresses due to short-time loading can be predicted with reasonable accuracy by this method.)

Working-stress design is based on the following assumptions:

Fig. 10-16 Stresses and strains in a beam with compression reinforcement, as assumed for working-stress design: (a) rectangular cross section of beam; (b) transformed section with twice the steel area, to allow for effects of creep of concrete; (c) assumed strains; (d) assumed distribution of stress in concrete.

1. A section plane before bending remains plane after bending. Strains therefore vary with distance from the neutral axis (Fig. 10-16c).

2. The stress-strain relation for concrete plots as a straight line under service loads within the allowable working stresses (Fig. 10-16c and d), except for *deep beams*.

3. Reinforcing steel takes all the tension due to flexure (Fig. 10-16a and b).

4. The modular ratio, $n = E_s/E_c$, where E_s and E_c are the moduli of elasticity of reinforcing steel and concrete, respectively, may be taken as the nearest whole number, but not less than 6 (Fig. 10-16b).

5. Except in calculations for deflection, n for lightweight concrete should be assumed the same as for normal-weight concrete of the same strength.

6. The compressive stress in the extreme surface of the concrete must not exceed $0.45f'_c$, where f'_c is the 28-day compressive strength of the concrete.

7. The allowable stress in the reinforcement must not be greater than the following:
Grades 40 and 50 .... 20 ksi
Grade 60 or greater .... 24 ksi
For ⅜-in or smaller-diameter reinforcement in one-way slabs with spans not exceeding 12 ft, the allowable stress may be increased to 50% of the yield stress but not to more than .... 30 ksi

8. For doubly reinforced flexural members, including slabs with compression reinforcement, an effective modular ratio of $2E_s/E_c$ should be used to transform the compression-reinforcement area for stress computations (Fig. 10-16b) to an equivalent concrete area. (This recognizes the effects of creep.) The allowable stress in the compression steel may not exceed the allowable tension stress.

Because the strains in the reinforcing steel and the adjoining concrete are equal, the stress in the tension steel f_s is n times the stress in the concrete f_c. The total force acting on the tension steel then equals nA_sf_c. The steel area A_s therefore can be replaced in stress calculations by a concrete area n times as large.

The transformed section of a concrete beam is a cross section normal to the neutral surface with the reinforcing replaced by an equivalent area of concrete (Fig. 10-16b). (In doubly reinforced beams and slabs, an effective modular ratio of $2n$ should be used to transform the compression reinforcement and account for creep and nonlinearity of the stress-strain diagram for concrete.) Stress and strain are assumed to vary with the distance from the neutral axis of the transformed section; that is, conventional elastic theory for homogeneous beams may be applied to the transformed section. Section properties, such as location of neutral axis, moment of inertia, and section modulus S, may be computed in the usual way for homogeneous beams, and stresses may be calculated from the flexure formula, $f = M/S$, where M is the bending moment at the section. This method is recommended particularly for T beams and doubly reinforced beams.

From the assumptions the following formulas can be derived for a rectangular section with tension steel only.

$$\frac{nf_c}{f_s} = \frac{k}{1-k} \quad (10\text{-}17)$$

$$k = \sqrt{2n\rho + (n\rho)^2} - n\rho \quad (10\text{-}18)$$

$$j = 1 - \frac{k}{3} \quad (10\text{-}19)$$

where $\rho = A_s/bd$ and b is the width and d the effective depth of the section (Fig. 10-16).
Compression capacity:

$$M_c = \tfrac{1}{2}f_c k j b d^2 = K_c b d^2 \quad (10\text{-}20a)$$

where $K_c = \tfrac{1}{2}f_c k j$.
Tension capacity:

$$M_s = f_s A_s j d = f_s \rho j b d^2 = K_s b d^2 \quad (10\text{-}20b)$$

where $K_s = f_s \rho j$.

Design of flexural members for shear, torsion, and bearing, and of other types of members, follows the strength design provisions of ACI 318-77, because allowable capacity by the alternate design method is an arbitrarily specified percentage of the strength.

10-46. Strength Design for Flexure.

Article 10-44 summarizes the basic assumptions for strength design of flexural members. The following formulas are derived from those assumptions.

The area A_s of tension reinforcement in a reinforced-concrete flexural member can be expressed as the ratio

$$\rho = \frac{A_s}{bd} \quad (10\text{-}21)$$

where b = beam width
d = effective beam depth = distance from the extreme compression surface to centroid of tension reinforcement

At ultimate strength of a critical section, the stress in this steel will be equal to its yield strength f_y, if the concrete does not first fail in compression. (See also Arts. 10-47 to 10-51 for additional reinforcement requirements.)

Singly Reinforced Rectangular Beams. For a rectangular beam, reinforced with only tension steel (Fig. 10-15), the total tension force in the steel at ultimate strength is

$$T = A_s f_y = \rho f_y b d \quad (10\text{-}22)$$

It is opposed by an equal compressive force

$$C = 0.85 f'_c b \beta_1 c \quad (10\text{-}23)$$

where f'_c = specified strength of the concrete
c = distance from extreme compression surface to neutral axis
β_1 = a constant (given in Art. 10-44)

Equating the compression and tension at the critical section yields:

$$c = \frac{\rho f_y}{0.85 \beta_1 f'_c} d \quad (10\text{-}24)$$

The criterion for compression failure is that the maximum strain in the concrete equals 0.003 in/in. In that case:

$$c = \frac{0.003}{f_s/E_s + 0.003} d \qquad (10\text{-}25)$$

where f_s is the steel stress, ksi, and $E_s = 29{,}000$ ksi is the steel modulus of elasticity.

Under balanced conditions, the concrete will reach its maximum strain of 0.003 in/in when the steel reaches its yield strength f_y. Then, c as given by Eq. (10-25) will equal c as given by Eq. (10-24). Also, the steel ratio for balanced conditions in a rectangular beam with tension steel only becomes:

$$\rho_b = \frac{0.85\beta_1 f'_c}{f_y} \frac{87{,}000}{87{,}000 + f_y} \qquad (10\text{-}26)$$

All structures should be designed to avoid sudden collapse. Therefore, reinforcement should yield before the concrete crushes. Gradual yielding will occur if the quantity of tensile reinforcement is less than the balanced percentage determined by strength design theory. To avoid compression failures, the ACI Building Code therefore limits the steel ratio ρ to a maximum of $0.75\rho_b$.

The Code also requires that ρ for positive-moment reinforcement be at least $200/f_y$ to prevent sudden collapse when the design positive-moment strength is equal to or less than the cracking moment. This requirement does not apply, however, if the reinforcement area at every section of the member is at least one-third greater than that required by the design moment.

For underreinforced rectangular beams with tension reinforcement only (Fig. 10-15) and a rectangular stress block with depth a ($\rho \leq 0.75\rho_b$), the flexural design strength may be determined from:

$$M_u = 0.90 bd^2 \rho f_y \left(1 - \frac{0.59 \rho f_y}{f'_c}\right) \qquad (10\text{-}27a)$$

$$= 0.90 A_s f_y \left(d - \frac{a}{2}\right) \qquad (10\text{-}27b)$$

$$= 0.90 A_s f_y jd \qquad (10\text{-}27c)$$

where $a = A_s f_y / 0.85 f'_c b$ and $jd = d - a/2$. (See Fig. 10-17 for determination of j.)

Tension-Steel Limitations. For flexural members of any cross-sectional shape, without compression reinforcement, the tension steel is limited by the ACI Building Code so that $A_s f_y$ does not exceed 0.75 times the total compressive force at balanced conditions. The total compressive force may be taken as the area of a rectangular stress block of a rectangular member; the strength of overhanging flanges or compression steel, or both, may be included. For members with compression reinforcement, the portion of tensile reinforcement equalized by compression reinforcement need not be reduced by the 0.75 factor.

Doubly Reinforced Rectangular Beams. For a rectangular beam with compression-steel area A'_s and tension-steel area A_s, the compression-steel ratio is

$$\rho' = \frac{A'_s}{bd} \qquad (10\text{-}28)$$

and the tension-steel ratio is

$$\rho = \frac{A_s}{bd} \qquad (10\text{-}29)$$

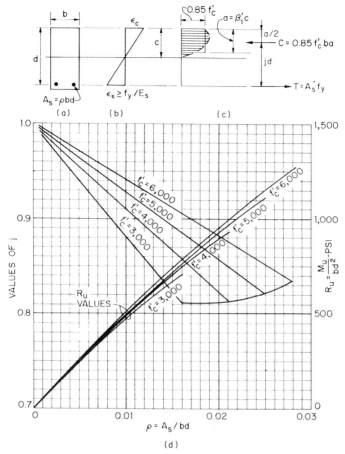

Fig. 10-17 For the rectangular beam shown in (*a*) reinforced, in tension only, with ASTM A615, Grade 60 bars: (*d*) moment-capacity chart based on assumed strain distribution in (*b*) and stress distribution in (*c*).

where b = width of beam
d = effective depth of beam

For design, ρ should not exceed

$$0.75\left(\rho_b - \rho'\frac{f'_s}{f_y}\right) + \rho'\frac{f'_s}{f_y} \qquad (10\text{-}30a)$$

for

$$\rho_b = \frac{0.85 f'_c \beta_1}{f_y} \frac{87,000}{87,000 + f_y} + \rho'\frac{f'_s}{f_y} \qquad (10\text{-}30b)$$

where f'_s = stress in compression steel, and other symbols are the same as those defined for singly reinforced beams. The compression force on the concrete alone in a cross section (Fig. 10-18) is

$$C_{1b} = 0.85 f'_c ba \qquad (10\text{-}31)$$

Fig. 10-18 Stress and strain in a rectangular beam with compression reinforcement at ultimate load: (a) section; (b) strain; (c) compression block stress; (d) compression steel stress.

where $a = \beta_1 c$ is the depth of the stress block and the compression steel carries $A'_s f'_s$. Forces equal in magnitude to these but opposite in direction stress the tension steel. The depth to the neutral axis c can be found from the maximum compressive strain of 0.003 in/in, or by equating the compression and tension forces on the section. (See also Art. 10-65.)

T Beams. When a T form is used to provide needed compression area for an isolated beam, flange thickness should be at least one-half the web width, and flange width should not exceed 4 times the web width.

When a T is formed by a beam cast integrally with a slab, only a portion of the slab is effective. For a symmetrical T beam, the effective flange width should not exceed one-fourth the beam span, nor should the width of the overhang exceed 8 times the slab thickness nor one-half the clear distance to the next beam. For a beam having a flange on one side only, the effective flange width should not exceed one-twelfth the span, 6 times the slab thickness, nor one-half the clear distance to the next beam.

The overhang of a T beam should be designed to act as a cantilever. Spacing of the cantilever reinforcing should not exceed 18 in or 5 times the flange thickness.

In computing the moment capacity of a T beam, it may be treated as a singly reinforced beam with overhanging concrete flanges (Fig. 10-19). The compression force on the web (rectangular beam) is

$$C_w = 0.85 f'_c b_w a \tag{10-32}$$

where b_w = width of web. The compression force on the overhangs is

$$C_f = 0.85 f'_c (b - b_w) h_f \tag{10-33}$$

Fig. 10-19 Stress and strain in a T beam at ultimate load: (a) section; (b) strain; (c) web compression block stress; (d) flange compression block stress.

where h_f = flange thickness
b = effective flange width of T beam

Forces equal in magnitude to these but opposite in direction stress the tension steel:

$$T_w = A_{sw}f_y \tag{10-34}$$
$$T_f = A_{sf}f_y \tag{10-35}$$

where A_{sw} = area of reinforcing required to develop compression strength of web
A_{sf} = area of reinforcing required to develop compression strength of overhanging flanges

The steel ratio for balanced conditions is given by

$$\rho_b = \frac{b_w}{b}\left[\frac{0.85f'_c\beta_1}{f_y}\frac{87{,}000}{87{,}000 + f_y} + \frac{A_{sf}}{b_w d}\right] \tag{10-36}$$

The depth to the neutral axis c can be found in the same way as for rectangular beams.

10-47. Shear in Flexural Members.

Design at a section of a reinforced-concrete flexural member with shear force V_u is based on

$$V_u \leq \phi V_n = \phi(V_c + V_s) \tag{10-37}$$

where ϕ = capacity reduction factor (given in Art. 10-44)
V_u = factored shear force at a section
V_c = nominal shear strength of concrete
V_s = nominal shear strength provided by reinforcement

Except for brackets, deep beams, and other short cantilevers, the section for maximum shear may be taken at a distance d from the face of the support when the reaction in the direction of the shear introduces compression into the end region of the member.

For shear in two-way slabs, see Art. 10-58.

For nonprestressed flexural members of normal-weight concrete without torsion, the nominal shear strength V_c carried by the concrete is limited to a maximum of $2\sqrt{f'_c}b_w d$, where b_w is the width of the beam web, d = depth of centroid of reinforcement, and f'_c is the specified concrete strength, unless a more detailed analysis is made. In such an analysis V_c should be obtained from

$$V_c = \left(1.9\sqrt{f'_c} + \frac{2{,}500\rho_w V_u d}{M_u}\right)b_w d \leq 3.5\sqrt{f'_c}\,b_w d \tag{10-38}$$

where M_u = design bending moment occurring simultaneously with V_u at the section considered, but $V_u d/M_u$ must not exceed 1.0
$\rho_w = A_s/b_w d$
A_s = area of nonprestressed tension reinforcement

For one-way joist construction, the ACI Building Code allows these values to be increased 10%.

For rectangular members with short cross-sectional dimension x and long cross-sectional dimension y for which the factored torsional moment T_u exceeds $0.5\phi\sqrt{f'_c}\,\Sigma x^2 y$, the nominal torsional strength that can be assigned to the concrete should be limited to the maximum value of V_c computed from

$$V_c = \frac{2\sqrt{f'_c}\,b_w d}{\sqrt{1 + (2.5C_t T_u/V_u)^2}} \tag{10-39}$$

where $C_t = b_w d/\Sigma x^2 y$.

For lightweight concrete, V_c should be modified by substituting $f_{ct}/6.7$ for $\sqrt{f_c'}$, where f_{ct} is the average splitting tensile strength of lightweight concrete, but not more than $6.7\sqrt{f_c'}$. When f_{ct} is not specified, values of $\sqrt{f_c'}$ affecting V_c should be multiplied by 0.85 for sand-lightweight concrete and 0.75 for all-lightweight concrete.

Shear Reinforcement. When V_u exceeds ϕV_c, shear reinforcement must be provided to resist the excess diagonal tension. The reinforcement may consist of stirrups making an angle of 45 to 90° with the longitudinal steel, longitudinal bars bent at an angle of 30° or more, or a combination of stirrups and bent bars. The nominal shear stress carried by the shear reinforcement V_s must not exceed $8\sqrt{f_c'}b_w d$.

Spacing of required shear reinforcement placed perpendicular should not exceed $0.5d$ for nonprestressed concrete, 75% of the overall depth for prestressed concrete, or 24 in. Inclined stirrups and bent bars should be spaced so that at least one intersects every 45° line extending toward the supports from middepth of the member to the tension reinforcement. When V_s is greater than $4\sqrt{f_c'}b_w d$, the maximum spacing of shear reinforcement should be reduced by one-half. (See Art. 10-111 for shear-strength design for prestressed concrete members.)

The area required in the legs of a vertical stirrup, in², is

$$A_v = \frac{V_s s}{f_y d} \tag{10-40a}$$

where s = spacing of stirrups, in
f_y = yield strength of stirrup steel, psi

For inclined stirrups, the leg area should be at least

$$A_v = \frac{V_s s}{(\sin \alpha + \cos \alpha) f_y d} \tag{10-40b}$$

where α = angle of inclination with longitudinal axis of member.

For a single bent bar or a single group of parallel bars all bent at an angle α with the longitudinal axis at the same distance from the support, the required area is

$$A_v = \frac{V_s}{f_y \sin \alpha}$$

in which V_s should not exceed $3\sqrt{f_c'}b_w d$.

See also Art. 10-66.

10-48. Torsion in Concrete Members.

Under twisting or torsional moments, a member develops normal (warping) and shear stresses. Torsional design may be based on

$$T_u \leq \phi(T_c + T_s) \tag{10-42}$$

where T_u = factored torsional moment
ϕ = capacity reduction factor = 0.85
T_c = nominal torsional strength provided by concrete
T_s = nominal torsional strength provided by torsion reinforcement

Torsional strength of the concrete may be computed from

$$T_c = \frac{0.8\sqrt{f_c'}\,\Sigma x^2 y}{\sqrt{1 + \left(\dfrac{0.4 V_u}{C_t T_u}\right)^2}} \tag{10-43}$$

where $C_t = b_w d/\Sigma x^2 y$
  x = shorter overall dimension of rectangular part of a cross section (Fig. 10-20)
  y = longer overall dimension of the rectangular part (Fig. 10-20)
For computations, it is usually convenient to divide a cross section into n rectangles with sides x_r and y_r, where r ranges from 1 to n. Calculation of $\Sigma x^2 y$ for flanged sections depends on how the rectangles are chosen. They can be selected to yield the greatest value for $\Sigma x^2 y$. But they should not overlap.

A rectangular box section may be taken as a solid section when wall thickness $h \geq x/4$. When $x/10 < h < x/4$, a solid section may be used, but $\Sigma x^2 y$ should be multiplied by $4h/x$. When $h < x/10$, divide the hollow section into separate rectangles. Fillets should be provided at all interior corners of box sections to reduce the effects of stress concentrations.

If T_u does not exceed $0.5\phi\sqrt{f_c'}\,\Sigma x^2 y$, where f_c' is the specified concrete strength, psi, torsional effects need not be included with those from shear and bending.

Fig. 10-20 Resolution of T beam into component rectangles for torsional shear calculations.

The maximum allowable torsional moment is

$$T_u \leq \phi 5 T_c \qquad (10\text{-}44a)$$

In a statically indeterminate system, where a reduction of the distributed torsional moment can occur, torsional cracking reduces the initial stiffness of the element in torsion. A redistribution of internal forces can be assumed, provided that ductility is maintained by torsional reinforcement to develop the cracking moment. For these conditions, the maximum factored torsional moment may be taken as

$$T_u = \phi(4\sqrt{f_c'}\,\Sigma x^2 y/3) \qquad (10\text{-}44b)$$

Members should be designed for the torsion at a distance d from the face of support, where d is the distance from extreme compression surface to centroid of tension steel of the members.

Stirrups. Torsion reinforcement, when required, should be provided in addition to other reinforcement for flexure, axial forces, or shear. For this purpose, the required area A_t for one leg of a closed stirrup can be computed from

$$A_t = (T_u - \phi T_c)\frac{s}{\alpha_t x_1 y_1 \phi f_y} \qquad (10\text{-}45)$$

where $\alpha_t = 0.66 + 0.33 y_1/x_1 \leq 1.50$
  s = spacing of torsion reinforcement measured parallel to longitudinal reinforcement
  x_1 = shorter center-to-center dimension of a closed rectangular stirrup
  y_1 = larger center-to-center dimension of a closed rectangular stirrup
  f_y = yield strength of stirrup steel

To control crack widths, ensure development of the ultimate torsional strength of the member, and prevent excessive loss of torsional stiffness after cracking, the spacing of closed-stirrup torsion reinforcement should not exceed $(x_1 + y_1)/4$ or 12 in, whichever is smaller.

Longitudinal Reinforcement. Steel reinforcement parallel to the axis of the member should be provided in each corner of closed-stirrup torsion reinforcement to assist in development of the torsional design strength without excessive cracking. The amount of longitudinal reinforcement A_l, in^2, for resisting torsion should be the greater of

$$A_l = 2A_t(x_1 + y_1)/s \qquad (10\text{-}46)$$

$$A_l = \left[\frac{400xs}{f_y}\left(\frac{T_u}{T_u + V_u/3C_t}\right) - 2A_t\right]\frac{(x_1 + y_1)}{s} \qquad (10\text{-}47)$$

except that A_l need not be greater than the values given by Eqs. (10-46) and (10-47) when $50b_w s/f_y$ is substituted for $2A_t$.

The spacing of longitudinal torsional reinforcement around the perimeter of closed stirrups should not exceed 12 in.

See also Art. 10-67.

10-49. Development, Anchorage, and Splices of Reinforcement.

Steel reinforcement must be bonded to the concrete sufficiently so that the steel will yield before it is freed from the concrete. Despite assumptions made in the past to the contrary, bond stress between concrete and reinforcing bars is not uniform over a given length, not directly related to the perimeter of the bars, not equal in tension and compression, and may be affected by lateral confinement. ACI Building Code requirements therefore reflect the significance of average bond resistance over a length of bar or wire sufficient to develop its strength (**development length**).

The calculated tension or compression force in each reinforcing bar at any section must be developed on each side of that section by a development length l_d, or by end anchorage, or both. Hooks can be used to assist in the development of tension bars only.

The critical sections for development of reinforcement in flexural members are located at the points of maximum stress and where the reinforcement terminates or is bent.

The following requirements of ACI 318-77 for the development of reinforcement were developed to help provide for *shifts* in the location of maximum moment and for *peak stresses* that exist in regions of tension in the remaining bars wherever adjacent bars are cut off or bent. In addition, these requirements help minimize any loss of shear capacity or ductility resulting from flexural cracks that tend to open early whenever reinforcement is terminated in a tension zone.

For All Flexural Reinforcement

Reinforcement should extend a distance of d or $12d_b$, whichever is larger, beyond the point where the steel is no longer required to resist stress, where d is the effective depth of the member and d_b is the reinforcement diameter. This requirement, however, does not apply at supports of simple spans and at the free end of cantilevers.

Continuing reinforcement should extend at least the development length l_d beyond the point where bent or terminated reinforcement is no longer required to resist tension.

Reinforcement should not be terminated in a tension zone unless *one* of the following conditions is satisfied:

1. Shear at the cutoff point does not exceed two-thirds of the shear permitted, including the shear strength of web reinforcement.

2. Stirrup area A_v not less than $60\ b_w s/f_y$ and exceeding that required for shear and torsion is provided along each terminated bar over a distance from the termination point equal to $0.75d$. (A_v = cross-sectional area of stirrup leg, b_w = width of member, and f_y = yield strength of stirrup steel.) The spacing should not exceed $d/8\beta_b$, where β_b is the ratio of the area of the bars cut off to the total area of bars at the cutoff section.

3. For No. 11 bars and smaller, continuing bars provide double the area required for flexure at the cutoff point, and the shear does not exceed three-fourths of that permitted.

For Positive-Moment Reinforcement

A minimum of one-third the required positive-moment reinforcement for simple beams should extend along the same face of the member into the support, and in beams, for a distance of not less than 6 in.

A minimum of one-fourth the required positive-moment reinforcement for continuous members should extend along the same face of the member into the support, and in beams, for a distance of at least 6 in.

For lateral-load-resisting members, the positive-moment reinforcement to be extended into the support in accordance with the preceding two requirements should be able to develop between the face of the support and the end of the bars the yield strength f_y of the bars.

Positive-moment tension reinforcement at simple supports and at points of inflection should be limited to a diameter such that the development length, in, computed for f_y with Eqs. (10-49) to (10-53) does not exceed

$$l_d = \frac{M_n}{V_u} + l_a \tag{10-48}$$

where M_n = nominal moment strength of a section, in-kips, assuming all reinforcement at the section stressed to f_y
V_u = factored design shear at the section, kips
l_a = embedment length, in, beyond center of support plus equivalent embedment length of any hook or mechanical anchorage, but not more than d or $12d_b$, whichever is greater
d = effective depth, in, of member
d_b = bar diameter, in

For Negative-Moment Reinforcement

Negative-moment reinforcement in continuous, restrained, or cantilever members should be developed in or through the supporting member.

Negative-moment reinforcement should have sufficient distance between the face of the support and the end of each bar to develop its full yield strength.

A minimum of one-third of the required negative-moment reinforcement at the face of the support should extend beyond the point of inflection the greatest of d, $12d_b$, or one-sixteenth of the clear span.

Computation of Development Length. The basic development length l_d for deformed reinforcing bars and deformed wire **in tension** is:

For No. 11 or smaller bars:

$$l_d = \frac{0.04 A_b f_y}{\sqrt{f_c'}} \geq 0.0004 d_b f_y \tag{10-49}$$

For No. 14 bars:

$$l_d = \frac{0.085 f_y}{\sqrt{f_c'}} \tag{10-50}$$

For No. 18 bars:

$$l_d = \frac{0.11 f_y}{\sqrt{f_c'}} \tag{10-51}$$

For deformed wire:

$$l_d = \frac{0.03 d_b f_y}{\sqrt{f'_c}} \tag{10-52}$$

Development length l_d, however, should be at least 12 in, except in calculation of lap splices and development length of web reinforcement.

For welded-wire fabric with deformed wires with at least one cross wire within the development length not less than 2 in from the point of critical section (Fig. 10-21):

$$l_d = \frac{0.03 d_b (f_y - 20{,}000)}{\sqrt{f'_c}} \geq \frac{0.2 A_w}{s_w} \left(\frac{f_y}{\sqrt{f'_c}} \right) \tag{10-53}$$

where f'_c = specified concrete strength, psi
f_y = steel yield strength, psi
s_w = spacing of the wires, in
A_w = area of one tension wire, in²

For smooth-wire fabric, the development length, in, should be the largest of the following: 6 in, one space plus 2 in, or

$$l_d = 0.27 \frac{A_w}{s_w} \frac{f_y}{\sqrt{f'_c}} \tag{10-54}$$

See Fig. 10-22.

The basic development lengths for tension bars with various concrete strengths are given in Table 10-7.

The basic development length in tension for top reinforcement, reinforcement with f_y greater than 60 ksi, reinforced lightweight concrete, reinforcement with a spacing of 6 in or greater, reinforcement in excess of that required, and bars enclosed within a spiral should be modified by multiplying the basic development length by the following factors:

For top reinforcement, horizontal bars with more than 12 in of concrete below: 1.4
For bars with $f_y > 60$ ksi: $2 - 60{,}000/f_y$
For reinforced, sand-lightweight concrete: 1.18
For reinforced all-lightweight concrete: 1.33
In lieu of the two preceding values for lightweight concrete: $6.7 \sqrt{f'_c}/f_{ct} \geq 1.0$, where f_{ct} = average splitting tensile strength of the concrete, psi

When reinforcement is at least 3 in from the side face of a member, is being developed in the length under consideration, and is spaced 6 in or more c to c: 0.80

For reinforcement in excess of that required: (A_s required/A_s provided), where A_s = cross-sectional area of reinforcement

Basic development length for deformed bars **in compression** is

$$l_d = \frac{0.02 d_b f_y}{\sqrt{f'_c}} \geq 0.0003 d_b f_y \geq 8 \text{ in} \tag{10-55}$$

When the reinforcement is enclosed by spirals at least ¼ in in diameter and with not more

Fig. 10-21 Minimum development length for deformed-wire fabric.

Fig. 10-22 Minimum development length for smooth-wire fabric.

TABLE 10-7 Development Length of Tension Reinforcement, l_d, in

Bar size no.	f'_c = 3,000 psi		f'_c = 3,750 psi		f'_c = 4,000 psi		f'_c = 5,000 psi		f'_c = 6,000 psi	
	Top‡ bars	Other bars	Top‡ bars	Other bars	Top‡ bars	Other bars	Top‡ bars	Other bars	Top‡ bars	Other bars
3	13	12	13	12	13	12	13	12	13	12
4	17	12	17	12	17	12	17	12	17	12
5	21	15	21	15	21	15	21	15	21	15
6	27	19	25	18	25	18	25	18	25	18
7	37	26	33	24	32	23	29	21	29	21
8	48	35	43	31	42	30	38	27	34	25
9	61	44	55	39	53	38	48	34	43	31
10	78	56	70	50	67	48	60	43	55	39
11	96	68	86	61	83	59	74	53	68	48
14	130	93	117	83	113	81	101	72	92	66
18	169	120	151	108	146	104	131	93	119	85

*Courtesy Concrete Reinforcing Steel Institute.
†1. For bars enclosed in standard column spirals, use 0.75 l_d.
 2. For bars, such as usual temperature bars, spaced 6 in. or more, use 0.8 l_d.
 3. Longer embedments for lightweight concrete are generally required, depending on the tensile splitting strength f_{ct}.
 4. Standard 90 or 180° end hooks may be used to replace part of the required embedment.
‡Horizontal bars with more than 12 in. of concrete below.

than a 4-in pitch, l_d can be reduced 25%. (See Table 10-8.) For general practice, use 20 bar diameters for compression embedment.

For **bundled bars,** the basic development length of each bar must be increased by 20% for a three-bar bundle and 33% for a four-bar bundle.

Hooks. For bars in tension, standard hooks (90 or 180°) can be used as part of the length required for development or anchorage of the bar. Table 10-9 gives the minimum tension embedment length E required with standard end hooks and Grade 60 bars to develop the specified yield strengths of bars.

TABLE 10-8 Minimum Compression Dowel Development Lengths l_d, in·†

Bar size no.	f'_c (Normal-weight concrete)			
	3,000 psi	3,750 psi	4,000 psi	Over 4,444 psi‡
3	8	8	8	8
4	11	10	10	9
5	14	12	12	11
6	17	15	14	14
7	19	17	17	16
8	22	20	19	18
9	25	22	22	20
10	28	25	24	23
11	31	28	27	25
14	37	33	32	31
18	50	44	43	41

* Courtesy Concrete Reinforcing Steel Institute.
† For embedments enclosed by spirals use 0.75 length given above but not less than 8 in.
‡ For $f'_c > 4{,}444$ psi, minimum embedment = $18d_b$.

TABLE 10-9 Minimum Tension Embedment E, in, for Standard End Hooks for Grade 60 Bars*

	Spacing s ≥ 6 in				Spacing s < 6 in			
	Other than top bars		Top bars		Other than top bars		Top bars	
Bar size no.	Enclosed bars†	Non-enclosed bars	Enclosed bars†	Non-enclosed bars	Enclosed bars†	Non-enclosed bars	Enclosed bars†	Non-enclosed bars
a. For f'_c = 4000 psi								
3	6	6	7	8	7	8	10	11
4	6	7	8	9	8	10	11	13
5	6	8	8	10	9	11	11	14
6	6	8	8	10	9	12	11	15
7	6	9	8	11	10	13	12	17
8	6	11	8	14	12	17	15	22
9	8	14	9	18	16	22	20	29
10	13	21	16	26	23	30	30	40
11	20	28	26	37	32	40	42	53
14	38	47	49	61	54	63	72	84
18	66	73	86	97	86	94	116	126
b. For f'_c = 3000 psi								
3	6	6	7	8	7	8	10	11
4	6	7	8	10	9	10	11	13
5	6	8	8	10	9	11	11	14
6	6	9	8	11	10	13	13	17
7	8	12	9	15	13	17	17	22
8	10	15	12	19	17	22	22	29
9	13	19	15	24	21	28	28	37
10	19	27	24	35	30	38	40	50
11	28	36	36	47	43	49	57	66
14	48	57	63	75	67	75	90	101
18	78	86	104	115	103	110	138	148

*For details of standard 90° and 180° hooks, see Table 10-5. The hook embedments in part *b* are based upon interpretations of ACI 318-77 requirements recommended by CRSI. Since there are at least eight possible permutations of such interpretations, other versions will be encountered. This table is quite conservative compared to latest recommendations based upon research (see "Suggested Development, Splice, and Standard Hook Provisions for Deformed Bars in Tension," ACI Committee 408, Concrete International, July 1979).

†Enclosure may consist of external concrete or internal closed ties, spirals, or stirrups or other tensile reinforcement transverse to the plane of the hooks.

Courtesy Concrete Reinforcing Steel Institute.

Web Reinforcement. Stirrups should be designed and detailed to come as close to the compression and tension surfaces of a flexural member as cover requirements and the proximity of other steel will permit. Ends of single-leg simple U stirrups or transverse multiple U stirrups must be anchored by one of the following means:

A standard hook plus half the development length l_d for tension reinforcement. The effective embedment length of a stirrup leg is the distance between middepth of the member ($d/2$ from the extreme compression surface) and the start of the hook (point of tangency).

Embedment, between middepth of the beam and the extreme compression surface, of the full l_d for tension reinforcement but not less than $24d_b$.

Bending around longitudinal reinforcement through at least 135° for stirrups consisting of No. 5 bars or D31 wire, or smaller sizes.

Between anchored ends, each bend in the continuous portion of a simple U or multiple U stirrup should enclose a longitudinal bar.

Stirrup Splices. Pairs of U stirrups or ties so placed as to form a closed unit can be considered properly spliced when the laps are $1.7l_d$. In members at least 18 in deep, such splices can be considered adequate for No. 3 bars of Grade 60 and Nos. 3 and 4 bars of Grade 40 if the legs extend the full available depth of the member.

Compression Lap Splices. Minimum lengths of lap in splices of compression reinforcement vary with bar diameter d_b and yield strength of steel f_y for all values of f'_c greater than 3000 psi. When f'_c is less than 3000 psi, the length of lap should be one-third greater than the value computed from the following appropriate splice equation.

The ACI Building Code defines three classes of compression splices and requires the following minimum laps for each bar size No. 11 and smaller:

1. Standard splice lap l_s equals the basic development length in compression [Eq. (10-55)], but not less than the largest of 12 in or the values computed from Eqs. (10-56a) and (10-56b).

$$l_s = 0.0005 f_y d_b \qquad f_y \leq 60{,}000 \text{ psi} \qquad (10\text{-}56a)$$
$$l_s = (0.0009 f_y - 24) d_b \qquad f_y > 60{,}000 \text{ psi} \qquad (10\text{-}56b)$$

Splices within column spirals must be lapped a minimum of 0.75 of the standard splice lap (use $24d_b$), but not less than 12 in.

2. Compression splices by bearing. In members with closed ties, closed stirrups, or spirals, compressive stress in a rebar may be transmitted by bearing of square-cut ends held in concentric contact by a suitable device. The ends must be fitted within 3° of full bearing after assembly.

3. Welded compression splices. The ACI Building Code requires butt-welded compression splices to develop at least 125% of the specified yield strength of the steel.

Tension Lap Splices. Bars No. 11 or less in size and deformed wire can be spliced by lapping. Full (butt) welded splices or full positive connections may be used for bars of all sizes. These splices must have a capacity equal to at least 125% of the specified f_y.

Tension lap splices are classified in four groups with the minimum lap length l_s expressed as a multiple of the basic development length l_d of tension reinforcement.

Class A Splices. Splices at sections away from maximum stress and where the steel design stress is less than $0.5f_y$, if not more than 75% of the bars at these sections are spliced within one Class A splice length of the section.

$$l_s = l_d \geq 12 \text{ in} \qquad (10\text{-}57)$$

Class B Splices. Splices where the design stress is less than $0.5f_y$ but where more than 75% of the bars at the section are spliced, or splices made at a section of maximum tensile stress where not more than 50% of the bars at the section are spliced.

$$l_s = 1.3 l_d \geq 12 \text{ in} \qquad (10\text{-}58)$$

Class C Splices. Splices at a section of maximum tensile stress where more than 50% of the bars at the section are spliced.

$$l_s = 1.7 l_d \geq 12 \text{ in} \qquad (10\text{-}59)$$

Table 10-10 lists splice lengths based on the preceding requirements for deformed bars and deformed wires. Splice lengths for welded-wire fabric are indicated in Figs. 10-23 and 10-24.

TABLE 10-10 Tension Lap Splice Lengths, in, for Grade 60 Bars*

Bar size no.	$f'_c = 3000$			$f'_c = 4000$			$f'_c = 5000$		
	A	B	C	A	B	C	A	B	C
a. For other than top bars†									
3	12	12	16	12	12	16	12	12	16
4	12	16	20	12	16	20	12	16	20
5	15	20	26	15	20	26	15	20	26
6	19	25	33	18	24	31	18	24	31
7	26	34	45	23	30	39	21	27	36
8	35	45	59	30	39	51	27	35	46
9	44	57	74	38	49	65	34	44	58
10	56	72	95	48	63	82	43	56	73
11	68	89	116	59	77	101	53	69	90
b. For top bars‡									
3	12	16	21	12	16	21	12	16	21
4	17	22	29	17	22	29	17	22	29
5	21	27	36	21	27	36	21	27	36
6	27	35	46	25	33	43	25	33	43
7	38	48	63	32	41	54	29	38	50
8	48	63	82	42	55	71	38	49	64
9	61	80	104	53	69	90	48	62	81
10	78	101	132	67	88	115	60	78	103
11	96	124	163	83	108	141	74	96	126

*Normal weight concrete. For general use where bar spacing is less than 6 in and side cover in the plane of the bars is less than 3 in.

†In standard spiral columns, use 0.75 basic lap lengths shown, but not less than 12 in.

‡Top bars are horizontal bars with more than a 12-in depth of concrete cast below the reinforcement. It is suggested that multiple horizontal bars in a single vertical plane, such as column ties or horizontal bars in walls, need not be considered top bars.

Courtesy Concrete Reinforcing Steel Institute.

Fig. 10-23 (*a*) Minimum splice length for deformed-wire fabric. (*b*) Slab reinforced with deformed-wire fabric.

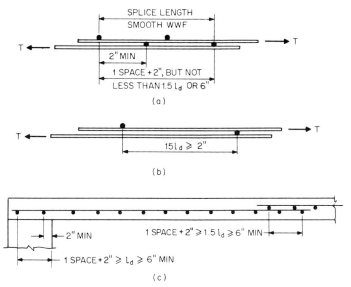

Fig. 10-24 Minimum splice length for smooth-wire fabric. Use the larger of the values shown in (*a*) and (*b*). In calculation of splice length, the computed value of development length l_d, not the minimum required value, should be used. (*a*) Splice length when steel area used is less than twice the required area. (*b*) Splice length when steel area used is 2 or more times the required area. (*c*) Slab reinforced with smooth-wire fabric providing twice the required reinforcement area.

Welded Tension Splices. These or other positive connections must be used for splices of Nos. 14 and 18 bars and may be used in lieu of lapped splices for No. 11 and smaller bars. Full (butt) welded splices or other full positive connections must develop at least 125% of the specified yield strength of the bars unless the splices are staggered at least 24 in and located to develop at every section at least twice the calculated tensile force at the section, but not less than 20,000 psi on the total cross-sectional area of all bars at the section.

10-50. Crack Control.

Because of the effectiveness of reinforcement in limiting crack widths, the ACI Building Code requires minimum areas of steel and limits reinforcement spacing, to control cracking.

Beams and One-Way Slabs. If, in a structural floor or roof slab, principal reinforcement extends in one direction only, shrinkage and temperature reinforcement should be provided normal to the principal reinforcement, to prevent excessive cracking. The additional reinforcement should provide at least the ratios of reinforcement area to gross concrete area of slab given in Table 10-11, but not less than 0.0014.

To control flexural cracking, tension reinforcement in beams and one-way slabs should be well distributed in zones of maximum concrete tension when the design yield strength of the steel f_y is greater than 40,000 psi. Spacing of principal reinforcement in slabs should not exceed 18 in nor 3 times the slab thickness, except in concrete-joist construction.

Where slab flanges of beams are in tension, a part of the main reinforcement of the beam should be distributed over the effective flange width or a width equal to one-tenth

TABLE 10-11 Minimum Shrinkage and Temperature Reinforcement

In slabs where Grade 40 or 50 deformed bars are used	0.0020
In slabs where Grade 60 deformed bars or welded-wire fabric, deformed or plain, are used (Table 10-16)	0.0018
In slabs reinforced with steel having a yield strength f_y exceeding 60,000 psi measured at a strain of 0.0035 in/in	$108/f_y$

This reinforcement should not be placed farther apart than 5 times the slab thickness or more than 18 in.

---

the span, whichever is smaller. When the effective flange width exceeds one-tenth the span, some longitudinal reinforcement should be provided in the outer portions of the flange. Also, reinforcement for one-way joist construction should be uniformly distributed throughout the flange.

To control flexural cracking in beams, reinforcement should be so distributed that:
For interior exposures,

$$z \le f_s \sqrt[3]{d_c A} \le 175 \text{ kips/in} \qquad (10\text{-}60)$$

For exterior exposures,

$$z \le f_s \sqrt[3]{d_c A} \le 145 \text{ kips/in} \qquad (10\text{-}61)$$

where d_c = thickness, in, or concrete cover measured from extreme tension surface to center of bar

A = effective tension area, in², of concrete per rebar. It can be computed by dividing the concrete area surrounding the main tension reinforcing bars and having the same centroid as that reinforcement by the number of bars. If bar sizes differ, the number of bars should be computed as the total steel area divided by the area of the largest bar used

f_s = calculated stress in reinforcement at service loads, ksi, but may be taken as $0.60 f_y$ in lieu of such calculations

The numerical limitations on z of 175 and 145 kips/in correspond to limiting crack widths of 0.016 and 0.013 in for interior and exterior exposures, respectively. For watertight slabs or severe exposures, z should be smaller.

For one-way slabs, tests indicate that z should not exceed 155 kips/in for interior exposures, or 129 kips/in for exterior exposures.

The z values can be transformed into the following expressions for the maximum spacing of Grade 60 bars to control flexural cracking:

Interior exposure	Maximum spacing, in
Beams	$57.4/d_c^2$
One-way slabs	$39.9/d_c^2$
Exterior exposure	Maximum spacing, in
Beams	$32.6/d_c^2$
One-way slabs	$22.9/d_c^2$

See Tables 10-12 and 10-20, Art. 10-68.

Two-Way Slabs. Flexural cracking in two-way slabs is significantly different from that in one-way slabs. For control of flexural cracking in two-way slabs, such as solid flat

TABLE 10-12 Maximum Spacing, in, of Grade 60 Bars for Control of Flexural Cracking

Bar size no.	Beams 2-in cover*		One-way slabs					
			Interior exposure			Exterior exposure		
	Interior exposure	Exterior exposure	Cover, in			Cover, in		
			¾	1	1½	1½	2	
3	...	...	18.0	18.0	...	8.0	...
4	...	...	18.0	18.0	...	7.5	...
5	10.7	6.1	18.0	18.0	...	6.9	...
6	10.2	5.8	18.0	18.0	...	...	4.1
7	9.7	5.5	18.0	18.0	...	...	3.9
8	9.2	5.2	18.0	17.7	...	...	3.7
9	8.7	5.0	18.0	16.3	...	...	3.5
10	8.3	4.7	18.0	14.9	...	...	3.3
11	7.8	4.5	17.9	13.7	...	...	3.1
14	7.1	4.0	...	...	7.2	...	...
18	5.9	...	...	...	5.8	...	...

*Cover to stirrups, if any, should be at least 1 1/2 in.

plates and flat slabs with drop panels, the ACI Building Code restricts the maximum spacing of tension bars to twice the overall thickness h of the slab but not more than 18 in. In waffle slabs or over cellular spaces, however, reinforcement should be the same as that for shrinkage and temperature in one-way slabs (see Table 10-12).

10-51. Deflection of Concrete Beams and Slabs.

Reinforced-concrete flexural members must have adequate stiffness to limit deflection to an amount that will not adversely affect the serviceability of the structure under service loads.

Beams and One-Way Slabs. Unless computations show that deflections will be small (Table 10-13), the ACI Building Code requires that the depth h of nonprestressed, one-way solid slabs, one-way ribbed slabs, and beams of normal-weight concrete—with Grade 60 reinforcement—be at least the fraction of the span L given in Table 10-14.

When it is necessary to compute deflections, calculation of short-time deflection may be based on elastic theory, but with an effective moment of inertia I_e.

For normal-weight concrete,

$$I_e = \left(\frac{M_{cr}}{M_a}\right)^3 I_g + \left[1 - \left(\frac{M_{cr}}{M_a}\right)^3\right] I_{cr} \leq I_g \tag{10-62}$$

where M_{cr} = cracking moment = $f_r I_g / y_t$
M_a = service-load moments for which deflections are being computed
I_g = gross moment of inertia of concrete section
I_{cr} = moment of inertia of cracked section transformed to concrete (for solid slabs, see Fig. 10-25)
f_r = modulus of rupture of concrete, psi = $7.5 \sqrt{f'_c}$
f'_c = specified concrete strength, psi
y_t = distance from centroidal axis of gross section, neglecting the reinforcement, to the extreme surface in tension

When structural lightweight concrete is used, f_r in the computation of M_{cr} should be

10-58 Concrete Construction

TABLE 10-13 Maximum Ratios of Computed Deflection to Span L for Beams and Slabs

Type of member	Deflection to be considered	Deflection limitation
Flat roofs not supporting or attached to nonstructural elements likely to be damaged by large deflections	Immediate deflection due to the live load	$L/180$*
Floors not supporting or attached to nonstructural elements likely to be damaged by large deflections	Immediate deflection due to the live load	$L/360$
Roof or floor construction supporting or attached to nonstructural elements likely to be damaged by large deflections	That part of the total deflection that occurs after attachment of the nonstructural elements (the sum of the long-time deflection due to all sustained loads and the immediate deflection due to any additional live load)†	$L/480$‡
Roof or floor construction supporting or attached to nonstructural elements not likely to be damaged by large deflections		$L/240$§

*This limit is not intended to safeguard against ponding. Ponding should be checked by suitable calculations of deflection, including the added deflections due to ponded water, and considering long-time effects of all sustained loads, camber, construction, tolerances and reliability of provisions for drainage.

†The long-time deflection may be reduced by the amount of deflection that occurs before attachment of the nonstructural elements.

‡This limit may be exceeded if adequate measures are taken to prevent damage to supported or attached elements.

§But not greater than the tolerance provided for the nonstructural elements. This limit may be exceeded if camber is provided so that the total deflection minus the camber does not exceed the limitation,

taken as $1.12 f_{ct} \leq 7.5 \sqrt{f'_c}$, where f_{ct} = average splitting tensile strength, psi, of the concrete. When f_{ct} is not specified, f_r should be taken as $5.6 \sqrt{f'_c}$ for all-lightweight concrete and as $6.4 \sqrt{f'_c}$ for sand-lightweight concrete.

For deflection calculations for continuous spans, I_c may be taken as the average of the values obtained from Eq. (10-62) for the critical positive and negative moments.

Additional long-time deflection for both normal-weight and lightweight concrete flex-

TABLE 10-14 Minimum Depths h of Reinforced-Concrete Beams and One-Way Slabs*

	One-way solid slabs	Beams and one-way ribbed slabs
Cantilever	$L/10 = 0.1000L$	$L/8\ \ = 0.1250L$
Simple span	$L/20 = 0.0500L$	$L/16\ \ = 0.0625L$
Continuous:		
End span	$L/24 = 0.0417L$	$L/18.5 = 0.0540L$
Interior span	$L/28 = 0.0357L$	$L/21\ \ = 0.0476L$

*For members with span L (Art. 10-41) not supporting or attached to partitions or other construction likely to be damaged by large deflections. Thinner members may be used if justified by deflection computations. For structural lightweight concrete of unit weight w, lb/ft^3, multiply tabulated values by $1.65 - 0.005w \geq 1.09$, for $90 < w < 120$. For reinforcement with yield strength $f_y > 60,000$ psi, multiply tabulated values by $0.4 + f_y/100,000$.

Fig. 10-25 Chart for determination of moment of inertia I_{cr} of transformed (cracked) section of one-way solid slab, given moment of inertia of gross section $I_g = bh^3/12$, reinforcement ratio $\rho = A_s/bd$, weight w of concrete, lb/ft^3, and ratio d/h of effective depth to thickness, for $f'_c = 4$ ksi.

ural members can be estimated by multiplying the immediate deflection due to the sustained load by:

$$2 - 1.2 \frac{A'_s}{A_s} \geq 0.6$$

where A'_s = area of compression reinforcement
A_s = area of tension reinforcement

The sum of the short-time and long-time deflections should not exceed the limits given in Table 10-13.

Two-Way Slabs. Unless computations show that deflections will not exceed the limits listed in Table 10-13, thickness h for nonprestressed two-way slabs with a ratio of long to short span not exceeding 2 should be at least the larger of the values computed from Eqs. (10-63) and (10-64).

$$h = \frac{L_n(800 + 0.005f_y)}{36,000 + 5000\beta[\alpha_m - 0.5(1 - \beta_s)(1 + 1/\beta)]} \qquad (10\text{-}63)$$

$$h = \frac{L_n(800 + 0.005f_y)}{36,000 + 5000\beta(1 + \beta_s)} \qquad (10\text{-}64)$$

But h need not be more than

$$h = \frac{L_n(800 + 0.005f_y)}{36,000} \qquad (10\text{-}65)$$

where L_n = clear span in long direction, in
α_m = average value of α for all beams along panel edges
α = ratio of flexural stiffness of beam section to flexural stiffness of a width of slab bounded laterally by the centerline of the adjacent panel, if any, on each side of the beam
β = ratio of clear span in long direction to clear span in short direction
β_s = ratio of length of continuous edges to total perimeter of slab panel

The thickness, however, should not be less than the following:

For slabs without beams or drop panels .... 5 in
For slabs without beams, but with drop panels .... 4 in
For slabs having beams on all four edges with α_m at least equal to 2.0 .... 3½ in

The minimum thicknesses for flat slabs with standard drop panels may be reduced 10%.

Unless edge beams with $\alpha \geq 0.8$ are provided at discontinuous edges, the minimum thicknesses of panels at those edges should be increased at least 10%.

The computed deflections of prestressed-concrete construction should not exceed the values listed in Table 10-13.

ONE-WAY REINFORCED-CONCRETE SLABS

A one-way reinforced-concrete slab is a flexural member that spans in one direction between supports and is reinforced for flexure only in one direction (Art. 10-64). If a slab is supported by beams or walls on four sides, but the span in the long direction is more than twice that in the short direction, most of the load will be carried in the short direction; hence, the slab can be designed as a one-way slab.

One-way slabs may be solid, ribbed, or hollow. (For one-way ribbed slabs, see Arts. 10-54 to 10-57.) Hollow one-way slabs are usually precast (Arts. 10-98 to 10-105). Cast-in-place, hollow one-way slabs can be constructed with fiber or cardboard-cylinder forms, inflatable forms that can be reused, or precast hollow boxes or blocks. One-way slabs can be haunched at the supports for flexure or for shear strength.

10-52. Analysis and Design of One-Way Slabs.

Structural strength, fire resistance, crack control, and deflections of one-way slabs must be satisfactory under service loads.

Strength and Deflections. Approximate methods of frame analysis can be used with uniform loads and spans that conform to ACI Code requirements (see Art. 10-41). Deflections can be computed as indicated in Art. 10-51, or in lieu of calculations the minimum slab thicknesses listed in Table 10-14 may be used. In Fig. 10-25 is a plot of ratios of moments of inertia of cracked to gross concrete section for one-way slabs. These curves can be used to simplify deflection calculations.

Strength depends on slab thickness and reinforcement and properties of materials used. Slab thickness required for strength can be computed by treating a 1-ft width of slab as a beam (Arts. 10-44 to 10-46).

Fire Resistance. One-way concrete slabs, if not protected by a fire-resistant ceiling, must have a thickness and a concrete cover around reinforcement that conforms to the fire-resistance rating required by the building code. Table 10-15 gives typical slab thickness and cover around reinforcement for various fire-resistance ratings for normal-weight and structural-lightweight-concrete construction.

TABLE 10-15 Typical Fire Ratings for Concrete Members*

Fire rating, hr	Slabs and joists				Cover, in., for beams and columns 12 in. or larger in section
	Lightweight concrete		Normal-weight concrete		
	Depth h, in.	Cover, in.	Depth h, in.	Cover, in.	
1	3	¾	3½†	¾	1½
2	4	¾	4½	¾	1½
3	4½	1	5½	1	1½
4	5	1	6½	1	1½

*From Uniform Building Code, for Grade A concrete containing less than 40% quartz, chert, or flint.
†May be reduced to 3 in. when limestone aggregate is used.

Reinforcement. Requirements for minimum reinforcement for crack control are summarized in Art. 10-50. Table 10-16 lists minimum reinforcement when Grade 60 bars are used. Reinforcement required for flexural strength can be computed by treating a 1-ft width of slab as a beam (Arts. 10-44 to 10-46).

Rebar weights, lb/ft² of slab area, can be estimated from Fig. 10-26 for one-way, continuous, interior spans of floor or roof slabs made of normal-weight concrete.

One-way reinforced concrete slabs with spans less than 10 ft long can be reinforced with a single layer of draped welded-wire fabric for both positive and negative moments. These moments can be taken equal to $wL^2/12$, where w is the uniform load and L is the span, defined in Art. 10-41, if the slab meets ACI Code requirements for approximate frame analysis with uniform loads.

For development (bond) of reinforcement, see Art. 10-49.

Shear. This is usually not critical in one-way slabs, but the ACI Building Code requires that it be investigated (see Art. 10-47).

TABLE 10-16 Minimum and Maximum Reinforcement for One-Way Concrete Slabs

Slab thickness, in.	Minimum reinforcement*			Maximum reinforcement†		
	Area A_s, sq in.	Bar size and spacing, in.	Weight, psf	Area A_s, sq in.	Bar size and spacing, in.	Weight, psf
4	0.086	No. 3 @ 12‡	0.30	0.555	No. 5 @ 9½	1.89
4½	0.097	No. 3 @ 13½	0.33	0.655	No. 6 @ 8	2.24
5	0.108	No. 3 @ 12½	0.36	0.750	No. 6 @ 7	2.56
5½	0.119	No. 3 @ 11	0.41	0.845	No. 6 @ 6	2.99
6	0.130	No. 4 @ 18	0.45	0.931	No. 7 @ 7½	3.26
6½	0.140	No. 4 @ 17	0.48	1.025	No. 7 @ 7	3.50
7	0.151	No. 4 @ 15½	0.53	1.110	No. 8 @ 8½	3.80
7½	0.162	No. 4 @ 14½	0.56	1.208	No. 8 @ 7½	4.30
8	0.173	No. 4 @ 13½	0.61	1.291	No. 9 @ 9	4.54
8½	0.184	No. 4 @ 13	0.63	1.382	No. 9 @ 8½	4.80
9	0.194	No. 4 @ 12	0.68	1.482	No. 9 @ 8	5.10

* For Grade 60 reinforcement. Minimum area $A_s \geq 0.0018bh$, where b = slab width and h = slab thickness.
† For f'_c = 3,000 psi and no compression reinforcement.
‡ This spacing for a 4-in. slab is for flexure and can be increased to 15 in. for temperature and shrinkage reinforcement.

Fig. 10-26 Weight of reinforcing steel required for an interior span of a continuous, one-way solid slab of 3000-psi concrete, weighing 150 lb/ft^3 and carrying 100-psf live load (170-psf design live load).

10-53. Embedded Pipes in One-Way Slabs.

Generally, embedded pipes or conduit, other than those merely passing through, should not be larger in outside dimension than one-third the slab thickness and should be spaced at least three diameters or widths on centers. Pipes that contain liquid, gas, or vapor should not be embedded in slabs if their temperature exceeds 150°F or the pressure is over 200 psi above atmospheric. Piping in solid one-way slabs is required to be placed between the top and bottom reinforcement unless it is for radiant heating or snow melting.

ONE-WAY CONCRETE-JOIST CONSTRUCTION

One-way concrete-joist construction consists of a monolithic combination of cast-in-place, uniformly spaced ribs (joists) and top slab (Fig. 10-27). (See also Art. 10-64.) The ribs are formed by placing rows of permanent or removable fillers in what would otherwise be a solid slab.

One-way joist construction was developed to reduce dead load. For long spans, the utility of solid-slab construction is offset by the increase in dead load of the slab. One-way concrete-joist construction provides adequate depth with less dead load than for solid slabs, and results in smaller concrete and reinforcement quantities per square foot of floor area.

Uniform-depth floor and roof construction can be obtained by casting the joists integral with wide, supporting band beams of the same total depth as the joists. This design eliminates the need for interior beam forms.

Fig. 10-27 Typical one-way reinforced-concrete joist construction.

10-54. Standard Sizes of Joists.

One-way concrete-joist construction that exceeds the dimensional limitations of the ACI Building Code must be designed as slabs and beams. These dimensional limitations are:

Maximum clear spacing between ribs—30 in
Maximum rib depth—3.5 times rib width
Minimum rib width—4 in
Minimum top-slab thickness with removable forms—2 in, but not less than one-twelfth the clear spacing of ribs
Minimum top-slab thickness with permanent forms—1½ in, but not less than one-twelfth the clear spacing of ribs

Removable form fillers can be standard steel *pans* or hardboard, corrugated cardboard, fiberboard, or glass-reinforced plastic. Standard removable steel pans that conform to "Types and Sizes of Forms for One-Way Concrete-joist Construction," NBS Voluntary Product Standard No. PS 16-69, include 20- and 30-in widths and depths of 6, 8, 10, 12, 14, 16, and 20 in. Standard steel square-end pans are available in 36-in lengths. Widths of 10 and 15 in and tapered end fillers are available as special items. For forms

10-64 Concrete Construction

20 and 30 in wide, tapered end forms slope to 16 and 25 in, respectively, in a distance of 3 ft.

Permanent form fillers are usually constructed from structural-clay floor tiles (ASTM C57) or hollow load-bearing concrete blocks (ASTM C90).

Structural-clay tiles are usually 12 in square and can be obtained in thicknesses varying from 3 to 12 in. The tiles are laid flat, or end to end, in rows between and at right angles to the joists. The usual clear distance between rows is 4 in, making the center-to-center spacing of the rows 16 in.

Hollow load-bearing concrete blocks are usually 16 in long and 8 in high. They can be obtained in thicknesses of 4, 6, 8, 10, and 12 in. The blocks are laid flat, end-to-end, in rows between and at right angles to the joists. The usual clear distance between rows is 4 in, making the center-to-center spacing of the rows 20 in.

If the clay-tile or concrete-block fillers have a compressive strength equal to that of the concrete in the joists, the vertical shells of the fillers in contact with the joist ribs can be included in the width of the joist rib.

10-55. Design of Joist Construction.

One-way concrete joists must have adequate structural strength, and crack control and deflection must be satisfactory under service loads. Approximate methods of frame analysis can be used with uniform loads and spans that conform to ACI Code requirements (see Art. 10-41). Table 10-14 lists minimum depths of joists to limit deflection, unless deflection computations justify shallower construction (Table 10-13). Load tables in the Concrete Reinforcing Steel Institute "Design Handbook" indicate when deflections under service live loads exceed specified limits.

Economy can be obtained by designing joists and slabs so that the same-size forms can be used throughout a project. It will usually be advantageous to use square-end forms for interior spans and tapered ends for end spans, when required, with a uniform depth.

Fire Resistance. Table 10-17 gives minimum top-slab thickness and reinforcement cover for fire resistance when a fire-resistant ceiling is not used.

Temperature and Shrinkage Reinforcement. This must be provided perpendicular to the ribs and spaced not farther apart than 5 times the slab thickness, or 18 in. The required area of Grade 60 reinforcement for temperature and shrinkage is 0.0018 times the concrete area (Table 10-18). For flexural reinforcement, see Art. 10-56. For shear reinforcement, see Art. 10-57.

Embedded Pipes. Top slabs containing horizontal conduit or pipes that are allowed by the ACI Code (Art. 10-53) must have a thickness of at least 1 in plus the depth of the conduit or pipe.

Bridging. Distribution ribs are constructed normal to the main ribs to distribute

TABLE 10-17 Required Top-Slab Thickness and Cover for Fire Resistance of One-Way Concrete Joist Construction*

Fire-resistance rating, hr	Normal-weight concrete		Structural lightweight concrete	
	Reinforcement cover, in.	Top-slab thickness, in.	Reinforcement cover, in.	Top-slab thickness, in.
1	¾	3½†	¾	3
2	1	4½	1	4
3	1¼	5½	1¼	4½

* From Uniform Building Code.
† Thickness can be reduced to 3 in. where limestones aggregate is used.

TABLE 10-18 Temperature and Shrinkage Reinforcement for One-Way Joist Construction

Top-slab thickness, in.	Required area of temperature and shrinkage reinforcement, sq in.	Reinforcement	Reinforcement weight, psf
2	0.043	WWF 4 × 12, W1.5/W1	0.19
2½	0.054	WWF 4 × 12, W2/W1	0.24
3	0.065	WWF 4 × 12, W2.5/W1	0.29
3½	0.076	No. 3 @ 17½ in.	0.26
4	0.086	No. 3 @ 15 in.	0.30
4½	0.097	No. 3 @ 13½ in.	0.33
5	0.108	No. 3 @ 12½ in.	0.36
5½	0.119	No. 3 @ 11 in.	0.41

concentrated loads to more than one joist and to equalize deflections. These ribs are usually made 4 to 5 in wide and reinforced top and bottom with one No. 4 or 5 continuous rebar. One distribution rib is usually used at the center of spans of up to 30 ft, and two distribution ribs are usually placed at the third points of spans longer than 30 ft.

Openings. These can be provided in the top slab of one-way concrete joist construction between ribs without significant loss in flexural strength. Header joists must be provided along openings that interrupt one or more joists.

10-56. Reinforcement of Joists for Flexure.

Reinforcement required for strength can be determined as indicated in Art. 10-46, by treating as a beam a section symmetrical about a rib and as wide as the spacing of ribs on centers.

Minimum Reinforcement. It is the opinion of the authors that all reinforcement (both positive and negative) with a yield strength f_y should have an area equal to or greater than $200/f_y$ times the concrete area of the rib $b_w d$, where b_w is the rib width and d = rib depth. Less steel can be used, however, if the areas of both the positive and negative reinforcement at every section are one-third greater than the amount required by analysis. (See also Art. 10-55.)

If bottom bars in continuous joists are not continuous through the support, top bars should meet the requirements for shrinkage and temperature reinforcement.

Maximum Reinforcement. Positive- and negative-moment steel ratios must not be greater than three-quarters of the ratio that produces balanced conditions (Art. 10-46). The positive-moment reinforcement ratio is based on the width of the top flange, and the negative-moment reinforcement ratio is based on the width of the rib b_w.

Reinforcement for one-way concrete-joist construction may consist of one straight and one-bent-up bar in each joist rib or straight top and bottom bars, cut off as required for moment. The straight-bar arrangement is preferred if crack control is essential at the top surface.

For top-slap reinforcement, see Art. 10-55 and Table 10-12. Straight top- and bottom-bar arrangements are more flexible in attaining uniform distribution of top bars to control cracking in the slab than straight and bent bars.

For development (bond) of reinforcement, see Art. 10-49.

Figure 10-28 shows rebar quantities, lb/ft^2 of floor or roof area, for continuous interior spans of one-way concrete-joist construction made with normal-weight concrete for superimposed factored live load of 170 psf.

Fig. 10-28 Weights of steel and concrete in flat-plate, flat-slab, and one-way joist construction, for preliminary estimates. *(Courtesy Concrete Reinforcing Steel Institute.)*

10-57. Shear in Joists.

The factored shear force V_u at a section without shear reinforcement should not exceed

$$V_u = \phi V_c = \phi(2.2\sqrt{f'_c}b_w d) \qquad (10\text{-}66)$$

where V_c = nominal shear strength of the concrete
ϕ = capacity reduction factor (Art. 10-44) = 0.85
d = distance, in, from extreme compression surface to centroid of tension steel
b_w = rib width, in

(Based on satisfactory performance of joist construction, the ACI Building Code allows the nominal shear strength V_c for concrete in joists to be taken 10% greater than for beams or slabs.) The width b_w can be taken as the average of the width of joist at the compression face and the width at the tension steel. The slope of the vertical taper of ribs formed with removable steel pans can safely be assumed as 1 in 12. For permanent concrete block fillers, the shell of the block can be included as part of b_w, if the compressive strength of the masonry is equal to or greater than the concrete.

If shear controls the design of one-way concrete-joist construction, tapered ends can be used to increase the shear capacity. The Concrete Reinforcing Steel Institute "Design Handbook" has comprehensive load tables for one-way concrete-joist construction that indicate where shear controls and when tapered ends are required for simple, end, and interior spans.

For joists supporting uniform loads, the critical section for shear strength at tapered ends is the narrow end of the tapered section. Shear need not be checked within the taper.

Reinforcement for shear must be provided when the factored shear force V_u exceeds the shear strength of the concrete ϕV_c. The use of single-prong No. 3 stirrups spaced at half depth, such as that shown in Fig. 10-29, is practical in narrow joists; they can be easily placed between two bottom bars.

Fig. 10-29 Stirrups for concrete-joist construction.

TWO-WAY CONSTRUCTION

A two-way slab is a concrete panel reinforced for flexure in more than one direction. (See also Art. 10-64.) Many variations of this type of construction have been used for floors and roofs, including flat plates, solid flat slabs, and waffle flat slabs. Generally, the columns that support such construction are arranged so that their center lines divide the slab into square or nearly square panels, but if desired, rectangular, triangular, or even irregular panels may be used.

10-58. Analysis and Design of Flat Plates.

The flat plate is the simplest form of two-way slab—simplest for analysis, design, detailing, bar fabrication and placing, and formwork. A flat plate is defined as a two-way slab of uniform thickness supported by any combination of columns and walls, with or without edge beams, and without drop panels, column capitals, and brackets.

Shear and deflection limit economical flat-plate spans to under about 30 ft for light loading and about 20 to 25 ft for heavy loading. While use of reinforcing-steel or structural-steel shear heads for resisting shear at columns will extend these limits somewhat, their main application is to permit use of smaller columns. A number of other variations, however, can be used to extend economical load and span limits (Arts. 10-61 and 10-62).

The ACI Building Code permits two methods of analysis for two-way construction: *direct design,* within limitations of span and load, and *equivalent frame* (Art. 10-42). Limitations on use of direct design are:

A minimum of three spans continuous in each direction
Rectangular panels with ratios of opposite sides less than 2
Successive clear span ratios not to exceed 2:3
Columns offset from centerlines of successive columns not more than 0.10 span in either direction
Specified ratio of live load to dead load not more than 3

The procedure for either method of design begins with selection of preliminary dimensions for review, and continues with six basic steps.

Step 1. Select a plate thickness expected to be suitable for the given conditions of load and span. This thickness, unless deflection computations justify thinner plates, should not be less than h determined from Eq. (10-65). With Grade 60 reinforcement, minimum thickness is, from Eq. (10-65), for an interior panel

$$h = \frac{L_n}{32.7} \geq 5 \text{ in} \qquad (10\text{-}67)$$

where L_n = clear span in direction moments are being determined. Also, as indicated in

Art. 10-51, for discontinuous panels, h computed from Eq. (10-65) or (10-67) may have to be increased at least 10%.

Step 2. Determine for each panel the total static design moment

$$M_o = 0.125 w L_2 L_n^2 \qquad (10\text{-}68)$$

where L_2 = panel width (center-to-center spans transverse to direction in which moment is being determined)
w = total design load, psf = $1.4D + 1.7L$, typically
D = dead load, psf
L = live load, psf

Step 3. Apportion M_o to positive and negative bending. In the direct-design method:

For interior spans, the negative design bending moment is

$$M_u = -0.65 M_o \qquad (10\text{-}69)$$

and the positive design bending moment is

$$M_u = 0.35 M_o \qquad (10\text{-}70)$$

For exterior spans (edge panels), the distribution of M_o is based on α_{ec}, the ratio of the flexural stiffness of the edge equivalent column K_{ec} to the flexural stiffness of the slab K_s.

$$\alpha_{ec} = \frac{K_{ec}}{K_s} \qquad (10\text{-}71)$$

Equivalent column consists of actual columns above and below the slab plus an attached torsional member transverse to the direction in which moments are being determined and extending to the centerlines on each side of the column of the bounding lateral panel. The torsional member should be taken as a portion of the slab as wide as the column in the direction in which moments are determined. (See also Art. 10-59.)

At the edge column, the negative design moment is

$$M_u = -0.65 M_o \frac{\alpha_{ec}}{\alpha_{ec} + 1} \qquad (10\text{-}72)$$

Also, in an edge panel the design positive moment is

$$M_u = M_o \left(0.63 - 0.28 \frac{\alpha_{ec}}{\alpha_{ec} + 1} \right) \qquad (10\text{-}73)$$

and at the first interior column, the negative design moment is

$$M_u = -M_o \left(0.75 - 0.10 \frac{\alpha_{ec}}{\alpha_{ec} + 1} \right) \qquad (10\text{-}74)$$

Step 4. Distribute panel moments M_u to column and middle strips.

Column strip is a design strip with a width of $0.25 L_2 \leq 0.25 L_1$ on each side of the column centerline, where L_1 is the center-to-center span in the direction in which moments are being determined (Fig. 10-30).

Middle strip is the design strip between two column strips (Fig. 10-30).
For flat plates without beams, the distributions of M_u become:

For positive moment, column strip 60%, middle strip 40%.
For negative moment at the edge column, column strip 100%.
For interior negative moments, column strip 75%, middle strip 25%.

A design moment may be modified up to 10% so long as the sum of the positive and

Fig. 10-30 Division of flat plate into column and middle strips.

negative moments in the panel in the direction being considered is at least that given by Eq. (10-68).

Step 5. Check for shear. Shear strength of slabs in the vicinity of columns or other concentrated loads is governed by the severer of two conditions: when the slab acts as a wide beam and when the load tends to punch through the slab. In the first case, a diagonal crack might extend in a plane across the entire width of the slab. Design for this condition is described in Art. 10-47. For the two-way action of the second condition, diagonal cracking might occur along the surface of a truncated cone or pyramid in the slab around the column.

The critical section for two-way action, therefore, should be taken perpendicular to the plane of the slab at a distance $d/2$ from the periphery of the column, where d is the effective depth of slab. The factored shear force V_u for punching action must be less than

ϕ times the nominal shear strength V_c of the concrete unless shear reinforcement is provided.

$$V_u \leq \phi V_c = (2 + 4/\beta_c)\sqrt{f'_c}b_o d \leq 4\sqrt{f'_c}b_o d \tag{10-75}$$

where ϕ = capacity reduction factor = 0.85
b_o = periphery of critical section
β_c = ratio of long to short side of shear periphery

Unless shear reinforcement is provided, V_u should not exceed $4\sqrt{f'_c}b_o d$, where f'_c is the specified concrete strength, psi. With shear reinforcement as for beams (Art. 10-47), V_u may be as high as $6\sqrt{f'_c}b_o d$. With steel-shape shearheads at interior columns meeting ACI Building Code requirements, V_u may be as high as $7\sqrt{f'_c}b_o d$.

Determine the maximum shear at each column for two cases: all panels loaded, and live load on alternate panels for maximum unbalanced moment to the columns. Combine shears due to transfer of vertical load to the column with shear resulting from the transfer of part of the unbalanced moment to the column by torsion. At this point, if the combined shear is excessive, steps 1 through 5 must be repeated with a larger column, thicker slab, or higher-strength concrete in the slab; or shear reinforcement must be provided where $V_u > \phi V_c$ (Art. 10-47).

Step 6. When steps 1 through 5 are satisfactory, select reinforcement.

10-59. Stiffnesses in Two-Way Construction.

The ACI Building Code prescribes a sophisticated procedure for computation of stiffness to determine α_{ec} (Art. 10-58). Variations in cross sections of slab and columns, drop panels, capitals, and brackets should be taken into account. Columns should be treated as infinitely stiff within the joint with the slab. The slab should be considered to be stiffened somewhat within the depth of the column.

With the equivalent-frame method, the stiffnesses of the equivalent column K_{ec} (Art. 10-58) and the slab K_s are employed in a straightforward elastic analysis. With the direct-design method, use of the value of α_{ec} for determination of moments for edge spans comprises a one-step moment-distribution analysis.

In the direct-design method, neglect of some of the refinements in computation of stiffnesses is permitted (see "Commentary" on ACI 318-77).

Column stiffness may be taken as

$$K_c = \frac{4E_c I_c}{L_c} \tag{10-76}$$

where E_c = modulus of elasticity of column concrete
I_c = column moment of inertia based on uniform cross section
L_c = story height

Slab flexural stiffness may be taken as

$$K_s = \frac{4E_{cs}I_{cs}}{L_1} \tag{10-77}$$

where E_{cs} = modulus of elasticity of slab concrete
I_{cs} = moment of inertia of slab based on gross, uniform section
L_1 = center-to-center span in direction in which moments are being determined

Equivalent column stiffness is given by

$$K_{ec} = \frac{\Sigma K_c}{1 + (1/K_t)\Sigma K_c} \tag{10-78}$$

where K_t = torsional stiffness of portion of slab (Art. 10-58).

The summation ΣK_c applies to the column above and the column below the slab.

$$K_t = \Sigma \frac{9E_{cs}C}{L_2(1 - c_2/L_2)^3} \quad (10\text{-}79)$$

where $C = \Sigma(1 - 0.63x/y)x^3y/3$
- c_2 = edge-column width transverse to direction in which moments are being determined
- L_2 = center-to-center span transverse to L_1
- x,y = dimensions of the section of flat plate in contact with side of column parallel to L_1, c_1 in width and h in depth ($x \leq y$)

In Eq. (10-79), the summation applies to the transverse spans of the slab on each side of the edge column.

Part of the unbalanced moment at the exterior-edge column is transferred to the column through torsional shear stresses on the slab periphery at a distance $d/2$ from the column, where d is the effective slab depth. The fraction of the unbalanced moment thus transmitted is

$$F = 1 - \frac{1}{1 + \frac{2}{3}\sqrt{\frac{c_1 + d/2}{c_2 + d}}} \quad (10\text{-}80)$$

For preliminary design, with square columns flush at edges of the flat plate, a rapid estimate of the shear capacity to allow for effects of torsion can be made by using uniform vertical load w only, with allowable strength for design load as follows:

For edge column, total shear $V_u = 0.5wL_2L_1$ and shear strength $V_c = 2\sqrt{f'_c}\,b_o d$
For first interior column, $V_u = 1.15wL_2L_1$ and shear strength $V_c = 4\sqrt{f'_c}\,b_o d$

where f'_c = specified concrete strength, psi.

Use of this calculation in establishing a preliminary design is a short cut, which will often avoid the need for repeating steps 1 through 5 in Art. 10-58, because it gives a close approximation for final design.

Figure 10-31 shows the minimum cantilever edge span with which all columns can be considered interior columns and the direct-design method can be employed without tedious stiffness calculations.

10-60. Bar Lengths and Details for Flat Plates.

The minimum bar lengths for reinforcing flat plates shown in Fig. 10-32, prescribed by the ACI Building Code, save development (bond) computations. The size of all top bars must be selected so that the development length l_d required for the bar size, concrete strength, and grade of the bar is not greater than the length available for development (see Table 10-7).

The size of top bars at the exterior edge must be small enough that the hook plus straight extension to the face of the column is larger than that required for full embedment (Table 10-7).

Middle-strip bottom bars in Fig. 10-32 are shown extended into interior columns so that they lap, and one line of bar supports may be used. This anchorage, which exceeds ACI minimum requirements, usually ensures ample development length and helps prevent temperature and shrinkage cracks at the centerline.

Crack Control. The ACI Code prescribes formulas for determination of maximum bar spacings based on crack width considered suitable for interior and exterior exposure (Art. 10-50). These requirements apply only to one-way reinforced elements. For two-

Fig. 10-31 Minimum cantilever span, for treatment of an exterior column as an interior column, equals $4L_n/15$. This result is obtained by equating the minimum cantilever moment at the exterior column to the minimum negative moment at the interior column.

Fig. 10-32 Reinforcing details for column and middle strips of flat plates.

way slabs, similar procedures are recommended. Bar spacing at critical sections, however, should not exceed twice the slab thickness, except in the top slab of cellular or ribbed (waffle) construction, where requirements for temperature and shrinkage reinforcement govern.

With Grade 60 bars, the suggested maximum spacings for control of flexural cracking in two-way flat plates are given in Table 10-12.

10-61. Flat Slabs.

A flat slab is a two-way slab generally of uniform thickness, but it may be thickened or otherwise strengthened in the region of columns by a drop panel, while the column top below the slab may be enlarged by a capital (round) or bracket (prismatic). If a drop panel is used to increase depth for negative reinforcement, the minimum side dimensions of this panel are $L_1/3$ and $L_2/3$, where L_1 and L_2 are the center-to-center spans in perpendicular directions. Minimum depth of a drop panel is $1.25h$, where h is the slab thickness elsewhere.

A **waffle flat slab** or **waffle flat plate** consists of a thin, two-way top slab and a grid of joists in perpendicular directions, cast on square dome forms. For strengthening

TABLE 10-19 Commonly Used Sizes of Two-Way Joist Forms

Depth, in.	Volume, cu ft per dome	Weight of displaced concrete, lb per dome	3-in. top slab		4½-in. top slab	
			Equiv. slab thickness, in.	Weight,* psf	Equiv. slab thickness, in.	Weight,* psf
30-in. wide domes						
8	3.85	578	5.8	73	7.3	92
10	4.78	717	6.7	83	8.2	102
12	5.53	830	7.4	95	9.1	114
14	6.54	980	8.3	106	9.9	120
16	6.54	1116	9.1	114	10.6	133
20	9.16	1375	10.8	135	12.3	154
			2½-in. top slab		4½-in. top slab	
19-in. wide domes						
6	1.09	163	5.2	66	7.2	90
8	1.41	211	6.3	78	8.3	103
10	1.90	285	6.8	85	8.8	111
12	2.14	321	8.1	101	10.1	126

* Basis: $w = 150$ pcf.

10-73

Fig. 10-33 Reinforcing details for column strips of flat slabs. Details for middle strips are the same as for middle strips of flat plates (Fig. 10-32).

around columns, the domes are omitted in the drop panel areas, to form a solid head, which also may be made deeper than the joists. Other variations of waffle patterns include various arrangements with solid beams on column centerlines both ways. Standard sizes of two-way joist forms are given in Table 10-19.

The main complication introduced in the design of flat slabs and waffle flat plates is the change in cross section of the section considered effective for torsional stiffness K_t (Art 10-59). The ACI Building Code permits this variation in depth to be neglected. This approximation results in an overestimation of flat-slab torsional stiffness, making it equal to that of a flat plate with thickness equal to that of the flat-slab drop panel or waffle-flat plate solid head. Except for the increase in negative moments at exterior columns, however, this assumption changes moments only slightly. Thus, the procedures outlined in Art. 10-58 can also be used for flat slabs.

The drop panel increases shear capacity. Hence, a solid flat slab can ordinarily be designed for concrete of lower strength than for a flat plate. Also, deflection of a flat slab is reduced by the added stiffness that drop panels provide.

The depth of drop panels can be increased beyond $1.25h$ to reduce negative-moment reinforcement and to increase shear capacity when smaller columns are desired. If this adjustment is made, shear in the slab at the edge of the drop panel may become critical. In that case, shear capacity can be increased by making the drop panel larger, up to about 40% of the span. See Fig. 10-33 for bar details (column strip).

Fig. 10-34 Reinforcing details for column strips of two-way waffle flat plates. $B = 24$ bar diameters or 12 in minimum. Details for middle strips are the same as for middle strips of flat plates (Fig. 10-32). *(Courtesy Concrete Reinforcing Steel Institute.)*

Waffle flat plates behave like solid flat slabs with drop panels. Somewhat higher-strength concrete, to avoid the need of stirrups in the joists immediately around the solid head, is usually desirable. If required, however, such stirrups can be made in one piece as a longitudinal assembly, to extend the width of one dome between the drop head and the first transverse joist. For exceptional cases, such stirrups can be used between the second row of domes also. See Fig. 10-34 for reinforcement details.

10-62. Two-Way Slabs on Beams.

The ACI Building Code provides for use of beams on the sides of panels, on column centerlines. (A system of slabs and beams supported by girders, however, usually forms rectangular panels. In that case, the slabs are designed as one-way slabs.)

Use of beams on all sides of a panel permits use of thinner two-way slabs, down to a minimum thickness $h = 3\frac{1}{2}$ in. A beam may be assumed to carry as much as 85% of the column-strip moment, depending on its stiffness relative to the slab (see Chapter 13, ACI 318–77). A secondary benefit, in addition to the direct advantages of longer spans, thinner slabs, and beam stirrups for shear, is that many local codes allow reduced live loads for design of the beams. These reductions are based on the area supported and the ratio of dead to live load. For live loads up to 100 psf, such reductions are usually permitted to a maximum of 60%. Where such reductions are allowed, the reduced total panel moment M_o (Art. 10-58) and the increased effective depth to steel in the beams offer savings in reinforcement to offset partly the added cost of formwork for the beams.

10-63. Estimating Guide for Two-Way Construction.

Figure 10-35 can be used to estimate quantities of steel, concrete, and formwork for flat slabs, as affected by load and span. It also affords a guide to preliminary selection of

Fig. 10-35 Weight of reinforcing steel in square interior panels of flat plates, flat slabs, and waffle flat slabs of 4000-psi concrete, with rebars of 60-ksi yield strength, carrying a superimposed load of 200 lb/ft², for preliminary estimates.

dimensions for analysis, and can be used as an aid in selecting the structural system most appropriate for particular job requirements.

BEAMS

Most ACI Building Code requirements for design of beams and girders refer to *flexural members*. When slabs and joists are not intended, the Code refers specifically to *beams* and occasionally to *beams* and *girders*, and provisions apply equally to beams and girders. So the single term, *beams*, will be used in the following.

10-64. Definitions of Flexural Members.

The following definitions apply for purposes of this section:

Slab—a flexural member of uniform depth supporting area loads over its surface. A slab may be reinforced for flexure in one or two directions.

Joist-slab—a ribbed slab with ribs in one or two directions. Dimensions of such a slab must be within the ACI Code limitations (see Art. 10-54).

Beam—a flexural member designed to carry uniform or concentrated line loads and to span in one direction. A beam may act as a primary member in beam-column frames, or may be used to support slabs or joist-slabs.

Girder—a flexural member used to support beams and designed to span between columns, walls, or other girders. A girder is always a primary member.

10-65. Flexural Reinforcement.

Nonprestressed beams should be designed for flexure as explained in Arts. 10-44 to 10-46. For preliminary design, Fig. 10-17 is useful in estimating bending-moment capacity of beams reinforced for tension only.

If the beam size desired is adequate for the design moment with tension reinforcement only, beam size selected from Fig. 10-17 may be used for final design for flexure. If capacity must be increased without increasing beam size, additional capacity may be provided by addition of compression bars and more tension-bar area to match the compression forces developable in the compression bars (Fig. 10-18). (Shear, torsion, crack control, and deflection requirements must also be met to complete the design. See Arts. 10-47 to 10-51 and 10-66 to 10-68.) Deflection need not be calculated for ACI Code purposes if the total depth h of the beam, including top and bottom cover, is at least the fraction of the span L given in Table 10-14.

A number of interdependent complex requirements (Art. 10-49) regulate the permissible cutoff points or bend points of bars within a span, based on various formulas and rules for development (bond). An additional set of requirements applies if the bars are cut off or bent in a tensile area. These requirements can be satisfied for cases of uniform load and nearly equal spans for the top bars by extending at least 50% of the top steel to a point in the span $0.30L_n$ beyond the face of the support, and the remainder to a point $0.20L_n$, where L_n = clear span. For the bottom bars, all requirements are satisfied by extending at least 40% of the total steel into the supports 6 in past the face, and cutting off the remainder at a distance $0.125L_n$ from the supports. Note that this arrangement does not cut off bottom bars in a tensile zone. Figure 10-36 shows a typical reinforcement layout for a continuous beam, singly reinforced.

The limit in the ACI Building Code on tension-reinforcement ratio ρ that it not exceed 0.75 times the ratio for balanced conditions applies to beams (Art. 10-46). Bal-

Fig. 10-36 Reinforcing details for uniformly loaded continuous beams. At columns, embed alternate bottom bars (at least 50% of the tension-steel area) a minimum of 6 in. to avoid calculation of development length at $0.125L_n$.

anced conditions in a beam reinforced only for tension exist when the tension steel reaches its yield strength f_y simultaneously with the maximum compressive strain in the concrete at the same section becoming 0.003 in/in. Balanced conditions occur similarly for rectangular beams, and for T beams with negative moment, that are provided with compression steel, or doubly reinforced. Such sections are under balanced conditions when the tension steel, with area A_s, yields just as the outer concrete surface crushes, and the total tensile-force capacity $A_s f_y$ equals the total compressive-force capacity of the concrete plus compression steel, with area A_s'. Note that the capacity of the compression steel cannot always be taken as $A_s' f_y$, because the straight-line strain distribution from the fixed points of the outer concrete surface and centroid of the tension steel may limit the compression-steel stress to less than yield strength (Fig. 10-18).

For design of doubly reinforced beams, the force $A_s f_y$ in the tension steel is limited to three-fourths the compression force in the concrete plus the compression in the compression steel at balanced conditions. For a beam meeting these conditions in which the compression steel has not yielded, the design moment capacity is best determined by trial and error:

1. Assume the location of the neutral axis.
2. Determine the strain in the compression steel.
3. See if the total compressive force on the concrete and compression steel equals $A_s f_y$ (Fig. 10-37).

Example. Design a T beam to carry a design negative bending moment of 225 ft-kips. The dimensions of the beam are shown in Fig. 10-37. Concrete strength $f_c' = 4$ ksi, the reinforcing steel has a yield strength $f_y = 60$ ksi, and capacity reduction factor $\phi = 0.90$.

Need for Compression Steel. To determine whether compression reinforcement is required, first compute

$$R_u = \frac{M_u}{b_w d^2} = \frac{225{,}000 \times 12}{15(12.5)^2} = 1150 \text{ psi} \tag{10-81}$$

A check with Fig. 10-17 indicates that, for the selected dimensions, compression reinforcement is needed, because the horizontal line through 1150 does not intersect the line from the lower left representing $f_c' = 4000$.

Compression on Concrete. (Trial-and-error solution.) Assume that the distance c from the neutral axis to the extreme compression surface is 5.1 in. The depth a of the portion of the concrete in compression then may be taken as $0.85c = 4.33$ in (Art. 10-44).

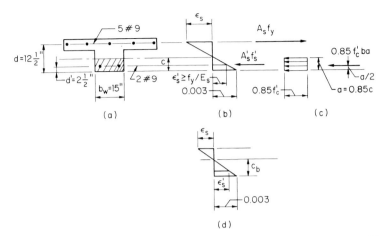

Fig. 10-37 Stress and strain in a T beam reinforced for compression: (a) section; (b) strain; (c) stress; (d) balanced strain.

For a rectangular stress distribution over the concrete, the compression force on the concrete is

$$0.85 f'_c b_w a = 0.85 \times 4 \times 15 \times 4.33 = 221 \text{ kips}$$

Selection of Tension Steel. To estimate the tension steel required, assume a moment arm $jd = d - a/2 = 12.5 - 4.33/2 = 10.33$ in. By Eq. (10-27b), the tension-steel force therefore should be about

$$A_s f_y = 60 A_s = \frac{M_u}{\phi jd} = \frac{225 \times 12}{0.9 \times 10.33} = 290 \text{ kips}$$

from which $A_s = 4.84$ in². Select five No. 9 bars, supplying $A_s = 5$ in². This yields a tensile-steel ratio

$$\rho = \frac{5}{15 \times 12.5} = 0.0267$$

The bars can exert a tension force $A_s f_y = 5 \times 60 = 300$ kips.

Stress in Compression Steel. For a linear strain distribution, the strain ϵ'_s in the steel 2½ in from the extreme compression surface can be found by proportion from the maximum strain of 0.003 in/in at that surface.

$$\epsilon'_s = \frac{5.1 - 2.5}{5.1} \times 0.003 = 0.0015 \text{ in/in}$$

With the modulus of elasticity E_s taken as 29,000 ksi, the stress in the compression steel is

$$f'_s = 0.0015 \times 29,000 = 43.5 \text{ ksi}$$

Selection of Compression Steel. The total compression force equals the 221-kip force on the concrete previously computed plus the force on the compression steel. If the total compression force is to equal the total tension force, the compression steel must resist a force

$$A'_s f'_s = A'_s (43.5 - 3.4) = 300 - 221 = 79 \text{ kips}$$

from which the compression-steel area $A'_s = 2$ in².

Check the Balance of Forces ($\Sigma F_c = \Sigma F_t$). For an assumed position of the neutral axis at 5.10 in, with five no. 9 tension bars and two no. 9 compression bars, the total compression force C is:

$$\begin{aligned}\text{Concrete: } 0.85 \times 4 \times 4.33 \times 15 &= 221 \text{ kips}\\ \text{Steel: } \quad 2(43.5 - 3.4) \quad &= \underline{\ 80\ \text{kips}}\\ C &= 301 \text{ kips}\end{aligned}$$

This compression force for practical purposes is equal to the total tension force: $5 \times 60 \times 1 = 300$ kips. The assumed position of the neutral axis results in a balance of forces within 1% accuracy.

Nominal Flexural Strength. For determination of the nominal flexural strength of the beam, moments about the centroid of the tension steel are added:

$$M_n = 0.85 f'_c ba\left(d - \frac{a}{2}\right) + A'_s f_y (d - d') \tag{10-82}$$

Substitution of numerical values yields:

$$\begin{aligned}M_n &= 0.85 \times 4 \times 15 \times 4.33(12.5 - 4.33/2) + 2(43.5 - 3.4)(12.5 - 2.5)\\ &= 221 \times 10.33 + 80 \times 10 = 257 \text{ ft-kips}\end{aligned}$$

Check Design Strength (ϕM_n).

$$\phi M_n = 0.90 \times 257 = 231 \text{ ft-kips} > M_u = 225 \text{ ft-kips}) \qquad \text{OK}$$

10-66. Reinforcement for Shear and Flexure.

Determination of the shear capacity of a beam is discussed in Art. 10-47. Minimum shear reinforcement is required in all beams with total depth greater than 10 in, or 2½ times flange (slab) thickness, or half the web thickness, except where the factored shear force V_u is less than half the design shear strength ϕV_c of the concrete alone. Torsion should be combined with shear when the factored loads cause a torsional moment T_u larger than $\phi(0.5\sqrt{f'_c})\,\Sigma x^2 y$ (see Art. 10-48).

Shear strength should be computed at critical sections in a beam from Eq. (10-37). Open or closed stirrups may be used as reinforcement for shear in beams; but closed stirrups are required for torsion. The minimum area for open or closed stirrups for vertical shear only, to be used where $0.5\phi V_c \leq V_u \leq \phi V_c$ and the factored torsional moment $T_u \leq \phi(0.5\sqrt{f'_c})\,\Sigma x^2 y$, should be calculated from

$$A_v = \frac{50 b_w s}{f_y} \tag{10-83}$$

where A_v = area of all vertical legs in the spacing s, in, parallel to flexural reinforcement, in²
b_w = thickness of beam web, in²
f_y = yield strength of reinforcing steel, psi

Note that this minimum area provides a capacity for 50-psi shear on the cross section $b_w s$.

Where V_u exceeds V_c, the cross-sectional area A_v of the legs of open or closed vertical stirrups at each spacing s should be calculated from Eq. (10-40). A_v is the total area of vertical legs, two legs for a common open U stirrup or the total of all legs for a transverse multiple U. Note that there are three zones in which the required A_v may be supplied by various combinations of size and spacing of stirrups (Fig. 10-38):

1. Beginning 1 or 2 in from the face of supports and extending over a distance d

Fig. 10-38 Required shear reinforcement in three zones of a beam between supports and midspan is determined by cross-hatched areas.

from each support, where d is the depth from extreme compression surface to centroid of tension steel (A_v is based on V_u at d from support).

2. Between distance d from each support and the point where $\phi V_s = V_u - \phi V_c = 50 b_w s$ (required A_v decreases from maximum to minimum).

3. Distance over which minimum reinforcement is required (minimum A_v extends from the point where $\phi V_s = 50\, b_w s$ to the point where $V_u = 0.5\phi V_c$).

10-67. Reinforcement for Torsion and Shear.

Any beam that supports unbalanced loads that are transverse to the direction in which it is subjected to bending moments transmits an unbalanced moment to the supports and must be investigated for torsion. Generally, this requirement affects all spandrel and other edge beams, and interior beams supporting uneven spans or unbalanced live loads on opposite sides. The total unbalanced moment from a floor system with one-way slabs in one direction and beams in the perpendicular direction can often be considered to be transferred to the columns by beam flexure in one direction, neglecting torsion in the slab. The total unbalanced moment in the other direction, from the one-way slabs, can be considered to be transferred by torsional shear from the beams to the columns. (Determination of the torsional capacity of a beam is discussed in Art. 10-48.)

The factored torque T_u for rectangular or T beams should be less than ϕT_n, where T_n is the nominal torsional strength [Eq. (10-42)]. If $T_u \leq \phi(0.5\sqrt{f'_c})\Sigma x^2 y$, T_u may be neglected. The portion of the torque carried by the concrete should not exceed T_c computed from Eq. (10-43).

The required area A_t of each leg of a closed stirrup for torsion should be computed from Eq. (10-45). Stirrup spacing should not exceed 12 in or $(x_1 + y_1)/4$, where x_1 and y_1 are the dimensions center to center of a closed rectangular stirrup.

Torsion reinforcement also includes the longitudinal bars shown in each corner of the closed stirrups in Fig. 10-39 and the longitudinal bars spaced elsewhere inside the perimeter of the closed stirrups at not more than 12 in. At least one longitudinal bar in each corner is required. [For required areas of these bars, see Eqs. (10-46) and (10-47).] If a beam is fully loaded for maximum flexure and torsion simultaneously, as in a spandrel beam, the area of torsion-resisting longitudinal bars A_l should be provided in addition to flexural bars.

For interior beams, maximum torsion usually occurs with live load only on a slab on one side of the beam. Maximum torsion and maximum flexure cannot occur simultaneously. Hence, the same bars can serve for both.

The closed stirrups required for torsion should be provided in addition to the stirrups required for shear, which may be the open type. Because the size of stirrups must be at

Fig. 10-39 Torsion reinforcement consists of closed stirrups and longitudinal bars within the perimeter of the stirrups. Required area of stirrup legs is the sum of that required for shear and that for torsion. If the same loading causes maximum moment, shear, and torsion, the required area of the top bars is the sum of that required for negative moment and that for torsion.

least No. 3 and maximum spacings are established in the ACI Building Code for both shear and torsion stirrups, a closed-stirrup size-spacing combination can usually be selected for combined shear and torsion. Where maximum shear and torsion cannot occur under the same loading, the closed stirrups can be proportioned for the maximum combination of forces or the maximum single force; whichever is larger (Fig. 10-40).

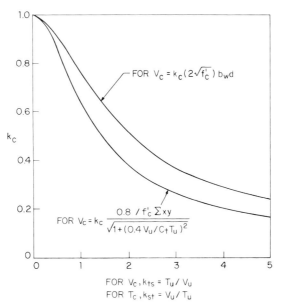

Fig. 10-40 Chart for determining torsion and shear carried by concrete.

TABLE 10-20 Minimum Clear Concrete Cover, in, for Beams and Girders

Bar size No.	Exterior exposure			Interior exposure		
	Cast-in-place concrete	Precast concrete	Prestressed concrete	Cast-in-place concrete	Precast concrete	Prestressed concrete
3, 4, and 5	1½	1¼	1½	1½	⅝	1½
6 through 11	2	1½	1½	1½	d_b*	1½
14 and 18	2	2	1½	1½	1½	1½
Stirrups	Same as above for each size				⅜	1

* d_b = nominal bar diameter, in.

(P. F. Rice and E. S. Hoffman, "Structural Design Guide to the ACI Building Code," Van Nostrand Rheinhold Company, New York.)

10-68. Crack Control in Beams.

The ACI Building Code contains requirements limiting reinforcement spacing to regulate crack widths when the yield strength f_y of the reinforcement exceeds 40,000 psi (Art. 10-50). Limits for interior and exterior exposures are prescribed. Since the computed crack width is made a function of the tensile area of concrete tributary to and concentric with each bar, the required cover is important to crack control. (See Tables 10-12 and 10-20.)

WALLS

Generally, any vertical member whose length and height are both much larger than the thickness may be treated as a wall. Walls subjected to vertical loads are called **bearing walls**. Walls subjected to no loads other than their own weight, such as panel or enclosure walls, are called **nonbearing walls**. Walls with a primary function of resisting horizontal loads are called **shear walls**. They also may serve as bearing walls. See Art. 10-91.

10-69. Bearing Walls.

Reinforced concrete bearing walls may be designed as eccentrically loaded columns or by an empirical method given in the ACI Building Code. The empirical method may be used when the resultant of the applied load falls within the middle third of the wall thickness. This method gives the capacity of the wall as

$$P_u \leq \phi P_{nw} = 0.55\phi f'_c A_g \left[1 - \left(\frac{kL_c}{32h}\right)^2\right] \quad (10\text{-}84)$$

where f'_c = specified concrete strength
ϕ = capacity reduction factor = 0.70
A_g = gross area of horizontal cross section of wall
h = wall thickness
L_c = vertical distance between supports
k = effective length factor

When the wall is braced against translation at top and bottom:

$k = 0.8$ for restraint against rotation at one or both ends
$k = 1.0$ for both ends unrestrained against rotation

When the wall is not braced against lateral translation, $k = 2.0$ (cantilever walls).

The allowable average design stress f_c for a wall with $k = 0.8$ is obtained by dividing P_u in Eq. (10-84) by A_g:

$$f_c = \frac{P_u}{A_g} = 0.385 f'_c \left[1 - \left(\frac{L_c}{40h} \right)^2 \right] \tag{10-85}$$

For given f'_c and ratio L_c/h, f_c can be obtained from Fig. 10-41.

Length. The effective length of wall for concentrated loads may be taken as the center-to-center distance between loads, but not more than the width of bearing plus 4 times the wall thickness.

Thickness. The minimum thickness of bearing walls for which Eqs. (10-84) and (10-85) are applicable is one-twenty-fifth of the least distance between supports at the sides or top, but at least 6 in for the top 15 ft of wall height and 1 in additional for each successive 25 ft, or fraction of it, downward. But bearing walls of two-story dwellings may be 6 in thick throughout their height. Exterior basement walls, foundation walls, party walls, and fire walls should be at least 8 in thick. Minimum thickness and reinforcement requirements may be waived, however, if justified by structural analysis.

Reinforcement. The area of horizontal steel reinforcement should be at least

$$A_h = 0.0025 A_{wv} \tag{10-86}$$

where A_{wv} = gross area of the vertical cross section of wall.

Area of vertical reinforcement should be at least

$$A_v = 0.0015 A_{wh} \tag{10-87}$$

where A_{wh} = gross area of the horizontal cross section of wall. For Grade 60 bars, No. 5 or smaller, or for welded-wire fabric, these steel areas may be reduced to $0.0020 A_{wv}$ and $0.0012 A_{wh}$, respectively.

Walls 10 in or less thick may be reinforced with only one rectangular grid of rebars.

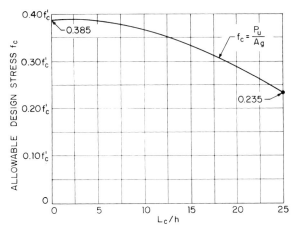

Fig. 10-41 Curve for determining allowable average design stress for bearing walls, given the ratio of unsupported height to thickness.

Thicker walls require two grids. The grid nearest the exterior wall surface should contain between one-half and two-thirds the total steel area required for the wall. It should have a concrete cover of at least 2 in, but not more than one-third the wall thickness. A grid near an interior wall surface should have a concrete cover of at least ¾ in, but not more than one-third the wall thickness. Minimum size of bars, if used, is No. 3. Maximum bar spacing is 18 in. (These requirements do not apply to basement walls, however. If such walls are cast against and permanently exposed to earth, minimum cover is 3 in. Otherwise, the cover should be at least 2 in for bar sizes No. 6 and larger, and 1½ in for No.5 bars or ⅝-in wire and smaller.)

At least two No. 5 bars should be placed around all window and door openings. The bars should extend at least 24 in beyond the corners of openings.

Design for Eccentric Loads. Bearing walls with bending moments sufficient to cause tensile stress must be designed as columns for combined flexure and axial load. Minimum reinforcement areas and maximum bar spacings are the same as for walls designed by the empirical method. Lateral ties, as for columns, are required for compression reinforcement and where the vertical bar area exceeds 0.01 times the gross horizontal concrete area of the wall. (For column capacity, see Arts. 10-83 and 10-85.)

Under the preceding provisions, a thin, wall-like (rectangular) column with a steel ratio less than 0.01 will have a greater carrying capacity if the bars are detailed as for walls. The reasons for this are: The effective depth is increased by omission of ties outside the vertical bars and by the smaller cover (as small as ¾ in) permitted for vertical bars in walls. Furthermore, if the moment is low (eccentricity less than one-sixth the wall thickness), so that the wall capacity is determined by Eq. (10-84), the capacity will be larger than that computed for a column, except where the column is part of a frame braced against sidesway.

10-70. Nonbearing Walls.

Nonbearing reinforced-concrete walls, frequently classified as panels, partitions, or cross walls, may be precast or cast in place. Panels serving merely as exterior cladding, when precast, are usually attached to the columns or floors of a frame, supported on grade beams, or supported by and spanning between footings, serving as both grade beams and walls. Cast-in-place cross walls are most common in substructures. Less often, cast-in-place panels may be supported on grade beams and attached to the frame.

In most of these applications for nonbearing walls, stresses are low and alternative materials, such as unreinforced masonry, when supported by beams above grade, or panels of other materials, can be used. Consequently, unless esthetic requirements dictate reinforced concrete, low-stressed panels of reinforced concrete must be designed for maximum economy. Minimum thickness, minimum reinforcement, full benefits of standardization for mass-production techniques, and design for double function as both wall and deep beam must be achieved.

Thickness of nonbearing walls of reinforced concrete should be at least one-thirtieth the distance between supports, but not less than 4 in.

The ACI Building Code, however, permits waiving all minimum arbitrary requirements for thickness and reinforcement where structural analysis indicates adequate strength and stability.

Where support is provided, as for a panel above grade on a grade beam, connections to columns may be detailed to permit shrinkage. Friction between base of panel and the beam can be reduced by an asphalt coating and omission of dowels. These provisions will permit elimination or reduction of horizontal shrinkage reinforcement. Vertical reinforcement is seldom required, except as needed for spacing the horizontal bars.

If a nonbearing wall is cast in place, reinforcement can be nearly eliminated except at edges. If the wall is precast, handling stresses will often control. Multiple pickup points with rigid-beam pickups will reduce such stresses. Vacuum pad pickups can eliminate nearly all lifting stresses.

Where deep-beam behavior or wind loads cause stresses exceeding those permitted on plain concrete, the ACI Code permits reduction of minimum tension-reinforcement [$A_s = 200bd/f_y$ (Art. 10-46)] if reinforcement furnished is one-third greater than that required by analysis. (For deep-beam design, see Art. 10-90.)

10-71. Cantilever Retaining Walls.

Under the ACI Building Code, cantilever retaining walls are designed as slabs. Specific Code requirements are not given for cantilever walls, but when axial load becomes near zero, the Code requirements for flexure apply.

Minimum clear cover for bars in walls cast against and permanently exposed to earth is 3 in. Otherwise, minimum cover is 2 in for bar sizes No. 6 and larger, and 1½ in for No. 5 bars or ⅝-in wire and smaller.

Two points requiring special consideration are analysis for load factors of 1.7 times lateral earth pressure and 1.4 times dead loads and fluid pressures, and provision of splices at the base of the stem, which is a point of maximum moment. The footing and stem are usually cast separately, and dowels left projecting from the footing are spliced to the stem reinforcement.

A straightforward way of applying Code requirements for strength design is illustrated in Fig. 10-42. Soil reaction pressure p and stability against overturning are determined for actual weights of concrete D and soil W and actual assumed lateral pressure of the soil H. The total cantilever bending moment for design of stem reinforcement is then based upon $1.7H$. The toe pressure used to determine the footing bottom bars is $1.7p$. And the top load for design of the top bars in the footing heel is $1.4(W + D_h)$, where D_h is the weight of the heel. The Code requires application of a factor of 0.9 to vertical loading that reduces the moment caused by H.

Where the horizontal component of backfill pressure includes groundwater above the top of the heel, use of two factors, 1.7 for the transverse soil pressure and 1.4 for the transverse liquid pressure, becomes confusing. Because the probability of overload is about the same for soil pressure and water pressure, use of a single factor, 1.7, seems logical. For sanitary structures where these conditions are common, ACI Committee 350 has recommended use of 1.7 for both soil and liquid pressure (see "Concrete Sanitary Engineering Structures," ACI 350R-77). Committee 350 also favored a more conservative approach for design of the toe. It is more convenient and conservative to consider 1.7 times the *entire* vertical reaction uniformly distributed across the toe as well as more nearly representing the actual end-point condition (Fig. 10-43).

The top bars in the heel can be selected for the unbalanced moment between the design forces on the toe and the stem, but need not be larger than for the moment of the

Fig. 10-42 Critical sections for design of cantilever retaining walls and design (factored) loads.

top loads on the footing (earth and weight of heel). For a footing proportioned so that the actual soil pressure approaches zero at the end of the heel, the unbalanced moment and the maximum moment in the heel caused by the top loads will be nearly equal.

Fig. 10-43 Loads for simplified strength design for toe of wall.

The possibility of an overall sliding failure, involving the soil and the structure together, must be considered, and may require a vertical lug extending beneath the footing, tie backs, or other provisions.

The base of the stem is a point of maximum bending moment and yet also the most convenient location for splicing the vertical bars and footing dowels. The ACI Code advises avoiding such points for the location of lap splices. But for cantilever walls, splices can be avoided entirely at the base of the stem only for low walls (8 to 10 ft high), in which L-shaped bars from the base of the toe can be extended full height of the stem. For high retaining walls, if all the bars are spliced at the base of the stem, a Class C tension lap splice is required (Art 10-49). If alternate dowel bars are extended one Class B lap-splice length and the remaining dowel bars are extended at least twice this distance before cutoff, Class B lap splices may be used. This arrangement requires that dowel-bar sizes and vertical-bar sizes be selected so that the longer dowel bars provide at least 50% of the steel area required at the base of the stem and the vertical bars provide the total required steel at the cutoff point of the longer dowels (Fig. 10-44).

10-72. Counterfort Retaining Walls.

In this type of retaining wall, counterforts (cantilevers) are provided on the earth side between wall and footing to support the wall, which essentially spans as a continuous one-way slab horizontally. Counterfort walls seldom find application in building construction. A temporary condition in which basement walls may be required to behave as counterfort retaining walls occurs though, if outside fill is placed before the floors are constructed. Under this condition of loading, each interior cross wall and end basement

Fig. 10-44 Splice details for cantilever retaining walls: (a) for low walls; (b) for high walls with Class C lap for dowels; (c) alternate detail for high walls, with Class B lap for dowels.

wall can be regarded as a counterfort. It is usually preferable, however, to delay the fill operation rather than to design and provide reinforcement for this temporary condition.

The advantages of counterfort walls are the large effective depth for the cantilever reinforcement and concrete efficiently concentrated in the counterfort. For very tall walls, where an alternative cantilever wall would require greater thickness and larger quantities of steel and concrete, the savings in material will exceed the additional cost of forming the counterforts. Accurate design is necessary for economy in important projects involving large quantities of material and requires refinement of the simple assumptions in the definition of counterfort walls. The analysis becomes complex for determination of the division of the load between one-way horizontal slab and vertical cantilever action.

(F. S. Merritt, "Standard Handbook for Civil Engineers," McGraw-Hill Book Company, New York.)

10-73. Retaining Walls Supported on Four Sides.

For walls more than 10 in thick, the ACI Building Code requires two-way layers of bars in each face. Two-way-slab design of this reinforcement is required for economy in basement walls or subsurface tank walls supported as vertical spans by the floor above and the footing below, and as horizontal spans by stiff pilasters, interior cross walls, or end walls.

This type of two-way slab is outside the scope of the specific provisions in the ACI Code. Without an "exact" analysis, which is seldom justified because of the uncertainties involved in the assumptions for stiffnesses and loads, a realistic design can be based on the simple two-way-slab design method of Appendix A, Method 2, of the 1963 ACI Building Code, as follows:

TABLE 10-21 Moment Coefficients C for Two-Way Slabs with Two Discontinuous Edges*

Location and type of moment	For short span when ratio of short span to long span is:						Long span, all values of m
	$m = 1.0$	$m = 0.9$	$m = 0.8$	$m = 0.7$	$m = 0.6$	$m = 0.5$	
Negative moment at continuous edge	0.049	0.057	0.064	0.071	0.078	0.090	0.049
Negative moment at discontinuous edge	0.025	0.028	0.032	0.036	0.039	0.045	0.025
Positive moment at midspan....	0.037	0.043	0.048	0.054	0.059	0.068	0.037

* Bending moments may be computed from $M = CwS^2$, where w = uniform load on slab and S = length of short span, the smaller of center-to-center distance between supports and clear span plus twice the slab thickness. From Appendix A, Method 2, "Building Code Requirements for Reinforced Concrete," ACI 318-63.

For the usual conditions of stiffness, the edges supported by a floor at the top of the wall and a footing below the wall can be considered "discontinuous." (Note that the following method provides for negative moments at "discontinuous" edges equal to 50% of those at "continuous" edges.) End walls or interior cross walls with separate footings can be considered continuous edge supports. Pilasters, unless very stiff and provided with separate footings, can be disregarded in the analysis for design of inner wall bars. Walls should be reinforced with at least a nominal amount of outer bars.

Fig. 10-45 Two-way retaining wall.

The wall panels should be divided into middle strips and column strips, each to be designed as one-way slabs (Fig. 10-30). Width of middle strips should be taken as half the panel width, in vertical and horizontal directions, respectively. Where the ratio of short span to long span is less than 0.5, however, the width of the middle strip in the short direction should be taken as the difference between the short and long spans.

Bending moments M at critical points in a two-way retaining wall may be computed with the aid of Table 10-21. It gives C for the computation of M for middle strips from

$$M = CwS^2 \qquad (10\text{-}88)$$

where $w = 1.7H$ or $1.4F$
H = lateral earth pressure
F = lateral fluid pressure
S = short span of wall

H can be converted into an equivalent uniform load, as indicated in Fig. 10-45 on the assumption that some load is resisted by cantilever action on outer-face dowels. The half-panel-wide column strips are assumed to resist two-thirds of the middle-strip moments. For the horizontal column strips one-quarter-panel wide, the moment per foot may be taken as one-third the middle-strip moment.

FOUNDATIONS

Building foundations should distribute wall and column loads to the underlying soil and rock within acceptable limits on resulting soil pressure and total and differential settlement. Wall and column loads consist of live load, reduced in accordance with the applicable general building code, and dead load, combined, when required, with lateral loads of wind, earthquake, earth pressure, or liquid pressure. These loads can be distributed to the soil near grade by concrete spread footings, or to the soil at lower levels by concrete piles or drilled piers.

10-74. Types of Foundations.

A wide variety of concrete foundations are used for buildings. Some of the most common types are illustrated in Fig. 10-46.

Spread wall footings consist of a plain or reinforced slab wider than the wall, extending the length of the wall (Fig. 10-46a). Plain- or reinforced-concrete **individual-column spread footings** consist of simple, stepped, or sloped two-way concrete slabs, square or rectangular in plan (Fig. 10-46b to d). For two columns close together, or an

Fig. 10-46 Common types of foundations for buildings.

exterior column close to the property line so that individual spread or pile-cap footings cannot be placed concentrically, a reinforced-concrete, spread **combined footing** (Fig. 10-46e) or a **strap footing** (Fig. 10-46f) can be used to obtain a nearly uniform distribution of soil pressure or pile loads. The strap footing becomes more economical than a combined footing when the spacing between the columns becomes larger, causing large bending moments in the combined footing.

For small soil pressures or where loads are heavy relative to the soil capacity, a reinforced-concrete **mat,** or **raft, foundation** (Fig. 10-46g) may prove economical. A mat consists of a two-way slab under the entire structure. Concrete cross walls or inverted beams can be utilized with a mat to obtain greater stiffness and economy.

Where sufficient soil strength is available only at lower levels, **pile foundations** (Fig. 10-46h) or **drilled-pier foundations** (Fig. 10-46i) can be used.

10-75. General Design Principles for Foundations.

The area of spread footings, the number of piles, or the number of drilled piers are selected by a designer to support actual unfactored building loads without exceeding *settlement limitations,* a safe soil pressure q_a, or a safe pile or drilled-pier load. A factor of safety from 2 to 3, based on the ultimate strength of the soil and its settlement characteristics, is usually used to determine the safe soil pressure or safe pile or drilled-pier load.

For foundations not subjected to overturning moment, the area of spread footings or the number of piles or drilled piers is determined simply by dividing the actual unfactored wall or column loads by the safe soil pressure q_a, or the safe pile or drilled-pier load. The weight of the footing or drilled pier can be neglected in determining the dead load, if the safe soil pressure is a net pressure in excess of that at the same level due to the surrounding surcharge.

Soil Pressures. After the area of the spread footing or the number and spacing of piles or drilled piers has been determined, the spread footing, pile-cap footing, or drilled pier can be designed. The strength-design method of the ACI Building Code (Art. 10-44) uses factored loads of gravity, wind, earthquake, earth pressure, and fluid pressure to determine *design soil pressure* q_s and design pile or pier load. The design loadings are used in strength design to determine design moments and shears at critical sections.

For concentrically loaded footings, q_s is usually assumed as uniformly distributed over the footing area. This pressure is determined by dividing the concentric wall or column design load P_u by the area of the footing. The weight of the footing can also be neglected in determining q_s, because the weight does not induce design moments and shears. The design pile load for concentrically loaded pile-cap footings is determined in a similar manner.

When individual or wall spread footings are subjected to overturning moment about one axis, in addition to vertical load, as with a spread footing for a retaining wall, the pressure distribution under the footing is trapezoidal if the eccentricity e_x of the resultant vertical load P_u is within the kern of the footing, or triangular if beyond the kern, as shown in Fig. 10-47. Thus, when $e_x < L/6$, where L is the footing length in the direction of eccentricity e_x, the pressure distribution is trapezoidal (Fig. 10-47a) with a maximum

$$q_{s1} = \frac{P_u}{BL}\left(1 + \frac{6e_x}{L}\right) \qquad (10\text{-}89)$$

and a minimum

$$q_{s2} = \frac{P_u}{BL}\left(1 - \frac{6e_x}{L}\right) \qquad (10\text{-}90)$$

where B = footing width. When $e_x = L/6$, the pressure distribution becomes triangular over the length L, with a maximum

$$q_s = \frac{2P_u}{BL} \qquad (10\text{-}91)$$

When $e_x > L/6$, the length of the triangular distribution decreases to $1.5L - 3e_x$ (Fig. 10-46b) and the maximum pressure rises to

$$q_s = \frac{2P_u}{1.5B(L - 2e_x)} \qquad (10\text{-}92)$$

Reinforcement for Bearing. The bearing stress on the interface between a column and a spread footing, pile cap, or drilled pier should not exceed the allowable stress f_b given by Eq. (10-93), unless vertical reinforcement is provided for the excess.

$$f_b = 0.85\phi f'_c \sqrt{\frac{A_2}{A_1}} \qquad \frac{A_2}{A_1} \le 4 \qquad (10\text{-}93)$$

where ϕ = capacity reduction factor = 0.70
f'_c = specified concrete strength
A_1 = loaded area of the column or base plate
A_2 = supporting area of footing, pile cap, or drilled pier that is the lower base of the largest frustrum of a right pyramid or cone contained wholly within the footing, with A_1 the upper face, and with side slopes not exceeding 2 horizontal to 1 vertical (Fig. 10-48)

If the bearing stress on the loaded area exceeds f_b, reinforcement must be provided by extending the longitudinal column bars into the spread footings, pile cap, or drilled pier or by dowels. If so, the column bars or dowels required must have a minimum area of 0.005 times the loaded area of the column, and at least four bars must be provided.

Fig. 10-47 Spread footing subjected to moment.

Fig. 10-48 Bearing stresses on column and footing.

The ACI Building Code requires that every column have a minimum tensile capacity in each face equal to one-fourth the area of the reinforcement in that face. This requirement applies throughout the full height of the column, from foundation up, and will sometimes control the dowel embedment for development of column verticals. If the same size and number of dowels are provided in the footing as rebars in the column and are fully developed in compression, they will usually develop the required minimum tension capacity required by the Code for the column.

Required compression-dowel embedment length cannot be reduced by hooks, nor by f'_c over 4444 psi. Compression dowels can be smaller than column reinforcement. They may be one bar size larger than the column bars.

If the bearing stress on the loaded area of a column does not exceed $0.85\phi f'_c$, column compression bars or dowels do not need to be extended into the footing, pile cap, or pier, if they can be developed within 3 times the column dimension (pedestal height) above the footing (Art. 10-49). It is desirable, however, that a minimum of one No. 5 dowel be provided in each corner of a column.

Footing Thickness. The minimum thickness allowed by the ACI Building Code for footings is 8 in for plain concrete footings on soil, 6 in above the bottom reinforcement for reinforced-concrete footings on soil, and 12 in above the bottom reinforcement for reinforced-concrete footings on piles. Plain-concrete pile-cap footings are not permitted.

Concrete Cover. The minimum concrete cover required by the ACI Code for reinforcement cast against and permanently exposed to earth is 3 in.

10-76. Spread Footings for Walls.

(See also Art. 10-75.) The critical sections for shear and moment for spread footings supporting concrete or masonry walls are shown in Fig. 10-49a and b. Under the soil pressure, the projection of footing on either side of a wall acts as a one-way cantilever slab.

Unreinforced Footings. For plain-concrete spread footings, the maximum permissible flexural tension stress, psi, in the concrete is limited by the ACI Building Code to $5\phi \sqrt{f'_c}$, where ϕ = capacity reduction factor = 0.65 and f'_c = specified concrete strength, psi. For constant-depth, concentrically loaded footings with uniform design soil pressure, the thickness h, in, can

Fig. 10-49 Critical sections for shear and moment in wall footings.

be calculated from

$$h = 0.08X \sqrt{\frac{q_s}{\sqrt{f_c'}}} \qquad (10\text{-}94)$$

where X = projection of footing, in
q_s = net design soil pressure, psf

Shear is not critical for plain-concrete wall footings. The maximum tensile stress, $\phi 5\sqrt{f_c'}$, due to flexure controls the thickness.

Because it is usually not economical or practical to provide shear reinforcement in reinforced-concrete spread footings for walls, V_u is usually limited to the maximum value that can be carried by the concrete, $\phi V_c = \phi 2\sqrt{f_c'}b_w d$.

Flexural Reinforcement. Steel area can be determined from $A_s = M_u/\phi f_y jd$ as indicated in Art. 10-46. Sufficient length of reinforcement must be provided to develop the full yield strength of straight tension reinforcement. The critical development length is the shorter dimension l_d shown in Fig. 10-49a and b. The minimum lengths required to develop various bar sizes are tabulated in Table 10-7 (Art. 10-49). Reinforcement at right angles to the flexural reinforcement is usually provided as shrinkage and temperature reinforcement and to support and hold the flexural bars in position. The ACI Code does not limit the amount of this reinforcement.

10-77. Spread Footings for Individual Columns.

(See also Art. 10-75.) Footings supporting columns are usually made considerably larger than the columns to keep pressure and settlement within reasonable limits. Generally, each column is also placed over the centroid of its footing to obtain uniform pressure distribution under concentric loading. In plan, the footings are usually square, but they can be made rectangular to satisfy space restrictions or to support rectangular columns or pedestals.

Under soil pressure, the projection on each side of a column acts as a cantilever slab in two perpendicular directions. The effective depth of footing d is the distance from the extreme compression surface of the footing to the centroid of the tension reinforcement.

Bending Stresses. Critical sections for moment are at the faces of square and rectangular concrete columns or pedestals (Fig. 10-50a). For round and regular polygon columns or pedestals, the face may be taken as the side of a square having an area equal to the area enclosed with the perimeter of the column or pedestal. For steel columns with steel base plates, the critical section for moment may be taken halfway between the face of the column and the edges of the plate.

(a) MOMENT SECTIONS

(b) SHEAR SECTIONS

Fig. 10-50 Critical sections for shear and moment in column footings.

For plain-concrete spread footings, the flexural tensile stress must be limited to a maximum of $5\phi\sqrt{f_c'}$, where ϕ = capacity reduction factor = 0.65 and $\sqrt{f_c'}$ = specified concrete strength, psi. Thickness of such footings can be calculated with Eq. (10-94). Shear is not critical for plain-concrete footings.

Shear. For reinforced-concrete spread footings, shear is critical on two different sections:

Two-way or punching shear (Fig. 10-50b) is critical on the periphery of the surfaces at distance $d/2$ from the column, where d = effective footing depth. Shear stress may be computed from Eq. (10-75).

One-way shear, as a measure of diagonal tension,

is critical at distance d from the column (Fig. 10-50b), and must be checked in each direction.

It is usually not economical or practical to provide shear reinforcement in column footings. So the shear ϕV_c that can be carried by the concrete controls the thickness required.

For one-way shear for plain or reinforced concrete sections,

$$V_c = 2\sqrt{f'_c}b_w d \qquad (10\text{-}95)$$

but Eq. (10-38) may be used as an alternative for reinforced concrete.

For two-way shear,

$$V_c = (2 + 4/\beta_c)\sqrt{f'_c}b_o d \leq 4\sqrt{f'_c}\,b_o d \qquad (10\text{-}96)$$

where b_w = width of section
b_o = perimeter of critical section
β_c = ratio of long side to short side of reaction area
d = effective depth of section

Flexural Reinforcement. In square spread footings, reinforcing steel should be uniformly spaced throughout, in perpendicular directions. In rectangular spread footings, the ACI Building Code requires the reinforcement in the long direction to be uniformly spaced over the footing width. Also, reinforcement with an area $2/(\beta + 1)$ times the area of total reinforcement in the short direction should be uniformly spaced in a width that is centered on the column and equal to the short footing dimension, where β is the ratio of the long to the short side of the footing. The remainder of the reinforcing in the short direction should be uniformly spaced in the outer portions of the footing. To maintain a uniform spacing for the bars in the short direction for simplified placing, the theoretical number of bars required for flexure must be increased by about 15%. The maximum increase occurs when β is about 2.5.

The required area of flexural reinforcement can be determined as indicated in Art. 10-46. Maximum-size bars are usually selected to develop the yield strength by straight tension embedment without end hooks. The critical length is the shorter dimension l_d shown for wall footings in Figs. 10-49a and b.

10-78. Combined Spread Footings.

(See also Arts. 10-74 and 10-75.) A combined spread footing under two columns should have a shape and location such that the center of gravity of the column loads coincides with the centroid of the footing area. It can be square, rectangular, or trapezoidal, as shown in Fig. 10-51.

Net design soil-pressure distribution can be assumed to vary linearly for most (rigid) combined footings. For the pressure distribution for flexible combined footings, see the report, "Suggested Design Procedures for Combined Footings and Mats," *ACI* 336.2R-66, and M. J. Haddadin, "Mats and Combined Footings—Analysis by the Finite Element Method," *ACI Journal Proceedings,* Vol. 69, No. 12, December 1971, pp. 945–949. A computer program based on this method is available from the Cement and Concrete Research Institute, Portland Cement Association, Old Orchard Road, Skokie, Ill. 60076.

If the center of gravity of the unfactored column loads coincides with the centroid of the footing area, the net soil pressure q_a will be uniform.

$$q_a = \frac{R}{A_f} \qquad (10\text{-}97)$$

where R = applied vertical load
A_f = area of footing

Fig. 10-51 Design conditions for combined spread footing of trapezoidal shape.

The factored design load on the columns, however, may change the location of the center of gravity of the loads. If the resultant is within the kern for footings with moment about one axis, the design soil pressure q_s can be assumed to have a linear distribution (Fig. 10-51b) with a maximum at the more heavily loaded edge:

$$q_{s1} = \frac{P_u}{A_f} + \frac{P_u ec}{I_f} \qquad (10\text{-}98)$$

and a minimum at the opposite edge:

$$q_{s2} = \frac{P_u}{A_f} - \frac{P_u ec}{I_f} \qquad (10\text{-}99)$$

where P_u = factored vertical load
e = eccentricity of load
c = distance from center of gravity to section for which pressure is being computed
I_f = moment of inertia of footing area

If the resultant falls outside the kern of the footing, the net design pressure q_s can be assumed to have a linear distribution with a maximum value at the more heavily loaded edge and a lengthwise distribution of 3 times the distance between the resultant and the pressed edge. The balance of the footing will have no net design pressure.

With the net design soil-pressure distribution known, the design shears and moments can be determined (Fig. 10-51d and e). Critical sections for shear and moment are shown in Fig. 10-51a. The critical section for two-way shear is at a distance $d/2$ from the columns, where d is the effective depth of footing. The critical section for one-way shear in both the longitudinal and transverse direction is at a distance d from the column face.

Maximum design negative moment, causing tension in the top of the footing, will occur between the columns. The maximum design positive moment, with tension in the bottom of the footing, will occur at the face of the columns in both the longitudinal and transverse direction.

Flexural reinforcement can be selected as shown in Art. 10-46. For economy combined footings should be made deep enough to avoid the use of stirrups for shear reinforcement. In the transverse direction, the bottom reinforcement is placed uniformly in bands having an arbitrary width, which can be taken as the width of the column plus $2d$. The amount at each column is proportional to the column load.

Size and length of reinforcing must be selected to develop the full yield strength of the steel between the critical section, or point of maximum tension, and the end of the bar (Art. 10-49).

10-79. Strap Footings.

(See also Arts. 10-74 and 10-75.) When the distance between two columns to be supported on a combined footing becomes large, cost increases rapidly, and a strap footing, or cantilever-type footing, may be more economical. This type of footing, in effect, consists of two footings, one under each column, connected by a strap beam (Fig. 10-52). This beam distributes the column loads to each footing to make the net soil pressure with unfactored loads uniform and equal at each footing, and with factored design loads uniform but not necessarily equal. The center of gravity of the actual column loads should coincide with the centroid of the combined footing areas.

The strap beam is usually designed and constructed to that it does not bear on the soil (Fig. 10-52b). The concrete for the beam is cast on compressible material. If the

Fig. 10-52 Strap footing.

concrete for the strap beam were placed on compacted soil, the resulting soil pressure would have to be considered in design of the footing.

The strap beam, in effect, cantilevers over the exterior column footing, and the bending will cause tension at the top. The beam therefore requires top flexural reinforcement throughout its entire length. Nominal flexural reinforcement should be provided in the bottom of the beam to provide for any tension that could result from differential settlement.

The top bars at the exterior column must have sufficient length to develop their full yield strength. Hence, the distance between the interior face of the exterior column and the property-line end of the horizontal portion of the top bar, plus the value of the hook, must provide tension development length (Art. 10-49).

Strength-design shear reinforcement [Eq. (10-40a)] will be required when $V_u > \phi V_c$, and minimum shear reinforcement [Eq. (10-83)] must be provided when $V_u > \phi V_c/2$, where V_c = shear carried by the concrete (Art. 10-47).

For the strap footing shown in Fig. 10-52, the exterior column footing can be designed as a wall spread footing, and the interior column footing as an individual-column spread footing (Arts. 10-76 and 10-77).

10-80. Mat Foundations.

A mat or raft foundation is a single combined footing for an entire building unit. It is economical when building loads are relatively heavy and the safe soil pressure is small. (See also Arts. 10-74 and 10-75.)

Weight of soil excavated for the foundation decreases the pressure on the soil under the mat. If excavated soil weighs more than the building, there is a net decrease in pressure at mat level from that prior to excavation.

When the mat is rigid, a uniform distribution of soil pressure can be assumed and the design can be based on a statically determinate structure, as shown in Fig. 10-53. (See "Suggested Design Procedures for Combined Footings and Mats," ACI 336.2R.)

If the centroid of the factored design loads does not coincide with the centroid of the mat area, the resulting nonuniform soil pressure should be used in the strength design of the mat.

Strength-design provisions for flexure, one-way and two-way shear, development length, and serviceability should conform to ACI Building Code requirements (Arts. 10-58 to 10-60).

Fig. 10-53 Rigid mat footing.

10-81. Pile Foundations.

Building loads can be transferred to piles by a thick reinforced-concrete slab, called a pile-cap footing. The piles are usually embedded in the pile cap 4 to 6 in. They should be cut to required elevation after driving and prior to casting the footing. Reinforcement should be placed a minimum of 3 in clear above the top of the piles. The pile cap is required by the ACI Building Code to have a minimum thickness of 12 in above the reinforcement. (See also Art. 10-75.)

Piles should be located so that the centroid of the pile cluster coincides with the center of gravity of the column design load. As a practical matter, piles cannot be driven exactly to the theoretical design location. A construction survey should be made to determine if the actual locations require modification of the original pile-cap design.

Pile-cap footings are designed like spread footings (Art. 10-77), but for concentrated pile loads. Critical sections for shear and moment are the same. Reaction from any pile with center $d_p/2$ or more inside the critical section, where d_p is the pile diameter at footing base, should be assumed to produce no shear on the section. The ACI Code requires that the portion of the reaction of a pile with center within $d_p/2$ of the section be assumed as producing shear on the section based on a straightline interpolation between full value for center of piles located $d_p/2$ outside the section and zero at $d_p/2$ inside the section.

For pile clusters without moment, the pile load P_u for strength design of the footing is obtained by dividing the design column load by the number of piles n. The design load equals $1.4D + 1.7L$, where D is the dead load, including the weight of the pile cap, and L is the live load. For pile clusters with moment, the design load on the pile ($n = 1$) is

$$P_{u(n=1)} = (1.4D + 1.7L)\left(\frac{1}{n} + \frac{eC_n}{\sum_1^n C_n^2}\right) \qquad (10\text{-}100)$$

where e = eccentricity of resultant load with respect to neutral axis of pile group
C_n = distance between neutral axis of pile group and center of nth pile

10-82. Drilled-Pier Foundations.

A drilled-pier foundation is used to transmit loads to soil at lower levels through end bearing and, in some situations, side friction. (See also Art. 10-75.) It can be constructed in firm, dry earth or clay soil by machine excavating an unlined hole with a rotating auger or bucket with cutting vanes and filling the hole with plain or reinforced concrete. Under favorable conditions, pier shafts 12 ft in diameter and larger can be constructed economically to depths of 100 ft and more. Buckets with sliding arms can be used to form bells at the bottom of the shaft with a diameter as great as 3 times that of the shaft (Fig. 10-54).

Some building codes limit the ratio of shaft height to shaft diameter to a maximum of 30. They also may require the bottom of the bell to have a constant diameter for the bottom foot of height, as shown in Fig. 10-54.

The compressive stress permitted on plain-concrete drilled piers with lateral support from surrounding earth varies with different codes. ACI limits the computed bearing stress based on service

Fig. 10-54 Bell-bottom drilled pier.

TABLE 10-22 Allowable Service (Unfactored) Loads on Drilled Piers, kips*

Shaft dia, ft	Shaft area, sq in.	$f'_c = 3,000$	$f'_c = 4,000$	$f'_c = 5,000$	$f'_c = 6,000$
1.5	254.5	229	305	382	458
2.0	452.4	407	543	679	814
2.5	706.9	636	848	1,060	1,272
3.0	1,017.9	916	1,221	1,527	1,832
3.5	1,385.4	1,247	1,662	2,078	2,494
4.0	1,809.6	1,629	2,172	2,714	3,257
4.5	2,290.2	2,061	2,748	3,435	4,122
5.0	2,827.4	2,545	3,393	4,241	5,099
5.5	3,421.2	3,079	4,105	5,132	6,158
6.0	4,071.5	3,664	4,886	6,107	7,329

Bell dia, ft	Bell area, sq ft	Safe allowable service-load bearing pressure on soil, psf					
		10,000	12,000	15,000	20,000	25,000	30,000
1.5	1.77	18	21	27	35	44	53
2.0	3.14	31	38	47	63	78	94
2.5	4.91	49	59	74	98	123	147
3.0	7.07	71	85	106	141	177	212
3.5	9.62	96	115	144	192	240	289
4.0	12.57	126	151	189	251	314	387
4.5	15.90	159	191	238	318	398	477
5.0	19.64	196	236	295	393	492	589
5.5	23.76	238	285	356	475	594	713
6.0	28.27	283	339	424	565	707	848
6.5	33.18	332	398	498	664	830	995
7.0	38.48	385	462	577	770	962	1,154
7.5	44.18	442	530	663	884	1,105	1,325
8.0	50.27	503	603	754	1,005	1,258	1,508
8.5	56.74	567	681	851	1,135	1,420	1,702
9.0	63.62	636	763	954	1,272	1,591	1,909
9.5	70.88	709	851	1,063	1,418	1,770	2,126
10.0	78.54	785	942	1,178	1,571	1,965	2,356
10.5	86.59	866	1,041	1,300	1,732	2,164	2,598
11.0	95.03	950	1,140	1,425	1,901	2,375	2,851
11.5	103.87	1,039	1,248	1,560	2,080	2,598	3,116
12.0	113.10	1,131	1,357	1,696	2,262	2,827	3,393
12.5	122.72	1,227	1,470	1,840	2,450	3,068	3,682
13.0	132.73	1,327	1,593	1,991	2,655	3,320	3,982
13.5	143.14	1,431	1,718	2,145	2,860	3,576	4,294
14.0	153.94	1,539	1,847	2,309	3,079	3,850	4,618
14.5	165.13	1,651	1,980	2,480	3,303	4,130	4,954
15.0	176.15	1,762	2,110	2,645	3,523	4,402	5,284
15.5	188.69	1,887	2,260	2,830	3,774	4,717	5,661
16.0	201.06	2,011	2,410	3,020	4,021	5,025	6,032
16.5	213.82	2,138	2,560	3,200	4,276	5,344	6,415
17.0	226.98	2,270	2,725	3,410	4,540	5,680	6,809
17.5	240.53	2,405	2,890	3,610	4,811	6,012	7,216
18.0	254.47	2,545	3,060	3,820	5,089	6,361	7,634

*$f_{c_1} = 0.30 f'_c$ (ACI 322-72 Alternate Design Method).

NOTE: Bell diameter preferably not to exceed 3 times the shaft diameter. Check shear stress if bell slope is less than 2:1. (Courtesy Concrete Reinforcing Steel Institute.)

loads to $0.30f'_c$, where f'_c is the specified concrete strength, and the computed bearing stress based on factored loads to $0.60f'_c$. ("Building Code Requirements for Structural Plain Concrete," ACI 322-72.)

Reinforced-concrete drilled piers can be designed as flexural members with axial load, as indicated in Art. 10-83.

Allowable unfactored loads on drilled piers with various shaft and bell diameters, supported by end bearing on soils of various allowable bearing pressures, are given in Table 10-22. For maximum-size bells (bell diameter 3 times shaft diameter) and a maximum concrete stress of $0.30f'_c$ for unfactored loads, the required concrete strength, psi, is 20.83% of the allowable soil pressure, psf.

COLUMNS

Column-design procedures are based on a comprehensive investigation reported by ACI Committee 105 ("Reinforced Concrete Column Investigation," *ACI Journal*, February 1933) and followed by many supplemental tests. The results indicated that basically the total capacity for axial load can be predicted, over a wide range of steel and concrete strength combinations and percentages of steel, as the sum of the separate concrete and steel capacities.

10-83. Basic Assumptions for Strength Design of Columns.

At maximum capacity, the load on the longitudinal reinforcement of a concentrically loaded concrete column can be taken as the steel area A'_s times steel yield strength f_y. The load on the concrete can be taken as the concrete area in compression times 85% of the compressive strength f'_c of the standard test cylinder. The 15% reduction from full strength accounts, in part, for the difference in size and, in part, for the time effect in loading of the column. Capacity of a concentrically loaded column then is the sum of the loads on the concrete and the steel.

The ACI Building Code applies a capacity reduction factor $\phi = 0.75$ for members with spiral reinforcement and $\phi = 0.70$ for other members. For small axial loads ($P_u \leq 0.10 f'_c A_g$, where A_g = gross area of column), ϕ may be increased proportionately to as high as 0.90. Capacity of columns with eccentric load or moment may be similarly determined, but with modifications. These modifications introduce the assumptions made for strength design for flexure and axial loads.

The basic assumptions for strength design of columns can be summarized as follows:

1. Strain of steel and concrete is proportional to distance from neutral axis (Fig. 10-55c).

2. Maximum usable compression strain of concrete is 0.003 in/in (Fig. 10-55c).

3. Stress, psi, in longitudinal bars equals steel strain ϵ_s times 29,000,000 for strains below yielding, and equals the steel yield strength f_y, tension or compression, for larger strains (Fig. 10-55f).

4. Tensile strength of concrete is negligible.

5. Capacity of the concrete in compression, which is assumed at a maximum stress of $0.85f'_c$, must be consistent with test results. A rectangular stress distribution (Fig. 10-55d) may be used. Depth of the rectangle may be taken as $a = \beta_1 c$, where c is the distance from the neutral axis to the extreme compression surface and $\beta_1 = 0.85$ for $f'_c \leq 4000$ psi and 0.05 less for each 1000 psi that f'_c exceeds 4000 psi, but β_1 should not be taken less than 0.65.

In addition to these general assumptions, design must be based on equilibrium and strain compatibility conditions. No essential difference develops in maximum capacity between tied and spiral columns, but spiral-reinforced columns show far more toughness before failure. Tied-column failures have been relatively brittle and sudden, whereas spiral-reinforced columns that have failed have deformed a great deal and carried a high percentage of maximum load to a more gradual yielding failure. The difference in behavior is reflected in the higher value of ϕ assigned to spiral-reinforced columns.

Additional design considerations are presented in Arts. 10-84 to 10-89. Following is an example of the application of the basic assumptions for strength design of columns.

Fig. 10-55 Stress and strain in reinforced-concrete column.

Example. Determine the capacity of the 20-in-square reinforced-concrete column shown in Fig. 10-55a. The column is reinforced with four No. 18 bars, with $f_y = 60$ ksi, and lateral ties. Concrete strength is $f'_c = 6$ ksi. Assume the factored load P_u to have an eccentricity of 2 in, and that slenderness can be ignored.

To begin, assume $c = 24$ in. Then, with $\beta_1 = 0.75$ for $f'_c = 6000$ psi, the depth of the compression rectangle is $a = 0.75 \times 24 = 18$ in. This assumption can be checked by computing the eccentricity $e = 12M_u/P_u$, where M_u is the moment capacity, ft-kips.

Since the strain diagram is linear and maximum compression strain is 0.003 in/in, the strains in the reinforcing steel are found by proportion to be 0.00258 and 0.00092 in/in (Fig. 10-55c). The strain at yield is 60/29,000 = 0.00207 < 0.00258 in/in. Hence, the stresses in the steel are 60 ksi and 0.00092 × 29,000 = 26.7 ksi.

The maximum concrete stress, which is assumed constant over the depth $a = 18$ in, is $0.85f'_c = 0.85 \times 6 = 5.1$ ksi (Fig. 10-55d). Hence, the compression force on the concrete is $5.1 \times 20 \times 18 = 1836$ kips and acts at a distance $20/2 - 18/2 = 1$ in from the centroid of the column (Fig. 10-55e). The compression force on the more heavily loaded pair of reinforcing bars, which have a cross-sectional area of 8 in², is 8×60 less the force on concrete replaced by the steel 8×5.1, or 439 kips. The compression force on the other pair of bars is $8(26.7 - 5.1) = 173$ kips (Fig. 10-55f). Both pairs of bars act at a distance of $20/2 - 3.375 = 6.625$ in from the centroid of the column.

The design capacity of the column for vertical load is the sum of the nominal steel and concrete capacities multiplied by a capacity reduction factor $\phi = 0.70$.

$$P_u \leq \phi P_n = 0.70(1836 + 173 + 439) = 1714 \text{ kips}$$

The capacity of the column for moment is found by taking moments of the steel and concrete capacities about the centroidal axis.

$$M_u \leq \phi M_n = 0.70 \left[1836 \times \frac{1}{12} + (439 - 173) \frac{6.625}{12} \right] = 209 \text{ ft-kips}$$

The eccentricity for the assumed value of $c = 24$ in is

$$e = \frac{209 \times 12}{1714} = 1.46 < 2 \text{ in}$$

If for a new trial, c is taken as 22.5 in, then $P_u = 1620$ kips, $M_u = 272$ ft-kips, and e checks out close to 2 in. If sufficient load-moment values for other assumed positions of the neutral axis are calculated, a complete load-moment interaction diagram can be constructed (Fig. 10-56).

The nominal maximum axial load capacity P_o of a column without moment equals the sum of the capacities of the steel and the concrete.

$$P_o = 0.85 f'_c (A_g - A_{st}) + f_y A_{st} \qquad (10\text{-}101)$$

where A_g = gross area of column cross section
  A_{st} = total area of longitudinal steel reinforcement

Fig. 10-56 Load-moment interaction diagram for determination of the capacity of a reinforced-concrete column.

10-84. Design Requirements for Columns.

The ACI Building Code contains the following principal design requirements for columns, in addition to the basic assumptions (Art. 10-83):

1. Columns must be designed for all bending moments associated with a loading condition.
2. For corner columns and other columns loaded unequally on opposite sides in perpendicular directions, biaxial bending moments must be considered.
3. All columns must be designed for an eccentricity, in, of at least $0.6 + 0.03h$, where h is the thickness, in, of the member in the direction of bending, and for maximum axial loads not larger than $0.80P_o$ for tied columns, or $0.85P_o$ for spiral columns, where P_o is given by Eq. (10-101).
4. The minimum ratio of longitudinal-bar area to total cross-sectional area of column A_g is 0.01, and the maximum ratio is 0.08. For columns with a larger cross section than required by loads, however, a smaller A_g, but not less than half the gross area of the columns, may be used for calculating both load capacity and minimum longitudinal bar area. This exception allows reuse of forms for larger-than-necessary columns, and permits longitudinal bar areas as low as 0.005 times the actual column area. At least four longitudinal bars should be used in retangular reinforcement arrangements, and six in circular arrangements.
5. The ratio of the volume of spiral reinforcement to volume of concrete within the spiral should be at least

$$\rho_s = 0.45 \left(\frac{A_g - A_c}{A_c} \right) \left(\frac{f'_c}{f_y} \right) \quad (10\text{-}102)$$

where A_g = gross cross-sectional area of concrete column, in^2
A_c = area of column within outside diameter of spiral, in^2
f'_c = specified concrete strength, psi
f_y = specified yield strength of spiral steel, psi (maximum 60,000 psi)

6. For tied columns, minimum size of ties is No. 3 for longitudinal bars that are No. 10 or smaller, and No. 4 for larger longitudinal bars. Minimum vertical spacing of sets of ties is 16 diameters of longitudinal bars, 48 tie-bar diameters, or the least thickness of the column. A set of ties should be composed of one round tie for bars in a circular pattern, or one tie enclosing four corner bars plus additional ties sufficient to provide a corner of a tie at alternate interior bars or at bars spaced more than 6 in from a bar supported by the corner of a tie.

TABLE 10-23 Minimum Clear Cover, in, for Column Reinforcement*

Type of construction	Reinforcement	Interior exposure†	Exposed to Weather*
Cast-in-place	Longitudinal	1 1/2	2
	Ties, spirals	1 1/2	1 1/2
Precast	Longitudinal	$5/8 \le d_b \le 1\ 1/2$‡	1 1/2
	Ties, spirals	3/8	1 1/4
Prestressed in-place	Longitudinal	1 1/2	1 1/2
	Ties, spirals	1	1 1/2
Prestressed, precast	Longitudinal	$5/8 < d_b < 1\ 1/2$‡	1 1/2§
	Ties, spirals	3/8	1 1/4

*From ACI 318–77.
†See local code; fire protection may require greater thickness.
‡d_b = nominal bar diameter, in.
§Increase these covers 50% where tensile stress exceeds $6\sqrt{f'_c}$, where f'_c = specified concrete strength, psi.

7. Splices of vertical bars from footing to roof must provide a minimum tensile capacity in each face of the column equal to 25% of the total vertical-bar area times the steel yield strength.

8. Minimum clear concrete cover required for reinforcement is listed in Table 10-23. See also Art. 10-86.

10-85. Design for Minimum Bending Moment.

The requirement for design of columns for a minimum eccentricity (Art. 10-84) means that the maximum design load $P_u = 1.4D + 1.7L$, where D is the dead load and L the live load, for cast-in-place tied columns more than 10 in thick is accompanied by a minimum bending moment $M_u = P_u (0.6 + 0.03h)$.

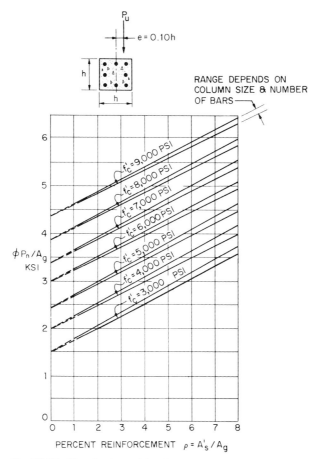

Fig. 10-57 Chart for determining the average design stress $\phi P_n/A_g$ in tied columns with eccentricity $e = 0.1h$, which corresponds closely with the conditions at a stress equal to $0.80P_o/A_g$, using a capacity reduction factor ϕ of 0.70, and reinforcement with 60-ksi yield strength. *(Courtesy Concrete Reinforcing Steel Institute.)*

Interior columns in buildings with reasonably uniform spans and usual loadings, when lateral deflection and moments due to wind are negligible or when lateral bracing is provided by walls or other elements, usually will be designed for this condition of maximum design axial load and minimum moment.

The universal column design chart in Fig. 10-57 gives the solution to this most common design problem directly for square columns with bars in four faces for the practical range of concrete strengths and standard Grade 60 bars. The slight spread of the curves for each value of specified concrete strength f'_c is a result of the variation in depth from face of concrete to centroid of various-size longitudinal bars and the practicable range of numbers of vertical bars. For rectangular columns with sides less than 2 to 1, round columns, or square columns with unknown, possibly larger bending moment, a preliminary size at a low steel ratio can be selected with confidence that the final design based on more accurate analysis can be completed for the size selected. Note that final design where large wind moment is a factor, where biaxial bending occurs, or where slenderness effects must be considered will usually require more steel and only occasionally a larger size. ("CRSI Handbook," Concrete Reinforcing Steel Institute).

10-86. Column Ties and Tie Patterns.

For full utilization, all ties in tied columns must be fully developed (for full tie yield strength) at each corner enclosing a vertical bar or, for circular ties, around the full periphery.

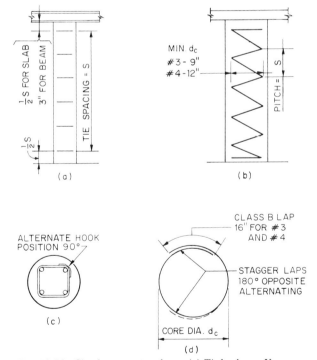

Fig. 10-58 Circular concrete column. (a) Tied column. Use separate ties when core diameter $d_c \leq s$. (b) Spiral-reinforced column. It may be used when $d_c > s$. (c) Rectangular tie for use in columns with four longitudinal bars. (d) Circular tie.

Splices. The ACI Building Code provides arbitrary minimum sizes and maximum spacings for column ties (Art. 10-84). No increase in size nor decrease in the spacings is required for Grade 40 materials. Hence, the minimum design requirements for splices of ties may logically be based on Grade 40 steel.

The ordinary closed, square or rectangular, tie is usually spliced by overlapping standard tie hooks around a longitudinal bar. Standard tie patterns require staggering of hook positions at alternate tie spacings, by rotating the ties 90 or 180°. ("Manual of Standard Practice," Concrete Reinforcing Steel Institute.) Two-piece ties are formed by lap splicing or anchoring the ends of U-shaped open ties. A Class C tension lap splice, with minimum lap of 20 in, must be provided, because it is not practicable to stagger such lap splices (Art. 10-49). Lapped bars should be securely wired together to prevent displacement during concreting.

Tie Arrangements. Commonly used tie patterns are shown in Figs. 10-58 to 10-60. In Fig. 10-59, note the reduction in required ties per set and the improvement in bending resistance about both axes achieved with the alternate bundled-bar arrangements. Bundles may not contain more than four bars, and bar size may not exceed No. 11.

Minimum tie sizes and maximum spacings per set of ties are listed in Table 10-24.

Drawings. Design drawings should show all requirements for splicing longitudinal bars, that is, type of splice, lap length if lapped, location in elevation, and layout in cross section. On detail drawings, dowel erection details should be shown if special large longitudinal bars, bundled bars, staggered splices, or specially grouped bars are to be used.

Fig. 10-59 Ties for square concrete columns. Additional single bars may be placed between any of the tied groups, but clear spaces between bars should not exceed 6 in.

Fig. 10-60 Ties for wall-like columns. Spaces between corner bars and interior groups of three bars, and between interior groups may vary to accommodate average spacing not exceeding 6 in. A single additional bar may be placed in any of such spaces if the average spacing does not exceed 6 in.

TABLE 10-24 Minimum Sizes and Laps and Maximum Spacing, in, for Column Ties

Longitudinal-bar size No.	Minimum size No. 3	Minimum size No. 4	Minimum size No. 5
5	10		
6	12		
7	14		
8	16	16	
9	18	18	
10	18	20	
11	...	22	22
14	...	24	27
18	...	24	30
Lap splice, in.	15	20	26

10-87. Biaxial Bending of Columns.

If column loads cause bending simultaneously about both principal axes of a column cross section, as for most corner columns, a biaxial bending analysis is required. For rapid preliminary design, Eq. (10-103) gives conservative results.

$$\frac{M_x}{M_{ox}} + \frac{M_y}{M_{oy}} \leq 1 \qquad (10\text{-}103)$$

where M_x, M_y = design moments about x and y axes, respectively

M_{ox}, M_{oy} = design capacities about x and y axes, respectively

For square columns with equal steel in all faces, $M_{ox} = M_{oy}$, and the relation reduces to:

$$\frac{M_x + M_y}{M_{ox}} \leq 1 \qquad (10\text{-}104)$$

Because $M_x = e_x P_u$ and $M_y = e_y P_u$, the safe biaxial capacity can be taken from uniaxial load-capacity tables for the load P_u and the uniaxial bending moment $M_u = (e_x + e_y)P_u$. Similarly, for round columns, the moment capacity is essentially equal in all directions, and the two bending moments about the principal axes may be combined into a single uniaxial design moment M_u which is then an exact solution.

$$M_u = \sqrt{M_x^2 + M_y^2} \qquad (10\text{-}105)$$

The linear solution always gives a safe design, but becomes somewhat overconservative when the moments M_x and M_y are nearly equal. For these cases, a more exact solution will be more economical for the final design. (Simple curves for direct solutions are available in the "CRSI Handbook," Concrete Reinforcing Steel Institute.)

10-88. Slenderness Effects on Concrete Columns.

The ACI Building Code requires that primary column moments be magnified to provide safety against buckling failure. Detailed procedures, formulas, and design aids are provided in the ACI Code and the "Commentary."

For most unbraced frames, an investigation will be required to determine the magnification factor to allow for the effects of sidesway and end rotation. The procedure for determination of the required increase in primary moments, after the determination that slenderness effects cannot be neglected, is complex. For direct solution, the requirements of Sec. 10, ACI 318-77 can be met by a P-Δ analysis. (See, for example, J. G. MacGregor and S. E. Hage, "Stability Analysis and Design of Concrete," *Journal of the Structural Division*, ASCE, Vol. 103, No. ST10, October 1977.)

The direct P-Δ method of MacGregor and Hage is based upon an equation for a geometric series that was derived for the final second-order deflection as a function of the first-order elastic deflection. This direct P-Δ analysis provides a very simple method for computing the moment magnifier δ when the stability index Q is greater than 0.04 but equal to or less than 0.22.

$$\delta = 1/(1 - Q) \qquad 0.04 < Q \leq 0.22 \qquad (10\text{-}106)$$

where $Q = P_u \Delta_u / (H_u h_s)$

P_u = sum of the factored loads in a given story
Δ_u = elastically computed first-order lateral deflection due to H_u (neglecting P-Δ effects) at the top of the story, relative to the bottom
H_u = total factored lateral force (shear) within the story
h_s = height of story, center-to-center of floors or roof

The approximate method of ACI 318-77 may also be used to determine the moment magnifier. This approximate method is a column-by-column correction based upon the stiffnesses of the column and beams, applied primary design column end moments, and consideration of whether the entire structure is laterally braced against sidesway by definition. (See ACI 318-77, Sec. 10 and 11.) Design aids are available (curves in "CRSI Handbook," Concrete Reinforcing Steel Institute; computer program, Portland Cement Association.)

The ACI Code permits slenderness effects to be neglected only for very short, braced columns, with the following limitations for columns with square or rectangular cross sections:

$L_u \leq 6.6h$ for bending in single curvature
$L_u \leq 10.2h$ for bending in double curvature with unequal end moments
$L_u \leq 13.8h$ for bending in double curvature with equal end moments

and for round columns, five-sixths of the maximum lengths for square columns, where L_u is the unsupported length and h the depth or overall thickness of column in the direction being considered.

These limiting heights are based on the ratio of the total stiffnesses of the columns to the total stiffnesses of the flexural members, $\Sigma K_c / \Sigma K_B = 50$, at the joint at each end of a column. As these ratios become less, the limiting heights can be increased. When the total stiffnesses of the columns and the floor systems are equal at each end of the column (a common assumption in routine frame analysis), the two ratios = 1.00, and the limiting heights increase about 30%. With this increase, the slenderness effects can be neglected for most columns in frames braced against sidesway.

A frame is considered braced when other structural elements, such as walls, provide stiffness resisting sidesway at least 6 times the sum of the column stiffnesses resisting sidesway in the same direction in the story being considered.

10-89. Economy in Column Design

Actual costs of reinforced-concrete columns in place per linear foot per kip carrying capacity vary widely. The following recommendations based on relative costs are generally applicable:

Formwork. Use of the same size and shape of column cross section throughout a floor and, for multistory construction, from footing to roof will permit mass production and reuse for economy. Within usual practicable maximum building heights, about 60 stories or 600 ft, increased speed of construction and saving in formwork will save more than the cost of the excess concrete volume over that for smaller column sizes in upper stories.

Concrete Strength. Use of the maximum concrete strength required to support the design loads with the minimum allowed steel area results in the lowest cost. The minimum size of a multistory column is established by the maximum concrete strength reliably available locally and the limit on maximum area of vertical bars. If the acceptable column size is larger than the minimum possible at the base of the multistory stack, the steel ratio can begin with less than the maximum limit (Art. 10-84). At successive stories above, the steel ratio can be reduced to the minimum, and thereafter, for additional stories, the concrete strength can be reduced. Near the top, as loads reduce further, a further reduction in the steel area to 0.005 may be made (Art. 10-84).

Steel. Grade 60 vertical bars offer the greatest economy. Minimum tie requirements can be achieved with four-bar or four-bundle (up to four bars per bundle) arrangements, or by placing an intermediate bar between tied corners not more than 6 in (clear) from the corner bars. For these arrangements, no interior ties are required; only one tie per set is needed. (See Fig. 10-59 and Art. 10-86.) With no interior ties, low-slump concrete can be placed and consolidated more easily, and the cost and time for assembly of column reinforcement cages are greatly reduced. Note that, for small quantities, the local availability of Nos. 14 and 18 bars should be investigated before they are specified.

Details of Column Reinforcement. Where Nos. 14 and 18 bars are used in compression only, end-bearing splices usually save money. If the splices are staggered 50%, as with two-story lengths, the tensile capacity of the columns will also be adequate for the usual bending moments encountered. For unusually large bending moments, where tensile splices of No. 10 bars and larger are required, tension couplers are usually least expensive in place. For smaller bar sizes, lap splices, tensile or compressive, are preferred for economy. Some provision for staggered lap splices for No. 8 bars and larger may be required to avoid Class C tension splices (Art. 10-49).

Where butt splices are used, it will usually be necessary to assemble the column reinforcement cage in place. Two-piece interior ties or single ties with end hooks for two bars (see Art. 10-86) will facilitate this operation.

Where the vertical bar spacing is restricted and lap splices are used, even with the column size unchanged, offset bending of the bars from below may be required. However, where space permits, as with low steel ratios, an additional saving in fabrication and erection time will be achieved by use of straight column verticals, offset one bar diameter at alternate floors.

SPECIAL CONSTRUCTION

10-90. Deep Beams.

The ACI Building Code defines deep beams as flexural members with overall span-depth ratios less than 2.5 for continuous spans and 1.25 for simple spans. Some types of building components behave as deep beams and require analysis for nonlinear stress distribution in flexure. Some common examples are long, precast panels used as spandrel beams; below-grade walls, with or without openings, distributing column loads to a continuous slab footing or to end walls; and story-height walls used as beams to eliminate lower columns in the first floor area.

Shear. When the span-depth ratio is less than 5, beams are classified as deep for shear reinforcement purposes. Separate special requirements for shear apply when span-depth ratio is less than 2 or between 2 and 5. The critical section for shear should be taken at a distance from face of support of $0.15 L_n \leq d$ for uniformly loaded deep beams, and of $0.50a \leq d$ for deep beams with concentrated loads, where a is the shear span, or distance from concentrated load to face of support, L_n the clear span, and d the distance from extreme compression surface to centroid of tension reinforcement. Shear reinforcement required at the critical section should be used throughout the span.

The nominal shear strength of the concrete can be taken as

$$V_c = 2 \sqrt{f'_c}\, b_w d \tag{10-107}$$

where f'_c = specified concrete strength, psi
b_w = width of beam web
d = distance from extreme compression fiber to the centroid of the tension reinforcement

The ACI Building Code also presents a more complicated formula that permits the concrete to carry up to $6 \sqrt{f'_c}\, b_w d$.

Maximum nominal shear strength when $L_n/d < 2$ should not exceed

$$V_n = V_c + V_s = 8 \sqrt{f'_c}\, b_w d \tag{10-108}$$

and when L_n/d is between 2 and 5 should not exceed

$$V_n = \frac{2}{3}\left(10 + \frac{L_n}{d}\right) \sqrt{f'_c}\, b_w d \tag{10-109}$$

Required area of shear reinforcement should be determined from

$$f_y d \left[\frac{A_v}{s}\left(\frac{1 + L_n/d}{12}\right) + \frac{A_{vh}}{s_2}\left(\frac{11 - L_n/d}{12}\right) \right] = V_u/\phi - V_c \tag{10-110}$$

where A_v = area of shear reinforcement perpendicular to main reinforcement within a distance s
s = spacing of shear reinforcement measured parallel to main reinforcement
A_{vh} = area of shear reinforcement parallel to main reinforcement within a distance s_2
s_2 = spacing of shear reinforcement measured perpendicular to main reinforcement
f_y = yield strength of shear reinforcement

Spacing s should not exceed $d/5$ or 18 in. Spacing s_2 should not exceed $d/3$ or 18 in. The area of shear reinforcement perpendicular to the main reinforcement should be a minimum of

$$A_v = 0.0015 bs \tag{10-111}$$

where b = width of beam compression face. Area of shear reinforcement parallel to main reinforcement should be at least

$$A_{vh} = 0.0025 b s_2 \qquad (10\text{-}112)$$

When $b_w > 10$ in, shear reinforcement should be placed in each face of the beam. If the beam has a face exposed to the weather, between one-half and two-thirds of the total shear reinforcement should be placed in the exterior face. Bars should not be smaller than No. 3.

Bending. The area of steel provided for positive bending moment in a deep beam should be at least

$$A_s = \frac{200 b_w d}{f_y} \qquad (10\text{-}113)$$

where f_y = yield strength of flexural reinforcement, psi. This minimum amount can be reduced to one-third more than that required by analysis.

A safe assumption for preliminary design is that the extreme top surface in compression is $0.25h$ below the top of very deep beams for computation of a reduced effective depth d for flexure (Fig. 10-61).

10-91. Shear Walls.

Cantilevered shear walls used for bracing structures against lateral displacement (sidesway) are a special case of deep beams. They may be used as the only lateral bracing, or in conjunction with beam-column frames. In the latter case, the lateral displacement of the combination can be calculated with the assumption that lateral forces resisted by each element can be distributed to walls and frames in proportion to stiffness. For tall structures, the effect of axial shortening of the frames and the contribution of shear to lateral deformation of the shear wall should not be neglected. Figure 10-62 indicates the forces assumed to be acting on a horizontal cross section of a shear wall.

Reinforcement required for flexure of shear walls as a cantilever should be proportioned as for deep beams (Art. 10-90). Shear reinforcement is usually furnished as a combination of horizontal and vertical bars distributed evenly in each story (for increment of load). For low shear (where the factored shear force V_u at a section is less than $0.5\phi V_c$, where V_c is the nominal shear permitted on the concrete), the minimum shear reinforcement required and its location in a wall are the same as for bearing walls (Art. 10-69). Maximum spacing of horizontal shear reinforcement, however, should not exceed $L_w/5$, $3h$, or 18 in, where L_w is the horizontal length of wall and h the overall wall thickness (Fig. 10-62). Maximum spacing of the vertical reinforcement should not exceed $L_w/3$, $3h$, or 18 in.

A thickness of at least $L_w/25$ is advisable for walls with high shear.

The factored horizontal shear force V_u acting on a section through the shear wall must not exceed the nominal shear strength V_n multiplied by $\phi = 0.85$.

$$V_u \leq (\phi V_n = \phi V_c + \phi V_s) \qquad (10\text{-}114)$$

where V_c = nominal shear strength of the concrete
V_s = nominal shear strength provided by reinforcement

The horizontal shear strength at any section should not be taken larger than

$$V_n = 10 \sqrt{f'_c} \, hd \qquad (10\text{-}115)$$

where f'_c = specified concrete strength, psi
d = effective depth of wall, but not to be taken larger than 80% of the wall length
h = wall thickness

Fig. 10-61 Reinforcing for deep beams. When beam thickness exceeds 10 in, a layer of vertical steel should be placed near each face of the beam.

Fig. 10-62 Shear and normal forces on a section of a shear wall.

Shear carried by the concrete should not exceed the smaller of the values of V_c computed from Eqs. (10-116) or (10-117).

$$V_c = 3.3 \sqrt{f'_c}\, hd + \frac{N_u d}{4L_w} \tag{10-116}$$

where N_u = factored vertical axial load on wall acting with V_u, including tension due to shrinkage and creep (positive for compression, negative for tension).

$$V_c = \left[0.6 \sqrt{f'_c} + \frac{L_w(1.25 \sqrt{f'_c} + 0.2N_u/L_w h)}{M_u/V_u - L_w/2} \right] hd \tag{10-117}$$

where M_u = design moment at section where V_u acts.

Alternatively, $V_c = 2\sqrt{f'_c}\, hd$ may be used if N_u causes compression. Shear strength V_c computed for a section at a height above the base equal to $L_w/2$ or one-half the wall height, whichever is smaller, may be used for all lower sections.

When $V_u > 0.5\phi V_c$, the area of horizontal reinforcement within a distance s_2 required for shear is given by

$$A_h = \frac{(V_u/\phi - V_s)s_2}{f_y d} \geq 0.0025 h s_2 \tag{10-118}$$

where s_2 = spacing of horizontal reinforcement (max $\leq L_w/5 \leq 3h \leq 18$ in)
f_y = yield strength of the reinforcement

Also, when $V_u > 0.5\phi V_c$, the area of vertical reinforcement with spacing s should be at least

$$A_{vh} = \left[0.0025 + 0.5 \left(2.5 - \frac{L_h}{L_w} \right) \left(\frac{A_h}{hL_h} - 0.0025 \right) \right] hs \geq 0.0025 hs \quad (10\text{-}119)$$

where L_h is the wall height. But A_{vh} need not be larger than A_h computed from Eq. (10-118).

10-92. Reinforced-Concrete Arches.

Arches are used in roofs for such buildings as hangars, auditoriums, gymnasiums, and rinks, where long spans are desired. An arch is essentially a curved beam with the loads, applied downward in its plane, tending to decrease the curvature. Arches are frequently used as the supports for thin shells that follow the curvature of the arches. Such arches are treated in analysis as two-dimensional, whereas the thin shells behave as three-dimensional elements.

The great advantage of an arch in reinforced concrete construction is that, if the arch is appropriately shaped, the whole cross section can be utilized in compression under the maximum (full) load. In an ordinary reinforced concrete beam, the portion below the neutral axis is assumed to be cracked and does not contribute to the bending strength. A beam can be curved, however, to make its axis follow the lines of thrust very closely for all loading conditions, thus virtually eliminating bending moments.

The component parts of a fixed arch are shown in Fig. 10-63. For a discussion of the different types of arches and the stress analyses required for each, see Arts. 5-88 to 5-91.

Because the depth of an arch and loading for maximum moments generally vary along the length, several cross sections must be chosen for design, such as the crown, springing, haunches, and the quarter points. Concrete compressive stresses and shear should be checked at each section, and reinforcement requirements determined. The sections should be designed as rectangular beams or T beams subjected to bending and axial compression, as indicated in Arts. 10-83 to 10-86.

When an arch is loaded, large horizontal reactions, as well as vertical reactions, are developed at the supports. For roof arches, tie rods may be placed overhead, or in or under the ground floor, to take the horizontal reaction. The horizontal reaction may also be resisted externally by footings on sound rock or piles, by reinforced concrete buttresses, or by adjoining portions of the structure, for example a braced floor or roof at springing level.

Hinged arches are commonly made of structural steel or precast concrete. The hinges simplify the arch analysis and the connection to the abutment, and they reduce the indeterminate stresses due to shrinkage, temperature, and settlements of supports. For cast-in-place concrete, hingeless (fixed) arches are often used. They eliminate the cost of

Fig. 10-63 Components of a fixed arch.

special steel hinges needed for hinged concrete arches and permit reduced crown thicknesses, to provide a more attractive shape.

Arches with spans less than 90 ft are usually constructed with ribs 2 to 4 ft wide. Each arch rib is concreted in a continuous operation, usually in 1 day. The concrete may be placed continuously from each abutment toward the crown, to obtain symmetrical loading on the falsework.

For spans of 90 ft or more, however, arch ribs are usually constructed by the alternate block, or voussoir, method. Each rib is constructed of blocks of such size that each can be completed in one casting operation. This method reduces the shrinkage stresses. The blocks are cast in such order that the formwork will settle uniformly. If blocks close to the crown section are not placed before blocks at the haunch and the springing sections, the formwork will rise at the crown, and placing of the crown blocks will then be likely to cause cracks in the haunch. The usual procedure is to cast two blocks at the crown, then two at the springing, and alternate until the complete arch is concreted.

In construction by the alternate block method, the block sections are kept separate by timber bulkheads. The bulkheads are kept in place by temporary struts between the voussoirs. Keyways left between the voussoirs are concreted later. Near piers and abutments where the top slopes exceed about 30° with the horizontal, top forms may be necessary, installed as the casting progresses.

If the arch reinforcement is laid in long lengths, settlement and deformation of the arch formwork can displace the steel. Therefore, depending on the curvature and total length, lengths of bars are usually limited to about 30 ft. Splicing should be located in the keyways. Splices of adjacent bars should be staggered (50% stagger), and located where tension is small.

Upper reinforcement in arch rings may be held in place with spacing boards nailed to props, or with wires attached to transverse timbers supported above the surface of the finished concrete.

Forms for arches may be supported on a timber falsework bent. This bent may consist of joists and beams supported by posts that are braced together and to solid ground. Wedges or other adjustment should be provided at the base of the posts so that the formwork may be adjusted if settlement occurs, and so that the entire formwork may be conveniently lowered after the concrete has hardened sufficiently to take its own load.

("Recommended Practice for Concrete Formwork," ACI 347, American Concrete Institute.)

10-93. Reinforced-Concrete Thin Shells.

Thin shells are curved slabs with thickness very small compared with the other dimensions. A thin shell possesses three-dimensional load-carrying characteristics. The best natural example of thin-shell behavior is that of an ordinary egg, which may have a ratio of radius of curvature to thickness of 50. Loads are transmitted through thin shells primarily by direct stresses—tension or compression—called membrane stresses, which are almost uniform throughout the thickness. Reinforced-concrete thin-shell structures commonly utilize ratios of radius of curvature to thickness about 5 times that of an eggshell. Because concrete shells are always reinforced, their thickness is usually determined by the minimum thickness required to cover the reinforcement, usually 1 to 4 in. Shells are thickened near the supports to withstand localized bending stresses in such areas. (See also Arts. 5-92 to 5-99.)

Shells are most often used as roofs for such buildings as hangars, garages, theaters, and arenas, where large spans are required and the loads are light. The advantages of reinforced-concrete thin shells may be summarized as follows:

Fig. 10-64 Cylindrical thin shell.

Most efficient use of materials.
Great freedom of architectural shapes.
Convenient accommodation of openings for natural lighting and ventilation.
Ability to carry very large unbalance of forces.
High fireproofing value due to lack of corners, thin ribs, and the inherent resistance of reinforced concrete.
Reserve strength due to many alternate paths for carrying load to the supports. One outstanding example withstood artillery fire punctures with only local damage.

Common shapes of reinforced-concrete thin shells used include cylindrical (barrel shells), dome, and hyperbolic paraboloid (saddle shape).

Cylindrical shells may be classified as long if the radius of curvature is shorter than the span, or as short (Fig. 10-64). Long cylindrical shells, particularly the continuous, multiple-barrel version which repeats the identical design of each bay (and permits reuse of formwork) in both directions, are advantageous for roofing rectangular-plan structures. Short cylindrical shells are commonly used for hangar roofs with reinforced-concrete arches furnishing support at short intervals in the direction of the span.

Structural analysis of these common styles may be simplified with design aids. ("Design of Cylindrical Concrete Shell Roofs," Manual No. 31, American Society of Civil Engineers; "Design Constants for Interior Cylindrical Concrete Shells," EB020D; "Design Constants for Ribless Concrete Cylindrical Shells," EB028D; "Coefficients for Design of Cylindrical Concrete Shell Roofs" (extension of ASCE Manual No. 31), EB035D; "Design of Barrel Shell Roofs," IS082D, Portland Cement Association.)

The ACI Building Code includes specific provisions for thin shells. It requires an elastic analysis and suggests model studies for important or unusual shapes, prescribes minimum steel, and prohibits use of the working-stress method for design, thus prescribing selection of all shear and flexural reinforcement by the strength-design method with the same load factors as for design of other elements. Figure 10-65 shows a typical reinforcing

Fig. 10-65 Reinforcement in a long cylindrical shell. Folded plates are similarly reinforced.

arrangement for a long cylindrical shell. (See also F. S. Merritt, "Standard Handbook for Civil Engineers," Sec. 8, "Concrete Design and Construction," McGraw-Hill Book Company, New York.)

10-94. Concrete Folded Plates.

Reinforced-concrete, folded-plate construction is a versatile concept applicable to a variety of long-span roof construction. Applications using precast, simple V folded plates include segmental construction of domes and (vertically) walls (Fig. 10-66). Inverted folded plates have also been widely used for industrial storage bins. ("Recommended Practice for Design and Construction of Concrete Bins, Silos and Bunkers for Storing Granular Materials," ACI 313, and "Commentary," ACI 313R, American Concrete Institute.) Determination of stresses in folded-plate construction is described in Arts. 5-102 and 5-103.

Formwork for folded plates is far simpler than that for curved thin shells. Precasting has also been a simpler process to save formwork, permit mass-production construction, and achieve sharp lines for exposed top corners (vees cast upside down) to satisfy aesthetic requirements. For very long spans, posttensioned, draped tendons have been employed to reduce the total depth, deflection, and reinforcing-steel requirements. The tendons may be placed in the inclined plates or, more conveniently, in small thickened edge beams. For cast-in-place, folded-plate construction, double forming can usually be avoided if the slopes are less than 35 to 40°.

Since larger transverse bending moments develop in folded plates than in cylindrical shells of about the same proportions, a minimum thickness less than 4 in creates practical problems of placing the steel. A number of areas in the plates will require three layers of steel and, near the intersections of plates, top and bottom bars for transverse bending will be required. Ratios of span to total depth are similar to those for cylindrical shells, commonly ranging from 8 to 15. (See also F. S. Merritt, "Standard Handbook for Civil Engineers," Sec. 8, "Concrete Design and Construction," McGraw-Hill Book Company, New York.)

10-95. Slabs on Grade.

Slabs on ground are often used as floors in buildings. Special use requirements often include heavy-duty floor finish (Art. 10-35) and live-load capacity for heavy concentrated (wheel) loads or uniform (storage) loads, or both.

Although slabs on grade seem to be simple structural elements, actual analysis is extremely complicated. Design is usually based on rules of thumb developed by experience. For design load requirements that are unusually heavy and outside common experience, design aids are available. Occasionally, the design will be controlled by wheel loads only, as for floors in hangars, but more frequently by uniform warehouse loadings. ("Design of Floors on Ground for Warehouse Loadings," Paul F. Rice, *ACI Journal*, August 1957, paper No. 54-7; "Design of Concrete Airport Pavement," EB050P, Portland Cement Association.)

Fig. 10-66 Typical folded-plate shapes.

A full uniform load over an entire area causes no bending moment if the boundaries of the area are simple construction joints. But actual loads in warehouse usage leave unloaded aisles and often alternate panels unloaded. As a result, a common failure of warehouse floors results from *uplift* of the slab off the subgrade, causing negative moment (top) cracking. In lieu of a precise analysis taking into account live-load magnitude, joint interval and detail, the concrete modulus of elasticity, the soil modulus, and load patterns, a quick solution to avoid uplift is to provide a slab sufficiently thick so that its weight is greater than one-fifth the live load. Such a slab may be unreinforced, if properly jointed, or reinforced for temperature and shrinkage stresses only. Alternatively, for very heavy loadings, an analysis and design may be performed for the use of reinforcement, top and bottom, to control uplift moments and cracking. ("Design of Floors on Ground for Warehouse Loadings," Paul F. Rice, *ACI Journal*, Aug., 1957, paper No. 54-7.)

Shrinkage and temperature change in slabs on ground can combine effects adversely to create warping, uplift, and top cracking failures with no load. Closely spaced joint intervals, alternate-panel casting sequence, and controlled curing will avoid these failures. Somewhat longer joint spacings can be employed if reinforcement steel with an area of about 0.002 times the gross section area of slab is provided in perpendicular directions.

With such reinforcement, warping will usually be negligible if the slab is cast in alternate lanes 12 to 14 ft wide, and provided with contraction joints at 20- to 30-ft spacings in the direction of casting. The joints may be tooled, formed by joint filler inserts, or sawed. One-half the bars or wires crossing the contraction joints should be cut accurately on the joint line. The warping effect will be aggravated if excess water is used in the concrete and it is forced to migrate in one direction to top or bottom of the slab, for example, when the slab has been cast on a vapor barrier or on a very dry subgrade. For very long slabs, continuous reinforcement, 0.006 times the gross area, has been used to eliminate transverse joints in more than 12,000 two-lane miles of highway and airport pavement. (Bulletin No. 1, "Design of Continuously Reinforced Concrete Pavement," and "Design and Construction," Continuously Reinforced Pavement Group, 180 N. LaSalle St., Chicago, Ill.; and "Suggested Specifications for Heavy-duty Concrete Floor Topping," IS021B; "Design of Concrete Floors on Ground," IS046B; "Suggested Specifications for Single-course Floors on Ground," IS070B, Portland Cement Association.)

10-96. Seismic-Resistant Concrete Construction.

The ACI Building Code contains special seismic requirements for design that apply only for areas where the probability of earthquakes capable of causing major damage to structures is high, and where ductility reduction factors for lateral seismic loads are utilized (ACI 318-77, Appendix A). The general requirements of ACI 318-77 for reinforced concrete provide sufficient seismic resistance for seismic zones where only minor seismic damage is probable and no reduction factor for ductility is applied to seismic forces. Designation of seismic zones is prescribed in general building codes, as are lateral force loads for design. (See also Art. 5-86.)

Special ductile-frame design is prescribed to resist lateral movements sufficiently to create "plastic" hinges and permit reversal of direction several times. These hinges must form in the beams at the beam-column connections of the ductile frame.

Shear walls used alone or in combination with ductile beam-column frames must also be designed against brittle (shear) failures under the reversing loads ("Commentary on ACI 318-77").

Ductility is developed in reinforced concrete by:

Conservative limits on the net tension steel ratio $\rho < 0.50\rho_b$, where ρ_b is the steel ratio for balanced conditions (Art. 10-46) to ensure underreinforced behavior.

Heavy confining reinforcement extending at joints through the region of maximum moment in both columns and beams, to include points where hinges may form. This confining reinforcement may consist of spiral reinforcing or heavy, closely spaced, well-anchored, closed ties with hooked ends engaging the vertical bars or the ties at the far face.

(ACI 315, "Details and Detailing of Concrete Reinforcement," American Concrete Institute.)

10-97. Composite Flexural Members.

Reinforced- and prestressed-concrete, composite flexural members are constructed from such components as precast members with cast-in-place flanges, box sections, and folded plates.

Composite structural-steel-concrete members are usually constructed of cast-in-place slabs and structural-steel beams. Interaction between the steel beam and concrete slab is obtained by natural bond if the steel beam is fully encased with a minimum of 2 in of concrete on the sides of soffit. If the beam is not encased, the interaction may be accomplished with mechanical anchors (shear connectors). Requirements for composite structural-steel-concrete members are given in the AISC "Specification for the Design, Fabrication and Erection of Structural Steel for Buildings."

The design strength of composite flexural members is the same for both shored and unshored construction. Shoring should not be removed, however, until the supported elements have the design properties required to support all loads and limit deflections and cracking. Individual elements should be designed to support all loads prior to the full development of the design strength of the composite member. Premature loading of individual precast elements can cause excessive deflections as the result of creep and shrinkage.

The design horizontal shear force for a composite member may be transferred between individual elements by contact stresses or anchored ties, or both. The factored shear force V_u at the section considered must be equal to or less than the nominal horizontal shear strenth V_{nh} multiplied by $\phi = 0.85$.

$$V_u \leq \phi \, V_{nh} \qquad (10\text{-}120)$$

When $V_u \leq 80b_v d$, where b_v is the section width and d the distance from the extreme compression surface to the centroid of tension steel, the design shear force may be transferred by contact stresses without ties, if the contact surfaces are clean and intentionally roughened with a full amplitude of about ¼ in. Otherwise, fully anchored minimum ties [Eq. (10-83)], spaced not over 24 in or 4 times the least dimension of the supported element, may be used.

When V_u is between $80b_v d$ and $350b_v d$ the design shear force may be transferred by shear-friction reinforcement placed perpendicular to assumed cracks. Shear force V_u should not exceed $800b_v d$ or $0.2f'_c \, b_v d$, where f'_c is the specified concrete strength. Required reinforcement area is

$$A_{vf} = \frac{V_u}{\phi f_y \mu} \qquad (10\text{-}121)$$

where f_y = yield strength of shear reinforcement
$\quad \mu$ = coefficient of friction
$\quad \ \ $ = 1.4 for monolithic concrete
$\quad \ \ $ = 1.0 for concrete cast against hardened concrete
$\quad \ \ $ = 0.7 for concrete cast against as-rolled structural steel (clean and without paint)

PRECAST-CONCRETE MEMBERS

Precast-concrete members are assembled and fastened together on the job. They may be unreinforced, reinforced, or prestressed. Precasting is especially advantageous when it permits mass production of concrete units. But precasting is also beneficial because it facilitates quality control and use of higher-strength concrete. Form costs may be greatly reduced, because reusable forms can be located on a casting-plant floor or on the ground at a construction site in protected locations and convenient positions, where workmen can move about freely. Many complex thin-shell structures are economical when precast, but would be uneconomical if cast in place.

10-98. Design Methods for Precast Members.

Design of precast-concrete members under the ACI Building Code follows the same rules as for cast-in-place concrete. In some cases, however, design may not be governed by service loads, because transportation and erection loads on precast members may exceed the service loads.

Design of joints and connections must provide for transmission of any forces due to shrinkage, creep, temperature, elastic deformation, gravity loads, wind loads, and earthquake motion.

("Suggested Design of Joints and Connections in Precast Structural Concrete," ACI 512-1R; "Precast Structural Concrete in Buildings," ACI 512.2R.)

10-99. Reinforcement Cover in Precast Members.

Less concrete cover is required for reinforcement in precast-concrete members manufactured under plant control conditions than in cast-in-place members because the control for proportioning, placing, and curing is better. Minimum concrete cover for reinforcement required by ACI 318-77 is listed in Table 10-25.

For special, thin-section, precast-concrete members, less cover for small-size reinforcement is acceptable (Table 10-26).

For all sizes of reinforcement in precast-concrete wall panels, minimum cover of ⅜ in is acceptable at nontreated surfaces exposed to weather and ⅜ in at interior surfaces.

TABLE 10-25 Minimum Reinforcement Cover for Precast Members, in

Concrete exposed to earth or weather:	
Wall panels:	
No. 14 and No. 18 bars	1½
No. 11 and smaller	¾
Other members:	
No. 14 and No. 18 bars	2
No. 6 through No. 11 bars	1½
No. 5 bars, ⅝-in. wire and smaller	1¼
Concrete not exposed to weather or in contact with the ground:	
Slabs, walls, joists:	
No. 14 and No. 18 bars	1¼
No. 11 and smaller	⅝
Beams, girders, columns:	
Principal reinforcement:	
Diameter of bar d_b but not less than ⅝ in. and need not be more than 1½ in.	
Ties, stirrups or spirals	⅜
Shells and folded-plate members:	
No. 6 bars and larger	⅝
No. 5 bars, ⅝-in. wire and smaller	⅜

TABLE 10-26 Minimum Reinforcement Cover for Thin-Section Precast Members, in

Concrete surfaces exposed to the weather:	
Main reinforcement in beams and columns	½
Slab reinforcement and secondary reinforcement in beams and columns	⅜
Concrete surfaces not exposed to weather, ground, or water:	
All reinforcement	⅜

10-100. Tolerances for Precast Construction.

Dimensional tolerances for precast members and tolerances on fitting of precast members vary for type of member, type of joint, and conditions of use. See "PCI Design Handbook," Prestressed Concrete Institute; "Precast Reinforced Concrete—Thin Sections," report of ACI Committee 525; "Details and Detailing of Concrete Reinforcement," ACI 315.

10-101. Accelerated Curing.

For strength and durability, precast concrete members require adequate curing. They usually are given some type of accelerated curing for economic reuse of forms and casting space. At atmospheric pressure, curing temperatures may be held between 125 and 185°F for 12 to 72 hr. Under pressure, autoclave temperatures above 325°F for 5 to 36 hr are applied for fast curing. Casting temperatures, however, should not exceed 90°F. See Fig. 10-6.

("Recommended Practice for Curing Concrete," ACI 308; "Low-pressure Steam Curing," ACI 517.1R; "High-pressure Steam Curing: Modern Practice, Properties of Autoclaved Products," ACI 516R; "Recommended Practice for Atmospheric-Pressure Steam Curing of Concrete," ACI 517.)

10-102. Precast Floor and Roof Systems.

Long-span, precast-concrete floor and roof units are usually prestressed. Short members, 30 ft or less, are often made with ordinary reinforcement. Types of precast units for floor and roof systems include solid or ribbed slabs, hollow-core slabs, single and double tees, rectangular beams, L-shaped beams, inverted-T beams, and I beams.

Hollow-core slabs are usually available in normal-weight or structural lightweight concrete. Units range from 16 to 96 in in width, and from 4 to 12 in in depth. Hollow-core slabs may come with grouted shear keys to distribute loads to adjacent units over a slab width as great as one-half the span.

Manufacturers should be consulted for load and span data on hollow-core slabs, because camber and deflection often control the serviceability of such units, regardless of strength.

("PCI Design Handbook," Prestressed Concrete Institute.)

10-103. Precast Ribbed Slabs, Folded Plates, and Shells.

Curved shells and folded plates have a thickness that is small compared with their other dimensions. Such structures depend on their geometrical configuration and boundary conditions for strength.

Thickness. With closely spaced ribs or folds, a minimum thickness for plane sections of 1 in is acceptable.

Reinforcement. Welded-wire fabric with a maximum spacing of 2 in may be used

10-120   Concrete Construction

for slab portions of thin-section members, and for wide, thin elements 3 in thick or less. Reinforcement should be preassembled into cages, using a template, and placed within a tolerance of +0 in or −⅛ in from the nearest face. The minimum clear distance between bars should not be less than 1½ times the nominal maximum size of the aggregate. For minimum cover of reinforcement, see Art. 10-99.

Compressive Strength. Concrete for thin-section, precast-concrete members protected from the weather and moisture and not in contact with the ground should have a strength of at least 4000 psi at 28 days. For elements in other locations, a minimum of 5000 psi is recommended.

Analysis. Determination of axial stresses, moments, and shears in thin sections is usually based on the assumption that the material is ideally elastic, homogeneous, and isotropic.

Forms. Commonly used methods for the manufacture of thin-section, precast-concrete members employ metal or plastic molds, which form the bottom of the slab and the sides of the boundary members. Forms are usually removed pneumatically or hydraulically by admitting air or water under pressure through the bottom form.

("Precast Reinforced Concrete—Thin Sections," Report by ACI Committee 525.)

10-104. Wall Panels.

Precast-concrete wall panels include plain panels, decorative panels, natural stone-faced panels, sandwich panels, solid panels, ribbed panels, tilt-up panels, loadbearing and nonloadbearing panels, and thin-section panels. Prestressing, when used with such panels, makes it possible to handle and erect large units and thin sections without cracking.

Forms required to produce the desired size and shape of panel are usually made of steel, wood, concrete, vacuum-formed thermoplastics, fiber-reinforced plastics, or plastics formed into shape by heat and pressure, or any combination of these. For complicated form details, molds of plaster, gelatin, or sculptured sand can be used.

Glossy-smooth concrete finish can be obtained with forms made of plastic. But, for exterior exposure, this finish left untreated undergoes gradual and nonuniform loss of its high reflectivity. Textured surfaces or smooth but nonglossy surfaces obtained by early form removal are preferred for exterior exposure.

Exposed-aggregate monolithic finishes can be obtained with horizontal-cast panels by initially casting a thin layer containing the special surface aggregates in the forms and then casting regular concrete backup. With a thickness of exposed aggregate of less than 1 in the panel can also be cast face up and the aggregate seeded over the fresh concrete or hand placed in a wet mortar. Variations of exposed surface can be achieved by use of set retardant, acid washes, or sandblasting.

Consolidation of the concrete in the forms to obtain good appearance and durability can be attained by one of the following methods:

External vibration with high-frequency form vibrators or a vibrating table.

Internal or surface vibration with a tamping-type or jitterbug vibrator.

Placing a rich, high-slump concrete in a first layer to obtain uniform distribution of the coarse aggregate and maximum consolidation, and then making the mix for the following layers progressively stiffer. This allows absorption of excessive water from the previous layer.

Tilt-up panels can be economical if the floor slab of the building can be designed for and used as the form for the panels. The floor slab must be level and smoothly troweled. Application of a good bond-breaking agent to the slab before concrete is cast for the panels is essential to obtain a clean lift of the precast panels from the floor slab.

If lifting cables are attached to a panel edge, large bending moments may develop at

TABLE 10-27 Joint Widths for Precast Wall Panels*

Recommended maximum panel dimensions, ft	Design normal joint width, in
5	3/8
18	1/2
30	3/4

*"Fabrication, Handling and Erection of Precast Concrete Wall Panels," ACI 533.3R.

the center of the wall. For high panels, three-point pickup may be used. To spread pickup stresses, specially designed inserts are cast into the wall at pickup points.

Another method of lifting wall panels employs a vacuum mat—a large steel mat with a rubber gasket at its edges to contact the slab. When the air between mat and panel is pumped out, the mat adheres to the panel, because of the resulting vacuum, and can be used to raise the panel. The method has the advantage of spreading pickup forces over the mat area.

Panels, when erected, must be temporarily braced until other construction is in place to provide required permanent bracing.

Joints. Joint sealants for panel installations may be mastics or elastomeric materials. These are extensible and can accommodate the movement of panels. Guide lines for joint width are given in Table 10-27.

Recommended maximum joint widths and minimum expansions for the common sealants are listed in Table 10-28.

The joint sealant manufacturer should be asked to advise on backup material for use with his sealant and which shape factor should be considered. A good backup material is a rod of sponge material with a minimum compression of 30%, such as foamed polyethylene, polystyrene, polyurethane, polyvinyl chloride, or synthetic rubber.

(ACI reports 533.1R, 533.2R, 533.3R, and 533.4R; "PCI Manual for Structural Design of Architectural Precast Concrete," Prestressed Concrete Institute.)

10-105. Lift Slabs.

Lift slabs are precast-concrete floor and roof panels that are cast on a base slab at ground level, one on top of the other, with a bond-breaking membrane between them. Steel collars are embedded in the slabs and fit loosely around the columns. After the slabs have cured, they are lifted to their final position by a patented jack system supported on the columns. The embedded steel collars then are welded to the steel columns to hold the lift slabs in place. This method of construction eliminates practically all formwork.

TABLE 10-28 Maximum Joint Widths for Sealants*

Type of sealant	Maximum joint width, in	Maximum movement, tension, and compression, %
Butyl	3/4	±10
Acrylic	3/4	±15–25
One-part polyurethane	3/4	±20
Two-part polyurethane	3/4	±25
One-part polysulfide	3/4	±25
Two-part polysulfide	3/4	±25

*"Fabrication, Handling and Erection of Precast Concrete Wall Panels," ACI 533.3R.

PRESTRESSED-CONCRETE CONSTRUCTION

Prestressed concrete is concrete in which internal stresses have been introduced during fabrication to counteract the stresses produced by service loads. The prestress compresses the tensile area of the concrete to eliminate or make small the tensile stresses caused by the loads.

10-106. Basic Principles of Prestressed Concrete.

In applying prestress, the usual procedure is to tension high-strength-steel elements, called tendons, and anchor them to the concrete, which resists the tendency of the stretched steel to shorten after anchorage and is thus compressed. If the tendons are tensioned before concrete has been placed, the prestressing is called **pretensioning**. If the tendons are tensioned after the concrete has been placed the prestressing is called **posttensioning**.

Prestress can prevent cracking by keeping tensile stresses small, or entirely avoiding tension under service loads. The entire concrete cross section behaves as an uncracked homogeneous material in bending. In contrast, in nonprestressed, reinforced-concrete construction, tensile stresses are resisted by reinforcing steel, and concrete in tension is considered ineffective. It is particularly advantageous with prestressed concrete to use high-strength concrete.

Loss of Prestress. The final compression force in the concrete is not equal to the initial tension force applied by the tendons. There are immediate losses due to elastic shortening of the concrete, friction losses from curvature of the tendons, and slip at anchorages. There are also long-time losses, such as those due to shrinkage and creep of the concrete, and possibly relaxation of the steel. These losses should be computed as accurately as possible or determined experimentally. They are deducted from the initial prestressing force to determine the effective prestressing force to be used in design. (The reason that high-strength steels must be used for prestressing is to maintain the sum of these strain losses at a small percentage of the initially applied prestressing strain.) (See also Art. 10-107.)

Stresses. When stresses in prestressed members are determined, prestressing forces can be treated as other external loads. If the prestress is large enough to prevent cracking under design loads, elastic theory can be applied to the entire concrete cross section (Fig. 10-67).

Prestress may be applied to a beam by straight tendons or curved tendons. Stresses at midspan can be the same for both types of tendons, but the net stresses with the curved tendons can remain compressive away from midspan, whereas they become tensile at the top fiber near the ends with straight tendons. For a prestressing force P_s applied to a beam by a straight tendon at a distance e_1 below the neutral axis, the resulting prestress in the extreme surface throughout is

$$f = \frac{P_s}{A_c} \pm \frac{P_s e_1 c}{I_g} \qquad (10\text{-}122)$$

where P_s/A_c is the compressive stress on a cross section of area A_c, and $P_s e_1 c/I_g$ is the bending stress induced by P_s (positive for compression and negative for tension), as indicated in Fig. 10-67. If stresses $\pm Mc/I_g$ due to moment M caused by external gravity loads are superimposed at midspan, the net stresses in the extreme fibers can become zero at the bottom and compressive at the top. Because the stresses due to gravity loads are zero at the beam ends, the prestress is the final stress there and the top surface of the beam at the ends is in tension.

If the tensile stresses at the ends of beams with straight tendons are excessive, the

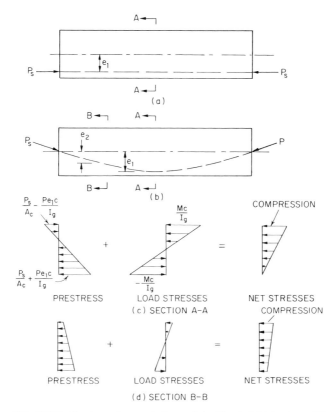

Fig. 10-67 Prestressed-concrete beam: (*a*) with straight tendons; (*b*) with curved tendons; (*c*) midspan stresses with straight or curved tendons; (*d*) stresses between midspan and supports with curved tendons. Net stresses near the supports become tensile with straight tendons.

tendons may be draped, or harped, in a vertical curve. Stresses at midspan will be substantially the same as with straight tendons (if the horizontal component of prestress is nearly equal to P_s), and the stresses at the beam ends will be compressive, because the prestressing force passes through or above the centroid of the end sections (Fig. 10-67). Between midspan and the ends, the cross sections will also be in compression.

10-107. Losses in Prestress.

Assumption in design of total losses in tendon stress of 35,000 psi for pretensioning and 25,000 psi for posttensioning to allow for elastic shortening, frictional losses, slip at anchorages, shrinkage, creep, and relaxation of steel usually gives satisfactory results. Losses greater or smaller than these values have little effect on the design strength but can affect service-load behavior, such as cracking load, deflection, and camber.

Elastic Shortening of Concrete. In pretensioned members, when the tendons are released from fixed abutments and the steel stress is transferred to the concrete by bond, the concrete shortens under the compressive stress. The decrease in unit stress in the tendons equals $P_s E_s / A_c E_c = n f_c$, where E_s is the modulus of elasticity of the steel, psi; E_c

the modulus of elasticity of the concrete psi; n the modular ratio, E_s/E_c; f_c the unit stress in the concrete, psi; P_s the prestressing force applied by the tendons; and A_c the cross-sectional area of the member.

In posttensioned members, the loss due to elastic shortening can be eliminated by using the members as a reaction in tensioning the tendons.

Frictional Losses. In posttensioned members, there may be a loss of prestress where curbed tendons rub against their enclosure. The loss may be computed in terms of a curvature-friction coefficient μ. Losses due to unintentional misalignment may be calculated from a wobble-friction coefficient K (per lin ft). Since the coefficients vary considerably, they should, if possible, be determined experimentally. A safe range of these coefficients for estimates is given in the "Commentary on ACI 318-77."

Frictional losses can be reduced by tensioning the tendons at both ends, or by initial use of a larger jacking force which is then eased off to the required initial force for anchorage.

Slip at Anchorages. For posttensioned members, prestress loss may occur at the anchorages during the anchoring. For example, seating of wedges may permit some shortening of the tendons. If tests of a specific anchorage device indicate a shortening δl, the decrease in unit stress in the steel is equal to $E_s \delta l/l$, where l is the length of the tendon. This loss can be reduced or eliminated by overtensioning initially by an additional strain equal to the estimated shortening.

Shrinkage of Concrete. Change in length of a member caused by concrete shrinkage results in a prestress loss over a period of time. This change can be determined from tests or experience. Generally, the loss is greater for pretensioned members than for posttensioned members, which are prestressed after much of the shrinkage has occurred. Assuming a shrinkage of 0.0002 in/in of length for a pretensioned member, the loss in tension in the tendons is $0.0002 E_s = 0.0002 \times 30 \times 10^6 = 6000$ psi.

Creep of Concrete. Change in length of concrete under sustained load induces a prestress loss proportional to the load over a period of time depending greatly on the aggregate used. This loss may be several times the elastic shortening. An estimate of this loss may be made with an estimated creep coefficient C_{cr} equal to the ratio of additional long-time deformation to initial elastic deformation determined by test. The loss in tension for axial prestress in the steel is, therefore, equal to $C_{cr} n f_c$. Values ranging from 1.5 to 2.0 have been recommended for C_{cr}.

Relaxation of Steel. A decrease in stress under constant high strain occurs with some steels. Steel tensioned to 60% of its ultimate strength may relax and lose as much as 3% of the prestressing force. This type of loss may be reduced by temporary overtensioning, which artificially accelerates relaxation, reducing the loss that will occur later at lower stresses.

(P. Zia et al., "Estimating Prestress Loss," *Concrete International,* June 1979, p. 32, American Concrete Institute; "PCI Design Handbook," Prestressed Concrete Institute.)

10-108. Allowable Stresses at Service Loads.

At service loads and up to cracking loads, straight-line theory may be used with the following assumptions:

Strains vary linearly with depth through the entire load range.

At cracked sections, the concrete does not resist tension.

Areas of unbonded open ducts should not be considered in computing section properties.

The transformed area of bonded tendons and reinforcing steel may be included in pretensioned members and, after the tendons have been bonded by grouting, in posttensioned members.

TABLE 10-29 Allowable Stresses for Prestressed Concrete

Concrete:	
Temporary stresses after transfer of prestress but before prestress losses:	
Compression	$0.60 f'_{ci}$
Tension in members without auxiliary reinforcement in tension zone	$3\sqrt{f'_{ci}}$*
Service-load stresses after prestress losses:	
Compression	$0.45 f'_c$
Tension in precompressed tensile zone	$6\sqrt{f'_c}$†
Prestressing steel:	
Due to jacking force	$0.80 f_{pu}$‡
Pretensioning tendons immediately after transfer or posttensioning tendons immediately after anchoring	$0.70 f_{pu}$‡

*Where the calculated tension stress exceeds this value, reinforcement should be provided to resist the total tension force on the concrete computed for assumption of an uncracked section.

† May be taken as $12\sqrt{f'_c}$ for members for which computations based on the transformed cracked section and on bilinear moment-deflection relationships show that immediate and long-time deflections do not exceed the limits given in Table 10-13.

‡f_{pu} = ultimate strength of tendons, but not to exceed the maximum value recommended by the manufacturer of the tendons or the strength of anchorages.

Flexural stresses must be limited to ensure proper behavior at service loads. Limiting these stresses, however, does not ensure adequate design strength.

In establishing permissible flexural stresses, the ACI Building Code recognizes two service-load conditions, that before and that after prestress losses. Higher stresses are permitted for the initial state (temporary stresses) than for loadings applied after the losses have occurred.

Permissible stresses in the concrete for the initial load condition are specified as a percentage of f'_{ci}, the compressive strength of the concrete, psi, at time of initial prestress. This strength is used as a base instead of the usual f'_c, 28-day strength of concrete, because prestress is usually applied only a few days after concrete has been cast. The allowable stresses for prestressed concrete, as given in ACI 318-77, are tabulated in Table 10-29.

Bearing Stresses. Determination of bearing stresses at end regions around posttensioning anchorages is complicated, because of the elastic and inelastic behavior of the concrete and because the dimensions involved preclude simple analysis under the St. Venant theory of linear stress distribution of concentrated loads.

Lateral reinforcement may be required in anchorage zones to resist bursting, horizontal splitting, and spalling forces. Anchorages are usually supplied by a prestressing company and are usually designed by test and experience rather than theory.

Concrete in the anchorage region should be designed to develop the guaranteed ultimate tensile strength of the tendons. The ACI Code formula for bearing stresses [Eq. (10-93)] does not apply to posttensioning anchorages, but it can be used as a guide by substitution of f'_{ci}, the compressive strength of the concrete at the time of initial prestress, for f'_c.

10-109. Design Procedure for Prestressed-Concrete Beams.

Beam design involves choice of shape and dimensions of the concrete member, positioning of the tendons, and selection of amount of prestress.

After a concrete shape and dimensions have been assumed, determine the geometrical properties—cross-sectional area, center of gravity, distances of kern and extreme surface from the centroid, moment of inertia, section moduli, and dead load of the member per unit of length.

Treat the prestressing forces as a system of external forces acting on the concrete.

Compute bending stresses due to service dead and live loads. From these, determine the magnitude and location of the prestressing force required at sections subject to maximum moment. The prestressing force must result in sufficient compressive stress in the concrete to offset the tensile stresses caused by the bending moments due to dead and live service loads (Fig. 10-67). But at the same time, the prestress must not create allowable stresses that exceed those listed in Table 10-29. Investigation of other sections will guide selection of tendons to be used and determine their position and profile in the beam.

After establishing the tendon profile, prestressing forces, and tendon areas, check stresses at critical points along the beam immediately after transfer, but before losses. Using strength-design methods (Art. 10-110), check the precentage of steel and the strength of the member in flexure and shear.

Design anchorages, if required, and shear reinforcement.

Finally, check the deflection and camber under service loads. The modulus of elasticity of high-strength prestressing steel should not be assumed equal to 29,000,000 psi, as for conventional reinforcement, but should be determined by test or obtained from the manufacturer as required by the ACI Building Code.

10-110. Flexural-Strength Design of Prestressed Concrete.

(See also Art. 10-109.) Flexural design strength should be based on factored loads and the assumptions of the ACI Building Code, as explained in Arts. 10-44 through 10-51. The stress f_{ps} in the tendons at design load ($1.4D + 1.7L$, where D is the dead load and L the live load), however, should not be assumed equal to the specified yield strength. High-strength prestressing steels lack a sharp and distinct yield point, and f_{ps} varies with the ultimate strength of the prestressing steel f_{pu}, the prestressing steel percentage ρ_p, and the concrete strength f'_c at 28 days. A stress-strain curve for the steel being used is necessary for stress and strain compatibility computations of f_{ps}. For unbonded tendons, successive trial-and-error analysis of tendon strain for strength design is straightforward but tedious. Assume a deflection at failure by crushing of the concrete (strain = 0.003 in/in). Determine from the stress-strain curve for the tendon steel the tendon stress corresponding to the total tendon strain at the assumed deflection. Proceed through successive trials, varying the assumed deflection, until the algebraic sum of the internal tensile and compressive forces equals zero. The moment of the resulting couple comprising the tensile and compressive forces times $\phi = 0.90$ is the design moment.

When such data are not available, and the effective prestress, after losses, is at least $f_{pu}/2$, the stress f_{ps} in bonded tendons may be obtained from

$$f_{ps} = f_{pu}\left(1 - 0.5\rho_p \frac{f_{pu}}{f'_c}\right) \tag{10-123}$$

where $\rho_p = A_{ps}/bd$
A_{ps} = area of tendons in tension zone
b = width of compression face of member
d = distance from extreme compression surface to centroid of tendons and other tension reinforcement, if used

and in unbonded tendons from

$$f_{ps} = f_{se} + 10,000 + \frac{f'_c}{100\rho_p} \leq f_{se} + 60,000 \tag{10-124}$$

where f_{se} = effective prestress in tendons, psi, after prestress losses; but f_{ps} should not be taken more than the specified tendon yield strength.

Nonprestressed reinforcement conforming to ASTM A185, A615, A616, A617, or A496, in combination with tendons, may be assumed equivalent, at design-load moment, to its area times its yield point stress, but only if

$$\omega_p \leq 0.30 \quad (10\text{-}125)$$
$$(\omega + \omega_p - \omega') \leq 0.30 \quad (10\text{-}126)$$
$$\omega_w + \omega_{wp} - \omega'_w \leq 0.30 \quad (10\text{-}127)$$

where $\omega = \rho f_y/f'_c$
f_y = yield strength of nonprestressed reinforcement
$\omega' = \rho' f_y/f'_c$
$\omega_p = \rho_p f_{ps}/f'_c$
$\rho = A_s/bd$ = ratio of nonprestressed tension reinforcement
A_s = area of nonprestressed tension reinforcement
$\rho' = A'_s/bd$ = ratio of nonprestressed compression reinforcement
A'_s = area of nonprestressed compression reinforcement
$\omega_w, \omega_{wp}, \omega'_w$ = reinforcement indices for flanged sections, computed as the ω, ω_p, and ω', except that b is the web width, and the steel area is that required to develop the compressive strength of the web only

Nominal strength in flexure (without the capacity reduction factor ϕ), as predicted by standard equations, does not correlate well with test results when ω exceeds 0.30.

Design and Cracking Loads. To prevent an abrupt flexural failure by rupture of the prestressing steel immediately after cracking without a warning deflection, the total amount of prestressed and nonprestressed reinforcement should be adequate to develop a design load in flexure of at least 1.2 times the cracking load, calculated on the basis of a modulus of rupture f_r. For normal-weight concrete, this modulus may be taken as

$$f_r = 7.5 \sqrt{f'_c} \quad (10\text{-}128)$$

and for lightweight concrete as

$$f_r = 1.1 f_{ct} \leq 7.5 \sqrt{f'_c} \quad (10\text{-}129)$$

where f_{ct} = average splitting tensile strength of lightweight concrete. When the value for f_{ct} is not available, the modulus of rupture of lightweight concrete can be computed for sand-lightweight concrete from

$$f_r = 6.375 \sqrt{f'_c} \quad (10\text{-}130)$$

and for all-lightweight concrete from

$$f_r = 5.625 \sqrt{f'_c} \quad (10\text{-}131)$$

("PCI Design Handbook," Prestressed Concrete Institute.)

10-111. Shear-Strength Design of Prestressed Concrete.

(See also Art. 10-109.) The ACI Building Code requires that prestressed concrete beams be designed to resist diagonal tension by strength theory. There are two types of diagonal-tension cracks that can occur in prestressed-concrete flexural members: flexural-shear cracks initiated by flexural-tension cracks, and web-shear cracks caused by principal tensile stresses that exceed the tensile strength of the concrete.

The factored shear force V_u computed from Eq. (10-37) can be used to calculate the diagonal-tension stress. The distance d from the extreme compression surface to the centroid of the tension reinforcement should not be taken less than 0.80 the overall depth h of the beam.

When the beam reaction in the direction of the applied shear introduces compression into the end region of the member, the shear does not need to be checked within a distance $h/2$ from the face of the support.

Minimum Steel. The ACI Code requires that a minimum area of shear reinforcement be provided in prestressed-concrete members, except where the factored shear force V_u is less than $0.5\phi V_c$, where ϕV_c is the assumed shear that can be carried by the concrete; or where the depth h of the member is less than 10 in, 2.5 times the thickness of the compression flange, or one-half the thickness of the web; or where tests show that the required ultimate flexural and shear capacity can be developed without shear reinforcement.

When shear reinforcement is required, the amount provided perpendicular to the beam axis within a distance s should be not less that A_v given by Eq. (10-83). If, however, the effective prestress force is equal to or greater than 40% of the tensile strength of the flexural reinforcement, a minimum area A_v computed from Eq. (10-132) may be used.

$$A_v = \frac{A_{ps} f_{pu}}{80 f_y} \frac{s}{d} \sqrt{\frac{d}{b_w}} \qquad (10\text{-}132)$$

where A_{ps} = area of tendons in tension zone
f_{pu} = ultimate strength of tendons
f_y = yield strength of nonprestressed reinforcement
s = shear reinforcement spacing measured parallel to longitudinal axis of member
d = distance from extreme compression surface to centroid of tension reinforcement
b_w = web width

The ACI Code does not permit the yield strength f_y of shear reinforcement to be assumed greater than 60,000 psi. The Code also requires that stirrups be placed perpendicular to the beam axis and spaced not farther apart than 24 in or $0.75h$, where h is the overall depth of the member.

The area of shear reinforcement required to carry the shear in excess of the shear that can be carried by the concrete can be determined from Eq. (10-40a).

Maximum Shear. For prestressed concrete members subjected to an effective prestress force equal to at least 40% of the tensile strength of the flexural reinforcement, the shear strength provided by the concrete is limited to that which would cause significant inclined cracking and, unless Eqs. (10-134) and (10-135) are used, can be taken as equal to the larger of $2\sqrt{f'_c} b_w d$ and

$$V_c = \left(0.60 \sqrt{f'_c} + 700 \frac{V_u d}{M_u}\right) b_w d \leq 5 \sqrt{f'_c} b_w d \qquad (10\text{-}133)$$

where M_u = design-load moment at section
V_u = design shear force at section

The design moment M_u occurs simultaneously with the shear V_u at the section. The ratio $V_u d/M_u$ should not be taken greater than 1.0.

If the effective prestress force is less than 40% of the tensile strength of the flexural reinforcement, or if a more accurate method is preferred, the value of V_c should be taken as the smaller of the shear forces causing inclined flexure-shear cracking V_{ci} or web-shear cracking V_{cw}, but need not be smaller than $1.7 \sqrt{f'_c} b_w d$.

$$V_{ci} = 0.6 \sqrt{f'_c} b_w d + V_d + V_i M_{cr}/M_{\max} \qquad (10\text{-}134)$$

$$V_{cw} = (3.5 \sqrt{f'_c} + 0.3 f_{pc}) b_w d + V_p \qquad (10\text{-}135)$$

where V_d = shear force at section caused by dead load
V_i = shear force at section occurring simultaneously with M_{max}
M_{max} = maximum bending moment at the section caused by externally applied design loads
M_{cr} = cracking moment based on the modulus of rupture (Art. 5-51)
b_w = width of web
f_{pc} = compressive stress in the concrete, after all prestress losses have occurred, at the centroid of the cross section resisting the applied loads, or at the junction of web and flange when the centroid lies within the flange
V_p = vertical component of effective prestress force at section considered

In a pretensioned beam in which the section $h/2$ from the face of the support is closer to the end of the beam than the transfer length of the tendon, the reduced prestress in the concrete at sections falling within the transfer length should be considered when calculating V_{cw}. The prestress may be assumed to vary linearly along the centroidal axis from zero at the beam end to the end of the transfer length. This distance can be assumed to be 50 diameters for strand and 100 diameters for single wire.
("PCI Design Handbook," Prestressed Concrete Institute.)

10-112. Bond, Development, and Grouting of Tendons.

Three- or seven-wire pretensioning strand should be bonded beyond the critical section for a development length, in, of at least

$$l_d = (f_{ps} - \tfrac{2}{3}f_{se})d_b \qquad (10\text{-}136)$$

where d_b = nominal diameter of strand, in
f_{ps} = calculated stress in tendons at design load, ksi
f_{se} = effective stress in tendons after losses, ksi

(The expression in parentheses is used as a constant without units.) Investigations for bond integrity may be limited to those cross sections nearest each end of the member that are required to develop their full strength under design load. When bonding does not extend to the end of the member, the bonded development length given by Eq. (10-136) should be doubled.

Minimum Bonded Steel. When prestressing steel is unbonded, the ACI Building Code requires that some bonded reinforcement be placed in the precompressed tensile zone of flexural members and distributed uniformly over the tension zone near the extreme tension surface. The amount of bonded reinforcement that should be furnished for beams, one-way slabs, and two-way slabs would be the larger of the values of A_s computed from Eqs. (10-137) and (10-138).

$$A_s = \frac{N_c}{0.5f_y} \qquad (10\text{-}137)$$

$$A_s = 0.004A \qquad (10\text{-}138)$$

where N_c = tensile force in the concrete under actual dead load plus 1.2 times live load
A = area of that part of the cross section between the flexural tension face and centroid of gross section
f_y = yield strength of bonded reinforcement, but not more than 60,000 psi

The minimum amount of bonded reinforcement for two-way slabs with unbonded tendons is the same as for one-way slabs, except that the minimum amount may be reduced when there is no tension in the precompressed tension zone under service loads.

Grouting of Tendons. When posttensioned tendons are to be bonded, a cement grout is usually injected under pressure (80 to 100 psi) into the space between the tendon and the sheathing material of the cableway. The grout can be inserted in holes in the anchorage heads and cones, or through buried pipes. To ensure filling of the space, the grout can be injected under pressure at one end of the member until it is forced out the other end. For long members, it can be injected at each end until it is forced out a vent between the ends.

Grout provides bond between the posttensioning tendons and the concrete member and protects the tendons against corrosion.

Members should be above 50°F in temperature at the time of grouting. This minimum temperature should be maintained for at least 48 hr.

Tendon Sheaths. Ducts for grouted or unbonded tendons should be mortar-tight and nonreactive with concrete, tendons, or filler material. To facilitate injection of the grout, the cableway or duct should be at least ¼ in larger than the diameter of the posttensioning tendon or large enough to have an internal area at least twice the gross area of the prestressing steel.

10-113. Application and Measurement of Prestress.

The actual amount of prestressing force applied to a concrete member should be determined by measuring the tendon elongation, also by checking jack pressure on a calibrated gage or load cell, or by use of a calibrated dynamometer. If the discrepancy in force determination exceeds 5%, it should be investigated and corrected. Elongation measurements should be correlated with average load-elongation curves for the particular prestressing steel being used.

10-114. Concrete Cover in Prestressed Members.

The minimum thicknesses of cover required by the ACI Building Code for prestressed and nonprestressed reinforcement, ducts, and end fittings in prestressed concrete members are listed in Table 10-30.

The cover for nonprestressed reinforcement in prestressed concrete members under plant control may be that required for precast members (Tables 10-25 and 10-26). When the general code requires fire-protection covering greater than that required by the ACI Code, such cover should be used.

("PCI Design Handbook," Prestressed Concrete Institute, 20 N. Wacker Drive, Chicago, Ill. 60606.)

TABLE 10-30 Minimum Concrete Cover, in, in Prestressed Members

Concrete cast against and permanently exposed to earth.	3
Concrete exposed to earth or weather:	
Wall panels, slabs, and joists.	1
Other members	1½
Concrete not exposed to weather or in contact with the ground:	
Slabs, walls, joists.	¾
Beams, girders, columns:	
Principal reinforcement	1½
Ties, stirrups, or spirals	1
Shells and folded-plate members:	
Reinforcement ⅝ in. and smaller.	⅜
Other reinforcement	d_b but not less than ¾

Section 11

Engineered Brick Construction

Alan H. Yorkdale
Vice President, Engineering and Research,
Brick Institute of America, McLean, Va.

Design of load-bearing brick structures may be based on rational engineering analysis instead of the empirical requirements for minimum wall thickness and maximum wall height contained in building codes and given in Sec. 12. Those requirements usually make bearing-wall construction for buildings higher than three to five stories uneconomical, and encourage use of other methods of support (steel or concrete skeleton frame). Since 1965, engineered brick buildings 10 or more stories high, with design based on rational structural analysis, have been built in the United States. This construction was stimulated by the many load-bearing brick buildings exceeding 10 stories in height constructed during the preceding two decades in Europe.

Design requirements for engineered brick structures given in this section are taken from "Building Code Requirements for Engineered Brick Masonry," promulgated by the Brick Institute of America, 11490 Commerce Park Drive, Reston, VA 22091.

11-1. Design Standards.

Working with a highly qualified Engineering Advisory Committee, the Structural Clay Products Institute developed in 1966 a rational design standard for engineered brick construction, based on laboratory research and historical performance data. The Institute, in collaboration with the Engineering Advisory Committee, also developed and published a second-generation standard, "Building Code Requirements for Engineered Brick Masonry," August, 1969. This newer edition incorporated the findings of additional research and the experience of architects and engineers in use of the 1966 standard.

Provisions of the first or second editions of the standard have been adopted by reference or incorporated in the masonry chapters of the BOCA Basic Building Code, Standard Building Code, National Building Code, and Uniform Building Code.

In addition to adoption by the model building codes, many local political subdivisions have also adopted, by reference or in principle, these provisions and requirements for engineered brick construction.

General Requirements of Design Standard. The 1969 standard, "Building Code Requirements for Engineered Brick Masonry," provides minimum requirements predicated on a general analysis of the structure and based on generally accepted engineering analysis procedures.

The standard contains a requirement for architectural or engineering inspection of the workmanship to ascertain, in general, if the construction and workmanship are in accordance with the contract drawings and specifications. Frequency of inspections should be such that an inspector can inspect the various stages of construction and see that the work is being properly performed. The standard requires reduced allowable stresses and capacities for loads when such architectural or engineering inspection is not provided.

Structural Requirements of Design Standard. Engineered brick bearing-wall structures do not require new techniques of analysis and design, but merely application of engineering principles used in the analysis and design of other structural systems. The method of analysis depends on the complexity of the building with respect to height, shape, wall location, and openings in the wall. A few conservative assumptions, however, accompanied by proper details to support them, can result in a simplified and satisfactory solution for most bearing-wall structures up to 12 stories high. More rigorous analysis for bearing-wall structures beyond this height may be required to maintain the economics of this type of construction.

11-2. Definitions of Design Terms.

Where terms are not defined in this section, they may be assumed to have generally accepted meanings. Unless otherwise noted in this section, terms listed in this article have the meanings given here. (For definitions of other commonly used masonry terms, see Art. 12-1.)

Bearing Wall. (See Load-Bearing Wall.)

Brick. A masonry unit having approximately the shape of a rectangular prism, made from fired clay or shale or a mixture of these.

Column. (See Art. 12-1.)

Eccentricity. The normal distance between the centroidal axis of a member and the component of resultant load parallel to that axis.

Effective Height. The height of a member to be assumed for calculating the slenderness ratio.

Effective Thickness. The thickness of a member to be assumed for calculating the slenderness ratio.

Lateral Support. Members such as cross walls, columns, pilasters, buttresses, floors, roofs, or spandrel beams that have sufficient strength and stability to resist horizontal forces transmitted to them may be considered lateral supports.

Load-Bearing Wall. A wall that supports any vertical load in addition to its own weight.

Prism. An assemblage of brick and mortar for the purpose of laboratory testing for design strength, quality control of materials, and workmanship. Minimum height for prisms is 12 in, and the slenderness ratio should lie between 2 and 5.

Slenderness Ratio. Ratio of the effective height of a member to its effective thickness.

Solid Masonry Unit. (See Art. 12-1.)

Solid Masonry Wall. A wall built of solid masonry units laid contiguously, with joints between units filled with mortar or grout.

Virtual Eccentricity. The eccentricity of resultant axial loads required to produce axial and bending stresses equivalent to those produced by applied axial and transverse loads.

11-3. Materials for Engineered Brick Construction.

Strength (compressive, shearing, and transverse) of brick structures is affected by the properties of the brick and the mortar in which they are laid. In compression, strength of brick has the greater effect. Although mortar is also a factor in compressive strength, its greater effect is on the transverse and shearing strengths of masonry. For these reasons, there are specific design requirements for and limitations on materials used in engineered brick structures.

Brick. These units must conform to the requirements for grade MW or SW, ASTM "Standard Specifications for Building Brick," C62. In addition, brick used in load-bearing or shear walls must comply with the dimension and distortion tolerances specified for type FBS of ASTM "Standard Specifications for Facing Brick," C216. Bricks that do not comply with these tolerance requirements may be used if the ultimate compressive strength of the masonry is determined by prism tests. Used or salvaged bricks are not permitted in engineered brick masonry construction.

Mortar. Most of the test data on which allowable stresses for engineered brick masonry are based were obtained for specimens built with portland cement–hydrated lime mortars. Three mortar types are provided for: M, S, and N, as described in ASTM C270 (see Table 12-1), except that the mortar must consist of mixtures of portland cement (type I, II, or III), hydrated lime (type S, non-air-entrained), and aggregate when the allowable stresses specified in "Building Code Requirements for Engineered Brick Masonry" are used. This standard provides, however, that "Other mortars ... may be used when approved by the Building Official, provided strengths for such masonry construction are established by tests ..."

11-4. Determination of Masonry Compressive Strength.

Allowable stresses are based, for the most part, on the ultimate compressive strength f'_m, psi, of masonry used. This strength may be determined by prism testing, or may be based on the compressive strength of the bricks and the type of mortar used.

Prism Tests. When the compressive strength of the masonry is to be established by preliminary tests, the tests should be made in advance of construction with prisms built of similar materials, assembled under the same conditions and bonding arrangements as for the structure. In building the prisms, the moisture content of the units at the time of laying, consistency of the mortar, thickness of mortar joints, and workmanship should be the same as will be used in the structure.

Prisms. Prisms should be stored in an air temperature of not less than 65°F and aged, before testing, for 28 days in accordance with the provisions of ASTM "Standard Method of Test for Compressive Strength of Masonry Assemblages," E447. Seven-day test results may be used if the relationship between 7- and 28-day strengths of the masonry has been established by tests. In the absence of such data, the 7-day compressive strength of the masonry may be assumed to be 90% of the 28-day compressive strength.

The value of f'_m used in the standard, "Building Code Requirements for Engineered Brick Masonry," is based on a height-thickness ratio h/t of 5. If the h/t of prisms tested is less than 5, which will cause higher test results, the compressive strength of the spec-

11-4 Engineered Brick Construction

imens obtained in the tests should be multiplied by the appropriate correction factor given in Table 11-1. Interpolation may be used to obtain intermediate values.

TABLE 11-1 Strength Correction Factors for Short Prisms

Ratio of height to thickness h/t	2.0	2.5	3.0	3.5	4.0	4.5	5.0
Correction factor	0.82	0.85	0.88	0.91	0.94	0.97	1.00

Brick Tests. When the compressive strength of masonry is not determined by prism tests, but the brick, mortar, and workmanship conform to all applicable requirements of the standard, allowable stresses may be based on an assumed value of the 28-day compressive strength f'_m computed from Eq. (11-1) or interpolated from the values in Table 11-2.

$$f'_m = A(400 + Bf'_b) \tag{11-1}$$

where A = coefficient (% without inspection and 1.0 with inspection)

B = coefficient (0.2 for type N mortar, 0.25 for type S mortar, and 0.3 for type M mortar)

f'_b = average compressive strength of brick, psi $\leq$ 14,000 psi

When there is no engineering or architectural inspection to ensure compliance with the workmanship requirements of the standard, the values in Table 11-2 under "Without inspection" should be used.

Compressive strength tests of brick should be conducted in accordance with ASTM "Standard Methods of Sampling and Testing Brick," C67.

TABLE 11-2 Assumed Compressive Strength of Brick Masonry, psi

Compressive strength of brick, psi	Without inspection			With inspection		
	Type N mortar	Type S mortar	Type M mortar	Type N mortar	Type S mortar	Type M mortar
14,000 plus	2,140	2,600	3,070	3,200	3,900	4,600
12,000	1,870	2,270	2,670	2,800	3,400	4,000
10,000	1,600	1,930	2,270	2,400	2,900	3,400
8,000	1,340	1,600	1,870	2,000	2,400	2,800
6,000	1,070	1,270	1,470	1,600	1,900	2,200
4,000	800	930	1,070	1,200	1,400	1,600
2,000	530	600	670	800	900	1,000

11-5. Allowable Stresses for Brick Construction.

Allowable stresses—compressive f_m, tensile f_t, and shearing f_v—for brick construction should not exceed the values shown in Tables 11-3 and 11-4. Allowable shears on bolts and anchors are given in Table 11-9. For allowable loads on walls and columns, see Art. 11-6.

For wind, blast, or earthquake loads combined with dead and live loads, the allowable stresses in brick construction may be increased by one-third, if the resultant section will not be less than that required for dead loads plus reduced live loads alone. Where the actual stresses exceed the allowable, the designer should specify a larger section or reinforced brick masonry.

TABLE 11-3 Allowable Stresses for Nonreinforced Brick Masonry, psi

Description		Without inspection	With inspection
Axial compression			
Walls	f_m	$0.20\,f'_m$	$0.20\,f'_m$
Columns	f_m	$0.16\,f'_m$	$0.16\,f'_m$
Flexural compression			
Walls	f_m	$0.32\,f'_m$	$0.32\,f'_m$
Columns	f_m	$0.26\,f'_m$	$0.26\,f'_m$
Flexural tension			
Normal to bed joints			
M or S mortar	f_t	24	36
N mortar	f_t	19	28
Parallel to bed joints			
M or S mortar	f_t	48	72
N mortar	f_t	37	56
Shear			
M or S mortar	v_m	$0.3\sqrt{f'_m} \le 40$	$0.5\sqrt{f'_m} \le 80$
N mortar	v_m	$0.3\sqrt{f'_m} \le 28$	$0.5\sqrt{f'_m} \le 56$
Bearing			
On full area	f_m	$0.25\,f'_m$	$0.25\,f'_m$
On one-third area or less	f_m	$0.375\,f'_m$	$0.375\,f'_m$
Modulus of elasticity	E_m	$1000\,f'_m \le 2{,}000{,}000$ psi	$1000\,f'_m \le 3{,}000{,}000$ psi
Modulus of rigidity	E_v	$400\,f'_m \le 800{,}000$ psi	$400\,f'_m \le 1{,}200{,}000$ psi

TABLE 11-4 Allowable Stresses for Reinforced Brick Construction, psi

Description		Without inspection	With inspection
Axial compression			
Walls	f_m	$0.25\,f'_m$	$0.25\,f'_m$
Columns	f_m	$0.20\,f'_m$	$0.20\,f'_m$
Flexural compression			
Walls and beams	f_m	$0.40\,f'_m$	$0.40\,f'_m$
Columns	f_m	$0.32\,f'_m$	$0.32\,f'_m$
Shear			
No shear reinforcement			
Flexural members	v_m	$0.43\sqrt{f'_m} \le 25$	$0.7\sqrt{f'_m} \le 50$
Shear walls	v_m	$0.3\sqrt{f'_m} \le 50$	$0.5\sqrt{f'_m} \le 100$
With shear reinforcement			
taking entire shear			
Flexural members	v	$1.2\sqrt{f'_m} \le 60$	$2.0\sqrt{f'_m} \le 120$
Shear walls	v	$0.9\sqrt{f'_m} \le 75$	$1.5\sqrt{f'_m} \le 150$
Bond			
Plain bars	u	53	80
Deformed bars	u	107	160
Bearing			
On full area	f_m	$0.25\,f'_m$	$0.25\,f'_m$
On one-third area or less	f_m	$0.375\,f'_m$	$0.375\,f'_m$
Modulus of elasticity	E_m	$1000\,f'_m \le 2{,}000{,}000$ psi	$1000\,f'_m \le 3{,}000{,}000$ psi
Modulus of rigidity	E_v	$400\,f'_m \le 800{,}000$ psi	$400\,f'_m \le 1{,}200{,}000$ psi

11-6. Allowable Loads on Walls and Columns.

Two stress-reduction factors are used in calculating allowable loads on walls and columns: slenderness coefficient C_s and eccentricity coefficient C_e. The eccentricity coefficient is used to reduce the allowable axial load in lieu of performing a separate bending analysis. The slenderness coefficient is used to reduce the allowable axial load to prevent buckling.

To determine these two factors, three constants are needed: end eccentricity ratio e_1/e_2, ratio of maximum virtual eccentricity to wall thickness e/t, and slenderness ratio h/t.

Eccentricity Ratio. At the top and bottom of any wall or column, a virtual eccentricity (Art. 11-2) of some magnitude (including zero) occurs. e_1/e_2 is the ratio of the smaller virtual eccentricity to the larger virtual eccentricity of the loads acting on a member. By this definition, the absolute value of the ratio is always less than or equal to 1.0. Where e_1 or e_2, or both, are equal to zero, e_1/e_2 is assumed to be zero. When the member is bent in single curvature (top and bottom virtual eccentricities occurring on the same side of the centroidal axis of a wall or column), e_1/e_2 is positive. When the member is bent in double curvature (top and bottom virtual eccentricities occurring on opposite sides of a wall or column centroidal axis), e_1/e_2 is negative (Fig. 11-1).

Eccentricity-Thickness Ratio. The ratio of maximum virtual eccentricity to wall thickness e/t is used in selecting the eccentricity coefficient C_e. Design of a nonreinforced member requires that e/t be less than or equal to ⅓. If e/t is greater than ⅓, the designer should specify a larger section, different bearing details for load transfer to the masonry, or reinforced brick construction.

Slenderness Ratio. This is the ratio of the unsupported height h to the wall thickness t. It is used in selecting the slenderness coefficient C_s.

The unsupported height h is the actual distance between lateral supports. Not always the floor-to-floor height, h may be taken as the distance from the top of the lower floor to the bearing of the upper floor where these floors provide lateral support.

Fig. 11-1 Axis of compression member with positive eccentricity ratio e_1/e_2 has single curvature, and with negative eccentricity ratio, double curvature.

The effective thickness t for nonreinforced solid masonry is the actual wall thickness, except for metal-tied cavity walls. In cavity walls, each wythe is considered to act independently, thus producing two different walls. When the cavity is filled with grout, the effective thickness becomes the total wall thickness.

Eccentricity Coefficients. C_e may be selected from Table 11-5, or calculated from Eqs. (11-2) to (11-4). Linear interpolation is permitted within the table.

$$C_e = 1.0 \qquad 0 < \frac{e}{t} \leq 0.05 \qquad (11\text{-}2)$$

$$C_e = \frac{1.3}{1 + 6e/t} + \frac{1}{2}\left(\frac{e}{t} - \frac{1}{20}\right)\left(1 - \frac{e_1}{e_2}\right) \qquad 0.05 < \frac{e}{t} \leq 0.167 \qquad (11\text{-}3)$$

$$C_e = 1.95\left(\frac{1}{2} - \frac{e}{t}\right) + \frac{1}{2}\left(\frac{e}{t} - \frac{1}{20}\right)\left(1 - \frac{e_1}{e_2}\right) \qquad 0.167 < \frac{e}{t} \leq 0.333 \qquad (11\text{-}4)$$

TABLE 11-5 Eccentricity Coefficients C_e

e/t	e_1/e_2								
	−1	−3/4	−1/2	−1/4	0	1/4	1/2	3/4	1
0–0.05	1.00	1.00	1.00	1.00	1.00	1.00	1.00	1.00	1.00
0.10	0.86	0.86	0.85	0.84	0.84	0.83	0.83	0.82	0.81
0.15	0.78	0.77	0.76	0.75	0.73	0.72	0.71	0.70	0.68
0.167	0.77	0.75	0.74	0.72	0.71	0.69	0.68	0.66	0.65
0.20	0.74	0.72	0.71	0.68	0.66	0.64	0.62	0.60	0.59
0.25	0.69	0.66	0.64	0.61	0.59	0.56	0.54	0.51	0.49
0.30	0.64	0.61	0.58	0.55	0.52	0.48	0.45	0.42	0.39
0.333	0.61	0.57	0.54	0.50	0.47	0.43	0.40	0.36	0.32

Slenderness Coefficient. C_s may be selected from Table 11-6, or calculated from Eq. (11-5). Linear interpolation is permitted within the table.

$$C_s = 1.20 - \frac{h/t}{300}\left[5.75 + \left(1.5 + \frac{e_1}{e_2}\right)^2\right] \leq 1.0 \qquad (11\text{-}5)$$

Allowable Axial Loads. Allowable loads, lb, on brick walls and columns can be computed from

$$P = C_e C_s f_m A_g \qquad (11\text{-}6)$$

where C_e = eccentricity coefficient
C_s = slenderness coefficient
f_m = allowable axial compressive stress, psi (Table 11-3 or 11-4)
A_g = gross cross-sectional area, sq in

For shear walls with eccentric loading, see Art. 11-16.

11-7. Shear-Wall Action.

The general design concept of a bearing wall should take into account the combined structural action of floor and roof systems with the walls. The floor and roof systems carry vertical loads and, acting as a diaphragm, lateral loads to the walls for transfer to the foundation. Some walls, called shear walls, are designed specifically to resist the components in their plane of lateral forces of wind and earthquake. These shear walls, by

TABLE 11-6 Slenderness Coefficients C_s

h/t	e_1/e_2								
	-1	$-3/4$	$-1/2$	$-1/4$	0	$1/4$	$1/2$	$3/4$	1
5.0	1.00	1.00	1.00	1.00	1.00	1.00	1.00	1.00	1.00
7.5	1.00	1.00	1.00	1.00	1.00	0.98	0.96	0.93	0.90
10.0	1.00	0.99	0.98	0.96	0.93	0.91	0.88	0.84	0.80
12.5	0.95	0.94	0.92	0.90	0.87	0.83	0.79	0.75	0.70
15.0	0.90	0.88	0.86	0.83	0.80	0.76	0.71	0.66	0.60
17.5	0.85	0.83	0.81	0.77	0.73	0.69	0.63	0.57	0.50
20.0	0.80	0.78	0.75	0.71	0.67	0.61	0.55	0.48	0.40
22.5	0.75	0.73	0.69	0.65	0.60	0.54	0.47	0.39	
25.0	0.70	0.67	0.64	0.59	0.53	0.47	0.39		
27.5	0.65	0.62	0.58	0.53	0.47	0.39			
30.0	0.60	0.57	0.52	0.47	0.40				
32.5	0.55	0.52	0.47	0.41					
35.0	0.50	0.46	0.41						
37.5	0.45	0.41							
40.0	0.40								

their resistance to shear and overturning, transfer the lateral loads to the foundation (Fig. 11-2).

Distribution of the lateral-load components to the shear walls parallel to them depends on the rigidities of the shear walls and the building components, such as floors and roofs, acting as diaphragms, that transmit the lateral loads to the shear walls (Art. 11-13).

Fig. 11-2 Vertical shear walls resist horizontal forces parallel to their plane.

11-8. Diaphragms.

Horizontal distribution of lateral forces to shear walls is achieved by the floor and roof systems acting as diaphragms (Fig. 11-3).

To qualify as a diaphragm, a floor or roof system must be able to transmit the lateral forces to the shear walls without exceeding a horizontal deflection that would cause distress to any vertical element. The successful action of a diaphragm also requires that it be properly tied into the supporting shear walls. Designers should ensure this action by appropriate detailing at the juncture between horizontal and vertical structural elements of the building.

Diaphragms may be considered analogous to horizontal (or inclined, in the case of some roofs) plate girders. The roof or floor slab constitutes the web; the joists, beams, and girders function as stiffeners; and the walls or bond beams act as flanges.

Diaphragms may be constructed of structural materials, such as concrete, wood, or metal in various forms. Combinations of such materials are also possible. Where a diaphragm is made up of units, such as plywood, precast-concrete planks, or steel decking, its characteristics are, to a large degree, dependent on the attachments of one unit to another and to the supporting members. Such attachments must resist shearing stresses due to internal translational and rotational actions.

The stiffness of a horizontal diaphragm affects the distribution of the lateral forces to the shear walls. For the purpose of analysis, diaphragms may be classified into three groups—rigid, semirigid or semiflexible, and flexible—although no diaphragm is actually infinitely rigid or infinitely flexible.

A rigid diaphragm is assumed to distribute horizontal forces to the vertical resisting elements in proportion to the relative rigidities of these elements (Fig. 11-4).

Semirigid or semiflexible diaphragms are diaphragms that deflect significantly under load, but have sufficient stiffness to distribute a portion of the load to the vertical elements in proportion to the rigidities of these elements. The action is analogous to a continuous beam of appreciable stiffness on yielding supports (Fig. 11-5). Diaphragm reactions are dependent on the relative stiffnesses of diaphragm and vertical resisting elements.

A flexible diaphragm is analogous to a continuous beam or series of simple beams spanning between nondeflecting supports. Thus, a flexible diaphragm is considered to distribute the lateral forces to the vertical resisting elements in proportion to the wall tributary areas (Fig. 11-6).

A rigorous analysis of lateral-load distribution to shear walls is sometimes very time-

Fig. 11-3 Floors of building distribute horizontal loads to shear walls (diaphragm action).

Fig. 11-4 Horizontal section through shear walls connected by a rigid diaphragm. R = relative rigidity and Δ_v = shear-wall deflection.

Fig. 11-5 Horizontal section through shear walls connected by a semirigid diaphragm. Δ_D = diaphragm horizontal deflection.

Fig. 11-6 Horizontal section through shear walls connected by a flexible diaphragm.

consuming, and frequently unjustified by the results. Therefore, in many cases, a design based on reasonable limits may be used. For example, the load may be distributed by first considering the diaphragm rigid, and then by considering it flexible. If the difference in results is not great, the shear walls can then be safely designed for the maximum applied load. (See also Art. 11-9.)

11-9. Torque Distribution to Shear Walls.

When the line of action of the resultant of lateral forces acting on a building does not pass through the center of rigidity of a shear-wall system, distribution of the rotational forces must be considered as well as distribution of the translational forces. If rigid or semirigid diaphragms are used, the designer may assume that torsional forces are distributed to the shear walls in proportion to their relative rigidities and their distances from the center of rigidity. A flexible diaphragm should not be considered capable of distributing torsional forces. See also Arts. 11-10 and 11-13.

11-10. Example of Torque Distribution to Shear Walls.

To illustrate load-distribution calculations for shear walls with rigid or semirigid diaphragms, Fig. 11-7 shows a horizontal section through three shear walls A, B, and C taken above a rigid floor. Wall B is 16 ft from wall A, and 24 ft from wall C. Rigidity of A is 0.33, of B 0.22, and of C 0.45 (Art. 11-13). A 20-kip horizontal force acts at floor level parallel to the shear walls and midway between A and C.

Fig. 11-7 Rigid diaphragm distributes 20-kip horizontal force to shear walls A, B, and C.

The center of rigidity of the shear walls is located, relative to wall A, by taking moments about A of the wall rigidities and dividing the sum of these moments by the sum of the wall rigidities, in this case 1.00.

$$x = 0.22 \times 16 + 0.45 \times 40 = 21.52 \text{ ft}$$

Thus, the 20-kip lateral force has an eccentricity of $21.52 - 20 = 1.52$ ft. The eccentric force may be resolved into a 20-kip force acting through the center of rigidity and not producing torque, and a couple producing a torque of $20 \times 1.52 = 30.4$ ft-kips.

The nonrotational force is distributed to the shear walls in proportion to their rigidities:

Wall A: $0.33 \times 20 = 6.6$ kips
Wall B: $0.22 \times 20 = 4.4$ kips
Wall C: $0.45 \times 20 = 9.0$ kips

For distribution of the torque to the shear walls, the equivalent of moment of inertia must first be computed:

$$I = 0.33(21.52)^2 + 0.22(5.52)^2 + 0.45(18.48)^2 = 313$$

Then, the torque is distributed in direct proportion to shear-wall rigidity and distance from center of rigidity and in inverse proportion to I.

Wall A: $30.4 \times 0.33 \times 21.52/313 = 0.690$ kips
Wall B: $30.4 \times 0.22 \times 5.52/313 \ = 0.118$ kips
Wall C: $30.4 \times 0.45 \times 18.48/313 = 0.808$ kips

The rotational forces should be added to the nonrotational forces acting on walls A and B, whereas the rotational force on wall C acts in the opposite direction to the nonrotational force. For a conservative design, the rotational force on wall C should not be subtracted. Hence, the walls should be designed for the following forces:

$$\text{Wall } A: 6.6 + 0.7 = 7.3 \text{ kips}$$
$$\text{Wall } B: 4.4 + 0.1 = 4.5 \text{ kips}$$
$$\text{Wall } C: 9.0 \text{ kips}$$

11-11. Relative Deflections of Shear Walls.

When shear walls are connected by rigid diaphragms (Art. 11-8) and horizontal forces are distributed to the vertical resisting elements in proportion to their relative rigidities, the relative rigidity of a shear wall is dependent on shear and flexural deflections. For the proportions of shear walls used in most highrise buildings, however, flexural deflections greatly exceed shear deflections. In such cases, only flexural rigidity need be considered in determination of relative rigidity of the shear walls (Art. 11-13).

Deflections can be determined by treating a shear wall as a cantilever beam. For a shear wall with a solid rectangular cross section, the cantilever flexural deflection under uniform loading may be computed from

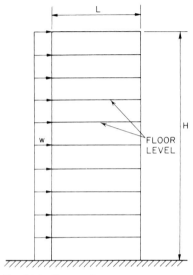

Fig. 11-8 Shear wall acts as vertical cantilever under uniformly distributed wind load w.

$$\delta_m = \frac{wH^4}{8E_m I} = 0.0015 \frac{wH}{f'_m t} \left(\frac{H}{L}\right)^3 \quad (11\text{-}7)$$

where w = uniform lateral load (Fig. 11-8)
H = height of wall
E_m = modulus of elasticity = $1000 f'_m$
f'_m = 28-day compressive strength of the masonry wall
I = moment of inertia of wall cross section = $tL^3/12$
t = wall thickness
L = length of wall

The cantilever shear deflection under uniform loading may be computed from

$$\delta_v = \frac{0.6 wH^2}{E_v A} = 0.0015 \frac{wH}{f'_m t} \left(\frac{H}{L}\right) \quad (11\text{-}8)$$

where E_v = modulus of rigidity of wall cross section = $400 f'_m$
A = cross-sectional area of wall = tL

The total deflection then is

$$\delta = \delta_m + \delta_v = 0.0015 \frac{wH}{f'_m t} \left[\left(\frac{H}{L}\right)^3 + \frac{H}{L}\right] \quad (11\text{-}9)$$

Table 11-7 lists relative flexural and shear deflections (δ_m/δ and δ_v/δ) for several height-length ratios.

For a cantilever wall subjected to a concentrated load P at the top (Fig. 11-9a), the flexural deflection is

$$\delta_m = \frac{PH^3}{3E_m I} \quad (11\text{-}10)$$

TABLE 11-7 Relative Deflections of Shear Walls

Wall ratio H/L	Relative flexural deflection δ_m/δ	Relative shear deflection δ_v/δ
1	0.50	0.50
2	0.80	0.20
3	0.90	0.10
4	0.94	0.06
5	0.96	0.04

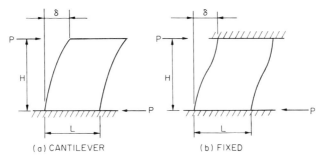

Fig. 11-9 Deflections δ of shear walls caused by horizontal load P applied at the top.

The shear deflection under P for the cantilevered wall is

$$\delta_v = \frac{1.2PH}{E_v A} \tag{11-11}$$

If E_m is taken as 3000 ksi, E_v as $0.4E_m$, t as 12 in, and P as 1000 kips, the total cantilever deflection is given by

$$\delta = 0.1112 \left(\frac{H}{L}\right)^3 + 0.0833 \frac{H}{L} \tag{11-12}$$

For a wall fixed at the top against rotation and subjected to a concentrated load P at the top (Fig. 11-9b), the flexural deflection is

$$\delta_m = \frac{PH^3}{12E_m I} \tag{11-13}$$

The shear deflection for the fixed wall is also given by Eq. (11-11). For the same values assumed for Eq. (11-12), the total deflection for the fixed wall is

$$\delta = 0.0278 \left(\frac{H}{L}\right)^3 + 0.0833 \frac{H}{L} \tag{11-14}$$

11-12. Diaphragm-Deflection Limitations.

As indicated in Art. 11-8, horizontal deflection must be considered in designing a horizontal diaphragm. Diaphragm deflection should be limited to prevent excessive stresses in walls perpendicular to shear walls. Equation (11-15) was suggested by the Structural Engineers Association of Southern California for allowable story deflection, in, of masonry or concrete building walls.

$$\Delta = \frac{h^2 f}{0.01 Et} \qquad (11\text{-}15)$$

where h = height of wall between adjacent horizontal supports, ft
t = thickness of wall, in
f = allowable flexural compressive stress of wall material, psi
E = modulus of elasticity of wall material, psi

This limit on deflection must be applied with engineering judgment. For example, continuity of wall at floor level is assumed, and in many cases is not present because of through-wall flashing. In this situation, the deflection may be based on the allowable compressive stress in the masonry, if a reduced cross section of wall is assumed. The effect of reinforcement, which may be present in a reinforced brick masonry wall or as a tie to the floor system in a nonreinforced or partly reinforced masonry wall, was not considered in development of Eq. (11-15). Note also that the limit on wall deflection is actually a limit on differential deflection between two successive floor, or diaphragm, levels.

Maximum span-width or span-depth ratios for diaphragms are usually used to control horizontal diaphragm deflection indirectly. Normally, if the diaphragm is designed with the proper ratio, the diaphragm deflection will not be critical. Table 11-8 may be used as a guide for proportioning diaphragms.

TABLE 11-8 Maximum Span-Width or Span-Depth Ratios for Diaphragms—Roofs or Floors*

Diaphragm construction	Masonry and concrete walls	Wood and light steel walls
Concrete	Limited by deflection	
Steel deck (continuous sheet in a single planc)	4:1	5:1
Steel deck (without continuous sheet)	2:1	2½:1
Cast-in-place reinforced gypsum roofs	3:1	4:1
Plywood (nailed all edges)	3:1	4:1
Plywood (nailed to supports only—blocking may be omitted between joists).	2½:1†	3½:1
Diagonal sheathing (special)	3:1†	3½:1
Diagonal sheathing (conventional construction)	2:1†	2½:1

* From California Administrative Code, Title 21, Public Works.

† Use of diagonal sheathed or unblocked plywood diaphragms for buildings having masonry or reinforced concrete walls shall be limited to one-story buildings or to the roof of a top story.

11-13. Shear-Wall Rigidity.

Where shear walls are connected by rigid diaphragms (Art. 11-8) so that they must deflect equally under horizontal loads, the proportion of total horizontal load at any level carried by a shear wall parallel to the load depends on the relative rigidity, or stiffness, of the wall in the direction of the load. Rigidity of a shear wall is inversely proportional to its deflection under unit horizontal load. This deflection equals the sum of the shear and flexural deflections under the load (Art. 11-11).

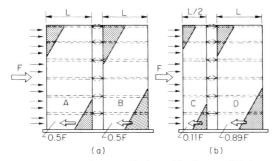

Fig. 11-10 Distribution of horizontal load to parallel shear walls: (a) walls with same length and rigidity share the load equally; (b) wall half the length of another carries less than one-eighth the load.

Where a shear wall contains no openings, computations for deflection and rigidity are simple. In Fig. 11-10a, each of the shear walls has the same length and rigidity. So each takes half the total load. In Fig. 11-10b, length of wall C is half that of wall D. By Eq. (11-7), C therefore receives less than one-eighth the total load.

Walls with Openings. Where shear walls contain openings, such as doors and windows, computations for deflection and rigidity are more complex. But approximate methods may be used.

For example, the wall in Fig. 11-11, subjected to a 1000-kip load at the top, may be treated in parts. The wall is 8 in thick, and its modulus of elasticity $E_m = 2400$ ksi. Its height-length ratio H/L is $12/20 = 0.6$. The wall is perforated by two, symmetrically located, 4-ft-square openings.

Fig. 11-11 Shear wall, 8 in thick, with openings.

Deflection of this wall can be estimated by subtracting from the deflection it would have if it were solid the deflection of a solid, 4-ft-deep, horizontal midstrip, and then adding the deflection of the three coupled piers B, C, and D.

Deflection of the 12-ft-high solid wall can be obtained from Eq. (11-12) modified, by proportion, for actual thickness t and E_m.

$$\delta = [0.1112(0.6)^3 + 0.0833 \times 0.6] \frac{3000}{2400} \times \frac{12}{8} = 0.138 \text{ in}$$

Rigidity of the solid wall then is

$$R = \frac{1}{0.138} = 7.22$$

Similarly, the deflection of the 4-ft-deep solid midstrip can be computed from Eq. (11-12), with $H/L = 4/20 = 0.20$.

$$\delta = [0.1112(0.2)^3 + 0.0833 \times 0.2] \frac{3000}{2400} \times \frac{12}{8} = 0.033 \text{ in}$$

Deflection of the piers, which may be considered fixed top and bottom, can be obtained from Eq. (11-14) modified for actual t and E_m, with $H/L = \frac{3}{4} = 1$. For any one of the piers, the deflection is

$$\delta' = [0.0278(1)^3 + 0.0833 \times 1]\frac{3000}{2400} \times \frac{12}{8} = 0.208 \text{ in}$$

The rigidity of a single pier is $1/0.208 = 4.81$, and of the three piers, $3 \times 4.81 = 14.43$. Therefore, the deflection of the three piers when coupled is

$$\delta = \frac{1}{14.43} = 0.069 \text{ in.}$$

The deflection of the whole wall, with openings, then is approximately

$$\delta = 0.138 - 0.033 + 0.069 = 0.174 \text{ in}$$

And its rigidity is

$$R = \frac{1}{0.174} = 5.74$$

11-14. Effects of Shear-Wall Arrangements.

To increase the stiffness of shear walls and thus their resistance to bending, intersecting walls or flanges may be used. Often in the design of buildings, Z-, T-, U-, L-, and I-shaped walls in plan develop as natural parts of the design (Fig. 11-12). Shear walls with these shapes have better flexural resistance than a single, straight wall.

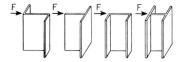

Fig. 11-12 Shear walls with flange walls have high resistance to lateral loads parallel to the web walls.

In "Building Code Requirements for Engineered Brick Masonry," a limit is placed on the effective flange width that may be used in calculating flexural stresses in shear walls. For symmetrical T or I sections, the effective flange width may not exceed one-sixth the total wall height above the level being analyzed. For unsymmetrical L or C sections, the width considered effective may not exceed one-sixteenth the total wall height above the level being analyzed. In either case, the overhang for any section may not exceed six times the flange thickness (Fig. 11-13).

The shear stress at the intersection of the walls should not exceed the permissible

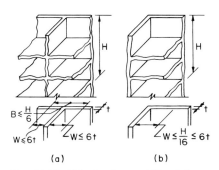

Fig. 11-13 Effective flange width of shear walls may be less than actual width: (a) limits for flanges of I and T shapes; (b) limits for C and L shapes.

shear stress. The allowable stress depends on the method used in bonding the two walls together. If the intersection is bonded with masonry, the allowable shear stress is that given in Table 11-3 or Table 11-4; if with metal anchors, the allowable stress is given in Table 11-9. See also Art. 11-15.

TABLE 11-9 Allowable Shear on Bolts and Anchors*

Dia, in.	Minimum embedment, in.†	Allowable shear, lb	
		Without inspection	With inspection
$1/4$	4	180	270
$3/8$	4	270	410
$1/2$	4	370	550
$5/8$	4	500	750
$3/4$	5	730	1,100
$7/8$	6	1,000	1,500
1	7	1,230	1,850
$1 1/8$	8	1,500	2,250

* From "Building Code Requirements for Engineered Brick Masonry," Brick Institute of America.

† Bolts and anchors should be solidly embedded in mortar.

11-15. Coupled Shear Walls.

Another method than that described in Art. 11-14 for increasing the stiffness of a bearing-wall structure and reducing the possibility of tension developing in shear walls under lateral loads is coupling of coplanar shear walls.

Figures 11-14 and 11-15 indicate the effect of coupling on stress distribution in a pair of walls under horizontal forces parallel to the walls. A flexible connection between the walls is assumed in Figs. 11-14a and 11-15a, so that the walls act as independent vertical cantilevers in resisting lateral loads. In Figs. 11-14b and 11-15b, the walls are assumed to be connected with a more rigid member, which is capable of shear and moment transfer. A rigid-frame-type action results. This can be accomplished with a steel, reinforced concrete, or reinforced brick masonry coupling. A plate-type action is indicated in Figs. 11-14c and 11-15c. This assumes an extremely rigid connection between walls, such as full story-height walls or deep rigid spandrels.

Fig. 11-14 Stress distribution in end shear walls: (a) with flexible coupling; (b) with rigid-frame-type action; (c) with plate-type action.

Fig. 11-15 Stress distributions in interior shear walls: (*a*) with flexible coupling; (*b*) with rigid-frame-type action; (*c*) with plate-type action.

11-16. Eccentrically Loaded Shear Walls.

"Building Code Requirements for Engineered Brick Masonry" also provides a basis for design of eccentrically loaded shear walls. The standard requires that, in a nonreinforced shear wall, the virtual eccentricity e_L about the principal axis normal to the length L of the wall not exceed an amount that will produce tension. In a nonreinforced shear wall subject to bending about both principal axes, $e_t L + e_L t$ should not exceed $tL/3$, where e_t = virtual eccentricity about the principal axis normal to the thickness t of the shear wall. Where the virtual eccentricity exceeds the preceding limits, shear walls should be designed as reinforced or partly reinforced walls.

Consequently, for a planar wall the virtual eccentricity e_L, which is found by dividing the overturning moment about an axis normal to the plane of the wall by the axial load, should not exceed $L/6$. In theory, any virtual eccentricity exceeding $L/6$ for planar shear wall will result in development of tensile stresses. If, however, intersecting walls form resisting flanges, the classic approach may be used where the bending stress Mc/I is combined with the axial stress P/A. If the bending stress exceeds the axial stress, then tensile stresses are present. (Note that some building codes require a safety factor against overturning. It is usually 1.5 for nonreinforced walls. But the codes state that, if the walls are vertically reinforced to resist tension, the safety factor does not apply.)

For investigation of biaxial bending, the eccentricity limitation may, for convenience, be placed in the form:

$$\frac{e}{t} = \frac{e_t}{t} + \frac{e_L}{L} \leq \frac{1}{3} \qquad (11\text{-}16)$$

11-17. Allowable Stresses for Shear Walls.

Allowable vertical loads on shear walls may be computed from Eq. (11-6). The effective height h for computing C_s should be taken as the minimum vertical or horizontal distance between lateral supports.

Allowable shearing stresses for shear walls should be taken as the sum of the allowable shear stress given in Table 11-3 or 11-4 and one-fifth the average compressive stress produced by dead load at the level being analyzed, but not more than the maximum values listed in these tables.

In computation of shear resistance of a shear wall with intersecting walls treated as flanges, only the parts serving as webs should be considered effective in resisting the shear.

When non-load-bearing shear walls are required to resist overturning moment only by their own weight, the design can become critical. Consequently, positive ties should be provided between shear walls and bearing walls, to take advantage of the bearing-wall loads in resisting overturning. The shear stress at the connection of the shear wall to the bearing wall should be checked. The designer should exercise judgment in assumption of the distribution of the axial loads into the non-load-bearing shear walls.

11-18. Bearing Walls.

Walls whose function is to carry only vertical loads should be proportioned primarily for compressive stress. Allowable vertical loads are given by Eq. (11-6). Effects of all loads and conditions of loading should be investigated. For application of Eq. (11-6), each loading should be converted to a vertical load and a virtual eccentricity.

In the lower stories of a building, the compressive stress is usually sufficient to suppress development of tensile stress. But in the upper stories of tall buildings, where the exterior walls are subject to high lateral wind loads and small axial loads, the allowable tensile stress may be exceeded occasionally and make reinforcement necessary.

When lateral forces act parallel to the plane of bearing walls that act as load-bearing shear walls, they must meet the requirements for both shear walls and biaxial bending (Arts. 11-16 and 11-17). Such walls must be checked, to preclude development of tensile stress or excessive compressive or shearing stresses. If analysis indicates that tension requirements are not satisfied, the size, shape, or number of shear walls must be revised or the wall must be designed as reinforced masonry.

11-19. Floor-Wall Connections.

Bending moments caused in a wall by floor loading depend on such factors as type of floor system, detail of floor-wall connections and sequence of construction. Because information available on their effects is limited, engineers must make certain design assumptions when providing for such moments. Conservative assumptions that may be used in design are discussed in the following.

For a floor system that acts hinged at the floor-wall connection, such as steel joists and stems from precast-concrete joists, a triangular stress distribution can be assumed under the bearing (Fig. 11-16). The moment in the wall produced by dead and live loads is then equal to the reaction times the eccentricity resulting from this stress distribution.

For precast-concrete-plank floor systems, which deflect and rotate at the time of placing, a triangular stress distribution similarly can be assumed to result from the dead load of the plank and to induce a moment in the wall. When the topping is placed as each level is constructed, a triangular stress distribution can still be assumed. The moment resulting from the dead load of the floor system, including the topping, then is that due to the eccentric loading. If, however, the topping is placed after the wall above has been built and the wall clamps the plank end in place, creating a restrained end condition, the moment in the wall will then be the sum of the moment due to the eccentric load of the plank itself and the fixed-end moment resulting from the superimposed loads of topping weight and live load (Fig. 11-17).

The degree of fixity and the resulting magnitude of the restrained end moments usually must be assumed. Full fixity of floors due to the clamping action of a wall under large axial loads in the lower stories of high- and medium-rise buildings appears a logical assumption. The same large axial loads that provide the clamping action in the lower stories also act to suppress development of tensile stresses in the wall at the floor-wall connection. Because axial loads are smaller in upper stories, however, the degree of fixity may be assumed reduced, with occurrence of slight rotation and elevation of the extreme

Fig. 11-16 Stress distribution in wall from joist reaction.

Fig. 11-17 Stress distribution in wall from plank reaction.

end of the slab. Based on this assumption, slight, local stress-relieving in connections in upper stories could take place. Regardless of the assumption, the maximum moment transferred to the wall can never be greater than the negative-moment capacity of the floor system.

When full fixity is assumed, the magnitude of the moment in the wall will be approximately the distribution factor times the initial fixed-end moment of the slab at the face of the wall. As an approximation for precast-concrete plank with uniform load w and span L, $wL^2/36$ may be conservatively assumed as the wall moment. [Preliminary test results have indicated about 80% moment transfer from the slab into the wall sections (40% to the upper and 40% to the lower wall section) with flat, precast plank penetrating the full wall thickness.]

Fig. 11-18 Stress distribution in wall with only joist stem embedded.

For a cast-in-place concrete slab, a fixed-end moment may be assumed for both dead and live loads, because usually the wall above the slab will be built before removal of shoring.

Because restrained end moments in a wall can become large, reduction of the eccentricity of the floor reaction is advantageous in limiting the moment in the wall. This may be accomplished by projecting only the stems of cast-in-place or precast-concrete systems into the wall (Fig. 11-18). In such cases, a bearing pad should be placed immediately under each stem.

11-20. Example—End Shear Wall, Concrete Joist Floors.

Design a nonreinforced, load-bearing shear wall between the second and third floors (Fig. 11-19) of a 10-story building, given the following data:

Nominal thickness is 8 in. Actual thickness t = 7.5 in.
Unsupported height h_u = 8 ft.
Precast-concrete-joist floor system with 4-in minimum bearing.
Story height = 8 ft 10 in.
Intersecting flange wall 3 ft long. Ratio of cross-sectional shear-wall area to total wall area = 16.6/18.6 = 0.892.

Example—End Shear Wall, Concrete Joist Floors

	Floor	Roof
Live loads LL, lb/ft²	40	30
Dead loads DL, lb/ft²	60	50
Total loads TL, lb/ft²	100	80

Maximum live-load reduction permitted is

$$R = \frac{100\,TL}{4.33\,LL} = \frac{100 \times 100}{4.33 \times 40} = 58\%$$

Lateral wind load = 20 lb/ft²
Parallel shear load = 22.5 kips
Parallel wind moment = 900 ft-kips
Wall dead load = 120 lb/ft³ = 74.4 lb/ft²

Fig. 11-19 Properties of end shear wall of 10-story building.

Required Masonry Strength. The allowable compressive stress for nonreinforced walls is $f_m = 0.20 f'_m$, where f'_m is the 28-day compressive strength of the masonry, psi. From Eq. (11-6), the required masonry strength is

$$f'_m = \frac{P}{0.20 A_g C_e C_s} \tag{11-17}$$

When gravity loads and wind loads are combined, f'_m given by Eq. (11-17) may be multiplied by 1/1.33.

Loads at Third Floor. Gravity loading per linear foot of wall from floors or roof is contributed by half the floor or roof span, a length of 14 ft. Loading from floors above is shared with the flange wall, so that the shear wall carries 0.892 of this loading.

The dead load just below the third floor equals the weight of eight stories, or $8 \times 8.333 = 70.7$ ft, of masonry, plus roof dead load, plus dead load of eight floors, seven of which are shared by the flange wall.

$$P_{DL} = 70.7 \times 74.4 + (14 \times 50 + 7 \times 14 \times 60)0.892 + 14 \times 60 = 11{,}970 \text{ lb/ft}$$

Similarly, the live load, without reduction is

$$P_{LL} = (14 \times 30 + 7 \times 14 \times 40)0.892 + 14 \times 40 = 4430 \text{ lb/ft}$$

The contributory floor area is large enough that the maximum reduction may be applied. Hence, the wall just below the third floor must carry $100 - 58 = 42\%$ of the total live load from the floors. The reduced live load is

$$P_{RLL} = (14 \times 30 + 0.42 \times 7 \times 14 \times 40)0.892 + 0.42 \times 14 \times 40 = 2080 \text{ lb/ft}$$

Loads at Second Floor. The dead load just above the second floor equals the weight of nine stories, or $9 \times 8.33 = 79.5$ ft, of masonry, plus roof dead load, plus dead load of eight floors.

$$P_{DL} = 79.5 \times 74.4 + (14 \times 50 + 8 \times 14 \times 60)0.892 = 12{,}555 \text{ lb/ft}$$

Similarly, the live load, without reduction, is

$$P_{LL} = (14 \times 30 + 8 \times 14 \times 40)0.892 = 4370 \text{ lb/ft}$$

and the reduced live load is

$$P_{RLL} = (14 \times 30 + 0.42 \times 8 \times 14 \times 40)0.892 = 2055 \text{ lb/ft}$$

Summation of Loads at Second Floor. The total dead load just above the second floor is

$$P'_{DL} = 14 \times 26(50 + 8 \times 60) + 79.5 \times 74.4(26.67 + 3.0) = 370{,}400 \text{ lb}$$

The total unreduced live load plus dead load is

$$P'_{TL} = 14 \times 26(80 + 8 \times 100) + 79.5 \times 74.4(26.67 + 3.0) = 497{,}700 \text{ lb}$$

Source of Moments. The bending moments to which the shear wall is subjected are moments acting about the longitudinal axis of the wall, caused by floor loads; moments about the longitudinal axis of the wall, caused by perpendicular wind loads acting between floor levels; and moments acting parallel to the wall, caused by wind loads.

Fig. 11-20 Bearing of concrete joist on end wall.

A conservative approach in selection of loading conditions to produce maximum stresses is to assume loadings such that total-load moment occurs at the top of the wall (third floor) and dead-load moment at the bottom (second floor). This arrangement makes e_1/e_2 small, thus giving conservative C_e and C_s values for use in Eq. (11-17). Note that end walls normally go into double curvature when moments are caused by floor loads (eccentricity ratio e_1/e_2 is negative).

Moments from Floor Loads. Assume 4-in bearing of joists on the wall and a triangular stress distribution under the bearing (Fig. 11-20). The eccentricity of the joist reaction then is

$$e_f = \frac{7.5}{2} - \frac{4}{3} = 2.42 \text{ in}$$

The dead-load and live-load moments due to this eccentricity are distributed in equal proportions to the wall sections above and below the floor, and equal

$$M_{DL} = \tfrac{1}{2} \times 40 \times 14 \times 2.42 = 675 \text{ in-lb/ft}$$
$$M_{LL} = \tfrac{1}{2} \times 60 \times 14 \times 2.42 = 1015 \text{ in-lb/ft}$$
$$\text{Total moment} = M_{TL} = 1690 \text{ in-lb/ft}$$

Example—End Shear Wall, Concrete Joist Floors

Perpendicular Wind Moment. For the 20-lb/ft² wind load, with the wall assumed to act as a fixed-end beam between floors, the bending moments at the floors are

$$M = \tfrac{1}{12} \times 20(8.83)^2 12 = 1560 \text{ in-lb/ft}$$

Parallel Wind Moment at Second Floor. This was given initially as $M' = 900$ ft-kips.

Overturn Check. To determine if tensile stress develops in the shear wall from overturning moment, the wall should be considered to be carrying only dead loads in addition to the wind parallel to the wall. The location of the neutral axis of the wall is found, relative to the flanged end, by dividing the sum of the moments of the wall areas about that end by the wall area $A = 18.6$ ft².

$$x = \frac{26.67}{2} \times 16.6 \times \frac{1}{18.6} = 11.91 \text{ ft}$$

The moment of inertia of the wall about this axis is $I = 1307$ ft⁴.

The allowable eccentricity e_L, which is the maximum for which no tension occurs on the wall cross section, is given by I/cA, where c is the maximum distance of a wall end from the neutral axis.

$$e_L = \frac{1307}{(26.67 - 11.91)18.6} = 4.76 \text{ ft}$$

The actual eccentricity is

$$e_L = \frac{M'}{P'_{DL}} = \frac{900{,}000}{370{,}400} = 2.43 < 4.76 \text{ ft}$$

Therefore, no tensile reinforcement is required.

Loading Conditions. Five cases are considered for checking eccentricities and stresses in the wall under gravity and wind loads:

Case I. Total loads on floors and roof plus wind parallel to the wall.
Case II. Dead loads on floors and roof plus wind parallel to the wall.
Case III. Total loads on floors and roof plus wind perpendicular to the wall.
Case IV. Dead loads on floors and roof plus wind perpendicular to the wall.
Case V. Dead loads plus reduced live loads.

Case I (TL + Parallel Wind). The wall is subjected to biaxial bending. Eccentricities must satisfy Eq. (11-16) and strength f'_m must satisfy Eq. (11-17).

The maximum eccentricity of floor load at the third floor is

$$e_t = \frac{1690}{4370 + 12{,}555} = 0.0997 \text{ in}$$

The virtual eccentricity of the 900-ft-kip overturning moment is

$$e_L = \frac{M'}{P'_{TL}} = \frac{900{,}000}{497{,}700} = 1.81 \text{ ft}$$

In accordance with Eq. (11-16) for biaxial bending,

$$\frac{e}{t} = \frac{0.0997}{7.5} + \frac{1.81}{26.67} = 0.081 < \frac{1}{3}$$

The eccentricity of floor load at the top of the wall is computed for floor total-load moment.

$$e_{\text{top}} = \frac{M}{P} = \frac{1690}{4430 + 11{,}970} = \frac{1690}{16{,}400} = 0.103 \text{ in}$$

The eccentricity of floor load at the bottom of the wall for floor dead-load moment is

$$e_{bot} = \frac{1015}{4370 + 12{,}555} = \frac{1015}{16{,}925} = 0.060 \text{ in}$$

The eccentricity ratio, negative because of double curvature of the wall, then is

$$\frac{e_1}{e_2} = -\frac{0.060}{0.103} = -0.582$$

For use in determining the slenderness coefficient C_s, the slenderness ratio is

$$\frac{h}{t} = \frac{8.0 \times 12}{7.5} = 12.8$$

From Eq. (11-5) or Table 11-6, $C_s = 0.92$. From Table 11-5, with $e/t = 0.081$, the eccentricity coefficient $C_e = 0.90$. For use in Eq. (11-17), the area of a linear foot of wall is $A_g = 7.5 \times 12 = 90$ in^2. From Eq. (11-17), the required masonry strength of Case I, with the one-third increase in stress allowed for combined gravity and wind loads, is

$$f'_m = \frac{16{,}926}{0.20 \times 90 \times 0.90 \times 0.92} \frac{1}{1.33} = 855 \text{ psi}$$

Case II (DL + Parallel Wind). This case is similar to Case I, but dead load is used instead of total load.

The eccentricity of the floor load at the bottom of the wall is

$$e_{bot} = \frac{1015}{12{,}555} = 0.081 \text{ in}$$

The virtual eccentricity of the 900-ft-kip overturning moment is

$$e_L = \frac{M'}{P'_{DL}} = \frac{900{,}000}{370{,}400} = 2.43 \text{ ft}$$

A check with Eq. (11-16) yields

$$\frac{e}{t} = \frac{0.081}{7.5} + \frac{2.43}{26.67} = 0.102 < \frac{1}{3}$$

The eccentricity of the floor dead load at the top of the wall is

$$e_{top} = \frac{1015}{11{,}970} = 0.085 \text{ in}$$

The eccentricity ratio for the floor loading is

$$\frac{e_1}{e_2} = -\frac{0.081}{0.085} = -0.950$$

With $h/t = 12.8$ as in Case I, $C_s = 0.94$. From Table 11-5, with $e/t = 0.102$, $C_e = 0.86$. From Eq. (11-17), the required masonry strength for Case II is

$$f'_m = \frac{12{,}555}{0.20 \times 90 \times 0.86 \times 0.94} \frac{1}{1.33} = 650 \text{ psi}$$

Case III (TL + Perpendicular Wind). This case differs from Case I in that the wall acts as a one-way slab, supported by floors and roof, in resisting the wind load. Bending due to wind and bending due to floor loads act about the same axis at the top and bottom of the wall in each story. The eccentricity, at the top, of the total floor loads and the perpendicular wind is

$$e_{top} = \frac{1690 + 1560}{16{,}400} = \frac{3250}{16{,}400} = 0.198 \text{ in}$$

Example—End Shear Wall, Concrete Joist Floors 11-25

A check of the eccentricity-thickness ratio yields

$$\frac{e}{t} = \frac{0.198}{7.5} = 0.026 < \frac{1}{3}$$

Because the transverse loading exceeds 10 lb/ft², e_1/e_2 is taken as unity. With $h/t = 12.8$, $C_s = 0.70$. From Table 11-5, with $e/t = 0.026$, $C_e = 1.0$. From Eq. (11-17), the required masonry strength for Case III is

$$f'_m = \frac{16{,}925}{0.20 \times 90 \times 1.0 \times 0.70} \frac{1}{1.33} = 1010 \text{ psi}$$

Case IV (DL + Perpendicular Wind). This case is similar to Case III, but dead load is used instead of total load.

$$e_{top} = \frac{1015 + 1560}{11{,}970} = \frac{2575}{11{,}970} = 0.215 \text{ in}$$

A check of the eccentricity-thickness ratio yields

$$\frac{e}{t} = \frac{0.215}{7.5} = 0.029 < \frac{1}{3}$$

Because the transverse loading exceeds 10 lb/ft², e_1/e_2 is taken as unity. With $h/t = 12.8$, $C_s = 0.70$. From Table 11-5, with $e/t = 0.029$, $C_e = 1.0$. From Eq. (11-17), the required masonry strength for Case IV is

$$f'_m = \frac{12{,}555}{0.20 \times 90 \times 1.0 \times 0.70} \frac{1}{1.33} = 750 \text{ psi}$$

Case V (DL + RLL). No wind is included in this case. The eccentricity at the top of the wall is

$$e_{top} = \frac{1690}{11{,}970 + 2080} = \frac{1690}{14{,}050} = 0.120 \text{ in}$$

The eccentricity at the bottom of the wall is

$$e_{bot} = \frac{1015}{12{,}555 + 2055} = \frac{1015}{14{,}610} = 0.069 \text{ in}$$

A check of the eccentricity-thickness ratio yields

$$\frac{e}{t} = \frac{0.120}{7.5} = 0.016 < \frac{1}{3}$$

The eccentricity ratio, with the wall in double curvature, is

$$\frac{e_1}{e_2} = -\frac{0.069}{0.120} = -0.575$$

With $h/t = 12.8$, $C_s = 0.93$. From Table 11-5, with $e/t = 0.016$, $C_e = 1.0$. From Eq. (11-17), the required masonry strength for Case V is

$$f'_m = \frac{14{,}610}{0.20 \times 90 \times 1.0 \times 0.93} = 885 \text{ psi}$$

Shear Check. Assume $f'_m = 1010$ psi and Type N mortar, without inspection. The allowable stress (Art. 11-17) is

$$v_m = 0.5\sqrt{f'_m} + \frac{f_{DL}}{5} = 0.5\sqrt{1010} + \frac{1}{5}\frac{370{,}400}{18.6 \times 144} = 44 > 28 \text{ psi}$$

Use 28 psi. The actual average shear stress over the wall web is

$$v_m = \frac{V}{A} = \frac{22{,}500}{16.6 \times 144} = 9.5 < 28 \text{ psi}$$

The 7.5-in-thick nonreinforced wall is satisfactory between the second and third floors if masonry with $f'_m \geq 1010$ psi is used.

11-21. Example—End Shear Wall, Concrete Plank Floors.

Design a shear wall for the same conditions as for the example in Art. 11-20, except that the floors are flat, precast-concrete plank with topping. All loadings remain the same for purposes of this example.

Sources of Moments. Assume 4-in bearing of the plank on the wall for dead loads (Fig. 11-21). The eccentricity of the dead-load reaction then is 2.42 in, as in Art. 11-20. For uniform live load w_{LL}, because of the restraint of the floor end imposed by the clamping action of the wall, $w_{LL}L^2/36$ is conservatively assumed as the live-load moment induced in the wall section above or below the floor. For the loads used in Art. 11-20, the moments in the wall are

Fig. 11-21 Bearing of concrete plank on end wall.

$M_{DL} = \frac{1}{2} \times 60 \times 14 \times 2.42 = 1{,}015$ in-lb/ft
$M_{LL} = \frac{1}{36} \times 40(28)^2 12 = 10{,}400$ in-lb/ft
Total moment $= M_{TL} = 11{,}415$ in-lb/ft

Case I (TL + Parallel Wind). This case is similar to Case I of Art. 11-20. The maximum eccentricity at the second floor is

$$e_t = \frac{11{,}415}{16{,}925} = 0.673 \text{ in}$$

As in Art. 11-20, the virtual eccentricity of the 900-ft-kip overturning moment is $e_L = 1.81$ ft. In accordance with Eq. (11-15) for biaxial bending,

$$\frac{e}{t} = \frac{0.673}{7.5} + \frac{1.81}{26.67} = 0.158 < \frac{1}{3}$$

The eccentricity of floor load at the top of the wall (third floor) is computed for the floor total-load moment.

$$e_{top} = \frac{11{,}415}{16{,}400} = 0.698 \text{ in}$$

The eccentricity of floor load at the bottom of the wall (second floor) for floor dead-load moment is

$$e_{bot} = \frac{1015}{16{,}925} = 0.060 \text{ in}$$

The eccentricity ratio, negative because of double curvature of the wall, then is

$$\frac{e_1}{e_2} = -\frac{0.060}{0.698} = -0.086$$

With $h/t = 12.8$, Table 11-6 gives $C_s = 0.88$. From Table 11-5, with $e/t = 0.158$, $C_e = 0.72$. From Eq. (11-17), the required masonry strength for Case I, with the one-third increase in stress allowed for combined gravity and wind loads, is

$$f'_m = \frac{16{,}925}{0.20 \times 90 \times 0.72 \times 0.88} \frac{1}{1.33} = 1115 \text{ psi}$$

Case II (DL + Parallel Wind). This case is the same as Case II in Art. 11-20, because there are no live-load moments.

Case III (TL + Perpendicular Wind). This case is similar to Case III, Art. 11-20.

$$e_{top} = \frac{11{,}415 + 1560}{16{,}400} = \frac{12{,}975}{16{,}400} = 0.790 \text{ in}$$

A check of the eccentricity-thickness ratio yields

$$\frac{e}{t} = \frac{0.790}{7.5} = 0.105 < \frac{1}{3}$$

Because the transverse loading exceeds 10 lb/ft^2, e_1/e_2 is taken as unity. With $h/t = 12.8$, $C_s = 0.70$. From Table 11-5, with $e/t = 0.105$, $C_e = 0.80$. From Eq. (11-17), the required masonry strength for Case III is

$$f'_m = \frac{16{,}925}{0.20 \times 90 \times 0.80 \times 0.70} \cdot \frac{1}{1.33} = 1270 \text{ psi}$$

Case IV (DL + Perpendicular Wind). This case is the same as Case IV, Art. 11-20, because there are no live-load moments.

Case V (DL + RLL). No wind is included in this case. The eccentricity at the top of the wall is

$$e_{top} = \frac{11{,}415}{14{,}050} = 0.815 \text{ in}$$

The eccentricity at the bottom of the wall is

$$e_{bot} = \frac{1015}{14{,}610} = 0.069 \text{ in}$$

A check of the eccentricity-thickness ratio yields

$$\frac{e}{t} = \frac{0.815}{7.5} = 0.119 < \frac{1}{3}$$

The eccentricity ratio, negative with the wall in double curvature, is

$$\frac{e_1}{e_2} = -\frac{0.069}{0.815} = -0.085$$

With $h/t = 12.8$, $C_s = 0.88$. From Table 11-5, with $e/t = 0.119$, $C_e = 0.79$. From Eq. (11-17), the required masonry strength for Case V is

$$f'_m = \frac{14{,}610}{0.20 \times 90 \times 0.79 \times 0.88} = 1170 \text{ psi}$$

The nonreinforced wall is satisfactory between the second and third floors if masonry with $f'_m \geq 1270$ psi is used.

11-22. Example—Interior Shear Wall, Concrete Joist Floors.

Design a nonreinforced, load-bearing, interior shear wall between the second and third floors of a 10-story building, loaded on both sides of the wall, given the same data as in the example in Art. 11-20, except with the changes shown in Fig. 11-22 and the following.

Parallel wind moment = 1500 ft-kips
Parallel shear = 37 kips

For an interior wall, there is no perpendicular wind load.

Fig. 11-22 Properties of interior shear wall of 10-story building.

Distribution of Axial Loads. The axial loads, as in the examples in Art. 11-20 and 11-21, are assumed to spread over the total cross-sectional area of the wall, which in this example is I-shaped.

Loads at the Third Floor. Gravity loading per linear foot of wall from floors or roof is contributed by half the floor or roof span on either side of the wall, lengths of 11.5 and 14.5 ft. Loading from floors above is shared with the flange walls, so that the shear wall carries $15.6/21.9 = 0.713$ of this loading.

The dead load just below the third floor, from eight stories of masonry, roof dead load, and dead load of eight floors, is

$$P_{DL} = 70.7 \times 74.4 + [(11.5 + 14.5)50 + 7 \times 26 \times 60]0.713 + 26 \times 60$$
$$= 15{,}550 \text{ lb/ft}$$

Similarly, the live load, without reduction, is

$$P_{LL} = (26 \times 30 + 7 \times 26 \times 40)0.713 + 26 \times 40 = 6800 \text{ lb/ft}$$

With the maximum permissible reduction, the wall need carry only 42% of the floor live load (Art. 11-20). The reduced live load is

$$P_{RLL} = (26 \times 30 + 0.42 \times 7 \times 26 \times 40)0.713 + 0.42 \times 26 \times 40$$
$$= 3180 \text{ lb/ft}$$

Loads at the Second Floor. The dead load just above the second floor, from nine stories of masonry, roof dead load, and dead load of eight floors, is

$$P_{DL} = 79.5 \times 74.4 + (26 \times 50 + 8 \times 26 \times 60)0.713 = 15{,}760 \text{ lb/ft}$$

Similarly, the live load, without reduction, is

$$P_{LL} = (26 \times 30 + 8 \times 26 \times 40)0.713 = 6500 \text{ lb/ft}$$

and the reduced live load is

$$P_{RLL} = (26 \times 30 + 0.42 \times 8 \times 26 \times 40)0.713 = 3050 \text{ lb/ft}$$

Example—Interior Shear Wall, Concrete Joist Floors

Summation of Load at the Second Floor. The total dead load just above the second floor is

$$P'_{DL} = 79.5 \times 120 \times 21.9 + 23.83 \times 26(30 + 8 \times 40) = 423{,}900 \text{ lb}$$

The total unreduced live load plus dead load is

$$P'_{TL} = 79.5 \times 120 \times 21.9 + 23.83 \times 26(80 + 8 \times 100) = 752{,}300 \text{ lb}$$

Source of Moments. To produce maximum total-load and reduced-live-load moments, assume checkerboard loading of floors and roof, with the live load occurring on the long span at the top of the wall (second floor) and on the short span at the bottom of the wall (third floor). This arrangement of loads will normally place the wall in single curvature, thus giving conservative (positive) values of e_1/e_2.

Moments from Floor Loads. Assume 4-in bearing of joists on the wall and a triangular stress distribution under each bearing (Fig. 11-23). The eccentricity of the joist reaction then is

Fig. 11-23 Bearing of concrete joist on interior wall.

$$e_f = \frac{7.5}{2} - \frac{4}{3} = 2.42 \text{ in}$$

The dead-load and total-load moments due to this eccentricity are distributed in equal proportions to the wall sections above and below the floor.

At the top of the wall:

$$M_{DL} = -\tfrac{1}{2} \times 60 \times 11.5 \times 2.42 = -835 \text{ in-lb/ft}$$
$$M_{TL} = \tfrac{1}{2} \times 100 \times 14.5 \times 2.42 = 1750 \text{ in-lb/ft}$$
$$\text{Moment in wall} = M = 915 \text{ in-lb/ft}$$

At the bottom of the wall:

$$M_{DL} = -\tfrac{1}{2} \times 60 \times 14.5 \times 2.42 = -1055 \text{ in-lb/ft}$$
$$M_{TL} = \tfrac{1}{2} \times 100 \times 11.5 \times 2.42 = 1390 \text{ in-lb/ft}$$
$$\text{Moment in wall} = M = 335 \text{ in-lb/ft}$$

Parallel Wind Moment. For the second floor, this was given initially for this example as $M' = 1500$ ft-kips.

Overturn Check. To determine if tensile stress develops in the shear wall from overturning moment, the wall should be considered to be carrying only dead loads in addition to the wind parallel to the wall. The location of the neutral axis of the wall is found relative to the narrow flange, by dividing the sum of the moments of the wall areas about that end by the wall area $A = 21.9$ ft^2.

$$x = \left[\frac{0.625(25)^2}{2} + 0.625 \times 6 \times 24.69 + \frac{0.46(5.5)^2}{2} \right] \frac{1}{21.9} = 13.40 \text{ ft}$$

The moment of inertia of the wall about this axis is $I = 1730$ ft^4.

The allowable eccentricity e_L, which is the maximum for which no tension occurs on the wall cross section, is given by I/cA, where c is the maximum distance of a wall end from the neutral axis.

$$e_L = \frac{I}{cA} = \frac{1730}{13.40 \times 21.9} = 5.90 \text{ ft}$$

The actual eccentricity is

$$e_L = \frac{M'}{P'_{DL}} = \frac{1,500,000}{423,900} = 3.55 < 5.90 \text{ ft}$$

Therefore, no tensile reinforcement is required.

Loading Conditions. Three cases are considered for checking eccentricities and stresses in the wall under wind and gravity loads:

Case I. Total load on floors and roof plus wind parallel to the wall.
Case II. Dead load on floors and roof plus wind parallel to the wall.
Case III. Dead load plus reduced live load.

Case I (TL + Parallel Wind). This case involves biaxial bending and is similar to Case I, Art. 11-20. The maximum eccentricity at the second floor is

$$e_t = \frac{915}{22,260} = 0.041 \text{ in}$$

The virtual eccentricity of the 1500-ft-kip overturning moment is

$$e_L = \frac{M'}{P'_{TL}} = \frac{1,500,000}{752,300} = 2.00 \text{ ft}$$

In accordance with Eq. (11-16) for biaxial bending,

$$\frac{e}{t} = \frac{0.041}{7.5} + \frac{2.00}{25.0} = 0.086 < \frac{1}{3}$$

The eccentricity of the floor load at the top of the wall (third floor) is computed for live load only on the long span.

$$e_{top} = \frac{915}{22,350} = 0.041 \text{ in}$$

The eccentricity of floor load at the bottom of the wall (second floor) for live load only on the short span is

$$e_{bot} = \frac{335}{22,260} = 0.015 \text{ in}$$

The eccentricity ratio, positive because of single curvature of the wall, is

$$\frac{e_1}{e_2} = \frac{0.015}{0.041} = 0.375$$

For use in determining the slenderness coefficient C_s, the slenderness ratio $h/t = 12.8$. From Eq. (11-5) or Table 11-6, $C_s = 0.81$. From Table 11-5, with $e/t = 0.086$, the eccentricity coefficient is $C_e = 0.87$. From Eq. (11-17), the required masonry strength for Case I, with the one-third increase in allowable stress for combined gravity and wind loads, is

$$f'_m = \frac{22,260}{0.20 \times 90 \times 0.87 \times 0.81} \frac{1}{1.33} = 1320 \text{ psi}$$

Case II (DL + Parallel Wind). This case is similar to Case I, but dead load is used instead of total load.

The eccentricity of the floor load at the bottom of the wall is

$$e_{bot} = \frac{1055 - 835}{15,760} = \frac{220}{15,760} = 0.014 \text{ in}$$

Example—Interior Shear Wall, Concrete Joist Floors   11-31

The virtual eccentricity of the 1500-ft-kip overturning moment is

$$e_L = \frac{M'}{P'_{DL}} = \frac{1,500,000}{423,900} = 3.55 \text{ ft}$$

A check with Eq. (11-16) yields

$$\frac{e}{t} = \frac{0.014}{7.5} + \frac{3.55}{25.0} = 0.144 < \frac{1}{3}$$

The eccentricity of the floor load at the top of the wall is

$$e_{top} = \frac{220}{15,550} = 0.014 \text{ in}$$

The eccentricity ratio, negative for the wall in double curvature, is

$$\frac{e_1}{e_2} = -\frac{0.014}{0.014} = -1$$

With $h/t = 12.8$, $C_s = 0.95$. From Table 11-5, with $e/t = 0.144$, $C_e = 0.80$. From Eq. (11-17), the required masonry strength for Case II is

$$f'_m = \frac{15,760}{0.20 \times 90 \times 0.80 \times 0.95} \frac{1}{1.33} = 860 \text{ psi}$$

Case III (DL + RLL). No wind is included in this case. The eccentricity at the top of the wall is, for live load only on the long span,

$$e_{top} = \frac{915}{15,550 + 3180} = \frac{915}{18,730} = 0.049 \text{ in}$$

The eccentricity at the bottom of the wall is, for live load only on the short span,

$$e_{bot} = \frac{335}{15,760 + 3050} = \frac{335}{18,810} = 0.018 \text{ in}$$

A check of the eccentricity-thickness ratio yields

$$\frac{e}{t} = \frac{0.049}{7.5} = 0.007 < \frac{1}{3}$$

The eccentricity ratio, with the wall in single curvature, is

$$\frac{e_1}{e_2} = \frac{0.018}{0.049} = 0.368$$

With $h/t = 12.8$, $C_s = 0.81$. From Table 11-5, with $e/t = 0.007$, $C_e = 1.0$. From Eq. (11-17), the required masonry strength for Case III is

$$f'_m = \frac{18,810}{0.20 \times 90 \times 1 \times 0.81} = 1300 \text{ psi}$$

Shear Check. Assume $f'_m = 1320$ psi and Type N mortar, without inspection. The allowable stress (Art. 11-17) is

$$v_m = 0.5\sqrt{f'_m} + \frac{f_{DL}}{5} = 0.5\sqrt{1320} + \frac{1}{5}\frac{423,900}{21.9 \times 144} = 46 > 28 \text{ psi}$$

Use 28 psi. The actual average shear stress over the wall web is

$$v_m = \frac{37,000}{15.6 \times 144} = 16.5 < 28 \text{ psi}$$

The nonreinforced wall is satisfactory between the second and third floors if masonry with $f'_m \geq 1320$ psi is used.

11-23. Example—Interior Shear Wall, Concrete Plank Floors.

Design a shear wall for the same conditions as given for the example in Art. 11-22, except that the floor is flat, precast-concrete plank with topping.

Moments from Floors. Assume conservatively that the wall above or below a floor receives a live-load moment from the floor of $w_{LL}L^2/36$, as in Art. 11-22. Because precast planks cannot lap each other, the bearing is assumed to be 3 in (Fig. 11-24), and the eccentricity of the plank reaction is

$$e_f = \frac{7.5}{2} - \frac{3}{3} = 2.75 \text{ in}$$

Fig. 11-24 Bearing of concrete planks on interior wall

With live load only on the long span, moments at the top of the wall are

$M_{DL} = -\frac{1}{2} \times 60 \times 11.5 \times 2.75 \qquad = -950$ in-lb/ft
$M_{TL} = \frac{1}{2} \times 60 \times 14.5 \times 2.75 + 40(29)^2 12/36$
$\qquad\qquad\qquad\qquad\qquad\qquad\qquad = 12{,}400$ in-lb/ft
Moment in wall $= M \qquad\qquad\qquad = 11{,}450$ in-lb/ft

With live load only on the short span, moments at the bottom of the wall are

$M_{DL} = -\frac{1}{2} \times 60 \times 14.5 \times 2.75 \qquad = -1200$ in-lb/ft
$M_{TL} = \frac{1}{2} \times 60 \times 11.5 \times 2.75 + 40(23)^2 12/36 = 8000$ in-lb/ft
Moment in wall $= M \qquad\qquad\qquad\qquad = 6800$ in-lb/ft

Case I (TL + Parallel Wind). This case is similar to Case I, Art. 11-22. The maximum eccentricity at the second floor is

$$e_t = \frac{11{,}450}{22{,}260} = 0.514 \text{ in}$$

As in Art. 11-22, the virtual eccentricity of the 1500-ft-kip overturning moment is $e_L = 2.00$ ft. In accordance with Eq. (11-16) for biaxial bending,

$$\frac{e}{t} = \frac{0.514}{7.5} + \frac{2.00}{25.0} = 0.148 < \frac{1}{3}$$

The eccentricity of floor load at the top of the wall is computed for live load only on the long span.

$$e_{top} = \frac{11{,}450}{22{,}350} = 0.512 \text{ in}$$

The eccentricity of floor load at the bottom of the wall for live load only on the short span is

$$e_{bot} = \frac{6800}{22{,}260} = 0.302 \text{ in}$$

The eccentricity ratio, positive because of single curvature, is

$$\frac{e_1}{e_2} = \frac{0.302}{0.512} = 0.590$$

With $h/t = 12.8$, $C_s = 0.77$. From Table 11-5, with $e/t = 0.148$, $C_e = 0.71$. From Eq. (11-17), the required masonry strength for Case I, with the one-third increase in stress allowed for combined gravity and wind loads, is

$$f'_m = \frac{22{,}260}{0.20 \times 90 \times 0.71 \times 0.77} \frac{1}{1.33} = 1700 \text{ psi}$$

Case II (DL + Parallel Wind). This case is the same as Case II, Art. 11-22, because there are no live-load moments.

Case III (DL + RLL). No wind is included in this case. The eccentricity at the top of the wall is

$$e_{\text{top}} = \frac{11{,}450}{18{,}730} = 0.613 \text{ in}$$

The eccentricity at the bottom of the wall is

$$e_{\text{bot}} = \frac{6{,}800}{18{,}810} = 0.360 \text{ in}$$

A check of the eccentricity-thickness ratio yields

$$\frac{e}{t} = \frac{0.613}{7.5} = 0.082 < \frac{1}{3}$$

The eccentricity ratio, positive with the wall in single curvature, is

$$\frac{e_1}{e_2} = \frac{0.360}{0.613} = 0.592$$

With $h/t = 12.8$, $C_s = 0.77$. From Table 11-5, with $e/t = 0.082$, $C_e = 0.87$. From Eq. (11-17), the required masonry strength for Case III is

$$f'_m = \frac{18{,}810}{0.20 \times 90 \times 0.87 \times 0.77} = 1560 \text{ psi}$$

The nonreinforced wall is satisfactory between the second and third floors if masonry with $f'_m \geq 1700$ psi is used.

11-24. Bibliography.

The following publications are available from the Brick Institute of America, McLean, Va.:

J. G. Gross, R. D. Dikkers, and J. C. Grogan, "Recommended Practice for Engineered Brick Masonry."
H. C. Plummer, "Brick and Tile Engineering."
"Building Code Requirements for Engineered Brick Masonry."
"Technical Notes on Brick and Tile Construction"—a series.

Section 12

Walls, Partitions, and Interior Finishes

Frederick S. Merritt
Consulting Engineer,
West Palm Beach, Fla.

This section discusses design and construction of systems generally used for enclosing buildings and the spaces within them. (Some such systems, such as roofs, foundations, doors, and windows, however, are treated in other sections, because of their special functions in addition to enclosure of spaces.) The subsystems covered in this section, as described in Art.1-7, include exterior walls; interior walls, or partitions; floors; and ceilings.

Each of these systems usually consists of one or more facing subsystems and a structural subsystem that supports them. The facing subsystems may be the surfaces of the structural subsystem or separate entities that enclose that subsystem. They serve esthetic purposes, provide privacy, and bar, or at least restrict, passage of people or other moving objects, water, air, sound, heat, and also often light.

Wood structural subsystems are discussed in Sec. 7, and concrete is discussed in Sec. 10. Basic principles of waterproofing building exteriors are presented in Art. 3-13. This section describes techniques applicable to unit masonry and curtain walls.

Floors not only provide a horizontal separation of interior building spaces but also a surface on which human activities can take place and on which materials and equipment can be stored. The structural subsystem usually consists of a slab or deck and also often of beams that support it. These are described in Secs. 7 through 10. This section discusses constructions used for the upper facing, or floor coverings, which serve esthetic purposes and act as a wearing surface. The bottom facing, or ceiling, may be the bottom surface of the slab or deck or a separate entity, such as a gypsum-plaster membrane, which is also discussed in this section, or acoustical tile, which is treated in Sec. 17.

MASONRY WALLS

Masonry comprises assemblages of nonmetallic, incombustible materials, such as stone, brick, structural clay tile, concrete block, glass block, gypsum block, or adobe brick. Unit masonry consists of pieces of such materials, usually between 4 and 24 in in length and height and between 4 and 12 in in thickness. The units are bonded together with mortar or other cementitious materials.

Walls and partitions are classified as load bearing and non-load bearing. Different design criteria are applied to the two types.

Minimum requirements for both types of masonry walls are given in ANSI Standard Building Code Requirements for Masonry, A41.1 and ANSI Standard Building Code Requirements for Reinforced Masonry, A41.2, American National Standards Institute; Building Code Requirements for Engineered Brick Masonry, Brick Institute of America, and ACI Standard Building Code Requirements for Concrete Masonry Structures, ACI 531, American Concrete Institute.

Like other structural materials, masonry may be designed by application of engineering principles (Sec. 11). In the absence of such design, the empirical rules given in this section and adopted by building codes should be used.

12-1. Masonry Definitions.

Following are some of the terms most commonly encountered in masonry construction:

Architectural Terra Cotta. (See Ceramic Veneer.)
Ashlar Masonry. Masonry composed of rectangular units usually larger in size than brick and properly bonded, having sawed, dressed, or squared beds. It is laid in mortar.
Bonder. (See Header.)
Brick. A rectangular masonry building unit, not less than 75% solid, made from burned clay, shale, or a mixture of these materials.
Buttress. A bonded masonry column built as an integral part of a wall and decreasing in thickness from base to top, though never thinner than the wall. It is used to provide lateral stability to the wall.
Ceramic Veneer. Hard-burned, non-load-bearing, clay building units, glazed or unglazed, plain or ornamental.
Chase. A continuous recess in a wall to receive pipes, ducts, conduits.
Column. A compression member with width not exceeding 4 times the thickness, and with height more than 3 times the least lateral dimension.
Concrete Block. A machine-formed masonry building unit composed of portland cement, aggregates, and water.
Coping. A cap or finish on top of a wall, pier, chimney, or pilaster to prevent penetration of water to masonry below.
Corbel. Successive course of masonry projecting from the face of a wall to increase its thickness or to form a shelf or ledge (Fig. 12-3f).
Course. A continuous horizontal layer of masonry units bonded together (Fig. 12-3).
Cross-Sectional Area. Net cross-sectional area of a masonry unit is the gross cross-sectional area minus the area of cores or cellular spaces. Gross cross-sectional area of scored units is determined to the outside of the scoring, but the cross-sectional area of the grooves is not deducted to obtain the net area.
Grout. A mixture of cementitious material, fine aggregate, and sufficient water to produce pouring consistency without segregation of the constituents.
Grouted Masonry. Masonry in which the interior joints are filled by pouring grout into them as the work progresses.
Header (Bonder). A brick or other masonry unit laid flat across a wall with end surface exposed, to bond two wythes (Fig 12-1b).
Height of Wall. Vertical distance from top of wall to foundation wall or other intermediate support.

Masonry Definitions 12-3

Fig. 12-1 Brick exterior walls: (a) brick veneer with metal-stud framing; (b) masonry-bonded hollow wall; (c) hollow wall with metal ties between wythes; (d) insulated cavity wall.

Hollow Masonry Unit. Masonry with net cross-sectional area in any plane parallel to the bearing surface less than 75% of its gross cross-sectional area measured in the same plane.
Masonry. A built-up construction or combination of masonry units bonded together with mortar or other cementitious material.
Mortar. A plastic mixture of cementitious materials, fine aggregates, and water.
Partition. An interior non-bearing wall one story or less in height.
Pier. An isolated column of masonry. A bearing wall not bonded at the sides into associated masonry is considered a pier when its horizontal dimension measured at right angles to the thickness does not exceed 4 times its thickness.
Pilaster. A bonded or keyed column of masonry built as part of a wall, but thicker than the wall, and of uniform thickness throughout its height. It serves as a vertical beam, column, or both.
Rubble:
Coursed Rubble. Masonry composed of roughly shaped stones fitting approximately on

level beds, well bonded, and brought at vertical intervals to continuous level beds or courses.

Random Rubble. Masonry composed of roughly shaped stones, well bonded and brought at irregular vertical intervals to discontinuous but approximately level beds or courses.

Rough or Ordinary Rubble. Masonry composed of nonshaped field stones laid without regularity of coursing, but well bonded.

Solid Masonry Unit. A masonry unit with net cross-sectional area in every plane parallel to the bearing surface 75% or more of its gross cross-sectional area measured in the same plane.

Stretcher. A masonry unit laid with length horizontal and parallel with the wall face (Fig. 12-3).

Veneer. A wythe securely attached to a wall but not considered as sharing load or adding strength to it (Fig. 12-1a).

Wall. Vertical or near-vertical construction, with length exceeding three times the thickness, for enclosing space or retaining earth or stored materials.

Bearing Wall. A wall that supports any vertical load in addition to its own weight.

Cavity Wall. (See *Hollow Wall* below.)

Curtain Wall. A non-load-bearing exterior wall.

Faced Wall. A wall in which the masonry facing and the backing are of different materials and are so bonded as to exert a common reaction under load.

Hollow Wall. A wall of masonry so arranged as to provide an air space within the wall between the inner and outer wythes (Fig. 12-1b, c, and d). A cavity wall is built of masonry units or plain concrete, or of a combination of these materials, so arranged as to provide an airspace within the wall, which may be filled with insulation, and in which inner and outer wythes are tied together with metal ties (Fig. 12-1d).

Nonbearing Wall. A wall that supports no vertical load other than its own weight.

Party Wall. A wall on an interior lot line used or adapted for joint service between two buildings.

Shear Wall. A wall that resists horizontal forces applied in the plane of the wall.

Spandrel Wall. An exterior curtain wall at the level of the outside floor beams in multistory buildings. It may extend from the head of the window below the floor to the sill of the window above.

Veneered Wall. A wall having a facing of masonry or other material securely attached to a backing, but not so bonded as to exert a common reaction under load (Fig. 12-1a).

Wythe. Each continuous vertical section of a wall one masonry unit in thickness (Fig. 12-1).

12-2 Quality of Materials for Masonry.

Materials used in masonry construction should be capable of meeting the requirements of the applicable standard of the American Society for Testing and Materials.

Second-hand materials should be used only with extreme caution. Much salvaged brick, for example, comes from demolition of old buildings constructed of solid brick in which hard-burned units were used on the exterior and salmon units as backup. Because the color differences that guided the original masons in sorting and selecting bricks become obscured with exposure and contact with mortar, there is a definite danger that the salmon bricks may be used for exterior exposure and may disintegrate rapidly. Masonry units salvaged from chimneys are not recommended because they may be impregnated with oils or tarry material.

12-3. Mortar.

For unit masonry, mortar should meet the requirements of the American Society for Testing and Materials Specifications C270 and C476. These define the types of mortar shown in Table 12-1. Each type is used for a specific purpose, as indicated in Table 12-2, based on compressive strength. However, it should not be assumed that higher-strength mortars are preferable to lower-strength mortars where lower strength is permitted for particular uses. The primary purpose of mortar is to bond masonry units together.

TABLE 12-1 Types of Mortar

Mortar type	Parts by volume				Min avg compressive strength of three 2-in cubes at 28 days, psi
	Portland cement	Masonry cement	Hydrated lime or lime putty	Aggregate measured in damp, loose condition	
M	1	1			2500
	1		1/4		
S	1/2	1			1800
	1		Over 1/4 to 1/2	Not less than 2 1/4 and not more than 3 times the sum of the volumes of the cements and limes used	
N		1			750
	1		Over 1/2 to 1 1/4		
O		1			350
	1		Over 1 1/4 to 2 1/2		
K	1		Over 2 1/2 to 4		75
PL	1		1/4 to 1/2		2500
PM	1	1			2500

TABLE 12-2 Mortar Requirements of Masonry

Kind of Masonry	Types of Mortar
Masonry in contact with earth:	
Footings	M or S
Walls of solid units	M, S, or N
Walls of hollow units	M or S
Hollow walls	M or S
Masonry above grade or interior:	
Piers of solid masonry	M, S, or N
Piers of hollow units	M or S
Walls of solid masonry	M, S, N, or O
Grouted masonry	PL or PM
Walls of hollow units; load-bearing or exterior, and hollow walls 12 in. or more in thickness	M, S, or N
Hollow walls less than 12 in. in thickness where assumed design wind pressure:	
a. Exceeds 20 psf	M or S
b. Does not exceed 20 psf	M, S, or N
Glass-block masonry	M, S, or N
Nonbearing partitions of fireproofing composed of structural clay tile or concrete masonry units	M, S, N, O, or gypsum
Gypsum partition tile or block	Gypsum
Firebrick	Refractory air-setting mortar
Linings of existing masonry	M or S
Masonry other than above	M, S, or N

Mortars containing lime are generally preferred because of greater workability. Commonly used:

For concrete block, 1 part cement, 1 part lime putty, 5 to 6 parts sand.
For rubble, 1 part cement, 1 to 2 parts lime hydrate or putty, 5 to 7 parts sand.
For brick, 1 part cement, 1 part lime, 6 parts sand.
For setting tile, 1 part cement, ½ part lime, 3 parts sand.

12-4. Allowable Stresses in Masonry.

In determining stresses in masonry, effects of loads should be computed on actual dimensions, not nominal. Except for engineered masonry, the stresses should not exceed the allowable stresses given in ANSI Standard Building Code Requirements for Masonry (A41.1), which are summarized for convenience in Table 12-3.

This standard recommends also that, in composite walls or other structural members composed of different kinds or grades of units or mortars the maximum stress should not exceed the allowable stress for the weakest of the combination of units and mortars of which the member is composed.

12-5. Construction of Masonry.

Compressive strength of masonry depends to a great extent on workmanship and the completeness with which units are bedded. Tensile strength is a function of the adhesion of mortar to a unit and of the area of bonding (degree of completeness with which joints are filled). Hence, in specifying masonry work, it is important to call for a full bed of mortar, with each course well hammered down, and all joints completely filled with mortar. To minimize the entrance of water through a masonry wall, follow the practices recommended in Art. 3-13.

In particular, in filling head joints, apply a heavy buttering of mortar on one end of the masonry, press the unit down into the bed joint, and push the unit into place so that the mortar squeezes out from the top and sides of the head joint. Mortar should correspondingly cover the entire side of a unit before it is placed as a header. An attempt to fill head joints by slushing or dashing will not succeed in producing watertight joints. Partial filling of joints by "buttering" or "spotting" the vertical edge of the unit with mortar cut from the extruded bed joint is likewise ineffective and should be prohibited. Where closures are required, the opening should be filled with mortar so that insertion of the closure will extrude mortar both laterally and vertically.

Mortar joints usually range from ¼ to ⅜ in in thickness.

Tooling of joints, if done properly, can help to resist penetration of water; but it is not a substitute for complete filling, or a remedy for incomplete filling of joints. A concave joint (Fig. 12-2) is recommended. Use of raked or other joints that provide horizontal water tables should be avoided. Mortar should not be too stiff at time of tooling, or compaction will not take place, nor should it be too fluid, or the units may move—and units should never be moved after initial contact with mortar. If a unit is out of line, it should be removed, mortar scraped off and fresh mortar applied before the unit is relaid.

Back-plaster, or parge, the back face of exterior wythes before backup units are laid. If the backup is laid first, the front of the backup should be parged. The mortar should be the same as that used for laying the masonry and should be applied from ¼ to ⅜ in thick.

The rate of absorption of water by unit masonry at the time of laying is important in determining the strength and resistance to penetration of water of a mortared joint. This rate can be reduced by wetting the unit before laying. Medium absorptive units may

TABLE 12-3 Allowable Stresses in Unit Masonry

	Allowable compressive stress on gross cross section, psi*				
Construction and grade of unit	Type M mortar	Type S mortar	Type N mortar	Type O mortar	Type PL or PM mortar
Solid masonry of brick or other solid units:					
8000+ psi	400	350	300	200	
4500 to 8000 psi	250	225	200	150	
2500 to 4500 psi	175	160	140	100	
1200 to 2500 psi	125	115	100	75	
Grouted solid masonry of brick or other solid units:					
8000+ psi					500
4500 to 8000 psi					350
2500 to 4500 psi					275
1200 to 2500 psi					225
Solid masonry of through-the-wall solid units:					
8000+ psi	500	400	300		
4500 to 8000 psi	350	275	200		
2500 to 4500 psi	275	200	150		
1200 to 2500 psi	225	175	125		
Masonry of hollow units:					
1000+ psi	100	90	85		
700 to 1000 psi	85	75	70		
Piers of hollow units, cellular spaces filled	105	95	90		
Hollow walls (cavity or masonry bonded):†					
Solid units:					
4500+ psi	180	170	140		
2500 to 4500 psi	140	130	110		
1200 to 2500 psi	100	90	80		
Hollow units:					
1000+ psi	80	70	65		
700 to 1000 psi	70	60	55		
Stone ashlar masonry:					
Granite	800	720	640	500	
Limestone or marble	500	450	400	325	
Sandstone or cast stone	400	360	320	250	
Rubble, coursed, rough or random	140	120	100	80	

*Allowable bearing stress directly under concentrated loads may be taken 50% larger than the tabulated values.

†On gross cross section of wall minus area of cavity between wythes. The allowable compressive stresses for cavity walls are based on the assumption that floor loads bear on only one of the two wythes. Increase stresses 25% for hollow walls loaded concentrically.

CONCAVE   WEATHERED   STRUCK   FLUSH   RAKED   EXTRUDED

Fig. 12-2 Types of masonry joints.

12-8 Walls, Partitions, and Interior Finishes

need to be thoroughly soaked with water. Highly absorptive units may require total immersion in water for some time before unit "suction" is reduced to the low limits needed.

The test for the rate of absorption of brick is described in ASTM Standard C67. In this absorption test, the brick, in a flat position, is partly immersed in water to a depth of ⅛ in for 1 min, and the absorption is calculated for a net equivalent area of 30 in^2. For water-resistant masonry, the suction (rate of absorption) of the brick should not exceed 0.35, 0.5, and 0.7 oz, respectively, for properly constructed all-brick walls or facings of normal 4-, 8-, and 12-in thickness.

The amount of wetting that the brick will require to control rate of absorption properly when laid should be known or determined by measurements made before the brick are used in the wall. Some medium absorptive brick may require only frequent wetting in the pile; others may need to be totally immersed for an hour or more. While the immersion method is more costly than hosing in the pile, it ensures that all brick are more or less saturated when removed from immersion. A short time interval may be needed before the brick are laid; but the brick are likely to remain on the scaffold, in a suitable condition to lay, for some time. Brick on the scaffold should be inspected and moisture condition checked several times a day.

In general, method of manufacture and surface texture of masonry do not greatly affect the permeability of walls. However, water-resistant joints may be difficult to obtain if the units are deeply scored, particularly if the mortar is of a dry consistency. Loose sand should be brushed away or otherwise removed from units that are heavily sanded.

Mortar to be used in above-grade, water-resistant brick-faced and all-brick walls should be of as wet a consistency as can be handled by the mason and meet requirements of ASTM Standard C270, Type N. Water retention of the mortar should not be less than 75%, and preferably 80% or more. For laying absorptive brick that contain a considerable amount of absorbed water, the mortars having a water retention of 80% or more may be used without excessive "bleeding" at the joints and "floating" of the brick. The mortar may contain a masonry cement meeting the requirements of ASTM Standard C91, except that water retention should not be less than 75 or 80%. Excellent mortar may also be made with portland cement and hydrated lime, mixed in the proportion of 1:1:6 parts by volume of cement, lime, and loose damp sand. The hydrated lime should be highly plastic. Type S lime conforming with the requirements of ASTM Standard C207 is highly plastic, and mortar containing it, in equal parts by volume with cement, will probably have a water retention of 80% or more.

Since capillary penetration of moisture through concrete and mortar is of minor importance, particularly in above-grade walls, the mortar need not contain an integral water repellent. However, if desired, water-repellent mortar may be advantageously used in a few courses at the grade line to reduce capillary rise of moisture from the ground into the masonry. The mortar should be of a type that does not stiffen rapidly on the board, except through loss of moisture by evaporation.

Mortar should be retempered frequently if necessary to maintain as wet a consistency as is practically possible for the mason to use. At air temperatures below 80°F, mortar should be used or discarded within 3½ hr after mixing; for air temperatures of 80°F or higher, unused mortar should be discarded after 2½ hr.

Cold-Weather Construction of Masonry Walls. Masonry should be protected against damage by freezing. The following special precautions should be taken:

Materials to be used should be kept dry. Tops of all walls not enclosed or sheltered should be covered whenever work stops. The protection should extend downward at least 2 ft.

Frozen materials must be thawed before use. Masonry units should be heated to at least 40°F. Mortar temperature should be between 40 and 120°F, and mortar should not be placed on a frozen surface. If necessary, protect the wall with heat and windbreaks for at least 48 hr. Use of mortars made with high-early-strength cement may be advantageous for cold-weather masonry construction.

12-6. Bond between Wythes in Masonry Walls.

When headers are used for bonding the facing and backing in solid masonry walls and faced walls, as shown in Fig. 12-3, not less than 4% of the wall surface of each face should be composed of headers, which should extend at least 4 in into the backing. These headers should not be more than 24 in apart vertically or horizontally (Fig. 12-3a and b). In walls in which a single bonder does not extend through the wall, headers from opposite sides should overlap at 4 in or should be covered with another bonder course overlapping headers below at least 4 in.

If metal ties (Figs. 12-3 e and 12-4) are used for bonding, they should be corrosion-resistant. For bonding facing and backing of solid masonry walls and faced walls, there should be at least one metal tie for each 4½ ft² of wall area. Ties in alternate courses

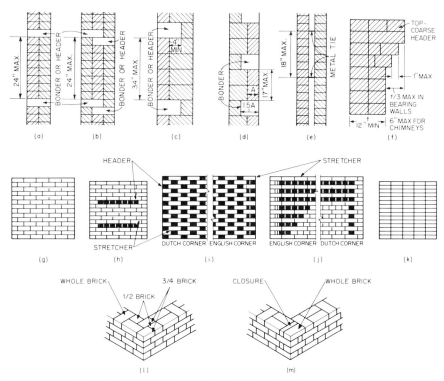

Fig. 12-3 Types of unit-masonry construction. Cross sections through walls show: (a) two-wythe solid wall with bonders; (b) three-wythe solid wall with bonders; (c) and (d) hollow-unit walls; (e) hollow or cavity wall; (f) corbeled wall. Elevations of walls show types of masonry courses: (g) running bond; (h) common, or header, bond with bonders every sixth course; (i) Flemish bond with bonders in every course; (j) English bond; (k) stack bond. Types of corner bond: (l) Dutch; (m) English.

12-10 Walls, Partitions, and Interior Finishes

Fig. 12-4 Types of metal ties for masonry walls: (*a*) rectangular; (*b*) ladder; (*c*) Z; (*d*) truss; (*e*) U.

should be staggered, the maximum vertical distance between ties should not exceed 18 in, and the maximum horizontal distance should not be more than 36 in.

In walls composed of two or more thicknesses of hollow units, stretcher courses should be bonded by one of the following methods: At vertical intervals up to 34 in, there should be a course lapping units below at least 4 in (Fig. 12-3*c*). Or at vertical intervals up to 17 in, lapping should be accomplished with units at least 50% thicker than the units below (Fig. 12-3*d*). Or at least one metal tie should be incorporated for each $4\frac{1}{2}$ ft^2 of wall area. Ties in alternate courses should be staggered; the maximum vertical distance between ties should be 18 in and maximum horizontal distance, 36 in. Full mortar coverage should be provided in both horizontal and vertical joints at ends and edges of face shells of the hollow units.

In ashlar masonry, bond stones should be uniformly distributed throughout the wall and form at least 10% of the area of exposed faces.

In rubble stone masonry up to 24 in thick, bond stones should have a maximum spacing of 3 ft vertically and horizontally. In thicker walls, there should be at least one bond stone for each 6 ft^2 of wall surface on both sides.

For bonding ashlar facing, the percentage of bond stones should be computed from the exposed face area of the wall. At least 10% of this area should be composed of uniformly distributed bond stones extending 4 in or more into the backup. Every bond stone and, when alternate courses are not full bond courses, every stone should be securely anchored to the backup with corrosion-resistant metal anchors. These should have a minimum cross section of $\frac{3}{16} \times 1$ in. There should be at least one anchor to a stone and at least two anchors for stones more than 2 ft long or with a face area of more than 3 ft^2. Larger facing stones should have at least one anchor per 4 ft^2 of face area of the stone, but not less than two anchors.

Cavity-wall wythes should be bonded with $\frac{3}{16}$-in-diameter steel rods or metal ties of equivalent stiffness embedded in horizontal joints. There should be at least one metal tie for each $4\frac{1}{2}$ ft^2 of wall area. Ties in alternate courses should be staggered, the maximum vertical distance between ties should not exceed 18 in (Fig 12-3*e*), and the maximum horizontal distance, 36 in. Rods bent to rectangular shape should be used with hollow masonry units laid with cells vertical. In other walls, the ends of ties should be bent to 90° angles to provide hooks at least 2 in long. Additional bonding ties should be provided at all openings. These ties should be spaced not more than 3 ft apart around the perimeter and within 12 in of the opening.

When two bearing walls intersect and the courses are built up together, the intersections should be bonded by laying in true bond at least 50% of the units at the intersection. When the courses are carried up separately, the intersecting walls should be regularly toothed or blocked with 8-in maximum offsets. The joints should be provided with metal anchors having a minimum section of $\frac{1}{4} \times 1\frac{1}{2}$ in with ends bent up at least 2 in or with cross pins to form an anchorage. Such anchors should be at least 2 ft long and spaced not more than 4 ft apart.

12-7. Grouted Masonry.

Construction of walls requiring two or more wythes of brick or solid concrete block, similar to the wall shown in Fig. 12-3a, may be speeded by pouring grout between the two outer wythes, to fill the interior joints. Building codes usually require that, for the wythes, the mortar be type M or S, consisting of portland cement, lime, and aggregate (Table 12-1). Also, they may require that, when laid, burned-clay brick and sand-lime units should have a rate of absorption of not more than 0.025 oz/in^2 over a 1-min period in the standard absorption test (Art. 12-5). All units in the two outer wythes should be laid with full head and bed joints.

Low-Lift Grouting. The vertical spaces between wythes, which are to be grouted, should be at least ¾ in wide. Masonry headers should not project into the gap. One of the outer wythes may be carried up 18 in before grout is poured. The other outer wythe is restricted to a height up to 6 times the grouting space, but not more than 8 in, before grout is poured. Thus, in this type of construction, grout is poured in lifts not exceeding 8 in. The grout should be puddled with a grout stick immediately after it has been poured. If work has to be stopped for an hour or more, horizontal construction joints should be formed by raising all wythes to the same level and leaving the grout 1 in below the top. A suitable grout for this type of construction consists of 1 part portland cement, 0.1 part hydrated lime or lime putty, and 2¼ to 3 parts sand.

High-Lift Grouting. This type of construction is often used where steel reinforcement is to be inserted in the vertical spaces between wythes; for example, in the cavity of the wall shown in Fig. 12-3e. Grout is poured continuously in lifts up to 6 ft high and up to 30 ft long in the vertical spaces. (Vertical barriers, or dams, of solid masonry may be built in the grout space to control the horizontal flow of grout.) Building codes may require each lift to be completed within one day. The grout should be consolidated by puddling or mechanical vibrating as it is placed and reconsolidated after excess moisture has been absorbed but before plasticity has been lost. A suitable grout for gaps 2 or more inches wide consists of 1 part portland cement, 0.1 part hydrated lime or lime putty, 2 to 3 parts sand, and not more than 2 parts gravel, by volume.

In construction of the wall, the wythes should be kept at about the same level. No wythe should lag behind the others more than 16 in in height. The masonry should be allowed to cure for at least 3 days, to gain strength, before grout is poured. The grout space should be at least 2 in wide. If, however, horizontal reinforcement is to be placed in the gap, it should be wide enough to provide ¼ in clearance around the steel, but not less than 3 in wide.

Cleanouts should be provided for every pour. This may be done by omitting every other unit in the bottom course of the wall section being poured. Before grout is placed, excess mortar, mortar fins, and other foreign matter should be removed from the grout space. A high-pressure water jet may be used for the purpose. After inspection but before placement of grout, the cleanout holes should be plugged with masonry units, which should then be braced to resist the grout pressure.

Wire ties should be inserted in the mortar joints between masonry courses and span across each grout space, to bond the wythes (Fig. 12-5). The ties should be formed into rectangles, 4 in wide and with a length 2 in less than the distance between outer faces of the wythes being bonded. The wire size should not be less than No. 9. Spacing of ties should not exceed 24 in horizontally. For running-bond masonry (Fig. 12-3f), vertical tie spacing should not exceed 16 in, and for stack-bond masonry (Fig. 12-3j), 12 in.

Fig. 12-5 Grouted masonry wall: (a) section at roof; (b) section at base of wall.

12-8. Flashing in Masonry Walls.

Flashing should be used to divert to the exterior of a building water that may penetrate or condense on the interior face of masonry walls. Accordingly, flashing should be installed in exterior walls at horizontal surfaces, such as roofs, parapets, and floors, depending on type of construction; at shelf angles; at openings, such as doors and windows (Fig. 12-6a and b); and at the bases of walls just above grade (Fig. 12-6c and e). The flashing should extend through a mortar joint to the outside face of the wall, where it should turn down to form a drip.

Flashing in tooled mortar joints, however, would trap water unless some means is provided to drain it to the outside. Consequently, flashing should be used in conjunction with weep holes, which should be formed in head joints immediately above the flashing (Fig. 12-6d). When the weep holes are left open, spacing should not exceed 24 in c to c. If wicks of glass-fiber or nylon rope, cotton sash cord, or similar materials are left in the holes, spacing should not exceed 16 in c to c.

Materials used for flashing include sheet copper, bituminous fabrics, plastics, or a combination of these. Copper may be selected for its durability, but cost may be greater than for other materials. Combinations of materials, such as cold-formed steel and plastic or bituminous coating, may yield a durable flashing at lower cost.

Fig. 12-6 Flashing in masonry walls: (*a*) over opening for window or door; (*b*) under window sill; (*c*) at base of wall; (*d*) and (*e*) below weep holes.

(J. H. Callender, "Time-Saver Standards for Architectural Design Data," McGraw-Hill Book Company, New York.)

12-9. Repair of Leaky Joints.

Leakage of wind-driven rain through the joints in permeable brick masonry walls can be stopped by either repointing or grouting the joints. It is usually advisable to treat all the joints, both vertical and horizontal, in the wall face. Some "tuck-pointing" operations in which only a few, obviously defective joints are treated may be inadequate and do not necessarily ensure that the untreated joints will not leak.

Repointing consists of cutting away and replacing the mortar from all joints to a depth of about ⅝ in. After the old mortar has been removed, the dust and dirt should be washed from the wall and the brick thoroughly wetted with water to near saturation. While the masonry is still very damp but with no water showing, the joints should be repointed with a suitable mortar. This mortar may have a somewhat stiff consistency to enable it to be tightly packed into place, and it may be "prehydrated" by standing for 1 or 2 hr before retempering and using. Prehydration is said to stabilize the plasticity and workability of the mortar and to reduce the shrinkage of the mortar after its application to the joints.

After repointing, the masonry should be kept in a damp condition for 2 or 3 days. If the brick are highly absorptive, they may contain a sufficient amount of water to aid materially in curing.

Weathering and permeability tests described in C. C. Fishburn, "Effect of Outdoor Exposure on the Water Permeability of Masonry Walls," *National Bureau of Standards BMS Report 76* indicate that repointing of the face joints in permeable brick masonry walls was the most effective and durable of all the remedial treatments against leakage that did not change the appearance of the masonry.

Joints are **grouted** by scrubbing a thin coating of a grout over the joints in the masonry. The grout may consist of equal parts by volume of portland cement and fine sand, the sand passing a No. 30 sieve.

The masonry should be thoroughly wetted and in a damp condition when the grout is applied. The grout should be of the consistency of a heavy cream and should be scrubbed into the joints with a stiff bristle brush, particularly into the juncture between brick and mortar. The apparent width of the joint is slightly increased by some staining of the brick with grout at the joint line. Excess grout may be removed from smooth-textured brick with a damp sponge, before the grout hardens; care should be taken not to remove grout from between the edges of the brick and the mortar joints. If the brick are rough-textured, staining may be controlled by the use of a template or by masking the brick with paper masking tape.

Bond of the grout to the joints is better for "cut" or flush joints than for tooled joints. If the joints have been tooled, they should preferably not be grouted until after sufficient weathering has occurred to remove the film of cementing materials from the joint surface, exposing the sand aggregate.

Grouting of the joints has been tried in the field and found to be effective on leaky brick walls. The treatment is not so durable and water-resistant as a repointing job but is much less expensive than repointing. Some tests of the water resistance of grouted joints in brick masonry test walls are described in *National Bureau of Standards BMS Report 76*.

The cost of either repointing or grouting the joints in brick masonry walls probably greatly exceeds the cost of the additional labor and supervision needed to make the walls water-resistant when built.

12-10. Lateral Support for Masonry Walls.

For unreinforced solid or grouted masonry bearing walls, the ratio of unsupported height to nominal thickness, or the ratio of unsupported length to nominal thickness, should not exceed 20. For hollow walls or walls of hollow masonry units, the ratio should be 18 or less. For cavity or stone walls, the ratio should not exceed 14. See ANSI Standard Building Code Requirements for Masonry, A41.1, American National Standards Institute.

In calculating the ratio of unsupported length to thickness for cavity walls, you can take the thickness as the sum of the nominal thickness of the inner and outer wythes. For walls composed of different kinds or classes of units or mortars, the ratio should not exceed that allowed for the weakest of the combinations. Veneers should not be considered part of the wall in computing thickness for strength or stability.

For nonbearing, unreinforced exterior walls, the thickness ratio should not exceed 20. For unreinforced partitions, the ratio should be 36 or less.

Cantilever walls and masonry walls in locations exposed to high winds should not be built higher than 10 times their thickness unless adequately braced or designed in accordance with engineering principles. Backfill should not be placed against foundation walls until they have been braced to withstand horizontal pressure.

In determining the unsupported length of walls, existing cross walls, piers, or buttresses may be considered as lateral supports, if these members are well bonded or anchored to the walls and capable of transmitting forces perpendicular to the plane of the wall to connected structural members or to the ground.

In determining the unsupported height of walls, floors and roofs may be considered as lateral supports, if they can resist a lateral force of at least 200 lb/lin ft and provision is made to transmit the lateral forces to the ground. Ends of floor joists or beams bearing on masonry walls should be securely fastened to the walls (Fig. 12-7). (See also Art. 12-13.) Interior ends of anchored joists should be lapped and spiked, or the equivalent, so as to form continuous ties across the building. When lateral support is to be provided by joists parallel to walls, anchors should be spaced no more than 6 ft apart and engage at least three joists, which should be bridged solidly at the anchors.

Unsupported height of piers should not exceed 10 times the least dimension. How-

Fig. 12-7 Anchorage of joists to bearing walls.

ever, when structural clay tile or hollow concrete units are used for isolated piers to support beams or girders, unsupported height should not exceed 4 times the least dimension unless the cellular spaces are filled solidly with concrete or either Type M or S mortar (see Table 12-1).

12-11. Masonry-Thickness Requirements.

Walls should not vary in thickness between lateral supports. When it is necessary to change thickness between floor levels to meet minimum-thickness requirements, the greater thickness should be carried up to the next floor level.

Where walls of masonry hollow units or bonded hollow walls are decreased in thickness, a course of solid masonry should be interposed between the wall below and the thinner wall above, or else special units or construction should be used to transmit the loads between the walls of different thickness.

The following limits on dimensions of masonry walls should be observed unless the walls are designed for reinforcement, by application of engineering principles (Sec. 11):

Bearing walls should be at least 12 in thick for the uppermost 35 ft of their height. Thickness should be increased 4 in for each successive 35 ft or fraction of this distance measured downward from the top of the wall. Rough or random or coursed rubble stone walls should be 4 in thicker than this, but in no case less than 16 in thick. However, for other than rubble stone walls, the following exceptions apply to masonry bearing walls:

Stiffened Walls. Where solid masonry bearing walls are stiffened at distances not greater than 12 ft by masonry cross walls or by reinforced-concrete floors, they may be made 12 in thick for the uppermost 70 ft but should be increased 4 in in thickness for each successive 70 ft or fraction of that distance.

Top-Story Walls. The top-story bearing wall of a building not over 35 ft high may be made 8 in thick. But this wall should be no more than 12 ft high and should not be subjected to lateral thrust from the roof construction.

Residential Walls. In dwellings up to three stories high, walls may be 8 in thick (if not more than 35 ft high), if not subjected to lateral thrust from the roof construction. Such walls in one-story houses and one-story private garages may be 6 in thick, if the height is 9 ft or less or if the height to the peak of a gable does not exceed 15 ft.

12-16 Walls, Partitions, and Interior Finishes

Penthouses and Roof Structures. Masonry walls up to 12 ft high above roof level, enclosing stairways, machinery rooms, shafts, or penthouses, may be made 8 in thick. They need not be included in determining the height for meeting thickness requirements for the wall below.

Plain Concrete and Grouted Brick Walls. Such walls may be 2 in less in thickness than the minimum basic requirements, but in general not less than 8 in—and not less than 6 in in one-story dwellings and garages.

Hollow Walls. Cavity or masonry bonded hollow walls should not be more than 35 ft high. In particular, 10-in cavity walls should be limited to 25 ft in height, above supports. The facing and backing of cavity walls should be at least 4 in thick, and the cavity should not be less than 2 in or more than 3 in wide.

Anchors for Masonry Facings. Support perpendicular to its plane may be provided an exterior masonry wythe, whether it is a veneer (non-load-bearing) or the outer wythe of a hollow wall, by anchoring it to construction capable of furnishing the required lateral support. Accordingly, a masonry veneer may be tied with masonry bonders or metal ties to a backup masonry wall that is given lateral support or the veneer may be anchored directly to structural framing. Methods of bonding wythes together are described in Art. 12-6. The following applies to anchorage of masonry walls to structural framing.

Several types of anchors are illustrated in Fig. 12-8. They should be corrosion resistant. Also, they should be able to resist tension and compression applied by forces acting perpendicular to the wall. Yet, the anchors should be flexible enough to permit, between walls and framing, small differential horizontal and vertical movements parallel to the plane of the wall. The anchors should be embedded at one end in the mortar of bed joints and extend almost to the face of the wall. The other end should be securely attached to framing providing lateral support. The type of anchor to use depends on the construction to which the wall is to be anchored.

Figure 12-8a shows a corrugated metal tie for attachment of masonry walls to wood studs. Such ties should be fastened to studs with corrosion-resistant nails that are driven through sheathing to penetrate at least 1½ in into the studs. The ties should have a thickness of at least 22 gage, width of ⅞ in, and length of 6 in.

Anchors shown in Fig. 12-8b and c may be used to attach masonry walls to metal studs. The wires of these anchors should be at least 9 gage. The anchor shown in Fig. 12-8d is suitable for tying masonry walls to structural steel framing, as illustrated in Fig. 12-9b.

Dovetail anchors for anchorage of masonry walls into dovetail slots in concrete framing are illustrated in Fig. 12-8e and f. Applications of the type shown in Fig. 12-8e are shown in Fig. 12-9c and d. Wires in these anchors should be at least 6 gage and should

Fig. 12-8 Anchors for use between masonry walls and structural framing: (*a*) corrugated metal for tie to wood studs; (*b*) and (*c*) wire ties for attachment to metal studs; (*d*) wire tie for anchorage to structural steel; (*e*) dovetail anchor for use with concrete; (*f*) corrugated metal for tie to cast-in-place concrete.

Fig. 12-9 Anchorage of walls to structural framing: (*a*) hollow wall ties to structural steel beam; (*b*) masonry veneer wall anchored to structural steel column; (*c*) masonry veneer wall attached to reinforced concrete beam; (*d*) hollow wall tied to concrete corner column.

be spread to a width of at least 4 in for embedment at least 2 in into bed joints in the wall. The flat-bar type (Fig. 12-8*f*) should have a minimum thickness of 16 gage and width of ⅞ in. The end to be embedded in a bed joint should be turned upward at least ¼ in.

Faced Walls. Neither the height of faced (composite) walls nor the distance between lateral supports should exceed that prescribed for masonry of either of the types forming the facing and the backing. Actual (not nominal) thickness of material used for facings should not be less than 2 in and in no case less than one-eighth the height of the unit.

Nonbearing Walls. In general, parapet walls should be at least 8 in thick and the height should not exceed 3 times the thickness. The thickness may be less than 8 in, however, if the parapet is reinforced to withstand safely earthquake and wind forces to which it may be subjected.

Nonbearing exterior masonry walls may be 4 in less in thickness than the minimum for bearing walls. However, the thickness should not be less than 8 in except that where 6-in bearing walls are permitted, 6-in nonbearing walls can be used also.

Nonbearing masonry partitions should be supported laterally at distances of not more than 36 times the actual thickness of the partition, including plaster. If lateral support

12-18   Walls, Partitions, and Interior Finishes

depends on a ceiling, floor, or roof, the top of the partition should have adequate anchorage to transmit the forces. This anchorage may be accomplished with metal anchors or by keying the top of the partition to overhead work. Suspended ceilings may be considered as lateral support if ceilings and anchorages are capable of resisting a horizontal force of 200 lb/lin ft of wall.

12-12. Corbeling.

Where a solid masonry wall 12 in or more thick must be increased in thickness above a specific level, the increase should be achieved gradually by corbeling. In this method, successive courses are projected from the face of the wall, as indicated in Fig. 12-3f.

The maximum corbeled horizontal projection beyond the face of a wall should not exceed one-third the wall thickness for walls supporting structural members. In any case, projection of any course of masonry should not exceed 1 in.

Chimneys generally may not be corbeled more than 6 in from the face of the wall. In the second story of two-story dwellings, however, corbeling of chimneys on the exterior of enclosing walls may equal the wall thickness.

12-13. Provisions for Dimensional Changes after Construction.

In design and construction of masonry walls, allowance should be made for relative movements of the masonry and contiguous construction. If this is not done, unsightly or troublesome cracking or even structural failure may result. In the past, such damage has occurred in masonry walls because of:

1. Restraint offered by contiguous construction to dimensional changes in the masonry. Such changes may be produced by temperature changes or by absorption of water by the masonry after construction.

2. Restraint offered by the masonry to movements of or dimensional changes in contiguous or bonded construction, such as concrete frames or backup walls. Such changes may be produced by drying shrinkage, elastic deformations under load, or creep of the concrete after construction.

To avoid such restraints, it is necessary to install in walls expansion joints with proper gaps, at appropriate intervals, and to break bond between the walls and construction that would restrain relative movements (Fig. 12-10).

Vertical expansion joints should be installed in masonry walls to permit horizontal movements of the masonry, and horizontal expansion joints, to permit vertical movements. In the absence of specific information on thermal and water-absorption properties, the unit strain may be assumed to be 0.0007 in/in in a brick wall when movement is restricted, for example, by bond to a concrete foundation. Thus, for 60-ft spacing of expansion joints in a straight brick wall, a joint width of $2 \times 60 \times 12 \times 0.0007$, or about 1 in, would be required. In general, spacing of vertical expansion joints should range between 50 and 100 ft, and a joint should be placed not more than 30 ft from wall intersections.

The width required for an expansion joint also depends on the maximum allowable strain of the sealant used to seal the gap. If the size of a joint is controlled by the elastic properties of the sealant, joint spacing should be adjusted to limit the joint size to accommodate the elasticity of the sealant. The sealant should be placed at the exterior wall face and inserted in the joint to a depth of at least ⅛ in but not deeper than one-half the joint width. This depth may be controlled by a backup material that is inserted in the joint before the sealant is applied and that will not adhere to the sealant.

Provisions for Dimensional Changes after Construction   12-19

Fig. 12-10 Cross sections show expansion joints in masonry walls: (a) types of fillers used; (b) expansion joint in brick veneer with control joint in concrete-block backup; (c) joint in hollow wall; (d) joint at anchorage to column; (e) joint at T intersection of masonry wall; (f) joints at offset in hollow wall; (g) joint below shelf angle.

In brick facades, horizontal expansion joints should be inserted directly under horizontal lintels that are supported on concrete frames (Fig. 12-10g). The joints should be sized for probable expansion of the masonry below the lintels plus probable shortening of the concrete frame produced by drying shrinkage, compressive loading, and creep. In the absence of specific information on the properties of the materials to be used, a relative vertical movement of 0.0014 in/in may be assumed. Thus, where a brick facade is supported on steel shelf angles, for example, spaced 15 ft apart vertically, a gap of 15 × 12 × 0.0014, or about ¼ in, would be required. As for vertical joints, a sealant and backup material should seal the horizontal expansion joint.

Slip joints should be provided where abrupt changes in wall dimensions occur; for example, at panels bounding or included between openings in a masonry wall, such as those for windows and doors. Bond should be prevented by insertion of sheet metal, building paper, or other material that would permit sliding when thermal movements occur.

Similarly, to permit relative horizontal movements, slip planes should be provided between cast-in-place concrete floors or roofs and masonry bearing walls that support them (Fig. 12-5a). Flexible anchors that permit sliding may be installed between the slabs and walls to prevent uplift. Such anchors, however, should not be installed within a distance from a slab corner of one-tenth the slab length. The reason for this is that such corners tend to curl upward when shrinkage occurs, in which case the anchorages would apply tension to and crack the walls.

Particular care should be taken to provide for relative movements when dissimilar materials are combined in a wall. Preferably, they should be separated at least ½ in and joined with flexible ties (Fig. 12-10b and d). In particular, because of different thermal movements of the materials, bond should be prevented between brick walls and contiguous concrete foundation walls below that are exposed to the weather (Fig. 12-5b). Flexible anchors should be provided between the walls and foundations, to permit horizontal sliding yet prevent uplift. Also, foundations should be made sufficiently stiff to prevent deflections that would crack the walls above. In addition, for the same purpose, the walls may be designed as deep beams or Vierendeel masonry trusses. (A Vierendeel truss does not consist solely of triangular configurations of members as do conventional trusses, and thus the members are subjected to a combination of axial forces and bending moments. Such a truss would be formed by a wall with openings for doors and windows.)

When bearing-wall construction is used for a building, differential movements of adjacent supports of horizontal structural members should be kept very small. For this reason, bearing walls of dissimilar materials should not be used in the same structure, inasmuch as they are likely to have physical properties that cause unequal deformations. For example, either load-bearing brick walls or concrete walls, but not both, should be used in a structure.

12-14. Glass Block.

For control of light that enters a building and for better insulation than obtained with ordinary glass panes, masonry walls of glass block are frequently used (Fig. 12-11). These units are hollow, 3⅞ in thick by 6 in square, 8 in square, or 12 in square (actual length and height ¼ in less, for modular coordination, to allow for mortar joints). Faces of the units may be cut into prisms to throw light upward or the block may be treated to diffuse light.

Glass blocks may be used as nonbearing walls and to fill openings in walls. The glass block so used should have a minimum thickness of 3 in at the mortar joint. Also, surfaces of the block should be satisfactorily treated for mortar bonding.

For exterior walls, glass-block panels should not have an unsupported area of more than 144 ft^2. They should be no more than 15 ft long or high between supports.

For interior walls, glass-block panels should not have an unsupported area of more than 250 ft^2. Neither length nor height should exceed 25 ft.

Exterior panels should be held in place in the wall opening to resist both internal and external wind pressures. The panels should be set in recesses at the jambs so as to provide a bearing surface at least 1 in wide along the edges. Panels more than 10 ft long should also be recessed at the head. Some building codes, however, permit anchoring small panels in low buildings with noncorrodible perforated metal strips.

Steel reinforcement should be placed in the horizontal mortar joints of glass-block panels at vertical intervals of 2 ft or less. It should extend the full length of the joints but not across expansion joints. When splices are necessary, the reinforcement should be lapped at least 6 in. In addition, reinforcement should be placed in the joint immediately below and above any openings in a panel.

The reinforcement should consist of two parallel longitudinal galvanized-steel wires, No. 9 gage or larger, spaced 2 in apart, and having welded to them No. 14 or heavier-gage cross wires at intervals up to 8 in.

Fig. 12-11 *(Top left)* First step in installation of a glass-block panel is to coat the sill with an asphalt emulsion to allow for movement due to temperature changes. Continuous-expansion strips are installed at jambs and head. *(Top right)* Blocks are set with full mortar joints. *(Bottom left)* Welded-wire ties are embedded in the mortar to reinforce the panel. *(Bottom right)* When all the blocks are placed, joints tooled to a smooth, concave finish and the perimeter of the panel calked, the block are cleaned.

Glass block should be laid in Type S mortar. Mortar joints should be from ¼ to ⅜ in thick. They should be completely filled with mortar.

Exterior glass-block panels should be provided with ½-in expansion joints at sides and top. These joints should be kept free of mortar and should be filled with resilient material (Fig. 12-11). An opening may be filled one block at a time, as in Fig. 12-11, or with a preassembled panel.

12-15. Openings, Chases, and Recesses in Masonry Walls.

Masonry above openings should be supported by arches or lintels of metal or reinforced masonry, which should bear on the wall at each end at least 4 in. Stone or other nonreinforced masonry lintels should not be used unless supplemented on the inside of the wall with structural steel lintels, suitable masonry arches, or reinforced-masonry lintels carrying the masonry backing. Lintels should be stiff enough to carry the superimposed load with a deflection of less than $\frac{1}{720}$ of the clear span.

In plain concrete walls, reinforcement arranged symmetrically in the thickness of the wall should be placed not less than 1 in above and 2 in below openings. It should extend at least 24 in on each side of the opening or be equivalently developed with hooks. Minimum reinforcement that should be used is one No. 5 bar for each 6 in of wall thickness.

In structures other than low residences, masonry walls should not have chases and recesses deeper than one-third the wall thickness, or longer than 4 ft horizontally or in horizontal projection. There should be at least 8 in of masonry in back of chases and recesses, and between adjacent chases or recesses and the jambs of openings.

Chases and recesses should not be cut in walls of hollow masonry units or in hollow walls but may be built in. They should not be allowed within the required area of a pier.

The aggregate area of recesses and chases in any wall should not exceed one-fourth of the whole area of the face of the wall in any story.

In dwellings not more than two stories high, vertical chases may be built in 8-in walls if the chases are not more than 4 in deep and occupy less than 4 ft^2 of wall area. However, recesses below windows may extend from floor to sill and may be the width of the opening above. Masonry above chases or recesses wider than 12 in should be supported on lintels.

Recesses may be left in walls for stairways and elevators, but the walls should not be reduced in thickness to less than 12 in unless reinforced in some approved manner. Recesses for alcoves and similar purposes should have at least 8 in of masonry at the back. They should be less than 8 ft wide and should be arched over or spanned with lintels.

If the strength of a wall will not be impaired, pipe or conduit may be passed horizontally or vertically through the masonry in a sleeve. Sleeves, however, should not be placed closer than three diameters center to center.

12-16. Support Conditions for Walls.

Provision should be made to distribute concentrated loads safely on masonry walls and piers. Heavily loaded members should have steel bearing plates under the ends to distribute the load to the masonry within allowable bearing stresses. Length of bearing should be at least 3 in. Lightly loaded members may be supported directly on the masonry if the bearing stresses in the masonry are within permissible limits and if length of bearing is 3 in or more.

Masonry should not be supported on wood construction.

12-17. Chimneys and Fireplaces.

Minimum requirements for chimneys may be obtained from local building codes or any model building code. (See also H. C. Plummer, "Brick and Tile Engineering," Brick Institute of America, McLean, Va.)

In brief, chimneys should extend at least 3 ft above the highest point where they pass through the roof of a building and at least 2 ft higher than any ridge within 10 ft. (For chimneys for industrial-type appliances with discharge temperatures between 1400 and 2000°F, minimum height above the roof opening or any part of the building within 25 ft should be 10 ft. For discharge temperatures over 2000°F, minimum height above any part of the building should be 20 ft.) Masonry chimneys should be constructed of solid masonry units or reinforced concrete and lined with firebrick or fire-clay tile. In dwellings, thickness of chimney walls may be 4 in. In other buildings, the thickness of chimneys for heating appliances should be at least 8 in for most masonry. Rubble stone thickness should be a minimum of 12 in. Cleanout openings equipped with steel doors should be provided at the base of every chimney.

When a chimney incorporates two or more flues, they should be separated by masonry at least 4 in thick.

In seismic zones where damage may occur, chimneys should be of reinforced masonry construction. They should be anchored to floors and ceilings more than 6 ft above grade and to roofs.

Fireplaces should have backs and sides of solid masonry or reinforced concrete, not less than 8 in thick. A lining of firebrick at least 2 in thick or other approved material should be provided unless the thickness is 12 in.

Fireplaces should have hearths of brick, stone, tile, or other noncombustible material supported on a fireproof slab or on brick trimmer arches. Such hearths should extend at least 20 in outside the chimney breast and not less than 12 in beyond each side of the fireplace opening along the chimney breast. Combined thickness of hearth and supporting construction should not be less than 6 in. Spaces between chimney and joists, beams, or girders and any combustible materials should be fire-stopped by filling with noncombustible material.

The throat of the fireplace should be not less than 4 in and preferably 8 in above the top of the fireplace opening. A metal damper (12-gage steel or thicker) extending the full width of the fireplace opening should be placed in the throat. The flue should have an effective area equal to one-twelfth to one-tenth the area of the fireplace opening. For more details, see H. C. Plummer, "Brick and Tile Engineering," Brick Institute of America, McLean, Va.; C. G. Ramsey and H. R. Sleeper, "Architectural Graphic Standards," John Wiley & Sons, Inc., New York.

STUD WALLS

Load-bearing walls in buildings up to three stories high and story-high partitions often are constructed of framing composed of thin pieces of wood or metal. When the main structural members of such walls are installed vertically at close spacing, the members are called **studs** and the walls are referred to as stud walls. Any of a wide variety of materials may be applied to the studs as facings for the walls.

Stud walls permit placement of insulation between studs, so that no increase in thickness is required to accommodate insulation. Also, pipe and conduit may be inexpensively hidden in the walls. Cost of stud construction is usually less than for all-masonry walls.

12-18. Stud-Wall Construction.

Load-bearing and non-load-bearing stud walls may be built of wood, aluminum, or cold-formed steel. Basic framing consists of vertical structural members, or **studs,** seated on a bottom, horizontal, bearing member, called a **sole plate,** and capped with a horizontal tie, called a **top plate** (Fig. 12-12). In addition, diagonal and horizontal bracing may be applied to the framing to prevent racking due to horizontal forces acting in the plane of the wall.

The studs usually are spaced 16 or 24 in on centers. Traditional surfacing materials are manufactured to accommodate these spacings; for example, panels to be attached to the framing usually come 48 in wide. (Inasmuch as the panels are fastened to each stud, panel thickness required, and hence cost, is determined by the stud spacing and generally is larger for 24-in spacing than for 16-in. Overall wall cost, however, may not be larger for the wider spacing, because it requires fewer studs.)

Wood stud walls are normally built of nominal 2×4-in lumber. This type of construction, usually used for residential buildings, is described in Art. 7-25. Advantages of wood construction include light weight and ease of fabrication and assembly, especially in the field.

Aluminum and cold-formed steel construction offer the advantages over wood of incombustibility and freedom from warping, shrinking, swelling, and attack by insects.

12-24 Walls, Partitions, and Interior Finishes

Fig. 12-12 Stud-wall construction.

Studs may be provided with punched openings, which not only reduce weight but also permit passage of pipe and conduit without the necessity of drilling holes in the field. Stud spacing usually is 24 in, rather than 16 in, to reduce the number of studs required.

Metal framing is not so easy to cut and fit in the field as wood. Hence, prefabrication of metal walls in convenient lengths is desirable.

Metal members are manufactured with a variety of widths, leg dimensions, lengths, and thicknesses. Steel studs, for example, are available as C shapes, channels and nailable sections; that is, attachments can be nailed to the flanges. Widths range from ½ to 6 in, and lengths, from 6 to 40 ft.

For partitions, a nonstructural interior finish, such as gypsum plaster, gypsumboard, fiberboard, or wood paneling, may be applied to both faces of stud-wall framing. For exterior walls, the interior face may be the same as for partitions, whereas the outer side must be enclosed with durable, weather-excluding materials, such as water-resistant sheathing and siding or masonry veneer.

For quick assembly, stud walls may be prefabricated. Figure 12-13 illustrates erection of a cold-formed-steel stud wall that has been preassembled with sheathing already attached.

12-19. Sheathing.

To deter passage of air and water through exterior stud walls, sheathing may be attached to the exterior faces of the studs. When sheathing in the form of rigid panels, such as plywood, is fastened to the studs so as to resist racking of the walls, it may be permissible to eliminate diagonal wall bracing, which contributes significantly to wall construction costs. Panels, however, may be constructed of weak materials, especially when the sheathing is also required to serve as thermal insulation, inasmuch as the sheathing is usually protected on the weather side by a facade of siding, masonry veneer, or stucco.

Materials commonly used for sheathing include plywood (see Art. 7-28), fiberboard, gypsum, urethanes, isocyanates, and polystyrene foams. Sheathing usually is available in 4-ft-wide panels, with lengths of 8 ft or more. Available thicknesses range from ½ to 2½ in. Some panels require a protective facing of waterproofing paper or of aluminum foil, which also serves as reflective insulation.

Fig. 12-13 Erection of preassembled steel-stud wall. *(U.S. Gypsum Co.)*

CURTAIN WALLS

With skeleton-frame construction, exterior walls need carry no load other than their own weight, and therefore their principal function is to keep wind and weather out of the building—hence the name curtain wall. Nonbearing walls may be supported on the structural frame of a building, on supplementary framing (girts or studs, for example) in turn supported on the structural frame of a building, or on the floors.

12-20. Functional Requirements of Curtain Walls.

Curtain walls do not have to be any thicker than required to serve their principal function. Many industrial buildings are enclosed only with light-gage metal. However, for structures with certain types of occupancies and for buildings close to others, fire resistance is an important characteristic. Fire-resistance requirements in local building codes often govern in determining the thickness and type of material used for curtain walls.

In many types of buildings, it is desirable to have an exterior wall with good insulating properties. Sometimes a dead-air space is used for this purpose. Sometimes insulating material is incorporated in the wall or erected as a backup.

The exterior surface of a curtain wall should be made of a durable material, capable of lasting as long as the building. Maintenance should be a minimum; initial cost of the wall is not so important as the life-cycle cost (initial cost plus maintenance and repair costs).

To meet requirements of the owner and the local building code, curtain walls may vary in construction from a simple siding to a multilayer-sandwich wall. They may be job-assembled or be delivered to the job completely prefabricated.

Walls with masonry components should meet the requirements of Arts. 12-2 to 12-16 or Sec. 11.

("Comparative Ultimate Cost of Building Walls," Brick Institute of America, McLean, Va.)

12-21. Wood Facades.

Wood is often applied on low buildings as an exterior finish in the form of siding, shingles, half timbers, or plywood sheets.

Siding may be drop, or novelty; lap, or clapboard; vertical boarding or horizontal flush boarding.

Drop siding can combine sheathing and siding in one piece. This type of siding consists of tongued-and-grooved individual pieces that are driven tightly up against each other when they are nailed horizontally in place to make the wall weathertight (Fig. 12-14a and b). It is not considered a good finish for permanent structures.

Lap siding or clapboard are beveled boards, thinner along one edge than the opposite edge, which are nailed horizontally over sheathing and building paper (Fig. 12-14c). Usually boards up to 6 in wide lap each other about 1 in; wider boards, more than 2 in. At the eaves, the top siding boards slip under the lower edge of a frieze board to make a weathertight joint.

Fig. 12-14 Types of wood siding: (a) and (b) drop siding; (c) lap siding.

When vertical or horizontal boards are used for the exterior finish, precautions should be taken to make the joints watertight. Joints should be coated with white lead in linseed oil just before the boards are nailed in place, and the boards should be driven tight against each other. Battens (narrow boards) should be applied over the joints if the boards are squared-edged.

In half-timber construction, timber may be used to form a structural frame of heavy horizontal, vertical, and diagonal members, the spaces between being filled with brick. This type of construction is sometimes imitated by nailing boards in a similar pattern to an ordinary sheathed frame and filling the space between boards with stucco.

Plywood for exterior use should be an exterior grade, with plies bonded with permanent waterproof glue (see Art. 7-28). The curtain wall may consist of a single sheet of plywood or of a sandwich of which plywood is a component. Also, plywood may be laminated to another material, such as a light-gage metal, to give it stiffness.

12-22. Wall Shingles.

Wood, asphalt, and asbestos cement are most frequently used for shingles over a sheathed frame. Shingles are made in a variety of forms and shapes and are applied in different ways. The various manufacturers make available instructions for application of their products.

12-23. Stucco.

Applied like plaster, stucco is a mixture of sand, portland cement, lime, and water. Two coats are applied to masonry, three coats on metal lath. The finish coat may be tinted by adding coloring matter to the mix or the outside surface may be painted with a suitable material.

The metal lath should be heavily galvanized. It should weigh at least 2.5 lb/yd^2, even though furring strips are closely spaced. When supports are 16 in c to c, it should weigh 3.4 lb/yd^2. (See Table 12-4 in Art. 12-37.) The lath sheets should be applied with long dimensions horizontal and should be tied with 16-gage wire. Edges should be lapped at least 1 in, ends 2 in.

The first, or scratch, coat should be forced through the interstices in the lath so as to embed the metal completely. In three-coat applications, the coat should be at least ½ in thick. Its surface should be scored to aid bond with the second, or brown, coat. That coat should be applied as soon as the scratch coat has gained sufficient strength to carry the weight of both coats, usually after about 4 or 5 hr from completion of the scratch coat. The second coat should be at least ⅜ in thick. It should be moist cured for at least 48 hr with fine sprays of water and then allowed to dry for at least 1 week. The finish coat should be at least ⅛ in thick. (When only two coats are used, for example, on a masonry base, the base coat should be a minimum of ⅜ in thick and the finish coat, ¼ in. Before application of the base coat, a bond coat, consisting of one part portland cement and one to two parts sand, should be dashed on the masonry with a stiff brush and allowed to set.)

For both the scratch and brown coats, the mix, by volume, may be 1 part portland cement to 3 to 5 parts sand, plus hydrated lime in amount equal to 25% of the volume of cement. Masonry cement may be used instead of portland cement, but without addition of lime, inasmuch as masonry cement contains lime. The finish coat may be a factory-prepared stucco-finish mix or a job mix of 1 part white portland cement, not more than ¼ part of hydrated lime, 2 to 3 parts of a light-colored sand, and mineral oxide pigment, if desired.

Ingredients should be thoroughly mixed dry. Then, water should be added and the materials mixed for at least 5 min in a power mixer. The first two coats usually are applied with a trowel. The finish coat may be sprayed or manually applied.

("Plasterer's Manual," EBO49M, Portland Cement Association.)

12-24. Metal, Plastic, and Asbestos-Cement Siding.

Either in flat sheets or corrugated form, light-gage metal, plastics or asbestos cement may be used to form a lightweight enclosure. Corrugated sheets are stiffer than the flat. If the sheets are very thin, they should be fastened to sheathing or closely spaced supports.

When corrugated siding is used, details should be planned so that the siding will shed water. Horizontal splices should be placed at supporting members and the sheets should lap about 4 in. Vertical splices should lap at least 1½ corrugations. Sheets should be held firmly together at splices and intersections to prevent water from leaking through. Consideration should be given to sealing strips at openings where corrugated sheets terminate against plane surfaces. The bottommost girt supporting the siding should be placed at least 1 ft above the foundation because of the difficulty of attaching the corrugated materials to masonry. The siding should not be sealed in a slot in the foundation because the metal may corrode or the asbestos cement may crack.

When flat sheets are used, precautions should be taken to prevent water from penetrating splices and intersections. The sheets may be installed in sash like window glass, or the splices may be covered with battens. Edges of metal sheets may be flanged to interlock and exclude wind and rain.

Pressed-metal panels, mostly with troughed or boxed cross sections, are also used to form lightweight walls.

Provision should be made in all cases for expansion and contraction with temperature changes. Allowance for movement should be made at connections. Methods of attachment vary with the type of sheet and generally should be carried out in accordance with the manufacturer's recommendations. For typical details, see C. G. Ramsey and H. R. Sleeper, "Architectural Graphic Standards," John Wiley & Sons, Inc., New York.

12-25. Precast-Concrete or Metal and Glass Facings.

In contrast to siding in which a single material forms the complete wall, precast concrete or metal and glass are sometimes used as the facing, which is backed up with insulation, fire-resistant material, and an interior finish. The glass usually is tinted and is held in a light frame in the same manner as window glass. Metal panels may be fastened similarly in a light frame, attached to mullions or other secondary framing members, anchored to brackets at each floor level, or connected to the structural frame of the building. The panels may be small and light enough for one man to carry or one or two stories high, prefabricated with windows.

Provision for expansion and contraction should be made in the frames, when they are used, and at connections with building members. Metal panels should be shaped so that changes in surface appearance will not be noticeable as the metal expands and contracts. Frequently, light-gage metal panels are given decorative patterns, which also hide movements due to temperature variations ("canning") and stiffen the sheets. Flat sheets may be given a slight initial curvature and stiffened on the rear side with ribs, so that temperature variations will only change the curvature a little and not reverse it. Or flat sheets may be laminated to one or more flat stiffening sheets, like asbestos cement or asbestos cement and a second light-gage metal sheet, to prevent "canning."

It may be desirable in many cases to treat the metal to prevent passage of sound. Usual practice is to apply a sound-absorbing coating on the inside surface of the panel. Some of these coatings have the additional beneficial effect of preventing moisture from condensing on this face.

Metal panels generally are flanged and interlocked to prevent penetration of water. A good joint will be self-flashing and will not require calking. Care must be taken that water will not be blown through weep holes from the outside into the building. Flashing and other details should be arranged so that any water that may penetrate the facing will be drained to the outside. (See also Art. 12-26.)

12-26. Sandwich Panels.

Walls may be built of prefabricated panels that are considerably larger in size than unit masonry and capable of meeting requirements of appearance, strength, durability, insulation, acoustics, and permeability. Such panels generally consist of an insulation core sandwiched between a thin lightweight facing and backing.

When the edges of the panels are sealed, small holes should be left in the seal. Otherwise, heat of the sun could set up sizable vapor pressure, which could cause trouble.

The panels could be fastened in place in a light frame, attached to secondary framing members (Fig. 12-15b), anchored to brackets at each floor level (Fig. 1-5c), or connected to the structural frame of the building. Because of the large size of the panels, special precautions should be taken to allow for expansion and contraction due to temperature changes. Usually such movements are provided for at points of support.

With metal curtain walls, special consideration also must be given to prevention of leakage, since metal and glass are totally nonabsorptive. It is difficult to make the outer face completely invulnerable to water penetration under all conditions; so a secondary defense in the form of an internal drainage system must be provided. Any water entering the wall must be drained to the exterior. Bear in mind that water can penetrate a joint through capillary action, reinforced by wind pressure, and running down a vertical sur-

Fig. 12-15 Methods for field installation of metal curtain walls: (a) stick; (b) mullion and panel; (c) panel system.

face, water can turn a corner to flow along a horizontal surface, defying gravity, into a joint.

When light frames are not used to support the units, adjoining panels generally interlock, and the joints are calked and sealed with rubber or rubberlike material to prevent rain from penetrating. Flashing and other details should be arranged so that any water that comes through will be drained to the outside. For typical details, see C. G. Ramsey and H. R. Sleeper, "Architectural Graphic Standards," John Wiley & Sons, Inc., New York; J.H. Callender, "Time-Saver Standards for Architectural Design Data," McGraw-Hill Book Company, New York; and "Tilt-up Concrete Walls," PA079.01B, Portland Cement Association.

Metal curtain walls may be custom, commercial, or industrial type. Custom-type walls are those designed for a specific project, generally multistory buildings. Commercial-type walls are those built up of parts standardized by manufacturers. Industrial-type walls are comprised of ribbed, fluted, or otherwise preformed metal sheets in stock sizes, standard metal sash, and insulation.

Metal curtain walls may be classified according to the methods used for field installation:

Stick Systems. Walls installed piece by piece. Each principal framing member, with windows and panels, is assembled in place separately (Fig. 12-15a). This type of system involves more parts and field joints than other types and is not so widely used.

Mullion-and-Panel Systems. Walls in which vertical supporting members (mullions) are erected first, and then wall units, usually incorporating windows (generally unglazed), are placed between them (Fig. 12-15b). Often, a cover strip is added to cap the vertical joint between units.

Panel Systems. Walls composed of factory-assembled units (generally unglazed) and installed by connecting to anchors on the building frame and to each other (Fig. 12-15c). Units may be one or two stories high. This system requires fewer pieces and fewer field joints than the other systems.

Ample provision for movement is one of the most important considerations in designing metal curtain walls. Movement continually occurs because of thermal expansion and contraction, wind loads, gravity, and other causes. Joints and connections must be designed to accommodate it.

When mullions are used, it is customary to provide for horizontal movement at each mullion location, and in multistory buildings, to accommodate vertical movement at each floor, or at alternate floors when two-story-high components are used. Common ways of providing for horizontal movements include use of split mullions, bellows mullions, batten mullions, and elastic structural gaskets. Split mullions comprise two channel-shaped components permitted to move relative to each other in the plane of the wall. Bellow mullions have side walls flexible enough to absorb wall movements. Batten mullions consist of inner and outer cap sections that clamp the edges of adjacent panels, but not so tightly as to restrict movement in the plane of the wall. Structural gaskets provide a flexible link between mullions and panels. To accommodate vertical movement, mullions are spliced with a telescoping slip joint.

When mullions are not used and wall panels are connected to each other along their vertical edges, the connection is generally made through deep flanges. With the bolts several inches from the face of the wall, movement is permitted by the flexibility of the flanges.

Slotted holes are unreliable as a means of accommodating wall movement, though they are useful in providing dimensional tolerance in installing wall panels. Bolts drawn up too tightly or corrosion may prevent slotted holes from functioning as intended. If slotted holes are used, the connections should be made with shoulder bolts or sleeves and Bellville or nylon washers, to provide light but positive pressure and prevent rattling.

Since metals are good transmitters of heat, it is particularly important with metal curtain walls to avoid thermal short circuits and metallic contacts between inner and outer wall faces. When, for example, mullions project through the wall, the inner face should be insulated, or each mullion should comprise two sections separated by insulation.

For more details on curtain walls, see W. F. Koppes, "Metal Curtain Wall Specifications Manual," National Association of Architectural Metal Manufacturers, 221 N. LaSalle St., Chicago, Ill. 60601; "Curtain Wall Handbook," U.S. Gypsum Co., Chicago, Ill. 60606.

PARTITIONS

Partitions are dividing walls one story or less in height used to subdivide the interior space in buildings. They may be bearing or nonbearing walls. (See also Art. 1-7.)

12-27. Types of Partitions.

Bearing partitions may be built of masonry or concrete or of wood or light-gage metal studs. These materials may be faced with plaster, wallboard, plywood, wood boards, plastic, or other materials that meet functional and architectural requirements. Masonry partitions should satisfy the requirements of Arts. 12-2 to 12-16 or Sec. 11. See also Art. 12-18.

Nonbearing partitions may be permanently fixed in placed, temporary (or movable) so that the walls may be easily shifted when desired, or folding. Since the principal function of these walls is to separate space, the type of construction and materials used may vary widely. They may be opaque or transparent; they may be louvered or hollow or solid; they may extend from floor to ceiling or only partway; and they may serve additionally as cabinets or closets or as a concealment for piping and electrical conduit.

Fire resistance sometimes dictates the type of construction. If a high fire rating is desired or required by local building codes, the local building official should be consulted for information on approved types of construction or the fire ratings given in the following should be used: "Fire Resistance Design Manual," Gypsum Association, 1603 Orrington Ave., Evanston, Ill. 60201; "Approval Guide," Factory Mutual System, 1151 Boston-Providence Turnpike, Norwood, Mass. 02062; "Fire Resistance Directory," Underwriters Laboratories, 207 East Ohio St., Chicago, Ill. 60611.

When movable partitions may be installed, the structural framing should be designed to support their weight wherever they may be placed.

Acoustics also sometimes affects the type of construction of partitions. Thin construction that can vibrate like a sounding board should be avoided. Depending on functional requirements, acoustic treatment may range from acoustic finishes on partition surfaces to use of double walls separated completely by an airspace or an insulating material.

Light-transmission requirements may also govern the selection of materials and type of construction. Where transparency or translucence is desired, the partition may be constructed of glass, or of glass block or plastic, or it may contain glass windows.

For installation of facings of ceramic wall tiles, see Arts. 12-29 and 12-30. For plaster or gypsumboard partitions, see Arts. 12-34 to 12-46.

Consideration should also be given to the necessity for concealing pipes, conduits, and ducts in partitions.

12-28. Structural Requirements of Partitions.

Bearing partitions should be capable of supporting their own weight and superimposed loads in accordance with recommended engineering practice and should rest in turn on adequate supports that will not deflect excessively. Masonry partitions should meet the requirements of Arts. 12-2 to 12-16 or Sec. 11.

Nonbearing partitions should be stable laterally between lateral supports or additional lateral supports should be added. Since they are not designed for vertical loads other than their own weight, such partitions should not be allowed to take loads from overhead beams that may deflect and press down on them. Also, the beams under the partition should not deflect to the extent that there is a visible separation between bottom of partition and the floor or that the partition cracks.

Folding partitions, in a sense, are large doors. Depending on size and weight, they may be electrically or manually operated. They may be made of wood, light-gage metal, or synthetic fabric on a light collapsible frame. Provision should be made for framing and supporting them in a manner similar to that for large folding doors (Art. 14-10.)

CERAMIC-TILE CONSTRUCTION

When a hard, incombustible, abrasion-resistant, easily cleaned finish is required for walls, ceilings, or floors, ceramic tile often is selected, although construction cost usually is higher than for other commonly used finishes. Ceramic tile is especially suitable where an attractive, water-resistant surface is desired, for instance, for shower enclosures or swimming pools.

12-29. Types of Ceramic Tile.

Ceramic tile is a surfacing unit, relatively thin with respect to surface area, with a body of clay or a mixture of clay and other ceramic materials that has been fired above red heat. The tile should meet the requirements of the Recommended Standard Specification for Ceramic Tile, TCA 137.1, of the Tile Council of America, Inc., P.O. Box 326, Princeton, N.J. 08540.

Ceramic tile is referred to as **glazed** if it is given a surface finish composed of ceramic materials that have been fused into the body of the tile and that may be impervious, vitreous, semivitreous, or nonvitreous. [Glazed tiles tend to be slippery, especially when wet. If this characteristic is undesirable (for example, for floors), then unglazed tile, which has a homogeneous composition, or tile incorporating a dispersed abrasive should be used.] **Impervious tile** is required to have water absorption of less than 0.5% when tested as described in ASTM C373. **Vitreous tile** is required to have water absorption between 0.5 and 3%.

Ceramic tile is classified as wall, ceramic mosaic, quarry, or paver tile. Mosaic, paver, and quarry tile are generally used as floor coverings.

Wall tile is glazed and usually nonvitreous (water absorption greater than 3% but less than 18%). It has a body that is suitable for interior applications, although it is not expected to withstand impact. Flat wall tile is nominally $\frac{5}{16}$ in. thick. Rectangular in shape, it generally comes in nominal sizes of $4\frac{1}{4} \times 4\frac{1}{4}$, $6 \times 4\frac{1}{4}$ and 6×6 in. Also, trim units are available with various shapes necessary for completing an installation. These units include bases, coves, caps, curbs, angles, and moldings.

Ceramic mosaic tile is made of porcelain or natural clay. It may be glazed or unglazed, and if glazed, may be either plain or contain a dispersed abrasive. Nominal thickness is $\frac{1}{4}$ in. Surface area is less than 6 in^2. Available in rectangular, hexagonal and other shapes, flat mosaic tiles come in nominal sizes of 1×1, 2×1, and 2×2 in. The tiles usually are mounted on sheets 1 ft by 2 ft, to facilitate tile setting.

Quarry tile is made by extrusion from natural clay or shale and is unglazed. It usually is $\frac{1}{2}$ to $\frac{3}{4}$ in. thick. Surface area generally exceeds 6 in^2.

Paver tile is similar to ceramic mosaic tile but is thicker, larger, and unglazed. Thickness usually is $\frac{3}{8}$ or $\frac{1}{2}$ in. Nominal sizes generally are 4×4, 6×6, and 8×4 in.

12-30. Tile Installation Methods.

Installation of ceramic tile should be in accordance with the appropriate specification in the A108 series of the American National Standards Institute, depending on the type of setting compound used. Specifications for materials commonly used for setting and grouting the tile are given in the series A118. These materials include portland-cement mortar, dry-set or latex portland-cement mortar, organic adhesive, and chemical-resistant, water-cleanable, tile-setting and grouting epoxy.

Tiles may be set into plastic mortar or adhesive or may be bonded to hardened mortar. (To ensure good alignment, the tile may be preassembled in groups, lightly adhered to sheets.) When the bond is sufficiently strong to prevent the tile from moving, the joints between tiles should be grouted. (Mounted tiles also are available pregrouted, usually with an elastomeric material.) The tiles may be face mounted, back mounted, or edge bonded. Face-mounted tiles are stuck face down on paper or other suitable material with a water-soluble adhesive so that the sheet can be removed easily after tile setting, before the joints are grouted. For back-mounted tiles, the back side, and for edge-bonded tiles, the edges, are adhered to perforated paper, fiber mesh, resin, or other suitable material, which remains in place after tile setting.

Quality and cost of tile installations are affected by the stability, permanence, and precision of installation of the backing or base material on which the tiles are set. The backing should be sound, clean, and dimensionally stable. Deflection of the supporting members under total load should not exceed $\frac{1}{360}$ of the span.

While the tile may not deteriorate when exposed to moisture, the backing may. For installations exposed to wetting, therefore, the backing should be concrete, masonry, or portland-cement mortar, or else a cleavage, or water-resistant, membrane should be inserted to protect the backing.

Backing for Walls and Ceilings. When portland-cement mortar is to be applied, for tile setting, to a gypsum-product backing, a membrane and metal lath should be applied first to cover the backing. In contrast, sound concrete or masonry may have tile directly applied, after the surface has been prepared by sandblasting, chipping, or scarifying to expose an uncontaminated surface. (Without such preparation, use of metal lath is desirable.) For installations with organic adhesive or dry-set or latex portland-cement mortar, which are applied as a thin coating, it is especially desirable that the backing used be dimensionally stable and have negligible variation from a plane.

In all types of installation, control joints or other effective means of preventing cracking should be placed at appropriate intervals.

Exterior walls should be constructed to prevent moisture from collecting behind the tile. For the purpose, flashing, coping, and vapor barriers should be installed, and weep holes draining to the outside should be provided at the base of the exterior walls (see Art. 12-8).

Backing for Floors. Crack-free concrete floors with a screed finish may have portland-cement mortar directly bonded to them as a backing for tile. Concrete floors with a steel-troweled finish, including precast concrete, should be covered with a cleavage membrane before placement of the mortar, and steel reinforcement should be incorporated in the mortar bed. When mortar or adhesive is the type that is applied as a thin coating, the backing should have negligible variation from a plane, or else, a portland-cement mortar bed at least 1¼ in thick should be placed atop the concrete slab.

Metal Lath. Installation of metal lath for reinforcing a mortar bed for tile should be in accordance with the Standard Specification for Installation of Interior Lathing and Furring, ASTM C841. The lath should be galvanized for protection against corrosion. It should be the flat-expanded type and weigh at least 2.5 lb/yd².

For application of metal lath to wood or steel studs or to wood furring, a cleavage membrane—15-lb roofing felt or 4-mil polyethylene film—should be attached first, with joints lapped at least 2 in. Then, the lath should be fastened in place on the side to be mortared (Fig. 12-16*i*). (See also Art. 12-37.) Where sheets of lath are joined, the lath should be lapped at least 2 in at sides and ends. (If wire fabric is used, it should be lapped at least one full mesh.)

Scratch and Leveling Coats. Over rough-surfaced backings, a level surface to receive tile may be built up with portland-cement mortar. For walls and ceilings, the mortar mix for the scratch and leveling coats for application to metal lath, concrete, or masonry should be 1 part portland cement and 3 parts dry sand or 4 parts damp sand, by volume. The ingredients should be mixed dry. Then, sufficient water should be added to make a stiff mix. The scratch coat should be cured for at least 24 hr.

A leveling coat is desirable when the surface of the scratch coat varies more than ¼ in in 8 ft from the required plane or when the required mortar-bed thickness exceeds ¾ in. As for the scratch coat, the leveling coat should be scratched and cured before the next bond coat is applied.

12-34 Walls, Partitions, and Interior Finishes

Fig. 12-16 Ceramic-tile construction. For interior floors: (*a*) portland-cement mortar bed on a concrete subfloor; (*b*) bond coat of dry-set or latex portland-cement mortar on a concrete subfloor; (*c*) epoxy bond coat on a concrete subfloor; (*d*) reinforced mortar bed on a wood subfloor. For an outdoor walkway: (*e*) portland-cement mortar bed on a concrete base. For roofs: (*f*) reinforced mortar bed on crushed stone. For interior walls: (*g*) cement mortar on masonry or concrete; (*h*) dry-set or latex portland-cement mortar on masonry or concrete; (*i*) cement mortar on wood studs or furring.

Mortars and Adhesives. For setting ceramic tile, portland-cement mortar is the only type recommended by the Tile Council of America for a thick bed. Other mortars and adhesives should be used only for thin beds. Setting materials such as epoxies and furans that do not contain portland cement offer special properties but are more expensive than cement-based mortars.

Portland-Cement Mortar. This is suitable for most surfaces and ordinary installations. For floors, the mix may be 1 part portland cement and about 6 parts sand. For walls, about ½ part of hydrated lime may be added to the floor mix. The bed may be ¾ to 1¼ in thick. Tiles may be set into the bed while the mortar is still plastic (ANSI A108.1). Or, after the mortar has cured, the tiles may be bonded to the mortar bed with a ¹⁄₁₆-in-thick coat of dry-set or latex portland-cement mortar or with neat cement. If a neat-cement bond coat is used and the tile is absorptive, it should be soaked before being set. (See Fig. 12-16*a*, *d*, *e*, *f*, *g*, and *i*.)

Dry-Set Mortar. This is a mix of portland cement and sand, with additives that impart water retentivity. The mortar should meet the requirements of ANSI A118.1 or A118.2. Applied as a coat as thin as $\frac{3}{32}$ in, the mortar is suitable for use on stable backings with true surfaces (Fig. 12-16*b* and *h*). Tile installation should be in accordance with ANSI A108.5 or A108.7.

Latex Portland-Cement Mortar. This is a mix containing portland cement, sand, and a latex additive, prepared to meet requirements of ANSI A118.4. Tile applications and installation method with this mortar are similar to those for dry-set mortar and should be in accordance with ANSI A108.5. (See Fig. 12-16*b* and *h*.) Early exposure of latex mortar to water inhibits development of full strength and increases sensitivity to water. Hence, for applications such as swimming pools and showers, the mortar should be cured at least 14 days and allowed to dry thoroughly before exposure to water.

Organic Adhesive. This is an organic material that cures or sets by evaporation. It should meet the requirements of ANSI A136.1. Applied in a thin layer with a notched trowel, organic adhesive is suitable for use on stable backings with true surfaces, but not for swimming pools or exterior applications. Tile installation should be in accordance with ANSI A108.4.

Epoxy Mortar. Suitable for use where chemical resistance, high bond strength, or high impact resistance is needed, this mortar contains epoxy resin and hardener and meets requirements of ANSI A118.3. It should be applied as a thin coat in accordance with ANSI A108.6 (Fig. 12-16*c*).

Furan Mortar. Suitable for use where chemical resistance is important, this mortar consists of furan resin and hardener. It should be used in accordance with the manufacturer's recommendations.

Grouting Materials for Tile. After tile has been set, the joints should be grouted. Most grouts have a portland-cement base but are modified to provide specific qualities, such as whiteness, uniformity, hardness, flexibility, or water retentivity. Damp curing of such grouts for about 72 hr is required.

A grout mix may be prepared in the field with portland cement and a fine graded sand, with up to ⅕ part lime, if desired. The ratio of portland cement to sand, by volume, may be as follows: 1:1 for joints up to ⅛ in wide; 1:2 for joints up to ½ in wide, and 1:3 for joints wider than ½ in. Prepared cement-based mixes include commercial portland-cement grout and dry-set grout. A latex portland-cement grout may be prepared by adding latex to any of the preceding mixes and offers the advantages of lower permeability and no curing, but it is less rigid than regular cement grout.

Grouts also are available with mastic, furan, epoxy, or silicone rubber, to provide such properties as flexibility, stain resistance, high impact resistance, and capacity to withstand below-freezing temperatures and hot, humid conditions. Such grouts, however, are more expensive than the cement-based ones.

Expansion Joints. These generally are required where tiles abut restraints, such as walls, curbs, columns, or pipes, and where joints occur in structural backings, such as concrete floors and walls. For interior tile floors covering large areas, expansion joints generally should be inserted every 24 to 36 ft if the mortars are cement based. For roofs and outdoor floors, joint spacing may range from 12 to 16 ft. For interior walls, expansion joints are generally required directly over masonry control joints, where there are changes in materials, or every 24 to 36 ft.

Joint width should be at least 4 times any expected movement. The width should be at least ⅜ in for exterior joints that are 12 ft apart and ½ in for those 16 ft apart. These widths should be increased $\frac{1}{16}$ in for each 15°F that the actual temperature range may exceed 100°F between summer high and winter low temperature. For interior joints, width should be at least ¼ in. In any case, if there is a structural joint in the backing, the tile expansion joint should be at least as wide as the structural joint.

The joint sealant should be a nonsagging type in vertical joints and a self-leveling type in horizontal joints. It should be inserted to a depth of not less than ⅛ in nor more than one-half the joint width. The depth should be controlled by insertion of a backup strip, which should be flexible and compressible, with a rounded surface where it contacts the sealant. Closed-cell foamed polyethylene or butyl rubber is a suitable material for the backup.

("Handbook for Ceramic Tile," Tile Council of America, Inc., P.O. Box 326, Princeton, N.J. 08540.)

PANEL FINISHES

For speedy installation of interior finishes on walls or ceilings, rigid or semirigid boards, some of which require no additional decorative treatment, may be nailed directly to studs or masonry or to furring. For acoustic materials, see Sec. 17. Gypsumboard is discussed in Arts. 12-41 to 12-46.

Joints between panels may be concealed or accentuated according to the architectural treatment desired. The boards may interlock with each other, or battens, moldings, or beads may be applied at joints.

When the boards are thinner than conventional lath and plaster, framing members may have to be furred out to conventional thicknesses unless stock doors and windows intended for use with panels can be used.

12-31. Plywood Finishes.

Available with a large variety of surface veneers, including plastics, plywood is nailed directly to framing members or furring. Small finish nails, such as 4d, should be used, with a spacing not exceeding 6 in. All edges should be nailed down, intermediate blocking being inserted if necessary. The plywood should also be nailed to intermediate members. When joints are not to be covered, the nailheads should be driven below the surface (set). Under wood battens, ordinary flat-headed nails can be used. Water-resistant adhesive may be applied between plywood and framing members for additional rigidity. (See also Art. 7-28.)

12-32. Asbestos-Cement Boards.

Like other types of boards, asbestos cement can be nailed directly to framing members or furring. Thin sheets usually are backed with plywood or insulation boards to increase resistance to impact. Joints generally are covered with moldings or beads. (See also Art. 4-38.)

12-33. Fiber and Pulp Boards.

Fabricated with a wide variety of surface effects, fiber and pulp boards also are available with high acoustic and thermal insulation values. Application is similar to that for plywood. Generally, adjoining boards should be placed in moderate contact with each other, not forced.

PLASTER AND GYPSUMBOARD

For walls or ceilings, an interior finish made of gypsum products may consist of materials partly or completely prepared in the field, or of prefabricated sheets (dry-type construction). Factors such as initial cost, cost of maintenance and repair, fire resistance, sound control, decorative effects, and speed of construction must be considered in choosing between them.

When field-prepared materials are used, the plaster finish generally consists of a base and one or more coats of plaster. When dry-type construction is used, one or more plies of prefabricated sheet may be combined to achieve desired results. For fire-resistance and sound-transmission ratings of plaster construction, see "Fire Resistance Design Manual," Gypsum Association, 1603 Orrington Ave., Evanston, Ill. 60201.

12-34. Plaster and Gypsumboard Construction Terms.

Absorption. The rate of absorption of water into gypsumboard, as determined by the test described in ASTM C473.

Accelerator. Any material added to gypsum plaster that speeds the set.

Acoustical Plaster. A finishing plaster that corrects sound reverberations or reduces noise intensity.

Adhesive Spreader. A notched trowel or special tool that aids in application of laminating adhesives.

Adhesive Wall Clips. Special clips or nails with large, perforated bases for mastic application to a firm surface.

Adhesives:

Contact. An adhesive that forms a strong, instantaneous bond between two plies of gypsumboard when the two surfaces are brought together.

Laminating. An adhesive that forms a slowly developing bond between two plies of gypsumboard or between masonry or concrete and gypsumboard.

Stud. An adhesive suitable for attaching gypsumboard to framing.

Admixture. Any substance added to a plaster component or plaster mortar to alter its properties. (See also Dope.)

Arris. A sharp edge forming an external corner at the junction of two surfaces.

Back Blocking. Support provided for gypsumboard butt joints that fall between framing members.

Back Clip. A clip attached to the back of gypsumboard and designed to fit into slots in framing to hold the board in place. Used for demountable partitions.

Back Plastering. Application of plaster to one face of a lath system after application and hardening of plaster applied to the opposite face (used for solid plaster partitions and curtain walls).

Backing (Backer) Board. A type of gypsumboard intended to serve as a base layer in a multilayer gypsumboard system or for application with adhesive of acoustical tile or panels.

Band. A flat molding.

Banjo Taper. A mechanical device that dispenses tape and taping compound simultaneously.

Base (Baseboard). A plastic, wood, or metal trim or molding applied to a wall at the floor line to protect the wall from damage.

Base Coat. The plaster coat or combination of coats applied before the finish coat.

Batten. A predecorated strip or joint covering used to conceal the junction between two boards. It is often used with demountable systems.

Bead. A strip of sheet metal usually with a projecting nosing, to establish plaster grounds, and two perforated or expanded flanges, for attachment to the plaster base, for use at the perimeter of a plaster membrane as a stop or at projecting angles to define and reinforce the edge.

Beaded Molding. A cast plaster string of beads set in a molding or cornice.

Beading. (See Ridging.)

Bed, or Bedding, Coat. The first coat of joint compound over tape, bead, or fastener heads.

Bed Mold or Bed. A flat area in a cornice in which ornamentation is placed.

Bench, Hangers'. A low scaffold used by workers to reach the ceiling.

Binder. A chemical added during formulation of the gypsumboard core, often starch, to improve bond between the core and paper facings.

Bleeding. A discoloration, usually at a joint, on a finished wall or ceiling of gypsumboard.

Blister. Protuberance on the finish coat of plaster caused by application over too damp a base coat, or troweling too soon, or a loose, raised spot on the face of a gypsumboard, usually due to an airspace or void in the core. Also denotes a bulge under joint reinforcing tape, usually caused by insufficient compound under the tape.

Blow. Separation of a large area of paper facing from the core during the manufacturing process and usually appearing as a large, puffy blister or a full loose sheet of paper.

Board Knife. A hand tool that holds a replaceable blade for scoring or trimming gypsumboards.

Board Saw. A short handsaw with very coarse teeth used to cut gypsumboards for framed openings for windows and doors.

Bond Plaster. A plaster formulated to serve as a first coat applied to monolithic concrete.

Boss. A Gothic ornament set at the intersection of moldings.

Broad Knife. A wide, flexible finishing knife for applying joint-finishing compound.

Brown Coat. Coat of plaster directly beneath the finish coat. In two-coat work, brown coat refers to the base-coat plaster applied over the lath. In three-coat work, the brown coat refers to the second coat applied over a scratch coat.

Bubble. A large void in the core of gypsumboard caused by entrapment of air while the core is in a fluid state during manufacture.

Buckles. Raised or ruptured spots in plaster that eventually crack, exposing the lath. Most common cause for buckling is application of plaster over dry, broken, or incorrectly applied wood lath.

Bull Nose. This term describes an external angle that is rounded to eliminate a sharp corner and is used largely at window returns and door frames.

Butterflies. Color imperfections on a lime-putty finish wall, caused by lime lumps not put through a screen, or insufficient mixing of the gaging.

Caging. Framing, usually of metal, used to enclose pipes, columns, beams, or other components to be concealed by gypsumboard.

Capital or Cap. The ornamental head of a column or pilaster.

Case Mold. Plaster shell used to hold various parts of a plaster mold in correct position. Also used with gelatin and wax molds to prevent distortions during pouring operation.

Casing Bead. A bead set at the perimeter of a plaster membrane or around openings to provide a stop or separation from adjacent materials.

Casts. (See Staff.)

Catface. Flaw in the finish coat of plaster comparable to a pock mark.

Ceilings:

Coffered. Ornamental ceilings composed of recessed panels between ribs.

Contact. Ceilings attached in direct contact with the construction above, without use of runner channels or furring.

Cross Furred. Ceilings applied to furring members attached at right angles to the underside of main runners or other structural supports.

Furred. Ceilings applied to furring members attached directly to the structural members of the building.

Suspended. Ceilings applied to furring members suspended below the structural members of the building.

Chamfer. A beveled corner or edge.

Chase. A groove in a masonry wall to provide for pipes, ducts, or conduits.

Check Cracks. Cracks in plaster caused by shrinkage, but the plaster remains bonded to its base.

Chip Cracks. Similar to check cracks, except the bond is partly destroyed. Also referred to as fire cracks, map cracks, crazing, fire checks, and hair cracks.

Circle Cutter. An adjustable scribe for cutting circular openings in gypsumboard for lighting fixtures and other devices.

Cockle. A crease-like wrinkle or small depression in gypsumboard paper facing, usually extending in the long direction.

Compounds:

All-Purpose. A joint treatment material that can be used for bedding tape, finishing, laminating adhesive, and texturing.

Joint. A cementitious material for covering joints, corners, and fasteners in finishing of gypsumboard installations, to produce a smooth surface.

Setting-Type. A joint compound that hardens by chemical reactions before drying and that is used for shortening the time required for patching and completing joint finishing.

Taping. A joint compound specially formulated for embedment of joint tape.

Topping. A joint compound specially formulated to serve as the final finishing coat for a joint, but not intended for embedment of tape.

Core. The gypsum structure between face and back papers of gypsumboard.

Coreboard. A gypsumboard, usually 24 in wide and up to 1 in thick, with square, rounded or tongue-and-groove edges, and homogeneous or laminated.

Corner Bead. A strip of sheet metal with flanges and a nosing at the junction of the flanges; used to protect arrises.

Corner Cracks. Cracks in joint of intersecting walls or walls and ceilings.

Corner Floating. (See Floating Angles.)

Cornerite. Reinforcement for plaster at a reentrant corner.

Cornice. A molding, with or without reinforcement.

Cove. A curved concave, or vaulted, surface.

Crown. A buildup of joint compound over a joint to conceal the tape over the joint.

Cure. Treatment, usually of a portland-cement plaster, to ensure hydration after application.

Dado. The lower part of a wall usually separated from the upper by a molding or other device.

Darby. A flat wood tool with handles about 4 in wide and 42 in long; used to smooth or float the brown coat; also used on finish coat to give a preliminary true and even surface.

Dentils. Small rectangular blocks set in a row in the bed mold of a cornice.

Dimple. The depression in the surface of gypsumboard caused by a hammer in setting a nail head slightly below the surface, to permit concealment of the nail with joint compound.

Dope. Additives put in any type of mortar to accelerate or retard set.

Double-up. Applications of plaster in successive operations without a setting and drying interval between coats.

Double Nailing. A method of applying gypsumboard to framing with pairs of nails at intervals of about 12 in along the framing, to ensure firm contact. (The nails in each pair are usually set about 2 in apart.)

Dry out. Soft chalky plaster caused by water evaporating before setting.

Drywall Construction. Application of gypsumboard. (This is basically a dry process rather than a wet process, such as lath and plaster.)

Edges:

Beveled. The factory-formed edge of gypsumboard that has been sloped so that, where two boards abut, a V-groove joint is created.

Chisel. A slanted factory edge on gypsumboard.

Feathered. A thin edge formed by tapering joint compound at a joint to blend with adjoining gypsumboard surfaces; also denotes the skived edge of joint tape.

Featured. A configuration of the paper-bound edges of gypsumboard that provides special design or performance characteristics.

Floating. An edge that does not lie directly over framing and that will be unsupported after installation of gypsumboard or plaster.

Hard. A special core formulation used along the paper-bound edges of gypsumboard to improve resistance to damage during handling and application of the board.

Skive. The outside edges of joint tape that have been sanded or chamfered to improve adhesion and reduce waviness.

Tapered. A factory edge on gypsumboard that is progressively reduced in thickness to allow for concealment of joint tape below the plane of the gypsumboard surface.

Efflorescence. White fleecy deposit on the face of plastered walls, caused by salts in the sand or backing; also referred to as "whiskering" or "saltpetering."

Egg and Dart. Ornamentation used in cornices consisting of an oval and a dart alternately.

Eggshelling. Plaster chip-cracked concave to the surface, the bond being partly destroyed.

Enrichments. Any cast ornament that cannot be executed by a running mold.

Expanded Metals. Sheets of metal that are slit and drawn out to form diamond-shaped openings.

Fat. Material accumulated on a trowel during the finishing operation of plaster and used to fill in small imperfections. Also denotes a mortar that is not too stiff, too watery, or oversanded.

Feather Edge. A beveled-edge wood tool used to straighten reentrant angles in the finish plaster coat.

Fines. Aggregate capable of passing through a No. 200 sieve.

Finish Coat. Last and final coat of plaster; also denotes a thin coat of joint treatment to reduce variations in surface texture and suction.

Finisher. A tradesman with skill in finishing of gypsumboard joints.

Fire Taping. Taping of gypsumboard joints without subsequent finishing coats, usually used where esthetics is not important.

Fisheyes. Spots in plaster finish coat about ¼ in in diameter, caused by lumpy lime because of age or insufficient blending of material.

Float. A tool shaped like a trowel, with a handle braced at both ends and wood base for blade, used to straighten, level, and texture finish plaster coats.

Floating Angles (Corner Floating). Unrestrained surfaces intersecting at about 90°, usually with fasteners omitted near the intersection (Fig. 12-28).

Foil Back. A gypsumboard with a reflective aluminum-foil composite laminated to its back surface.

Furring. Strips that are nailed over studs, joists, rafters, or masonry to support lath or gypsumboard. This construction permits free circulation of air behind the plaster or gypsumboard.

Gaging. Mixing of gaging plaster with lime putty to acquire the proper setting time and initial strength. Also denotes type of plaster used for mixing with the putty.

Green Board. A gypsumboard with a tinted face paper, usually light green or blue, to distinguish special types of board; also denotes gypsumboard that is damp.

Green Plaster. Wet or damp plaster.

Grounds. A piece of wood, metal, or plaster attached to the framing to indicate the thickness of plaster to be applied.

Gypsum. Fully hydrated calcium sulfate (calcium sulfate dihydrate).

Gypsum Base. Gypsum lath used as a base for veneer plasters.

Gypsumboard. A noncombustible board with gypsum core enclosed in tough, smooth paper.

Type X. A gypsumboard specially formulated with high fire resistance for use in fire-rated assemblies.

Hardwall. Gypsum neat base-coat plaster.

Joint Treatment. Concealing of gypsumboard joints, usually with tape and joint compound.

Joints:

Butt. Joints in which gypsumboard ends with core exposed (usually in the direction of the board width) are placed together.

Crown (High or Hump). Protrusion of joint compound from gypsumboard surface at a joint.

Floating. (See Edges, Floating.)

Keene's Cement. A dead-burned gypsum product that yields a hard, high-strength plaster.

Lamination:

Sheet. A ply of gypsumboard attached to another ply with adhesive over the entire surface to be bonded.

Strip. A ply of gypsumboard attached to another ply by parallel strips of adhesive, usually 16 or 24 in apart.

Lath. A base to receive plaster.

Lime. Oxide of calcium produced by burning limestone. Heat drives out the carbon dioxide leaving calcium oxide, commonly termed "quicklime." Addition of water to quicklime yields hydrated or slaked lime.

Lime Plaster. Base-coat plaster consisting essentially of lime and aggregate.

Lime Putty. Thick paste of water and slaked quicklime or hydrated lime.

Marezzo. An imitation marble formed with Keene's cement to which colors have been added.

Mud. (See Compounds.)

Nail Popping. Protrusion above the face of gypsumboard of a nail used to attach the board to framing; usually caused by shrinkage due to drying of inadequately cured wood framing.

Nail Spotter. A small, box-type applicator used to cover with joint compound the heads of nails in gypsumboard.

Neat Plaster. A base-coat plaster to which sand is added at the job.

Niche. A small recess in a wall.

Ogee. A curved section of a molding, partly convex and partly concave.

Papers:

Calendered. Papers with a high glossy finish.

Cream (Ivory or Manila). Highly sized and calendered papers used as the face papers on gypsumboard.

Gray. Unsized, uncalendered papers used on the back side of regular gypsumboard and as the face and back papers of backing boards.

Sized. Paper treated with a sealant to equalize suction for paint and prevent rise of nap.

Perimeter Relief. A construction arrangement that permits building movements; also denotes gaskets that relieve stresses at intersections of walls and ceilings.

Pinhole. A small hole that appears in a plaster cast because of excess water in preparation of the plaster; also denotes a small perforation in gypsumboard paper or paper joint tape.

Plasterboard. (See Gypsumboard.)

Prefill. An application method used in preparation of joints of tapered- or beveled-edge gypsumboard to receive tape and joint treatment with the objective of reducing the possibility of ridging or beading.

Primer. A base coat of paint used to improve the bond and appearance of the finish coat of paint; usually referred to as an undercoat when tinted. (See also Sealer and Sizing.)

Punch out. A hole made in gypsumboard to fit closely around pipe that passes through the board.

Putty Coat. A smooth, troweled-finish coat containing lime putty and a gaging material.

Quicklime. (See Lime.)

Relief. Ornamental figures above a plane surface.

Retarder. Any material added to gypsum plaster that slows its set.

Return. The terminal of a cornice or molding that takes the form of an external miter and stops at the wall line.

Reveal. The vertical face of a door or window opening between the face of the interior wall and the window or door frame.

Ridging. A linear surface protrusion along treated joints.

Ripper. A narrow strip of gypsumboard used for soffits, window reveals, and finished openings.

Runner. A metal or wood track or strip placed at floor and ceiling to receive framing members for partitions.

Sanding. Smoothing a joint treatment for gypsumboard with sandpaper. In wet sanding, the joint is smoothed with a coarse wet sponge, so that less dust is produced than in dry sanding.

Scagliola. An imitation marble, usually precast, made with Keene's cement.

Score. A groove cut in the surface of a board with a sharp blade to expedite manual breaking of the board.

Scratch Coat. First coat of plaster in three-coat work.

Screeds. Long, narrow strips that serve as guides for plastering; also denotes tools, such as straightedges, used to shape an unhardened surface.

Sealer. A base coat of paint used to seal a surface and equalize differences in surface suction, to improve the bond and appearance of the finish coat. (See also Primer and Sizing.)

Seam. A treated gypsumboard joint.

Sheathing, Gypsum. A gypsumboard formulated for use as a enclosure for an exterior wall and a base for siding or other exterior facings, but not intended for long-time direct exposure to the weather.

Ship Lap. An offset lamination of two layers of gypsumboard.

Shoulder. The area between the tapered edge and the face of a gypsumboard.

Sizing. A surface sealant used to equalize suction of paint when applied to gypsumboard paper and prevent rising of the nap.

Skim Coat. (See Finish Coat.)

Skip Trowel. A method of texturing a surface that results in a rough *Spanish Stucco* effect.

Slaking. Adding water to hydrate quicklime into a putty.

Soffit. The underside of an arch, cornice, bead, or other construction.

Soffit Board. A gypsumboard formulated for use on the underside of exterior overhangs, carport ceilings, and other areas protected from the weather.

Splay Angle. An angle of more than 90°.

Spray Texture. A mechanically applied material, which may contain aggregates to produce various effects, used to form decorative finishes.

Staff (Casts). Plaster casts made in molds and reinforced with fiber; usually wired or nailed into place.

Stucco. Plaster applied to the exterior of a building.

Substrate. A surface capable of receiving additional finish or decoration; also denotes the base or concealed layer of gypsumboard in a composite assembly.

Suspension System. Construction, usually incorporating heavy-gage hanging wire, for supporting a ceiling set below structural floor framing, roof, subfloor, or floor deck.

Sweat out. A soft, damp wall area of plaster caused by poor drying conditions.

Swirl Texturing. A method of applying texturing material in a decorative circular pattern.

Tape:

Dry. A tape applied over gypsumboard joints with adhesive other than conventional joint compound.

Joint. A paper tape or fiber mesh for reinforcing joint compound to conceal and reinforce the joints of gypsumboard.

Tape Creaser. A hand-held tool for folding joint tape for use in reentrant corners.

Temper. Mixing of plaster to a workable consistency.

Template. A gage, pattern, or mold used as a guide to produce arches, curves, and various other shapes.

Veneer Plasters. Gypsum plasters meeting requirements of ASTM C587, and that may be applied in one or more coats to a maximum thickness of ¼ in.

Wadding. The act of hanging staff by fastening wads made of plaster of paris and excelsior or fiber to the casts and winding them around the framing.

Wainscot. The lower 3 or 4 ft of an interior wall when it is finished differently from the remainder of the wall.

Wallboard:

Gypsum. A gypsumboard used primarily as an interior surface.

Laminated. Two or more layers of gypsumboard held together with an adhesive.

Predecorated A gypsumboard with a finished surface, such as paint, texturing material, vinyl film, or printed paper coverings, applied before the board is delivered to the building site.

White Coat. A gaged lime-putty troweled-finish coat.

12-35. Plaster Finishes.

Prepared on the building site, plaster finishes are classified as wet-type construction. They take longer to complete than dry-type construction, because they are field mixed with water and require curing before decorative materials, such as paint or wallpaper, can be applied. Plaster finishes are selected nevertheless because they are hard, abrasion resistant, rigid, incombustible, and provide a monolithic (unseamed) surface, even at corners. They are relatively brittle, however, and must be properly applied to avoid cracking when movements due to drying shrinkage or thermal changes are restrained.

A plaster finish consists of a supporting base, such as masonry or lath, and one or more coats of plaster or mix of plaster and other ingredients with water that is troweled or machine sprayed over the base.

The principal ingredient of plaster usually is gypsum but may be portland cement.

(Portland-cement plaster, or stucco, is discussed in Art. 12-23.) Gypsum plasters generally are formulated to meet the requirements of American Society for Testing and Materials "Standard Specification for Gypsum Plasters," ASTM C28, or "Standard Specification for Gypsum Veneer Plaster," ASTM C587.

C28 gypsum plasters include ready-mixed, neat, wood-fiber and gaging plasters. They may be applied over a masonry or lath base, generally in two or more coats, with a total thickness exceeding ½ in. They are required to contain 66% or more by weight of $CaSO_4 \cdot \frac{1}{2}H_2O$.

Veneer plasters must be applied over a special gypsum base, meeting the requirements of ASTM C588, and are limited in thickness to a maximum of ¼ in. They are usually selected because of low cost, rapid installation (permitting application of decorative materials 24 hr after plastering), and high resistance to cracking, nail popping, and impact and abrasion failure. The finishes are, however, not so rigid as conventional lath and plaster. Also, veneer plasters are more susceptible than gypsum plasters to ridging and cracking at joints when dried too rapidly because of low humidity, high temperature, or exposure to drafts.

Application of gypsum and veneer plasters should meet the requirements of the following ASTM specifications:

C841. Installation of Interior Lathing and Furring
C842. Application of Interior Gypsum Plaster
C843. Application of Gypsum Veneer Plaster
C844. Application of Gypsum Base to Receive Gypsum Veneer Plaster

See also Art. 12-46.

Plaster Mixes. Plaster coats other than veneer plasters are generally composed of gypsum plaster, lime, an aggregate (sand, vermiculite, perlite), and water.

Sisal or synthetic fibers, such as nylon, may be added to some scratch-coat plasters for application to metal lath, to limit the amount of plaster that passes through the lath meshes to what is needed for good bond. Fibering, however, adds no strength.

Sand should comply with ASTM C35. It should be clean, free of organic material, more than about 5% clay, silt, or other impurities, and should not contain salt or alkali. The proportion of sand in the plaster has an important bearing on the characteristics of the product. Oversanding results in considerable reduction in strength and hardness. A mix as lean as 4:1 by weight should never be used.

The Gypsum Association suggests that a 1-ft³ measuring box be used for preparing mixes. Some plasterers use a No. 2 shovel, which holds about 16 lb of moist sand, for maintaining proper proportions. Thus, with each 100-lb bag of plaster, a 1:2 mix requires 12 shovels of sand and a 1:3 mix 18 shovels ("Manual of Gypsum Lathing and Plastering").

Water should be clean and free of substances that might affect the rate of set of the plaster. It is not advisable to use water in which plasterers' tools have been washed because it might change the set. Excessive water is undesirable in the mix, because when the water evaporates, it leaves numerous large voids, which decrease the strength of the plaster. Hence, manufacturers' recommendations should be observed closely in determining water requirements.

Perlite and vermiculite are manufactured lightweight aggregates that are used to produce a lightweight plaster with relatively high fire resistance for a given thickness. Both aggregates should conform with ASTM C35, and the mix should be prepared strictly in accordance with manufacturers' recommendations.

Mixing Plaster. A mechanical mixer disperses the ingredients of a mix more evenly and therefore is to be preferred over box mixing. Recommended practice is as follows: (1) Place the anticipated water requirements in the mixer; (2) add about half the required sand (or all required perlite or vermiculite); (3) add all the plaster; (4) add the rest of the sand; (5) mix at least 2 min, but not more than 5 min, adding water, if necessary, to obtain proper workability; and (6) dump the entire batch at once.

The mixer should be thoroughly cleaned when it is not in use. If partly set material is left in it, the set of the plaster might be accelerated. For this reason also, tools should be kept clean.

For hand mixing, first sand and plaster should be mixed dry to a uniform color in a mixing box, water added, and the plaster hoed into the water immediately and thoroughly mixed. Undermixed plaster is difficult to apply and will produce soft and hard spots in the plastered surface.

Plaster should not be mixed more than 1 hr in advance. Nor should a new mix, or gaging, be mixed in with a previously prepared one. And once plaster has started to set, it should not be remixed or retempered.

Plaster Drying. A minimum temperature of 55°F should be maintained in the building where walls are to be plastered when outdoor temperatures are less than 55°F, and held for at least 1 week before plaster is applied and 1 week after the plaster is dry.

In hot, dry weather, precautions should be taken to prevent water from evaporating before the plaster has set. Plastered surfaces should not be exposed to drafts, and openings to the outside should be closed off temporarily. After the plaster sets, the excess moisture it contains evaporates. Hence, the room should be adequately ventilated to allow this moisture to escape.

("Architect Data Book—Construction Products and Systems," Gold Bond Building Products, a National Gypsum Division, 2001 Rexford Road, Charlotte, N.C. 28211; "Gypsum Products Design Data," Gypsum Association, 1603 Orrington Ave., Evanston, Ill. 60201; "Gypsum Construction Handbook," United States Gypsum, 101 South Wacker Drive, Chicago, Ill. 60606.)

12-36. Gypsum Bases for Plaster.

One commonly used base for plaster is gypsum lath. This is a noncombustible sheet generally 16 × 48 in by ⅜ or ½ in thick, or 16 × 96 in by ⅜ in thick. It is composed principally of calcined gypsum that has been mixed with water, hardened, dried, and then sandwiched between two paper sheets (ASTM C37). Insulating gypsum lath is made by cementing shiny aluminum foil to the back of plain lath. It is used for vapor control, and for insulation against heat loss or gain. Also available is gypsum lath with a core specially formulated with minerals for high resistance to fire.

Gypsum lath has the advantage over metal lath that less plaster is required, because the first, or scratch, coat is applied only over the lath surface. Also, when used for suspended ceilings or hollow partitions, plaster on gypsum lath is less susceptible to cracking than on metal lath.

Installation of gypsum lath should meet the requirements of ASTM C841, "Installation of Interior Lathing and Furring." The lath should be applied to studs with long dimension horizontal, and vertical joints should be staggered (Fig. 12-17). In ceilings, the long sides should span supports. Ends should rest on or be nailed to framing, headers, or nailing blocks. Each lath should be in contact with adjoining sheets, but if spaces more than ½ in wide are necessary, the plaster should be reinforced with self-furring metal lath that is stapled or tied with wire to the gypsum lath.

When nailed to wood members, ⅜-in-thick gypsum lath should be attached with four nails, and ½-in lath with five nails, to each framing member covered. Nails should be blued gypsum lath nails, made of 13-gage wire, 1⅛ in long, with a ¹⁹⁄₆₄-in-diam flat head. They should be driven until the head is just below the paper surface without breaking the paper. Lath also may be attached to wood framing with four or five 16-gage staples,

7/16 in wide, with 3/8-in divergent legs. With metal framing, screws should be used, as recommended by the lath manufacturer. Clips, however, are a suitable alternative for use with wood or metal framing. Fasteners should be driven at least 3/8 in away from ends and edges. Clips must secure the lath to framing at each intersection with the framing.

Studs or ceiling members supporting lath may be spaced up to 16 in c to c with 3/8-in gypsum lath, and up to 24 in c to c with 1/2-in lath.

Except at intersections that are to be unrestrained, reentrant corners should be reinforced with cornerite stapled or tied with wire to the gypsum lath. Exterior-angle corners should be finished with corner beads set to true grounds and nailed or tied with wire to the structural frame or to furring. Casing beads should be used around wall openings and at intersections of plaster with other finishes and of lath and lathless construction.

Gypsum Base for Veneer Plasters. Special gypsum lath meeting requirements of ASTM C588 is required as a base for veneer plasters. It is formulated to provide the strength and absorption necessary for proper application and performance of these thin coatings. The lath comes in thicknesses of 3/8 (for two-coat systems), 1/2, and 5/8 in, the last permitting 24-in spacing of wood framing. Installation should meet the requirements of ASTM C844. In general, gypsum base should be applied first to the ceiling, then to the walls. Maximum spacing of nails is 7 in on ceilings and 8 in on walls. Screw spacing should not exceed 12 in for wood framing 24 in c to c or for steel framing or for wood ceiling framing, or 16 in for wood studs apaced 16 in c to c.

12-37. Metal Lath.

A metal base often is used in plaster construction because it imparts strength and resists cracking. Plaster holds to metal lath by mechanical bond between the initial coat of plaster and the metal. So it is important that the plaster completely surrounds and embeds the metal.

Basic types commonly used are expanded-metal, punched sheet-metal, and paper-backed welded-wire lath. Woven-wire lath may be used as a supplemental reinforcement over solid plaster bases, but not as the primary base for gypsum plaster. Wire lath should be made of galvanized, copper-bearing steel. The other types should be made of copper-bearing steel with a protective coat of paint or of galvanized steel.

Expanded-metal lath is fabricated by slitting sheet steel and expanding it to form a mesh. Several types are available:

Diamond-mesh lath, with more than 11,000 meshes/yd^2, is an all-purpose lath, suitable as a base for flat or curved plaster surfaces (Fig. 12-18). The small meshes are helpful in reducing droppings of plaster during plastering. A self-furring type also is availa-

Fig. 12-17 Two coats of plaster applied to perforated gypsum lath attached to wood studs.

Fig. 12-18 Three coats of plaster applied to metal lath attached to steel studs.

ble. When it it attached to a backing, it is separated from the backing by at least ¼ in. Thus, self-furring lath is convenient for use as exterior stucco bases and bases for column fireproofing and replastering over old surfaces.

Flat-rib expanded-metal lath comes with smaller openings than diamond mesh, and it has ribs parallel to the length of the sheet that make it more rigid. Flat-rib lath is generally preferred for nailing to wood framing and tying to framing for flat ceilings, but it is not suitable for contour lathing.

High-rib expanded-metal lath is used when greater rigidity is desired, for example, for spacing supports up to 24 in c to c and for solid, studless plaster partitions. The lath has a herringbone mesh pattern and V-shaped ribs running the length of each sheet. For ⅜-in rib lath, ⅜-in-deep ribs are spaced 4½ in c to c, alternating with inverted ³⁄₁₆-in ribs. For ¾-in rib lath, ¾-in-deep ribs are spaced 6 in c to c. This type of high-rib lath may be used as a form and reinforcement for concrete slabs, but its thickness makes it generally unsuitable for plaster construction. The ⅜-in rib lath may also be used as a concrete form, but its rigidity makes it unsuitable for contour lathing.

Welded-wire lath should be made of wire 16 gage or thicker, forming 2 × 2-in or smaller meshes, stiffened continuously parallel to the long dimension of the sheet at intervals not exceeding 6 in. The paper backing should comply with Federal Specifications UU-B-790, "Building Paper, Vegetable Fiber (Kraft, Waterproofed, Water Repellant, and Fire Resistant)." Acting as a base to which plaster can adhere while hardening about the wire, the backing should permit full embedment in at least ⅛ in of plaster of more than half the total length and weight of the wires.

Table 12-4 lists limiting spans for various types and weights of metal lath for ceilings and walls.

Tying and Nailing. Installation of metal lath should meet the requirements of ASTM C841. Attachments of metal lath to supports should not be spaced farther apart than 6 in along the supports.

TABLE 12-4 Limiting Spans for Metal Lath, in

Type of lath	Min wt., lb/yd²	Vertical supports			Horizontal supports	
			Metal			
		Wood	Solid partitions*	Other	Wood or concrete	Metal
Diamond mesh (flat expanded)	2.5	16	16	12	12	12
metal lath	3.4	16	16	16	16	13½
Flat- (1/8-in) rib expanded	2.75	16	16	16	16	16
metal lath	3.4	19	24	19	19	19
3/8-in rib expanded metal lath†	3.4	24	24†	24	24	24
	4.0	24	24†	24	24	24
3/4-in rib expanded metal lath	5.4		†	24	36‡	36‡
Sheet-metal lath†	4.5	24	†	24	24	24
Welded-wire lath	1.16§	16	16	16	16	16
	1.95¶	24	24	24	24	24

* For paper-backed lath, only absorbent, perforated, or slotted paper separator should be used.
† Permitted for studless solid partitions.
‡ Permitted only for contact or furred ceilings.
§ Welded 16-gage wire, paper-backed lath.
¶ Paper-backed lath with welded wire, face wires 16 gage, every third back wire parallel to line dimension of lath 11 gage.
SOURCE: Based on "Uniform Building Code," International Conference of Building Officials, Inc.

When metal framing or furring is used, metal lath should be tied to it with 18-gage, or heavier, galvanized soft-annealed wire. Rib lath, however, should be attached to open-web steel joists with single loops of 16-gage, or heavier, wire, or double loops of 18-gage wire, with the ends of each loop twisted together. Also, rib lath should be tied to concrete joists with loops of 14-gage, or heavier, wire or with wire hangers not less than 10 gage.

With wood supports, diamond-mesh, flat-rib, and welded-wire lath should be attached to horizontal framing with 1½-in, 11 gage, $\frac{7}{16}$-in-head, barbed, galvanized, or blued roofing nails, driven full length. For vertical wood supports the following may be used: 4d common nails; 1-in, 14-gage wire staples driven full length; and 1-in roofing nails driven at least ¾ in into the supports. Common nails should be bent over to engage a rib or at least three strands of lath. Alternatives of equal strength also may be used.

Metal lath should be applied with long sides of sheets spanning supports. Each sheet should underlap or overlap adjoining sheets on both sides and ends. Expanded-metal and sheet-metal lath should be lapped ½ in along the sides, or have edge ribs nested, and 1 in along the ends. Welded-wire lath should be lapped one mesh at sides and ends. All side laps of metal lath should be fastened to supports and tied between supports at intervals not exceeding 9 in.

Wherever possible, end laps should be staggered, and the ends should be placed at and fastened to framing. If end laps fall between supports, the adjoining ends should be laced or securely tied with 18-gage, galvanized, annealed steel wire.

Normally, metal lath should be applied first to ceilings. Flexible sheets may be carried down 6 in on walls and partitions. As an alternative, preferable for more rigid sheets, sides and ends of the lath may be butted into horizontal reentrant angles and the corner reinforced with cornerite. But for large ceilings (length exceeding 60 ft in any direction, or more than 2400 ft^2 in area) and other cases in which restraint should be avoided, and for portland-cement plaster ceilings, cornerite should not be used. Instead, the abutting sides and ends should terminate at a casing bead, control joint, or similar device that will isolate the ceiling lath and plaster from the walls and partitions.

Similar considerations govern installation of metal lath at vertical reentrant corners. Between partitions, flexible sheets may be bent around vertical corners and attached at least one support away from them; more rigid sheets may be butted and the corner reinforced with cornerite. But where restraint is undesirable at reentrant corners, for example, where partitions meet structural walls or columns, or where load-bearing walls intersect, cornerite should not be used. Instead, the walls, partitions, and structural members should be isolated from each other, as described for ceiling-wall corners.

12-38. Masonry Bases for Plaster.

Gypsum partition tile has scored faces to provide a mechanical bond as well as the natural bond of gypsum to gypsum plaster. The 12 × 30-in faces of the tile present an unwarped plastering surface because the tile is dried without burning; so a mechanic can lay a straighter wall than with other types of units. The Gypsum Association's "Manual of Gypsum Lathing and Plastering" recommends that only gypsum plaster be applied to gypsum partition tile, since lime and portland cement do not bond adequately. Also, only gypsum mortar should be used for laying tile.

Brick and clay tile can be used as a plaster base if they are not smooth-surfaced or of a nonporous type. If the surface does not provide sufficient suction, it should offer a means for developing a mechanical bond, such as does scored tile.

Plaster should not be applied directly to exterior masonry walls because dampness may damage the plaster. It is advisable to fur the plaster at least 1 in in from the masonry.

Properly aged concrete block may serve in walls as a plaster base, but for block ceilings, a bonding agent or a special bonding plaster should be applied first.

For precast or cast-in-place concrete with smooth dense surfaces, a bonding agent or a special bonding plaster should be used first. But if a plaster thickness of more than ⅜ in is required for concrete ceilings, or ⅝ in for concrete walls, metal lath should be secured to the concrete before plastering, in which case sanded plaster can be used.

12-39. Plaster Base Coats.

The base coat is the portion of the plaster finish that is applied to masonry or lath bases and supports the finish coat (Fig. 12-19).

Except for veneer plasters, plaster applications may be three-coat (Fig. 12-18) or two-coat (Fig. 12-17). The former consists of (1) a scratch coat, which is applied directly to the plaster base, cross-raked after it has "taken up," and allowed to set and partly dry; (2) a brown coat, which is surfaced out to the proper grounds, darbied (float-finished), and allowed to set and partly dry; and (3) the finish coat. Three-coat plaster is required over metal lath, ½-in gypsum lath spanning horizontal supports more than 16 in c to c, all gypsum lath attached by clips providing only edge support, and ⅜-in perforated gypsum lath on ceilings.

The two-coat application is similar, except that cross-raking of the scratch coat is omitted and the brown coat is applied within a few minutes to the unset scratch coat. Three coats are generally preferred, because the base coat thus produced is stronger and harder.

Veneer plaster applications, 1/16 in thick, may be one-coat or two-coat, both applied to special gypsum base (Arts. 12-35 and 12-36). The single coat is composed of a scratch coat without cross-raking and a double-up coat immediately applied, then worked to a smooth or textured finish.

See also Art. 12-40 and "Manual of Gypsum Lathing and Plastering," Gypsum Association.

Fig. 12-19 Application of a base coat of plaster to gypsum lath. *(Gold Bond Building Products, a National Gypsum Division)*

Plaster Grounds. Except for veneer plasters, thickness of base-coat plaster should be controlled with grounds—wood or metal strips applied at the perimeter of all openings and at baseboards or continuous strips of plaster applied at intervals along a wall or ceiling, to serve as screeds. Plaster screeds should be used on all plaster surfaces of large area.

Minimum Thicknesses. Grounds should be set to provide a minimum plaster thickness of ½ in over gypsum lath and gypsum partition tile; ⅝ in over brick, clay tile, or other masonry; and ⅝ in from the face of metal lath. A thickness of $\frac{1}{16}$ in is included for the finish coat.

Gypsum Base-Coat Plasters. Three types of gypsum base-coat plasters are in general use: gypsum neat plaster, gypsum ready-mixed plaster, and veneer plasters, which may be used in thin one-coat or two-coat systems. In two-coat veneer-plaster systems, the base-coat veneer plaster should be applied to a thickness of $\frac{1}{16}$ to $\frac{3}{32}$ in, and left with a rough surface to receive the finish coat. Veneer plasters should meet the requirements of ASTM C587 and application should be in accordance with ASTM C843.

Gypsum neat plaster, sometimes called hardwall or gypsum-cement plaster, is sold in powder form and mixed with an aggregate and water at the construction site. Mixed with no more than 3 parts sand by weight, it makes a strong base coat at low cost. Scratch coats generally consist of 1 part plaster powder to 2 parts sand by weight, fibered or unfibered; the base coat in two-coat work usually is a 1:2½ mix; brown coats are 1:3 mixes. With perlite or vermiculite instead of sand, a 1:2 mix may be used.

Gypsum ready-mixed plaster requires the addition only of water at the site, since it is sold in bags containing the proper proportions of aggregate and plaster. It is specified when good plastering sand is high cost or not available, or to avoid the possibility of oversanding. It costs a little more than neat plaster because of the extra cost of transporting the sand.

The water ratio for base coat neat and ready-mixed plasters should be such that slump does not exceed 4 in when tested with a $2 \times 4 \times 6$-in cone at the mixer, for mixes with sand proportions not exceeding those given for gypsum neat plaster.

Application of gypsum plaster should meet the requirements of ASTM C842. The scratch coat (Figs. 12-17 and 12-18) applied to lath should be laid on with enough pressure to form a strong clinch or key. The coat should cover the lath to a thickness of ¼ in. For two-coat systems, the double-up brown coat is applied immediately. For three-coat systems, after the surface has been trued, the scratch coat should be scratched horizontally and vertically with a toothed tool to form a good bonding surface, then left to dry partly. When the surface is so hard that the edges of the scoring do not yield easily under the pressure of a thumbnail, the brown coat may be applied. Hardening may take at least 1 day, and sometimes as long as 1 week, depending on drying conditions.

The brown coat not only forms the base for the finish coat, but is also the straightening coat. The plaster should be laid on with a steel float, trued with rod or darby, and left rough in preparation for the finish coat.

12-40. Finish Plaster Coats.

Several types of plasters are available for the finish coat (Figs. 12-17 and 12-18). Usually, lime is an important ingredient, because it gives plasticity and bulk to the coat.

Gaging plasters are coarsely ground gypsum plasters, which are available in quick-setting and slow-setting mixtures; so it is not necessary to add an accelerator or a retarder at the site. Gaging plasters also are supplied as white gaging plaster and a slightly darker local gaging plaster. Finish coats made with these plasters are amply hard for ordinary usages and are the lowest-cost plaster finishes. However, they are not intended for ornamental cornice work or run moldings, which should be made of a finer-ground plaster. Gypsum gaging plasters should conform with ASTM C28. Application should conform with ASTM C842.

Typical mixes consist of 3 parts lime putty to 1 part gaging plaster, by volume. If a harder surface is desired, the gaging content may be increased up to 1 part gaging to 2 parts lime putty.

The lime is prepared first, being slaked to a smooth putty, then formed on the plasterer's board into a ring with water in the center. Next, gaging plaster is gradually sifted into the water. Then, aggregates, if required, are added. Finally, all ingredients are thoroughly mixed and kneaded. Alternatively, materials, including Type S hydrated lime, may be blended in a mechanical mixer.

The lime-gaging plaster should be applied in at least two coats, when the brown coat is nearly dry. The first coat should be laid on very thin, with sufficient pressure to be forced into the roughened surface of the base coat. After the first coat has been allowed to draw a few minutes, a second or leveling coat, also thin, should be applied.

The base coat draws the water from the finish coats; so the finished surface should be moistened with a wet brush as it is being troweled. Pressure should be exerted on the trowel to densify the surface and produce a smooth hard finish. Finally, the surface should be dampened with the brush and clean water. It should be allowed to stand at least 30 days before oil paints are applied.

Prepared gypsum trowel finishes also are available that require only addition of water at the site. The resulting surface may be decorated as soon as dry. The plaster is applied in the same manner as lime-gaging plaster, but the base coat should be dry and, because the prepared plaster has a moderately fast set, it should be troweled before it sets. For best results, three very thin coats should be applied and water should be used sparingly.

Sand float finishes are similar to gypsum trowel finishes, except that these float finishes contain a fine aggregate to yield a fine-textured surface and the final surface is finished with a float. The base coat should be firm and uniformly damp when the finish coat is applied. These finishes have high resistance to cracking.

Molding plaster, intended for ornamental work, is made with a finer grind than other gaging plasters. It produces a smooth surface, free from streaks or indentations as might be obtained with coarser-ground materials. Equal parts of lime putty and molding plaster are recommended by the Gypsum Association for cornice moldings.

Veneer plasters, applied to a thickness of only $\frac{1}{16}$ to $\frac{3}{32}$ in, develop hard abrasion-resistant surfaces that can be decorated the day after application. Factory prepared, these plasters are easy to work, have high plasticity, and provide good coverage. The finish may be applied in one-coat systems over special gypsum base (Art. 12-36), or in two-coat systems over a veneer-plaster or sanded gypsum base coat. They require only addition of water on the job. Veneer plasters should meet the requirements of ASTM C587 and application should conform with ASTM C843.

12-41. Gypsumboard Finishes.

To avoid construction delays due to the necessity of mixing plaster ingredients with water on the building site and, after application, waiting for the plaster to cure, dry-type construction with gypsum products may be used. (For discussions of other dry-type interior finishes, see Arts. 12-31 to 12-33.) For the purpose, gypsumboard is factory fabricated and delivered to the site ready for application.

Gypsumboard is the generic name for a variety of panels, each consisting of a noncombustible core, made primarily of gypsum, and a bonded tough paper surfacing over the face, back, and long edges. Different types of paper are used for specific purposes. Gypsumboard with a factory-applied, decorative face or with an aluminum-foil back for insulation purposes also is available.

The panels may be attached directly to framing, such as studs or joists (Fig. 12-20). While gypsumboard also may be attached directly to unit masonry or concrete, use of furring strips between the panels and backing is desirable because of the possibilities of interference from surface irregularities in the backing or of moisture penetration through exterior walls. Successive panels are applied with edges or ends abutting each other. Depending on esthetic requirements, the joints may be left exposed, covered with battens, or treated to present an unseamed, or monolithic, appearance. Gypsumboard also can be used to construct self-supporting partitions spanning between floor and ceiling.

Gypsumboard generally comes in widths of 4 ft, lengths of 8 to 16 ft, and thicknesses of ¼ to 1 in. Panels ½ in or more thick are suitable for 16- to 24-in spacing of supports.

The types of gypsumboard generally available include wallboard, backing board, coreboard, fire-resistant gypsumboard, water-resistant gypsumboard, gypsum sheathing, and gypsum formboard.

Fig. 12-20 Application of wallboard to wood studs. *(Gold Bond Building Products, a National Gypsum Division)*

Gypsum Wallboard. This type is used for the surface layer on interior walls and ceilings. Regular gypsum wallboard comes with gray liner paper on the back and a special paper covering, usually cream-colored, on facing side and edges. This covering provides a smooth surface suitable for decoration. Foil-backed gypsum wallboard has aluminum foil bonded to the liner paper to serve as a vapor barrier and, when contiguous to an airspace, as thermal insulation. Predecorated gypsum wallboard does not require decorative treatment after installation because it comes with a finished surface, often a decorative vinyl or paper sheet. Wallboard should conform with ASTM C36.

Wallboard usually is available in the following thicknesses:

¼ in—for covering and rehabilitating old walls and ceilings.
⅜ in—mainly for the outer face in two-layer wall systems.
½ in—for single-layer new construction with supports 16 to 24 in c to c.
⅝ in—for better fire resistance and sound control than ½ in provides.

Standard edges are rounded, beveled, tapered, or square.

Backing Board. This type is used as a base layer in multi-ply construction, where several layers of gypsumboard are desired for high fire resistance, sound control, and strength in walls. It has gray liner paper on front and back faces. Also available is backing board with aluminum foil bonded to the back face. Gypsum backing board should conform with ASTM C442.

Gypsum Coreboard. To save space, this type is used as a base in multi-ply construction of self-supporting (studless) gypsum walls. Coreboard may be supplied as 1-in-thick, solid backing board or as two factory-laminated, ½-in-thick layers of backing board.

Type X Gypsumboard. For use in fire-rated assemblies, Type X may be gypsum wallboard, backing board, or coreboard with core made more fire resistant by addition of glass fiber or other reinforcing materials.

Water-Resistant Gypsum Backing Board. This type comes with a water-resistant gypsum core and water-repellant face paper. It may be used as a base for wall tile in baths, showers, and other areas subject to wetting. The board should conform with ASTM C630.

Gypsum Sheathing. This type is used as fire protection and bracing of exterior frame walls. It must be protected from the weather by an exterior facing. Sheathing should conform with ASTM C79.

Gypsum Formboard. This type is used as a permanent form in the casting of gypsum-concrete roof decks.

("Architect Data Book—Construction Products and Systems," Gold Bond Building Products, a National Gypsum Division, 2001 Rexford Road, Charlotte, N.C. 28211; "Gypsum Products Design Data," Gypsum Association, 1603 Orrington Ave., Evanston, Ill. 60201; "Gypsum Construction Handbook," United States Gypsum, 101 South Wacker Drive, Chicago, Ill. 60606.)

12-42. General Application Procedures for Gypsumboard.

Gypsumboards may be used in single-ply construction (Fig. 12-22) or combined in multi-ply systems (Fig. 12-25). The latter are preferred for greater sound control and fire resistance. Application and finishing should conform with ASTM C840.

Precautions. When outdoor temperatures are less than 55°F, the temperature of the building interior should be maintained at a minimum of 55°F for at least 24 hr before installation of gypsumboards. Heating should be continued until a permanent heating system is in operation, or outdoor temperatures stay continuously above 55°F. In warm or cold weather, gypsumboards should be protected from the weather.

Green lumber should not be used for framing or furring gypsumboard systems. Moisture content of the lumber should not exceed 19%, to avoid defects caused by shrinkage as the wood dries. For the same reason, dry lumber should be kept dry during storage and erection and afterward.

To prevent damage from structural movements or dimensional changes, control and expansion joints should be provided and floating-angle construction used, as described in Art. 12-46.

Many building components may be affected by the decision to use gypsumboard construction. For example, window and door frames should have the appropriate depth for the wall thickness resulting from use of wallboard. Therefore, walls and partitions and related installations, including mechanical and electrical equipment, should be carefully planned and coordinated.

Application. Wallboard preferably should be applied to studs with long dimension horizontal (Fig. 12-22), and vertical joints should be staggered. In ceilings, the long sides should span supports (often referred to as horizontal, or across, application). Board ends and edges parallel to framing or furring should be supported on those members, except for face layers of two-ply systems. Otherwise, back blocking should be used to reinforce the joints.

Ceiling panels should be installed first, then the walls. Adjoining boards should be placed in contact, but not forced against each other. Tapered edges should be placed next to tapered edges, square ends in contact with square ends. (Joints formed by placing square ends next to tapered edges are difficult to conceal.)

Furring. Supplementary framing, or furring, should be used when framing spacing exceeds the maximum spacing recommended by the gypsumboard manufacturer for the thickness of board to be used (Table 12-5), or when the surface of framing or base layer is too far out of alignment.

At gypsumboard joints, wood furring should be at least $1\frac{1}{2}$ in wide, and metal furring $1\frac{1}{4}$ in wide, to provide adequate bearing surface and space for attachment of the gypsumboard. The furring should be aligned to receive the board, and securely fastened to framing or masonry or concrete backing. For rigidity, lumber used for furring should be at least 2×2 in when nails are used and 1×3 in when screws are used for attachment of gypsumboard. Furring on masonry or concrete may be as small as $\frac{3}{8} \times 1\frac{1}{2}$ in.

Fastening. Gypsumboards may be held in place with various types of fasteners or adhesives, or both. Special nails, staples, and screws are required for attachment of gypsumboards, because ordinary fasteners may not hold the boards tightly in place or countersink neatly, to be easily concealed. Clips and staples may be used only to attach the base layer in multi-ply construction.

To avoid damaging edges or ends, fasteners should be placed no closer to them than $\frac{3}{8}$ in (Figs. 12-23 and 12-24). With a board held firmly in position, fastening should start at the middle of the board and proceed outward toward the edges or ends. Nails should be driven with a crown-headed hammer until a uniform depression, or dimple, not more than $\frac{1}{32}$ in deep is formed around the nail head. Care should be taken in driving not to tear the paper or crush the gypsum core.

Nails. For use in attachment of gypsumboard, nails should conform with ASTM C514. (Some fire-rated systems, however, may require special nails recommended by the gypsumboard manufacturer.) For easy concealment, nail heads should be flat or slightly concave and thin at the rim. Diameter of nail heads should be $\frac{1}{4}$ or $\frac{5}{16}$ in. Length of smooth-shank nails should be sufficient for penetration into wood framing of at least $\frac{7}{8}$ in, and length of annular-grooved nails, for $\frac{3}{4}$-in penetration.

Screws. Three types of screws, all with cupped Phillips heads, are used for attachment of gypsumboard. Type W, used with wood framing, should penetrate at least $\frac{5}{8}$ in into the wood. Type S, used for sheet metal, should extend at least $\frac{3}{8}$ in beyond the inner gypsumboard surface. Type G, used for solid gypsum construction, should penetrate at least $\frac{5}{8}$ in into supporting gypsumboard. These screws, however, should not be used to attach wallboard to $\frac{3}{8}$-in backing board, because sufficient holding power cannot be developed. Nails or longer screws should be driven through both plies.

Staples. Used for attaching base ply to wood framing in multi-ply systems (Fig. 12-21), staples should be made of flattened, galvanized, 16-gage wire. The crown should be at least $\frac{7}{16}$ in wide. Legs should be long enough to permit penetration into supports of at least $\frac{5}{8}$ in, and should have spreading points.

When the face ply is to be laminated with adhesive to the base ply, the staples should be spaced 7 in c to c. When the face ply is to be nailed, the staples should be placed 16 in c to c.

Adhesives. These may be used to attach gypsumboard to framing or furring, or to existing flat surfaces. Nails or screws may also be used to provide supplemental support.

Adhesives used for bonding wallboard may be classified as stud, laminating, or contact or modified contact.

Stud adhesives are used to attach wallboard to wood or steel framing or furring. They should conform with ASTM C557. They should be applied to supporting members in continuous, or nearly so, beads.

Laminating adhesives are used to bond gypsumboards to each other, or to suitable masonry or concrete surfaces. They are generally supplied in powder form, and water is added on the site. Only as much adhesive should be mixed at one time as can be applied within the period specified by the manufacturer. The adhesive may be spread over the entire area to be bonded, or in parallel beads or a pattern of large spots, as recommended by the manufacturer. Supplemental fasteners or temporary support should be provided the face boards until sufficient bond has been developed.

Contact adhesives are used to laminate gypsumboards to each other, or to attach wallboard to metal framing or furring. A thin, uniform coat of adhesive should be applied to both surfaces to be joined. After a short drying time, the face board should be applied to the base layer and tapped with a rubber mallet, to ensure overall adhesion. Once contact has been made, it may not be feasible to move or adjust the boards being bonded.

Modified contact adhesives, however, permit adjustments, often for periods of up to ½ hr after contact. Also, they generally are formulated with greater bridging ability than contact adhesives. Modified adhesives may be used for bonding wallboard to all kinds of supporting construction.

See also Arts. 12-43 to 12-46.

12-43. Single-Ply Gypsumboard Construction.

(See also Art. 12-42.) A single-ply system consists of one layer of wallboard attached to framing, furring, masonry, or concrete (Fig. 12-22). This type of system is usually used for residential construction and where fire-rating and sound-control requirements are not stringent. Maximum spacing of supports should not exceed the limits specified in Table 12-5.

Nail Attachment. Spacing of nails generally is determined by requirements for fire resistance and for firmness of contact between wallboard and framing necessary to avoid surface defects. Spacing normally used depends on whether single nailing or double nailing is selected. Double nailing provides tighter contact, but requires more nails. In either method, one row of nails is driven along each support crossed by a board or on which a board end or edge rests, and the spacing applies to the center-to-center distance between nails in each row.

In the single-nailing method, nails should be spaced not more than 7 in apart for ceilings and 8 in apart for walls (Fig. 12-23).

In the double-nailing method, pairs of nails are driven 12 in c to c, except at edges or ends, in a special sequence (Fig. 12-24). First, one nail of each pair is placed, starting at

Fig. 12-21 Gypsumboard base ply attached to wood supports with staples.

Fig. 12-22 Single-ply application of gypsumboard.

TABLE 12-5 Maximum Spacing, in, of Framing for Single-Ply Gypsumboard

Board thickness, in	Orientation	Spacing c to c, in
a. Applications in ceilings		
3/8*†	Across	16
1/2*	Across	24
1/2*	Parallel	16
5/8	Across	24
5/8	Parallel	16
b. Applications in walls		
3/8	Across or parallel	16
1/2	Across or parallel	24
5/8	Across or parallel	24

*Gypsumboard for ceilings to receive a water-base-spray texture finish should be applied only across (perpendicular to) framing. For 16-in spacing of framing, board thickness should be increased from 3/8 to 1/2 in. For 24-in spacing of framing, board thickness should be increased from 1/2 to 5/8 in.
†Should not support thermal insulation.

Fig. 12-23 Wallboard attached to wood framing with single nailing.

the middle of the board and then proceeding toward edges and ends. Next, the second nail is driven 2 in from the first. Finally, the first set of nails placed should be given an extra hammer blow to reseat them firmly. Single-nailing spacing should be used for edges or ends at supports.

Screw Attachment. Fewer screws than nails are needed for attachment of wallboard. With wood framing, Type W screws should be spaced 12 in c to c for ceilings. Spacing should not exceed 16 in for studs 16 in c to c, or 12 in for studs 24 in c to c. With metal framing, spacing of Type S screws should not exceed 12 in for walls or ceilings.

Adhesive Nail-on Attachment. With this method, at least 50% fewer nails are required in the middle of the wallboard than with conventional nailing. Also, a stiffer assembly results.

The first step is to apply stud adhesive to framing or furring in beads about ¼ in in diameter. Each bead should contain enough adhesive so that it will spread to an average width of 1 in and thickness of ⅟₁₆ in when wallboard is pressed against it, but there should

Fig. 12-24 Wallboard attached to wood framing with double nailing. One man nails rows 1 to 4, and a second man, rows 2A to 4A (solid dots), starting at the middle of the board. They then drive the second set of nails (open dots), again starting at the middle of the board. Finally, they reset the first set of nails with a blow on each.

not be so much adhesive that it will squeeze out at joints. If joints are to be treated, an undulating or zigzag bead should be applied to supports under joints. If joints are to be untreated, as is likely to be the case for predecorated wallboards, two parallel beads should be applied near each edge of the framing. Along other supports a single, continuous, straight bead is adequate. Adhesive, however, should not be applied to members, such as diagonal bracing, blocking, and plates, not required for wallboard support.

The number and spacing of supplemental fasteners needed with adhesives depend on adhesive properties. Stud adhesives generally require only perimeter fasteners for walls. One fastener should be driven wherever each edge or end crosses a stud. For edges or ends bearing on studs, fasteners should be spaced 16 in c to c. The same perimeter fastening should be used for ceilings, but in addition, fasteners should be spaced 24 in c to c along all framing crossed by the wallboard.

Where perimeter fasteners cannot be used, for example, with predecorated wallboard for which joint treatment is unnecessary or undesirable, prebowing or temporary bracing should be used to keep the wallboard pressed against the framing until the adhesive develops full strength. Bracing should be left in place for at least 24 hr. Prebowing bends a wallboard so that the finish side faces the center of curvature. The arc is employed to keep the board in tight contact with the adhesive as the board is pressed into place starting at one end.

12-44. Multi-Ply Gypsumboard Construction.

(See also Art. 12-42.) A multi-ply system consists of two or more layers of gypsumboard attached to framing, furring, masonry, or concrete (Fig. 12-25). This type of system has better fire resistance and sound control than can be achieved with single-ply systems, principally because of greater thickness, but also because better insulation can be used for the base layer.

Face layers usually are laminated (glued), but may be nailed or screwed, to a base layer of wallboard, backing board, or sound-deadening board. Backing board, however, often is used for economy. When adhesives are used for attaching the face ply, some fasteners generally are also used to ensure bond.

Maximum support spacing depends principally on the thickness of the base ply and its orientation relative to the framing, as indicated in Table 12-6, and is the same for wood or metal framing or furring.

Nails, screws, or staples may be used to attach base ply to supports. For wood framing, when the face ply is to be laminated to the base ply, staples should be spaced 7 in c to c (Fig. 12-21), and nails and screws should be single-nailed as recommended for single-ply construction (Art. 12-43). If the face ply is to be nailed, nails and staples for the base ply should be driven 16 in c to c, and screws 24 in c to c. For metal framing and furring, Type S screws for the base ply should be spaced 12 in c to c if the face ply is to laminated, and 16 in c to c if the face ply is to be attached with screws.

Fig. 12-25 Double-ply application of wallboard.

For attachment of the face ply with adhesive, sheet, strip, or spot lamination may be used. In **sheet lamination,** the entire back of the face ply is covered with adhesive, usually applied with a notched spreader. In **strip lamination,** adhesive is applied by a special spreader in grouped parallel beads, with groups spaced 16 to 24 in c to c. In **spot lamination,** adhesive is brushed or daubed on the back of the face ply at close intervals. Partitions with strip or spot lamination provide better sound control than those with sheet lamination.

Supplemental fasteners or temporary bracing is required to ensure complete bond between face and base plies. If a fire rating is not required, temporary fasteners may be placed at 24-in intervals. For fire-rated assemblies, permanent fasteners generally are required, and spacing depends on that used in the assembly tested and rated. Nails should penetrate at least 1⅛ in into supports. With sound-deadening base plies, fastener spacing should be as recommended by the base manufacturer.

TABLE 12-6 Maximum Spacing of Framing for Two-Ply Gypsumboard

Base thickness, in	Face thickness, in	Orientation Base	Orientation Face	Spacing, c to c, in Fasteners only	Spacing, c to c, in Adhesive between plies
		a. Application in ceilings			
3/8	3/8*	Across	Across	16	16
3/8	3/8*	Across	Parallel	†	16
1/2	3/8*	Parallel	Across	16	16
1/2	1/2	Parallel	Across	16	16
1/2	1/2*	Across	Across	24	16
5/8	1/2*	Across	Across	24	24
5/8	1/2*	Parallel	Across or parallel	†	24
5/8	5/8	Across	Across	24	24
5/8	5/8	Parallel	Across or parallel		24
		b. Application in walls			
3/8,1/2,5/8	3/8	Across or parallel		16	24
3/8,1/2,5/8	1/2,5/8	Across or parallel		24	24

*Gypsumboard for ceilings to receive a water-base-spray texture finish should be applied only across (perpendicular to) framing. For 16-in spacing of framing, board thickness should be increased from 3/8 to 1/2 in. For 24-in spacing of framing, board thickness should be increased from 1/2 to 5/8 in.
†Not recommended.

In placing face ply on base ply, joints in the two layers should be offset at least 10 in. Face ply may be applied horizontally or vertically. Horizontal application (long sides of sheet perpendicular to supports) usually provides fewer joints, but vertical application may be preferred for predecorated wallboard that is to have joints trimmed with battens.

12-45. Finishing Procedures for Gypsumboard.

The finish surface of wallboard may come predecorated or may require decoration. Predecorated wallboard may require no further treatment other than at corners, or may need treatment of joints and concealment of fasteners. Corner and edge trim are applied for appearance and protection, and battens often are used for decorative concealment of flush joints. Other types of wallboard require preparation before decoration can be applied.

While trim can be applied to undecorated wallboard, as for predecorated panels, usually, instead, joints are made inconspicuous.

Joint-Treatment Products. Materials used for treatment of wallboard joints to make them inconspicuous should meet the requirements of ASTM C475. Application of joint treatments should conform with ASTM C840. These materials include:

Joint tape, a strip of strong paper reinforcement with feathered edges, for embedment in joint compound. Sometimes supplied with small perforations, the tape usually is about 2 in wide and $\frac{1}{16}$ in thick.

Joint compound, and adhesive, with or without fillers, for bonding and embedding the tape. Two types may be used. One, usually referred to as joint, or taping, compound, is applied as an initial coat for filling depressions at joints and fasteners and for adhering joint tape. The second, called topping, or finishing, compound, is used to conceal the tape and for final smoothing and leveling at joints and fasteners. As an alternative, an all-purpose compound that combines the features of taping and topping compounds may be used for all coats. Compounds may be supplied premixed, or may require addition of water on the job.

Flush Joints and Fasteners. For concealing fastener heads and where wallboards in the same plane meet, at least three coats of joint compound should be used. The first coat, of taping compound, should be used for adhering tape at edges and ends of the boards and to fill all depressions over fastener heads and at tapered edges. Joint tape should be centered over each joint for the length or width of the wall or panel and pressed into the compound, without wrinkling, with a tape applicator, a broad knife. Excess compound should be redistributed as a skim coat over the tape. A skim coat of joint compound applied immediately after tape embedment reduces the possibility of edge wrinkling or curling, which may result in edge cracking.

A second, or bedding, coat should be applied with topping compound at fastener heads and joints after the taping compound has dried. The waiting period may be 24 hr if regular taping compound is used, and about 3 hr if a quick-setting compound is used. This bedding coat should be thin. At joints, it should completely cover the tape and should be feathered out 1 to 2 in beyond each edge of the tape at tapered edges, and over a width of 10 to 12 in at square edges.

When the coat has dried, a second, thin coat of the topping compound should be applied for finish at fasteners and joints (Fig. 12-26). At joints, it should completely cover the preceding coat and should be feathered out 1 to 2 in beyond its edges at tapered edges, and over a width of 18 in at square edges. Irregularities may be removed by light sanding after each coat has hardened. But if joint treatment is done skillfully, such sanding should not be necessary.

Fig. 12-26 Application of joint compound to conceal a wallboard joint.

Reentrant Corners. Treatment of interior corners differs only slightly from that of flush joints. Joint compound should first be applied to both sides of each joint. Joint tape should be creased along its center, then embedded in the compound to form a sharp angle. After the compound has dried, finishing compound should be applied, as for flush joints.

Trim. Exposed corners and edges should be protected with a hard material. Wood or metal casings may be used for protection and to conceal the joints at door and window frames. Baseboard may be attached to protect wallboard edges at floors. Corner beads may be applied at exterior corners when inconspicuous protection is desired. Casing beads may be similarly used in other places.

At least three-coat finishing with joint compound is required to conceal fasteners and obtain a smooth surface with trim shapes.

Decoration. Before wallboard is decorated, imperfections should be repaired. This may require filling with joint compound and sanding. Joint compound should be thoroughly dry before decoration starts.

The first step should be to seal or prime the wallboard surface. (Glue size, shellac, and varnish are not suitable sealers or primers.) An emulsion sealer blocks the pores and reduces suction and temperature differences between paper and joint compound. If wallpaper is used over such a sealer, the paper can be removed later without damaging the wallboard surface and will leave a base suitable for redecorating. A good primer provides a base for paint and conceals color and surface variations. Some paints, such as high-quality latex paints, often have good sealing and priming properties.

12-46. Isolation and Control Joints.

Plaster and gypsumboard construction have low resistance to stresses induced by structural movements. They also are subject to dimensional changes because of changes in temperature and humidity. If proper provision is not made in the application of plaster and gypsumboard systems for these conditions, cracking or chipping may occur or fasteners may pop out.

Generally, such defects may be prevented by applying these systems so that movement is not restrained. Thus, gypsumboard or lath and plaster surfaces should be isolated by expansion or control joints, floating angles, or other means where a partition or ceiling abuts any structural element, except the floor, or meets a dissimilar wall or partition assembly, or other vertical penetration. Isolation also is advisable where the construction changes in the plane of the partition or ceiling. In addition, expansion or control joints should be inserted where expansion or control joints, respectively, occur in structural components of the building and along the intersection of wings of L-, U-, and T-shaped ceilings.

In long walls or partitions, control joints should be inserted not more than 30 ft apart. In large ceiling areas, control joints should not be spaced more than 30 ft apart for plaster and 50 ft apart for gypsumboard. Control joints may be conveniently located along lighting fixtures, heating vents, or air-conditioning diffusers, where continuity would ordinarily be broken. Door frames extending from floor to ceiling may serve as control joints. With shorter frames, however, control joints should be inserted from the top corners of the frames to the ceiling.

Installation of Joints. Where a control joint is to be provided, the continuity of gypsumboard or lath in plaster installations should be interrupted (Fig. 12-27). The gap for gypsumboard should be at least ½ in and for lath at least ¼ in. A framing member should be inserted on each side of the gap to support the edges of the boards or lath. (Steel studs, however, should be fastened on only one side of the joint.)

The joint should be covered at the surface with a flexible material to prevent entrance of dirt into the wall. For the purpose, plastic or noncorrodible-metal, bellows-type stops, called expansion joints or control joints, may be used (Fig. 12-27). For plaster, the flanges of the control joint should be stapled or wire tied to the lath, and the surface of the gap should be protected with tape. Then, plaster should be applied over the lath, flush to grounds. After the finished surface has been completed, the protective tape should be

Fig. 12-27 Control joints: (*a*) expansion joint in a plastered, metal-stud partition; (*b*) control joint in a wallboard, metal-stud wall; (*c*) control joint in a plastered, suspended ceiling.

12-62 Walls, Partitions, and Interior Finishes

removed. For gypsumboard, similarly, the control-joint flanges should be stapled to the board and the gap protected with tape. Next, the joint should be treated with joint compound or veneer plaster. Then, the protective tape should be removed. Where sound control or fire ratings are important, a sealant should be installed behind the control joints.

Floating Interior Angles. A floating angle is desirable at reentrant intersections of ceilings and walls or partitions (Fig. 12-28a). It is constructed by installing gypsum lath or gypsumboard first in the ceiling. The first line of ceiling fasteners should be spaced as shown in Fig. 12-28b and c. Normal fastener spacing should be used in the rest of the ceiling away from the wall. Gypsum lath or gypsumboard should then be applied to the wall with firm contact at the ceiling line to support the edges of the ceiling panels. The first line of fasteners for the wall panels also should be spaced as shown in Fig. 12-28b and c. At wall intersections, fasteners should be omitted or inserted as indicated in Fig. 12-28d.

("Architect Data Book—Construction Products and Systems," Gold Bond Building Products, a National Gypsum Division, 2001 Rexford Road, Charlotte, N.C. 28211; "Gypsum Products Design Data," Gypsum Association, 1603 Orrington Ave., Evanston, Ill. 60201; "Gypsum Construction Handbook," United States Gypsum, 101 South Wacker Drive, Chicago, Ill. 60606.)

Fig. 12-28 Floating interior angles with gypsumboard construction (gypsum lath similar): (a) omission of nails at intersections; (b) wall-ceiling intersection with ceiling boards applied parallel to joists; (c) wall-ceiling intersection with ceiling boards applied perpendicular to joists; (d) reentrant corner at walls.

FLOOR COVERINGS*

For proper selection of floor coverings, designers should take into account many factors, including use of lightweight-concrete slabs, subfloors in direct contact with the ground, radiant heating, air conditioning, possible necessity for decontamination, dustlessness, traffic loads, and maintenance costs—all of which have an important bearing on flooring selection. Consideration must be given to current standards of styling, comfort, color, and quietness.

The primary consideration of the designer of a flooring system is to select a floor covering that can meet the maximum standards at reasonable cost. To avoid the dissatisfaction that would arise from failure to select the proper flooring, designers must consider all the factors relevant to flooring selection.

This section contains information that can provide a guide toward this end. It summarizes the characteristics of the major types of floor coverings and describes briefly methods for the proper installation of these materials. (For ceramic-tile installations, see Arts. 12-29 and 12-30.)

12-47. Asphalt Tiles.

These tiles are intended for use on rigid subfloors, such as smooth-finished or screeded concrete, structurally sound plywood, or hardboard floors not subject to excessive dimensional changes or flexing. The tiles can be satisfactorily installed on below-grade concrete subject to slight moisture from the ground.

Low cost and large selection of colors and designs make asphalt tile an economically desirable flooring.

Asphalt tile is composed of asbestos fibers, mineral coloring pigments, and inert fillers bound together. For dark colors the binder is Gilsonite asphalt; for intermediate and light colors, the binder may consist of resins of the cumarone indene type or of those produced from petroleum. Tiles most commonly used are 9×9 in and $\frac{1}{8}$ in thick.

Colors are classified into groups A, B, C, and D, graded from black and dark red (A) to cream, white, yellow, blue, and bright red (D). Cost is generally lower for the darker colors.

To avoid permanent indentations in asphalt tiles, contact surfaces of furniture or equipment should be smooth and flat to distribute the weight. This is particularly necessary for installations over radiant-heated floors and on areas near windows exposed to sun.

Never use on asphalt tiles waxes containing benzene, turpentine, or naphtha-type solvents and free fats or oils. Avoid strong detergents or cleaning compounds containing abrasives or preparations not readily soluble in water. These may soften the tiles and cause colors to bleed. Grease, oils, fats, vinegar, and fruit juices allowed to remain in contact with asphalt tiles will stain and soften them. Because of these restrictions, asphalt tiles are not recommended for use in kitchens or bathrooms.

See also Art. 12-54.

12-48. Cork Tiles.

Cork flooring is intended for use on rigid subfloors, such as smooth-finished or screeded concrete supported above grade and free of moisture, or on structurally sound plywood or hardboard. Cork tile is not recommended for application below grade. When it is installed at grade, moisture-free conditions must be ensured.

---

*Excerpted with revisions from Sec. 14, Floor Coverings, by Thomas H. Boone and John E. Fitzgibbons, of the "Building Construction Handbook, 3d ed."

Cork tile is manufactured by baking cork granules with phenolic or other resin binders under pressure. Four types of finishes are produced: natural, factory-prefinished wax, resin-reinforced wax, and vinyl cork tile (Art. 14-8). The tiles are generally 6×6, 6×12, 9×9, 12×12, 12×24, or 36×36 in and $\frac{1}{8}$, $\frac{3}{16}$, $\frac{5}{16}$, or $\frac{1}{2}$ in thick.

Natural cork tile must be sanded (to level), sealed, and waxed immediately after installation.

Unless the exposed surface of cork floors is maintained with sealers and protective coatings, permanent stains from spillage and excessive soiling by heavy traffic will result.

Cork tiles are particularly suitable for areas where quiet and comfort are of paramount importance.

See also Art. 12-54.

12-49. Vinyl Flooring.

Flooring of this type is unbacked. For a discussion of backed vinyl, see Art. 12-51. It is intended for use on rigid subfloors, such as smooth-finished or screeded concrete supported above grade, or structurally sound plywood or hardboard floors. Vinyl floors are not recommended for use below grade. They must be applied with an alkaline, moisture-resistant adhesive when used at grade.

Vinyl mats or runners may be laid without adhesive over relatively smooth surfaces. Large mats generally are installed in a recess in the concrete floor at building entrances. The mats are ribbed or perforated for drainage.

Vinyl flooring consists predominantly of polyvinyl chloride resin as a binder, plasticizers, stabilizers, extenders, inert fillers, and coloring pigments. Because of its unlimited color possibilities and opaqueness to transparent effects, it is widely used. Common thicknesses are 0.080, $\frac{3}{32}$, and $\frac{1}{8}$ in.

Since vinyl resins are tough synthetic polymers, vinyl flooring can withstand heavy loads without indentation, and yet is resilient and comfortable under foot. It is practically unaffected by grease, fat, oils, household cleaners, or solvents. But it is easily scratched and scuffed. The lustrous surface of new tile is not retained under constant foot traffic unless adequately protected with a floor polish.

See also Art. 12-54.

12-50. Vinyl-Asbestos Tiles.

These are intended for use on rigid subfloors, such as smooth-finished or screeded concrete, or structurally sound plywood or hardboard subfloors not subjected to excessive dimensional change or flexing. Like asphalt tiles, vinyl-asbestos tiles can be installed satisfactorily on below-grade concrete subject to slight moisture from the ground.

Vinyl-asbestos tiles are composed of asbestos fibers, mineral coloring pigments, and inert fillers, bound together with vinyl resin and plasticizers. They have brighter and clearer colors than asphalt tiles, which have a similar composition except for the binder. Tiles usually come 9×9 in. Common thicknesses are $\frac{1}{16}$ and $\frac{1}{8}$ in.

Vinyl-asbestos tile has a duller and harder surface than vinyl tile, and therefore its appearance is not damaged as much by foot traffic. Resistance to oils, grease, and household reagents is a little less than that of vinyl.

See also Art. 12-54.

12-51. Backed Vinyl.

The family of backed-vinyl flooring comprises vinyl wearing surfaces from 0.02 to 0.050 in thick, laminated to many different backing materials. In some products, the vinyl surfaces are unfilled transparent films placed over a design on paper, cork, or degraded vinyl. Filled vinyl surfaces with a 34% vinyl resin binder are placed over plastic composition backing or asphalt-saturated or resin-impregnated felt. The asphalt-felt type may be used in moist areas. Foamed rubber or plastic is incorporated in some of these materials to increase comfort and decrease impact noise.

Asphalt-felt-backed vinyl materials may be applied with a moisture-resistant adhesive on concrete at or below grade. Installation of other backed vinyl flooring is similar to that of linoleum.

See also Art. 12-54.

12-52. Rubber.

Rubber flooring is intended for use on rigid subfloors, such as smooth-finished or screeded concrete supported above grade, or on structurally sound plywood or hardwood subfloors. Rubber is not recommended for use below grade. When used at grade, it must be applied with an alkaline, moisture-resistant adhesive.

Rubber mats or runners may be laid without adhesive over relatively smooth surfaces. Large mats generally are installed in a recess in the concrete floor at building entrances. The mats are ribbed or perforated for drainage.

Most rubber flooring is produced from styrene-butadiene rubber. Reclaimed rubber is added to some floorings. The flooring also contains mineral pigments and mineral fillers, such as zinc oxide, magnesium oxide, and various clays. Another synthetic-rubber flooring, chlorosulfonated polyethylene (Hypalon), also is available.

Rubber floorings can be obtained in thicknesses of $\frac{3}{32}$, $\frac{1}{8}$, or $\frac{3}{16}$ in. They have excellent resistance to permanent deformation under load. Yet they are resilient and quiet under foot.

See also Art. 12-54.

12-53. Linoleum.

Linoleum basically is made from drying oils, such as linseed, natural and synthetic resins, fillers, and pigments similar to those used in paints. The filler is largely wood flour. The flooring usually has a backing or carrier of burlap or rag felt.

Since the burlap and rag felt backings of linoleum are susceptible to moisture and fungus attack, and the wearing surface is damaged by alkaline solutions, linoleum should never be used on floors where moisture can accumulate between the flooring and subfloor. Like wood and cork floors, linoleum should not be used on concrete at or below grade.

With burlap backing (commercial linoleum), thickness is $\frac{1}{8}$ or $\frac{3}{16}$ in. With felt backing (household linoleum), overall thickness may be 0.070 in (light gage) or 0.090 in (standard gage), with thickness of linoleum wearing surface of 0.03 or 0.05 in.

Linoleum is available as sheet or tile. When properly maintained, it performs outstandingly in residential and commercial buildings.

See also Art. 12-54.

12-54. Installation of Thin Coverings.

Most manufacturers and trade associations make available instructions and specifications for installation and maintenance of their floorings.

The most important requirement for a satisfactory installation of a thin floor covering is a dry, even, rigid, and clean subfloor.

Protection from moisture is a prime consideration in applying a flooring over concrete. Moisture within a concrete slab must be brought to a low level before installation begins. Moisture barriers, such as 6-mil polyethylene, 55-lb asphalt-saturated and coated roofing felt, or $\frac{1}{32}$-in butyl rubber, should be placed under concrete slabs at or below grade, and a minimum of 30 days' (90 in some cases) drying time should be allowed after placement of concrete before installation of the flooring.

Particular care should be taken with installations on lightweight concrete. It always has a higher gross water requirement than ordinary concrete and therefore takes longer to dry. So a longer drying period should be allowed before installation of flooring. Flooring manufacturers provide advice and sometimes also equipment to test for moisture.

An indication of dryness at any given time is no assurance that a concrete slab at or below grade will always remain dry. Therefore, protection from moisture from external sources must be given considerable attention. (See also Art. 3-13.)

Some concrete curing agents containing oils and waxes may cause trouble with adhesive-applied flooring. Commercial curing agents that have been shown to be satisfactory include styrene-butadiene copolymer or petroleum-hydrocarbon resin dissolved in a hydrocarbon solvent. These products dry to a hard film in 24 hr. Parting agents, used as slab separators in lift-slab and tiltup construction, may cause bond failure of adhesive-applied flooring if they contain nonvolatile oils or waxes. Concrete surfaces that have been treated with or have come in contact with oils, kerosene, or waxes must be cleaned as recommended by the adhesive manufacturer.

All concrete surfaces to receive adhesive-applied, thin flooring must be smooth, and free from serious irregularities that would "telegraph" through the covering and be detrimental to appearance and serviceability. For rough or uneven concrete floors, a troweled-on underlayment of rubber latex composition or asphalt mastic is recommended. It can be applied from a thickness of $\frac{1}{4}$ in to a featheredge. Small holes, cracks, and crevices may be filled with a reliable cement crack filler.

Wood subfloors, of sufficient strength to carry intended loads without deflection, should be covered with plywood or hardboard underlayments. These should be nailed 6 in c to c in perpendicular directions over the entire area with ringed or barbed nails. The nails should be driven flush with the underlayment surface without denting it around the nail head.

Adhesives commonly used to attach flooring to concrete and underlayment are listed in Table 12-7.

12-55. Carpets.

Extending from wall to wall, carpets are frequently used as floor coverings in residences, offices, and retail stores. They are often selected for the purpose because they offer foot comfort and, being available in many colors, patterns, and textures, attractive appearance. Rugs, often used as an alternative in residences, differ from carpets chiefly in being single pieces of definite shape and usually not covering an entire floor between walls.

A carpet is a thick, heavy fabric that is usually piled but could be woven or felted. Pile consists of closely placed loops of fiber, or tufts, that produce a raised surface on a backing to which they are locked. The tufts may be sheared to produce a soft, velvety surface with a wide variety of patterns and textures.

Sheared or unsheared, the piled fabric is very resilient, thus contributing to foot comfort. Nevertheless, thin pads, generally of foam rubber or plastic, sometimes are placed under carpets, to improve resilience. They offer the additional advantage of absorbing minor irregularities in the floor surface that otherwise would cause rapid wear in local areas.

TABLE 12-7 Adhesives Used to Install Flooring*

Flooring	Concrete below grade	Concrete on grade	Concrete above grade	Plywood or hardboard
Asphalt and vinyl-asbestos tiles	Asphalt, cutback Asphalt, emulsion	Asphalt, cutback Asphalt, emulsion	Asphalt, cutback Asphalt, emulsion	Asphalt, cutback Asphalt, emulsion
Rubber and vinyl	Chemical set Latex	Latex	Latex	Latex
Linoleum, cork, and vinyl backed with felt or cork	Do not install	Do not install	Linoleum paste	Linoleum paste
Vinyl backed with asbestos felt	Latex	Latex	Latex Linoleum paste	Linoleum paste
Laminated wood block	Do not install	Asphalt, hot melt Asphalt, cutback Rubber base	Asphalt, hot melt Asphalt, cutback Rubber base	Asphalt, hot melt Asphalt, cutback Rubber base
Solid unit wood block	Do not install	Asphalt, hot melt Asphalt, cutback	Asphalt, hot melt Asphalt, cutback	Asphalt, hot melt Asphalt, cutback

* The adhesives listed are intended for each type of flooring; i.e., asphalt adhesives used for asphalt and vinyl-asbestos tiles are not the same as asphalt adhesives used for wood blocks. There are a number of adhesives having special properties and characteristics, such as heat resistance, resistance to water spillage, and brush-on types not included in this table.

Fibers used for tufting indoor carpets include wool, acrylic, polyester, continuous-filament or heat-set spun nylon and nylon with antistatic treatment for high resistance to soiling. Pile weight generally ranges between 15 and 40 oz/yd^2. The primary backing, to which the pile is tufted and which provides dimensional stability to the carpet, usually is polypropylene weighing 3.5 oz/yd^2. A secondary backing, which generally is attached to the primary backing with a latex adhesive, is used to protect the underside of the carpet and improve other characteristics, such as resilience and noise absorption. The secondary backing usually is woven jute weighing 7 oz/yd^2, nonwoven polypropylene weighing 3.5 oz/yd^2, or $3/16$-in-thick, high-density foam rubber weighing 38 oz/yd^2, which eliminates the need for an underlying pad.

In selection of carpeting, consideration should be given to the intensity of traffic to which the covering will be subjected; availability of desired colors, patterns, and textures; colorfastness; resistance to crushing and matting; soil resistance; cleanability; resistance to fuzzing, beading, and pilling, as measured by bundle wrap and latex penetration on the underside of the primary backing; subfloor conditions; and installed cost of the carpet. Colorfastness may be judged by performance in an 80-hr xenon-arc fadeometer test and in wet-method cleaning in accordance with Federal Specification DDD-C-001559 and by negligible crocking (color transfer to a white cloth). Preference should be given to carpets that meet the following requirements:

Flammability: flame spread rating (ASTM E84) of 75 or less; flame propagation index (floor chamber test) of 4.0 or less (see Underwriters Laboratories Subject 992); and corrected maximum specific optical density of 450 or less in the National Bureau of Standards smoke-density-chamber test.

Static propensity: 3.0 kV or less.

Acoustical properties: impact noise rating of 30 or more; impact insulation class 81 or more; NC 30 masking level of 30 or less; and noise reduction coefficient, measuring airborne sound absorption, of 0.5 or more.

Tuft bind: average force of 10 lb or more required to pull out tufts from the face of the carpet (ASTM D1335).

Installation of Carpet. Carpet is supplied usually in widths of 12 or 15 ft in rolls in long lengths. It may be cut to desired sizes and shapes with a carpet knife. Strips of carpet are laid side by side to extend the covering from wall to wall. Joints may be stitched or taped.

Installation should conform to recommendations of the carpet manufacturer. In general, carpet may be laid on any firm, smooth floor.

Carpets with jute or synthetic secondary backing generally should be stretched over a good-quality pad, to eliminate bulges, and anchored at the walls with tacks or tackless strips. With this type of installation, carpets may be removed easily when replacement is necessary. However, they must still be cleaned in place, may require restretching, and can be difficult to repair. (Power stretchers should be used for carpets with synthetic secondary backing.) Alternatively, this type of carpet may be directly cemented to subfloors, eliminating an underlying pad and future restretching. But wearability may be lower and there may be a greater tendency to soil under heavy traffic.

Carpets with high-density foam-rubber backing also may be cemented directly to subfloors. Such carpets, however, are not suitable for carrying heavy traffic and may be difficult to remove when replacement is necessary.

In all cases, use of chair pads under castered chairs is desirable.

12-56. Terrazzo.

A Venetian marble mosaic, with portland cement matrix, terrazzo is composed of two parts marble chips to one part portland cement. Color pigments may be added. Three methods of casting in place portland-cement terrazzo atop structural concrete floor slabs are commonly used: sand cushion, bonded, and monolithic.

Sand-cushion (floating) terrazzo is used where structural movement that might injure the topping is anticipated from settlement, expansion, contraction, or vibration. This topping is at least 3 in thick. First, the underlying concrete slab is covered with a $\frac{1}{4}$- to $\frac{1}{2}$-in bed of dry sand. Over this is laid a membrane, then wire-fabric reinforcing. The terrazzo underbed is installed to $\frac{5}{8}$ in below the finished floor line. Next, divider strips are placed and finally, the terrazzo topping.

Bonded terrazzo has a minimum thickness of $1\frac{3}{4}$ in. After the underlying concrete slab has been thoroughly cleaned and soaked with water, the surface is slushed with neat portland cement to ensure a good bond with the terrazzo. Then, the underbed is laid, divider strips are installed, and terrazzo is placed.

Monolithic terrazzo is constructed by placing a $\frac{5}{8}$-in topping as an integral part of a green-concrete slab. Adhesive-bonded monolithic terrazzo with an epoxy resin adhesive also has been used successfully, with a topping thickness of only $\frac{3}{8}$ in.

Terrazzo may be precast. It generally is used in this form for treads, risers, platforms, and stringers on stairs.

Because of the large variety of color and surface textures that can be attained with terrazzo, it is used extensively as an exterior and interior decorative flooring. Portland-cement terrazzo, however, should not be used in areas subject to spillage, such as might be encountered in kitchens.

Details on selecting the proper type of terrazzo, marble-chip sizes, methods of applying, and finishing of the surface may be obtained from the National Terrazzo and Mosaic Association, 2-A West Loudon St., Leesburg, Va. 22075. Also, specification sheets for terrazzo are published by the American Institute of Architects.

Other matrix materials used with marble chips include rubber latex, epoxy, and polyesters. Suppliers should be consulted for installation details.

12-57. Concrete Floors.

A concrete topping may be applied to a concrete structural slab before or after the base slab has hardened. Integral toppings usually are ½ in thick; independent toppings about 1 in.

It is well to remember in specifying or installing a concrete floor that the difference between good and poor concrete lies in selection and grading of aggregates, proportioning of the mix, and the care with which the vital operations of placing, finishing, and curing are carried out. (See also Sec. 10.)

(ACI302.1, "Guide for Concrete Floor and Slab Construction," American Concrete Institute.)

12-58. Wood Floors.

Both hardwoods and softwoods are used for floors. Hardwoods most commonly used are maple, beech, birch, oak, and pecan. Softwoods are yellow pine, Douglas fir, and western hemlock. The hardwoods are more resistant to wear and indentation than softwoods.

Hardwood strip floorings are available in thicknesses of $\frac{11}{32}$, $\frac{15}{32}$, and $\frac{25}{32}$ in and in widths of 1½, 2, 2¼, and 3¼ in.

Softwood strip flooring usually is $\frac{25}{32}$ in thick and can be obtained in widths of 2⅜, 3¼, and 5$\frac{3}{16}$ in.

Solid-unit wood blocks for floors are made from two or more units of strip-wood flooring fastened together with metal splines or other suitable devices. A block usually is square. Tongued and grooved, either on opposite or adjacent sides, it is held in place with nails or an asphalt adhesive (Table 12-7). Blocks $\frac{25}{32}$ in thick are made up of multiples of 1½- or 2¼-in strips; blocks ½ in thick are composed of multiples of 1½- or 2-in strips.

A laminated block is formed with plywood comprising three or more plies of wood glued together. The core or cross bonds are laid perpendicular to the face and back of the block. Usually square, the block is tongued and grooved on either opposite or adjacent sides. The most common thickness is ½ in, but other thicknesses used are ⅜, $\frac{7}{16}$, ⅝, and $\frac{13}{16}$ in. Laminated blocks are installed with adhesives (Table 12-7).

Average moisture content of wood flooring at time of installation should be 6% in dry southwestern states, 10% in damp southern coastal states, and 7% in the rest of the United States ("Moisture Content of Wood in Use," U.S. Forest Products Laboratory Publication No. 1655, Madison, Wis.). The U.S. Forest Products Laboratory also recommends heating the building before installing any type of wood flooring, except when outdoor temperatures are high. Bundles of flooring should be opened and stored in the building before installation. Leave at least 1 in of expansion space at walls and columns.

Specifications for wood floors on concrete, gymnasium floors, or other special designs or conditions may be obtained from associations of wood flooring manufacturers and installers.

12-59. Industrial Floors.

The primary purpose of a floor covering in an industrial building is to protect the structural floor from foot and truck traffic and from corrosive effluents. The flooring also must provide a durable surface that will not be detrimental to plant operations.

12-70 Walls, Partitions, and Interior Finishes

A good-quality concrete floor is adequate if factory conditions are dry and the surface has to withstand only foot and truck traffic. A cast-iron grid filled with concrete on ¼-in-thick steel plates may be used in areas where heavy equipment or materials are dragged continually. End-grain wood blocks may be used in areas of heavy trucking of heavy castings, to protect both the structure and castings from impact damage. Asphalt mastic floors also are used in areas with heavy truck traffic.

If the factory process is wet and corrosive, the designer may have to choose between a less-resistant flooring that has to be replaced periodically and a flooring with low maintenance costs but high initial cost. Table 12-8 rates the relative resistance to liquids of some floor toppings and bedding and jointing materials for clay tiles and bricks. In addition to selecting the most suitable materials, the designer should provide adequate slope and drainage, a liquid-tight layer below the finished floor as a second line of defense, and a nonslip surface. Selection and laying of chemical-resistant materials are a specialized operation on which specialists should be consulted.

12-60. Conductive Flooring.

Where explosive vapors are present, sparks resulting from accumulation of static electricity constitute a hazard. The most effective of several possible ways of mitigating this hazard is to keep electrical resistance so low that dangerous voltages never are attained.

Most objects normally rest or move on the floor and therefore can be electrically connected through the floor. Flooring of sufficiently low electrical resistance (conductive flooring) thus is of paramount importance in elimination of electrostatic hazards. But the electrical resistance must be high enough to eliminate electric shock from faulty electrical wiring or equipment.

Electrical conductivity of flooring is improved by addition of acetylene black (carbon). In ceramic, linoleum, rubber, and vinyl floors, the carbon is finely dispersed in the material during manufacture; in latex terrazzo, concrete terrazzo, and setting-bed

TABLE 12-8 Resistance of Floor Toppings and of Bedding and Jointing Material to Various Liquids*

Material	Water	Organic solvents	Oils	Acids	Alkalies
Floor topping:					
Portland cement concrete	VG	G	F	VP	G
High-alumina cement concrete	G	G	F–G	VP	F
Pitch (coal tar) mastic	VG	F	F	F–G	G
Asphalt mastic	VG	VP	P	F–G	G
Epoxy	VG	G	G	G	G
Polyester	VG	G	G	G	F
Bedding and jointing materials for clay tiles and bricks:					
Portland cement mortar		As above for concrete			
High-alumina cement mortar		As above for concrete			
Silicate cement	F	G	G	G	VP
Sulfur cement	VG	G	F	G	G
Epoxy			As above		
Furan	VG	VG	VG	VG	VG

VG = very good; G = good; F = fair; P = poor; VP = very poor.

*This table is intended for broad comparisons, as the items under various headings may need qualifications for particular conditions, such as temperature, and concentration of the liquids and acids that are oxidizing agents.

cement for ceramic tile, the carbon is uniformly dispersed in the dry powder mixes, placed in containers, and shipped for on-the-job composition ("Conductive Flooring for Hospital Operating Rooms," Monograph 11, National Bureau of Standards, Washington, D.C. 20234).

For linoleum flooring, brass seam connectors with projecting points are used for electrical intercoupling between sheets. For vinyl tiles, copper foil is placed between the adhesive and tile for electrical intercoupling.

When flammable gases are in use, everyone and everything must be electrically intercoupled via a static conductive floor at all times. To ensure this, constant vigilance, inspection by testing, and a high standard of housekeeping are required. Wax or dirt accumulation on the floor and grounding devices can provide high resistance to electrical flow and cancel the effect of conductive flooring.

12-61. Specifications and Standards for Flooring.

Federal specifications are written to establish minimum standards for purposes of competitive bidding and to ensure that government agencies will receive quality material. Some flooring materials, however, cannot be classed under these specifications because only one manufacturer makes the products. Prices of the Federal specifications may be obtained from the General Services Administration, Washington, D.C. 20405, or the nearest GSA regional office.

Commercial Standards are specifications that establish quality levels for manufactured products in accordance with the principal demands of the trade. The Office of Commodity Standards, National Bureau of Standards, Washington, D.C. 20234, will supply a price list.

Section 13

Windows

Phillip M. Grennan
Consulting Engineer, Rockville Centre, N.Y.

Selection of a proper window for a building entails many considerations. Functional requirements are usually most important. However, size, type, arrangement, and materials very often establish the character of the building, and this external show of fenestration is an integral part of design and expression.

13-1. How Much Window Area?

Distribution and control of daylight, desired vision of outdoors, privacy, ventilation control, heat loss, and weather resistance are all-important aspects of good design principles. Most building codes require that glass areas be equal to at least 10% of the floor area of each room. Nevertheless, it is good practice to provide glass areas in excess of 20% and to locate the windows as high in the wall as possible to lengthen the depth of light penetration. Continuous sash or one large opening in a room provides a more desirable distribution of light than separated narrow windows, and eliminates the dark areas between openings.

Location, type, and size of window are most important for natural ventilation. The pattern of air movement within a building depends to a great extent on the angle at which the air enters and leaves. It is desirable, particularly in summer, to direct the flow of air downward and across a room at a low level. However, the type and location of windows best suited to ventilation may not provide adequately for admission of light and clear vision, or perhaps proper weather protection. To arrive at a satisfactory relationship, it may be necessary to compromise with these functional requirements.

While building codes may establish a minimum percentage of glass area, they likewise may limit the use of glass and also require a fire rating in particular locations. In hazardous industrial applications subject to explosions, scored glass is an added precaution for quick release of pressure.

Heat transmittance is of economic importance and can also affect comfort. Weather stripping or the use of windows with integral frame and trim can minimize air infiltration. Double glazing or heat-absorbing glazing makes large glass areas feasible by mate-

13-2 Windows

rially reducing heat transmittance. Solar heating may be achieved by placement of glass areas so that the rays of the sun can be admitted during the winter.

13-2. Window Materials.

Window units are generally made of wood or metal. Plastics have been employed for special services.

Use of metal in window construction has been highly developed, with aluminum and steel most commonly used. Bronze, stainless steel, and galvanized steel are also extensively used for specific types of buildings and particular service. Use of metal windows is very often dictated by fire-resistance requirements of building codes.

13-3. Wood Windows.

These can be used for most types of construction and are particularly adaptable to residential work. The most important factor affecting wood windows is weather exposure. However, with proper maintenance, long life and satisfactory service can be expected. Sash and frames have lower thermal conductivity than metal.

The kinds of wood commonly used to resist shrinking and warping for the exposed parts of windows are white pine, sugar pine, ponderosa pine, fir, redwood, cedar, and cypress. The stiles against which a double-hung window slides should be of hard pine or some relatively hard wood. The parts of a window exposed to the inside are usually treated as trim and are made of the same material as the trim. (See also Art. 13-4 and Figs. 13-1 to 13-3.)

Fig. 13-1 Relation of modular standard double-hung wood windows to grid opening in a brick wall.

13-4. Definitions for Wood Windows.

Sash. A single assembly of stiles and rails made into a frame for holding glass, with or without dividing bars. It may be supplied either open or glazed.

Window. One or more single sash made to fill a given opening. It may be supplied either open or glazed.

Window Unit. A combination of window frame, window, weather strip, balancing device, and at the option of the manufacturer, screen and storm sash, assembled as a complete and properly operating unit.

Measurements:

Between Glass. Distance across the face of any wood part that separates two sheets of glass.

Face Measure. Distance across the face of any wood part, exclusive of any solid mold or rabbet.

Finished Size. Dimension of any wood part overall, including the solid mold or rabbet.

Fig. 13-2 Dimensions for multiple modular openings for double-hung wood windows.

Wood Allowance. The difference between the outside opening and the total glass measurement of a given window or sash.
Stiles. Upright, or vertical, outside pieces of a sash or screen.
Rails. Cross, or horizontal, pieces of the framework of a sash or screen.
Check Rails. Meeting rails sufficiently thicker than the window to fill the opening between the top and bottom sash made by the check strip or parting strip in the frame. They are usually beveled and rabbeted.
Bar. Either vertical or horizontal member that extends the full height or width of the glass opening.
Muntin. Any short light bar, either vertical or horizontal.
Frame. A group of wood parts so machined and assembled as to form an enclosure and support for a window or sash.
Jamb. Part of frame that surrounds and contacts the window or sash the frame is intended to support.
Side Jamb. The upright member forming the vertical side of the frame.
Head Jamb. The horizontal member forming the top of the frame.
Rabbeted Jamb. A jamb with a rectangular groove along one or both edges to receive a window or sash.
Sill. The horizontal member forming the bottom of the frame.
Pulley Stile. A side jamb into which a pulley is fixed and along which the sash slides.
Casing. Molding of various widths and thicknesses used to trim window openings.

13-4 Windows

Fig. 13-3 Details of installation of double-hung wood windows. *(National Woodwork Manufacturers Assoc.)*

Parting Stop. A thin strip of wood let into the jamb of a window frame to separate the sash.

Drip Cap. A molding placed on the top of the head casing of a window frame.

Blind Stop. A thin strip of wood machined so as to fit the exterior vertical edge of the pulley stile or jamb and keep the sash in place.

Extension Blind Stop. A molded piece, usually, of the same thickness as the blind stop, and tongued on one edge to engage a plow in the back edge of the blind stop, thus increasing its width and improving the weathertightness of the frame.

Dado. A rectangular groove cut across the grain of a frame member.

Jamb Liner. A small strip of wood, either surfaced four sides or tongued on one edge, when applied to the inside edge of a jamb, increases its width for use in thicker walls.

13-5. Steel Windows.

These are, in general, made from hot-rolled structural-grade new billet steel. However, for double-hung windows, the principal members are cold-formed from new billet strip steel. The principal manufacturers conform to the specifications of The Steel Window Institute, which has standardized types, sizes, thickness of material, depth of sections, construction, and accessories.

The life of steel windows is greatly dependent on proper shop finish, field painting, and maintenance. The more important aspects of proper protection against corrosion are as follows:

Shop Paint Finish. All steel surfaces should be thoroughly cleaned and free from rust and mill scale. A protective treatment of hot phosphate, cold phosphate-chromate, or similar method such as bonderizing or parkerizing is necessary to protect the metal and provide a suitable base for paint. All windows should then be given one shop coat of rust-inhibitive primer, baked on.

Field Painting. Steel windows should receive one field coat of paint either before or immediately after installation. A second coat should then be applied after glazing is completed and putty set.

Hot-Dipped Galvanized. Assembled frames and assembled ventilators should be galvanized separately. Ventilators should be installed into frames after galvanizing and bonderizing. Specifications must guard against distortion by the heat of hot dipping, abrasion in handling, and damage to the zinc galvanizing by muriatic acid used to wash brickwork. Painting is optional.

13-6. Aluminum Windows.

Specifications set up by the Aluminum Window Manufacturers' Association are classified as residential, commercial, and monumental. These specifications provide for a minimum thickness of metal and regulate stiffness of the component parts, as well as limiting the amount of air infiltration.

The manufacturers offer about the same window types that are available in steel. However, in addition to these standard types, substantial progress has been made in the development of aluminum windows as an integral part of the walls of a building. With the exception of ventilator sections, which are shop-fabricated, some windows are completely assembled at the job. The wide range of extruded aluminum sections now manufactured has made it feasible to treat large expanses of glass as window walls. Picture windows with many combinations of ventilating sash are also furnished in aluminum. Many features are available, such as aluminum trim and casing, weather stripping of stainless steel or Monel metal, combination storm sash and screens, sliding wicket-type screens, and metal bead glazing.

Aluminum windows should be protected for shipment and installation with a coating of clear methacrylate lacquer or similar material able to withstand the action of lime mortar.

13-7. Stainless-Steel and Bronze Windows.

These are high-quality materials used for durability, appearance, corrosion resistance, and minimum maintenance. Stainless steel is rustproof and extremely tough, with great strength in proportion to weight. Weather stripping of stainless steel is quite often used for wood and aluminum sash.

Bronze windows are durable and decorative and used for many fine commercial and monumental buildings.

13-8. Window Sizes.

Manufacturers have contributed substantially toward standardization of window sizes and greater precision in fabrication. The American National Standards Institute has established an increment of sizing generally accepted by the building industry ("Basis for the Coordination of Building Materials and Equipment," A62.1).

The basis for this standardization is a nominal increment of 4 in in dimension. However, it will be noted in Fig. 13-1, that for double-hung wood windows the 4-in module applies to the grid dimensions, whereas the standard window opening is 4 in less in width and 6 in less in height than the grid opening. Also, to accommodate 2- and 6-in mullions for multiple openings in brick walls, it is necessary to provide a 2-in masonry offset at one jamb (Fig. 13-2).

Use of modular planning effects maximum economies from the producer and simplification for the building industry. At the same time, a given layout does not confine acceptable products to such a limited range.

13-9. Window Types.

The window industry has kept well apace with modern advances in building design and has made available a wide range of window types to satisfy the ever-increasing scope of requirements. Windows are basically functional, but at the same time, they are usually employed as a means of architectural expression. There are many materials, sizes, arrangements, details, and specific features from which to choose.

Types of windows (see symbols Fig. 13-4) include:

Pivoted windows, an economical, industrial window used where a good tight closure is not of major importance (Figs. 13-5 and 13-8). Principal members of metal units are usually 1⅜ in deep, with frame and vent corners and muntin joints riveted. However, when frame and vent corners are welded, principal members may be 1¼ in deep. Vents are pivoted about 2 in above the center. These windows are adaptable to multiunits, both vertical and horizontal, with mechanical operation. Bottom of vent swings out and top swings in. No provision is made for screening. Putty glazing is placed inside.

Commercial projected windows, similar to pivoted windows (Figs. 13-6 and 13-8). These are an industrial-type window but also are used for commercial buildings where economy is essential. Vents are balanced on arms that afford a positive and easy control of ventilation and can be operated in groups by mechanical operators. Maximum opening is about 35°. Factory provision is made for screens.

Security windows, an industrial window for use where protection against burglary is important, such as for factories, garages, warehouses, and rear and side elevations of stores (Figs. 13-7 and 13-8). They eliminate the need for separate guard bars. They consist of a grille and ventilated window in one unit. Maximum grille openings are 7×16 in. The ventilating section, either inside or outside the grille, is bottom-hinged or projected. Muntins are continuous vertically and horizontally. Factory provision is made for screening. Glazing is placed from the inside.

Fig. 13-4 Symbols for common window types (viewed from outside).

Fig. 13-5 Pivoted window.

Fig. 13-6 Commercial projected window.

Fig. 13-7 Security windows.

Fig. 13-8 Details of installation of commercial projected, pivoted, and security windows.

13-8 Windows

Basement and utility windows, economy sash designed for use in basements, barracks, garages, service stations, areaways, etc. (Figs. 13-9 and 13-10). Ventilator opens inward and is easily removed for glazing or cleaning, or to provide a clear opening for passage of materials. Center muntin is optional. Screens are attached on the outside.

(a)   (b)

Fig. 13-9 Basement and utility windows.

Architectural projected windows, a medium-quality sash used widely for commercial, institutional, and industrial buildings (Figs. 13-11 and 13-12). Ventilator arrangement permits cleaning from inside and provides a good range of ventilation control with easy operation. Steel ventilator sections are usually 1¼ in deep, with corners welded. Frames may be 1⅜ in deep when riveted or 1¼ in deep when welded. Screens are generally furnished at extra cost. Glazing may be either inside or outside.

Intermediate projected windows, high-quality ventilating sash used for schools, hospitals, commercial buildings, etc. (Figs. 13-13 and 13-14). When made of metal, corners of frames and ventilators are welded. Depth of frame and vent sections varies from "intermediate" to "heavy intermediate," depending on the manufacturer. Each vent is balanced on two arms pivoted to the frame and vent. The arms are equipped with friction shoes arranged to slide in the jamb sections. Screens are easily attached from inside. All cleaning is done from the inside. Glazing is set from the outside. Double insulating glass is optional.

Psychiatric projected windows, for use in housing for mental patients, to provide protection against exit but minimizing appearance of restraint (Fig. 13-15). Ventilators open in at the top, with a maximum clear opening of 5 to 6 in. Glazing is set outside. Screens also are placed on the outside but installed from inside. Metal casing can be had com-

Fig. 13-10 Details of installation of basement and utility windows.

pletely assembled and attached to window ready to install. Outside glass surfaces are easily washed from the inside.

Detention windows, designed for varying degrees of restraint and for different lighting and ventilating requirements. The guard type (Fig. 13-16) is particularly adaptable to jails, reform schools, etc. It provides security against escape through window openings. Ventilators are attached to the inner face of the grille and can be had with a removable key-locking device for positive control by attendants. Screens are installed from inside between vent and grille. Glazing is done from outside, glass being omitted from the grille at the vent section.

Residential double-hung windows (Figs. 13-1 to 13-3 and 13-17), available in different designs and weights to meet various service requirements for all types of buildings. When made of metal, the frame and ventilator corners are welded and weathertight. These windows are also used in combination with fixed picture windows for multiple window openings. They are usually equipped with weather stripping, which maintains good weathertightness. Screens and storm sash are furnished in either full or half sections. Glazing is done from outside.

Fig. 13-11 Architectural projected window.

Residence casements, available in various types, sizes, and weights to meet service requirements of homes, apartments, hotels, institutions, commercial buildings, etc. (Fig. 13-18). Rotary or lever operators hold the vent at the desired position, up to 100% opening. Screens and storm sash are attached to the inside of casement. Extended hinges on vents permit cleaning from inside. Glazing is done from outside.

Intermediate combination windows, with side-hinged casements and projected-in ventilators incorporated to furnish flexibility of ventilation control, used extensively for apartments, offices, hospitals, schools, etc., where quality is desired (Fig. 13-19). They

Fig. 13-12 Details of installation of architectural projected windows.

are available in several weights with rotary or handle operation. When made of metal, corners of vents and frames are welded. Factory provision is made for screens. All cleaning is done from inside. Glazing is set from outside. Special glazing clips permit use of double insulating glass.

Intermediate casements, a heavier and better quality than residence casements, used particularly for fine residences, apartments, offices, institutions, and similar buildings (Fig. 13-20). Frames and ventilators of metal units have welded corners. Easy control of ventilation is provided by rotary or handle operation. Extended hinges permit safe cleaning of all outside surfaces from inside. Screens are attached or removed from inside. Single or double glazing is set from outside.

Fig. 13-13 Intermediate projected window.

Awning windows, suitable for residential, institutional, commercial, and industrial buildings (Fig. 13-21), furnishing approximately 100% opening for ventilation. Mechanical operation can provide for the bottom vent opening prior to the other vents, which open in unison. This is desirable for night ventilation. Manual operation can be had for individual or group venting. All glass surfaces are easily cleaned from inside through the open vents. Glazing is set from outside, storm sash and screens from inside.

Jalousie windows (Fig. 13-22) combine unobstructed vision with controlled ventilation and are used primarily for sunrooms, porches, and the like where protection from the weather is desired with maximum fresh air. The louvers can be secured in any position. Various kinds of glass, including obscure and colored, often are used for privacy or decoration. They do not afford maximum weathertightness but can be fitted with storm sash on the inside. Screens are furnished interchangeable with storm sash.

Ranch windows, particularly suited to modern home design and also used effectively in other types of buildings where more light or better view is desired (Fig. 13-23). When made of metal, corners of frames and vents are welded. Depth of sections vary with manufacturers. These windows are designed to accommodate insulating glass or single panes. Screens are attached from inside.

Fig. 13-14 Details of installation of intermediate projected window.

Fig. 13-15 Psychiatric projected window.

Fig. 13-16 Guardtype detention window.

Fig. 13-17 Double-hung windows.

Fig. 13-18 Residence casement window.

Fig. 13-19 Intermediate combination window.

13-11

Fig. 13-20 Intermediate casement window.   **Fig. 13-21** Awning window.   **Fig. 13-22** Jalousie window.

Continuous top-hung windows, used for top lighting and ventilation in monitor and sawtooth roof construction (Fig. 13-24). They are hinged at the top to the structural-steel framing members of the building and swing outward at the bottom. Two-foot lengths are connected end to end on the job. Mechanical operators may be either manual or motor-powered, with lengths of runs as specified in Table 13-1. Sections may be installed as fixed windows. Glazing is set from outside.

Additional types of windows (not illustrated) include:

Vertical pivoted, which sometimes are sealed with a rubber or plastic gasket.

Horizontal sliding sash usually in aluminum or wood for residential work.

Vertical folding windows, which feature a flue action for ventilation.

Double-hung windows with removable sash, which slide up and down, or another type which tilts for ventilation but does not slide.

Austral type, with sash units similar to double-hung but which operate in unison as projected sash. The sash units are counterbalanced on arms pivoted to the frame, the top unit projecting out at bottom and the bottom unit projected in at top.

Picture windows, often a combination of fixed sash with or without auxiliary ventilating units.

FIXED LIGHTS 36"X 24" GLASS.
VENTS MAY BE PLACED AS DESIRED.

Fig. 13-23 Ranch windows.

Fig. 13-24 Continuous top-hung windows.

TABLE 13-1 Maximum Run for Mechanical Operators, ft

Operators for Pivoted Windows

Type of operator	Lever (1-in. pipe)				Rack and pinion (1-in. pipe)				
	Manual				Manual				Electric oil encl.
Control	Dust*		Oil†		Dust*		Oil†		
	Chain	Rod	Chain	Rod	Chain	Rod	Chain	Rod	
Sash open 16-in.	100	80	150	120	160	140	200	180	240

Operators for Projected Windows

Type of operator	Rack and pinion				
	Manual				
Control	Dust*		Oil†		Electric
	Chain	Rod	Chain	Rod	
	100	90	140	120	200

Operators for Continuous Top-hung Windows

Type of operator			Rack and pinion	Tension	
Control			Manual or electric	Manual	Electric
	Sash	Height, ft	1-in. pipe	1-in. pipe	1-in. pipe
Limits	Vertical	3, 4, 5, 6	150	200	300
	30° slope	3 4 5 6	120 100 80 60	160 140 120 100	240 210 180 150

* Dust-protected—cast gearing.
† Oil-enclosed—cut gearing.

13-14 Windows

Store-front construction, usually semicustom-built of stock moldings of stainless steel, aluminum, bronze, or wood.

13-10. Windows in Wall-Panel Construction.

In the development of thin-wall buildings and wall-panel construction, windows have become more a part of the wall rather than units to fill an opening. Wall panels of metal and concrete incorporate windows as part of their general makeup; a well-integrated design will recognize this inherent composition.

Manufacturers of metal wall panels can furnish formed metal window frames specifically adapted to sash type and wall construction.

In panel walls of precast concrete, window frames can be cast as an integral part of the unit or set in openings as provided (Fig. 13-25). The cast-in-place method minimizes air infiltration around the frame and reduces installation costs on the job. Individual units or continuous bands of sash, both horizontal and vertical, can be readily adapted by the proper forming for head, jamb, and sill sections.

Fig. 13-25 Window installed in precast-concrete sandwich wall panel.

13-11. Mechanical Operators for Windows.

Mechanical operation for pivoted and projected ventilators is achieved with a horizontal torsion shaft actuated by an endless hand chain or by motor power. Arms attached to the torsion shaft open and close the vents.

Two common types of operating arms are the lever and the rack and pinion (Fig. 13-26). The lever is used for manual operation of small groups of pivoted vents where rapid opening is desirable. The rack and pinion is used for longer runs of vents and can be motor powered or manually operated. The opening and closing by rack and pinion are slower than by lever arm.

Fig. 13-26 Levers, rack-and-pinion, and tension-type window operators.

Usually one operating arm is furnished for each vent less than 4 ft wide and two arms for each vent 4 ft or more wide. Mechanical operators should be installed by the window manufacturer before glazing of the windows.

Continuous top-hung windows are mechanically controlled by rack-and-pinion or tension-type operators (Fig. 13-26). Maximum length of run is given in Table 13-1.

13-12. Weather Stripping and Storm Sash.

Weather stripping is a barrier against air leakage through cracks around sash. Made of metal or a compressible resilient material, it is very effective in reducing heat loss.

Storm sash might be considered an overall transparent blanket that shields the window unit and reduces heat loss. Weather stripping and storm sash also reduce condensation and soot and dirt infiltration, in addition to decreasing the amount of cold air near the floor.

Storm sash and screens are often incorporated in a single unit.

13-13. Window Glass.

Various types and grades of glass are used for glazing:

Clear Window Glass. This is the most extensively used type for windows in all classes of buildings. A range of grades, as established by Federal Government Standard DD-G-451c, classifies quality according to defects. The more commonly used grades are A and B. A is used for the better class of buildings where appearance is important, and B is used for industrial buildings, some low-cost residences, basements, etc.

With respect to thickness, clear window glass is classified as "single-strength" about $\frac{3}{32}$ in thick; "double-strength," about $\frac{1}{8}$ in thick; and "heavy-sheet," up to $\frac{7}{32}$ in thick. Maximum sizes are as follows: single-strength, 40 × 50 in; double-strength, 60 × 80 in; and heavy sheet, 76 × 120 in. Because of flexibility, single strength and double strength should never be used in areas exceeding 12 ft^2, and for appearance's sake areas should not exceed 7 ft^2.

Plate and float glass have, in general, the same performance characteristics. They are of superior quality, more expensive, and have better appearance, with no distortion of vision at any angle. Showcase windows, picture windows, and exposed windows in offices and commercial buildings are usually glazed with polished plate or float glass. Thicknesses range from $\frac{1}{8}$ to $\frac{7}{8}$ in. There are two standard qualities, *silvering* and *glazing,* the latter being employed for quality glazing.

Processed glass and rolled figured sheet are general classifications of obscure glass. There are many patterns and varying characteristics. Some provide true obscurity with a uniform diffusion and pleasing appearance, while others may give a maximum transmission of light or a smoother surface for greater cleanliness. The more popular types include a clear, polished surface on one side with a pattern for obscurity on the other side.

Obscure wired glass usually is specified for its fire-retarding properties, although it is also used in doors or windows where breakage is a problem. It should not be used in pieces over 720 in^2 in area (check local building code).

Polished wired glass, more expensive than obscure wired glass, is used where clear vision is desired, such as in school or institutional doors.

There are also many special glasses for specific purposes:

Heat-absorbing glass reduces heat, glare, and a large percentage of ultraviolet rays, which bleach colored fabrics. It often is used for comfort and reduction of air-conditioning loads where large areas of glass have a severe sun exposure. Because of differential temperature stresses and expansion induced by heat absorption under severe sun exposure, special attention must be given to edge conditions. Glass having clean-cut edges is particularly desirable, because these affect the edge strength, which, in turn, must resist the central-area expansion. A resilient glazing material should be used.

Corrugated glass, corrugated wired glass, and corrugated plastic panels are used for decorative treatments, diffusing light, or as translucent structural panels with color.

Laminated glass consists of two or more layers of glass laminated together by one or more coatings of a transparent plastic. This construction adds strength. Some types of laminated glass also provide a degree of security, sound isolation, heat absorption, and glare reduction. Where color and privacy are desired, fadeproof opaque colors can be included. When fractured, laminated glass tends to adhere to the inner layer of plastic and, therefore, shatters into small splinters, thus minimizing the hazard of flying glass.

Bullet-resisting glass is made of three or more layers of plate glass laminated under heat and pressure. Thicknesses of this glass vary from ¾ to 3 in. The more common thicknesses are 1³⁄₁₆ in, to resist medium-powered small arms; 1½ in, to resist high-powered small arms; and 2 in, to resist rifles and submachine guns. (Underwriters Laboratories lists materials having the required properties for various degrees of protection.) Greater thicknesses are used for protection against armor-piercing projectiles. Uses of bullet-resisting glass include cashier windows, bank teller cages, toll-bridge booths, peepholes, and many industrial and military applications. Transparent plastics also are used as bullet-resistant materials, and some of these materials have been tested by the Underwriters Laboratories. Thicknesses of 1¼ in or more have met UL standards for resisting medium-powered small arms.

Tempered glass is produced by a process of reheating and sudden cooling that greatly increases strength. All cutting and fabricating must be done before tempering. Doors of ½- and ¾-in-thick tempered glass are commonly used for commercial building. Other uses, with thicknesses from ⅛ to ⅞ in, include backboards for basketball, showcases, balustrades, sterilizing ovens, and windows, doors, and mirrors in institutions. Although tempered glass is 4½ to 5 times as strong as annealed glass of the same thickness, it is breakable, and when broken, disrupts into innumerable small fragments of more-or-less cubical shape.

Tinted and coated glasses are available in several types and for varied uses. As well as decor, these uses can provide for light and heat reflection, lower light transmission, greater safety, sound reduction, reduced glare, and increased privacy.

Transparent mirror glass appears as a mirror when viewed from a brightly lighted side, and is transparent to a viewer on the darker opposite side. This one-way-vision glass is available as a laminate, plate or float, tinted, and in tempered quality.

Plastic window glazing, made of such plastics as acrylic or polycarbonate, is used for urban school buildings and in areas where high vandalism might be anticipated. These plastics have substantially higher impact strength than glass or tempered glass. Allowance should be made in the framing and installation for expansion and contraction of plastics, which may be about 8 times as much as that of glass. Note also that the modulus of elasticity (stiffness) of plastics is about one-twentieth that of glass. Standard sash, however, usually will accommodate the additional thickness of plastic and have sufficient rabbet depth.

Suspended glazing (Fig. 13-27) utilizes metal clamps bonded to tempered plate glass at the top edge, with vertical glass supports at right angles for resistance to wind pressure. These vertical supports, called stabilizers, have their exposed edges polished. The joints between the large plates and the stabilizers are sealed with a bonding cement. The bottom edge or sill is held in position by a metal channel, and sealed with resilient waterproofing. Suspended glazing offers much greater latitude in use of glass and virtually eliminates visual barriers.

Fig. 13-27 Typical details of suspended glazing *(F. H. Sparks Co., Inc., New York, N.Y.)*

Safety glazing. A governmental specification Z-97, adopted by many states, requires entrance-way doors and appurtenances glazed with tempered, laminated, or plastic material.

Thickness. Figure 13-28 can be used as a guide to estimate thickness of plate, float, or sheet glass for a given glass area, or the maximum area for a given thickness to withstand a specified wind pressure. Based on minimum thickness permitted by Federal Specification DD-G-451c, the wind-load chart provides a safety factor of 2.5. It is intended for rectangular lights with four edges glazed in a stiff, weathertight rabbet.

For example, determine the thickness of a 108 × 120-in (90-ft^2) light of polished plate glass to withstand a 20-lb/ft^2 wind load. Since the 20-lb/ft^2 and 90-ft^2 ordinates intersect below the ⅜-in glass thickness line, the thickness to use is ⅜ in.

The correction factors in Table 13-2 also allow Fig. 13-28 to be used to determine the thickness for certain types of fabricated glass products. The table, however, makes no allowance for the weakening effect of such items as holes, notches, grooves, scratches, abrasion, and welding splatter.

TABLE 13-2 Resistance of Glass to Wind Loads

Product	Factor*
Plate, float, or sheet glass	1.0
Rough rolled plate glass	1.0
Sand-blasted glass	2.5
Laminated glass†	2.0
Sealed double glazing:‡	
Metal edged, up to 30 ft^2	0.667
Metal edged, over 30 ft^2	0.556
Glass edged	0.5
Heat-strengthened glass	0.5
Fully tempered glass	0.25

*Enter Fig. 13-28 with the product of the wind load in psf multiplied by the factor.

†At 70°F or above, for two lights of equal thickness laminated to 0.015-in-thick vinyl. At 0°F, factor approaches 1.

‡For thickness, use thinner of the two lights, not total thickness.

Fig. 13-28 Wind load chart indicates maximum glass area for various nominal thicknesses of plate, float or sheet glass supported on four sides to withstand specified wind pressures, with design factor of 2.5.

Fig. 13-29 Sound-isolation glazing.

The appropriate thickness for the fabricated glass product is obtained by multiplying the wind load, lb/ft² by the factor given in Table 13-2. The intersection of the vertical line drawn from the adjusted load and the horizontal line drawn from the glass area indicates the minimum recommended glass thickness. Glass producers should be consulted for more accurate thickness recommendations.

Table 13-3 gives the overall thermal conductance, or U factor, and weight of glass for several thicknesses. Table 13-4 presents typical sound-reduction factors for glass, and for comparison, for other materials. As a sound barrier, glass, inch for inch, is about equal to concrete and better than most brick, tile, or plaster. Double glazing (Fig. 13-29) is particularly effective where high resistance to sound transmission is required. Unequal glass thicknesses minimize resonance in this type of glazing, and a nonreflective mounting reduces sound transmission. For optimum performance, sound-isolation glazing should be carefully detailed and constructed, and should be mounted in airtight, heavy walls with acoustically treated surfaces.

TABLE 13-3 Overall Conductance and Weight of Glass

Thickness, in.	U factor	Weight, psf
1/8	1.14	1.65
1/4	1.13	3.29
3/8	1.12	4.94
1/2	1.11	6.58
1-in. insulating glass	0.55	7.00

TABLE 13-4 Sound Reduction—Glass and Other Materials*

Material thickness, in.	Sound transmission classification rating
Glass:	
0.087 (single strength)	26
0.118 (double strength)	29
1/8	29
3/16	29
1/4	27
5/16	29
3/8	30
1/2	33
1	32
1/4 laminated (up to 0.045 plastic layer)	33
1-in. sealed double glazing, 1/2-in. air space	27
Other materials:	
1/4 plywood	22
1/2 fiberboard	21
1/4 lead	45
2 solid vermiculite plaster on metal lath	30
Hollow wall, 2 × 4 studs, 1/2-in. gypsum plaster on metal lath on two sides	33

* Courtesy Libby-Owens-Ford Company, Toledo, Ohio.

13-14. Glazing Compounds.

Putty, glazing compound, rubber, or plastic strips and metal or wood molds are used for holding glass in place in sash. Metal clips for metal sash and glazing points for wood sash also are employed for this purpose.

Putty and glazing compounds are generally classified in relation to the sash material and should be used accordingly for best results.

Bedding of glass in glazing compound is desirable, because it furnishes a smooth bearing surface for the glass, prevents rattling, and eliminates voids where moisture can collect (Commercial Standards, U.S. Department of Commerce). Glazing terminology relative to putty follows:

Face Puttying (Fig. 13-30). Glass is inserted in the glass rabbet and securely wedged where necessary to prevent shifting. Glazing points are also driven into the wood to keep the glass firmly seated. The rabbet then is filled with putty, the putty being beveled back against the sash and muntins with a putty knife.

Back Puttying (Fig. 13-30b). After the sash has been face-puttied, it is turned over, and putty is run around the glass opening with a putty knife, thus forcing putty into any voids that may exist between the glass and the wood parts.

Bedding (Fig. 13-30c). A thin layer of putty or bedding compound is placed in the rabbet of the sash, and the glass is pressed onto this bed. Glazing points are then driven into the wood, and the sash is face-puttied. The sash then is turned over, and the excess putty or glazing compound that emerged on the other side is removed.

Wood-Stop or Channel Glazing (Fig. 13-30d). A thin layer of putty or bedding compound is placed in the rabbet of the sash, and the glass is pressed onto this bed. Glazing points are not required. Wood stops are securely nailed in place. The sash then is turned over, and the excess putty or glazing compound that emerged on the other side is removed.

Glazing Beads (Fig. 13-30f). These are designed to cover exterior glazing compound and improve appearance.

Continuous Glazing Angles (Fig. 13-30g). These angles or similar supports for glass usually are required for "labeled windows" by the National Board of Fire Underwriters.

Table 13-5 lists some commonly used glazing compounds and indicates the form and method in which they usually are used. Other compounds used for special service include silicone rubber, butadiene-styrene (GR-S) rubber, polyethylene, nitrile rubbers, polyurethane rubbers, acrylics, and epoxy.

TABLE 13-5 Commonly Used Glazing Compounds

Compound	Type of material	Final form	Type of glazing
Vegetable oil base (skin-forming type)	Gun, knife	Hard	Channel, bedding, face
Vegetable oil-rubber or nondrying oil blends (polybutene) etc. . .	Gun, knife	Hard, plastic	Face, channel, bedding
Nondrying oil types.	Gun, knife, tacky tape	Plastic	Channel, bedding
Butyl rubber or polyisobutylene	Tacky tape	Plastic, rubber	Bedding, channel
Polysulfide rubber	Gun	Plastic, rubber	Bedding, channel
Neoprene (polychloroprene). . .	Gun, preformed gasket	Rubber	Channel
Vinyl chloride and copolymers .	Preformed gasket	Hard, plastic	

13-15. Factory-Sealed Double Glazing.

This is a factory-fabricated, insulating-glass unit composed of two panes of glass separated by a dehydrated airspace. This type of sash is also manufactured with three panes of glass and two airspaces, providing additional insulation against heat flow or sound transmission. (See Fig. 13-30e.)

Fig. 13-30 Glazing details.

Heat loss and heat gain can be substantially reduced by this insulated glass, permitting larger window areas and added indoor comfort. Heat-absorbing glass often is used for the outside pane and a clear plate or float glass for the inside. However, there are many combinations of glass available, including several patterned styles. Thickness of glass and airspace between can be varied within prescribed limitations. In the selection of a window for double or triple glazing, accommodation of the overall glass thickness in the sash is an important consideration.

13-16. Gaskets for Glazing.

Structural gaskets, made of preformed and cured elastomeric materials, may be used instead of sash for some applications to hold glass in place. Gaskets are extruded in a single strip, molded into the shape of the window perimeter and installed against the glass and window frame. A continuous locking strip of harder elastomer is then forced into one side of the gasket as a final sealing component. Fit and compression of the gasket determine weathertightness. With proper installation, calking is not ordinarily required; but should it be necessary, sealants are available that are compatible with the material of which the gasket is made.

There are many variations in the design of lockstrip gaskets, some of which are shown in Fig. 13-31. The two most commonly used types are the H and reglet types. The H gasket fits over a flange on the surrounding frame (Fig. 13-31c and d), whereas the reglet gasket fits into a recess in the frame (Fig. 13-31a and b). When gaskets are used in concrete wall panels, the entire opening should be cast within one panel. In all cases, the smoothness, tolerance, and alignment of the contact surfaces are very important. The

Fig. 13-31 Structural gasket glazing: (a) and (b) gasket fitted into groove; (c) and (d) H shape locked to continuous metal fin; (e) and (f) weepholes for factory-sealed double-glazing installations. *(Architectural Aluminum Manufacturers Association, Chicago, Ill.)*

gaskets shown in Fig. 13-31e and f for use with double glazing require weepholes to allow water that may collect in the gaskets to drain.

The maximum glass area recommended for glazing with standard H-type lockstrip gaskets may be obtained from Fig. 13-32. Loads for the chart are based on minimum glass thicknesses allowed in Federal Specification DD-G-451c. Glass producers should be consulted to obtain the latest thickness recommendations.

Fig. 13-32 Maximum permissible areas of plate, float, or sheet glass, with design factor of 2.5, for various wind loads on vertical windows supported on four sides and having standard H-type lockstrip gaskets. Loads apply to glass lights having a length-to-width ratio of 3 or less. *(Architectural Aluminum Manufacturers Association, Chicago, Ill.)*

Section 14

Doors

Frederick S. Merritt
Consulting Engineer,
West Palm Beach, Fla.

The purpose of doors is to close openings that are needed in walls and partitions for access to building interior spaces, when closing is required to prevent trespass by unauthorized persons, to provide privacy, to protect against weather, drafts and noise, and to act as a barrier to fire and smoke. Many types of doors are available for this. They may be hinged on top or sides to swing open and shut, they may slide horizontally or vertically, or they may revolve about a vertical axis in the center of the opening.

A large variety of materials also is available for door construction. Wood, metal, glass, plastics, and combinations of these materials with each other and with other materials, in the form of sandwich panels, are in common use.

Selection of a type of door and door material depends as much on other factors as on the primary function of serving as a barrier. Cost, psychological effect, fire resistance, architectural harmony, and ornamental considerations are but a few of the factors that must be taken into account.

Doors may be classified as ordinary or special purpose. Ordinary doors are those used to protect openings up to about 12 ft high or wide. Often, such doors are available in stock or standard sizes. Larger doors may be considered special purpose, because they are generally custom designed and built and require specially designed framing.

In either case, a door system consists of a door proper, hardware for control of door movement (Sec. 16), a door frame around the wall opening to support the door and trim the opening, and structural framing, such as a lintel, to support the wall and other building components directly over the opening.

ORDINARY DOORS

Ordinary doors may be classified as exterior or interior doors. Exterior doors are those that are installed in an opening in a wall that separates an interior space from outdoors. Serving as an entrance or an exit door, or both, these doors must be capable of excluding weather but usually need not be fire rated. Interior doors are those installed in an opening in a wall or partition between two interior spaces. Such doors are not required to exclude weather but may be required to bar passage of fire and smoke.

14-1. Exterior Doors.

Entrance and exit doors in exterior walls should be suitably selected and located for smooth traffic flow and safety. Choice of door movement, for example, should take into account in which direction persons are likely to turn immediately before or after they enter a doorway. Proper swing of door will not only smooth traffic flow but also reduce wear and tear on the doors by decreasing impact loading on the hardware from persons who lean on it to change directions.

Also, selection of entrance doors should take not only aesthetics into account but also traffic volume. The heavier traffic volume will be, the more rugged the doors should be to withstand wear.

Water Exclusion at Exterior Doors. Exterior doors are subjected to all the effects of natural forces, as are walls and windows, including solar heat, rain, and wind. For ordinary installations, closed doors cannot be expected to completely exclude water or stop air movement under all conditions. One important reason for this is that clearances must be provided around each door. These are necessary to permit easy operation and thermal expansion and contraction of doors.

Exterior doors, however, can be made less vulnerable to water penetration by setting them back from the building face, or by providing overhead protection, such as a canopy, marquee, or balcony. These measures also will help reduce collection of snow and ice at door thresholds. In addition, provision of an entrance vestibule is desirable, because it can serve as a weather barrier. Also, weatherstripping around doors assists in preventing passage of water and air past closed doors.

At the bottom of a door opening for an exterior door is a sill (Fig. 14-1), which forms a division between the finished floor on the inside and the outside construction. The sill generally also serves as a step, for the door opening usually is raised above exterior grade to prevent rain from entering. The top of the sill is sloped to drain water away from the interior. Also, it may have a raised section in the plane of the door or slightly to the rear, so that water dripping from the door will fall on the slope. The raised section may be integral with the sill or a separate threshold. In either case, the rear portion covers the joint between sill and floor. In addition, all joints should be sealed.

Control of Air Movements at Entrances. In design of entrances to buildings, consideration should be given to the effects in those areas of outdoor winds. These may impose pressures or suctions on exterior doors. As a result, doors may be held partly open or may be violently opened beyond their design limits and severely damaged. To avoid such occurrences, doors should be set back from the building face or provided with a wind shield. Also, design of the building facade should be checked, if necessary by wind-tunnel tests, to ensure that it will not create objectionable air movements.

Also, in tall buildings, there is likely to be a difference in air pressure between the inside and outside at entrances. This pressure results from air movements through stairways and shafts. When the building is heated in cold weather, warm air rises in the vertical passageways. When the building is cooled in warm weather, cool air flows downward through these passageways. The resulting pressure differential varies with building height and difference between indoor and outdoor temperatures. In many cases, this stack effect causes large airflows through entrance doors, significantly increases heating and cooling loads, and may make entrance-door operation difficult.

A commonly used means of reducing such loads is provision of an entrance vestibule with one set of doors leading outdoors and another set leading to the inside. In some cases, however, traffic flow is so great that at least one door of both sets of doors may be partly open simultaneously, permitting air to flow through the vestibule between the inside and outside of the building. Thus, the effectiveness of the vestibule is decreased. The loss of effectiveness may be reduced, however, by venting the vestibule to outdoors and providing compensating heating or cooling.

Another means of reducing air movements due to stack effect is installation of revolving doors (Art. 14-9). While in continuous contact with its enclosure, a revolving door permits entry and exit without extra force to offset the differential pressure between indoors and outdoors. (Building codes, however, require some swinging doors in conjunction with revolving doors.)

Where stack effects make operation of swinging doors difficult, operation may be made easier by hanging the doors on balanced pivots or equipping the doors with automatic operators. In some cases, it may be advantageous to replace the doors with automatic, horizontally sliding doors.

Automatic Entrances. Power-operated doors may be used to speed traffic flow at entrances to buildings with heavy traffic volume, such as retail stores, hospitals, and public buildings. Not only are such doors convenient for persons carrying packages or pushing carts, but they also are advantageous for physically handicapped persons. The doors are activated to open fully by persons approaching when their weight is applied to a floor mat or when they are detected by microwave or optical-electronic sensors. The doors are kept open until the persons are safely past, even if traffic stops or moves slowly. (ANSI A156.10, "Power-Operated Pedestrian Doors," American National Standards Institute.)

As an alternative to solid doors, an automatic entrance can be formed with an air curtain. At such an entrance, air is blown upward past a floor grate to the top of the door opening, to keep outdoor air from entering the building interior and prevent loss of interior air. Yet, human traffic can readily penetrate the curtain.

Safety at Entrances. Because of the heavy flow of traffic at building entrances, safety provisions at entrance doors are an important design consideration. Account must be taken of the location of such doors in the building faces, flow of outdoor traffic, and type and volume of traffic generated by the building. Following are some design recommendations:

Arc of a door swing should exceed 90°.

An entrance should always be set back from the building face.

Hinge jambs of swinging doors should be located at least 6 in from a wall perpendicular to the building face.

If hinge jambs for two doors have to be placed close together, there should be enough distance between them to permit the doors to swing through an arc of 110°.

If several doors swinging in the same direction are placed close together in the same plane, they should be separated by center lights and should also have sidelights, to enable the doors to swing through 110° arcs.

If doors hung on center pivots are arranged in pairs, they should be hinged at the side jambs and not at the central mullion.

The floor on both sides of an exit door should be substantially level for a distance on each side equal to at least the width of the widest single leaf of the door. If, however, the exit discharges to the outside, the level outside the door may be one step lower than inside, but not more than 7½ in lower.

14-2. Traffic Flow and Safety.

Openings in walls and partitions must be sized for their primary function of providing entry and exit to or from a building or its interior spaces, and doors must be sized and

14-4 Doors

capable of operating so as to prevent or permit such passage, as required by the occupants of the building. In addition, openings must be adequately sized to serve as an exit under emergency conditions. (See Art. 3-22.) In all cases, traffic must be able to flow smoothly through the openings.

To serve these needs, doors must be properly selected for the use to which they are to be put, and properly arranged for maximum efficiency. In addition, they must be equipped with suitable hardware for the application. (See also Sec. 16 and Art. 14-3.)

Safety. Exit doors and doors leading to exit passageways should be so designed and arranged as to be clearly recognizable as such and to be readily accessible at all times. A door from a room to an exit or to an exit passageway should be the swinging type, installed to swing in the direction of travel to the exit.

Code Limitations on Door Sizes. To ensure smooth, safe traffic flow, building codes generally place maximum and minimum limits on door sizes. Typical restrictions are as follows:

No single leaf in an exit door should be less than 28 in wide or more than 48 in wide. Minimum nominal width of opening should be at least:

36 in for single corridor or exit doors.
32 in for each of a pair of corridor or exit doors with central mullion.
48 in for a pair of doors with no central mullion.
32 in for doors to all occupiable and habitable rooms.
44 in for doors to rooms used by bedridden patients and single doors used by patients in such buildings as hospitals, sanitariums, and nursing homes.
32 in for toilet-room doors.

Jambs, stops, and door thickness when the door is open should not restrict the required width of opening by more than 3 in for each 22 in of width.

Nominal opening height for exit and corridor doors should be at least 6 ft 8 in. Jambs, stops, sills, and closures should not reduce the clear opening to less than 6 ft 6 in.

For a specific type of occupancy, the number of exit doors required in each story of a multistory building and the minimum width permitted for each door can be determined from the maximum capacity listed in Table 3-5 in Art. 3-22. The maximum sizes of openings permitted in fire barriers is given in Table 14-2.

14-3. Fire and Smokestop Doors.

Building codes require fire-resistant doors in critical locations to prevent passage of fire. Such doors are required to have a specific minimum fire-resistance rating, and are usually referred to as fire doors. The codes also may specify that doors in other critical locations be capable of limiting passage of smoke. Such doors, known as smokestop doors, should have a fire-resistance rating of at least 20 min and should be tight fitting.

Fire protection of an opening in a wall or partition depends on the door frame and hardware, as well as on the door. All these components must be "labeled" or "listed" as suitable for the specific application. (See NFPA 80, "Fire Doors and Windows," National Fire Protection Association, 470 Atlantic Ave., Boston, Mass. 02210.)

All fire doors must be self-closing or should close automatically when a fire occurs. In addition, they should be self-latching, so that they remain closed. Push-pull hardware should not be used. Exit doors for places of assembly for more than 100 persons usually must be equipped with fire-exit hardware capable of releasing the door latch when pressure of 15 lb, or less, is applied to the device in the direction of exit. Combustible materials, such as flammable carpeting, should not be permitted to pass under a fire door.

Fire-door assemblies are rated, in hours, according to ability to withstand a standard fire test, such as that specified in American Society for Testing and Materials Standard

E152. They may be identified as products qualified by tests by a UL label, provided by Underwriters Laboratories, Inc.; an FM symbol of approval, authorized by Factory Mutual Research Corp.; or by a self-certified label, provided by the manufacturer (not accepted by National Fire Protection Association and some code officials).

TABLE 14-1 Typical Fire Ratings Required for Doors

Door use	Rating, hr*
Doors in 3- or 4-hr fire barriers	3†
Doors in 2- or 1 1/2-hr fire barriers	1 1/2
Doors in 1-hr fire barriers	3/4
Doors in exterior walls:	
Subject to severe fire exposure from outside the building	3/4
Subject to moderate or light fire exposure from outside the building	1/3
Doors to stairs and exit passageways	1 1/2
Doors in 1-hr corridors	1/3
Other corridor doors	0‡
Smokestop doors	1/3§

*Self-closing, swinging doors. Normally kept closed.
†A door should be installed on each side of the wall.
‡Should be noncombustible or 1 3/4-in solid wood-core doors. Some codes do not require self-closing for doors in hospitals, sanitariums, nursing homes, and similar occupancies.
§May be metal, metal covered, or 1 3/4-in solid wood-core doors (1 3/8 in in buildings less than three stories high), with 1296-in^2 or larger, clear, wire-glass panels in each door.

Openings in walls and partitions that are required to have a minimum fire-resistance rating must have protection with a corresponding fire-resistance rating. Typical requirements are listed in Table 14-1.

This table also gives typical requirements for fire resistance of exit doors, doors to stairs and exit passageways, corridor doors, and smokestop doors.

In addition, some building codes also limit the size of openings in fire barriers. Typical maximum areas, maximum dimensions, and maximum percent of wall length occupied by openings are given in Table 14-2.

Smokestop doors should be of the construction indicated in the footnote to Table 14-1. They should close openings completely, with only the amount of clearance necessary for proper operation.

["Standard for Fire Doors and Windows," NFPA No. 80; Life Safety Code, NFPA No. 101; "Fire Tests of Door Assemblies," NFPA No. 252, National Fire Protection Association, 470 Atlantic Ave., Boston, Mass. 02210.]

TABLE 14-2 Maximum Sizes of Openings in Fire Barriers

Protection of adjoining areas	Max area, ft^2	Max dimension, ft
Unsprinklered	120*	12†
Sprinklers on both sides	150*	15*
Building fully sprinklered	Unlimited*	Unlimited*

*But not more than 25% of the wall length or 54 ft^2 per door if the fire barrier serves as a horizontal exit.
†But not more than 25% of the wall length.
SOURCE: Based on National Building Code, American Insurance Association, New York.

14-6   Doors

"Fire Tests of Door Assemblies," Standard UL 10(b); "Fire Door Frames," Standard UL 63; "Building Materials List" (annual, with bimonthly supplements), Underwriters Laboratories, Inc., 333 Pfingsten Road, Northbrook, Ill. 60062.

"Factory Mutual Approval Guide," Factory Mutual Research Corp., 1151 Boston-Providence Turnpike, Norwood, Mass. 02062.

"Hardware for Labeled Fire Doors," Door and Hardware Institute, 1815 N. Fort Myer Drive, Arlington, Va. 22209.]

14-4. Door Materials.

Wood is used in several forms for doors. When appearance is unimportant and a low-cost door is required, it may be made of boards nailed together. When the boards are vertical and held together with a few horizontal boards, the door is called a batten door. Better-grade doors are made with panels set in a frame or with flush construction.

Fig. 14-1 Typical paneled wood door.

Paneled doors consist of solid wood or plywood panels held in place by verticals called **stiles** and horizontals known as **rails** (Fig. 14-1). The joints between panels and supporting members permit expansion and contraction of the wood with atmospheric changes. If the rails and stiles are made of a single piece of wood, the paneled door is called solid. When hardwood or better-quality woods are used, the doors generally are veneered; rails and stiles are made with cores of softwood sandwiched between the desired veneer.

Tempered glass or plastic may be used for panels. In exterior doors, the lights should be installed to prevent penetration of water, especially in veneered doors. One way is to insert under the glass a piece of molding that extends through the door and is turned down over the outside face of the door to form a drip. Another way is to place a sheet-metal flashing under the removable outer molding that holds the glass. The flashing is turned up behind the inside face of the glass and down over the exterior of the door, with only a very narrow strip of the metal exposed.

Flush doors also may be solid or veneered. The veneered type has a core of softwood, while the flat faces may be hardwood veneers. When two plies are used for a face, they are set with grain perpendicular to each other.

Flush doors, in addition, may be of the hollow-core type. In that case, the surfaces are made of plywood and the core is a supporting grid. Edges of the core are solid wood boards.

Metal doors generally are constructed in one of three ways: cast as a single unit or separate frame and panel pieces; metal frame covered with sheet metal; and sheet metal over a wood or other type of insulating core. (See "ANSI Standard Nomenclature for Steel Doors and Steel Door Frames," A123.1, and "Recommended Standard Details, Steel Doors and Frames," Steel Door Institute, 712 Lakewood Center North, 14600 Detroit Ave., Cleveland, Ohio 44107.)

Cast-metal doors are relatively high-priced. They are used principally for monumental structures.

Fig. 14-2 Typical steel flush door. *(Courtesy V. C. Braun, Amweld Building Products, Niles, Ohio.)*

Hollow metal doors may be of flush or panel design, with steel faces having a thickness of at least 20 gage. Flush doors incorporate steel stiffeners; or an insulation core, such as hydrous calcium silicate, polyurethane foam, or polystyrene foam; or a honeycomb core as a lightweight support for the faces (Figs. 14-2 and 14-3). Voids between stiffeners may be filled with lightweight insulation. Panel doors may be of stile-and-rail or stile-and-panel construction with insulated panels. Also, a light-duty, 24-gage, steel-faced door with an insulated core is available.

Metal-clad (Kalamein) doors are of the swinging type only. They may be of flush or panel design, with metal-covered wood cores for stiles and rails and insulated panels covered with steel 24 gage or lighter.

Other Materials. Doors may be made wholly or partly transparent or translucent. Lights may be made of tempered glass or plastic. Doors made completely of glass are pivoted at top and bottom because the weight makes it difficult to support them with hinges or butts.

Sliding doors of the collapsible accordion type generally consist of wood slats or a light steel frame covered with textile. Plastic coverings frequently are used.

14-5. Structural Requirements for Openings and Doors.

A wall or partition above a door or window opening must be adequately supported by a structural member. In design of such a member, stiffness as well as strength must be taken into account. Excessive deflection could interfere with door operation.

14-8 Doors

It is common practice to install a frame along the perimeter of a door opening, to support the door. Anchored to the wall, the frame usually also serves as trim around the opening. In design of door framing, wind loads are generally more critical than dead loads imposed by the wall or partition above. Wind load should be taken as at least 15 lb/ft². Under this loading, maximum deflection of framing members should not exceed ¾ in or ¹⁄₁₇₅ the clear span. The design should take into account the fact that wind forces acting outward may be larger than those acting inward. Door framing should generally be made independent of other framing.

14-6. Swinging Doors.

These are doors hinged near one edge to rotate about a vertical axis. Swinging doors are hung on butts or hinges (Arts. 16-6 to 16-9). The part of a doorway to which a door is hinged and against which it closes is known as the door frame. It consists of two verticals, commonly called jambs, and a horizontal member, known as the header (Figs. 14-1 and 14-2). **Single-acting doors** can swing 90° or more in only one direction; **double-acting doors** can swing 90° or more in each of two directions.

To stop drafts and passage of light, the header and jambs have a stop, or projection, extending the full height and width, against which the door closes. The projection may be integral with the frame, or formed by attaching a stop on the surface of the frame, or inset slightly.

Door frames for swinging wood doors generally are fastened to rough construction members known as rough bucks, and the joints between the frame and the wall are covered with casings, or trim. For steel door frames, the trim is often integral with the frame, which is attached to the wall with anchors. ("Recommended Standard Details, Steel Doors and Frames," SDI 111, Steel Door Institute, 712 Lakewood Center North, 14600 Detroit Ave., Cleveland, Ohio 44107.)

Swinging doors are constructed in a variety of ways: usually flush (Fig. 14-2); stile and rail, with one or more recessed panels (Fig. 14-1); or stile and panel, with a wide center panel between hinge and lock stiles. Flush doors may be solid or hollow. Hollow doors usually are braced internally. Some of the types of bracing used for steel doors are shown in Fig. 14-3. Those with steel stiffeners (Fig. 14-3a to c) normally are thermally or acoustically insulated with a bat-type material or sprayed insulation. The core types (Fig. 14-3d and e) may have heavy-paper honeycomb, rigid foamed insulation, solid structural mineral blocking, or other bracing laminated to the door facings. (See also Art. 14-4.)

Fig. 14-3 Some types of bracing used for hollow steel doors: (a) horizontal stiffeners; (b) vertical stiffeners; (c) grid of horizontal and vertical stiffeners; (d) rigid insulation core; (e) honeycomb core.

Fig. 14-4 Recommended mounting heights for hardware for ordinary swinging doors.

Wood or steel swinging doors and frames for them are available for ordinary applications in several stock sizes. Standard door thicknesses for these are 1¾ in and 1¾ in. The doors are fabricated to fit in openings with heights of 6 ft 8 in, 7 ft, or 7 ft 2 in. Standard opening widths for single doors are 24, 28, 30, 32, and 36 in. Single 1¾-in stock doors are also available for opening widths of 40, 42, 44, and 48 in and heights of 8 ft. Standard opening widths for pairs of doors are twice as large as for single doors. (See SDI 100, "Recommended Specifications for Standard Steel Doors and Frames," Steel Door Institute.)

Nonstandard sizes are obtainable on special order, but certain precautions should be observed: Size and number of butts or hinges and offset pivots should be suitable for the door size. For rail-and-stile doors, check with the manufacturer to ensure that face areas will not be excessive for the stile width. Large doors should not be used where they will be exposed to strong winds. To prevent spreading of stiles of tall doors, push bars or other intermediate bar members should be attached to both stiles of doors over 8 ft high.

Package entrance doors are available from several manufacturers. The package includes a single door or pair of doors, door frame, and all hardware. Such doors offer the advantages of integrated design, assumption by the manufacturer of responsibility for satisfactory performance, usually quick delivery, and often cost savings.

Recommended mounting heights for hardware for swinging doors are shown in Fig. 14-4. For specific installations, refer to hardware templates supplied by door manufacturers.

14-10   Doors

Fig. 14-5 Swing of doors. KS indicates key, or locking, side.

The direction of swing, or hand, of each door must be known and specified in ordering such hardware as latches, locks, closers, and panic hardware, because the hand determines the type of operation required of them. The hand is determined with respect to the outside or key, or locking, side, following the conventions illustrated in Fig. 14-5.

To weatherproof the joint between the bottom of an exterior door and the threshold, a weather strip in the form of a hooked length of metal often is attached to the underside of the door. When the door is closed, the weather strip locks into the threshold to seal out water. Other types of weather stripping, including plastic gaskets, generally are used for steel doors and may be installed on header and jambs ("Recommended Weatherstripping for Standard Steel Doors and Frames," SDI 111-E, Steel Door Institute).

Fig. 14-6 Horizontally sliding door: (*a*) horizontal section through opening; (*b*) vertical section through opening. *(Reprinted with permission from F. S. Merritt, "Building Engineering and Systems Design," Van Nostrand Reinhold, New York.)*

14-7. Ordinary Horizontally Sliding Doors.

This type of door may roll on a track in the floor and have only guides at the top, or the track and rollers may be at the top and guides at the bottom (Fig. 14-6). Some doors fold or collapse like an accordion, to occupy less space when open. A pocket must always be provided in the walls on either or both sides, to receive rigid-type doors; with the folding or accordion types, a pocket is optional.

Horizontally sliding doors are advantageous for unusually wide openings, and where clearances do not permit use of swinging doors. Operation of sliding doors is not hindered by windy conditions or differences in air pressure between indoors and outdoors. Sliding doors, however, are not accepted as a means of egress.

14-8. Ordinary Vertically Sliding Doors.

This type of door may rise straight up (Fig. 14-7), may rise up and swing in, or may pivot outward to form a canopy. Sometimes, the door may be in two sections, one rising up, the other dropping down. Generally, all types are counterweighted for ease of operation.

In design of structures to receive sliding doors, the clearance between the top of the doors and the ceiling or structural members above should be checked—especially the clearance required for vertically sliding doors in the open position. Also, the deflection of the construction above the door opening should be investigated to be certain that it will not cause the door to jam.

To keep out the weather, the upper part of a sliding door either is recessed into the wall above, or the top part of the door extends slightly above the bottom of the wall on the inside. Similarly, door sides are recessed into the walls or lap them and are held firmly against the inside. Also, the finished floor is raised a little above outside grade. Very large sliding doors require special study (Arts. 14-10 and 14-11).

Fig. 14-7 Vertically sliding door: (a) vertical section through opening; (b) elevation of door. *(Reprinted with permission from F. S. Merritt, "Building Engineering and Systems Design," Van Nostrand Reinhold, New York.)*

14-9. Revolving Doors.

This type of door is generally selected for entranceways carrying a continuous flow of traffic without very high peaks. They offer the advantage of keeping interchange of inside and outside air to a relatively small amount compared with other types of doors. They usually are used in combination with swinging doors because of the inability to handle large groups of people in a short period of time.

Revolving doors consist of four leaves that rotate about a vertical axis inside a cylindrical enclosure. Diameter of the enclosure generally is at least 6 ft 6 in, and the opening to the enclosure usually is between 4 and 5 ft.

Building codes prohibit use of revolving doors for some types of occupancies; for example, for theaters, churches, and stadiums, because of the limited traffic flow in emergencies. Where they are permitted as exits, revolving doors have limitations imposed on them by building codes. The National Fire Protection Association "Life Safety Code" allows only one-half unit of exit width per revolving door in computation of exit capacity, but some codes permit one unit per door. Also, revolving doors may not provide more than 50% of the required exit capacity at any location. The remaining capacity must be supplied by swinging doors within 20 ft of the revolving doors. Rotation speed must be controlled so as not to exceed 15 rpm. Each wing should be provided with at least one push bar and should be glazed with tempered glass at least $\frac{7}{32}$ in thick. Some codes also require the doors to be collapsible.

Fig. 14-8 Horizontally sliding, telescoping door: (*a*) horizontal section through opening; (*b*) vertical section through opening; (*c*) door fully open; (*d*) door partly open; (*e*) door closed; (*f*) and (*g*) alternative arrangements of top rollers.

SPECIAL-PURPOSE DOORS

Large-size doors, such as those for hangars, garages, and craneway openings in walls and for subdividing gymnasiums and auditoriums, often have to be designed individually. As distinguished from those made on a production basis, special-purpose doors require careful design of the main structure to support loads and to allow space for the open door and its controls.

Manufacturers classify special-purpose doors as horizontally sliding, vertically sliding, swinging, and top or horizontally hinged.

14-10. Large Horizontally Sliding Doors.

Door leaves in the horizontally sliding type are equipped with bearing-type bottom wheels and ride rails in the floor while top rollers operate in overhead guides. Two variations are in common use—telescoping and folding.

Telescoping doors (Fig. 14-8) are frequently used for airplane hangars. Normally composed of several leaves, they may be center parting or open to one side only, as illustrated in Fig. 14-8. Open doors are stacked in pockets at ends of the opening. They are built of wood, steel, or a combination of the two.

Telescoping doors are frequently operated by motors located in the end pockets. The motors drive an endless chain attached to tops of closing leaves. Remaining leaves are moved by a series of interconnecting cables attached to the powered leaf and arranged

Fig. 14-9 Wood folding door: (*a*) elevation of door nearly closed; (*b*) vertical section through opening; (*c*) detail at fixed end of door; (*d*) detail at strike jamb. *(Reprinted with permission from F. S. Merritt, "Building Engineering and Systems Design," Van Nostrand Reinhold, New York.)*

so that all leaves arrive at open or closed position simultaneously. Motor size ranges from 1 to 10 hp, travel speed of leaves from 45 to 160 ft/min.

The weight of the leaves must be taken by footings below the rails. Provision also must be made to take care of wind loads transmitted to the top guide channels by the doors and to carry the weight of these channels.

Folding doors (Fig. 14-9) are commonly used for subdividing gymnasiums, auditoriums, and cafeterias and for hangars with very wide openings. This type of door is made up of a series of leaves hinged together in pairs. Leaves fold outward, and when the door is shut, they are held by automatic folding stays. Motors that operate biparting doors usually are located in mullions adjacent to the center of the opening. The mullions are connected by cables to the ends of the opening, and when the door is to be opened, the mullions are drawn toward the ends, sweeping the leaves along. Travel speed may vary from 45 to 160 ft/min.

Chief advantage of folding over telescoping types is that only two guide channels are required, regardless of width. Thus, less metal is required for guide channels and rails, and less material for the supporting members above. Also, since wind loading is applied to leaves that are always partly in folded position, the triangular configuration gives the door considerable lateral stiffness. Hence, panel thickness may be less for folding than for telescoping doors, and a lighter load may be used for designing footings.

14-11. Large Vertically Sliding Doors.

When space is available above and below an opening into which door leaves can be moved, vertically sliding doors are advantageous. They may be operated manually or electrically. Leaves may travel at 45 to 60 ft/min. About 1½ ft in excess of leaf height must be provided in the pockets into which the leaves slide. So the greater the number of leaves the less space needed for the pockets. Figure 14-10 illustrates a vertically sliding, telescoping door.

Vertically sliding doors normally are counterweighted. About 15 in of clear space back of the jamb line is needed for this purpose, but on the idler side only 4 in is required.

Vertical loads are transmitted to the door guides and then by column action to the footings. Lateral support is provided by the jamb and building walls.

Fig. 14-10 Vertically sliding, telescoping door: (*a*) door components; (*b*) door in open position; (*c*) door partly open; (*d*) door closed.

14-12. Large Swinging Doors.

When there is insufficient space around openings to accommodate sliding doors, swinging types may be used. Common applications have been for firehouses, where width-of-building clearance is essential, and railway entrances, where doors are interlocked with the signal system.

Common variations include single-swing (solid leaf with vertical hinge on one jamb), double-swing (hinges on both jambs), two-fold (hinge on one jamb and another between folds and leaves), and four-fold (hinges on both jambs and between each pair of folds).

The more folds, the less time required for opening and the smaller the radius needed for swing. Tighter swings make doors safer to open and allow material to be placed closer to supports.

14-13. Horizontally Hinged Doors.

Effective use of horizontally hinged swinging doors is made in such applications as craneway entrances to buildings. Widths exceeding 100 ft at or near the top of the building can be opened to depths of 4 to 18 ft. Frequently, horizontally sliding doors are employed below crane doors to increase the opening. If so, the top guides are contained in the bottom of the crane door; so the sliding door must be opened before the swinging door.

Top-hinged swinging doors are made of light materials, such as structural steel and exterior-grade plywood. The panel can be motor-operated to open in 1 min or less.

14-14. Radiation-Shielding Doors.

These are used as a barrier against harmful radiation and atomic particles across openings for access to "hot" cells, and against similar radioactive-isotope handling arrangements and radiation chambers of high-energy X-ray machines or accelerators. Usually, they must protect not only personnel but also instruments even more sensitive to radiation than people.

Shielding doors usually are much thicker and heavier than ordinary doors, because density is an important factor in barring radiation. Generally, these special-purpose doors are made of steel plates, steel-sheathed lead, or concrete. To reduce thickness, concrete doors may be of medium-heavy (240 lb/ft^3) or heavy (300 lb/ft^3) concrete, often made with iron-ore aggregate.

The heavy doors usually are operated hydraulically or by electric motor. Provision must be made, however, for manual operation if the mechanism should break down.

Common types of shielding doors include hinged, plug, and overlap. The hinged type is similar to a bank vault swinging door. The plug type, flush with the walls when closed, may roll on floor-mounted tracks or hang from rails. Overlap doors, surface-mounted, also may roll or hang from rails. In addition, vertical-lift doors sometimes are used.

Section 15

Roof Coverings

J. C. Gudas
Editor-Publisher,
American Roofer & Building Improvement Contractor,
Downers Grove, Ill.

The term *roofing* or *roof covering* is generally considered to refer to the covering material installed in a building over the roof deck, which in turn is erected over the building frame. Roofing does not normally include rafters, sheathing, or purlins, but it does include the roofing felt placed over the roof deck. The type of covering used on a building depends on the roofing system specified to weatherproof the structure properly.

Primarily, design of a roofing system is the responsibility of an architect or engineer. The designer often relies on specifications developed by roofing trade associations and materials manufacturers.

It is of extreme importance that plans and specifications call for a stable, structurally sound building, with a minimum of movement. Roofing materials cannot withstand the three-way movement of a poorly designed building, which often results from a misguided attempt to reduce costs.

The importance of proper design of roofs was expressed by an architect who stated that the roof, representing 2% of the total construction cost of a new building, creates 98% of the structural problems.

15-1. Roof Slopes.

Roofs are constructed in a variety of shapes, of which the more common are gable or shed, hip, gambrel, mansard, sawtooth, and flat. The slope of a roof is generally referred to as the pitch and is expressed as the ratio of the rise of the roof to the horizontal span.

If a roof is dead flat, there is no rise. If a roof that has a constant slope and spans a building 20 ft wide has the peak 10 ft above the eaves, the pitch is 10/20 or **half-pitch** (also expressed as 6 in on 12 in, or 6:12). Common roof pitches are 1/3, 1/2, and occasionally 5/8 or 3/4.

The greater the pitch, the more roofing material is required to cover the roof. With added allowances for waste, pitches result in the following increase in roofing-material area over a flat area: 1/4, 12%; 1/3, 20%; 1/2, 42%; 5/8, 60%; and 3/4, 80%. (See also Table 15-1.)

TABLE 15-1 Variation of Roof Area with Slope

Classification	Incline		Inclined area per sq ft horizontal area	Percentage increase in area over flat roof
	In. per ft horizontal	Angle with horizontal		
Flat roofs	1/8	0°36'	1.000	0.0
	1/4	1°12'	1.000	0.0
	3/8	1°47'	1.000	0.0
	1/2	2°23'	1.001	0.1
	5/8	2°59'	1.001	0.1
	3/4	3°35'	1.002	0.2
	1	4°46'	1.003	0.3
	1 1/8	5°21'	1.004	0.4
	1 1/4	5°57'	1.005	0.5
	1 1/2	7°8'	1.008	0.8
	1 3/4	8°18'	1.011	1.1
	2	9°28'	1.014	1.4
Steep roofs	2 1/4	10°37'	1.017	1.7
	2 1/2	11°46'	1.021	2.1
	2 3/4	12°54'	1.026	2.6
	3	14°2'	1.031	3.1
	3 1/4	15°9'	1.036	3.6
	3 1/2	16°16'	1.042	4.2
	3 3/4	17°21'	1.048	4.8
	4	18°26'	1.054	5.4
	4 1/4	19°30'	1.061	6.1
	4 1/2	20°34'	1.068	6.8
	5	22°37'	1.083	8.3
	6	26°34'	1.118	11.8
	7	30°16'	1.158	15.8
	8	33°42'	1.202	20.2
	9	36°52'	1.250	25.0
	10	39°48'	1.302	30.2
	11	42°31'	1.356	35.6
	12	45°0'	1.414	41.4
Extra-steep roofs	14	49°24'	1.537	53.7
	16	53°8'	1.667	66.7
	18	56°19'	1.803	80.3
	20	59°2'	1.943	94.3
	22	61°23'	2.088	108.8
	24	63°26'	2.235	123.5

15-2. Bituminous and Plastic Roofing Materials.

Expensive errors in roof application have resulted from failure of architects, engineers, builders, roofing contractors, and property owners to agree on trade terminology. As many as three or four different meanings have been conveyed by a term or phrase in a contract. The generally accepted definitions of bituminous materials are as follows:

Bitumen. A generic term used to indicate either asphalt or coal-tar pitch.
Asphalt. A by-product of the refining processes of petroleum oils.
Coal-Tar Pitch. A by-product of crude tars derived from coking of coal. The crude tars are distilled to produce coal-tar pitch.

Asphalt and coal-tar pitch serve in a variety of waterproofing and water-shedding applications. In their solid state, coal-tar pitch and asphalt resemble each other in appearance. There are, however, major differences: Coal-tar pitch has a lower melting point, and as a result is self-sealing and melts faster. Because of the low melting point, the top layer of a coal-tar-pitch roof must be restrained by a covering of gravel or slag from flowing caused by solar heat. Asphalt usually does not require such a covering, but it does not have the self-sealing features of coal-tar pitch. Nevertheless, the range of use of asphalt is greater, because it can be used on steeper slopes.

Both asphalt and pitch are used as a saturant to produce the felts and prepared roofings in common usage. These products can be classified into four basic groups: saturated felts, coated felts, prepared roofing, and plastic roll materials.

For the first three categories, two types of felts are used in the manufacturing process: organic felts, manufactured from rags, paper, wood pulp, or a combination of these; and inorganic felts, composed of asbestos or glass fibers. Saturated and coated felts, delivered in rolls, have bitumen on both sides and are surfaced with a mineral release agent on either side or both sides, to prevent adhesion in the roll. Prepared roofings have a surfacing material, such as ceramic granules, sand, mica, or talc, added on the top side for decorative purposes.

The fourth category, plastic roll materials, includes sheets of such plastics as polyethylene, vinyl, hypalon, and rubber. These materials are used as roof coverings, as well as for flashing, expansion joints, and vapor barriers.

15-3. Types of Roof Coverings.

While a multitude of materials are utilized as roof coverings, there are basically two main groups into which they are classified, single-unit roof coverings and multiunit roof coverings.

Single-unit roof coverings are any roof surfaces, flat or sloped, that become, after application, a single entity. Included in this category are:

Built-up roofs (often referred to as *flat roofs*), with little or no slope, consisting of a number of layers (plies) of roofing felts bonded together with asphalt or pitch (Fig. 15-1).

Fig. 15-1 Plan views of a built-up roof showing steps in application over (*a*) concrete nonnailable deck and (*b*) steel deck, with one layer of insulation mechanically fastened and a second layer, with joints staggered, asphalt adhered. *(Courtesy Johns-Manville Corp.)*

Flat-seam metal roofings.

Single-ply roof systems comprising products installed as a single sheet but which may consist of several layers of felts and adhesives.

Fluid-applied products, which are sprayed or brushed on and are manufactured from various components, such as plastic, rubber, epoxy compounds, and bitumen emulsions.

Multiunit roof coverings are materials suitable for application on steep roofs (slopes of 4 in or more) and consist of many pieces installed individually, usually partly overlapped. This category includes such products as roofing shingles manufactured from asphalt-coated felts; tiles and shingles of plastic or glass compounds; asbestos-cement, aluminum, steel, and wood shingles; and clay, concrete, and slate tiles. Glass-fiber felts may be used as a shingle base.

Attachment to Decks. For monolithic roof decks, such as cast-in-place concrete and poured gypsum, felts can be cemented directly to the roof deck. For roof decks made of separate units that permit nailing, the first two plies of felt are nailed to the deck, sealing the joints between the units from leakage and tying the roof covering to the deck. For multiunit roof decks that do not permit nailing, such as precast-concrete slabs, the first ply of felt is cemented to the deck by a strip-mopping operation, the deck surface being mopped with bitumen to within 4 in of its perimeter.

Surface Coatings. Where the slope of a roof deck is 2 in on 12 in or more, it is difficult to retain slag or gravel surfacing applied to bitumen roofing to prevent flowing. The surfacing has a tendency to slide or to be blown off the roof. This has resulted in specification of top-layer, or cap, felts with a mineral surfacing embedded in the exposed face of the felts. Minimum weight of mineral-surfaced cap sheets is 55 lb/square (100 ft^2). Maximum specifications require the cap sheets to be applied over two 15-lb felts, one mopped and the first one nailed.

Use of smooth-surfaced roofs on suitable low slopes has resulted in specifications that eliminate the heavy surface coating in favor of a 30-lb felt nailed to the deck and followed by three 15-lb felts and a final coat of asphalt, applied at the rate of 10 lb/square. The thin coating is preferable to a heavy coating, because the thin film is not so subject to **alligatoring** (a wrinkling of membranes with an alligator-skin appearance). Also, loss of volatile oils under solar radiation is less than with thick films.

Slag, however, can be used on roofs with slopes up to 4 in/ft if special precautions are taken in its application.

Mineral-Surfaced Cap Sheets. These also may be used on slopes up to 4 in/ft. These cap sheets are furnished in three forms. The simplest is a heavy sheet of saturated felt with its upper surface entirely covered with granules. The second type is similar, except for an unsurfaced lap portion along one side. The third has about one-half its width surfaced with granules, and is applicable to two-ply construction in the final assembly. Known as double-coverage or 19-in selvage-edge roofing, this type should be applied so that each sheet laps the unsurfaced portions, which can be nailed to the roof deck or to wood nailers embedded in the deck.

Flat Roofs. Because of the tendency of a flat deck to retain water, application of built-up roofing on a level roof requires greater mechanical skill and judgment than for steep roofs. Gravity helps even a badly constructed steep roof to shed water.

15-4. Selection of Roof Coverings.

The type of roofing material used to waterproof a structure depends largely on shape and slope of the roof, compatibility of materials to be installed together, geographical location, end use of the building, desired esthetic effect, and cost.

For steep roofs, a multiunit roof covering generally is advisable. In selection of the type, location of the building is an important factor, because climatic differences dictate the choice. For example, in the southern United States, clay or concrete tiles are usually selected to provide insulation from the semitropical heat and to protect the deck from rot and mildew. In Alaska, extreme temperature changes create severe stress on built-up bituminous roofs. Hence, an elastomeric coating might be advisable, even though costlier initially. Asphalt shingles are a practical, inexpensive choice for North American houses. Slate is used extensively in England, but less frequently in the United States because of high cost.

Apartment, commercial (stores and offices), and industrial (manufacturing and storage) buildings, schools, and hospitals are generally covered with a flat, single-unit roof. Occasionally, combinations of single-unit and multiunit roofings are specified, for example, for a mansard roof, flat on top, with a part of the top exterior of the building framed out to form a base for decorative shingles.

Roofs with sweeping curves, thin-shell domes, hyperbolic-paraboloid shells, folded-plate roofs, and other unconventional surfaces call for a single-unit covering that can be applied over difficult-to-waterproof shapes. Elastomeric or plastic fluid-applied compounds are often specified for the purpose.

Compatibility of materials used in the construction of a roof is of extreme importance and requires careful study before a decision is made in the selection of products. Chemical reaction between two incompatible products can create undesirable softening, hardening, bleeding, cracking, or peeling. Differing thermal contraction or expansion of roof deck, insulation, and covering can result in splitting and cracking of the roof membrane. For example, foamed-in-place polyurethane insulation, with its exceptionally high coefficient of expansion, when used with the wrong coatings, will crack and split, permitting moisture to enter.

Structures such as laundries, with a high interior moisture content, or cold-storage buildings, with a sharp contrast between interior and exterior temperatures, require vapor barriers to safeguard the roof membrane against infiltration of moisture and eventual deterioration of the roof covering.

SINGLE-UNIT COVERINGS

15-5. Built-Up Roofs.

The term built-up roof describes the roof covering that is applied on a deck with relatively little or no slope and that is composed of a sandwich of alternate layers (or plies) of saturated felt and moppings of bitumen. An optional top pouring of pitch or asphalt, in which slag or gravel may be embedded, completes the built-up roof. The felt membranes may be of the organic or inorganic type, and may be over an insulated or uninsulated deck.

The roof sandwich may consist of a few or many of the following products, depending on the specifications: vapor barrier; asbestos base sheet or coated base sheet; sheathing paper (over wood-fiber or wood decks); primer (over concrete and precast-concrete decks); insulation (board, urethane foam, or insulating concrete); coated or saturated felts, No. 15, 20, 30, or 40, or a combination, according to specifications; steep asphalt; low-slope asphalt; slag or gravel topping; nails or fasteners; No. 11 glass-fiber sheet; glass-fiber cap sheet or mineral cap sheet; glass-fiber tape to seal insulation; flashings.

Manufacturers' or roof-association specifications, as selected by the architect or engineer, must be closely observed to ensure a serviceable roof covering.

The five-ply built-up roof specification is the oldest of all built-up roofing specifications. With the introduction of cast-in-place roof decks, the four-ply specification came into wide use, because it required one ply of felt fewer than a wood-sheathed deck or a metal deck that has a multiplicity of side and end joints.

While one ply of felt may be dispensed with for a cast-in-place concrete deck, one extra layer of bitumen is required to compensate for omission of nailing of the first two felts. Thus, a four-ply built-up roof over concrete is equal to a five-ply over wood. (See Figs. 15-1 and 15-2.)

Fig. 15-2 Gravel spreader distributes gravel on a built-up roof. Operator wears safety clothing: hard hat, long-sleeved shirt, gloves, long trousers with bottoms tucked into above-the-ankle work shoes, and safety glasses.

All subsequent specifications are modifications of the standard specifications for four- and five-ply roofing to take care of varying conditions. Minimum specifications have been achieved by reducing these standards by one or two plies of felt and one or two layers of bitumen, or equal.

Many factors must be considered to obtain a successful built-up roof application. For example: In multiacre-sized roofs, restraint of relative movement of roofing creates stresses never meant for the roofing to withstand. Dimensionally unstable insulation can cause ridges that weaken and split the membrane. Water in insulation not properly stored and handled can vaporize and blister the roof. A concrete deck, shrinking 1 in per 100 ft in 16 years, will split a roof covering that otherwise would have performed for 25 years. Equipment, such as heating and air-conditioning units installed on top of a building, creates an additional source of roof-covering problems.

("The NRCA Built-up Roofing Manual," National Roofing Contractors Association, 1515 N.Harlem Ave., Oak Park, Ill. 60302.)

15-6. Dependence of Built-Up Roofing on Roof Decks.

The roof deck is the foundation or base on which the whole roofing system depends. The deck must satisfy considerably more than the engineering requirement that the deck be designed to support live loads (rain, snow, equipment, and workmen) and dead loads (weight of the deck itself and the roofing assembly). The roof deck must also provide proper drainage of water from the roof with enough slope so that water does not collect in the low areas between roof-forming members, and so that the roof area is completely dry within about 48 hr after it stops raining. The deck must, in addition, provide a foundation that is smooth, free of humps, depressions, and offsets at joints (Fig. 15-3), and that is stiff enough to support the equipment and materials needed to apply and maintain the roof system.

Many roof failures can be traced to poorly designed roof decks, faulty preparation of the deck to receive the roofing, and roofing applied before the concrete or gypsum deck has properly cured and dried. The following test may be used to determine deck dryness before start of a roofing application:

Pour on the surface of the deck 1 pt of bitumen that is specified for membrane, heated to 350 to 400°F. If the bitumen foams, the deck is not dry enough to roof. After the bitumen has cooled, try to strip it from the deck. If the bitumen strips clean from the deck, the deck is not dry enough to roof.

15-7. Types of Roof Decks.

Following are brief descriptions of commonly used roof decks.

Steel Decks. This type is composed of a number of pieces of sheet metal formed with ribs, to give strength and rigidity. The sheets are attached by welding or clips to the roof framing system.

Precast, Prestressed Concrete Decks. Individual units of concrete may be cast away from the building site and erected after they harden. They may be T-shaped, double tees, hollow-core slabs, flat plates, or channel slabs. Prestress applied with high-strength steel offsets in some part stresses that will be imposed on the unit by live and dead loads. Camber differentials between adjacent members after erection must not be more than ¼ in. Lightweight fill or grout over the slab or between adjacent members may be used to correct differentials in the deck before the roofing application begins.

Structural Cement-Fiber Roof Decks. These are assembled of panels made with treated wood fibers bonded with cementitious binders. The panels are rated noncombustible and possess insulating qualities. Additional insulation may be required when metal supports are within ½ in of the deck top.

Wood Decks. A wood deck should be of sound, well-seasoned lumber, connected by tongue and groove, or shiplapped or splined together at the side joints. Because of the moisture content of lumber, it is important that a slip sheet be installed between the deck and roofing membrane to permit dimensional changes to take place in the wood.

Gypsum Concrete Decks. This type is comprised of slabs or planks made of a mixture of gypsum, wood chips or shavings, mineral aggregate, and water, and reinforced with steel, or a galvanized woven-wire fabric. Manufacturers' specifications must be adhered to rigidly. Gypsum concrete must not be poured in rain or in freezing temperatures. If material freezes before it is cured, it must be replaced completely before roofing commences. This type of deck will not prove satisfactory if the building will have a high-moisture occupancy, such as frozen-food storage, laundry, or restaurant.

Insulating-Concrete Decks. These are made of lightweight aggregates, such as vermiculite or perlite, mixed with portland cement and water, or by foaming the concrete with chemical agents. If this material is used with galvanized steel decks, underside venting should be used. This type of deck will not give satisfactory service in areas or over surfaces and forms where complete evaporation of moisture is not possible. (See also Arts. 15-12 and 15-21.)

Reinforced Concrete Decks. These are made of stone or crushed gravel and sand mixed with portland cement and water and reinforced with steel bars or welded-steel fabric.

Thermosetting Insulating Fill. This is a mixture of asphalt and vermiculite or perlite compacted in place, to provide insulation and slope for drainage. The fill must be applied over a rigid subbase that will not deflect under pressure from the compacting roller. If the building will have a high-moisture occupancy, venting precautions should be taken, and a vapor barrier should be installed on the deck under the insulation. If moisture from rain or other sources wets the fill before the roofing is applied, it is essential to thoroughly dry the fill. (See also Arts. 15-12 and 15-21.)

15-8. Vapor Barriers for Built-Up Roofing.

Unless precautions are taken, water vapor in the interior of a building, especially if it has a high-moisture occupancy, will condense on the cold underside of roofing felts, saturating the insulation and reducing its effectiveness, or will drip back into the building, staining the ceiling or wetting the floor. In some cases, the moisture will remain in the felts, freezing in cold weather and rupturing the roof membrane.

A solid, unbroken vapor barrier placed below insulation installed on the underside of the roofing felts can, however, prevent such condensation.

The following general rules will assist in determining whether or not to install a vapor seal under built-up roofing:

A vapor seal should generally be applied on all roof decks in cold climates, and on all decks of heated buildings where the midwinter temperature may fall below 45°F.

Vapor barriers should be used on all roof decks in temperate climates wherever conditions of high inside humidity exist, such as in textile mills, laundries, canning factories, creameries, and breweries.

Vapor barriers should be used on all roof decks where the roof deck itself contains an appreciable amount of moisture. A vapor barrier should be used in all cases on top of concrete and poured gypsum decks.

Perm Ratings. The effectiveness of a vapor barrier is measured by its perm rating, which is a measure of porosity of material to passage of water vapor. Perm ratings are established by ASTM procedures. To be classified as a vapor barrier, the material should have a perm rating between 0.00 and 0.50 perms.

A perm rating for a material is the number of grains of water vapor (7000 grains equal 1 lb) that will pass through 1 ft^2 of the material in 1 hr when the vapor-pressure differential between the two sides of the material equals 1 in of mercury (0.49 psi).

Vapor-Barrier Materials. Coated base sheets, saturated felts, laminated kraft papers, and polyvinyl chloride (PVC) sheets are commonly used as roof vapor barriers.

PVC sheets are generally specified when a noncombustible roof rating is required. PVC has a very low perm rating, about 0.10 to 0.50 perms. The sheets are attached to the deck with mechanical fasteners that pass through the insulation (on top of the vapor barrier), or with a compatible adhesive.

Coated base sheets may be used with lightweight, insulating roof fills, such as foamed, cellular, perlite, or vermiculite concretes, which have a high moisture content. Vents, for example, a vent stack or edge venting, should be provided to release moisture from the deck itself. Coated base sheets have proved to be a satisfactory vapor barrier when the material has been properly stored and handled before application, and if the installation includes venting.

Kraft laminates, vapor-barrier sheets made up of two layers of special heavy-duty kraft paper laminated together, have a low perm rating, about 0.25 perm. Used with compatible adhesives or mechanical fasteners through the insulation, these laminates will result in a UL or FM acceptable fire-retardant system.

Saturated felts are sometimes used as vapor barriers, but coated base sheets are preferable.

15-9. Insulation with Built-Up Roofing.

Insulation is a material or product that retards transfer of heat through itself. In a roof system, heat is transferred primarily by conduction, the movement of heat from a warm surface to a cold surface of a material. Insulation usually is placed below the roof deck, but it may instead be placed above (Art. 15-13).

Roof insulations are rated by their coefficients of thermal conductance C or resistance R, rather than by thickness. See Art. 18-4.

The principal types of roof insulations are:

Cellular-glass insulation. It has a C value of 0.39 for 1 in thickness. Because it has a perm rating of 0.0, a vapor barrier is not required.

Glass-fiber insulation. It has a nominal k value of 0.25.

Perlite insulation. Consisting of expanded volcanic glass combined with mineral fibers and binders, it has a C value of 0.36 for a 1-in nominal thickness.

Cork insulation. It has a k value of 0.26 for densities of 6.4 lb/ft^3.

Fiberboard insulation. Made primarily of wood fibers and a binder, it has a k value of 0.36.

Urethane rigid-board insulation or urethane foamed-in-place insulation. These are made of a closed-cell plastic foamed under pressure. The rigid-board insulation is sometimes sandwiched between two layers of coated roofing felts. Foamed-in-place urethane (a term used interchangeably with polyurethane) is sprayed directly on the roof. The board insulation is usually covered with a standard built-up roof, whereas the foamed-in-place insulation requires a liquid roof coating formulated for the purpose. Urethane insulation has a C value of about 0.15 for a 1-in nominal thickness.

Installation of Insulation. There are six methods by which roof insulation is generally installed. These are:

1. *Standard application* with the insulation attached to the roof deck either with mechanical fasteners, nails, adhesives, or asphalts.
2. *Lightweight insulating concrete,* which becomes an integral part of the roof deck.
3. Attached to the *underside* of the deck.
4. As a part of the ceiling, which is suspended below the deck and separated by a vented airspace.
5. The inverted roof system (Art. 15-13), where the insulation is installed above the membrane and ballasted with stone.
6. *Foamed-in-place polyurethane roof system,* fabricated with foaming equipment on the job site.

Regardless of which method is used to install the insulation, it must be protected from adverse weather, and at the end of a work day, it is vital that edges be sealed against moisture. Fastening methods vary with the type of roof deck and specifications. Insulation may be mechanically fastened, nailed, or adhered with asphalt or an adhesive, especially if a vapor retardant or vapor barrier is used. The term *retardant* is more accurate, but *vapor barrier* is commonly used. Adhesives are often used as a fastening agent with steel or concrete decks. But the time-tested method, mopping with hot asphalt, generally is the preferred installation method.

Often, installation of insulation starts with a "sprinkle" of hot asphalt on the deck

15-10   Roof Coverings

into which insulation boards are laid. (Channel mopping is another term used to describe the application. Solid mopping refers to covering of the entire roof surface with hot, steep asphalt.)

15-10. Built-Up Roofing Failures.

The primary cause of roofing failures may be summarized as follows:

Faulty design of the entire structure.
Faulty design of the roof deck.
Poorly designed flashing details.
Improperly installed flashings.
Application of roofing materials during inclement weather.
Improper storage of roofing materials before installation.
Improper use of materials.
Faulty and careless workmanship. (See, for example, Fig. 15-3.)

Fig. 15-3 Poor installation of insulation on metal deck. Irregularities (1) in the top flanges of the deck prevent adherence of the insulation. Top bends (2) of the ribs provide the only support for the insulation. The end joint (3) of the insulation board is cantilevered, because the top flange is not level, as a result of which the board may fracture under foot traffic.

Movement of roofing, which often causes roofing failures, is difficult to avoid, because of differences in dimensional changes that occur in connected materials as temperatures change or moisture content varies. These failures result basically from a design error caused by any of the following factors:

Deck or building movement.
Thermal movement of the substrate, such as insulation.
Thermal movement of the built-up roofing.
Movement of the built-up roofing due to alternate wetting and drying as condensation occurs in the roofing plies and evaporates. Prevention requires careful attention to detail in preparing the specifications, with full knowledge of the properties of materials to be installed.

See also Art. 15-12.

15-11. Temporary Roof Coverings.

When weather conditions are likely to be adverse to application of built-up roofing, or on roofs where trades other than roofers have further work, a temporary roof covering may be applied and application of the permanent roof covering delayed to a date when more favorable conditions prevail. Use of temporary coverage helps extend the construction season through the full year; allows use of the roof as a storage area during construction; permits installation on the roof of equipment, such as ventilating fans, signs, air conditioners, and heating units; and allows foot traffic on the roof.

A temporary roof covering may consist of a coated base sheet with 4-in side and end laps. If extra-heavy traffic is expected, double coverage should be employed.

The temporary roof is installed exactly as a conventional roof, but will have served its purpose when the final complete roof is installed. In most instances, the temporary covering should be removed, especially if extensive damage has occurred to it. If damage is light, the temporary roof can be repaired, to serve as a vapor barrier under the roof insulation or as the deck for the new application.

15-12. Good-Practice Recommendations for Built-Up Roofing.

Careful adherence to the basic conditions for good application of a built-up roofing system will result in an economical long life for the roof. Often, however, when the simple rules are overlooked, expensive failures requiring premature roof replacement occur. The basic requirements for good roofing are:

Structurally sound buildings and stable roof decks.
Insulation with acceptable thermal dimensional stability.
Acceptable vapor barrier, whenever required.
Dry insulation and roofing materials.
Uniform and positive adhesion between materials, including adhesion of the vapor barrier to the deck, insulation to the vapor barrier, the first ply of roofing felt to the insulation, and between all plies of felt.
Application of asphalt or pitch in quantities not less than those established in the specifications.
Prevention of moisture in any form from being sealed into the roofing system during its application.

The roofer should exercise good quality control and insist on good workmanship throughout the roofing application.

The roofing contractor must have the cooperation of the architect or engineer, general contractor, and roofing materials manufacturers, so that the application of the roofing system may be coordinated under the most favorable weather and working conditions.

The architect and the general contractor should avoid the lowest bid price if it appears to be far below the price structure within a particular area. An abnormally low price for roofing initially establishes the primary reasons for a roof failure before construction of the project is even started.

Recommendations for Installation of Insulation. During the past few decades, use of insulation on roofs came into general practice. Previously, most structures were uninsulated. But construction of multiacre-sized roofs resulted in many premature roofing failures because of thermal stresses from expansion and contraction. New lessons had to be learned by building designers and roofing contractors. Among these are:

1. The need for new construction specifications for roofs. Where insulation is installed under the roof deck, the insulation, in cold weather, interrupts the flow of warm air from the building interior to the deck and roof membrane. Consequently, the membrane gets cold and may split, crack, blister, or develop other defects. Such conditions should be corrected with appropriate construction specifications.

2. Closer spacing of expansion joints may be advisable.

3. Added membrane plies may be required for increased strength to withstand thermal strains.

4. Supplemental nailing or mechanical attachments of the insulation to the deck may be required, especially at the roof perimeter to prevent blow-offs.

5. Reflective gravel or coatings that deflect the summer sunshine lower surface temperatures substantially and prolong roof life.

6. If the C-value of the roof insulation is 0.79 or less, it is advisable that the insulation be installed in two layers. The first layer should be mechanically fastened. The second layer should be installed with staggered joints between the layers and adhered with hot asphalt to the first layer.

7. When preparing specifications and in calculating heat flow, the architect or engineer should consider the air film that exists on every surface. Each element affecting the efficiency of insulation should be taken into account. (See Art. 18-4.)

15-13. Inverted Built-Up Roofing System.

Also known as upside-down roof assembly, an inverted built-up roofing system consists of several layers installed in a sequence different from the usual one in which insulation is placed below the roof deck. First, standard built-up roofing is applied directly to the deck. Then, rigid insulation that is impervious to moisture, such as extruded polystyrene foam, is bonded to the top of the built-up roofing with a mopping of steep asphalt (Fig. 15-4). A layer of ¾-in crushed stone (1000 lb/square) or paving blocks or structural concrete on top of the insulation completes the assembly. Gravel or slag should not be used, because the sharp edges would damage the bare insulation underneath.

The theory is that the insulation, set above the roofing, both insulates the building and protects the built-up roofing from the harmful effects of thermal cycling, ultraviolet degradation, weathering, and roof traffic. Most common defects caused by these elements, such as blistering, ridging, cracking, alligatoring, and wrinkling, are virtually eliminated.

The roofing can be installed over a variety of decks: wood, structural wood-fiber boards, cast-in-place concrete, precast concrete, poured gypsum, lightweight fills, or light-gage metal (with a rigid board installed as a base).

Fig. 15-4 Inverted built-up roofing (patented), with insulation placed above the deck and roofing.

Note that the inverted system specifies twice as much stone per square as usually applied. Thus, this roof system is about 300 to 400 lb heavier per square, and consequently requires a stronger roof deck.

15-14. Foamed-in-Place Polyurethane Roof System.

Urethane (or polyurethane) board insulation, factory produced, is often included in built-up roofing. Foamed-in-place urethane, however, is usually specified when a roof is difficult to waterproof with conventional built-up roofing. Foamed-in-placed urethane insulation is produced on the job site and foamed directly onto the roof deck. The urethane immediately conforms to the shape of the substrate and seals all cracks and crevices.

Chemically, polyurethane foam is not a single material, but a family of plastics based on urethane polymer. Flexible, rigid, and semirigid foams are produced by the reaction of a number of components when mixed together in differing ratios, depending on the end use of the resulting product. Roofing insulation is usually foamed with truck-mounted, portable equipment. Materials used include a polyisocyanate and a polyhydroxyl, which are reacted under carefully controlled conditions at a low viscosity. Other ingredients include a blowing agent and surfactants to control cell structure, and a catalyst to control speed of reaction. Formulations may also include flame retardants, fillers, extenders, and pigments.

Flexible foam derives its physical properties from cells that are broken and open. Semirigid or rigid foams are produced by adding functional resins to conventional foam formulations. The purpose is to keep the cells closed, retaining the gas and giving the foam a rigid or semirigid structure.

The insulation produced in this manner rises (somewhat like bread dough) to 30 times its original volume. Then, it sets into a plastic compound of uniform cross-linked cells (more than 200,000 cells/ft^3). Actually, the material is 97% gas and is better insulation than still air.

The foam, uniformly sprayed over the surface to be insulated, rises to its full height and hardens within a few minutes. A protective coating should be applied as soon as possible thereafter, to avoid moisture from rain or dew condensing on the surface and preventing good adherence of a top covering.

Except in cold weather, deck preparation is minimal. All that is required is that the deck be clean, dry, and free of debris. On structures being reroofed, major patching should be done in advance. Because the foam adheres to any surface, it is self-flashing. Urethane foam has been used successfully on sloping decks and on particularly troublesome decks with unusual design features.

Cold-weather applications present special problems. Although it is possible to heat the isocyanate and the resin and to insulate spraying lines, the substrate to be sprayed is still cold. Adhesion to the deck will be poor, and the entire roof may crack and shear away from its base. Foamed-in-place urethane may be used, however, if the roof is temporarily enclosed and the enclosure is heated.

15-15. Metal Roof Coverings.

One of the most important factors in selection of metal roof coverings is their ability to conform readily to a wide variety of roof shapes and slopes. Principal metals used include steel, aluminum, copper, lead, and alloys, such as terneplate, stainless steels, and Monel metal. In addition, multimetal systems, new alloys, dead-soft metals, together with fast installation methods, such as those used for lock-in or snap-in panel systems and the zipperlike zip-rib roofing, offer many design possibilities.

Multimetal systems employ two or more metals to obtain properties not available in a single metal. The metals are joined in a permanent molecular bond by heat and pressure. Usually, one metal is bonded to both sides of a base metal that provides over 50% of the thickness of the system. Cladding, however, need be done only on one side for some applications, with any ratio of thickness desired.

The multimetals used in roof coverings include:

Terne-coated stainless steel, which is a nickel-chrome stainless steel clad on both sides with a terne alloy (80% lead, 20% tin).

Copper-clad stainless steel, fully annealed and provided with a bright surface. More oxide-free than commercial copper, the surface can be lacquered to preserve its brilliance. Stainless steel clad on only one side with copper is frequently laminated to a stiff substrate, such as plywood. This product has a low coefficient of thermal expansion.

Brass-clad stainless steel consists usually of 85-15 brass clad to a 400-series, ferritic stainless steel.

Galvanized steel.

Nickel stainless steel.

Roofing panels are manufactured from either 24- or 26-gage steel, usually with a protective coating on both sides. Color pigment is often added by manufacturers, who offer 15- and 20-year guarantees on the color fastness.

Dead-Soft Stainless Steel. This metal roofing handles easily, conforming to any desired shape with little or no springback. It is easy to solder and can be joined by all the standard methods. It also may be used for flashing, expansion joints, fascia, and copings.

Corrosion Prevention. Corrosion or gradual deterioration is a phenomenon encountered in the use of practically all commercial metals. This disintegration occurs because each metal possesses a potential force that tends to make it revert to the more stable state of chemical combination in which it is frequently found in nature.

Dissimilar metals, when in contact or connected by an electrolyte, set up galvanic action, resulting in the destruction of the metal lower in the electromotive-force series. Aluminum, high in the series, for example, will attack zinc, next lowest in the series. Roofing metals, in the order of highest potential, list as follows: copper, lead, tin-nickel (Monel), iron (steel), aluminum, and zinc. Because current flows from copper to steel, for example, these metals should not be in contact.

A metal lower in the series, however, may be used for sacrificial protection of another metal. For example, an unbroken zinc coating (galvanizing) will protect steel against corrosion.

Metals near each other in the series suffer little damage from each other's contact. The farther apart two metals in the series, the greater the damage may be from contact. For this reason, a copper roof should not be applied directly over a wood roof deck fastened with steel nails. Instead, the copper should be separated from the deck by insertion of an asphalt felt to insulate the copper from attack by the steel nails in the roof deck. Also, aluminum sheets should not rest directly on steel purlins, but should be insulated from the steel with felt strips.

As far as possible, nails or other fasteners used in securing metal roof coverings should be of the same metal or alloy as the roofing.

Thermal Movements. All building materials are in constant, although imperceptible, state of motion because of temperature changes, and metal sheets move much more than other roofing materials. Table 15-2 indicates the relative thermal movements of several roofing metals.

Because the amount of expansion and contraction is proportional to the dimensions, large roofing sheets undergo larger movements than small sheets. A metal roof laid in large sheets that are permanently fastened to the deck, consequently, would buckle and split at seams and joints.

To allow for expansion and contraction, four types of metal-roof installations are often used: corrugated metal, standing seam, batten seam, and flat seam.

TABLE 15-2 Change, in, in 8-ft Sheet for 150°F Temperature Change

Steel	$6/64$
Monel metal	$7/64$
Copper	$9/64$
Aluminum	$12/64$
Lead	$15/64$
Zinc	$16/64$

Corrugated-metal roofing, as the name implies, consists of sheets with corrugations, which absorb thermal movements in the direction across the undulations. In the other direction, connections between sheets must permit thermal movements.

Standing-seam construction employs narrow panels interconnected along their length by raised folded edges that provide for expansion and contraction, while looselocked seams allow longitudinal motion. (This is also true of batten-seam construction, a type of roofing not used as frequently as standing seam.)

Batten-seam roofing has a metal cap, or batten, installed over raised seams.

Flat-seam construction employs a series of small sheets that are interconnected by solder on all four edges. The sheets are so small that expansion and contraction are negligible and are amply provided for by a slight buckling in the center of each sheet.

Except for corrugated metal, metal roofing sheets should not be fastened directly to structural framing, because the metal must be free to move. Connections to structural framing may be made with small metal clips for standing-seam and flat-seam roofing. In all three cases, the clips should be locked into the laps or folds of the roofing sheets.

Where metal sheets are fastened directly to the framing, although provisions are made for expansion and contraction, the consequent movement of the sheets makes it virtually impossible for any of the fastening devices used to remain tight. This has led to use of lead or Neoprene washers, which take up the slack caused by loosening of bolts, friction of moving sheets, and enlargement of holes drilled for fasteners. Note that interior and exterior air pressures on this type of roof should also be reckoned with in providing for sheet movement.

Special caps shaped to the angle of the roof should be used where metal-roofing sheets meet at a ridge. These caps should have pockets on both sides to house the ends of the sheets.

15-16. Cold-Process Roof-Covering Systems.

These are used primarily for maintenance and repair rather than for new construction, but they are not limited to these applications. Many systems are available, some involving chemical combinations with bituminous products. One cold-process roofing system requires a heavy-duty glass mesh to be embedded in a high-viscosity, asphalt-based emulsion which is brushed or sprayed on a roof deck, and a finish coat is added for a smooth surface. Another system combines polyurethane with coal tar to produce a sprayed-on waterproofing membrane.

To produce workable liquidity, bituminous materials are thinned with solvents or emulsified with water.

Cutbacks. Solvent-thinned bitumens are known as cutbacks. Solvents include kerosene, in the manufacture of cutback asphalt, and toluene, in the manufacture of cutback tar. With cutbacks, the cementing process of a roof is achieved by evaporation of the solvents, which leaves a film of bituminous material.

Emulsions. These are produced by melting bitumen and dispersing it into tiny particles suspended in water by means of soap or a chemical emulsifying agent. On a roof, the water evaporates, leaving a solid film of bitumen on the surface.

Felts and Fabrics. Plies used with cold-process bitumen differ from those used with hot. Since cementing takes place on evaporation of solvents or emulsion water, it is necessary that the drying-out process be speeded. For the purpose, porous mats made of glass fibers often are used, because they have the advantage that the open weave allows water vapor to travel through it, a breathing process that eliminates blisters. Glass fibers, however, require a thick mastic-type emulsion, because light films are not strong enough to hold the fibers in place.

Application. Cold-process materials are applied by brush or spray. But first, the roof surface must be prepared, to eliminate cracks, ruptures, holes, fish mouths, and other undesirable characteristics of weathered roof decks.

In selection of cold-processing roofing materials, designers should beware of unprincipled producers who promote doubtful products by making extravagant advertising claims. Cold-process roofing materials, when properly selected and applied, can add years of life to roofs, but they are not cure-alls and will not add much life to roof coverings that have disintegrated. They will not bridge gaps in surfaces and will prolong life only in materials in which life exists.

15-17. Elastomeric Roof Coverings.

The term elastomer refers to a synthetic polymer with rubberlike properties. This type of material gained rapid acceptance for use on free-form roof shapes, such as large domes, barrel roofs, folded plates, and hyperbolic paraboloids. Also, elastomers are advantageous for cable-supported roof decks, which require a roofing membrane with minimal weight and an elasticity that withstands continuous movement. Furthermore, liquid-applied Neoprene-hypalon systems weighing only 20 to 50 lb/square are often specified for very large industrial buildings rather than conventional built-up roofs weighing up to 600 lb/square.

Elastomeric roofing systems fall into two general categories:

Fluid-applied systems, applied by roller, pressure-fed roller, spray gun, brush, or squeegee in single or multiple coats.

Sheet-applied systems of the one-ply type or those requiring a top coat for color.

The fluid-applied system has the advantages of conforming readily to any shape of substrate without material waste due to pattern-cutting requirements, as is necessary with the sheet system, and better adhesion. Fluid application also forms a continuous membrane, free of joints. The elastomer will not, however, fill voids and requires continuous substrate junctures and expansion joints. Insulated-covered decks present special problems, except for sprayed-in-place polyurethane insulation, which offers a continuous unbroken surface to be covered by the elastomer.

One-ply sheet systems often have material, such as asbestos, laminated to the back. The single ply is provided in roll form and is installed with an adhesive applied to the roof deck and at the joints where rolls lap. The product is lightweight, easy to install, and forms a long-lasting watertight roofing, with excellent expansion-contraction properties. The one-ply system also helps to resist the high stress caused by roof-deck movements when air conditioning is used in the building interior. Furthermore, this roofing will withstand light foot traffic. Repairs are made with patches, cutouts, or sealants.

Sheet systems should not be used for steep roofs or roofs with intricate designs that would involve extensive cutting, fitting, and seaming. Sheet systems will, however, tolerate greater deck defects. If large thermal movements may occur, sheet elastomers can be permitted to "float," remaining unbonded over some areas.

A number of compounds fall into the elastomer family, including:

Silicone. This is a group name for polymers with a structure of alternate silicon and oxygen atoms, with various organic groups attached to the silicon. The elastomer is compounded with a suitable filler and vulcanizing agent to produce a liquid-applied film. This coating has higher dirt retention than most other elastomers.

Butyl Rubber. This is a synthetic rubber, a liquid formed by copolymerization of isobutylene with a small proportion of isoprene or butadiene. It is compounded in black only, and is somewhat difficult to adhere under normal field conditions.

Polyurethane Rubber. This elastomer should not be confused with polyurethane insulation, which is an entirely different product. The polyurethane rubber formulation is produced by addition of a diisocyanate to polyesters. Fire retardants can be added to increase fire resistance. This rubber can be made with a wide range of physical properties. It has low vapor transmission and excellent resistance to thermal exposure and weathering. Fewer coats are required in application than for other elastomers, thus reducing labor costs.

Neoprene. This is a family of elastomers made by polymerization of chloroprene and formed into liquid, sheet, or puttylike consistency. The product will not support combustion and has excellent adhesive properties. It is normally compounded in dark colors.

Hypalon. This is made of chlorosulfonated polyethylene, in liquid, sheet, or puttylike consistency. It will not support combustion, and has even greater resistance to thermal movement and weathering than Neoprene. It is formulated in a variety of subtle colors.

Roof Decks. To be successful, elastomeric roofing must be applied over a satisfactory roof deck, carefully selected and specified. Lightweight-aggregate concrete decks or foamed-concrete fills, for example, should not be used with liquid elastomeric coatings. Air and moisture trapped in such substrates will eventually blister the elastomeric film. Structural concrete, properly finished and cured for a minimum of 30 days, makes an excellent foundation for liquid roofing. Plywood provides an excellent surface for Neoprene and hypalon coatings. Closed-cell polyurethane insulation serves as an excellent substrate for elastomers, and where cost is not an obstacle, elastomers are the preferred covering for the foamed-in-place product.

See also Art. 15-18.

15-18. Single-Ply Roofing Systems.

Once a part of the elastomeric family, these systems have multiplied until dozens of different products are marketed as single-ply roofing systems. This name, however, does not imply that the actual membrane consists of one-ply only, but rather that a number of layers have been bonded to form a single, thick sheet (Fig. 15-5). It may or may not be an elastomeric material; it could be any of many compounds, rubberized materials, etc. Some membranes may have a smooth asphaltic surfacing, a granular topping, or perhaps a thin metal sheeting, such as copper. Some are thick and some are very thin. Or a membrane may come with a protective peel-back paper over an adhesive film that adheres to the deck with light pressure. Some membranes have asphalt compounds on

15-18 Roof Coverings

Fig. 15-5 Single-ply compositions used for roofing: (a) polyethylene surfacings; (b) polyethylene bottom surface and embossed aluminum top, which provides a reflective surface; (c) bituminous ply with bituminous aggregate topping; (d) bituminous ply with paving blocks on top; (e) detail at a drain.

the underside, which are torch-heated with special equipment to soften the coating so that it adheres to the substrate.

The thinner *plastic-like* membranes are often loosely laid by unrolling the required lengths side by side, slightly overlapped. These overlaps are then adhered into a single sheet by a heat-weld, or a solvent-type material to soften edges so they stick together, or a compatible adhesive (Fig. 15-6). Gravel or stone ballast is placed on top to prevent uplift by the wind. Roof walkways are usually installed on top of the stone ballast to provide protection to the surfacing and permit access to roof equipment.

Fig. 15-6 Application of a single-ply roofing system. Shown is a variation calling for a combination of fastened and free-floating application, in which a neoprene sheet is fastened along its edges with a prepunched metal strip and mechanical fasteners. In between, the membrane is free floating. An adhesive is used for seams and terminations at flashings and edges.

Application. Manufacturers' specifications should be followed very carefully in all applications. A typical application of the liquid coatings might require (1) a neoprene primer, (2) two body coats of elastomer or other coating, and (3) two finishing coats of color elastomer.

One of the sheet systems might require (1) a water-dispersed deck adhesive; (2) the sheet or roofing laminate product, cut to size and shape according to the roof design and then prerolled before application; (3) brooming or pushing out of the air pockets or blisters, and pressing the material into place before the adhesive sets; (4) *flashing tape,* caps, and seals for field joints, vents, and roof edges; and (5) often, a top coating of a liquid membrane to cover any taping used to seal laps of membranes and provide a seamless surface.

15-19. Corrugated Asbestos-Cement Sheets.

These are applied somewhat like corrugated metal. The same type of fastener is employed. Because asbestos cement is a more easily punctured material, however, the sheets are placed over welded studs on purlins, and the corrugations are malleted until the studs pierce the material. Rubber mallets are used to deliver enough force without damaging the sheets or the studs.

Asbestos-cement roofing differs in shape from metal roofing. Asbestos-cement sheets are molded so that the valleys of the first and last corrugations lie downward, to allow the minimum side lap to be one corrugation instead of 1½.

Two methods of application are employed, the step method and the miter method. Both prevent occurrence of four-thickness end and side laps, a feature of corrugated-metal roofing that would be unsightly as well as difficult to waterproof with the thicker asbestos-cement roofing. In the step method, alternate courses of the sheets are started with half sheets. In the miter method, no effort is made to break joints, but with a miter at one corner of each sheet, end and side laps have uniform thickness.

15-20. Rigid Vinyl Roofing.

Polyvinyl chloride in flat, corrugated, and ribbed sheets is available for roofing in standard and special lengths. Standard lengths range from 8 to 20 ft. Longer sheets can be ordered, however, to reduce the number of end laps required.

The sheets are fastened to structural framing with machine screws, bolts, or wood screws, depending on whether the frame is metal or wood. Fastening may be made less laborious by use of fasteners that seal both top and underside when thrust through a hole drilled from the top. These fasteners include "pop" rivets and machine screws with a Neoprene cap and jacket.

Vinyl sheets can be cut quickly on the job with portable power saws with abrasive cutting wheels. Where possible on large jobs, sawing, cutting, and drilling of vinyl sheets should be done in the shop. The sheets can be nested and drilled and cut a half-dozen at a time. In addition, much of the sawing can be speeded with stationary machines.

As with all sheet roofing material, application starts at the eaves opposite the prevailing wind direction. While the sheeter has no jurisdiction over purlin spacing, the sheeting should be arranged to end over a purlin. Overhangs at eaves should not exceed 6 in. Each of the second and succeeding sheets placed alongside each other should cover one corrugation of the previous sheets and should be fastened along the lap (in the valley) at 12-in intervals. Sheets laid at the ends of these, and succeeding courses, should cap the lower ones at least 8 in. At a ridge, corrugated closures should be placed to ensure a neat finish and adequate sealing. Where full weatherproofing is required, filler strips may be inserted at eaves, and calking cement employed between end laps.

For economy, sheets should extend from eaves to ridge. This cuts down the number of fasteners, amount of accessories, and amount of material, because end laps are eliminated. For many arched roofs, however, sufficiently long sheets are not available, not because manufacturers cannot produce them, but because of shipping difficulties.

15-21. Venting Lightweight-Concrete Roof Fills.

Lightweight-concrete roofs with a high moisture content have been responsible for many premature roof failures. Entrapped moisture emitted by the fills and transmitted to the roof membrane has caused splitting, cracking, blistering, and other sources of leaks. Blisters as large as 200 ft in diameter have been observed. Nevertheless, lightweight concrete offers many advantages, including thermal insulation, light weight, and ease of surfacing for taper to drains. Consequently, it is worthwhile to overcome the disadvantages, which can be done with proper precaution during application, especially by installation of roof vents. But primarily, use of this type of construction should be avoided where it is not suitable. For example, if the building interior will have a high moisture content, lightweight-concrete fills are unsuitable, because they will absorb moisture from the interior. Also, suspended ceilings should not be used in such cases, because condensation may occur on the bottom of the roof deck.

Fills cast over vented galvanized steel roof decks or wire fabric may not require top venting. Underside venting, however, may be provided if the architect and roofing contractor deem it necessary.

On a small building, effective venting can be obtained through use of perimeter or edge venting, with an open-type metal gravel guard or fascia. As an additional safeguard, this can be supplemented by metal roof vent stacks. On decks more than 50 ft wide, vent stacks, whether turbine, free-spin, or stationary, with a minimum diameter of 6 in, should be used to vent each 1000 ft^2 of fill. On smaller decks, at least two vents should be installed.

The space under each vent should be free of concrete down to the structural deck. The space can be formed either before or after placement of the concrete. Before concreting, the space can be blocked out and the blocking removed after the concrete has hardened. Or the space can be created after concreting by removal of the fill down to the deck. This gap may then be filled with loose, granular, or fibrous insulating material, and the metal vent should be installed over such areas on top of the fill. If desired, the vents may be removed after 18 months if the fill has dried to the satisfaction of the architect and the roofing contractor and if the built-up roofing is performing well.

15-22. Roof Moisture-Detection Surveys.

Moisture within a roof membrane system tends to migrate from its point of entry. Thus, blisters, cracks, and splits caused by the moisture are often located at a distance from the source of a leak. Several methods are available for locating such sources. The equipment utilizes infrared photographic techniques, or nuclear or electronic devices, or variations of these systems. Each method has advantages and disadvantages but can be effective if properly used.

In the method employing an infrared roof scanner, an infrared photograph is taken of a roof at night from a helicopter. Because wet insulation emits vapors and loses heat, it can be detected by the scanner and photographed.

The electronic and nuclear methods use differing means for locating roof moisture, but the procedures are similar. First, a drawing to scale of the roof is made, showing all projections, roof equipment, etc. A grid is marked on the drawing, then physically reproduced on the roof with ropes or paint. Specially trained operators take a reading, with an instrument that can detect moisture, at every grid intersection and record the observation. At the same time, the roof is visually inspected and observations are carefully noted. The data collected then are evaluated by a specialist and adjusted for the effects of up to nine different factors, such as gravel depth, roof construction, etc. The information is entered on a final map of the roof, to provide a picture of water-damaged areas, under-membrane water-flow patterns, percentage of moisture in the roof, and locations of moisture. From this map, it is possible to determine where replacement is necessary, where repair will be sufficient, and what to budget for future work that may be needed to maintain the roof or whether to replace it section by section as required. Such moisture detection surveys are valuable tools for saving energy and for management planning.

15-23. Air- and Tension-Supported Roofs.

Buildings with roofs and sidewalls held aloft by pressurized air continuously pumped into the interior have often been constructed at exhibitions and fairs. The application has also been extended to more permanent and elaborate structures, such as stadiums, covers for industrial parks and shopping malls, and office buildings (Fig. 15-7). Other applications include industrial storage facilities and covers for construction-site weather protection, to extend the work seasons.

The concept of air-supported roofs should be easy to comprehend because their construction is similar to that of a gas-filled balloon or an air cushion. Basically, the construction process is simple. For example, a storage area may be constructed in a few hours after the area and shape of the structure have been laid out and a floor has been formed, either of earth, wood, or concrete. A peripheral compression ring, exactly fitting the area to be enclosed, is made ready to secure the material to be used as walls and roof. This material may be laid out flat over the floor, secured to the compression ring and then raised by air pressure. Or the material may be suspended from one or more masts

Fig. 15-7 Air-supported dome over Syracuse University stadium covers 6.5 acres.

and then filled with air pressurized sufficiently to support it. The building is now ready for use. After occupation, air lost through entrances and exits is small and can be replaced by operation of an air compressor.

In the case of a structure that must be kept open frequently on one or more sides, a variation of the preceding technique is used. A virtual air cushion is constructed and filled with air. The cushion comprises an upper and a lower membrane, stretched inside a circumferential compression ring. Pressurized air is pumped between the two membranes, which confine it and become inflated, to form the roof. Walls are suspended from the roof and secured at grade. The space below the membranes need not be pressurized.

Membrane structures under tension demonstrate high load-bearing properties and permit use of economical amounts of material. Weight is inversely proportional to strength, and the greater the strength, the thinner the membrane can be. (See Arts. 5-100 and 5-101.)

Canvas was the first fabric to be used as a membrane and the structures were simple. But with the development of more sophisticated, translucent, lightweight, colorful, waterproof products, such as glass-fiber fabrics coated with Teflon, architects produced more elaborate designs, with steel cables to give a variety of shapes to the buildings. Materials such as stainless steel give permanence to the structures while providing significant cost savings. Estimated life expectancy of most air-supported construction is shorter than that for more conventional structures, but with changing needs of owners, prolonged life of a building is not always a very important consideration.

MULTIUNIT ROOF COVERINGS

15-24. Asphalt Shingles.

A typical asphalt shingle is produced from five major compatible components: a smooth base felt, saturating asphalt, coating asphalt reinforced with mineral stabilizers, back coating for dimensional stability and as a barrier against moisture from the building interior, and opaque mineral granules embedded in the heavy coating of asphalt for texture and color.

Asphalt shingles are widely used in preference to traditional slate, tile, or thatched roofs. Many types of asphalt shingle are available. They range from a low-cost, lightweight type to a heavyweight shingle weighing from 230 to 350 lb/square. Giant individual shingles and square-butt strip shingles, with or without cutouts for design tabs, provide labor-saving advantages. Self-sealing shingles, with a factory-applied adhesive activated by the sun's rays (roof temperature 60°F or over), offer resistance to wind uplift. Another design features a slit or tab that matches an engaging device in the adjacent shingle and interlocks the shingles into a secure roof covering.

To prevent wind damage, self-sealing or interlocking shingles should be used on roofs with less than 4 on 12 pitch in all wind zones. In medium-wind areas, roofs with slopes of 4 on 12 or steeper should have sealed tabs if the shingle weight is less than 275 lb/square. Secured tabs are required in areas where winds reach 70 mi/hr or more.

Also, a large variety of asphalt roofing styles is available, in a wide range of colors, textures, and patterns. The style of the building will reflect on the type of shingles selected to complement it.

Application. With all asphalt-shingle materials, the first step in application is to fasten an underlay of No. 15 asphalt felt to the roof deck, with a starting strip at the low end of the roof from which work proceeds. The felt should be lapped 6 in at the end and sides.

Because many backup leaks at eaves are caused by moisture traveling upward from the eaves beneath the shingles, through capillary attraction, it is a wise precaution to cement the starter material to the roof deck, rather than nail it. In winter, water may dam up in a gutter and, backing up under the shingles, follow a nail puncture through the roof into the building interior.

Faulty shingle applications on wood roof decks can be traced in part to too few nails, but many defective roofs can also be attributed to nailing shingles too high. If the shingle is nailed ½ in above the slots, as in the case of square-butt shingles, each shingle will be held to the roof not just by the four nails called for in specifications, but by eight nails—four driven to attach the shingle and four others driven to attach the shingle on top. Each nail travels through two thicknesses of shingle. If the shingle is nailed 2 in or more above the slots, however, it is held by only four nails at the top, and it is more likely to be pulled loose by winds.

Nails should be at least 1¼ in long on new work, and 1¾ in long on reroofing. Hot-dipped, zinc-coated iron or steel nails, with barbed stems, usually should be used. But ordinary steel nails give good service, and if the stems rust slightly, the effect will be beneficial, because the holding power of a rusty nail is excellent.

While shingle tabs are not required to be fastened, it is a useful precaution in high-wind areas to place a small dab of asphalt cement under each tab, which then should be pressed firmly onto the concealed portion of the underlying shingle.

Special Treatments. Color and form may be used imaginatively in special applications to dramatize or accent a roof, or to stress the horizontal portions of a building (Fig. 15-8). Special treatments most frequently employed are ribbon courses, two-tone rib courses, single shingle applications, and complementary color courses.

Ribbon Courses. A rugged, horizontal roof texture can be achieved by applying a triple thickness of asphalt shingles every third, fourth, or fifth course. This presents eye appeal and texture and is especially suited for the large roof areas of a long one-story building.

Two-Tone Rib Courses. A colorful approach to roof styling, similar in appearance to ribbon courses, utilizes two colors or tones on the roof. The boldness of the design depends on the colors of the shingles selected for the entire roof.

Every three or four courses, one course of 9-in black starter strip is applied. The next course is applied with only 1 in of starter strip exposed to the weather. A black band or rib results. The lighter the main roof color, the greater the contrast with the black and the bolder the styling. This design is particularly effective on square one-story buildings.

Single Shingle Applications. Visual appeal can be achieved by application of heavyweight shingles to obtain a rough-textured individual shingle look. Application is simple with a no-tab, two-tab, or three-tab shingle.

15-24 Roof Coverings

Fig. 15-8 Asphalt roofing shingles create a rustic appearance by combining an overlay texture with a staggered butt edge. *(Courtesy Certainteed Corp.)*

The first course is applied the standard way. Shingle sections 4, 6, and 8 in wide are cut from additional bundles of shingles and are nailed over the cutouts and the butted ends of the shingles. For a three-tab shingle, a 4-in section is cut off each end, and each remaining section is cut in half. If no-tab shingles are used, the cut sections are placed at random intervals between butted joints. The sections are extended about ¼ in below the shingle course, depending on the amount of texture desired. Care should be used so that no two sections of the same size are applied next to one another.

Complementary Color Courses. Complementary color courses take advantage of the variety of colors asphalt shingles offer. This application is suited to commercial buildings, such as restaurants and motels. Different-color shingles are used in alternate courses up the roof. The shingles can be a contrasting or complementary color. For example, the different courses may be dark green and a lighter green, or red and brown, or slate and black.

15-25. Wood Shingles.

These may be obtained machine-sawn or hand-split. Fire-retardant-treated shingles also are available for both varieties.

An undereaves course is required under the first course of shingles on the roof; but no cant strip is required, because the tapered wood shingles fit closely to the roof.

Because of water drip at the eaves, eaves shingles receive more abuse than the other shingles on the roof. To prevent any possibility of leakage at eaves or cornice, a triple layer of shingles at those areas is a wise and comparatively inexpensive precaution, adding little in material and labor to the average job. If a triple layer is used, the joints in all three layers should be laid so as not to coincide.

The standard exposure of wood shingles is 5 in. But depending on the length of shingles, they may be laid at exposures ranging from 3½ to 12 in.

As the shingles are applied to alternate courses, with joints broken in relation to courses above and below, the desired exposure is maintained, except where it is adjusted to line up with dormers or with the ridge line.

Because a sharp hatchet and saw are used for instant trimming, there is no necessity for special shingles on wood shingle roofs. At gables, joints are broken by use of wide and narrow shingles in alternate courses. To prevent water dripping down the gables, it is a good plan to nail a length of beveled siding from the eaves to ridge along the gable and nail shingles over it. This forces water onto the main body of the roof, and the slight elevation is imperceptible from the ground.

Although it is difficult to avoid lining up an occasional joint, matching joints of shingles in any three courses should be avoided. This is important because a leak is unavoidable if a shingle splits on a line between the joints in alternate courses.

To avoid such splits, nails should never be driven into the center of a shingle, no matter how wide the shingle. There is the additional hazard that a center-nailed shingle may have a joint above it, exposing the nail to the weather, to corrosion, and to an eventual leak.

Only the widest shingles should be used on valley sides. For ridges and hips, the shingles themselves are employed, the best construction being a modified Boston hip. Hip and ridge shingles should be at least 5 in wide. To avoid waste, shingles of this width should be selected from the random widths available. Hip and ridge shingles should be applied horizontally and allowed to project above the ridge shingle on the other side of the roof, the projection being chipped and shaved off with the shingler's hatchet. The hip shingles should be applied so that the shingles are woven in a weathertight pattern.

("The NRCA Steep Roofing Manual," National Roofing Contractors Association, 1515 N. Harlem Ave., Oak Park, Ill. 60302.)

15-26. Asbestos-Cement Shingles.

Composed of 15 to 25% asbestos fiber and 75 to 85% portland cement, asbestos-cement products might be described as lightweight sheets of reinforced concrete. They are produced in a variety of shapes and sizes for numerous specific applications. Principal products are roofing shingles, flat sheets, siding shingles, and corrugated sheets.

Basically, there are four standard types of asbestos-cement roofing shingles, known as American-method, multiple-unit, Dutch-lap, and hexagonal.

American-method shingles (Fig. 15-9a) are so called because of their shape and finish. Laid in a rectangular pattern and having a simulated wood grain, they present an appearance similar to conventional wood shingles.

Fig. 15-9 Application of asbestos-cement shingles begins at the eaves with a cant or starter strip. Ridge shingles or ridge rolls can be used at the peak of sloping roofs.

Multiple-unit shingles are produced in large units, each of which covers an area equal to that of two to five standard-size shingles. Made in styles and sizes that vary with manufacturers, they retain the same general appearance, when installed, as the small units.

Dutch-lap shingles (Fig. 15-9b), sometimes called *Scotch-lap*, are larger than conventional shingles and are lapped at the top and one side, a method that effects savings in material and labor.

Hexagonal shingles (Fig. 15-9c), also known as shingles laid in the *French method*, are nearly square. They are laid in a diamond pattern with an overlap at the top and bottom. The effect is a hexagonal pattern.

Sizes of asbestos-cement roof shingles vary from 16×16 in to 14×30 in.

Asbestos-cement shingles are highly resistant to water. But if, before installation, they are not stored under tarpaulins or under cover during inclement weather, seeping moisture may discolor them.

Special eaves starters are manufactured to aid in application of the first course of shingles (Fig. 15-9). A cant strip or furring strip is required along the eaves. Furring strips are also required at hips and ridges to enable ridge and hip shingles to cover adjacent material at the correct angle. Specially designed hip and ridge shingles and terminals are manufactured and should be used for finishing these areas.

Most asbestos-cement shingles require a minimum roof pitch of 5 in/ft. American-method shingles, however, can be applied (because of the small exposure) on roofs with a 4-in pitch.

15-27. Roofing Tiles.

Clay roofing tiles offer considerable latitude in design and range of choice. The tiles are manufactured from special clays. Shale, a claylike rock, which when fired becomes as enduring as stone, is often preferred.

While a variety of patterns is available, all tiles fall into two classifications, roll and flat. Roll tiles are furnished in semicircular shape, reverse-curve S shape, pan and cover, and flat-shingle design.

To lay tiles successfully, vertical and horizontal chalk lines should be struck as guide lines. In laying out these lines, it must not be assumed that the manufacturer's standard tile measurements coincide with the tile size delivered to the job. Although variations may be small, contractions occurring in the burning of the tile can cause enough difference to warrant checking of the delivered tile.

To ensure a full-length tile at the ridge, and to avoid needless cutting at dormers, chimneys, and other projections, the horizontal guide lines should be adjusted, always in favor of increasing the lap. With interlocking tile, however, because the tile exposure and lap are governed by the locking device, it is impossible to make any adjustments.

Except for Spanish mission tile, an undereaves course plus a cant strip are required before the first course of tile is laid. Tile application generally starts from the bottom right-hand corner of a roof and proceeds from right to left. Tiles adjacent to chimneys, vents, and other projections are left loose so that flashing material can be inserted.

Tile joints are broken, as with other types of roofing. But because there is an element of risk in fastening a half tile at such exposed locations as gables, manufacturers make special tile-and-a-half units. These not only save time by eliminating cutting, but make the gable windtight.

These large-width tiles can be used effectively at valleys and hips. Where cutting must be done, as around chimneys, the tile may be trimmed on a slater's stake, the cutting line being tapped with a hammer until the unwanted portion breaks off.

Tiles that verge along the hips must be fitted tight against the hip board with mastic or cement colored to match the tile.

("The NRCA Steep Roofing Manual," National Roofing Contractors Association, 1515 N. Harlem Ave., Oak Park, Ill. 60302.)

15-28. Roofing Slate.

This is one of the older mechanical arts, one in which skills are disappearing as craftsmen become scarce.

Slate roofs may be divided into two classifications, standard commercial slate and textural slate, or random slating. The latter is the older form, in which slates are delivered to the job in a variety of sizes and thicknesses, to be sorted by slaters. The longer and heavier slates are placed at the eaves, medium sized at the center, and the smallest at the ridge.

Standard commercial slating results from grading of slates at the quarries and costs less in place. Slates are graded not only by length and width but also by thickness, the latter being about ¼ in.

In manufacture, slate is split, resulting in a smooth and a rough side. The slates are applied on the roof with the smooth side down.

A slate should be laid with a lap of at least 3 in over the second slate beneath it. Slate exposure, or margin, is found by deducting the lap from the total length of the slate and dividing the result by 2. For example, the exposure of a 20-in slate with a 3-in lap is $(20 - 3)/2 = 8\frac{1}{2}$ in.

Application starts with an undereaves course of slate fastened over a batten. The purpose of the batten is to cant the first course of slate to coincide with the angular projection of the other courses.

Slate is applied so that all joints are broken. It is nailed through countersunk holes with noncorroding nails, which are not driven home but level with the slate. Slating nails, with a large, thin head designed to fit within the countersunk holes, should be used.

15-29. Roof Finishing Details.

The juncture of chimneys, walls, and roof projections with the edge of a roofing membrane is the most vulnerable area of the whole structure, because of the possibility of poor flashing. Flashings usually join two different materials, roof to wall, with different thermal characteristics. Resulting relative movement is likely to cause tears or cracks in the fabric or felt, and even in the roof membrane.

Whenever a sharp angle is introduced into roof or wall construction—in fact, at almost all intersecting planes—flashing is necessary to waterproof the angle. Because movement is to be expected at intersecting planes, flashing should be made of fairly elastic material, or fashioned so that there is room for movement.

All flashing material is affected by the expansion and contraction of the material itself, and by similar forces in the material with which it is in contact. These forces tend to pull the flashing away from the wall, leaving a gap through which water can leak past the flashing into the wall.

Bituminous flashings have the ability to hug tight against the wall when applied properly. Metal flashings, on the other hand, require added protection in the shape of cap flashing installed above and covering the top edge of the base flashing. If the flashing pulls away from the wall, there will be no gap through which water can pass to the rear. Plastic flashing sheets are useful in sealing the junctions of vents and pipes with roof decks.

On flat roof decks, installation of flashing is fairly simple, base flashing of bituminous materials or metal being applied so that at least 6 in of the material is secured to the roof deck and 8 in to the wall. Counterflashing should be installed above it. Not less than 1½ in of the counterflashing should enter a raggle cut into a mortar line between the nearest row of bricks above the base flashing, and not less than 4 in of the flashing should hang vertically down the wall.

On steep roofs, recourse must be made to step flashing. This entails cutting short pieces of flashing metal, bending them at right angles, and sandwiching one flange between every roofing unit adjacent to the intersecting surface, the remaining flange being in contact with the vertical plane. In application of the flashing, care must be taken that each unit laps the unit below by at least 2 in. Thus, water traveling downward will be conducted from one unit to another, and should water seep under the shingles or other roofing materials, the sandwiched metal will conduct it safely back to the surface. Counterflashing is installed above these base flashings in a series of steps dictated by the slope of the roof and the mortar-line raggles into which the counterflashing is inserted.

Crickets. Chimneys projecting through one side of a sloping roof require the building of a cricket, or flashing saddle, behind the chimney. Ordinary methods of flashing are not suitable in such locations, where water propelled by a 15-mi/hr wind can build up to a depth of several inches within minutes.

Saddle is an apt description of the sheet-metal flashing designed for chimney backs. It takes the form of an equestrian saddle or a miniature roof with a ridge and two valleys built to divide the streams of water traveling down the roof and conduct them away from the chimney, instead of pocketing behind the chimney.

The saddle should be securely soldered to a base step flashing behind the chimney. Because of the severe punishment they take, saddles should be fabricated of heavy-gage metals, highly resistant to corrosion.

Valleys. Where roofs intersect, special precautions are necessary to design a waterproof roof. It is obvious that, where the intersecting planes meet at the aptly named valley, the area receives twice as much rainwater as flows down the main roof area.

Bituminous valley materials should be laid in two plies at this vulnerable point, an 18-in-wide base sheet followed by a 35-in-wide mineral-surfaced top sheet.

Valley metal should be at least 12 in wide, preferably much wider, and should bridge all junctures to make the intersection watertight. To guard against water building up in the valley, metal should be folded back 1 in to act both as a barrier to infiltrating water and as a device in which fastening clips can be hooked. Metal valleys, because of expansion and contraction, should not be fastened directly to the deck, but through the clips. The roof covering should overlap the folded edges of the valley metal, to guide rainwater into the valley.

Section 16

Builders' Hardware

Richard A. Hudnut
Product Standards Coordinator,
Builders Hardware Manufacturers Association, Inc.
New York, N.Y.

Hardware is a general term covering a wide variety of fastenings and devices for operating or controlling the operation of movable building components, such as doors and windows. By common usage, the term builders' hardware generally covers only finishing hardware, but some rough hardware is discussed in this section.

In point of cost, the finishing hardware for a building represents a relatively small part of the finished structure. But the judicious selection of suitable items of hardware for all the many conditions encountered in construction and use of any building can mean a great deal over the years in lessened installation and maintenance costs and general satisfaction.

To make the best selection of hardware requires some knowledge of the various alternates available and the operating features afforded by each type. In this section, pertinent points relating to selecting, ordering, and installing some of the more commonly used builders' hardware items are discussed briefly.

16-1. Finishing and Rough Hardware.

Finishing hardware consists of items that are made in attractive shapes and finishes and that are usually visible as an integral part of the completed building. Included are locks, hinges, door pulls, cabinet hardware, door closers and checks, door holders, exit devices, and lock-operating trim, such as knobs and handles, escutcheon plates, strike plates, and knob rosettes. In addition, there are push plates, push bars, kick plates, door stops, and flush bolts.

Included in rough hardware are utility items not usually finished for attractive appearance, such as fastenings and hangers of many types, shapes, and sizes—nails, screws, bolts, studs secured by electric welding guns, studs secured by powder-actuated cartridge guns, inserts, anchor bolts, straps, expansion bolts and screws, toggle bolts, sash balances, casement and special window hardware, sliding-door and folding-door supports, and fastenings for screens, storm sash, shades, venetian blinds, and awnings.

16-2. When to Select Hardware.

An important point to bear in mind in connection with selection of hardware for a new building is that, in many instances, it is one of the earliest decisions that should be made, particularly when doors and windows are to be of metal. This is true despite the fact that the finishing hardware may not actually be applied until near the end of the construction period.

Metal door and window frames, for example, are among the first things that are needed on the job after construction gets above the first floor. These frames cannot be designed until the exact hardware to be applied to them has been decided on in detail, and precise template information has been obtained from the hardware manufacturer. Even after all the shop drawings for these items have been completely detailed, revised, and approved by the architect, it usually takes a considerable period of time to have them manufactured and delivered to the building site. Furthermore, the particular type of finishing hardware selected may also require considerable time to manufacture. Yet, if the frames are not available on the job during the early stages, work on the exterior walls and interior partitions cannot proceed. The delay would, of course, set back progress on the entire building. So it is wise to select the hardware as soon as possible.

A good deal of time and expense can sometimes be saved on these metal frames by specifying those of standard design. If any fancy trim is desired, it is often advisable to have this separately made and applied later. Such trim may well be of wood; its fabrication will not delay delivery of the frames that must be built into the walls and partitions as they progress.

16-3. Effects of Codes and Regulations on Hardware.

Building codes, NFPA 80, "Standard for Fire Doors and Windows," and the "Life Safety Code" promulgated by the National Fire Protection Association contain many regulations directly affecting the use of builders' hardware. Some of these regulations permit tradeoffs in fire protection. For example, one regulation permits reduction or elimination of compartmentation of large floor areas in exchange for fully sprinklering a building.

Some of the major considerations of such regulations are life safety, security, and handicapped people (see Sec. 3). Life-safety requirements deal primarily with provision of safe egress in emergencies; requirements for handicapped people deal with ease of circulation and use of facilities in buildings, and security requirements aim at making both unauthorized ingress and egress difficult. These requirements make conflicting demands on hardware. For example, local security ordinances often conflict with applicable building-code rules and may contain a clause exempting doors within means of egress from security requirements. Such an exemption, however, weakens security, because most doors are within a means of egress. The difficulty of designing or selecting hardware is further exacerbated when provisions for the handicapped are superimposed on building-code and security ordinances.

Efforts are being made to find compromise solutions to the conflicting regulations. In the meantime, design professionals should be especially careful to keep abreast of the latest edition of all codes, regulations, and legislation governing the areas where their buildings are being erected.

16-4. Function of Hardware Consultants.

In view of the wide variety of types, styles, materials, and finishes found in the numerous items comprising finishing hardware, it is advisable to call in a hardware consultant or manufacturer's representative before specifications are made in too great detail. Such consultants are in the best position to streamline the requirements, to obtain the desired result in the simplest, most economical and satisfactory manner. In this way, one can often avoid unnecessary delay and complications. The consultant's knowledge of codes affecting the selection of hardware may also be of great assistance (see Art. 16-3).

It is well to remember that what was standard a short time ago may now be out of regular production. A hardware consultant should be able to advise on the selection of stock items.

Advance study with a representative of the manufacturer usually has another distinct advantage. The representative can often point out how definite savings can be achieved by making slight changes in the specifications. In many instances, these changes may mean little or nothing to the owner of the building and may be just as acceptable as what was originally specified. But these minor changes may mean a great deal in what can contribute to faster progress and lower ultimate cost of the completed work.

16-5. Factors Affecting Selection of Hardware.

The operating characteristics of hardware items govern their selection, according to the particular requirements in each case. Then, the question of material, such as plastics, brass, bronze, aluminum, or steel, can be settled, as well as the finish desired. Selection of material and finish depends on the architectural treatment and decorative scheme.

Wrought, forged, and stamped parts are available for different items. Finishes include polished, satin, and oxidized. When solid metal is desired rather than a surface finish that is plated on metal, the order should definitely so specify.

National standards defining characteristics, sizes, dimensions, and spacings of holes, materials, and finishes for many items can assist greatly in identification and proper fit of hardware items to the parts on which they will be mounted.

16-6. Template and Nontemplate Hardware.

For hardware items that are to be fastened to metal parts, such as jambs or doors, so-called template hardware is used. Template items are made to a close tolerance to agree exactly with drawings furnished by the manufacturer. The sizes, shapes, location, and size of holes in this type of hardware are made to conform so accurately to the standard drawings that the ultimate fit of all associated parts is assured. In the case of hinges or butts, holes that are template-drilled usually form a crescent-shaped pattern (Fig. 16-4).

Hardware for stock may be nontemplate; however, certain lines are all template-made, whether for stock or for a specific order. Nontemplate items may vary somewhat and may not fit into a template cutout.

For use on wood or metal-covered doors, template drilling is not necessary; nontemplate hardware of the same type and finish can be used. Generally, there is no price difference between template and nontemplate items.

FINISHING HARDWARE

16-7. Applicable Standards for Finishing Hardware.

Several American National Standards present performance requirements for builders' hardware. These standards are available from the American National Standards Institute, Inc. (ANSI), 1430 Broadway, New York, N.Y. 10018. The Builders Hardware Manufacturers Association, Inc., is the sponsor of the A156 series and submits additional standards or revisions of existing standards to ANSI for approval on a regular basis. The latest edition of the following should be consulted:

ANSI A156.1. Butts and Hinges
ANSI A156.2. Locks and Lock Trim
ANSI A156.3. Exit Devices
ANSI A156.4. Door Controls (Closers)
ANSI A156.5. Auxiliary Locks and Associated Products
ANSI A156.6. Architectural Door Trim
ANSI A156.7. Template Hinge Dimensions
ANSI A156.8. Door Controls (Overhead Holders)
ANSI A156.9. Cabinet Hardware
ANSI A156.10. Power-Operated Pedestrian Doors
ANSI A156. 11. Cabinet Locks
ANSI/ASTM F 476. Standard Test Methods for Security of Swinging Door Assemblies
ANSI/ASTM F 571. Standard Practice for Installation of Exit Devices in Security Areas

In addition, there are a series of ANSI standards for the preparation of standard steel doors and frames to receive hardware—the A115 series, sponsored by the Door and Hardware Institute.

16-8. Hinges and Butts.

A hinge is a device permitting one part to turn on another. In builders' hardware, the two parts are metal plates known as leaves. They are joined together by a pin, which passes through the knuckle joints.

When the leaves are in the form of elongated straps, the device is usually called a hinge (Figs. 16-1 and 16-2). This type is suitable for mounting on the surface of a door.

When the device is to be mounted on the edge of a door, the length of the leaves must be shortened, because they cannot exceed the thickness of the door. The leaves thus retain only the portion near the pin, or butt end, of the hinge (Figs. 16-3 to 16-5). Thus, hinges applied to the edge of a door have come to be known as butts, or butt hinges.

Butts are mortised into the edge of the door. They are the type of hinge most commonly used in present-day buildings.

Sizes of butt hinges vary from about 2 to 6 in, and sometimes to 8 in. Length of hinge is usually made the same as the width; but they can be had in other widths. Sometimes, on account of projecting trim, special sizes, such as 4½ × 6 in, are used.

For the larger, thicker, and heavier doors receiving high-frequency service, and for doors requiring silent operation, bearing butts (Fig. 16-4) or butts with Oilite bearings or other antifriction surfaces are generally used. It is also customary to use bearing butts wherever a door closer is specified.

Plain bearings (Fig. 16-5) are recommended for residential work. The lateral thrust of the pin should bear on hardened steel.

Fig. 16-1 Heavy strap hinge. *(C. Hager & Sons Hinge Mfg. Co.)*

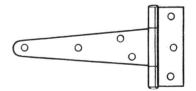

Fig. 16-2 Heavy tee hinge. *(C. Hager & Sons Hinge Mfg. Co.)*

(a) LEFT HAND LOOSE JOINT HINGE (b) RIGHT HAND LOOSE JOINT HINGE (c) LEFT HAND OLIVE KNUCKLE HINGE (d) RIGHT HAND OLIVE KNUCKLE HINGE

Fig. 16-3 Special types of butt hinges that are handed. For hand of doors, see Fig. 16-13.

Fig. 16-4 Bearing butt hinge. *(Builders Hardware Manufacturers Association.)*

Fig. 16-5 Plain bearing butt hinge. *(Builders Hardware Manufacturers Association.)*

Unusual conditions may dictate the use of extra-heavy hinges or ball bearings where normal hinges would otherwise be used. One such case occurred in a group of college dormitories where many of the doors developed an out-of-plumb condition that prevented proper closing. It was discovered that the students had been using the doors as swings. Heavier hinges with stronger fastenings eliminated the trouble.

Two-bearing and four-bearing butt hinges should be selected, as dictated by weight of doors, frequency of use, and need for maintaining continued floating, silent operation. Because most types of butt hinges may be mounted on either right-hand or left-hand doors, it should be remembered that the number of bearing units actually supporting the thrust of the vertical load is only one-half the bearing units available. With a two-bearing butt, for example, only one of the bearings carries the vertical load, and with a four-bearing butt, only two carry the load. It should be noted, however, that some hinges (Fig. 16-3), are "handed" and must be specified for use on either a right-hand or left-hand door.

When butts are ordered for metal doors and jambs, "all machine screws" should be specified.

Location of Hinges. One rule for locating hinges for ordinary doors is to allow 5 in from rabbet of head jamb to top edge of top hinge and 10 in from finished floor line to bottom hinge. The third hinge should be spaced equidistant between top and bottom hinges. Another location for hinges is that of the standard steel frame and door (Fig. 14-4). The location varies a little among door manufacturers, but each is perfectly satisfactory from a functional standpoint.

Types of Hinge Pins. A very important element in the selection of hinges is the hinge pin. It may be either a loose pin or a fast pin.

Loose pins are generally used wherever practicable, because they simplify the hanging of doors. There are four basic pin types:

1. Ordinary loose pins
2. Nonrising loose pins
3. Nonremovable loose pins
4. Fast (or tight) pins

The **ordinary loose pin** can be pulled out of the hinge barrel so that the leaves may be separated. Thus, the leaves may be installed on the door and jamb independently, with ease and "mass-production" economy. However, these pins have a tendency—with resulting difficulties—of working upward and out of the barrel of the hinge. This "climbing" is caused by the constant twisting of the pin due to opening and closing the door. Present-day manufacture of hinges has tended to drift away from this type of pin, which is now found only in hinges in the lowest price scale.

The **nonrising loose pin** (or self-retaining pin) has all the advantages of the ordinary loose pin; but at the same time, the disadvantage of "climbing" is eliminated. The method of accomplishing the nonrising features varies with the type of hinge and its manufacture.

The **nonremovable loose pin** is generally used in hinges on entrance doors, or doors of locked spaces, which open out and on which the barrel of the hinge is therefore on the outside of the door. If such a door were equipped with ordinary loose-pin hinges or nonrising loose-pin hinges, it would be possible to remove the pin from the barrel and lift the door out of the frame and in so doing overcome the security of the locking device on the door. In a nonremovable loose-pin hinge, however, a setscrew in the barrel fits into a groove in the pin, thereby preventing its removal. The setscrew is so placed in the barrel of the hinge that it becomes inaccessible when the door is closed. This type of hinge offers the advantage of the ordinary loose-pin type plus the feature of security on doors opening out. Some manufacturers achieve the same results using other means, such as a safety stud.

The **fast (or tight) pin** is permanently set in the barrel of the hinge at the time of manufacture. Such pins cannot be removed without damaging the hinge. They are regularly furnished in hospital- or asylum-type hinges. The fact that the leaves of this type of hinge cannot be separated, however, makes the installation more difficult and costly. However, the difficulty is not too great, because with this type of hinge it is only necessary to hold the door in position while the screws for the jamb leaf are being inserted.

Ends of pins are finished in different ways. Shapes include flat-bottom, ball, ovalhead, modern, cone, and steeple. They can be chosen to conform with type of architecture and decoration. Flat-button tips are generally standard and are supplied unless otherwise specified.

16-9. How to Select Hinges.

One hinge means one pair of leaves connected with a pin. The number of hinges required per door depends on the size and weight of the door, and sometimes on conditions of use. A general rule recommends two butt hinges on doors up to 60 in high; three hinges on doors 60 to 90 in high; and four hinges on doors from 90 to 120 in high.

Table 16-1 gives general recommendations covering the selection of suitable hinges. Figure 16-6 indicates how hinge width is determined when it is governed by clearance.

The proper operating clearance between the hinged edge of a door and the jamb is taken care of in the manufacture of the hinges by "swaging" or slightly bending the leaves of the hinge near the pin. Since the amount of such bending required is determined by whether one or both leaves are to be mortised or surface-mounted, it is important in ordering hinges to specify the type of hinge needed to satisfy mounting conditions. If hinge leaves are to be mortised into both the edge of the door and the jamb, a full-mortise hinge is required (Fig. 16-7a, c, and h). If leaves are to be surface-applied to

TABLE 16-1. Hinges for Doors

Door thickness, in.	Door width, in.	Minimum hinge height, in.
7/8 or 1	Any	2 1/2
1 1/8	To 36	3
1 3/8	To 36	3 1/2
1 3/8	Over 36	4
1 3/4	To 41	4 1/2
1 3/4	Over 41	4 1/2 heavy
1 3/4 to 2 1/4	Any	5 heavy*

* To be used for heavy doors subject to high-frequency use or unusual stress.

both the side of the door and the face of the jamb, a full-surface hinge is needed (Fig. 16-7d). If one leaf is to be surface-applied and the other mortised, the hinge is a half-surface hinge or a half-mortise hinge, depending on how the leaf for the door is to be applied—half-surface if applied to door surface (Fig. 16-7b, e, and f) and half-mortise if mortised in the door end (Fig. 16-7g).

Fig. 16-6 When determined by clearance, width of door hinge equals inset (typically 1/8 in), plus hinge backset (typically 1/4 in for doors up to 2 1/4 in thick and 3/8 in for thicker doors), plus twice the door thickness, plus clearance.

16-8 Builders' Hardware

Fig. 16-7 Hinge-mounting classification depends on where and how hinge is fastened: (a), (c), and (h), full mortise; (d) full surface; (b), (e), and (f), half surface; and (g) half mortise.

Exterior doors should have butts of nonferrous metal. Although chromium plating does not tarnish, it is not considered to be satisfactory on steel for exterior use. Interior doors in rooms where dampness and steam may occur should be of nonferrous metal and should also have butts of nonferrous metal. Butts for other interior work may be of ferrous metal.

Ferrous-metal butts should be of hardened cold-rolled steel. Where doors and door frames are to be painted at the job site, butts should be supplied with a prime-coat finish. For doors and trim that are to be stained and varnished, butts are usually plated.

Other types of hinges include some with a spring that closes the door. They may be either single- or double-acting. The spring may be incorporated in a hinge mounted on the door in the usual manner, or it may be associated with a pivot at the bottom of the door. In the latter case, the assembly may be of the type that is mortised into the bottom of the door, or it may be entirely below the floor.

16-10. Door-Closing Devices.

These include overhead closers, either surface-mounted or concealed, and floor-type closers. These are some of the hardest-worked items in most buildings. To get the most satisfactory operation at low first cost and low maintenance cost, each closer should be carefully selected and installed to suit the particular requirements and conditions at each door.

Most of these devices are a combination of a spring—the closing element—and an oil-cushioned piston, which dampens the closing action, inside a cylinder (Fig. 16-8). The piston operates with a crank or a rack-and-pinion action. It displaces the fluid through ports in the cylinder wall, which are closed or open according to the position of the piston in the cylinder. Opening of the door energizes the spring, thus storing up closing power. Adjustment screws are provided to change the size of the ports, controlling flow of fluid. This arrangement makes the closer extremely responsive to the conditions of service at each individual door and permits a quiet closure, which at the same time ensures positive latching of the door.

While the fluid type of closer is preferred, pneumatic closers are also used, particularly for light doors, like screen doors.

Overhead door closers are installed in different ways, on the hinge side of the door or on the top jamb on the stop side of the head frame or on a bracket secured to the door frame on the stop side. Various types of brackets are available for different conditions. Also, when it is desired to install a closer between two doors hung from the same frame, or on the inside of a door that opens out, an arrangement with a parallel arm makes this possible. Other types of closers may be mortised into the door or housed in the head above the door.

Fig. 16-8 Door closer-spring closes the door, while hydraulic mechanism (cylinder with piston) keeps the door from slamming. *(Builders Hardware Manufacters Association.)*

Closers may be semiconcealed or fully concealed. Total concealment greatly enhances appearance but certain features of operation are limited.

An exposed-type closer should be mounted on the hinge side or stop side of the door unless there is real need for a bracket or parallel-arm mounting.

Whereas the use of brackets reduces headroom and may become a hazard, a parallel-arm closer mounted on the door rides out with the door, leaving the opening entirely clear.

When surface-applied door closers are used, careful consideration should be given to the space required in order that doors may be opened at least 90° before the closers strike an adjacent wall or partition.

Semiconcealed door closers are recommended for hollow metal doors. These closers are mortised into the upper door rail.

Various hold-open features also are available in different closer combinations to meet specific requirements. Another available feature is delayed action. This allows plenty of time to push a loaded vehicle through the opening before the door closes and also permits handicapped people time to maneuver through a door opening.

When floor-type checking and closing devices are used, floor conditions should be carefully determined in order that there will be sufficient unobstructed depth available for their installation.

To get maximum performance from any door closer, it must be of ample size to meet the conditions imposed on it. If abnormal conditions exist, such as drafts or severe traffic, a closer of larger than the normal capacity should be employed. Installing too small a closer is an invitation to trouble. Manufacturers' charts should be used to determine the proper sizes and types of closers to suit door sizes and job conditions.

It is very important that door closers be installed precisely as recommended by the manufacturer. Experience has shown that a large percentage of troubles with closers results from disregard of mounting instructions.

In response to various code requirements for room-to-corridor protection, closer manufacturers have produced products that provide automatic door closing but still allow flexibility. These units permit the door to be held open in many locations of hold-open and yet upon a signal from a smoke detector, the hold-open mechanism disengages and the closer causes the door to close. A similar type of unit incorporates a device allowing the door to swing free as though it were not equipped with a door closer. As with the multiple-point hold-open device just described, the "swing-free" model, upon a signal from a smoke detector, cancels the closer nullifying device and the closer causes the door to close. These units are suitable in a variety of occupancies, particularly institutional.

Barrier-free provisions for handicapped people dictate certain minimum opening resistance of doors. Depending upon the jurisdiction and location of doors, these forces range from a 5-lb force maximum to a 15-lb force maximum. There will be instances, because of the door size or air-pressure conditions, when the necessary closing force exerted by the closer to overcome these conditions will create opening resistance in excess of what is permitted. Power-operated doors or doors with a power assist specially made to solve this problem are recommended as a practical solution.

16-11. Locks and Latches.

The function of locks and latches is to hold doors in a closed position. Those known as rim locks or latches are fastened on the surface of the door. The ones that are mortised into the edge of the door are known as mortise locks (Fig. 16-9) or latches.

When the locking bolt is beveled, the device is usually referred to as a latch bolt; such a bolt automatically slides into position when the door is closed. A latch is usually operated by a knob or lever. Sometimes it may be opened with a key on the other side. Night latches came to be so called because they are generally used at night with other ordinary locks to give additional security.

Latches must of course take into account the hand of the door so the bevel will be right. A large percentage of latches are "reversible"—that is, they may be used on a right- or left-hand door (Art. 16-12). It is well, however, when ordering any lock to specify the hand of the door on which it is to be used.

When the locking bolt is rectangular in shape, it does not slide into position automatically when the door is closed; it must be projected or retracted by a thumb turn or key. This type of bolt is referred to as a dead bolt, and the lock as a dead lock. It may be worked with a key from one or both sides. Such a bolt is often used in conjunction with a latch. When the latch bolts and dead bolts are combined into one unit, it is known as a lock.

For keyed locks that do not have dead bolts, it is desirable that the latch bolt be equipped with a dead latch (Fig. 16-12). This is a small plunger or an auxiliary dead latch (Fig. 16-10) that is held depressed when the door is closed and "deadlocks" the latch bolt so that it cannot be retracted by a shim, card, or similar device inserted between latch and door frame.

Various combinations of latches, dead bolts, knobs and keys, and locking buttons are applied to all types and kinds of doors. The exact combination most suitable for any given door is determined by a careful analysis of the use to which the door is to be put.

Fig. 16-9 Mortise lock. *(Builders Hardware Manufacturers Association.)*

Fig. 16-10 Unit lock. *(Builders Hardware Manufacturers Association.)*

A point to bear in mind is that a uniform size should be selected, if practicable, for a project, no matter what the individual functions of the different locks may be. This makes possible the use of standard-size cutouts or sinkages on each installation. When this is done, not only is the cost of installation reduced, but any changes that may be made in the drawings as the job progresses, or any changes that may have to be made later are simplified, and special hardware is avoided.

Unit locks (Fig. 16-10) are complete assemblies that eliminate most of the adjustments during installation that would otherwise be necessary. These locks have merely to be slipped into a standard notch cut into a wood door or formed in a metal door.

Bored-in locks are another type that can be installed by boring standard-size holes in wood doors or by having uniform circular holes formed in metal doors. These bored-in locks are often referred to as tubular-lock sets or cylindrical-lock sets depending on how the holes have to be bored to accommodate them.

Tubular locks have a tubular case extending horizontally at right angles to the edge of the door. This type of case permits a horizontal hole of small diameter to be bored into the door at right angles to the vertical edge of the door; another small hole is required at right angles to the first hole to take care of the latching mechanism. Most tubular locks now produced fit the standard bores in steel doors conforming to ANSI A115.2 and A115.3, just as do cylindrical locks.

Cylindrical locks (Fig. 16-11), the other type of bored-in lock, have a cylindrical case requiring a relatively large-diameter hole in the door, bored perpendicular to the face of the door. This hole accommodates the main body of the lock. A hole of smaller diameter to take the latch bolt must be bored at right angles to the edge of the door.

When bored-in locks are used in hollow metal doors, a reinforcing unit is required in the door. This unit is generally supplied by the door manufacturer.

Exit devices are a special series of locks required by building codes and the National Fire Protection Association "Life Safety Code" on certain egress doors in public

Fig. 16-11 Cylindrical lock. *(Builders Hardware Manufacturers Association.)*

16-12 Builders' Hardware

buildings. On the egress side, there is a horizontal bar running a minimum of one-half the width of the door. When pressed against, the bar releases the locking or latching mechansim, allowing the door to open. These devices are required to be labeled for safe egress by a nationally recognized independent testing laboratory. When used on fire doors, they must carry an additional label showing that the devices have also been investigated for fire. They then bear the name "fire exit hardware." These devices are available with various functions, including arrangements for having them locked from the outside and openable with a key. For all functions, however, they must be openable from the inside by merely depressing the cross bar.

Keys. Locks are further classified according to the type of key required to retract the bolts. On all but the cheaper installations the principal type of key used is the cylinder key, which operates a pin-tumbler cylinder.

In the preceding locks, the locking cylinder, which is the assembly that supplies the security feature of the lock, is a cylindrical shell with rotatable barrel inside. The barrel has a longitudinal slot or keyway formed in it. The cross section of the keyway for every lock has a shape requiring a similarly designed key. Several holes (usually five or six) are bored through the shell and into the barrel at right angles to the axis of the barrel (Fig. 16-12). In each hole is placed a pair of pins—a driver pin in the shell end and a tumbler pin in the barrel end. Pins vary in length, and the combination of lengths differs from that of pins in other locks. A spring mounted in the shell end of each hole behind each driver pin forces these pins into the hole in the barrel, so that normally they are partly

Fig. 16-12 Mechanism of cylindrical lock. *(Kwikset Locks.)*

in the barrel portion of the hole and partly in the shell portion. Thus, the barrel is prevented from rotating in the shell to operate the bolt, and the door remains locked.

Keys have notches spaced along one side to correspond with the spacing and length of the pins. When a key is inserted in the slot in the barrel, each notch forces a tumbler pin back, against pressure from the spring. When the proper key is inserted in the slot, each notch pushes its corresponding pair of pins just far enough back into the shell to bring the junction point of that pair exactly at the circumference of the barrel. The barrel then can be rotated within the shell by merely turning the key.

Turning the key operates a cam attached to the end of the barrel. The cam withdraws the bolt from its locked position.

The security feature of the cylinder lock results from the fact that only one series of notches in the key will correspond to the respective lengths of the several individual pins. With as many as five or six tumblers, it is apparent that many combinations are available.

By using split pins, different keyways, and various sizes and arrangements of the pins, master keying and grand master keying of cylinder locks are made possible. Thus one master key may be made to open all the separate locks on each floor of a building, while a grand master key will open any lock in any floor of that building. In case there are a number of such buildings in a group, a great-grand master key can be made that will open any door on any building in the group.

Security Precautions. The security of a building from the standpoint of unauthorized entry and life safety is affected by the proper selection of hardware. Although locks are important for security, the total system, including the building design, must be considered. A lock installed in a door that is loosely fitted in the frame may be ineffective. Hollow metal and aluminum frames that are not reinforced and anchored properly are easily "spread," and undetectable entry can result.

In high levels of master keying, more split pins must be used, and thus more "shear lines" are established. This makes the cylinder easier to pick. Hence, unnecessarily complicated master key systems should be avoided.

A good security lock is available with a dead bolt and a deadlocking latch bolt. Both bolts are retractable in one operation, merely by turning the inside knob. This satisfies requirements for security against unauthorized entry and life safety.

Hardware requirements for life safety and fire doors are covered by two National Fire Protection Association Standards, "Life Safety Code" and "Fire Doors and Windows," Both are incorporated in many building codes.

16-12. Hand of Doors.

As you look at a door, the hinges can be either on your right or left, and the door can open either away from or toward you. Because of the fact that it is from the outside of a door that the lock is operated by a key, it is customary to speak of the hand of a door as viewed from the outside—the side on which protection is needed against unauthorized entry. In a series of connecting rooms, the outside of each door is considered to be the side you first approach as you proceed into the rooms from the entrance.

If when you are standing outside the door the hinges are to your right and the door opens away from you, it is a right-hand, regular-bevel door; if the door opens toward you, it is a right-hand, reverse-bevel door. Similarly, if when you are outside the door, the hinges are to your left and the door swings away from you, it is a left-hand, regular-bevel door; if the door opens toward you, it is a left-hand, reverse-bevel door (Fig. 16-13).

In ordering hardware, the hand of the door must be given; also, whether the beveled face of the latch bolt has to be regular-beveled or reverse-beveled. In some cases, the face is reversible, but it is best to state the bevel definitely anyway. For closet and cabinet doors, which never open away from you, it is sufficient to state merely whether the hand of these doors is right or left and the fact that they are closet or cabinet doors. It is then understood that reverse-beveled latch bolts are required.

For casement sash, the hand is stated a little differently; it is safest to specify whether the hinges are on the right or on the left as viewed from the room side, and also whether the sash opens in or out.

16-13. Casement-Sash Operators.

There are many types of devices for operating casement sash. Some of these require that screens have openings to make the operating handle accessible. Other types, which operate independently of the screens, have an interior crank that actuates a worm gear. Various arrangments of arms actuated by the gear are used to move the sash.

Fig. 16-13 Conventional method of determining hand of doors and bevel.

16-14. Sliding Hardware.

Hardware for interior and exterior sliding doors and screens ranges from light to heavy. Hangers, tracks, and wheels are of various designs to serve different requirements. Ball bearings and composition or nylon wheels are generally used. The tracks and hangers are often made of aluminum, especially for the lighter installations. Because of the wide choice, a careful examination of the features peculiar to each design is well worthwhile before a decision is made.

16-15. Folding-Door Hardware.

Folding doors require special supports and tracks because of their wide range of movement and their various weights. Hardware is available for folding doors weighing from 30 to 250 lb each.

16-16. Locking Bolts (Doors and Windows).

Various types of bolts and rods are fastened to doors and windows for the purpose of securing them in closed position or to other doors and windows. Flush bolts and surface bolts, manual or automatic, are often used.

Top and bottom vertical bolts operated by a knob located at convenient height between them are known as **cremorne bolts**.

ROUGH HARDWARE

16-17. Window Hardware.

Sash balances are commonly used with double-hung windows as counterweights instead of weights and pulleys. Some balances have tape or cable with clock-type springs, which coil and uncoil as the sash is raised and lowered. Another type employs torsion springs with one end fixed to the side jamb of the window and the other arranged to turn as the sash goes up or down. The turning device in this type of balance may be a slide working in a rotatable spiral tube, or it may be a slotted bushing attached to the free end of the spring and fitted around a vertical rod attached to the sash. This rod (Fig. 16-14) is a flat piece of metal twisted into a spiral shape. The up-and-down movement of the sash causes the slotted bushing to revolve on its spiral sliding rod, thus winding and unwinding the spring. Still another type utilizes a vertical tension spring of the ordinary coil variety. One end of the spring is fixed and the other is fastened to the sash; the spring stretches or compresses in a vertical direction as the sash is moved up or down.

Fig. 16-14 Constant-tension sash balance. *(Unique Balance Co.)*

Some sash balances combine weather stripping with the balances. Others have friction devices to hold the sash in the desired position.

One patented type of sash balance known as the Unique sash balance incorporates a clever counterbalancing feature by making a change in the degree of pitch of the spiral rod from top to bottom, thereby controlling the increase and decrease of spring tension (Fig. 16-14). The pitch varies from 30 to 80°. As the spring turns, the changing spring tension is automatically compensated for by the variable pitch of the spiral rod sliding through the slotted bushing on the end of the spring. Thus the tension is equalized at every point of operation. This automatically prevents the sash from creeping or dropping of its own accord, without the necessity of introducing friction devices that interfere with easy operation.

Two sash balances are used per sash (one on each side) or four balances per double-hung window.

Other window hardware—locks, sash pulls, sash weights, and pulleys—are simple items, supplied as standard items with specific windows.

16-18. Inserts, Anchors, and Hangers.

Metal inserts of various types are cast into concrete floor slabs to serve as hangers or connectors for other parts of the structure that will be supported from the floor system. These inserts include electric conduit and junction boxes, supports for hung ceilings, slots for pipe hangers, fastenings for door closers to be set under the floor, and circular metal shapes for vertical pipe openings.

Metal anchors of various types are used in building construction. Each type is specifically shaped, according to the purpose served. These include anchors for securing

16-16 Builders' Hardware

⟨────────⟩	CASING HEAD WOOD SIDING NAIL
⟨────────⟩	SINKER HEAD WOOD SIDING NAIL
⟨────────⟩	GENERAL PURPOSE FINISH NAIL
⟨────⟩	ROOFING NAIL
⟨─────⟩	WOOD SHINGLE NAIL
⟨─────⟩	WOOD SHAKE NAIL
⟨──────⟩	GYPSUM LATH NAIL
⟨───────⟩	INSULATED SIDING NAIL
⟨∿∿∿∿⟩	SPIRAL-THREADED ROOFING NAIL WITH NEOPRENE WASHER
⟨▬▬▬▬⟩	ANNULAR-RING ROOFING NAIL WITH NEOPRENE WASHER
⟨∿∿∿∿⟩	SPIRAL-THREADED ROOFING NAIL FOR ASPHALT SHINGLES AND SHAKES
⟨▬▬▬▬⟩	ANNULAR-RING ROOFING NAIL FOR ASPHALT SHINGLES AND SHAKES
⟨∿∿∿∿∿∿⟩	SPIRAL-THREADED CASING HEAD WOOD SIDING NAIL
⟨▬▬▬⟩	ANNULAR-RING PLYWOOD SIDING NAIL FOR APPLYING ASBESTOS SHINGLES AND SHAKES OVER PLYWOOD SHEATHING
⟨▬▬⟩	ANNULAR-RING PLYWOOD ROOFING NAIL FOR APPLYING WOOD OR ASPHALT SHINGLES OVER PLYWOOD SHEATHING
⟨∿∿⟩ ⟨▬▬▬⟩	SPIRAL-THREADED ⎫ ASBESTOS ANNULAR-RING ⎭ SHINGLE NAILS
⟨▬▬▬⟩	ANNULAR-RING GYPSUM BOARD DRYWALL NAIL
⟨∿∿∿∿⟩	SPIRAL-THREADED INSULATED SIDING FACE NAIL

Fig. 16-15 Nail heads and points.

stone facings to masonry walls, anchors for fastening marble slabs in place, column anchor bolts set in foundations, and anchor bolts for fastening sills to masonry.

Joist hangers are used for framing wood joists into girders and for framing headers into joists around stairwells and chimneys.

16-19. Nails.

Wire nails, made of mild steel, are commonly used for most nailing purposes. Cylindrical in shape, they are stronger for driving than cut nails and are not so liable to bend when driven into hardwoods.

Cut nails, sheared from steel plate, are flat and tapered. They have holding power considerably greater than wire nails. They are usually preferred to wire nails for fastening wood battens to plaster; also in places where there is danger that the nails may be drawn out by direct pull. They are frequently used for driving into material other than wood. They are generally used for fastening flooring.

When driven with width parallel to grain of wood, they have less tendency to split wood than wire nails.

The length of cut and wire nails is designated by the unit "pennies." Both cut and wire nails come in sizes from 2-penny (2d) which are 1 in long, to 60-penny (60d), which are 6 in long. (See Table 7-13.)

Above 6 in, the fasteners are called **spikes.** They run in 7-, 8-, 9-, 10-, and 12-in lengths (Table 7-13).

Various gages or thicknesses of nails, and different sizes and shapes of heads and points (Figs. 16-15 and 16-16), are available in both wire and cut nails. Certain types have distinguishing names. For examples, the term **brads** is applied to nails with small heads, suitable for small finish work.

Common brads are the same thickness as common nails but have different heads and points. **Clout nails** have broad flat heads. They are used mostly for securing gutters and metalwork.

Casing nails are about half a gage thinner than common wire nails of the same length; **finishing nails,** in turn, are about half a gage thinner than casing nails of the same length. **Shingle nails** are half a gage to a full gage thicker than common nails of the same length.

Certain manufacturers have developed nails for special purposes that hold tighter and longer. Among these are threaded nails (Fig. 16-16), which combine the ease of driving of the ordinary nail with much greater holding power.

One type is a spiral-threaded flooring nail. These nails turn as they drive, minimizing splitting of the tongues of the floor boards. These nails are said to actually grip more firmly with the passage of time. A nailing machine is available for driving these nails. In one operation, it starts the nail, drives the joint between the flooring strips tight, and drives and sets the nail. With this machine, mashed tongues and marred edges of the wood are avoided.

Nails of aluminum and stainless steel are particularly useful in exposed locations where rust or corrosion of steel nails might cause unsightly stains to form on exterior finished surfaces. These nails are now made in most of the usual sizes, including special spiral-threaded nails.

Galvanized nails are used for fastening slate and shingles. These nails are sometimes used in exposed locations as protection against corrosion.

To satisfy the special requirements and varying conditions in each case, the proper nails to use for fastening various sizes and types of materials must be carefully and intelligently chosen by men experienced in this kind of work.

Fig. 16-16 Typical nails.

16-20. Screws.

These are used for applying hardware of all descriptions; also for panel work, cabinet work, all types of fine finish work, and support of electric and plumbing fixtures. Screws have greater holding power than nails and permit easy removal and replacement of parts without injury to the wood or finish. Screws avoid danger of splitting the wood or marring the finish, when the screw holes are bored with a bit.

Screws are made in a large variety of sizes and shapes to suit different uses (Fig. 16-17) and they are made of different metals to match various materials. A much-used type of head, other than the ordinary single-slot type, is the **Phillips head,** which has two countersunk slots at right angles to each other. The head keeps the screwdriver exactly centered during driving and also transmits greater driving power to the screw, while holding the screwdriver firmly on the head. Phillips heads are smoother at the edges, because the slots in the head do not extend to the outer circumference.

Steel screws for wood vary in length from ¼ to 6 in. Each length is made in a variety of thicknesses. Heads may be ordinary flat (for countersinking), round, or oval.

Parker screws for securing objects to thin metal are self-threading when screwed into holes of exactly the correct size.

Sizes of screws are given in inches of length and the gage of the diameter. Lengths vary by eighths of an inch up to 1 in, by quarters from there up to 3 in, and by halves from 3 to 5 in. Unlike wire gages, the smallest diameter of a screw gage is the lowest number; the larger the number, the greater is the diameter in a screw gage. Gage numbers range from 0 to 30 (Table 7-16).

Lag screws are large, heavy screws used for framing timber and ironwork. Lengths vary from 1½ to 12 in and diameters from 5/16 to 1 in. Two holes should be bored for lag screws, one to take the unthreaded shank without binding and the other (a smaller hole) to take the threaded part. This smaller hole is usually somewhat shorter in length than the threaded portion. Lag screws usually have square heads and are tightened with a wrench. (See Table 7-17.)

Fig. 16-17   Typical screws.

1. FLAT-HEAD WOOD SCREW
2. SHEET-METAL SCREW
3. OVAL-HEAD WOOD SCREW
4. ROUND-HEAD WOOD SCREW
5. PYRAMID-HEAD WOOD SCREW
6. PHILLIPS-HEAD WOOD SCREW
7. SPANNER-HEAD WOOD SCREW
8. FLAT-HEAD MACHINE SCREW
9. ROUND-HEAD MACHINE SCREW
10. OVAL-HEAD MACHINE SCREW
11. PHILLIPS-HEAD MACHINE SCREW
12. SPANNER-HEAD MACHINE SCREW
13. SLOTTED TWIN-HEAD MACHINE SCREW
14. OVAL-HEAD MACHINE SCREW WITH GROMMET NUT
15. LAG SCREW
16. CARRIAGE BOLT
17. FLAT-HEAD STOVE BOLT

16-21. Welded Studs.

Studs electrically welded to the steel framework of a building are often used as the primary element for securing corrugated siding and roofing, insulation, metal window frames, ornamental outer skins, anchorages for concrete, and other items. The welded studs thus form an integral part of the basic structure.

Many types of studs or fasteners are available, each one being designed for a particular purpose. Most of the studs have threads formed on them, either externally or internally. Some of the studs are designed to have the material impaled over them and riveted to them.

The studs are designed so as to project the exact distance desired after they have been welded. Special sealing washers and nuts are usually placed on each stud over the flat sheet of material being fastened. Tightening the nut or expanding the head of the stud with a riveting hammer then makes the fastening complete, weathertight, and secure.

Studs are usually cadmium-plated mild steel or stainless steel. The latter is recommended for corrosive atmospheric conditions. Flux to assure a good weld is contained in the center of each stud at the welding end.

Equipment required for stud welding includes a stud-welding gun, a control unit for adjusting the amount of welding current fed to the gun, and a power source. The source of welding current may be a direct-current generator, a rectifier, or a battery unit. When a welding generator is used, the minimum National Electrical Manufacturers Association rating should be 400 A.

The welding gun usually has a chuck for holding the stud in position for welding (Fig. 16-18), and a leg assembly holding an adjustable-length extension sleeve into which the necessary arc-shielding ferrule is inserted. Expendable ceramic arc ferrules are generally used to confine the arc and control the weld fillet. After each weld, the arc ferrule is broken and removed by a light tap with any convenient metal object. In some cases where the required finished stud is short, an extra length is provided on the stud as furnished, for proper chucking in the gun. A groove is provided so the extra length can be easily broken off after the stud is welded.

The operation of fastening corrugated or flat asbestos-cement roofing and siding to steel girts is typical of how these stud fastenings are installed on buildings. First, the sheet of asbestos cement is placed in the correct position on the girts or purlins. If the sheets have not already been predrilled, a carboloy-tip drill is then used to drill a hole of proper diameter through the sheet. Next, an end-mill drill is inserted through the hole in the material to remove any scale, rust, paint, or dirt and to provide a clean welding surface. Then, the stud is inserted in the gun chuck, the ceramic arc shield or ferrule is put into place in the adjustable-length device, and the tubular assembly is inserted through the hole in the asbestos-cement siding. The gun is held perpendicular to and firmly against the purlin or girt. The stud is end welded in a fraction of a second by pressing the gun trigger. Finally, the ferrule is knocked off.

The sealing collar and nut are then placed

Fig. 16-18 Steps in stud welding: (1) press welding end of gun against work plate; (2) press trigger, creating an electric arc between stud and plate and melting portions of each; (3) stud is automatically plunged into molten pool; (4) remove gun and knock off ferrule.

on the stud and tightened sufficiently to make a positive and weathertight seal between the collar and the asbestos-cement sheet. A similar fastening, but with corrugated-metal sheet, is shown in Fig. 16-19.

The above method permits application of siding and roofing entirely from the outside. A worker is not required to be on the inside because there are no clips or fasteners on the interior. The expense of an interior scaffold is thus saved.

In securing corrugated metal, it is sometimes desirable to weld the studs to the steel frame in advance of placing predrilled sheets. In these cases, templates for quickly marking sheet and stud locations may be employed, as desired. Stud welding is applicable to any steel frame composed of standard structural steel. Steels of the high-carbon variety such as rerolled rail stock are not suitable for stud welding.

Fig. 16-19 Corrugated metal impaled on and fastened in place with welded stud.

The manufacturer's recommendations should be followed in selecting the best type and size of stud for each specific installation. The leading manufacturers have direct field representatives in all areas who can supply valuable advice as to the best procedures to follow.

16-22. Powder-Driven Studs.

For many applications requiring the joining of steel or wood parts, or fastenings to concrete, steel, and brick surfaces, powder-actuated stud drivers are found to decrease costs because of their simplicity and speed. These drivers use a special powder charge to drive a pin or stud into relatively hard materials. The key to their efficiency is the proper selection of drive pins and firing charges. Because of the high velocity, the drive pin, in effect, fuses to the materials and develops the holding power.

Pullout tests of these driven studs prove remarkable holding power. Average pullout in 3500-psi concrete exceeds 1200 lb for 10-gage studs, and 2400 lb for $\frac{1}{4}$-in-diameter studs. In steel plate the pullout resistance is still greater.

All that is required is a stud driver, the correct stud, and the correct cartridge, as recommended by the manufacturer for each specific set of fastening conditions. There are about a half dozen different strengths of cartridge, each identified by color, and some two dozen varieties of studs. Some studs have external threads, others internal threads. A plastic coating protects the threads from damage while driving.

The drivers will force studs into steel up to about $\frac{1}{2}$ in thick, into concrete through steel plates up to about $\frac{1}{4}$ in thick, into concrete through various thicknesses of woods, and into steel through steel up to $\frac{1}{4}$ in thick.

Powder-driven studs should never be driven into soft materials or into very hard or brittle materials, such as cast iron, glazed tile, or surface-hardened steel. Neither should they be used in face brick, hollow tile, live rock, or similar materials.

In driving studs a suitable guard must be used for each operation, and the driver must be held squarely to the work. If the driver is not held perpendicular, a safety device prevents firing of the charge. There also must be sufficient backup material to absorb the full driving power of the charge. Studs cannot safely be driven closer than $\frac{1}{2}$ in from the edge of a steel surface, or closer than 3 in from the outside edge of concrete or brick surfaces.

16-23. Miscellaneous Anchoring Devices.

Various types of expansion bolts, screw anchors, and toggle bolts are available for securing fixtures, brackets, and equipment to solid material, such as masonry, brick, concrete, and stone. Anchors also can be had for fastening to hollow walls, such as plaster on metal lath in furred spaces and hollow tile. The best device to use depends on the requirements in each case.

For use with practically any materials, including soft and brittle ones, such as composition board, glass, and tile, fiber screw anchors with a hollow metal core find a universal application. These plugs with braided metal cores possess many advantages: They can be used with wood screws and with lag screws. The flexible construction permits the plug to conform to any irregularities. Because of this elastic compression, the fibers are compressed as the screw enters. Screws can be unscrewed and replaced. Shock and vibration have no effect on gripping power. The plugs come in about 40 sizes to fit anything from a No. 6 screw to a ⅝-in lag screw. In practice, a hole is drilled first, the plug is driven into it, and then the screw is inserted, expanding the plug against the sides of the hole.

For some fastenings, one-piece drive bolts are hammered like a nail into prepared holes in masonry or concrete. Other types of expansion bolts have expansion shields or are calked into place. The expansion shield is expanded in the hole by a tapered sleeve forced against a cone-shaped nut by the tightening of a bolt threaded into the cone. These types are not recommended for soft or brittle materials.

For thin hollow walls, toggle bolts equipped with spring-actuated wings are used. The wings will pass into the hole in folded position. After entering the hollow space, the wings open out, thereby obtaining a secure hold.

For fastening lightweight materials to nailable supports, stapling machines are extensively used. In one patented system, for example, staples secure acoustic tile to wood furring strips. In this system, invisible fastenings are obtained by using a full-spline suspension for the kerfed pieces of tile and a special stapling machine adjusted to function at the proper distance below the furring strips. The machine staples the splines (at the joints of the tiles) to the supporting strips, giving a speedy, economical, and secure fastening.

16-24. Bolts.

The number of sizes and varieties of this basic type of fastening found in buildings is great. One of the more significant applications is in making field connections for structural-steel framing. The bolts are known as high-tensile-strength bolts (Art. 8-55). For other types of bolts used with structural steel, see Arts. 8-54 and 8-56.

In timber construction, bolts generally are used with timber connectors—patented metal rings, grids, shear plates, framing anchors, and clamping plates, which increase the strength of joints (Art. 7-16).

Section 17

Acoustics

Lyle F. Yerges
Consulting Engineer, Downers Grove, Ill.

Acoustics, derived from a Greek word meaning to hear, refers to the entire science of sound, including its generation, transmission, absorption, and control.

Acoustics has a long history. In Roman times, Vitruvius wrote extensively on the acoustics of rooms, particularly theaters and performing spaces. Some of his observations were correct, some incorrect. During the latter part of the nineteenth century, excellent scientific work in acoustics was done by Helmholz and Lord Rayleigh. But it was the pioneer work of Wallace Clement Sabine at Harvard University during the first two decades of the twentieth century that founded the modern science of acoustics.

Sabine established the relationship between reverberation and intelligibility within a room. He also discovered the relationship of reverberation time to the acoustical absorption present within a room.

Although acoustics remains part science and part art, modern studies are stripping away much of its mystique. As a result, it is possible to plan and predict the acoustics of finished spaces with reasonable certainty.

17-1. Sound Production and Transmission.

Sound is a vibration in an elastic medium. It is a simple form of mechanical energy, and can be described by the mathematics associated with the generation, transmission, and control of energy.

Almost any moving, vibrating, oscillating, or pulsating object is a potential sound source. Usually, though, sound sources of interest to us radiate enough energy to be audible to humans or felt by them.

An understanding of the nature of "elastic matter" is essential to an understanding of the mechanism of sound transmission. If a single molecule of a quantity of matter subject to sound motion were observed, it would be found to move first to one side of its normal, or rest, position to some maximum displacement, then back through normal to an opposite maximum position, and back to normal (Fig. 17-1). If the motion is regular and repetitive, it is called a **vibration**.

While each molecule or particle moves only an infinitesimal distance to either side of its rest position, it bumps the particles adjacent to it and imparts motion and energy to these particles. Thus, the impulse or disturbance progresses rapidly and travels great distances, but the medium in which it travels only vibrates. Figure 17-2 shows what the molecules in a medium look like when a sound wave passes through the medium. The impulse travels in the same direction as the movement of the particles; in other words, it is a longitudinal or "squeeze" wave motion, not a transverse wave like light.

Fig. 17-1 Motion of a molecule during a single cycle of vibration.

Fig. 17-2 An elastic medium subjected to sound.

Sound waves in air (or other gases or fluids) travel outward from the source like a rapidly expanding soap bubble, although any particular group of molecules behaves like a pulsating balloon, moving only slightly, while the wave progresses swiftly to great distances.

17-2. Nomenclature for Analysis of Sound.

Cycle. A complete single excursion of a vibrating molecule (Fig. 17-1).

Frequency. The number of cycles of vibration in a given unit of time, usually cycles per second (cps) or hertz (Hz).

Sound Wave. The portion of a sound between two successive compressions or rarefactions.

Wavelength. The distance between two successive rarefactions or compressions in a sound wave (Fig. 17-2).

Amplitude. Maximum displacement, beyond its normal, or rest, position, of a vibrating element. In most audible sounds, these excursions are very small, although low-frequency sound may cause large excursions (as would be observed in the motion of a loudspeaker cone reproducing very low frequency sounds at audible level). Amplitude of motion is related to the increased pressure created in the medium, and to the intensity of the energy involved.

Velocity. Speed at which a sound impulse travels (not the speed of movement of any particular molecule). In a given medium, under fixed conditions, sound velocity is a constant. Therefore, the relationship between velocity, frequency, and wavelength can be expressed by the equation:

$$\text{Velocity} = \text{frequency} \times \text{wavelength} \qquad (17\text{-}1)$$

Because velocity is constant in air, low-frequency sound has long wavelengths, and high-frequency sound has short wavelengths. This is important to remember in any acoustical design considerations.

In addition to the longitudinal (compression or "squeeze") waves by which sound travels, there is another type of vibrational motion to which most building materials and construction systems are subjected, a **transverse wave** (Fig. 17-3). This is the familiar motion of a vibrating string or reed. This type of wave, too, transmits energy that can be felt or heard, or both.

Sheets or panels, studs and joists, hangers and rods, and similar slender and somewhat flexible members are particularly apt to vibrate in a transverse mode and to transmit sound energy along their length, as well as radiating it from their surfaces to the surrounding air.

Fig. 17-3 Transverse waves.

17-3. Elastic Media.

The elasticity and the mass of all materials determine their acoustical performance. All matter has mass; that is, inertia and weight. (For most purposes, we can think of weight and mass as interchangeable in our gravitational system.)

Sound does not travel faster in dense media than in less dense media. In fact sound velocity is inversely proportional to the density of the medium, as can be seen from Eq. (17-2).

$$\text{Velocity} = k \sqrt{\frac{\text{modulus of elasticity}}{\text{density}}} \qquad (17\text{-}2)$$

where k is a constant. Usually, however, dense materials often have a much larger modulus of elasticity than less dense materials, and the effect of the modulus more than offsets the effects of density. Table 17-1 lists the velocity of sound in some common materials.

Another characteristic of material, very important to its acoustical performance, is acoustical **impedance**. This value is determined by multiplying the velocity of sound in the material by the density of the material. An examination of the units resulting from this operation will show that this is a way of measuring the rate at which the material will accept energy. Table 17-2 lists the acoustical impedance of several typical materials used in building construction.

Sound must travel via some path—the elastic media. Hard, rigid, dense materials make excellent transmission paths, because they can accept much energy readily.

TABLE 17-1 Velocity of Sound in Various Media

Material	Approximate Sound Velocity, ft per sec
Air	1,100
Wood	11,000
Water	4,500
Aluminum	16,000
Steel	16,000
Lead	4,000

TABLE 17-2 Acoustical Impedance of Various Materials

Material	Acoustical Impedance, Psi per Sec
Rubber	100
Cork	165
Pine	1,900
Water	2,000
Concrete	14,000
Glass	20,000
Lead	20,500
Cast iron	39,000
Copper	45,000
Steel	58,500

17-4. Hearing.

In most cases of interest, a human is the "receiver" in the source-path-receiver chain. Hearing is the principal subjective response to sound. Within certain limits of frequency and energy levels, sound creates a sensation within the auditory equipment of humans and most animals.

At very low frequencies or at very high energy levels, additional sensations, ranging from pressure in the chest cavity to actual pain in the ears, are experienced.

Our sense of feeling is also affected by vibrations in building structures, a subject closely related to ordinary building acoustics but outside the scope of this section.

From a strictly mechanical standpoint, the ear responds in a relatively predictable manner to physical changes in sound. Table 17-3 relates the objective characteristics of sound to our subjective responses to those characteristics.

Loudness is the physical response to sound pressure and intensity. It is influenced somewhat by the frequency of the sound.

Pitch is the physical response to frequency. Low frequencies are identified as low in pitch, high frequencies as high in pitch. Middle C on the piano, for example, is 261 Hz; 1 octave below is 130 Hz; 1 octave above is 522 Hz.

An **octave** represents a 2:1 ratio of frequencies.

For practical purposes, humans can be considered to have a hearing range from about 16 Hz to somewhat less than 20,000 Hz, and to be significantly deaf to low frequencies and very high frequencies. Human loudness response to sounds of identical pressure but varying frequency plots something like the curve in Fig. 17-4.

Any measure of loudness must, in some way, specify frequency as well as pressure or intensity to have any real significance in human acoustics. As shown in Art. 17-7, measuring equipment and scales used are all modified to recognize this. Table 17-4 lists some of the significant frequency ranges.

TABLE 17-3 Subective Responses to Characteristics of Sound

Objective (Sound)	Subjective (Hearing)
Amplitude, Pressure, Intensity	Loudness
Frequency	Pitch
Spectral distribution of acoustical energy	Quality

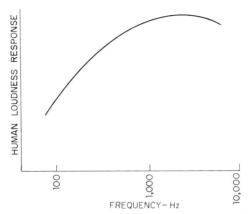

Fig. 17-4 Human response to sounds of identical level but varying frequency.

Low frequencies do not affect humans very strongly, while those from 500 to 5000 Hz are very important.

Sounds generated by mechanical equipment may encompass the entire frequency range of human hearing. Jet aircraft, for example, have significant output throughout the entire range, and large diesel engines produce substantial energy from 30 to 10,000 Hz.

Wanted sound (signal), whatever its nature, communicates information to us. *Unwanted sound,* whatever its nature, is *noise.* When the signal becomes sufficiently louder than the noise, the signal can be detected and the information becomes available. Thus, the distinction between noise and communication is completely subjective, completely a matter of human desires and needs of the moment.

While very weak sounds can be heard in a very quiet background, the level of wanted sound may be increased to override the unwanted background (or noise); but there is a significant noise level above which even loud (shouting) speech is scarcely intelligible. Such a level, if long continued, can prove damaging to the hearing mechanism. At still higher levels, pain can be experienced, and immediate physical damage (often irreversible) may occur.

Complete silence (if such a thing could occur) would be most unpleasant. Humans would quickly become disoriented and distressed. Somewhere between silence and the uncomfortably loud, however, is a wide range that humans find acceptable.

TABLE 17-4 Significant Frequency Ranges

	Approximate Frequency Range, Hz
Range of human hearing	16–20,000
Speech intelligibility (containing the frequencies most necessary for understanding speech)	600–4,800
Speech privacy range (containing speech sounds that intrude most objectionably into adjacent areas)	250–2,500
Typical small table radio	200–5,000
Male voice	350*
Female voice	700*

* Frequency at about which energy output tends to peak.

17-5. Force and Energy Concepts.

In acoustics, the fundamental concept is the well-known:

$$\text{Force} = \text{mass} \times \text{acceleration} \tag{17-3}$$

All materials possess mass. This mass, when moved with change in direction and velocity, which any elastic, oscillating motion must involve, must be accelerated. This process requires force; and force, acting over displacement, is energy. Thus, for sound, the amount of motion of the particles in the medium depends on the sound pressure (force) or the intensity (amount) of the sound energy involved.

The energy concepts of importance are kinetic energy (energy of motion), potential energy (stored energy), and energy conversion.

Transmission (flow), or prevention of transmission of sound, and conversion of sound energy to a nonaudible form are the functions of so-called **acoustical materials.**

17-6. Units and Dimensions for Sound Measurements.

Generally, absolute numbers obtained from measurements have little significance in acoustics. Instead, measurements are almost always compared with some base or reference, and they are usually quoted as **levels** above or below that reference level. The levels are usually ratios of observed values to the reference level, such as 2:1, 1:2, 1:10, because rarely are simple, linear relationships found between stimuli and effects in humans.

Zero level of sound pressure is not a true physical zero; that is, the absence of any sound pressure at all. Rather, zero level is something of an average threshold of human hearing, of sound at about 1000 Hz. The physical pressure associated with this threshold level is very small, 0.00002 Pa (N/m^2).

Changes in human response tend to occur according to a ratio of the intensity of the stimuli producing the response. In acoustics, the ratio of 10:1 is called a bel, and one-tenth of a bel, a decibel (dB). Thus, power and intensity levels, in dB, are computed from

$$\text{Level} = 10 \log_{10} \frac{\text{quantity measured}}{\text{reference quantity}} \tag{17-4}$$

Intensity level IL, dB, for example, represents the ratio of the intensity being measured to some reference level, and is given by

$$\text{IL} = 10 \log_{10} \frac{I}{I_o} \tag{17-5}$$

where I = intensity measured, W/cm^2
I_o = reference intensity = 10^{-16} W/cm^2

Since intensity varies as the square of the pressure, intensity level also is given by

$$\text{IL} = 10 \log_{10} \frac{p^2}{p_o^2} = 20 \log_{10} \frac{p}{p_o} \tag{17-6}$$

where p = pressure measured, Pa
p_o = reference pressure = 0.00002 Pa

Sound pressure level, dB, to correspond with intensity level, is defined by

$$\text{SPL} = 20 \log_{10} \frac{p}{p_o} \tag{17-7}$$

Sound power level refers to the power of a sound source relative to a reference power of 10^{-12} W. (*Note:* At one time, 10^{-13} W was used; thus, it is imperative that the reference level always be explicitly stated.)

TABLE 17-5 Subjective Effect of Changes in Sound Characteristics

Change in Sound Level, dB	Change in Apparent Loudness
3	Just perceptible
5	Clearly noticeable
10	Twice as loud (or ½)
20	Much louder (or quieter)

The ear responds in a roughly logarithmic manner to changes in stimulus intensity, but approximately as shown in Table 17-5.

Another comparison, which gives more meaning to various levels, is shown in Table 17-6.

17-7. Sound Measurements and Measurement Scales.

Most measurements or evaluations of sound intensity or level are made with an electronic instrument that measures the sound pressure. The instrument is calibrated to read pressure levels in decibels (rather than volts). It can measure the overall sound pressure level throughout a frequency range of about 20 to 20,000 Hz, or within narrow frequency bands (such as an octave, third-octave, or even narrower ranges). Usually, the sound level meter contains filters and circuitry to bias the readings so that the instrument responds more like the human ear—"deaf" to low frequencies and most sensitive to the midfrequencies (from about 500 to 5000 Hz). Such readings are called A-scale readings. Most noise level readings (and, unless otherwise specifically stated, most sound pressure levels with no stated qualifications) are A-scale readings (often expressed as dBA). This means that actual sound pressure readings have been modified electrically within the instrument to give a readout corresponding somewhat to the ear's response (Fig. 17-4).

For various measurements and evaluations of performance for materials, constructions, systems, and spaces, see Art. 17-17.

TABLE 17-6 Comparison of Intensity, Sound Pressure Level, and Common Sounds

Relative intensity	SPL, dBA*	Loudness
100,000,000,000,000	140	Jet aircraft and artillery fire
10,000,000,000,000	130	Threshold of pain
1,000,000,000,000	120	Threshold of feeling
100,000,000,000	110	
10,000,000,000	100	Inside propellor plane
1,000,000,000	90	Full symphony or band
100,000,000	80	Inside automobile at high speed
10,000,000	70	Conversation, face-to-face
1,000,000	60	
100,000	50	Inside general office
10,000	40	Inside private office
1,000	30	Inside bedroom
100	20	Inside empty theater
10	10	
1	0	Threshold of hearing

SPL as measured on A scale of standard sound level meter.

17-8. Sound and Vibration Control.

This is a process, not a product. It consists of:

A. Acoustical analysis
   1. Determining the use of the structure—the subjective needs.
   2. Establishing the desirable acoustical environment in each usable area.
   3. Determining noise and vibration sources inside and outside the structure.
   4. Studying the location and orientation of the structure and its interior spaces with regard to noise and noise sources.
B. Acoustical design
   1. Designing shapes, areas, volumes, and surfaces to accomplish what the analysis indicates.
   2. Choosing materials, systems, and constructions to achieve the desired result.

Sound and vibration sources are usually speech and sounds of normal human activity—music, mechanical equipment sound and vibration, traffic, and the like. Characteristics of these sound sources are well known or easily determined. Therefore, the builder or designer is usually most interested in the transmission paths for sound and vibration. These are gases (usually air); denser fluids (water, steam, oil, etc.); and solids (building materials themselves).

The mechanism for energy transmission via these paths is discussed in Art. 17-1. During this transmission process, some of the sound energy is absorbed or dissipated, some is reflected from various surfaces or constructions, and some is transmitted through the constructions.

Sound control is accomplished by means of barriers and enclosures, acoustically absorbent materials, and other materials, constructions, and systems properly shaped and assembled. Vibration control is accomplished by means of various resilient materials and assemblies, and by damping materials (viscoelastic materials of various types).

Airborne and structure-borne energy are controlled by somewhat different techniques, described in the following articles.

17-9. Airborne Sound Transmission.

A sound source in a room sets the air into vibration. The vibrating air causes any barrier it touches (partitions, floors, ceilings, etc.) to vibrate; the vibrating barrier, in turn, sets into vibration the air on the opposite side of the barrier (Fig. 17-5).

The barrier, like any other body, resists motion because of its inherent inertia. Because it takes more force, and therefore more energy, to move a heavier (more massive) barrier, sound transmission through the barrier depends directly on the mass of the barrier.

The loss in energy level between the original signal striking the barrier and the level

Fig. 17-5 Sound transmission through a barrier.

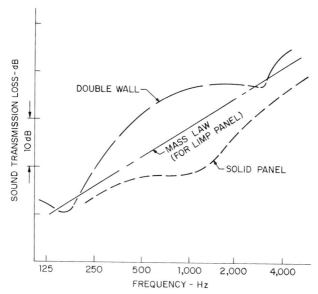

Fig. 17-6 Sound transmission loss of barriers.

of the energy transmitted to the opposite side is called the **sound transmission loss** of the barrier. The more effective the barrier, the greater the sound transmission loss.

Figure 17-6 shows graphically how a truly limp barrier of infinite width and height responds to sound energy of varying frequency (the *mass law*, straight-line variation). The straight line rises at the rate of 6 dB per octave. At 800 Hz, for example, such a barrier moves back and forth twice as fast as at 400 Hz (1 octave lower), and one-half as fast as at 1600 Hz (1 octave higher). The moving barrier has kinetic energy KE, which is related to mass M and velocity v by

$$KE = \tfrac{1}{2}Mv^2 \tag{17-8}$$

As the velocity of movement (frequency) doubles (1 octave higher), the energy of motion quadruples. Therefore, the energy or intensity level increases by 6 dB, inasmuch as Eq. (17-5) gives $10 \log_{10} 4 \approx 6$ dB.

Likewise, as the mass of the barrier doubles, the force or pressure required to maintain the same frequency doubles, and because energy is proportional to the square of the pressure, the energy level quadruples. By Eq. (17-6), the increase in intensity level is $10 \log_{10} 4 \approx 6$ dB.

In practice, however, this simple mathematical relationship rarely exists. No barrier is truly limp; a wall, floor, or ceiling always has a finite stiffness. Therefore, at low frequencies, barriers tend to have higher sound transmission losses than the mass law predicts. In some region of the spectrum, however, barriers tend to have lower sound transmission losses (pass sound more readily) than the mass law indicates (see curve for solid panel in Fig. 17-6).

This latter phenomenon results because the barrier, in addition to its back-and-forth motion, as shown in Fig. 17-5, moves in a simultaneous shear wave, like a rope being shaken (Fig. 17-3). At some frequency, the velocity of this shear wave in the barrier coincides with the velocity of the impinging sound wave in the air. Then, the partition is quite transparent to sound, and a deep coincidence dip in the sound transmission loss curve occurs (see curve for double wall in Fig. 17-6). At frequencies above the coincidence frequency, the curve tends to recover its 6 dB per octave slope.

There are few practical limp materials useful for ordinary building panels. Soft sheet lead is occasionally used for specialized barriers, as described in Art. 17-16, or is applied to other panel material to increase the mass of the composite without increasing its stiffness. In addition, damping materials (Art. 17-12) can be applied to the panels to damp out their vibrations quickly and to increase the energy loss as the panels vibrate.

Figure 17-6 shows another characteristic of barriers that further complicates their performance. While single, solid panels have somewhat smaller transmission losses than the mass law predicts, a double wall, a barrier comprising separated wythes or leaves, of the same total weight produces greater transmission losses than the mass law predicts.

If the mass of a single, solid panel is divided into two separate, unconnected layers, the only energy transfer between them occurs via the air between the layers. Air is not stiff; it can sustain little of the shear wave, and it is not an efficient transmitter of the back-and-forth motion of one layer to the other (except at certain resonant frequencies). As a result, particularly in the midfrequencies, the sound transmission loss actually exceeds the mass-law values, often considerably. The greater the distance between layers, the better the performance. Theoretically, the sound transmission loss should increase about 6 dB for each doubling of the width of the airspace. In practice, however, the increase in decibels is somewhat less than this.

If a porous, sound-absorbent blanket is inserted in the void between layers, the standing waves in the air layers are minimized. Futhermore, such a blanket has the effect of reducing the stiffness of the air, as well as absorbing some of the energy of the air as it pumps back and forth through the blanket. The result is an additional increase in sound transmission loss.

Fig. 17-7 Structural flanking: (a) poor isolation at a partition; (b) adequate isolation at a partition; (c) poor isolation at an exterior wall (balloon framing); (d) adequate isolation (platform framing).

The performance of various barriers, and rating systems for their performance, are shown in Table 17-7.

17-10. Bypassing of Sound Barriers.

Rarely is a sound barrier the sole transmission path for the acoustic energy reaching it. Some energy invariably travels via the connecting structures (floors, ceilings, etc.), or through openings in or around the barrier.

Structural flanking via edge attachments and junctions of walls, partitions, floors, and ceilings can seriously degrade the performance of a barrier. Figure 17-7 shows some typical flanking paths and how to avoid such flanking.

Even trivial openings through a sound barrier seriously degrade its performance. An opening with an area of only about 1 in^2 transmits as much acoustical energy as a 6-in thick concrete block wall of 100-ft^2 area. Figure 17-8 shows the effect of openings of various sizes on walls of various effectiveness. (*Note:* Sound transmission class *STC* is a rating system described in Art. 17-16.) The better the wall, the more serious the effect of openings or leaks.

The more common points of leakage through or around barriers include perimeter of pipes, ducts, or conduits penetrating the barriers; relief grilles for return air; perimeter of doors or glazing; shrinkage or settlement cracks at partition heads or sills; joints between partitions and exterior curtain-wall mullions; joints around or openings through back-to-back electrical outlet boxes, medicine cabinets, etc.; common supply or return

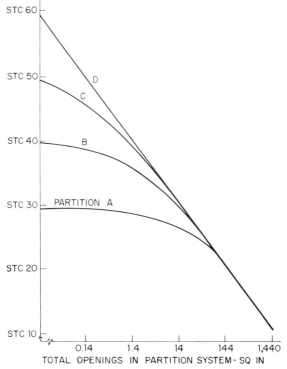

Fig. 17-8 Effect of leaks on STC of partitions (for 100-ft^2 test wall).

ducts with short, unlined runs between rooms; and operable windows opening to a common court. Such openings should be avoided, when possible, and all cracks, joints, and perimeters should be calked and sealed.

In some spaces (an open-plan school, "landscaped" office, etc.), screens, or partial-height barriers are used. Their effectiveness is much less than that of full-height partitions or walls. When partial-height partitions are contemplated in building design, the services of experienced acoustical consultants should always be used, because experts are needed to deal with the many factors that enter into the acoustical design of open-plan spaces.

See also Art. 17-15.

17-11. Structure-Borne Sound Transmission.

Acoustical energy transmission from sound sources to distant parts of the structure via the structure itself is often a major problem in building acoustics. One reason, as discussed in Art. 17-10, is that structural flanking can seriously degrade the performance of sound barriers. Additional transmission paths include pipes, ducts, conduits, and almost any solid, continuous, rigid member in the building. Impacts and vibrations can be transmitted through the floor-ceiling assemblies quite readily, if proper precautions are not taken.

Normally, in ordinary construction, a good carpet over a good pad will provide sufficient impact isolation against the sounds of footfalls, heel clicks, and dropped objects. Where carpet is not feasible, or in critical spaces, special floor-ceiling assemblies can be used. (See Art. 17-16 for various constructions and their impact isolation performance.)

Isolation against vibration or impact produced by machinery or other vibrating equipment is usually provided by use of special resilient or mounting systems. Springs, elastomeric pads, and other devices are used in such work. For a detailed discussion of such isolation, refer to the current ASHRAE "Guide," American Society of Heating, Refrigerating and Air-Conditioning Engineers, or other references on vibration and shock isolation.

It is important to remember that any moving, rotating, oscillating, or vibrating equipment, rigidly attached to the building structure, will transmit some of its energy to the structure. Isolating against such transmission is always advisable, even in ordinary construction, and is imperative in critical buildings and where the equipment is large or of high power.

Noise and vibration transmitted via plumbing and heating pipes, ducts, and conduits can be objectionable unless proper precautions are taken. Resilient connectors for rigid members, acoustically absorptive duct linings, and similar approaches are described in detail in the ASHRAE "Guide."

17-12. Damping of Vibrations.

As a panel or object vibrates, it radiates acoustical energy to the air surrounding it and to solid surfaces touching it or attached to it. If the energy of vibration could be dissipated, the radiation would be reduced and the sound and vibration levels lowered. One way to do this is to attach firmly to a vibrating panel certain "lossy" substances (those with high internal friction or poor connections between particles) or viscoelastic materials (neither elastic nor completely viscous, such as certain asphaltic compounds). These damp the vibrations by absorbing the energy and converting it to heat.

In the assembly of barriers, damping can be accomplished with proper connections and attachments, use of viscoelastic adhesives, proper attachment of insulating materials, and similar means. Special viscoelastic materials for adhesive attachment or brush or spray application to panels are available.

17-13. Sound Absorption.

The best-known acoustical materials are acoustical absorbents (although actually all materials are acoustical). Generally, these absorbents are lightweight, porous, "fuzzy" types of boards, blankets, and panels.

Acoustical absorbents act as energy transducers, converting the mechanical energy of sound into heat. The conversion mechanism involves either pumping of air contained within the porous structure of the material, or the flexing of thin panels or sheets. Most materials employ the first principle.

The internal construction of most absorbents consists of a random matrix of fibers or particles, with interconnected pores and capillaries (Fig. 17-9). It is necessary that the air contained within the matrix be able to move sufficiently to create friction against the fibers or capillaries. Nonconnected-cell or closed-cell porous materials are not effective absorbents.

Tuned chambers, with small openings and a restricted neck into the chambers, also are used for sound absorption; but they are somewhat specialized in design and function, and their use requires expert design in most instances.

The surface of absorbents must be sufficiently porous to permit the pressures of impinging sound waves to be transferred to the air within the absorbent. Very thin, flexible facings (plastic or elastomeric sheets) stretched over panels and blankets do not interfere significantly with this pressure transfer, but thick, rigid, heavy coatings (even heavy coatings of paint) may seriously restrict the absorption process. Perforated facings, if sufficiently thin and with sufficient closely spaced openings, do not appreciably degrade the performance of most absorbents.

The visible surface of absorbent panels and tiles may be smooth or textured, fissured or perforated, or decorated or "etched" in many ways. Figure 17-10 shows a typical commercial tile. Figure 17-11 illustrates an example of the use of both acoustical tiles and panels in a ceiling.

Fig. 17-9 Cross section of sound absorbent (greatly enlarged).

Absorbents are normally produced from vegetable or mineral fibers, porous or granular aggregates, foamed elastomers, and other products, employing either added binders or their own structure to provide the structural integrity required. Because of their lightweight, porous structure, most such products are relatively fragile and must be installed where they are not subject to abuse; or they may be covered with sturdy perforated or porous facings to protect them.

Absorbents are chosen for their appearance, fire resistance, moisture resistance, strength, maintainability, and similar characteristics. Their performance ratings in respect to these characteristics are usually published in advertising materials and in various bulletins and releases of trade associations.

The acoustical property of absorbents most important to designers and builders is absorptive efficiency. [Sound transmission through most absorbents takes place readily. They are very poor in this respect and should never be used to attempt to improve the airborne sound isolation of a barrier. Sound transmission over the top of a partition, through a lightweight, mechanically suspended acoustical panel ceiling, is frequently a serious problem in buildings (Art. 17-15).] The absorptivity of a tile or panel is usually expressed as the fraction or percentage of acoustical energy absorbed from an impinging plane wave. If a perfectly absorptive plane surface represents 1.00 or 100%, the ratio of absorption of a given product to this perfect absorber is called the **sound absorption coefficient** of the product.

Fig. 17-10 Fissured-surface acoustical tile with the appearance of travertine stone.

The absorptivity of a material varies with its thickness, density, porosity, flow resistance, and other characteristics. Further, absorptivity varies significantly with the frequency of the impinging sound. Usually, very thick layers of absorptive material are required for good absorption of low-frequency sound, while relatively thin layers are effective at higher frequencies. Little or no absorption, however, can be obtained with thin layers or flocked surfaces, textured but nonporous surfaces, or other products often mistakenly thought to be absorptive or claimed to break up the sound.

Generally, there is an optimum density and flow resistance for any particular family of materials (particularly fibrous materials). Usually, absorptivity increases with thickness of the material.

Performance data for absorbent materials are usually readily available from manufacturers and trade associations. Performance of some typical products is shown in Table 17-10. See also Arts. 17-14 to 17-16.

17-14. Control of Reflection and Reverberation.

Normally, acoustical absorbents are used to prevent or minimize reflections of sound from the surfaces of rooms or enclosures. Distinct reflections—**echoes**—are usually objectionable in any occupied space. Rapid, repeated, but still partly distinguishable echoes, such as occur between parallel sidewalls of a corridor—**flutter**—are also objectionable.

Reverberation comprises very rapid, repeated, jumbled echoes, blending into an indistinct but continuing sound after the source that created them has ceased.

Fig. 17-11 Library ceiling comprising both acoustical panels and acoustical tiles, integrated with lighting panels. *(U.S. Gypsum Co.)*

Usually, reverberation is one of the major causes of poor intelligibility of speech within a room; but, within limits, it may actually enhance the sound of music within a space. Reverberation control is a necessary and important aspect of good acoustic design, but it is often greatly overemphasized. Good room proportions and configuration, control of echoes, and absorption of noise usually assure an acceptable reverberation time within a space. Where careful determination and control of reverberation are required in a room, the services of a competent acoustical consultant are always advisable.

Reflections from strategically located and properly shaped room surfaces may be highly desirable, because such reflections may strongly enhance the source signal. But excessively delayed or highly persistent reflections are usually undesirable. (For most purposes, and within the normal frequency range of importance to human hearing, it can be assumed that sound waves, like light waves, reflect from a surface at the same angle of incidence. Because of the enormously longer wavelength of sound compared with light, this assumption is inexact but acceptable for most design work.)

In most rooms, absorption of most of the acoustical energy inpinging on many of the surfaces (the floor, distant walls, etc.) is desirable to prevent buildup or increase of unintelligible or useless sound. For this purpose, sound absorbents may be placed on some or all of the surfaces. The difference in sound pressure level (or noise level) caused within a space by the introduction of absorbents can be calculated simply. From such a calculation, it is possible to determine how effective such treatment will be.

Noise reduction NR, dB, provided by adding acoustical absorbents in a space can be determined from

$$\text{NR} = 10 \log_{10} \frac{A_o + A_a}{A_o} \tag{17-9}$$

where A_o = original acoustical absorption present
A_a = added acoustical absorption

Acoustical absorption equals the sum of the products of each area (in consistent units) in the space times the absorption coefficient of the material constituting the surface of the area; for example, the floor area, ft^2 × its absorption coefficient, plus ceiling area, ft^2 × its absorption coefficient, plus total wall area, ft^2 × its absorption coefficient.

[*Note:* An anomalous but useful term is often used in advertising data, the **noise reduction coefficient** NRC. This is the arithmetic average of the sound absorption coefficients of a material as determined at 250, 500, 1000, and 2000 Hz (Art. 17-13). Since these frequencies include the most significant speech and intelligibility ranges, such a figure is a reasonably good means of comparing similar materials; that is, materials with absorption characteristics not differing widely from one another within this frequency range. Often, NRC is used, instead of the absorption coefficients at various frequencies, to determine an average noise reduction from Eq. (17-9).]

Equation (17-9) indicates that the more absorption present originally the less the improvement provided by added absorption. Thus, in a very "hard," bare room, addition of acoustical (sound-absorbent) tile to a full ceiling significantly reduces the noise level. But addition of the same ceiling tile in a room with a thick carpet, upholstered furniture, and heavy draperies would make little change.

Heavy carpet, upholstered furniture, heavy draperies, and similar materials are very effective absorbers. In residences, for example, rarely is additional acoustical absorption required in bedrooms, living rooms, and similar spaces. In kitchens, bathrooms, or recreation rooms, with normal hard floors and few additional furnishings or fabrics, however, an acoustical tile ceiling is helpful. This is equally true of offices and similar spaces. In theaters and auditoriums, the large expanse of upholstered seating and aisle carpets is normally adequate for most noise control, but added absorption on some wall surfaces may be required for control of echoes.

Reverberation Time. The reverberation within a space is usually expressed as the time required for a sound pulse to decay 60 dB (to one-millionth of its original level). For most purposes, reverberation time T, s, can be calculated from the simple Sabine formula:

$$T = \frac{0.049 V}{A} \qquad (17\text{-}10)$$

where V = volume of the space, ft^3
A = total acoustical absorption in the space

Equation (17-10) assumes a smooth, steady, logarithmic decay; random distribution of sound within the room, with the wave front striking every surface quickly and within the decay time; and no standing waves between surfaces that could support a persistent mode. These are idealized conditions and never exist, but the formula is sufficiently accurate for most purposes.

Because the absorption of an absorbent material varies with frequency of sound, it is necessary to calculate T for each significant frequency. For most reverberation calculations, determinations at 500 Hz are adequate. For concert halls, and critical spaces, calculations are usually made 2 octaves above and 2 octaves below 500 Hz as well.

Fig. 17-12 Recommended reverberation time, indicated by shaded area, varies with size of room.

Fig. 17-13 Typical suspended acoustical panels.

Optimum reverberation time for a room is a subjective determination, governed by speech intelligibility and the fullness and richness of musical sound desired. Figure 17-12 shows, within the shaded area, the acceptable range of reverberation times for normal spaces of varying volume. For critical spaces (radio studios, concert halls, auditoriums, etc.), it is advisable to obtain the advice of competent acoustical consultants, because experts are needed to evaluate and provide for many additional factors that are equally important.

It is imperative that designers understand that a reverberation-time determination is not an acoustical analysis, and that, in many instances, reverberation time is a trivial part of an acoustical study.

17-15. Installation of Absorbents.

The amount, location, and installation method for absorbents are all significant in a room. As discussed in Art. 17-14, the material is located on surfaces from which reflections are undesirable and where it is reasonably free from damage. The amount of material required can be determined from noise-reduction and reverberation-time calculations. (For most simple spaces, a full ceiling treatment is acceptable.)

Acoustical absorbents can be cemented directly to a smooth, solid surface, nailed or stapled to furring strips, or suspended by any of a number of mechanical systems, such as that shown in Fig. 17-13.

While acoustical absorbents are usually installed as, or on, flat, horizontal surfaces, various other configurations are used. Coffers, grids of hanging panels, and similar arrangements are often employed. The acoustical performance of each configuration must be tested to determine its effectiveness.

Sprayed-on and trowel-applied acoustical products are also employed, although their use is limited.

The structural, fire-resistance, and acoustical performance of most acoustical panels and tiles are significantly affected by the installation method. It is imperative that performance data be explicitly related to the specific installation method.

Partition Bypassing. When partitions or other sound barriers are constructed to (but not through) a suspended acoustical tile or panel ceiling, sound transmission through the tile from one space, over the top of the partition, into an adjacent space can be a serious problem. Figure 17-14 shows how sound can bypass a partition, and alternative methods of preventing this.

Similarly, structure-borne flanking transmission along the continuous metal runners of some suspension systems may seriously degrade the performance of an acoustical ceiling as a sound barrier. Performance data for most typical commercial ceilings are published in various bulletins and advertising matter.

Fig. 17-14 Prevention of sound transmission over partitions.

17-16. Performance Data.

To simplify and standardize evaluation of the acoustical performance of materials, systems, and constructions, various rating systems have been adopted. The best known and most widely used are those published by the American Society for Testing and Materials (ASTM) 1916 Race St., Philadelphia, Pa. 19103.

Partitions, Floor-Ceiling Assemblies, and Barriers. Insulation (or isolation) of airborne sound provided by a barrier is usually expressed as its **sound transmission class** STC. For a specific construction, STC is determined from a sound-transmission-loss curve obtained from a standardized test of a large-scale specimen in a recognized laboratory. This curve is compared with a standard contour, and a numerical rating is assigned to the specimen (ASTM E90 test procedure and ASTM E413, Determination of Sound Transmission Class).

Table 17-7 lists typical STC ratings of several partition, wall, and floor-ceiling components or assemblies. Published data for almost any type of construction can be obtained from various sources.

It is important to remember that a difference of one or two points between two similar constructions is rarely significant. Normally, constructions tend to fall into groups or classes with their median values about five points apart. In Table 17-8, the italic number represents the median of a performance group which includes the numbers on either side of the median.

The impact insulation (or isolation) provided by floor-ceiling assemblies is usually expressed as their impact noise rating INR or impact insulation IIC. Like STC, INR and IIC values are obtained by comparing the curve of the sound spectrum obtained in a test with a standard contour (except that for INR and IIC the sound pressure level is measured in the room below the noise source). The entire procedure is controversial and far from widely accepted; but its use is so widespread that designers and builders should be aware of it. (See ASTM RM 14-4.)

Table 17-9 lists the impact isolation provided by several types of construction. Note particularly the enormous effect of certain floor coverings on the performance of the construction. (While IIC values are shown in Table 17-9, they can be converted to INR values by subtracting 51 points; for example, IIC 60 = INR 9, and IIC 45 = INR −6, etc.)

Acoustical Absorbents. Sound absorption coefficients and noise reduction coefficients of acoustical absorbents, including carpets and other furnishings, are usually readily available from manufacturers. The coefficients are normally obtained from laboratory

TABLE 17-7 STC of Various Constructions

Construction	STC
¼-in. plate glass	26
¾-in. plywood	28
½-in. gypsum board, both sides of 2 × 4 studs	33
¼-in. steel plate	36
6-in. concrete block wall	42
8-in. reinforced-concrete wall	51
12-in. concrete block wall	53
Cavity wall, 6-in. concrete block, 2-in. air space, 6-in. concrete block	56

TABLE 17-8 STC Performance Groups

30	31	*32*	33	34
35	36	*37*	38	39
40	41	*42*	43	44
45	46	*47*	48	49
50	51	*52*	53	54
55	56	*57*	58	59

TABLE 17-9 IIC of Various Floor Constructions

Construction	IIC
Oak flooring on ½-in. plywood subfloor, 2 × 10 joists, ½-in. gypsum board ceiling	23
With carpet and pad	48
8-in. concrete slab	35
With carpet and pad	57
2½-in. concrete on light metal forms, steel bar joists	27
With carpet and pad	50

TABLE 17-10 Performance of Commonly Used Sound Absorbents

Absorbent	Thickness, in.	Density, lb per cu ft	Noise reduction coefficient
Mineral or glass fiber blankets	½-4	½-6	0.45-0.95
Molded or felted tiles, panels, and boards	½-1⅛	8-25	0.45-0.90
Plasters (porous)	⅜-¾	20-30	0.25-0.40
Sprayed-on fibers and binders	⅜-1⅛	15-30	0.25-0.75
Foamed, open-cell plastics, elastomers, etc	½-2	1-3	0.35-0.90
Carpets	Varies with weave, texture, backing, pad, etc.		0.30-0.60
Draperies	Varies with weave, texture, weight, fullness		0.10-0.60

Absorbent	Absorption coefficient per sq ft of floor area at frequencies, Hz:				
	125	250	500	1,000	2,000	4,000
Seated audience	0.60	0.75	0.85	0.95	0.95	0.85
Unoccupied upholstered (fabric) seats	0.50	0.65	0.80	0.90	0.80	0.70

tests of panel specimens of about 72- to 80-ft^2 area, tested according to ASTM C423. It is imperative that the test specimens be as nearly identical as possible, in construction system and detail, with actual field installations, since construction details enormously affect acoustical performance.

Table 17-10 lists performance ranges of typical absorbent materials. For specific data on specific materials or systems, always refer to specific tests by accredited test agencies.

The sound transmission loss through an acoustical ceiling (up and over a barrier) when the ceiling is used as a continuous membrane is often an important rating. As might be expected, the effectiveness of acoustical absorbents as acoustical barriers is limited, and supplementary barriers or isolation are frequently required in normal construction.

Other Acoustical Materials. Performance ratings of damping materials, duct linings, vibration isolation materials and devices, etc., are available from various sources. Use of such materials is somewhat complex and specialized.

17-17. Acoustical Criteria.

Increasingly, acoustical performance criteria and environmental specifications are becoming a part of building contract documents. Governmental agencies, lending institutions, owners, and tenants frequently require objective standards of performance.

To a large degree, acoustical criteria are subjective or are based on subjective response to acoustic parameters. This complicates attempts to provide objective specifications, but long experience has permitted acoustical experts to determine broad classes or ranges of criteria and standards that will produce satisfaction in most instances.

It is important to remember that one-point differences are normally insignificant in acoustical criteria. Usually, a tolerance of $\pm 2\frac{1}{2}$ points from a numerical value is acceptable in practice.

Tables 17-11 to 17-13 list some of the more common criteria for ordinary building spaces. They should, however, be used only as a guide; always attempt to obtain specific data or requirements whenever possible.

Acceptable Background Noise Levels. Steady, constant, unobtrusive sound levels that normally occur in typical rooms and that are acceptable as background noise are indicated in Table 17-11. For specific applications and in acoustically critical spaces, specific requirements should always be determined.

[*Note:* Levels are given in dBA, easily obtained numbers with simple measuring equipment (Art. 17-7). When noise criterion NC values are specified, they can be determined by subtracting about 7 to 10 points from dBA values; that is, dBA 50 = NC 40 to NC 43.]

Use of electronically generated *masking sound*, usually by means of speakers located above the ceiling, has become commonplace in open-plan spaces as a means of establishing a uniform background level. However, it is an expensive and complex procedure and should be handled only by acoustics experts.

Sound-Transmission-Loss Requirements. Acoustical performance requirements of sound barriers separating various occupancies may vary widely, depending on the particular needs of the particular occupants. Typical requirements for common occupancies in normal buildings are shown in Table 17-12.

In highly critical spaces, or where a large number of spaces of identical use are involved (as in a large hotel or multiple-dwelling building), it is always advisable to obtain expert acoustical advice.

Impact Isolation Requirements. Because the only standard test method available is controversial, impact isolation performance specifications are only broad, general suggestions, at best. The values in Table 17-13, however, are reasonably safe and economically obtainable with available construction systems.

TABLE 17-11 Typical Acceptable Background Levels

Space	Background Level, dBA
Recording studio	25
Suburban bedroom	30
Theater	30
Church	35
Classroom	35
Private office	40
General office	50
Dining room	55
Computer room	70

TABLE 17-12 Sound Isolation Requirements between Rooms

Room	Between and Adjacent area	Sound isolation requirement, STC
Hotel bedroom	Hotel bedroom	47
Hotel bedroom	Corridor	47
Hotel bedroom	Exterior	42
Normal office	Normal office	33
Executive office	Executive office	42
Bedroom	Mechanical room	52
Classroom	Classroom	37
Classroom	Corridor	33
Theater	Classroom	52
Theater	Music rehearsal	57

TABLE 17-13 Impact Isolation Requirements between Rooms

Room	Between and Room below	Impact isolation requirement, IIC
Hotel bedroom	Hotel bedroom	55
Public spaces	Hotel bedroom	60
Classroom	Classroom	47
Music room	Classroom	55
Music room	Theater	62
Office	Office	47

Acoustical Absorption Requirements. It is ironic that requirements for the most widely used acoustical materials are the least well defined. So-called optimum reverberation time requirements (Fig. 17-12) are a reasonably safe specification for most rooms, but many additional requirements are often involved.

In general, where noise control is the significant requirement, the equivalent absorption of a full ceiling of acoustical tile providing a noise-reduction coefficient of about 0.65 to 0.70 is adequate. This absorption may be provided by carpet, furnishings, or other

TABLE 17-14 Available Options in Noise Control

Objectives of sound-control efforts	Sound-control procedures*					
	Quiet the source†	Barriers or enclosures	Vibration isolation or damping	Absorption	Masking	Personal protection
Reduce the general noise level to:						
Improve communication	X		X	X‡		
Increase comfort	X		X	X‡		
Reduce risk of hearing damage	X		X	X‡		
Reduce extraneous, intruding noise to:						
Increase privacy		X‡			X	
Increase comfort		X‡	X			
Improve communication		X‡	X			
Protect many persons against localized source producing damaging levels	X	X‡	X	X		X
Protect many persons against many distributed sources producing damaging levels	X	X	X	X		X‡
Protect one person against localized source producing damaging levels	X	X§	X			X‡
Protect a few persons against many distributed sources producing damaging levels		X§				X‡
Eliminate echoes and flutter				X¶		
Reduce reverberation				X‡		
Eliminate annoying vibration			X‡			

* No mention has been made of another option, reinforcement, because its purpose is to increase levels. It is the only available option when the signal level must be increased.
† Always the best and simplest means of eliminating noise, if practical and economical.
‡ Indicates the most likely solution(s).
§ A closed booth or small room for the person(s) is often feasible.
¶ Assumes that the configuration of the reflecting surface(s) cannot be modified.

materials, as well as by acoustical tile. In many instances, however, no absorption at all should be applied to the ceiling of a room, because the ceiling may be a necessary sound reflector.

For important projects, always obtain the advice of competent acoustical consultants.

17-18. Helpful Hints for Noise Control.

A building encloses a myriad of activities and the equipment used in those activities. It is imperative that designers and builders consider the control of noise and vibration associated with such activities. While it is not practical to give here detailed solutions to the many potential acoustical problems that may arise in the design of even one building, the more common problems encountered and the most likely solutions to such problems are indicated in Tables 17-14 and 17-15.

TABLE 17-15 Typical Acoustical Problems and Likely Solutions

Problem	Possible causes	Solution
"It's so noisy, I can't hear myself think"	High noise levels Excessive reverberation Excessive transmission Excessive vibration Focusing effects	Absorption Absorption Sound isolation Vibration isolation Eliminate cause of focusing effects
"Speech and music are fuzzy, indistinct"	Excessive reverberation	Absorption
"Little sounds are most distracting"	Background level too low Room too "dead"	Masking Optimum reverberation
"There's an annoying echo"	Echo Flutter Focusing effects Excessive reverberation	Proper room shape Proper room shape Eliminate cause of focusing effects Absorption
"I can hear everything the fellow across the office says"	Room too "dead" Background level too low Focusing or reflection	Optimum reverberation Masking Eliminate focusing or reflection
"It doesn't sound natural in here"	Flutter Distortion caused by improper sound system Selective absorption Room too "dead", low reverberation time	Alter room shape Add absorption Proper sound system Proper type and amount of absorption
"It feels oppressive"	Reverberation time too low Background sound level too low	Proper amount of absorption Use of background and masking sound
"It's not loud enough at the rear of the room"	Room too large Improper shape Lack of reflecting surfaces Poor distribution Too much absorption	Electronic amplification Alter room shape Add reflecting surfaces Eliminate absorption on surfaces needed for reflection
"There are dead spots in the room"	Poor distribution Improper shape Echoes	Reflecting surfaces Eliminate cause of focusing effects Alter room shape
"Sound comes right through the walls of these offices"	Sound leaks Sound transmission Vibration Receiving room too quiet Poor room location	Eliminate leaks Sound isolation Vibration isolation Masking Proper room layout
"I can hear machine noises and people walking around upstairs"	Vibration Sound transmission Poor room location	Vibration isolation Sound isolation Proper room layout
"Outside noises drive me crazy"	Poor room location Sound leaks Sound transmission	Proper acoustical environment Eliminate leaks Sound isolation

Source-Path-Receiver. Sound originates at a source and travels via a path to a **receiver.** Sound control consists of modifying or treating any or all of these three elements in some manner.

The most effective control measures often involve eliminating noise at the source. For example:

Balancing moving parts, lubricating bearings, improving aerodynamics of duct systems, etc.
Modifying parts or processes.
Changing to a different, less noisy process.

The most common sound control measures usually involve acoustical treatment to absorb sound; but equally important are:

Use of barriers to prevent airborne sound transmission.
Interruption of the path with carefully designed discontinuities.
Use of damping materials to minimize radiation from surfaces.

Another important approach involves reinforcing the direct sound with controlled reflections from properly designed reflective surfaces.

Often, the most simple and effective approach involves protecting the receiver, enclosing the person within adequate barriers, or equipping the person with personal protection devices (ear plugs or muffs), rather than trying to enclose or modify huge sources or an entire room or building.

Table 17-14 indicates the usual options available in noise control, and the most likely solutions to problems.

In existing buildings, identifying problems, determining their causes, and deciding on the most economical solutions are frequently the tasks of designers and builders. Table 17-15 indicates typical acoustical problems, their possible causes, and likely solutions to the problems.

17-19. Bibliography.

L. L. Beranek, "Noise and Vibration Control," McGraw-Hill Book Company, New York.

P. D. Close, "Sound Control and Thermal Insulation of Buildings," Van Nostrand Reinhold Company, New York.

M. D. Egan, "Concepts in Architectural Acoustics," McGraw-Hill Book Company, New York.

C. M. Harris, "Handbook of Noise Control," McGraw-Hill Book Company, New York.

P. M. Morse and K. U. Ingard, "Theoretical Acoustics," McGraw-Hill Book Company, New York.

L. F. Yerges, "Sound, Noise and Vibration Control," Van Nostrand Reinhold Company, New York

Section 18

Heating, Ventilation, and Air Conditioning

Frank C. Yanocha
Chief Mechanical Engineer, HVAC
Sanders and Thomas, Inc.
Pottstown, Pa.

The necessity of heating, ventilation, and air-conditioning (HVAC) control of environmental conditions within buildings has been well established over the years as being highly desirable for various types of occupancy and comfort conditions as well as for many industrial manufacturing processes. In fact, without HVAC systems, many manufactured products produced by industry that are literally taken for granted would not be available today.

Significance of Design Criteria. The desired performance of any HVAC system, whether designed for human comfort or industrial production or industrial process requirements are significantly related to the development of appropriate and accurate design criteria.

Some of the more common items that are generally considered are as follows:

1. Outside design temperatures:
   Winter and summer
   Dry bulb (DB), wet bulb (WB)
2. Inside design temperatures:
   Winter: heating °F DB and relative humidity
   Summer: cooling °F DB and relative humidity
3. Filtration efficiency of supply air
4. Ventilation requirements
5. Exhaust requirements
6. Humidification
7. Dehumidification
8. Air-change rates
9. Positive-pressure areas
10. Negative-pressure areas

11. Balanced-pressure areas
12. Contaminated exhausts
13. Chemical exhausts and fume hoods
14. Energy conservation devices
15. Economizer system
16. Enthalpy control system
17. Infiltration
18. Exfiltration

Outline of Design Procedure. Design of an HVAC system is not a simple task. The procedure varies considerably from one application or project to another, and important considerations for one project may have little impact on another. But for all projects, to some extent, the following major steps have to be taken:

1. Determine all applicable design conditions, such as inside and outside temperature and humidity conditions for winter and summer conditions, including prevailing winds and speeds.

2. Determine all particular and peculiar interior space conditions that will be maintained.

3. Estimate, for every space, heating or cooling loads from adjacent unheated or uncooled spaces.

4. Carefully check architectural drawings for all building materials used for walls, roofs, floors, ceilings, doors, etc. and determine the necessary thermal coefficients for each.

5. Establish values for air infiltration and exfiltration quantities, for use in determining heat losses and heat gains.

6. Determine ventilation quantities and corresponding loads for heat losses and heat gains.

7. Determine heat or cooling loads due to internal machinery, equipment, lights, motors, etc.

8. Include allowance for effects of solar load.

9. Total the heat losses requiring heating of spaces and heat gains requiring cooling of spaces, to determine equipment capacities.

18-1. Definitions of Terms for Heating, Ventilation, and Air Conditioning (HVAC).

Adiabatic Process. A thermodynamic process that takes place without any heat being added or subtracted and at constant total heat.

Air, Makeup. New, or fresh, air brought into a building to replace losses due to exfiltration and exhausts, such as those from ventilation and chemical hoods.

Air, Return (Recirculated). Air that leaves a conditioned space and is returned to the conditioning equipment for treatment.

Air, Saturated. Air that is fully saturated with water vapor (100% humidity), with the air and water vapor at the same temperature.

Air, Standard. Air at 70°F and standard atmospheric pressure (29.92 in of mercury) and weighing about 0.075 lb/ft^3.

Air Change. The complete replacement of room air volume with new supply air.

Air Conditioning. The process of altering air supply to control simultaneously its humidity, temperature, cleanliness, and distribution to meet specific criteria for a space. Air conditioning may either increase or decrease the space temperature.

Air Conditioning, Comfort. Use of air conditioning solely for human comfort, as compared with conditioning for industrial processes or manufacturing.

Air Conditioning, Industrial. Use of air conditioning in industrial plants where the prime objective is enhancement of a manufacturing process rather than human comfort.

Baseboard Radiation. A heat-surface device, such as a finned tube with a decorative cover, used in lieu of a room's baseboard trim.

Blow. Horizontal distance from a supply-air discharge register to a point at which the supply-air velocity reduces to 50 ft/min.

Boiler. A cast-iron or steel container fired with solid, liquid, or gaseous fuels to generate hot water or steam for use in heating a building through an appropriate distribution system.

Boiler-Burner Unit. A boiler with a matching burner whose heat-release capacity equals the boiler heating capacity less certain losses.

Boiler Heating Surface. The interior heating surface of a boiler subject to heat on one side and transmitting heat to air or hot water on the other side.

Boiler Horsepower. The energy required to evaporate 34.5 lb/hr of water at 212°F, equivalent to 33,475 Btu/h.

Booster Water Pump. In hot-water heating systems, the circulating pump used to move the heating medium through the piping system.

British Thermal Unit (Btu). Quantity of heat required to raise the temperature of 1 lb of water 1°F at or near 39.2°F, which is its temperature of maximum density.

Central Heating or Cooling Plant. One large heating or cooling unit used to heat or cool many rooms, spaces or zones or several buildings, as compared to individual room, zone, or building units.

Coefficient of Performance. For machinery and heat pumps, the ratio of the effect produced to the total power of electrical input consumed.

Comfort Zone. An area plotted on a psychrometric chart to indicate a combination of temperatures and humidities at which, in controlled tests, more than 50% of the persons were comfortable.

Condensate. Liquid formed by the condensation of steam or water vapor.

Condensers. Special equipment used in air conditioning to liquefy a gas.

Condensing Unit. A complete refrigerating system in one assembly, including the refrigerant compressor, motor, condenser, receiver, and other necessary accessories.

Conductance, Thermal C. Rate of heat flow across a unit area (usually 1 ft^2) from one surface to the opposite surface under steady-state conditions with a unit temperature difference between the two surfaces.

Conduction, Thermal. A process in which heat energy is transferred through matter by transmission of kinetic energy from particle to particle, the heat flowing from hot points to cooler ones.

Conductivity, Thermal. Quantity of heat energy, usually in Btu, that is transmitted through a substance per unit of time (usually 1 hr) from a unit area (usually 1 ft^2) of surface to an opposite unit surface per unit of thickness (usually 1 in) under a unit temperature difference (usually 1°F) between the surfaces.

Convection. A means of transferring heat in air by natural movement, usually a rotary or circulatory motion caused by warm air rising and cooler air falling.

Cooling. A heat-removal process usually accomplished with air-conditioning equipment.

Cooling, Evaporative. Cooling effect produced by evaporation of water, the required heat for the process being taken from the air. (This method is widely used in dry climates with low wet-bulb temperatures.)

Cooling, Sensible. Cooling of a unit volume of air by a reduction in temperature only.

Cooling Effect, Total. The difference in total heat in an airstream entering and leaving a refrigerant evaporator or cooling coil.

Cooling Unit, Self-Contained. A complete air-conditioning assembly consisting of a compressor, evaporator, condenser, fan motor, and air filter ready for plug-in to an electric power supply.

Cooling Tower. A mechanical device used to cool water by evaporation in the outside air. Towers may be atmospheric or induced- or powered-draft type.

Damper. A plate-type device used to regulate flow of air or gas in a pipe or duct.

Defrosting. A process used for removing ice from a refrigerant coil.

Degree Day. The product of 1 day (24 hr) and the number of degrees Fahrenheit the daily mean temperature is below 65°F. It is frequently used to determine heating-load efficiency and fuel consumption.

Dehumidification. In air conditioning, the removal of water vapor from supply air by condensation of water vapor on the cold surface of a cooling coil.

Direct Expansion. A means of air conditioning that uses the concept of refrigerant expansion (through a thermostatic expansion valve) in a refrigerant coil to produce a cooling effect.

Ductwork. An arrangement of sheet-metal ducts to distribute supply air, return air, and exhaust air.

Efficiency. Ratio of power output to power input. It does not include considerations of load factor or coefficient of performance.

Emissivity. Ratio of radiant energy that is emitted by a body to that emitted by a perfect black body. An emissivity of 1 is assigned to a perfect black body. A perfect reflector is assigned an emissivity of 0.

Enthalphy. A measure of the total heat (sensible and latent) in a substance and which is equal to its internal energy and its capacity to do work.

Entropy. The ratio of the heat added to a material or substance to the absolute temperature at which the heat is added.

Environment. A general term used to define the many conditions of atmosphere, water, and soil including acceptable or unacceptable conditions.

Evaporator. A cooling coil in a refrigeration system in which the refrigerant is evaporated and absorbs heat from the surrounding fluid (airstream).

Exfiltration. Unintentional loss of conditioned supply air by leakage from ductwork, rooms, spaces, etc., that is to be considered a load on the air-conditioning system.

Film Coefficient (Surface Coefficient). Heat transferred from a surface to air or other fluid by convection per unit area of surface per degree temperature difference between the surface and the fluid.

Furnace, Warm-Air. Heating system that uses a direct- or indirect-fired boiler to produce warm air for heating.

Grille. A metal covering, usually decorative, with openings through which supply or return air passes.

Head. Pressure expressed in inches or feet of water. A head of 12 in, or 1 ft, of water is the pressure from a column of water 12 in, or 1 ft, high. See also Inch of Water below.

Heat, Latent. Heat associated with the change of state (phase) of a substance, for example, from a solid to a liquid (ice to water) or from a liquid to a gas (water to steam vapor).

Heat, Sensible. Heat associated with a change in temperature of a substance.

Heat, Specific. Ratio of the thermal capacity of a substance to the thermal capacity of water.

Heat, Total. Sum of the sensible and latent heat in a substance above an arbitrary datum, usually 32°F or 0°F.

Heat Capacity. Heat energy required to change the temperature of a specific quantity of material 1°.

Heat Pump. A refrigerant system used for heating and cooling purposes.

Heat Transmission Coefficient. Quantity of heat (usually Btu in the United States) transmitted from one substance to another per unit of time (usually 1 hr) through one unit of surface (usually 1 ft^2) of building material per unit of temperature difference (usually 1°F).

Heater, Direct-Fired. A heater that utilizes a flame within a combustion chamber to heat the walls of the chamber and transfers the heat from the walls to air for space heating, as in a warm-air heater.

Heater, Unit. A steam or hot-water heating coil, with a blower or fan and motor, used for space heating.

Heating. The process of transferring heat from a heat source to a space in a building.

Heating, District. A large, central heating facility that provides heat from steam or hot water to a large number of buildings under different ownership.

Heating, Radiant. Heating by ceiling or wall panels, or both, with surface temperatures higher than that of the human body in such a manner that the heat loss from occupants of the space by radiation is controlled.

Heating, Warm-Air. A heating system that uses warm air, rather than steam or hot water, as the heating medium.

Heating Surface. Actual surface used for transferring heat in a boiler, furnace, or heat exchanger.

Heating System, Automatic. A complete heating system with automatic controls to permit operation without manual controls or human attention.

Heating System, Hot-Water. A heating system that utilizes water at temperatures of about 200°F.

Humidity. Water vapor mixed with dry air.

Humidity, Absolute. Weight of water vapor per unit volume of a vapor-air mixture. It is usually expressed in grains/ft^3 or lb/ft^3.

Humidity, Percent. Ratio of humidity in a volume of air to the maximum amount of water vapor that the air can hold at a given temperature, expressed as a percentage.

Humidity, Relative (RH). Ratio of the vapor pressure in a mixture of air and water vapor to the vapor pressure of the air when saturated at the same temperature.

Humidity, Specific (Humidity Ratio). Ratio of the weight of water vapor, grains, or pounds, per pound of dry air, at a specific temperature.

Hygrometer. A mechanical device used to measure the moisture content of air.

Hygroscopic. Denoting any material that readily absorbs moisture and retains it.

Hygrostat. A mechanical device that is sensitive to changes in humidity and used to actuate other mechanical devices when predetermined limits of humidity are reached.

Inch of Water. A unit of pressure intensity applied to low-pressure systems, such as air-conditioning ducts. It is equivalent to 0.036136 psi.

Infiltration. Leakage into an air-conditioned area of outside air (usually unwanted), which becomes a load on the air-conditioning system.

Insulation, Thermal. Any material that slows down the rate of heat transfer (offers thermal resistance) and effects a reduction of heat loss.

Louvers. An arrangement of metal blades to provide air slots that will permit passage of air and exclude rain or snow.

MBH. 1000 Btu/hr (Btu/h).

Micron. 0.001 mm. It is frequently used to designate particle sizes of dust and the efficiency of filtration by air-conditioning filters.

Modulating. Process of making incremental adjustments, usually by an automatic device operating a valve or damper motor.

Pressure, Absolute. Pressure above an absolute vacuum. Absolute pressure equals the sum of gage and barometric pressures.

Pressure, Atmospheric. Air pressure indicated by a barometer. The standard atmospheric pressure is 760 mm (29.92 in) of mercury, or 14.696 psi.

Pressure, Head. Condensing pressure, often considered as the refrigerant compressor-discharge pressure.

Pressure, Saturation. The pressure that corresponds to a specific temperature that will permit simultaneous condensation and evaporation.

Pressure, Suction. The pressure in the suction line of a refrigeration system.

Pressure Head. See Head above.

Psychrometry. A branch of physics that concerns itself with the measurement and determination of atmospheric conditions, with particular emphasis on moisture mixed in the air.

Psychrometer. A mechanical device utilizing a wet-bulb and dry-bulb thermometer and used to determine the humidity in an air-water vapor mixture, such as room air.

Psychrometric Chart. A chart used in air-conditioning design and analysis that indicates various properties of an air-water vapor mixture along with various relevant mathematical values.

Radiation. Transfer of energy in wave form, from a hot body to a colder body, independent of any matter between the two bodies.

Radiation, Equivalent Direct. Rate of steam condensation at 240 Btu/(h)(ft^2) of radiator surface.

Refrigerant. A substance that will accept large quantities of heat, that will cause boiling and vaporization at certain temperatures, and that can be utilized in air-conditioning systems.

Resistance, Thermal. The thermal quality of a material that resists passage of heat. Also, the opposite of conductance.

Resistivity, Thermal. The reciprocal of conductivity.

Split System. A separation of air-conditioning components, such as location of an air-blower-evaporator coil far from the compressor-condenser unit.

Steam. Water in gas or vapor form.

Steam Trap. A mechanical device that allows water and air to pass but prevents passage of steam.

Subcooling. Cooling at constant pressure of a refrigerant liquid to below its condensing temperature.

Suction Line. The low-temperature, low-pressure refrigerant pipe from an evaporator to a refrigerant compressor.

Sun Effect. The effect or quantity of heat from the sun that tends to increase the internal temperature of a space or building.

Temperature, Absolute. Temperature measured on a scale for which zero is set at $-273.16°C$, or $-459.69°F$ (presumably the temperature at which all molecular motion stops in a gas under constant pressure). The scale is called Kelvin, and $1°K = 1°C = 9/5°F$.

Temperature, Design. An arbitrary design criterion used to determine equipment size to produce air conditioning, heating, or cooling capable of maintaining the designated temperature.

Temperature, Dew Point. Temperature of air at which its wet-bulb temperature and dry-bulb temperature are identical and the air is fully saturated with moisture. Condensation of water vapor begins at this temperature and will continue if the temperature is reduced further.

Temperature, Dry-Bulb. Temperature measured by a conventional thermometer. It is used to determine the sensible heat in air.

Temperature, Effective. A single or arbitrary index that combines into a single value the effects of temperature, humidity, and air motion on the sensation of comfort. This value is that of the temperature of still, saturated air that will induce an identical feeling of comfort.

Temperature, Wet-Bulb. Air temperature as indicated by a thermometer with a wet bulb. This temperature is less than the dry-bulb temperature, except when the air is fully saturated with water vapor, or at 100% relative humidity, when wet-bulb and dry-bulb temperatures will be equal.

Ton, Refrigeration. Refrigeration effect equivalent to 200 Btu/min, or 12,000 Btu/h.

Vapor. The gaseous state of water and other liquid substances.

Vapor Barrier. An impervious material used to prevent the passage of water vapor through the walls or roofs of a building and to prevent the subsequent condensation within those spaces.

Velocity Pressure. The pressure caused by a moving airstream, composed of both velocity pressure and static pressure.

Ventilation. The process of supplying air to any space within a building without noticeable odors and without objectionable levels of contaminants, such as dusts and harmful gases, and of removing stale, polluted air from the space. Outside air is generally used as an acceptable source of ventilation air.

Ventilator, Unit. A type of unit heater with various modes of operation and degrees or percentages of outside air (frequently used for heating classrooms).

Volume, Specific. Volume, ft^3/lb, occupied by a unit weight of air.

Water, Makeup. Generally the water supplied to a cooling tower to replace the cooling water lost by evaporation or bleedoff.

Water Vapor. A psychrometric term used to denote the water in air (actually low-pressure, superheated steam) that has been evaporated into the air at a temperature corresponding to the boiling temperature of water at that very low pressure.

18-2. Heat and Humidity.

Man has always struggled with the problem of being comfortable in his environment. First attempts were to use fire directly to provide heat through cold winters. It was only in recent times that man's interest and technology permitted development of greater understanding of heat and heating, and substantial improvements in comfort were made. Comfort heating now is a highly developed science and, in conjunction with air conditioning, provides comfort conditions in all seasons in all parts of the world.

As more was learned about humidity and the capacity of the air to contain various amounts of water vapor, greater achievements in environmental control were made. Control of humidity in buildings now is a very important part of heating, ventilation, and air conditioning, and in many cases is extremely important in meeting manufacturing requirements. Today, it is possible to alter the atmosphere or environment in buildings in any manner, to suit any particular need, with great precision and control.

Thermometers and Scales. Energy in the form of heat is transferred from one material or substance to another because of a temperature difference that exists between them. When heat is applied to a material or substance, there will be an increase in average velocity of its molecules or electrons, with an increase in their kinetic energy. Likewise, as heat is removed, there will be a decrease in the average molecular velocity and, therefore, also the electron or molecular kinetic energy.

A thermometer is used to measure the degree of heat in a substance or material. The thermometer includes an appropriate graduated scale to indicate the change in temperature of the substance. The change in temperature as read on a thermometer is a measure of heat transferred to or from the substance. A unit of temperature is called a degree and is equivalent to one graduation on the scale.

By convention, the scale is an interval scale. The Celsius thermometer is a metric system of measuring temperature; 0°C is assigned to the temperature at which water freezes and 100°C to the temperature at which water boils at normal atmospheric conditions. Hence, on a Celsius thermometer, there are 100 intervals or graduations, called degrees, between the freezing and boiling temperatures. Each interval or degree is called 1 Celsius degree.

In the Fahrenheit system, 32°F is used to designate the freezing temperature of water and 212°F the boiling temperature at normal atmospheric pressure. Hence, on the Fahrenheit scale, a degree is equal to 1/180 of the distance on the scale between the freezing and boiling temperatures. Conversion formulas used for each scale are as follows:

$$°F = 1.8 \times °C + 32 \qquad (18\text{-}1)$$
$$°C = \tfrac{5}{9}(°F - 32) \qquad (18\text{-}2)$$

Thermal Capacity and Specific Heat. The thermal capacity of a substance is indicated by the quantity of heat required to raise the temperature of 1 lb of the substance 1°F. In HVAC calculations, thermal capacity is usually expressed by the British thermal unit (Btu).

One Btu is the amount of heat that is required to increase the temperature of 1 lb of water 1°F at or near 39.2°F, which is the temperature at which water has its maximum density. Conversely, if 1 Btu is removed from 1 lb of water, its temperature will be reduced by 1°F.

Various quantities of heat will produce changes of 1°F per pound of substances other than water. Thus, thermal capacity is entirely dependent on the specific heat of the substances.

The specific heat of a substance is the ratio of the heat content or thermal capacity of a substance to that of water. And by definition, the specific heat of water is unity.

It is customary in HVAC calculations to use specific heat in lieu of thermal capacity, because of the convenience of using the Btu as a unit of heat quantity without conversions. Specific heats of air and some common building materials are shown in Table 18-1. Data for other substances may be obtained from tables in the "Handbook of Fundamentals," American Society of Heating, Refrigeration and Air Conditioning Engineers (ASHRAE). An examination of Table 18-1 indicates that the specific heat of these

TABLE 18-1. Specific Heats—Common Materials

Substance	Specific heat, Btu/(lb)(°F)
Air at 80°F	0.24
Water vapor	0.49
Water	1.00
Aluminum	0.23
Brick	0.20
Brass	0.09
Bronze	0.10
Gypsum	0.26
Ice	0.48
Limestone	0.22
Marble	0.21
Sand	0.19
Steel	0.12
Wood	0.45–0.65

materials is less than unity and that, of all common substances, water possesses the largest specific heat and the largest thermal capacity.

Sensible Heat. When heat energy is added to or taken away from a substance, the resulting changes in temperature can be detected by the sense of touch, or sensibly. Therefore, this type of heat is called sensible heat. Since sensible heat is associated with a change in temperature, the quantity of sensible heat energy transferred in a heat exchange is usually calculated from

$$Q = Mc(t_2 - t_1) \tag{18-3}$$

where Q = sensible heat, Btu, absorbed or removed
M = mass, lb, of the substance undergoing the temperature change
c = specific heat of the substance
$(t_2 - t_1)$ = temperature difference of the substance, where t_2 is the final temperature after the heat exchange and t_1 is the temperature of the material before the heat exchange

Laws of Thermodynamics. The application of the laws of thermodynamics to HVAC calculations are usually limited to two well-known laws. These laws can be expressed differently, but in equivalent ways. A simplification of these laws as follows will permit an easier understanding.

The first law of thermodynamics states that when work is performed, heat must also be produced, and the quantity of the heat produced is proportional to the work performed. And conversely, when heat energy is utilized to perform work, the quantity of the heat dissipated is proportional to the work performed. Work, ft-lb, is equal to the product of the force, lb, acting on the body for a distance, ft, that the body moves in the direction of the applied force.

Hence, this first law of thermodynamics can be expressed mathematically by the following equation:

$$W = JQ \tag{18-4}$$

where W = work, ft-lb
J = Joule's constant = mechanical equivalent of heat
Q = heat, Btu, generated by the work

Experiments have shown that the mechanical equivalent of heat, known as Joule's constant, is equivalent to 778 ft-lb/Btu. The first law is also known as the law of conservation of energy.

The second law of thermodynamics states that is is impossible for any machine to transfer heat from a substance to another substance at a higher temperature (if the machine is unaided by an external agency). This law can be interpreted to imply that the available supply of energy for doing work in our universe is constantly decreasing. It also implies that any effort to devise a machine to convert a specific quantity of heat into an equivalent amount of work is futile.

Entropy is the ratio of the heat added to a substance to the absolute temperature at which the heat is added.

$$S = \frac{dQ}{T_a} \tag{18-5}$$

where S = entropy
dQ = differential of heat (very small change)
T_a = absolute temperature

The second law of thermodynamics can be expressed mathematically with the use of the entropy concept.

18-10 Heating, Ventilation, and Air Conditioning

Suppose an engine, which will convert heat into useful mechanical work, receives heat Q_1 from a heat source at temperature T_1 and delivers heat Q_2 at a temperature T_2 to a heat sink after performing work. By the first law of thermodynamics, the law of conservation of energy, Q_2 is less than Q_1 by the amount of work performed. And by the second law of thermodynamics, T_2 is less than T_1. The universe at the start of the process loses entropy $\Delta S_1 = Q_1/T_1$ and at the end of the process gains entropy $\Delta S_2 = Q_2/T_2$. Hence, the net change in the entropy of the universe because of this process will be $\Delta S_2 - \Delta S_1$.

Furthermore, this law requires that this net change must always be greater than zero and that the entropy increase is and must always be an irreversible thermodynamic process.

$$\Delta S_2 - \Delta S_1 > 0 \qquad (18\text{-}6)$$

Because of the irreversibility of the process, the energy that has become available for performing work is

$$Q_u = T_2(\Delta S_2 - \Delta S_1) \qquad (18\text{-}7)$$

Absolute Temperature. The definition of entropy given above involves the concept of absolute temperature measured on a ratio scale. The unit of absolute temperature is measured in degrees Kelvin (°K) in the Celsius system and in degrees Rankine (°R) in the Fahrenheit system. Absolute zero or zero degrees in either system is determined by considering the theoretical behavior of an ideal gas, and for such a gas,

$$P_a V = mRT_a \qquad \text{or} \qquad P_a v = RT_a \qquad (18\text{-}8)$$

where P_a = absolute pressure on the gas, psf
V = volume of the gas, ft^3
v = specific volume of the gas, ft^3/lb = reciprocal of the gas density
m = mass of the gas, lb
T_a = absolute temperature
R = universal gas constant

For a gas under constant pressure, the absolute temperature theoretically will be zero when the volume is zero and all molecular motion ceases. Under these conditions, the absolute zero temperature has been determined to be nearly $-273°C$ and $-460°F$. Therefore,

$$\text{Kelvin temperature } °K = \text{Celsius temperature } + 273° \qquad (18\text{-}9)$$
$$\text{Rankine temperature } °R = \text{Fahrenheit temperature } + 460° \qquad (18\text{-}10)$$

In the Rankine system, the universal gas constant R equals 1545.3 divided by the molecular weight of the gas. For air, $R = 53.4$, and for water vapor, $R = 85.8$.

Latent Heat. The sensible heat of a substance is associated with a sensible change in temperature. In contrast, the latent heat of a substance is always involved with a change in state of a substance, such as from ice to water and from water to steam or water vapor. Latent heat is very important in HVAC calculations and design, because the total heat content of air almost always contains some water in the form of vapor. The concept of latent heat may be clarified by consideration of the changes of state of water.

When heat is added to ice, the temperature rises until the ice reaches its melting point. Then, the ice continues to absorb heat *without a change in temperature* until a required amount of heat is absorbed per pound of ice, at which point it begins melting to form liquid water. The reverse is also true: if the liquid is cooled to the freezing point, this same quantity of heat must be removed to cause the liquid water to change to the solid (ice) state. This heat is called the **latent heat of fusion** for water. It is equal to 144 Btu and will convert 1 lb of ice at 32°F to 1 lb of water at 32°F. Thus,

$$\text{Latent heat of fusion for water} = 144 \text{ Btu/lb} \qquad (18\text{-}11)$$

If the pound of water is heated further, say to 212°F, then an additional 180 Btu of heat must be added to effect the 180°F sensible change in temperature. At this temperature, any further addition of heat will not increase the temperature of the water beyond 212°F. With the continued application of heat, the water experiences violent agitation, called boiling. The boiling temperature of water is 212°F at atmospheric pressure.

With continued heating, the boiling water absorbs 970 Btu for each pound of water *without a change in temperature* and completely changes its state from liquid at 212°F to water vapor, or steam, at 212°F. Therefore, at 212°F,

$$\text{Latent heat of vaporization of water} = 970 \text{ Btu/lb} \tag{18-12}$$

Conversely, when steam at 212°F is cooled or condensed to a liquid at 212°F, 970 Btu per pound of steam (water) must be removed. This heat removal and change of state is called **condensation**.

When a body of water is permitted to evaporate into the air at normal atmospheric pressure, 29.92 in of mercury, a small portion of the body of water evaporates from the water surface at temperatures below the boiling point. The latent heat of vaporization is supplied by the body of water and the air, and hence both become cooler. The amount of vapor formed and that absorbed by the air above the water surface depends on the capacity of the air to retain water at the existing temperature and the amount of water vapor already in the air.

Table 18-2 lists the latent heat of vaporization of water for various air temperatures and normal atmospheric pressure. More extensive tables of thermodynamic properties of air, water, and steam are given in the "Handbook of Fundamentals," American Society of Heating, Refrigerating and Air Conditioning Engineers (ASHRAE).

Enthalpy. Enthalpy is a measure of the total heat (sensible and latent) in a substance and is equivalent to the sum of its internal energy plus its ability or capacity to perform work, or PV/J, where P is the pressure of the substance, V its volume, and J its mechanical equivalent of heat. Specific enthalpy is the heat per unit of weight, Btu/lb, and is the property used on psychrometric charts and in HVAC calculations.

The specific enthalpy of dry air h_a is taken as zero at 0°F. At higher temperatures, h_a is equal to the product of the specific heat, about 0.24, multiplied by the temperature, °F. (See Table 18-2.)

The specific enthalpy of saturated air h_s, which includes the latent heat of vaporization of the water vapor, is indicated in Table 18-2. The specific enthalpy of the water vapor or moisture at the air temperature may also be obtained from Table 18-2 by subtracting h_a from h_s.

Table 18-2 also lists the humidity ratio of the air at saturation for various temperatures (weight, lb, of water vapor in saturated air per pound of dry air). In addition, the specific enthalpy of saturated water vapor h_g, Btu/lb, is given in Table 18-2 and represents the sum of the latent heat of vaporization and the specific enthalpy of water at various temperatures.

The specific enthalpy of unsaturated air is equal to the sensible heat of dry air at the existing temperature, with the sensible heat at 0°F taken as zero, plus the product of the humidity ratio of the unsaturated air and h_g for the existing temperature.

Cooling by Evaporation. Evaporation of water requires a supply of heat. If there is no external source of heat, and evaporation occurs, then the water itself must provide the necessary heat of vaporization. In other words, a portion of the sensible heat in the liquid will be converted into the latent heat of vaporization. As a result, the temperature of the liquid remaining will drop. Since no external heat is added or removed by this process of evaporation, it is called **adiabatic cooling**.

Human beings are also cooled adiabatically by evaporation of perspiration from skin surfaces. Similarly, in hot climates with relatively dry air, air conditioning is provided

TABLE 18-2. Thermal Properties—Dry and Saturated Air at Atmospheric Pressure

Air temperature, °F	Specific volume, ft³/lb — Dry air	Specific volume, ft³/lb — Saturated air	Pounds of water in saturated air per pound of dry air (humidity ratio)	Latent heat of vaporization of water, Btu/lb	Specific enthalpy of dry air h_a,* Btu/lb	Specific enthalpy of saturated air h_s,† Btu/lb	Specific enthalpy of saturation vapor h_g, Btu/lb
0	11.58	11.59	0.0008		0	0.84	1075.2
32	12.39	12.46	0.0038	1075.2	7.69	11.76	1076.5
35	12.46	12.55	0.0043	1073.5	8.41	13.01	1078.7
40	12.59	12.70	0.0052	1070.6	9.61	15.23	1080.9
45	12.72	12.85	0.0063	1067.8	10.81	17.65	1083.1
50	12.84	13.00	0.0077	1065.0	12.01	20.30	1085.2
55	12.97	13.16	0.0092	1062.2	13.21	23.22	1087.4
60	13.10	13.33	0.0111	1059.3	14.41	26.46	1089.6
65	13.22	13.50	0.0133	1056.5	15.61	30.06	1091.8
70	13.35	13.69	0.0158	1053.7	16.82	34.09	1094.0
75	13.47	13.88	0.0188	1050.9	18.02	38.61	1096.1
80	13.60	14.09	0.0223	1048.1	19.22	43.69	1098.3
85	13.73	14.31	0.0264	1045.2	20.42	49.43	1100.4
90	13.85	14.55	0.0312	1042.4	21.62	55.93	1102.6
95	13.98	14.80	0.0367	1039.6	22.83	63.32	1104.7
100	14.11	15.08	0.0432	1036.7	24.03	71.73	1125.8
150	15.37	20.58	0.2125	1007.8	36.1	275.3	1145.8
200	16.63	77.14	2.295	977.7	48.1	2677	1150.4
212				970.2			

*Enthalpy of dry air is taken as zero for dry air at 0°F.
†Enthalpy of water vapor in saturated air = $h_s - h_a$, including sensible heat above 32°F.

by the evaporation of water into air. And refrigeration is also accomplished by the evaporation of a refrigerant.

Heating by Condensation. Many thermal processes occur without addition or subtraction of heat from the process. Under these conditions, the process is called **adiabatic.**

When a volume of moist air is cooled, a point will be reached at which further cooling cannot occur without reaching a fully saturated condition, that is, 100% saturation or 100% relative humidity. With continued cooling, some of the moisture condenses and appears as a liquid. The temperature at which condensation occurs is called the **dew point temperature.** If no heat is removed by the condensation, then the latent heat of vaporization of the water vapor will be converted to sensible heat in the air, with a resultant rise in temperature.

Thus, an increase in temperature is often accomplished by the formation of fog, and when rain or snow begins to fall, there will usually be an increase in temperature of the air.

Psychrometry. The measurement and determination of atmospheric conditions, particularly relating to the water vapor or moisture content in dry air, is an important branch of physics known as psychrometry. Some psychrometric terms and conditions have already been presented in this article. Many others remain to be considered.

An ideal gas follows certain established laws of physics. The mixture of water vapor and dry air does behave at normal atmospheric temperatures and pressures almost as an ideal gas. As an example, air temperatures, volumes, and pressures may be calculated by use of Eq. (18-8), $Pv = RT$.

Dalton's law also applies. It states:

When two or more gases occupy a common space or container, each gas will fill the volume just as if the other gas or gases were not present. Dalton's law also requires:

1. That each gas in a mixture occupy the same volume or space and also be at the same temperature as each other gas in the mixture.

2. That the total weight of the gases in the mixture equal the sum of the individual weights of the gases.

3. That the pressure of a mixture of several gases equal the sum of the pressures that each gas would exert if it existed alone in the volume enclosing the mixture.

4. That the total enthalpy of the mixture of gases equal the sum of the enthalpies of each gas.

An excellent example of the application of Dalton's law of partial pressures is the use of a liquid barometer to indicate atmospheric pressure. The barometer level indicates the sum of the partial pressure of water vapor and the partial pressure of the air.

Partial pressures of air and water vapor are of great importance in psychrometry and are used to calculate the degree of saturation of the air or relative humidity at a specific dry-bulb temperature.

Relative Humidity and Specific Humidity. Relative humidity is sometimes defined by the use of mole fractions, a difficult definition for psychrometric use. Hence, a more usable definition is desired. For the purpose, relative humidity may be closely determined by the ratio of the partial pressure of the water vapor in the air to the saturation pressure of water vapor at the same temperature, usually expressed as a percentage.

Thus, dry air is indicated as 0% relative humidity and fully saturated air is termed 100% relative humidity.

Computation of relative humidity by use of humidity ratios is also often done, but with somewhat less accuracy. Humidity ratio, or specific humidity W_a, at a specific temperature is the weight, lb, of water vapor in air per pound of dry air. If W_s represents the humidity ratio of saturated air at the same temperature (Table 18-2), then relative humidity can be calculated approximately from the equation

$$RH = \frac{W_a}{W_s} \times 100 \qquad (18\text{-}13)$$

Dalton's law of partial pressures and Eq. (18-6) may also be used to calculate humidity ratios:

$$W_a = 0.622 \frac{p_w}{P - p_w} \qquad (18\text{-}14)$$

where P = barometric pressure, atmospheric, psi
p_w = partial pressure of water vapor, psi

It is difficult to use this equation, however, because of the difficulty in measuring the partial pressures with special scientific equipment that is required and rarely available outside of research laboratories. Therefore, it is common practice to utilize simpler types of equipment in the field that will provide direct readings that can be used to convert into humidity ratios or relative humidity.

A simple and commonly used device is the wet- and dry-bulb thermometer. This device is a packaged assembly consisting of both thermometers and a sock with scales. It is called a **sling psychrometer.** Both thermometers are identical, except that the wet-bulb thermometer is fitted with a wick-type sock over the bulb. The sock is wet with water, and the device is rapidly spun or rotated in the air. As the water in the sock evaporates, a drop in temperature occurs in the remaining water in the sock, and also in the wet-bulb thermometer. When there is no further temperature reduction and the temperature remains constant, the reading is called the wet-bulb temperature. The other thermometer will simultaneously read the dry-bulb temperature.

A difference between the two thermometer readings always exists when the air is less than saturated, at or less than 100% relative humidity. Inspection of a psychrometric chart will indicate that the wet-bulb and dry-bulb temperatures are identical only at fully saturated conditions; that is, at 100% relative humidity. Commercial psychrometers usually include appropriate charts or tables that indicate the relative humidity for a wide range of specific wet- and dry-bulb temperature readings. These tables are also found in books on psychrometry and HVAC books and publications.

Dew-Point Temperatures. Dew is the condensation of water vapor. It is most easily recognized by the presence of droplets in warm weather on grass, trees, automobiles, and many other outdoor things in the early morning. Dew is formed during the night as the air temperature drops, and the air reaches a temperature at which it is saturated with moisture. This is the dew-point temperature. It is also equal to both the wet-bulb temperature and dry-bulb temperature. At the dew-point temperature, the air is fully saturated, that is, at 100% relative humidity. With any further cooling or drop in temperature, condensation begins and continues with any further reduction in temperature. The amount of moisture condensed is the excess moisture that the air cannot hold at saturation at the lowered temperature. The condensation forms drops of water, frequently referred to as *dew.*

Dew-point temperature, thus, is the temperature at which condensation of water vapor begins for any specific condition of humidity and pressure as the air temperature is reduced.

The dew-point temperature can be calculated, when the relative humidity is known, by use of Eq. (18-13) and Table 18-2. For the temperature of the unsaturated air, the humidity ratio at saturation is determined from Table 18-2. The product of the humidity ratio and the relative humidity equals the humidity ratio for the dew-point temperature, which also can be determined from Table 18-2. As an example, to determine the dew-point temperature of air at 90°F and 50% relative humidity, reference to Table 18-2 indicates a humidity ratio at saturation of 0.0312 at 90°F. Multiplication by 0.50 yields a humidity ratio of 0.0156. By interpolation in Table 18-2 between humidity ratios at saturation temperatures of 65 and 70°F, the dew-point temperature is found to be 69.6°F.

A simpler way to determine the dew-point temperature and many other properties of air-vapor mixtures is to use a psychrometric chart. This chart graphically relates dry-bulb, wet-bulb, and dew-point temperatures to relative humidity, humidity ratio, and specific volume of air. Psychrometric charts are often provided in books on psychrometrics and HVAC handbooks.

Refrigeration Ton. A ton of refrigeration is a common term used in air conditioning to designate the cooling rate of air-conditioning equipment. A ton of refrigeration indicates the ability of an evaporator to remove 200 Btu/min or 12,000 Btu/h. The concept is a carry-over from the days of icemaking and was based on the concept that 200 Btu/min had to be removed from 32°F water to produce 1 ton of ice at 32°F in 24 hr. Hence,

$$1 \text{ ton refrigeration} = \frac{2000 \frac{\text{lb}}{\text{day}} \times 144 \frac{\text{Btu}}{\text{lb}}}{24 \frac{\text{hr}}{\text{day}}}$$

$$= 288{,}000 \text{ Btu/day} \qquad (18\text{-}15)$$
$$= 12{,}000 \text{ Btu/h}$$
$$= 200 \text{ Btu/min}$$

("Handbook of Fundamentals," American Society of Heating, Refrigerating and Air-Conditioning Engineers, 1791 Tully Circle, N. E., Atlanta, Ga. 30329.)

18-3. Heat Flow and Human Comfort.

This article presents the necessary concepts for management of heat energy and aims at development of a better understanding of its effects on human comfort. The concepts must be well understood if they are to be applied successfully to modification of the environment in building interiors, computer facilities, and manufacturing processes.

Temperatures Determined by Heat Balances. In cold weather, comfortable indoor temperatures may have to be maintained by a heating device. It should provide heat to the space at the same rate as the space is losing heat. Similarly, when cooling is required, heat should be removed from the space at the same rate that it is gaining heat. In each case, there must be a heat-balance between heat in and heat out when heating and the reverse in cooling. Comfortable inside conditions can only be maintained if this heat balance can be controlled or maintained.

The rate at which heat is gained or lost is a function of the difference between the inside air temperature to be maintained and the outside air temperature. Such temperatures must be established for design purposes in order to properly size and select HVAC equipment that will maintain the desired design conditions. Many other conditions that also affect the flow of heat in and out of buildings, however, should also be considered in selection of equipment.

Design Criteria Accuracy. Some engineers apply much effort to determination of design conditions with great accuracy. This is usually not necessary, because of the great number of variables involved in the design process. Strict design criteria will increase the cost of the necessary machinery for such optimum conditions and may be unnecessary. It is generally recognized that it is impossible to provide a specific indoor condition that will satisfy every occupant at all times. Hence, HVAC engineers tend to be practical in their designs and accept the fact that the occupants will adapt to minor variations from ideal conditions. Engineers also know that human comfort depends on the type and quantity of clothing worn by the occupants, the types of activities performed, environmental conditions, duration of occupancy, ventilation air, and closeness of and number of people within the conditioned space and recognize that these conditions are usually unpredictable.

Methods of Heat Transfer. Heat always flows from a hot to a cold object, in strict compliance with the second law of thermodynamics (Art. 18-2). This direction of heat flow occurs by conduction, convection, or radiation and in any combination of these forms.

Thermal conduction is a process in which heat energy is transferred through matter by the transmission of kinetic energy from molecule to molecule or atom to atom.

Thermal convection is a means of transferring heat in air by natural or forced movements of air or a gas. Natural convection is usually a rotary or circular motion caused by warm air rising and cooler air falling. Convection can be mechanically produced (forced convection), usually by use of a fan or blower.

Thermal radiation transfers energy in wave form from a hot body to a relatively cold body. The transfer occurs independently of any material between the two bodies. Radiation energy is converted energy from one source to a very long wave form of electromagnetic energy. Interception of this long wave by solid matter will convert the radiant energy back to heat.

Thermal Conduction and Conductivity. Thermal conduction is the rate of heat flow across a unit area (usually 1 ft^2) from one surface to the opposite surface for a unit temperature difference between the two surfaces and under steady-state conditions. Thus, the heat-flow rate through a plate with unit thickness may be calculated from

$$Q = kA(t_2 - t_1) \tag{18-16}$$

where Q = heat flow rate, Btu/h
k = coefficient of thermal conductivity for a unit thickness of material, usually 1 in
A = surface area, normal to heat flow, ft^2
t_2 = temperature, °F, on the warm side of the plate
t_1 = temperature, °F, on the cooler side of the plate

The coefficient k depends on the characteristics of the plate. The numerical value of k also depends on the units used for the other variables in Eq. (18-16). When values of k are taken from published tables, units given should be adjusted to agree with the units of the other variables.

In practice, the thickness of building materials often differs from unit thickness. Consequently, use of a coefficient of conductivity for the entire thickness is advantageous. This coefficient, called **thermal conductance,** is derived by dividing the conductivity k by the thickness L, the thickness being the length or path of heat flow.

Thermal Conductance C and Resistance R. Thermal conductance C is the same as conductivity, except that it is based on a specific thickness, instead of 1 in as for conductivity. Conductance is usually used for assemblies of different materials, such as cast-in-place concrete and concrete block with an airspace between. The flow of heat through such an assembly is very complex and is determined under ideal test conditions. In such tests, conductance is taken as the average heat flow from a unit area of surface (usually 1 ft^2) for the total thickness of the assembly. In the case of 9-in-thick concrete, for example, the conductance, as taken from appropriate tables, would be 0.90 Btu/(h)(ft^2)(°F). (It should be understood, however, that conversion of the conductance C to conductivity k by dividing C by the thickness will produce significant errors.)

Conductance C is calculated from

$$Q = CA(t_2 - t_1) \tag{18-17}$$

where $t_2 - t_1$ is the temperature difference causing the heat flow Q and A is the cross-sectional area normal to the heat flow.

Values of k and C for many building materials are given in tables in the ASHRAE "Handbook of Fundamentals" and other publications on air conditioning.

Thermal resistance, the resistance to flow of heat through a material or an assembly of materials, equals the reciprocal of the conductance:

$$R = \frac{1}{C} \tag{18-18}$$

Thermal resistance R is used in HVAC calculations for determining the rate of heat flow per unit area through a nonhomogenous material or a group of materials.

Air Films. In addition to its dependence on the thermal conductivity or conductance of a given wall section, roof, or other enclosure, the flow of heat is also dependent on the surface air films on each side of the constructions. These air films are very thin and cling to the exposed surface on each side of the enclosures. Each of the air films possesses thermal conductance, which should always be considered in HVAC calculations.

The indoor air film is denoted by f_i and the outdoor film by f_o. Values are given in Table 18-3 for these air films and for interior or enclosed air spaces of assemblies. In this

TABLE 18-3 Thermal Conductance of Air, Btu/(h)(ft²)(°F)

f_i	for indoor air film (still air)	1.5
	Vertical surface, horizontal heat flow	
	Horizontal surface	
	Upward heat flow	1.6
	Downward heat flow	1.1
f_o	for outdoor air film, 15-mi/h wind (winter)	6.0
f_o	for outdoor air film, 7.5-mi/h wind (summer)	4.0
C	for vertical air gap, 3/4 in or more wide	1.1
C	for horizontal air gap, 3/4 in or more wide	
	Upward heat flow	1.2
	Downward heat flow	1.0

table, the effects of air films along both enclosure surfaces have been taken into account in developing the air-film coefficients. Additional data may be obtained from the ASHRAE "Handbook of Fundamentals."

Air-to-Air Heat Transfer. In the study of heat flow through an assembly of building materials, it is always assumed that the rate of heat flow is constant and continues without change. In other words, a steady-state condition exists. For such a condition, the rate of heat flow in Btu per hour per unit area can be calculated from

$$Q = UA(t_2 - t_1) \tag{18-19}$$

where U = coefficient of thermal transmittance.

Coefficient of Thermal Transmittance U. The coefficient of thermal transmittance U, also known as the overall coefficient of heat transfer, is the rate of heat flow under steady-state conditions from a unit area from the air on one side to the air on the other side of a material or an assembly when a steady temperature difference exists between the air on both sides.

In calculation of the heat flow through a series of different materials, their individual resistances should be determined and totaled to obtain the total resistance R_t. The coefficient of thermal transmittance is then given by the reciprocal of the total resistance:

$$U = \frac{1}{R_t} \tag{18-20}$$

Tables of U values for various constructions are available in the ASHRAE "Handbook of Fundamentals," catalogs of insulation manufacturers, and other publications.

Computation of R and U. An assembly that is constructed with several different building materials with different thermal resistances R_1, R_2, R_3, ..., R_n provides a total thermal resistance

$$R_t = \frac{1}{f_i} + R_1 + R_2 + R_3 + \cdots + R_n + \frac{1}{f_o} \tag{18-21}$$

including the indoor and outdoor air film resistances f_i and f_o. The U value, coefficient of thermal transmittance, is then determined by use of Eq. (18-20). This coefficient may be substituted in Eq. (18-19) for calculation of the steady-state heat flow through the assembly.

As a typical example, consider an exterior wall section that is constructed of 4-in face brick, 4-in cinder block, ¾-in airspace, and lightweight ¾-in lath and plaster. The wall is 8 ft 6 in high and 12 ft long. The inside air temperature is to be maintained at 68°F, with an outdoor air temperature of +10°F and a 15-mi/hr prevailing wind. What will be the total heat loss through this wall?

From Table 18-3, the indoor air film conductance is 1.5. Its resistance is equal to $1/1.5 = 0.67$. The outdoor air-film conductance for a 15-mi/hr wind is 6.0. Its resistance is equal to $1/6.0 = 0.17$. Conductivity of the 4-in face brick is 5.0. Conductance of the 4-in cinder block is 0.90; of the ¾-in airspace, 1.1, and of the ¾-in lath and lightweight plaster, 7.70. The total resistance of the wall is then:

$$\begin{aligned} R_t &= \frac{1}{6.0} + 4 \times \frac{1}{5.0} + \frac{1}{0.90} + \frac{1}{1.1} + \frac{1}{7.70} + \frac{1}{1.5} \\ &= 0.17 + 0.8 + 1.11 + 0.91 + 0.13 + 0.67 \\ &= 3.79 \end{aligned}$$

and the coefficient of thermal transmittance is

$$U = \frac{1}{3.79} = 0.264$$

The heat flow rate through the entire wall will be, from Eq. (18-19),

$$Q = 0.264 \times (8.5 \times 12.0)(68 - 10) = 1562 \text{ Btu/h}$$

Convection. Heating by natural convection is very common, because air very easily transfers heat in this manner. As air becomes warmer, it becomes less dense and rises. As it leaves the proximity of the heating surface, other cooler air moves in to replace the rising volume of heated air. As the warm air rises, it comes in contact with cooler materials, such as walls, glass, and ceilings. It becomes cooler and heavier and, under the influence of gravity, begins to fall. Hence, a circulatory motion of air is established, and heat transfer occurs.

When a heating device called a convector operates in a cool space, heat from the convector is transmitted to the cooler walls and ceiling by convection. The convection process will continue as long as the walls or ceiling are colder and the temperature difference is maintained.

Heating of building interiors is usually accomplished with convectors with hot water or steam as the heating medium. The heating element usually consists of a steel or copper pipe with closely spaced steel or aluminum fins. The convector is mounted at floor level against an exterior wall. The fins are used to increase greatly the area of the heating surface. As cool room air near the floor comes in contact with the hot surfaces of the convector, the air quickly becomes very warm and rises rapidly along the cold wall surface above the convector. Additional cold air at floor level then moves into the convector to replace the heated air. In this manner, the entire room will become heated. This process is called heating by natural convection.

Radiation. The most common form of heat transfer is by radiation. All materials and substances radiate energy and absorb radiation energy.

The sun is a huge radiator and the earth is heated by this immense source of radiated energy, which is often called solar energy (sunshine). Solar-collector devices are used to collect this energy and transfer it indoors to heat the interior of a building.

When radiation from the sun is intercepted by walls, roofs, or glass windows, this heat is transmitted through them and heats the interior of the building and its occupants. The reverse is also true; that is, when the walls are cold, the people in the space radiate their body heat to the cold wall and glass surfaces. If the rate of radiation is high, the occupants will be uncomfortable.

Not all materials radiate or absorb radiation equally. Black- or dark-body materials radiate and absorb energy better than light-colored or shiny materials. Materials with smooth surfaces and light colors are poor absorbers of radiant energy and also poor radiators.

Much of the radiation that strikes the surface of window glass is transmitted to the interior of the building as short-wave radiation. This radiation will strike other objects in the interior and radiate some of this energy back to the exterior, except through glass, as a longer wavelength of radiation energy. The glass does not efficiently transmit the longer wavelengths to the outside. Instead, it acts as a check valve, limiting solar radiation to one-way flow. This one-way flow is desirable in winter for heating. In summer, however, it is not desirable, because the longer-wavelength energy eventually becomes an additional load on the air-conditioning system.

The rate of radiation from an object may be determined by use of the **Stefan-Boltzmann law** of radiation. This law states that the amount of energy radiated from a perfect radiator, or a blackbody, is proportional to the fourth power of the absolute temperature of the body. Because most materials are not perfect radiators or absorbers, a proportionality constant called the hemispherical emittance factor is used with this law. Methods for calculating and estimating radiation transfer rates can be found in the ASHRAE "Handbook of Fundamentals."

The quantity of energy transferred by radiation depends on the individual temperatures of the radiating bodies. These temperatures are usually combined into a **mean radiant temperature** for use in heating and cooling calculations. The mean radiant temperature is the uniform temperature of a black enclosure with which a solid body (or occupant) would exchange the same amount of radiant heat as in the actual nonuniform environment.

Thermal Criteria for Building Interiors. There are three very important conditions to be controlled in buildings for human comfort. These important criteria are dry-bulb temperature, relative humidity, and velocity or rate of air movement in the space.

Measurements of these conditions should be made where average conditions exist in the building, room, or zone and at the breathing line, 3 to 5 ft above the floor. The measurements should be taken where they would not be affected by unusually high heat sources or heat losses. Minor variations or limits from the design conditions, however, are usually acceptable.

The occupied zone of a conditioned space does not encompass the total room volume. Rather, this occupied zone is generally taken as that volume bounded by levels 3 in above the floor and 6 ft above the floor and by vertical planes 2 ft from walls.

Indoor design temperatures are calculated from test data compiled for men and women with various amounts of clothing and for various degrees of physical exertion. For lightly clothed people doing light, active work in relatively still room air, the design dry-bulb temperature can be determined from

$$t = 180 - 1.4 t_r \quad (18\text{-}22)$$

where t = dry-bulb temperature, °F

t_r = mean radiant temperature of the space or room (between 70° and 80°F)

When temperatures of walls, materials, equipment, furniture, etc., in a room are all equal, $t = t_r = 75°F$. With low outside temperature, the building exterior becomes cold, in which case the room temperature should be maintained above 75°F to provide the necessary heat that is being lost to the cold exterior. In accordance with Eq. (18-22), the design dry-bulb temperature should be increased 1.4°F for each 1°F of mean radiant temperature below 75°F in the room. In very warm weather, the design temperature should be decreased correspondingly.

Humidity is often controlled for human comfort. Except in rare cases, relative humidity (RH) usually should not exceed 60%, because the moisture in the air may destroy wood finishes and support mildew. Below 20% RH, the air is so dry that human nostrils become dry and wood furniture often cracks from drying out.

In summer, a relative humidity of 45 to 55% is generally acceptable. In winter, a range of 30 to 35% RH is more desirable, to prevent condensation on windows and in walls and roofs. When design temperatures in the range of 75°F are maintained in a space, the comfort of occupants who are inactive is not noticeably affected by the relative humidity.

Variations from the design criteria are generally permitted for operational facilities. These variations are usually established as a number of degrees above or below the design point, such as 75°F DB ± 2°F. For relative humidity, the permitted variation is usually given as a percent, for example, 55% RH ± 5%.

Design conditions vary widely for many commercial and industrial uses. Indoor design criteria for various requirements are given in the "Applications" volume of the ASHRAE "Guide and Data Book."

Outdoor Design Conditions. The outdoor design conditions at a proposed building site are very important in design of heating and cooling systems. Of major importance are the dry-bulb temperature, humidity conditions, and prevailing winds.

Outside conditions assumed for design purposes affect the heating and cooling plant's physical size, capacity, electrical requirements, and of considerable importance, the estimated cost of the HVAC installation. The reason for this is that in many cases, the differences between indoor and outdoor conditions have a great influence on calculated heating and cooling loads, which determine heating and cooling equipment capacities. Since in most cases the design outdoor air temperatures are assumed, the size of equipment will be greatly affected by assumed values.

Extreme outside air conditions are rarely used to determine the size of heating and cooling equipment, since these extreme conditions may occur, in summer or winter, only once in 10 to 50 years. If these extreme conditions were used for equipment selection, the results would be greatly oversized heating and cooling plants and a much greater installed cost than necessary. Furthermore, such oversized equipment will operate most of the year at part load and with frequent cycling of the machines. This results in inefficient operation and, generally, consumption of additional power, because most machines operate at maximum efficiency at full load.

On the other hand, when heating or cooling equipment is properly sized for more frequently occurring outdoor conditions, the plants will operate with less cycling and greater efficiency. During the few hours per year when outside conditions exceed those used for design, the equipment will run continuously in an attempt to maintain the intended interior design conditions. If such conditions persist for a long time, there will probably be a change in interior conditions from design conditions that may or may not be of a minor extent and that may produce uncomfortable conditions for the occupants.

Equipment should be selected with a total capacity that includes a safety factor to cover other types of operation than under steady-state conditions. In the midwest, for instance, the outdoor air temperature may fall as much as 45°F in 2 hr. The heating capacity of a boiler in this case would have to be substantially larger than that required for the calculated heat loss alone. As another example, many heating and cooling systems are controlled automatically by temperature control systems that, at a predetermined time, automatically reset the building temperature downward to maintain, say, 60°F at night for heating. At a predetermined time, for example, 7:30 a.m., before arrival of occupants, the control system instructs the boiler to bring the building up to its design temperature for occupancy. Under these conditions, the boiler must have the additional capacity to comply in a reasonable period of time before the arrival of the occupants.

Accordingly, design outdoor conditions should be selected in accordance with the manner in which the building will be used and, just as important, to obtain reasonable initial cost and low operational costs. Outdoor design conditions for a few cities are shown in Tables 18-4 and 18-12. Much more detailed data are presented in the ASHRAE "Handbook of Fundamentals."

TABLE 18-4 Recommended Design Outdoor Summer Temperatures

State	City	Dry-bulb temp, °F	Wet-bulb temp, °F	State	City	Dry-bulb temp, °F	Wet-bulb temp, °F
Ala........	Birmingham	95	78	Miss.......	Vicksburg	95	78
Ariz.......	Flagstaff	90	65	Mo........	St. Louis	95	78
Ariz.......	Phoenix	105	75	Mont.......	Helena	95	67
Ark........	Little Rock	95	78	Nebr.......	Lincoln	95	78
Calif.......	Los Angeles	90	70	Nev........	Reno	95	65
Calif.......	San Francisco	85	65	N.H.......	Concord	90	73
Colo.......	Denver	95	65	N.J........	Trenton	95	78
Conn......	Hartford	95	75	N.Mex.....	Albuquerque	95	70
D.C.......	Washington	95	78	N.Y.......	New York	95	75
Fla........	Jacksonville	95	78	N.C.......	Greensboro	95	78
Fla........	Miami	95	79	N.Dak.....	Bismarck	95	73
Ga........	Atlanta	95	76	Ohio.......	Cincinnati	95	75
Idaho......	Boise	95	65	Okla.......	Tulsa	100	77
Ill.........	Chicago	95	75	Ore........	Portland	90	68
Ind........	Indianapolis	95	75	Pa.........	Philadelphia	95	78
Iowa.......	Des Moines	95	78	R.I........	Providence	95	75
Kans......	Topeka	100	78	S.C.......	Charleston	95	78
Ky........	Louisville	95	78	S.Dak.....	Rapid City	95	70
La........	New Orleans	95	80	Tenn......	Nashville	95	78
Maine.....	Portland	90	73	Tex.......	Dallas	100	78
Md........	Baltimore	95	78	Tex.......	Houston	95	78
Mass......	Boston	95	75	Utah......	Salt Lake City	95	65
Mich......	Detroit	95	75	Vt.........	Burlington	90	73
Minn......	Minneapolis	95	75	Va.........	Richmond	95	78

The use of outdoor design conditions does not yield accurate estimates of fuel requirements or operating costs, because of the considerable variations of outdoor air temperature seasonally, monthly, daily, and even hourly. These wide variations must be taken into account in attempts to forecast the operating costs of a heating or cooling system.

Since most equipment capacities are selected for calculated loads based on steady-state conditions, usually these conditions will not provide acceptable estimates of annual operating costs. (Wide fluctuations in outside temperatures, however, may not always cause a rapid change in inside conditions as outdoor temperatures rise and fall. For example, in buildings with massive walls and roofs and small windows, indoor temperatures respond slowly to outdoor changes.) Hence, forecasts of fuel requirements or operating costs should be based on the average temperature difference between inside and outside air temperatures on an hourly basis for the entire year. Such calculations are extremely laborious and are almost always performed by a computer that utilizes an appropriate program and local weather tapes for the city involved. Many such programs are currently available from various sources.

18-4. Thermal Insulation.

Article 18-3 indicates that the magnitudes of heat losses and heat gains of a building exterior are directly related to the materials used for construction. Since the heat-loss or heat-gain rate through a building exterior usually represents a major portion of the total heating and cooling load, design of the HVAC systems and the architectural design of the building should be coordinated. Failure to do so will result in oversized heating and cooling equipment, excessively high initial costs, and exorbitant operating costs.

A substantial reduction in heating and cooling loads can be made by the judicious use of thermal insulation in wall and roof construction. Addition of insulation results in an increase in thermal resistance R, or a reduction in the coefficient of heat transfer U of the walls and roof.

Any material with high resistance to flow of heat is called insulation. Many kinds of insulation materials are used in building construction.

Types of Thermal Insulation. All materials have some resistance to heat flow, but high cost and weight of structural materials make it uneconomical to use walls of stone, brick, or other heavy structural materials in thicknesses that provide sufficient resistance to heat flow for economical heating and cooling. For this reason, new and more economical insulations for structures have been and are being developed. These insulating materials generally add no strength to the structure; they are employed only for the purpose of resisting flow of heat. In every case, the cost and weight of these materials are less than those of a sufficient thickness of structural materials with the same resistance to heat flow.

Building insulation may be made of any of a wide variety of materials—glass fibers, glass foam, mineral fibers, wood or other organic fibers, metal foil, foamed plastic, or lightweight concrete. Proper selection depends generally first on cost and then on the physical characteristics.

Most types of insulation have a thermal conductivity k of less than 1, and the best, less than 0.25 $Btu/(h)(ft^2)(°F)$ per inch of thickness.

Two types of insulation are in common use. One is a reflective type, usually a shiny metal plate or foil (aluminum). This type of insulation is used in steel mills and other structures in which extremely high temperatures are encountered. The second type of insulation, which is more widely used, comprises soft and porous materials with numerous trapped air spaces. Typical examples are cork, cellular glass, foamed plastics, vermiculite, perlite, and pumice. For most insulation of this type, thermal conductance is inversely proportional to thickness.

Insulation of the second type is available as blankets, bats, boards, board and slab, and loose fill. Information on thermal conductance of insulation is usually available from the manufacturers.

Heat Flow across an Airspace. The best practical heat insulation at our disposal is still air. An airspace of about ¾ in has been found to be best. Airspaces, however, have a limitation because it is impossible to keep the air motionless.

Surfaces of all materials exposed to air retain a film of air held fast by molecular attraction. This film also is an excellent insulation, but its thickness and effectiveness are diminished by any air motion wiping the surface.

In a vertical airspace, such as may be constructed in a wall, with a higher temperature on one side than on the other, the air within the gap transfers heat from one side to the other by convection (see Art. 18-2). The amount of heat transferred is the same for the same temperature difference, regardless of which side is warm or cold.

Consider now a horizontal airspace, such as may exist between a flat roof and a furred ceiling. In summer sunshine, the roof slab becomes very hot. The air in contact, in the space below, becomes heated, expands, and becomes lighter. On the other hand, the air in contact with the cooler ceiling below the airspace is cooled, contracts, and becomes heavier. Since the light air is at the top of the airspace it cannot rise, and the cooler, heavier air at the bottom of the airspace cannot sink. Hence air motion due to temperature differences, namely convection, is at a minimum, and so is transfer of heat.

This condition is desirable for summer. But in winter the situation is less desirable. The roof slab is cold, and the air below is cooled, tending to fall—which it is free to do. The air above the ceiling is warmed and tends to rise—which it is free to do. So a thermal circulation is set up. This again is convection, a conveyor belt taking heat from the ceiling below and transporting it to the roof slab above.

Airspaces in pitched roofs fall somewhere in between vertical and horizontal airspaces.

Insulations that depend on airspaces for their resistance to heat flow incorporate within them a great many tiny cells. Not only do they contain a large number of air films in the path of heat flow, but the minute size of the air spaces tends to keep the air still and thus considerably reduces convection.

Airspaces often are used for insulation purposes in building construction. For example, walls are built with one or more airspaces within to improve overall thermal resistance. Also, window sash is available with one or more airspaces between two or more sheets of glass.

Weather Stripping. Weather stripping utilizes various devices to eliminate infiltration of outside air into a building and thus prevent this air from applying a substantial heat load. Hence, the effect of weather stripping is similar to insulation in that it reduces the heat loss or heat gain of a structure and also conserves energy.

Weather stripping is available in many forms, such as foam rubber or foam plastic strips, which are resilient and compressible. They are usually placed in a door frame or window frame in such a manner that the door or window compresses the material and thereby eliminates air infiltration.

In many cases where weather stripping is not possible, such as minute cracks and crevices of construction, calking is usually applied. It utilizes a plastic compound usually furnished in tubes and applied by a calking gun.

In all cases, the installation of weather stripping and calking of joints reduces the heat loss of a building in winter and the heat gain in summer and thus reduces the cost of operation for the HVAC systems. This saving will more than pay for the cost of the weather stripping and calking.

Installation of Insulation. Insulations must be properly installed if the intended results are to be realized. Poor installation practices and disregard for the integrity of the vapor barrier that is usually installed with insulation (see Art. 18-5) will produce undesirable results. Specialists in installation of insulation and vapor barriers are available.

Insulations that are flexible or semirigid and are supplied as blankets or bats vary in compressibility and flexibility. Also, they are available with or without a vapor barrier. These insulations generally come in rolls that fit between wood wall studs, ceiling joists, or roof rafters spaced 16 in on centers.

The insulation blankets or bats should be installed so that the vapor barrier is as close to the warm side of the wall (usually the inside surface) as possible. Built-in flaps on the insulation backing or the vapor barrier permit stapling to the studs, joists, or rafters. A finish-type gypsumboard or lath and plaster are then applied over the studs to form the finished inside surface. Many kinds of flexible insulation blankets are also made for HVAC ductwork with and without vapor barriers.

Loose-fill insulations include granular perlite, vermiculite, and pumice. They are used as blow-in type insulations for existing structures or installed in masonry construction in one or more insulation spaces. In many cases, also, these insulations are used with insulating cements or pastes and are troweled or blown on a surface. When dry, they serve as insulation. This type is often used on irregular surfaces, such as hot-water pumps and boilers, tanks, and vessels.

Rigid insulation of many kinds are used in roofs as sheets, boards, slabs, or blocks. For use as roof insulation, the boards are laid flat on a metal or concrete deck. They are held in place by a waterproof, bituminous adhesive mopped in place on the decking.

Pipe insulation is available as rigid curved sections and is made of glass fiber, calcium silicate, or other materials with low conductance. Thickness used depends on the temperature range of the construction to be insulated. Calcium silicate is generally used for high-temperature applications.

U Values. Note that the maximum overall conductance U encountered in building construction is 1.5 Btu/(h)(ft^2)(°F). This would occur with a sheet-metal wall. The metal has, for practical purposes, no resistance to heat flow. The U value of 1.5 is due entirely to the resistance of the inside and outside air films. Most types of construction have U factors considerably less than 1.5.

The minimum U factor generally found in standard construction with 2 in of insulation is about 0.10.

Since the U factor for single glass is 1.13, it can be seen that windows are a large source of heat gain, or heat loss, compared with the rest of the structure. For double glass, the U factor is 0.45. For further comparison, the conductivity k of most commercial insulations varies from about 0.24 to about 0.34.

Calculation of Heat Flow. It is not always possible to find in tables the U factor corresponding to the actual type of construction and choice of materials desired. In that case, it is necessary to calculate U by first finding the thermal conductivity k or thermal conductance C for each material. These are found in tables, such as those in the "Handbook of Fundamentals" (American Society of Heating, Refrigerating and Air-Conditioning Engineers). The thickness in inches must also be known for each material.

Fig. 18-1 Section through a brick-veneer wall.

For example, the thermal conductivity k for brick is 9.2. The conductance of a 4-in thickness of brick veneer is then 9.2/4, or 2.3. Note that the value 2.3 might be listed in tables as thermal conductance C for 4-in veneer brick, in which case no calculation is needed.

Before the thermal characteristics of any wall, roof, etc., can be calculated it is necessary to know the heat conductance of the air films and possible airspaces encountered. These are generally accepted as given in Table 18-3 (Art. 18-3).

In all tables, conductivities or conductances are listed, but to find the overall conductance U of, say, a built-up wall, it is necessary to compute the total resistance, which is the sum of the resistances R corresponding to the thickness of each material employed.

Resistance is the reciprocal of conductance. If, for example, the heat conductivity of 1 in of insulation is 0.25, then the resistance is 1/0.25 or 4.

There are formulas for calculating U [see, for example, Eqs. (18-20) and (18-21)], but a simpler method is as follows: Assume the hypothetical wall in Fig. 18-1.

Item	K	Thickness, in.	C	$R = 1/C$
Outside film	...	...	6	0.166
Brick	9.2	4	2.30	0.434
Air space	...	...	1.10	0.910
Wallboard	...	½	1.00	1.000
Inside film	...	...	1.65	0.606
Total resistance	...	...	...	3.116

$$\text{Overall conductance } U = \frac{1}{3.116} = 0.32$$

Assume we wish to find out what the overall conductance would be for the wall in Fig. 18-1 with 1-in insulation ($k = 0.25$).

Item	K	Thickness, in.	C	$R = 1/C$
Wall	...	...	0.32	3.116
Insulation	0.25	1	0.25	4.000
Total resistance	...	...	...	7.116

$$\text{Overall conductance } U = \frac{1}{7.116} = 0.14$$

The last calculation indicates that, if the overall conductance of a construction is known and it is desired to add to or subtract a material from this wall, it is not necessary to recalculate completely, but as above, use the existing resistance and add or subtract the resistance of the material to be varied. Then, the reciprocal is the conductance sought.

A simpler device for finding the overall conductance resulting from addition of insulation to a known wall is as follows:

Using the above example, we have 0.32 for the overall conductance of the wall, and the conductivity of the desired 1-in insulation is 0.25. We then multiply the two conductances together, and divide by their sum:

$$\frac{0.32 \times 0.25}{0.32 + 0.25} = \frac{0.08}{0.57} = 0.14$$

If we wish to find the overall conductance of the above wall by the addition of 2 in of insulation ($C = 0.25/2$) instead of 1 in, we have

$$\frac{0.32 \times 0.125}{0.32 + 0.125} = \frac{0.04}{0.445} = 0.09$$

Effect of Insulation. There is a very simple way of evaluating insulation without going into details and theoretical considerations. Consider, for example, a wall in which we are thinking of adding, say, 1 in of a certain insulation. From tables, we can get the heat conductance of the insulation and the overall conductance of the wall. If, for example, the conductance of the insulation is one-third that of the wall, then the overall conductance of the wall with the insulation added is one-fourth of what it would be without insulation. The heat flow stopped is three-fourths.

Table 18-5 will make this simple method very clear. (All terms are relative heat conductances, with the heat conductance of the uninsulated wall taken as unity.)

TABLE 18-5 Effect of Insulating a Typical Wall

Relative conductance of insulation added	Resultant conductance of insulated wall	Decrease in heat flow
$1/2$	$1/3$	$2/3$
$1/3$	$1/4$	$3/4$
$1/4$	$1/5$	$4/5$
$1/5$	$1/6$	$5/6$
$1/n$	$1/(n+1)$	$n/(n+1)$

It can be seen that the savings in heat are considerable. Nevertheless, increasing the thickness of insulation provides diminishing returns. Considering the effect of insulation alone, if the thickness is doubled, say, from 1 in to 2 in, the heat flow becomes 50%, or 50% is saved. If another 1 in is added, the heat flow compared with the original 1-in thickness is 33%. With another 1-in addition, the heat flow is reduced to 25%. So doubling the original thickness reduces the heat flow by 50%; tripling the thickness reduces heat flow another 17%; and quadrupling the thickness reduces heat flow only another 8%. So for each increase in the thickness of insulation, the effect is less, and hence an economical balance can be calculated.

In any case, once installed, insulation not subject to abuse needs no repair or maintenance. Hence, first cost is the final cost, which in a new structure is offset by a considerable reduction in the required heating or cooling plant. Furthermore, in any structure, insulation makes possible a continuous reduction in operating expense for heating or cooling. These savings can generally pay for the cost of insulation in anywhere from 3 to 10 years, depending on the structure. Any investment having a return of 10 to 30% is attractive. Also, added comfort due to the resulting more uniform inside temperatures is another advantage.

Economics of Insulation. Generally, construction materials almost always require some insulation. The question, therefore, is not whether to insulate but rather what thickness and type of insulation to use.

Select the type of insulation by the installed cost involved. This should be the one whose prevailing cost per square foot divided by its resistance R_i gives the lowest figure, all values being based on 1-in thickness.

Figure the total wall area to be insulated for each building exposure, as for example A_n (north wall), A_s (south wall), A_e (east wall), A_w (west wall), ft². Calculate winter heat loss and summer heat gain for each wall exposure without insulation. Do not neglect summer solar gains.

Determine the effect of adding the resistance of 1 in of insulation to the resistance of the uninsulated wall for each case: $R_w + R_{1i}$, where R_{1i} is the resistance of 1 in of insulation. The insulated wall heat gain or loss will be that of the uninsulated wall times $R_w/(R_w + R_{1i})$, and for 2-in insulation, the heat gain or loss of the uninsulated wall times $R_w/(R_w + R_{2i})$. The effect of 3 or more inches of insulation can be similarly determined.

As an example, the resistance R_w of the wall shown in Fig. 18-1 is 3.116. Addition of 1 in of insulation with a resistance $R = 1/C = 1/0.25 = 4.0$ increases the total resistance to 7.116. Thus, the reduction in heat gain or loss due to 1 in of insulation is $R_w/(R_w + R_{1i}) = 3.116/7.116 = 0.438$, or 43.8%.

With 2 in of insulation, the total resistance is $3.116 + 2 \times 4.0 = 11.116$. Therefore, reduction in heat gain or loss is $3.116/11.116 = 0.28$, or 28%.

Bear in mind that the cost of insulation material is proportional to the thickness installed. Installation labor costs, however, do not rise proportionately. Therefore, the total installed cost of insulation is not proportional to thickness but increases to a lesser extent than the cost of increase in thickness.

Note also that insulation conserves energy and permits a reduction in capacity of the heating and cooling plants, thereby decreasing the initial cost of these high-price items. This cost reduction offsets to some extent the cost of increased insulation.

The preceding calculations can be continued to cover the reductions with 3, 4, 5, and 6 in of insulation. With costs of fuel and power for heating and cooling known, the designer can then compare the resulting savings in operating cost with investment in installed cost of insulation. The savings in operating cost provided by insulation are worthwhile if they prove greater than interest and amortization charges of the added cost of the insulation that brings about the savings.

Thickness of insulation is generally limited in residential use to 3 in in walls and 6 in in roof or ceilings, because of the depth available between 2×4 studs and 2×6 rafters. In commercial and industrial construction, thickness of insulation is usually limited by the distance columns and beams or trusses extend within the structure. In any case, the preceding calculations indicate the optimum thickness of insulation for any type of structure.

As an adjunct to the above considerations, the savings due to double or triple fenestration should not be overlooked.

18-5. Prevention of Damage from Condensation.

If reduction of heat flow were all that had to be considered in the structural application of insulation, the whole matter of design would be merely a simple matter of economics—comparing the cost of insulation with the cost saved in heating or cooling, as well as the reduction in size of heating or cooling plant in new structures.

Unfortunately for insulation, air almost always contains water vapor. When insulation is not correctly installed, this water vapor may, under certain conditions, destroy the insulating value, as well as the structure itself. This by-product nuisance of insulation is the greatest single cause of insulation failure. It is essential, therefore, to the proper installation of insulation that the behavior of water vapor be investigated, and then its effect on insulation. Here it can truthfully be stated: "It is not the heat but the humidity."

The problem involved is more easily grasped if the following two statements are kept in mind:

1. For practical purposes, water vapor is a gas and behaves exactly like air. It goes wherever air goes, penetrating voids, porous materials, etc.
2. Generally, water vapor always flows from a region of high temperature to a region of lower temperature.

When a certain volume of air contains all the water vapor it can, it is saturated, and the so-called relative humidity is 100%. The higher the temperature, the more vapor it can contain. Conversely, the lower the temperature, the less vapor it can hold.

Whatever the temperature, if the air is saturated, any drop in temperature will immediately cause some of the water to condense.

Assume air containing a certain amount of vapor less than saturation. As the temperature drops, it takes less and less vapor to saturate the air, until a temperature is reached at which the air is saturated with vapor. It has reached the dew point. Any further cooling will condense out moisture.

Assume a room with air at 70°F and 30% relative humidity. The dew point is approximately 38°F. If the outside temperature is, say, 0°F, the temperature in the outside wall of the room will vary from 0°F at the exterior air film up to 70°F at the inside air film. Therefore, at some point within the wall the temperature will be 38°F, the dew point.

Since almost all building materials are porous, the vapor permeating the pores will tend to condense into water near the section where the dew point occurs, and the higher temperature inside the room will cause more vapor to penetrate the wall, repeating the process.

The dew point often occurs within the insulation and may in time saturate it, drastically reducing its insulating value. Furthermore, the moisture may rot or rust the structure, or stain interior finishes. Under severe conditions, the moisture may freeze, and in expanding, as ice always does, it may cause cracks in the structure.

Condensation within a structure is generally a winter problem. The greater the relative humidity existing within the building, the more serious the problem becomes.

The relative humidity maintained within a structure depends on the occupancy of the building. If the conditions are intermittent, for example, as in a church, condensation that may occur for a short period will reevaporate in between such periods. On the other hand, an indoor swimming pool will maintain a continuous comparatively high inside relative humidity; so unless condensation is prevented, the process is accumulative during any cold weather.

Buildings often are maintained at lower temperatures (50 to 60°F) during the night when outside temperatures are at their minimum. The entire wall or roof is then at an even lower temperature. So condensation may occur at night, even if not during the day. On the other hand, though condensation may occur at night, it may not exceed the amount that would evaporate during daytime occupancy.

Generally, there is always danger of condensation within a wall or roof, etc., if the temperature on the cold side is below the dew point of the air on the warm side. Whenever condensation occurs on a window, it can be assumed that it is also occurring within the structure, although to a lesser degree.

Since vapors flow from the warm side of a wall or roof to the cold side, the exterior surface should be as porous as possible or vented and yet offer protection against penetration of rain. This is particularly important with "blown-in" insulation as applied to frame houses, for which a vapor barrier generally cannot be installed. This type of insulation also involves another principle, which, if ignored, frequently is the cause of peeling of paint and leads to unnecessary repair of rain gutters that do not leak.

"Blown-in" insulation is sprayed into the spaces between the studs of frame construction. The interior surface is generally lath and plaster, or wallboard—both porous. The exterior is generally wood sheathing, with shingles, clapboards, or stucco. The heat resistance of the insulation is such that during the winter the location of the dew point falls within the insulation. Theoretically, the resulting condensation should occur within the insulation. This, however, does not occur. Condensation, when it does happen, does not form at the location of the dew point within the insulation, but on the inside surface of the sheathing.

The principle involved is this: Whenever the dew point occurs within a material, condensation will not occur until the flow of water vapor encounters the surface of another material of greater resistance to the flow of water vapor. That is, as long as the air can keep on moving, it will carry the moisture along with it and will not deposit the moisture until it reaches a surface that resists its flow and is colder than the dew point.

The problem inherent in blown-in insulation can be solved by "cold-side venting." In applying blown-in insulation, an opening usually is drilled through the exterior wall surface between each pair of studs. These holes should never be sealed, only covered with porous water-repellent material for protection against the weather. Then, whatever water vapor flows through the inside porous finish can escape to the cold air outside without condensing. With clapboard construction, "toothpick" wedges may be driven under the lower edge of each clapboard to provide the required openings for breathing.

Condensation during Summer Cooling. The preceding discussions of winter condensation seem to contradict summer requirements when the warm and cold sides of a construction are the reverse of what they are in winter. In most parts of the United States, however, cooling seldom results in maintenance of inside temperatures more than 15°F below outside conditions, whereas in winter, inside temperatures are generally maintained at 60 to 75°F above outside conditions. So in winter, the prevailing maximum temperature differences are from 4 to 5 times what they are in summer. Furthermore, in summer very little cooling is required during the night. Hence, as far as insulation is concerned, summer condensation is so intermittent that it can be completely disregarded for the average structure and average occupancy.

It should be mentioned, however, that in low-temperature work, such as cold-storage rooms and low-temperature test cells, special conditions arise for which it is best to refer to a specialist.

Ventilation to Get Rid of Condensation. In addition to cold-side venting, ventilation is used in other circumstances to prevent damage from condensation. For example, where insulated ceilings exist below an attic, or an airspace below a roof, louvers or some other provision for admission of outside air should be provided.

Under winter conditions, outside air, being very cold, has little capacity for holding moisture. Yet, outside air is the only source for inside air, and inside any structure there are many sources from which moisture is absorbed—people, shower baths, kitchens.

Hence, an occupied and heated building contains air with more moisture than in the outside air. The moisture tends to flow from the higher temperatures inside to the lower temperatures in the attic or airspace under the roof. If these spaces are not ventilated with air capable of absorbing the moisture, it will condense and cause rust, rot, leaks, stains, etc.

In other words, any unheated space above heated spaces should be ventilated. Where prevailing winds are not dependable, one opening should be high, and another low, to take advantage of the tendency of heated air to rise.

As a general rule, the upward movement of heated air should be provided for in all airspaces above insulation placed over warmer spaces.

Ventilation is required because of the lack of a perfect vapor barrier and is workable only because the drier outside air is capable of absorbing more moisture. In general, vent area should total about $1/300$ of the horizontal projection of the roof area.

Use of Vapor Barriers to Prevent Condensation. Whenever insulation is installed in a wall, roof, or slab, its resistance to the flow of heat is so much greater than that of the other elements of the construction that the dew point and resulting condensation may occur within the insulation.

Since water vapor flows from regions of high temperature to regions of low temperature, a simple solution to condensation is to stop the flow of water vapor by means of some surface material impervious to moisture—provided this surface is always maintained above the dew point. Such a surface is called a **vapor barrier**. It must always be applied on the warm side.

Because condensation is generally most severe during the heating season, vapor barriers should be installed on the interior side of walls and roofs. From a practical standpoint, this means that the vapor barrier should be next to and part of the insulation.

One of the best and most economical vapor barriers is aluminum foil. Some insulations come equipped with this foil attached to one surface. However, unless reinforced with kraft paper or some other strong material, the foil is easily ripped or punctured, and so is of little value as a barrier.

Since vapor behaves as a gas, a vapor barrier, to be effective, should be airtight, or as nearly so as possible. But this is often an impractical requirement. For example, consider a roof with the insulation above the deck and between a vapor barrier and waterproof roofing. Unless the insulation is of a firm material, such as cellular glass, the heat of the sun will cause the air within the insulation to expand, forming bubbles under the waterproofing. During the coolness of the night, the bubbles will contract. After a series of sunny days and cool nights, the bending back and forth of the surface may destroy the roofing. One way to prevent this is to side-vent the roof insulation so the contained air can freely expand and contract. The side vents must, however, be protected from driving rain.

Vapor barriers can be made of other materials besides aluminum foil. There are aluminum paints, plastic paints, some plastic films, asphalt paints, rubber-base paints, asphalt, and foil-laminated papers. It must be remembered that water-repellent surfaces are not necessarily vapor barriers, that is, airtight.

To evaluate a vapor barrier, a unit known as the **perm** is used. It is defined as a vapor-transmission rate of 1 grain of water vapor through 1 ft^2 of material per hour when the vapor-pressure difference is equal to 1 in of mercury (7000 grains equals 1 pound). A material having a vapor-transmission rate of 1 perm or less is considered a good vapor barrier. The corresponding unit for permeance of 1-in thickness is **perm-inch**.

Resistance to vapor transmission is the reciprocal of the permeance.

To sum up: Vapor barriers, or as much resistance as possible to vapor flow (or air) should be provided on the warm side of walls and roofs. Openings or porous materials—as little resistance as possible to vapor flow—should be provided on the cold side.

If vapor barriers were perfect, cold-side venting would not be required. Unfortunately, vapor barriers are not perfect; therefore, cold-side venting is worthwhile insurance against failure of insulation in all cases.

18-6. Ventilation.

Ventilation is utilized for many different purposes, the most common being control of humidity and condensation. Other well-known uses include exhaust hoods in restaurants, heat removal in industrial plants, fresh air in buildings, odor removal, and chemical and fume hood exhausts. In commercial buildings, ventilation air is used for replacement of stale, vitiated air, odor control, and smoke removal. Ventilation air contributes greatly to the comfort of the building's occupants. It is considered to be of such importance that many building codes contain specific requirements for minimum quantities of fresh, or outside, air that must be supplied to occupied areas.

Ventilation is also the prime method for reducing employee exposure to excessive airborne contaminants that result from industrial operations. Ventilation is used to dilute contaminants to safe levels or to capture them at their point of origin before they pollute the employees' working environment. The Occupational Safety and Health Act (OSHA) standards set the legal limits for employee exposures to many types of toxic substances.

Methods of Ventilation. Ventilation is generally accomplished by two methods: natural and mechanical. In either case, ventilation air must be air taken from the outdoors. It is brought into the building through screened and louvered or other types of openings, with or without ductwork. In many mechanical ventilation systems, the outside air is brought in through ductwork to an appropriate air-moving device, such as a centrifugal fan. With a network of ductwork, the supply air is distributed to areas where it is needed. Also, mechanical ventilation systems are usually designed to exhaust air from the building with exhaust fans or gravity-type ventilators in the roof or a combination-type system.

Many mechanical ventilation systems are installed for fire protection in buildings to remove smoke, heat, and fire. The design must be capable of satisfying the provisions of the National Fire Protection Association "Standard for Installation of Air-Conditioning Systems," NFPA 90-A. The standard also covers installation provisions of air intakes and outlets.

Natural ventilation in buildings is caused by the temperature difference between the air in the building and the outside air and by openings in the outside walls or by a combination of both. With natural ventilation, there should be some means for removing the ventilation air from the building, such as roof-mounted gravity vents or exhaust fans.

Minimum Ventilation Requirements. There are many codes and rules governing minimum standards of ventilation. All gravity or natural-ventilation requirements involving window areas in a room as a given percentage of the floor area or volume are at best approximations. The amount of air movement or replacement by gravity depends on prevailing winds, temperature difference between interior and exterior, height of structure, window-crack area, etc. For controlled ventilation, a mechanical method of air change is recommended.

Where people are working, the amount of ventilation air required will vary from one air change per hour where no heat or offensive odors are generated to about 60 air changes per hour.

At best, a ventilation system is a dilution process, by which the rate of odor or heat removal is equal to that generated in the premises. Occupied areas below grade or in windowless structures require mechanical ventilation to give occupants a feeling of outdoor freshness. Without outside air, a stale or musty odor may result. The amount of fresh air to be brought in depends on the number of persons occupying the premises, type of activity, volume of the premises, and amount of heat, moisture, and odor generation. Table 18-6 gives the recommended minimum amount of ventilation air required for various activities.

TABLE 18-6 Minimum Ventilation Air for Various Activities

Type of Occupancy	Ventilation Air, Cfm per Person
Inactive, theaters	5
Light activity, offices	10
Light activity with some odor generation, restaurant	15
Light activity with moderate odor generation, bars	20
Active work, shipping rooms	30
Very active work, gymnasiums	50

The amount of air to be handled, obtained from the estimate of the per person method, should be checked against the volume of the premises and the number of air changes per hour given in Eq. (18-23).

$$\text{Number of air changes per hour} = \frac{60Q}{V} \qquad (18\text{-}23)$$

where Q = air supplied, ft^3/min
V = volume of ventilated space, ft^3

When the number of changes per hour is too low (below one air change per hour), the ventilation system will take too long to create a noticeable effect when first put into operation. Five changes per hour are generally considered a practical minimum. Air changes above 60 per hour usually will create some discomfort because of air velocities that are too high.

Toilet ventilation and locker-room ventilation are usually covered by local codes—50 ft^3/min per water closet and urinal is the usual minimum for toilets and six changes per hour minimum for both toilets and locker rooms.

Heat, Odor, and Moisture Removal. Removal of heat by ventilation is best carried out by locating the exhaust outlets as close as possible to the heat source. Where concentrated sources of heat are present, canopy hoods will remove the heat more efficiently.

Figure 18-2 shows a canopy-hood installation over a kitchen range. Grease filters reduce the frequency of required cleaning. When no grease is vaporized, they may be eliminated.

Greasy ducts are serious fire hazards and should be cleaned periodically. There are on the market a number of automatic fire-control systems for greasy ducts. These systems usually consist of fusible-link fire dampers and a means of flame smothering—CO_2, steam, foam, etc.

Figure 18-3 shows a double hood. This type collects heat more efficiently; i.e., less exhaust air is required to collect a given amount of heat. The crack area is arranged to yield a velocity of about 1000 ft/min.

A curtain of high-velocity air around the periphery of the hood catches the hot air issuing from the range or heat source. Canopy hoods are designed to handle about 50 to 125 ft^3/min of exhaust air per square foot of hood. The total amount of ventilation air should not yield more than 60 changes per hour in the space.

Fig. 18-2 Canopy hood for exhausting heat from a kitchen range.

Fig. 18-3 Double hood for exhausting heat.

Where hoods are not practical to install and heat will be discharged into the room, the amount of ventilation air may be determined by the following method:

Determine the total amount of sensible heat generated in the premises—lights, people, electrical equipment, etc. This heat will cause a temperature rise and an increase in heat loss through walls, windows, etc. To maintain desired temperature conditions, ventilation air will have to be used to remove heat not lost by transmission through enclosures.

$$q_v = 1.08 Q(T_i - T_o) \tag{18-24}$$

where q_v = heat, Btu/h, carried away by ventilation air
Q = flow of ventilation air, ft^3/min
T_i = indoor temperature to be maintained
T_o = outdoor temperature

With Eq. (18-24), we can calculate the amount of ventilation air required by assuming a difference between room and outdoor temperatures or we can calculate this temperature gradient for a given amount of ventilation air.

The same method may be used to calculate the air quantity required to remove any objectionable chemical generated. For example, assume that after study of a process we determine that a chemical will be evolved in vapor or gas form at the rate of X lb/min. If Y is the allowable concentration in pounds per cubic foot, then $Q = X/Y$, where Q is the ventilation air needed in cubic feet per minute.

Where moisture is the objectionable vapor, the same equation holds, but with X as the pounds per minute of moisture vaporized, Y the allowable concentration of moisture in pounds per cubic foot above outdoor moisture concentration.

Once the amount of ventilation air is determined, a duct system may be designed to handle it, if necessary.

Ventilation air may be provided by installing either an exhaust system, a supply system, or both.

In occupied areas where no unusual amounts of heat or odors are generated, such as offices and shipping rooms, a supply-air system may be provided, with grilles or ceiling outlets located for good distribution. When the building is tight, a relief system of grilles or ducts to the outside should be provided. But when the relief system is too extensive, an exhaust fan should be installed for a combination supply and exhaust system.

In spaces where concentrated heat, odor, or other objectionable vapors are generated, it is more practical to design an exhaust distribution system with hoods. The closer the hood to the source, the more efficient the system.

All air exhausted from a space must be replaced by outside air either by infiltration through doors and windows or by a fresh-air makeup system. Makeup air systems that have to operate during the winter season are often equipped with heating coils to temper the cold outside air.

Natural Ventilation. Natural ventilation in buildings is accomplished by use of windows, louvers, skylights, roof ventilators, roof monitors, jalousies, intake hoods, etc. They should be located to admit fresh air only and not near sources of smoke, dust, odors, or polluted air from adjacent sources. Discharge vents should also be provided to eliminate vitiated air from the building. The outlet locations must not discharge toward other fresh-air intakes of the building or its neighbors. In multifloor buildings, vertical vent shafts, or risers, are used to supply ventilation air throughout the building.

Mechanical Ventilation. Mechanical ventilation is almost always preferred over natural ventilation because of reliability and the ability to maintain specific design requirements, such as air changes per hour and face velocities for exhaust hoods. Natural ventilation permits wide variations in ventilation-air quantities and uncertain durations of ventilation. (In critical areas, such as in carcinogenic research laboratories, natural ventilation is never relied upon.) For this reason, mechanical ventilation systems are almost always used where ventilation requirements are critical and must be highly reliable.

Mechanical ventilation is often required by various building codes for various applications as follows:

1. Control of contaminants in the work area for health protection and compliance with OSHA standards for achieving the legal limits set on employee exposure to specific toxic and hazardous substances
2. Fire and explosion prevention for flammable vapors
3. Environmental protection
4. Reuse of valuable industrial materials
5. Human comfort—removal of heat, odors, and tobacco smoke
6. Humidity control
7. Corrosive fumes and noxious gases

Mechanical ventilation may be a single system without heating, cooling, filtration, humidification, dehumidification, etc., or it may include various combinations of these functions. In other words, the systems can be heating-ventilating units or heating–ventilating–air-conditioning (HVAC) units.

In many complex and specialized buildings, certain functional areas will be required to have various degrees of positive pressurization, negative pressurization, or balanced atmospheric conditions. The ventilation air as a part of the system supply air is used to provide the positive and balanced pressures. An exhaust system is utilized to maintain the negative-pressure areas. In many designs, air lost by pressurization is exfiltrated from the system and does not become part of the return-air stream.

Recirculation of ventilation air is prohibited from certain areas, such as toilets, bathrooms, biology labs, chemistry labs, hospital operating rooms, mortuary rooms, isolation rooms, and rooms with flammable vapors, odors, dust, and noxious gases.

In all ventilation systems, a quantity of air equal to the ventilation air should leave the building. If this is not accomplished, then the building will become pressurized, and the ventilation air will exfiltrate through available doors, windows, cracks, crevices, relief vents, etc. Since in many cases this is undesirable and unreliable, exhaust systems are usually employed. The exhaust, in many cases, may be part of a complete HVAC system.

Types of Ventilation Fans. Inasmuch as most ventilation systems are designed as mechanical ventilation systems that utilize various kinds of fans, a knowledge of the types of fans in use will be of value in selection of ventilation fans. Fans are used to create a pressure differential that causes air to flow in a system. They generally incorporate one of several types of impellers mounted in an appropriate housing or enclosure. An electric motor usually drives the impeller to move the air.

Two types of fans are commonly used in air-handling and air-moving systems: axial and centrifugal. They differ in the direction of airflow through the impeller.

Centrifugal fans are enclosed in a scroll-shaped housing, which is designed for efficient airstream energy transfer. This type of fan has the most versatility and low first cost and is the workhorse of the industry. Impeller blades may be radial, forward-curved, backward-inclined, or airfoil. When large volumes of air are moved, airfoil or backward-inclined blades are preferable because of higher efficiencies. For smaller volumes of air, forward, curved blades are used with satisfactory results. Centrifugal fans are manufactured with capacities of up to 500,000 ft^3/min and can operate against pressures up to 30 in water gage.

Axial-flow fans are versatile and sometimes less costly than centrifugal fans. The use of axial fans is steadily increasing, because of the availability of controllable-pitch units, with increased emphasis on energy savings. Substantial energy savings can be realized by varying the blade pitch to meet specific duty loads. Axial fans develop static pressure by changing the velocity of the air through the impeller and converting it into static pressure. Axial fans are quite noisy and are generally used by industry where the noise level can be tolerated. When used for HVAC installations, sound attenuators are almost always used in series with the fan for noise abatement. Tubeaxial and vaneaxial are modifications of the axial-flow fan.

Propeller-type fans are also axial fans and are produced in many sizes and shapes. Small units are used for small jobs, such as kitchen exhausts, toilet exhausts, and air-cooled condensers. Larger units are used by industry for ventilation and heat removal in large industrial buildings. Such units have capacities of up to 200,000 ft^3/min of air. Propeller-type fans are limited to operating pressures of about ½ in water gage maximum, and are usually much noisier than centrifugal fans of equal capacities.

Vaneaxial fans are available with capacities up to 175,000 ft^3/min and can operate at pressures up to 12 in water gage. Tubeaxial fans can operate against pressures of only 1 in water gage with only slightly lower capacities.

In addition to the axial and centrifugal fan classifications, a third class for special designs exists. This classification covers tubular centrifugal fans and axial-centrifugal, power roof ventilators. The tubular centrifugal type is often used as a return-air fan in low-pressure HVAC systems. Air is discharged from the impeller in the same way as in standard centrifugal fans and then changed 90° in direction through straightening vanes. Tubular centrifugal fans are manufactured with capacities of more than 250,000 ft^3/min of air and may operate at pressures up to 12 in water gage.

Power roof ventilators are usually roof mounted and utilize either centrifugal or axial blade fans. Both types are generally used in low-pressure exhaust systems for factories, warehouses, etc. They are available in capacities up to about 30,000 ft^3/min. They are, however, limited to operation at a maximum pressure of about ½ in water gage. Powered roof ventilators are also low in first cost and low in operating costs. They can provide positive exhaust ventilation in a space, which is a definite advantage over gravity-type exhaust units. The centrifugal unit is somewhat quieter than the axial-flow type.

Fans vary widely in shapes and sizes, motor arrangements and space requirements. Fan performance characteristics (variation of static pressure and brake horsepower) with changes in the airflow rate (ft^3/min) are available from fan manufacturers and are presented in tabular form or as fan curves.

Dampers. Dampers are mechanical devices that are installed in a moving airstream in a duct to reduce the flow of the stream. They, in effect, purposely produce a pressure drop (when installed) in a duct by substantially reducing the free area of the duct.

Two types of dampers are commonly used by HVAC designers, parallel blade and opposed blade. In both types, the blades are linked together so that a rotation force applied to one shaft simultaneously rotates all blades. The rotation of the blades opens or closes the duct's free area from 0 to 100% and determines the flow rate.

Dampers are used often as opening and closing devices. For this purpose, parallel dampers are preferred.

When dampers are installed in ducts and are adjusted in a certain position to produce a desired flow rate downstream, opposed-blade dampers are preferred. When dampers are used for this purpose, the operation is called **balancing.**

Once the system is balanced and the airflows in all branch ducts are design airflows, the damper positions are not changed until some future change in the system occurs. However, in automatic temperature-control systems, both opening-and-closing and balancing dampers are commonly used. In complex systems, dampers may be modulated to compensate for increased pressure drop by filter loading and to maintain constant supply-air quantity in the system.

Filters. All air-handling units should be provided with filter boxes. Removal of dust from the conditioned air not only lowers building maintenance costs and creates a healthier atmosphere but prevents the cooling and heating coils from becoming blocked up.

Air filters come in a number of standard sizes and thicknesses. The filter area should be such that the air velocity across the filters does not exceed 350 ft/min for low-velocity filters nor 550 ft/min for high-velocity filters. Thus, the minimum filter area in square feet to be provided equals the airflow, ft^3/min, divided by the maximum air velocity across the filters, ft/min.

Most air filters are of either the throwaway or cleanable type. Both these types will fit a standard filter rack.

Electrostatic filters are usually employed in industrial installations, where a higher percentage of dust removal must be obtained. Check with manufacturers' ratings for particle-size removal, capacity, and static-pressure loss; also check electric service required. These units generally are used in combination with regular throwaway or cleanable air filters, which take out the large particles, while the charged electrostatic plates remove the smaller ones.

Duct Design. After air discharge grilles and the air handler, which consists of a heat exchanger and blower, have been located, it is advisable to make a single-line drawing showing the duct layout and the air quantities each branch and line must be able to carry.

Of the methods of duct design in use, the equal-friction method is the most practical. It is considered good practice not to exceed a pressure loss of 0.15 in of water per 100 ft of ductwork by friction. Higher friction will result in large power consumption for air circulation. It is also considered good practice to stay below a starting velocity in main ducts of 900 ft/min in residences; 1300 ft/min in schools, theaters, and public buildings; and 1800 ft/min in industrial buildings. Velocity in branch ducts should be about two-thirds of these and in branch risers about one-half.

Too high a velocity will result in noisy and panting ductwork. Too low a velocity will require uneconomical, bulky ducts.

The shape of ducts usually installed is rectangular, because dimensions can easily be changed to maintain the required area. However, as ducts are flattened, the increase in perimeter offers additional resistance to airflow. Thus a flat duct requires an increase in cross section to be equivalent in air-carrying capacity to one more nearly square.

A 12 × 12-in duct, for example, will have an area 1 ft^2 and a perimeter of 4 ft, whereas a 24 × 6-in duct will have the same cross-sectional area but a 5-ft perimeter and thus greater friction. Therefore, a 24 × 7-in duct is more nearly equivalent to the 12 × 12. Equivalent sizes can be determined from tables, such as those in the ASHRAE "Handbook of Fundamentals," where rectangular ducts are rated in terms of equivalent round ducts (equal friction and capacity). Table 18-7 is a shortened version.

Charts also are available in the ASHRAE handbook giving the relationship between duct diameter in inches, air velocity in feet per minute, air quantity in cubic feet per minute, and friction in inches of water pressure drop per 100 ft of duct. Table 18-8 is based on data in the ASHRAE handbook.

In the equal-friction method, the equivalent round duct is determined for the required air flow at the predetermined friction factor. For an example illustrating the method of calculating duct sizes, see Art. 18-11.

(H. E. Bovay, Jr., "Handbook of Mechanical and Electrical Systems for Buildings," McGraw-Hill Book Company, New York.)

TABLE 18-7 Diameters of Circular Ducts in Inches Equivalent to Rectangular Ducts

Side	4	8	12	18	24	30	36	42	48	60	72	84
3	3.8	5.2	6.2									
4	4.4	6.1	7.3									
5	4.9	6.9	8.3									
6	5.4	7.6	9.2									
7	5.7	8.2	9.9									
12	...	10.7	13.1									
18	...	12.9	16.0	19.7								
24	...	14.6	18.3	22.6	26.2							
30	...	16.1	20.2	25.2	29.3	32.8						
36	...	17.4	21.9	27.4	32.0	35.8	39.4					
42	...	18.5	23.4	29.4	34.4	38.6	42.4	45.9				
48	...	19.6	24.8	31.2	36.6	41.2	45.2	48.9	52.6			
60	...	21.4	27.3	34.5	40.4	45.8	50.4	54.6	58.5	65.7		
72	...	23.1	29.5	37.2	43.8	49.7	54.9	59.6	63.9	71.7	78.8	
84	...	....	....	39.9	46.9	53.2	58.9	64.1	68.8	77.2	84.8	91.9
96	...	....	....	....	49.5	56.3	62.4	68.2	73.2	82.6	90.5	97.9

18-7. Heat Losses.

Methods and principles for calculation of heat losses are presented in Art. 18-3. These methods provide a rational procedure for determination of the size and capacity of a heating plant.

Heat loads for buildings consist of heat losses and gains. Heat losses include those from air infiltration, ventilation air, and conduction through the building exterior caused by low temperatures of outside air. Heat gains include those due to hot outside air, solar radiation, electrical lighting and motor loads, and heat from miscellaneous interior equipment. These loads are used to determine the proper equipment size for the lowest initial cost and for operation with maximum efficiency.

Walls and Roofs. Heat loss through the walls and roofs of a building constitutes most of the total heat loss in cold weather. These losses are calculated with Eq. (18-19), $Q = UA(t_2 - t_1)$, with the appropriate temperature differential between inside and outside design temperatures.

Architectural drawings should be carefully examined to establish the materials of construction that will be used in the walls and roofs. With this information, the overall coefficient of heat transmittance, or U factor, can be determined as described in Art. 18-3. Also, from the drawings, the height and width of each wall section should be determined to establish the total area for each wall or roof section for use in Eq. (18-19).

Heat Loss through Basement Floors and Walls. Although heat-transmission coefficients through basement floors and walls are available, it is generally not practicable to use them because ground temperatures are difficult to determine because of the many variables involved. Instead, the rate of heat flow can be estimated, for all practical purposes, from Table 18-9. This table is based on groundwater temperatures, which range from about 40 to 60°F in the northern sections of the United States and 60 to 75°F in the southern sections. (For specific areas, see the ASHRAE "Handbook of Fundamentals.")

Heat Loss from Floors on Grade. Attempts have been made to simplify the variables that enter into determination of heat loss through floors set directly on the ground. The most practical method breaks it down to a heat flow in Btu per hour per linear foot of edge exposed to the outside. With 2 in of edge insulation, the rate of heat loss is about

TABLE 18-8 Sizes of Round Ducts for Airflow*

Friction, in per 100 ft	0.05		0.10		0.15		0.20		0.25		0.30	
Airflow, cfm	Diam, in	Velocity, ft/min	Diam, in	Velocity, ft/min	Diam, in	Velocity, ft/min	Diam, in	Velocity, ft/min	Diam, in	Velocity, ft/min	Diam, in	Velocity, ft/min
50	5.3	350	4.6	450	4.2	530	3.9	600	3.8	660	3.7	710
100	6.8	420	5.8	550	5.4	640	5.1	720	4.8	780	4.7	850
200	8.7	480	7.6	650	6.9	760	6.6	860	6.3	940	6.1	1020
300	10.2	540	8.8	730	8.2	850	7.7	960	7.3	1050	7.1	1120
400	11.5	580	9.8	770	9.0	920	8.5	1040	8.2	1130	7.8	1200
500	12.4	620	11.8	820	9.8	970	9.3	1080	8.8	1160	8.6	1270
1,000	15.8	730	13.7	970	12.8	1140	12.0	1280	11.5	1400	11.2	1500
2,000	20.8	870	18.0	1150	16.6	1370	15.7	1520	15.0	1660	14.5	1780
3,000	24.0	960	21.0	1280	19.7	1500	18.3	1680	17.5	1850		
4,000	26.8	1050	23.4	1360	21.6	1600	20.2	1800				
5,000	29.2	1100	25.5	1463	23.7	1700	22.2	1900				
10,000	37.8	1310	33.2	1770	30.3	2000						

*Based on data in ASHRAE "Handbook of Fundamentals," American Society of Heating, Refrigerating and Air-Conditioning Engineers.

50 in the cold northern sections of the United States, 45 in the temperate zones, 40 in the warm south. Corresponding rates for 1-in insulation are 60, 55, and 50. With no edge insulation the rates are 75, 65, and 60 Btu/(h)(ft).

TABLE 18-9 Below-Grade Heat Losses

Ground water temp, °F	Basement floor loss,* Btu per hr per sq ft	Below-grade wall loss, Btu per hr per sq ft
40	3.0	6.0
50	2.0	4.0
60	1.0	2.0

* Based on basement temperature of 70°F.

Heat Loss from Unheated Attics. Top stories with unheated attics above require special treatment. To determine the heat loss through the ceiling, we must calculate the equilibrium attic temperature under design inside and outside temperature conditions. This is done by equating the heat gain to the attic via the ceiling to the heat loss through the roof:

$$U_c A_c (T_i - T_a) = U_r A_r (T_a - T_o) \tag{18-25}$$

where U_c = heat-transmission coefficient for ceiling
U_r = heat-transmission coefficient for roof
A_c = ceiling area
A_r = roof area
T_i = design room temperature
T_o = design outdoor temperature
T_a = attic temperature

Thus

$$T_a = \frac{U_c A_c T_i + U_r A_r T_o}{U_c A_c + U_r A_r} \tag{18-26}$$

The same procedure should be used to obtain the temperature of other unheated spaces, such as cellars and attached garages.

Air Infiltration. When the heating load of a building is calculated, it is advisable to figure each room separately, to ascertain the amount of heat to be supplied to each room. Then, compute the load for a complete floor or building and check it against the sum of the loads for the individual rooms.

Once we compute the heat flow through all exposed surfaces of a room, we have the heat load if the room is perfectly airtight and the doors never opened. However, this generally is not the case. In fact, windows and doors, even if weather-stripped, will allow outside air to infiltrate and inside air to exfiltrate. The amount of cold air entering a room depends on crack area, wind velocity, and number of exposures, among other things.

Attempts at calculating window- and door-crack area to determine air leakage usually yield a poor estimate. Faster and more dependable is the air-change method, which is based on the assumption that cold outside air is heated and pumped into the premises to create a static pressure large enough to prevent cold air from infiltrating.

The amount of air required to create this static pressure will depend on the volume of the room.

If the number of air changes taking place per hour N are known, the infiltration Q in cubic feet per minute can be computed from

$$Q = \frac{VN}{60} \qquad (18\text{-}27)$$

where V = volume of room, ft^3. The amount of heat q in Btu per hour required to warm up this cold air is given by

$$q = 1.08QT \qquad (18\text{-}28)$$

where Q = ft^3/min of air to be warmed
T = temperature rise

18-8. Heat Gains.

These differ from heat losses only by the direction of the heat flow. Thus, the methods discussed in (Art. 18-7) for heat losses can also be used for determining heat gains. In both cases, the proper inside and outside design conditions and wet-bulb temperatures should be established as described in Art. 18-3.

Heat gains may occur at any time throughout the year. Examples are heat from electric lighting, motor and equipment loads, solar radiation, people, and ventilation requirements. When heat gains occur in cold weather, they should be deducted from the heat loss for the space.

Ventilation and infiltration air in warm weather produce large heat gains and should be added to other calculated heat gains to arrive at the total heat gains for cooling-equipment sizing purposes.

To determine the size of cooling plant required in a building or part of a building, we determine the heat transmitted to the conditioned space through the walls, glass, ceiling, floor, etc., and add all the heat generated in the space. This is the cooling load. The unwanted heat must be removed by supplying cool air. The total cooling load is divided into two parts—sensible and latent.

Sensible and Latent Heat. The part of the cooling load that shows up in the form of a dry-bulb temperature rise is called sensible heat. It includes heat transmitted through walls, windows, roof, floor, etc.; radiation from the sun; and heat from lights, people, electrical and gas appliances, and outside air brought into the air-conditioned space.

Cooling required to remove unwanted moisture from the air-conditioned space is called latent load, and the heat extracted is called latent heat. Usually, the moisture is condensed out on the cooling coils in the cooling unit.

For every pound of moisture condensed from the air, the air-conditioning equipment must remove about 1050 Btu. Instead of rating items that give off moisture in pounds or grains per hour, common practice rates them in Btu per hour of latent load. These items include gas appliances, which give off moisture in products of combustion; steam baths, food, beverages, etc., which evaporate moisture; people; and humid outside air brought into the air-conditioned space.

Design Temperatures for Cooling. Before we can calculate the cooling load, we must first determine a design outside condition and the conditions we want to maintain inside.

For comfort cooling, indoor air at 80°F dry bulb and 50% relative humidity is usually acceptable.

Table 18-4 gives recommended design outdoor summer temperatures for various cities. Note that these temperatures are not the highest ever attained; for example, in New York City, the highest dry-bulb temperature recorded exceeds 105°F, whereas the design outdoor dry-bulb temperature is 95°F. Similarly, the wet-bulb temperature is sometimes above the 75°F design wet-bulb for that area.

Heat Gain through Enclosures. To obtain the heat gain through walls, windows, ceilings, floors, etc., when it is warmer outside than in, the heat-transfer coefficient is multiplied by the surface area and the temperature gradient.

Radiation from the sun through glass is another source of heat. It can amount to about 200 Btu/(h)(ft^2) for a single sheet of unshaded common window glass facing east and west, about three-fourths as much for windows facing northeast and northwest, and one-half as much for windows facing south. For most practical applications, however, the sun effect on walls can be neglected, since the time lag is considerable and the peak load is no longer present by the time the radiant heat starts to work through to the inside surface. Also, if the wall exposed to the sun contains windows, the peak radiation through the glass also will be gone by the time the radiant heat on the walls gets through.

Radiation from the sun through roofs may be considerable. For most roofs, total equivalent temperature differences for calculating solar heat gain through roofs is about 50°F.

Roof Sprays. Many buildings have been equipped with roof sprays to reduce the sun load on the roof. Usually the life of a roof is increased by the spray system, because it prevents swelling, blistering, and vaporization of the volatile components of the roofing material. It also prevents the thermal shock of thunderstorms during hot spells. Equivalent temperature differential for computing heat gain on sprayed roofs is about 18°F.

Water pools 2 to 6 in deep on roofs have been used, but they create structural difficulties. Furthermore, holdover heat into the late evening after the sun has set creates a breeding ground for mosquitoes and requires algae-growth control. Equivalent temperature differential to be used for computing heat gain for water-covered roofs is about 22°F.

Spray control is effected by the use of a water solenoid valve actuated by a temperature controller whose bulb is embedded in the roofing. Tests have been carried out with controller settings of 95, 100, and 105°F. The last was found to be the most practical setting.

The spray nozzles must not be too fine, or too much water is lost by drift. For ridge roofs, a pipe with holes or slots is satisfactory. When the ridge runs north and south, two pipes with two controllers would be practical, for the east pipe would be in operation in the morning and the west pipe in the afternoon.

Heat Gains from Interior Sources. Electric lights and most other electrical appliances convert their energy into heat.

$$q = 3.42W \qquad (18\text{-}29)$$

where q = Btu/h developed
W = watts of electricity used

Where fluorescent lighting is used, add 25% of the lamp rating for the heat generated in the ballast. Where electricity is used to heat coffee, etc., some of the energy is used to vaporize water. Tables in the ASHRAE "Handbook of Fundamentals" give an estimate of the Btu per hour given up as sensible heat and that given up as latent heat by appliances.

Heat gain from people can be obtained from Table 18-10.

Heat Gain from Outside Air. The sensible heat from outside air brought into a conditioned space can be obtained from

$$q_s = 1.08Q(T_o - T_i) \qquad (18\text{-}30)$$

where q_s = sensible load due to outside air, Btu/h
Q = ft^3/min of outside air brought into conditioned space

TABLE 18-10 Rates of Heat Gain from Occupants of Conditioned Spaces*

Degree of activity	Typical application	Total heat, adults, male, Btu/h	Total heat adjusted, Btu/h†	Sensible heat, Btu/h	Latent heat, Btu/h
Seated at rest	Theater, movie	400	350	210	140
Seated, very light work, writing	Office, hotel, apartment	480	420	230	190
Seated, eating	Restaurant‡	520	580‡	255	325
Seated, light work, typing	Office, hotel, apartment	640	510	255	255
Standing, light work, or walking slowly	Retail store, bank	800	640	315	325
Light bench work	Factory	880	780	345	435
Walking, 3 mi/hr, light machine work	Factory	1040	1040	345	695
Bowling§	Bowling alley	1200	960	345	615
Moderate dancing	Dance hall	1360	1280	405	875
Heavy work, heavy machine work, lifting	Factory	1600	1600	565	1035
Heavy work, athletics	Gymnasium	2000	1800	635	1165

*Tabulated values are based on 78°F room dry-bulb temperature. For 80°F room dry bulb, the total heat remains the same, but the sensible-heat value should be decreased by approximately 8% and the latent-heat values increased accordingly. All values are rounded to the nearest 10 Btu/h.

†Adjusted total heat gain is based on normal percentage of men, women, and children for the application listed, with the postulate that the gain from an adult female is 85% of that for an adult male, and that the gain from a child is 75% of that for an adult male.

‡Adjusted total heat gain for eating in a restaurant includes 60 Btu/h for food per individual (30 Btu sensible and 30 Btu latent).

§For bowling, figure one person per alley actually bowling, and all others as sitting (400 Btu/h) or standing and walking slowly (790 Btu/h).

Reprinted from "ASHRAE Handbook of Fundamentals," Chap. 25, 1977, American Society of Heating, Refrigerating and Air-Conditioning Engineers, with permission.

T_o = design dry-bulb temperature of outside air
T_i = design dry-bulb temperature of conditioned space

The latent load due to outside air in Btu per hour is given by

$$q_l = 0.67Q(G_o - G_i) \tag{18-31}$$

where Q = ft³/min of outside air brought into conditioned space
G_o = moisture content of outside air, grains per pound of air
G_i = moisture content of inside air, grains per pound of air

The moisture content of air at various conditions may be obtained from a psychrometric chart.

Miscellaneous Sources of Heat Gain. In an air-conditioning unit, the fan used to circulate the air requires a certain amount of brake horsepower depending on the air quantity and the total resistance in the ductwork, coils, filters, etc. This horsepower will dissipate itself in the conditioned air and will show up as a temperature rise. Therefore, we must include the fan brake horsepower in the air-conditioning load. For most low-pressure air-distribution duct systems, the heat from this source varies from 5% of the sensible load for smaller systems to 3½% of the sensible load in the larger systems.

Where the air-conditioning ducts pass through nonconditioned spaces, the ducts must be insulated. The amount of the heat transmitted to the conditioned air through the insulation may be calculated from the duct area and the insulation heat-transfer coefficient.

METHODS OF HEATING BUILDINGS

The preceding articles present the basic knowledge necessary for accurate determination of the heat losses of a building. Such procedures make possible sizing and selection of heating equipment that will provide reliable and satisfactory service.

18-9. General Procedure for Sizing a Heating Plant.

The basic procedure used in sizing a heating plant is as follows: We isolate the part of the structure to be heated. To estimate the amount of heat to be supplied to that part, we must first decide on the design indoor and outdoor temperatures. For if we maintain a temperature of, say, 70°F inside the structure and the outside temperature is, say, 0°F, then heat will be conducted and radiated to the outside at a rate that can be computed from this 70° temperature difference. If we are to maintain the design inside temperature, we must add heat to the interior by some means at the same rate that it is lost to the exterior.

Recommended design inside temperatures are given in Table 18-11. Recommended design outdoor temperatures for a few cities are given in Table 18-12. (More extensive data are given in the ASHRAE "Handbook of Fundamentals," American Society of Heating, Refrigerating and Air-Conditioning Engineers.)

TABLE 18-11 Recommended Design Indoor Winter Temperatures

Type of Building	Temp, °F
Schools:	
Classrooms	72
Assembly rooms, dining rooms	72
Playrooms, gymnasiums	65
Swimming pool	75
Locker rooms	70
Hospitals:	
Private rooms	72
Operating rooms	75
Wards	70
Toilets	70
Bathrooms	75
Kitchens and laundries	66
Theaters	72
Hotels:	
Bedrooms	70
Ballrooms	68
Residences	72
Stores	68
Offices	72
Factories	65

Note that the recommended design outdoor winter temperatures are not the lowest temperatures ever attained in each region. For example, the lowest temperature on record in New York City is $-14°F$, whereas the design temperature is $0°F$. If the design indoor temperature is $70°F$, we would be designing for $(70-0)/(70+14)$, or 83.3%, of the capacity we would need for the short period that a record cold of $-14°F$ would last.

Once we have established for design purposes a temperature gradient (indoor design temperature minus outdoor design temperature) across the building exterior, we obtain the heat-transmission coefficients of the various building materials in the exterior con-

TABLE 18-12 Recommended Design Outdoor Winter Temperatures

State	City	Temp, °F	State	City	Temp, °F
Ala.	Birmingham	10	Miss.	Vicksburg	10
Ariz.	Flagstaff	−10	Mo.	St. Louis	0
Ariz.	Phoenix	25	Mont.	Helena	−20
Ark.	Little Rock	5	Nebr.	Lincoln	−10
Calif.	Los Angeles	35	Nev.	Reno	−5
Calif.	San Francisco	35	N.H.	Concord	−15
Colo.	Denver	−10	N.J.	Trenton	0
Conn.	Hartford	0	N.Mex.	Albuquerque	0
D.C.	Washington	0	N.Y.	New York	0
Fla.	Jacksonville	25	N.C.	Greensboro	10
Fla.	Miami	35	N.Dak.	Bismarck	−30
Ga.	Atlanta	10	Ohio	Cincinnati	0
Idaho	Boise	−10	Okla.	Tulsa	0
Ill.	Chicago	−10	Ore.	Portland	10
Ind.	Indianapolis	−10	Pa.	Philadelphia	0
Iowa	Des Moines	−15	R.I.	Providence	0
Kans.	Topeka	−10	S.C.	Charleston	15
Ky.	Louisville	0	S.Dak.	Rapid City	−20
La.	New Orleans	20	Tenn.	Nashville	0
Maine	Portland	−5	Tex.	Dallas	0
Md.	Baltimore	0	Tex.	Houston	20
Mass.	Boston	0	Utah	Salt Lake City	−10
Mich.	Detroit	−10	Vt.	Burlington	−10
Minn.	Minneapolis	−20	Va.	Richmond	15

struction for computation of the heat flow per square foot. These coefficients may be obtained from the manufacturers of the materials or from tables, such as those in the ASHRAE "Handbook of Fundamentals." Next, we have to take off from the plans the areas of exposed walls, windows, roof, etc., to determine the total heat flow, which is obtained by adding the sum of the products of the area, temperature gradient, and heat-transmission coefficient for each item (see Art. 18-3).

Choosing Heating-Plant Capacity. Total heat load equals the heat loss through conduction, radiation, and infiltration.

If we provide a heating plant with a capacity equal to this calculated heat load, we shall be able to maintain design room temperature when the design outside temperature prevails, if the interior is already at design room temperature. However, in most buildings the temperature is allowed to drop to as low as 55°F during the night. Thus, theoretically, it will require an infinite time to approach design room temperature. It, therefore, is considered good practice to add 20% to the heating-plant capacity for morning pickup.

The final figure obtained is the minimum heating-plant size required. Consult manufacturers' ratings and pick a unit with a capacity no lower than that calculated by the above method.

On the other hand, it is not advisable to choose a unit too large, because then operating efficiency suffers, increasing fuel consumption.

With a plant of 20% greater capacity than required for the calculated heat load, theoretically after the morning pickup, it will run only 100/120, or 83⅓%, of the time. Furthermore, since the design outdoor temperature occurs only during a small percentage of the heating season, during the rest of the heating season the plant would operate intermittently, less than 83⅓% of the time. Thus it is considered good practice to choose a heating unit no smaller than required but not much larger.

If the heating plant will be used to produce hot water for the premises, determine the added capacity required.

18-10. Heating-Load-Calculation Example.

As an example of the method described for sizing a heating plant, let us take the building shown in Fig. 18-4.

Fig. 18-4 First-floor and second-floor plans of a two-story building.

A design outdoor temperature of 0°F and an indoor temperature of 70°F are assumed. The wall is to be constructed of 4-in brick with 8-in cinder-block backup. Interior finish is metal lath and plaster (wall heat-transmission coefficient $U = 0.25$).

The method of determining the heat load is shown in Table 18-13.

Losses from the cellar include 4 Btu/(h)(ft²) through the walls [column (4)] and 2 Btu/(h)(ft²) through the floors. Multiplied by the corresponding areas, they yield the total heat loss in column (6). In addition, some heat is lost because of infiltration of cold air. One-half an air change per hour is assumed, or 71.2 ft³/min [column (3)]. This causes a heat loss, according to Eq. (18-28), of

$$1.08 \times 71.2 \times 70 = 5400 \text{ Btu/h}$$

To the total for the cellar, 20% is added to obtain the heat load in column (7).

Similarly, heat losses are obtained for the various areas on the first and second floors. Heat-transmission coefficients [column (4)] were obtained from the "Handbook of Fundamentals," American Society of Heating, Refrigerating and Air-Conditioning Engi-

TABLe 18-13 Heat-Load Determination for Two-Story Building

Space	Heat-loss source	Net area or cfm infiltration	U or coefficient	Temp gradient, °F	Heat loss, Btu per hr	Total plus 20%
(1)	(2)	(3)	(4)	(5)	(6)	(7)
Cellar................	Walls	1,170	4		4,680	
	Floor	1,000	2		2,000	
	½ air change	71.2	1.08	70	5,400	14,500
First-floor store.........	Walls	851	0.25	70	14,900	
	Glass	135	1.13	70	10,680	
	Doors	128	0.69	70	1,350	
	1 air change	170	1.08	70	12,850	47,600
Vestibule.............	Glass	48	1.13	70	3,800	
	2 air change	12	1.08	70	905	5,650
Office................	Walls	99	0.25	70	1,730	
	Glass	21	1.13	70	1,660	
	½ air change	9	1.08	70	680	4,900
Men's room...........	Walls	118	0.25	70	2,060	
	Glass	6	1.13	70	48	
	½ air change	3	1.08	70	226	2,800
Ladies' room...........	Walls	60	0.25	70	1,050	
	Glass	6	1.13	70	48	
	½ air change	3	1.08	70	226	1,590
Second-floor office No. 1	Walls	366	0.25	70	6,400	
	Glass	120	1.13	70	9,500	
	Roof	606	0.19	70	8,050	
	½ air change	50	1.08	70	3,780	33,300
Office No. 2...........	Walls	207	0.25	70	3,620	
	Glass	72	1.13	70	5,700	
	Roof	234	0.19	70	3,120	
	½ air change	19	1.08	70	1,435	16,650
Men's room...........	Walls	79	0.25	70	1,380	
	Glass	6	1.13	70	475	
	Roof	20	0.19	70	266	
	½ air change	3	1.08	70	226	2,820
Ladies' room...........	Walls	44	0.25	70	770	
	Glass	6	1.13	70	475	
	Roof	20	0.19	70	266	
	½ air change	3	1.08	70	226	2,080
Hall.................	Walls	270	0.25	70	4,720	
	Door	21	0.69	70	1,015	
	Roof	75	0.19	70	1,000	
	2 air change	40	1.08	70	3,020	12,585
						144,475

neers. These were multiplied by the temperature gradient (70 − 0) to obtain the heat losses in column (6).

The total for the building, plus 20%, amounts to 144,475 Btu/h. A heating plant with approximately this capacity should be selected.

18-11. Warm-Air Heating.

A warm-air heating system supplies heat to a room by bringing in a quantity of air above room temperature, the amount of heat added by the air being at least equal to that required to counteract heat losses.

A gravity system (without a blower) is rarely installed because it depends on the difference in density of the warm-air supply and the colder room air for the working pressure. Airflow resistance must be kept at a minimum with large ducts and very few elbows. The result usually is an unsightly duct arrangement.

A forced-warm-air system can maintain higher air velocities, thus requires smaller ducts, and provides much more sensitive control. For this type of system,

$$q = 1.08Q(T_h - T_i) \tag{18-32}$$

where T_h = temperature of air leaving grille
T_i = room temperature
q = heat added by air, Btu/h
Q = ft^3/min of air supplied to room

Equation (18-32) indicates that the higher the temperature of the discharge air T_h, the less air need be handled. In cheaper installations, the discharge air may be as high as 170°F and ducts are small. In better systems, more air is handled with a discharge temperature as low as 135 to 140°F. With a room temperature of 70°F, we shall need $(170 - 70)/(135 - 70) = 1.54$ times as much air with the 135°F system as with a 170°F system.

It is not advisable to go much below 135°F with discharge air, because drafts will result. With body temperature at 98°F, air at about 100°F will hardly seem warm. If we stand a few feet away from the supply grille, 70°F room air will be entrained with the warm supply air and the mixture will be less than 98°F when it reaches us. We probably would complain about the draft.

Supply grilles should be arranged so that they blow a curtain of warm air across the cold, or exposed, walls and windows. (See Fig. 18-5 for a suggested arrangement.) These grilles should be placed near the floor, since the lower-density warm air will rise and accumulate at the ceiling.

Return-air grilles should be arranged in the interior near unexposed walls—in foyers, closets, etc.—and preferably at the ceiling. This is done for two reasons:

1. The warm air in all heating systems tends to rise to the ceiling. This creates a large temperature gradient between floor and ceiling, sometimes as high as 10°F. Taking the return air from the ceiling reduces this gradient.

2. Returning the warmer air to the heating plant is more economical in operation than using cold air from the floor.

Duct sizes can be determined as described in Duct Design, Art. 18-6. To illustrate the procedure, duct sizes will be computed for the structure with floor plans shown in Figs. 18-4 and 18-5. The latter shows the locations chosen for the discharge grilles and indicates that the boiler room, which contains the heating unit, is in the basement.

Table 18-13 showed that a heating unit with 144,475 Btu/h capacity is required. After checking manufacturers' ratings of forced-warm-air heaters, we choose a unit rated at 160,000 Btu/h and 2010 ft^3/min.

If we utilized the full fan capacity and supply for the rated 160,000 Btu/h and applied Eq. (18-32), the temperature rise through the heater would be

$$\Delta T = \frac{q}{1.08Q} = \frac{160,000}{1.08 \times 2010} = 73.6°F$$

If we adjust the flame (oil or gas) so that the output was the capacity theoretically required, the temperature rise would be

$$\Delta T = \frac{q}{1.08Q} = \frac{144,475}{1.08 \times 2010} = 66.5°F$$

Fig. 18-5 Layout of a duct system for warm-air heating of the basement, first floor, and second floor of the building shown in Fig. 18-4. The boiler room is in the basement.

In actual practice, we do not tamper with the flame adjustment in order to maintain the manufacturer's design balance. Instead, the amount of air supplied to each room is in proportion to its load. (See Table 18-14, which is a continuation of Table 18-13 in the design of a forced-air heating system.) Duct sizes can then be determined for the flow

TABLE 18-14 Air-Supply Distribution in Accordance with Heat Load

Space	Load, Btu per hr	% of Load	Cfm (% × 2,010)
Cellar	14,500	10.05	200
First-floor store	47,600	33.00	660
Vestibule	5,650	3.91	80
Office	4,900	3.39	70
Men's room	2,800	1.94	40
Ladies' room	1,590	1.10	25
Second-floor office No. 1	33,300	23.01	460
Office No. 2	16,650	11.51	230
Men's room	2,820	1.95	40
Ladies' room	2,080	1.44	30
Hall	12,585	8.70	175
	144,475	100.00	2,010

indicated in the table (see Fig. 18-5). In this example, minimum duct size for practical purposes is 12 × 3 in. Other duct sizes were obtained from Table 18-8, for a friction loss per 100 ft of 0.15, and equivalent rectangular sizes were obtained from Table 18-7 and shown in Fig. 18-5.

Humidification in Warm-Air Heating. A warm-air heating system lends itself readily to humidification. Most warm-air furnace manufacturers provide a humidifier that can be placed in the discharge bonnet of the heater. This usually consists of a pan with a ball float to keep the pan filled.

Theoretically, a building needs more moisture when the outside temperature drops. During these colder periods, the heater runs more often, thus vaporizing more water. During warmer periods less moisture is required and less moisture is added because the heater runs less frequently.

Some manufacturers provide a woven asbestos-cloth frame placed in the pan to materially increase the contact surface between air and water. These humidifiers have no control.

Where some humidity control is desired, a spray nozzle connected to the hot-water system with an electric solenoid valve in the line may be actuated by a humidistat.

With humidification, well-fitted storm windows must be used, for with indoor conditions of 70°F and 30% relative humidity and an outdoor temperature of 0°F, condensation will occur on the windows. Storm windows will cut down the loss of room moisture.

Control of Warm-Air Heating. The sequence of operation of a warm-air heating system is usually as follows:

When the thermostat calls for heat, the heat source is started. When the air chamber in the warm-air heater reaches about 120°F, the fan is started by a sensitive element. This is done so as not to allow cold air to issue from the supply grilles and create drafts.

If the flame size and air quantity are theoretically balanced, the discharge air will climb to the design value of, say, 150°F and remain there during the operation of the heater. However, manual shutoff of grilles by residents, dirty filters, etc., will cause a reduction of airflow and a rise in air temperature above design. A sensitive safety element in the air chamber will shut off the heat source when the discharge temperature reaches a value higher than about 180°F. The heat source will again be turned on when the air temperature drops a given amount.

When the indoor temperature reaches the value for which the thermostat is set, the heat source only is shut off. The fan, controlled by the sensitive element in the air chamber, will be shut off after the air cools to below 120°F. Thus, most of the usable heat is transmitted into the living quarters instead of escaping up the chimney.

With commercial installations, the fan usually should be operated constantly, to maintain proper air circulation in windowless areas during periods when the thermostat is satisfied. If residential duct systems have been poorly designed, often some spaces may be too cool, while others may be too warm. Constant fan operation in such cases will tend to equalize temperatures when the thermostat is satisfied.

Duct systems should be sized for the design air quantity of the heater. Insufficient air will cause the heat source to cycle on and off too often. Too much air may cool the flue gases so low as to cause condensation of the water in the products of combustion. This may lead to corrosion, because of dissolved flue gases.

Warm-Air Perimeter Heating. This type of heating is often used in basementless structures, where the concrete slab is laid directly on the ground. The general arrangement is as follows: The heater discharges warm air to two or more under-floor radial ducts feeding a perimeter duct. Floor grilles or baseboard grilles are located as in a conventional warm-air heating system, with collars connected to the perimeter duct.

To prevent excessive heat loss to the outside, it is advisable to provide a rigid waterproof insulation between the perimeter duct and the outside wall.

Air Supply and Exhaust for Heaters. Special types of packaged, or preassembled, units are available for heating that include a direct oil- or gas-fired heat exchanger complete with operating controls. These units must have a flue to convey the products of combustion to the outdoors. The flue pipe must be incombustible and capable of withstanding high temperatures without losing strength from corrosion. Corrosion is usually caused by sulfuric and sulfurous acids, which are formed during combustion and caused by the presence of sulfur in the fuel.

Gas-fired heaters and boilers are usually provided with a draft hood approved by building officials. This should be installed in accordance with the manufacturer's recommendations. Oil-fired heaters and boilers should be provided with an approved draft stabilizer in the vent pipe. The hoods and stabilizers are used to prevent snuffing out of the flame in extreme cases and pulling of excessive air through the combustion chamber when the chimney draft is above normal, as in extremely cold weather.

Flues for the products of combustion are usually connected to a masonry type of chimney. A chimney may have more than one vertical flue. Where flue-gas temperatures do not exceed 600°F, the chimney should extend vertically 3 ft above the high point of the roof or roof ridge when within 10 ft of the chimney. When chimneys will be used for higher-temperature flue gases, many codes require that the chimney terminate not less than 10 ft higher than any portion of the building within 25 ft.

Many codes call for masonry construction of chimneys for both low- and high-temperature flue gases for low- and high-heat appliances. These codes also often call for fireclay flue linings that will resist corrosion, softening, or cracking from flue gases at temperatures up to 1800°F.

Flue-pipe construction must be of heat-resistant materials. The cross-sectional area should be not less than that of the outlet on the heating unit. The flue or vent pipe should be as short as possible and have a slope upwards of not less than ¼ in/ft. If the flue pipe extends a long distance to the chimney, it should be insulated to prevent heat loss and the formation of corrosive acids by condensation of the combustion products.

All combustion-type heating units require air for combustion, and it must be provided in adequate amounts. Combustion air is usually furnished directly from the outside. This air may be forced through ductwork by a fan or by gravity through an outdoor-air louver or special fresh-air intakes. If outside air is not provided for the heating unit, unsatisfactory results can be expected. The opening should have at least twice the cross-sectional area of the vent pipe leaving the boiler.

(H. E. Bovay, Jr., "Handbook of Mechanical and Electrical Systems for Buildings," McGraw-Hill Book Company, New York.)

18-12. Hot-Water Heating Systems.

A hot-water heating system consists of a heater or furnace, radiators, piping systems, and circulator.

A gravity system without circulating pumps is rarely installed. It depends on a difference in density of the hot supply water and the colder return water for working head. Airflow resistance must be kept to a minimum, and the circulating piping system must be of large size. A forced circulation system can maintain higher water velocities, thus requires much smaller pipes and provides much more sensitive control.

Three types of piping systems are in general use for forced hot-water circulation systems:

One-pipe system (Fig. 18-6). This type has many disadvantages and is not usually recommended. It may be seen in Fig. 18-6 that radiator No. 1 takes hot water from the supply main and dumps the colder water back in the supply main. This causes the supply-water temperature for radiator No. 2 to be lower, requiring a corresponding increase in radiator size. (Special flow and return fittings are available to induce flow through the radiators.) The design of such a system is very difficult, and any future adjustment or balancing of the system throws the remainder of the temperatures out.

Two-pipe direct-return system (Fig. 18-7). Here all radiators get the same supply-water temperature, but the last radiator has more pipe resistance than the first. This can

Fig. 18-6 One-pipe hot-water heating system.

Fig. 18-7 Two-pipe direct-return hot-water heating system.

be balanced out by sizing the pump for the longest run and installing orifices in the other radiators to add an equivalent resistance for balancing.

Two-pipe reversed-return system (Fig. 18-8). The total pipe resistance is about the same for all radiators. Radiator No. 1 has the shortest supply pipe and the longest return pipe, while radiator No. 3 has the longest supply pipe and the shortest return pipe.

Supply design temperatures usually are 180°F, with a 20°F drop assumed through the radiators; thus the temperature of the return riser would be 160°F.

When a hot-water, heating system is designed, it is best to locate the radiators, then calculate the water flow in gallons per minute required by each radiator. For a 20°F rise

$$q = 10,000Q \tag{18-33}$$

where q = amount of heat required, Btu/h
Q = flow of water, gal/min

A one-line diagram showing the pipe runs should next be drawn, with gallons per minute to be carried by each pipe noted. The piping may be sized, using friction-flow charts and tables showing equivalent pipe lengths for fittings, with water velocity limited to a maximum of 4 ft/s. (See ASHRAE "Handbook of Fundamentals.") Too high a water velocity will cause noisy flow; too low a velocity will create a sluggish system and costlier piping.

The friction should be between 250 and 600 milinches/ft (1 milinch = 0.001 in). It should be checked against available pump head, or a pump should be picked for the design gallons per minute at the required head.

It is very important that piping systems be made flexible enough to allow for expansion and contraction. Expansion joints are very satisfactory but expensive. Swing joints as shown in Fig. 18-9 should be used where necessary. In this type of branch takeoff, the

Fig. 18-8 Two-pipe reversed-return hot-water heating system.

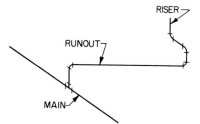

Fig. 18-9 Swing joint permits expansion and contraction of piping.

runout pipe, on expansion or contraction, will cause the threads in the elbows to screw in or out slightly, instead of creating strains in the piping.

(W. J. McGuinness et. al., "Mechanical and Electrical Equipment for Buildings," John Wiley & Sons, Inc., New York.)

Hot-Water Radiators. Radiators, whether of the old cast-iron type, finned pipe, or other, should be picked for the size required, in accordance with the manufacturer's ratings. These ratings depend on the average water temperature. For 170°F average water temperature, 1 ft^2 of radiation surface is equal to 150 Btu/h.

Expansion Tanks for Hot-Water Systems. All hot-water heating systems must be provided with an expansion tank of either the open or closed type.

Figure 18-10 shows an open-type expansion tank. This tank should be located at least 3 ft above the highest radiator and in a location where it cannot freeze.

The size of tank depends on the amount of expansion of the water. From a low near 32°F to a high near boiling, water expands 4% of its volume. Therefore, an expansion tank should be sized for 6% of the total volume of water in radiators, heater, and all piping. That is, the volume of the tank, up to the level of its overflow pipe, should not be less than 6% of the total volume of water in the system.

Figure 18-11 is a diagram of the hookup of a closed-type expansion tank. This tank is only partly filled with water, creating an air cushion to allow for expansion and contraction. The pressure-reducing valve and relief valve are often supplied as a single combination unit.

Hot-Water Heating Systems

Fig. 18-10 Open-type expansion tank.

Fig. 18-11 Closed-type expansion tank.

The downstream side of the reducing valve is set at a pressure below city watermain pressure but slightly higher than required to maintain a static head in the highest radiator. The minimum pressure setting in pounds per square inch is equal to the *height in feet* divided by 2.31.

The relief valve is set above maximum possible pressure. Thus, the system will automatically fill and relieve as required.

Precautions in Hot-Water Piping Layout. One of the most important precautions in a hot-water heating system is to avoid air pockets or loops. The pipe should be pitched so that vented air will collect at points that can be readily vented either automatically or manually. Vents should be located at all radiators.

Pipe traps should contain drains for complete drainage in case of shutdown.

Zones should be valved so that the complete system need not be shut down for repair of a zone. Multiple circulators may be used to supply the various zones at different times, for different temperature settings and different exposures.

Allow for expansion and contraction of pipe without causing undue stresses.

All supply and return piping should be insulated.

In very high buildings, the static pressure on the boiler may be too great. To prevent this, heat exchangers may be installed as indicated in Fig. 18-12. The boiler temperature and lowest zone would be designed for 200°F supply water and 180°F return. The second lowest zone and heat exchanger can be designed for 170°F supply water and 150°F return, etc.

Control of Hot-Water Systems. The

Fig. 18-12 Piping layout for tall building, including heat exchangers.

control system is usually arranged as follows: An immersion thermostat in the heater controls the heat source, such as an oil burner or gas solenoid valve. The thermostat is set to maintain design heater water temperature (usually about 180°F). When the room thermostat calls for heat, it starts the circulator. Thus, an immediate supply of hot water is available for the radiators. A low-limit immersion stat, usually placed in the boiler and wired in series with the room stat and the pump, is arranged to shut off the circulator in the event that the water temperature drops below 70°F. This is an economy measure; if there is a flame failure, water will not be circulated unless it is warm enough to do some good. If the boiler is used to supply domestic hot water via an instantaneous coil or storage tank, hot water will always be available for that purpose. It should be kept in mind that the boiler must be sized for the heating load plus the probable domestic hot-water demand.

High-Temperature, High-Pressure Hot-Water Systems. Some commercial and industrial building complexes have installed hot-water heating systems in which the water temperature is maintained well above 212°F. This is made possible by subjecting the system to a pressure well above the saturation pressure of the water at the design temperature. Such high-temperature hot-water systems present some inherent hazards. Most important is the danger of a leak, because then the water will flash into steam. Another serious condition occurs when a pump's suction strainer becomes partly clogged, creating a pressure drop. This may cause steam to flash in the circulating pump casing and vapor bind the pump.

These systems are not generally used for heating with radiators. They are mostly used in conjunction with air-conditioning installations in which the air-handling units contain a heating coil for winter heating. The advantage of high-temperature hot-water systems is higher rate of heat transfer from the heating medium to the air. This permits smaller circulating piping, pumps, and other equipment.

(H. E. Bovay, Jr., "Handbook of Mechanical and Electrical Systems for Buildings," McGraw-Hill Book Company, New York.)

18-13. Steam-Heating Systems.

A steam-heating system consists of a boiler or steam generator and a piping system connecting to individual radiators or convectors.

A one-pipe heating system (Fig. 18-13) is the simplest arrangement. The steam-supply pipe to the radiators is also used as a condensate return to the boiler. On startup, as the steam is generated, the air must be pushed out of the pipe and radiators by the steam. This is done with the aid of thermostatic air valves in the radiators. When the system is cold, a small vent hole in the valve is open. After the air is pushed out and steam comes in contact with the thermostatic element, the vent hole automatically closes to prevent escape of steam.

Where pipe runouts are extensive, it is necessary to install large-orifice air vents to eliminate the air quickly from the piping system; otherwise the radiators near the end of the runout may get steam much later than the radiators closest to the boiler.

Air vents are obtainable with adjustable orifice size for balancing a heating system. The orifices of radiators near the boiler are adjusted smaller, while radiators far from the boiler will have orifices adjusted for quick venting. This helps balance the system.

The pipe must be generously sized so as to prevent gravity flow of condensate from interfering with supply steam flow. Pipe capacities for supply risers, runouts, and radiator connections are given in the "ASHRAE Guide" (American Society of Heating, Refrigerating and Air-Conditioning Engineers). Capacities are expressed in square feet of

equivalent direct radiation (EDR):

$$1 \text{ ft}^2 \text{ EDR} = 240 \text{ Btu/h} \tag{18-34}$$

Where capacities are in pounds per hour, 1 lb/h = 970 Btu/h.

Valves on radiators in a one-pipe system must be fully open or closed. If a valve is throttled, the condensate in the radiator will have to slug against a head of steam in the pipe to find its way back to the boiler by gravity. This will cause water hammer.

A **two-pipe system** is shown in Fig. 18-14. The steam supply is used to deliver steam to the supply end of all radiators. The condensate end of each radiator is connected to

Fig. 18-13 One-pipe steam-heating system. Condensate returns through supply pipe.

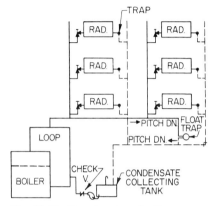

Fig. 18-14 Two-pipe steam-heating system. Different pipes are used for supply and return purposes.

the return line via a thermostatic drip trap (Fig. 18-15). The trap is adjusted to open below 180°F and close above 180°F. Thus when there is enough condensate in the radiator to cool the element, it will open and allow the condensate to return to the collecting tank. A float switch in the tank starts the pump to return this condensate to the boiler against boiler steam pressure.

There are many variations in combining the two-pipe system and one-pipe system to create satisfactory systems.

In the **two-pipe condensate pump return system,** the pressure drop available for pipe and radiator loss is equal to boiler pressure minus atmospheric pressure, the difference being the steam gage pressure.

A **wet-return system,** shown in Fig. 18-16, will usually have a smaller head available for pipe loss.

Fig. 18-15 Drip trap for condensate end of steam radiator.

It is a self-adjusting system depending on the load. When steam is condensing at a given rate, the condensate will pile up in the return main above boiler level, creating a hydraulic head that forces the condensate into the boiler. The pressure above the water level on the return main will be less than the steam boiler pressure. This pressure difference—boiler pressure minus the pressure that exists above the return main water level—is available for pipe friction and radiator pressure drop.

On morning start-up when the air around the radiators is colder and the boiler is fired harder than during normal operation in an effort to pick up heat faster, the steam side of the radiators will be higher in temperature than normal. The air-side temperature of the radiators will be lower. This increase in temperature differential will create an increase in heat transfer causing a faster rate of condensation and a greater piling up of condensate to return the water to the boiler.

In laying out a system, the steam-supply runouts must be pitched to remove the condensate from the pipe. They may be pitched back so as to cause the condensate to flow against the steam or pitched front to cause the condensate to flow with the steam.

Since the condensate will pile up in the return pipe to a height above boiler-water level to create the required hydraulic head for condensate flow, a check of boiler-room ceiling height, steam-supply header, height of lowest radiator, height of dry return pipe, etc., is necessary to determine the height the water may rise in the return pipe without flooding these components.

Fig. 18-16 Wet-return two-pipe steam-heating system.

The pipe may have to be oversized where condensate flows against the steam (see ASHRAE "Handbook of Fundamentals"). If the pitch is not steep enough, the steam may carry the condensate along in the wrong direction, causing noise and water hammer.

Fig. 18-17 Diagram of a vacuum pump for a vacuum steam-heating system. The eductor (1) maintains the desired vacuum in the lower tank.

A **vacuum heating system** is similar to a steam pressure system with a condensate return pump. The main difference is that the steam pressure system can eliminate air from the piping and radiators by opening thermostatic vents to the atmosphere. The vacuum system is usually operated at a boiler pressure below atmospheric. The vacuum pump must, therefore, pull the noncondensables from the piping and radiators for discharge to atmosphere. Figure 18-17 shows diagrammatically the operation of a vacuum pump.

This unit collects the condensate in a tank. A pump circulates water through an eductor (1), pulling out the noncondensable gas to create the required vacuum. The discharge side of the eductor nozzle is above atmospheric pressure. Thus, an automatic control system allows the noncondensable gas to escape to atmosphere as it accumulates, and the excess condensate is returned to the boiler as the tank reaches a given level.

Vacuum systems are usually sized for a total pressure drop varying from ¼ to ½ psi. Long equivalent-run systems (about 200 ft) will use ½ psi total pressure drop to save pipe size.

Some systems operate without a vacuum pump by eliminating the air during morning pickup by hard firing and while the piping system is above atmospheric pressure. During the remainder of the day, when the rate of firing is reduced, a tight system will operate under vacuum.

(H. E. Bovay, Jr., "Handbook of Mechanical and Electrical Systems for Buildings," McGraw-Hill Book Company, New York.)

18-14. Unit Heaters.

Large open areas, such as garages, showrooms, stores, and workshops, are usually best heated by unit heaters placed at the ceiling.

Figure 18-18 shows the usual connections to a steam unit-heater installation. The thermostat is arranged to start the fan when heat is required. The surface thermostat strapped on the return pipe prevents the running of the fan when insufficient steam is available. Where hot water is used for heating, the same arrangement is used, except that the float and thermostatic trap are eliminated and an air-vent valve is installed. Check manufacturers' ratings for capacities in choosing equipment.

Where steam or hot water is not available, direct gas-fired unit heaters may be installed. However, an outside flue is required to dispose of the products of combustion properly (Fig. 18-19). Check with local ordinances for required flues from direct gas-fired

Fig. 18-18 Steam unit heater. Hot-water heater has air vent, no float and trap.

Fig. 18-19 Connections to outside for gas-fired unit heater.

equipment. (Draft diverters are usually included with all gas-fired heaters. Check with the manufacturer when a draft diverter is not included.) For automatic control, the thermostat is arranged to start the fan and open the gas solenoid valve when heat is required. The usual safety pilot light is included by the manufacturer.

Gas-fired unit heaters are often installed in kitchens and other premises where large quantities of air may be exhausted. When a makeup air sytem is not provided, and the relief air must infiltrate through doors, windows, etc., a negative pressure must result in the premises. This negative pressure will cause a steady downdraft through the flue pipe from the outdoors into the space and prevent proper removal of the products of combustion. In such installations, it is advisable to place the unit heater in an adjoining space from which air is not exhausted in large quantities and deliver the warm air through ducts. Since propeller fans on most unit heaters cannot take much external duct resistance, centrifugal blower unit heaters may give better performance where ductwork is used. Sizes of gas piping and burning rates for gas can be obtained from charts and tables in the ASHRAE "Handbook of Fundamentals" for various capacities and efficiencies.

The efficiency of most gas-fired heating equipment is between 70 and 80%.

18-15. Radiant Heating.

Radiant heating, or panel heating as it is sometimes referred to, consists of a warm pipe coil embedded in the floor, ceiling, or walls. The most common arrangement is to circulate warm water through pipe under the floor. Some installations with warm-air ducts, steam pipes, or electric-heating elements have been installed.

Warm-air ducts for radiant heating are not very common. A modified system normally called the perimeter warm-air heating system circulates the warm air around the perimeter of the structure before discharging the air into the premises via grilles.

Pipe coils embedded in concrete floor slabs or plaster ceilings and walls should not be threaded. Ferrous pipe should be welded, while joints in nonferrous metal pipe should be soldered. Return bends should be made with a pipe bender instead of fittings to avoid joints. All piping should be subjected to a hydrostatic test of at least 3 times the working pressure, with a minimum of 150 psig. Inasmuch as repairs are costly after construction is completed, it is advisable to adhere to the above recommendations.

Construction details for ceiling-embedded coils are shown in Figs. 18-20 to 18-23. Floor-embedded coil construction is shown in Fig. 18-24. Wall-panel coils may be installed as in ceiling panels.

Electrically heated panels are usually prefabricated and should be installed in accordance with manufacturer's recommendations and local electrical codes.

The piping and circuiting of a hot-water radiant heating system are similar to hot-water heating systems with radiators or convectors, except that cooler water is used. However, a 20°F water-temperature drop is usually assumed. Therefore, charts used for

Fig. 18-20 Pipe coil embedded in a structural concrete slab for radiant heating.

Fig. 18-21 Pipe coil embedded in a suspended plaster ceiling for radiant heating.

Fig. 18-22 Pipe coil attached to joists or studs and embedded in plaster for radiant heating.

Fig. 18-23 Pipe coil installed above lath and plaster ceiling for radiant heating.

Fig. 18-24 Pipe coil embedded in a floor slab on grade for radiant heating.

the design of hot-water piping systems may be used for radiant heating, too. (See ASHRAE "Handbook of Fundamentals.") In radiant heating, the radiant coils replace radiators. Balancing valves should be installed in each coil, as in radiators. One wise precaution is to arrange coils in large and extensive areas so as not to have too much resistance in certain circuits. This can be done by using high-resistance continuous coils (Fig. 18-25) or low-resistance grid (Fig. 18-26) or a combination of both (Fig. 18-27).

One of the advantages of this system is its flexibility; coils can be concentrated along or on exposed walls. Also, the warmer supply water may be routed to the perimeter or exposed walls and the cooler return water brought to the interior zones.

Panel and Room Temperatures. Heat from the embedded pipes is transmitted to the panel, which in turn supplies heat to the room by two methods: (1) convection and (2) radiation. The amount of heat supplied by convection depends on the temperature difference between the panel and the air. The amount of heat supplied by radiation depends on the difference between the fourth powers of the absolute temperatures of

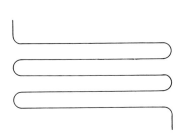
Fig. 18-25 Continuous pipe coil for radiant heating.

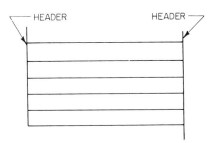
Fig. 18-26 Piping arranged in a grid for radiant heating.

panel and occupants. Thus, as panel temperature is increased, persons in the room receive a greater percentage of heat by radiation than by convection. Inasmuch as high panel temperatures are uncomfortable, it is advisable to keep floor panel temperatures about 85°F or lower and ceiling panel temperatures 100°F or lower. The percentage of radiant heat supplied by a panel at 85°F is about 56% and by one at 100°F about 70%.

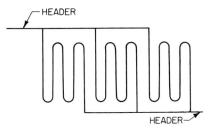

Fig. 18-27 Combination of grid arrangement and continuous pipe coil.

Most advocates of panel heating claim that a lower than the usual design inside temperature may be maintained because of the large radiant surface comforting the individual; i.e., a dwelling normally maintained at 70°F may be kept at an air temperature of about 65°F. The low air temperature makes possible a reduction in heat losses through walls, glass, etc., and thus cuts down the heating load. However, during periods when the heating controller is satisfied and the water circulation stops, the radiant-heat source diminishes, creating an uncomfortable condition due to the below-normal room air temperature. It is thus considered good practice to design the system for standard room temperatures (Table 18-11) and the heating plant for the total capacity required.

Design of Panel Heating. Panel output, Btu per hour per square foot, should be estimated to determine panel-heating area required. Panel capacity is determined by

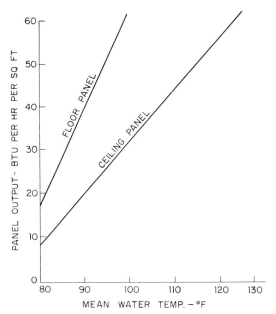

Fig. 18-28 Output of a radiant-heating panel for various panel temperatures and mean radiant temperatures of premises. *(Reprinted with permission from "ASHRAE Guide," American Society of Heating, Refrigerating and Air-Conditioning Engineers.")*

pipe spacing, water temperature, area of exposed walls and windows, infiltration air, insulation value of structural and architectural material between coil and occupied space, and insulation value of structural material preventing heat loss from the reverse side and edge of the panel. For other than ordinary panel-heating design, it is best to leave such design to a specialist or refer to "Systems, ASHRAE Guide and Data Book." For the usual installation with pipe spacing between 6 and 9 in and installed as indicated in Figs. 18-22, 18-23, and 18-24, the curves shown in Fig. 18-28 may be used.

Example. Figure 18-29 illustrates a room 10 × 10 ft with two walls exposed, two walls unexposed, ceiling exposed, and floor panel heated.

First, obtain the heating-load requirements for this room, as discussed in Art. 18-9. Assume storm windows with an overall glass factor of 0.65 Btu/(h)(ft²)(°F) are used. Assume also that exposed walls and ceiling are insulated to provide an overall factor of 0.1 Btu/(h)(ft²)(°F). Then, the heating load comes to 7,320 Btu/h. The panel output for a floor panel with 85°F average water temperature, from Fig. 18-28, is 28.5 Btu/(h)(ft²).

Fig. 18-29 Room to be heated by radiant panel.

$$\text{Required panel area} = \frac{\text{heat load}}{\text{panel output}} = \frac{7320}{28.5} = 256 \text{ ft}^2$$

Inasmuch as the total floor area is only 100 ft², this panel output will be insufficient to maintain comfortable conditions. If this panel were to be designed for 95°F average water temperature (not recommended), the panel capacity would be 54 Btu/(h)(ft²). See Fig. 18-28. The panel area required would be 135 ft², again exceeding 100 ft². Thus, the total heating job in this room cannot be done with panel heating alone. The floor, however, can provide 100 × 28.5 = 2850 Btu/h with panel heating. The remainder, 4470 Btu/h, can be taken care of with radiators set under the large windows.

Where conditions are such that only part of the floor area is required, the pipe grid should be placed near the exposed windows and the interior area left blank.

18-16. Snow Melting.

Design of a snow-melting system for sidewalks, roads, parking areas, etc., involves the determination of a design amount of snowfall, sizing and layout of piping, and selection of heat exchanger and circulating medium. The pipe is placed under the wearing surface, with enough cover to protect it against damage from traffic loads, and a heated fluid is circulated through it.

If friction is too high in extensive runs, use parallel loops (Art. 18-15). All precautions for drainage, fabrication, etc., hold for snow-melting panels as well as interior heating panels.

Table 18-15 gives a design rate of snowfall in inches of water equivalent per hour per square foot for various cities.

Table 18-16 gives the required slab output in Btu per hour per square foot at a given circulating-fluid temperature. This temperature may be obtained once we determine the

TABLE 18-15 Water Equivalent of Snowfall

City	In. per hr per sq ft	City	In. per hr per sq ft
Albany, N.Y.	0.16	Evansville, Ind.	0.08
Asheville, N.C.	0.08	Hartford, Conn.	0.25
Billings, Mont.	0.08	Kansas City, Mo.	0.16
Bismarck, N.Dak.	0.08	Madison, Wis.	0.08
Boise, Idaho.	0.08	Minneapolis, Minn.	0.08
Boston, Mass.	0.16	New York, N.Y.	0.16
Buffalo, N.Y.	0.16	Oklahoma City, Okla.	0.16
Burlington, Vt.	0.08	Omaha, Nebr.	0.16
Caribou, Maine.	0.16	Philadelphia, Pa.	0.16
Chicago, Ill.	0.06	Pittsburgh, Pa.	0.08
Cincinnati, Ohio.	0.08	Portland, Maine.	0.16
Cleveland, Ohio.	0.08	St. Louis, Mo.	0.08
Columbus, Ohio.	0.08	Salt Lake City, Utah.	0.08
Denver, Colo.	0.08	Spokane, Wash.	0.16
Detroit, Mich.	0.08	Washington, D.C.	0.16

TABLE 18-16 Heat Output and Circulating-Fluid Temperatures for Snow-Melting Systems

Rate of showfall, in/(h)(ft²) of water equivalent		Air temp, 0°F			Air temp, 10°F			Air temp, 20°F			Air temp, 30°F		
		Wind velocity, mi/h											
		5	10	15	5	10	15	5	10	15	5	10	15
0.08	Slab output, Btu/(h)(ft²)	151	205	260	127	168	209	102	128	154	75	84	94
	Fluid temp, °F	108	135	162	97	117	138	85	97	110	70	75	79
0.16	Slab output, Btu/(h)(ft²)	218	273	327	193	233	274	165	191	217	135	144	154
	Fluid temp, °F	142	169	198	120	149	170	117	129	142	100	105	109
0.25	Slab output, Btu/(h)(ft²)	292	347	401	265	305	346	235	261	287	203	212	221
	Fluid temp, °F	179	206	234	165	186	206	151	163	176	134	139	144

NOTE: This table is based on a relative humidity of 80% for all air temperatures and construction as shown in Fig. 18-30.

rate of snowfall and assume a design outside air temperature and wind velocity. The table assumes a snow-melting panel as shown in Fig. 18-30.

Once the Btu per hour per square foot required is obtained from Table 18-16 and we know the area over which snow is to be melted, the total Btu per hour needed for snow melting can be computed as the product of the two. It is usual practice to add 40% for loss from bottom of slab.

The circulating-fluid temperature given in Table 18-16 is an average. For a 20°F rise, the fluid temperature entering the panel will be 10°F above that found in the table, and the leaving fluid temperature will be 10°F below the average. The freezing point of the fluid should be a few degrees below the minimum temperature ever obtained in the locality.

Check manufacturers' ratings for antifreeze solution properties to obtain the gallons per minute required and the friction loss to find the pumping head.

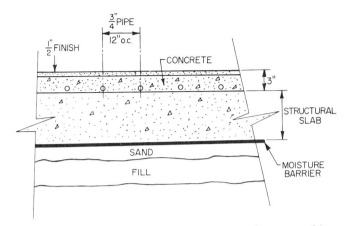

Fig. 18-30 Pipe coils embedded in outdoor concrete slab for snow melting.

When ordering a heat exchanger for a given job, specify to the manufacturer the steam pressure available, fluid temperature to and from the heat exchanger, gallons per minute circulated, and physical properties of the antifreeze solution.

18-17. Radiators and Convectors.

In hot-water and steam-heating systems, heat is released to the spaces to be warmed by radiation and convection. The percentage transmitted by either method depends on the type of heat-dispersal unit used.

A common type of unit is the tubular radiator. It is composed of a series of interconnected sections, each of which consists of vertical tubes looped together. Steam-radiator sections are attached by nipples only at the base, whereas hot-water sections are connected at both top and bottom. Steam radiators should not be used for hot-water heating because of the difficulty of venting.

Pipe-coil radiators are sometimes used in industrial plants. The coils are usually placed on a wall under and between windows and are connected at the ends by branch trees or manifolds. Sometimes, finned-pipe coils are used instead of ordinary pipe. The fins increase the area of heat-transmitting surface.

Since radiators emit heat by convection as well as radiation, any enclosure should permit air to enter at the bottom and leave at the top.

Convectors, as the name implies, transmit heat mostly by convection. They usually consist of finned heating elements placed close to the floor in an enclosure that has openings at bottom and top for air circulation.

Baseboard units consist of continuous heating pipe in a thin enclosure along the base of exposed walls. They may transmit heat mostly by radiation or by convection. Convector-type units generally have finned-pipe heating elements. Chief advantage is the small temperature difference between floor and ceiling.

18-18. Heat Pumps.

A heat-pump cycle is a sequence of operations in which the heat of condensation of a refrigerant is used for heating. The heat required to vaporize the refrigerant is taken from ambient air at the stage where the normal refrigeration cycle (Art. 18-22) usually rejects the heat. In the summer cycle, for cooling, the liquid refrigerant is arranged to flow to the cooling coil through an expansion valve, and the hot gas from the compressor is condensed as in the standard refrigeration cycle (Fig. 18-32). During the heating season, the refrigerant gas is directed to the indoor (heating) coil by the use of multiport electric valves. The condensed liquid refrigerant is then directed to the "condenser" via an expansion valve and is evaporated. This method of heating is competitive with fuel-burning systems in warmer climates where the cooling plant can provide enough heat capacity during the winter season and where electric rates are low. In colder latitudes, the cooling plant, when used as a heat pump, is not large enough to maintain design indoor temperatures and is therefore not competitive with fuel-burning plants.

Other heat sources for heat pumps are well water or underground grid coils. These installations usually call for a valve system permitting warm condenser water to be piped to the air-handling unit for winter heating and the cold water from the chiller to be pumped to the air-handling unit for summer cooling. In winter, the well water or the water in the ground coil is pumped through the chiller to evaporate the liquid refrigerant, while in the summer this water is pumped through the condenser.

Packaged units complete with controls are available for both air-to-air and air-to-water use. Small residential air-to-air units are available in all-electric modes, with capacities up to 9 tons of cooling capacity. Units are made as roof-top or grade-mounted units. Special built-up systems with large capacity have also been constructed.

18-19. Solar Heating.

Solar radiation may be used to provide space heating, cooling, and domestic hot water, but the economics of the application should be carefully investigated. Returns on the initial investment from savings on fuel costs and from tax credits may permit a payback for solar domestic hot-water systems in about 6 to 10 years. Heating and cooling systems will take much longer. The advantage of solar heat is that it is free. Therefore, use of solar heat will overcome the continuing high cost of energy from other sources and conserve those fuels that are in limited supply.

There are many disadvantages in using solar radiation. For one thing, it is available only when the sun is shining and there are hourly variations in intensity, daily and with the weather. Also, the energy received per square foot of radiated surface is small, generally under 400 Btu/(h)(ft^2). This is a very low energy flux. It necessitates large areas of solar collectors to obtain sufficient energy for practical applications and also provide a reasonable payback on the investment.

Solar heating or cooling is advantageous only when the cost of the solar energy produced is less than the cost of energy produced by the more conventional methods. In general, the cost of solar systems may be reduced by obtaining minimum initial costs, favorable amortization rates for the equipment required, and governmental investment tax credits. Also, continuous heat loads and an efficient heating-system design will keep costs low. Bear in mind that the efficiency of solar-system designs for heating or cooling depends on the efficiency of solar collectors, the efficiency of the conversion of the solar radiation to a more useful form of energy, and the efficiency of storage of that energy from the time of conversion until the time of use.

The simplest method of collecting solar energy is by use of a flat-plate-type collector (Fig. 18-31b). The collector is mounted in a manner that allows its flat surface to be held normal, or nearly so, to the sun's rays, thereby, in effect, trapping the solar radiation within the collector. Flat-plate-type collectors are used only in low-temperature systems (70°F to 180°F). Evacuated-tube focusing/concentrating collectors generate much higher temperatures by minimizing heat losses and concentrating sunlight on a reduced absorber surface. Evacuated-tube-type collectors operate in a range of 185° to 250°F, while the concentrating-type collectors operate in an even higher range of 250° to 500°F, or more.

Fig. 18-31 Example of a solar-heating system for a building: (*a*) heat absorbed from solar radiation by a glycol solution is transferred to water for heating a building and producing domestic hot water; (*b*) solar collector for absorbing solar radiation; (*c*) detail of batten assembly of the solar collector. (*Reprinted with permission from F. S. Merritt, "Building Engineering and Systems Design," Van Nostrand Reinhold Company, New York.*)

Precautions should be taken in design and installation of solar collectors on low-sloped roofs to prevent excessive roof deflections or overloads. Also, care should be taken to avoid damaging the roofing or creating conditions that would cause premature roofing failures or lead to higher roofing repair, maintenance, or replacement costs. In particular:

1. Solar collectors should not be installed where ponding of rainwater may occur or in a manner that will obstruct drainage of the roof.
2. Solar collector supports and the roof should be designed for snow loads as well as for wind loads on the collectors, including uplift.
3. The installation should not decrease the fire rating of the roof.
4. The roofing should be protected by boards or other means during erection of the collector. If the roofing membrane must be penetrated for the supports or piping, a roofer should install pipe sleeves, flashing, and other materials necessary to keep out water. If bitumen is used, it should not be permitted to splatter on the collector cover plates.
5. At least 24-in clearance should be provided between the bottom of the collector frames and the roof, to permit inspection, maintenance, repair, and replacement of the roof. Similarly, clearance should be provided at the ends of the frames.
6. At least 14-in clearance should be provided between thermal pipes and the roof surface for the preceding reason. The pipes should be supported on the collector frames, to the extent possible, rather than on the roof.

A typical solar heating and cooling system consists of several major system components: collectors, heat storage, supplemental heat source, and auxiliary equipment.

Many liquid solutions are used for heat transmission in solar systems. A typical solution may be a mixture of water and 10% propylene-glycol antifreeze solution. It is pumped through the collector tubes in a closed-loop circuit to a heat exchanger, where the solution transfers its energy to a water-heating circuit (Fig. 18-31a). The cold glycol solution is then pumped back to the collectors.

The heated water travels in its own closed circuit to a hot-water storage tank and back to the exchanger. Hot water from the storage tank is withdrawn, as required, to satisfy building heating requirements.

The hot water may be pumped from the storage tank to another heat exchanger to produce hot water for domestic use. Hot water from the storage tank may also be pumped through another closed-loop circuit to a heating coil mounted in a central heating blower unit for space heating. In both cases, the cooled water is returned to the storage tank or to the heat exchanger served by the glycol solution.

For use when insufficient solar heat is available, an auxiliary water heater should be added to the system. Providing hot water for heating and domestic purposes, the auxiliary heater may be electrically operated or gas, oil, or coal fired.

When solar heat is to be used for cooling, the hot-water loop must be modified. A three-way valve is placed in the hot-water line to direct the hot water to an absorption-type water chiller. The chilled water discharge is then pumped to a cooling coil in the main air-handling unit.

18-20. Integrated Systems.

The proliferation of mechanical installations in new structures has resulted in the use of hitherto dead space for wireways, pipe runs, air supply or air return, etc. Raised floor space (as in computer rooms), hung-ceiling space, hollow structural columns (round, rectangular, or square tubing) may be used. The basic principles in the design of air requirements for the various spaces do not change.

Once the air quantities have been determined, it is then necessary to check the areas available to move these air quantities and the required horsepower for the job. Utilizing otherwise unused spaces will require reference to basic design manuals for relationship of velocity, friction factors, etc., and should be left to a specialist.

METHODS OF COOLING AND AIR CONDITIONING

The methods for establishing accurate heat losses for heating are also applicable for heat gains and air conditioning. It is mandatory that such procedures be used in order that the necessary cooling equipment can be sized and selected with the lowest first cost that will provide reliable and satisfactory service. (See Arts. 18-7 and 18-8.)

18-21. Sizing an Air-Conditioning Plant.

Consider the building shown in Fig. 18-4 (p. 18-45). The first and second floors only will be air conditioned. The design outdoor condition is assumed to be 95°F DB (dry-bulb) and 75°F WB (wet-bulb). The design indoor condition is 80°F DB and 50% relative humidity.

The temperature gradient across an exposed wall will be 15°F (95°F − 80°F). The temperature gradient between a conditioned and an interior nonconditioned space, such as the cellar ceiling, is assumed to be 10°F.

Exterior walls are constructed of 4-in brick with 8-in cinder backup; interior finish is plaster on metal lath. Partitions consist of 2×4 studs, wire lath, and plaster. First floor has double flooring on top of joists; cellar ceiling is plaster on metal lath.

Lights may be assumed to average 4 W/ft^2. Assume 50 persons will be in the store, 2 in the first-floor office, 10 in the second-floor office No. 1, and 5 in second-floor office No. 2.

Cooling Measured in Tons. Once we obtain the cooling load in Btu per hour, we convert the load to tons of refrigeration by Eq. (18-15)..

Air Requirements for First-Floor Store (Table 18-17a).

$$\text{Load in tons} = \frac{86{,}951}{12{,}000} = 7.25 \text{ tons}$$

If supply air is provided at 18°F differential, with fresh air entering the unit through an outside duct, the flow required for sensible heat is [from Eq. (18-30)]:

$$Q = \frac{55{,}241}{1.08 \times 18} = 2842 \text{ ft}^3/\text{min}$$

If no fresh-air duct is provided and all the infiltration air enters directly into the premises, the flow required is computed as follows:

$$\text{Store volume} = 841 \times 12 = 10{,}100 \text{ ft}^3$$

$$\text{Infiltration} = \frac{10{,}100}{60} \times 1.5 = 252 \text{ ft}^3/\text{min}$$

Sensible load from the infiltration air [Eq. (18-30)] = $252 \times 1.08 \times 15 = 4090$ Btu/h

Internal sensible load plus infiltration = $55{,}241 + 4090 = 59{,}331$ Btu/h

$$\text{Flow with no fresh-air duct} = \frac{59{,}331}{1.08 \times 18} = 3052 \text{ ft}^3/\text{min}$$

Air Requirements for First-Floor Office (Table 18-17b).

$$\text{Load in tons} = \frac{4{,}172}{12{,}000} = 0.35 \text{ ton}$$

TABLE 18-17a First-Floor Store Load

	Area, ft^2		$U(T_o - T_i)$, °F	Heat load, Btu/h
Heat gain through enclosures:				
North wall	92	×	0.25 × 15	344
North door	28	×	0.46 × 15	193
South wall	412	×	0.25 × 15	1,540
South glass	154	×	1.04 × 15	2,405
East wall	480	×	0.25 × 15	1,800
West wall	228	×	0.25 × 15	854
Partition	263	×	0.39 × 10	1,025
Partition door	49	×	0.46 × 10	225
Floor	841	×	0.25 × 10	2,100
Ceiling	112	×	0.25 × 10	280
Load due to conduction				10,766
Sun load on south glass, 154 ft^2 × 98				15,100
Occupants (sensible heat) from Table 18-10, 50 × 315				15,750
Lights [Eq. (18-29)] 841 ft^2 × 4 W/ft^2 × 3.42				11,500
				53,116
Fan hp (4% of total)				2,125
Total internal sensible heat load				55,241
Fresh air (sensible heat) at 10 ft^3/min per person [Eq. (18-30), 500 ft^3/min ×1.08 ×15]				8,100
Total sensible load				63,341
Occupants (latent load) from Table 18-10, 50 × 325				16,250
Fresh air (latent heat) Eq. (18-31), 500 × 0.67 (99-77)				7,360
Total latent load				23,610
Total load				86,951

TABLE 18-17b First-Floor Office Load

	Btu/h	
Heat gain through enclosures:		
North wall, 102 × 0.25 × 15	382	
North glass, 18 × 1.04 × 15	281	
Partition, 30 × 0.39 × 10	117	
Floor, 79 × 0.25 × 10	198	
	978	
2 occupants (sensible load) at 255	510	
Lights, 79 × 4 × 3.42	1080	
	2568	
Fan hp 4%	103	
Total internal sensible load		2671
Fresh air (sensible load) at 2 changes per hour, 32 × 1.08 × 15		518
Total sensible load		3189
Occupants (latent load), 2 at 255	510	
Fresh air (latent load), 32 × 0.67(99 − 77)	473	
Total latent load		983
Total load		4172

$$\text{Supply air required with fresh-air duct} = \frac{2{,}671}{1.08 \times 18} = 137 \text{ ft}^3/\text{min}$$

$$\text{Supply air required with no fresh-air duct} = \frac{3{,}189}{1.08 \times 18} = 164 \text{ ft}^3/\text{min}$$

Air Requirements for Second-Floor Office No. 1 (Table 18-17c).

$$\text{Load in tons} = \frac{41{,}888}{12{,}000} = 3.5 \text{ tons}$$

$$\text{Supply air required with fresh-air duct} = \frac{33{,}768}{1.08 \times 18} = 1737 \text{ ft}^3/\text{min}$$

$$\text{Supply air required with no fresh-air duct} = \frac{36{,}678}{1.08 \times 18} = 1887 \text{ ft}^3/\text{min}$$

Air Requirements for Second-Floor Office No. 2 (Table 18-17d).

$$\text{Load in tons} = \frac{13{,}362}{12{,}000} = 1.13 \text{ tons}$$

$$\text{Supply air required with fresh-air duct} = \frac{9{,}697}{1.08 \times 18} = 499 \text{ ft}^3/\text{min}$$

$$\text{Supply air required with no fresh-air duct} = \frac{10{,}947}{1.08 \times 18} = 563 \text{ ft}^3/\text{min}$$

Cooling and supply-air requirements for the building are summarized in Table 18-18.

TABLE 18-17c Second-Floor Office No. 1

	Btu/h	
Heat gain through enclosures:		
South wall, $130 \times 0.25 \times 15$	388	
South glass, $120 \times 1.04 \times 15$	1,870	
East wall, $260 \times 0.25 \times 15$	975	
West wall, $40 \times 0.25 \times 15$	150	
Partition, $265 \times 0.39 \times 10$	1,035	
Door in partition, $35 \times 0.46 \times 10$	161	
Roof, $568 \times 0.19 \times 54$	5,820	
	10,399	
Sun load on south glass, 120×98	11,750	
Occupants (sensible load), 10×255	2,550	
Lights, $568 \times 4 \times 3.42$	7,770	
	32,469	
Fan hp 4%	1,299	
Total internal sensible load		33,768
Fresh air (sensible load), $180 \times 1.08 \times 15$		2,910
Total sensible load		36,678
Occupants (latent load), 10×255	2,550	
Fresh air (latent load), $180 \times 0.67(99 - 77)$	2,660	
Total latent load		5,210
Total load		41,888

18-70 Heating, Ventilation, and Air Conditioning

TABLE 18-17d Second-Floor Office No. 2

	Btu/h	
Heat gain through enclosures:		
North wall, 103 × 0.25 × 15	386	
North glass, 72 × 1.04 × 15	1,110	
East wall, 132 × 0.25 × 15	495	
Partition, 114 × 0.39 × 10	445	
Door in partition, 18 × 0.46 × 10	83	
Roof, 231 × 0.19 × 54	2,370	
	4,889	
Occupants (sensible load), 5 × 255	1,275	
Lights, 231 × 4 × 3.42	3,160	
	9,324	
Fan hp 4%	373	
Total internal sensible load		9,697
Fresh air (sensible load), 77 × 1.08 × 15		1,250
Total sensible load		10,947
Occupants (latent load), 5 × 255	1,275	
Fresh air (latent load), 77 × 0.67(99 − 77)	1,140	
Total latent load		2,415
Total load		13,362

TABLE 18-18 Cooling-Load Analysis for Building in Fig. 18-4 (p. 18-45)

Space	Tons	Flow with fresh-air duct, ft³/min	Flow with no fresh-air duct, ft³/min
First-floor store	7.25	2842	3052
First-floor office	0.35	137	164
Second-floor office No. 1	3.50	1737	1887
Second-floor office No. 2	1.13	499	563
	12.23	5215	5666

18-22. Refrigeration Cycles.

Figure 18-32 shows the basic air-conditioning cycle of the **direct-expansion type.** The compressor takes refrigerant gas at a relatively low pressure and compresses it to a higher pressure. The hot gas is passed to a condenser where heat is removed and the refrigerant liquefied. The liquid is then piped to the cooling coil of the air-handling unit and allowed to expand to a lower pressure (suction pressure). The liquid vaporizes or is boiled off by the relatively warm air passing over the coil. The compressor pulls away the vaporized refrigerant to maintain the required low coil pressure with its accompanying low temperature.

Chilled-Water Refrigeration Cycle. In some systems, water is chilled by the refrigerant and circulated to units in or near spaces to be cooled (Fig. 18-33), where air is cooled by the water.

In formerly used water-cooled and belt-driven air-conditioning compressors, motor winding heat was usually dissipated into the atmosphere outside the conditioned space (usually into the compressor rooms). For average air-conditioning service 1 hp could produce about 1 ton of cooling. When sealed compressor-motor units came into use, the motor windings were arranged to give off their heat to the refrigerant suction gas. This

Fig. 18-32 Direct-expansion air-conditioning cycle.

Fig. 18-33 Chilled-water air-conditioning cycle.

heat was therefore added to the cooling load of the compressor. The result was that 1 hp could produce only about 0.85 ton of cooling.

Because of water shortages, many communities restrict the direct use of city water for condensing purposes. As an alternative, water can be cooled with cooling towers and recirculated. But for smaller systems, cooling towers have some inherent disadvantages. As a result, air-cooled condensers have become common for small- and medium-sized air-conditioning systems. Because of the higher head pressures resulting from air-cooled condensers, each horsepower of compressor-motor will produce only about 0.65 ton of cooling. Because of these developments, air-conditioning equipment manufacturers rate their equipment in Btu per hour output and kilowatts required for the rated output, at conditions standardized by the industry, instead of in horsepower.

18-23. Air-Distribution Temperature for Cooling.

The air-distribution system is the most critical part of an air-conditioning system. If insufficient air is circulated, proper cooling cannot be done. On the other hand, handling large quantities of air is expensive in both initial cost and operation. The amount of cool air required increases rapidly the closer its temperature is brought to the desired room temperature.

If, for example, we wish to maintain 80°F DB (dry bulb) in a room and we introduce air at 60°F, the colder air when warming up to 80°F will absorb an amount of sensible heat equal to q_s. According to Eq. (18-30), $q_s = 1.08Q_1(80 - 60) = 21.6Q_1$, where Q_1 is the required airflow in cubic feet per minute. From Eq. (18-30), it can be seen also that, if we introduce air at 70°F, with a temperature rise of 10°F instead of 20°F, $q_s = 10.8Q_2$, and we shall have to handle twice as much air to do the same amount of sensible cooling.

From a psychrometric chart, the dew point of a room at 80°F DB and 50% relative humidity is found to be 59°F. If the air leaving the air-conditioning unit is 59°F or less, the duct will sweat and will require insulation. Even if we spend the money to insulate the supply duct, the supply grilles may sweat and drip. Therefore, theoretically, to be safe, the air leaving the air-conditioning unit should be 60°F or higher.

Because there are many days when outside temperatures are less than design conditions, the temperature of the air supplied to the coil will fluctuate, and the temperature of the air leaving the coil may drop a few degrees. This will result in sweating ducts. It is good practice, therefore, to design the discharge-air temperature for about 3°F higher than the room dew point.

Thus, for 80°F DB, 50% RH, and dew point at 59°F, the minimum discharge air temperature would be 62°F as insurance against sweating. The amount of air to be handled may be obtained from Eq. (18-30), with a temperature difference of 18°F.

18-24. Condensers.

If a water-cooled condenser is employed to remove heat from the refrigerant, it may be serviced with city water, and the warm water discharged to a sewer. Or a water tower (Fig. 18-34) may be used to cool the condenser discharge water, which can then be recirculated back to the condenser.

Where practical, the water condenser and tower can be replaced by an evaporative condenser as in Fig. 18-35.

The capacity of water savers, such as towers or evaporative condensers, depends on the wet-bulb temperature. The capacity of these units decreases as the wet-bulb temperature increases.

Such equipment should be sized for a wet-bulb temperature a few degrees above that used for sizing air-conditioning equipment.

As an example, consider an area where the design wet-bulb temperature is 75°F. If we size the air-conditioning equipment for this condition, we shall be able to maintain design inside conditions when the outside conditions happen to be 75°F WB. There will be a few days a year, however, when the outside WB may even attain 79 or 80°F WB. During the higher wet-bulb days, with the air-conditioning equipment in operation, we shall balance out at a relative humidity above design. For example, if the design relative humidity is 50%, we may balance out at 55% or higher.

Fig. 18-34 Water-tower connection to condenser of an air-conditioning system.

Fig. 18-35 Evaporative condenser for an air-conditioning system.

However, if the water towers and evaporative condensers are designed for 75°F WB, then at a wet-bulb temperature above 75, the water-saver capacity would be decreased and the compressor head pressure would build up too high, overload the motor, and kick out on the overload relays. So we use 78°F WB for the design of water savers. Now, on a 78°F WB day, the compressor head pressure would be at design pressure, the equipment will be in operation, although maintaining room conditions a little less comfortable than desired. At 80°F WB, the compressor head pressure will build up above design pressure and the motor will be drawing more than the design current but usually less than the overload rating of the safety contact heaters.

Water condensers are sized for 104°F refrigerant temperature and 95°F water leaving. When condensers are used with city water, these units should be sized for operation with the water at its warmest condition.

The amount of water in gallons per minute required for condensers is equal to

$$Q = \frac{\text{tons of cooling} \times 30}{\text{water-temperature rise}} \qquad (18\text{-}35)$$

This equation includes the heat of condensation of the compressor when used for air-conditioning service. In many cities, the water supplied is never above 75°F. Thus 75°F is used as a condenser design condition.

When high buildings have roof tanks exposed to the sun, it will be necessary to determine the maximum possible water temperature and order a condenser sized for the correct water flow and inlet-water temperature.

When choosing a condenser for water-tower use, we must determine the water temperature available from the cooling tower with 95°F water to the tower. Check the tower manufacturer for the capacity required at the design WB, with 95°F water to the sprays, and the appropriate wet-bulb approach. (Wet-bulb approach is equal to the number of degrees the temperature of the water leaving the tower is above the wet-bulb temperature. This should be for an economic arrangement about 7°F.) Thus, for 78°F, the water leaving the tower would be 85°F, and the condenser would be designed for a 10°F water-temperature rise; i.e., for water at 95°F to the tower and 85°F leaving the tower.

Evaporative condensers should be picked for the required capacity at a design wet-bulb temperature for the area in which they are to be installed. Manufacturers' ratings should be checked before the equipment is ordered.

For small- and medium-size cooling systems, air-cooled condensers are available for outdoor installation with propeller fans or for indoor installation usually with centrifugal blowers for forcing outdoor air through ducts to and from the condensers.

18-25. Compressor-Motor Units.

Compressor capacity decreases with increase in head pressure (or temperature) and fall in suction pressure (or temperature). Therefore, in choosing a compressor-motor unit, one must first determine design conditions for which this unit is to operate; for example:

Suction temperature—whether at 35°F, 40°F, 45°F, etc.

Head pressure—whether serviced by a water-cooled or air-cooled condenser.

Manufacturers of compressor-motor units rate their equipment for Btu per hour output and kilowatt and amperage draw at various conditions of suction and discharge pressure. These data are available from manufacturers, and should be checked before a compressor-motor unit is ordered.

Although capacity of a compressor decreases with the suction pressure (or back pressure), it is usual practice to rate compressors in tons or Btu per hour at various suction temperatures.

For installations in which the latent load is high, such as restaurants, bars, and dance halls, where a large number of persons congregate, the coil temperature will have to be brought low enough to condense out large amounts of moisture. Thus, we find that a suction temperature of about 40°F will be required. In offices, homes, etc., where the latent load is low, 45°F suction will be satisfactory. Therefore, choose a compressor with capacity not less than the total cooling load and rated at 104°F head temperature, and a suction temperature between 40 and 45°F, depending on the nature of the load.

Obtain the brake horsepower of the compressor at these conditions, and make sure that a motor is provided with horsepower not less than that required by the compressor.

A standard NEMA motor can be loaded about 15 to 20% above normal. Do not depend on this safety factor, for it will come in handy during initial pull-down periods, excess occupancy, periods of low-voltage conditions, etc.

Compressor manufacturers have all the above data available for the asking so that one need not guess.

Sealed compressors used in most air-conditioning systems consist of a compressor and a motor coupled together in a single housing. These units are assembled, dehydrated, and sealed at the factory. Purchasers cannot change the capacity of the compressor by changing its speed or connect to it motors with different horsepower. It therefore is necessary to specify required tonnage, suction temperature, and discharge temperature so that the manufacturer can supply the correct, balanced, sealed compressor-motor unit.

18-26. Cooling Equipment—Central Plant Packaged Units.

For an economical installation, the air-handling units should be chosen to handle the least amount of air without danger of sweating ducts and grilles. In most comfort cooling, an 18°F discharge temperature below room temperature will be satisfactory.

If a fresh-air duct is used, the required fresh air is mixed with the return air, and the mix is sent through the cooling coil; i.e., the fresh-air load is taken care of in the coil, and discharge air must take care of the internal load only. Then, from Eq. (18-30), the amount of air to be handled is

$$Q = \frac{q_s}{1.08 \, \Delta T} \qquad (18\text{-}36)$$

where q_s = internal sensible load

ΔT = temperature difference between room and air leaving coil (usually 18°F)

If a fresh-air duct is not installed, and the outside air is allowed to infiltrate into the premises, we use Eq. (18-30) with the sensible part of the load of the infiltration air added to q_s, because the outside air infiltrating becomes part of the internal load.

Once the amount of air to be handled is determined, choose a coil of face area such that the coil-face velocity V would be not much more than 500 ft³/min. Some coil manufacturers recommend cooling-coil-face velocities as high as 700 ft³/min. But there is danger of moisture from the cooling coil being carried along the air stream at such high velocities.

$$V = \frac{Q}{A_c} \qquad (18\text{-}37)$$

where Q = air flow, ft³/min
A_c = coil face area, ft²
V = air velocity, ft/min

The number of rows of coils can be determined by getting the manufacturer's capacity ratings of the coils for three, four, five, six, etc., rows deep and choosing a coil that can handle not less than the sensible and latent load at the working suction temperature.

Multizone air-handling units (Fig. 18-36) are used to control the temperature of more than one space without use of a separate air-handling unit for each zone. When a zone thermostat calls for cooling, the damper motor for that zone opens the cold deck dampers

Fig. 18-36 Multizone air-handling unit.

and closes or throttles the warm deck dampers. Thus, the same unit can provide cooling for one zone while it can provide heating for another zone at the same time.

18-27. Zoning.

In large and complex buildings, there will be many spaces that have different inside design conditions. The reason for this is that it is very difficult, in many cases, to provide an air-handling unit that discharges supply air at a given set of conditions to satisfy the various spaces with different design conditions. Common practice is to combine areas with similar design requirements into a zone. Each zone is then served by a separate air-conditioning unit independently of the other zones.

In some cases, a zone may be satisfied by use of reheat coils to satisfy the zone requirements. However, reheating the low-temperature supply air consumes energy. Hence, this practice is in disfavor and in many cases prohibited by building codes and governmental policies. Other formerly well-established systems, such as multizone and dual-duct systems, are also considered energy consumers and are also in disfavor. Exceptions that permit these systems by codes and statutes are for specific types of manufacturing or processing systems, or for areas where the cooled space must be maintained at very specific temperatures and humidities, such as computer rooms, libraries, operating rooms, paper and printing operations, etc.

(H. E. Bovay, Jr., "Handbook of Mechanical and Electrical Systems for Buildings," McGraw-Hill Book Company, New York.)

18-28. Packaged Air-Conditioning Units.

To meet the great demand for cheaper air-conditioning installations, manufacturers produce packaged, or preassembled, units. These vary from 4000 to 30,000 Btu/h for window units and 9000 to 1,200,000 Btu/h for commercial units. Less field labor is required to install them than for custom-designed installations.

Packaged window units operate on the complete cycle shown in Fig. 18-32, but the equipment is very compactly arranged. The condenser is air cooled and projects outside the window. The cooling coil extends inside. Both the cooling-coil fan and condenser fan usually are run by the same motor.

These packaged units require no piping, just an electric receptacle of adequate capacity. Moisture that condenses on the cooling coil runs via gutters to a small sump near the condenser fan. Many manufacturers incorporate a disk slinger to spray this water on the hot condenser coil, which vaporizes it and exhausts it to the outside. This arrangement serves a double purpose:

1. Gets rid of humidity from the room without piping.
2. Helps keep the head pressure down with some evaporative cooling.

Floor-type and ceiling-type package units also contain the full air-conditioning cycle (Fig. 18-32). The air-cooled packaged units are usually placed near a window to reduce the duct runs required for the air-cooled condenser.

Roof-type packaged units are available for one-story structures or top-floor areas, such as industrial spaces and supermarkets. These units contain a complete cooling cycle (usually with an air-cooled condenser) and a furnace. All controls are factory prewired, and the refrigeration cycle is completely installed at the factory. Necessary ductwork, wiring, and gas piping are supplied in the field.

A combination packaged unit including an evaporative condenser is also available.

Most packaged units are standardized to handle about 400 ft^3/min of air per ton of refrigeration. This is a good average air quantity for most installations. For restaurants, bars, etc., where high latent loads are encountered, the unit handles more than enough air; also, the sensible capacity is greater than required. However, the latent capacity may be lower than desired. This will result in a somewhat higher relative humidity in the premises.

For high sensible-load jobs, such as homes, offices, etc., where occupancy is relatively low, air quantities are usually too low for the installed tonnage. Latent capacity, on the other hand, may be higher than desired. Thus, if we have a total load of 5 tons—4½ tons sensible and ½ ton latent—and we have available a 5-ton unit with a capacity of 4 tons sensible and 1 ton latent, it is obvious that during extreme weather the unit will not be able to hold the dry-bulb temperature down, but the relative humidity will be well below design value. Under such conditions, these units may be satisfactory. In the cases in which the latent capacity is too low and we have more than enough sensible capacity, we can set the thermostat below design indoor dry-bulb temperature and maintain a lower dry-bulb temperature and higher relative humidity to obtain satisfactory comfort conditions. Where we have insufficient sensible capacity, we have to leave the thermostat at the design setting. This will automatically yield a higher dry-bulb temperature and lower relative humidity—and the premises will be comfortable if the total installed capacity is not less than the total load.

As an example of the considerations involved in selecting packaged air-conditioning equipment, let us consider the structure in Fig. 18-4 and the load analysis in Table 18-18.

If this were an old building, it would be necessary to do the following:

First-floor store (load 7.25 tons). Inasmuch as a 5-ton unit is too small, we must choose the next larger size—a 7½-ton packaged unit. This unit has a greater capacity than needed to maintain design conditions. However, many people would like somewhat lower temperature than 80°F dry bulb, and this unit will be capable of maintaining such conditions. Also, if there are periods when more than 50 occupants will be in the store, extra capacity will be available.

First-floor office (load 0.35 tons). Ordinarily a 4000-Btu/h window unit would be required to do the job. But because the store unit has spare capacity, it would be advisable to arrange a duct from the store 7½-ton unit to cool the office.

Second-floor offices (No. 1, load 3.50 tons; No. 2, load 1.13 tons). A 5-ton unit is required for the two offices. But it will be necessary to provide a fresh-air connection to eliminate the fresh-air load from the internal load, to reduce air requirements to the rated flow of the unit. Then, the 2000 ft^3/min rating of the 5-ton packaged unit will be close enough to the 2236 ft^3/min required for the two offices.

Inasmuch as the total tonnage is slightly above that required, we will balance out at a slightly lower relative humidity than 50%, if the particular packaged unit selected is rated at a sensible capacity equal to the total sensible load.

A remote air-conditioning system would be more efficient, because it could be designed to meet the needs of the building more closely. A single air-handling unit for both floors can be arranged with the proper ductwork. However, local ordinances should be checked, since some cities have laws preventing direct-expansion systems servicing more than one floor. A chilled water system (Fig. 18-33) could be used instead.

Split Systems. Split systems differ from other packaged air-conditioning equipment in that the noisy components of the system, notably the refrigerant compressor and air-cooled condenser or cooling tower fans, are located outdoors, away from the air-handling unit. This arrangement is preferred for apartment buildings, residences, hospitals, libraries, churches, and other buildings for which quiet operation is required. In these cases, the air-handler is the only component located within the occupied area. It is usually selected for low revolutions per minute to minimize fan noise. For such installations, chilled-water piping or direct-expansion refrigerating piping must be extended from the cooling coil to the remote compressor and condensing unit, including operating controls.

Packaged Chillers. These units consist of a compressor, water chiller, condenser, and all automatic controls. The contractor has to provide the necessary power and condensing water, from either a cooling tower, city water, or some other source such as well or river. The controls are arranged to cycle the refrigeration compressor to maintain a given chilled-water temperature. The contractor need provide only insulated piping to various chilled-water air-handling units.

Packaged chillers also are available with air-cooled condensers for complete outdoor installation of the unit, or with a condenserless unit for indoor installation connected to an outdoor air-cooled condenser.

The larger packaged chillers, 50 to 1000 tons, are generally powered by centrifugal compressors. When the units are so large that they have to be shipped knocked down, they are usually assembled by the manufacturer's representative on the job site, sealed, pressure-tested, evacuated, dehydrated, and charged with the proper amount of refrigerant.

18-29. Absorption Units for Cooling.

Absorption systems for commercial use are usually arranged as packaged chillers. In place of electric power, these units use a source of heat to regenerate the refrigerant. Gas, oil, or low-pressure steam is used.

In the absorption system used for air conditioning, the refrigerant is water. The compressor of the basic refrigeration cycle is replaced by an absorber, pump, and generator. A weak solution of lithium bromide is heated to evaporate the refrigerant (water). The resulting strong solution is cooled and pumped to a chamber where it absorbs the cold refrigerant vapor (water vapor) and becomes dilute.

Condensing water required with absorption equipment is more than that required for electric-driven compressors. Consult the manufacturer in each case for the water quantities and temperature required for the proper operation of this equipment.

The automatic-control system with these packaged absorption units is usually provided by the manufacturer and generally consists of control of the amount of steam, oil, or gas used for regeneration of the refrigerant to maintain the chilled water temperature properly. In some cases, condenser-water flow is varied to control this temperature.

When low-cost steam is available, low-pressure, single-stage absorption systems may be more economical to operate than systems with compressors. In general, steam consumption is about 20 lb/hr per ton of refrigeration. Tower cooling water required is about 3.7 gal/min per ton with 85°F water entering the absorption unit and 102°F water leaving it.

As an alternative, high-pressure steam, two-stage absorption units are available that operate much more efficiently than the low-pressure units. The high-pressure units use about 12 lb of steam per ton of refrigeration at full load. This is a 40% savings in energy.

18-30. Ducts for Air Conditioning.

In designing a duct system for air conditioning, we must first determine air-outlet locations. If wall grilles are used, they should be spaced about 10 ft apart to avoid dead spots. Round ceiling outlets should be placed in the center of a zone. Rectangular ceiling outlets are available that blow in either one, two, three, or four directions.

Manufacturers' catalog ratings should be checked for sizing grilles and outlets. These catalogs give the recommended maximum amount of air to be handled by an outlet for the various ceiling heights. They also give grille sizes for various lengths of blows. It is obvious that the farther the blow, the higher must be the velocity of the air leaving the grille. Also the higher the velocity, the higher must be the pressure behind the grille.

When grilles are placed back to back in a duct as in Fig. 18-37, be sure that grille A and grille B have the same throw; for if the pressure in the duct is large enough for the longer blow, the short-blow grille will bounce the air off the opposite wall, causing serious drafts. But if the pressure in the duct is just enough for the short blow, the long-blow grille will never reach the opposite wall. Figure 18-38 is recommended for unequal blows because it allows adjustment of air and buildup of a higher static pressure for the longer blow.

Fig. 18-37 Grilles placed back to back on an air-supply duct with no provisions for adjustment of discharge.

Fig. 18-38 Vanes placed in duct to control airflow to back-to-back grilles, useful when air throws are unequal.

In some modern buildings perforated ceiling panels are used to supply conditioned air to the premises. Supply ductwork is provided in the plenum above the suspended ceiling as with standard ceiling outlets. However, with perforated panels, less acoustical fill is used to match the remainder of the hung ceiling.

After all discharge grilles and the air-handling unit are located, it is advisable to make a single-line drawing of the duct run. The air quantities each line and branch must carry should be noted. Of the few methods of duct-system design in use, the equal-friction method is most practical. For most comfort cooling work, it is considered good practice not to exceed 0.15 in friction per 100 ft of ductwork. It is also well to keep the air below 1500 ft/min starting velocity.

If a fresh-air duct is installed, the return-air duct should be sized for a quantity of air equal to the supply air minus the fresh air.

It is advisable, where physically possible, to size the fresh-air duct for the full capacity of the air-handling unit. For example, a 10-ton system handling 4000 ft^3/min of supply air—1000 ft^3/min fresh air and 3000 ft^3/min return air—should have the fresh-air duct sized for 4000 ft^3/min of air. A damper in the fresh-air duct will throttle the air to 1000 ft^3/min during the cooling season; however, during an intermediate season, when the outside air is mild enough, cooling may be obtained by operating only the supply-air fan and opening the damper, thus saving the operation of the 10-ton compressor-motor unit.

As an example of the method for sizing an air-conditioning duct system, let us determine the ductwork for the first floor of the building in Fig. 18-39. Although a load analysis shows that the air requirement is 2979 ft^3/min, we must design the ducts to handle the full capacity of air of the packaged unit we supply. Handling less air will unbalance the unit, causing a drop in suction temperature, and may cause freezing up of the coil. If a 7½-ton packaged unit is used, for example, the ducts should have a capacity at least equal to the 3000 ft^3/min at which this unit is rated.

Table 18-19 shows the steps in sizing the ducts. The 3000 ft^3/min is apportioned to the various zones in the store in proportion to the load from each, and the flow for each segment of duct is indicated in the second column of the table. Next, the size of an equivalent round duct to handle each airflow is determined from Table 18-8 with friction equal to 0.15 in per 100 ft. The size of rectangular duct to be used is obtained from Table 18-7.

Fig. 18-39 Ductwork for cooling a store.

The preceding example of ductwork design falls into the category of low-pressure duct systems. This type of design is used for most air-distribution systems that are not too extensive, such as one- or two-floor systems, offices, and residences. In general, the starting air velocity is below 2000 ft/min, and the fan static pressure is below 3 in of water.

For large multistory buildings, high-velocity air-distribution duct systems often are used. These systems operate at duct velocities well above 3000 ft/min and above 3-in static pressure. Obvious advantages include smaller ducts and lower building cost, since smaller plenums are needed above hung ceilings. Disadvantages are high power consumption for fans and need for an air-pressure-reducing valve and sound attenuation box for each air outlet, resulting in higher power consumption for operation of the system.

Some of the more elaborate heating and air-conditioning installations consist of a high-pressure warm-air duct system and a high-pressure cold-air duct system. Each air outlet is mounted in a sound attenuation box with pressure-reducing valves and branches from the warm- and cold-air systems (Fig. 18-40). Room temperature is controlled by a thermostat actuating two motorized volume dampers. When cooling is required, the thermostat activates the motor to shift the warm-air damper to the closed or throttled position and the cold-air damper to the open position.

TABLE 18-19 Duct Calculation for 7½-ton Package Unit for Store (Fig. 18-39)

Duct	Cfm	Equivalent round duct (friction = 0.15 in. per 100 ft, max velocity = 1,500 fpm), diam, in.	Rectangular duct, in.
A-B	3,000	19.7	28 × 12
B-C	2,700	18.4	24 × 12
C-D	1,900	16.1	18 × 12
D-E	1,100	13.2	12 × 12

18-31. Built-Up Air-Conditioning Units.

Built-up air-conditioning units differ from packaged units in that built-up units are assembled at the site, whereas components of packaged units are preassembled in a factory. Built-up units are usually limited to the larger-size units, with capacity of 50 tons

Fig. 18-40 Double-duct air-distribution system.

and over. They provide cooling air in summer and heated air in winter. In intermediate seasons, the units may provide 0 to 100% outside air for ventilation, with economy control or enthalpy control. The units come complete with prefilters, final filters, return-air fan, dampers, and controls, as required. The units may be installed outdoors at grade or on rooftop, or indoors as a central-plant unit.

A built-up unit usually is enclosed with sandwich-type, insulated panels, which incorporate thermal insulation between steel or cement-asbestos sheets. Special details have been developed for connections of enclosure panels to provide rigidity and to make them weatherproof or airtight. Access doors should be provided between major in-line components.

Built-up units can be made as a complete system with return-air fan, refrigerant compressors and air-cooled condensers for outside installations. Other types without these components are available for split systems that utilize remote refrigerant compressors and condensing units. In chilled-water systems, chilled water is pumped to the cooling coil from the remote chilled-water unit. Heating water is also piped to the heating coil from a remote heat exchanger but with its own separate pumping system.

18-32. All-Air Systems.

HVAC systems for large commercial and institutional buildings are chiefly all-air systems. There are five predominate types: double-duct, dual-conduit, multizone, single-duct with terminal reheat, and variable-volume single duct. These units all have a combination of preheat coils, cooling coils, final heating coils, prefilters, final filters, fan section, return-air duct, and outside-air duct with dampers. A central plant with a hot-water or steam boiler along with a chilled-water refrigeration machine supplies chilled water to the cooling coils and the steam or hot water from the boiler for heating.

The return air is often returned by a return-air fan. This air is mixed with fresh air for ventilation in a mixing chamber prior to the prefilter and filter. If large quantities of outside air are used, a tempering coil may be provided in the outdoor-air stream prior to mixing.

After the air has been filtered, fans move it to the air-handler and the air-distribution equipment. Design of these varies greatly for the different types of air systems mentioned previously.

Double-Duct System. This system utilizes for air distribution two ducts—a hot-air duct and a cold-air duct. At the air handler, the supply-airstream is divided internally to flow over a hot deck (for heating) and a cold deck (for cooling). Pressure from the fan section forces each airstream into the hot and cold distribution ducts for transfer throughout the building. At a specific room or zone to be conditioned, the two airstreams enter a mixing box, which proportions the hot and cold air to produce a specific mix temperature as required to meet the thermal conditions of the space or zone. Mixing or proportioning the supply air in the mixing box is controlled with thermostats located in the conditioned area.

This system is more expensive than other systems because of the necessity of using two distribution ducts and their mixing boxes and controls. It also requires a considerable plenum above the ceiling for concealment. The system is often used in high-rise commercial office buildings, hospitals, laboratories and similar buildings where great diversity of conditions occurs in individual rooms and other methods of zoning would be ineffective.

This system is not usually considered a terminal reheat system but does, in fact, operate essentially on similar principles. Air is cooled down to 50 to 55°F and then is reheated at the mixing box, at the point of use, by being mixed with the warm air. Hence, it inherently wastes energy. Its use is currently in disfavor and is, in many cases, limited by building codes or governmental policies.

High-Velocity Dual-Duct System. This system also employs two main distribution ducts—one for hot air and the other for cold air. The major difference is that the airstreams are delivered at relatively high pressures and inside of specially constructed high-pressure ductwork. With the high-pressure, high-velocity air, much noise is created and sound attenuators must be included in the design. One duct carries the primary supply air and the other, a secondary air supply (Fig. 18-41). A thermostat in the room or zone proportions the amount of hot and cold air to be mixed and supplied to that zone to maintain a constant temperature, irrespective of changes in heating or cooling requirements. Each mixing unit has a constant-volume device so that, under any conditions, the same amount of air will be supplied to the zone.

Fig. 18-41 Example of a high-velocity, dual-duct air-conditioning system.

The secondary airstream, held at 50 to 55°F, is supplied directly to the conditioned spaces to handle the internal load of lights, people, equipment, and sun. A thermostat controls the rate of flow. The primary airstream is heated to between 55°F and 115°F in proportion to the heat load of the building perimeter. A constant volume of this primary air is supplied to each space to be conditioned.

Initial installation and operating costs are lower than those of the double-duct system. The system is applicable to buildings with shallow above-ceiling plenums and for which overhead heating is acceptable, particularly under mild seasonal conditions.

The high-velocity, dual-duct system, like the double-duct systems, is not generally considered a terminal reheat system but nevertheless operates essentially as one. Temperature control is maintained by the mixing of a hot and cold airstream with the hot air being used as reheat. As a result, the system wastes energy. Like the double-duct system, its use is currently in disfavor, and is, in many cases, limited by building codes or governmental policies.

Multizone System. A multizone central-station air handler for heating and cooling provides a number of individual ducts for conducting supply air to an equal number of zones. The air handler contains internally a means to divide the airflow through a hot (for heating) and cold (for cooling) deck. In each deck, there are several motor-operated, modulating dampers, each individually controlled by a zone thermostat. Each zone damper responds to the thermostat's need for heating or cooling by regulating the mix of hot and cold air from each deck as required to satisfy the thermostat. This mixed-air supply is distributed throughout the zone, as required, and is typical for each zone. This system has higher initial and operating costs than other systems because of the multiplicity of individual duct runs and automatic temperature controls.

The multizone system also is a form of reheat system and is considered as an energy waster. Another disadvantage is that, in many units, the zone dampers leak large quantities of hot and cold air, and the system attempts to compensate for the additional hot and cold air, thus wasting additional energy. Innovations have been made, however, to reduce the leakage of zone dampers. In addition, a control device has been included to reset the leaving conditions of the air for both the hot and cold deck in accordance with changes in outside-air temperature from minimum to maximum design conditions. A unit so equipped will limit the amount of reheating to the absolute minimum. Various other methods have also been developed by manufacturers, such as bypassing condenser heat to the heating coil, to reduce further the energy consumption of multizone units.

Single-Duct Reheat System. This system employs a single supply-air duct for distributing cold air at about 55°F continuously. Heating coils at each room or zone, under the control of the room or zone thermostat, reheat the cold supply air, as required, to maintain the room or zone conditions. The reheat device may be steam, hot-water, or electric coils (Fig. 18-42a).

This system provides excellent control of spaces, rooms, or zones with wide variations in loadings. It also provides very good conditions for heating or cooling for perimeter areas and wall exposures, and very close control for process or specific types of industrial manufacturing requirements. This system is often utilized for hospitals, research laboratories, office buildings, and similar facilities that will experience large deviations in load or occupancy.

The room terminal units may be of the straight reheat types mentioned above, induction reheat, or reheat induction type. The induction reheat unit (Fig. 18-42b) utilizes cold supply air that discharges through a special type of nozzle, called an induction nozzle. The discharge from the nozzle induces an equivalent quantity of room air to flow through a reheat coil into the terminal unit and mix with the supply air. This reheated mixture is then discharged into the room.

In a reheat induction unit (Fig. 18-42c), the reheat coil is located upstream of the induction nozzles. The supply air is heated, then mixed with the induced room air for

Fig. 18-42 Room terminal units for all-air central-plant air-conditioning systems: (a) straight reheat; (b) induction reheat; (c) reheat induction.

discharge into the room. This system provides good air circulation in the room because of low resistance through the unit.

Of the various types of systems, the terminal reheat types, when properly designed, provide the best temperature and humidity control; but operating costs will be high. Like the previously described all-air systems, the terminal reheat system wastes energy, because the air is cooled to 50 to 55°F and then reheated to maintain the desired room or zone temperature. As a consequence, terminal reheat systems are prohibited by many building codes.

Older existing systems were often installed to control dehumidification. If dehumidification is no longer required and zone control can be adequately maintained (or dehumidification can be eliminated), then the system can be operated with temperature control only, and the reduction in energy consumption will be considerable.

Variable-Volume, Single-Duct System. The most popular heating and air-conditioning system is the variable-volume system. It consumes relatively little energy because it eliminates reheating at room outlets.

In a variable-volume, single-duct system, as the room thermostat calls for heating or cooling, the volume of supply air is varied as required to satisfy the room load. This variation in air volume is accomplished by use of *variable air-volume boxes.*

As the airflow varies, the system static pressure also tends to vary. Inasmuch as this is undesirable, static pressure regulators are utilized to maintain constant static pressure in the system and at the supply-air fans. The variation in air quantity at the fan is accomplished with fan discharge dampers or with vortex-inlet fan guide vanes, or with both. Operation of these devices are under the control of the system static pressure regulator.

Air volume at the fan is not controlled in some designs of variable-volume systems. In such cases, when the maximum system volume is not required by the conditioned areas, the excess supply air is "dumped" into the return-air plenums and returned to the main air-handling unit. This "dump" system is very important when the cooling coil in the air handler is a direct-expansion refrigerating coil, for which it is mandatory that 100% of the supply-air quantity pass through the coil. This type of system is not quite as efficient as the fan-controlled type, but it saves energy by reducing the system static pressure.

Variable-volume systems are suitable for office buildings, hotels, apartment buildings, schools, hospitals, etc. Their use results in the lowest installed and operating costs of the all-air type systems previously described, and temperature and humidity control is highly acceptable.

(H. E. Bovay, Jr., "Handbook of Mechanical and Electrical Systems for Buildings," McGraw-Hill Book Company, New York.)

18-33. Air-Water Systems.

The only alternative to the air-to-air systems described in Art. 18-32, air-water systems furnish chilled water from a remote chiller or central plant to the room terminal devices. These contain a cooling coil or a heating coil, or both. Room temperature is maintained by varying the flow of chilled water or heating fluid through the coils with valves that respond to the thermostat. Ventilation air is provided from a separate central plant directly to the room or the terminal device.

Two-pipe or four-pipe systems are used for distribution of chilled water and hot water to the room terminals from a central plant. In a two-pipe system, the supply pipe may carry either chilled water or hot water and the second pipe is used as a return. The four-pipe system provides two pipes for chilled-water supply and return and two pipes for hot-water supply and return. The installed cost of the two-pipe system is less than that of the four-pipe system, but the versatility is less. The major disadvantage of the two-pipe system is its inability to provide both heating and cooling with a common supply pipe on days for which both heating and cooling are desired. The four-pipe system has a major drawback in loss of temperature control whenever a changeover from cooling to heating is desired. To overcome this, thermostats are used that permit selection of either cooling or heating by a manual changeover at the thermostats.

Terminal devices for air-water systems are usually of the fan-coil or induction types.

Fan-Coil Terminal Units. A fan-coil terminal device consists of a fan or blower section, chilled-water coil, hot-water heating coil or electric-resistance heating elements, filter, return-air connection, and a housing for these components with an opening for ventilation air. The electric-resistance heating coil is often used with two-pipe systems to provide the performance of a four-pipe system without the cost of the two extra pipes for hot water, insulation, pumps, etc. Fan-coil units may be floor mounted, ceiling mounted-exposed or ceiling mounted-recessed, or ceiling mounted-recessed with supply- and return-air ductwork. When furnished with heating coils, the units are usually mounted on the outside wall or under a window, to neutralize the effects of perimeter heat losses.

Built-in centrifugal fans recirculate room air through the cooling coil. Chilled water circulating through the coil absorbs the room heat load. Ventilation air that is conditioned by another remote central plant is ducted throughout the building and supplied directly to the room or room terminal devices, such as a fan-coil unit. A room thermostat varies the amount of cooling water passing through the cooling coil, thus varying the discharge temperature from the terminal unit and satisfying the room thermostat.

Induction Terminal Units. These units are frequently used in large office buildings. The units are served by a remote air-handling unit that provides high-pressure conditioned air, which may be heated or cooled and is referred to as primary air. It is distributed to individual induction units that are located on the outside walls of each room or zone. At the terminal induction unit, a flow of high-pressure primary air through several nozzles induces a flow of room air through the heating or cooling coil in the unit. Room air mixes with the primary air to provide a mixed-air temperature that satisfies the thermal requirements of the space. In most systems, the ratio of induced air to primary air is about 4 to 1.

The induction system is a large energy consumer because of the extra power required to maintain the high pressure necessary to deliver the primary air to the room induction nozzles and induce room air to flow through the unit coils. Also, the induction terminal unit operates simultaneously with heating and cooling, wasting energy as in a terminal-reheat type of operation.

Air-water systems generally have substantially lower installed and operating costs than all-air systems. They do not, however, provide as good control over room temperature, humidity, air quality, air movement, and noise. The best control of an air-water system is achieved with a fan-coil unit with supplemental ventilation air from a central, primary-air system that provides ventilation air.

(H. E. Bovay, Jr., "Handbook of Mechanical and Electrical Systems for Buildings," McGraw-Hill Book Company, New York.)

18-34. Control Systems for Air Conditioning.

One commonly used system has a thermostat wired in series with the compressor holding-coil circuit (Fig. 18-43). Thus the compressor will stop and start as called for by room conditions. The high-low pressure switch in series with the thermostat is a safety device that stops the compressor when the head pressure is too high and when the suction pressure approaches the freezing temperature of the coil. A liquid solenoid will shut off the flow of refrigerant when the compressor stops, to prevent flooding the coil back to the compressor during the off cycle. This valve may be eliminated when the air-handling unit and compressor are close together.

A second type of control is the pump-down system (Fig. 18-44). The thermostat shuts off the flow of the refrigerant, but the compressor will keep running. With the refrigerant

Fig. 18-43 Control circuit for an air-conditioning system in which the thermostat controls compressor operation.

Fig. 18-44 Pump-down system for control of air conditioning.

supply cut off, the back pressure drops after all the liquid in the coil vaporizes. Then, the high-low pressure switch cuts off the compressor.

Either of these two systems is satisfactory. However, the remainder of this discussion will be restricted to the pump-down system.

Additional safety controls are provided on packaged units to reduce compressor burnouts and increase the average life of these units. The controls include crankcase heaters, motor-winding thermostats, and nonrecycling timers. These controls are usually pre-wired in the factory. The manufacturer supplies installation and wiring instructions for interconnecting the various components of the air-conditioning system.

Where city water is used for condensing purposes, an automatic water-regulator valve is supplied (usually by the manufacturer). Figure 18-45 shows a cross section of such a valve. The power element is attached to the hot-gas discharge of the compressor. As the head pressure builds up, the valve is opened more, allowing a greater flow of water to the condenser, thus condensing the refrigerant at a greater rate. When the compressor is shut off, the head pressure drops, the flow of water being cut off by the action of a power spring working in opposition to the power element. As the water temperature varies, the valve responds to the resulting head pressure and adjusts the flow automatically to maintain design head pressure.

Fig. 18-45 Water-regulating valve for condensing in an air-conditioning system.

When a water tower is used, the automatic water-regulating valve should be removed from the circuit, because it offers too much resistance to the flow of water. It is not usually necessary to regulate the flow of water in a cooling-tower system. For when the outside wet-bulb temperature is so low that the tower yields too low a water temperature, then air conditioning generally is not needed.

Figure 18-46 shows a simplified wiring diagram for a tower system. Because of the interconnecting wiring between the magnetic starters, the tower fan cannot run unless the air-conditioning fan is in operation. Also, the circulating pump cannot run unless the

Fig. 18-46 Control circuit for a cooling-tower-type air-conditioning system.

tower fan is in operation, and the compressor cannot run unless the circulating pump is in operation.

When the system is started, the air-conditioning fan should be operated first, then the tower via switch No. 1, the pump via switch No. 2, and the compressor via switch No. 3. When shutting down, switch No. 3 should be snapped off, then switch No. 2, then switch No. 1, and then the air-conditioning fan.

In buildings where one tower and pump are used to provide water for many units, pressure switches may be used in series with the holding coil of each compressor motor starter. No interconnecting wiring is necessary, for the compressor will not be able to run unless the pump is in operation and provides the necessary pressure to close the contacts on the pressure switch.

Some city ordinances require that indirect-expansion, or chilled-water, systems be installed in public buildings, such as theaters, night clubs, hotels, and depots. This type of system is illustrated in Fig. 18-33. The refrigerant is used to cool water and the water is circulated through the cooling coil to cool the air. The water temperature should be between 40 and 45°F, depending on whether the occupancy is a high latent load or a low latent load. Because the cost of equipment is increased and the capacity decreased as water temperature is lowered, most designers use 45°F water as a basis for design and use a much deeper cooling coil for high-latent-load installations.

The amount of chilled water in gallons per minute to be circulated may be obtained from

$$Q = \frac{24 \times \text{tons of refrigeration}}{\Delta T} \qquad (18\text{-}38)$$

where ΔT is the temperature rise of water on passing through the cooling coil, usually 8 or 10°F. The smaller the temperature change, the more water to be circulated and the greater the pumping cost, but the better will be the heat transfer through the water chiller and cooling coil.

Coil manufacturers' catalogs usually give the procedure for picking the number of rows of coil necessary to do the required cooling with the water temperatures available. The water chiller should be sized to cool the required flow of water from the temperature leaving the cooling coil to that required by the coil at a given suction temperature.

Assuming 45°F water supplied to the cooling coil and a temperature rise of 8°F, the water leaving the coil would be at 53°F. The chiller would then be picked to cool the required flow from 53 to 45°F at 37°F suction. The lower the suction temperature, the smaller will be the amount of heat-transfer surface required in the chiller. Also, the lower will be the compressor capacity. In no case should the suction temperature be less than 35°F, since the freezing point of water—32°F—is too close. A frozen and cracked chiller is expensive to replace.

A simplified control system for a chilled-water system with an evaporative condenser is shown in Fig. 18-47.

18-35. Heating and Air Conditioning.

Most manufacturers of air-conditioning equipment allow space in the air-handling compartment for the installation of a humidification unit and a heating coil for hot water or steam. These make it possible to humidify and heat the air in cold weather. Before a decision is reached to heat through the air-conditioning duct system, it is important to consider the many pitfalls present:

When a unit provides air conditioning in a single room, a heating coil may do the job readily, provided the Btu-per-hour rating of the coil is equal to or greater than the heat-

Fig. 18-47 Control circuit for a chilled-water system with evaporative condenser.

ing load. The fact that the supply grilles are high is usually no disadvantage; since winter heating is designed for the same amount of air as summer cooling, a small temperature rise results, and the large air-change capacity of the air-handling unit creates enough mixing to prevent serious stratification. Difficulties usually are encountered, however, in a structure with both exposed and interior zones. In the winter, the exposed zones need heating, while the interior zones are warm. If the heating thermostat is located in the exposed zone, the interior zones will become overheated. If the thermostat is located in the interior zone, the exposed zones will be too cold. Where some heat is generated in the interior zones, the system may require cooling of the interior and heating of the exterior zones at the same time. Thus, it is impossible to do a heating and cooling job with a single system in such structures.

Where heating and cooling are to be done with the same duct system, the air-handling equipment should be arranged to service individual zones—one or more units for the exposed zones, and one or more units for the interior zones.

When a heating system is already present and an air-conditioning system is added, a heating coil may be used to temper the outside air to room temperature. A duct-type thermostat may be placed in the discharge of the unit to control the steam or hot-water valve. When a room thermostat is used, the spare coil capacity may be used for quick morning pickup. Later in the day, the system may be used for cooling the premises with outside air if the building-heating system or other internal heat sources overheat the premises. In buildings with large window areas, it is advisable to place under the windows for use in cold weather some radiation to supply heat in addition to that from the heating coil, to counteract the down draft from the cold glass surfaces.

The heating-coil size should be such that its face area is about equal to the cooling-coil face area. The number of rows deep should be checked with manufacturers' ratings. When the unit is to be used for tempering only, the coil need be sized for the fresh-air load only [Eq. (18-30)].

When large quantities of outside air are used, it is usual practice to install a preheat coil in the fresh-air duct and a reheat coil in the air-conditioning unit. Install the necessary filters before the preheat coil to prevent clogging.

Fig. 18-48 Circuit for close control of temperature and humidity.

18-36. Industrial Air Conditioning.

Certain manufacturing processes call for close control of temperature and relative humidity. A typical control system is shown in Fig. 18-48. The humidistat may be of the single-pole, double-throw type.

On a rise in room humidity, the refrigeration compressor cuts in for the purpose of dehumidification. Since the sensible capacity of the cooling coil usually is much higher than the sensible load, whereas the latent capacity does not differ too much from the latent load, undercooling results. The room thermostat will then send enough steam or hot water to the reheat coil to maintain proper room temperature.

A small drop in room relative humidity will cut off the compressor. A further drop in room relative humidity will energize the water or steam solenoid valve to add moisture to the air. The thermostat is also arranged to cut off all steam from the reheat coil when the room temperature reaches the thermostat setting. However, there may be periods when the humidistat will not call for the operation of the refrigeration compressor and the room temperature will climb above the thermostat setting. An arrangement is provided for the thermostat to cut in the compressor to counteract the increase in room temperature.

When calculating the load for an industrial system, designers should allow for the thermal capacity and moisture content of the manufactured product entering and leaving the room. Often this part of the load is a considerable percentage of the total.

18-37. Chemical Cooling.

When the dew point of the conditions to be maintained is 45°F or less, the method of controlling temperature and humidity described in Art. 18-36 cannot be used because the coil suction temperature must be below freezing and the coil will freeze up, preventing flow of air. For room conditions with dew point below 45°F, chemical methods of moisture removal, such as silica gel or lithium chloride, may be used. This equipment is arranged to have the air pass through part of the chemical and thus give up its moisture, while another part of the system regenerates the chemical by driving off the moisture previously absorbed. The psychrometric process involved in this type of moisture removal is adiabatic or, for practical purposes, may be considered as being at constant wet-bulb temperature. For example, if we take room air at 70°F and 30% relative humidity and pass the air through a chemical moisture absorber, the air leaving the absorber

Fig. 18-49 Controls for cooling with a chemical moisture absorber.

will be at a wet-bulb temperature of 53°F; i.e., the conditions leaving may be 75°F and 19% RH or 80°F and 11% RH, etc.

It may be noticed that the latent load here is converted to sensible load, and a refrigeration system will be necessary to do sensible cooling. See Fig. 18-49 for the control of such a system.

18-38. Year-Round Air Conditioning.

Electronic-computer rooms and certain manufacturing processes require temperature and humidity to be closely controlled throughout the year. Air-conditioning systems for this purpose often need additional controls for the condensing equipment.

When city water is used for condensing and the condensers are equipped with a water regulating valve, automatic throttling of the water by the valve automatically compensates for low water temperatures in winter.

When cooling towers or evaporative condensers are used, low wet-bulb temperatures of outside air in winter will result in too low a head pressure for proper unit operation. A controller may be used to recirculate some of the humid discharge air back to the evaporative-condenser suction to maintain proper head pressure. With cooling towers, the water temperature is automatically controlled by varying the amount of air across the towers, or by recirculating some of the return condenser water to the supply. Figure 18-50 shows the piping arrangement.

Fig. 18-50 Piping layout for control of temperature of water from a cooling tower.

When air-cooled condensers are used in cold weather, the head pressure may be controlled in one of the following ways: for a single-fan condenser, by varying the fan speed; for a multifan condenser, by cycling the fans or by a combination of cycling and speed variation of fans for closer head-pressure control.

Some head-pressure control systems are arranged to flood refrigerant back to the condenser coil when the head pressure drops. Thus, the amount of heat-transfer surface is reduced, and the head pressure is prevented from dropping too low.

Section 19

Plumbing—Water-Supply, Sprinkler, and Wastewater Systems

Tyler G. Hicks
International Engineering Associates,
New York, N.Y.

This section treats the major subsystems for conveyance of liquids and gases in pipes within a building. The pipes generally extend beyond the building walls to a supply source or a disposal means, such as a sewer.

19-1. Health Requirements for Plumbing.

Building codes, many of which adopt by reference the National Plumbing Code, A40.8, American National Standards Institute, place strict constraints on plumbing installations in the interests of public health. Following are typical basic provisions:

All buildings must be provided with potable water in quantities adequate for the needs of their occupants. Plumbing fixtures, devices and appurtenances should be supplied with water in sufficient volume and at pressures adequate to enable them to function properly. The pipes conveying the water should be of sufficient size to provide the required water without undue pressure reduction and without undue noise under all normal conditions of use.

Plumbing should be designed and adjusted to use the minimum quantity of water consistent with proper performance and cleansing of fixtures and appurtenances.

Devices for heating and storing water should be designed, installed and maintained to guard against rupture of the containing vessel because of overheating.

The wastewater system should be designed, constructed and maintained to guard against fouling, deposit of solids and clogging. It should be provided with adequate cleanouts. A **cleanout** is an opening that provides access to a pipe, either directly or through a short branch, to permit cleaning the pipe. The opening is kept plugged, until the plug has to be removed for a cleansing operation.

Provision should be made in every building for conveying stormwater to a sewer if one is available.

Pipes, joints and connections in the plumbing system should be gastight and watertight for the pressure required by tests. Tests should be made to discover any leaks or defects in the system.

Plumbing fixtures should be located in ventilated enclosures and should be readily accessible to users, but they should not interfere with normal operation of windows or doors or other exits.

Plumbing fixtures should be made of smooth, nonabsorbent materials. They should not have concealed fouling surfaces. Plumbing fixtures, devices and appliances should be protected to prevent contamination of food, water, sterile goods and similar material by backflow of wastewater. Indirect connections with the building wastewater system should be provided when necessary.

Every fixture directly connected to the wastewater system should be equipped with a **liquid-seal trap.** This is a fitting so constructed that passage of air or gas through a pipe is prevented while flow of liquid through the pipe is permitted. An example is a vertical *J* bend in a pipe. When water is not flowing in the pipe, water trapped in the bend of the J blocks passage of gases. When flow in the pipe resumes, the pressure moves the water through the bend.

Foul air in the wastewater system should be exhausted to the outside, through vent pipes. These should be located and installed to minimize the possibility of clogging and the return of foul air to the building.

If a wastewater system is subject to backflow of sewage from a sewer, suitable provision should be made to prevent overflow of the sewage into the building.

The structural safety of a building should not be impaired in any way as a result of the installation, alteration, renovation or replacement of a plumbing system or any of its parts. Pipes should be installed and supported to prevent stresses and strains that would cause malfunction of the system. Provision should be made for expansion and contraction of the pipes due to temperature changes and for structural settlements that might affect the pipes.

Where pipes pass through a construction that is required to have a fire-resistance rating, the space between the pipe and the opening or a pipe sleeve should not exceed ½ in. The gap should be completely packed with rope asbestos, mineral wool or equivalent noncombustible material and closed off with close-fitting metal escutcheons on both sides of the construction.

Pipes, especially those in exterior walls or underground outside a building, should be protected, with insulation or heat, against freezing. Underground pipes should be placed deep enough so as not to be damaged by heavy traffic. Pipes subject to external corrosion should be protected with coatings, wrappings or other means that will prevent corrosion. Dissimilar metals should not be connected to each other, because corrosion will result.

WATER SUPPLY

Enough water to meet the needs of occupants must be available for all buildings. Further water needs for fire protection, heating, air conditioning, and possibly process use must also be met. This section provides specific data on all these water needs except the last—process use. Water needs for process use must be computed separately because the demand depends on the process served.

19-2. Water Quality.

Sources of water for buildings include public water supplies, groundwater, and surface water. Each requires careful study to determine if sufficient quantities are available for the building being designed.

Water for human consumption must be of suitable quality to meet local requirements. Public water supplies generally furnish suitably treated water to a building, eliminating the need for treatment in the building. However, ground and surface waters may require treatment in the building prior to distribution for human consumption. Useful data on water treatment are available from the American Water Works Association.

Useful data on water supply for buildings are available in the following publications: American Society of Civil Engineers, "Glossary—Water and Sewer Control Engineering"; E. W. Steel, "Water Supply and Sewerage," McGraw-Hill Book Company, New York; G. Fair, J. C. Geyer, and D. A. Okun, "Water and Wastewater Engineering," John Wiley & Sons, Inc., New York; E. Nordell, "Water Treatment for Industrial and Other Uses," Van Nostrand-Reinhold, New York; American Society for Testing and Materials, "Manual on Industrial Water." The last-mentioned publication contains extensive data on process-water and steam requirements of a variety of industries. Data on water for fire protection are available from the National Board of Fire Underwriters (NBFU).

Transmission and distribution of water for buildings is generally in pipes, which may be run underground or aboveground. Useful data on pipeline sizing and design are given in: J. Church, "Practical Plumbing Design Guide," and C. E. Davis and K. E. Sorenson, "Handbook of Applied Hydraulics," both McGraw-Hill Book Company, New York. NBFU has a series of publications on water storage tanks for a variety of services.

Characteristics of Water. Physical factors to be considered in raw water are temperature, turbidity, color, taste, and odor. All but temperature are characteristics to be determined in the laboratory by qualified technicians working on carefully procured samples.

Turbidity, a condition due to fine, visible material in suspension, is usually due to presence of colloidal soils. It is expressed in parts per million (ppm) of suspended solids. It may vary widely with discharge of relatively small streams of water. Larger streams or rivers tending to be muddy are generally muddy all the time. The objection to turbidity in potable supplies is its ready detection by the drinker; 10-ppm turbidity shows up as cloudiness in a glass of water. Therefore, the usual limit is 5 ppm.

Color, also objectionable to the drinker, is preferably kept down to 10 ppm or less. It is measured, after all suspended matter (turbidity) has been centrifuged out, by comparison with standard hues.

Tastes and odors due to organic matter or volatile chemical compounds in the water should of course be removed completely from drinking water. But slight, or threshold, odors due to very low concentrations of these compounds are not harmful—just objectionable. Perhaps the most common source of taste and odor is due to decomposition of algae.

Chemical Content. Chemical constituents commonly found in raw waters and measured by laboratory technicians if the water is to be considered for potable use include hardness, pH, iron, and manganese, as well as total solids. Total solids should not exceed 1000 ppm and in some places are kept below 500 ppm. Other limits on individual chemical constituents generally are as given in Table 19-1.

Hardness, measured as calcium carbonate, may be objectionable as a soap waster with as little as 150 ppm of $CaCo_3$ present. But use of synthetic detergents decreases its significance and makes even much harder waters acceptable for domestic uses. Hardness is of concern, however, in waters to be used for boiler feed, where boiler scale must be avoided. Here, 150 ppm would be too much hardness; the water would require softening (treatment for decrease in hardness).

Hydrogen-ion concentration of water, commonly called pH, can be a real factor in corrosion and encrustation of pipe and in destruction of cooling towers. A pH under 7 indicates acidity; over 7 indicates alkalinity; 7 is neutral. Color indicators commonly used can measure pH to the nearest tenth, which is near enough.

TABLE 19-1 Maximum Quantities of Chemicals Permitted in Water

	Ppm
Lead	0.1
Arsenic	0.05
Copper	0.2
Zinc	5.0
Magnesium	125
Iron	0.3
Chlorides	250
Sulfates	250
Phenol (in compounds)	0.001
Fluorine	1.0 (optimum concentration for prevention of tooth decay)

Iron and manganese when present in more than 0.3-ppm concentrations may discolor laundry and plumbing. Their presence and concentration should be determined. More than 0.2 ppm is objectionable for most industrial uses.

Organic Content. Bacteriological tests of water must be made on carefully taken and transported samples. A standard sample is five portions of 10 cm^3, each sample a different dilution of the water tested. A competent laboratory will use standard methods for analyses.

Organisms other than bacteria, such as plankton (free-floating) and algae, can in extreme cases be real factors in the design of water treatment; therefore, biological analyses are significant. Microscopic life, animal and vegetable, is readily tracked down under a high-powered microscope.

Maintenance of Quality. It is not sufficient that potable water be delivered to a building. The quality of the water must be maintained while the water is being conveyed within the building to the point of use. Hence, the potable-water distribution system must be completely sealed to prevent contamination.

No cross connections may be made between this system and any portion of the wastewater-removal system. Furthermore, the potable-water distribution system should be completely isolated from parts of plumbing fixtures or other devices that might contaminate the water. Check valves or air gaps may be used to prevent backflow or back-siphonage.

Backflow is the flow into the distribution piping of any substance from any source other than the intended water-supply source, such as a public water main.

Back-siphonage is the flowing back into the distribution piping of used or polluted water from a plumbing fixture because of a pressure differential.

19-3. Water Treatment.

To maintain water quality within the limits given in Art. 19-2, water supplied to a building usually must undergo some treatment. Whether treatment should be at the source or after transmission to the point of consumption is usually a question of economics, involving hydraulic features, pumping energies and costs, and possible effects of raw water on transmission mains.

In general, treatment, in addition to disinfection, should be provided for water averaging over 20 coliform group organisms per 100 mL, if the water is to be used for domestic purposes. Treatment methods include screening, plain settling, coagulation and sedimentation, filtration, disinfection, softening, and aeration. It is the designer's problem to choose the method that will take the objectionable elements out of the raw water in the simplest, least expensive manner.

Softening of water is a process that must be justified by its need, depending upon the use to which the water will be put. It has been stated that, with a hardness in excess of about 150 ppm, cost of softening will be offset partly by the saving in soap. When synthetic detergents are used instead of soap, this figure may be stretched considerably. But when some industrial use of water requires it, the allowable figure for hardness must be diminished appreciably.

Since corrosion can be costly, corrosive water must often be treated in the interest of economics. In some cases, it may be enough to provide threshold treatment that will coat distribution lines with a light but protective film of scale. But in other cases—boiler-feed water for high-pressure boilers is one—it is important to have no corrosion or scaling. Then, deaeration and pH control may be necessary. (The real danger here is failure of boiler-tube surfaces because of overheating due to scale formation.)

(American Water Works Association, "Water Quality and Treatment," McGraw-Hill Book Company, New York; G. M. Fair, J. C. Geyer, and D. A. Okun, "Elements of Water Supply and Wastewater Disposal," John Wiley & Sons, Inc., New York.)

19-4. Water Quantity and Pressures.

Quantity of water supplied must be adequate for the needs of occupants and processes to be carried out in the building. The total water demand may be calculated by adding the maximum flows at all points of use and applying a factor less than unity to account for the probability that only some of the fixtures will be operated simultaneously (see Art. 19-7).

In addition, the pressure at which water is delivered to a building must lie within acceptable limits. Otherwise, low pressures may have to be increased by pumps and high pressures decreased with pressure-reducing valves. Table 19-2 lists minimum flow rates and pressures generally required at various water outlets. The pressure in Table 19-2 is the pressure in the supply pipe near the water outlet while the outlet is wide open and water is flowing.

In delivery of water to the outlets, there is a pressure drop in the distribution pipes because of friction. Therefore, water supplied at the entrance to the distribution system must exceed the minimum pressures required at the water outlets by the amount of the pressure loss in the system. But the entrance pressure should not exceed 85 psi, to prevent excessive flow and damage to fixtures. Velocity of water in the distribution system should not exceed 8 ft/s.

TABLE 19-2 Rate of Flow and Required Pressure during Flow of Different Fixtures*

Fixture	Flow pressure,† psi	Flow rate, gal/min
Ordinary basin faucet	8	3.0
Self-closing basin faucet	12	2.5
Sink faucet, 3/8 in	10	4.5
Sink faucet, 1/2 in	5	4.5
Bathtub faucet	5	6.0
Laundry-tub cock, 1/2 in	5	5.0
Shower	12	5.0
Ball cock for closet	15	3.0
Flush valve for closet	10–20	15–40‡
Flush valve for urinal	15	15.0
Garden hose, 50 ft and sill cock	30	5.0

*From National Plumbing Code, ASME.

†Flow pressure is the pressure in the pipe at the entrance to the particular fixture considered.

‡Wide range due to variation in design and type of flush-valve closets.

A separate supply of water must be reserved for fire fighting. This supply must be of the most reliable type obtainable. Usually, this requirement can be met with water from a reliable municipal water supply. Otherwise, pumps or storage in an elevated water tank must be provided to supply fire-fighting water. Generally, such water must be provided at a pressure of at least 15 psi at the highest level of fire sprinklers, while flow at the base of the supply riser is a minimum 250 gal/min for light-hazard occupancies and 500 gal/min for ordinary-hazard occupancies.

If a building is so located that it cannot be reached by a fire department with 250 ft of hose, a private underground water system may have to be installed. Hydrants should be placed so that all sides of a building can be reached with hoses.

19-5. Water Distribution in Buildings.

Cold water may be conveyed to plumbing fixtures under the pressure of a water source, such as a public water main, by pumps, or by gravity flow from elevated storage tanks.

Hot water may be delivered at a temperature between 120 and 160°F. It may be distributed by pressure of a water source, by pumps, or by temperature differences. In the last case, the hot water rises above a heater or hot-water storage tank because it is lighter in weight than cold water. If the water is in a closed loop that is heated at the bottom, the water will rise in one leg and drop in the other leg on cooling, because of the difference in pressure in the two legs. As a result, there will be continuous circulation of water in the loop. Hot water can be withdrawn from either leg of the loop.

The water-distribution system should be so laid out that, at each plumbing fixture requiring both hot and cold water, the pressures at the outlets for both supplies should be nearly equal. This is especially desirable where mixing valves may be installed, to prevent the supply at a higher pressure from forcing its way into the lower-pressure supply when the valves are opened to mix hot and cold water. Pipe sizes and types should be selected to balance loss of pressure head due to friction in the hot- and cold-water pipes, despite differences in pipe lengths and sudden large demands for water from either supply.

Horizontal pipe runs should not be truly horizontal. They should have a minimum slope of about ¼ in/ft toward the nearest drain valve.

Upfeed Water Distribution. To prevent rapid wear of valves, such as faucets, water is usually supplied to building distribution systems at pressures not more than about 50 psi. This pressure is large enough to raise water from four to six stories upward and still retain desired pressures at plumbing fixtures (see Table 19-2). Hence, in low buildings, cold water can be distributed by the upfeed method (Fig. 19-1), in which at each story plumbing fixtures are served by branch pipes from risers that carry water upward under pressure from the water source.

In Fig. 19-1, cold water is distributed under pressure from a public water main. The hot-water distribution is by a discontinuous system. Hot water rises from the water heater in the basement to the upper levels under pressure from the cold-water boiler fill. This type of distribution has the disadvantage that, when the system is not in use, hot water in the risers cools and, when a faucet is opened, cooled-off water flows until hot water reaches the faucet.

Downfeed Water Distribution For buildings more than about four stories high, it is conventional to pump water to one or more elevated storage tanks, from which pipes convey the water downward to plumbing fixtures and water heaters. Water in the lower portion of an elevated tank often is reserved for fire-fighting purposes (see Fig. 19-2). Generally, also, the tank is partitioned to provide independent, side-by-side chambers, each with identical piping and controls. During hours of low demand, a chamber can be emptied, cleaned and repaired, if necessary, while the other chamber supplies water as

Fig. 19-1 Upfeed water-distribution system for two-story apartment building. *(Reprinted with permission from F. S. Merritt, "Building Engineering and Systems Design," Van Nostrand Reinhold Company, New York.)*

needed. Float-operated electric switches in the chambers control the pumps supplying water to the tank. When the water level in the tank falls below a specific elevation, the switches start a pump, and when the water level becomes sufficiently high, the switches stop the pump.

Usually, at least two pumps are installed to supply each tank. One pump is used for normal operation. The other is a standby, for use when the first pump is inoperative. For fire fighting, a pump must be capable of refilling the tank at a rate of at least 65 gal/min.

When a pump operates to supply a tank, it may draw so much water from a public main that the pressure in the main would be considerably reduced. To avoid such a condition, water often is stored in a suction tank at the bottom of the building for use by the pumps. The tank is refilled automatically from the public main. Because refilling can take place even when the pumps are not operating, water can be drawn from the public main without much pressure drop.

Figure 19-2 is a simplified schematic diagram of a downfeed distribution system of a type that might be used for buildings up to 20 stories high.

Tall buildings may be divided into zones, each of which is served by a separate downfeed system. (The first few stories may be supplied by an upfeed system under pressure from a public main.) Each zone has at its top its own storage tank, supplied water by its own set of pumps in the basement. All the pumps draw on a common suction tank in the basement. Also, each zone has at its base its own water heater and a hot-water circulation system. In effect, the distribution in each zone is much like that shown in Fig. 19-2.

All water-supply and distribution piping must be designed so there is no possibility of backflow at any time. The minimum required air gap—distance between the fixture outlet and the flood-level rim of the receptable—should be maintained at all times. Before any potable-water piping is put into use it must be disinfected with chlorine, using a procedure approved by the local code.

Pipe Materials. Pipes and tubing for water distribution may be made of copper,

Fig. 19-2 Downfeed water-distribution system for a tall building. *(Reprinted with permission from F. S. Merritt, "Building Engineering and Systems Design," Van Nostrand Reinhold Company, New York.)*

brass, lead, plastics, cast iron, wrought iron, or steel, provided they are approved by the local code. When choosing materials for potable-water piping, care should be taken to see there is no possibility of chemical action, or any other action that might cause a toxic condition.

Fittings. These are used to change the direction of water flow (because it usually is not practical to bend pipe in the field), to make connections between pipes, and to plug openings in pipes or close off the terminal of a pipe. In a water-supply system, fittings and joints must be capable of containing pressurized water flow. Fittings should be of quality equal to that of the pipes connected.

Standard fittings are available and generally may be specified by reference to an American National Standards Institute or a Federal specification. Fitting sizes indicate the diameters of the pipes to which they connect. For threaded fittings, the location of the thread should be specified: a thread on the outside of a pipe is called a **male thread**, whereas an internal thread is known as a **female thread**.

Cast-iron pipe is generally available with standard cast-iron or malleable-iron flanged or threaded fittings. Brass or bronze fittings for copper or brass pipe also may be flanged or threaded. Flanges are held together with bolts. In some cases, to make connections watertight, a gasket may be placed between flanges, whereas in other cases, the flanges may be machine-faced. Threaded fittings often are made watertight by coating the threads with a pipe compound, made with graphite, lead or iron, or by wrapping the threads with lamp wick before the fittings are screwed onto the pipe.

Valves. These are devices incorporated in pipelines to control the flow into, through and from them. Valves are also known as faucets, cocks, bibs, stoppers, and plugs. The term **cock** is generally used with an adjective indicating its use; for example, a sill cock (also called a hose bib) is a faucet used on the outside of a building for connection with a garden hose. A **faucet** is a valve installed on the end of a pipe to permit or stop withdrawal of water from the pipe.

Valves usually are made of cast or malleable iron, brass, or bronze. Faucets in bathrooms or kitchens are usually faced with nickel-plated brass or white alloy for aesthetic reasons.

The types of valves generally used in water-supply systems are gate, globe, angle, and check valves.

Gate valves control flow by sliding a disk perpendicular to the water flow to fit tightly against seat rings when a hand wheel is turned. This type of valve is usually used in locations where it can be left completely open or closed for long periods of time.

Globe valves control the flow by changing the size of the passage through which water can flow past the valves. Turning a hand wheel moves a disk attached at the end of the valve stem to vary the passage area. When the valve is open, the water turns 90° to pass through an orifice enclosed by the seat and then turns 90° again past the disk, to continue in the original direction. Flow can be completely stopped by turning the hand wheel to compress the disk or a gasket on it against the seat. This type of valve usually is used in faucets.

Angle valves are similar to globe valves but eliminate one 90° turn of the water flow. Water is discharged from the valves perpendicular to the inflow direction.

Check valves are used to prevent reversal of flow in a pipe. In the valves, water must flow through an opening with which is associated a movable plug. When water flows in the desired direction, the plug automatically moves out of the way; however, a reverse flow forces the plug into the opening, to seal it.

Pipe Supports. When standard pipe is used for water supply in a building, stresses due to ordinary water pressure are well within the capacity of the pipe material. Unless the pipe is supported at short enough intervals, however, the weight of the pipe and its contents may overstress the pipe material. Generally, it is sufficient to support vertical pipes at their base and at every floor. Maximum support spacing for horizontal pipes depends on pipe diameter and material. Building codes may set limits such as the following:

Threaded pipe, 1 in or less in diameter	8 ft
Threaded pipe, 1¼ in or more in diameter	12 ft
Copper tubing, 1¼ in or less in diameter	6 ft
Copper tubing, 1½ in or more in diameter	10 ft

While the supports should be firmly attached to the building, they should permit movement of the pipe caused by thermal dimensional changes or differences in settlement of building and pipe. Risers should pass through floors preferably through sleeves and transfer their load to the floors through tight-fitting collars. Horizontal pipe runs may be carried on rings or hooks on metal hangers attached to the underside of floors. The hangers and anchors used for plumbing piping should be metal and strong enough to prevent vibration. Each should be designed and installed to carry its share of the total weight of the pipe.

Expansion and Contraction. To provide for expansion and contraction, expansion joints should be incorporated in pipelines. Such joints should be spaced not more than 50 ft apart in hot-water pipe. While special fittings are available for the purpose, flexible connections are a common means of providing for expansion. Frequently, such connections consist of a simple U bend or a spiral coil, which permits springlike absorption of pipe movements.

Meters. These are generally installed on the service pipe to a building to record the amount of water delivered. The meters may be installed inside the building, for protection against freezing, or outside, in a vault below the frost line. Meters should be easily accessible to meter readers. Meter size should be determined by the maximum probable water flow, gal/min.

Water Hammer. This is caused by pressures developing during sudden changes in water velocity or sudden stoppage of flow. The result is a banging sound. It frequently results from rapid closing of valves, but it also may be produced by other means, such as displacing air from a closed tank or pipe from the top.

Water hammer can be prevented by filling a closed tank or pipe from the bottom while allowing the air to escape from the top. Water hammer also can be prevented by installing on pipelines air chambers or other types of water-hammer arresters. These generally act as a cushion to dissipate the pressures.

19-6. Plumbing Fixtures.

The water-supply system of a building distributes water to plumbing fixtures at points of use. Fixtures include kitchen sinks, water closets, urinals, bathtubs, showers, lavatories, drinking fountains, laundry trays, and slop (service) sinks. To ensure maximum sanitation and health protection, most building codes have rigid requirements for fixtures. These requirements cover items like materials of construction, connections, overflows, installation, prevention of backflow, flushing methods, types of fixtures allowed, and inlet and outlet sizes. Most codes and the National Plumbing Code also list the minimum number of each type of fixture that must be installed in buildings of various occupancies (Table 19-3).

The plumbing fixtures are at the terminals of the water-supply system and the start of the wastewater system. To a large extent, the flow from the fixtures determines the quantities of wastewater to be drained from the building.

Traps. Separate traps are required by almost every code on most fixtures not fitted (a)　　　　　　　(b)　　　　　　　(c)　　　　　　　(d)　　　　　　　(e)

Fig. 19-3 Typical water closets: (a) siphon-vortex; (b) siphon-jet; (c) reverse trap; (d) washdown; (e) blowout.

with an integral trap. The trap should be installed as close as possible to the unit served. More than one fixture may be connected to a trap if certain code regulations are observed. For specific requirements, refer to the governing code. Table 19-5 lists the minimum trap size recommended by the National Plumbing Code for various fixtures.

A water seal of at least 2 in, and not more than 4 in, is generally required in most traps. Traps exposed to freezing should be suitably protected to prevent ice formation in the trap body. Cleanouts of suitable size are required on all traps except those made integral with the fixture or those having a portion which is easily removed for cleaning of the interior body. Most codes prohibit use of traps in which a moving part is needed to form the seal. Double trapping is also usually prohibited.

Water Closets. These consist of a bowl and integral trap, which always contain water, and a tank or a flushometer valve, which supplies water for flushing the bowl (Fig. 19-3). The passage through the trap to the discharge usually is large enough to pass a solid ball 2 to 3 in in diameter. Flushometer valves generally require a pressure of at least 15 psi for operation. Tanks, in contrast, are raised above the water level in the bowl so that gravity provides sufficient pressure for flushing.

The cleansing action of water flow in the bowl may be achieved in any of several different ways. One method is illustrated by the **siphon jet** in Fig. 19-3b. The tank discharges water around the rim and also jets water into the upleg of the trap. As a result, the contents of the bowl are siphoned out of the downleg of the discharge pipe. Other types of action include the **reverse trap** (Fig. 19-3c), which is similar to the siphon-jet type but smaller; the **siphon vortex** (Fig. 19-3a), in which water from the rim washes the bowl, creates a vortex, becomes a jet and discharges by siphonage; the **washdown** (Fig. 19-3d), in which pressure buildup causes the upleg to overflow and create a discharge siphon; and the **blowout** (Fig. 19-3e), used with a flushometer valve, which projects a strong jet into the upleg to produce the discharge.

Air Gaps. To prevent backflow of wastewater into the water supply at plumbing fixtures, an air gap must be provided between the fixture water-supply outlet and the flood-level rim of the receptacle. Building codes usually require a minimum gap of 1 to 2 in for outlets not affected by a nearby wall and from 1½ to 3 in for outlets close to a wall. Table 19-4 lists minimum air gaps usually used.

In addition to the usual drain at the lowest point, receptacles generally are provided with a drain at the flood-level rim to prevent water from overflowing. The overflow should discharge into the wastewater system on the fixture side of the trap.

19-7. Water Demand and Fixture Units.

For each fixture in a building, a maximum requirement for water flow, gal/min, can be anticipated. Table 19-2 indicates the minimum flow rate and pressure that building codes require must be provided. The maximum flow may be considerably larger. Branch pipes to each fixture should be sized to provide the maximum flow and minimum pressure the fixture will require. Mains serving these branches, however, need not be sized to handle the sum of the maximum flows for all branches served. It is generally unlikely

TABLE 19-3 Minimum Facilities Recommended by National Plumbing Code

Type of building or occupancy[b]	Water closets		Urinals	Lavatories	Bathtubs or showers	Drinking fountains[c]
Dwelling or apartment houses[d]	1 for each dwelling or apartment unit		....	1 for each apartment or dwelling unit	1 for each apartment or dwelling unit	
Schools:[e]	Male	Female				
Elementary	1 per 100	1 per 35	1 per 30 male	1 per 60 persons		1 per 75 persons
Secondary	1 per 100	1 per 45	1 per 30 male	1 per 100 persons		1 per 75 persons
Office or public buildings	No. of persons 1–15 16–35 36–55 56–80 81–110 111–150 1 fixture for each 40 additional persons	No. of fixtures 1 2 3 4 5 6	Wherever urinals are provided for men, one water closet less than the number specified may be provided for each urinal installed[b] except that the number of water closets in such cases shall not be reduced to less than 2/3 of the minimum specified	No. of persons 1–15 16–35 36–60 61–90 91–125 1 fixture for each 45 additional persons	No. of fixtures 1 2 3 4 5	1 for each 75 persons
Manufacturing, warehouses, workshops, loft buildings, foundries and similar establishments[f]	No. of persons 1–9 10–24 25–49 50–74 75–100 1 fixture for each additional 30 employees	No. of fixtures 1 2 3 4 5	Same substitution as above	1–100 persons, 1 fixture for each 10 persons Over 100, 1 for each 15 persons[g,h]	1 shower for each 15 persons exposed to excessive heat or to skin contamination with poisonous, infectious, or irritating material	1 for each 75 persons
Dormitories[i]	Male: 1 for each 10 persons Female: 1 for each 8 persons Over 10 persons, add 1 fixture for each 25 additional males and 1 for each 20 additional females		1 for each 25 men Over 150 persons, add 1 fixture for each additional 50 men	1 for each 12 persons. (Separate dental lavatories should be provided in community toilet rooms. Ratio of dental lavatories for each 50 persons is recommended.) Add 1 lavatory for each 20 males, 1 for each 15 females	1 for each 8 persons. In the case of women's dormitories, additional bathtubs should be installed at the ratio of 1 for each 30 females Over 150 persons, add 1 fixture for each 20 persons	1 for each 75 persons

TABLE 19-3   Minimum Facilities Recommended by National Plumbing Code *(Continued)*

Type of building or occupancy[b]	Water closets			Urinals		Lavatories		Bathtubs or showers	Drinking fountains[c]
	No. of persons	No. of fixtures		No. of persons	No. of fixtures	No. of persons	No. of fixtures		
		Male	Female	(Male)		1-200	1		
Theaters, auditoriums	1-100	1	1	1-200 201-400 401-600 Over 600; 1 for each additional 300 males	1 2 3	201-400 401-750 Over 750, 1 for each additional 500 persons	2 3		1 for each 100 persons
	101-200	2	2						
	201-400	3	3						
	Over 400, add 1 fixture for each additional 500 males and 1 for each 300 females								

[a] The figures shown are based upon one fixture being the minimum required for the number of persons indicated or any fraction thereof.

[b] Building category not shown on this table will be considered separately by the administrative authority.

[c] Drinking fountains shall not be installed in toilet rooms.

[d] Laundry trays—one single compartment tray for each dwelling unit or 2 compartment trays for each 10 apartments. Kitchen sinks—1 for each dwelling or apartment unit.

[e] This schedule has been adopted (1945) by the National Council on Schoolhouse Construction.

[f] As required by the American Standard Safety Code for Industrial Sanitation in Manufacturing Establishments (ANSI Z4.1-1935).

[g] Where there is exposure to skin contamination with poisonous, infectious, or irritating materials, provide 1 lavatory for each 5 persons.

[h] 24 lin in. of wash sink or 18 in. of a circular basin, when provided with water outlets for such space, shall be considered equivalent to 1 lavatory.

[i] Laundry trays, 1 for each 50 persons. Slop sinks, 1 for each 100 persons.

General. In applying this schedule of facilities, consideration must be given to the accessibility of the fixtures. Conformity purely on a numerical basis may not result in an installation suited to the need of the individual establishment. For example, schools should be provided with toilet facilities on each floor having classrooms.

Temporary workingmen facilities:
1 water closet and 1 urinal for each 30 workmen
24-in. urinal trough = 1 urinal.   48-in. urinal trough = 2 urinals.
36-in. urinal trough = 2 urinals.   60-in. urinal trough = 3 urinals.   72-in. urinal trough = 4 urinals.
Courtesy of American Society of Mechanical Engineers.

that all the fixtures would be supplying maximum flow simultaneously or even that all the fixtures would be operating at the same time. Consequently, the diameters of the mains need be sized only for the probable maximum water demand.

In practice, the probable flow is estimated by weighting the maximum flow in accordance with the probability of fixtures being in use. The estimate is based on the concept of **fixture units.**

A fixture unit is the average discharge, during use, of an arbitrarily selected fixture, such as a lavatory or toilet. Once this value is established, the discharge rates of other types of fixtures are stated in terms of the basic fixture. For example, when the basic fixture is a lavatory served by a 1¼-in trap, the average flow during discharge is 7.5 gal/min. So a bathtub that discharges 15 gal/min is rated as two fixture units (2×7.5). Thus, a tabulation of fixture units can be set up, based on an assumed basic unit.

A specific number of fixture units, as listed in Table 19-5, is assigned to each type of plumbing fixture. These values take into account:

1. The anticipated rate of water flow from the fixture outlet, gal/min
2. The average duration of flow, minutes, when the fixture is used
3. The frequency with which the fixture is likely to be used

TABLE 19-4 Minimum Air Gaps for Generally Used Plumbing Fixtures*

Fixture	Minimum air gap	
	When not affected by near wall†	When affected by near wall‡
Lavatories with effective openings not greater than ½-in. diameter....................	1.0	1.50
Sink, laundry trays, and goose neck bath faucets with effective openings not greater than ¾-in. diameter .	1.5	2.25
Overrim bath fillers with effective openings not greater than 1-in. diameter...............	2.0	3.00
Effective openings greater than 1-in.	¶	§

* From National Plumbing Code, ASME.

† Side walls, ribs, or similar obstructions do not affect the air gaps when spaced from inside edge of spout opening a distance greater than three times the diameter of the effective opening for a single wall, or a distance greater than four times the diameter of the effective opening for two intersecting walls (see figure).

‡ Vertical walls, ribs, or similar obstructions extending from the water surface to or above the horizontal plane of the spout opening require a greater air gap when spaced closer to the nearest inside edge of spout opening than specified in note † above. The effect of three or more such vertical walls or ribs has not been determined. In such cases, the air gap shall be measured from the top of the wall.

¶ 2 × effective opening.
§ 3 × effective opening.

The ratings in fixture units listed in Table 19-5 represent the relative loading of a water-distribution system by the different types of plumbing fixtures. The sum of the ratings for any part or all of a system is a measure of the load the combination of fixtures would impose if all were operating. The probable maximum water demand, gal/min, can be determined from the total number of fixture units served by any part of a system by use of Figs. 19-4 and 19-5.

The demand obtained from these curves applies to fixtures that are used intermittently. If the system serves fixtures, such as air-conditioning units, lawn sprinklers, or hose bibs, that are used continuously, the demand of these fixtures should be added to the intermittent demand.

Fixtures not listed in Table 19-5 should be estimated in accordance with Table 19-6. For a continuous or semicontinuous flow into a drainage system, such as from a pump, pump ejector, air-conditioning system, or similar device, two fixture units should be used for each gallon per minute of flow. When additional fixtures are to be installed in the future, pipe and drain sizes should be based on the ultimate load, not on the present load.

TABLE 19-5 Fixture Units per Fixture or Group*

Fixture type	Fixture-unit value as load factors	Min size of trap, in.	
1 bathroom group consisting of water closet, lavatory and bathtub or shower stall	Tank water closet, 6 Flush-valve water closet, 8		
Bathtub† (with or without overhead shower)............................	2		$1\frac{1}{2}$
Bathtub†............................	3		2
Bidet...............................	3	Nominal	$1\frac{1}{2}$
Combination sink and tray............	3		$1\frac{1}{2}$
Combination sink and tray with food-disposal unit......................	4	Separate traps	$1\frac{1}{2}$
Dental unit or cuspidor..............	1		$1\frac{1}{4}$
Dental lavatory.....................	1		$1\frac{1}{4}$
Drinking fountain...................	$\frac{1}{2}$		1
Dishwasher, domestic................	2		$1\frac{1}{2}$
Floor drains‡......................	1		2
Kitchen sink, domestic...............	2		$1\frac{1}{2}$
Kitchen sink, domestic, with food-waste grinder..........................	3		$1\frac{1}{2}$
Lavatory¶.........................	1	Small P.O.	$1\frac{1}{4}$
Lavatory¶.........................	2	Large P.O.	$1\frac{1}{2}$
Lavatory, barber, beauty parlor........	2		$1\frac{1}{2}$
Lavatory, surgeon's..................	2		$1\frac{1}{2}$
Laundry tray (1 or 2 compartments).....	2		$1\frac{1}{2}$
Shower stall, domestic................	2		2
Showers (group) per head.............	3		
Sinks:			
Surgeon's........................	3		$1\frac{1}{2}$
Flushing rim (with valve)...........	8		3
Service (trap standard).............	3		3
Service (P trap)....................	2		2
Pot, scullery, etc...................	4		$1\frac{1}{2}$
Urinal, pedestal, siphon jet, blowout.....	8	Nominal	3
Urinal, wall lip.....................	4		$1\frac{1}{2}$
Urinal stall, washout.................	4		2
Urinal trough (each 2-ft section)........	2		$1\frac{1}{2}$
Wash sink (circular or multiple) each set of faucets........................	2	Nominal	$1\frac{1}{2}$
Water closet, tank-operated...........	4	Nominal	3
Water closet, valve-operated..........	8		3

* From National Plumbing Code, ASME.
† A shower head over a bathtub does not increase the fixture value.
‡ Size of floor drain shall be determined by the area of surface water to be drained.
¶ Lavatories with $1\frac{1}{4}$- or $1\frac{1}{2}$-in. trap have the same load value; larger P.O. (plumbing orifice) plugs have greater flow rate.

19-8. Water-Pipe Sizing.

Pipe diameters should be determined by application of the principles of hydraulics. While economy dictates use of the smallest sizes of pipe permitted by building-code requirements, other factors often make larger sizes advisable. These factors include:

  1. Pressure at the water-supply source, usually the public main, psi
  2. Pressure required at the outlets of each fixture, psi

Fig. 19-4 Estimate curves for domestic water-demand load. *(National Plumbing Code, ASME.)*

Fig. 19-5 Enlargements of low-demand portion of Fig. 19-4. *(National Plumbing Code, ASME.)*

TABLE 19-6 Other Fixture-Unit Values*

Fixture Drain or Trap Size, In.	Fixture-unit Value
1¼ in. and smaller	1
1½	2
2	3
2½	4
3	5
4	6

* From National Plumbing Code, ASME.

3. Loss of pressure because of height of outlets above the source and friction of water flow in pipe

4. Limitations on velocity of water flow, ft/s, to prevent noise and erosion

The National Plumbing Code recommends the following steps in sizing cold-water distribution piping systems:

1. Sketch all the proposed risers, horizontal mains, and branch lines, indicating the number and the type of fixtures served, together with the required water flow.

2. Compute the demand weights of the fixtures, in fixture units, using Table 19-5.

3. Using Fig. 19-4 or 19-5, and the total number of fixture units, determine the water demand in gallons per minute.

4. Compute the equivalent length of pipe for each stack in the system, starting from the street main.

5. Obtain by test or from the water company the average minimum pressure in the street main. Determine the minimum pressure needed for the highest fixture in the system.

6. Compute the pressure loss in the piping, using the equivalent length found in item 4.

7. Choose the pipe sizes from a chart like that in Fig. 19-6 or 19-7, or from the charts given in the National Plumbing Code.

(See also V. T. Manas, "National Plumbing Code Handbook," McGraw-Hill Book Company, New York.)

Effects of Pressure. Rate of flow, ft^3/s, in a pipe is determined by

$$Q = AV \tag{19-1}$$

where A = pipe cross-sectional area, ft^2
V = water velocity, ft/s

Fig. 19-6 Chart for determination of flow in copper tubing and other pipes that will be smooth after 15 to 20 years of use.

In general, V should be kept to 8 ft/s or less to prevent noise and reduce erosion at valve seats. Hence, pipe area should be at least the flow rate Q divided by 8.

The minimum pressures at plumbing fixtures generally required by building codes are listed in Table 19-2. These pressures are those that remain when the pressure drop due to height of outlet above the water source and the pressure lost by friction in pipes are deducted from the pressure at the water source. The pressure loss due to height can be computed from

$$p = 0.433h \qquad (19\text{-}2)$$

where p = pressure, psi
h = height or pressure head, ft

The total head H, ft, on water at any point in a pipe is given by

$$H = Z + \frac{p}{w} + \frac{V^2}{2g} \qquad (19\text{-}3)$$

where Z = elevation, ft, of the point above some arbitrary datum
p/w = pressure head, ft
w = specific weight of water = 62.4 lb/ft^3

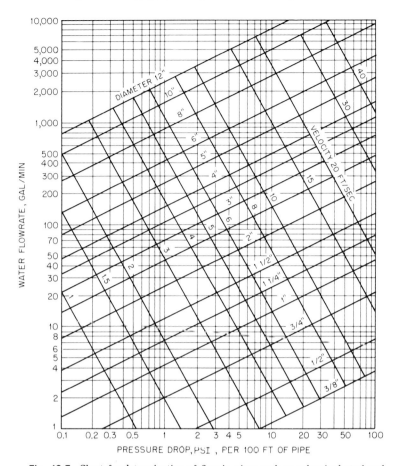

Fig. 19-7 Chart for determination of flow in pipes such as galvanized steel and wrought iron that will be fairly rough after 15 to 20 years of use.

$V^2/2g$ = velocity head, ft
g = acceleration due to gravity, 32.2 ft/s^2

When water flows in a pipe, the difference in total head between any two points in the pipe equals the friction loss h_f, ft, in the pipe between the points.

Any of several formulas may be used for estimating h_f. One often used for pipes flowing full is the Hazen-Williams formula:

$$h_f = \frac{4.727}{D^{4.87}} L \left(\frac{Q}{C_1}\right)^{1.85} \qquad (19\text{-}4)$$

where Q = discharge, ft^3/s
D = pipe diameter, ft
L = length of pipe, ft
C_1 = coefficient

The value of C_1 depends on the roughness of the pipe, which, in turn, depends on pipe material and age. A new pipe has a larger C_1 than an older one of the same size and

Water-Pipe Sizing

TABLE 19-7 Allowances for Friction Losses in Valves and Fittings, Expressed as Equivalent Length of Pipe, ft*

Fitting size, in	Standard ells 90°	Standard ells 45°	90° tee Side branch	90° tee Straight run	Coupling	Gate valve	Globe valve
3/8	0.5	0.3	0.75	0.15	0.15	0.1	4
1/2	1	0.6	1.5	0.3	0.3	0.2	7.5
3/4	1.25	0.75	2	0.4	0.4	0.25	10
1	1.5	1.0	2.5	0.45	0.45	0.3	12.5
1 1/4	2	1.2	3	0.6	0.6	0.4	18
1 1/2	2.5	1.5	3.5	0.8	0.8	0.5	23
2	3.5	2	5	1	1	0.7	28
2 1/2	4	2.5	6	1.3	1.3	0.8	33
3	5	3	7.5	1.5	1.5	1	40
3 1/2	6	3.5	9	1.8	1.8	1.2	50
4	7	4	10.5	2	2	1.4	63
5	9	3	13	2.5	2.5	1.7	70
6	10	6	15	3	3	2	84

*Allowances are for streamlined soldered fittings and recessed threaded fittings. For threaded fittings, double the allowances shown in the table. (Courtesy Copper Development Association.)

material. Hence, when pipe sizes are being determined for a new installation, a future value of C_1 should be assumed to ensure adequate flows in the future. Design aids, such as charts (see Figs. 19-6 and 19-7) or nomograms, are often used to evaluate Eq. (19-4).

In addition to friction loss in pipes, there also are friction losses in meters, valves, and fittings. These pressure drops can be expressed for convenience as equivalent lengths of pipe of a specific diameter. Table 19-7 indicates typical allowances for friction loss for several sizes and types of fittings and valves.

The pressure reduction caused by pipe friction depends, for a given length of pipe and rate of flow, on pipe diameter. Hence, a pipe size can be selected to create a pressure drop in the pipe to provide the required pressure at a plumbing fixture, when the pressure at the water source is known. If the pipe diameter is too large, the friction loss will be too small and the pressure at the fixture will be high. If the pipe size is too small, the friction loss will be too large and the pressure at the fixture will be too small.

Minimum Pipe Sizes. Table 19-8 lists the minimum sizes for fixture-supply pipes, as recommended by the National Plumbing code. These sizes apply to both hot- and cold-water fixtures.

Sizes of pipes for small buildings, such as single-family houses, can usually be deter-

TABLE 19-8 Minimum Sizes for Fixture-Supply Pipes*

Type of Fixture or Device	Pipe Size, In.	Type of Fixture or Device	Pipe Size, In.
Bath tubs	½	Shower (single head)	½
Combination sink and tray	½	Sinks (service, slop)	½
Drinking fountain	⅜	Sinks, flushing rim	¾
Dishwasher (domestic)	½	Urinal (flush tank)	½
Kitchen sink, residential	½	Urinal (direct flush valve)	¾
Kitchen sink, commercial	¾	Water closet (tank type)	⅜
Lavatory	⅜	Water closet (flush valve type)	1
Laundry tray, 1, 2 or 3 compartments	½	Hose bibs	½
		Wall hydrant	½

* From National Plumbing Code, ASME.

mined from experience of the designer and applicable building-code requirements, without extensive calculations. For short branches to individual fixtures, for example, the minimum pipe diameters listed in Table 19-8 generally will be satisfactory. Usually also, the following diameters can be used for the mains supplying water to the fixture branches:

½ in for mains with up to three ⅜-in branches
¾ in for mains with up to three ½-in or five ⅜-in branches
1 in for mains with up to three ¾-in or eight ½-in or fifteen ⅜-in branches

The adequacy of these sizes, however, depends on the pressure available at the water source and the probability of simultaneous use of the plumbing fixtures. In domestic installations, each person will use a total of 20 to 80 gal/day, depending on family size, garden requirements, etc. For apartment houses, a consumption of 50 gal per person per day is generally assumed in system design, while 40 gal per person per day is often used for housing projects.

19-9. Domestic-Water Heaters.

Domestic-water heaters may be direct-fired or unfired. Two types of unfired heaters are in common use, storage or instantaneous.

Storage heaters (Fig. 19-8) generally use steam or hot water as the heating medium. The tank stores hot water for future use. In general, the less uniform the demand for hot water, the greater the storage capacity needed in the tank. When the water demand is fairly uniform, a smaller storage capacity is allowable, but the capacity of the heating coil must be greater. However, a large storage capacity is used whenever possible, because this reduces the load on the boiler and allows use of a smaller heating coil. Industrial, office, and school-building hot-water demands are usually nonuniform with peak loads occurring during very short periods. Loads in hotels, apartment houses, and hospitals are usually more uniform.

Instantaneous heaters (Fig. 19-9) have V-shaped or straight tubes to heat the water

Fig. 19-8 Storage heater for domestic hot water.

Fig. 19-9 Instantaneous heater for domestic hot water.

when steam or hot water is the heating medium. Alternatively, electricity may be used. The water for the building is heated as needed; there is no storage section in the heater.

The hot-water load for a given building is computed in a manner similar to that described in Art. 19-7 but with Table 19-9 and the tabulated demand factor for the particular building type. The heating-coil capacity of the heater must at least equal the maximum probable demand for hot water.

For storage-type heaters, the storage capacity is obtained by multiplying the maximum probable demand by a suitable factor, which usually varies from 1.25 for apartment houses to 0.60 for hospitals. Table 19-10 lists hot-water temperatures for various services.

Example. An industrial plant has 9 showers, 200 private lavatories, 20 slop sinks, 20 public lavatories, and 1 dishwasher. What is the hourly water consumption and what storage and heating capacities are required for this load?

From Table 19-9:

9 showers at 225 gal/h	= 2025 gal/h
200 private lavatories at 2 gal/h	= 400
20 slop sinks at 20 gal/h	= 400
20 public lavatories at 12 gal/h	= 240
1 dishwasher at 250 gal/h	= 250
Total hourly demand	= 3315 gal/h

For the demand factor from Table 19-9, maximum probable demand is

$$0.40 \times 3315 = 1326 \text{ gal/h}$$

Heating-coil capacity must at least equal this maximum probable demand. Storage capacity should be $1.00 \times 3315 = 3315$ gal.

WASTEWATER PIPING

Human, natural, and industrial wastes resulting from building occupancy and use must be disposed of in a safe, quick manner if occupant health and comfort are to be safe-

TABLE 19-9 Hot-Water Demand per Fixture for Various Building Types
(Based on average conditions for types of buildings listed, gal/h of water per fixture at 140°F)

Type of fixture	Apartment House	Hospital	Hotel	Industrial plant	Office building
Basins, private lavatories	2	2	2	2	2
Basins, public lavatories	4	6	8	12	6
Showers	75	75	75	225	
Slop sinks	20	20	30	20	15
Dishwashers (per 500 people)	250	250	250	250	250
Pantry sinks	5	10	10		
Factors					
Demand factor	0.30	0.25	0.25	0.40	0.30
Storage factor	1.25	0.60	0.80	1.00	2.00

TABLE 19-10 Hot-Water Temperatures for Various Services, °F

Cafeterias (serving areas)	130
Lavatories and showers	130
Slop sinks (floor cleaning)	150
Slop sinks (other cleaning)	130
Cafeteria kitchens	130 + steam

guarded. So design of an adequate plumbing system requires careful planning and adherence to state or municipal regulations governing these systems.

The National Plumbing Code, published by ASME, gives basic goals in environmental sanitation and is a useful aid in designing plumbing systems for all classes of buildings. It is also helpful for areas having antiquated codes or none at all. Much of the material in this section is based on this code.

19-10. Wastewater Disposal.

There are three main types of wastewater: domestic, storm, and industrial. Separate plumbing systems are generally required for each type.

Domestic wastewater is primarily spent water from the building water supply, to which is added wastes from bathrooms, kitchens, and laundries. It generally can be disposed of by discharge into a public sanitary sewer, if one is available.

Storm water is primarily the water that runs off the roof of the building. The water usually is led to roof drains or gutters. These then feed the water to drainpipes, which convey it to a storm sewer. Special conditions at some building sites, such as large paved areas or steep slopes, may require capture of land drainage in inlets. From these, the storm water is conveyed to storm sewers. Discharge into sanitary sewers is objectionable because the large flows interfere with effective wastewater treatment and increase treatment costs. If kept separate from other types of wastewater, storm water usually can be safely discharged into a large body of water. Raw domestic wastewater and industrial wastes, on the other hand, have objectionable characteristics that make some degree of treatment necessary before they can be discharged.

There are numerous treatment methods available for domestic sewage after its removal from a building. Some treatments are relatively simple, such as those afforded by septic tanks, Imhoff tanks, sewage lagoons, or sedimentation tanks (clarifiers). Some treatments are complex, requiring many steps. These include filtration, activated-sludge and aeration methods. The degree of treatment necessary generally depends on the assimilation potential of the body of water to receive the effluent, primarily the ability of the body to dilute the impurities and to supply oxygen for decomposition of organic matter present in the wastewater.

Industrial waste may present special problems because (1) the flow volume may be beyond the public sewer capacity, and (2) local regulations may prohibit the discharge of industrial waste into public sewers. Furthermore, many recent pollution regulations prohibit discharge of industrial waste into streams, lakes, rivers, and tidal waters without suitable prior treatment. Industrial wastes generally require treatments engineered to remove the specific elements injected by industrial processes that make the wastes objectionable. Often, these treatments cannot be carried out in public sewage treatment plants. Special treatment plants may have to be built for the purpose. Treatment methods for a variety of industrial wastes are discussed in E. B. Bellelievre, "The Treatment of Industrial Wastes," McGraw-Hill Book Company, New York. Specific design procedures for sewers, drains, and sewage treatment, with accompanying numerical examples, are given in T. G. Hicks, "Standard Handbook of Engineering Calculations," McGraw-Hill Book Company, New York.

19-11. Sewers.

A sewer is a conduit for water carriage of wastes. For the purpose of this section any piping for wastewater inside a building will be considered plumbing or process piping; outside the building, wastewater lines are called sewers.

Sewers carry **sewage**. And a system of sewers and appurtenances is **sewerage**. (See "Glossary—Water and Sewer Control Engineering," American Society of Civil Engineers.)

Sanitary sewers carry domestic wastes or industrial wastes. Where buildings are located on large sites or structures with large roof areas are involved, a storm sewer is used for fast disposal of rain and is laid out to drain inlets located for best collection of runoff.

For figuring rates of runoff to determine storm-sewer requirements, the rational method may be used. The so-called **rational method** for determination of runoff involves the formula $Q = AIR$, where Q is the maximum rate of runoff in cubic feet per second; A the watershed area in acres; I the imperviousness ratio of the watershed area; and R the rate of rainfall on the area in inches per hour. The imperviousness ratio I is the ratio of water running off the land to water precipitating on it. It will range from 0.7 to 0.9 for built-up areas and paved surfaces; 0.05 to 0.30 for unpaved surfaces, depending on slopes. In storm-sewer design, however, it is necessary to know not only rate of runoff and total runoff, but also at what point in time after the start of a storm the rate of runoff reaches its peak. It is this peak runoff for which pipe must be sized and sloped. It should be noted that the water necessary to fill the drainage system—storm sewers and manholes—is, in effect, storage. Such storage is a sort of buffer tending to decrease slightly the magnitude of theoretical max-

Fig. 19-10 Hydraulic properties of circular pipe for various depths of flow.

imum flow. Its effect can be approximated by figuring about how much flow it would take to fill the system before it can begin to flow as designed (Fig. 19-10).

Determining Sewer Size. Sanitary sewers or lines carrying exclusively industrial wastes from a building to disposal must be sized and sloped according to best hydraulic design. The problem is generally one of flow in a circular pipe. (C. V. Davis and K. E. Sorenson, "Handbook of Applied Hydraulics," McGraw-Hill Book Company, New York.) Gravity flow is to be desired; but pumping is sometimes required.

Pipe should be straight and of constant slope between manholes, and manholes should be used at each necessary change in direction, slope, elevation, or size of pipe. Manholes should be no farther apart than 500 ft, and preferably as close as 200 ft where pipe used is of a diameter too small to enter.

The sewer from a building must be sized to carry out all the water carried in by supply mains or other means. Exceptions to this are the obvious cases where losses might be appreciable, such as an industrial building where considerable water is consumed in a process or evaporated to the atmosphere. But, in general, water out about equals water in, plus all the liquid and water-borne solid wastes produced in the building.

Another factor to consider in sizing a sewer is infiltration. Sewers, unlike water mains, often flow at less pressure than that exerted by ground water around them. Thus they are more likely to take in ground water than to leak out sewage. An infiltration rate of 2000 to 200,000 gal/(day)(mi) might be expected. It all depends on diameter of pipe (which fixes length between joints), type of soil, groundwater pressure, and workmanship.

In an effort to keep infiltration down, sewer-construction contracts are written with a specified infiltration rate. Weir tests in a completed sewer can be used to check the contractor's success in meeting the specification; but unless the sewer is large enough to go into, prevention of excessive infiltration is easier than correction. In addition to groundwater infiltration through sewer-pipe joints, the entry of surface runoff through manhole covers and thus into sewers is often a factor. Observers have gaged as much as 150 gal/min leaking into a covered manhole.

Size and slope of a sanitary sewer depend lastly—and vitally—on a requirement that velocity when flowing full be kept to at least 2 ft/s to keep solids moving and prevent clogging. In general, no drain pipe should be less than 6 in in diameter; an 8-in minimum is safer.

Sewer-Pipe Materials. Vitrified clay, concrete, asbestos-cement, cast iron, plastics, bituminized fiber, or steel may be used for pipe to carry sewage and industrial wastes. Vitrified clay up to about 36 in diameter and steel or concrete in larger sizes are used most generally; bituminized fiber or plastics are used for the smaller diameters.

Choice of pipe material depends on required strength to resist load or internal pressure; corrosion resistance, which is especially vital in carrying certain industrial wastes; erosion resistance in sewers carrying coarse solids; roughness factor where flat slopes are desirable; and cost in place.

Vitrified-clay pipe requires some care in handling to minimize breakage. Concrete pipe must be made well enough or protected to withstand effects of damaging sewer gas (hydrogen sulfide) or industrial wastes. Asbestos-cement pipe comes in greater lengths, therefore has fewer joints. It is smooth, noncorrosive, and easily jointed. Cast-iron sewer pipe is good under heavy loads, exposed as on bridges, in inverted siphons, or in lines under pressure. Bituminized-fiber pipe in smaller diameters (to 8 in) for drains is lighter than but about as strong as clay or concrete. It may deteriorate in sunlight and must be handled with reasonable care. Steel is used chiefly for its strength or flexibility. Corrugated-steel pipe with protective coatings is made especially for sewer use; its long lengths and light weight ease handling and jointing. Plastic pipes are used because of corrosion resistance, light weight, and low cost installed.

19-12. Wastewater System Elements.

The usual steps in planning a plumbing system are: (1) secure a sewer or waste-disposal plan of the site; (2) obtain preliminary architectural and structural plans and elevations of the building; (3) tabulate known and estimated occupancy data, including the number of persons, sex, work schedules, and pertinent details of any manufacturing process to be performed in the building; (4) obtain copies of the latest edition of the applicable code; (5) design the system in accordance with code recommendations; and (6) have the design approved by local authorities before construction is begun.

The typical plumbing layout in Fig. 19-11 shows the major elements necessary in the usual plumbing system. **Fixtures** (lavatories, water closets, bathtubs, showers, etc.) are located as needed on each floor of the structure (Art. 19-6).

Each fixture is served by a **soil** or **waste stack,** a **vent** or **vent stack,** and a **trap,** Fig. 19-11. Vertical soil or waste stacks conduct waste solids and liquids from one or more fixtures to a sloped **house** or **building drain** in the building cellar, crawl space, or wall.

Fig. 19-11 Wastewater-removal system for a multistory building. *Reprinted with permission from F. S. Merritt, "Building Engineering and Systems Design," Van Nostrand Reinhold Company, New York.)*

Each vent stack extends to a **stack vent** that projects above the building roof and may or may not have **branch vents** connected to it. Vents and vent stacks permit the entrance of fresh air to the plumbing system, diluting any gases present and balancing the air pressure in various branches. See also Art. 19-19.

Traps on each fixture provide a water seal, which prevents sewer gases from entering the working and living areas. The house drain delivers the discharge from the various stacks to the **house** or **building trap** (Fig. 19-11), which is generally provided with a separate vent. Between the house trap and **public sewer,** or other main sewer pipe, is the **building sewer.** The building sewer is outside the building structure, while the building trap is just inside the building foundation wall.

Where the building drain is below the level of the public sewer line, some arrangement for lifting the sewage to the proper level must be provided. This can be done by allowing the house drain to empty into a suitably sized **sump pit.** The sewage is discharged from the sump pit to the public sewer by a pneumatic ejector or motor-driven pump.

Pipe Supports. Pipes of wastewater-removal systems should be supported and braced in the same way as pipes of water-supply systems (see Art. 19-5). Vertical pipes generally should be supported at every floor. Horizontal pipes should be supported at intervals not exceeding the following: cast-iron soil pipe, 5 ft and behind every hub; threaded pipe, 12 ft; copper tubing, 10 ft. Supports also should be provided at the bases of stacks.

Consideration should be given to the possibility of building settlement and its effects on vertical pipes and also to thermal expansion and contraction of pipes, especially when the pipes are made of copper.

Cleanouts. In horizontal drainage lines, at least one cleanout is required per 50 ft of 4-in or smaller pipe. For larger sizes, a cleanout is required for each 100 ft of pipe. In underground drainage lines, the cleanout must be extended to the floor or grade level to allow easier cleaning. Cleanouts should open in a direction opposite to that of the flow in the pipe, or at right angles to it.

In pipes up to 4 in, the cleanout should be the same size as the pipe itself. For pipes larger than 4 in, the cleanout should be at least 4 in in diameter but may be larger, if desired. When underground piping over 10 in in diameter is used, a manhole is required at each 90° bend and at intervals not exceeding 150 ft.

Combined Drainage and Vent Systems. These offer the possibility of considerable cost savings over the separate drainage and venting systems previously described.

A combination waste and vent system, introduced by the Western Plumbing Officials' Association, is one such system. It employs horizontal wet venting of one or more sinks or floor drains by means of a common waste and vent pipe adequately sized to provide free movement of air above the flow line of the pipe. Relief vents are connected at the beginning and end of the horizontal pipe. The traps of the fixtures are not individually vented. Some building codes permit such a system only where structural conditions preclude installation of a conventional system and, where it can be used, require larger than normal waste pipes and traps in the system.

The **Sovent system,** invented by Fritz Sommer, a Swiss, and promoted in the United States by the Copper Development Association, Inc., is another type of combination system. It requires drainage branches and soil stacks, with special fittings, but no individual and branch vents and no vent stacks.

The system has four basic parts: a soil or waste stack, a Sovent aerator fitting on the stack at each floor, horizontal branches, and a Sovent deaerator fitting on the stack at its base and at horizontal offsets (see Fig. 19-12). The aerator and deaerator provide means for self-venting the stack. In a conventional drainage system, a vent stack is installed to supply air to vent pipes connected to the drainage branches and to the soil or waste stack to prevent destruction of the trap seals. In the Sovent system, however,

Fig. 19-12 Copper single-stack Sovent plumbing system. *(Courtesy Copper Development Association, Inc.)*

the vent stack is not needed because the aerator, deaerator, and stack vent avoid creation of a strong suction.

The aerator does the following: It reduces the velocity of both liquid and air in the stack. It prevents the cross section of the stack from filling with a plug of water. And the fitting mixes the wastewater from the drainage branches with the air in the stack.

The deaerator separates the airflow in the stack from the wastewater. As a result, the wastewater flows smoothly into a horizontal offset or building drain. Also, air pressure preceding the flow at 90° turns is prevented from rising excessively by a pressure relief line between a deaerator and a stack offset or the building drain to allow air to escape from the deaerator.

An aerator is required on the stack at each level where a horizontal soil branch or a waste branch the same size as or one tube size smaller than the stack discharges to it. Smaller waste branches may drain directly into the stack. At any floor where an aerator fitting is not required, the stack should have a double inline offset, to decelerate the flow (see Fig. 19-12). No deaerators are required at stack offsets of less than 60°.

19-13. Waste-Pipe Materials.

Cast iron is the most common piping material for systems in which extremely corrosive wastes are not expected. In recent years, copper has become increasingly popular because it is lightweight and easy to install. Vitrified-clay, steel, brass, wrought-iron, plastic, and lead pipe are also used in plumbing systems.

No matter what material is used for plumbing-system piping, it should conform to one or more of the accepted standards approved by the National Plumbing Code, or the code used in the area in which the building is located.

In cast-iron pipe, the fitting joints are calked with oakum or hemp and filled with molten lead not less than 1 in deep. This provides some flexibility without the danger of leaks. Clay pipe is calked with oakum, cement grout, and mortar. Copper pipe is commonly brazed, while steel and wrought-iron pipe have screwed or flanged connections.

When planning a plumbing system, check with the applicable code before specifying the type of joint to be used in the piping. Joints acceptable in some areas may not be allowed in others.

19-14. Layout of Waste Piping.

Wastewater-removal systems should be sized and laid out to permit use of the smallest-diameter pipes capable of rapidly carrying away the wastewater from fixtures without clogging the pipes, without creating annoying noises, and without producing excessive pressure fluctuations at points where fixture drains connect to soil or waste stacks. Such pressure changes might siphon off the liquid seals in traps and force sewer gases back through the fixtures into the building. Positive or negative air pressure at the trap seal of a fixture should never be permitted to exceed 1 in of water.

The drainage system is considered a nonpressure system. The pipes generally do not flow full. The discharge from a fixture drain is introduced to a stack through a stack fitting, which may be a long-turn tee-wye or a short-turn or sanitary tee. The fitting gives the flow a downward, vertical component. As the water accelerates downward under the action of gravity, it soon forms a sheet around the stack wall. If no flows enter the stack at lower levels to disrupt the sheet, it will remain unchanged in thickness and will flow at a terminal velocity, limited by friction, to the bottom of the stack. A core of air at the center of the stack is dragged downward with the wastewater by friction. This air should be supplied from outdoors through a stack vent (see Fig. 19-11), to prevent creation of a suction that would empty trap seals.

When the sheet of wastewater reaches the bottom of the stack, a bend turns the flow 90° into the building drain. Within a short distance, the wastewater drops from the upper part of the drain and flows along the lower part of the drain.

Slope of Horizontal Drainage Pipes. Building codes generally require that horizontal pipes have a uniform slope sufficient to insure a flow with a minimum velocity of 2 ft/s. The objective is to maintain a scouring action to prevent fouling of the pipes. Codes therefore often specify a minimum slope of ¼ in/ft for horizontal piping 3 in in diameter or less and ⅛ in/ft for larger pipe.

Because flow velocity increases with slope, greater slopes increase pipe carrying capacity. In branch pipes, however, high velocities can cause siphonage of trap seals. Therefore, use of larger-size pipes is preferable to steeper slopes for attaining required capacity of branch pipes.

See also Art. 19-19.

19-15. Interceptors.

These are devices installed to separate and retain grease, oil, sand, and other undesirable matter from the sewage, while permitting normal liquid wastes to discharge to the sewer or sewage-treatment plant.

Grease interceptors are used for kitchens, cafeterias, restaurants, and similar establishments where the amount of grease discharged might obstruct the pipe or interfere with disposal of the sewage. Oil separators are used where flammable liquids or oils might be discharged to the sewer. Sand interceptors are used to remove sand or other solids from the liquid sewage before it enters the building sewer. They are provided with large cleanouts for easy removal of accumulated solids.

Other types of plants in which interceptors are usually required include laundries, beverage-bottling firms, slaughterhouses, and food-manufacturing establishments.

19-16. Piping for Indirect Wastes.

Certain wastes like those from food-handling, dishwashing (commercial), and sterile-materials machines should be discharged through an indirect waste pipe; i.e., one that is not directly connected with the building drainage pipes but handles waste liquids by discharging them into a plumbing fixture or receptacle from where they flow directly to the building drainage pipes. Indirect-waste piping is generally required for the discharge from rinse sinks, laundry washers, steam tables, refrigerators, egg boilers, iceboxes, coffee urns, dishwashers, stills, sterilizers, etc. It is also required for units that must be fitted with drip or drainage connections but are not ordinarily regarded as plumbing fixtures.

An air gap is generally required between the indirect-waste piping and the regular drainage system. The gap should be at least twice the effective diameter of the drain it serves. A common way of providing the required air gap is by leading the indirect-waste line to a floor drain, slop sink, or similar fixture that is open to the air and is vented or trapped in accordance with the governing code. To provide the necessary air gap, the indirect-waste pipe is terminated above the flood level of the fixture.

Where a device discharges only clear water, such as water from engine-cooling jackets, sprinkler systems, or overflows, an indirect-waste system must be used. Or, if desired, devices discharging only clear water may be emptied by leading their waste lines to the building roof. The roof provides the air gap and the waste flows off it into the building storm drains.

Hot water above 140°F and steam pipes usually must be arranged for indirect connection into the building drainage system or into an approved interceptor.

To prevent corrosion of plumbing piping and fittings, any chemicals, acids, or corrosive liquids are generally required to be automatically diluted or neutralized before being discharged into the plumbing piping. Sufficient fresh water for satisfactory dilution, or a neutralizing agent, should be available at all times. A similar requirement is contained in most codes for liquids that might form or give off toxic or noxious fumes.

19-17. Rainwater Drainage.

Exterior sheet-metal **leaders** for rainwater drainage are not normally included in the plumbing contract. Interior leaders or storm-water drains, however, are considered part of the plumbing work. Depending on the size of the building and the code used in the locality, rainwater from various roof areas may or may not be led into the building drains. Where separate rainwater leaders or storm drains are used, the building drains are then called **sanitary drains** because they convey only the wastes from the various plumbing fixtures in the building.

Interior storm-water drain pipes may be made of cast iron, steel, or wrought iron. All joints must be tight enough to prevent gas and water leakage. It is common practice to insert a cast-iron running trap between the storm drain and the building drain to maintain a seal on the storm drain at all times. However, not every code requires this type of protection. Use of storm drains does not eliminate the need for separate drains and vents for sewage. All codes prohibit use of storm drains for any type of sewage.

Water falling on the roof may be led either to a gutter, from where it flows to a leader, or it may be directed to a leader inlet by means of a slope in the roof surface. Many

Fig. 19-13 Elements of a storm-drainage system: (a) roof gutter, exterior leader, and splash pan; (b) roof drain and top portion of interior leader; (c) piping to a storm sewer; (d) piping to a combined sewer. *(Reprinted with permission from F. S. Merritt, "Building Engineering and Systems Design," Van Nostrand Reinhold Company, New York.)*

different designs of leader inlets are available for different roof constructions and stormwater conditions. When vertical leaders are extremely long, it is common practice to install an expansion joint between the leader inlet and the leader itself. Figure 19-13c shows an example of a connection of a building storm drain to a storm sewer. When the drain must be connected to a combined sanitary-storm sewer, a trap should be installed before the connection to the sewer (Fig. 19-13d).

Sizes of vertical leaders and horizontal storm drains depend on the roof area to be drained. Table 19-11 indicates the maximum horizontal projection of roof area permitted for various sizes of leaders and horizontal storm drains.

Semicircular gutters are sized on the basis of the maximum projected roof area served. Table 19-12 shows how gutter capacity varies with diameter and pitch. Where maximum rainfall is either more than, or less than, 4 in/hr, refer to the National Plumbing Code for suitable correction factors.

Drains for building yards, areaways, and floors may be connected to the inlet side of an interior rain-leader trap. Where this is not possible, these drains may be run to a municipal sewer, or to a dry well. When a dry well is used, only the drains may discharge to it, other sewage being prohibited by most codes.

TABLE 19-11 Sizes of Vertical Leaders and Horizontal Storm Drains*

Vertical Leaders	
Size of leader or conductor,† in.	Max projected roof area, sq ft
2	720
2½	1,300
3	2,200
4	4,600
5	8,650
6	13,500
8	29,000

Horizontal Storm Drains			
Diam of drain, in.	Max projected roof area for drains of various slopes, sq ft		
	⅛-in. slope	¼-in. slope	½-in. slope
3	822	1,160	1,644
4	1,880	2,650	3,760
5	3,340	4,720	6,680
6	5,350	7,550	10,700
8	11,500	16,300	23,000
10	20,700	29,200	41,400
12	33,300	47,000	66,600
15	59,500	84,000	119,000

* From National Plumbing Code, ASME.
† The equivalent diameter of square or rectangular leader may be taken as the diameter of that circle which may be inscribed within the cross-sectional area of the leader.

TABLE 19-12 Sizes of Gutters*

Diam of gutter,† in.	Max projected roof area for gutters of various slopes, sq ft			
	¹⁄₁₆-in. slope	⅛-in. slope	¼-in. slope	½-in. slope
3	170	240	340	480
4	360	510	720	1,020
5	625	880	1,250	1,770
6	960	1,360	1,920	2,770
7	1,380	1,950	2,760	3,900
8	1,990	2,800	3,980	5,600
10	3,600	5,100	7,200	10,000

* From National Plumbing Code, ASME.
† Gutters other than semicircular may be used provided they have an equivalent cross-sectional area.

19-18. Waste-Pipe Sizing.

There are two ways of specifying the pipe size required for a particular class of plumbing service: (1) directly in terms of area served, as in roof-drainage service (Table 19-11) and (2) in terms of "fixture units" (Table 19-13).

As can be seen from these tables, the capacity of a leader or drain varies with the pitch of the installed pipe. The greater the pitch per running foot of pipe, the larger the capacity allowed, in terms of either the area served or the number of fixture units. This is because a steeper pitch increases the static head available to produce flow through the pipe. As static head increases, so does the amount of liquid which the pipe can handle.

Fixture Units. These are described in Art. 19-7. The steps in determination of pipe sizes by means of fixture units are: (1) list all fixtures served by one stack or branch; (2) alongside each fixture list its fixture unit; (3) add the fixture units and enter the proper table (Tables 19-13, 19-14, or 19-15) to determine the pipe size required for the stack or the branch.

Fixture branches connecting one or more fixtures with soil or waste stack are usually sized on the basis of the maximum number of fixture units for a given size of pipe or trap (Table 19-15). Where a large volume of water or other liquid may be contained in a fixture, such as in bathtubs, slop sinks, etc., an oversize branch drain may be provided to secure more rapid emptying.

19-19. Venting.

Waste pipes are vented to the outside to balance the air pressure in various branches and to dilute any gases present. The availability of air prevents back pressure and protects traps against siphonage.

The **main vent** is the principal artery of the venting system. It supplies air to **branch vents,** which, in turn, convey it to individual vents and wastewater pipes. See also Combined Drainage and Vent Systems, Art. 19-12.

Every building should have at least one main **vent stack.** It should extend undiminished in size and as directly as possible from outdoor air at least 6 in above the roof to

TABLE 19-13 Sizes of Building Drains and Sewers*

Diam of pipe, in.	Max number of fixture units that may be connected to any portion† of the building drain or the building sewer			
	Fall per ft			
	1/16 in.	1/8 in.	1/4 in.	1/2 in.
2	....	....	21	26
2½	....	....	24	31
3	....	20‡	27‡	36‡
4	....	180	216	250
5	....	390	480	575
6	....	700	840	1,000
8	1,400	1,600	1,920	2,300
10	2,500	2,900	3,500	4,200
12	3,900	4,600	5,600	6,700
15	7,000	8,300	10,000	12,000

* From National Plumbing Code, ASME.
† Includes branches of the building drain.
‡ Not over two water closets.

TABLE 19-14 Size and Length of Vents*

Size of soil or waste stack, in.	Fixture units connected	Diam of vent required, in.								
		$1\frac{1}{4}$	$1\frac{1}{2}$	2	$2\frac{1}{2}$	3	4	5	6	8
		Max length of vent, ft								
$1\frac{1}{4}$	2	30								
$1\frac{1}{2}$	8	50	150							
$1\frac{1}{2}$	10	30	100							
2	12	30	75	200						
2	20	26	50	150						
$2\frac{1}{2}$	42		30	100	300					
3	10		30	100	200	600				
3	30			60	200	500				
3	60			50	80	400				
4	100			35	100	260	1,000			
4	200			30	90	250	900			
4	500			20	70	180	700			
5	200				35	80	350	1,000		
5	500				30	70	300	900		
5	1,100				20	50	200	700		
6	350				25	50	200	400	1,300	
6	620				15	30	125	300	1,100	
6	960					24	100	250	1,000	
6	1,900					20	70	200	700	
8	600						50	150	500	130
8	1,400						40	100	400	120
8	2,200						30	80	350	110
8	3,600						25	60	250	80
10	1,000							75	125	100
10	2,500							50	100	50
10	3,800							30	80	35
10	5,600							25	60	25

* From National Plumbing Code, ASME.

the building drain. The main vent should be so located as to provide a complete loop for circulation of air through the wastewater-removal system. As an alternative to direct extension through the roof, a vent stack may be connected with a stack vent, if the connection is made at least 6 in above the flood-level rim of the highest fixture.

A **stack vent** is the extension of a soil or waste stack above the highest horizontal drain connected to the stack. This vent terminates above the roof.

Generally, a group of fixtures, consisting of one bathroom group and a kitchen sink or combination fixture, may be installed without individual fixture vents in a one-story building or on the top floor of a building—but only if each fixture drain connects independently to the soil stack and the water-closet and bathtub drains enter the stack at the same level. In that case, all the fixtures are vented through the stack extension, or stack vent.

An **individual,** or **back, vent** is a pipe installed to vent a fixture trap and connected with the venting system above the fixture served or terminated outdoors. To ensure that the vent will adequately protect the trap, building codes limit the distance downstream that the vent opening may be placed from the trap. This distance generally ranges from 2½ ft for a 1¼-in fixture drain to 10 ft for a 4-in fixture drain, but not less than two pipe

TABLE 19-15 Horizontal Fixture Branches and Stacks*

Diam of pipe, in.	Max number of fixture units that may be connected to			
	Any horizontal† fixture branch	One stack of 3 stories in height or 3 intervals	More than 3 stories in height	
			Total for stack	Total at one story or branch interval
$1\frac{1}{4}$	1	2	2	1
$1\frac{1}{2}$	3	4	8	2
2	6	10	24	6
$2\frac{1}{2}$	12	20	42	9
3	20‡	30¶	60¶	16‡
4	160	240	500	90
5	360	540	1,100	200
6	620	960	1,900	350
8	1,400	2,200	3,600	600
10	2,500	3,800	5,600	1,000
12	3,900	6,000	8,400	1,500
15	7,000			

* From National Plumbing Code, ASME.
† Does not include branches of the building drain.
‡ Not over two water closets.
¶ Not over six water closets.

diameters. The vent opening should be located above the bottom of the discharge end of the trap. See Fig. 19-11.

To reduce the amount of piping required, two fixtures may be set back to back, on opposite sides of a wall, and supplied air by a single vent (**common vent**). In that case, however, the fixtures should discharge wastewater separately into a double fitting with inlets at the same level.

A **branch vent** is a pipe used to connect one or more individual vents to a vent stack or to a stack vent.

A **wet vent** is a pipe that serves both as a vent and as a drainage pipe for wastes other than those from water closets. This type of vent reduces the amount of piping from that required with individual vents. For example, a bathroom group of fixtures may be vented through the drain from a lavatory, kitchen sink or combination fixture if such a fixture has an individual vent (see Fig. 19-14a).

A battery of fixtures is any group of similar fixtures that discharges into a common horizontal waste or soil branch. A battery of fixtures should be vented by a circuit or loop vent. (Building codes usually set a limit on the number of fixtures that may be included in a battery.)

A **circuit vent** is a branch vent that serves two or more traps and extends from the vent stack to a connection to the horizontal soil or waste branch just downstream from the farthest upstream connection to the branch (see Fig. 19-14b).

A **loop vent** is like a circuit vent but connects with a stack vent instead of a vent stack (see Fig. 19-14c). Thus, air can circulate around a loop.

Soil and waste stacks with more than 10 branch intervals should be provided with a relief vent at each tenth interval installed, starting with the top floor. A **branch interval** is a section of stack at least 8 ft high between connections of horizontal branches. A **relief vent** provides circulation of air between drainage and venting systems. The lower end of a relief vent should connect to the soil or waste stack, through a wye, below the horizontal branch serving a floor where the vent is required. The upper end of the relief

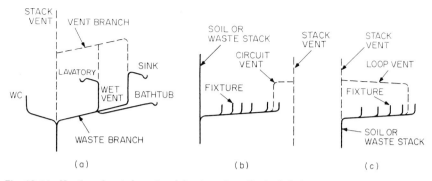

Fig. 19-14 Venting of waste branches: (a) wet venting of bathtub drainage pipe; (b) circuit venting and (c) loop venting of a group of plumbing fixtures.

vent should connect to the vent stack, through a wye, at least 3 ft above that floor. Such vents help to balance the pressures that are continuously changing within a plumbing system.

Slopes and Connections for Vent Pipes. While the venting system is intended generally to convey only air to and from the drainage system, moisture may condense from the air onto the vent pipe walls. To remove the condensation from the venting system, all individual and branch vent pipes should be so sloped and connected as to conduct the moisture back to soil or waste pipes by gravity.

Sizing of Vent Pipes. The fixture unit (Art. 19-4) is also used for sizing vents and vent stacks (Table 19-14). In general, the diameter of a branch vent or vent stack is one-half or more of the branch or stack it serves, but not less than 1¼ in. Smaller diameters are prohibited, because they might restrict the venting action.

19-20. Plumbing-System Inspection and Tests.

Plans for plumbing systems must usually be approved before construction is started. After installation of the piping and fixtures is completed, both the new work and any existing work affected by the new work must be inspected and tested. The plumber or plumbing contractor is then informed of any violations. These must be corrected before the governing body will approve the system.

Either air or water may be used for preliminary testing of drainage, vent, and plumbing pipes. After the fixtures are in place, their traps should be filled with water and a final test made of the complete drainage system.

When testing a system with water, all pipe openings are tightly sealed, except the highest one. The pipes are then filled with water until overflow occurs from the top opening. Using this method, either the entire system or sections of it can be tested. In no case, however, should the head of water on a portion being tested be less than 10 ft, except for the top 10 ft of the system. Water should be kept in the system for at least 15 min before the inspection starts. During the inspection, piping and fixtures must be tight at all points; otherwise approval cannot be granted.

An air test is made by sealing all pipe outlets and subjecting the piping to an air pressure of 5 psi throughout the system. The system should be tight enough to permit maintaining this pressure for at least 15 min without the addition of any air.

The final test required of plumbing systems uses either smoke or peppermint. In the smoke test, all traps are sealed with water and a thick, strong-smelling smoke is injected into the pipes by means of a suitable number of smoke machines. As soon as smoke appears at the roof stack outlets, they should be closed and a pressure equivalent to 1 in of water should be maintained throughout the system for 15 min before inspection begins. For the peppermint test, 2 oz of oil of peppermint are introduced into each line or stack.

GAS PIPING

As with piping and fixtures for plumbing systems, gas piping and fixtures are subject to code regulations in most cities and towns. Natural and manufactured gases are widely used in stoves, water heaters, and space heaters of many designs. Since gas can form explosive mixtures when mixed with air, gas piping must be absolutely tight and free of leaks at all times. Usual codes cover every phase of gas-piping size, installation, and testing. The local code governing a particular building should be carefully followed during design and installation.

19-21. Gas Supply.

The usual practice is for the public-service gas company to run its pipes into the building cellar, terminating with a brass shutoff valve and gas meter inside the cellar wall. From this point the plumbing contractor or gas-pipe fitter runs lines through the building to the various fixture outlets. When the pressure of the gas supplied by the public-service company is too high for the devices in the building, a pressure-reducing valve can be installed near the point where the line enters the building. This valve is usually supplied by the gas company.

Besides municipal codes governing the design and installation of gas piping and devices, the gas company serving the area will usually have a number of regulations that must be followed. In general, gas piping should be run in such a manner that it is unnecessary to locate the meter near a boiler, under a window or steps, or in any other area where it may be easily damaged. Where multiple meter installations are used, the piping must be plainly marked by means of a metal tag showing which part of the building is served by the particular pipe. When two or more meters are used in a building to supply separate consumers, there should be no interconnection on the outlet side of the meters.

19-22. Gas-Pipe Sizes.

Gas piping must be designed to provide enough gas to appliances without excessive pressure loss between them and the meter. It is customary to size gas piping so the pressure loss between the meter and any appliance does not exceed 0.3 in of water during periods of maximum gas demand. Other factors influencing the pipe size include maximum gas consumption anticipated, length of pipe and number of fittings, specific gravity of the gas, and the diversity factor.

19-23. Estimating Gas Consumption.

Use the manufacturer's Btu rating of the appliances and the heating value of the gas to determine the flow required in cubic feet per hour. When Btu ratings are not immediately available, the values in Table 19-16 may be used for preliminary estimates. The average heating value of gas in the area can be obtained from the local gas company, but when this is not immediately available, the values in Table 19-17 can be used for preliminary estimates.

Example. A building has two 8 gal/min hot-water heaters and 10 domestic ranges, each of which has four top burners and one oven burner. What is the maximum gas consumption that must be provided for if gas with a net heating value of 500 Btu/ft^3 is used?

TABLE 19-16 Heat Input to Common Appliances

Unit	Approx Input, Btu per Hr
Water heater, side-arm or circulating type	25,000
Water heater, automatic instantaneous:	
4 gpm	150,000
6 gpm	225,000
8 gpm	300,000
Refrigerator	2,500
Ranges, domestic:	
4 top burners, 1 oven burner	62,500
4 top burners, 2 oven burners	82,500

TABLE 19-17 Typical Heating Values of Commercial Gases

Gas	Net Heating Value, Btu per Cu Ft
Natural gas (Los Angeles)	971
Natural gas (Pittsburgh)	1,021
Coke-oven gas	514
Carbureted water gas	508
Commercial propane	2,371
Commercial butane	2,977

Solution. Heat input, using values from Table 19-16, is

$$2(300,000) + 10(62,500) = 1,225,000 \text{ Btu/h}$$

Maximum gas consumption is therefore $1,225,000/500 = 2,450 \text{ ft}^3/\text{h}$. The supply piping would be sized for this flow, even though all appliances would rarely operate at the same time.

19-24. Gas-Pipe Materials.

Materials recommended for gas piping in buildings are wrought iron or steel complying with American Standard B36.10. Malleable-iron or steel fittings should be used, except for stopcocks and valves. Above 4-in nominal size, cast-iron fittings may be used.

Fig. 19-15 Drips for gas piping.

Some local codes permit the use of brass or copper pipe of the same sizes as iron pipe if the gas handled is not corrosive to the brass or copper. Threaded fittings are generally required with these two materials.

Since the usual gas supplied for heating and domestic cooking generally contains some moisture, all piping should be installed so it pitches back to the supply main, or drips should be installed at suitable intervals. Neither unions nor bushings are permitted in gas pipes because of the danger of leakage and moisture trapping. To permit moisture removal, drips are installed at the lowest point in the piping, at the bottom of vertical risers, and at any other places where moisture might accumulate. Figure 19-15 shows typical drips for gas piping.

19-25. Testing Gas Piping.

Once installation of the gas piping in a building is complete, it must be tested for leaks. Portions of the piping that will be enclosed in walls or behind other permanent structures should be tested before the enclosure is completed.

The usual code requirement for gas-pipe testing stipulates a test pressure of about 10 psig for 15 min, using air, city gas, or some inert gas. Use of oxygen in pressure tests is strictly prohibited. Prior to connecting any appliances, the piping should be purged, using the regular gas supply.

SPRINKLER SYSTEMS

Well-proved as effective in preventing the spread of fires in buildings of all types, sprinkler systems for fire protection are in wide use. Essentially (Fig. 19-16), they consist of a series of horizontal parallel pipes placed near the building ceiling. Vertical risers (Fig. 19-17) supply the liquid—usually water—used for extinguishing fires. The horizontal pipes contain sprinklers (Fig. 19-18) preset to open automatically and discharge liquid when a fire is detected.

19-26. Sprinkler System Standards and Approvals.

The most widely accepted standard for sprinkler system design is NFPA 13, "Installation of Sprinkler Systems," supplemented when necessary by NFPA 24, "Outside Protec-

Fig. 19-16 Water-supply piping for sprinklers.

Fig. 19-17 Typical layouts of sprinkler piping.   **Fig. 19-18** Sprinklers above and below ceiling.

tion," NFPA 15, "Water Spray Fixed Systems," and NFPA 20, "Centrifugal Fire Pumps," National Fire Protection Association, 470 Atlantic Ave., Boston, Mass. 02210. Other Standards will be required for special hazards. These standards must be used with discretion when they are incorporated in specifications, because of the many references to the "authority having jurisdiction." Thus, before these standards are incorporated in a specification, rulings must be obtained from the "authority having jurisdiction" over the project at hand. The owner's insurance carrier can furnish the necessary information.

If the owner's insurance carrier is the Factory Mutual System, the FM standards will apply, and Factory Mutual will be the authority to be consulted.

In many cases the local building code will be the governing factor, and an insurance requirement a secondary matter. In such cases, conflicts must be reconciled to avoid a penalty for the owner and ambiguity for the construction bidders.

19-27. Types of Sprinkler Systems.

The type of system that should be used depends chiefly on temperature maintained in the building, damageability of contents, expected propagation rate of a fire, and total fire load.

Wet-pipe systems are so called because there is always water in the piping. Consequently, building temperature must be maintained above freezing. Wet-pipe systems are the simplest and most economical type that can be used. Other than a gate valve and an alarm valve, there are no other devices between the water supply and the sprinkler heads. Except for periodic tests of the alarm, no maintenance is required.

An alarm can be provided to indicate when a sprinkler head has started to operate and water is flowing in the piping. This is generally done by installing, in the main supply pipe, an alarm-check valve designed to distinguish between a momentary pressure surge and a steady flow. An alarm-check valve (Fig. 19-19) is essentially a swing check valve with an interior orifice that will admit water to a retard chamber that must fill completely to open a valve to the alarm supply pipe. The retarding chamber can also admit water to a pressure-actuated switch that can transmit a local or remote alarm electrically. In cases in which only a local alarm is considered adequate, water is admitted directly to a water-motor-driven gong, which requires no outside energy. Where no water-motor

Fig. 19-19 Alarm-check valve.

Fig 19-20 Curves determine sprinkler discharge for design purposes, for buildings containing 20-ft-high rack storage of Class I commodities, in double-row racks, with conventional pallets. Curve application:

A, 8-ft aisles with 286°F ceiling sprinklers and 165°F in-rack sprinklers
B, 8-ft aisles with 165°F ceiling sprinklers and 165°F in-rack sprinklers
C, 4-ft aisles with 286°F ceiling sprinklers and 165°F in-rack sprinklers
D, 4-ft aisles with 165°F ceiling sprinklers and 165°F in-rack sprinklers
E, 8-ft aisles with 286°F ceiling sprinklers
F, 8-ft aisles with 165°F ceiling sprinklers
G, 4-ft aisles with 286°F ceiling sprinklers
H, 4-ft aisles with 165°F ceiling sprinklers.
(Reproduced with permission from NFPA 231C, "Rack Storage of Materials," 1975, National Fire Protection Association.)

gong is needed or required, a vane-type water-flow indicator can be inserted in the supply pipe to give notification electrically of water flow.

Pipe sizing for wet-pipe systems, as well as for all other types of systems, is based on the degree of hazard covered. Pipe-sizing schedules are provided in "Installation of Sprinkler Systems," National Fire Protection Association, for light-hazard, ordinary-hazard, and extra-hazard systems. These schedules apply to all systems having sealed heads. A pipe-sizing schedule is also given for deluge systems, but this is only a guide. Determination of the schedule to be used must be made by the insurance or code authority. Frequently, a single building can require the use of all the schedules.

For buildings where the fire load is unusually heavy, or where stock is piled higher than 12 ft, or where deluge systems are used, the authority can require that the schedule be disregarded and the sizing be calculated for a specific rate of discharge over a specific area. In this case, the authority will indicate the total area over which a specified discharge rate must be made. The area indicated will have no relation to the total area of the building or the total number of heads subject to the same fire. For example, in a 25,000-ft^2 building the authority might consider that the sprinkler system will bring a fire under control within an area of 2000 ft^2 with a water application rate (density) of 0.35 gal/min per square foot of floor area. This means that the maximum anticipated discharge of the sprinkler system would be 2,000 × 0.35, or 700 gal/min. With heads spaced for a coverage of 100 ft^2 each, this means that each head would have to discharge 35 gal/min. With the discharge rate per head known, and the total area known, the pressure required to deliver this volume of water at the most distant point from the water supply can be calculated. Calculations are made in the manner described in NFPA 13, and NFPA 15, "Water Spray Fixed Systems."

In determining the area and density required, the authority considers the contents, type of building construction, heat venting, and type of system, among other factors. If high piling of stock is involved, the authority takes into account the pile size, pile height, aisle width, and combustibility of contents. Based on full-scale fire tests, a standard, NFPA 231C, describes the protection required for stock piling from 12 to 20 ft, in racks. Storage heights less than 12 ft are covered by NFPA 13.

Figure 19-20 presents sprinkler system design curves for 20-ft-high rack storage, with conventional pallets and with Class I commodities (noncombustible products on wood pallets or in ordinary paper cartons on wood pallets, such as noncombustible foodstuffs or beverages).

Dry-pipe systems are used in locations and buildings where it is impractical to maintain sufficient heat to accommodate a wet-pipe system. The system of piping is the same, but it is normally empty, containing only air under pressure. A normally high water pressure is held by a normally low air pressure by use of a **dry-pipe valve.** This valve generally employs a combined air and water clapper (Fig. 19-21) where the area under air pressure is about 16 times the area subject to water pressure. When a sprinkler head operates, air is released to the atmosphere, allowing the water to overcome the pressure differential and enter the piping. Air pressure must be checked periodically, but devices are available to maintain pressure automatically and are equipped with an alarm that goes off in case of a malfunction allowing air pressure to drop too low.

In dry-pipe systems of large capacity, the relatively slow drop in air pressure when a single head or a few heads operate is overcome by use of an accelerator or exhauster. The former is a device, installed near the dry-pipe valve, to sense a small drop in pressure and transmit the system air pressure to a point under the air clapper. The water then can open the clapper quickly and push its way into the system without requiring a further pressure drop. An exhauster accomplishes the opening by sensing a small pressure drop and exhausting the pressurized air to the atmosphere through a large opening.

The dry-pipe valve must be installed in a heated area, or in a heated enclosure, because there is water in the piping up to the valve, and priming water in the valve itself.

Deluge systems are those in which all heads open to discharge water simultaneously. The water is controlled by a **deluge valve,** which is operated by a temperature-sensitive detection system installed throughout the same area as the piping and heads. The detecting devices may be electrical, pneumatic, or mechanical, and may be connected to the deluge valve by wiring, tubing, or piping. Some valves are adaptable to all three means of operation.

A deluge system may be used for a few heads for a small, isolated industrial hazard, or in an entire, huge structure, such as an airplane hangar requiring several thousand heads. Deluge systems are also used to advantage in chemical plants where process vessels and tanks contain flammable materials. The purposes of such systems are to isolate a fire and to cool exposed equipment not originally involved in the fire.

Fig. 19-21 Differential dry-pipe valve: (a) air pressure keeps clapper closed; (b) venting of air permits clapper to open and water to flow.

Piping in deluge systems must be sized by calculation. Consideration must be given to the probability of simultaneous operation of adjacent systems, if any are involved. In addition, authorities frequently require that provision be made in the water supply system for operation of two or more fire-hose streams while the deluge system is operating. In a multisystem installation, each system must be of manageable size, and good practice requires that no system be larger than one 6-in deluge valve can supply.

An electrically controlled system may employ fixed-temperature thermostats, rate-of-rise units, smoke detectors, combustion products detectors, or infrared or ultraviolet detectors. The nature and extent of the hazard determine the kind of detector required.

Operation of a hydraulically or pneumatically controlled deluge valve can be by means of a pilot line of small-diameter pipe on which are spaced automatic sprinkler heads at suitable intervals. These heads can be augmented, when necessary, by use of a mechanical air- or water-release device, which operates on the rate-of-rise principle as well as the fixed temperature of the sprinkler heads. Other penumatic means include small copper air chambers, sensitive to rate-of-rise conditions, connected by small-diameter copper tubing to the release mechanism of the valve.

Most deluge valves are essentially check valves with a clapper that is normally latched in the closed position. The actuating system unlatches the valve, allowing water to enter the system and flow out the heads.

For application of water from a deluge system, a variety of nozzles may be needed. If general area coverage is required, a standard sprinkler head, without a fusible element, may be used. A standard sprinkler orifice is ½ in, but larger or smaller orifices may be required. For specific equipment protection, orifices may be directional, or with a wide or narrow included angle, and with varying degrees of droplet size, including atomizing nozzles.

Preaction systems employ the same components as deluge systems, but the sprinkler heads are sealed and the pipe sizing is in accordance with wet- and dry-pipe systems. A separate heat-detecting system with sealed heads is employed, because the detecting system is designed to operate before a head opens and at the same time to sound an alarm. If the fire can be extinguished before any heads open, the total damage from the fire and water may be less. In addition, if any part of the piping is damaged, water will not issue unless heat is present also. The sealed-head piping system is usually supervised by air at low pressure, 1 or 2 psi, the loss of which will sound a trouble alarm.

This system lends itself to use where highly damageable contents are involved. It is also a substitute for a dry-pipe system when the additional expense of the detecting system is justified.

Combined dry-pipe and preaction systems are suitable for large-area structures, such as piers, where more than one dry-pipe system would be required and where a water-supply pipe would be required in an unheated area.

In this arrangement (Fig. 19-22), the heat-detecting system is the same as for a stan-

Fig. 19-22 Typical piping layout for combined dry-pipe and preaction sprinkler systems. See Fig. 19-23 for supply header and Fig. 19-24 for exhaust header. *(Reproduced by permission from NFPA 13, "Installation of Sprinkler Systems," 1978, National Fire Protection Association.)*

Fig. 19-23 Supply header for sprinkler system in Fig. 19-22. *(Reproduced by permission from NFPA 13, "Installation of Sprinkler Systems," 1978, National Fire Protection Association.)*

dard preaction system, but two dry-pipe valves in parallel are used instead of deluge valves. The system uses compressed air, as in a dry-pipe system. Valve arrangement at the supply end is shown in Fig. 19-23, and the exhauster system at the outboard end in Fig. 19-24.

When a fire is detected, the detecting system sounds an alarm and operates the exhausters at the valves and at the outboard end. The exhausters hasten the entry of water into the sprinkler pipes. Rubber-faced check valves on each subsystem avoid the need for exhausting air in the subsystems that are not involved in the fire. If there is a failure in the detecting system, it will still operate as a dry-pipe system, and if both dry-pipe valves fail, at least an alarm has been given.

In a large system, a number of dry-pipe valves, each with expensive heated enclosures, can be eliminated, as well as wrapping and heat-tracing of the supply line and the consequent maintenance of these items.

Firecycle system is a recycling sprinkler system (patented by Viking Corporation, Hastings, Mich.) that will detect a fire, extinguish it with standard sprinklers, and turn off the water automatically when the fire is out (Fig. 19-25). If the fire is not completely out, or if it rekindles, the system will start again and repeat the performance as many times as necessary. This system can be valuable when there is highly damageable stock, or if it is likely to be a long time before there is a response to the alarm.

The basic design of the system is similar to the preaction system. It employs self-resetting, rate-of-rise, electrically operated detectors connected with conductors not subject to fire damage. The detector circuit operates a flow-control valve that is a modified form of the Viking deluge valve. As soon as the detecting system senses the fire, an alarm is actuated and the valve is opened, admitting water to the system. As ceiling temperature increases, sprinklers operate and discharge water over the fire area. This discharge

Fig. 19-24 Exhaust header for sprinkler system in Fig. 19-22. *(Reproduced by permission from NFPA 13, "Installation of Sprinkler Systems," 1978, National Fire Protection Association.)*

will continue until the ceiling temperature is reduced to below the operating temperature of the detector. At that time, the detector circuit closes and a timing cycle starts. At the end of the timing cycle, the flow-control valve closes, and the system is ready for a repeat operation if the ceiling temperature increases again.

Outside sprinklers for exposure protection may be installed outside a building on any side exposed to an adjacent combustible structure or other potential fire hazard. Such systems have open nozzles directed onto the wall, windows, or cornices to be protected. The water supply may be taken from a point below the inside-sprinkler-system control valve if the building is sprinklered, otherwise from any other acceptable source,

Fig. 19-25 General arrangement of Firecycle system. *(The Viking Corporation, Hastings, Mich.)*

with the controlling valve accessible at all times. The system is usually operated manually by a gate valve but can be made automatic by use of a deluge valve actuated by suitable means on the exposed side of the building. The distribution piping is usually installed on the outside of the wall, with nozzles provided in sufficient numbers to wet the entire surface to be protected.

See also Art. 3-19.

19-28. Sprinkler Piping.

The local fire-prevention authorities and the fire underwriters usually have specific requirements regarding the type, material, and size of pipe used in sprinkler systems. They should be consulted prior to the design of any system. Table 19-18 lists pipe-size schedules for the occupancy classifications met in various structures.

NFPA 13 requires that sprinkler pipes be made of wrought steel (ASTM A120), which preferably should be galvanized, or wrought iron (ASTM A72). The pipe should be capable of withstanding a pressure of at least 175 psi. (Standard-wall pipe may be used for pressures up to 300 psi.) NFPA 13 also contains tables that give the minimum diameters of pipe that may be used with specific numbers of sprinklers for various types of building occupancies.

Fittings on sprinkler pipes should be capable of resisting the working pressures in the pipes, but at least 175 psi. Fittings may be made of cast or malleable iron.

Connections should be made with screwed or flanged fittings. Joints in risers and feed mains, however, may be welded.

TABLE 19-18 Pipe-Size Schedule for Typical Sprinkler Installations.

Occupancy and Pipe Size, In.	No. of Sprinklers
Light Hazard	
1	2
1¼	3
1½	5
2	10
2½	40
3	No limit
Ordinary Hazard	
1	2
1¼	3
1½	5
2	10
2½	20
3	40
3½	65
4	100
5	160
6	250
Extra Hazard	
1	1
1¼	2
1½	5
2	8
2½	15
3	27
3½	40
4	55
5	90
6	150

Area Protected by Sprinklers. In structures of fire-resistive construction and light-hazard occupancy, the area protected by one sprinkler should not exceed 200 ft². In this type of construction with ordinary-hazard occupancy, the area protected per sprinkler should not ordinarily exceed 130 ft². For extra-hazard occupancy, the area protected by each sprinkler should not exceed 100 ft². Local fire-protection codes and underwriters' requirements cover other types of construction, including mill, semimill, open-joist, and joist type with a sheathed or plastered ceiling.

Sprinkler Position. In all usual applications, the sprinklers should be installed in the upright position (Fig. 19-18), where they are protected from drainage by vehicles and other objects passing beneath them. When it is necessary to install sprinklers in a pendent position, they must be of a design approved for that position.

Sprinkler deflectors should be parallel to the building ceilings and roof, or the incline of stairs when installed in a stairwell. Deflectors for sprinklers in the peak of a pitched roof should be horizontal.

Sprinkler Alarms. These are often required, depending on what other types of alarms or fire-notification devices are installed in the structure. The usual sprinkler alarm is designed to sound automatically when the flow from the system is equal to or greater than that from a single automatic sprinkler. Dry-pipe systems use a water-actuated alarm; preaction and deluge systems use electric alarm attachments actuated independently of the water flow in the system (see Art. 19-27).

Sprinkler Layout. Horizontal branch pipes with sprinklers are usually laid out in parallel lines in each story (Figs. 19-16 and 19-17). In buildings with light- and ordinary-hazard occupancies, the maximum distance permitted between horizontal branch pipes and between sprinklers on those pipes is 15 ft. For high-piled storage, however, maximum spacing generally is limited to 12 ft. For extra-hazard occupancies, maximum spacing also is 12 ft. The distance from a wall to an end branch pipe or to an end sprinkler on a branch pipe should not exceed one-half the allowable spacing for the pipes or sprinklers, respectively.

Locations of sprinklers should be staggered on successive branch pipes in buildings with extra-hazard occupancies. Sprinklers also should be staggered in ordinary-hazard buildings when sprinkler spacing on branch pipes exceeds 12 ft or when the sprinklers are placed under solid beams spaced 3 to 7½ ft on centers. In areas where aesthetics is important, however, the sprinklers need not be staggered.

Sprinklers should be located to prevent structural members from interfering with the discharge pattern. Where branch pipes are installed transverse to lines of beams, the deflectors of sprinklers, which direct the discharge, should preferably be located above the bottom of the beam. If the branch pipes are installed under beams, sprinkler deflectors should be placed 1 to 4 in below the beams, but not more than 14 in below combustible ceilings nor more than 16 in below noncombustible ceilings. Under smooth-ceiling construction, deflectors of upright sprinklers should be at least 3 in and of pendent sprinklers at least 2½ in below the ceiling; but they should be not more than 10 in below combustible ceilings nor more than 12 in below noncombustible ceilings.

Drainage of Sprinkler Systems. Provision should be made for draining all or any part of a sprinkler system. For the purpose, air-inlet valves or plugged fittings should be installed at high points, and valve-controlled outlets at low points. A drain and valve connection should be provided for each group of up to 20 sprinkler heads. To ensure drainage, so-called horizontal pipes should have a slope of at least ¼ in/ft.

For 4-in diam and larger supply-pipe risers, drip or drain pipes should be at least 2 in in diameter. For risers between 1½ and 4 in in diameter, drain pipes should be at least 1½ in in diameter, and they should be the same size as the risers when these are smaller than 1½ in in diameter.

Test Pipes. Sprinkler systems should be tested periodically to ensure that they will function when needed. For the purpose, water should be drained from the sprinkler systems through test or drain connections at least ¾ in in diameter; the higher the pressure head, the larger the diameter of the connection should be. The discharge should be visible. The test pipe for the main water supply should be at least 2 in in diameter.

Approvals of Sprinkler-System Design. Before a sprinkler system is installed or remodeled, the applicable drawings should be submitted to the local municipal agency and the underwriter of the building. Since beneficial reductions in insurance rates may be obtained by suitable installation of sprinkler systems, it is important that the underwriter have sufficient time for a full review of the plans before construction begins. Likewise, municipal approval of the sprinkler-system plans is usually necessary before the structure can be occupied. In actual construction, the piping contractor installing the sprinkler system generally secures the necessary municipal approval. However, existence of the approval should always be checked before construction is begun.

19-29. Water Supplies for Sprinklers.

These must be of the most reliable type obtainable. An example of a satisfactory supply is a connection to a reliable municipal water system. Where this is not possible, storage or pumping facilities must be provided. An elevated water tank is acceptable for this purpose and requires no outside energy source. An automatic fire pump supplied from a storage reservoir is usually used as a secondary supply, augmenting a municipal water connection or a gravity tank, but such a supply is generally acceptable as a primary supply if it is supervised by a central-station system (Art. 19-31). A hydropneumatic storage tank is also acceptable as a primary supply. It is generally augmented by an automatic fire pump. National Fire Protection Association standard NFPA 20 applies to fire pumps, NFPA 22 to water tanks.

In general, gravity tanks should have a capacity of at least 5000 gal with the tank bottom not less than 35 ft above the underside of the roof for light- and ordinary-hazard occupancies. Pressure tanks should have at least 2000 and 3000 gal of water available, respectively, for these occupancies. Where penstocks, flumes, rivers, or lakes supply the water, approved strainers or screens must be provided to prevent solids from entering the piping.

The absolute minimum water supply requirement is to provide for a pressure of 15 psi at the highest line of sprinklers concurrent with a flow at the base of the riser of 250 gal/min for light-hazard occupancy, and 500 gal/min for ordinary-hazard occupancy. This requirement is based on the availability of a municipal fire department that will respond to alarms set off by a system in the building. Water supply requirements are given in municipal building codes. If, however, NFPA 13, "Installation of Sprinkler Systems," is the governing standard, the rules are indefinite and the local authority must be consulted.

Where the occupancy requires a system with an unusual water demand, this must be met. In a calculated system, the maximum demand is known within limits, and at most must be supplemented with a reasonable flow from hose streams. Where there are deluge systems, attention must be given to the possible simultaneous operation of more than one system.

A property located so that buildings

Fig. 19-26 Fire department tie-in pit.

cannot be reached by a fire department with 250 ft of hose may require installation of a private, underground water-distribution system. Figure 19-26 shows a fire-department tie-in pit. If a municipal water system is available within a reasonable distance, this can form the primary water source. Water demand may be such that this would have to be supplemented by a private elevated tank, or automatic fire pump, or both.

Under such circumstances, the distribution system would have to provide private hydrants to reach all sides of all buildings. Supplies for automatic sprinklers and standpipes would be taken directly from this underground system. Standard for the installation of such systems is NFPA 24.

19-30. Standpipes.

Hoses supplied with water from standpipes are the usual means of manual application of water to interior fires in buildings higher than six stories or 75 ft, or in 4- or 5-story buildings with an area of more than 7500 ft^2 per floor. Standpipes are vertical pipes in which water is stored under pressure, or which can be rapidly supplied with water under pressure, primarily for use by a fire department, but they also may be used by building fire fighters.

Riser Sizes. Risers up to 150 ft high, measured to the highest hose outlet, should be at least 4 in in diameter. Taller risers should be at least 6 in in diameter. The local building code should be consulted for exact requirements.

Number of Risers. Sufficient risers should be installed in a building so that every point of every floor can be reached by a 20-ft stream from a nozzle attached to 125 ft or less of hose connected to a riser outlet valve.

Maximum Pressure. The standpipe system should be zoned by the use of gravity tanks, automatic fire pumps, pressure tanks, and street pressure so that the maximum gage pressure at the inlet of any hose valve does not exceed 160 psi. Building codes often require that the hose valve be equipped with an adjustable-type, pressure-reducing valve to limit the pressure on the downstream side of the hose valve. Usually, this pressure should be less than 50 psi for a discharge of 200 gal/min from a 1-in hose nozzle.

Hose Stations. Standpipes should be placed within stairway enclosures, or as near such an enclosure as possible, for safe access to hose outlets by fire fighters. Hose stations should be located at the standpipes. The hoses should be stored on hose racks.

Where standpipes are required for fire department use, 2½-in hose valves should be provided. Where it is intended and permitted that building fire fighters use hose, 1½-in hose may be specified. Some codes provide for both, in which case the valves should be 2½ in and equipped with easily removable adapters. The valves should be readily accessible from a floor or a stairway landing and should be placed about 5 or 6 ft above that level.

Water Supply. Standpipes can be supplied from municipal water mains when adequate pressure and volume are available. When necessary, this source can be supplemented with water from a tank or automatic fire pump.

NFPA requires that the minimum supply be 500 gal/min for one riser, and an additional 250 gal/min for each additional riser, up to a maximum of 2500 gal/min. A pressure of 65 psi is required at the highest outlet for all risers.

Standpipes with no water supply other than a fire-department pumper connection are permitted in open-air parking garages.

All risers should be cross connected at, or below, the street entrance-floor level. In buildings that have zoned water-distribution systems, standpipes in each zone should be cross connected below, or in, the story with the lowest hose outlets from the water source in each zone. Cross connections should have a diameter at least as large as that of the largest riser supplied by the cross connection. The cross connection between standpipe risers at the street entrance-floor level and at the bottom of each zone should extend to the exterior walls of the building and terminate in a siamese connection on the outside, 18 to 36 in above grade. A siamese connection is a wye to which two fire hoses can be connected. The hoses, when connected by fire fighters to a street hydrant or to a mobile pump, can supply the standpipes with water in an emergency. Building codes may require a siamese connection for each 300 ft of building wall facing on a street or public place.

19.31. Central-Station Supervisory Systems.

Any mechanical device or system is more reliable if it is tested and supervised regularly. Sprinkler systems are designed to be rugged and dependable, as shown by their satisfactory performance record of 96.5% over a period exceeding 75 years. Performance, however, is improved where systems have been connected to an approved central-station supervisory service that provides continuous monitoring of alarms, gate valves, pressures, and other pertinent functions.

A central station supplies the necessary attachments to a system or water supply and transmits appropriate signals over supervised circuits to a constantly monitored location. When any signal is received at the station, at whatever hour, suitable action should be taken by dispatch of investigators, notification of the fire department, or whatever is necessary. Notification of a closed gate valve, for example, should be investigated immediately to avoid a possible loss. Any special extinguishing system, such as a carbon dioxide system, can also be supervised as to pressure, operation, or other vital condition.

Such services are available in most cities and are arranged by contract, usually with an installation charge and an annual maintenance charge. Requirements for such systems are in National Fire Protection Association standard NFPA 71. Where no such service is available, a local or proprietary substitute can be provided when such a facility is desirable. Standards for such installations are NFPA 72A, "Local Protective Signaling Systems," NFPA 72B, "Auxiliary Protective Signaling Systems," NFPA 72C, "Remote Station Protective Signaling Systems," and NFPA 72D, "Proprietary Protective Signaling Systems."

Section 20

Electric Power

Charles J. Wurmfeld
Consulting Engineer,
Wurmfeld Associates, P. C.
New York, N.Y.

In many ways, transmission of electricity in buildings is analogous to water-supply distribution. Voltage is equivalent to pressure drop; wire resistance, to pipe friction; and electric current, or flow of electrons, to water droplets.

For water flow, mathematical formulas have been developed to measure the relationship between pressure, measured in feet of water or psi; flow, gal/min or ft^3/s; and pipe friction f, an experimentally determined number that varies with the material, shape, and smoothness of pipe. For electricity, formulas have been developed to measure the relationship between voltage, measured in volts; current, in amperes; and wire resistance, in ohms, an experimentally determined number that varies with the metal, shape, and temperature of the conductor.

The hydraulic analogy is limited to only very elementary applications with electric flow like direct current, which always flows in the same direction. The analogy does not hold for alternating current, which reverses flow many times per second without apparent inertia drag. Direct-current systems are simple two-wire circuits, whereas alternating current uses two, three, or four wires and the formulas are more complex. Any attempt to apply the hydraulic analogy to alternating currents would be more confusing than helpful. The mathemetical concepts are the only guides that remain true over the whole area of application.

Ampere (abbreviated A) is the basic unit for measuring flow of current. The unit flowing is an electric charge called a **coulomb**. An ampere is equivalent to a flow of one coulomb per second.

One source of direct current is the battery, which converts chemical energy into electric energy. By convention, direct current flows from the positive terminal to the negative terminal when a conductor is connected between the terminals. The voltage between battery terminals depends on the number of cells in the battery. For a lead-plate–sulphuric-acid battery, this voltage is about 1.5 to 2 V per cell.

For high voltages, a generator is required. A generator is a machine for converting mechanical energy into electrical energy. The basic principle involved is illustrated by the simple experiment of moving a copper wire across the magnetic field between a north pole and south pole of a magnet. In a generator, the rotor is wound with coils of wire and the magnets are placed around the stator in pairs, two, four, six, and eight. When the coil on the rotor passes through the magnetic field under a south pole, current flows in one direction. When the same coil passes through the north-pole field, the current reverses. For this reason, all generators produce alternating current. If direct current is required, the coils are connected to contacts on the rotor, which transfer the current to brushes arranged to pick up the current flowing in one direction only. The contacts and brushes comprise the commutator. If the commutator is omitted, the generator is an alternator, producing alternating current.

20-1. Direct-Current Systems.

Resistance of flow through a wire, measured in units called **ohms**, depends on the wire material. Metals like copper and aluminum have low resistance and are classified as conductors.

Resistance for a given material varies inversely as the area of the cross section and directly as the length of wire. The unit *mil* designates the area of a wire 1 mm in diameter.

Ohm's law states that the voltage E required to cause a flow of current I through a wire with resistance R is given by

$$E = IR \tag{20-1}$$

Power P is measured in watts and is the product of volts and amperes:

$$P = EI = (IR)I = I^2R \tag{20-2}$$

Large amounts of power are measured in kilowatts (kW), a unit of 1000 W, or megawatts (MW), a unit of 1,000,000 W.

Ohm's law can be used to solve problems involving simple two-wire circuits. For more complex circuits, such as flow through a wire consisting of sections with different resistances [series circuit (Fig. 20-1a)] or flow through a junction point where the current divides into several different conductors [parallel circuits (Fig. 20-1b)] with varying characteristics, other formulas, known as Kirchhoff's laws, have been developed.

Electric Energy. The energy expended in a circuit equals the product of watts and time, expressed as watt-seconds or watt-hours (Wh). For large amounts of energy, a unit of 1000 kilowatt-hours, kWh, is used.

Charges for electric use are usually based on two separate items. The first is total energy used per month, kWh, and the second is the maximum kW required over any short period during the month, usually 15 to 30 min.

Power Transmission. Power is usually transmitted at very high voltages to minimize the power loss over long distances. This power loss results from the energy consumed in heating the transmission cables and is equal to the square of the current flowing I, times a constant representing the resistance r of the wires, Ω/ft, times the length L, ft, of the wires. Measured in watts,

$$\text{Heat loss} = I^2 rL \tag{20-3}$$

Equation (20-3) is strictly correct only for direct currents. With alternating currents (ac), the variation in magnetic fields surrounding the wires introduces additional power losses. A multiplying factor greater than 1.0 is applied to the heat loss, W, in Eq. (20-3) (see Table 20-4). This factor varies with the frequency of the alternating current.

Kirchhoff's laws, together with Ohm's law [Eq. (20-1)], can be used to determine currents and voltages in direct-current circuits. The basic circuits are parallel and series circuits and combinations of these two types.

Fig. 20-1 Types of electric circuits: (a) series; (b) parallel.

Series Circuits. A series circuit is by definition one in which the same current I flows through all parts of the circuit (Fig. 20-1a). In such a circuit, the resistance R of each part is the resistance per foot times the length, ft. Also, by Ohm's law, for each part of the circuit, the voltage drop is

$$E_1 = IR_1 \quad E_2 = IR_2 \quad \ldots \quad E_n = IR_n \tag{20-4}$$

Kirchhoff's law for series circuits states that the total voltage drop in a circuit is the sum of the voltage drops:

$$\begin{aligned} E_T &= E_1 + E_2 + E_3 + \ldots + E_n \\ &= IR_1 + IR_2 + IR_3 + \ldots + IR_n \\ &= I(R_1 + R_2 + R_3 + \ldots + R_n) \end{aligned} \tag{20-5}$$

By Ohm's law the total resistance in a series circuit then is

$$R = R_1 + R_2 + R_3 + \ldots + R_n \tag{20-6}$$

Parallel Circuits. These are, by definition, circuits in which the same voltage drop is applied to each circuit (Fig. 20-1b). The resistance of each circuit is obtained by multiplying the resistance per foot by the length, ft.

Kirchhoff's law for parallel circuits states that the total current in a circuit is equal to the sum of the currents in each part:

$$\begin{aligned} I_T &= I_1 + I_2 + I_3 + \ldots + I_n \\ &= \frac{E}{R_1} + \frac{E}{R_2} + \frac{E}{R_3} + \ldots + \frac{E}{R_n} \\ &= E\left(\frac{1}{R_1} + \frac{1}{R_2} + \frac{1}{R_3} + \ldots + \frac{1}{R_n}\right) \end{aligned} \tag{20-7}$$

By Ohm's law, then, the total resistance R in a parallel circuit is given by

$$\frac{1}{R} = \frac{1}{R_1} + \frac{1}{R_2} + \frac{1}{R_3} + \ldots + \frac{1}{R_n} \tag{20-8}$$

It is sometimes convenient to use **conductance** G, siemens, which is the reciprocal of resistance R:

$$G = \frac{1}{R} \tag{20-9}$$

By Ohm's law,

$$I_T = EG_1 + EG_2 + EG_3 + \ldots + EG_n \tag{20-10}$$

Series circuits are usually used in street and subway lighting. Most building systems use parallel circuits for both motor and lighting distribution.

Network systems consisting of a combination of series and parallel circuits are used for municipal distribution. With this type of system, a given building can be served from two different sides. This provides a second source of power, which is often accepted in lieu of an emergency generator.

20-2. Alternating-Current Systems.

Any change in flow of current, such as that which occurs in alternating current, produces a magnetic field around the wire. With steady flow, as in direct current, there is no magnetic field.

One common application of magnetic fields is for solenoids. These are coils of wire, with many turns, around a hollow cylinder in which an iron pin moves in the direction of the magnetic field generated by the current in the coil. The movement of the pin is used to open or close electric switches, which start and stop motors. The pin returns to a normal position, either by gravity or spring action, when the current in the coil is stopped.

The motion of the pin can be predicted by the *right hand rule*. If the fingers of the right hand are curled around the solenoid with the fingers pointing in the same direction as the current in the coil, the thumb will point in the direction of the magnetic field, or the direction in which the pin will move.

With direct current, a magnetic field exists only as the flow changes from zero to steady flow. Once steady flow is established in the wire, the magnetic field collapses. For this reason, all devices and machines that rely on the interaction of current and magnetic fields must use alternating current, which changes continuously. This equipment includes transformers, motors, and generators.

Transformers. These are devices used to change voltages. A transformer comprises two separate coils, primary and secondary, that wind concentrically around a common core of iron (Fig. 20-2). A common magnetic field consequently cuts both the primary and secondary windings. When alternating current (ac) flows in the primary coil, the changing magnetic field induces current in the secondary coil. The voltage resulting in each winding is proportional to the number of turns of wire in each coil. For example, a transformer with twice the number of turns in the secondary coil than in the primary will have twice the primary voltage across the secondary coil.

Generators and Motors. Just as changes in current flowing in a wire produce a magnetic field, movement of a wire through a magnetic field produces current in the wire. This is the principle on which electric motors and generators are built.

In these machines, a rotating shaft carries wire coils wound around an iron core, called an armature. A stationary frame, called the stator, encircling the armature, also carries iron cores around which are wound coils of wire. These cores are arranged in pairs opposite each other around the stator, to serve as poles of magnets. The windings are so arranged that if a north pole is produced in one core a south pole is produced in the opposite core. Current flowing in the stator, or field, coils creates a magnetic field across the rotating armature.

When the machine is to be used as a motor, voltage is applied across the armature windings, and the reaction with the magnetic field produces rotary motion of the shaft. When the machine is to be used as a generator, mechanical energy is applied to rotate the shaft, and the rotation of the armature windings in the mag-

Fig. 20-2 Transformer.

netic fields produces current in the armature windings. The current varies in magnitude and reverses direction as the shaft rotates.

For direct-current motors and generators, the ends of each armature coil are connected to bare copper terminals, which transmit the current to or from an electrical circuit through carbon brushes. As an armature coil rotates through 180°, the current reverses direction, but the brushes transmit the current, which varies in magnitude, always in the same direction. This flow is called variable direct current.

Sine-Type Currents and Voltages. In generation of alternating current, rotation of the armature of a generator produces a current that starts from zero as a wire enters the magnetic field of a pole on the stator and increases as the wire moves through the field. When the wire is directly under the magnet, the wire is cutting across the field at right angles and the maximum flow of current results. The wire then moves out of the field and the current decreases, passing through zero at 180° from the start. The wire next moves into the magnetic field of the opposite pole, and the process repeats, except that the current now flows in the opposite direction in the wire. This variation from zero to a maximum in one direction (positive direction), down to zero, then continuing down to a maximum in the opposite direction (negative direction), and back to zero is called a **sine wave**.

The number of complete cycles per second of the wave is called the **frequency** of the current. This is usually 60 Hz (cycles per second) in the U.S.; 50 Hz in many European countries.

If P is the number of poles on the stator of a generator, the frequency of the alternating current equals $P \times RPM/120$, where RPM is the revolutions per minute of the armature. This relationship also holds for ac motors. Hence, for a frequency of 60 Hz, RPM = $60 \times 120/P = 7200/P$. This indicates that theoretically a standard four-pole motor would run at 1800 rpm, and a two-pole motor at 3600 rpm. Because of slippage, however, these speeds are usually 1760 and 3400 rpm, respectively.

Phases. Two currents or voltages in a circuit may have the same frequency but may pass through zero at different times. This time relationship is called phase. As explained in the preceding, the variation of the current (or voltage) from zero to maximum is a result of the rotation of a generator coil through 90° to a pole and back to zero in the subsequent 90°. The particular phase of a current is therefore given as angle of rotation from the zero start. If current (or voltage) 1 passes through its zero value just as another current (or voltage) 2 passes through its maximum, current 2 is said to *lead* current 1 by 90°. Conversely, current 1 is said to *lag* current 2 by 90°.

Effective Currents. The instantaneous value of an alternating current (or voltage) is continuously varying. This current has a heating effect on wire equal to the *effective current I* times the resistance R. Mathematically, the effective current is 0.707 times the maximum instantaneous current of the sine wave.

Ohm's law, $E = IR$, can be used in alternating circuits with E as the *effective voltage*, I, the effective current, and R the resistance, Ω.

Inductive Reactances and Susceptance. When alternating current flows through a coil, a magnetic field surrounds the coil. As the current decreases in instantaneous value from maximum to zero, the magnetic field increases in strength from zero to maximum. As the current increases in the opposite direction, from zero to maximum, the magnetic-field strength decreases to zero. When the current starts to decrease, a new magnetic field is produced that is continuously increasing in strength but has changed direction.

The magnetic field, in changing, induces a voltage and current in the wire, but the phase, or timing, of the zero and maximum values of this induced voltage and current are actually 90° behind the original voltage and current wave in the wire.

The induced voltage and current are proportional to a constant called the **inductive**

reactance of the coil. This constant, unlike resistance, which depends on the material and cross-sectional area of the wire, depends on the number of turns in the coil and the material of the core on which the coil is wound. For example, a simple coil wound around an airspace has less inductive reactance than a coil wound around an iron core. Inductive voltage E_L and inductive current I_L are related by

$$E_L = I_L X_L \qquad (20\text{-}11)$$

where X_L is the constant for inductive reactance of the circuit, Ω. The reciprocal $1/X_L$ is called **inductive susceptance**.

When an inductive reactance is wired in a series circuit with a resistance, the inductive reactance does not draw any power (or heating effect) from the circuit. This is because the induced current I_L is 90° out of phase with the applied voltage E. In the variation of the instantaneous value of applied voltage, power is taken from the circuit in making the magnetic field. Then, as the magnetic field collapses, the power is returned to the circuit.

Capacitive Reactance and Susceptance. An electrostatic condenser, or capacitor, consists basically of two conductors, for example, flat metal plates, with an insulator between. Another familiar form in laboratory use is the arrangement of two large brass balls with an airspace between. Electrostatic charges accumulate on one plate when voltage is applied. When the voltage is high enough, a spark jumps across the air space. With direct current, the discharge is instantaneous and then stops until the charge builds up again. With alternating current, as one plate is being charged, the other plate is discharging, and the flow of current is continuous. In this case, the circuit is called capacitive.

$$E_C = I_C X_C \qquad (20\text{-}12)$$

where X_C is the constant for **capacitive reactance,** Ω. The reciprocal $1/X_C$ is called **reactive susceptance**. The current I_C reaches its peak when the impressed voltage E is just passing through zero. Capacitive current is said to lead the voltage by 90°.

Impedance and Admittance. A circuit can have resistance and inductive reactance, or resistance and capacitive reactance, or resistance and both inductive and capacitive reactance. Resistance is present in all circuits. When there is any inductive or capacitive reactance, or both, in a circuit, the relation of the voltage E and current I is given by

$$E = IZ \qquad (20\text{-}13)$$

where Z is the **impedance,** Ω, the vector sum of the resistance, and the inductive and capacitive reactances. The reciprocal $Y = 1/Z$ is called the **admittance**.

A vector may be represented by a line and arrowhead. In electrical calculations, the length of the line is made proportional to the value of E or I, and the arrowhead indicates plus or minus. In resistance circuits, the vector E is indicated by a horizontal line with the arrowhead at the right:

$$\xrightarrow{E = IR}$$

In a circuit that contains inductive reactance, the current I_L lags behind the voltage E by 90°. This is indicated by vectors as follows:

The vector sum of these voltages is indicated by vectors as follows:

The diagram indicates that

$$E = IZ \cos \theta_L \qquad (20\text{-}14)$$

where θ_L is the phase angle between voltage and current.

The relation between resistance R and inductance L is indicated by a similar vector addition:

The diagram indicates that

$$R = Z \cos \theta_L \qquad (20\text{-}15)$$

In a similar way, capacitive reactance in the circuit is indicated by:

and the vector sum is shown as follows:

The diagrams indicate that

$$E = IZ \cos \theta_C \qquad (20\text{-}16)$$
$$R = Z \cos \theta_C \qquad (20\text{-}17)$$

Note that the vectors for inductance and capacitance are in opposite directions. Thus, when a circuit contains both inductance and capacitance, they can be added algebraically. If they are equal, they cancel each other, and the Z value is the same as R. If the inductance is greater, Z will be in the upper quadrant. A greater value of capacitance will throw Z into the lower quadrant.

The diagram indicates that

$$\tan \theta = \frac{L - C}{R} \qquad (20\text{-}18)$$

Kirchhoff's laws are applicable to alternating current circuits containing any combinations of resistance, inductance, and capacitance by means of vector analysis:

In a **series circuit,** the current I is equal in all parts of the circuit, but the total voltage drop is the vector sum of the voltage drops in the parts. If the circuit has resistance R, inductance L and capacitance C, the voltage drops must be added vectorially as described in the preceding. Equations (20-14) to (20-18) hold for ac series circuits. To find the voltage drop in each part of the circuit, compute

$$E_R = IR \qquad E_L = IX_L \qquad E_C = IX_C$$

$$E_Z = E_R + E_L - E_C \qquad \text{(vectorially)} \tag{20-19}$$

In a **parallel circuit,** the voltage E across each part is the same and the total current I_Z is the vector sum of the currents in the branches,

$$\frac{E}{R} = I_R \qquad \frac{E}{X_L} = I_L \qquad \frac{E}{X_C} = I_C$$

For parallel circuits, it is convenient to use the reciprocals of the resistance and reactances, or susceptances, respectively S_R, S_L, and S_C. To find the current in each branch then, compute

$$ES_R = I_R \qquad ES_L = I_L \qquad ES_C = I_C$$
$$I_Z = I_R + I_L - I_C \qquad \text{(vectorially)} \tag{20-20}$$

Power in AC Circuits. Pure inductance or capacitance circuits store energy in the magnetic field and when the magnetic field declines to zero, this energy is restored to the electric circuit.

Power is consumed in an ac circuit only in the resistance part of the circuit and equals E_R, the effective voltage across the resistance, times I_R, the effective current. E_R and I_R are in phase. In a circuit with impedance, however, the total circuit voltage E_Z is out of phase with the current by the phase angle θ. In a series circuit, the current I is in phase with E_R; the voltage E_Z, on the other hand, is out of phase with E_R by the angle θ. In parallel circuits, the voltage E is in phase with E_R, but the current I_Z is out of phase with E_R. In both circuits, the power P is given by

$$P = E_R I_R \tag{20-21}$$

In series circuits, $E_R = E \cos \theta$ and $P = (E \cos \theta) I_R$. in parallel circuits, $I_R = I \cos \theta$ and $P = EI \cos \theta$. In any circuit with impedance angle θ, therefore, the power is given by

$$P = EI \cos \theta \tag{20-22}$$

Power Factor. The term $\cos \theta$ in Eq. (20-22) is called the power factor of the circuit. Because it is always less than 1, it is usually expressed as a percentage.

Low power factor results in high current, which requires high fuse, switch, and circuit-breaker ratings and larger wiring. Induction motors and fluorescent lamps are a common cause of low power factor. Since they are both inductive reactances (coils), the low power factor can be corrected by inserting capacitive reactances in the circuit to balance the inductive effects. This can be done with capacitors that are available commercially in standard kilovolt-ampere capacities.

For example, a 120-V, 600-kVA circuit with a 50% power factor has a current of 5000 A. The actual power expended is only 300 kW, but the wire, switches, and circuit breakers must be sized for 5000 A. If a capacitor with a 300-kVA rating is wired into the circuit, the current is reduced to 2500 A, and the wiring, switches and circuit breakers may be sized accordingly.

Conversion of AC to DC. Alternating current has the advantage of being convertible to high voltages by transformers. High voltages are desired for long-distance transmission. For these reasons, utilities produce and sell alternating current. However, many applications requiring accurate speed control need direct-current motors, for example, building elevators and railroad motors, including subways. In buildings, ac may be converted to dc by use of an ac motor to drive a dc generator, which, in turn, provides the power for a dc motor. The ac motor and dc generator are called a **motor-generator set.**

Another device used to convert ac to dc is a rectifier. This device allows current to flow in one direction but cuts off the sine wave in the opposite direction. The current obtained from the motor-generator set described previously is a similar unidirectional current of varying instantaneous value. The only nonvarying direct current is obtained from batteries.

Two-Phase and Three-Phase Systems. A single-phase ac circuit requires two wires, just like a dc circuit. One wire is the **live wire** and the other is the **neutral,** so called because it is usually grounded (Fig. 20-3a). Both the generator frame and any motor frames used on this circuit must also be grounded, to prevent shock to persons coming in contact with the machines. For this reason, the neutral is connected to the grounded frames of electric machines.

A voltage commonly used in the United States is 240 V, which is obtained from the two terminals of the secondary coil of pole transformers on utility high-voltage transmission lines. If a third wire is connected to the midpoint of the secondary coil as a neutral, the voltage between either of the two terminal wires and the neutral will be 120 V, which also is a commonly used voltage (Fig. 20-3b).

The currents in the two terminal wires are 180° apart in phase. The neutral current from each is also 180° apart. These two currents, traveling in the same neutral wire, offset each other because of the phase difference. If the load currents in the two terminal wires are equal, the current in the neutral will become zero.

Fig. 20-3 Examples of circuit wiring: (a) single-phase, two-wire circuit; (b) two-phase, three-wire circuit; (c) three-phase, four-wire circuit; (d) current I_n in the neutral wire of the three-phase circuit is the vector sum of the currents in the phase legs.

In a similar way, three-phase electric service can be obtained directly from the utility company with three live wires and a grounded neutral (Fig. 20-3c). The currents in the three live wires, as well as their respective return flows in the neutral, are 120° apart in phase. If the currents are equal in the three live wires, the current in the neutral will be zero.

If the phase currents are not equal, the current in the neutral will be the vector sum of the phase currents (Fig. 20-3d).

In many two-phase or three-phase systems, it is necessary therefore to balance the single-phase loads on each wire as much as possible. When the current in the neutral is zero, there is no voltage drop in the return circuit. Any voltage drop in the neutral subtracts from the voltage on the single-phase wires and affects the loads on these circuits. The voltage drop times the current flowing in the neutral times the cosine of the phase angle is the power consumed in the neutral wire, and this adds to the total metered power on the utility bill.

20-3. Electrical Loads.

Electric services in a building may be provided for several different kinds of loads: lighting, motors, communication, equipment. These loads may vary in voltage and times of service, as for example, continuous lighting or intermittent elevator motors. Motors have high instantaneous starting currents, which can be twice or four times the running current.

It is highly improbable that all of the intermittent loads will occur at once. For determination of the probable maximum load, a demand factor is applied to the loads. **Demand factor** is the ratio of probable maximum demand for power to the total connected load. Demand factors are set by building codes or may be obtained from the National Fire Protection Association "National Electrical Code" (see Art. 20-7).

Lighting Loads. The minimum, and often the maximum, watts per square foot of floor area to be used in design are specified by building codes for various uses of the floor area. Electrical engineers may exceed the minimum wattages if the proposed use requires more. For example, lighting may be designed to give a high intensity of illumination, which will require more watts per square foot than the code minimum. (Recommended lighting levels are given in the Illuminating Engineering Society "Lighting Handbook.") Maximum wattages are set to conserve energy.

Power Loads. In residential, office, and commercial buildings, the power loads are mainly the elevators and air-conditioning equipment. Certain computers and communication equipment, however, also require considerable power.

Most process equipment must be counted at 100% connected load. Some small machinery and portable tools carry a code demand factor that may be used at the option of the designer. Codes also establish demand factors for domestic ranges.

20-4. Electrical Conductors and Raceways.

All metals conduct electricity but have different resistances. Some metals, like gold or silver, have very low resistance, but they do not have the tensile strength required for electrical wire and are too costly. Consequently, only two metals are used exclusively in buildings as conductors, copper and aluminum.

Conductors may be solid round wire, strands, or bus bars of rectangular cross section. Usually, conductors are wrapped in insulation of a type that prevents electric shock to persons in contact with it. The type of insulation also depends on the immediate environment surrounding the wire in its proposed use; dry or moist air, wetness, buried in earth, temperature, and exposure to mechanical or rodent damage.

Each size of commercial wire with a particular insulation is given by building codes a safe current-carrying capacity in amperes, called the **ampacity** of that wire. The code ampacity is based on the maximum heating effect that would be permitted before damage to the insulation.

The codes also require that wires installed in a building be protected from mechanical damage by encasement in steel pipes, called conduits, or other metal and nonmetallic enclosures, termed raceways.

Safety Regulations. The safety regulations for use of wire in buildings are given in local building codes, which are usually based on the National Fire Protection Association "National Electric Code" (see Art. 20-7). These codes are revised frequently; hence, the latest editions should be used.

Another agency, Underwriters Laboratories, Inc., tests electric materials, devices and equipment. If approved, the item carries a UL label of approval.

Major Distribution Conductors. Large buildings, apartment houses or groups of apartment houses, factories, and other commercial types can have their own power plant, but most buildings receive their electrical power supply through service conductors from the street mains and transformers of a public utility.

The service conductors may be underground or above ground if taken from a utility-system pole. If a building is set back a great distance from the street poles, additional poles can be installed on the customer's property or the service conductors may be placed in an underground pipe extending to the building from the street pole. At the pole, the pipe should be extended 10 ft up to receive the service conductors.

At the building end, the service conductors come into a steel entrance box mounted on the building wall and then are brought to a service switch or circuit breaker. Service switches are commercially available up to 6000 A. Where the service load is greater, two or more service switches can be installed, up to a limit of six main service switches, called *drops*, on each service.

For large buildings, where six drops are not sufficient, the utility will install additional services with six drops available on each service.

Each service switch feeds a distribution center or groups of distribution centers, called panels. The connection between switch and center is called a **feeder.** These main distribution panels consist of several circuit breakers or fused switches. Each of these breakers or switches feeds a load, either a motor or another remote panel or group of panels. The panels, in turn, serve branch circuits connected to lighting, wall receptacles, or other electrical devices.

Distribution systems in buildings are usually three-phase, four-wire. The final branch circuits are generally single-phase, two-wire. One wire in each circuit is grounded. The grounded-wire insulation in a feeder is colored white or natural gray, in accordance with the color code of the "National Electrical Code." This is commercially available up to No. 6 size wire. Larger feeders may be identified by white markings at the connections.

In a **branch circuit,** the grounding conductor is colored green. When several grounded conductors are in one feeder raceway, one of the grounded conductors should be colored white or gray. The other grounded conductors should have a colored stripe (but not green) over the white or gray, and a different color should be used for the stripe on each wire. For four-wire systems, the colors for ungrounded conductors are usually blue, black, red, and white.

Conductor ampacity depends on the accumulative heating effect of the IR power loss in the wire. This loss is different for a given size wire with different insulations and depends on whether the wire is in open air and can dissipate heat or confined in a closed conduit with other heat-producing wires. Tables in the "National Electrical Code" give the safe ampacity for each type of insulation and the derated ampacity for more than three live wires in a raceway.

Open wiring is used for temporary wiring in construction. Single wires mounted in knobs or cleats must be ½ to 1 in away from the building surfaces and at least 2½ in apart for voltages below 300; for voltages between 300 and 600, the spacing should be increased to 4 in.

Types of Insulated Conductors. Following is a list of the various types of insulated conductors rated in the National Electrical Code:

Type MI. Mineral-insulated cable sheathed in a watertight and gastight metallic tube. Cable is completely incombustible and can be used in many hazardous locations and underground.

Type MC. No. 4 wire and larger, sheathed in an interlocking metal tape or a close-fitting, impervious tube. With lead sheath or other impervious jacket, Type MC may be used in wet locations.

Type AC. (Also known as BX cable.) This has an armor of flexible metal tape with an internal copper bonding strip in close contact with the outside tape for its entire length. This provides a grounding means at outlet boxes, fixtures, or other equipment. Type AC cable may be used only in dry concealed locations.

Type ACL. In addition to insulation and covering as for Type AC, Type ACL has lead-covered conductors. This makes this type suitable for wet or buried locations.

Type ACT. Only the individual conductors have a moisture-resistant fibrous covering.

Type NM or NMC. Nonmetallic-sheathed cables (also known as Romex). This type may be used in partly protected areas. The New York City Code permits BX (Type AC) but does not allow Romex because it is not rodentproof and is subject to nail damage in partitions.

Type SNM. The conductors are assembled in an extruded core of moisture-resistant, flame-resistant, nonmetallic material. The core is then covered with an overlapping metal tape and wire shield and sheathed in extruded nonmetallic material impervious to flames, moisture, oil, corrosion, fungus, and sunlight. This type may be used in hazardous locations.

Type SE or USE. Service-entrance cable has a moisture resistant, fire-resistant insulation with a braid over the armor for protection against atmospheric corrosion. Type USE is the same as Type SE, except that it has a lead covering for underground uses.

Type UF. This type is factory assembled in a sheath resistant to flames, moisture, fungus, and corrosion, suitable for direct burial in the earth. The assembly may include an uninsulated grounding conductor. Cables may be buried under 18 in of earth or 12 in of earth and a 2-in concrete slab.

Nonmetallic Extensions. Two insulated conductors within a nonmetallic jacket or extruded thermoplastic cover may be used for surface extensions on walls or ceilings or as overhead cable with a supporting steel cable made part of the assembly. Extensions may be used in dry locations within residences or offices.

Aerial cables may be used only for industrial purposes. At least 10 ft should be provided above the floor as clearance for pedestrians only, 14 ft for vehicular traffic.

Busways. These are bare conductors of rectangular cross section, which are assembled in a sheet-metal trough. The conductors are insulated from the enclosure and each other. Busways must be exposed for heat dissipation. They are arranged with access openings for plug-in and trolley connections.

Cable Bus. For heavy current loads, such as services, several insulated cables may be mounted in parallel, at least one diameter apart, within a ventilated metal enclosure with access facilities. Cable bus costs less than bus bars for the same load but generally takes up more space. Use is limited to dry locations.

Electrical Connections. It is often necessary to join two or more wires together. For the purpose, copper wire may be soldered. Aluminum wiring may be welded.

Also, there are a variety of devices commercially available for joining and squeezing two wires together. One type, a pressure connector, called a wire nut, may be screwed over two or three copper wires twisted together. Another type consists of end lugs attached to wires by soldering or welding. The lugs have a flat extension with a bolt hole for connection by bolts to a switch or another end lug.

All metal electrical connectors must be wrapped with insulating tape and should be enclosed in a metal box with cover. Several connections properly insulated can be enclosed in the same metal box if the box is adequate in size. The number of spliced conductors in a box is limited by building codes.

Raceways. A raceway is a general term used to describe the supports or enclosures of wires. For most power distribution systems in buildings, rigid conduit or tubing is used. The dimensions of such conduit or tubing and the number of wires of each size permitted is fixed by tables in the "National Electrical Code." Three or more conductors may not occupy more than 40% of the interior area, with some exceptions for lead-sheathed cable. All metallic raceways must be continuously grounded.

One wide use of **rigid steel conduit,** galvanized, is for branch circuits buried in the concrete slabs of multistory buildings.

Electrical metallic tubing is a thin-walled tube that is permitted by codes in locations where the raceway is not subject to physical damage.

For economy in industrial installations, a continuous, rigid structure may be designed to carry both power and signal wiring. This structure may be in the form of a trough, a ladder run, or a channel. It is limited in use to certain cables specifically approved by Underwriters Laboratories for such use.

Flexible metallic conduit, also known as Greenfield, is a continuous winding of interlocking metal stripping similar to that used for Type AC metal-clad cables (BX). These conduits are often used in short lengths at the terminal connection of a feeder to a motor. For wet locations, a watertight-type is available.

Surface raceways are usually oval shaped and flat. When painted the same color as the wall or ceiling, they are less conspicuous than round pipe conduit.

Underfloor raceways are ducts placed under a new floor in office spaces where desks and other equipment are frequently moved. Laid in parallel runs 6 to 8 ft apart, with separate ducts for power, signal, and telephone wires, these raceways may have flat-plate outlet covers spaced 4 to 6 ft along each run (Fig. 20-4). Large retail stores also find these installations a great convenience. The alternative is feeder runs above the hung ceiling of the story below, with *poke-through* construction to reach new outlets above the floor.

Underfloor raceways may be single level (Fig. 20-5) or two level (Fig. 20-6). In steel-frame buildings, with cellular steel decking, single-level raceways may be included in the structure of the floor itself. A concrete header across the cellular runs provides the means of entering from the finished floor. A similar arrangement can be used in cellular precast-concrete decks, with metal headers for connections. Wireways to carry large numbers of conductors carrying light-current signal or control circuits are commercially available in fixed lengths.

20-5. Power System Apparatus.

Most buildings, commercial, industrial, and residential, receive their power from a public utility. Usually, the customer is given a choice of voltages. For example, 240/120-V single-phase, three-wire service is very common in suburban and rural areas. This service comes from a single-phase, 240-V transformer on a utility pole, with one wire from each

Fig. 20-4 Underfloor distribution of electric power through ducts.

Fig. 20-5 Electric ducts in a concrete floor.

Fig. 20-6 Cellular steel decking serves as underfloor electric ducts. Wires in headers distribute power to wires in the cells.

end of the secondary coil and with the neutral from the midpoint of its secondary coil. The voltage between the end terminal connections is 240 V and between each end wire and the neutral, 120 V (Fig. 20-3b).

In large cities, the service to large buildings is 208/120 V, three-phase, four-wire, with 208 V available between phase wires and 120 V between a phase wire and the neutral (Fig. 20-3c). Another choice is 460/277 V, three-phase, four-wire, with 277 V available between a phase leg and the neutral. It is more economical to use the higher voltage,

460/277 V, for motors and industrial lighting. The lower voltage 208/120 V is required for residential or commercial lighting and appliances.

In some areas, such as New Jersey, the utility will provide both voltage services on separate meters to a large building. But in other areas, such as New York City, the customer must choose one or the other voltage from only one meter and then use transformers to provide the second voltage service.

For example, the engineer for a 50-story apartment building may elect to receive 460/277-V service in the cellar meter room. Some service cables will go through a large service switch and the owner's meter and then to the owner's main distribution panel. This will feed all the large motors in the building and the garage and corridor lighting. For the areas that require 208/120 V, transformers will be provided in the cellar and in roof spaces. Those areas include the apartments, which require 208/120-V service, usually with a three-wire meter. This provides 208 V for air-conditioning units and 120 V for lighting and small appliances. Because only one service drop has been used for the owner's metered load, the remaining service cables are assigned to five additional service switches, or drops. Each of these switches will feed a remote transformer that will reduce the voltage to 208/120, to feed a certain number of apartment meters. With three-wire meters, it is possible to combine service to three apartments on one common neutral, picking up three apartments in a vertical line and then extending horizontally to a meter closet. This makes it convenient to locate meter closets on every third floor of the building and the remote transformers in the corresponding space on the floor below the meter closet.

Power losses in the feeders to the transformers and in the transformers themselves occur before the apartment meters, and the costs are borne by the utility company. Transformer losses are generally limited to 2% and feeder drops to 1%. For this reason, the utility company has an interest in the wiring design inside the building and will require the customer to obtain its approval for designs that include transformers and meters remote from the point of service entry.

Transformers may be dry or liquid-immersed type. The wet type is used for large installations. It requires special precautions for fire protection.

All transformers are rated in kVA, with primary and secondary voltages. Taps can be specified to provide up to 2.5% above or below rated secondary voltages. The manufacturer can also make available to the engineer the reactance and resistance of the coils and the noise ratings. Noise can be minimized by use of vibration isolation mountings.

The power losses in a transformer create heat, which must be dissipated. Dry-type transformers are cooled by circulating air in the spaces enclosing the transformers. For liquid-filled transformers, which usually have very high capacity, the liquid may be circulated through coolers to transfer heat from the coils. Average losses in transformers used in buildings are about 2% of the rated capacity.

Meters. Consumption of electrical energy is measured by watt-hour meters. Utilities also include another charge, for demand, based on the maximum amount of power used in a specified time interval, usually about 15 to 30 min.

Three-wire meters are generally used for residences, either 208 V or, in some areas, 230/240 V. The 208-V service is usually taken from a three-phase, four-wire street or pole main. The voltage therefore differs 120° in phase from the current. There is a 120-V difference between the third, or neutral, wire and the phase leg. For the three 120-V, single-phase circuits, the total power, W, is computed from

$$P = 3EI \cos \theta \tag{20-23}$$

where E = voltage between phase legs and neutral
I = current, A
$\cos \theta$ = power factor

Industries and commercial installations with large motors require three-phase, four-

Fig. 20-7 (*a*) Snap switch. (*b*) Blade switch. *(Reprinted with permission from F. S. Merritt, "Building Engineering and Systems Design," Van Nostrand Reinhold Company, New York.)*

wire meters. Distribution can be over one of three different types of circuits: 208 V, three phase (motors); 208 V, single phase (motors, appliances); or 120-V, single phase (lighting, motors, appliances).

Meters for services supplied by a utility are provided and installed by the utility. The meter pans and current transformers must be provided by the customer in accordance with the utility's requirements.

All the service to one building may be measured by one meter, usually called a master meter. Apartment buildings may have one meter for the owner's load and individual meters for each apartment.

Switches. These are disconnecting devices that interrupt electric current. Toggle switches (Fig. 20-7*a*) or snap switches are used for small currents like lighting circuits. They employ pressure contacts of copper to copper. Knife switches (Fig. 20-7*b*) are used for larger loads. A single-phase knife switch employs a movable copper blade hinged to one load terminal. To close a circuit, the blade is inserted between two fixed copper blades connected to the other terminal. The ground leg is usually continuous and unswitched, for safety reasons. For multiphase circuits, one hinged blade is used for each phase; thus, the switch may be double pole (Fig. 20-8*a*) or three-pole (Fig. 20-8*b*), as the case may be.

Fig. 20-8 (*a*) Double-pole, single-throw switch. (*b*) Triple-pole, single-throw switch. (*c*) Single-pole, double-throw switches used for remote control of lights from two locations.

A switch may be single throw (Fig. 20-7b), as described, or double throw (Fig. 20-8c). A double-throw switch permits the choice of connecting the load (always on the movable blade) to two different sources of power, each connected to opposite, fixed blades.

Once the blades of a switch are in solid contact, the heating effect at the contact surface is minimized. Opening and closing the switch, though, draws a hot arc, which burns the copper. This may cause an uneven surface of contact, with continuing small arcs across the separated points, and result in continual weakening of the contact switch and eventual breakdown.

Switches are carefully rated for load and classified for use by the National Electric Manufacturer's Association (NEMA) and the Underwriters Laboratories (UL). For example, a motor-circuit switch, which carries a heavy starting current, is rated in maximum horsepower allowed for connection.

Many types of service-entrance switches are available to meet the requirements of utility companies. They may be classified as fuse pull switch, externally operated safety switch, bolted pressure contact-type switch, or circuit breaker.

An isolating switch may not be used to interrupt current. It should be opened only after the circuit has been interrupted by another general-use switch. Since isolating switches are very light, an arc will create high temperatures and can severely burn the operator.

For control of large, separate loads, the live copper blades of the various switches are concealed in steel enclosures, and the movable blades are operated by insulated levers on the front of the board. The equipment is called a dead-front switchboard.

Protective Devices for Circuits. In an electrical distribution system for a building, each electric service should have a means of disconnection, but it should not consist of more than six service switches. Each service switch may disconnect service to a panel from which lighter feeders extend to other distribution points, up to the final branch circuits with the minimum size wire, No. 12. This panel contains switches with lower disconnecting ratings than that of the service switch and that serve as disconnecting means for the light feeders. The rating and type of each switch must correspond to the size and kind of load and the wire size.

There must also be in every circuit some protective device to open the circuit if there is an unexpected overload, such as a short circuit or a jammed motor that prolongs a high inrush current. These protective devices may be fuses combined with knife switches, or circuit breakers, which provide both functions in one device.

Fuses in lighting and appliance circuits with loads up to 40 A may be the screwed plug type, with a metallic melting element behind a transparent top. For each rating, plug fuses are given different colors and the screw size is made intentionally different, to prevent errors.

Cartridge fuses are another type of fuse. They are cylindrical in form and are available in any size. They are classified for special purposes, and the rating is clearly marked on the cylinder. So-called HI-CAP fuses (high capacity) are used for fused service switches that have the capacity to interrupt very high short-circuit currents.

Short circuits in the heavy copper conductor immediately following a service switch can be very high, because of the high power potential of the street transformers. The currents may be 25,000 to 100,000 amp. Upon inquiry, the utility will advise as to the maximum short-circuit current for an installation. The service switch fuse should be selected to suit this capacity.

Circuit breakers operate on a different principle. The points of contact are blocks of carbon, which are brought together in a wiping or sliding motion. This grinds the two surfaces into close contact. The blocks are held in position by an arm curved in the form of a segment of a circle. The arm is composed of layers with different heating and expansion coefficients. The current flows through this arm. Any overload tends to expand the metals and straighten the arm, thus breaking contact between the blocks. A spring provides for fine adjustment.

Fuses are destroyed on overload and must be replaced. Circuit breakers, on the other hand, can be reset after the cause of the overload has been removed. Circuit breakers, however, are not available with the high interruptable capacity required at the service entrance. For this reason, the service disconnect on all large-power installations are fused service switches with HI-CAP fuses. It is also good practice to design the first panel immediately after the service switch for fused switches and then use circuit breakers on subsequent panels where the loads are lighter.

Current limiters may also be used at the service connection. These are strips of metal that have a high rate of increase in electrical resistance when heated. If a short circuit occurs, this limits the flow of short-circuit current even before the short is cleared.

Current-Limiting Reactors. A coil with high inductive reactance may be placed in series with the service. If a heavy short-circuit current occurs, the impedance limits the current by temporarily storing energy in the magnetic field.

Protective Relays. There are several kinds of faults that may occur in an electric service. Excessive variations in voltage and phase reversal are two types. Relays that consist of a solenoid and switch mechanism may be used to indicate such faults and open circuit breakers in the control circuits. Such relays serve to disconnect heavy equipment before costly damage occurs.

Switchgear and Switchboards. The service switches and main distribution panels in large buildings are usually assembled in a specially designed steel frame housed in a separate room in the cellar. The assembly is usually referred to as switchgear for large power units and switchboards for smaller assemblies.

A switchboard is defined in the National Electrical Code as a large single panel, frame, or assembly of panels, on which are mounted, on the face or back or both, switches, overcurrent and other protective devices, buses, and usually instruments. Switchboards are generally accessible from the rear as well as from the front and not intended to be installed in cabinets.

Switchboards are commonly divided into the following types:

1. Live-front.
2. Dead-front.
3. Safety enclosed switchboards.
    a. Unit or sectional.
    b. Draw-out.

Live-front switchboards have the current-carrying parts of the switch equipment mounted on the exposed front of the vertical panels and are usually limited to systems not exceeding 600 V. They generally are installed in restricted areas.

Dead-front switchboards have no live parts mounted on the front of the board and are used in systems limited to a maximum of 600 V for dc and 2500 V for ac.

Unit safety-type switchboard is a metal-enclosed switchgear consisting of a completely enclosed self-supporting metal structure, containing one or more circuit breakers or switches.

Draw-out-type switchboard is a metal-clad switchgear consisting of a stationary housing mounted on a steel framework and a horizontal draw-out circuit-breaker structure. The equipment for each circuit is assembled on a frame forming a self-contained and self-supporting mobile unit.

Metal-clad switchgear consists of a metal structure completely enclosing a circuit breaker and associated equipment such as current and potential transformers, interlocks, controlling devices, buses, and connections.

Because the switchboard room contains a heavy concentration of power, these rooms have special building code requirements for ventilation and safety. The safety rules may require two exits remote from each other and minimum clear working spaces at front, top, back, and sides of the equipment. Also, the rules may prohibit overhead piping and ducts that may leak.

Substations. These are arrangements of transformers and switchgear used to step down voltages and connect to or disconnect from the mains. A master substation is used to transform from 13,800 V to 2300 V and a load-center substation, to reduce to 600 V or less for customer use. The load-center substation may be located outside the building in an underground vault or on a surface pad. Where street space is limited, utilities sometimes permit inside substations in the cellar of the customer adjacent to the switchboard room. These inside vaults must comply with strict rules for ventilation and drainage set by the utility, and access should be available from the street through doors that are normally locked.

Panelboards. These are distribution centers that follow the service switches and switchgear. Whereas the latter have fused switches with a high current-interrupting capacity, panelboards contain circuit breakers or, for large loads, fused knife switches.

A panelboard is a single panel or a group of panel units designed for assembly in the form of a single panel in which are included buses and perhaps switches and automatic overcurrent protective devices for control of light, heat, or power circuits of small capacity. It is designed to be placed in a cabinet or cutout box placed in or against a wall or partition and accessible only from the front. In general, panelboards are similar to but smaller than switchboards.

A panelboard consists of a set of copper mains from which the individual circuits are tapped through overload protective devices or switching units.

Panelboards are designed for dead-front or live-front construction. In dead-front construction, no live parts are exposed when the door of the panelboard is opened. In the live-front type, the current-carrying parts of the fuses or switches are exposed when the door of the panelboard is opened. Panelboards preferably should be dead-front construction.

Panelboards also are designed for flush, semiflush, or surface mounting. They fall into two general classifications, those designed for medium loads, usually required for lighting systems, and those for heavy-duty industrial-power-distribution loads.

Distribution panelboards are designed to distribute current to lighting panelboards and power loads and panelboards.

Lighting panelboards are generally used for distribution of branch lighting circuits and fall into the following types: plug-fuse branches; single-pole tumbler switch, with one-fuse branches, plug or cartridge type; double-pole tumbler switch, with two-fuse branches, plug or cartridge type; single-pole quick-lag circuit breaker branches; single-pole quick-make and quick-break circuit-breaker branches.

Power panelboards fall into the following types:

1. Live-front, fused cutouts in branches.
2. Live-front, fused knife switches in branches.
3. Safety dead-front, pull switch in branches.
4. Safety dead-front, circuit breaker in branches.

Since motors fed from power panelboards vary in sizes, the switches and breakers in a power panelboard are made of several different sizes corresponding to the rating of the equipment.

Panelboards are designed with mains for distribution systems consisting of:

1. Three-wire, single-phase 120/240-V, solid-neutral, alternating current.
2. Three-wire, 120/240-V, solid-neutral, direct current.
3. Four-wire, three-phase, 120/208-V, solid-neutral, alternating current.

The mains in the panelboard may be provided with lugs only, fuses, switch and fuses, or circuit breakers.

A single-phase, three-wire panel consists of two copper busbars set vertically in the center of the panel and horizontal strip connections on each side for branch circuit breakers or switches. The third wire, or neutral, is connected to a copper plate at the top of the panel with several bolted studs. Neutral wire connectors from each circuit are connected to that plate. When the capacity of the panel bus is 200 A per phase or greater, a disconnect switch is required at the line connection.

A similar construction is used for three-phase, four-wire panelboards, which are required for motor circuits, but with three, instead of two, copper busbars set vertically in the center of the board. Single-phase circuits may be taken from both types of panels, but it is important to balance the loads as closely as possible on the two- or three-phase legs, to minimize the current in the neutral.

The following items should be taken into consideration in determining the number and location of panelboards:

1. No lighting panelboard should exceed 42 single-pole protective devices.
2. Panelboard should be located as near as possible to the center of the load it supplies.
3. Panelboard should always be accessible.
4. Length of run from panelboard to the first outlet should not exceed 100 ft.
5. Panelboard should be located so that the feeder is as short as possible and have a minimum number of bends and offsets.
6. Spare circuit capacity should be provided at the approximate rate of one spare to every five circuits originally installed.
7. At least one lighting panelboard should be provided for each floor of a building.

Motor Control Centers. These are an assembly, in one location, of motor controllers, devices that start and stop motors and protect them against overloads, and of disconnect switches for the motors. For safety reasons, the National Electrical Code requires that a disconnect switch be located within sight of the motor and its controller.

The controller may be a manual or an electromagnetic switch. The electromagnetic switch is a surface-contact switch operated by a solenoid valve and returned to its normal position by a spring. Such switches may be normally closed or normally open, depending on the automatic function required. Usually, an on-off automatic push button switch is installed to bypass the electromagnetic switch and allow manual operation of the motor for testing purposes and then return to automatic for normal functioning. Overload protection is incorporated in the electromagnetic switch. For the purpose, thermal, heat-operated relays are provided for motors larger than 1 hp. A reset button is pushed to close the switch after the overload has been removed. For motors 1 hp and smaller, overload protection may be provided by a fused switch or a circuit breaker.

Lightning Protection. The design of lightning protection systems should conform with the standards of the American National Standards Institute, the National Fire Protection Association (Bulletin No. 78, "Lightning Code") and Underwriters Laboratories (Standard UL96A "Master-Labeled Lightning Protection Systems").

Protection of a building is accomplished by projecting several metallic rods above high points such as chimneys. These rods are electrically bonded together and to a heavy copper cable that extends down to a good electrical ground. The entire circuit including the ground must have very low electrical resistance and should be run without sharp bends to avoid arcs jumping to nearby objects or structures.

The electric wiring system forms an especially attractive receptor for lightning discharges. Protection for it may be provided by special valve devices, which are connected to the ungrounded circuits and which are open for normal voltages but pass high-voltage discharges to the ground. Another type, an expulsion arrestor, contains a gas-producing material that quenches the arc formed by a heavy discharge.

20-6. Communication Systems.

Circuits that convey information or control other circuits are classified as communication systems. They are usually low voltage (commonly 12 to 24 V), in contrast to power systems, which start at 120 V. Communication systems include all types of signal systems—telephone, telegraph, and alarm—as well as remote control for power systems. The National Electrical Code divides these circuits into three classes and provides separate requirements for each.

Class 1 circuits are similar to power circuits in that they are permitted to carry up to 600 V and, in general, are subject to the same installation requirements. One exception is that use of No. 18 wire is permitted for Class 1 circuits. Special insulation and overcurrent protection devices are also permitted. Thermostats and other sensing devices controlling remote motor starters, available at 120 V, may be incorporated in Class 1 circuits and may use No. 18 wire, but with overcurrent protection not exceeding 20 A.

Class 2 and 3 circuits are limited to a maximum of 150 V. The conductors may not be placed in the same raceway as power circuits. Class 2 wire should provide protection against both fire and electric shock, whereas Class 3 wire should offer protection primarily against fire.

Small alarm circuits for bell, buzzer, and annunciator panels operate at about 20 V and use a maximum of 5 A. (Such systems often use battery power.) A small transformer on a 120-V power system can step down the voltage for a good bell-ringing circuit. Larger communication systems use a rectifier connected to a power circuit and transformer.

Alarm Systems. These systems cover security, fire detection, and fire alarm. Fire-protective signaling systems are classified as non-power-limited or power-limited. Power-limited circuits are limited to currents less than 5 A and may use No. 22 wire.

Non-power-limited circuits may carry up to 600 V and may use No. 18 wire if overcurrent protection does not exceed 20 A.

All fire-protective systems should comply with the general requirements of National Fire Protection Association Standards (NFPA 71, 72A, 72B, 72C, 72D, 72E, or 74).

Telephone Systems. The utility company pulls all the wire for telephone service in a building, provides the equipment, and makes all final connections. In new construction, the builder must provide the required empty conduits with a pull wire inside.

Television Systems. Large apartment buildings generally provide a master antenna and amplifier systems. For buildings up to 20 stories high, the amplifiers may be located at the roof. A distribution system then feeds down risers, with coaxial cables, to each apartment. The risers may be ¾ in in diameter. One coaxial cable, about ⅜ in in diameter, may feed 20 outlets. In very tall buildings, the amplifying equipment may be located at midheight and serve upfeed and downfeed risers.

20-7. Electrical Distribution in Buildings.

The National Fire Protection Association "National Electrical Code" is the basic standard for electrical design for buildings in the United States and has been adopted by reference in many building codes. In some cases, however, local codes may contain more restrictive requirements. The local ordinance should always be consulted.

The "National Electrical Code," or the "National Electrical Code Handbook," which explains provisions of the code, may be obtained from NFPA, 470 Atlantic Ave., Boston, Mass. 02210.

The American Insurance Association sponsors the Underwriters Laboratories, Inc.,

20-22 Electric Power

which passes on electrical material and equipment in accordance with standard test specifications. The UL also issues a semiannual List of Inspected Electrical Appliances, which can be obtained from the UL at 207 East Ohio Street, Chicago, Ill. 60611.

Electrical codes and ordinances are written primarily to protect the public from fire and other hazards to life. They represent minimum safety standards. Strict application of these codes will not, however, guarantee satisfactory or even adequate performance. Correct design of an electrical system, over these minimum safety standards, to achieve a required level of performance, is the responsibility of the electrical designer.

Electrical Symbols. Table 20-1 illustrates the graphic symbols commonly used for electrical drawings for building installations. ANSI Y32.2, American National Standards Institute, contains an extensive compilation of such symbols.

TABLE 20-1 Electrical Symbols*

Wall	Ceiling	
⊢○	○	Outlet
-ⓑ	ⓑ	Blanked outlet
	ⓓ	Drop cord
-ⓔ	ⓔ	Electrical outlet—for use only when circle used alone might be confused with columns, plumbing symbols, etc.
-ⓕ	ⓕ	Fan outlet
-ⓙ	ⓙ	Junction box
-ⓛ	ⓛ	Lamp holder
-ⓛ$_{PS}$	ⓛ$_{PS}$	Lamp holder with pull switch
-ⓢ	ⓢ	Pull switch
ⓥ	-ⓥ	Outlet for vapor-discharge lamp
-ⓧ	ⓧ	Exit-light outlet
-ⓒ	-ⓒ	Clock outlet (specify voltage)
⊖		Duplex convenience outlet
⊖$_{1,3}$		Convenience outlet other than duplex. 1 = single, 3 = triplex, etc.
⊖$_{WP}$		Weatherproof convenience outlet
⊖$_R$		Range outlet
⊖$_S$		Switch and convenience outlet
⊖[R]		Radio and convenience outlet
△		Special-purpose outlet (designated in specifications)
⊙		Floor outlet

Wall	Ceiling	
S		Single-pole switch
S_2		Double-pole switch
S_3		Three-way switch
S_4		Four-way switch
S_D		Automatic door switch
S_E		Electrolier switch
S_K		Key-operated switch
S_P		Switch and pilot lamp
S_{CB}		Circuit breaker
S_{WCB}		Weatherproof circuit breaker
S_{MC}		Momentary-contact switch
S_{RC}		Remote-control switch
S_{WP}		Weatherproof switch
S_F		Fused switch
S_{WF}		Weatherproof fused switch
⊖ a,b,c etc. ○ a,b,c etc. S a,b,c etc.		Any standard symbol as given above with the addition of a lower-case subscript letter may be used to designate some special variation of standard equipment of particular interest in a specific set of architectural plans. When used, they must be listed in the key of symbols on each drawing and if necessary further described in the specifications
■		Lighting panel
▨		Power panel
———		Branch circuit; concealed in ceiling or wall
—·—		Branch circuit; concealed in floor
----		Branch circuit; exposed
→→		Home run to panel board. Indicate number of circuits by number of arrows. NOTE: Any circuit without further designation indicates a two-wire circuit. For a greater number of wires indicate as follows: ─///─(three wires), ─////─(four wires), etc.
———		Feeders. NOTE: Use heavy lines and designate by number corresponding to listing in feeder schedule

TABLE 20-1 Electrical Symbols *(Continued)*

Wall	Ceiling	
⇌⬜⇌		Underfloor duct and junction box. Triple system. NOTE: For double or single systems eliminate one of two lines. This symbol is equally adaptable to auxiliary-system layouts
Ⓖ		Generator
Ⓜ		Motor
Ⓘ		Instrument
Ⓣ		Power transformer (or draw to scale)
⊠		Controller
⎕⊢		Isolating switch
⊡		Push button
⎕╱		Buzzer
⎕◯		Bell
⊢◇		Annunciator
⊢◀		Outside telephone
⊢◁		Interconnecting telephone
◁⊳		Telephone switchboard
Ⓣ		Bell-ringing transformer
⎕D⎕		Electric door opener
⎕F⎕◯		Fire-alarm bell
⎕F⎕		Fire-alarm station
◼		City fire-alarm station
⎕FA⎕		Fire-alarm central station
⎕FS⎕		Automatic fire-alarm device
⎕W⎕		Watchman's station
⎕W⎕		Watchman's central station
⎕H⎕		Horn
⎕N⎕		Nurse's signal plug
Ⓣⓥ		Television antenna outlet

Wall	Ceiling	
R		Radio outlet
SC		Signal central station
◻		Interconnection box
ı|ı|ı|		Battery
—·—·—		Auxiliary-system circuits. NOTE: Any line without further designation indicates a two-wire system. For a greater number of wires designate with numerals in manner similar to 12—No. 18W-¾″-C., or designate by number corresponding to listing in schedule.
☐a,b,c		Special auxiliary outlets. Subscription letters refer to notes on plans or detailed description in specifications.

*Standard electrical symbols are compiled in ANSI Y32.2, American National Standards Institute.

Building Wiring Systems. The electrical load in a building is the sum of the loads, in kilowatts (kW), for lighting, motors, and appliances.

In computing the size of feeders for some types of buildings, the designer may apply a demand factor to the lighting load, since all lights may not be on at the same time. Table 20-2 gives recommended minimum lighting loads based on floor areas and the allowed demand factors for several types of buildings. These minimum loads should be used only when the actual lighting-fixture installation is not known; for example, in apartment building design. For other building types, such as factories or commercial buildings, where illumination is designed, the installed-fixture load should be used.

Motor and appliance loads usually are taken at full value. Household and kitchen appliances, however, are exceptions. The National Electrical Code lists demand factors for household electric ranges, ovens, and clothes dryers. Some municipal codes allow the first 3000 W of apartment appliance load to be included with lighting load and therefore to be reduced by the demand factor applied to lighting.

For factories and commercial buildings, the electrical designer should obtain from the mechanical design the location and horsepower of all blowers, pumps, compressors, and other electrical equipment, as well as the load for elevators, boiler room, and other machinery. The load in amperes for running motors is given in Tables 20-9 and 20-10.

Plans. Electrical plans should be drawn to scale, traced or reproduced from the architectural plans. Architectural dimensions may be omitted except for such rooms as meter closets or service space, where the contractor may have to detail his equipment to close dimensions. Floor heights should be indicated if full elevations are not given. Locations of windows and doors should be reproduced accurately, and door swings shown, to facilitate location of wall switches. For estimating purposes, feeder or branch runs may be scaled from the plans with sufficient accuracy.

Indicate on the plans by symbol the location of all electrical equipment (Table 20-1). Show all ceiling outlets, wall receptacles, switches, junction boxes, panel boxes, telephone and interior communication equipment, fire alarms, television master-antenna connections, etc.

A complete set of electrical plans should include a diagram of feeders, panel lists, service entrance location, and equipment. Before these can be shown on the plans, how-

TABLE 20-2 Code Demand Factors for Lighting*

Type of occupancy	Unit load per ft^2, watts	Load to which demand factor applies, watts	Demand factor, %
Armories and auditoriums	1	Total wattage	100
Banks	2	Total wattage	100
Barber shops and beauty parlors	5	Total wattage	100
Churches	1	Total wattage	100
Clubs	2†	Total wattage	100
Court rooms	2	Total wattage	100
Dwellings (other than hotels)	3†	3000 or less	100
		Next 117,000	35
		Over 120,000	25
Garages—commercial (storage)	1/2	Total wattage	100
Hospitals	2	50,000 or less	40‡
		Over 50,000	20
Hotels and motels, including apartment houses without provisions for cooking by tenants	2†	20,000 or less	50‡
		Next 80,000	40
		Over 100,000	30
Industrial commercial (loft) buildings	2	Total wattage	100
Lodge rooms	1 1/2	30,000 or less	100
		Over 30,000	70
Office buildings	5	Total wattage	100
Restaurants	2	Total wattage	100
Schools	3	Total wattage	100
Stores	3	Total wattage	100
Warehouses, storage	1/4	12,500 or less	100
		Over 12,500	50
In any of above occupancies except single-family dwellings and individual apartments of multifamily dwellings:			
Assembly halls and auditoriums	1	Total wattage as specified for the specific occupancy	
Halls, corridors	1/2		
Closets, storage spaces	1/4		

*Reproduced by permission from "National Electrical Code," 1978, National Fire Protection Association.
In view of the trend toward higher-intensity lighting systems and increased loads due to more general use of fixed and portable appliances, each installation should be considered as to the load likely to be imposed and the capacity should be increased to ensure safe operation.
Where electric-discharge lighting systems are to be installed, high power-factor type should be used, or the conductor capacity may need to be increased.
†For general illumination in dwelling occupancies, it is recommended that one 15-A branch circuit be installed for each 375 ft^2 of floor area.
‡For subfeeders to areas in hospitals and hotels where entire lighting is likely to be used at one time, e.g., in operating rooms, ballrooms, dining rooms, a demand factor of 100% shall be used.

ever, wire sizes should be computed in accordance with procedures outlined in the following paragraphs.

Where there is only one panel box in an area, and it is clear that all circuits in that area connect to that box, it is not necessary to number the panel other than to designate it as, for example, "apartment panel." In larger areas, where two or more panel boxes may be needed, each should be labeled for identification and location; for example, L.P. 1-1, L.P. 1-2 ... for all panels on the first floor; L.P. 2-1, L.P. 2-2 ... for panels on the second floor.

Branch Circuits.   It is good practice to limit branch runs to a maximum of 50 ft by installing sufficient panels in efficient locations.

Connect each outlet with a branch circuit and show the home runs to the panel, as indicated in Table 20-1. General lighting branch circuits with a 15-A fuse or circuit breaker in the panel usually are limited to 6 to 8 outlets, although most codes permit 12. No more than two outlets should be connected in a 20-A appliance circuit.

It is good practice to use wire no smaller than No. 12 in branch circuits, though some codes permit No. 14. Special-purpose individual branch circuits for motors or appliances should be sized to suit the connected load.

Electric Services. For economy, alternating current is transmitted long distances at high voltages and then changed to low voltages by step-down transformers at the point of service.

Small installations, such as one-family houses, usually are supplied with three-wire service. This consists of a neutral (transformer midpoint) and two power wires with current differing 180° in phase. From this service, the following types of interior branch circuits are available:

Single-phase two-wire 230-V—by tapping across the phase wires.
Single-phase two-wire 115-V—by tapping across one phase wire and the neutral.
Single-phase three-wire 115/230-V—by using both phase wires and the neutral.

For larger installations, the service most widely used is the 120/208-V three-phase four-wire system. This has a neutral and three power wires carrying current differing 120° in phase. From this service, the following types of interior branch circuits are available:

Single-phase two-wire 208-V—by tapping across two phase wires.
Single-phase two-wire 120-V—by tapping across one phase wire and the neutral.
Two-phase three-wire 120/208-V—by using two phase wires and the neutral.
Three-phase three-wire 208-V—by using three phase wires.
Three-phase four-wire 120/208-V—by using three phase wires and the neutral.

20-8. Circuit and Conductor Calculations.

The current in a conductor may be computed from the following formulas, in which
I = conductor current, A
W = power, W
f = power factor, as a decimal
E_p = voltage between any two phase legs
E_g = voltage between a phase leg and neutral, or ground

Single-phase two-wire circuits:

$$I = \frac{W}{E_p f} \quad \text{or} \quad I = \frac{W}{E_g f} \tag{20-24}$$

Single-phase three-wire (and balanced two-phase three-wire) circuits:

$$I = \frac{W}{2E_g f} \tag{20-25}$$

Three-phase three-wire (and balanced three-phase four-wire) circuits:

$$I = \frac{W}{3E_g f} \tag{20-26}$$

When circuits are balanced in a three-phase four-wire system, no current flows in the neutral. When a three-phase four-wire feeder is brought to a panel from which single-phase circuits will be taken, the system should be designed so that under full load the load on each phase leg will be nearly equal.

TABLE 20-3 Properties of Conductors*

AWG or kcmil	Area, circular mils	Concentric lay stranded conductors		Bare conductors		DC resistance, ohms per 1000 ft at 25°C, 77°F		
		No. of wires	Diam each wire, in	Diam, in	Area,† in²	Copper Bare cond.	Copper Tin'd. cond.	Aluminum
18	1,624	Solid	0.0403	0.0403	0.0013	6.51	6.79	10.7
16	2,583	Solid	0.0508	0.0508	0.0020	4.10	4.26	6.72
14	4,107	Solid	0.0641	0.0641	0.0032	2.57	2.68	4.22
12	6,530	Solid	0.0808	0.0808	0.0051	1.62	1.68	2.66
10	10,380	Solid	0.1019	0.1019	0.0081	1.018	1.06	1.67
8	16,510	Solid	0.1285	0.1285	0.0130	0.6404	0.659	1.05
6	26,250	7	0.0612	0.184	0.027	0.410	0.427	0.674
4	41,740	7	0.0772	0.232	0.042	0.259	0.269	0.424
3	52,640	7	0.0867	0.260	0.053	0.205	0.213	0.336
2	66,370	7	0.0974	0.292	0.067	0.162	0.169	0.266
1	83,690	19	0.0664	0.332	0.087	0.129	0.134	0.211
0	105,500	19	0.0745	0.373	0.109	0.102	0.106	0.168
00	133,100	19	0.0837	0.418	0.137	0.0811	0.0843	0.133
000	167,800	19	0.0940	0.470	0.173	0.0642	0.0668	0.105
0000	211,600	19	0.1055	0.528	0.219	0.0509	0.0525	0.0836
250	250,000	37	0.0822	0.575	0.260	0.0431	0.0449	0.0708
300	300,000	37	0.0900	0.630	0.312	0.0360	0.0374	0.0590
350	350,000	37	0.0973	0.681	0.364	0.0308	0.0320	0.0505
400	400,000	37	0.1040	0.728	0.416	0.0270	0.0278	0.0442
500	500,000	37	0.1162	0.814	0.520	0.0216	0.0222	0.0354
600	600,000	61	0.0992	0.893	0.626	0.0180	0.0187	0.0295
700	700,000	61	0.1071	0.964	0.730	0.0154	0.0159	0.0253
750	750,000	61	0.1109	0.998	0.782	0.0144	0.0148	0.0236
800	800,000	61	0.1145	1.031	0.835	0.0135	0.0139	0.0221
900	900,000	61	0.1215	1.093	0.938	0.0120	0.0123	0.0197
1,000	1,000,000	61	0.1280	1.152	1.042	0.0108	0.0111	0.0177
1,250	1,250,000	91	0.1172	1.289	1.305	0.00863	0.00888	0.0142
1,500	1,500,000	91	0.1284	1.412	1.566	0.00719	0.00740	0.0118
1,750	1,750,000	127	0.1174	1.526	1.829	0.00616	0.00634	0.0101
2,000	2,000,000	127	0.1255	1.631	2.089	0.00539	0.00555	0.00885

*Reproduced by permission from "National Electrical Code," 1978, National Fire Protection Association.

†Area given is that of a circle having a diameter equal to the overall diameter of a stranded conductor.

The values given in the table are those given in NBS Circ. 31 except that those shown in the eighth column are those given in Specification B33 of the American Society for Testing and Materials.

The resistance values given in the last three columns are applicable only to direct current. When conductors larger than No. 4/0 are used with alternating current, the multiplying factors in Table 20-4 should be used to compensate for skin effect.

Voltage-Drop Calculations. Voltage drop in a circuit may be computed from the following formulas, in which

V_d = voltage drop between any two phase wires, or between phase wire and neutral when only one phase wire is used in the circuit

I = current, A

L = one-way run, ft

R = resistance, Ω/mil-ft

c.m. = circular mils

Single-phase two-wire (and balanced single-phase three-wire) circuits:

$$V_d = \frac{2RIL}{\text{c.m.}} \qquad (20\text{-}27)$$

Balanced two-phase three-wire, three-phase three-wire, and balanced three-phase four-wire circuits:

$$V_d = \frac{\sqrt{3}\,RIL}{\text{c.m.}} \qquad (20\text{-}28)$$

Equations (20-27) and (20-28) contain a factor R that represents the resistance in ohms to direct current of 1 mil-ft of wire. The value of R may be taken as 10.7 for copper and 17.7 for aluminum. Table 20-3 gives the resistance, ohms per 1000 ft, for various sizes of conductors. For small wire sizes, up to No. 3, resistance is the same for alternating and direct current. But above No. 3, ac resistance is larger, and a correction factor should be applied (Table 20-4).

Voltage drops used in design may range from 1 to 5% of the service voltage. Some codes set a maximum for voltage drop of 2.5% for combined light and power circuits from service entry to the building to point of final distribution at branch panels.

When this voltage drop is apportioned to the various parts of the circuit, it is economical to assign the greater part, say 1.5 to 2%, to the smaller, more numerous feeders, and only 0.5 to 1% to the heavy main feeders between the service and main distribution panels. Table 20-5 gives the maximum allowable current for each wire size for copper

TABLE 20-4 AC/DC Resistance Ratio at 60 Hz*

Size, AWG or kcmil	Multiplying factor			
	For nonmetallic sheathed cables in air or nonmetallic conduit		For metallic sheathed cables or all cables in metallic raceways	
	Copper	Aluminum	Copper	Aluminum
Up to 3	1	1	1	1
2	1	1	1.01	1.00
1	1	1	1.01	1.00
0	1.001	1.000	1.02	1.00
00	1.001	1.001	1.03	1.00
000	1.002	1.001	1.04	1.01
0000	1.004	1.002	1.05	1.01
250	1.005	1.002	1.06	1.02
300	1.006	1.003	1.07	1.02
350	1.009	1.004	1.08	1.03
400	1.011	1.005	1.10	1.04
500	1.018	1.007	1.13	1.06
600	1.025	1.010	1.16	1.08
700	1.034	1.013	1.19	1.11
750	1.039	1.015	1.21	1.12
800	1.044	1.017	1.22	1.14
1000	1.067	1.026	1.30	1.19
1250	1.102	1.040	1.41	1.27
1500	1.142	1.058	1.53	1.36
1750	1.185	1.079	1.67	1.46
2000	1.233	1.100	1.82	1.56

*Reproduced by permission from "National Electrical Code," 1978, National Fire Protection Association.

TABLE 20-5 Allowable Current-Carrying Capacities of Copper Conductors, Amperes*

(Not More Than Three Conductors in Raceway, Cable, or Direct Burial)
[Based on Room Temperature of 30°C (86°F)]

Size, AWG or kcmil	Temperature rating of conductor							
	60°C (140°F) Types RUW, T, TW, UF	75°C (167°F) Types RH, RHW, RUH, THW, THWN, XHHW, USE, ZW	85°C (185°F) Types V, MI	90°C (194°F) Types TA, TBS, SA, AVB, SIS, FEP, FEPB, RHH, THHN, XHHW	110°C (230°F) Types AVA, AVL	125°C (257°F) Types AI, AIA	200°C (392°F) Types A, AA, FEP, FEPB	250°C (482°F) Type TFE (nickel or nickel-coated copper only)
18				21				
16			22	22				
14	15	15	25	25	30	30	30	40
12	20	20	30	30	35	40	40	55
10	30	30	40	40	45	50	55	75
8	40	45	50	50	60	65	70	95
6	55	65	70	70	80	85	95	120
4	70	85	90	90	105	115	120	145
3	80	100	105	105	120	130	145	170
2	95	115	120	120	135	145	165	195
1	110	130	140	140	160	170	190	220
1/0	125	150	155	155	190	200	225	250
2/0	145	175	185	185	215	230	250	280
3/0	165	200	210	210	245	265	285	315
4/0	195	230	235	235	275	310	340	370
250	215	255	270	270	315	335		
300	240	285	300	300	345	380		
350	260	310	325	325	390	420		
400	280	335	360	360	420	450		
500	320	380	405	405	470	500		
600	355	420	455	455	525	545		
700	385	460	490	490	560	600		
750	400	475	500	500	580	620		
800	410	490	515	515	600	640		
900	435	520	555	555				
1000	455	545	585	585	680	730		
1250	495	590	645	645				
1500	520	625	700	700	785			
1750	545	650	735	735				
2000	560	665	775	775	840			

Correction factors for ambient temperatures over 30°C

Ambient temp.									
°C	°F								
40	104	0.82	0.88	0.90	0.91	0.94	0.95		
45	113	0.71	0.82	0.85	0.87	0.90	0.92		
50	122	0.58	0.75	0.80	0.82	0.87	0.89		
55	131	0.41	0.67	0.74	0.76	0.83	0.86		
60	140		0.58	0.67	0.71	0.79	0.83	0.91	0.95
70	158		0.35	0.52	0.58	0.71	0.76	0.87	0.91

Correction factors for ambient temperatures over 30°C

Ambient temp. °C	°F						
75	167	0.43	0.50	0.66	0.72	0.86	0.89
80	176	0.30	0.41	0.61	0.69	0.84	0.87
90	194			0.50	0.61	0.80	0.83
100	212				0.51	0.77	0.80
120	248					0.69	0.72
140	284					0.59	0.59
160	320						0.54
180	356						0.50
200	392						0.43
225	437						0.30

*Allowable current-carrying capacities of aluminum wire are 84% of the values in this table. Reproduced by permission from "National Electrical Code," 1978, National Fire Protection Association.

wire; use 84% of these values for aluminum. Table 20-3 gives the area, in circular mils, to be used in the voltage-drop formulas.

First, select the minimum-size wire allowed by the building code, and test it for voltage drop. If this drop is excessive, test a larger size, until one is found for which the voltage drop is within the desired limit. This trial-and-error process can be shortened by first assuming the desired voltage drop, and then computing the required wire area with Eqs. (20-27) and (20-28). The wire size can be selected from Table 20-3.

For circuits designed for motor loads only, no lighting, the maximum voltage drop may be increased to a total of 5%. Of this, 1% can be assigned to branch circuits and 4% to feeders.

Tables 20-6 and 20-7 give dimensions of trade sizes of conduit and tubing and permissible numbers of conductors that can be placed in each size.

Wiring for Motor Loads. Motors have a high starting current that lasts a very short time. But it may be 2 to 3 times as high as the rated current when running. Although motor windings will not be damaged by a high current of short duration, they cannot take currents much greater than the rated value for long periods without excessive overheating and consequent breakdown of the insulation.

Overcurrent protective devices, fuses and circuit breakers, should be selected to protect motors from overcurrents of long duration, and yet permit short-duration starting currents to pass without disconnecting the circuit. For this reason, the National Electrical Code permits the fuse or circuit breaker in a motor circuit to have a higher ampere rating than the allowable current-carrying capacity of the wire. Table 20-8 gives the overcurrent protection for motors allowed by the Code. Time-delay fuses are available that permit smaller fuse holders for a given-size motor than with standard fuses (Tables 20-9 and 20-10).

The National Electrical Code requirements for motor circuit conductors and overcurrent protection are as follows:

Branch Circuits (One Motor). Conductors shall have an allowable current-carrying capacity not less than 125% of the motor full-load current. Overcurrent protection, fuses or circuit breakers, must be capable of carrying the starting current of the motor. Maximum rating of such protection is given in Table 20-8, which shows that the maximum rating varies with the type, starting method, and locked-rotor current of the motor. For the great majority of motor applications in buildings, conductor and fuse protection may be selected more readily from Tables 20-9 and 20-10.

TABLE 20-6 Maximum Number of Conductors of Same Size in Trade Sizes of Conduit or Tubing*

Type letters	Conductor size, AWG or kcmil	Conduit or tubing trade size, in										
		1/2	3/4	1	1 1/4	1 1/2	2	2 1/2	3	3 1/2	4	4 1/2	5	6
TW, T, RUH, RUW, XHHW (14 thru 8)	14	9	15	25	44	60	99	142	171				
	12	7	12	19	35	47	78	111	131				
	10	5	9	15	26	36	60	85	131	176				
	8	2	4	7	12	17	28	40	62	84	108			
RHW and RHH (without outer covering), THW	14	6	10	16	29	40	65	93	143	192			
	12	4	8	13	24	32	53	76	117	157			
	10	4	6	11	19	26	43	61	95	127	163		
	8	1	3	5	10	13	22	32	49	66	85	106	133	
TW, T, THW, RUH (6 thru 2), RUW (6 thru 2),	6	1	2	4	7	10	16	23	36	48	62	78	97	141
	4	1	1	3	5	7	12	17	27	36	47	58	73	106
	3	1	1	2	4	6	10	15	23	31	40	50	63	91
	2	1	1	2	4	5	9	13	20	27	34	43	54	78
	1		1	1	3	4	6	9	14	19	25	31	39	57
FEPB (6 thru 2), RHW and RHH (without outer covering)	0		1	1	2	3	5	8	12	16	21	27	33	49
	00		1	1	1	3	5	7	10	14	18	23	29	41
	000		1	1	1	2	4	6	9	12	15	19	24	35
	0000			1	1	1	3	5	7	10	13	16	20	29
	250				1	1	2	4	6	8	10	13	16	23
	300			1	1	1	2	3	5	7	9	11	14	20
	350				1	1	1	3	4	6	8	10	12	18
	400				1	1	1	3	4	5	7	9	11	16
	500				1	1	1	2	3	4	6	7	9	14
	600						1	1	3	4	5	6	7	11
	700						1	1	2	3	4	5	7	10
	750					1	1	1	2	3	4	5	6	9

Type letters	Conductor size, AWG or kcmil	Conduit or tubing trade size, in										
		1/2	3/4	1	1 1/4	1 1/2	2	2 1/2	3	3 1/2	4	4 1/2	5	6
THWN, THHN, FEP (14 thru 2), FEPB (14 thru 8),	14	13	24	39	69	94	154							
	12	10	18	29	51	70	114	164						
	10	6	11	18	32	44	73	104	160					
	8	3	5	9	16	22	36	51	79	106	136			
	6	1	4	6	11	15	26	37	57	76	98	154	154	137
	4	1	2	4	7	9	16	22	35	47	60	75	94	116
	3	1	1	3	6	8	13	19	29	39	51	64	80	97
	2	1	1	3	5	7	11	16	25	33	43	54	67	72
	1		1	1	3	5	8	12	18	25	32	40	50	
XHHW (4 thru 500 kcmil)	0		1	1	3	4	7	10	15	21	27	33	42	61
	00		1	1	2	3	6	8	13	17	22	28	35	51
	000		1	1	1	3	5	7	11	14	18	23	29	42
	0000		1	1	1	2	4	6	9	12	15	19	24	35
	250			1	1	1	3	4	7	10	12	16	20	28
	300			1	1	1	3	4	6	8	11	13	17	24
	350			1	1	1	2	3	5	7	9	12	15	21
	400				1	1	1	3	5	6	8	10	13	19
	500					1	1	2	4	5	7	9	11	16
	600				1	1	1	1	3	4	5	7	9	13
	700					1	1	1	3	4	5	6	8	11
	750					1	1	1	2	3	4	6	7	11
XHHW	6	1	3	5	9	13	21	30	47	63	81	102	128	185
	600				1	1	1	1	3	4	5	7	9	13
	700					1	1	1	3	4	5	6	7	11
	750					1	1	1	2	3	4	6	7	10

TABLE 20-6 Maximum Number of Conductors of Same Size in Trade Sizes of Conduit or Tubing* (Continued)

Type letters	Conductor size, AWG or kcmil	Conduit or tubing trade size, in											
		1/2	3/4	1	1 1/4	1 1/2	2	2 1/2	3	3 1/2	4	4 1/2	5	6
RHW,	14	3	6	10	18	25	41	58	90	121	155			
	12	3	5	9	15	21	35	50	77	103	132			
	10	2	4	7	13	18	29	41	64	86	110			
	8	1	2	4	7	9	16	22	35	47	60	75	94	137
RHH	6	1	1	2	5	6	11	15	24	32	41	51	64	93
(with	4	1	1	1	3	5	8	12	18	24	31	39	50	72
outer	3	1	1	1	3	4	7	10	16	22	28	35	44	63
covering)	2		1	1	3	4	6	9	14	19	24	31	38	56
	1		1	1	1	3	5	7	11	14	18	23	29	42
	0			1	1	2	4	6	9	12	16	20	25	37
	00			1	1	1	3	5	8	11	14	18	22	32
	000			1	1	1	3	4	7	9	12	15	19	28
	0000			1	1	1	2	4	6	8	10	13	16	24
	250				1	1	1	3	5	6	8	11	13	19
	300				1	1	1	3	4	5	7	9	11	17
	350				1	1	1	2	4	5	6	8	10	15
	400				1	1	1	1	3	4	6	7	9	14
	500					1	1	1	3	4	5	6	8	11
	600					1	1	1	2	3	4	5	6	9
	700					1	1	1	1	3	3	4	5	8
	750						1	1	1	3	3	4	5	8

*Reproduced by permission from "National Electrical Code," 1978, National Fire Protection Association. See also Table 20-7.

TABLE 20-7 Dimensions of Conductors and Areas of Conduit or Tubing Permitted to Be Occupied by Conductors*

a. Occupiable area in conduit or tubing, in^2

Trade size	Internal diameter, in	Total 100%	Not lead-covered			Lead-covered			
			1 cond. 53%	2 cond. 31%	3 cond. and over 40%	1 cond. 55%	2 cond. 30%	3 cond. 40%	4 cond. 38%	Over 4 cond. 35%
1/2	0.622	0.30	0.16	0.09	0.12	0.17	0.09	0.12	0.11	0.11
3/4	0.824	0.53	0.28	0.16	0.21	0.29	0.16	0.21	0.20	0.19
1	1.049	0.86	0.46	0.27	0.34	0.47	0.26	0.34	0.33	0.30
1 1/4	1.380	1.50	0.80	0.47	0.60	0.83	0.45	0.60	0.57	0.53
1 1/2	1.610	2.04	1.08	0.63	0.82	1.12	0.61	0.82	0.78	0.71
2	2.067	3.36	1.78	1.04	1.34	1.85	1.01	1.34	1.28	1.18
2 1/2	2.469	4.79	2.54	1.48	1.92	2.63	1.44	1.92	1.82	1.68
3	3.068	7.38	3.91	2.29	2.95	4.06	2.21	2.95	2.80	2.58
3 1/2	3.548	9.90	5.25	3.07	3.96	5.44	2.97	3.96	3.76	3.47
4	4.026	12.72	6.74	3.94	5.09	7.00	3.82	5.09	4.83	4.45
5	5.047	20.00	10.60	6.20	8.00	11.00	6.00	8.00	7.60	7.00
6	6.065	28.89	15.31	8.96	11.56	15.89	8.67	11.56	10.98	10.11

b. Dimensions of rubber-covered and thermoplastic-covered conductors

Size, AWG or kcmil	Types RFH-2, RH, RHH, RHW, SF-2		Types TF, T, THW, TW, RUH, RUW		Types THHN, THWN, TFN		Types FEP, FEPB, FEPW, TFE, PF, PFA, PFAH, PGF, PTF	
	Approx. diam., in	Approx. area, in^2	Approx. diam., in	Approx. area, in^2	Approx. diam., in	Approx. area, in^2	Approx. diam., in	Approx. area, in^2
18	0.146	0.0167	0.106	0.0088				
16	0.158	0.0196	0.118	0.0109				
14	2/64 in 0.171	0.0230	0.131	0.0135	0.105	0.0087	0.105 0.105	0.0087 0.0087
14	3/64 in 0.204	0.0327						
14			0.162	0.0206				
12	2/64 in 0.188	0.0278	0.148	0.0172	0.122	0.0117	0.121 0.121	0.0115 0.0115
12	3/64 in 0.221	0.0384						
12			0.179	0.0251				
10	0.242	0.0460	0.168	0.0224	0.153	0.0184	0.142 0.142	0.0159 0.0159
10			0.199	0.0311				
8	0.311	0.0760	0.228	0.0408	0.201	0.0317	0.189 0.169	0.0280 0.0225
8			0.259	0.0526				
6	0.397	0.1238	0.323	0.0819	0.257	0.0519	0.244 0.302	0.0467 0.0716
4	0.452	0.1605	0.372	0.1087	0.328	0.0845	0.292 0.350	0.0669 0.0962
3	0.481	0.1817	0.401	0.1263	0.356	0.0995	0.320 0.378	0.0803 0.1122
2	0.513	0.2067	0.433	0.1473	0.388	0.1182	0.352 0.410	0.0973 0.1316
1	0.588	0.2715	0.508	0.2027	0.450	0.1590		
0	0.629	0.3107	0.549	0.2367	0.491	0.1893		
00	0.675	0.3578	0.595	0.2781	0.537	0.2265		
000	0.727	0.4151	0.647	0.3288	0.588	0.2715		
0000	0.785	0.4840	0.705	0.3904	0.646	0.3278		
250	0.868	0.5917	0.788	0.4877	0.716	0.4026		
300	0.933	0.6837	0.843	0.5581	0.771	0.4669		
350	0.985	0.7620	0.895	0.6291	0.822	0.5307		
400	1.032	0.8365	0.942	0.6969	0.869	0.5931		
500	1.119	0.9834	1.029	0.8316	0.955	0.7163		
600	1.233	1.1940	1.143	1.0261				
700	1.304	1.3355	1.214	1.1575				
750	1.339	1.4082	1.249	1.2252				
800	1.372	1.4784	1.282	1.2908				
900	1.435	1.6173	1.345	1.4208				
1000	1.494	1.7531	1.404	1.5482				
1250	1.676	2.2062	1.577	1.9352				
1500	1.801	2.5475	1.702	2.2748				
1750	1.916	2.8895	1.817	2.5930				
2000	2.021	3.2079	1.922	2.9013				

*Reproduced by permission from "National Electrical Code," National Fire Protection Association.

TABLE 20-8 Overcurrent Protection for Motors*

Full-load current rating of motor, A	For running protection of motors		Maximum allowable rating or setting of branch-circuit protective devices							
	Maximum rating of nonadjustable protective devices, A	Maximum setting of adjustable protective devices, A	With code letters: Single-phase, squirrel-cage, and synchronous. Full voltage, resistor or reactor starting, code letters F to V inclusive. Without code letters: Same as above		With code letters: Single-phase, squirrel-cage, and synchronous. Full voltage, resistor or reactor start, code letters B to E inclusive. Autotransformer start, code letters F to V inclusive. Without code letters: (Not more than 30 A) Squirrel-cage and synchronous, autotransformer start, high-reactance cage†		With code letters: Squirrel-cage and synchronous autotransformer start, code letters B to E inclusive. Without code letters: (More than 30 A) Squirrel-cage and synchronous autotransformer start, high-reactance squirrel-cage†		With code letters: All motors code letter A. Without code letters: D-c and wound-rotor motors	
			Fuses	Circuit breakers (nonadjustable overload trip)	Fuses	Circuit breakers (nonadjustable overload trip)	Fuses	Circuit breakers (nonadjustable overload trip)	Fuses	Circuit breakers (nonadjustable overload trip)
1	2	1.25	15	15	15	15	15	15	15	15
2	3	2.50	15	15	15	15	15	15	15	15
3	4	3.75	15	15	15	15	15	15	15	15
4	6	5.0	15	15	15	15	15	15	15	15
5	8	6.25	15	15	15	15	15	15	15	15
6	8	7.50	20	15	15	15	15	15	15	15
7	10	8.75	25	20	20	15	15	15	15	15
8	10	10.0	25	20	20	20	20	20	15	15
9	12	11.25	30	30	25	20	20	20	15	15
10	15	12.50	30	30	25	20	20	20	15	15
11	15	13.75	35	30	30	30	25	30	20	20
12	15	15.00	40	30	30	30	25	30	20	20
13	20	16.25	40	40	35	30	30	30	20	20
14	20	17.50	45	40	35	30	30	30	25	30
15	20	18.75	45	40	40	30	30	30	25	30
16	20	20.00	50	40	40	40	35	40	25	30

Col1	Col2	Col3	Col4	Col5	Col6	Col7	Col8	Col9	Col10	Row
30	30	40	35	40	45	50	60	21.25	25	17
30	30	40	40	40	45	50	60	22.50	25	18
30	30	40	40	40	50	50	60	23.75	25	19
30	30	40	40	40	50	50	60	25.00	25	20
40	35	50	45	50	60	70	70	27.50	30	22
40	40	50	50	50	60	70	80	30.00	30	24
40	40	70	60	70	70	70	80	32.50	35	26
50	45	70	60	70	70	70	90	35.00	35	28
50	45	70	60	70	80	100	90	37.50	40	30
50	50	70	70	70	80	100	100	40.00	40	32
70	60	100	70	100	90	100	110	42.50	45	34
70	60	100	80	100	90	100	110	45.00	45	36
70	60	100	80	100	100	100	125	47.50	50	38
100	70	100	90	100	100	125	125	50.00	50	40
100	70	100	90	100	110	125	125	52.50	50	42
100	70	100	90	100	110	125	125	55.00	60	44
100	90	125	110	125	125	125	150	57.50	60	46
100	90	125	125	125	125	125	150	60.00	60	48
100	90	125	125	125	125	125	150	62.50	60	50
125	90	125	125	125	150	150	175	65.00	70	52
125	100	125	125	125	150	150	175	67.50	70	54
125	100	150	150	150	150	150	175	70.00	70	56
125	100	150	150	150	150	150	175	72.50	70	58
125	110	150	150	150	150	150	200	75.00	80	60
150	110	150	150	150	175	175	200	77.50	80	62
150	125	150	150	150	175	175	200	80.00	80	64
150	125	175	175	175	175	175	200	82.50	80	66
150	125	175	175	175	175	175	225	85.00	90	68
125	125	150	150	150	175	175	225	87.50	90	70
125	125	150	150	150	200	200	225	90.00	90	72
125	125	175	150	175	200	200	225	92.50	90	74
150	125	175	175	175	200	200	250	95.00	100	76
125	125	175	175	175	200	200	250	97.50	100	78
125	125	175	175	175	200	200	250	100.00	100	80
125	125	175	175	175	225	225	250	102.50	110	82
150	150	175	175	175	225	225	250	105.00	110	84
150	150	175	175	175	225	225	300	107.50	110	86
150	150	200	200	200	225	225	300	110.00	110	88
150	150	200	200	200	225	225	300	112.50	125	90
150	150	200	200	200	250	250	300	115.00	125	92

20-37

TABLE 20-8 Overcurrent Protection for Motors* (Continued)

Full-load current rating of motor, A	For running protection of motors		Maximum allowable rating or setting of branch circuit protective devices							
	Maximum rating of nonadjustable protective devices, A	Maximum setting of adjustable protective devices, A	With code letters: Single-phase, squirrel-cage, and synchronous. Full voltage, resistor or reactor starting, code letters F to V inclusive / Without code letters: Same as above		With code letters: Single-phase, squirrel-cage, and synchronous. Full voltage, resistor or reactor start, code letters B to E inclusive. Autotransformer, start, code letters F to V inclusive / Without code letters: (Not more than 30 A) Squirrel-cage and synchronous autotransformer start, high-reactance squirrel-cage†		With code letters: Squirrel-cage and synchronous autotransformer start, code letters B to E inclusive / Without code letters: (More than 30 A) Squirrel-cage and synchronous autotransformer start, high-reactance squirrel-cage†		With code letters: All motors code letter A / Without code letters: D-c and wound-rotor motors	
			Fuses	Circuit breakers (nonadjustable overload trip)	Fuses	Circuit breakers (nonadjustable overload trip)	Fuses	Circuit breakers (nonadjustable overload trip)	Fuses	Circuit breakers (nonadjustable overload trip)
94	125	117.50	300	250	250	200	200	200	150	150
96	125	120.00	300	250	250	200	200	200	150	150
98	125	122.50	300	250	250	200	200	200	150	150
100	125	125.00	300	250	250	200	200	200	150	150
105	150	131.50	350	300	300	225	225	225	175	175
110	150	137.50	350	300	300	225	225	225	175	175
115	150	144.00	350	300	300	250	250	250	175	175
120	150	150.00	400	300	300	250	250	250	200	200
125	175	156.50	400	350	350	250	250	250	200	200
130	175	162.50	400	350	350	300	300	300	200	200
135	175	169.00	450	350	350	300	300	300	225	225
140	175	175.00	450	350	350	300	300	300	225	225
145	200	181.50	450	400	400	300	300	300	225	225
150	200	187.50	450	400	400	300	300	300	225	225
155	200	194.00	500	400	400	350	350	350	250	250
160	200	200.00	500	400	400	350	350	350	250	250

HP										
165	225	206.00	500	500	450	350	350	350	250	250
170	225	213.00	500	500	450	350	350	350	300	300
175	225	219.00	500	500	450	350	350	350	300	300
180	225	225.00	600	500	450	400	400	400	300	300
185	250	231.00	600	500	500	400	400	400	300	300
190	250	238.00	600	500	500	400	400	400	300	300
195	250	244.00	600	500	500	400	400	400	300	300
200	250	250.00	600	500	500	400	400	400	300	300
210	250	263.00	800	600	600	500	500	500	350	350
220	300	275.00	800	600	600	500	500	500	350	350
230	300	288.00	800	600	600	500	500	500	350	350
240	300	300.00	800	600	600	500	500	400	400	400
250	300	313.00	800	700	800	500	500	500	400	400
260	350	325.00	800	700	800	600	600	600	400	400
270	350	338.00	1000	700	800	600	600	600	450	450
280	350	350.00	1000	700	800	600	600	600	450	500
290	350	363.00	1000	800	800	600	600	600	450	500
300	400	375.00	1000	800	800	600	600	600	450	500
320	400	400.00	1000	800	800	700	700	800	500	500
340	450	425.00	1200	800	1000	700	700	800	600	600
360	450	450.00	1200	...	1000	800	800	800	600	600
380	500	475.00	1200	...	1000	800	800	800	600	600
400	500	500.00	1200	...	1000	800	800	800	600	600
420	600	525.00	1600	...	1200	1000	800	800	800	700
440	600	550.00	1600	...	1200	1000	...	...	800	700
460	600	575.00	1600	...	1200	1000	...	...	800	700
480	600	600.00	1600	...	1200	1000	...	...	800	800
500	...	625.00	1600	...	1600	1000	...	...	800	800

SOURCE: Table 17-21 from W. T. Stuart, Wiring Design—Commercial and Industrial Buildings, in D. G. Fink and J. M. Carroll (eds.), "Standard Handbook for Electrical Engineers," 10th ed., McGraw-Hill Book Co., 1968. Used by permission.

* High-reactance squirrel-cage motors are those designed to limit the starting current by means of deep-slot secondaries or double-wound secondaries and are generally started on full voltage.

TABLE 20-9 Protection of Single-Phase Motors and Circuits

Size of motor		Branch-circuit protection*		Motor-running protection†				Size of fused switch or fuse holder		Minimum size of starter: NEMA size	Minimum size and type of wire: AWG or MCM	Minimum size of conduit: diameter, in.
Horse-power	Ampere rating	Maximum size fuse permitted by the Code	Time-delay-cartridge or low-peak fuse that can be used	Size of time-delay-cartridge or low-peak fuse				Maximum size switch that can be used	Size that can be used with time-delay-cartridge or low-peak fuses			
				Ordinary service	Heavy service	Maximum size						
						40°C motor	All other motors					
115 volts												
1/6	4.4	15	7	4½	5	5 6/10	5 6/10	30	30	00	14 R	½
1/4	5.8	20	10	5 6/10	6¼	8	7	30	30	0	14 R	½
1/3	7.2	25	12	7	8	9	9					
1/2	9.8	30	17½	10	12	12	12					
3/4	13.8	45	25	15	17½	17½	17½	60	30	0	12 R	½
1	16	50	30	17½	20	20	20	60	30			
1½	20	60	30	20	25	25	25	60	30			
2	24.0	80	40	25	30	30	30	100	60	1	10 R	¾
230 volts												
1/6	2.2	15	3½	2¼	2½	2 8/10	2 8/10	30	30	00	14 R	½
1/4	2.9	15	5	2 8/10	3 2/10	4	3½					
1/3	3.6	15	6	3½	4	4½	4½					
1/2	4.9	15	8	5	5 6/10	6¼	6¼					
3/4	6.9	25	12	7	8	9	8	30	30	00	14 R	½
1	8	25	15	8	9	10	10	30	30			
1½	10	30	17½	10	12	12	12	30	30			
2	12	40	20	12	15	15	15	60	30	0	14 R	½
3	17	60	25	17½	20	20	20	60	30	1	10 R	¾
5	28	90	45	30	35	35	35	100	60	2	8 R	¾
7½	40	125	60	40	45	50	50	200	60	2	6 R	1
10	50	150	80	50	60	70	60	200	100	3	4 R	1¼

* These do not give motor-running protection.
† On normal installations these also give branch-circuit protection.

TABLE 20-10 Protection of Three-Phase 208-V Motors and Circuits

Size and class of motor			Branch circuit protection*		Motor-running protection†				Size of fused switch or fuse holder		Minimum size of starter: NEMA size	Minimum size and type of wire: AWG or MCM	Minimum size of conduit: diameter, in.
Horsepower	Ampere rating	Class	Maximum size fuse permitted by the Code	Time-delay-cartridge or low-peak fuse that can be used	Size of time-delay-cartridge or low-peak fuse		Maximum size		Maximum size switch that can be used	Size that can be used with time-delay-cartridge or low-peak fuses			
					Ordinary service	Heavy service	40°C motor	All other motors					
½	2.1	Any	15	5	2¼	2½	2⁸⁄₁₀	2½	30	30	00	14 R	½
¾	3	Any	15	8	3	3½	4	3½					
1	3.7	Any	15	8	4	4½	5	4½					
1½	5.3	Any	15	10	5⁶⁄₁₀	6¼	7	6¼					
2	6.9	1	25	12	7	8	9	8	30	30	0	14 R	½
		2	20	12	7	8	9	8					
		3-4	15	12	7	8	9	8					
3	9.5	1	30	15	10	12	12	12	30	30	0	14 R	½
		2	25	15	10	12	12	12					
		3	20	15	10	12	12	12					
		4	15	15	10	12	12	12					
5	15.9	1	50	25	17½	20	20	20	60	30	1	12 R	½
		2	40	25	17½	20	20	20	60	30			
		3	35	25	17½	20	20	20	60	30			
		4	25	25	17½	20	20	20	30	30			
7½	23.3	1	80	35	25	30	30	30	100	60	1	10 R	¾
		2	60	35	25	30	30	30	60	60			
		3	50	35	25	30	30	30	60	60			
		4	40	53	25	30	30	30	60	60			

20-41

TABLE 20-10 Protection of Three-Phase 208-V Motors and Circuits (Continued)

Size and class of motor			Branch circuit protection*		Motor-running protection†					Size of fused switch or fuse holder		Minimum size of starter: NEMA size	Minimum size and type of wire: AWG or MCM	Minimum size of conduit: diameter, in.
Horsepower	Ampere rating	Class	Maximum size fuse permitted by the Code	Time-delay-cartridge or low-peak fuse that can be used	Size of time-delay-cartridge or low-peak fuse					Maximum size switch that can be used	Size that can be used with time-delay-cartridge or low-peak fuses			
					Ordinary service	Heavy service	Maximum size		All other motors					
							40°C motor							
10	28.6	1	90	45	30	35	40		35	100	60	2	8 R	¾
		2	70	45	30	35	40		35	100	60			
		3	60	45	30	35	40		35	60	60			
		4	45	45	30	35	40		35	60	60			
15	42.3	1	125	60	45	50	50		50	200	60	2	6 R	1
		2	110	60	45	50	50		50	200	60			
		3	90	60	45	50	50		50	100	60			
		4	70	60	45	50	50		50	100	60			
20	55	1	175	90	60	70	70		70	200	100	3	4 R	1¼
		2	150	90	60	70	70		70	200	100			
		3	110	90	60	70	70		70	200	100			
		4	90	90	60	70	70		70	100	100			
25	68	1	225	100	70	80	90		80	400	100	3	2 R	1¼
		2	175	100	70	80	90		80	200	100			
		3	150	100	70	80	90		80	200	100			
		4	110	100	70	80	90		80	200	100			
30	83	1	250	125	90	100	110		100	400	200	3	1 R	1½
		2	225	125	90	100	110		100	400	200			
		3	175	125	90	100	110		100	200	200			
		4	125	125	90	100	110		100	200	200			

HP	Amps	No.									No.	Wire	Conduit
40	110	1	350	175	110	125	150	150	400	200	4	0 RH	2
		2	300	175	110	125	150	150	400	200			
		3	225	175	110	125	150	150	400	200			
		4	175	175	110	125	150	150	200	200			
50	132	1	400	200	150	175	175	175	400	200	4	00 RH	2
		2	350	200	150	175	175	175	400	200			
		3	300	200	150	175	175	175	400	200			
		4	200	200	150	175	175	175	200	200			
60	159	1	500	250	175	200	200	200	600	400	5	000 RH	2
		2	400	250	175	200	200	200	400	400			
		3	350	250	175	200	200	200	400	400			
		4	250	250	175	200	200	200	400	400			
75	196	1	600	300	200	225	250	250	600	400	5	250 RH	2½
		2	500	300	200	225	250	250	600	400			
		3	400	300	200	225	250	250	400	400			
		4	300	300	200	225	250	250	400	400			
100	260	1		400	250	300	350	300		400	5	400 RH	3
		2		400	250	300	350	300		400			
		3	600	400	250	300	350	300	600	400			
		4	400	400	250	300	350	300	400	400			
125	328	1–3	500	500	350	400	450	400	600	600	6	2 sets‡ 0000 RH	2½‡
		4		500	350	400	450	400		600			
150	381	1–3	600	600	400	450	500	450	600	600	6	2 sets‡ 250 RH	2½‡
		4		600	400	450	500	450		600			

* These do not give motor-running protection.
† On normal installations these also give branch-circuit protection.
‡ Indicates two sets of multiple conductors and two runs of conduit.

Feeder Circuits (More than One Motor on a Conductor). The conductor should have an allowable current-carrying capacity not less than 125% of the full-load current of the largest motor plus the sum of the full-load currents of the remaining motors on the same circuit. The rating of overcurrent protection, fuses or circuit breakers, shall not be greater than the maximum allowed by the code for protection of the largest motor plus the sum of the full-load currents of the remaining motors on the circuit.

If the allowable current-carrying capacity of the conductor or the size of the computed overcurrent device does not correspond to the rating of a standard-size fuse or circuit breaker, the next larger standard size should be used.

Amp-Traps and Hi-Caps are high-interrupting-capacity current-limiting fuses used in service switches and main distribution panels connected near service switches. This type of fuse is needed here because this part of the wiring system in large buildings consists of heavy cables or buses and large switches that have very little resistance. If a short circuit occurs, very high currents will flow, limited only by the interrupting capacity of the protective device installed by the utility company on its own transformers furnishing the service. Ordinary fuses cannot interrupt this current quickly enough to avert damage to the building wiring and connected electrical equipment. The interrupting-capacity value needed can be obtained from the utility company.

Fuses in service switches and connected main panels should have current-time characteristics that will isolate only the circuit in which a short occurs, without permitting the short-circuit current to pass to other feeders and interrupt those circuits too. The electrical designer should obtain data from manufacturers of approved fusing devices on the proper sequence of fusing.

Service-Entrance Switch and Metering Equipment. Fused switches or circuit breakers must be provided near the entrance point of electrical service in a building for shutting off the power. The National Electrical Code requires that each incoming service in a multiple-occupancy building be controlled near its entrance by not more than six switches or circuit breakers.

Metering equipment consists of a meter pan, meter cabinet, current transformer cabinet, or a combination of these cabinets, depending on the load requirements and other characteristics of the specific project. The meters and metering transformers for recording current consumed are furnished by the utility company. Unless otherwise permitted by the utility company, meters must be located near the point of service entrance. Sometimes, the utility company permits one or more tenant meter rooms at other locations in the cellar of an apartment house to suit economical building wiring design. Tenant meter closets on the upper floors, opening on public halls, also may be permitted. The most common form of tenant meters used is the three-wire type, consisting of two phase wires and the neutral, taken from a 208/120-V three-phase four-wire service.

The service switch and metering equipment may be combined in one unit, or the switch may be connected with conduit to a separate meter trough. For individual metering, the detachable-socket-type meter with prongs that fit into the jaws of the meter-mounting trough generally is used.

Switchboards and Panelboards. For low-capacity loads, wiring may be taken directly to a panelboard. For larger loads, wiring may be brought first to a switchboard and then to panelboards. This equipment is described in Art. 20-5. Branch circuits extend from panelboards to the various loads.

Sample Calculations for Apartment-building Riser. A diagram of a light and power riser for a nine-story apartment building is shown in Fig. 20-9.

1. *Typical Meter Branch to Apartment Panel.* Note that the meters are three-wire type, and three apartments are connected to the same neutral. Under balanced conditions, when each of the three identical apartments is under full load, no current will flow in the neutral. But at maximum unbalance, current in the neutral may be twice the cur-

Fig 20-9 Diagram of a typical apartment-building electrical riser.

20-45

rent in the phase wire for any one apartment. The neutral wire must be sized for this maximum current, though the usual practice is to compute voltage drops for the balanced condition.

Assume that the apartment area is 900 ft^2. The one-way run from meter to apartment panel (apartment A) is 110 ft.

$$\text{Apartment lighting load} = 900 \times 3 \text{ W/ft}^2 = 2700 \text{ W}$$
$$\text{Apartment appliance load} = \underline{3000}$$
$$\text{Total} = 5700 \text{ W}$$

The electric service is three-phase four-wire 208/120 V. Thus, the voltage between phase wires is 208, and between one phase and neutral 120 V. Assume a 90% power factor. From Eq. (20-25):

$$\text{Current per phase} = \frac{5700}{2 \times 120 \times 0.9} = 26.4 \text{ A}$$

The local electrical code requires the minimum size of apartment feeder to be No. 8 wire. According to Table 20-5, the allowable current in No. 8 RH wire is 45 A; so this wire would be adequate for the current, but it still must be checked for voltage drop. The neutral must be sized for the maximum unbalance, under which condition the current in the neutral will be $2 \times 26.4 = 52.8$ A. This will require No. 6 wire.

The voltage drop between phase wires can be obtained from Eq. (20-28), with the area of No. 8 wire taken as 16,510 circular mils (Table 20-3) and length of wire as 110 ft:

$$V_d = \frac{\sqrt{3} \times 10.7 \times 26.4 \times 110}{16,510} = 3.24 \text{ V}$$

$$\% \text{ voltage drop} = \frac{3.24 \times 100}{208} = 1.56\%$$

2. *Feeder to Meter Bank on Sixth Floor.*

$$\text{Total load (24 apartments)} = 24 \times 5700 \text{ W} = 136,800 \text{ W}$$
$$\text{Demand load: first 15,000 W at 100\%} = 15,000$$
$$\text{Balance, 121,800 W at 50\%} = \underline{60,900}$$
$$\text{Total demand load} = 75,900 \text{ W}$$

Assume a power factor of 90% and apply Eq. (20-26):

$$\text{Current per phase} = \frac{75,900}{3 \times 120 \times 0.9} = 234 \text{ A}$$

According to Table 20-5, minimum-size type RH wire for this current is 250 kcmil. This has to be tested for voltage drop. For a one-way run of 150 ft from service switch to sixth-floor meter bank and an area of 250,000 circular mils, Eq. (20-28) gives:

$$V_d = \frac{\sqrt{3} \times 10.7 \times 234 \times 150}{250,000} = 2.60 \text{ V}$$

From Table 20-4, the correction factor for ac resistance of 250-kcmil wire is 1.06. Application of this factor yields a corrected voltage drop of $1.06 \times 2.60 = 2.76$ volts.

$$\% \text{ voltage drop} = \frac{2.76 \times 100}{208} = 1.33\%$$

Then, the total voltage drop from service switch to apartment panel A is

$$1.56 + 1.33\% = 2.89\%$$

which exceeds the 2.5% maximum voltage drop allowed by the local code. It is necessary, therefore, to increase the size of the meter bank feeder over the minimum size required for the current.

To bring the total drop down to 2.5% the meter bank feeder drop must be reduced to 0.94% or 1.96 volts. The required wire size may be found by proportion:

$$\text{Required area} = \frac{2.76}{1.96} \times 250{,}000 = 352{,}000 \text{ circular mils}$$

According to Table 20-5, the nearest larger wire size is 400 kcmil.

For computing the size of the neutral for carrying 234 A, the local code allows a demand factor of 70% on lighting loads over 200 A. Hence, the net current in the neutral is 200 + 34 × 0.70 = 223.8 A. From Table 20-5, minimum-size wire is No. 4/0 RH. And with Table 20-7b, the size of conduit required for the feeder is computed as follows:

Three 400-kcmil RH wires = 3 × 0.8365 = 2.5095 in²
One 0000 RH wire   = 0.4840
Total area   = 2.9935 in²

According to Table 20-7a, permissible raceway fill for four conductors is 40%. Hence the minimum area required for the conduit is 2.9935/0.40 = 7.484 in². And from Table 20-7a, the required conduit size is 3½ in.

3. Feeder to Elevator and Roof Fans Panel. The total load on this feeder is

8 roof fans at ½ hp = 16.8 A/phase
8 roof fans at ¼ hp = 17.4
2 elevators at 15 hp = 84.6
Total   = 118.8 A/phase

The minimum current-carrying capacity of the feeder must be 125% of the rated full-load current of the largest motor, an elevator motor, 42.3 A, plus the sum of the rated load currents of the other motors. Thus, the capacity must be:

$$1.25 \times 42.3 + 42.3 + 16.8 + 17.4 = 129.4 \text{ A}$$

From Table 20-5, the minimum-size wire that can be used is No. 1 RH, with an area of 83,690 circular mils.

Check for voltage drop with a run of 150 ft:

$$V_d = \frac{\sqrt{3} \times 10.7 \times 118.8 \times 150}{83{,}690} = 3.95 \text{ V}$$

$$\% \text{ voltage drop} = \frac{3.95 \times 100}{208} = 1.90\% \quad \text{OK}$$

Section 21

Lighting

Jonathan S. Fink
Electrical Engineer,
Sanders & Thomas, Inc.,
Pottstown, Pa.

Lighting is part of the environmental control system, which also includes sound control and heating, ventilation, and air conditioning (HVAC), within a building. The prime purpose of a lighting system is to provide good visibility for execution of the tasks to be performed within the building. With good visibility, occupants can execute their tasks comfortably, efficiently, and safely.

Lighting also is desirable for other purposes. For example, it can be used to develop color effects for pleasure or accident prevention. It can be used to decorate select spaces or to accent objects. It can produce effects that influence human moods. And it can serve an emergency egress system and as part of a security system.

Good lighting requires good quality of illumination (Art. 21-2), proper color rendering (Art. 21-3), and an adequate quantity of light (Art. 21-4). This result, however, cannot be achieved economically solely by selection and arrangement of suitable light sources. Lighting effects are also dependent on other systems and factors such as the characteristics of surrounding walls, floor, and ceiling; nature of tasks to be illuminated; properties of the backgrounds of the tasks; age and visual acuity of occupants; and characteristics of the electrical system. Design of a lighting system, therefore, must take into account its interfacing with other systems. Also, lighting design must take into account the influence of lighting requirements on other systems, including architectural systems; heating and cooling effects of windows provided for daylighting; energy supply required from the electrical system; and loads imposed by electric lighting on the HVAC system.

Sources of light within a building may be daylight or artificial illumination. The latter can be produced in many ways, but only the most commonly used types of electric lighting are discussed in this section.

Like other building systems, lighting design is significantly affected by building codes. These generally contain minimum requirements for illumination levels, for the safety and health of building occupants. In addition, electric-lighting equipment and electrical distribution must conform to safety requirements in building codes and the National Electrical Code, which is promulgated by the National Fire Protection Association, and to standards of the Underwriters Laboratories, Inc. Also, the Illuminating Engineering Society has developed standards and recommended practices to promote good lighting design.

In the interests of energy conservation, Federal and state government agencies have set limits on the amount of energy that may be expended (energy budget) for operation of buildings. These limits may establish maximum levels of illumination for specific purposes in buildings.

Because of the importance of good lighting, the need to control lighting costs and to conserve energy, and the multiplicity of legal requirements affecting lighting design, engagement of a specialist in lighting design is advisable for many types of buildings.

21-1. Light and Sight.

A light source produces light by converting energy to electromagnetic waves. Light sources used in practical applications emit waves with a broad spectrum of frequencies or wavelengths. Light consists of those waves that the human eye normally perceives. A normal eye interprets the wavelengths as colors, the shortest wavelengths being recognized as blue, the longest wavelengths as red, and intermediate wavelengths as green, yellow, and orange. The eye also recognizes differences in intensity of light, or levels of illumination.

The eye sees an object because it receives light emitted by the object (if it is a light source) or reflected from it (if the object is not a light source). In the latter case, the eye can see the object if it reflects light received from a light source directly or from surfaces reflecting light. The total light reflected from the object equals the sum of the light from all sources that strike the object and is not absorbed. Thus, an object shielded from light sources will be revealed to the eye by light reflected from other surfaces, such as walls, floor, ceiling or furniture. The amount of detail on the object that the eye can recognize, however, depends not only on the intensity of light that the eye receives from the object but also on the intensity of that light relative to the intensity of light the eye receives from the background (field of vision behind the object). Consequently, the eye can readily recognize details on a brightly lit object set against a darker background. But the eye perceives little detail or considers the object dark (in shadow) if the background is much brighter than the object.

Inverse Square Law. Consider now a point source radiating luminous energy equally in all directions and located at the common center of two transparent spheres of unequal diameter. Because each sphere receives the same amount of luminous energy from the light source, the quantity of light per unit of area is less on the larger sphere than that on the smaller one. In fact, the quantity per unit area varies inversely as the areas of the spheres, or inversely as the square of the radii. These considerations justify the inverse square law, which states:

> Illumination level at any point is inversely proportional to the square of distance from the point light source.

For large light sources, the law holds approximately at large distances (at least 5 times the largest dimension of the sources) from the sources.

Light Source Power. Analogous to a pump in a water system or a battery in an electrical system, a light source emits luminous power. The unit used to measure this power is **candlepower** (cp), or **candela** (cd) (metric unit). (At one time, 1 candlepower was assumed equivalent to the luminous intensity of a wax candle, but now a more precise definition based on radiation from a heated black body is used.) The unit used to measure luminous power at a distance from the light source is the lumen (lm).

A **lumen** is the luminous power on an area of 1 ft^2 at a distance of 1 ft from a 1-cp light source.

Luminous efficacy is the unit used to measure the effectiveness of light sources. It is calculated by dividing the total lumen output of a light source by the total input, watts (W), and thus is measured in lm/W.

Level of Illumination. A major objective of lighting design is to provide a specified **illuminance**, or level of illumination, on a task. For design purposes, the task often is taken as a flat surface, called a **work plane**. If the task is uniformly illuminated, the level of illumination equals the lumens striking the surface divided by the area. The unit used to measure illuminance is the **footcandle** (fc). In accordance with the inverse square law, the illuminance on a work plane normal to the direction to a point light source is given by

$$\text{fc} = \text{cp}/D^2 \qquad (21\text{-}1)$$

where D = distance, ft, from work plane to light source
  cp = candlepower of light source
For a work plane at an angle θ to the direction of the light source,

$$\text{fc} = \text{cp} \sin \theta /D^2 \qquad (21\text{-}2)$$

A **luminaire** is a lighting device that consists of one or more lamps, or light sources, a fixture that positions and shields them, components that distribute the light, and elements that connect the lamps to the power supply. In general, luminaires do not radiate light of equal intensity in all directions, because of the characteristics of the lamps or the geometry of the fixtures. The actual illuminance around a single luminaire is an important design consideration. This environment may be characterized by the candlepower distribution curve of the luminaire. A typical such curve for a light source symmetrical about two vertical, perpendicular planes is shown in Fig. 21-1.

To produce the curve, illuminance, fc, is measured, with a ganiophotometer, on a plane that is placed at various points at equal distances from the light source and that at each point is normal to the direction to the source. From Eq. (21-1), the candlepower corresponding to the illuminance at each point is computed. In practice, the candlepower is usually calculated for points in vertical planes along longitudinal and transverse axes through the luminaire. The results for each plane are then plotted with polar or rectangular coordinates relative to the light source. Such plots show the variation in illuminance with direction from the light source. For example, Fig. 21-1 shows in polar coordinates that the light source investigated directs light mainly downward. Because the source is symmetrical, only one curve is needed in this case to represent illuminance in longitudinal, transverse, and diagonal planes.

Equivalent Spherical Illumination. If proper care is not taken in positioning observer and task relative to primary light sources, one or more of these light sources may be reflected into the field of vision as from a mirror. The resulting glare reduces visibility of the task; that is, effective illuminance, fc, is less than the raw (actual) footcandles. For measurement of the effective illuminance of lighting installations, the concept of equivalent spherical illumination, ESI fc, was introduced. It is based on a standard sphere lighting that produces equal illumination from all directions (almost glarefree). In

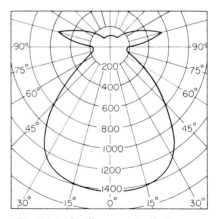

Fig. 21-1 Candlepower distribution curve indicates variation in lighting intensity with direction from a light source.

employment of the ESI concept, an actual lighting installation is assumed replaced by sphere illumination with task visibility equivalent to that of the actual installation.

Brightness. As mentioned previously, an observer sees an object because of light reflected from it. The observer interprets the intensity of the sensation experienced as brightness. The sensation of brightness usually is partly attributable to the general luminous environment, which affects the state of adaptation of the eye, and partly attributable to the intensity of light emanating from the object. The latter component is called luminance, or photometric brightness.

Luminance is the luminous power emitted, transmitted, or reflected by a surface in a given direction, per unit of area of the surface projected on a plane normal to that direction. The unit of measurement of luminance is the footlambert.

A **footlambert** (fL) of luminance in a given direction is produced by one lumen per square foot emanating from a surface in that direction. Thus, a self-luminous surface emitting 10 lm/ft^2 has a luminance of 10 fL. For surfaces that reflect or transmit light, however, luminance depends both on the illuminance of light incident on the surface and characteristics of the surface.

For a reflecting surface, luminance is determined from

$$fL = fc \times \text{reflectance} \tag{21-3}$$

where fc = footcandles of incident light. A mirror (specular reflector) may give almost 100% reflection, whereas a black surface absorbs light and therefore has negligible reflectance. Most materials have an intermediate value of reflectance.

For a transmitting surface, luminance is determined from

$$fL = fc \times \text{transmittance} \tag{21-4}$$

Clear glass (transparent material) may have a transmittance of about 90%, whereas an opaque material has no transmittance. Transmittances of other transparent materials may be about the same as that for clear glass, while transmittances of translucent materials may be 50% or less. Light incident on a surface and not reflected or transmitted is absorbed by it.

In general, visibility improves with increase in brightness of a task. Because increase in brightness is usually accomplished at increase in operating cost caused by consumption of electric power, it is neither necessary nor desirable, however, to maintain levels of illumination higher than the minimum needed for satisfactory performance of the task. For example, tests show that speed of reading and comprehension are nearly independent of illuminance above a minimum level. This level depends on several factors, such as difficulty of the task, age of observers, duration of the task, and luminance relation between task and its surroundings. The more difficult the task, the older the occupants, and the longer the task, the higher the minimum level of illumination should be.

High brightness also is useful in attracting visual attention and accenting texture. For this reason, bright lights are played on merchandise and works of art.

Contrast. This is created when the brightness of an object and its surroundings are different. The effects of contrast on visibility depend on several factors but especially on the ratio of brightness of object to that of its background. Ideally, the brightness of a task should be the same as that of its background. A 3:1 brightness ratio, however, is not objectionable; it will be noticed but usually will not attract attention. A 10:1 brightness ratio will draw attention, and a brightness ratio of 50:1 or more will accent the object and detract attention from everything else in the field of vision.

High background brightness, or low brightness ratios, may have adverse or beneficial effects on visibility. Such high contrast is undesirable when it causes glare or draws attention from the task or creates discordant light and dark patterns (visual noise). On the other hand, high contrast is advantageous when it helps the observer detect task details; for example, read fine print. High contrast makes the object viewed appear dark so that its size and silhouette can be readily discerned. But under such circumstances, if surface detail on the object must be detected, object brightness must be increased at least to the level of that of the background.

The reason for this is that the eye adapts to the brightness of the whole field of vision and visualizes objects in that field with respect to that adaptation level. If background brightness is disturbing, the observer squints to reduce the field of vision and its brightness and thus increase the brightness of the task. The need for squinting is eliminated, however, by increasing the illumination on the task.

Color Rendering. A black body is colorless. When increasing heat is applied to such a body, it eventually develops a deep red glow, then cherry red, next orange, and finally blue-white. The color of the radiated light is thus related to the temperature of the heated body. This phenomenon is the basis for a temperature scale used for the comparison of the color of light from different sources. For example, the light from an incandescent lamp, which tends to be yellowish, may be designated 2500° Kelvin (K), whereas a cool white fluorescent lamp may be designated 4500°K.

Light used for general illumination is mainly white, but white light is a combination of colors and some colors are more predominant than others in light emitted from light sources commonly used. When light other than white is desired, it may be obtained by selection of a light source rich in the desired hue or through use of a filter that produces that hue by absorbing other colors.

Color rendering is the degree to which a light source affects the apparent color of objects. **Color rendering index** is a measure of this degree relative to the perceived color of the same objects when illuminated by a reference source for specified conditions. The index actually is a measure of how closely light approximates daylight of the same color temperature. The higher the index, the better is the color rendering. The index for commonly used light sources ranges from about 20 to 99. See also Art. 21-3.

Color of light affects the color of an object, because the surface of the object absorbs light of certain frequencies and reflects light of other frequencies. An object appears red because it reflects only red light and absorbs other hues. If a light source emits light that is only blue-green, the color complementary to red, the red object will reflect no light and will therefore appear to be gray. The eye, however, can, to some degree, recognize colors of objects despite the color of the illuminant; that is, the eye can adapt to colored light. It also becomes sensitive to the colors that would have to be added to convert the illuminant to white light.

Apparent color is also affected by high levels of illumination. For example, all colors appear less bright, or washed out, under high illumination. But brightness of color also depends on the hue. Under the same illumination, light colors appear brighter than dark colors. Thus, lower levels of illumination are desirable with white-, yellow-, and red-colored (warm-colored) objects than with black-, blue-, and green-colored (cool-colored) objects. (Warm-colored objects also have the psychological effect of appearing to be closer than they actually are, whereas cool-colored objects tend to recede. In addition, cool colors create a calm and restful atmosphere, conducive to meditation, but are not flattering to skin colors or food, whereas warm colors produce opposite effects.)

See also Art. 21-10, Bibliography.

21-2. Quality of Light.

A good lighting system not only provides adequate quantities of light for safe, efficient visual performance but also good quality of light. Quality determines the visual comfort of building occupants and contributes to good visibility. In accordance with the relationships between light and sight described in Art. 21-1, quality and quantity therefore should be considered together in lighting design. For ease of presentation, however, the factors affecting these characteristics of light are discussed separately. Color rendering is treated in Art. 21-3 and quantity of light, in Art. 21-4.

The characteristics of a luminous environment that determine quality are contrast, diffusion, and color rendering. Contrast, created by shadows or by relatively bright areas in a field of vision, affects visibility, mood, comfort, and eyestrain (Art. 21-1). Diffusion, the dispersion of light in all directions, may be produced by transmission or reflection. Diffuse transmission occurs when light from a bright source passes through a material that disperses the incident light, with consequent reduction in brightness. Translucent materials or specially constructed lenses are often employed for this purpose in lighting fixtures. Diffuse reflection occurs when incident light on a surface is reflected almost uniformly in all directions by tiny projections or hollows. Such surfaces appear nearly equally bright from all viewing angles. Diffusion tends to reduce contrast and promote uniformity of lighting.

Brightness Ratios. For good quality of lighting, the degree of contrast of light and dark areas in the field of vision should be limited to provide for viewing angle changes. The reason for this is that the eye adapts to the luminance of a task after a period of time. When the eye leaves the task and encounters a field of different brightness, the eye requires an appreciable time to adapt to the new condition, during which eyestrain or visual discomfort may be experienced. To promote quick, comfortable adaptation, brightness ratios in the visual environment should be kept small.

For example, in offices, the ratio of brightness of task to that of darker immediate surroundings should not exceed 3:1, and to that of darker, more remote surroundings, 5:1. Similarly, the ratio of brightness of lighter, more remote surroundings to that of the task should be less than 5:1. (See "Office Lighting," RP 1, Illuminating Engineering Society of North America, 345 E. 47th St., New York, N.Y. 10017.)

Direct Glare. When background luminance is much greater than that of the task, glare results. It may take the form of direct glare or reflected glare. Direct glare is produced when bright light sources are included in the field of vision and cause discomfort or reduced visibility of the task. The intensity of glare depends on the brightness, size, and relative position of the light sources in the field of vision, and on the general luminance of the field of vision.

The brighter the light source, the greater will be the glare, other factors remaining substantially constant. Similarly, the larger the light source, the greater will be the glare. A small bright lamp may not be objectionable; in fact, in some cases, it may be desirable to provide sparkle and relieve the monotony of a uniformly lit space, whereas a large, bright luminaire in the field of vision would cause discomfort. In contrast, glare decreases as the distance of the light source from the line of sight increases. Also, glare decreases with increase in general luminance of the visual environment, or level of eye adaptation.

The Illuminating Engineering Society has established standard conditions for determining a criterion, called **visual comfort probability** (VCP), for rating discomfort glare. VCP indicates the percentage of observers with normal vision who will be visually comfortable in a specific environment. Tables of VCP values for various luminaires are available from their manufacturers. VCP values should be applied with caution, because they may not be applicable under conditions that depart significantly from the IES standard.

In general, direct glare should not be troublesome if all of the following conditions are satisfied for an overhead electric lighting system:

1. VCP is about 70 or more.

2. The ratio of maximum luminance of each luminaire to its average luminance is 5:1 (preferably 3:1) or less, at 45, 55, 65, 75, and 85° with respect to the vertical, when viewed lengthwise and crosswise.

3. The maximum luminance of each luminaire, when viewed lengthwise and crosswise, does not exceed the values given in Table 21-1 for the specified angles.

Reflected Glare. Also called veiling reflection because of the effect on visibility, reflected glare results when incident light from a bright light source is reflected by the task into the eyes of the observer and causes discomfort or loss of contrast. Occurrence of glare depends on the brightness of the light source, overall luminance of the task, reflectance of the task surface, and relative positions of light source, task, and observer. When a bright light source is reflected into an observer's eyes, it casts an apparent veil over the image of the task. The result is a loss of contrast that would otherwise be useful in perception of task size and silhouette details; for example, print that would be readily legible without reflected glare would become difficult to read in the presence of a veiling reflection.

Several techniques have been found useful in maintaining an adequate quantity of light while limiting loss of contrast. These include the following:

Observers, tasks, and light sources should be positioned to reduce reflected glare. If there is only a single light source, positional change should remove it from the field of vision. When daylighting is used, occupants should be faced parallel to or away from windows, rather than toward them. When overhead luminaires are used, they should be positioned on either side of and behind the occupants, instead of in the general area above and forward of them. When continuous rows of linear luminaires are used, the occupants should be positioned between the rows with the line of sight parallel to the longitudinal axes of the luminaires.

Increase in luminance of the task will offset the loss of contrast, if the added illumination is provided at nonglare angles.

Decrease in the overall brightness of light sources or the brightness at angles that cause glare also is helpful. When windows are the sources of glare, it can be reduced or eliminated by tilting blinds. When luminaires are used, brightness of the source should be kept low to minimize glare. Low-brightness sources with large areas usually cause less glare than small sources with high luminance. As an alternative, the whole ceiling can be used as an indirect light source, which reflects light directed onto it by luminaires. These fixtures should be suspended at least 18 in below the ceiling to prevent high-brightness spots from being created on it. The ceiling should be white and clean and have a matte finish to diffuse the light.

Reflected glare also can be reduced by use of luminaires that distribute light mostly at nonoffending angles. Tasks are usually observed in a downward direction at angles with the vertical of 20 to 40°. Consequently, light incident on the task at those angles may cause reflected glare. Therefore, selection of a luminaire that produces little or no light at angles less than 40° cannot cause glare. Figure 21-2a illustrates the candlepower distribution curve for a luminaire that directs little light downward. (Note that it also directs little light near the horizontal, to prevent direct glare.) The curve is called a batwing, because of its shape. As indicated in Fig. 21-2b, a high percentage of the light from this luminaire cannot produce veiling reflections.

TABLE 21-1 Recommended Maximum Luminances of Light Sources

Angle with vertical, degrees	Luminance, fL
45	2250
55	1605
65	1125
75	750
85	495

Fig. 21-2 Example of control of light distribution to reduce veiling reflections: (a) batwing candlepower distribution curve for a luminaire; (b) arrangement of fluorescent lighting fixtures with distribution shown in (a), task and observer to limit veiling reflections.

Specularity of the task can be a major cause of reflected glare. Consequently, high-gloss surfaces in the field of vision should be avoided. Often, tilting the task can reduce or eliminate veiling reflections.

Use of low-level illumination throughout a space, supplemented by local lighting on the task, offers several advantages, including adequate light with little or no glare and flexibility in positioning the local light source.

See Art. 21-10, Bibliography.

21-3. Color Rendering of Light.

Color rendering of light affects the perceived colors of objects (Art. 21-1). Hence, for good lighting, light sources with an appropriate color rendering index should be chosen to produce desired effects.

Generally, the color rendering of light should be selected to enhance color identification of an object or surface. This is especially important in cases where color coding is used for safety purposes or to facilitate execution of a task. Color enhancement is also important for stimulating human responses; for example, in a restaurant, warm-colored light would make food served appear more appetizing, whereas cool-colored light would have the opposite effect.

Sources producing white light are generally used. Because of the spectral energy distribution of the light, however, some colors predominate in the illumination. For example, for daylight, north light is bluish, whereas direct sunlight at midday is yellow-white; and light from an incandescent lamp is high in red, orange, and yellow. The color composition of the light may be correlated with a color temperature (Art. 21-1). For a specific purpose, a source with the appropriate color characteristics should be chosen. Lamp manufacturers provide information on the color temperature and color rendering index of their products.

Colored light, produced by colored light sources or by filtering of white light, is sometimes used for decorative purposes. Colored light also may be used to affect human moods or for other psychological purposes, as indicated in Art. 21-1. Care must be taken in such applications to avoid objectionable reactions to the colored light; for example, when it causes unpleasant changes in the appearance of human skin or other familiar objects.

Perceived color of objects also is affected by the level of illumination (Art. 21-1).

When brilliant color rendition is desired and high-intensity lighting is to be used, the color saturation of the objects should be high; that is, colors should be vivid. Also, a source that would enhance the colors of the objects should be chosen.

See Art. 21-10, Bibliography.

21-4. Quantity of Light.

As indicated in Art. 21-2, quantity and quality of light are actually inseparable in contributing to good lighting, although they are treated separately, for convenience of presentation, in this section. Illumination should meet the requirements of visual tasks for safe, efficient performance, aesthetic reasons, and the purpose of attracting attention. Factors that affect visual performance of a task include:

Luminance, or brightness, of the task
Luminance relation between task and surroundings
Color rendering of the light
Size of details to be detected
Contrast of the details with their background
Duration and frequency of occurrence of the task
Speed and accuracy required in performance of the task
Age of workers

The influence of these factors on visibility is described in Art. 21-1.

Dim lighting is sometimes desirable for mood effects. For merchandising, however, a pattern of brightness is depended on to capture the attention of potential customers. For task lighting, sufficient lighting must be provided on the work area if the task is to be executed without eyestrain and fatigue. Higher than the minimum required level of illumination usually improves visibility but often with greater energy consumption and increased life-cycle costs both for lighting and cooling the building. Consequently, for task lighting, illumination should be kept to the minimum necessary for maintenance of adequate quantity and quality of lighting.

For many years, single values for minimum illuminance, or level of illumination, fc, developed by the Illuminating Engineering Society of North America (IES) and listed in tables in the "IES Lighting Handbook," have been widely used in the United States. In 1979, however, the IES revised its criteria to a recommended range of target illuminances. These recommended values take into account many of the factors listed above.

To determine the target illuminance, start by ascertaining the type of activity, or illuminance category, for the space to be illuminated. (The "IES Lighting Handbook" contains a detailed table correlating specific areas or activities with categories labeled A to I. For some of these categories, the effects of veiling reflections should be evaluated, for which purpose equivalent-sphere-illumination, ESI, calculations may be used, as indicated in Art. 21-1.) The first column of Table 21-4, which is based on descriptions in "Selection of Illuminance Values for Interior Lighting Design," IES RP-15A, gives a general description of these categories.

Next, determine the appropriate weighting factors W from Table 21-2 to adjust for loss of visual acuity with age, for importance of speed and accuracy in performing tasks, and for reflections of task background. For categories A to C, for which there are no task activities, use $W = 0$ as the weight for speed and accuracy. Then, add the weighting factors. From Table 21-3, determine the visual condition number corresponding to ΣW and the category. Decrease the condition number by one if the task is of short duration or occurs infrequently. Finally, select the target illuminance, fc, from Table 21-4 corresponding to the category and visual condition.

The IES recommends that if information is available on the effect of changes in illu-

TABLE 21-2 Weighting Factors for Determining Visual Conditions*

Task and worker characteristics	Weight W
Workers' ages	
Under 40	-1
40 to 55	0
Over 55	$+1$
Importance of speed or accuracy†	
Not important	-1
Important (errors costly)	0
Critical (errors unsafe)	$+1$
Reflectance of task background	
Greater than 70%	-1
30 to 70%	0
Less than 30%	$+1$

*Based on "Selection of Illuminance Values for Interior Lighting Design," RP-15A, Illuminating Engineering Society of North America. Calculate ΣW by adding the weighting factors W for the specific environment or required performance and determine the corresponding visual condition from Table 21-4.

†Use $W = 0$ for environment categories A to C.

minance or equivalent sphere illumination, ESI, on task performance, the data may be used to determine if a variation from the design illuminance or ESI will be meaningful in terms of increased or decreased productivity, which may be used in a cost-benefit analysis.

See Art. 21-10, Bibliography.

21-5. Lighting Methods.

Interior lighting may be accomplished with natural or artificial illumination, or both. Natural illumination is provided by daylight. Artificial illumination usually is produced by consumption of electric power in various types of lamps and sometimes by burning candles or oil or gas in lamps. Usually, electric lighting for building spaces is produced by lighting devices called **luminaires,** which consist of one or more lamps, a fixture in which the lamps are held, lenses for distributing the light and parts for supplying electricity. Fixtures may be portable or permanently set in or on ceilings or walls.

To meet specific lighting objectives, the following lighting methods may be used alone or in combination:

General Lighting. This provides uniform and, often, diffuse illumination throughout a space. This type of lighting is useful for performing ordinary activities and for reducing the relative luminance of surroundings when local lighting is applied to a work area.

Local or Functional Lighting. This provides a high level of illumination on the relatively small area in which a task is to be performed, such as reading, writing, or operation of tools.

Accent Lighting. This actually is a form of local lighting, but it has the objective of creating focal points for observers, to emphasize objects on display.

Decorative Lighting. This employs color or patterns of light and shadow to attract attention, hold interest, produce visual excitement or a restful atmosphere, or create aesthetic effects.

TABLE 21-3 Visual Condition Number for Determining Recommended Illuminances

Category	ΣW*						
	3	2	1	0	-1	-2	-3
A to C		4	4	3	2	2	
D to I	4	4	3	3	3	2	2

* ΣW = sum of weighting factors given in Table 21-3 for a specific environment or required performance.

TABLE 21-4 Recommended Illuminances for Interior Lighting, fc[a]

Category of environment or required performance	Visual condition[b]			
	1. Short exposure[c]	2. Moderate	3. Ordinary	4. Severe
A. Public areas with dark surroundings[d]		2	3	5
B. Simple orientation for short visits[d]		5	7.5	10
C. Working spaces for infrequent tasks[d]		10	15	20
D. Tasks with high contrast or large size[e,f]		20	30	50
E. Tasks with medium contrast or small size[e,g]	30	50	75	100
F. Tasks with low contrast or very small size[e,h]	75	100	150	200
G. Tasks with low contrast, very small size and long duration[i,j]		200	300	500
H. Very prolonged and exacting tasks[i,k]		500	750	1000
I. Tasks with extremely low contrast and small size[i,l]		1000	1500	2000

[a]Based on "Selection of Illuminance Values for Interior Lighting Design," RP-15A, Illuminating Engineering Society of North America.
[b]Visual condition depends on workers' ages, importance of speed and accuracy in performance of task, and reflectance of task background. For tasks that occur infrequently or are of short duration and if conditions 2, 3, or 4 would be applicable, use conditions 1, 2, or 3, respectively. Where no value is given for condition 1, use condition 2. See text and Tables 21-2 and 21-3.
[c]Based on British IES "Code for Interior Lighting," 1977.
[d]General lighting throughout the building space.
[e]Illuminance on task.
[f]For example, reading printed material, typed originals, handwriting in ink or good xerography; rough bench or machine work; ordinary inspection; or rough assembly.
[g]For example, reading medium pencil handwriting, poorly printed or reproduced material; medium bench or machine work; difficult inspection, or medium assembly.
[h]For example, reading handwriting in hard pencil on poor-quality paper or very poor reproductions, or highly difficult inspections.
[i]Illuminance on task, obtained by combining general and local (supplementary) lighting.
[j]For example, fine assembly, unusually difficult inspection, or fine bench or machine work.
[k]For example, the most difficult inspection, extra-fine bench or machine work, or extra-fine assembly.
[l]For example, surgical procedures.

Illumination may be classified as indirect, semiindirect, diffuse or direct-indirect, semidirect, or direct.

For indirect lighting, about 90 to 100% of the illumination provided in a space is directed at the ceiling and upper walls, and nearly all of the light reaches the task by reflection from them. The resulting illumination is, therefore, diffuse and uniform, with little or no glare.

For semiindirect lighting, about 60 to 90% of the illumination is directed at the ceiling and upper walls; the remaining percentage in generally downward directions. When overhead luminaires are used, the downward components should be dispersed by passage through a diffusing or diffracting lens to reduce direct glare. The resultant illumination on a task is diffuse and nearly glarefree.

General diffuse or direct-indirect lighting is designed to provide nearly equal distribution of light upward and downward. General-diffuse luminaires enclose the light source in a translucent material to diffuse the light and produce light in all directions. Direct-indirect luminaires give little light near the horizontal. Quality of the resulting illumination from either type depends on the type of task and the layout of the luminaires.

For semidirect lighting, about 60 to 90% of the illumination is directed downward, the remaining percentage upward. Depending on the eye adaptation level, as determined by overall room luminance, the upward component may reduce glare. Diffuseness of the lighting depends on reflectance of room enclosures and furnishings.

For direct lighting, almost all the illumination is directed downward. If such luminaires are spread out, reflections from room enclosures and furnishings may diffuse the light sufficiently that it can be used for general lighting, for example, in large offices. A concentrated layout of these luminaires is suitable for accent, decorative, or local lighting. Because direct lighting provides little illumination on vertical surfaces, provision of supplementary perimeter lighting often is desirable.

Lighting Distribution. Luminaires are designed for a specific type of lamp to distribute light in a way that will meet design objectives. For this purpose, luminaires incorporate various shapes of reflectors and various types of lenses (Fig. 21-3). Also, size and shape of openings through which light is emitted is controlled.

A luminaire may provide symmetrically or asymmetrically distributed light. With symmetrical distribution (Fig. 21-1), the level of illumination, fc, on a work plane is nearly the same at the same angle with the vertical and at equal distances from the light source. With asymmetrical distribution (Fig. 21-3c), the luminaire concentrates light in a specific direction. Symmetrical distribution is appropriate for general lighting. Asymmetrical distribution is advantageous for accent lighting.

See Art. 21-10, Bibliography.

21-6. Daylight.

Use of natural light has the advantages, compared with artificial illumination, of not consuming fuel and not having associated operating costs. But daylight has the disadvantages of being dependent on the availability of windows and on the absence of light-blocking obstructions outside the windows, of not being available between sunset and sunrise, and of providing weak light on cloudy days and around twilight and dawn. When lighting is needed within a building at those times, it must be provided by artificial illumination. Also, for parts of rooms at large distances from windows, where adequate daylight does not reach, artificial illumination is needed to supplement the daylight. When supplementary lighting is required, initial, maintenance, and replacement costs of lamps and fixtures are not saved by use of daylight, although costs of power for lighting can be reduced by turning off lamps not needed when daylight is available. In addition, design

Fig. 21-3 Examples of recessed fixtures (*a*) for direct widespread lighting, (*b*) for direct, narrow-beam lighting, (*c*) for asymmetrical direct lighting, and (*d*) for direct diffuse lighting.

for daylighting should be carefully executed so as not to introduce undesirable effects, for example, glare, intensive sunlight, or excessive heat gain.

Elements can be incorporated in building construction to control daylight to some extent to provide good lighting within short distances from windows. For example, to prevent glare, windows may be shielded, by blinds or by outside overhangs, against direct sunlight. To reduce heat gain, windows may be glazed with reflecting, insulating, or heat-absorbing panes. Ceiling, floor, and walls with high reflectance should be used to diffuse light and reflect it into all parts of rooms. To illuminate large rooms, daylight should be admitted through more than one wall and through skylights or other roof openings, if possible. In rooms that extend a long distance from windows, the tops of the windows should be placed as close to the ceiling as possible, to permit daylight to penetrate to the far end of the room.

Design procedures for daylighting are presented in "Recommended Practice for Daylighting," RP-5, Illuminating Engineering Society of North America.

21-7. Characteristics of Lamps.

Selection of the most suitable lamp consistent with design objectives is critical to performance and cost of a lighting system. This decision should be carefully made before a fixture for the lamp is selected. Fixtures are designed for specific lamps.

Lamps are constructed to operate at a specific voltage and wattage, or power consumption. In general, the higher the wattage rating of a specific type of lamp, the greater will be its **efficacy,** or lumen output per watt.

Greatest economy will be secured for a lighting installation through use of a lamp with the highest lumen output per watt with good quality of illumination. In addition to lumen output, however, color rendering and other characteristics, such as lighting distribution, should also be considered in lamp selection. Information on these characteristics can be obtained from lamp manufacturers. Latest data should be requested, because characteristics affecting lamp performance are changed periodically. The following information usually is useful:

Lamp life, given as the probable number of hours of operation before failure.

Lamp efficacy, measured by the lamp output in lumens per watt of power consumed.

Lamp lumen depreciation, as indicated by tests. Curves are plotted from data to show the gradual decrease in light output with length of time of lamp operation. The decrease occurs because of both aging and dirt accumulation. The latter can be corrected with good maintenance, but the possible effects nevertheless should be considered in lighting design.

Lamp warm-up time, which is significant for some fluorescent lamps and all high-intensity-discharge lamps, for which there is a delay before full light output develops.

Lamp restart time, or time that it takes some lamps to relight after they have been extinguished momentarily. The lamps may go out because of low voltage or power interruption.

Color rendering index and color acceptability, which are, respectively, measures of the degree to which illumination affects the perceived color of objects and the human reaction to perceived colors (see Art. 21-3).

Voltages and frequency of current at which lamps are designed to operate. In most buildings in the United States, electricity is distributed for light and power as an alternating current at 60 Hz. Lamp output generally increases at higher frequencies but capital investment for the purpose is higher. Voltage often is about 120, but sometimes, especially for industrial and large commercial buildings, voltages of 208, 240, 277, or 480 are used, because of lower transmission losses and more efficient operation of electrical equipment. Direct current from batteries often is used for emergency lighting, when the prime ac source fails.

Noise, which is significant for some types of lamp applications, such as fluorescent and high-intensity-discharge lamps. These depend for operation on a device, called a ballast, which may hum when the lamp operates. Whether the hum will be annoying under ordinary circumstances depends on the ambient noise level in the room. If the ambient level is high enough, it will mask the hum. Before a combination of lamp, ballast, and fixture is selected, ballast noise rating should be obtained from the luminaire manufacturer.

Ambient temperature, or temperature around a lamp when it is operating, which may affect lamp life, lumen output, and color rendering. If the ambient temperature exceeds the rated maximum temperature for the luminaire, the life of incandescent lamps may be considerably reduced. Consequently, lamps rated at a wattage greater than that recommended by the fixture manufacturer should not be used. Also, provision should be made for dissipation of the heat produced by lamps. Low ambient temperature slows starting of fluorescent and high-intensity-discharge lamps. Low temperature also reduces lumen output and changes the color of fluorescent lamps.

Maintenance of Lamp Output. The efficiency of a lighting system decreases with time because of dirt accumulation, decrease in lumen output as lamps age, lamp failures, and deteriorating lighting fixtures. Depending on type of fixtures, cleanliness of the environment, and time between cleanings of lamps and fixtures, lumen losses due to dirt may range from 8 to 10% in a clean environment to more than 50% under severe conditions. Also, the longer lamps operate, the dimmer they become; for example, a fluorescent lamp at the end of its life will yield only 80 to 85% of its initial lumen output. And when one or more lamps fail and are not replaced, the space being illuminated may suffer a substantial loss in light. Furthermore, in the case of lights operating with ballasts, the lamps, before burning out, overload the ballast and may cause it to fail. Consequently, a poorly maintained lighting installation does not provide the illumination for which it was designed and wastes money on the power consumed.

The design illumination level may be maintained by periodic cleaning and prompt replacement of aged or failed lamps. Relamping may be carried out by the spot or group methods. Spot relamping, or replacement of lamps one by one as they burn out, is more inefficient in use of labor and more costly. Group relamping calls for scheduled replacement of lamps at intervals determined from calculations based on expected lamp life and variation of lumen output with time. This method reduces labor costs, causes fewer work interruptions, maintains higher levels of illumination with no increase in cost of power, prevents the appearance of the lighting system from deteriorating and reduces the possibility of damage to auxiliary equipment, such as ballasts, near the end of lamp life.

Because of the decrease in lamp output with time, in design calculations initial lamp output should be multiplied by a lamp lumen depreciation factor to correct for the effects of aging. This product equals the output after a period of time, usually the interval between group relampings. In addition, the output should be multiplied by a dirt depreciation factor. Both factors are less than unity. The two factors may be combined into a single maintenance factor M.

$$M = \text{LLD} \times \text{LDD} \tag{21-5}$$

where LLD = lamp lumen depreciation factor
LDD = dirt depreciation factor

After a period of operation, lamp output then is given by

$$L = L_i M \tag{21-6}$$

where L_i = initial lamp output, lm

Lamp Control. Incoming power for a lamp normally is by a local switch in the power circuit. The switch turns the lamps on or off by closing or opening the circuit.

An alternative method may be used that is more economical when the lamps are distant from the switch and that reduces the possibility of personal injury or short-circuit damage at the switch. In this method, the main power circuit is opened and closed by a relay located near the lamps. The relay, in turn, may be activated by low-voltage power controlled by a remote switch. Control-system voltages usually range between 6 and 24 V, obtained by stepping down the normal distribution voltage with transformers.

For control of light output from a luminaire, the switch may be replaced with a dimmer. For incandescent lamps, this device can vary the voltage across the lamps from zero to the rated value and thus can be used to adjust the level of illumination. For fluorescent and high-intensity-discharge lamps, the dimmer is coordinated with the lamp ballast.

Types of Lamps. Lamps that are commonly used may be generally classified as incandescent, fluorescent, or high-intensity-discharge (HID). HID lamps include mercury-vapor, metal-halide, low-pressure sodium, and high-pressure sodium lamps. See Tables 21-5 and 21-6.

Incandescent Lamps. These lamps generate light by heating thin tungsten wires until they glow. The filaments are enclosed in a sealed glass bulb from which air is evacuated or that is filled with an inert gas, to prevent the heated tungsten from evaporating. In a tungsten-halogen incandescent lamp, for prolonged life, the filler gas contains halogens (iodine, chlorine, fluorine, and bromine), which restores to the filaments any metal that may evaporate.

Incandescent lamps are available in a variety of shapes. They also come with a variety of bases, making it necessary to ensure that selected fixtures provide sockets that can accommodate the desired lamps.

Incandescent lamps produce light mainly in the yellow to red portion of the spectrum. Color depends on the wattage at which the lamp is operated. Generally, the higher the

TABLE 21-5 Characteristics of Lamps Often Used for General Lighting

Lamp characteristic	Incandescent	Tungsten-halogen incandescent	Fluorescent	Clear mercury	Clear metal halide	Clear high-pressure sodium	Clear low-pressure sodium
Efficacy, lm/W	15–25	6–23	25–84	30–63	68–125	77–140	137–183
Initial lumens	40–33,600	40–33,600	96–15,000	1,200–63,000	12,000–155,000	5,400–140,000	4,800–33,000
Lumen maintenance, %	80–90	75–97	75–91	70–86	73–83	90–92	75–90
Wattage range	5–1,500	6–1,500	4–215	40–1,000	175–1,500	70–1,000	35–180
Life, hr	750–1,000	750–8,300	9,000–20,000	16,000–24,000	1,500–15,000	20,000–24,000	18,000
Color temperature, K	2,400–3,400	2,800–3,400	2,700–6,500	3,300–5,900	3,200–4,700	2,100	1,780
Color rendering index	89–92	95–99	55–95	22–52	65–70	21	0
Color acceptability	Good	Good	Good	Fair	Good	Fair	Poor
Light control	Excellent	Excellent	Poor	Fair	Fair to good	Good	Poor
Initial cost per lamp	Low	Low	Moderate	Moderate	High	High	Moderate
Energy cost	High	High	Moderate	Moderate	Low	Low	Low

Courtesy of Sylvania Lighting.

TABLE 21-6 Comparison of Lamps Often Used for General Lighting*

Type of lamp	Lamp characteristics																								
	Color rendering			Initial efficacy, lm/W			Lumen maintenance (mean lm)			Rated average life, hr			Degree of light control			Input power required for equal light				System operating cost for equal light			Initial equipment cost for equal light			Total owning and operating cost		
	Very important	Important	Unimportant	Highest (80 up)	Medium (50–80)	Lowest (15–50)	Highest (85% up)	Medium (75–85%)	Fair (65–75%)	Shortest (5,000 or less)	Intermediate (5,000–15,000)	Longest (15,000–25,000)	Highest	Intermediate	Lowest	Highest	High	Intermediate	Lowest	Highest	Intermediate	Lowest	Highest	Intermediate	Lowest	Highest	Intermediate	Lowest
	Relative rating of lamps																											
Incandescent	•					•		•		•			•			•				•					•	•		
Tungsten-halogen	•					•	•			•			•			•				•					•	•		
Fluorescent	•				•			•				•			•			•			•			•			•	
Clear mercury		•			•			•				•	•				•				•			•			•	
Coated mercury		•			•			•				•		•			•				•			•			•	
Clear metal halide	•			•					•		•		•					•				•		•				•
Coated metal halide			•	•					•		•			•				•				•		•				•
Clear high-pres. sodium			•	•					•			•	•						•			•	•					•
Coated high-pres. sodium			•	•					•			•		•					•			•		•				•

*Dot indicates that the light source exhibits the listed characteristics. Courtesy of Sylvania Lighting.

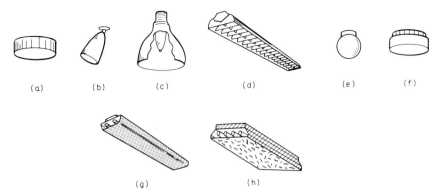

Fig. 21-4 Examples of ceiling-mounted fixtures: (a) opaque-side drum for direct diffuse lighting; (b) spot lighting with an incandescent lamp; (c) direct, widespread lighting with an HID lamp; (d) fluorescent fixture for direct diffuse lighting; (e) globe fixture with incandescent lamp for direct diffuse lighting; (f) small diffusing drum; (g) fluorescent fixture for semidirect diffuse lighting; (h) fluorescent fixture with diffusing lens for direct lighting.

wattage, the whiter is the color of light produced. A reduction in wattage or voltage results in a yellower light. See Tables 21-5 and 21-6 for other characteristics of incandescent lamps.

Often, the glass bulbs of these lamps are treated to obtain special effects. Usually, the effect desired is diffusion of the emitted light, to soften glare. For the purpose, the glass may be frosted, or etched, or silica coated (white bulb). Light diffusion and other effects can also be achieved with control devices incorporated in the fixtures, such as lenses or louvers (Figs. 21-3 to 21-6). Also, lamps may be treated with coatings or filters to produce any of a variety of colors.

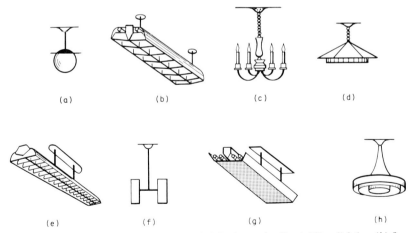

Fig. 21-5 Examples of pendent fixtures: (a) globe fixture for direct diffuse lighting; (b) fluorescent fixture for general diffuse lighting; (c) exposed-lamp fixture for direct lighting; (d) direct downlight fixture; (e) fluorescent fixture for semidirect lighting; (f) fixture for direct-indirect lighting; (g) fluorescent fixture for semiindirect lighting; (h) fixture with high-intensity-discharge lamp for indirect lighting.

Fig. 21-6 Examples of wall-mounted fixtures: (*a*) small diffuser type; (*b*) linear type, for example, four-lamp incandescent or a fluorescent lamp; (*c*) bullet-type directional fixture, for accent lighting; (*d*) fixture for directional night lighting; (*e*) exposed-lamp fixture for direct lighting.

Reflectorized incandescent lamps are made, with standard or special shapes, with a reflective aluminized or silver coating applied directly to part of the inside bulb surface. Such lamps are widely used for spot or flood lighting. Type PAR lamp has a parabolic shape to focus the light beam. Types EAR and ER lamps have elliptical shapes that cause the light beam to concentrate near the front of the lens, then to broaden into the desired pattern, yielding more usable light than other types of lamps, with little glare.

Fluorescent Lamps. These lamps are sealed glass tubes coated on the inside surface with phosphors, chemicals that glow when bombarded by ultraviolet light. The tubes are filled with an inert gas, such as argon, and low-pressure mercury vapor. Passage of an electric arc through the mercury vapor causes it to emit the ultraviolet rays that activate the phosphors to radiate visible light. The electric arc is started and maintained by cathodes at the ends of the glass tubes.

The high voltage needed to form the arc is provided initially by a device called a **ballast.** After the arc has been formed, the ballast limits the current in the arc to that needed to maintain it. Ballasts also may be designed to decrease the stroboscopic effect of the lamp output caused by the ac power supply and to keep the variation in current nearly in phase with the variation in voltage, thus maintaining a high power factor.

Fluorescent lamps generally are available as linear, bent U, or circular tubes, and fixtures are designed to be compatible with the selected shape.

Fluorescent lamps may be classified as preheat, rapid start, or instant start. They differ in the method used to decrease the delay in starting after a switch has been thrown to close the electrical circuit. The preheat type requires a separate starter, which allows current to flow for several seconds through the cathodes, to preheat them. For a rapid-start lamp, the cathodes are electrically preheated much more rapidly without a starter. For an instant-start lamp, high voltage from a transformer forms the arc, without the necessity of preheating the cathodes.

Instant-start lamps may be of the hot-cathode or cold-cathode type, depending on cathode shape and voltage used. Efficacy of cold-cathode lamps is lower than that of hot-cathode lamps. Both types are more expensive than rapid-start lamps and are less efficient in lumen output. Instant-start lamps, however, come in sizes that are not available for rapid-start lamps and can operate at currents that are not feasible for rapid-start lamps. Also, instant-start lamps can start at lower temperatures, for instance, below 50°F.

The life of most types of fluorescent lamps is adversely affected by the number of lamp starts. Cold-cathode lamps, however, have a long life, which is not greatly affected by the number of starts.

Fluorescent lamps last longer and have a higher efficacy than incandescent lamps. (See Tables 21-5 and 21-6.) Hence, fluorescent lamps cost less to operate. Initial cost of fixtures, however, may be higher. Fluorescent lamps require larger fixtures, because of tube length, and special equipment, such as ballasts and transformers (Figs. 21-3 to 21-6). Also associated with such lamps is ballast hum and possible interference with radio reception. Pattern control of light is better with incandescent lamps, but lamp brightness is low with fluorescent lamps, so that there is less likelihood of glare, even when the lamps are not shielded.

Color rendering of light emitted by a fluorescent lamp depends on the phosphors used in the tube. The best color rendering for general use may be obtained with the deluxe cool white (CWX) type. Check with manufacturers for color characteristics of lamps currently available, because new lamps designed with color rendering appropriate for specific purposes are periodically introduced.

High-Intensity-Discharge (HID) Lamps. These lamps generate light by passage of an electric arc through a metallic vapor, which is contained in a sealed glass or ceramic tube. The lamps operate at pressures and electrical current densities sufficient to produce desired quantities of light within the arc. Three types of HID lamps are generally available: mercury vapor, metal-halide, and high-pressure sodium. Major differences between them include the material and type of construction used for the tube and the type of metallic vapor. In performance, the lamps differ in efficacy, starting characteristics, color rendering, lumen depreciation, price, and life. (See Tables 21-5 and 21-6.) HID lamps are available with lumen outputs considerably greater than those of the highest-wattage fluorescent lamps available. HID lamps require ballasts that function like those for fluorescent lamps and that should be coordinated with the type and size of lamp for proper operation.

Each time an HID lamp is energized from a cold start, the lamp produces a dim glow initially and there is a time interval called *warm-up time* until the lamp attains its full lumen output. Warm-up time for metal-halide and sodium vapor lamps may range from 3 to 5 min and for mercury vapor lamps from 5 to 7 min. When the power to the lamp is interrupted, even momentarily, the lamp extinguishes immediately; and even if the power is restored within a short time, while the lamp is still hot, there is a delay, until the lamp cools to provide a condition that will permit the arc to restrike. Sodium vapor lamps have the fastest restart time, 1 min, compared with 3 to 6 min for mercury vapor lamps and as much as 10 min for metal-halide lamps. Because of this condition, it is desirable to employ supplemental lighting, usually incandescent, to provide minimal illumination during these intervals.

The color of clear mercury lamps tends to be bluish. This type of lamp, however, also is available coated with phosphors that improve color rendering. The color of clear metal-halide lamps is stark white, with subtle tints ranging from pink to green. These lamps also may be coated for color correction. Light from clear sodium lamps, however, is yellowish. It strengthens yellow, green, and orange, but grays red and blue and turns white skin complexions yellow. Sodium lamps also may be coated to improve color rendering, but as color rendering is improved, efficacy decreases somewhat.

With respect to annual cost of light, high-pressure sodium lamps with ceramic (aluminum-oxide) arc tubes, with relatively small size, high efficacy, long life, and excellent lumen maintenance, appear to be the most economical HID type. Some variations of this type of lamp also offer improved color rendering.

HID lamps require special fixtures and auxiliaries. Some of these fixtures will accept replacement HID lamps of any of the three types in specific wattages. Others will accept only one type.

See Art. 21-10, Bibliography.

21-8. Characteristics of Lighting Fixtures.

A lighting fixture is that component of a luminaire that holds the lamps, serves as a protective enclosure, or housing, delivers electric power to the lamps, and incorporates devices for control of emitted light. The housing contains lampholders and usually also reflective inside surfaces shaped to direct light out of the fixture in controlled patterns. In addition, a fixture also incorporates means of venting heated air and houses additional light-control equipment, such as diffusers, refractors, shielding, and baffles. The power component consists of wiring and auxiliary equipment, as needed, such as starters, ballasts, transformers, and capacitors. The light-control devices include louvers, lenses, and diffusers. Fixture manufacturers provide information on construction, photometric performance, electrical and acoustical characteristics, installation, and maintenance of their products.

Some luminaires are sealed to keep out dust. Some are filtered and vented to dissipate heat and prevent accumulation of dust. Also, some are designed as part of the building air-conditioning system, which removes heat from the lamps before it enters occupied spaces. In some cases, this heat is used to warm spaces in the building that require heating.

Safety Requirements. Construction and wiring of fixtures should conform with local building codes and the National Electrical Code (NEC), recommendations of the National Electrical Manufacturers Association (NEMA) and "Standard for Lighting Fixtures," Underwriters Laboratories, Inc.

The NEC requires that fixtures to be installed in damp or wet locations or in hazardous areas containing explosive liquids, vapors, or dusts be approved by Underwriters Laboratories for the specific application. Auxiliary equipment for fluorescent and HID lamps should be enclosed in incombustible cases and treated as sources of heat.

The NEC specifies that fixtures that weigh more than 6 lb or are larger than 16 in in any dimension not be supported by the screw shell of a lampholder. The code permits fixtures weighing 50 lb or less to be supported by an outlet box or fitting capable of carrying the load. Fixtures also may be supported by a framing member of a suspended ceiling if that member is securely attached directly to structural members at appropriately safe intervals or indirectly via other adequately supported ceiling framing members. Pendent fixtures should be supported independently of conductors attached to the lampholders.

The NEC also requires that fixtures set flush with or recessed in ceilings or walls be so constructed and installed that adjacent combustible material will not be exposed to temperatures exceeding 90°C. (Thermal insulation should not be installed within 3 in of a recessed fixture.) Fire-resistant construction, however, may be exposed to temperatures as high as 150°C if the fixture is approved for such service. Screw-shell-type lampholders should be made of porcelain.

Lenses may be made of glass or plastic. In the latter case, the material should be incombustible and a low-smoke-density type. It should be stable in color and strength. The increase in yellowness after 1000 hr of testing in an Atlas FDA-R Fade-Ometer in accordance with ASTM G23 should not exceed 3 IES-NEMA-SPI units. Acrylics are widely used.

Considerations in Fixture Selection. Because fixtures are designed for specific types of lamps and for specific voltage and wattage ratings of the lamps, a prime consideration in choosing a fixture is its compatibility with lamps to be used. Other factors to consider include:

Conformance with the chosen lighting method (see Art. 21-5)

Degree to which a fixture assists in meeting objectives for quantity and quality of light through emission and distribution of light

Luminous efficiency of a fixture, the ratio of lumens output by the fixture to lumens produced by the lamps

Esthetics—in particular, coordination of size and shape of fixtures with room dimensions so that fixtures are not overly conspicuous

Durability
Ease of installation and maintenance

Light distribution from fixtures, to summarize, may be accomplished by means of transmission, reflection, refraction, absorption, and diffusion. Reflectors play an important role. Their reflectance, consequently, should be high—at least 85%. The shape of a reflector—spherical, parabolic, elliptical, hyperbolic—should be selected to meet design objectives; for example, to spot or spread light in a building space or to spread light over a fixture lens that controls light distribution. (The need for a curved reflector, which affects the size of the fixture, can be avoided by use of a Fresnel lens, which performs the same function as a reflector. With this type of lens, therefore, a smaller fixture is possible.) Light control also is affected by shielding, baffles, and louvers that are positioned on fixtures to prevent light from being emitted in undesirable directions.

A wide range of light control can be achieved with lenses. Flat or contoured lenses may be used to diffuse, diffract, polarize, or color light, as required. Lenses composed of prisms, cones, or spherical shapes may serve as refractors, producing uniform dispersion of light or concentration in specific directions.

Types of Installations. Luminaires may be classified in accordance with type and location of mountings, as well as with type of lighting distribution: flush or recessed (Fig. 21-3), ceiling mounted (Fig. 21-4), pendent (Fig. 21-5), wall mounted (Fig. 21-6) or structural.

Structural lighting is the term applied to lighting fixtures built into the structure of the building or built to use structural elements, such as the spaces between joists, as parts of fixtures. Structural lighting offers the advantage of a lighting system conforming closely to the architecture or interior decoration of a room. Some types of structural lighting are widely used in residences and executive offices. For the purposes of accent or decorative lighting, for example, cornices, valences, coves, or brackets are built on walls to conceal fluorescent lamps. For task lighting, fixtures may be built into soffits or canopies. For general lighting, large, low-brightness, luminous panels may be set flush with or recessed in the ceiling.

Lighting objectives can be partly or completely met with portable fixtures in some types of building occupancies. For the purpose, a wide variety of table and floor lamps are commercially available. Because the light sources in such fixtures are usually mounted at a relatively low height above the floor, care should be taken to prevent glare, by appropriate placement of fixtures and by selection of suitable lamp shades.

Number and Arrangement of Luminaires. With the type of lamp and fixture and the required level of illumination known, the number of luminaires needed to produce that lighting may be calculated and an appropriate arrangement selected. The **lumen method of calculation,** which yields the average illumination in a space, is generally used for this purpose.

The method is based on the definition of footcandle (Art. 21-1), in accordance with which the level of illumination on a horizontal work plane is given by

$$\text{fc} = \frac{\text{lumens output}}{\text{area of work plane, ft}^2} \quad (21\text{-}7)$$

Lamp manufacturers provide data on initial lumen output of lamps, but these values cannot be substituted directly in Eq. (21-7), because of light losses in fixtures and building spaces and the effects of reflection.

To adjust for the effects of fixture efficiency, distribution of light by fixtures, room proportions and surface reflectances, and mounting height and spacing of fixtures, the design lamp output, lm, is multiplied by a factor CU, called **coefficient of utilization,** to obtain lumens output for Eq. (21-7). (CU is the ratio of lumens striking the horizontal work plane to the total lumens emitted by the lamps. It may be obtained from tables available from fixture manufacturers.) Thus,

$$\text{fc} = \frac{\text{design lamp output, lm} \times \text{CU}}{\text{area of work plane, ft}^2} \quad (21\text{-}8)$$

To adjust for decreasing illumination with time, initial lamp output, lm, is multiplied by a light loss factor LLF to obtain the design lamp output. (LLF is the ratio of the lowest level of illumination on the work plane just before corrective action is taken, such as relamping and cleaning, to the initial level of illumination, produced by lamps operating at rated initial lumens. Maintenance of lighting installations thus is taken into account in LLF.) Substitution of this product in Eq. (21-8) yields

$$\text{fc} = \frac{\text{initial lamp output, lm} \times \text{CU} \times \text{LLF}}{\text{area of work plane, ft}^2} \quad (21\text{-}9)$$

Factors contributing to LLF include ballast performance, voltage to luminaires, luminaire reflectance and transmission changes, lamp outages, luminaire ambient temperature, provisions for removal of heat from fixtures, lamp lumen depreciation with use, and luminaire dirt depreciation.

The initial lamp output equals the product of the number of lamps by the rated initial lumens per lamp. Substitution in Eq. (21-9) and rearrangement of terms gives

$$\text{Number of lamps required} = \frac{\text{fc} \times \text{area}}{\text{initial lm per lamp} \times \text{CU} \times \text{LLF}} \quad (21\text{-}10)$$

$$\text{Number of luminaires required} = \frac{\text{number of lamps required}}{\text{lamps per luminaire}} \quad (21\text{-}11)$$

Layout of luminaires depends on architectural and decorative considerations, size of space, size and shape of fixtures, mounting height, and the effect of layout on quality of lighting. Different types of fixtures may be used in a space; for example, one type to provide general lighting, other types to provide supplementary local lighting, and still other types to produce accent or decorative lighting.

Details of the lumen method of calculation are given in books on lighting (see Art. 21-10, Bibliography).

Manufacturers list the maximum permissible spacing for each type of luminaire in photometric reports on their products. This spacing depends on the mounting height, relative to the work plane, for direct, semidirect, and general-diffuse luminaires, and relative to the ceiling height for indirect and semidirect luminaires. Spacings closer than the maximum improve uniformity of lighting and reduce shadows. Perimeter areas, however, require much closer spacing, depending on location of tasks and on the reflectance of the walls; generally, the distance between luminaires and the wall should not exceed half the distance between luminaires, and in some cases, supplementary lighting may have to be added.

21-9. Systems Design of Lighting.

The objectives of and constraints on lighting systems and the interrelationship of lighting and other building systems are summarized at the start of this section and treated in this article. To design a lighting system for specific conditions, it is first necessary for the designer to determine the nature of and lighting requirements for the activities to be carried out in every space in the building. Also, the designer should cooperate with architects, interior designers, and structural, electrical, and HVAC engineers, as well as with the owner's representatives, to establish conditions for optimization of the overall building system. For example, where feasible, reflectances for ceiling, walls, and floor for each space may be selected for high lighting efficiency and visual comfort. Also, HVAC may be designed to remove and utilize heat from luminaires.

With tasks known, the designer should establish, for every space, criteria for illumination levels for task performance, safety, and visual comfort and also determine luminance ratios and light-loss factors. (For establishment of the light-loss factors, maintenance of the lighting system should be planned with the owner's representatives.) Based on the criteria and the lighting objectives, the designer can then decide how best to use daylighting and artificial lighting and select lamps and fixtures, luminaire mounting and layout, and lighting controls, such as switches and dimming. Because quality, color rendering and quantity of light are interrelated, they should be properly balanced. This should be checked in an appraisal of the lighting system, which also should include comparisons of alternatives and studies of life-cycle costs and energy consumption. The analysis should compare alternatives not only for the lighting system but also for the other building components that affect or are affected by the lighting system.

Value analysis should examine illumination levels critically. A quantity of light that is sufficient for functional purposes is essential; but more light does not necessarily result in better lighting, higher productivity, or greater safety. Furthermore, higher illumination levels are undesirable, because they increase costs of lamps and fixtures, of lighting operation, of the electrical installation, and of HVAC installation and operation. Consequently, lighting should be provided, at the levels necessary, for visual tasks, with appropriate lower levels elsewhere, for example, for circulation areas, corridors, and storage spaces. Provision preferably should be made, however, for relocation or alteration of lighting equipment where changes in use of space can be expected.

The various types of lamps differ in characteristics important in design, such as color rendering, life, size, and efficacy. For each application, the most efficient type of lamp appropriate to it should be chosen. Consequently, prime consideration should be given to fluorescent and high-intensity-discharge lamps, which are highly efficient. Also, consideration should be given to high-wattage lamps of the type chosen, because the higher the wattage rating, the higher the lumen output per watt. Furthermore, in selection of a lamp, much more weight should be placed on life-cycle costs than on initial purchase price. Cost of power consumed by a lamp during its life may be 30 or more times the lamp cost. Consequently, use of a more efficient, though more expensive, lamp can save money because of the reduction in power consumption.

Similar consideration should also be given to luminaire selection. Efficient luminaires can produce more light on a task with less power consumption. Additional consideration, however, should be given to ease of cleaning and relamping, to prevention of direct glare and veiling reflections, and to removal of heat from the luminaires.

Control of lighting should be flexible. Conveniently located, separate switches or dimmers should be installed for areas with different types of activities. It should be easy to extinguish lights for areas that are not occupied and to maintain minimal emergency lighting, for safety.

Where feasible, daylighting should be used, supplemented, as needed, with artificial lighting. Light from windows can be reflected deeply into rooms, with venetian blinds or glass block. Glare and solar heat can be limited with blinds, shades, screens, or low-transmission glass. Provision should be made for decreasing or extinguishing supplementary artificial lighting when there is adequate daylight. Consideration should be given to use of photoelectric-cell sensors with dimmers for control of the artificial lighting.

Such sensors are also useful in maintaining preset illumination levels when only electric lighting is used, with considerable savings in operating costs. These savings are made possible because initial illumination levels are much higher than design illumination levels to allow for light losses with time (Art. 21-8). The light-control system can be set to decrease power input below lamp wattage rating initially, thus decreasing initial lumen output to maintain the preset illumination level. As the lumen output deteriorates with time, the light-control system increases power input to the lamps and increases lumen output. Thus, excessive illumination and power consumption are avoided, and required illumination levels are maintained between relampings. An alternative to use of dimmers for light control is switching on or off of fixtures, individually or as a group, under computer control, with commands issued by building operators, clocks, or photoelectric sensors.

Maximum use of daylighting and other energy conservation measures are often essential, not only to meet the owner's construction budget but also to satisfy the energy budget, or limitations on power consumption, set by building codes or state or Federal agencies.

21-10. Bibliography.

H. E. Bovay, Jr., "Handbook of Mechanical and Electrical Systems for Buildings," McGraw-Hill Book Company, New York.

D. G. Fink and H. W. Beaty, "Standard Handbook for Electrical Engineers," McGraw-Hill Book Company, New York.

J. E. Flynn and A. W. Segil, "Architectural Interior Systems: Lighting, Air Conditioning, Acoustics," Van Nostrand Reinhold Company, New York.

"Lighting Handbook," Illuminating Engineering Society of North America, 345 E. 47 St., New York, N.Y. 10017.

W. J. McGuinness, B. Stein, and J. S. Reynolds, "Mechanical and Electrical Equipment for Buildings," John Wiley & Sons, Inc., New York.

Section 22

Vertical Circulation

Frederick S. Merritt
Consulting Engineer,
West Palm Beach, Fla.

Circulation, as usually applied in architecture, is the movement of people and goods between interior spaces in buildings and to entrances and exits. Safe, convenient, rapid circulation is essential for all buildings under both normal and emergency conditions. Such circulation may be channeled through any of several different types of passageways, such as lobbies, corridors, ramps, stairways and elevator hoistways. General requirements for these have been discussed in previous sections. This section presents in more detail design and construction considerations in provision of means of vertical circulation, the movement of people and goods between floors of multistory buildings.

Vertical circulation systems may be divided into two classes. Class I systems are intended for movement of both people and goods and include ramps, stairs, escalators, and elevators. Class II systems, including dumbwaiters and vertical conveyors, in contrast, may not be used for movement of people.

Class I systems may be divided into two subclasses, A and B. Class IA systems can be used by people both under normal and emergency conditions as a means of egress. This class includes ramps, stairs, and escalators (powered stairs) that meet requirements for means of egress specified in building codes or the National Fire Protection Association "Life Safety Code" (see Art. 3-22). Systems not acceptable as an emergency means of egress comprise Class IB. (Such systems nevertheless may be used for emergency evacuation of a building, but the capacity of Class IA systems alone must be sufficient for rapid, safe evacuation of the maximum probable building population.)

Vertical circulation of traffic in a multistory building is the key to successful functioning of the design, both in normal use and in emergencies. In fact, location of elevators or stairs strongly influences the floor plan. So in the design of a building, much thought should be given to the type of vertical circulation to be provided, number of units needed, and their location, arrangement, and design.

Traffic may pass from level to level in a multistory building by ramps, stairs, elevators, or escalators. The powered equipment is always supplemented by stairs for use when power is shut off, or there is a mechanical failure, or maintenance work is in prog-

22-2   Vertical Circulation

ress, or in emergencies. In addition to conventional elevators, other types of man lifts are occasionally installed in residences, factories, and garages. For moving small packages or correspondence between floors, dumbwaiters, chutes, or vertical conveyors also are installed in certain types of buildings. Ladders may be used for occasional access to attics or roofs.

22-1. Ramps.

When space permits, a sloping surface, or ramp, can be used to provide an easy connection between different levels or floors (Fig. 22-1). In some garages, to save space, every floor serves as a ramp; each one is split longitudinally, each section sloping gradually in opposite directions to meet the next level above and below.

Ramps are especially useful when large numbers of people or vehicles have to be moved from floor to floor. Therefore, they are frequently adopted for public buildings, such as railroad stations, stadiums, and exhibition halls. And they are either legally required or highly desirable for all buildings, especially to accommodate persons in wheelchairs. In all cases, ramps should be constructed with a nonslip surface.

Ramps have been built with slopes up to 15% (15 ft in 100 ft), but 8% is a preferred maximum. Some idea of the space required for a ramp may be obtained from the following: With the 8% maximum slope and a story height of, say, 8 ft, a ramp connecting two floors would have to be 100 ft long (Fig. 22-1a). The ramp need not be straight for the whole distance, however. It can be curved, zigzagged (Fig. 22-1b), or spiraled. Level landings, with a length of at least 44 in in the direction of travel, should be provided at door openings and where ramps change slope or direction abruptly. Ramps and landings should be designed for a live load of at least 100 lb/ft^2. Railings should be designed for a load of 200 lb applied downward or horizontally at any point of the handrail or for a horizontal thrust of 50 lb/ft at top of rail. Guards higher than the minimum required guard height of 42 in should be designed for 50 lb/ft applied 42 in above the floor.

Inside Ramps.   Local building codes and the National Fire Protection Association "Life Safety Code" contain general requirements for acceptability of a ramp as a means of emergency egress (see Art. 3-22). Egress ramps are classified as Class A or Class B. The latter may be as narrow as 30 in, whereas Class A must be at least 44 in wide. (This

Fig. 22-1 Types of ramps: (*a*) straight ramp; (*b*) zigzag ramp.

width can accommodate two adults abreast.) Also, the "Life Safety Code" restricts Class A ramps to slopes of 10% or less and Class B ramps to slopes of not more than 1 in 8. In addition, for Class B, the vertical distance between landings may not exceed 12 ft, but no limit is placed on this distance for Class A. Building codes usually require Class A ramps only for places of assembly of more than 1000 persons. For other types of occupancy, codes may permit the choice of ramp class to be based on emergency exit capacity required.

The capacity, in persons per 22-in unit of ramp width, may be taken as 100 in the downward direction and 60 in the upward direction for Class B ramps. For Class A ramps, the capacity may be taken as 100 persons per unit of width in either direction.

To be acceptable as a means of egress, a ramp inside a building more than three stories high or a building of noncombustible or fire-resistant construction is required to be of noncombustible construction. The ramp also should be protected by separation from other parts of the building in the same way as other means of egress. There should be no enclosed usable space, such as closets, under the ramp, nor should the open space under the ramp be used for any purpose. (Other enclosed ramps, however, are permitted to be located under the ramp.)

For all inside ramps, guards—vertical protective barriers—should be provided along the edges of ramps and along the edges of floor openings over ramps, to prevent falls over the open edges. Requirements for type of construction and minimum height for such barriers are the same as those for stairs (see Art. 22-2). Handrails are required only for Class B ramps.

Outside Ramps. A ramp permanently installed on the outside of a building is acceptable as a means of egress if the life-safety requirements for inside egress ramps are met. For outside ramps more than three stories high, however, guards along ramp edges should be at least 4 ft high. Also, for such ramps, provision should be made to prevent accumulations of snow or ice.

Powered Ramps. In some buildings, such as air terminals, in which pedestrians have to be moved speedily over long distances, traffic may be transported on a moving walk, a type of passenger-carrying powered device on which passengers stand or walk. The moving floor, or treadway, guards and handrails are continuous and travel parallel to the direction of motion, which may be horizontal or on a slope up to 15°. (For greater slopes, an escalator should be used. See Art. 22-7.) Moving walks generally transport passengers at speeds up to 180 ft/min.

Inclined moving walks are classified as powered ramps. Such ramps are acceptable as a means of egress if they meet the egress requirements of stationary ramps. (Moving walks are acceptable if they meet the requirements for exits. See Art. 3-22.) Moving transporters, however, must be incapable of operation in the direction opposite to normal exit travel.

The installation of moving walks and powered ramps should meet the requirements of the American National Standard Safety Code for Elevators, Dumbwaiters, Escalators and Moving Walks, ANSI A17.1.

Basically, moving walks and powered ramps consist of a grooved treadway moved by a driving machine; a handrail on each side of the treadway that moves at the same speed as the treadway; balustrades, or guards, that enclose the treadway on each side and support the handrails; brakes; control devices; and threshold plates at the entrance to and the exit from the treadway. The purpose of the threshold plates is to facilitate smooth passage of passengers between treadway and landing. The plates are equipped with a comb, or teeth, that mesh with and are set into grooves in the treadway in the direction of travel, so as to provide firm footing and to prevent things from becoming trapped between the treadway and the landing.

The treadway may be constructed in one of the following ways:

22-4　Vertical Circulation

1. **Belt type**—a power-drive continuous belt.
2. **Pallet type**—a series of connected, power-drive pallets. (A pallet is a short, rigid platform, which, when joined to other pallets, forms an articulated treadway.)
3. **Belt pallet type**—a series of connected, power-drive pallets to which a continuous belt is fastened.
4. **Edge-supported belt type**—a belt supported near its edges by rollers in sequence.
5. **Roller-bed type**—a treadway supported throughout its width by rollers in sequence.
6. **Slider-bed type**—a treadway that slides on a supporting surface.

Powered ramps resemble escalators in construction (see Art. 22-7). For example, both types of transporters are supported on steel trusses. The driving machine may be connected to the main drive shaft by toothed gearing, a coupling, or a chain. Movement of the treadway and handrails can be halted by an electrically released, mechanically applied brake, located either on the driving machine or on the main drive shaft and activated automatically when a power failure occurs, when the treadway or a handrail breaks or when a safety device is actuated. For moving walks and ramps, safety devices required include switches for starting, emergency stopping, and maintenance stopping and a speed governor that will prevent the treadway speed from exceeding 40% more than the maximum design speed.

Balustrades should be at least 30 in high, measured perpendicular to the treadway. Hand or finger guards should be installed where the handrails enter the balustrades. The handrails should extend at normal height at least 12 in beyond each end of the exposed treadway, to facilitate entry and exit of passengers from or onto a level landing.

Information on passenger capacity of a moving walk or powered ramp should be obtained from the manufacturer. The capacity depends on treadway width and speed. Standard widths are 24, 32, and 40 in. With level entry and exit, ramp speeds generally are a maximum of 180 ft/min for slopes up to 8° and 140 ft/min for slopes between 8° and 15°.

("Life Safety Code," National Fire Protection Association, 470 Atlantic Ave., Boston, Mass. 02210.

"American National Standard Safety Code for Elevators, Dumbwaiters, Escalators and Moving Walks," A17.1, and "Making Buildings and Facilities Accessible to and Usable by the Physically Handicapped," A117.1, American National Standards Institute, New York, N.Y. 10018.)

22-2. Stairs.

Less space is required for stairs than for ramps, because steeper slopes can be used. Maximum slope of stairs for comfort is estimated to be about 1 on 2 (27°), but this angle frequently is exceeded for practical reasons. Exterior stairs generally range in slope from 20 to 30°, interior stairs from 30 to 35°.

Types of Stairs. Generally, stairs are of the following types: straight, circular, curve, or spiral, or a combination.

Straight stairs are stairs along which there is no change in direction on any flight between two successive floors. There are several possible arrangements of straight stairs. For example, they may be arranged in a **straight run** (Fig. 22-2a), with a single flight between floors, or a series of flights without change in direction (Fig. 22-2b). Also, straight stairs may permit a change in direction at an intermediate landing. When the stairs require a complete reversal of direction (Fig. 22-2c), they are called **parallel stairs**. When successive flights are at an angle to each other, usually 90° (Fig. 22-2d), they are called **angle stairs**. In addition, straight stairs may be classified as **scissors**

Fig. 22-2 Arrangements of straight stairs: (a) a single flight between floors; (b) a series of flights without change in direction; (c) parallel stairs; (d) angle stairs; (e) scissors stairs.

stairs when they comprise a pair of straight runs in opposite directions and are placed on opposite sides of a fire-resistive wall (Fig. 22-2e).

Circular stairs when viewed from above appear to follow a circle with a single center of curvature and large radius.

Curved stairs when viewed from above appear to follow a curve with two or more centers of curvature, such as an ellipse.

Spiral stairs are similar to circular stairs except that the radius of curvature is small and the stairs may be supported by a center post. Overall diameter of such stairs may range from 3 ft 6 in to 8 ft. There may be from 12 to 16 winder treads per complete rotation about the center.

Stairway Components. Among the principal components of a stairway are:

Flight. A series of steps extending from floor to floor, or from a floor to an intermediate landing or platform.

Guard. Protective vertical barrier along edges of stairways, balconies, and floor openings.

Landings (Platforms). Are used where turns are necessary or to break up long climbs. Landings should be level, as wide as the stairs, and at least 44 in long in the direction of travel.

Step. Combination of a riser and the tread immediately above.

Rise. Distance from floor to floor.

Run. Total length of stairs in a horizontal plane, including landings.

Riser. Vertical face of a step. Its height is generally taken as the vertical distance between treads.

Tread. Horizontal face of a step. Its width is usually taken as the horizontal distance between risers.

Nosing. Projection of a tread beyond the riser below.

Soffit. Underside of a stair.

Header. Horizontal structural member supporting stair stringers or landings.

Carriage. Rough timber supporting the steps of wood stairs.

Stringers. Inclined members along the sides of a stairway. The stringer along a wall is called a wall stringer. Open stringers are those cut to follow the lines of risers and treads. Closed stringers have parallel top and bottom, and treads and risers are supported along their sides or mortised into them. In wood stairs, stringers are placed outside the carriage to provide a finish.

Railing. Framework or enclosure supporting a handrail and serving as a safety barrier.

Balustrade. A railing composed of balusters capped by a handrail.

Handrail. Protective bar placed at a convenient distance above the stairs for a handhold.

Baluster. Vertical member supporting the handrail in a railing.

Newel Post. Post at which the railing terminates at each floor level.

Angle Post. Railing support at landings or other breaks in the stairs. If the angle post projects beyond the bottom of the stringers, the ornamental detail formed at the bottom of the post is called the **drop**.

Winders. Steps with tapered treads in sharply curved stairs.

Headroom. Minimum clear height from a tread to overhead construction, such as the ceiling of the next floor, ductwork, or piping.

Vertical Clearance. Ample headroom should be provided not only to prevent tall people from injuring their heads, but to give a feeling of spaciousness. A person of average height should be able to extend his hand forward and upward without touching the ceiling above the stairs. Minimum vertical distance from the nosing of a tread to overhead construction should never be less than 6 ft 8 in and preferably not less than 7 ft.

Stairway Width. Width of a stairway depends on its purpose and the number of persons to be accommodated in peak hours or emergencies. Generally, the minimum width that can be used is specified in the local building code. For example, for interior stairs, clear width may be required to be at least 36 in in one- and two-family dwellings, and 44 in in hotels, motels, apartment buildings, industrial buildings, and other types of occupancy.

Number of Stairways Required. This is usually controlled by local building codes. This control may be achieved by setting a minimum of two exits per floor, a restriction on the maximum horizontal distance from any point on a floor to a stairway, or a limitation on the maximum floor area contributory to a stairway. In addition, codes usually have special provisions for assembly buildings, such as theaters and exhibition halls. Restrictions usually also are placed on the maximum capacity of a stairway. For example, the National Fire Protection Association "Life Safety Code" sets a maximum capacity for stairways of 60 persons per 22-in unit of width, up or down.

Step Sizes. Risers and treads generally are proportioned for comfort, although sometimes space considerations control or the desire to achieve a monumental effect, particularly for outside stairs of public buildings. Treads should be 11 to 14 in wide, exclusive of nosing. Treads less than 11 in wide should have a nosing of about 1 in. The most comfortable height of riser is 7 to 7½ in. Risers less than 4 in and more than 8 in high should not be used. The steeper the slope of the stairs, the greater the ratio of riser to tread. Among the more common simple formulas generally used with the preceding limits are:

1. Product of riser and tread must be between 70 and 75.
2. Riser plus tread must equal 17 to 17.5.
3. Sum of the tread and twice the riser must lie between 24 and 25.5.

In design of stairs, account should be taken of the fact that there is always one less tread than riser per flight of stairs. No flight of stairs should contain less than three risers.

Curved Stairways. Winders should be avoided when possible, because the narrow width of tread at the inside of the curve may cause accidents. Sometimes, instead, **balanced steps** can be used. Instead of radiating from the center of the curve, like winders, balanced steps, though tapered, have the same width of tread along the line of travel as the straight portion of the stairs. (Line of travel in this case is assumed to be about 20 in from the rail on the inside of the curve.) With balanced steps, the change in angle is spread over a large portion of the stairs.

Loading. Stairs and landings should be designed for a live load of 100 lb/ft^2 or a concentrated load of 300 lb placed to produce maximum stresses.

Guards. To prevent people from falling over edges of stairs and landings, barriers, called guards, should be placed along all edges and should be at least 42 in high. These should support 2-in-diameter handrails, which should be set 30 to 34 in above the intersections of treads and risers at the front of the steps.

Interior stairs more than 88 in wide should have intermediate handrails that divide the stairway into widths of not more than 88 in, preferably into a nominal multiple of 22 in. Handrails along walls should have a clearance of at least 1½ in.

Guards should be designed for a horizontal force of 50 lb/ft, applied 42 in above the floor, or for the force transmitted by the handrail, whichever is greater. Handrails should be capable of supporting a load of at least 200 lb, downward or horizontally.

See also Arts. 22-3 to 22-7.

22-3. Emergency Egress Stairways.

In many types of buildings, interior exit stairways must be enclosed with walls having a fire-resistance rating, to prevent spread of smoke and flames. The walls should have a 1-hr rating for stairways serving up to three stories and a 2-hr rating for stairways serving four or more stories. Openings in the walls should be protected by approved, self-closing fire doors. Stairs in buildings required by the code to be of fire-resistant construction should be completely made of noncombustible materials. Open space under stairs to be used as a means of egress should not be used for any purpose, including closets, except for another flight of stairs.

In buildings requiring such egress stairways, an alternative type of construction, called a **smokeproof tower,** may be used. A smokeproof tower is a continuous, vertical, fire-resistant enclosure protecting a stairway from fire or smoke that may develop elsewhere in a building. The intent is to limit the entrance into the stairway of products of combustion so that during a 2-hr period the tower air will not contain smoke or gases with a volume exceeding 1% of the tower volume. All components of the tower should be made of noncombustible materials, and the enclosure should have a 2-hr fire rating. Walls between the stairs and the building interior should not have any openings. If the exterior wall of the tower will not be subjected to a severe fire-exposure hazard, however, that wall may incorporate fixed or automatic fire windows.

Access to a smokeproof tower should be provided in each story through vestibules open to the outside on an exterior wall, or from balconies on an exterior wall, neither exposed to severe fire hazards. Doors should be at least 40 in wide, self-closing, and provided with a viewing window of clear, wired glass not exceeding 720 in^2 in area. It also is wise to incorporate some means of opening the top of the shaft, either with a thermally operated device or with a skylight, to let escape any heat that might enter the tower from a fire. Exits at the bottom of smokeproof towers should be directly to the outdoors, where people can remove themselves quickly to a safe distance from the building.

Stairs outside a building are acceptable as a required fire exit instead of inside stairs, if they satisfy all the requirements of inside stairs. Where enclosure of inside stairs is required, however, outside stairs should be separated from the building interior by fire-resistant walls with fire doors or fixed wired glass windows protecting openings. Some building codes limit the height of outside stairs to a maximum of six stories or 75 ft.

Fire escapes, outside metal-grating stairs, and landings attached to exterior walls with unprotected openings were acceptable at one time as required exits, but usually now are not accepted as such for new construction.

See also Art. 3-22.

("Life Safety Code," National Fire Protection Association, Boston, Mass. 02210.)

22-4. Wood Stairs.

(See also Art. 22-2.) In wood-frame buildings, low nonfireproof buildings, and one- and two-family houses, stairs may be constructed of wood (Fig. 22-3). They may be built in place or shop fabricated.

Construction of a built-in-place stair starts with cutting of carriages to the right size and shape to receive the risers and treads (Fig. 22-3b). Next, the lower portion of the wall stringer should be cut out at least ½ in deep to house the steps (Fig. 22-3c). The stringer should be set in place against the wall with the housed-out profile fitted to the stepped profile of the top of the carriage. Then, treads and risers should be firmly nailed to the carriages, tongues at the bottom of the risers fitting into grooves at the rear of the treads. Nosings are generally finished on the underside with molding.

If the outer stringer is an open stringer (Fig. 22-3c) it should be carefully cut to the same profile as the steps, mitered to fit corresponding miters in the ends of risers, and nailed against the outside carriage. Ends of the treads project beyond the open stringer.

If the outer stringer is a curb or closed stringer it should be plowed out in the same way as the wall stringer to house the steps. Ends of the treads and risers should be wedged and glued into the wall stringer (Fig. 22-3d).

(Ramsey and Sleeper, "Architectural Graphic Standards," John Wiley & Sons, Inc., New York; J. D. Wilson and S. O. Werner, "Simplified Stair Layout," Delmar Publishers, Albany, N.Y.)

22-5. Steel Stairs.

(See also Art. 22-2.) Cold-formed-steel or steel-plate stairs generally are used in fire-resistant buildings. They may be purchased from various manufacturers in stock patterns.

Fig. 22-3 Typical construction for wood stairs.

Escalators 22-9

Fig. 22-4 Types of metal stairs: (a) stairs made of cold-formed steel; (b) stairs made of steel plate. (Reprinted with permission from F. S. Merritt, "Building Engineering and Systems Design," Van Nostrand Reinhold, New York.)

The steel sheets are formed into risers and subtreads or pans, into which one of several types of treads may be inserted (Fig. 22-4). Stringers usually are channel-shaped. Treads may be made of stone, concrete, composition, or metal. Most types are given a nonslip surface.

("Metal Stairs Manual," National Association of Architectural Metal Manufacturers, 1033 South Blvd., Oak Park, Ill. 60302.)

22-6. Concrete Stairs.

(See also Art. 22-2.) Depending on the method of support provided, concrete stairs may be designed as cantilevered or inclined beams and slabs (Fig. 22-5). The entire stairway may be cast in place as a single unit, or slab or T beams formed first and the steps built up later. Soffits formed with plywood or hardboard forms may have a smooth enough finish to make plastering unnecessary. Concrete treads should have metal nosings to protect the edges. Stairs also may be made of precast concrete.

22-7. Escalators.

Escalators, or powered stairs, are used when it is necessary to move large numbers of people from floor to floor. They have the advantage of continuous operation without the need for operators. They have large capacity with low power consumption. Large department stores provide vertical-transportation facilities for one person per hour for every 20 to 25 ft^2 of sales area above the entrance floor, and escalators generally carry 75 to 90% of the traffic, elevators the remainder.

22-10   Vertical Circulation

Fig. 22-5 Reinforced concrete stairs. *(Reprinted with permission from F. S. Merritt, "Building Engineering and Systems Design," Van Nostrand Reinhold, New York.)*

An escalator resembles a powered ramp in construction (Art. 22-1). The major difference is that a powered ramp has a continuous treadway for carrying passengers, whereas the treadway of an escalator consists of a series of moving steps. As for a powered ramp, the installation of powered stairs should conform with the requirements of the American National Standard Safety Code for Elevators, Dumbwaiters, Escalators and Moving Walks, ANSI A17.1.

Basically, an escalator consists of articulated, grooved treads and risers attached to a looped belt moved by a driving machine and supported by a steel trussed framework (Fig. 22-6). The installation also includes a handrail on each side of the steps that moves at the same speed as the steps; solid balustrades, or guards, that enclose the steps on each side and support the handrails; brakes; control devices, and threshold plates at the entrance to and the exit from the treadway. The purpose of the threshold plates is to facilitate smooth passage of passengers between the treadway and landing. The plates are equipped with a comb, or teeth, that mesh with and are set into grooves in the tread-

Fig. 22-6 Escalator details. *(Courtesy of Otis Elevator Co.)*

way in the direction of travel, so as to provide firm footing and to prevent things from becoming trapped between treadway and the landing.

Each step is formed by a pair of pallets that are hinged together and to adjoining steps. One member of each pair forms a tread and stays level while it is visible to passengers. The other member of the pair forms a riser while traveling along the incline. The riser pallet, however, rotates to the horizontal at the top and bottom of the escalator so that the step is converted to a level platform for safe entry of passengers to and exit from the powered stairs. Carried on resilient rollers, the steps ride on a set of tracks attached to the trussed framework.

ANSI A17.1 sets the following limitations on escalator steps (Fig. 22-7):

Minimum depth of tread in direction of travel—15¾ in
Maximum rise between treads—8½ in
Minimum width of tread—16 in
Maximum width of tread—40 in
Maximum clearance between tread and adjacent skirt panel—⅜ in
Maximum distance between handrail centerlines—width between balustrades plus 6 in with not more than 3 in on either side of the escalator (see Fig. 22-7b)

The escalator width is measured on the incline between balustrades, as indicated in Fig. 22-7b. It should be at least as wide as the step but not more than 13 in wider than the step.

To provide a firm footing, treads should be grooved in the direction of travel. The grooves should mesh with the combs or teeth of the threshold plates at top and bottom of the escalator.

The handrails, which move in synchronization with the steps, should be between 30 and 34 in above the treads. The handrails should extend at normal height at least 12 in beyond the line of points of the combplate teeth. The balustrades carrying the handrails and acting as a guard to prevent passengers from falling off the sides of the moving steps should be designed to resist simultaneous application of a horizontal load of 40 lb/ft and a vertical load of 50 lb/ft, both applied to the top of the balustrades.

Fig. 22-7 Limitations on escalator dimensions.

The driving machine may be connected to the main drive shaft by toothed gearing, a coupling, or a chain. Movement of the steps can be halted by an electrically released, mechanically applied brake, located either on the driving machine or on the main drive shaft. The brake should be activated automatically when a power failure occurs, when the treadway or a handrail breaks or when a safety device is activated.

Safety devices required for escalators include switches for starting, emergency stopping, and maintenance stopping and a speed governor that will prevent the step speed from exceeding 40% more than the maximum design speed. An emergency stop button, protected against accidental activation, should be set in the right-hand (when facing the escalator) newel at top and bottom landings. The button, however, should not be capable of starting the driving machine.

Information on passenger capacity should be obtained from the escalator manufacturer. The capacity depends on step width and speed.

Escalators usually can operate at 90 or 120 ft/min, as needed for peak traffic, and are reversible in direction. Slope of the stairs is standardized at 30°.

Standard escalator widths are 32 and 48 in, with corresponding capacities at 90 ft/min of 5000 and 8000 persons per hour. At 120 ft/min, a 48-in escalator can carry as many as 10,000 persons per hour.

Planning for Escalators. Location of powered stairs should be selected only after a careful study of traffic flow in the building. They should be installed where traffic is heaviest and where convenient for passengers, and their location should be obvious to people approaching them. In stores, escalators should lead to strategic sales areas.

In the design of a new building, adequate space should be allotted for powered stairs. Structural framing should be made adequate to support them.

For an escalator installation in an existing building, careful study should be made to determine the necessary alterations to assure adequate space and supports. When the driving machine cannot be housed inside the trussed framework, building beams must be cut to permit passage of the drive chain; these beams must be reinforced.

Floor-to-floor height should be taken into account in determining loads on supporting members. If the total rise is within certain limits, the stairs need be supported only at top and bottom; otherwise, they may have to be supported at intermediate points also. A structural frame should be installed around the stair well to carry the floor and wellway railing. The stairway should be independent of this frame.

Installation. Design of escalators permits a vertical variation of ½ in in the level of the supporting beams from the specified floor-to-floor height. The escalator is shimmed to bring it level. If variations in elevation exceed ½ in, installation is difficult and much time may be lost. Also, when upper and lower beams are too far apart horizontally, there will be delay, because an extension must be added to the trussed framework.

Trusses generally are brought to the job in three sections. There, they are assembled and raised into position with chain hoists, either through an elevator shaft or on the outside of the building.

Either the owner or the manufacturer may furnish the exterior treatment of the balustrade and wellway railings.

Escalator Arrangements. Escalators usually are installed in pairs—one for carrying traffic up and the other for moving traffic down. The units may be placed parallel to each other in each story (Fig. 22-8), or crisscrossed (Fig. 22-9). Crisscrossed stairs generally are preferred because they are more compact, reducing walking distance between stairs at various floors to a minimum.

Fire Protection. Escalators may be acceptable as required means of egress if they comply with the applicable requirements for exit stairs (Art. 22-3). Such escalators must be enclosed in the same manner as exit stairs. Escalators capable of reversing direction, however, may not qualify as required means of egress.

Fig. 22-8 Parallel arrangement of up and down escalators.

Fig. 22-9 Crisscross arrangement of escalators.

An escalator not serving as a required exit must have its floor openings enclosed or protected as required for other vertical openings. Acceptable protection, as an alternative, is afforded in buildings completely protected by a standard supervised sprinkler system by any of the following:

Sprinkler-vent method, a combination of an automatic fire or smoke detection system, automatic air-exhaust system, and an automatic water curtain.

Spray-nozzle method, a combination of an automatic fire or smoke detection system and a system of high-velocity water-spray nozzles.

Rolling shutter method, in which an automatic, self-closing, rolling shutter is used to enclose completely the top of each escalator.

Partial enclosure method, in which kiosks, with self-closing fire doors, provide an effective barrier to spread of smoke between floors.

Escalator trusses and machine spaces should be enclosed with fire-resistant materials. But ventilation must be provided for machine and control spaces.

("Life Safety Code," National Fire Protection Association, Boston, Mass. 02210; "American National Standard Safety Code for Elevators, Dumbwaiters, Escalators, and Moving Walks," A17.1, American National Standards Institute; G. R. Strakosch, "Vertical Transportation: Elevators and Escalators," John Wiley & Sons, Inc., New York.)

22-8. Elevator Installations.

An **elevator** is a hoisting and lowering mechanism equipped with a car or platform that moves along guides in a shaft, or hoistway, in a substantially vertical direction and that transports passengers or goods, or both, between two or more floors of a building. **Passenger elevators** carry persons in addition to an operator or freight loaders or unloaders (if such personnel are necessary). **Hospital elevators** employ special cars, suitable in size and shape for transportation of patients in stretchers or standard hospital beds and of attendants accompanying them. **Freight elevators** carry freight, which may be accompanied only by an operator and persons necessary for loading and unloading it.

Elevators are desirable in all multistory buildings for providing vertical transportation of passengers or freight and may be required by local building codes for buildings more than four stories high or for transportation of physically handicapped persons. Elevators, however, are not usually accepted as a means of egress, because there is a

22-14   Vertical Circulation

possibility that they may not operate properly, if at all, in an emergency. Nevertheless, elevators are useful for emergency evacuation of building occupants (although they should not be permitted to stop for passengers other than firefighters at floors at which fire or smoke is present). It also is desirable that elevators be available at street level for use of fire fighters when a fire occurs in the building.

Elevators usually are either electric or hydraulic powered. Electric elevator cars are suspended on wire ropes that are moved by electric driving machines (Fig. 22-10a and b) and are used exclusively in tall buildings and most low buildings (Art. 22-12). Hydraulic elevator cars (Fig. 22-10c) are raised and lowered by liquid under pressure in a cylinder equipped with a plunger or piston (Art. 22-13). They generally are used for low-rise freight service and may be used for low-rise passenger service where low initial cost is desired.

Elevator installations should meet the requirements of the American National Standard Safety Code for Elevators, Dumbwaiters, Escalators and Moving Walks, ANSI A17.1. It is desirable that car sizes and shapes be in accord with NEI standards, such as "Elevator Engineering Standard Layouts" and "Suggested Minimum Passenger Elevator Requirements for the Handicapped," National Elevator Industry, Inc., 600 Third Ave., New York, N.Y. 10016.

Owner's Responsibilities. Elevators and related equipment, such as machinery, signal systems, ropes, and guide rails, are generally supplied and installed by the manufacturer. The general contractor has to guarantee the dimensions of the shaft and its freedom from encroachments. The owner's architect or engineer is responsible for the design and construction of components needed for supporting the plant, including buffer supports, machine-room floors, trolley beams, and guide-rail bracket supports. Magnitudes of loads generally are supplied by the manufacturer with a 100% allowance for impact.

For design of machinery and sheave beams and floor systems, unit stresses should not exceed 80% of those allowed for static loads in the design of usual building structural members. Where stresses due to loads, other than elevator loads, supported on the beams

Fig. 22-10 Types of elevators: (a) electric elevator with driving machine at top of hoistway; (b) electric elevator with driving machine in basement; (c) hydraulic elevator.

or floor system exceed those due to the elevator loads, 100% of the allowable unit stresses may be used. Unit stresses in a steel guide rail or its reinforcement, caused by horizontal forces, calculated without impact, should not exceed 15,000 psi, and deflection should not exceed ¼ in. Guide-rail supports should be capable of resisting horizontal forces with a deflection of not more than ⅛ in.

(G. R. Strakosch, "Vertical Transportation: Elevators and Escalators," John Wiley & Sons, Inc., New York.)

22-9. Definitions of Elevator Terms.

(See also Figs. 22-10 to 22-14.)

Annunciator. An electrical device that indicates, usually by lights, the floors at which an elevator landing signal has been registered.

Buffer. A device for stopping a descending car or counterweight beyond its bottom terminal by absorbing and dissipating the kinetic energy of the car or counterweight. The absorbing medium may be oil, in which case the buffer may be called an **oil buffer**, or a spring, in which case the buffer may be referred to as a **spring buffer**.

Bumper. A device other than a buffer for stopping a descending car or counterweight beyond its bottom terminal by absorbing the impact.

Car. The load-carrying element of an elevator, including platform, car frame, enclosure, and car door or gate.

Car-Door Electric Contact. An electrical device for preventing normal operation of the driving machine unless the car door or gate is closed.

Car Frame. The supporting frame to which the car platform, guide shoes, car safety, and hoisting ropes or hoisting-rope sheaves, or the plunger of a hydraulic elevator are attached.

Car Platform. The floor of the car.

Car Switch. A manual operating device in a car by which an operator actuates the control.

Control. The system governing the starting, stopping, direction of motion, acceleration, speed, and retardation of the car.

> **Generator-field control** employs an individual generator for each elevator, with voltage applied to the driving-machine motor adjusted by varying the strength and direction of the generator field.
>
> **Multivoltage control** impresses successively on the armature of the driving-machine motor various fixed voltages, such as those that might be obtained from multicommutator generators common to a group of elevators.
>
> **Rheostatic control** varies the resistance or reactance of the armature or the field circuit of the driving-machine motor.
>
> **Single-speed, alternating-current control** governs a driving-machine induction motor that runs at a specified speed.
>
> **Two-speed alternating-current control** governs a two-speed driving-machine induction motor, with motor windings connected to obtain various numbers of poles.

Dispatching Device. A device that operates a signal in a car to indicate when the car should leave a designated floor or to actuate the car's starting mechanism when the car is at a designated floor.

Driving Machine. See Machine.

Emergency Stop Switch. A car-located device that, when operated manually, causes the car to be stopped by disconnecting electric power from the driving-machine motor.

Hoistway. A shaft for travel of one or more elevators. It extends from the bottom of the pit to the underside of the overhead machine room or the roof. A **blind hoistway**

is the portion of the shaft that passes floors or other landings without providing a normal entrance.

Hoistway Access Switch. A switch placed at a landing to permit car operation with both the hoistway door at the landing and the car door open.

Hoistway-Unit System. A series of hoistway-door interlocks, or electric contacts, or a combination of such contacts and mechanical locks installed to prevent operation of a driving machine by the normal operating device unless all hoistway doors are closed.

Hoistway-Door Electric Contact. An electrical device for preventing normal operation of the driving machine unless the hoistway door is closed.

Hoistway-Door Locking Device. A device for preventing the hoistway door or gate from being opened from the landing side unless the car has stopped within the landing zone.

Leveling Device. A mechanism for moving a car that is within a short distance of a landing toward the landing and stopping the car there. An **automatic maintaining, two-way, leveling device** will keep the car floor level with the landing during loading and unloading.

Machine (Driving Machine). The power unit for raising and lowering an elevator car.

Electric driving machines include an electric motor and brake, driving sheave or drum, and connecting gearing, belts, or chain, if any. A **geared-drive machine** operates the driving sheave or drum through gears. A **traction machine** drives the car through friction between suspension ropes and a traction sheave. A **gearless traction machine** has the traction sheave and the brake drum mounted directly on the motor shaft. A **winding-drum machine** has the motor geared to a drum on which the hoisting ropes wind. A **worm-geared machine** operates the driving sheave or drum through worm gears.

Hydraulic driving machines raise or lower a car with a plunger or piston moved by a liquid under pressure in a cylinder.

Nonstop Switch. A device for preventing a car from making registered landing stops.

Operating Device. The car switch, pushbutton, lever, or other manual device used to actuate the control.

Operation. The method of actuating the control.

Automatic operation starts the car in response to operating devices at landings, or located in the car and identified with landings, or located in an automatic starting mechanism, and stops the car automatically at landings. **Group automatic operation** starts and stops two or more cars under the coordination of a supervisory control system, including automatic dispatching means, with one button per floor in each car and up and down buttons at each landing. **Nonselective collective automatic operation** has one button per floor in each car and only one button at each landing. **Selective collective automatic operation** is a form of group automatic operation in which car stops are made in the order in which landings are reached in each direction of travel after buttons at those landings have been pressed. **Single automatic operation** has one button per floor in each car and only one button per landing, so arranged that after any button has been pressed, pushing any other button will have no effect on car operation until response to the first button has been completed.

Car-switch operation starts and stops a car in response to a manually operated car switch or continuous-pressure buttons in a car.

Preregister operation is one in which signals to stop are registered in advance by buttons in a car or at landings and then, at the proper point in car travel, are given to an operator in the car, who initiates the stop, which is completed automatically.

Signal operation starts and stops a car automatically as landings are reached, in response to actuation of buttons in cars or at landings, irrespective of direction of car travel or sequence in which buttons are pressed, but the car can be started only by a button or starting switch in the car.

Parking Device. A device for opening from the landing side the hoistway door at any landing when the car is within the landing zone.

Pit. Portion of a hoistway below the lowest landing.

Position Indicator. Device for showing the location of a car in the hoistway.

Rise. See Travel.

Rope Equalizer. A device installed on a car or counterweight to equalize automatically the tensions in the hoisting ropes.

Runby. The distance a car can travel beyond a terminal landing without striking a stop.

Safety. A mechanical device attached to the counterweight or to the car frame or an auxiliary frame to stop or hold the counterweight or the car, whichever undergoes overspeed or free fall, or if the hoisting ropes should slacken.

Safety Bulkhead. In a cylinder of a hydraulic elevator, a closure, at the bottom of the cylinder but above the cylinder head, with an orifice for controlling fluid loss in case of cylinder-head failure.

Signal Registering Device. A button or other device in a car or at a landing that causes a stop signal to be registered in a car.

Signal Transfer Device. A device used in automatic elevator operation for automatically transferring a signal registered in one car to the next car if the former passes, without stopping, a floor for which a signal has been registered.

Signal Transfer Switch. A manually operated switch for accomplishing the same function as a signal transfer device.

Slack-Rope Switch. A device that automatically disconnects electric power from the driving machine when the hoisting ropes of a winding-drum machine become slack.

Starter's Control Panel. An assembly of devices with which an elevator starter can control the way in which one or more elevators function.

Terminal Speed-Limiting Device (Emergency). A device for reducing automatically the speed of a car approaching a terminal landing, independently of the car-operating device and the normal terminal stopping device if the latter should fail to slow the car as intended.

Terminal Stopping Device. Any device for slowing or stopping a car automatically at or near a terminal landing, independently of the car-operating device. A **final terminal stopping device,** after a car passes a terminal landing, disconnects power from the driving apparatus, independently of the operating device, normal terminal stopping device, or emergency terminal speed-limiting device. A **stop-motion switch,** or **machine final terminal stopping device,** is a final terminal stopping device operated directly by the driving machine.

Transom. One or more panels that close an opening above a hoistway entrance.

Travel (Rise). The vertical distance between top and bottom terminal landings.

Traveling Cable. A cable containing electrical conductors for providing electrical connections between a car and a fixed outlet in a hoistway.

Truck Zone. A limited distance above a landing within which the truck-zoning device permits movement of a freight-elevator car with its door or the hoistway door open.

Truck-Zoning Device. A device that permits a car operator to move, within a specified distance above a landing, a freight-elevator car with its door or the hoistway door open.

Waiting-Passenger Indicator. An indicator that shows at which landings and for which direction stop-or-signal calls have been registered and have not been served by an elevator car.

22-10. Elevator Hoistways.

A hoistway is a shaft in which an elevator travels. To provide access to an elevator car, the shaft enclosure has openings, protected by doors with safety devices, at landings. In a pit at the bottom of the hoistway, buffers or bumpers must be installed to stop a descending car or counterweight beyond its normal limit of travel, by storing or by absorbing and dissipating its kinetic energy (Fig. 22-10). Construction of the hoistway and installation of the associated equipment should meet the requirements of the American National Standard Safety Code for Elevators, Dumbwaiters, Escalators and Moving Walks, ANSI A17.1.

The code permits up to four elevators to operate in a hoistway (multiple hoistway). If, however, four or more elevators serve the same portion of a building, they should be installed in at least two hoistways, with not more than four elevators in any of them.

Hoistway Enclosure. The code requires that hoistways be enclosed throughout their height with fire-resistant construction, except possibly for private residences and for cases where no solid floors are penetrated. The enclosure should have a 2-hr fire rating, and hoistway doors and other opening protective assemblies should have a 1½-hr rating. Where fire-resistant construction is not required, laminated-glass curtain walls or unperforated metal, such as 18-gage sheet steel, should enclose the hoistway to a height of 6 ft above each floor and above the treads of adjacent stairways. Openwork enclosures may be used above that level, if openings are less than 2 in wide or high.

At the top of a hoistway, a metal or concrete floor should be provided (but it is not necessary below secondary and deflection sheaves of traction-type driving machines located over the hoistway). If a driving machine is installed atop the hoistway, the floor should be level with or above the top of the beams supporting the machine. Otherwise, the floor should be set under the overhead sheaves. The floor should cover the entire top of the hoistway if its area would be 100 ft^2 or less. For larger hoistway cross-sectional areas, the floor should extend from the entrance to the machine space, at or above the level of the platform, for a distance at least 2 ft beyond the general contour of the driving machine, sheaves or other equipment. In such cases, exposed floor edges should be protected with a toe board at least 4 in high and a railing at least 42 in high and conforming to the requirements of the American National Standard Safety Code for Floor and Wall Openings, Railings and Toe Boards, ANSI A12.1.

Venting of Hoistways. For elevators serving four or more stories, hoistways should be provided with means for venting smoke and hot gases to the outdoors in case of fire. (Such venting, however, is not required for hoistways equipped with automatic sprinklers responsive to smoke as well as heat, if the building does not contain overnight sleeping quarters.) Vents, with ductwork, if necessary, must be installed for the purpose, since hoistway enclosures are not permitted to have windows. Such vents may be inserted directly below the floor at the top of the hoistway or in the walls or roof of the overhead machine space above the roof of the building.

ANSI A17.1 requires that the vent area total 3.5% of the hoistway cross-sectional area but not less than 3 ft^2 per elevator car. The code also specifies that at least one-third of this vent area should be permanently open or automatically openable by a damper. The remaining vent area may be provided by windows or skylights, with metal frames and sash, and glazed with plain glass. The glass should not be more than ⅛ in thick, to permit it to be broken easily for venting in emergencies. To protect this thin glass against accidental breakage, a wire-mesh or expanded-metal screen should be placed above each skylight. A screen also should be installed underneath, to prevent broken glass from falling into the hoistway.

Machine Rooms. Construction of enclosures of spaces containing machines, control equipment and sheaves should be equivalent to that used for the hoistway enclosure. To dissipate machinery heat, the spaces should be ventilated.

If the driving machine is located at the top of the hoistway, other machinery and equipment for building operation may also be installed in the machine room, but separated from the elevator equipment by a substantial metal grille at least 6 ft high. The entrance to the elevator machine room should be guarded by a self-closing, self-locking door. If, however, the driving machine is not at the top of the hoistway, only elevator equipment is permitted in the machine room.

In machine rooms at the top of the hoistway, headroom of at least 7 ft above the floor should be provided. For spaces containing only overhead, secondary or deflecting sheaves, headroom may be only 3½ ft, but 4½ ft is required if the spaces also contain speed governors, signal machines, or other equipment.

Hoistway Doors. Each opening in a hoistway enclosure for access to elevator cars must be protected with a 1½-hr fire-rated door for the full width and height of the opening. ANSI A17.1 lists types of doors that may be used and gives requirements for their openings. Generally, however, single-section swinging doors or horizontally sliding doors are used for hoistways for freight and passenger elevators and vertically sliding doors are used for freight elevators.

Horizontally sliding or swinging doors for automatic elevators should be equipped with door closers. They should close open doors automatically if the car leaves the landing zone. (A **landing zone** is the space 18 in above and below a landing.) A horizontally sliding hoistway door may be kept open while a car is at a landing, but only while the car is being loaded and unloaded or when the door is under the control of an operator or an automatic elevator dispatching system.

For safety reasons, normal operation of the elevator car in a hoistway should not be possible unless all hoistway doors are closed and preferably also locked. For the purpose, the doors should be equipped with door-locking devices, hoistway access switches and parking devices. A **locking device** holds a door closed, preventing it from being opened on the landing side, except for repair, maintenance, or emergency purposes. An **access switch** is placed at a landing and operated to permit movement of a car with the car door and the hoistway door at the landing open, for access to the top of the car or to the hoistway pit. A **parking device** is used to open or close a hoistway door from the landing side at any landing if a car is within the landing zone. Unless all hoistway doors automatically unlock as a car enters their landing zones, at least one landing of the hoistway should be provided with a parking device.

ANSI A17.1 also requires that hoistway doors be equipped with hoistway-unit-system interlocks. These consist of electric contacts or mechanical locks, or a combination of these devices, that prevent operation of the elevator driving machine by the normal operating device unless all hoistway doors are closed and locked.

Hoistway doors should be openable by hand from the hoistway side from a car within the interlock unlocking zone, except when the doors are locked out of service. (Doors at the main-entrance landing or at the top or bottom terminal landing should be incapable of being locked out of service, so that some means of access to the hoistway is always available.) Automatic fire devices controlled by heat should not lock any hoistway door so that it cannot be opened manually from inside the hoistway nor lock any exit leading from a hoistway door to the outside.

Vision panels of clear wired glass or laminated glass, with an area between 25 and 80 in^2, may be inserted in any type of hoistway door and car door, to enable passengers in a car to see if passengers at landings are waiting to enter. ANSI A17.1, however, specifically requires such a vision panel to be installed in all horizontally swinging hoistway doors and in manually operated, self-closing, sliding hoistway doors for elevators with automatic or continuous-pressure operation. But the code does not require a vision panel at landings for automatic elevators provided with a device that indicates the location of the car in the hoistway (hall position indicator).

Guide Rails. The paths of elevator cars and of counterweights, if used, are controlled by vertical guide rails installed in the hoistway. The rails usually are T shaped in cross section and have smooth guiding surfaces along which the car wheels roll. A rail is installed on each side of the shaft to guide the car (Fig. 22-11). When a counterweight is used with electric elevators, to reduce power requirements, a second pair of rails is placed along one wall of the shaft, to guide the counterweight.

The elevator manufacturer usually supplies and installs the guide rails. The owner is responsible for supplying beams to support them.

Buffers and Bumpers. Energy-absorbing devices are required at the bottom of a hoistway to absorb the impact from a car that descends below its normal limit of travel (Fig. 22-10a). ANSI A17.1 specifically requires buffers under cars and counterweights in hoistways over accessible spaces. The code also requires buffers or solid bumpers in the pits for passenger elevators with speeds up to 50 ft/min and for freight elevators with speeds up to 75 ft/min but buffers are required for greater speeds. Oil buffers may be used under elevators at any rated speed, but spring buffers may be used only for speeds up to 200 ft/min.

Solid bumpers, which are permissible only for slow elevators, may be made of wood or other resilient material. As the name implies, spring buffers use springs, whereas oil buffers use the hydraulic pressure of oil against a plunger contacted by a descending car, to bring the car to a gradual stop.

Hoistway Dimensions. The National Elevator Industry standard, "Elevator Engineering Standard Layouts," lists clear inside dimensions required for hoistways for elevator cars covered by the standard. Cars supplied by various manufacturers, however, may differ somewhat in sizes from those recommended in the standard. Consequently, it is advisable to obtain recommended hoistway inside dimensions from the car manufacturer or elevator installer for a specific installation.

For proper elevator operation and for safety, both maximum and minimum clearances between hoistways and cars and other moving equipment, as recommended by elevator manufacturers and ANSI A17.1, should be provided. The clearance between a car and the hoistway enclosure, for example, should be at least ¾ in, except on the sides used for loading and unloading. The clearance between the car-platform sill and vertically sliding hoistway doors or the hoistway edge of the landing sill should be at least ½ in where side door guides are used and ¾ in where corner guides are used, but not more than 1½ in. Maximum clearance between the loading side of the car platform and the hoistway enclosure generally is 5 in but may be as much at 7½ in when vertically sliding hoistway doors are used.

Clearance between a car and its counterweight should be at least 1 in. Between the counterweight and other components, the clearance should be at least ¾ in.

In multiple hoistways, a minimum of 2-in clearance should be provided between moving equipment.

22-11. Elevator Cars.

A car consists basically of a platform for transporting passengers and goods that is raised or lowered by wire ropes or a hydraulic piston or plunger. For safety reasons, the car is required to be completely enclosed with light metal or laminated glass or a combination. (For light loads, platforms and their structural members may be constructed of wood.) To provide access to the car, openings protected by doors are provided in one or two of the car walls. In addition, the platform and the car enclosure are supported on a structural steel frame (Fig. 22-10).

For electric elevators, the wire ropes that move the car are attached to the frame or wound around a sheave connected to it. For hydraulic elevators, the frame is seated on the piston. The frame also supports the upper and lower wheels that roll along the vertical guide rails in the hoistway (Fig. 22-11). In addition, the frame carries safety devices for preventing the car from falling in case of failure of wire ropes or hydraulic piston.

Design, construction, and installation of elevator cars should meet the requirements of the National Standard Safety Code for Elevators, Dumbwaiters, Escalators and Moving Walks, ANSI A17.1.

Door Controls. Car doors may be horizontally or vertically sliding. They usually are power operated. For safety, they should be equipped with devices that prevent them from opening while the car is moving or is outside the landing zones, the space 18 in above and below a landing. Also, ANSI A17.1 requires safety devices that will keep the car from moving while the doors are open. When doors open under automatic control, they should remain open for at least 3 s.

Additional devices are needed for power-operated doors to reopen the car and hoistway doors when they start to close on a passenger or other object. The National Elevator

Fig. 22-11 Electric traction passenger elevator. *(Courtesy Otis Elevator Co.)*

Industry standard, "Suggested Minimum Passenger Elevator Requirements for the Handicapped," recommends that the devices be capable of sensing a person or object in the path of a closing door, without requiring contact for activation, at a nominal 5 and 29 in above the floor. Also, the doors should be kept open for at least 20 s after reopening. Still other devices should be installed for other safety reasons, for example, to prevent car and hoistway doors from closing and the car from moving when it is overloaded. Furthermore, these doors should be so installed that they can be opened manually from inside the car when power operation of the doors fails, if the car is not moving.

Car Equipment. The interior of the car should be ventilated and illuminated with at least two electric lamps. Lighting provided at the landing edge of the car platform should be at least 5 fc for passenger elevators and 2.5 fc for freight elevators. In addition, an emergency electric-lighting power source should be installed, to operate immediately after failure of the normal power source. For a period of at least 4 hr, this system should maintain at least 0.2 fc at a level 4 ft above the car floor and about 1 ft in front of a car station.

The car should also house, in a convenient location for operation by passengers or by an elevator operator, an alarm button, telephone, or other means for two-way communication with persons outside the hoistway, an emergency stop switch, and a signal registering device, usually a set of push buttons, that controls the stopping of the car at selected landings. To be available for use by persons in wheelchairs, the alarm button, which, when pressed, should sound an alarm outside the hoistway, and emergency stop switch should be about 35 in above the platform. The height of the highest push button or of a telephone should not exceed 54 in. A handrail should be provided about 32 in above the floor along the rear car wall.

It is also desirable that the car contain a car position indicator, located above the push buttons or the door. It should indicate by a light the number of the floor that the car is passing or at which it has stopped. Also, when the car is stopping or passing a floor that is served by the elevator, an audible signal should be given to advise the passengers of the occurrence.

Similarly, a visual and audible signal should be given at each hoistway door to indicate in the hall, or lobby, that the car is stopping at that floor in response to a call. The audible signal should sound once for the up direction and twice for the down direction of car travel. Call buttons for summoning cars should be located in the elevator lobbies, or halls, about 42 in above the floor. A lamp should light when a call is registered and go out when the call has been answered by a car.

An emergency exit should be provided in the roof of each car. Also, means should be available for operating the car from its top during inspection, maintenance, and repair. In addition, an electric light and convenience outlet should be installed on the roof, with a switch near the fixture.

See also Art. 22-12.

Car Capacities and Sizes. Cars are rated in accordance with their load-carrying capacity. For passenger elevators, capacities generally range from 1500 lb for use in apartment buildings to 5000 lb for use in department stores and hospitals. (Approximate capacity in passengers can be estimated by dividing the rated capacity, lb, by 150.) Capacities of freight elevators usually range from 1500 lb for light duty to 10,000 lb for general-purpose work or 20,000 lb for heavy duty.

The National Elevator Industry standard, "Elevator Engineering Standard Layouts," lists standard car platform sizes for various rated capacities for electric and hydraulic passenger, hospital, and freight elevators. The sizes give clear inside width and depth of the cars. To obtain the outside dimensions of a car, add 4 in to the clear width (parallel to car door) and the following to the clear depth:

11 in for passenger elevators with center-opening doors or a single sliding door
12½ in for passenger and hospital elevators with two-speed or two-speed, center-opening doors at one end only
21 in for hospital elevators with two-speed front and rear doors
7 in for freight elevators with front doors only
10 in for freight elevators with front and rear doors

Cars supplied by various manufacturers, however, may differ somewhat in sizes from those recommended in the standard. Consequently, it is advisable to obtain recommended car sizes from the car manufacturer or elevator installer for a specific installation.

22-12. Electric Elevators.

An electric-elevator installation requires, in addition to the car described in Art. 22-11 and the hoistway components described in Art. 22-10, wire ropes for raising and lowering the car and for other purposes, a driving machine, sheaves for controlling rope motion, control equipment for governing car movements, a counterweight, and safety devices (Fig. 22-11).

Driving Machines. Components of an electric driving machine include an electric motor, a brake, a drive shaft turned by the motor, a driving sheave or a winding drum, and gears, if used, between the drive shaft and the sheave or drum. The brake operates through friction on the drive shaft to slow or halt car movement. Hoisting-rope movement is controlled by the driving sheave or the winding drum around which the ropes are wound.

Traction machines are generally used for electric elevators. These machines have a motor directly connected mechanically to a driving sheave, with or without intermediate gears, and maintain and control motion of the car through friction between the hoisting ropes and the driving sheave. Also called a traction sheave, this wheel has grooves in its metal rim for gripping the ropes.

Geared-traction machines, used for slow elevators, have gears interposed between the motor and the driving sheave. The gearing permits use of a high-speed ac or dc motor with low car speeds, for economical operation. (AC motors usually are used for slow freight elevators.)

Gearless traction machines, in contrast, are used with dc motors for elevators that operate at speeds of 35 ft/min or more. Intermediate gearing is not needed for this type of driving machine because the traction sheave and the brake drum are mounted directly on the motor shaft. As a result of this type of mounting, however, the traction sheave must turn at the same speed as the motor. This arrangement is satisfactory for high-speed elevators, but for slow elevators, the motor speed would have to be too slow for efficient operation.

A **winding-drum machine** gear-drives a grooved drum to which the hoisting ropes are attached and on which they wind and unwind. Because of the limitations of this arrangement, this type of machine is usually used only for freight elevators with speeds not exceeding 50 ft/min and with a travel, or rise, not exceeding 40 ft. The installation does not require counterweights, which are generally used with traction machines.

Elevator Control. The system governing starting, stopping, direction of motion, speed, and acceleration and deceleration of the car is called control. Multivoltage control (also known as variable-voltage control) or rheostatic control is generally used for electric elevators.

Multivoltage control usually is used with driving machines with dc motors. For elevator control, the voltage applied to the armature of the motor is varied. Because buildings usually are supplied with ac power, the variable voltage generally is obtained from a motor-generator set that converts ac to dc. This type of control commonly is used for passenger elevators because it combines smooth, accurate speed regulation with efficient motor operation and permits rapid acceleration and deceleration and accurate car stops, with low power consumption and little maintenance. But multivoltage control costs more initially than rheostatic control.

Rheostatic control is achieved by varying the resistance or the reactance of the driving machine motor to change its speed. This type of control is often used with an ac motor for low-rise, slow elevators that are used infrequently (less than five trips per hour).

Car Leveling at Landings. Elevator installations should incorporate equipment capable of stopping elevator cars level with landings within a tolerance of ½ in under normal loading and unloading conditions. Because changing car loads vary the stretch of the hoisting ropes, provision should be made to compensate for this variation and keep the car platform level with the landing. Either of two types of leveling equipment may be used: inching or automatic leveling.

Inching permits an operator to move a car with continuous-pressure push buttons or with levers to floor level from within a limited zone (about 9 in above or below) with the car door open.

Automatic leveling controls the driving motor to level the car. In the best types of elevator installation, a two-way, automatic-maintaining leveling device is used, to correct the car level on both overrun and underrun at a landing and hold the car level with the landing during car loading and unloading.

Terminal Stopping Devices. For safety, provisions should be made to control car movement as it approaches a terminal landing and to keep it from passing the terminal. For the purpose, special speed-limiting and stopping devices are needed.

An **emergency terminal speed-limiting device** is required to reduce car speed automatically as the car approaches the terminal landing. This should be done independently of the functioning of the operating device, which actuates the elevator control, and of the normal terminal stopping device if it should fail to slow the car down as intended.

The **normal terminal stopping device** slows down and stops the car at or near a terminal landing independently of the functioning of the operating device. It should continue to function until the final terminal stopping device operates.

The **final terminal stopping device** is required to interrupt automatically the electric power to the driving-machine motor and brake after the car has passed a terminal landing. But this device should not operate when the car has been stopped by the normal terminal stopping device. When the final terminal stopping device has been actuated, normal car operating devices should be rendered incapable of moving the car.

Car and Counterweight Safeties. A safety is a mechanically operated device that is capable of stopping and supporting the weight of an elevator car and its load when the device is actuated by a car-speed governor. The safety should be actuated when the car travels at more than 15% above its rated speed.

At least one safety should be attached to the car frame and one should be placed within or below the bottom members of the frame (safety plank). When tripped, the safeties should squeeze the guide rails hard enough to stop the car. (When a hoistway is located above an accessible space, safeties such as those used for cars should also be provided on the counterweight frame.) The safeties should be capable of being released by upward motion of the car.

The governor may be conveniently located in the machine space, where the device will not be struck by the car or the counterweight if either should overtravel. The governor may measure car speed from the rotation of a sheave around which is wound a wire rope connected to the car and held under tension. When the car goes too fast, the governor trips jaws that grip a wire rope connected through linkages to a safety and release a spring to actuate the safety. Also, electrical switches on the governor and the safety are opened to remove power from the driving machine and apply a friction brake to the drive shaft.

Counterweights. Power requirements of the driving machine for moving the car may be reduced by hanging a counterweight on the hoisting ropes. Use of a counterweight also is advantageous for maintaining traction between the hoisting ropes and the driving sheave. The weight of the counterweight usually is made equal to the weight of the unloaded car and the ropes plus about 40% of the rated load capacity of the car (Fig. 22-12).

A counterweight usually is made up of blocks of cast iron set in a steel frame. Moving up as the car moves down and down when the car moves up, the counterweight is kept in a fixed vertical path by upper and lower wheels that are attached to its frame and roll along a pair of guide rails.

Roping for Elevators. The American National Standard Safety Code for Elevators, Dumbwaiters, Escalators and Moving Walks, ANSI A17.1, requires that a car be suspended from at least three hoisting ropes for traction-type machines and two ropes for winding-drum machines. At least two ropes are needed for a counterweight. All these ropes must be at least ½ in in diameter.

A wire rope for an elevator installation should be made up of a group of steel strands laid helically around a hemp core. Each strand, in turn, should consist of steel wires placed helically around a central wire and should have a symmetrical cross section.

For a given weight of car and load, the method of roping an elevator has a considerable effect on car speed and loading on the hoisting ropes, machine bearings, and building structural members. The simple arrangement of hoisting ropes, car and counterweight shown in Fig. 22-12a, for example, is called **1:1 roping,** because car speed equals rope speed. The ropes are attached to the top of the car frame, wind around the driving sheave, bend around a deflector sheave and then extend downward to the top of the counterweight. This rope arrangement is also known as **single-wrap roping,** because the ropes pass over the driving sheave only once between the car and the counterweight. The 1:1 single-wrap roping often is used for high-speed passenger elevators.

Fig. 22-12 Types of roping for electric traction elevators. Rope tension and loads imposed on sheaves and supports depend on type of roping, car weight W and car capacity C.

For single-wrap roping, the rim of the driving sheave is given wedge-shaped or undercut grooves, to obtain sufficient traction. The sheave grips the ropes because of a wedging action between the sides of the grooves and the ropes. The pinching, however, tends to shorten rope life.

For good traction with less rope wear, **double-wrap roping** (Fig. 22-12b) is frequently used for high-speed passenger elevators instead of single-wrap. For double-wrap roping, the ropes are attached to the top of the car frame, wind twice around the driving sheave and an idler sheave, and are then deflected down to the counterweight by the idler sheave. Because of the double wrap, less grip is needed at the driving sheave, as a result of which its rim may be given U-shaped or round-seat grooves, which cause less rope wear. The 1:1 double-wrap roping shown in Fig. 22-12b, however, applies twice the load to the driving sheave for the same weight of car and counterweight as does the single-wrap roping in Fig. 22-12a and requires a heavier design for affected components.

For the double-wrap roping shown in Fig. 22-12c, rope speed is twice the car speed. The arrangement, called **2:1 roping,** is suitable for heavily loaded, slow freight elevators. For this arrangement, a higher-speed, less costly motor can be used for a given car speed than with 1:1 roping. The ropes in this case are not attached to the car and counterweight as for 1:1 roping but instead wind around idler sheaves on them, and the ends of the ropes are anchored at the top of the hoistway at beams. As a result, the load on the driving and idler sheaves is only about one-half that for 1:1 roping.

In most buildings, driving machines are located in a penthouse. When a machine must be placed in a basement (Fig. 22-12d), the load on the overhead supports is increased, rope length is tripled and additional sheaves are needed, adding substantially to the cost. Other disadvantages include higher friction losses and a larger number of rope bends, requiring greater traction between ropes and driving sheave for the same elevator loads and speeds; higher power consumption; greater rope wear; and, consequently, larger operating expenses.

Figure 22-12e shows a type of 2:1 roping suitable for slow, low-rise elevators. In contrast to the roping in Fig. 22-12c, only one end of the hoisting ropes is dead-ended at the top of the hoistway. The other end is attached to the counterweight. Also, the ropes pass around idler sheaves at the car that are placed on the underside of the car frame.

Elevator Operating Systems. The method of actuating elevator control is called elevator operation. Many types of operation are available and some are complex and sophisticated. These may cost more than the simpler systems for installation and operation, but the sophisticated systems accomplish more automatically and handle traffic more efficiently. Following are descriptions of several types of operation:

Car-Switch Operation. With a manually operated car switch or continuous-pressure buttons in the car, an operator controls movement and direction of travel of the car. To ensure that the operator controls car movement, the handles of lever-type operating devices should return to the stop position and latch there automatically when the operator's hand is removed. In automatic car-switch, floor-stop operation, the operator releases the lever or button to stop the car at a landing. Slowing and stopping are then accomplished automatically.

Preregister Operation. As in car-switch operation, the car is controlled by an operator in it. The operator initiates stops, when advised by passengers in the car or by an audible or visible signal when the car approaches a stop point, by pressing and releasing a push button. The stop is accomplished automatically. Stop signals are preregistered by passengers pushing a button **(hall button)** in the lobby, or hall, at a landing.

Signal Operation. The car can be started only by an operator pushing a start button in the car. The operator can register stops in advance by pressing and releasing a push button in the car corresponding to the predetermined floor number. Persons calling for the elevator can similarly register a stop by pushing an up or down hall button. The car automatically stops at landings for which signals were registered, regardless of the direction of car travel or of the sequence in which buttons at various floors were pressed. When a landing is served by two or more elevators, the first available car approaching the floor in the specified direction makes the stop automatically.

Automatic Operation. An operator is not needed; starts and stops are signaled by passengers in the car or by hall buttons or by an automatic operating mechanism. The car starts either in response to this mechanism or when a passenger presses a car or hall button. Responding to signals from car or hall buttons, the car travels to and stops at the signaled landings, and car and hoistway doors open automatically. Following, in order of increasing sophistication, are descriptions of several types of automatic operation.

Single Automatic Operation. The car starts when a passenger presses and releases a car button corresponding to a landing. The car then travels to that floor and stops. The car also starts when a hall button is pressed, travels to that landing and stops. After any button has been pressed, depression of any other button has no effect on car movement until the stop signaled by the first button has been made.

Nonselective Collective Automatic Operation. Signaling is similar to that for single automatic operation. In this case, however, all stops signaled are made, regardless of the number of buttons that were pressed or the sequence in which the signals were registered. Also, regardless of the direction of travel, the car stops in the order in which landings are reached. This type of operation has the disadvantage that passengers waiting at an intermediate landing who may, for example, wish to descend may be served by an elevator going up with passengers.

Selective Collective Automatic Operation. This eliminates the preceding disadvantage of nonselective operation. When a car button corresponding to a landing is pressed, the car travels to that floor but also, on the way, makes other stops signaled. Hall calls are answered in the order in which landings are reached in each direction of travel, regardless of the sequence in which signals were received. All up calls are served when the car is traveling upward, and all down calls are answered when the car is traveling downward.

Group Automatic Operation. This is an extension of selective collective operation to a group of cars serving the same landings. A supervisory control system automatically dispatches cars to answer calls and coordinates the operation of the group. A call is answered by the first car to approach a landing in the proper direction. In response to a timer, the cars leave a terminal at predetermined intervals, which can be varied in accordance with traffic requirements. Group operation with automatic dispatching increases the number of passengers the elevators can carry in a given time.

Elevator Group Supervision. The supervisory control system for group automatic operation of elevators should be capable of making adjustments for varying traffic conditions. It should control car motion so that cars in the best location for answering calls do so. Also, the system should dispatch cars from upper and lower terminals in a predetermined, regulated order.

For example, for an office building, the supervisory control system may schedule trips as follows:

At the start of the work day, when traffic moves predominantly upward, cars are dispatched from the bottom terminal either at the end of a specific time interval or when they are loaded to 80% of capacity, as indicated by a load-weighing device in the car platform. As cars finish serving the highest call signaled, they immediately return to the bottom terminal.

In tall buildings with a very heavy up peak, the building may be divided into high, low, and, if necessary, intermediate zones, to reduce average round-trip time. One elevator then is assigned to each zone and acts as an express, bypassing floors, outside its zone.

22-28 Vertical Circulation

At the end of the work day, when traffic moves predominantly downward, parking at the bottom of the terminal of cars waiting to load is eliminated. Cars start upward immediately when a call is registered and answer up calls normally. As cars finish serving the highest call signaled, they start down, answering downward calls in the order in which landings are approached. When the cars become fully loaded, however, they automatically bypass remaining stop calls. As may be done for the up peak, the building may be divided into zones to improve round-trip time during the down peak in tall buildings.

To reduce passenger waiting time and car travel between up and down peak periods, when traffic is moderate and generally a mixture of up and down travel, **multiple zoning operation** may be used. In this case, the supervisory control system parks, in a convenient location in each zone, a car for which no calls have been registered from that zone. When a call is received within the zone, the parked car will be near the floor to be served. When demand builds up, for example, at lunchtime, the supervisory system dispatches cars at rapid intervals and shifts, when necessary, to up-peak or down-peak operation.

When traffic is very small, as usually is the case in office buildings at night, weekends, and holidays, cars not required for operation are parked at the bottom terminal. The power supply to them may be shut off (operation of ac-dc electronic rectifiers or motor-generator sets stopped). As the cars are needed, the power supply is restored automatically. Also, in buildings not equipped with a freight elevator, one or more of the elevators may be used during off hours as a service elevator to hoist deliveries.

Additional Safety Devices. Automatic elevators require several safety devices in addition to those usually installed in elevators with operators. These devices include:

An automatic load weigher to prevent doors from closing and the car from starting when it is overloaded

Car and hall buttons that passengers can push to stop the doors from closing and to hold them open

Means for preventing doors from closing when the entrance is obstructed

Emergency power system that is activated as soon as the primary system fails

Lights to indicate landings for which calls have been registered

Two-way communication with a supervisor outside the hoistway

(G. R. Strakosch, "Vertical Transportation: Elevators and Escalators," John Wiley & Sons, Inc., New York.)

22-13. Hydraulic Elevators.

For low-rise elevators, hydraulic equipment may be used to supply the lift. The car sits atop a plunger, or piston, which operates in a pressure cylinder (Fig. 22-13). Oil serves as the pressure fluid and is supplied through a motor-driven positive-displacement pump, actuated by an electric-hydraulic control system.

To raise the car, the pump is started, discharging oil into the pressure cylinder and forcing the plunger up. When the car reaches the desired level, the pump is stopped. To lower the car, oil is released from the pressure cylinder, returning to a storage tank.

Single-bearing cylinders (Fig. 22-14a) are a simple type that operate like a hydraulic jack. They are suitable for elevator and sidewalk lifts where the car is guided at top and bottom, preventing eccentric loading from exerting side thrust on the cylinder bearing. A cylinder of heavy steel usually is sunk in the ground as far as the load rises. The plunger, of thick-walled steel tubing polished to a mirror finish, is sealed at the top of the cylinder with compression packing. Oil is admitted under pressure near the top of the cylinder, while air is removed through a bleeder.

A different cylinder design should be used where the car or platform does not operate in guides. One type capable of taking off-balance loads employs a two-bearing plunger (Fig. 22-14b). The bearings are kept immersed in oil.

Another type, suitable for general industrial applications, has a movable bearing at the lower end of the plunger to give support against heavy eccentric loads (Fig. 22-14c). At the top of the cylinder, the plunger is supported by another bearing.

For long-stroke service, a cage-bearing type can be used (Fig. 22-14d). The cage bearing is supported by a secondary cylinder about 3 ft below the main cylinder head. Oil enters under pressure just below the main cylinder head, passes down through holes in the bearing, and lifts the plunger.

When the car or platform is not heavy enough to ensure gravity lowering, a double-acting cylinder may be used (Fig. 22-14e). To raise the plunger, oil is admitted under pressure below the piston; to lower it, oil is forced into the cylinder near the top, above the piston, and flows out below. Jack plunger sizes for the various types range from 2½ in in diameter for small low-capacity lifts to 18 in for large lifts, operating at 150 to 300 psi.

Hydraulic elevators have several advantages over electric elevators: They are simpler. The car and its frame rest on the hydraulic plunger that raises and lowers them. There are no wire ropes, no overhead

Fig. 22-13 Hydraulic elevator.

equipment, no penthouse. Without heavy overhead loads, hoistway columns and footings can be smaller. Car safeties or speed governors are not needed, because the car and its load cannot fall faster than normal speed. Speed of the elevators is low; so the bumpers need be only heavy springs.

Capacity of hydraulic passenger elevators usually ranges from 1200 to 5000 lb at speeds from 75 to 200 ft/min. With gravity lowering, down speed may be 1.5 to 2 times up speed. So the average speed for a round trip can be considerably higher than the up speed. Standard hospital elevators have capacities of 3500 to 5000 lb at speeds of 75 to 200 ft/min.

Capacity of standard freight elevators ranges from 2500 to 8000 lb at 50 to 125 ft/min, but they can be designed for much greater loads.

(G. R. Strakosch, "Vertical Transportation: Elevators and Escalators," John Wiley & Sons, Inc., New York.)

22-14. Planning for Passenger Elevators.

The number of passenger elevators needed to serve a building adequately depends on their capacity, the volume of traffic, and the interval between cars.

Traffic is measured by the number of persons handled in 5-min periods. For proposed buildings, peak traffic generally can be estimated from comparisons with existing buildings of the same class and occupancy located in the same or a similar neighborhood. Or the peak traffic may be estimated in terms of the probable population. The maximum 5-min traffic flow can be counted (in existing buildings) or estimated from the population of the building, which in turn may be computed from the net rentable area.

Dividing the peak 5-min traffic flow by the 5-min handling capacity of an elevator gives the minimum number of elevators required. The 5-min handling capacity of an elevator is determined from the round-trip time.

$$HC = P\frac{300}{T} \tag{22-1}$$

Fig. 22-14 Jacks commonly used for hydraulic elevators: (*a*) single-bearing plunger for guided loads; (*b*) two-bearing plunger for off-balance loads; (*c*) movable-bearing plunger for heavy service; (*d*) cage bearing for long-stroke service; (*e*) double-acting plunger.

where HC = handling capacity of car, persons in 5 min or 300 s
P = car capacity, persons
T = round-trip time of car, s

The minimum number of elevators required can then be computed from

$$n = \frac{V}{\text{HC}} = \frac{VT}{300P} \tag{22-2}$$

where V = peak traffic, persons in 5 min. Equation (22-2) indicates that the minimum number of elevators required is directly proportional to the round-trip time for a car and inversely proportional to the car capacity. Consequently, space requirements for elevators in a building, being generally dependent on the number of elevators needed, can be reduced by use of large, fast cars. For economy, however, unduly large or fast cars may be undesirable. Large cars cost more than small cars, not only initially but also for operation, because cars often travel unloaded or only partly loaded. Fast cars may not be economical because the short time between stops may not allow sufficient time for the car to accelerate to full speed. For these reasons, an optimum car is not likely to be the largest or fastest car available.

In addition to handling capacity, there is another factor to be considered in determining the number of elevators for adequate service—**interval,** the average time between elevators leaving the ground floor. As an indication of the length of time passengers must wait for an elevator, it is a significant measure of good service.

After the number of elevators has been computed on the basis of traffic flow, the average interval should be checked. It is obtained by dividing the round-trip time by the number of elevators.

Round-trip time is composed principally of the time for a full-speed round-trip run without stops, time for accelerating and decelerating per stop, time for leveling at each stop, time for opening and closing gates and doors, time for passengers to move in and out, reaction time of operator, lost time due to false stops, and standing time at top and bottom floors.

Opening and closing of doors may contribute materially to lost time unless the doors are properly designed. A 3-ft 6-in opening is good, because two passengers may conveniently enter and leave a car abreast. A slightly wider door would be of little advantage. Department stores, hospitals, and other structures served by larger passenger elevators (4000 lb and over) usually require much bigger openings.

Center-opening doors, preferred for power operaton, are faster than either the single or two-speed type of the same width. The impact on closing is smaller with the center-opening door; hence, there is less chance of injuring a passenger. Also, transfer time is less since passengers can move out as the door starts to open.

Another factor affecting passenger-transfer time is the shape of the car. The narrower and deeper a car, the greater is the time required for passengers to leave likely to be.

(See also Elevator Operating Systems and Elevator Group Supervision in Art. 22-12.)

Elevators in General-Purpose Buildings. For a proposed diversified-tenancy, or general-purpose, office building, peak traffic may be estimated from the probable population computed from the net rentable area (usually 65 to 70% of the gross area). Net rentable area per person may range from 90 to 135 ft^2. However, when several floors are occupied by a single organization with a large clerical staff, the population density may be much higher.

Diversified-tenancy office buildings usually have important traffic peaks in the morning, at noon, and in the evening, and very little interfloor traffic. The 5-min morning peak generally is the controlling factor, because if the elevators can handle that peak satisfactorily, they can also deal with the others. In a well-diversified office building, the 5-min peak will be about one-ninth of the population; but it can rise to one-eighth if the tenancy is not sufficiently diversified.

22-32 Vertical Circulation

For busy, high-class office buildings in large cities, time intervals between elevators may be classified as follows: 20 s, excellent; between 20 and 25 s, good; between 25 and 30 s, fair; between 30 and 35 s, poor; and over 35 s, unsatisfactory. In small cities, however, intervals of 30 s and longer may be satisfactory.

When elevators are divided into local and express service, it is desirable to equalize the intervals. Often, this cannot be done because the necessary number of express elevators would not be economically justified. Intervals for high-use elevators of 30 to 35 s, therefore, may be considered acceptable.

Car speeds used vary with height of building: 4 to 10 stories, 200 to 500 ft/min; 10 to 15 stories, 600 to 700 ft/min; 15 to 20 stories, 700 to 800 ft/min; 20 to 50 stories, 900 to 1200 ft/min; and over 50 stories, 1200 to 1800 ft/min.

Elevators should be easily accessible from all entrances to a building. For maximum efficiency, they should be grouped near the center. Except in extremely large buildings, two banks of elevators located in different parts of the structure should not serve the same floors. Generally, traffic cannot be equalized on the two banks, and the bank near the principal exit or street will handle most of the traffic.

Elevators should not serve two lower terminal floors, unless double-deck cars are used. (Double-deck cars are sometimes used in very tall buildings to reduce space requirements for hoistways by increasing car-handling capacity by substitution of two platforms for a single, large platform.) The extra stop increases the round-trip time and decreases the handling capacity. If there is sufficient traffic between the two lower floors, escalators should be installed.

In laying out an elevator group, not more than four elevators should be placed in a straight line. Elevators placed in alcoves off the main corridor make a good arrangement. Such a layout eliminates interference between elevator and other traffic, makes possible narrow corridors, saves space in the upper floors, decreases passenger-transfer time, and reduces walking distances to the individual elevators. Width of alcove may be 10 to 12 ft.

It is necessary to divide elevator groups into local and express banks in tall buildings, especially those with setbacks and towers, and in low buildings with large rental areas. In general, when more than eight elevators are needed, consideration should be given to such an arrangement. One or two of the elevators of the high-use express bank should be provided with openings at all floors for Sunday and holiday service. In addition to improving service, the division into local and express banks has the advantage that corridor space on the floors where there are no doors can be used for toilets, closets, and stairs.

Unless the building is very large (500,000 ft^2 of net rentable area), separate service elevators are not needed for a diversified-tenancy office building. One of the passenger elevators can be detached from the bank for freight work when passenger traffic is light.

Elevators in Single-Purpose Buildings. Elevator requirements and layouts are similar in general for both single-purpose and diversified-tenancy office buildings, but several different factors should be taken into consideration: Single-purpose buildings are occupied by one large organization. Generally, the floors that are occupied by the clerical staff are not subdivided into many offices; the net rentable area is about 80% of the gross area. Population densities are higher than for general-purpose buildings. Depending on the kind of business to be carried on, population density varies from 40 ft^2 per person for some life-insurance companies to about 100 for some telephone companies.

While traffic peaks occur at the same periods as in the diversified-tenancy type, the morning peak is very high, unless working hours are staggered. The maximum 5-min periods may be one-eighth to one-half the population, depending on the type of occupancy. Because of this volume of traffic, elevators should have a capacity of 4000 lb or more, to keep the number required to a minimum. Furthermore, interfloor traffic is large in this type of building and must be considered in designing the elevator system. Usually, so many elevators are required for handling traffic that the interval is satisfactory.

Elevators in Government Buildings. Municipal buildings, city halls, state office buildings, and other government office buildings may be treated the same as single-purpose office buildings. Population density often may be assumed as one person per 100 ft^2 of net area. The 5-min maximum peak occurs in the morning, and may be as large as one-third the population.

Professional-Building Elevators. Population cannot be used as the sole basis for determining the number of elevators needed for buildings occupied by doctors, dentists, and other professional people, because of the volume of patient and visitor traffic. Peaks may occur in the forenoon and midafternoon, with the maximum taking place when reception hours coincide. Traffic studies indicate that the maximum peak varies from two to six persons per doctor per hour up and down.

Since crowding of incapacitated patients is inadvisable, elevators should be of at least 3000-lb capacity. If the building has a private hospital, then one or two of the elevators should have stretcher-size platforms, 5 ft 4 in by 7 ft 8 in.

Hotel Elevators. Hotels with transient guests average 1.3 to 1.5 persons per sleeping room. They have pronounced traffic peaks in morning and early evening. The 5-min maximum occurs during checkout hour and is about 10% of the estimated population, with traffic moving in both directions.

Ballrooms and banquet rooms should be located on lower floors and served by separate elevators. Sometimes it is advisable to provide an express elevator to serve heavy roof-garden traffic. Passenger elevators should be of 3000-lb capacity or more to allow room for baggage carriers. Intervals for passenger elevators should not exceed 50 s.

Service elevators are very important in hotels. They should have a capacity of at least 4000 lb, with platforms about 8 ft by 9 ft 11 in. The number required usually is determined by comparison with existing hotels.

Apartment-Building Elevators. Multistory residential buildings do not have peaks so pronounced as other types of buildings. Generally, the evening peak is the largest. Traffic flow at that time may be 6 to 8% of the building population in a 5-min period. Building population may be estimated at 2.6 persons per apartment.

One elevator rated at 2000 lb and 200 ft/min usually is sufficient for a six-story apartment building with 50 to 75 units. For taller buildings, two elevators operating at 500 to 700 ft/min are generally adequate. One of these elevators can be used for service work at times.

Department-Store Elevators. Department stores should be served by a coordinated system of escalators and elevators. The required capacity of the vertical-transportation system should be based on the transportation or merchandising area and the maximum density to which it is expected to be occupied by shoppers.

The **transportation area** is all the floor space above or below the first floor to which shoppers and employees must be moved. Totaling about 80 to 85% of the gross area of each upper floor, the transportation area includes the space taken up by counters, showcases, aisles, fitting rooms, public rooms, restaurants, credit offices, and cashiers' counters but does not include kitchens, general offices, accounting departments, stockrooms, stairways, elevator shafts, or other areas for utilities.

The **transportation capacity** is the number of persons per hour that the vertical-transportation system can distribute from the main floor to the other merchandising floors. The ratio of the peak transportation capacity to the transportation area is called the density ratio. This ratio is about 1 to 20 for a busy department store. So the required hourly handling capacity of a combined escalator and elevator system is numerically equal to one-twentieth, or 5%, of the transportation area. The elevator system generally is designed to handle about 10% of the total.

22-34   Vertical Circulation

The maximum peak hour usually occurs from 12 to 1 p.m. on weekdays and between 2 and 3 p.m. on Saturdays.

The type of elevator preferred for use with moving stairs is one with 3500-lb capacity or more. It should have center-opening, solid-panel, power-operated car and hoistway doors, with at least a 4-ft 2-in opening and a platform 7 ft 8 in by 4 ft 7 in.

22-15. Hospital Elevators.

Traffic in a hospital is of two types: (1) medical staff and equipment and (2) transient traffic, such as patients and visitors. Greatest peaks occur when visitor traffic is combined with regular hospital traffic. Waiting rooms should be provided at the main floor and only a limited number of visitors should be permitted to leave them at one time, so that the traffic peaks can be handled in a reasonable period and to keep corridors from getting congested. In large hospitals, however, pedestrian and vehicular traffic should be separated.

For vehicular traffic or a combination of vehicular and pedestrian traffic, hospital elevators should be of stretcher size—5 ft 4 in to 6 ft wide and 8 or 9 ft deep, with a capacity of 4000 to 5000 lb. Speeds vary from 100 to 700 ft/min for electric elevators, depending on height of building and load. The elevators should be self-leveling, to avoid bumping patients on and off. For staff, visitors, and other pedestrian traffic, passenger-type elevators, with wide, shallow platforms, such as those used for office buildings, should be selected (see Art. 22-14).

Elevators should be centrally located and readily accessible from the main entrance. Service elevators can be provided with front and rear doors and, if desired, so located that they can assist the passenger elevators during traffic peaks.

22-16. Planning for Freight Elevators.

In low-rise buildings, freight elevators may be of the hydraulic type (Art. 22-13), but in taller buildings (higher than about 50 ft) electric elevators (Art. 22-12) generally will be more practical. Fig. 22-15 shows the components of an electric freight elevator.

In planning for freight elevators, the following should be considered:

1. Building characteristics, including the travel, number of floors, floor heights and openings required for a car. Also, structural conditions that may influence the size, shape, or location of the elevator should be studied.

2. Units to be carried on the elevator—weight, size, type, and method of loading.

3. Number of units to be handled per hour.

4. Probable cycle of operation and principal floors served during the peak of the cycle.

5. Freight elevators are not permitted to carry passengers.

For low-rise, slow-speed applications, especially where industrial trucks will be used, rugged hydraulic freight elevators generally will be more economical than electric freight elevators.

Classification of Freight Elevators. The American National Standard Safety Code for Elevators, Dumbwaiters, Escalators and Moving Walks, ANSI A17.1, defines three classes of freight elevators. Class A applies to general freight loading. This is defined as a distributed load that is loaded or unloaded manually or by hand truck and no unit of which, including loading equipment, weighs more than one-quarter the rated load of the elevator. Class B elevators may handle only motor vehicles. Class C elevators may be subjected to heavy concentrated loads and fall into one of three subclasses. Class C1 applies to elevators that carry industrial trucks, Class C2 to elevators for which indus-

Fig. 22-15 Electric traction freight elevator. *(Courtesy of Otis Elevator Co.)*

trial trucks are used only for loading and unloading the cars and are not carried by them, and Class C3 to elevators carrying heavy concentrated loads other than trucks.

Electric-Elevator Controls. Two types of control equipment are generally used for freight elevators—multivoltage and ac rheostatic (see Art. 22-12).

With **multivoltage,** the hoisting motor is dc operated. An ac-dc rectifier or motor-generator set is provided for each elevator, and the speed and direction of motion of the car are controlled by varying the generator field. This type of elevator permits the most accurate stops, the most rapid acceleration and deceleration, and minimum power consumption and maintenance for an active elevator. Automatic leveling, which is required for Class C1 and C3 freight elevators, compensates for cable stretch or other variations from floor level and is an inherent part of multivoltage equipment.

Standard speeds with multivoltage geared freight elevators range from 50 to 200 ft/min. Generally recommended are 100 ft/min for 2 or 3 floors, 150 ft/min for 4 or 5 floors, and 200 ft/min for 6 to 10 floors. If one of the stories exceeds 20 ft, however, the next higher speed should be used.

The **ac rheostatic type** is often chosen to keep initial cost down when the elevator is to be used infrequently (less than five trips per hour on a normal business day). If the ordinary landing accuracy of this type will not be acceptable for the loading methods expected, either of two leveling devices may be installed:

1. Inching, which may be added to the controller for a single-speed ac motor at nominal cost. It permits an operator to move the car with continuous-pressure switches to floor level within a limited zone (about 9 in) with the doors open.

2. Automatic leveling, which is achieved with a two-speed motor and control apparatus. It is available only with selective-collective or single-automatic operation (Art. 22-12) at speeds up to 100 ft/min for capacities up to 8000 lb, and at 75 ft/min for 10,000-lb elevators.

Standard speeds for ac rheostatic-type geared freight elevators are 50, 75, and 100 ft/min. Generally recommended are 50 ft/min for two floors, 75 ft/min for three or four floors, and 100 ft/min for five to eight floors. If one of the stories is more than 20 ft high, the next higher speed should be used.

Car Capacity. The size of car to be used for a freight elevator is generally dependent on the dimensions of the freight package to be carried per trip and the weight of the package and loading equipment. Power trucks, for example, impose severer strains on the entire car structure and the guide rails than do hand trucks. As a power truck with palletized load enters an elevator, most of the weight of the truck and its load are concentrated at the edge of the platform, producing heavy eccentric loading. Maximum load on an elevator should include most of the truck weight as well as the load to be lifted, since the truck wheels are on the elevator as the last unit of load is deposited.

The carrying capacity per hour of freight elevators is determined by the capacity or normal load of the elevator and the time required for a round trip. Round-trip time is composed of the following elements:

1. Running time, which may be readily calculated from the rated speed, with due allowance for accelerating and decelerating time (about $2\frac{1}{4}$ s for ac rheostatic with inching, $1\frac{3}{4}$ s for multivoltage), and the distance traveled.

2. Time for operation of the car gate and hoistway doors (manual 16 s, power 8 s).

3. Loading and unloading time (hand truck 25 s, power truck 15 s). Wherever practical, a study should be made of the loading and unloading operations for a similar elevator in the same type of plant.

Operation. The most useful and flexible type of operation for both multivoltage and ac rheostatic freight elevators is selective-collective with an annunciator. When operated with an attendant, the car automatically answers the down calls as approached when moving down and similarly answers up calls when moving up. The elevator attendant, when present, has complete control of the car and can answer calls indicated by the annunciator by pressing the corresponding car button.

The single-automatic type, available for both multivoltage and ac rheostatic controls, is a somewhat simpler, less expensive type for operation without an attendant. But it has the disadvantage that corridor calls cannot be registered when the car is in motion, nor can the car respond to more than one call at a time.

If an attendant will always operate the elevator, use may be made of car-switch releveling, which is a simple, inexpensive means of operating with variable-voltage control. The car will travel in either direction, or stop, in response to the movement of a switch, leveling automatically at each floor.

Also, a car-switch type of operation with ac rheostatic control may be used for freight elevators. The car is leveled to a mark on the hoistway door by the attendant using the normal car switch.

The least expensive type of operation is the continuous-pressure type. It is available for light-service ac rheostatic elevators. Buttons are installed in the corridor and car and, when pressed, will cause the car to move in a selected direction. The car will stop when the button is released. This control is not recommended for travel of more than 70 ft.

The standard hoistway door is the vertical biparting metal-clad wood type. For active elevators and openings wider than 8 ft, doors should be power-operated.

Automatic freight elevators can be integrated into material-handling systems for multistory warehouse or production facilities. On each floor, infeed and outfeed horizontal conveyors may be provided to deliver and remove loads, usually palletized, to and from the freight elevator. The elevator may be loaded, transported to another floor, and unloaded—all automatically.

22-17. Dumbwaiters.

These may be used in multistory buildings to transport small loads between levels. Generally too small to carry an operator or passenger, dumbwaiters are cars that are raised or lowered like elevators. They may be powered—controlled by pushbuttons—or manually operated by pulling on ropes. Powered dumbwaiters available can automatically handle from 100 to 500 lb at speeds from 45 to 150 ft/min. They also are available with special equipment for automatic loading and unloading and also designed for floor-level loading suitable with cart-type conveyances.

The American National Standard Safety Code for Elevators, Dumbwaiters, Escalators and Moving Walks, ANSI A17.1, contains safety requirements for dumbwaiters. Powered dumbwaiters may be constructed like electric elevators with winding-drum or traction driving machines or like hydraulic elevators, but many of the safety requirements for elevators are waived for dumbwaiters. Standard heights for a dumbwaiter are 3, 3½, and 4 ft. ANSI A17.1 restricts platform area to a maximum of 9 ft^2. Rated load capacity usually ranges from 20 to 500 lb. As for elevators, hoistways should have fire-resistant enclosures.

22-18. Vertical Conveyors and Pneumatic Tubes.

When there is a continuous flow of materials to be distributed throughout a multistory building, vertical conveyors may be the most economical means. In some installations, 200 lb or more of paper work and light supplies are circulated per minute.

A typical conveyor installation is similar to an escalator (Art. 22-7). A continuous roller chain is driven by an electric motor. Engaging sprockets at top and bottom, the chain extends the height of the building, or to the uppermost floor to be served. Carriers spaced at intervals along it transport trays from floor to floor at a speed of about 70 ft/min.

Like elevators, however, vertical conveyors should be enclosed in fire-resistant shafts. Generally, the only visual evidences of the existence of the installation are the wall cutouts for receiving and dispatching trays at each floor. Short gravity-control conveyor sections may lead from the receiving station into a work area.

In event of fire, vertical sliding doors, released by fusible links, should snap down over the openings, sealing off the conveyor shaft at each floor.

Operation of vertical conveyors is simple. When the trays are ready for dispatch, the attendant sets the floor-selector dial or presses a button alongside the dispatch cutout. As the trays are placed on the loading station, they are automatically moved into the path of the traveling carriers. Each tray rides up and around the top sprocket and is automatically discharged on the downward trip at the preselected floor. It takes only 4½ min for a 26-story delivery of interoffice correspondence.

The best place to install a vertical conveyor is in a central location, next to other vertical shafts, to minimize horizontal runs in collecting and distributing correspondence at each level. An installation alongside a stairway is advantageous in providing access to the rear maintenance platform of the conveyor, for servicing equipment.

Vertical conveyors are made and shipped in sections about 10 ft long, which are assembled in the shaft. Principal components include the head sprocket and drive unit, the foot sprocket and take-up units, and intermediate sections consisting of braced steel shapes and tubular track guides. Generally, one or more walls of the shaft are left open until the conveyor is installed. A removable floor plate is provided above the shaft for access to the drive machinery.

Pneumatic Tubes. These also may be used to transport small loads within buildings. Units are moved through tubes under air pressure or suction, or both. Items to be transported are carried inside cylinders slightly smaller in diameter than the tubes.

In choosing between vertical conveyors and pneumatic tubes, the designer's first consideration should be the number of floors to be served. An arterial system of pneumatic tubes would satisfy the requirements of a predominantly horizontal building, whereas a vertical conveyor generally would be more advantageous in a tall building. The amount of paper work to be distributed is a second consideration. A vertical conveyor is more efficient when the volume of items to be moved is large.

22-19. Mail Chutes.

Used in multistory buildings for gravity delivery of mail from the various floors to a mailbox in the main lobby, a mail chute is simply a vertical, unpressurized, rectangular tube. With permission of the Post Office, one or more chutes may be installed in office buildings more than four stories high and in apartment buildings with more than 40 apartments.

Usually made of 20-gage cold-formed steel, with a glass front, and supported by vertical steel angles, a chute is about 3×8 in in cross section. In the front of the chute, available in a lobby in each story, a slot is provided for insertion of flat mail into the chute.

The mailbox usually is 20 in wide, 10 in deep, and 3 ft high. It should be placed with its bottom about 3 ft above the floor. The mailbox should be placed within 100 ft of the building entrance.

Section 23

Surveying for Building Construction

Reginald S. Brackett
*Formerly Architectural Superintendent,
Smith, Smith, Haines, Lundberg & Waehler,
New York, N.Y.*

Surveying for building construction can be divided into three classifications:

   1. *Preliminary.* So the architect can prepare plans for the building, information should be obtained on:
      a. Boundary lines and general topography (Art. 23-2).
      b. Streets, curbs, and pavements.
      c. Utilities—sewers, water mains, gas lines, electric and steam service.
      d. Buildings on and adjacent to the site.
      e. Subsurface conditions—from test holes and borings.
      f. Soil tests.
      g. Quantity survey (engineers' estimate).
   2. *Construction.* Condition survey of adjacent structures and establishing lines and grades. Needed are:
      a. Survey of conditions of adjacent structures that are subject to damage from new construction (Art. 23-3).
      b. Elevation readings on existing structures, including utility appurtenances and street pavements (Art. 23-4).
      c. General layout for construction—base lines, offsets, and bench marks (Art. 23-5).
   3. *Possession Survey.* Made after the building is completed.

23-1. Units of Measurement.

In land surveying, both horizontal and vertical distances are measured in feet and decimals of a foot. In building construction, however, horizontal distances are given in feet

23-2 Surveying for Building Construction

and inches, while as a rule, elevations are noted in feet and decimals of a foot. Conversion is easily made by remembering that there are 96 eighth inches in 1 ft, or ⅛ in equals 0.01 ft (more accurately 0.0104) and 1 in equals 0.0833 ft.

For measurements in the metric system, meters or millimeters may be used (see Appendix).

23-2. Survey of General Topography.

For a preliminary survey, property lines can be established by any competent person, provided the locality is properly monumented. In many cities and towns where monu-

Fig. 23-1 Building site in open country.

ments are not established, the local surveyor is the only one who has the knowledge of landmarks referred to in deeds for land in the vicinity. In such cases, it is always advisable to have the main corners of property staked out by a local licensed surveyor.

One or two bench marks are also required to establish the datum, or reference point for elevation. It is a good policy when the excavation goes below groundwater level to call water level + 100, so as to eliminate minus elevations.

The new survey should show any difference between the present standard and that used in the original survey. Surveys will differ when a city has been monumented long after buildings are in place. The differences should be noted so that the architect can

Fig. 23-2 Map of city block.

23-4 Surveying for Building Construction

make corrections before the new plans are started. It is confusing to find too late that 100 ft in the deed is only 99.85 ft.

After property lines have been staked out, record with practical accuracy all pertinent information required to make detail drawings, such as slope of ground and location of any existing buildings or excavated areas.

In general, there are two types of site—open country and built-up city plots. A map for the first would be similar to Fig. 23-1, showing contours, and a map for the second would be like Fig. 23-2, showing existing buildings. The contours can be drawn using either a cross-sectional grid system (Fig. 23-3) or the stadia method, laying off angles and distances (Fig. 23-4). The latter is perhaps a little quicker but probably less accurate.

Fig. 23-3 Contours plotted from a grid system.

(R. E. Davis, F. S. Foote, J. M. Anderson and E. M. Mikhail, "Surveying: Theory and Practice," McGraw-Hill Book Company, New York.)

The type of building to be constructed and its details will determine the frequency with which elevations and dimensions are required. Walls of existing buildings should be plumbed when on the property line.

For street curbs and pavements in this preliminary work, only the location, elevation, and condition are required.

Sizes and condition of sewers should be recorded, giving invert elevations at manholes. Other utilities can be arrived at from hydrants and other manhole covers.

Fig. 23-4 Stadia method of plotting contours.

23-6　Surveying for Building Construction

Existing buildings on the site should be shown. The number of stories, kind of material, and whether there is a cellar or basement should be noted as in Fig. 23-2.

Unless the character of the soil on which the foundation is to rest is known, test holes or borings should be made (Sec. 6).

An estimate should be made of excavation quantities. This involves plotting cross sections from elevations obtained from a survey and grading for the new building. Volumes can be figured by either the end-area formula [Eq. (23-1)] or the prismoidal formula [Eq. (23-2)].

$$V = \tfrac{1}{2}L(A_1 + A_2) \tag{23-1}$$
$$V = \tfrac{1}{6}L(A_1 + 4A_m + A_2) \tag{23-2}$$

where V is the volume in cubic feet of a prismoid with end bases or cross sections having areas in square feet A_1 and A_2 and a midsection A_m and with a length in feet of L.

23-3. Survey of Nearby Buildings.

It is always possible that adjacent buildings might be damaged by construction operations for a new building. So an examination, both inside and out, of all buildings close enough to the new construction to be affected should be made. The survey should be carried out, prior to start of construction, by a party representing the architect, contractor, and owners of the building to be inspected.

The record of this survey should be made accurately and in detail, to be available for comparison, if needed, to prove or disprove any change in appearance. The exterior can be examined, where beyond range of unaided vision, with high-powered field glasses. Record all defects or irregularities of masonry, which if noted later could be attributed to construction of the new building. It is a good policy to mention all sides where exposed, whether defective or not. If various elements of existing structures are in good condition, just state no comment.

For example, the record of an examination could read as follows: "Empire Building on the north side of North St.—South elevation, granite watertable requires pointing, no cracks in stone, a few spalls at corners, limestone above to 3rd floor, no comment. The entablature at 3rd floor has a large crack, starting at the top of the 5th column from the west corner of the building and extending diagonally upward through the architrave and frieze to the cornice, open at the top about the thickness of a mortar joint. Above the 3rd floor . . ."

The interior inspection should cover all rooms from the lowest floor to the roof. Cracks developing from settlement will generally carry from floor to floor. If a crack in a wall does not appear in rooms above and below it, it is safe to say that it did not come from settlement of the foundation.

In describing rooms, describe the floor, walls, and ceiling. Thus: "Room 203—Floor covered with asphalt tile. South wall has a diagonal crack in plaster, starting in the upper left-hand corner and running down to a point one foot above the floor and 3 ft from the right corner. Starts as $\tfrac{1}{32}$ in closing to zero. East wall has one vertical crack in center, extending from floor to ceiling, open $\tfrac{1}{16}$ in for full height. North wall—no comment. West wall has some hairline crazing near ceiling. Ceiling is acoustical gypsum tile, two pieces missing. Door to room binds at top on the lock side. Door trim split at top on the hinge side . . ."

Sewers and pavement should be examined by the architect, contractor, and a city representative. As with the building survey, photographs contribute an important part of the record.

23-4. Establishment of Elevations before Construction.

The examination described in Art. 23-3 must be supported by readings of elevations at the foundations of the various buildings to determine any settlement or lateral movement. The number and type of such reference marks will be controlled by the kind of new foundation. In general, reference points should be located and recorded in such a way that settlement or horizontal motion can be detected.

A reference grid can be laid out on street pavements and sidewalks. (Spikes driven into the pavement will aid in making future readings.) The water table, window sills, or mortar joints on existing buildings can be used as reference points to observe settlement. These have an advantage over marks placed on walls at some convenient arbitrary elevation, inasmuch as readings on them will show up any irregular previous settlement.

Establish levels inside the buildings too, if they are close to the excavation. Readings on utility castings, hydrants, valve boxes, etc., will tie in subsurface pipelines.

It is generally accepted that most buildings settle during construction as the loading increases; but it also must be remembered that settlement continues for many years. If any of the adjacent buildings are comparatively new, then differential settlement would have to be watched for. After work starts, keep a record of readings taken at regular intervals. Let the readings give warning; do not wait for cracks to appear.

23-5. General Layout for Construction.

This differs, in general, for private dwellings, low industrial buildings, and large commercial buildings. However, they all require building lines at the start as a reference for measurements.

Small Buildings. Because it is not practical to work to a masonry line without disturbing it, offset lines are used. Distances to control points—building corners, wall faces, footings, etc.—are measured from these lines. The simplest layout (Fig. 23-5) requires principally setting of batter or corner boards nailed to posts or other supports. The boards should be long enough to enable the building and offset lines to be marked, and the top should be set to a known elevation, such as level of the first floor. The posts should be driven first and grade nails set so as to determine the elevation of the boards before the lines are marked. A piece of shiplap, laid with the tongue projecting up, is generally used for the board. The step in the shiplap top is set at the desired elevation, and a saw cut the depth of the tongue allows a string to rest on the step at the correct level.

Fig. 23-5 Offset line for building layout determined with batter boards.

Layouts for low industrial buildings follow the same pattern, except that the corner boards are farther apart, and additional intermediate batter boards are needed. The position of the batter boards will vary with the depth and method used in excavation.

Large Buildings. The layout for large buildings, although basically the same as for small buildings, has to be expanded. Offset lines have to be established with the accuracy of a base line and will be called base lines in the following discussion.

Temperature correction for tape measurements is very important for layouts of large buildings; so it is necessary to have a standardized tape, which is kept in the field office

23-8   Surveying for Building Construction

and is used for this kind of work only. (The temperature at which any tape is standard can be found by comparison with a standard tape. For this purpose, the tape may be sent to the National Bureau of Standards, Washington, D.C.) A 100-ft steel tape will change ¾ in when the temperature changes 100°F.

Work that requires extreme accuracy should be done on a cloudy day, in the early morning or evening. It is very difficult to determine the temperature of a steel tape in the sun. One expedient is to measure two lines, each 100 ft long—one that will remain in the sun and the other in the shade all day, where they will be readily accessible for checking a tape at any time. This will automatically give the correction to be applied to tape measurements.

Fig. 23-6 Base-line layout for a large commercial building.

Figure 23-6 shows a typical layout of main reference lines. The offset traverse or base line, which is wholly outside the building site, can vary in distance from the building line, but 6 ft is good because the line will clear the conventional swing scaffold. The sidewalk bridge, which is necessary for protection, prevents sighting above it, unless the transit is set up outside the curb line at either end. The base line, therefore, must be produced well beyond the limits of the sidewalk bridge, so as to be of use above it.

After the base line is finished and checked, all points of use in layout should be projected to a permanent location on nearby existing buildings. Control points for base lines can in many cases be obtained by sighting on some architectural ornamentation a short distance away. Those for the inside building-line offset and column offset lines should be located as high as possible, so as to be seen from the bottom of the excavation. The roof cornice in Fig. 23-7 is ideal. Targets at such locations can be used to establish column lines by "bucking in" or double centering between control points on opposite sides.

The inside building line is useful in aligning foundation walls after the general excavation is completed. Placing removable windows in the sidewalk fence, which is part of the sidewalk bridge, permits transit lines to be dropped into the excavation.

Steel Framing. The centerlines of columns can be marked on batter boards set with the top a given number of feet above subgrade, footings, or pile caps. If piles are used, they can be staked out with a tape by measuring from the centerlines. Before columns are set, transit lines can be run on the centerlines, but after they are in place, the offset lines have to be used.

There is a certain satisfaction in being able to sight directly on the line to be used. As soon as some concrete is placed, get a permanent bench mark set in it. In a steel-frame building, an anchor bolt may be used; if not, set a special rod for this purpose.

When all the walls and columns are in place to ground level, set at least two bench marks some distance apart, preferably at each end of the building where they can be used to make vertical measurements as the building rises. Stair wells or elevator shafts are good locations for these control points.

Most mechanics work from a level mark 4 ft above each finished floor. Therefore, such marks should be placed on each column, or about 25 ft apart, in every story. They must

Fig. 23-7 Control points fo base-line are set high upon an existing building to permit sighting by an instrument in the excavation.

be tied in accurately between floors, which is the reason for the vertical base lines. Two marks must be seen from the same level setup before the 4-ft marks are established on the columns. With an extension-leg tripod, set the horizontal hair of the level directly on the taped 4-ft marks. On the original run do not set above or below the mark and read a rule. Between setting the rule and reading the backsight, and reading the rule to set the foresight, errors can be made, especially when about ½ in from it. If the rule is turned the wrong way, the error will be twice the amount of the reading.

Since the building starts to settle shortly after loading, do not try to check with an outside bench mark. Hold and watch the original vertical base lines, and from time to time take a quick check on the preestablished 4-ft marks on the lower floors to spot a differential settlement. If there is any, set the 4-ft marks on the upper floors so that, when a particular column that is slow or fast in settling catches up, all the marks will be level. In some factory buildings, settlement effects are unimportant, but when the interior finish is marble, trouble may be encountered if settlement is not taken care of in advance.

Extreme care must be used in setting columns around elevator shafts. Check each column independently with a heavy plumb bob. (It can consist merely of a 4-in pipe, 8 in long, filled with lead and suspended by piano wire.) Swinging can be reduced by suspending the bob in a pail of oil.

As the building rises above ground level, the exterior base line is the most important control line. The interior column offset lines, when used, must always be checked with the building line.

Accuracy of alignment of a steel-framed building stems from the anchor bolts and billets. It is important to have all horizontal measurements correct so that the lower section will go together without difficulty. Each tier of framing above depends on the accuracy of shop fabrication.

The elevation of the column base plates controls for the full height of the building; so it is obvious that extreme care must be used in setting them. If this is done, column splices, as a rule, will be as close to the true elevation as if established with a tape. This is very important so that the beams and girders will be in the correct vertical location. Clearances for mechanical work also could be affected by variation from the true location.

As columns are loaded, a shortening takes place. So vertical measurements must be made and tied in before this happens. Any total overall measuring done after loading will have to be corrected by calculated deformations based on the modulus of elasticity of the column material.

Fireproofing and the rough floor are located with respect to the steel. This is done because it is not practical to use an instrument on the frame until it is braced by some kind of deck, whether concrete or steel.

The frame is plumbed by the steel erector with a heavy bob and checked by the engineer using the base line produced beyond the sidewalk bridge. When the temperature in the field differs from that of the fabrication shop, the frame will be longer or shorter, an effect to be considered when checking. (Shop temperature is always a little above tape standard.)

After the structural floors are in place, the centerline is marked on each exterior column in each story for the mason and other trades to work from. This also is done from the base-line extension.

A straight piece of 1 × 1-in white pine, about 8 ft long, should be used for an offset rod, with a 6-in piece fastened to the end, forming a tee. Two nails are driven into the head of the tee to sight on. The transit man will know the rod is at right angles to his line when the nails line up.

The column marks should be set accurately, but the mechanic building the walls should be instructed to set his masonry line ¼ in inside the finish line. This is advisable to compensate for the slight variations usually found in masonry work.

The column offset lines can be plumbed from the opposite side of the street to the floor edge at both ends or sides of the building. The lines can be carried through the building by setting up on the floor and "bucking in" between the end marks.

Concrete Framing. A concrete frame requires a layout for each floor. A level mark to work from is put on a column splice rod 1 ft above the finish floor. Exterior column centers are set from the base line. Intermediate columns may be located by taping.

After column forms are stripped, 4-ft marks are established as is done with a steel frame. Column centers for masonry also are established as for a steel frame. The mason uses the 4-ft marks to determine window head and sill heights.

A competent mason foreman checks to the level mark on the floor above before he gets there.

23-6. Surveying Instruments.

For survey and layout work for building construction, generally transits and levels, supplemented by tapes and stadia and level rods, can perform all required tasks satisfactorily. Surveys for remote sites and very large lots, especially over rough terrain, can be expedited with electronic devices such as geodimeters, which use the speed of light, tellurometers, which employ microwaves, and electrotapes, which utilize radio-frequency signals.

Transits. These instruments principally provide a telescope, a scale for reading horizontal angles, a scale for reading vertical angles, and a plumb bob attachment for positioning the telescope over a point below, plus a tripod support. The telescope contains two perpendicular hairs, which when the instrument has been leveled become horizontal and vertical, for determination of line of sight, and two additional hairs on opposite sides of and parallel to the horizontal hair, for stadia measurement of distance. Transits are used mainly for property surveys, setting base lines and other reference lines, alignment, and other observations requiring measurement of horizontal or vertical angles. In choosing a transit, always select one with a telescope supported by standards high enough to permit a 360° vertical swing of the telescope and with a prism eyepiece in place. A transit equipped with an optical plummet is a time saver in high-wind setups.

Levels. These instruments consist principally of a telescope and a leveling device, a means for ensuring that the line of sight is horizontal. The telescope contains a central hair, which is horizontal when the instrument has been leveled. Levels are used mainly to determine or set elevations.

A hand level consists of a telescope designed to be held in the hand of an observer or supported on a rod. It is suitable for rough readings and for sights up to about 50 ft.

An engineer's level consists of a telescope with a level-indicating bubble and leveling screw, supported on a tripod. It may be of the wye or Dumpy type. Either is capable of measuring elevations to within 0.01 ft. When the instrument has been leveled, the telescope rotates only about a vertical axis, and the lines of sight always lie in the same plane.

For faster, more accurate sighting, a tilting level or a self-leveling level may be used. With a conventional engineer's level, the bubble must be checked before each sight. Care must be taken to observe the bubble length, which changes with temperature. This is automatically compensated for with tilting or self-leveling levels.

A tilting level is a Dumpy level with a telescope that can be titled a few degrees through a vertical angle by a micrometer screw at the eyepiece end. When a tilting level is to be set up at a new station, a preliminary setup is first made with the aid of a bullseye level, similar to that used on some cameras. When a reading is taken, the telescope bubble is reflected by three mirrors and observed through a small window at the left side of the telescope. The mirrors are arranged so as to split the image and enable the observer to see both ends of the bubble simultaneously. The observer, while viewing the split image, can line up the bubble ends with the tilting screw and thus level the telescope.

A self-leveling level, like a tilting level, has a bulls-eye level for making the initial setup. With the instrument nearly level, a self-leveling device takes over. The line of sight through the telescope is kept horizontal with the aid of gravity and three prisms. One or more of these prisms are suspended as a pendulum in the telescope. Even if the telescope is not level, gravity tends to make the pendulums plumb, a braking arrangement stopping their swing. With the pendulums plumb, the line of sight becomes horizontal. Functioning of the device can be checked by turning one of the bulls-eye screws and observing if the line of sight moves up and down. These instruments can also be obtained with stadia hairs and an azimuth scale.

(R. E. Davis, F. S. Foote, J. M. Anderson and E. M. Mikhail, "Surveying: Theory and Practice," and P. Kissam, "Surveying Practice," McGraw-Hill Book Company, New York; J. B. Evatt, "Surveying," John Wiley & Sons, Inc., New York.)

Section 24

Estimating Building Construction Costs

E. D. Lowell
Formerly Chief Estimator, Kaiser Engineers,
Oakland, Calif.

Scoping, planning, scheduling, estimating, and economic evaluation of complex engineering and construction projects require the specialized skills of many persons. Estimators, planners, and schedulers, along with architects, engineers, economists, managers, and other professionals, are very important members of the team.

Because building construction involves millions of dollars, accurate estimating is critical to the success of these projects. A poor estimate by a construction company can inflict serious losses on the company and could well threaten its solvency.

Plant capital-cost estimating has developed specialty application; that is, an estimator of commercial facilities, such as banks, hospitals, and office buildings, tends to stay in this field and further expand his knowledge and expertise in commercial construction. The broad categories of specialization are generally:

Commercial facilities
Industrial complexes
Chemical and refinery plants
Power plants—fossil and nuclear
Dams, tunnels, and highways

The following discussions, illustrations, and applications are addressed to all of these fields of specialization, with recognition, however, of the fact that each field has its own unique methods, procedures, and presentation relative to capital-cost estimating.

24-1. Estimating before, during, and after Design.

Project cost estimates are prepared by different firms for different purposes and are of varying degrees of accuracy. A general contractor may prepare lump-sum bid estimates from completed or semicompleted plans and specifications. His methods and techniques are designed for this type of estimating, which, in many cases, depends on his having firm lump-sum quotations from specialty subcontractors on 85 to 100% of the facility cost. At the other end of the spectrum, an owner's management may decide that a new facility having a certain approximate area or annual capacity is needed. Based on prior plant construction-cost data, a reasonable unit price for the area or capacity may be obtained, from which a total value for the plant, including engineering, administrative, financing, and start-up costs, can be calculated. Between these two extremes lies a broad band of estimating methods and techniques.

The general contractor bidding a lump sum usually has the advantage of completed drawings and specifications, as well as lump-sum quotations for most of the work. The architect-engineer, on the other hand, relies on his in-house conceptual estimating capability to estimate construction costs as design proceeds.

Monitoring Estimates. An architect-engineer estimator is involved in a project from its inception. The scope of his activities encompasses the entire project. During the development of a project and through detail design and construction, the estimator is involved in making continuous *monitoring estimates* to assure design-cost compliance and, if necessary, to evaluate alternative methods. (The tradeoff study, value engineering described in Art. 1-8, is an outgrowth of this. So are other evaluation techniques, which include complete reviews of life-cycle cost, scope, plan, and schedule, at least at the one-third and two-thirds points of design progress. Participants in these reviews should include the owner and the planning, management, and engineering staff, as well as the estimator.) Consequently, these monitoring estimates should be in sufficient depth to assure understanding of the project by all concerned, and should yield reliable costs commensurate with the data available and properly correlated with original concepts.

Cost Control. A related aspect of estimating, cost control of a project, requires particular attention during the conceptual and detail design stages to assure compliance with the amount appropriated for construction and with the engineering and architectural requirements of the project for quality and quantity of product. It is less costly and time-consuming to revise a "runaway" project during the detail design phase than after construction has commenced, when few alternatives or options are left. Estimating during the detail design phase is critical, and is dependent on the expertise of the conceptual estimators.

Planning-Stage Estimating. From the inception of a facility, financial studies should be continually developed, evaluated, and reevaluated on the basis of increasing in-depth studies. During these stages, for one reason or another, projects may be estimated, reestimated, and re-reestimated; many "tradeoff" studies are required as the scope of the project continually changes. This is the period in which the reliability of the estimate is a direct function of the architect-engineer conceptual estimator's knowledge and experience, over and above the quality and quantity of information available.

The form of the estimates and the estimating methods used are generally unique to the individual estimator, facility, and conditions. The estimator must spend a great deal of time in determining the scope and general parameters of the job and then translating these into meaningful, measurable quantities with reasonable units of cost that are all-encompassing and defendable. For example, for concrete construction, instead of estimating item by item the form costs, the rebar density and cost, and the concrete purchase, placing, and finishing costs, the estimator may choose to use an all-inclusive unit cost, such as $85 per cubic yard. The reason is that at this stage it is more important to determine the quantity of concrete accurately than to detail unit prices.

The estimator must involve himself in the overall project plan, its economics, its phases, and its schedule, to evaluate properly the intangibles that must always be reflected in construction costs. The scope and schedule of the project are so closely related to the estimate that isolated or inadequate analysis of scope or schedule can negate the reliability of the estimate.

24-2. Types of Estimates.

Estimates may be defined by at least two criteria, quality and end use.

Quality. The quality of a cost estimate is directly proportional to the quality and quantity of the engineering available, the expertise of the estimators, and the methods used for estimate preparation. The simplified dot chart (Fig. 24-1a) is an approach to quality definition of a cost estimate. This chart indicates that estimates may range from a doorknob estimate (Type 1), which is based on facility quantity and unit price, to an engineer's estimate (Type 5), which is an in-house cost estimate based on completed detail plans and specifications. A bid estimate (Type 6) is the same as an engineer's estimate, except that it incorporates an evaluation of specialty contractors' competitive bids.

The degree of accuracy of these estimates is indicated in Fig. 24-1b and c by the magnitude of the required contingency allowance, which may range from a probable 25% for Type 1 estimates to 5% for Type 6. No one estimate conforms completely with one type, because usually an estimate is a montage of varying qualities.

End Use. Types of cost estimates differ in accordance with the end use of the estimate. Doorknob estimates, for example, have been used for studying the economic feasibility of a project, as well as for funding or financing a project. Any quality of estimate may be prepared for any one or more of the following end uses:

Studies
Comparative appraisals
Checking
Budgeting purposes
Control
Feasibility studies
Financing
Proposals
Bids
Value engineering
Scoping efforts

Doorknob and Magnitude Estimates. Figure 24-1a indicates that the data available for preparing Type 1 and 2 estimates are very general in nature. These estimates therefore cannot be prepared by the accepted methods of quantity takeoff and pricing. These types of estimates are normally prepared by a highly skilled conceptual estimator, who from his own knowledge and experience applies ratio and comparison techniques to determining the cost of the major features. The techniques and methods employed are unique to the estimator and unique for the project.

Often, such estimates are merely the proposed plant's annual (or daily) capacity times a unit cost, or ratios of total plant costs from prior job histories factored to the proposed capacity, by use of the sixth-tenth factor or some other method of cost-capacity adjustments. This type of estimating is extremely demanding of an estimator and important in the early feasibility study stage.

Preliminary, Definitive, and Engineer's Estimates. Type 3, 4, and 5 estimates may be further subdivided according to the progress of the project. For instance, an engineer's estimate, Type 5, may be made when detail design is 30% complete, 60% complete, and 100% complete. These estimates are usually prepared by architect-engineer conceptual estimators and are based on quantity takeoffs and pricing of partly or fully

	TYPES OF ESTIMATES					
	DOOR-KNOB	MAGNITUDE	PRELIM'RY.	DEFINITIVE	ENGINEER'S	BID
INFORMATION REQUIRED	TYPE 1	TYPE 2	TYPE 3	TYPE 4	TYPE 5	TYPE 6
PRODUCT, CAPACITY, & LOCATION	●	●	●	●	●	●
FACILITY DESCRIPTION	●	●	●	●	●	●
PLANT LAYOUT	●	●	●	●	●	●
TIME FOR ESTIMATE PREPARATION	●	●	●	●	●	●
LIST OF EQUIPMENT, MAJOR EQUIPMENT PRICED		●	●	●	●	●
GENERAL ARRANGEMENTS, APPROVED BY CLIENT		●	●	●	●	●
OUTLINE SCOPE, GENERAL PLANT FEATURES		●	●	●	●	●
ELECTRICAL MOTOR LIST, WITH HP			●	●	●	●
PIPE & INSTRUMENT DIAGRAMS			●	●	●	●
SINGLE-LINE ELECTRICAL DRAWINGS			●	●	●	●
DRAWINGS OF PIPING-SYSTEM RUNS				●	●	●
PRELIMINARY DESIGN DRAWINGS				●	●	●
DETAIL EQUIPMENT, LIST PRICED				●	●	●
DETAIL # SCOPE OF WORK				●	●	●
DETAIL CONSTRUCTION DRAWINGS					●	●
DETAIL SPECIFICATIONS					●	●
SUBCONTRACT AND VENDOR-FIRM LUMP-SUM QUOTES						●

Fig. 24-1 (*a*) Types of estimates and information required for preparation of each type (indicated by dots). Basic data needed include craft wage rates and fringe benefits; payroll taxes and insurance; local sales, use, and other taxes; design and construction schedule; and insurance requirements. (*b*) Probable contingency of each type of estimate when prepared without historic cost data. (*c*) Probable contingency with historic cost data.

designed portions of the facility to be built. As design progresses, the margin of error due to quantity variances decreases. Consequently, estimate reliability increases, while the contingency requirement decreases. From 50 to 70% of the time consumed in preparing these estimates is spent on quantity takeoff, with the balance for pricing, summarizing, and comparing.

Bid Estimates. Estimates for preparation of a competitive bid, Type 6, which are based on completed construction drawings, technical specifications, and general and special conditions, require the estimator to take off and price only the work to be performed by the general contractor. Specialty subcontractor's competitive quotations are received for the balance of the work, which may cover from 80 to 100% of the job. The proper evaluation of local conditions, productivity, competitiveness of the market, and volume of work the general contractor has underway, as well as the quality and completeness of the plans and specifications, play a significant role in determination of the final price of the facility. An award to a low bidder does not relieve the architect-engineer of the responsibility for informing the client of the probable final cost based on his best judgment. This estimate may be in the form of the low bid plus a contingent reserve for change orders, changes in conditions, and other unforeseen events best evaluated by the estimator.

24-3. Preliminary Steps in Estimating.

Certain basic information is required for each cost estimate. A standard checklist should be available to form a guide and reminder, and should include at least the following:

Abstract of the job
Local craft union agreements and wages
Climatic data
Labor availability and relative productivity
Insurance and tax requirements
Proposed schedule for preparation, review, and submittal of the estimate
Man-hours assigned for estimate preparation
Approximate engineering and construction schedule
Outline and summary page format for the estimate

Prior to preparing a cost estimate, the estimator, to plan and schedule the estimate properly, should establish a list of job task and manpower requirements. A general estimator who estimates all the work other than mechanical and electrical and assembles the completed estimate should be allocated sufficient time to prepare the estimate.

There are at least two good bases for establishing manpower requirements for estimating:

1. By actual drawing count and, from past experience, the average man-hour effort per drawing. The latter generally ranges from 3 to 6 hr per drawing. Examples of drawing distribution expressed as percentage of the total drawings might be as follows:

	Commercial construction, %	Industrial construction, %
Pipe drawings	25	15
Electrical drawings	25	35
Other drawings	50	50
Total drawings	100	100

2. By relative dollar value of the major disciplines or crafts; for example:

	Commercial construction, %	Industrial construction, %
General construction	60	40
Piping	20*	10
Electrical	10	10
Engineered equipment	10	40
Total	100	100

* Includes plumbing, heating, ventilation, air conditioning, etc.

Whatever method is used, note that electrical and mechanical takeoffs represent from 20 to 50% of the effort, and should be allocated adequate time for proper preparation of the estimate.

The estimate summary format, with general outlines of each division, should be established so that takeoff sheets, pricing sheets, and supporting data can be properly organized. In summarizing an estimate, it is advisable to adopt the organization of one of the following:

Job specifications
Outline of prior estimate, if one or more exists
Standard specifications, if job specifications do not exist
Process flow of industrial-type plants

24-4. Quantity Takeoffs.

Clearly defined, accurate, and well-organized quantities required for a project are prerequisite to good estimates. Special forms usually are used for recording takeoff. They vary with type of material being surveyed, individuality of the particular contractor, architect, or engineer, and requirements of the particular job. Figure 24-2 shows a sample form applicable to general constructions items of work. Figure 24-3 is a piping takeoff and pricing sheet developed for manual pipe estimating and incorporating an extensive checklist across the column headings. Figure 24-4 is a takeoff and pricing sheet generally

Fig. 24-2 Sample form for quantity takeoff of general construction items.

Fig. 24-3 Sample takeoff and pricing sheet for piping estimating.

used by electrical estimators. The quantities must be clearly defined with proper references to their location (drawing number, specification reference, facility name, project title, date of takeoff, and name of person doing the takeoff), and recorded in clean, clear, legible lettering and readily understandable computation sequences.

The extent of architectural and engineering data available will determine the technique best suited for quantity surveying. Estimators working only with preliminary scope information generally do not have the time or information available for takeoff of detail quantities. These must be developed by one of the following means:

Ratio basis (for example, past experience may indicate that 1½ yd^3 of concrete is needed per ton of installed equipment, or 150 lb of reinforcing steel per cubic yard of concrete).

Comparison basis (for example, records may show that a certain-size pump required 150 yd^3 of concrete for foundations; thus, the pump for the project being estimated, being smaller, might take only 100 yd^3).

Use of good judgment and experience.

Engineers' estimates based on 75% design completion, however, require definite quantities to justify the quality of the estimate. Detailed quantity surveying may represent 50 to 70% of the estimating effort.

Fig. 24-4 Sample takeoff and pricing sheet for electrical estimating.

Hurried quantity surveys with inadequate checks and poor summarization and definition are the basic cause of poor-quality estimates.

For industrial construction, properly prepared and priced *capital equipment lists* are essential. Generally, a process engineer will define equipment requirements during the conceptual stages of a program and will obtain either judgment prices or telephone-quote prices. As the facility becomes better defined, more refined practices are employed. In such cases, the engineer defines each piece of equipment and solicits reliable quotations. Figure 24-5 is a form that minimizes the repetitive writing effort by simultaneously serving for equipment definition, equipment pricing, and equipment installation estimating.

Recording Requirements. There are four basic requirements for quantity surveys:

1. **Neatness.** Write clearly and simply and do not crowd your work.
2. **Exactness.** Define the item being surveyed and give the location and drawing reference.
3. **Accuracy.** Check all computations and check the quantities for rationality.
4. **Utility.** Be sure quantities taken off are the proper quantities for the pricing structure of the job.

24-5. Unit Prices.

Pricing techniques vary with the type of estimate, estimator, construction company, and data available.

Material Costs. Current material unit prices often can be obtained by telephone. Otherwise, material prices based on prior job records, properly adjusted to reflect current costs, competitive nature of the market, and specific job locations and quantities, have to be extracted from company records. Such prices are probably the easiest to obtain and the most reliable for preliminary-type estimates. Piping and electrical material prices are available as published list prices which must be adjusted to reflect market experience.

Process Equipment Unit Prices. These are important inasmuch as equipment may represent 40% of the estimated value of an industrial project. Sizing and pricing of equipment are the most critical steps in preparation of an industrial-type estimate; hence, care and good judgment must be used. The best sources of price data are the vendors of the equipment. They are usually agreeable to giving, for estimating purposes, budget prices based on minimum specifications.

It is important that all material and equipment prices be quoted FOB job site and, if required, include applicable sales, use, and other taxes.

Labor Costs. Estimating labor unit prices is another major task of estimators. It can be done by various methods, such as:

Crew size and experienced productivity
Experienced man-hours per unit of work
Experienced unit costs

In all cases, the costs should be well defined and may include all or a portion of the following:

Bare craft wage rate
Craft fringe benefits
Statutory taxes and insurance
Small tools and consumable supply allowances
Construction equipment costs
Contractors' overhead and profit

Fig. 24-5 Sample form for equipment listing, pricing, and installation estimating.

24-9

24-10   Estimating Building Construction Costs

In the crew system of estimating, a first step is establishment of the crew size and crew rate. For example:

One each carpenter foreman:	8 hr/day @ $15.35 =	$ 122.80 per day
Eight each carpenter journey men:	64 hr/day @ $14.39 =	920.96 per day
Four each construction laborers:	32 hr/day @ $11.92 =	381.44 per day
Crew rate:	104 hr @ $13.70	= $1425.20 per day

From job experience, the estimator knows that this crew will build and erect, complete with screeds, headers, keyways, and the like, about 500 ft^2 of average industrial constructions forms in one 8-hr day, at about 80 to 85% productivity. If the job calls for 1000 ft^2 of forms, building and erecting them will take 2 days and cost 2 × $1425.20 = $2850.40 for labor only.

If the method of experienced man-hours per units were to be used, the estimator would refer to prior job records. For the preceding example, the records may show that this type of form takes 0.21 hr/ft^2 to build and erect. Based on the previously calculated average hourly rate of $13.70, the cost for 1000 ft^2 of forms in place would be, for labor only: 0.21 × 1000 × $13.70 = $2877.

The unit-cost method is very broad, and is used for preliminary estimating only. For instance, in unit-cost estimating for the preceding example, the estimator would have to know the concrete forms would cost $2.85 per sq ft for labor only. Then, the cost of 1000 sq ft of forms in place would be 1000 × $2.85 = $2850 for labor only.

While estimators hope for as much detail regarding equipment and materials as is possible, often the quantities and units are on a much broader scale. For instance: What is the cost of an integrated steel plant with modest finishing facilities for 1,000,000 short tons annually? Records would indicate that construction cost of the plant should range from $700 to $800 per ton, or from $700,000,000 to $800,000,000, plus the owner's costs, which could include some or all of the following:

Escalation
Spare parts
Land
Owners' engineering and administrative costs
Interest during construction
Technical assistance and training
Operating inventory
Working capital
Start-up expenses
Financing costs

Quantities and unit prices, be they for a cubic yard of concrete or a complete facility, must be clearly defined and limited so that costs based on them represent as accurate a prediction of future costs as is possible to obtain. Simplicity and clarity of presentation cannot be overstressed.

24-6. Allowances.

The alternative to detailed quantities and unit prices (Arts. 24-4 and 24-5) is use of allowances. These are generally used in lieu of detailed unit prices when the scope or definition of the work is incomplete. But their use should be held to a minimum and established by experienced estimators who have developed a feel for costs. For instance, the electrical subcontract value, excluding equipment, for an industrial plant may be from 10 to 13% of the total costs; process and utility piping may be 6%. During the early stages of an estimate, this method may be more realistic than an attempt to detail the quantities.

Before an allowance is used for any element in the estimate, all sources of information should be explored, such as data on similar installations in other plants, or perhaps a sketch by process engineers as to the intent of the undefined element. In presenting the estimate to management and the client, the estimator should identify and list the value of each allowance.

24-7. Estimate Details.

The details of an estimate vary with the type of estimate (see Art. 24-2), information available, and time allocated for preparing the estimate. A magnitude estimate (Type 2), such as the one for the 1,000,000-ton integrated steel mill in Art. 24-5, may require only a one-page tabulation of major plant facilities, listing annual capacity of each facility and an all-inclusive experienced unit cost per annual unit of capacity. On the other hand, a definitive estimate (Type 4) for a process plant provides more details and is further subdivided into plant facilities and subfacilities. The definitive estimate would have construction quantity and pricing detail for all elements of the work. Figure 24-6 is a form for preparing such an estimate. The depth of detail varies with the item being estimated, the amount of engineering available, and the time available. Care should be taken not to overdetail the item of work being estimated.

Fig. 24-6 Sample form for preparation of a definitive estimate.

Bid (Type 6) estimates, which are prepared by estimators for general contractors and specialty subcontractors, are usually based on completed plans and specifications. Such estimates are developed in greater detail than definitive estimates, and are prepared only for the work the contractor intends to do himself. The estimate outline and resulting summarization will invariably conform to the construction specification divisions.

The cost estimate, as presented, does not include all the supporting studies and detail that an estimator develops to establish his unit prices and quantities. It is good practice, however, for the estimator to assemble an appendix to the estimate, incorporating worksheets properly classified and identified, to complete the file of detail support information. The extra time and effort required to obtain, create, and organize all papers used in the preparation of the estimate are well spent, because the data may be invaluable at a later date when the inevitable extra-work orders and claims develop.

24-8. Evaluations of General Cost Items.

After quantities have been taken off and priced (Arts. 24-4 and 24-5), the engineered equipment has been defined and priced for purchase and installation, and all such costs have been properly summarized, the next step involves judgment evaluation of escalation, contingency, fee, overhead, and engineering costs.

The labor, material, and equipment prices used in an estimate are usually defined as those in effect at the day of the estimate. Future price increases, either known or evaluated, must be taken into account to develop a valid estimate.

24-12 Estimating Building Construction Costs

Escalation. This applies to increases in costs for material and labor and does not take into consideration changes in craft-labor productivity, design innovations, and improved construction techniques. These should be reflected in the body of the estimate, or as suggested alternative schemes.

One method of evaluating escalation is shown for illustrative purposes in Fig. 24-7, and consists of four steps:

The first step involves preparation of a schedule showing the design and construction periods in simple bar chart form. The time scale is selected to match that of physical progress curves drawn later below the bar chart.

Next, the craft wage rates for the five major crafts (carpenters, laborers, ironworkers, pipefitters, and electricians) are summarized directly underneath the bar chart using the same time scale. The X to the right of each craft indicates the expiration date of the wage contract with the craft union, and the dollar amount shows the wage and benefit amounts defined by the contract for the period indicated.

Fig. 24-7 Chart illustrating evaluation of cost escalation for an industrial building. Bar chart at the top illustrates schedule for design and construction. Journeymen hourly rates are given immediately below, with date of expiration of wage contract. The two curves at the bottom indicate, respectively, design and construction progress. Vertical lines show when union contracts expire and are labeled with anticipated amount of hourly wage increase. Cost escalation with data from the chart is computed as follows:

Labor, including fringe benefits (10% annual increase):
3,200,000 hr @ 100% @ $1.65	= $ 5,280,000
3,200,000 hr @ 75% @ $1.82	= $ 4,368,000
	$ 9,648,000
Payroll taxes and insurance (15% total increase):	$ 1,447,000
Material and other field costs (10% annual increase, 21-month period:)	
33,000,000 × 10%/12 × 21	$ 5,775,000
Engineered equipment:	
Firm quotation: $41,000,000 @ 0%	....
Balance (10% annual rate): $17,500,000 × 10%/12 × 21	= $ 3,062,000
Engineering: $10,500,000 @ 10%	= $ 1,050,000
Total escalation exposure	= $20,982,000

The third step involves preparation of curves indicating physical progress of the project. The construction-progress curve is based on on-site construction-labor man-hours and represents an average indicated by experience. The vertical lines with +1.65, +1.82, and +2.00 indicate actual or estimated craft wage increase at appropriate effective dates.

The final step involves calculating the expected cost escalation, which is divided into five areas:

1. Labor. For this size job and based on the cost estimate, the estimator determines that there will be 3,200,000 on-site construction man-hours. Since the cost estimate is based on wage rates in effect at the time of the estimate, it is necessary to allow for future increases. Because construction did not commence until mid-1980, all man-hours (100%) received the $1.65 estimated increase for mid-1980, and 75% received the $1.82 increase in 1981. The curve indicates that the $2.00 increase will have no effect if the project stays on schedule.

2. Payroll taxes and burden. An estimated 15% is added to the labor increase to allow for the increase in statutory wage costs.

3. Material. The estimate indicates the value of material purchases required. The example in Fig. 24-7 shows $33,000,000 of such purchases. Records show that material costs have been and probably will increase at an annual rate of 10%. The center of these purchases is taken at the 50% physical completion date, which is 21 months from the date of the estimate. This results in the calculation shown in Fig. 24-7 for this group of costs.

4. Engineered equipment. An analysis of the project indicates that $41,000,000 of equipment is based on firm quotations, leaving $17,500,000 subject to escalation. With the same formula for equipment as was used for material, the estimator can calculate the escalation exposure for equipment.

5. Engineering. The estimate allows $10,500,000 for engineering. An estimated 10% is added for the probable increase in these costs over the duration of the project.

Total. The sum of labor and other escalation is the total estimated escalation exposure for the project.

Contingency. This is an evaluation based on experience, and is included as part of the estimated cost to compensate for accidental events; errors; imperfections in estimating; quality and quantity of engineering data available; omissions; and unknown events. With the addition of an evaluated contingency, the estimator intends that the resulting estimate precisely represents the final cost of the facility. Man is not perfect, however, and the best estimator usually is only able to keep the variation between total estimated cost and actual cost figures within a very small percentage.

An example of a contingency evaluation for an estimate might be as follows:

Labor costs	$ 9,297,000 @ 15% =	$1,395,000
Construction equipment	1,932,000 @ 15% =	290,000
Material purchase	6,125,000 @ 10% =	612,000
Subcontracts		
Firm lump sum	13,250,000 @ 2% =	265,000
Unit price	3,073,000 @ 10% =	307,000
Estimated	2,150,000 @ 15% =	322,000
Engineered equipment		
Firm lump sum	22,351,000 @ 3% =	670,000
Estimated	5,625,000 @ 10% =	563,000
Field overhead		
General conditions	3,632,000 @ 15% =	545,000
Escalation	8,250,000 @ 15% =	1,238,000
Total construction cost	$75,685,000	
Total contingency		$6,207,000

This calculation indicates that the average contingency for the job is about 8%. The estimator who prepared the estimate is the best judge of contingent requirements, because he is intimately associated with all facets of the estimate and the detail of information available.

Overhead. Contractor's field overhead or job general conditions include project costs not readily or equitably assignable to individual direct-cost items of work. There are many standard check sheets available for use in defining the costs to be included in this category. Individual company accounting practices, nature of work and estimating-department definitions should be the bases for a standard and comprehensive checklist that reflects the individuality of the company and gives continuity of estimate definition from job to job.

Contractor's Fee. Profit or fee evaluation is management's prerogative. Estimators, however, should be aware of their management's considerations and evaluating techniques, so that the proper bare figures for computation are readily available. Individual company policies vary so widely in the definition of profit and fees that a discussion would require extreme detail analysis. Each company should establish clear definitions compatible with management's policies.

24-9. Computer Estimating.

For estimates that are prepared from minimum engineering data, where basic quantities are nonexistent and where maximum use of experience, ratios, and factors, as well as the knowledge of a conceptual estimator are required, computerization is improbable, because the methods and techniques used are not amenable to strict mathematical solution. Where design has progressed sufficiently to permit reasonable quantity surveying, there is a definite place in estimating for the use of computers, especially where large volumes of data are used, as in piping and electrical estimating. The cost of maintaining up-to-date material prices, however, is a major cost factor that must be considered, because an outdated material data base will adversely affect the estimate.

Computer equipment availability and cost are also important considerations in developing and using estimating systems. Programs developed for a specific computer, for example, require availability of that model on a continuing basis. Service-center operation with, if need be, terminal equipment at the home office is practical and indeed desirable. But establishment of files and procedures for using the computer at outlying job locations or under bidding conditions requires planning and careful implementation.

Because of the large volume of detail unit material and labor prices in pipe estimating, an outline of a computer approach to this subject is given in Art. 24-10.

24-10. Pipe Estimating by Computer.

Development and implementation of a detailed piping estimating program require an in-depth study of all possible alternatives and requirements. For instance, the program should be able to sort and summarize according to various items and subjects, such as takeoff sheet number, bill of material number, commodity number, cost-account number, drawing number, line or system number, and specification number. These may not all be developed and used in the initial program, but consideration should be given to them and space allotted for them during the establishment of the program, because they may be developed at a later date.

Data to be included in the computer memory bank must be defined in detail to assure consistency from estimate to estimate. Unit material costs may be defined as list prices; unit labor man-hours may be defined as craft journeyman-hours; and unit subcontract costs for insulation, painting, etc., may be the current competitive subcontract unit price for medium-size jobs. The weight of each item, while not essential, does provide useful information for checking overall costs and determining shipping weights and discounts.

Introduction of adjustment factors permits the estimator to make reasonable judgment decisions to revise the standard units in the files, and is necessary for any practical program. Adjustment factors are generally classified in three major groups, as follows:

Labor wage-rate adjustments, including fringe benefits, statutory insurances, and taxes, which have to be inserted for each project because of the variation in these costs from time to time and from place to place.

Standard unit man-hours adjustments to take into account the individuality of the particular system and circumstances. These adjustments compensate for completeness of drawings, adequacy of specifications, craft supervisory requirements, availability and quality of crafts, area productivity, weather, adequacy of working space, job size, complexity of installation, crew size, area congestion, and job schedule.

Adjustments to list prices of materials, for area pricing and available discounts.

Test runs on piping estimates indicate that little cost savings may be realized by use of a computer. It does, however, reduce the estimator's time by 30% or more, thus freeing him for other activities for which he is skilled.

Human judgment is still the prime ingredient of good estimates. But with continuing development, study, and objective approach, computer estimating will improve and incorporate more sophisticated judgment factoring. Figures 24-8 and 24-9 contain samples of quantity input sheets and computer output sheets for pipe estimates.

24-11. Cost Estimate and Construction Schedule.

The cost of a project is very much dependent on the project schedule. The relationship is best described by means of an optimum time-cost curve (Fig. 24-10). This curve indicates that construction cost for a project is at a minimum for a specific construction time,

Fig. 24-8 Sample input sheet for computerized estimating of piping costs.

24-16 Estimating Building Construction Costs

```
FILE UPDATE NO.         1                      ********************************       PAGE         1
FILE REVISE DATE        13JAN2                 *   ESTIMATE BY TAKE-OFF SHEET  *       DATE         13JAN2
                                               ********************************

JOB TITLE               ROCK PLANT BRIDGEPORT                                          JOB NO. 3001

TAKE-OFF DESCRIPTION    SCREENING PLANT PLANT WATER                                    TAKE-OFF SHEET    2
BY                      EDL
DRAWING NO./REVIS
COST ACCOUNT            1363.635
TAKE-OFF DATE           31AUG1


                                        DESC    RATING                  MATERIAL         LABOR         SUBCON  WEIGHT
  SPEC CODE   ITEM            SIZE      CODE SCH PRES WALL   QUANT UN   $ STD  $ ADJ   MH STD  MH ADJ  $ STD    LBS

               SPRAY NOZZLES  1/4       *Z01                 672 EA       0      0      54      67      0        0
               DRILL HOLES    1/4       *Z02                 672 EA       0      0      67      84      0        0
   A CELEST    PIPE           2         120A  40             143 LF     117     70      30      38      0      522
   A CELEST    PIPE           3         53XA  40              34 LF      73     44       8      10      0      257
                                                    ～～～～～～～～～～～～～～～～
   A CELEST    HANGER ASSEMBLY 16       Z203                  1 EA       29     18       3       4      0       33
                                                                        ********************************************
* SUMMARY *                                                  TOTALS    4286   2572     540     673      0     8218

TOTAL COST (ADJ)           $ 2572
TOTAL PIPE LENGTH            419 LF
EQUIV PIPE DIAMETER         5.67 INCHES
TOTAL COST/LENGTH          $ 6.14/LF
TOTAL LBS/LENGTH            19.61/LF

STD MH/(TOTAL LENGTH X EQUIV DIAM)           0.23
STD MATERIAL $/(TOTAL LENGTH X EQUIV DIAM)   1.81
```

Fig. 24-9 Sample computer output for computer estimate of piping costs.

increases rapidly when construction is speeded, and also rises, although not so fast, when project time is lengthened. The cost estimate, therefore, is affected by the schedule, which should be well thought out to obtain an optimum project cost. It should be recognized, however, that an optimum project schedule resulting in best costs may not be economically sound as far as the overall project is concerned. For example, if an increase of $100,000 in construction cost will bring the plant on-line 2 months earlier, the additional product profit may far outweigh the construction cost penalty. A cost estimate prepared for too short or too long a construction period also simulates the time-cost curve in reflecting project value.

Preparation of a cost estimate should start with a plan and schedule of accomplishment indicating major milestones. These schedules range from a detailed bar chart to a relatively complex critical-path-method (CPM) diagram. First, the estimator should ascertain the time available and the manpower required for preparing the estimate. With this information, a simple bar chart indicating estimating activities and dates can be prepared. The chart should allow ample time for management review and approval. Because most estimates are due at a specific time and date, the plan for accomplishing the major tasks should establish well-defined milestone dates for easy monitoring.

Fig. 24-10 Curve indicates the variation in construction cost of a project with construction duration.

Methods for scheduling construction vary from a simple bar chart showing design, procurement, and construction times to complex CPM diagrams requiring computer solutions. To support and define the estimate properly, the estimator should establish and submit a schedule of project activities with the cost estimate.

If the project is in the design stage, adequate detail of design accomplishment should be included as restraints on procurement and construction activities. The scheduling of the detail design effort is an extremely complex task with many interrelations and dependencies. The drawing release schedule, however, could be the basis for determining initiation of procurement activities and commencement of field activities.

There are various rules of thumb for predicting duration of design and construction activities. The relatively simple, but sometimes unreliable, relationship between dollars and time offers a basic check of schedule. Table 24-1, which applies to industrial construction projects, may be used as a guide:

TABLE 24-1 Costruction Times for Industrial Buildings

Construction value, millions of dollars	Design period, months	Construction period, months	Overall period, months
10	12	13	20
15	12	15	20
30	14	18	24
50	16	20	26
75	20	24	30
100	20	27	36
150	24	30	36

Detail Control. With development of the detailed scope of work and the definitive cost estimate, a detailed project schedule showing the interrelationship between design, procurement, and construction, with proper resource allocation and reasonable durations, should be established. This schedule, when reviewed and approved, should be updated periodically and used for monitoring physical accomplishment relative to dollars expended, as a check against the estimate and as a method of highlighting cost variances.

How much detail is required for proper schedule definition? The answer is not simple. A small job requires fewer activities than a large job, and a complex job requires more activities than a simple job. The detail required therefore should be only that needed to assure proper definition and sequencing of the salient job activities.

Monitoring. A project manager uses project schedules and cost estimates for evaluating his team accomplishments. The estimate and schedule should be developed so that periodic updatings are as free from partial or slanted input data as possible. All input should be factual. It is the responsibility of the project manager and his staff to interpret these facts and initiate the required action.

24-12. Presentation of the Estimate.

The time consumed in and cost of preparation of cost-estimate details certainly justify allowing adequate time for assembling the worksheets and data that support the estimate.

During the assembly phase of an estimate, two very important checks should be made:

1. Mathematical computations should be verified by a competent clerk knowledgeable in the estimating field. Mathematical errors are inexcusable.

2. The plans, specifications, scope, equipment list, and other supporting details should be reviewed thoroughly to make sure that all cost items are included and are in their proper place.

The estimate in its entirety should then include at least the following documents: summary, criteria sheet, scope or job abstract, comparisons with other estimates, and supporting quantity and pricing detail.

Summary. The summary page should be consistent with prior jobs. One method of accomplishing this is adoption of a standard summary-page format for presentation of all cost estimates. Standard summary pages consistent with the type of estimating being done should be established within a company for ready understanding and ease of recognizing project elements. The summary-page format in Table 24-2 is typical for a turnkey industrial project.

Another approach for summarization would be to follow the outline of the specifications. As a substitute in the case of commercial work, the format established by the Construction Specifications Institute, Inc. for building specifications is excellent (Art. 26-7). Estimate summaries should be confined to a single page, and present as tersely as possible the basic dollar facts about the job.

Criteria Sheet. Probably one of the most significant documents that can be prepared is the estimate criteria sheet (Table 24-3), which outlines in estimator's language the overall financial limitations of the estimate. A criteria sheet should be prepared for each job to include at least the following:

The schedule expressed in starting and completing months and years for both design and field construction.

The work shift—standard or otherwise—on which the cost estimate is based.

The date for labor and material costs and an indication of the inclusion or exclusion of evaluated escalation exposure.

The statement that added costs, if required, have or have not been included for: hiring and training of special workers; recruiting and expediting costs due to shortage of skilled craftsmen or material; and implementation of certain new or unusual governmental regulatory acts.

The statement that costs have or have not been included for any or all of the following: land, owner's engineering and administration, training of operating personnel, fur-

TABLE 24-2 Typical Summary-Page Format for Industrial Building

1.0	**Construction Costs**
1.1	**Direct costs**
1.11	Yard improvements and development
1.12	Buildings and structures
1.13	Crushing and grinding
1.14	Screening and classifying
1.15	Concentrating
1.16	Finish-product processing
1.17	Storage and shipping
1.18	Escalation
	Total direct costs (1.1)
1.2	**Indirect costs** (general conditions)
1.21	Contractor's field overhead
1.22	Construction plant and equipment
	Total indirect costs (1.2)
	Total construction costs (1.0)
2.0	**Engineering, Supervision, and Procurement**
3.0	**Engineer-contractor Fee**
	Subtotal
4.0	**Contingency**
	Total project

TABLE 24-3. Example of Estimate Criteria for an Industrial Building March 10, 1980

Estimate Criteria

The estimate of cost for the XYZ Plant is based on the following assumptions:

1. The estimate is based on the construction program starting January 1980, and being completed in 14 months.
2. The estimate is based on labor, material, and equipment prices as of October 1979. An escalation allowance has been included as a line item for future increases.
3. The demolition of existing structures that are in the way of construction has been included.
4. Rough grading of the plant site has been included.
5. Prices for certain pieces of equipment to be furnished by the owner have been obtained from the owner and are included.
6. The estimate does not include additional monies, if required, for hiring and training of minority workers.
7. The cost of ecological studies and the costs due to these studies, such as added work or construction delays, are not included.
8. Added costs, if any, due to the Occupational Safety and Health Act are not included.
9. The estimate is based on engineering and construction being performed on a one-shift standard work week.
10. The Arizona Gross Receipts Tax (sales tax) is not included.
11. The cost of spare parts is not included.
12. Special tools or equipment, such as zinc furnace and vulcanizer, are not included.
13. Rerouting of utility lines within the plant area is by others.
14. The costs for plant start-up are not included.
15. Owner's engineering, administration, financing, and other such costs are not included.

niture and furnishing, spare parts, interest during construction, production materials, and financing.

A list of major process or facility exclusions, if any, that may include owner-furnished material or equipment.

It should be stressed that the criteria sheet is not to be worded as a legal document, but is the medium through which an estimator, in his own language, outlines major considerations in the estimate.

Scope. Inclusion of a general scope or descriptive abstract of the facility with a general arrangement plan of the plant and, if available, a material-flow diagram, will aid considerably in defining the estimate limitations and afford easy reference during review meetings with management, client, and financial representatives. The statements should be clear, concise, and factual for the plant description and for the engineering data available.

Estimate Comparisons. A page should be included in the estimate package with data from the cost estimate for comparison with prior estimates for the same facility with a brief analysis of the differences, if any, and comparisons with costs for similar installations that have been constructed or estimated. For instance, if an estimate of $20,000,000 for an 8000-short-ton-per-day copper concentrator can be compared with four or five similar plants within the range of $2000 to $3000 per short ton per day, it would appear that the estimate being reviewed is within reason. Comparisons in detail or in totals are good checks and a basis for evaluating the estimate at hand with other experience data. Well-documented and readily accessible historical information makes such comparisons easy and reliable.

Details. Estimate details should be organized in a loose-leaf binder with proper tabs, so that answers to questions and supporting documents are readily accessible. Neat, orderly worksheets may not improve the estimate quality (many people believe that they do, however), but they do improve the ability to sell the cost estimate. Estimating is the application of personal knowledge, experience, and understanding on a judgment basis to the logical flow of plant and material. Clarity of thought and logical presentation of detail will materially enhance the quality of the estimate.

After the estimate with its details has served its purpose, it should be given an identification symbol, placed in fixed binders, and properly filed for future references.

24-13. Reference Material.

Job cost data derived from experience must be properly analyzed and organized to be readily available for estimators' use. There are many fine books available for aiding in detail and overall cost estimating of mechanical, electrical, and brick-and-mortar activities to supplement job records. The sources of prices on engineered mechanical equipment are the manufacturers, who are generally willing to give prices that range from list prices to firm quotations. The following is a sampling of the many good cost guide books available. Be sure to read and understand the introductions of these books, to determine the particular qualifications and areas of specialization of each:

"Marshall Valuation Service," Marshall and Swift Publication Company, Los Angeles, Calif., updated periodically.

"Building Construction Cost Data," Robert Snow Means Company, Inc., Duxbury, Mass., annual editions.

"Construction Pricing and Scheduling Manual," McGraw-Hill Information Systems Company, New York, annual editions.

"Commercial-Industrial Estimating & Engineering Standards," Richardson Engineering Service, Inc., Solana Beach, Calif., 3 volumes, updated periodically.

Series of publications by John S. Page and James G. Nation, published by Gulf Publishing Company: "Estimator's Piping Man-Hour Manual"; "Estimator's Man-Hour Manual on Heating, Air Conditioning, Ventilating and Plumbing"; "Estimator's General Construction Man-Hour Manual"; "Estimator's Electrical Man-Hour Manual."

24-14. Cost Indices.

These are invaluable for updating past estimates and prices, as well as for plotting trends to aid in forecasting future costs. The composition and intent of an index should be studied and thoroughly understood before it is used.

The various indices published periodically by *Engineering News-Record* are probably the most widely known. Other indices that may be helpful are the building cost index published in *Architectural Record;* the "Nelson Refinery Construction Cost (Inflationary and True Cost) Index," and the "Chemical Engineering Plant Cost Index." These represent only a few of the indices available. It is sometimes advantageous, however, for the estimator to develop his own index for specific building types.

Section 25

Construction Management

Robert F. Borg
President, Kreisler Borg Florman Construction Company
Scarsdale, N.Y.

Construction management of a building project encompasses organizing the field forces and backup personnel in administrative and engineering positions necessary for supervising labor, awarding subcontracts, purchasing materials, record keeping, and financial and other management functions to ensure profitable and timely performance of the job. The combination of managerial talents required presupposes training and experience, both in field and office operations of a construction job. Proper construction management will spell the difference between a successful building or contracting organization and a failure.

This section outlines practical considerations in construction management based on the operations of a functioning general contracting organization. Wherever possible, in illustrations given, the forms are from actual files for specific jobs. These forms, therefore, not only illustrate various management techniques, but also give specific details as they apply to particular situations.

25-1. Types of Construction Companies.

The principles of construction management, as outlined in this article, apply equally to those engaged in subcontracting and those engaged in general contracting.

Small Renovation Contractors. These companies generally work on jobs requiring small amounts of capital and the type of work that does not require too great a knowledge of estimating or construction organization. They usually perform home alterations or small commercial and office work. Many small renovation contractors have their offices in their homes and perform the "paper work" at night or on weekends after working with the tools of their trade during the day. The ability to grow from this type of contractor to a general contractor depends mainly on the training and business ability of the individual. Generally, if one is intelligent enough to be a good small renovation contractor, that person may be expected to eventually move into the field of larger work.

General Contractors. These companies often are experts in either new buildings or alteration work. Many building contractors subcontract a major portion of their work, while alteration contractors generally perform many of the trades with their own forces.

Some general contractors specialize in public works. Others deal mainly with private and commercial work. Although a crossing of the lines by many general contractors is common, it is often in one or another of these fields that many general contractors find their niche.

Owner-Builder. The company that acts as an owner-builder is not a contractor in the strict sense of the word. Such a company builds buildings only for its own ownership, either to sell on completion, or to rent and operate. Examples of this type of company include giants in the industry, and many of them are listed on the various stock exchanges. Many owner-builders, on occasion, act in the capacity of general contractor or as construction manager (see below) as a sideline to their main business of building for their own account.

Real Estate Developer. This is a type of owner-builder who, in addition to building for personal ownership, may also build to sell before or after completion of the project. One- and two-family home builders are included in this category.

Professional Construction Manager. A professional construction manager may be defined as a company, an individual, or a group of individuals who perform the functions required in building a project for an owner, as the agent of the owner, as if the owner were performing the job with his own employees. The construction management organization usually supplies all the personnel required for the various duties the owner would need done were he doing them himself. Such personnel include construction superintendents, expediters, project managers, and accounting personnel.

The manager sublets the various portions of the construction work in the name of the owner and does all the necessary office administration, requisitioning, paying of subcontractors, payroll reports, and other work on the owner's behalf. Generally, construction management is performed without any risk of capital to the construction manager. All the financial obligations are contracted in the name of the owner by the construction manager. (See Art. 25-9.)

Package (Turnkey) Builders. Such companies take on a contract for both design and construction of a building. Often these services, in addition, include acquisition of land and financing of the project. Firms that engage in package building usually are able to show prospective clients prototypes of similar buildings completed by them for previous owners. From an inspection of the prototype and discussion of possible variations or features to be included, an approximate idea is gained by the prospective owner of the cost and function of the proposed building.

Package builders often employ their own staff of architects and engineers, as well as construction personnel. Some package builders subcontract the design portion to independent architects or engineers. It is important to note that, when a package builder undertakes design as part of the order for a design-construction contract, the builder must possess the necessary professional license for engineering or architecture, which is required in most states for those performing that function.

Sponsor-Builder. In the field of government-aided or subsidized building, particularly in the field of housing, a sponsor-builder may be given the responsibility for planning design, construction, rental, management, and maintenance. A sponsor guides a project through the government processing and design stages. The sponsor employs attorneys to deal with the various government agencies, financial institutions, and real estate consultants, to provide the know-how in land acquisition and appraisal. On signing the contract for construction of the building, the sponsor assumes the building role, and in this sense functions very much as an owner-builder would in building for its own account.

25-2. Construction Company Organization.

How a construction company organizes for its work depends on number and size of projects, project complexity, and geographical distribution of the work.

Sole Proprietor. This is a simple form of organization for construction contractors. It is often used by subcontractors, including those licensed in plumbing or electrical work. The advantage of operating as a sole proprietorship is that taxes on profits are much lower for individual owners. But there is the disadvantage of having the personal exposure to potential debts associated with a disastrous job.

Partnership. This is the joint ownership and operation of a company by two or more persons. Each partner, however, is personally liable for all the debts of the partnership. Profits and losses are shared in some manner predetermined by the partners. A partnership comes to an end with the death of one of the partners. (For typical provisions to be included in a partnership agreement, see Richard H. Clough, "Construction Contracting," John Wiley & Sons, Inc., New York.)

Corporation. This is the most common form of organization used by general contractors. A corporation is an entity that has the power to act as a separate body and enter into contracts. It has perpetual life and is owned by stockholders, each of whom has a share in the profits and losses of the corporation. An important advantage of the corporate form of ownership for general contractors is the absence of personal liability of the stockholders. This is desirable because of the risks of the contracting business, and is more than recompense for the additional burden of taxes that those taking part in corporate ownership must bear. (Small corporations can obtain some relief from Federal taxes, however.)

Corporations formed in one state must obtain, as a foreign corporation, a certificate of authority to do business in other states. This is important when bidding jobs in locations other than the home state of the general contractor.

Some general contractor corporations are large enough to find it advantageous to raise capital by becoming public corporations, with shares sold over the counter or on the various stock exchanges. Such corporations publish financial reports yearly for the benefit of the stockholders, as required by law. A study of such reports is often helpful for those engaged in the contracting business.

Joint Venture. Often when an individual job is too large to be undertaken by one company, or the risks involved are too great for one company to want to assume (although it may be capable of doing so), a joint venture is formed. This is an association between two or more contracting firms for a particular project. It joins the resources of the venturers, who share the financing and management of the job and the profits or losses in some predetermined manner.

Generally, there are specific reasons for the formation of a joint venture between specific companies. For example, one may possess the equipment and the other the know-how for a particular job. Or one may possess the financing and the other the personnel required to perform the contract. Joint ventures do not bind the members to any debts of the coventurers other than for those obligations incurred for the particular jobs undertaken.

Staff Organization. An organization chart for a typical medium-size general building contracting company is shown in Fig. 25-1. The organization shown is for a company that subcontracts most of its work and is engaged mainly in new construction. For an organization with district offices, see the chart in Paul G. Gill, "Systems Management Techniques for Builders and Contractors," McGraw-Hill Book Company, New York.

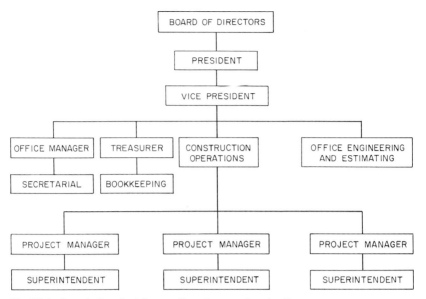

Fig. 25-1 Organization chart for a medium-size general contracting company.

25-3. Contractors' Business Consultants.

Construction contractors often have to engage experts from various disciplines to advise and assist them in conducting their business. In addition to architectural and engineering consultants, contractors usually consult the following:

Accountants. The accountant selected by a construction company should preferably be one who has had experience in contracting and construction accounting. This accountant will know the generally accepted principles of accounting applying to construction projects, such as costs, actual earnings, and estimated earnings on uncompleted construction contracts. In addition, the accountant should have an understanding of management's role to help formulate the financial picture of the firm. This role of management involves estimating the probable earnings on uncompleted jobs and the amounts of reserves that should be provided for by the accountant for contingencies on jobs that have already been completed but for which final settlements have not yet been made with all the subcontractors and suppliers.

Attorney. A construction company may find that it needs more than one attorney to handle all of its affairs. For example, it will in all likelihood have an attorney who will be retained for most of the routine matters of corporate business, such as formation of the corporation, registration by the corporation in other states, routine contract advice, and legal aid in the general affairs of the business. In addition to this, however, many construction companies require different attorneys for such phases of their activities as claims, personal affairs and estate work, real-estate and tax-shelter matters, and government programs and processing. Many of these functions are performed by attorneys who specialize in that type of work and can offer the most up-to-date advice.

Insurance and Bonding. An insurance broker who has a fairly sizable volume of business is the preferred source of insurance. Such a broker will have the greatest amount of leverage with insurance companies when questions arise about claims for losses or about requirements by insurance companies for premium deposits on policy renewals.

Most general insurance brokers with a fairly large clientele should be able to handle construction company insurance.

When it comes to bonding, however, it may be advisable to deal with a firm that specializes in general contractors and their bonding problems. Often, general insurance brokers do not have much experience in this field. While surety companies, who issue bonds, are generally subsidiaries of insurance companies, bonding and insurance are guided by entirely different principles. A broker who has many clients for whom he obtains performance and payment bonds is to be preferred for meeting bonding requirements. Such a broker will be able to offer advice on the bonding companies best suited for the contractor's specific business. Also, the broker should be able to aid the contractor and the contractor's accountant in preparation of financial statements so as to show the contractor's financial position in the most favorable light for bonding purposes.

Banking. Bank accounts needed for the company's affairs should preferably be divided between two or more banking institutions. A banking relationship with more than one bank could be advantageous at times. For example, it could facilitate a request for banking references for credit; for a bank to be available for temporary loans or for writing letters of credit; for equipment loans, and for advice on investments or for custodian accounts for the handling of surplus funds of the contractor.

Trade Associations. Many contractors find it advantageous to become a member of a local trade association or local branches of national organizations, such as the Associated General Contractors of America. Before doing so, however, the contractor should investigate and be thoroughly acquainted with the rules of such associations, particularly with regard to labor bargaining.

In most cases, membership in a local trade association binds the contractor to permit that association to do all of its labor-contract bargaining with local labor unions and the various construction trades. Furthermore, if there is a strike by any of the local labor organizations, the members must obey the dictates of the association and may be required to join in the association's stand against the strikers. On the other hand, if the contractor is not a member of the local association, he is free to bargain individually with the trade unions involved and can sign their contract as an *independent.* Often, labor unions, during a strike, will be willing to offer an interim contract to independent contractors under which terms the contractor will be able to resume work on agreeing to be bound by the terms of the labor agreement consummated to settle the strike, effective retroactively.

25-4. Sources of Business.

For continuity of operation, a construction organization needs a supply of new projects to build. After a company has been in existence for a long time and built up a reputation, new business may come to it with less effort. But most companies must work hard at obtaining new jobs. Furthermore, work that happens to come in may not be of a type that the organization prefers. To find that type requires serious skillful efforts.

To be successful, a contracting company should have a person specifically assigned to attract new business. This person might be the proprietor of the construction company. In large firms, a complex organization with sales and public relations personnel, backed up by engineers and cost estimators, is used. The organization should be geared to follow up on all possible sources of new business.

Public Works. The following sources for leads to new jobs and submitting proposals can be used for public work bids:

Dodge Bulletin, or other construction industry newsletters.
Engineering News-Record—"Pulse" and "Official Proposals" sections.

Bid invitations, as a result of requests to be placed on bid-invitation mailing lists of various government agencies.

Newspaper announcements and articles.

Official publications of government agencies that contain advertisements of contracts to be bid.

Private Contracts. All the sources for public works.

Contacts with and letters to architects.

Contacts with and letters to owners.

Personal recommendation.

Sponsorship. Applying for sponsorship of any of the following:

Government-encouraged housing programs.

Urban renewal.

Purchase of land, with financing of building construction to be provided by various government programs.

Owner-Builder/Developer.

Construction and rental of apartment buildings.

Construction and rental of commercial and office facilities.

Construction and leasing of post offices and other government buildings.

Professional Construction Manager.

Application to city and state agencies of large corporations awarding this type of contract.

All the sources for obtaining private contracts.

Uses of Dodge Bulletin. From a typical *Dodge Bulletin* (McGraw-Hill Information Systems Company, New York), a subscriber to this daily information bulletin can gain the following information:

Contracts for which general contract and prime bids for mechanical and electrical work, etc., are being requested. A contractor interested in any of these types of work can obtain the plans and specifications from owners or designers, whose names and addresses are given, and submit a bid.

For contracts awarded, lists of names of contractors and amounts of contracts. Subcontractors or material suppliers who are interested in working for the contractors can communicate with those who have received the awards.

Lists of jobs being planned and estimates of job costs. Contractors and subcontractors who are interested in jobs in those locations and of the sizes indicated can communicate directly with the owner.

Additional information that may be obtained from the *Dodge Bulletin* includes lists of subcontractors and suppliers being employed by general contractors on other jobs that are already under way, and tabulation of the low bidders on jobs bid and publicly opened.

25-5. What Constitutes the Contract Documents?

Generally, plans and specifications for a project are completed before issuance to contractors. This, however, is not always the case. For example, during the course of design, the architect may distribute progress plans and specifications for review by government agencies or for pricing by contractors. Sometimes, the owner may be anxious to enter into a contract or start a job before the plans and specifications have been totally completed or approved. Therefore, to the construction manager, the most important thing to be alert to regarding plans and specifications is: What constitutes the agreed-upon plans and specifications and other contract documents that pertain to the project?

It is surprising how often this question is neglected by those entering into construction contracts and by attorneys and others concerned with signing of contracts. A clear understanding of what constitutes the contract documents and the revisions, if any, is one of the most urgent aspects of construction management. Furthermore, a precise list of what constitutes the contractor's obligation under the contract is essential to proper performance of the contract by the contractor.

In general, the contract documents should be identified and agreed to by both parties. A listing of the contract documents should be included as part of every contract. Contract documents, which are described in Sec. 26, generally include the following:

Plans (list each plan and revision date of each plan, together with title)
Specifications, properly identified, a copy initialed by each party and in the possession of each party
The agreement, or contract (Art. 25-8)
General conditions
Soil borings
Existing site plan
Special conditions
Original proposal (if it contains alternates and unit prices, and these are not repeated in the contract)
Invitation (if it contains data on completion dates or other information that is not repeated in the contract)
Addenda (if any)

The contractor should repeat and list the contract documents in each subcontract and purchase order (Arts. 25-11 and 25-12).

It is essential to have a properly drawn and understood list of the contract documents agreed to by all parties if construction management is to proceed smoothly and if changes and disputes are to be handled in an orderly manner (Art. 25-20). To prevent misunderstandings and doubts as to which documents are in the possession of various subcontractors and suppliers for estimating, a properly drafted transmittal form should accompany each transmission of such documents (Fig. 25-2).

25-6. Major Concerns with Building Codes.

Contractors should have a working knowledge of a variety of building codes. In most cities and municipalities, there is a local building code. It also may be the same as the state code.

Local and state codes usually govern most of the construction activities of a contractor, in addition to the requirements of the plans and specifications. The contractor's task of satisfying code regulations, however, is made more difficult because most building codes continually undergo significant changes. Consequently, contractors should be alert to such changes and have sources that provide them with information on the new regulations.

In addition, local and state building codes provide means for enforcing compliance of contractors with code requirements. Usually, such enforcement takes the form of inspections by building inspectors, affidavits by architects and engineers that construction has been performed consistent with the code, and submissions of plans, specifications, and details to building officials for their approval before construction starts.

TRANSMITTAL

Kreisler Borg Florman
CONSTRUCTION COMPANY

TO: Lipsky & Rosenthal, Inc.
155 Utica Avenue
Brooklyn, N.Y. 11213

97 Montgomery Street, Scarsdale, New York 10583/Telephone (914) SC 5-4600

ATT: Mr. D. G.
Mr. A. H.

DATE February 25, 1982

RE: J. 47 Combined School and Apartment, New York, N.Y.

WE ARE SENDING YOU THIS DATE THE FOLLOWING:

- [X] BLUEPRINTS
- [ ] SPECS
- [ ] SAMPLES
- [ ] BOOKLETS
- [ ] CATALOGUE CUTS
- [ ] LETTERS
- [ ] TEST REPORTS
- [ ]

SENT FOR THE FOLLOWING REASON:

- [ ] YOUR APPROVAL
- [ ] YOUR FINAL APPROVAL
- [ ] APPROVED
- [ ] APPROVED AS NOTED
- [ ] DISAPPROVED
- [ ] RESUBMIT
- [X] YOUR INFORMATION
- [ ] YOUR USE
- [ ]

NUMBER COPIES	DRAWING NUMBER	PREPARED BY	DESCRIPTION
1 Set	A-1 thru A-10 Dated 2/18/82	Carl Puchall & Assoc.	School Preliminary Plans
1	7 of 12 Rev. 10/18/82	" " "	Typical Apartment Floor Plan
1	9 of 12 Rev. 1/17/82	" " "	Sub. Cellar Garage

REMARKS: We would appreciate your budget figure for referenced project, as soon as possible. Price must be broken down to School portion and Residential portion.

Thank you.

Kreisler Borg Florman
CONSTRUCTION COMPANY

BY _____

TITLE __Estimator_____

Fig. 25-2 Form for transmittal of contract documents to subcontractors and suppliers.

It is only after all code requirements have been complied with and all required approvals have been obtained that a contractor will receive a Certificate of Occupancy for a completed building. Before this, however, approval of the work may also have to be obtained from other local officials: Fire Department for sprinklers, fire alarms, and exit and elevator safety; Health Department for food-handling and kitchen equipment;

Environmental Protection Agencies for boiler and incinerator emissions; Police Department for fire and sprinkler alarrms.

See also Art. 1-10.

25-7. Estimating, Bidding, and Costs.

Methods of preparing cost estimates for building construction are described in Sec. 24. It is advisable to have the routine to be followed in preparing cost estimates and submitting bids well established in a contractor's organization.

Particular attention should be given to the answers to the question: For whom is the estimate being prepared and for what purpose? The answers will influence the contractor as to the amount of time and effort that should be expended on preparation of the estimate, and also indicate how serious the organization should be about attempting to negotiate a contract at the figure submitted. Decision on the latter should be made at an early date, even before the estimate is prepared, so that the type of estimate can be decided.

Bid Documents. The documents should be examined for completeness of plans and specifications, and for the probable accuracy that an estimate will yield from the information being furnished. For example, sometimes contract documents are sent out for bid when they are only partly complete and the owner does not seriously intend to award a contract at that stage but merely wishes to ascertain whether construction cost will be within his budget.

Preparation of the Top Sheet. This is usually based on an examination of the specifications table of contents. If there are no specifications, then the contractor should use as a guide top sheets (summary sheet showing each trade) from previous estimates for jobs of a similar nature, or checklists.

Subcontractor Prices. Decide on which trades subbids will be obtained, and solicit prices from subcontractors and suppliers in those trades. These requests for prices should be made by postcard, telephone, or personal visit.

Decide on which trades work will be done by the contractor's own forces, and prepare a detailed estimate of labor and material for those trades.

Pricing. Use either unit prices arrived at from the contractor's own past records, estimates made by the members of the contractor's organization, or various reference books that list typical unit prices ("Building Construction Cost Data," Robert Snow Means Co., Inc., P.O. Box G, Duxbury, Mass. 02332). Computerized banks of unit prices for various types of work on different structures may be maintained by a contractor. These can be updated electronically with new wage and material costs, depending on the program used, so that prices can be applied nearly automatically.

Hidden Costs. Carefully examine the general conditions of the contract and visit the site, so as to have a full knowledge of all the possible hidden costs, such as special insurance requirements, portions of site not yet available, etc.

Final Steps. Receive prices for materials and subcontracts.

Review the estimate and carefully go over exclusions and exceptions in each subcontract bid and in material quotations. Fill in with allowances or budgets those items or trades for which no prices are available.

Decide on the markup. This is an evaluation that should be made by the contractor, weighing factors such as the amount of extras that may be expected, the reputation of the owner, the need for work on the part of the general contractor, and a gage as to the contractor's overhead.

Finally, and most importantly, the estimate must be submitted in the form requested by the owner. The form must be filled in completely, without any qualifying language or exceptions, and must be submitted at the time and place specified in the invitation to bid.

25-8. Types of Bids and Contracts.

Contractors usually submit bids for a lump-sum contract or a unit-price-type contract, when based on complete plans and specifications.

When plans and specifications have not yet advanced to a stage where a detailed estimate can be made, the type of contract usually resorted to is a cost-plus-fee type. The bid may be based on either a percentage markup the contractor will receive over and above costs, or may include a lump-sum fee that the contractor charges over and above costs. Sometimes an incentive fee is also incorporated, allowing owner and contractor to share, in an agreed-on ratio, the cost savings achieved by the contractor, or to reward the contractor for completing the project ahead of schedule.

Evaluation of bids by an owner may take into consideration the experience and reputation of the contractor, as a result of which awards may not be made on a strictly low-bid basis.

Budget Estimate. Often, an owner in preparing plans and specifications will want to determine the expected construction cost while the plans are still in a preliminary stage. He will then ask contractors for a cost estimate for budget purposes. If the estimate from a specific contractor appears to be satisfactory to an owner, and if the owner is desirous of establishing a contractual relationship with the contractor early in the planning stage so as to benefit from the contractor's suggestions and guidance, a contract may be entered into after the submission of the estimate.

On the other hand, the owner may refrain from formally entering into the contract, but may treat the contractor as the "favored contractor." When requested, this contractor will assist the architect and engineers with advice and cost estimates and will expect to receive the contract for construction on completion of plans and specifications, if the cost of the project will lie within the budget estimate when plans and specifications have been completed.

Separate Prime Contracts. Sometimes an owner has the capability for managing construction projects and will take on some of the attributes of a general contractor. One method for an owner to do this is to negotiate and award separate prime contracts to the various trades required for a project. Administration of these trades will be done either by the owner's own organization or by a construction manager hired by the owner (Art. 25-9).

Sale Lease-Back. This is a method used by some owners and government agencies to obtain a constructed project. Prospective builders are asked to bid not only on cost of construction, but also on supplying a completed building and leasing it to the prospective user for a specified time. This type of bid requires a knowledge of real estate analysis and financing, as well as construction. Contractors who bid may have to associate with a real estate firm to prepare such a bid.

Developer/Sponsor-Builder. In this type of arrangement (Art. 25-1), the contractor may not only have to prepare a construction-cost estimate but may also require knowledge of real estate, to be prepared to act as owner of the completed project, in accordance with the terms of a sponsor-builder agreement with a government agency, or government-assisted neighborhood or nonprofit group.

The following types of contracts are used for general construction work:

Letter of Intent. This is used where a quick start is necessary and where there is not sufficient time for drafting a more detailed contract. A letter of intent also may be used where an owner wishes material ordered before the general contract is started, or where the commitment of subcontractors requiring extensive lead time must be secured immediately.

Lump-Sum Contract. (For example, Document A-101, American Institute of Architects, 1735 New York Ave., N.W. Washington, D.C. 20006.) Basis of payment is a stipulated sum. Progress payments, however, are made during the course of construction.

Cost-plus-Fixed-Fee Contract. (For example, A-111, American Institute of Architects.) This type of agreement is used generally with an "up-set" price. The contractor guaranteees that the total cost plus a fee will not exceed a certain sum. (See types of bids, preceding.) Generally, there are provisions for auditing of construction costs by the owner.

Cost-plus-Percentage-of-Fee Contract. Similar to cost-plus-fixed-fee contracts, but the fee paid, instead of being a lump sum in addition to the cost, is a percentage of the costs.

Unit-Price Contract. This type of agreement is used where the type of work involved is subject to variations in quantities and it is impossible to ascertain the total amount of work when the job is started. Bids will be submitted by the contractors on the basis of estimated quantities for each classification of work involved. On the basis of the unit prices submitted and the estimated quantities, a low bidder will be chosen for award of the contract. After the contract has been completed, the final amount paid to the contractor will be the sum of the actual quantities encountered for each class of work multiplied by the unit prices bid for that class.

Design-Construction Contracts. This type of agreement is used for turnkey projects and by package builders (Art. 25-1). Cost estimates must be prepared from preliminary plans or from similar past jobs. Preparation of these estimates requires high skill and knowledge of construction methods and costs, because the usual methods for preparing cost estimates do not apply.

25-9. Professional Construction Managers.

A professional construction manager (CM) is an individual or organization specializing in construction management or practicing it on a particular project as part of a project management team that also includes an owner and a design organization. A prime construction contractor or a funding agency may also be a member of the team. (See Art. 1-13.)

As the primary construction professional on the project management team, the CM provides the following services or such portion thereof as may be appropriate to the specific project:

1. The CM works with the owner and the design organization from the beginning of design through completion of construction, providing leadership to the construction team on all matters relating to construction, keeping the project management team informed, and making recommendations on design improvements, construction technology, schedules, and construction economies.

2. The CM proposes construction and design alternatives to be studied by the project management team during the planning phase and analyzes the effects of these alternatives on the project cost and schedule.

3. After the project budget, schedule, and quality requirements have been established, the CM monitors subsequent development of the project to ascertain that those targets are not exceeded without knowledge of the owner.

4. The CM advises on and coordinates procurement of equipment and materials and the work of all contractors on the project. Also, the CM may monitor payments to contractors, changes ordered in the work, contractors' claims for extra payments, and inspection for conformance with design requirements. In addition, the CM provides current cost and progress information as the work proceeds and performs other construction-related services required by the owner.

The CM does not normally perform significant design or construction work with his own forces but may provide some or all of the site facilities and services specified in the General Conditions of the construction contract.

The advantages of engaging a CM over conventional construction with a general contractor (see Art. 1-4) are as follows:

The CM treats project planning, design, and construction as integrated tasks, which are assigned to a project management team. The team works in the owner's best interests from the beginning of design to project completion. The contractual relationships between the members of the team are intended to minimize adversary relationships and to contribute to greater responsiveness within the team. In this way, the project can be completed faster and at lower cost. (See also Art. 2-19.)

(D. Barry and B. C. Paulson, "Professional Construction Management," Journal of the Construction Division, American Society of Civil Engineers, September 1976.)

25-10. Contract Administration.

Administration of construction contracts requires an intimate knowledge of the relationship of the various skills required for construction, which involves labor, material suppliers, and subcontractors. Feeding into the job are all of the life-supplying services. This

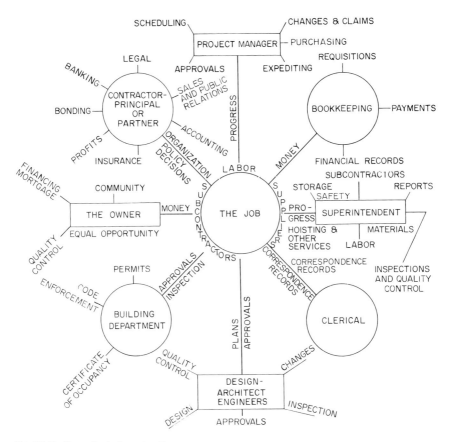

Fig. 25-3 Borg wheel of construction progress.

is illustrated in the "Borg wheel of construction progress" (Fig. 25-3). Like any wheel, it is dependent on all the spokes being in place if it is to move. Whether the contractor combines one or more of the jobs shown in more than one person in his organization is immaterial. But all the spokes shown in the wheel are necessary.

In the upper left of Fig. 25-3, the task of the contractor, principal or partner, shows his relationship to the wheel of construction progress. As indicated, this person must be familiar with and has responsibility for legal, bonding, insurance, and banking requirements of the firm. He feeds into the job necessary organization and policy decisions. This contribution, when added to what is fed into the job by the project manager (progress), bookkeeping (money), superintendent (progress), clerical (correspondence and records), architect and engineers (plans and approvals), building department (approvals and inspection), and the owner (money), is essential for job progress.

Planning of the job is dealt with in Art. 25-17. Profit and loss of the job are controlled in the manner described in Art. 25-24.

25-11. Subcontracts.

General contractors usually obtain subcontract bids as well as material-price solicitations during the general-contract bidding stage. Sometimes, however, general contractors continue shopping after award of the general contract to attain budget goals for the work that may have been exceeded during the initial bidding stages. In such cases, additional bids from subcontractors are solicited after the award of contract.

Purchasing Index. Contractors would find it advantageous to approach purchasing of subcontracts with a purchasing index (Fig. 25-4). The index should list everything necessary to be purchased for the job, together with a budget for each of the items. As subcontracts are awarded, the name of the subcontractor is entered, and the amount of the subcontract is noted in the appropriate column. Then, the profit or loss on the purchase is later entered in the last column; thus, a continuous tabulation is maintained of the status of the purchases.

Priority numbers are given to the various items, in order of preference in purchasing.

		KREISLER BORG FLORMAN CONSTRUCTION CO.				
		97 Montgomery Street				
		Scarsdale, New York 10583				
PRIORITY NO.		ITEM	ESTIMATE	SUBCONTRACTOR	AMOUNT	PROFIT
1	2A	Excavation	77,000			
1	2B	Foundations	INCL.			
1	2H	Piling	INCL.			
1	3A	Concrete Superstructure	242,000			
2	4	Masonry	104,400			
3	4B	Cast Stone	INCL.			
3	4C	Ext. Cement Ash Sills	INCL.			
6	12A	Venetian Blinds	2,200			
2	14A	Elevators	29,600			

Fig. 25-4 Purchasing index. Numbers in first column indicate priority in purchasing subcontracts, materials, and equipment.

The contractor should concentrate his efforts on those subcontracts that must be awarded first. Those that follow in due course are given priority numbers that are appropriate. These priorities are indicated in Fig. 25-4 by the numbers in the left column.

Bid Solicitation. Bids are generally solicited through notices in trade publications, such as *Dodge Bulletin,* or from lists of subcontractors that the contractor maintains. Solicitation also can be by telephone call, letter, or postcard to those invited to bid. Where the owner or the law requires use of specific categories of subcontractors, bids have to be obtained from qualified members of such groups.

After subcontractor bids have been received, careful analysis and tabulation are needed for the contractor to compare bids fairly (Fig. 25-5).

In a complicated trade, such as Miscellaneous Iron, shown in Fig. 25-5, it is necessary to tabulate, from answers obtained by questioning each of the bidders, the exact items that are included and excluded. In this way, an evaluation can intelligently be made, not only of the prices submitted but also as to whether or not the subcontractors are offering a complete job of the section of work being solicited. Where an indication in the subcontractor's proposal as in Fig. 25-5 shows that a portion of the work is being omitted, it is necessary to cross-check the specifications and other trades to be purchased to ascertain that the missing items are covered by other subcontractors.

Subcontract Forms. Various subcontract forms are available for the written agreements. A commonly used form is the standard form of agreement between contractor and subcontractor (Contractor-Subcontractor Agreement, A401, American Institute of Architects). A short form of subcontract with all the information appearing on two sides of one sheet is shown in Fig. 25-6a and b. Changes may be made on the back of the printed form with the permission of both parties to the agreement.

Important, and not to be neglected, is a subcontract rider (Fig. 25-7), which is tailored

	KREISLER BORG FLORMAN CONSTRUCTION CO.
FOR: C.I. 17	97 Montgomery St., Scarsdale, N.Y. 10583
MISC. IRON SUB CHECK SHEET - SEC. 5A	

ITEMS	Lieb Iron Works	Premier Structural Steel	Ment Bros.	REMARKS
Plans & Specs. C.I. 17 - Base Bid	13,180.	13,987.	14,900.	
No Sales Tax	✓	✓	✓	
Shop Drawings & Samples	✓	✓	✓	
Painting - one coat	✓	✓	✓	
field	NO	NO	NO	
two coat	NO	NO	NO	shelf angle only
Attached Lintels - F & I	✓	✓	✓	
Loose Lintels - Furnish	✓	✓	✓	
Site Bridge (no crane)	+ 2,350.	NO	NO	
Exterior Saddles furn. & install	+ 1,360.	NO	NO	
Alts.				
Site Bridge Crane	+ 1,000.	—	—	

Fig. 25-5 Analysis of subcontractor bids.

Kreisler Borg Florman
CONSTRUCTION COMPANY (INC.)
97 Montgomery Street, Scarsdale, New York 10583
Telephone (914) SC 5-4600   DATE August 31, 1985

TO: Brisk Waterproofing Co Inc
720 Grand Avenue
Ridgefield, New Jersey

PROJECT: East Midtown Plaza Stage II, Project #HRB 66-14B
24th & 25th Sts betw. 1st & 2nd Aves, New York, NY

SUBJECT TO CONDITIONS ON REVERSE SIDE HEREOF, YOU ARE AUTHORIZED TO PROCEED WITH THE WORK AS DESCRIBED BELOW IN CONNECTION WITH THE ABOVE PROJECT, THE COST OF WHICH IS TO BE $ 4,000.00

·············· FOUR THOUSAND DOLLARS ···

DESCRIPTION

HYDROLITHIC IRON TYPE WATERPROOFING (Specification Division 5, Section 35) and all such work shown on the Plans, Specification, Specification Addenda 1, 2, 3 and 4, Construction Contract, General Conditions of the Contract, Invitation, Contractor's Loan Agreement, Bid Form, Performance and Payment Bond, and Subcontract Rider. If Performance and Payment Bond is required premium is to be paid by Contractor.

As a condition precedent to the duty of performance of the work hereunder the Contractor shall be awarded and execute a contract for the work of the project.

Subcontractor agrees to include at no additional cost any further details or corrections to the plans and specifications resulting from requirements of the Building Department or Housing and Development Administration and agrees to be bound by final plans and specifications when they are completed and approved.

This work includes the H.I.T. waterproofing of two elevator pits, 1 ejector pit and laundry troughs and curbs as indicated on plans. In connection with this work, light, water, heat, pumping and power to be furnished to us without charge. Floor slabs to be left raked by others.

PLEASE SIGN EXTRA COPY OF THIS ORDER AND RETURN TO KREISLER-BORG AT ONCE. PLEASE SEND US CERTIFICATES OF INSURANCE FOR WORKMEN'S COMPENSATION, PUBLIC LIABILITY AND PROPERTY DAMAGE INSURANCE.

ALL OF THE ABOVE MATERIALS TO BE DELIVERED OR THE WORK COVERED BY THIS ORDER ARE TO BE COMPLETED IN ACCORDANCE WITH PLANS, SPECIFICATIONS AND ALL CONTRACT REQUIREMENTS BETWEEN US AND THE OWNER, BY ALL OF WHICH YOU AGREE TO BE BOUND UPON ACCEPTANCE OF THIS ORDER.

Kreisler Borg Florman (CONTRACTOR)
CONSTRUCTION COMPANY (INC.)

BY ....
Robert F. Borg, President

ACCEPTED:
BRISK WATERPROOFING CO INC
SUBCONTRACTOR

BY ....
R. W. Ehrenberg, V.P.

DATE: 9/6/85

(over)

Fig. 25-6a Front side of short form for subcontract.

---

for each job. Only one page of the rider is shown in Fig. 25-7. The rider takes into account modifications required to adapt the standard form to the specific project. The rider, dealing with such matters as options, alternates, completion dates, insurance requirements, and special requirements of the owner or lending agency, should be attached to all copies of the subcontract and initialed by both parties.

CONDITIONS OF CONTRACT

Within five days after the date of this contract and before commencement of the work, the subcontractor agrees to furnish the contractor with a certificate showing that he is properly covered by Workmen's Compensation Insurance as required by the law of the State where the work is to be performed and with such other insurance that may be required by Contractor, the specifications and terms of contract between Contractor and the Owner.

Where the order covers the furnishing of labor and materials on a time and material basis, it is distinctly understood the subcontractor will furnish daily vouchers for verification and signature to an authorized representative of Contractor showing labor used and materials installed in the work. A copy of these signed vouchers to be presented with invoice, together with duplicate bills for materials furnished. Contractor shall have the right to examine all records of the subcontractor relating to said charges.

Where this order is issued to a subcontractor, and purports to cover labor and materials in addition to the original contract, it is given with the express understanding that, should it subsequently be proven that the work covered herein is in the subcontractor's original contract, this order becomes null and void.

Should the subcontractor or material dealer at any time refuse or neglect to supply an adequate number of properly skilled workmen or sufficient materials of the proper quality, or fail in any respect to prosecute the work with promptness and diligence, the Contractor shall in its exclusive opinion be at liberty after three days' notice to the subcontractor or material dealer to provide any such labor or materials in accordance with such notice and to deduct the cost thereof from any money then due or to become due to the subcontractor or material dealer under this contract, any excess cost to Contractor will be immediately paid by the subcontractor or material dealer.

Where this order covers materials only, it is agreed that the materials will be delivered F.O.B. job unless otherwise ordered, in such quantity and at such times as may be authorized by this company's representative. It is further understood that in all cases, quantities of materials mentioned herein are approximate only and the dealer agrees that deliveries will be based upon actual needs and requirements of the work.

It is understood that no claims for extra work performed or additional materials furnished shall be made unless ordered in writing by an officer or authorized representative of the Contractor.

TERMS OF PAYMENT: Payments to subcontractors will be made monthly to the amount of 85% of the value of the work and materials incorporated in the building during the previous calendar month. Final balance to become due and payable within sixty days after the subcontractor has completed his work to the entire satisfaction of the architects, engineers, other representatives of the Owner and Contractor.

All payments covering subcontractor's work and/or material shall be payable only after the Contractor has received corresponding payments from the Owner.

The subcontractor hereby accepts exclusive liability for the payment of all taxes now or which may hereafter be enacted covering the labor and material to be furnished hereunder, and any contributions under the New York State Unemployment Insurance Act, The Federal Social Security Act, and all legislation enacted either Federal, State or Municipal, upon the payroll of employees engaged by him or materials purchased for the performance of this contract, and agrees to meet all the requirements specified under the aforesaid acts or legislation, or any acts or legislation which may hereafter be enacted affecting said labor and/or materials. The subcontractor will furnish to the Contractor, any records the Contractor may deem necessary to carry out the intent of said acts or legislation and hereby authorizes the Contractor to deduct the amount of such taxes and contributions from any payments due the subcontractor and to pay same direct or take any such precaution as may be necessary to guarantee payment.

Samples and details are to be submitted to Contractor, if requested, and approval must be secured before proceeding with the work.

If inferior work or material is installed or furnished and allowed to remain, the Owner and/or Contractor at its option, may reject such work and/or material or are to be allowed the difference in value between cost of work and/or material installed or furnished and cost of work and/or material specified or ordered.

The subcontractor will furnish all labor, materials, tools, scaffolds, rigging, hoists, etc. required to carry on the work in the best and most expeditious manner and furnish protection for his and other work, and will do all necessary cutting, and patching and also remove and replace any interfering work, for the proper installation of his work. The subcontractor will remove all rubbish from premises in connection with his work. The subcontractor agrees to perform work in a safe and proper manner and save the Owner and Contractor harmless against all penalties for violation of governing ordinances and all claims for damages resulting from negligence of the subcontractor, or his employees, or accidents in carrying on his work.

The subcontractor agrees to repair, replace or make good any damages, defects or faults resulting from defective work, that may appear within one year after acceptance of work or for such additional period as may be required by Owner or by the specifications relating to same.

Time is of the essence of this agreement.

All labor employed shall be Union labor of such type and character as to cause no Union or jurisdictional disputes at the site of the work.

The subcontractor shall furnish all labor, material and equipment and permits and pay all fees and furnish all shop drawings, templates, and field measurements incidental to the work hereby let to it that the Contractor is required to perform and furnish under the General Contract, and whatever the Contractor is required to do or is by the General Contractor bound in and about the work hereby let to the Subcontractor shall be done by the Subcontractor without any extra charge.

Any controversy or claim arising out of or relating to this contract, or breach thereof, shall be settled by arbitration in the City of New York in accordance with the Rules of the American Arbitration Association, and judgment upon the award rendered by the Arbitrator(s) may be entered in any Court having jurisdiction thereof.

The Subcontractor expressly covenants and agrees to file no lien of any nature or kind for any reason whatsoever arising out of this contract for matters and things related thereto and does hereby expressly and irrevocably constitute the Contractor as its agent to discharge as a public record any lien which may have been filed by it for any reason or cause whatsoever.

Fig. 25-6b Back side of short form for subcontract.

One additional clause that should appear in all subcontracts is the following:

This subcontract shall be subject to all the terms and conditions of the prime contract between the general contractor and the owner insofar as that contract shall be applicable to the work required to be performed under this subcontract, it being the intention that the subcontractor shall assume to the general contractor all obligations of the general contractor to the owner insofar as the work covered by this subcontract is concerned and that the subcontractor shall be bound by all rulings, determinations, and directions of the architect and/or owner to the same extent the general contractor is so bound.

B
            Canaan IV Apartments
            115th St. and Lenox Ave.

The printed part of this contract is hereby modified and supplemented as follows. Wherever there is any conflict between this rider and the printed part of the subcontract, the provisions of this rider are paramount and the subcontract shall be construed accordingly:

31. *ARTICLE 3 shall be amended by the addition of the following:*
    The reference herein to payments to the contractor for the owner shall be deemed to be absolute conditions precedent to the contractor's obligation to pay the subcontractor herein, it being understood distinctly that said provision shall not be deemed to relate solely to the timing of the payments due from the contractor to the subcontractor, but, shall be deemed to relate to the contractor's liability therefor. The contractor shall not be liable to the subcontractor until and unless it receives the said payments from the owner.

32a *ARTICLE 10 OF STANDARD FORM OF SUBCONTRACT, eliminate and replace by:*
    The subcontractor agrees to indemnify and save harmless the Owner and the Contractor against loss or expense by reason of the liability imposed by law upon the Owner, Contractor, their partners, representatives, agents, employees, officers, and directors, for damage because of bodily injuries, including death at any time resulting therefrom accidentally sustained by any person or persons or on account of damage to property arising out of this work, whether such injuries to person or damage to property are due or claimed to be due to any negligence of the subcontractor, the Owner, Contractor, their partners, representatives, agents, employees, officers and directors, or any other person. The foregoing, however, shall not be deemed a covenant, promise, agreement, or understanding that the subcontractor shall indemnify and save harmless the Owner, Contractor, their partners, representatives, agents, employees, officers, and directors, for damage resulting from their sole negligence.

32b The subcontractor agrees to obtain and maintain at his own expense and until the completion of this project the following insurance:
    (I) Workmen's Compensation with Employer's Liability coverage as required by State Statute.
    (II) Comprehensive General Liability covering the Subcontractor's operations, sublet operations, completed operations and contractual [as described in 10a above]. This insurance shall be in the minimum amount of:
        With respect to bodily injury $3,000,000 each occurrence
        With respect to property damage $1,000,000 each occurrence
        With respect to property damage, coverage for explosion, collapse and damage to underground utilities shall be included.
    (III) Fire, extended coverage, theft insurance, etc. on subcontractor's equipment, forms, and temporary facilities.

Fig. 25-7. Subcontract rider.

Also, the list of drawings (Fig. 25-8) should not be neglected. The content of the exact contract drawings should never be left in doubt. Without a dated list of the drawings that both parties have agreed to have embodied in the subcontract, disputes may later arise.

See also Arts. 25-12 and 25-20.

25-12. Purchase Orders.

Issuance of purchase order differs from issuance of a subcontract (Art. 25-11). A purchase order is issued for material on which no labor is expected to be performed in the field. A subcontract, on the other hand, is an order for a portion of the work for which the sub-

25-18 Construction Management

```
                    East Midtown Plaza Stage II
                24th & 25th Sts. between 1st & 2nd Aves.
                            New York, New York

DRAWING
  NO.      TITLE                                    DATE     REV.       REV.

T-1        Title & Drawing List
T-2        Notes, Symbols & Abbreviations
Z-1        Zoning
Z-2        Zoning
A-1        Cellar Plan - Parts G, GH & H
A-2        First Floor Plan - Parts G, GH & H

A-22       Elevators
A-23       Doors

                        STRUCTURAL
S-1        Foundation Plan - Bldgs. G & H

S-11       Typical Floor Framing Plan-Bldg. H          (R. B.)
           West half.                        5/1/82
                                                      PLEASE INITIAL
```

Fig. 25-8 List of drawings for construction of a project.

contractor is expected not only to furnish materials but also to perform labor in the field. An example of a purchase order form, front and back, is shown in Fig. 25-9a and b.

For the specific project, a purchase order rider (like Fig. 25-7) and list of contract drawings (Fig. 25-8) should be appended to the standard purchase order form. The rider describes special conditions pertaining to the job, options or alternates, information pertaining to shop drawings, or sample submissions and other particular requirements of the job.

Material-price solicitations are handled much in the same manner as subcontract-price solicitations (Art. 25-11). Material bids should be analyzed for complicated trades in the same manner as for subcontracts (Fig. 25 5).

To properly administer both the subcontract and the purchase orders, which may number between 40 and 60 on an average building job, it is necessary to have a purchasing log (Fig. 25-10) in which is entered every subcontract and purchase order after it has been sent to the subcontractor or vendor. The entry in the log is copied from the file copy of the typed document, and an initial in the upper right-hand corner of the file copy indicates that entry has been made. The log serves as a ready cross reference, not only to names of subcontractors and vendors but also to the amounts of their orders and the dates the orders were sent.

Computers in Purchasing. On a large-scale construction project, which costs many millions of dollars, thousands of items may have to be procured. Also, information, shop drawings, and part lists may have to be distributed and revised, redistributed, approved, and distributed again. Systems have been developed to use computers to keep track of all equipment and materials and related purchase information, such as specifications, quotations, final orders, shipment, and delivery dates.

The systems are based on the concept of critical-path items, and the various tasks that must be performed are assigned due dates. For example, one such system is the Martin-Marietta Data Systems Technical Requirements Management System (TRMS), which operates on an IBM 370/158 computer linked to terminals and remote printers in various parts of the contracting-organization home office and in the field. Information about any item, or group of items, can be read from cathode-ray-tube screens or the printout from remote printers. For example, a computer report could be by project and show all open purchase-order items for one project, or by buyer name, with all open purchase-order items for each buyer, including all projects.

In negotiating and awarding either a subcontract or a material purchase, the contractor should take into account the scope of the work, list inclusions properly, note exceptions or exclusions, and, where practicable, record unit prices for added or deleted work.

KBF N° 3002   PURCHASE ORDER   **Kreisler Borg Florman**

97 Montgomery Street, Scarsdale, New York 10583 Telephone (914) SC 5-4600 CONSTRUCTION COMPANY, INC.
(herein referred to as "Purchaser")

TO Construction Products Co.
   Route #7   Att: Mr. Alan Fishkin
   Brookfield, Connecticut

DATE 180
(referred to herein as "Vendor")   JOB NO.
Please enter our order for the following materials, subject to the terms and conditions and all of the provisions set forth herein and in accordance with all pertainent provisions of the contract dated __6/9/81__ entered into between purchaser and __Construction for Progress, Inc.__
(referred to herein as "The Owner") for the construction of: (Name of project and shipping address)
   13-Story Apartment Building, 170th Street to 172nd Street and 93rd
   Avenue, Jamaica New York
(Which contract, together with all plans, specifications and all provisions and documents incorporated or referred to therein is referred to herein as the "Principal Contract").

QUANTITY	DESCRIPTION	PRICE (Unit or Lump Sum)
	REFUSE CHUTE HOPPER DOORS, ACCESS DOOR AND SPARK ARRESTOR (Specification Division 23, Section 10), furnished, fabricated, and delivered, and all such work shown on the Plans, Specifications, AIA General Conditions, Modifications to AIA General Conditions, General Conditions, Contract and Rider.	
	This order is based on the following:	
	Twelve (12) 15"x18" hopper doors and frames One (1) 12"x12", 1-1/2 hr. fireproof self-closing access door and sleeve One (1) explosive vent	
	Five (5) hoppers to have sprinkler heads installed.	
	The price for this material FOB job site will be	...............$1,350.00
	Shop drawings and sample hopper door to be submitted immediately.	

The price(s) set forth above shall constitute payment in full for the prompt and proper furnishing of the materials, hereunder in accordance with the terms of this Purchase Order. If unit prices are set forth above, Vendor agrees to be bound by the quantities of such items which may be certified by the Owner as having been properly furnished under the terms of the Principal Contract, it being understood in such event that unit quantities set forth above are estimated and used only for convenience in determining the approximate amount of this Purchase Order and that the actual quantities required may substantially vary, upward or downward from the estimated quantity.
If no price is set forth above, it is agreed that Vendor's price will be the lowest prevailing market price and in no event is the order to be filled at higher prices than last previously quoted or charged without Purchaser's consent..
This order shall not become effective unless within ten (10) days from the date of execution hereof by Purchaser as set forth below, Purchaser receives the acknowledgment copy hereof, unconditionally executed by an officer of Vendor. If Purchaser does not so receive the executed acknowledgment hereof, the offer contained herein is withdrawn and shall thereafter be renewed only by written statement to such effect made by Purchaser.

TIME OF DELIVERY: THE ABOVE IS HEREBY ACCEPTED   TERMS OF PAYMENT: Regular
approximately six   VENDOR _____ PROD. CO. INC   **Kreisler Borg Florman**
to eight weeks.   BY _____ VICE PRES   CONSTRUCTION COMPANY, INC.
                            DATE 9/28/82   BY _____
                            (Name and Title)   (Name and Title) President

1. ACKNOWLEDGMENT MUST BE SIGNED AND RETURNED BEFORE INVOICES WILL BE PAID.   2. PURCHASE ORDER NUMBER MUST APPEAR ON ALL INVOICES.
                        3. THERE MUST BE A SEPARATE INVOICE FOR EACH PURCHASE ORDER.

VENDOR

Fig. 25-9 *a*   Front side of purchase order.

Construction Management

TERMS AND CONDITIONS

The purchase orders is subject to the following terms and conditions and by accepting this Purchase Order, Vendor agrees to be bound by each and every of said terms and conditions.

1. Acceptance of this Purchase Order by Vendor in time and manner hereinbefore specified shall constitute the only possible manner of acceptance hereof and Purchaser will in no way be responsible or indebted to Vendor for any materials or samples thereof furnished by Vendor or for any work or services performed by Vendor in the absence of such acceptances, it being understood that Vendor shall in such event make no claim against the Purchaser.

2. This Purchase Order may only be accepted by Vendor unconditionally and exactly as written. Any additional or different terms or conditions stated by Vendor are hereby objected to and rejected and the statement by Vendor of such additional terms and conditions shall be deemed to constitute a rejection of the offer contained therein, even if accompanied by the acknowledgment copy hereof executed by Vendor.

3. The principal contract has been, is and will be available for review by Vendor at Purchaser's office during normal business hours. By accepting this order, Vendor acknowledges that it has read the Principal Contract and is fully aware of each and every of its provisions which in any way bear upon or relate to the materials to be furnished hereunder and Vendor understands and agrees that all of such provisions of the Principal Contract are applicable hereto and made a part hereof as if expressly set forth herein in full, including but not limited to the general provisions of the Principal Contract dealing with the subject project as a whole as well as the special provisions and parts thereof specifically relating to the materials to be furnished hereunder and Vendor agrees to furnish said materials as if it were the contractor named in the principal contract, with Vendor hereby assuming the same obligations, undertakings, warranties, guarantees, risk and liabilities to the Purchaser as Purchaser has assumed to the Owner in addition to any other obligations, undertakings, warranties, guarantees, risks and liabilities provided for herein or otherwise provided as a matter of Law. In the event of any conflict between the Provisions of this Purchase Order and of the Principal Contract, this Purchase Order shall control. It is understood that if the Principal Contract affords to the Owner the right to approve a Vendor, then the Purchase Order is conditioned upon such approval and in the event the Owner does not so approve, then this Purchase Order shall be null and void and of no force or effect and Vendor shall have no rights whatsoever against Purchaser and of all the materials hereunder and/or any samples thereof, if any, which have previously been furnished by Vendor, may be returned to Vendor at Vendor's sole risk and expense and Vendor shall have no claim of any sort against Purchaser.

4. All materials hereunder shall be furnished in strict accordance with the Principal Contract and to the satisfaction of both Purchaser and the Owner. Both the Owner and Purchaser have the right to inspect the materials hereunder at Vendor's place of manufacture and/or at any time prior to delivery to the job and/or after delivery to the job and/or after installation in the project and in the event that the materials are rejected in whole or in part by the Owner and/or Purchaser for not conforming to the requirements hereof, such materials may be returned to Vendor at Vendor's sole risk and expense. In such event all costs incurred by Purchaser in testing and/or inspecting the materials shall be for the account of Vendor. Purchaser reserves the right to elect to keep non-conforming materials or some portion thereof rather than returning them to Vendor. In such event, in addition to all other liabilities provided for hereunder and under the law otherwise, Vendor shall be charged for the difference between the value of the delivered non-conforming materials and the value of such materials would have had had they conformed to the requirements of this purchase order. Notwithstanding anything to the contrary hereunder or under the Law otherwise, Purchaser is not required to inspect and/or reject the materials hereunder and/or any particular period of time and delay in such inspection and/or rejection shall in no way limit Vendors obligation to furnish materials which strictly conform to the requirements hereof and shall in no way constitute an acceptance of non-conforming goods, nor shall payment to Vendor for or utilization by Purchaser of any of all of the materials hereunder be deemed an acceptance thereof. Vendor shall, at its own expense, arrange for an assist in all testing of materials hereunder as required under the Principal Contract or when requested by Purchaser and/or the Owner.

5. Unless hereinbefore expressly stated to the contrary, Purchaser shall, within ten (10) days after receiving payment therefor from the Owner, pay to the Vendor eighty-five (85%) percent of the amount due in accordance with the prices set forth above, for materials furnished and delivered by Vendor and Vendor shall not be so entitled to said payment or any portion thereof unless and until Purchaser is paid by the Owner for such furnishings. The remaining fifteen (15%) percent of said payment shall be paid to Vendor within ten (10) days after Purchaser receives payment from the Owner of all retained percentages under the terms of the Principal Contract. Before any partial of final payment, Vendor shall furnish written proof satisfactory to Purchaser that all labor and all suppliers of material, equipment and/or services in connection with the materials hereunder have been paid in full and that there are no claims outstanding against it in connection with the materials hereunder.

6. Vendor shall be entitled to no price increases or escalators of any kind in connection with the furnishing of the materials hereunder nor shall Vendor be entitled to reimbursement for boxing or packing costs or cartage charges or taxes of any type (including any Federal, State or local taxes and including but not limited to Sales or Use Taxes), all of such costs and expenses and all other costs and/or expenses which might be incurred by Vendor in connection with the furnishing of the materials hereunder being included in the consideration set forth on the reverse side hereof, payment of which consideration shall constitute payment in full for the furnishing of the materials hereunder. Vendor represents that the price charged for the items or services covered by this Purchase Order, is the lowest price charged by Vendor to buyers of a class similar to Purchaser under conditions to those specified in this Purchase Order and that prices comply with applicable Government regulations in effect at time of quotation, sale or delivery. Vendor agrees that any price reduction made in merchandise covered by this order subsequent to the placement of this Purchase Order will be applicable to the Purchase Order.

7. The number of this Purchase Order must appear on all invoices and if more than one purchase order has been issued by Purchaser to Vendor, there must be separate invoices for each order. Written evidence (such as delivery receipts) of delivery of the materials hereunder to Purchaser must be attached to all invoices.

8. Unless a specific delivery date or period of delivery is expressly specified herein, Vendor shall within five (5) days after being notified so to do by Purchaser, deliver the subject material or any portion thereof in such amounts and at such times as Purchaser may direct. It is understood that such time of delivery is of the essence and shall be determined by the requirements of the Principal Contract and that Purchaser shall be the sole judge thereof. Delivery shall be made FOB job site. Vendor shall unload, uncrate and stock pile in the places and manner requested by Purchaser.

9. Purchaser is under no obligation to accept any delivery exceeding the quantity specified in this Purchase Order and reserves the right to return, at Vendor's expense, part or all of any amount so delivered. Unless the entire order can be filled, Purchaser is under no obligation to accept delivery of any part thereof, unless previously consented to in writing and Purchaser reserves the right to return, at Vendor's expense, part or all of any such delivery. Material must be delivered as per instructions and if not so delivered, any extra handling costs incurred by Purchaser will be billed back to Vendor.

10. Vendor shall prepare and submit all samples and drawings and any other back-up data when and as required by the Principal Contract and/or when and as otherwise required by Purchaser or Owner, and Vendor shall perform any and all other matters necessary in order to have the materials hereunder approved by the Owner. All of the preparation, submissions and other work, materials and/or services under or in connection with this paragraph shall be at the sole expense of Vendor. Unless otherwise agreed in writing, all special dies, molds, patterns, jigs, fixtures, and any other property furnished to the Vendor by the Purchaser, or specifically paid for by the Purchaser for use in the performance of this Purchase Order, shall be and remain the property of Purchaser, shall be subject to removal upon Purchaser's instructions; shall be used only in filling orders from Purchaser; shall be held at Vendor's risk, and shall be kept insured by Vendor at the Vendor's expense while in its custody or control in an amount equal to the replacement cost thereof, with loss payable to Purchaser. Copies of policies or certificates of such insurance will be furnished to Purchaser on demand.

11. Vendor agrees to employ workmen and to use materials and work practices and delivery methods which shall not in any manner cause or contribute to strikes or other labor disturbances by workmen employed by Purchaser and/or by subcontractors of Purchaser or by any other contractors involved in the subject project. Union labels shall be affixed to the materials hereunder if Purchaser so requires. Vendor will not discriminate against any employee or applicant for employment because of race, creed, color, or national origin. Vendor will take affirmative action to insure that applicants are employed, and that employees are treated during employment, without regard to their race, creed, color, or national origin.

12. If the Owner makes or causes to be made changes or issues or causes to be issued orders or directions, any of which effect the materials to be furnished hereunder, this Purchase Order shall be deemed to be modified accordingly. If such modification causes an increase or a decrease in the cost of furnishing the materials hereunder, then in lieu of all claims and demands which vendor might otherwise have or make, Vendor shall accept in full satisfaction and discharge of all such claims and demands, the amounts, if any, as may be allowed by the Owner to the Purchaser on account of Vendor's claims and demands. If such modification causes a decrease in the cost of furnishing the materials hereunder, Vendor shall accept and be bound by the Owner's determination as to the amount of money to be deducted on account thereof. Vendor agrees to accept the Owner's determination, as aforesaid, in full and complete discharge and release of Purchaser's liability to Vendor by reason of such modification and agrees to be bound and concluded thereby and agrees never to look to Purchaser on account thereof, except for such monies, if any, as the Owner may pay to Purchaser in satisfaction of Vendor's claims and demands. Vendor agrees to immediately comply with all orders and directions of Purchaser in any way bearing upon the material to be furnished hereunder, including but not limited to, orders or directions modifying the materials hereunder, irrespective of whether Vendor shall dispute same in any particular (including but not limited to Vendor claiming additional reimbursement therefor), it being understood that the furnishing of the materials hereunder is not to be held up pending the resolution of such dispute.

13. In addition and without prejudice to all other warranties expressed or implied by Law, Vendor warrants that all materials covered by this Purchase Order will strictly conform to the requirements set forth in the Principal Contract and will be of the best quality and workmanship and free from defects and fit for the intended purposes. All warranties, both expressed and implied, asis constitute conditions and shall survive inspection, acceptance and payment and shall inure to Purchaser and the Owner. Without limitation of any rights by reason of any breach of warranty or otherwise, material or articles which are not so warranted may at any time be returned to Vendor at Vendor's expense for credit, correction or replacement, as Purchaser may direct. This warranty shall remain in effect for one (1) year from the date of acceptance of all such materials under the Principal Contract unless a longer period of warranty is contained in the Principal Contract, in which event such longer period shall govern.

14. Vendor shall observe and comply with all Federal, State, Municipal and Local Laws, orders, rules, ordinances and regulations and those of all pertinent bureaus, agencies and departments, which in any way relate to the furnishing of the materials hereunder.

15. Vendor undertakes and agrees to defend at Vendor's own expense and to indemnify and hold Purchaser, its bonding company and the Owner harmless from and in connection with all claims, suits, actions, and proceedings in which Purchaser, and/or Purchaser's bonding company and/or the Owner are made defendants for actual, or alleged infringement of any U.S. or foreign letters patent or copyright or trade secret or unfair competition resulting from the use or sale of the materials purchased hereunder and Vendor further agrees to pay and discharge any and all judgments or decrees which may be rendered in any such suit, action or proceeding against such defendants. The provisions of this paragraph shall apply whether or not Purchaser furnishes any specification to Vendor.

16. Vendor shall promptly pay for all labor, materials, equipment, services and all other charges it incurs in connection with the furnishing of the materials hereunder. All payments received by Vendor hereunder shall be held by Vendor as trust funds for the payment of such labor, materials, equipment, services and all other charges incurred by Vendor in connection with the furnishing of the materials hereunder and shall not be applied to any other purposes until all of Vendor's obligations in connection with the materials hereunder are satisfied. Vendor agrees that it shall file no lien on connection herewith and that it shall file no notice or claims with the Owner or with Purchaser's bonding company. Vendor further agrees that if a lien is filed by any third person or entity for whom Vendor shall be liable, Vendor shall at its own expense and immediately after being notified by Purchaser of the existence of such lien, take all steps necessary to discharge or vacate such lien including but not limited to, furnishing a lien bond if necessary to discharge such lien. If such lien is filed by a third party and is not discharged or vacated within five (5) days after Vendor is notified as to its existence, Purchaser may take the steps necessary to discharge or vacate such lien and charge to Vendor all costs in connection therewith, including attorney's fees. Further to the above, Vendor agrees to assume the entire responsibility and liability for and defense of and to indemnify and hold Purchaser and Purchaser's bonding company harmless from any and all demands, claims, liens and/or charges which may be filed, made and/or asserted by any third person or entity, including but not limited to demands, claims, liens and/or charges for payment for labor, equipment, materials or services as well as those for personal injury (including death) and/or property damage in any way related to the materials hereunder. If at any time there should be any such lien, demand, claim or charge, Purchaser shall have, in addition to all other rights, the right to retain out of any payment then due or thereafter might become due to Vendor, an amount sufficient in Purchaser's opinion to fully indemnify it against any loss in connection therewith. If Purchaser so elects, it may make payment directly to such third persons or entities in order to resolve such third party 'claim and such payment shall be for the account of Vendor.

17. Vendor agrees that it shall not delegate or sublet or assign this Purchase Order, or any of its obligations or rights hereunder or any portion of such right or obligations, without prior written permission of Purchaser. Any such delegation, subletting or assignment without prior written permission of Purchaser shall vest no rights against Purchaser and any third party.

18. In the event that any provisions of the Bankruptcy Act of the United States of America shall be invoked by or against Vendor, or in the event Vendor shall become insolvent, or make an assignment for the benefit of creditors, or a Trustee or Receiver shall be appointed of his property, or any judgment is taken against Vendor and execution issued thereon, or any writ of attachment or the like is served upon Purchaser because of any pending litigation of Vendor, or if the Vendor shall in the opinion of Purchaser violate any of the terms of this Purchase Order, then upon the happening of any of the foregoing, Purchaser shall have the right to terminate the employment of Vendor under this Purchase Order, and in such event, may either employ its own forces or those of others to complete the work remaining to be performed under this Purchase Order either in whole or in part, and shall be entitled to have and use without cost to it, either through its own forces or those of others, any materials so employed by it, any plant, equipment or materials, whether on the premises or otherwise provided by Vendor in the performance of the work called for herein and the completion thereof (the right to possession of which plant, equipment and materials Vendor hereby assigns and transfers to Purchaser), and Vendor shall receive no further payment until the work under the Purchase Order shall be finished, completed, accepted, and final payment made by the Owner. Then, if the unpaid balance that would be due under this Purchase Order on the cost to Purchaser of completing same, including Purchaser's overhead, and any damages or loss which the Purchaser may sustain by reason of Vendor's fault such excess shall be paid to Vendor, but if such cost plus all such damages shall exceed such unpaid balance, Vendor shall pay the difference to Purchaser. In addition to having the right to terminate this Purchase Order for cause, as aforesaid, Purchaser may at any time, terminate this Purchase Order in whole or in part, for Purchaser's own convenience. Upon such termination for convenience, the principals contained in Section 8-706 of the Armed Service Procurement Regulations relating to termination of subcontracts (as such is in effect as of the date of execution by Purchaser hereof) shall determine the rights of Vendor and Vendor shall have no other claims or rights against Purchaser.

19. None of the rights and remedies conferred upon or reserved by Purchaser under this Purchase Order or the liabilities or obligations hereby imposed or assumed by Vendor, shall be deemed exclusive, or impose any limitation upon or be in derogation of any right or remedy of Purchaser now existing or which hereafter may exist, at law, in equity or by statute. Each and every right and remedy of Purchaser under this Purchase Order shall be cumulative and in addition to all of the rights and remedies of Purchaser now existing or hereafter to exist, at law, in equity or by statute. In the event that Vendor fails to perform any of its obligations hereunder, then in addition to all other obligations and/or liabilities imposed upon Vendor hereunder and under the Law otherwise, Vendor shall hold Purchaser harmless from and indemnify Purchaser for any and all losses, costs, damages and expenses (including, but not limited to, all incidental and consequential damages due to delay or otherwise) incurred by Purchaser as a result of any of such failures, it being understood by Vendor that any failure on its part may subject Purchaser to an extent in excess of the dollar amount of this Purchase Order.

20. Failure on the part of Vendor to perform any of the obligations called for hereunder shall be deemed to be a substantial breach of the within Purchase Order. It is understood that this Purchase Order constitutes an entire and invisible agreement. No waiver by Purchaser of any breach of this Purchase Order shall be deemed to be a waiver of any other or any subsequent breach. Acquiescence by Purchaser in a course of performance by Vendor contrary to the terms of this Purchase Order shall not be construed as a waiver or modification of any term hereof. A waiver or modification of any term hereof may be accomplished only be a writing signed by Purchaser as states its intent.

21. It is understood and agreed that each and every provision of this Purchase Order, including any alleged breach thereof, shall be interpreted in accordance with the Law of the State of New York.

22. This Purchase Order constitutes the entire agreement between Purchaser and Vendor and shall supersede and merge all prior understandings, whether oral or in form of correspondence heretofore had between said parties in connection with any and all matters pertaining to this purchase order and the performance thereof, and it is specifically agreed that the within Purchase Order shall be subject to modification only by a writing signed by an officer of Purchaser. Nothing contained in this Purchase Order is intended to inure to the benefit of any third party.

23. Vendor agrees that valid only if signed by an authorized officer of the Purchaser.

Fig. 25-9b Back side of purchase order.

Consideration should be given to the time of performance of units of work and availability of men and materials, or equipment for performing the work. Purchase orders should contain a provision for field measurements by the vendor, if this is required, and should indicate whether delivery and transportation charges and sales taxes are included in the prices.

PURCHASING LOG

KREISLER BORG FLORMAN CONSTRUCTION CO.
97 Montgomery Street, Scarsdale, N. Y.
FOR: 170 Street, Jamaica, N.Y. — TURNKEY PROJECT

SPEC	ITEM	NAME	DATE	AMOUNT
4.00	Excavation and Grading	Dedona Contracting	4/27/81	$25,000.-
5.00	Concrete Foundations	AD&M General Cont.	4/26/81	28,000.-
30.00	Heating & Ventilation	John A. Jones	6/11/81	150,000.-
29.18, 19 &22	Asphalt Paving & Steel Curb	Strada Contracting	6/26/82	4,500.-
22.0	Lath & Plaster	A. Palmese & Son	7/20/82	1,600.-
12.00	Resilient Flooring	Staples Floorcraft	7/20/82	20,000.-

Fig. 25-10 Purchasing log for recording subcontract and purchasing orders.

25-13. Insurance*

Insurance policies are contracts under which an insurance company agrees to pay the insured, or a third party on behalf of the insured, should certain contingencies arise.

No business is immune to loss resulting from ever-present risks. It is imperative, therefore, that a sound insurance program be designed and that it be kept up to date.

Because few businesses can afford the services of a full-time insurance executive, it is important that a competent agent or broker be selected to: (1) prepare a program that will provide complete coverage against the hazards peculiar to the construction business, as well as against the more common perils; (2) secure insurance contracts from qualified insurance companies; (3) advise about limits of protection; and (4) maintain records necessary to make continuity of protection certain. While a responsible executive of the business should interest himself in the subject of insurance, much of the detail can be eliminated by utilizing the services of a competent agent or broker.

There are three kinds of insurance carriers—stock, mutual, and reciprocal. Nevertheless, policy forms are rather well standardized. Rates are established by the rating organizations that serve the member companies of different groups of carriers, but generally, basic rates are very nearly the same.

It is necessary, of course, that the responsibility for providing protection be placed upon an insurance company whose financial strength is beyond doubt.

Another important point for the buyer of insurance is the service that the company selected may be in a position to render. Frequently, construction operations are conducted at a considerable distance from city facilities. It is necessary that the company charged with the responsibility of protecting the construction operations be in a position to render "on-the-job" service from both a claim and an engineering standpoint.

The interests of contractors, subcontractors, and building owners are very closely allied. Particular attention should be given to the definition of these respective interests in all insurance policies. Where the insurable interest lies may depend upon the terms of the contract. Competent advice is frequently needed in order that all policies protect all interests as required.

While the forms of protection purchased and the adequacy of limits are of great importance to a prime or general contractor, it is also of great importance to him to see that the insurance carried by subcontractors is written at adequate limits and is broad enough to protect against conditions that might arise as a result of their acts. Also, the policies should include the interests of the prime or general contractor insofar as such interests should be protected.

*Extracted from Sec. 28, Insurance and Bonds, by C. S. Cooper, "Building Construction Handbook," 3d ed.

TABLE 25-1 Typical Insurance Needs of Construction Contractors

Type	Purpose	Suggested limits for small-to-medium-size company	Basis for premium
Builders Risk Fire Insurance with extended coverage, vandalism, and malicious mischief	Protect building, fire and lightning, explosion, windstorm, etc. All-risk coverage is the broadest and most expensive	Amount of contract less demolition, construction below lowest basement slab and landscaping	Dollars per hundred
Comprehensive Liability			
1a. Premises, operations of contractor	Third-party bodily injury and property damage	$300,000 B.I. $100,000 P.D.	Payroll classification
1b. Independent contractors contingent public liability and property damage	Protect contractor against subcontractor's negligence	Same as above	Amount of subcontract
1c. Completed operations insurance	Claims by third parties on projects already completed	Same as above	Amount of coverage
1d. (Hold-harmless) contractual liability	Contractually assumed liability	Same as above or as per contract	Amount of coverage
2. Owner's protective liability insurance	Third-party liability	As per contract	Amount of contract
3. Umbrella liability	Third-party liability	$3,000,000	Various
Owned and nonowned motor-vehicle insurance	Third-party liability and physical damage insurance	B.I., P.D., collision, fire and theft (no fault)	Number of cars and experience
Contractor's equipment insurance	Protects equipment, tools, etc.	$50,000	Coverage amount
Material floater	Losses to materials before installation in job (may be included in Builders Risk)	$50,000	Coverage amount
Workmen's compensation and employers liability	Employee protection	Set by state	Payroll
Employers liability insurance	Employee suits outside of Workmen's Compensation	Unlimited in some states or $100,000	Payroll
Disability benefits insurance	(Where required by the state)	Set by the state	Payroll
Fidelity insurance	Embezzlement of funds	$100,000	Number of employees
Valuable papers	Loss of valuable papers	$10,000	Amount of coverage
Key man insurance	Life insurance policy to cover corporate officer in case of death or disability	$100,000, or amount to be determined	Amount of coverage
Group hospitalization	Employees' protection (Blue Cross)	120-day extended coverage	Number of employees and coverage
Major medical	Employees' protection	Limits beyond Blue Cross	Same as above
Group life insurance	All eligible employees	6 mo. to year salary	Age and number of employees

TABLE 25-1 Typical Insurance Needs of Construction Contractors (*Continued*)

Type	Purpose	Suggested limits for small-to-medium-size company	Basis for premium
Technically not insurance			
Pension and/or profit sharing plan	Employee retirement	25% salary when combined with Social Security	Salary of employees
Unemployment insurance	(Where required by the state)	Set by state and federal government	Payroll
Social Security	Employees' welfare	Set by federal government	Payroll
Street Obstruction Bond	Where required by municipality	As required	As required
Architects and engineers insurance			
Architects and engineers liability insurance	Damage to persons and property	$300,000 B.I. $100,000 P.D.	Amount of coverage
Errors and omissions insurance	Errors or omissions from the plans	Highest available	Amount of coverage
Owner insurances			
Owner's protective liability insurance	Public liability and property damage as a result of construction operations	$500,000 $1,000,000	
Property fire insurance	When construction is being done on an existing building, fire, etc., general contractor and subcontractors	Full insurance value	
Loss of use insurance	Losses in property as result of fire or other loss	Approx. loss of income	
Steam boiler	Explosion, etc.	$500,000	
Nuclear incident	Nuclear reactor loss (if required)	High	

This section merely outlines those forms of insurance that may be considered fundamental (Table 25-1). It includes brief, but not complete, descriptions of coverages without which a contractor should not operate. It is not intended to take the place of the advice of experienced insurance personnel.

Fire Insurance. Fire insurance policies are now well standardized. They insure buildings, contents, and materials on job sites against direct loss or damage by fire or lightning. They also include destruction that may be ordered by civil authorities to prevent advance of fire from neighboring property. Under such a policy, the fire insurance company agrees to pay for the direct loss or damage caused by fire or lightning and also to pay for removal of property from premises that may be damaged by fire.

Particular attention should be given to the computation of the amount of insurance to be applied to property exposed to loss. In addition, the cost of debris removal should be taken into consideration if the property could be subject to total loss. Under no circumstances will the amount paid ever exceed the amount stated in the policy. If the fire insurance policy has a coinsurance clause, the problem of valuation and adequate amount of insurance becomes even more important.

The form of fire insurance particularly applicable in the construction industry is known as Builders Risk Insurance. The purpose of this form is to insure an owner or contractor, as their interests may appear, against loss by fire while buildings are under construction. Such buildings may be insured under the Builders Risk form by the following methods:

1. The rarely used reporting form, under which values are reported monthly. Reports must be made regularly and accurately. If so, the form automatically covers increases in value.

2. The commonly used completed value form under which insurance is written for the actual value of the building when it is completed. This is written at a reduced rate because it is recognized that the full amount of insurance is not at risk during the entire term of the policy. No reports are necessary in connection with this form.

3. Automatic Builders Risk Insurance, which insures the contractor's interest automatically in new construction, pending issuance of separate policies for a period not exceeding 30 days. This form generally is used for contractors who are engaged in construction at a number of different locations.

There are certain hazards which, though not quite so common as fire and lightning, are nevertheless real. The contractor should insist that these be included in the insurance, by endorsement. A few of the available endorsements are:

1. Extended coverage endorsement, which insures the property for the same amount as the basic fire policy against loss or damage caused by windstorm, hail, explosion, riot, civil commotion, aircraft, vehicles, and smoke.

2. Vandalism and malicious mischief endorsement, which extends the protection of the policy to include loss caused by vandalism or malicious mischief. There is a special extended coverage form that provides coverage on an all-risk basis and includes the peril of collapse.

Other forms that extend and make fire insurance protection more complete also are available. Special conditions and specific exposures should be studied to determine the advisability of the purchase of certain of these additional protections.

Note that completed-value and reporting forms treat foundations of a building in the course of construction as a part of the Builders Risk Value for insurance purposes. But all work below the lowest basement slab, site work, and demolition are not covered by Builders Risk Insurance. The value of these items should be deducted from the amount of the policy. This will result in a saving of insurance premium. Builders' machinery and equipment should be specifically insured as a separate item if coverage is desired under the policy.

Under the terms of the AIA General Conditions, the owner is to provide the Builders Risk Insurance and pay for it. Under those circumstances, the contractor should request that the following clause appear as an endorsement to the policy:

Additional insureds under this policy: The Contractor and its Subcontractors, as their interests may lie.

This endorsement will assure the contractor and its subcontractors of a share in the insurance proceeds in case of claim. It further protects the contractor and subcontractors from suit by the insurance company under its subrogation clause, should either of these parties have caused or contributed to the loss.

Insurance Required for Contractor's Equipment. The so-called Inland Marine insurance market is the place to look for many coverages needed by contractors. From the insurance point of view, each contractor's problem is considered separately. The type of operation, the nature of equipment, the area in which the contractor works, and other pertinent factors are all points considered by an Inland Marine underwriter in arriving at a final form and rate.

Obvious contractors' equipment—mechanical shovels, hoists, bulldozers, ditchers, and all other mobile equipment not designed for highway use—is the primary subject of the Contractors Equipment Floater. Such protection is necessary because of the size of the investment in such equipment and of the multiplicity of perils to which the equipment is exposed.

Some companies will write the Contractors Equipment Floater Policy only on a named-perils basis, which ordinarily includes fire, collision, or overturning of a transporting conveyance, and sometimes theft of an entire piece of equipment. Other companies will write certain kinds of contractors' equipment on the so-called all-risk basis. Certain perils, such as collision during use, are subject to a deductible fixed amount. The rate for this broader insurance generally is higher than that for the named-perils form. In the all-risk form, the customary exclusions, such as wear and tear, the electrical exemption clause, strikes, riots, and other similar exclusions, are present. Because of increased exposure to nuclear hazards, all Inland Marine policies that insure against fire must carry a nuclear exclusion clause that provides that the company shall not be liable for loss by nuclear reaction or radiation or radioactive contamination, whether controlled or uncontrolled and whether such loss is direct or indirect, approximate or remote.

In addition to the equipment, there is a need to provide protection for the materials and supplies en route to or from the site. If these materials and supplies are transported at the risk of a contractor and are in the custody of a common carrier, a Transportation Floater should be obtained. If, on the other hand, these materials and supplies are moved on the contractor's own trucks, a "Motor Truck Cargo-Owners Form" should be obtained. The premium for the transportation form is usually based on the value of shipments coming under the protection of the policy. The coverage is usually on an all-risk basis. The Motor Truck Cargo-Owners Form is generally on a named-perils basis at a flat rate applied against the limit of liability required by the insured's needs.

Occasionally, a contractor may be responsible for machinery, tanks, and other property of that nature until such time as they are completely installed, tested, and accepted. Exposures of this kind are usually covered under an "Installation Floater," which would provide insurance to the site as well.

The Installation Floater Form is generally on a named-perils basis, including perils of loading and unloading, at a rate for the exposure deemed adequate by the underwriter. Large contractors should have such insurance written on a monthly reporting form to reflect increasing values as installation progresses. Small contractors generally can provide for the coverage under a stated amount on an annual basis subject to coinsurance.

There is also a form of policy known as a "Riggers Floater" that is designed for contractors doing that type of work. This policy is usually a named-perils form at rates based on the nature of the rigging operation.

Neither the forms nor rates in any of these classes are standard among the companies writing them, although in general they are all similar.

Some contractors building bridges might be required to take a Bridge Builders Risk Form. This is an exception to most of the insurance provided to contractors, in that it is required to be rated in accordance with forms and rates filed in most of the states and administered by a licensed rating bureau.

Motor-Vehicle Insurance. Loss and damage caused by or to motor vehicles should be separately insured under specific policies designed to cover hazards resulting from the existence and operation of such vehicles. Bodily injury or property damage sustained by the public as a result of the operation of contractor's motor vehicles or other self-propelled equipment is insured under a standard policy of insurance. To secure complete protection, a Comprehensive Automobile policy should be obtained to provide protection, in addition to the preceding, for hired cars, employers nonownership, and any newly acquired motor vehicles or self-propelled equipment during the term of the policy. Damage to owned vehicles can be added to the same policy on an automatic basis to provide Comprehensive Coverage and Collision Insurance.

All vehicles should be covered, for high limits.

Protection furnished by automobile liability and property damage insurance serves in two ways: (1) the insurance company agrees to pay any sum for bodily injury and property damage for which the insured is legally liable; (2) the policy agrees to defend the insured. It is important, therefore, that the limits be adequate to guarantee that the insured obtain full advantage of the service available. For example, suppose only $25,000 and $50,000 limits are carried, and action is brought against the contractor in the amount of $500,000. It may be necessary that he employ an attorney to safeguard his interests for the amount of the action that exceeds the limits of the policy.

It is presumed that no contractor will operate any type of motor vehicle without insurance. Automatic coverage should be provided to include all motor vehicles owned or acquired. The cost for highest available limits is reasonable.

Damage to owned motor vehicles may be insured under a fire, theft, and collision policy. Comprehensive motor-vehicle protection covers physical damage sustained by motor vehicles because of fires, theft, and other perils, including glass breakage. Collision insurance insures against loss from collision or upset. While the latter is available on a full-coverage basis, it is generally written on a deductible basis (loss less a fixed sum).

To cover a contractor for liability arising from the use by employees of their own automobiles while on his business, nonownership or contingent liability coverage is necessary. This may be included in the policy by endorsement.

Frequently, contractors have occasion to hire trucks or other vehicles. Liability and property damage insurance to cover the contractor's liability when using hired vehicles should be included at the time automobile insurance is arranged.

Boiler and Machinery Insurance. Boilers and other pressure vessels and machinery require the protection provided by boiler and machinery insurance. These policies cover loss resulting from accidents to boilers or machinery, and in addition, cover contractor's liability for damage to the property of others. Policies may also include liability arising from bodily injuries sustained by persons other than employees. This is needed because of the exposure that many contractors have as a result of the interest of the public in construction work.

The service rendered by boiler and machinery companies is of great help. Nearly all insurance companies that write this form have staffs of competent and experienced inspectors whose job it is to see that boilers, pressure vessels, and other machinery are adequately maintained.

Liability Policies Covering Contractor's Operations. Anyone who suffers bodily injuries or whose property is damaged as a result of the negligence of another person can recover from that person, if the latter is legally liable. Every business should protect itself against claims and suits that may be brought against it because of bodily injuries or property damage suffered by third parties.

Maintenance of an office or yard, as well as the conduct of a construction job, presents exposures to the public. There may be no negligence, and consequently no legal liability on the part of the contractor, but should claim be brought or suit instituted for an injury, without insurance coverage the contractor will require trained personnel to investigate the claim and negotiate a settlement or defend a lawsuit if the claim goes to court.

Public liability insurance is expressly designed to serve contractors by providing insurance that will pay for bodily injuries and property damage suffered by third parties if the contractor is legally liable, but the policy will also serve by defending the interests of the contractor in court. Sometimes the problem in court is one of amount of damage; but frequently, it is one in which the contractor being sued is not legally liable for the injuries or damage. It is fundamental, therefore, that a policy be obtained at substantial limits for both bodily injuries and property damage, that the policy cover all existing exposures and also provide for protection against exposures that may not exist or be contemplated on the inception date of the policy. The scope of operations conducted by most contractors is such that it is frequently difficult to visualize all the hazards that may exist or come about simply by being in the construction business.

The changes that have taken place in public-liability protection have been very rapid and in the best interests of the insurance-buying public. While at one time it was necessary to obtain several policies to cover the numerous public-liability hazards, it is possible now to obtain a **comprehensive liability policy** that will include the legal liability resulting from known operations and hazards, as well as those from operations and hazards not contemplated at the time the policy is written. The comprehensive general-liability policy that covers all liability of the insured, except that resulting from the use of automobiles, is a standard form available in all states at rates regulated by law.

The comprehensive general-liability policy protects the contractor under one insuring clause and with one limit against claims that formerly had to be covered specifically under a schedule liability policy or by a number of different policies and endorsements. Blanket coverage is provided by this policy at a premium based upon actual exposures disclosed by an audit at the end of the policy term. As required by this policy, every contractor must make provisions for maintaining accurate records of payrolls, value of sublet work, dollar amount of sales, and other factors that will be important at the time an audit is made.

Under the bodily injury liability clause of this policy the insurance company agrees to pay on behalf of the insured "all sums that the insured shall become legally obligated to pay as damages because of bodily injury, sickness or disease, including death at any time resulting therefrom, sustained by any person and caused by an occurrence." This is a very broad insuring clause. It obviously includes the entire business operations of the insured.

Property-damage liability is an optional coverage but should be taken. The coverage will apply to the entire risk with one or two exceptions.

There are certain exclusions in the policy that should be noted. The policy does not include:

1. Ownership, maintenance, or use (including loading or unloading) of water craft away from premises owned, rented, or controlled by the insured; automobiles while away from the premises or the ways immediately adjoining; and aircraft under any condition. However, this exclusion does not apply to operations performed by independent contractors or to liability assumed by any contract covered by the policy.
2. Bodily injury sustained by employees while engaged in the employment of the insured.
3. Liability for damage to property occupied, owned, or rented to the insured or in his care, custody, or control.

The contractor must be familiar with the provision of the comprehensive policy pertaining to contractual liability. The policy automatically provides coverage for the following written contracts: lease of premises; easement agreement, except in connection with construction or demolition operations on or adjacent to a railroad; undertaking to indemnify a municipality; side-track agreement; or elevator maintenance agreement. A premium is charged for such agreements as may be disclosed by audit.

There is no protection for the liability assumed in some very common types of agreements that include service, delivery, and work contracts. Many of these contracts are signed without full realization of the liability assumed. Each such agreement should be submitted to the insurance company at the time the policy is written in order that a premium charge may be computed and the agreement covered under the policy.

The most important contractual liability that may be assumed by the contractor is the so-called hold-harmless clause. A common type of hold-harmless clause as written by a general contractor to a subcontractor is illustrated in the subcontract rider in Fig. 25-7, Paragraph 32a. A similar clause written by the owner for inclusion in the general contract would be slightly modified and would substitute in appropriate places the word "Contractor" for "subcontractor" and the word "Owner" for "Contractor."

Hold-harmless clauses can be automatically included in a general contractor's Comprehensive Liability Policy. However, the general contractor should include the previously mentioned subcontractor hold-harmless clause in each subcontract. Subcontractors will probably be required by their insurance companies to pay an additional premium for this coverage.

The comprehensive general-liability policy includes complete and automatic products-liability insurance, including completed-operations protection. The one exception to this complete coverage is that the policy does not include liability for damage to the work or to the goods themselves, such as the obligation of the contractor to repair or replace if there are defects. While the policy provides coverage, it is in fact an optional protection, which may be deleted. However, every contractor should take advantage of this coverage.

Most building contractors use elevators or hoists during construction. The comprehensive policy automatically covers elevators, hoists, and other such hazards. Escalators may be covered for an additional premium.

The breadth of public-liability protection available, the numerous hazards to which a contractor may be exposed, both known and unknown, and the necessity for having complete coverage at all times indicate the need for the advice of trained insurance representatives.

Workmen's Compensation. Every state requires an employer to secure a policy of workmen's compensation to provide for an injured employee the benefits of the workmen's compensation law of that state. An insurance company that has had extensive experience in the workmen's compensation field is best suited to meet the requirements of most contractors.

The problem of acquiring and keeping labor may be troublesome. It is in the employer's, as well as the employee's, best interests to see that the company entrusted to provide workmen's compensation insurance is equipped to provide loss-prevention service, prompt first aid, and to settle compensation claims fairly and speedily.

Some states impose on contractors liability for injury to subcontractors and their employees unless insurance is specifically provided for subcontractors and their employees. Check the law of the state in which construction is to be performed.

An injured employee may believe that it suits his purpose to waive his benefits and sue his employer for negligence or failure to maintain a safe place to work. Since benefits under workmen's compensation are limited by statute, the employee may believe that recovery in a lawsuit may be substantially higher. Insurance coverage for such suits is provided by Employers Liability Insurance. Another coverage provided by the Comprehensive Liability Policy is the Independent Contractors Contingent Public Liability and Property Damage protection, which insures the general contractor against subcontractors' negligence.

The law of the state in which the work is done is the one that will control.

Subcontractors' insurance should be carefully examined and made complete where it is deficient. Certificates of insurance should be required, including coverage for contractual liability (hold-harmless). Because many complex situations arise, it is important that the advice of a qualified agent or broker be obtained.

Money and Securities Protection. Every contractor has cash, securities, a checking account, and payrolls that are vulnerable to attack by dishonest people, both on and off the payroll. The same hazards present in every business are present also in the construction business. And no contractor is immune to dishonesty, robbery of payroll, burglary of materials, or forgery of his signature on a check.

Employee dishonesty may be covered on a blanket basis, either under a Primary Commercial or Blanket Position Form of Bond. The fact that contractors generally entrust the maintenance of payroll records and the payment of employees to subordinates demonstrates the necessity for blanket dishonesty protection.

While many contractors maintain an organization on a year-round basis, some may not. For those contractors who do have a permanent staff, the bond may be written on a 3-year basis at a saving. It is important that adequate limits be purchased. A blanket bond in an amount equal to 5% of the gross sales is desirable.

General funds, securities, and payroll funds should be covered on the broadest basis now available that protects against burglary, robbery, mysterious disappearance, and destruction, on and away from any premises. The general funds may be covered in an amount sufficient to protect against the maximum single exposure. Payroll funds may be insured specifically and in a different amount.

Contractors who maintain inventories of materials should insure them against burglary and theft. It should be noted, however, that insurance companies are not willing to insure against loss by burglary or theft unless materials are under adequate protection. Insurance is not available to cover property on open sites or in yards, but only while within buildings that are completely secured when not open for business.

Every business that maintains a checking account, however small the balance may be, should insure against loss caused by the forgery of the maker's name or by the forgery of an endorsement of checks issued.

The modern policy to cover all these hazards is the Comprehensive Dishonesty, Disappearance, and Destruction Policy. Its several insuring agreements include employee-dishonesty coverage, broad-form money and securities protection, on and off the premises, and forgery. Other coverages to provide burglary and theft protection for merchandise and materials may be added by endorsement. Optional coverages available are numerous, and the contract may be designed specifically for all of a contractor's exposures.

Employee Group Benefits. Interest in group benefits for employees has grown appreciably in recent years. There are many forms of group insurance. Included in the following is a brief description of those forms of "group insurance benefits" that are of greatest current interest.

Group Life Insurance. A form of term life insurance written to cover in a specified amount a group of employees of a single employer. Frequently, the amount of insurance is related to an employee's earnings and increases as the earnings increase. However, the amount of insurance may be a flat sum. Usually, a specified number of employees must participate.

Group Disability Insurance. If an employee is away from work as a result of a nonoccupational disability, this insurance provides a continuing income during his period of absence. The amount of benefit is usually a percentage of earnings. Many states have adopted compulsory disability laws.

Group Hospitalization and Surgical Benefits. These forms of group protection are designed to protect an employee from the results of high hospital expense or the expense of a surgical operation. The protection may be written to cover an employee solely or it may be written to cover an employee and dependents.

The allowance for hospital room and board charges may be set at a figure intended to provide semiprivate care. Certain extras, such as the use of operating room, drugs, X-rays, and electrocardiograms, may be included either on a service basis or on a basis that provides a maximum amount that will be paid for such service.

The period during which benefits are payable will vary depending on the contract. Some plans provide for payment of full charges for a comparatively short period and part charges for a longer specified period.

Surgical benefits are usually based on a schedule of amounts set forth in the policy. The maximum provided will, of course, depend upon the premium.

Group Coverage for Major Medical Expenses. Increased medical expense has brought about a demand for protection against catastrophic medical costs. Rates vary greatly, as do forms of policy. The coverage is usually provided on a deductible basis (medical expense less a fixed sum). Generally, the amount that the insured receives under a group hospitalization and surgical policy is deducted. Frequently, a coinsurance feature is included so that the individual protected by such insurance bears part of the cost.

Amounts of Insurance. For a construction manager, the limits of insurance that should be carried by the contractor and the limits that should be carried by the subcontractors are of fundamental importance.

Usually, the contractor's limits are set forth in the agreement with the owner. When the owner's insurance limits are not high enough to afford the contractor full protection for the exposure the contractor anticipates, it is often worth the additional cost of increasing these limits. For example, insurance with limits of $500,000 to $1,000,000 for public liability and $500,000 for property damage often may be less expensive in the long run than lower limits that may be the maximum required by the owner. Furthermore, if the general contractor requires subcontractors to carry insurance with the same limits, it may be provident for the general contractor to pay the additional cost to the subcontractor for the increase of limits above what the subcontractor normally carries. When insurance limits are raised from $100,000 to $500,000, or from $50,000 to $500,000, the increase in cost is not proportional to the increase in limits.

25-14. Construction Contract Bonds.*

Agencies at all levels of government generally obtain competitive bids for all construction. Awards are made to the lowest responsible bidder, who is required to furnish performance and payment bonds provided by a qualified corporate surety. In addition, all construction projects financed or insured by the Federal government, such as FHA-insured projects, require such bonds. Also, many private owners require bonds of contractors.

Generally, bidders are required to post a certified check or furnish a bid bond. A bid bond assures that, if a contract is awarded to the contractor, he will, within a specified time, sign the contract and furnish bond for its performance. If the contractor fails to furnish the performance bond, the measure of damage is the smaller of the following: the penalty of the bid bond or the amount by which the bid of the lowest bidder found to be responsible and to whom the contract is awarded exceeds the initial low bid.

It should be noted that most surety companies follow the practice of authorizing a bid bond only after the performance bond on a particular contract has been underwritten

---

*Extracted from Sec. 28, Insurance and Bonds, by C. S. Cooper, "Building Construction Handbook," 3d ed.

and approved. For this reason, contractors are well advised against depositing a certified check with a bid unless there are assurances that the performance bond on that particular contract has been underwritten and approved.

There is no standard form of construction contract bond. The Federal government and each state, county, or municipal government has its own form. Private owners generally use the bond form recommended and copyrighted by the American Institute of Architects. Surety companies have developed a very broad form of bond, which is available to owners of private construction. Whatever form is used, the surety generally has a twofold obligation:

1. To indemnify the owner against loss resulting from the failure of the contractor to complete the work in accordance with the contract.

2. To guarantee payment of all bills incurred by the contractor for labor and materials.

Usually, two bonds are furnished, one for the protection of the owner, and another to protect exclusively those who perform labor or furnish materials. If one bond is furnished, the owner has prior rights.

Sureties underwrite construction contract bonds carefully. They are interested in determining whether a contractor has the capital to meet all financial obligations, the equipment to handle the physical aspects of the particular undertaking, and the construction experience to fulfill the terms of the contract.

A Performance and Payment Bond is not an insurance policy that will pool the premiums from those contractors who receive bonds and will pay the losses out of that pool, as is done with insurance. Rather, the bond essentially is a credit guarantee by the bonding company, and, as for any credit guarantee given in business, the bonding company expects to be reimbursed if this guarantee is enforced. Therefore, before providing a bond to a contractor, the surety requires the contractor to sign an Application for Surety Bond. This application, among other things, includes an agreement by the contractor to reimburse the surety for any losses that the surety may sustain as a result of having written the bond. The contractor, in order to receive the bond, therefore is indemnifying the surety. Cost of the bond is added by the contractor to the construction contract price.

It is essential that contractors understand all the information that a surety will require and that is necessary to the underwriting of a contract bond. It is also important that a contractor take advantage of the services of a competent agent or broker who has close affiliations with a surety company that has the capacity to meet all the contractor's needs. Handling of construction contract bonds is rather highly specialized; consequently, agents and brokers must be experienced in the insurance requirements of contractors, and they can be of valuable assistance in many ways to principal and obligee.

Complete records should be kept. Accurate information about the amount of work on hand, value of equipment, value of materials, and records of past performances will prove necessary.

Outlined below are several items of information that will be required by the surety:

1. A complete balanced financial statement with schedules of the principal items. This is a condition precedent to the approval of any contract bond. Sureties have forms on which contractors may furnish financial information. However, when such forms are used, they should be completed by the individual responsible for the financial operations of the company and the data taken directly from the company's books. It is preferable to have the financial statement prepared and certified by a public accountant. Contractors will find that they will save time and money if the services of qualified accountants are obtained.

2. A report on the contractor's organization. The surety is interested in knowing the length of time the contractor has been in business, whether he operates as an individual, a partnership, or a corporation, and certain specific details depending upon the form of organization.

3. A report on the technical qualifications and experience of the individuals who will be in charge of work to be performed.

4. A report on the type of work undertaken in the past, together with information regarding jobs successfully completed.

5. An inventory of equipment, noting value and age of each piece and any existing encumbrance. An inventory of materials will also be helpful.

From the construction management point of view, the most important question involving bonding is: What avenues of business are open to the contractor who lacks sufficient bonding capacity to do bonded work?

In Art. 25-3 various sources of business are described, and in Art. 25-1 various types of construction companies are discussed. Many of these types of business and construction companies do not require bonds for their work. For example, it is very rare that a bond is required in a construction management contract. When a contractor lacks capacity for bonding, it is well to pursue the lines of work described in those articles for which a bond will not be required.

The second question confronting a construction manager is whether or not to require a bond of subcontractors. The problem arises when a contemplated subcontractor award is to a firm that is deficient in either financial capability or experience for the job on hand. Usually, the question answers itself when it arises. If a firm has questionable financial capability and questionable experience, so that the general contractor believes that additional protection afforded by a subcontract bond is necessary, then generally the bonding companies will look with disfavor on issuing a bond to such a subcontractor. In general, therefore, if the financial capability or experience of a subcontractor is sufficiently doubtful as to require bonding, the job should not be awarded to that company. Exceptions to this can be made to assist young companies in starting and gaining experience.

There are alternatives to subcontract bonds. These alternatives include the following:

Personal guarantees by the principals of the subcontracting company

Personal guarantees of other individuals of substantial worth unconnected with the subcontracting company

Posting of a sum of money or of a security until performance of the subcontractor's work has been completed by the subcontractor

25-15. Scheduling and Expediting.

Two common methods of scheduling construction projects are by means of the Gantt (bar) chart (Fig. 25-11) and by means of the critical path method (Fig. 25-12).

Bar Chart. The bar chart is preferred by many contractors because of its simplicity, ease in reading, and ease of revision. The bar chart can show a great deal of information besides expected field progress. It can also show actual field progress; dates for required delivery; fabrications and approvals; percent of completion, both planned and actual; and time relationships of the various trades. Copies of bar charts may be distributed to subcontractors, trade foremen, and in many cases laymen, with the expectation that it will be easily understood.

Steps in preparation of a bar chart should include the following:

1. On a rough freehand sketch, lay out linearly and to scale horizontally, the amount of time contemplated for the total construction of the job, based on either the contract or past experience.

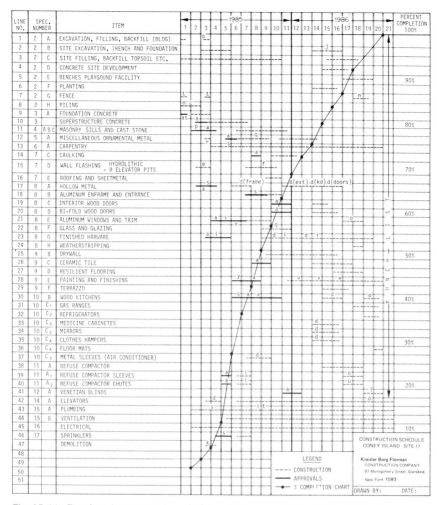

Fig. 25-11 Bar chart for construction scheduling. Letters associated with bars indicate: *a*, approvals; *b*, backfill; *c*, contract piles; *d*, delivery; *e*, elevator pits, hydrolithic; *f*, spandrel waterproofing; *g*, glazing; *h*, handrails; *i*, installation; *j*, site work; *k*, taxpayers; *l*, hardware schedule; *m*, load tests; *n*, permanent; *o*, adjustments; *p*, brick panels; *q*, lintel; *r*, saddles, wedge inserts; *s*, shop drawing; *t*, temporary; *u*, tape; *v*, prime, ceiling spray; *w*, finish; *x*, fabrication; *y*, finish schedule.

2. List in the first column all the major trades and items of work to be performed by the contractor for the job.

3. Based on past experience or on previous bar charts that give actual times of completion for portions of past jobs, block in the amount of time that will be needed for each of the major trades and items of work, and indicate their approximate starting and completion date in relationship to the other trades on the job.

4. On completion of Step 3, reexamine the chart in its entirety to ascertain whether the total amount of time being allocated for completion is realistic.

5. After adjusting various trades and times of completion and starting and completion dates on the chart, work backward on dates for those trades requiring fabrication of material off the site and for those trades that require approval and submission of shop drawings, samples, or schedules.

6. Block in the length of time necessary for fabrication and approval of items requiring those steps.

7. Using the contractor's trade-payment breakdown or schedule of payments for each trade and month-by-month analysis of which trades will be on the job, sketch in the percent-completion graph across the face of the chart.

8. After the chart has been completely checked and reviewed for errors, draw the chart in final form.

Critical Path Method (CPM). This is favored by some owners and government agencies because it provides information on the mutually dependent parts of a construction project and the effect that each component has on the overall completion of the project and scheduling of other components. CPM permits a more realistic analysis of the daily problems that tend to delay work than does the customary bar chart. Strict adherence to the principles of CPM will materially aid builders to reduce costs substantially and to enhance their competitive position in the industry.

Briefly, CPM involves detailing, in normal sequence, the various steps to be taken by each trade, from commencement to conclusion. The procedure calls for the coordination of these steps with those of other trades with contiguous activities or allied or supplementary operations, to ensure completion of the tasks as scheduled, so as not to delay other work. And CPM searches out those trades that control the schedule. This knowledge enables the contractor to put pressure where it will do the most good to speed a project and to expedite the work at minimum cost.

An important, but not essential, element of CPM is a chart consisting of a network of arrows and nodes. Each arrow represents a step or task for a particular trade. Each node, assigned a unique number, represents the completion of the steps or tasks indicated by the arrows leading to it and signifies the status of the project at that point. The great value of this type of chart lies in its ability to indicate what tasks can be done concurrently and what tasks follow in sequence. This information facilitates expediting, and produces a warning signal for future activities.

The actual critical path in the network is determined by the sequence of operations requiring the most time or that would be considered the most important parts of the contract on which other trades would depend. This path determines the total length of time for construction of the project. Usually, it is emphasized by heavier or colored lines, to remind the operating staff that these tasks take precedence (see Fig. 25-12). Also, it is extremely important that the contractor be guided by the knowledge that a task leaving a node cannot be started until all tasks entering that node have been completed.

Another benefit of CPM scheduling is the immediate recognition of float time. Associated with noncritical activities, float time is the difference between time required and time available to execute a specific item of work. This information often enables a supervisor to revise his thinking and scheduling to advantage.

Float is determined in two steps, a forward and a backward pass over the network. The forward pass starts with the earliest begin time for the first activity. Addition of the duration of this task to the begin time yields the earliest complete time. This also is the earliest begin time for the next task or tasks. The forward pass continues with the computation of earliest complete times for all subsequent tasks. At nodes where several arrows meet, the earliest begin time is the largest of the earliest complete times of those tasks. The backward pass starts with the earliest complete time of the final task. Subtraction of the duration of that activity yields the latest allowable begin time for it. The backward pass continues with the computation of the latest allowable begin times for all

Fig. 25-12 Network for critical path method of scheduling comprises arrows representing steps, or tasks, and numbered nodes representing completion of those tasks. Heavy line marks critical path, the sequence of steps taking longest to complete.

preceding tasks. At nodes from which several arrows take off, the latest allowable complete time is the smallest of the latest allowable begin times, and the latest allowable begin times of preceding tasks are found by subtracting their duration from it. Float is the difference between earliest and latest begin (or complete) times for each task.

The computations can be done manually or with a high-speed electronic computer. The latter is desirable for projects involving a large number of tasks, or for which frequent updating of the network is required.

Because of its complexity, the critical path method of scheduling construction has not been widely adopted by contractors. Two reasons for this are that computerized schedules often do not reflect the expectations or abilities of those who are actually doing the work, and the computer data lack simplicity and ease of interpretation and understanding.

As an alternative, the time-scale arrow diagram with the activities shown to scale may be used. Compared with bar charts, such arrow diagrams are almost as easy to read and understand. (S. S. Pinnell, "Critical Path Scheduling: An Overview and Practical Alternative," *Civil Engineering,* July 1980, pp. 66 to 69.) See also Art. 25-17.

Expediting. The task of keeping a job on schedule should be assigned to an expediter. The expediter must be alert to and keep on schedule the following major items: letting of subcontracts; securing of materials; and expediting of shop drawings, sample approvals, fabrication and delivery of materials, and building department and government agency submissions and approvals.

Records must be kept by the expediter of all these functions, so that the work can be properly administered. One method of record keeping for the expediting of shop drawings, change orders, samples, and other miscellaneous approvals is illustrated in Fig. 25-13. Information entered includes name of subcontractor, description of work being done, and dates of submissions and approvals received. Space is provided for shop-drawing and sample submissions, approvals and distributions, and other miscellaneous information, as well as contract and change orders. A page similar to the one shown in Fig. 25-13 should be used for every subcontractor and every supplier on the job.

Expediting requires detail work, follow-up work, and awareness of what is happening and what will be needed on the job. Its rewards are a completed job on or ahead of schedule.

Fig. 25-13 Record for expediting approvals.

25-16. Fast Tracking.

When time of completion is the most essential element in a construction job, superseding even the requirements for total coordination of the plans and specifications during design, and when competitive bids from general contractors are not essential, then a method of design and construction known as fast tracking may be utilized.

On a job that is fast tracked, design and construction begin almost simultaneously with groundbreaking and take place even while the designers are working on the foundations of the structure. Thereafter, work proceeds in the field directly after applicable portions of the design leave the drafting boards. This means that the normal time for design of the project and later bidding is telescoped by a procedure that, in effect, sets design and construction on separate tracks parallel to each other, rather than in consecutive sequence.

The advantage of fast tracking is that a considerable amount of time is saved if construction can indeed proceed simultaneously with, and in conjunction with, the design. Presumably, job completion will take place shortly after completion of the design or, certainly, much earlier than it otherwise would have.

The disadvantages of fast tracking are that in coordination of the work, the necessary input from various consultants will be largely lacking or after the fact. This may necessitate redoing certain work, or at best, repairing deficiencies that are discovered only after portions of the construction are in place.

Another disadvantage of fast tracking is that much less control exists over costs than in a job where designs are complete when prices are solicited. This disadvantage can be partly overcome, however, if the work is given to reputable, carefully prequalified subcontractors and suppliers. Certain rules of thumb from past jobs certainly will give an indication of what the final cost will be. But because of the expected loss of efficiency and redoing of certain parts of the work, costs can be expected to be higher. Nevertheless, such cost increases could be offset by resultant savings in interest on construction loans, revenues accruing from earlier use of the building, and avoidance of the effects of monetary inflation on the cost of the job if it had been begun at a considerably later date.

Fast tracking lends itself particularly to professional construction management as a form of contract, although the cost-plus-fixed-fee or cost-plus-percentage-fee contract is applicable as well. (See also Art. 2-18.)

25-17. Project Management.

In a small contracting organization, project management is generally the province of the proprietor. In larger organizations, an individual assigned to project management will be responsible for one large job or several small jobs.

A project manager is responsible in whole or in part for the following:

Progress schedule (Fig. 25-11 or 25-12).
Purchasing (Arts. 25-11 and 25-12).
Arranging for surveys and layout.
Obtaining permits from government agencies from start through completion of the job.
Familiarity with contract documents and their terms and conditions.
Dealing with changes and extras.
Submission of and obtaining approvals of shop drawings and samples, and material certifications.
Conducting job meetings with the job superintendent and subcontractors and following up decisions of job meetings. Job meetings should result in assignments to various individuals for follow-up of the matters discussed at the meetings. Minutes kept of the plans of action discussed at each meeting should be distributed as soon as possible to all attending. At the start of the following job meeting, these minutes can be checked to verify that follow-up has been properly performed. Figure 25-14 illustrates typical job-meeting minutes.

Construction Productivity. One of the principal tasks of the construction manager is coping with decreased labor availability and productivity coupled with fewer working hours per day. Added to this is the increased complexity of modern construction, particularly in the structural and mechanical phases. These increasingly complex factors, however, can be kept within the contractor's control by efficient management tools, including scheduling and monitoring techniques.

On a complex construction project for which thousands of items are to be procured, approved, manufactured, delivered, and installed, satisfactory use of computers can be made. An effective system for monitoring the time and cost of a project from design through procurement and installation, consisting of one data bank from which four reports are drawn, is as follows:

Purchasing/Cost Report. This report lists the various items to be procured and sets target dates for bidding and award of contracts. It keeps track of the budget and actual cost for each item. A summary prepared for top management provides totals in each category and indicates the status of the purchasing.

KBF JOB MEETING

Kreisler Borg Florman
Date: August 19, 1983

UNITED CEREBRAL PALSY

Present: Robert F. Borg, John V. Cricco, John A. Nowak, Robert W. Hewet, Barbara Brandwein

Spec. Section	Item of Work	Remarks	RWH	JVC	JAN	BB	GR
2A	Preliminary Work						
2B	Excavating, Bckfl., Grading/HOWNOR:	PH check subs & suppliers paid before check issued.					
2C	Site Improvement LA STRADA:	see page 3					
2D	Topsoil, Seed, Plant.	RWH push for decision w/Vella.	X				
3A	Concrete & Cement HOWNOR:	Make open items/omissions list.					X
3B	Precast Con. Plank KOEHLER:						
4A	Masonry KOEHLER:	see page 3					
5A	Misc. Metal UNITED:	One section stair rail still remains					X
5B	Struct. Steel UNITED:		---	---	---	---	---
6A	Carpentry BE-MI:	RWH call Bellona re: pulls. Install. bi-fold heads & shelves. Expedite 6th & 1st	X				X
6A	Millwork METALOC	fl. bi-folds. Measure handrail.					X
6B	Kitchen Cabinets BOISE CASCADE:	Follow up approval of pull.				X	
7A	Roofing & Sheet Metal/ COLONIAL:		---	---	---	---	--
7B	Dampproof KINGS COUNTY:	Waterprf. lobby flr. at main entrance					X
7C	Wall Wtpf. KINGS COUNTY:		---	---	---	---	...
7D	Caulking & Sealants KINGS COUNTY:	Make open items list.					
8A	HM Door Frames WILLIAMSBURG:	Expedite 3 missing frames					X
8B	HM Doors WILLIAMSBURG:	Expedite 6 missing doors.					X
8C	Alum. Entrances LOXCREEN:	Expedite install. Q.T. & entrances.					X
8D	Alum. Windows LOXCREEN						
8E	Glass & Glazing LOXCREEN:	Make open items list.					X
8F	Finish Hardware SEECO:	Make open items list.					X

Fig. 25-14 Minutes of a job meeting.

Expediting/Traffic Report. This report picks up the items once they are purchased and gives a continual update of delivery dates, shop drawing and approval status, shipping information, and location of the material when stored either on or off the site.

Equipment List. This report, which is normally used when the job involves a process or refinery, can also be used for lists of equipment in a complex building, such as a hospital. The report describes all the utility information for each piece of equipment, its size, functions, intent, characteristics, manufacturer, part number, location in the finished job, and guarantees, as well as information relating to its source, procurement, price, and location or drawing number of the plan it appears on.

Accounting System. The system consists of a comprehensive series of accounting reports, including a register for each supplier, and shows all disbursements. This information is used in preparing requisitions for progress payments. It also can be used to report costs of the job to date and to make predictions of probable costs to complete.

All of the above reports can be integrated from start to finish and synchronized with each other. (M. Kupperman, "Improving Productivity, the Construction Phase," *Consulting Engineer,* December 1977, pp. 52 to 55.)

Project Time Management. After construction operations commence, there should be a continual comparison of field performance with the established schedule. When the schedule is not being met, some method should be used to make management aware of this and correct schedule lags. The corrective actions and rescheduling phases are known as project time management.

The monitoring phase of time management involves periodic measurement of actual job progress and comparison with the planned objectives. The only way that this can be done positively is to determine the work quantities put into place and to report this information for comparison with work quantities anticipated in the job schedule. Then, a determination can be made of the effect of the current status of the job on the completion date for the project. Any corrective actions necessary can then be planned and implemented, after which the schedule can be updated.

Computer-calculated network activities provide a convenient basis for measuring progress and for issuance of reports (Art. 25-15). The network diagram should be corrected as needed so that the current job schedule reflects actual job status.

(R. H. Clough and G. A Sears, "Construction Project Management," John Wiley & Sons, Inc., New York.)

Computerized Project Management Control System. This system combines project scheduling with cost controls, resource allocation controls, and a contract-progress statistical reporting system so as to yield total control over time, cost, resources and statistics.

Time. The time aspect of the system is designed to produce, through project scheduling, a set of time objectives, a visual means of presenting these objectives, and the devising and enforcing of a corrective method of adhering to the objectives in order that the desired results will be achieved, as described previously (Project Time Management).

Cost. These are summary costs monitored by budget reports, produced monthly and distributed to the owner. In addition, detailed reports for construction company management list costs under each class of construction activity and are used by project managers and field, purchasing, and top management personnel. A report on probable total cost to complete the project is intended for all levels of construction company personnel but is used primarily by those responsible for corrective action.

Resource Allocation. For the purpose of resource allocation, a graphical summary should be prepared of projected monthly manpower for individual activities and also of the estimated quantities of work to be in place for all trades on a cumulative basis. An update of these charts monthly will indicate which trades have low work quantities in place, so that these lagging trades may be augmented with the proper number of men to permit them to catch up with and adhere to the schedule.

Statistics. From the information received from the preceding reports, an accurate forecast can be made of the probable construction completion date and total cost of the project.

Software and programs for computerized preparation of all the preceding reports can be developed by company personnel or by outside consultants.

25-18. Field Supervision.

A field superintendent has the most varied duties of anyone in the construction organization. His responsibilities include the following: field office (establishment and maintenance); fencing and security; watchmen; familiarity with contract documents; ordering out, receiving, storing, and installing materials; ordering out and operation of equipment and hoists; daily reports; assisting in preparation of the schedule for the project; maintenance of the schedule; accident reports; monitoring extra work; drafting of back-charges; dealing with inspectors, subcontractors, and field labor; punch-list work; and safety.

Familiarity with contract documents and ability to interpret the plans and specifications are essential for performance of many of the superintendent's duties. (The importance of knowing the contract documents is discussed in Art. 25-5.) Should the superintendent detect that work being required of him by the architect, owner, or inspectors exceeds the requirements of the contract documents, he should alert the contractor's office. A claim for pay for extra work, starting with a change-order proposal, may result (Art. 25-20).

The daily reports from the superintendent are a record that provides much essential information on the construction job. From these daily reports, the following information is derived: names of men working and hours worked; cost code amounts; subcontractor operations and description of work being performed; materials received; equipment received or sent; visitors to the job site; other remarks; temperature and weather; accidents or other unusual occurrences. Figure 25-15 shows a typical daily report.

Back-Charges. Frequently, either at the request of a subcontractor or because of the failure of a subcontractor to perform, work must be done on behalf of a subcontractor and his account charged. If the work performed and the resultant back-charge is at the request of the subcontractor, then obtaining the information and the agreement of all parties to a back-charge order (Fig. 25-16) is easy.

If, however, there is a dispute as to whether or not the work is part of the obligation of the subcontractor, then the task becomes more complicated. Back-charges in this situation should be used sparingly. Sending a back-charge to a subcontractor under circumstances that are controversial is at best only a self-serving declaration, and in all likelihood will be vigorously disputed by the subcontractor.

25-19. Who Pays for the Unexpected?

A well-drafted construction contract between contractor and owner should always contain a changed-conditions clause in its general conditions. (See "General Conditions of the Contract for Construction," AIA A201, American Institute of Architects, 1735 New York Ave., NW, Washington, D.C. 20006.) This clause should answer the question:

If the contractor encounters subsurface conditions different from what might normally be expected or is described in plans, specifications, or other job information provided by the owner or his agents, resulting in increased cost, should the contractor absorb the increased cost or should the owner pay it?

A properly drafted changed-conditions clause should contain the following elements:

A requirement that the owner pay for the unexpected.

Arrangements so that the owner will not be the arbiter of whether the unexpected has occurred.

Recognition that the contract documents have been based on an assumed, described set of facts.

DAILY REPORT

Kreisler Borg Florman
CONSTRUCTION COMPANY (INC.)

97 Montgomery Street, Scarsdale, New York 10583
Telephone (914) SC 5-4600

JOB No. __182__ SHEET __1__ OF __1__
JOB NAME __Coney Island Site 17__ WEATHER __Fair__
SUPTS NAME __H. L.__ TEMP. A.M. __38°__ P.M. __49°__

KREISLER BORG FLORMAN OPERATIONS

	CLASS	EMPLOYEE'S NAME	HRS.	COST CODE	EXPLANATION	TOTAL HOURS	RATE	AMOUNT
1	Proj. Mgr.	J. C.						
2	Secty	M. R.						
3	Secty	P. W.			UTSCO deliv. 500' of .018 S/S base flashing			
4	Supt.	H. L.			for Sepia			
5	Asst.	D. W.						
6	Lab.	J. F.	7		Sweeping & scraping 10th & 11th fl.			
7	Lab.	V. C.	7		Sweeping & cleaning floors for bathroom layout			
8	Lab.	B. M.	8		on the 17th & 18th fls. light salamanders and			
9	Lab.	C. T.	7		change propane cylinder on 2nd, 3rd & 4th fls.			
10	Lab.	E. M.	7	(1)	Cleaning "C" type window for installation of hopper frame on the 6th & 7th floors			

SUBCONTRACTOR OPERATIONS

SUB'S NAME & WORK FORCE	WORK PERFORMED
Island Security (4) 4 Watchmen	2 shifts, 3PM to 5AM
Alwinseal 2 Ironworkers	setting 19th floor windows; installing hopper frames on 6th fl.
Lieb 2 Ironworkers (1)	Setting roof level lintels; Stair platform frame level Bldg. "C" at Mech. Rm.
Lipsky 1-15 3-17 Plb (1)(1)(1)	C.I. test 13th fl. thru roof Bldg B; installing gas risers 16th, 17th fls Blds. A,B,C; setting tubs 16th fl; maint. temp. water
Lefferts Waterproofer	Spandrel waterproofing south end "C" east side of B & A at roof level & mech.room slab level at bldg C.
Public Improvement 2-12-8 elec. (1)(3)	Installing conduits & corrective work 2nd & 3rd fl. working with sheetrock crew; misc. corrective work on 6th flr; pulling wire on 10th flr. bldgs B & C; nippling on 14th & 15th fl; layout for nippling on 15th & 16th flr; fabrication work; maint. temp. lights
Star-Circle 2-26 Carp (4)	Sheetrock 2nd flr; framing stairs 2 to 3 & 3 to 4 flr; framing kit. 3 fl; sheetrock bathrms 3 fl; installing insul & sheetrock back of stairs 4 fl; fire cod'g duct risers 5 flr & framing 5 flr. apts; layout of bathrooms on 17 & 18 flrs.
Bafill 26-2-27 Brklyrs (7) (2) 1-19 M. Tender 1 Hst. Egr. (7)	Laying face brick and backup block to roof level at Building A West side of Building B. Brickwk. at bulkhead to mech. rm. slab level at building C Block work at elev. shafts 7 & 8 flr. of buildings A,B & c
Dic 2 carp. 4 lab.	Framing & grouting pipe & duct openings on 10th flr; cleaning inserts at roof level

REMARKS

Reid - 3 engr.	Took settlement readings with Raamot's Engr.
Active - 2 carp.	Installing storefront frames at 1st flr. of Bldgs B & C
Tropey	Continue excavating for Con Edison vault at 24th St near Bldgs B & C

MATERIAL RECEIVED	VISITORS
	F. B - UDC
153 men on jobsite	F. F - Bafill
40 minority	

DATE __Jan. 16__, 19__82__ DAY __Tuesday__
PROJ. MGR. INITIAL _____
REPORT No. __266__
OVER ☐ SUPT. SIGNATURE _____
Superintendent

Fig. 25-15 Daily report, prepared by superintendent.

BACK CHARGE ORDER

Kreisler Borg Florman
CONSTRUCTION COMPANY
97 Montgomery Street, Scarsdale, New York 10583
Telephone (914) SC 5-4600

PAGE 1 OF 1

NAME OF COMPANY TO BE BACKCHARGED: Universal Ductwork Corporation
LOCATION: Bronx Municipal Hospital Center, Pelham Parkway So & Eastchester Rd, Bronx, NY (JOB NAME)
JOB NO. 173
PERFORMED BY: General Contractor (INSERT GENERAL CONTRACTOR OR NAME OF SUBCONTRACTOR)
DATE: January 22, 1985

LABOR

NAME OF EMPLOYEE	OCCUPATION	HOURS WORKED	RATE OF PAY	AMOUNT
John M	Laborer	7		

MATERIALS, SUPPLIES AND/OR EQUIPMENT

QUANTITY	DESCRIPTION	UNIT PRICE	AMOUNT
2	Ballpoints	$7.50	

COMPLETE DESCRIPTION OF WORK DONE THIS DATE

Chopped holes in existing wall for new ducts, Room 301

(SIGNED) _____ (SUBCONTRACTOR'S SUPERINTENDENT OR TRADE FOREMAN)

THIS BACK CHARGE WORK HAS BEEN PERFORMED AND THE COST HAS BEEN DEDUCTED FROM THE MONEY DUE YOU ON THIS JOB

UNIVERSAL DUCTWORK CORP.
(NAME OF SUBCONTRACTOR)
BY _____ (SIGNATURE OF K-B REPRESENTATIVE)

AT YOUR REQUEST ☒
BECAUSE OF FAULTY WORK BY YOU ☐
BECAUSE OF YOUR FAILURE TO DO THE WORK DESCRIBED ☐
OTHER, AS FOLLOWS ☐

January 22, 1985 (DATE)   Superintendent (TITLE)

NUMBER 173-6

UNLESS WE RECEIVE YOUR WRITTEN PROTEST WITHIN 3 DAYS IT WILL BE ASSUMED THAT YOU ACCEPT THE CHARGE.

Fig. 25-16 Back-charge order, for work done by general contractor on behalf of a subcontractor.

Indication that a changed condition can exist because of unanticipated difficulty of performance.

A requirement that the owner be made aware of a changed condition when it occurs. Indication that changed conditions can include obstructions.

A requirement that the contractor stop work in the area of a changed condition until ordered to proceed.

Indication that either party can claim changed conditions.

Recognition that the method of procedure for handling changed conditions is provided for in the contract.

An arbitration clause or effective means of appeal other than to the courts.

Because of the comprehensiveness of the elements and because of the lack of uniformity in changed-conditions clauses, the American Society of Civil Engineers Committee on Contract Administration drafted the following recommended changed-conditions clause:

> The contract documents indicating the design of the portions of the work below the surface are based upon available data and the judgment of the Engineer. The quantities, dimensions, and classes of work shown in the contract documents are agreed upon by the parties as embodying the assumptions from which the contract price was determined.
>
> As the various portions of the subsurface are penetrated during the work, the Contractor shall promptly, and before such conditions are disturbed, notify the Engineer and Owner, in writing, if the actual conditions differ substantially from those which were assumed. The Engineer shall promptly submit to Owner and Contractor a plan or description of the modifications which he proposes should be made in the contract documents. The resulting increase or decrease in the contract price, or the time allowed for the completion of the contract, shall be estimated by the Contractor and submitted to the Engineer in the form of a proposal. If approved by the Engineer, he shall certify the proposal and forward it to the Owner with recommendation for approval. If no agreement can be reached between the Contractor and the Engineer, the question shall be submitted to arbitration as provided elsewhere herein. Upon the Owner's approval of the Engineer's recommendation, or receipt of the ruling of the arbitration board, the contract price and time of completion shall be adjusted by the issuance of a change order in accordance with the provisions of the sections entitled, "Changes in the Work" and "Extensions of Time."

A contractor would be well advised not to enter into agreement for construction without a changed-conditions clause.

25-20. Changes, Extras, and Claims.

Most contracts provide means by which the owner can order changes in the work or require extra work. These changes or extras may be priced in any of the following ways:

Unit Prices. At the time of either the bid or the signing of the contract, unit prices are listed by the contractor for various classes of work that may be subject to change. Usually, unit prices are easily administered for such trades as excavation, concrete, masonry, and plastering. The task of the purchaser is to obtain unit prices from subcontractors for various classes of work for the same trades that are in the contract. Although usually the same unit price is agreed for both added and deducted work, occasionally the unit prices for deducted work will be agreed to be 10% less than those for added work.

Cost of Labor and Materials, plus Markup. Another method of computing the value of changes or extra work is by use of actual certified costs, as derived from record keeping as the project proceeds. Rates for wages and fringe benefits must be verified, and the amount of percentage markup must be agreed to either in the contract or before the work is started. Usually, the general contractor is allowed a markup over and above subcontractors' costs and markup, but the general contractor's markup is less than the subcontractor's allowance in such cases.

When work is done on a cost-plus-markup basis, the contractor must maintain an accurate daily tabulation of all field costs. This document should be agreed to and signed by all parties responsible for the record keeping for the change. It will form an agreed-on certification that the work has been performed and of the quantities of labor and material used. A daily work-report certification is shown in Fig. 25-17.

Negotiation by Lump Sum. If the owner desires to have changes or extra work performed and does not want it done on a cost-plus or unit-price basis, owner and contractor may negotiate a lump-sum payment. In this situation, a cost estimate is prepared by the contractor or subcontractor involved, and a breakdown of costs is submitted, together with the estimate total. If the owner accepts the lump sum, he or the architect writes a change order, and the work is performed. Such a change order is shown in Fig. 25-18 as issued to a subcontractor.

Extra Work. An order from either the architect or the owner's representative may sometimes result in work that is not part of the original contract documents (Art. 25-5). Usually, there is acknowledgment on the part of the person requesting the work that it is not the contractor's obligation to perform it. This acknowledgment will generally be in the form of an authorization to proceed with the extra work or with a change order to the contractor.

Claims. Sometimes, however, a dispute may arise as to whether or not the work is part of the contract documents and hence the obligation of the contractor. Such a dispute may result in a claim on the part of the contractor. In making a claim, the contractor may request that a change order be issued prior to proceeding with the work. Or the contractor may nevertheless proceed with the work so as not to delay the job, but request that a change order be issued to him. Such a request is made in the form of a change-order proposal (Fig. 25-19). Depending on the size of the claim and the attitude of the owner or architect, the contractor may decide whether to continue with the extra work

Kreisler Borg Florman
CONSTRUCTION COMPANY

97 Montgomery Street, Scarsdale, New York 10583
Telephone (914) SC 5-4600

DESCRIPTION OF WORK relocation site drainage line   NO. 3   PAGE 1   OF 1
LOCATION Apartment House, 170th Street, Jamaica, New York   JOB NO.
PERFORMED BY   General Contractor   DATE January 23, 1985

LABOR

NAME OF EMPLOYEE	OCCUPATION	HOURS WORKED	RATE OF PAY	AMOUNT
George M	Laborer	7	$7.50	

MATERIALS, SUPPLIES AND/OR EQUIPMENT

QUANTITY	DESCRIPTION
60 LF	4" Clay Drainage Pipe

COMPLETE DESCRIPTION OF WORK DONE THIS DATE

Relocated site drainage to catch basin B as shown on revised sketch A-301 dated January 10, 1983

THE INFORMATION CONTAINED ABOVE IS CORRECT. ALL WORK SUBJECT TO TERMS OF SUBCONTRACT AND PRIME CONTRACT.

Kreisler Borg Florman SUBCONTRACTOR   OWNER'S OR REPRESENTATIVE   CONSTRUCTION FOR PROGRESS, INC.
CONSTRUCTION COMPANY

BY   BY   BY
TITLE Superintendent   TITLE   TITLE Engineer

Fig. 25-17 Daily work-report certification by general contractor for changes made or extra work performed.

CHANGE ORDER

Kreisler Borg Florman
CONSTRUCTION COMPANY
97 Montgomery Street, Scarsdale, New York 10583
Telephone (914) SC 5-4600

DATE: March 2, 1982

TO: National Tile & Marble Corporation
300 West 102nd Street
New York, New York 10025

RE: 13-Story Apartment Building
170th Street, Jamaica, New York

WE ARE INCREASING (DECREASING) YOUR SUBCONTRACT AMOUNT BY THE SUM OF $1100.00 FOR

Change tile in lobby and vestibule in accordance with letter from Clarence Lilien and Associates dated January 26, 1982.

Reference your proposal dated February 15, 1982.

ALL WORK MUST COMPLY WITH THE PLANS, SPECIFICATIONS AND ALL OTHER CONDITIONS AND REQUIREMENTS OF THE GENERAL CONTRACT. ALL TERMS AND CONDITIONS OF THE ORIGINAL SUBCONTRACT BETWEEN US SHALL APPLY AND REMAIN IN FULL FORCE.

KINDLY SIGN AND RETURN ONE COPY OF THIS ORDER.

ACCEPTED:

NATIONAL TILE & MARBLE CORPORATION

BY _____ DATE _____

OWNER'S CHANGE ORDER NO. 4

BASED ON CHANGE ORDER PROPOSAL NO. _____

Kreisler Borg Florman
CONSTRUCTION COMPANY

BY _____

TITLE ____Vice-President____

Fig. 25-18 Change order.

or to press for a decision on the claim, through either arbitration (Art. 25-21), or some other remedy available to him either under the contract or at law.

25-21. Arbitration and Mediation.

One of the favored ways of settling disputes in the construction industry is by arbitration. Parties to a contract, either at the time of entering into the contract or when a dispute arises, agree to submit the facts of the dispute to impartial third parties who will hear a presentation of the claims by both parties and render a decision.

If the parties have agreed to submit a matter to arbitration, the decision of the arbitrators is binding on both parties. The decision can be enforced in any court of law having jurisdiction over the party against whom a claim is made.

Because of the large number of construction arbitration cases filed, the American Arbitration Association has, in conjunction with representatives from the construction industry, drafted special construction-industry arbitration rules ("Construction Contract Disputes—How They May Be Resolved under the Construction Industry Arbitration Rules," American Arbitration Association, 140 West 51st St., New York, N.Y. 10020). These rules provide information for proceeding with construction arbitration.

Parties may agree beforehand to submit disputes to arbitration by inclusion of the standard arbitration clause in their argeement:

> Any controversy or claim arising out of or relating to this contract, or the breach thereof, shall be settled by arbitration in accordance with the Construction Industry Arbitration Rules of the American Arbitration Association, and judg-

CHANGE ORDER PROPOSAL NO.

Kreisler Borg Florman
CONSTRUCTION COMPANY
97 Montgomery Street, Scarsdale, New York 10583
Telephone (914) SC 5-4600

TO Bond-Ryder Associates, Inc.
    101 Central Park North
    New York, New York 10026

DATE: January 15, 1983
RE: Lionel Hampton Houses
    UDC #29
    New York, New York

WE SUBMIT OUR ESTIMATE IN THE AMOUNT OF $ 325.40 + Mark up FOR PERFORMING THE FOLLOWING extra WORK:

Furnishing and installing labor and material to remove and or eliminate receptacles that are directly over kitchen sinks and install blank cover plates in Buildings "A", "B", and "C".

Attached find back-up information from Meyerbank Electric Co., Inc., dated 12/13/82 and letter of authorization to proceed from Mr. R. Germano, Project Manager, Urban Development Corporation dated December 11, 1982.

This proposal is our request for a Change Order to reimburse us for Change Order No. 56 issued to Meyerbank Electric Co., Inc., a copy of which we enclose.

This work is not part of the completed plans dated September 1, 1982, nor is it part of the Addenda, as agreed to by contractor. We are proceeding with this work so as not to delay the job.

~~BEFORE STARTING ANY WORK ON THIS CHANGE WE ARE AWAITING YOUR DECISION. IF YOU WISH US TO PROCEED WITH THIS WORK~~ PLEASE SIGN ONE COPY OF THIS PROPOSAL INDICATING YOUR APPROVAL. ALL OTHER CONDITIONS OF THE CONTRACT BETWEEN US SHALL REMAIN IN FULL FORCE AND EFFECT. ~~IN ACCORDANCE WITH THE APPLICABLE PROVISIONS OF THE CONTRACT WE REQUEST AN EXTENSION OF OUR CONTRACT COMPLETION DATE BECAUSE OF THIS CHANGE. A PROMPT DECISION IS REQUESTED.~~

APPROVED AS CHANGE ORDER NO._____ TO THE CONTRACT
BY _____ DATE:_____

DISAPPROVED; PROCEED IN ACCORDANCE WITH CONTRACT PLANS & SPECIFICATIONS.
BY _____ DATE:_____

Kreisler Borg Florman
CONSTRUCTION COMPANY

CC: Urban Development Corporation

BY _____
TITLE ____Project Manager____

Fig. 25-19 Change-order proposal

ment upon the award rendered by the Arbitrator(s) may be entered in any Court having jurisdiction thereof.

To the standard arbitration clause, it is sometimes best to add the words "in the city of _____" immediately after "shall be settled by arbitration," to specify where the arbitration hearings should be held.

Mediation. In addition to arbitration, there is available through the services offered by the American Arbitration Association a system of mediation of disputes. Mediation differs from arbitration in that not only is it entered into voluntarily by the parties, but the results are nonbinding, as opposed to arbitration where the results are binding on both parties. When a dispute is submitted to mediation, a trained mediator, or several mediators, will assist the parties in reaching a settlement of the controversy.

The mediator participates impartially in the negotiations and advises and consults the various parties involved. The result of the mediation should be an agreement that the parties find acceptable. The mediator cannot impose a settlement and can only seek to guide the parties to a settlement.

The American Arbitration Association administers the mediation process by providing panels of trained and acceptable mediators. In addition, it will provide a place to hold the mediation hearings and will maintain records of cases and issues involved, the names and types of mediators appointed, and the results of the mediation efforts.

The National Roofers Association, under a separate set of rules, mediates disputes between roofers and owners under industry and American Arbitration Association auspices.

The American Arbitration Association general mediation rules and roofing mediation rules are available in AAA booklets, "Construction Industry—Mediation Rules" and "Roofing Industry Mediation."

When drawing up a contract, the parties to the contract may, by mutual agreement, wish to insert into the contract the following mediation clause:

> If a dispute arises out of or relating to this contract, or the breach thereof, and if said dispute cannot be settled through direct discussions, the parties may agree to endeavor first to settle the dispute in an amicable manner by mediation under the Voluntary Construction Mediation Rules of the American Arbitration Association, before having recourse to arbitration or a judicial forum.

25-22. Trade Payment Breakdowns and Payments.

The Contractor's Trade Payment Breakdown (Fig. 25-20) is usually prepared by the contractor long in advance of his having actually purchased his materials or subcontracts. Therefore, the amounts shown may vary from the actual costs of the various items. In addition to this, the contractor usually has start-up costs for such expenses as mobilization, temporary structures, temporary utilities and layout, which may not be reflected completely in the breakdown. To offset both of these conditions, the breakdowns may be unbalanced by giving greater value than true costs to the work of some of the trades done early than to the work of the trades done later. This unbalancing will enable the contractor to receive funds for these early start-up costs and thus to be compensated for having to pay various subcontractors and suppliers greater amounts than anticipated at the time the breakdown was prepared. The unbalancing will also make funds available for paying certain subcontractors and suppliers their retained percentages prior to receipt by the general contractor of his retained percentage. The reason why such payments are made earlier is that direct labor costs and many purchase commitments made by contractors, such as lumber, hardware, millwork, doors, and frames, are not subject to retained percentage. These materials, which are merely delivered and not contracted for to be both delivered and installed, must be paid for in full. Since there is a retained percentage being held on all of the work of the general contractor, funds must be available for the part of the retainage that is paid out by the contractor. The unbalancing of the Trade Payment Breakdown will serve this purpose.

Payments to subcontractors and suppliers are made on the basis of payments received by the general contractor on monthly requisitions. The general contractor should pay to these subcontractors and suppliers only the same percentage as approved on his requisition to the owner. In addition, the contractor should pay only extras for which he has received corresponding payments from the owner. In case of disputed extras, the con-

25-48 Construction Management

```
                    U. S. DEPARTMENT OF HOUSING AND URBAN DEVELOPMENT
                              FEDERAL HOUSING ADMINISTRATION
                              CONTRACTOR'S and/or MORTGAGOR'S
```

Date: August 23, 1978
Sponsor: 80-86 Houses
Name of Project: 80-86 Houses.
COST BREAKDOWN (SCHEDULES OF VALUES)
Project No. 012-38024-LP
Building Identification
Location: 80-86 Green Ave Brooklyn N.Y.

LINE	DIV	TRADE ITEM	COST	TRADE DESCRIPTION
1	3	Concrete	530,000	Foundations, superstructure plank, slab-on-grade
2	1	Masonry	410,000	Brick and block masonry and caulking
3	5	Metals	70,000	Lintels, stairs, railings
1	6	Rough carpentry	10,000	Block and rough framing
5	6	Finish carpentry	60,000	Millwork, material, and labor
38	2	Site improvements	5,000	Benches, etc.
39	2	Lawns & planting	10,000	Topsoil, seeding, plants
10	2	Unusual site conditions		
11		Total land imprvts.	130,000	
12		Tot. struct. & land imprvts.	2,860,000	
13	1	General requirements	240,000	
14		Subtotal (lines 41 and 42)	3,100,000	
15		Builder's overhead	62,000	
16		Builder's profit		
17		Subtotal (lines 44 thru 46)	3,162,000	
18				
19		Other fees		
50		Bond premium		
51		Total for all improvements		
52		Builder's profit paid by means other than cash		
53		Total for all improvements less line 52		

NONRESIDENTIAL AND SPECIAL EXTERIOR LAND IMPROVEMENT (costs included in item breakdown)

DESCRIPTION	EST. COST

OFFICE COSTS (costs not included in item breakdown)

DESCRIPTION	EST. COST

TOTAL $
OTHER FEES
TOTAL $
DEMOLITION (costs not included in trade item breakdown)

DESCRIPTION	EST. COST

TOTAL $ _____ TOTAL $ _____

Mortgagor: Kreigler Borg Florman   By _____ Date _____
Contractor: Construction Company (Inc.)   By (signed) Samuel C. Florman, VP   Date August 23, 1978
FHA _____ Processing Analyst   (Date) _____ (Chief, Cost Branch or Cost Analyst) _____ (Date) _____
FHA _____ (Chief Underwriter) _____ Date _____

Fig. 25-20 Contractor's Trade Payment Breakdown.

tractor may have to pay a portion of the disputed amount to the subcontractor to keep the job running smoothly, despite the fact that the owner has not paid anything on these disputed claims.

Subcontractors should be paid only when the contractor gets paid for the subcontractors' portions of the work. Furthermore, subcontractors should be paid only after their subcontracts are signed and insurance certificates have been received. All subcontractors should be required to use the same requisition form. A typical Application for Payment on Account of Contract is illustrated in Fig. 25-21. Both the front and back must be filled out by the subcontractor and signed. Payments to a subcontractor should be shown cumulatively on the subcontractor's financial folder, which should be separately set up for each subcontractor.

After the first payment is made to a subcontractor, the contractor should verify that the subcontractor is paying his suppliers. If suppliers are not being paid, the general contractor should insist on joint checks to suppliers and the subcontractor and require that the subcontractor endorse the check so that the supplier can receive the funds.

In making final payments to subcontractors and suppliers, the contractor may be able

Trade Payment Breakdowns and Payments   25-49

Kreisler Borg Florman
CONSTRUCTION COMPANY (INC.)
97 Montgomery Street, Scarsdale, New York 10583
Telephone (914) SC 5-4600

APPLICATION FOR PAYMENT ON ACCOUNT OF CONTRACT

ALL REQUISITIONS MUST BE <u>MAILED</u> AND IN THIS OFFICE ON OR BEFORE THE LAST DAY OF THE MONTH. BOTH SIDES OF THIS REQUISITION MUST BE COMPLETED AND SIGNED.

SUBCONTRACTOR Modern Pollution Control, Inc.   PERIOD FROM Start   TO September 30, 1985

FOR 170th St @92rd Ave, Jamaica, New York   WORK   REQ. No. 1

PROJECT Apartment Building

ITEM	AMOUNT	DO NOT WRITE IN THIS SPACE
AMOUNT OF ORIGINAL CONTRACT	9,825	
APPROVED EXTRAS TO DATE (LIST ON REVERSE SIDE)	250*	
TOTAL CONTRACT AND EXTRAS	10,075	
CREDITS TO DATE (LIST ON REVERSE SIDE)	nil	
NET CONTRACT TO DATE	10,075	
VALUE OF WORK PERFORMED TO DATE	TOTAL	
LESS RESERVE OF ____% AS PER CONTRACT	1,475	
BALANCE	8,600	
LESS PREVIOUS PAYMENTS	nil	
AMOUNT OF THIS REQUISITION		

```
        DO NOT WRITE IN THIS SPACE
   JOB _____  DISTRIBUTION _____
   CHECKED: QTY. _____  PRICE _____ EXT. _____
   AMOUNT PAID _____  DATE _____
   CHECK NO. _____  FOLIO NO. _____
   ENTER _____  PAY _____
                                          (OVER)
```

Fig. 25-21a Front side of application for payment submitted by subcontractor.

to get a reduction in their claims and extras. In addition to this, the records of backcharges and other charges for use of hoists, etc. should be shown to them. A close check should be kept on credits due. If a subcontractor omits portions of the work and if this is known, it is essential that a credit be requested of the subcontractor before final payment. All omitted work and shortcuts, especially in alteration work, should be recorded.

25-50 Construction Management

THE FOLLOWING IS A FULL AND COMPLETE LIST OF ANY AND ALL PERSONS, FIRMS OR ENTITIES OF EVERY NATURE OR KIND AND DESCRIPTION WHO FURNISHED WORK, LABOR (OTHER THAN ON DIRECT PAYROLL OF APPLICANT), SERVICE AND/OR MATERIAL (INCLUDING BUT NOT LIMITED TO INSURANCE PREMIUMS, WATER, GAS, POWER, LIGHT, HEAT, OIL, GASOLINE, TELEPHONE SERVICE OR RENTAL OF EQUIPMENT AS WELL AS UNION WELFARE, PENSION AND OTHER FRINGE BENEFIT PAYMENTS AND FEDERAL, STATE & LOCAL TAXES) IN EXCESS OF $100.00 ARISING OUT OF OR IN CONNECTION WITH THE JOB FOR WHICH THIS PAYMENT IS REQUESTED TOGETHER WITH ANY AND ALL AMOUNTS NOW DUE AND OWING TO THEM AS WELL AS THE DATE THAT THEY LAST PERFORMED ANY WORK, LABOR, SERVICE OR DELIVERED ANY MATERIAL:

NAME	ADDRESS	DATE	AMOUNT DUE
NONE DUE AND OWING			

ALL CLAIMS FOR EXTRAS NOT HEREIN LISTED ARE WAIVED BY THE SUBCONTRACTOR

EXTRAS TO DATE					CREDITS TO DATE			
Order No.	Description	Date	Amount	Order No.	Description	Date	Amount
	Damages on electrical control box		$250.00				

THE FOREGOING REPRESENTATIONS ARE MADE IN ORDER TO INDUCE YOU TO MAKE AN INTERIM OR FINAL PAYMENT TO US, WELL KNOWING THAT YOU ARE RELYING ON THE TRUTH THEREOF

MODERN POLLUTION CONTROL INC
APPLICANT'S NAME

36-50 38th St, L.I.C., New York
ADDRESS

SIGNATURE

President
TITLE

Fig. 25-21b Back side of application for payment.

After final settlement with the subcontractor of all his claims, and while final payment is being awaited from the owner, the subcontractor may often be willing to discount his final payment for immediate receipt of the money, rather than having to wait.

25-23. Cost Records.

Segregation of costs for each job is essential for proper construction management. Only with such records can profits or losses for each job be calculated and predictions made for future work costs. After award of subcontracts, in order for records to be up to date as to any extras or credits that are being claimed by or awarded to the subcontractors for changes (Art. 25-20), a monthly update of all anticipated costs for each subcontract is essential. This can be done by means of an Application for Payment Form (Fig. 25-21) from each subcontractor.

On this form, the approved extras are shown on the second line on the front of the sheet under Amount (Fig. 25-21a). Also, all extras claimed by the subcontractor must be listed on the back (Fig. 25-21b), as shown on the portion of the form Extras to Date.

Payrolls. From the daily reports received from the field (Art. 25-18), weekly payrolls can be prepared for each job. Labor for each job must be segregated and tabulated in a payroll report (Fig. 25-22). This report also provides statistics on tax withholding to the government, as well as other payroll information. In addition, the report gives the gross wages week by week.

Monthly Cost Report. This report summarizes subcontracts and extras, material purchases, and labor costs encountered to date and expected to be encountered until completion. A monthly report prepared in a manner similar to Fig. 25-23 will yield information to the contractor long before the job has been completed for calculating anticipated profit for the job, and will offer a method of monitoring job progress and costs to ascertain whether the anticipated profit is being maintained.

25-24. Accounting Methods.

The contractor together with his accountant should select a method of accounting best suited for the contractor's operations. The method selected will determine, to a great degree, the amount of taxes paid as well as the accuracy of the information that the contractor will be able to furnish to his bonding company.

One of two bases is generally used: the cash basis or accrual basis. The accrual basis is subdivided into (1) straight accrual basis; (2) completed-contract basis; (3) percentage-of-completion basis.

The cash basis is used mainly by small contracting companies. The gross income is reported on the basis of cash receipts. Contract costs are reported on the basis of actual expenses paid. This method has the advantages of simplicity, control of net income by the timing of requisitions and receipts, and payments of taxes *after* profits have been *earned* and *collected*.

The disadvantages of the cash basis are as follows: It is not an accurate reflection of the company's financial condition. It does not show monies earned and not collected, or debts incurred and not paid. It cannot be used for surety purposes.

The *straight-accrual method* can be used for both short-term and long-term contracts. In this method, the gross income is recorded when earned and expenses are recorded when incurred, regardless of when the cash is received or disbursed. It has the advantages that statements of income reflect actual operations during the period, all receivables and payables are recorded as incurred so that the balance sheet is more useful, and if there are more payables than receivables, less tax has to be paid.

The disadvantages are as follows: The gross income may not be accurate. It may not be the same as the gross profit because of advance billings (caused by unbalanced requisitioning, or front-loading) or job-site inventories. There is less flexibility in tax planning than in other methods. If there are more receivables than payables, the accrual method will result in a higher tax.

The *completed-contract basis* is used for long-term contracts. It is also very good for joint ventures. The income of each long-term contract is recognized only when that contract has been completed or substantially completed.

The advantages are as follows: Taxes are effectively deferred until completion of the project, thus augmenting the contractor's working capital. Contract profits are figured

Kreisler Borg Florman
CONSTRUCTION COMPANY (INC.)
97 MONTGOMERY STREET, SCARSDALE, N.Y., 10583, SC 5-4600

PAYROLL SHEET NO. 1 OF 1

PROJECT: TWIN PARKS NORTHWEST

PERIOD FROM 1/17/85 TO 1/23/85

Badge Number	Name of Employee	Position	B.T./O.T.	Hours Worked Daily (Wed/Thu/Fri/Sat/Sun/Mon/Tue)	Total Hours	Rate	Amount Earned (Straight / Premium)	Gross Wages	W.T. Exemp	Deductions (O.A.B. / W.T. / N.Y.S. W.T. / N.Y.C. W.T. / Other)	Net Payment
1	J.R.							200.00		11.70 / 26.30 / 6.90 / 2.30 /	152.80
2	V.L.	Lab.						262.50		15.36 / 46.40 / 12.60 / 4.00 /	184.14
3	P.F.	Lab.						262.50		15.36 / 30.50 / 8.70 / 3.05 /	204.89
4											
5											
6											
7											
8											
9											
10											
11											
12											
13											
14											
15											
16											
17											
18											
	TOTAL	RB						725.00		42.42 / 103.20 / 28.20 / 9.35 /	541.83

AMOUNT OF PREVIOUS WEEK'S PAYROLL: $755.00

DEPT. HEAD APPROVAL _____ SIGNATURE OF SUPERINTENDENT _____

CODE:
A — ABSENT, NOT PAID FOR   P — VACATION
B — ABSENT, PAID FOR   S — SICK
H — HOLIDAY   W — WITHOUT PAY
V — VACATION WITH PAY

Fig. 25-22 Payroll report.

Accounting Methods 25-53

KREISLER BORG FLORMAN CONSTRUCTION CO.

JOB: _____
LOCATION: _____
ARCHITECT: _____

PERIOD ENDING FEB. 28, 1984
PREPARED BY S. F.
CHECKED BY J. F.

Cost Analysis

NO	SUBCONTRACTS & MAJOR PURCHASE ORDERS ITEM	NAME	COST OR COMMITMENT TO DATE	COST TO COMPLETE	ANTICIPATED TOTAL COST	REMARKS
1	EXCAVATION	AAA EXCAVATORS	23000.00	*500.00	23500.00	*MAINTAINING TEMPORARY ROADS
2	CONCRETE-MATERIAL	BBB READY-MIX	8349.21	4650.79	*13000.00	*MAXIMUM PURCHASE AMOUNT
3	REINFORCING-MATERIAL	CCC STEEL CO.	2100.00	—	2100.00	
4	" -INSTALLATION	DDD ERECTORS	1850.00	—	1850.00	
5	STEEL & MISC. IRON	EEE FABRICATORS	42000.00	*600.00	42600.00	*CLAIM, CORRECTING BASE PLATES
6	MASONRY	FFF MASON CO.	73000.00	1100.00	74100.00	OPTION CATCH BASINS-1100.*
7	LATH & PLASTER	GGG PLASTERING CO.	9000.00	—	9000.00	
8	HOLLOW METAL	HHH STEEL DOOR	1750.00	—	1750.00	
9	ROOFING	III ROOFING	12000.00	*200.00	12200.00	*CLAIM-FLASHING EXTRA OPENINGS
10	WINDOWS	JJJ CO.	16000.00	—	16000.00	
11	GLAZING	NOT LET	—	*8300.00	8300.00	*ANTICIPATED COST MAXIMUM
12	ROUGH LUMBER	LLL LUMBER CO.	6914.63	2085.37	*9000.00	*PURCHASE AMOUNT
13	MILLWORK	MMM MILLWORK	28500.00	*(260.00)	28240.00	*BACKCHARGE REPAIRING CABINETS
14	HARDWARE	NNN CO.	9100.00	400.00	9500.00	*ALLOWANCE 9500.
15	ASPHALT TILE	OOO FLOOR CO.	8400.00	*500.00	8900.00	*OPTION WAXING 500.
16	PAINTING	NOT LET	—	*23000.00	23000.00	*ANTICIPATED COST
17	PLUMBING & HEATING	QQQ HEATING CORP.	110000.00	—	110000.00	
18	ELECTRICAL	RRR ELECTRICAL CO.	49000.00	—	49000.00	ALTERNATE
	TOTAL SUBCONTRACTING & MAJOR P.O.		400963.84	41076.16	442040.00	STAGE LIGHTING 2200.*
	MISCELLANEOUS MATERIALS & EQUIPMENT					
19	CONCRETE ACCESSORIES, ETC.		743.21	200.00	943.21	
20	NAILS & ROUGH HARDWARE		612.73	400.00	1012.73	
21	CRANE & PUMP RENTAL		1350.00	500.00	1850.00	
	TOTAL MISC. MATERIALS & EQUIPMENT		2705.94	1100.00	3805.94	
	GENERAL CONDITIONS					
22	INSURANCE & WELFARE		8353.20	8920.00	17273.20	APPROXIMATELY 20% OF LABOR
23	SURVEYS		350.00	—	350.00	
24	SHANTIES & SUPPLIES		642.15	500.00	1142.15	
25	TEMPORARY FACILITIES		2315.10	600.00	2915.10	
26	GLASS BREAKAGE		—	200.00	200.00	
27	RUBBISH TRUCKING		160.00	120.00	280.00	
28	TRAVEL & MISC. EXPENSE		629.32	750.00	1379.32	
29	GLASS & ALUMINUM CLEANING		—	1000.00	1000.00	
	TOTAL GENERAL CONDITIONS		12449.77	12090.00	24539.77	
	LABOR					
30	SUPERVISION		6755.00	9000.00	15755.00	
31	HAND EXCAVATION		5787.65	600.00	6387.65	
32	CONCRETE		9930.36	10000.00	19930.36	
33	CARPENTRY		16355.34	22000.00	38355.34	
34	CLEANUP & MISC.		2937.66	3000.00	5937.66	
	TOTAL LABOR		41766.01	44600.00	86366.01	
	SUMMARY					
35	SUBCONTRACTS & MAJOR PURCHASE ORDERS		400963.84	41076.16	442040.00	
36	MISC. MATERIALS & EQUIPMENT		2705.94	1100.00	3805.94	
37	GENERAL CONDITIONS		12449.77	12090.00	24539.77	
38	LABOR		41766.01	44600.00	86366.01	
39	FUTURE COSTS ON APPROVED CHANGES		—	2100.00	2100.00	
40	CONTINGENCIES		—	2000.00	2000.00	
	ANTICIPATED TOTAL COST		457885.56	102966.16	560851.72	
	CONTRACT PRICE				593252.00	
	APPROVED CHANGE ORDERS				6941.16	
	PENDING CHANGES FOR WORK DONE				1159.74	
	ADJUSTED CONTRACT PRICE				601352.90	
	ANTICIPATED TOTAL COST				560851.72	
	ANTICIPATED GROSS PROFIT				40501.18	

Fig. 25-23 Monthly cost report. *(Courtesy Samuel C. Florman.)*

on the basis of actual results, rather than on estimates that could overlook unforeseen future costs or delays. All receivables and payables are recorded as they occur, making a more accurate balance sheet. Taxable income can be regulated between years by deliberately hastening or deferring contract completion. Profits as accrued can be invested and used to earn interest before taxes are paid. And payment of taxes has a better relationship to cash flow from accounts receivable and retained percentages.

The disadvantages of the completed-contract basis are that it does not reflect current performance when a contract spans more than one fiscal year and unincorporated contractors may be penalized, since net income may be irregular and taxed at higher rates one year and at lower rates in another year.

The *percentage-of-completion method* recognizes the gross income on each contract as work progresses. Annual income for each contract should be the same percentage of total income as contract costs to date are of the estimated total contract costs. In computation of contract costs, all or portions of costs may be temporarily excluded during the early stages—for example, start-up costs—if the exclusion would result in a more accurate allocation of income. The advantages of this method are: It is the most realistic annual determination of income. Receivables and payables are recorded when earned or incurred. Liability for taxes arises in the fiscal year when it is actually incurred. And it provides consistent data with very little distortion.

The disadvantages are: It is dependent on estimates from management of percentage of completion and cost to complete. Retained percentages and accounts receivable not yet billed are taken as income although they may not be collected until a much later date. It does not permit the deferral of income taxes, a procedure that may be available with the cash or completed-contract method.

25-25. Safety.

Responsibility for job safety rests initially with the superintendent. Various safety manuals are available giving recommended practice for all conceivable types of construction situations; for example, see "Manual of Accident Prevention in Construction," Associated General Contractors of America, Inc., 1957 E St., N. W., Washington, D.C. 20006.

Because of wide diversity in state safety laws, the Federal government in 1970 passed the Occupational Safety and Health Act (OSHA) (Title 29—Labor Code of Federal Regulations, Chapter XVII, Part 1926, U.S. Government Printing Office). Compared with state safety laws of the past, the Federal law had much stricter requirements. For example, in the past, a state agency had to take the contractor to court to penalize him for illegal practices. In contrast, the Occupational Safety and Health Administration can impose fines on the spot for violations, despite the fact that inspectors ask the employers to correct their deficiencies. OSHA enforcement, however, may eventually be taken over by the states if they develop state regulations as strict as those of OSHA.

The most frequently cited violations of OSHA, in order, have been: guardrails, handrails, covers; scaffolding; ladders; gas welding and cutting; grounding and bonding; cranes and derricks; and housekeeping. Also cited have been handling of flammable and combustible liquids; general electrical installation and maintenance; fire protection; trenching; motor vehicles; head protection; and material hoists, personnel hoists, and elevators. Other violations include medical services and first aid; stairways; general requirements for hand and power tools; excavation; sanitation; personal protective equipment; eye and face protection; arc welding and cutting; and safety nets.

An essential aspect of job safety is fire prevention. In aiding in this endeavor, superintendents and project managers often have the advice of insurance companies who perform inspection of the jobs, free of charge. Such inspections by insurance companies often result in reports with advice on fire-prevention procedures. Contractors will benefit from adoption of these recommendations.

To deal most effectively with safety in the contractor's organization, the contractor should assign one man responsibility for safety. He should be familiar with all Federal and state regulations in the contractor's area. This man should also instruct superintendents and foremen in safety requirements and, on his visits to job sites, be constantly alert for violations of safety measures. A file containing all the necessary records relative to government regulations should also be the responsibility of this person. He should obtain a copy of Record-Keeping Requirements under the Occupational Safety and Health Act, Occupational Safety and Health Administration (U.S. Department of Labor, Washington, D.C.). Management should hold frequent conferences with this individual and with the insurance company to review the safety record of the firm and to obtain advice for improving this safety record.

25-26. Community Relations.

Community concerns with the results of new construction or disturbances from construction operations materially affect the construction industry. In the past, many communities merely helped shape projects that were being planned for construction in their environs. This aid consisted of recommendations from community advisory boards and localization of planning. More recently, communities have assumed a more vigorous role in regulating construction, including the power of veto or costly delay over many projects.

Some of the areas of community relations that must be dealt with in construction management are discussed in the following:

Employment of Local Labor. Because many construction projects are built in inner city, or core areas, where there is much unemployment, communities may insist on utilization of unemployed local labor. This may be done in accordance with local plans or in the form of an Equal Opportunity Program of nondiscrimination, or by recruitment of local labor for employment on the job site. Additional equal opportunity must be given by contractors to women and the handicapped in both office and field positions.

Utilization of Local Subcontractors. Various government agencies may require or give preference to Federal, state, or municipal employment of local subcontractors for construction work. In many cases, general contractors may be required to enter into written understandings with a government agency specifying goals to be set for local subcontractor employment on a job. As a result of such actions, poorly capitalized subcontractors have been able to make initial employment gains in fields requiring small capital investment, such as watchmen services and painting. Payrolls for such subcontractors, however, in a number of these trades do present difficulties. In many instances, it may be necessary for the general contractor to make special arrangements for interim payments to these subcontractors, prior to the regular payment date, for work performed.

Among the routes used to bring about employment of subcontractors short of capital on construction jobs are the following:

Awarding of a subcontract and orders to such firms.

Subdividing of work into manageable-size subcontracts.

Encouragement of subcontractors to enter into joint ventures with better-financed subcontractors.

Awarding of sub-subcontracts to subcontractors by better-financed subcontractors who hold a large subcontract.

Awarding of a pilot contract for a small job to a subcontractor, for example, tiling of one or two bathrooms, just to create an opportunity to begin to function.

Recruitment of local community labor and local subcontractors for a project requires maintenance by the contractor of an active program for the purpose. When the job is started, if there is a community group strongly organized and vocal in the area, the leaders of this group should be approached. If a request is made by the community group for employment of a member of the group as a community liaison man or organizer, this request should be given earnest consideration. With a salaried liaison between the contractor's organization and the community, many pitfalls can be avoided.

Up-to-date lists of community subcontractors should be maintained by the contractor's office. These lists should be frequently updated, or they will rapidly become obsolete as these small subcontractors either expand or phase out. The contacts thus made with local firms are important, because an acquaintanceship with local conditions is essential in obtaining and executing contracts and dealing with communities.

Public Interest Groups. These also express their opinions and ask for a voice in planning and construction of proposed projects. They can promote projects they favor or seriously delay or cause to be canceled projects they oppose, by lobbying, court actions, presentation of arguments at public hearings, or influencing local officials in a position to regulate construction.

Environmental Impact Statements. An environmental impact statement is an analysis of the effect that proposed construction will have on the environment of the locality in which the project is to be built. The statement should take into consideration, among other things, the following factors: effect on traffic; potential noise, sound, and air pollution; effect on wild life and ecology; effect on population and community growth; racial characteristics; economic factors; and esthetics and harmony with the appearance of the community.

For each project, contractors should ascertain whether an environmental impact statement is required. If such a statement is required, it should be begun early in the construction planning stage, if it is required to be drafted by the contractor. If the impact statement for a project is to be drafted by a government agency, the contractor should ascertain that the agency has drafted the statement and that it has been filed and approved.

25-27. Relations with Public Agencies in Executing Construction Operations.

A contractor must deal with numerous public agencies. In some localities, to obtain the necessary permits for start and approval of completed construction, a contractor, for example, may have business with the following agencies: building department, highway department, fire department, police department, city treasurer or controller, sewer and water department, and various government agencies providing the financing, such as Federal Housing Administration, State Division of Housing, or a public-interest corporation formed by the state or municipality for undertaking construction work.

The contractor must be familiar with the organization of the agency and the division of functions, that is, which portion of the agency does design, which does construction-cost approval, and which does inspection. The contractor should also determine whether financing is provided for completed construction, or merely by providing or guaranteeing a mortgage. In addition, the contractor should be knowledgeable on construction-code enforcement; source of payments, whether through a capital construction budget or merely a building mortgage; permits required, methods of obtaining them, and fees; record keeping; and methods of obtaining information from the agency's files.

The contractor should do the following to deal most effectively with public agencies. First, various members of the contractor's staff should concentrate their efforts and become experts on dealing with one or more agencies. For instance, the person in charge of field operations should be the one to deal with the building department and building inspectors. He should make it his business to be familiar with the organizational structure of the building department and to know the inspectors. Others in the organization should be versed in dealing with city treasurers or comptrollers, and still others with state or Federal agencies.

Second, up-to-date files and information should be kept and segregated on agency regulations and procedures. For example, building department codes and regulations should be obtained, and revisions of these should be maintained in the contractor's offices.

Also, it is very important that a personal relationship be established between the contractor's personnel and the members of the agencies with which the personnel deal. The contractor too should visit the agencies and introduce himself. He should explain his problems to agency personnel. Most of the personnel are willing to help when approached on a frank and open basis.

25-28. Labor Relations.

Proper labor relations on a construction job involve many facets of a contractor's ability. Such relations often are affected by the type of labor organization involved.

Most craft labor on jobs in large cities and in the industrialized portions of the country is unionized. Most unionized employees on construction are members of American Federation of Labor building crafts unions. Open-shop contractors, however, often are able to perform work on a nonunion basis. In a few cases, local labor unions of an industrialized type perform construction work.

Strikes. Construction labor strikes after expiration of a labor contract can place a contractor in a difficult position. If the contractor is a member of a contractor's association, the association will do the bargaining for him. If not an association member, he has one or two alternatives. He can sign up as an independent contractor on condition that the final terms of the agreement settled at the conclusion of the strike will apply to his agreement retroactively. Or he can continue working around the affected trade and hope that the job will not be delayed or advanced too far from the normal job sequence.

Jurisdictional Disputes. Jurisdictional disputes between two crafts often make a contractor an innocent bystander. Appeals may be made to the national headquarters of the crafts involved, and machinery exists in such cases for resolution of such disputes in the National Joint Board for the Settlement of Jurisdictional Disputes and the National Appeals Board in Washington, D.C. A simpler way is for the contractor to file a charge with the nearest office of the National Labor Relations Board.

Standby Pay. In many union contracts, a requirement is imposed on contractors for standby pay for men who may not actually be engaged in installing materials or performing work because of a requirement that members of the union be assigned to stand by for certain purposes. Examples of standby pay are pay for temporary water (plumbers), for temporary electric standby to maintain temporary power and lighting (electricians), and for steamfitters or electricians to maintain temporary heat. Standby pay is usually considered a normal part of the construction process, if in the union agreement, and must be allowed for by the contractor in his cost estimates.

Prevailing Wages. Contracts being done for city, state, or Federal agencies often require that the labor on a job be paid prevailing wages. These are defined as the wages received by persons normally performing that trade in the locality where the job is being performed. Prevailing wages in cities or localities where there is a strong unionized work force usually are the union wages paid to labor in that area. Also considered part of prevailing wages are all fringe benefits and allowances usually paid in addition to hourly wages. If a contractor enters into an agreement that provides for payment of prevailing wages, he may also be responsible for a subcontractor who fails to pay prevailing wages to his labor.

Labor Recruitment. A contractor's labor recruitment takes many forms. Most importantly, the work force comes from a following of labor that has previously worked with the contractor. These men are summoned by telephone or by word of mouth when they are needed on the job site. In instances in which the contractor's usual labor force must be vastly expanded, he can resort to advertising in newspapers or recruitment by advising local labor-union officials that men are needed.

Men are assigned to tasks on the basis of their trade. In strong union jurisdictions, crossing-over by labor from one trade to another is prohibited when the union labor is organized along craft lines.

In times of labor shortage, when other contractors may be competing for his labor, the contractor may have to resort to overtime work to attract and hold capable men, with the expectation that overtime pay will prevent others from hiring the labor from him. In instances in which a great many jobs, or one tremendous job, will be under construction in a locality with a relatively small labor force, other problems may arise. Labor may have to be brought in from the outside, and to do this the contractor may find it necessary to pay travel time or living allowances for such labor.

Termination. When a project ends, generally all labor is discharged, except the contractor's key personnel, who may be moved to another job. Subcontractors, however, frequently, through judicious timing, are able to employ entire labor groups by moving them from one project to another.

Efficiency. Keeping production efficiency high is generally the chief task of the labor superintendent or foreman. The means used may involve careful scheduling and planning of materials and equipment availability, weeding out of inefficient men, training and instruction of those who may not be totally familiar with the work, and most importantly, skilled supervision.

25-29. Social and Environmental Concerns in Construction.

Construction project managers should seriously consider the social and environmental effects of their construction operations, job safety operations (Art. 25-26), and for proper community relations, use of environmental impact statements and the opinions of public interest groups (Art. 25-27). How the constructor operates and whether or not he is prepared to act on these possible effects depend to a great extent on his awareness and knowledge of how his operations interfere with or become the concern of others. It is well therefore that builders be familiar with operations that have resulted in criticism and restraints, so that they can avoid pitfalls and operate within desirable guidelines.

The Committee on Social and Environmental Concerns of the Construction Division of the American Society of Civil Engineers has defined the three main areas of concern for constructors as follows:

Social. These areas cover:

Land usage, such as the visual aspect of the construction project, including housekeeping and security; avoidance of landscape defacement, such as needless removal of trees; prevention of earth cuts and borrow pits that would deface certain areas for a long time; protection of wildlife, vegetation, and other ecological systems; and visual protection of surrounding residential areas through installation of proper fencing, plantings, etc.

Historical and archaeological, including preservation of historical and archaeological items of an irreplaceable nature.

Crime. The construction process often creates temporary negative impacts on a community, resulting in a crime increase. This can include local crime as well as fraud and bribing of public officials.

Economics, including impact of a project on the economics of a region, such as a rapidly increased demand for labor far in excess of supply, with a negative effect on the wage structure in the area and economic harm to the area after construction has been completed.

Community involvement, including hiring practices and dealing with the leadership in the community, whether they be of different ethnic groups, income levels, or organizational affiliations (Art. 25-26); union hiring practices; and training programs and foreign language programs.

Safety (Art. 25-25).

Physical Media. The effects of construction on land, air, water, and of release of pollutants and toxics are the next broad area of concern. Water is often altered in its purity and temperature, and wildlife often is destroyed on land and water by construction of such projects as dams, power plants, and river and harbor facilities.

Energy Conservation, Vibration, and Noise. Vibration has become of increasing concern through the increasingly frequent use in buildings of light construction materials. These are usually flexible and prone to vibrate. In addition, construction machinery has become larger and more powerful, with the result that vibration of this machinery requires strict control.

Identification of noise-producing construction operations and equipment, and control of building construction noise must also be the concern of contractors. Noise-abatement codes for construction exist in many municipalities. Unless the provisions of these codes are properly understood and enforced, they may result in prohibiting of two- or three-shift construction work and delaying of work that requires overtime. Also, some Federal agencies have promulgated regulations requiring noise readings on projects. In accordance with such readings, local officials may place a construction project in one of the following categories:

Unacceptable: Noise levels exceed 80 dB for 1 hr or more per 24 hr, or 75 dB for 8 hr per 24 hr.

Normally unacceptable (discretionary): Noise exceeds 65 dB for 8 hr per 24 hr, or loud repetitive noises on site.

Normally acceptable (discretionary): Noise does not exceed 65 dB for more than 30 min per 24 hr.

Acceptable: Noise does not exceed 45 dB for more than 30 min per 24 hr.

Equipment may not be permitted on Federal building projects and generally should not be used on other types of projects if it produces a noise level 50 ft away exceeding the limits in Table 25-2.

Energy conservation in construction projects is part of the overall problem of the conservation of energy resources of the nation as a whole. Contractors should be alert to and aware of any ways to bring this about.

25-30. Systems Building.

The term systems building is used to define a method of construction in which use is made of integrated structural, mechanical, electrical, and architectural systems. The ultimate goal is integration of planning, designing, programming, manufacturing, site operation, scheduling, financing, and management into a disciplined method of mechanized production of buildings. Application of these systems should be controlled by an engineer-construction management firm rather than by use of prevailing contract building-management procedures.

TABLE 25-2 Limits on Noise Levels of Construction Equipment

Equipment	Maximum Noise Level at 50 Ft, dB(A)
Earthmoving	
Front loader	75
Backhoes	75
Dozers	75
Tractors	75
Scrapers	80
Graders	75
Trucks	75
Pavers	80
Materials handling	
Concrete mixers	75
Concrete pumps	75
Cranes	75
Derricks	75
Stationary	
Pumps	75
Generators	75
Compressors	75
Impact	
Pile drivers	95
Jackhammers	75
Rock drills	80
Pneumatic tools	80
Other	
Saws	75
Vibrators	75

Housing Production. The greatest concentration of effort has occurred in the realm of structural framing, leading to development of mass-production methods. These have, in general, been of the following types:

Panel type, consisting of floors and walls that are precast on site or at a factory and stacked in a house-of-cards fashion to form a building.

Volumetric type, consisting of boxes of precast concrete or preassembled steel, aluminum, plastic, or wood frames, or combinations of these, which are erected on the site after being produced in a factory.

Component type, consisting of individual members of precast-concrete beams and columns or prefabricated floor elements, which are brought to the site in volume, or mass produced.

These systems have displayed inherent disadvantages that came about through lack of opportunity to use the entire systems-building process, and because the systems often were not able to attain sufficient volume production to pay for the many fixed costs and start-up costs for the component factories that were built.

Experience has shown, however, that housing can be built efficiently and economically from standardized, modular designs. Some examples are as follows:

Mobile homes manufactured, complete with floor, roof, walls, electric wiring, plumbing, and cabinetry, by techniques similar to those used by automobile manufacturers. Built in factories, mobile homes are trucked to the sites and then installed and finished by on-site builders.

Packaged homes, also known as prefabricated or panelized. They are factory subassembled and, with an assortment of other building components, are delivered to the site for final assembly.

Modular homes, also known as sectional homes, manufactured in off-site factories, usually on an assembly line similar to that used by mobile-home manufacturers. Completely furnished, three-dimensional sections, including mechanical systems, of one or more rooms are factory assembled and then delivered to the sites by truck, railroad or barge.

(A. D. Bernhardt, "Building Tomorrow: The Mobile/Manufactured Housing Industry," The MIT Press, Cambridge, Mass.)

The lessons learned from past experiences offer the following guidelines for construction managers:

Early commitment must be made to use of a specific system in the design stage.
The systems builder must control the design.
Shop drawings should be started early and refined during the design.
Preparations of schedules, cost estimate, and bids must involve all the members of the team, that is, the designers, owner, and contractor, because there will be many last-minute proposed changes and details that will have to be challenged and modified.

Construction must follow an industrialized-building sequence, rather than a conventional bar chart or CPM diagram. Unlike the procedure for conventional construction, the method of scheduling and monitoring for industrialized building requires that all trades closely follow the erection sequence and that every trade match the speed of erection. Thus, if the erection speed is eight apartments per day, every trade must automatically fall on the critical path. It is necessary that each of the subcontractors work at the rate of eight apartments per day, otherwise the scheduled date of occupancy will not be met.

School Construction. The second major area in which systems building has been used appreciably is school construction. In this case, 8 or 10 prefabricated subsystems have been brought together for assembly on the site. The technique has been most effective in California school construction.

Metal Building Systems. These systems, produced by several manufacturers, have come into widespread use for many different types of low-rise buildings. The systems coordinate design and construction of buildings whose primary structure is a structural steel frame of standard, modular sizes; whose secondary members are mass-produced, cold-formed steel shapes; whose roof is light-gage metal; and whose walls may be built of any material.

Metal building systems generally cost less than other construction systems because of in-plant automation, use of precut and prepunched components, and quick erection on the site. Furthermore, costs are predictable. Metal buildings usually are purchased under design-build contracts. In addition, these systems offer flexibility, in that expansion can be readily achieved by removal of end walls, erection of new modular framing, and addition of matching walls and roof.

(Metal Building Dealers Association and Metal Building Manufacturers Association, "Metal Building Systems," MBMA, 1230 Keith Building, Cleveland, Ohio 44115.)

Types of Systems. In general, there are two kinds of system in use.

A *closed system* comprises one set of major components designed to integrate exclusively with each other. Such a system requires linked bids and coordinated product research and development by the manufacturers. In a closed system, the buyer has no alternative but to purchase the components specified. The California school construction system, for example, is virtually a closed system.

An *open system* is one in which many different components manufactured by different suppliers are compatible. Where the closed system will accept only a specific item, a specified ceiling lighting component for example, the ceiling lighting components in an open system can be met by three, four, or more components manufactured by different companies.

(See also Art. 2-16.)

25-31. Basics of Successful Management.

As outlined in this chapter, construction management, which will result in a profitable, on-time job, involves the organization and interplay of many talents. Engineers, accountants, field supervisors, construction labor, suppliers and subcontractors all aided by attorneys, insurance and bonding underwriters and the design professionals, must be organized and carefully monitored.

Those who succeed in this complex and difficult business are the ones who familiarize themselves thoroughly with the daily operations of their jobs. They are constantly learning by reading the latest literature and professional journals; they are alert and openminded about new ideas.

Hard work, planning, and diligent study of plans and specifications and thorough familiarity with other contract documents all combine, when properly applied, to yield a successful, profitable project.

Section 26

Specifications

Joseph F. Ebenhoeh, Jr.
Architect, Senior Associate
Chief, Architectural Specifications Division,
Albert Kahn Associates, Inc., Architects and Engineers,
Detroit, Mich.

26-1. Role of Specifications in Construction.

Architects or engineers prepare contract documents for a project for the purpose of obtaining competitive bids for construction and to guide the successful contractor during construction. To the extent these documents, such as The Agreement between the Owner and the Contractor and certain aspects of the general conditions, involve legal matters, the role of the architects and engineers is to assist the owner's attorney. The contract documents consist of

    The Agreement between the Owner and the Contractor
    General Conditions (and Special Conditions)
    Drawings and Specifications

The specifications are a written description and the drawings a diagrammatic presentation of the construction project. The drawings and specifications are complementary.

Although explanatory notes are included on drawings, these notes are short, general, and describe a type of construction, its location and quantity required, whereas specifications expand on the characteristics of the materials involved and the workmanship desired in installing them. A note on the drawings may read "Suspended acoustic ceiling." The specifications will completely describe the ceiling and establish such details as whether the acoustic material is prefinished perforated metal with sound-absorbing pads, mineral fiber, or vegetable fiber; the size of the units; whether the metal grid supporting the acoustic units is concealed or exposed, and numerous other conditions.

Specifications are addressed to the prime contractor. They present a written description of the project in an orderly and logical manner. They are organized into divisions and sections representing, in the opinion of the specification writer, the trades that will be involved in construction (Art. 26-7). Proper organization of the specifications will facilitate estimating and aid in preparation of bids.

26-2 Specifications

Workmanship required should be detailed in the specifications. Contractors study specifications to determine the sequence of work, quality of workmanship, and appearance of end product. From this information, they determine costs of the various skills and labor required. If workmanship is not determined properly, unrealistic costs will result.

For example, specifications will describe workmanship for architectural concrete in terms of forms, surface finish, curing methods, and weather protection. These items, representing large costs, will affect the bids substantially. Contractors will not include in their bids the cost of any item that is not specified or install it. If the form material (wood, metal, or plastic) and finished surface of the concrete are not defined in the specifications, contractors will select the form material that will give what they consider a satisfactory finish with the greatest number of reuses. The end product may or may not satisfy the building designer. A clearly written and comprehensive specification is the only means of indicating to contractors what must be done to produce an acceptable building.

Whenever a dispute arises, the contracting parties will scrutinize the specifications for instructions or requirements spelled out in writing before consulting any other part of the construction documents. So specifications must be specific and accurate in describing the requirements of the project.

A good specification will expand or clarify drawing notes, define quality of materials and workmanship, establish the scope of work, and spell out the responsibilities of the prime contractor.

26-2. Types of Specifications.

These include performance, descriptive, reference, proprietary, and base-bid specifications.

Performance specifications define the work by the results desired. For example, curtain-wall specifications will establish: that drawings will govern the design, character, and arrangement of the various components; materials and finishes; wind loads the assembled units will have to resist; that provisions should be made for expansion and contraction; acceptable minimum air infiltration; tests that must be conducted; acceptable heat-transmission factors; method of operation for ventilators (projected, pivoted, sliding); type of glazing (gaskets, metal bead with dry-seal gaskets, or putty glazing); and that all connections shall be made with concealed fasteners.

This type of specification gives the contractor complete freedom to employ his knowledge and experience to provide the itemized results. He will engineer the end product, design internal reinforcing and bracing, and assemble the various components to comply with the specification.

As another example: The engineer knows that numerous manufacturers are capable of producing fans with various characteristics. So he will write a performance specification that will establish the rotation, discharge, type, size, and capacity of the fan that will meet his requirements. Also, he will require the manufacturer to submit test data certifying the fan's performance. Under the performance specification, the manufacturer is responsible for establishing the physical and chemical properties of the metal used to build the fan and housing and the dimensions of the various parts, to ensure that the fan will perform as specified.

The terms of the contract should obligate each prime contractor to execute and deliver to the architect or engineer and owner a written guarantee that all labor and materials furnished and the work performed are in accordance with the requirements of the contract. Should any defects develop from use of inferior materials, equipment, or workmanship during the guarantee period (one year or more from the date of final completion of the contract or full occupancy of the building by the owner, whichever is earlier), the contractor should be required by the contract to put all guaranteed work in satisfactory condition. Also, he should make good all damage resulting from the inferior work and restore any work, material, equipment, or contents disturbed in fulfilling the guarantee.

The prime contractor should get guarantees from fabricators, manufacturers, and subcontractors who furnish products. Therefore, these firms would have to make good any defects, even to the extent of furnishing an entirely new product, to comply with performance defined in the specifications.

Descriptive specifications describe the components of a product and how they are assembled. The specification writer specifies the physical and chemical properties of the materials, size of each member, size and spacing of fastening devices, exact relationship of moving parts, sequence of assembly, and many other requirements. The contractor has the responsibility of constructing the work in accordance with this description. The architect or engineer assumes total responsibility for the function and performance of the product. Usually, architects and engineers do not have the resources, laboratory, or technical staff capable of conducting research on the specified materials or products. Therefore, unless the specification writer is very sure the assembled product will function properly, he should avoid the use of descriptive specifications.

Reference specifications employ standards of recognized authorities to specify quality. Among these authorities are the American Society for Testing and Materials, American National Standards Institute, National Bureau of Standards, Underwriters Laboratories, Inc., American Association of State Highway Officials, American Institute of Steel Construction, and American Concrete Institute.

Here is an example of a reference specification: Cement shall be portland cement conforming to ASTM C150-78a, "Specification for Portland Cement," using Type I or Type II for general concrete construction.

Reputable companies state in their literature that their products conform to specific recognized standards and will furnish independent laboratory reports supporting their claims. The buyer is assured that the products conform to minimum requirements and that he will be able to use them repeatedly and expect the same end result.

Reference specifications generally are used in conjunction with one or more of the other types of specifications.

Proprietary specifications call for materials, equipment, and products by trade name, model number, and manufacturer. This type of specification simplifies the specification writer's task because commercially available products set the standard of quality acceptable to the architect or engineer.

Use of these specifications is dangerous, because manufacturers reserve the right to change their products without notice, and the product that is incorporated in the project may not be what the writer thought he would get.

For example, on one project, a product was specified by model number and manufacturer, in accordance with usual practice. A catalog showed that a door on the product was mounted on a stainless-steel piano hinge. But after the specification was completed, the manufacturer's agent delivered a new catalog. This one showed the door mounted on three hinges. The specification writer asked to see a sample and discovered that the door actually was mounted on three bright metal hinges that distracted from the esthetic value of the product. Yet, this was the standard product identified by model number of the specified manufacturer that, because of the proprietary specification, became the standard of quality for the project. The specification was promptly changed to instruct the manufacturer to provide a continuous stainless-steel piano hinge, which was still available but at extra cost.

A disadvantage of proprietary specifications is that they may permit the use of alternative products that are not equal in every respect. Therefore, the specifier should be familiar with the products and their past performance under similar use and should know whether they have had a history of satisfactory service. He should also take into consideration the reputation of the manufacturers or subcontractors for giving service and their attitude toward making good on defective or inferior work on previous projects.

Under a proprietary specification, the architect or engineer is responsible to the owner for the performance of the material or product specified and for checking the installation to see that it conforms with the specification. The manufacturer of the product specified by model number has the responsibility of providing the performance promised in his literature.

In general, the specification writer has the responsibility of maintaining competition between manufacturers and subcontractors, to keep costs in line. Naming only one supplier may result in a high price. Two or more names should be supplied for each product to ensure competition.

Use of "or equal" should be avoided. It is not satisfactory in controlling quality of materials and equipment, though it saves time in preparing the specification, since only one or two products need be investigated and research time on other products is postponed.

Base-bid specifications establish acceptable materials and equipment by naming one or more (three preferred) manufacturers and fabricators. The bidder is required to prepare his proposal with prices submitted from these suppliers. Usually, base-bid specifications will permit the bidder to submit substitutions or alternates for the specified products. When this is done, the bidder should state in his proposal the price to be added to or deducted from the base bid and include the name, type, manufacturer, and descriptive data for the substitutions. Final selection rests with the owner.

Base-bid specifications provide the greatest control of quality of materials and equipment. But there are many pros and cons for the various types of specifications, and there are many variations of them.

26-3. Components of Specifications.

A properly arranged specification comprises four major parts—bidding requirements, general conditions (and special conditions), contract forms, and technical specifications. Each major part has a function of grouping all pertinent information for easy reference. See also Arts. 26-4 to 26-7.

26-4. Bidding Requirements.

These inform contractors of all provisions for submission of proposals.

The usual documents prepared for bidders are called advertisement for bids, invitation to bid, instructions to bidders, and the proposal form.

The advertisement for bids generally is prepared by the specification writer for projects financed with public funds. But it may be used on private projects if the owner elects to solicit bids in this manner. The advertisement for bids is a printed notice in newspapers or other periodicals. It is used to give public notice that proposals will be received for construction and completion of a structure.

The architect or engineer selects newspapers and other periodicals most likely to reach the largest number of contractors and other interested parties. Statute will determine the number of consecutive times the advertisement must appear to satisfy the requirements of public notice.

The advertisement for bids will give:

Name of owner
Location of project
Name of the prime contract or contracts

Time and place for receiving bids
Brief description of the project
Time and place the contract documents (which actually are not contract documents officially until an agreement is signed by both parties, but for simplicity they will be called contract documents here) can be obtained
Amount of deposit for the documents
Place where interested parties may examine them
Information regarding bid guarantees
Information governing performance and payment bonds

An invitation to bid contains practically the same information as an advertisement for bids. Both are addressed to bidders and signed by the owner or his duly authorized representative.

Invitations to bid generally are prepared for projects financed with private funds, but they may be used for public work when permitted by statute. The invitations are sent directly to qualified contractors selected for experience, qualifications, and financial ability to perform under the terms of the contract. The document usually is prepared in letter form.

Instructions to bidders describe conditions governing submission of proposals that must be established to protect the financial interests of the owner and the bidders. Each has a sizable sum of money invested in the project by the time bids are opened. The conditions usually are in accordance with acceptable industry standards mutually established by such organizations as the American Institute of Architects, American Consulting Engineers Council, and Associated General Contractors. Some of these conditions must be in accordance with statutes.

The instructions to bidders will describe in detail all requirements governing:

Preparation and submission of bids
Receipt and opening of bids
Withdrawal and rejection of bids
Interpretation of and addenda to the drawings and specifications
Bidders' visits to the site to ascertain pertinent local conditions

In addition, for public works or projects where the owner elects to advertise for bids, the specification writer will have to include other requirements. These will describe the qualification of bidders, submission and return of bidders' guarantee, taxes, and wage rates. After careful review of the project, the specification writer should add other pertinent information. For example, he may wish the bidders to submit a list of subcontractors and material suppliers. If so, he should define the terms of acceptance or rejection of these.

The proposal form is a means of obtaining uniformity in presentation of bidders' proposals. Preparation of this form by the architect or engineer is a widely accepted procedure and recommended in the interest of good practice.

The bidder is requested to complete the proposal form by filling in all blank spaces with the appropriate information. The proposal is addressed to the owner and signed by the bidder. When he signs, the bidder acknowledges having familiarized himself with the general and special conditions, drawings, and specifications; examined the site; and received addenda, written instructions affecting the work to be performed. Further, the bidder agrees to furnish all materials and perform the work for the stated price, to complete the work in the stipulated time, and to execute a contract for the project if the proposal is accepted within the established time.

The proposal form should provide blank space for prices that may be added to or deducted from the sum quoted for the complete project if the owner elects to make the changes described under the heading of "alternates." The general conditions will establish the rules governing changes in the contract. They also will itemize various methods that may be employed to determine the value of changes in the work. Usually, among the items included in the list of allowed costs is the contractor's fee stated as a percentage. Constituting all his charges for supervision, overhead, and profit, this fee varies, depending on whether the work is performed by his own forces or by a subcontractor. The proposal form should have blank spaces for the bidder to fill in the fees to be applied to the estimated cost of labor, materials, and equipment.

Another method used to determine the costs of changes is to compute the value of the work on the basis of unit prices stated in the contract. The usual method of obtaining unit prices is to require the bidder to state them in his proposal. The proposal should describe the work to which the unit price will apply and the unit of measure, and the form should provide blank spaces in which the bidders can place the prices.

Each paragraph containing blank spaces should define the scope of the item being quoted in clear, concise, and simple language.

The proposal form should be as simple and short as possible. It should not repeat general-condition and technical specification items, unless necessary for identification of quoted prices. Other items than those previously mentioned, however, can be included. Some of these are subcontractors' names and prices; subdivision of the bid total into prices for different types of work, or, if the project has more than one building, into prices for each of the buildings; statement on bidders' security; and space for the bidder to itemize substitutions.

The proposal form should have at the end blank spaces where the bidder can insert his full firm name and business address, his official signature, typewritten or printed name of signing party or parties, and his business form—individual, partnership, or corporation.

Proposals that are not signed by the individuals submitting them should be accompanied by a power of attorney establishing the authority of the signer to do so. Proposals made by a partnership should be signed by all the partners or an agent. If an agent signs, the partners should execute a power of attorney. Proposals submitted by a corporation should have the correct corporate name and the signature of the president or other authorized officer written below the corporate name following "By _____" and the impression of the corporate seal. If an official other than the president signs, a certified copy of a resolution of the Board of Directors authorizing the official to do so should be attached to the proposal.

26-5. General Conditions and Special Conditions.

The general conditions contain provisions that should be included in the contract documents for every project. Primarily, these provisions define the relationship and duties of the owner and the contractor.

The standard form, General Conditions of the Contract for Construction, American Institute of Architects, is general enough to apply to all building projects. This document should be used either by reference or by physical inclusion (binding) in the specifications. Experienced designers and some owners, however, may prepare a set of general conditions of their own. These often will serve better the practice of the designer or owner.

The specification writer should always use the general conditions adopted by his employer and never make revisions unless all parties concerned concur. This principle ensures that all members of the office and field staff will be thoroughly familiar with all provisions and that contractors will be familiar with all provisions governing costs, a factor tending to reduce estimating costs and thus to yield better prices to the owner.

The standard AIA general conditions comprise many articles assembled under 14 headings. The author's office has found through long experience that additional articles are necessary to protect the interests of all parties concerned. Certain of these articles, such as use of the site by the contractor, are modified to suit the individual project.

When general conditions prepared by owners or agencies are used on a project, the author's office prepares a document titled special conditions. These modify and supplement the general conditions to the extent the architects and engineers consider necessary to protect the parties concerned.

26-6. Contract Forms.

Since all parties executing a contract should be familiar with all provisions of the contract, the contract forms should be included in the specifications to give bidders the opportunity to review them before submitting their proposals. Then when a bidder is awarded a contract, he will be acquainted with its provisions.

Usual documents comprising the contract forms are the form of agreement, bid bond, labor and material payment bond, performance bond, sworn statements, waiver of lien, guarantee, arbitration agreement, and form of subcontract.

Standard contract forms, such as those prepared by the American Institute of Architects and the American Consulting Engineers Council, can be used with confidence. They have been prepared under the direction of counsel and with the help of consultants thoroughly familiar with the subject.

Generally, the architect or engineer will assist the owner's attorney in the final formulation of the general conditions and contract forms. Government agencies, insititutions, and large corporations, however, will prepare contract forms best suited to serve their interests, because the standard forms are too general.

26-7. Technical Specifications.

Difficult and time-consuming to prepare, the technical specifications give a written description of the project, lacking only a portrayal of its physical shape and its dimensions. The specifications describe in detail every piece of material, whether concealed or exposed, in the project and every piece of fixed equipment needed for the normal functioning of the project. If they are properly prepared, well organized, comprehensive, and indexed, any type of work, kind of material, or piece of equipment in a project can be easily located.

Usually, the specification writer will receive a set of prints for a project when the drawings are from 50 to 75% completed. The ideal, but not always practical, time to start the specifications is at the time when drawings are 95% or more complete.

The technical specifications cover the major types of work—architectural, civil, structural, mechanical, and electrical. Each of these types is further divided and subdivided in the specifications and given a general title that describes work performed by specific building trades or technicians, such as plasterers, tile setters, plumbers, carpenters, masons, and sheet-metal workers.

The specification writer studies the drawings, reports of meetings, and correspondence, and then assigns all pertinent work to the trade best qualified to perform it. Though the writer should make a sincere effort to assign work to the subcontractors for these trades, there is no rule requiring such assignments.

The prime contractor has the responsibility to perform all work, to furnish all materials, and to complete the project as rapidly as possible. The contractor, therefore, has the right to select subcontractors or do the work with the contractor's own forces. In recognition of this, each specification should contain a statement either in the general conditions or in the special conditions, that, regardless of the subdivision of the technical

26-8 Specifications

specifications, the contractor shall be responsible for allocation of the work to avoid delays due to conflict with local customs, rules, and union jurisdictional regulations and decisions.

Standard forms for technical specifications can be obtained from the Construction Specifications Institute. The CSI publishes a Master List of Section Titles and Numbers, which is a generally accepted standard. (See CSI Document MP-2-1, "Masterformat," The Construction Specifications Institute, 601 Madison St., Alexandria, VA 22314.) In it, technical specifications are organized into 16 divisions, each having titles that identify a major class of work. Each division contains basic units of work, called sections, related to the work described by the division title. Following is the division format developed by CSI:

1. General Requirements
2. Sitework
3. Concrete
4. Masonry
5. Metals
6. Wood and Plastics
7. Thermal and Moisture Protection
8. Doors and Windows
9. Finishes
10. Specialties
11. Equipment
12. Furnishings
13. Special Construction
14. Conveying Systems
15. Mechanical
16. Electrical

In preparing technical specifications, the specifier relies heavily on background information developed from years of experience. This information is recorded in specifications written for previous projects, checklists, master specifications, interoffice correspondence, letters replying to contractors' inquiries, and research notes. Inexperienced specification writers should use as a guide specifications of the Construction Specifications Institute accepted by many professionals. These basic documents set the framework for the trade sections, but the initiative of the specifier still is required to complete the technical specifications.

The form of each section should be based on the work to be described. But there is a similarity in structure of sections. Table 26-1 shows examples taken from two separate projects.

TABLE 26-1 Comparison of Forms of Technical Specifications

Ceramic tile	Movable metal partitions
Special note	Special note
Work required	Work required
Work included under other sections	Work included under other sections
Samples	Shop drawings
Manufacturers	Manufacturers
Materials	Types of partitions
Standard specifications	Partitions
Setting methods	Doors and frames

TABLE 26-1 (*Continued*)

Ceramic tile	Movable metal partitions
Mortar	Hardware
Tile work-general	Metal base
Application of wall tile	Metal ceilings
Application of floor tile	Hose cabinets
Protection	Glass and glazing
Completion	Finish
Extra tile for maintenance	Protection

The special note at the beginning of each section should make the general conditions (and special conditions) part of the technical specifications. The work-required paragraph following should clearly describe the scope of work.

Items of work not to be performed by the trade contractor should be described in the work-included-under-other-sections paragraph.

The paragraph on shop drawings should state that these are to be provided. Since general-condition items should not be repeated in the trade sections, it should refer to the applicable articles in the general and special conditions that describe the provisions governing the submission of the drawings.

The other subheadings are general. They were selected because they represent classes or subdivisions for which requirements usually would be presented in these sections.

When preparing technical specifications, the specifier should place himself in the role of the estimator, general contractor, fabricator, architect's project representative, and subcontractor, so as to provide information required by each.

Language should be clear and concise. Good specifications contain as few words as necessary to describe the materials and the work.

Use "shall" when specifying the contractor's duties and responsibilities under the contract. Use "will" to specify the owner's or architect's responsibilities. (See also Art. 26-8.)

Do not use "as directed by the architect," " ... to the satisfaction of the architect," or " ... approved by the architect." The specification should be comprehensive and adequate in scope to eliminate the necessity of using these phrases. "Approved by the architect" may be used, however, if it is accompanied by a specification that can be used to determine what the architect would consider in his evaluation. "By others" is not clear or definite and, when used, has resulted in many extras. Never use "any" when "all" is meant.

26-8. Automated Specifications.

This refers to a process that manipulates filed and new text and prints the text for a project specification. The process requires master specifications, data processing equipment, and editing techniques.

The master specification is a file of text covering every possible condition that would be needed. This text is stored in the memory bank of the data processing system, ready to be used on command. Use of master specifications gives the continued assurance that comprehensive and accurate specifications are produced.

The project specification is the written description of a specific project. To be useful, the master specification file must be periodically updated to include supplemental text for new and revised building components and systems and text reflecting changing office experience. The automated specification process must permit updating without affecting text that specification writers intend to be preserved.

Preservation of master text during the manipulation of text required during the production of project specifications is another demand placed on the process.

Data processing equipment performing these operations include the automatic typewriter and the electronic computer.

Automatic Typewriter. This is a machine resembling a standard typewriter, except for a few modifications consistent with its function. The automatic typewriter, manually operated, produces a typed original and simultaneously records on tape the material being typed. Later, when the need arises, the machine can be used to prepare automatically a typed copy of the text that was recorded on tape. Changes can be made by stopping the running of the tape and typing out the change. The machine simultaneously records the revised text on a second tape. The second tape then becomes the file copy of a project specification, or a new master specification.

A specification writer converts the master text into a project specification by reading each line of the printed copy taken off the automatic typewriter and editing as required. Changes are indicated by drawing a line through the text, or using a caret indicating the location of text written in the margins. Thus, the edited copy serves as instructions to be followed by the operator of the automatic typewriter. This system is well suited for small and some medium-size offices.

Computer. The most advanced data processing system employed in the automated specification process, the computer has a tape or disk storage bank, core memory bank, and high-speed printer, all run by complex integrated mechanical, electrical, and electronic systems. The tape or disk storage bank holds the master specifications. The core memory bank holds among other things the program containing the basic instructions for operating the computer to manipulate or edit the text in the tape or disk storage bank. The high-speed printer produces a copy of the text brought up from the storage bank and edited by the program. The computer records the revised text in the form of a project specification which is preserved and later printed on media acceptable for reproduction. At the same time, the master text is preserved and may be used for the production of another project specification.

Computers are available through direct in-house and time-sharing plans.

The direct in-house plan allows for the leasing or purchasing of computer equipment and the purchasing of programs available from numerous sources.

The time-sharing plan is a method of computer usage that allows two or more firms to use a computer system in such a way that each can operate as if the entire system were under the firm's exclusive control. The computer system is usually located at a data center remote from the user's location. The user has in the office a teletype or typewriter terminal that is tied to the remote data center through a telephone circuit. Master and project specifications are prepared, edited, and printed out at the user's terminal. Many offices find this plan economical. Normally, the printing of the specification is done at the console of the teletype, which is a slow printer. But high-speed printers are available to the user at the data center. The user, however, must establish a technique for delivery of the specification printout by United States mail or special messenger.

Programs. Success of the computer system depends on a program and a file of master specifications. A variety of programs, though, is available. The program and the editing techniques adapted are closely related.

The program is the set of provisions or instructions in the core memory unit that directs the computer to operate in a predetermined manner on input data to produce required output data. The simplest of programs for specifications translates instructions into the action of deleting lines or inserting text.

A more advanced program permits selection of large blocks of text related to common subject matter from the master specification. This program directs the computer to write a project specification from a large volume of master specification text. The specification printed as a result of the automatic editing process resembles very closely the final project specification. This project specification is read, and additional editing, if needed, is done by line or word deletions and insertions. Hand editing is kept to a minimum, thus releasing the text for early reproduction.

The advanced program requires development of editing techniques. It also requires that text be assigned in the master specification to common related blocks. This is done by giving each block a group name (corresponding to a family name) and each subdivision of that group a member name (like the given name of a family member). Also, phrases that change frequently can be assigned a number, and optional phrases can be given an alternate number. The advanced program is able to identify the options and select or delete them from the master text.

Because the goal of the advanced program is to select automatically from the file text applicable to a project, a system of editing techniques to decide which option is to be included must be developed by the office. Thus, when the responsible parties have selected the basic materials and systems to be included in the project, the decision is recorded in terms recognized by the program and master file. One method is to use questionnaires related to basic materials and systems, with questions requiring a simple yes or no answer. The computer is used to translate each yes answer into action, which consists of calling up from the master file the optional text identified with the selected materials and systems and then printing copy with the high-speed printer. This project specification can be further edited by adding or deleting text.

When final approval is given to the project specification, the computer uses another program to assign section numbers and page numbers, format text, drop out all the optional names assigned to blocks of text, and omit notes to the specification writer that are customarily recorded in the master file.

26-9. Covers and Binding for Specifications.

The cover page, title page, table of contents, and binding also are the responsibility of the specification writer.

The cover page should be made of heavy durable paper or plastic. Usually, the title of the project, location, name of owner or agency, name of architect or engineer, date of issue, and project number are printed on the cover page or title page or both.

The first page after the cover or title page should be the table of contents. It should list all parts, divisions, and sections included in the book.

Binding at the left side with staples, screw bolts, plastic binders, or equivalent fasteners is permitted. The binding method allowing the book to be laid flat is preferred. Some agencies, for legal reasons, require the stapling method.

26-10. Construction Contracts.

The specification writer must know, before starting the specifications, what construction contracts will be awarded on any given project. The kind and number of contracts will be based on size of project, usual practice of owner or architect, time allowed to complete the project, necessity of purchasing long-delivery items, and many other considerations.

An owner may award a single contract for construction of a project or several contracts, each covering portions of the work. When several contracts are awarded, each will have a complete set of contract documents. Usual practice is to divide the project into the major classes—civil, architectural, structural, mechanical, and electrical. Construction contracts may be awarded for the following:

Architectural work	Primary and secondary switchgear
Structural steel	Precast-concrete work
Plumbing and heating	Track work
Ventilating and air conditioning	Demolition
Fire-protection work	Landscaping
Electrical work	Compressors
Site preparation work	Kitchen equipment
Paving	Hardware
Foundations	Stage equipment
Transportation equipment	Laboratory equipment
Elevators	Boilers
Escalators	Interior decorating

Individual preference will dictate whether a single contract or separate contracts will be awarded for the major classes of work. Some architects and engineers prefer to receive separate proposals for each class but to award one contract. They inform the contractor submitting a proposal for the architectural work to include in his proposal a fee for managing the work of the contractors selected by the owner. In submitting a proposal, the contractor for the architectural work agrees to sign contracts with the other contractors. Similarly, they, on submitting a proposal, agree to sign with the contractor selected for the architectural work. The owner reviews all bids and accepts them independently.

The total contract amount with separate contracts is the sum total of the individual bids accepted, including a fee for managing the work. The fee may be a lump sum or a percentage of the total of accepted bids.

Reasons frequently given favoring separate contracts for architectural, mechanical, and electrical work include: (1) Elimination of bid shopping for the major subcontracts. (Bid shopping is a practice of some contractors to force subcontractors to submit a lower bid by informing them of lower prices submitted by others.) (2) Owner retains control of selection of major contractors. (3) Some construction may be started while contract documents are being prepared for other work. This results in an early occupancy date.

A single contract usually is preferred because it permits the general contractor to obtain bids from and award contracts to subcontractors with whom he knows he can work efficiently. Also, a single contract enables the contractor to manage all the work, including requests for approval of materials and equipment, shop drawings, tests, changes in the work, application of payments, establishing a detailed construction schedule, and expediting the work.

When separate contracts are awarded, it is very important that the specifications define the responsibilities of each contractor. For example, who will employ the excavator when the work requires excavating, backfilling, compacting, and grading for building foundations, underground piping, and underground electrical service? Building foundations usually are specified as part of the architectural work; piping is part of the mechanical, and electrical service obviously is the responsibility of the electrical contractor. The specifications must tell each of the bidders and subsequently the contractors who will employ the excavator to complete the earthwork.

Also, the specifications will tell each of the contractors how the work constructed under one separate contract is to be prepared to receive and connect to work performed under other separate contracts. For example, consider the case of a wall with face brick set on a ledge on a concrete foundation wall and anchored to concrete backup, a steel column anchored to the concrete, and dampproofing laid under the brick, down the foundation wall and up the backup specified distances. Assume that contract documents will be prepared for foundations, structural steel, and superstructure work. Then, the foundations specifications will establish that the work includes construction of foundations and concrete backup; column anchor bolts will be furnished by the structural steel contractor and installed by the foundation contractor; face brick, dampproofing, and metal anchors will be furnished and installed by the superstructure contractor; and dovetail slots to receive the metal anchors will be furnished and installed by the foundation contractor. The specifications for structural steel and for superstructure work will each establish the responsibilities of the contractors concerned in accordance with the assignment of work in the foundations specifications.

26-11. Revisions of Contract Documents.

Revisions made after issuance of contract documents to the bidders must be in writing.

Revisions of the contract documents made before receipt of bids should be issued to all parties concerned as *addenda*. Revisions issued after award of contract to the successful bidder should be described in a bulletin, with or without accompanying drawings, and made part of the contract on execution of a *change order*. The specification writer prepares the written description for addenda and bulletins, sometimes also the change order, but rarely the field, or work, orders.

Addenda inform bidders of revisions and modifications of work that will affect the quotations. Bulletins require the contractor to state a price for described revisions and modifications, but do not authorize a change in the contract. The contract is changed only when the price quoted is accepted by the owner, and contractor and owner sign the change order. Field orders describe a change in work that will take effect immediately after signing by architect or engineer, owner, and contractor.

26-12. Aids for Specification Writers.

In the course of preparing specifications, specification writers have to review all drawings pertaining to the work, determine the various components detailed, then prepare specifications for each of these components, and supplement the drawings by specifying work that is not detailed. Each experienced specification writer has his own system of recording information pertaining to each trade section thus collected. The more experienced the writer, the fewer the notes that will be jotted down.

The specifier is expected to establish order out of the confusion resulting from numerous instructions and complex drawings. His usual procedure for accomplishing this is to prepare an outline. Some suggestions follow:

The writer should purchase a spiral notebook, or staple together several blank pages at the left-hand margin, to serve as a specification notebook or outline. The first page should be a list of divisions and trade sections he has assigned to the project. This list also serves as an index for the notebook.

Place the title of each division and section at the top of a blank sheet. Then, review the drawings and correspondence for the project. Decide, as each bit of information, item of material or equipment is discovered, under what trade the item should be specified. Use a short note to record the item under the selected section. Verbal instructions received also should be recorded. Assignment of materials and work to trade sections should continue until all drawings and correspondence have been reviewed. After that, commence writing the sections.

If the outline is complete and thorough, the trade sections will be comprehensive and can be readily prepared. Table 26-2 shows an example taken from an outline.

26-14 Specifications

TABLE 26-2   Movable Metal Partitions

Report of meeting 5/6, p. 4.
Type—Flush, with recessed base, ceiling high
Dwg. Sht. 4—Sliding doors
Dwg. Sht. 18—Metal acoustic ceiling over shop offices
Dwg. Sht. 18—Metal filler above partition—Section A8
Dwg. Sht. 14—6-in.-high metal base, detail 58/14. Is this by MMP supplier?
Specify dry-seal glazing
Fire hose cabinets
Sliding-door hardware by MMP. Other hardware specified under hardware section

The final specification for movable partitions was prepared, a few weeks after the notes were recorded, from a master specification The specifier had only to delete paragraphs that were not applicable and to modify others.

When personal experience with performance of materials and equipment is lacking, or it is necessary to specify unknown materials and equipment, the specification writer can rely on products conforming with the minimum standards established by recognized authorities.

One such authority is the American Society for Testing and Materials. Its committees have developed over 3500 standard specifications and methods of test for materials. Each specification writer should have available those that are frequently used in preparing construction specifications.

Many industries produce construction materials complying with the physical and chemical properties specified in ASTM standards. Lack of an ASTM standard for a material should be a warning to a specifier that the material may be new, special (made on order only), or supplied by one manufacturer and more costly than alternatives.

Another important authority is the American National Standards Institute. It promulgates ANSI Standards, which define products, processes, or procedures with reference to one or more of the following: nomenclature, composition, construction, dimensions, tolerances, safety, operating characteristics, performance, quality, rating, certification, testing, and the service for which defined. Like ASTM specifications, these standards are subject to periodic review. They are reaffirmed or revised to meet changing economic conditions and technological progress. Users of ASTM specifications and ANSI Standards should avail themselves of the latest editions.

Federal specifications are the most comprehensive and competition-encouraging specifications available. They establish minimum requirements for various types of work. And often, manufacturers will submit test reports certifying that their product complies with a certain Federal specification. Therefore, specifiers will find it helpful to maintain a file of Federal specifications pertaining to construction. (See "Index of Federal Specifications and Standards" and supplements, Superintendent of Documents, U.S. Government Printing Office, Washington, D.C. 20402.)

26-13. Trade Associations and Societies.

The following publishers, trade associations, and societies publish reliable information and standards useful to specification writers:

American National Standards Institute, 1430 Broadway, New York, N.Y. 10018
American Society for Testing and Materials, 1916 Race St., Philadelphia, Pa. 19103
Architectural Aluminum Manufacturers Association, 35 East Wacker Drive, Chicago, Ill. 60606.

Architectural Woodwork Institute, 2310 South Walter Reed Drive, Arlington, Va. 22206.
Asphalt Institute, Asphalt Institute Building, College Park, Md. 20740.
Brick Institute of America, 11490 Commerce Park Drive, Reston, VA 22091.
Building Stone Institute, 420 Lexington Avenue, New York, N.Y. 10017.
Construction Specifications Institute, 601 Madison St., Alexandria, VA 22314

Copper Development Association, Inc., 405 Lexington Ave., New York, N.Y. 10017.
Door and Hardware Institute, 1815 N. Fort Myer Drive, Arlington, Va. 22209.
Facing Tile Institute, P.O. Box 8880, Canton, Ohio 44711.
Factory Mutual System, 1151 Boston-Providence Turnpike, Norwood, Mass 02062.
Gypsum Association, 1603 Orrington Avenue, Suite 1210, Evanston, Ill. 60201.
Indiana Limestone Institute of America, P.O. Box 72, Bedford, Ind. 47421.
Industrial Fasteners Institute, 1505 East Ohio Building, 1717 E. 9th St., Cleveland, Ohio 44114.
Industrial Risk Insurers, 85 Woodland, Hartford, Conn. 06102.
International Masonry Institute, Inc., 823 15th St., N.W., Washington, D.C. 20005.
Maple Flooring Manufacturers Association, 1800 Pickwick Ave., Glenview, Ill. 60025.
Marble Institute of America, 60 Plato Blvd., St. Paul, Minn. 55107.
Metal Lath/Steel Framing Association, 221 N. LaSalle Street, Chicago, Ill. 60601.
National Association of Architectural Metal Manufacturers, Suite 2026, 221 North LaSalle Street, Chicago, Ill. 60601.
National Concrete Masonry Association, P.O. Box 135, McLean, Va. 22101.
National Electrical Manufacturers Association, 2101 L St., N.W., Washington, D.C. 20037.
National Fire Protection Association, 470 Atlantic Ave., Boston, Mass. 02210.
National Forest Products Association, 1619 Massachusetts Avenue, N.W., Washington, D.C. 20036.
National Institute of Building Sciences, 1015 15th St., N.W., Washington, D.C. 20005.
National Terrazzo and Mosaic Association, Inc., 2-A West Loudoun Street, Leesburg, Va. 22075.
Portland Cement Association, Old Orchard Road, Skokie, Ill. 60076.
Sheet Metal and Air Conditioning Contractors National Association, Inc., 8224 Old Court House Rd., Vienna, Va. 22180.
Steel Deck Institute, P.O. Box 3812, St. Louis, Mo. 63122.
Steel Door Institute, 14600 Detroit Ave., Suite 712, Cleveland, Ohio 44107.
Steel Joist Institute, 1205 48th Ave. North, Myrtle Beach, SC 29577.
Steel Structures Painting Council, 4400 Fifth Ave., Pittsburgh, Pa. 15213.
Superintendent of Documents, U.S. Government Printing Office, Washington, D.C. 20402.
Sweet's Division, McGraw-Hill Information Systems Company, 1221 Avenue of the Americas, New York, N.Y. 10020.
Tile Council of America, Inc., P.O. Box 326, Princeton, N.J. 08540.
Underwriters Laboratories, Inc., 35 East Wacker Drive, Chicago, Ill. 60601.
Western Wood Products Association, Yeon Building, Portland, Oreg. 97204.
Wire Reinforcement Institute, 5034 Wisconsin Ave., N.W., Washington, D.C. 20016.
Zinc Institute, Inc., 292 Madison Ave., New York, N.Y. 10017.

Appendix

Factors for Conversion to the Metric System (SI) of Units

Frederick S. Merritt
Consulting Engineer
West Palm Beach, Fla.

Congress committed the United States to conversion to the metric system of units when it passed the Metric Conversion Act of 1975. This Act states that it shall be the policy of the United States to change to the metric system in a coordinated manner and that the purpose of this coordination shall be to reduce the total cost of the conversion. While conversion has already taken place in some industries and in some engineering disciplines, conversion is taking place in short steps at long time intervals in building design and construction. Consequently, conventional units are used throughout the preceding portion of this handbook. The metric system is explained and factors for conversion to it are presented in this Appendix, to guide and assist those who have need to apply metric units in design or construction.

The system of units that is being adopted in the United States is known as the International System of units, or SI, an abbreviation of the French *Le Système International d'Unités*. This system, intended as a basis for worldwide standardization of measurement units, was developed and is being maintained by the General Conference on Weights and Measures (CGPM).

For engineering, the SI has the advantages over conventional units of being completely decimal and of distinguishing between units of mass and units of force. With conventional units, there sometimes is confusion between use of the two types of units. For example, lb or ton may represent either mass or force.

SI units are classified as base, supplementary or derived units. There are seven base units (Table A-1), which are dimensionally independent, and two supplementary units (Table A-2), which may be regarded as either base or derived units.

Derived units are formed by combining base units, supplementary units and other derived units in accordance with algebraic relations linking the corresponding quantities. Symbols for derived units represent the mathematical relationships between the com-

A-2 Factors for Conversion to the Metric System (SI) of Units

TABLE A-1 SI Base Units

Quantity	Unit	Symbol
Length	meter	m
Mass	kilogram	kg
Time	second	s
Electric current	ampere	A
Thermodynamic temperature	kelvin	K
Amount of substance	mole	mol
Luminous intensity	candela	cd

TABLE A-2 Supplementary SI Units

Quantity	Unit	Symbol
Plane angle	radian	rad
Solid angle	steradian	sr

ponent units. For example, the SI unit for velocity, meter per second, is represented by m/s or m·s^{-1}; that for acceleration, meters per second per second, by m/s^2, and that for bending moment, newton-meters, by N·m. Figure A-1 indicates how units may be combined to form derived units.

As indicated in Fig. A-1, some of the derived units have been given special names; for example, the unit of energy, N·m, is called joule and the unit of pressure or stress, N/m^2, is called pascal. Table A-3 lists and defines derived SI units that have special names and symbols approved by CGPM. Some derived units used in building design and construction are given in Table A-4; others are listed with the conversion factors in Table A-6.

Prefixes. Except for the unit of mass, the kilogram (kg), names and symbols of multiples of SI units by powers of 10, positive or negative, are formed by adding a prefix to base, supplementary, and derived units. Table A-5 lists prefixes approved by CGPM. For historical reasons, kilogram has been retained as a base unit. Nevertheless, for units of mass, prefixes are attached to gram, 10^{-3}kg. Thus, from Table A-5, 1 Mg = 10^3 kg = 10^6 g.

The prefixes should be used to indicate orders of magnitude without including insignificant digits in whole numbers or leading zeros in decimals. Preferably, a prefix should be chosen so that the numerical value associated with a unit lies between 0.1 and 1000. Preferably also, prefixes representing powers of 1000 should be used. Thus, for building construction, units of length should be millimeters, mm; meters, m; and kilometers, km. Units of mass should be milligrams, mg; gram, g; kilogram, kg; and megagram, Mg.

When values of a quantity are listed in a table or when such values are being compared, it is desirable that the same multiple of a unit be used throughout.

In the formation of a multiple of a compound unit, such as of velocity, m/s, only one prefix should be used, and, except when kilogram occurs in the denominator, the prefix should be attached to a unit in the numerator. Examples are kg/m and MJ/kg; do not use g/mm or kJ/g, respectively. Also, do not form a compound prefix by juxtaposing two or more prefixes; for example, instead of Mkm, use Gm. If values outside the range of approved prefixes should be required, use a base unit multiplied by a power of 10.

To indicate that a unit with its prefix is to be raised to a power indicated by a specific exponent, the exponent should be applied after the unit; for example, the unit of volume, mm^3 = (10^{-3}m)^3 = 10^{-9}m^3.

TABLE A-3 Derived SI Units with Special Names

Quantity	Unit	Symbol	Formula	Definition
Celsius temperature	degree Celsius	°C	$K - 273.15$	The *degree Celsius* is equal to the kelvin and is used in place of the kelvin for expressing Celsius temperature (symbol t) defined by the equation $t = T - T_o$ where T is the thermodynamic temperature and $T_o = 273.15$ K by definition.
Electric capacitance	farad	F	C/V	The *farad* is the capacitance of a capacitor between the plates of which there appears a difference of potential of one volt when it is charged by a quantity of electricity equal to one coulomb.
Electric conductance	siemens	S	A/V	The *siemens* is the electric conductance of a conductor in which a current of one ampere is produced by an electric potential difference of one volt.
Electric inductance	henry	H	Wb/A	The *henry* is the inductance of a closed circuit in which an electromotive force of one volt is produced when the electric current in the circuit varies uniformly at a rate of one ampere per second.
Electric potential difference, electromotive force	volt	V	W/A	The *volt* (unit of electric potential difference and electromotive force) is the difference of electric potential between two points of a conductor carrying a constant current of one ampere, when the power dissipated between these points is equal to one watt.
Electric resistance	ohm	Ω	V/A	The *ohm* is the electric resistance between two points of a conductor when a constant difference of potential of one volt, applied between these two points, produces in this conductor a current of one ampere, this conductor not being the source of any electromotive force.
Energy	joule	J	N·m	The *joule* is the work done when the point of application of a force of one newton is displaced a distance of one meter in the direction of the force.
Force	newton	N	$kg \cdot m/s^2$	The *newton* is that force which, when applied to a body having a mass of one kilogram, gives it an acceleration of one meter per second squared.
Frequency	hertz	Hz	1/s	The *hertz* is the frequency of a periodic phenomenon of which the period is one second.

TABLE A-3 Derived SI Units with Special Names *(Continued)*

Quantity	Unit	Symbol	Formula	Definition
Illuminance	lux	lx	lm/m^2	The *lux* is the illuminance produced by a luminous flux of one lumen uniformly distributed over a surface of one square meter.
Luminous flux	lumen	lm	$cd \cdot sr$	The *lumen* is the luminous flux emitted in a solid angle of one steradian by a point source having a uniform intensity of one candela.
Magnetic flux	weber	Wb	$V \cdot s$	The *weber* is the magnetic flux which, linking a circuit of one turn, produces in it an electromotive force of one volt as it is reduced to zero at a uniform rate in one second.
Magnetic flux density	tesla	T	Wb/m^2	The *tesla* is the magnetic flux density given by a magnetic flux of one weber per square meter.
Power	watt	W	J/s	The *watt* is the power which gives rise to the production of energy at the rate of one joule per second.
Pressure or stress	pascal	Pa	N/m^2	The *pascal* is the pressure or stress of one newton per square meter.
Quantity of electricity	coulomb	C	$A \cdot s$	The *coulomb* is the quantity of electricity transported in one second by a current of one ampere.

TABLE A-4 Some Common Derived Units of SI

Quantity	Unit	Symbol
Acceleration	meter per second squared	m/s^2
Angular acceleration	radian per second squared	rad/s^2
Angular velocity	radian per second	rad/s
Area	square meter	m^2
Density, mass	kilogram per cubic meter	kg/m^3
Energy density	joule per cubic meter	J/m^3
Entropy	joule per kelvin	J/K
Heat capacity	joule per kelvin	J/K
Heat flux density	watt per square meter	W/m^2
Irradiance	watt per square meter	W/m^2
Luminance	candela per square meter	cd/m^2
Magnetic field strength	ampere per meter	A/m
Moment of force	newton-meter	$N \cdot m$
Power density	watt per square meter	W/m^2
Radiant intensity	watt per steradian	W/sr
Specific heat capacity	joule per kilogram kelvin	$J/(kg \cdot K)$
Specific energy	joule per kilogram	J/kg
Specific entropy	joule per kilogram kelvin	$J/(kg \cdot K)$
Specific volume	cubic meter per kilogram	m^3/kg
Surface tension	newton per meter	N/m
Thermal conductivity	watt per meter kelvin	$W/(m \cdot K)$
Velocity	meter per second	m/s
Viscosity, dynamic	pascal second	$Pa \cdot s$
Viscosity, kinematic	square meter per second	m^2/s
Volume	cubic meter	m^3

Factors for Conversion to the Metric System (SI) of Units

TABLE A-5 SI Prefixes

Multiplication factor		Prefix	Symbol
1 000 000 000 000 000 000	= 10^{18}	exa	E
1 000 000 000 000 000	= 10^{15}	peta	P
1 000 000 000 000	= 10^{12}	tera	T
1 000 000 000	= 10^{9}	giga	G
1 000 000	= 10^{6}	mega	M
1 000	= 10^{3}	kilo	k
100	= 10^{2}	hecto*	h
10	= 10^{1}	deka*	da
0.1	= 10^{-1}	deci*	d
0.01	= 10^{-2}	centi*	c
0.001	= 10^{-3}	milli	m
0.000 001	= 10^{-6}	micro	μ
0.000 000 001	= 10^{-9}	nano	n
0.000 000 000 001	= 10^{-12}	pico	p
0.000 000 000 000 001	= 10^{-15}	femto	f
0.000 000 000 000 000 001	= 10^{-18}	atto	a

*To be avoided where practical.

Fig. A-1 How SI units of measurement may be combined to form derived units.

A-6   Factors for Conversion to the Metric System (SI) of Units

Units in Use with SI. Where it is customary to use units from different systems of measurement with SI units, it is permissible to continue the practice, but such uses should be minimized. For example, for time, while the SI unit is the second, it is customary to use minutes (min), hours (h), days (d), etc. Thus, velocities of vehicles may continue to be given as kilometers per hour (km/h). Similarly, for angles, while the SI unit for plane angle is the radian, it is permissible to use degrees and decimals of a degree. As another example, for volume, the cubic meter is the SI unit, but liter (L), mL, or μL may be used for measurements of liquids and gases. Also, for mass, while Mg is the appropriate SI unit for large quantities, short ton, long ton, or metric ton may be used for commercial applications.

For temperature, the SI Unit is the kelvin, K, whereas degree Celsius, °C (formerly centigrade) is widely used. A temperature interval of 1°C is the same as 1 K, and °C = $K - 273.15$, by definition.

Loads. Building codes, at present, usually give minimum uniformly distributed gravity loads for design in lb/ft^2 and minimum concentrated gravity loads in lb. These loads are caused by the weight of objects to be sustained, and weight in commerce generally means mass. Hence, one method of specifying minimum design gravity loads in SI units might be use of kg/m^2 and kg. In that case, for computation of stress, the forces of gravity acting on the masses would have to be computed as the product of the masses by the acceleration g due to gravity. Hence, kg/m^2 and kg would be multiplied by g to convert to design forces in kN/m^2, or kPa, and kN, respectively. The standard international value for for g is 9.806650 m/s^2, whereas it actually ranges between 9.77 and 9.83 m/s^2 over the surface of the earth.

Dimensional Coordination. The basic concept of dimensional coordination is selection of the dimensions of the components of a building and installed equipment so that sizes may be standardized and the items fitted into place with a minimum of cutting in the field. One way to achieve this is to make building components and equipment to fit exactly into a basic cubic module or multiples of the module, except for the necessary allowances for joints and manufacturing tolerances. For the purpose, a basic module of 4 in is widely used in the United States. Larger modules often used include 8 in, 12 in, 16 in, 2 ft, 4 ft, and 8 ft.

For modular coordination in the SI, Technical Committee 59 of the International Standards Organization has established 100 mm (3.937 in) as the basic module. In practice, where modules of a different size would be more convenient, preferred dimensions have been established by agreements between manufacturers of building products and building designers. For example, in Great Britain, the following set of preferences have been adopted:

1st preference	300 mm (about 12 in)
2d preference	100 mm (about 4 in)
3d preference	50 mm (about 2 in)
4th preference	25 mm (about 1 in)

Accordingly, for a dimension exceeding 100 mm, the first preference would be a multimodule of 300 mm. Second choice would be the basic module of 100 mm.

The preferred multimodules for horizontal dimensioning are 300, 600 (about 2 ft), and 1200 (about 4 ft) mm, although other multiples of 300 are acceptable. Preferred modules for vertical dimensioning are 300 and 600 mm, but increments of 100 mm are acceptable up to 3000 mm. The submodules, 25 and 50 mm, are used only for thin sections.

Some commonly used dimensions, such as the 22 in used for unit of exit width, cannot be readily converted into an SI module. For example, 22 in = 558.8 mm. The nearest larger multimodule is 600 mm (23 ⅝ in), and the nearest smaller multimodule is 500 mm (19 ¹¹⁄₁₆ in). The use of either multimodule would affect the sizes of doors, windows, stairs,

TABLE A-6 Factors for Conversion to SI Units of Measurement

To convert from	to	multiply by
acre foot	cubic meter, m^3	1.233 489 E + 03
acre	square meter, m^2	4.046 873 E + 03
ampere hour	coulomb, C	3.600 000*E + 03
angstrom	meter, m	1.000 000*E − 10
are	square meter, m^2	1.000 000*E + 02
astronomical unit	meter, m	1.495 979 E + 11
atmosphere (standard)	pascal, Pa	1.013 250*E + 05
atmosphere (technical = 1 kgf/cm^2)	pascal, Pa	9.806 650*E + 04
bar	pascal, Pa	1.000 000*E + 05
barrel (for petroleum, 42 gal)	cubic meter, m^3	1.589 873 E − 01
board-foot	cubic meter, m^3	2.359 737 E − 03
British thermal unit (International Table)	joule, J	1.055 056 E + 03
British thermal unit (mean)	joule, J	1.055 87 E + 03
British thermal unit (39°F)	joule, J	1.059 67 E + 03
British thermal unit (59°F)	joule, J	1.054 80 E + 03
British thermal unit (60°F)	joule, J	1.054 68 E + 03
Btu (International Table)·ft/ (h)(ft^2)(°F) (k, thermal conductivity)	watt per meter kelvin, W/(m·K)	1.730 735 E + 00
Btu (International Table)·in/ (h)(ft^2)(°F) (k, thermal conductivity)	watt per meter kelvin, W/(m·K)	1.442 279 E − 01
Btu (International Table)·in/ (s)(ft^2)(°F) (k, thermal conductivity)	watt per meter kelvin, W/(m·K)	5.192 204 E + 02
Btu (thermochemical)·in/ (s)(ft^2)(°F) (k, thermal conductivity)	watt per meter kelvin, W/(m·K)	5.188 732 E + 02
Btu (International Table)/h	watt, W	2.930 711 E − 01
Btu (International Table)/s	watt, W	1.055 056 E+ 03
Btu (International Table)/ (h)(ft^2)(°F) (C, thermal conductance)[16]	watt per square meter kelvin, W/(m^2·K)	5.678 263 E + 00
Btu (thermochemical)/ (h)(ft^2)(°F) (C, thermal conductance)[16]	watt per square meter kelvin, W/(m^2·K)	5.674 466 E + 00
Btu (International Table)/ (s)(ft^2)(°F)	watt per square meter kelvin, W/(m^2·K)	2.044 175 E + 04
Btu (International Table)/lb	joule per kilogram, J/kg	2.326 000*E + 03
Btu (International Table)/ (lb)(°F) (c, heat capacity)	joule per kilogram kelvin, J/(kg·K)	4.186 800*E + 03
Btu (International Table)/ft^3	joule per cubic meter, J/m^3	3.725 895 E + 04
bushel (U.S.)	cubic meter, m^3	3.523 907 E − 02
calorie (International Table)	joule, J	4.186 800*E + 00
calorie (mean)	joule, J	4.190 02 E + 00
calorie (15°C)	joule, J	4.185 80 E + 00
calorie (20°C)	joule, J	4.181 90 E + 00
cd/in^2	candela per square meter, cd/m^2	1.550 003 E + 03
centimeter of mercury (0°C)	pascal, Pa	1.333 22 E + 03
centimeter of water (4°C)	pascal, Pa	9.806 38 E + 01
centipoise	pascal second, Pa·s	1.000 000*E − 03
chain	meter, m	2.011 684 E + 01

TABLE A-6 Factors for Conversion to SI Units of Measurement *(Continued)*

To convert from	to	multiply by
circular mil	square meter, m^2	5.067 075 E − 10
day	second, s	8.640 000*E + 04
day (sidereal)	second, s	8.616 409 E + 04
degree (angle)	radian, rad	1.745 329 E − 02
degree Celsius	kelvin, K	$T_K = t_c + 273.15$
degree Fahrenheit	degree Celsius	$t_C = (t_F - 32)/1.8$
degree Fahrenheit	kelvin, K	$T_K = (t_F + 459.67)/1.8$
degree Rankine	kelvin, K	$T_K = T_R/1.8$
(°F)(h)(ft²) Btu (International Table) (R, thermal resistance)	kelvin square meter per watt, $K \cdot m^2/W$	1.761 102 E − 01
(°F)(h)(ft²)/(Btu (International Table)·in) (thermal resistivity)	kelvin meter per watt, $K \cdot m/W$	6.933 471 E + 00
dyne	newton, N	1.000 000*E − 05
dyne cm	newton meter, $N \cdot m$	1.000 000*E − 07
dyne/cm²	pascal, Pa	1.000 000*E − 01
fathom	meter, m	1.828 804 E + 00
fluid ounce (U.S.)	cubic meter, m^3	2.957 353 E − 05
foot	meter, m	3.048 000*E − 01
foot(U.S. survey)	meter, m	3.048 006 E − 01
foot of water (39.2°F) (pressure)	pascal, Pa	2.988 98 E + 03
ft²	square meter, m^2	9.290 304*E − 02
ft²/h (thermal diffusivity)	square meter per second, m^2/s	2.580 640*E − 05
ft²/s	square meter per second, m^2/s	9.290 304*E − 02
ft³ (volume or section modulus)	cubic meter, m^3	2.831 685 E − 02
ft³/min	cubic meter per second, m^3/s	4.719 474 E − 04
ft³/s	cubic meter per second, m^3/s	2.831 685 E − 02
ft⁴ (area moment of inertia)	meter to the fourth power, m^4	8.630 975 E − 03
ft/h	meter per second, m/s	8.466 667 E − 05
ft/min	meter per second, m/s	5.080 000*E − 03
ft/s	meter per second, m/s	3.048 000*E − 01
ft/s²	meter per second squared, m/s^2	3.048 000*E − 01
footcandle	lux, lx	1.076 391 E + 01
footlambert	candela per square meter, cd/m^2	3.426 259 E + 00
ft·lbf	joule, J	1.355 818 E + 00
ft·lbf/h	watt, W	3.766 161 E − 04
ft·lbf/min	watt, W	2.259 697 E − 02
ft·lbf/s	watt, W	1.355 818 E + 00
ft-poundal	joule, J	4.214 011 E − 02
free fall, standard g	meter per second squared, m/s^2	9.806 650*E + 00
Gallon (Canadian liquid)	cubic meter, m^3	4.546 090 E − 03
gallon (U.K. liquid)	cubic meter, m^3	4.546 092 E − 03
gallon (U.S. dry)	cubic meter, m^3	4.404 884 E − 03
gallon (U.S. liquid)	cubic meter, m^3	3.785 412 E − 03
gallon (U.S. liquid) per day	cubic meter per second, m^3/s	4.381 264 E − 08
gallon (U.S. liquid) per minute	cubic meter per second, m^3/s	6.309 020 E − 05
grad	degree (angular)	9.000 000*E − 01
grad	radian, rad	1.570 796 E − 02
grain	kilogram, kg	6.479 891*E − 05
gram	kilogram, kg	1.000 000*E − 03
g/cm³	kilogram per cubic meter, kg/m^3	1.000 000*E + 03

To convert from	to	multiply by
gf/cm²	pascal, Pa	9.806 650*E + 01
hectare	square meter, m²	1.000 000*E + 04
horsepower (550 ft·lbf/s)	watt, W	7.456 999 E + 02
horsepower (boiler)	watt, W	9.809 50 E + 03
horsepower (electric)	watt, W	7.460 000*E + 02
horsepower (water)	watt, W	7.460 43 E + 02
horsepower (U.K.)	watt, W	7.457 0 E + 02
hour	second, s	3.600 000*E + 03
hour (sidereal)	second, s	3.590 170 E + 03
inch	meter, m	2.540 000*E − 02
inch of mercury (32°F) (pressure)	pascal, Pa	3.386 38 E + 03
inch of mercury (60°F) (pressure)	pascal, Pa	3.376 85 E + 03
inch of water (39.2°F) (pressure)	pascal, Pa	2.490 82 E + 02
inch of water (60°F) (pressure)	pascal, Pa	2.488 4 E + 02
in²	square meter, m²	6.451 600*E − 04
in³ (volume or section modulus)	cubic meter, m³	1.638 706 E − 05
in³/min	cubic meter per second, m³/s	2.731 177 E − 07
in⁴ (area moment of inertia)	meter to the fourth power, m⁴	4.162 314 E − 07
in/s	meter per second, m/s	2.540 000*E − 02
in/s²	meter per second squared, m/s²	2.540 000*E − 02
kelvin	degree Celsius	$t_c = T_K - 273.15$
kilogram-force (kgf)	newton, N	9.806 650*E + 00
kgf·m	newton meter, N·m	9.806 650*E + 00
kgf·s²/m (mass)	kilogram, kg	9.806 650*E + 00
kgf/cm²	pascal, Pa	9.806 650*E + 04
kgf/m²	pascal, Pa	9.806 650*E + 00
kgf/mm²	pascal, Pa	9.806 650*E + 06
km/h	meter per second, m/s	2.777 778 E − 01
kWh	joule, J	3.600 000*E + 06
kip(1000 lbf)	newton, N	4.448 222 E + 03
kip/in²(ksi)	pascal, Pa	6.894 757 E + 06
knot (international)	meter per second, m/s	5.144 444 E − 01
lambert	candela per square meter, cd/m	3.183 099 E + 03
liter	cubic meter, m³	1.000 000*E − 03
maxwell	weber, Wb	1.000 000*E − 08
mho	siemens, S	1.000 000*E + 00
microinch	meter, m	2.540 000*E − 08
micron	meter, m	1.000 000*E − 06
mil	meter, m	2.540 000*E − 05
mile (international)	meter, m	1.609 344*E + 03
mile (U.S. statute)	meter, m	1.609 347 E + 03
mile (international nautical)	meter, m	1.852 000*E + 03
mile (U.S. nautical)	meter, m	1.852 000*E + 03
mi² (international)	square meter, m²	2.589 988 E + 06
mi² (U.S. statute)	square meter, m²	2.589 998 E + 06
mi/h (international)	meter per second, m/s	4.470 400*E − 01
mi/h (international)	kilometer per hour, km/h	1.609 344*E + 00
mi/min (international)	meter per second, m/s	2.682 240*E + 01
mi/s (international)	meter per second, m/s	1.609 344*E + 03
millibar	pascal, Pa	1.000 000*E + 02

TABLE A-6 Factors for Conversion to SI Units of Measurement (*Continued*)

To convert from	to	multiply by
millimeter of mercury (0°C)	pascal, Pa	1.333 22 E + 02
minute (angle)	radian, rad	2.908 882 E − 04
minute	second, s	6.000 000*E + 01
minute (sidereal)	second, s	5.983 617 E + 01
ounce (avoirdupois)	kilogram, kg	2.834 952 E − 02
once (troy or apothecary)	kilogram, kg	3.110 348 E − 02
ounce (U.K. fluid)	cubic meter, m^3	2.841 307 E − 05
ounce (U.S. fluid)	cubic meter, m^3	2.957 353 E − 05
ounce-force	newton, N	2.780 139 E − 01
ozf·in	newton meter, N·m	7.061 552 E − 03
oz (avoirdupois)/gal (U.K. liquid)	kilogram per cubic meter, kg/m^3	6.236 021 E + 00
oz (avoirdupois)/gal (U.S. liquid)	kilogram per cubic meter, kg/m^3	7.489 152 E + 00
oz (avoirdupois)/in^3	kilogram per cubic meter, kg/m^3	1.729 994 E + 03
oz (avoirdupois)/ft^2	kilogram per square meter, kg/m^2	3.051 517 E − 01
oz (avoirdupois)/yd^2	kilogram per square meter, kg/m^2	3.390 575 E − 02
perm (0°C)	kilogram per pascal second meter, kg/(Pa·s·m)	5.721 35 E − 11
perm (23°C)	kilogram per pascal second meter, kg/(Pa·s·m)	5.745 25 E − 11
perm·in (0°C)	kilogram per pascal second meter, kg/(Pa·s·m)	1.453 22 E − 12
perm·in (23°C)	kilogram per pascal second meter, kg/(Pa·s·m)	1.459 29 E − 12
pint (U.S. dry)	cubic meter, m^3	5.506 105 E − 04
pint (U.S. liquid)	cubic meter, m^3	4.731 765 E − 04
poise (absolute viscosity)	pascal second, Pa·s	1.000 000*E − 01
pound (lb avoirdupois)[22]	kilogram, kg	4.535 924 E − 01
pound (troy or apothecary)	kilogram, kg	3.732 417 E − 01
lb·ft^2 (moment of inertia)	kilogram square meter, kg·m^2	4.214 011 E − 02
lb·in^2 (moment of inertia)	kilogram square meter, kg·m^2	2.926 397 E − 04
lb/ft·h	pascal second, Pa·s	4.133 789 E − 04
lb/ft·s	pascal second, Pa·s	1.488 164 E + 00
lb/ft^2	kilogram per square meter, kg/m^2	4.882 428 E + 00
lb/ft^3	kilogram per cubic meter, kg/m^3	1.601 846 E + 01
lb/gal (U.K. liquid)	kilogram per cubic meter, kg/m^3	9.977 633 E + 01
lb/gal (U.S. liquid)	kilogram per cubic meter, kg/m^3	1.198 264 E + 02
lb/h	kilogram per second, kg/s	1.259 979 E − 04
lb/in^3	kilogram per cubic meter, kg/m^3	2.767 990 E + 04
lb/min	kilogram per second, kg/s	7.559 873 E − 03
lb/s	kilogram per second, kg/s	4.535 924 E − 01
lb/yd^3	kilogram per cubic meter, kg/m^3	5.932 764 E − 01
poundal	newton, N	1.382 550 E − 01
pound-force (lbf)[23]	newton, N	4.448 222 E + 00
lbf·ft	newton-meter, N·m	1.355 818 E + 00
lbf·ft/in	newton-meter per meter, N·m/m	5.337 866 E + 01
lbf·in	newton-meter, N·m	1.129 848 E − 01
lbf·in/in	newton-meter per meter, N·m/m	4.448 222 E + 00
lbf·s/ft^2	pascal second, Pa·s	4.788 026 E + 01
lbf·s/in^2	pascal second, Pa·s	6.894 757 E + 03
lbf/ft	newton per meter, N/m	1.459 390 E + 01
lbf/ft^2	pascal, Pa	4.788 026 E + 01
lbf/in	newton per meter, N/m	1.751 268 E + 02

To convert from	to	multiply by
lbf/in² (psi)	pascal, Pa	6.894 757 E + 03
quart (U.S. dry)	cubic meter, m³	1.101 221 E − 03
quart (U.S. liquid)	cubic meter, m³	9.463 529 E − 04
rod	meter, m	5.029 210 E + 00
second (angle)	radian, rad	4.848 137 E − 06
second (sidereal)	second, s	9.972 696 E − 01
square (100 ft²)	square meter, m²	9.290 304*E + 00
ton (assay)	kilogram, kg	2.916 667 E − 02
ton (long, 2240 lb)	kilogram, kg	1.016 047 E + 03
ton (metric)	kilogram, kg	1.000 000*E + 03
ton (refrigeration)	watt, W	3.516 800 E + 03
ton (register)	cubic meter, m³	2.831 685 E + 00
ton (short 2000 lb)	kilogram, kg	9.071 847 E + 02
ton (long)/yd³	kilogram per cubic meter, kg/m³	1.328 939 E + 03
ton (short)/yd³	kilogram per cubic meter, kg/m³	1.186 553 E + 03
ton-force (2000 lbf)	newton, N	8.896 444 E + 03
tonne	kilogram, kg	1.000 000*E + 03
Wh	joule, J	3.600 000*E + 03
yard	meter, m	9.144 000*E − 01
yd²	square meter, m²	8.361 274 E − 01
yd³	cubic meter, m³	7.645 549 E − 01
year (365 days)	second, s	3.153 600*E + 07
year (sidereal)	second, s	3.155 815 E + 07
year (tropical)	second, s	3.155 693 E + 07

*Exact value.
From "Standard for Metric Practice," E380, American Society for Testing and Materials.

etc. For conversion of SI to occur, building designers and product manufacturers will have to agree on preferred dimensions.

Conversion Factors. Table A-6 lists factors with seven-digit accuracy for conversion of conventional units of measurement to SI units. To retain accuracy in a conversion, multiply the specified quantity by the conversion factor exactly as given in Table A-6, then round the product to the appropriate number of significant digits that will neither sacrifice nor exaggerate the accuracy of the result. For the purpose, a product or quotient should contain no more significant digits than the number with the smallest number of significant digits in the multiplication or division.

In Table A-6, the conversion factors are given as a number between 1 and 10 followed by E (for exponent), a plus or minus, and two digits that indicate the power of 10 by which the number should be multiplied. For example, to convert lbf/in² (psi) to pascals (Pa), Table A-6 specifies multiplication by $6.894\ 757 \times 10^3$. For conversion to kPa, the conversion factor is $6.894\ 757 \times 10^3 \times 10^{-3} = 6.894\ 757$.

["Standard for Metric Practice," E 380, and "Practice for Use of Metric (SI) Units in Building Design and Construction," E 621, American Society for Testing and Materials, 1916 Race St., Philadelphia, Pa. 19103; "NBS Guidelines for Use of the Metric System," NBS LC 1056, Nov., 1977, and "The International System of Units (SI)," NBS Specification Publication 330, 1977, Superintendent of Documents, Government Printing Office, Washington, D.C. 20402.]

Index

Abutments, **5**-115, **5**-118, **10**-112
Acceleration, gravity, **5**-90
ACI (*see* Associations, American Concrete Institute)
Acoustical absorption, **17**-16, **17**-21
Acoustical materials:
 absorption coefficients for, **17**-13, **17**-16, **17**-18
 absorption criteria for, **17**-21
 in ceilings, 1-14, 1-16, **9**-54, **9**-55, **9**-63, **17**-17, **17**-18, **21**-13
 impact insulation class of, **17**-18, **17**-20, **17**-21
 impact noise ratings of, **17**-18
 impedance of, **17**-3, **17**-4
 installation of, **17**-17
 lossy, **17**-12
 noise reduction coefficients for, **17**-15, **17**-18
 purposes of, **17**-6
 sound-absorbing, **17**-10, **17**-13 to **17**-17
 sound transmission classes of, **13**-19, **17**-11, **17**-18 to **17**-21
 sound transmission losses in, **17**-8 to **17**-10, **17**-18 to **17**-20
 viscoelastic, **17**-12
Acoustics:
 analysis in, **17**-8
 bibliography for, **17**-24
 criteria for, **17**-20 to **17**-22, **25**-59, **25**-60
 defined, **17**-1
 design for, **17**-8
 early studies of, **17**-1
 mass law in, **17**-8, **17**-9
 methods of (*see* Sound, control of)
 objectives of, 1-18, 1-19
 units of measurement for, **17**-6, **17**-7
 (*See also* Acoustical materials; Noise; Sound)
Acrylics, 4-63, 4-64, 4-76
Addenda, **2**-13, **26**-13

Adhesives:
 for ceramic tile, **12**-34, **12**-35
 for flooring, **12**-67
 for gypsumboard, **12**-54, **12**-55
 for wood, 4-33, 4-34, **7**-3, **7**-14, **7**-15, **7**-20, **7**-27, **7**-28, **7**-47, **7**-48
Adiabatic process, **18**-2, **18**-11, **18**-13
Admiralty metal, 4-54, 4-55, 4-57
Admixtures:
 for concrete (*see* Concrete admixtures)
 for plaster, **12**-37, **12**-39, **12**-42
Agreements (*see* Contracts)
Aggregates, 4-7 to 4-9, **10**-3 to **10**-5
Air:
 exfiltration of, **18**-4, **18**-23
 infiltration of, **18**-5, **18**-23, **18**-33, **18**-39, **18**-40
 makeup, **18**-2, **18**-33
 return, **18**-2, **18**-79, **18**-81
 saturated, **18**-2
 specific enthalpy of, **18**-11, **18**-12
 specific heat of, **18**-8
 specific volume of, **18**-12
 standard, **18**-2
 supply: for cooling, **18**-67 to **18**-70, **18**-72, **18**-75, **18**-82 to **18**-85
  for heating, **18**-47 to **18**-49
 thermal conductance of, **18**-4, **18**-17
 (*See also* HVAC; Ventilation)
Air changes, **18**-2, **18**-31, **18**-32, **18**-40
Air conditioners:
 absorption, **18**-78
 air distribution for (*see* Air conditioning)
 air-handling units for, **18**-75 to **18**-77, **18**-83
 built-up, **18**-80, **18**-81
 central plant, **18**-3, **18**-75, **18**-76, **18**-80
 with chemical cooling, **18**-90, **18**-91
 chilled-water, **18**-70 to **18**-72, **18**-78, **18**-81, **18**-85

1

Air conditioners (*Cont.*):
    compressor-motor units for, **18**-70 to
      18-72, **18**-74
    condensers for, **18**-3, **18**-71 to **18**-74,
      18-91
    defrosting of, **18**-4
    direct-expansion, **18**-4, **18**-70, **18**-71,
      18-76, **18**-77, **18**-81
    evaporators for, **18**-4
    heat-pump, **18**-4, **18**-63, **18**-64
    operating costs of, **18**-22
    packaged (self-contained), **18**-4, **18**-75
      to **18**-78
    rating of, **18**-72
    sizing of, **18**-20, **18**-21, **18**-67 to **18**-70
    split-system, **18**-6, **18**-77, **18**-78
Air conditioning:
    air distribution for: in cellular floors,
      9-61 to **9**-63
    double-duct, **18**-80 to **18**-83
    ducts for, **18**-78 to **18**-80
    grilles (registers) for, **18**-3, **18**-78,
      18-79
    high-velocity, **18**-80, **18**-82,
      18-83
    via perforated ceilings, **18**-79
    supply air for, **18**-67 to **18**-70, **18**-72,
      18-75
    temperatures for, **18**-72, **18**-75
    with terminal reheat, **18**-83, **18**-84
    in unused spaces, **18**-66, **18**-67
    comfort zone for, **18**-3
    control systems for, **18**-86 to **18**-89
    cooling loads for, **18**-15, **18**-37, **18**-40 to
      18-42
    defined, **18**-2, **18**-3
    with evaporative cooling, **18**-3, **18**-11,
      18-13
    fan-coil terminal units for, **18**-85
    induction terminal units for, **18**-83 to
      18-86
    industrial, **18**-90
    single-duct heating and cooling for,
      18-88, **18**-89
    standards for, **18**-31
    variable volume, **18**-85, **18**-86
    year-round, **18**-91
    zoned, **18**-75, **18**-76, **18**-83
    (*See also* Air conditioners; Cooling;
      Heating; HVAC; Ventilation)
Air curtains, **14**-3
Air-stabilized structures:
    applications of, **15**-21
    construction of, **5**-130 to **5**-132, **15**-21,
      15-22
    membranes for, **4**-68, **15**-22
    shapes of, **5**-130, **5**-131
    stresses in, **5**-132
AISC (*see* Associations, American
    Institute of Steel Construction)

Alarms, **3**-36, **19**-39, **19**-40, **19**-46, **20**-21,
    22-22
Alkyds, **4**-78
Alloys:
    aluminum, **4**-47 to **4**-51
    copper, **4**-52 to **4**-58
    lead, **4**-56, **4**-58, **4**-59
    nickel, **4**-41 to **4**-44, **4**-54, **4**-55, **4**-57,
      4-59, **4**-60
    steel, **4**-38, **4**-40 to **4**-45
Aluminum:
    alloys of, **4**-47 to **4**-51
    applications of, **4**-47
    bibliography for, **4**-52
    brazing of, **4**-51
    clad, **4**-48
    corrosion of, **4**-48, **4**-50, **4**-51
    finishes for, **4**-51
    foil, **18**-22, **18**-30
    hardened, **4**-48 to **4**-50
    heat-treated, **4**-48 to **4**-50
    in paint, **4**-78
    specific heat of, **18**-8
    tempers of, **4**-49, **4**-50
    weight of, **5**-5
    welding of, **4**-51
Ampere, **20**-1
Anchorage:
    of arch bases, **7**-62, **8**-2, **10**-112
    of column bases, **7**-50
    of reinforcing steel (*see* Steel
      reinforcement)
    of tendons, **6**-55, **10**-124, **10**-125
Anchors:
    dead-men, **6**-53 to **6**-55
    fiber screw, **16**-21
    government, **8**-6
    masonry, **8**-6, **9**-2, **9**-6, **11**-17, **12**-10,
      12-11, **12**-14 to **12**-17, **16**-15, **16**-16,
      16-21
    tendon, **6**-54, **6**-55
    types of, **16**-15, **16**-16, **16**-21
Angle post, for stairs, **22**-6, **22**-9
Angles:
    floating, **12**-40, **12**-62
    splay, **12**-43
Annunciators, **22**-15
ANSI (*see* Associations, American
    National Standards Institute)
Arbitration, **25**-45, **25**-46
Arches:
    components of, **5**-118, **10**-112
    concrete, **10**-112, **10**-113
    defined, **5**-115, **10**-112
    elastic center of, **5**-118, **5**-119
    fixed-end, **5**-118 to **5**-120, **8**-12, **8**-13,
      10-112, **10**-113
    reactions of, **5**-115 to **5**-120
    segmented, **7**-61, **7**-62
    steel, **8**-12, **8**-13

Arches (*Cont.*):
  stresses in, 5-116, 5-120
  supports for, 5-115, 5-116
  three-hinged, 5-116, 8-12, 8-13
  tied, 5-115, 5-116, 7-61
  Tudor, 7-59 to 7-61
  two-hinged, 5-116, 5-117, 8-13
  types of, 7-59, 7-60, 8-12, 8-13
  wood, 7-4, 7-23, 7-25, 7-59 to 7-63, 7-87
  (*See also* Beams, curved)
Architects:
  business administration by, 2-1 to 2-3
  collaboration with engineers, 1-4, 2-17
  construction duties of, 1-6, 2-14 to 2-16
  consultants of, 1-4, 2-7, 2-17
  cost estimating by, 1-31, 1-32, 2-10, 2-12, 24-2 to 24-5
  design activities of, 1-2, 1-4, 1-5, 2-6 to 2-13
  ethics of, 2-4 to 2-6
  fees for, 2-2
  insurance for, 25-23
  legal liabilities of, 2-16, 2-17
  licensed, 2-4
  practice standards of, 2-5, 2-6
  profession of, 1-2, 1-4, 2-1, 2-3 to 2-6
  project management by, 2-2, 2-3
  relations of: with contractors, 2-5, 2-12, 2-14 to 2-16
    with employees, 2-1, 2-5
    with owners, 1-7, 2-7, 2-10 to 2-12, 2-14
  selection of, 2-6, 2-7
  services provided by, 1-4, 1-5, 1-7, 2-6 to 2-19
Architecture:
  art of, 2-3, 2-4
  business of, 2-1 to 2-3
  defined, 1-2
  economic influences on, 2-19
  effects of construction changes on, 2-19
  principles of, 1-1, 1-2
  (*See also* Building design; Systems; Systems design)
Areas:
  of sections, 5-28
  variations of, with slope, 15-1, 15-2
Arris, 12-37
Asbestos-cement:
  components of, 4-23
  finishes of, 12-36
  properties of, 4-23, 4-24
  roofs of, 4-24, 15-19, 15-20, 15-25
  shingles of, 4-24, 5-9, 12-26, 15-25, 15-26
  siding of, 4-24, 5-9, 12-27, 12-28
Ashlar, 5-9, 12-2, 12-7, 12-10
ASHRAE (*see* Associations, American Society of Heating, Refrigerating and Air-Conditioning Engineers)

Asphalt:
  applications of, 4-74, 4-75, 15-2, 15-3
  block, 5-8
  defined, 15-2
  in flooring, 4-75, 5-8, 12-63 to 12-67, 12-70
  grout, 4-75
  mastics of, 4-75, 12-70
  pavements of, 4-75
  in roofing, 4-74, 4-75, 5-9, 15-22 to 15-24
  shingles of, 4-75, 5-9, 12-26, 15-22 to 15-24
  siding of, 4-75
  tile, 5-8, 12-63, 12-65 to 12-67
  types of, 4-74, 4-75
  weight of, 5-5
Associations:
  Aluminum Association, 4-51, 4-52
  American Concrete Institute (ACI), 3-29
  American Institute of Architects, 2-4 to 2-6, 25-10
  American Institute of Steel Construction (AISC), 8-3, 8-112, 8-113
  American Institute of Timber Construction, 7-18, 7-47, 7-59
  American Iron and Steel Institute, 4-47, 8-108, 8-112, 8-113, 9-12
  American National Standards Institute (ANSI), 3-2, 16-3, 22-13, 22-14, 26-14
  American Plywood Association, 7-66, 7-82
  American Society of Heating, Refrigerating and Air-Conditioning Engineers (ASHRAE), 2-9, 18-15
  American Society for Metals, 4-47, 4-59
  American Society for Testing and Materials (ASTM), 17-18, 26-14, A-11
  American Welding Society, 8-39, 8-113
  Architectural Aluminum Manufacturers Association, 26-14
  Architectural Woodwork Institute, 26-15
  Asphalt Institute, 26-15
  Associated General Contractors of America, Inc., 25-54
  Brick Institute of America, 11-1, 26-15
  Building Stone Institute, 26-15
  Concrete Reinforcing Steel Institute (CRSI), 10-22
  Construction Specifications Institute, 26-8, 26-15
  Copper Development Association, Inc., 4-58, 26-15

Associations (*Cont.*):
  Door and Hardware Institute, 14-6, 26-15
  Facing Tile Institute, **26**-15
  Gypsum Association, 1-16, 8-112, **12**-31, **12**-37, **12**-45, **12**-53, **26**-15
  Illuminating Engineering Society, 21-6, 21-25
  Indiana Limestone Institute of America, **26**-15
  Industrial Fasteners Institute, **26**-15
  International Masonry Institute, Inc., **26**-15
  Maple Flooring Manufacturers Association, **26**-15
  Marble Institute of America, **26**-15
  Metal Building Dealers Association, 9-66
  Metal Building Manufacturers Association, 9-66
  Metal Lath/Steel Framing Association, 8-112, **26**-15
  National Association of Architectural Metal Manufacturers, 12-30, **26**-15
  National Association of Home Builders, 7-66
  National Electrical Manufacturers Association, **26**-15
  National Elevator Industry, Inc., **22**-14
  National Fire Protection Association, 3-29, 14-4, 14-5, 19-39, 20-21, **22**-4, **26**-15
  National Forest Products Association, 4-35, 7-18, 7-45, 7-50, 7-56, 7-66, **26**-15
  National Hardwood Lumber Association, 4-35
  National Institute of Building Sciences, **26**-15
  National Roofing Contractors Association, 15-6
  National Terrazzo and Mosaic Association, Inc., **12**-68, **26**-15
  Perlite Institute, 8-112
  Portland Cement Association, 10-93, **26**-15
  Prestressed Concrete Institute, 10-130
  Steel Deck Institute, 9-58, **26**-15
  Steel Door Institute, 14-6, 14-8, **26**-15
  Steel Joist Institute, 9-3, 9-4, **26**-15
  Steel Structures Painting Council, 8-103, **26**-15
  Tile Council of America, Inc. **12**-32, **12**-36, **26**-15
  Western Wood Products Association, 4-35, 7-46, 7-50, 7-56
  Wire Reinforcement Institute, **26**-15
  Zinc Institute, Inc. **26**-15

ASTM (*see* Associations, American Society for Testing and Materials)
Back-charge orders, **25**-40, **25**-42
Balusters, **22**-6
Balustrade, **22**-4, **22**-5, **22**-11
  (*See also* Guard)
Band, **12**-37
Bar charts (Gantt), **24**-12, **25**-32 to **25**-34
Barometer, 18-13
Bars:
  flexibility of, 5-166
  impact stresses in, 5-96 to 5-98
  steel: chemical composition of, 4-37
    dimensional limits for, 9-9
  (*See also* Steel reinforcement; Tie rods)
  stiffness of, 5-166
  window, 13-3
  (*See also* Tendons)
Basalt, 4-20, 4-21, 5-6
Base lines, **23**-7 to **23**-9
Baseboard, **12**-37
Basements, 1-9, 1-10, 3-19 to 3-26
  (*See also* Floors; Foundations)
Bathtubs, **19**-5, **19**-12, **19**-13, **19**-15, **19**-19
Battens, **12**-26, **12**-27, **12**-36, **12**-37
Bead, **12**-37
  casing, **12**-38, **12**-46, **12**-48, **12**-60
  corner, **12**-39, **12**-46, **12**-60
Beading (ridging), **12**-42
Beams:
  allowable stresses for (*see specific materials, for example:* Steel; Wood)
  bending axis of, 5-38
  bending moments in (*see* Bending moment)
  bracing of (*see* Bracing; Bridging)
  buckling of (*see* Buckling)
  camber for, 7-56, 7-57, 8-2, 8-3
  cantilever, 5-20, 5-21, 5-35, 5-36, 7-53, 7-56, 11-12, 11-13
  (*See also* suspended-span, *below*)
  carry-over factors for, 5-62, 5-71, 5-72
  cold-formed-shape, 9-20, 9-28 to 9-32, 9-41
  (*See also* Cold-formed shapes)
  compact, 8-28, 8-29
  composite: of precast and cast-in-place concrete, 10-117
    steel-concrete, 8-20, 8-41 to 8-45, 9-3, 9-61, 9-63, 9-64, 10-117
    steel-wood, 7-82, 7-83
  concrete (*see* Concrete beams)
  conjugate, 5-32
  connections of (*see* Connections; Framing)
  continuous: algebraic equations for, 5-58 to 5-59, 5-74 to 5-76

Beams, continuous (*Cont.*):
  defined, 5-21
  influence lines for, 5-57
  moment-distribution methods for, 5-62 to 5-76
  as restrained simple beams, 5-57, 5-58
  slope-deflection equations for, 5-76
  support movements of, 5-66 to 5-69, 5-76, 6-25
  (*See also* carry-over factors for, *above;* fixed-end, *below*)
  cover-plated, 8-9, 8-46
  curvature of, 5-30 ·
  curved, 5-39 to 5-42
  (*See also* Arches)
  defined, 1-10, 5-20, 10-76
  deflections of (*see* Deflections; elastic curves for, *below*)
  dynamic models of, 5-90, 5-91, 5-98, 5-106 to 5-108
  elastic curves for, 5-30 to 5-32, 5-34 to 5-36
  fixed-end: defined, 5-59
    end moments in, 5-59 to 5-61, 8-57, 8-58
  flexibility of, 5-167, 5-168
  flexure formula for, 5-27, 5-38, 5-39
  hung-span (*see* suspended-span, *below*)
  lateral support of (*see* Bracing; Bridging)
  lintel, 1-10, 1-13, 1-15, 8-5, 8-6, 8-102, 8-103, 12-13
  minimum depths of (*see* Deflections, limits on)
  neutral axis (surface) of, 5-22, 5-27
  with overhang, 5-20 to 5-22, 5-35, 5-36
  in plastic range, 5-30, 5-156 to 5-159, 8-56 to 8-58
  propped-cantilever, 5-20, 5-21, 5-67
  reactions for (*see* Reactions)
  reinforced concrete (*see* Concrete beams)
  resisting moments in, 5-22
  (*See also* Bending moment)
  restraint of, 5-71
  rotation (slope) of, 5-34, 5-35, 5-54, 5-67, 5-70, 5-74, 5-76
  shear in (*see* Shear)
  sheet or strip steel (*see* cold-formed-shape, *above*)
  simple: defined, 5-20
    end rotations (slopes) of, 5-34, 5-35, 5-54, 5-70
    moments in, 5-23 to 5-26, 5-34, 5-35
    (*See also* Bending moment)
    reactions of (*see* Reactions)
    shear in, 5-34, 5-35
    (*See also* Shear)
  span of, 5-20, 5-62

Beams (*Cont.*):
  spandrel, 1-10, 8-7, 8-8, 8-35
  splices in, 7-2
  statically determinate, 5-21
    (*See also* cantilever; with overhang; simple, *above*)
  statically indeterminate, 5-21, 5-50, 5-57
    (*See also* continuous; fixed-end; propped-cantilever, *above*)
  steel (*see* Girders, steel; Structural steels, shapes of)
  stiffness of, 5-61, 5-62, 5-167, 5-168
  stresses in, 5-22, 5-27, 5-29, 5-30
    (*See also* Buckling; Cold-formed shapes; Concrete beams; Girders, steel; torsion in; web crippling in, *below*)
  supports for, 5-20, 5-21
  suspended-span, 5-51, 7-53, 7-56, 8-15
  torsion in, 8-35, 8-36
  ultimate strength of, 5-156 to 5-159
    (*See also* Concrete beams)
  unsupported length of, 7-23, 7-24
  unsymmetrical, 5-38, 5-39
  unsymmetrical bending of, 5-37
  vibration of, 5-90 to 5-95, 5-100, 5-107, 5-108
  web crippling in, 8-26, 8-27, 9-29 to 9-32
  wood (*see* Wood beams)
  (*See also* Framing; Girders; Joists)
Bearing stress, 5-14
  (*See also specific materials, for example:* Concrete; Masonry; Structural steels)
Bed mold, 12-38
Bel, 17-6
Belts, rubber, 4-73
Bench, hangers', 12-38
Bending (*see* Beams; Girders; Joists)
Bending moment:
  defined, 5-23
  diagrams of: for concentrated loads, 5-23, 5-24, 5-34, 5-35, 5-59, 5-60
    defined, 5-23
    by graphic statics, 5-45, 5-46
    plotting of, 5-23, 5-24
    for uniform loads, 5-23, 5-24, 5-34, 5-36, 5-59
  energy of, 5-16
  fixed-end, 5-59 to 5-61
  influence factors for, 5-66, 5-68
  influence lines for, 5-25, 5-26, 5-56, 5-57
  maximum: in continuous beams, 5-64 to 5-66
    location of, 5-24, 5-26, 5-27
  for moving loads, 5-26, 5-27
  for uniform load, 5-24

Bending moment (*Cont.*):
  plastic, 5-157 to 5-160
  redistribution of, 5-157
  sign conventions for, 5-23, 5-59
  in statically determinate beams, 5-23 to 5-26
  in statically indeterminate beams (*see* fixed-end; maximum, in continuous beams, *above*)
Bending moment-shear relationships, 5-23, 5-25
Bending stress (*see* Stress, bending)
Bents (*see* Frames)
Bidders:
  instructions to, 2-13, 26-5
  proposals from, 2-15, 26-5, 26-6
Bidets, 19-15
Bids:
  advertisement for, 2-13, 26-4, 26-5
  analysis of, 25-10, 25-14
  contingency allowances in, 24-3 to 24-5, 24-12, 24-13
  estimates for (*see* Cost estimates)
  invitations for, 2-13, 26-5
  material, 25-18
  reception of, 2-13
  solicitation of, 2-12, 2-13, 25-14, 26-4, 26-5
  subcontract, 25-13, 25-14
  submission of, 2-13, 26-5, 26-6
Bins, 10-115
Bitumen, 15-2, 15-16
Boards:
  backing, 12-37, 12-53
  batter, 23-7
  corner, 23-7
  lumber, 4-30, 7-5
  particle, 4-35
  (*See also* Fiberboard; Gypsumboard; Hardboard; Pulp board)
Bolts:
  allowable stresses for, 7-37, 7-38, 8-39 to 8-41, 9-49, 11-17
  anchor, 7-44
  applications of, 16-21
  carriage, 16-18
  clearances for, 8-76, 9-49
  edge distances for, 7-15, 7-37, 7-40, 7-42, 8-78, 8-79, 9-49
  expansion, 16-21
  high-strength, 4-38, 8-39 to 8-41, 8-56, 8-57, 8-69 to 8-71, 8-81, 8-82
  holes for, 7-25, 7-26, 7-37, 8-68, 8-71, 8-72
  Huckbolts, 8-71
  identification of, 8-70
  lag (*see* Screws, lag)
  locking (*see* Locking bolts)
  locking devices for, 8-69

Bolts (*Cont.*):
  in plastic design, 8-56, 8-57
  ribbed, 8-71
  spacing for, 7-15, 7-25, 7-32, 7-37, 8-79, 9-49
  specifications for, 8-39, 8-68, 8-69, 9-49
  steels for, 4-38, 4-39
  stove, 16-18
  symbols for, 8-68
  tightening of, 8-68, 8-70, 8-71
  toggle, 16-21
  unfinished (machine), 4-39, 7-37, 8-39 to 8-41, 8-56, 8-57, 8-68, 8-69
  washers for, 7-44 to 7-46, 8-68 to 8-70, 8-72
  for wood construction, 7-15, 7-32, 7-37 to 7-46
  (*See also* Connections, bolted)
Bond, of concrete (*see* Concrete; Steel reinforcement)
Bonder, masonry, 12-2, 12-9
Bonds (in contracts):
  alternatives to, 25-32
  bid, 2-13, 25-30 to 25-32
  brokers for, 25-5, 25-31
  information required for, 25-31, 25-32
  interests protected by, 25-31
  money and security protection, 25-22, 25-29
  performance, 25-30 to 25-32
Boss, plaster, 12-38
Boussinesq stress distribution, 6-25, 6-26
Bow's notation, 5-49
Box system, 5-111
Bracing:
  batter-pile, 6-37, 6-53
  of beams, 7-22 to 7-24, 7-50, 8-18, 8-33, 8-56, 8-58 to 8-61, 9-6, 9-28
  (*See also* Bridging)
  capacity required for, 7-88, 8-61, 9-35
  of columns, 8-24, 8-25, 8-58, 8-59, 9-35
  connections of, to stiffeners, 8-47
  of crane structures, 8-67
  in erection, 7-87, 7-88, 8-58, 9-6
  of excavations, 6-53, 6-54
  of footings, 3-16
  of formwork, 10-20, 10-21
  functions of, 1-11, 3-9
  K, 3-12
  knee, 3-12, 7-58, 8-58, 8-62, 8-65
  methods of, 3-9 to 3-13, 3-16, 5-110 to 5-112, 7-58, 8-62 to 8-67
  of piles, 3-16, 6-37, 6-38
  for plastic design, 8-56
  of rigid frames, 8-56, 8-67
  (*See also* for plastic design, *above*; wind, *below*)
  shear-wall (*see* Shear walls)

Bracing (*Cont.*):
  of trusses, 7-58, 7-87, 7-88, 8-58
    (*See also* wind, *below*)
  V, 3-12
  for walls, 7-61, 7-63, 7-64, 11-2, 12-23, 12-24
  wind, 3-9 to 3-13, 8-61 to 8-67
    (*See also* Connections; Rigid frames; Shear walls)
  X, 3-9, 3-10, 3-12, 3-13, 5-153, 7-58, 8-62, 8-65 to 8-67
    (*See also* Buildings, drift of; Stiffeners)
Brackets (*see* Connections, eccentrically loaded)
Brads, 16-17
Brass, 4-52 to 4-57, 5-6, 18-8
Brick:
  absorption test for, 12-8
  allowable stresses for, 8-34, 11-18, 12-7
  concrete (*see* Concrete block)
  defined, 11-2, 12-2
  dimensions of, 4-16
  engineered construction with:
    allowable loads on, 11-6, 11-7
    allowable stresses for, 11-3 to 11-5
    bibliography for, 11-33
    compressive strength of, 11-3, 11-4
    inspection of, 11-2, 11-4
    mortar for, 11-3
    in shear walls, 11-7 to 11-9
    standards for, 11-1 to 11-3
      (*See also* Masonry walls, standard code requirements for)
    tolerances for, 11-3
  expansion of, 4-16
  laying of, 12-6, 12-8 to 12-12
  modulus of elasticity of, 11-5
  modulus of rigidity of, 11-5
  moisture limits for, 4-16
  as plaster base, 12-48
  prism tests of, 11-2 to 11-4
  salmon, 4-16, 12-4
  secondhand, 12-4
  specific heat of, 18-8
  specifications for, 3-22, 4-16, 11-3
  strength of, 4-16, 11-4
  thermal conductivity of, 4-16
  weight of, 5-8
  wetting of, 12-6, 12-8
    (*See also* Masonry; Masonry walls)
Bridges:
  highway, 10-38
  temporary, 1-6
Bridging:
  for cold-formed beams, 9-28
  for concrete joists, 10-64, 10-65
  for steel joists, 8-18, 9-6, 9-28
  for wood beams, 7-22 to 7-24, 7-50
British thermal unit (Btu), 18-3, 18-8

Bronze, 4-53 to 4-58, 5-6, 18-8
Buckling:
  column, 5-82 to 5-86, 8-23, 8-24, 8-59
  local, 5-85
    (*See also* Cold-formed shapes)
  steel-beam, 8-23, 8-26, 8-58
  wood-beam, 7-22 to 7-24
Bucks, door, 14-6, 14-8
Budget, cost, 2-10, 2-11
Buffers, 22-14, 22-15, 22-18, 22-20, 22-21, 22-35
Builders:
  designer, 2-24, 2-25
  home, 1-6, 2-24, 25-2
  package, 2-24, 25-2
    (*See also* Construction managers; Contractors; Owner-builder; Sponsor-builder)
Building codes (*see* Codes)
Building construction (*see* Construction)
Building design:
  competency for, 1-2
  defined, 1-1, 1-2
  for emergency conditions, 1-5, 3-1, 3-11, 3-12, 3-29 to 3-31, 3-43 to 3-47, 3-50, 5-1
    (*See also* Fire; Fire protection; Floodproofing; Lightning protection; Security measures; Seismic design; Wind design)
  for energy conservation, 2-9, 2-10, 25-59
  for handicapped persons, 1-5, 1-19, 16-2, 16-9, 16-10, 22-2, 22-13, 22-22
  principles of, 1-2
  procedure for, 1-4, 1-5, 1-26, 1-27, 2-10 to 2-16
  program for, 1-4, 1-20, 2-7 to 2-10
  for security (*see* Security measures)
    (*See also* Architecture; Systems; Systems design)
Building team, 1-37, 2-17, 2-18
Buildings:
  air-supported (*see* Air-stabilized structures)
  area limitations for, 1-34 to 1-36, 3-33, 3-34
  circulation in, 22-1, 22-2
    (*See also* Corridors; Doors; Dumbwaiters; Egress, emergency; Elevators; Escalators; Passageways, exit; Ramps; Stairs)
  construction classes of, 1-34, 3-31, 3-33 to 3-35, 7-31
  control center for, 3-42, 3-47, 3-51, 19-49
  defined, 1-1
  design life of, 3-2
  drift of, 3-9 to 3-13, 3-16
    (*See also* Rigid frames, sidesway of)

Buildings (*Cont.*):
    fire-resistance classification of, 3-34 to 3-35
    height limitations for, 1-35, 3-33, 3-34
    nuclear containment, 10-37
    occupancy classes of, 1-34, 3-31, 3-34, 3-35
    overturning of, 3-8, 3-9
    preengineered, 1-30, 2-21, 9-66, 25-61
    sanitary engineering, 10-38, 10-85
    sliding of, 3-8, 3-9
    systems in, 1-3, 2-20
    (*See also* Systems)
Bull nose, 12-38
Bumpers, 22-15, 22-18, 22-20, 22-29
Burglar protection, 3-50, 3-51, 13-6, 13-7
Buttress, 12-2
Butts, 16-4 to 16-8, 16-13, 16-14
Butyl rubber, 4-71, 4-76, 15-17

Cable-supported structures:
    network, 5-146, 5-151 to 5-153
    simple, 5-141 to 5-146
    stayed-girder, 5-141
    truss, 5-146 to 5-150
    (*See also* Cables)
Cables:
    applications of, 5-141
    catenary, 5-142, 5-143
    characteristics of, 5-141, 5-146
    defined, 5-140, 5-141
    deflections of, 5-145
    elongation of, 5-144
    length of, 5-142, 5-144
    parabolic, 5-143 to 5-145
    shapes of, 5-141
    stresses in, 5-141 to 5-145
    vibration of, 5-147, 5-148
    (*See also* Cable-supported structures)
Caissons:
    drilled, 6-25, 6-47
    drilled-in, 6-43, 6-44
    nested, 6-47
    pneumatic, 6-46, 6-47
    sinking of, 6-46, 6-47
Calcimine, 4-77
Calcium chloride, 10-10, 10-11
Calking, 3-17, 3-27, 18-23
Calking compounds, 4-75, 4-76
Camber (*see* Beams, camber for; Trusses, camber for)
Candlepower (candela), 21-2
Cantilever method, 5-155
Cantilevers, 5-20, 5-21, 5-35, 5-36, 7-53, 7-56, 11-12, 11-13
Capital recovery, 1-24
Capitals, column, 10-72, 12-38
Carbon dioxide, 3-37, 3-38
Carbon equivalent, 4-46

Carpets, 12-66 to 12-68, 17-12, 17-18 to 17-20
Carriage, 22-5, 22-8
Carry-over factors, 5-62, 5-71, 5-72
Casings, 13-3, 14-6, 14-8
Cast iron (*see* Iron, cast)
Castigliano's theorems, 5-52
Catenaries, 5-142, 5-143
Cathodic protection, 4-47
Ceilings:
    acoustical, 1-14, 1-16, 9-54, 9-55, 9-63, 17-17, 17-18, 21-13
    construction of, 12-38, 12-39
    defined, 1-16, 12-1
    as fire protection, 1-15, 1-16, 3-33, 8-109 to 8-112
    gypsumboard (*see* Gypsumboard)
    openings permitted in, 8-111, 8-112
    panel-heating, 18-58, 18-59
    plaster (*see* Plaster)
    suspended, 1-15, 1-16, 12-39, 12-43, 12-61, 15-20, 17-17, 21-13
    weights of, 5-9
    (*See also* Floors)
Cellars (*see* Basements)
Cellulosics, 4-63, 4-66, 4-67, 4-80
Cement:
    aluminous, 4-4
    classes of, 4-1, 4-2
    defined, 4-1
    Keene's, 12-41
    masonry, 4-2, 4-7, 12-8
    natural, 4-4, 10-11
    oxychloride, 4-2, 4-6, 4-7
    portland: air-entraining, 4-3, 4-4
        blast-furnace slag, 10-2
        color variations from, 10-2
        components of, 4-2, 4-3
        high-early-strength, 4-2, 4-3
        low-heat, 4-3
        manufacture of, 4-2
        paint of, 3-26 to 3-28
        plaster (*see* Coatings, cementitious)
        pozzolan, 10-2
        properties of, 4-2 to 4-4
        shrinkage-compensating, 10-2
        specifications for, 4-2, 4-3, 10-1, 10-2
        sulfate-resistant, 4-2, 4-3
        types of, 4-2, 4-3, 10-2
        weight of, 4-9, 5-6
    Sorel, 4-2, 4-6, 4-7
Centering (*see* Formwork)
Ceramic veneer, 4-17, 12-2
Chamfer, 12-39
Change orders, 2-15, 25-44 to 25-46, 26-13
    (*See also* Contracts, construction, changes in)
Charcoal, 5-6

Chases, 12-2, 12-22, 12-39
Chimneys, 12-18, 12-22, 18-50
Cinders, 5-6, 5-9, 6-49
Circuit breakers, 20-17, 20-18, 20-36 to 20-39
Clapboard, 12-26
Clay, 4-8, 5-6, 6-1, 6-6, 6-25 to 6-29, 6-49, 6-52, 6-53, 6-56
Clerk of the works, 1-6
Clients, building (*see* Owners)
Coal, 5-6, 6-49
Coatings:
 bibliography for, 4-80
 bituminous, 3-22 to 3-27, 3-29, 4-74
 cementitious, 3-22, 3-23, 3-26 to 3-29
 chemical, 4-47
 colorless, 3-27, 3-28
 for concrete walls (*see* Membrane, waterproofing)
 elastomeric, 15-3, 15-13, 15-16, 15-17
 fire-retardant, 8-111
 metallic, 4-47
 organic, 3-27 to 3-29
 porcelain-enamel, 4-70, 4-71
 rubber, 4-73
 water-repellent, 3-28
 waterproof, 3-21 to 3-26
 (*See also* Concrete, sprayed; Paint)
Cocks, 19-5, 19-9
Codes:
 building, 1-33, 1-34, 2-7, 3-1, 3-30, 25-7, 25-8
 of ethics, 2-4 to 2-6
 zoning, 1-34 to 1-36, 2-7, 3-33
Cofferdams, 6-46
 (*See also* Sheeting)
Cold-formed shapes:
 allowable loads for welds of, 9-44 to 9-46
 allowable stresses for: axial and bending, 9-38
  basic bending, 9-20, 9-41
  basic tension, 9-20
  bolted connections of, 9-49 to 9-51
  compression, 9-21 to 9-23, 9-33, 9-34, 9-39
  seismic and wind, 9-20
  shear, 9-29
  shear and bending, 9-29
  web crippling, 9-29 to 9-32
 applications of, 9-7, 9-12
 as beams, 9-20, 9-28 to 9-32, 9-41
 bolting of, 9-49 to 9-51
 coatings for, 9-12, 9-54
 cold-work-strengthened, 9-13, 9-14
 as columns, 9-20, 9-31 to 9-38
 compression elements of, 9-18 to 9-29, 9-31 to 9-38
 corrugated, 9-39 to 9-41, 9-59, 9-60
 curved, 9-20

Cold-formed shapes (*Cont.*):
 defined, 9-7
 depth-thickness limits for, 9-29
 edge stiffeners for, 9-18
 effective width of, 9-23 to 9-27
 fasteners for, 9-49 to 9-53
 fire ratings for, 9-59, 9-60, 9-65, 9-66
 flat-width ratios of: as buckling criteria, 9-19 to 9-27, 9-31
  defined, 9-16, 9-18
  maximum permissible, 9-27, 9-28
  selection of, 9-22, 9-23
 as floor deck, 8-19, 8-20, 8-44, 8-45, 9-61 to 9-66
 local buckling of, 9-16, 9-19, 9-20, 9-27, 9-28, 9-31, 9-38
 materials for, 9-8 to 9-12
 multiple stiffened, 9-18
 riveting of, 9-51, 9-52
 as roof deck, 9-53 to 9-60, 15-7, 15-14, 15-15
 screw connections for, 9-52, 9-53
 section properties of, 9-14 to 9-17, 9-19, 9-23 to 9-27, 9-39 to 9-41
 (*See also* compression elements of, *above*)
 shear lag in, 9-27
 stainless steel, 9-12, 9-41, 9-42
 thicknesses of, 9-13
 torsional-flexural buckling of, 9-32 to 9-37
 tubular, 9-38, 9-39
 types of, 9-13 to 9-16
 as wall studs, 5-9
 web crippling in, 9-29 to 9-32
 welding of, 9-42 to 9-48
 wide-flange, 9-27
 (*See also* Sheet steel; Strip steel)
Columns:
 bracing of, 8-24, 8-25, 9-35
 buckling of, 5-82 to 5-86, 8-23, 8-24, 8-59
 cold-formed-shape, 9-20, 9-31 to 9-38
 (*See also* Cold-formed shapes)
 composite, 8-15, 8-59
 concrete (*see* Concrete columns)
 curves for, 5-84, 5-85
 defined, 5-82
 end-restraint effects on, 5-84, 5-85
 Engesser equation for, 5-84
 Euler (critical) load for, 5-83, 5-84
 Euler equation for, 5-84, 8-24
 masonry, 11-5 to 11-7, 12-2
 pipe, 8-28
 purpose of, 1-11
 sheet and strip steel (*see* cold-formed-shape, *above*)
 slenderness ratios of, 5-84, 5-85, 8-23 to 8-25, 8-50
 structural steel: capacity of, 8-23 to

Columns, structural steel (*Cont.*):
8-26, 8-55
  effective length of, 8-24 to 8-26
  shapes for, 8-28, 8-58
  shear in, 8-56
  splices of, 8-64, 8-93 to 8-96
  thickness limits for, 8-50
  tubular, 8-28
  web crippling of, 8-27, 8-28
  wood, 7-18, 7-48 to 7-51, 7-86
  (*See also* Compression)
Communication systems, 20-21
  (*See also* Alarms; Signal and alarm systems)
Composite beams (*see* Beams, composite)
Composite columns, 8-15, 8-59
Composite decking, 8-19, 8-44, 8-45
Compression:
  in cold-formed shapes, 9-18 to 9-29, 9-31 to 9-38
  stresses, 5-2, 5-13, 5-14, 5-16
  (*See also* Columns; Concrete columns; Studs)
Compressors, for cooling, 18-70 to 18-72, 18-74
Concrete:
  air in, 4-10, 4-13, 10-6, 10-10, 10-13, 10-14
  alkali-aggregate expansion of, 10-11
  batching ingredients of, 10-9
  bearing stresses on, 8-34, 10-39, 10-90 to 10-91, 10-125
  bonding of: to concrete, 10-31
    (*See also* joints in, *below*)
  to steel (*see* Steel reinforcement, development length of)
  calcium chloride in, 10-10, 10-11
  cement in, 10-1, 10-2, 10-6
    (*See also* Cement; water-cement ratios for, *below*)
  characteristics of, 4-10 to 4-14, 10-7, 10-9
  coatings for (*see* Coatings)
  coefficient of expansion of, 4-12
  coefficients of friction for, 10-117
  cold-weather precautions for, 10-32
  color variation of, 10-2
  compressive strength of: for dynamic loads, 5-88
    effects of age on, 4-11, 10-3
    effects of aggregates on, 4-8, 4-13, 4-14, 10-4
    effects of air entrainment on, 10-10
    effects of cement on, 4-11, 10-2 to 10-4
    effects of curing on, 4-11, 10-16, 10-17
    effects of water on, 4-11, 10-3, 10-4
    for mix design, 10-12, 10-13, 10-18
    relationship to modulus of elasticity, 10-8

Concrete, compressive strength of (*Cont.*):
  tests of, 4-11, 10-6, 10-7, 10-15 to 10-18
  for watertightness, 10-14
consistency (plasticity) of, 10-6
consolidation of, 10-28, 10-29, 10-120
crack control in, 10-33, 10-34, 10-55 to 10-57, 10-82
crack width in, 10-56
creep (plastic flow) of, 4-12, 10-40, 10-124
curing of, 4-11, 4-12, 10-16, 10-17, 10-21, 10-32, 10-33, 10-119, 12-66
defined, 4-10, 10-1
ductility development of, 5-116, 5-117
expanded, 4-13
for fire protection of steel, 8-108, 8-109
flexural members of, 10-76
  (*See also* Concrete beams; Concrete floors)
foamed, 4-13
formwork for (*see* Formwork)
framing (*see* Concrete beams; Concrete floors; Rigid frames, concrete; Shear walls, concrete)
heavyweight, 4-8, 4-14
hot-weather precautions for, 10-2, 10-11, 10-32
insulating, 4-8, 4-9, 4-13, 4-14, 15-7
joints in, 3-21, 3-22, 10-33, 10-34, 10-116, 10-121
laitance on, 10-29
lightweight, 4-8, 4-9, 4-13, 4-14, 8-109, 15-7, 15-20, 18-22
mix proportioning for, 10-3 to 10-5, 10-12 to 10-15
mixing of, 5-9, 5-10
modular ratio of, 10-40
modulus of elasticity of, 4-12, 10-7, 10-8
modulus of rupture of, 10-7 to 10-9, 10-127
nailable, 4-14
permeability of, 3-18
placement of: in horizontal forms, 10-30, 10-32, 10-34
  inspection of, 10-35
  methods for, 10-28 to 10-31
  precautions in, 10-28, 10-30 to 10-33
  with preplaced aggregates, 10-29
  underwater, 10-28, 10-29
  in vertical forms, 10-30, 10-34
as plaster base, 12-49
polymer, 4-68
precast, 3-26, 3-28, 10-85, 10-118 to 10-121, 12-28, 12-49, 13-14, 15-7
prestressed: allowable stresses for, 10-125
  behavior of, 10-122, 10-123
  defined, 10-122

Concrete, prestressed (*Cont.*):
  deflections of, **10**-58, **10**-60
  prestress losses in, **10**-122 to **10**-124
  prestress measurement for, **10**-130
  stresses in, **10**-122 to **10**-125
  tendons for (*see* Tendons)
  (*See also* Concrete beams, prestressed)
 pullout tests of, **10**-17
 pumped, **10**-9, **10**-28
 quality control of, **10**-9, **10**-12 to **10**-18
 reinforcement for (*see* Steel reinforcement)
 repair of, **4**-70
 setting of, **4**-11
 shrinkage of, **4**-12, **10**-32, **10**-124
 slump of, **4**-13, **10**-3, **10**-5, **10**-6, **10**-15
 slump tests for, **10**-6
 sprayed, **3**-22, **3**-27, **10**-10, **10**-28
 standards for, **10**-1
 strength-design principles for, **10**-38, **10**-39
  (*See also* Concrete beams; Concrete columns; Concrete floors)
 stress-design principles for (*see* Concrete beams)
 tensile strength of, **10**-7 to **10**-9
 tensile-strength tests of, **10**-7 to **10**-9, **10**-15, **10**-16
 torsion design of, **10**-39, **10**-45 to **10**-48, **10**-79 to **10**-81
 vibration of, **10**-29, **10**-120
 volume changes in, **4**-12
 water in, **10**-5, **10**-6, **10**-15, **10**-29
  (*See also* water-cement ratios for *below*)
 water-cement ratios for, **4**-10, **4**-11, **10**-3, **10**-4, **10**-13, **10**-14
  (*See also* cement in; water in, *above*)
 water-resistant, **3**-18, **3**-22, **3**-23, **3**-26 to **3**-29, **10**-11, **10**-14
 weight of, **4**-11, **4**-13, **4**-14, **5**-8, **5**-9
 workability of, **4**-10, **4**-13, **10**-3, **10**-5, **10**-6, **10**-10
 yield calculations for, **10**-4, **10**-5, **10**-15
 (*See also* Aggregates; Cement; Concrete admixtures; Steel reinforcement)
Concrete admixtures:
 accelerators, **10**-10, **10**-11
 air-entraining, **4**-13, **10**-10
 bonding agents, **10**-11
 cement-replacement, **10**-11
 coloring agents, **10**-11, **10**-12
 defined, **10**-1
 expansion agents, **10**-11
 fly ash, **10**-11
 lime, **4**-4, **4**-5, **10**-11
 natural cement, **4**-4, **10**-11
 pozzolans, **10**-11

Concrete admixtures (*Cont.*):
 retarders, **10**-11
 water-repellent, **3**-18, **3**-22, **3**-23, **10**-11
Concrete beams:
 concreting of, **10**-31
  (*See also* Concrete, placement of)
 continuous, **10**-35 to **10**-37, **10**-76 to **10**-79
 deep, **10**-40, **10**-109 to **10**-111
 deflections of, **6**-25, **10**-57 to **10**-59
  (*See also* Deflections)
 effective depth of, **10**-41
 effective flanges of, **10**-44
 fire ratings for, **10**-61
 minimum depth of, **10**-58
 plastic flow (creep) in, **4**-12, **10**-40
 precast, **10**-119
 prestressed: deflections of, **10**-58, **10**-60
  design procedure for, **10**-125, **10**-126
  reinforcement for, **10**-122, **10**-127, **10**-129
  (*See also* Tendons)
  strength design of, **10**-126, to **10**-129
  (*See also* Concrete, prestressed)
 rectangular: design of, **10**-41 to **10**-44
  (*See also* strength design of; stress design of, *below*)
 reinforcing for (*see* Steel reinforcement; Stirrups; Tendons)
 shear in: capacity reduction factor for, **10**-39
  maximum design, **10**-45, **10**-66, **10**-79, **10**-109, **10**-110, **10**-127 to **10**-129
  (*See also* Steel reinforcement, for shear; Stirrups)
 strength design of: assumptions for, **10**-38, **10**-39, **10**-76, **10**-77
  capacity reduction factors for, **10**-39
  design loads for, **10**-38, **10**-39
  load factors for, **10**-38, **10**-39
  reinforcement limitations for, **10**-42, **10**-43
  steps in, **10**-76 to **10**-82
  (*See also* rectangular, *above;* T section, *below*)
 stress design of: allowable stresses for, **10**-40
  assumptions for, **10**-39 to **10**-41
  (*See also* rectangular, *above*)
 T section: design example for, **10**-77 to **10**-79
  reinforcement distribution in, **10**-44, **10**-55, **10**-56
  strength design of, **10**-44, **10**-45
  (*See also* strength design, *above*)
 torsion in (*see* Concrete, torsion design of)
 transformed section for, **8**-42, **10**-40
  (*See also* Beams)

Concrete block:
  allowable stresses for, 12-7
  curing of, 4-14, 4-15
  defined, 12-2
  as floor form filler, 10-64
  moisture limits for, 4-15
  as plaster base, 12-49
  shrinkage prevention for, 3-27
  sizes of, 4-14
  specifications for, 3-22, 4-15
  strength of, 4-15
  weight of, 5-8
  (*See also* Masonry; Masonry walls)
Concrete columns:
  biaxial bending of, 10-106, 10-107
  concrete strengths for, 10-108
  concreting of, 10-30, 10-34
  design loads for, 10-39
  equivalent, 10-37
  fire ratings for, 10-61
  formwork for (*see* Formwork)
  precast, 10-118
  reinforcement for, 10-91, 10-99 to 10-106
  (*See also* Steel reinforcement)
  spiral-reinforced, 10-99, 10-100, 10-102, 10-104
  stiffness of, 10-70, 10-71
  strength design of: assumptions for, 10-99, 10-100
    capacity reduction factors for, 10-39, 10-99
    charts for, 10-101, 10-103, 10-104
    constraints on, 10-102, 10-103
    example of, 10-100, 10-101
    for slenderness, 10-107, 10-108
  tied, 10-99 to 10-106
Concrete floors:
  arch type, 8-17
  basement, 3-21, 3-25, 5-7
  Cofar, 8-20
  concrete topping for, 5-8, 10-31, 12-69, 12-70
  concreting of, 10-30, 10-31, 10-34, 10-116
  drying time for, 12-66
  epoxy toppings for, 4-70, 12-70
  flat-plate, 8-15, 8-20, 10-60, 10-61, 10-67 to 10-72
  (*See also* two-way-slab, *below*)
  flat-slab, 10-60, 10-61, 10-72 to 10-75
  (*See also* flat-plate, *above*; two-way-slab, *below*)
  formwork for (*see* Formwork)
  on ground, 3-19 to 3-21, 10-115, 10-116
  heavy-duty finishes for, 10-31, 12-70
  joints in (*see* Concrete, joints in)
  joist-slab, 10-76
  (*See also* one-way-joist; waffle, *below*)

Concrete floors (*Cont.*):
  lift-slab, 8-15, 8-16, 10-121
  over metal deck, 9-63, 9-64
  minimum reinforcing for, 10-55 to 10-57
  one-way-joist: advantages of, 8-17, 8-18, 10-62
    bridging for, 10-64, 10-65
    defined, 10-62
    deflections of, 10-64
    design of, 10-64 to 10-66
    dimensions of, 10-63
    filler forms for, 10-63, 10-64
    fire ratings for, 10-61, 10-64
    minimum depth of, 10-64
    reinforcement for, 10-55, 10-56, 10-64 to 10-66
    shear in, 10-45, 10-66, 10-67
    uniform depth construction, 10-62
  one-way-slab: defined, 10-60
    deflections of, 10-57 to 10-59
    design of, 10-60 to 10-62
    fire ratings for, 10-60, 10-61
    minimum depth of, 10-57, 10-58
    moments in, 10-35 to 10-37, 10-61
    reinforcing in, 9-6, 10-55 to 10-57, 10-61, 10-62
    shear in, 10-61
    types of, 10-60
    (*See also* Rigid frames, concrete)
  openings in, 10-65
    (*See also* Floors, openings in)
  pan-formed (*see* one-way-joist, *above*; waffle, *below*)
  pipes in, 10-62
  precast, 5-119, 5-121, 8-20
  prestressed, 10-58, 10-60
  reshoring of, 10-21
  ribbed (*see* one-way-joist, *above*; waffle, *below*)
  smooth-ceiling (*see* flat-plate, *above*)
  two-way-joist (*see* waffle, *below*)
  two-way-slab: analysis of, 10-37, 10-67 to 10-71, 10-74, 10-75
    on beams, 10-75
    column arrangements for, 10-67
    defined, 10-67
    deflections of, 10-58, 10-59
    design guide for, 10-75, 10-76
    fire ratings for, 10-61
    minimum depth of, 10-60
    reinforcing in, 10-56, 10-57, 10-71, 10-72, 10-74, 10-75
    shear in, 10-69, 10-70, 10-74, 10-75
    stiffness of, 10-70, 10-71
    (*See also* flat-plate; flat-slab, *above*)
  waffle, 10-57, 10-72 to 10-75
    (*See also* flat-plate; flat-slab; two-way-slab, *above*)
  weight of, 5-8

Concrete girders (*see* Beams; Concrete beams; Girders)
Concrete piles, **6**-39 to **6**-41
Concrete roofs, **10**-35 to **10**-37, **10**-119, **15**-6 to **15**-8
(*See also* Concrete floors)
Concrete shells, **10**-31, **10**-113 to **10**-115, **10**-119, **10**-120
(*See also* Shells)
Concrete slab, **10**-76
(*See also* Concrete floors)
Concrete stairs, **22**-9, **22**-10
Concrete walls:
 basement, **3**-21, **3**-22, **3**-25, **3**-26
 bearing, **10**-82 to **10**-84, **12**-2, **12**-4
 capacity reduction factor for, **10**-82
 coatings for (*see* Coatings; Membrane, waterproofing)
 concreting of, **10**-30
 defined, **10**-82
 deflection limitations for, **11**-14
 effective length of, **10**-82, **10**-83
 finishes for, **10**-120
 formwork for, **10**-19 to **10**-21
 joints in (*see* Concrete, joints in)
 nonbearing, **10**-82, **10**-84, **10**-85
 precast, **3**-26, **3**-28, **10**-85, **10**-120, **10**-121, **12**-28, **13**-14
 reinforcement for (*see* Steel reinforcement)
 retaining: cantilever, **10**-85, **10**-86
  counterfort, **10**-86, **10**-87
  earth pressures on, **6**-49 to **6**-52, **10**-85
  two-way-slab, **10**-87, **10**-88
 support for: lateral, **11**-14
  at openings, **1**-13, **12**-21, **12**-22
 thickness requirements for, **10**-83, **10**-84, **10**-110, **12**-15, **12**-16
 tilt-up, **10**-120, **10**-121
 (*See also* precast, *above*)
 water-resistant, **3**-26 to **3**-29
 (*See also* basement, *above*; Concrete, water-resistant)
(*See also* Masonry walls; Walls)
Condensation, **15**-8, **18**-3, **18**-11, **18**-14, **18**-20, **18**-27 to **18**-30
Condensers:
 air-conditioning, **18**-3, **18**-71 to **18**-74, **18**-91
 electric, **20**-6
Conductance:
 electric, **20**-3
 thermal, **13**-19, **15**-9, **18**-3, **18**-16 to **18**-18
Conduction, thermal, **18**-3, **18**-16 to **18**-18
Conductivity, thermal, **15**-9, **18**-3, **18**-16, **18**-22
Connections:
 at arch bases, **7**-62, **8**-2, **10**-112

Connections (*Cont.*):
 beam, **7**-51, **7**-52, **8**-84 to **8**-93, **9**-6
 (*See also* bolted; end-plate; riveted; welded; wind, *below*)
 bolted: bearing-type, **8**-39 to **8**-41, **8**-68 to **8**-71, **8**-77 to **8**-81, **9**-49 to **9**-51
  bracket-type, **8**-62, **8**-63
  framed, **8**-78, **8**-79, **8**-84, **8**-86
  friction-type, **8**-69 to **8**-71, **8**-81, **8**-82
  seated, **8**-84 to **8**-86
  specifications for, **8**-69, **9**-51
  tolerances for, **7**-25, **7**-26
  for wood members, **7**-15, **7**-32, **7**-37 to **7**-46, **7**-51, **7**-52, **7**-62, **7**-63
  (*See also* Bolts; splices, *below*)
 with combinations of fasteners, **8**-68, **8**-74, **8**-75
 corrosion protection of, **7**-32
 crown, **7**-62
 detailing of, **8**-77
 eccentrically loaded, **8**-79 to **8**-83
 electrical, **20**-13
 end-plate, **8**-87 to **8**-89
 fixity of, **8**-90 to **8**-93
 flange-web, **8**-45
 floor-wall, **8**-5 to **8**-7, **11**-19, **11**-20
 hinged, **7**-62, **8**-2, **8**-13, **8**-14, **10**-112
 moment (*see* rigid; semirigid; wind, *below*)
 one-sided, **8**-89
 prying action in, **8**-81
 rigid (moment), **3**-10, **7**-63, **8**-62, **8**-63, **8**-90, **8**-92, **8**-93
 (*See also* wind, *below*)
 riveted, **8**-67, **8**-68
 selection of fasteners for, **8**-67
 semirigid (moment), **8**-90, **8**-93
 (*See also* wind, *below*)
 simple, **8**-90, **8**-91, **8**-93
 (*See also* bolted, *above*; welded, *below*)
 skewed, **8**-89
 splices: of arches, **7**-61, **7**-63
  in beams, **8**-96
  in columns, **8**-64, **8**-93 to **8**-96
  finger-joint, **7**-26, **7**-27
  in frames, **7**-2, **8**-15
  glued, **7**-26, **7**-27
  in piles, **6**-41
  scarf, **7**-26
 (*See also* bolted, *above*; welded, *below*)
 web-plate, **8**-89, **8**-90
 welded: bracket-type, **8**-62, **8**-63
  framed, **8**-86, **8**-87
  seated, **8**-87
  stresses in, **8**-82, **8**-83, **8**-86 to **8**-88
  (*See also* Welding; Welds)
 wind, **3**-9, **3**-10, **5**-153, **8**-62, **8**-63
 (*See also* rigid; semirigid, *above*)

Connections (*Cont.*):
  wood: beam, 7-51, 7-52
    glued, 7-46, 7-47
    nailed, 7-32 to 7-34, 7-63, 7-78 to 7-81
    screwed, 7-34 to 7-37
    spiral-doweled, 7-37, 7-51
    Z type, 8-89, 8-90
Connectors:
  shear, 8-42 to 8-44, 10-117
  wood, 7-25, 7-26, 7-37, 7-39 to 7-45
Construction:
  accelerated, 1-8, 2-22, 2-23, 25-36, 25-37
  approvals needed for, 1-33, 1-35, 1-36, 25-7 to 25-9
  defined, 1-2
  dry-wall, 12-36, 12-40
    (*See also* Gypsumboard)
  expediting of, 25-35, 25-36
  methods of (*see specific systems, for example:* Floors; Framing; Roofs; Walls)
  procedures for, 1-5 to 1-8
  productivity control in, 25-37 to 25-39, 25-58
  progress wheel for, 25-12, 25-13
  safety practices in, 25-54, 25-55
  social and environmental concerns in, 25-56, 25-58, 25-59
  supervision of, 1-6, 1-7, 1-33, 2-14, 25-7, 25-40
  by systems building, 1-30, 2-20, 2-21, 9-66, 25-59 to 25-62
  (*See also* Bids; Bidders; Cost estimates; Contractors; Contracts)
Construction budget, 1-8, 1-9, 25-10
Construction costs:
  components of, 1-8
  for contract changes, 25-43 to 25-45
  control of, 24-2
  data published on, 24-20, 25-9
  effects of project duration on, 24-15 to 24-17
  indices for, 24-20
  life-cycle, 1-8, 2-10
  (*See also* Cost estimates)
Construction managers, 1-6, 1-8, 2-23, 2-24, 25-2, 25-11, 25-12, 25-37 to 25-40
  (*See also* Builders; Contractors; Owner-builder; Sponsor-builder)
Construction program managers, 1-6
Construction scheduling:
  with bar charts, 24-12, 25-32 to 25-34
  for cost estimates, 24-12, 24-15 to 24-17
  with critical path method, 25-34, 25-35
Contractors:
  accounting bases for, 25-51, 25-53, 25-54

Contractors (*Cont.*):
  basic principles for, 25-62
  bidding instructions for, 2-13, 26-5
  bonding of (*see* Bonds)
  building-code concerns of, 25-7 to 25-9
  claims for extra work by, 25-44 to 25-47
  community relations of, 25-55, 25-56
  consultants to, 25-4, 25-5
  environmental concerns of, 25-56, 25-58, 25-59
  fees for, 24-14, 25-9 to 25-11
  general: bid shopping by, 25-13
    cost estimates by (*see* Cost estimates)
    functions of, 1-2, 1-5 to 1-8
    management personnel of, 25-37, 25-40, 25-54
    new-business sources for, 25-5, 25-6
    organization of, 25-3, 25-4
    responsibility for work allocation, 25-58, 26-7, 26-8
    work done by, 25-2
  insurance for (*see* Insurance carriers; Insurance policies)
  joint-venture, 25-3
  labor relations of, 25-5, 25-57, 25-58
  management duties of, 25-1, 25-12, 25-13, 25-37 to 25-40
  in multiple-contract construction, 1-5, 2-21, 2-22
  owner payments to, 2-15, 25-47
  payments to subcontractors by, 25-47 to 25-50
  prime, 1-5, 2-21, 2-22
  relationships: with architects, 2-5, 2-12, 2-14 to 2-16
    with public agencies, 25-56, 25-57
  renovation, 25-1, 25-2
  reports of: accounting, 25-39
    approval, 25-35, 25-36
    computer-aided, 25-39
    cost, 25-37, 25-50 to 25-53
    daily, 25-40, 25-41
    equipment, 25-38, 25-39
    expediting/traffic, 25-38
    job meeting, 25-37, 25-38
    OSHA, 25-55
    payroll, 25-51, 25-52
    purchasing, 25-18, 25-20, 25-37
    trade payment breakdown, 25-47, 25-48
    work certification, 25-43, 25-44
  responsibility of: for formwork, 10-18
    for safety, 25-54, 25-55
  social concerns of, 25-58, 25-59
  sponsor-builder, 25-2, 25-3, 25-10
  turnkey, 1-6, 2-25, 25-2
  (*See also* Builders; Construction managers; Owner-builder; Subcontractors)

Contracts:
    construction: arbitration clause in, 25-45, 25-46
        award of, 2-12, 2-13
        changed-conditions clause in, 25-40, 25-42, 25-43
        changes in, 2-15, 25-43 to 25-45, 26-6, 26-13
        cost-plus-fee, 25-11
        documents in, 1-5, 2-12, 2-13, 25-6 to 25-8, 26-1, 26-7
        general (single), 1-6, 2-21, 26-11, 26-12
        general conditions in, 2-13, 26-1, 26-6, 26-7
        guarantee clause in, 26-2, 26-3
        hold-harmless clauses in, 25-28
        interpretation of, 2-14
        letter of intent, 1-6, 25-10
        lump-sum, 25-10
        management, 25-11, 25-12
        mediation clause in, 25-47
        multiple, 1-5, 2-21, 2-22, 25-10, 26-11 to 26-13
        negotiated, 2-12, 2-13
        preparation of, 1-4, 26-1, 26-7
        special (supplementary) conditions in, 26-1, 26-7
        unit-price, 25-11, 26-6
        work covered in, 26-12
    design-build, 2-24, 2-25, 25-11
    sale lease-back, 25-10
    sponsor-builder, 25-2, 25-3, 25-10
    structural steel, 8-96
    subcontractor, 25-7, 25-13 to 25-18
    turnkey, 1-6, 2-25, 25-2
    (*See also* Drawings; Specifications)
Control center, 3-42, 3-47, 3-51, 19-49
Convection, 18-3, 18-16, 18-18, 18-19
Convectors, 18-18, 18-19, 18-63
Conversion:
    boiler hp to Btu/h, 18-3
    gage to inches, 9-13
    grains to pounds, 15-8, 18-30
    inches to feet, 23-2
    to metric (SI) units, A-1 to A-11
    of temperature scales, 18-6, 18-8, 18-10
    tons to Btu/h, 18-7, 18-15
    watts to Btu/h, 18-41
Conveyors:
    belts for, 4-73
    vertical, 22-1, 22-37, 22-38
Cooling:
    chemical moisture absorption with, 18-90, 18-91
    defined, 18-3
    design temperatures for, 18-20, 18-21, 18-40
    evaporative, 18-3, 18-11, 18-13
    latent, 18-40

Cooling (*Cont.*):
    by roof sprays, 18-41
    sensible, 18-3, 18-40
    with solar heat, 18-66
    by ventilation, 18-32, 18-33, 18-79
    (*See also* Air conditioners; Air conditioning; HVAC)
Cooling effect, total, 18-3
Cooling unit (*see* Air conditioners)
Coping, 12-2
Copper:
    alloys of, 4-52 to 4-58
    bibliography for, 4-58
    deoxidized, 4-52, 4-53
    electrolytic tough pitch, 4-52, 4-53
    lead-coated, 4-52
    properties of, 4-52, 4-53
    water stops of, 3-17, 3-22, 3-26
    weight of, 5-6, 5-9
Corbeling, 12-2, 12-9, 12-18
Coreboard, 12-39, 12-53
Cork, 5-6, 5-9, 12-63 to 12-67, 15-9
Cornerite, 12-39, 12-46, 12-48
Cornice, 12-37 to 12-40
Corridors, 3-43, 3-46
    (*See also* Ramps)
Corrosion:
    chemical, 4-50
    by dissimilar metals, 4-50, 4-51, 15-14
    electrolytic, 4-16, 4-50, 4-51
    of ferrous metals, 4-46, 4-47, 10-11, 8-4, 8-103, 8-104
    protection against, 4-47, 7-32, 8-104, 15-14, 19-2
Corrugated sheets:
    asbestos-cement, 15-19, 15-20
    metal, 9-39 to 9-41, 9-59, 9-60, 12-27, 12-28, 15-15
    vinyl, 15-20
Cost estimates:
    accuracy variation in, 24-1 to 24-5
    allowances in, 24-10, 24-11
    basic information for, 24-5
    of capital equipment, 24-8, 24-9
    checking of, 24-17
    comparative, 1-21 to 1-25, 24-19
    (*See also* Value analysis)
    with computers, 24-14, 24-15
    construction-schedule effects on, 24-15 to 24-17
    contingency evaluations for, 24-3 to 24-5, 24-13, 24-14
    of contractor's fees, 24-14, 25-9
    cost sources for, 24-8, 24-20, 25-9
    criteria sheet for, 24-18, 24-19
    definitive, 24-3 to 24-5, 24-11
    by designers, 1-31, 1-32, 2-10, 2-12, 24-2 to 24-5
    detail needed for, 24-11, 25-9
    door-knob, 24-3, 24-4

Cost estimates (*Cont.*):
electrical, 24-7
elements of, 24-18 to 24-20
end uses of, 24-3
engineer's, 24-3 to 24-5, 24-7
escalation allowances in, 24-11 to 24-13
final, 24-4, 24-5, 24-11, 25-9
hidden costs for, 25-9
from incomplete documents, 24-3, 24-5, 24-10, 24-11, 25-9
of labor, 24-8, 24-12, 24-13, 24-15
magnitude, 24-3, 24-4, 24-11
monitoring, 24-2
of overhead, 24-14
of piping, 24-7, 24-14 to 24-16
planning-stage, 1-31, 1-32, 2-10, 2-12, 24-2, 24-3, 25-10
preliminary, 24-3 to 24-5, 24-7
preparation manpower for, 24-5, 24-6
preparation records for, 24-8, 24-11, 24-19, 24-20
probable error in, 24-4
in project management, 24-17
quantity takeoffs for, 24-6 to 24-9
scope statements in, 24-19
specialists for, 24-1
subcontractor prices for, 25-9
summary format for, 24-6, 24-18, 25-9
types of, 24-1 to 24-5
unit prices in, 24-9, 24-10
(*See also* Construction costs)
Costs (*see* Construction costs; Cost estimates)
Coulomb damping, 5-104, 5-106
Coulomb electrical unit, 20-1
Coulomb pressures, 6-49, 6-52
Counterforts, 10-86, 10-87
Cove, 12-39
Crane girders, 8-51
Cranes:
in buildings, 8-11, 8-13
construction, 7-86, 8-97
Creosote, 4-33, 7-28 to 7-31
Cricket, chimney, 15-28
Cripple, 1-13
Critical path method, 25-34, 25-35
Crown, 5-116, 5-118, 8-13, 10-112
CRSI (*see* Associations, Concrete Reinforcing Steel Institute)
Culverts, 6-52
Cupronickel, 4-59
Current (*see* Electric currents)
Cutbacks, 15-16

Dado, 12-39, 13-5
Dalton's law, 18-13
Dampers, 18-4, 18-35, 18-36; 18-75, 18-76
Darby, 12-39

Decibel, 17-6
Decking (*see* Floors; Roofs)
Deflections:
cable, 5-145
of concrete beams and slabs, 6-25, 10-57 to 10-60, 10-64
conjugate-beam method for, 5-32
diagrams (elastic curves) for, 5-30, 5-31, 5-34 to 5-36
dummy unit-load method for, 5-53 to 5-56
of elastic beams, 5-30 to 5-36, 5-53 to 5-57, 5-78 to 5-82
of horizontal diaphragms, 11-13, 11-14
influence coefficients for, 5-78, 5-79
influence lines for, 5-56, 5-57
limits on: for concrete beams and slabs, 9-64, 10-57 to 10-60
for door frames, 14-8
for open-web joists, 9-5
for roofs, 7-56, 8-51, 9-7
for steel beams, 8-50, 8-51, 8-54
for walls, 11-14
for wood beams, 7-56
modal static, 5-100
moment-area method for, 5-31, 5-32
of plywood panels, 7-82
reciprocal relationships for, 5-33
shear, 5-33
shear-wall, 11-12 to 11-16
of steel beams, 6-25
of steel joists, 9-5
support, 5-66 to 5-69, 5-76, 6-25, 12-20
(*See also* Buildings, drift of; Foundations, settlement of; Rigid frames, sideway of)
of trusses, 5-54 to 5-56
of wood beams, 7-56
of wood decking, 7-68
Deformation, 5-2, 5-14
(*See also* Deflections; Strain)
Degree, thermal, 18-6 to 18-8, 18-10
Degree day, 18-4
Degree of freedom (vibration), 5-90
Dehumidification, 18-4
Dentils, 12-39
Derricks, 8-97
Developers, real estate, 25-2
Deviation:
beam tangential, 5-31
standard, 10-12, 10-13
Dew point, 18-6, 18-13 to 18-15, 18-28, 18-29
Diabase, 4-20
Diaphragm action, 3-16, 5-112, 7-78, 7-79, 9-64, 11-9, 11-10, 11-13, 11-14
Diorite, 4-20
Dishwashers, 19-15, 19-19
Displacements:
matrix representation of, 5-162

Displacements (*Cont.*):
  resistance to (*see* Beams, restraint of;
    Flexibility; Modulus of elasticity;
    Resistance, to displacement;
    Spring constant; Stiffness)
  vector representation of, 5-161, 5-162
  in vibration (*see* Vibration)
  virtual, 5-50 to 5-52
  (*See also* Deflections; Deformation)
Domes, 5-121, 5-127, 7-4
  (*See also* Shells)
Doors:
  air pressure on, 14-2, 14-3
  automatic operators for, 14-3
  batten, 14-6
  bevel of, 16-10, 16-13, 16-14
  bucks for, 14-6, 14-8
  capacity of, 3-46
  closing devices for, 16-8 to 16-10
  components needed for, 14-1
  concrete, 14-15
  corridor, 14-5
  elevator (*see* Elevators)
  exit, 3-44
  exterior, 14-1 to 14-3, 14-5
  fire, 3-44, 14-4, 14-15, 16-12, 16-13, 22-7
  flush, 14-7, 14-8
  folding, 14-7, 14-10, 14-13
  frames for, 1-13, 1-17, 14-6 to 14-8
  glass, 4-22, 14-6, 14-7
  hand (swing) of, 14-10, 16-13, 16-14
  hardware for, 1-17, 14-4, 14-9, 16-4 to 16-14
  horizontally hinged, 14-12, 14-14, 14-15
  horizontally sliding, 14-3, 14-10, 14-12, 14-13
  interior, 14-1, 14-2
  Kalamein, 14-7
  lead, 14-15
  locks for (*see* Locks)
  materials for, 14-1, 14-6, 14-7
  metal, 14-6 to 14-8, 14-15
  paneled, 14-6
  purposes of, 1-16, 1-17, 14-1
  radiation-shielding, 14-15
  revolving, 14-3, 14-11
  safety requirements for, 14-3 to 14-5
  size limitations for, 14-4
  smokestop, 14-4, 14-5
  to stairs, 14-5
  swinging, 14-3, 14-8 to 14-10, 14-12, 14-14, 16-5, 16-13, 16-14
  telescoping, 14-12 to 14-14
  types of, 14-1, 14-12
  vertically sliding, 14-10 to 14-12, 14-14
  water exclusion at, 14-2, 14-6
  wind loads on, 14-2
  wood, 14-6, 14-7

Dope, 12-39
Dormers, 1-12
Dowels, 7-37, 7-51, 10-50, 10-51, 10-86, 10-91
Drainage:
  at grade, 3-19
  roof, 3-19, 7-53, 7-56, 15-7, 15-18, 19-29 to 19-31
  subgrade, 3-19, 3-20, 6-53, 6-54, 7-66
Drawings:
  contract, 1-2, 1-4, 1-5, 2-12, 2-13, 26-1
  shop, 2-14, 2-15, 25-35, 25-36
Drinking fountains, 19-12, 19-13, 19-19
Drip caps, 13-4
Drips, 19-37, 19-38
Drop:
  electric, 20-11, 20-15
  for stairs, 22-6, 22-9
Ductility:
  defined, 3-16, 5-155
  under dynamic loads, 3-16, 5-88
Ductwork:
  air flow in, 18-36, 18-38
  circle-rectangle equivalent, 18-36, 18-37
  defined, 18-4
  design of, 18-36 to 18-38, 18-47 to 18-50, 18-78 to 18-80
  fire protection of, 18-32
  (*See also* Air conditioning; Heating, warm-air; Ventilation)
Dummy unit-load method, 5-53 to 5-56
Dumbwaiters, 22-1, 22-37
Dynamic load factors, 5-98 to 5-101, 5-105, 5-106
Dynamic magnification factor, 5-106
Dynamics, 5-88 to 5-115

Earth (*see* Soils)
Earth pressure:
  active, 5-7, 6-49 to 6-53
  on culverts, 6-52
  passive, 6-49, 6-51, 6-54, 6-55, 7-84
  on poles, 7-84
  at-rest, 6-49, 6-52
  surcharge, 5-7, 6-51
  wedge theory of, 6-49 to 6-51
Earthquakes:
  characteristics of, 3-13 to 3-15
  design for (*see* Cantilever method; Portal method; Seismic design)
  design loads for, 3-2, 3-3, 3-15, 5-3, 5-11, 5-108 to 5-114
  intensity scales for, 3-14, 3-15
  intensity zones for, 5-108, 5-109
  seismograms of, 3-14
Eccentricity, 11-2, 11-3
Echoes, 17-14 to 17-16
Efficiency, 18-4

Efflorescence, 12-40
Egg and dart, 12-40
Egress, emergency, 3-43 to 3-47, 22-1 to 22-3, 22-7, 22-12
Elastic limit, 5-13
Elastic theory:
  for arches, 5-115 to 5-120
  for bars, 5-13 to 5-17
  for beams, 5-27 to 5-42, 5-57 to 5-76
  for cables, 5-143 to 5-153
  for columns, 5-82 to 5-84
  for finite elements, 5-161 to 5-173
  for folded plates, 5-132 to 5-140
  general, 5-50 to 5-57
  for grid framing, 5-76 to 5-82
  for rigid frames, 5-57 to 5-76
  for trusses, 5-54 to 5-56
Elasticity, modulus of, 5-13, 5-88
  shearing modulus of, 5-15
  (*See also* Structural steels, modulus of elasticity of; Wood, modulus of elasticity of)
Elastomers (*see* Plastics; Rubber)
Electric admittance, 20-6
Electric apparatus:
  batteries, 20-1, 20-9
  capacitors (condensers), 20-6
  generators, 20-1, 20-2
  meters, 20-15, 20-16, 20-44
  motor control centers, 20-20
  motor-generator sets, 20-9
  motors (*see* Motors, electric)
  panelboards, 20-19, 20-20, 20-44
  panels, 20-11, 20-17 to 20-20
  rectifiers, 20-9
  substations, 20-19
  switchboards (switchgear), 20-17 to 20-19, 20-44
  switches (*see* Switches)
  transformers, 20-4, 20-15
Electric circuits (*see* Electric wiring)
Electric conductance, 20-3
Electric currents:
  alternating: calculation of, 20-8, 20-9, 20-27 to 20-31, 20-46
    cycles of, 20-5
    defined, 20-1
    effective, 20-5
    frequency of, 20-5
    phase of, 20-5
    production of, 20-2, 20-4, 20-5
  conversion of ac to dc, 20-9
  direct: calculation of, 20-2, 20-3
    defined, 20-1
    flow direction of, 20-1
    production of, 20-1, 20-2, 20-4, 20-5
  hydraulic analogy of, 20-1
  short-circuit, 20-17
Electric energy, 20-2
Electric impedance, 20-6, 20-7

Electric lighting (*see* Lamps; Lighting; Luminaires)
Electric outlets, 20-22, 20-27
Electric power:
  in ac circuits, 20-8, 20-10
  in dc circuits, 20-2
  demand factors for, 20-10, 20-25, 20-26
  distribution of, 1-19, 20-2, 20-11, 20-13 to 20-15, 20-21, 20-22, 20-26, 20-27
  (*See also* Electric wiring)
  emergency, 3-47, 20-4
  loads on, 20-10, 20-25, 20-26
  power factor for, 20-8
Electric reactance, 20-5, 20-6
Electric resistance, 20-2, 20-3, 20-28, 20-29
Electric susceptance, 20-5, 20-6
Electric voltages (*see* Voltage)
Electric wiring:
  ampacity of, 20-11, 20-29 to 20-31
  areas of, 20-28, 20-35
  bus bar, 20-12
  circuits: analysis of, 20-2 to 20-4, 20-8, 20-27 to 20-31
    branch, 20-11, 20-17, 20-26, 20-27, 20-31
    for communication, 20-21
    multiphase, 20-9, 20-10, 20-27, 20-29
    network, 20-4
    parallel, 20-3, 20-4, 20-8
    protection of, 20-17, 20-18, 20-20, 20-21, 20-31, 20-36 to 20-44
    series, 20-3, 20-4, 20-8
  color coding for, 20-11
  components, of, 20-10
  conduit for, 20-11 to 20-13, 20-31 to 20-35, 20-40 to 20-43
  connection of, 20-13
  design example for, 20-44 to 20-47
  feeder, 20-11, 20-29, 20-31, 20-44, 20-46
  grounding, 20-11
  insulation for, 20-10 to 20-12
  maximum voltage drops for, 20-29, 20-31
  neutral (ground), 20-9 to 20-11
  open, 20-12
  power losses in, 20-2, 20-10, 20-15
  raceways for, 20-11, 20-13, 20-14
  regulations for, 20-11, 20-21
  service conductor, 20-11, 20-12
  sizes of: determination of, 20-31
    minimum permissible, 20-17, 20-21, 20-27, 20-40 to 20-43, 20-46
    standard, 20-28, 20-35
  types of, 20-12
  underfloor, 1-14, 1-16, 8-19, 9-61 to 9-63, 20-13, 20-14
  underground, 20-12
Electrical plans, 20-25, 20-26

Electrical symbols, 20-22 to 20-25
Elevators:
   allowable stresses for, 22-14
   automatic, 22-16, 22-27, 22-28
   cars for: capacities of, 22-22, 22-32 to
      22-34, 22-36
     clearances for, 22-20
     communication with, 22-22
     components of, 22-15, 22-20 to 22-22
     doors for, 22-21, 22-22, 22-31, 22-37
     double-deck, 22-32
     hospital, 22-34
     position indicators for, 22-17
     sizes of, 22-22, 22-23
     (*See also* freight; passenger, *below*)
   definitions of terms for, 22-13, 22-15 to
     22-17
   dispatching devices for, 22-15
   electric traction: applications of, 22-14
     components of, 1-19, 1-20, 22-21,
       22-23 to 22-25, 22-35
     controls for, 22-15 to 22-17, 22-23,
       22-24, 22-35, 22-36
     counterweights for, 22-21, 22-24,
       22-25, 22-35
     driving machines for, 22-16, 22-23
     roping for, 22-17, 22-25, 22-26
   emergency use of, 3-47
   fire protection of, 3-33, 22-18, 22-19
   freight: capacity of, 22-36
     classification of, 22-34, 22-35
     defined, 22-13
     in material-handling systems, 22-37
     operation of, 22-36, 22-37
     selection of, 22-34
     speeds of, 22-35, 22-36
     truck-zoning devices for, 22-17
     (*See also* geared-drive; hydraulic,
       below)
   geared-drive, 22-16, 22-23, 22-35, 22-36
   guide rails for, 22-19 to 22-21, 22-29,
     22-35
   hoistways for: doors of, 22-19, 22-31,
     22-37
     enclosures, of, 22-18, 22-20
     venting of, 22-18
   hospital: cars for (*see* cars, *above*)
     defined, 22-13
     planning for, 22-34
   hydraulic: advantages of, 22-29
     applications of, 22-14
     capacities of, 22-29
     components of, 1-19, 1-20, 22-28 to
       22-30
     driving machines for 22-16, 22-28
     operation of, 22-28
     safety bulkheads for, 22-17
     speeds of, 22-29
   installation of, 22-14
   intervals for, 22-31 to 22-33

Elevators (*Cont.*):
   landing zones for, 22-19, 22-21
   leveling devices for, 22-16, 22-24, 22-35
     to 22-37
   machine rooms for, 22-14, 22-18, 22-19
   operation systems for, 22-15 to 22-17,
     22-19, 22-26 to 22-28
   parking devices for, 22-17, 22-19
   passenger: for apartment buildings,
     22-33
     arrangements of, 22-32, 22-33
     defined, 22-13
     for department stores, 22-33, 22-34
     for general-purpose buildings, 22-31,
       22-32
     for government buildings, 22-33
     for hotels, 22-33
     number required, 22-29, 22-31
     for professional buildings, 22-33
     for single-purpose buildings, 22-32,
       22-33
     speeds of, 22-32, 22-33
     zoning of, 22-28
   pits for, 22-17
   planning for, 22-13, 22-14, 22-29, 22-31
     to 22-34
   round-trip time for, 22-31, 22-36
   runby of, 22-17
   safety devices for, 22-15 to 22-17, 22-19
     to 22-22, 22-24, 22-28
   service, 22-32 to 22-34
   signaling for, 22-17, 22-22, 22-26,
     22-27
   standards for, 22-14, 22-18, 22-21,
     22-22, 22-25, 22-34
   travel (rise) of, 22-17
   traveling cables for, 22-17, 22-21
   waiting-passenger indicator for, 22-17,
     22-22
   winding-drum, 22-23
Emissivity, 18-4
Emulsions, 15-16
Enamel, 4-70, 4-71, 4-78
Endurance limit, 5-88, 5-89
Energy, 5-15 to 5-17, 5-51 to 5-53
Energy budget levels, 2-9, 2-10
Energy conservation, 2-9, 2-10, 25-59
Engesser equation, 5-84
Engineers:
   collaboration with architects, 1-4, 2-17,
     2-18
   insurance for, 25-23
   profession of, 1-2, 1-4, 2-17
   resident, 1-6
   responsibility for formwork, 10-18,
     10-19
   selection of, 2-17
Enthalpy, 18-4, 18-11, 18-12
Entropy, 18-4, 18-9, 18-10
Environment, 18-4

Environmental impact, 2-8 to 2-10, 25-56, 25-59
Epoxies, 4-62, 4-63, 4-69, 4-70, 4-76, 4-78
Equilibrium:
 dynamic, 5-90, 5-91
 laws of, 5-11, 5-12, 5-21
Equilibrium (funicular) polygons, 5-43 to 5-46
Escalators:
 advantages of, 22-9
 arrangements of, 22-12, 22-13
 capacity of, 22-9, 22-12
 components of, 1-20, 22-10 to 22-12
 dimensional limitations for, 22-11
 exit, 3-44, 22-1
 fire protection of, 3-33, 22-12, 22-13
 installation of, 22-12
 planning for, 22-12
 safety devices for, 22-12
 slope of, 22-12
 speeds of, 22-12
 standards for, 22-10
Estimates (*see* Cost estimates)
Euler load, 5-84
Evaporators, 18-4
Excavations, 6-52 to 6-55, 23-6
Exits (*see* Doors; Egress, emergency; Stairs)
Expansion, coefficient of, 5-15
Extrados, 5-118, 5-121, 10-112

Factory Mutual Systems, 3-30, 12-31, 14-6, 26-15
Fans, 18-34, 18-35, 18-42
Fast track (accelerated building), 1-8, 2-22, 2-23, 25-36, 25-37
Fat, 12-40
Fatigue, 5-88, 5-89, 8-36, 8-37
Faucets, 19-5, 19-9
Feather-edge tool, 12-40
Fees (*see* Architects; Contractors)
Felts, roofing (*see* Membrane, waterproofing; Roofing, felts for)
Fiberboard, 15-9
Fills:
 earth, 6-16 to 6-18
 floor, 5-9
Filters, air, 18-36
Finisher, 12-40
Finite-element methods:
 classification of, 5-161
 displacement (stiffness), 5-161, 5-164 to 5-166
 elements in: axially loaded bar, 5-166
  beam, 5-167, 5-168
  connections between, 5-161
  triangular, 5-169 to 5-173
  types of, 5-161
 force (flexibility), 5-161 to 5-164

Fire:
 building classifications for (*see* Buildings)
 classes of, 3-35
 severity of, 3-32
Fire detectors, 3-39 to 3-42
Fire divisions, 3-33, 3-34, 14-5
Fire escapes, 22-7
Fire extinguishers:
 carbon dioxide, 3-37, 3-38
 dry chemical, 3-38
 dry powder, 3-38, 3-39
 foam, 3-36, 3-37
 Halon, 3-38
 portable, 3-48
 (*See also* Fire, classes of; Hoses; Sprinklers; Standpipes)
Fire loads, 3-31, 3-32
Fire protection:
 with chemicals (*see* Fire extinguishers)
 in construction, 3-47, 3-48, 25-54
 of conveyors, 22-37
 for elevators, 3-33, 22-18, 22-19
 for escalators, 3-33, 22-12, 22-13
 methods for, 3-29, 7-30 to 7-32, 8-108 to 8-111
 (*See also* Doors; Ramps; Stairs; Ventilation)
 of openings, 3-33, 8-110 to 8-112, 22-7
 (*See also* Doors, for elevators; for escalators, *above*; for stairs, *below*; Windows)
 planning for, 3-29, 7-30 to 7-32
 smoke control in, 3-33, 3-41, 18-31
 with sprinklers (*see* Sprinklers)
 for stairs, 3-33, 3-43, 3-44, 22-7
 standards for, 3-29 to 3-31
 with standpipes, 3-36, 3-47, 19-48, 19-49
 of steel decking, 8-20, 8-110, 8-111
 for steel framing, 1-12, 1-14 to 1-16, 8-107 to 8-112
 for wood framing, 1-12, 1-16, 4-33, 7-30 to 7-32
 (*See also* Fire; Fire ratings; Framing, fire protection for)
Fire ratings:
 for concrete members, 10-60, 10-61, 10-64
 for doors, 14-5
 required, 3-32, 14-5
 for steel floor deck, 9-65, 9-66
 for steel framing, 8-108, 8-112
 for steel joists, 9-7, 9-8, 9-60
 for steel roof deck, 9-59, 9-60
 tests for, 1-16, 3-32
Firebrick, 12-5, 12-23
Fireplaces, 12-23
Firestops, 3-33, 7-63, 7-64

Flashing, 3-27, 4-52, 8-106, 12-12, 12-13, 15-27, 15-28
Flat-plate construction (*see* Concrete floors)
Flat-slab construction (*see* Concrete floors)
Flexibility, 5-162, 5-166
   (*See also* Matrix, flexibility)
Flexure formula, 5-27, 5-38, 5-39
Float tools, 12-40
Floodproofing, 3-16, 3-17
Floor coverings:
   asphalt block, 5-8
   asphalt tiles, 5-8, 12-63, 12-65 to 12-67
   carpets, 12-66 to 12-68, 17-12, 17-18 to 17-20
   ceramic tiles, 4-17, 5-8, 12-31 to 12-36
   concrete toppings, 5-8, 10-31, 12-70
   conductive, 12-70, 12-71
   cork, 5-6, 5-9, 12-63 to 12-67
   installation of, 12-65 to 12-67
      (*See also* Concrete floors; Wood floors; Tile, ceramic)
   linoleum, 5-8, 12-65 to 12-67
   mastic, 4-75, 12-70
   purposes of, 1-16
   resistance to liquids of, 12-70
   rubber, 12-65 to 12-67
   selection of, 12-63
   specifications for, 12-71
   terrazzo, 5-8, 12-68, 12-69
   underlayments for, 7-79, 12-66, 12-68
   vinyl, 12-64 to 12-67
   wood, 4-31, 5-8, 12-67, 12-69
Floors:
   area per occupant, 1-34, 3-46
   basement, 3-21, 3-25, 5-7
   components of, 1-9, 1-11, 1-14, 1-16
   concrete (*see* Concrete floors; Floor coverings, concrete toppings)
   connections to walls, 11-19, 11-20
   defined, 1-9, 12-1
   diaphragm action of, 3-16, 5-112, 5-113, 9-64, 11-9, 11-10, 11-13, 11-14
   electrified, 1-14, 1-16, 8-19, 9-61 to 9-63
   fills in, 5-9
   framing of (*see* Framing)
   industrial, 10-31, 12-70
   loads on, 8-16
      (*See also* Loads)
   metal deck, 8-19 to 8-21, 8-44, 8-45, 9-61 to 9-66
   minimum thicknesses of, (*see* Deflections, limits on)
   open-web joists in, 8-18, 9-2, 9-4 to 9-6
   openings in, 1-10, 1-19, 3-33, 9-4, 10-65, 22-3, 22-7
   panel-heating, 18-58 to 18-63
   sound control for (*see* Sound, control of)

Floors (*Cont.*):
   wood (*see* Wood floors)
   (*See also* Floor coverings; Framing; Subfloors)
Flues, 18-50, 18-57
Fluoroplastics, 4-63, 4-65, 4-68, 4-80
Flutter, 17-14
Fly ash, 10-11
Foam, 3-36, 3-37
Folded plates:
   analysis of, 5-133 to 5-140
   concrete, 10-114, 10-115
   defined, 5-132
Footcandle, 21-3, 21-22
Footings:
   bearing on, 10-90, 10-91, 10-99
   bracing of, 3-16
   capacity reduction factors for, 10-90 to 10-92
   effective depth of, 10-92
   loads on, 6-33, 10-85, 10-86, 10-88, 10-90
   minimum thickness of, 10-91
   pile, 1-10, 6-37 to 6-41, 10-89, 10-97
      (*See also* Piles)
   purposes of, 1-9
   safety factors for, 10-89
   shaft (*see* Shafts, foundation)
   spread: allowable bearing capacity of, 6-31, 6-32
      (*See also* Foundations, bearing capacity of; safety factors for)
   cantilever (*see* strapped, *below*)
   column, 1-10, 6-30, 10-88 to 10-93
   combined, 6-36, 10-88, 10-89, 10-93 to 10-95
   design of, 6-33 to 6-36
      (*See also* Foundations, design steps for)
   mat, 1-10, 6-36, 6-37, 10-89, 10-96
   pedestal, 10-89
   pressures on, 6-33 to 6-35
   raft (*see* mat, *above*)
   settlement of, 6-21, 6-25 to 6-30, 6-32, 6-33
   sloped, 10-89
   stepped, 10-89
   strapped, 6-37, 10-89, 10-95, 10-96
   wall (strip), 1-10, 10-88 to 10-92
   types of, 1-10
   (*See also* Foundations)
Footlambert, 21-4
Force-displacement relationships, 5-161 to 5-166
Force polygons, 5-43 to 5-45
Forces:
   addition of, 5-43
   concurrent, 5-42, 5-43
   coplanar, 5-42 to 5-44
   equilibrant of, 5-43

Forces (*Cont.*):
  equilibrium (funicular) polygon for, 5-43 to 5-46
  matrix representation of, 5-161
  parallelogram of, 5-42, 5-43
  resolution of, 5-43
  resultant of, 5-43
  vector representation of, 5-42, 5-43, 5-161
  (*See also* Loads)
Foremen, construction, 1-6
Formaldehydes, 4-62 to 4-64, 4-80
Formwork:
  for arches, 10-113
  bracing for, 10-20, 10-21
  for concrete columns, 10-19, 10-108
  for floors, 8-17 to 8-20, 8-44, 9-6, 10-19 to 10-21, 10-63, 10-64, 10-73, 10-74
  for folded plates, 10-115
  gypsumboard, 12-53
  inspection of, 10-21, 10-22
  metal-lath, 9-6, 12-47
  plywood, 7-69
  for precast concrete, 10-120
  pressures on, 10-19 to 10-21
  removal of, 10-18, 10-21
  responsibility for, 10-18, 10-19
  sheet-steel, 8-17 to 8-20, 8-44, 9-6
  specifications for, 10-18, 10-19
  for uncommon structures, 10-21
Foundations:
  bearing capacity of, 6-22 to 6-25
  bibliography for, 6-57
  deep, 6-22 to 6-25
  design steps for, 6-21, 6-22
  drainage of, 3-19, 3-20, 6-53, 6-54, 7-66
  drilled-pier (*see* Caissons, drilled-in; Shafts, foundation)
  earth pressures on (*see* Earth pressure)
  failures of, 6-21, 6-32, 6-56, 6-57
  floating (*see* Footings, spread, mat)
  pressures under, 6-25 to 6-27
  purposes of, 1-9, 1-10, 6-1, 10-88
  safety factors for, 6-21
  settlement of, 6-21, 6-25 to 6-30, 6-32, 6-33, 6-56, 6-57
  shallow, 6-22, 6-23
  in unexpected subsoils, 25-40, 25-42, 25-43
  vibration control at, 5-115
  wall (*see* Concrete walls, basement; Masonry walls, basement; Walls, foundation)
  (*See also* Caissons; Footings; Piles; Shafts)
Frames:
  aseismic design of, 5-108 to 5-114
  box-system, 5-111
  braced, 8-54, 8-62 to 8-64, 10-108

Frames (*Cont.*):
  combination, 7-63
  continuous (*see* Beams, continuous; Rigid frames)
  door (*see* Doors)
  force-displacement relationships for, 5-161 to 5-166
  moment-resisting (*see* portal; space, *below;* Rigid frames)
  portal, 8-62
  rigid (*see* Rigid frames; space, *below*)
  space, 5-110 to 5-112, 8-11
  splices in (*see* Connections, splices)
  sway, 8-66
  unbraced, 8-55
  window (*see* Windows)
  (*See also* Framing)
Framing:
  balloon, 7-63, 7-64
  beam-and-column, 8-7 to 8-9
  braced-timber (eastern), 7-63
  bracing of (*see* wind bracing, *below;* Bracing)
  combination steel-concrete, 8-15, 8-16
    (*See also* Beams, composite; Columns, composite)
  concrete (*see* Concrete beams; Concrete columns; Concrete floors; Concrete walls)
  connections in (*see* Connections; Joints)
  erection layout of, 23-9 to 23-11
  fire protection for, 1-12, 1-14, 1-16, 4-33, 7-30 to 7-32, 8-107 to 8-112
    (*See also* Fire ratings; Steel reinforcement, protective cover for; Wood, preservatives for)
  grid, 5-76 to 5-82, 8-11
  lift-slab, 8-15, 8-16, 10-121
  open-web-joist, 1-10, 1-15, 1-16, 8-18, 9-2, 9-4 to 9-6
  P-Δ effect on, 8-34
  platform (western), 7-63, 7-64
  seismic bracing of (*see* Shear walls; wind bracing, *below*)
  skeleton, 1-10, 1-11, 8-7 to 8-16
    (*See also* beam-and-column, *above;* Frames; Rigid frames)
  steel: elastic-theory design of, 8-21
    for floors, 8-16, 8-17
    with intermediate beams, 8-17, 8-20
    long-span, 8-8 to 8-15
    plastic-theory design of, 8-21, 8-54 to 8-58
    (*See also* Cold-formed shapes; Columns, structural steel; Girders, steel; Structural steels)
  tube-type, 3-11, 3-12, 8-64
  wall-bearing, 1-11, 8-5 to 8-7
    (*See also* Connections, floor-wall)
  wind bracing, 3-9 to 3-13, 8-61 to 8-67

Framing, wind bracing (*Cont.*):
 (*See also* Connections; Rigid frames;
  Shear walls)
 wood: combination, 7-63
  economical designs for, 7-2 to 7-4
  fire protection of, 1-16, 4-33, 7-30 to
   7-32
  for houses, 1-13, 1-14, 7-61, 7-63,
   7-64
  (*See also* Floors; Frames; Loads; Rigid
   frames; Roofs)
Frequency:
 of alternating currents, 20-5
 natural, 5-91, 5-148
 natural circular, 5-91 to 5-95, 5-107
 of sound waves: defined, 17-2
  human response to, 17-4, 17-5
  in intensity measurements, 17-7
  relation of, to wave length, 17-3
Friction, 6-3, 6-4
Frieze, 12-26
Funicular polygons, 5-43 to 5-46
Furan mortar, 12-35, 12-70
Furring, 12-41, 12-54
Fuses, 20-17, 20-18, 20-36 to
 20-44

g, 5-90, A-6
Gage:
 of fasteners, 8-22
 inch equivalence of, 9-13
Gaging, 12-41
Galvanizing, 4-47, 15-14
Gas:
 heating, 19-36, 19-37
 ideal, 18-10, 18-13
 meters for, 19-36
 piping for, 19-36 to 19-38
Gaskets, glazing, 4-75, 4-77, 13-21 to
 13-23
Gasoline, 5-5
Gin pole, 8-97
Girders:
 composite (*see* Beams, composite)
 concrete (*see* Beams; Concrete beams;
  Girders)
 defined, 1-10, 8-45, 10-76
 in grid framing, 5-76 to 5-82
 ring, 5-122, 5-129
 steel: allowable stresses for (*see*
  Structural steels, allowable
  stresses for)
  cover plates for, 8-45, 8-46
  crane, 8-51
  deflections of, 6-25
   (*See also* Deflections)
  design of, 8-45 to 8-49, 8-59, 8-60
  flange limitations on, 8-42, 8-44,
   8-46, 8-47, 8-50, 8-55

Girders, steel (*Cont.*):
 hybrid, 8-9, 8-29, 8-46, 8-47
 lateral supports of (*see* Bracing;
  Structural steels, allowable
  stresses for)
 shear capacity of, 8-23, 8-47 to 8-49,
  8-56
 splices in (*see* Connections, splices)
 stiffeners for (*see* Stiffeners)
 thickness limits for, 8-32, 8-33, 8-46,
  8-55
 types of, 8-8, 8-9
 (*See also* Beams; Framing)
Girts, 1-10, 7-63, 8-21
Glass:
 bullet-resisting, 13-16
 cellular (foamed), 5-9, 15-9, 18-22,
  18-24
 characteristics of, 4-22, 4-23, 13-15 to
  13-19
 clear window, 13-15
 coated, 13-16
 corrugated, 13-16
 in double glazing, 13-1, 13-2, 13-18 to
  13-23, 18-24
 float, 13-15, 13-18, 13-23
 grades of, 13-15
 heat-absorbing, 13-1, 13-2, 13-15,
  13-16, 13-21
 installation of, 13-16 to 13-23
 laminated, 4-22, 13-16 to 13-19
 maximum areas of, 13-15, 13-17, 13-18,
  13-23
 obscure, 13-15
 one-way-vision (transparent mirror),
  13-16
 plate, 13-15, 13-18, 13-23
 processed, 13-15
 rolled figured sheet, 13-15
 sound reduction with, 13-19
 suspended, 13-16, 13-17
 tempered, 13-16 to 13-18
 thermal conductance of, 13-19, 18-24
 thickness classes of, 13-15, 13-18,
  13-19, 13-23
 tinted, 13-16
 weight of, 5-9, 13-19
 wired, 13-15
 (*See also* Doors; Glass block; Glass
  fiber; Walls; Windows)
Glass block, 4-23, 5-8, 12-5, 12-20, 12-21
Glass fiber:
 as insulation, 5-9, 15-9, 18-22, 18-24
 reinforcing, 4-67
 in roofing, 15-4, 15-5
 as sound absorbent, 17-19
Glazing, 1-15, 13-15 to 13-19
 (*See also* Glass; Plastics)
Glazing compounds, 13-19, 13-20
Glazing methods, 13-16 to 13-23

Glues, 4-33, 4-34
  (*See also* Adhesives)
Gold, 5-6
Grains, 15-8
Granite, 4-20, 4-21, 5-9, 12-7
  (*See also* Masonry; Masonry walls)
Graphic statics, 5-42 to 5-50
Gravel, 4-8, 5-5, 5-6, 6-1, 6-6, 6-49
  (*See also* Aggregates)
Grilles (registers), 18-3, 18-4, 18-47,
  18-48, 18-78, 18-79
Grounds, plaster, 12-41, 12-50
Grout:
  asphalt, 4-75
  epoxy, 4-70
  portland cement, 3-20, 3-28, 12-2,
    12-11 to 12-14, 12-35
GR-S rubber, 4-71
Guard (balustrade), 22-2 to 22-5, 22-7,
  22-11
Gunite (*see* Concrete, sprayed)
Gutters, 19-29 to 19-31
Gypsum:
  applications of, 4-2, 12-36, 12-37
  block, 4-22, 5-8, 8-109, 12-5
  calcined (plaster of paris), 4-6
  defined, 12-41
  formboard, 12-53
  lath, 4-22, 12-41, 12-45, 12-46
  plank, 4-22, 15-7
  plaster (*see* Plaster)
  roofs of, 4-22, 15-7
  sheathing, 4-22, 5-9, 7-63, 7-64, 12-42,
    12-53
  specific heat of, 18-8
  tile, 4-22, 12-5, 12-48
  weight of, 5-5
  (*See also* Gypsumboard)
Gypsum base, 12-41, 12-44, 12-46
Gypsumboard:
  absorption of, 12-37
  advantages of, 12-51
  back blocking for, 12-37, 12-54
  backing boards for, 12-37, 12-53
  binder in, 12-38
  bleeding of, 12-38
  blow on, 12-38
  bubbles in, 12-38
  caging for 12-38
  cockles in, 12-39
  composition of, 4-21, 12-51, 12-52
  core of, 12-39, 12-51
  coreboard, 12-39, 12-53
  corner treatment of, 12-60
  decoration of, 12-59, 12-60
  defined, 12-41
  definitions of terms for, 12-37 to 12-43
  dimpling of, 12-39, 12-54
  edge treatments for, 12-40, 12-52
  for fire protection of steel, 8-109

Gypsumboard (*Cont.*):
  floating angles (corner floating) of,
    12-40, 12-62
  foil-back, 12-40, 12-52
  framing for, 12-52 to 12-58
  gluing of, 12-37, 12-54 to 12-58
  green board, 12-41
  installation of, 12-37 to 12-42, 12-52 to
    12-62
  joints of: compound for, 12-38, 12-39,
    12-59, 12-60
    control, 12-53, 12-60 to
      12-62
    isolation, 12-60 to 12-62
    mating of, 12-52, 12-54
    ridging (beading) of, 12-42
    treatment of, 12-37, 12-39 to 12-43,
      12-59, 12-60
    types of, 12-41
  laminations of: with adhesives, 12-37,
    12-55, 12-58, 12-59
    ship lap, 12-42
    types of, 12-41
  multi-ply construction with, 12-53,
    12-57 to 12-59
  (*See also* laminations of, *above*)
  nailing of, 12-39 to 12-41, 12-54 to
    12-58, 16-16
  paper facings for, 12-41, 12-42
  perimeter relief for, 12-42
  punch out in, 12-42
  ripper, 12-42
  screws for, 12-54, 12-56, 12-58
  single-ply construction with, 12-53,
    12-55 to 12-57
  sizes of, 12-52
  sizing on, 12-42
  soffit board, 12-43
  stapling of, 12-54, 12-55, 12-58
  type X, 12-41, 12-53
  types of, 12-51 to 12-53
  wallboard, 4-21, 4-22, 12-43, 12-52,
    12-55 to 12-59
  weight of, 5-9
  (*See also* Gypsum; Gypsum base)

Half timbers, 12-26
Halon, 3-38
Handrails, 22-3, 22-4, 22-6 to 22-9, 22-11,
  22-22
  (*See also* Railing)
Hangers, 7-45
  (*See also* Tension members)
Hankinson formula, 7-41
Hardboard, 4-35
Hardpan, 6-31
Hardware:
  consultants for, 16-2, 16-3
  defined, 1-17, 16-1

Hardware, (*Cont.*):
 finishing: defined, **16**-1
  door closers, **16**-8 to **16**-10
  for doors, **1**-17, **14**-4, **14**-9, **16**-4 to **16**-14
  finishes for, **16**-3, **16**-8
  fire exit, **16**-11, **16**-12
  hinges, **16**-4 to **16**-8, **16**-13, **16**-14
  locks (*see* Latches; Locking bolts; Locks)
  nontemplate, **16**-3
  standards for, **16**-3, **16**-4
  template, **16**-3
  window operators, **16**-14
  for windows, **1**-16, **16**-14, **16**-15
 legal requirements for, **16**-2
 rough: anchors, **16**-15, **16**-16, **16**-21
  bolts (*see* Bolts)
  defined, **16**-1
  nails (*see* Nails)
  powder-driven studs, **16**-20
  screws (*see* Screws)
  welded studs, **4**-46, **8**-19, **8**-20, **8**-44, **15**-19, **16**-18 to **16**-20
  for windows, **1**-16, **16**-14, **16**-15
 selection of, **16**-2, **16**-3, **16**-6 to **16**-8, **16**-11, **16**-13
Harmonic motion, **5**-91 to **5**-95, **5**-99
Haunches, **5**-118, **10**-112
Hazen-Williams formula, **19**-18
Headers:
 door, **14**-6 to **14**-8
 masonry, **12**-2, **12**-9
 structural, **1**-10, **1**-13, **1**-19, **7**-64, **9**-4, **22**-5, **22**-8, **22**-9
 window, **1**-13, **1**-15
Heat:
 from air conditioners, **18**-42
 balancing of, **18**-15
 gains of (*see* Loads, cooling)
 latent, **18**-4, **18**-10 to **18**-12, **18**-40, **18**-42
 from lighting, **18**-41, **21**-21
 losses of (*see* Loads, heating)
 mechanical equivalent of, **18**-9
 from people, **18**-41, **18**-42
 sensible, **18**-4, **18**-9, **18**-40, **18**-41
 specific, **18**-4, **18**-8, **18**-9
 from sun, **13**-2, **18**-6, **18**-19, **18**-41, **18**-64 to **18**-66
 total, **18**-4
Heat capacity, **18**-4, **18**-8
Heat flow, **18**-15 to **18**-22
Heat transmission:
 by air, **18**-33, **18**-40, **18**-47
 calculation of, **18**-16 to **18**-18, **18**-24 to **18**-26, **18**-72
 cause of, **18**-7
 coefficient *U* for, **18**-5, **18**-17, **18**-18, **18**-24 to **18**-26

Heat transmission (*Cont.*):
 by conduction, **18**-3, **18**-16 to **18**-18
 by convection, **18**-3, **18**-16, **18**-18, **18**-19
 direction of, **18**-16
 film (surface) coefficient for, **18**-4, **18**-17
 by radiation, **18**-6, **18**-16, **18**-19
 resistance to, **15**-9, **18**-6, **18**-17, **18**-18, **18**-25
 (*See also* Insulation)
 by water, **18**-52, **18**-88
Heating:
 baseboard, **18**-3, **18**-63
 by condensation, **18**-13
 by convectors, **18**-18, **18**-19
 defined, **18**-5
 design temperatures for, **18**-20, **18**-21, **18**-43, **18**-44
 district, **18**-5
 hot-water, **18**-5, **18**-51 to **18**-54
  high-temperature, **18**-54
 loads for, **18**-15, **18**-37 to **18**-40
 panel (radiant), **18**-5, **18**-58 to **18**-63
  snow melting, **18**-61 to **18**-63
 solar: active, **18**-19, **18**-64 to **18**-66
  passive, **13**-2, **18**-19
 steam, **18**-54 to **18**-57
 vacuum, **18**-56, **18**-57
 warm-air: control for, **18**-49, **18**-50
  defined, **18**-5
  duct design for, **18**-47 to **18**-50
  (*See also* Ductwork)
  function of, **18**-46, **18**-47
  gravity, **18**-47
  grilles for, **18**-3, **18**-4, **18**-47, **18**-48
  humidification for, **18**-49
  perimeter, **18**-50, **18**-58
  supply air for, **18**-47
 (*See also* Air conditioning; Heating plants; HVAC)
Heating plants:
 automatic, **18**-5
 boilers, **18**-3, **18**-50, **18**-51, **18**-54
 central, **18**-3
 combustion air for, **18**-51
 direct-fired, **18**-5, **18**-50, **18**-51, **18**-57, **18**-58
 flues for, **18**-50, **18**-57
 heat-pump, **18**-4, **18**-63, **18**-64
 operating costs of, **18**-22
 operation of, **18**-49, **18**-50, **18**-54
 sizing of, **18**-20, **18**-21, **18**-43 to **18**-46
 unit-heater, **18**-5, **18**-7, **18**-57, **18**-58
 warm-air furnace, **18**-4, **18**-50, **18**-51
 (*See also* Heating; Water heaters)
Hertz, **17**-2
Hinges:
 arch, **7**-62, **8**-2, **10**-112
 door, **16**-4 to **16**-8, **16**-13, **16**-14
 plastic structural, **5**-157 to **5**-159, **8**-56
 window, **16**-14

Home builders, 1-6, 2-24, 25-2
Hoods, canopy, 18-32
Hooke's law, 5-13
Hoses, 4-73, 19-5, 19-19, 19-48
Housing (*see* Construction)
Humidifiers, 18-49
Humidity, 18-5, 18-11 to 18-15, 18-20
Hurricanes, 3-1, 3-6
HVAC:
   air-water systems for, 18-85, 18-86
   all-air systems for, 18-81 to 18-85
   definitions of terms for, 18-2 to 18-7
   design criteria for, 18-1, 18-2, 18-15, 18-16, 18-19 to 18-22
   design procedure for, 18-2
   objectives of, 1-18, 18-1, 18-7
   (*See also* Air conditioning; Heating; Ventilation)
Hydrants, 19-6, 19-48
Hygrometer, 18-5
Hygrostat, 18-5
Hypalon, 4-68, 15-17

Ice, 5-5, 18-8
Illumination (*See* Lighting)
Impact, 5-2, 5-3, 5-95 to 5-98
Impedance:
   acoustical, 17-3, 17-4
   electrical, 20-6, 20-7
Inconel, 4-59, 4-60
Inertia:
   moment of, 5-27 to 5-29
   product of, 5-29
Influence lines, 5-25, 5-26, 5-56, 5-57
Inspection (*see* Construction, supervision of)
Institutes (*see* Associations)
Insulation:
   air, 18-23
   electrical, 20-10 to 20-12
   thermal: blown-in, 18-24, 18-28, 18-29
      conductivity of, 15-9, 18-22, 18-24
      defined, 15-9, 18-5, 18-22
      economics of, 18-26, 18-27
      effects on heat flow of, 18-26
      failure of, 18-27, 18-28
      under floors, 3-20
      installation of, 1-14, 15-8 to 15-12, 18-23, 18-24
      materials used for, 15-9, 18-22 to 18-24
      pipe, 18-24, 19-2
      purposes of, 1-13, 18-22
      reflective, 18-22
      resistance of, 15-9
      roof, 9-54, 9-61, 15-8 to 15-13, 18-24
      wall, 18-24 to 18-26
      weight of, 5-9
      (*See also* Condensation; Heat flow)

Insulators, vibration, 4-73
Insurance carriers, 25-4, 25-5, 25-21
Insurance policies:
   agents and brokers for, 25-4, 25-5, 25-21
   for architects and engineers, 25-23
   boilers and machinery, 25-26
   bridge builders risk form, 25-25
   builders machinery and equipment, 25-22, 25-25
   builders risk, 25-22 to 25-24
   Contractor's Equipment Floater, 25-22, 25-25
   coverage limits for, 25-22, 25-23, 25-30
   defined, 25-20
   dishonesty, disappearance, and destruction, 25-22, 25-29
   employee group, 25-22, 25-29, 25-30
   extended coverage, 25-22, 25-24
   fire, 25-22 to 25-24
   Inland Marine, 25-25
   Installation Floater, 25-25
   interests protected by 25-21
   materials coverage, 25-22, 25-25
   motor-truck cargo-owners form, 25-25
   motor-vehicle, 25-22, 25-25, 25-26
   products liability, 25-28
   public liability, 25-22, 25-23, 25-26 to 25-28
   rigging, 25-25
   transportation floater, 25-25
   vandalism, 25-22, 25-24
   workmen's compensation, 25-22, 25-28, 25-29
Intrados, 5-118, 5-121, 10-112
Invar, 4-41
Iron:
   cast, 4-36, 5-5
   ingot, 4-36
   malleable, 4-36
   weight of, 5-5, 5-9
   wrought, 4-36, 4-45
Iron-carbon diagram, 4-35, 4-36
Ironworkers, 8-1

Jamb liner, 13-5
Jambs:
   door, 14-6 to 14-8
   window, 13-3 to 13-5
Japan, 4-78
Joint ventures, 25-3
Joints:
   in ceramic tile, 12-35, 12-36
   in concrete, 3-21, 3-22, 10-33, 10-34, 10-116, 10-118, 10-121
   in gypsumboard (*see* Gypsumboard)
   in masonry walls (*see* Masonry walls)
   in panel walls, 12-27 to 12-30
   in pipe (*see* Pipe, joints in)

Joints (*Cont.*):
  in plaster construction, 12-48, 12-60 to 12-62
  sealants for, 4-75 to 4-77
  (*See also* Connections)
Joists:
  concrete (*see* Concrete beams; Concrete floors)
  defined, 1-10
  junior beam, 8-18, 8-19
  open-web: allowable stresses for, 9-3
    applications of, 8-11, 8-18, 9-1, 9-2
    composite, 9-3
    defined, 1-10, 9-1
    deflections of, 9-5
    fabrication of, 9-1 to 9-4
    fire ratings for, 9-7, 9-8, 9-60
    in floors, 8-18, 9-2, 9-4 to 9-6
    painting of, 9-4
    in roofs, 1-15, 9-6, 9-7
    sizing of, 9-4, 9-5
    specifications for, 9-1, 9-3, 9-4
    supports for, 9-2, 9-6
    types of, 9-3
  tail, 9-4
  wood, 1-13, 5-8, 7-50, 7-59, 7-61, 7-63 to 7-65
Joule's constant, 18-9

Kern, 10-90
Keys, 16-12, 16-13
Kilowatthours, 20-2
Kip, 5-12
Kirchhoff's laws, 20-2, 20-3, 20-8

Labor:
  bargaining with, 25-5, 25-57
  costs of, 24-8, 24-12, 24-13, 24-15, 25-40, 25-41, 25-50, 25-51, 25-57
  jurisdictional disputes of, 25-57
  unions of, 1-6, 25-57
Lacquer, 4-78
Ladders, 22-2
Laitance, 10-29
Lamps:
  color performance of, 21-14, 21-16, 21-17, 21-20
  control of, 21-15, 21-24
  costs for, 21-16, 21-17, 21-19, 21-20
  current frequency for, 21-14
  fluorescent, 21-16, 21-17, 21-19, 21-20
  high-intensity discharge (*see* mercury; metal-halide; sodium, *below*)
  incandescent, 21-15 to 21-19
  life of, 21-14, 21-16, 21-17, 21-19
  light control by, 21-16, 21-17, 21-19

Lamps (*Cont.*):
  lumen depreciation of, 21-14
  lumen output of, 21-15 to 21-17, 21-20, 21-22
  luminous efficacy of, 21-3, 21-13, 21-14, 21-16, 21-17
  maintenance of, 21-14 to 21-17
  mercury, 21-16, 21-17, 21-20
  metal-halide, 21-16, 21-17, 21-20
  selection of, 21-13, 21-14, 21-24
  sodium, 21-16, 21-17, 21-20
  temperature effects on, 21-14
  voltages for, 21-14
Lanterns, in shells, 5-122
Latches, 16-10
Lath:
  back plastering of, 12-37
  defined, 12-41
  fastening of, 12-45 to 12-48
  gypsum, 4-22, 12-41, 12-44 to 12-46, 12-49
  metal, 12-27, 12-33, 12-46 to 12-48
Laundry tubs, 19-5, 19-15
Lavatories:
  fixture units for, 19-15
  hot water for, 19-22
  minimum pipe sizes for, 19-19
  number required, 19-5, 19-12, 19-13
Lead, 4-58, 4-59, 5-5, 13-19
  alloys of, 4-56, 4-58, 4-59
Leaders, rain, 19-29 to 19-31
Letter of intent, 1-6, 25-10
Lift-slabs, 8-15, 8-16, 10-121
Lighting:
  bibliography for, 21-25
  color rendering of, 21-5, 21-8, 21-9
  design of, 21-22 to 21-25
  diffusion of, 21-6, 21-18
  effects: brightness, 21-4 to 21-9
    contrast, 21-2, 21-4 to 21-6
    of other systems on, 21-1, 21-7, 21-12, 21-13, 21-24
    on other systems, 21-1, 21-9, 21-24
    on people, 21-1, 21-4, 21-5, 21-8, 21-9
  electric: heat from, 18-41, 21-21
    lamps for (*see* Lamps)
    methods of, 21-7, 21-10, 21-12, 21-13
    power requirements for, 20-10, 20-25, 20-26
  for elevators, 22-22
  equivalent spherical illumination of, 21-3, 21-4
  fixtures for (*see* Luminaires)
  glare, 21-3, 21-4, 21-6 to 21-8, 21-13
  for good visual performance, 21-1, 21-4 to 21-10
  illuminance of: calculation of, 21-3
    defined, 21-3
    recommended, 3-45, 21-9 to 21-11, 21-24
  inverse square law for, 21-2

Lighting (*Cont.*):
  lumen method of calculation for, **21**-22, **21**-23
  luminance of, **21**-4
  natural, **1**-18, **13**-1, **21**-12, **21**-13, **21**-24
  objectives of, **1**-18, **21**-1
  perception of, **21**-2, **21**-4 to **21**-10
  quality of, **21**-5 to **21**-8
  reflectance for, **21**-4
  standards for, **21**-1, **21**-2
  temperature scale for, **21**-5, **21**-16
  transmittance for, **21**-4
  veiling reflection of, **21**-7, **21**-8
Lightning protection, **3**-48 to **3**-50, **20**-20, **20**-21
Lime, **4**-1, **4**-2, **4**-4 to **4**-6, **5**-5, **5**-6, **12**-5, **12**-6, **12**-8, **12**-41
  (*See also* Putty, lime; Quicklime)
Limestone, **4**-19 to **4**-21, **5**-6, **5**-9, **12**-7, **18**-8
  (*See also* Masonry; Masonry walls)
Linings (*see* Coatings)
Linoleum, **5**-8, **12**-65 to **12**-67
Lintels, **1**-10, **1**-13, **1**-15, **8**-5 to **8**-8, **8**-102, **8**-103, **12**-13, **12**-21, **12**-22
Load factors:
  dynamic, **5**-98 to **5**-101, **5**-105, **5**-106
  static, **3**-2, **8**-21, **8**-54, **10**-39
Loads:
  axial, **5**-2
  axial and bending, **5**-33, **5**-37
  concentrated, **5**-2
  cooling, **18**-15, **18**-37, **18**-40 to **18**-42
  critical buckling, **5**-83, **5**-84
  dead, **5**-2, **5**-3, **5**-8, **5**-9
  defined, **5**-2
  design (service), **1**-11, **5**-3 to **5**-11, **5**-108 to **5**-114
  dynamic, **5**-2, **5**-88, **5**-89, **5**-95 to **5**-108
  (*See also* earthquake, *below*)
  earth-pressure (*see* Earth pressure)
  earthquake, **3**-2, **3**-3, **3**-15, **5**-3, **5**-11, **5**-108 to **5**-114
  eccentric, **5**-2, **5**-33, **5**-37
  electrical (*see* Electric power)
  fire, **3**-31, **3**-32
  heating, **18**-15, **18**-37 to **18**-40
  hydrostatic, **5**-7, **6**-50
  ice, **5**-7
  impact, **5**-2, **5**-3, **5**-95 to **5**-98
  lateral (*see* Bracing)
  live, **5**-2 to **5**-7, **5**-10, **5**-11
  moving, **5**-25, **5**-26
  rain, **5**-7
  recurrence intervals for, **5**-3
  repeated, **5**-2, **5**-11, **5**-88, **5**-89
  seismic, **3**-2, **3**-3, **3**-15, **5**-3, **5**-11, **5**-108 to **5**-114
  sinusoidal, **5**-105, **5**-106
  snow, **3**-2, **5**-3, **5**-7, **5**-10, **5**-11

Loads (*Cont.*):
  static, **5**-2, **5**-88
  for strength design (*see* Load factors)
  sudden, **5**-96, **5**-98, **5**-108
  torsional, **5**-2, **5**-86
  ultimate, **5**-156 to **5**-160
  uniformly distributed, **5**-2
  wind, **3**-2 to **3**-8, **5**-3, **5**-5, **5**-7, **5**-50, **13**-17 to **13**-19, **13**-23
  (*See also* Forces)
Loam, **6**-49
Locking bolts:
  beveled, **16**-10
  cremorne, **16**-14
  dead, **16**-10
Locks, **16**-10 to **16**-13
Louvers, **18**-5, **18**-29
Lumber:
  boards, **4**-30, **7**-5
  common, **4**-31
  connections of (*see* Connections, bolted; splices; wood)
  dimension, **4**-30, **7**-5
  dressed, **4**-30
  dressed sizes of, **7**-3, **7**-5 to **7**-7
  dry, **7**-5
  factory and shop, **4**-30, **4**-31
  glued-laminated: adhesives for, **4**-33, **4**-34, **7**-3, **7**-14, **7**-15, **7**-20, **7**-27, **7**-28
    advantages of, **7**-14, **7**-15
    allowable stresses for (*see* Wood, allowable unit stresses for)
    beam camber of, **7**-56, **7**-57
    beam capacity of, **7**-51, **7**-53 to **7**-55
    beam types in, **7**-51, **7**-53
    curving of, **7**-22
    decking of, **7**-66
    fabrication of, **7**-26 to **7**-28
    sectional properties of, **7**-3, **7**-10 to **7**-13
    sizes of, **7**-3, **7**-10 to **7**-13, **7**-26
  grades of, **4**-29 to **4**-31
  green, **7**-5, **7**-8
  hardwood, **4**-31
  matched, **4**-30
  nailed-laminated, **7**-66
  patterned, **4**-30
  rough, **4**-30
  select, **4**-30
  shiplapped, **4**-30
  sizes of, **4**-30
  softwood, **4**-29 to **4**-31
  structural, **4**-30, **4**-31
  weight per foot of, **7**-6, **7**-7
  worked, **4**-30
  yard, **4**-30, **4**-32
  (*See also* Framing; Timber; Wood)
Lumen, **21**-3
Luminaires:

Luminaires (*Cont.*):
candlepower distribution curves for, 21-3, 21-7, 21-8
ceiling-mounted, 21-18, 21-22
components of, 21-3, 21-10
fixtures in, 21-20 to 21-22
layout of, 21-22, 21-23
light control by, 21-7, 21-8, 21-12, 21-22
maximum luminances for, 21-7
pendent, 21-18, 21-21, 21-22
recessed, 21-13, 21-21, 21-22
selection of, 21-24
visual comfort probability for, 21-6
wall-mounted, 21-19, 21-22

Magnesium alloys, 5-6
Mail chutes, 22-38
Management (*see* Construction managers)
Maps, topographical, 23-2 to 23-5
Marble, 4-18 to 4-21, 5-9, 18-8
imitation, 12-41, 12-42
(*See also* Masonry; Masonry walls)
Marezzo, 12-41
Masonry:
allowable stresses for, 8-34, 11-3 to 11-5, 12-6, 12-7
anchors for (*see* Anchors, masonry)
area of, 12-2
ashlar, 5-9, 12-2, 12-7, 12-10
brick (*see* Brick)
concrete unit (*see* Concrete, precast; Concrete block)
courses of, 12-2, 12-9
defined, 4-1, 12-1, 12-3
hollow unit, 12-3, 12-5
mortar for (*see* Mortar)
permeability of, 3-18
as plaster base, 12-48, 12-49
quality requirements for, 12-4
rubble, 12-3, 12-4, 12-6, 12-7, 12-10, 12-15
solid unit, 12-4, 12-5
stretcher, 12-4, 12-9
unit, 1-13, 12-1, 12-2
(*See also* Masonry walls)
weights of, 5-8, 5-9
(*See also* Brick; Concrete block; Masonry walls; Stone, building)
Masonry cement (*see* Cement, masonry)
Masonry walls:
basement, 3-22, 3-23
bearing, 11-1, 11-2, 11-7, 11-19, 12-2, 12-4, 12-15
(*See also* Brick, engineered construction with)
bonding: of courses, 12-6, 12-9
of facing and backing, 12-9 to 12-12

Masonry walls, bonding (*Cont.*):
of intersecting walls, 12-10, 12-11
building of, 12-6, 12-8 to 12-12
chases in, 12-2, 12-22
coatings for (*see* Coatings; Membrane, waterproofing)
corbeled, 12-2, 12-9, 12-18
defined, 12-4
deflection limitations on, 11-13, 11-14
dimensional changes in, 12-18
eccentricity coefficients for, 11-6, 11-7
eccentricity ratio of, 11-6
eccentricity-thickness ratio of, 11-6
grouted, 12-2, 12-5, 12-11 to 12-14, 12-16
height of, 11-2, 11-6, 11-18, 12-2, 12-14, 12-15
joints in, 3-22, 3-27, 12-6 to 12-8, 12-11 to 12-14, 12-18 to 12-21
load distribution on, 11-19, 11-20, 11-22, 11-26, 11-29, 11-32, 12-22
leakproofing of, 3-26, 3-27, 12-13, 12-14
mortar for (*see* Mortar)
nonbearing, 12-2, 12-17, 12-18
recesses in, 12-22
repointing of, 12-13
slenderness coefficients for, 11-6 to 11-8
slenderness ratio of, 11-2, 11-6
solid, 11-3, 12-9, 12-10
standard code requirements for, 12-2
(*See also* Brick, engineered construction with)
support for: lateral, 11-2, 12-14 to 12-18
at openings, 12-21, 12-22
vertical, 12-20, 12-22
surface treatments for (*see* Coatings)
thickness of, 11-2, 11-7, 12-14, 12-15
thickness requirements for, 12-15 to 12-18
(*See also* Brick, engineered construction with)
tuck pointing of, 12-13
water-resistant, 3-22, 3-23, 3-26 to 3-29
(*See also* basement, *above*)
weights of, 5-8, 5-9
winter construction of, 12-8, 12-9
(*See also* Brick; Masonry; Shear walls; Stone, building; Walls)
Mastics, 3-28, 4-75
Matrix:
flexibility: for bars, 5-166
for beams, 5-167, 5-168
computation of, 5-162, 5-163, 5-166 to 5-170
defined, 5-162, 5-166
for rotated axes, 5-170
for triangular elements, 5-169, 5-170
inverse of, 5-161

Matrix (*Cont.*):
  notation for, **5**-161
  stiffness: for bars, **5**-166
    for beams, **5**-167, **5**-168
    computation of, **5**-162, **5**-164 to **5**-168, **5**-170 to **5**-173
    defined, **5**-162, **5**-166
    for rotated axes, **5**-173
    for triangular elements, **5**-171 to **5**-173
  transpose of, **5**-163, **5**-165
Maxwell diagram, **5**-48 to **5**-50
Mediation, **25**-46, **25**-47
Membrane, waterproofing, **3**-20, **3**-23 to **3**-26, **4**-74, **4**-75, **5**-9, **7**-66
  (*See also* Roofing, felt)
Membrane construction (*see* Air-stabilized structures; Shells)
Membrane theory, **5**-121 to **5**-128
Mercalli scale, **3**-14, **3**-15
Metals:
  Admiralty, **4**-54, **4**-55, **4**-57
  expanded, **12**-40, **12**-46
  ferrous, **4**-35 to **4**-47
  Muntz, **4**-53, **4**-56
  (*See also specific metals, for example:* Iron; Steel)
Metric system, **A**-1 to **A**-11
Micron, **18**-5
Milinch, **18**-52
Modal method, **5**-99, **5**-100
Models, **1**-30 to **1**-32
Modular ratio ($n = E_s/E_c$), **10**-40
Modular sizes, **13**-2, **13**-3, **13**-6, **A**-6
Modulating, **18**-5, **18**-36
Modulus of elasticity, **5**-13, **5**-88
  shearing, **5**-15
  (*See also* Structural steels, modulus of elasticity; Wood, modulus of elasticity)
Modulus of resilience, **5**-16, **5**-95
Modulus of rigidity, **5**-15
Modulus of rupture, **5**-30
Mohr's circle, **5**-19, **5**-20
Molding:
  band, **12**-37
  base, **12**-37
  beaded, **12**-37
  ogee, **12**-41
  wood, **4**-31
Moment (*see* Bending moment)
Moment distribution, **5**-62 to **5**-76
Moment-distribution check, **5**-74 to **5**-76
Moment of inertia, **5**-27 to **5**-29
  polar, **5**-86, **5**-87
Monel, **4**-59, **4**-60
Monitors, **1**-12, **3**-41
Mortar:
  cement for (*see* Cement, masonry; Coatings, cementitious)

Mortar (*Cont.*):
  defined, **4**-9, **12**-3
  high-bond, **4**-9
  lime in, **4**-5, **4**-6, **12**-5, **12**-6, **12**-8
  mixes for, **3**-23, **4**-5, **4**-6, **4**-9, **12**-5, **12**-6, **12**-8, **12**-33, **12**-34
  polymer, **4**-68
  properties of, **4**-9, **4**-10
  retempering of, **12**-8
  for severe exposures, **3**-22
  specifications for, **4**-9, **11**-3, **12**-5, **12**-8
  for tile, **12**-5, **12**-32 to **12**-35
  types of, **11**-3, **12**-5
  weight of, **4**-9, **5**-6
Mosely principle, **6**-36
Motors, electric:
  ampere ratings of, **20**-31, **20**-40 to **20**-43
  circuit protection for, **20**-31, **20**-36 to **20**-43
  components of, **20**-4, **20**-5
  speeds of, **20**-5
  starter sizes for, **20**-40 to **20**-43
Moving walks, **22**-3, **22**-4
Mud (joint compound), **12**-38, **12**-39, **12**-59, **12**-60
Muller-Breslau principle, **5**-56
Mullions, **12**-30, **13**-3
Muntin, **13**-3
Muntz metal, **4**-53, **4**-56

Nails:
  aluminum, **16**-17
  brad, **16**-17
  casing, **16**-16, **16**-17
  clout, **16**-17
  corrosion of, **7**-32
  corrosion-resistant, **7**-65, **7**-66
  cut, **16**-16, **16**-17
  deformed-shank (threaded), **7**-32, **16**-16, **16**-17
  dimensions of, **7**-32
  finishing, **16**-16, **16**-17
  flooring, **16**-17
  galvanized, **16**-17
  for gypsumboard, **12**-54, **16**-16
  hardened, **7**-32
  heads for, **16**-16, **16**-17
  holes for, **7**-33
  for lath, **12**-45, **12**-48, **16**-16
  load capacity of, **7**-33
  minimum spacing of, **7**-33
  penetration required of, **7**-33, **7**-78, **7**-79
  points for, **16**-16, **16**-17
  roofing, **16**-16
  shingle, **15**-23, **16**-16, **16**-17
  siding, **16**-16
  for slate roofing, **15**-27

Nails (*Cont.*):
  stainless steel, **16**-17
  for underlayments, **12**-66
  wire, **7**-32, **7**-33, **16**-16, **16**-17
  wood shake, **16**-16
Neoprene, **4**-68, **4**-71, **15**-17, **15**-19
Newel post, **1**-19, **22**-6, **22**-8
Newton's law of motion, **3**-14, **5**-89
Niche, **12**-41
Nickel, **4**-59, **4**-60
  alloys of, **4**-41 to **4**-44, **4**-54, **4**-55, **4**-57, **4**-59, **4**-60
Nickel silver, **4**-54, **4**-55, **4**-57
Nitril rubber, **4**-71
Noise:
  acceptable levels of, **17**-20, **25**-59, **25**-60
  causes of, **17**-22 to **17**-24
  defined, **17**-5
  effects on humans, **17**-5
  from electric lamps, **21**-14
  reduction of, **17**-15, **17**-22 to **17**-24
  (*See also* Sound; Vibration)
Nosing, **22**-5, **22**-8
Notice to contractors, **1**-6, **2**-13
Nuts, **4**-39, **8**-69
Nylon, **4**-63, **4**-66

Oakum, **4**-77
Occupancy, certificate of, **1**-7, **1**-33, **25**-8
Occupant loads, **1**-34, **3**-46
Occupational Safety and Health Act, **1**-36, **2**-7, **3**-30, **18**-31, **25**-54, **25**-55
Octave, **17**-4
Offset lines, **23**-7
Ogee, **12**-41
Ohm, **20**-2
Ohm's law, **20**-2, **20**-3, **20**-5
Owner-builder, **25**-2
Owners, **1**-4, **1**-7, **1**-8, **2**-7, **2**-10 to **2**-12, **2**-14

Paint:
  aluminum, **4**-78
  for corrosion protection, **4**-47, **8**-104 to **8**-106
  for fire protection, **4**-33
  fire-retardant, **8**-111
  for masonry, **3**-28, **3**-29
  oil, **4**-77, **4**-78
  pigments for, **4**-79
  primer, **8**-105, **8**-107, **12**-42, **12**-60
  resins for, **4**-79, **4**-80
  sealer, **12**-42, **12**-60
  sizing for, **12**-42
  for steel joists, **9**-4
  vehicles for, **4**-78
  water, **4**-77

Paint (*Cont.*):
  for wood, **7**-31
  (*See also* Coatings)
Panels:
  drop, **10**-72, **10**-74, **10**-75
  wall (*see* Walls, curtain; nonbearing)
Parapets, **3**-33, **12**-17
Partitions:
  bearing, **1**-16, **12**-30, **12**-31
  defined, **1**-16, **12**-3, **12**-30
  folding, **12**-31
  (*See also* Doors, folding)
  gypsum tile or block, **12**-5
  masonry (*see* Masonry; Masonry walls)
  movable, **12**-31
  nonbearing, **1**-16, **1**-17, **12**-17, **12**-18, **12**-30, **12**-31
  studless gypsumboard, **12**-52, **12**-53
  weights of, **5**-3, **5**-9
  (*See also* Concrete walls; Walls)
Passageways, exit, **3**-43 to **3**-46
Pavements, **4**-75
Pentachlorphenol, **4**-33, **7**-29
Performance, coefficient of, **18**-3
Perlite, **4**-6, **4**-8, **15**-9, **18**-22, **18**-24
Perm, **15**-8, **18**-30
Permeability, **3**-17, **3**-18, **6**-5
Permits, **1**-6, **1**-8, **1**-33, **1**-36
Pervious materials, **3**-18
pH, **19**-3
Phenolics, **4**-33, **4**-62, **4**-63, **4**-69, **4**-80
Piers:
  drilled bell (*see* Shafts, foundation)
  masonry, **12**-3, **12**-5, **12**-7, **12**-14, **12**-15
Pilaster, **8**-6, **10**-87, **12**-3
Piles:
  bearing, **6**-25, **6**-43, **6**-44
  bracing of, **3**-16, **6**-37, **6**-38
  caps for, **6**-37, **10**-97
  concrete, **6**-39 to **6**-41
  defined, **1**-10, **6**-37
  deterioration of, **6**-44, **6**-45
  displacement-type, **6**-40, **6**-41
  drilled-in caisson, **6**-43, **6**-44
  drilled injection, **6**-40
  driving of, **6**-45
  driving formulas for, **6**-24
  embedment of: in cap, **10**-97
    in soil, **6**-38
  Franki, **6**-40, **6**-41
  friction, **6**-24, **6**-44
  hammers for, **6**-45
  jetting of, **6**-45
  load tests of, **6**-42, **6**-43
  loads on: allowable, **6**-39, **6**-42 to **6**-44
    applied, **6**-42
    transfer to soil, **6**-37
    ultimate, **6**-23 to **6**-25, **6**-43, **10**-97
  minimum sizes of, **6**-38
  points (shoes) on, **6**-39

Piles (*Cont.*):
  reused, **6**-40, **6**-41
  soldier, **6**-54
  spacing of, **6**-38
  splices for, **6**-41
  steel, **6**-39
  timber, **6**-38, **6**-39
  tolerances for, **6**-45
  in unconsolidated soils, **6**-37, **6**-38
  (*See also* Footings, pile; Sheetpiles)
Pins, **8**-33
Pipe:
  corrosion protection for, **19**-2
  fittings for, **19**-9, **19**-19
  gas, **19**-36 to **19**-38
  hydraulic properties of, **19**-23
  insulation of, **18**-24, **19**-2
  joints in, **18**-52, **18**-58, **19**-2, **19**-9, **19**-10, **19**-28, **19**-45
  materials for: for fire sprinklers, **19**-45
    for heating gas, **19**-37, **19**-38
    for roof drainage, **19**-29
    for wastewater, **19**-27, **19**-28
    for water supply, **4**-58, **4**-59, **19**-7, **19**-9
  protection of openings for, **19**-2
  rubber-lined, **4**-73
  sizing of: for heating gas, **19**-36
    for hot-water heating, **18**-52
    for roof drainage, **19**-30, **19**-31
    for sprinklers, **19**-39, **19**-42, **19**-45
    for steam heating, **18**-54 to **18**-56
    for venting, **19**-35
    for wastewater, **19**-28, **19**-32 to **19**-34
    for water supply, **19**-6, **19**-11, **19**-13, **19**-15 to **19**-20
  support of, **19**-2, **19**-9, **19**-10, **19**-26
  underground, **19**-2, **19**-26
  (*See also* Sewers)
Pitch:
  coal-tar, **4**-74, **5**-5, **15**-2, **15**-3
  of fasteners, **8**-22
Plaster:
  accelerators in, **12**-37
  acoustical, **12**-37
  admixtures for, **12**-37
  advantages of, **12**-43
  application of, **12**-37 to **12**-46, **12**-49 to **12**-51, **12**-62
  base coats for, **12**-37, **12**-49, **12**-50
  bases for: concrete, **12**-49
    gypsum, **4**-22, **12**-41, **12**-44 to **12**-46, **12**-49
    masonry, **12**-48, **12**-49
    (*See also* Lath)
  blisters on, **12**-38
  bond, **12**-38
  brown coat of, **12**-38, **12**-46, **12**-50
  buckles in, **12**-38
  butterflies on, **12**-38

Plaster (*Cont.*):
  case mold, **12**-38
  casts (staff), **12**-42
  catface, **12**-38
  composition of, **12**-43, **12**-44
  cracks in, **12**-39, **12**-40
  curing of, **12**-39
  definitions of terms for, **12**-37 to **12**-43
  double-up applications of, **12**-39
  dry out of, **12**-40
  drying of, **12**-45, **12**-51
  efflorescence on, **12**-40
  eggshelling of, **12**-40
  enrichments of, **12**-40
  fibered, **12**-44
  finish coat of, **12**-40, **12**-46, **12**-50, **12**-51
  for fire protection of steel, **8**-109 to **8**-111
  fisheyes in, **12**-40
  floating angles (corner floating) of, **12**-40, **12**-62
  gaging, **12**-41, **12**-44, **12**-50, **12**-51
  green, **12**-41
  grounds for, **12**-41, **12**-50
  gypsum-cement (*see* neat, *below*)
  joints in, **12**-48, **12**-60 to **12**-62
  Keene's cement, **12**-41
  lath for (*see* Lath)
  lime, **12**-41
  mixes for, **12**-44, **12**-50, **12**-51
  mixing of, **12**-43 to **12**-45, **12**-51
  molding, **12**-51
  neat (hardwall), **4**-6, **12**-41, **12**-44, **12**-50
  perimeter relief for, **12**-42, **12**-48
  perlite, **4**-6, **12**-44
  pinholes in, **12**-42
  portland-cement (*see* Mortar; Stucco)
  properties of, **4**-6
  putty coat of, **12**-42
  ready-mixed, **12**-44, **12**-50
  retarder for, **12**-42
  scratch coat of, **12**-42, **12**-46, **12**-50
  screeds for, **12**-42, **12**-50
  skim coat of (*see* finish coat of, *above*)
  as sound absorbent, **17**-20
  specifications for, **12**-44, **12**-50, **12**-51
  stucco (*see* Stucco)
  sweat out of, **12**-43
  tempering of, **12**-43
  (*See also* water in, *below*)
  textures for, **12**-42, **12**-43, **12**-51
  thickness of, **12**-44, **12**-49 to **12**-51
  veneer, **4**-22, **12**-43, **12**-44, **12**-46, **12**-49 to **12**-51
  vermiculite, **4**-6, **12**-44, **13**-19
  water in, **12**-44, **12**-45, **12**-50, **12**-51
  weight of, **5**-9
  white coat, **12**-43, **12**-50, **12**-51

Plaster of paris, 4-6
Plasterboard (*see* Gypsum, lath; Gypsumboard)
Plastic theory, 5-30, 5-100 to 5-104, 5-155 to 5-160, 8-21, 8-54 to 8-58
Plastics:
  bibliography for, 4-70
  composition of, 4-60 to 4-62
  defined, 4-60
  elastomeric, 15-16, 15-17
  fabrication of, 4-61, 4-62
  foamed, 4-61
  as glazing, 13-16, 13-17
  glazing-compound, 13-19, 13-20
  laminates of, 4-62, 4-67, 13-16 to 13-18
  mechanical properties of, 4-60, 4-62 to 4-68
  reinforced, 4-62, 4-67 to 4-69
  thermoplastic, 4-60, 4-63 to 4-66
  thermosetting, 4-60, 4-62 to 4-64
  (*See also* Floor coverings; Roofing, plastic)
Plate girders (*see* Girders, steel)
Plates:
  folded (*see* Folded plates)
  shear, 7-39 to 7-45
  sole, 7-64, 12-23, 12-24
  steel: bearing, 8-6
    black, 9-9
    butt, 8-94, 8-95
    classified by thickness, 9-9
    cover, 8-45, 8-46
    top (wall), 7-64, 12-23, 12-24
Plumbing:
  air gaps in, 19-4, 19-11, 19-14, 19-29
  back-siphonage in, 19-4, 19-32
  backflow in, 19-2, 19-7, 19-11
  cleanouts in, 19-1, 19-11, 19-25, 19-26
  combination waste and vent, 19-26, 19-27
  drains in, 19-11, 19-15, 19-16, 19-25, 19-29, 19-30
  health requirements for, 19-1 to 19-4, 19-7
  for indirect wastes, 19-29
  interceptors in, 21-28, 21-29
  piping for, 19-2, 19-6 to 19-10
    (*See also* Pipe; Wastewater; Water supply)
  purposes of, 1-18
  Sovent, 19-26, 19-27
  stacks in, 19-25, 19-28, 19-32 to 19-35
  standards for, 19-1, 19-22
  sump pits for, 19-26
  testing of, 19-35, 19-36, 19-38
  traps in: building, 19-25, 19-26
    fixture units for, 19-16
    location of, 19-2, 19-10, 19-11, 19-26, 19-29, 19-30
    maximum air pressure at, 19-28

Plumbing, traps in (*Cont.*):
  minimum size of, 19-11, 19-15
  purpose of, 19-2, 19-26
  requirements for, 19-11
  water seal in, 19-11
  venting of, 19-2, 19-25 to 19-27, 19-32 to 19-35
  (*See also* Pipe; Plumbing fixtures; Sprinklers; Storm water; Wastewater; Water supply)
Plumbing fixtures:
  bathtubs, 19-5, 19-12, 19-13, 19-15, 19-19
  connections to, 19-2
  dishwashers, 19-15, 19-19
  drinking fountains, 19-12, 19-13, 19-19
  faucets, 19-5, 19-9
  fixture units for, 19-13 to 19-16, 19-32 to 19-35
  laundry tubs, 19-5, 19-15
  lavatories (*see* Lavatories)
  minimum required, 19-12, 19-13
  overflow protection for, 19-11
  requirements for, 19-2, 19-10
  showers (*see* Showers)
  sinks, 19-5, 19-11, 19-15, 19-19, 19-22
  types of, 19-10
  urinals (*see* Urinals)
  water closets (*see* Water closets)
Plywood:
  allowable loads on, 7-76 to 7-79, 7-81 to 7-83
  allowable stresses for, 7-74, 7-75
  characteristics of, 4-34, 4-35, 7-75
  composition of, 4-34, 7-68
  deflections of, 7-82
  finishes of, 12-36
  for formwork, 7-69
  in foundation walls, 7-64 to 7-66
  grade stress levels of, 7-70, 7-71, 7-75
  grades of, 4-34, 4-35, 7-69 to 7-71
  identification stamp on, 7-69
  section properties of, 7-72 to 7-75
  sheathing, 7-4, 7-63, 7-64, 7-76 to 7-81, 12-24
  siding, 7-80, 7-81, 12-26
  sound-transmission rating of, 13-19
  span rating of, 7-69, 7-74 to 7-77, 7-79, 7-80
  stiffness groups of, 7-68, 7-69
  subfloors of, 5-8
  waterproof, 4-34
  weight of, 5-8, 7-72, 7-73
Pneumatic construction (*see* Air-stabilized structures)
Pneumatic tubes, 22-38
Poisson's ratio, 5-15
Polar moment of inertia, 5-86, 5-87
Poles, 7-83 to 7-86
Polycarbonate, 4-63, 4-65

Polyesters, 4-62, 4-63, 4-69
Polyethylene, 4-63 to 4-65, 4-76, **15**-18
Polypropylene, 4-63, 4-65
Polystyrene, 4-63, 4-66, **5**-9
Polysulfides, 4-71, 4-76
Polytetrafluoroethylene, 4-65
Polyurethane (*see* Urethane)
Polyvinyls (*see* Vinyls)
Porcelain enamel, 4-70, 4-71
Portal method, **5**-154, **5**-155
Portals (*see* Frames; Rigid frames)
Posttensioning, **10**-122
  (*See also* Concrete, prestressed)
Pozzolans, **10**-11
Prefabrication, **1**-30, **25**-60, **25**-61
  (*See also* Buildings, preengineered;
    Concrete, precast)
Present worth, **1**-23 to **1**-25
Pressure:
  absolute, **18**-5
  atmospheric, **18**-6
  Dalton's law for, **18**-13
  head, **18**-6
  saturation, **18**-6
  suction, **18**-6
  velocity, **18**-7
Pressure head, **18**-4, **18**-5, **19**-17
Pretensioning, **10**-122
  (*See also* Concrete, prestressed)
Prices (*see* Bids; Construction costs; Cost estimates)
Product of inertia, **5**-29
Program, building, **1**-4, **1**-20, **2**-7 to **2**-10
Proportional limit, **5**-13, **5**-155
Proposals, **2**-15, **25**-44, **25**-46, **26**-5, **26**-6
Psychrometer, **18**-6, **18**-14
Psychrometric chart, **18**-6, **18**-14, **18**-15
Pulp board, **12**-36
Pumice, 4-8, 4-9, **18**-22, **18**-24
Pumps:
  booster water (circulator), **18**-3, **18**-53, **18**-54
  fire, **19**-47
  heat, **18**-4, **18**-63, **18**-64
Purchase orders, **25**-7, **25**-17 to **25**-20
Purchasing index, **25**-13, **25**-14
Purchasing log, **25**-18, **25**-20
Purlins:
  defined, **1**-10
  open-web, **9**-6, **9**-7
  steel, **8**-9, **8**-21
  wood, **7**-59
  (*See also* Beams; Joists; Rafters)
Putty:
  glazing, **13**-19, **13**-20
  lime, 4-5, **12**-41

Quartzite, 4-20, 4-21
Quicklime, 4-5

Radiation:
  defined, **18**-6
  equivalent direct, **18**-6, **18**-55
  solar, **13**-2, **18**-6, **18**-19, **18**-41, **18**-64
Radiators, **18**-16, **18**-19, **18**-51, **18**-52, **18**-54 to **18**-56, **18**-63
Radius of gyration, **5**-28, **5**-37, **5**-83
Rafters, **1**-10, **7**-61, **7**-64, **9**-6, **9**-7
  (*See also* Beams; Joists; Purlins)
Railing, **22**-2 to **22**-9, **22**-23
Rails, **13**-2, **13**-3, **14**-6
  (*See also* Handrails)
Rain (*see* Storm water)
Ramps:
  powered, **3**-44, **22**-3, **22**-4
  stationary, **1**-19, **22**-1 to **22**-3
  (*See also* Corridors)
Ranges, domestic, **19**-37
Rayleigh method, **5**-92, **5**-93
Reactions:
  arch (*see* Arches)
  beam: cold-formed, **9**-30 to **9**-32
    for concentrated loads, **5**-34, **5**-35
    for distributed loads, **5**-24, **5**-34, **5**-36
    for dynamic loads, **5**-108
    from equilibrium laws, **5**-21
    by graphic statics, **5**-44, **5**-45
    from influence lines, **5**-25, **5**-26, **5**-56, **5**-57
    by virtual work, **5**-51
  defined, **5**-20
  truss, **5**-12, **5**-44, **5**-45
Rebars, defined, **10**-22
  (*See also* Steel reinforcement)
Reciprocal theorem, **5**-56, **5**-57
Refrigerant, **18**-6
Refrigeration (*see* Air conditioners; Air conditioning)
Refrigerators, **19**-37
Refuge areas, **3**-44
Registers (grilles), **18**-3, **18**-4, **18**-47, **18**-48, **18**-78, **18**-79
Relief, **12**-42
Resilience, modulus of, **5**-16, **5**-95
Resistance:
  to displacement, **5**-100 to **5**-102, **5**-107
  (*See also* Beams, restraint of;
    Flexibility; Modulus of elasticity;
    Spring constant; Stiffness)
  electric, **20**-2, **20**-3, **20**-28, **20**-29
  thermal, **15**-9, **18**-6, **18**-17, **18**-18, **18**-25
Resonance, **5**-104, **5**-105, **5**-147, **5**-148
Retaining walls (*see* Concrete walls;
    Masonry walls)
Return, **12**-42
Reveal, **12**-42
Ribband, **7**-61, **7**-63, **7**-64

Richter scale, 3-14
Ridging, **12**-42
Rigid frames:
　bracing of, **8**-67
　concrete: analysis of, **10**-35 to **10**-38
　　direct design method for, **10**-37, **10**-67 to **10**-71
　　ductile, **10**-37, **10**-116, **10**-117
　　equivalent frame method for, **10**-37, **10**-67 to **10**-71
　　high-rise, **10**-37
　defined, 1-11, **3**-9, **3**-10
　drift of (*see* sidesway, *below*)
　dynamic models of, **5**-90, **5**-91, **5**-98, **5**-106 to **5**-108
　moment distribution in: by converging approximations, **5**-62 to **5**-66
　　single-cycle, **5**-72, **5**-73
　purposes of, 1-11, **3**-9
　seismic forces on, **5**-108 to **5**-112
　sidesway of, **3**-10, **5**-67 to **5**-70, **8**-24, **8**-25
　steel: characteristics of, **8**-13 to **8**-15
　　knees of, **8**-23
　　splices in, **8**-14, **8**-15
　ultimate load of, **5**-156 to **5**-160
　vibration of, **5**-90 to **5**-93, **5**-98 to **5**-100, **5**-106 to **5**-108
　wood, **7**-59 to **7**-61
　(*See also* Frames)
Rigidity, modulus of, **5**-15
(*See also* Stiffness)
Risers:
　electrical, **20**-44 to **20**-47
　stair, 1-19, **22**-5, **22**-6, **22**-8, **22**-9
Rivets:
　holes for, **8**-68
　in plastic design, **8**-56, **8**-57
　spacing of, **8**-79
　specifications for, **8**-68
　stresses for, **8**-39 to **8**-41
　(*See also* Connections, riveted)
Rock (*see* Soils; Stone)
Rockers, **8**-34
Rollers, expansion, **8**-34
Roof sprays, **18**-41
Roofing:
　alligatoring, **15**-4
　area variation with slope, **15**-1, **15**-2
　asbestos-cement, **4**-24, **15**-19, **15**-20
　asphalt, **4**-74, **4**-75, **5**-9, **15**-3
　built-up: coatings on, **15**-4, **15**-6
　　composition of, **15**-3 to **15**-6
　　deck conditions for, **15**-6, **15**-7
　　installation of, **15**-3 to **15**-6
　　inverted, **15**-12, **15**-13
　　mineral-surfaced, **15**-4
　　weight of, **5**-9
　classification of, **15**-3
　cold-process bituminous, **15**-15, **15**-16

Roofing (*Cont.*):
　copper, **4**-52
　for curved roofs, **15**-5, **15**-16
　deck preparation for, **15**-4, **15**-7, **15**-16, **15**-19
　defined, 1-12, **15**-1
　elastomeric (*see* plastic, *below*)
　expansion of, **15**-10, **15**-14, **15**-15
　failures of, **15**-1, **15**-4 to **15**-8, **15**-10, **15**-11
　felts for, **4**-74, **15**-3 to **15**-6
　(*See also* Membrane, waterproofing)
　fluid-applied, **15**-4, **15**-16
　foamed urethane, **15**-13
　glass fiber in, **15**-4, **15**-5, **15**-15, **15**-16
　moisture-detection surveys for, **15**-21
　plastic, **15**-3, **15**-13, **15**-16 to **15**-20
　prepared, **4**-75, **15**-3
　at ridges, **15**-15, **15**-25, **15**-26
　selection of, **15**-4, **15**-5
　sheet-applied, **15**-4, **15**-16, **15**-17
　shingle, **4**-31, **5**-9, **15**-22 to **15**-26
　(*See also* Shingles)
　single-ply, **15**-4, **15**-16 to **15**-19
　slate, **4**-19 to **4**-21, **5**-9, **15**-5, **15**-27
　specifications for, **15**-1
　square of, **4**-74, **15**-4
　for steep roofs, **15**-4, **15**-5, **15**-23 to **15**-28
　temporary, **15**-10, **15**-11
　tile, **12**-34, **15**-26, **15**-27
　(*See also* Tile)
　in valleys, **15**-25, **15**-26, **15**-28
　venting of, **15**-8, **15**-20, **15**-21, **18**-30
　weights of, **5**-9
　(*See also* Roofs)
Roofs:
　air-supported (*see* Air-stabilized structures)
　cable-supported (*see* Cable-supported structures; Cables)
　cement-fiber, **15**-7
　components of, 1-12, 1-13, 1-15, 1-16
　concrete (*see* Concrete floors; Concrete roofs)
　drainage of, **3**-19, **7**-53, **7**-56, **15**-7, **15**-18, **19**-29 to **19**-31
　dryness test of, **15**-7
　expansion of, **15**-10, **15**-14, **15**-15
　fill over, **15**-8, **15**-20, **15**-21
　flashing for (*see* Flashing)
　flat, 1-12, **15**-2 to **15**-13
　(*See also* ponding on, *below*)
　framing for, **8**-9 to **8**-15, **8**-21
　(*See also* Air-stabilized structures; Cable-supported structures; Floors)
　gypsum, **4**-22, **15**-7
　insulation of, **9**-54, **9**-61, **15**-8 to **15**-13, **18**-24
　(*See also* Condensation; Heat transmission; Insulation)

Roofs (*Cont.*):
    loads on (*see* Loads)
    metal, **5**-9, **8**-21, **9**-53 to **9**-60, **15**-4, **15**-7, **15**-13 to **15**-15
    minimum thicknesses of (*see* Deflections, limits on)
    moisture venting of, **15**-8, **15**-20, **15**-21
    ponding on, **3**-18, **3**-19, **7**-56, **7**-57, **8**-51, **9**-7, **10**-58
    sheathing for (*see* Sheathing)
    slope (pitch) of, **7**-56, **7**-57, **9**-7, **15**-1, **15**-2
    thin-shell (*see* Shells)
    types of, **1**-12, **15**-1
    upside-down assembly of, **15**-12, **15**-13
    vapor barriers for (*see* Vapor barriers)
    wood, **7**-4, **7**-66 to **7**-68, **15**-7
    (*See also* Roofing)
Rubber, **4**-71 to **4**-74, **12**-65 to **12**-67, **15**-17
Rubble, **12**-3, **12**-4, **12**-6, **12**-7, **12**-10, **12**-15
Rugs, **12**-66
Runners, **12**-42

Sabine formula, **17**-17
Safety factor, **3**-2, **5**-13, **8**-21
Sand, **4**-8, **5**-6, **5**-9, **6**-1, **6**-6, **6**-49
    (*See also* Aggregates)
Sandstone, **4**-20, **4**-21, **5**-6, **5**-9, **12**-7
    (*See also* Masonry; Masonry walls)
Sash, **1**-15, **13**-2
    storm, **13**-15
Saxe seats, **8**-93
Scagliola, **12**-42
Scarf, **7**-26
Schedules (*see* Construction scheduling)
Schmidt orthogonalization, **5**-93
Scholten nomograph, **7**-41, **7**-42
Scientific method, **1**-20
Screeds, plaster, **12**-42, **12**-50
Screws:
    applications of, **16**-18
    for gypsumboard, **12**-54, **12**-56, **12**-58
    lag, **7**-34 to **7**-37, **7**-63, **16**-18
    for thin metal (Parker), **9**-52, **9**-53, **16**-18
    types of, **16**-18
    for wood, **7**-34 to **7**-37, **16**-18
Sealants, **4**-75 to **4**-77, **10**-121, **12**-18, **12**-19, **12**-35, **12**-36
    (*See also* Calking compounds)
Section modulus:
    elastic, **5**-28, **5**-29
    plastic, **8**-55, **8**-56
Section properties, geometric, **5**-28, **7**-3, **7**-5 to **7**-7, **7**-10 to **7**-13
Security measures, **3**-50, **3**-51, **13**-6, **13**-7, **16**-2, **16**-13

Seismic coefficient, **5**-108, **5**-110, **5**-112, **5**-113
Seismic design, **3**-1 to **3**-3, **3**-13 to **3**-16, **5**-108 to **5**-114
    (*See also* Cantilever method; Loads, seismic; Portal method)
Serpentine, **4**-20, **4**-21
Service core, **3**-11, **8**-16
Sewage (*see* Wastewater)
Sewers:
    building, **19**-25, **19**-26, **19**-32
    defined, **19**-23
    materials for, **19**-24
    minimum flow velocity in, **19**-24
    sanitary, **19**-22 to **19**-24
    sizing of, **19**-23, **19**-24, **19**-32
    storm, **19**-22, **19**-23
    (*See also* Pipe)
Shafts:
    elevator (*see* Elevators, hoistways for)
    foundation: drilled, **6**-47, **10**-89, **10**-97 to **10**-99
        excavation methods for, **6**-47, **6**-48
        open, **6**-46, **6**-47
        (*See also* Caissons; Cofferdams; Sheeting)
    smoke-vent, **3**-42
Shale, **4**-8, **5**-6
Shear:
    in beams: defined, **5**-22, **5**-45
        stresses from, **5**-29, **5**-30
        (*See also* Concrete beams; Girders, steel; Steel reinforcement)
    in brackets, **5**-13, **5**-14
    in composite beams, **8**-42, **8**-43, **10**-117
    in concrete members (*see* Concrete beams; Concrete floors)
    defined, **5**-2
    diagrams of: for concentrated loads, **5**-22
        defined, **5**-22
        by graphic statics, **5**-45, **5**-46
        plotting of, **5**-22, **5**-23
        for uniform loads, **5**-23
    energy of, **5**-16
    in fasteners, **8**-38, **8**-39, **8**-78, **8**-82
    in footings, **10**-91 to **10**-94, **10**-96
    influence lines for, **5**-25, **5**-26, **5**-56, **5**-57
    maximum: at a point, **5**-19
        at a section, **5**-25, **5**-26, **5**-29, **5**-30
    on perpendicular planes, **5**-17
    pure, **5**-19
    seismic (*see* Loads, seismic)
    on shear connectors, **8**-42, **8**-43
    in steel beams, **8**-22, **8**-23, **8**-56
    strain, **5**-14, **5**-15
    unit stress: in beams, **5**-29, **5**-30
        constant, **5**-14
        from torsion, **5**-86 to **5**-88

Shear (*Cont.*):
  in walls, **10**-110, **10**-111
  in welds, **8**-38, **8**-82
Shear center, **5**-2, **5**-38, **5**-39, **5**-86
Shear connectors, **8**-42 to **8**-44, **10**-117
Shear flow, **5**-87
Shear lag, **9**-27
Shear-moment relationships, **5**-23, **5**-25
Shear span, **10**-109
Shear walls:
  allowable stresses for, **11**-18
  arrangements of, **1**-11, **3**-9 to **3**-12,
    8-16, **8**-64, **11**-16
  brick, **11**-7 to **11**-19
  concrete, **10**-110 to **10**-112
  coupled, **11**-17, **11**-18
  defined, **3**-9, **5**-153, **10**-82, **12**-4
  deflections of, **11**-12 to **11**-16
  design examples for, **11**-20 to **11**-33
  eccentrically loaded, **11**-18
    (*See also* design examples for, *above*)
  effective length of, **11**-16
  load distribution to, **11**-8 to **11**-12,
    10-110
  overturning moments in, **11**-18, **11**-19,
    11-23, **11**-29
  purpose of, **1**-11, **3**-9 to **3**-12, **5**-153,
    8-16, **8**-64, **10**-110, **11**-7, **11**-8
  rigidity (stiffness) of, **11**-11, **11**-12,
    11-14 to **11**-16
  seismic design of, **5**-111
  torque distribution to, **11**-11, **11**-12
    (*See also* Concrete walls; Masonry
    walls)
Shearheads, **8**-15, **8**-20, **10**-70
Sheathing:
  gypsum, **4**-22, **5**-9, **12**-42, **12**-53
  plywood, **7**-4, **7**-63, **7**-64, **7**-76 to **7**-81,
    12-24
  roof, **1**-13, **5**-9, **7**-4, **7**-76 to **7**-79
  wall, **1**-13, **5**-9, **7**-63, **7**-64, **7**-80, **7**-81,
    12-24
Sheet steel, **4**-38, **9**-1, **9**-9, **9**-10, **9**-12,
  9-13
  (*See also* Cold-formed shapes)
Sheeting, **6**-52 to **6**-55
Sheetpiles, **4**-39, **6**-54
Shellac, **4**-78, **4**-79
Shells:
  advantages of, **10**-113, **10**-114
  bending theory for, **5**-121, **5**-123, **5**-128,
    5-129
  concrete, **10**-31, **10**-113 to **10**-115,
    10-119, **10**-120
  conical, **5**-121, **5**-127, **5**-128
  cylindrical (barrel), **5**-121, **5**-129,
    5-130, **10**-114, **10**-115
  defined, **5**-121, **10**-113
  design of, **5**-121 to **5**-123, **5**-129
    (*See also* concrete, *above*)

Shells (*Cont.*):
  dome, **5**-121, **5**-127
  elliptical-paraboloid, **5**-124, **5**-125
  hyperbolic-paraboloid, **5**-121, **5**-124,
    5-125
  membrane theory for, **5**-121 to **5**-128,
    10-113
  prismatic (folded plate) (*see* Folded
    plates)
  of revolution, **5**-125 to **5**-128
  ribbed, **5**-121
  stresses in, **5**-121, **5**-129, **5**-130
  supports for, **5**-122
  thickness of, **5**-121, **10**-113
  unit forces in, **5**-123 to **5**-128
Sherardizing, **4**-47
Shingles:
  asbestos-cement, **4**-24, **5**-9, **12**-26,
    15-25, **15**-26
  asphalt, **4**-75, **5**-9, **12**-26, **15**-22 to
    15-24
  wood, **4**-31, **5**-9, **12**-26, **15**-24, **15**-25
  (*See also* Roofing, shingle)
Shoring (*see* Formwork)
Shotcrete, **3**-22, **3**-27, **10**-10, **10**-28
Showers:
  fixture units for, **19**-15
  minimum flow for, **19**-5, **19**-22
  minimum pipe size for, **19**-19
  number required, **19**-12
Siamese connections, **19**-49
Siding:
  asbestos-cement, **4**-24, **5**-9, **12**-27, **12**-28
  asphalt, **4**-75
  metal, **5**-9, **12**-27, **12**-28
  plastic, **12**-27, **12**-28
  wood, **7**-80, **7**-81, **12**-26
Signal and alarm systems, **1**-19, **19**-49
Silicones, **3**-28, **4**-63, **4**-64, **4**-71, **4**-72,
  4-76, **4**-80, **15**-17
Sill:
  door, **14**-2, **14**-6, **14**-7
  wall, **7**-63, **7**-64
  window, **12**-13, **13**-3, **13**-4
Silt, **5**-6, **6**-1, **6**-6
Silver, **5**-6
Sinks, **19**-5, **19**-11, **19**-15, **19**-19, **19**-22
Skewback, **5**-118, **10**-112
Skylights, **1**-12
Slabs:
  concrete (*see* Concrete floors)
  steel (*see* Plates)
Slag, **4**-8
Slaking, **4**-5, **12**-43
Slate, **4**-18, **4**-19 to **4**-21, **5**-6, **5**-8, **5**-9,
  15-5, **15**-27
Slope-deflection equation, **5**-76
Smoke:
  constriction of, **3**-33
  detectors of, **3**-39 to **3**-42

Smoke (*Cont.*):
venting of, 3-41, **18**-31
(*See also* Fire protection)
Snow:
design loads for, **3**-2, **5**-3, **5**-7, **5**-10, **5**-11
melting of, **18**-61 to **18**-63
rate of fall of, **18**-62
Soapstone, **4**-21
Societies (*see* Associations)
Soffit, **5**-118, **12**-43
Soil mechanics, **6**-1
(*See also* Soils)
Soils:
aerial surveys of, **6**-10
angle of internal friction of, **6**-4, **6**-22, **6**-23, **6**-49 to **6**-52
angle of repose of, **6**-3
Atterberg limits of, **6**-6, **6**-15
bearing capacity of, **6**-22 to **6**-25
bibliography for, **6**-57
classification of **6**-1, **6**-5 to **6**-9
coarse-grained, **6**-6
(*See also* Gravel)
coefficient of friction for, **6**-4, **6**-22, **6**-23, **6**-49 to **6**-52
cohesion in, **6**-4
compressibility of, **6**-3 to **6**-5
compression strength of, **6**-2, **6**-15
consolidation of, **6**-25 to **6**-29
defined, **6**-1
densification of (*see* improvement of, *below*)
drainage of, **3**-19, **3**-20, **6**-53, **6**-54
elastic modulus of, **6**-4, **6**-29, **6**-30
fine-grained, **6**-6, **6**-7
(*See also* Clay; Sand; Silt)
friction in, **6**-3, **6**-4
heaving of, **6**-56
identification of, **6**-2, **6**-5 to **6**-9
improvement of: applicability of, **6**-18, **6**-19
by compaction, **6**-3, **6**-4, **6**-16 to **6**-20
by draining, **6**-21
with fills, **6**-16 to **6**-18
by grouting, **6**-21
with piles, **6**-18
with surcharges, **6**-19 to **6**-21
by Terra-Probe, **6**-18
by vibroflotation, **6**-18, **6**-19
liquid limit of, **6**-7
organics in, **6**-1, **6**-5
permeability of, **6**-5
plastic limit of, **6**-8
plasticity index of, **6**-7, **6**-15
residual, **6**-1, **6**-2
rock, **6**-12, **6**-14
sampling of, **6**-12
sedimentary, **6**-1, **6**-2
seismic effects on, **3**-14, **3**-15

Soils (*Cont.*):
shear strength of, **6**-2, **6**-15
shearing of, **6**-2 to **6**-4
states of, **6**-2, **6**-3
subsurface explorations of: by borings, **6**-11 to **6**-14
electrical, **6**-10, **6**-11
in situ, **6**-11 to **6**-14
geophysical, **6**-10, **6**-11
planning of, **6**-10, **6**-11
by probing, **6**-11
purposes of, **6**-10
remote-sensing, **6**-10, **6**-11
seismic, **6**-10
by soundings, **6**-12
with test pits or trenches, **6**-11
symbols for, **6**-6, **6**-7
tests of: density, **6**-4
dilatancy, **6**-8
dry-strength (crushing), **6**-8
laboratory, **6**-15, **6**-26 to **6**-29
penetrometer, **6**-12, **6**-29
plate loading, **6**-31, **6**-32
pressure-meter, **6**-12
Proctor (moisture-density), **6**-16, **6**-17
standard penetration, **6**-12, **6**-29, **6**-30
toughness, **6**-8
vane, **6**-12
water effects on, **6**-3 to **6**-5
weight of, **5**-6, **6**-49
(*See also* Earth pressure; Footings; Foundations)
Solders, **4**-59
Solenoid, **20**-4
Sound:
A-scale readings of, **17**-7
in communication, **17**-5
control of: by absorption, **17**-10, **17**-12 to **17**-14, **17**-16, **17**-17, **17**-22 to **17**-24
with barriers, **17**-10, **17**-12, **17**-22, **17**-24
with carpet, **17**-12, **17**-16
by damping, **17**-12, **17**-22, **17**-24
by isolation (insulation), **17**-10, **17**-12, **17**-22 to **17**-24
with low partitions, **17**-12
by masking noise, **17**-5, **17**-20
methods for, **1**-18, **1**-19, **17**-8, **17**-22 to **17**-24
by personnel protection, **17**-22, **17**-24
by reinforcement, **17**-22, **17**-23
by room configuration, **17**-15 to **17**-17, **17**-23
by source elimination, **17**-22, **17**-24
(*See also* Acoustical materials)
defined, **17**-1
echoing of, **17**-14 to **17**-17
effects on humans, **17**-4, **17**-5

Sound (*Cont.*):
  intensity of, **17**-4, **17**-6, **17**-7
  loudness of, **17**-4, **17**-5, **17**-7
  measurements of, **17**-6, **17**-7
  pitch of, **17**-4
    (*See also* Frequency)
  power level of, **17**-6
  pressure of, **17**-4, **17**-6, **17**-7
  reflection of, **17**-14 to **17**-17
  reverberation of, **17**-14 to **17**-17
  sources of, **17**-1, **17**-8
  transmission of: in elastic medium, **17**-1 to **17**-3, **17**-8
    through barriers, **17**-8 to **17**-11, **17**-18
    by bypassing barriers, **17**-10 to **17**-13, **17**-17
    via structure, **17**-10 to **17**-12
  (*See also* Acoustical materials; Acoustics; Noise; Sound waves; Vibration)
Sound waves:
  amplitude of, **17**-2
  defined, **17**-2
  frequency of (*see* Frequency)
  length of, **17**-2
  motions of, **17**-1 to **17**-3
  spectral distribution of, **17**-4
  velocity of, **17**-2, **17**-3
  (*See also* Noise; Sound; Vibration)
Spandrels (*see* Beams, spandrel; Walls, spandrel)
Specifications:
  automated, **26**-9 to **26**-11
  base-bid, **26**-4
  bidding requirements in, **26**-4 to **26**-6
  binding for, **26**-11
  competition objective of, **26**-4
  components of, **26**-4 to **26**-9
  computer programs for, **26**-10, **26**-11
  contract inclusion in, **26**-7
  defined, **1**-2, **1**-4, **26**-1
  descriptive, **26**-3
  "or equal," **26**-4
  federal, **26**-14
  information provided in, **26**-1 to **26**-4, **26**-12, **26**-13
  information sources for, **26**-14, **26**-15
  legal responsibilities of, **2**-12, **26**-1
  master, **26**-9 to **26**-11
  with multiple contracts, **26**-11 to **26**-13
  organization of, **26**-1, **26**-4, **26**-8
  outline, **2**-11, **2**-12
  performance, **26**-2, **26**-3
  proprietary, **26**-3, **26**-4
  reference, **26**-3
  technical: defined, **26**-7
    divisions of, **26**-8
    preparation of, **26**-7 to **26**-11, **26**-13, **26**-14

Specifications (*Cont.*):
  work covered in, **26**-7 to **26**-9, **26**-12, **26**-13
  (*See also specific materials and building components, for example:* Floor coverings; Steel reinforcement; Wood)
Spikes, **7**-32 to **7**-34, **16**-17
Split rings, **7**-25, **7**-26, **7**-37, **7**-39 to **7**-44
Sponsor-builder, **25**-2, **25**-3, **25**-10
Spring constant, **5**-90, **5**-98
Springing line, **5**-118, **10**-112
Sprinklers:
  alarms with, **3**-36, **19**-39, **19**-40, **19**-46
  applicability of, **3**-35, **3**-36
  area protected by, **19**-46
  components of, **3**-36, **19**-38
  deluge, **19**-41, **19**-42
  design approval of, **19**-47
  drains for, **19**-46, **19**-47
  dry-pipe, **19**-41 to **19**-43
  Firecycle, **19**-43, **19**-44
  heads for, **19**-38, **19**-39, **19**-42
  height and area limits with, **3**-34
  orifices for, **19**-42
  outside, **19**-44, **19**-45
  piping for, **19**-38, **19**-39, **19**-42 to **19**-46
  positioning of, **19**-46
  preaction, **19**-42, **19**-43
  spacing of, **19**-46
  standards for, **19**-38, **19**-39
  testing of, **19**-46, **19**-47
  water requirements for, **19**-40, **19**-41, **19**-47, **19**-48
  wet-pipe, **19**-39 to **19**-41
  (*See also* Fire; Fire detectors; Fire protection; Pipe)
Square, of roofing, **4**-74, **15**-4
Stainless steel (*see* Steel, stainless)
Stairs:
  angle, **22**-4, **22**-5
  capacity of, **3**-46, **22**-6
  circular, **22**-5, **22**-6
  components of, **1**-19, **1**-20, **22**-5, **22**-6
  concrete, **22**-9, **22**-10
  curved, **22**-5, **22**-6
  exit, **3**-33, **3**-43, **3**-44, **22**-1, **22**-7
  fire protection of (*see* exit, *above*)
  flight of, **22**-4, **22**-5
  guards for, **22**-5, **22**-7
  headroom for, **22**-6
  landings for, **1**-19, **22**-5, **22**-7
  loads on, **22**-7
  number required, **22**-6
  parallel, **22**-4, **22**-5
  powered (*see* Escalators)
  railings for, **22**-5 to **22**-9
  rise of, **22**-5
  risers for, **1**-19, **22**-5, **22**-6, **22**-8, **22**-9
  run of, **22**-4, **22**-5

Stairs (*Cont.*):
  scissors, **22**-4, **22**-5
  slopes of, **22**-4
  spiral, **22**-5, **22**-6
  steel, **22**-8, **22**-9
  steps for, **1**-19, **22**-5, **22**-6
  straight, **22**-4, **22**-5
  treads for, **1**-19, **22**-5, **22**-6, **22**-8, **22**-9
  width required for, **22**-6
    (*See also* capacity of, *above*)
  wood, **22**-8
Standards (*see* Specifications)
Standpipes, **3**-36, **3**-47, **19**-48, **19**-49
Staples:
  for acoustical tile, **16**-21
  for gypsumboard, **12**-54, **12**-55
  for lath, **12**-45, **12**-46
Steel:
  alloys of, **4**-38, **4**-40 to **4**-45
  bibliography for, **4**-47
  for bolts, **4**-38, **4**-39
  cast, **4**-39
  coefficient of expansion of, **15**-15
  cold-formed (*see* Cold-formed shapes)
  corrosion of (*see* Corrosion)
  hardening of, **4**-37, **4**-40
  heat treatment of, **4**-37, **4**-40
  hypereutectoid (tool), **4**-36, **4**-45
  hypoeutectoid (carbon), **4**-36, **4**-45
  light-gage (*see* Cold-formed shapes; Sheet steel; Strip steel)
  pipe columns, **8**-28
    (*See also* Columns; Pipe)
  prestressing (*see* Tendons)
  reinforcing (*see* Steel reinforcement)
  sheet (*see* Sheet steel)
  specific heat of, **18**-8
  stainless, **4**-41, **4**-44, **4**-45, **9**-12, **9**-41, **9**-42, **15**-14
  standard, **4**-36, **4**-37
  strip (*see* Strip steel)
  structural (*see* Structural steels)
  tendons (*see* Tendons)
  for tubular sections, **8**-28
  weight of, **5**-6
  welding of (*see* Welding; Welds)
Steel construction:
  cold-formed (*see* Floors, metal deck; Roofs, metal; Walls, metal-panel; sandwich-panel)
  structural (*see* Beams; Floors, open-web joists in; Framing; Girders; Rigid frames; Trusses)
Steel piles, **6**-39
  (*See also* Piles)
Steel reinforcement:
  allowable stresses for, **10**-40
  anchorage of, **10**-48 to **10**-52, **12**-21, **12**-22
  areas of, **10**-22, **10**-24
  for bearing, **10**-90, **10**-91

Steel reinforcement (*Cont.*):
  bending of, **10**-22, **10**-23, **10**-26, **10**-27
  bundled, **10**-26, **10**-51, **10**-105
  compression, **10**-44, **10**-77, **10**-78, **10**-99 to **10**-101, **10**-108
  corrosion of, **10**-10, **10**-11
  cut-off points for, **10**-48, **10**-49, **10**-72, **10**-74, **10**-76, **10**-77
  in deep beams, **10**-109 to **10**-111
  delivery of, **10**-25, **10**-26
  detailing of, **10**-25
  development length of: defined, **10**-48
    in flexural members, **10**-48 to **10**-52, **10**-71, **10**-76, **10**-92
    with hooks, **10**-48, **10**-51, **10**-52
  dowel, **10**-50, **10**-51, **10**-86, **10**-91
  fabrication of, **10**-25, **10**-26
  for glass block, **12**-20
  grades of, **4**-39, **10**-22
  hooks for, **10**-23, **10**-26, **10**-48, **10**-51, **10**-52
  inspection of, **10**-27, **10**-28
  lath, **12**-47
  lifts of, **10**-25, **10**-26
  maximum: in beams, **10**-42, **10**-43, **10**-76, **10**-77
    in columns, **10**-102
    in joists, **10**-65
    in one-way slabs, **10**-61
  maximum spacing of: in beams, **10**-56, **10**-57, **10**-82, **10**-109, **10**-111
    in columns, **10**-105
    in one-way slabs, **10**-55 to **10**-57
    in two-way slabs, **10**-57, **10**-71, **10**-72
    in walls, **10**-84, **10**-106, **10**-110
  mechanical properties of, **4**-39
  minimum: in beams, **10**-42, **10**-128
    (*See also* for shear, *below*)
    in columns, **10**-102
    in joists, **10**-65, **10**-128
    in one-way slabs, **10**-55, **10**-56, **10**-61
    in walls, **10**-83 to **10**-85, **12**-22
  minimum spacing of, **10**-102, **10**-120
  modulus of elasticity of, **10**-42
  perimeters of, **10**-22
  placement of, **10**-26, **10**-27
  in precast concrete, **10**-119, **10**-120
  prestressed (*see* Tendons)
  protective cover for: in beams, **10**-57, **10**-61, **10**-82, **10**-118, **10**-119, **10**-130
    in columns, **10**-61, **10**-102, **10**-103, **10**-118, **10**-119, **10**-130
    in footings, **10**-91
    in joists, **10**-61, **10**-118, **10**-119, **10**-130
    in precast concrete, **10**-82, **10**-118, **10**-119
    in prestressed concrete, **10**-82, **10**-130
    in shells, **10**-118, **10**-130

Steel reinforcement, cover for (*Cont.*):
  in slabs, **10**-57, **10**-60, **10**-61, **10**-64, **10**-118, **10**-121
  tolerances for, **10**-26, **10**-120
  in walls, **10**-84, **10**-85, **10**-118, **10**-130, **12**-21
  rusty, **10**-28
  for shear, **10**-46, **10**-66, **10**-67, **10**-79 to **10**-81, **10**-110 to **10**-112, **10**-128
  shipping limitations for, **10**-26
  for shrinkage control, **10**-55 to **10**-57, **10**-64, **10**-65, **10**-116
  sizes of, **10**-22, **10**-24
  specifications for, **4**-39, **10**-22, **10**-24, **10**-25
  spiral, **10**-53, **10**-102, **10**-104
  splices of, **10**-53 to **10**-55, **10**-86, **10**-91, **10**-103, **10**-105, **10**-108
  stirrup (*see* Stirrups)
  supports for, **10**-27
  for temperature, **10**-55 to **10**-57, **10**-64, **10**-65, **10**-116
  tension, **10**-40 to **10**-42, **10**-45, **10**-76 to **10**-78
  ties, **10**-23, **10**-102, **10**-104 to **10**-106
  for torsion, **10**-47, **10**-48, **10**-79 to **10**-81
  weights of, **10**-22, **10**-62, **10**-65, **10**-66, **10**-75
  welding of, **10**-22, **10**-27, **10**-55
  wire-fabric, **10**-24, **10**-25, **10**-50, **10**-54, **10**-55
  yield strengths of, **4**-39, **10**-22
Steel stairs, **22**-8, **22**-9
Steel strand, **4**-37, **4**-39, **4**-40, **10**-25, **22**-25
Steel wire, **4**-37, **4**-39, **4**-40, **10**-24, **10**-25
Stefan-Boltzmann law, **18**-19
Stiffeners:
  columns, **8**-27, **8**-28
  girder: bearing, **8**-26, **8**-33, **8**-47
    transverse, **8**-47 to **8**-49
  at plastic hinges, **8**-56
Stiffness:
  actual beam, **5**-70, **5**-71
  of concrete members, **10**-70, **10**-71
  fixed-end beam, **5**-61, **5**-62
  of shear walls, **11**-11, **11**-12, **11**-14 to **11**-16
  in vibration, **5**-107
  (*See also* Matrix, stiffness)
Stiles, **13**-2, **13**-3, **14**-6
Stirrups:
  hooks for, **10**-23
  minimum size of, **10**-80, **10**-81
  protective cover for, **10**-57
  for shear, **10**-46, **10**-52, **10**-53, **10**-66, **10**-67, **10**-79, **10**-80, **10**-109, **10**-110
  for torsion, **10**-47, **10**-79 to **10**-81
Stodola-Vianello method, **5**-92
Stone:
  building: allowable stresses for, **12**-7
  characteristics of, **4**-18 to **4**-21

Stone, building (*Cont.*):
  weights of, **5**-6, **5**-9
  (*See also* Masonry; Masonry walls)
  for concrete (*see* Aggregates)
Stops:
  blind, **13**-4, **13**-5
  door, **14**-6 to **14**-8
  parting, **13**-4
Storm water:
  disposal of, **19**-2, **19**-22
  drainage of, **19**-29 to **19**-31
Strain, **5**-13, **5**-14, **5**-17, **5**-18
  (*See also* Deflections; Deformation)
Strain energy, **5**-15 to **5**-17, **5**-51 to **5**-53
  stress-strain relationship (*see* Hooke's law; Stress-strain relationship)
Strand, **4**-37, **4**-39, **4**-40, **22**-25
Strength:
  ultimate: under dynamic loads, **5**-88
    under static loads (*see* Beams, in plastic range; Concrete beams, strength design of; Columns; *specific materials, for example:* Structural steels; Wood)
  yield: under dynamic loads, **5**-88
    under static loads, **5**-13
  (*See also* Structural steels)
Stress:
  allowable, **5**-12, **5**-13
  (*See also specific materials, for example:* Structural steels; Wood)
  arch, **5**-116
  bearing, **5**-14
  bending, **5**-27, **5**-39 to **5**-41
  (*See also* Beams; Cold-formed shapes; Concrete beams; Girders)
  cable (*see* Cable-supported structures; Cables)
  combined: bending-and-axial, **5**-33, **5**-37, **5**-41, **5**-42
    bending-and-shear, **5**-30
    normal-and-shear, **5**-18 to **5**-20
    (*See also* allowable stresses for *under specific materials*)
  components of, **5**-17 to **5**-20
  compression (*see* Cold-formed shapes, as columns; Columns; Compression; Concrete columns)
  defined, **5**-2, **5**-13
  diagonal tension (*see* Shear, in concrete beams)
  dynamic, **5**-91
  (*See also* impact, *below*; Vibration)
  of eccentric loads, **5**-33, **5**-37
  fatigue, **5**-88, **5**-89
  impact, **5**-96 to **5**-98
  membrane, **5**-132
  (*See also* Membrane theory)
  normal, **5**-17 to **5**-20
  notation for, **5**-17, **5**-18
  principal, **5**-18 to **5**-20

Stress (*Cont.*):
  proof, **5**-13
  reversed, **5**-88, **5**-89
  shear (*see* Shear)
  shell, **5**-121, **5**-129, **5**-130
  tension (*see* Tension)
  thermal, **5**-15
  torsion (*see* Torsion)
  wind, **5**-154, **5**-155
  working (*see* allowable, *above*)
  yield (*see* Yield point)
Stress-strain relationship, **5**-13, **5**-85
Stretcher, **12**-4, **12**-9
Stringers, stair, **1**-10, **22**-5, **22**-8, **22**-9
  (*See also* Beams)
Strip steel, **4**-38, **9**-1, **9**-9, **9**-10, **9**-12, **9**-13
  (*See also* Cold-formed shapes)
Structural design, **5**-1, **5**-13
  (*See also* Systems, structural; *specific materials, for example:* Brick; Concrete; Structural steels; *specific structural members, for example:* Arches; Beams; Columns)
Structural steels:
  allowable stresses for: bearing, **8**-26, **8**-27, **8**-33, **8**-34, **8**-39, **8**-40
    bending, **8**-28 to **8**-33
    bending and compression, **8**-34, **8**-35
    bending and shear, **8**-49
    bending and tension, **8**-35
    bolts, **8**-39 to **8**-41
    compression, **8**-23 to **8**-26
    plate girders, **8**-46
    rivets, **8**-39 to **8**-41
    seismic, **8**-35 to **8**-37
    shear, **8**-22, **8**-23
    tension, **8**-21, **8**-22
    threaded parts, **8**-39
    web crippling, **8**-26 to **8**-28
    welds, **8**-37, **8**-38, **8**-56, **8**-57
    wind, **8**-35 to **8**-37
  alloy, **4**-38, **4**-40 to **4**-43, **8**-4, **9**-11
  beams of (*see* Beams; Girders, steel)
  bendability of, **9**-11
  coefficient of expansion of, **15**-15
  columns of (*see* Columns, structural steel)
  concrete-embedded, **8**-41, **8**-42, **8**-59, **8**-106, **8**-107
    (*See also* Beams, composite; Columns, composite)
  connections for (*see* Connections; Framing)
  corrosion resistance of, **8**-4, **8**-103, **8**-104
    (*See also* Corrosion)
  defined, **8**-1
  design conditions for, **8**-21
  elongation of, **4**-38, **9**-11
  erection of: clearances for, **8**-76, **8**-77, **8**-86, **8**-94, **8**-97 to **8**-99

Structural steels, erection of (*Cont.*):
  equipment for, **8**-97
  planning of, **8**-6, **8**-7, **8**-15
  sequence of, **8**-99
  tolerances for, **8**-102
  fabrication of: defined, **8**-1
    tolerances for, **8**-100 to **8**-102
  fatigue in, **8**-36, **8**-37
  fire protection of, **8**-107 to **8**-112
  framing of (*see* Floors; Framing; Roofs)
  girders of (*see* Girders, steel)
  heat effects on, **8**-99, **8**-100, **8**-102, **8**-107
  identification of, **8**-4
  items furnished as, **8**-96
  mill tolerances for, **8**-2
  minimum thicknesses of, **8**-1
  modulus of elasticity of, **8**-23, **8**-108
  painting of, **8**-3, **8**-97, **8**-103 to **8**-107
  plates of (*see* Plates, steel)
  shapes of, **8**-1 to **8**-3
  shipping limitations for, **8**-14, **8**-15
  specifications for, **4**-38, **8**-3 to **8**-5, **9**-11
  tensile strength of, **4**-38, **8**-3, **8**-4, **9**-11
  weathering, **8**-3, **8**-104, **8**-105
  weldable types of, **4**-45, **4**-46, **8**-3, **8**-4
  welding of (*see* Welding; Welds)
  yield strength of, **4**-38, **8**-4, **9**-11
  (*See also* Steel; Framing, steel)
Structural theory, **5**-1
  (*See also* Strains; Stress; Systems, structural; *specific materials, for example:* Brick; Concrete; Wood; *specific structural members, for example:* Arches; Beams; Columns)
Stucco, **3**-28, **12**-26, **12**-27, **12**-43
Studs:
  metal wall, **5**-9, **12**-23 to **12**-25
  powder-driven, **16**-20
  spacing for plaster, **12**-46, **12**-47
  welded, **4**-46, **8**-19, **8**-20, **8**-44, **15**-19, **16**-18 to **16**-20
  wood (*see* Wood walls)
Subcontractors:
  back-charge orders to, **25**-40, **25**-42
  bonding of, **25**-32
  function of, **1**-5
  organization of, **25**-3
  payments to, **25**-47 to **25**-50
  (*See also* Contractors)
Subcontracts, **25**-7, **25**-13 to **25**-18
  (*See also* Contracts)
Subfloors, **1**-16, **5**-8, **7**-50, **7**-64, **7**-79
  (*See also* Floor coverings; Floors)
Substrate, **12**-43
Substructures, **1**-10
Subsystems, defined, **1**-3
  (*See also* Systems; Systems design)
Sun effect, **18**-6, **18**-19, **18**-41, **18**-64
Superintendents, **1**-6
Superposition, principle of, **5**-16

Superstructures, **1**-10
Surveying instruments, **23**-11, **23**-12
Surveys:
　of building movements, **23**-7
　construction, **1**-6, **23**-1, **23**-7 to **23**-11
　of existing buildings, **1**-6, **23**-6
　possession, **23**-1
　preliminary, **1**-6, **23**-1 to **23**-7
　quantity, **24**-6 to **24**-9
　units of measurement for, **23**-1, **23**-2
Switches:
　dimmer, **21**-15, **21**-24, **21**-25
　electromagnetic, **20**-20
　fused, **20**-40 to **20**-44
　limitations on, **20**-17
　service-entrance, **20**-11, **20**-15, **20**-17, **20**-44
　types of, **20**-16, **20**-17, **20**-20
Syenite, **4**-20
Systems:
　analysis of, **1**-3, **1**-4, **1**-28, **1**-30, **1**-33
　　(*See also* Systems design)
　appraisal of (*see* analysis of, *above*; Value analysis)
　building-enclosure, **1**-9, **1**-12 to **1**-15, **12**-1
　　(*See also* Foundations; Roofs; Shells; Walls)
　comparisons of, **1**-21 to **1**-25
　　(*See also* Value; Value analysis)
　criteria for, **1**-29
　defined, **1**-3
　design of (*see* Systems design)
　electrical communication, **20**-21
　　(*See also* Alarms)
　electric power, **1**-19
　　(*See also* Electric apparatus; Electric energy; Electric power; Electric wiring)
　finish, **1**-7
　floor-ceiling, **1**-11, **1**-12, **1**-14 to **1**-16, **8**-16 to **8**-21, **8**-110
　HVAC (*see* Air conditioning; Cooling; Heating; HVAC; Ventilation)
　interior-enclosure, **1**-16, **1**-17
　lighting (*see* Lighting)
　models of, **1**-30 to **1**-32
　packaged, **2**-20
　plumbing, **1**-17, **1**-18
　protective signaling, **1**-19, **19**-49
　roof-ceiling, **1**-11, **1**-12, **1**-14 to **1**-16, **8**-21
　standards for, **1**-29
　structural, **1**-10 to **1**-12
　　(*See also* Foundations; Framing; Roofs; Shells; Walls)
　synthesis of (*see* Systems design)
　vertical-circulation, **1**-19, **1**-20
　　(*See also* Dumbwaiters; Escalators;

Systems (*Cont.*):
　　Elevators; Ramps; Stairs)
　(*See also specific types of systems, for example:* Electric wiring; Fire protection; Floors; Framing; Lighting; Plumbing)
Systems building, **1**-30, **2**-20, **2**-21, **9**-66, **25**-59 to **25**-62
Systems design:
　advantages of, **1**-1, **1**-3
　aims of, **1**-27, **1**-29, **1**-30
　application of, **1**-37
　constraints in, **1**-29
　defined, **1**-2
　goals in, **1**-27, **1**-29
　objectives in, **1**-29
　optimization in, **1**-31, **1**-32
　organization for, **1**-37, **2**-17, **2**-18
　procedure for, **1**-2, **1**-3, **1**-27 to **1**-33
　(*See also* Building design; Systems; Value analysis)

Tanks:
　expansion, **18**-52, **18**-53
　rubber-lined, **4**-73
　seismic loads for, **5**-111 to **5**-113
　storage requirements for, **19**-6, **19**-7, **19**-47
Tapes, surveying, **23**-7, **23**-8
Tar, **5**-5
Telephone systems, **20**-21
Television systems, **20**-21
Temper, **12**-43
Temperatures:
　design, **18**-6, **18**-20 to **18**-22, **18**-40, **18**-43, **18**-44
　dew-point, **18**-6, **18**-13 to **18**-15, **18**-28, **18**-29
　effective, **18**-6, **18**-7
　for heat balance, **18**-15
　mean radiant, **18**-19, **18**-20
　scales for, **18**-6 to **18**-8, **18**-10
Template, **12**-43
Tendons:
　allowable stresses for, **10**-125
　anchorage of, **6**-55, **10**-124, **10**-125
　curved, **10**-122, **10**-123
　design stress in, **10**-126
　development length of, **10**-129
　function of, **10**-122
　grouting of, **10**-129, **10**-130
　properties of, **4**-37, **4**-39, **4**-40, **10**-25
　protective cover for, **10**-130
　relaxation of, **10**-124
　sheaths for, **10**-130
　specifications for, **10**-25
　strength of, **10**-25
　transfer length of, **10**-129
　types of, **10**-25

Tension, **5**-2, **5**-13, **5**-14, **5**-16
Tension members:
  steel: allowable stresses for, **8**-21, **8**-22
    net section for, **8**-22
    slenderness ratio of, **8**-50, **8**-67
  wood, **7**-37
Terneplate, **4**-47
Terra cotta, **4**-17, **12**-2
Terra-Probe, **6**-18
Terrazzo, **5**-8, **12**-68, **12**-69
Thermodynamics, laws of, **18**-9, **18**-10
Thermometers, **18**-6 to **18**-8, **18**-14
Threshold, **14**-2, **14**-6
Tie rods, **7**-45, **7**-47, **7**-61, **7**-62
  (*See also* Bars, steel)
Ties:
  for arches and rigid frames, **7**-45, **7**-61, **7**-62, **8**-14, **10**-112
  for concrete columns (*see* Concrete columns, tied; Steel reinforcement, ties)
  for masonry, **12**-9 to **12**-11, **12**-16, **12**-17
Tile:
  acoustical, **16**-21, **17**-13 to **17**-15, **17**-19
  asphalt, **5**-8, **12**-63, **12**-64, **12**-67
  ceramic, **4**-17, **5**-8, **12**-31 to **12**-36
  cork, **5**-6, **5**-9, **12**-63 to **12**-67
  drain, **3**-19
  expansion joints for, **12**-35, **12**-36
  grouting of, **12**-35
  gypsum, **4**-22, **12**-5, **12**-48
  mortar for, **12**-5, **12**-32 to **12**-35, **12**-48
  rubber, **12**-65 to **12**-67
  structural clay, **3**-22, **4**-17, **4**-18, **10**-64, **12**-5, **12**-48
  vinyl, **12**-64 to **12**-67
  weights of, **5**-8
  (*See also* Roofing, tile)
Timber:
  defined, **4**-30
  dressed sizes of, **7**-5
  half, **12**-26
  laminated (*see* Lumber)
  (*See also* Framing; Wood)
Tin, **5**-6, **5**-9
Ton:
  refrigeration, **18**-7, **18**-15
  weight, **5**-12
Tornadoes, **3**-1, **3**-4
Torsion:
  angular twist of, **5**-87
  in concrete, **10**-39, **10**-45 to **10**-48, **10**-79 to **10**-81
  defined, **5**-2
  energy of, **5**-16
  rotation center in, **5**-2, **5**-38, **5**-39, **5**-86
  on shear walls, **11**-11, **11**-12
  simple, **5**-86
  in steel beams, **8**-35, **8**-36

Torsion (*Cont.*):
  stresses from, **5**-86 to **5**-88
Towers:
  cooling (water), **18**-4, **18**-72 to **18**-74, **18**-87, **18**-88, **18**-91
  smokeproof, **3**-44, **22**-7
Trade associations (*see* Associations)
Transformers, **20**-4, **20**-15
Transoms, **22**-17
Traps:
  plumbing (*see* Plumbing, traps in)
  steam, **18**-6, **18**-55
Treads, **1**-19, **22**-3 to **22**-6, **22**-8 to **22**-11
Trimmers, **1**-13, **1**-19, **9**-4, **22**-8
Trusses:
  Bow's notation for, **5**-49
  bracing of (*see* Bracing)
  camber for, **8**-51
  chords in, **5**-46
  continuous, **5**-76 to **5**-82
  defined, **1**-10, **5**-46, **8**-49
  deflections of, **5**-54 to **5**-56
  graphical analysis of, **5**-49, **5**-50
  indeterminate, **5**-50, **5**-52, **5**-53, **5**-57
    (*See also* continuous, *above*)
  loads on, **5**-46, **5**-50
  method of joints for, **5**-48 to **5**-50
  method of sections for, **5**-47
  panel points of, **8**-49
  reactions for, **5**-12, **5**-44, **5**-45
  steel, **8**-9 to **8**-13, **8**-49, **8**-50, **8**-58
    (*See also* Joists, open-web)
  types of, **5**-46, **7**-58, **8**-9 to **8**-13
  Vierendeel, **8**-10, **12**-20
  web members of, **5**-46
  wood, **7**-4, **7**-57 to **7**-59, **7**-87
Tubes, structural, **8**-28, **9**-38, **9**-39
Turnkey developers, **1**-6, **2**-25, **25**-2

U factor, **13**-19
Underwriters' Laboratories, **3**-30, **12**-31, **14**-6, **20**-11, **20**-21, **20**-22, **26**-15
Unions, labor (*see* Labor)
Urethane, **4**-63, **4**-64, **4**-76, **4**-78, **5**-9, **15**-9, **15**-13, **15**-17
Urinals:
  fixture units for, **19**-15
  minimum flow for, **19**-5
  minimum pipe size for, **19**-19
  number required, **19**-12, **19**-13

Value:
  defined, **1**-20
  future, **1**-23, **1**-24
  measures of, **1**-21 to **12**-24
  salvage, **1**-23, **1**-24
Value analysis:
  aims of, **1**-20, **1**-21
  application of, **1**-26, **1**-33

Value analysis (*Cont.*):
  comparisons in, **1**-21 to **1**-25
  defined, **1**-21
  procedure for, **1**-25, **1**-26
Value engineering, **1**-20 to **1**-26
Valves, **19**-4, **19**-9, **19**-19, **19**-39 to **19**-45
Vandal protection, **3**-50, **3**-51, **13**-6, **13**-7
Vapor barriers, **3**-23 to **3**-28, **15**-5, **15**-8, **15**-9, **18**-7, **18**-30
Variation, coefficient of, **10**-12
Varnish, **4**-78, **4**-80
Veneer:
  ceramic, **4**-17, **12**-2
  masonry, **12**-4
  wood (*see* Plywood)
Ventilation:
  air exhaust from, **18**-34
  air recirculation for, **18**-34
  air required for, **18**-31 to **18**-33
  defined, **18**-7
  for heat removal, **18**-32, **18**-33, **18**-79
  mechanical, **18**-31 to **18**-36
  for moisture removal, **18**-29, **18**-32, **18**-33
  natural, **13**-1, **18**-31, **18**-33
  for odor removal, **18**-32, **18**-33
  purposes of, **1**-18, **18**-30, **18**-31
  for smoke removal, **3**-41, **18**-31
  with windows, **13**-1, **18**-31, **18**-33
Ventilators:
  power roof, **18**-35
  unit, **18**-7
Vents:
  for fire-heat release, **3**-41, **22**-18
  for flue gases, **18**-50, **18**-57
  plumbing, **19**-2, **19**-25 to **19**-27, **19**-32 to **19**-35
  for roof moisture, **15**-8, **15**-20, **15**-21, **18**-30
Vermiculite, **4**-6, **4**-8, **5**-9, **18**-22, **18**-24
Vertical transportation (*see* Conveyors; Dumbwaiters; Egress, emergency; Elevators; Escalators)
Vestibules, **14**-2, **14**-3
Vibration:
  characteristic shape in, **5**-92 to **5**-95, **5**-100, **5**-107
  of concrete, **10**-29
  cycle of, **17**-2
  damage from, **5**-115
  damping of, **5**-104 to **5**-106, **17**-12
  defined, **17**-1
  elimination of, **5**-104 to **5**-106, **5**-114
  forced, **5**-98 to **5**-108
  frequency of (*see* Frequency)
  human sensitivity to, **5**-115
  insulators for, **4**-73
  isolation of, **17**-12
  natural period of, **5**-88 to **5**-95, **5**-107, **5**-110
  normal modes of, **5**-91 to **5**-95

Vibration (*Cont.*):
  resonance in, **5**-104, **5**-105, **5**-147, **5**-148
  steady-state response, **5**-106
  transient response, **5**-105, **5**-106
  (*See also* Noise; Sound)
Vibroflotation, **6**-18, **6**-19
Vinyl, **4**-63, **4**-65, **4**-66, **4**-68, **4**-78, **15**-20
Vinyl floor coverings, **12**-64, **12**-65, **12**-67
Voltage:
  calculation of, **20**-2, **20**-3, **20**-8, **20**-28, **20**-29, **20**-31, **20**-46, **20**-48
  defined, **20**-2
  service, **20**-27
Volume:
  formulas for, **23**-6
  specific, **18**-7, **18**-12

Wadding, **12**-43
Wainscot, **12**-43
Walks, moving, **3**-44
Wallboard (*see* Asbestos-cement; Fiberboard; Gypsumboard; Plywood; Pulp board)
Walls:
  basement (*see* foundation, *below*)
  bearing, **1**-11, **1**-15, **1**-16, **8**-5 to **8**-7, **10**-82
  (*See also* Brick, engineered construction with; Concrete walls, bearing; Masonry walls, bearing; Wood walls, stud)
  bracing for, **7**-61, **7**-63, **7**-64, **11**-2, **12**-23, **12**-24
  brick (*see* Brick, engineered construction with; Masonry walls)
  cavity (*see* hollow, *below*)
  ceramic-tile, **12**-31 to **12**-36
  coatings for (*see* Coatings)
  cold-side venting of, **18**-29, **18**-30
  combination, **1**-14
  concrete (*see* Concrete walls; Shear walls, concrete)
  connections to floors, **11**-19, **11**-20
  curtain, **1**-14, **1**-15, **8**-7, **10**-84, **10**-85, **12**-25 to **12**-30
  (*See also* metal-panel; nonbearing; siding, *below*)
  defined, **10**-82, **12**-4
  exterior, **1**-13 to **1**-15, **12**-3
  (*See also specific materials, for example:* Concrete walls; Masonry walls; metal-panel, *below*; Wood walls)
  faced, **1**-14, **12**-4, **12**-17
  fire-barrier, **3**-33, **3**-34, **14**-5
  fire protection of openings in, **14**-4, **14**-5
  flashing for (*see* Flashing)

Walls (*Cont.*):
   foundation, 1-10, 5-7, 6-49 to 6-55, 7-64 to 7-66, 10-88, 10-89
      (*See also* Concrete walls, basement; Foundations; Masonry walls, basement)
   framed, 1-13, 12-26
      (*See also* stud, *below*; Wood walls)
   glass, 12-28
   glass block, 4-23, 5-8, 12-5, 12-20, 12-21
   half-timber, 12-26
   hollow, 3-27, 11-7, 12-3 to 12-5, 12-7, 12-9 to 12-12, 12-14, 12-16, 12-17
   insulation of (*see* Condensation; Heat transmission; Insulation)
   interior finishes for: gypsumboard (*see* Gypsumboard)
      masonry (*see* Masonry; Masonry walls; Tile)
      paneling, 12-36
      plaster (*see* Plaster)
      requirements for, 1-17
   masonry (*see* Masonry walls)
   metal-panel, 3-26, 12-27 to 12-30, 13-14
   nonbearing, 1-14, 1-15, 8-7
      (*See also* Concrete walls; curtain, *above*; Masonry walls; Partitions; siding for, *below*)
   panel, 1-13, 3-26
      (*See also* bearing; curtain; metal-panel, *above*)
   parapet, 3-33, 12-17
   partition (*see* Partitions)
   party, 12-4
   plywood, 12-26
   sandwich-panel, 12-28 to 12-30
   shear (*see* Shear walls)
   sheathing for (*see* Sheathing)
   shingle, 4-31, 5-9, 12-26
   siding for, 5-9, 12-26 to 12-28
   slurry trench, 6-47, 6-48
   spandrel, 12-4
   stucco, 3-28, 12-26, 12-27, 12-43
   stud, 1-13, 1-16, 5-9, 12-3, 12-23, 12-24
      (*See also* Wood walls)
   supports for openings in, 10-84, 12-21, 14-7
      (*See also* Beams, spandrel; Lintels)
   tilt-up (*see* Concrete walls)
   vapor barriers for (*see* Vapor barriers)
   veneered, 12-3, 12-4, 12-13, 12-14, 12-17
   waterproofing membrane for (*see* Membrane, waterproofing)
   weights of, 5-8, 5-9
   wood (*see* Wood walls)
      (*See also* Concrete walls; Masonry walls)

Washers, for bolts, 7-44 to 7-46, 8-68 to 8-70, 8-72
Wastewater:
   disposal of, 19-21 to 19-23
   industrial, 19-22, 19-23
   minimum flow velocity for, 19-28
   piping for, 19-25 to 19-29
      (*See also* Pipe; Sewers)
   quantity of, 19-10
   treatment of, 19-22, 19-23, 19-29
   types of, 19-22
      (*See also* Plumbing; Sewers; Storm water)
Water:
   characteristics of, 19-3, 19-4
   latent heat of fusion for, 18-10
   latent heat of vaporization of, 18-11, 18-12
   makeup, 18-7
   specific heat of, 18-8
   weight of, 5-6
      (*See also* Water supply)
Water closets:
   fixture units for, 19-15
   functioning of, 19-11
   minimum flow for, 19-5
   minimum pipe size for, 19-19
   number required, 19-12, 19-13
   types of, 19-11
Water hammer, 19-10
Water heaters, 18-54, 19-20, 19-21
Water-repellent materials, 3-18
Water-resistant materials, 3-18
Water supply:
   biliography for, 19-3, 19-5
   contamination prevention for, 19-4, 19-7
   for fire fighting, 19-6, 19-47 to 19-49
   hot, 19-6, 19-7, 19-20 to 19-22
   maximum probable flow for, 19-13 to 19-16
   maximum velocity for, 19-5, 19-17
   meters for, 19-10
   piping for, 19-6 to 19-10, 19-19, 19-20
      (*See also* Pipe)
   pressures for, 19-5, 19-6
   pumping for 19-7
   quality requirements for, 19-2 to 19-4
   quantity requirements for, 19-1, 19-2, 19-5, 19-6, 19-11, 19-13 to 19-16, 19-20, 19-22
   sources for, 19-2
   for sprinklers, 19-47, 19-48
   storage of, 19-6 to 19-9
   treatment of, 19-4, 19-5
      (*See also* Plumbing)
Waterproof materials, 3-18, 7-66
Waterproofing, 3-16 to 3-29, 5-9, 7-66
   (*See also* Coatings; Floodproofing)
Waterstops, 3-17, 3-22, 3-26, 10-34
Watt, 20-2

Weatherstripping, 3-17, 13-1, 13-15, 14-2, 14-10, **18**-23
Weep holes, 3-27, **12**-12, **12**-13
Weights, material, 4-21, 5-5, 5-8, 5-9, **6**-49
(See also specific materials, for example: Aggregates; Cement; Concrete; Steel*)*
Welding:
  advantages of, 8-72
  of aluminum, 4-51
  clearances for, 8-76, 8-77
  of ferrous metals: base-metal temperatures for, 8-73, 8-74
    electrodes for, 4-45, 4-46, 8-38
    processes for, 4-45, 4-46, 8-72, **9**-42 to **9**-47
  field procedures for, 8-99, 8-100, **9**-42
  specifications for, 4-45, **9**-44
  stud, 4-46, 8-19, 8-20, 8-44, **15**-19, **16**-18 to **16**-20
  (See also Welds*)*
Welds:
  allowable loads or stresses for, 8-37, 8-38, 8-56, 8-57, **9**-44 to **9**-46
  for butt joints *(see* groove, *below)*
  for cold-formed steel, **9**-43 to **9**-48
  fillet, 8-37, 8-38, 8-72 to 8-74, 8-82, **9**-43 to **9**-45, **9**-48
  groove, 8-37, 8-38, 8-72 to 8-74, 8-82, **9**-43, **9**-45
  in plastic design, 8-56, 8-57
  symbols for, 8-75
  (See also Beams, splices in; Columns, structural steel; Connections, welded; Welding*)*
Westergaard stress distribution, **6**-25, **6**-26
Wind connections, 3-9, 3-10, 5-153
  (See also Connections, rigid (moment); semirigid*)*
Wind design, 3-1 to 3-13
  (See also Cantilever method; Dynamics; Portal method; Wind connections; Wind loads; Wind stresses*)*
Wind loads, 3-2 to 3-8, 5-3, 5-5, 5-7, 5-50, **13**-17 to **13**-19, **13**-23
Wind stresses, 5-154, 5-155
Wind-tunnel tests, 3-4
Winders, **22**-6
Windows:
  air infiltration at, **13**-1
  aluminum, **13**-5
  architectural effects of, **13**-1
  architectural projected, **13**-8, **13**-9, **13**-13
  area required for, **13**-1
  austral, **13**-12
  awning, **13**-10, **13**-12
  basement, **13**-8

Windows *(Cont.)*:
  bronze, **13**-6
  casement, **13**-9 to **13**-12, **16**-14
  casings for, **13**-3
  commercial projected, **13**-6, **13**-7
  components of, 1-15, **13**-2 to **13**-5
  continuous top hung, **13**-12, **13**-13
  defined, **13**-2
  detention, **13**-9, **13**-11
  double-glazed, **13**-1, **13**-2, **13**-18 to **13**-23
  double-hung, **13**-2 to **13**-6, **13**-9, **13**-11, **13**-12, **16**-14, **16**-15
  frames for, 1-13, 1-15, **13**-3
  glass for *(see* Glass*)*
  hand (swing) of, **13**-7, **16**-14
  hardware for, 1-16, **16**-14, **16**-15
  in hazardous locations, **13**-1
  heat transmission of, **13**-1, **13**-2
  horizontal sliding, **13**-12
  intermediate combination, **13**-9 to **13**-11
  intermediate projected, **13**-8, **13**-10
  jalousie, **13**-10, **13**-12
  for light, 1-18, **13**-1, **21**-12, **21**-13
  materials for, **13**-2 *(See also* Glass*)*
  measurements for, **13**-2, **13**-3
  mechanical operators for, **13**-13 to **13**-15
  modular, **13**-2, **13**-3, **13**-6
  monitor, 1-12, 3-41
  picture, **13**-12, **13**-15
  pivoted, **13**-6, **13**-7, **13**-12, **13**-13
  psychiatric projected, **13**-8, **13**-9, **13**-11
  purposes of, 1-16, **13**-1
  ranch, **13**-10, **13**-12
  sash balances for, **16**-14, **16**-15
  security, **13**-6, **13**-7
  stainless steel, **13**-6
  steel, **13**-5
  storefront, **13**-14
  storm, **13**-15
  symbols for, **13**-7
  utility, **13**-8
  for ventilation, **13**-1, **18**-31, **18**-33
  vertical folding, **13**-12
  vertical pivoted, **13**-12
  wind on, **13**-17 to **13**-19, **13**-23
  wood, **13**-2 to **13**-5 *(See also* Glass*)*
Wire, concrete reinforcing, 4-37, 4-39, **10**-24, **10**-25
  (See also Electric wiring; Steel reinforcement; Tendons*)*
Wire-fabric reinforcement, **10**-24, **10**-25, **10**-50, **10**-54, **10**-55
  (See also Steel reinforcement*)*
Wood:
  adhesives for *(see* Adhesives, for wood*)*
  advances in construction with, **7**-1, **7**-2
  advantages of, **7**-2
  allowable unit stresses for, 4-25, **7**-9, **7**-14, **7**-15, **7**-18 to **7**-25

Wood (*Cont.*):
  basic unit stresses for, **7**-9
  bibliography for, **4**-35
  buckling factor for, **7**-23
  characteristics of, 4-25 to 4-29, **7**-2, **7**-3
  checks in, **7**-15, **7**-18
  chemical resistance of, **7**-30, **7**-32
  commercial grades of, 4-29 to 4-31
  in composite construction, **7**-82, **7**-83
  composition of, 4-25, **7**-30
  curvature effects on, **7**-21, **7**-22
  decay resistance of, 4-31, 4-32, **7**-28
  erection of, **7**-84, **7**-86 to **7**-88
  fabrication of, **7**-25 to **7**-28
  fire resistance of, **7**-30 to **7**-32
  fire-resistant treatments for, 1-16, 4-33, **7**-31, **7**-32
  grading of, **7**-9, **7**-14
  job-site storage of, **7**-84, **7**-86, **7**-87
  laminated (*see* Lumber)
  load-duration effects on, **7**-19
  modulus of elasticity of, 4-26, 4-27, **7**-2, **7**-19, **7**-75, **7**-85
  modulus of rigidity of, **7**-75
  modulus of rupture of, **7**-85
  moisture effects on, 4-28, 4-29, **7**-2, **7**-14 to **7**-18, **7**-25
  painting of, **7**-31
  preservatives for, 4-31 to 4-33, **7**-3, **7**-20, **7**-28 to **7**-30, **7**-65
  radial-stress effects on, **7**-21
  section properties of, **7**-3, **7**-5 to **7**-7, **7**-10 to **7**-13
  size effects on, **7**-10 to **7**-13, **7**-20, **7**-21
  slenderness effects on, **7**-22
  species for construction, **7**-8
  specific gravity (weight) of, 4-26, 4-27, **7**-3, **7**-8
  specific heat of, **18**-8
  specifications for, **7**-9, **7**-14, **7**-15, **7**-18, **7**-19
  strength of 4-25 to 4-27, **7**-9, **7**-14, **7**-15, **7**-18 to **7**-22
  strength tests of, **7**-3, **7**-9
  temperature effects on, **7**-2, **7**-19, **7**-20
  (*See also* Connections; Framing; Lumber; Timber)
Wood arches, **7**-4, **7**-23, **7**-25, **7**-59 to **7**-63, **7**-87
Wood beams:
  buckling factor for, **7**-23
  camber for, **7**-56, **7**-57
  curvature factor for, **7**-21, **7**-22
  design of, **7**-50, **7**-51, **7**-53 to **7**-56
  effective length of, **7**-23, **7**-24
  lateral support of, **7**-22 to **7**-24
  (*See also* Bracing; Bridging)
  load capacities of, **7**-51, **7**-53 to **7**-55
  prestressed, **7**-83

Wood beams (*Cont.*):
  radial stress in, **7**-21
  section properties of, **7**-6, **7**-10 to **7**-13
  shear strength of, 4-26, 4-27, **7**-18
  size factors for, **7**-10 to **7**-13, **7**-20, **7**-21
  slenderness factors for, **7**-22
  spans of, **7**-4
  types of, **7**-51, **7**-53
  (*See also* Beams; Deflections)
Wood block, 5-8, **12**-67, **12**-69
Wood clapboard, **12**-26
Wood columns, **7**-18, **7**-48 to **7**-51, **7**-86
  (*See also* Columns)
Wood decking (*see* Wood floors; Wood roofs)
Wood doors, **14**-6, **14**-7
Wood floors, 4-31, 5-8, **7**-4, **7**-24, **7**-66 to **7**-68, **12**-67, **12**-69
Wood framing (*see* Framing)
Wood joists, 1-13, 5-8, **7**-50, **7**-59, **7**-61, **7**-63 to **7**-65
Wood piles, **6**-38, **6**-39
Wood planks (*see* Wood floors)
Wood poles, **7**-83 to **7**-86
Wood rigid frames, **7**-59 to **7**-61
Wood roofs, **7**-4, **7**-66 to **7**-68, **15**-7
Wood sheathing (*see* Sheathing)
Wood shingles, 4-31, 5-9, **12**-26, **15**-24, **15**-25
Wood siding, **12**-26
Wood stairs, **22**-8
Wood subfloors, 1-16, 5-8, **7**-50, **7**-64
Wood trim, 4-31, **16**-2
Wood trusses, **7**-4, **7**-57 to **7**-59, **7**-87
Wood walls:
  facings for, **12**-13, **12**-26 to **12**-28
  (*See also* Brick; Concrete block; Gypsumboard; Masonry; Plaster)
  foundation, **7**-64 to **7**-66
  sheathing for, 1-13, 5-9, **12**-24
  stud, 1-13, 1-16, 5-9, **7**-61, **7**-63 to **7**-65, **12**-23, **12**-24
  (*See also* Partitions; Walls)
Work:
  defined, **18**-9
  least, 5-52, 5-53
  reciprocal theorem for, **5**-56, **5**-57
  virtual, 5-50 to 5-52
  (*See also* Energy)
Workmen's compensation, **25**-22, **25**-28, **25**-29
Wythes, **12**-4

Yield point (strength), 5-13, **5**-88
  (*See also under* Steel reinforcement; Structural steels)
Yitzhaki method, 5-133 to 5-140

Zinc, **5**-6
Zinc chloride, **7**-20, **7**-29
Zoning, 1-34 to 1-36, 3-33

About the Author

Frederick S. Merritt, formerly Senior Editor of *Engineering News-Record,* is a consulting engineer with many years' experience in building and bridge design, structural analysis, and construction management. He is a Fellow of the American Society of Civil Engineers and a member of both the American Concrete Institute and the American Society for Testing and Materials. He is the author of *Building Engineering and Systems Design,* which was chosen by the Professional and Scholarly Publishing Division of the Association of American Publishers as the best engineering book published in 1979. He also is the author of *Mathematics Manual, Applied Mathematics in Engineering Practice, Modern Mathematical Methods in Engineering,* and editor of the *Standard Handbook for Civil Engineers* and the *Structural Steel Designers' Handbook.*

WITHDRAWN

JUN 0 5 2024

DAVID O. McKAY LIBRARY
BYU-IDAHO